A GAZETTEER OF THE WORLD

PEAK TENERIFFE.

A GAZETTEER OF THE WORLD

A Complete Repertory of Knowledge

PHYSICAL, STATISTICAL, HISTORICAL
&
ETHNOGRAPHICAL GEOGRAPHY

Compiled from :
The Most Recent Authorities

Vol. VII
AA-BRAZEY

GYAN PUBLISHING HOUSE
NEW DELHI-110 002

GYAN PUBLISHING HOUSE
5, Ansari Road,
New Delhi 110 002

India : 1997

ISBN : 81-212-0127-6
81-212-0134-9

Published by **Gyan Publishing House** and
Printed by **Goyal Offset Printers**, Delhi.

GAZETTEER OF THE WORLD

TA, or LADY'S ISLAND (LOUGH), a sea-lough or lagoon in co. Wexford, connected with the ocean by a narrow and brief strait, 1½ m. WNW of Carnsore-point. It penetrates the land nearly 2 m. NNE, with an extreme breadth of about 7 furl.

TAAIF, or TAYEF, a town of Arabia, 60 m. SE of Mecca, situated in a mountainous but fertile and well-cultivated district, whence Jidda and Mecca are supplied with fruits, particularly almonds.

TAAL, a town on the SE coast of the island of Luçon, in the Philippine group, S of Manila.

TAAPAN, a town on the W coast of Illano bay, in the island of Mindanao.

TAARS, a village on the NW coast of the Danish island of Laaland. Pop. 750.

TAAS, or TAAES, a city of Yemen, in Arabia, on the road from Mokha to Sana, and 48 m. ENE of Mokha. It is encompassed with a wall from 16 to 30 ft. thick, flanked with several towers, and composed of bricks dried in the sun, but with a facing of burnt bricks. The garrison consists of about 600 men. It is surrounded by mountains, which are in the possession of independent sheikhs.

TAASINGE, or THORSENGE, an island of Denmark, about 8 m. long and 4 m. broad, lying between Funen and Langeland. The soil is tolerably fertile, and the inhabitants, 3,750 in number, are employed in agriculture, navigation, and fishing. The island has a small town of the same name.

TAAWIRRY, an island in the S. Pacific, situated within the reef of the island of Otaheite, and on the E side of the main island.

TAB, a river of Persia, the ancient *Arosis*, which rises in the mountains of Fars, in two head-streams, the Harharas and the Rudkhonisk-Shir, which unite about 20 m. WNW of Shiraz. It runs SW, dividing Fars from Khuzistan; and falls into the Persian gulf near Indian, after a course of 180 m. At its mouth it is 80 yds. wide, and is navigable for boats of 20 tons up to Indian.

TAB, a village of the Punjab, on the l. bank of the Chenab, in N lat. 32° 32′.

TABA ISLANDS, four small islands in the Eastern seas, near Saint-Anthony's point, on the E coast of Borneo, in N lat. 2° 6′.

TABABELLAH, a town of Hindostan, in the prov. of Malwah, in N lat. 23° 16′.

TABAFRA, a village on the Ivory coast of Africa, 15 m. E of Drewin.

TABAGO, or TABOGA, an island in the Pacific, near the coast of Mexico, 18 m. S of Panama about [illegible] m. long and 2 m. broad. It is mountainous, and on the N side declines with a gentle descent to the sea. It produces plantains and bananas. Before it is a good road, about 1 m. from the shore, where ships may ride safely in 16 or 18 fath.

TABAGO. See TOBAGO.

TABAI, an island of the S. Pacific, in the Society group, in S lat. 23° 40′.

TABAJANA, a village of Wulli, in Western Africa, 12 m. WSW of Medina.

TABAJO, a river of Brazil, in the prov of São-Paulo, a tributary of the Paranapanema.

TABAK, a town of Bessarabia, 30 m. NNW of Ismail.

TABANERA, a village of Spain, in the prov. and 52 m. N of Palencia. Pop. 300.

TABANIEH, a village of Lower Egypt, on the E branch of the Nile, 6 m. SW of Mansurah.

TABARCAH, a small port of Algeria, situated in the midst of extensive woods, at the mouth of the Wad-el-Barba, opposite to a small island, famous for its coral fishery. It was long in possession of the Lomellines, a noble Genoese family, from whom it was taken in 1740.

TABARCA-NUEVA, an island of Spain, in the Mediterranean, off the coast of Valencia. It is almost destitute of trees and water, but is inhabited by the descendants of a colony of Spaniards who were redeemed from slavery in Barbary by the government, and settled here in 1771.

TABARIAH, TABARIEH, TABARIYEH, or TIBERIAS, a town of Syria, in the pash. and 48 m. ESE of Acre, on the W shore of the lake of Tiberias, a little to the S of its centre. It is built in the form of an irregular crescent, and is enclosed toward the land by a thick wall about 20 ft. in height, flanked with circular towers. Some of the houses on its E front are almost washed by the sea. Its S wall approaches almost to the beach; but the NW angle of the northern wall, being on a rising ground, recedes some little distance from the water, and thus gives an irregular form to the enclosure. The whole does not appear to exceed 1 m. in circuit, and cannot contain more than 500 separate dwellings. Its pop. has been estimated at 4,000. On the W it has a small plain fit for cultivation, whence the land rises into lofty hills which almost overhang the town. The interior presents few objects of interest beside the ordinary habitations, which are small and mean. There is a mosque with a dome and minaret, and another with an octangular tower in ruins. There are also two synagogues near the centre of the town, and one

Christian place of worship, called 'the House of Peter,' near the southern quarter, close to the water's edge. This last, which is thought by some to be the oldest place of Christian worship now extant in Palestine, is a vaulted room, 30 ft. by 15 ft., and about 15 ft. in height. T. is one of the four holy cities of the Talmud; the other three being Szaffad, Jerusalem, and Hebron. It is esteemed holy ground, because Jacob is supposed to have resided here, and because it is situated on the lake of Gennesareth, from which, according to the generally received opinion of the Talmud, the Messiah is to rise. The greater part of the Jews who reside in these holy places do not engage in mercantile pursuits; but are a society of religious persons occupied solely with sacred duties. Three or four missionaries are sent out from the four sacred cities every year; one to the coasts of Africa from Damietta to Mogadore, another to the coasts of Europe from Venice to Gibraltar, a third to the Archipelago, Constantinople, and Anatolia; and a fourth through Syria. Those who go abroad as missionaries are allowed 10 per cent. upon all alms collected, besides their travelling expenses. The Jewish devotees pass the whole day in the schools or the synagogue, reciting the Old Testament and the Talmud.

The lake of T., or the sea of Gennesareth, occupies the mountain-basin formed within the sweep of the two great mountain-chains of Syria as they run southward, forming the extreme N part of the singularly depressed tract called El-Ghor. It is in fact an expansion of the river El-Urdun or Jordan, of an irregular oval form, 12 m. in length from N to S, and upwards of 5 m. in its greatest width, near Tabariyeh. It is everywhere surrounded by rocky, bare, brown-coloured mountains.

TABARINSKAIA, a town of Asiatic Russia, in the gov. and 150 m. SW of Tobolsk, on the r. bank of the Tafda.

TABARISTAN, a province of Persia, lying between the parallels of 35° 35′ and 36° 15′ N; and bounded by Mazanderan on the N; by Khorasan on the E; on the SE by Kohistan; on the S by Irak-Ajemi; and on the W by Ghilan. It is a mountainous region, being intersected by the great Elburz chain, and several of its spurs. Its cap. is Demavend.

TABAS, a city of Khorasan, on the road from Herat to Yezd, 337 m. from the former, and 150 m. from the latter. It is situated amid a range of mountains, and contains a pop. of about 20,000.—Also a village of Anatolia, in Asiatic Turkey, 34 m. S of Degnizlu.

TABASCO, a province of Mexico, bounded on the N by the gulf of Mexico, along which it extends from the river Coazacualco on the W, to the Lago-dos-Terminos on the E. On the E it has Yucatan; on the S, the provs. of Chiapa and Oaxaca; and on the W, Vera-Cruz. The soil is fertile, and the surface is in general flat, marshy, and filled with small lagoons or lakes. The coast is subject, from September to March, to dreadful storms; the northerly gales prevailing during that period rend r navigation dangerous and difficult. In February, March, and April, the heat is insupportable; and infinite swarms of mosquitoes and other venomous insects, add to the annoyances of the season. There are, however, numerous good farms, well stocked with cattle, in this prov. The natives reap from three to four harvests of maize in the year, and rear rice, barley, European garden-herbs, fruits, cacao, sugar-cane, and cotton.—The capital of the state is San-Juan-Bautista.

TABASCO, a river of Mexico, which rises on the frontiers of Guatimala, and flowing WNW, and then N and NNE, intersects Chiapan and the state of Tabasco, and flows into the bottom of the gulf of Mexico, in N lat. 18° 34′, W long. 92° 38′. Its bar has from September to March in general 11 ft. water; but vessels are sometimes detained a couple of months inside for want of water. At the W side of the entrance to the river is the Isla-del-Buey or Great Ox island; and ¾ m. to the N of it lies the smaller Ox island. About 1 m. above the entrance, on the E bank, is a small fort, and the pilot establishment. Between this fort and the town of Tabasco or La Frontera, is a thickly wooded island about 1½ m. long. The river abreast of the fort and at Frontera, is nearly ¾ m. wide, but above the town its breadth greatly diminishes. About 8 m. above the town, the Tabasquillo branches off to the SW; and 3 m. above the mouth of this arm, at Los-Tres-Brazos, the San-Pedro-el-Chiquito, and the Usumasinta, join the T. At 9 leagues above La Frontera, the Chilapa enters the T. on the r. bank; and 4 leagues above the latter point of confluence, the T. is joined by the Chilapilla. The town of Villa-Hermosa-de-San-Juan-Bautista, the cap. of the state of Tabasco, is situated on the W or l. bank of the T., about 24 leagues above La Frontera.

TABASCO (La Frontera de), a village of the Mexican state of Tabasco, built on the E bank of the Tabasco river, in N lat. 18° 31′. It is a straggling assemblage of bamboo and adobe huts. Its principal inhabitants are the agents of the logwood-cutters. It is the port of San-Juan-Bautista, and is chiefly frequented by vessels from Campeachy, the United States, and the Havannah. Its chief exports are logwood, cocoa, and fustic.

TABASSERAN, a town and small district of Shirvan, in the Russian prov. of Daghestan, 20 m. WSW of Derbend.

TABATINGA, a town of Brazil, in the prov. of Para, 60 m. S of Olivenza, on the l. bank of the Amazon, in S lat. 4° 12′, at the confluence of the Rio-Jabari.

TABERG, a village of Oneida co., New York. U. S., 110 m. WNW of Albany.

TABERNA, a town of Spain, in the prov. and 18 m. NE of San-Felipe, on the great road leading along the coast in the direction of Catalonia. Pop. 4,500.

TABERNAS-Y-TURRILLAS, a town of Spain, in the prov. and 15 m. NE of Almeria. Pop. 5,500. Coal, lead, iron, and copper are or have been wrought in the vicinity.

TABIO, a settlement of New Granada, in the prov. of Cundinamarca, 15 m. NW of Santa Fé.

TABLADA, a settlement of New Granada, in the prov. of Santa-Marta, on the shore of the river Magdalena, on an island formed by an arm of this river, 90 m. SSE of Mampox.

TABLAS, one of the Philippine islands, due S of Luçon and N of Panay. It is about 30 m. long, by 3 m. in average breadth.

TABLAT, or Sant-Fiden, a town of the Swiss cant. of St. Gall, 7 m. SW of Rorshach. Pop. 2,500.

TABLE (La), a river of Savoy, 3 m. NE of La Rochette, on the r. bank of the Jelon. Pop. 1,300.

TABLE-BAY, a bay on the E coast of Labrador, in N lat. 53° 44′.—Also a bay on the W coast of the Cape of Good Hope. See that article.

TABLE-CAPE, a cape on the E coast of New Zealand, in S lat. 39° 7′.—Also a steep rocky point of land on the N coast of Van Diemen's Land, in S lat. 40° 56′.

TABLE-ISLAND, a small flat island of the Eastern seas, in Gaspar's strait.—Also a small island in the S. Pacific, in S lat. 18° 54′.—Also a small island near the coast of Spitzbergen, in N lat. 80°

57′.—Also one of the New Hebrides in the S. Pacific, in S lat. 15° 38′.

TABLE-MOUNTAIN. See CAPE OF GOOD HOPE.

TABLE-MOUNTAIN, a mountain of S. Carolina, U. S., near the NW border of the state, 26 m. NW of Greenville, 3,168 ft. higher than the surrounding country, and about 4,000 ft. above the level of the sea. It presents on one side a tremendous precipice of solid rock.—Also a mountain of California, in Sierra co., 151 m. NE of Vallejo, said to exceed 8,000 ft. in alt.

TABLE-MOUNTAIN, one of the central alps of co. Wicklow. Its summit forms part of the water-shed between the river-systems of the Slaney and the Ovoca; and has an alt. above sea-level of 2,312 ft. But so nearly fused is this cloud-cleaving summit into the great general ridge of the Slaney and Ovoca watershed, that the 'gap' which affords a path of communication between Glenmalure and the glen of Imail, and which is situated about ½ m. S of the summit, has an alt. above sea-level of 2,266 ft.

TABLE-POINT, or TAFEL-HOECK, the S extremity of Bali island, in the Eastern seas, and the eastern boundary of the S entrance into Bali straits, in S lat. 3° 50′.

TABLE-RIVER, a river of Louisiana, which runs into the Mississippi, in N lat. 37° 12′.

TABLEY (NETHER and OVER), two townships of Cheshire, 2 m. WNW of Nether Knutsford. Pop. of former in 1851, 123; of latter, 537.

TABO, or LITTLE DIEPPE, a river and port on the Grain coast of Africa, 52 m. NE of Cape Palmas.

TABOAÇO, a village of Portugal, in the prov. of Beira, comarca of Samego.

TABOCAS, a river of Brazil, in the prov. of Goyaz, which descends from the E flank of the Cordillera-Grande, and flowing ENE, joins the Tocantins on the l. bank, after a course of 90 m.

TABOGA. See TABAGO.

TABOLEOO, a river of Chili, which runs E and NNE, and enters the Biobio.

TABOR, a circle in the SE of Bohemia; bounded on the N by the circle of Czaslau; on the E by Moravia; and on the S and W by the circle of Budweis. Its extent is 1,270 sq. m.; its pop. in 1843, 207,824. Its soil is fertile, and adapted both to tillage and pasturage.—Its capital, of the same name, situated on a hill on the river Luschnitz, 88 m. W by N of Brunn, and 49 m. SSE of Prague, is fortified, and naturally strong, but has been frequently taken. It contains 4,000 inhabitants, whose principal occupation is weaving. It was built by Zisca, the Hussite general, and called by the Hussites, Hradistie-Hory-Tabor, or 'the Camp of Mount Tabor;' and from it they take the name of Taborites.

TABOR, or THABOR, the *Ithabarius* of the ancients, and JEBEL-TUR of the Arabs, a mountain of Palestine, situated to the SW of the lake of Tiberias, over which it commands a most extensive prospect. It is of a conical form, and limestone formation; and contains on its summit a plain of great extent, and highly cultivated. Its rounded sides are richly clothed with the valonia oak, the myrtle, and wild pistachio. Mr. Spencer, who recently ascended it, thus describes the view from its summit: "Notwithstanding the desolation of T., compared with its glory once, in other days, we beheld there what man has neither given nor can take away—the glorious view of the surrounding country, which is unsurpassed in all Palestine. I would that I possessed the power of graphic description, that I might tell you of what we saw from the summit of this noble mount; how grandly loomed up in the far distance, to the N, the snow-crowned Jebel-es-Sheikh, the Hermon of Scripture! How picturesque appeared the hills and mountains to the NE and the E, beyond and on this side the silvery Jordan, which springs out of their very bosom! How lovely seemed that lake, of all others most interesting to the Christian's heart, the lake of Tiberias, a part of which we could plainly see! How noble toward the S, looked the valley of the Jordan, Gilead, Gilboa, the Little Hermon, and the charming vales between! But how surpassingly beautiful, which I cannot find words rightly to express, was the scene in the west, as, at this commanding elevation, some 1,500 ft. above the plain, we looked down upon Esdraelon, in all its glory and magnificent verdure, its extent, its fertility, its loveliness, its surrounding hills, its streams and rivulets, its river, the Kishon and its many, many points of attractiveness! Believe me, I stood as it were entranced on the steep brow of T., and beheld this scene with emotions too deep for utterance." Colonel Wildenbruch estimates the alt. of the plain at the foot of Mount T., extending towards the lake of Tiberias, at 845 ft. above the level of the Mediterranean; but the American survey makes it only 652 ft.

TABORAO, a river of Brazil, in the prov. of Sante Catharina. Its banks are said to produce the best flax in the prov.

TABRABOUCHA, a swamp of New South Wales in the co. of Roxburgh, near the head of Cunningham creek, and 131 m. from Sydney.

TABREK, a mountain of Persia, in the prov. of Irak-Ajemi, and district of Teheran, near Rei. On its side is a castle of the same name.

TABRIZ, or TEBRIZ, a city of Persia, the cap. of the prov. of Azerdbijan, in the centre of which it is situated, in N lat. 46° 8′ 30″, E long. 38° 3′ 59 [*Monteith*]. It is the principal commercial depot of Northern Persia. Chardin estimated its pop. in 1686 at 500,000; at present it does not exceed 20,000. It contains about 8,000 brick houses, several good bazaars and mosques, and an extensive *maidan* or square. Its SW extremity is occupied by the ark or citadel. "The towering pile of the ark, being the only building to be seen in the distance above its embosoming forest of fruit-trees, one might readily fancy the place to be a huge orchard stretching along the base of the red and barren mountain-wall which bounds it on the south. The area thus covered by the town itself, and these miles of surrounding gardens," continues a very recent visitor, "is, I am told, nearly equal to that of London, T. being by far the largest and most important city in Persia. The town proper is surrounded by a crumbling brick wall and a fosse, constructed with such utter disregard to any principles of scientific engineering, as to have been, even in their best condition, but a poor protection against any attacking force provided with artillery. At present these fortifications are in many places a mere heap of ruins. Inside, the streets are of the worst description; and, excepting in the bazaars, are in reality little more than mud-wall avenues, broken at irregular intervals by doorways of burnt bricks, which lead—usually through small gardens—to the quiet and well-hidden dwelling-houses within. Excepting the ark, an old tower within the citadel, where cannon are cast, and bored by machinery driven by a pair of buffaloes, and the splendid remains of the old blue mosque destroyed about 80 years ago by an earthquake, T. contains no public buildings worth mentioning. Its bazaars and caravansaries, however, are unequalled in Persia: the former are filled with European goods, especially those of England; and the latter with the costly products of India, Cashmere, and Arabia, *en route* for Constantinople, to be there exchanged for the manufactures of the West. The inhabitants themselves weave silk goods—the raw material of which comes from Mazanderan and Ghilan, on the Caspian—that command extensive sales throughout Asia Minor, in Constantinople, and very largely in Russia." A few European merchants have establishments in this city. Our journalist adds that the present pop. of T. (including the suburban villages) numbers about 180,000; but this estimate must include the entire district administratively attached to the city. T. was founded A. D. 787. It has frequently been devastated by earthquakes. In 1792, upwards of 80,000 inhabitants perished by one of these fearful convulsions. It has also suffered greatly from the constant struggles of the Persians and the Turks.

TABUA, a village of Brazil, in the prov. of Minas-Geraes, and comarca of Rio-de-Jequitinhon a, on a mountain of the same name.

TABUENCA, a town of Spain, in the prov. and 39 m. W of Zaragoza, and partido of Borja. Pop. 780.

TABURUHINA, a river of Brazil, in the prov. of Mato-Grosso, which descends from the N side of the Cordilheira-dos-Parecis, runs NE, and joins the Juruena, on the r. bank. A tribe of the Mambare Indians, who inhabit its banks, bear the same name.

TABY, or GABANIT, a mountain of Northern Africa, in the territory of Bertat, about 30 m. from the l. bank of the Bahr-el-Azrek, in N lat. 11° 20′. It is partly covered with wood, and detached huts inhabited by negro idolaters.

TACALAYO, a settlement of Laplata, in the prov. and 105 m. N of Salta.

TACALOA, a town of New Granada, in the dep. of the Magdalena, prov. and 84 m. SE of Carthagena, on the l. bank of the Cauca, a little above its confluence with the Magdalena.

TACAMBARO, a settlement of Mexico, in the state of Michoacan, and 33 m. SE of Valladolid. Pop., Spanish and Creoles, 2,000.

TACAMES. See ATACAMES.

TACAMOCHO, a settlement of New Granada, in the dep. of the Magdalena, and prov. of Carthagena, on the Magdalena, near the mouth of the Cauca.

TACANA, a town of Guatimala, in the dep. of Quezaltenango and Soconusco. Pop. 2,600.

TACANHUNA, TOCAHUNOS, or TARACUNA, a river of Brazil, in the prov. of Para, which has its source in N lat. 8° 10′, and W long. 51° 40′, on the NE confines of the prov. of Mato-Grosso; takes at first the name of Paraupaba; runs NE; forms several sinuosities; and after a course of about 330 m., joins the Tocantins on the l. bank near Itabocas. It derives its name from a tribe of the Tupinambas Indians, who inhabit its banks.

TACARATU, or TARCAVATU, a town of Brazil, in the prov. of Alagoas, 135 m. NW of Penedo, and 33 m. NNE of the fall of Paulo-Affonso, formed by the São-Francisco. It has a chapel frequented by several Indian tribes.

TACARIGUA. See VALENCIA (LAKE OF).

TACAZZE′, or TAKKAZEE, a river of Abyssinia, second in importance to the Abai or Nile. It rises in about 11° 30′ N lat., and 39° E long.; and first runs N, receiving the Tzelari and the Bellegas in this early part of its course. It then sweeps round to the W, enclosing the lofty plateau of Samen within its bend, and after running some distance nearly W, about the meridian of 37°, and nearly under the parallel of 14°, turns NW, receives the Sireh and the Setif on its r. bank, and the Angrab and the Guangue on its l. After the junction of the latter stream, on the N frontier of the Shangalla territory, it assumes the name of the Atbara, and flows in a NNW direction to the Nile, which it joins at Addemer, the extreme N point of the island of Meroe, as the large region enclosed by the Nile on the W, and the T. on the E, is named. Its principal tributaries in Nubia are the Mareb and the Mogren. Dr. Ruppell states that a descent of 2,400 ft. from the plains of Temben, led to the banks of the T., which he found rushing like a torrent through a deep valley, on the W side of which the mountains of Simen or Samen rear their rugged summits nearly to the limits of perpetual snow. The absolute height of the river where he crossed it was 2,900 ft., and as its course is extremely rapid, we may infer that the plains in which, under the name of Atbara, it joins the Nile 450 m. lower down, cannot be raised much more than 2,000 above the level of the sea, which is but half of the elevation assigned to them by Rennell, Humboldt, and other speculative geographers. Salt says, that the Abyssinians in general shudder at the thought of bathing in the river, on account of the crocodiles which infest it.

TACCORARAY, a town of Upper Guinea, on the Gold Coast, in the kingdom of Ashantee, 27 m. NE of Cape Three-Points, on a rocky headland. The Dutch had formerly a fort here.

TACENO, a village of Austria, in Lombardy, in the prov. and 21 m. NNE of Como, in the valley of Sasina, near the r. bank of a river of the same name.

TACHAU, TACHOW, or DRZEWNOW, a town of Bohemia, in the circle and 36 m. W of Pilsen, at the foot of the Böhmerwald, on the l. bank of the Misa. Pop. 2,932, of whom 266 are Jews. It has a castle, and a Franciscan convent. It has a forge and a copper-mine. In the vicinity is the glass-manufactory of Ströbl, the most ancient in Bohemia.

TACHBROOK (BISHOP'S), a parish in Warwickshire, 2¾ m. SE of Warwick, watered by a branch of the Avon. It comprises the hamlet of Tachbrook-Mallory. Area 3,446 acres. Pop. 663.

TA-CHING-HEEN, a district and town of China, in the prov. of Chih-le, and dep. of Shun-teen-fu, in N lat. 38° 48′, and E long. 116° 40′.

TA-CHU, a district and town of China, in the prov. of Sze-chuen, and div. of Shun-king-fu, in N lat. 29° 50′, and E long. 105° 51′.

TACKLEY, a parish in Oxfordshire, 3 m. NE by N of Woodstock, on the road from Oxford to Deddington. Area 2,850 acres. Pop. in 1851, 558.

TACLATUR, or TAEKLAGUR, a town of Northern Hindostan, in Nepal, on the W side of the Gogra or Sarjon, 85 m. N of Jemlah, at an alt. of 14,500 ft. above sea-level. It is a dependency on Gurdon, a Chinese station and carries on an active trade with Tartary in wool, woollen fabrics, gold, tea, wax, borax, and a species of grain locally named awajou The latter is carried N to the banks of lake Rawans hrad, where it forms during winter the chief sustenance of the horses.

TACNA, a city of Peru, in the prov. of Arequipa, situated in S lat. 18° 1′ 50″, on the Barrosa, about 25 m. from the sea, and 120 m. direct distance SW by W of La Paz. It stands in a small valley from 2 to 3 m. wide, on the interior edge of a broad barren plain, extending from Arica on the coast, at an alt. of 1,795 ft. above sea-level, with moderate elevations on the N and S, and the snowy-crested Cordillera on the E. In the general plan of the arrangement of its streets and walks, in the paving and the style of building, one is reminded of Lima, but in several respects T. is decidedly superior; the shops are larger and more commodious, and the houses of the better classes are larger and finished exteriorly in decidedly better style,—some of them being built of hewn stone, and many of them with a balcony to the second floor, while the interior is in many instances finished elegantly and furnished in a style in keeping with the wealth of the proprietors. This is accounted for by the fact, that the prominent merchants of the place are English, that is, they were born in England, and educated there, though most of them have been long resident in T., and a majority have married Spanish ladies. The merchants of T. boast of the finest club-room in South America. It is very spacious, and is occupied as a reading-room, and supplied with the American, French, and English papers; a good library is also attached to the establishment. Several attempts have been made to adorn the place by the erection of a cathedral, but it has twice been partially thrown down by earthquakes, and the walls and towers which are yet standing are rent and weakened. The city has suffered greatly from these commotions. Twice since the commencement of the present century it has been destroyed, and the inhabitants are raising the inquiry whether they may not secure themselves against so great liability of disaster from this cause, by the use of iron to a considerable extent in the framework of their buildings. A large proportion of the primary instruction. here is private, yet there is a primary school supported by the government and free to the children of every citizen. In these schools, whether public or private, it is said that nearly all the children are taught to read and write, and many study the elements of geography and arithmetic; many of the slaves even receive instruction thus far. For more advanced instruction there are two colleges, so called—one for males, and one for females—designed to meet the wants not of T. only, but of the whole southern province of Peru, of which this is the principal city. These are institutions of recent origin, have no funds, no library or apparatus of any amount, and with the exception of a small gratuity from the government, depend on tuition fees for support. Pupils who board and lodge at home pay about 100 d. per annum; such as are furnished everything pay about 250 d. In the college for boys a corps of 5 instructors is employed. The matriculation list in 1853 showed a list of 39 pupils in the regular course, and 36 in the primary course.—A large proportion of the business by which Bolivia is connected with the exterior world is done here. Bolivia has a small seaport, Cobija, and between that and Arica there are for local purposes some others; but access to every part of Bolivia is less difficult and less expensive from Arica than from any other port. The great business of Bolivia—though the whole is but a trifle—centres here. T. is a part of Peru only in its political relations; by treaty-stipulations, goods pass in bond to the Peruvian frontier free of duty, and the Bolivian duties are paid there as if it were a port-of-entry. This city, therefore, sustains to Bolivia precisely the same relations that Lima does to Peru. Inconsiderable in size compared with Lima, having a pop. of only about 8,000, but in several respects they are much alike. The merchants realize fully that the proper place for a commercial city is the sea-port where goods are shipped and landed, and more than one effort has been made to effect a removal to Arica. Between T. and its sea-port—a distance of 40 m.—except in the immediate vicinity of Arica, there is not a blade of grass or a drop of water; the transmission of freight was therefore a matter of great difficulty, and performed entirely by mules and donkeys; but a railroad between the two places is now being constructed under a concession from government for 99 years, which will diminish the cost of labour and of living in T., and greatly increase its business.—The environs of T. are very barren. One indeed is surprised that vegetation can subsist at all in a region apparently of such absolute sterility. But irrigation has developed considerable fertility. The valley near the river consists of a bed of clay overlying a deep bed of pebbles. This clayey surface on each side of the city, and for miles above and below it, is divided into lots of a few acres each, called *chacras* or farms. Anything that will allow of hilling, as corn or potatoes, is planted on low ridges, and the water is let into the intervening ditches long enough each alternate day to furnish the roots with moisture. The *alfalfa*, much like red clover, and upon which mules, horses, and cattle are chiefly fed, is sown in beds a rod square. Each bed is surrounded with a slight ridge of earth. The water is let into each bed till it covers all the surface, and in sufficient quantity to give, when absorbed, abundant moisture to the soil. The pear, peach, grape, &c., are planted near the permanent water-courses, so that they need no care of irrigation. Many other shrubs are allowed to grow along these runs, but are turned to no account: among others the cotton-plant grows spontaneously, producing an article of excellent quality, but it is never gathered.

TACOARA, or TAGUARY, a river of Brazil, in the prov. of Matto-Grosso, which rises in the district of Camapuan, and runs WSW to the Paraguay, which it joins after a course of about 300 m.

TACOLNESTON, or TACLESTON, a parish in Norfolk, 10 m. SW by S of Norwich. Area 1,580 acres. Pop. in 1831, 486; in 1851, 501.

TACORA (NEVADO-DE), a snowy summit on the exterior edge of the Bolivian Andes. It rises above the pass of Las Guahllas, which runs between it and the Nevado of Niuta, and probably exceeds 17,000 ft. in alt. Near its base is a village and convent of Franciscan monks, at an alt. of 14,275 ft. above sea-level.

TACORONTA, a town on the N coast of the island of Teneriffe. Pop. 3,500.

TACOTALPE, a settlement of Mexico, in the state of Tabasco, 47 leagues SE of Vera Cruz.

TACUBA, or TALCOPAN, a town of Mexico, the capital of a district of the same name, 6 m. WNW of the city of Mexico, with which it is connected by a beautiful causeway of cut stone, being the same by which Cortes made his way into the capital. Pop. 2,500, chiefly Indians.

TACUBAYA, a town of Mexico, 4 m. SW of the city of Mexico. Pop. 2,000. The archbishop of Mexico has a fine palace here.

TACULUM, a town of Hindostan, in the Carnatic, in N lat. 13° 4′.

TACUMSHANE, a parish on the S coast of co. Wexford, 2 m. SW of Broadway. Area 3,153 acres. Pop. in 1831, 907; in 1851, 875.

TACUNGA, a province of Ecuador; bounded on the N by the prov. of Quito; on the E by the valley of Vicioso; on the S by the prov. of Ambato; on the SW by Chimbo; and on the NW by Esmeraldas and Guayaquil. It abounds in excellent pastures, and produces wheat, barley, and rye.—The capital, of the same name, is situated in an extensive plain to the S of Quito, near the cordillera of the Andes. It is a large and well-arranged town, with wide streets, and houses built chiefly of a light porous lava. It was destroyed in 1698 by an earthquake, when the greater part of the inhabitants perished. A similar catastrophe was repeated in 1743 and 1757. The inhabitants are computed at 17,000, the greater part being Spaniards and Mestizoes.

TACURUCA, an island of Brazil, off the coast of the prov. of Porto-Seguro, near Santa-Cruz.

TADCASTER, a parish and market-town of Yorkshire, 9 m. SW of York, on the river Wharfe, and within 3 m. of the North Midland railway.

The p. includes the townships of Stutton with Hazlewood, Catterton, Oxton, and T. Area 6,010 acres. Pop. in 1801, 2,072; in 1831, 2,855; in 1851, 2,979. The town, built in the form of a cross, is pleasantly situated on the river Wharfe, which is here spanned by a bridge of 9 arches, and is navigable up to the town. The houses are neat and well-built. No manufactures are carried on in the town. In the neighbourhood are several stone-quarries, and corn or flour mills. T. was the *Calcaria* of the Romans. In the civil wars of England it was always esteemed a post of considerable importance.

TADDINGTON, a chapelry in Bakewell p., in Derbyshire, 4 m. S by W of Tideswell. Pop. 460.

TADEN, a canton and village of France, in the dep. of Cotes-du-Nord, 2 m. NE of Dinan, near the Rance. Pop. 1,200.

TADICOMBO, a town of Hindostan, in the Carnatic, 9 m. N of Dondigol.

TADLA, a town of Marocco, on the l. bank of the Morbeya, 85 m. NE of Marocco.

TADLEY, a township of Southamptonshire, 6½ m. NNW of Basingstoke. Area 2,047 acres. Pop. in 1851, 876.

TADLOW, a parish of Cambridgeshire, 6 m. S of Caxton. Area 1,717 acres. Pop. in 1851, 189.

TADMARTON (Great), a parish of Oxfordshire, in which is situated the hamlet of Little Tadmarton, 4½ m. WSW of Banbury. Area ?,500 acres. Pop. in 1831, 355; in 1851, 450.

TADMOR. See Palmyra.

TADOUSAC, a settlement of Lower Canada, at the mouth of the Saguenay, on the l. bank of the St. Lawrence. It has a fort.

TADVAN, a village of Persia, on a small bay of Lake Van, 10 m. E by N from Betlis. It is inhabited by about 40 Armenian families.

TAE-CHU, a division and town of China, in the prov. of Shan-se. The div. comprises 3 districts. The town is 96 m. NE of Tae-yuen-fu, near the great wall, in N lat. 39° 3′ 50″, and E long. 112° 18′.—Also a division and town in the prov. of Kan-suh. The div. comprises 7 districts.

TAE-CHU-FU, a division and town of China, in the prov. of Che-keang. The division comprises 6 districts. The town is 120 m. SE of Hang-chu-fu in N lat. 24° 54′, and E long. 121° 9′ 24″. It is situated on an affluent of the Yellow sea, 30 m. from the coast, which here abounds with a species of ray, the skin of which is manufactured into sheaths for cutlasses, and forms an important article of export to Japan and all parts of the empire.

TAE-CHUN, a district and town of China, in the prov. of Che-keang, and div. of Wan-chu-fu, 120 m. S of Hang-chu, in N lat. 27° 34′ 48″, and E long. 119° 50′ 20″.

TAE-GAN-FU, a division and town of China, in the prov. of Shan-tung. The div. comprises 7 districts. The town is 39 m. SSW of Tse-nan-fu, in N lat. 36° 14′ 15″, and E long. 117° 15′ 30″. It is scantily peopled. In the principal street is a triumphal arch, and at the extremity a large handsome pagoda.

TAE-HANG-SHAN, a chain of mountains in China, which detaches itself from the In-shan mountains in Mongolia; penetrates China proper; enters the dep. of Tae-tung-fu in Shan-se, and that of Seuen-hwa-fu in Chih-le, and running S to the confines of Honan, forms a line of separation between these two former provinces.

TAE-HING-HEEN, a district and town of China, in the prov. of Keang-su, and div. of Tung-chu, in N lat. 32° 12′, and E long. 120°.

TAE-HO, a district and town of China, in the prov. of Keang-se, and div. of Keih-gan-fu, 150 m. SSW of Nan-chang, in N lat. 26° 46′, and E long. 114° 54′. It is pleasantly situated on the l. bank of the Kan-keang. At its entrance is a tower 9 stories in height.

TAE-HO-HEEN, a district and town of China, in the prov. of Gan-hwuy, and div. of Ying-chu-fu, in N lat. 33° 10′, and E long. 115° 43′.—Also a district and town of China, in the prov. of Yun-nan, and div. of Ta-le-fu.

TAE-HO-SHIN, a river of China, in the prov. of Kwang-tung, and island of Haï-nan. It issues from the S side of the Wei-chishan; runs first SE, then E; passes 3 m. N of Leen-chu-fu, and along the base of a mountain which gives rise to the Pokishin; divides soon after into two branches, forming Cotton-tree island, and throws itself into the bay of Shwuy-kaou-keang.

TAE-HU-HEEN, a district and town of China, in the prov. of Gan-hwuy, and div. of Gan-king-fu, in N lat. 30° 30′, and E long. 116° 20′.—Also a new town in the prov. of Kiang-nan, and dep. of Su-chu-fu, in an island of a lake of the same name.

TAE-KANG, a district and town of China, in the prov. of Ho-nan, and div. of Chin-chu-fu, in N lat. 34° 7′, and E long. 114° 54′.

TAE-KUH, a district and town of China, in the prov. of Shan-se, and div. of Tae-yuen-fu, in N lat. 37° 25′, and E long. 112° 33′.

TAE-LAO, an island of China, in the prov. of Kwang-tung; at the entrance to the roadstead of Macao, in N lat. 22° 7′, and E long. 113° 39′.

TAENSAPAVA, a river of West Florida, which runs into the Ibberville, in N lat. 30° 19′.

TAE-PE-SHAN, a mountain of China, in the prov. of Shen-se, and dep. of Fung-seang-fu, in N lat. 33° 55′, and E long. 107° 42′.—Also a mountain in the prov. of Shan-se, in the dep. of Tae-tung-fu, in N lat. 39° 20′, and E long. 111° 59′.—Also a mountain in the prov. of Kan-suh, in N lat. 32° 46′, and E long. 105° 3′. These mountains are all covered with perpetual snow.

TAE-PING, a district and town of China, in the prov. of Che-keang, and div. of Tae-chu-fu, in N lat. 28° 26′, and E long. 121° 25′, or according to Klaproth, 120° 42′.—Also a district and town in the prov. of Shan-se, and div. of Ping-yang-fu, in N lat. 35° 46′, and E long. 111° 18′.—Also a district and town of China, in the prov. of Sze-chuen, and div. of Kwei-chu-fu, in N lat. 32° 8′, and E long. 108° 14′.

TAE-PING-FU, a division and town of China, in the prov. of Gan-hwuy. The div. comprises 3 districts. The town is 120 m. NE of Gan-king-fu, in N lat. 31° 38′ 38″, and E long. 118° 31′ 45″, on the r. bank of the Yang-tse-keang, which here receives 3 tributaries. It has a considerable trade.—Also a division and town in the prov. of Kwan-se. The div. comprises 7 districts. The town is 69 m. SW of Nan-ning-fu, on the r. bank of the Le-keang, on a peninsula formed by that river, in N lat. 22° 25′ 12″, and E long. 107° 7′ 10″. It is defended landwards by a wall. The surrounding district is the most fertile and populous in the prov., and contains numerous forts erected for the defence of its frontier towards An-nam.

TAE-PING-HEEN, a district and town of China, in the prov. of Gan-hwuy, and div. of Ning-kwo-fu, in N lat. 30° 25′, and E long. 118° 8′.

TAERENDO-ELF, a branch of the river Tornea, in Lapland, which is said to communicate with the river Calix, though it afterwards takes a contrary direction, and flows N to the Frozen ocean.

TAE-TEEN-HEEN, a district and town of China, in the prov. of Fo-keen, and div. of Yung-chun-chu, in N lat. 25° 40′, and E long. 18° 8′.

TAE-TSANG-CHU, a division and town of

China, in the prov. of Keang-su. The div. comprises 4 districts. The town is 150 m. ESE of Nanking-fu, in N lat. 31° 30′, and E long. 120° 52′.

TAE-TUNG-FU, a division and town of China, in the prov. of Shan-se. The div. comprises 10 districts. The town is 180 m. NNE of Tae-yuen-fu, near the great wall, in N lat. 40° 5′ 42″, and E long. 113° 16′ 30″. It stands amid mountains, in a locality exposed to the incursions of Tartars, but it is well-fortified, and has a numerous garrison. The streets are narrow, and the houses confined, but tolerably well-built. There are several triumphal arches built of wood, some of them of great antiquity. It has a large pop., and possesses an active trade in furs. The environs afford marble, porphyry, lapis-lazuli, jasper, and many kinds of rare medicinal plants.

TAE-TUNG-HEEN, a district and town of China, in the prov. of Shan-se, and div. of Tae-tung-fu.

TAE-WAN-FU. See Formosa.

TAE-YO, an island of China, in the prov. of Kwang-tung, at the entrance to the roadstead of Macao, formerly called Lan-tao, in N lat. 22° 15′, and E long. 113° 52′. It has a town of the same name.

TAE-YUEN, a district and town of China, in the prov. of Shan-se, div. and 12 m. SW of Tae-yuen-fu, near the r. bank of the Fuen-ho.

TAE-YUEN-FU, a division and town of China, in the prov. of Shan-se. The div. comprises 11 districts. The town is 279 m. SW of Pe-king, in N lat. 37° 53′ 30″, and E long. 112° 33′. It is 9 m. in circumference. It is well fortified, and populous, although considerably decayed in importance since it ceased to be an imperial residence. The palace is rapidly falling into ruins. It has manufactories of fine earthen vessels, of various kinds of iron-ware, and of varieties of woollen fabrics, inclusive of carpets. The environs afford musc and azure stone. On the adjacent mountains are the sepulchres of several princes, some of them of great beauty.

TAF, or Taafe, a river of Wales, which rises among the hills in Brecknockshire, in two streams, the Greater and Lesser Taf, which unite their waters below the village of Coed-y-Cymmer, on their entrance into Glamorganshire; and thence proceed by Merthyr-Tydvil. About 12 m. below the latter town, the T. receives the Bangoid-Taf, a mountain-stream which flows into it from the E. Lower down it is joined by the Cynon from the W, and a few miles lower by the united waters of the two Rhonddas. It then flows nearly S. to Llandaff, and afterwards passing Cardiff, enters the small bay of Penarth. In dry weather the T. contains but little water; when swollen by the floods from the mountains, it rolls over its rocky bed in an impetuous torrent. It is navigahle for small craft as far as Cardiff, to which the tide ascends.

TAFALISGA, a town of Gallam, in Central Africa, situated at the junction of the Falemé with the Senegal.

TAFALLA, a small but ancient town of Spain, in the prov. of Pampeluna, on the river Cidacos, 27 m. N of Tudela, and 19 m. S of Pampeluna. Pop. 4,100. It was formerly the residence of the kings of Navarre, and the seat of. a university. It has several churches and convents.

TAFARA, a walled village of Bambarra, in Central Africa, 120 m. WSW of Sego.

TAFELBERG, a town on the E coast of the island of Ceram, in S lat. 3° 20′.

TAFELFICHTE, or Tafelstein, a mountain of Prussia, situated at the point of meeting of the Bohemian, Silesian, and Lusatian ranges. Its elevation is 3,370 ft., and the prospect from its summit, on the side of Bohemia, is immense.

TAFELNEH (Cape), a promontory on the coast of W. Africa, 27 m. SSW of Mogador. It rises to an alt. of 780 ft.

TAFFERS, a commune and village of Switzerland, in the cant. and 3 m. ENE of Friburg. Pop. 2,600.

TAFILET, a large district situated to the SE of the chain of Atlas, tributary to the empire of Marocco. It has the prov. of Draha on the SW, and Fez on the N; and consists of a vast plain, presenting an unvaried surface like the sea. It is traversed by two rivers running in opposite directions, one of which loses itself in the desert of Angad, the other in the loose sands of the Sahara. Water, though brackish, is everywhere to be found at the depth of 12 ft. Dates are its chief produce, though wheat and barley are cultivated on the banks of the rivers. There are mines of antimony and lead, which are carefully worked for the preparation of the composition called *kahol fileli*, used all over the East for blackening the eyebrows. The inhabitants possess numerous herds of sheep and goats, from the wool of which the women manufacture stuffs and carpets which are held in considerable estimation. Sigilmessa was once the most remarkable place in this region, but the town called Tafilelt or Tafilet, on the r. bank of the Zin, and 230 m. ESE of the city of Marocco, has supplanted it.

TAFT, a town of Persia, in the prov. of Fars, 21 m. SW of Yezd, on the Debala. Pop. 6,000.

TAFU, a town of Upper Guinea, on the Gold coast, 12 m. N of Coomassi.

TAFYLE, a town of Syria, in the pash. of Damascus, near the frontiers of Arabia, inhabited by about 3,000 Howalat Arabs.

TAGADEMPT, a village of Algiers, 106 m. SW of Algiers, round which are the remains of a very large city, supposed to be the ancient *Vaga*.

TAGAI, a town of Russia, in the gov. and 46 m. W of Simbirsk, on the Tagaika, an affluent of the Sviiaga. Pop. 1,200.

TAGAL. See Taggal.

TAGALUN, one of the Fox islands, in the N. Pacific, in N lat. 53° 30′.

TAGAMA, a district in the African desert, to the south of Fezzan, inhabited by Tuaricks, remarkable for the comparative whiteness of their complexion.

TAGANROG, a town of European Russia, in the gov. of Ekaterinoslav, near the N extremity of the sea of Azof, in N lat. 47° 13′. It is built on the side of a lofty promontory, nearly opposite the embouchure of the Don, and conducts through the medium of that river an extensive traffic. It has a harbour and a fortress, maritime and commercial courts, a naval hospital, and a lazaretto. It trades with Kertsh, Caffa, Odessa, Constantinople, Smyrna, and some Italian ports. Its exports are corn, Siberian iron, timber, pitch, tar, potash, linseed, sailcloth, wax, isinglass, leather, fish, and caviar; it imports Greek wines, fruit, and manufactured articles. The vessels which arrived at this port in 1846 were 1,287; among which were 49 English, 3 French, and 16 American. The chief exports during the same period were, 293,436 chetwerts of grain, 13,392 chetwerts of linseed, 3,090 chetwerts of rye, 21,257 poods of wool, and 125,263 poods of tallow. T. was fortified by Peter I. in 1697, but dismantled after the treaty of Pruth, and given up to the Turks, in whose possession it remained till 1768, when it was seized by Catharine II., to whom it was finally ceded in 1774. Alexander I. died here in December 1825. The environs are extremely fertile.—The bay of T. extends from Point Armianskaia to Cape T., 5 m. S of the former.

TAGAPOLA, a small island in the Philippine group, 25 m. W of Samar.

TAGARABUENA, a village of Spain, in the prov. and 18 m. ESE of Zamora.

TAGASA, or TAGASTA, a town of Marocco, in the prov. of Fez, situated on a river about 5 leagues from the Mediterranean, 20 m. N of Melilla.

TAGAVOST, a town of Marocco, in the prov. of Susa, on the l. bank of the Messa, 24 m. W of Talent.

TAGAZOUTE, a village of Algiers, 45 m. SE of Oran.

TAGERFELDEN, a village of Switzerland, in the cant. of Aargau, 3 m. S of Zurzach. Pop. 750.

TAGGAH, a village of Algiers, 50 m. SW of Constantina, in a fertile plain. In the vicinity are some ancient Roman ruins.

TAGGAL, or TEGAL, a town of the island of Java, on the N coast, in S lat. 6° 44', E long. 109° 12½', 2° 20' E of Batavia, prettily situated on a broad river. It has a church and a small fort, and presents altogether a very neat appearance. The resident's house is a commodious and very handsome building. The country around is fertile; and this part of Java is the rice granary not only for the supply of Batavia but for exportation to the East-India isles.—Also an island in the strait of Sunda, 15 m. NE of Mahitam.

TAGGART, an inhabited island of Lough-Strangford, co. Down, on the W side of the lough, nearly m. NNE of the town of Killyleagh. It measures m. in length from N to S.—Also an islet near the head of Clew bay, co. Mayo, 3⅜ m. WNW of Westport.

TAGGIA, a town of the Sardinian states, in the prov. of Genoa, 6 m. NE of San-Remo.

TAGHADOE, TEAGHDOE, or TAPTOO, a parish in Kildare, 2 m. SSW of Maynooth. Area 4,126 acres. Pop. in 1831, 467; in 1851, 377.

TAGHANIE, a township of Columba co., in the state of New York, U. S., 40 m. S of Albany. Pop.

TAGHATU, or TATAU, a river of Persian Armenia, one of the principal feeders of Lake Urumiyah. It is formed by the junction of two considerable streams, one flowing N from the mountains of Kurdistan, the other called the Saruk, flowing W from the Takht-i-Soleiman. These streams unite E of M...del, under the parallel of 10° 15' N, and the united stream pursues a NNW course to the extremity of the lake, into which it flows at a point to the SSW of Binab.

TAGHBOY, or TAUGHBOY, a parish 5⅜ m. S by E of Athleague, partly in co. Galway, but chiefly in co. Roscommon. Area 13,996 acres. Pop. 2,773.

TAGHEEN, or TAUGHEEN, a parish in co. Mayo, m. NNE of Hollymount. Area 6,837 acres. Pop. in 1831, 2,561; in 1851, 2,051.

TAGHKANIC MOUNTAINS, a range partly in the E part of the state of New York, U. S., and partly in Massachusetts, a spur of the Green mountains. It is about 50 m. in length, and attains an alt. of 3,000 ft. above sea-level.

TAGH KASI, a populous village of Asiatic Turkey, in the district of Kaisariyah, on a rivulet which flows from the N foot of Mount-Arjish to the Sasak.

TAGHMACONNEL, a parish in co. Roscommon, m. NNE of Ballinasloe. Area 18,826 acres, of which 215 are water. Pop. in 1851, 3,048.

TAGHMON, a parish in co. Westmeath, 5 m. NE of Mullingar. Area 3,452 acres. Pop. in 1831, 922; in 1851, 798. It contains the hamlets of Rathcloghrin, Taghmon-bridge, and Crookedwood.—Also a parish in co. Wexford. Area 10,132 acres. Pop. in 1831, 2,803; in 1851, 3,119. The market-town, and formerly a parl. borough of T., stands at the intersection of the road from Enniscorthy to Bannow, 6½ m. W by S of Wexford.

TAGIL, a river of Asiatic Russia, which rises about 40 m. NW of Yekaterinburg, on the E flank of the Urals; runs N, and then ENE, and joins the Tura on the r., 60 m. SE of Verkhoturi, after a course of 230 m.

TAGLIACOZZO, a town of the kingdom of Naples, in Abruzzo-Ultra 2da, 20 m. SSW of Aquila, near the r. bank of the Imele. Pop. 4,000. It has several fine churches and a ducal palace.

TAGLIAMENTO, a large river of Austrian Lombardy, in the gov. of Venice, which rises on Monte-Mauro, in the prov. of Udine; flows E past Tolmezzo; then turns S, and passing Spielenberg and Mandusco, flows into the Adriatic, in N lat. 45° 38', after a course of 120 m. It is navigable from the small town of Latisana. Its banks were the scene of military operations in the spring of 1797.

TAGLIO, a river of Italy, in the prov. of Genoa, which flows into the Mediterranean, 4 m. E of San Remo.—Also a town of Austrian Lombardy, 6 m. NE of Ariano.

TAGLO-BAY, a bay on the S coast of the island of Mindanao, in N lat. 6° 8'.

TAGOAT, a village in the p. of Roslare, co. Wexford, 1½ m. W of the nearest part of St. George's channel.

TAGOLANDA, a small island, about 20 m. in circumf., situated off the NE extremity of Celebes.

TAGRIVELT (RAS), or CAPE SIM, a low sandy point on the W coast of Africa, 8½ m. SW of Mogador. The coast between this point and Mogador is a continued line of bare sand-hills sloping from a height of 70 ft. to the beach.

TAGUS, the TAJO of the Spaniards, and TEJO of the Portuguese, a magnificent river which issues from a spring in the mountains of Albarracin, on the frontiers of the prov. of Cuença, in N lat. 40° 38', W long. 2° 46', at the distance of little more than 100 m. from the Mediterranean. Pursuing its course first NW, and then in a W direction, a little inclined to the S, it passes the palace of Aranjuez, the cities of Toledo, Talavera, and Alcantara in Spain, and Abrantes, Santarem, and Lisbon in Portugal, and flows into the sea, 7 m. below the capital of Portugal, in N lat. 38° 45', W long. 10° 2', after a course of 510 m., of which 375 m. are in Spain. Its principal Spanish affluents on the r. are the Jarama, the Guadarrama, the Alberche, the Tietar, and the Alagan; and in Portugal, the Elga, the Ponsal, and the Zezere. On the l. it is joined by the Guadiela, the Guadarranque, and the Solor, in Spain; and by the Sever, the Saraya, the Zatas, and the Cunha, in Portugal. Like the Guadalquivir, and other great rivers of Spain, it absorbs the waters collected between two long parallel chains of mountains. Long before reaching Lisbon it receives the tide, and becomes expanded into an estuary. Its volume of water is large throughout; but in so mountainous a country as Spain, river navigation is very limited, and that of the Tagus is not at present carried farther than Abrantes. It has often been proposed to clear it of its rocks, and to prolong the navigation eventually by canals; but as yet nothing of this nature has been decisively attempted.

TAGYPEEL, a small island in the Eastern seas, near the NE coast of Borneo, in N lat. 6° 29'.

TAGZA, a village of Algiers, 12 m. SSE of Constantina.

TAHA. See OTAHA.

TAHA-EL-MODAIN, a village of Egypt, 5 m. SW of Somalut.

TA-HEEN, a district and town of China, in the prov. of Sze-chuen and div. of Suy-ting-fu, in N lat. 31° 18′, E long. 107° 37′ 30″.

TAHEJ, a town of Hindostan, in the prov. of Cutch, of which, in the 16th cent., it was the capital, but it is now dwindled to a small village, in N lat. 23° 17′.

TAHIM, a parish of Brazil, in the prov. of São-Pedro-do-Rio-Grande, on a river of the same name, an outlet of Lake Mangueira, by which the p. is bordered in its entire extent from N to S. Pop. 900.

TAHIRAN. See Tehran.

TAHIRBEGY, a village of Afghanistan, 80 m. E of Merochak, and near the confines of Bokhara.

TAHITI, or Otaheite, an island of the S. Pacific, the principal in the Society archipelago, in S lat. 17° 40′, E long. 150° 30′, about 130 m. SE of Raiatea. It is an elongated high ridge of land, composing two peninsulas joined by a low narrow isthmus, each rising gradually from the shore to an alt. of 7,000 ft. above sea-level, and covered to the summit with the most luxuriant foliage. The one section is in length about 25 m., and nearly the same in breadth; the other is about 20 m. in length, by 15 m. in breadth. "From the great central peaks of the larger peninsula—Orohena, Aorai, and Pirohitee—the land radiates on all sides to the sea in sloping green ridges. Between these are broad and shadowy valleys, in aspect each a Tempe, watered with fine streams and thickly wooded. Unlike many of the other islands there extends nearly all round T. a belt of low alluvial soil, teeming with the richest vegetation. Here, chiefly, the natives dwell. Seen from the sea, the prospect is magnificent. It is one mass of shaded tints of green, from beach to mountain-top; endlessly diversified with valleys, ridges, glens, and cascades. Over the ridges here and there, the loftier peaks fling their shadows, and far down the valleys. At the head of these, the waterfalls flash out into the sunlight as if pouring through vertical bowers of verdure. Such enchantment, too, breathes over the whole, that it seems a fairy world, all fresh and blooming from the hand of the Creator. Upon a near approach the picture loses not its attractions. It is no exaggeration to say, that to a European of any sensibility who for the first time wanders back into these valleys, away from the haunts of the natives, the ineffable repose and beauty of the landscape is such that every object strikes him like something seen in a dream; and for a time he almost refuses to believe that scenes like these should have a commonplace existence. No wonder that the French bestowed upon the island the appellation of the New Cytheren. 'Ofton,' says De Bougainville, 'I thought I was walking in the garden of Eden.'"—[*Melville.*]

Climate, &c.] The climate is fine, and the soil is a fertile blackish mould, producing spontaneously a great variety of edible fruits, roots, and vegetables; of these the most common and most useful are the *artocarpus*, or bread-fruit, of which there are above fifty varieties, the cocoa-nut, the [illegible] thirteen sorts of bananas, the arum, [illegible], sugar-cane, and the paper-mulberry of which the inhabitants make their cloth. The guava, though only introduced about thirty years ago, now composes the greater part of the woodlands and bush in the lower lands. Dogs, hogs, and rats are numerous. Saddle-horses are imported from S. America; and the missionaries possess a few hundred head of cattle. The birds are of various kinds, and the natives rear the domestic poultry common in Europe; there are also wild ducks, turtle-doves of a green colour, large pigeons, small paroquets, king-fishers, cuckoos, and herons. Of the small birds which abound in the forests, some are remarkable for the melody of their notes. The ants are troublesome, but no species of serpent has yet been found; and frogs, toads, and scorpions, if they exist at all, are rare. The sea on the coast abounds with fish, among which is the sea-snake of which the bite is mortal.

Inhabitants.] Mr. Ellis estimated the pop. of this island at about 10,000; Mr. Williams, at from 18,000 to 20,000. There is no question but that depopulation has taken place here, as in the other islands, to a considerable extent; although for a season, under the renovating and genial influence of Christianity, it seemed to be rapidly increasing. The natives are well-made. Their mode of life—which affords them exercise, without subjecting them to severe labour—imparts to their bodies much of that strength and elegance which the human frame exhibits under good training. Some of the other inhabitants of the islands of the Pacific, and among the rest those of the Friendly islands, may be more hardy and robust, and may display a greater appearance of muscular strength; but none of them seem so generally to unite that strength and elegance of shape which is often seen in the natives of T. Their colour is tawny,—of a shade much lighter than that of the inhabitants of the Friendly islands; sometimes, particularly in those of the higher ranks, it approaches to what we call fair, but the cheeks never display that glow by which the northern Europeans are distinguished. The hair is generally black, but sometimes brown, red or flaxen, and frizzled, but never woolly. The eyes are black, the nose flat, the mouth large, and the teeth white and even. The women are distinguished from the men by greater delicacy of form; and their skin is delicate and soft, which is procured by their frequent bathing, and copious use of the oil of the cocoa. The chiefs—like the mandarins in China—used to be distinguished by long nails. The practice of tatooing was nowhere more prevalent than in T.; and the custom was common to both sexes. The dress of the Otaheitans is formed of a kind of cloth resembling paper, made of the bark of certain trees, particularly of the paper-mulberry; but as this substance cannot long resist the rain, the garments formed of it are thrown aside during wet weather, and their place is supplied by others formed of a kind of matting. The missionaries introduced the culture of the cotton-plant, and taught the natives to manufacture a coarse cotton-cloth, and dress in a becoming manner. The climate of this island renders houses almost superfluous, and as the inhabitants require little shelter, they have been proportionally careless in the construction of their habitations. These generally consist only of a shed or roof, resembling the roof of a barn in Europe, supported by three rows of pillars, one in the centre, and one on each side. They are thatched with palm-leaves; and the floor is strewed with hay and covered with mats. A few blocks of wood serve as stools by day, and by night as pillows. The food commonly consists of the bread-fruit, bananas, plantains, yams, apples, and a sort of sour fruit, which is used as a relish to the roasted bread-fruit. The most general dish is *popoi*, which is made of the mountain plantain beaten up to a paste or jelly, with cocoa-nut milk. Every kind of fish is eagerly devoured. Their small canoes are formed of a single tree, and carry from 2 to 6 men. Their larger canoes are formed of several planks sewed together, and are either fitted with an outrigger, or two of them are fixed together with planks. The seams are calked with rushes, and the whole is covered with a gummy substance, which serves the purpose of pitch.

Civil condition.] A rivalry of proselytism has long existed between the Protestant and Catholic missionaries for upwards of ten years in the South seas. In 1827 two Catholic missionaries landed for the first time in the Sandwich islands. They found the country ruled by Protestant missionaries: the laws, both political and religious, the administration, commerce, taxes, sales, cultivation, all emanated from them or their schoolmasters. Nevertheless, the progress of Catholicism was rapid, and MM. Bachelot and Short were, in spite of the energetic protests of the English and American consuls, transported at the end of 1831 to the coast of California. Hearing of a change in the government of those isles in consequence of the death of the Queen Kaahumanu, these two priests attempted a second visit in 1837; but had scarcely set foot on shore at Woahu than they were obliged once more to embark on board their vessel, and retrace their voyage to California. These events are nearly an exact counterpart of what recently took place at T. About the end of 1836 MM. Laval and Caret left the isle of Mangavera and arrived at T. They were admitted into the presence of Queen Pomare, and received with kindness and condescension, but were soon after ordered to take their departure. The two strangers refused to depart, and placed themselves under the protection of the American consul; but were ultimately, by order of the queen, taken by force on board an English vessel, and conducted to Valparaiso. This act was resented by the French government, which eventually assumed the protectorate of the island. The French seem to have introduced many stringent municipal regulations into the government of the island; and, it is to be feared, their presence has not contributed to the improvement of Tahitian morals. The pop. of T., which was between 200,000 and 300,000 at the time of Cook's discovery, has dwindled, it is said, to 6,000; and we believe on the best authority that the rate of decrease, during the last five years, is in enormous excess of all previous periods.

Commerce.] The foreign trade of the island is confined to the exportation of pearl-shell, pearls, sugar, cocoa-nut oil, and arrow-root, and is entirely in the hands of Americans and Europeans. The port-dues, however, and trade in supplying shipping, especially the English and American whale-ships, yield the natives considerable emolument.—T. offers every facility for provisioning ships. The harbour is beautiful, and a ship can lie alongside of the dock and be supplied with water in a very few hours. The French authorities are extremely anxious for the establishment of a line of steamers. The following is the time which a well-furnished steamer would take to reach England from Sydney in Australia, by way of T:—From Sydney to T., 11 days, 12 hours; from T. to Panama, 16 days; to cross the isthmus, 7 hours; from Aspinwall to England, 18 days; total, 45 days, 19 hours, with 3 days' detention in coaling at T. Returning it would take 40 days, with 3 days' detention at T. in coaling. T. may be regarded as under the full influence of the SE trade-winds.

TAHLEQUAH, the capital of the Cherokee Indians, in the state of Arkansas, U. S., 18 m. E of Fort-Gibson, and 53 m. NW of Van Buren, on the Arkansas Illinois river. It became the seat of government of the Cherokee nation in 1839, after the reunion of the eastern and western branches of the Cherokee family. The location of the town is central and beautiful, and combines advantages of good health, excellent spring water, and a plentiful supply of timber for firewood and purposes of building. The surrounding country presents a diversity of mountain, woodland, and prairie scenery. After T. became the seat of government, a number of log cabins were thrown up about the place, without, however, much regard to order, as they were designed for the temporary accommodation of those engaged in the transaction of public business; but a regular town has been laid off, a number of lots sold to citizens of the nation, and a few houses have been erected, and others are in contemplation. The supreme court holds its annual session in a commodious brick court-house.

TAHMUR, a town of Hindostan, in the prov. of Oude, 30 m. NE of Keyr-abad.

TAHOUATA. See OHITAHOU.

TAHRAH, a town and fortress of Hindostan, in the prov. of Kutch, 12 m. SE of Luckput-Bunder. The fortress is an irregular building, flanked with round towers. On the N and E of it are extensive basins of water, which render it difficult of access on these points. The town itself is quite undefended. The inhabitants are chiefly Hindus.

TA-HU, or TAI-HO, [*i.e.* 'Great Lake,'] a lake of China, between the provinces of Keang-su and Chekeang. It is 150 m. in circumf., and is enclosed by a chain of picturesque mountains. It is a place of resort to the inhabitants of the adjacent towns. Its waters abound with fish. The great canal passes, at a point to the N of Kia-hing and W of Song-kiang, through a broad sheet of water of considerable extent which probably is part of—or at least joins—this celebrated lake. "This part of the lake is very shallow; and much of it is quite covered with the *Trapa bicornis*—a plant called *ling* by the Chinese. It produces a fruit of a very peculiar shape, resembling the head and horns of a bullock more than anything else,—and is highly esteemed as a vegetable by the Chinese in all parts of the empire. I have seen," adds Mr. Fortune, "three very distinct species or varieties of this plant,—one of which has fruit of a beautiful red colour. Women and boys were sailing about on all parts of the lake in tubs of the same size and form as our common washing-tub, gathering the fruit of the *ling*. I do not know of any contrivance, rude as it appeared, which would have answered the purpose better than these tubs; for they held the fruit as it was gathered, as well as the gatherers, and at the same time were easily propelled through the masses of the *ling* without doing the slightest injury to the plants. Nevertheless, the sight of a number of people swimming about on the lake each in his washing-tub had, in my eyes, something very amusing and ludicrous about it. After leaving this lake, the banks of the canal—and indeed the greater part of the country—are covered with the mulberry. Silk is evidently the staple production in this part of China."

TAHURA, a small island of the N. Pacific, in the group of the Sandwich islands, to the SW of the island of Oni-hao, in N lat. 21° 37′, and W long. 160° 30′.

TAHURAWI, one of the smallest of the Sandwich islands, in the N. Pacific, 9 m. SW of the island of Mawi, in N lat. 20° 30′, and W long. 156° 35′.

TAHYS, two lakes of Brazil, in the prov. of Rio-de-Janeiro, between S. Salvador and S. Joao-de-Praya, near the r. bank of the Parahiba. The largest is 3 m. in length.

TAIABANA, an extensive parish of Brazil, in the prov. of Parahiba and district of Pilar, on the r. bank of the Rio-Parahiba. Pop. 1,400. The locality is noted for its cotton.

TAIABATE, a river of Brazil, in the prov. of São-Paulo, which joins the Parahiba a little above Jacarehi.

TAIARBU, a peninsula dependent on Tahiti.

TAIBAREISKI, a tribe of the Samoïdes, in Russia in Europe, in the gov. of Arkhangel and district of Mezen, to the S of the island of Varander.

TAIDSI, a district of Japan, in the island of Nifon, and prov. of Monts.

TAIFARA, a district of Japan, in the island of Nifon and prov. of Tutomi.

TAIFE. See TEFFE.

TAIKANY, a district of Japan, in the island of Nifon and prov. of Sangami.

TAILLADE, a village of France, in the dep. of the Vaucluse, cant. and 3 m. E of Cavaillon, at the foot of Mont Leberon. Pop. 320. It has several madder-mills and iron-works.

TAILLANT (LE), a village of France, in the dep. of the Gironde, cant. and 2 m. WSW of Blanquefort, on a hill. Pop. 820.

TAILLEBOURG, a commune of France, in the dep. of the Charente-Inferieure, cant. and 4 m. SSE of St. Savinien, partly on the summit, and partly on the slope of a hill, crowned with the ruins of an ancient castle, and the base of which is washed by the Charente. Pop. 1,106. This town was formerly fortified. It is noted for a victory over the English won by St. Louis in 1242.

TAILLES, a department and commune of Belgium, in the prov. of Luxembourg and arrond. of [illegible]nfchateau. Pop. of dep. 446; of com. 334.

[illegible]LLEVILLE, a village of France, in [illegible] the Calvados, cant. and 2 m. NW of Douvres. [illegible] It has the remains of an ancient [illegible]

[illegible]-LE-HAMEAU, a village of France, in the dep. ot the Ardennes, cant. and 6 m. E of Buzancy. Pop. 587. It has a blast-furnace, several lime-kilns, an oil-mill, and a distillery.

TAIMBURNY, a town of Hindostan, in the prov. of Aurungabad and district of Solopur, 105 m. ESE of Poona.

TAIMELA, a tribe who inhabit the vicinity of the coast of Abyssinia, near the Danakil or Dankali territory.

TAIMYR, a gulf of the Arctic ocean, on the coast of Russia in Asia, in the gov. of Jeniseisk and district of Touroukhansk, between Capes Sievére-Vostochnoi and Taïmour. It is 210 m. in length from N to S, and 60 m. in medium breadth.—Also a lake in the same gov. It is several leagues in length, and discharges itself into the gulf of T. by a river of the same name, which traverses its W part, and has a considerable volume of water far above the point where it enters the lake.

TAIN, a parish on the N border of Ross-shire; bounded on the N by the Dornoch frith. The surface consists of three well-defined districts; a belt of low flat plain along the coast, a broad sheet of land, of middle character, between a terrace and a hanging plain, receding from bank or escarpment of about 50 ft. above the level of the plain; and a ridge or series of gentle uplands along the exterior frontier, sending up their loftiest summit in the hill of Tain to an alt. of nearly 800 ft. above sea-level. The fishing village of Inver, with a pop. of about 150, stands in the extreme E, 4 m. from the burgh of T. Pop. of p. in 1831, 3,078; in 1851, 3,570.

TAIN, a small town, about ¾ m. from the Dornoch frith, in the centre of the above parish, 1¾ m. N by E of Invergordon, and 47 m. N by E of Inverness. Pop. in 1851, 2,588. The only establishments connected in any sense with manufacture, are an iron-foundry, a brewery, and mills respectively for grinding, sawing, carding, and dyeing. Some trade by sea is conducted in the import of coal and lime, and the export of fir-props for coal-pits; but, as T. enjoys no better facility than a dry berth for vessels on the broad belt of sand between tide-marks, it very generally gives place to a land-communication with Cromarty and Invergordon. T. unites with Wick, Dingwall, Cromarty, Dornoch, and Kirkwall, in sending a member to parliament. Constituency in 1850, 86.

TAIN, a canton, commune, and town of France, in the dep. of the Drome and arrond. of Valence. The cant. comprises 10 com. Pop. in 1831, 11,008; in 1846, 12,512. The town is 11 m. NNW of Valence, on the l. bank of the Rhône, opposite Tournon, with which it is connected by an iron suspension-bridge. Pop. in 1846, 2,541. It lies at the foot of Hermitage hill, so named from a hermitage on its summit, noted for the wines to which it gives its name, and for the finest granite in France. It has several silk-mills, and carries on an active trade also in wool and almonds. The plain between Tain and the Isere is noted for the victory of Q. Fabius over the Allobroges.

TAINIERES-SUR-HON, a commune of France, in the dep. of the Nord, cant. and 2 m. NE of Bavay. Pop. 1,296. It has a distillery of gin, a brewery, and a marble-work. In this com. is the hamlet of Malplaquet, noted for the battle to which it gave its name, fought on the 11th Sept. 1709.

TAINTIGNIES, a department and commune of Belgium, in the prov. of Hainault and dep. of Tournai. Pop. 2,160.

TAINTREUX, a village of France, in the dep. of the Vosges, cant. and 4 m. SW of Diey, in a valley of the Vosges, near the r. bank of the Taintroné. Pop. 1,250.

TAINTRONE', a river of France, in the dep. of the Vosges, which has its source to the S of Taintreux; runs NNE, and after a course of 9 m. throws itself into the Meurth, on the l. bank, 1½ m. below St. Diey.

TAIPALE, a town of Russia in Europe, in Finland, in the gov. and 60 m. ESE of Kuopio, and district of Oefre-Karelens, on the S coast of a considerable lake.

TAIPU, a village of Brazil, in the prov. of Parahiba, 3 m. E of Pilar. Pop. 1,200.—Also a headland of the island of Engua-Guaçu, which with Point-Manduba, to the S of the island of Guahíbe, or Santo Amaro, forms the entrance to the bay of Santos. Its summit is in S lat. 24° 1′ 11″.

TAIPU. See ITAIPU.

TAISERO, a town of Japan, in the island of Kiusiu, and prov. of Tsikousen, 30 m. NE of Sanga. Pop. 2,500.

TAITZA. See JAITZE.

TAIVERAM, a town of Hindostan, in the Carnatic, 117 m. N of Cape Comorin.

TAIZE-AIZIE, a village of France, in the dep. of the Charente, cant. and 3 m. NE of Ruffec, on the r. bank of the Charente, opposite the confluence of the Bioussac. Pop. 824. It has several iron-works.

TAJACANTES, or TAKANTES, a Moorish tribe who inhabit the SW of the Sahara.

TAJAHY, or TAJUCA. See ITAJAHI.

TAJARDI, a town of Turkey in Europe, in Rumelia, in the sanj. of Gallipoli, 13 m. NNW of Ghumurdjina, in a narrow valley, on the r. bank of the Karatch, at the confluence of the Slino.

TAJO, or TAYO, a mountain of Brazil, on the confines of the provinces of São-Paulo and Santa Catherina, in S lat. 26° 50′, and W long. 50° 10′. It gives rise to the Itajahi, and contains gold mines.

TAJO. See TAGUS.

TAJO, a river of the island of Luzon, in the Philippine archipelago, which runs from S to N, and after a course of about 180 m., throws itself into the ocean near New Segovia.

TAJOVE, a village of Hungary, in the comitat of Sohl, 5 m. W of Neusohl, at the foot of a mountain. It has a copper foundry.

TAJPUR, a town of Hindostan, in the prov. of Malwa, 8 m. from Oojein. Pop. in 1820, 2,500.

TAJUMULEV, a town of Guatimala, in the dep. of Guezaltenango and Seconusco. Pop. 2,566.

TAJUNA, a river of Spain, which has its source in the mountains of Solorio, in the prov. of Guadalajara; passes Brihuega, and Loranca-de-Tajuna; waters the SE part of the prov. of Madrid, and joins the Jarama, on the l. bank, at a point 21 m. SSE of the capital.

TAJURAH, or TAJURRA, a town on the Adel coast of Eastern Africa, in N lat. 11° 46′ 35″, E long. 43° 0′ 20″ [*Harris*], on the N shore of an extensive bay from 32 to 34 leagues long, and 6 or 7 wide. This bay, whose entrance is obstructed by an infinite number of small islands, is studded throughout its whole extent with reefs at 7, 5, and even 3 ft. only below the surface of the water, which causes it to be a very dangerous harbour; besides which, it is exposed to impetuous winds from W by S, and N by W. The islands of Mushakh in this bay have been acquired by the British government from the sultan of T. The NW end of the NE island is in N lat. 11° 43′, E long. 43° 19′. The town or village of T., at the bottom of the channel, consists of about 300 wretched wooden huts, and 2 mosques, with a pop. of nearly 1,500. The spot itself and its environs is a scene of aridity and desolation. It conducts some trade with Hodeidah, Mokha, and Aden, to which it sends wood, sheep, goats, and butter. From T. to the kingdom of Shoa is 129 leagues, which can be traversed only when the rains have filled the natural reservoirs met with on the road: at any other season the traveller runs the risk of dying of thirst. The vast desert which forms the country of Adel is an upheaved volcanic formation, rarely susceptible of cultivation, and still more rarely cultivated. It is traversed in all directions by chains of hills of moderate height, bearing the impress of subterranean fire. The quantity of lava met with is prodigious. About 21 leagues from T. there is a lake, which formerly was part of the bay of Tajurah, but which is now separated from it by a valley 4 leagues long and 9 leagues wide. The lava occurring here is of various character, and in some places is from 130 to 140 ft. thick. From T. to Shoa M. Rochet met with 23 thermal springs, whose temp. varied from 151·25 F. to the boiling point. In the interior of the country, to the distance of 50 leagues from the sea, there is a stratum of siliceo-calcareous clay, containing an immense quantity of fossil shells. From T. to the banks of the Hawash the desert surface of the country is wandered over by various animals, antelopes, gazelles, wild asses, ostriches, and pintadoes; but the animal met with in the greatest numbers is the spotted hyena. The vegetation is very limited; there are a few gum-bearing trees and aloes. The Dankali language is spoken at T.

TAK, a district of the Derajat, on the W side of the Indus; bounded by the Buttunni branch of the Suliman range on the N; Puharpur on the E; Kolachi on the S; and the Viziri and Shirani hills on the W. It is irrigated by the Zam and the Gomul. Its town and fort of the same name are 40 m. direct distance NW of Dera-Ismail-Khan. The fort is an immense pile about 250 yds. square, with walls 4 yds. thick, and a ditch 5 yds. broad. The town in 1847, contained about 300 Hindu and Mahommedan families, but was fast falling into decay. It is famed for its fruits. It yielded a revenue of 100,000 rupees to the Sikhs.—Also a town of Chinese Turkestan, in N lat. 36° 48′, and E long. 82° 42′, 3 m. E of a small river which has its source in Mount Khera-Tak, 24 m. S of the town.

TAK (JEBEL), a mountain chain on the frontiers of Turkey in Asia, and Persia. It runs from NE to SW, between the pash. of Bagdad on the W, and the Persian prov. of Kurdistan on the E, and joins the Bisuton on the NE. This chain of mountains is crossed by a pass of the same name, said to have been used by the Romans. It is cut out of the solid rock, 12 ft. in breadth, and at the height of 20 ft. is surmounted by an arch. The entrance is ornamented with pilasters of architectural form.

TAKA, a district of Japan, in the island of Nifon, and prov. of Farima.—Also a district in the prov. of Simosa.—Also a town in the island of Sikokf, and prov. of Sanoki, on the Mi-tsou-sima-nada channel.

TAKA, a prov. of Nubia, to the E of the Tacazze, between 16° and 17° N lat. Near Soderab there is a narrow pass in the mountains whence in the rainy season issues a large volume of water, by which a considerable proportion of the prov. is regularly inundated to the depth of several feet. The soil is generally very fertile. It affords excellent dourra, and pastures large herds of cattle.

TAKABOKOUSIMA, or POPENBERG, an island of Nagasaki bay, in Japan, near the coast of the island of Kiu-siu.

TA-KAE-SINE-SHAN, a mountain of China, in the prov. of Sze-chuen, in N lat. 30° 27′, E long. 101° 45′. It is covered with perpetual snow.

TAKAGA-VOUROUNG, a mountain in the E part of the island of Java. Its prolongation forms Cape Sedano, and presents a series of perpendicular pillars of basaltic breccia.

TAKAI-IZ, a district of Japan, in the island of Nifon and prov. of Yamatto.

TAKAKAKKAN, a small island of the Celebes sea, near the E coast of the island of Borneo, in N lat. 3° 8′, E long. 116° 51′.

TAKAKI, a district of Japan, in the island of Kiu-siu, and prov. of Satsoumi.

TAKAKIDA, a district of Japan, in the island of Nifon and prov. of Yetsisen.

TAKAKOU, a district of Japan, in the island of Kiu-siu and prov. of Fizen.

TAKAKOUSA, a town of Japan, in the island of Nifon and prov. of Inaba.

TAKALA, TAKEL, or TAKLE, a district of Nigritia, in Kordofan, 100 m. SSE of Obeit. It is inhabited by a tribe of Mahommedan Nubas, who have no form of government. Maize is its only production, and forms the chief sustenance of the inhabitants. They suffer much from the ravages of the agents of the king of Sennaar, by whom numbers are annually carried off into slavery.

TAKAMIJA, a town of Japan, in the island of Nifon and prov. of Aki.

TAKAMIYA, a district of Japan, in the island of Nifon and prov. of Aki.

TAKANO, a district of Japan, in the island of Nifon and prov. of Mouts.

TAKAREE, a pass of Beluchistan, across the mountains of the same name, in the Hala range, running from E to W, between Cutch-Gundava and Saravan.

TAKASANGO, a town of Japan, in the island of Nifon and prov. of Nizen, on a bay at the mouth of a river. Pop. 2,000.

TAKASSIMA, a district of Japan, in the island of Nifon and prov. of Omi.

TAKATA, a town of Japan, in the island of Nifon and prov. of Idsu.

TAKATTA, a district of Japan, in the island of Nifon and prov. of Aki.

TAKATU, a river of Guayana, which rises in N

lat. 1° 50′, 19 m. W of Pirara; runs N between the Rio-Branco and the Rupununi; and joins the Mahu in N lat. 3° 35′. Its width a little above the point of junction is 192 yds. It abounds in kaimans.

TAKAY, a town of Japan, in the island of Nifon and prov. of Sinano.

TAKDE, a district of Nigritia, in Kordofan, about 90 m. S of Obeit. It is inhabited by a tribe of Mahommedan Nubas.

TAKEAH, a village of Bhawlpur, on the l. bank of the Ghera, 45 m. NE of Bhawlpur.

TA-KEANG, a river of China, in the prov. of Kwang-tung and island of Hai-nan. It has its source in a mountain named Wei-chi-shan; runs N till it reaches a mountain named Na-Khian-ling, to the SE of which it divides into two arms, of which that to the r. bears the name of Gung-kao-keang; runs N, and falls into the bay of Thun-tsu-keang. The other, retaining the name of Ta-keang, takes a W course; forms two islands; passes to the N of Tan-chu, and throws itself into the bay of Gang-fu-keang.

TA-KEANG, or Hung-shui-keang, a river of China, which has its source in the E part of the prov. of Yun-nan; traverses that of Kwang-se, and in the dep. of Wei-chu, at the junction of the Kwei-keang takes the name of Se-keang. A little below it enters the prov. of Kwang-tung; flows to the W of Canton; and after a course of upwards of 600 m., joins the Pe-keang to form the Chu-keang. Its principal affluents are the Nyo-yu-keang and Keen-keang on the r.; and on the l. the Lieu-keang and Kwei-keang.

TAKEEA, a village of Afghanistan, 30 m. NE of Ghuzni.

TAKEL. See Takala.

TAKELEY, a parish in Essex, 14 m. NW of Chelmsford. Area 3,154 acres. Pop. in 1851, 991.

TAKENO, a district of Japan, in the island of Kiu-siu and prov. of Tsikougo.

TAKERN, a lake of Sweden, in the prefecture of Linkoeping, communicating by a river with Lake Wetter.

TAKEYSON, a town of Caffraria, in the Betjuanas territory, on an affluent of the Gariep.

TAKHTALU-DAGH, a mountain of Turkey in Asia, in the S part of Anatolia, on the confines of the sanj. of Meis and Teke-eli, near the W coast of the gulf of Satalleh, and to the SW of Ighidler. Its naked summit rises to the height of 7,700 Parisian ft. above sea-level. The base, consisting of fragile rock, is broken with deep and irregular ravines.

TAKHT-I-SULIMAN, a summit of the Suliman range, in Afghanistan, in N lat. 31° 35′, rising to an alt. of 12,000 ft. above sea-level.

TAKHT-I-SULIMAN, an elevated mountain of Persia, in N lat. 36° 30′, E long. 47° 10′. Its summit is crowned by a grey hoary mass of crumbling walls and buildings, which Major Rawlinson is inclined to identify with the *Ecbatana* of Herodotus, the capital of *Media Atropatene*, the *Gaza* of some later geographers. The hill appears at first as if it were isolated, but this is not strictly the case. The brow of the hill is crowned by a wall, the most perfect part of which is along the S face, and the most ruinous upon the W. There are the remains of 37 bastions, and the circuit of the wall, measured from point to point of these bastions, is 1,330 paces, or a little more than ¾ m. At a few points only is the line of wall perfect; but where it is, the masonry is shown to be most excellent. Above the gateway, and extending from one bastion to the other, are a line of blocks, each carved with a rude representation of an arch, which thus form a sort of ornamental frieze to the portal, and offer the only specimen of ancient sculpture to be found upon the walls. Passing through the gateway, Major Rawlinson says, "I found myself within the precincts of the deserted city; the first object that attracted my attention was the lake. I found this to be an expanse of water on the highest point of the hill, irregularly shaped, and about 300 paces in circuit; the rocky banks that surround it are formed of a deposit of carbonate of lime, of which the water holds vast quantities in solution, and there can be no doubt but that they are daily narrowing as the calcareous deposit continues; a very short distance from the surface they recede inwards, thus forming a huge incurvated basin for the lake. Sir R. K. Porter states his belief, that the hill has been formed entirely by deposition from the water, and this, in very remote antiquity, would seem to have been the case, for the depth of the water, recently determined by repeated experiments of the Afshar chief at 47 Persian yards, agrees, as near as possible, with the height of the hill, ascertained by myself."

TAKHT-SULIMAN, or Och, a town of Independent Tartary, in the khanate and 195 m. ESE of Kokan. It is governed by a hakim, is well peopled, and is supplied with good water. This town is noted for the tomb of Azef-Barkhia, vizier of Suliman. It is of great dimensions. The throne of Suliman, a low hill to the W, which gives name to the place, is surmounted by a building with a dome. Large numbers of pilgrims from the surrounding districts resort hither in the spring to visit this tomb, and bring with them various kinds of merchandise for sale and barter. In the hot season the town is much infested with musquitoes.

TAKI, a district of Japan, in the island of Nifon and prov. of Mino.—Also a district in the prov. of Tango.

TAKIAN, a village of Turkey in Asia, in the sanj. of Bija, on the coast opposite the island of Tenedos. In its vicinity are the ruins of *Alexandria Troas*, the modern Eski-Stambul.

TAKIMA, a kingdom of Upper Guinea, on the Gold coast, tributary to the state of Ashanti, by which it is bounded on the S. The capital, which bears the same name, is 102 m. N of Kumassi.

TAKINOS, a lake of Turkey in Europe, in Rumelia, in the beglik and 6 m. S of Seres, and 3 m. from the gulf of Orphano. It forms an irregular oval, and is 18 m. from NW to SE, and 5 m. in extreme breadth. It receives several rivers, amongst others the Kara-su, which enters at the NW extremity, and issues on the SE to throw itself into the sea. Except on the N and NE, in which direction it extends into the great marshy plain of Seres, it is bordered on all sides by mountains. On the W bank is a small village of the same name, inhabited entirely by Greeks.

TAKIPUR, a village of Kashmir, 20 m. N of Sopur, on the road thence to Solab. It has a thermal spring.

TAKKAL, a village of Afghanistan, in the district and 6 m. W of Peshawur, on the road through the Khyber pass.

TAK-KUM-TA, a halting-place in Abyssinia, on the confines of the kingdom of Tigre, and of the gov. of Coast districts, and at the foot of Mount Taranta. It is one of the most picturesque localities in the country, and abounds with water.

TAKLI (Cape), a headland of Russia in Europe, in the Crimea, in the district of Kertch, and at the W side of the entrance to the strait of that name, in N lat. 45° 6′ 22″, E long. 36° 27′ 29″, 4 m. NE of Cape Opuk. It has a light-house. About ½ m. WNW is a village named Takel.

TAKRURI, a town of Nubia, about 18 m. from the l. bank of the Tacazzé

TAK-RUSTAN, or TAK-KHOSRU, a mountain of Persia, in Kurdistan, to the E of Kermanshah. It rises precipitously to a considerable height, and exhibits numerous sculptures in bas relief, and several excavations which appear to have been the ancient sepulchres of Persian kings and heroes. In the latter, which are difficult of access, urns and antique sarcophagi have been found. The mountain derives its name from Rostan or Rustan, the *Hercules* of the Orientals.

TA-KUNG-TING, a district and town of China, in the prov. of Kwei-chu and div. of Chin-yuen-fu.

TA-KWAN-TING, a district and town of China, in the prov. of Yun-nan and div. of Tung-chuen-fu.

TAL, a river of Afghanistan, which has its source in the Laspissor mountains; flows SW; and after a course of about 120 m., joins the Panjkore or Lundye. On its l. bank, 25 m. NNE of Panjkore, is a village of the same name.

TALA, a town of India beyond the Ganges, in the prov. and 45 m. SE of Aracan, near the gulf of Bengal.

TALA, a lake of New South Wales, in the district of Murrumbidgee, and near the river of that name.

TALABO, a headland of the island of Celebes, on the E coast, in S lat. 0° 15′, E long. 123° 57′.

TALABOR, or TEREBLA, a river of Hungary, in the comitat of Marmaros, formed by the junction of the Raditna and Ozorna-Reka, which descend from the S side of the Carpathian chain. It runs first WSW, then SE, and afterwards in a generally SSW direction; divides into two branches, distinguished as Kis-Talabor and Nagy-Talabor; and after a total course of about 45 m., joins the Theiss on the r. bank, a little to the WNW of Tecso.

TALACH-DDU, a parish in Breconshire, 4 m. NE of Brecon. Area 1,818 acres. Pop. 187.

TALACRE, a small port in the p. of Llanasaph, co. Flint, within 1 m. of the point of Ayr, at the mouth of the estuary of the Dee.

TALAFRU, a village of the Punjab, on the road from Doda to Chumba.

TALAGIR, a small island of the Philippine archipelago, 27 m. W of Samar.

TALAHIGUA, a town of New Granada, in the dep. of the Magdalena, prov. and 27 m. NNW of Mompox, on the l. bank of the Magdalena, a little above the embouchure of the Cauca.

TALAI-HYI, a town of Mongolia, in N lat. 44° 19′ 12″, E long. 121° 15′ 40″.

TALAIN-PYI. See TALONG.

TALAK, a town of India beyond the Ganges, in the prov. and 70 m. SSE of Aracan, on a small river of the same name. It consisted in 1827 of about 100 huts, occupied chiefly by Burmese, who carried on an active trade in cotton, yarn, khut, silk fabrics, bees-wax, and ivory, which they bartered for British piece-goods, tobacco, betel-nuts, guapee, and balachang.—About 4 m. NE of this town is Phoongee-Dong, a lofty mountain summit, over which the pass runs by which the Burmese army invaded Aracan in 1825. The surrounding country is unhealthy but fertile. It abounds with elephants, and produces tobacco, cotton, ginger, and pepper in great abundance.

TALALHO-KARA-PALGASIN, a town of the Chinese empire, in the Khalkas country, in N lat. 47° 32′ 24″, and E long. 106° 6′.

TA-LA-MA, or TARAMA, an island of the Eastern sea, in the Majicosima group, to the NE of the island of Formosa.

TALAMAN, or HALAMON, a village of Asia Minor, 10 m. N of Cape Artemisium, about 4 m. E of a river of the same name, which Mr. Hoskyn supposes may be the ancient *Indus*.

TALAMANCA, a town of Spain, in the prov. and 30 m. NNE of Madrid, and partido of Colmen-ar-Viejo, in a flat district, near the l. bank of the Jarama. Pop. 490. It is enclosed by old walls.

TALAMONA, a village of Austria, in Lombardy, in the prov. of the Valteline, district and 1 m. E of Morbegno, near the l. bank of the Adda. Pop. 2,657.

TALAMONE, a town of Tuscany, in the prov. of Sienna, podesteria and 9 m. NNW of Orbitello, on the Tyrrhenian sea.

TALANCE, a village of France, in the dep. of the Gironde, cant. and 2 m. SSW of Bordeaux. Pop. 1,489.

TALANDA, TALANTI, or ATALANTI (STRAIT OF), an arm of the Grecian Archipelago, extending between the dioceses of Locri and Thebes, in the NE part of Livadia and the island of Eubœa. On the NW it forms the gulf of Zeitoun, and opens thence by the channel of Trikeri into the Archipelago. On the SE it is connected by the strait of Egripo with the gulf of Negropont. Its total length from NW to SE is about 48 m.; in breadth it varies from 3 to 15 m. On the W side are several considerable bays, the principal of which bears the same name. The banks are bordered with lofty hills, from which numerous torrents descend to the channel. This channel is the *Euripus* of the ancients. In the NW part of the gulf of T. is a small island of the same name.

TALANDA, a town in the diocese of Locri, 18 m. NNE of Livadia, on the N of Mount Talanda or Klomo, and 4 m. from the shore of the gulf of the same name. Pop. 1,000. It is defended on the S by a castle, and has considerable trade. This town is the *Opus* of the ancients.

TALANGE, a village of France, in the dep. of the Moselle, cant. and 8 m. N of Metz, near the l. bank of the Moselle. Pop 534. It has manufactories of beet-root sugar.

TA-LANG-TING, a district and town of China, in the prov. of Yun-nan and div. of Tsinunh-fu.

TALANT, a village of France, in the dep. of the Cote-d'Or, cant. and 1½ m. WNW of Dijon, on the summit of a mountain. Pop. 580. This place was formerly a town, the mayor of which had the right of entrance to the states of Burgundy. It had a fortress belonging to the dukes of B., in which the Duke de Bar was held prisoner in 1431, and which was demolished by Henry IV.

TALARN, a town of Spain, in the prov. and 45 m. NNE of Lerida, and partido of Tremp, on the r. bank of the Noguera-Pallaresa. Pop. 1,450. It is the residence of a civil and military governor, and has a convent and an hospital. It has manufactories of woollen fabrics, and carries on an active trade in butter, cheese, and meat.

TALARUBIAS, a town of Spain, in the prov. and 66 m. E of Badajoz, and partido of Herrera-del-Duque, in a fertile plain, near the l. bank of the Guadiana. Pop. 2,690. It has several churches and convents, an hospital, a custom-house, and a tannery. It has extensive manufactories of woollen fabrics.

TALAS, a river of Independent Tartary, in the khanat of Kokan, which has its sources in the Alatau mountains; runs WNW, and after a course of 300 m. falls into Lake Sikirlik. On the N its bed is separated from that of the Tchui or Tzui, by the Archatou or Kubakai mountains.

TALASANI, a town of Corsica, in the cant. of Tavagna, 23 m. S of Bastia. Pop. 453.

TALAVAN, a town of Spain, in the prov. and 21 m. NNE of Caceres, and partido of Garrovillas, on an inclined plain, 1½ m. from the l. bank of the Tagus. Pop. 1,160. It has a parish-church, a custom-house, and several convents.

TALAVERA-DE-LA-REINA, an ancient town of Spain, in the prov. of Toledo, 63 m. WSW of Madrid, on the r. bank of the Tagus, over which there is here a fine but dilapidated bridge of 35 arches, 1,200 ft. in length. It stands in a fertile plain of vast extent, intersected by the river. Part of the old ramparts are still in preservation, but the town is, in a military sense, altogether an open place. It contains several well-built churches, in particular that of the Hieronymites; has two public walks, one on the N, the other on the S of the town; but nothing can be poorer than the general appearance of the place. The dwelling-houses are seldom more than one story in height; the streets are badly paved, and crossed by a number of narrow lanes; the pavement is wretched, and the whole place is full of pools after a heavy fall of rain. Its pop., including that of the suburbs, is about 6,400; and it is the seat of several provincial *bureaux*, has four hospitals, a classical and a theological school. Silk manufactures, established about the year 1748, are still carried on in the town, and in the neighbouring village of Cervera, as well as manufactures of soap, hats, and earthen-ware.—T. is a place of great antiquity, and contains many Roman monuments. It is the *Talabriga* of ancient times; but some identify it with the ancient *Ebora*, *Libora*, or *Elbora*. It received the adjunct of La-Reina from having been bestowed by Alonzo XI. on his queen, Dona Maria. It fell into the hands of the Moors in 714, continued long in their possession, and when the Christians succeeded in re-occupying the inland provinces, was repeatedly taken and retaken. In the beginning of the 12th cent., it was sacked by the Moors, and its walls demolished. This place has given birth to several men of eminence, of whom the best known is Mariana the historian. In the present age it is memorable for a battle fought on 27th and 28th July 1809, between a French army, amounting to 34,000 men, under Victor, Jourdan, and Joseph, and an allied force, in which there were 16,000 British, and between 30,000 and 40,000 Spaniards. The French, after making several desperate attacks on the British position, were repulsed, with the loss of 20 cannon and 10,000 men.

TALAVERA-LA-REAL, a town of Spain, in the prov. and partido and 9 m. ESE of Badajoz, on the l. bank of the Guadiana. Pop. 2,820. It is tolerably well-built, and has a convent, an hospital, and manufactories of silk, cord, and leather.

TALAVERA-LA-VIEGA, a town of Spain, in the prov. and 51 m. NE of Caceres, and partido of Navalmoral-de-la-Mata, on the l. bank of the Tagus, in a fertile locality. Pop. 450. This town is supposed by some to mark the site of the *Elvora* or *Eura* of the ancients. It contains the ruins of two temples, and some other fine monuments of antiquity.

TALAVO, a canton of Corsica, at the SE extremity of the arrond. of Ajaccio. Pop. 4,335. Zicavo is its chief place.

TALAWAH, a mountain of New South Wales, in the county of Gloucester, to the N of the Wollomba river.

TALAYUELA, a town of Spain, in the prov. and 48 m. NE of Caceres, and partido of Navalmoral-de-la-Mata. Pop. 290.

TALBAILA, a town of Spain, in the prov. and 29 m. WNW of Ivria, and partido of Burgo-de-Osma, in the midst of pine-forests. Pop. 226.

TALBENNY, a parish in the co. of Pembroke, 7 m. WSW of Haverford-West. Area 1,425 acres. Pop. in 1801, 145; in 1831, 222; in 1851, 235.

TALBERT (POINTE DE), a cape on the coast of Brittany, in the English channel, in N lat. 48° 52′.

TALBERT'S ISLAND, a small island in the Atlantic, on the coast of Georgia, U. S., in N lat. 30° 44′.

TALBINGO, a mountain of New South Wales, in the district of Murrumbidgee, on the E bank of the Doomut.

TALBOLTON, a village of Talbot co., in the state of Georgia, U. S., 73 m. WSW of Milledgeville. Pop. in 1840, 800; in 1850, 600.

TALBOT, a river of Western Australia, in the co. of Gork, which flows SE, and joins the Dale river, on the l. bank.—Also a mountain of Australia, in the district of Wimmera, between the river of that name and the Glenelg.

TALBOT, a county in the W part of the state of Georgia, U. S., comprising an area of 457 sq. m., drained by branches of Flint river, by which it is bounded on the E and NE. The surface is hilly, but the soil is fertile, and is well adapted to the culture of cotton. It is intersected by the Muscogee railway. Pop. in 1840, 15,627. Its capital is Talbottan.—Also a county in the NE part of the state of Maryland, comprising an area of 336 sq. m., drained by several affluents of Choptank river, by which it is bounded on the E, and of Chesapeak bay. On the NW it is bounded by the Wye. Pop. in 1840, 12,090; in 1850, 13,811. Its capital is Easton.

TALBOT DISTRICT, a large district of Upper Canada, bounded by the Niagara and Gore districts on the NE, and Lake Erie on the S, and watered by Big-creek and Otter-creek. Its cap. is Simcoe.

TALBRAGAR, or POOLABATTA, a river of New South Wales, which has its source in the co. of Bligh, near Turee; flows along the confines of the county of Bligh, and after receiving the Coolaburragundy, joins the Erskine, an affluent of the Macquarie.

TALCA, or ST. AUGUSTINE, a town of Chili, in the prov. of Maule, of which it is the cap., in S lat. 35° 13′, on the r. bank of the Maule, 193 m. NNE of La Concepcion. Its pop. is considerable, owing to the mines of gold in the adjacent mountains, and to the low price of provisions. It contains a parish-church, two monasteries, and a college built by the Jesuits. To the NE of this place is a small hill consisting almost entirely of amethysts; and in its vicinity is another hill which furnishes a species of cement known by the name of Talca sand.

TALCAGUANA (PUNTA), a projection of land on the coast of Chili, which bounds the bay of Concepcion towards the W.

TALCAGUANO, or TALCAHUANO, a port of Chili, within the bay of La Concepcion, in S lat. 36° 42′, W long. 73° 10′. It is much frequented by small vessels, being completely sheltered from the N winds. A steamer has recently been built in the harbour of T., and finished these three months, destined for the navigation of the Itata. All the agricultural product of Maule, Nuble, and even of Concepcion, will be brought by this steamer to T. at a saving of 300 per cent. upon the present transportation. Large quantities of grain, which now remain in the interior of the country for want of means of transportation, will be brought to the mills at a price not exceeding 12½ cents per fanega, or a little less than 6 cents per bushel.

TALCAN, a town of Independent Tartary, 100 m. SE of Termed.

TA-LE-FU, a division and town of China, in the prov. of Yun-nan. The div. comprises 7 districts. The town is 171 m. WNW of Yun-nan-fu, on the W bank of Lake Shang-koen, in N lat. 25° 44′ 24″, E long. 100° 21′ 50″. It is large and populous, and is noted for its manufacture of ornaments of the fine jasper marble, with which the adjacent mountains abound.

TALGARTH, a parish and village in co. Brecon, 8 m. NE by E of Brecknock. The p. includes the hamlets of Forest, Groyne-Vawr, Groyne-Vechan, Pwll-y-Wrach, and Trevecca. Area 16,900 acres. Pop. in 1831, 1,473; in 1851, 1,325.

TALGAU, or THALGAU, a town of the archd. of Austria, 9 m. E of Salzburg, on the l. bank of the Fushl. Pop. 2,000.

TALIANGIERI (CAPE), a headland on the coast of the Black sea, in N lat. 41° 33′.

TALISCAYAN, a settlement of Mexico, in the prov. of Vera-Cruz, 24 m. S of Vera-Cruz.

TALISHIN, a khanate of Asiatic Russia, in the prov. of Shirvan. Its chief place is Astara. It produces silk, cotton, rice, tobacco, and wine.

TALIDJ MOUNTAINS, a chain in Asiatic Russia, in the prov. of Shirvan, which runs NW and SE, and is connected with the Elburz.

TALISSE, a small island in the Eastern seas, near the N coast of the island of Celebes, in N lat. 1° 40′.

TALKAN, a village of Persia, in the prov. of Irak, 30 m. E of Sultaniyeh.

TALKHAN. See TALCAN.

TALK-O'-TH'-HILL, a chapelry in Audley p., co. of Stafford, 5 m. NNW of Newcastle-under-Lyne. Pop. in 1831, 1,196; in 1851, 1,979.

TALKIN, a township in Hayton p., Cumberland, 11 m. E by N of Carlisle. Pop. in 1831, 376; in 1851, 311. There are several collieries here, and quarries of slate and free-stone on the river Gelt.

TALLA, a village of Tuscany, in the prov. of Florence, 7 m. SSE of Poppi.

TALLAGHT, a parish in co. Dublin, containing the villages of Greenhills and T. Area 21,868 acres. Pop. in 1831, 4,646; in 1851, 4,367. It is watered along the E side by the rivulet Dodder. The principal summits, together with their respective alts. above sea-level, are Kippure, 2,475 ft.; Slieve-Bane, 2,128 ft.; Mount-Seskin, 1,049 ft.; Killakee-hill, 1,271 ft.; and the hill of Tallaght, 1,263 ft.

TALLAND, a parish in Cornwall, on the English channel, 2 m. SW by W of East Looe. Area 2,665 acres. Pop. in 1831, 1,434; in 1851, 1,605.

TALLANSTOWN, a parish in co. Louth, 2 m. S of Louth. Area 3,210 acres. Pop. in 1851, 637.

TALLANTINE, a hamlet of Cumberland, 3 m. NNW of Cocker-mouth.

TALLARD, a canton, commune, and town of France, in the dep. of the Hautes-Alpes. The cant. comprises 9 com. Pop. in 1831, 4,899; in 1846, 4,812. The town is on the r. bank of the Durance, 7 m. S of Gap. Pop. 1,140.

TALLAPOOSA, or OAKFUSKEE, a river of the United States, which rises in Georgia; enters Alabama; flows SW, and unites with the Coosa 3 m. SW of Fort-Jackson, to form the Alabama. It is navigable, except in dry seasons, to the great falls, about 35 m.

TALLATON, a parish in Devon, 3½ m. NW by N of Ottery. Area 2,365 acres. Pop. in 1851, 443.

TALLEGHANY, a town of New South Wales, in the co. of Gloucester, on the banks of the Karuah.

TALLENTIRE, a township in Bride-Kirk p., Cumberland, 3 m. NW by N of Cockermouth. Area 1,913 acres. Pop. in 1831, 237; in 1851, 250.

TALLERATH, TELLEROUGHT, or TULLYRATH, a parish of co. Wexford, 3¾ m. SSE of New Ross. Area 1,653 acres. Pop. in 1831, 468; in 1851, 323.

TALLEY (LOWER and UPPER), a parish in Carmarthenshire, 7 m. N of Llandilo-Fawr. Area 7,167 acres. Pop. in 1831, 1,058; in 1851, 1,005.

TALLEYRAND-BAY, a bay of Australia Felix, to the SW of Port Philip, between Capes Petton and Albany-Otway.

TALLIKO, a town of Senegambia, in the kingdom of Futatoro, and 81 m. SE of Sedo.

TALLINGTON, a parish in Lincolnshire, 3 m. WSW of Market-Deeping, on the N bank of the Welland. Area 690 acres. Pop. in 1851, 267.

TALLMADGE, a township of Ottawa co., in the state of Michigan, U. S., on Grand river, 64 m. W by N of Lansing. Pop. in 1840, 139; in 1850, 534.—Also a township of Summit co., in the state of Ohio, 109 m. NE of Columbus, intersected by the Akron branch railroad and the Pennsylvania and Ohio canal. Pop. in 1840, 2,134; in 1850, 2,456.

TALLOIRES, a town of Sardinia, in the div. of Savoy, prov. of Genevois, mand. and 6 m. SE of Annecy, finely situated on the E bank of the lake of Annecy. Pop. 1,115. A noble convent, an hospital, and an infirmary were founded here in 1020, by Rodolph I., king of Burgundy.

TALLOW, a parish, containing a town of the same name, in co. Waterford. Area 5,026 acres. Pop. in 1831, 4,716; in 1851, 3,884.—The town, formerly a parliamentary borough, stands on the r. bank of the river Bride, 4½ m. SW by S of Lismore. The Bride is navigable to the vicinity for barges of 40 tons. Pop. in 1831, 2,998; in 1851, 1,986.

TALLOW-BRIDGE, a village in the p. of Lismore and Mocollop, co. Waterford, on the l. bank of the river Bride, a ½ m. NNE of Tallow.

TALLU, a town of the island of Celebes, in the kingdom and 21 m. from Macassar. The inhabitants manufacture cotton fabrics, white and coloured, which the Dutch export to the Molucca and Philippine islands and to the coasts of India. Fabrics of raw silk, in imitation of those of China, are also manufactured here.—Also a port of the Society islands, on the N coast of the island of Eimeo, in S lat. 17° 30′, and W long. 150°. It is 3 m. in length, and 2 m. in breadth, and has a considerable depth of water, but a reef at the entrance renders it difficult of ingress.

TALLULAH, a river of the state of Georgia, U. S., which has its source in Rabun co., in the NE corner of the state; flows SE, and joins the Chatuga, one of the two head-streams of the Savannah.

TALLUYERS, a village of France, in the dep. of the Rhône, cant. and 3 m. NE of Mornant. Pop. 558.

TALLVIG, a village of Norway, in the diocese of Nordland, in Finmark, on the Alten-fiord, 54 m. SSW of Hammerfest.

TALL-WATER, a rivulet of co. Armagh, which rises near Hamilton's-Bawn; and flows N and W to the river Callan, a short distance above its confluence with the Blackwater. Its course is about 9 m.

TALLYA, a town of Hungary, in the comitat and 32 m. SW of Zemplin, at the foot of the Sâtor, near the l. bank of the Kacsonta. Pop. 3,700. It has a castle, and two churches, a Catholic and a Reformed. It is noted for its wine.

TALMAS, a commune of France, in the dep. of the Somme, cant. and 10 m. ESE of Domart. Pop. in 1841, 1,912. The houses consist of only a single story, are ill-built, and covered with thatch.

TALMAY, a commune and town of France, in the dep. of the Côte-d'Or, cant. and 4 m. NNE of Pontaeillier, on the Vingeanne. Pop. 1,230.

TALMONT, a canton, commune, and town of France, in the dep. of the Vendee, and arrond. Les Sables. The cant. comprises 9 com. Pop. in 1831, 9,498; in 1846, 10,750. The town is 9 m. ESE of Les Sables-d'Olonne, on a hill, near the Guy-Chatenay. Pop. in 1846, 3,363. It is small and ill built. In the environs are several salt marshes. A benedictine abbey was founded here by William-the-Bald in 1046.—Also a town in the dep. of the Charente-Inferieure, cant. and 5 m. SW of Coxe, in a peninsula, on the r. bank of the Gironde. Pop. 500. It has a small port, and carries on an active trade in wine. This town was formerly fortified and defended by a citadel flanked with towers, the ruins of which still exist.

TALNERE, a fortress of Hindostan, in the prov. of Candeish, on the r. bank of the Tapti, 90 m. W of Burhanpur. This place was the ancient capital of the sultans of the dynasty of Adil Shah. At the dissolution of the Mogul empire it fell into the hands of the Maharattas, and was ceded by the Holcar family to the British.

TALOKOVI-GORY, a range of mountains in Russia in Asia, in the gov. of Irkutsk and district of Kirensk.

TALONG, TALAIN-PYR, or PEGU-PROPER, one of the three provinces of the kingdom of Pegu, in the Burman empire, of which it forms the NE part. Its capital is Pegu. See article PEGU.

TALOOLY, a town of New South Wales, in the co. of Bligh, on Bylong creek.

TALORO, a river of Sardinia, in the div. of Cape Cagliari, formed in the prov. of Nuoro by several streams, amongst which is the Gavoi, which descends from the NW side of the mountains of Corno-di-Bue. It directs its course first SE, then W; receives on the l. the Pietre-fitte and the Tiana; enters the prov. of Busachi; waters the plain of Ottana; and, after a course of about 29 m., throws itself into the Oristano, on the l. bank, opposite Sedilo, and 10 m. NNE of Busachi.

TALOUI, a town of Russia in Europe, in the gov. of Voronetz, district and 24 m. WSW of Bogutshar, on a river of that name.

TALOVSKAIA, a town of Russia in Asia, in the gov. of Orenburg, district and 120 m. ENE of Tcheliabinsk, in the vicinity of several small lakes.

TALSEN, a town of Russia in Europe, in the gov. of Kourland, 32 m. NW of Toukoum.

TALT (LOUGH), a lake in the p. of Kilmacteigue, co. Sligo, among the Ox mountains, 6¾ m. WNW of Tobbercurry. It extends a little upwards of 1 m. SE, and has a mean breadth of about 3 furl. Its surface elevation above sea-level is 455 ft.

TALVENDE-LE-GRAND, a commune of France, in the dep. of the Calvados, cant. and 2 m. W of Vire. Pop. in 1846, 3,284.

TALVENDE-LE-PETIT, or ST. GERMAIN-DE-TALUANDE, a village of France, in the dep. of the Calvados, cant. and 3 m. S of Vire, on a plateau. Pop. 3,152.

TALYLLYNN, a parish in Merionethshire, 6 m. S by W of Dolgelly. Area 15,182 acres. Pop. in 1831, 767; in 1851, 1,123. The steep cliffs of Cader-Idris add much to the surrounding scenery.

TALYRANG-PEAK, a summit of New South Wales, in the co. of Murray, in the Gurock range.

TALYSCH. See TALISHIN.

TAM (EL), a town of Sistan, on the Helmund, 25 m. E of Zareng.

TAMA, a small river of Brazil, in the prov. of Para, which runs NNW, and enters the arm of the Amazon which forms the island of Marayo.

TAMA, a central county of Iowa, U. S., intersected by the Iowa river. Area 720 sq. m. Pop. in 1850, 8.

TAMAGA, or TAMEGA, a river of Spain, which rises in the Sierra-de-San-Mamed, in the prov. and 25 m. SE of Orense; flows S through the valley of Laza, and then SSW; enters Portugal, and joins the Doura, on the r. bank, 12 m. SSW of Amarante, after a course of 110 m., of which about one-third is in Spain.

TAMAHU, a small island in the Eastern seas, near the W coast of Borneo, in N lat. 0° 7′.

TAMALAMEQUE, a town of New Grenada, in the prov. of Santa-Marta, situated on the r. bank of the Magdalena, 158 m. S of Santa-Marta.

TAMAMES, a town of Spain, in the prov. and 32 m. SSW of Salamanca. Pop. 780. There are mineral springs in the vicinity.

TAMAN, an island of European Russia, in the gov. of Taurida, formed by the straits of Kertch, separating it from the E extremity of the Crimea, the sea of Azof on the N, and the Black sea on the S. On the E it is connected with the Caucasian provinces by a narrow strip of land, but the large lagunes of Ak-tanez and Kuban nearly complete its isolation by water on that side also. It is of very irregular form, being deeply indented on the W side by Taman bay; and on the E by the two lagunes and their offsets. Its extreme N point, Cape Kamennoi, is in N lat. 45° 26′, E long. 36° 54′. Its S point, Point Kishla, is in N lat. 45° 8′, E long. 36° 43′. Its extreme length from Peresip, on the NE, to Panagia, on the SW, is 30 m. Measured along the parallel of Cape Kamennoi it is 22 m. in length. Part of it lies low, but it contains also a number of small hills from which volcanic eruptions occasionally take place. In 1853, M. Begitschef, a Russian *savant*, while crossing over the strait of T. from Kertch, early in the morning of the 6th of August, saw a flame accompanied by a thick vapour rising from the summit of the mountain of Korabetoff, situated about 3 m. from the town of T. In a few minutes the column of fire and smoke had reached a great height, and remained in that state for five or six minutes. Two other eruptions followed at short intervals, but with less violence than the first. This volcano had remained tranquil for at least 35 years. Immediately on landing, M. Begitschef proceeded towards the mountain, which he reached at 10 o'clock. At about 700 yds. from the foot of it, he found a mass of black mud spread on the ground to a considerable depth, which had been thrown out by the eruption. The crater made a whistling noise, similar to that made by an engine when letting off its steam. The ground was cracked in different directions, and round the fissures the grass appeared perfectly calcined, and in several parts had caught fire. The eruptions lasted altogether about three hours. Wood is scarce, but pasturage is abundant throughout the island. The inhabitants were formerly Crim-Tartars, but in 1793 a Cossack tribe settled here, and continue to occupy the island. Before the Russians took possession of it, it was called ZMUTARAKAN. It is now frequently called FANAGORIA, from its chief town, or fortress rather, which is situated on the S coast of Taman bay, 1 m. E of the village of Taman.

TAMANDALAHI, a river of Brazil, in the prov. of São-Paulo, which flows near the capital, and with the Hinhangabahu, throws itself into the Rio Tiete.

TAMANDARE', a river of Brazil, in the prov. of Pernambuco, which, after a short course ESE, throws itself into the Atlantic, 30 m. SSW of Cape Augustine, by an embouchure which forms one of the best harbours on the coast. It has a depth at the entrance of from 4 to 5 fath. to 18 fath. It is defended by a fort.

TAMANDUA, a town of Brazil, in the prov. of Minas-Geraes, and comarca of Rio-Grande, 90 m. WNW of Villa-Rica, at the foot of the Sierra-Nigra, between two of the head-streams of the Lambari. It has a church and three chapels. The district comprises three parishes. Pop. 8,000. Agriculture and mining form the chief objects of local industry.

TAMAR, a river of England, which rises in the NW part of Devonshire, on the borders of Cornwall, about 3 m. from the sea; and on being joined by the Tavy, and passing near Saltash, forms the noble harbour of Hamoaze, below Plymouth, falling into the sea at Mount-Edgecombe.

TAMAQUE, a village of Schuylkill co., in Pennsylvania, U. S., 79 m. NE of Harrisburg. Pop. in 1840, 465; in 1850, 600. It is situated in a wild mountainous district, rich in anthracite coal.

TAMAR, a river of Van Diemen's Land, formed by the union of the North and South Esk, at Launceston, about 30 m. above Port Dalrymple, into which the united stream flows through a deep and narrow valley. It has more the appearance of a chain of lakes, than of a regularly formed river; and such, according to Flinders, it probably was, until, by long undermining, a passage was forced out to the sea.

TAMAR (CAPE), the NW point of a large aby and harbour on the N shore of the straits of Magalhaens, in S lat. 52° 51′. The SE point of this bay is Cape Providence.

TAMARA, a village of Marocco, on the coast of the Atlantic, 30 m. W of Tarodant.

TAMARA, or TAMARIDA, a port on the NE part of the island of Socotra, in the Indian ocean.

TAMARA, a settlement of New Granada, in the prov. of San-Juan-de-los-Llanos, at the foot of the mountains of Bogota, 126 m. NE of Santa-Fe, on a small affluent of the Meta.

TAMARA, a town of Spain, in the prov. and 15 m. NNE of Palencia.

TAMARACA. See ITAMARACA.

TAMARARE, or TAMARE, a tribe of Indians of Brazil, who inhabit the strait between the Sao-Simao and Jamari rivers, affluents of the Guapore, in the prov. of Matto-Grosso. Fishing and the chase form their chief means of subsistence.

TAMAREVA, a village of St. Clair co., in Illinois, U. S., 114 m. S by W of Springfield, on the Kaskaskia.

TAMARITE, a town of Spain, 20 m ESE of Balbastro, in the prov. of Huesca. Pop. 3,200.

TAMARIZ, a village of Spain, in the prov. and 24 m. NNW of Valladolid, near the l. bank of the Rio-Seco. Pop. 450.

TAMARO, a small river of New Granada, in the prov. of Maracaybo, which rises near the sea-coast, runs S, and flows into Lake Maracaybo.

TAMARON, a village of Spain, in the prov. and 15 m. WSW of Burgos.

TAMAROVKA, a town of Russia, in the gov. of Kursk, 15 m. WNW of Belgorod.

TAMARUGAL (PAMPA-DE), a flat tract from 3,000 to 3,500 ft. above sea-level, stretching N and S at the foot of the Andes in Bolivia, between the arid mountains of the coast and the outlying ranges of the E slope of the Cordillera, running N into the prov. of Arica, and S into the desert of Atacama. It has an average width of 30 m. It takes its name from the *tamaruga* or tamarisk tree, which grows wherever water reaches the pampa from the ravines on the E. Vast deposits of nitrate of soda occur on the W side of the pampa, commencing immediately where the level plain ceases, and on the sides of some of the ravines running from the pampa towards the coast, and in some of the hollows of the mountains. The nitrate has not been found nearer to the coast than 18 m., and looks as if it gradually transferred itself into salt as it approached the coast. The deposits commence about Tiliviche, and extend S near to Quilliagua, with interruptions of deposits of common salt. The nitrate caliche grounds vary in breadth; the average may be 500 yds., and is in some places 7 to 8 ft. thick. In the ravines and hollows the nitrate is found on their shelving sides; the hollows look like dried-up lakes, and are covered with salt from 2 to 3 ft. thick. On the margins of these deposits there is nitrate of soda, often going down to some depth; in other cases there is a hard crust upon it occasionally of 4 ft. thick. The nitrate caliche found under this crust is in thin layers, and very solid and pure. There are several varieties of the nitrate of soda, and the pampa of T. contains sufficient nitrate for the consumption of Europe for ages. On the practical question, how to obtain the produce of this fruitful mine at a cheap rate, Mr. Pusey says, "The existence of cubic saltpetre in Tarapaca has been known in Europe about a century, but none was sent to England until 1820; guano had been known for 200 years to be accumulated in the same parched neighbourhood, yet did not reach Europe till some years later. In 1820, however, as Mr. Bollaert informs us, some nitrate was sent to England, but the duty being too high, it was thrown overboard. Ten years afterwards, in 1830, a cargo was sent to the United States, but found unsaleable there; a part of it was therefore forwarded to Liverpool, but returned unsaleable from Liverpool also. Such is the risk of dealing in a new article; yet in the year following another cargo sold in England for £35 per ton; and up to 1850, 239,860 tons of the nitrate were exported from the port of Iquique alone, making a return of towards £5,000,000. The market-rate has since settled down to £17 or £16 per ton, but even at that figure the price, owing to the excessive cost of production, must be greatly too high. For, according to Mr. Darwin, the chief expense in producing it is the transfer from the quarry to the sea-coast. Now the distance as the crow flies is not more than 10 m., and by the circuitous track (road there is none, even for this large traffic), that traveller reached the works from the port, though mounted upon a mule, in a single day. The nitrate also is brought down on the backs of mules, and Mr. Darwin found the desert interval strewed with the bones and dried skins of the many beasts of burden which had perished on it from fatigue. The only living animal was the vulture, which preys on the carcases." Mr. Pusey suggests that instead of taking English coals to the nitre-pits in Peru, the nitrates might be brought to this country quite fit for agricultural purposes without being refined at all, and might be sold at £6 per ton.

TAMASI, a town of Hungary, 70 m. SSW of Pest, and 15 m. WSW of Simontornya.

TAMATAM, an island of the Pacific, in the archipelago of the Carolina, in N lat. 7° 32′.

TAMATAVE, a town on the E coast of Madagascar, 90 m. SSW of Saint Mary's island. It has a pop. of about 2,000, and is surrounded by palisades. It was taken possession of by the French in 1829.

TAMAULIPA. See SANTANDER.

TAMAULIPAS. See TAMPICO.

TAMBACH, a village of Bavarian Franconia, on the borders of Saxony, 4 m. W of Coburg.—Also a village in the principality of Saxe-Coburg-Gotha, 10 m. NE of Schmalcalden, on the Apfelstedt.

TAMBACUNDA, a town of Western Africa, in the country of Neola, 52 m. W of Baniserile.—Also a town of Wulli, in Western Africa, 30 m. ENE of Medina.

TAMBAH, a town of Hindostan, in the prov. of Bejapur, in N lat. 17° 28′.

TAMBAOHU, a village of Brazil, in the prov. of Parahiba, on the shore of the Atlantic, 6 m. E of Parahiba. It has an hospital and a convent.

TAMBAROURA, a creek in New South Wales, between the Turon and the Pyramul, and parallel to both, which disembogues itself into the Macquarie several miles below the junction of the Turon. This place has lately taken an important position among the diggings for richness and extent. The diggings are situated chiefly on table-land.

TAMBEKAN, a town of Hindostan, in the prov. of Nepal, in N lat. 27° 25′, celebrated for the copper mines in its vicinity.—Near to it is a celebrated pass through the mountains, which commands the road leading to the coast.

TAMBERTCHERY, a town of Hindostan, in the prov. of Malabar, 18 m. NE of Calicut. The country in the vicinity of this place produces fine teak timber, and abundance of cocoa-nuts.

TAMBI, or ITAMBI, a parish of Brazil, in the prov. of Rio-de-Janeiro and district of Santo-Antonio-de-Sa, near the r. bank of the Aldeia. It has a church substantially built of stone. It produces sugar, rice, millet, and mandioc.

TAMBO, a town of Peru, in the prov. of Calca, situated on the r. bank of the Quillabamba, 45 m. NNW of Cusco. In the vicinity is a mountain, on which are the remains of a fortress said to have belonged to the Incas. It is built of large wrought stones beautifully fitted together. It has its bulwarks, gate, and small squares, arranged with singular art; and is entered by long, wide flights of steps, with several landing-places. At a small distance from this fortress are two strong towers, which may have served as advanced posts.—Also a settlement of Chili, in Coquimbo, near the source of the river Choapa.—Also a town of Bolivia, in the prov. of Atacama, on the shore of a small river which enters the Quillabamba.—Also a river of Peru, which, rising on the W flank of the Andes, in S lat. 16° 50′, runs SSE to Moquehua, and enters the Pacific ocean, in the bay of Quilca, after a course of about 100 m.—It is also the name of several other inconsiderable settlements in South America.

TAMBOR, a settlement of New Granada, in the prov. of Popayan, 13 m. W of Popayan.

TAMBORETES, a group of islets of Brazil, in the N part of the prov. of Santa-Catharina, and near the island of São-Francisco.

TAMBOV, or TAMBOF, a large province in the central part of European Russia, lying between the parallels of 51° 15′, and 55° 10′ N; and bounded by the govs. of Vladomir and Nijni-Novgorod on the N; by Penza and Saratov on the E; by Voronetz on the S; and by Orel, Tula, and Riazan on the W. Its territorial extent is 58,170 sq. versts. Its pop. in 1817 was returned at 1,135,000; in 1846, 1,750,900, Russians, Tartars, and Mordwines. Like the greater part of European Russia, it is level; the whole country from the river Khoper to the borders of Saratov, is occupied by extensive steppes. Many tracts are woody and marshy, but in general a sandy soil pervades the N, and a black and fertile mould the S. It is well watered both by rivers and small lakes. The Oka intersects it on the NW, and here receives the Moksha; in the S, the Vorona, the Savala, and the Takai, are tributary to the Khoper; the W is traversed by the Don and the Voronetz. Winter has here all the severity of the Russian climate, but in summer the heat is such that the Polish cochineal and the Spanish fly are common. The mineral products are iron, sulphur, and saltpetre. The prov. is divided into 12 circles.

TAMBOV, the capital of the above prov., is situated on the river Zna, in N lat. 52° 44′, E long. 41° 45′, 286 m. SE of Moscow. Pop. 10,700. It is the see of a Greek bishop, and has a large monastery in which are two churches, besides 14 other churches or chapels. It has manufactories of woollen, canvass, linen, and alum, and a considerable carrying trade; but the chief employment of the inhabitants is agriculture. An extensive fair is annually held in the vicinity of this town on an extensive steppe or down, about ¾ m. off. On this down a village, at the period of the fair, is erected of wooden booths, in which shops are opened for the sale of all kinds of goods, especially every article necessary for winter-clothing. A number of fire-engines are stationed round the booths, to be useful not only in the event of fire, but as assistants to the police in keeping order, since any engine playing into the midst of a mob speedily disperses the crowd.

TAMBRE, a river of Spain, which rises near Sobrado, in the mountains of Coruna; flows WSW; and enters the Atlantic at Noya, after a course of nearly 60 m.

TAMBUCO, or TABUCO, a town on the E coast of the island of Celebes, situated in a bay, to which it gives name, in S lat. 3° 50′.

TAME, a river of England, which rises near Winslow, in Buckinghamshire, and runs into the Thames below Dorchester.—Also a river in Staffordshire, which rises near Dudley; flows SE; and falls into the Trent, near Eadinghall, about 7 m. above Burton.

TAME, a settlement of New Granada, in the prov. of San-Juan-de-los-Llanos, 58 m. S of Pampeluna, on an affluent of the Casanara.

TAMEGA. See TAMAGA.

TAMERTON (NORTH), a parish in Cornwall, 8 m. N by W of Launceston. Area 5,261 acres. Pop. in 1831, 517; in 1851, 516.

TAMERTON-FOLLIOTT, a parish on a creek of the river Tamar, in Devonshire, 4 m. N by W of Plymouth. Area 5,150 acres. Pop. in 1851, 1,147.

TAMERVILLE, a village of France, in the dep. of La Manche, 2 m. NNE of Valognes. Pop. 1,240.

TAMIAGUA, a village of Mexico, in the state of Vera-Cruz, situated on a narrow isthmus on the W bank of the lake or lagune of the same name, 68 m. SE of Tampico, at the point where a channel of about 1 m. in width discharges a portion of the waters of the lagune over the bar of Tanquijo. It is built in a straggling manner, and presents only adobe houses. In 1830 its pop. did not exceed 200.

TAMIAGUA (LAGUNA DE), a lake of Mexico, in the state of Vera-Cruz, which joins the sea between the river Tampico and the Punta-Delgada, under the parallel of 21° 50′ N. It is about 60 m. in length, and 26 m. in greatest breadth. It has a long and narrow outlet towards the S, running parallel with the shore. Its waters are fresh and clear, and abound in fish and alligators.

TAMIEH, a village of the prov. of Fayum, in Egypt, on a canal which forms a communication between the Nile and the Birket-el-Karun, 12 m. NE of Fayum.

TAMIHAK, a town of India, in the Punjab, in N lat. 33° 17′, between Attock and Rhotas.

TA-MING-FU, a division and town of China, in the prov. of Chih-le. The div. comprises 7 districts. The town is 270 m. SSW of Pe-king, in N lat. 36° 21′ 4″, E long. 115° 22′. It is situated in a fertile locality, and is well-peopled.

TAMISE, or THEMESCHE, a town of Belgium, in the prov. of E. Flanders, on the Scheldt, 8 m. N of Dendermonde. Pop. 5,000. It has manufactories of linen and lace, and enjoys, like most other towns in this level country, the advantages of inland navigation.

TAMLAGHT, a parish 3½ m. SSE of Moneymore, partly in co. Londonderry, partly in co. Tyrone. Area 4,954 acres. Pop. in 1831, 2,854;

in 1851, 2,489.—Also a village in the p. of Tamlaght-O'Crilly, co. Londonderry, 3¾ m. SSW of Kilrea. Pop. in 1831, 188; in 1841, 211.

TAMLAGHTFINLAGAN, a parish in co. Londonderry, containing the villages of Ballykelly, Crindle, and Moys. Area 17,402 acres. Pop. 5,647.

TAMLAGHT-O'CRILLY, a parish, partly in co. Coleraine, but chiefly in co. Londonderry, containing the villages of Tamlaght and Innisrush, and part of the town of Portglenone. Area of the Coleraine section, 1,334 acres; of the Londonderry section, 15,504 acres. Pop. of the whole, 8,301.

TAMMARO, or TAMARO, a small river of Naples, in the prov. of Molise. It rises near Campobasso; runs E, then SSE, S, and SW; and falls into the Calore after a course of 50 m.

TAMMERFORS, a town of European Russia, in the gov. of Finland, 42 m. NNW of Tavasthus. Pop. 600.

TAMOYOS, a tribe of Indians, who formerly inhabited the coast of Brazil from Cape Frio to the prov. of Sao-Paulo, and who resisted the efforts of the Jesuits to establish themselves in their territory.

TAMPA-BAY, a large bay in the gulf of Mexico, which sets up NE from the gulf into Hillsboro' co., in the state of Florida, U. S. It is about 40 m. in length, and is in one place 35 m. in breadth. It has from 15 to 20 ft. water on the bar, and affords safe anchorage. It receives five small creeks, of which the Hillsboro' and the Manata are the principal.—Little T. bay is an elliptical inlet about 10 m. in axis, in the N part of the larger bay.

TAMPICO, a town of Mexico, in the state of Vera-Cruz, near the sea, on a rising ground between the river Panuco and the Laguna-del-Carpentero, about 214 m. NNW of Vera-Cruz. It has a few good houses both of brick and stone, and a pretty large square. Its pop. has recently been estimated at about 7,000. Improvements are in progress both with regard to this town and the navigation of its river. New buildings are continually erecting, and the centre of the town is fast filling-up; the principal streets are being paved, and side-walks are completing all over the town. The commerce of T. has suffered much in consequence of the extensive trade in smuggling carried on at Matamoras and the whole of the northern frontier, principally induced and supported by the prohibitory Mexican tariff, which the late revolution has in a measure abolished. At present the communication of T. with the interior is most imperfect; the roads are only suited for mule traffic, goods from T. to San-Luis-Potosi having to be conveyed on mules' backs exposed to sun and rain for 28 or 30 days. The junta of San-Luis-Potosi have in contemplation the formation of a cart-road, on their side, down a branch of the river Panuco called Famina, where steamboats can arrive from T. in 20 hours, by which means goods will reach San-Luis from T. in 8 days. The most important improvement of all is the erection of a light on the N side of the entrance at T. bar, which can be seen 10 or 12 m. off in fine weather.—The old town of T., situated on the borders of the Laguna-de-Tampico, chiefly consists of a few adobe houses, with a pop. of 1,500.

TAMPISCO, a river of Nicaragua, which runs SSE into the gulf of Nicoya. It has a course of 30 m.

TAMRUCK, a small fortress of European Russia, in the gov. of Taurida, on the coast of the sea of Azof, between the mouths of the river Kuban, 156 m. SSW of Azof.

TAMSWEG, a village of Upper Austria, in the circle of Salzburg, 58 m. SSE of Salzburg, on the l. bank of the Minho. Pop. 2,200.

TAMUJA, a river of Spain, which rises in the prov. of Caceres, and flows NNW to the Rio-del-Monte.

TAMWORTH, a parish and market-town, partly in the co. of Stafford, partly in that of Warwick; at the confluence of the rivers Thame and Anker, and intersected by the Birmingham and Derby junction railway, which has a station here, distant by railway, 23½ m. NNE of Birmingham; 24¼ m. SSW of Derby; and 114¼ m. NW of London. The railway at this point is carried across the Anker by a viaduct of 18 arches. The Coventry canal passes about 1 m. S of the town. The parish includes the town of T., the chapelries of Fazeley, Wiggington, Wilnecote, Amington, and Hopewas, the townships of Biddescote, Bonehill, Bole-hall-with-Glascote, and the liberties of Tamworth castle and Syerscote. Area 12,420 acres. Pop. in 1801, 3,870; in 1831, 7,182; in 1851, 8,655. The church, an ancient edifice, with a tower, is supposed to occupy the site of a nunnery founded by St. Editha. The town is finely situated in the midst of a tract of rich meadow-land, watered by the rivers Thame and Anker, over which there are here two bridges. It has a neat and handsome appearance. The ancient castle stands on the SE side of the town, upon a lofty artificial mount, and is kept in tolerable repair. Drayton-manor, the seat of Sir Robert Peel, Bart., is an elegant and spacious mansion in this vicinity. In the 16th cent. the town appears to have declined, and ceased to be regarded as a corporation. It was incorporated, however, anew, by Queen Elizabeth. The income of the corporation, in 1840, was £377; in 1850, £384. It returns 2 members to parliament. The boundaries of the parl. borough were enlarged by the reform act, and include the whole of the parish. The electors registered in 1837, 491; in 1852, 307. The trade of the town is chiefly confined to supplying the inhabitants with the usual articles of consumption and wear. There are breweries and tan-works of considerable extent, and extensive cotton-mills; there are also two large wool-stapling establishments, and in different parts of the parish, corn-mills, dyeing-houses, and bleaching-works.

TAMWORTH, a township of Strafford co., New Hampshire, U. S., 43 m. NNE of Concord, on Bearcamp river. Pop. in 1840, 1,717; in 1850, 1,786.

TAMWORTH, a town of New South Wales, on the Peel river, 154 m. W from Maitland.

TAN, a district and town of China, in the prov. of Shan-tung and div. of Tsaou-chu-fu, in N lat. 34° 57′, E long. 116° 18′.

TANA-ELV, a large river of Lapland, which forms the boundary between Russia and Norway for 150 m.; traverses the NE part of Finmark; and falls into the Tana-fiord, a gulf of the Arctic ocean, in N lat. 71°, E long. 31° 30′, after a course of 200 m.

TANAGA, one of the Fox islands, in the N. Pacific, to the E of Kanaga. It is about 40 m. in circuit, and rises into high snowy peaks, some of which emit smoke.

TANAH, a town of Hindostan, in the prov. of Gujerat, in N lat. 21° 21′.

TANAHPILEH, a town of Sumatra, the cap. of the state of Jambi, about 70 m. from the coast. Pop. 4,000.

TANAIS. See DON.

TANAK POINT, a cape on the N coast of Java, in S lat. 6° 24′.

TANAKEKE, a small island about 12 m. in circumf., surrounded by a cluster of smaller ones, off the SW extremity of Celebes, in S lat. 5° 30′.

TANALITZKAIA, a fortress of Asiatic Russia, in the gov. and 120 m. E of Orenburg, on the r. bank of the Ural, which here receives the Tanalik

TANAMO, a port on the N coast of the island of Cuba, in N lat. 20° 42′.

TANAR, a river of Aberdeenshire, which rises at the foot of Mount Battock, and falls into the river Dee near the church of Aboyne. It gives the name of Glentanar to the district through which it runs.

TANARO, a river of the north of Italy, which rises among the Apennines, and after passing by Ceva, Cherasco, Alba, Asti, and Alessandria, falls into the Po at Bassignana, 3 m. E of Valenza, after a course of 150 m. Its principal affluents are the Corsaglia, Ellero, Pesco, Stura, Borbo, and Versa on the l.; and the Belbo and Bormida on the r.

TANASSERIM. See Tenasserim.

TANAT, a river of Wales, in the counties of Montgomery and Denbigh. It forms part of the boundary between these cos., and runs into the Severn on the NE point of Montgomeryshire.

TAN-CHU, a town of China, in the prov. of Kwang-tung and dep. of Keun-che-fu, or island of Hainan, in N lat. 19° 32′, E long. 109° 20′.—Also a district and town of China, in the prov. of Hu-pih and div. of Hwang-chu-fu.

TANCITARO, a village of the Mexican state of Michoacan, 60 m. ENE of Colima, in N lat. 19°.

TANCONVILLE, a village of France, in the dep. of La Meurthe, cant. and 6 m. SW of Lorquin. Pop. 500.

TANCOOK (Great), an island off the SE coast of Nova Scotia, at the entrance of Mahone bay. It is about 3 m. in length.—The little T. lies to the NE of it.

TANCOS, a town of Portugal, in Estremadura, on the r. bank of the Tagus, 68 m. WNW of Lisbon. Pop. 700.

TANDAH, a town of Hindostan, in the prov. and 9 m. SE of Oude, on the S bank of the Gogra.

TANDERAGEE — anciently Tawnatclee — a market-town in the p. of Ballymore, co. Armagh, on the river Cusher, 1¼ m. W of Newry canal, and 2 m. SW of Guilford. In the vicinity of the town are flour, meal, and flax mills. Considerable quantities of flax, and of all kinds of agricultural produce, are sold at the weekly markets. Pop. in 1851, 1,496.

TANDRIDGE, a parish in Surrey, 9¾ m. SSE of Croydon. Area 3,944 acres. Pop. in 1851, 594.

TANEGASIMA, an island of Japan, to the S of Kiu-siu. It is about 30 m. in length, and is well cultivated.

TANEPANTLA, a town of Mexico, in the state and 10 m. N of the city of Mexico.

TANETE, a town and small principality on the island of Celebes, situated half-way between Fort Rotterdam and the bay of Sorian, in S lat. 4° 14′.

TANEY, Tawney, or Churchtown, a parish in co. Dublin, containing the villages of Dundrum and Windymills. Area 4,571 acres. Pop. 3,929.

TANEY, or Tawney, a village in the p. of Clondevaddock, co. Donegal, in the peninsula of Fannat, 6¾ m. N of Millford.

TANEYTOWN, a township of Frederic co., Maryland, U. S.

TANFIELD, a chapelry in Chester-le-Street parish, co.-palatine of Durham, 6½ m. SW of Gateshead. It comprises the townships of Beamish and Lintz-Green. Area 7,072 acres. Pop. in 1831, 2,498; in 1851, 3,480. Here are iron and coal mines. The collieries are very extensive.

TANFIELD (West), a parish in the N. R. of Yorkshire, 6 m. NW by N of Ripon, on the N bank of the Ouse. Area 3,139 acres. Pop. in 1851, 628.

TANG, a district and town of China, in the prov. of Chih-le and div. of Paon-ting-fu, in N lat. 38° 44′, E long. 115°.—Also a district and town in the prov. of Ho-nan and div. of Nan-gan-fu, in N lat. 32° 47′, E long. 112° 53′.—Also a district and town in the prov. of Kwan-se and div. of Wu-chu-fu, in N lat. 23° 26′, E long. 110° 21′.

TANGALA, a small island in the Eastern seas, near the S coast of Java, in S lat. 8° 20′.

TANGALLA, a town on the S coast of the island of Ceylon, 21 m. ENE of Matura.

TAN-GANG, a town of China, in the prov. of Keang-su and div. of Chu-keang-fu, in N lat. 32° 4′, E long. 119° 32′.

TANGE, a small river of E. Prussia, which falls into the Kurische-haff, at Memel, where it is of sufficient depth to be navigated.

TANGER, a small river of Prussia, in the Old Mark of Brandenburg, which falls into the Elbe at Tangermünde.

TANGERANG, a village of the island of Java, about 15 m. W of Batavia, on a river of the same name. It is a considerable place; and before the neighbouring territory was ceded to the Dutch, it was a large military station of the state of Bantam. A weekly bazaar is held here, to which the produce of the adjacent country is brought, and thence carried to Batavia, by means of a canal which communicates with the river Tangerang or Tjidanee, and then runs parallel to the carriage-road the whole way to the city of Batavia.

TANGERMUNDE, a small town of Prussia, in the Old Mark of Brandenburg, on the Elbe, 32 m. NNE of Magdeburg, at the confluence of the Tanger. It has 3,000 inhabitants, employed in linen weaving, brewing, and cultivating the neighbouring district. It has an ancient castle, separated from the rest of the town.

TANG-GANG-HEEN, a district and town of China, in the prov. of Hu-pih and div. of King-mun-chu, in N lat. 30° 45′, E long. 111° 36′.

TANG-HEEN, a district and town of China, in the prov. of Shan-tung and div. of Yen-chu-fu, in N lat. 35° 15′, E long. 117° 24′.

TANGIERS, or Tanger, a considerable port of the prov. of Fez, in Marocco, situated at the W entrance of the straits of Gibraltar, to the E of Cape Spartel, and 38 m. WSW of Gibraltar. It has a pop. of 9,000, which subsists chiefly by supplying the British garrison of Gibraltar, and the cities of Cadiz and Lisbon, with cattle, fowls, fruit, and vegetables. It is a place of considerable strength. Its old walls, crenulated and flanked with round and square towers, are perhaps capable of a greater resistance than their appearance indicates. Several bastioned towers have been repaired and put in a state of defence within the last few years. In itself it is of little importance, but from its being the residence of the European consuls. Its commerce is inferior to that of some other ports, notwithstanding its proximity to the European coast. The houses of the consuls form the only ornament of the town. It is situate on the declivity of a calcareous mountain, part of which is not built upon, and presents a naked and unpleasant *coup d'œil.* Like most other places in Marocco, it is surrounded by a wall flanked by round and square towers, and enclosed by a ditch 2,200 metres in extent, but which has no counterscarp. It is protected by a kasbah or fortress of imposing appearance, and by a fort of Portuguese construction, bastioned in the modern style, but half in ruins. The kasbah is mounted by 12 pieces of cannon, which command the straits of Gibraltar. The ditches round the city are filled with trees, and cultivated as kitchen gardens. The rampart opposite the sea is remarkable for having two tiers of embankments with embrasures mounted by guns. On the N side the city is defended by steep rocks.

which prevent the approach of a besieging force. The principal defences have been raised in front of the landing-place near the marine gate. They consist of two batteries, mounting 60 pieces of heavy calibre, and 8 mortars all bearing on the port. The landing-place is flanked on either side by two batteries. Besides these defences of the port, the bay of T. is defended by six batteries in masonry, one built on Cape Malabata, and the other on the ruins of Old Tangier. These batteries mount together 40 guns. The two batteries which flank the roadstead at its two extremities are raised on small hills, about 150 ft. above the level of the sea; the others are at the water's edge. From 18 to 20 days is required to send a despatch, and get an answer, between T. and Fez, and from Marocco requires from 45 to 50 days. The bay of T. is now encumbered by the ruins of the old mole and fortification, and is not very safe during winter in W winds. The best anchorage for frigates and large vessels is at the E point.—T. was known under the name of *Tingis* or *Tinja* to the Romans. On the invasion of the Saracens, it was surrendered to them by Count Julian. In modern times, T. has often been a subject of contest between the Moors and the Portuguese. In 1437 it was besieged by Prince Ferdinand; but his army was completely defeated. In 1471, Alonzo, king of Portugal, succeeded in obtaining possession of it; and about two centuries after, in 1662, it became the property of England, being ceded to Charles II. as a marriage-portion with the princess Catherine of Portugal. The English, however, abandoned it in 1684, destroying the fortifications. T. became afterwards a distinguished station of piracy, for which its situation at the mouth of the straits gave it great advantages.

TANGIER-ISLANDS, several islands of the Chesapeake, near the coast of Maryland, U. S., opposite the mouth of the Potomac.

TANGIPAO, a river of the United States, which rises in Mississippi, crosses the E part of Louisiana, and flows into Lake Pontchartrain, 10 m. NE of the pass of Manchac.

TANG-KEU-EUL, a small but populous town on the N frontiers of the Chinese prov. of Kan-su. It is an animated place, full of business and bustle, being a rendezvous of caravans from Tibet. Once a-year the Eastern Tibetians quit the slopes of the Bayan-Kara mountains, in large caravans, and bring here furs, butter, and wild fruits for sale.

TANGLEY, a parish in Southamptonshire, 5 m. NNW of Andover. Area 1,561 acres. Pop. in 1831, 283; in 1851, 278.

TANG-MAO, an island of China, in the prov. of Kwang-tung and gulf of Canton, in N lat. 22° 2', E long. 113° 31'.

TANGMERE, a parish in Sussex, 3 m. N by E of Chichester. Area 734 acres. Pop. in 1851, 221.

TANGO, a town of Nifon, in Japan, 65 m. SW of Meaco.

TANGOLOTANGO, a port of Mexico, in the prov. of Oaxaca, in the gulf of Tehuantepec, in N lat. 15° 45', 100 m. SSE of Oaxaca.

TANGRU-MOUNTAINS, a chain in Mongolia, which commences on the frontiers of Siberia, in a spur of the Little Altai, and runs SE towards the Hongaer mountains.

TANG-SHAN, a district and town of China, in the prov. of Chih-le and div. of Shun-tih-fu, in N lat. 37° 22', E long. 114° 48'.—Also a district and town in the prov. of Keang-su and div. of Seu-chu-fu, in N lat. 34° 28' 30", E long. 116° 40' 55".

TANG-TU-HEEN, a district and town of China, in the prov. of Gan-hwuy and div. of Tae-ping-fu.

TANGUEY, or TONGUEY, a bay of Chili, on the coast of the Pacific, in S lat. 30° 16'.

TANG-YANG, a lake of China, in the prov. of Keang-su, 36 m. N of Hwae-gan-fu. It is about 30 m. in circumf.

TANG-YIH, a district and town of China, in the prov. of Shan-tung and div. of Tung-chang-fu, in N lat. 36° 35', E long. 115° 58'.

TANG-YIN, a district and town of China, in the prov. of Ho-nan and div. of Chang-tih-fu, 78 m. N of Kae-fung-fu, in N lat. 35° 36', E long. 114° 33'.

TANG-YUE-CHU, a district and town of China, in the prov. of Yun-nan and div. of Yung-chang-fu, on the confines of Burmah, in N lat. 24° 58', E long. 98° 45'. It is an important military station.

TANILA, a river of Mexico, which runs into the gulf of Mexico, in N lat. 18° 10'.

TA-NING, a district and town of China, in the prov. of Sze-chuen and div. of Kew-chu-fu, in N lat. 31° 37', E long. 109° 38'.

TANINGE, a town of the Sardinian states, in the prov. of Faucigny, situated in a fertile valley, on the river Feron, 22 m. E by S of Geneva. Pop. 2,700.

TA-NING-HEEN, a district and town of China, in the prov. of Chih-le, div. and 3 m. S of Ta-ning-fu, in N lat. 36° 18', E long. 115° 20'.—Also a district and town of China, in the prov. of Shan-se and div. of Sih-chu, in N lat. 36° 30', E long. 110° 43'.

TANJORE, an extensive, populous, and well cultivated district of the south of Hindostan, in the Carnatic, lying between the parallels of 9° and 12° S. It is bounded on the N by the Colerun; on the E by the sea; on the S by the zemindaries of Ramnad and Shevagunga; and on the W by Trichinopoly. Its length from N to S is 115 m.; its breadth 75 m. Area 8,625 sq. m. Pop. in 1831, 1,128,730, chiefly Hindus. The principal rivers are the Cavery, the Colerun, and the Venaar. About one-half of the surface is a flat alluvial delta, intersected by various branches of the Cavery and the Colerun. In the month of January, the whole face of the country is one continuous sheet of paddy ground, here and there interspersed with villages. The total number of villages paying revenue is 4,700. The number of landholders, 40,600. The revenue in 1837-8 was 47,38,607 rupees.—Next to the cap. the chief towns are Negapatam, Combaconum, and Manargudy. For a considerable period this district constituted a small principality which was never completely subdued by the Mahommedan arms. It was, however, conquered by the Mahratta chief Eccojee, about the middle of the 17th cent., and ultimately became tributary to the nabob of the Carnatic. In consequence of the cession of the nabob's territories, in 1799, this district is now in possession of the British.

TANJORE, the capital of the above-mentioned district, is situated in N lat. 10° 45', E long. 79° 11', 205 m. from Madras. It is, with its straggling suburbs, nearly 6 m. in circumf. There are two forts, one of which comprehends the palace and other public buildings; the other contains one of the handsomest temples in the south of India, and a college, formerly a celebrated place of Hindu learning.

TANJORE, a town of the island of Java, situated on high ground, 73 m. SSE from Batavia.

TANKARDSTOWN, a parish in co. Limerick, 2 m. WNW of Kilmallock. Area 1,710 acres. Pop. in 1831, 613; in 1851, 344.—Also a parish partly in Queen's co., and partly in co. Kildare, 3½ m. SSE of Athy. Area 8,350 acres. Pop. 1,447.

TAN-KEANG-TING, a district and town of China, in the prov. of Kwei-chu and div. of Tu-yun-fu.

TANKEL, an island in the strait of Sunda, to the NE of Tambikel.

TAN-KEO-HEEN, a district and town of China, in the prov. of Shan-se and div. of Keang-chu.

TANKERSLEY, a parish in the W. R. of Yorkshire, 4½ m. S of Barnesley. Area 8,404 acres. Pop. in 1831, 1,596; in 1851, 1,928.

TANKROWAL, a town of Western Africa, on the l. bank of the Gambia, in N lat. 13° 25′. The English African company had once a factory here, now abandoned. Its chief trade is in wax.

TANKSAL, a town of Hindostan, in the prov. of Delhi, in N lat. 30° 51′.

TANLAY, a village of France, in the dep. of Yonne, 6 m. E of Tonnerre. Pop. 750.

TAN-LING-HEEN, a district and town of China, in the prov. of Sze-chuen and div. of Mei-chu, in N lat. 30° 5′, E long. 103° 35′.

TANN, a town of Bavarian Franconia, on the small river Ulster, 9 m. ENE of Fulda. Pop. 100.

TANNA, a town of Saxony, in the Voigtland, 6 m. SSE of Schleitz.

TANNA (Cape), a headland of Java, in S lat. 6° 30′, 14 m. NW of Cheribon.

TANNA, an island in the S. Pacific ocean, one of the New Hebrides group, in S lat. 19° 32′, discovered by Cook in 1774. It is about 22 m. in length, and 10 m. in breadth. The inhabitants are of middle size, rather slender than otherwise, and of a very dark colour. The island contains a considerable volcano, and some hot springs, which raised the thermometer to 170°, and in one place to 202°. Cook named the harbour where he lay Port Resolution.

TANNA-BALLU, a small island in the Eastern seas, near the E coast of Borneo, in N lat. 4° 52′.

TANNAH, a town and fortress of Hindostan, in the prov. of Aurungabad, district of Bombay, on the E side of the island of Salsette, in N lat. 19° 11′. The fort, which is very strong, commands the passage between the island and the mainland, which is only about 200 yds. broad. It was taken from the Mahrattas by the British in December 1773, after an obstinate resistance. The town is straggling, but not large, although it contains several Portuguese churches. The Great Indian peninsular railway from Bombay to Callian, after sweeping across Sienmarsh, and giving off a branch to Machim, bends to the right; enters the island of Salsette, and has a station at Bhandup. From this point for the next 14 m. the line is perfectly level, the rails being laid along the surface of the ground, with merely so much embanking as to save them from the risk of flooding during the rains. To the l. are the low rocky wooded ridges of Salsette, woodlands and richly cultivated fields, hamlets, and cottages filling up the intervening space; on the r., parallel to, and close beside, the railway for about 8 m. is the salt water creek called the Tannah river, and beyond are the magnificent ghauts. The railway, on approaching T., becomes embowered under magnificent trees. On entering the village it turns quickly round towards the viaduct, by which it crosses the river, which is here divided into two by a long and rocky island. The channel on the T. side is shallow, and the arches are of moderate span; on the mainland side, the channel is deep and narrow, and an iron bridge, somewhat on the tubular plan, 86 ft. in span, permits the shipping to pass under it. The whole viaduct from shore to shore, including the part over the river, is about 1,000 ft. in length, the ferry-way about 40 ft. above the high water mark. From this the line runs along the base of the hill, at a distance of about 3 m. to the projecting part of Parsick-point, which it penetrates by a channel about 100 yds. in breadth. On the other side the traveller finds himself passing for about a mile along the margin of the Callian river, surrounded by some of the most magnificent scenery in the world, when a second tunnel is encountered, and no further difficulty presents itself till the village of Callian is attained. The first sod of this line was turned on the 31st of October 1850, and the line was completed to Callian on 1st May 1854. One line is now laid down complete to T., and the second is in rapid progress. —Also a fortress of Bengal, situated on the W bank of the Hugli river, about 2 m. below Calcutta. It was taken by the British in 1687, but was afterwards restored to the nabob. It was again taken by the British in 1756; but the erection of the fortress of Fort William having rendered T. unnecessary, the fortifications have been allowed to decay.

TANNADICE, a parish a little N of the centre of Forfarshire. Area 38,400 acres. The South Esk, ploughing its way between romantic and beautiful banks, forms, over a great distance, the W and S boundary-line. The Noran has most of its fine course either in the interior or along the eastern boundary. The uplands are to a large extent heathy and almost wholly pastoral. Pop. 1,517.

TANNA-MERA, a small island near the E coast of Borneo, in N lat. 3° 45′.

TANNAY, a town of France, dep. of the Nievre, 9 m. S by E of Clamecy, near the l. bank of the Yonne. Pop. 1,400.

TANNE, a village of Brunswick, in the principality of Blankenburg, near Hasselfeld, on the l. bank of the Bode. The village is insignificant, but there are large iron-works in the neighbourhood.

TANNENBURG, a village of E. Prussia, 54 m. S by W of Königsberg.

TANNERRE, a town of France, in the dep. of the Yonne, 21 m. SSW of Ivigny.

TANNER'S CREEK, a river of Indiana, U. S., which runs into the Ohio, 2 m. below Laurenceburg. It is 30 yds. wide at its mouth.

TANNER'S HILL, a village of Newbury district, S. Carolina, U. S.

TANNESAR, or Thanasir, a town of Hindostan, in the prov. of Delhi, in N lat. 29° 55′. This place formerly contained the celebrated temple of Jug-Sum, which was held in the highest estimation by the Hindus. Its riches and celebrity attracted the cupidity of Mahmud of Ghizni, who, in 1012, marched against it and captured it.

TANNESERIM. See Tenasserim.

TANNEWANG, a river on the S coast of the island of Celebes, which runs into the sea, 5 m. W of Bonthain.

TANNHAUSEN, a village of Prussian Silesia, in the principality of Schweidnitz, near the borders of Bohemia, 9 m. SE of Waldenburg.

TANNINGTON, a parish in Suffolk, 3½ m. NW of Framlington. Area 1,602 acres. Pop. in 1831, 264; in 1851, 244.

TANON (Point), a cape on the S coast of the island of Sibu, in N lat. 9° 52′, E long. 123° 18′.

TANORE, a port of Malabar, 24 m. SE of Calicut, in N lat. 10° 55′. It was formerly a place of considerable note, but is now reduced to a mere village.

TANQUA, a river of Southern Africa, which descends from the Wittemberg, and running WNW, joins the Little Doorn, on the r. bank, after a course of 120 m.

TANQUE, a village on the coast of Chili, on the S side of Tongey bay.

TANSEY, a river of America, which rises in the Rocky mountains, and falls into the Missouri near the great falls. Its general course is from E to W.

through wide valleys, well supplied with both the long and broad leafed cotton wood.

TAN-SHWUY, a district and town of China, in the prov. of Hu-pih and div. of Hwang-chu-fu.

TAN-SHWUY-CHING, a town of China, in the prov. of Fo-keen and dep. of Tae-wan-fu, on the NW coast of the island of Formosa, on a bay of the same name, in N lat. 25° 7′ 10″, E long. 121° 11′.

TAN-SHWUY-KI (LOWER), a river of China, in the prov. of Fo-keen and dep. of Tae-wan-fu, or island of Formosa, which runs to the SE of Fung-shan, and throws itself into the bay of that name.

TAN-SHWUY-KI (UPPER), a river of China, in the prov. of Fo-keen and dep. of Tae-wan-fu, or island of Formosa, which has its source in the lofty mountains on the SE frontier of the Chinese portion of the island; runs NW; receives the Pa-lang-seuen, and throws itself into the bay of the same name. Its banks are finely shaded with wild palm-trees. It is navigable a considerable distance, and its waters abound with a species of fish locally named the *hung-sin-hu*, which averages a length of nearly 10 ft.

TANSIF. See TENSIF.

TANSLEY, a township in Crich parish, Derby, 1½ m. E of Matlock. Area 1,150 acres. Pop. in 1831, 507; in 1851, 593.

TANSOR, or TANSOVER, a parish in Northamptonshire, 2 m. NNE of Oundle, on the E bank of the Nen. Area 2,050 acres. Pop. in 1851, 256.

TANTOYUCA, a settlement of Mexico, in the district of Tampico, 60 leagues NE by N of Mexico.

TANTUMQUERI, a port on the Gold coast of Africa, 18 m. E of Cormantin.

TANWORTH, a parish in the co. of Warwick, 4 m. NW of Henley-in-Arden. Area 9,400 acres. Pop. in 1831, 2,201; in 1851, 1,892.

TAN-YANG, a town of China, in the prov. of Keang-su and div. of Chin-keang-fu, in N lat. 32° 4′, and E long. 119° 32′.

TAOLI-MONASTIR. See MONASTIR.

TAONG-SU, a nomade people of the Burman empire, who frequent the country between the Thanluyn and Setang.

TAORMINA, a town of Sicily, in the prov. and 33 m. SW of Messina, and district of Castroreale, on one of the heights of Mount Taurus, in a magnificent situation, near the shore of the Ionian sea, in N lat. 37° 48′ 15″, and E long. 15° 17′ 25″. It is partly enclosed by walls, and is commanded on the NW by two forts. Pop. in 1831, 3,929. It is ill-built and dirty, but contains some handsome churches and convents, an hospital, and several fine ruins. The trade consists chiefly in wine and hemp. This town, the *Tauromenium* of the ancients, was destroyed by the Saracens in 968. About 1½ m. to the SW are the ruins of the ancient *Naxus*. T. gives its name to a bay enclosed on the NE by Cape S. Andrea, and on the SW by Point Pietragala, 3 m. distant from one another. It has a depth of about 2 m.

TAORO. See OROTAVA.

TAOS, a county in the NE of the state of New Mexico, between the parallels of 36° and 38° N, traversed by the Rio-Grande, and by numerous small streams. Pop. in 1850, 9,507.

TAOS (PUEBLO DE), a village of New Mexico, situated in N lat. 36° 23′, W long. 105° 47′, on the r. bank of a small affluent of the Rio-del-Norte, 52 m. NNE of Santa-Fe, immediately at the foot of the W side of the first ridge of the Rocky mountains. The village possesses much singularity, combining great strength, with bold conception in the architecture, and enjoying the most complete utility in its honey-comb formation. It is formed of two distinct departments, the rivulet dividing it in the centre. The buildings are exactly similar. The first, or basement story, occupies a large area; upon this is reared another story, smaller than the first by 10 ft.;—each successive story falling off in the same ratio, until the pagoda becomes 9 stories high. In the basement story there are no doors, the entrance being through a round hole in the offset which forms the roof; a pole being planted in the aperture, with a ladder attached, by which they ascend and descend. The buildings resemble pyramids, only being square. In the formation of the village an equal number of the heads of families divided themselves off and laid two foundations precisely similar, and reared upon them two equal structures. The original inhabitants of the valley of Taos are Indians, and known by the name of Pueblos. They were in possession of this country when the Spaniards first visited Mexico, and they have remained until the present time. The patron saint of the village is St. Jerome; the Pueblos having adopted the Catholic religion through the powerful influence of the missionaries from Spain. The day of St. Jerome is celebrated by selecting thirty of the choicest youths from each side of the stream, to compete with one another in manly sports. Their honesty and industry is proverbial, and well-cultivated fields and well-stocked farms give token of much thrift and independence.

TAOU-CHU, a district and town of China, in the prov. of Hu-nan, and dep. of Yung-chu-fu, 210 m. SSW of Chan-sha-fu, on a river of the same name, an affluent of the Seang-keang, in N lat. 25° 32′ 29″, E long. 111° 28′ 10″.

TAOU-CHU-TING, a district and town of China, in the prov. of Kan-suh and dep. of Kung-chang-fu, in N lat. 34° 35′, and E long. 103° 31′.

TAOUD, a village of Upper Egypt, in the prov. of Thebes, and near Esneh. It contains the remains of a temple now occupied by fellahs and their cattle. It is the *Taphnim* of the ancients.

TAOU-HING-TENG-SHAN, a mountain of China, in the prov. of Kwei-chu, and dep. of Szenan-fu. It is covered with perpetual snow.

TAOU-SU-SHAN, a mountain of China, in the prov. of Chih-le, and div. of Ssun-wha-chu, in N lat. 39° 52′, and E long. 114° 45′. It is covered with perpetual snow.

TAOU-YUEN, a district and town of China, in the prov. of Keang-su, and div. of Hwae-gan-fu, in N lat. 33° 43′, and E long. 118° 48′.—Also a district and town in the prov. of Hu-nan, and div. of Chang-tih-fu, in N lat. 28° 52′ 10″, and E long. 111° 11′ 9″.

TAPACHULA, a town of Guatimala, in the state of that name and dep. of Quezaltenango and Soconusco.

TAPACOAS, a tribe of Indians in Brazil, who inhabit the banks of the Tocantins and Somno, in the N part of the prov. of Goyaz. They subsist by fishing and the chase, but are much addicted to rapine.

TAPACORA, a village of Brazil, in the prov. of Rio-de-Janeiro, and district of São-João-d'Itaborahi.—Also a mountain in the same district.—Also a river of the prov. of Pernambuco, which passes the town of Santo-Antão, and throws itself into the Rio-Capibaripe.

TAPADA, a lake of Brazil, in the prov. of Espirito-Santo, between the Doce and São-Matheos rivers. It is of little extent, but abounds with fish.

TAPADO, a river of Brazil, in the prov. of Pernambuco, 6 m. N of Olinda.

TAPAGI, a headland of Brazil, in the prov. of Caara, and district of Granja, to the E of the Acaracu.

TAPAGIPE, or ITAPAGIPE, a parish of Brazil, in the prov. and to the N of Bahia, on a peninsula of the bay of Todos-os-Santos, and at the mouth of a river of the same name.

TAPAGONIA, a comarca of Brazil, in the prov. of Para, bounded on the N by the Amazon, on the S by the comarca of Dos Arinos; on the W by that of Mundrucania, from which it is separated by the Tapajos; and on the E by the comarca of Xingutania. It is 300 m. in length from N to S, and 180 m. in medium breadth, and is watered by the Xingu, Tapajos, Guajara and Urancu rivers. It is extremely fertile, producing spontaneously cloves, nutmeg, cacao, and varieties of medicinal plants. It is inhabited chiefly by Indians. Its principal towns are Souzel and Santarem.

TAPAJOS, a river of Brazil, formed by the junction, in the prov. of Mato-Grosso, of the Rio-dos-Arinos and Jaruena, the former of which has its sources in the Sierra-Pany, the latter in the Sierra-Parecis, and which unite in S lat. 9° 10′, and E long. 57° 36′. Running NE it enters the prov. of Para, in which it bathes the districts of Para and Rio-Negro; receives on the r. the Azevedo, one of its most important tributaries; and, after a total course of 600 m., throws itself, by two branches, into the Amazon, on the r. bank, a little above Santarem. This river is navigable throughout a great part of its course.—Also a tribe of Indians, who inhabit the banks of the above named river.

TAPAJOS, or SANTAREM, a town of Brazil, in the prov. of Para and comarca of Rio-de-Tapajos, on the S bank of the Amazon, a little below the confluence of the Tapajos. Its trade consists chiefly in cacao and medicinal plants. Pop. of district 10,000, of whom the majority are whites.

TAPALGUEN, a town, or rather a collection of oven-shaped Indian huts, in the state and 170 m. SSW of the city of Buenos-Ayres, on a river of the same name, which here flows through the pampa in a NE direction to the Rio-Flores, a branch of the Salado.

TAPALGUEN (SIERRA), a low range of quartz hills, in the state of Buenos-Ayres, which terminates in Cape Corrientes.

TAPANHUACANGA, TAPAHUARANGA, or JOSE DE TAPAHUANCANGA, a village of Brazil, in the prov. of Minas-Geraes, 12 m. SSW of Villa-do-Principe, in a valley at the E base of the Serra-do-Espinhaco. The church, standing on a platform, commands the houses, which are small, low, and covered with tiles. There were formerly gold mines in its vicinity.

TAPANHUNA, a tribe of Indians of Brazil, who inhabit the banks of the Rios dos Arinos and Jaruena, in the prov. of Matto-Grosso.

TAPARI, a fort of Brazil, in the island of Maranhao, founded in 1612.

TAPARICA. See ITAPARICA.

TAPAXANAS, a tribe of Indians who inhabit the banks of the Amazon in Brazil, in the prov. of Para and district of the Rio-Negro.

TAPE′, a village of Hungary, in the comitat of Csongrad, 2 m. ENE of Szegedin, on the r. bank of the Theiss, and in the midst of extensive marshes. It has extensive manufactories of mats.

TAPERA, a village of Brazil, in the prov. of Minas Geraes, on the l. bank of the Chopoto.—Also an island in the bay of Nitherohi or Rio-de-Janeiro.

TAPERI, a lake of Brazil, in the prov. of Ceara and district of Mecejana.

TAPES, a numerous tribe of Indians of Brazil, who inhabit the NW part of the prov. of Sao-Pedro-do-Rio-Grande, between the chain of the Cochilla-Grande-de-los-Tapes, and the N bank of the Uruguay.

TAPIA, a town of New Granada, in the prov. and 45 m. SSW of Mompox, on the r. bank of the Cauca,

TAPIAN, a headland of the island of Mindanao. on the W coast, in N lat. 7°, and E long. 124° 30′.

TAPIAN, a town of Prussia, in the prov. of E. Prussia, regency and 24 m. ESE of Königsberg, and circle of Wylau, on the r. bank of the Pregels, at the junction of the Deime canal. Pop. in 1843, 3,226. It has a royal castle, now used as a house-of-correction, an asylum for the poor, and possesses manufactories of carpets, hosiery, and several tanneries and breweries.

TAPINO, a mountain of the Papal States, in the NW part of the comarca of Rome, to the N of Lake Bracciana, and 3 m. SSE of Sutri.

TAPIOLES, a town of Spain, in the prov. of Zamora, partido and 14 m. ESE of Benevento, in a fertile plain, watered by the Valderaduey. Pop. 660. Wine is cultivated in the environs.

TAPIRAPE, or TAPIRAQUE, an Indian tribe of Brazil, who inhabit the banks of a river of the same name, in the prov. of Matto-Grosso and comarca of Tapiraquia. The river T. joins the Araguaia, on the l. bank, some leagues below the island of Bananal.

TAPIRAPOAN, a mountain of Brazil, in the prov. of Matto-Grosso. It forms a ramification of the Cordelheira-dos-Parecis. The Preto, the first important affluent of the Paraguai, and also the Pari and Barvados, take their rise in this mountain.

TAPIRAQUIA, a comarca of Brazil, at the NE extremity of the prov. of Mato-Grosso, bounded on the N by the prov. of Para, on the r. by the Rio-Xingu, by which it is separated from the Com-dos-Arinos, on the S by the comarca of Bororonia, and on the E by the Uruguay, by which it is separated from the prov. of Goyaz. It is 330 m. in length, from N to S, and 210 m. in breadth. With the exception of the E part, this com. is little known. It is inhabited by several Indian tribes, of whom the principal are the Guapindayas, Tapiraques, Ximbinas, and Aracys. It is very fertile, well-watered, and abounds with game, and is said to contain gold and silver-mines.

TAPISI, a river which has its source in Peru, in the mountains of Conomamas, and runs NNW to the Ucayle, which it joins in S lat. 5° 20′.

TAPITAY, a town of Brazil, in the prov. and 465 m. W of São-Paulo, on the l. bank of the Parana.

TAPLOW, a parish in Bucks, 5 m. NW of Eton, on the E bank of the Thames, and in the line of the Great Western railway. Area 1,920 acres. Pop. 704.

TAPO. See TARMA.

TAPOLCSAN (KIS), or MALE-TOPOLCANY, a town of Hungary, in the comitat of Bars, 11 m. W of Konigsberg, in the midst of mountains. Pop. 1,800. It has a castle, with a superb park, a Catholic church, and an orphan's asylum. It has manufactories of cloth, and in the environs is a mineral spring.

TAPOLCSAN (NAGY), or WELKE-TAPOLCANY, a town of Hungary, in the comitat and 19 m. NNE of Neutra, near the r. bank of the Neutra, a little below the confluence of the Hetenye. Pop. 2,547.

TAPOLCZA, a hamlet of Hungary, in the comitat of Borsod, in the midst of woody mountains, near the source of the Hejo. It has several mineral baths.—Also a town in the comitat of Szalad, 28 m. SW of Veszprim, at the foot of the mountains of Bakony.

TAPOLCZA (KUN). See KEPLITZ.

TAPOLLAMA (SERRA-DA), a mountain-chain of Brazil, in the E part of the prov. of Rio-Grande, and S part of that of Santa-Catharina. It is about 180 m. in length, and runs from NNE to SSW, parallel to and at the distance of about 36 m. from the shore of the Atlantic.

TAPPA, a small island of the Asiatic archipelago, in the group of the Molucca islands, to the NE of the island of Latta-Latta, from which it is separated by a narrow channel. It has a small port on the E side, in which vessels can find safe anchorage.

TAPPA. See NYFFE.

TAPPAHANNOCK, a town of Essex co., in Virginia, U. S., on the SW side of the Rappahannock, about 50 m. above its entrance into Chesapeake bay. It has a good port on the river.

TAPPANTOWN, a village in Rockland co., in the state of New York, U. S., 2 m. W of the Hudson river, and 108 m. S of Albany.

TAPPANULY, a bay of the island of Sumatra, on the SW coast, in the Battas territory, in N lat. 1° 45′, E long. 98° 40′. This bay, with the island of Mansalar, forms one of the finest harbours in the world. It is so spacious and well sheltered, that vessels of the largest size can find safe harbourage within it from all winds, and so numerous are its bays and ramifications that a large vessel might effectually elude a laborious research. Its banks are inhabited by Battas who carry on an active barter trade with the vessels which visit the harbour. This bay is noted for the kima cockle, a species of mollusca which attains the size of 3 ft. 3 in. in length, and 2 ft. 1 in. in breadth. The shell is white and several inches in thickness, and the fish when detached sometimes exceeds 25 lbs. in weight. The bay abounds also with corallines and madrepores. On the N bank of the bay is a village of the same name, originally a British settlement, but ceded to the Dutch in 1818.

TAPTI, a river of Hindostan, which has its source in the W part of the prov. of Gundwana; separates the prov. of Berar and Kandeish; traverses the latter prov. and the S part of that of Gujerat; and after a course in a generally W direction of 450 m., throws itself into the gulf of Cambay, between Domus and Swally. Its principal affluents are the Goul and Anheir on the r.; and the Boaree, Guirna, and Panzeera on the l. Burhanpur and Surat are the chief places on its banks.

TA-PU, a district and town of China, in the prov. of Kwang-tung and div. of Chaou-chu-fu, in N lat. 24° 40′, E long. 116° 43′.

TAPUERCA, or ATAPUERCA, a town of Spain, in the prov. and partido and 9 m. E of Burgos, near the source of the Pico.

TAPUL, a small island of the Sulu archipelago, to the S of Sulu island. It abounds with excellent water, is well-cultivated, and pastures large herds of cattle and goats.

TAQUA, a town of Upper Guinea, on the Gold coast, in the state of Aowin, 84 m. SW of Kumassi.

TAQUARAÇU, a village of Brazil, in the prov. of Minas Geraes, on the r. bank of the Rio Guaicuhi or das Velhas, and 24 m. N of Sabara.

TAQUARAL, a river of Brazil, in the prov. of Goyaz, which joins the Vermelho on the l. bank, below Goyaz.—Also a river of the prov. of Mato-Grosso, which joins the Peixe, an affluent of the Rio-das-Mortes.—Also two rivers in the prov. of Rio-de-Janeiro.—Also a mountain in the prov. of Mato-Grosso and district of Cuiaba.

TAQUARATINGA, a village of Brazil, in the prov. of Pernambuco and district of Limoeiro, on a mountain of the same name. This village is the oldest in the prov. The mountain has an alt. of 3,000 ft. above sea-level.

TAQUAREMA, a river of Brazil, in the prov. of São-Pedro-do-Rio-Grande and district of Alegrette, which runs W; receives the Jaguari, and joins the Santa-Maria, an affluent of the Paraguay.

TAQUARI, or TEBICUARI, a river of Brazil, in the prov. of São-Pedro-do-Rio-Grande and district of Triumpho. It has its source in the mountains of Geral, in S lat. 18°, to the E of the source of the Jacuhi, and after a course in a generally S direction of 105 m., joins that river on the l. bank. Its principal affluents are the Antas and Taquari-Mirim, the latter of which it receives on the r.—Also a river of the prov. of Rio-de-Janeiro, an affluent of the bay of Angeo-dos-Reis.

TAQUARI-MIRIM, a river of Brazil, in the prov. of São-Pedro-do-Rio-Grande, an affluent of the Taquari.—Also a river of the prov. of Mato-Grosso, an affluent of the Coxim, which it joins on the l. bank, 3 m. above its confluence with the Taquari.

TAQUARITIBA, a village of Brazil, in the prov. of Parahiba and district of Pianco, on the Piranhas. Millet, mandioc, and cotton are extensively cultivated in the environs.

TAQUASO. See CHUQUNAQUE.

TAQUENE, TACUNA, or TAQUILE, an island of Bolivia, in the dep. of Cuzco, district and 33 m. NNE of Chucuito, in Lake Titicaca or Chucuito, near the W bank. It is 9 m. in circumference, and is covered with gardens, orchards, and meadows. In its highest part is a plateau, on which are the ruins of an Indian town, the buildings of which are of stone, are vaulted, and exhibit great regularity.

TAR, or PAMLICO, a river of the state of North Carolina, U. S., which has its source in Person co., 8 m. SE of Roxboro; waters the cos. of Granville, Franklin, Nash, Edgecombe, Pitt, Beaufort, and Hyde; and after a course in a generally ESE direction of 180 m., flows by a wide estuary into Pamlico sound. Its principal affluent is Fishing creek.

TARA, a river of Russia in Asia, which has its source in the gov. of Tomsk and district of Kainsk; runs W into the gov. of Tobolsk; waters the district of the same name; and after a course of about 180 m., joins the Irtish on the r. bank, near the town of Tara.—Also a town in the gov. and 240 m. SE of Tobolsk, on the Arkarka, which, ¾ of a mile below, joins the Irtish opposite the confluence of the Tara. Pop. 6,000. It consists of two parts, one of which, on a mountain, is enclosed by an earthen rampart, and possesses a fort. The lower part extends along the river into the plain. It contains two churches, a mosque, and manufactories of hats, vitriol, and leather. It has an active trade with Bokara and the Chinese empire. The lower part of the town is inhabited chiefly by Tartars. In the vicinity is the steppe of Baraba, in which there is a colony of exiles.

TARA, a river of Australia Felix, in Gipp's Land, an affluent of Port Albert.

TARA, a village of Sind, 12 m. SW of Tattah.

TARA, a small bay on the E coast of co. Down, opening 1½ m. N by E of Ballyquintin point.—Also a rivulet of co. Armagh, and co. Louth, which rises to the E of the town of Newtown-Hamilton; runs nearly parallel to the Newtown-Hamilton river, and falls into the Castletown or Dundalk river.—Also a hill in co. Wexford, with an alt. of 826 ft. above sea-level, situated 1⅛ m. W of the nearest part of the coast, and 2¾ m. NE of Gorey.

TARA, TARAH, or TARAGH, a parish, containing a village and a celebrated hill of the same name, in co. Meath. Area 3,364 acres. Pop. in 1831, 641; in 1851, 422. A prominent feature, not only of the parish, but of the vast rich plain in the midst of which it lies, is the far-famed hill of Tara, a verdant, moundish, flowingly-outlined mass, about ¾ m. in length from N to S, and rather less than a ½ m. in extreme breadth,—possessing a wavy, tumulated, tabular summit,—lifting up a large, solitary standing stone or monumental pillar on the crown of one of its tumuli,—sharing with the hill of Skreen, 1¾ m. to the E, and 507 ft. in alt., the power and interest of relieving the monotony of the vast central expanse of the plain of Meath.

TARABA, a town of Arabia, in the prov. of Sabya, 150 m. E of Mecca. It is well-fortified.

TARABEL, a village of France, in the dep. of the Haute-Garonne, cant. and 4 m. SSE of Lauta. Pop. 340.

TARABOLUS. See TRIPOLI.

TARABUZUN, a pashalik of Asiatic Turkey, extending southwards from the coast of the Black sea to the frontier of Sivas, and westwards to the river Joruk or Tchoruk. It is a beautiful and richly varied district of mountain-slopes, valleys, and plains, and comprises a large portion of the ancient *Pontus*. On its coast are the ports of Khotz, Sololer, Riza, Eski, Tarabuzun, Kopa, Makral, Terabolis, Kerasun, Ordu, Fatsa, and Unieh.—The cap., of the same name, is situated on the lower slope of the mountain-chain which rises abruptly from the shores of the Black sea. It is divided into two parts, of which the more ancient probably marks the site of the ancient *Trapezus*. With the exception of a small harbour on the E side of the city, it has only an open roadstead; yet the journey into Persia from this port is shorter by several days than that by Sivas.

TARACENA, a town of Spain, in the prov. and 5 m. ENE of Guadalajara, near the l. bank of the Henares. Pop. 460.

TARACHTCHA, a town of Russia in Europe, in the gov. and 68 m. S of Kiev, on a small affluent of the Ros.

TARADELL, a town of Spain, in the prov. and 30 m. NNE of Barcelona and partido of Vich, on a rugged locality. Pop. 1,600. The streets are sloping, but there is a handsome square ornamented with a fine fountain. It has manufactories of woollen fabrics.

TARAGHUR, a fortress and town of Hindostan, in the prov. of Ajmir. The fortress stands on a hill, and is nearly 2 m. in circumf. It is well supplied with water, and is to a great extent bomb-proof.—Also a fortress of the Punjab, in N lat. 32° 7′, E long. 76° 30′.

TARAGUDO, a town of Spain, in the prov. and 15 m. NNE of Guadalajara, and partido of Brihueja, on a hill, in a fine and salubrious situation. Pop. 150. In its vicinity is a palace of the duke of Infantado, with a fine park, orangery, and orchards.

TARAGUES, an Arab tribe of Nubia, in the prov. of Shendy. They inhabit El-Boeydah, and employ themselves in the salt-works of that town.

TARAI, or TARIYANI, the name given to a tract of highly absorbent land in India, skirting the S edge of the outlying ranges of the Himalaya, between the hills and the plains of India, and to the E of the Ganges, with a breadth of about 10 m., usually covered by forest, and remarkable for its utter want of water.

TARAK, a village of Persia, in the prov. of Irak-Ajemi, in the district and vicinity of Ispahan. It has manufactories of articles in ivory and ebony.

TARAKLI, or TEREKLI, a town of Turkey in Asia, in Anatolia, and sanj. of Sultanieh, on the r. bank of a river of the same name, an affluent of the Sakaria, and 48 m. SE of Isnikmid. It contains 400 houses, and seems a place of considerable prosperity. It has manufactories of gold and wood ware, and several dye-works. The vine and mulberry are extensively cultivated in the environs. This is probably the *Heraclia* of the ancients.

TARALGA, a creek of New South Wales, in the co. of Argyle, an affluent of Guinecor creek.

TARAMANDI, a lake of Brazil, in the E part of the prov. of Rio-Grande-do-Sul. It is 30 m. in length from N to S, and 12 m. in breadth, and discharges itself by a river of the same name into the Atlantic.

TARANCON, a town of Spain, in the prov. of Cuenca and partido of Ucles, 45 m. SE of Madrid, in a rugged locality, near the r. bank of the Rianzaros. Pop. 4,790. The streets are irregular and ill-paved; the houses, although of stone, are poorly built, and the fountains are in bad taste. The parish-church, originally Gothic, was restored by Philip II., and has a fine front in the Ionic order. There are also several convents and an hospital. Linen and common cloth are the chief manufactures of the town, but the locality is noted for its wine. Coal is also found in the vicinity.

TARANNON, a river of Montgomeryshire, which runs into the Severn, near Corfe-castle, about 5 m. W of Newton.

TARANOVKA, a town of Russia in Europe, in the gov. and 27 m. SSW of Kharkov, and district of Zmiev.

TARANSAY, an Hebridean island of the Harris group, Inverness-shire, off the entrance of West Loch-Tarbert, 1¼ m. from Ru-Grodernish. Its length from E to W is 4½ m.; its extreme breadth upwards of 2 m. It consists of two peninsulated hills, probably 800 ft. in height, connected by a narrow sandy isthmus.

TARANTA, a mountain near Arkiko, on the N frontier of Tigre, in Abyssinia, on the road from that city into the interior. On its summit is the village of Halai. Ruppell ascertained the height of this mountain to be 8,600 ft. A small stream creeps from the mountain-plain westwards to the valley of the Mareb; all the other streams southwards to Ategerat, it appears, flow towards the coast.

TARANTA, a town of Naples, in Abruzzo-Citra, 18 m. SSW of Lanciano, near the l. bank of the Avent no.

TARANTAISE (THE), or TARANTASIA, a province of the Sardinian states, in the E of Savoy, between Faucigny on the N, and Aosta on the S. Its superficial extent is about 780 sq. m.; its pop. 50,000. Its surface is covered with mountains and rocks, ranges and spurs of the Cottian Alps, and is little susceptible of culture. By the industry of the inhabitants, however, the least sterile parts of it have been brought into cultivation, and made to produce corn, saffron, fruit, and chestnuts; and the pastures nourish a fine race of cattle. Its principal river is the Isere, which rises on Mount Iseran in the SE part of the prov.

TARANTASCA, a village of the Sardinian states, in the prov. and 10 m. N of Coni, on the l. bank of the Grana. Pop. 1,200.

TARANTO, or TARENTO, an ancient town of the kingdom of Naples, in the prov. of Otranto, 58 m. WNW of Lecce. It is built on a small island at the N extremity of the great bay called the gulf of Taranto, in N lat. 40° 28′, E long. 17° 35′, and has several advantages as a maritime position, having behind it, towards the N and E, an inlet called Il Mare Piccolo, which extends a considerable way into the interior, while it communicates with the sea by two channels, one on each side of the island, which is occupied by the town. A castle of some strength protects the harbour. Pop. 18,000. The surrounding country is fertile, but the trade of the town and neighbourhood is inferior to what it might be rendered in a populous and industrious country. Some intercourse is carried on with other ports on the Mediterranean; but the chief support of the inhabitants of T. is derived from fishing. T., at present a town of little interest, filled a conspicuous place in ancient history. It was founded 703 B. C. Its inhabitants, descended from a colony of Greeks, kept up their connection with the mother country, and on the approach of the Roman arms after the conquest of Samnium, about 280 B. C., called to their aid Pyrrhus, king of Epirus. The Tarantines, abandoned by Pyrrhus, and unable to defend themselves, ultimately called in the Carthaginians; a step which, coinciding with the collisions between that nation and the Romans in Sicily, was the cause of the first Punic war.

TARANTO (Gulf of), a spacious bay of the Mediterranean, formed between the parallels of 39° 21′ and 40° 30′ N, by the two extremities of the Italian peninsula. Point Alice on the SW, and Cape Maria-de-Leuca on the NE, form its points of entrance. The rivers Brandano, Bisiento, Agri, Sinno, and Cosale, flow into this gulf.

TARAPACA, a province of Southern Peru, in the dep. of Arequipa, lying between the parallels of 19° and 21° 30′ S, and the meridians of 68° 15′ and 70° 22′ W; and bounded on the N by Arica; on the E by Bolivia; on the S by the desert of Atacama; and on the W by the Pacific. It is subdivided into the 4 *curatos* of T., Peca, Sibuya, and Camina; and has a mixed pop.* of about 10,000, chiefly Aymara Indians. Its cap. of the same name is situated in 19° 56′ S lat., and 69° 35′ W long., at the mouth of a ravine which runs down from the cordillera of Lerima. The cultivation in the vicinity consists of maize, wheat, lucern, fruits, and a few vegetables. See articles Atacama, Iqueque, and Tamarugal.

TARAPIA, or Therapia, a town of Turkey, in the sanj. of Visa, 12 m. NNE of Constantinople, in N lat. 41° 8′.

TARAPUR, a town of Hindostan, in the prov. of Bahar, in N lat. 25° 7′.—Also a town and fortress of Hindostan, situated on a high point of land on the coast between Bombay and Surat, in N lat. 19° 30′.

TARARE, a town of France, in the dep. of the Rhone, on the l. bank of the river Tardine, 16 m. SW of Villefranche, in a valley at the foot of a mountain to which it gives name. Pop. of com. in 1846, 9,690. It has manufactories of muslin and printed cottons, cotton-thread, leather, and pottery ware. The cant., comprising 16 coms., had a pop. of 26,291 in 1846.

TARASCON, a town of France, in the dep. of Bouches-du-Rhone, 9 m. N of Arles, on the Rhone, opposite Beaucaire, with which it communicates by a bridge of boats. Pop. in 1846, 11,968. Of public buildings, the principal is the castle, now a prison, a massive structure of hewn stone, surmounted by a platform affording a view of the adjacent country and of the Rhone, which here approaching to its mouth, rolls along a broad and rapid volume of water. Some of the churches are likewise handsome buildings. The inhabitants manufacture woollens, silks, and stockings, and conduct an export trade in wine, brandy, olive-oil, almonds, and other products of the south of France.

TARASCON-SUR-ARIEGE, a town of France, on the l. bank of the river Ariège, 9 m. S of Foix. Pop. 1,551. In the neighbourhood are the iron-mines of La Rancié.—The cant. comprises 24 coms. Pop. in 1846, 17,775.

TARASORKA, a town of Russia, in the gov. of Charkov, 25 m. from Bogodukhof.

TARASP, a village of Switzerland, in the cant. of the Grisons, 2 m. S of Schuls, near the r. bank of the Inn.

TARAVO, a river of Corsica, which runs SW into the gulf of Valinco, after a course of 30 m.

TARAZONA, a town of Spain, in the prov. and 55 m. SE of Logrono, at the foot of the hill of Moncayo. It is a bishop's see, and has three churches and seven monasteries. Pop. 10,000, whose chief employment is the manufacture of coarse brown cloth. The environs produce large quantities of excellent fruit.

TARAZONA-DE-LA-MANCHA, a town of Spain, in the prov. and 54 m. SSE of Cuença, in a fertile plain, watered by the Jucar. Pop. 7,000, who carry on a trade in the products of the country, particularly wine and oil.

TARBAGATAI, the chief place of a division of Sungaria, at the E foot of Mount Takhta, in N lat. 46° 8′, E long. 82° 38′, on the river Imily. It is surrounded by a strong wall, and contains 6,000 inhabitants.

TARBAT, a parish in the extreme NE of the cos. of Ross and Cromarty, bounded on the N by the Dornoch frith. Area 5,081 acres. Along the coast are six natural harbours, and a number of small creeks. The chief harbour is at the village of Portmaholmack. Pop. in 1831, 1,809; in 1851, 2,151.

TARBATNESS, the point of the peninsula of the above parish, a headland which splits the waters of the Dornoch frith from those of the Moray frith, sweeping round to form the frith of Cromarty. On this narrow and far-projecting point has lately been erected a lighthouse in N lat. 57° 51′, E long. 3° 48′.

*** Tarbert, or Tarbet, a name applied to numerous places in Scotland, chiefly where there are such narrow necks of land between opposite indentations of sea or lake, as afforded in the primitive state of Scottish navigation an easy portage for curraghs and boats.

TARBERT, a port-town in the p. of Kilnaughtin, co. Kerry, on the coast road from Limerick to Tralee, 3 m. W by S of Glinn, at the head of a small bay, opposite Clonderalaw bay, in co. Clare, which bears the name of Tarbert bay. The steamers which ply between Limerick and Kilrush usually call off T. both in upward and in downward transit; and they achieve an average passage hence to Limerick in 4 hours. Considerable quantities of pigs, butter, and corn are shipped here for Limerick. Pop. in 1831, 956; in 1851, 996.

TARBERT (East and West), two sea-lochs on opposite sides of the southern peninsula of Argyleshire, approaching each other to within 1 m., and, together with the narrow isthmus between them, separating Kintyre from Knapdale. The W loch opens 13 m. due E of Ardmore-point in Islay; extends, in nearly a straight line, NE; and measures 11 m. in length, and about ¾ m. in mean breadth. At its head is a pier or quay for the accommodation of the Islay steam-packet.—The E loch is of small size,—only about 1 m. long, and nowhere more than about 3 furl. broad; but it is a curious and singularly safe and landlocked natural harbour, and is entered by a narrow and circling passage between low ridges of naked rock. On its S side, near the head, is a steam-boat quay. Behind the quay, and around the head of the loch, stands the neat, picturesque, and crowded village of Tarbert. T. is the grand seat of the celebrated Loch-Fyne herring-fishery; and, additional to its own pop., it is the resort, during the fishing-season, of several hundreds of stranger fishermen.

TARBES, a town of France, the capital of the dep. of Hautes-Pyrenees, situated on the l. bank of the Adour, 22 m. ESE of Pau, and 50 m. SW of Auch, in N lat. 43° 13′. It is surrounded with a wall, and defended by an old castle, and forms the key to communication with the roads conducting through the Pyrenees. The streets are tolerably broad, and well paved; the houses, though low, are constructed of brick or grey marble, and covered with slate. The only public edifices worth notice are the modern cathedral, the churches, the palace of the bishop, the theatre, and an hospital. Pop. in 1821, 8,035; in 1846, 13,521. There are here manufactories of linen, handkerchiefs, paper, knives, small copper articles, and leather. On 20th March 1814, the French army under Soult were forced from their position here by Wellington.

TARBET (EAST and WEST), two deep indentations of the sea, on the opposite coasts of Harris, approaching each other to within a ¼ m. See HARRIS. East Loch-Tarbet is 5½ m. long, and from 4½ to nearly 2 m. broad; it forks at the head into two slender bays; and it embosoms several islets, and has the considerable island of Scalpa at its entrance. West Loch-Tarbet is 6 m. long, and diminishes in breadth from 4½ m. to nearly a point. It is screened from the W winds by Taransay; and it is overhung by lofty mountains, which stoop precipitously down to its margin. At its head stands the solitary village of Tarbet, the seat of a mission on the Royal bounty.

TARBET, or TURBOT ISLAND, a small inhabited island in the p. of Omey, co. Galway, off the N side of Ardbear harbour. It measures about 1 m. in length, and 3 furl. in breadth. It is well cultivated, and has about 100 inhabitants.

TARBO, a town of Upper Guinea, on the Gold coast, on the r. bank of the Ancobra, 9 m. above its embouchure.

TARBOLTON, a parish in Kyle, Ayrshire. Its length is between 7 and 8 m., and its breadth about 6½ m. It lies about 5 m. from the sea-coast, and has an aggregate elevation above sea-level higher than the medium height of the co. The rivulet Faile flows through the interior to the Ayr. Pop. in 1831, 2,274; in 1851, 2,824.—Tarbolton, a considerable village, stands near the centre of the cognominal parish, on the r. bank of the rivulet Faile, 4 m. W of Mauchline. Weaving, in various departments of cotton, woollen, and silk, is carried on to a considerable extent here.

TARBOROUGH, a township of Edgecombe co., N. Carolina, U. S., on the Tar river, 85 m. above its mouth, and 63 m. E by N of Raleigh. Pop. in 1850, 700. Large quantities of beef, pork, Indian corn, tobacco, &c. are collected here for exportation.

TARBRET, a handsome village of co. Kerry, on the Shannon, 24 m. WSW of Limerick.

TARCHIN, a town of Bosnia, in the sanj. and 34 m. SE of Trawnik, on the l. bank of the Lepenitza.

TARCZAL, a town of Hungary, 2 m. W of Tokay. The wine produced at this place can scarcely be distinguished from Tokay.

TARCZYN, a village of Poland, in the obwodie and 21 m. SSW of Warsaw.

TARDAGIS, a village of Spain, in the prov. and 6 m. WSW of Burgos. Pop. 780.

TARDEBIGG, a parish in Worcestershire, 2½ m. E by S of Bromsgrove, crossed by the Worcester and Birmingham canal. It includes the hamlets of Bentley, Redditch, and Wibheath-Yields, with Tutnal and Cobley. Area 8,994 acres. Pop. 6,423.

TARDES, a river of France, in the dep. of Creuse, which rises 3 m. ENE of Croce; runs SW; and flows into the Cher on the l. bank, after a course of 45 m.

TARDETS, a village of France, in the dep. of Basses-Pyrenees, 30 m. WSW of Pau, on the r. bank of the Soison. Pop. 500.

TARDIERE (LA), a village of France, in the dep. of La Vendee, cant. and 2 m. NNE of La Chataigueraye. Pop. 1,200.

TARDOIRE, a river of France, which rises near Chalus, in the dep. of Haute-Vienne; runs NW, then S, and then NW; and after a course of 48 m. becomes spent in the vicinity of Coulgens.

TAREIRI, a river of Brazil, in the prov. of Rio-Grande-del-Norte, which runs into the Atlantic, in S lat. 6°, after a course of 18 m.

TAREM. See TARUN.

TARENA, a river of New Granada, in the prov. of Ystmo, which rises in the central ridge of the isthmus; runs ENE, collecting the waters of many other tributary streams; and enters the Atlantic by four mouths, forming three large islands in the gulf of Darien. This river also forms a large lake of the same name, at some distance from its entrance into the sea.

TARENT, an island on the W shore of the Persian gulf, immediately opposite Ketif. It is about 7 m. in length and in breadth, is well supplied with good water, and produces abundance of fruit.

TARENT, or TARRANT, a river of Dorsetshire, which rises near Studhampton, and gives name to several villages through which it passes in its way to where it falls into the Stour, 3 m. SE of Blandford.

TARENTO. See TARANTO.

TARF, a beautiful rivulet of Kirkcudbrightshire, formed by the union of Anstool-burn and Glengap-burn. Their united stream, afterwards augmented by seven or eight small burns, runs S 11 m. to the Dee, 1¾ m. above Kirkcudbright.—Also a rivulet of Athole forest, in Blair-Athole, Perthshire, which rises on the W side of Benvrackie, and flows 9 m. E to a point 2 m. S of Loch-Tilt, where it is joined from the N by the brief stream, misnamed the Tilt, and from the E by a stream of 5 m. run called Glenmorewater. The whole course of the Tarf is among the savage but sublime scenery of the most towering and impervious part of the great central mountain-range of Scotland.—Also a rivulet of about 7 m. length of course, in Inverness-shire, which rises near the great glen between Lochs Oich and Ness, and circles round the SW end of Stratherrick to the head of Loch-Ness at Fort-Augustus.

TARGON, a town of France, dep. of the Gironde, 17 m. NW of La Reole. Pop. 1,200.

TARGOWICA, or TERGOWICE, a town of Russian Poland, in the gov. of Podolia, on the river Simicha, 76 m. ESE of Breslaw. A confederation was formed here in 1791, by some Polish noblemen, which vainly attempted to support the new constitution of Poland.

TARHAR, a small district of Hindostan, in the prov. of Allahabad, bounded on the N by the river Jumna, near its confluence with the Ganges. Its inhabitants are chiefly Hindus.

TARIEGO, a village of Spain, in the prov. and 7 m. SSE of Palencia, on the l. bank of the Pisuerga.

TARIFA, a town of Spain, situated on a small bay on the N side of the straits of Gibraltar, in N lat. 36° 0′, W long. 5° 35′, 17 m. WSW of Gibraltar. It is fortified with a wall and towers; and in 1811 a British force lodged in it baffled all the efforts of the French to take it after a long siege. It was the *Julia Traducta* of the Romans, and received its present name from the Moors.

TARIFFVILLE, a village of Hartford co., in Connecticut, U. S., 12 m. NNW of Hartford.

TARIJA, a district and town of Bolivia, in the dep. of Potosi, stretching along a river of the same name, which descends from the Tacsora mountains, and flows ESE to the Vermejo. The serene sky and a fine temperature produce wheat, maize, the herb of Paraguay, the cocoa, the vine, and flax, which is cultivated merely for the sake of its seed. A vast number of cattle and sheep also find pasturage in this district.—The town is situated in S lat. 21° 44′, 190 m. N of Salta.

TARISSA, or TARCZA, a river of Hungary, which descends from the E flank of Mount Bringh, in the com. of Zips; and flows SE to the Hernath, which it joins after a course of 70 m.

TARKA, a river of S. Africa, which joins the Grand-Brakke, after a SW course of 80 m.

TARKHAN, a town of Russia, in the gov. of

Taurida, on the W coast of the Crimea, in N lat. 45° 21′.

TARKHU, a town of Asiatic Russia, in the prov. of Daghestan, 72 m. SSE of Kizliar, 2½ m. from the Caspian sea, on the slope of a mountain on the crest of which is the fortress of Burnu. Pop 9,000.

TARLAND-AND-MIGVIE, a parish, consisting of four detached districts, in the SW division of Aberdeenshire. The entire area is about 22 sq. m. The village of T. stands nearly in the centre of the subdivision of Mar, called Cromar, 9 m. NE of Ballater. Pop. in 1831, 1,074; in 1851, 1,197.

TARLETON, a parish of Lancashire, 8 m. N by E of Ormskirk. Area 5,405 acres. Pop. 1,945.

TARLETON, a township of Pickaway co., Ohio, U. S., 17 m. NE of Chillicothe. Pop. in 1850, 456.

TARMA, a province of Peru, comprehending several minor districts, and bounded by the prov. of Truxillo on the N, the Apurimac on the E, and the prov. of Lima and Huancavelica on the S. It is very productive in maize, and has abundance of cattle, the wool of which is here manufactured into cloth. It has also numerous mines of silver and quicksilver, which are worked to considerable profit.—Its capital, of the same name, is situated on the N shore of the river Chanchamayo, a branch of the Paro, in S lat. 11° 36′, in a deep narrow valley. Pop. 5,000, chiefly Creoles, Mestizoes, and Indians. The adjoining district is fertile; but the climate is unhealthy, as the surrounding mountains prevent a free circulation of air. Quicksilver, antimony, and silver are worked in the vicinity; and in several pits they dig saltpetre of an excellent quality.

TARMON, a rivulet of cos. Donegal and Fermanagh, which rises on the NW side of Crockinnagoe, near Lough Derg, and flows about 5 m. S to Lower Lough-Erne.—Also a hamlet and fishing harbour in the p. of Kilmore-Erris, co. Mayo, on the W side of Blacksod bay.

TARMONBARRY, or Termonbarry, a parish 7 m. E by S of Strokestown, in co. Roscommon, containing part of the village of Rooskey. Area 9,295 acres. Pop. in 1831, 4,048; in 1851, 3,044.

TARN, a river of France, which, rising in the SE of the dep. of Lozère, flows SW into the dep. of Aveyron; turns NW and then WSW, and joins the Garonne on the r. bank after a course of 220 m. Its principal affluents are the Dourbre, the Dourdon, the Rance, and the Agout, on the l.; and the Tescoa and Aveyron on the r.

TARN, a department in the S of France, formed of part of the old province of Languedoc, and bounded by the dep. of Aveyron on the N and E; by Herault on the SE; by Aude on the S; and by Haute-Garonne and Tarn-et-Garonne on the W. It has an area of 574,414 hectares; with a pop. in 1801 of 270,908; in 1841 of 351,656; in 1851, of 363,073, of whom above 50,000 are Protestants. Its surface is an undulating plain, traversed by several chains of small hills, which on the S and SE are connected with the Cevennes. Its principal river is the Tarn, here a large stream flowing from the NE to the Garonne, and navigable to Alby. The next river in importance is the Agout. In the N are the Aveyron and the Viaur. The soil is in general fertile; the climate temperate. The productions are wheat, barley, maize, hemp, flax, vines, chestnuts, fruit, coriander and anise seeds, and saffron. The culture of the vine is extensive. Mines of iron and coal are wrought. The department is divided into the 4 arrondissements of Alby, Castres, Lavaur, and Gaillac, which are subdivided into 35 cantons, and 319 communes.

TARN-ET-GARONNE, a department in the S of France, formed so lately as 1808, of portions of the deps. of Lot and the Haute-Garonne, Lot-et-Garonne, Gers, and Aveyron. It is intersected on the N by the dep. of Lot; by Aveyron on the NE; by Tarn on the E; by Haute-Garonne on the S; Gers on the SW; and Lot-et-Garonne on the NW. It has an area of 371,367 hectares; with a pop. in 1821 of 238,143; in 1841, 239,297; in 1851, 237,553, of whom above 40,000 are Protestants. Its surface is a plain traversed by three chains of hills, the highest of which is said not to exceed 1,200 ft. Its principal rivers are the Tarn, which here receives the Tescon, the Aveyron, and the Gimone; and the Garonne. Its soil is in general fruitful, and its climate mild. Its productions are wheat, barley, maize, hemp, flax, beetroot, vines, chestnuts, almonds, and figs. The pastures, from the general deficiency of rain, are of comparatively limited extent. Manufacturing industry is of small extent. The department is divided into the 3 arrondissements of Montauban, Moissac, and Castel-Sarrasin; which are subdivided into 23 cantons, and 193 communes.

TARNOCZ, or Tarnowce, a town of Hungary, in the com. of Liptau, 40 m. W of Kesmark. Pop. 1,100.

TARNOE, an island in the Baltic, off the SE coast of Sweden, in the group called the Skæres of Carlscrona.

TARNOGROD, a town of Poland, 52 m. WSW of Balcz. Pop. 1,700.

TARNOPOL, a circle of Austrian Gallicia, bordering on Russia on the E and N. Its territorial extent is 66·9 German sq. m. Pop. in 1843, 211,248. It was ceded in 1810 to Russia, as a recompense for the services which she had rendered to France in the campaign of the preceding year, but was restored to Austria at the congress of Vienna.—The capital, of the same name, is situated on the river Sireth, 84 m. ESE of Lemberg. Pop. 13,000, of whom a large proportion are Jews. The principal manufacturing establishments are tanneries, and woollen and linen manufactories.

TARNOS, a village of France, in the dep. of Landes, cant. and 3 m. N of Sant-Esprit. Pop. 1,300.

TARNOW, a circle of Austrian Gallicia, bounded on the N by the Vistula, and by the river Dunajec on the W. Its area is 71·1 German sq. m.; and its pop. in 1843, 240,000. It is in general a level country, with the exception of some hills in the S.—Its capital, of the same name, is situated on the river Biala, an affluent of the Dunajec, 47 m. E of Cracow, in N lat. 59° 0′. Pop. 4,400. It is a bishop's see, and has a gymnasium and high school. It has pleasant environs; but is on the whole ill-built.

TARNOWITZ, a town of Prussian Silesia, on the confines of Poland, 42 m. SE of Oppeln. Pop. 2,200. Productive mines of silver, iron, lead, and calamine are wrought in the vicinity.

TARO, a river which has its source in the Apennines, in Sardinia, in the div. of Genoa and prov. of Chiavari. It runs first NE; enters the state of Parma; bathes Compano; passes to the W of Parma; and 14 m. NNW of that town throws itself into the Po by the l. bank. It receives the Zeno on the l., and has a total course of about 75 m. Under the French empire, this river gave its name to a dep. of which Parma was the capital.

TARODANT, or Terodant, a town of Marocco, in the prov. of Susa, on the r. side of the river of that name. Pop. 10,000. It is built of stone, and is enclosed by walls 25 ft. in height, crenelated in all their extent, and of prodigious thickness. It has manufactories of saltpetre, and salt-works, and several tanneries and dye-works. The surrounding country is well-cultivated, and produces maize, water-melons, dates, olives and grapes of the largest size.

TAROLI, a small town of Hindostan, in the prov. of Agra, 26 m. E by N of Jansi.

TARONCA, a town of Portugal, in the prov. of Beira, comarca and 12 m. S of Lamego. Pop. 1,689.

TAROUSSA, or TARUSA, a town of Russia in Europe, in the gov. of Kaluga, 42 m. NNW of Tula, on the l. bank of the Oka, at the confluence of the Taroussa. Pop. 1,000. It has still some remains of an earthen rampart, by which it was formerly enclosed, three churches, and several shops. Sail-cloth is its chief article of manufacture. T. was formerly a small principality belonging to the younger princes of Tchernigov.

TARPA, a town of Hungary, in the comitat of Szathmar, 10 m. NNE of Matocz, in the midst of a woody marsh, near the r. bank of the Theiss.

TARPAULIN COVE, a harbour of the state of Massachusetts, U. S., on the S side of Naushon island, in Vineyard sound. It has a lighthouse on its W side.

TARPERON, or TARPEZON, a village of France, in the dep. of the Côte-d'Or, cant. and 1½ m. NW of Aignay. Pop. 180.

TARPORLEY, a parish and market-town in the co.-palatine of Chester, 10 m. ESE of Chester, comprising the townships of Eaton, Rushton, Utkinton, and T. Area 6,057 acres. Pop. in 1801, 1,866; in 1831, 2,391; in 1851, 2,632. The town, which is a place of considerable antiquity, is pleasantly situated on the London road to Chester, and within one m. of the Nantwich and Chester canal. It is neatly built, and consists chiefly of one long street, well-paved. The inhabitants are principally occupied in the manufacture of stockings, and in agricultural pursuits. About 2 m. to the S are the extensive ruins of Beeston-castle, a fortress, the strength of which was once proverbial.

TARQUI, a village of Ecuador, in the dep. of Assuay, prov. and 7 m. S of Cuenca, in the midst of lofty mountains, and near the source of the Paute. This village was visited early in the last century by La Condamine. A battle took place in its vicinity between the Peruvians and Columbians in 1829.

TARQUINO, or TARGUINO (SIERRA DE), a chain of mountains in the E part of the island of Cuba. It attaches itself on the E to the Sierra-de-Cobre, and terminates on the W in Cape Cruz. Its S base is bathed by the Atlantic. Its highest summit bears the same name.

TARRAFEL, a bay on the SW coast of the island of Santa-Antonia, Cape Verd archipelago, at the mouth of a river. It affords good anchorage.

TARRAGO, a lagune of New South Wales, in the co. of Argyle, between Breadalbane plains and Lake George.

TARRAGONA, an administrative province and a town of Spain, in Catalonia.—The prov. contains 7 partidos, and had a pop. of 253,477 in 1834; and 290,000 in 1849.—The town and port of T. is situated near the mouth of the river Francoli, in N lat. 41° 9', 49 m. WSW of Barcelona. Pop. 11,000. It is built on a hill, and is surrounded by walls with turrets. T. is the see of a bishop, and has a large cathedral in the Gothic style. It was under the Romans the chief town of the province called *Tarraconensis*, and in 516 was the seat of a church council. It was occupied by the British in the beginning of the 18th cent., with an intention of rendering it a naval station, for which, however, the harbour was not well-calculated, and on the acquisition of Gibraltar in 1704, the design was abandoned. In 1810, it was taken by the French; and an attempt to retake it, in June 1813, by a force under Sir John Murray, was not successful.

TARRAKAI CHANNEL OF), or GULF OF TARTARY, an arm of the sea, which extends NNE between the island of Sagalin and Mandshuria. It forms the N part of the sea of Japan, and is 450 m. in length from N to S, and 180 m. in extreme breadth. On the N it communicates by a narrow strait with the gulf of Sagalin.

TARRAMUNGALA, a mountain of New South Wales, in the co. of Argyle, 4 m. N of Lake George.

TARRANARKI, a district of New Zealand, in the W part of the prov. of New Ulster. Near its centre is a volcanic mountain, named Egmont or Taranaki, which has an alt. of about 8,840 ft. above sea-level. In its N part are Port Eliot and New Plymouth.

TARRANO, a village of Corsica, in the cant. of Alesani, 48 m. NE of Ajaccio, at the foot of lofty mountains, near the l. bank of the Alesani. Pop. 394.

TARRAS, a small but romantic river in Eskdale, Dumfries-shire, which rises on the S side of Hartsgarth, and runs 9 m. S and SSW to the Esk, 2½ m. below the town of Langholm.

TARRASA, a judicial partido and town of Spain, in the prov. of Barcelona. The partido comprises 32 pueblos. The town is 18 m. NNW of Barcelona, in a mountainous locality. Pop. 5,118. It has a collegiate and two other churches, a convent, and an hospital. It has manufactories of fine woollens.

TARRATA, a mountain of New South Wales, in the Wellington district, near the Kalingalungaguy river.

TARRAY, or DYUNBURS, a river of Australia-Felix, in the district of Western Port, an affluent of the Loddon.

TARREGA, a town of Spain, in the prov. and partido and 30 m. ENE of Lerida, in a pleasant and fertile plain. Pop. 3.915. Its streets are spacious and well-paved, and the houses well-built. It has three convents, an hospital, and cavalry barracks. It possesses several distilleries, and carries on an active trade in grain, wine, and oil.

TARRENGOWER, a mountain of Australia-Felix, in the district of Western Port, near the Loddon.

TARRETBURN, a township in Bellingham p., Northumberland, 4½ m. NW of Bellingham. Pop. in 1831, 265; in 1851,

TARRINGTON, a parish in Herefordshire, 6½ m. WNW of Ledbury. Area 2,224 acres. Pop. in 1831, 540; in 1851, 534.

TARRING (WEST), a parish in Sussex, 1½ m. NW by W of Worthing. Pop. in 1851,

TARSET (WEST QUARTER), a township in Thorneyburn p., Northumberland, 18 m. NW of Hexham. Area 17,408 acres. Pop. in 1831, 149; in 1851, 173.

TARSIA, a town of Naples, in the prov. of Calabria-Citra, district and 15 m. SSE of Castrovillari, in a fertile valley, near the l. bank of the Crati. Pop. 1,140. It has two parish churches.

TARSUS, a city of Asiatic Turkey, the ancient capital of *Cilicia*, situated in a fertile plain on the r. bank of the Cydnus, 9 m. above its embouchure. The houses are interspersed by gardens and orchards; they seldom exceed one story in height, are flat-roofed, and mostly constructed of hewn stone derived from the demolition of the ancient edifices. The castle is said to have been built by Bajazet; and the town is partly surrounded by a wall probably the remains of that erected by Harun-al-Raschid. The city contains public baths, a number of mosques, several handsome caravanserai, and a small ancient church. The land in the neighbourhood is exceedingly fertile, yielding wheat, barley, sesame, and cotton, which are exported to Malta, and thence to Spain and Portugal. Copper from Maden, and gallnuts from the mountains, are also staple commodities. The imports consist chiefly of rice, sugar, and coffee. The pop. during winter is stated at 60,000. The Turkmen migrate with their families in summer to the uplands of Tarsus. There are a few families from Cyprus, and a few Europeans. Mr. Barker reports, that the air of T. is unhealthy during the months of July and August, when the town and its environs are subject to exhalations productive of putrid and intermittent fevers. "The principal cause of this evil is a stagnant lake, about 30 m. in circumf., now a few miles from T., which formerly communicated with the sea, but which is now separated from it by a sand-bank. This is the harbour mentioned by Strabo, which he says was the port of T. Indeed, its position leads us to infer that the sea once came up to T.; but as the alluvium of the river has raised the ground considerably, it would be easy to drain this lake, while the ground reclaimed would be well adapted to the cultivation of sesame, cotton, and wheat, and its incomparable fertility the first year would no doubt repay a thousand-fold all expenses." T. certainly possesses a very high antiquity, and is said by Arrian and Strabo to have been founded by Sardanapalus on the same day with Anchiale, while others are of opinion that it was a Grecian colony founded by Triptolemus. It was much favoured by Augustus, as well as Adrian, and rose to such celebrity as to rival Athens, Antioch, and Alexandria, in wealth and grandeur, as well as in the cultivation of literature and science. It was also adorned with a number of magnificent temples, as well as with a gymnasium and theatre. It has, however, been subject to so many vicissitudes, and has been so often taken and plundered, as to retain scarcely a vestige of its former magnificence. Hardly a single inscription, or any monument of beauty or art, can now be discovered.

TARSUL, a village of France, in the dep. of the Côte-d'Or, and cant. of Is-sur-Tille, on the Jignon. Pop. 317.

TARTAKOW, a town of Galicia, in the circle and 33 m. NE of Zolkiew, in a marshy locality, on a small affluent of the Bug. It has manufactories of wood ware.

TARTARAGHAN, a parish of co. Armagh, 3¾ m. NNE of Loughgall, containing the villages of Maghery and Milltown. Area 11,612 acres, of which 1,917 acres are in Lough Neagh. Pop. in 1831, 6,321; in 1851, 6,724.

TARTARAS, a village of France, in the dep. of the Loire, cant. and 3 m. NE of Rive-de-Gier, on a hill, near the l. bank of the Gier and of the Givors canal. Pop. 700. Coal is wrought in the environs.

TARTARO, a river of Austria in Lombardy, which has its source in the prov. of Verona, district and 1½ m. ESE of Bardolina; runs first S; then SE, passes Villa-Franca; forms afterwards to some extent the boundary-line of the provinces of Mantua and Polesina; turns ESE into the latter prov.; waters Trecenta; and, after a total course of about 66 m., joins the Castagnaro, on the l. bank, to form the Blanco canal. In the lower part, which is navigable, its banks are generally marshy.

TARTARY,

A name vaguely given to a very extensive region comprehending all the tract between the Caspian sea and the great plateau of Central Asia, and extending from the frontiers of Persia and Afghanistan on the S, to those of Asiatic Russia on the N. The geography of this tract is only known in a very general way; so obscure is it that it has been justly characterized as being "chiefly conjectural," and as remaining, to the disgrace of science, "in a wretched state of imperfection." For our knowledge of it we are chiefly indebted to the Oriental historians and geographers, Abulfeda, Ebn-Haukel, Abulghazi, and others, and to some notices from Jenkinson, Thompson, and Hanway. These imperfect accounts have been followed of late years by more accurate information obtained by Mr. Elphinstone, in his mission to Peshawer, by Lieut. Abbott in his mission to Khiva, by Burnes, by the Russians, in their several embassies to Kokan, Khiva, and Bokhara, and from the memoirs of Baber, a native of this region, and sultan of Kokan, previous to his expulsion by the Usbeks, the present possessors. —Independent Tartary has the Caspian sea and the river Jaik or Ural on the W; the Ural-Tau, and the Algydim-Shalo, on the N and NE; Sungaria and Eastern Turkistan on the E; on the SE, the basin of the Upper Indus, or what we now denominate Western Tibet; and on the S, Persia and Afghanistan. From S to N it extends 16 degrees, or from 35° to 51° N lat., and from W to E 20 degrees, or from 54° to 74° E long., that is, from the Caspian sea to the summit of the Belur-Tagh. It must be observed, however, that as the eastern limit of this vast tract has never been astronomically determined, it is impossible to be precise on this point. The maps generally carry the longitude not farther than 70° and 72° E, and even Malte Brun fixes the eastern frontier, at the summit of the Belur, in only 69° E long., adopting the opinion of Rennel that the range of the Belur has been placed 5 degrees too far E by D'Anville. Independent of the Kirghissian portion of this tract, Balbi has estimated the total superficies of the Usbek and Turkoman portion at 555,000 British sq. m. thus:

The khanate of Bokhara,	230,000 sq. m.
" Khiva, . .	193,000
" Kokan, . .	132,000
	555,000

In this estimate the surface of the Aral lake is included, but whether he includes the tract S of the Oxus or Jihun, is not quite certain. Malte Brun makes the whole superficies, exclusive of the steppe of Ishim, only 460,000 sq. m., which is certainly too low. If the medium breadth of this tract be 15 degrees of longitude by 16 degrees of meridional length, then the surface, including the tract S of the Oxus, will be found to be 843,600 sq. m.; and this will not appear too great, if we compare it with the statements of Klaproth and Balbi. The former states the superficies of the tract occupied by the Lesser and Middle Kirghis hordes, lately incorporated with the Russian empire, at 220,000 sq. m., which, added to Balbi's statement of that of the Usbek khanates, makes 775,000 sq. m. To this must be added the large and extensive province of Bactria, extending more than 8 degrees of longitude by 2 degrees of medial breadth.—In compliance with long-established usage, we apply the customary appellation of Independent Tartary to this region. The name of Tartar has been for six centuries applied as a generic appellation to all the nomadic tribes of Northern and Middle Asia, in the same way as the ancients applied the term *Scythiæ* to all the erratic nations of which they had any knowledge whether in Europe or Asia. All the vagrant tribes to the N and E of the Danube, as far as geographical knowledge then extended, were anciently so called, from the time of Herodotus downwards; and the modern appellation Tartar, has succeeded by common consent to that of Scythian, ever since the middle of the 13th cent. So firmly has this ethnographic term taken root amongst all European writers, that it is now no more possible to eradicate it, than it would be to abolish the generic term *Indian*, universally applied to all the aboriginal natives of the American continent ever since its discovery. The first writer who introduced the term *Tartar* was the political missionary, Carpini, in 1246, but he applied it solely to the four great Mongolian tribes, who, he says, all spoke the same language, a strong proof that they belonged to the same parent stock. Rubruquis, in 1254, used it in the same sense, when he visited the court of Sartakh-Khan, grandson of Jenghis-Khan; but the Mongolian princes were indignant at being called Tartars, and told him that they were Mongols and not Tartars, who were a different and a vanquished tribe, and that they did not choose to be denominated from the name of a vassal horde. Still, however, the name prevailed, in spite of Mongolian remonstrance, and has been, and still is, applied not merely to the Mongolians, but also, most improperly and erroneously, by almost all modern writers, to designate exclusively, all the tribes of Turkish extraction, although it is certain these are a very different and distinct race. Hence the modern division of Great Tartary amongst Tartars, Mongols, and Mandshurs. The two latter names are just and proper, as designating distinct races, and as recognized by these races themselves; but the first appellation has never been acknowledged by any of the numerous tribes to whom it is applied. They all speak dialects of the same language, and know themselves only by the particular name of their own tribe, or by the general name of Turks. As all these tribes have the best right to fix their own name, it is unjust to call them by one they have never acknowledged; and if the name Tartar be at all applicable to any of the great races, it belongs to the Mongols,—one of whose tribes the ancient Tartars were,—with much greater propriety than to either of the others. By the Arab conquerors of Asia, and by the Arab and Persian geographers, the name of Turks was applied to all the nomadic hordes, Mongols as well as others, of which they had any knowledge, and they divided them into two great branches, Western and Eastern Turks, the former extending to the Black sea, and the latter as far E as China. In this they were more correct than succeeding European writers: both nations knew the people and their language. Their error lay in the too great extension of the name, in applying it to the whole of the nomadic races, instead of limiting it to one. As the real proper name of the race is *Turks*, so the whole region inhabited by the numberless tribes speaking Turkish, even as far E as Hami, at the eastern extremity of the Lesser Bukharia, should have been called *Turkistan*, instead of receiving the name *Tartary*. It is probable that all the Scythian tribes known to the ancients since the days of Herodotus were of Turkish origin, and spoke dialects of the same language. We shall, therefore, divide it into three great divisions, Southern, Middle, and Northern Turkistan, all comprehended under the general name of Western Turkistan, to distinguish it from the extensive region to the E of the Belur, called Chinese or Eastern Turkistan.

Divisions.] This region may be regarded as divided into

I. Southern Turkistan, S of the Amur or Oxus, which is subdivided into 1. Turkman Desert and Khiva; 2. Balkh; 3. Badakshan.

II. Middle Turkistan, which, N of the Amur, comprises 1. Khoilan; 2. Karrategin; 3. Hissar or Saganian; 4. Kesh; 5. The vale of Sogd, comprehending Bokhara and Samarcand; and in the vale of the Syr or Jaxartes: 1. Kokan and Nemingan; 2. Tashkent; 3. Uratippa; 4. Ghaznah or the desert of Ural; 5. Ilak or Iestan; and 6. Turkistan Proper.

III. Northern Turkistan, or the Kirghissian region, which is divided amongst the three hordes of the Lesser, Middle, and Great Kirghissian hordes.

Physical features.] All the W and N part of this region is occupied with immense desert plains, whilst the S and E parts, or the basins of the Syr and Amur, are of a different description, having fine tracts of land defended by inaccessible mountains and barren deserts, and watered by numerous streams. But even to this there are exceptions, the fertile plain of Kharizm being a mere basis surrounded by moving sands, and several steppes even occur here and there on both sides of the Syr and Oxus. The surface is very varied in the S and SE parts, consisting of numerous valleys, vast mountains, and extensive plains. The E shores of the Caspian present nothing but a long and gloomy succession of rocks and arid downs.—The southern chain of the Hindu-kush divides this region from Persia and Afghanistan. A number of lateral ranges, enclosing extensive well-watered valleys, run N toward the Amur; and the descent is much greater on the N side than on the S of the Hindu-kush. The other great ranges are the Belur-Tagh, the Muz-Tagh, the Asfera range, and the Kynder-Tagh, besides a vast number of subordinate ranges running in various directions from these chains. The Belur-Tagh, the great range which on the E divides this region from Central Asia, is called Thsoung-ling, or 'the Azure mountains,' by the Chinese, and has been long celebrated in the history of Tartary as the great culminating point where the waters flow to the Aral-Nor on the W, and the desert of Shamo on the E. The sources of these waters, running in opposite directions, are estimated by the Chinese geographers at 1,000 li., or 300 geo. m. of elevation; and, according to the geographers of the Ming dynasty, this range is many thousand Chinese feet in height. In the Mongolian and Oigurian languages, it is called the Bolur range; and in the Persian, Bellor Koh, or 'the shining mountain,' from the transparency of the quartz, and sometimes the Belut-Tagh, or 'the dark mountain,' from the perpetual clouds which overhang it. This chain is at least 500 m. in length from SSW to NNE, or from E long. 71° and N lat. 34° 30′, to where it meets the Muz-Taghin, in 41° N lat. and 71° E long., according to Waddington's map. From this point it passes N to 42° N lat. where it is intersected by the Alak-Tagh coming from the E. In this part of its course it is called Khashghar Divan. This chain is covered with ever-during snow, and so lofty, abrupt, and precipitous, that the only known passes are those of Badakshan and the source of the Syr. The Asfera range, otherwise called the Pamer mountains, form the S boundary of Firghana or Kokan, running E and W. It is a very broad chain of mountains rising from an elevated land, and probably is a continuation westward of the Muz-Tagh as far as the vicinity of Khojund. The fact that the only communication between Kokan and Bokhara is by the pass of Khojund, between the extremity of this range and the river Syr, is a strong evidence of its magnitude and impracticability.—From the Asfera range, in 67° 30′ E long., are detached the Ak-Tagh, or 'White mountains.' On approaching Uratippa, these are again subdivided into two branches, the most western of which terminates in 63° 30′ E long., and forms the N boundary of the vale of Sogd, whilst the Ak-Tagh, properly so called, separates Bokhara from Yar-Ailak, and terminates in two branches at Jizzukh and Joupar.—The Kara-Tagh, or 'Black mountains,' are another lateral range projected from the S side of the Asfera range, and runs S and SW near 400 m. towards the Amur. Next to the Asfera, it is the most lofty, rugged, and precipitous range in Usbek-Turkistan.—The Samarcand mountains, which form the S boundary of the valley of Sogd, are a branch of this range, projected westward as far as 66° E long. It separates the vale of the Sogd from the district of Kesh, where Tamerlane was born, and is called the hill of Kesh by Sherifeddin, his biographer, and the mountain of Zarkah by Ebn-Haukel. The Ak-Tau ridge is called Al-Botom by Abulfeda.—The Kinder-Tau is the range which bounds the country of Kokan and the vale of the Syr on the N, westwards as far as 65° E long., when it expires in the desert of Aral. It is another offset from the great range of the Belur, and has a longitudinal extent of eight degrees, or 400 m. It is of prodigious elevation, as we are informed by Nazaroff, and is covered with perpetual snow. It is also called the Ming-Bulak mountains, and corresponds to the Arga-Tau or Argjun, or Arka-ula, of Strahlenberg. From this range a lateral one runs S to the Syr, and to the W of Akhsikat, and divides the district of Kokun from that of Tashkant. It appears to be of no great elevation. To the N of the Kinder-Tau the country is little known, but seems not to be so mountainous as the territory of Usbek Turkistan. It is said that, at the NE angle of the Kinder-Tau, it is joined to a range of mountains running far to the E, and thus connecting it with the Ulugh-Tagh. This is probably the Kichuck-Tagh. But we are comparatively ignorant of the directions of the numerous mountain ridges which traverse the country of the great Kirghissian horde. The fact is, the whole of Eastern Usbek Turkistan is full of mountains and difficult both of access and description.—The whole of Usbek Turkistan may be considered as a large basin hollowed out by the waters descending from the Paropamisan and Hindu-kush mountains on the S, and those of the Belur and Kinder-Tagh on the E and N; but formed into two divisions by the Asfera mountains, thus making the two great valleys of the Syr and Amur. The geography of this region will be found further detailed in the articles Balkh, Badakshan, Bokhara, Khiva, Kokan, Kirghiz-Kazaks, and Turkistan.

TARTAS, a canton, commune, and town of France, in the dep. of the Landes and arrond. of Saint Sever. The cant. comprises 18 coms. Pop. in 1831, 16,325; in 1846, 17,104. The town is 13 m. WSW of Saint Sever, on the Midouse, 3 m. above its confluence with the Adour. Pop. in 1846, 3,039. It is divided by the river into two parts, distinguished as the upper and lower, and is generally well-built. The upper town, which occupies the slope of a hill, commands a magnificent view. T. has manufactories of linseed oil, and tanneries, and carries on an active trade in wine, grain, timber, saffron, and resin, fruit, hams, &c., the produce of the environs. It is also noted for its bread. Tortoises and red partridges are common in the vicinity. This town was taken from the English by Charles VII. It had a fortress which was long in the possession of the Huguenots.

TARTH, a small river, a tributary of the Lyne, in Peebles-shire, which, during the first 5 m. of its course, is identical with the Medwin. At Garwaldfoot, the Medwin, which had flowed chiefly on the boundary between Lanarkshire and Peebles-shire, splits into two streams, the one of which goes off into Lanarkshire and becomes tributary to the Clyde, and the other passes on to the Lyne, and through it, to the Tweed. The latter, though it continues for a brief space to be called the Medwin, and again is sometimes called near its mouth Newlands-water, is the Tarth, and undisputedly bears that name over the greater part of its course. Its length of run, measured in straight lines from Garwaldfoot, is only 7 m.; and over that distance it flows SE. Its tributaries, though 16 or 17 in number, are all inconsiderable. Compared to the Lyne, it is a deep, dull, and muddy stream.

TARTLAN, Tartlen, or Prasmar, a town of Transylvania, in the Saxon territory, district and 10 m. ENE of Cronstadt, on the Tartel. Pop., Saxon, 4,000. It has two churches, a Greek and a Lutheran, one of which was formerly fortified. Grain, flax, and tobacco are cultivated in the environs.

TARTOW, a town of Poland, in the gov. and obwod of Sandomir, 15 m. from Opatow, near the Vistula. Pop. 1,531.

TARTZMANSDORF, or Tarcsa, a village of Hungary, in the comitat of Eisenburg, 15 m. WSW of Guns, in the midst of mountains. It has several mineral springs.

TARUN, or Tarem, a town of Persia, in Farsistan, and district of Laristan, 210 m. SE of Shiraz. Pop. 3,000. It is enclosed on all parts, except the W, by mountains, and is surrounded by a lofty crenulated wall, flanked with towers, and a deep ditch. It is entered by only one gate. It has three mosques, a caravanserai, a bazaar, and an aqueduct. The houses are closely packed together, and the bad quality of the water, with which the town is supplied, renders it extremely unhealthy. It has a considerable trade with Bunder-Abbas in corn, sugar,coffee, and Indian goods.

TARUT, or Tirhut, an island in the W part of the Persian gulf, near the coast of Arabia, in the prov. of Lahsa, and opposite El Kalif. It is about 11 m. in length from NW to SE; is well-watered and adorned with fine gardens.

TARVES, a parish some miles NE of the centre of Aberdeenshire. Area about 30 sq. m. The village of T. stands nearly in the centre of the parish, 6 m. W of Ellon. Pop. in 1851, 2,469.

TARVIN, or Tarven, a parish in the co.-palatine of Chester, 6 m. E by N of Chester, comprising the townships of Ashton, Bruen-Stapleford, Burton, Clotton-Hoofield, Dudden, Foulk-Stapleford, Hockenhull-Stapleford, Horton-with-Peele, Kelsall, Mouldsworth, and T. Area 2,007 acres. Pop. 1,282.

TARVIS, a town of Illyria, in the circle and 19 m. SW of Villach, in a deep valley, near the l. bank of the Gailitz. About 4 m. S is Lake Raibler, in which the Gailitz has its source, and in the vicinity of which are mines of lead and calamine. T. formerly belonged to the bishop of Bamberg, and bore the rank of an archducal town.

TARVOERNE, a group of islands in the Atlantic, near the W coast of Norway, in the bail. of South Drontheim-Huusoee. The largest of the group is in N lat. 63° 48′, E long. 9° 23′.

TARWIN, a river of South Australia, in the district of Western Port, which has its source in the Strzelecki range, and flows into the ocean to the NW of Cape Liptrap.

TARYSSA, a district and town of Russia in Europe, in the gov. of Kalouga, on the Oka, at the confluence of a river of the same name.

TA-SANG, a town of the Corea, in the prov. of Hwang-hai, 90 m. NW of Han-yang.

TASAU, a town of Moravia, in the circle and 18 m. ESE of Iglau, on the slope of a hill, near the l. bank of the Aslawa.

TASBURGH, or Tasborough, a parish in Norfolk 8 m. S by W of Norwich. Area 916 acres. Pop. 475

TASCHITZ. See Daschitz.

TASCO, a town of New Granada, in the deps. of Boyaca and Tunja. Pop. 180.

TASCO, or TLACHCO, a town of Mexico, in the state and 75 m. SSW of Mexico, near the r. bank of the Zacatula. It is delightfully situated at an alt. of 915 toises above sea-level, and has a fine parish-church. The environs abound with fruit trees. Veins of silver are found in the locality.

TASCOFFIN, or TISCOFFIN, a parish in co. Kilkenny, 3½ m. NNW of Gowran. Area 4,708 acres. Pop. in 1831, 1,283; in 1851, 1,122.

TAS-EL-NAMLI, a cataract formed by the Tigris, between Mosul and Teksit, in Turkey in Asia.

TASGAON, a town of Hindostan, in the prov. of Bejapur, on an affluent of the Kistna, 90 m. WNW of Bejapur. It is well-fortified.

TA-SHAN, a range of mountains in China, which traverse the island of Formosa from N to S, and the summits of which are covered with snow as early as November. The principal summit, Mu-kang-shan, sometimes gives its name to the entire chain.

TASHAN-DEGH, a range of mountains in Turkey in Asia, in the NW part of the pash. of Sivas, on the confines of the sanjaks of Djanik and Amasia.

TASH-BURIK, TASH-BALIG, or TASH-BALIK, a small town of Chinese Turkistan, on the l. bank of the Yaman-yar, 42 m. SW of Kashgar, in N lat. 39° 6′, E long. 73° 34′ 20″.

TASHEM, ID, or ROWO, a volcano in the E part of the island of Java, and prov. of Balembang. At its last eruption, which took place in 1796, a large volume of water issued from its N side, charged with large quantities of sulphuric acid and particles of flint.

TASHEN-SEE, a lake of Bavaria, in the circle of Upper Bavaria, 15 m. NW of Salzburg. It is 8 m. in length from NNW to SSE, and from ¾ to 1 m. in breadth, and discharges itself by a small river into the Salza.

TASHINNY, or TAGHSHINNY, a parish in co. Longford, 3 m. NE of Ballymahon, containing the village of Barry. Area 4,880 acres. Pop. 2,333.

TASHKAND, or TASHKANT, a town of Independent Tartary, formerly an independent state, and now making part of that of Kokan, on the Tcherchek, a little above its confluence with the Sir or Sihon, and 150 m. NW of Kokan. Pop. 100,000. The greater part of the town is built in a valley, and is enclosed for 14 m. by a lofty brick wall, with 12 gates. In the interior, skirting the wall, are numerous gardens and orchards. Water from the river is conveyed into the town by means of canals, and is distributed to numerous fountains. Each house is supplied besides with a reservoir or bath for the use of the inmates. The mosques are without roofs, but many of the old temples are surmounted with antique cupolas. The suburbs contain numerous gardens. The constant passage of caravans through this town renders it a place of great importance. The inhabitants are indolent, but polite in their manners, and great lovers of music and dancing. It contains few artisans. They profess Islamism and keep their women in strict seclusion. Near T. is a fortress containing a garrison of 10,000 men, defended in the direction of Kokan by lofty walls and two deep ditches, and on the side towards Tashkand, by a wall at the foot of which is a deep canal. In the midst of these fortifications, on the site of the palace of the ancient Khans, is a castle, also surrounded by lofty walls and by ditches, and forming the residence of the governor. This town was formerly called Shah. The climate of the locality is pleasant, and the soil highly productive, affording every necessary of life, and in great abundance grapes, pomegranates, oranges, peaches, and figs. The banks of the river are finely lined with poplars.

TASH-KULVIN, a town of Asiatic Turkey in Anatolia, in the sanj. and 24 m. NE of Kastamim, in a plain, on the r. bank of the Kara-su, which is here crossed by a fine stone-bridge constructed to a great extent of the ruins of ancient architecture. Pop. 400. It contains 13 mosques, a khan, and baths, and has manufactories of stuffs and leather. This town occupies the site of the ancient *Pompeiopolis*.

TASHLIDGE, TASHLIZZA, or PLEVLE, a town of Bosnia, in the sanj. of Novi-Bazar, 70 m. SE of Bosna-Serai, on the slope of a mountain, near the r. bank of the Oschockina. Pop. 4,000. It has a public fountain, on the basin of which is an inscription in honour of Adrian. The environs are stony, but tolerably well cultivated.

TASIEEVA. See OUDA.

TASIMA, a prov. of Japan, in the W part of the island of Nifon, to the W of the provinces of Tango and Tamba, to the N of that of Farima, and E of the prov. of Inaba. On the N it is bathed by the sea of Japan. Yoki is its chief place. Its inhabitants find their chief employment in fishing and the culture of silk.

TASMANIA. See VAN DIEMEN'S LAND.

TASMAN'S HEAD, the most southerly point of Brune Island, in the co. of Buckingham, Van Diemen's Land, in N lat. 43° 46′.

TASMAN'S ISLAND, or THE PILLAR, an island of Van Diemen's Land, in the co. of Pembroke, to the S of Cape Pillar, the SE extremity of Tasman's peninsula.

TASMAN'S PENINSULA, a peninsula of Van Diemen's Land, in the SE part of the island, and co. of Pembroke, extending between 42° 56′ and 43° 12′ 31″, S lat. and between 147° 44′ and 148° 7′ E long., and bounded on the N by Norfolk bay, on the E and S by the ocean, and on the W by Storm bay. It is 22 m. in extreme length from N to S, and about 17 m. in breadth, and is connected with Forestier's peninsula, by a narrow isthmus on which is a military station. Its principal headlands are Capes Raoul and Pillar. Between these is Port Arthur; on the E coast is Fortesque bay, and on the W Wedge bay and Burnett harbour. It was first visited by Tasman in 1642.

TASNAD, TOSNAD, TRESSENBERG, TRESTENBURY, or TESNADU, a town of Transylvania, in the comitat of Kosop Szolnok, on the slope of a hill. Pop. 3,068. It has a fortress in ruins. The vine is cultivated in the environs.

TA-SOW HEEN, a district and town of China, in the prov. of Keang-se and div. of Nan-gan-fu.

TASQUE, a village of France, in the dep. of the Gers, cant. and 3 m. NNW of Plaisance, on the l. bank of the Sarros. Pop. 458. It had a Benedictine abbey.

TASSAI, a district of Japan, in the island of Kiusiu, and prov. of Tsikonsen.

TASSENIERE, a village of France, in the dep. of the Jura, cant. and 6 m. SE of Chaussin, in a woody and marshy locality. Pop. 400.

TASSISUDON, a town of Northern Hindostan, capital of Bhotan, 180 m. SW of Lassa, in a well-cultivated valley, about ¾ of a mile in breadth, and watered by the Tchin-tchien, the banks of which are lined with willows. It consists chiefly of the sovereign's palace, the residences of the government officers, and a long range of sheds in which forges are constantly employed in the manufacture of bronze idols and ornaments for the temples. It has also extensive manufactories of paper manufactured from the bark of a tree named deah, which abounds in the environs, and the filaments of which are sufficiently strong to admit of their being woven into textures. The palace, which is surrounded by a wall 30 ft. in height, forms a kind of citadel. It consists of 7 stories, each from 15 to 20 ft. in height.

TASWELL, or ST. AUGUSTINE, an island of the South Pacific, or more properly two islands on one reef, in S lat. 5° 37′, and E long. 176° 9′. It was discovered in 1781.

TAT, an insular rock in the Baltic, forming part of the group of the Erd-Holmer, and lying about 700 ft. to the N of Græsholm.

TATA. See DOTIS.

TATALISGA, a small town of Gallam, in Western Africa, 60 m. W of Gallam.

TATARA PASS, a route leading through the Khyber mountains in Afghanistan, in N lat. 34° 10′, to the N of the Khyber pass.

TATAR-BAZAR, or TATAR-BAZARGIK, a considerable town of European Turkey, in Bulgaria, on the l. bank of the Maritza, 16 m. WNW of Philippopoli, and on the great road from Constantinople to Belgrade. It is said to contain several mosques, baths, and other good buildings, with about 10,000 inhabitants.

TATARBUNAR, a town of Bessarabia, 70 m. S of Bender. It is said to have been once a populous city; but is now almost deserted. Near it is a lake, partly dried up in summer, which communicates with the Black sea.

TATE, a township of Clermont co., in Ohio, U. S. Pop. in 1840, 2,364.

TATENHILL, a parish in Staffordshire, 3¼ m. WSW of Burton-upon-Trent, containing the chapelries of Barton-under-Needwood and Wichnor, with the township of Dunstall. Area 9,408 acres. Pop. in 1831, 2,180; in 1851, 2,329.

TATESVILLE, a village of De Soto co., in Mississippi, U. S., 161 m. N of Jackson.

TATHAM, a parish in the co.-palatine of Lancaster, 11½ m. NE by E of Lancaster, watered by the Wenning, comprising the township of Ireby. Area 8,501 acres. Pop. in 1851, 654.

TATHWELL, a parish in Lincolnshire, 2½ m. S by W of Louth. Area 4,314 acres. Pop. in 1831, 338; in 1851, 429.

TATIBSCHEVSKAIA, a fortress of Asiatic Russia, in the gov. of Ufa, on the r. bank of the Ural, 28 m. W of Orenburg.

TATING, a town of Denmark, in the duchy of Sleswick, 9 m. WNW of Tonningen, on the r. bank of the Eider. Pop. 1,000.

TA-TING-FU, a division and town of China, in the prov. of Kwei-chu. The div. comprises 5 districts. The town is in N lat. 27° 5′, E long. 103° 33′.

TATMAGUCHE, or TATAMAGUCHE, a port of Nova Scotia, on a short bay which sets up southerly from the straits of Northumberland, about 25 m. from Onslow. It has a good road for vessels.

TATNALL, a county in the SW part of Georgia, U. S. Area 1,023 sq. m. Pop. in 1840, 2,724; in 1850, 3,227. It is watered by the Great Ohoopee, a tributary of the Ocmulgee. It cap. is Reidsville.

TATNAM (CAPE), the E point of Haye's river, in Hudson's bay, in N lat. 57° 35′.

TATRA MOUNTAINS, that part of the Carpathian chain which lies in the Hungarian counties of Zyps, Liptau, and Arva, and on the S frontier of Gallicia. It contains the highest summits of the chain, amongst which are the Krivan and the Lomnitz, rising to the height of at least 8,500 ft. above the level of the sea.

TA-TSEEN-TU-TING, a district, town, and fortress of China, in the prov. of Sze-chuen and div. of Ya-chu-fu, 150 m. WSW of Ching-tu-fu, on the Tung-po-ho, in N lat. 30° 8′ 24″, E long. 101° 49′ 50″. The walls are of free-stone. It is inhabited by Chinese and Tibetians, and forms an important egress from China of troops for Tibet, and of exports of tea under government inspection. Its inhabitants are Buddhists. The surrounding country is cold, bristled with mountains, rocks, and precipices, and is intersected by the Lou-ho. It formerly belonged to Han-chao.

TATSFIELD, a parish in Surrey, 8½ m. SE of Croydon. Area 1,276 acres. Pop. in 1851, 182.

TA-TSUH, a district and town of China, in the prov. of Sze-chuen, and div. of Chun-king-fu, in N lat. 29° 50′, and E long. 105° 51′.

TATTA, a village of Barbary, 240 m. SE of Marocco, in the prov. of Draha, on the skirts of the Sahara.

TATTA, an extensive district and a town of Sind. The district comprises the whole of the delta of the river Indus, 150 m. in length, by 50 m. in breadth. It is in general sandy and barren; but it produces rice and salt, and the numerous rivers abound with fish. This prov. was invaded by the Arabs in the beginning of the 8th cent., and may be considered as having been subject to the Mahommedans from that period. It was taken possession of by Akbar in 1590, and acknowledged the Mogul authority till the dissolution of that empire. Its remaining history is identical with that of Sind.—Its former capital was Brahminabad; its present capital, of the same name as the prov., is situated about 3 m. W of the Indus, and 130 m. from the sea, in a fertile valley formed by the low range of the Minkali hills, which, during the freshes of the river, is frequently inundated. Some of the houses are built of brick and mortar, but the greater number are constructed of mud and timber. The circumf. of the modern town is 4 m., and it is supposed to contain 15,000 inhabitants, but some carry it as high as 40,000. Barnes states that it was once 30 m. in circuit. It was formerly famous for its commerce and manufactures; but its wealth and trade have greatly declined, although it still continues to carry on a considerable traffic, and to manufacture large quantities of a rich cloth, called *lungis*, and coarse cottons. The art of cotton-printing is skilfully practised here. The modern city of T. was founded by Jam-Mundel in 1445, the 14th of the Someah dynasty, A. D. 1585, and was plundered by the Portuguese in 1555. Robertson is of opinion that T. was the *Pattala* of the Greeks.

TATTARAN, a small island in the Sulu archipelago, in N lat. 6° 10′.

TATTENHALL, a parish in the co.-palatine of Chester, 4½ m. SW by W of Taporley, watered by a small branch of the Dee, comprising the townships of Golborn-Bellow, Newton by Tattenhall, and T. Area 4,134 acres. Pop. in 1831, 1,080; in 1851, 1,204.

TATTENHOE, or TOTTENHOE, a parish in Bucks, 3¾ m. W of Fenny-Stratford. Area 690 acres. Pop. 55.

TATTERFORD, a parish in Norfolk, 4 m. W by S of Fakenham, watered by the Wensum. Area 959 acres. Pop. in 1831, 75; in 151, 86.

TATTERSET, or GATESEND, a parish in Norfolk, 5 m. W of Fakenham. Area 1,759. Pop. in 1851, 189.

TATTERSHALL, a parish and town in the co. of Lincoln, 9 m. SSW of Horncastle, on the river Bane, near its junction with the Witham, and intersected by the Horncastle canal. The church, which was made collegiate in the reign of Henry VI., is a spacious and elegant cruciform edifice. SW of the town stand the remains of T. castle, a stately edifice erected about the year 1440. Area of p. 4,580 acres. Pop. in 1831, 599; in 1851, 987.

TATTINGSTONE, or TADINGSTON, a parish of

Suffolk, 5½ m. S by W of Ipswich, in the line of the Eastern Counties railway. Area 1,637 acres. Pop. in 1831, 356; in 1851, 597.

TATTORA, a town of Hindostan, in the prov. of Bejapur, in N lat. 17° 53′.

TATTUBT, the remains of a considerable town in Algiers, 25 m. S of Constantina.

TATTYMOLE, a mountain in co. Tyrone, 3½ m. SSW of Fintona, connected by hill ranges with the great mountain-group of both Tyrone and Fermanagh. Its summit has an alt. of 1,031 ft. above the level of the sea.

TA-TUNG, a district and town of China, in the prov. of Kan-suh and div. of Se-ning-fu, in N lat. 36° 50′, E long. 102° 50′.

TATUTCHE, a barren island on the NW coast of North America, at the entrance of Juan-de-Fuca's straits. It is of small extent, and is surrounded with breakers in every direction.

TAUBATE, or THAUBATE, a town of Brazil, in the prov. and district and 90 m. NE of São-Paulo, on the r. bank of the Paraiba, a little above the confluence of a small river, in S lat. 22° 54′ 12″, W long. 45° 20′. It is one of the principal places in the prov., and has a church, two chapels, and two convents, and possesses manufactories of matting and vases. The district produces tobacco, sugar, coffee, cotton, millet, and pulse. Pop. 10,000.

TAUBER, a river which has its source in Würtemberg, in the circle of the Jaxt, bail. and 11 m. NE of Gerabronn; runs first SE past Hausen, turns afterwards E into Bavaria; runs N and NNW; passes Rotenburg; re-enters Würtemberg, but soon returns to Bavaria; bends W into the circle of the Lower Main; bathes Rottingen; again re-enters Würtemberg; waters Weikersheim, Markelsheim, Mergentheim, and Edelfingen; takes a NW direction; penetrates the grand-duchy of Baden; passes Konigshofen, Landa, Distelhausen, and Bischofsheim: and after a total course of about 84 m., joins the Main on the l. bank, at Wertheim.

TAUBER-BISCHOFSHEIM. See BISCHOFSHEIM.

TAUBRATH, a village of Austria, in Bohemia, in the circle of Elnbogen and district of Frais. It has mines of cobalt and of iron.

TAUCHA, a town of Saxony, on the small river Partha or Parde, 6 m. ENE of Leipsic. It was the scene of sharp fighting at the battle of Leipsic, on 18th October 1813. Pop. 1,300.

TAUCRA, or TEUKERA, a town of Tripoli, in Barca, 21 m. SW of Ptolemeta, on the Mediterranean. This town, the *Teuchira* of antiquity, still retains a portion of its ancient walls, strongly built of free-stone, defended at short intervals by quadrangular towers. It has also the remains of several monuments. The sepulchral grottoes of T. bear a close resemblance to those of Ptolemais.

TAUDA, a river of Asiatic Russia, formed by the junction of the Sosva and the Losva, which falls into the Tobol, 40 m. S of Tobolsk.

TAUDENY, a large village of the Sahara, on the caravan route from Marocco to Timbuctu. A supply of excellent water gives fertility to the district, and enables it to afford refreshment to the travellers crossing the desert. In the neighbourhood also are numerous beds of salt, an article of considerable demand throughout the countries on the Niger. These beds are from 5 to 6 ft. deep, and from 20 to 30 yds. in circumf. The salt, which is partly red, comes up in large lumps mixed with earth.

TAUDICUMBU, a town of India, in the district of Dindigul, in N lat. 10° 24′.

TAUGON-LA-RONDE, a village of France, in the dep. of Charenté-Inferieure, 18 m. NE of La Rochelle. Pop. 1,000.

TAUJEPUR, a town of Bengal, in the district of Purneah, in N lat. 25° 45′.

TAULE′, a town of France, dep. of Finistere, 3 m. NW of Morlaix. Pop. 2,800.

TAULIGNAN, a town of France, dep. of the Drôme, 14 m. ESE of Montelimart. Pop. 1,800, employed partly in the manufacture of silk.

TAULLAH, or JALLAH-MHUKI, a town of Hindostan, in the prov. of Lahore, district of Nadone, in N lat. 32° 5′. It contains a temple held in high estimation by the Hindus, on account of a volcanic flame which issues from the side of a mountain in its vicinity.

TAUMAGO, an island in the Pacific, in S lat. 10° 0′, E long. 169° 25′, discovered by Quiros in 1606, and is about 24 or 25 m. in circumf.; and abounds with bananas, cocoa-trees, palms, and many kinds of nutritious roots.

TAUNDA. See TANDA.

TAUNSK, or TAUISK, a small fortress of Russia in Asia, in the district and 240 m. ENE of Okotsk, at the confluence of the Tain with the Kova, and at the distance of about 30 m. from the shore of the gulf of Taunskaia-Gouba.

TAUNSKAIA-GOUBA, a gulf of Russia in Asia, in the district of Okotsk and N part of the see of that name. It is 135 m. in length from E to W, and 60 m. in breadth. Its principal affluent is the Tain, which descends from the Stannovoi mountains, and flows into the NW part of the gulf.

TAUNTON, a town and parl. borough in the co. of Somerset, on the navigable river Tone, and in the line of the Bridgewater canal, and the Bristol and Exeter railway, on which there is here a principal station 45 m. SW of Bristol, and 22 m. SSW of Bridgewater. The two parishes of Taunton St. James, and Taunton St. Mary Magdalene, contain 2,755 acres. Pop. in 1801, 5,794; in 1831, 11,139; in 1851, 13,519.—The town, which has extended beyond the old borough limits, into the parishes of Bishops-Hull and Wilton, is a large well-built place. The main streets are spacious, well-paved, and lighted with gas. There are various smaller streets, lanes, and courts, inhabited by the poorer classes, and here popularly called 'colleges;' but on the whole the town is considered both airy and salubrious. The country around, in the large and fertile vale of Taunton, abounds with pleasant orchards and villages, and renders the situation of this town one of peculiar beauty. T. returns 2 members to parliament. Pop. of parl. borough in 1851, 14,176. The number of electors registered in 1837, was 864; in 1852, 790. T. is also the principal place of election for the western division of the county. The principal manufacture of T. has been that of silk, in which about 1,800 hands have been at one time employed; but this manufacture has been on the decline of late years. There was formerly a considerable woollen manufacture, which has also gradually declined.—T. was the residence of some of the West Saxon kings, especially of their lawgiver, Ina, who held in it the great council of his kingdom.

TAUNTON, a township and one of the capitals of Bristol co., Massachusetts, U. S., on the river Taunton, 33 m. S of Boston, in N lat. 41° 54′ 11″, W long. 71° 05′ 55″. It is a pleasant and handsome town, and contains a court-house, a town-house, and ten churches. It has paper-mills, nail manufactories, rolling and slitting mills, and cotton manufactories, and several cloth-printing works. Pop. in 1820, 4,520; in 1850, 10,431.

TAUNTON, a river of the United States, which empties into Narraganset bay, at Tiverton, opposite the N end of Rhode Island. It is formed by several streams which rise in Plymouth co., Massachusetts.

Its course is about 50 m. from NE to SW, and it is navigable for small vessels to Taunton, about 20 m.

TAUNTON-DEAN, a district of England, in the co. of Somerset, extending about 30 m. along the course of the river Tone, and noted for its remarkable fertility and produce.

TAUNUS, a mountain-ridge of W. Germany, in the duchy of Nassau, which extends SW to the borders of the Rhine, and on the NE is attached to the Vogelsgebirge. It forms the water-stead between the Lahn and the Rhine. Its highest point, the Feldberg, has an alt. of 2,605 ft. above sea-level.

TAUPONT, a village of France, in the dep. of Morbihan, cant. and 1 m. NW of Ploermel. Pop. 2,300.

TAURANO, a town of Naples, in the prov. of Terra-de-Lavoro, 6 m. ESE of Nola. Pop. 1,280.

TAURASI, a town of Naples, in the prov. of Principato-Ultra, 12 m. NW of Santo-Angelo, near the r. bank of the Calore.

TAURAT, a settlement in the island of Cuba, 38 m. NNE of St. Jago.

TAURICASTRO, or TAURO-CASTRO, a town of Greece, in Livadia, opposite to Negroponte, 20 m. NNE of Athens.

TAURIDA, a government in the south of European Russia, between the parallels of 44° 26′, and 47° 43′ N; bounded on the N by the govs. of Kherson and Yekaterinoslav; on the E by the sea of Azof and the strait of Jenikale; on the S and W by the Black sea. It is composed of the following territories: 1st, The peninsula of the Crimea; 2d, a considerable track to the N of that peninsula, between the Dnieper and the Berda; 3d, the island of Taman or Tmutarakan; 4th, the land of the Tschernomorski or Black Sea Cossacks, lying to the E of the Crimea. All these, except the second, are described under their respective articles in the course of the present work. They are combined into one government, which takes the name of Taurida, from its principal part called by the ancients the *Taurica Chersonesus*. The superficial area of this prov. is 56,280 sq. versts, or 1163·38 German sq. m. of 15 to a degree; but the inhabitants are thinly scattered, their number being returned at 572,200 in 1846. This gov. contains some fertile tracts, particularly in the Crimea, but has also in its N part immense steppes extending to the Dnieper, the Kaskaia, and the Berda, many of them almost entirely unproductive, and all deficient in water, though others are covered with the finest grass. The summer is mild, but the winter, though short, is very severe. The mountains of Taurida, a high range of limestone mountains, precipitous on the S face, but sinking gradually into the plain on the N, divide the Crimea into two parts, remarkable for difference of climate; the northern, by much the larger, being neither pleasant nor healthy, while the southern, which is properly a stripe, may be said to resemble in its degree of heat, and in its vegetable products, the most favoured parts of Asia Minor. The highest points are the Tchatyr-Dagh, 4,700 ft., and the Babugan-Jaila, of about equal height. Between these a cross valley runs with a pass at an elevation of 2,000 ft. "The tract lying between the steep side of the mountains and the low sandy sea shore of the Crimea is peculiarly circumstanced. It is entirely protected from all the rough winds of the N, while it is quite open to all the warm breezes blowing from the S shores of the Black sea. It consequently enjoys an exceedingly mild climate, which allows the vine, olive, pomegranate, and all the south European fruits, to come to perfection. The principal culture is that of the vine. The vine-yards extend at present into a number of the northern valleys of the Tauric chain, but the wine is much poorer than that of the south coast. Fruit is, however, largely cultivated in the northern valleys, with an exclusive privilege of supplying Moscow and St. Petersburg. The southern declivity of the ridge is covered with forests of the Corsican pine, the region extending from 600 to 3,000 ft.; on the N side, which is much colder, they are replaced by beech. *Arbutus Andrachne* occurs only on the S side, from the coast up to 1,200 ft., but usually solitary, and is supposed to have been derived from seeds brought by birds-of-passage from Anatolia." [*Henfrey.*] Situated between the Black sea and the sea of Azof, that is to say, between the Danube on the W, the Dnieper on the N, and the Kuban on the E, all grand commercial affluents of the eastern portion of the European continent, and of Southern Russia, as likewise of the Caspian basin, this region holds a most favourable position for commerce and trade. The Crimean district is specially favoured in its interior by the mildness of its climate and by the fertility of a large portion of its territory, which is susceptible of every culture. Corn, wine, cattle, wool, pelts and furs, hides, hemp, honey, oil, salt, and fisheries, constitute the chief elements composing the wealth of the country, and a transit-trade also exists in corn and grain, oleaginous seeds, tallow and grease, tobacco, and silk tapestry, which are brought for barter with the stuffs, sugar, hardware, and other articles wrought in Europe, more especially in Russia itself. Corn constitutes the bulk of the exports from the Crimean harbours—these harbours being adjuncts and almost dependents on the harbour of Odessa, that granary of the Levant, or rather of Southern Europe. According to the official reports for 1851 from the government of Taurida, the corn-harvest had increased to 2,568,497 hectolitres: 10 years before it was hardly 1,000,000 h. It is particularly in the district of Berdiansk, peopled in part by foreign settlers, that the culture of the cereals is most developed, and, it is thought, that the entire region of the Crimea, with that of the sea of Azof, may supply commerce annually with 5,000,000 or 6,000,000 hectolitres of corn. Moreover, the Crimea, in 1851, was found to possess nearly 2,000,000 sheep, half of which were fine-wooled, 248,260 head of horned cattle, and 85,700 horses. The salt-mines of Perekop and Eupatoria have some celebrity, and although inadequately worked, are a valuable source of wealth to the country. The entire vintage of the Crimea—the greater part of which is consumed in the country, and the remainder of which is sold to customers in the provinces of Southern Russia—may amount, it is said, to about 160,000 hectolitres. The wines exported from Taurida are in general of secondary quality, and are chiefly used, like those from the Caucasus, for mixing with other wines or with other preparations. M. Tegoborski says, that Taurida possessed, in 1848, 35,577,000 vines, a number six times larger than what grew there sixteen years before. The Russian government has at all times made great efforts to develop the culture of the vine in the Crimea, and to say the truth, it is almost the only culture which has acquired there any importance. Manufactures are at the lowest ebb. Two or three factories for the weaving of common cloth, a few tanneries, and a few yards for making morocco leather seem to constitute the amount of manufacturing means possessed by this gov. As for the value of the exchanges carried on in the entire basin of the Crimea and the sea of Azof, the *Annales du Commerce Extérieur* reports, that in 1841 the imports and exports were as follow:—

	Imports.	Exports.	Total.
Ports in the Crimea.	780,000f.	2,308,000f.	3,088,000f.
Ports in the sea of Azof,	5,208,000f.	22,088,000f.	27,296,000f.

Ten years later, in 1851, the value of the traffic of the Crimea was only 1,747,000f. a result showing a great diminution, and for the ports in the sea of Azof, 34,084,000f., showing a great increase. Kertch, placed at the straits separating the Crimea from the Transcaucasian provinces, and Taganrok, situated quite at the bottom of the sea of Azof, conduct much of this commerce. They alone exported in 1851, corn to the value of 7,564,000f.—a sum almost equal to the aggregate amount from all the other ports. We must not, however, measure the commercial activity in the ports of the Crimea and the sea of Azof simply by the results of the foreign trade: the coasting trade, which is there extremely active, would give almost an equal value of exchanges. The home trade is also of some importance in the Crimea, and it may be judged of by remarking that there are 79 fairs held in this district every year. Goods to the value of 2,494,000 rubles, or £75,000, were brought to them in 1851; and what is remarkable is the fact, that, with the exception of the two fairs at Simpheropol, all of them are held in the northern districts, which are almost exclusively peopled with Christian agriculturists. To sum up, the foreign trade of the two seas, in 1851, employed 1,561 ships, carrying 400,000 tons; and the coasting-trade may well have been three times larger. The coasts of the Crimea offer a large number of harbours that in all times have been eminently useful to ships frequenting these difficult and sometimes dangerous seas. The chief harbours are Eupatoria, Theodosia, or Kaffa, Kertch, and Sebastopol, to which we must add, as belonging to the same sphere of commercial activity, the ports in the sea of Azof, viz. Berdiansk, Mariopol, Rostoff, and Taganrok.—The Genoese thoroughly understood the importance of such a line of coast when, towards the end of the 13th century, they purchased, or rather took, from the Mongol-Tartars the ancient *Theodosia*, spread their colonies over all Taurida, covered with their ships the shores of the Euxine, and founded Kaffa, which soon became the principal centre of Europe's commerce with Asia Minor, Persia, and the Indies. Two centuries later the Crimea was for a long time blighted, as it were, with sloth and sterility; its cultures, its commerce pined away more and more through atrophy, and the yoke imposed on it by the Muscovites in 1740 was little calculated to restore it; but by the franchise granted by the Empress Catherine to its ports subsequently, the peninsula saw its prosperity rapidly return. Unfortunately, the Czar Paul thought he ought to protect the commerce of Taurida by cancelling this franchise, and replacing it by an oppressive system of customs, with all their restrictive regulations. The inhabitants are Tartars of three or four different tribes, Cossacks, Russians, Armenians, Jews, Gypsies, and foreign colonists, chiefly of German descent. Schnitzler, in 1835, estimated the Tartar pop. at only 100,000, and reported it to be rapidly decreasing. The prov. is divided into 6 circles, besides the isle of Taman, and the land of the Tchernomorski Cossacks. See articles Azof (Sea of), Crimea, Odessa, and Sebastopol.

TAURINES, a village of France, in the dep. of Aveyron, 15 m. SSW of Rhodez, near the Seor, an affluent of the Viaur.

TAURION, or Thorion, a river of France, which rises in the dep. of Creuse, cant. and 1 m. S of Paillier, and flows into the Vienne, on the r. bank, at St. Priest, 7 m. NE of Limoges, after a course of 52 m.

TAURIS. See Tabriz.

TAURISANO, a town of Naples, in the Terra-d'Otranto, 15 m. SE of Gallipoli. Pop. 1,300.

TAUROGEN, a large town of Russia in Europe, in the gov. and 135 m. W of Vilna, and district of Rossiena, on a small affluent of the Niemen.

TAUROGINA, a town of Russia in Europe, in the gov. and 60 m. NNE of Vilna and district of Vidzoni, on a small lake.

TAURUS, a lofty chain of mountains, situated in the eastern part of Asia Minor, where it borders on Syria. It commences on the r. bank of the Euphrates, and runs from E to W through the pash. of Marash; then turns SW, separating Caramania from the pash. of Itchil; then turns NW to the source of the Nabis, near which it forks into two branches, one of which runs NW towards the straits of Constantinople; the other SW, terminating at Cape Arbora, a little S of the embouchure of the Buyuk-Mendereh. The chain is very lofty, and approaches so near to the Mediterranean as in some places to leave only narrow passes, the most celebrated of which is that of Issus, where the battle was fought between Darius and Alexander. These mountains are in many places covered with vast pine forests. They are traversed in summer by Turcoman shepherds, who in winter descend and take up their residence in the towns. See article Asia Minor.

TAUSHAN, or Rabbit Islands, a group of small islands in the Archipelago, 6 m. N of Tenedos, near the coast of Turkey in Asia. They are mountainous and rocky, but well-wooded.

TAUSS, or Drastow, a town of Bohemia, 15 m. W of Klatau. Pop. 4,600. It is surrounded with a wall, and has large manufactures of thread and linen.

TAUSTE, a town of Spain, in the prov. and 27 m. NW of Saragossa, on the small river Riguel, near its influx into the Ebro. Pop. 3,500. It is situated in a fruitful district. A canal from this place forms the great canal of Aragon, and promotes the internal trade of the country.

TAUTENBURG, a village in the grand-duchy of Saxe-Weimar, 1 m. ESE of Dornburg.

TAUTEREN, an island on the coast of Norway, in N lat. 62° 41'.

TAUVES, a town of France, in the dep. of Puy-de-Dome, on the small river Mourlagne, 25 m. SW of Clermont. Pop. 1,200.

TAUXIGNY, a village of France, in the dep. of Indre-et-Loire, cant. and 9 m. NW of Loches, on the Echandon. Pop. 1,400.

TAVAI. See Tavoy.

TAVAI-POENAMMON. See New Zealand.

TAVALY, one of the Molucca group, to the SW of Gilolo, in S lat. 0° 20'.

TAVANDA, a river of Bohera, in the prov. of Ibarra, which rises in the desert of Cayamburu, and passing through the town of Ibarra, turns E, and enters the Mira.

TAVANNES, a village of Switzerland, in the cant. and 21 m. NW of Berne, and 1 m. NNE of the Pierre-Pertuis passage. Pop. in 1851, 672.

TAVARA, a village of Spain, in the prov. and 27 m. NNW of Zamora. Pop. 2,000. There is a fine palace here of the Dukes de l'Infantado.

TAVASTEHUS, or Tavastland, a district of Russian Finland, bounded by E. Bothnia on the N, and by the gulf of Finland on the S. Its length from N to S is about 180 m.; its breadth from E to W varies from 35 to 100 m. Pop. about 350,000, chiefly Fins. The N part is hilly, and covered with wood, but the rest is a level country, diversified with arable and meadow land, and well-watered with rivers and lakes. The exports comprise barley, pease, beans, flax, hemp, dried fish, cattle, leather, tallow, butter, lime, and bark. Its capital is Helsingfors.

TAVASTEHUS, or KRONEBORG, a small town of Russian Finland, formerly the capital of the district of the same name, 80 m. ENE of Abo, in N lat. 61° 3′. It is situated among marshes, and built of wood, but its streets are broad and irregular. It has a castle, with an arsenal and magazines.

TAVDA. See TAUDA.

TAVE, a river of Carmarthenshire, which runs into the Severn at St. Clear, near Laugharn.

TAVE. See TAF.

TAVEL, a village of France, in the dep. of Gard, 15 m. E of Uzes. Pop. 800.

TAVENNA, a town of Naples, in the prov. of Sannio, 9 m. NW of Larino. Pop. 1,500, speaking an Illyrian dialect.

TAVERHAM, a parish of Norfolk, 5¾ m. NW of Norwich, watered by the Wensum. Area 2,099 acres. Pop. in 1831, 191; in 1851, 207.

TAVERNA, a town of Naples, in Calabria-Ultra, 12 m. ENE of Nicastro, near the l. bank of the Alli. Pop. 2,400. It has coarse woollen manufactures. It was greatly devastated by an earthquake in 1783.

TAVERNA, a village of Switzerland, in the cant. of Tessin, in the valley of the Agno, 4 m. NNW of Lugano. Pop. 512.

TAVERNES, a town of France, in the dep. of the Var, 20 m. N of Brignoles. Pop. 1,400.

TAVERNIER KAY, a small island on the N coast of Cuba, one of the Tortugas, 2 m. from the SW end of Kay Largo.

TAVERNILLA, a village of New Granada, in the prov. of Ystmo, on the line of railway from Chagres to Panama, 21 m. from Chagres, and at an alt. of about 50 ft. above the level of the river Chagres.

TAVERNY, a village of France, in the dep. of Seine-et-Oise, cant. and 4 m. NW of Montmorency. Pop. 1,500.

TAVIANO, a town of Naples, in the prov. of Terra-d'Otranto, 9 m. SE of Gallipoli, within 3 m. of the E of the gulf of Torento. Pop. 1,500.

TAVIGLIANO, a village of the Sardinian states, in the prov. and 6 m. N of Bielle. Pop. 1,200.

TAVIGLIONE, a mountain of Switzerland, in the cant. of Vaud, 6 m. E of Aigle, having an alt. of 7,000 ft. above sea-level.

TAVIGNANO, a river of Corsica, which rises to the W of Monte Rotondo; runs NE, and then SE, and flows into the sea in N lat. 42° 13′, after a winding course of 50 m.

TAVINSK, a town of Russia, in the gov. of Orenburg, 30 m. NNE of Sterlitamak, on the r. bank of the Bielaya.

TAVIRA, or TAVILA, a port of Portugal, in Algarva, at the mouth of the river Seca, which divides it into two, and which is here crossed by a bridge of 7 arches, 140 m. SSE of Lisbon, in N lat. 37° 7′. Pop. 8,700. It is surrounded by a wall, and defended by a castle: there are also two small forts at the mouth of the harbour. The palace of the governor of Algarva, who resides here, is an elegant structure, and the town is, on the whole, comparatively well-built. It has two churches, an hospital, and six convents. The entrance of the harbour is obstructed by a sand-bank; but the export trade, in figs, almonds, and other fruit, is considerable.

TAVISTOCK, a parish and parl. borough in the co. of Devon, 34 m. WSW of Exeter, on the river Tavy, which is here crossed by three bridges, and in the line of the Tavistock canal, which unites it with the navigation of the river Tamar. Area of the p. 10,700 acres. Pop. in 1831, 5,602; in 1851, 8,147. The town stands on the NW bank of the Tavy, in a valley surrounded by hills, and possessed of many local attractions. Many of the houses are old, but in general the place is well-built, though the streets are for the most part narrow and indifferently paved. It was formerly of greater note than at the present time, but it still returns 2 members to parliament. Electors registered, in 1837, 329; in 1852, 349. The limits of the borough are now extended so as to coincide with the whole parish. The richest tin and copper mines in Devon are in this vicinity: see article DEVONSHIRE. Lead and manganese are also found here, and occasionally the magnetic ore or loadstone. Many of the inhabitants are occupied in these mines; others in the manufacture of serges and coarse linens.

TAVO, an island of the gulf of Bothnia, 30 m. SW of Uleaborg, in N lat. 64° 49′.

TAVOLARA, a small island off the NE coast of Sardinia, in front of the entrance of the harbour of Terra-Nova, in N lat. 40° 54′. The only occupants of this island are wild goats.

TAVORA, a town of Portugal, in the prov. of Beira, 6 m. E of Lamego.

TAVOY, TAVAY, or DHA-WAY, a town of the Trans-Gangetic British prov. of Tenasserim, situated on the l. bank of a river of the same name, 35 m. above its mouth, but only 8 m. distant from the sea in an E direction, in the midst of a rich alluvial valley, which is bounded on the W, at a distance of about 3 m., by a range of precipitous and densely wooded hills from 200 to 400 ft. in height, and extending along the river to its embouchure, and on the E, at the distance of 7 or 8 m., by another range of similar elevations, above which towers at the distance of 20 m. the lofty chain which intersects the whole peninsula. The town contains 20,000 inhabitants. The town is scattered, most of the houses having a court round them planted with trees. The houses are in general roomy and comfortable; but a few are built of bamboo-matting set on wooden frames and raised several feet from the ground. The Tavoyers are a different people from the Burmans, and may be said to speak a distinct language. They had the reputation under the Burman government of being a riotous and rebellious set, and were kept with a strict hand: when they went to other towns they were obliged to quarter themselves in a particular spot, whence they were not allowed to be absent between sunset and sunrise. Hence a suburb of Maulmein is still called Tavoy-zu. This is the only town on the Tenasserim coast that manufactures silk cloths in any quantity. The raw material is brought from China. Coarse cotton-pieces are also woven by the women. The principal trade of T. is carried on by Chinese junks and Burman boats, which carry the staple produce, rice, to Penang; there is also a small trade with Rangoon and Maulmein. T. is only 15 days' journey, partly by water, from Bancock, the capital of Siam; yet from the jealousy of the Siamese government in excluding foreigners, little or no trade is carried on with that country. A few Siamese, however, yearly bring to T. brass cups, gold leaf, and false hair, which the Burman women wear at the back of the head. T. is celebrated for the manufacture of Burman musical instruments, the chief of which are the harp, the *mee-gyoung*, (an instrument in the shape of a crocodile, with three strings along the back) the *patala*, a sort of harmonicon made of bamboos and played with sticks, and a three-stringed fiddle, precisely the same in form as the European violin, which must have been borrowed from the Portuguese who settled on this coast more than two centuries ago. The hills E of the town contain tin.

which was formerly used for coining by the Burman government; the mines are not now worked. T. formerly belonged to the king of Siam, but was taken by the Birmans in 1785. It was besieged the following year by the Siamese, but was so well defended that it remained in possession of the conquerors till 1790, when bribery caused the gates to be opened, and restored it to the monarch of Siam. It was again taken by the Birmans in 1792, and confirmed to them by the treaty of peace of 1793. In 1826 it came under British sway.

TAVROVKAIA, a town of Russia, in the gov. and 9 m. S of Voronetz. Pop. 1,000.

TAW, a river of Devon, which rises near the centre of the county, about 3 m. SE of Oakhampton; flows to Barnstaple; then turns W, and joins the Torridge, at its mouth, in the Bristol channel.

TAWALLY, one of the Gilolo islands, in S lat. 0° 21′. It is 35 m. long from N to S, and 6 m. in average breadth.

TAWI-TAWI, the chief of a cluster of islands, 56 in number, composing part of the Sulu archipelago. It is situated in N lat. 5° 15′, and E long. 120° 10′. It is 36 m. in length. The other islands are of various size, some merely rocks; and all have inhabitants, though but thinly peopled.

TAWSTOCK, a parish in Devon, 2 m. S of Barnstaple, on the W bank of the Taw. Area 6,582 acres. Pop. in 1831, 1,348; in 1851, 1,383.

TAWTON (Bishop's), a parish in Devon, 2½ m. S by E of Barnstaple, on the E bank of the Taw. Area 4,263 acres. Pop. in 1851, 2,004.

TAWTON (North), a parish in Devonshire, 6½ m. NE of Okehampton, watered by the river Taw. Area 5,814 acres. Pop. in 1851, 1,906.

TAWTON (South), a parish in Devon, 3½ m. E by S of Okehampton, on the river Taw. Area 10,879 acres. Pop. in 1831, 1,937; in 1851, 2,758.

TAWY, a river of Brecknockshire and Glamorganshire, which runs into the Bristol channel at Swansea.

TAXALL, a parish in the co.-palatine of Chester, 8¼ m. NE by E of Macclesfield, on the river Goyt, and crossed by the Cromford railway. Area 3,718 acres. Pop. in 1851, 205.

TAXAMALCA, a town of Mexico, 60 m. S of Mexico.

TAXAMARCA, a town of Mexico, in the prov. and 40 m. E of Mechoacan.

TAXIMAROA, a settlement of Mexico, in the state of Mechoacan, 15 m. S of Valladolid.

TAY, a river chiefly of Perthshire and partly of Fife and Forfar, the largest and the most exquisitely scenic of all the Scottish streams, and one which pours more water into the ocean than any other river of Great Britain. The Tay of common topography is the southern one of two streams which unite 7½ m. above Dunkeld, and includes no more of even that stream than the portion below Loch-Tay; while the Tay of correct geography must either be the northern one of the two great streams, as both the larger and the longer, or must at least include the southern stream up to the source of its remotest head-water. The northern stream has three successive names,—the Gauir, the Rannoch, and the Tummel; and, in its progress, expands itself into three great lakes,—Lydoch, Rannoch, and Tummel: which see. It rises at a point about 18 or 20 m. NNW of the source of the Fillan, the remotest of the head-waters which find their way into Loch-Tay; and, including its progress through its lakes, but excluding its sinuosities, it performs an entire separate run of about 60 m. The southern of the two great streams, part of which is recognised as the popular Tay, performs, from its highest source, and measured in straight lines and through lakes, to its junction with the Tummel, a run of about 57 m. Its origin is on the side of Benloy, 7 m. NNW of the head of Loch-Lomond. It expands itself when about third way into Loch-Dochart, and bears respectively before and after this expansion, the names of the Fillan and the Dochart. From the lower end of Loch-Tay, to its junction with the Tummel, is a distance of about 14 or 15 m. From its junction with the Tummel to its junction with the Earn, where it begins to expand into an estuary, it achieves, irrespective of sinuosities, a distance of 32 m., and has on its r. bank Little Dunkeld, Kinclaven, Auchtergaven, Redgorton, Perth, and Rhynd, —and, on its l. bank, Logierait, Dunkeld, Caputh, Cargill, St. Martins, Scone, Kinnoul, Kinfauns, and St. Madoes. As an estuary, it extends 26 m. from the mouth of the Earn to Buddonness, the point where it becomes quite lost in the German ocean; has for 16 m. a breadth of from three-fourths of a mile to 3 m., and the direction of NE by E; has, over the other 10 m., a prolonged contraction of from 2 m. to less than 1 m. in breadth, and then an expansion, down to St. Andrew's-bay, of 9 or 10 m. in breadth, and, in both places, a prevailing easterly direction. Its entire length of course, jointly as a river, and as an estuary, is thus, if measured in straight lines from the head of the Gauir, 118 m.; if measured in the same way from the head of the Fillan, 115 m.; and, if measured along its channel from either of the remote sources, between 160 and 200 m. The chief tributaries of the northern great head-branch are the Ericht, which falls into Loch-Rannoch, the Gary, and the Tilt, which falls into the Tummel a little below Killiecrankie. Those of the southern great head-branch are the Lochy and the Lyon. Those of the united stream are the Bran, the Isla, the Ericht, the Shochie, the Almond, and the Earn. The estuary in general is shallow, and receives such an amount of debris from the steady and large current of the river, as it has been apprehended, must eventually drive shipping from its waters. Though it cannot compare in spaciousness and some other properties with the Forth, it is at present not a little commodious, and all the way from Buddonness to past the mouth of the Earn, may be considered as a continuous harbour. Vessels of 500 tons can go up to Newburgh; and vessels drawing 9 ft. of water can go up, though with difficulty, and in critical states of tide and freshet, to Perth. The tide from the ocean flows to a point about 1 m. above Perth, and rises at Perth harbour to a height of 10 ft. above the bed of the stream. The extent of surface drained by the Tay and its tributaries is variously computed at 2,396, and 2,750 sq. m.; that of the Spey, the entirely Scottish river next to it in size, at 1,300 and 1,298⅔ sq. m. The geographic positions and character of the district, whence most of the waters are drawn, being in the case of the two rivers very similar, the Tay may be supposed to discharge about twice as much water as the Spey. Dr. Anderson determined the quantity of water which, in the mean state of the river, flows through a section of it opposite Perth, to be at the rate of 3,640 cubic ft. per second; but it was reduced, in the course of the summer of 1819, to 457 cubic ft., and at the close of the summer of 1835, to a still smaller volume. Much of the country which now forms the sea-board of the estuary, and especially the whole of the carse of Gowrie, and the lower part of Strathtay, exhibit evidence of having, at a comparatively recent period, lain under the sea, and been gradually raised above its level by depositions from the Tay.

TAY (Loch), a magnificent sheet of water, in

Breadalbane, Perthshire, upwards of 15 m. in length, and averaging from 15 to 100 fath. in depth. The united streams of the Dochart and Lochy pour their waters into the head of the loch, a little below the village of Killin; while the noble Tay issues from its lower end at Kenmore. Ben-Lawers rises from the N margin of the lake to the height of 4,015 ft. above the level of the sea. Near the E end of the lake, and 200 yds. from the N shore, is a small island of a circular form, about 400 yds. in circumf. containing the ruins of a priory erected in 1122. The waters of this lake have frequently, without any apparent cause, undergone singular and violent agitations, occasioned, it may be presumed, from some subterraneous impulse.

TAYA, a small island in the Eastern seas, situated off the E coast of Sumatra, in S lat. 0° 48′.

TAYABANNA, a town of Brazil, in the prov. and 52 m. WSW of Parahiba, on the l. bank of the Parahiba.

TAYABO, a town on the E coast of the island of Celebes, on Gunong-Tellu bay, in S lat. 1° 10′.

TA-YAOU, a district and town of China, in the prov. of Yun-nan, and div. of Tsoo-heung-fu, in N lat. 25° 46′, and E long. 101° 30′.

TA-YAY, a district and town of China, in the prov. of Hu-pih, and div. of Wu-chang-fu, in N lat. 30° 6′, and E long. 114° 57′.

TA-YIH-HEEN, a district and town of China, in the prov. of Sze-chuen, and div. of Kung-chu, in N lat. 30° 37′, and E long. 103° 33′.

TAYLOR, a county in the W of Iowa, U. S., drained by Modaway and La Platte rivers. Area 522 sq. m. Pop. in 1850, 204.—Also a co. in the central part of Kentucky. Area 208 sq. m. Pop. 7,250. Its cap. is Campbellsville.—Also a co. in the NW of Virginia, watered by the Monongahela. Area 214 m.

TAYLOR'S ISLES, three small islands on the E coast of New Holland, between Thistle island and the shore, from which they are about 2 m. distant.

TAYLORSVILLE, a village in Spencer co., Kentucky, U. S., on the N side of Salt river, 20 m. WSW of Frankfort.—Also a village in Bucks co., in Pennsylvania, 106 m. E of Harrisburg.—Also the cap. of Hanover co., in Virginia, 20 m. N of Richmond.

TAYNUILT, a small ill-built village in Argyleshire, on the S coast of Loch-Etive, about 6 m. from Bunawe.

TAYOMAYO, a small island of Spain, in the Mediterranean, off the NE coast of the island of Majorca.

TAZE, a river of Asiatic Russia, which rises from two lakes in the N part of the gov. of Tobolsk, and after a considerable course from S to N, falls into the Tazovskaia gulf, in the Frozen ocean.

TAZEWELL, a county in the SW part of Virginia, U. S. Area 1,056 sq. m. Pop. in 1850, 9,942. Its cap., of the same name, is situated on the S side of the N fork of Cliesh river.—Also a co., near the centre of Illinois, U. S. Area 704 sq. m. Pop. in 1850, 12,052. Its cap. is Fremont.

TAZOVSKAIA, a gulf or bay in the Obskaia gulf, on the coasts of the Russian govs. of Tobolsk and Jenisesk, formed by the waters of several rivers of Siberia, and joined to the Obskaia gulf. It is about 140 m. in length, and 3 m. in breadth.

TAZRI, a village of Persia, in the prov. of Laristan, 15 m. NE of Tarem.

TCHABLAR-SAGHI, CHEBLER-SAGHI, or CHAMOULA (CAPE), a headland of Turkey in Europe, in Bulgaria, in the sanj. of Silistria, 16 m. S of Mangalia, on the Black sea, in N lat. 43° 32′ 33″, E long. 28° 38′ 37″. A tower of pyramidal form used as a light-house formerly existed here.

TCHACHNIKI, a town of Russia in Europe, in the gov. of Vitebsk, district and 15 m. E of Novoi-Lepel.

TCHAD. See CHAD (LAKE).

TCHADYR, CHADIR-DAGH, or PALATE-GORA ('Tent Mount'), a mountain of Russia in Europe, in the gov. of the Taurida district, and 18 m. SSE of Simferopol, and 20 m. E of Bakchi-Serai, between 44° 46′ and 44° 48′ N lat., and between 34° 17′ and 34° 18′ E long. It has an alt. of 5,135 ft. above sea-level, and is the highest summit of the Crimea. It consists of an immense mass of calcareous rock, rising precipitously, especially on the E and W, and flattened on the top, whence in clear weather nearly the entire peninsula is visible. It presents numerous fissures, some of which, inaccessible to the sun's rays, are filled with perpetual snow. Except here and there, a few alpine plants, and the base which is covered with fine forests of beech, it is almost entirely arid. This mountain gives rise on the E and N to several of the head-streams of the Salghir; on the W, to the Kuise, an affluent of the Alma; and on the S to the Alushta. It was named by the Greeks, *Trapezos*.

TCHADOBETZ, a river of Russia in Asia, in the gov. and district of Jeniseisk, which runs W, and after a course of about 240 m., joins the Verkhniaia-Toungouska on the r. bank, near Tchadobsko.

TCHADOBSKO, a town of Russia in Asia, in the gov. and district and 375 m. ENE of Yeniseisk, on the Verkhniaia-Toungouska, near the confluence of the Tchadobetz.

TCHADOSY, a town of Russia in Europe, in the gov. of Vilna, district and 66 m. NNE of Vilkomir.

TCHADRIN, or YADRIN, a town of Russia in Europe, capital of a district, in the gov. and 120 m. W of Kazan, on the l. bank of the Sura. This district is in the E part of the gov., and is populous and well-cultivated.

TCHAGAINY, or SAGAINY, a town of Burmah, in the Mranma, on the r. bank of the Irawady, opposite Amarapura. It occupies the slope of several hills, and contains numerous temples, the greater number of which are adorned with spires and gilt domes, presenting from a distance an imposing effect. On closer inspection, however, it affords very different impressions. The houses are built of wood and covered with tiles. The fort, formerly one of the most important points of defence in the empire, is now in ruins. This town is the great entrepot for the cotton trade with China. In its vicinity are quarries of fine white marble. From 1760 to the death of Namdojy-Pra, in 1764, this town was the capital of the Burman empire.

TCHAGANI, or GLOUKHOI (CAPE), a headland of the Crimea, 8½ m. ESE of Cape Kazantep or Kettle, with which it forms the entrance to a bay of that name.

TCHAGATAI. See TARTARY.

TCHAGHIRKHAN, a village of Turkey in Asia, in Anatolia, in the sanj. of Khodebendkiai, 12 m. NW of Brosissa, near the r. bank of the Ufer-su. It has four thermal baths.

TCHAGHOURI, a fortress of Afghanistan, in the prov. of Bamian, on a mountain of the same name.

TCHAGLAIK, a town of Turkey in Europe, in Rumelia, in the sanj. of Gallipoli, 17 m. NE of Le Cavale, at the foot of a lofty mountain, on the r. bank of the Kara-su.

TCHAGLASA. See THSIANGLODZIAN.

TCHAGODA, a river of Russia in Europe, which has its source in the gov. of Tver, district and 18 m. E of Bejetzk; runs W through the N part of the district of Kaschin; enters the gov. of Jaroslav; turns N, and joins the Mologa on the r. bank, 21

m. NW of the town of that name. It is navigable by boats of shallow draught.

TCHAGRA, a river of Russia in Europe, in the gov. of Saratov, and district of Khvalinsk, which has its source 45 m. SSW of Samara; runs generally WSW, partly on the confines of the gov. of Simbirsk; and after a sinuous course of upwards of 90 m., throws itself into the Volga on the l. bank, 6 m. SE of Khvalinsk. It flows through an uninhabited though fertile country.

TCHAHAN-SOUBARKAN-KHOTON, a town of Mongolia Proper, about 180 m. NNE of Pe-king, in a fertile valley to the N of the Great Wall, and to the W of Lan-ho. With the exception of the houses of the mandarins, it consists of miserable cabins, and yet in its vicinity is a fine palace and magnificent park, in which the emperor habitually spends a portion of the summer. This town is connected by a fine road with the capital.

TCHAHASU, a town of China, in Mandshuria, 270 m. NE of Tsitsikar, near the l. bank of the Amour.

TCHAHDARRA, a village of Hindostan, in the prov. and 3 m. N of Lahore, near the r. bank of the Ravy. It contains two fine mausoleums.

TCHAIDAM, a river of China, in the central plateau of the territory of the Koko-Nor. It has its sources in the Alak-Nor and Toson-Nor; runs NW; receives the Gedergou, Oulan-oussom, Bouloungir; and after a course of about 240 m., throws itself into the Dabsoun-Nor, a lake of the desert which has no outlet.

TCHAI-KING-KOEN, a town of China, in the prov. of Sze-chuen, 100 m. SW of Tching-le.

TCHAINAT. See CHAINAT.

TCHAINITZA, TCHENITZA, or CENIZZA, a town of Turkey in Europe, in Bosnia, in the sanj. of Novi-Bazar, 45 m. SE of Bosna-Serai, in a fertile plain, on a torrent, of which a portion of the waters passes under the single arch of a lofty stone-bridge, and the remainder is conducted to supply power to the manufactories of blankets and coarse cloth in the vicinity. It has a castle, rectangular in form, with a tower at each angle, in which the pasha of Bosnia maintains a garrison, and a portion of his household.

TCHAIRLI-DAGH, a mountain of Turkey in Asia, in Anatolia, in the sanj. of Kodja-Ili, a little to the S of the gulf of Ismid, and to the NE of Lake Isnik.

TCHAISSI, a fortress of Russia in Asia, in Mingrelia, on a lofty mountain, 12 m. NE of Anakria.

TCHAKA-KHOTON, a town of Mongolia Proper, in N lat. 43° 59′, E long. 117° 54′.

TCHAKHARS, or TSAKHARS, a district in the S part of Mongolia Proper, between the Nishan mountains and China, comprising a tract 300 m. in extent, generally mountainous, and watered by several small rivers. It is in some parts capable of culture, and affords good pasture. It is inhabited by a tribe of the same name. They are divided into eight banners. T. signifies in the Mongol language a border town. Under the Ming dynasty it bore the name of Tsagan or Tchagan. The Tchakars formed one of the 8 corps of the Manchu army, by whom China was conquered in 1644.

TCHAKON-TOWN, a group of islands, eight in number, in the gulf of Saghalien, to the NE of Mandshuria, between the island of Saghalien and the continent.

TCHAKTCHUR, a town of Turkey in Asia, in the pash. and 120 m. NE of Diarbekir.

TCHALA, a town of Russia in Asia, in the gov. of Georgia, in the Kartalinia, to the ESE of Gori. It is enclosed by walls. In the vicinity is the convent of Samtavissi.

TCHALATDERE, SALATDERE, or BOKLU, a river of Turkey in Asia, in Anatolia, in the sanj. of Biga, which has its source in Kaz-dagh, a summit of Mount Ida; runs NNE; receives the Ustvola on the r.; and after a total course of 48 m., throws itself into the sea of Marmara. This river is celebrated in history for the victory of Alexander the Great over the Persians.

TCHALBAH, TCHEMBACHE, or TCHELBASI, a river of Russia in Europe, in the gov. of the Caucasus. It takes its rise in the district and 60 m. NW of Stavropol; runs NW; enters the Cossack territory; turns afterwards W; forms several lakes, amongst others that of Hadkoi; and after a course of upwards of 150 m., throws itself into the gulf of Beisongski, in the sea of Azof.

TCHALBUKHA, a town of Russia in Asia, in the gov. of Irkutsk and district of Nertchinsk, on a river of the same name.

TCHALDIR, or TCHELDER-DAGH, a range of mountains in Turkey in Asia, a ramification of the Taurus chain, on the confines of the pash. of Akalzik and Erzerum.

TCHA-LING-CHU, a district and town of China, in the prov. of Hu-nan, dep. and 105 m. SSE of Chan-sha-fu, in N lat. 26° 53′ 40″, E long. 113° 23′.

TCHALIT, a territory and tribe of Mongolia Proper. The tribe forms a banner of the r. wing of the Kort-shin.—The territory is 120 m. in length from N to S, and 18 m. in breadth.

TCHALKHA, a district of Russia in Europe, in Circassia, on the Shadghir. The sterility of the soil occasioned the emigration of a large portion of its inhabitants to the banks of the Kumbalei, the capital of which, also named Tchalkha, contains about 200 families.

TCHALLITCHAKA, a lake of Little Tibet, near Roudok. It affords borax.

TCHALU, a village of Tibet, in the prov. of Thsang, 39 m. NNW of Tissisudon, between two lakes, one of which is held in high veneration by the Tibetians.

TCHAM-HALEKA-TANDA, a fortress of Sind, district of Rona, 18 m. NE of Hydrabad. It has a considerable suburb.

TCHAN, a river of China, in the prov. of Kiang-si, which flows into the Kan-kiang, on the l. bank, after an easterly course of 120 m.

*** TCHAN and TCHANG. For words with these prefixes, see CHAN and CHANG.

TCHARDAH, a town of Turkey, in the sanj. and 56 m. NW of Sophia, near the Nissara.

TCHARDAKLI, a village of Russia, in the gov. of Yekaterinoslav, 18 m. N of Marionpol, on the river Kaleti.

TCHARTORISK, a town of Russia, in the gov. of Volhynia, 36 m. NNE of Lutsk, on the l. bank of the Styr. Pop. 1,600.

TCHARYTCH, a river of Asiatic Russia, which rises in the Little Altai; runs first NW, and then NE; and joins the Obi, on the l. bank at Bijsk, after a winding course of about 200 m.

TCHAVDAR (CAPE), a low sandy point on the coast of the Black sea, in N lat. 45°, E long. 35° 51′.

TCHEBOKSARI, a town of Russia, in the gov. and 74 m. WNW of Kasan, on the r. bank of the Volga, at the confluence of a river of the same name. Pop. 5,000. It has a traffic in tallow, wine, wax, and honey.

TCHECKMEH, or TEHISME, a port of Asiatic Turkey, in the sanj. of Saghala, 42 m. W of Smyrna, in N lat. 38° 24′, opposite to the island of Chios. Pop. 5,000. It is defended by a large citadel. In its neighbourhood are hot springs. It occupies the site of the ancient *Cyssos*.

TCHE-KIANG. See CHE-KIANG.

TCHELGA, a town of Abyssinia, in the prov. of Amhara, 18 m. NW of Gondar.

TCHELIABINSK, a town of Asiatic Russia, in the gov. of Orenburg, 52 m. N of Troitskaia, on the Mijas. It is fortified with a wall and palisades, and is generally garrisoned by a few Cossack troops and Bashkirs.

TCHEMBUL. See CHUMBUL.

TCHENAB. See CHENAB.

TCHEPELOVO, a town of Albania, in the sanj. and 18 m. NNE of Joanina.

TCHEPTZA, a river of Russia, which rises on the E frontier of the gov. of Viatka; flows N, and WNW; passes Glazov; and joins the Viatka on the l., after a course of 250 m.

TCHERDIN, a town of Russia, in the gov. of Perm, on the r. bank of the Kolva, near the confluence of the Vischera. It is a place of considerable antiquity.

TCHEREMSHANE, a river of Russia, which rises about 30 m. SW of Bugulma, in the gov. of Orenburg; flows NW, and then SW; and joins the Volga, on the l. bank, 10 m. SSE of Singlei, after a course of 150 m.

TCHERKASK. See CHERKASK.

TCHERKESSES. See CIRCASSIA.

TCHERNIGOV, a government in the S of Russia, lying between the parallels of 50° 20′, and 53° 10′ N; and separated on the W by the Dnieper from the govs. of Minsk and Kiev. It has an area of 48,379 sq. versts, with a pop. in 1846 of 1,430,000. Its principal river is the Desna. Its surface is level and soil fertile.—Its cap. of the same name, situated on the Desna, 80 m. NNE of Kiev, has a pop. of 8,000.

TCHERNOILAR, or TCHERNOIARSK, a fortified town of Russia, in the gov. of Astrakhan, on the r. bank of the Volga, 154 m. NW of Astrakhan. It conducts an important traffic in cattle and fish.

TCHIKISLA, a village of Asiatic Turkey, in the sanj. of Konieh, 36 m. SSW of Nigdeh, on the site of the ancient *Nazianzum*.

TCHILKA. See CHILKA.

TCHIMBURSK (CAPE), a headland on the coast of the sea of Azoff, 9½ m. S of Cape Otchakof.

⁂ TCHIN For words with this prefix, see CHIN or CHING.

TCHORLU, a town of Turkey, in the sanj. and 39 m. SW of Viza, near the l. bank of a river of the same name, which flows into the sea of Marmora, 3 m. E of Rodosto, after a course of 45 m. Pop. 4,000.

TCHOUGOUEV, a town of Russia, in the gov. and 27 m. SE of Charkof, on the r. bank of the Sevarnoi-Donetz. Pop. 10,000. It is celebrated for its extensive trade in tanned sheep-skins and in leather.

TCHUI. See CHUI.

TCHUKTSHES, or TSCHUKTSCHI, a people of Asiatic Russia, who, according to Wrangell, inhabit the NE part of Asia, extending from Tschaun bay to Behring's straits in one direction, and in the other from the Anadyr, and the upper country of the Aniuj, to the Polar sea. Their neighbours to the S are the Koraki, and to the W the Tschuwanzi and Jukahiri of the Aniuj. "They formerly occupied a more extensive territory, before the Cossacks from the Lena subdued the country through which the Kolyma flows. Though still in great measure a nomade race, they have less of the characteristics which usually accompany such a mode of life, than the wandering Tungusi; they are less cheerful, and more careful; they lay up stores for the future, and in general do not remove their dwellings without an object, but only when it becomes necessary to seek fresh pasture for their reindeer. They are more covetous and more saving than belongs to the character of genuine nomade races. Their dress differs greatly from that of the Tungusi, which is tight and close-fitting, and well-adapted to an active wandering life, whereas the clothing of the Tschuktschi is large, loose, and cumbrous. They wear long wide trousers made of fur, and an ample kuchlanka. Formerly all the T. lived on the produce of their reindeer; but those among them who lost their herds by sickness or other causes settled by degrees along the coast, where they kill whales, seals, and walruses. The walrus is almost as useful to the settled, as the reindeer is to the nomade T. Their principal weapons are different kinds of spears, and particularly the batags. Iron being scarce, they sometimes employ walrus' tusks instead."

TCHULIM, a river of Asiatic Russia, formed by the confluence of the Great and Little Yuis, to the W of the Sagaitsis mountains. It flows in a winding, but prevailingly N course to its junction with the Kemptchuga; after which it flows WNW to the Obi, which it joins on the r. bank, 100 m. NNW of Tomsk.

TCHUMBUL. See CHUMBUL.

TEA, a river of Spain, in the prov. of Vigo, which has its source on the S side of Monte Taro, 15 m. SE of Pontevedra; runs S; receives several streams; and after a course of about 27 m., joins the Minho on the r. bank, opposite the Portuguese town of Valencia.

TEAKIKI, or THEAKIKI, a river which has its source at the N extremity of the state of Indiana, U. S., on the confines of the territory of Michigan; runs WNW through the country of the Pottowato rivers; enters afterwards into the state of Illinois; bathes the counties of Clark and Edgar; and after a course of about 105 m., joins the Plein river on the l. bank, 51 m. SW of Chicago, to form the Illinois.

TEALBY, a parish of Lincolnshire, 4 m. ENE of Market-Raisen. Area 3,050 acres. Pop. 862.

TEALING, a parish in the S part of the Sidlaw district of Forfarshire. Area 4,300 imperial acres. Its kirktown is 5⅔ m. N of Dundee. Pop. 844.

TEAN, a village of Staffordshire, 2 m. S of Cheadle. Pop. 1,300.

TEANA, a town of Naples, in the prov. of Basilicata, district and 23 m. E of Lagonegro, cant. and 4 m. WNW of Chiaromonte, on the summit of a hill, near the r. bank of the Serapotamo. Pop. 1,200. It has two alms-houses and a fort.

TEANO, a town of Naples, in the prov. of the Terra-di-Lavoro, district and 20 m. NW of Caserta, on a hill. Pop. 6,000. It has a cathedral, designed by Vaccaro, and erected on 16 granite columns, five parish churches, two of which are collegiate, five convents, a seminary, an alms-house, and the ruins of an extensive castle.

TEATH. See TEITH.

TEBA, a town of Spain, in the prov. and 39 m. NW of Malaga, in the midst of the Sierra-Camorra. Pop. 4,500. It has an ancient Moorish castle, which was rendered defensible by the French in 1810, a parish-church, two convents, and an hospital. It contains some Roman antiquities. Basket-ware is its chief manufacture.

TEBALEH, a town of Arabia, in Yemen, 180 m. SE of Mecca. It is a large place, and is defended by a castle, situated on a steep rock and well fortified.

TEBAR, a town of Spain, in the prov. and 45 m. S of Cuenca, in a stony but fertile locality, near the r. bank of the Xucar. Pop. 1,560. Wine, oil, and saffron are the chief productions of the environs.

TEBBES, a town of Persia, capital of a district in the prov. of Kuhistan, on the road from Herat to Yezd, 285 m. WSW of the former. Pop. 10,000. It possesses, from its situation, an active transit trade.

TEBELHELT, a town of Marocco, on the confines of the Sahara, and in the prov. of Tafilet, to the SW of the lake into which the Ziz discharges itself.

TEBENIATZKAIA, a town of Russia in Asia, in the gov. of Tobolsk and district of Kourgan, 17 m. SE of Rofailova, on the l. bank of the Tebeniak, a small affluent of the Tobol.

TEBESSA, a town of Algeria, 72 m. SE of Constantina. Pop. 9,000.

TEBIQUARIGUAZU, a river of Paraguay, which descends from the S side of the mountains of Villarica, 48 m. ESE of the town of that name; runs first SSW, then W, forming numerous sinuosities; and after a course of about 225 m., throws itself into the Paraguay on the l. bank, 21 m. SSW of Agatape.

TEBRIS. See TAURIS.

TEBUA, or HOOD, an island of the S. Pacific, in the Mindanao archipelago, in S lat. 9° 26′, W long. 138° 52′. It is small and inhospitable looking, but possesses a considerable population. It was discovered by Cook in 1774.

TEBUK, a village of Arabia, of about 60 houses, on the high road of the Syrian pilgrims, 4 days from Maan, and the like distance from Muweilah. The surrounding plain is one of the most insecure parts of the Desert. There is a small castle here.

TECALI, a town of Mexico, in the state and 18 m. SE of La Puebla. Pop. 1,000.

TECAMACHALCO, a town of Mexico, in the state of that name. Pop. 2,350.

TECAPA, a volcano of Guatemala, in the state of St. Salvador, to the SE of the volcano of Bastlan.

TECH, a river of France, in the dep. of the Eastern Pyrenees, on the Spanish frontier, in the cant. and 8 m. W of Prats-de-Mollo; runs first E. then NE, passing Prats-de-Mollo and Arles; and a little to the N of Ceret, and after a course of about 48 m., throws itself into the Mediterranean, 3 m. NE of Argeles.

TECHA, TIOSCHA, or TEJE, a river of Russia in Europe, which has its source in the gov. of Nijnii-Novgorod, in the district and 8 m. SSW of Loukoianov; waters that town and Arzamas; passes a little to the N of Ardatov; enters afterwards into the gov. of Vladimir; and after a total course in a generally WNW direction of about 126 m., joins the Oka on the r. bank, nearly opposite Murom.

TECHADA, a town of Spain, in the prov. and 33 m. WNW of Toledo, now nearly abandoned from the insalubrity of the situation.

TECHE (BAYOU), a river of the state of Louisiana, U. S., forming a branch of Crocodile Bayou, from which it detaches itself on the r., in the co. and 5 m. NE of Opelousas. It runs SE through the parishes of St. Martin and St. Mary, and joins the Atchafalaya on the r. bank, 15 m. above the embouchure of that river in the gulf of Mexico. It has a very tortuous course of about 165 m., of which 100 are navigable. Martinsville, New Iberia, and Franklin, are the chief places on its banks.

TECHE, a village of France, in the dep. of the Isere, cant. and 4 m. NE of St. Marcellin, on the r. bank of the Vezy. Pop. 800.

TECHEIN, or TECHAN, a town of Turkey in Europe, in Bosnia, in the sanj. and 39 m. NE of Trawnik, on a mountain torrent.

TECHOU-LOUMBOU, a town of Tibet, in the prov. of Thsang, a little to the W of Jika-dze, and near the r. bank of the Yarou-dzangbo-tchu, in N lat. 29° 4′ 20″, E long. 84° 27′. It is the ordinary residence of the Techou-Lama, and is in reality a large monastery, consisting of about 400 houses inhabited by *gelongs* or priests, and comprising numerous temples, mausoleums, the pontifical palace, in which all the ecclesiastical and civil officers attached to the court reside, and the interior of which is very handsome.. An extensive exchange trade in gold and silver is maintained between this town and Lhassa.

TECKENDORF, TEKE, or TYAKA, a town of Transylvania, in the lower circle of the comitat of Klausenburg, and 17 m. S of Bistricz, on the r. bank of the Sajo. It has a salt lake.

TECKLENBURG, a town of Prussia, cap. of a circle in the prov. of Westphalia, regency and 20 m. NNE of Munster, in the midst of the Teutoburgerwald. Pop. 1,200. It has a poor's-house, and the ruins of an old castle. Coal is wrought in the environs.—The circle comprises an area of 99 sq. m.

TECLA (SANTA), a town of the republic of Monte-Video, in the Banda-Oriental, 129 m. WNW of Villa-Guarda, on the confines of the Brazilian prov. of Rio-Grande-do-Sul, in the midst of the mountains of Cochilla-Grande-de-los-Tapes, and near the source of the Rio-Negro.

TECOANTEPEC. See TEHUANTEPEC.

TECOLINE, a parish in Queen's co., 4½ m. SSE of Stradbally. Area 1,022 acres. Pop. 144.

TECOLOTLAN, a town of Mexico, in the state of Xalisco, 75 m. SW of Guadalajara.

TECOMACHALCO, a river of Mexico, in the state of that name, which has its source in the mountains to the W of the capital; runs W; and after a short course, throws itself into the lakes of the valley of Mexico.

TECSO, TACOWA, or TECSEU, a town of Hungary, in the comitat of Marmaros, 15 m. WNW of Szigeth, on the r. bank of the Theiss, at the confluence of a small river of the same name.

TECUMSEH, a township in the co. of Lenawee, in the state of Michigan, U. S., 38 m. WNW of Monroe, on the r. bank of Raisin river. Pop. 2,600.

TECUMSETH, a township of Upper Canada, in the Simcoe district, intersected by the Nottawasaga. Pop. in 1842, 2,491.

TEDAVNET, a parish of co. Monaghan, containing the villages of Bellanode and Scotstown. Area 26,502 acres. Pop. in 1831, 11,352; in 1851, 8,442. The surface descends from the culminating point of the Slieve-Beagh, to within a ½ m. of the town of Monaghan. The western and northern districts are, to a large extent, identical with a portion of the Slieve-Beagh mountains; and their highest ground has an alt. of 1,254 ft. above the level of the sea. The chief of about 19 lakes within the limits are Loughs Meenish, Mullaghmore, North Mullaghinshigo, South Mullaghinshigo, Carrowhatta, Sheetrim, Lamb, and Slack's-Grove.

TEDBURN (SAINT MARY), a parish of Devon, 5 m. SSW of Crediton. Area 4,433 acres. Pop. 861.

TEDDINGTON, a parish and village of Middlesex, 12 m. WSW of London. Area of p. 1,120 acres. Pop. in 1841, 1,199; in 1851, 1,146. Wax-candles are largely made here.

TEDELES (CAPE), a headland of Algiers, in the prov. of Constantine, between Bougia and Algiers, in N lat. 36° 54′ 30″, E long. 4° 14″.

TEDIF, or TEDEF, a town of Turkey in Asia, in Syria, in the pash. and 21 m. E of Aleppo. It has a synagogue. In an adjacent mountain are several tombs and an aqueduct cut in the rock.

TEDLA. See TADLA.

TEDLES. See DELLYS.

TEDMENT, a town of Tripoli, in the S part of Fezzan, on the confines of the Sahara. Senna is grown in great quantities in the environs.

TEDNEST, a town of Marocco, 70 m. WNW of Marocco, on an affluent of the Tensift. Pop. 4,000, many of whom are Jews.

TEDSI, a town of Marocco, 25 m. ENE of Tarodant, in N lat. 30° 15′. Pop. 12,000. Its vicinity is well cultivated and productive.

TEDSTONE (DELAMERE), a parish of Herefordshire, 4 m. NE of Bromyard. Area 1,677 acres. Pop. in 1851, 193.

TEDSTONE (WAFER), a parish of Herefordshire. adjoining above. Area 683 acres. Pop. in 1851, 69.

TEDVAN, a town of Turkey in Asia, in the

pash. and 90 m. WNW of Van, and near the NW extremity of the lake of that name.

TEDZEN, or TEDJEN, a river which has its source in Persia, in Khorassan, to the NE of Meshed; flows thence into Independent Tartary, where it waters the SW part of the Kharism; and after a total course in a NW direction of about 300 m., throws itself into a marshy gulf, in the E part of the gulf of Balkan, a branch of the Caspian sea.

TEELIN, a small estuarial harbour on the coast of co. Donegal. It opens 1¾ m. E of Carrigan-head, and penetrates the land about 1¾ m. northward, but has a very inconsiderable width, and is properly the estuary of a rivulet called the Glen river.

TEELIN-HEAD, a sublime cape in the p. of Glencollumbkill, co. Donegal, the most south-westerly or seaward portion of the series of magnificent cliffs which commences at Carrigan-head.

TEEN-TSIN, or TIEN-SING, a city of China, in the prov. of Pe-che-li, at the junction of the Imperial canal and the Pei-ho, 60 m. SE of Pekin. It stretches several miles along the Grand canal, and is the seat of an active internal trade in salt. This salt is formed in vats near the sea-shore, and is thence transported to Taku, where it is kept for some time piled up on hillocks of sand. It is finally transported to Teen-tsin, and there again stored for several years before it is fit for sale.

TEES, a river which rises in the mountains of Cumberland and Westmoreland, and pursues a serpentine course along the S margin of the county of Durham, which it separates from Yorkshire throughout its whole extent. After emerging from Milbourn forest, this fine river flows through the romantic vale of Teesdale, where it receives several tributary streams, and after passing Barnard-castle, Sockburn, and Yarm, falls into the German ocean near Stockton. It is tidal to Yarm, and navigable for vessels of 60 tons burden up to Stockton. In its approach to the ocean from Portrack, the river expands into a large bay, about 3 m. across, which is slightly contracted at its mouth by a tongue of land called Seaton-Snook, from which a bar of sand stretches across the estuary to the Cleveland coast. The depth of water on the bar at spring-tides, is about 10 or 12 ft. at low water, and from 26 to 28 ft. at high water. The estuary of the Tees is a place of great safety for vessels in stormy weather. The principal fishery on the river is that of salmon.

TEESTA, a river of Asia, which rises on the N flank of the Himalaya, near the great peak of Chumulari, in several head-streams which unite under the parallel of 27° 30′, and in E long. 88°. The united stream flows S, piercing the Himalaya, and forming the line of demarcation between the territories of Butan and Sikkim, and finally joins the Ganges, under the name of Atri, about 40 m. WNW of Dacca, after a total course of 360 m.

TEFANE, a town of Asiatic Turkey, in the sanj. of Kutahiyah, 30 m. WNW of Estenas, and not far from the l. bank of the Sultanimir-Shai.

TEFFE, or JEPE, a river of Brazil, in the prov. of Para, which running NNE, joins the Amazon, on the r. bank, under the parallel of 3° 30′, after a course of 550 m.—The district which lies between this river and the Amazon, and is bounded by the Coary on the E, is fertile in rice, sugar, manioc, cacao, and honey.

TEFFE, a village of Nubia, on the W bank of the Nile, 30 m. SSW of Assuan.

TEFFONT (EIRAS), a parish of Wilts, 7 m. W of Wilton. Area 742 acres. Pop. in in 1851, 177.

TEFFONT (MAGNA), a parish of Wilts, 6 m. E of Hindon. Area 1,460 acres. Pop. in 1851, 281.

TEFLENAZ, a town of Syria, in the pash. and 18 m. SW of Aleppo.

TEFLIS, or TIFLIS, [in Georgian, THILIS-KALAKHI, *i. e.* 'the Town of Hot waters'], the capital of Russian Georgia, situated in a narrow valley on the river Kur, in N lat. 41° 41′, E long. 44° 50′. It is composed of an upper and lower town, and a citadel; the whole of which are encompassed by a wall with 6 gates. The lower town, which is situated on the bank of the river, is the principal seat of trade, and is inhabited chiefly by Armenians. Its streets are narrow, irregular, and filthy in the extreme. The upper town, in which are the official buildings and residences of the Russian part of the pop., presents a few broad streets and open squares. This town, says a very recent visitor, "has a peculiar aspect. On the side from which we entered, the quarter inhabited by the Russians, it has a perfectly European look: straight streets, rows of modern houses, elegant shops, milliners, apothecaries, even a bookseller, with cafés, public buildings, a government palace, churches with cupolas and towers, the various Russian military uniforms with French paletots and frock-coats, quite transported us back to Europe. But where this European town ends, one of a perfectly Asiatic character begins, with bazaars, caravansaries, and long streets, in which the various trades are carried on in open shops. In one part is seen a row of smithies, the men all hammering away at their anvils, heedless of the crowds of passers-by. Then follows another row of houses, where tailors are seated at work, in precisely the same fashion, and with the same gesticulations and agility, as with us. After these succeed shoemakers, furriers, &c. The pop. is no less varied and interesting; here Tartars, in the costume from which the so-called Polish dress is evidently derived; in another part thin, sunburnt Persians, with loose flowing dresses; Kurds, with a bold and enterprising look; Lesghis and Circassians, engaged in their traffic of horses; lastly, the beautiful Georgian women, with long flowing veils and high-heeled slippers; nearly all the population displaying a beauty of varied character, which no other country can exhibit,—an effect heightened by the parti-coloured, picturesque, and beautiful costumes. In no place are both the contrasts and the connecting links between Europe and Asia found in the same immediate juxtaposition as in Tiflis." This city has long been celebrated for its baths, which are situated at one extremity of the bazaar. At this point, a small bridge is thrown over a deep ravine, at the bottom of which flows a mountain-stream, which, pure and cold at its fountain-head, mingles here with the hot springs which take their rise in the adjacent heights, and becomes warm. Over this steaming flood the public baths are erected, which form not only a resource in sickness to the natives, but are the daily resort of both sexes as places of luxury and amusement. On one side of the bridge stand those appropriated to the men, and on the other, immediately below the gloomy walls of the citadel, the range intended for the women. The water which supplies these distinct bath-houses is strongly impregnated with sulphur. Its heat ranges from 15° to 36° of Reaumur in the several basons. At the source of the hot stream it is about 42°. The basons are excavated in the solid rock; and these are divided under one immense vaulted roof, into different apartments, whence the day-light is excluded, and which are merely rescued from total darkness by the faint glimmerings of a few twinkling lamps struggling with the vapours from the stream. The citadel is situated on the summit of a high promontory, which forms the termination of the mountain that overshadows the town on its SW side. There are several fine churches of different Christian persuasions, in T.; that which is dedicated to the Roman Catholic mode of worship is one of the most beautiful; the cathedral of Holy Sion, the great Armenian church, is more extensive, but does not equal its rival in richness and grace of architecture.—The government of Teflis, under the Russian administrative arrangements organized in 1846, is bounded on the N by Caucasus; on the NE by Derbend; on the E by Shamaki; on the S by the Araxes, forming the boundary with Persia; on the SW and W by Asiatic Turkey; and on the NW by Kutais; and embraces the circles of Elisabetpol, Erwan, Nakshivan, Alexandropol, Siknash, Telav, Gori, and Teflis.

TEGERNSEE, a village of Bavaria, in the circle of Upper Bavaria, 7 m. SW of Miesbach, finely situated on the E bank of a lake of the same name, at the foot of the Alps, and at the entrance to the valley of Achen. It formerly possessed a rich and ancient abbey, which was secularised in 1802, and converted into a hunting lodge. The town contains about 80 houses. Surrounding it are extensive forests, and the adjacent mountains abound with game. They contain also beds of bituminous lime and quarries of fine marble.—The lake of T. is 6 m. long, and about 1½ m. in breadth. The Mangfall, an affluent of the Inn, issues from its NE extremity.

TEGHERY, or Toghara, a town of Tripoli, in Fezzan, 45 m. S of Gatrone, on the road from Murzuk to Kuka, and on the banks of a salt lake. It has a castle, and is surrounded by a double wall, the interior of which is pierced with loop-holes, and within the town are wells of good water. The environs produce excellent dates. The situation of T. is extremely pleasant, and the adjacent salt lakes abound in wild ducks and geese. The inhabitants are black, but differ in features from the Negroes.

TEGLA, a chain of mountains in the E part of Nigritia, to the S of Kordofan. It runs from E to W.

TEGLAS, a town of Hungary, in the com. of Zabolcs, near Hedhuz. Pop. 1,500.

TEGNA, a village of Switzerland, in the cant. of Ticino, district of Locarno. Pop. 240.

TEGRA, a village of France, in the dep. of the Lot, cant. and 3 m. NE of Gramat, at the foot of a mountain. Pop. 000.

TEGUANTEPEC. See Tehuantepec.

TEGUAYO, a lake in the N part of Mexico, to the SSW of Lake Timpanogos, in N lat. 39°. It receives the Rio San Buenaventura, on the NE, and on the E the Rio Salado.

TEGUCIGALPA, a town of Honduras, 60 m. ESE of Comayagua. This town, which is one of the most flourishing in the state, contains a fine church, two convents, and barracks, and in the vicinity are mines of gold and silver. The climate of the locality is mild and salubrious, and the soil productive.

TEGUISE, a town of the Canary islands, the capital of the island of Lancerote, situated near its centre, and at the foot of a mountain, the summit of which is surmounted by the castle of Santa-Barbara. Pop. 2,800. In its vicinity is a large pond, named Mareta, which in winter forms a general reservoir for the inhabitants of the island. The church is one of the finest in the Canaries. The palace of the ancient marquisate is now in ruins.

TEHALLON, or Tehollands, a parish in co. Monaghan, 2½ m. NE by E of Monaghan. Area 5,949 acres. Pop. in 1831, 4,846; in 1851, 2,906.

TEHAMA, a flat tract of land, extending with irregular breadth along the coast of the Red sea, in Arabia, for the distance of about 600 m. between 16° and 20° N lat. It is generally desert and barren, and its coast is lined with coral reefs.

TEHERAN, a city of Persia, the present capital of the kingdom and of the beglerbeglik of the same name, in the prov. of Irak-Ajemi, 215 m. NNW of Ispahan, and 1,350 ESE of Constantinople, in N lat. 35° 41′ 50″, and E long. 51° 21′. It lies in a sandy plain possessing little fertility, and exposed in summer to intense heat; bounded on the N and E by the Elburz mountains, and by the peak of Demavend; and on the W by a fertile and well-cultivated plain. Its outline is an oblong square, about 4 m. in circuit, and it is enclosed by thick walls, flanked with towers and preceded by a ditch. The gates, of which there are four in number, are ornamented with figures of tigers and other animals. Within the walls are numerous gardens and vacant spaces, and the general aspect of the houses, bazaars, king's palace, and other edifices suggests the idea of a town of modern erection. The royal palace, situated in the N part of the town, occupies upwards of a fourth part of its entire extent. It is square in form, is well-fortified, and possesses gardens of great extent and beauty. In one of the latter is the seraglio, enclosed by lofty walls, and defended by a guard. The dome of one of the mosques is covered with plates of gold. The distance of T. from the great commercial routes, limits its trade to articles of local consumption. In winter the pop. fluctuates from 50,000 to 60,000. In summer upwards of 7-10ths of the inhabitants, inclusive of the monarch, quit the town on account of its insalubrity, and encamp in the plains of Sultanieh. On an adjacent hill, are a palace and magnificent gardens belonging to the king. This town under the Sophis was not a place of any importance. It was taken by the Afghans after the battle of Salman-abad, and almost entirely destroyed by them. It was rebuilt by Kerim-Khan, and has been frequently fortified and embellished by his successors. Early in the 18th century it was constituted the capital of the kingdom. Tehran, says Southgate, "appears to have been a place of very little importance until the middle of the last cent., when Agha Mohammed Shah, the founder of the present dynasty, removed the seat of government thither, from Isfahan, in order to be nearer his own tribe, the Kujars of Mezanderan, on them he chiefly depended for the support of his throne: before this the town seems to have been little known. In the days of the old travellers it was a village, and the great route to the south did not pass through it. It had, from the first, no local advantages, aside from its proximity to Mezanderan, for which it could have been chosen as the seat of royalty, and now, after fifty years, it has hardly any other recommendation. The place presents, in outward appearance, none of the features of a royal city. Its bazaars are extensive, and are roofed with tile, so as to present a succession of small domes; they are filthy, however, and less attractive in every respect than those of Tebriz; they are thronged with beasts as well as men, which makes a walk through them no easy nor pleasant matter. The streets are peculiarly bad, for the most part destitute of pavements, narrow, irregular, encumbered with filth, and full of dangerous holes. The houses are extraordinarily mean, even for an Eastern town; and unsightly ruins, covering, in some instances, extensive areas, frequently meet the eye. There are no magnificent structures or outward marks of grandeur to affect the general meanness of its appearance. The palace of the Russian minister is an extensive range of very plain buildings. The British residency, though unfavourably situated in a low position near the southern wall, is a neat edifice, with an open piazza, and a range of columns in front. It has also pleasant gardens, with paved walks, both in front and rear, which makes it altogether a delightful spot. But the chief attraction is 'the Ark,' or royal residence. It consists of a great number of buildings, courts, and gardens, covering a large area, and enclosed within a high wall, which separates it from the rest of the city. Just without the wall, on the side where I entered, is a public square, in the centre of which is a large cannon, said to have been used by Nadir Shah. It is now converted into the more pacific and sacred use of a sanctuary for criminals: whoever takes refuge beneath it is safe."

TEHINTCHIEW, a river of Asia, which rises on the S frontiers of Tibet; flows S, passing Tassisuden; and joins the Brahmaputra, after a course of about 150 m.

TEHRUT, a town of Persia, in the prov. of Kerman, 54 m. NW of Banim, on the Uchku, in a fertile and richly cultivated locality.

TEHUACAN, a town of Mexico, in the state and 69 m. SE of La Puebla, on the road from that town to Oaxaca, in a rich plain. Pop. 200. It has three convents, and an hospital, and some well-built houses, and carries on an active trade in flour. This was one of the most venerated of the sacred places of the Aztecs.

TEHUANTEPEC, a town of Mexico, in the state and 158 m. ESE of Oaxaca, on a river of the same name, about 11 m. above its mouth. It is a place of considerable trade, and has manufactures of salt, and of coarse cottons. The pop. has recently been estimated at 14,000, mostly Indians, some half-breeds, and a few Castilians. It has 16 churches, and a college was established here in 1850. Its manufactures consist of leather, cotton cloths, silk sashes, shoes, hats, mats, silverware, saddlery, pottery, and soap.

TEHUANTEPEC (Isthmus of), that portion of Mexican territory, comprising the eastern portions of the states of Vera Cruz and Oaxaca, which lies between the gulf of Mexico and the Pacific ocean, where the two seas approach nearest to each other. From the mouth of the Coatzacoalcos, which discharges itself into the gulf, in N lat. 18° 8′ 20″, and W long. 94° 32′ 50″, to the harbour of Ventosa, on the Pacific, in N lat. 16° 11′ 45″, W long. 95° 15′ 40″, the direct distance is 143½ m. The coast-lines on either side have a general direction nearly E and W. A belt of level country, of some 40 or 50 m. in breadth, lies contiguous to the gulf coast. The chain of the Cordilleras traverses the central portion of the isthmus, but here exhibits a sudden depression or breach of continuity, nearly at a point directly in the line of shortest communication between the two oceans. The plains on the Pacific side average about 20 m. in breadth, from the base of the mountains to the coast, and have a slope varying from 10 to 15 ft. in the mile. The principal river on the Atlantic side of the isthmus, is the Coatzacoalcos, which rises in the sierra to the E of Santa-Maria-Chimalapa, and has a general direction of NNE and SSW. Its mouth has a width of 350 ft., and a depth of 13 ft.; and at 7 m. from the gulf, a depth of 40 ft. The Chicapa and the Tehuantepec are the principal rivers on the Pacific side. In the rainy season the latter river has a depth of 12 ft. In March 1853, a convention was entered into between the United States of America, and the United Mexican States, to facilitate the construction and insure the maintenance of a transit way across this isthmus. The importance of such a work is second only to that noticed under the head Miguel (Gulf of San), of forming a ship-canal across the isthmus of Darien, of which it is a rival. The latter scheme, though by far the more important to the mercantile interests and general progress of the world, is just now in abeyance, and its necessity promises to be superseded by the former. The plan proposed by Dr. Cullen and Mr. Gisborne, of making a ship-canal across the isthmus of Darien, is a project of surpassing promise to the commerce of the world; but, in consequence of the large amount of capital required to carry it into execution, its projectors are not likely to bring it speedily to a successful issue. The other undertaking, therefore, that of forming both a plank road and a railroad across the isthmus of T., becomes of more importance to the commercial interests of this and other countries than it at one time appeared probable would be the case. It appears that after much loss of time in negotiation, legislation, and arbitration, all obstacles to the opening of this route have at length been removed. The treaty between the American and Mexican government contains, among other stipulations, the following articles:

Art 1st. The United Mexican States solemnly engage that the inter-oceanic transit-way across that part of its territory which is denominated the isthmus of Tehuantepec shall be open and free for all the nations of the globe.

Art. 2d. and 3d. The United States of America and the United Mexican States engage to protect the persons engaged and property employed in the construction of the said work, from the commencement thereof to its completion against all confiscation, spoliation or violence of whatsoever nature. The contracting parties further engage to extend their protection to the transit-way when completed, during the continuance of the contract, and also that they will guarantee the neutrality thereof, so that the conveyance of persons and property thereon shall at all times be uninterrupted, and the capital invested therein entirely secure. Neither passports nor letters of security will be required of persons crossing the isthmus, and not remaining in the country.

Art 6th. No interest in the said transit-way, nor in the proceeds thereof, shall be sold, or otherwise transferred to any foreign government, corporation or body politic, nor shall its benefits be extended to any other nation, or to its citizens or subjects, on any terms more favourable than those accorded to the United States and its citizens.

Art. 8th. Armed forces not being permitted to pass over the said transit-way, without the express authorization of the Mexican government, it is agreed that the said government shall either give this permission directly itself, or will empower in proper form the chief military officer of the isthmus, or his substitute, to permit the passage over the said transit-way of the troops which the United States may find it necessary from time to time to transmit from one part of its territory to another lying on opposite sides of the continent, so that the said troops shall meet with no impediment.

A contract has been signed between the Tehuantepec railroad company and Messrs. Sykes and Co., of England, for the construction of a plankroad and a railway across the isthmus. The former is to begin at Suchil, the head of navigation on the Coatzacoalcos, and terminate at Ventosa bay, on the Pacific—a distance of 96 m.—and is to be completed in November 1855. The railway is to begin at Minatilla, only 20 m. from the mouth of the Coatzacoalcos, and terminates also at Ventosa bay. It is to be completed within four years, and will be 166 m. in length. This will be the shortest route between New York and San-Francisco. From New York to San Francisco, *via* the Panama route, the distance is 4,992 m.; *via* Nicaragua, 4,531 m.; *via* Tehuantepec, 3,804 m. From England to San Francisco, *via* Panama, the distance is 7,502 m.: *via* Nicaragua, 7,041, and *via* Tehuantepec, 6.671 m. The distance from New York to San-Francisco, is *via* Tehuantepec, 1,100 m. nearer than Panama, and 727 m. nearer than Nicaragua. The distance from England to San-Francisco, *via* Tehuantepec, is 831 m. less than by Panama, and 370 m. less than by Nicaragua. New Orleans is more favourably located in relation to this route than any port of England, or the port of New York, in the distance saved. From New Orleans to San-Francisco, *via* Tehuantepec, the distance is 1,801 m. less than by

Panama, and 1,063 m. less than by Nicaragua. The geographical position of New Orleans is, therefore, not only more favourable as regards total distance, but as regards distance saved. Whilst the Panama route will monopolise the trade and traffic of the South Pacific, the Tehuantepec route will probably command almost the entire travel and traffic of California and the North Pacific. "Nobody doubts that we shall soon see steamers cross from the W coasts of the British islands to Halifax in less than a week. The Grand Trunk railway from that city to Detroit on Lake Michigan —1,100 m.—is under contract with Jackson, Peto, Brassey and Betts, of European renown, and £11,000,000 cash capital under British management insure the vigorous prosecution and early completion of this stupendous work. This road connects the N and S shores of the river St. Lawrence, at Montreal, by a tubular bridge upwards of 2 m. in length, now constructing under the superintendence of Robert Stephenson. From Detroit—the terminus of the Grand Trunk railway through Canada—the great railroad line to the gulf of Mexico will be shortly completed, much of it being already made. It will, therefore, be but a short time before the English traveller will find himself on the Pacific side of the Tehuantepec isthmus, steaming towards Australia, in 12 days from England, having traversed the Atlantic, the British American provinces, the heart of the American continent through the Mississippi valley, the gulf of Mexico, and the Tehuantepec isthmus, where he looks on the Pacific ocean, and has passed the last obstacle which separated him from a vast and golden hemisphere which his country owns on the other side of the globe. This sounds like fiction, and seems like a Utopian dream; but within five years it will have grown as familiar to us as the nursery tales of good King Alfred, and not half so romantic."

TEHUELS, a tribe of Indians, who inhabit the district of Patagonia, extending between the Andes and desert. They are said to be of great stature, but to be mild and pacific in disposition.

TEHUILOTEPEC, a village of Mexico, in the state and 60 m. S of the town of that name, and on the road thence to Acapulco.

TEICHEL, a town of the principality of Schwarzburg-Rudolstadt, bail. and 6 m. NNW of Rudolstadt, in a fine valley. Pop. 400.

TEICHMANSDORF, a village of the principality of Schwarzburg-Rudolstadt, 4 m. E of Stadt-Ilm. It has a fortress.

TEIFY, or Tivy, a river, which, rising in the mountains of Cardiganshire, separates it from the counties of Carmarthen and Pembroke. After flowing S to Tregaron, it enters the plain in which Llanbedr is situated, and, as it approaches Newcastle-in-Emlyn, gradually confines itself within steep banks fringed with wood. Soon afterwards it becomes suddenly ingulfed within two piles of high rocks, from which it acquires the rapidity of a cataract. Augmented by the tide, it proceeds in a broad majestic stream, winding between the bases of two lofty ridges, on one of which stand the ruins of Cilgerran-castle, till it falls into the sea below Cardigan. It is navigable to Newcastle-in-Emlyn.

TEIGHSHINOD, a parish, 3¼ m. N by E of Ballymahon, co. Longford, Leinster. Area 2,713 acres. Pop. in 1831, 2,553; in 1851, 1,517.

TEIGNGRACE, a parish in Devon, 12¼ m. S by W of Exeter. Area 1,329 acres. Pop. in 1851, 187.

TEIGNMOUTH (East), a parish in Devon, 12½ m. S by E of Exeter. Area 745 acres. The town has been greatly improved of late years, and is much frequented in the bathing season. Pop. 1,760.

TEIGNMOUTH (West), a parish, sea-port, and market-town, in Exminster hund., Devon, 12 m. S by E of Exeter, at the mouth of the Teign, and separated from East T. by a small rivulet, called the Tame. Area of p., 493 acres. Pop. in 1801, 1,528; in 1831, 2,878; in 1851, 3,389.—The town, though irregularly built, contains many good houses. With its quay and dock-yard, it stands on a curve formed by the sudden expansion of the river. The harbour is safe and capacious, and a new wharf has been erected. Vessels belonging to this port are engaged in the Newfoundland trade. The chief imports are coal and culm; the exports, granite and potters' clay. Of the latter many thousand tons are annually shipped to the Staffordshire and other potteries. A bridge, constructed over the Teign, here, and opened in 1827, consists of 34 arches, and a swing bridge, which opens in two parts, so that vessels of 400 tons may pass.

TEIGNTON (Bishop's), a parish in Devon, 4 m. ENE of Newton-Bushel. Area 4,748 acres. Pop. in 1851, 1,119.

TEIGNTON-DREWS, a parish in Devon, 8¾ m. SW of Crediton. Area 6,937 acres. Pop. 1,232.

TEIL (Le), a commune of France, in the dep. of the Ardeche, cant. and 5 m. N of Viviers, on the r. bank of the Rhône. Pop. in 1846, 2,397. It has manufactories of silk fabrics, and several tile and brick-kilns.—Also a village in the dep. of the Ille-et-Vilaine, cant. and 2 m. WNW of Retiers, in a mountainous and marshy locality. Pop. 1,400. It has a ferruginous spring. See also Thiel (Le).

TEILLE', a village of France, in the dep. of the Loire Inferieure, cant. and 5 m. S of Riaille, near the r. bank of the Havre. Pop. 1,200.—Also a village in the dep. of the Sarthe, cant. and 2 m. W of Ballon, near the l. bank of the Sarthe. Pop. 1,020.

TEILLET, or Teulet, a village of France, in the dep. of the Tarn, cant. and 8 m. SW of Alban. Pop. 230.

TEILLEUL (Le), a canton, commune, and town of France, in the dep. of the Manche, and arrond. of Mortain. The cant. comprises 8 com. Pop. in 1831, 8,063; in 1846, 8,350. The town is 8 m. SSE of Mortain, in a marshy plain. Pop. in 1846, 2,566.

TEILLOU, a village of France, in the dep. of the Deux-Sevres, cant. and 5 m. NW of Chef-Boutonne, on a hill near the r. bank of the Sompt, a small affluent of the Boutonne. Pop. 536. It has a considerable trade in grain and sheep, and rears large numbers of goats.

TEINITZ, a village of Austria, in Bohemia, in the circle and 26 m. SE of Beraun, on the l. bank of the Sazawa. It has a manufactory of stone vases.

TEINITZ, or Tynetz, a town of Austria, in Moravia, in the circle and 36 m. SE of Brunn, on the r. bank of the March. Pop. 575.

TEINITZ (Bischof), or Horssow Teyn, a town of Bohemia, in the circle and 18 m. NW of Klattau. Pop. 2,403. It has a fine castle and gardens belonging to the prince of Trautmansdorf.

TEINITZ, (Elbe), or Tegnice-Nad-Labein, a town of Bohemia, in the circle of Chrudim, on the r. bank of the Elbe. Pop. 1,830.

TEINITZ, (Hrochow), or Hrochowa-Tegnice, a town of Bohemia, in the circle and 4 m. E of Chrudim. Pop. 964.

TEIONES (Cape), a headland of Tripoli, on the coast of Barca, to the SW of Benghazy, in N. lat. 31° 50′ 30″, and E long. 19° 52′ 20″.

TEISBACH, a village of Bavaria, 13 m. WSW of Landau, on the r. bank of the Isar. Pop. 480.

TEISENDORF, or Deisendorf, a village of Bavaria, 14 m. WNW of Salzburg, on the Sur. Pop. 640.

TEITH, or Teath, a river of Perthshire, whose

head-sources are not far from a point at which the cos. of Argyle, Perth, and Dumbarton meet, or from another point whence waters are shed respectively to the Tay, the Forth, and the lower Clyde. They rise respectively 1 m. and 2 m. due E of the head of Loch-Lomond. The northern stream, starting from the N side of the height called Mealpalan, runs 6½ m. E by N to Loch-Doine. Issuing from Loch-Doine, it proceeds east, till it is lost for 3½ m. in Loch-Voil. On egressing from the latter lake, it takes the name of the Balvag; sweeps past Balquidder; and flows 3 miles sinuously southward to Loch-Lubnaig. Issuing from Loch-Lubnaig, it begins to tumble along the wild romantic Pass of Leny, and having flowed 1 m. S, and 2 m. E, it unites with its sister stream to form the Teith. Its entire length of course, including its lacustrine expansions, is about 23 m.—The southern stream makes a run of 4 m. from its origin due SE to the head of Loch-Katrine; and is identified for 8 m. with the lake of second rank for scenic brilliance in Scotland. Emerging from its SE extremity, it begins to traverse the bosky, bristling, tumultuously wooded Trossachs; and debouching from among them, is lost for 1⅞ m. in Loch-Achray; egressing thence, it makes a run of only 1½ m. eastward, till it is ingulphed by Loch-Vennacher. On issuing from the latter lake, it careers 1½ m. ENE to a junction with the other great head-water of the Teith. Its entire length of run, including its course through lakes, is about 22 m.—The Teith, or united stream, wants the grand and boldly romantic features of its head-waters; and subsides as to scenery into simple yet often thrilling beauty. Its entire course is SE; and, measured in a straight line, extends only to between 11 and 12 m. Its tributaries are numerous; but, excepting the Keltie, they are all inconsiderable. The T., if either its volume of water or its length of course had been made the ground of decision, would have been regarded as the parent-stream, and the Forth which joins it as the tributary. The point at which they unite is the bridge of Drip, 2¼ m. NW of Stirling. The T. is a clear stream, and, for the most part, rapid. It abounds in salmon and trout; and has some small beds of a peculiar sort of mussels which yielded valuable pearls.

TEIXEIRA, a town of Portugal, in the prov. of Beira, comarca and 11 m. W of Lamego, near the l. bank of the Douro.

TEJA. See Teza.

TEJADA, a town of Spain, in the prov. and 29 m. SSE of Burgos, in a fertile valley, near the r. bank of the Esgueva. Pop. 250.

TEJADO, a town of Spain, in the prov. and 15 m. SE of Soria, near the l. bank of the Rituerto, an affluent of the Duero. Pop. 250.

TEJARES, a town of Spain, in the prov. and 2 m. SW of Salamanca, on the road from that town to Ciudad-Rodrigo, and near the left bank of the Tormes, which is here crossed by a bridge. Pop. 157.

TEJEDA, a town of Spain, in the prov. and 47 m. NNE of Caceres, near the r. bank of the Tietar. —Also a town in the prov. and 33 m. SSW of Salamanca. Pop. 250. It has a convent.

TEJO. See Tagus.

TEK-DAGH, a range of lofty mountains in Turkey in Asia, in the pash. and to the SE of Erzerum.

TEKE-DEREH, or Kawakti-direh, a river of Turkey in Europe, in Rumelia, in the sanj. of Kirk-Kilissia. It has its source in the E part of the sanj., on the W side of the S mountains, 6 m. SSW of Binarhissar, on the confines of the sanj. of Viza; runs first WNW through a narrow valley, then SSW; passes to the SE of Kirk-Kilissia and Eski-Baba, where it is crossed by a fine stone-bridge; receives soon afterwards the Sarajiala; forms for a short distance the boundary line of the sanj. of Gallipoli; and after a total course of about 75 m. joins the Erkene, on the r. bank, 9 m. WSW of Eski-Baba.

TEKE-ILI, a sanjak of Turkey in Asia, in the S part of Anatolia, bounded on the N by the sanj. of Hamid-ili, on the E by Karmania-Itshil, on the W by the sanjaks of Mentesha and Meis, and on the S by the gulf of Satalieh. It is 105 m. in length from E to W, and 81 m. in extreme breadth. It is surrounded by lofty mountains, of which the principal is the Taktalu-dagh on the SW. The Duden-su, an affluent of the Estenas, the Ak-chaï and Kapri are its principal streams. Springs of waters are generally scarce. The mountains are covered with fine forests, and in the plains are grown wine, cotton, and other varieties of fruit. This sanj. occupies a large portion of *Pisidia* and *Pamphylia* of the ancients.

TEKES, a river of China, in Sungaria, in the military div. of Ili. It joins the Khounghes at Kash, and thence the united stream takes the name of Ili.

TEKIR-DAGH, or Tekiri-dagh, a range of mountains in Turkey in Europe, in Rumelia, in the E part of the sanj. of Gallipoli, between the l. bank of the Maritza and the sea of Marmora, on the shore of which it terminates, 11 m. SSW of Rodosto. It has an extent of about 78 m., runs in a generally ESE direction, and sends forth a ramification in a SW direction into the peninsula of Enos. It gives rise to numerous streams, the greater number of which belong to the basin of the Maritza. Of these the principal are the Kizil-nahar, Ipsala-su, Ainajik, and Kawa-chaï—the latter flows into the gulf of Saros. From the proximity of the sea of Marmora, the streams which it sends thither are mere mountain torrents. See also Rodosto.

TEKIS, a maritime town of Japan, in the island of Kiu-siu, and prov. of Fizen.

TEKIYEH, a town of Turkey in Asia, in the pash. and 99 m. WNW of Erzerum, and 6 m. E by N of Chorum.

TEKMAN, a small town of Turkey in Asia, in the pash. and to the NE of Erzerum.

TEKOA, a village of Turkey in Asia, in the pash. of Damascus, 9 m. S of Bethlehem. It occupies the site of an ancient town.

TEKRIT, a town of Turkey in Asia, in the pash. and 90 m. NNW of Bagdad, on a hill, near the r. bank of the Tigris, in N lat. 34° 35′. Pop. 1,000. It is built of earth; but is a comparatively clean place. The environs are fertile, and have several naphtha springs. This town occupies the site of the ancient *Bertha*, which sustained a long siege against Tamerlane, and which in the 11th century was the residence of a Jacobite primate. The ruins of the ancient town are of great extent.

TEKROVA. See Fionda.

TEKTEK, a mountain of Turkey in Asia, in the pash. of Reha, ENE of the town of that name.

TEKUTSH, a town of Turkey in Europe, capital of a district in Moldavia, 33 m. NNE of Rimnik, and 101 m. S of Jassy, near the r. bank of the Byrtal.

TEL-AKBERIN, a town of Turkey in Asia, in Syria, in the pash. and 30 m. W of Aleppo.

TELAV. See Thelavi.

TELDOM-KHOTON, a town of China, in Mandshuria, in the prov. of He-lung-keang, on the Amour, in N lat. 49° 56′, and E long. 129° 53′.

TELEGD (Mezo), or Tyelegd, a town of Hungary, in the comitat of Behar, 13 m. E of Gross-Wardein, on the l. bank of the Sebes-körös.

TELEGU, a village of Wallachia, celebrated for its salt mines. The salt is cut out with great regularity. Long parallel lines are first deeply traced with the axe; cross cuts are then made at regular intervals, after which a single stroke of the tool is generally sufficient to raise each block from its bed. The larger mine is worked exclusively by convicts, numbering about 200 in 1854, of whom 60 are condemned for life, and are only brought up to the surface for a few hours once every month: the others come up every evening. The majority of the whole are Gipsies. The crimes of which they have been guilty are mostly petty thefts and robberies with violence; several of those condemned for life to this terrible punishment have committed no greater offence than a violation of the quarantine on the Turkish frontier! "It was well known that the object of Russia was to make the line of separation between the two countries, both commercially and politically, as broad as possible—to destroy the influence of Turkey, by rendering her a *terra incognita*, and attach the principalities more closely than ever to the 'sacred soil of holy Russia.' Russia is likewise the author and promoter of the law which condemns the first poor peasant who, in a moment of impatience, endeavours to escape 96 hours of durance vile, to a life of hard labour in cold and darkness and a poisonous atmosphere." There are three salt mines in Wallachia, all belonging to the government, and all worked mainly by convicts. Their united annual product is 60,000,000 okas, which is delivered at the mouth of the mine at 8 paras the oka, or rather less than one penny English, which will give a total annual revenue at about £250,000. From this, however, has to be deducted the expenses of working and management, which are trifling, and the amount abstracted by the various officials. The whole of these 60,000,000 okas, minus what is needed for home consumption, passes into Turkey and Servia. Wallachian salt is prohibited both in Austria and Russia.

TELEKHANOUI, a town of Russia in Europe, in the gov. and 129 m. SW of Minsk, and district of Pinsk, on the Oginskoi canal.

TELEMBI, a river of New Grenada, in the dep. of the Cauca, which has its source in the Andes, near the confines of Ecuador, runs first N; then NW; waters Barbacoas; and after a course of about 90 m., joins the Patia, on the l. bank, a little above the entrance of that river into the Pacific.

TELEMBI (San-Luis-de), a town of New Granada, in the dep. of the Cauca, prov. and 66 m. WNW of Pasto, on the l. bank of the Telembi.

TELEMSAN. See Tlemecen.

TELEOUTES, or Telengoutes, a Tartar people who inhabit the S part of the gov. of Tomsk and Yeniseisk, in Russia in Asia, and called by the Russians, White Kalmucks. They derive their name from lake Telengoul in the Altaï mountains. They did homage to Russia for the first time in 1609, but did not become incorporated with the empire till about the middle of last century.

TELESE, a village of Naples, in the prov. of Terra-di-Lavoro, 15 m. NE of Caserta, near a small lake of the same name, and on the site, it is supposed, of the ancient *Samnium*.

TELETSKOE, a lake of Tartary, in the Kirghiz territory, near the r. bank of the Syr-Deria, and to the W of Otrar, into which a stream called the Sarasu flows, but which has no apparent outlet.

TELFAIR, a central county of the state of Georgia, U. S., watered by the Ocmulgee. Area 950 sq. m. Pop. in 1840, 2,763; in 1850, 2,126. Its cap. is Jacksonville.

TELFS, a village of Tyrol, in the circle of Imst, on the l. bank of the Inn.

TELGET, or Telgte, a village of Prussia, in the reg. and 6 m. ENE of Munster, on the l. bank of the Ems. Pop. 2,150. It has potteries and flour-mills, and a trade in cattle.

TELGRUE, a village of France, in the dep. of Finistere, cant. and 6 m. ESE of Crozun, near the coast. Pop. of com. 1,750.

TELIGUL, a river of Russia, which rises on the frontier of Podolia, and flows into the Black sea, at a point 25 m. NE of Odessa, after a SSE course of 90 m. It gives name to a small gulf or bay at its embouchure.

TELIORMAN, a river of Upper Wallachia, which descends from the SE flank of the Arjish range; runs SSE; and joins the Veda, on the l. bank, 16 m. NNE of Sistova, after a course of about 90 m.

TELISKOF, a town of Asiatic Turkey, in the pash. and 30 m. NNW of Mosul, in the centre of a vast plain, extending from the Tigris to the W, to Mount Zagros on the E, on which Alexander won the battle of Arbela, and some centuries afterwards Heraclius quelled the power of Khosru-Perviz. The town has a pop. of about 4,000.

TELKI-BANYA, a village of Hungary, in the com. of Abaujvar, 4 m. ENE of Gönez, on an affluent of the Hernad. There are chalybeate springs in the vicinity.

TELLARNOUGHT, a parish in co. Wexford. Area 1,654 acres. Pop. in 1841, 323.

TELLICHERRY, a town and port of Hindostan, on the Malabar coast, in N lat. 11° 44′, E long. 75° 36′, 6 m. NW of Mahe. Its commerce, especially in the articles of sandal-wood, pepper, and spices, has somewhat declined, but is still considerable.

TELLINGSTEDT, a village of Denmark, in the duchy of Holstein, 18 m. WSW of Rendsburg.

TELLIO, a village of Austrian Lombardy, in the Valteline, 6 m. SW of Terano, near the r. bank of the Adda. Pop. 1,500.

TELLISFORD, a parish of Somersetshire, 5 m. NNE of Frome. Area 757 acres. Pop. in 1841, 150; 1851, 124.

TELLO, a village on the W coast of Celebes, 30 m. NE of Macassar, in S lat. 5° 5′.

TELLO-DELANO, a port on the S coast of the island of Nias, off the SW coast of Sumatra.

TELLORE, an islet off the SW coast of Sumatra, in N lat. 0° 7′, E long. 99° 20′.

TELLTOWN, or Killalton, a parish, 3½ m. SE by E of the town of Kells, co. Meath. Area 4,266 acres. Pop. in 1831, 1,308; in 1851, 1,128.

TELMA, a town of Asiatic Russia, in the gov. and 15 m. NW of Irkutsk. It consists of two rows of log houses, erected on the sides of a log-road covered with smooth planks. A stone church, in the Italian style, and spacious barracks, give the place an air of importance. "The workhouse," says M. Ermann, "a fabric of two stories, is no doubt the largest and finest specimen of architecture in North Asia. The front of it has a length of 364 ft., and is adorned with massive columns, between which, in two rows, are windows of the purest plate glass. The lower story is divided into three apartments, in which are carried on the manufacture of cloth. Above dwell the officers who manage the institution for the crown. Stone warehouses, and mills of different kinds, are situated along the banks of the stream which drives the machinery of the workhouse. The advantages of the locality were discovered a century ago, by private speculators, since which time, T. has been famous for its cloth manufactory. More recently, glass, paper, and linen have been added to its productions. The inhabitants are about 2,000 in number, of whom 800 find employment in the manufactories. They are persons exiled for crimes. They are supplied gratuitously with meal, and receive besides an amount of wages, proportioned in each instance to the value of the labour. The wool required for the manufactory at T. is procured chiefly from the Buraets and Tungúses, who wander with their flocks over the southern borders of Siberia. The machinery for combing and spinning the wool was originally procured from England, and was afterwards made in Siberia, according to the English model. T. produces annually about 50,000 yds. of woollen cloth, and half that quantity of linen. The former is sold at a price not exceeding half-a-crown a yard. Pains are taken to improve the wool; and in 1830 a flock of 480 Spanish sheep were driven from Moscow to Irkutsk; and, notwithstanding the length of the journey, and the plagues of the Barabinskian steppes, 300 of them reached their destination in safety."

TELOCHE, a village of France, in the dep. of Sarthe, cant. and 4 m. N of Ecomoy, on a small affluent of the Sarthe. Pop. of com. 1,400.

TELSCOMBE, a parish of Sussex, 3 m. NW of Newhaven. Area 1,349 acres. Pop. in 1851, 176.

TELSHA, a town of Russia, in the gov. and 150 m. NW of Wilna, on the NW bank of a small lake, which discharges itself into the Windau.

TELTOW, or TELTAW, a town of Prussia, in the reg. and 9 m. E of Potsdam, on the Dolger-see. Pop. 1,400, chiefly linen-weavers.

TELTSH, a town of Moravia, in the circle and 15 m. SSW of Iglau, on the l. bank of the Thaya-Morava. Pop. 3,500. It has considerable manufactories of fine woollens.

TELU-SAMAVA, a port on the N coast of Sumatra, 150 m. ESE of Achim.

TELYCH, a hamlet in Llandigat p., Carmarthenshire; adjacent to Llandovery. Pop. 270.

TEMACIN, a town of Northern Africa, in N lat. 33° 20′, about 12 m. SSW of Tuggurt, the capital of a small independent district.

TEMASCALTEPEC, a town of Mexico, in the state and 70 m. WSW of Mexico. Pop. 2,500, chiefly miners and cotton-weavers.

TEMBLEQUE, a town of Spain, in the prov. and 21 m. SE of Toledo. Pop. 3,000. It has manufactories of saltpetre and chocolate, and some trade in corn.

TEMBO, or PIO-DE-NEIGE, a mountain of Switzerland, in the cant. of the Grisons, in the Rhintal, rising to an alt. of 6,815 Parisian ft. above sea-level. Its summit commands a view embracing Lake Como on the S, and the lake of Constanz on the N.

TEMBUCTU. See TIMBUCTU.

TEMBY, a river of Eastern Africa, a branch of English river, which is navigable from Delagoa bay, upwards of 40 m.

TEME, a river rising in the mountains which divide Radnorshire from Montgomeryshire, and flowing eastward by Knighton, through the pleasant vale of Brampton-Bryan, receiving in its course a number of tributaries, as it proceeds, in a placid stream, to Tenbury, where it enters Worcestershire, and, after traversing a romantic glen, falls into the Severn near Powick, about 3 m. from Worcester. It is celebrated for grayling and trout.

TEMENDFUS. See MATIFU.

TEMERIN, a town of Hungary, in the com. and 30 m. E of Bacs. Pop. 6,300. It is surrounded by a marshy district.

TEMESH, or TEMES, a river of Hungary, which rises on the NE flank of Mount Szemenik, in the Banat; runs ESE, and then NNW; and passing Karansebes, receives the Sebes on the r.; passes Lugoz; turns SE, and then SSE; and joins the Danube on the l. bank, a little below Pancsova, after a winding course of above 300 m. Its principal tributaries are the Bisztra, the Bega, and the Temessecz on the r.; and the Boganicz on the l.—The comitat of T. is bounded by that of Arad on the N; by Krassova on the E and SE; by the Wallachia-Illyrian regimental district on the S; by the German banat on the SW; and by the com. of Torantal on the W. The Maros defines its whole northern frontier; the Karash, its SE. It is a level marshy district; fertile in soil, and producing maize, rice, tobacco, saffron, fruit, and wine. Its inhabitants, about 360,000 in number, are chiefly Wallachians and Servians. It is administratively divided into the four marches of St. Andras, Lippa, Temeswar, and Vershitz.

TEMESVAR, or TEMESCHWAR, in Wallachian TIMISIVARA, a town of Hungary, the cap. of the com. of Temesh, situated in a vast marshy plain, on the Bega canal, in N lat. 45° 42′, E long. 21° 14′, 74 m. NNE of Belgrade. Pop. in 1846, 19,000. It is a well-built place, and is strongly fortified. It has a handsome Gothic cathedral, and several other large public edifices, besides extensive military arsenals and barracks. It has manufactories of woollens, paper, and hardware; and conducts an active trade in grain and the other agricultural produce of the district. In 1551, it was taken by the Turks, from whom, in 1716, Prince Eugene wrested it. In 1849, it resisted the Hungarian insurgents for 107 days, until relieved by Haynau. D'Anville identifies it with the ancient *Thybiscus*. T., with a pop. of 24,000, the capital of Southern Hungary, and the principal seat of its civilization, is of comparatively modern and French aspect. In situation, says a recent writer, it is not much better than Szegedin as regards salubrity, and it was in the time of the Turks entirely surrounded by marshes except to the N. The river Bega, which now forms an ellipse to the S of the town, then flowed through the middle of it, forming numerous islands, and even now fevers abound. If we ascend to the observatory tower a wide champaign prospect of cultivated but far from perfectly drained fields and villages is seen all around; no hills are visible, except on a clear day to the eastward in the direction of Transylvania, where the last spurs of the Carpathians appear like a cloud on the horizon. The town itself, when handed over to Prince Eugene on the 13th of Oct. 1716, consisted of four parts; the inner town of wooden houses—only the mosques and the powder magazine being of stone—surrounded with a wall and a ditch; then the castle of the Hunyady family, a Middle age fortress, connected with the town by a draw-bridge and forming the citadel; and to the N, E, and W, the so called great and small Palanka, not walled but pallisaded. The inner town has entirely disappeared. The palaces and symmetrical streets of T. are in the style of Louis Quatorze. Even the mosques which were built of stone have disappeared: two of them, immediately after the conquest reconsecrated as churches, were pulled down, and the modern church of the Franciscans and the Bishops' seminary occupy their place. The principal feature of T. is the great square, in which are the Catholic and Greek cathedrals; the former an extensive edifice, built during the government of our countryman, Count Andrew Hamilton, who was president and commander-in-chief in the Banat between 1734—8, and distinguished himself by great activity in building and other improvements.—At right angles with the cathedral is the principal edifice of the town—the palace of the government, which occupies the space between two streets. A triple line of fortifications, according to the rules of Vauban, encircles the town. Beyond each curtain is the ravelin; beyond each bastion the contregarde; and an envelope of solid masonry forms the third and outer line of defence. The great defect of the fortifications is, that in the lapse of time they have sunk at various places, on account of the marshy land, so that the relative gradation of the outer to the inner works is, in several places, disturbed to such an extent that the former are not sufficiently dominated by the latter. The principal gates, three in number, are named from the directions in which they lead—Peterwardein, Transylvania, and Vienna. Without the Peterwardein gate, and beyond the rayon of the fortress is the Josephstadt suburb, intersected by the Bega canal, cut in 1745—60, to connect T. with the SW part of the Banat in connexion with the Theiss, Danube, and Save. Here is a crowd of canal craft; here are the large magazines of the Banat wheat, and a constant bustle of loading and unloading. The old castle of Hunyady, which is within the fortifications, and is now the armoury, was built by John Hunyady, Count of Temes, in 1442. It resisted the recent siege-operations better than most of the houses in the town, but its towers, rising high above the bastion, are now quite unroofed. Passing the large infantry barracks—the lower part of which is bomb proof, and forms a section of the fortification—we go out at the gate of Transylvania and come upon the principal suburb of T., called the Fabrik. This suburb received the surname of Fabrik from the manufactures of metals, cloths, paper, hats, &c., which Count Mercy, the first governor of the Banat after the conquest, attempted to establish.

TEMIMEH, or NAHIL, a river of Tripoli, in Barca, which runs NE, and flows into the gulf of Bomba, under the parallel of 47° 20′.

TEMISCAMING, a lake of Canada, which discharges itself by a narrow channel into the Ottawa. It is 30 m. in length from N to S, and 10 m. in medium width, and contains several small islands. Its banks are inhabited by a tribe of Indians of the same name.

TEMISCOUATA, a lake of Lower Canada, 125 m. NE of Quebec. It is 22 m. in length, and from 1 to 2 m. in breadth.

TEMISSA, a town of Tripoli, in Fezzan, 120 m. NE of Murzuk, on a hill. The ruins by which it is surrounded indicate its having formerly been of greater extent. The environs yield large quantities of dates.

TEMMEN, TEMA, or TREMINI, a town of Upper Guinea, on the Gold coast, in the state of Inkran, 48 m. WSW of the embouchure of the Volta. It has a Dutch establishment.

TEMNIKOV, a town of Russia in Europe, in the gov. and 156 m. NNE of Tambov, on the right bank of the Moksha. Pop. 3,200. It contains six churches, numerous shops, and some manufactories of sailcloth. A small trade is carried on in corn. The era of the foundation of this town is unknown, but it was rebuilt on its present site in 1536.

TEMNOLESKAIA, a fortified town of Russia in Europe, in the prov. of the Caucasus, district and 54 m. W of Alexandrof, on a mountain, near the r. bank of the Kuban. It makes part of the military line of the Caucasus.

TEMOAL, or TIMOEL (CAPE), a headland of the island of Celebes, in N lat. 0° 1′, and E long. 119° 32′.

TEMOURTU. See TUZ-KUL.

TEMPE′, a valley of Turkey in Europe, in the NE part of Thessaly. It lies between the chain of Olympus on the N, and that of Ossa on the S, and is about 6 m. in length. It is traversed by the Selembria, the *Peneus* of the ancients, which, a little below, throws itself into the gulf of Salonica. Between Baba, where it is narrowest, and the point at which it begins to widen, the valley bears the name of the Boghaz or gorge. This valley, so celebrated for its beauty by the ancient poets, is described by some travellers as consisting merely of a deep and narrow gorge, remarkable for the wild sterility of its aspect, while others distinguish its SW part as extremely picturesque and pleasing. See PENEUS.

TEMPELBURG, a town of Prussia, in the prov. of Pomerania, regency and 48 m. S of Köslin, and circle of Neu-Stettin, at the SW extremity of Lake Drazig. Pop. in 1843, 3,368. It is enclosed by walls, and has three gates, two suburbs, and two churches, a Catholic and Lutheran. It has manufactories of cloth, breweries, and several distilleries of brandy. This town was founded in the 12th century by the Templars.

TEMPERANCEVILLE, a settlement of Upper Canada, in the township of Malahide, on Catfish creek or river Barbu, 10 m. E of St. Thomas. Pop. 100.

TEMPIO, a town of Sardinia, in the div. of Cape Sassari, prov. and 27 m. N of Ozieri, on a mountain, a little to the W of the Limbara mountains. Pop. 7,057. It is well-built and fortified, and has a college and three convents. It is noted for salted meats and hams. Wine is cultivated in the environs, and large numbers of cattle are reared, and form an important branch of contraband trade with Corsica.

TEMPLE, a parish in Cornwall, 6 m. NE of Bodmin. Area 843 acres. Pop. in 1851, 24.

TEMPLE, a township of Hillsborough co., in the state of New Hampshire, U. S., 44 m. SSW of Concord, drained by branches of Souhegan river. Pop. in 1840, 567; in 1850, 579.—Also a township of Franklin co., in the state of Maine, 40 m. NNW of Augusta, bordered on the E by Sandy creek. Pop in 1840, 955; in 1850, 785.

TEMPLE (BAY), a bay of New South Wales, on the NE coast, between 11° 58′ and 12° 25′ S lat., and in 143° 5′. It is enclosed on the N by Cape Granville, and on the SE by Cape Fear, and is 33 m. in length, and 12 m. in breadth. The surrounding coast is generally low and sandy.

TEMPLE (LE), a village of France, in the dep. of the Loire Inferieure, cant. and 4 m. N of St. Etienne-de-Montluc, on an elevated plateau. Pop. 350.

TEMPLE (LE), or TEMPLE-SUR-LOT (LE), a commune and town of France, in the dep. of the Lot-et-Garonne, cant. and 4 m. WSW of Sainte-Livrade, near the l. bank of the Lot. Pop. 551.

TEMPLEBEG, a parish in co. Tipperary, 4⅘ m. SSW of Borris-o'-leagh. Area 3,427 acres. Pop. in 1831, 1,064; in 1851, 902.

TEMPLEBODANE, a parish in co. Cork, 5 m. SE by S of Rathcormack. Area 4,736 acres. Pop. in 1831, 1,337; in 1851, 924.

TEMPLEBOY, a parish in co. Sligo, 2 m. E by S of Dunmore-West. Area 9,112 acres. Pop. in 1831, 3,787; in 1851, 2,048.

TEMPLEBREADY, a parish in co. Cork, 5½ m. SSE of Passage. Area 2,654 acres. Pop. 1,587.

TEMPLEBREDEN, a parish in co. Tipperary, 2½ m. N of Emly. Area 2,455 acres. Pop. 1,001.

TEMPLEBRIEN, a parish in the E division of the barony of East Carbery, co. Cork. Area 1,189 acres. Pop. in 1831, 496; in 1851, 319.

TEMPLECARNE, a parish, partly in co. Fermanagh, and partly in co. Donegal, containing part of the town of Pettigoe. Area of the mainland of the Fermanagh section, 7,719 acres, of which 4,399 acres are in Lower Lough Erne, of the Donegal section, 38,149 acres, of which 2,140 acres are in Lough Derg, and 1,072 acres in small lakes. Pop. of the whole, in 1831, 5,461; in 1851, 4,926. Only about one-sixth of the whole land in the p. can be used for tillage; the greater part of the enormous remainder is so upland, moorish, and heathy as to be of no other service than for coarse summer pasturage to black cattle.

TEMPLECORRAN, a parish in co. Antrim, 4½ m. NE of Carrickfergus, containing the village of Ballycarry. Area 4,744 acres. Pop. 1,472.

TEMPLECRONE, a parish on the coast of co. Donegal, containing the town of Dunglo. Area 52,921 acres,—of which 4,353 acres are in the island of Arran. Pop. in 1831, 8,198; in 1851, 9,592. The mainland district extends southward from Guidore-bay to Trawenagh-bay, and is identical over all the N and the centre with the wild and unique tract of country called the Rosses.

TEMPLEDERRY, a parish in co. Tipperary, 6½ m. SE of Nenagh. Area 6,998 acres. Pop. 1,457.

TEMPLEDOWNEY, a parish in co. Tipperary, consisting of two detached districts, the larger one N of Toomavara, and the smaller one S of that place. Area of the whole, 1,850 acres. Pop. 348.

TEMPLE-ERRY, or TEMPLE-REE, a parish in the barony of Ikerrin, co. Tipperary. Area 4,241 acres. Pop. in 1831, 1,415; in 1851, 1,189.

TEMPLEHARRY, a parish in King's co., 2½ m. NNW of Moneygall. Area 4,589 acres. Pop. in 1831, 1,156; in 1851, 939.

TEMPLEKELLY, or TEMPLEJEHALLY, a parish in co. Tipperary, containing Ballina, or the eastern suburb of Killaloe. Area 10,038 acres,—of which 1,009 acres are in Lough Derg. Pop. in 1831, 4,722; in 1851, 2,098. About one-half of the length extends along Lough Derg; and the remainder along the Shannon.

TEMPLEKIERAN, a parish in co. Meath, 3¾ m. SE of Navan. Area 1,067 acres. Pop. 113.

TEMPLEMALEY, a parish in co. Clare, 2½ m. N by E of Ennis. Area 4,648 acres. Pop. 1,014.

TEMPLEMARTIN, a parish in co. Cork, 3½ m. N of Bandon. Area 7,515 acres. Pop. in 1831, 2,730; in 1851, 1,550. The surface is part of the N screens of the valley of the Bandon river.—Also a parish 2½ m. E of Kilkenny. Area 782 acres. Pop. in 1851, 224.

TEMPLEMICHAEL, a parish in co. Longford, containing the whole of the town of Longford. Area of the Ardagh section, 9,115 acres. Pop. of the whole, in 1831, 8,323; in 1851, 9,106.—Also a parish on the E border of co. Tipperary, containing the village of Ninemilehouse. Area 2,869 acres. Pop. in 1831, 796; in 1851, 775.—Also a parish in co. Waterford, on the r. bank of the Blackwater, 2¾ m. NNW of Youghal. Area 8,215 acres. Pop. in 1831, 2,573; in 1851, 1,645.—Also a parish in co. Wicklow, 2¾ m. N of the town of Arklow. Area 1,325 acres.—Also a parish in co. Cork, 2½ m. E by N of Innishannon. Area 2,064 acres. Pop. in 1831, 764; in 1851, 550.

TEMPLEMOLOGGA, a parish in co. Cork, 3½ m. NW of Mitchellstown. Area 4,396 acres. Pop. in 1831, 1,876; in 1851, 1,180.

TEMPLEMORE, a parish, containing the chief part of the city of Londonderry, and identical with the North-west Liberties of the City of Londonderry. Area 12,615 acres. Pop. in 1831, 19,620; in 1851, 23,772 "The parish," says the *Ordnance Memoir*, "generally presents the appearance of a rich and cultivated country, eminently diversified in its picturesque features, and imposing from the grandeur of its undulating outlines. Its chief beauties are, however, connected with the broad and navigable Foyle."—Also a parish in co. Tipperary, containing the town of Templemore. Area 8,472 acres. Pop. in 1831, 5,218; in 1851, 5,699.—The town stands on the r. bank of the Suir, on the road from Roscrea to Cashel, 9 m. N of Thurles.

TEMPLEMOYLE, a hamlet, and a celebrated agricultural school, in the p. of Faughanvale, co. Londonderry, 1½ m. SW by S of Muff, on the road to Strabane. The agricultural school stands 7 furl. SSW of Muff, and 5 m. ENE of Londonderry, upon a farm of 172 acres. The buildings cost £2,400; and are capable of accommodating not less than 70 boarders. The annual expense of board and instruction amounts to £10 a-year. Parents or landlords pay this for the largest proportion (the parents paying for about one-third), the Drapers' and Grocers' companies, the Irish society, and the Commissioners of the Woods and Forests for the remainder. The lowest age for admission is 14. The school is open to all classes and persuasions, and is frequented from all parts of Ireland. It is divided into two equal divisions; one remains in school, while the other is occupied abroad. The class in school pursue, under the head-master, reading English, English grammar, writing, arithmetic, geography, construction of maps, book-keeping in reference to their special occupations, farmers' accounts, &c. Euclid's elements, trigonometry, in its application to heights and distances, land-surveying, together with the use of the water-level, theodolite, chain, &c. The other division works the farm and garden, under the head-farmer, divided into bands, each superintended by its respective monitor, who is responsible for the conduct of his band. The situation of the establishment is beautiful. The house stands near to the top of a steep hill, looking down upon a wooded glen, and abroad over the rich levels stretching to the lough, and over the lough to the mountains of Donegal and the grand Coleraine rocks. Besides feeding the whole establishment, the produce brings in a yearly increasing profit. Within ten years, there have been additions of new dormitories, an infirmary, washing-rooms, a museum of models of farming implements, an improved cow-house, and an excellent house for sheep, the introduction of which, with all modern improvements in the management of them, is an important new feature in the education given. The land is divided into nine portions, five of which are regularly tilled on the five-shift rotation, and the other on the four-shift. Every part of the work is, sooner or later, done by the hands of each pupil, the only help hired being for the drudgery, which would be mere waste of time when once learned. From the first attempt to plough a furrow or set a fence to the highest skill in judging of stock at fairs and markets, the pupils are exercised in the whole of their art.

TEMPLEMURRIG, a parish in co. Mayo, 2½ m. NNW of Killala. Area 2,240 acres. Pop. 514.

TEMPLENECARRIGA, a parish in co. Cork, 4¾ m. N by W of Middleton, containing the village of Walshtown. Area 5,208 acres. Pop. 991.

TEMPLENEIRY, a parish in co. Tipperary, 3¾ m. ESE of Tipperary, containing the village of Bansha. Area 12,840 acres. Pop. in 1851, 2,677.

TEMPLE-NEWSHAM, a township in Whitkirk p., Yorkshire, 4 m. E by S of Leeds. Area 3,097 acres. Pop. in 1831, 1,458; in 1851, 1,690.

TEMPLENOE, a parish in co. Kerry, 5½ m. SW by W of Kenmare. Area 32,428 acres. Pop. in 1831, 3,882; in 1851, 3,011.—Also a parish in co. Tipperary, 2¾ m. NE of Tipperary. Area 2,729 acres. Pop. in 1831, 1,158; in 1851, 1,238.

TEMPLEOMALUS, a parish in co. Cork, 2¾ m. SE by E of Clonakilty. Area 1,931 acres. Pop. in 1831, 1,352; in 1851, 913.

TEMPLEORAN, or FORAN, a parish in co. Westmeath, 5½ m. NW of Mullingar. Area 5,188 acres. Pop. in 1831, 1,298; in 1851, 961.

TEMPLEOUTRAGH, or UPPERCHURCH, a parish in co. Tipperary, 6½ m. NW by W of Holycross. Area 12,902 acres. Pop. in 1851, 2,350.

TEMPLEPATRICK, a parish in co. Antrim, containing the village of Templepatrick. Area 14,190 acres. Pop. in 1851, 3,925.

TEMPLEPETER, a parish in co. Carlow, 4¾ m. SW of Tullow. Area 1,045 acres. Pop. 234.

TEMPLEPLACE, a village in the p. of Donaghcamper, co. Kildare, on the r. bank of the river Liffey, 5 furl. S by W of Celbridge. Pop. 310.

TEMPLEPORT, a parish in co. Cavan, 3¼ m. WSW of Ballyconnel. Area 42,171 acres. Pop. in 1831, 10,758; in 1851, 9,788. The summit of Slievenakilla, at the NW extremity, has an alt. above sea-level of 1,793 ft.

TEMPLEQUINLAN, a parish in co. Cork, 2¾ m. ENE of Clonakilty. Area 2,270 acres. Pop. in 1831, 1,042; in 1851, 870.

TEMPLEROAN, a parish in co. Cork, 3 m. E of Doneraile. Area 3,866 acres. Pop. in 1851, 1,154.

TEMPLEROBIN, a parish in co. Cork, containing the greater part of the town of Cove; and consisting of the islands of Spike and Hawlbowline, and the eastern division of Great Island, all in the upper part of Cork harbour. Area 3,594 acres. Pop. in 1831, 8,220; in 1851, 15,332.

TEMPLESCOBBIN, or TEMPLESCOBY, a parish in co. Wexford, 1½ m. WSW of Enniscorthy. Area 1,707 acres. Pop. in 1831, 405; in 1851, 431.

TEMPLESHAMBO, or TEMPLESHANBOUGH, a parish in co. Wexford, 4½ m. S by W of Newtonbarry. Area 19,516 acres. Pop. in 1831, 4,198; in 1851, 3,688. The summit of Mount Leinster, at the NW extremity of the p., has an alt. above sea-level of 2,610 ft.

TEMPLESHANNON, a parish in co. Wexford, containing part of the town of Enniscorthy. Area 4,982 acres. Pop. in 1831, 2,743; in 1851, 2,851.

TEMPLE-SOWERBY. See SOWERBY.

TEMPLETENNY, a parish of co. Tipperary, 5 m. W of Clogheen, containing the village of Ballyporeen. Area 18,181 acres. Pop. in 1831, 5,786; in 1851, 4,667. The surface extends southward from the summit-line of the Galtee mountains, across the valley of the Tar rivulet and the Mitchellstown Caves, up and over the Knockmeledown mountains to the bottom of the vale or to the stream of the Arraglin.

TEMPLETHAY, or TEMPLE-ETNEY, a parish in co. Tipperary, 4 m. NE by E of Clonmel. Area 6,677 acres. Pop. in 1831, 1,437; in 1851, 1,050. The summits of Slievenaman, Carrickabrock, and another height, all on the N boundary, have altitudes of respectively 2,364, 1,859, and 1,589 ft. above sea-level.

TEMPLETOGHER, a parish 3½ m. W by N of the village of Ballymoe, co. Galway. Area 13,705 acres. Pop. in 1831, 4,881; in 1851, 4,042.

TEMPLETON, a parish of Devon, 5¼ m. W by N of Tiverton. Area 1,895 acres. Pop. in 1831, 222; in 1851, 218.—Also an ancient village in Narberth p., co. of Pembroke, 2 m. SSE of Narborth.

TEMPLETON, a township of Worcester co., in the state of Massachusetts, U. S., 61 m. WNW of Boston. It has a hilly surface, and is watered by branches of Miller's and Chicapee rivers. Pop. in 1840, 1,776; in 1850, 2,173.

TEMPLETON'S CREEK, a river of Australia Felix, in the district of Murray, an affluent of Goulburn river.

TEMPLETOWN, a parish in co. Wexford, 2¼ m. SW by W of Fethard. Area 4,156 acres. Pop. in 1831, 1,387; in 1851, 1,442.—Also a village in the p. of Carlingford, co. Louth, near the extremity of Carlingford peninsula.—Also a locality in co. Antrim, which gives the title of Baron and Viscount in the peerage of Ireland to the noble family of Upton.

TEMPLETRINE, a parish of co. Cork, 5 m. SW of Kinsale. Area 4,784 acres. Pop. in 1851, 1,019.

TEMPLETUOHY, or BALLINSIN, a parish in co. Tipperary, containing the village of Templetuohy. Area 8,460 acres. Pop. in 1851, 2,561.

TEMPLEUDIGAN, or ST. PETER'S, a parish in co. Wexford, 5 m. NNE of New Ross. Area 8,177 acres. Pop. in 1831, 2,434; in 1851, 1,723.

TEMPLEUVE, a department, commune, and town of Belgium, in the prov. of Hainault, arrond. and 6 m. NW of Tournay, on the line of railway to Courtray. Pop. 4,000. It has manufactories of linen, and carries on an active trade in honey, wax, and butter.

TEMPLEUVE - EN - PEWELL, a town of France, in the dep. of the Nord, cant. and 4 m. SSW of Cysoing. Pop. in 1846, 3,180. It has manufactories of furniture stuffs, oil-mills, and sugar factories.

TEMPLEUSQUE, a parish in co. Cork, 5 m. NE of Cork, containing the village of Riverstown. Area 4,603 acres. Pop. in 1831, 1,289; in 1851, 1,128.

TEMPLEUX - LE - GUERARD, a village of France, in the dep. of the Somme, cant. and 3 m. ENE of Roisel. Pop. 650. It has manufactories of embroidered muslin.

TEMPLIN, a circle and town of Prussia, in the prov. of Brandenburg, and regency of Potsdam. The circle comprises an area of 216 sq. m., and contains 38,000 inhabitants.—The town is 23 m. SE of New Strelitz, on Lake Dolgen. Pop. in 1843, 3,352. It is enclosed by walls, and has four gates, an hospital, and four churches. It has manufactories of hosiery, and distilleries, and carries on a considerable trade in wood.

TEMPO, a rivulet of co. Fermanagh, which rises among the mountains on the boundary of cos. Tyrone and Fermanagh, and runs SSW past the village of Tempo, to a confluence with the Colebrook river, at a point 1¼ m. S of Maguire's-bridge.—Also a village in the p. of Enniskillen, co. Fermanagh, 4½ m. NE by N of Lisbellaw. Pop. in 1851, 407.

TEMPSFORD, a parish of Bedfordshire, 6 m. NW of Biggleswade. Area 2,350 acres. Pop. 622.

TEMPZIN, a village of the grand-duchy of Mecklenburg-Schwerin, and duchy of Mecklenburg-Güstrow, on a small lake, 1½ m. NNW of Brisel, and 15 m. NE of Schwerin. It has a tile-work.

TEMRIOUK, a fortified town of Russia in Europe, in the prov. of the Caucasus, and territory of the Cossacks of the Black sea, 30 m. N of Anapa, on a narrow tongue of land, extending into a gulf of the same name, in the SE part of the sea of Azof, in N lat. 45° 20′. It makes part of the Kuban line of defence, and contains a new church, and about 120 houses, of which a half are outside the ramparts. Under the Turkish government this place was one of much greater extent and importance.

TEMTHELS, or TEMMELS, a village of Prussia, in the prov. of the Rhine, regency and 10 m. SW of Trèves, and circle of Sarrebury, on the r. bank of the Moselle. Pop. 500.

TENA, a valley of Spain, in the N part of the prov. of Huesca, on the S side of the Pyrenees, and between two ramifications of these mountains. It is 15 m. in length from N to S, and 9 m. in extreme breadth, and is watered by the Gallego and several other streams. This valley, which is one of the finest in Aragon, produces corn, barley, oats, millet, flax, and hemp, and pastures large herds of cattle, and is noted for its mineral wells and baths. In the adjacent mountains are a silver and several iron mines. The valley contains 11 villages, of which Panticosa is the principal.

TENA, a town of New Grenada, in the prov. of Bogota, 24 m. from Santa-Fe-de-Bogota. Pop. 800.

TENAILLE (LA), a hamlet of France, in the dep. of Charente-Inferieure and cant. of Mirambeau, 6 m. WNW of Jouzac, at the source of the Seudre. It had formerly an abbey.

TENANGO, a town of Mexico, in the intendancy

and 32 m. SW of Mexico. A great annual fair is held here.

TENANT'S BAY, an indentation of the SE coast of Nova Scotia, 12 m. E of the bay of Mascaret. It is about 5 m. in length, and 1½ m. in medium breadth, and is divided into two parts connected by a narrow neck. It affords safe anchorage, but its banks are of the wildest description.

TENASSERIM PROVINCES, a district of Trans-Gangetic India, lying between Siam on the E, and Pegu on the W, formerly a part of the Burmese empire, but ceded to the East India company by the treaty of Yandabu, on the termination of the war in 1826. The provinces, comprising Martaban, Ye, Tavoy, and Mergui, are situated on the eastern shores of the bay of Bengal; and though the boundaries have never been very accurately defined, may be said to extend from 10° to 17° N lat., and from 97½° to 99° E long., forming a narrow strip of territory about 420 m. in length, and from 50 to 70 m. in breadth, except towards the southern extremity, where the breadth diminishes to 10 or 12 m. On the N, this territory is bounded by a lofty range of mountains, branching from the table-land of Yunnan, and clothed with dense forests, which after attaining the height of 5,300 ft., makes a sweep to the southward, and runs in a long continuous chain, but with diminished altitude, between the eastern extremity of the provinces and the kingdom of Siam. The peninsula of Malacca forms the S boundary, while on the W the bay of Bengal extends for the distance of more than 300 m. along the coast, to the embouchure of the Saluen, which river forms the rest of the boundary in that direction, and separates our possessions from the Pegu territories. The entire ex-extent of coast is bordered by islands. The characteristic feature of this extensive territory, particularly in the N province, is that of a vast alluvial plain partially enclosed by the high mountain-range above referred to, and diversified by a double line of hills from 200 to 800 ft. in height, forming in some parts a narrow continuous chain; in others, starting into sharp precipitous eminences which present a most singular and picturesque aspect. These are principally of limestone formation, but along the course of the Saluen, from Maulmein to its embouchure, there runs a ridge of sandstone from 200 to 300 ft. in height. On the sea-coast also, the ground is elevated and rocky, but being densely wooded, and for the most part unexplored, little is known of its geological formation. The space between the hills, as well as between the sea-coast and the mountain-boundary of the interior, consists of a succession of extensive valleys or plains, which though open in some parts for many miles, are more generally covered with dense and almost impenetrable jungle. The whole country is intersected by numerous rivers and streams, which pour an immense volume of water to the sea. On the N the Saluen or Thaluen has a course of nearly 1,000 m., penetrating beyond the confines of China; the Attayen and the Gyne, though of more limited extent, are of very considerable depth, and intersect the N division of the province with numerous windings navigable for boats. Opposite the cantonment of Maulmein, where these three rivers form a junction, the breadth of their united stream is from 1 to 2 m., and the depth sufficient for vessels of several hundred tons burden. To the S, the country is equally well watered by the Ye, the Tavoy, and the Tenasserim, all rivers of considerable magnitude and navigable for many miles. Owing to the level character of the surface too, numerous creeks, branching from these rivers, run through the provinces in various directions, and tend considerably to facilitate the communications. All these sources of outlet, however, are insufficient to carry off the immense fall of rain during the wet season, when the low grounds are so completely covered as to present the appearance of an extensive lake; and even during the first two months of the dry season, they are left in a very marshy state. It is computed that in the province of Amherst or Martaban alone, at least 3,000 sq. m. of plain are thus regularly inundated, and thereby fertilized like the delta of Egypt by the overflowing of the Nile. As may be supposed, under such circumstances, the soil is almost universally alluvial, and covered in many parts with the remains of decayed vegetation to the depth of more than 3 ft. Beyond the reach of the inundation the surface is clothed with forests of the most stately timber, particularly teak, which grows in great abundance as far S as the parallel of 16°, while long coarse grass or a mass of jungle overspreads the low grounds. The durian, mangosteen, pine-apple, sugar-cane, and various spices and fruits, are abundant; and dye-woods, storax, gamboge, sandal-wood, and cajeput are produced in the forests. The banks of the rivers, as far as the influence of the tide extends, give birth to a dense barrier of mangrove bushes and rank weeds. Indeed, the exuberant fertility of the soil produces upon a most gigantic scale every species of tropical vegetation, wherewith the whole face of the country is so overrun as to present one of the principal obstacles to its being, for many years to come, completely explored. The animals found in these provinces are elephants, rhinoceroses, tigers, panthers, leopards, buffaloes, and wild hogs.—Coal has been discovered in Mergui, and iron-ore in Ye and Tavoy.

Climate.] The principal characteristic of the climate is the extreme length and severity of the wet season, when the quantity of rain is greater than in any British colony except Western Africa. It begins to fall in April, and continues almost incessantly till October, during which upwards of 200 inches have been measured; the quantity, however, varies materially in different years. As in Ceylon, the seasons are principally influenced by the SW and NE winds or monsoons, the former extends from April to Oct., and constitutes the wet season, during which the wind, coming direct from the ocean, is laden with moisture. The other continues during nearly all the rest of the year, and blowing over the immense extent of land to the NE is just as remarkable for its dryness. For about a month of the intervals between the monsoons, the winds are variable, and the setting in of each is always accompanied by terrific storms of thunder and lightning. The seasons on this coast are divided by the natives into the hot, the rainy, and the cold, the first commencing with the full moon in February, the second with the full moon in June, and the third with the full moon in October. The mere range of the thermometer, however, would not lead to the inference that there was any great difference in the degree of heat at these seasons, as will be seen from the following table:—

	1829.		1831.	
	Highest.	Lowest	Highest.	Lowest.
January,	86°	61°	89°	53°
February,	92½	65	96	71
March,	93½	72	98	72
April,	95	74	102	75
May,	89	74	98	75
June,	88	74½	94	76
July,	85½	74	89	74
August,	88	73½	92	74
September,	86	73	92	75
October,	90½	72	94	71
November,	89	62	92	53
December,	89	60	"	"

The temperature, as shown by this table, will be found much the same as at Colombo, in Ceylon, ex cept that it falls about 10° lower in the cold, and rises from 5° to 10° higher in the hot season; the difference between the maximum and minimum is also twice as great throughout the year.

Population.] Notwithstanding this amazing fertility, and the variety of its productions, the whole of this immense territory has for many centuries past been almost a desert, the population, at the time it was delivered over to the British government, not having exceeded 30,000 souls on a space comprising 33,000 sq. m.; and with the exception of the inhabitants of the towns of Tavoy and Mergui, these consisted of wandering tribes subsisting chiefly on the produce of the forests, and yielding only a nominal subjection to any government. On the British taking possession, however, many thousands from Rangoon and its vicinity took refuge under our flag, and every succeeding year has brought such accessions to their number that, on the l. bank of the Saluen alone, the pop. at present exceeds 100,000 souls, and it was computed that altogether there were in 1839 nearly 112,000 throughout the provinces.

Burmese and Talaings,	92,242
Siamese,	1,425
Chinese,	1,499
Malays,	533
Kayens and Toung-thoos,	13,503
Natives of India,	2,517
Miscellaneous,	686
	112,405

The chief towns, when the British took possession, were Tavoy and Mergui, the former about 150, the latter 250 m. S of the nearest part of the exiting Burmese frontier. These being at too great a distance to afford effectual protection to the N province, it was determined to station most of the troops opposite to the Burmese town of Martaban, on a peninsular piece of ground at the junction of the Attayen, the Gyne, and the Saluen, three rivers which afford the principal means of access to the interior. Accordingly, in January, 1827, the cantonment of Maulmein or Moelmyne, now the principal station of the troops, and the chief seat of the pop. in these provinces, was founded on the l. bank of the Saluen, about 30 m. from its embouchure, on a spot said to have been occupied as a military position by the Portuguese several centuries ago. See articles AMHERST, MERGUI, MAULMEIN, and TAVOY.

TENASSERIM, or TANATHARE, a river of India beyond the Ganges, in the prov. of the same name. It has its source in the mountains to the NE of Tavoy; runs through a narrow valley to the parallel of the town of that name; thence it directs its course to the E till it approaches Mergui, when it turns W, and flows by two branches into the sea. The southern embouchure is open to ships, but the other cannot with safety be navigated. It has a total course of about 240 m.—The prov. of T. forms a narrow tract extending along the E side of the bay of Bengal, between 11° and 19° N lat., between Tavoy on the N, and the isthmus of Kra, by which the peninsula of Malacca is connected with the continent on the S, and separated from Siam on the E by a range of mountains.

TENAY, a commune of France, in the dep. of the Ain, cant. and 4 m. SE of St. Rambert on the Albarine. Pop. 1,130. It has manufactories of linen and cashmere, a silk-mill, and bleacheries, and carries on an active trade in linen.

TENBURY, a parish and market-town in Worcestershire, 22 m. NW by W of Worcester, on the S bank of the Teme, near the Leominster canal. The p. comprises the hamlets of Berrington, Sutton, and Tenbury-Foreign. Area 5,060 acres. Pop. in 1831, 1,768; in 1851, 1,786. The town is indifferently built, but contains some good houses. It occupies a low position, close to the Teme, in consequence of which it is often liable to sudden inundations. The Kyre, a small rapid stream, flows into the Teme at the upper end of the principal street. Both rivers are here crossed by handsome bridges. The trade of the town is chiefly in hops and cider, great quantities of which are produced in the neighbourhood. Tanning and glove-making also employ some hands. In 1839, a spring of strongly impregnated saline water was discovered here. T. is a polling-place in the election of members for the W division of the county.

TENBY, a borough and small sea-port town in the p. of St. Mary, co. of Pembroke, 12 m. E of Pembroke, on one of the bays on the N side of the Bristol channel. Pop. in 1831, 2,128; in 1851, 3,208. T. contributes with Pembroke, Wiston, and Milford, in returning 1 member to parliament. It is also a polling-place for the co. The town is very picturesquely situated on a ridge of mountain limestone, which curving gently towards the E, forms a small bay, which has been converted into a commodious and well-sheltered harbour, skirted by a bold amphitheatre of rocks and houses. The town has been greatly improved of late years. The houses in general are neatly built, and the streets are clean and well-lighted with gas. Besides the ecclesiastical edifices, the public buildings are the town-hall, assembly-rooms, theatre, spacious baths, and a market-house. The trade of the town is inconsiderable. The harbour consists of two piers, which shoot into a corner of the bay, and nearly encircle a small but safe spot for the anchorage of vessels. The imports are chiefly articles of domestic consumption. The exports consist of butter, corn, coal, and culm. Some fishing vessels from Plymouth and Broxham make T. their station during the fishing-season. The oyster-beds here constitute a source of considerable profit to the fisherman. Tenby, however, is important principally as a watering-place, for which it is singularly well adapted, by the great beauty of its situation, and the protection from rough weather which it receives from the headlands close to it. The lodging-houses are numerous and excellent; and the beach is hard and smooth. Steam-packets from Dublin and Bristol touch here in the summer season. The ruins of Tenby-castle are still considerable, though mostly in a very dilapidated state.

TENBY, a town of Van Diemen's Land, in the co. of Pembroke, 49 m. from Hobart Town.

TENCE, a canton, commune, and town of France, in the dep. of the Haute-Loire and arrond. of Yssingeaux. The cant. comprises 4 coms. Pop. in 1831, 13,323; in 1846, 14,071.—The town is 9 m. ESE of Yssingeaux, on the r. bank of the Lignon, at the confluence of the Serigoul. Pop. in 1846, 6,158. It has an alms-house; and possesses extensive manufactories of black and white lace, and of felt-hats, a silk and a paper-mill.

TENCH'S ISLAND, an island of the S. Pacific, in the archipelago of New Ireland, in S lat. 1° 39', E long. 151° 31'. It is about 2 m. in circumf., is low and entirely covered with trees, of which a large number are cocoa palms. The inhabitants, who are copper coloured, are tall, well-made, and manage their canoes with great dexterity. They are about 1,000 in number. This island was so named by Bale, who visited it in 1790.

TENCIN, a commune of France, in the dep. of the Isère, cant. and 2 m. SSW of Goncelin, on the r. bank of the Theys, near its confluence with the Rhone. Pop. 1,040.

TENCOA, a small town of Honduras, in the dep. and 75 m. NW of Comayagua.

TENDA, a town of Sardinia, capital of a mandemento, in the prov. and 32 m. NE of Nice, and 25 m. NNE of Sospello, on the flank of a steep mountain near the r. bank of the Roya. Pop. 1,500. The houses are ill-built, and the general aspect of the place extremely poor. It gives its name to a pass of the Alps leading from Piedmont to Nice, and is noted as the place from which the family of the Lascaris sprung, one member of which, named Theodore, was emperor of Constantinople.

TENDA, a state of Senegambia, on the banks of the Gambia, in N lat. 13°. It is covered with woody mountains.

TENDA (Col-de), a pass in the chain of the Maritime Alps, a little to the W of its junction with the Apennines, in Sardinia, on the confines of the divisions of Nice and Coni, in N lat. 44° 8′, E long. 7° 42′. It is 6,160 ft. in perpendicular height, and is crossed by a fine road, practicable for carriages, constructed by Victor-Amedeo III. The ascent on the S side occupies three hours, and about two-thirds up is a vast excavation, commenced by order of Anne, duchess of Savoy, with the intention of penetrating the mountain, but remaining still unfinished. The descent on the other side takes about 1½ hour. The summit of the col is covered with snow and exposed to violent storms. It is surrounded on all sides, except the S, in which the Mediterranean opens to view, by mountains of the wildest description. This pass was long infested by the Barbets, a kind of mountain militia, notorious in defiance of their professed character for their depredations upon travellers.

TENDA-MAIE, a district of Southern Senegambia, enclosed by a bend of the Río-Grande. It is of small extent, and is flat and sandy, but very fertile, and produces in large quantities maize, millet, and rice. It abounds with deer and wild oxen, and is noted for its iron. Its inhabitants are gentle, but indolent, poor, and little disposed to hospitality.

TENDERBAR, a town of Senegambia, in the state of Kaen, on the l. bank of the Gambia, 51 m. E of Bathurst.

TENDILLA, a town of Spain, in the prov. and 15 m. ESE of Guadalajara, in a country covered with vines and olives, and near the l. bank of the Tajuna. Pop. 1,135.

TENDON, a commune of France, in the dep. of the Vosges, cant. and 18 m. NE of Remiremont. Pop. 1,291.

TENDRA, or Tenter, an island of the Black sea, in Russia in Europe, near the NW coast of the gov. of Kherson, and district of Aleshki, near the mouths of the Dnieper, and to the W of the gulf of Odessa, in N lat. 46° 21′ 40″, E long. 31° 29′ 5″. It is 26 m. in length from WNW to ESE, but is very narrow, forming a tongue of land, separated from the continent by only a narrow channel. It has a lighthouse on its N point, 32 m. ESE of Odessa.

TENDRE (Mont), a mountain of Switzerland, in the cant. of Vaud, 10 m. W of Cossonay, above Montricher, in a ramification of the Jura. It has an alt. of 4,030 Parisian feet above the level of the lake of Geneva, and is crossed by a road at the height of 3,860 ft.

TENDRING, a parish in Essex, 5¾ m. SSE of Manningtree. Area 2,827 acres. Pop. 953.

TENE. See Fatima.

TENEBRON (El), a town of Spain, in the prov. and 48 m. SSW of Salamanca. Pop. 150.

TENEDOS, or Bokhtsha-Adassis, an island of Turkey in Asia, in the Grecian archipelago, near the coast of Anatolia, 12 m. SSW of the SW entrance to the straits of the Dardanelles, in N lat 39° 50′, E long. 26° 3′. Pop. 7,000. It is about 8 m. in length from E to W, and 3 m. in breadth. Although covered with rocky mountains—of which the principal, St. Elias, appears to have been volcanic—it is generally fertile, and produces in considerable quantities, grain, fruit, cotton, and sesame; but its wealth consists chiefly in the culture of the vine, the annual revenue from which is estimated at 30,000 piastres. It has no port, and vessels are here subjected to all the hazards of an open coast. On its NE coast is a town of the same name, with a pop. of 4,500. It is built semicircularly in a valley, and on the slope of two hills, and is commanded by another pyramidal-shaped hill, which rises to some height above. On the S it is defended by a small fort, and on the N by a citadel. It conducts an active trade with Constantinople. An immense grain magazine was erected here by Justinian, for storing the cargoes of vessels detained here on their route from Egypt, but it long since fell into disuse, and has been suffered to fall to ruins. This island is mentioned by Homer, and its position at the mouth of the Hellespont has rendered it at all times a place of importance. It fell early under Ottoman domination. In 1656, it was taken by the Venetians, but the death of their general the following year again threw it into the hands of the Turks, in whose power it has ever since continued.

TENEMBER, an island of the Molucca sea, in S lat. 6° 30′, E long. 132° 45′. It is 12 m. long, and 3 m. broad.

TENEN, a town of Russia in Europe, in the gov. of Vilna, district and 50 m. W of Rossiena.

TENERIFFE, or Tenerife, the largest island in the group of the Canaries, off the western coast of Africa, situated between the parallels of 28° 0′ and 28° 35′ N; and intersected by the meridian of 16° 25′ W. It is of a triangular form, each side being about 36 m. in length, and has an area of about 1,000 sq. m. The NE extremity of the island bears 15½ leagues NW by W from the NE point of Canaria. It is of volcanic formation, and in the peak of Teyde, of the sloping sides of which the island actually consists, it attains an alt. of 3,706 metres, or 12,158 ft. above sea-level. The height to which it rises, the distance from which it is perceived at sea, and the volcanic eruptions which continually issue from its sides, have long rendered it an object of curiosity to naturalists. The coasts are everywhere cliffy and precipitous; the cliffs being of a dark brown basalt. By the rapidity of its rise, it presents, within a short distance, every variation of temperature, from the colder climates of Europe to those of the equinoctial regions. Humboldt, after having traversed the banks of the Orinoco, the Cordilleras, and the most beautiful valleys of Mexico, declares that he never beheld a landscape more agreeable, more harmonious, and more attractive than that which the Laguna or NE coast of this island offers. There, orange, myrtle, and cypress trees, entwine the chapels reared on the eminences; the declivities are covered with vines, and cultivated like a garden; a perpetual spring prevails, and in the summer evenings the breeze from the sea bears with it delicious coolness. In this part of the island the date-tree, the plantain, the sugar-cane, the Indian fig, the *Arum colocasia*, the root of which furnishes the lower class with a nutritive meal, the olive-tree, all the fruit trees of Europe, the vine, and wheat, maize, pulse, and potatoes, are cultivated. Wheat is reaped from the end of March to the beginning of May; the culture of the bread-fruit tree, of cinnamon, coffee, and cocoa, have been tried with success. Above the fertile level track rises what is called the region of laurels, forming an extensive border all round the island. These trees are fed by a vast number of springs which rise up amid a turf covered with perpetual verdure. Extensive plantations of chestnut occur in the lower part of this zone, above which several species of laurel, an oak resembling that of Tibet, and other trees flourish. The underwood in the lower part consists of arborescent heath; in the upper part, of ferns. Above this region, we enter a vast forest of fir and pine. The prevailing species of fir has the appearance of the Scotch fir, with long and stiff leaves sprouting by two or oftener by three in one sheath. Passing through this belt, we enter a vast plain, like a sea of sand, covered with yellow pumice dust which continually fills the air, and embellished with tufts of the beautiful *Retama blanca* growing to the height of nine feet, and loaded with odoriferous flowers. On the skirts of this plain the rich verdure of the island terminates, as well as all appearance of habitation; the traveller ascends afterwards through a complete solitude. Above this sandy plain are the *malpays*, a name which the Spaniards, in all their volcanic districts, apply to a ground destitute of vegetable mould, and covered with loose and broken fragments of lava. The ascent here is steep and extremely fatiguing, over blocks of lava which easily roll from beneath the feet. At the extremity of the *malpays* is a small plain called the *Rambleta*, from the centre of which, the *Piton* or sugar-loaf summit of Teyde rises in N lat. 22° 16′ 24″, W long. 16° 39′ 0″. Here are found those spiracles which are called by the natives 'the Nostrils of the Peak,' consisting of watery and heated vapours, which issue at intervals from crevices in the ground. The Piton is of a rugged, conical, truncated form, and 852 ft. in height; its ascent is steep, and rendered still more difficult by the loose ashes with which it is covered. At the top there is scarcely room to stand, and the crater is enclosed by a wall so steep that it could not be entered, were there not a breach in one spot. The sides of the crater slope gently to a depth of about 106 ft. The crater has long ceased to emit flames, but gives vent to aqueous vapours with a peculiar buzzing noise. The view from the top of the peak is characterised by peculiar beauty. The traveller, placed on the summit of such colossal mountains, sees usually only their own barren steeps; while the plains, covered with rich vegetation, appear in the immensity of distance; but the slender form and rapid rise of this mountain causes the cultivated and wooded parts of the island to be seen in very close proximity. From the summit of this solitary region the eye hovers over an inhabited world; the steep and naked declivities above are contrasted by the smiling aspect of the country beneath; the transparent atmosphere enables the spectator to distinguish even the houses, the sails of the vessels, and the trunks of the trees. Beyond all lies a vast extent of ocean studded with the whole archipelago of the Fortunate islands. Prince Adalbert of Prussia thus describes the scene from a point a little way up the black cone. "Looking down on the long violet-coloured ridge of the Cumbra, which with its indentations formed bays and promontories, right and left, in the white sea of fleecy clouds, whilst it descended gently towards the Llano-de-las-Retamas. On the right, a long wall of rock joined these acclivities of the Cumbra,—a continuation of the margin of the pumice-stone plain. Ascending between the obsidians of the cone, I saw at every step this colossal wall, nearly 2,000 ft. high, [?] curving more and more round the base of the peak, whilst its upper contour, quitting the straight line, began to form single cupolas. I recognized the Circus—the mighty, semicircular portion of the elevation-crater, which alone had braved all the later eruptions of the volcano—the Circus, with its horizontal strata, which at times resemble narrow terraces, or rather small steps, whilst high up along its wall is seen the narrow, white, trachytic band, the silver stripe of Angostura. I looked down: my eye glided quickly along the steep declivity on which I was standing, over the huge, black mass of pointed glassy blocks of obsidian scattered deep below in wild confusion, upon two rounded cupolas, which rose close to the base of the peak from the subjacent plain of pumice-stone, passing from yellow into red, and contrasting strongly with the edge of the black obsidian field. This lower surface, the bottom of the elevation-crater, from which the Teyde arose, is covered with black masses and red blocks of lava; on every side rise up ridges and cones, between which wind single streams of lava. Below on the right, close under the highest precipice of the Circus, stand the dark-red waves of a mighty stream, arrested and stiffened in its course, the force of whose current is still perceptible. Above this fearful scene of volcanic action, this gigantic wall, which once rose out of the depths of ocean—high above all these scenes and objects, the giant peak, like a colossal pyramid purpled by the setting sun, flung his shadow over the mists which covered the ocean." The question has been agitated, but seems yet undecided, whether it is possible to perceive from this point Cape Bojador, on the coast of Africa, 90 m. to the NW. The summit of the peak may be considered as a *solfatara* or extinguished volcano, having remained tranquil during many ages, and presenting no symptom threatening a new eruption. From its flanks, however, several violent eruptions have taken place in the course of the present century. In 1704 there occurred one in the district of Guimar, which buried several valleys, and approached within a short distance of the port of Orotava on the NW coast. Two years afterwards the lava, issuing forth in a different quarter, in the space of two hours buried the port of Garachico, then the finest and most frequented harbour in the island. The volcanic power remained dormant for nearly a century, till 1796, when the mountain of Chahorra, to the W of Teyde, which had always been considered as an extinguished volcano, began pouring out vast torrents of lava. It continued in action for three months and six days, but being fortunately in an uncultivated part of the island, no serious injury was done.—The commercial importance of T. depends chiefly on its wine, which, though of an inferior quality to that of Madeira, yet being afforded at a cheaper rate, is in considerable demand. From 20,000 to 30,000 pipes are annually exported. The price on the island runs from £12 to £20 the pipe of 100 gallons. The consumption, we believe, has considerably increased. About 650,000 lbs. of cochineal were exported from the island in 1853. T. exports also orchilla-weed, rose-wood, and a few other trifling articles. The island derives great advantage, in consequence of its capital and port of Santa-Cruz, on the NE coast, in N lat. 28° 27′ 54″, W long. 16° 15′ 18″, forming a great place of refreshment, or, as Humboldt terms it, a grand caravansery, between Spain and the Indies. For this purpose it affords beef and fish in plenty, and excellent water. A considerable trade is also carried on between this island and the Spanish West Indies.—The pop. is about 85,000.—The climate of Teneriffe is dryer than that of Madeira, though in other respects very similar. In certain pulmonary affections it will, therefore, be found preferable, so far as atmosphere is concerned; but there are not the accommodations for invalids which Madeira possesses, and a very limited English society. Prices, however, are proportionately low.

TENERIFFE, a town of New Granada, in the prov. and 96 m. SSW of Santa-Marta, founded on the shore of the Magdalena, in 1536. It was formerly a large and commercial town; but is now reduced to a miserable village.

TENES, TENEZ, or TENNIS, a headland of Algeria, in the prov. of Mascara, on the Mediterranean, NNE of a town of the same name, in N lat. 36° 34′. The town is 105 m. WSW of Algiers, in the prov. of Mascara, on the Mediterranean. It forms the port of the newly founded French town of Orleansville.

TENGALLE. See TANGALLA.

TENGGAR, a mountainous district of Java, to the SE of the prov. of Surabaya, and forming the finest and most picturesque portion of the island. The summits and slopes of the mountains are covered with pines, firs, and many plants common to Southern Europe. It contains 40 villages, with about 1,200 inhabitants, who in manners and customs differ remarkably from the Javanese of other portions of the island. They have preserved their independence and maintain strict separation from other tribes. They profess Hinduism. Each village is under the government of a petinghi or chief elected by the inhabitants.

TENGHI, a village of Afghanistan, on the l. bank of the Lundye, 30 m. N of Peshawur.

TENGHISTUN, a town of Persia, in the prov. of Farsistan, 2 m. from the shore of the Persian gulf, on the road from Abu-shehr to Firuz-abad. It is enclosed by walls flanked with towers, and by a deep ditch, and is defended at the entrance by two small cannons. It contains about 500 families, and has a considerable trade in grain.

TENGHIZ, or TENIS, a lake of Russia in Asia, in the prov. and 105 m. S of Omsk, about 90 m. from the l. bank of the Irtish.

TENJO, a town of New Granada, in the dep. of Cundinamarca, and prov. of Bogota. Pop. 300.

TENKIRI-NOR, a lake of Tibet, lying under the parallel of 31° 15′ N, about 9 days NW of Lhassa. It is a large sheet of water, nearly 60 m. in length, and is bordered on the N by snowy mountains. It receives from the SE the Tarkit-Tsangbo river.

TENNEBERG, a bail. of the duchy of Saxe-Coburg-Gotha, and principality of Gotha. Pop. 6,300. It has a castle of the same name, 9 m. WSW of Gotha.

TENNESSEE, one of the United States of North America, in the basin of the Mississippi, between the parallels of 35° and 36° 40′ N; bounded on the N by Kentucky; on the E by N. Carolina and Virginia; on the S by Georgia, Alabama, and Mississippi; and on the W by the Mississippi river, dividing it from the states of Arkansas and Missouri. It has an area of 45,600 sq. m. This state is divided by that section of the Alleghany chain called the Cumberland mountains, into divisions, East Tennessee and West Tennessee; but a third administrative division is known as Central T. It is washed for 160 m. by the great river Mississippi on the W; and the fine rivers Tennessee and Cumberland, both tributaries of the Ohio, pass through it in serpentine courses. The western part is undulating, some of it level; in the middle it is hilly and rolling. East Tennessee abounds in lofty mountains, presenting scenery peculiarly grand and picturesque. Of these the Cumberland or great Laurel ridge is the most remarkable. Stone, Yellow, Iron, Bald, Smoky, and Unika mountains, join each other, and form, in a direction nearly NE and SW, and under the name of the Kittatiny ridge, the eastern boundary of the state. To the NW of these, and separated from each other by valleys of from 5 to 15 m. wide, are Bay's mountain, Copper-ridge, Clinch-mountain, Powell's-mountain, and Welling's-ridge. The last four terminate N of Tennessee river. None of the summits of these mountains exceed 2,000 ft. in alt. above sea-level.—The geological formation of this state is wholly secondary, except a small portion of the eastern part, which is transition, and numerous spots on the banks of rivers, which are alluvial. A considerable portion of the state lies on limestone. The produce of iron in 1852 was 44,500 tons. Copperas, alum, nitre, and lead, are among the minerals which chiefiy occur in the eastern and middle divisions. Copper and silver have been found; and coal is supposed to be plentiful. Saltpetre is so abundant as to form a great article of commerce. There are several mineral springs, and many valuable salt springs.—The principal rivers are the Tennessee, Cumberland, Holston, Clinch, French-Broad, Nolichucky, Hiwassee, Tellico, Duck, Reelfoot, Obian, Forked-Deer, and Wolf. The Cumberland has about 250 m. of its course within T.; the Tennessee, about 200 m.

Climate.] The climate is generally mild and genial. In East T. the air is so tempered by the mountain-air on one side, and by refreshing breezes from the gulf of Mexico on the other, that this part of the state has one of the most desirable climates in North America. The middle part resembles Kentucky in climate. The winter in T. resembles the spring in New England. Snow seldom falls to a greater depth than 10 inches, or lies longer than 10 days. Cumberland river has seldom been frozen over since the country was settled. Cattle are rarely sheltered in winter. In the western parts there are some low bottoms on which the inhabitants are subject to bilious fevers, and to fever and ague in the autumn.

Soil and productions.] The soil is very various in quality. The western part of the state has a rich black soil; in the middle is a considerable extent of excellent land; in the eastern part of the mountains are many fertile valleys lying on a calcareous soil. Oak of different species, black and white walnut, beech, red cedar, black and honey locust, ash, elm, mulberry, dogwood, sassafras, maple, sugar-tree, papaw, cherry, hornbeam, juniper, and cucumber-tree, are among the forest-trees. In the eastern district there is a species of pitch pine, useful for boards, timber, and tar. Cane, on the low lands, grows to the height of 20 ft. The wild strawberry is of a delicious flavour; and the wild grape vine yields tolerable grapes. Of plants, the following are indigenous in this state: wild hop, ginseng, Virginia and the Seneca snake-root, angelica, ginger, and sweet anise, wild rye, wild oats, clover, and buffalo grass.—The improved farm lands in 1850 were returned at 5,175,172 acres; the unimproved at 13,809,849 acres. The agricultural productions are the same as those of Kentucky, with the exception of cotton, which, in the western parts of T., forms a staple commodity. Wheat, barley, oats, rye, buckwheat, Indian corn, flax, hemp, tobacco, indigo, rice, and cotton, thrive here luxuriantly. The limestone lands, which are well adapted to the culture of cotton, are in many parts deficient in water, which escapes through fissures in the beds of the streams. Lands of the first and second quality produce Indian corn and hemp, but for wheat the soil is too rich, unless reduced by two or three crops of maize, hemp, tobacco, or cotton. The third quality bears every kind of grain which is cultivated on the dry grounds of the Atlantic states. The produce of wheat in 1840 was 4,569,692 bushels; in 1850, 1,619,386 b.; of Indian corn in 1840, 44,986,188 b.; in 1850, 52,279,223 b. On Cumberland river, the common produce of Indian corn is from 60 to 70 bushels; that of cotton is usually 800 lbs. to the acre. Fruit trees succeed well. The farmers in Upper T. grow little artificial grass, but they have potatoes, carrots, and turnips. They have generally each a herd of pigs, which roves through the woods with the cows; and the latter have a bell strapped round their necks, as a means of finding them. The live stock in 1850 amounted to 270,636 horses, 75,303 asses and mules, 250,546 milk cows, 86,255 working oxen, 414,051 other cattle, 811,591 sheep, 3,114,111 swine. The wool raised in 1840 amounted to 1,060,832 lbs.; in 1850, to 1,864,378 lbs. The animals are such as are generally found in other parts of the United States. The large herds of bisons often seen after the first white settlements in T. were formed have now disappeared. The elk and moose still inhabit some of the mountainous parts, but are not numerous; the deer also, constantly pursued by the hunter, have become scarce, except on the mountains. Bears, panthers, wild cats, and wolves, are yet seen in the forests, but seldom visit cultivated places; the beaver, otter, and musk-rat, are still occasionally seen on the upper branches of the Cumberland.

Exports.] The exports of this state consist of cotton, tobacco, hemp, horses, live cattle, Indian corn, pork, fowls, potatoes, flour, saltpetre, flax, deer skins, ginseng, lumber, and iron. The great staple productions are saltpetre, tobacco, cotton, hogs, and cattle. The imports chiefly consist of dry goods and groceries imported in waggons to East Tennessee from Philadelphia and Baltimore, and to West Tennessee by land to Pittsburg, and thence down the Ohio and up the Cumberland river. Orleans sugar, and some articles of groceries, are imported thence by the Mississippi. This state also supplies Kentucky, Ohio, &c. with cotton for inland manufactures; and from East T. considerable numbers of cattle are sent to the seaports on the Atlantic. In 1850, the total number of manufacturing establishments was 2,789; the value of the cotton goods produced, 510,644 dollars. New avenues to commerce have been opened by means of roads and railroads. The central points of the railway system in this state are Knoxville and Nashville. T. has no canals; but much of the produce seeks a northern outlet by way of Pittsburg and the lakes.

Population.] The pop. and decennial increase of this state have been as follows in the periods undernoted:—

	White.	Coloured.	Total.	Increase per cent.
1790,	32,013	3,778	35,791	
1800,	91,709	13,893	105,602	195·0
1810,	215,875	45,852	261,727	147·8
1820,	339,927	82,886	422,813	61·5
1830,	535,646	146,258	681,904	61·2
1840,	640,627	188,583	829,210	21·6
1850,	756,893	245,732	1,002,725	20·9

Of the coloured pop. in 1850, 239,461 were returned as slaves. The same census returned 5,740 of the inhabitants as of foreign birth. The population, originally consisting chiefly of emigrants from the Carolinas, Virginia, and Georgia, from the New England states, and Europe, has scarcely any uniform character. They are said to be somewhat rough in their manners, but high spirited and hospitable. They are all good horsemen, and expert at the rifle. Their stockings, clothes, and bedding, and even their candles and shoes, are generally of domestic manufacture. The number of paupers in 1850 was 1,005. The number of children attending school in 1848 was 272,000. The number of churches in 1850 was 1,939, of which 831 were Methodist, 611 Baptist, and 357 Presbyterian.—Nashville is the present seat of government. The other most considerable towns in West Tennessee are Franklin, Fayetteville, Shelbyville, Columbia, Clarksville, Carthage, and Gallatin; in East Tennessee, Knoxville, Jonesborough, Greenville, and Rogersville.

The legislature is composed of a senate of 25 members and a house-of-representatives consisting of 75 members. The members of each, together with the governor, are chosen biennally by every free white male citizen 21 years of age, and resident in the co. in which he votes for 6 months. The judiciary is vested in a supreme court, a court of chancery, and circuit courts. The state returns 10 representatives to congress.—The biennial revenue of the state, in 1850 and 1851, was 1,004,004 dollars. The state debt in 1851 was 3,651,856 d.

History.] This country, which formed a part of Carolina, according to the second charter of Charles II. was inhabited by the Cherokee Indians, by whom the first colonists, consisting of above 60 families, in the year 1754, were nearly destroyed. Their settlements were not renewed till 1774, when the Indians, refusing to join the British standard, were attacked and driven towards the Kenhawa. The country then belonged to N. Carolina; and delegates, in 1776, were sent from this district to the convention held for the purpose of forming a state constitution. In 1789, it was ceded by Carolina to the United States, and in 1796, was received into the federal union. The military force consists of a militia, which amounted, in the winter of 1812 to 20,193; in 1840 to 71,252. The inhabitants of this state, active inured to the chace, familiar with the rifle, and proud of their rights, form a militia which no regular army could long despise.

TENNESSEE, a large and navigable river of the United States, which rises in the mountains of Virginia, in the Holston and Clinch rivers, which unite at Kingston in Tennessee; traverses the western parts of Tennessee; crosses the NE corner of Alabama; bends NW; again enters Tennessee, flowing N; and after flowing 60 m. through Kentucky, joins the Ohio at Paducah, 57 m. from the Mississippi, by an outlet 600 yds. wide. It has a total course of 1,100 m., and drains a basin of 41,600 sq. m. It is navigable for the largest bow-boats as far as the Mussel-shoals, 250 m. from its mouth; and thence to its passage through the Cumberland mountains, about an equal distance, there is depth of water sufficient for boats of 40 or 50 tons. Of the two head branches of this river, the one known by the name of the Clinch or Pelison river, is navigable for boats 200 m. from its outlet, which is 150 yds. in width. The other, called the Holston, runs a course of 200 m., and is navigable for boats of 25 tons upwards of 100 m. It has several branches, the most considerable of which are Watauga and French-Broad river. The Hiwassee, Chickamago, and other streams, run into the T. from the N parts of Georgia. The Elk and other streams flow from the S parts of Tennessee through Alabama, into the Tennessee at the Mussel-shoals. The whole descent of the river is 1,700 ft.

TENNIS, a ruined city of Lower Egypt, situated on an island in Lake Menzaleh, 28 m. SE of Damietta.

TENNIS. See TENIS.

TENNSTADT, a town of Prussia, in the prov. of Saxony, regency and 16 m. NNW of Erfurt, and circle of Langensalza, on the Schambach, in a pleasant locality. Pop. in 1843, 2,984. It has three churches, sulphureous springs and baths, and an hospital; and possesses manufactories of linen and cloth, and a considerable trade in yarn.

TENQUIN (GROS), a commune of France, in the dep. of the Moselle and arrond. of Sarreguemines. The cant. comprises 25 com. Pop. in 1831, 16,863; in 1846, 17,578. The village is 20 m. SW of Sarreguemines. Pop. 1,335.

TENRIU, a river of Japan, in the island of Ni-fou, which has its source in the prov. of Sinani; traverses that of Tutomi, from N to S; and after a course of about 120 m., throws itself by several branches into the Pacific, a little to the W of Totomina bay.

TENSA, a town of New Grenada, in the dep. of Boyaca, prov. and 27 m. SE of Tunja, in the midst of the Andes. Pop. 500.

TENSAS, a parish of the state of Louisiana, U. S., in the E district, and comprising an area of 619 sq. m., bounded on the E by the Mississippi, and drained by branches of that river, and by a river of the same name. It has a level surface containing several small lakes, and is generally fertile. Pop. in 1850, 9,040.—Also a river of the same state, which has its source in Chicot co., in the state of Arkansas; flows S by W nearly parallel to the Mississippi, and falls into the Washita river.

TENSAW, a river of the state of Alabama, U. S., forming the E outlet of Mobile river, and flowing S, enters Mobile river by several mouths.

TENSIFT, a river of Marocco, which descends from the NW side of the Great Atlas, from a point about 40 m. E of the city of Marocco; flows at their base, and about 4 m. to the N of the capital; and after a course of about 240 m., in a generally WNW direction, throws itself into the Atlantic, 15 m. SSW of Safi. It is a rapid stream, but fordable in almost all places, except in spring.

TENTER. See Tendra.

TENTERDEN, a borough, parish, and market-town, in the cinque-port liberty of Hastings, Kent, 18 m. SE of Maidstone, and 8 m. S of the line of the South-Eastern railway. Area of p., 8,620 acres. Pop. in 1801, 2,370; in 1831, 3,177; in 1851, 3,901. The ancient limits of the borough are co-extensive with those of the hundred of Tenterden, which comprises the whole of the p. of that name, and part of the p. of Ebony. The town, which occupies a gentle eminence in the midst of a rich agricultural country, consists chiefly of one street about a mile in length. It contains some good houses; but is not remarkable either for the regularity or elegance of its buildings. The inhabitants are principally engaged in agricultural pursuits.

TENT-HILL, a summit of New South Wales, in Laidley Plains, in the co. of Stanley.

TENTOLI, a town of the island of Celebes, on the NW coast, in N lat. 1° 20′, at the head of a roadstead, to which it gives its name.

TENTUGAL, a town of Portugal, in the prov. of Beira, comarca and 12 m. WNW of Coimbra, in a fertile plain, near the r. bank of the Mondego. Pop. 1,788. It has a convent and an alms-house.

TENTYRA. See Denderah.

TENU. See Achenau.

TENYAH. See Faleme.

TEOCOCUILCO, a town of Mexico, in the state and 87 m. NE of Oaxaca, in the midst of the Andes, in N lat. 17° 24′, and W long. 100° 33′.

TEODO, a small district in Dalmatia, in the circle of Cattaro, extending between the Pettane and Saline, and remarkable for its fertility. Pop. 800. It contains numerous villas, belonging to the inhabitants of Persagno and Cattaro. Its wines are said to rival those of Spain.

TEOFIPOL, a town of Russia in Europe, in the gov. of Volhynia, district and 27 m. SSW of Zaslav, and 45 m. NE of Tarnopol.

TEOGE. See Tioughe.

TEOLO, a town of Austrian Lombardy, in the prov. and 11 m. SW of Padua. Pop. 1,630.

TEOMABAL, a small island of the Sulu archipelago, in N lat. 6° 15′, and E long. 120° 55′.

TEOPIXCA, a town of Mexico, in the prov. of Chiapa, 21 m. SE of Ciudad-Real. It is inhabited by Indians, who are noted for their horsemanship.

TEORA, a town of Naples, in the prov. of the Principato-Ultra, district and 7 m. SSE of San Angelo-de-Lombardi, on a hill, between the Ofanto and Sele. Pop. 3,300. It has a palace remarkable for an ancient round tower.

TEOTITMACON, a village of Mexico, in the state and 27 m. NE of Mexico, near a small river of the same name, which throws itself into lake Tezcuco. Pop. 1,000. It has a Franciscan convent. Between this place and San-Bartolomeo are the remains of two ancient pyramids, erected by the Aztecs, and dedicated to the sun and moon.

TEPAN-GUATIMALA, a town of Guatimala, in the state of that name, dep. of Sacatepec and Chimaltenango, and district of Xilotepeque. Pop. 4,958. It was formerly, as its name denotes, the residence of the kings of Guatimala.

TEPEACA, a town of Mexico, in the dep. and 15 m. SE of La Puebla, on a river which winds around it. Pop. 3,000. It is regularly built, and adorned with gardens and terraces. Its public square is spacious, and in its centre is a fort, into which it is said Cortez retired after his retreat from Mexico. In the outskirts is a Franciscan convent, and on the opposite side a fine church. It has manufactories of woollen fabrics, which, with corn and flour, form its chief articles of trade. This town formerly bore the name of Segura-de-la Frontera.

TEPELEN, or Tepelini, a town of Turkey in Europe, in Albania, in the sanj. and 33 m. ESE of Valona, in a valley enclosed by sterile mountains, on the l. bank of the Voïoussa. Pop. 2,000. It has a fortress.

TEPEXE, a town of Mexico, in the dep. and 45 m. S of La Puebla. Pop. 2,000.

TEPIC, a town of Mexico, in the state of Xalisco, 134 m. WNW of Guadalajara, and near the l. bank of the Rio Grande or Tololotlan, in N lat. 21° 36′, and W long. 104° 45′. Pop. 10,000. Next to Guadalajara, this town is the largest in the state. It is built with uniformity, and has terraced gardens, fine promenades, a Franciscan convent, and a small theatre. It is the residence during the rainy season of the merchants connected with the unhealthy port of San Blas.

TEPININSKAIA, a town of Russia in Europe, in the gov. of the Don Cossacks, 78 m. E of Pavlovsk, on the Khoper.

TEPITITAN, a town of the Mexico state of Tabasco, situated on the l. bank of a river of the same name, about 40 leagues above Frontera. Pop. 1,000. Rum is largely distilled here.—The T. river is the upper part of the Chilapa, and is nearly 100 fath. wide opposite the town.

TEPL, Tepla, Tepel, or Topel, a town of Bohemia, in the circle and 31 m. NW of Pilsen, on a small river of the same name, which after a course in a generally E and N direction of about 36 m. joins the Eger, on the r. bank, 5 m. NE of Elnbogen. Pop. 1,734. It is noted for its abbey; and has manufactories of linen and woollen fabrics, and a considerable trade in corn. In the vicinity are several forges, quarries of millstone and serpentine, and mineral springs, of which the most celebrated are those of Marienbad.

TEPLICA (Kunowa). See Teplitz.

TEPLICE. See Toplitz.

TEPLICKA. See Teplitscha.

TEPLICZ, or Teplice, a village of Hungary, in the comitat and 6 m. ENE of Trentsin, in a valley, on the Tepliczka, a small affluent of the Vag.

TEPLIK, a town of Russia in Europe, in the gov. of Podolia, district and 26 m. SE of Gaisin.

TEPLITSCHA, Teplicz (Kis), or Teplicka, a

village of Hungary, in the comitat of Lyptau, 20 m. SW of Kaysmark, at the foot of Mount Königsberg.

TEPLITZ. See TOPLITZ.

TEPLITZ, KUN-TAPOLCZA, or KUNOWA-TEPLICA, a village of Hungary, in the comitat and 12 m. NNE of Gomor, on the r. bank of the Csetnek, a small affluent of the Sajo. It has warm baths.

TEPLOVKA, a town of Russia in Europe, in the gov. and 129 m. NW of Poltova, and district of Piriatin.

TEPOSCOLULA, a town of Mexico, in the state and 60 m. NW of Oaxaca. Pop. 4,405. It has several parish churches, and a fine convent. It possesses manufactories of linen. Cochineal is found in large quantities in the environs.

TEPTIARIS, a people of Russia in Europe, in the gov of Orenburg, who inhabit the territory of the Bashkirs, at the E base of the Ural mountains. They number upwards of 35,000, who pay in common with the Russian peasantry a capitation tax, in addition to delivering 800.000 pounds of salt, which they are required to convey from the Ilek to the banks of the Belaia, whence it is embarked, and to an annual contingent of 400 pioneers for the maintenance of the Orenburg line of fortifications. They are courageous hunters, good labourers, and have adopted from the Bashkirs their excellent method of raising bees. In manners and customs they are similar to that people, but more deficient in cleanliness. They are chiefly Mahommedans, but have introduced into their worship some pagan usages adopted from the Finns. The name of this people signifies, in the Tartar language, a man who is not in a condition to pay the capitation. At the period of the destruction of the kingdom of Kazan, by the Czar Ivan Vasilivitch, a great number of Tartars, Tsheremisses, and Tshuvashes, fearing to be forced to adopt the religion of their conquerors, found refuge among the Bashkirs, and thus formed a race remarkable for the admixture of idiom religion and usages still preserved amongst them.

TEQUAMENON, a river of Michigan, U. S., which flows in an E by N course, into a bay of the same name, near the outlet of Lake Superior.

TEQUENDAMA (SAUT DE), a cataract of New Grenada, in the prov. of Bogota, a little to the W of Santa-Fe-de-Bogota. It is formed by the Bogota, which after collecting its waters in the valley of the same name, precipitates itself through a rocky chasm, 650 Parisian ft. in height.

TEQUIA, a town of New Grenada, in the dep. of Boyaca, 120 m. NNE of Tunja.

TEQUILA, a town of Mexico, in the state of Xalisco, 18 m. W of Guadalajara, near the l. bank of the Rio Grande.

TER, a river of Spain, which descends from the S side of the Pyrenees, on the confines of the French dep. of the Pyrenees Orientales, and at the NW extremity of the prov. of Gerona. It flows first S, past Camprodon; enters the prov. of Barcelona; bends abruptly ENE, returns into that of Gerona; passes the town of that name, and a little below Torroela-de-Mongri, and 30 m. E of Gerona, throws itself into the Mediterranean, opposite the Medas islands, and after a course of about 102 m. Except when flooded, it is fordable throughout almost its entire course. A large portion of its waters is employed in irrigation. This river is the *Alba* of the Romans.

TERA, a river of Spain, in the prov. of Zamora. It issues from the E extremity of Lake San-Martin-de-Castaneda; flows ESE, passing near Sanabria; and after a course of 60 m., joins the Esla, on the r. bank, 6 m. SSW of Benavento. In summer it is fordable throughout its entire course.—The lake S. Martin-de-Castaneda, formed amid the Sierra Segundera by several mountain streams, has a circumference of 6,900 yds. Its waters are clear, and abound with excellent trout.

TERABINS, a tribe of Bedouin Arabs, in Egypt. They are governed by a sheik, and number about 150 foot-soldiers.

TERAMO, a district, canton, and town of Naples, in the prov. of Abruzzo-Ultra. The district comprises 10 cants. The town is 15 m. SSE of Ascoli, on a plateau, at an alt. of 900 ft. above sea-level, and near the confluence of the Tordino and Vezzola. Pop. 9,500. It is about 4 m. in circumf., and has several long, spacious, and well-paved streets, and some fine edifices of medieval architecture. The cathedral is a handsome building. It has numerous convents, an orphan's and two other asylums, a seminary, and a provincial college. With the exception of the manufacture of hats, and of cream of tartar, this town possesses no branches of industrial employment. It was the *Interamnia* or *Interrana-Prœcutia* of the Romans, and contains the ruins of several ancient edifices.

TERAPEA. See TARAPIA.

TERBORGH, a village of Holland, in the prov. of Guelderland, 18 m. SE of Zutphen. Pop. 600.

TERCEIRA, an island near the coast of Africa, forming the central one of the group of the Azores. In consequence of its central situation, and of the safety of the roadstead at Angra, it has been made the seat of government. Its E extremity is in N lat. 38° 46'. It is about 54 m. in circumf., and is surrounded with steep cliffs, which render it inaccessible, except at a few points which are strongly fortified. The interior, which rises by gentle slopes towards a mountainous centre, is agreeable: the summits of the mountains consisting, for the most part, of beautiful and fertile plains, and being destitute of those craters and conical points which, in St. Michael's and the other islands of the group, visibly mark the operation of volcanic agency. Though abundantly supplied with fine water, it contains no boiling or mineral springs. Vines are cultivated, but the wine—of which about 4,000 pipes are annually produced—is by no means excellent. Oranges and lemons are largely exported; and almost every European and tropical plant may be grown. Agriculture and pasturage are the chief employments, and are carried on with great success; wheat, Indian corn, yams, pulse, and French beans, are raised to such an extent as to afford a surplus for the market of Lisbon. The yearly produce of Indian corn, wheat, and barley, is about 720,000 bushels. The pop. is about 40,000, and is reckoned superior in civilization to those of the other islands; the police is also better enforced, and the manners of Lisbon generally prevail.—The only places of consequence in the island are Angra, Praya, San-Salastio, and New Town. Fine roads lead to every part of the island.

TERCEIRAS. See AZORES.

TERCERO, a river of the La Plata confederation, which rises in the mountains to the SW of the city of Cordova; runs E, collecting the waters of a smaller river called the Saladillo, and enters the river Parana just above the town of Rosario, and 73 m. S of Santa-Fe, after a course of 300 m. It is navigable for barges as high as the pass of Fereira, about 80 m. below Cordova.

TERCHIZ, a town of Persia, in Khorasan, 90 m. SW of Medshid.

TERDOPPIO, a river of Italy, which rises near the S extremity of Lago-Maggiore, and falls into the Po, 12 m. ESE of Lumello, after a course of 50 m.

TEREBES, or TREBISSOW, a town of Hungary, in the com. of Zemplin, 22 m. ESE of Kaschau, on the river Bodrog. Pop. 2,900.—Also a village in the com. of Ugocz, on the river Thur.

TEREBOLI. See TRIPOLI.

TEREK, a river of Asiatic Russia, in the gov. of Caucasus, which rises amid the snow-covered heights of the Kasbek, in the loftiest part of the Caucasian chain, and descending in a rapid course and northerly direction, to the plain, turns eastward, passes Mosdok, flows along the foot of the mountains in a NE direction, and after passing Kizliar, divides into three branches, and falls into the Caspian. Its banks in the upper part of its course are adorned with fine forests of oak, while the lower produce vines, mulberries, and other fruit trees. It has a total course of 300 m. Its principal affluents are the Suja, Aksai, Baksu, and Malka.

TERENA, a town of Portugal, in the prov. of Alentejo, 26 m. SW of Elvas.

TERERY, a river of Brazil, in the prov. of Mato-Grosso, which descends from the Sierra Amambay, and runs W to the Paraguay, which it joins on the l. bank, after a course of 90 m.

TERESA, a town of Spain, in the prov. of San-Felipe, 10 m. W by N of Segorbe. Pop. 1,800.

TERESA (SANTA), a town of New Granada, at the conflux of the Meta and Orinoco, 300 m. E of Santa-Fe-de-Bogota.—Also a town of New Mexico, in the prov. of Cinaloa, 95 m. NE of Cinaloa.—It is also the name of several other inconsiderable settlements in South America.

TERESHKA, a river of Russia, which rises 20 m. SE of Kusnetz, in the gov. of Saratof; flows ESE, and then turns SW, and joins the Volga, 27 m. NE of Saratof.

TERESPOL, a town of Poland, 3 m. SSW of Brzesc, and 103 m. E by S of Warsaw, on the l. bank of the Bug. Pop. 1,000, chiefly the descendants of German settlers.

TERETTE, a town of Naples, in the prov. of Terra-de-Lavoro, 15 m. SE of Sora, on a lofty mountain. Pop. 1,800.

TERFUA, a village of Tunis, 120 m. S of Tunis.

TERFUI, a watering place in the desert of Nubia, 150 m. S of Syene.

TERGIL, a village of Diyarbekir, in Asiatic Turkey, 70 m. NNE of Diyarbekir.

TERGLU, one of the highest mountains of the Julian Alps, in the NW of Carniola. It has an alt. of 10,855 ft. above sea-level, and is covered with continual snow on its N side.

TERGOVISTE, or TERGOWITZ, a town of Wallachia, in the district of Zara-de-Suss, on the river Jalomnitza, 64 m. SE of Hermanstadt. It has several churches, two convents, a number of large but ruinous houses, and a pop. of 5,000.

TERHEYDEN, a town of Holland, in the prov. of N. Brabant, 3 m. N of Breda.

TERI, a town of Beluchistan, in N lat. 29° 51', 70 m. N of Shawl.

TERIM, a village of Arabia, in the prov. of Hadramaut, 230 m. E of Sanaa.

TERKAT, a village of Asiatic Turkey, in the pash. of Sivas, 20 m. NW of Tokat.

TERKI, a fortress of Russia, at the mouth of the Tumenka, which forms one of the branches by which the Staroi-Terek falls into the Caspian. The place bearing this name was formerly of importance; but after the Persian conquests, it ceased to be a frontier-town, and was razed at the peace of 1736. The present fort was afterwards built on a different spot.

TERKUL, a river of Asiatic Russia, which falls into the Ural at Uralsk.

TERLING, a parish of Essex, 3½ m. W of Witham. Area 3,205 acres. Pop. in 1851, 900.

TERLIZZI, a town of Naples, in the prov. of Bari, on the high road leading from Bari on the coast, across the Ofanto, 18 m. W of Trani. Pop. 9,000.

TERMIGNON, or THERMIGNON, a town of Savoy, in the district of Maurienne, near the river Arc, 12 m. ENE of St. André. Pop. 1,100. It has iron forges, and a manufacture of anchors.

TERMINATION ISLAND, an island in the S. Pacific, so named by Vancouver, as being the termination of his researches on the SW coast of New Holland, near which it lies, in S lat. 34° 32'.

TERMINI, a considerable port on the N coast of Sicily, in the Val-di-Mazzara, situated at the mouth of the river Termini, on a rocky eminence, 23 m. ESE of Palermo, with a well cultivated and productive country in its vicinity. It is surrounded by a wall, and has, on the side of the sea, an old but strong castle. The inhabitants, amounting to about 8,500, are employed partly in trade, partly in an active fishery which supplies the market of Palermo. The port is good, and corn, olive oil, and other products are pretty largely exported from it. It has long been celebrated for its hot saline springs, the *Thermæ Himerenses* of antiquity, which are still much frequented by visitors from Palermo.—The river rises 6 m. NNE of Bivona; flows NNE through the Val-di-Mazzara; and enters the sea a little to the NW of the above port, after a course of 50 m.

TERMINIER, a village of France, in the dep. of Eure-et-Loir, cant. and 5 m. SSE of Orgeres. Pop. 950.

TERMINILLO (MONTE), a mountain of Naples, in Abruzzo-Ultra 2da, 7 m. NNE of Civita-Ducale, having an alt. of 2,143 metres = 7,030 ft. above sea-level.

TERMINOS (LAGUNA-DE), a large inlet of the gulf of Mexico, on the coast of Tabasco and Yucatan, intersected by the parallel of 18° 30', and the meridian of 93° 30', and separated from the sea chiefly by the Isla-del-Carmen. The Rio-Palisado flows into its SW extremity.

TERMOLI, a town of Naples, in the prov. of Sarmio, on the Adriatic, 40 m. SE of Ortona. Pop. 2,000.

TERMONAMUNGAN, a parish in co. Tyrone, 3¾ m. SW of Castle-Derg. Area 45,399 acres. Pop. in 1831, 7,253; in 1851, 6,890. It forms the extremity of the W wing of the county, and is prevailingly wild, rocky, heathy, sterile, and repulsive; diversified principally by the glen of the river Derg.

TERMONCARRA, a village in the p. of Kilmore-Erris, co. Mayo, 5 furl. N of the marine inlet of Portnafranka. Pop. in 1851, 133.

TERMONEENY, a parish in co. Londonderry, 2 m. SE of Maghera. Area 4,773 acres. Pop. 1,875.

TERMONFECKAN, or TORFECKAN, a parish on the coast of co. Louth, containing the villages of Termonfeckan and Baltray. Area 6,382 acres. Pop. in 1831, 4,784; in 1851, 2,862.—The village stands on the Termonfeckan rivulet, 2 m. SSW of Clogher. Pop. in 1831, 470; in 1851, 329.

TERMONMAGUIRK, a parish in co. Tyrone, 4½ m. W of Pomeroy, containing the villages of Six-mile-cross and Termonrock. Area 41,078 acres. Pop. in 1831, 10,307; in 1851, 10,043. About one-third of the area is arable land; about 2,000 acres are reclaimable upland; all the remainder is either deep bog or sheer mountain, partly pastoral but chiefly waste.

TERMONROCK, a village in the p. of Termonmaguirk, co. Tyrone, 3⅜ m. NE of Six-mile-cross. Pop. in 1841, 113; in 1851, 121.

TERMONDE. See DENDERMONDE.

TERNA, a small town of Austrian Italy, in the delegation of Como, which gives name to a small lake in the neighbourhood, 30 m. NW of Milan.

TERNANT, a village of France, in the dep. of Rhone, cant. and 3 m. NW of Bois-d'Oingt, on the r. bank of the Oyngue. Pop. 600.

TERNATE, the northernmost of a chain of islands off the W coast of Gilolo, in the Indian archipelago, in N lat. 0° 48', E long. 127° 18', and formerly the seat of sovereignty over all the adjacent Molucca islands. On the expulsion of the Portuguese from the Molucca isles by the Dutch, in 1607, the princes of T. dwindled away into a state of political insignificance and dependence; the king of T., however, is still the most powerful of the Molucca princes, possessing the N part of Gilolo, with the Bachian group and part of Papua, whence he receives a tribute of gold, amber, and birds of paradise. T. is of small extent, being only about 10 m. in diam.; but it contains a peak 5,480 ft. in alt., which sometimes discharges flames. At present it produces only a little rice, tobacco, and cotton: though the trade it formerly carried on with New Guinea and the Chinese was very considerable.

TERNAY, a village of France, in the dep. of Isère, cant. and 3 m. SW of Saint-Symphorien-d'Ozon, on the l. bank of the Rhone. Pop. 1,200.

TERNERA, a settlement of New Granada, in the prov. of Carthagena, 6 m. N of Carthagena.

TERNES (Les), a village of France, in the dep. of Cantal, cant. and 5 m. WSW of St. Flour.

TERNEUSE, a town of Holland, in the prov. of Zieland, on the banks of the Hondt or Western Scheldt, 12 m. SE of Flushing.

TERNI, a town of the States of the Church, situated in a valley, between two branches of the river Nera, in the deleg. and 12 m. SSW of Spoleto. Its pop. is about 9,000; and it has a cathedral and several churches, but no other public building of consequence. It is, however, a place of considerable trade, and has large woollen and cotton factories. The cascades of the Velino, called the Caduta-delle-Marmora, about 4 m. E of T., are considered among the finest in the world. They consist of several streams, which, after running with great impetuosity, precipitate themselves from the brink of a rock into the Nera. "Imagine," says Shelley, "a river 60 ft. in breadth, with a vast volume of waters, the outlet of a great lake among the higher mountains, falling 300 ft. into a sightless gulf of snow-white vapour, which bursts up for ever and for ever from a circle of black crags, and thence leaping downwards, make five or six other cataracts, each 50 or 100 ft. high, which exhibit, on a smaller scale, and with beautiful and sublime variety, the same appearances. But words (and far less could painting) will not express it. Stand upon the brink of the platform of cliff, which is directly opposite. You see the ever-moving water stream down. It comes in thick and tawny folds, flaking off like solid snow gliding down a mountain. It does not seem hollow within, but without it is unequal, like the folding of linen thrown carelessly down; your eye follows it, and it is lost below; not in the black rocks which gird it around, but in its own foam and spray, in the cloud-like vapours boiling up from below, which is not like rain, nor mist, nor spray, nor foam, but water, in a shape wholly unlike anything I ever saw before. It is as white as snow, but thick and impenetrable to the eye. The very imagination is bewildered in it. A thunder comes up from the abyss wonderful to hear; for, though it ever sounds, it is never the same, but modulated by the changing motion, rises and falls intermittingly. The surrounding scenery is, in its kind, the loveliest and most sublime that can be conceived."

TERNOE, a small island in the Baltic, off the coast of Blekingen, in Sweden, near Carlshafen.

TERNOISE, a river of France, in the dep. of Pas-de-Calais, which joins the Canche at Hesdin, after a course of 22 m.

TERNOVA, or Tirnova, an inland town of European Turkey, in Bulgaria, in the sanj. and 52 m. SSW of Rustchuk, on the Jantra. It was formerly into comparative decay, though its pop. is estimated at 12,000. It lies in a gorge of the Balkan in an extremely picturesque site, the houses rising above one another, in the form of an amphitheatre, from the edge of the river.

TERRA-AUSTRALIS, the ancient name of New Holland, revived by Flinders, who applies it to the strongly fortified, and was the residence of the prince or governor of the country, but is now fallen whole continent, in contradistinction to the names New Holland and New South Wales, which he applies to different divisions of that extensive island.

TERRA-AUSTRALIS-DEL-ESPIRITU SANTO. See Espiritu-Santo.

TERRACINA, a town of the States-of-the-Church, situated at the SE extremity of the Pontine marshes, 25 m. SSW of Frosinone. It stands partly on the declivity of a rugged eminence, partly on low ground adjoining the shore, in a picturesque situation. Pop. 5,000. It was anciently a town of note, but is indebted to Pius VI. for its modern improvements. In the prosecution of his grand object, the draining of the Pontine marshes, that pontiff caused an elegant mansion to be erected at T., where he resided during spring and autumn to urge on his works. The cathedral is a dark and gloomy pile, and most of the buildings have a decayed appearance. The ancient castle of Theodoric is in ruins, but holds a commanding situation on the summit of a lofty rock. Traces are still visible of the ancient port of Antoninus Pius, and ruins of the ancient *Anxur* cover a neighbouring eminence.

TERRA-DEL-FUEGO. See Tierra-del-Fuego.

TERRA-DE-LAVORO. See Lavoro.

TERRA-FIRMA, a name applied by our older navigators to that portion of South America which comprehends the provinces of Darien, Veragua, and Ystmo or Terra-Firma proper.

TERRANEH, or Teraneh, a town of Lower Egypt, in the prov. of Baheireh, on the l. bank of the principal western branch of the Nile, 6 m. WSW of Menouf, and 36 m. NW of Cairo. It is a small, scantily peopled, earth-built town, its importance being chiefly derived from the natron lakes, 40 m. to the W, for the transit of the produce of which caravans arrive here annually at the commencement of winter. In the vicinity are the ruins of the ancient *Terinuthis*.

TERRANOVA, a town of Naples, in the prov. of Basilicata, district and 33 m. ESE of Lagonegro, and cant. of Noja, at the foot of a mountain, near the l. bank of the Sermento. Pop. 1,100. It has a convent.—Also a town of the prov. of Calabria-Citra, district and 14 m. SSE of Castrovillari, near the l. bank of the Crati, in a plain. Pop. 2,180. It has two parish churches, and four convents.—Also a town of Calabria-Ultra, district and 11 m. ESE of Palmi, and cant. of Casalnuovo, in a salubrious plain. Pop. 450. It has two churches. This town, founded in the 9th century, was one of the finest in the prov. previous to the earthquake of 1783, by which a great portion of it was overthrown.—Also a village of Sardinia, in the div. of Capo-Sassari, prov. and district and 25 m. E of Templo, at the head of a gulf of the same name, in an unhealthy plain. Pop. 1,600. It has a fine old cathedral. The port, wanting in depth, and consequently practicable by light vessels only, is sheltered from all winds, and has a considerable trade in corn and cattle. In the vicinity are several salt-works. The adjacent coast is uninhabited, but has some good roadsteads by which an active contraband traffic is carried on. T. was originally a Roman colony named *Olbia*. It became a town of some importance about the time of St. Gregory, and then bore the name of *Pausa-*

nias. After its destruction it retained the appellation of Civita.—Also a town of Sicily, capital of a district, in the prov. and 35 m. SSE of Caltanisetta, at the extremity of a desert shore, and near the entrance of a river of the same name into the Mediterranean. Pop. 9,200. It has a fortress, several fine churches, numerous convents, and an hospital. The port is good, and during summer the bay affords excellent anchorage. The trade consists chiefly in corn, wine, fruit, sulphur, and soda. This town was founded towards the end of the 13th century by Frederick of Aragon. The district comprises 4 cantons.

TERRANOVA-FOSSACECA, a town of Naples, in the prov. of the Principato-Ultra, district and 9 m. NNW of Avellino, and cant. of Altavilla, in a pleasant and fertile locality. Pop. 1,120. It has three churches, and an alms-house.

TERRA-NUOVA, a town of Tuscany, in the prov. and 31 m. SE of Florence, and vicariat of S. Giovanni-in-Valdarno, on the slope of a hill, near the r. bank of the Bregna or Ciaffenna, near its junction with the Arno. Pop. in 1846, 6,472. It is enclosed by walls.

TERRA-PILATE, a plain of Sicily, in the prov. and a little to the E of Caltanisetta. It has an area of 73,000 sq. yds. The soil consists of whitish clay, and is covered with the carbonate of lime, and a variety of other substances. In its centre are several boiling springs and jets-d'eau, the latter of which are attended with a loud report.

TERRAQUE, a commune of Belgium, in the prov. of Hainault, and dep. of Wannebecq. Pop. 136.

TERRASSE (LA), a commune of France, in the dep. of the Isere, and cant. of Le Touvet, 15 m. NE of Grenoble. Pop. 1,287.—Also a hamlet in the dep. of the Haute-Loire, cant. and 5 m. SSE of Puy, and com. of Coubon. In an adjacent mountain are several curious grottoes.

TERRASSON, a canton, commune, and town of France, in the dep. of the Dordogne, and arrond. of Sarlat. The cant. comprises 17 com. Pop. in 1831, 14,607; in 1846, 15,050.—The town is 18 m. NNE of Sarlat, on the l. bank of the Vezere. Pop. in 1846, 3,137. It has several tanneries, tile and brick-kilns, and extensive quarries of freestone, containing carbonated copper. The coal basin of T. extending into the dep. of the Correze, covers an area of 2,355 hect., and yields an annual produce of

TERRAUBE, a commune and town of France, in the dep. of the Aube, cant. and 4 m. SW of Lectoure, on a mountain. Pop. 1,138.

TERRAZOS, a village of Spain, in the prov. and 24 m. NE of Burgos, and partido of Briviesca.

TERRE (SAINTE), a village of France, in the dep. of the Gironde, cant. and 4 m. WSW of Castillon, and near the r. bank of the Dordogne. Pop. 2,040.

TERRE-BONNE, a parish in the S part of the state of Louisiana, comprising an area of 1,850 sq. m., bounded on the S by the gulf of Mexico, and drained by Terre-Bonne, Petite, and Grand-Bayou-Cailloux, and Bayou-de-Large rivers. It has a low and level surface, and is to a great extent unfit for cultivation. Its principal productions are sugar, cotton, rice, and Indian corn. Pop. in 1840, 4,410; in 1850, 7,724. Its capital is Houma.

TERRE-COUPER, a village of St. Joseph co., in the state of Indiana, U. S., 135 m. N by W of Indianapolis, and intersected by the Michigan Southern and Northern Indiana railroad.

TERREGLES, a parish of Kirkcudbrightshire, stretching along the Nith, to the W of Dumfries. Area 8 sq. m. Pop. in 1841, 564; in 1851, 566.

TERRE-HAUTE, a town of Vigo co., in the state of Indiana, U. S., on the l. bank of the Wabash river, 73 m. W of Indianapolis. This is one of the most important places in the state, being situated on the Wabash and Erie canal, and at the point of junction of the Terre-Haute and Richmond, the Atlantic and Mississippi, the Terre-Haute and Alton, and Evansville and Illinois railroads. Pop. in 1840, 2,000; in 1850, 4,900.

TERRE-NAPOLEON, the name given to a tract of coast on the S shore of Australia, comprised between the parallels of 37° 36′ and 35° 40′, and the meridians of 138° 58′ and 140° 10′ E, first explored by the French navigator, Baudin.

TERRENEGRE, a fortress of France, in the dep. of the Charente-Inferieure, cant. and 3 m. WNW of Royan, and com. of St. Palais.

TERRESSA, an island of the Nicobar archipelago, in the gulf of Bengal, to the NW of Camorta, in N lat. 8° 20′, and E long. 93° 36′. It lies in a SE direction, rises to a considerable height, and is covered with wood. It is scantily inhabited.

TERRIA, or TELLIA-GURHY, a village of Hindostan, situated in N lat. 25° 15′, in the pass which divides the provinces of Bengal and Bahar, formed by a range of hills, approaching very near the Ganges.

TERRIBLE (MONT), a chain of mountains in what was once the bishopric of Bale, but is now part of the Swiss canton of Bern, which brings the Jura mountains into connection with the Vosges. It gave name for some years to a dep. of France, consisting of the bishopric of Bale, small portions of Switzerland, and the county of Montbeliard.

TERRICCIOLA, a town of Tuscany, 18 m. SE of Pisa, in the Val-d'Era. Pop. 3,000.

TERRIER-ROUGE (LE), a town of the island of Hayti, 7 m. SW of Fort-Liberte.

TERRINCHES, a village of Spain, in the prov. and 52 m. SE of Ciudad-Real. Pop. 850.

TERRINGTON, a parish in the N. R. of Yorkshire, 8 m. W by S of New Malton. Area 3,630 acres. Pop. 753.

TERRINGTON (ST. CLEMENT'S), a parish of Norfolk, 5 m. W by N of Lynn-Regis. Area 34,236 acres. Pop. in 1831, 1,466; in 1851, 2,250.

TERRINGTON (ST. JOHN'S), a parish in the above county, adjoining T. St. Clements, and included in its area. Pop. in 1851, 784.

TERRIORE, a town and fortress of Hindostan, in the Carnatic, in N lat. 11° 12′.

TERROIG, or ATERICK, a lake on the mutual border of cos. Galway and Clare, among the Slieve-Baughta mountains, 6 m. N of Scariff. It has a surface-elevation above sea-level of 450 ft., measures about 1¼ m. in length, and sends off its superfluent waters in a head-stream of the Scariff river.

TERROWEH, a town and fortress of Hindostan, in the prov. of Allahabad, district of Bundelcund, 60 m. WNW of Allahabad.

TERRYGLASS, a parish in co. Tipperary, 4½ m NW of Borris-o'-kane. Area 9,761 acres. Pop. in 1831, 2,573; in 1851, 1,986.

TERSA, a river of Russia, which rises 18 m. ESE of Balachev, in the gov. of Saratov, which joins the Medvieditza, after a winding but prevailingly SE course of 130 m.

TERSATZ, a town and fort of Bosnia, 30 m. W of Novi, on the r. bank of the Korana.

TERSCHELLING. See SCHELLING.

TERSDORF, a village of the archd. of Austria, 10 m. NNE of Neustadt, on the l. bank of the Triesting.

TERSTENIK, a town of Servia, in the sanj. and 19 m. W of Kirschovatz, near the Western Morava.

TERTENIA, a village of Sardinia, in the prov. and 15 m. SSE of Lanusei.

TERTZENA, or TREZNA, a town of Greece, in the Morea, 12 m. N of Dimiczana.

TERUEL, a province of Spain, in the S of Aragon; bounded on the N by the provs. of Saragossa and Huesca; on the E and S by Tarragona, Castilan-de-la-Plana, and Valencia; and on the W by Cuenca and Guadalajara. Area 2,363 geog. sq. m. It is traversed by a chain of mountains which give rise to the Guadalaviar, the Guadalope, and the Xiloco. Pop. in 1849, 250,000.—The cap. of the same name is situated at the confluence of the rivers Guadalaviar and Alhambra, 13 m. ESE of Albarracin. It stands on a hill, and is defended by a citadel. It is a gloomy ill-built place, with narrow streets, but contains a number of churches and convents, and a theatre. The environs are fertile. This town was destroyed by the Moors, and after being a long time deserted, was rebuilt by Alphonso II. in 1171.

TERVAY, a village of France, in the dep. of Jura, cant. and 4 m. NE of Montmiry-le-Chateau.

TERVERE, or VEERE, a town of Holland, in the prov. of Zieland, 4 m. NE of Middleburg, on the N coast of the island of Walcheren. Pop. 1,300.

TERVES, a village of France, in the dep. of Deux-Sevres, cant. and 3 m. SSW of Bressuire.

TERVISO, a small town of Austrian Illyria, 8 m. W of Mitterburg.

TERVUREN, a town of Belgium, in the prov. of S. Brabant, 6 m. ESE of Brussels, with a fine chateau. Pop. 1,800.

TERWICH, a parish of Sussex, near Midhurst. Area 718 acres. Pop. in 1851, 97.

TERWOLDE, a village of Holland, in the prov. of Guelderland, 20 m. NNE of Arnheim.

TERZA (LA), a town of Naples, in the Terra-d'Otranto, 25 m. W of Taranto.

TESCHEN, a circle of Austrian Silesia, comprehending the eastern part of that province, and separated from Prussia by the Oder and the Vistula. It has an area of 740 sq. m. It is hilly on the S side, where the Carpathian chain commences; but the N is flat and marshy. The inhabitants are mostly of Slavonian origin. The greater part of this circle is formed of the ancient duchy of Teschen, which belonged to the emperors in their quality of kings of Bohemia.—Its cap. of the same name is situated on the r. bank of the river Olsa, 62 m. ENE of Olmutz. It has one Lutheran and several Catholic churches, a gymnasium, a high school, two convents, and two ducal castles. Pop. 6,400. It has manufactories on a small scale of broad cloth, kerseymeres, leather, and muskets; also a considerable traffic in leather, wool, cloth, wine from Hungary, honey, and wax. The dispute about the succession to the electorate of Bavaria was terminated by a treaty concluded here in 1779.—Also a town of Bohemia, 59 m. SSE of Prague, and 14 m. SE of Tabor. Pop. 800. In the vicinity are hot springs.

TESCOU, a river of France, in the dep. of Tarn, which joins the Tarn, on the l. bank, after a W and NW course of 30 m.

TESCUCO. See TEZCUCO.

TESEGDELT, a village of Marocco, situated on a lofty rock, supposed to be impregnable, 20 m. S of Mogodor.

TESENI, a village of Asiatic Turkey, 34 m. SE of Degnizlu, in a valley in the Taurian chain.

TESORO, a small island near the coast of South America, in N lat. 10° 8′.

TESSAN, a town of Bosnia, in the sanj. and 38 m. NNE of Trawnik, near the r. bank of the Ussora.

TESSARAGH, TISRARA, or TAUGHSRARA, a parish of co. Roscommon, 2 m. SSW of Athleague. Area 8,482 acres. Pop. in 1831, 3,492; in 1851, 2,031.

TESSAURAN, TISERAN, or KILGALLY, a parish in King's co., 2¼ m. NW by W of Cloghan. Area 7,316 acres. Pop. in 1831, 2,032; in 1851, 1,576.

TESSENDERLOO, a village of Belgian Limburg, 15 m. NW of Hasselt.

TESSERMUIT, an island near the SW coast of East Greenland, in N lat. 59° 59′.

TESSIN. See TICINO.

TESSIN, a walled town of Mecklenburg-Schwerin, on the Recknitz, 21 m. NNE of Gustrow. Pop. 2,200. Its chief manufactures are linen and nails.

TESSIURSAK, an island near the W coast of W. Greenland, in N lat. 61° 10′.

TESSON, a village of France, in the dep. of Charente-Inferieure, cant. and 4 m. NNE of Gemozac. Pop. 700.

TESSOWA, a considerable town of Fezzan, 100 m. ESE of Murzuk.

TESSY, a town of France, dep. of La Manche, 12 m. S of St. Lo.

TESTACCIO, a town of the Neapolitan island of Ischia, 3 m. SW of Ischia. Pop. 1,200.

TESTERTON, a parish of Norfolk, 2½ m. SE by S of Fakenham. Area 613 acres. Pop. in 1851, 13.

TESTIGOS (LOS), a cluster of small islands off the coast of Venezuela, 40 m. NW of Cape Mala-Pascua. They are 7 in number, besides several rocks. The principal island is 2½ m. in length.

TESTON, a parish of Kent, 4 m. W by S of Maidstone. Area 491 acres. Pop. in 1851, 268.

TET, a river of France, which rises in the Pic-de-Prique, in the Pyrenees; runs SE, and then ENE, passing Perpignan; and 8 m. below the latter town, enters the Mediterranean, after a course of 60 m.

TETBURY, a parish and town of Gloucestershire, 17 m. SSE of Gloucester. Area of p. 4,582 acres. Pop. in 1841, 2,982; in 1851, 3,325.—The town, pleasantly situated on a rising ground, near the source of the river Avon, consists of four principal streets meeting in the centre. T. has manufactures of woollen cloth, and is a considerable thoroughfare, the direct road from Bath and Bristol to Cirencester passing through the town, as also the road from Gloucester and Stroud to Bath, Devizes, Salisbury, and Southampton.

TETCOTT, a parish of Devonshire, 5 m. S by W of Holsworthy. Area 2,181 acres. Pop. 289.

TETE, a town of Mozambique, 110 m. NW of Senna, on the Zambese river. The surrounding territory is highly fertile, and the town itself is comparatively healthy, being built on an elevated site, with the river flowing beneath it. Boats ascending from Senna usually employ 6 weeks in surmounting the rapid current of the river.

TETE-HOIRE, a pass in the Swiss cant. of Valais, leading, at an alt. of 4,000 ft., into the Val-d'Orsine.

TETE-DE-BACH (LA), a village of France, in the dep. of Gironde, 32 m. WSW of Bordeaux, on the S side of the bay of Arcachon, and connected by railway with Bordeaux. It is now a place of seaside resort in summer.

TETELA, a town of Mexico, in the intendancy of Puebla, 60 m. SE of Mexico.—Also a town in the intendancy of Mexico, 108 m. N of Mexico.—There are several other small settlements of the same name in Spanish America.

TETENY, a town of Hungary, in the com. and 7 m. SSW of Pest.

TETEREW, a river of Russian Poland, which rises in the SW of Volhynia, and flows NE to a junction with the Dnieper, after a course of 150 m. It is navigable to Jitomir.

TETEROA, a group of six or seven islets, not many feet above the level of the sea, lying in the S. Pacific, in S lat. 17° 1′, 24 m. NW of Point Venus, in Tahiti. The inhabitants, originally about 3,000 in number, are chiefly employed in catching fish, which they exchange for bread-fruit.

TETEROW, a town of Germany, in Mecklenburg-Schwerin, 15 m. E of Gastrow. Pop. 3,750.

TETFORD, a parish of Lincolnshire, 6 m. NE by E of Horncastle. Area, 2,210 acres. Pop. in 1851, 799.

TETH, a village of Hungary, 14 m. SSW of Raab.

TETI. See CHIETI.

TETIUSCHI, a town of European Russia, in the gov. and 60 m. SSW of Kasan, on the Volga. Pop. 1,700.

TETNANG, a town of Würtemberg, 19 m. E of Constanz, on the Muhlenbach. Pop. 1,200. Having been formerly one of the four towns of Suabia entitled to coin money, it was notorious for the quantity of debased pieces issued under its stamp.

TETNEY, a parish of Lincolnshire, 10½ m. N by E of Louth. Area 3,825 acres. Pop. 869.

TETON, a hamlet of Northamptonshire, 8 m. NNW of Northampton.

TETRECHENNE, a village of France, in the dep. of Moselle, cant. and 6 m. NE of Boulay. Pop. 750.

TETSCHEN, DACZIN, or WARTA, a town of Bohemia, on the Elbe, 18 m. N of Leitmeritz, and 48 m. N by E of Prague. It is finely situated at the foot of a lofty sandstone ridge, on which stands an old castle.

TETSHA, a river of Asiatic Russia, which rises on the E flank of the Urals, 30 m. NW of Chebbiansk, and flows ENE to a junction with the Iset, after a course of 120 m.

TETSO, a town of Hungary, in the com. of Marmaros, 40 m. N of Nagy-Banya. Pop. 1,350.

TETSWORTH, a parish of Oxfordshire, on the river Thames, 12½ m. ESE of Oxford. Area 1,172 acres. Pop. in 1851, 512.

TETTE. See TETE.

TETTENHALL-REGIS, a parish of Staffordshire, 2 m. NW of Wolverhampton. Area 7,600 acres. Pop. in 1841, 3,143; in 1851, 3,396. Locks, hinges, nails, and bolts are largely manufactured here. The village, situated at the bottom and on the declivities of a lofty eminence which rises above the Smestall rivulet, contains many good houses, chiefly occupied by genteel families and persons engaged in the trade of Wolverhampton.

TETTENS, a village of Oldenburg, 6 m. N of Jever. Pop. 200.

TETTUA-MOTU, a cape on the E coast of New Zealand, forming the NE point of Poverty bay, in S lat. 38° 36′.

TETUAN, a port of Marocco, in the prov. of Fez, on the Mediterranean, immediately within the straits of Gibraltar, in N lat. 35° 57′, and 33 m. SE of Tangier, at the mouth of the Martil river. A branch of the Atlas, passing through the province of Tedla, approaches within a few miles of the town. The environs are laid out in vineyards and gardens. The inhabitants, about 15,000 in number, consist of Moors and Jews, who, from their intercourse with Spain, almost all speak the language of that country. T. is a purely Moorish town, and is divided into two parts,—the Jews' quarter and the Moors' quarter. A strong gate separates the two quarters; and on certain occasions this gate is kept by Moorish soldiers, and the ordinary Jewish pop. is not permitted to pass into the Moorish quarter.—The Jew's house is built in the style of the Moors, a fashion of building which always makes a dull-looking street; but in proportion as the house is cold and forbidding in its outside appearance, it is interesting and animating within. Every room opens into the *patio* or court, and it would be possible to traverse the whole of the Jewish quarter of the city on the roofs of the houses. No sooner, however, have we passed the gate and entered the great square of T. than we find a life and bustle that contrast strangely with the lifeless city out of which we have just come. Here in the *fabricas* of swords, guns, and pistols, mats, and devices in coloured woods, slippers, &c., we find abundant evidences of activity and industry. A considerable communication is maintained from this port with Gibraltar, whence ships often repair, when the wind is unfavourable for making Tangier.

TETWORTH, a parish of Huntingdonshire, 6 m. S by E of St. Neot's. Area 1,446 acres. Pop. 221.

TEUCHERN, a town of Prussia, in the prov. of Saxony, regency and 17 m. S of Merseburg, and circle of Weissenfels, on the Rippach. Pop. in 1843, 1,558. It has manufactories of pottery.

TEUDITZ, a village of Prussia, in the prov. of Saxony, regency and circle and 8 m. SE of Merseburg. Pop. 360. It has a salt work.

TEUFELSBERG, a mountain of the chain of the Bohmerwald, in Bohemia, at the SW extremity of the circle of Pilsen. It is of great height, and is crossed by the road from Pilsen into Bavaria.—Also a mountain of Russia in Europe, in the gov. of Livonia, near Laitzen. It has an alt. of 860 Parisian ft. above the level of Riga.

TEUFEN, or TEUFFEN, a town of Switzerland, in the cant. and 5 m. N of Appenzell, in the Ausser Rhoden. Pop. 3,867. It has a poor's house and mineral baths. It contains extensive manufactories of calico, muslin, and brocade.

TEUFENBACH, a village of Styria, in the circle and 14 m. SSW of Grätz, near Staniz.

TEUGMO, a small island of Russia in Europe, in the gulf of Bothnia, near the W coast of Finland, in the gov. and 24 m. NNE of Wasa, in N lat. 63° 19′, and E long. 21° 58′.

TEULADA, a town of Spain, in the prov. and 39 m. NE of Alicante, and partido of Denia, in a plain scantily watered, but remarkable for its fertility. Pop. 1,638. It has a parish-church, a customhouse, and a public granary, and the streets are generally spacious and substantially built of stone. Wine and silk are the chief productions of the locality.

TEULADA, a market-town of Sardinia, in the div. of Capo Cagliari, prov. and 26 m. SW of Cagliari and district of Domus-de-Maria, in the midst of mountains, on a small river of the same name, an affluent of the gulf of Teulada. Pop. 2,700. In the vicinity are found white marble and porphyry, and near the coast is a bed of magnetic iron.

TEULADA, or TOLARO (CAPE), a headland of Sardinia, in the div. of Capo Cagliari, and prov. of Iglesias, in N lat. 38° 52′, and E long. 18° 36′. It is the most southerly point of the island, and forms, with Cape Spartivento, 12 m. ENE the entrance to the gulf of Tuelada. This gulf, which is partly in the prov. of Iglesias, and partly in that of Cagliari, is about 5 m. in depth, and encloses several considerable bays. It contains the small island of Bossa, near which is good anchorage. The traffic from this coast consists chiefly in faggots for Cagliari, Malta, and Marseilles.

TEULET. See TEILLET.

TEUPITZ, a town of Prussia, in the prov. of Brandenburg, regency and 32 m. SE of Potsdam and circle of Teltow-Storkow, on the SE bank of a lake of the same name. Pop. 388. It has an ancient castle. The lake is 3 m. in length from NE to SW.

TEURTEVILLE-AU-BOCAGE, a village of France, in the dep. of the Manche, cant. and 4 m. WNW of Queltehon, in a fine valley. Pop. 1,800.

TEURTEVILLE-HAGUE, a village of France, in the dep. of the Manche, cant. and 5 m. SW of Octeville, near the l. bank of the Dirette. Pop. 1,200.

TEUSACA, a village of New Granada, in the prov. of Bogota. It was formerly a place of considerable importance.

TEUSCHNITZ, a town of Bavaria, in the circle of Upper Franconia, and 32 m. N of Bayreuth, on a mountain. Pop. 800. It has a castle.

TEUSING. See Theysing.

TEUTENWINKEL, a village of the grand-duchy of Mecklenburg-Schwerin, and duchy of Mecklenburg-Gustrow. Pop. 800.

TEUTOTLAN, a town of Mexico, in the state and 60 m. ENE of Oaxaca. Pop. 1,285. Cochineal is abundant in the environs.

TEUTOBURGER-WALD, or Teutschburger-wald, a chain of hills in Germany, which extends in a NW direction through Hesse-Cassel, Rhenish Prussia, Lippe, and Hanover, from the vicinity of the source of the Lippe, to the r. bank of the Ems, near the confluence of the Hopster-Aa, a distance of about 120 m. The Ems, its affluent the Haase, and the Werre, an affluent of the Weser, are the principal streams to which it gives rise. It is covered with fine wood. These hills are noted for the defence here made by Arminius against the legions of Varus.

TEVA, a town of Spain, in the prov. and 30 m. NW of Malaga.

TEVAI, an island of the S. Pacific, in the archipelago of Santa-Cruz. It lies in a bay of the same name, on the E side of Vanikoro island. On its NE shore is a village named Vanikoro.

TEVERE. See Tiber.

TEVERONE, a river of the Papal States, which has its source at the N extremity of the deleg. of Frosinone, on the confines of the kingdom of Naples, in Mount Tarino; runs first S, then NW into the comarca of Rome; passes Subiaco; turns afterwards WSW; bathes Vicovano and Tivoli, and at the latter place forms several fine cascades; and after a total course of about 66 m., falls into the Tiber on the l. bank, 3 m. N of Rome. This river, the *Anio* of the Romans, is noted in Roman history. It was here that Tarquin the elder put the Sabines to flight, that Camillus defied the Gauls, 387 B. C., and that Manlius Torquatus, in 361, attacked and killed a gigantic Gaul in single combat.

TEVERSALL, a parish of Nottinghamshire, 3¾ m. W by N of Mansfield. Area 2,820 acres. Pop. in 1831, 400; in 1851, 373.

TEVERSHAM, a parish in Cambridgeshire, 2¾ m. E of Cambridge. Area 1,200 acres. Pop. 298.

TEVESAN, a town on the W coast of the island of Celebes, in S lat. 2° 2′, E long. 119° 20′.

TEVIOT, a beautiful river of Roxburghshire, the largest tributary of the Tweed, and a stream of so much importance as popularly to impose its name on the whole of the county in which it flows. It rises in several short head-waters in the S extremity of the parishes of Cavers and Hawick, within a ½ m. of the sources of the Megget and the Stennis, tributaries of the Dumfries-shire and Cumberland Esk. The direction of its course for 30 m., or till it enters the parish of Eckford, is, with the exception of constant but brief distances, uniformly and almost due NE; for 5¼ m. further, it is N, or N by E; and for the last ¾ m. it is E, and nearly parallel to that of the noble river into which it falls. Its principal tributaries on its l. bank are Hislop-burn and Borthwick-water in Hawick, and Ale-water in Ancrum; and on its r. bank, Frostly-burn in Upper Cavers, Allan-water between Upper Cavers and Hawick, Slitrig-water at the town of Hawick, Rule-water between Lower Cavers and Bedrule, Jed-water in Jedburgh, Oxnam-water in Crailing, and Kail-water in Eckford.

TEW (Great), a parish in Oxfordshire, 5½ m. ENE of Chipping-Norton. Area 3,140 acres. Pop. in 1831, 616; in 1851, 541.

TEW (Little), a chapelry in the above p. Area 1,600 acres. Pop. in 1831, 216; in 1851, 237.

TEWIN, a parish in Hertfordshire, 4½ m. W by N of Hertford. Area 2,615 acres. Pop. in 1851, 532.

TEWKESBURY, a parish and parl. borough in the lower division of the hund. of Tewkesbury, co. of Gloucester, in the vale of Evesham, on the eastern bank of the Avon, near its junction with the Severn, 10 m. NNE of Gloucester, and 2 m. W of the Birmingham and Gloucester railway, with which it is united by a branch line. The parish includes the hamlets of Mythe and Southwick. Area of the p. 2,333 acres. Pop. in 1801, 4,199; in 1831, 5,780; in 1851, 5,878. The town is advantageously situated a little to the S of the Upper Avon, which has been diverted from its natural channel by a cut which meets the Severn lower down. Two other streams, called the Carron and the Swilgate, here flow into the Avon, one just above the town, and the other a short distance below it. The Birmingham and Gloucester branch-railway terminates in the High-street of the town. The town itself consists chiefly of three principal streets with several lanes and alleys. The principal, or High-street, running north and south, is of considerable length, and is spacious and handsome. The abbey church presents an interesting example of early Norman architecture, mixed with specimens of other kinds. It is built in the cruciform cathedral style, and consists of a nave, choir, transept, and massive anglo-Norman or Saxon central tower, with several chapels.—Most of the old buildings have been replaced by houses of more modern form; but a few specimens of the ancient style of building still exist. For the accommodation of the river navigation there are quays on the E side of High-street. The tides in the Severn reach Tewkesbury. The principal manufactures latterly carried on have been the stocking-frame work, knitting in cotton and lambs' wool, but particularly the former; the bobbinet lace manufacture, the manufacture of nails and leather, and the business of malting.—T. returns 2 members to parliament. The parliamentary boundaries have been extended so as to coincide with those of the parish. The number of electors registered in 1837, was 394; in 1851, 370. T. is also a polling-place in the elections for the E division of the county.—The most interesting and important military event in the annals of T. was the last decisive battle between the houses of York and Lancaster, known as the battle of Tewkesbury, which was fought in a field about a ½ m. S of the town, still named the Bloody meadow.

TEWKESBURY, a township of Middlesex co., in the state of Massachusetts, U. S., 21 m. W of Boston. It has a level surface, bounded on the NW by the Merrimac, and by Concord river on the W, and intersected by the Shawsheen. It is traversed by the Salem and Lowell, and Lowell and Lawrence railroads. Pop. in 1840, 906; in 1850, 1,048.—Also a township of Hunterdon co., in the state of New Jersey, 14 m. NE of Flemington. It has a hilly surface, and is drained by Rockaway creek and its branches, and by Lamington river. The soil, consisting chiefly of clay and loam, is fertile and well cultivated. Pop. in 1840, 1,944.

TEXAR (SAN SEBASTIAN DEL), a district and town of Guatimala, in the state of that name, and dep. of Sacatepec and Chimaltenango. Pop. of district 7,280; of town, 1,260.

TEXAS, one of the United States of North America, formerly a province of Mexico, and more recently an independent republic, situated between the 26th and 37th parallels of N lat., and the meridians of 93° 30′ and 110° W long.; and bounded on the N by the Red river and the Arkansas, which separate it from the Indian territory; on the E by the Sabine river and a conventional line, which divides it from Louisiana; on the S by the gulf of Mexico; and on the W by the Rio-Grande-del-Norte or Bravo, and by New Mexico. The extreme length of this state from N to S is 700 m.; and the breadth from E to W about 800 m. It has an area of 237,321 sq. m. The boundaries recently claimed by Texas, included portions of New Mexico and a large tract of country between the Nueces and Rio-Grande, which formerly belonged to the Mexican states of Tamaulipas, Nueva-Leon, and Coahuila; but the Texan claim has not been accorded by congress.

Physical features.] The face of this country exhibits three essentially different physical regions,—the level, the undulating, and the mountainous. Along the coasts, and inland from 30 to 60 m., the surface is a low level, free from swamps, and composed of excellent arable and pasture lands, and prairies. The soil here is a deep black mould, and of inexhaustible fertility. The bottoms along the rivers are well-wooded, and have a rich red soil of great depth. Over this vast plain immense droves of wild horses and cattle roam, in the enjoyment of liberty, except near the settlements, where they are fast diminishing in number. In the interior and to the N, the lands are considerably elevated, and the surface undulating. The prairies of this section are covered with forests of vast extent. The W and NW section of T. is mountainous, and spanned crosswise with the several ranges of sierras which divide T. from New Mexico. The mountains are clothed with forests of oak, pine, cedar, and a great variety of trees and shrubs, and they enclose extensive alluvial valleys, most of which are susceptible of irrigation and culture. The surface, for the most part, is covered with a fine growth of grasses, comprising, with the common prairie-grass, the *gama*, *musquite*, wild clover, wild rye, &c., and affording excellent pasturage. The natural growth of woods comprises many of the most valuable descriptions. Among the remarkable features of the country, the 'Cross-Timbers' of northern T. are conspicuous. These are a continuous series of forests, varying in width from 5 to 50 m., and extending in a direct line about the longitude of 97° W, from the woody region, at the sources of the Trinidad, northward to the Arkansas river. They appear at a distance as an immense wall of wood; and seen from the W, such is their linear regularity, that they look as if planted by the hand of man. Here the western prairies terminate. The country between the Nueces and the Rio-Grande has little soil susceptible of cultivation; and, except on the margins of the streams, and near the gulf of Mexico, is here a barren wilderness, overgrown with cacti and brush-wood. Several lakes, some of which are strongly impregnated with salt, are found in this region. "There is no region in the world," says M. Maillard, "that possesses more fertile land than T., and where Nature has raised so many obstacles in the way of its cultivation and the development of its fertility. The prairies are swamps for two-thirds of the year, and the timber of the higher regions cannot be obtained without a considerable outlay of capital and labour, the latter of which will at all times be extremely expensive, and more or less difficult to obtain. The surface soil throughout the country is a mixture of lime, sand, and marl, the proportion of the latter being equal to the two former: its colour determines that of the soil, as it appears in different sections. In the elevated tracts to the north-west it is whitish; in the prairies it is black; and in the cane prairies and woodlands it is red. Every where the land is rich, and adapted to the cultivation of different tropical staples. The substratum is still more diversified than the surface soil. In the elevated regions it is a white marl, which, on exposure to the atmosphere, becomes as hard as marble; in the woodlands and cane prairies it is rock sand. In the alluvial country, generally, it is a black greasy clay, interlaid at intervals of about 10 ft. with a stiff bluish marl. The alluvial region extends along the coast from the Sabine river to the Rio-Grande, and is about 80 m. inland. The physical constitution of the atmosphere, and other natural causes, render the prairies in the lower part of T., along the coast, worse than useless for two-thirds of the year, and the prospective productiveness of the upper sections of the country somewhat problematical."

Climate.] The climate of T., except in the more northern parts, is essentially tropical. The average temp. is about 80°. A wet and a dry season divide the year. The former lasts from December to March; the latter from March to the end of November. The great heats of summer are tempered by constantly-recurring breezes, which commence soon after sunrise, and last until 3 or 4 o'clock, P.M. The nights throughout the year are cool. From March to October little rain falls, though thunder-storms are frequent; during the rest of the year rain predominates, and the rivers inundate the country. Snow, except on the mountains, is seldom seen. In the river bottoms and along the coasts, intermittent fevers, and in the summer season yellow fever, prevail to an alarming extent. "Nine-tenths of the country," says Mr. Kennedy, "are considered healthier than the most healthy parts of the United States." But M. Maillard says "that the suddenness of the atmospherical changes is most injurious to the health. In the beginning of spring (March), the weather is wet and cold, the therm. seldom rising above 45°: but towards the close of the spring (May), it suddenly becomes sultry, rising to 95°. In summer the therm. ranges from 95° to 105°; and the rapid progress of the vegetation appears almost supernatural, while the air teeming with insects renders life intolerable. This is the season of epidemics, yellow and bilious fevers (in 1839, 65 per cent. of the pop. were carried off by yellow fever). In autumn, the therm. gradually falls to 60°, and the weather becomes more congenial for a short time, when the rainy season sets in, and continues throughout the winter and the greater part of the spring, accompanied by awful thunder, lightning, and northerly winds, which search the frame with an acuteness almost peculiar to themselves. The consequence of the climate on vegetation is, that though the earth is as rich as any in the world, every thing is no sooner above ground than it is matured and vanishes. The flowers are of a single day's creation, showing themselves in all their primitive and exquisite beauty in the evening, but the next day reduces them to a cinder, or they are inundated with water. At the close of the rainy season the waters do not disappear, but on the contrary; for the sun coming forth, thaws the snow off the mountains in the north, and the water which results is kept pent up in the mountain valleys by the strong SSE winds which prevail at this season, till the wind veering to the north permits of their escape, when they pour down in torrents, not unfrequently sweeping every thing before them, but never failing to continue the inundation of the lower country from about the 1st of April to the middle of June."

Rivers.] In addition to a coast-line presenting numerous harbours, bays, and inlets, no country is better supplied with rivers than Texas. The principal, proceeding from E to W, are the Sabine or Neches, Trinidad, Brazos, Colorado, Guadalupe, San-Antonio, Nueces, and the Rio-Grande. All these fall into the gulf of Mexico, and they are all navigable for steamboats for 100 or 200 m. from their outlets. The Red river and the Arkansas wind along the northern frontier and must eventually become the commercial highways of that section of the state. Among the minor streams are the Angelina, San-Jacinto, San-Bernard, Navidad, and Nueces. The bays and lagunes are formed by a range of long and narrow islands, which line the coasts. Galveston bay, into which the Trinidad, San-Jacinto, &c., fall, is by far the finest on the coast, being 35 m. long from N to S, and from 12 to 18 m. broad from E to W. Its average depth is from 9 to 10 ft, but in the channel the depth varies from 18 to 30 ft. The entrance, however, is obstructed by bars.

Productions.] In many parts of the rolling-prairie region, coal of a superior quality, and iron ore, exist abundantly; and it has been supposed that beds of these valuable minerals extend over the greater portion of the country. The precious metals are found in the mountains to the NW, and silver mines were worked in these regions until the miners were driven off by the Comanche Indians. The silver mine of San-Saba is one of the richest in the world; and gold has recently been obtained in many localities in Northern T. Saltpetre abounds in the E, and salt is obtained from numerous lakes and springs; and bitumen in several places. Granite, limestone, gypsum, shale, &c., are abundant, except in the low alluvial region.—The rearing of live stock has long been the principal and favourite occupation of the Texan settlers, and many of the prairies are covered with a valuable breed of oxen, which scarcely require, and certainly do not receive much more care or attention than the prairie-deer. These have increased enormously, and a profitable trade in cattle has been opened with New-Orleans and the West-India Islands. The rearing of horses and mules is also pretty extensively pursued. Sheep do not thrive well, except in the northern counties, and even there the wool is inferior and the mutton coarse. Hogs form a considerable item in the economy of the settler, and bees, which are in great numbers, might, with a little care, be made productive, there being an extensive demand for wax and honey for the northern market. The value of the live stock in 1850, was 10,266,880 dollars.—Vast herds of buffaloes and wild horses wander over the prairies, and deer are everywhere abundant. Bears, couguars, panthers, peccaris, wolves, foxes, and raccoons, are common; and most of the farmers are obliged to keep packs of large and powerful dogs, to preserve their stock from destruction. Most of the birds known to the United States are common to T., and the bays and lagunes abound with excellent fish and oysters, and other testacea. Alligators inhabit the Red river and its tributaries; and turtles and tortoises, the estuaries and mouths of the rivers flowing into the gulf. Reptiles and insects, some of the most venomous kinds, are prolific, and alike dangerous and troublesome.—Cotton, the great staple of T., is of a superior quality, producing more largely, and of a finer staple, than the same article in the Atlantic states. The product of 1848 was 39,742 bales. Superior cotton-lands, it is said, produce from 1½ to 2 bales per acre. Its cultivation is principally confined to the rich alluvial soils of the Red river, the Brazos, the Colorado, and Trinidad. The cotton is either shipped to New Orleans from Galveston, or is brought down the Red river and the Mississippi in steam-boats, which ply regularly on those rivers. Planting begins in February, and the cotton is picked in June. The grains chiefly cultivated are maize and wheat. The production of 1847 was as follows:—Wheat, 1,110,000 bushels; Indian corn, 500,000 bushels. The potato crop yielded 200,000 bushels. Rye, barley, oats, &c., are suited for the upper country; rice grows best near the river estuaries. A double crop of Indian corn is frequently obtained within the year. Sugar-cane grows well, and yields heavily throughout the level region. The crop of 1847 was 20,000 lbs. Cane plantations, however, are as yet very limited, and hitherto the culture has been altogether experimental. Tobacco will probably become a staple. The mulberry-tree grows vigorously, and the climate is well-suited for the silk-worm. The vanilla plant grows wild; nopal is abundant; and the *yaupan* is said to furnish a tea little inferior to the common black tea of China. Among fruits, the fig, peach, nectarine, quince, pine-apple, orange, and lemon, ripen well.

Texas has no manufactures except those of ordinary domestic kinds, or tho: which are made in every family. The tanning of leather, however, and the making of soap, &c., are beginning to create some attention. The value of home-made manufacture in 1850 was 255,719 ds. The direct foreign commerce is scarce; but the coasting-trade is considerable.

Population.] The pop. of T. in 1850 was 212,592, of whom 154,100 were whites, and 58,161 slaves. The provision made for education is much on the same basis as in the new states in the NW. The constitution also provides that one-tenth of the annual revenue of the state derived from taxation, shall be set aside as a permanent fund for the support of free public schools; and that all public lands granted for such schools shall not be alienated in fee, nor be leased for more than twenty years. The Roman Catholics, Methodists, and Baptists are the most numerous religious denominations. The number of periodicals published within the state in 1850 was 37.

Government.] The present constitution was adopted in convention at Austin, 27th August, 1845, and ratified by the people on 13th October of the same year. Every white male inhabitant, 21 years of age, being a citizen of the United States or of T., who has resided in the state one year, and six months in the district, county, city or town where he offers his vote, is invested with suffrage. The General assembly consists of a senate and house-of-representatives. The senate is composed of not less than 17 nor more than 33 members. Senators are elected for four years, one-half going out every two years. They must be qualified voters, at least 30 years of age, and have lived in the state three years, the last year thereof in the district for which they are chosen. Representatives—of whom there shall not be less than 45, nor more than 90—are chosen for two years. They must be qualified voters, having lived in the state two years, and the last year thereof in the place where they are chosen. The legislature meets biennally. The members of the legislature receive each 3 dollars per diem, and 3 d. for every 25 m. of travel to and from the capital. The governor and lieutenant-governor are chosen by a plurality of the popular vote for two years, and are ineligible to serve for more than four years out of any six years. The governor must have the same qualifications as a senator, and receives 2,000 dollars per annum. The lieutenant-governor is *ex-officio* president of the senate. The secretary-of-state is nominated by the governor, and his nomination confirmed by the senate for two years. The state-treasurer and comptroller are chosen biennally by joint vote of both houses. The governor may *veto* a bill, but by a two-thirds vote of both houses in its favour, it becomes law.—The Judiciary consists of a supreme court, with a chief-justice and 2 associate-justices, appointed for six years. There are eight district courts, with one judge to each district. These courts have original jurisdiction both in civil and criminal suits. In criminal cases, if the punishment be not specifically determined by law, the jury determine it. An attorney-general is appointed for two years by the governor, and confirmed by the senate, and district attorneys are chosen by joint ballot of both houses of the legislature, also for two years. In equity causes, either party may demand a jury. The constitution has several peculiar provisions. It renders duellists ineligible to public offices; forbids grants of money for any purposes than the ordinary expenses of government, except by a two-thirds vote; provides for the revision of the laws, civil and criminal, every ten years; makes a wife's property, real and personal, her separate property, not liable for the husband's debts; protects the homestead to the value of 2,000 d.; forbids the creation or extension of corporations for banking or other purposes; forbids the creation of a debt of more than 100,000 d., except in case of war, insurrection or invasion, &c., &c. The provisions with regard to slavery are similar to those of the other southern states. No law of emancipation shall be passed without the owner's consent, and by paying full compensation for the slaves liberated; nor shall emigrants to T. be prevented from bringing their slaves with them, but bringing them as merchandize is forbidden. Slaves are entitled to a jury in criminal suits.

Finances.] The annexed statement exhibits the amount and condition of the debt of T., according to the official report of the comptroller, dated March, 1846.

Total amount of audited drafts,	156,905 68 d.
„ „ of promissory notes,	2,674,447 10
„ „ of 8 per cent. bonds,	811,000 00
„ „ of funded debt,	1,675,800 00
„ „ of interest on liabilities,	1,873,804 92
„ „ of outstanding claims,	822,000 00
Aggregate of domestic debt,	8,013,957 65 d.
Aggregate of foreign debt,	1,935,049 40
Aggregate of indebtedness,	9,949,007 05 d.

The public debt of T., according to this statement, amounted in 1848 to about 10,000,000 dollars, in round numbers.—According to the state-treasurer's report the revenues and expenditures of the state, in available means, from the 31st of October, 1851, to the 31st of October, 1853, were as follows:—

RECEIPT.

From ordinary sources of revenue, such as taxes, &c.,	156,370 27 d.
From government of the U. S. in 5 per cent. bonds,	5,000,000 00
Premium and interest on the above bonds to 1st Jan. 1853,	510,050 10
Refunded by government of the U. S. on account of the advances of the state,	8,094 12
Means in the treasury on the 31st of Oct. 1851,	25,890 68
Total,	5,701,205 17

EXPENDITURE.

On account of expenditures of the state,	584,387 51
On account of debt of the late republic,	1,113,144 64
Transferred from state revenue to county revenue,	1,371 08
Amount of state bonds destroyed and United States bonds substituted as per act of Feb. 1852,	36,000 00
Leaving on the 31st of Oct. 1853, a balance of	3,965,301 94

The following was the estimate of receipts and expenditures for the fiscal year ending October 31, 1854:

RECEIPT.

Amount of taxes, &c., due prior to 1852,	48,000 00
Interest on United States bonds now in the treasury, from Jan. 1853 to Oct. 31, 1854,	327,708 33
	375,708 33
Balance in the treasury on the 31st October, 1853, as above shown,	3,965,301 94
Total estimated means for the support of the government for the year ending October 31, 1854,	4,341,010 27

EXPENDITURE.

Legislative department,	75,000 00		
Executive, „	40,500 00		
Judicial, „	37,850 00		
Miscellaneous expenses,	25,364 00		
		178,714 00	
Due on appropriations for support of the state for 1853 and previous years,		98,623 37	
Due for the payment of the public debt,		348,685 08	
Amount of public debt now audited, including that which will probably be audited up to the 1st of Feb. 1854, under act of Feb. 1853,		100,000 00	
			726,022 35
Leaving in the treasury on the 31st of October 1854, an estimated balance of			3,614,987 92 d.

The taxable property throughout the state has increased in value for the last eight years, but more especially for the past three years. The increase in this period has been at a rate seldom known, being almost 100 per cent. The following table presents the aggregate property for each of the last eight consecutive years, commencing with 1846 and terminating with 1853, with the increase per cent. for each year:—

Year.	Aggregate taxable Property.	Increase taxable Property.	Increase each Year.	Per cent.
1846,	34,391,175			
1847,	37,563,505	3,171,330	1847 over 1846	8¼
1848,	43,812,537	6,250,032	1848 „ 1847	16⅔
1849,	46,241,589	2,429,052	1849 „ 1848	5½
1850,	51,814,615	5,573,026	1850 „ 1849	12 2-5
1851,	69,739,581	17,924,966	1851 „ 1850	33⅓
1852,	80,754,094	11,014,513	1852 „ 1851	60½
1853,	99,155,114	18,401,020	1853 „ 1852	23

Counties.] The counties into which T. was recently divided are 78 in number, viz.: Anderson, Angelina, Austin, Bastrop, Bexar, Bowie, Brazoria, Brazos, Burleson, Caldwell, Calhoun, Cameron, Cass, Cherokee, Collin, Colorado, Comal, Cook, Dallas, Denton, De-Witt, Fannin, Fayette, Fort-Bend, Galveston, Gillespie, Goliad, Gonzales, Grayson, Grimes, Guadalupe, Harris, Harrison, Hays, Henderson, Hopkins, Houston, Hunt, Jackson, Jasper, Jefferson, Kaufman, Lamar, Lavacca, Leon, Liberty, Limestone, Matagorda, Medina, Milam, Montgomery, Nacogdoches, Navarro, Newton, Nueces, Panola, Polk, Red-River, Refugio, Robertson, Rusk, Sabine, San-Augustine, San-Patricio, Shelby, Smith, Starr, Titus, Travis, Tyler, Upshur, Van-Zandt, Victoria, Walker, Washington, Webb, Wharton, and Williamson.

The city of Austin, the capital of the state, is situated on the N bank of the Colorado, in N lat. 30° 24′, and W long. 95° 42′. It consists chiefly of frame houses, and a few brick buildings, occupied by the wealthier classes and tradesmen. The country in the neighbourhood is becoming settled, and quite a number of farm houses have been built. Pop. about 5,000. Sabine, at the outlet of the Sabine lake, is a port of entry. Galveston, the chief port in T., situated at the E extremity of the island of the same name, is the depôt of a vast and fertile region watered by the Trinidad and other rivers. The commerce of this port has increased to something considerable, and its prospects are, at the present time, more encouraging than those of any others of the sea-board towns. Galveston, has at the present period about 4,600 inhabitants. Houston, the former capital, is situated at the head of navigation of Buffalo-bayou, a small tributary of Galveston bay, and is a place of little consideration. Velasco, on the N bank of the Brazos, at its mouth, is a pleasant town, and said to be very healthy. Matagorda, at the mouth of the Colorado, near the head of Matagorda bay, is a considerable town. Goliad is an old Spanish town on the N bank of the San Antonio, 40 m. from its mouth. San-Antonio de-Bexar, also an old Spanish town, is situated higher up the river, and is celebrated in history as the scene of several battles and the gallant defence of the Alamo, by Col. Travis, for two weeks. Zodiac is a new town on the Pedernales and 4 m. from Fredericksburg, the seat of Gillespie county, and 60 m. N of San Antonio. Germans and Mormons are the principal settlers. This is the most northerly settlement in Western T.

History.] The first settlements in T. were made by the French at Matagorda, but the settlers were soon after expelled by the Spaniards, who, in 1690, extended their cordons from New Spain over the whole territory. Previous to this time, the Indians held undisputed sway over this fine country. Spanish forces were stationed at Goliad, Bexar, Nacogdoches, &c., and held military possession until the revolution which overthrew the Spanish power in Mexico. On the consummation of Mexican independence, T. became, in connection with the adjoining state of Coahuila, a state of the republic. Soon after this period, several colonies of Americans, who had been invited into the territory by the Mexicans, settled in the eastern section, and were for a long time secured from the onerous burdens of Mexico proper. For some time previous to 1835, however, serious complaints against the central government were uttered, and in the following year open rebellion was rife. The Mexican president invaded T. with a large army, and was successively victorious and defeated in a number of skirmishes. The American settlers declared their independence in March, 1836, and the defeat of the Mexicans and the capture of Santa Anna at San-Jacinto (21st April 1836), secured to the 'patriot party' the acknowledgment of their independent position by the Mexican president, and the whole country E of the Rio-Grande was acceded to them. The acts of Santa Anna, however, were never confirmed by the senate, and of course their authority was, to say the least, equivocal. No serious attempt having been made by the Mexicans to regain T., for eight or nine years, the political nationality of the country was considered as consolidated, and in 1845, under this supposition, the United States of America admitted the young republic as a state of the Union. The consequence of this act was war with Mexico—a war which the United States assumed by the annexation of T. at a time when that country was at war with Mexico. Of this transaction, the great statesman, Daniel Webster, thus speaks: "In 1841-2-3, T. was an independent government, so nominally, so practically, so recognised by our own and other governments. Mexico had no ground of complaint in the annexation of T.; we are the party to complain; we did not want T. But she did take offence; she persisted in claiming T. as her province, long as it had been an independent territory, and broke off terms of amity with our government, calling home her ministers, and refusing to receive our own. But this was not war—very far from it. I contend that the precise object of the war was simply this—Paredes became president in March 1845. Santa Anna was exiled to Cuba in June of the same year. Through 1845 this state of things continued; matters between the two governments were unsettled and unsatisfactory. Mexico held back and was warlike in her tone. In January, 1846, the United States army was ordered to advance to the Rio-Grande, in order, says Mr. Buchanan, to act as congress should authorise in case of the rejection of Mr. Slidell, then in Mexico. At this time came forth rumours that Santa Anna was more favourable to peace, that is, more willing to accommodate on the annexation difficulty, than Paredes; and on that very day in January, a message was sent to Commodore Conner, of the Gulf squadron, saying that if Santa Anna came that way to let him into Mexico. How came it to be supposed that he was likely to come that way? About the same time, Alexander Slidell, brother to the minister to Mexico, was sent to Cuba by the government. It was a matter of general conversation in Cuba that Santa Anna had been invited to return to Mexico by the United States authorities. The president said Paredes was the obstacle in the way of settling our differences: the inference is plain—strong. On the 11th of May war actually commenced. On the 1st of June a proclamation, prepared at Washington, was sent to General Taylor, to be distributed throughout Mexico as fast as he progressed into the country. This purported to give the causes of the war to the Mexicans. It set forth that we are friends of Mexico; that the war was waged on account of the debts, and the rejection of Mr. Slidell by the then existing government, and a supposed declaration of Paredes, of which I never heard or could ever learn any other authority. The fact is not stated, intimated, or alluded to, that was in the act of war. It was said that we came as friends, and that we desired to keep the Mexican government from falling into the hands of a European prince, a proposition to which effect, it was thrown out, was on foot. Some time in the same month (June) Santa Anna went to Vera Cruz, passed into the interior, issued his proclamation or pronunciamento. This, too, said he came as a friend to put down tyrants and avert the impending monarchy. The ideas and language are strikingly like those set forth through General Taylor. Whether Santa Anna borrowed from Mr. Polk, or Mr. Polk borrowed from Santa Anna, or whether it was an extraordinary coincidence of thought on the part of two great geniuses, I will not attempt to say. Sufficient that they are remarkably similar in language as well as character. General Salas, under Santa Anna's pronunciamento, overthrew Paredes, and he in turn was banished. Santa Anna became the head of the nation. The sequel all know. With equal mortification and regret did I read, in the last message of the president the admission that his object in assisting Santa Anna to return to Mexico was to overthrow the existing government." The history of this war is recent and well known: the arms of the United States were everywhere victorious, and when peace returned, not T. alone, but New Mexico and Upper California, were parcelled off to the Americans, and for ever lost to the Mexican republic. The resolutions of Congress admitting T. into the Union, were signed by the president of the United States, 1st March, 1845, and ratified by the Congress of T. on the 4th of July, 1845. The boundary between New Mexico and T. was adjusted in 1850, as follows: The state of T. agrees that her boundary on the N shall commence at the point at which the meridian of 100° W from Greenwich is intersected by the parallel of 36° 30′ N lat., and shall run from said point due W to the meridian of 103°; thence her boundary shall run due S to the 32° N lat.; thence on the said parallel of 32° N lat. to the Rio Bravo-del-Norte; and thence with the channel of said river to the gulf of Mexico. The state of T. cedes to the United States all her territory exterior to her limits and boundaries, which she agrees to establish by the first article of this agreement; and relinquishes all claim upon the United States for liability of the debts of T., and for compensation or indemnity for the surrender to the United States of her ships, forts, arsenals, custom-houses, custom-house revenue, arms, and munitions of war, and public buildings, with their sites, which became the property of the United States at the time of the annexation. The United States, in consideration of said reduction of boundaries, cession of territory, and relinquishment of claims, pay to the state of T. the sum of 10,000,000 of dollars, in a stock bearing 5 per cent. interest, and redeemable at the end of 14 years, the interest payable half-yearly at the treasury of the United States.

TEXAS, a county in the S part of the state of Missouri, U. S., comprising an area of 1,247 sq. m., generally undulating, and drained by the head branches of Current river and the Big Piney Fork of Gasconade river. It produces wheat, maize, and tobacco, and abounds with wood. Pop. in 1850, 2,312. Its capital is Houston.—Also a township of Wayne co., in the state of Pennsylvania, drained by Saxawaxen creek. Pop. in 1840, 1,094.—Also a township of Kalamazu co., in the state of Michigan. Pop. 294.

TEXEIRA, or Teixeira, a mountain of Brazil, in the prov. of Parahiba. It forms a ramification of the Borborema chain.

TEXEL, an island of the North sea, off the N coast of Holland, in the prov. of North Holland, and arrond. of Alkmaer, of which it forms a canton, at the entrance to the Zuider-zee, 2 m. SSW of the island of Vlieland, and 2 m. N of Helda. Pop. 9,000. It is 14 m. in length from NNE to SSW, and 8 m. in extreme breadth, and of little elevation, but is defended from irruptions of the sea, on the W, by extensive downs, and on the E, which is generally marshy, has a good port, in which vessels often take refuge during their detention by unfavourable winds. The currents and sand banks, however, by which it is surrounded, render approach to its shores at all times difficult. The Texel-stroom, or W inlet to the Zuider-zee, lies between the mainland and the S end of T. island. Kykduin lighthouse, on the S side of the Texel-stroom, is in N lat. 52° 57′ 5″, E long. 4° 43′ 37″. The island affords pasturage to large herds of cattle, and produces in considerable quantities butter, cheese, and wool. Little more than an immense sandbank, it is, however, noted for the richness and fineness of the wool produced from the sheep that are grazed upon it. The shepherds who tend these sheep are original and unique specimens of the pastoral character: it is impossible to conceive any thing more primitive than the language, manners, and customs of those poor people. Here, too, on the mainland, and opposite the island, is to be found by far the most remarkable of the 'dykes' now existing in Holland. The sea is here deeper, and the danger of incursion greater than elsewhere, though the tempests are not so severe: accordingly, the dyke is composed of a stronger substance than the straw-knit materials with which the embankments enclosing the *polders* are formed.' It consists of enormous blocks of granite, brought from Norway, and here forming a rampart of 2 leagues in length, and nearly 50 ft. in breadth, and at certain points projecting some hundreds of yards into the sea. Nothing in Holland is more calculated to give one a high idea of the perseverance of the people than this dyke. Large quantities of sea-birds' eggs are found on its N coast, whence its name of Egerland. It contains several villages, of which the principal is also named Texel. The culture of the soil, boat-building, fishing, and the manufacture of woollen and linen fabrics form the chief branches of local industry. A violent storm in 1825 submerged a large portion of this island, and swept great numbers of cattle into the sea. It is noted for the naval engagement which took place near its shores on the 8th August 1653, between the English and Dutch, in which the famous admiral Tromp lost his life; and it was also in the vicinity of this island, that in January 1794, a detachment of French cavalry took possession of an ice-locked Dutch fleet.

TEXERINA, a town of Spain, in the prov. and 33 m. NE of Leon, near one of the sources of the Cea. Pop. 117.

TEXUTLA, or TEXTLA, a town of Guatimala, 240 m. NW of the town of that name. Pop. 2,000. It has a manufactory of tobacco.

TEY (GREAT), a parish in Essex, 3¾ m. NE of Great Coggeshall. Area 2,503 acres. Pop. 735.

TEY (LITTLE), a parish in Essex, 3 m. E by N of Great Coggeshall, in the line of the Eastern Counties railway. Area 486 acres. Pop. in 1851, 74.

TEYA, a town of Spain, in the prov. and 11 m. NE of Barcelona, near the Mediterranean. Pop. 3,637, engaged chiefly in fishing.

TEYDE (PEAK of). See TENERIFFE.

TEY-MARKS, or TEY-AT-THE ELMS, a parish in Essex, 4 m. E by N of Great Coggeshall, intersected by the Eastern Counties railway. Area 1,214 acres. Pop. in 1831, 363; in 1851, 473.

TEYNGA, or TEYNIA, a small island of the Mindoro sea, near the SW coast of Mindanao, in N lat. 6° 50′, and E long. 121° 42′.

TEYNHAM, a parish in Kent, 4 m. WNW of Faversham. Area 2,648 acres. Pop. in 1851, 842.

TEYSING. See THEYSING.

TEYSSIEN, a village of France, in the dep. of the Lot, cant. and 6 m. E of Bretenoux, at the foot of a high hill. Pop. 700.

TEYUPURU, or TAGIPOM, an arm of the Amazon in Brazil, in the prov. and district of Para, which detaches itself from that river on the r., 50 m. ENE of Gurupu; runs first S, then NE, receives the Annapu, Pacajas and Jacundas, and with the Rio-Tocantins, which it joins in W long. 49°, forms the island of Joannes or Maraja. Its bed is obstructed by islands, and is to a great extent unnavigable.

TEZA, a town of Marocco, in the prov. and 90 m. E of Fez, at the foot of the Atlas mountains, on a rock, the base of which is washed by a small stream. It is tolerably well-built, and has a fine mosque, and the environs are adorned with numerous gardens.

TEZA, TESA, or TESSE, a river of Russia in Europe, which has its source in the gov. of Kostroma, a little to the SSW of Ples; runs S into the gov. of Vladimir; passes Chonia; bends SE; bathes the districts of Kovrov and Viazniki, and after a tortuous course of about 120 m., joins the Kliazma on the l. bank, 23 m. NNW of Viazniki. It was formerly navigable, but a great portion of the volume of its waters is now employed by mills.

TEZANILLOS, a town of Spain, in the prov. of Santander, and partido of Villacarriedo. Pop. 560.

TEZCUCO, a town of Mexico, in the intendancy of Mexico, situated on the E bank of a lake of the same name, in N lat. 19° 31′, 15 m. ENE of the city of Mexico. When Cortez invaded Mexico, it was the capital of a district rich and populous; it now exhibits numerous remains of Mexican antiquity. The modern town has some handsome edifices, and a pop. of 5,000. It formerly possessed considerable cotton manufactories, which have suffered, however, by the competition of those of Queretaro.—The lake of T. is the largest, and the lowest in elevation, of the five lakes in the valley of Mexico. It is not easy to determine the exact boundaries of this lake, the ground around it being so uniform in level, that the difference of level is not more than 7½ in. in a mile, and in consequence, when the E winds blow with any violence, the water withdraws towards the W bank of the lake, and sometimes leaves an extent of more than a ½ m. dry on the opposite side. It is not in general more than from 9 to 16 ft. deep; in some parts only 2 ft. Hence the commerce of the small town of T., situated on its banks, suffers much in the dry months of January and February. The water is more brackish than that of the four other lakes in its vicinity. Its extent is about 10 sq. leagues, but it has been gradually diminishing since the conquest of the country by the Europeans. This is ascribed by Humboldt to the thinning of the woods by the Spaniards, who cut down the trees both of the plain and the mountains, and thus exposed the country, without any protection, to the sun's rays: the dry soil imbibes a greater quantity of moisture than before, and a smaller portion remains for the supply of the lakes. The waters of Lake T. are also drawn off by means of a *desague* or drain begun in 1607, in the form of a subterranean tunnel, through the surrounding mountains. This drain has not only reduced within narrower limits the two lakes in the northern part of the valley, viz. the lakes of Zumpango and San Christoval, but it has also prevented their waters in the rainy season from flowing into the lake of T. The bed of the lake is also progressively rising, in consequence of the loose earth carried down by the torrents and deposited in it. If the report of a survey made in the 16th cent. be correct, the great square of Mexico was then elevated 43 ft. above the level of the lake of T., but the latter is now only 47 inches lower than the former. The lakes have therefore decreased both in extent and depth; and the city of Mexico, which in the days of Cortez stood environed by water, is now 3 m. from the lake of T., which receives much less water by infiltration than in the 16th cent.

TEZZUTE, an assemblage of magnificent ruins, in the S part of the prov. of Constantina, in Algiers, in N lat. 35° 24′, E long. 5° 58′. *Lambese* being mentioned in 'the Itinerary' as the most considerable city in this quarter, and that name being found on one of the inscriptions, leaves no doubt of its being the original of these noble ruins.

THAGARA, a town and fortress of Hindostan, in the prov. of Aurungabad. The fort is on the summit of a hill, and is surrounded by several stone walls, the lowest of which encloses the town. It is a place of great strength, but has notwithstanding been repeatedly taken by different bodies of Mahommedans. At the beginning of the 14th century, the Afghan emperor Mahammed III., endeavoured to make it his capital, and gave it its present name. It is now comprised in the territory of the Nizam, but since the foundation of Aurungabad, it in common with many other fortified eminences in India, has declined in importance. It is still however considered the key to the Deccan.

THAHO. See Tajo.

THAIF. See Tayef.

THAILFINGEN (Neckar), a town of Würtemberg, in the circle of the Schwarzwald, bail. and 9 m. E of Bahlingen, on the Schmich. Pop. in 1840, 1,129. Hosiery is manufactured here.

THAI-NGUYEN, or Xu-thai, a prov. of Annam, in Tonquin, on the confines of China.

THAINI, a town of Tunis, 9 m. SSW of Sfax, on a small stream of the same name, near the sea, in a dry, rocky locality. In its vicinity are the ruins of *Thoena*.

THAI-PING-SHAN, Tafusan, or Typinsan, the largest and most easterly of the Madjicosima islands in the Eastern sea. On its N coast is the temple of Miako. It is surrounded by numerous reefs.

THAKEHAM, a parish in Sussex, 6½ m. NW of Steyning. Area 2,980 acres. Pop. in 1851, 631.

THAL, a village of Switzerland, in the cant. and 11 m. ENE of St. Gall, and district of Rheinthal. Pop. 200.—Also a village of the duchy of Saxe-Coburg-Gotha, to the W of Waltershausen. Pop. 295. It contains the ruins of the castle of Scharfenberg.

THALABA, a parish of New South Wales, in the co. of Gloucester, bounded on the W by William's river.—Also a town in the co. of Durham, at the confluence of the Williams and Wangi, or Thalaba rivers.

THALE, or Thaale, a village of Prussia, in the prov. of Saxony, regency and 39 m. SW of Magdeburg, and circle of Aschersleben, at the foot of the Harz mountains, on the l. bank of the Bode. Pop. 1,250. It has a large iron-work.

THALERN, a village of the archduchy of Austria, in the prov. of Lower Austria, 13 m. NE of St. Polten, in the midst of hills. It has mines of coal, and manufactures of alum, sulphur, and vitriol.

THALFANG, a village of Prussia, in the prov. of the Rhine, regency and 17 m. E of Treves, and circle of Bernkastel. Pop. 315. It has extensive iron-mines.

THALFINGEN, a village of Bavaria, in the circle of Swabia. Pop. 400.—Also a village of Würtemberg, to the NE of Ulm. It has mineral baths.

THALGAN. See Talgan.

THALGUT, a village of Switzerland, in the cant. and 8 m. SE of Bern, bail. of Seftegen, near the l. bank of the Aar.

THALHEIM, a village of Switzerland, in the cant. of Argau, and circle of Beltheim. Pop. 840. It is noted for its wine.—Also a village of Würtemberg, in the circle of the Schwarzwald, bail. and 8 m. WNW of Tuttlingen, on a small affluent of the Danube. Pop. 1,125.—Also a village to the SW of Pfullingen. Pop. 1,050.—Also a village in the circle of the Neckar, and bail. of Heilbronn, on the Schotzach. Pop. 1,330. Also a town in the circle of the Jaxt, bail. and 8 m. E of Hale, on a small affluent of the Bühler. Pop. 300.

THALITTER, a village of the grand-duchy of Hesse-Darmstadt, and prov. of Upper Hesse, on the Itter, 32 m. WSW of Cassel. Pop. 260. It has a copper-mine, and manufactories of copper and tin.

THALMASSING, a town of Bavaria, in the circle of the Upper Pfalz, 9 m. SSE of Hilpoltstein, on the Thallach. Pop. 850. It has three churches.

THALWEIL, a fine village of Switzerland, in the cant. and 8 m. S of Zurich, and bail. of Wadenschweil, near the W bank of the lake of Zurich, on a height commanding a fine view of the lake and its environs. Pop. of com. 1,150. It has manufactories of cotton fabrics. The vine is cultivated in the environs.

THALWITZ, a village of Saxony, in the circle and 15 m. NE of Leipzig, and bail. of Wurtzen, on the Lossa.

THAMAN. See Taman.

THAME, a river of Oxfordshire, rising on the E side of Buckinghamshire, and uniting with the Thames at Dorchester: see Thames.

THAME, or Tame, a parish and market-town in Oxfordshire, 13 m. E by S of Oxford. Area of p., 5,310 acres. Pop. in 1801, 2,293; in 1831, 2,885; in 1851, 3,259. The church is a large and handsome structure in the early English style. The town, which is of Roman origin, stands on the S bank of the river Thame, which is navigable from this place. It consists principally of one long and spacious street, in the centre of which are a market-place and town-hall. The lace-manufacture is carried on to a limited extent in the town.

THAMES, the chief river in Britain, and, in a commercial view, the most important in the world, formed by the junction of the Thame and the Isis near Dorchester in Oxfordshire. The Thame rises from three sources, two of which are near Ivinghoe in Bucks, the other near Tring in Herts. The two principal sources unite to the W of Aylesbury, and the Thame thence runs SW by the town of Thame, where it becomes navigable, to Dorchester. The Isis is the principal source of the Thames, or Thame-Isis, as it was originally named. The principal head of the Isis, or Thames-head, as it is also called, is near Cirencester in Gloucestershire. Thence from an elevation of about 376 ft. above the sea, it runs southwards into Wilts; but afterwards divides these counties, receiving the Churnet from the N of Cirencester, and proceeding eastwardly by Cricklade, above which it is joined by the Swill from the N; and by Lechlade, above which it receives the Colne from the N, the Cole from the S, and the Thames and Severn canal from the W, and becomes navigable. From Lechlade, where its bed is 242 ft. above sea-level, it flows E by N, dividing Oxfordshire from Berkshire, and receiving the waters of the Wainrush and the united waters of the Evenlode and Glynn from the NW; after which it turns SE along the border of Oxfordshire, by Oxford, where it receives the Oxford and Warwick canal, and the united streams of the Charwell and the Bay. Proceeding still south-eastward it deviates by Abingdon, where it receives the Ock, and then accomplishes its junction with the Thame a little below Dorchester. From Dorchester the Thames continues to divide Oxfordshire from Berks, flowing south-eastward by Wallingford and Reading, at the latter of which places it receives the navigable river Kennet, and then turns NNE by Henley, being previously joined by the Loddon. Quitting Oxfordshire, it then divides Berks from Bucks, flowing first E by Marlow, and afterwards SE in a winding course by Windsor to Staines, where it is joined by the Colne from the N, and quits Berks and Bucks to divide Surrey from Middlesex. Still winding in its course, it runs SE by Chertsey, below which it is joined by the Wey from the S; then E by Kingston, above which it is joined by the Mole; and N by Teddington, where it becomes tidal, and by Richmond, to Brentford, where it is joined by the Grand Junction canal with the waters of the Brent. From Brentford it winds, in a more easterly course, by Chelsea and Westminster, through the metropolis, dividing Lambeth from Westminster, and Southwark from London; till, winding round the Isle of Dogs,—on the W side of which it is joined by the Regent's canal, and on the E by the Lea navigation, both on the Middlesex side,—it passes Greenwich, Woolwich, and Gravesend, dividing Kent from Essex, till after a winding course of 215 m., or of 124 m. in direct distance, it opens into its noble estuary at the Nore, previous to which it is joined by the Roding, and other Essex streams on the N, and by the Darent, the Medway, and other Kent streams on the S. The Medway issues into it at Sheerness in the Isle of Sheppey, which the Medway forms, immediately above the expansion of the Thames into its estuary at the Nore.

The Thames, with its tributaries, drains about 6,000 sq. m., with a mean velocity of 2 m. an hour. The fall of the main stream from Oxford to Maidenhead is about 25 ft. in every 10 m.; from Maidenhead to Chertsey-bridge, 22 ft. in every 10 m.; from Chertsey-bridge to Mortlake, 16 ft. in every 10 m.: afterwards the fall diminishes more gradually till the river unites with the sea. At London-bridge, 146½ m. below Lechlade, its height above sea-level is only 4 ft. 3 in., and its fall from this point to the Nore does not exceed 1 inch per mile. Its depth from London-bridge to Greenwich, in the fair way, is from 12 to 13 ft. The mean depth of high water at the entrance of the London docks is 21·3 ft., at spring-tides 22·8 ft. The tides flow up the channel of the T. nearly 80 m.; and occur twice in 24 hours. Occasionally they rise to so great a height as to overflow the banks of the river, and do considerable damage, especially to the metropolis, a grievance of recent as well as of ancient, though by no means of frequent, occurrence, and which has been attributed to such a combination of untoward circumstances, as the swelling by long-continued rains of the numerous rivers flowing into the T., simultaneously with the beating up of the tides through the swollen river by high winds off the sea.

The T. is navigable for vessels of almost any burthen as far as Deptford, although it has been customary for Indiamen to take out a part of their cargoes in Longreach. To the Pool it is navigable for vessels of 400 tons; to London-bridge for those of 200 tons; and to nearly 130 m. above London-bridge for large barges. And as it then communicates with the Severn, through the Thames and Severn canal, the navigation is in fact continued quite across the country from the E coast to the W. Other immense fields of inland navigation, N and S of its course, are opened up by its various navigable tributaries, and by the important and extensive chains of canal navigation with which it communicates.

The T. traverses some of the richest districts of England; but it is one of those rivers which rather derive their peculiar character from the countries through which they flow, than impose distinctive features on the landscape by the boldness and rapidity of their course. Even in the most hilly districts, where its current is naturally strongest, the T. can at no point be properly called a rapid stream; yet it is by no means a sluggish river, and is throughout distinguishable for the majestic progress of its pure and silvery stream, which generally fills its verdant banks, and is rarely discoloured with mud except after great floods. To these, independently of its tidal floods, it is also occasionally liable; and their influence has been very extensive as well as of long continuance in the level parts of its tract, making the whole country appear like a sea, and many of the towns and villages on its banks like islands, when viewed from the several eminences commanding the plain. Till it passes Oxford, the Isis is not a very considerable river; but it increases much in dignity and width after it has formed its junction with the Thame and assumed its proper title of the Thames: it also suffers a manifest enlargement from the accession of the Kennet. That part of its course which leads from Wallingford to Reading, Henley, Marlow, and Maidenhead-bridge, is most distinguished by romantic scenery and natural beauty leading through valleys bounded by hills richly clothed with beech woods, and finely embellished by the magnificent seats of many of our highest nobility and gentry. For many miles above the metropolitan vicinity, the river is enlivened with a multitude of small commercial craft, and latterly with river steamers running to and from the metropolis; but here it becomes a great and noble tide river, full of vessels of every description in the bold sweeps which it continues to make through increasing marshes to the sea, preserving to the end that air of placid dignity and imposing consequence which so eminently distinguish this monarch of the British rivers. Its breadth at London-bridge is 290 yds. The aspect of the T. in its course through the metropolis, to which it has certainly been the chief source of wealth and greatness, is more particularly noticed under article LONDON: which see. The different bends of the river between London-bridge and the sea are denominated *reaches*, and are 18 in number. As a fishing stream, the T. is much used, both commercially, and by the citizens of the metropolis for amusement. Besides the fish usually found in large rivers, there is one species which is not obtained in many of them, namely, the lamprey; which is supposed to ascend rivers from the sea for the purpose of spawning, as they are only found in the summer months. These fish are highly appreciated by epicures, but produce a better price, it is alleged, when sold to the Dutch fishermen to be used as bait in cod-fishing: they are exported in tanks of fresh-water during the winter. See article LONDON.

THAMES, a river of the state of Connecticut, U. S., formed by the junction of the Shetucket and Yantic rivers, which unite at Norwich; and after a course in a generally S direction of 14 m., fall into Long-Island sound, at New London. Its embouchure forms one of the best harbours in the United States.

THAMES, or LA FRANCHE, a river of Upper Canada, formed by the confluence of two principal streams, distinguished as the N and E branches, the former of which has its source in Huron district, and the latter in Easthope, and which unite at London; thence the united stream flows SW, and after a sinuous course of 115 m. in a generally SW direction, throws itself by a wide embouchure into Lake St. Clair. Its waters abound with fish, especially trout, and on its banks are numerous mills. It is navigable for steam-boats to Louisville, a distance of 30 m. from its mouth.

THAMES-DITTON. See DITTON-THAMES.

THAMEVILLE, a settlement of Upper Canada, in Camden West, on the Thames, 15 m. E of Chatham. Pop. 50.

THAMISE. See TAMISE.

THAMSBRUCH, or THOMASBRUCKEN, a town of Prussia, in the prov. of Saxony, regency and 21 m. NW of Erfurt, and circle of Langensalza, on the l. bank of the Unstrut. Pop. 1,000. It has manufactories of linen, and has several spinning-mills.

THANA, a town of Hindostan, in the Punjab, on the road from Lahore to Kashmir through the Pir-Panjal pass, at an alt. of 5,000 ft. above sea-level. It is finely situated amid plantations of walnut and mulberry trees, on a height which rises precipitously above the Rajawur river, and is intersected by numerous small streams. It has a large serai. Its inhabitants, about 400 in number, find their chief employment in the manufacture of shawls.

THANASIR, a town of Hindostan, in the prov.

and 93 m. NNW of Delhi. It is of considerable extent, and, on account of the famous temple of Juggsum which it contains, is held in high veneration by the Hindus.

THANATO, a town of Turkey in Europe, in Thessaly, in the district and 6 m. NNE of Hagia, on a mountain, at some distance from the Archipelago. Pop. 1,500. The environs are noted for their wines, and produce also silk, cotton, and grain.

THANET (ISLE OF), a district of Kent, forming the NE angle of the county, from the remainder of which it is isolated, on the S, by the river Stour, and on the W by the water-course called the Nethergong, while in every other direction it is surrounded by the sea. In ancient times it was divided by a wide estuary, capable of receiving vessels of considerable size, and was actually used as a short cut in the ordinary passage between France and London. The length of the isle is 9 m. from E to W; its breadth, 5 m. from N to S. A balk or strip of unploughed land, named St. Mildred's lynch, divides it into two manors—Minster and Monkton. The soil is fertile, and in a high state of cultivation. The whole isle comprises nearly 23,000 acres of arable land, and 3,500 acres of good pasture and marsh land. At North Foreland, its extreme E point, there is a lighthouse. Between this point and Margate, near Foreness-point, a harbour-of-refuge has been proposed. The Reculvers forms the limit of the Isle of Thanet on the W side. The popular watering-places of MARGATE, RAMSGATE, and BROADSTAIRS, [which see,] are on this island.

THANINGTON, a parish in Kent, 1¾ m. SW by W of Canterbury, on the river Stour. Area 1,212 acres. Pop. in 1831, 316; in 1851, 385.

THANN, a canton, commune, and town of France, in the dep. of the Haut-Rhin and arrond. of Belfort. The cant. comprises 12 coms. Pop. in 1831, 13,731; in 1846, 18,177.—The town is 17 m. NE of Belfort, on the r. bank of the Thurr, near the foot of the Vosges, at the entrance to the valley of St. Amarin, and at an alt. of 364 yds. above sea-level. Pop. in 1846, 5,872. Its principal objects of interest are the church of St. Theobald, a fine Gothic structure, founded in 1430 and finished in 1516, the tower of which, constructed on the model of the spire of Strasburg cathedral, is 300 ft. in height; the castle of Enguelbert, the ruins of which crown the summit of an adjacent mountain; and the promenade of Bonguert. It has several spinning-mills, manufactories of cotton, woollen, and silk fabrics, printed muslins, brocades, hats, and other articles in felt, combs, chemical substances, and starch, several dye-works, tanneries, breweries, &c. The environs are noted for their wine. This town, founded in the 13th century, originally belonged to the house of Austria. It was taken by the Swedes in 1632.—Also a town of Bavaria, in the circle of Lower Bavaria, 9 m. SE of Eggenfelden. Pop. 900. It has an hospital and manufactories of cloth.

THANNGELSTADT, a village of Saxe-Weimar, on the Schwarza, to the SE of Tanrode. Pop. 250.

THANNHAUSEN, a town of Bavaria, capital of a seignory of the same name, in the circle of Swabia, on the Mindel, 1½ m. NE of Ursberg. Pop. 1,500.

THANNHEIM, a village of the grand-duchy of Baden, in the circle of the See, to the SW of Villingen. Pop. 680.—Also a village of Würtemberg, in the circle of the Danube, bail. and 14 m. NNE of Leutkirch, near the l. bank of the Iller. Pop. 500.

THARAND, or GRANATEN, a town of Saxony, in the circle of the Erzgebirge and bail. of Grillenberg, 8 m. SW of Dresden, to the N of a forest of the same name, on the Shlozbach, and at an alt. of 646 ft. above sea-level. Pop. in 1843, 1,733. It has a celebrated agricultural school, library, museum, mineral baths, tanneries, and manufactories of linen. In the vicinity are the ruins of an ancient fortress of the same name.

THARSTON, a parish in Norfolk, 9½ m. S by W of Norwich. Area 1,582 acres. Pop. in 1851, 374.

THASSO, or TASSO, a beautiful island of the Archipelago, near the SE coast of Turkey in Europe, E of the gulf of Contessa, and separated from the continent by a channel 5 m. in width, in N lat. 40° 42′ 2″, E long. 24° 42′ 30″. It is in form an irregular oval, 19 m. in length from N to S, and 15 m. in breadth, and is covered with mountains clothed with wood, and giving rise to numerous streams. The soil, which is extremely fertile, produces corn, fruit, oil, and wine; and with wax and honey, forming important objects of commerce. It has quarries of marble, and is said to have formerly contained silver mines. It contains a small town, named Volgaro, and three villages, one of which bears the name of the island, and has a good port. The inhabitants are chiefly Greeks. This island was known to the ancients under the appellations, *Thasos*, *Æthria*, and *Aëria*.

THATCHAM, a parish in Berks, 3 m. E by N of Newbury, intersected by the river Kennet, and the Kennet and Avon canal. Area, including the chapelries of Greenham and Midgham, 12,259 acres. Pop. in 1831, 3,912; in 1851, 4,293.

THAU (ETANG DE), a lagune of France, in the dep. of the Herault, separated from the Mediterranean by a narrow tongue of land, 14 m. in length from SW to NE, and 5 m. in extreme breadth. It receives on the SW the Canal-du-Midi, and communicates on the NE with the Etang-de-Maguelonne. The Canal-de-Cette connects it with the sea. Although salt on the surface, its waters at a certain depth are fresh and salubrious. At the SW extremity is a species of natural jet-d'eau, which is in constant action, and throws the water in large volumes to the height of several feet. This lagune forms an important medium of communication between localities on the coast. Its waters abound with fish, and in summer its surface is literally covered with water-fowl. The principal places on its banks are Marseillan, Mire, and Balaruc.

THAUBATE. See TAUBATE.

THAUHA. See JOAM-DO-PRINCIPE.

THAUMACO, a town of Turkey in Europe, in Thessaly, 19 m. NW of Zeitun, on a mountain.

THAUMIER, a village of France, in the dep. of the Cher, cant. and 7 m. N of Charenton, near the Auron, in the midst of lagunes. Pop. 1,045.

THAUM-PE, a district of the Birmese empire, about 25 or 30 days NNE of Tongho, on the borders of Siam, which, when conquered by the Birmans, received from them the appellation of Tongsu. Its chief town, bearing the name of the district, is situated about 40 m. from the hills, in 19° N lat. The prov. is governed by a Birman chief, who resides at the capital, which is stockaded, and contains about 5,000 inhabitants. The face of the country is flat, and tolerably clear. Rice is cultivated to an extent sufficient for the consumption of the district. The inhabitants possess numerous herds of cattle, and a considerable number of small horses; a few buffaloes are employed in agriculture. Several kinds of cotton—one of which appears to be the brown or nankeen cotton—are grown; the tea-plant is also cultivated, and the leaves are pickled. Stick-lac is brought for sale in considerable quantities to Penang from this district; and the silk-worm is reared, being fed on the leaf of a plant called puja. Gold is found in the sands of the mountain-streams. Iron is abundant, and is smelted and wrought into swords, knives, and other implements. Tin, after disappearing to the north of Tavoy, again presents itself here in considerable quantities, in the beds of rivers, in the form of a fine black sand. The most productive mines, however, are those of lead. The ore is obtained in lumps, but in what state of combination we are not informed. From these sources the annual exports to Rangoon are estimated at 120,000 rupees, and might, no doubt, be much extended. The natives carry back from Rangoon and other ports, salt, areca-nuts, salt fish, broad cloth, woollens, piece goods, crockery, and spices. A commercial intercourse is also maintained between T. and China. Traders from the frontier districts of the latter country bring spices, including the clove and nutmeg, silk, cloth, woollens, paints, papers, cutlery, and other articles, and take back the products of the country. They come annually in a caravan, consisting sometimes of 1,000 persons well-armed; the merchandise is transported by asses and horses.—The Plau, as the inhabitants of this district are called, are a distinct people from both the Siamese and Birmans, and also from the neighbouring tribes, in language, features, and character. They are shorter and less robust than the Birmans, and bear a greater resemblance to the Chinese than to any other people. Their dress partakes also of the Chinese costume. They wear their hair twisted into a knot like the Birmans, and are tattooed like those people and the Laos; like the former also they thrust small cylinders of wood or silver through holes made in the lobes of their ears. Their clothes are usually quilted, which they say is necessary owing to the frigidity of their climate. They are a lively, simple race, addicted to agricultural and commercial pursuits, and have readily been reduced to subjection by the Birmans and Peguers, for whom they, nevertheless, entertain a profound contempt, and from whose rule, whenever it becomes very irksome or oppressive, they withdraw into the thick forests, and the mountains in their vicinity. The Plau profess the faith of Buddha, and, like all Buddhists, burn their dead.

THAXTED, a parish in Essex, 17 m. NNW of Chelmsford, on the high road to Cambridge. Area 6,219 acres. Pop. in 1831, 2,293; in 1851, 2,556. The church, erected in the time of Edward II., presents one of the finest specimens of architecture in the co. The town, which is of great antiquity, is situated on the river Chelmer, near its source. The cutlery business, once carried on here to a great extent, has declined.

THAYA, Theya, or Taya, a river of Austria, formed by the junction of the Deutsche Thaya, and Mohrische or Bömische Thaya, the former of which has its source 12 m. ENE of Zwetter, and waters Altensteig, Windischsteig, Waidhofen, Thaya, Dobersberg, and Carlstein; and the latter in the mountains of the Moravian circle of Iglau, and bathes Teltsch and Daschitz, which unite in the archduchy of Austria, a little to the W of Raps. Thence the united stream flows NE past Drosendorf; enters Moravia, and forms to a considerable extent the line of separation between the circle of Znaym and the archduchy of Austria; it then takes first a SE, and afterwards a NE direction; waters Wottau, Frain, Hardeck, Znayn, and Laa; enters the circle of Brunn; passes Dürnholz and Muschan, where it is joined on the l. by the Iglawa, with its tributary the Schwarza; it then turns SE; and after a total course of about 123 m., throws itself into the Morawa, on the r. bank, 2 m. ENE of Rabensburg. This river, which is remarkable for the number of its fords and bridges, is frequently confined within lofty hills, while at other parts it divides into numerous branches, forming woody islands. Throughout the greater part of its course, its banks, from their low marshy character, are extremely insalubrious. At the junction of the Iglawa, it has a width of 100 yds., and a depth of upwards of 20 ft. It was crossed near Laa by the French after the battle of Wagram.—Also a town in the upper circle of the Manhartsberg, 3 m. N of Waidhofen, on the bank of the Deutsche Thaya,

THAYINGEN, or Tayngen, a town of Switzerland, in the cant. and 5 m. NE of Schaffhausen, near the confines of the grand-duchy of Baden. Pop. 1,252. The environs are fertile, and produce corn, hemp, fruit, and wine.

THAZA, a town of Algeria, in the prov. and 90 m. SW of Algiers, near the source of the Oued Dourdour.

THEAKI. See Ithaca.

THEBERTON, a parish in Suffolk, 4 m. NE of Saxmundham. Area 1,954 acres. Pop. 591.

THEBES, a prov. of Upper Egypt, extending along both banks of the Nile; bounded on the N by the prov. of Girgeh, and on the S by Nubia. It is 180 m. in length, and comprises an area of 249 sq. m. Pop. about 90,000. Its chief towns are Gheneh, its capital, Coptos, Ghous, Esneh, Edfu, and Assuan. It is noted for the ruins of the ancient city of the same name, as well as those of Denderah, and other places of less importance.—The ruins of Thebes are situated 30 m. NNE of Esneh, and 300 m. SSE of Cairo, in a valley formed by the Arabian and Libyan mountains. They occupy a space on both banks of the Nile of about 9 m. in length, and 8 m. in breadth, and the circumference of the ancient city must have been about 30 m. The principal ruins are on the E side of the Nile, and of these the most remarkable is the temple of Karnak. It is oblong in form, of great extent, with a double colonnade at either extremity, and covered over the entire surface of its walls and columns with hieroglyphics. See Karnak. On the same side is the temple of Memnon, and at Luxor is a palace with two fine obelisks of single blocks of red granite. See Luxor. On the W bank of the Nile are also numerous ruins and avenues lined with rows of sphynxes. Leading from the valley of Thebes to the mountains is the pass of Biban-el-Melek, and at its extremity are the famous caverns supposed to have been the tombs of the Egyptian kings of the 18th and 19th dynasties. The pacha is now making extensive excavations at T.

THEBES, Thiva, or Thibai, a town of Greece, capital of a district of the same name, 30 m. NW of Athens. Pop. 5,000. It is a poorly-built place, and presents few traces of its ancient magnificence, having been quite ruined in the war of independence. In August 1853, it suffered severely from a series of earthquakes. The locality affected by the shocks was the extent of country terminated on the N by Mount Atlante, on the E by the gulf of Eubœa, on the S by the river Asopus, and on the W by Lake Copais. The central point of the impulse, or that at which it was felt with greatest force, was Mount Ptoo; under which is a gulf of Lake Copais.

THEBO, a town of Burmah, in the territory of Cochan-pu, 114 m. E of Ava.

THEDINGHAUSEN, a town of the duchy of Brunswick, capital of a bail. enclosed within the kingdom of Hanover, and bounded on the N by the Weser. The town is about 2 m. from the l. bank of the Weser, and 5 m. S of Achim. Pop. 1,680. It has manufactories of linen, and tanneries, and carries on a considerable trade in linen-yarn.

THEDIRAC, a village of France, in the dep. of the Lot, cant. and 6 m. SSE of Salviac, at the foot of a high hill. Pop. 820.

THEGONEC (Saint), a canton and commune of France, in the dep. of Finistere and arrond. of Morlaix. The cant. comprises 4 coms. Pop. in 1831, 12,190; in 1846, 12,754. The village is 7 m. SW of Morlaix, on a rising ground. Pop. in 1846, 3,962. It has manufactories of fine linen.

THEIL (Le), a commune of France, in the dep. of the Manche, cant. and 5 m. SW of St. Pierre-Eglise, on a hill. Pop. 1,146.—Also a canton and commune in the dep. of the Orne and arrond. of Mortagne. The cant. comprises 10 coms. Pop. in 1831, 12,038; in 1846, 11,629. The town is 26 m. SSE of Mortagne, near the r. bank of the Huisne. Pop. 870.

THEIL (Le Gros), a commune of France, in the dep. of the Eure, 15 m. W of Louviers, in the cant. of Amfreville-la-Campagne. Pop. 1,145. It has several lime-kilns.

THEILLAY, or Tilley-le-Pailleux, a village of France, in the dep. of the Loir-et-Cher, cant. and 8 m. S of Salbris. Pop. 1,390.

THEISS, or Tisza, a large river of Hungary, which rises under the parallel of 48° 3′ N, in the com. of Marmaros, on the NE frontier of the kingdom, in two streams called the Black and the White Theiss, descending from the Carpathians. These soon unite, and after flowing above 100 m. in a W direction, and passing Szigeth, Tecsö, and Ujlak, a little above Nemeny, the united stream receives the Szamos, and turns N, and then W to Tokay, where it receives the Bodrog. It then pursues a S course, and flows into the Danube at Salankemen, below Titul, after a course of above 700 m. measured by the windings, but not exceeding 260 m. in direct distance. It receives a great number of rivers, some of them of large size, such as the Szamos, the Bodrog, the Sajo, the Zagyva, the Korosch, the Maros, the Aranka, and the Bega. It flows through so level a country as to be navigable to within a few miles of its source. All along the T. the soil is of the richest quality; in some places, as in the case of the valley of the Nile, the alluvial soil formed by a river traversing a sandy plain is not of great breadth, but in others the fat land extends to many miles on either side of the river. The present state of the valley of the T. is a double evil. Its pestilential atmosphere is fatal to colonization from the more densely populated and more industrious and civilized parts of the empire, while at the same time it creates a military barrier forming a focus for future rebellion. The desiccation of the T. would therefore be attended with a double benefit. Several millions of acres of the richest wheat land could be created by a systematic lateral canalization of the basin of the river; and in a military point of view, while the Carpathian arc of the military position, forming a rampart to the eastward, would remain, the chord separating the regions beyond the T. from the other western parts of the monarchy would in a great measure disappear. No river in Europe is more abundant in fish. It is the *Tibiscus* of Ptolemy, the *Tibisis* of Herodotus, and the *Pathissus* of Latin writers.

THEIX, a village of France, in the dep. of Morbihan, 6 m. SE of Vannes.

THELBRIDGE, a parish in Devonshire, 7 m. E by S of Chulmleigh. Area 2,240 acres. Pop. 260.

THELNETHAM, a parish in Suffolk, 14 m. NE of Bury St. Edmund's. Area 1,773 acres. Pop. in 1851, 552.

THELO (Saint), a village of France, in the dep. of Cotes-du-Nord, 4 m. SSW of Uzel.

THELVERTON, or Thelton, a parish of Norfolk, 1½ m. N by E of Scole. Area 1,050 acres. Pop. in 1851, 154.

THELWALL, a hamlet of Cheshire, 9½ m. NW by N of Nether Knutsford.

THEMAR, a town of the duchy of Saxe-Coburg, on the r. bank of the Werra, 11 m. SE of Meiningen. Pop. 1,200. It is the chief place of a bailiwic, long the common property of the houses of Saxe-Coburg and Saxe-Gotha.

THEMILTHORPE, a parish of Norfolk, 3 m. WNW of Reepham. Area 652 acres. Pop. 93.

THENELLES, a village of France, in the dep. of Aisne, cant. and 3 m. N of Ribemont. Pop. 950.

THENEZAY, a town of France, in the dep. of Deux-Sevres, 14 m. NE of Parthenay. Pop. 1,500.

THENFORD, a parish of Northamptonshire, 5½ m. NW by W of Brackley. Area 890 acres. Pop. in 1851, 132.

THENINGEN, a village of Baden, on the Elz, 10 m. N by W of Freyburg. Pop. 1,300.

THENNENBRONN, a village of Baden, 7 m. SE of Hornberg, on the Schiltach. Pop. 1,200.

THENON, a town of France, in the dep. of the Dordogne, 6 m. NW of Montignac.

THEODORA, an islet off the N coast of Candia, in N lat. 35° 31′, E long. 23° 54′.

THEODOSIA. See Kaffa.

THEODOSIA (Saint), a small island in the N. Pacific, in N lat. 62° 20′.

THEODUNA, a town of Albania, in the sanj. and 30 m. SSE of Janina.

THEOG, a small village on the range of the Himalaya, immediately S of the Sutledge, 22 m. E by N of Simla, on the road to Mattiana. Barley ripens here in June at an alt. of 8,000 ft. above sea-level.

THEOPOLIS, a missionary village of S. Africa, on the l. bank of the Kasuga, 10 m. SW of Bathurst.

THEOTSAK, a town of Bosnia, in the sanj. and 65 m. ENE of Trawnik, on an affluent of the Drin.

THERAD, a town of Hindostan, in the prov. of Gujerat, and district of Neyer, 60 m. NW of Patan, in N lat. 24° 37′. It contains nearly 3,000 houses, and is surrounded by a wall and dry ditch. The palace and temple are lofty buildings; but the houses are seldom more than one story high. The district attached to this town contains 32 villages; but water is scarce, and the cultivation is scanty.

THERAIN, a river of France, which rises in the dep. of Seine-Inferieure, 6 m. E of Forges, and joins the Oise, on the r. bank, after a SE course of 50 m.

THERALPHENE, a village of Belgium, in S. Brabant, 12 m. WNW of Brussels.

THERESA (Santa), a river of Brazil, which descends from the Cordillera-Grande, in the prov. of Goyaz, and joins the Tocantins 30 m. WSW of Natividad, after a course of 180 m.

THERESIENFELD, a town of Lower Austria, 4 m. N of Neustadt.

THERESIENSTADT, or Marien-Theresienstadt, or Theresianopel, a town of Hungary, in the palatine of Bacs, 27 m. SW of Debreczin. It is an assemblage of villages, consisting of upwards of 3,000 cottages, inhabited by 22,000 inmates, partly of Servian, partly of Rascian descent, and depending for their support chiefly on the extensive townlands, amounting to 340 sq. m. The town is open, but has large barracks. It has a pretty active traffic in cattle, horses, wool, and hides. There are no collective manufacturing establishments, but a number of individuals work at their own homes, chiefly, if not exclusively, for the use of their townsmen. Of these, weavers and tanners are the most numerous; the women make linen, carpets, and aprons, also soap.

THERESIENSTADT, a town of Bohemia, situated in the angle formed by the Elbe and the Egra, at their junction, 31 m. NNW of Prague. It is to be considered rather as a fortress than a town, the civil inhabitants, only 1,000 in number, depending chiefly for subsistence on the garrison.

THERESIOPEL, a town of Hungary, in the com. of Temesh, 20 m. N of Temesvar, inhabited by a Bulgarian colony.

THERFIELD, a parish of Hertfordshire, 2½ m. SW by S of Royston. Area 4,761 acres. Pop. 1,335.

THERMEH, a village of Asiatic Turkey, in the pash. of Sivas, 130 m. W by N of Trebizond, on the l. bank of a river of the same name, which falls into the Black sea, 3½ m. W of Cape Tchaldi, after a N course of 120 m.

THERMIA, or CYTHNOS, an island of European Turkey, in the archipelago of the Cyclades, situated about 40 m. NW of Paros, and 25 m. SSE of Cape Colonna, the S point of Attica, in N lat. 37° 26′. It is 13 m. in length from NNE to SSW, and about 5 m. in breadth. Its surface is less mountainous than that of the surrounding islands, and the cultivated tracts are of considerable fertility, producing silk, cotton, wheat, barley, wine, honey, and wax. Its inhabitants, amounting to 6,000, are Greeks. It takes the name of Thermia from its hot springs, and has a village of the same name.

THERMOPYLÆ, a pass in the NE of Greece, at the N foot of Mount Œta, on the S side of the gulf of Zeitun, near the embouchure of the Hellada or Sperchius, in N lat. 38° 52′. It consists of a narrow passage between high cliffs on one side, and the sea or an impassable marsh formed by alluvial deposits, on the other. It extends on the whole not less than 5 m.; but the breadth of the passable ground, only 50 or 60 paces in the time of the ancient Greeks, is now nearly doubled, from the retiring of the sea and the deposit of alluvial soil. The cliffs overhanging the pass are in general from 400 to 600 ft. in height, and the situation would be admirable for defence, were it the sole entrance into the N of Greece; but there are paths over the mountains which were used subsequently by the Gauls, the Romans, and even by the Huns. "This passage," says Dr. Wordsworth, "was the scene of numerous struggles at various periods of Greek history. It was defended by the Phocians against the Thessalonians; subsequently, by Leonidas and his 300 Spartans, against Persia; again by the Ætolians against Philip, by Antiochus against the Romans, and by the Greeks against Brennus and the Gauls. In the three latter instances, the same manœuvre—namely, the detachment on the parts of the aggressors of a force which, having scaled the heights of Anopæa or Callidromus, was to fall on the rear of the defenders of the pass—was uniformly resorted to, as it had been employed by the Persians, and with the same success. The pass of T. was never stormed by main force: its conqueror, and its only one, has been Nature. So great is the change which has been effected by her means in the character and features of the place, that it has ceased to be an object of military importance. While the river Sperchеius has brought down in its channel a copious supply of alluvial deposit on the coast, the waters of the Malian gulf have retired so far to the NE as to extend what was once a narrow defile of a few yards into a broad and swampy plain. When such a revolution has been wrought in the grander features of this remarkable place—when the rivers which flowed through the pass of T. have formed for themselves new beds—when fields of rice, and salt pits, occupy the place which was once sea, it is agreeable to observe that the smaller objects which were characteristic of the spot in the time of Leonidas, are still visible here, to call to the mind of the traveller that he is treading the soil of Thermopylæ. The hot springs which supplied a name to the place, and which are connected with the history both of Hercules and Leonidas. still flow from the earth and expand their streams into pools of the clearest blue, as they did in the ages of the demigod and of the king, while the broad Spercheius has wandered from its course, and while it is no longer possible to trace upon the spot the ancient coastline of the Malian sea."

THEROUANNE, a town of France, in the dep. of Pas-de-Calais, 7 m. S of St. Omer. It was a fortress so lately as 1553, when it was taken by the Austrians and demolished.

THERY, a village of Hindostan, in the prov. of Malwah, in N lat. 24° 45′.

THESSALONICA. See SALONICHI.

THESSALY, or TRIKALA, a province of European Turkey, extending in an oblong form from N to S; and bounded on the N by Macedonia; on the E by that part of the Ægean which forms the gulf of Salonica; on the S by Greece; and on the W by the high Pindus range. It has an area of 5,500 sq. m.; and while mountainous, presents many fertile plains watered by streams descending from the mountains on the W, and flowing to the Ægean. These plains, equally renowned in ancient times for pasture and tillage, still continue highly fertile, producing maize, wheat, rice, vines, silk, cotton, and tobacco, and the cultivation extends a considerable way along the acclivities of the mountains. The chain of Pindus, traversing the W part of the prov., forms the watershed between the Ionian sea and the Archipelago. The high lying tracts nourish large flocks of sheep, and wool forms a considerable article of export. Of the rivers, the most remarkable is the Salembria or *Peneus*, a full and rapid stream flowing from W to E, and traversing towards its mouth the celebrated vale of Tempe. The Aspropotamos or *Achelous* traverses the W part of the prov. The pop. is about 250,000, of whom scarcely one-fifth are Turks, two-fifths are Albanians, and the remainder Greeks and Vlacks. A recent well-informed writer, to whose pen we are mainly indebted for the only accurate details of the recent insurrectionary movement in Epirus and Thessaly, says: "It is not everybody who knows how much the Greek element in Epirus and Thessaly owes to Ali Pasha of Janina. This extraordinary man, who had conceived the idea of creating an empire for himself, saw from the beginning the necessity of severing the ties which connected the provinces under his sway with the Ottoman empire, by raising the subject populations from their degradation and gaining their support. Being himself an Albanian, he knew well enough how unruly the Albanian spirit is, and how difficult to tame; he crushed it, therefore, mercilessly, while he did his utmost to promote Greek nationality in Epirus and T., and finished by making the Greek the official language, which it remains until this day, in these provinces. This latter measure had an incredible influence on the spreading of the Greek language. It has now become the only written language in the country, and even Albanian chiefs and the Turkish authorities use it among themselves. Nobody except the pashas, who have to correspond with other pashalics or with Stambul, have Turkish secretaries; all the rest employ Greek *grammatikoi*. Since the establishment of the kingdom of Greece the opportunity for strengthening the Greek element in Epirus and T. has much increased, owing to the constant communication which was kept up with Greece, to the great number of Greek subjects established in these provinces for the sake of commerce, and lastly, to the schools in Greece. The Christians of Epirus and T. having but very few and badly organized schools at home, an evil for which their rapacious bishops are more to be blamed than the Turks, send their children, whenever they can afford it, to Greece to study there and learn a handicraft. These youths, when they return, become the most active propagandists of Hellenism, and they had considerable share in preparing their countrymen for the late invasion. In this way the Greek element in Epirus and T. increased daily in strength, and exerted also considerable influence over the Christian populations of other races. The large Slavonic pop. which occupied once a great part of these countries has mostly yielded to this influence; it has been nearly entirely assimilated, and it is only here and there, in more retired spots—for instance, in the mountains of Zagori—that one finds traces of it. The Vlacks and Christian Albanians have escaped this fate, but they could not help being influenced by the movement of the Greek spirit, and being to a certain point carried away by its impulse. These Vlacks, who inhabit the central and upper ranges of the Pindus, form part of the numerous Vlack colonies scattered all over the Turkish empire south of the Danube down into Greece Proper, and designated by the Moldo-Wallacks as *kutzo* or 'lame Vlacks,' a term very likely indicating that they are considered as degenerate scions of the Rumani race. They themselves disclaim all relationship with the Moldo-Wallacks, and pretend to be the descendants of the Roman garrisons and the natives of the countries they inhabit. They are partly nomade and partly settled, the first occupying the highest chain of the Pindus, the second spreading down through the lower ranges towards the plains of T. and Macedonia. Both speak their own language, similar to that spoken in the principalities, although many Vlacks know Greek too. They have likewise preserved their ancient customs, and are, curiously enough, the only people in those parts who have something like a feeling of their nationality. This is so much the more interesting as they form so small a portion of the population (80,000 to 100,000,) and as the Greeks and Albanians of Epirus and T. have not the slightest feeling of that kind. This feeling of nationality is principally strong among the nomade Vlacks. Occupying the most out-of-the-way plateau of the Pindus, little or not at all frequented even by their lowland neighbours, they have formed there a little world of their own. Every village forms a separate (nearly independent) community, paying a small tribute to the Turkish government, but governed by its own *kodjabashis* or 'elders,' according to its old customs. These nomade Vlacks are for the most part employed as shepherds, cattle breeders, and muleteers, and wander about in those qualities over all the Turkish empire. Many of them settle for a while as merchants in towns, but they retain their affection for their native mountains, and return to them nearly always after having made their fortune, in order to pass there the remainder of their lives."

THETFORD, a chapelry in Stretham p., Cambridgeshire, 2¾ m. S by W of Ely. Area 1,630 acres. Pop. in 1831, 257; in 1851, 366.

THETFORD, a borough and market-town, chiefly in the hund. of Shropham, Norfolk, but partly in Suffolk, on the Little Ouse river, which divides the counties, 28½ m. SW by W of Norwich, and 80 m. NNE of London, on the railway from London to Norwich. It comprises three parishes, namely, St. Cuthbert, St. Peter, and St. Mary the Less. Area 8,270 acres. Pop. in 1801, 2,246; in 1831, 3,462 in 1851, 4,075. It is irregularly built and widely spread, the houses being in many parts separated by gardens, small pastures, and other enclosures. The Little Ouse in its course through the town is here joined by the Thet, a small tributary stream, the name of which appears to have been given to the town. The environs and surrounding country are open, and the latter is thinly peopled and divided into immense farms, with sheep-walks and corn-fields. There is a mineral-spring in the neighbourhood of the town, similar to those of Tunbridge and Toplitz, and a suite of baths has been erected. A trade is carried on in barges through the Ouse navigation, by way of Lynn, chiefly in the import of coals, and the export of corn, wool, and other agricultural produce. Coarse woollen cloth, paper, and iron are made, and there are four large breweries. T. at one period had 20 parish churches and 8 monasteries, besides other religious and charitable foundations; in virtue of which it was called in past ages, *Hierapolis* or *Monachopolis*. The borough jail and house-of-correction is of considerable extent. The income of the old corporation, in 1833–4, was £1,054; in 1849, £713. T. has returned 2 members to parliament ever since 1 Edw. VI. The number of electors in 1837, was 155; in 1852, 200.

THEUREN, a village of the duchy of Saxe-Meinengen, bail. and 3 m. NE of Schakau, on the Grümpen. Pop. 135.

THEUX, a department, commune, and town of Belgium, in the prov. of Liege, and arrond. of Verviers. Pop. of dep. 4,472. The town is 6 m. SW of Verviers, in a hollow, on a stream of the same name. Pop. 1,202. It has a fulling and several spinning mills, manufactories of cloth, and iron fineries. It is said that Louis-le-Debonnaire and his son Lothaire had a palace in this locality.

THEVE, a river of France, which issues from a lagune in the NW extremity of the Seine-et-Marne, a little to the N of Dammartin, runs first N; enters the dep. of the Oise; bends, and after a course of about 18 m., joins the Oise on the l. bank, 4 m. ESE of Beaumont-sur-Oise.

THEVENARD, an island of the Pacific, off the NW coast of Australia, in S lat. 21° 26', and E long. 114° 54'.

THEYS, a commune of France, in the prov. of the Isere, cant. and 4 m. SSE of Goncelin, on a river of the same name, a small affluent of the Isere. Pop. in 1846, 2,518. It has quarries of fine marble.

THEYSING, Theusing, or Teysing, a town of Bohemia, in the circle and 30 m. NW of Pilsen, on the Strzela. It has manufactories of vitriol, alum, and cloth.

THEYSS. See Theiss.

THEZAC, a village of France, in the dep. of the Charente Inferieure, cant. and 7 m. NNE of Cozes. Pop. 560. It has a church, now partly in ruins.

THEZAN, a village of France, in the dep. of the Herault, cant. and 2 m. SE of Murviel, on a mountain. Pop. 1,050.

THEZE, a canton and commune of France, in the dep. of the Basses Pyrenees, and arrond. of Pau. The cant. comprises 21 com. Pop. in 1831, 7,818; in 1846, 7,707. The village is 13 m. N of Pau, on a mountain, near the r. bank of the Luy-de-France. Pop. 517.

THIAGO-DE-CACEM (San), a town of Portugal, in the prov. of Alemtejo, comarca and 36 m. NW of Ourique, at the foot of a hill, crowned with a castle, and on the N bank of the Pera lagune, a gulf of the Atlantic, which forms a small port. Pop. 2,050. It has an hospital, an alms-house, and a Latin school.

THIAGUR, or TIAGUR, a fortress of Hindostan, in the prov. of the Carnatic, situated on a mountain, in N lat. 11° 45′, E long. 79° 10′, 52 m. WSW of Pondicherry. It was taken by the French in 1760, and retaken by the British in the following year.

THIAN, a town of the Corea, in the prov. of King-ki-tao, 18 m. NNE of Han-yang.

THIAN-CHAN, a town of the Corea, in the prov. of Hwang-hai, 18 m. SE of the embouchure of the Ya-lu-keang.

THIAN-CHAN-NANLU, or TEEN-CHAN-NANLU. See TURKESTAN.

THIAN-CHAN-PELU, or TEEN-CHAN-PELU. See SUNGARIA.

THIAN-CHANG, or TEEN-CHANG, a district and town of China, in the prov. of Gan-hwuy, and dep. of Sze-chu, in N lat. 32° 41′, and E long. 118° 55′.

THIAN-CHIN, a district and town of China, in the prov. of Shan-se, div. and 54 m. NE of Tae-tung-fu, near the Great wall, in N lat. 40° 28′ 30″, and E long. 114° 3′. It is of considerable extent, and is enclosed by lofty brick walls, but has few inhabitants.—Also a district and town in the prov. of Kwei-chu, and div. of Chin-yuen-fu, in N lat. 26° 45′, and E long. 108° 58′.

THIANETHI, a town of Russia in Asia, in Georgia, in the prov. of Kakhethi, district and 30 m. WNW of Thelavi, on the Jora.

THIANGES, a village of France, in the dep. of the Nievre, cant. and 6 m. NNE of Decize. Pop. 328. It is noted for its coals.

THIAN-HI-CHAN, a mountain of China, in the prov. of Yun-nan, and div. of Chu-ning-fu, on the r. bank of the Lan-sang, in N lat. 23° 50′, and E long. 100° 22′. It is covered with perpetual snow.

THIAN-HO-HEEN, a district and town of China, in the prov. of Kwan-si, div. and 27 m. N of King-yuen-fu, in N lat. 24° 48′, and E long. 108° 20′, on the l. bank of the Lung-keang.

THIAN-KEANG, a district and town of China, in the prov. of Sze-chuen, and div. of Chung-chu, in N lat. 30° 30′, and E long. 107° 24′.

THIAN-KEOU-CHAN, TEN-HOU-SAN, or IGOUCH-KUND, a mountainous island of China, near the W coast of the Great Lu-chu island, in N lat. 26° 43′, and E long. 128° 4′. It is visible from a distance of 90 m., and presents on all sides the same aspect. It is covered from the base upwards to a third of its height with houses, and resembles a garden placed in the midst of the sea.

THIAN-MEN-CHAN, a mountain of China, in the prov. of Kan-suh, and dep. of Kwang-chung, in N lat. 34° 32′, and E long. 104° 32′. It is covered with perpetual snow.

THIAN-MUN, a district and town of China, in the prov. of Hu-pih, and div. of Gan-luh-fu, in N lat. 30° 40′, and E long. 113° 5′.

THIAN-PAON-HEEN, a district and town of China, in the prov. of Kwan-se, and div. of Chin-gan-fu.

THIAN-PIH, a district and town of China, in the prov. of Kwang-tung, div. and 36 m. SE of Kaou-chu-fu, in N lat. 21° 30′, and E long. 110° 22′, on the China sea. It has a spacious port, well defended by several forts, but barred at the entrance by sand-banks.

THIAN-PO, or TEM-PO, a town of China, in the Great Lu-chu island, and prov. of Chung-chan.

THIANS, a commune of France, in the dep. of the Seine, cant. and 2 m. SSE of Villejuif, in a fine plain, surrounded by hills. Pop. 1,035. It is well-built, and is surrounded by numerous country houses.

THIAN-SHAN, SIU-SHAN, or TENGRI-TAGH, (Snowy mountains), a chain of mountains, in the Chinese empire, on the N confines of Chinese Turkestan, and to the S of Sungaria, and extending E from the meridian of 80° to that of 95°, or the W frontier of the prov. of Kansu. It forms the NW buttress of the great central plateau of Asia. On its N side, in 42° 25′ N lat., between Korgos and Koutche, is the volcano of Pe-shan (White mountain), now said to be dormant. On the same side is the solfatara of Urumtsi, the crater of which is 15 m. in circumference, and is in active operation. and near it is another active volcano. All these afford large quantities of sal-ammoniac. On the S side is the volcano of Turfan or Ho-chu, near which was a town of the same name. The highest portion of this chain consists of three summits covered with perpetual snow, and known as the Bokdaoola or Holy mountain.

THIAN-SING. See TEEN-TSIN.

THIAN-SUN, a lofty mountain peak in China, in the prov. of Yun-nan, div. of Ta-le-fu. It is noted for its marble.

THIANT, a village of France, in the dep. of the Nord, cant. and 5 m. SW of Valenciennes, on the l. bank of the Ecaillon. Pop. 933. It has a church adorned with fine bas-reliefs.

THIAN-TAE, a district and town of China, in the prov. of Che-keang, and div. of Tae-chu-fu, in N lat. 29° 10′, and E long. 120° 3′.

THIAN-TSEUEN-CHU, a district and town of China, in the prov. of Sze-chuen, and div. of Ya-chu-fu, in N lat. 30°, and E long. 102° 20′.

THIAUCOURT, a town of France, in the dep. of the Meurthe, on the small river Madive, 25 m. NW of Nancy. Pop. 1,400.

THIBAUD-DE-CUZE (SAINT), a village of Savoy, 9 m. NNE of Echelles. Pop. 1,100.

THIBERVILLE, a town of France, in the dep. of the Eure, on the river Arve, 11 m. E of Lisieux. Pop. 1,300.

THIBERY (SAINT), a village of France, in the dep. of Herault, cant. and 4 m. S of Pezenas. Pop. 1,600.

THIBET. See TIBET.

THICKLEY, a hamlet in the county of Durham, 4 m. SE of Bishops Auckland.

THICOURT, a village of France, in the dep. of the Moselle, cant. and 4 m. SSE of Fauquemont. Pop. 400.

THIECOURT, a village of France, in the dep. of Oise, cant. and 2 m. SE of Lassigny. Pop. 1,250.

THIEL, a town of Holland, in the prov. of Gelderland, situated in a fertile district, near the river Waal, 20 m. SE of Utrecht. It has 4,100 inhabitants, employed partly in manufactures of linen and salt, partly in a traffic in corn, fruit, and provisions.

THIELE, or ZIHL, a river of Switzerland, which rises on the E flank of the Jura, and after a SW course of 18 m. enters the lake of Neufchatel, whence it issues at the NE extremity, and soon after expands into the lake of Bienne, a few leagues below which it joins the Aar, on the l. bank, 9 m. NNE of Aarberg, after a course of about 60 m.

THIELRODE, a village of Belgium, in the prov. of E. Flanders, 6 m. NNE of Dendermonde, on the l. bank of the Durme.

THIELT, a town of Belgium, in the prov. of W. Flanders, situated at the foot of an eminence, 18 m. WSW of Ghent. Pop. 12,000. It has considerable manufactories of linen and lace, also of soap, glue, leather, and hats, and numerous distilleries and bleachfields.—Also a village in the prov. of Brabant, 24 m. ENE of Brussels.

THIEMBRONNE, a village of France, in the dep. of Pas-de-Calais, 3 m. NW of Fauquembergue. Pop. 1,200.

THIEMENDORF, a village of Prussian Silesia, 3 m. E of Lauban. Pop. 1,200.

THIENGEN, a town of Baden, formerly the capital of the district called the Clettgau, 18 m. W of Schaffhausen, and 3 m. E of Waldshut, on the r. bank of the Wutach. Pop. 900.

THIENNES, a village of France, in the dep. of Nord, 6 m. SSW of Hazebruck. Pop. 1,100.

THIERACHEREN, a village of Switzerland, in the cant. and 15 m. SSE of Bern. Pop. 803.

THIERGARTEN, a village of Hungary, in the com. and 30 m. NNW of Gömör, on the Gran near its source.

THIERMAUPTEN, a village of Bavaria, on the r. bank of the Acha, 15 m. N of Augsburg.

THIERS, a town of France, in the dep. of Puy-de-Dome, on the declivity of a hill, which is skirted by the small river Durolle, 23 m. E by N of Clermont. Pop. 8,800. The buildings in the outskirts of the town have a pleasing appearance; but the interior presents dark, crowded, and winding streets, bordered with gloomy and ill-built houses. It is, however, a place of considerable activity, being inhabited entirely by artizans and manufacturers. The principal branch of its industry is hardware, and particularly cutlery, which employs here and in the neighbourhood, upwards of 20,000 workmen. It has likewise extensive tanneries and paper manufactories.

THIERSHEIM, a village of Bavaria, 6 m. NE of Wunsiedel. Pop. 700.

THIERSTEIN, a town of Bavaria, in the principality of Baireuth, 25 m. SSE of Hof.

THIETREVILLE, a village of France, in the dep. of Seine-Inferieure, 12 m. NW of Yvetot.

THIEULAIN, a village of Belgium, in the prov. of Hainault, 24 m. NW of Mons.

THIL, a village of France, in the dep. of Landes, cant. and 9 m. ESE of Pouillon. Pop. 1,300.

THILEUR, a village of Belgium, in the prov. and 3 m. SW of Liege, on the l. bank of the Meuse. Pop. 600.

THILOUZE, a village of France, in the dep. of Indre-et-Loire, cant. and 6 m. ESE of Azay-le-Rideau. Pop. 1,200.

THIMBDA, a village of Tunis, 8 m. SW of Deserta.

THIMBLEBY, a parish of Lincolnshire, 1½ m. NW of Horncastle. Area 1,770 acres. Pop. 492.

THIMBLE ISLANDS, a group of small islands, near the coast of Connecticut, U. S., in N lat. 41° 11′.

THINGWALL, a township and village of Cheshire, 5½ m. N by W of Great Neston. Area of t., 370 acres. Pop. in 1851, 96.

THINUSTERRE, a town of Belgium, in the prov. and 14 m. E of Liege. Pop. of cant. 2,800.

THIONVILLE, a town of France, the capital of an arrondissement, in the dep. of Moselle, situated on the Moselle, 19 m. N of Metz. Pop. 4,000. It is well-built; and though a small place, is of considerable strength, and has stood several sieges. It has manufactories of stockings, hats, and glass.

THIRD (Point), a cape on the E coast of Sumatra, in the straits of Banca, in S lat. 2° 24′.

THIRKLEBY, a township in the E. R. of Yorkshire, 10 m. E by S of New Malton. Area 1,343 acres. Pop. in 1851, 50.—Also a parish in the N. R., 4½ m. SE of Thirsk. Area 2,597 acres. Pop. in 1851, 300.

THIRLBY, a township of Felixkirk p., Yorkshire, 5 m. ENE of Thirsk. Area 575 acres. Pop. in 1851, 113.

THIRLEMERE-LAKE, or Leathes-Water, a narrow, irregular sheet of water, skirting the immense base of Helvellyn for about 4 miles, and receiving a variety of torrents from the sides of that huge mountain, in the parish of Crossthwaite, Cumberland. This lake, situated between Keswick and Grasmere, has an elevation of 500 ft. above sea-level. Its shores are indented, and, near the middle, it becomes so contracted that it is spanned by a few rude arches for the convenience of the neighbouring population.

THIRLWALL, a township of Haltwhistle p., Northumberland, near the borders of Cumberland. Area 7,944 acres. Pop. in 1831, 382; in 1851, 425.

THIRNE, or Thorne, a township of Thornton-Watlas p., Yorkshire, 4 m. SW by W of Bedale. Area 413 acres. Pop. in 1831; 142; in 1851, 127.

THIRSK, a borough and parish in the N. R. of Yorkshire, 23 m. NW by N of York, and within 1½ m. of the Great North of England railway. It is divided by the river Codbeck into two parts, which are connected by two small but handsome stone-bridges. The p. includes the chapelries of Carlton-Islebeck or Miniott, Sandhutton, and Sowerby, besides the township of T. Area of p. 8,354 acres. Pop. in 1801, 2,916; in 1831, 3,829; in 1851, 4,704.—The old town is chiefly composed of one long street. The vicinity is entirely agricultural, but the neat village of Sowerby is within ¾ m. of the town, while those of Miniott and Sandhutton lie about 1 m. to the W. The manufacture of coarse linen and sacking has been carried on in this parish. This borough returned 2 members to parliament till the passing of the Reform act, when its representation was limited to one. The number of electors registered in 1837, was 302; in 1852, 357. The new borough comprises the p. of Thirsk, the township of S. Kilvington in the p. of that name, and the township of Bagby in the p. of Kirby Knowle. T. is a polling-place at the elections for the north riding of the county.

THIRSTON (East and West), a township in Felton p., Northumberland, 9½ m. S by E of Alnwick, on the S bank of the river Coquet. Area 1,161 acres. Pop. in 1831, 307; in 1851, 281.

THIRSTY-SOUND, a bay on the E coast of Australia, in S lat. 22° 6′.

THISSELT, a town of Belgium, in the prov. and 13 m. S of Antwerp. Pop. 1,740. It has breweries, salt-refineries, and flour-mills.

THISTED, a town of Denmark, in Jutland, on the NW coast of the Lymfiord, 48 m. NW of Viborg. Pop. 2,200. It has a small port, which, in 1842, was entered by 268 vessels.

THISTLE-ISLAND, an island on the S coast of Australia, at the mouth of Spencer's gulf, in S lat. 34° 56′. It is about 12 m. long, and from 1 to 3 m. in breadth.

THISTLETON, a parish in Rutlandshire, 7¼ m. NE by N of Oakham. Area 1,420 acres. Pop. 142.

THIVA. See Thebes.

THIVIERS, a town of France, dep. of the Dordogne, 20 m. NW of Perigueux. Pop. 1,400. It has manufactures of paper, leather, and stone-ware, and a trade in cheese and truffles.

THIXENDALE, a township in Wharram-Percy p., Yorkshire, 8½ m. SE of New Malton. Area 3,697 acres. Pop. in 1831, 207; in 1851, 266.

THIZY, a town of France, dep. of the Rhone, situated on an eminence, 20 m. W by N of Villefranche. Pop. 1,600. It has manufactories of linen and calicoes.

THOARD, a town of France, dep. of Basses-Alps, near the small river Raduge, 6 m. NW of Digne. Pop. 1,000.

THOCKRINGTON, or Tockerington, a parish in Northumberland, 10 m. N by E of Hexham, including the townships of Little Barrington, Cary

Coats, Sweethope, and T. Area 6,943 acres. Pop. in 1831, 203; in 1851, 173.

THOIRAS, a village of France, in the dep. of Gard, 18 m. ENE of Vigan. Pop. 500.

THOIRE, a village of France, in the dep. of Sarthe, cant. and 4 m. NNE of Chateau-du-Loir. Pop. 1,000.

THOIRY, a village of Savoy, 6 m. ENE of Chambery. Pop. 1,350.

THOISSEY, a town of France, dep. of Ain, on the small river Chalaronne, 11 m. S of Macon.

THOLEN, or Ter Tholen, a town of Holland, in the prov. of Zieland, on the island of Tholen, 4 m. NW of Bergen-op-Zoom. It contains about 1,900 inhabitants, who are partly employed in the spinning of yarn, a branch of industry carried here to great nicety and perfection. The surrounding country is appropriated to the culture of flax and madder.—The island of T., situated near the mouth of the Scheldt, is about 12 m. long and 6 m. broad. It is separated from Brabant by an arm of the East Scheldt, and is fertile and well cultivated.

THOLEY, a village of Prussia, in the prov. of the Rhine, 26 m. SE of Treves. Pop. 700.

THOLLEMBECQUE, a village of Belgium, in the prov. of S. Brabant, 18 m. SW of Brussels. Pop. 2,530. It has breweries and distilleries.

THOLLON, a town of Savoy, 29 m. ENE of Geneva, at the N foot of the Dent-d'Hoche. Pop. 1,200.

THOLTHORPE, a township in Alne p., Yorkshire, 4¼ m. SW of Easingwold. Area 1,690 acres. Pop. in 1831, 265; in 1851, 303.

THOLY (Le), a village of France, in the dep. of Vosges, cant. and 9 m. NE of Remiremont. Pop. 1,200.

THOMAR, a town of Portugal, in Estremadura, near the r. bank of the river Nabão, an affluent of the Zezere, 17 m. NNW of Abrantes. Pop. 3,700. It is regularly built, and has an establishment for spinning cotton, a silk-mill, some soap-works, and a little trade in olive oil.

THOMAS CREEK, a river of S. Carolina, U. S., which runs into the Great Pedee.

THOMAS (Saint), a considerable island in the gulf of Guinea, off the W coast of Africa, about 100 m. W from the mouth of the Rio-Gabon, in N lat. 0° 14′ 42″. It appears to be about 25 m. in length from NNE to SSW; and is entirely of volcanic formation, attaining in one of its peaks an alt. of 7,020 ft. The face of the low country is undulating and adorned with luxuriant verdure. Situated immediately under the line, the excess both of heat and moisture renders it extremely pernicious to the health of European residents. The Dutch took possession of it in 1641; but were soon induced by its unhealthiness to abandon it. The soil is clayey, frequently coarse, and mixed with chalk. Such is its fertility, that a plantation, when once abandoned, is almost entirely overgrown with luxuriant vegetation. Rice and millet are successfully cultivated, though yams, sweet cassava, and other roots are chiefly used by the inhabitants. Fish, turtle, poultry, hogs, sheep, and goats, are abundant. The only important article raised for exportation is sugar, for which the soil is extremely well fitted. The inhabitants, from 15,000 to 20,000 in number, consist partly of Portuguese, the descendants chiefly of outlaws and banished persons, sent hither on the first discovery of the island; and partly of Negroes, either native or brought over from the coast of Angola. On the E part of the island, facing the continent, is built the town of Povoacon or Saint T., consisting of 700 houses, and defended by several forts. The safest anchorage is in Man-of-War bay, on the NE side of the island.—Off the SW end of the island is a small islet named Ilha-das-Rolas.

THOMAS (Saint), the principal of the Virgin islands, in the West Indies, in N lat. 18° 20′, W long. 64° 55′, belonging to Denmark. It is 12 m. in length from E to W, and about 3 m. in average breadth. Its surface is rugged and elevated. The soil is sandy. It produces potatoes, millet, manioc, fruits, sugar, and tobacco, but is extremely infested with mosquitoes and other troublesome vermin. About one-half of the cultivation is sugar-cane. It has been adopted as a principal station of the West India mail-steamers from Southampton. Most of the houses of its capital, Charlotte-Amalie, are of brick, and built and tiled in the Dutch fashion. The streets of the town are clean and well regulated, and though Negroes are to be met with on all sides, they are decently dressed, respectful in their demeanour, and all seem to have some honest method of earning their daily bread. The harbour is defended by several fortifications of ancient date, and rather defective construction. The military, however, preserve admirable order in the town. If two people quarrel, they are immediately taken to the fort, which is prison as well as fortification. Every offence seems to be punished by fine, and this method touching the inhabitants in the tenderest spot of all—the pocket—is found the most effectual. Fighting, thieving, and other minor offences are all punished in this way, and in consequence there is but little crime.—St. T. is a free port, and a depot for all vessels visiting these seas. In 1688, the Danish factory here was attacked and plundered by the French buccaneers. In March 1801, it was taken by the British. It was given up at the peace of Amiens; but was again taken in the course of the subsequent war, and was restored to Denmark at the peace of Paris in 1814. The pop. is about 14,000.—St. T. is 3,622 m. sailing distance from Southampton; 1,031 m. from Havanna; 151 m. from St. Kitts; and 860 m. from Nassau.

THOMAS (Saint), a village of Lower Canada, on the S bank of the St. Lawrence, where it is joined by the Riviere-du-Sud.

THOMAS (Saint), a parish in Devon, a ½ m. S by W of Exeter, containing the town of St. Thomas, which is united with Exeter by a bridge across the Exe, and within the boundaries of the borough of Exeter. Area 3,700 acres. Pop. in 1831, 4,203; in 1851, 4,577.—Also a parish adjoining Launceston borough, in Cornwall. Area 1,817 acres. Pop. in 1831, 626; in 1851, 1,005.

THOMASTON, a township of Lincoln co., in the state of Maine, U. S., on the E side of George's river, and 36 m. SE of Augusta. Pop. in 1840, 6,227; in 1850, 2,732.—Also a village of Upton co., in the state of Georgia, 69 m. W by S of Milledgeville. Pop. in 1840, 150; in 1850, 180.

THOMASTOWN, a parish in co. Kildare, 2 m. SE of Rathangan. Area 853¼ acres. Pop. in 1831, 139; in 1851, 97.—Also a hamlet in the p. of Relickmurry, co. Tipperary, 2 m. W by S of Golden-Bridge. Pop. in 1851, 74.—Also a parish and town in co. Kilkenny. Area of p. 2,041 acres. Pop. in 1831, 3,182; in 1851, 2,180.—The town, formerly a parl. borough, stands on the river Nore, 3¾ m. NW of Innistiogue. The river Nore divides it into two sections, placing the larger section on the l. bank. A handsome stone-bridge of 5 arches connects the two sections of the town. Pop. in 1851, 1,779.

THOMASVILLE, a village of Thomas co., in the state of Georgia, U. S., 169 m. S by W of Milledgeville.

THOME (Saint), a village of France, in the dep.

of the Ardeche, cant. and 4 m. WNW of Viviers, at the confluence of the Escontay and Negue. Pop. 472.

THOME (SAINT), or MAILAPORAM, a town of Hindostan, in the presidency and 6 m. S of Madras, in the Karnatic, on a small bay at the extremity of a fine plain. It consists of about 100 detached houses, intermingled with numerous ruins, and has two Roman Catholic churches. Its inhabitants are chiefly Portuguese, Hindus, and Mussulmans. It has some calico manufactures. This town was built amid the ruins of Mailaporam, which was taken by the Portuguese in 1545, and their chief establishment on the Coromandel coast. It fell successively into the hands of the French and Dutch, and ultimately into those of the English, in 1749.

THOME (SAO). See THOMAS (SAINT).

THOME (SAO), or ANNA-DE-CHAVES. See THOMAS (SAINT).

THOME-DE-LA-GUAYANA (SAN). See ANGOSTURA.

THOMERY, a commune of France, in the dep. of the Seine-et-Marne, cant. and 3 m. NNW of Moret, on the l. bank of the Seine. Pop. 782. It has two fine castles, and is noted for its fruit, and especially for a fine species of grape.

THOMOND—originally TUAIDH-MUIN, or NORTH MUNSTER—an ancient district of Ireland, of fluctuating extent and changeful political character, identical, in a general view, with the present NW of the prov. of Munster. In 1800, Murrough, fifth Earl of Inchiquin, was created Marquis of Thomond, in the peerage of Ireland. In 1826, the second Marquis of Thomond was created Baron Tadcaster, in the peerage of Great Britain.

THOMPSON, a township of Wyndham co., in the state of Connecticut, U. S., 46 m. ENE of Hartford, drained by French, Quinnebaug, and Five-Mile rivers, and intersected by the Norwich and Worcester railroad. Pop. in 1840, 3,536; in 1850, 4,638.—Also a township of Susquehanna co., in the state of Pennsylvania, 184 m. NNE of Harrisburg. Pop. in 1840, 325; in 1850, 400.—Also a township of Geauga co., in the state of Ohio, 190 m. NE of Columbus. Pop. in 1840, 1,028; in 1850, 1,211.—Also a township of Seneca co., in the same state, 84 m. N of Columbus, and intersected by the Mad river and Sandusky railroad. Pop. in 1840, 1,404; in 1850, 1,668.—Also a township of Delaware co., in the same state, on the W side of Scioto river. Pop. in 1840, 660.—Also a village of Warren co., in the state of Georgia, 38 m. W of Augusta, on the Georgia railroad.—Also a township of Pike co., in the state of Arkansas. Pop. in 1840, 390.

THOMPSON, a river of Australia Felix, in the district of Gipp's Land, an affluent of Lake King.—Also a mountain of New South Wales, in the co. of Stanley, to the E of Mount Hallen.

THOMPSON, or TOMPSON, a parish in Norfolk, 2½ m. S by E of Watton. Area 2,890 acres. Pop. 496.

THOMPSONTOWN, a village of Juniata co., in the state of Pennsylvania, U. S., 36 m. NW of Harrisburg, on the Juniata river and Pennsylvania canal. Pop. in 1840, 350.

THOMPSONVILLE, a village of Hartford co., in the state of Connecticut, U. S., 20 m. N of Hartford, at the junction of Freshwater and Connecticut rivers. Pop. in 1840, 800.—Also a village of Thompson township, Sullivan co., in the state of New York, 105 m. SSW of Albany. Pop. in 1840, 150.

THON, a department and commune of Belgium, in the prov. of Namur, on the Meuse. Pop. of dep. 695; of com. 515. It has several iron works and quarries of stone.

THONANCE-LES-JOINVILLE, a village of France, in the dep. of the Haute-Marne, cant. and 2 m. NE of Joinville, on the Montreuil, a little above its junction with the Marne. Pop. 1,000. It has a blast furnace.

THONES, a town of Sardinia, capital of a mandemento, in the prov. of Guervois, 7 m. ESE of Annecy, near the confluence of the Fier and Nom, and at the junction of three great valleys. Pop. 2,500. It is triangular in form, and has three gates. The houses are built on low arcades. The public square is spacious, and in its centre is the church. It has extensive manufactories of agricultural implements, and of kirchwasser and saffron; several silk-mills and tanneries; and carries on an active trade with the adjacent valleys.

THONG (NETHER), a chapelry in Almondbury p., in Yorkshire, 4½ m. S of Huddersfield. Area 850 acres. Pop in 1851, 1,207.

THONG (UPPER), a township and village in Almondbury p., in Yorkshire, 5 m. S of Huddersfield. Area of t. 710 acres. Pop. in 1831, 1,648; in 1851, 2,463. The village is well-built, and has several woollen mills.

THONHAUSEN, a town of the duchy of Saxe-Altenburg, 8 m. SE of Ronneburg. Pop. 680.

THONIN BAY, an indentation of the E coast of Van Diemen's land, on the E side of Freycinet's peninsula, in the co. of Glamorgan.

THONIS. See TONIS.

THONISBERG. See ANTONISBERG (ST.).

THONLAGEE, or TONLAGEE, a mountain in co. Wicklow, 2¼ m. N of Wicklow-Gap, with an alt. above sea-level of 2,307 ft.

THONNE-LE-THIL, a village of France, in the dep. of the Meuse, cant. and 4 m. N of Montmedy, on the l. bank of the Thonne, a small affluent of the Chiers. Pop. 540. In the vicinity are several iron mines.—At the distance of 3 m. ESE is the village of Thonne-la-Long.

THONNELLES, a village of France, in the dep. of the Meuse, cant. and 2 m. N of Montmedy, on the Thonne. Pop. 320. It has iron-works.

THONON, a town of Sardinia, capital of the prov. of Chablais, 14 m. SW of Lausanne, and 21 m. NE of Geneva, on the S bank of the lake of that name, which here forms a bay also named Thonon, and at an alt. of 482 yds. above sea-level. Pop. 4,000. It is irregularly built, and consists of two parts distinguished as the upper and lower town. In the former, which is the larger, is the principal church, the college, and town-house. The lower town consists of only two streets, one of which terminating at the lake, crosses the other at right angles. A small esplanade, planted with trees and adorned with a fountain surmounted by an obelisk of grey marble, occupies the site of a fortress destroyed in the 14th century, by the Bernese and Vallaisans, and affords a fine view of the town, lake, and cant. of Vaud. It has manufactories of pottery, and carries on an active trade. The environs are fertile and well-cultivated. About 2 m. SW is a mineral spring, and a little to the NNE is the castle of Ripaille, an ancient Augustine convent, founded by Amedeus, eighth Duke of Savoy. Thonon is of great antiquity. It was destroyed by the Burgundians in the 5th cent., and a portion of the town and the castle were destroyed in the 14th cent. by the Bernese. It was the cap. of the French dep. of the Leman.

THOORA, a town of Sind, on the road from Cutch to Hyderabad, and 55 m. S of the latter place.

THOR (DAS HOHE), a mountain of Upper Austria, in the circle of Salzburg, having an alt. of 7,720 ft. above the level of the sea.

THOR (LE), a commune of France, in the dep. of

Vaucluse, cant. and 3 m. WNW of L'Isle, on the Sorgue. Pop. in 1846, 3,881. It has manufactories of madder, and a considerable trade in grain and hay. Silk is cultivated in the environs.

THORAISE, a village of France, in the dep. of the Doubs, cant. and 1 m. N of Roussieres, on the l. bank of the Doubs, and at the foot of a hill. Pop. 228.

THORAS, a village of France, in the dep. of the Haute-Loire, cant. and 7 m. S of Saugues, on the l. bank of a stream of the same name, an affluent of the Ance. Pop. 1,100.

THORDA, or THORENBURG, a county of Transylvania, nearly in the centre of the principality, having the co. of Clausenburg to the N, and that of Weissenburg to the S. Its area of 1,600 sq. m. is very hilly, but presents many pleasant valleys, and is watered by the Maros, the Görgeny, the Lucz, and the Aranyos. It has productive mines of salt and of iron, and extensive forests. The pop. in 1841 was 68,000.—Its capital of the same name is a considerable town, 17 m. SSE of Clausenburg, situated near the river Aranyos, and divided by a smaller stream called the Salzbach, into an old and new town. It contains about 8,000 inhabitants, whose chief source of subsistence is the salt trade, there being rich salt-mines in the neighbourhood. These mines were wrought in the time of the Romans, who had here a town called *Salinæ*, the ruins of which are still visible on an eminence close by the town. The Hungarian language is spoken here in great purity.

THORE', a river of France, which rises 6 m. SW of Saint-Pons-de-Thomières, in the dep. of Herault, and after a WNW course of 30 m., falls into the Agout on the l. bank.—Also a village of France, in the dep. of Loir-et-Cher, cant. and 4 m. W of Vendome, near the l. bank of the Loir.

THORENS-SALES, a town of Savoy, 16 m. SSE of Geneva, on the Fillieve. Pop. 2,500. It has large glass-works.

THORESBY (NORTH), a parish of Lincolnshire, 9 m. N by W of Louth. Area 2,485 acres. Pop. in 1851, 733.

THORESBY (SOUTH), a parish of Lincolnshire, 3½ m. W by N of Alford. Area 933 acres. Pop. in 1851, 156.

THORESWAY, a parish of Lincolnshire, 4 m. SE of Caistor. Area 2,730 acres. Pop. 135.

THORGANBY, a parish of Lincolnshire, 6 m. ESE of Caistor. Area 1,568 acres. Pop. in 1851, 120.—Also a parish in the E. R. of Yorkshire, 10 m. SE by S of York. Area 3,190 acres. Pop. in 1851, 388.

THORIGNY, a town of France, dep. of La Manche, near the river Vire, 9 m. SE of St. Lö. Pop. 2,200.—Also a village of France, in the dep. of Yonne, 9 m. NE of Sens. Pop. 700.—Also a village in the dep. of Seine-et-Marne, 1 m. NE of Lagny.

THORINGTON, a parish of Suffolk, 3 m. SE of Halesworth. Area 1,800 acres. Pop. in 1851, 157.

THORLEY, or THORNLEY, parish of Hertfordshire, 2 m. SW by S of Bishop's Stortford. Area 1,516 acres. Pop. in 1851, 402.—Also a parish in the Isle of Wight, 1 m. ESE of Yarmouth. Area 1,574 acres. Pop. in 1851, 154.

THORMANBY, a parish of the N. R. of Yorkshire, 4 m. NW by N of Easingwold. Area 958 acres. Pop. in 1851, 154.

THORN, a town of West Prussia, in the gov. and 50 m. SSW of Marienwerder, pleasantly situated on the r. bank of the Vistula, about 90 m. from the mouth of that river, over which it has a wooden bridge of nearly a ½ m. in length. It consists of an old and new town, which are separated from each other by a wall and ditch, but are both surrounded by one mound and moat. It was formerly considered a place of great strength. Its population in 1846 was 12,687, partly Protestants, partly Catholics. It has manufactories of woollens, linen, hats, soap, leather, gloves, and starch; and exports corn and wood, but the extent of its trade is less considerable than formerly, the Vistula having here become more shallow, so that vessels of burden can no longer come up to the town. Its best edifices are of old date. It enjoyed a considerable share of prosperity until the 18th cent.; but in 1708 and 1710 it was ravaged by the plague. In 1793 it was taken possession of by a Prussian garrison. In the end of 1806 it suffered from the invasion of the French, and remained in their hands until 1812. The well known Copernicus was born at Thorn in 1742.

THORN, a small town of Dutch Limburg, 7 m. SSW of Ruremonde, formerly the seat of a richly endowed abbey. Pop. 1,162.

THORN, a township of Perry co., in the state of Ohio, U. S. Pop. in 1840, 2,007.

THORNABY, a chapelry in Stainton p., Yorkshire, 3½ m. N by E of Yarm, on the E bank of the Tees. Area 1,230 acres. Pop. in 1851, 1,759.

THORNAGE, a parish in Norfolk, 2¾ m. WSW of Holt. Area 1,266 acres.. Pop. in 1851, 349.

THORNBOROUGH, a parish in Bucks, 3½ m. E by S of Buckingham. Area 2,530 acres. Pop. 754.

THORNBURG, a village of Spottsylvania co., in the state of Virginia, U. S., 76 m. NNW of Richmond, on both sides of Ta river.

THORNBURY, a township of Chester co., in the state of Pennsylvania, U. S., 20 m. SW of Philadelphia. It has a level surface, and is drained by Chester creek and Brandywine river. The soil is chiefly sandy loam. Pop. in 1840, 230.—Also a township of Delaware co., in the same state, 20 m. W of Philadelphia. It has a hilly surface, and is drained by Chester creek. The soil consists of sandy loam. Pop. in 1840, 675.

THORNBURY, a parish in Devon, 5 m. NE of Holsworthy, on the river Waldon. Area 2,772 acres. Pop. in 1831, 546; in 1851, 489.

THORNBURY, a parish and market-town in the co. of Gloucester, on the banks of a small rivulet in Berkeley vale, 24 m. SW by S of Gloucester, and 3½ m. E of the mouth of the Severn. The p. comprises the chapelries of Falfield, Oldbury-upon-Severn, and Rangeworthy, with the tythings of Kington and Moorton, and has an area of 15,732 acres. Pop. in 1801, 2,488; in 1831, 4,375; in 1851, 4,614. The church is a spacious and handsome structure in the cathedral form, with a lofty tower. The town consists principally of three streets, disposed in the form of the letter Y; or, "having one longe street," as Leland remarks, "and two hornes goyne out of it." The principal object of notice in this place is the remains of a magnificent palatial castle, begun by Stafford, Duke of Buckingham; but, in consequence of his attainder and death, in the reign of Henry VIII., never finished.—Also a parish in Herefordshire, 4 m. NNW of Bromyard. Area 2,130 acres. Pop. in 1851, 196.

THORNBY, or THURNBY, a parish in Northamptonshire, 10 m. NW of Northampton. Area 1,212 acres. Pop. in 1831, 198; in 1851, 220.

THORNCOMBE, a parish in Devon, 5⅝ m. NE by E of Axminster. Area 4,896 acres. Pop. 1,317.

THORNCOTE, a hamlet in Northill p., Bedfordshire, 3¼ m. W by N of Biggleswade. Pop. 208.

THORNDON, a parish in Suffolk, 2¼ m. S by W of Eye, and within the parliamentary boundaries of the borough of Eye. Area 2,680 acres. Pop. 725.

THORNDYKE, a township of Waldo co., in the

state of Maine, U. S., 42 m. NE of Augusta. It has a finely diversified surface, and is drained by a branch of Sebasticook river. The soil is generally fertile. Pop. in 1840, 897.

THORNE, a parish and market-town in the W. R. of Yorkshire, 26 m. S by E of York, and intersected by the river Don, and the Stainforth and Keadby canal. Area of p. 11,900 acres. Pop. in 1831, 3,779; in 1851, 3,484. The town is tolerably well-built, and consists principally of three nearly parallel streets. The rivers Don, Aire, and Ouse, communicating with each other, and with the Stainforth and Keadby canal, have greatly contributed to the increase of the carrying trade of this place. Coasting-vessels are built here and floated to Hull.

THORNE-COFFIN, a parish in Somerset, 2½ m. NW by W of Yeovil. Area 410 acres. Pop. in 1831, 101; in 1851, 102.

THORNE-FALCON, a parish in Somersetshire, 3½ m. E by S of Taunton. Area 814 acres. Pop. in 1831, 273; in 1851, 229.

THORNE-GRAFTON, a township in Haltwhistle p., Northumberland, 10 m. W by N of Hexham.

THORNE-ST.-MARGARET, a parish in Somersetshire, 3 m. W by N of Wellington, crossed by the Western canal, and near the Bristol and Exeter railway. Area 805 acres. Pop. in 1851, 143.

THORNER, a parish and village in the W. R. of Yorkshire, 5½ m. S by W of Wetherby, comprising the townships of Scarcroft, Shadwell, and T. The village consists of one long street. Area 4,380 acres. Pop. in 1831, 1,220; in 1851, 1,530.

THORNES, a chapelry in Wakefield p., Yorkshire, ¾ m. S by W of Wakefield. The village of T. is included within the boundaries of the parl. borough of Wakefield.

THORNEY, a parish and market-town in the Isle of Ely, co. of Cambridge, 10 m. NW by W of Marsh. Area of p. 17,590 acres. Pop. in 1831, 2,055; in 1851, 2,174. The church consists of part of a celebrated abbey of Anchorites founded here in 972. Many of the inhabitants are descendants of French Protestants.—Also a parish in Notts, 8 m. E by N of Tuxford, comprising the hamlets of Broadholme and Wiggesley. Area 4,140 acres. Pop. in 1831, 308; in 1851, 412.

THORNEY (WEST), or THORNEY ISLAND, a parish in Sussex, 6½ m. W by S of Chichester, near the centre of the estuary called Chichester harbour. Area 3,005 acres. Pop. in 1851, 111.

THORNEYBURN, a parish in Northumberland, 4½ m. NW by W of Bellingham. Area 20,133 acres. Pop. in 1831, 331; in 1851, 340.

THORNFORD, a parish in Dorset, 3½ m. SW by S of Sherborne. Area 1,407 acres. Pop. 410.

THORN-GUMBALD, a chapelry in Paul p., Yorkshire, 10 m. E by S of Kingston-upon-Hull. Area 1,450 acres. Pop. in 1831, 266; in 1851, 278.

THORNHAM, or THURNHAM, a parish in Kent, 3⅓ m. N by E of Maidstone. Area 3,319 acres. Pop. in 1831, 571; in 1851, 511.—Also a township in Middleton p., Lancashire, 3½ m. S by W of Rochdale. Area 1,998 acres. Pop. in 1831, 1,455; in 1851, 1,510.—Also a parish in Norfolk, 6¼ m. W by N of Burnham-Westgate. Area 2,934 acres. Pop. in 1831, 668; in 1851, 792.

THORNHAM (GREAT), a parish in Suffolk, 3½ m. WSW of Eye. Area 1,324 acres. Pop. 322.

THORNHAM (LITTLE), a parish in Suffolk, 2¾ m. W by S of Eye. Area 676 acres. Pop. 155.

THORNHAUGH, a parish in Northamptonshire, 9 m. WNW of Peterborough. Area 1,796 acres. Pop. in 1831, 271; in 1851, 266.

THORNHILL, a township in Hope p., co. of Derby, 5¾ m. NE by N of Tideswell. Area 589 acres. Pop. in 1831, 135; in 1851, 131.—Also a tything in Stalbridge p., Dorset, 2 m. S of Stalbridge. Pop. in 1831, 257; in 1851, 266.—Also a parish in the W. R. of Yorkshire, 5½ m. W by S of Wakefield, on the S bank of the Calder, and crossed by the Manchester and Leeds railway. It comprises the chapelry of Flockton, and the townships of Shitlington, Lower Witley, and T. Area 7,997 acres. Pop. in 1831, 6,271; in 1851, 6,858. The church consists of a nave, chancel, and side aisles, with a tower. Manufactures of woollens, chemical and glass-works, and large collieries, exist in this p. The v. was formerly a place of importance, and adorned by a magnificent castellated mansion, belonging to the Thornhills and Savilles, which was demolished by the parliamentarians in the civil wars.

THORNHILL, a village in the p. of Morton, Dumfries-shire, on the post-road between Glasgow and Dumfries, 8 m. ENE of Minnyhive, and 12 m. SE by S of Sanquhar, on the summit of a terrace-formed rising ground, between 200 and 300 ft. high, and from 2½ to 5 fur. E of the Nith. Its streets are spacious, and the houses are, for the most part, uniform one-story cottages. An extensive manufacture of hosiery was, for a series of years, conducted in workshops; but has declined. A tannery, a brewery, and a few remaining stocking-looms alone prevent its manufactures from being entirely domestic. The village is wholly under the superiority of the dukes of Buccleuch, and, since 1827, has been very greatly and liberally improved by the present duke. Pop. in 1817, 750; in 1831, 1,373; in 1851, 1,658.

THORNHILL, a village of Upper Canada, 11 m. from Toronto, on a branch of the river Don. Pop. 700.

THORNHOLME, a township in Burton-Agnes p., Yorkshire, 5 m. SW by W of Bridlington. Area 1,324 acres. Pop. in 1831, 93; in 1851, 100.

THORNLEY-WITH-WHEATLEY, a township in Chipping p., co.-palatine of Lancaster, 8 m. W by S of Clitheroe. Area 3,179 acres. Pop. 491.

THORNLIEBANK, a village in the p. of Eastwood, Renfrewshire, about 1½ m. S of Pollockshaws, upon the rivulet called Auldhouse-burn. It originated in manufactories and other works established here about the end of the 18th cent. Calico-printing, cotton-spinning, weaving by power and hand-loom, and bleaching, are carried on. The pop. amounts to about 1,600.

THORNSETT, a township in Glossop p., Derby, 5 m. NW of Chapel-in-le-Frith. Pop. in 1851, 869.

THORNTHWAITE, a chapelry and township in Crossthwaite p., Cumberland, 4 m. NW by W of Keswick. Pop. in 1831, 174; in 1851, 275.

THORNTHWAITE-WITH-PADSIDE, a chapelry and township in Hampsthwaite p., Yorkshire, 12 m. W by N of Knaresborough. Area 1,960 acres. Pop in 1831, 304; in 1851, 303.

THORNTON, a parish in Bucks, 4 m. ENE of Buckingham, crossed by the river Ouse, and the Buckingham canal. Area 1,332 acres. Pop. in 1831, 94; in 1851, 103.—Also a township in Norham p., Northumberland, 4½ m. SW of Berwick-upon-Tweed. Area 1,390 acres. Pop. in 1831, 190; in 1851, 169.—Also a township in Poulton p., co.-palatine of Lancaster, 1¾ m. N of Poulton, on the Wyre and Preston railway. Pop. in 1831, 842; in 1851, 1,013.—Also a township in Sephton p., co.-palatine of Lancaster, 6¾ m. N of Liverpool. Area 773 acres. Pop. in 1831, 342; in 1851, 298.—Also a parish and village in the co. of Leicester, 4¾ m. NE by E of Market-Bosworth, intersected by the Leicester and Swannington railway, and containing the chapelry of Bagworth, and the township of Stanton-under-Bardon with Horsepool. Area 3,930 acres. Pop. in 1831, 1,078; in 1851, 1,350.—Also a parish in Lincolnshire, 1¾ m. SW of Horncastle. Area 1,755 acres. Pop. in 1831, 216; in 1851, 240. —Also a parish in the E. R. of Yorkshire, 4 m. SW of Pocklington, including the townships of Melbourne and Storwood. Area 6,540 acres. Pop. in 1831, 791; in 1851, 816.—Also a chapelry and township in Bradford p., W. R. of Yorkshire, 3½ m. W of Bradford. Area 4,390 acres. Pop. in 1831, 2,246; in 1851, 8,051. The woollen manufacture is extensively carried on here.

THORNTON, a township of Grafton co., in the state of New York, U. S., 52 m. N by W of Concord. It has a hilly surface, and is watered by Pemigewasset river and its affluent Mad river. Pop. in 1840, 1,012; in 1850, 1,045.—Also a township of Cook co., in the state of Illinois, 192 m. NNE of Springfield, on Thorn creek. Pop. in 1850, 400.

THORNTON-LE-BEANS, a township in North Otterton p., Yorkshire, 3½ m. SE of North Allerton. Area 1,351 acres. Pop. in 1831, 219; in 1851, 247. —Also a chapelry and township in Ripon p., Yorkshire, 5½ m. SW by S of Ripon. Area 3,027 acres. Pop. in 1831, 614; in 1851, 691.

THORNTON-CHILDER, a township in Eastham p., co.-palatine of Chester, 8 m. NNW of Chester. Area 723 acres. Pop. in 1831, 296; in 1851, 319.

THORNTON-LE-CLAY, a township in Foston p., Yorkshire, 11½ m. NNE of York. Area 1,270 acres. Pop. in 1831, 205; in 1851, 266.

THORNTON-IN-CRAVEN, a parish in the W. R. of Yorkshire, 6 m. WSW of Skipton, crossed by the Leeds and Liverpool canal. Area 6,710 acres. Pop. in 1831, 2,246; in 1851, 2,202. Thornton-force is a picturesque cataract, issuing from an aperture in a high rock, and having a fall of about 90 ft. in one sheet of water 12 ft. wide.

THORNTON-CURTIS, or THORNETON-UPON-HUMBER, a parish in the co. of Lincoln, 5 m. SE by E of Barton-upon-Humber. Area 4,610 acres. Pop. in 1831, 362, in 1851, 497.

THORNTON-DALE, a parish in the N. R. of Yorkshire, 2½ m. E by S of Pickering, comprising the townships of Farmanby and T. Area 1,066 acres. Pop. in 1831, 1,368; in 1851, 927.

THORNTON-HOUGH, a township in Neston p., co.-palatine of Chester, 2½ m. NNE of Great Neston. Area 1,481 acres. Pop. in 1831, 144; in 1851, 164.

THORNTON-IN-LONSDALE, a parish in the W. R. of Yorkshire, 11½ m. NW by W of Settle, comprising the chapelry of Black-Burton and the township of T. Area 11,432 acres. Pop. 1,130.

THORNTON-LE-MOOR, a township in North Otterington p., Yorkshire, 4½ m. NW by N of Thirsk. Area 1,492 acres. Pop. in 1831, 337; in 1851, 339.—Also a parish in the co.-palatine of Chester, 5½ m. WSW of Frodsham, containing the townships of Dunham, Elton, Hapsford, T., and Wimbolds-Trafford. Area 4,866 acres. Pop. in 1831, 914; in 1851, 942.—Also a parish in Lincolnshire, 6 m. SW of Caistor. Area 1,503 acres. Pop. in 1831, 99; in 1851, 108.

THORNTON-RUST, a township in Aysgarth p., Yorkshire, 9 m. W of Middleham. Area 1,923 acres. Pop. in 1831 and 1851, 158.

THORNTON-STEWARD, a parish in the N. R. of Yorkshire, 3½ m. E by S of Middleham. Area 2,079 acres. Pop. in 1831, 310; in 1851, 304.

THORNTON-LE-STREET, a parish in the N. R. of Yorkshire, 3 m. NNW of Thirsk, comprising the townships of North Kilvington and T. Area 2,750 acres. Pop. in 1831, 226; in 1851, 234.

THORNTON-WATLAS, a parish in the N. R. of Yorkshire, 3 m. SW of Bedale, comprising the townships of Clifton-upon-Ure, Rookwith, Thirn, and T. Area 3,783 acres. Pop. in 1831, 418; in 1851, 421.

THORNTOWN, a village of Boone co., in the state of Indiana, U. S., on the N side of Prairie creek, and on the Lafayette and Indianapolis railroad, 37 m. NW of Indianapolis. Pop. 550.

THORNVILLE, a village of Perry co., in the state of Ohio, U. S., 35 m. E by S of Columbus, on the Scioto and Hocking Valley railroad. Pop. 200.

THOROE, a small island and parish of Denmark, in the Great Belt, bail. and 3 m. ESE of Svendborg, to the NE of the island of Taasing. Pop. 450. It is 3 m. in length, and 2 m. broad.

THOROLD, a township of Upper Canada, separated by Welland river, on the south from the township of Crowland, and intersected by the Welland canal. Pop. in 1840, 2,284. It has a village of the same name, situated on a mountain, 4 m. from St. Catherines, and near the Welland canal. Pop. in 1826, 1,000.

THORON, an island of the Baltic, near the E. coast of Sweden, in N lat. 58° 49′, E long. 17° 48′.

THOROTON, a parish in Notts, 7½ m. SSW of Newark. Area 730 acres. Pop. in 1851, 177.

THOROUGHFARE, a village of Prince William co., in the state of Virginia, U. S., 88 m. W by N of Richmond, and intersected by the Manassasgap railroad.

THORPE, a parish in Derbyshire, 3 m. NW by N of Ashborne, on the E bank of the Dove. Area 1,400 acres. Pop. in 1831, 189; in 1851, 188. In this vicinity is the celebrated rocky vale called Dovedale.—Also a parish in Notts, 3¾ m. SW of Newark. Area 698 acres. Pop. in 1831, 39; in 1851, 115.—Also a parish in Surrey, 2 m. NW by W of Chertsey. Area 1,495 acres. Pop. in 1851, 555.

THORPE-ABBOT'S, a parish in Norfolk, 4½ m. NNE of Eye. Area 1,122 acres. Pop. in 1851, 258.

THORPE-ACHURCH, a parish in Northamptonshire, 4½ m. NNE of Thrapstone, on the river Nene. Area 1,580 acres. Pop. in 1831, 240; in 1851, 242.

THORPE-ACRE-WITH-DISHLEY, a parish in Leicestershire, 1¾ m. WNW of Loughborough. Area 890 acres. Pop. in 1831, 366; in 1851, 260. The hamlet of Dishley was the residence of the celebrated agriculturist Bakewell, and is still in high repute for its breed of cattle and sheep.

THORPE-ARCH, a parish in Yorkshire, 2¾ m. SE by E of Wetherby, on the river Wharfe, which is here crossed by a handsome bridge of 4 arches. Area 1,607 acres. Pop. in 1831, 316; in 1851, 315.

THORPE-ARNOLD, a parish in Leicestershire, 1½ m. ENE of Melton-Mowbray. Area 1,742 acres. Pop. in 1831, 117; in 1851, 122.

THORPE-AUDLING, a township in Badsworth p., Yorkshire, 4 m. SSE of Pontefract. Area 1,260 acres. Pop. in 1831, 355; in 1851, 315.

THORPE-IN-BALNE, a township in Barnby-upon-Don p., Yorkshire, 5 m. N by E of Doncaster. Area 1,507 acres. Pop. in 1831, 121; in 1851, 130.

THORPE-BASSET, a parish in the E. R. of Yorkshire, 4 m. E by N of New Malton. Area 1,792 acres. Pop. in 1831, 206; in 1851, 207.

THORPE-BOCHART, a parish in Nottinghamshire, 10 m. S by E of Nottingham. Area 930 acres. Pop. in 1831, 39; in 1851, 33.

THORPE-CONSTANTINE, a parish in the co. of Stafford, 4 m. NE by E of Tamworth. Area 953 acres. Pop. in 1831, 49; in 1851, 58.

THORPE-MALZOR, or MALSOVERES, a parish in Northamptonshire, 2½ m. W by N of Kettering. Area 680 acres. Pop. in 1831, 297; in 1851, 287.

THORPE-MANDEVILLE, a parish in Northamptonshire, 7 m. NW by N cf Brackley. Area 1,230 acres. Pop. in 1831, 175; in 1851, 151.

THORPE-MARKET, a parish in Norfolk, 4½ m. NW by N of North Walsham. Area 1,309 acres.

Pop. in 1831, 254; in 1851, 221. The village is small, but handsome and picturesque.

THORPE-MORIEUX, a parish in Suffolk, 17 m. WNW of Ipswich. Area 2,457 acres. Pop. 414.

THORPE-NEXT-NORWICH, a parish in Norfolk, 2½ m. E of Norwich, comprising the hamlets of T. and Pockthorpe. Area 2,525 acres. Pop. in 1831, 2,151; in 1851, 3,000. The village is situated on the declivity of a hill which overlooks the city of Norwich on the opposite or W side of the Yare, which is here enlivened by the frequent passage of steam-packets, pleasure-boats, and wherries. The natural beauty of the vicinity has led to the erection of several elegant mansions or villas.

THORPE-SALVIN, a parish in the W. R. of Yorkshire, 13 m. SW of Bawtry, crossed by the Chesterfield canal. Area 2,180 acres. Pop. 313.

THORPE-SATCHVILLE, a chapelry and township in Twyford p., Leicestershire, 5½ m. S by W of Melton-Mowbray. Pop. in 1851, 159.

THORPE-LE-SOKEN, a parish in Essex, 8 m. SE by S of Manningtree. Area 3,203 acres. Pop in 1831, 1,173; in 1851, 1,294.

THORPE-SUPER-MONTEM, a parish in Lincolnshire, 6 m. SW by W of Lincoln. Area 1,820 acres. Pop. in 1831, 273; in 1851, 379.

THORPE-UNDERWOODS, a township in Little Ouseburn p., Yorkshire, 6½ m. SE of Aldborough. Area 2,185 acres. Pop. in 1831, 144; in 1851, 163.

THORPE (West), or Thorpe-in-the-Fallows, a parish in Lincolnshire, 7½ m. NW by N of Lincoln. Area 640 acres. Pop. in 1831, 62; in 1851, 56.

THORPE-WILLOUGHBY, a township in Brayton p., Yorkshire, 2½ m. WSW of Selby, intersected by the Leeds and Selby railway. Area 850 acres. Pop. in 1831, 148; in 1851, 161.

THORSENGE. See Taasing.

THORSHALLA, a village of Sweden, in the laen of Nykoping, near the SW shore of Lake Mälar. Pop. 500.

THORSHAVN, a town of the Faroe islands, on the SE coast of Stromoe. Pop. 500. It is the chief place in the archipelago, and has a Latin school and an hospital. It has a good port defended by fortifications, and possessing an active trade. The houses, about 100 in number, are small and built of wood.

THORVERTON, a parish in Devon, 6 m. E by N of Crediton. Area 4,036 acres. Pop. 1,511.

THOSOE, an island of the North sea, near the W coast of Norway, in the dio. of Drontheim and bail. of North Drontheim, and a little to the S of the island of Hotteren, in N lat. 63° 37′, E long. 8° 27′.

THOSTE, a village of France, in the dep. of the Cote-d'Or, cant. and 6 m. NW of Precy-sous-Thil, on a height, near the l. bank of the Serain. Pop. 400.

THOU, a village of France, in the dep. of the Loiret, cant. and 9 m. SE of Briare. Pop. 390.

THOU (Le), a village of France, in the dep. of the Charente-Inferieure, cant. and 3 m. SSE of Aigrefeuille. Pop. 700. It has the remains of the ancient castle of Sigogne.

THOUARCE′, a canton and commune of France, in the dep. of the Maine-et-Loire and arrond. of Angers. The cant. comprises 20 coms. Pop. in 1831, 18,957; in 1846, 19,244.—The village is 15 m. S of Angers, on the r. bank of the Leyon. Pop. 1,671. The locality affords good wine.

THOUARE, a town of France, in the dep. of the Loire-Inferieure, cant. and 4 m. SE of Carquefou, in a marshy locality, near the r. bank of the Loire. Pop. 800.

THOUARS, a canton, commune, and town of France, in the dep. of the Deux-Sevres and arrond. of Bressuire. The cant. comprises 24 coms. Pop. in 1831, 15,355; in 1846, 15,460.—The town is 17 m. NE of Bressuire, on the slope of a hill, near the r. bank of the Thoue. Pop. in 1846, 2,397. It is enclosed by walls partly in ruins, and has two fine churches, a communal college, two hospitals, fine promenades, one of which is in the centre of the town, and a castle. It possesses manufactories of woollen fabrics, linen, hats, cutlery, and tanneries; and carries on an active trade in corn, horses, mules, cattle, goats, fowls, hemp, nuts, almonds, and brandy. This town is of great antiquity. It was taken and fortified by Pepin in 758, and at a later period was ceded to the English, and rendered by them one of the places in Poitou; but in 1372, was taken by capitulation by Duguesclin. Towards the end of the 16th cent., it was erected into a duchy in favour of the lords of La Tremouille. The revocation of the edict of Nantes proved fatal to its prosperity. In 1793, it fell into the hands of the Vendeens.

THOUARSAIS, a village of France, in the dep. of La Vendee, cant. and 8 m. SW of Chataigneraye. Pop. 1,100.

THOUCHETHI, a canton of Russian Georgia, watered by the Upper Alasan. Its mountains are inhabited by the Tchouchi, a pastoral people, who profess the Greek religion.

THOUE, or Thouet, a river of France, which has its source in the dep. of the Deux-Sevres, at the village of La Pointerie; waters Secondigny, Parthenay, St. Loup, Airvault and Thouars; enters the dep. of the Maine-et-Loire; passes Montreuil-Bellay, and adjacent to Saumur, and a little below the latter town, throws itself into the Loire, on the l. bank at St. Hilaire, and after a course first in a NE direction, then N, and afterwards NNW direction of about 81 m. Its transports consist chiefly in grain, brandy, wine, and forage. Its principal affluents on the l. are the Cebron, Thouaret, and Argenton, and on the r. the Dive.

THOUIN, a village of Military Croatia, in the generalat and 21 m. SW of Cronstadt.

THOULOURENC, a river of France, which has its source in the mountain of Peyre-Belle, in the dep. of the Drôme; runs first S, then WNW; enters the dep. of the Vaucluse; flows through a narrow valley, at the foot of Mont Ventoux, and after a course of about 24 m. joins the Ouveze, on the l. bank, 4 m. E of Vaison.

THOUNE. See Thun.

THOUNY, a nomade Arab tribe, who inhabit the S part of the Persian prov. of Khorassan. They number about 15,000.

THOUNZA, a town of Burmah, in the territory of Cochanpri, 75 m. E of Ava, near the Myynguyamyyt.

THOUR. See Tor.

THOURIE, a commune of France, in the dep. of the Ille-et-Vilaine, and cant. of Rhetiers. Pop. 1,047.

THOUROUT, a department, commune, and town of Belgium, in the prov. of West Flanders, and arrond. of Bruges. Pop. of dep. 8,204. The town is 11 m. SSW of Bruges. Pop. 2,505. It has manufactories of hats, linen, carding machines, earthenware, and chicory, and several oil-mills. It returns a deputy to the provincial states. This town is of great antiquity. An abbey is said to have been founded here by Dagobert I., which was given in 830 by Louis-le-Debonnaire to the bishop of Hambourg, and ultimately destroyed by the Normans.

THOURY, a village of France, in the dep. of the Eure-et-Loir, cant. and 2 m. E of Janville. Pop. 1,232. It has manufactories of beet-root sugar, and hosiery, has a considerable trade in horses, asses, mules, and cattle. Also a village in the dep. of the

Ille-et-Vilaine, cant. and 6 m. SW of Retiers. Pop. 1,059.

THOURY-FEROLLES, a village of France, in the dep. of the Seine-et-Marne, cant. and 5 m. NNE of Lorrez-le-Bocage, on the slope of a hill, near the l. bank of the Oroanne. Pop. 570.

THOURY-SUR-ABRON, a village of France, in the dep. of the Nievre, cant. and 4 m. ENE of Dorne. Pop. 326.

THOUSAND ISLANDS, a numerous group of islands in the St. Lawrence river, which is here from 2 to 9 m. broad, a little below Lake Ontario. The American portion of them belong to Jefferson co., in the state of New York, and the remainder to Canada. They are rocky, and generally covered with cedar and hemlock to the water's edge. Their number is estimated at 1,500, within an area of 27 m., and in many instances so close together as scarcely to allow passage to a canoe. The largest belonging to Canada are Grand and Howe islands, and to the United States, Carlton, Grindstone, and Wells. See St. Lawrence.

THOYDON-BOIS, a parish in Essex, 3 m. S by W of Epping. Area 2,176 acres. Pop. in 1851, 591.

THOYDON-GARNON, or Coopershall, a parish in Essex, 2 m. SE by S of Epping. Area 3,161 acres. Pop. in 1831, 841; in 1851, 1,237.

THOYDON-MOUNT, a parish in Essex, 3 m. SE by E of Epping. Area 1,500 acres. Pop. 194.

THRACE. See Romelia.

THRANDESTON, a parish in Suffolk, 2¾ m. NW of Eye. Area 1,375 acres. Pop. in 1851, 397.

THRAPSTON, a parish and market-town in Northamptonshire, 20 m. NNE of Northampton, on the river Nen, over which there is here a good stone-bridge. Area of p. 990 acres. Pop. in 1801, 675; in 1831, 1,014; in 1851, 1,183. The town, which consists of four streets disposed in the form of a cross, is in general well-built, and occupies a cultivated valley on the E banks of the Nen, which is navigable from Lynn to Northampton, and has greatly facilitated the trade of this place.

THRECKINGHAM, a parish in Lincolnshire, 2¼ m. N by E of Folkingham. Area 2,270 acres. Pop. in 1831, 191; in 1851, 180.

THREE-BROTHERS, a group of mountains in New South Wales, in the co. of Macquarie. The most northerly is in S lat. 31° 43′, and E long. 152° 45′, and is visible at the distance sea-ward of about 60 m.

THREE-HILLS ISLAND, one of the New Hebrides, in the S. Pacific ocean, in S. lat. 17° 6′. It is about 12 m. in circumf.

THREE-HUMMOCK ISLAND, an island in the W part of Bass strait, in the group of Hunter Isles, between King's Island and the N extremity of Van Diemen's Land. It was discovered in 1798 by Flinders.

THREE-ISLANDS, small islands in the Eastern seas, near the E coast of Bintang, in S lat. 1° 10′.

THREE-ISLAND HARBOUR, a bay on the coast of Patagonia, in the straits of Magalhaen, 8 m. NNW of Bachelor's river.

THREE-ISLES, a group off the NE coast of Australia, in S lat. 15°.

THREE-KINGS, three small islands in the S. Pacific, in S lat. 34° 13′. They are all uninhabited, but the largest is said to have both pigs and goats running wild upon it.

THREE-POINTS CAPE, the N head of the entrance into Broken bay, on the E coast of Australia, in S lat. 33° 32½′.—Also a cape on the E coast of America, in the bay of Honduras, in N lat. 15° 56′.—Also a cape on the E coast of Patagonia, in S lat. 46° 50′, forming the S extreme of St. George's gulf.

THREE-RIVERS, a village of St. Joseph's co., in the state of Michigan, U. S., on St. Joseph's river, near the junction of Portage river and Rocky creek.—Also a village of Palmer township, Hampden co., in the state of Massachusetts, 76 m. WSW of Boston, on Chicapee river, near the junction of Ware and Swift rivers, and intersected by the Amherst and Belcherton railway. Pop. in 1840, 100.

THREE-RIVERS, a town of Lower Canada, situated on the NW side of the river St. Maurice, at its confluence with the St. Lawrence, 90 m. from Quebec. It derives its name from the entrance into the former river being here separated by two islands lying in its mouth, into three channels. The town stands on a light and sandy soil, which extends also over the environs. Towards the St. Maurice the ground rises considerably, but in the opposite direction it sinks almost to a level with the river. The principal public buildings are the Ursuline convent, the Protestant and Catholic churches, the court-house, jail, and barracks. The major part of the private dwelling-houses are built of wood, the oldest of them only one story high, having small gardens about them; those of more recent date are in a better style. The trade carried on here is chiefly in British manufactured goods, which from hence are distributed through the middle district of the province. The exports consist of wheat, timber, and its iron foundry, and a small quantity of peltry. Several pot and pearl ash manufactories, and two or three breweries, and brick manufactories, considerably increase the general trade of the place. As a shipping port it is conveniently situated, there being a sufficient depth of water for ships of large tonnage to lie close to the wharfs, and receive or discharge their cargoes by a temporary stage from their gangways. The pop. in 1854 was 6,500.

THREE-RIVERS HARBOUR, a bay of Nova Scotia, on the E coast of Prince Edward's Island, in the gulf of St. Lawrence, in N lat. 46° 8′, and W long. 66° 50′.

THRELKELD, a chapelry and township in Greystock p., Cumberland, 4 m. ENE of Keswick. Area 4,710 acres. Pop. in 1831, 320; in 1851, 387.

THRESHFIELD, a township in Linton p., Yorkshire, 8 m. N of Skipton. Area 2,644 acres. Pop. in 1831, 212; in 1851, 271.

THREXTON, a parish in Norfolk, 2¼ m. W by S of Watton. Area 1,097 acres. Pop. in 1851, 61.

THRIALETHI, a district of Russia in Europe, in the gov. of Georgia. Its chief town is Tsalki.

THRICKBY, or Thrigby, a parish in Norfolk, 6 m. WNW of Yarmouth. Area 575 acres. Pop. 49.

THRINGSTONE, a township in Whitwick p., co. of Leicester, 5 m. E by N of Ashby-de-la-Zouch. Pop. in 1831, 1,267; in 1851, 1,298.

THRINTOFT, a township in Ainderby-Steeple p., Yorkshire, 2¾ m. W by S of Northallerton. Area 1,183 acres. Pop. in 1831, 170; in 1851, 160.

THRIPLOW, a parish in Cambridgeshire, 7½ m. S of Cambridge. Area 2,489 acres. Pop. 521.

THROCKING, or Thorking, a parish in Hertfordshire, 1½ m. WNW of Buntingford. Area 903 acres. Pop. in 1831, 83; in 1851, 35.

THROCKLEY, a township in Newburn p., Northumberland, 6 m. NNW of Newcastle-upon-Tyne. Area 1,268 acres. Pop. in 1831, 208; in 1851, 159.

THROCKMORTON, a township and chapelry in Fladbury p., Worcestershire, 3½ m. NE of Pershore. Area 1,560 acres. Pop. in 1831, 159; in 1851, 153

THROG'S POINT, a headland of Westchester township and co., in the state and 16 m. NE of New York. It has a fort erected for the defence of the entrance to Long Island Sound.

THRONSE, Troenses, or Trouses, a town of Den-

THUN.

mark, in the diocese of Tyen, bail. and 2 m. SSE of Svendborg, on the NE coast of the island of Taasinge, of which it is the capital, and opposite the island of Thoroe. Pop. 1,000. It has a marine school, a well organised alms-house, building-docks, &c., and possesses an active trade in fruit, fuel, &c.

THROOPSVILLE, a village of Mentz township, Cayuga co., in the state of New York, U. S., 3 m. NW of Auburn, on Owasco inlet. Pop. 150.

THROPTON, a township in Rothbury p., Northumberland, 13 m. SW of Alnwick, near the confluence of the Snitter-burn and the Coquet. Area 827 acres. Pop. in 1831, 218; in 1851, 248.

THROWLEY, a parish in Devon, 6 m. ESE of Oakhampton. Area 1,943 acres. Pop. in 1831, 460; in 1851, 395.—Also a parish in Kent, 4 m. SW by S of Faversham. Area 3,180 acres. Pop. 614.

THRUMPTON, a parish in Notts, 7 m. SW of Nottingham. Area 1,080 acres. Pop. in 1851, 133.

THRUSHELTON, a parish in Devon, 10 m. SW by W of Oakhampton. Area 3,714 acres. Pop. 535.

THRUSSINGTON, a parish in Leicestershire, 8½ m. NNE of Leicester, on the river Wreak, and in the line of the Melton canal and the Roman Foss road. Area 2,200 acres. Pop. in 1851, 544.

THRUXTON, a parish in Herefordshire, 6 m. SW of Hereford. Area 437 acres. Pop. in 1831, 59; in 1851, 71.—Also a parish in Hants, 5½ m. W of Andover. Area 1,865 acres. Pop. in 1851, 267.

THRYBERGH, or THRIBERG, a parish in the W. R. of Yorkshire, 3¼ m. NE of Rotherham. Area 1,624 acres. Pop. in 1831, 332; in 1851, 331.

THSANG. See TIBET.

THSIANGABRANG, a town of Tibet, in the prov. of Thsang, at some distance from the l. bank of the Yaru-dzang, 24 m. WSW of Thsianglodzian.

THSIANGAMRING, a town of Tibet, in the prov. of Thsang, on the Chang, 180 m. WSW of Lhassa.

THSIANGLODZIN, a town of Tibet, in the prov. of Thsang, on the r. bank of the Yaru-dzangbo, and 240 m. WSW of Lhassa.

THUBŒUF, or TUBOEUF, a commune of France, in the dep. of the Mayenne, cant. and 5 m. NE of Massy. Pop. 1,197.

THUEYT'S, a canton, commune, and town of France, in the dep. of the Ardeche and arrond. of Largentiere. The cant. comprises 9 coms. Pop. in 1831, 14,220; in 1846, 15,225.—The town is 18 m. N of Largentiere. Pop in 1846, 2,876.

THUILE-ST.-BERNARD (LA), a village of Sardinia, in the div. and prov. and 19 m. W of Aosta, and mand. of Morgex, at the foot of Little St. Bernard, surrounded by meadows which are irrigated by a rivulet that bears its name. Hence the road leads up to a *col* and plain on which are evidences of the encampment of Hannibal, and a circle of stones round the most elevated part of the plain is called *Cirque d'Annibal*. The circle is 260 paces in circumf., and the rude blocks of stone which mark it are rather more than three paces apart, and somewhat buried in the soil.

THUILLES, a department and commune of Belgium, in the prov. of Hainault and arrond. of Charleroi. Pop. of dep. 1,270; and of village, 855. It has manufactories of soap.

THUILLEY-AUX-GROSEILLES, a village of France, in the dep. of the Meurthe, cant. and 7 m. NE of Colombey, in a gorge. Pop. 300. It has a castle, and on a plateau of an adjacent village are the ruins of the royal residence of Tussey. This village, which was once an important place, was ruined in the wars of the 17th cent., and by the pestilence of 1709.

THUIN, a department, commune, and town of Belgium, in the prov. of Hainault and arrond. of Charleroi. Pop. of dep. 3,865. The town is 10 m. SW of Charleroi, in a deep hollow on the r. bank of the Sambre, a little below the confluence of the Biemelle. It has a college, and possesses manufactories of woollen and linen fabrics, iron-ware, and leather. It returns a deputy to the provincial states. In the vicinity are the important iron-mines of Hourbes. This town owes its origin to a fortress which belonged to the adjacent abbey of Lobbes. It was erected into a town and fortified in 972. In 1053, it was destroyed by Baudouin V. count of Flanders. In 1466, it was taken by Charles-le-Hardi, duke of Burgundy.

THUIR, a canton, commune, and town of France, in the dep. of the Pyrenees and arrond. of Perpignan. The cant. comprises 20 coms. Pop. in 1831, 8,564; in 1846, 9,400.—The town is 9 m. SW of Perpignan, on a canal which issues from the r. bank of the Tet. It has manufactories of pottery, a tannery, oil and paper mills, and distilleries of brandy. Bees are extensively reared in the locality.

THUIT-SIGNOL (LE), a village of France, in the dep. of the Eure, cant. and 3 m. NNE of Tourville-la-Campagne, in a valley. Pop. 1,050.

THULIN, a department, commune, and town of Belgium, in the prov. of Hainault and arrond. of Mons. Pop. of dep. 2,050. The town is 10 m. WSW of Mons, on the Haine. Pop. 1,485. It has a spinning mill, and is noted as the scene of an engagement between the French and Austrians, in 1792.

THULLADA, a village of the Punjab, on the road from Ferozpur to Moultan.

THUM, a town of Saxony, in the circle of the Erzgebirge, bail. and 6 m. WNW of Wolkenstein. Pop. in 1843, 2,095. It has manufactories of hosiery, parchment, and lace; and in the vicinity are mines of silver, tin, and cobalt.

THUMA. See TUMA.

THUMAIDE, a department and commune of Belgium, in the prov. of Hainault, and dep. of Tournai. Pop. of dep. 788.

THUMERIES, a village of France, in the dep. of the Nord, cant. and 4 m. SW of Pont-a-Marcq. Pop. 765. It has manufactories of beet-root sugar, a tannery, and fine nursery gardens.

THUN, a village of France, in the dep. of the Nord, cant. and 3 m. NNE of St. Amand, on the l. bank of the Scarpe. Pop. 900. It formerly had a fine abbey. Hay is its chief article of trade.

THUN, or THOUNE, a town of Switzerland, in the cant. and 17 m. S of Berne, in a picturesque situation, near the NW extremity of the Thuner-see, on the Aar, which here divides into two branches, one of which runs through the town. Pop. 3,379. It has a fine parish-church in a commanding situation, an hospital, several good schools, an orphan's asylum, a public library, a federal military school, and on an adjacent mountain, a castle. It has a considerable trade in linen and in cheese.—The Thuner-see or Lake Thun, which is 12 m. in length, and 2 m. in extreme breadth, is formed by the Aar, which enters its SW extremity and issues from it on the NW. It is enclosed by lofty rocky eminences, amongst which the most remarkable are the Niesen, Stockhorn, and Morgenberghorn, and has a level of 606 yds. above that of the Mediterranean. It has a depth in some places of 720 ft.; and its waters abound with excellent fish.

THUNDER-BAY, an indentation of the Canadian coast, formed by Lake Huron, between the bay of Saganaw and the strait of Michilimakinak. It receives a river, and contains several small islands of the same name. It derives its name from the frequency of thunder in the locality.

THUNDERIDGE, a parish in Hertfordshire, 2 m. N by E of Ware. Area 2,200 acres. Pop. 572.

THUNDERSLEY, a parish in Essex, 2¾ m. SW of Rayleigh, including the hamlet of Thundersley. Area 2,499 acres. Pop. in 1831, 526; in 1851, 492.

THUNGEN, a market-town of Bavaria, in the circle of Lower Franconia, 5 m. ESE of Carlstadt, on the Werrin. Pop. 880. It has two castles.

THUNGERSHEIM, a village of Bavaria, in the circle of Lower Franconia, 9 m. NNW of Wurtzburg, on the r. bank of the Main. Pop. 1,300. The locality affords good wine.

THUNOE, an island of Denmark, in the dio. of Sieland and bail. of Holbek, in the Cattegat, 3 m. from the NW coast of the island of Samsoe, in N lat. 55° 58', E long. 10° 17'. It is 2 m. in length from WNW to ESE, and 1½ m. in breadth, and is to a great extent covered with furze. It forms a small parish.

THUN-SAINT-MARTIN, a village of France, in the dep. of the Nord, cant. and 5 m. NNE of Cambray, on the r. bank of the Schelde. Pop. 859. It has a manufactory of beet-root sugar.

THUNSTETTIN, a village of Switzerland, in the cant. of Berne and bail. of Aarwangen. Pop. 1,774. It has a fine castle.

THUR, a river of Switzerland, which has its source in the cant. of St. Gall, in the district of Haut-Tockenburg, near Wildhausen, 12 m. SSW of Appenzel; runs first NNW, then NE; enters the cant. of Thurgovia; passes Bischofszell, where it is joined by the Sitter on the r.; turns afterwards WNW; receives the Murg on the l.; waters the N part of the cant. of Zurich; and after a course of 72 m., throws itself into the Rhine on the l. bank, 7 m. SSW of Schaffhausen. It is navigable only a short distance.

THUR, or TUR, a river of Hungary, which has its source at the NE extremity of the comitat of Szathmar, 5 m. W of Visk; runs first S, then WNW; waters the S part of the comitat of Ugocz; re-enters that of Szathmar; traverses extensive marshes; and after a sinuous course of about 84 m., joins the Theiss on the l. bank, a little above the confluence of the Szamos.

THURCASTON, a parish and village in Leicestershire, 3¾ m. S by W of Mount-Sorrell, comprising the chapelry of Anstey and the township of Cropston. Area of p. 2,960 acres. Pop. in 1831, 1,241; in 1851, 1,102. The inhabitants are chiefly employed in framework-knitting.

THURDA. See THORENBURG.

THURE, a river of France, in the dep. of the Haut-Rhin, which descends from the Grand-Ventron in the Vosges, on the confines of the dep. of that name; runs first SE through the fine valley and town of St. Amarin, Thun and Cernay; directs its course afterwards NE, then N; bathes the cantons of Ensisheim, Soultz, Rouffach, Wintzenheim, and Colmar; and 2 m. S of the latter town, joins the Lauch on the r. bank, and after a total course of about 45 m.—Also a commune in the dep. of the Vienne, cant. and 5 m. WNW of Chatellerault, at the foot of hills. Pop. 1,631.

THUREN, a village of Prussia, in the prov. of W. Prussia, regency and circle and 3 m. SW of Gumbinnen. Pop. 100. It has a mineral spring.

THURET, a commune of France, in the dep. of the Puy-de-Dome, cant. and 5 m. SE of Aigueperse. Pop. 1,884.

THURGARTON, a parish in Norfolk, 4½ m. SW by S of Cromer. Area 961 acres. Pop. in 1831, 247; in 1851, 272.—Also a parish in Notts, 9½ m. NE of Nottingham, and 2 m. W of the Trent. Area 2,770 acres. Pop. in 1831, 329; in 1851, 385.

THURGAU, or THURGOVIA, a canton in the NE of Switzerland, bounded on the N by the Rhine, which separates it from Baden and Schaffhausen; on the E by the lake of Constanz, separating it from Bavaria, Wurtemberg, and Baden; on the S by the cant. of St. Gall; and on the W by Zurich and Schaffhausen. Its area is 270 sq. m.; its pop. in 1850, 88,908, of whom one-fourth only are Catholics. The surface is comparatively level, the hills, where they occur, not exceeding a height of 1,400 ft. above the lake of Constanz. They run NE and divide the cant. into three valleys, of which the central and largest is traversed by the Thur. The productions of this cant. are wheat, barley, oats, rye, flax, potatoes, hemp, and vines; the pastures are extensive; and there are numerous and large orchards. Cotton and silk goods are manufactured here; but the staple article is linen. The cant. is divided into eight bailiwics, and is governed by two *landrammans* chosen annually, and a council of 100 members chosen by all citizens above 25 years of age.

THURGOLAND, a township in Silkstone p., Yorkshire, 5 m. SW of Barnesley. Area 2,080 acres. Pop. in 1831, 1,147; in 1851, 1,548.

THURINGERWALD, a hilly and woody tract of country in the centre of Germany, extending from the sources of the Werra, NW to near Eisenach. It is a part of the ancient Hercynian forest, and has a length of about 70 m., with a breadth of from 9 to 16 m. It is thinly peopled, but is rich in metals, particularly iron and cobalt. Its highest peaks vary in height from 2,000 to 3,200 ft. It is covered with wood in almost every direction, and is traversed by only two great roads. It gives rise to a number of streams which flow into the adjacent plain, and eventually into the Main, the Weser, and the Elbe.

THURINGIA, the name formerly borne by an extensive tract of country in the central part of Germany, having Franconia on the W, and the country of Meissen on the E. It long gave the title of landgrave to the elector of Saxony, though, from its extent, it was subdivided among a number of petty princes. The name occurs in history so early as the 6th cent., and became gradually disused after the incorporation of the territory with other states. Its capital was Erfurt.

THURLASTON, a hamlet in Dunchurch p., Warwickshire, 1 m. W of Dunchurch. Pop. in 1831, 281; in 1851, 287.—Also a parish in Leicestershire, 6 m. NE by E of Hinckley, including the liberty of Newparks, with the hamlet of Normanton-Turville. Area 2,980 acres. Pop. in 1851, 797.

THURLBY, a parish in Lincolnshire, 8 m. SW of Lincoln, on the river Witham. Area 1,802 acres. Pop. in 1831, 145; in 1851, 156.

THURLBY-WITH-NORTHOPE, a parish in Lincolnshire, 5½ m. NNW of Market-Deeping. Area 5,070 acres. Pop. in 1831, 632; in 1851, 799.

THURLEBERE, a parish in Somersetshire, 3¾ m. SE of Taunton. Area 949 acres. Pop. in 1851, 212.

THURLEIGH, or THURLEY, a parish in Bedfordshire, 6 m. N of Bedford. Area 3,480 acres. Pop. in 1831, 538; in 1851, 498.

THURLES, a parish, containing a town of the same name, in co. Tipperary. Area of p. 8,268 acres. Pop. in 1831, 10,459; in 1851, 10,546. The surface is watered southward through the centre by the river Suir, and consists wholly of a section of the rich valley of that river.—The town stands on the Suir, at the intersection of the road from Clonmel to Roscrea, 3 m. NE by N of Holycross. The ruins of several ecclesiastical and castellated buildings are objects of interest in themselves, and monuments of the town's antiquity and comparative early importance. Among the modern public edifices are a neat church, an infantry barrack, a bridge, a bridewell, a court-house, a market-house, a fever-hospital, and a poor-law union workhouse. But Thurles derives its distinguishing character, both architectural and social, from its being the seat of the Roman Catholic archb. of Cashel, and, with the exception of Galway, the most decidedly Roman Catholic town in Ireland. The archbishop's residence is a beautiful and commodious dwelling. The cathedral is a simple structure, large enough to accommodate from 7,000 to 8,000 persons, with a magnificent altar. At either side of the cathedral stands a convent, one of Ursuline, and the other of Presentation nuns. T. is the market for the sale of the agricultural produce of an extensive circumjacent district of country; and the depot for the supply of that district with all the innumerable goods of a general retail trade. Pop. in 1851, 5,908.

THURLESTONE, a parish in Devon, 4¼ m. W by S of Kingsbridge. Area 1,898 acres. Pop. 460.

THURLOW (Great), a parish in Suffolk, 6½ m. NW by W of Clare. Area 2,023 acres. Pop. 431.

THURLOW (Little), a parish in Suffolk, 7 m. NW by W of Clare. Area 1,470 acres. Pop. 449.

THURLOW'S ISLAND, an island of the N. Pacific, near the coast of N. America, in N lat. 50° 24'. It is about 24 m. in length from E to W.

THURLOXTON, a parish in Somersetshire, 5 m. SSW of Bridgewater. Area 551 acres. Pop. in 1831, 229; in 1851, 192.

THURLSTON, a township and village in Penistone p., Yorkshire, 9½ m. W by S of Barnesley. Area 7,740 acres. Pop. in 1831, 1,599; in 1851, 2,018.

THURLTON, a parish in Norfolk, 10 m. SSW of Great Yarmouth. Area 1,170 acres. Pop. 445.

THURM, a village of Saxony, on a small affluent of the Mulda, 5 m. NNE of Zwickau. Pop. 1,100.

THURMASTON (North and South), a chapelry and village, partly in the p. of Barkby and partly in that of Belgrave, Leicestershire, 3½ m. NNE of Leicester, on the E bank of the Soar, and in the line of the Midland Counties railway. Area of South T. 1,220 acres. Pop. of both in 1851, 1,251.

THURNAU, a town of Bavarian Franconia, 6 m. SW of Culmbach, the chief place of the lordship of Giech, of 84 sq. m. Pop. 1,425. The town has a lyceum, a castle, and two churches.

THURNBY. See Thornby.

THURNBY, a parish of Leicestershire, 4 m. E by S of Leicester Area 2,740 acres. Pop. 374.

THURNE, a parish of Norfolk, 3½ m. N by E of Acle. Area 660 acres. Pop. in 1851, 330.

THURNEN, a village of Switzerland, in the cant. and 12 m. SE of Berne. Pop. in 1850, 284.

THURNHAM, a township of Lancashire, in the p. and 5 m. SSW of Lancaster. Area 1,470 acres. Pop. in 1831, 526; in 1851, 748.

THURNING, a parish of Huntingdon and Northampton shires, 7½ m. SW by W of Stilton. Area 1,000 acres. Pop. in 1851, 211.—Also a parish of Norfolk, 4½ m. N by W of Reepham. Area 1,504 acres. Pop. in 1851, 212.

THURNSCOE, a parish of the W. R. of Yorkshire, 8 m. E of Barnesley. Area 1,665 acres. Pop. in 1831, 223; in 1851, 298.

THUROTZ, or Thurocz-var-Megye, a palatinate in the NW of Hungary, lying between those of Trentsin, on the NW, and Sohl on the SE. It has a territorial extent of 450 sq. m., with 45,000 inhabitants, chiefly Slowaks. It is very hilly, being traversed by ridges of the Carpathians. The principal grain raised is buckwheat. Much of the surface is covered with forests. The Waag crosses the N part, and is joined by the Thurocz from the S. The chief town is Szent-Martony.

THURROCK (Gray's), a town and parish of Essex, situated on a small creek of the Thames, which is here navigable for hoys and vessels of 300 tons burden, 4 m. NW of Gravesend. The p. has an area of 1,634 acres. Pop. in 1831, 1,248; in 1851, 1,713. The town consists principally of one street.

THURROCK (Little), a parish of Essex, 1 m. E by S of Gray's Thurrock. Area 1,495 acres. Pop. in 1831, 302; in 1851, 308.

THURROCK (West), a parish of Essex, 1 m. W of Gray's Thurrock. Area 3,607 acres. Pop. 754.

THURSBY, or Thor's Town, a parish and village of Cumberland, 6½ m. SW of Carlisle. Area 2,984 acres. Pop. in 1831, 564; in 1851, 591. The p. is said to derive its name from the god Thor. At Woodrigs, a ½ m. NW of the church, stood a building supposed to have been a temple dedicated to that god, the foundation of which was dug up about 80 years since.

THURSCROSS, a township of Fewston p., Yorkshire, 14 m. W by N of Knaresborough. Area 6,340 acres. Pop. in 1831, 601; in 1851, 339, a decrease attributed to the stoppage of factories.

THURSFORD, a parish of Norfolk, 3½ m. SE by E of Little Walsingham. Area 1,350 acres. Pop. in 1831, 392; in 1851, 346.

THURSLEY, a parish of Surrey, 5½ m. SW by W of Godalming. Area 4,348 acres. Pop. 756.

THURSO, a river of Caithness-shire, which rises among the mountains in the SW corner of Halkirk, near the boundary with Sutherlandshire, and flows 12½ m. NE, and 14 m. N to the head of its cognominal bay, at the town of Thurso. Its tributaries are neither numerous nor large; and its greatest breadth is about 300 ft. Its banks are almost everywhere destitute of wood.—Also a parish on the N coast of Caithness-shire, having an area of about 35 sq. m. The surface gently rises from the shore, and is throughout a slightly inclined plane. The headland of Brimsness flanks the estuary of the Forss; and the magnificent promontory of Holborn-head runs out on the W side of Thurso-bay, forming a twin object to Dunnet-head, 7 m. to the NE. The soil is principally clay and loam lying on rock. The arable and the untilled lands bear the proportion to each other of 6 to 5. Pop. in 1851, 5,096.

THURSO, a town at the mouth of Thurso-water, in the above parish, 20 m. W by S of John-o'-Groat's-house, and 20¾ m. NW of Wick. It consists of an old and a new town: the latter occupying a pleasant and elevated situation on the SW; and, if completed on the plan on which it was originally designed, would be a very handsome place. The pop. in 1851 was 2,908. Linen, woollen, leather, and herring-net manufactures employ about 230 persons; and the manufacture of straw-plait employs about 60 females. The fisheries in the bay are extensive; and, along with the dressing of flags in the neighbouring quarries, afford the chief employment.—The harbour, when a bar at its entrance has been crossed, is abundantly safe.—The bay of Thurso lies exactly at the W entrance of the Pentland firth, and is in the direct course of the North Sea trade with our American possessions. It is a capacious and secure roadstead in all weathers, and naturally adapted for the construction of a complete and convenient harbour.

THURSTANTON, a parish of Cheshire, 5 m. NW by N of Great Neston. Area 2,762 acres. Pop. in 1831, 92; in 1851, 98.

THURSTON, a parish of Suffolk, 5½ m. E by N of St. Edmund's Bury. Area 2,200 acres. Pop. 759.

THURSTONLAND, a township and village of Yorkshire, 5 m. S by E of Huddersfield. Pop. in 1851, 1,320.

THURSTON-MERE, a river of Cumberland and Lancashire, which runs into the Fosse near Crakeford.

THURTON, a parish of Norfolk, 8½ m. SE of Norwich. Area 771 acres. Pop. in 1851, 242.

THURWASTON, a township of Derbyshire, 7½ m. WNW of Derby.

THURY, a village of France, in the dep. of Cote-d'-Or, 15 m. W of Beaune. Pop. of com. 1,800.—Also a village in the dep. of Yonne, 5 m. SE of Saint-Sauveur.

THURY-HARCOURT. See HARCOURT.

THUSIS, or TOSSANA, a town of the Swiss cant. of the Grisons, in the valley of the Domleschg, near the confluence of the Albula and the Hinter Rhine, 10 m. SSW of Coire. Pop. in 1850, 769. It has cattle and grain markets, and is the centre of transit trade between Italy and Germany.

THUXTON, or THURSTON, a parish of Norfolk, 5½ m. SE by S of East Dereham. Area 1,102 acres. Pop. in 1831, 83; in 1851, 133.

THWAITE, a parish of Norfolk, 4½ m. N by E of Aylesham. Area 676 acres. Pop. in 1851, 138.—Also a parish in the same co., 12 m. SE by S of Norwich. Area 500 acres. Pop. 130.—Also a parish of Suffolk, 4½ m. SW by S of Eye. Area 832 acres. Pop. 179.

THWAITES, a township of Cumberland, 10 m. SE of Ravenglass. Pop. in 1851, 374.

THWING, a parish in the E. R. of Yorkshire, 8½ m. WNW of Bridlington. Area 4,040 acres. Pop. in 1831, 350; in 1851, 444.

TIAGER, a town and fortress of Hindostan, in the Carnatic, in N lat. 11° 45′.

TIAHUANACU, or TIAGUANUCO, a village of Bolivia, 38 m. WNW of La Paz, and to the E of Lake Titicaca. It is celebrated for the remains of edifices supposed to have been the palaces of the Peruvian emperors, which still excite astonishment by the immensity of their stones.

TIAMUCHU, a river of Bolivia, in the prov. of Moxos, which runs from SW to NE, and enters the Marmore, in S lat. 14° 13′, by two branches.

TIANA, a village of Spain, in the prov. and 9 m. NE of Barcelona, near the coast. Pop. 1,600. About 1 m. to the W of this v. stood the beautiful monastery of Montalegre, which was sacked in the civil commotions of 1835.

TIASMIN, a river of Russia, which rises in the prov. of Kherson, 27 m. NNW of Elizabetgrad; flows ENE, and then SSE, and joins the Don on r. bank near Krilov.

TIBACUI, a settlement of New Granada, 30 m. W of Santa-Fe.

TIBAGY, a river of Brazil, which rises on the W flank of the Serra-do-Espinhaço; runs NNW; and joins the Paranapamena, on the l. bank after a course of 330 m.

TIBBERAGHNY. See TIPPERAGHNEY.

TIBBERMORE, or TIPPERMUIR, a parish in the SE division of Perthshire, with a superficial extent of about 4,670 acres. Pop. in 1851, 1,495.

TIBBERTON, a parish of Gloucestershire, 3½ m. SE of Newent. Area 1,400 acres. Pop. in 1831, 307; in 1851, 362.—Also a parish of Worcestershire, 4 m. ENE of Worcester. Area 1,320 acres. Pop. in 1831, 337; in 1851, 329.—Also a parish of Herefordshire, 10 m. W of Hereford. Area 1,111 acres. Pop. in 1831, 118; in 1851, 141.—Also a hamlet of Salop, 4½ m. W by N of Newport.

TIBBUS, a native African people of the Berber race, inhabiting the vast desert tract which extends along the S frontier of Fezzan, thence eastward towards Egypt, and along the N of Bornu. They are small in stature, have quick eyes, thick lips, a small turned-up nose, and well-shaped limbs. Their complexion is not perfectly black, and their hair is long and less curled than that of the Negroes. Sheep-skins form their usual dress, but the more opulent wear large blue shirts, wrapping their heads so closely in dark blue cloth as to leave nothing visible but the eyes. Their language derives a peculiar sound from the frequency of the consonants *l* and *s*, and from the rapidity with which it is pronounced, whence both ancients and moderns have compared it to the whistling of birds. The Rock Tibbus, who inhabit a mountainous district situated on the SE of Fezzan, are peculiarly rude and ferocious, and harass their neighbours with frequent hostilities. Though they carry on a trade with Fezzan for senna and camels, they frequently molest the caravans of that country in their journey to Bornu. The Tibbus of Bilma, inhabiting the tract between Fezzan and Bornu, are a milder race, and carry on most of the trade between the two countries.

TIBENHAM, a parish of Norfolk, 5 m. WSW of St. Mary Stratton. Area 3,286 acres. Pop. 727.

TIBER, a celebrated river of Italy, which rises in the Apennines, in the Tuscan prov. of Florence; flows first S, and then SSE; receives a number of smaller streams; passes by Rome, and falls into the Mediterranean, about 15 m. below that city, by two branches, the Fiumicino and the Fiumare. The length of its course is about 250 m., but it is navigable only for small vessels, and little more than 90 m. It has considerable depth and rapidity of current. See article ROME.

TIBERIAS. See TABARIAH.

TIBERON, or TIBURON, a village on the SW coast of the island of Hayti, on a bay formed by a cape of the same name on the NW, and by Point-Burgau on the SE. The cape is in N lat. 18° 20′ 30″.

TIBESTY, a district in the Sahara of Africa, to the SE of Fezzan, belonging to the Rock Tibbus. It is represented as a mountainous country, which, though suffering from the want of rain, abounds in springs and excellent pasturage. Senna of superior quality is raised here for the market of Fezzan, and the camels are highly esteemed. The huts of the natives are of a circular form, and composed of stakes covered with brush-wood.

TIBET, or THIBET, a region of Central Asia, lying between the parallels of 27° and 37° N, and the meridians of 72° and 105° E, or from the E frontiers of Badakshan on the NW to the W frontiers of the Chinese prov. of Sech-wen on the SE. The name, however, is generally confined by geographers to that portion of this region extending from the sources of the Indus and Sutledge to the W frontier of Sech-wen. This tract has the Gobi or Great desert on the N, separating it from the eastern part of Chinese Turkestan; the territory of the Eluths of Kokonor on the NE; the Sifan or Tufan, and the Chinese prov. of Sech-wen on the E; the prov. of Yun-nan on the SE; the Burman dominions, the valley of Assam, Butan, Nepal, and British India on the S; and Little Tibet on the W. These are its boundaries generally speaking, but its particular limits we cannot accurately determine for want of materials. Until within these few years, we had almost nothing but Chinese information, communicated through the channel of the Jesuits to guide us on the subject of the geography of this region, in addition to a few gleanings from Marco Polo, and such Romish missionaries as have visited it for the purpose of converting the natives to the Roman Catholic faith, as Fathers Andrada, Desideri, Horace-de-la-Penna, and the monk Cassiano. Horace spent 18 years in Lassa, and might consequently be supposed qualified to give us a tolerable account of Tibet; his account of the mission to T. was printed at Rome in 1742. A long and curious account of T., in 900 4to pages by Father Giorgi, was printed at Rome in 1762, under the title *Alphabetum Tibetianum;* but Cassiano is the real author of this book. All these productions, however, throw little light upon the geography of T., the attention of its missionaries being naturally much more engrossed with the religion than the geography of the country. Not above 19 or 20 pages are devoted in Giorgi's large work, to the latter subject, but an itinerary from Katmandu to Lassa is given by him, which slightly illustrates the geography of the intervening space. We have, in addition to these, an account of two embassies from the British East India Company to the Grand Lama in 1774 and 1783, the former by Mr. George Bogle is given in the Philosophical Transactions; the latter by Captain Turner, was published in 4to, 1800. But these concern only that part of T. which lies between Butan and Lassa. We have preserved in Kircher a short itinerary of Fathers Grœber and D'Orville from Sining to Lassa and from thence to Katmandu in Nepal; but it is so meagre that little satisfaction is got from the perusal. A Chinese account of T. was published at St. Petersburgh in 1828, in 2 vols. 8vo, by Father Hyacinth, long resident with the Russian mission at Pekin. The first of these vols. contains the geography of T.; the second its history and conquests, religion, manners, and customs. In this publication is inserted a chart of the road from Tschendu, or Chingtu in Sech-wen to H'lassa or Lassa. The orthography of the names of places, as given in this work, is so different from what has been usually employed in Du Halde's maps and other works as to render it a very difficult business to identify them. These, in conjunction with the Lama's map of T. were all our materials for a description of the country until within these twenty years. Klaproth has observed that the maps of the Chinese empire, made by the missionaries under Kanghi, were very imperfect extracts of the Chinese and Manshu originals, and that the proper names had been translated by a person but little versed in the Chinese language; and these materials were put into D'Anville's hands in order that he might reduce and superintend their publication. The maps made by order of Kienlung, he remarks, differ materially from those of the missionaries under Kanghi, so far as respects T. both in the longitudes and the latitudes, as in the case of the exit of the Sanpu from Tibet. He remarks farther that the greater the distance from the meridian of Pekin, the more erroneous are the longitudes in the Jesuits' maps of T. Of late years considerable information has been obtained respecting that portion of T. which borders on the NE confines of British India; and the French missionaries Huc and Gabet have made us somewhat acquainted with Eastern T.

Names and Divisions.] Among the natives, *Bhodt* and *Bhodtiul*, or *Bodyul*, i. e. 'the kingdom of Bhodt or Boodh,' are the most ordinary denominations of that portion of this region with which Europeans are most familiar; and the most usual designation of the people is *Bhotiyas.* The name *Thebeth*, *Thibet*, or *Tobbhote*, known to the Arabians and Persians as early as the 10th cent., is not used by the natives themselves. According to father Giorgi, Tibet [pronounced *Tibbet* in Bengal] is a corruption of the Tibetian appellation *Tenlt*, or 'kingdom of Boodh.' According to Hyacinth and Klaproth, the Tibetians add the word *ba*, signifying 'man,' to the word *bhot*, and call themselves and their country *Bhotba* and *Bhotas.* The Mongols use the term *tu* in place of *ba*, and call *Bhba*, *Tubat*, of which Europeans have made Tibet. According to this etymon, the name *Tibet* is originally Mongolian, and passed from the Mongols to the Persians and Arabs, and there can be no doubt that Marco Polo got the name from the Mongols when residing at the court of Kublai-Khan. It is also called *Poy*, or *Puë-Koachin*, *i. e.* 'the region of snow.' The Chinese sometimes call it *Si-fan*, or 'Western Fan,' though that name is now restricted to the region between the Tibetian prov. of Amdoa and Sech-wen. They also occasionally denominate Tibet by the name of *Chlassa*, and at other times by *Shan*, or *Sihan*, *i. e.* 'the western part of Shan.' The name *Shan*—or as Klaproth writes it, *Dzang*—is also given to the city of Lassa. *Tangut* is a Mongolian word, by which is designed all the region which touches the western frontier of China, and which is inhabited by Tibetians; but the Tangut, conquered by Jenghis-Khan, was distinct from T., and seems to have comprehended the Sifaun or Tufan, the region of Ko-konor to the W of Shen-si, the district of Kansu, part of Shen-si, and the territory of the Ortu-Mongols. Tangut is known in the Chinese annals under the names of the kingdom of *Si asp* and *Hya*, and *Si-Hya*, or 'Hya of the west.' *Oui Dzang*, or *Shan*, an ancient name of T., is compounded of the two words, *oui* and *dzang.* The Chinese transcribe this name by *Ou-szu-thsang;* and the Mongols render it by *Barohn-djao*, *i. e.* 'the right side,' or 'the West.'

The geographical divisions of T. are involved in obscurity, and the terms High, Middle, and Lower Tibet, are very vague. Marco Polo found it divided into eight kingdoms or provinces. Father Giorgi gives it twelve provinces: viz. Latak, now known to be a distinct sovereignty from T.; Nagari, which seems to correspond to the NW part of T.; Hor, with the lake of Terkiri; Kiang, Daum, and the principality of Kahang, in the N; Amdoa on the E; Brediong or Bramasciong on the S; together with Takbo and Congbo, and in the centre the provinces of Ou and Chang, the Oui and Dzang of Klaproth. Of these, Bramasciong seems to be to the N of Assam, and near the source of the Brahmaputra. On inspection of the Lama's map, we find the tract watered by the Sanpu, E and SE of Lassa, to be divided into *poys* or *poois*,—a term signifying a province, region, or district. Thus the tract to the S and SW of the Sanpu, betwixt it and the Omchu, is called Tak-poy or Tak-pooi, *i. e.* 'the prov. of Tak,' which comprehends a considerable number of inferior divisions, and which does not correspond to the Butan on the side of Bengal, as Malte Brun imagined, but is clearly the Tac-po of Giorgi. On the opposite, or NE side of the San-pu is a prov. called Konk-poo-i in the same map, also subdivided into a number of districts: this is clearly the Congbo of Giorgi; and to the SE of this is the prov. of Ken-pu-i, between the Sanpu and the Nu-kiang. Further E of this, is the country or district of Dsanclo, between the Lat-san-kiang and the Kin-sha-kiang. Amdoa is considered by Malte Brun to correspond to the Ardandam of Marco Polo; but this is merely putting the difficulty a little out of sight; for a reader will ask, where is Ardandam? Malte Brun indeed tells us that Ardandam is the SE part of Tibet, and the same with the Amdoa of Father Giorgi; Amdoa consequently is the SE part of T. Unfortunately for this conjecture, both Marsden in his notes on Marco Polo, and after him Klaproth, have shown that the Ardandam of that noted traveller makes no part of T, but corresponds to the SW part of the province of Yun-nan in China, of which Yong-chang, called Unchian by Marco Polo, is this capital. The name of this district, besides, is not Ardandam, or Arcladam, as Malte Brun has written it, after Muller's edition of Marco Polo; nor is it Kardandam, as Marsden has it; but Zar-dandan, a Persian appellation, signifying 'teeth of gold;' because in that district gold is abundant. Malte Brun has also identified Caridi, as he writes it, with Ardandam, but the proper reading is Carazan, a corruption of Karadjang, or, 'the Black country,' the NW part of Yun-nan, so called from the colour of the people, called Ou-man, or 'Black barbarians,' by the Chinese. The capital of this district is the modern Ta-li-fu, the Dai-leiou of the Chinese, and the capital of an independent state till destroyed by the Mongols in 1255, subsequent to the conquest of T. Marco Polo says, he took five days' journey on horseback W from Ta-li-fu to the prov. of Zardandan. The Caraiam of Marco Polo was certainly not Assam, as Malte Brun imagined, but lay to the W and N of Karadjang, and was called Karayan by the Mohammedan writers, and Thsuonman by the Chinese, who call the people Carains. The large lake of Eul-hai, or Tali, separates Karayan from Karazan or Karadjang. Klaproth places Caraiam to the SE of Karazan, and makes it the S part of Yun-nan, but Caraiam extended N and W as far as Caindhu of Marco Polo, on the Bruis, or Kin-sha-kiang—nor the Sanpo, as Klaproth dreams, or his Irrawaddy,—but the genuine Kin-sha-kiang of Tibet, and the NW part of Yun-nan. From what we have been able to gather from these confused and imperfect accounts of the divisions of T. proper are the following: 1st, Nagari, Nari, or Dzang, towards the chain that divides the sources of the Indus and Sutledge from Upper Tibet, and which contains the sources of the Yaru-Sanpu, the Kiang-ku, and other large rivers; 2d, Kam, probably the Kahang of Giorgi, called by him a principality, SW of Amdoa; 3d, Karra-Tœbet, on the Yarkia-Sanpu; 4th, Hor, on the N; 5th, Kongbo, on the l. of the Sanpu; 6th, Tac-po, on the r. of that river, and both SE of Lassa; 7th, Bregiong, to the W of Kongbo; 8th and 9th, Oui or Wei, and Chang or Tsang, both comprehended, according to Hyacinth, in the province of Chlassa or Lassa, and called Uchang; 10th, Jiamdo; 11th, Zanba, or Tishulumbu; and 12th, Amdoa, on the borders of Sech-wen. The district of Kiang lies to the N of Wei.

Physical features.] On an inspection of the Lama's map, T. presents a confused assemblage of mountains, valleys, lakes, and rivers, resembling a piece of net-work, of which to attempt a verbal description would be to make confusion worse confounded; and as no triangulation of this vast region was ever made, we cannot depend on the accuracy of the relative positions, or the directions of the interior chains. The declination of Eastern T., as indicated by the courses of the rivers, seems to be chiefly to the SE. From China to the E and N frontiers T. is a very sensible ascent the whole way. The mountains, which are very numerous, are represented as being higher above the horizon on the side towards China than on that towards T.; but when once they are passed, the climate becomes more mild, and the country less wild and savage, than it is on the borders of the Chinese provs. of Sech-wen and Yun-nan. According to the account of a Chinese officer, who served in the war of 1792 against Nepal, the army took 72 days to march from the frontiers of China to the vicinity of Lassa, 12 days more to Upper Tibet, and 15 days more to the W range which separates T. from Nepal. The passes, in his account, are spoken of with horror, some of them requiring a whole day to cross, and being totally impervious to cavalry. On the side of Hindostan the descent is much more rapid than on that of China, as being much shorter. When once the pass of Phari is passed, the contrast between Butan and T. is striking: the latter region appears to the eye as one of the least favoured countries under heaven, and seems in a great measure incapable of cultivation, exhibiting only low rocky hills, without any visible vegetation, or extensive arid plains of the most stern and stubborn aspect, promising fully as little as they produce. The very gradual descent of 70 m., from this pass to the Sanpo, was all that Turner saw of Tibet. The following description is given of the oriental plateau, or table-land, of T., by an eminent physical geographer. "The table land of T. is only 4,000 ft. above the sea towards the N, but it rises in Little T. to between 11,000 and 12,000 ft. According to Moorcroft the sacred lake Manasarowar, in Great T., and the surrounding country is 17,000 ft. above the sea, which is 1,270 ft. higher than Mont Blanc. In this elevated region the sheltered valleys and the borders of the streams alone are available for agriculture; and as the summer sun is powerful, wheat and barley grow and many of the fruits of Southern Europe ripen. There are no trees in this country, and the ground in cultivation bears a small proportion to the grassy steppes which extend in endless monotony grazed by thousands of the short wool goats, sheep, and cattle."

Mountains.] The mountain ranges in the SW and S of T., towards Hindostan, are the only ranges which have been geographically described to us by such of our travellers and surveyors as have reached them,—as Bogle, Turner, Webb, Colebrooke, Crawford, Hooker, Thomson, Strachey, and others; and the elevations of some of the passes and peaks towards India have been ascertained. The mountains, on the side of Kumaon, Nepal, and Butan, correspond to the *Emodus* of the ancients, or the Sanscrit Himadri. There are, properly speaking, three district ranges, the Northern, Central, and Southern Emodus; the first of which, invisible from the plains of Hindostan, and which bounds the valley of the Sanpo to the S, is not perforated by rivers. The central range is that immediately to the N of Nepal, and is perforated by several rivers, as the Gogra, the Gandak, the Arun, and the Tista—all branches of the Ganges; but amidst such tremendous precipices, and by such narrow gaps, as to render their openings generally totally impracticable. It was through the pass of the Arun, by Lungercote, that the Nepalese army retreated from Tishulumbu; but another division, which took the pass of Kuti, to the W, lost 2,000 men amidst the snow. The pass of Langur, or Lungercote, is represented by the Jesuit fathers as the most tremendous and precipitous they had encountered all the way from Sining in Shen-si. This central range is upwards of 40 m. in horizontal depth, and at the source of the Tista it is denominated *Khaœa Karpola*, or 'the Mountain white with snow.' The southern range is that called by Kirkpatrick the Kuchar Alps, immediately N of Katmandu; and there its summits are not, as he imagined, covered only with patches of snow, but with perennial snow to a great extent. The distance from Katmandu to Lassa is 536 British, or 462 geog. m., by the road, according to Father Giorgi, but not above 300 geog. m. in direct distance, supposing Lassa to be in 29° 35′ N lat., as in the Lama's map; so that more than one-half the space traversed is occupied by the windings of the mountain-road. According to Giorgi, Mount Langur is 50 m. beyond Mount Rimda, in the southern Himalaya, and abounds in suffocating exhalations, which increase as it is ascended by the pass; but 25 m. beyond it lies the beautiful alpine valley of Tingri,—an earthly paradise in every respect but the sharpness of the air. At the S foot of Mount Langur lies the fortress and town of Tankya, the first place recognised in the Lama's map; and 90 m. beyond this is Zuenga on the Bontsu. From hence two roads lead to Lassa,—the most northern by Sgigatche and Ringbu,—and the other by Kiangtsi. Kiangtsi is a fine city and fortress, with a convent so very extensive and magnificent as to seem another city of itself. About 50 m. beyond this, to the NE, and within three days' journey of Lassa, is the famous lake of Palti or Jango, said to be so large by the natives as to be 18 days' journey in circumf.; but in the Lama's map its circumf. is only 150 m. The road from Kiangsi to Lassa lies along the N side of this lake, a day and a half's journey. Between this lake and the Sanpo is a very lofty range called Kambala, which travellers must cross in the way to Lassa. From the summit of this elevated range is seen, towards the N, a still higher range covered with during snow. At 7 m. beyond the base of Kambala the Sanpo is struck and 12 m. further on is crossed by a bridge of iron chains stretched from side to side, with planks or logs laid across. This chain bridge, composed of 500 links each a foot long, is laid across the narrowest part of the stream, which here is deep and rapid. To the NE of this crossing-place of the Sanpo is the famed city of Lassa, 24 m. distant by the road. So far Giorgi's itinerary of the route from Katmandu to Lassa is correct in its details. Turner in his journey to Tishulumbu, entered Tibet at the pass of Phari, called Paridrong and Paridsong in the Lama's map, which was ascertained to be placed a full degree too far S in the Lama's map, as Rennel justly suspected, being in 27° 58′ N lat., and 89° 1′ E long. of Greenwich. The pass ascends steep for 12 m. up the S face of the dreary Sumunang till the summit is scaled, which is found to bear a long row of small flags, fixed in rude cairns of stones, and fluttering in the wind, marking the boundary between T. and Butan. To the NE of this, a few miles is the noted peak of Chumulari, visible at a direct distance of 232 m. At the foot of the pass is the sterile vale and fortress of Phari. E of the Lantang mountang mountains, which separate Assam from the valley of the Seri-Serhit, the mountains run all the way to the frontiers of Yun-nan, a distance of 100 m. at least, nearly in the parallel of 27° 30′ in an unbroken course of tremendous elevation, skirted at its base by a large tract of rugged subalpine country, impracticable even for the hardy mountaineers themselves, and barring all communication with the N. The range of Koiran seen from the summit of Mount Kambala is esteemed the highest in T., and runs to the S of the lake of Terkiri or Tengri-nor; but the only reason assigned for its superiority of elevation to the other chains is merely the circumstance of its central position; no conclusive proof. The range to the W communicating with the Muz-Taugler to the N and the Himalaya to the S, is called Kentaisse or Kantes-han. The Sanscrit Rimola, applied to the dividing range between T. and India, has been corrupted into Morial, in the geographical maps of the 17th cent. Respecting the interior ranges, we cannot say whether they are groups or ranges.

Lakes.] This very mountainous region contains a host of lakes, a great number of which have no outlet, especially in the N part. The Terkiri or Tengri is 70 m. long and 25 m. broad, and contains a superficies of 2,300 sq. m. Malte Brun observes, that if a line be drawn from the Terkiri lake 220 m. N, and another W of the same 470 m., we shall find 23 other lakes which have no outlet or which flow the one into the other. These masses of stagnant water are the result of the configuration of the Tibetian plateau, which is itself a collection as it were of smaller plateaus encircled by mountains, or so many concave basins, whose waters, having no outlets, descend to the bottoms of their respective hollows, where they either form lakes, or find them already prepared for the reception of their waters. Fifteen days' journey from Tishulumbu is a lake 20 m. in circumf., in which *lincal*, or crude borax, is formed or deposited in its bed near the bank; from the deeper parts rock-salt is procured; and during one-half of the year this expanse of saline fluid is covered with a smooth sheet of ice.

Rivers.] Though many of the streams which arise on this most elevated of all terrestrial convexities are lost in the numerous lakes which here and there occupy the hollows of the circular depressions of the mountainous surface, yet several of the longest and largest rivers of the Eastern hemisphere originate in this region. The Tachok-Sangpo has been identified with the main trunk of the Brahmaputra; and the Sangi-Sangpo, with the chief source of the Indus. The edict of the emperor Kanghi, published in 1721, and given by Klaproth in his memoir on the sources of the Brahmaputra and Irrawaddy, published at Paris in 1828, affords a curious and particular description of the Tibetian rivers, according to accurate data furnished by the superior lamas. The origin of the Whang-ho or Hoang-ho is, in this document, said to be without the frontier of Sining. The real name of the incipient stream is Altun-kol, or 'the Golden river,' as it is denominated by the Mongols. It is about 3 ft. deep, and rises 2° W of the Tsing-su-hai lakes. Its source is placed in 35° N lat. and 21° W of Pekin. The tract watered by its upper course is called Moma in Tibetian, and Thokan in Chinese. The Hoang-ho enters Shen-si, near Ho-chu, 10 days' journey from its source in direct distance, by a narrow pass between two vast steep rocks.—The next river of T. mentioned by Kanghi is the Min-kiang; but it is a Chinese river, its course being almost wholly within Sech-wen.—The Yalong-kiang forms, for 400 m. in a SE course, the SW frontier of the Sifan, and for 140 m. more, the W frontier of Sech wen, dividing it from T. After a course of 160 m. more, through the SW angle of Sech-wen, it finally joins the Kincha-kiang, coming NE from Yun-nan, thus accomplishing a comparative course of 700 m., and receiving in its broad and deep channel the waters of the Sifan on the W of Sech-wen. From its source to its entrance into Sech-wen, in 29° 54′ N lat., it bears the appellation of the Sachu-Tsitsirhana, and then of the Yalong.—The Kiaung-ku, according to the emperor's memoir, rises in the NE of the states of the Dalailama, runs SE into T., and then enters Sech-wen. Farther on, the Kiang passes Kwei-chu-fu, enters Hu-kwang, waters King-chu-fu, and joins the Han-kiang before Wu-chang-fu.—The Han-kiang comes from the Po-chung-shan, a mountain of Shen-si, in the district of Nin-kiang-chu, and bears at its source the name of Shih-yang-shwi. It runs to the eastward, and at Nan-ching-hun enters Hu-kwang, and joins the Great Kiang near Han-yang-hun. In the map of the Lamas, the Kincha-kiang rises in Ngari, or Upper T., [not in the Sifan] in 35° N lat., and 90° 27′ E long. from a lake called Pahaton-kol. It runs E to a place called Hurha, where it is joined by a stream from the S. At 2° E of its source it receives the Piti-Muran, a large stream from the S, and a little beyond a still larger called the Aktam. At 4½° E of its source, where it is separated by Mount-Kulkun from the parent-stream of the Hoang-ho, it turns to the S and SE, passing by Tsitsirkhana, a ruined city. It is called in this part of its course the Porsic-ho. In 31° N lat. it enters the country of Laton, and runs almost due S till at Tachinquan, in 27° 32′ N lat., and 16° 40′ W long. of Pekin, it enters Yun-nan, after a comparative course of 1,000 m. through T. Its course through China is at least double that distance. Its volume of water is the largest of any river in the Eastern hemisphere, discharging, it is calculated, 464,400 cubical ft. per second of time into the sea. Including sinuosities, its length of course is to that of the Thames as 21½ to 1, or as 3,780 to 180 m.; and its basin to that of the Thames as 138 to 1, or 760,000 sq. m., as it comprehends all the central part of China and the eastern part of T., including the Sifan.—The Lantsan-kiang rises in 34° 30′ N lat., and 22° W long. of Pekin, according to the Lama's map, and enters Yun-nan under the name of the Lak-chu, whence it proceeds to the country of the Northern Laos. To the W of this river runs another called Kara-ussu, the modern Lu-kiang of Yun-nan; and to the W of this is the Long-chuen-kiang. We are not exactly of the opinion of Klaproth with respect to the geographical merits of the edict on which some of these details appear to rest. He appears to have used it mainly for the purpose of contradicting our learned countrymen of Calcutta, and the German journalists of Weimar, who happened to coincide with them in their opinion, that the Seri-Serhit of the Bor-Khampti country is the genuine Irrawadi of Ava; whilst the Sanpo of T. on the one hand, and the Nu kiang of Yun-nan on the other, are both denied to be the Irrawadi.

Climate.] In respect of climate—to reason analogically—Eastern T. must bear a great resemblance to Western T.; our actual knowledge of its climate is confined to that part which lies to the S of the Sanpo. In the temp. of the seasons a remarkable uniformity prevails in T., both in their periodical duration and return: the same division of these takes place as in Bengal. The spring is from March to May, with a variable atmosphere and heat, thunder-storms, and occasional showers. From June to September is the humid season, when heavy and continued rains swell the rivers. From October to March a clear and uniform sky succeeds, seldom obscured by fogs or clouds. For three months of this season a degree of cold, far greater than is known in Europe, prevails. M. Huc, in travelling from the Koko-Nor to Lassa, had the little cakes of barley which he carried with him, close to his person and under three robes of skin, continually frozen. On the confines of T. its extreme severity has been felt by such as have crossed its mountainous frontier, whether to Tishulumbu or Lassa. On this rest eternal snows and permanent congelation; and its vicinity is at all times remarkable for the violence and dryness of the winds. "In summer the sun is powerful at mid-day, the air is of the purest transparency, and the azure of the sky so deep that it seems black as in the darkest night. The rising moon does not enlighten the atmosphere; no warning radiance announces her approach till her limb touches the horizon; and the stars shine with the distinctness and brilliancy of suns. In Southern T. the verdure is confined to the favoured spots—the black mountains and high plains are sternly gloomy—a scene of barrenness not to be conceived: solitude reigns in those dreary wastes, where there is not a tree nor even a shrub to be seen of more than a few inches high. The scanty short-lived verdure vanishes in October; the country looks as if fire had passed over it, and cutting dry winds blow with irresistible fury, howling in the bare mountains, whirling the snow through the air, and freezing to death the unfortunate traveller benighted in their defiles." It is remarkable that the snow-line is considerably higher in T. than on the S declivity of the Himalaya. Barley is grown in T. at an alt. of 14,700 ft.; wheat at 12,000 ft.; and small shrubs are found vegetating at 17,000 ft. above sea-level in Southern T. Strachey fixes the snow-line on the S side of the Himalayas at 15,500 ft.; but at from 19,000 to 20,000 ft. on the great plateau of T.

Soil and productions.] In such a lofty region, in a climate where the inhabitants are obliged to seek refuge from the intensity of the cold in the valleys and hollows, and where, from the glare of light reflected from the snow, they are subjected to ophthalmia and blindness, we cannot expect much fertility of soil, or abundance of vegetable and farinaceous produce. Yet in some of the valleys grain is abundant, and numerous clusters of villages meet the eye. *Ua* or black barley is the grain chiefly cultivated; little wheat is raised. Beans, turnips, and cabbage, are among the vegetables raised. Large herds of sheep and cattle are reared in Eastern T. The musk-deer is found in great numbers on the mountains of T. These animals are about the size of a moderately sized hog, which they resemble in figure. They have a small head, a thick and round hind quarter, no scut, and extremely delicate limbs. The greatest singularity in this animal is the sort of hair with which it is covered, which is prodigiously copious, and grows erect all over the body, between two and three inches long, lying smooth only where it is short, on the head, legs, and ears. Upon examination, it partakes in its texture, less of the nature of hair than feathers; or rather, it resembles the porcupine's quill, at the same time it is thin, flexible, and not straight, but undulated. The colour at the base is white, in the middle black, and brown at the points. The musk is a secretion formed in a little bag or tumor, resembling a wen, situated at the navel, and found only in the male. Among the valuable and useful animals of T., the breed of sheep merits a distinguished rank. A peculiar species is marked almost invariably by black heads and legs. They are of small size; their wool is soft, and their flesh the finest mutton in the world. They are fed principally upon the short herbage peculiar to the sides of eminences, and bleak, exposed plains. They are occasionally employed as beasts of burden; and numerous flocks of them may be seen in motion, laden with salt and grain, each carrying from 12 to 20 pounds. Their wool is usually fabricated into a narrow cloth resembling frieze, or a thick coarse blanket. When slaughtered, their skins are most commonly cured with the wool on, and form a most excellent winter garment for the peasant and the traveller. The skins of lambs are cured also with the wool on, and constitute a valuable article of traffic. The autumn being clear and serene, the farmer spreads his corn on the ground to dry, and then employs oxen to tread it. The variety and quantity of wild fowl, game, and beasts of prey, are astonishing. Deer, antelopes, wild asses, wild goats, lynxes, ounces, and bears, are numerous. Turner met with no wild animal in Butan but the monkey, and amid all his travels through it, he saw no game except a few pheasants near Chuka. It is not till near the source of the Patchieu, at the foot of Somunang, that wild animals begin to appear. Herons, ducks, geese, gulls, and tern frequent the great lakes. Lizards, grasshoppers, crickets, spiders, and bees occur.—Gold, rock-salt, and *lapis lazuli*, are among the mineral productions.

Inhabitants.] This large and mountainous region contains a number of distinct tribes all classed under the common name of Tibetians, and branches of the great Mongolian family. Besides the aborigines of the country, we know of two other classes of pop. in T., the Thorpo and the Hor. The former of these tribes dwell to the N, between the Yarkea-Sanpo on the S, and the Gobi on the N, and to the W of the sources of the Kiang-Ku, between the eastern frontier of Khotan and the tract through which the road passes from Lassa to Sining. They speak a distinct language both from the Kalmuks and the Tibetians. Moorcroft confounded them with the Eluths who are themselves Kalmuks, though at the same time he distinguishes them from the Hor or Sogpo, who are an Eluth tribe. They are the descendants of the Oigur tribes, who dwelt in the same tract during the time of the Ywen dynasty, and who then bore the name of Kara-Oigurs. They are the undoubted representatives of the Oigurs who made such a figure in the history of Jenghis-Khan, and who seemed to have vanished quite out of sight, like their ancient neighbours the Hyongnu. Their country is called Kara-Tibet, and they are divided into Upper and Lower Oigurs. They were subdivided in 1573 by one of the Mongolian chiefs, to the N of the Great wall. The Hor are a branch of the Eluth stem, who roam to the N of the Kara-Nor. *Hor* or Hor-pa, is the Tibetian name for the Mongolian race generally, who are called by the appellation Ghia-Hor, or 'the people of Hor,' whilst the name of this tribe in Mongolian is Siraigol or Karagol. They are also called by the Tibetians Sogh-po or wanderers, and Ghia Sogh or 'the people of the prairies,' because they wander in the steppes. In Carey's Tibetian lexicon the names Hor and Sogh-po are rendered by the word Tartars, and their country, Sogh-tul, is rendered Tartary. We know of no other distinct tribes in T., though doubtless there are more, as several languages or at least dialects are spoken. Whether the priesthood are a different stock from the shepherds and goatherds it is impossible to determine, but there is ground to suspect so, and that they are of Hindu origin; that they came from Hindostan, and importing hither the system of Budh, by means of it, and their superior learning and science, obtained a complete ascendency over the simple and ignorant aborigines. Although the age of Budha himself, or of the author of the system which bears his name, was at least ten centuries anterior to our era, yet his system was not introduced into T. until a period comparatively modern, although it is impossible to fix the date of its introduction. It seems certain, however, that the present Tibetian language and literature are of Hindu origin. The priesthood—who are possessed of whatever literature exists in T.—point to Benares as the source whence all their learning has been derived; and Moorcroft, in a communication from Cashmere, in 1823, to the Asiatic society of Calcutta, gives a sketch of the language of T. illustrated by drawings of the various alphabets used there. According to this account, not less than 10 varieties of character are employed in that country for familiar and religious purposes.—T. is the chief seat of Buddhism and of its incarnate head, the Grand lama. The influence of this spiritual lord extends over the whole of Central Asia, and he formerly united in his own person the regal as well as the sacerdotal character. The latter he still preserves in its fullest extent, in his own name and by means of spiritual vicars, who reside in different parts of his vast spiritual domain; but his regal power has of late been exercised by the emperor of China, who acts in his name, and has got military possession of all T. The Tibetians assemble in chapels, and unite together in prodigious numbers, to perform their religious service, which they chant in alternate recitative and chorus, accompanied by an extensive band of loud and powerful instruments; so that it is mentioned by Turner, when he heard these congregations at worship, they forcibly recalled to his recollection the solemnity and sound of the Roman Catholic mass. The instruments made use of were all of an enormous size; trumpets above six feet long; drums stretched over a copper cauldron, such as are termed nowbut in Hindostan; the gong, a circular Chinese instrument of thin hammered bell-metal, capable of producing a surprising sound; cymbals, hautboys, and a double drum, shallow, but of great circumference, mounted upon a tall, slender pedestal, which the performer turns with great facility, striking either side with a long curved iron, as the piece requires a higher or a lower tone. These, together with the human tibia and sea conch, compose, for the most part, their religious band. Harsh as these instruments, individually taken, might sound to a musical ear, when joined together in unison, with the voices of two or three hundred boys and men, managed with varying modulation, from the lowest and softest cadence to the loudest swell, they produce an effect extremely grand.

Commerce and Manufactures.] Hair and woollen stuffs, velvets of remarkable beauty of colour, elegant tobacco pipes and bowls, and jewellery of considerable taste, are manufactured by the Tibetians. The principal commerce is with China, to which T. exports gold-dust, incense, woollens, wax, fleeces, and European goods. Formerly a considerable commerce was carried on with Bengal through Nepal, but since 1792 the timid jealousy of the celestial court has greatly impeded the development of this traffic. To Cashmere, T. sends shawl-wool, rock salt, turquoises, musk, tea, and borax. No money is coined in T., being forbidden by the principles of their religion, and a very adulterated coin is the common medium of exchange.

Government.] The administration of affairs is managed by four *ghylongs*, or priests, each of whom has the administration of one-fourth of T. Under these functionaries, the civil and military affairs of T. are conducted by the Chinese who dwell at Lassa, and whose nomination must be approved by the Dalai lama. The Tibetian army is composed, according to father Hyacinth's Chinese author, of 60,000 men, of whom 15,000 are cavalry. The levy is said to be made with great impartiality, one man out of 10 is ordinarily taken. This, if correct, would give 600,000 men fit to bear arms; and this number, if reckoned as representing one-fourth of the pop., would give 2,400,000 as the pop. of T.; but to this must be added all those who belong to the religious orders, whether male or female, so that the pop. may perhaps amount to 3,000,000.

Manners and Customs.] The houses of the peasantry resemble brick-kilns, being built of rough stones heaped on each other, with a few apertures to admit light, and a flat terrace for the roof surrounded with a small parapet. In their food, the Tibetians prefer raw undressed meat; and at their feasts the table is seen spread with joints of raw fresh mutton as well as boiled, the former being most esteemed. They have no occasion to salt their meat during winter, as it will remain fresh during the whole season, from the coldness and dryness of the air. Tea is a favourite beverage amongst them. The milk of the yak is much used as food. This most useful animal gives abundance of lacteal fluid, rich, and yielding excellent butter, which is easily preserved in skins or bladders excluding the air. It keeps in this climate during all the year, so that after some time tending their flocks, when a sufficient stock is accumulated, they have only to load the yaks and drive them to market with their own produce, which constitutes, to the utmost verge of Tartary, a material article of merchandise.—"The yak," says Lieut. Wood, "is to the inhabitants of T. and Pamir, what the reindeer is to the Laplander in Northern Europe. Where a man can walk a kash-gow may be ridden. Like the elephant he possesses a wonderful knowledge of what will bear his weight. If travellers are at fault, one of these animals is driven before them, and it is said that he avoids the hidden depths and chasms with admirable sagacity. His footing is sure. Should a fall of snow close a mountain pass to man and horse, a score of yaks driven a-head answer the purpose of pioneers, and make, as my informant expresses it, 'a king's highway.' In this case, however, the snow must have recently fallen; for when once its surface is frozen, and its depth considerable, no animal can force its way through it. Other cattle require the provident care of man to subsist them through the winter. The most hardy sheep would fare but badly without its human protection, but the kash-gow is left entirely to itself. He frequents the mountain-slopes and their level summits. Wherever the mercury does not rise above zero, is a climate for the yak. If the snow on the elevated flats lie too deep for him to crop the herbage, he rolls himself down the slopes and eats his way up again. When arrived at the top, he performs a second summerset, and completes his meal as he displaces another groove of snow in his second ascent. The heat of the summer sends the animal to what is termed the old ice, that is to the regions of eternal snow; the calf being retained below as a pledge for the mother's returning, in which she never fails. In the summer, the women, like the pastoral inhabitants of the Alps, encamp in the higher valleys, which are interspersed among the snowy mountains, and devote their whole time to the dairy. The men remain on the plain, and attend to the agricultural part of the establishment, but occasionally visit the upper stations, and all speak in rapture of these summer wanderings. The kash-gows are gregarious, and set the wolves, which here abound, at defiance. Their hair is clipped once a-year in the spring. The tail is the well-known *Chowry* of Hindostan; but in this country, its strong, wiry, and pliant hair is made into ropes, which, for strength, do not yield to those manufactured from hemp. The hair of the body is woven into mats, and also into a strong fabric which makes excellent riding trousers. The milk of the yak is richer than that of the common cow, though the quantity it yields be less. The *kurut* made from it is considered to be first-rate, even superior to the produce of the Kohistan of Kabul, which has great celebrity in Afghanistan. The Kirghiz never extract the butter."—The Tibetians are said to be polyandrists; one woman having several husbands. The eldest brother of the family has the privilege of choosing his wife, but she becomes the common property of all the brothers however numerous. The general mode of disposing of the dead is like that of the Parsis of Bombay: they are exposed in the open air, and left to be devoured by carnivorous birds. A place set apart for this purpose was seen by Turner and his suite, when descending into the plain of Phari. But a fate far different is reserved for the body of the sovereign lamas, the Dalai lama, and the Tishu lama. As soon as the soul of Budha has left the body of the grand lama, it is placed upright in an attitude of devotion, the legs being folded under the body, with each thigh resting on the instep, and the soles of the feet turned upwards; in this posture they are deposited in shrines. The inferior lamas have their remains usually burned, and their ashes deposited in small metallic idols. Over the shrines of the deceased sovereign lamas, splendid pyramidal mausoleums are built. Sunk as the Tibetians are in the most abject subjection, they are comparatively an amiable, mild, humane race, and free from many of the cruel and sanguinary customs of the Hindus. There is no selling of female infants in T.; no putting them to death or exposure on trees or on the banks of rivers; nor funeral piles whereon widows are burnt to accompany the manes of their deceased husbands.

Cities.] The greater number of places marked on the map of T., as given by the lamas, seem, as Malte Brun very justly remarks, to be nothing more than villages or groups of cabins, each surrounding some temple. According to the Dai-Syn-itoundchi, there are but 16 cities in all T. Of these Lassa, the capital of Wei-tsang or Eastern T., and Tishulumbu, the capital of Western T., are the principal. See these articles.

Historical notice.] Like many other pagan countries of Asia, Hindostan not excepted, T. has no historical annals, at least none have yet been discovered: we only know its history by its connection with and vicinity to China. Beyond 790 of the Christian era, we have nothing but tradition—a most uncertain and precarious guide; but according to it, T. was peopled 1340 years before Christ, by Prasinpo and Prasrimno. Some 300 years later, Budha or Sacyo, was born of a virgin, having descended from the skies to restore a purer system of faith. The first king of T. was Guiathritz Bengo, son of Macchiaba, a Hindu sovereign. His capital was Jarlon, and he died 1102 years before our era. The people, weary of anarchy, voluntarily submitted to China in 790, A.D. From this period some light begins to dawn on Tibetian history. From the Chinese writers we learn that the Tibetians were a powerful people in the 8th cent., and possessed of all the country from the sources of the Sanpo to the mountains of Cashmere, and the frontiers of Tokharistan to the W of the Belur. Jenghiz-Khan and Timur successively conquered T. In 1727 the supremacy of China was finally submitted to.

TIBET (Little). See Bulti.

TIBET (Middle). See Ladak.

TIBI, a town of Spain, in the prov. and 18 m. NW of Alicante, and partido of Jijona, on the l. bank of the Ibi, in a fertile locality. Pop. 1,571. It has a parish church, and a custom-house, and in the environs is a fine reservoir.

TIBITINGA. See Tabatinga.

TIBOHINE, or Taughboyne, a parish in co. Roscommon, containing the town of Frenchpark and the village of Loughglynn. Area 44,092 acres. Pop. in 1831, 16,460; in 1851, 13,718.

TIBSHELF, a parish in Derbyshire, 4¼ m. NE by N of Alfreton. Area 2,280 acres. Pop. in 1801, 661; in 1831, 759; in 1851, 806.

TIBTHORPE, a township in Kirk-Burn p., Yorkshire, 5¼ m. WSW of Great Driffield. Area 2,780 acres. Pop. in 1831, 227; in 1851, 271.

TIBURON. See Tiberon.

TIBURON, an uninhabited island of the gulf of California, near the coast of the Mexican prov. of Sonora. It is about 54 miles in length from NNW to SSE, and 24 m. in medium breadth. A little to the N of this island is a pearl fishery.—Also a headland of New Granada, in the dep. of Ystmo, and prov. of Panama, on the W coast of the gulf of Darien.

TIBURONES, a group of islands, two in number, surrounded by rocks, near the Mosquitia coast, in N lat. 15° 10′, and W long. 82° 8′.

TICAO, an island of the Philippine archipelago, a little to the NE of the island of Masbate, and to the S of the SE peninsula of the island of Luzon, in N lat. 12° 30′, and E long. 123° 35′. It is 33 m. in length, and nearly 9 m. in medium breadth. It is generally mountainous. On its E coast is Port S. Jacinto, and on the NW that of S. Miguel. It is well-watered and very productive, and is frequently touched at for provisions by Spanish vessels on their voyages to Acapulco.

TICEHURST, a parish in Sussex, 3 m. ESE of Wadhurst. Area 8,202 acres. Pop. in 1851, 2,850. The village occupies a gentle eminence surrounded by hills and valleys of great beauty and fertility.

TICHANSKAIA, a town of Russia in Europe, in the gov. of the Don Cossacks, 75 m. ENE of Bogotchar, on the Khoper.

TICHAU, a village of Prussia, in the prov. of Silesia, regency and 63 m. SE of Oppeln, and circle of Pless. Pop. 1,650.

TICHBOURNE, or TITCHBOURNE, a parish in Hants, 2½ m. SW of New Alresford. Area 3,060 acres. Pop. in 1831, 363; in 1851, 378.

TICHET, or TAGAZZA, a town in the SW part of the Sahara, about 600 m. E of St. Louis-de-Senegal. The houses are constructed of rock-salt, and covered with camels' skins. The surrounding district is completely desert. It contains mines of fine salt, remarkable for the uniformity of its crystals. T. is inhabited by Moors.

TICHFIELD, a parish and market-town in Hants, 2½ m. W of Fareham, and 9 m. ESE of Southampton, on a small stream falling into Southampton water, with which it communicates by a navigable cut, and near the Gosport branch of the South-Western railway. Area of p. 17,512 acres. Pop. in 1831, 3,712; in 1851, 3,956.—The town consists principally of one good street running nearly N and S. It has breweries and corn-mills.

TICHITHOCKTHNOCK, a small island of the Atlantic, near the E coast of Labrador, in N lat. 55° 50′, and W long. 59° 42′.

TICHMARSH, a parish in Northamptonshire, 1¾ m. NNE of Thrapston. Area 4,480 acres. Pop. in 1831, 843; in 1851, 308.

TICINETTO, a town of Sardinia, in the div. of Alexandria, prov. and 7 m. SE of Casale, and mand. of Frassinetto-di-Po, in a fine plain. Pop. 1,000.

TICINO, or TESSIN, a considerable river of Italy, which has its source in Switzerland, on Mount St. Gothard, and after traversing the Val-Levantina, in the cant. of Ticino, turns SW, and flows into the Lago-Maggiore. Issuing from that lake, at its southern extremity, it flows SSE, and forms the boundary between Austrian Italy and the Sardinian states, until its junction with the Po below Pavia, after a course of 120 m., of which 45 m. are above the Lago-Maggiore. It becomes navigable at its efflux from the lake.

TICINO, or TESSIN, a canton in the S of Switzerland, situated between the central cantons and the frontier of Lombardy; and bounded on the N by the Grisons and Uri; on the E and S by Lombardy; on the W and SW by Piedmont. It was formerly called the Italian bailiwics, and is a tract of mountainous territory acquired long since by the Swiss, but inhabited by Italians, and governed by temporary deputies from the respective cantons until 1815, when it was formed into an independent canton, divided into the 8 following districts:—

	Pop. in 1815.	Pop. in 1850.
Mendrisio,	12,000	17,372
Lugano or Lanis,	26,700	36,473
Valmaggia or Maynthal,	6,000	7,482
Locarno or Luggarus,	17,400	22,362
Bellinzona or Bellenz,	8,000	11,582
Riviera,	3,000	4,445
Blegno or Blenio,	6,300	7,687
Levantina,	9,600	10,331
	89,000	117,759

The area of the whole, 1,030 sq. m., is equal to a middle-sized English county. It is watered in all its extent by the Ticino, and is extremely mountainous, the ramifications of the Alps dividing it into more than 20 distinct valleys. On its N frontier rise the Splugen, Mount St. Gothard, and Mount St. Bernard. The whole cant. belongs to the basin of the Po. The soil is fertile, and the climate mild. Figs, almonds, oranges, citrons, the myrtle, the pomegranate, and the olive flourish in the S part. The inhabitants, however, are represented as comparatively ignorant, indolent, and unacquainted with manufactures, the men generally emigrating, like the Savoyards, to large towns in Italy or France, and leaving to the women the care, not only of the cattle, but of tillage. Wine is cultivated and exported in small quantities; and the quantity of silk made is large. The exports take place chiefly to Italy, and though small in amount, comprise a diversity of articles, viz. silk, fruits, cheese, skins, marble, crystals, and timber; also the only manufacture of the country, straw hats. The government is vested in a grand council, chosen by all citizens 25 years of age, and possessing property to the value of £12 in life-rent.

TICKARRY, a town of Hindostan, in the prov. and district of Bahar, and 51 m. WSW of the town of that name. It is advantageously situated at the junction of 7 roads, and was formerly fortified.

TICKENCOTE, a parish in Rutlandshire, 9 m. E of Oakham. Area 1,256 acres. Pop. in 1851, 98.

TICKENHAM, a parish in Somersetshire, 9 m. W by S of Bristol. Area 1,627 acres. Pop. 424.

TICKHILL, a parish in the W. R. of Yorkshire, 4¼ m. W of Bawtry, comprising the townships of T. and Stancill with Wellingley and Wilsick. Area 6,514 acres. Pop. in 1831, 2,084; in 1851, 2,159. The church, built in the reign of Richard II., is a very perfect specimen of the 14th century.—The town is straggling, and covers a large space of ground. An extensive trade in malt was formerly carried on at this place, but it has much decreased.

TICKILLEN, a parish in co. Wexford, 3¼ m. NW of Wexford. Area 2,866 acres. Pop. in 1851, 1,135.

TICKMACREVAN, a parish in co. Antrim, containing the town of Glenarm, and part of the village of Carnlough. Area 20,506 acres. Pop. in 1831, 3,859; in 1851, 3,903. The summit of Collin-Top, a little beyond the NW extremity of the p. has an alt. above sea-level of 1,419 ft. The hamlet of Straidkelly, on the coast between Glenarm and Carnlough, had, in 1831, a pop. of 172.

TICKNALL, a parish in Derbyshire, 8½ m. S of Derby, and in the line of a railway connected with the Ashby-de-la-Zouch canal. Area 1,867 acres. Pop. in 1831, 1,278; in 1851, 1,241.

TICONDEROGA, a township and village of Essex co., in the state of New York, U. S., on both sides of the outlet of Lake George, and 87 m. N by E of Albany. It has a hilly surface, but near the lakes is very fertile. Pop. in 1840, 2,169; in 1850, 2,669.

TICON, or TOKU, a group of islets in the Indian ocean, near the W coast of Sumatra, in S lat. 0° 6′, and E long. 99° 50′. The nearest of the group is not more than 1½ m. from the coast. They are well-wooded, and inhabited. Establishments were formed here by the English and Dutch for the pepper trade, but were abandoned in 1621.

TICUNAS, a tribe of Indians who inhabit the tract of Ecuador, extending between the Tutumayo or Ica and Amazon.

TIDAN, a river of Sweden, in the prefecture of Skareborg, which issues from the N extremity of a small lake, on the confines of the prefecture of Joenkoeping, 15 m. NW of the town of that name; runs first N, then W; forms Lake Osten; takes afterwards a NNW direction; and after a sinuous course of about 51 m. throws itself near Lake Mariestadt into Lake Wener.

TIDCOMBE, a parish in Wilts, 5¾ m. NNE of Ludgershall. Area 2,321 acres. Pop. in 1851, 218.

TIDDINGTON, a hamlet in Albury p., Oxfordshire, 3 m. NW of Tetsworth. Pop. in 1851, 174.

TIDDY, a town of New South Wales, in the district of Darling Downs, on the Condamine, near Peel Plains.

TIDENHAM, a parish in Gloucestershire, 14 m. SW by S of Newnham, situated between the rivers Severn and Wye. Area 9,527 acres. Pop. 1,753.

TIDESWELL, a parish and market-town in Derbyshire, 32 m. NNW of Derby. The parish comprises the township of T., the chapelry of Wormhill, and the hamlets of Litton and Whestone. Area 10,350 acres. Pop. in 1831, 2,807; in 1851, 3,411.—The town is but indifferently built, but appears to have once been a place of more consequence than it is at present.

TIDI, a river of Cornwall, which rises near Liskeard, and, uniting with the Lynher, falls into the Tamar.

TIDMARSH, a parish in Berks, 6 m. WNW of Reading, on a branch of the Thames. Area 779 acres. Pop. in 1831, 143; in 1851, 145.

TIDMINGTON, a parish in Worcestershire, but located in the co. of Warwick, 1½ m. S of Shipston-upon-Stour. Area 754 acres. Pop. in 1851, 53.

TIDONE, a river which has its source in the div. of Genoa, and prov. of Bobbio, 7 m. E of Varzi; runs first N, then NW; passes Zavatarello; turns afterwards NE; enters the state of Parma, in which it waters the W part of the duchy of Piacentia; and after a course of about 39 m., joins the Po, on the r. bank, 7 m. NW of Piacentia. It is noted for a victory over the Sardo-German forces by a French and Spanish army in 1746.

TIDONG, or WAPEN-VAN-HOORN, a small island of the Sunda archipelago, in the Java sea, at the E entrance to the Sunda strait, in S lat. 5° 41′, and E long. 106° 24′.

TIDOR, an island of the Asiatic archipelago, in the group of the Molucca islands, in N lat. 45°, and E long. 127° 25′, near the W coast of the island of Gilolo, and 9 m. S of that of Ternate, from which it is separated by a channel affording good and safe anchorage. This island is about 21 m. in circumference, and is well-watered by several streams, which descend from the lofty mountains by which a great part of its surface is covered, and one of which is an active volcano. Spices form its chief productions. Its inhabitants are Mahommedans, and are subject to a sovereign, under the protection of the Dutch. This island was discovered by the Spanish navigator, Magalhaen, and a factory was established by him on its shores. In 1527 it was taken by the Portuguese, and in 1607 by the Dutch.

TIDWORTH (NORTH), a parish in Wilts, 2 m. SW by W of Ludgershall. Area 3,069 acres. Pop. in 1831, 392; in 1851, 385.

TIDWORTH (SOUTH), a parish in Hants, 7¾ m. WNW of Andover. Area 2,175 acres. Pop. 230.

TIEB. See DIBBIE.

TIECHOBUS, a village of Austria, in Bohemia, in the circle and 15 m. NE of Tabor. It has a glass-work.

TIEDRA-VIEGA, a town of Spain, in the prov. of Valladolid, and partido of Rioseca, in a valley of the Sacada, on a steep height. Pop. 1,695. It has an old castle, four parish churches, a priory depending upon the Bernardine convent, and an hospital.

TIEFENBACH, a village of Würtemberg, in the circle of the Neckar, NE of Maulbronn. Pop. 420.

TIEFENBRONN, a village of the grand-duchy of Baden, in the circle of the Middle Rhine, on the Würm, SW of Pforzheim. Pop. 850.

TIEFENDORF, a village of Saxony, in the circle of Lusatia. It has several mineral springs.

TIEFENGRUBEN, a village of the grand-duchy of Saxe-Weimar, in the circle of Weimar-Sena and bail. of Berka. Pop. 200. In its vicinity are extensive beds of gypsum, alabaster, potter's clay, fuller's earth, and rock crystal.

TIEFENKASTEN, a village of Switzerland, in the cant. of the Grisons, on the l. bank of the Albula, on the road into Italy by the Septim pass. Pop. 200.

TIEFENORT, a market-town of the duchy of Saxe-Weimar, in the circle and 13 m. SW of Eisenach, on the Werra. Pop. 1,264. It has manufactories of woollen fabrics. In its vicinity are the ruins of the castle of Craynberg.

TIEFFURTH, a village of the grand-duchy of Saxe-Weimar, in the circle of Weimar-Sena, 4 m. NE of Weimar, on the Ilm. Pop. 261. It has a ducal castle and an agricultural school.

TIEFHARTMANSDORF, a village of Prussia, in the prov. of Silesia, regency and 23 m. SW of Liegnitz and circle of Schonau. Pop. 1,270. It has manufactories of linen damask.

TIEGENHOF, a town of Prussia, in the prov. of W. Prussia, regency and 24 m. SE of Danzig, and circle of Marienburg, on the Tiege. Pop. 1,800. It has manufactories of linen and of printed calico, several dye-works, a brewery, and a vinegar manufactory.

TIEGHEM, a department and commune of Belgium, in the prov. of W. Flanders and arrond. of Courtrai. Pop of dep. 2,567. The village is 11 m. E of Courtrai and cant. of Avelghem, on a hill. It has several flax mills.

TIEHELERY, a commune of Belgium, in the prov. of Limburg and dep. of Nieuwerkerken. Pop. 224.

TIELMES, a town of Spain, in the prov. and 27 m. SE of Madrid and partido of Chinchon, on the r. bank of the Tajuna. Pop. 830.

TIEMBLO (EL), a town of Spain, in the prov. and 21 m. SSE of Avila and partido of Cebreros, in a marshy and woody locality, near the r. bank of the Alberche. Pop. 1,373. It has a parish-church, a convent, a custom-house, and a public granary, and possesses manufactories of woollen and linen fabrics.

TIENE, a town of Austrian Lombardy, in the prov. and 12 m. NNW of Vicenza, in a fertile and salubrious plain. Pop. 5,000. It has an hospital, and possesses manufactories of cloth, hats, linen, and silk.

TIERACHE, a commune of Belgium, in the prov. of Hainault and dep. of Momigines. Pop. 350.

TIERE'E, a commune of France, in the dep. of the Maine-et-Loire, cant. and 4 m. NNE of Briollay, near the l. bank of the Sarthe. Pop. 2,023. It has quarries of slate and free-stone.

TIERMAS, a town of Spain, in the prov. and 72 m. N of Zaragoza and partido of Sos, on a pine-covered mountain, and near the r. bank of the Aragon. Pop. 570. It has numerous thermal springs.

TIERNEY, a parish and town of Van Diemen's Land, in the co. of Somerset. The p. is bounded on the W by the Lake river, and on the E by the Isis. The town is on the Lake river.

TIERPS, or TIERP, a village and parish of Sweden, in the prefecture and 33 m. NNW of Upsala, on the l. bank of a river of the same name. The p. comprises the island of Saertsoe, formed by the Dalelf, and contains numerous iron forges.—The river T. has its source in Lake Skersioen, at the NE extremity of the prefecture of Westeras; runs first ESE, then NNE; forms Lake Tennaren; waters the N part of the prefecture of Upsala; and after a tortuous course of about 60 m., falls into the gulf of Bothnia, near Karlholm.

TIERRA-AUSTRAL-DEL-ESPIRITU-SANTO. See ESPIRITU-SANTO.

TIERRA-BOMBA, an island of the Caribbean sea, near the N coast of New Grenada, in the dep. and 5 m. SSW of Cartagena, at the entrance to a bay formed on the S by the island of Bara. It is about 6 m. in length, and 9 m. in breadth, and near its S extremity are two forts for the defence of the Boca-Chica, the entrance to the bay.

TIERRA-DEL-FUEGO. See FUEGO.

TIERRA-WETTE, TERA-WITTE, or POLIWERO (CAPE), a headland of New Zealand, forming the SW extremity of the island of Eaheinomauwe or New Ulster, in Cook's strait, in S lat. 41° 21′, E long. 175° 48′. It was discovered by Cook in 1769.

TIERS, a commune of Belgium, in the prov. of Liege and dep. of Chenee. Pop. 145.

TIERSHANAGHAN, a village in the p. of Ballyheigue, co. Kerry. Pop. in 1841, 222.

TIESCHTIN, or TISCHTIN, a town of Moravia, in the circle of Olmutz, 11 m. SSW of Tobitschau.

TIESI, a village of Sardinia, in the div. of Cape Sassari, prov. and 18 m. ESE of Alghero, in an unhealthy plain, near the source of the Turritano. Pop. 2,120. The environs, which are partly level, afford grain and pasture in great abundance.

TIETAR, a river of Spain, which has its source on the W side of Mount Pinos, in the prov. of Toledo, 12 m. N of Talavera-de-la-Reina; runs first W; enters the prov. of Caceres; turns SW; and after a course of about 90 m., joins the Tagus on the r. bank, 24 m. NNE of Caceres. In the upper part of its course it flows through a sandy district of little fertility, but forms afterwards high and nearly inaccessible banks.

TIETE, or ANHEMBI, a river of Brazil, which has its source in the Serra-Cubitao, in the district and 30 m. E of São-Paulo; runs first W, then NW; and after a total course of about 450 m., joins the Parana on the l. bank, in S lat. 20° 36′, W long. 51° 50′. Its principal affluents are on the r., the Jundiahi, Capibari, Piracicaba, Jacarepipira-Mirim, Jacarepipira-Acu or Pipira; and on the l. the Sorocaba and Lancoes. Near the junction of the Jundiahi, the T. forms a magnificent cataract, and throughout all the lower part of its course its navigation is obstructed by numerous cataracts.

TIETZ, TUETZ, or TUCZNO, a town of Prussia, in the prov. of W. Prussia, regency of Marienwerder, circle and 15 m. SW of Deutsch-Krone, in the midst of a group of lakes. Pop. 1,000. It has a manufactory of cloth and a saw-mill.

TIEUDENBERG, a commune of Belgium, in the prov. of W. Flanders and dep. of Westroosebeke. Pop. 340.

TIEVAS, a town of Spain, in the prov. of Navarra and partido of Aoiz, 9 m. SE of Pamplona, at the foot of the Alaiz mountains. Pop. 238. It has a fortress in which the most important state documents were kept by Philippe-le-Beau, and which was destroyed in 1378.

TIEWINKEL, a commune of Belgium, in the prov. of Limburg and dep. of Lummen. Pop. 475.

TIFEH, or TIFFECH, a town of Algiers, in the prov. and 35 m. E of Constantine, on the Hamiz. It is of small extent, and possesses little strength. It occupies the site of the *Theveste* of the Romans.

TIFERNO, or BIFERNO, a river of Naples, in the prov. of Sannio, which has its source in the district of Isernia, a little to the W of Bojano; runs first E, then N; enters the district of Campobasso; bends NE; traverses the district of Larino; and after a course of about 63 m., throws itself into the Adriatic, 2 m. SSE of Termoli. Its waters abound with trout and eels.

TIFESCH, or TEFESSAD, a town of Algiers, in the prov. and 40 m. WSW of the town of that name, in a fertile locality. It is the *Tipasa* of the Romans, and contains numerous ruins.

TIFFANGES, a commune and town of France, in the dep. of La Vendee, cant. and 8 m. WNW of Mortagne-sur-Sevre, on the l. bank of the Sevre-Nantaise. Pop. 847. It has a paper-mill. This town is said to have been the capital of the Teiphaliens, a people of Scythian origin, brought by the Romans to garrison the towns of Poictou.

TIFFIELD, a parish in Northamptonshire, 2 m. N of Towcester. Area 2,530 acres. Pop. 154.

TIFFIN, a village of Seneca co., in the state of New York, U. S., 76 m. N of Columbus, on Sandusky river, and intersected by the Mad river and Lake Erie railway. Pop. in 1840, 600; in 1850, 2,728.—Also a township of Adamas co., in the state of Ohio. Pop. in 1840, 1,540.—Also a township of Williams co., in the same state. Pop. 322.

TIFFINS, a river which has its source in the state of Michigan, U. S., flows thence into the state of Ohio, and joins the Maumee river.

TIFLIS. See TEFLIS.

TIFLISKOI, a fortress of Russia in Europe, in the prov. of the Caucasus, district and 93 m. WNW of Stavropol, on the r. bank of the Kouban.

TIFORE, a small low island of the Molucca sea, 90 m. W of that of Gilolo, in N lat. 1° 6′, E long. 126° 8′.

TIGA, a small island in the Eastern seas, near the NW coast of Borneo, in N lat. 6° 25′.—Also an islet in Lampong bay, on the coast of Sumatra.

TIGAON, an island in the Indian sea, near the NW coast of Borneo, in N lat. 6° 10′.

TIGER, an island of Guayana, at the mouth of the river Essequibo. It is in a high state of cultivation, and produces excellent coffee and sugar.

TIGER, a small island in the Spanish Main, near the coast of Darien, in N lat. 8° 35′, W long. 77° 30′.

TIGITAS, a small river of Venezuela, which rises to the S of Conception-de-Pao, and unites itself with the Galamotal.

TIGLAGANTI, a river of New Granada, in the prov. of Darien, which enters the sea in the gulf of Atrato.

TIGLIOLE, a town of Piedmont, 9 m. W by N of Asti. Pop. 2,500.

TIGNALE, a village of Lombardy, in the prov. of Brescia, near the W side of the Lago-de-Garda. Pop. 1,150.

TIGNES, a village of Savoy, 12 m. SE of St. Maurice, on the Isère.

TIGRE, a river of Ecuador, which rises in the prov. of Quixos; flows SE; and having been joined by the Nahuapo, after a course of 300 m., falls into the Amazon, in S lat. 4° 53′.

TIGRE, a province of Abyssinia, which has given name to almost all the NE districts of that country. It lies between the parallels of 12° and 15° 30′ N, and is composed in great measure of an elevated plateau supported by steep and lofty mountains, and intersected with fertile valleys and plains. The Tacazze, which intersects it in a N and W course, receives nearly all its streams. On the E it reaches in part to the sea, from which, however, it is to a great extent separated by a salt and sandy plain. Its principal subdivisions are Sereh, T. Proper, Tembren, Enderta, Wojerat, Adana, Agami, Serawi, Samen, Sulo, Woggam, Waldabba, and Walkayt. See article ABYSSINIA.

TIGRE, or AMAPALLA, a small island of the Pacific, in the gulf of Fonseca, on the S coast of Honduras. It is considered one of the best anchoring-places in the Pacific, and would be of considerable value as an intermediate coaling-station for steamers on their route to and from San Francisco, and also as a depôt for merchandise, for the markets of Central America. See FONSECA.

TIGRILLO-BAY, a bay on the coast of Venezuela, between Punta-de-Tigrillo and Punta-Gorda. At its farther extremity is a canal communicating with the harbour of Mochima.

TIGRIS, a large and celebrated river of Western Asia, which rises in two head-streams in Central Armenia, on the southern slope of the Anti-Taurus, near the source of the Euphrates. These two great rivers flow parallel, but at a varying distance from each other to their point of junction at Kornah, in N lat. 31°, about 100 m. above the Persian gulf, and after a course of 1,150 m. See article EUPHRATES. The western branch of the T. rises under the mountain of Hagar-Baha, 10 m. S of the beautiful little lake of Ghuiljik, near the southern limits of the Taurus. The little torrent having passed through the wild scenery of the mountain, emerges, increased by numerous rivulets and springs, into the high plain of Diyarbekir, through which it flows in a SSE direction, and sweeps under the city of that name, a shallow stream, but 250 yds. wide when flooded. At Osman-Keui the western branch, after a course of 150 m., is joined by the eastern branch of the T., whose head-streams rise on the sides of Ali-Tagh, the ancient *Niphates*, and which has had a course of 100 m. to this point. For nearly 100 m. it bends to the E, through an open, undulating country, bare of wood in the vicinity of the river, and possessing little to attract attention or admiration. The stream runs slowly here, and frequent shallows and little rapids show the impossibility of any but the lightest barks being borne on the waters of this branch of the T. No remains of interest are to be found on this part of its course. After the junction of the Batman-su, or Sert river, that flows into the T. from the N, the river becomes more deep and rapid; the banks change to steep cliffs, on one hand, and a rich sloping country on the other, alternately, until after a few miles, it is winding in short turns between high cliffs in a SE direction. The romantic beauty of numerous Christian villages pitched on the cliffs, and the dark ravines here and there admitting a passage from the river to the open country, have a fine effect, and add much to the scenery of this part of the river. Numerous shallows and rapids occur in every part of its course, rendering it unfit for the purposes of communication, except downwards. About 40 m. below the Batman-su, is the noble bridge, and on the r. bank of the river the fine old ruins of Hussain-Keif. About 15 m. further on the Ridgwan river falls into it on the l. bank: at 20 m. further on, the Tigris receives the Buhtan-chai, from the E; it then turns ESE and flows through a beautiful country about 50 m. to Jezireh, and passes through the Jebel-Judi, the "overhanging cliffs" of Xenophon. The country between Hussain-Keif and Jezireh is thronged with Christian villages, spread over a high uneven rocky plain, which produces the finest grapes in abundance. About 4 m. above Jezireh the river emerges into a more open country. The Khabur, or Zucho river, flows into the Tigris 23 m. below Jezireh, from the eastward, at a point which must have been that where the Greeks turned from the banks of the T. to seek a passage over the Carduchian mountains by the road that now leads across the plain of Zucho, and over the Jebel-Judi. The Khabur is not fordable near where it joins the T.; but it carries into the T. a much smaller body of water than the great stream bearing the same name which joins the Euphrates below Deir. At Jezireh, a fine old castle commands the river, and the bridge of boats that now crosses it. The remains of a splendid stone bridge cross it about 2 m. below the castle. The river below Jezireh, as far as Mosul runs through a high undulating plain; on both banks, hills are seen in the vicinity of the river, those on the l. bank converging towards it until they nearly touch, just below the confluence of the Khabur. The road along the l. bank lies over this high plain about 7 m. from the river, crossing the hills into the Zucho plain, through which the Khabur runs. Its course is SE to Eski-Mosul, but it curves round so as to approach Mosul in a SSE direction. Capt. Lynch looked with much attention for a ford over the Tigris above Mosul, where Alexander might have crossed before the battle of Arbil, but no such ford exists at the present day. On the l. bank, both above and below Mosul are the ruins of Nineveh. See article MOSUL. Below Mosul about 28 m. by the river occurs the first *bund* or dyke of solid masonry called Zikru-l'ahwaz; and 12½ m. below this point the Zab enters the T. Flowing in a course to the E of S, the T. passes Tekrit, below which it gives off a canal which rejoins it near Bagdad. After receiving the Adhim, the T. bends SSE, and then S by W to Bagdad, between which city and Mosul, it has run with an average current of 4½ m. per hour, and a breadth of 200 yds. At 5 m. below Bagdad, the canal of Saklawiyah or Isa brings during the season of floods, a portion of the waters of the Euphrates into the Tigris, from a spot 6 m. below Felujah. At 21 m. below the Saklawiyah, the Diala or Diyalah brings into the T. a vast body of water. Below the confluence of this stream, the T. pursues a very winding course, with an inclination westward, until its waters and its name are lost in those of the Euphrates at Kornah, 232 m. from Kut-el'Amrah by the windings of the stream, and 144½ m. in direct distance.—Mr. Rich, while consul at Bagdad, obtained information from the natives that where the Eastern Tigris rises, it bursts at once from a cavern in the mountain, and in its earliest course is a much more considerable river than the Diyarbekir branch. The distinguished traveller and archæologist, Mr. Layard, thus describes that part of his journey which refers to the country between Lake Wan and the caverned sources of the Eastern T. in 1850:—"We had now left the lake of Wan, and our track led up a deep ravine, which gradually became more narrow as we drew nigh to the high mountains that separated us from the unexplored districts of Mukus and Buhtan. On both sides of the ravine were villages and ruined castles. Numerous streams from the hill sides irrigated plots of cultivated ground. After climbing up a steep declivity of loose stones, like the moraine of a Swiss glacier, and dragging our horses with much difficulty after us, we found ourselves amidst eternal snow, over which we toiled for nearly two hours, until we reached the crest of the mountain, and looked down into the deep valley of Mukus. This is considered one of the highest passes in Kurdistan. At the foot of the pass is a small Armenian church, called Khurus-Kilissa, or "the Church of the Cock,' because a black cock is said to warn the traveller when the snowdrifts hide the mountain tracks. There was no other pathway down the rocky ravine than the dry bed of the torrent. Leaving the laden horses to find their way over the stones and through the torrent, we passed soon after a natural cavern, from which burst, white with foam and struggling through a bed of pink flowers, a most abundant spring. This was one of the principal sources of the eastern branch of the T." It is supposed by Mr. Ainsworth that this subterranean source of the Eastern T. is lower than Lake Wan, and that it is not at all improbable that its waters may be derived from that lake, and that thus the Lake Wan is in reality the source of the Eastern Tigris. He remarked that there is nothing in the elevation of the Buhtan-Chai—the ancient *Centrites*—or of its tributaries, and no tradition of waterfalls in the course of the Mukus river, to discountenance the fact of a much diminished elevation in the bed of the T. and its feeders, and the upland lake of Wan, which, it is estimated by the editor of the *Journal of the Royal Geographical Society* in 1838, has an alt. of 5,000 ft. above the level of the sea. The T. begins to rise in November, but its permanent rise does not usually begin till the middle of March, nor does it reach its greatest height till the middle of May, when it has a mean velocity of 7·33 ft. per second.

TIGRIS. See Bocca-Tigris.

TIGUA, a river of Ecuador, in the prov. of Esmeraldas, which runs NW, and enters the Toachi, in S lat. 19°.

TIGUIL, a river of Kamtchatka, which rises in the district of Nizni-Kamtchatka, and falls into the Penjinskaia gulf. On its banks, about 20 m. from its mouth, is built the fort of Tiguilskoi, the strongest in Kamtchatka, where a garrison is maintained, to keep in awe the Koriaks and the Tchutski.

TIHANG, or Tihony, a town of Hungary, in the com. of Szalad, 23 m. E of Tapolcza, on the E bank of a peninsula extending into the N part of Lake Balaton.

TIHANGE, a department and commune of Belgium, in the prov. of Liege, and arrond. of Huy. Pop. 872. It has a distillery and several breweries.

TIHARUH, a town of Hindostan, in the prov. of Delhi, 66 m. WNW of Sirhind.

TIH-CHU, a district and town of China, in the prov. of Shan-tung, div. and 75 m. NW of Tse-nan-fu, on the Imperial canal, in N lat. 37° 32′ 20″, and E long. 116° 24′ 54″. It has numerous shops, and has manufactories of straw bonnets. It is well-peopled. The locality is noted for its horses.

TIH-GAN, a district and town of China, in the prov. of Keang-se, and div. of Kew-keang-fu, 54 m. NNW of Nan-chang-fu, in N lat. 29° 19′, and E long. 115° 43′. In its vicinity is a fine pagoda, consecrated to Fo.

TIH-GAN-FU, a division and town of China, in the prov. of Hu-pih. The div. comprises 5 districts. The town is 66 m. NW of Wu-chang-fu, in a fertile locality, on an affluent of the Yang-tse-keang, in N lat. 31° 18′, and E long. 113° 37′ 40″. It has manufactories of wax-candles.

TIH-HING-HEEN, a district and town of China, in the prov. of Keang-se, and div. of Faou-chu-fu, in N lat. 28° 54′ 50″, and E long. 117° 42′ 8″.

TIH-HWA-HEEN, a district and town of China, in the prov. of Keang-se, and div. of Kew-keang-fu.—Also a district and town in the prov. of Fo-keen, div. of Yung-chun-chu, and 90 m. SW of Fuh-chu-fu, in N lat. 25° 26′, and E long. 118° 16′.

TIH-KING-CHU, a district and town of China, in the prov. of Kwang-tung, div. of Chaou-king-fu, and 105 m. W of Canton, in N lat. 23° 13′ 42″, and E long. 111° 13′ 10″.

TIH-LING, a district and town of China, in the prov. of Leaou-tung, or Shing-king, in N lat. 42° 25′, and E long. 123° 45′.

TIH-PING, a district and town of China, in the prov. of Shan-tung, and div. of Tse-nan-fu, in N lat. 37° 34′, and E long. 117° 4′.

TIH-TAOU-TING, a district and town of China, in the prov. of Kan-suh, and div. of Lan-chu-fu.

TIH-TSIN, a district and town of China, in the prov. of Che-keang, and div. of Hu-chu-fu, in N lat. 30° 37′, and E long. 119° 56′. This place is remarkable for the art of pearl-making, which is here practised. In the month of May or June large quantities of the pearl-mussel [*Mytilus cygnus*] are brought in baskets from the Táhú, a lake in Kiang-sú, about 30 m. distant, the largest amongst the full grown being specially selected. The shell is gently opened, and the free portion of the mollusc is carefully separated from one surface of the shell with an iron probe; the foreign bodies are then successively introduced at the point of a bifurcated bamboo stick, and placed in two parallel rows upon the mantle or fleshy surface of the animal; a sufficient number having been placed on one side, the operation is repeated on the other; stimulated by the irritating bodies, the suffering animal spasmodically presses against both sides of its testaceous skeleton, keeping the matrices in place. This being done the mussels are deposited one by one in canals, or streams, or pools connected therewith, five or six inches apart, at depths of from two to five feet, in lots of from 5,000 to 50,000.

TI-HWA, Tyhoua, or Ouroumtsi, a division and town of Mongolia, in the prov. of Kan-suh. The div. comprises three districts. The town is 320 m. E of Ele, at the foot of a ramification of the Thiai-Chan chain, in a fertile and well-watered locality. Its streets are spacious, and from the activity of its commerce, it presents a generally animated aspect. The circumjacent mountains contain coal, and are in some parts volcanic. A plain 33 m. in circumf. is covered with light cinders, which, whenever disturbed, emit a flame, and produce rapid combustion.

TIH-YANG-HEEN, a district and town of China, in the prov. of Sze-chuen, and div. of Meen-chu, in N lat. 31° 10′, and E long. 104° 22′.

TIIS-SOE, a lake of Denmark, in the diocese and W part of the island of Sieland, bail. and 19 m. SW of Holbeck. It is nearly oval in form, and is 4 m. in length from N to S, and 3 m. in breadth. It is traversed by the Hallebye-aae, which issues from it on the W, and flows into the Great Belt.

TIJARAFE, a village of the Canary islands, in the N part of the island of Palma, and partido of Santa-Cruz-de-la-Palma, on a hill which is ascended by a flight of steps cut in the rock. Pop. 2,216. It has a parish-church and a customhouse.

TIJELENKING, or Chillingching, a Malay village, on the N coast of the island of Java, 10 m. E of Batavia. It is traversed by a river which affords good fishing, and has several salines. The English landed here in 1811.

TIJOLA, a town of Spain, in the prov. and 33 m. N of Almeria, and partido of Purchena, in a fine plain, watered by the Seron and Bacares. Pop. 1,690. It has an hospital, and flax and oil mills.

TIJUCO (SANTO-ANTONIO-DE), a town of Brazil, in the prov. of Minas-Geraes, district of Serro-do-Frio, and 24 m. NNE of Villa-do-Principe, in the centre of the diamond district, of which it is the chief depot. It rises amphitheatrally, on a mountain of the Serra-Espinhaco, near the source of the Jequitinhonha, and is a large and flourishing place. Some of the houses are built of stone or of mud; but the greater number of wood. It has a church, five chapels, several convents, three hospitals, an almshouse, and commodious barracks. It is well supplied with good water, and the climate, although cold, is salubrious. The working of diamonds in this locality was at first a private speculation, but in 1772 was committed to royal administration.

TIKHAIA-SOSNA, a river of Russia in Europe, in the gov. of Voronej, which has its source 12 m. W of Livensk, in the district of Biriontch, the capital of which it waters; flows thence through the district, and bathes the town of Astrogojsk; and after a course, in a generally NE direction, of about 105 m., joins the Don on the r. bank, 8 m. E of Korotvïak.

TIKHINETZ, a town of Russia in Europe, in the gov. and 54 m. SSW of Mohilev, and district of Rogatchev, near the r. bank of the Droutz.

TIKHVIN, a town of Russia in Europe, in the gov. and 57 m. NE of Novgorod, on the l. bank of the Tikhvina. Pop. 5,000. It has eight churches, two convents, extensive magazines for grain, salt, and brandy, tallow founderies, and tanneries.

TIKHVINA, or TIKHVINKA, a river of Russia in Europe, in the gov. of Novgorod, and district of Tikhvin, which issues from Lake Ozerskoïe, 33 m. E of the town of that name; runs first ESE, then S, and afterwards WNW; passes Tikhvin, when it becomes navigable for small boats, and 15 m. WNW of that town, joins the Sias, on the r. bank, and after a total course of about 90 m.

TIKOPIA. See TUCOPIA.

TIKSA, a lake of Russia in Europe, in the gov. of Arkhangel, in N lat. 66° 26′, and E long. 31° 30′. It is 14 m. in extent from NW to SE, and presents an extremely irregular outline. It discharges itself into Lake Pija, by the Chadra, which issues from its NW extremity. It contains several islands.

TIKVESH, a town of Turkey in Europe, in Rumelia, in the sanj. and 84 m. SW of Ghiustendil, on the Vetitcha, an affluent of the Vardar.

TILAFAR, a town of Asiatic Turkey, in the pash. and 36 m. W of Mosul. Pop. 1,000.

TILBORG, a town of Holland, in the prov. of N. Brabant, cap. of a cant., in the arrond. and 15 m. SW of Bois-le-Duc, near the l. bank of the Ley. Pop. 14,000. It has three churches, a chapel, and a fine castle, and contains extensive manufactories of cloth and other woollen fabrics. It returns three deputies to the provincial states.

TILBROOK, a parish in Bedfordshire, 14 m. N by E of Bedford. Area 1,680 acres. Pop. 350.

TILBURY-JUXTA-CLARE, a parish in Essex, 2¾ m. S of Clare. Area 940 acres. Pop. 401.

TILBURY (EAST), a parish in Essex, 13 m. ESE of Romford, on a creek near the Thames. Area 3,237 acres. Pop. in 1831, 245; in 1851, 401. This is believed to have been the place where the emperor Claudius crossed the Thames in pursuit of the Britons, as related by Dion Cassius.

TILBURY (EAST), a township of Upper Canada, in the Western district, bounded on the N by the Thames, by several branches of which it is intersected, and on the SE by Lake Erie. Pop. in 1844, 540, of whom nearly a half are French Canadians.

TILBURY (WEST), a parish in Essex, 26 m. E of London, on the N side of the Thames. Area 2,086 acres. Pop. in 1831, 276; in 1851, 519. Here are still some traces of Queen Elizabeth's camp, formed in 1588, when the kingdom was threatened by the Spanish Armada.—T. fort stands on the bank of the Thames, opposite Gravesend. It was originally erected by Henry VIII., as a blockhouse; but was enlarged and made a regular fortification by Charles II. after the Dutch fleet had sailed up the river, in the year 1667, and burnt 3 English man-of-war ships at Chatham. The esplanade is large, and the bastions are faced with brick and surrounded with a double ditch or moat. On the land side are two small redoubts of brick; but on this side the chief strength of the fort consists in its being able to lay the whole level under water. On the river side is a strong curtain, with a ditch.

TILBURY (WEST), a township of Upper Canada, in the co. of Kent, bounded on the N by Lake St. Clair. It is to a great extent marshy and unfit for cultivation. Pop. in 1844, 437, of whom two-thirds are French Canadians.

TIL-CHATEL, a commune and town of France, in the dep. of the Cote-d'Or, and cant. of Is-sur-Tille, at the confluence of the Tille and Ignon. Pop. 916. It has a blast furnace and several founderies.

TILDIZ-DAGH, a mountain of Asiatic Turkey, 10 m. S of Tokat.

TILEHURST, or TYLEHURST, a parish in Berks, 2¾ m. W of Reading. Area 5,164 acres. Pop. 519.

TILESWORTH, a parish in Bedfordshire, 3⅓ m. NW by W of Dunstable. Area 1,510 acres. Pop. 1831, 275; in 1851, 337.

TILFF, a village of Belgium, in the prov. and 3 m. S of Liege.

TILGEN, a village of Saxony, 4 m. ESE of Glaucha, on an affluent of the Mulda.

TILGMAN'S ISLAND, an island of Maryland, U. S., at the mouth of the Choptank, containing about 1,720 acres.

TILGUN, a village of Asiatic Turkey, 36 m. ENE of Akshehr.

TILHAC, a village of France, in the dep. of Gers, 7 m. SE of Marciac. Pop. 650.

TILKERODE, a village of the principality of Anhalt-Bernburg, in the bail. of Harzgerode, with large iron-works.

TILL, a river of Northumberland, which rises among the Cheviot-hills; runs S to Chillingham; thence NW to the Tweed, 4 m. below Coldstream.

TILL, a town of Prussia, in the reg. of Dusseldorf, 5 m. SE of Cleves. Pop. 1,680.

TILLANCHONG, one of the Nicobar islands, in the gulf of Bengal, in N lat. 8° 30′. It is high and rugged.

TILLE, a river of France, in the dep. of Cote-d'Or, which rises 3 m. W of Grancey, and running SE, falls into the Saone, on the r. bank, after a course of 40 m.—Also a town of France, in the dep. of Oise, 3 m. NNE of Beauvais. It has iron-works.

TILLEDA, a village of Prussian Saxony, 8 m. WSW of Sangerhausen, at the E base of the Kyffhausen. Pop. 960.

TILLENDORF, a village of Prussian Silesia, 1 m. W of Bunzlau, on the opposite side of the Bobar.

TILLICOULTRY, a parish in Clackmannanshire, marching over one-half of its boundary with Perthshire. Its superficial extent is upwards of 6,000 Scottish acres. About two-thirds of this area lies among the Ochil hills, and includes some of their highest summits. Bencleuch, the loftiest of the heights, rises 2,300 ft. above sea-level. A rich variety of ironstone, and veins of iron-ore of the kidney kind, are in sufficient quantity to be an object of attention to the Devon company. The village of T. is situated on the r. bank of the Devon, 2 m. E of Alva. It has considerable manufactories of shawls, plaidings, and other woollen fabrics, and, seated at the foot of the Ochils, with a plentiful supply of coals and pure water, has facilities for conducting them. Pop. of p. in 1851, 3,217.

TILLIE'RES, a town of France, in the dep. of Eure, on the river Arve, 5 m. NE of Verneuil.

TILLIERS, a village of France, in the dep. of Maine-et-Loire, 4 m. NW of Montfaucon. Pop. of com. 1,150.

TILLINGHAM, a parish in Essex, 18 m. E by S of Chelmsford. Area 7,235 acres. Pop. 1,048.

TILLINGTON, a township in Burghill p., Herefordshire, 5 m. NW of Hereford. Pop. 509.—Also a parish in Sussex, 1 m. W by N of Petworth. Area 3,766 acres. Pop. in 1851, 982.

TILLOUTA, a town of Hindostan, in the prov. of Bahar, district of Rotas, pleasantly situated on the N or l. bank of the Sone, in N lat. 24° 48′.

TILLY, a seigniory of Lower Canada, in the co. of Buckingham, on the S side of the St. Lawrence.

TILLY-SUR-SEULE, a village of France, in the dep. of Calvados, 7 m. SSE of Bayeux. Pop. 900.

TILMANSTONE, a parish in Kent, 5 m. W by S of Deal. Area 1,124 acres. Pop. in 1851, 447.

TILNEY-WITH-ISLINGTON, a parish in Norfolk, 3 m. WSW of Lynn-Regis. Area, including that of the following p., 7,511 acres. Pop. 273.

TILNEY-ALL-SAINTS, a parish in Norfolk, 4 m. W by S of Lynn-Regis. Pop. in 1851, 575.

TILNEY-ST.-LAWRENCE, a parish in Norfolk, 6 m. SW of Lynn-Regis. Pop. in 1851, 573.

TILSHEAD, a parish in Wilts, 14 m. NNW of Salisbury. Area 3,990 acres. Pop. in 1851, 523.

TILSIT, a town of Prussian Lithuania, in the reg. of Gumbinnen, 56 m. ENE of Konigsberg, at the confluence of the Little Tilse, a small stream separating the town from the castle, and the Niemen, which flows past the town on the N, and over which is a bridge of boats. The inhabitants, 13,271 in number in 1846, are partly Catholics, but chiefly Protestants. They carry on a trade, by means of the Niemen, with Konigsberg on the one side, and the interior of Poland on the other, of which the chief articles are corn, wax, salt, salted provisions, hats, and leather. T. gives name to a district which extends to the great maritime inlet of the Curische-haff, a level and fertile tract, noted for its produce of barley, butter, and cheese. The castle of T. is said to have been built in 1289. A treaty of peace was concluded here on 1st July 1807, between France, Russia, and Prussia.

TILSTON, a parish in Cheshire, 12¼ m. SSE of Chester, comprising the townships of Carden, Grafton, Horton, Stretton, and T. Area 3,644 acres. Pop. in 1831, 873; in 1851, 837.

TILSTON-FERNHALL, a township in Bunbury p., Cheshire, 2¾ m. SE of Tarporley. Area 864 acres. Pop. in 1831, 170; in 1851, 155.

TILT, a romantic stream in Blair-Athole, Perthshire, which issues from Loch-Tilt, on the boundary with Aberdeenshire; and after running 2 m. S, receives from the E Glenmore-water, and from the W Tarf-water, each of which greatly excels it both in volume and in length of run. Thus augmented, it flows 9 m. SW, and 4 m. S to the Garry, a little below the village of Blair-Athole. The glen through which it flows is distinguished from every other in the Highlands at once by its extreme depth, narrowness, and prolongation, and by the bold contrast of excessive wildness at the upper end, and ornamental beauty at the other.

TILTON-ON-THE-HILL, a parish in Leicestershire, 10½ m. E by N of Leicester, comprising the townships of Halstead and Marefield, with part of the chapelry of Whatborough. Area 3,112 acres. Pop. in 1831, 342; in 1851, 410.

TILTY, a parish in Essex, 3 m. S by W of Thaxted, watered by the Chelmer. Area 1,040 acres. Pop. in 1831, 82; in 1851, 101.

TILWARA, a town of Hindostan, in the state of Jodpur, on the l. bank of the Loni, here a ¾ m. wide when in flood. Immediately opposite to T. is Mulinathka-Than, the site of a celebrated shrine, which has all the privileges of a sanctuary. A large cattle fair is held here annually, commencing 12 days after the *halee*, and lasting for half-a-month, during which time 80,000 people are said to be collected with large quantities of camels, horses, and bullocks, upon the sale of which fees of 5 rupees, 3 rupees, and 8 annas per head, respectively, are paid to the hakim of Sewanna, who attends the fair on the part of the Jodpur government. The prices of the best horses and mares sold at this fair are from 400 to 600 rupees; riding camels bring 120 r., and baggage camels 60 r. The Loni is here passed.

TIM, a town of Russia, in the gov. of Kursk, on the river Tim, 39 m. E of Kursk.—The river runs about 8 m. W of the town, and flows into the Sosna after a winding course of 80 m.

TIMA, or Al-Ablak, a town of Nedjed, in Arabia, 180 m. NNE of Medina.

TIMAAN, or Timoan, a small island, high and woody, off the E coast of the Malay peninsula, in N lat. 2° 52′. It has several others still smaller lying off it to the westward.

TIMAHOE, a parish in co. Kildare, 3¾ m. N by W of Robertstown. Area 9,896 acres. Pop. 1,052.—Also a village in the p. of Fossy or Timahoe, Queen's co., 4 m. SSW of Stradbally.

TIMANA, a town of New Granada, in the prov. and 81 m. E of Popayan. Pop. 2,000, chiefly cultivators of cocoa.

TIMANOVKA, a town of Russia, in the gov. of Podolia, 21 m. S of Bratzlav.

TIMAVO, a river of Italy, which rises on the confines of Istria, 8 m. S of Goritz, and after a short course, falls into the bay of Trieste, 1 m. NW of Diuno.

TIMBERLAND, a parish in Lincolnshire, 8 m. NE by N of Sleaford, comprising the hamlet of Martin, with the township of Thorpe-Tinby. Area 9,190 acres. Pop. in 1831, 1,278; in 1851, 1,638.

TIMBERSCOMBE, a parish in Somersetshire, 3 m. WSW of Dunster. Area 1,902 acres. Pop. 442.

TIMBO, a town of Senegambia, the capital of Futa-Jallon, in N lat. 10° 33′. Pop. about 19,000. It stands on a rolling plain. N. of it a lofty mountain range rises at the distance of 10 or 15 m. and sweeps eastwardly to the horizon. The landscape which declines from these slopes to the S, is in many places bare: yet fields of plentiful cultivation, groves of cotton-wood, tamarind, and oak, thickets of shrubbery, and frequent villages, stud its surface, and impart an air of rural comfort to the picturesque scene. The town presents narrow streets, low houses, mud walls, cul-de-sacs, and a few mosques. It has no fine bazaars, market-places, or shops. The chief wants of life are here supplied by pedlars. Platters, jars, and baskets of fruit, vegetables, and meat, are borne around twice or thrice daily. "Horsemen," says a French traveller, who has recently given an account of a visit to this place, "dashed about on beautiful steeds towards the fields in the morning, or came home at night-fall at a slower pace. I never saw man or woman bask lazily in the sun. Females were constantly busy over their cotton and spinning-wheels when not engaged in household occupations; and often I have seen an elderly dame quietly crouched in her hovel at sunset, reading the Koran. Nor are the men of T. less thrifty. They weave cotton, work in leather, fabricate iron from the bar, engage diligently in agriculture, and, whenever not laboriously employed, devote themselves to reading and writing, of which they are excessively fond." One cannot but hope that our author who writes anonymously, but calls himself 'a World-wide Tourist,' describes what he has really seen, not merely imagined. "The native villages through which I passed on this excursion," he continues, "manifested the great comfort in which these Africans live throughout their prolific land, when unassailed by the desolating wars that are kept up for the slave-trade. It was the height of the dry season, when everything was parched by the sun, yet I could trace the outlines of fine plantations, gardens, and rice-fields. Everywhere I found abundance of peppers, onions, garlic, tomatoes, sweet potatoes, and cassava, while tasteful fences were garlanded with immense vines and flowers. Fowls, goats, sheep, and oxen, stalked about in innumerable flocks, and from every domicile depended a paper, inscribed with a charm from the Koran to keep off thieves and witches. My walks through T. were promoted by the constant efforts of my entertainers to shield me from intrusive curiosity. Whenever I sallied forth, two townsfolk in authority were sent forward to warn the public that the Furtu desired to promenade without a mob at his heels. These lusty criers stationed themselves at the corners with an iron triangle, which they rattled to call attention to the king's command; and in a short time the highways were so clear of people, who feared a *bastinado*, that I found my loneliness rather disagreeable than otherwise. Every person I saw shunned me. When I called the children or little girls, they fled from me. My reputation as a slaver in the villages, and the fear of a lash in the town, furnished me much more solitude than is generally agreeable to a sensitive traveller."

TIMBUCTU, a town of Central Africa, on the skirts of the Sahara, the geographical position of which, as ascertained by Dr. Barth, is nearly 18° 4′ N lat., and 1° 45′ W long. The existence of this great city in the interior of Africa was long known; but, although it had been approached from the W, the N, and the S, within a few days' journey, by different European travellers, no authentic account of it reached Europe until M. Caillié gave the result of his travels to the world. That enterprising traveller, on having disembarked from the Joliba or Niger, at Cabra, set out from this point at ½ past 3, and reached T. just as the sun was touching the horizon. Of the place itself, he says it presents at first sight nothing but a mass of ill-looking houses built of earth; that it is surrounded on all sides by immense plains of quicksand, of a yellowish-white colour, where not even the warbling of a bird could be heard; and that he thinks "the river formerly flowed close to T., though at present it is 8 m. to the N of that city, and 5 m. from Cabra, in the same direction." This last paragraph appears to make the distance between T. and Cabra far less than a day's journey, independent of its being at issue with every previous account, which concur in placing T. to the N of the river, instead of having the river on its N side. The inhabitants are said by Caillié to be negroes of the Kissur nation, but zealous Mahommedans, and their king a negro of the name of Osman; he adds, "many Moors also reside there." According to the same authority, the government is mild and patriarchal, and the slaves are kindly treated. The city he represents as in outline a sort of triangle, about 3 m. in circuit; as having 7 mosques, 2 of which are large, and each surmounted by a brick tower; and as not closed by any barrier, but open on all sides, and containing at most about 10,000 or 12,000 inhabitants. He also represents the water used there as being brought from deep reservoirs dug in the sands. Leo Africanus, who was himself at T., states that the branch river of T. flows to the W, and joins the Niger at the port of Cabra, 12 m. distant. Sidi Hamet says, that the river which runs by T. is a small one, which occasionally dries up; so that the natives are then obliged to go to the southward, two hours distant, where there is a large river called Zolibib. This Sidi Hamet describes a much larger city, called Wassanah, on the Niger, 60 days to the S of T. Our late traveller, Richardson, while at Ghat, was informed that T. is situated upon the N flats of the Niger, at about a half-day's distance from it in summer, but only 3 hours in winter, the difference arising from the increase of the water of the river. When Dr. Barth, in September 1852, says his zealous friend Mr. Petermann—whose details we follow—lost his only companion, Dr. Overweg, and saw himself reluctantly compelled to abandon, for the time, his contemplated journey across the continent towards the Indian ocean, he resolved to undertake alone a journey to T., which, though greatly less in magnitude than a journey to the Indian ocean, yet was looked on as a most difficult and dangerous attempt. "As the sole survivor of the mission," so wrote Dr. Barth before his departure from Kuka, "the completion of its objects now devolving entirely on me, I feel my powers doubled, and my mind all the more determined single-handed to follow up the results already obtained. My means consist of a tolerable supply of presents, in addition to 200 dollars, four camels, and four horses; my health is in the best condition; and with five trustworthy long-tried servants, well-armed and having plenty of powder and shot with us, I shall with fresh and redoubled courage, and with full confidence of success, start on my journey to Timbuktu." He left Kuka about the end of November, 1852, and proceeded first to Sakatu, by way of Zinder and Kashna, the route by Kano being impracticable on account of the war in that region between the Bornuese and Fellatas. It appears that the general direction of his route from Sakatu to T. was at first WNW, and that he crossed the Kowara, Quorra, or Niger, at Say, an important place situated in about 14° N lat. and 3° 45′ E long. from Greenwich, 180 geog. m. WNW from Sakatu. Both from this place and Libtako he despatched letters to Europe by way of Sakatu. From Libtako in about N lat. 14° 10′, and 0° 39′ E long., the general direction of Dr. Barth's course was NW, till he reached Saraiyamo, a large town 60 m. S of T., and situated on a tributary or branch of the Kowara or Quorra, running towards the latter place. On the former river he embarked on the 1st of September. At first it presented a fine sheet of water 300 yds. in width, but afterwards a most intricate system of narrow meandering channels, partly overgrown with reeds and grass. At a distance of 40 m., in a straight line, from Saraiyamo, after a very tedious winding navigation, he entered the main stream, the Kowara, on the 4th of Sept., and arrived at Kabara or Cabra on the following day. Kabara he describes as a small town of 400 houses and huts, but it has attained great celebrity as the port of T. It scarcely, however, he thinks, deserves that distinction, as it is only approachable by water during four months of the year at an average, and at most during five months, when the floods are unusually high. The creek on which it is situated is of so inconsiderable a size and depth that even at the time of Dr. Barth's visit, which was during the rainy season, the boat with himself and his effects had to be dragged up to the place with great difficulty. The creek measured about 15 ft. across, and the water scarcely reached up to the boatmen's knees. The docks of Kabara—as an artificial large handsome basin, close to the town, may be called—contained but a few boats at the time of Dr. Barth's arrival. Koromeh, a place situated at the junction of the creek of Kabara with the main stream, and the islands of Day, between it and Kabara, have greater claims to be considered the port of Timbuktu. There the Kowara presented a magnificent aspect, covered as it was with a numerous fleet of vessels and boats of various sizes. On the 7th September, 1853, Dr. Barth entered T. in grand style, escorted by the brother of the Sheikh-el-Bakay, the ruling chief of that city, and by a splendid suite on horseback, on camels, and on foot, welcomed and saluted by the festive multitudes of the inhabitants. The latter had been made to believe that the stranger was a messenger from the Great sultan of Stambul! the real character of Dr. Barth being only known to the sheikh himself, whose protection and goodwill the intrepid traveller had been fortunate enough to obtain, and who considered it advisable that he should assume that character on account of the very fanatical disposition of the people; but even under this character Dr. Barth considered himself not entirely free from danger, owing to the complicated character of the political powers which exercise sovereign sway over T.—the inhabitants being composed of various nationalities. There are, first, the Sonray, forming the great bulk of the people; then Arabs of various tribes, Fellatas and Tuaricks, together with a few tribes of Bambara and Mandingo. The accomplishment of the journey from Lake Tsad to T., which in linear extent—taking into account the windings of the road—amounts to at least 2,000 m., might well be a trying task for the physical powers of any man, from its extent only; but when to this are added a preceding three years' travel and toil, the difficulties arising from the rainy season, with its swollen rivers, floods and inundations, during which, partly at least, the journey to T. was accomplished, together with the harassing difficulties and dangers arising from the fanatical character of the inhabitants he had to pass through, it will scarcely be wondered at that Dr. Barth should have reached Timbuctu in a rather exhausted condition. Such were the exertions of the journey that out of six camels two had died on the road, and the remainder were rendered unfit for further use. Dr. Barth describes T. as being closely built of houses mostly of clay and stone, many with handsome and tasteful fronts, the arrangement of the interior being similar to that of Agadez. The pop. he estimated at 20,000. He found the market, which is celebrated as the centre of the North African caravan-trade, to be of less extent than that of Kano, but the merchandise of a superior quality and greater value. He has obtained a complete *imana* from the Sheikh for any English traders that may wish to visit Timbuctu. The Tuaricks and their neighbours had expressed their anxious desire to see English trade extended up the Kowara as far as T., for which all possible advantages and protection were held out. If but the navigation of the river to T. be found practicable, these people, who are masters of the middle course of the Kowara, would, in Dr. Barth's opinion, be sure to protect any English vessel that might proceed in that direction. At present, the market of T. is supplied by two very tedious channels, one coming from the N, from the Barbary states, importing European manufactures, such as guns and cutlery; the other from Nufi and other Koussa states lying near the lower course of the Kowara, not going up the river, but by a circuitous northerly route, also through the desert. The merchandise supplied by the latter channel consists mostly of fine black woollen and silk robes, chiefly manufactured in Nufi (or Nyffi), of which alone 2,000 camel-loads, to the value of 300,000 or 400,000 dollars, are brought to T. annually, and disposed of at a great profit. Sugar is extremely in request. The common medium for trading is salt, which is brought from Taodenni, a place in the desert N of T., and without which article the latter city would not exist. Provisions are very cheap at T., particularly poultry and other animal food; 100 pigeons, for example, may be bought for about 3s. Dr. Barth while in T. found that the river did not attain its *maximum* before the middle of December, so that the river may be presumed to be high during a great portion of the year, the rise in the lower course of the Kowara commencing as early as July. He estimated the rise of the river at T. to exceed greatly that of the Nile.

Historical notice.] Leo Africanus, who visited T. in 1500, was the first traveller who gave to Europe any account of this city, though it was known long before his time as one of the grand emporiums of the interior trade of the African continent. According to him, T. was founded in the year of the Hegira 610 [1215 A. D.], by a king called Mense-Suliman. Under his successor, named Izchia, T. extended its dominion over all the neighbouring states, of which the principal were Ghinea or Genni, Cassina, Guber, Zanfara, and Cano; in his time also it seems to have acquired that commercial prosperity for which it has ever since been distinguished. The city, when visited by Leo Africanus, contained many shops of artizans and merchants, and numerous manufacturers of cotton cloth. There were in it many persons of great opulence, particularly foreign merchants, two of whom were reckoned so considerable by their wealth, that the king had given them his daughters in marriage. The surrounding country abounded both in grain and in pasture; yet neither garden nor orchard, he says, was cultivated round the city. Salt was brought from Tegazza, a distance of 500 m., and sold at an enormous price. The king possessed an ample treasure, and held a splendid and well-regulated court. Great care was taken to keep his stud in good condition. The army consisted of 3,000 cavalry, and a numerous infantry, many of whom used poisoned arrows. The houses of the ordinary inhabitants appear to have been built in a somewhat humble style, in the form of bells; the walls being composed of stakes or hurdles, and the roofs of interwoven reeds; stone, however, had been used in the construction of the principal mosque, and of the royal palace, the latter of which, according to our traveller, had been designed by an artist from Granada. The city was extremely exposed to fire; Leo, in one of his visits, had seen half of it consumed by a single conflagration. Water was supplied from numerous wells; besides which, the waters of the Niger, when they overflowed, were conveyed through the town by sluices. Such is the description given of T. during the period probably of its greatest glory; for the wide dominion which it then appears to have possessed over the surrounding states, has now ceased; yet it continues still to be an emporium of Central Africa, and a medium of communication with the northern parts of that continent. In this character it attracted the notice of the Portuguese, when they carried their career of discovery along the western coast of Asia. It is described by De Barros, as situated three leagues to the N of the Niger; as a great mart for gold; and as being resorted to by merchants from Cairo, Tunis, Marocco, Fez, and all the kingdoms of Northern Africa. The nation who, after the Portuguese, chiefly directed their exertions to penetrate into the interior of Africa, were the French, who, from their settlements on the Senegal, then generally believed to be the same river with the Niger, entertained sanguine hopes of penetrating to T. From the native merchants who came through Bambarra, they learned the existence of the lake Dibbie, under the name of Maberia; and of the well-watered territory of Ghingala or Jinbala; and the position also of T. on the river beyond it. A large caravan of white men, with firearms, justly conjectured to be Moors from Barbary, were stated to arrive annually at T. for the purposes of trade. It was added, that large barks with masts had been seen in the river near T.: these were erroneously supposed to belong to the Tripoli merchants; but no French mission ever extended beyond the banks of the Senegal, or attained the frontier of Bambarra; and the earliest efforts of the English, made by the Gambia, then supposed to be also a branch of the Niger, were productive of still less important results than those made by the rival nation. Jobson and Stibbs could not reach higher than the neighbourhood of Tenda; and their reports went only to discourage the idea of penetrating into the interior by this channel. Park, under the auspices of the African association, penetrated nearly a thousand miles beyond all the former limits of French or English discovery; but the pressure of disastrous circumstances compelled him to stop considerably short of T. He, however, learned that the kingdom of T. lay beyond that of Masina, which extended along Lake Dibbie, and bordered on the E frontier of Bambarra. The city, Park was informed, lay about a day's journey to the N of the Niger; while its port, Cabra, was situated at the point of junction between two branches of that river which issue out of Lake Dibbie. The government was said to be, in the hands of the Moors. It is stated that some years after Park's journey, the king of Bambarra conquered T., and established there a Negro government; the Moors, however, were not only permitted to reside and carry on their trade, but the internal police of the city was left in their hands. The report of Adams, an American sailor, who described himself as having been shipwrecked near Cape Blanco, carried as a slave to T., and detained there during six months, describes this city as standing on nearly the same extent of ground as Lisbon, though the pop. was probably much less, as the houses were built in a very scattered manner. The king and all his principal officers were Negro. No Moors were allowed to reside in, or even to enter the city, unless in small parties. This circumstance, which contrasts singularly with Park's account, may, however, be connected with the frequent revolutions which have taken place in the political destinies of T. Adams does not describe T. in very lofty terms. The king's palace, called the *kusba*, consisted merely of a square space enclosed by a mud wall, and containing eight small apartments on the ground floor; the houses of the chief citizens were built of wooden cases filled with clay and sand; while the huts of the poorer class consisted merely of branches of trees bent in a circle, covered with a matting of palmeto, and overlaid with earth. The king and queen wore dresses of blue nankeen, profusely ornamented with gold and ivory. Adams describes T. as situated immediately upon a river called the Mar Zarah, about ¾ m. wide, and flowing, as he believed, to the SW. About 2 m. S of the town it passes between high mountains, where its breadth is contracted to a ½ m.—A still more recent account of T., though not that of an eye-witness, is contained in the narrative of Riley, an American captain, who in 1815 suffered shipwreck on the coast of the Sahara. His information, however, was entirely derived from Sidi Hamet, an African merchant, by whom he was purchased and brought to Mogodor. According to him T. is a large town, six times as populous as Mogodor. The pop. is entirely Negro. He states, however, what seems to have entirely escaped the notice of Adams, that there was a distinct town, divided from the other by a strong partition wall, which formed the residence of those Mussulmans who were permitted to remain at T. Captain Lyon during his residence in Fezzan, was informed by the merchants, that T. was not so large a town as had been supposed; some represented it as not more extensive than Murzuk. The immense pop. which some have ascribed to it, is accounted for by supposing that they included the kafilas, who arrived there in great numbers, and being often obliged to remain during the rainy season, erected temporary huts. The first European traveller who succeeded in penetrating to this mysterious city, was Major Laing, who reached it in 1826, but was barbarously murdered on his return homewards. An enterprising Frenchman, M. Caillié, was more fortunate. Setting out on the 19th of April, 1827, from Kakondy, the tomb of Peddie and Campbell, he crossed the Senegal at Bafila, and after passing Lake Debo, the Debbie of former writers, he arrived at Cabra on the Joliba, and disembarked at a point 13 m. distant from Timbuctu, which he says lies to the S of the river. Of the city, he says, "I found it neither so large nor so populous as I had expected. Its commerce is not so considerable as fame has reported There was not, as at Jenné, a concourse of strangers from all parts of the Sudan. I saw in the streets of T. only the camels which had arrived from Cabra laden with the merchandise of the flotilla, a few groups of the inhabitants sitting on mats, conversing together, and Moors lying asleep in the shade before their doors: in a word, every thing had a dull appearance. I was surprised at the inactivity, I may even say indolence, displayed in the city. Some colat-nut venders were crying their goods in the streets as at Jenne. About 4 in the afternoon, when the heat had diminished, I saw several Negro traders, all well-clothed, and mounted on good horses, richly harnessed, go out to ride. Prudence forbids them to venture far from the city, for fear of the Tuariks, who would make them pay dearly for their excursions. In consequence of the oppressive heat, the market is not held until 3 in the afternoon. There were few strangers to be seen except the Moors of the neighbouring tribe of Zawat, who often come hither; but in comparison with Jenne, the market is a desert At T. it is very unusual to see any other merchandise except what is brought by the vessels, and a few articles from Europe, such as glass wares, amber, coral, sulphur, paper, &c. I saw three shops kept in small rooms, well-stored with stuffs of European manufacture. The merchants put out at their doors cakes of salt for sale, but they never exhibit them in the market. Such as do business at the market have stalls made of stakes, covered with mats, to protect them against the heat of the sun. My host, Sidi-Abdallahi, was obliging enough to show me over one of his magazines, in which he stowed his European merchandise. I observed there many double-barrelled guns, with the mark of Saint Etienne, and other manufactories. In general French muskets are much prized, and sell at a higher rate than those of other nations. I also saw some beautiful elephants' teeth. My host told me that he procured some from Jenne, but the larger ones had been bought at T. They are brought hither by the Tuariks or Surgus, the Kissurs, and the Dirimans, who inhabit the banks of the river." There are, he says, "many Moors in T., and they occupy the finest houses in the city. They soon become rich in trade, and receive consignments of merchandise from Adrar, Tafilet, Tawât, Ardamas, Tripoli, Tunis, and Algiers. They receive from Europe, tobacco and other articles, which they send by canoes to Jenne and elsewhere. T. may be regarded as the principal *entrepôt* of this part of Africa. All the salt obtained from the mines of Toudeyni is brought hither on camels. The Moors of Marocco and other countries who travel to the Sudan, remain six or eight months at T. to sell their goods and get their camels re-laden." T., he says, "contains 7 mosques, two of which are large; each is surmounted by a brick tower. This mysterious city, which has been an object of curiosity for so many ages, and of whose population, civilization, and trade with the Sudan, such exaggerated notions have prevailed, is situated in an immense plain of white sand, having no vegetation but stunted trees and shrubs, such as the *Mimosa ferruginea* which grows no higher than three or four ft. The city is not closed by any barrier, and may be entered on any side. Within the town are seen some of the *Balanitis Ægyptiaca*, and in the centre is a palm-tree. T. may contain at most about 10,000 or 12,000 inhabitants; all are engaged in trade. The pop. is at times augmented by the Arabs, who come with the caravans, and remain a while in the city. In the plain, several species of grass and thistles afford food for the camels. Fire-wood is very scarce, being all brought from the neighbourhood of Cabra; it is an article of trade, and the women sell it in the market-place: it is only burnt by the rich; the poor use camel dung for fuel. Water is also sold in the market-place; the women give a measure containing about half a pint for a cowrie. T., though one of the largest cities I have seen in Africa, possesses no other resources but its trade in salt, the soil being totally unfit for cultivation. The inhabitants procure from Jenne every thing requisite for the supply of their wants, such as millet, rice, vegetable butter, honey, cotton, Sudan cloth, preserved provisions, candles, soap, allspice, onions, dried fish, pistachios, &c. If the vessels from Cabra should chance to be stopped by the Tuariks, the inhabitants of T. would be reduced to famine: to obviate this misfortune, they take care to have their warehouses always amply stored with every kind of provision. I saw the magazines of Sidi-Abdallahi full of great sacks of rice, which keeps better than millet. For these reasons the vessels which come down the river to Cabra are deterred from making any resistance to the Tuariks, notwithstanding the burden of their exactions. Sidi-Abdallahi informed me that there was no traffic or communication by water between T. and the country of Houssa; because, said he, the navigation of the river ceases at Cabra. The Negroes and Moors devote their attention exclusively to trade: they possess but limited ideas of geography. All to whom I applied for information respecting the course of the river, to the E and ESE of T., agreed in stating that it runs to Houssa, and empties itself into the Nile. I was unable to obtain any more accurate information on this point; and the great problem of the issue of the Dhioliba into the ocean will thus be left to the demonstration of some more fortunate traveller; but if I may be permitted to hazard an opinion as to the course of the river, I should say that it probably empties itself by several mouths into the gulf of Benin." The poeple of T. are gentle and complaisant to strangers; in trade they are industrious and intelligent; and the traders are, generally, wealthy, and have many slaves. The men are of the ordinary size, well-made, upright, and walk with a firm step. Their colour is a fine deep black. Their noses are a little more aquiline than those of the Mandingoes, and, like them, they have thin lips and large eyes. I saw some women who might be considered pretty. They are all well-fed: their meals, of which they take two a-day, consist of rice, and coocoosoo made of small millet, dressed with meat or dried fish. Those Negroes who are in easy circumstances, like the Moors, breakfast on wheaten bread, tea, and butter made from cows' milk. Those of inferior condition use vegetable butter. They are exceedingly neat in their dress and in the interior of their dwellings. Their domestic articles consist of calabashes and wooden platters. Their furniture merely consists of mats for sitting on; their beds are made by fixing four stakes in the ground at one end of the room, and stretching over them a few mats or a cow-hide. The women of the richer class have always a great number of glass beads about their necks and in their ears. Like the women of Jenne, they wear nose-rings; and the female who is not rich enough to procure a ring substitutes a bit of red silk for it: they wear silver bracelets, and ankle-rings of plated steel, the latter of which are made in the country. The amount of this traveller's information regarding the mysterious Niger is much less ample than his details respecting T. All that we can gather from him is that the Niger at its source is called the Tombia, or the Ba, and afterwards, until it reaches Sego, the Joliba. To the N of Sego, it divides into two branches, which, according to one authority, bear the names of Ba-niou and Ba-leo, or, as they have been interpreted, 'the White river,' and 'the Black river.' These two branches unite at Isaca, and from that town to Cabra, the Joliba—which is there known by the name of Couara only—proceeds due N to Cabra. At Cabra it turns E, and soon after is again divided into two branches, one of which the Yeou is supposed to run eastward into Lake Chad; the other and larger was traced by Park as far as Boussa, whence, it has now been ascertained, it proceeds to the bight of Benin.

TIME', a town of Nigritia, 300 m. ESE of Timbo on the W side of the Granite Peaks. Pop., chiefly Mandingoes, 600.

TIMELKAM, a town of Upper Austria, on the river Ager, 3 m. W of Vocklabruck.

TIMELKAMIN, a town of the archduchy of Austria, in the prov. of Upper Austria, and circle of Hansruck, 2 m. W of Wöcklabruck, on the r. bank of the Wöckla.

TIMERYCOTTA, a town and fortress of Hindostan, in the Carnatic, in N lat. 16° 35'. Six miles distant from this place is a cataract 60 ft. high, from which the water falls into a basin 120 ft. in breadth, on the banks of which are a number of small temples.

TIMIMUN, a town of the Sahara of Africa, in the oasis of Touat, in N lat. 27° 50', consisting of between 500 and 600 houses, and surrounded by a wall and ditch.

TIMISCOUATA. See Temiscouta.

TIMKOVITCHI, a town of Russia in Europe, in the gov. and 69 m. SSW of Minsk, and district of Sloutzk.

TIMMERINGUE, or Dimeringue, a village of France, in the dep. of the Bas-Rhin, cant. and 5 m. N of Drulingue, on the slope of a hill. Pop. 950. It has manufactories of hosiery, cordage, earthenware and tiles. In the environs are several salines and quarries.

TIMOAN (Pulo), or Tioman, an island of the China sea, near the E coast of the peninsula of Malacca, in N lat. 2° 52'. It is high and woody, abounds with the cabbage palm, and produces considerable quantities of rice. This island lies a little to the E of the bent route towards China, during the strength of the monsoon. It in common with the adjacent islands is inhabited by Malays.

TIMOCHI, Riochery, or Pic-de-Langle, an island of the sea of Japan, to the NW of the island of Jeso, in N lat. 45° 10', and E long. 141°. It is circular in form, and rises to the height of 7,000 Parisian ft. above sea-level.

TIMOFIEEVKA, a town of Russia in Europe, in the gov. and 71 m. NW of Kharkov, and district of Soumoui.

TIMOGUE, a parish in Queen's co., 2 m. SSW of Stradbally. Area 2,490 acres. Pop. in 1851, 243.

TIMOK, a river of Turkey in Europe, which descends from the S side of mount Haiduchki, in the Balkan chain, in Servia, sanj. of Semendria, and 5 m. W of a town of the same name; runs first ESE, then NNE to the confines of Bulgaria; runs along these till it reaches the Danube, which it joins on the r. bank, 20 m. NW of Widdin, and after a sinuous course of about 120 m. Its principal affluents are the Kutchuk-Timok on the r., and on the l. the Krivivir, Tukhovatz and Kreszno.—The town of Timok, 84 m. SE of Semendria, is the capital of a jurisdiction. It lies in a narrow valley, at the E base of mount Haiduchki, on a small affluent of the river of the same name.

TIMOLEAGUE, a parish in co. Cork, Munster. Area 2,873 acres. Pop. in 1831, 1,824; in 1851, 1,000.—The ancient and once important but now decayed town of T. stands on the r. bank of the Arigadeen estuary, 3 m. above its expansion into Courtmacsherry bay, and 4 m. ENE of Clonakilty. Pop. in 1831, 720; in 1851, 393.

TIMOLIN, a parish of co. Kildare, containing the villages of Ballytore and Timolin. Area 2,289 acres. Pop. in 1831, 2,941; in 1851, 1,265. The village of T. is 1¾ m. S of Ballytore. Pop. in 1851, 137.

TIMONHA, a mountain of Brazil, in the prov. of Ceara, to the N of the Serra-Araripe. It runs from N to S, a distance of several leagues, and is said to contain considerable quantities of gold.

TIMOR, the southernmost and largest of the Molucca islands, in the Eastern seas. Its extent in a NE and SW direction, is nearly 300 m.; its breadth from 30 to 60 m. Its S extremity is in S lat. 10° 24', E long. 123° 32'; its NE, in S lat. 8° 20', E long. 127° 10'. The interior is intersected, in the direction of the island's length, by a chain of mountains, some of which nearly equal the peak of Teneriffe in elevation; and various ranges of hills of limestone formation occur in different parts of the island. Its mountains exhibit sea-shells at an elevation of 800 ft.; they frequently present a conical shape. It is not known whether any volcanoes exist among them; but the whole island is subject to frequent earthquakes. The valleys are generally narrow with steep sides, but in a few instances open into plains of considerable extent. The enthusiasm of navigators, who have visited this island immediately after leaving the monotonous shores of New Holland, has probably created some exaggeration in their descriptions of the fertility of this island; yet it is certainly a pleasing region. The shores on the SE side are low, and overrun with mangroves. The streams are generally of short course, and few of them continue to flow during the dry season. Maize, cotton, tobacco, sago, indigo, pine-apples, and a variety of fruit, sandal-wood, honey, and wax, are among the productions of the island. Maize is the principal article of food, but the natives depend for a great part of their food on the sugar of the lontar-palm, and the produce of the sago-palm. The use of the plough is unknown; a wooden hoe and sharp-pointed stick are the only implements used in the hill-cultivation. The average annual crop of paddy is 70 fold. Cocoas and areca palms are scarce; but the lontar is abundant, and small quantities of sugar-cane are raised. Fish can scarcely be considered as an article of subsistence, as there are scarcely any of the natives who will trust themselves in a canoe. The bee is not domesticated here, nor indeed in any of the islands in this quarter; but the vegetation supports myriads of wild bees. An article of export is *beche-de-mer*, which is partly procured near the island, and partly fished on the coast of Australia, and brought here for sale. The bread-fruit tree grows in great luxuriance, and seems to thrive as well here as in Otaheite. Sandal-wood, bees-wax, honey, and slaves, are exported; and rice, arrack, sugar, tea, coffee, betel-nut, and the manufactures of China, with some from India and Europe, received in return. The original inhabitants, who are black, but whose hair is not woolly, inhabit the mountainous parts, to which they appear to have been driven by the Malays who are mostly in possession of the sea-coast. There were formerly several Portuguese establishments on the N side of the island, of which Dielli, in S lat. 8° 34', and E long. 125° 37', Gedeh in S lat. 8° 57', and E long. 124° 55', and Liefouw, in S lat. 9° 12', and E long. 124° 24', still remain. The principal Dutch settlement, that of Koepang, is on the W side of the river, in S lat. 10° 10', E long. 123° 35'. The Dutch have made some attempts to establish Christianity, but with little success, the natives mostly remaining in their original ignorance. The Dutch and Portuguese claim between them the entire sovereignty of this island: Fort Concordia being the seat of the government of the former, and Dielli of the latter. But the power of both is so much decreased at the present day, that their authority is only acknowledged by such of the native chiefs as need their assistance against their more powerful neighbours. The nominal boundary of the two governments cannot be formed by a line drawn in any direction, as some of the petty states near Dielli are under Dutch protection, while others, near Koepang, are under the Portuguese. It is, however, considered that the whole of the country to the E of Dielli belongs to the Portuguese; and the whole of the S coast to the Dutch. Along the NW coast the two governments are completely mixed. It is impossible to form any correct estimate of the pop. The inhabitants are said to be numerous in the interior and along the S coast; but very few villages are to be seen on the N coast, and these consist only of a few huts. It appears, however, the general custom of the island not to form themselves into large communities. The natives are generally of a dark colour, with frizzled bushy hair, but less inclining in appearance to the Papuans than those of Ende. They are below the middle size, and rather slight in their figure. In countenance they more nearly resemble the South-sea islanders than any of the Malay tribes. The peasants of both sexes wear only a cloth wrapped round their loins; the rajahs generally wear *bajus* of silk or chintz, with five or six handkerchiefs of different colours wrapped loosely round their heads. Their ornaments chiefly consist of arm-rings of gold, silver, or ivory; the women wear arm and ankle-rings of earthenware. There does not appear to be any system of laws amongst them; the will of the sovereign being in all cases supreme. The religion of the island is pagan; most of the princes indeed profess Christianity, but are entirely guided by their pagan priests and customs. The feudal system seems to exist throughout the island; for every man capable of bearing arms is obliged to attend the call of his feudal lord.

TIMOR-LAUT, an island in the Eastern seas, about 70 m. long, by 25 m. in average breadth, situated between the 7th and 8th parallels of S lat. and the 132d and 133d meridians of E long.

TIMPERLEY, a township in Bowden p., co.-palatine of Chester, 8 m. NNE of Nether Knutsford. Area 1,628 acres. Pop. in 1831, 752; in 1851, 1,008.

TIMSBURY, a parish in Somersetshire, 6½ m. SW by W of Bath. Area 1,148 acres. Pop. in 1831, 1,367; in 1851, 1,639.—Also a parish in Hants, 2 m. N by W of Romsey; crossed by the Anter or Test river. Area 1,411 acres. Pop. in 1851, 194.

TIMUR-HISSAR. See DEMIR-HISSAR.

TIMWORTH, a parish in Suffolk, 3½ m. N of St.-Edmunds-Bury. Area 1,358 acres. Pop. 251.

TINACO, a river of Venezuela, which enters the Portuguesa. It has a settlement of the same name on its banks, 45 m. SW of Valencia.

TINACORO, a volcanic islet of the S. Pacific, in S lat. 10° 23′, E long. 165° 49′.

TINAJAS, a village of Spain, in the prov. and 24 m. NW of Cuenca. Pop. 650.

TINAQUILLO, a settlement of Venezuela, situated on the river Caxede, to the S of Valencia.

TINCHEBRAY, a town of France, in the dep. of the Orne, on the river Noireau, 14 m. N of Domfront. Pop. 1,916. It has manufactories of linen and thread, also tanneries, and nail-works. A battle was fought here between the English and Normans in 1106.

TINCULEN, or TINZULEN, a village of Marocco, in the prov. and 120 m. SW of Tafilet.

TINDARO (CAPE), a headland on the N coast of Sicily, in N lat. 38° 7′, 4 m. NE of Patti. The ruins of the ancient *Tyndaris* lie near it.

TINDFIELD, a mountain range of Norway, under the 60th parallel, connected on the NW with the Hardanger-field, and stretching SE about 45 m.

TINDSIŒ, a lake of Norway, in N lat. 59° 45′ E long. 8° 50′. It is 21 m. in length from NNW to SSE, and discharges itself into the Nordsöe.

TINEH, or THINEH, a ruined city of Lower Egypt, situated upon the Pelusiac branch of the Nile, in N lat. 30° 55′.

TINEH, a small town of Tripoli, situated on a river which falls into the gulf of Sidra or Syrtis, in N lat. 30° 5′.

TINEHELY, a village in the p. of Kilcommon, co. Wicklow, on the rivulet Derry, 5¾ m. N by E of Carnew. It occupies a romantic site among the broken grounds skirting the base of the mountains, which extend from Glen-Aughrim to the vicinity of Newtownbarry; and it is overhung on the W, by a summit which has an alt. of 1,312 ft. above the level of the sea. Pop. in 1831, 575; in 1851, 562.

TING-CHU, a division and town of China, in the prov. of Chih-le. The div. comprises two districts. The town is 38 m. SW of Paou-ting-fu, near the r. bank of the Tam-ho, in N lat. 38° 32′ 30″, and E long. 115° 9′.

TING-CHU-FU, a division and town of China, in the prov. of Shan-tung. The div. comprises 10 districts. The town is 240 m. ENE of Tse-nan-fu, on the Yellow sea, in N lat. 37° 48′ 26″, and E long. 121° 4′ 30″. It has a commodious port, defended by a numerous garrison, and by several vessels of war. —Also a division and town in the prov. of Fo-keen. The div. comprises 7 districts. The town is 195 m. W of Fuh-chu-fu, in N lat. 25° 44′ 54″, and E long. 116° 29′ 35″. It is enclosed by mountains of considerable height, and containing mines of silver. The environs are fertile, but extremely unhealthy.

TINGEWICK, a parish in Bucks, 2 m. W by S of Buckingham. Area 2,290 acres. Pop. 877.

TING-FAN-CHU, a district and town of China, in the prov. of Kwei-chu, div. and 33 m. SSW of Kwei-yang-fu, on the r. bank of the Leen-keang, in N lat. 26° 6′, and E long. 106° 32′.

TING-GAN, a district and town of China, in the prov. of Kwang-tung and div. of Keun-che-fu, island of Hai-nan. The town is near the r. bank of the Li-mu-keang or Keen-keang, 23 m. S of Kiung-chu. Pop. in 1819, 54,955. It is enclosed by substantial walls, and is of considerable extent. Its streets are straight and well-paved, and it has an extensive library founded in 1736.

TING-HAE, a district and town of China, in the prov. of Che-keang, and div. of Ning-po-fu. The town is 39 m. E of Ning-po-fu, in N lat. 30° 0′ 24″, and E long. 122° 6′ 35″. Pop. 35,000. It lies on the S coast of the island of Chusan, in the Yellow sea. The walls by which it is surrounded are 18 ft. in height, and 15 ft. thick, and are surmounted by another wall 14 ft. 16 in. high, and 2 ft. thick. It is traversed by numerous canals, across which are thrown lofty bridges connecting the streets, which are narrow and paved with flag-stones. The houses are generally only a single story in height. The general aspect of the town is that of activity and industry. It was taken by the British in 1841.

TING-HING, a district and town of China, in the prov. of Chih-le, div. and 33 m. NNE of Paou-ting-fu, in N lat. 39° 17′, and E long. 115° 56′.

TING-JANG-HEEN, a district and town of China, in the prov. of Shan-se and div. of Hin-chu, in N lat. 38° 32′, and E long. 112° 59′.

TING-KE-HEEN, a district and town of China, in the prov. of Kwei-chu and div. of Sze-chu-fu.

TING-NAN-TING, a district and town of China, in the prov. of Keang-se, and div. of Kan-chu-fu, in N lat. 24° 45′, and E long. 114° 45′.

TING-PEEN, a district and town of China, in the prov. of Shen-se, and div. of Yen-gan-fu.—Also a district and town in the prov. of Fo-keen, and div. of Tsu-heung-fu, on the r. bank of the Ta-shan, in N lat. 25° 6′, and E long. 100° 43′.

TING-PWAN-CHU, a district and town of China, in the prov. of Kwei-chu, and div. of Kwei-yang-fu.

TINGRITH, or TYNGRAVE, a parish in Bedfordshire, $4\frac{1}{4}$ m. E by S of Woburn. Area 946 acres. Pop. in 1831, 162; in 1851, 196.

TING-TAOU, a district and town of China, in the prov. of Shan-tung, div. and 15 m. SE of Tsaou-chu-fu, in N lat. 35° 11′ 18″, and E long. 115° 44′ 30″.

TING-YUEN, a district and town of China, in the prov. of Sze-chuen, div. and 32 m. SSE of Chun-ting-fu, on the r. bank of the Kia-cing-keang, in N lat. 30° 25′, and E long. 106° 20′.

TING-YUEN-CHU, a district and town of China, in the prov. of Shen-se, and div. of Han-chung-fu.

TING-YUEN-HEEN, a district and town of China, in the prov. and 27 m. S of Fung-yang-fu, in N lat. 32° 32′ 46″, and E long. 117° 32′ 47″. It contains numerous shops of mediocre quality, and has two triumphal arches.—Also a district and town in the prov. of Yun-nan, div. and 18 m. N of Tsu-heung-fu, on a small affluent of the Lang-chuen-ho, in N lat. 25° 22′, and E long. 101° 44′.

TINGWALL, WHITENESS, and WEESDALE, a united parish a little S of the centre of the mainland of Shetland. Its greatest length is between 18 and 20 m., and its greatest breadth is 10 m.; but it is so much indented with *voes*, that no part of it is more than 2 m. from the sea. Several small islands belong to it; the chief of which are Oxnor, Trousa, Hildessie, and Linga, all inhabited. The principal voes or harbours are those of Wadbaster, Laxforth, and Deal, on the E, and Weesdale, Stromness, Ustaness, and Scalloway, on the W. Pop. 2,874.

TINHOSA, an island in the China sea, in N lat. 18° 40′.

TINISCHT, a town of Bohemia, 73 m. E of Prague, on the r. bank of the Adler. Pop. 1,000.

TINJULIEN, a town of Marocco, on the Wad-Drah, 105 m. SE of Marocco.

TINKLETON, a parish in Dorset, $5\frac{1}{2}$ m. E by N of Dorchester. Area 885 acres. Pop. in 1851, 176.

TINLEYSVILLE, a village of Goochland co., Virginia, U. S., 45 m. WNW of Richmond.

TINMOUTH, a township of Rutland co., Vermont, U. S., watered by Otter creek, 10 m. S of Rutland. Pop. 1,000.

TINNEHINCH, a hamlet in the p. of Rearymore, Queen's co., on the rivulet Barrow, 2 m. E by S of Clonaslee, at an elevation above sea-level of 352 ft. —Also a village in the p. of St. Mullins, co. Carlow, on the l. bank of the Barrow, directly opposite to Graiguenamanagh, so as to form a suburb of that town, 5 m. S by W of Borris. Pop. in 1851, 401.

TINNEVELLY, an extensive district of Hindostan, in the Southern Carnatic, occupying the SE extremity of the peninsula, between the parallels of 8° and 10° N, and separated from Ceylon by the gulf of Manaar. Generally speaking, this district, the area of which is 5,700 sq. m., may be called an open and level country, with large tracts of waste and jungle, although it contains some woods and several hills. It does not contain any river of magnitude, but is watered by numerous streams flowing from the western mountains; and in favourable seasons, yields abundant crops of rice and cotton. The banks of the Tamburapurni are very fertile. Towards the S and E extremities are numerous salt marshes. Its principal towns are Palamcotta, Manapur, and Tinnevelly; and its best seaports, Tutacorin and Coilpatam. The pop., about 850,000 in number, consists chiefly of Hindus, who retain much of their primitive customs and manners. During the first wars in which the British were engaged in this country, from 1740 to 1760, it was in the hands of a number of petty chiefs called *polygars*, who were constantly quarrelling with each other, and plundering the country. In this disorderly state the district continued till 1799, when the British reduced to obedience a number of these chieftains. During the war with Tippu Sultan, the polygars took advantage of the absence of the army, and broke into rebellion; but as soon as a sufficient number of troops could be spared, a large force was sent against them; and in 1801, they were all subdued. The revenue, amounting to 70,000 pagodas or about £23,000 per annum, in 1804, is now 24 lacs, or £240,000. Cotton, silks, and iron wares are manufactured.—The capital, of the same name, stands in N lat. 8° 48′, about 25 m. E of a range of mountains, and is large and populous. On three sides it is surrounded by extensive rice-fields; and on the W by rocky ground.

TINNIS, a rivulet of Liddesdale, Roxburghshire, which rises between the heights of Tinnis and Lockknowe, and runs SE to the Liddel at Burnmouth.

TINOS, an island of the Grecian archipelago, to the SE of Andros, from which it is separated by a narrow channel. It is about 60 m. in circuit. It is very mountainous, but fertile, producing wheat, barley, silk, olives, figs, wine, racki, wax and unrivalled honey, hares, wild pigeons, and doves, sheep, pigs, and poultry. It has two or three small rivers, countless fountains of excellent water, valuable marble and stone quarries, three or four towns, nearly 70 villages, innumerable convents and churches, millions of snakes, a few jackalls, and about 30,000 inhabitants. Strong gloves, stockings, and stuffs, are manufactured in the island, and small marble figures are occasionally sculptured here in a manner that would do credit to Italy. Its trade in [illegible]e, oil, and brandy is considerable.

TINOSO (CAPE), a cape of Spain, on the coast of Murcia, in N lat. 37° 30′.

TINSLEY, a chapelry in Rotherham p., Yorkshire, $2\frac{3}{4}$ m. SW by W of Rotherham. Area 1,623 acres. Pop. in 1831, 368; in 1851, 613.

TINTAGELL-AND-BOSSINEY, a parish and disfranchised borough in Cornwall, $5\frac{1}{2}$ m. NW of Camelford. Area 4,350 acres. Pop. in 1831, 1,006; in 1851, 1,084. It returned 2 members to parliament from the 7th of Edward VI. till 1832, when it was disfranchised by the Reform act. T. is the reputed birth-place of King Arthur.

TINTERN, or KINNEAGH, a parish in co. Wexford, 3 m. N of Fethard, containing the village of Saltmills. Area 6,863 acres. Pop. in 1851, 2,163.

TINTERNE-ABBEY, a celebrated ruin in Chapel-hill p., co. of Monmouth, $4\frac{1}{2}$ m. N of Chepstow, on the W bank of the river Wye. It was a Cistertian abbey, founded in 1131. Its ruins are justly esteemed the most beautiful and picturesque objects on the Wye.

TINTERNE-PARVA, a parish in Monmouthshire, 5 m. N of Chepstow. Area 827 acres. Pop. in 1831, 313; in 1851, 370.

TINTINHULL, a parish in Somersetshire, 2 m. SW of Ilchester. Area 1,828 acres. Pop. 529.

TINTO, a river of Spain, in the prov. of Seville, which rises in Huelva; flows SE; and runs into the Atlantic, to the W of the Guadalquivir, near the town of Moguer, after a total course of 65 m. It derives its name from the colour of its water, which in the early part of its course is a yellowish green, and so bitter that no animal except goats will drink of it. This arises probably from a metallic infusion received at or near its source; for it disappears after its stream has been increased by other rivers; and at the town of Niebla it has hardly any properties distinct from those of common water. At its mouth it forms a considerable bay, about 7 m. in breadth; and large vessels sail up as far as San-Juan-del-Puerto, about 12 m. inland.

TINTO, a lofty mountain at the head of Clydesdale, in Lanarkshire, stretching above 2 m. from E to W, and elevated 2,312 ft. above the level of the sea, and 1,740 ft. above the Clyde. The upper part of this mountain consists of a flesh-coloured felspar, resting on conglomerate.

TINTO, a river of Central America, which flows into the bay of Honduras, 20 leagues E of Cape Honduras, after a course of about 120 m.

TINTWISTLE, or TINGETWISSEL, a chapelry in Mottram p., Cheshire, 9½ m. NE by E of Stockport, on the N bank of the Etherow, and in the line of the Manchester and Sheffield railway, which here enters a tunnel 3 m. in length, by which it is conducted to the Yorkshire side of the ridge: the elevation of the SW entrance of the tunnel above sea-level is 875 ft. Area 14,120 acres. Pop. 3,027.

TINWALD, a parish lying between Nithsdale and Annandale, but belonging mainly to Dumfries-shire. Its superficial extent is about 15 sq. m. Pop. in 1831, 1,220; in 1851, 1,080.

TINWELL-AND-INGTHORP, a parish in Rutlandshire, 10½ m. ENE of Uppingham. Area 1,651 acres. Pop. in 1831, 262; in 1851, 287.

TIOGA, a county on the N side of the state of Pennsylvania, U. S. Area 1,056 sq. m. Pop. in 1840, 15,498; in 1850, 23,987. Chief town, Wellsborough.—Also a central co. in New York. Area 456 sq. m. The surface is considerably broken and hilly. Pop. in 1840, 20,527; in 1850, 24,880. Its cap. is Owego.—Also a post township of Tioga co., Pennsylvania, 110 m. N by W of Harrisburg.—Also a river of the United States, which rises among the Alleghany mountains, Pennsylvania; runs N, entering the state of New York; after which it turns to the E, and joins the Couhocton.

TIONE, a small town of the Tyrol, on the river Sarca, 19 m. W of Trent. Pop. 1,000.

TIORN, an island of Sweden, 18 m. N of Gottenburg. It is about 30 m. in circumf., and has good pasturage. The inhabitants export butter, cheese, and hops.

TIOUGHE, or TEOGE, a recently discovered river of S. Africa, the principal feeder of Lake Ngami, into the NW extremity of which it falls. It has recently been ascended by Mr. Anderson for 13 days in a NNW direction; but so strangely tortuous is its course that he only made one degree of northing in all that time. Here he reached a large village, to which he had been directed by the chief of Lake Ngami, with the assurance that fresh boats and men should be supplied from it for his further progress. However, on his arrival, he found that the whole male population had decamped with their canoes, professedly on a seacow hunt, but in reality to deprive him of all means of further progress. He waited a week, when, as matters continued the same, and as the sickly season was approaching, he had no alternative but to return. The T., in that portion reached by Mr. Anderson, does not exceed 40 yds. across; it is deep and full of reeds, and runs at about 2½ m. per hour. Its banks are low and reedy, and are flooded for many miles during the rainy season. Here and there they rise somewhat higher; and then clumps of fine trees entwined with parasitical plants, and presenting a varied and luxuriant vegetation, cover them. Seacows are numerous in the river, and much dreaded by the natives, frequently attacking their canoes. A party of Griquas, in 1852-3, had travelled in waggons to the lake, and thence to the NNW for 19 days, when they succeeded in reaching the town Libébè, which is the cap. of the Bavicko country, and situated high up on the T. It is the centre of a great inland traffic, and the Mambari, the Ovapangari, and the Ovahanyama are among the nations that visit it. The Griquas returned by a different route, and then fell in with the *tsetse* fly, which committed such sad havoc with their cattle that they were but just able to get back to the lake before the very last of their horses and oxen died. Half of the Griquas themselves had perished during this short excursion of malignant fever. Strong evidence is adduced by Mr. Anderson to prove that an important branch of the great Cunene river, which runs towards the W coast, rises near to where the T. falls into the lake. The dry upper part of this branch was crossed by the above-mentioned Griquas. See article NGAMI.

TIOUGHNIOGA, a river of New York, U. S., which rises in the S part of Onondaga co., and flows into the Chenango, in the SE part of Broome co.

TIPERAH, an extensive district of Bengal, situated on the E side of the Brahmaputra or Megna, between the 22d and 24th parallels of N lat. On the E it is divided from Ava by a range of mountains and impervious woods, abounding with wild elephants and other ferocious animals. On the S it is bounded by Chittagong. The district is supposed to contain nearly 5,000 sq. m.; but of this space a considerable portion is wild and uncultivated, and inhabited only by a scanty population. The W division is productive, and yields abundant crops of rice and cotton. A quantity of salt is also manufactured near the coast, on account of government. It is so celebrated for the growth of betel-nut, that the Birman merchants often make advances to the growers, in order to secure the whole crop. Its principal manufacture is that of coarse cotton cloth, denominated *baftah*, and some cossas. Its cap. is Comillah. Its other towns are Odeypore, the ancient capital, Luckipore, Doudcandy, and Colinda.—This district appears to have been long the seat of an independent Hindu principality. In our old maps it is described as the kingdom of Tipra. It retained its independence till the last cent., although frequently invaded by the Mahommedans, after they had gained possession of the rest of Bengal. By Mahommedan historians it is denominated the country of Jagenagur, which was probably the name of its principal port, subsequently known as Alumgeernagur. It was invaded in 1279, by Toghril, the Afghan governor of Bengal. During the government of Sultan Shuja, about the middle of the 17th cent., it was invaded, and some of the lands near the river Megna annexed to Bengal. Along with the rest of Bengal, it devolved to the British in 1765. The pop. has been estimated at 1,400,000, in a nearly equal proportion of Hindus and Mahommedans.

TIPITAPA, a river of Nicaragua, which flows from the SW extremity of the lake of Leon into the NW extremity of the lake of Nicaragua. See article NICARAGUA (LAKE OF).

TIPPAH, a county in the N of the state of Mississippi, U. S. Area 1,037 sq. m. Pop. 20,741.

TIPPECANOE, a river of Indiana, U. S., which joins the Wabash, about 420 m. from its mouth, after a course of about 170 m. Several of its branches issue from lakes, swamps, and ponds, some of which have double outlets running into the St. Joseph's, a branch of the Miami of the Lakes.—It gives name to a county of Indiana. Area 504 sq. m. Pop. in 1850, 19,377. Capital, La Fayette.

TIPPER, a parish in co. Kildare, 1½ m. E by S of the town of Naas. Area 3,288 acres. Pop. 661.

TIPPERAGHNEY, or TYBUROUGHNEY, a parish in co. Kilkenny, 1¼ m. WSW of Pilltown. Area 1,147 acres. Pop. in 1831, 283; in 1841, 278.

TIPPERARY, an inland county of the prov. of Munster, in Ireland, constituting all the NE and the middle-eastern portion of that prov., and ranking with Cork and Galway as one of the three greatest counties of Ireland. It is bounded on the NW by co. Galway; on the N by King's co.; on the E by Queen's co. and the co. of Kilkenny; on the S by co. Waterford; on the SW by co. Cork; and on the W by cos. Limerick and Clare. Its greatest length, southward from the mouth of the Little Brosna to the Knockmeledown mountains at the junction point with the counties of Waterford and Cork, is 50 m.; its greatest breadth westward is 29½ m.; but its breadth over 10 m. at the N extremity nowhere exceeds 9¾ m. Its area comprises 843,887 acres of arable land, 178,183 of uncultivated land, 23,779 of continuous plantations, 2,359 of towns, and 13,523 of water,—in all, 1,061,731 acres. In proportion to its extent, this co. contains less unimproved land than any other in the S of Ireland.

Surface.] The SE termination of the Slievebloom mountains slightly touches the extremity of the wing of the co. situated on the E side of the King's co. indentation. The magnificently grand mountain-range of the Galtees come in from the SW extremity of the co. of Limerick, and extend, within Tipperary, 9 m. E to the valley of the Suir. They have their culminating-point in Galteemore, 3,015 ft. of alt. above sea-level, on the boundary with Limerick; they possess within Tipperary an extreme basis-breadth of 5½ m., and everywhere a basis-breadth of considerable expansion. The Knockmeledown mountains extend strictly parallel with the Galtees, and belong rather more to Waterford than to T., and partially also, under the name of the Kilworth mountains, to Cork. The portion or rather side of these mountains within Tipperary has a basis-breadth of from 1¾ m. to 3¾ m. Various groups and ridges, and congeries of mountains, impart strength and brilliance to the landscapes of all parts of the co. The broad, spreading, fertile, and luscious valley of the Suir flanks all the E side of the Devil's Bit and the Bilboa mountains; the broad, beautiful, and most fructiferous dingle, called the Golden Vale, extends from the valley of the Suir E to the Slievenaman mountains. The district which constitutes the NE and the E borders, from the vicinity of Roscrea S to the vicinity of the Slievenaman mountains, are first a sheet of sheer bog about 23½ m. in length, and next an expanse of low, spreading, moorish hills, whose loftiest summits acquire an alt. of from 300 to 600 ft. above the level of the adjacent plain.

Waters.] All the lakes of the co. of T., excepting its portion of Lough Derg, are small and unimportant. The principal are Loughs Curra, Diheen, and Boheen. Excepting these lakes, all the 13,600 acres of water exhibited in the Ordnance survey as belonging to T., are in Lough Derg, and the rivers Suir and Shannon. All the district W of the summit-line of the Devil's Bit and the Bilboa mountains, and also a small portion of the Golden Vale around Emly, belong to the basin of the Shannon, and are drained to that river, or to its great expansion of Lough Derg, by a number of inconsiderable rivulets, the chief of which are the Brosna, the Nenagh, the Kilmastulla, and the Mulkern. Of the remainder of the co. nearly all belongs to the basin of the Suir. The only navigations of the co. are that of the Shannon along all the boundary with Clare and Galway, and that of the Suir up the S boundary of the co. to Clonmel.

Minerals.] A formation of clay slate, greywacke, and greywacke slate, constitutes the greater portion of the Keeper mountains, the Bilboa mountains, and the Lough Derg mountains, a small part of the Knockmeledown mountains, and a considerable part of the Galtees mountains, the Devil's Bit mountains, and the Slievenamuck mountains. A formation of old conglomerate, with red, purple, and grey clay slate, constitutes most of the Knockmeledown mountains, rather more than one-half of the Galtees, and a considerable portion of the Slievenamuck group. A formation of yellow sandstone and sandstone conglomerate constitutes the northern part of the Devil's Bit range, a broad belt round the Keeper and Bilboa congeries, and narrow belts round the Lough Derg mountains, the Galtees, the Knockmeledown mountains, and the Slievenamuck mountains. A formation of old red sandstone and sandstone conglomerate, constitutes a portion of the centre of the Keeper mountains, the western declivities of these mountains, the district around Newport-Tip, and thence to the Shannon, a narrow belt between the clay slate and the yellow sandstone of the Lough Derg mountains, and the central and eastern portions of the Slievenamuck range. A coal formation, partially containing workable seams of coal, constitutes the district of moorish hill on the eastern border of the county, the district westward thence till within a few miles of Cashel, a tiny district closely adjacent to Cashel, and another tiny district a little NNW of Clonmel. Wings and ramifications of the great central floetz limestone plain of Ireland constitute nearly all the remainder of the co.

Agriculture.] In the barony of Middlethird, which was selected by the Irish Poor Commissioners as the field of inquiry into the agricultural and economical condition of T., and which may probably be regarded as a fair average specimen of the co., about one-third or one-fourth of the surface was pasturage, and most of the remainder in tillage. Leases were much less in use than at a recent period; and such as still existed were usually for 31 years or three lives in the case of farms, and for 21 years in the case of cottage and glebe lands. Competition for small holdings was so great that, on occasion of almost every vacancy, persons offered more than the land was worth, and entered on its possession on terms quite incompatible with their obtaining more from it than the most miserable subsistence. This competition had been a prolific source of crime. Potatoes, wheat, and oats, formed the series, which was repeated if the land would bear it. After this rotation of crops had been taken once or twice, or on some of the best lands three or four times, the ground was left to grass for 6 or 8 years. The con-acre system was prevalent; and the quantity of land usually taken by a family was from a rood to an acre. Very little cheese was made; and even butter was not produced in large quantity. In 1841, there were within the civic districts of the co., 230 farms of from 1 acre to 5 acres, 110 of from 5 to 15 acres, 22 of from 15 to 30 acres, and 21 of upwards of 30 acres; and within the rural districts, 13,032 of from 1 acre to 5 acres, 12,787 of from 5 to 15 acres, 4,938 of from 15 to 30 acres, and 2,960 of upwards of 30 acres. In 1851, wheat occupied 55,110 acres; oats 93,088; barley and rye 22,764; potatoes 43,348; and turnips 30,203 acres.—The Limerick or common Irish breed of black cattle are the most common, in consequence of being thought the most hardy. In 1841, there were within the co. 33,215 horses and mules, £265,720; 7,076 asses, £7,076; 87,162 cattle, £556,553; 129,848 sheep, £142,833; 105,333 pigs, £131,666; and 586,783 poultry, £14,669.—The plantations within the co. in 1841, occupied 29,602 acres.

Divisions and Towns.] The county was recently divided into two ridings, N and S. Its baronies are Lower Ormond, in the N part upon the Shannon; Upper Ormond, in the central part upon the Shannon; Owney and Arra, in the S part upon the Shannon; Ikerrin, in the NE; Eliogurty, in the N part of the central E; Slieveardagh, in the S part of the central E; East Iffa and Offa, in the SE; West Iffa and Offa, in the SW; Clanwilliam, in the S part of the central W; Lower Kilnemanagh, in the N part of the central W; Upper Kilnemanagh, in the NW centre, between Lower Kilnemanagh and Upper Ormond; and Middlethird, in the SE centre, between Slieveardagh and Clanwilliam. The towns and chief villages are Borris-o'-kane, Goatstown, Puskawn, Lorrha, Ballyloughnane, Cloghjordan, and part of Nenagh, in Lower Ormond; Toomavarra, Silvermines, and part of Nenagh, in Upper Ormond; Portroe, Newport, and Ballina, in Owney and Arra; Roscrea and Templetuohy, in Ikerrin; Thurles, Moyne, Loughmoe, Templemore, Littleton, and Two-mile-Borris, in Eliogurty; Mullinahone, Killenaule, Ballingarry, New Birmingham, Nine-mile-house, and Maradyke, in Slieveardagh; Carrick-on-Suir, Abbey, Ballyclarahan, Kilcash, Tobberaheena, Kilsheelan, and the chief part of Clonmel, in East Iffa and Offa; Cahir, Clogheen, Ardfinnan, Ballyporeen, Newcastle, and Burncourt, in West Iffa and Offa; Tipperary, Emly, Cullen, Golden, Thomastown, and Banshaw, in Clanwilliam; Ballagh, in Lower Kilnemanagh; Borris-o'-leagh and Cappaghwhite, in Upper Kilnemanagh; and Cashel, Fethard, Drangan, Holycross, and New-Inn, in Middlethird.

Statistics.] In 1824, according to Protestant returns, the number of schools within the co. of T. was 657, of scholars 33,676. At the close of 1843, the National board had in full operation within the co. 118 schools, conducted by 95 male and 47 female teachers, and attended by 9,288 male and 7,129 female scholars. The cost of maintaining the entire constabulary force of the co. during 1843, amounted to £39,577. The number of electors registered Feb. 1, 1844, was 2,674; of whom 814 were £50 freeholders; the number in 1851 was 6,760. The pop. of the co. in 1792 was 169,000; in 1821, 346,896; in 1831, 402,363; in 1841, 435,553; in 1851, 331,487. The following statistics are all of the year 1841:—Males, 216,650; females, 218,903; families, 74,570. Inhabited houses, 66,384. First-class inhabited houses, 1,722; second-class, 13,653. Families residing in first-class houses, 21,123; in second-class houses, 16,699. Families employed chiefly in agriculture, 52,805; in manufactures and trade, 13,512; in other pursuits, 8,253. Families dependent chiefly on property and professions, 1,607; on the directing of labour, 25,978; on their own manual labour, 45,034.

History.] The *Coriondi* and the *Udiæ* or *Uodiæ* of Ptolemy are supposed to have occupied the country which now constitutes the co. of Tipperary, and the counties adjoining it in the W and SW. In the settlement which followed the Anglo-Norman conquest, a principal part of the territory which now constitutes co. T. seems to have continued as a tributary toparchy, in the possession of Donald O'Brien, the native prince of Thomond and Ormond. In 1210, T. was erected into a county by King John, during his expedition to Ireland. Almost at the first blush of the great rebellion of 1642, Cashel, Clonmel, Fethard, Carrick-on-Suir, and all the other towns of T. were seized by the insurgents. In 1649, after Lord Inchiquin had made common cause with the Earl of Ormond, when Cromwell invaded Ireland, he found himself opposed by both royalists and confederates in this co. At the period of the Restoration, Clonmel was one of the towns in possession of the royalists. In the war of the Revolution, and after the battle of the Boyne, Clonmel was abandoned by the Jacobites on the advance of William; and it formed the retreat and asylum of the latter's army, on occasion of his relinquishing the siege of Limerick, and embarking at Duncannon for England. The county of T. was not involved in the rebellion of 1798; but it has figured with painful and ignominious prominence in many an agrarian disturbance.

TIPPERARY, a parish in co. Tipperary, containing part of the town of the same name. Area 4,362 acres. Pop. in 1831, 7,996; in 1851, 7,601.—Also a town in the parishes of Tipperary, Cordangan, and Corroge, co. Tipperary, on the rivulet Arra, 10 m. W by S of Cashel, and 87 m. SW of Dublin. It is what a mercantile traveller would call 'a good little town.' There is no town westward nearer than Limerick, and there is consequently a busy retail trade. Pop. in 1831, 6,972; in 1851, 7,001.

TIPPERKEVIN, a parish of co. Kildare, 2 m. N by E of Ballymore. Area 3,751 acres. Pop. 512.

TIPTON, a parish and v. in the co. of Stafford, 2 m. NE of Dudley. Area of p., with the vs. of Dudley-port, Horseley-heath, and Great-Bridge, 3,020 acres. Pop. in 1831, 14,950; in 1851, 24,872, an increase mainly due to the coal and iron-works here.

TIPTON, a central co. of the state of Indiana, U. S., comprising an area of 264 sq. m., drained by Cicero and Duck creeks, and intersected by the Indianapolis and Peru, and the Newcastle and Logansport railroads. It is level and well-timbered. Pop. in 1850, 3,532. It has a village of the same name.—Also a county in the W part of the state of Tennessee, comprising an area of 682 sq. m., drained by branches of Big Hatchee and Loosahatchee rivers, affluents of the Mississippi. Pop. in 1840, 6,800; in 1850, 8,887. Its capital is Covington. It has a generally level surface, and is fertile in cotton and tobacco.—Also a village of Cedar co., in the state of Indiana, 1½ m. W of Sugar creek, and 20 m. E by S of Iowa city. Pop. in 1840, 50.

TIPUNA, a town and port of New Zealand, near the NE extremity of Eaheinomauwe or North Island, and on the N side of the Bay of Islands.

TIPUTINI, a river of Ecuador, in the prov. of Imbambura, which runs E, and joins the Napo, on the r. bank, in S lat. 1° 30'.

TIQUE, a river of Venezuela, in the prov. of Cumana, which has its source on the confines of Barcelona; runs ENE; expands into a lake; issues thence by 3 outlets, which conjoin about 30 m. below, and thence taking the name of Caripe, flow into Cayo-Manamao, the most westerly arm of the Orinoco, and a little above its entrance into the gulf of Paria. The T. is connected in its course with the Marichal on the S, and the Cayo-Cacao on the N.

TIQUIRA, or Itiqueira, a river of Brazil, which has its source in a mountain of the same name, and joins the Pequiri, an affluent of the São Lourenço.

TIRA, or Shah Jehanpur, a town and stronghold of Hindostan, in the NE part of the Punjab, on a rocky plateau, at an alt. of about 100 ft. above the Beas, which runs at its base. Pop. 5,000.

TIRANA. See Terran.

TIRANGES, a village of France, in the dep. of the Haute-Loire, cant. and 6 m. W of Bas, at the foot of a mountain near the l. bank of the Ance, here crossed by two old stone-bridges. Pop. 1,680.

TIRANO, a town of Austrian Lombardy, in the gov. of Milan, delegation and 17 m. ENE of Sondrio, on the l. bank of the Adda. Pop. 4,000. Adjacent is the village of Madona-di-Tirano, noted for a magnificent church dedicated to the Virgin.

TIRASPOL, a district and town of Turkey in Europe, in the gov. of Kherson. The town is 6 m. E of Bender, on the l. bank of the Dniester, by which it is separated from Bessarabia. Pop. 5,100. Its streets are straight and spacious, and the houses regularly built. It has two churches, and in its vicinity is a mud fortress well garrisoned, designed to defend the passage of the river. This town was fortified in 1793, when, after the taking of Otchakov, the Russians formed a line of fortified ports along the Dniester.

TIRAXANA, or Tirajana (San-Bartoleme), a town of the Canary islands, in the island of Gran Canaria, and partido of Las Palmas. Pop. 3,656. It has a parish church, three convents, and a custom-house. Its industry is chiefly agricultural.

TIREE, one of the Inner Hebrides, 15 m. W of Mull, in N lat. 56° 20′, W long. 6° 53′. It is about 13 m. in length, from SW to NE, and from 1 to 5 m. in breadth. It consists chiefly of gneiss. Its surface is low, and in few places fertile. Pop. in 1851, 3,709.

TIREH, TIRRA, or TYRIA, a town of Turkey in Asia, in the sanjak of Aïdin, 36 m. SE of Smyrna, at the N base of a lofty mountain of the Kestane-dagh, and 3 m. from the l. bank of the Kutchuk-Mendere, which is here crossed by a stone bridge. Pop. 20,000. It has 14 mosques, two Greek churches, numerous houses intermingled with gardens and plantations, and a suburb, and possesses manufactories of carpets and calico. About 36 m. to the W are the ruins of Ephesus.

TIRGO, a town of Spain, in the prov. and 30 m. WNW of Logrono, and partido of Haro-y-sus-Barrios, on a flat but fertile locality, on the r. bank of the Tiron. Pop. 400.

TIRGOSCHYL, a town of Turkey in Europe, in Little Wallachia, capital of the district of the Upper Schyl, and 42 m. NE of Tchernetz, on the slope of a mountain, near the l. bank of the Schyl.

TIRGU-FORMOS, a town of Turkey in Europe, in Moldavia, in the district and 24 m. NW of Jassy, in a valley, on the l. bank of the Baglinez.

TIRGUSHINA, or ETELS, a town of Turkey in Europe, in Albania, in the sanj. of Scutari, pash. and 50 m. NNW of Prisvend, at the base of the Djamus Dagh.

TIRHUT, an extensive district of Hindostan, in the prov. of Bahar, situated principally between the 25th and 27th parallels of N lat., and bounded by Nepal on the N; the district of Purneah on the E; Bhagulpore on the SE; the Ganges on the S; and Sarum on the W. Area 7,400 sq. m. Although not hilly, the surface of this district is more elevated, the soil drier, and the climate healthier, than in Bengal. It is, generally speaking, well cultivated, and productive of cotton, grain, sugar, indigo, tobacco, opium, and saltpetre; of late years the breeding of horses and cattle has received encouragement here from government. Towards the N frontier are extensive forests, abounding with valuable timber, but unfortunately the rivers are not of sufficient depth to float it into the Ganges. Its principal towns are Mozufferpur, Durbungah, Nagorbussy, Jenjapur, and Turki. Its chief rivers are the Little Gunduck, the Gogary, and the Bhagmutty. In the remote eras of Hindu antiquity, this region formed a part of the kingdom of Maithila, the boundaries of which were the Gunduck and Cosa rivers. Within these territories a peculiar language was spoken, still named the Maithila or Tirhuty: it also gave name to a division of Brahmins, who pride themselves on their purity. It appears to have continued an independent principality until 1237, when it was invaded by Toghan-Khan, the Afghan governor of Bengal. It was finally subdued by the Mahommedans about 1325, by the emperor Ala-Addin, who annexed it to the throne of Delhi. Along with the rest of the prov., it came into possession of the British in 1765. The pop. is estimated at 2,400,000, in the proportion of three Hindus to one Mahommedan.

TIRIOLO, a town of Naples, in the prov. of Calabria-Ultra, and district of Catanzaro, on a summit of the Apennines, between the Corace and Lamato. Pop. 2,500. It has two parish churches, and a convent. The environs, which are extremely fertile, are noted for their medicinal plants, and contain mines of iron and of coal. T. is of great antiquity, and contains numerous ancient remains.

TIRIX, a town of Spain, in the prov. and 27 m. N of Castellon-de-la-Plana, and partido of Albocacer, amid mountains. Pop. 677.

TIRKCHLE′, a town of Russia in Europe, in the gov. of Vilna, district and 26 m. NNE of Telch.

TIRLEMONT, a department, commune, and town of Belgium, in the prov. of Brabant, and arrond. of Louvain. Pop. of dep. 8,160. The town is 12 m. SE of Louvain, on the Ghete. It is enclosed by a wall, 5 m. in circuit, comprising numerous gardens, orchards, &c. Pop. 7,221. It has a large square, in which are the town-house, and the principal church, a college, an academy and a savings' bank, and possesses manufactories of cotton and woollen fabrics, hosiery, soap, oil, pottery, several salt refineries, distilleries and breweries. Its trade consists chiefly in grain, wool, cattle, and pigs. It returns two deputies to the provincial states. This town was formerly one of the most important in the Brabant, but has frequently suffered much from war. During the revolt of the Netherlands it was taken by the Spaniards, in 1635 by the French and Dutch, and again in 1792 by the French. The following year it fell into the hands of the Austrians, but in 1794 was regained by the French.

TIRMEZ, or TERMEZ, a town of Independent Tartary, in the khanat and 250 m. SE of Bokara, on the r. bank of the Jihun, which here becomes navigable.

TIRON, a river of Spain, which has its source in the mountains of San Antonio, on the confines of the provinces of Logrono and Burgos; flowing N for some distance between these provs., it passes Belorado; thence taking a NE course, it enters Logrono, waters Angunciana and Haro, and a little below the latter town joins the Ebro, on the r. bank, and after a total course of 39 m.

TIRON, or TIRAN, an island of the Arabian gulf, near the coast of Hedjaz, in Arabia, and at the entrance to the gulf of Akaba, in N lat. 27° 55′. It is surrounded by sand banks.

TIROON, a district of Borneo, on the E coast, extending between the 3d and 4th degrees of N lat. It is generally flat, and is watered by numerous streams, the principal of which is the Solowang. It abounds with sago trees and birds' nests, and produces also sugar, wax, honey, beche-de-mer, rattans, &c. Gold is found in some parts. The inhabitants who are supposed to be a subdivision of the Idaan, or Horaforas, are said to be addicted to cannibalism.

TIRPIED, a village of France, in the dep. of the Manche, cant. and 5 m. W of Brecey, on the r. bank of the See. Pop. 1,350.

TIRRE. See TIREH.

TIRSCHENREUTH, a presidial and town of Bavaria, in the circle of the Upper Pfalz. Pop. of presidial, 13,350. The town is 17 m. S of Eger, on the Waldnab. Pop. 2,300. It is enclosed by walls, and has three churches. It has manufactories of cloth and linen, and productive fisheries.

TIRSCHTIEGEL, or TRZCIEL, a town of Prussia, in the prov. and regency and 45 m. W of Posen, and circle of Meseritz, between two lakes, on the Obra, by which it is divided into two parts distinguished as the Old and New town. Pop. in 1843, 2,398. It has a Lutheran and two Catholic churches, and possesses manufactories of cloth and tobacco, and distilleries of brandy.

TIRSI. See ORISTANO.

TIRSO (SAN), a village of Portugal, in the prov. of Minho, comarca and 15 m. N of Oporto, near the l. bank of the Ave. Pop. 300. There is a superb convent here.

TIRTAPURI, a town of Tibet, in the prov. of Ndzang, to the S of the Cailas mountains, near the source of the Sutlej, and 50 m. SSE of Gortope. It stands on a plateau 200 ft. in height, and is overhung by lofty limestone crags. It is the residence of a lama and of several gylnogs. It contains numerous *maths* or diminutive temples, and nearly opposite the lama's house is a wall 450 ft. in length, and 12 in breadth, covered with stones, on which prayers have been inscribed by pilgrims who have visited the locality. About 2 m. to the W of T. are two hot springs.

TIRUPMEN, a town of Burmah, in the prov. of Pegu, near the l. bank of the Irawaddy, 30 m. SSE of Prome.

TIRVIA, a town of Spain, in the prov. and 81 m. NNE of Lerida, and partido of Sort, in a mountainous locality, near a small affluent of the Noguera-Pallaresa. Pop. 420.

TIRWAN-ANGADY, or TIRAVANA-ANGADI, a town of Hindostan, in the prov. of Malabar, 21 m. SSE of Calicut, between an affluent of the Irnada and a lagune. It is inhabited chiefly by a Mahommedan tribe named Moplays. A decisive victory was here gained by the English over the Sultan Tippu in 1790.

TISBURY, a parish and village of Wilts, 3½ m. SE of Hindon. The church is a spacious building of great antiquity. Area 7,355 acres. Pop. of East T. 945; of West T. 680. Various members of the Arundel family have monuments here.

TISBURY, a township of Duke's co., Massachusetts, U. S., on the N side of Martha's Vineyard, 8 m. W of Edgartown, and 85 m. S of Boston. Pop., including that of the Elizabeth islands, 1,520.

TISCHINGEN, a town of Würtemberg, 10 m. NNE of Dillingen. Pop. 1,000. In the neighbourhood is the castle of Trugenhofen, the residence of the prince of Tour and Taxis, to whom the town also belongs.

TISCHNOWITZ, a town of Moravia, 15 m. NW of Brunn, on the l. bank of the Schwarzawa. Pop. 1,200. It has a woollen manufactory.

TISQUI, a river of Ecuador, in the prov. of Esmeraldas, which runs NW into the Blanco, opposite the mountain of Quindiu, in N lat. 21° 30′.

TISSINGTON, a parish of Derbyshire, 4½ m. N of Ashborne. Area 2,316 acres. Pop. in 1851, 344.

TISTED (EAST), a parish of Southamptonshire, 5 m. S by W of Alton. Area 2,602 acres. Pop. 229.

TISTED (WEST), a parish in the same co., 5 m. SE by E of New Alresford. Area 2,268 acres. Pop. in 1851, 268.

TISZA. See THEISS.

TITA (SAINT), a small island in the N. Pacific, in N lat. 63° 51′.

TITAN, a small island of France, in the Mediterranean, on the coast of Provence, the most eastern of the Hyeres group.

TITCHFIELD. See TICHFIELD.

TITCHWELL, a parish of Norfolk, 5 m. W by N of Burnham Westgate. Area 4,480 acres. Pop. in 1851, 163.

TITEL, a village of Hungary, in the com. of Bacs, at the confluence of the Theiss and the Danube, 22 m. ESE of Peterwardein. Pop. 2,840.

TITICACA, a lake of South America, on the NW frontiers of Bolivia, situated in the great alpine valley, between the two cordilleras of the Andes, between the parallels of 15° 15′, and 16° 35′ S. It is the most considerable of all the lakes of South America. Its figure is irregular, but inclining to oval, with its principal line of direction lying NNW and SSE 170 m. Its greatest breadth is 70 m. The great chain of the Andes, which forms a single and continuous ridge from the most southern extremity of the American continent to the neighbourhood of the tropic of Capricorn, separates into two great ridges near the city of Potosi, in S lat. 19° 35′. These, after enclosing a vast alpine plain raised 12,000 to 13,000 ft. above the level of the sea, reunite in about 14° S lat., where they form the Andes of Vilcanote and Cusco. The western of these ridges runs parallel to the shores of the Pacific, and is called the Cordillera of the Coast; the eastern, or that of the interior, is called the Cordillera Real. The intermediate plain is the basin of the celebrated lake T., the physical features of which are scarcely less extraordinary than its history is interesting. The lake is of vast size, having an area, according to Mr. Pentland, of above 5,000 sq. m.; and its depth is not less remarkable, having been sounded to 120 fath., and probably greatly exceeding this depth. It has only one outlet, the Desaguadero, which, after running about 150 m. towards the SW, is lost in a small lake. The heights in the Western Cordillera are chiefly dome or bell-shaped; those in the Eastern are peaked, giving the range generally a serrated form. The descent of both, E and W, is rapid; but that of the Western into the basin of T. is less so than that of the Eastern. The breadth of the former is about 100 m.; that of the latter it is less easy to determine, in consequence of its throwing out many lateral chains on its eastern side, the length of which may be considered portions of the breadth of the main ridge. Excluding these, however, this may be estimated at from 35 m. where narrowest, or under the parallel of 17° 58′ S to above 70 m. where widest, under the parallel of 16° 50′. The entire width of the two ridges, including that also of the basin of Lake T., varies from 200 to 300 m., exclusive of the projecting chains: including them it approaches 500 m.; and the length of this portion of the Andean chain, bounded, as we have already stated, by the 14th and 20th parallels, is nearly 400 m. All the waters from the E face of the Western Cordillera, and all those from the W face of the Eastern one, up to the height of 14,000 ft., flow into Titicaca, and its outlet, the Desaguadero, whence they are carried off by absorption and evaporation, there being no visible outlet for them on either side. But above 14,000 ft. on the Eastern Cordillera, the waters are turned aside by some low ridges of hills, not improbably alluvial deposits, being only a few hundred feet high, and are directed by them to the S; whence they issue to the E in about 16° S lat., constituting the Mapiri, one of the largest affluents of the great river Beni, itself, with the Mamore, forming the Madera, one of the most considerable affluents of the Amazon. The drainage of a portion of the W face of the Cordillera-Real, is thus to the eastward. On the E shore of the lake, Mr. Pentland found a large agricultural pop. almost exclusively composed of Indians speaking the Aymara language. The lake is navigated by ships; but is subject to storms and tremendous gusts of wind descending from the lofty mountains by which it is surrounded. Ten or twelve rivers, the chief of which are the Asangaro and the Laganillas, and a number of small streams, empty themselves into it; but its only outlet is the Desaguadero already mentioned. Its water, though neither bitter nor brackish, is turbid, and, from its nauseous taste, not drinkable: yet it abounds with fish, particularly two kinds, one large and palatable, called *suchis*, and the other small, insipid, and bony, termed *boyas*. Immense flocks of water-fowl frequent it; and its shores are covered with flags and rushes which serve many purposes of domestic manufacture. On the shores of this inland sea Mango-Capac first appeared to the Peruvians; and, on a few islands near its SE extremity, were accumulated the richest and most sacred objects of the superstition on which he founded his powerful and highly civilized empire.

At the Spanish conquest the greater part of these were thrown into the lake by the despairing but still hostile Peruvians; and Garcilasso de la Vega, himself a descendant of the Incas, gives an astonishing, but not, perhaps, altogether an improbable account of the treasures thus sacrificed.

TITLEY, a parish of Herefordshire, 3 m. NE by E of Kington. Area 1,876 acres. Pop. 375.

TITLINGTON, a hamlet of Northumberland, 6½ m. W by N of Alnwick.

TITLIRBERG, a mountain-range of Switzerland, in the cant. of Unterwalden, on the frontiers of Uri, which attains an alt. of 11,414 ft. above sea-level.

TITMANING, or DITMANING, a town of Bavaria, on the Salza, 23 m. NNW of Salzburg. Pop. 1,200.

TITSCHEN, or NEU-TITSCHEIN, a town of Moravia, 12 m. ENE of Weisskirchen, near an affluent of the Oder. It is a comparatively well-built place, with active manufactures of woollens. Pop. 5,600.—About 3 m. to the SW of it is the village of Alt-Titschein.

TITSEY, a parish of Surrey, 5 m. NE by E of Godstone. Area 1,936 acres. Pop. in 1851, 154.

TITTENHANGER, a hamlet of Hertfordshire, 2½ m. SE by E of St. Alban's.

TITTENSOR, a hamlet of Staffordshire, 4 m. NW by N of Stone.

TITTERIE, the central province of the kingdom of Algiers, under the regime of the dey. It is about 60 m. long by 40 m. broad, extending from the coast of the Mediterranean to the plain of the Bled-el-Jerid, which separates it from the Sahara, and from the river Masaffran on the W, to the river Buberak on the E.

TITTESWORTH, a township in the p. of Leeke, Staffordshire. Pop. in 1851, 606.

TITTING, or DIETTING, a town of Bavaria, 8 m. N of Eichstadt.

TITTLESHALL, or TITLESHALL, a parish of Norfolk, 6 m. SSW of Fakenham. Area 3,364 acres. Pop. in 1851, 615.

TITUL. See TITEL.

TIUMEN, a town of Asiatic Russia, in the gov. and 100 m. SW of Tobolsk. Pop. 8,000, chiefly Tartars. It is a place of considerable trade.

TIVDIJA, a small river of European Russia, which flows into Lake Onega on the W side.

TIVERNO. See BIFERNO.

TIVERTON, a parish and parliamentary borough of Devonshire, situated at the confluence of the Axe and the Loman rivers, from which circumstance it was anciently called Twyfordtown, or Twofordtown. It stands on the slope of a hill, rising gently towards the N, and consists of four principal streets, which form a quadrangle, enclosing an area of gardens, in the centre of which is a bowling-green. The principal public buildings are the church, the townhouse, assembly-rooms, and the free grammar-school. The ancient ruined castle stands on the W side of the town, on an eminence well adapted for defence. It appears to have been nearly of a quadrangular form, enclosing an area of about an acre. The manufacture of bobinet-lace, serges, druggets, and other woollen goods, employs a large proportion of the pop. Area of p. 17,650 acres. Pop in 1801, 6,505; in 1831, 9,766; in 1851, 11,144. The burgh returns 2 members to parliament. Registered electors in 1837, 496; in 1852, 651.

TIVERTON, a township of Cheshire, 2 m. S of Tarporley. Area 1,657 acres. Pop. in 1851, 747.

TIVERTON, a township of Newport co., Rhode island, U. S., 1,000 ft. long, on the mainland, opposite to Portsmouth, with which it is connected by a bridge, 18 m. SE by S of Providence. Pop. in 1850, 3,183; 1840, 4,699.

TIVETSHALL, ST. MARGARET, and ST. MARY, two adjacent parishes of Norfolk, 5½ m. NE of Diss. Area of St. Margarets, 1,668 acres. Pop. in 1851, 355. Area of St. Mary, 1,125 acres. Pop. 352.

TIVIOT. See CHEVIOT.

TIVISSA, a town of Spain, in the prov. and 28 m. W of Tarragona. Pop. 3,787. It has manufactories of oil, linen, charcoal, wax-candles, and rush-work.

TIVOKEA, an island in the S. Pacific ocean, in S lat. 14° 28′, W long. 144° 56′. It is low and sandy, of an elliptic form, 18 m. in its longest diameter. A lagune in the centre is entered from the SW end of the island. The inhabitants are stout made, and of a dark brown colour, and tatoo themselves with the figures of fishes. Their language approaches the Otaheitan dialect, but is more guttural.

TIVOLI, the *Tibur* of the ancients, a considerable town of the Papal states, about 18 m. E by N of Rome, delightfully situated on an eminence sheltered on one side by Monte-Castali and a circular range of the Sabine mountains, and on the other commanding an extensive prospect over the Campagna-di-Roma. It has a population of nearly 17,000, and, without being in general well-built, contains a good cathedral and several other churches. The neighbourhood affords quarries of excellent stone, of which several of the public edifices of Rome are built; and the waters of the Teverone form by their deposit a kind of stone of no slight interest in the eyes of the naturalist. Not far distant is the small but remarkable lake of Solfatara. The great attraction of the town now, as in former ages, consists in the falls of the Teverone, the ancient *Anio*, which glides gently through the town, till reaching the brink of a rock over which it precipates itself nearly 100 ft. in one mass, and after boiling up in its narrow channel, rushes through a chasm of the rock into a cavern below. On the summit of the steep bank of the Teverone stands a temple of the Corinthian order, supposed to have been dedicated to Vesta. The remains of several Roman villas, in particular of the *Villa Adriana*, remarkable for its extent and magnificence, at tract the antiquary in visiting this place.

TIVY, or TOWY. See TEIFY.

TIXALL, a parish of Staffordshire, 3½ m. E by S of Stafford. Area 2,352 acres. Pop. in 1851, 221.

TIXOVER, a parish of Rutlandshire, 7½ m. E by N of Uppingham. Area 1,080 acres. Pop. 115.

TIZE, or TIZ, a port of Mekran, in Beluchistan, supposed to represent the *Tiza* of Ptolemy. It lies in a valley, 10 m. S of Chinbar, surrounded with steep hills, except on the S, where it is open to the sea.

TIZOIRAS, a river of Brazil, which descends from the W flank of the Cordillera-Grande, and running WNW, flows into the Araguay, on the r. bank, after a course of 180 m.

TIZZANA, a commune and town of Tuscany, in the prov. and 13 m. WNW of Florence, near the r. bank of the Ombrone. Pop. 7,700. It has a trade in silk and wine, the produce of the surrounding fertile district.

TIZZANO, a town of the duchy and 24 m. S of Parma.

TJANJOR, a town of the island of Java, 52 m. SE of Batavia. The surrounding territory is fertile and picturesque.

TJERINGKIN, a considerable fishing village of Java, on the W coast, 102 m. W of Batavia. Pop. 1,800. It is situated at the mouth of a river which is navigable a long way up for small prows. It was formerly a military post, with a block-house, which was burnt by the British.

TJIDOVEAN, a river of Java, which runs a N course, and falls into the Java seas in the bay of Bantam. During the rains it is rapid, and scarcely passable.

TKASHIKA, a village of Austrian Gallicia, 45 m. S of Chennowitz.

TLACOTALPAN, a town of Mexico, 60 m. SE of Vera Cruz, on an island formed by the Alvarado.

TLAJOMULCO, a town of Mexico, in the state and 9 leagues S of Guadalaxara.

TLALPUJAGUA, or TLAPUXAHUA, a town of Mexico, in the state of Mechoasan, 52 m. ENE of Valladolid. It is the place where the mine duties of the district are paid, and is inhabited by 500 families of Spaniards, Mestizoes, and Mulattoes, with many families of Indians who labour in the mines.

TLAMANALCO, a settlement of Mexico, in the state of Mexico, 6 m. E of Chalco.

TLAMATH, a river of Oregon and California, which rises in a lake of the same name, and flows W into the Pacific.

TLAPA, a settlement of Mexico, in the state of La Puebla, 138 m. SSE of Mexico.

TLAXCALLA, or TLASCALLA, a district and town of the Mexican confederacy. The district, now embraced in the state of La Puebla, is bounded on the N by the state of Vera Cruz; on the E by Oaxaca or Guaxaca; on the S by the Pacific; and on the W by the state of Mexico; and is about 354 m. in length, and from 40 to 150 m. in breadth. It is so fertile in maize, that it received its name from that produce of the soil, *Tlaxcalla* signifying 'The Land of bread.'—The town is situated on the side of a high mountain, 64 m. E of Mexico, on the banks of a small river which runs into the Pacific, in N lat. 19° 19′. When the Spaniards first arrived here, it is said to have contained 300,000 inhabitants; and Acosta affirms that it had a market-place large enough to hold 30,000 buyers and sellers, and that in the shambles were seldom less than 15,000 sheep, 4,000 oxen, and 2,000 hogs. But Gemelli, who visited T. in 1698, says, it was then become an ordinary village, with a parish-church in which hung a picture of the ship which brought Cortez to Vera-Cruz. It is now so much reduced that it scarcely contains 5,000 inhabitants, among whom there are not more than 900 Indians of pure extraction.

TLAZASALCA, a town of Mexico, the capital of a jurisdiction of the same name, in the intendancy of Valladolid, 70 m. W by N of Mexico.

TLEMSEN, TLEMSAN, or TREMECEN, the most westerly of the three provinces into which the kingdom of Algiers was divided under its deys. On the N it had the Mediterranean; on the E the prov. of Titterie; on the S the Atlas chain; on the W the river Mulavia separating it from Fez. Besides the capital of the same name, it contained the important sea-port of Oran.—The city of T. is situated in the modern prov. and 60 m. SSW of Oran. It was once an important place, and the metropolis of a kingdom; but in 1670, on occasion of a revolt of the inhabitants, Hassan, dey of Algiers, laid it in ruins. It is placed on a rising ground, below some rocky precipices whence issue a number of springs. Its streets are narrow, and the houses seldom of more than one story. It forms an entrepot between Marocco and Fez, and has considerable traffic in consequence. The pop. in 1850 was returned at 9,443. The citadel is a large building. The town still presents remains which can be traced to a Roman origin. In the western part is a large square basin of Moorish workmanship, 200 yds. long, and about half as broad.

TLETCH, a town of Asiatic Russia, on the Irtish, 72 m. ESE of Tobolsk.

TLUMASCHAU, or TLUMASZOW, a small town of Moravia, on the river March, 12 m. E of Hradisch and 39 m. E of Brunn. Pop. 1,100.

TMAJE, or TMI-EL-EMDID, a village of Lower Egypt, 12 m. SE of Mansura.

TMLAIA, a river of Russia, which rises in N lat. 48° 30′, and flows into the Donetz, on the l. bank, after a course of 120 m.

TMULGA, a village of Algiers, 10 m. E of Sinab.

TMUTORAKAN. See TAMAN.

TOA, a river of the island of Porto-Rico, which runs into the harbour of Porto-Rico.

TOBA, a small island in the Eastern seas, near the W coast of Aru, in S lat. 5° 8′.

TOBA (LA), a village of Spain, in the prov. and 24 m. S of Guadalaxara. Pop. 600.

TOBACCO KEY, a small island in the bay of Honduras, near the coast of Yucatan, in N lat. 16° 45′.

TOBACCO (POINT), a cape on the coast of Maryland, U. S., in the river Potomack, 37 m. SSW of Annapolis.

TOBACCO-PORT, a village of Stewart co., in the state of Tennessee, U. S., 60 m. WNW of Nashville.

TOBAGO, one of the British West India islands, in the Windward group. It is about 30 m. in length from SE to NW, and about 12 m. in its greatest breadth; and has its N point in N lat. 11° 25′, W long. 60° 32′.

Physical features.] The surface of the island is agreeably diversified; no part of it is rugged or impassable, though its NW extremity is mountainous. The prevailing slope is towards the SW. The highest ground has an alt. of 900 ft. above sea-level. The hills are not broken and rugged, as those of St. Vincent's, but regular though steep in ascent and descent. "The scene of nature is on an extensive scale, and gives the idea of a continent rather than an island. It is not alone its vicinity to the Spanish main that suggests this idea. The appearance of the island fully warrants this assumption; and the contiguity of South America only more fully marks its being torn therefrom, and of its being in old times the southern point, or bold promontory of the vast bay of Mexico." Its soil is of different kinds, but in general the mould is rich and black, and fit for producing whatever crops are raised in other parts of the West Indies. The abundance of springs upon the island contributes to its healthfulness, and its bays and creeks are so disposed as to be commodious for all kinds of shipping. Man-of-War bay, Courland bay, Sandy-point, and King's bay, are calculated for vessels of the largest size. Tyrrel's bay, Bloody bay, Mangrove bay, Englishman's bay, Castera's bay, and La Guira's bay, have good anchorage for vessels of 150 tons and under. Halifix bay is fit for ships of 250 tons.—The climate of T. is more temperate than could be expected in an island that is but 11° N of the equator. Another favourable circumstance in its position is, that it lies out of the tract of those hurricanes which so often prove destructive in the other West India islands. Besides producing the different kinds of wood that are to be found in the other West India islands, the Dutch affirm, that both the true nutmeg-tree and the cinnamon-tree, with that which produces the real gum copal, grows upon this island. Indian corn, Guinea corn, pease, beans, French beans, figs, pine-apples, pomegranates, oranges, lemons, limes, plantains, bananas, grapes, guavas, tamarinds, prickly pears, papaws, and a variety of other fruits which are not to be found in Europe, are freely grown. The cocoa-tree attains great perfection; and musk-melons, water-melons, gourds, cucumbers, and pompions, are abundantly raised. Wild hogs at one time so greatly abounded upon T., that at least 20,000 of them were annually killed without their number being sensibly diminished. Horses, cows, asses, sheep, deer, goats, and rabbits, probably introduced by the Dutch, have multiplied exceedingly. The birds indigenous to T., or which frequent the coasts, are wild ducks, water-hens, wood-pigeons, turtle and Virginian doves, three varieties of humming birds, blackbirds of yellow and black colours, thrushes, white woodcocks, herons, the pouched pelican, and flamingoes. The sea is stored with excellent fish, particularly turtle of every kind, and mullet.

Commerce, &c.] The principal articles of export from T. are sugar, molasses, and rum. The exports in 1810 were returned at £70,787; in 1845, at £83,412; in 1847, at £119,691; and in 1849, at £53,051. The imports in 1845 were £36,966; in 1847, £34,856; in 1849, £22,925. Nearly the whole trade is with the United Kingdom; but the Azores and Madeira, the United States of America, and the northern states of South America, conduct some commerce with it. In cultivating the soil, the *metayer* system has recently been adopted, and with success, on several estates.—T. is governed by a lieut.-governor, assisted by a legislative council and a house-of-assembly of 16 members. The pop. in 1812 was returned at 470 whites, 250 free coloured persons, and 17,000 slaves; in 1852, it was 13,027.

History.] This island was first discovered by Columbus, in 1498. William, Earl of Pembroke, in 1628, obtained a grant of this island, with that of Barbuda. About 1632, some Zealanders trading with the West India islands, made such a favourable report of T. in particular, that a company of merchants undertook to settle it, and gave it the name of New Walcheren, from one of the islands of their prov. The Spaniards exterminated the new colony. James, duke of Courland, sent a colony to T. who settled upon Great Courland bay, and erected a small fort, with a town. The duke's title was confirmed by a grant from Charles II., but disputed by the Dutch. Upon the extinction of the Kettler family, dukes of Courland, the fief of the island reverted to the crown of England in 1737, and by the treaty concluded at Paris in 1763, it was ceded in full right to Great Britain. In May 1781, a small squadron of ships from Martinique appeared before this island with 1,200 troops, under the command of M. Blanchelande, governor of St. Vincent's, and the island was surrendered on favourable terms. In 1793 it was retaken by the British, by whom it was retained at the peace of Amiens.

TOBAGO (Little), a small island near the NE coast of Tobago. It is about 2 m. long, and 1 m. broad.

TOBAK, a town of Russia, in Bessarabia, 34 m. NNW of Ismail. In 1789 the Russians were defeated here by the Turks, and the place was almost entirely destroyed.

TOBALINILLA, a village of Spain, in the prov. of Burgos and partido of Villarcayo, 39 m. WSW of Vitoria, in the valley of Tobalina, on the r. bank of the Ebro. Pop. 30.

TOBAR, a town of Spain, in the prov. and 15 m. NW of Burgos, near the l. bank of the Hormazuela. Pop. 280. It consists of two parts, distant about 500 paces from one another, and has a parish-church and a priory.

TOBARUELA, a town of Spain, in the prov. and 21 m. NNE of Jaen, and partido of Baeza, in a vast plain covered with vineyards and olive groves. Pop. 105.

TOBAS, an Indian tribe of La Plata, who inhabit the mountains of Tocsora. in the SE part of the prov. of Chaco, between the Pilcomayo, Vermejo, and Paraguay.

TOBAT, a district of the island of Sumatra, in the Battas country. Near its centre is a large lake.

TOBATI, a settlement of Paraguay, 30 m. E of Assumption.

TOBBER, a parish of co. Dublin, $1\frac{1}{4}$ m. NE of Dunlavin. Area 1,422 acres.—Also a village in the p. of Kilcroan, co. Galway, 1 m. W of the river Suck.

TOBBERAHEENA, a village in the p. of Abbey, co. Tipperary, near the river Suir, $2\frac{1}{2}$ m. SW of Clonmel. Pop. in 1851, 395.

TOBBERCURRY, a small town in the p. of Achonry, co. Sligo, 3 m. ENE of Banada. Pop. 755.

TOBBERDONY, or Tobbereendoney, a village in the p. of Beagh, co. Galway, $4\frac{3}{4}$ m. SSW of Gort. Pop. in 1851, 103.

TOBBERMORE, a village in the p. of Kilcronaghan, co. Londonderry, $2\frac{1}{4}$ m. SSW of Maghera. Pop. in 1831, 672; in 1851, 445.

TOBBERSCANAVAN, a village in the p. of Ballysadere, co. Sligo, $1\frac{1}{3}$ m. SSE of Collooney. Pop. in 1851, 155.

TOBEL, a bailiwick, parish, and village of Switzerland, in the cant. of Thurgau, 7 m. SE of Frauenfeld. Pop. (Cath.) 1,279. It has a house-of-correction, formerly a rich commandery of Malta.

TOBERMORY, a port on the NE coast of the island of Mull, in N lat. 56° 38′, at the head of a sheltered bay, $3\frac{1}{2}$ m. SW of Auliston-point, where Loch-Sunart forks off from the sound of Mull, and 9 m. SE of Ardnamurchan-point. It was built in 1788, at the same time as Ullapool, by the British fishing company, as the site of a fishing establishment, and the rendezvous of the herring-vessels. The harbour or bay is spacious, and almost completely land-locked; and is covered across the entrance, and at a brief distance, by Calve island. There are two excellent quays. As the only town in Mull, and in a large circumjacent district, both Hebridean and continental, it possesses much provincial importance, and is the seat of a considerable domestic trade. Pop. in 1851, 1,547.

TOBI. See Scombi.

TOBIA, a town of Spain, in the prov. and 24 m. SW of Logrono, and partido of Najera, on the l. bank of a small river of the same name, an affluent of the Najarilla. Pop. 187. It has manufactories of common woollen and linen fabrics.

TOBITSCHAU, or Towacow, a town of Austria, in Moravia, in the circle of Olmutz, 8 m. WSW of Prerau, on a mountain, near the r. bank of the March. Pop. 1,100, of whom about 100 are Jews.

TOBLACH, or Dobbiaco, a town of Austria, in the Tyrol, in the circle of Pusterthal, 11 m. ESE of Prunecken, a little to the N of a small lake of the same name, on the Rienz.

TOBOL, a river of Russia in Asia, which has its source on the frontier of Independent Tartary, on the E side of the S branch of the Ural chain, in the gov. of Orenburg; runs NE; enters the gov. of Tobolsk; and after a course of 600 m., joins the Irtish on the l. bank, near Tobolsk. Its principal affluents are the Abouga on the r.; and on the l. the Wei, at the confluence of which it becomes navigable, the Iset, Toura, and Tavda. Its waters near its head are saltish.

TOBOLSK, one of the governments into which Asiatic Russia is divided, forming the western part of Siberia. On the NW the chain of the Urals separates this gov. from European Russia; on the N it is bounded by a vast extent of the Northern ocean, broken into deep bays; on the E the govs. of Yeniseisk and Tomsk; on the S it has the gov. of Omsk. This gov., which includes the most of the cultivated part of Asiatic Russia, except along the frontiers, consists almost entirely of a vast and uniform plain, which, however, varies greatly in point of fertility. From the shores of the Frozen ocean, as far south as the parallel of 58°, the surface presents an aspect of dismal sterility: a few pines and stunted shrubs are the only produce of the soil, which as we proceed N is found at length to yield nothing but bare moss. Fishing, and the chase of fur-bearing animals, afford the only sources of subsistence or profit. A great portion, on the other hand, of the southerly districts, consists of steppes covered with saline lakes and marshes. The only truly agricultural districts are those on the Tobol and its tributaries, in the lower part of their course, which yield luxuriant crops, and supply with grain not only the tracts to the north, but the govs. of Perm and Orenburg. In respect of mineral wealth, few regions rival the mountains of its western frontier. See article URALS. The area of this gov. as now defined is nearly 425,000 geog. sq. m. By the enumeration of 1801, its fixed inhabitants did not exceed 622,422. Of these, 2,017 were returned as Mongols, 17,236 Ostiaks, 273 Tchuwashes, 2,936 Bokharians, 1,993 Tunguses, 2,581 Yakuts, 3,302 Samoyedes, 308 Yakagirs. Besides these were the Cossacks, who were reckoned at about 22,000, and several tribes of wandering Tartars. In 1850 the pop. was estimated at 600,000. The gov. is administratively divided into the 9 circles of T., Berezov, Jalutorovsk, Johun, Kurgansk, Tarsk, Tuimen, Tukalinsk, and Turinsk. See article SIBERIA.

TOBOLSK, the capital of the above government, and of Asiatic Russia in general, is situated on the river Irtisch, close to its junction with the Tobol, in N lat. 58° 12′, E long. 63° 20′. Pop. 20,000. The Russians, when they first took possession of this country in 1587, built here a mere *ostrog* or wooden fort, with the view of keeping the natives in subjection. In 1643, this was burnt to the ground, and they began to erect the present city. It is composed of two parts, the high and the low town. The former is built on an elevated plateau, running parallel to the Irtisch, and at a little distance from it; the latter fills the level space between it and the river. The high town, which occupies the site of the ancient fort, and forms what is properly called the city, contains the residence of the governor, the tribunals, public offices, seminaries, hospital, and the magazines of foreign merchandise. These, with the churches and convents, are all the edifices composed of stone; the rest are of wood. This part of the city is exempt from the danger of inundation; but its inhabitants have to procure water by going to the foot of the hill. The low town, on the contrary, is plentifully supplied with water, but seldom a year passes without its being overflowed. Connected with the lower town is a large suburb, inhabited by Tartars, who are described as in general a quiet orderly race, honest in their dealings, courteous to strangers, and free from those habits of intoxication which form the prevailing vice of the Russian inhabitants. The other residents are in a great measure the descendants of exiles sent hither by the Russian government. The largest colony ever transported hither consisted of the Swedish officers made prisoners at the battle of Pultawa. The most important branch of trade of the city arises from the immense extent of land and river carriage, between European Russia and the frontier of China, which annually concentrates here. The merchants from Europe arrive in spring, with their commodities destined for that distant market; and at the end of summer, the boats return with their cargoes to be transported to Moscow and Petersburg. Merchants from Tartary and Bokhara arrive in the beginning of winter, spend that season at T., and return in spring. The manufactures consist of linens, silk ribbons, reindeer and otter skin dresses, soap, candles, and glass.

TOBOSO (EL), a town of Spain, in the prov. and 63 m. SE of Toledo, on a plateau slightly elevated above surrounding marshes. Pop. 2,817. Its insalubrity, and the indolence of its inhabitants have reduced this place, rendered famous by the genius of Cervantes, to one of the most miserable description. It has manufactories of vases for wine, and some wind-mills; but the environs possess little cultivation, and with the exception of a little grain, wine, saffron and barilla are altogether unproductive.

TOBOUAI-MANOU, or TOUBAI, an island of the South Pacific, in the group of Society islands, in S lat. 17° 28′, and W long. 150° 40′. It is about 6 m. in length, and rises in the centre to a considerable height. It appears fertile, but is scantily populated. It was discovered by Wallis in 1797.

TOCA, a town of New Granada, in the dep. of Boyaca, prov. and 9 m. SSW of Tunja, on the road from that town to Santa-Fe-de-Bogota. Pop. 2,500, of whom a large proportion are Indians.

TOCACHI. See TOACHI.

TOCAHUNOS. See TACANHUNAS.

TOCAIGH, or TOWAIHAE, a bay of the W coast of the island of Hawai-i or Owhyhee, Sandwich islands, in N lat. 20°, and W long. 156°. It is exposed to NW winds, which strike with great violence upon the rocks and reefs by which its banks are surrounded.

TOCALON, a town of New Granada, in the dep. of the Magdalena, and prov. of Carthagena, on the Magdalena.

TOCANCIPA, a town of New Granada, in the dep. of Cundinamarca, and prov. of Bogota, 21 m. N of Santa-Fe-de-Bogota. Pop. 1,100.

TOCANTINS, a large river of Brazil, formed by the confluence of the Almas with its affluent the Maranhão and the Paranatinga, which descend from the Sierra Escalvada and Pyreneos respectively, and unite in S lat. 12°. The united stream directs its course first NNE, then NW, and afterwards N, through the district of Paranna; and between those of Tocantins and Nova Beira, at São-João-das-duas-Barras, it enters the prov. of Para, and separates the districts of Xinjutania and Para; passes Villa-Vicosa; receives the Rio-dos-Bocas; and swelling into a wide estuary, runs between the island of Marajo and the continent, into the Atlantic, which it enters to the E of the embouchure of the Amazon, and between Capes Maguari and Tijiocca, by a mouth upwards of 35 m. in width, and after a total course of about 1,050 m. On reaching Para, 75 m. above its junction with the ocean, it takes the name of that town. Its principal affluents besides those already named are the Manoel, Alvez, and Capim on the r.; and on the l. the Uraguay, and Taracuna.—Also a modern town in the prov. of Goyaz, on the l. bank of the Bacalhao.—Also a town in the prov. of Para, recently erected with the view of promoting the civilization of the Indians.

TOCANTINS-PEQUENO, a river of Brazil, in the prov. of Goyaz, which has its source in the Serra-dos-Viadeiros, and joins the Maranhão.

TOCAYMA, a town of New Granada, in the dep. of Cundinamarca, and prov. of Bogota, 45 m. WSW of Santa-Fe-de-Bogota, and near the r. bank of the Magdalena. Pop. 1,000. It is noted for its thermal springs.

TOCCIA, or TOSA, a river of Sardinia, in the div. of Novare, which has its source in Monte Griesio, on the confines of the Swiss cant. of Tecino, and N extremity of the prov. of Ossola; runs S through the valleys of Formazza, Antigorio, and Ossola; passes Domo-d'Ossola; directs its course afterwards SE, enters the prov. of Pallanza; and 4 m. WNW of the town of that name, throws itself into Lago-Maggiore, on the W bank at the head of a bay which bears the same name. It has a course of about 48 geog. m., and forms several fine cascades. It is crossed several times by the road from Geneva to Milan by the Simplon. At the entrance to the bay of T. are the Borromean islands. Many smaller torrents, on either hand, join it in its course. On the r., the Diveria, descending from Monte-Sempio, and bordered by the Val-di-Vedro, swells it with its waters near to Crevola; at Domo-d'Ossola it receives the Bogna, issuing from the valley of Bugnanco; next, the Ovesca or Mezzavalle, flowing through the valley of Antrona; and finally, the Anza, which has its origin in a small lake on the N side of Monte-Rosa, and descends through the wild and picturesque valley of Anzazca. On the l., the Isorno, rising in Orana-Rossa, forms the valley which bears its name, and issues in the T. at Crevola; and the Trontano, watering the rich valley of Vegezzo, falls into it at Masero. From the lake of Maggiore the royal post-road from Turin and Milan, leading to the pass of the Simplon, ascends the valley of the T., parallel with the river, till it has crossed a fine bridge over the Diveria, near to Crevola; when it turns to the left through the wild scenes of Val-Vedro. From this point an inferior road continues, in line with the T., directly N to Monte Griesio, and a branch also to San-Gottardo. In the descent from Monte-Griesio, the T. forms one of the finest cataracts in all the Alpine regions of Italy. If it fell direct, the perpendicular height would exceed 600 ft.; but its descent is by several leaps, in almost instantaneous succession, from precipice to precipice, until it finally plunges into a deep rocky abyss; its whole course forming one continuous foaming volume, to the extent, probably, of 1,000 ft. The body of water is immense; and, in still weather, its roar is echoed afar, from rock to rock, and from mountain to mountain. At the head of the fall, the embouchure of the torrent is 76 ft. wide; but the gigantic masses that oppose its downward race expand its frantic volume; and looking from below, it presents a triangular form, the width of its base being vastly increased.

TOCCO, a town of Naples, in the prov. of Abruzzo-Citra, district and 18 m. SW of Chieti, and cant. of S. Valentino, near the r. bank of the Pescara, on the slope, and at the foot of a hill. Pop. 3,040. It has five convents, and an hospital.—Also a town in the prov. of the Principato-Ultra, district and 19 m. NW of Avellino, and cant. of Vitolano, on a rock, at the N base of Mount Taburno. Pop. 1,200.

TOCINA, a town of Spain, in the prov. and 15 m. NNE of Seville, and partido of Lora-del-Rio, on the l. bank of the Guadalquiver. Pop. 1,610. It has manufactories of woollen fabrics.

TOCKENBURG. See Toggenburg.

TOCKHOLES, a chapelry and township in Blackburn p., co.-palatine of Lancaster, 4 m. SSW of Blackburn. Area 2,050 acres. Pop. in 1851, 939.

TOCKINGTON (Lower), a tything in Almondsbury p., Gloucestershire, 3¾ m. S by E of Thornbury. Pop. in 1831, 327; in 1851, 443.

TOCKINGTON (Upper), a tything in Olveston p., Gloucestershire, 3 m. S by W of Thornbury. Pop. in 1831, 729; in 1851, 880.

TOCKWITH, a village in Bilton p., Yorkshire, 9 m. W of York. Area 1,610 acres. Pop. 506.

TOCOIOS, a village of Brazil, in the prov. of Minas-Geraes, on the r. bank of the Jequitinhonha, about 100 m. N of Minas-Novas.

TOCONOA, a village of Bolivia, in S lat. 23° 20′, W long. 68° 10′, 10 leagues SE of Atacama.

TOCQUEVILLE, a village of France, in the dep. of the Manche, cant. and 3 m. E of St. Pierre-Eglise. Pop. 800.

TOCUJUS, a large uninhabited island of Brazil, near the coast of the prov. of Para.

TOCUYO, a river of Venezuela, which has its source in the Sierra-Rosas, in the S part of the dep. of Barquisimento; flows NNW to the junction of the Corora, when it takes a NE direction, and after a tortuous course of about 240 m., flows into the Caribbean sea, 20 m. NW of the gulf of Triste. It becomes navigable at the confluence of the Baragua, a distance of 120 m. from its embouchure. Its banks are covered with fine timber.—Also a town in the dep. and 60 m. SW of Barquisimento, pleasantly situated in a fine plain, enclosed on all sides by lofty mountains, and bathed by the river of the same name. Pop. 7,000. It has two churches, a Franciscan and a Dominican convent, and a college, and straight and spacious streets. It possesses manufactories of woollen fabrics and tanneries, and carries on an active trade in salt, grain, and flour. It is also noted for its bread and fruit. The locality is salubrious, the intensity of the sun's rays being modified by the clouds which frequently overcast its sky.

TODBERE, a parish in Dorset, 5 m. SW by W of Shaftesbury. Area 384 acres, Pop. in 1851, 119.

TODD, a county in the SW part of the state of Kentucky, U. S., comprising an area of 362 sq. m., drained by Elk fork of Red river, and by tributaries of Green river. Pop. in 1840, 9,991; in 1850, 12,268. Its capital is Elkton.—Also a township of Huntingdon co., in the state of Pennsylvania, 56 m. W of Harrisburg. Pop. in 1850 about 900.

TODDENHAM, a parish in Gloucestershire, 3 m. SSW of Shipston-on-Stour. Area 2,477 acres. Pop. in 1831, 481; in 1851, 462.

TODDIN, a village of the grand-duchy of Mecklenburg-Schwerin, SW of Hagenau.

TODDINGTON, a parish in Gloucestershire, 2¾ m. N by E of Winchcombe. Area 1,857 acres. Pop. in 1831, 290; in 1851, 189.—Also a parish and market-town in Bedfordshire, 5 m. N by W of Dunstable, including the hamlet of Chalton. Area 5,390 acres. Pop. in 1831, 1,926; in 1851, 2,438. The manufacture of straw-plait affords employment to a considerable number of the inhabitants.

TODES. See Dotis.

TODI, a town of the Papal states, capital of a district, in the delegation and 24 m. S of Perugia, on an isolated mountain between a small river of the same name and the Naja, a little above its confluence with the Tiber. Pop. 2,410. It has a cathedral, 12 parish churches and 8 convents. It is the *Tudertum* of the ancients, in the neighbourhood of which Naises defeated and slew Totila, king of the Goths.

TODMOOS, or Todtmoos, a village of the grand-duchy of Baden, in the circle of the Upper Rhine, to the NE of Schopfheim. Pop. 1,684. It has manufactories of vitriol and alum.

TODMORDEN, a chapelry and market-town in Rochdale p., co.-palatine of Lancaster, 7 m. NNE of Rochdale, including the township of Walsden. The chapelry is intersected by the Rochdale canal, and by the Manchester and Leeds railway, which passes here through the celebrated Summit tunnel. Pop. in 1831, 6,054; in 1851, 7,699. The town, which contains some good houses, is situated in one of the most picturesque valleys in the north of England, surrounded by lofty and precipitous mountains, and watered by the river Calder, which rises on the margin of the forest of Rossendale, at the NW extremity of the township. The houses are irregularly built, being scattered up and down the valley in a straggling manner. The cotton manufacture is extensively carried on at this place, which possesses, in an eminent degree, the three great requisites for manufacturing prosperity—building materials, coal, and water-communication.

TODOROVO, a town and fort of Turkey in Europe, in Bosnia, in Croatia, 23 m. W of Novi, on a mountain.

TODOS-OS-SANTOS, a river of Brazil, in the prov. of Minas-Geraes, and comarca of Serro, which has its sources in the Serra-das-Safiras, and joins the Mucuri on the r. bank.

TODOS-OS-SANTOS, See BAHIA.

TODOS-SANTOS, a bay of California, on the Pacific, in N lat. 31° 47′, W long. 116° 27′, and enclosed on the S by Cape Grajero.

TODTENHAUSEN, a village of Prussia, in the prov. of Westphalia, regency and circle and 4 m. N of Minden, near the l. bank of the Weser.

TODTNAU, or TODNAU, a village of the grand-duchy of Baden, in the circle of the Upper Rhine, 5 m. NE of Schonau, in the midst of the Schwarzwald, on the Wiesen. Pop. 1,480. It has manufactories of wooden ware, matches, straw-hats, and paper, and carries on a considerable trade. In the environs are mines of silver and lead.

TODWICK, a parish in the W. R. of Yorkshire, 7½ m. SE by S of Rotherham. Area 1,860 acres. Pop. in 1831, 224; in 1851, 660.

TOE, a small river of Afghanistan, which has its source in the mountains near Kohat; flows SW, and after a rapid course of about 40 m., falls into the Indus, on the r. side, opposite the Soheila rocks.

TOE-HEAD, a cape on the SW coast of Lewis, 42 m. SW of Stornoway, in N lat. 57° 50′.

TOE-HEAD, a cape in co. Cork, projecting southward from the W side of the entrance of Castlehaven, 11½ m. NE of Cape Clear.

TOEJO ISLANDS, a group of 7 high and woody islands off the N shore of Banca, in the Eastern sea, in S lat. 1° 6′.

TOEPLITZ. See TOPLITZ.

TOFFALA, an island in the Baltic, off the SW coast of Finland, in N lat. 60° 33′.

TOFT, a parish in Cambridgeshire, 4 m. E by S of Caxton. Area 1,242 acres. Pop. in 1831, 279; in 1851, 360.—Also a township in Knutsford parish, co.-palatine of Chester, 2 m. S by E of Nether Knutsford. Area 1,298 acres. Pop. in 1831, 200; in 1851, 241.—Also a hamlet in the p. of Witham-on-the-Hill, Lincolnshire, 3 m. SSW of Bourne. Pop. in 1831, 194; in 1851, 231.

TOFT-MONK'S, a parish in Norfolk, 11 m. SW of Great Yarmouth. Area 2,238 acres. Pop. 421.

TOFT-NEXT-NEWTON, a parish in Lincolnshire, 4½ m. W by S of Market-Raisen. Area 1,293 acres. Pop. in 1831, 74; in 1851, 77.

TOFTES, or TOFT-REES, a parish in Norfolk, 2¼ m. SW of Fakenham. Area 1,184 acres. Pop. 68.

TOFTS (WEST), a parish in Norfolk, 6¾ m. NNW of Thetford. Area 3,051 acres. Pop. in 1851, 191

TOGA, a village of Spain, in the prov. and 21 m. WNW of Castellon-de-la-Plana, on the r. bank of the Millares.

TOGDA, or TODGA, a town and district of Western Africa, 50 m. W of Sigilmessa.

TOGEBAT, a village of Persia, in the prov. of Irak-Arabi, 81 m. N of Ispahan.

TOGGENBURG (UNTER and OBER), a long valley in the N of Switzerland, lying between the cantons of Appenzel and Zurich. In length it exceeds 50 m., and is traversed by the river Thur, from which it sometimes takes the name of the Thur-thal. The population, above 50,000 in number, are all Protestants, and employ themselves partly in agriculture and pasturage, partly in weaving linen and muslin, and spinning cotton for the larger establishments in the towns. The cap. of this district is the small town of Lichtensteig. See THUR.

TOGHER, a village in the p. of Clonmore, co. Louth, 4½ m. E by N of Dunleer.

TOGING, a village of Bavaria, on the l. bank of the Altmuhl, 1 m. W of Dietfurt.

TOGOMI, a town of Nifon, in Japan, 80 m. NW of Meaco.

TOHOTCHIE-HOTUN, a town of Chinese Tartary, in the country of Hami, 30 m. NW of Hami-Hotun.

TOIKO, a town of Nifon, in Japan, 80 m. ESE of Yedo.

TOIRAC, a village of France, in the dep. of Lot, 7 m. NE of Cajara.

TOIRAN, or DOIRAN, a town of Turkey, in the sanj. and 72 m. S of Ghiustendil.

TOIRANO, a town of the Sardinian states, in the prov. and 5 m. N of Albenga. Pop. 1,416.

TOIROS, a port of Brazil, in the prov. of Rio-Grande-do-Norte, 16 m. N of Natal. Salt, hides, and leather are largely exported from this place.

TOK, a river of Russia, which rises in the Obshtshi-syrt, 46 m. NNW of Orenburg; and flowing first NW, and then SW, falls into the Samara, on the r. bank, after a course of 135 m.

TOKA, a town of Hindostan, in the prov. of Aurungabad, in N lat. 19° 25′, at the confluence of the Pera with the Godavery.

TOKAT, a large and commercial city of Asiatic Turkey, situated in the interior of Asia Minor, in the pash. of Sivas, in N lat. 40° 7′. It rises in the form of an amphitheatre, in a deep valley on the banks of the Kizil-Irmak, the ancient *Iris*, flanked by high limestone hills. The streets, though frequently steep, from the unevenness of the ground, are well-paved, but the houses are poor, and chiefly built of mud-dried bricks. The surrounding territory is mountainous, and covered with forests of fir and pine. The chief employment of the place consists in the manufacture of vessels of copper, kettles, candlesticks, &c. with which it supplies the neighbouring provinces; a considerable quantity of blue morocco and silk stuffs is also manufactured here, and there are dyeing and calico-printing establishments. T. is, or rather was, the centre of a very extensive inland trade to and from all parts of Asia Minor. Caravans from Diarbekir arrive in 18 days; from Sinope in 6; from Brusah in 20; from Smyrna in 27. The pop. is stated by Kinneir at 60,000; by some other travellers it has been estimated at 100,000. The majority are Turks, a considerable number Armenians, who have seven churches, and a few Greeks, who have only one church.—The port of T. is Eunieh, 9 m. W by N of Cape Karajik, in N lat. 41° 8′, E long. 37° 18′.

TOKAY, a town of Hungary, in the com. of Theiss, 114 m. ENE of Pest, in N lat. 48° 7′, at the confluence of the rivers Bodrog and Theiss. Pop. 5,700. This town has attained great celebrity for its wine, which is the finest and most expensive in Europe. The vines are raised on a range of low hills called the Hegyallya, about 20 m. in extent. Throughout this district, the grapes are plucked one by one, after they are perfectly ripe, instead of being gathered green, ripe, and rotten, without distinction, and thrown into the press with the stalks, as in most other parts of Hungary. The wine is of three sorts: the essence, or that which runs from the grapes when put into a cask, without artificial pressure. The second sort, called *Ausbruch*, is obtained by applying a slight pressure to the same grapes; lastly comes the *Maslas*, obtained by greater pressure, but still superior to common wine.

TOKEN-BESSEYS, or TOUCAMBASO ISLANDS, a cluster of small rocky islands, lying off the E coast of the island of Bouton, in the Eastern seas, in about 5° 17′ S lat. They are all of them either connected with, or surrounded by rocky shoals, between which rapid currents set strongly to the eastward, and render the navigation hazardous.

TOKIS, a town of Nifon, in Japan, 40 m. NNE of Meaco.

TOKMAK, a town of Russia, in the gov. of Taurida, 21 m. S of Oriekhov.

TOKOZ, an insular district in the Hungarian com. of Raab, formed by the Raab on the S, and the Rabnitz on the N, commencing to the W of the town of Raab. It is 18 m. in length from E to W.

TOKTONAI, a river of Mongolia, which joins the Murui-Ussu after a course from W to E of 180 m.

TOLA, or TULA, a river of Mongolia, which rises on the W flank of the Little-Kentai chain; flows WSW, and then NW; and joins the Orkhon on the r. bank, after a course of 240 m.

TOLAGO-BAY, a bay on the NE coast of the N island of New Zealand, in S lat. 38° 22′, discovered by Cook in 1769. It has from 7 to 13 fath., with a clean sandy bottom and good anchorage, and is sheltered from all winds except the NE.

TOLATOLA, a town on the N coast of the island of Celebes, near Cape Donda. It is a considerable place, with a fine harbour.

TOLBANOS, a village of Spain, in the prov. and 42 m. SE of Burgos.

TOLCSVA, a village of Hungary, in the com. of Zemplin, 12 m. NNE of Tokay.

TOLEDO, a central province of Spain, in New Castile, situated chiefly to the S of the Tagus, between the parallels of 39° 3′, and 40° 54′. It is bounded on the N by the provs. of Madrid and Avila; on the E by Cuenca; on the S by Ciudad-Real; on the W by Caceres. It has an area of 4,215 geog. sq. m. Pop. in 1849, 330,000. Its surface consists partly of mountain-tracts, and partly, towards the centre, of elevated and extensive plains, the soil of which, however, is frequently sandy or chalky, so that water is scarce, and few trees enliven the prospect. Like the rest of Castile, this prov. is mountainous. The Tagus traverses the prov., and the Guadiana approaches its S part. Pease, lentiles, kidney-beans, potatoes, flax, saffron, melons, and oranges, are abundantly grown. Sheep are not numerous, but their wool is of the best quality. Vines, silk, honey, wax, and fruits, are the chief products of this prov. Gold, silver, lead, iron, and copper are mined. The manufactures are trifling, though some articles, such as woollens, silks, soap, earthenware, and sword-blades, are exported.

TOLEDO, an ancient city of Spain, the chief town of the preceding prov., and the see of an archbishop, is situated on the sides and top of a steep hill, bathed by the Tagus, on the E, S, and W sides, in a narrow valley surrounded by lofty mountains, 37 m. WSW of Madrid, and at an alt. of 2,400 ft. above sea-level. Pop. 14,000. The environs are unproductive, and the surrounding hills, by concentrating the sun's rays, render the heat in summer excessive. The houses are generally massive and lofty structures, but crowded; the streets are narrow and steep. The Alcazar is a large structure built with solidity, and decorated with statues. Its central gate, vestibule, court, and subterranean apartments, are entitled to admiration; but being no longer wanted as a princely residence, it has been converted into a factory or working establishment. The cathedral is of great antiquity. It is a Gothic building, which, from its magnitude, would be magnificent, were not its front too low, and its interior so much divided and subdivided. Its painted windows are of great beauty; and a valuable library is attached to it. There are upwards of 20 churches, besides hospitals, monasteries, and convents in T. The hospital of St. Cruz is an elegant building of the 15th cent., that of St. John, built in the 16th, is equally rich. Among the educational establishments is a military school attended by about 600 cadets, and an ecclesiastical seminary. The university of T., formerly in great repute, had long lost its estimation, and was entirely suppressed in 1807.—The Tagus is said to have been navigable as far as T. in times when the exports of the town were of importance. These chiefly consisted of woollens, stockings, and sword-blades. Woollens, linen, silks, serges, taffeta, church-robes, oil, chocolate, and leather, are still manufactured; and the sword-blades, so noted throughout Spain, till excluded a century ago from fashionable dress by French swords, are now made in a large building on the banks of the Tagus, 2 m. SW of the city. The secret of tempering them is said to have been recovered, and they still fetch a high price.—T. is a place of great antiquity, and was successively the seat of government under the Goths, the Moors, and the kings of Castile. The Gothic kings fixed their residence here in 567. In 711 it was taken by the Moors, and became the abode, first of a viceroy, and eventually of an independent prince. In 1085 it fell into the hands of the Christians, and became anew the residence of their kings. It was besieged by the Moors in 1109, 1114, and 1127, but without success; at a subsequent date it was less fortunate, having been besieged and taken in 1467 and in 1641. Great part of the town was burned on each occasion, which, with the removal of the government to Madrid, has been the main cause of its decline. T. was formerly the seat of several meetings of the Cortes, and of a number of national church councils.

TOLEDO, a port-of-entry of Lucas co., in the state of Ohio, U. S., on the W bank of the Maumee, 5 m. above its entrance into Maumee bay, and 117 m. N by W of Columbus. Pop. in 1840, 1,222; in 1853, 6,512. The rapid growth of this place has been chiefly owing to its connection with the Wabash and Erie canal, which extends from a point 4 m. below Toledo, to Evansville on the Ohio river, in the state of Indiana, 467 m. The total tonnage arriving at and departing from T. by this canal in 1851, was 250,000 tons, chiefly wheat, flour, and iron.

TOLEGNO, a village of the Sardinian states, in the prov. and 2 m. NNW of Bielle. Pop. 1,050.

TOLENTINO, a town of the Papal states, situated on a rising ground bathed by the river Chienti, 12 m. SW of Macerata. Pop. of district 9,500. It is remarkable for a treaty of peace concluded here between Bonaparte and Pius VI. in February, 1797; also for some actions between the Austrians and Neapolitans in the beginning of May 1815.

TOLETHORPE, a hamlet of Rutlandshire, 12½ m. E of Oakham.

TOLFA, a town of the Papal states, 10 m. ENE of Civita-Vecchia. Pop. 1,100. Alum is extensively wrought in the vicinity.

TOLGEN, a village and parish of Norway, in the bail. of Hedemarken, 15 m. SW of Roraas, on the r. bank of the Glommen-elv.

TOLJAPUR, a town of Hindostan, in the district of Aurungabad, in N lat. 18° 17′.

TOLKA, a village in the p. of Finglass, co. Dublin, on the Tolka rivulet, ¾ m. SSE of the village of Finglass. Pop. in 1851, 223.

TOLKEMIT, a town of W. Prussia, near the

Frische-haff, 12 m. NE of Elbing, and 44 m. SW of Konigsberg. Pop. 2,300.

TOLLAND, a parish of Somersetshire, 3 m. NNE of Wiveliscombe. Area 824 acres. Pop. 147.

TOLLAND, a county in the N of Connecticut, U. S. Area 337 sq. m. Pop. in 1840, 17,980; in 1850, 19,946. Its cap., of the same name, 18 m. NE by E of Hartford, had a pop. of 1,410 in 1850.—Also a post-town of Hampden co., Massachusetts, 104 m. WSW of Boston. Pop. 600.

TOLLARD-ROYAL, a parish of Wiltshire, 11½ m. S by E of Hindon. Area 2,807 acres. Pop. 574.

TOLLEN-SEE, a lake in the grand-duchy of Mecklenburg-Strelitz, formed by a river of the same name, which rises 6 m. NNW of New-Strelitz, and after its issue from the lake, flows N and NW, by New Brandenburg and Treptow, and falls into the Peene near Demmin, after a course of 50 m. The lake or expansion of the river is 7 m. in length from NNE to SSW.

TOLLER-FRATRUM, a parish in Dorset, 8 m. E by S of Beaminster. Area 500 acres. Pop. 54.

TOLLER-PORCORUM, a parish in Dorset, 7 m. SE by E of Beaminster. Area 3,143 acres. Pop. 527.

TOLLERTON, a parish in Notts, 4 m. SE of Nottingham. Area 1,240 acres. Pop. in 1831, 149; in 1851, 157.—Also a township in Alne p., Yorkshire, 4¼ m. SSW of Easingwold, crossed by the Great North of England railway. Area 2,340 acres. Pop. in 1831, 529; in 1851, 551.

TOLLESBURY, a parish in Essex, 7½ m. ENE of Maldon. Area 10,638 acres. Pop. in 1851, 1,193.

TOLLESHUNT-DARCY, a parish in Essex, 6 m. NE by E of Maldon. Area 3,371 acres. Pop. 792.

TOLLESHUNT-KNIGHTS, a parish in Essex, 7 m. NE of Maldon. Area 2,079 acres. Pop. 371.

TOLLESHUNT-MAJOR, or BECKINGHAM, a parish in Essex, 5 m. NE by E of Maldon. Area 2,344 acres. Pop. in 1831, 428; in 1851, 479.

TOLLO, a town of Naples, in the prov. of Abruzzo-Citra, 8 m. E of Chieti. Pop. 2,000.

TOLMEIN, a town of Illyria, 40 m. N of Trieste, near the l. bank of the Isonzo. Pop. 500.

TOLMEZZO, a town of Austrian Italy, in the deleg. and 25 m. NNW of Udina. Pop. 2,000. It has considerable manufactures of linen, and a cotton-spinning mill.

TOLNA, a comitat or county of Hungary, lying to the W of the Danube, between the parallels of 46° 7′ and 46° 50′, between the coms. of Veszprim and Stuhlweissenburg on the N; and Baranya on the S. It has an area of nearly 1,400 sq. m., with 180,000 inhabitants, descended from Magyars, Rascians, Bohemian Slavonians, and Germans. With the exception of a few low ridges or hills chiefly in the S, it consists entirely of level ground, and is traversed by the Sarviz, and the canal of that name. Its chief products are corn, wine, tobacco, and fruit; and it is in general fertile. The chief town of the co. has for some time been Simontornya.—Its former cap. of the same name, is situated on the r. bank of the Danube, 139 m. SE of Presburg, and 73 m. S of Buda. Pop. 3,000.

TOLO-BAY, a large bay on the E coast of the island of Celebes, between the parallels of 1° 30′ and 3° 5′ S, very broad at its entrance, but growing narrower towards the bottom.

TOLOCZYN, or TOLOTSTIN, a town of Russia, in the gov. and 42 m. NW of Mohilev, near the l. bank of the Droutz.

TOLOMETA, a village of Tripoli, on the coast of Barcah, in N lat. 32° 48′, 72 m. NE of Bengazi, the representative of the ancient *Ptolemeta*.

TOLOSA, a town of Spain, the chief place of Guipuzcoa, in the prov. and 15 m. S of San-Sebastian. Pop. 8,000. It is situated in a pleasant valley at the confluence of the rivers Oria and Arages. Its streets are tolerably wide and well-paved, and its houses large and comparatively well-built. It has manufactures of nails, horse-shoes, kitchen-utensils, agricultural implements, and sword-blades; also some tanneries.—Also a village of Portugal, in the prov. of Alemtejo, 12 m. NNW of Crato.

TOLOX, a town of Spain, in the prov. and 30 m. W of Malaga, at the foot of the Sierra-Nieve.

TOLPUDDLE, a parish in Dorset, 7 m. ENE of Dorchester. Area 2,039 acres. Pop. in 1851, 354.

TOLSTONOSKOI, a fort of Asiatic Russia, in the gov. of Tobolsk, on the Yenisei, 280 m. NNW of Turuchansk.

TOLTEN, a river of Chili, which has its rise in a lake at the W base of the volcano of Villarica; runs N, and then W; and enters the Pacific, after collecting the waters of several other rivers, 7 leagues W of Port Imperial, in S lat. 39° 11′.

TOLU, a town of New Granada, in the prov. and 60 m. S of Cartagena, on the E coast of the gulf of Morosquillo. It has a good harbour; and the country around abounds in grain, and in the tree from which is extracted the celebrated balsam of Tolu.

TOLUCA, a town of Mexico, situated at the foot of the Nevado-de-Toluca, in N lat. 19° 16′ 19″, 28 m. SW of Mexico, in a valley 8,800 ft. above the level of the sea.—The Nevado has an alt. of 2,370 toises = 5,051½ yds.

TOLVA, a village of Spain, in the prov. and 45 m. ESE of Huesca. Pop. 790.

TOLVE, a town of Naples, in the prov. of Basilicata, 8 m. NE of Potenza. Pop. 3,400.

TOLZ, a town of Bavaria, on the Isar, 26 m. S of Munich. Pop. 3,100. It has iron and saltpetre works, and a considerable trade in agricultural produce.

TOM, a river of Russia in Asia, in the gov. of Tomsk, which has its source in the district of Kouznetzk; runs NW; passes Kuznetzk and Tomsk; and after a course of upwards of 300 m., joins the Obi on the l. bank, 27 m. from Tomsk. Its waters abound with fish.

TOMAC, a creek of Australia Felix, in the district of Western Port, which flows into the Great Swamp.

TOMACHPOL, a town of Russia in Europe, in the gov. of Podolia, district and 27 m. NNE of Tzekinovka, on the r. bank of the Rouzava.

TOMAGAN, a town of New South Wales, in the co. of St. Vincent, on a creek of the same name, an affluent of Broulee harbour.

TOMAH, a mountain of New South Wales, in the co. of Cork, in the Carmarthen chain. It is visible from the Sydney lighthouse.

TOMAHAWK, an island near the N coast of Van Diemen's Land, in Ringaroome bay, co. of Dorset.

TOMAKOVKA, a town of Russia in Europe, in the gov. and 51 m. SSW of Yekaterinoslav, and district of Alexandrovsk, on a river of the same name, which, after a course in a generally SSW direction of 45 m., joins the Dnieper, on the r. bank.

TOMALPIN, a mountain of New South Wales, in the co. of Northumberland.

TOMANISI, a district and town of Japan, in the island of Nifon, and prov. of Mimasaka.

TOMARES, a village of Spain, in the prov. and 3 m. NW of Sevilla, on the r. bank of the Guadalquiver. Pop. 172.

TOMARI, a headland of New South Wales, in the co. of Gloucester. It consists of a line of lofty cliffs, and forms the SE point of Port-Stephens.

TOMARIN, a town of Japan, in the island of Kiu-siu, and the prov. of Satsuma.

TOMAS, a town of Nubia, in the Barabras territory, on the l. bank of the Nile, opposite Derr. It consists of an assemblage of detached houses, intermingled with plantations of date-palms.

TOMASZOW, a town of Poland, in the gov. and 69 m. SE of Lublin, and obwod of Hrubieszow. Pop. 2,900. It has extensive manufactories of porcelain and earthenware, and carries on an extensive trade in wiszniak and matinnik liqueurs, prepared from honey and fruit.—Also a town in the gov. of Masovie, obwod and 23 m. SSW of Rawa, on an elevated and salubrious plateau, in the midst of extensive forests, and to the l. of the Pitica, a navigable river which forms a communication between the Vistula and Baltic, and at the confluence of the Wolborka. Pop. 5,000. It is small, but well-built, and has two churches, a Catholic and a Protestant, and a synagogue. It possesses manufactories of calico and fine cloth, yarn-mills, dye-works, printing-mills, extensive iron-works, &c., and carries on an active trade. In the environs are mines of iron, and quarries of lime and free-stone. The advantageous situation of this town renders it one of rapidly increasing importance.

TOMB, or TOOME, a parish 5½ m. SW of Gorey, in co. Wexford. Area 6,979 acres. Pop. 1,768.

TOMBARO. See NEW IRELAND.

TOMBE (LA), a commune of Belgium, in the prov. of Hainault and dep. of Kain. Pop. 550.

TOMBEBŒUF, a village of France, in the dep. of the Lot-et-Garonne, cant. and 6 m. NNW of Monclar, on a mountain. Pop. 2,900.

TOMBEBOUE, a village of France, in the dep. of the Lot-et-Garonne, cant. and 3 m. SSE of Ste. Livrade. Pop. 1,455.

TOMBEL, a commune of Belgium, in the prov. of Hainault and dep. of Ellezelles. Pop. 127.

TOMBERGEN, a commune of Belgium, in the prov. of Brabant and dep. of Lennich-Saint-Marlin. Pop. 415.

TOMBES, a commune of Belgium, in the prov. of Namur and dep. of Mozet. Pop. 366.

TOMBIGBEE, a river of the United States, formed by the confluence of several streams in Tishamingo co., in the state of Mississippi; flows S through Lowndes co.; enters the state of Alabama, and, after a course of 60 m., is joined by the Black Warrior, a river of nearly equal size; thence it pursues a meandering course to the S, and joins the Alabama, when the united stream takes the name of Mobile. It has a total course of nearly 500 m., and for 9 months in the year is navigable for steam-boats to Columbus.

TOMBLAINE, a village of France, in the dep. of the Meurthe, cant. and 2 m. ESE of Nancy, on the r. bank of the Meurthe, which is here crossed by a ferry. Pop. 534. It has a cotton-spinning, and a fulling-mill, a manufactory of cloth, and tile and lime-kilns.

TOMBORO, a volcano of the island of Sumbava, in the Sunda archipelago. A terrible eruption of this mountain occurred in 1815, when 12,000 persons were destroyed. The report of its outbreak was audible at the distance of nearly 900 m.

TOMBOS, an island of the Nile, in the Mahas territory, a little to the SE of Hannek. It is well-cultivated. On the r. bank of the river opposite are several ruins.

TOMBROUK, a commune of Belgium, in the prov. of West Flanders and dep. of Limogne. Pop. 487.—Also a commune in the prov. of West Flanders and dep. of Monscron. Pop. 414.

TOMBUCTOO. See TIMBUCTU.

TOME-DO-MAR (SAN), a town of Spain, in the prov. and 11 m. WNW of Pontevedra, at the entrance of the Umia into the bay of Arosa. Pop. 855.

TOME'-DEL-PIE-DEL-PUERTO (CONCEJO-DE-SANTO), a town of Spain, in the prov. of Segovia, and partido of Sepulveda. Pop. 567. It has a parish-church and a customhouse.

TOMELLOSA, a village of Spain, in the prov. and 18 m. E of Guadalajara, and partido of Brihuega, on the l. bank of the Tajuna. It is well-built, and has a fine public square. Pop. 650. It has manufactories of linen, and an oil-mill.

TOMELLOSO, a town of Spain, in the prov. and 54 m. ENE of Ciudad-Real, and partido of Alcazar-de-San-Juan, in an arid plain. Pop. 1,020. It is well-built, but depends upon wells for its supply of water. Baskets are its chief article of manufacture.

TOMEPENDA, a settlement of Ecuador, in the prov. of Jaen-de-Bracamoros, on the l. bank of the Tunguragua. The environs are fertile in tobacco and cotton.

TOMFINLOUGH, a parish in co. Clare, containing the town of Newmarket-on-Fergus, and the village of Bohererroan. Area 6,736 acres. Pop. in 1831, 4,053; in 1851, 3,182.

TOMGRANEY, or TOMGRENI, a parish in co. Clare, containing the villages of Scariff and Tomgraney. Area 14,181 acres. Pop. in 1831, 5,568; in 1851, 6,673. The greater part of Lough O'Grady lies within the p.; and the surface-elevation of this lake has a height of 122 ft. above the level of the sea.

TOMHAGGARD, a parish of co. Wexford, 4¼ m. W by S of Broadway. Area 2,180 acres. Pop. in 1831, 723; in 1851, 826.

TOM-HO, a river of China, which rises on the frontier of Shen-si, and flows SSW through Sechwen to the Kia-ling, which it joins on the l. bank, after a course of 100 m.

TOMIEH, a small town of Upper Egypt, 15 m. S of Siut.

TOMINA, a district of Bolivia, in the dep. of Charcas, bordering eastward on the Chiriguanos, a nation of independent Indians. It is a mountainous country, in the valleys of which are some sugar plantations; its higher parts afford pasturage to cattle and horses. The rivers which water the district flow into the Cochabamba, and it is separated from Santa-Cruz-de-la-Sierra by the Rio-Grande, which joins the Mamore.—Its capital, of the same name, is an inconsiderable place, 55 m. E of Chuquisaca.

TOMISCANING. See TEMISCAMING.

TOMISVAR, a small port of European Turkey, in the sanj. of Silistria, situated on an inlet of the Black sea. It is supposed to be the representative of the ancient *Tomi*, the scene of Ovid's exile.

TOMLUK, a town of Hindostan, on the r. bank of the Rupnaran, 42 m. SW of Calcutta. In the vicinity are large salt-works.

TOMOGUY, an island in the Eastern seas, in S lat. 0° 15', E long. 127° 4'. It is shaped like a horse-shoe, the hollow forming a bay; and is about 2½ m. in circuit.

TOMOR, or TOMOROS, a mountain of Albania, in the sanj. of Avlona, in N lat. 40° 37'.

TOMORINDSHA, a town of European Turkey, situated in the mountains between the towns of Prespa and Elbassan. It is inhabited by Albanians.

TOMOS, a village and pass of the Carpathians, between Transylvania and Wallachia, 6 m. S of Cronstadt. It is defended by a redoubt, and is watered by a small rivulet, an affluent of the Wiedenbach.

TOMPKINS, a central town of the state of New York, U. S., comprising an area of 580 sq. m., watered by Fall and Six-mile creeks and Cayuga inlet, affluents of Cayuga lake, and intersected by the Cayuga and Ithaca railroad. The surface is hilly, but fertile in grain and grass. Pop. in 1840, 37,948; in 1850, 38,738. Its cap. is Ithaca.—Also a township of Jackson co. in the state of Michigan, 24 m. S of Lansing, drained by Grand river and Sandstone creek. Pop. in 1840, 209; in 1850, 623. —Also a township of Delaware co., in the state of New York, 22 m. SW of Delaware. It has a hilly surface, is drained by Mohawk and W branch of Delaware river and its tributaries, and by tributaries of the Popacton branch of the Delaware, and intersected by the New York and Erie railroad. Pop. in 1840, 2,035.

TOMPKINSVILLE, a village of Castleton township, Richmond co., U. S., in the state and 6 m. S of New York, on the SE part of Staten island. It is noted as the quarantine station of New York, and contains an extensive lazaretto. Pop. in 1840, 1,400.—Also a village of Monroe co., in the state of Kentucky, 153 m. SSW of Frankfort, on a branch of Big Barren river. Pop. in 1850, 240.

TOMREGAN, a parish of co. Cavan. Area 10,677 acres. Pop. in 1831, 4,118; in 1851, 3,405. The summit of Legavagra in this p. has an alt. above sea-level of 1,279 ft. The considerable lakes Annagh and Kilwilly lie on the E boundary of the Cavan section; Lough Togher lies on the S.

TOMSEH, a village of Asiatic Turkey, on the Euphrates, 70 m. W of Diyarbekir.

TOMSK, a government of Asiatic Russia, lying between the parallels of 49° and 60°; and bounded on the NW and N by Tobolsk; on the NE and E by Yeniseisk; on the S by China; and on the W by the gov. of Omsk. The Little Altai chain and its ramifications cover the S part of this gov., which is wholly within the basin of the Obi, and is watered by the Biia, the Alei, the Tom, the Shulym, the Chany, and the Sumy. The N part of this gov. is altogether a wild and desolate region; towards the S cultivation commences. The pop. is about 350,000, Russians, Cossacks, Tartars, and Ostiaks.

TOMSK, the capital, is situated on the r. bank of the Tom, about 25 m. from its junction with the Obi, in N lat. 56° 29′ 39″. Nothing can be more irregularly constructed than the old and lower part of the town; but the upper and more modern section contains some good houses. The kremlin, a fortress constructed in the 17th cent., is now almost in ruins; within its circuit, however, are the cathedral church, the tribunals, the treasury, and the magazines. There are nine churches, two monasteries, and several mosques. The greater part of the inhabitants, about 10,000 in number, subsist by commerce, for which the place is very advantageously situated, being on the great line of rivers which connect Tobolsk with the Chinese frontier and the eastern part of Siberia. It is the centre also of the trade in brandy, which is brought hither from distilleries on the Tobol and the Iset, and thence distributed to the countries to the eastward. Besides Russians, the place contains a great number of Tartar, Bucharian, and Kalmuck merchants.

TOMU, a port of Nifon, in Japan, 115 m. SW of Meaco.

TOMYSL (NEU), a village of Prussia, in the reg. and 34 m. WSW of Posen.

TON, or TON-SHIN, a city of China, of the second rank, in Yun-nan, in N lat. 26° 39′.

TONALE, a mountain at the NE extremity of the Austro-Italian prov. of Bergamo, on the frontiers of Tyrol, in N lat. 46° 5′. A new road was commenced in 185[illegible], which is to connect Lombardy with the Tyrol, through the defile of Tonale, situated at 6,250 ft. above the level of the sea. It is to supersede the old road over the Stilfs, which is upwards of 8,000 ft. above the level of the sea, and is often impassable for months together on account of the snows.

TONAMY, a town of Nifon, in Japan, 155 m. NW of Jedo.

TONARA, a town of the Sardinian states, 12 m. E of Busachi. Pop. 2,240.

TONAWANDA, a village of New York, U. S., 12 m. N of Buffalo.

TONCO, a village of the Sardinian states, in the division and 22 m. WNW of Alessandria, near the l. bank of the Versa. Pop. 2,000.

TONDA, or TONRA, an ancient town of Bengal, situated opposite to the ruins of Gour, but divided from them by the Ganges, in N lat. 24° 49′, E long. 88° 15′. In 1564, Soliman-Shah, of the Shere-Shah dynasty, fearing the unhealthy situation of Gour, made this place the cap. of his united kingdom of Bengal and Bahar. It retained this dignity till 1592, when Man-Sing transferred the seat of government to Rajmahal. It has now dwindled to a mere village, frequently confounded with Taunda, a manufacturing town of Oude.

TONDELLA, or BESTEIROS, a town of Portugal, in the prov. of Beira, comarca and 12 m. SW of Viseu. Pop. 2,400.

TONDER, or TONDERN, a town of Denmark, in the duchy of Sleswig, 30 m. S of Ribe, in a low territory on the Widaa, which is here crossed by a bridge. Pop. 2,800. It is enclosed by old walls. There are here several schools, an hospital, and a house-of-detention. The principal manufacture is that of lace; bonnets, hosiery, woollen and linen fabrics, and tobacco are also manufactured.

TONDI, a town and port of Hindostan, in the Carnatic and district of Dindigul and Madura, at the mouth of the Vaigam and opposite the NW coast of Ceylon.

TONDIMAN, a zemindary of Hindostan, in the presidency of Madras, and prov. of the Carnatic, bounded on the N by the prov. of Tanjore, and on the E by the bay of Bengal. It is to a great extent covered with jungle, and, except towards the S possesses little fertility. It is governed by a chief who takes the name of Tondiman, and has at all times maintained the strictest adherence to the British government. Its chief place is Podocatta.

TONDJA, a river of Turkey in Europe, in Rumelia, which has its source in the sanj. of Sophia 44 m. N of Philippopoli, on the S side of the Balkan chain, and near the confines of Bulgaria; runs first E, then SSE, through the sanj. of Tchirmen; and after a course of about 165 m., joins the Maritza, on the l. bank, a little below Adrianople. The principal places on its banks, besides Adrianople, are Kezanlik, Janboli, and Jenidge-Kizilaghadj.

TONDO, a town of the Philippine island of Luçon, on the Pasig, nearly opposite to Manila. Pop. 15,000.

TONDON, a town of Mandshuria, in the prov. and 510 m. ESE of Sagalin-Oola, on the r. bank of the Amour, at the confluence of a river of the same name.

TONG, a chapelry and village of Yorkshire, 6 m. WSW of Leeds. Pop. in 1851, 511.—Also a township of Birstall p., Yorkshire. Area 2,644 acres. Pop. in 1851, 2,797.

TONGA. See FRIENDLY ISLANDS.

TONGATABU, an island in the S. Pacific, one of the southern group of the Friendly islands, in S lat. 21° 4′, W long. 175° 28′. It is 20 m. in length, by 10 m. wide at its broadest part, and is very low in surface; the only part which can be seen from a ship in the offing, is the NE point; and it is only

30 ft. above the level of the sea. The soil is prolific, and consists of a fine rich mould, free from stones, except near the beach where coral rocks appear above the surface. At a distance the surface seems entirely clothed with trees, amongst which the tall cocoa-palms raise their tufted heads above all the rest. Of cultivated fruits, the principal are plantains, of which there are fifteen varieties. Several plants grow here which are not known at Otaheite, especially shaddocks. The only quadrupeds, besides hogs, are a few rats and dogs. The pop. was reported to be about 9,000 in 1850, of whom 5,000 were Protestants, and 600 Roman Catholics. In 1844, on the death of the king of the Tonga group, George, king of the Haabai or Havai group, became king of the three groups composing the Friendly islands. This island was discovered in 1643, by Tasman, a Dutch navigator. It has since been visited by different navigators; by Cook in 1773, and in 1777 by Perouse; in 1787, by Edwards; in 1791, by Messrs. D'Entrecasteaux and Huon; by the missionary ship, Duff, in 1797; by the Favourite, in 1840; and by Meander, in 1850.

TONGATAVA, an island of the S. Pacific, in the group of Friendly islands. It is the principal of a cluster of islands to the S of the island of Pylstaert. It is low and surrounded by coral reefs. It was discovered by Depps.

TONGE, a parish and village of Kent, 1½ m. E of Milton. Area 1,883 acres. Pop. in 1831, 226; in 1851, 242.—Also a hamlet of Lancashire, 2 m. NE of Great Bolton. Pop. in 1811, 1,226; in 1851, 3,831.—Also a township of Lancashire, 5½ m. NNE of Manchester. Area 367 acres. Pop. in 1851, 3,831.—Also a hamlet of Leicestershire, 5½ m. NE of Ashby-de-la-Zouch.—Also a parish of Salop, 4 m. E by S of Shiffnall. Area 3,464 acres. Pop. 511.

TONGE (Oude), a village of Holland, in the prov. of South Holland, cant. and 14 m. SE of Goedereede, in the island of Over-Flakke. Pop. 1,500. About 3 m. NW is the village of Nieuwe Tonge. Pop. 1,200.

TONGERLOO, a department and commune of Belgium, in the prov. and 26 m. ESE of Antwerp, and arrond. of Turnhout, watered by the Wimpe. Pop. of dep. 1,558. It has distilleries and breweries. —Also a department and commune in the prov. of Limburg and arrond. of Tongres. Pop. 585.

TONGORA, a small port of Peru, in the dep. of Truxillo, 15 m. WSW of Sechura, in S lat. 5° 41′.

TONGOUZLEK. See Degnizli.

TONGOY. See Tanguey.

TONGRE-NOTRE-DAME, a department and commune of Belgium, in the prov. of Hainault, and arrond. of Mons, watered by the Hunelle. Pop. of dep. 808; of com. 606.

TONGRES, or Tongeren, a department, commune, and town of Belgium, in the prov. of Limburg, and arrond. of the same name. Pop. of dep. 5,981. The town is 12 m. SW of Maestricht, on the Jaar. It was formerly enclosed by walls, of which some remains still exist; and has a college and several tanneries. The trade consists chiefly in pigs and grain. In the vicinity is a mineral spring noted by Pliny. This town returns two deputies. T. is supposed to have derived its origin from an ancient fortress in the territory of the Eburi, named by Augustus, *Tongri*. After rising to a place of considerable importance, it was ruined by Attila in 498. It was subsequently devastated by the Normans in 881, and after numerous changes of masters, fell finally into the hands of the French, in 1677.

TONGRES-SAINT-MARTIN, a department and commune of Belgium, in the prov. of Hainault, and arrond. of Mons, on the Blanche. Pop. 172.

TONGRINNE, a department and commune of Belgium, in the prov. and arrond. of Namur. Pop. of dep. 1,011. The village is 7 m. WNW of Namur, near the l. bank of the Ligne. Pop. 400.

TONG-SEU. See Plau.

TONGUE, a parish on the north coast of Sutherlandshire; bounded on the N by the North sea; on the E, SE, and S, by Farr; on the SW by Edderachylis, having a length from NNE to SSW of 20 m., and an area of about 120 sq. m. The coast, measured in a straight line, extends 10 m., and is in general high and rocky. The Kyle-of-Tongue strikes off from the sea at Eilean-na-roan, and penetrates the parish 8½ m. SSW, with a mean breadth of about 1½ m. A lofty semicircular range of hills rises boldly and suddenly from the ocean, and sweeps quite round the Kyle-of-Tongue, forming the enclosed valley into a stupendous amphitheatre. On its W side it commences with Ben-Hutig, 1,345 ft. above sea-level; runs along in the rugged, trackless, boggy, mountain waste of the Moin; and terminates not far from the head of the Kyle, in the sublime Ben-Hope. Wood, natural and planted, covers about 700 acres. The lands in tillage comprise only about 1,000 acres. The village of Tongue or Kirkiboll is pleasantly situated on the slope of a hill, about 1 m. S of Tongue-house, and 44 m. from Thurso. Pop. of p. in 1801, 1,348; in 1821, 1,736; in 1851, 2,041. Houses in 1851, 372.

TONIAK, a river of Japan, in the island of Nifon, and prov. of Musasi, which runs S, and after a course of about 90 m., throws itself into the bay of Yedo, at the town of that name.

TONIKAKI, an island of the N. Pacific, near the N coast of the island of Celebes, in N lat. 5° 31′, E long. 117° 17′.

TONK, a town of Hindostan, in the prov. of Ajmir, 54 m. S of Jeypur. It stands in a triangular hollow, of which two sides are formed by hills, and the third by a substantial wall. It is of considerable extent, and is well-built of stone.

TONKER, or Dounggar-Dzoung, a town of Tibet, in the prov. of Kam, near the E bank of Lake Napatsu, near its outlet into the Kin-cha-keang.

TONKIN. See Tonquin.

TONMARUP, a parish of Sweden, in the prefecture of Christianstad, and haerad of Jaeresta. It has quarries of black marble.

TONNA, or Grafentonna, an amt or bailiwick and market-town of the duchy of Saxe-Coburg-Gotha, principality and 11 m. N of Gotha, on a river of the same name, near the Unstrut. Pop. 1,300. It has two castles, an old and a new, and possesses manufactories of tobacco, and a mineral spring. Pop. of amt 4,130.

TONNAY-BOUTONNE, a canton, commune, and town of France, in the dep. of the Charente-Inferieure, and arrond. of St. Jean-d'Angely. The cant. comprises 9 coms. Pop. in 1831, 4,456; in 1846, 4,759.—The town is 9 m. WNW of St. Jean-d'Angely, on the r. bank of the Boutonne. Pop. 1,142. It has an active trade in grain, wine, and brandy. On an adjacent hill are the ruins of an ancient fortress.

TONNAY-CHARENTE, or Charente, a canton, commune, and town of France, in the dep. of the Charente-Inferieure, and arrond. of Rochefort. The cant. comprises 10 coms. Pop. in 1831, 9,714; in 1846, 10,624.—The town is 3 m. E of Rochefort, pleasantly situated on the r. bank of the Charente, which is here crossed by a ferry. Pop. in 1846, 3,304. It is the residence of several foreign consuls, and has an exchange, and a well-frequented port.

It possesses manufactories of home-grown sugar, and several tile and brick kilns. Its trade consists chiefly in brandy, wine, grain and seeds, flax, salt, timber, tartar, minium, iron, coal, coal-tar, wax, &c. Its principal imports from England are coal, wrought steel, sharping stones, and fine wool. The port is capable of affording safe harbourage, even at low water, to 200 vessels of the large size.

TONNDORF, a village of Saxe-Weimar, 3 m. W of Berka. Pop. 540.

TONNEINS, a town of France, in the dep. of Lot-et-Garonne, situated on the Garonne, which is here crossed by a suspension-bridge, 10 m. SE of Marmande. Pop. 4,468. It is a well-built place, with a few handsome public buildings. Its manufactures are of snuff and tobacco, and articles of woollens and cordage. It has a considerable transit trade by the river in hemp, corn, wine, and brandy.

TONNERRE (Mont), or Donnersberg, a mountain of Bavaria, on the l. bank of the Rhine, at the N extremity of the Vosges chain, 12 m. NE of Kaiserslautern. It has an alt. of 682 metres = 745 yds. above the level of the Rhine. The French gave the name of this mountain to a department which comprehended the greater part of the electorate of Mentz, the Lower palatinate, the bishoprics of Spires and Worms, and several counties and lordships. At the congress of Vienna, it was divided between Austria and Hesse-Darmstadt; the latter receiving the N part, the former the southern or larger division, which was afterwards ceded to Bavaria.

TONNERRE, a town of France, in the dep. of the Yonne, situated on the Armençon, 20 m. E by N of Auxerre. It is well-built, and contains a magnificent hospital. Its manufactures consist of glass, pottery, and hats; and it has a traffic in the wine of the vicinity, known by the name of Vin-de-Tonnerre.

TONNING, or Tonningen, a town of Denmark, in the duchy of Sleswick, on the r. bank of the Eyder, 30 m. SW of Sleswick, and 11 m. from the mouth of the river, in N lat. 45° 19'. Pop. 2,700. It has become, since the opening of the canal of Kiel, a place of activity: many vessels from the Baltic are also laid up here for the winter. The depth of water in the harbour is only 10 ft. at high water. The town has also an active traffic in corn, and other country-produce raised in the surrounding district called the Eiderstedt. In August, 1854, a railway was completed between this port and Flensburg, a distance of about 40 m. By this line, T., situated in the extreme W of Denmark, is brought into connexion with the Baltic. The line runs past Husum, containing about 5,000 inhabitants, principally occupied in cattle-feeding. From Husum a branch extends to Rendsburg, formerly a border-fortress between Holstein and Sleswick, where it forms a junction with the Kiel and Altona line, thus becoming the direct route to Hamburg. Until lately, Denmark and Northern Europe could only be reached from London by the long rail and sea route *via* Hamburg, or the still longer one *via* Hull to Copenhagen, a distance of 735 m., the voyage alone occupying 3½ days at least; while to reach Hamburg, the traveller being then distant from Copenhagen nearly 300 m., usually takes about 55 hours. On the other hand, from London to T. can be accomplished in from 30 to 34 hours; and from thence to Hamburg in six hours; while to Copenhagen it will be reduced to not more than 150 m.

TONNOMAIA, one of the smaller Friendly islands, 11 m. S of Annamuka.

TONNOY, a village of France, in the dep. of Meurthe, 12 m. SSE of Nancy. Pop. 600.

TONQUIN, Tunquin, or Tungquin, an extensive region of South-Eastern Asia, at present comprehended in the Cochin-Chinese empire. According to the best authorities, it lies between the parallels of 17° and 23° N lat. and the meridians of 101° and 108° E long.; and is 420 m. from N to S, and from 400 to 450 m. from E to W. It is bounded by Laos on the W; by China on the N; on the E by the Chinese sea; and on the S by Cochin-China. The name *Tonquin* signifies 'the Court of the east.' The country known in Europe by the name of Tonquin, is properly named Nuoc Annam. Cochin-China is sometimes named *Dangtrong*, which signifies 'the Internal kingdom;' while Tonquin is called *Dangnay*, or 'the External kingdom.' The region properly called Tonquin anciently formed part of the Chinese empire, but on the Mogul invasion of China in the 13th cent., the Chinese governors of the south took the opportunity of setting up the standard of independence, and in this manner several distinct kingdoms were created, the sovereigns of which, however, continued to acknowledge a nominal vassalage to the throne of China. Among these, the Tonquinese princes gradually assumed a greater degree of independence; and about 1553, are said to have subdued Cochin-China. For some time before and after the above era, the sovereigns of T. were assisted by a succession of hereditary prime-ministers named *chovas*, similar to the Mahratta peshwas, or 'the Mayors of the Palace' in France under the second dynasty. T. was finally conquered by the Cochin-Chinese sovereign, about 1800, and has ever since been ruled by a viceroy delegated from the seat of government.—It is divided into 11 provinces, four of which are named provinces of the east, west, north, and south, according to their bearings from the capital, which is situated in the middle of them. The others are Hu-Tien-Kuong, Hu-Huong-Hoa, which border on China; Hu-Then, Hu-Kasbang, Hu-Thank, Hu-Nghe, one part of which forms the frontier to Cochin-China, and another to Laos, and the province of Jen-Quang, which, properly speaking, is only a part of the province of the east under a different name. The province of Xunam forms the centre of the kingdom, and is situated in the middle of those named after the four cardinal points. Ba-king or Ki-ho, the capital, is situated about 40 leagues from the sea, upon the banks of the great river, San-Koy, in 21° N lat., and is said to contain about 40,000 inhabitants. Among the other towns are Hanring, Tranhanc, Causang, Vihoang, and Hunnam, in which latter town the Dutch had formerly a factory.

Physical features.] On the N and NW this country is mountainous. On the side next China, a ridge of high mountains runs along the frontiers of Kang-su and Yun-nan, in a winding course of more than 500 m., forming a natural boundary, and presenting only one pass which is fortified with a high wall. On the frontiers of Cochin-China and Laos, the country is also hilly, but not so much so as in other quarters. The lower part of T. is a level country, so much so that the principal part of the maritime provinces have been gained from the sea. The land still continues to gain upon the sea in this quarter, which may be accounted for by the immense quantities of mud and other substances brought down by the inundations from the mountainous districts, where the heavy rains carry off the soil. The western provinces are watered by seven large rivers, which unite their waters about 40 m. above the capital, and form a large lake or inland sea, whence the water again issues in numerous branches, and enters the bay of Tonquin by several mouths. The parent-stream of this vast body of water is the Hoti-Kiang or Song-koi of the Tonquinese, which rises in the NW part of Yunnan, and which, after running a S course of more than 500 m., falls into the gulf of T., forming a delta of which the Domea is the chief branch.

Climate and productions.] Although T. is situated within the tropics, yet its vicinity to the sea on the one hand, and being surrounded on the N and E with high mountains, render the temperature comparatively cool and pleasant. During the months of January and February, the cold is very perceptible; for then the winds blow from the northward, and, issuing from the frozen deserts of Siberia, bear the cold of the frigid zone into regions within the tropics. From September to March the air is pure and healthy; from March to September it is less so, particularly in the months of April, May, June, and July. During this part of the year the monsoons blow; though these winds are by no means so regular here as in the more westerly parts of India. The only difference of the seasons is wet and dry. The rains are most violent from May till August, because the sun then approaches the tropic of Cancer, and brings along with him, in his advance, dark clouds which descend in dreadful torrents. The inundations caused by the heavy rains are often sudden and destructive; in the upper and mountainous districts they sweep away rocks, soil, trees, and every thing within the reach of their current; but on the low lands the waters spread, and do no damage beyond covering the country with water, at which time a rich soil is deposited on the ground which greatly fertilizes it. On the low grounds the soil is rich and fertilized by the inundations, which render the cultivation easy; and the return is always such as to reward the husbandman richly for his labour. Where there is a constant supply of water—which is generally the case throughout the low country—two crops are raised in the year; towards the mountainous districts, where the land is poorer, only one crop is obtained. Rice is the chief article of produce, and the principal object of agriculture. This valuable plant, after having been made to shoot in troughs placed in the houses, is sown very thick, in well-watered beds, which are afterwards levelled as exactly as possible. It soon springs up; and, in six weeks, is transplanted stem by stem, into other fields. In about three months from the time it is transplanted, it is ripe, and fit to be cut. Maize is abundant, and different species of yams and leguminous plants. Sugar-cane also attains perfection in this country; but the Tonquinese are not so skilful refiners as the Cochin Chinese. T. produces all the fruits common to the tropical countries of India; bananas, plantains, pine-apples, guavas, passas, and sapadillas. Mulberry-trees are abundant, and are here valuable as they supply food for the silk-worm. The only European fruit-trees that thrive are the plum, the pomegranate, the citron, and the orange. The tea plant is said to be plentiful, but of an inferior quality.—Among the domestic animals are oxen and buffaloes, which are only used for labour; the people never kill or eat them, though not prohibited doing so by religious views. Hogs are very numerous, and poultry is plentiful. In the forests, stags, deer, and wild boars abound; tigers are also numerous, and of great strength, and wild elephants. White termites are very numerous and destructive.—Peacocks, quails, a species of partridge peculiar to the country, and a few others, form the leading members of the ornithological catalogue. Musquitoes are numerous and troublesome, particularly near the sea-coast; but during the northern monsoons, the country is in a great measure cleared of these tormentors.

Commerce, &c.] The Chinese at present conduct the principal part of the commerce of T. Formerly the Portuguese and Spaniards, and the Dutch and English, had a considerable share; but the trade of Europeans with this country has long been nearly annihilated. The articles of interior commerce are the nuts and fruit of the areca, fruits, cotton in the pod, spun cotton, linen, and cloth made from bark. The exports are raw or wrought silk, such as gauzes, grograms, or strong stuffs, linen made from the bark of trees, which is a very fine and neat material, different works in mother-of-pearl and ratan, all kinds of small furniture, ebony, ivory, tortoise-shell, cinnamon, copper, cotton, and various other valuable articles. Gold and silver occur in this country, and might be extensively wrought if the inhabitants were possessed of knowledge and industry sufficient to explore and conduct the mines. The only coin which they have in circulation, are a few pieces of gold which they obtain from China and Japan, and a copper coin from the former country. They receive in return for them exports from the Chinese, various medicinal drugs, fine tea, porcelain, fine silk stuffs, linen, sugar powdered and candied, wheat and barley flour, iron and copper kitchen utensils, iron, spices, cloves, nutmegs, cinnamon, dried preserves, hemp, flax, silk, wax, and cotton, glass-ware, iron-ware, small looking-glasses, telescopes, white glass bottles, and glass or paste beads. From Europe, or the European settlements in the East, they receive tanned ox-hides, which, as leather is very scarce, is in great demand, woollen stuff cloths, which are in much esteem, painted linens of India, broad linen, and cotton cloths. T. is most advantageously situated for commerce; and might supply a great part of Europe with articles much sought after from the East. It also borders upon the richest provinces of China, by which means it might command some of the finest productions of that vast empire, and in return, the manufactures of Europe might find a wider market.

Population.] The total pop. of T. has been computed by the missionary Bissachere at 18,000,000, but certainly this estimate greatly exceeds the reality, and he furnishes no account of the facts upon which he grounds his opinion. It is calculated, that one-tenth of the inhabitants of Lower T. live constantly on the water. The men are well-made and healthy; but there seems to be a general defect in the eye-sight, which is weak. They bear a strong resemblance, in physical characteristics, to the Chinese, from whom, no doubt, many of them are descended. The inhabitants of many of the mountainous districts are represented as little better than savages. The Tonquinese nation, though subjects of the same empire, are in fact composed of different races of men, whose habits and dispositions are widely dissimilar. The houses of the more civilized are raised about 7 ft. from the ground, and placed on posts; under the floor, which is made of canes, the cattle are lodged. The only fire-place, or kitchen, is a square of about four ft. covered with clay; there is no chimney, nor any other ventiduct to let out the smoke, which the people endure without any inconvenience, and even consider a benefit to their dwellings. Fishermen and those employed in the inland navigation live on the canals and rivers, in a manner similar to the Chinese, in cabins built on boats, and fastened to the shore. Polygamy is tolerated and common in T., but the woman who is first married, or whose parents are of the highest and wealthiest rank, always retains the precedence. Each province has its capital; a city of the first class, named *tow;* one of the second class, or *chew;* and one of the third class, or *hiuen.* Each burgh with the dignity of *xa*, conveying a certain local jurisdiction, has its limits marked upon a tablet, which is exposed to public view. Each xa is divided into villages; each village into quarters; and if the place is considerable, these are again divided into other subdivisions. Three villages compose a xa, and three xas a *tong*, the chief of which is elected by a plurality of voices of the xas. Each xa may indeed be considered as a kind of popular republic, in which women have no part, nor young men below 20 years of age: after that age the latter have a vote and right of suffrage in the assemblies. Criminal affairs are adjudged by the king only. Sentence of death is seldom passed; and the few executions which take place are always performed in the capital.—The more respectable Tonquinese follow the tenets of Confucius. They acknowledge one supreme deity, who directs, governs, and protects every thing; they believe in the immortality of the soul, and in transmigration. At an early period the Jesuits sent missionaries to T., and had made considerable progress, until being suspected of carrying on a secret political correspondence with the Cochin-Chinese, they were expelled. Missionaries were subsequently sent into T. by Louis XIV. under the character of commercial agents. Numerous obstacles, however, still present themselves to the propagation of the Christian religion in T., among which is the obligation imposed on every subject of contributing to the support and worship of the national idols, and to appear at certain festivals which have both a civil and religious character. The Jesuits tolerated their usages, from which it is not easy to detach the natives, but the court of Rome disapproved of this indulgence. See article Annam.

TONS, a village of Spain, in the prov. and 30 m. SW of Valencia, on the l. bank of the Xucar. Pop. 1,200.

TONSBERG, a port of Norway, situated on a bay of the Baltic, in N lat. 59° 15′, 43 m. SSW of Christiania. It is a very ancient town, but is ill-built, and consists only of wooden houses. Its harbour, though difficult of access, is capable of receiving large vessels. Its trade consists in the export of timber, and in a retail traffic with the surrounding country. In 1536, it was laid in ashes by the Swedes, and it has never recovered this disaster.

TONSE, or Tonsa, a river of Hindostan, which rises in the mountains of Bogilcund, and running E and NE, falls into the Ganges, 20 m. below Allahabad, after a course of 100 m.

TONYN'S ISLANDS, a cluster of small islands in the Eastern seas, near the SW coast of Celebes, in S lat. 5° 31′.

TOOM, or Toemverig, a parish of co. Tipperary, 6 m. N by W of Tipperary. Area 12,276 acres. Pop. in 1831, 3,566; in 1851, 3,521.

TOOM, a parish of co. Wexford. Area 5,980 acres. Pop. in 1831, 1,805; in 1851, 1,768.

TOOMAVARRA, a village in the p. of Aghnameadle, co. Tipperary, 3½ m. SW of Moneygall. Pop. in 1831, 790; in 1851, 419.

TOOME, or Toome-Bridge, a village in the p. of Duneane, co. Antrim, on the r. bank of the Bann, 5 m. W by N of Randalstown. A noble bridge of 9 arches here, takes the thoroughfare across the Bann, and gives the village its name.

TOOMIES, or Tomies, a mountain in co. Kerry, forming the N part of the western screen of the Lower lake of Killarney. It rises precipitously from the edge of the lake, is clothed over the base and the lower declivities with noble masses of wood, and shakes down from one of its sides the superb waterfall called O'Sullivan's Cascade.

TOPACURO, a small river of Brazil, in the province of Ceara, which runs N, and enters the Atlantic, between the Josavi and the Iguarasu.

TOPAHOT, a village of Asiatic Turkey, in the pash. of Sivas, 9 m. W of Inzghat.

TOPANFALVA, a town of Transylvania, on the l. bank of the Aranyos, 30 m. NW of Karlsburg.

TOPAS, a village of Spain, in the prov. and 15 m. N of Salamanca.

TOPAYOS. See Tapayos.

TOPCLIFFE, a parish and village of the N. R. of Yorkshire, situated on the banks of the Swale, 5½ m. SSW of Thirsk. Area of p. 15,565 acres. Pop. in 1811, 1,327; in 1851, 473. The church is a large building; and the village lying on the great N road, has, or rather had, several good inns.

TOPCROFT, a parish of Norfolk, 4½ m. E by S of St. Mary-Stratton. Area 1,875 acres. Pop. 477.

TOPDALS-ELV, a river of Norway, which rises in the bail. of Nedenaes; runs SE, and then SSW; and flows into a small bay to the E of Christiansand, after a course of 75 m.

TOPERTZ, Toporcz, or Toporec, a village of Hungary, in the comitat of Zips, 2 m. NW of Podolincz, at the foot of the Carpathian chain, on a small affluent of the Poprad.

TOPESFIELD, a parish in Essex, 5 m. S by W of Clare. Area 3,320 acres. Pop. in 1851, 1,051.

TOP-GALLANT ISLANDS, a cluster in Investigator group, South Australia, to the E of Flinder's Island. They are little more than mere rocks.

TOPINO, a river of the Papal states, which descends from Monte Santo in the N part of the delegation of Spoleto; runs first N; enters the delegation of Perugia; turns N; waters Foligno, and after a course of about 36 m. joins the Tiber, on the l bank, 6 m. SSE of Perugia. It receives the Calcignolo and Chiascio, on the r., and on the l. the Timia.

TOPL, Topla, Toplya, or Topolya, a river of Hungary, which descends from the Carpathian chain on the frontier of Galicia, in the comitat of Saros; runs first E past Bartfeld; then SSE; separates Kurmia from the comitat of Zemplin; traverses the S part of that comitat, divides the comitats of Ujhely and Mihaly; and after a course of about 84 m., throws itself into the Ondeva, on the r. bank, 9 m. N of Zemplin.

TOPLICZA, a village of Transylvania, in the comitat of Hunyad, 7 m. N of Deva, in the midst of mountains, on a small affluent of the Maros.

TOPLIKA, or Toplicze, a town of Civil Croatia, in the comitat and 9 m. SSE of Warasdin, on the l. bank of the Bednya. It has a glass-work. It is noted for its thermal springs, the *Aquæ-Jasæ* of the Romans.

TOPLITZ, Teplitz, or Teplice, a town of Bohemia, in the circle and 17 m. NW of Leitmeritz, at the foot of the Erzgebirge, in a fine valley, at an alt. of 728 Parisian ft. above sea-level. Pop. 2,750. It has a castle with fine gardens, and possesses manufactories of linen, woollen, and cotton fabrics. It is noted for its mineral baths, varying from 60° to 120° in temp. In the environs are quarries of limestone and coal-mines.

TOPLITZ, or Toeplitz, a village of Austrian Illyria, in the gov. and 32 m. SE of Laibach, and circle of Neustadtl, on a small affluent of the Gurk. It has two thermal springs.

TOPLITZ, or Neuhaus, a village of Styria, in the circle of Cilly, on the San. Pop. 60. It is noted for its thermal springs and baths.

TOPLITZA, a river of Turkey in Europe, in Servia, which has its source in the sanj. of Kruchovatz, on the confines of Albania, 12 m. N of Pristina; passes near Krattovo, and a village of the same name; thence flows NNE through a fertile plain, and in the environs of Kurchumba and Orkup; forms for a short distance the W frontier of Bulgaria; and after a course of 84 m. joins the Eastern Morava, on the l. bank, 9 m. W of Nissa.—Also a town in the sanj. of Semendria, 16 m. E of Valliero.

TOPLYA. See Tepl.

TOPOLI, a town of Russia in Europe, in the gov. and 80 m. E of Kharkov, and district of Koupiansk, on the r. bank of the Oskol.

TOPOLIAS. See Copais.

TOPOLNITZA, a town of Turkey in Europe, in Little Wallachia, in the district of Mehenditzi, 8 m. NE of Neu-Orsova, on a small river of the same name, an affluent of the Danube.

TOPONAR, a town of Hungary, in the comitat of Schimegh, 4 m. NE of Kaposvar, between two small affluents of the Kapos.

TOPOREZ, or Toporec. See Topertz.

TOPOROW, a town of Galicia, in the circle and 23 m. NNW of Zloczow, on an island of the Styr. It has two churches, a Catholic and a United Greek.

TOPOZERO, a lake of Russia in Europe, in the W part of the gov. of Arkhangel. It is 53 m. in length from NW to SE, and 9 m. in medium breadth. It forms several bays, especially on the NE, where an extensive peninsula runs in a NW direction. It contains several islands.

TOPRA-KALEH, or Turba-Kaleh, a town of Armenia, 69 m. ESE of Erzerum, on the slope of a hill, commanded by a rock on which is an earthen fortress, flanked with towers, and armed with several small pieces of cannon. It is ill-built, and contains about 500 families, chiefly Turkish and Armenian.

TOPSFIELD, a township and village of Essex co., in the state of Massachusetts, U. S., 20 m. N by E of Boston. It is hilly, and is watered by the Ipswich river. Pop. in 1840, 1,059; in 1850, 1,171.—Also a township of Washington co., in the state of Maine, 235 m. NE of Augusta, bordered on the N by a large lake. Pop. in 1840, 188; in 1850, 268.

TOPSHAM, a port of Devonshire, 4 m. SE of Exeter, at the head of the estuary of the river Exe, at its junction with the Clyst. Area 1,740 acres. Pop. in 1801, 2,748; in 1831, 3,184; in 1851, 3,377. The town, which consists of one long street of irregular breadth, is almost surrounded by the rivers Clyst and Exe, the latter of which here suddenly widens its bed, and is navigable for ships of several hundred tons. The quays and wharfs are spacious and convenient. A considerable trade is still carried on here, and steamers ply regularly between T. and London; but nearly the whole of the Exeter shipping now proceeds through the canal navigation direct to Exeter. There are bonded warehouses here for all foreign goods except wine and spirits, East India goods, and tobacco. Ship-building and the manufacture of ropes and chain cables are carried on to some extent.

TOPSHAM, a township, and with Wiscasset and Warren, capital of Lincoln co., in the state of Maine, U. S., 28 m. SSW of Augusta, bounded on the E by the Kennebec, and on the S and W by the Androscoggin river, and intersected by the Kennebec and Portland railroad. Pop. in 1840, 1,883; in 1850, 2,010.—Also a township of Orange co., in the state of Vermont, 23 m. SE of Montpelier. It has a diversified surface, with little fertility, and is watered by the head streams of Wait's river. Pop. in 1840, 1,745; in 1850, 1,668.

TOR, or Tur, a port of Arabia, situated near the head of the gulf of Suez, in N lat. 28° 14', E long. 33° 41'. It was in former times a place of great importance, in the trade between Syria and India; but since Suez became the emporium of the Red sea, T. has sunk into a village; and only such vessels as are prevented by strong N winds from reaching Suez, unlade their cargoes at T., whence they are conveyed by land to their destination. The harbour is formed by a reef of coral rocks, and to the northward by a low point of land, on which is placed a beacon. The place is inhabited by Greeks and Bedouin Arabs; and the monks of Sinai have a convent here. This is the nearest point from which a traveller can proceed from the gulf to Mount Sinai, a journey of 2 days on camels. The width of the gulf opposite T. is 17 m. The best water on the coast of the Red sea is obtained here.

TOR, a village of Spain, in the prov. and 90 m. NE of Lerida.

TORA, a village of Egypt, on the Nile, 8 m. S of Cairo.—Also a village of Spain, in the prov. and 50 m. NE of Lerida. Pop. 900.—Also a town of Naples, in the prov. of Terra-di-Lavoro, 24 m. NE of Gaeta. Pop. 1,200.

TORAK, a town, or rather two contiguous villages of Hungary, in the com. of Torontal, on the Bega.

TORAL-DE-LOS-GUZMALA, a village of Spain, in the prov. and 27 m. S of Leon, on the r. bank of the Esla. Pop. of partido, 2,000.

TORANO, a village of Naples, in Calabria-Citra, 15 m. NNE of Cosenza.—Also a village of Abruzzo-Ultra 1ma., 12 m. NNE of Teramo.

TORBALA, a town of Hindostan, in the prov. of Lahore, on the l. bank of the Dur river, near its junction with the Indus, in N lat. 34° 12'.

TORBALI, a town of Asiatic Turkey, in the sanj. and 52 m. SW of Boli, situated in a valley between two high mountains, on the Terekli-su. Pop. 2,800.

TORBAY, a bay on the E coast of Newfoundland, in N lat. 47° 48'.—Also a town and bay on the SE coast of Nova Scotia, in N lat. 45° 8'.—Also a bay on the coast of W Australia, in S lat. 35° 5'.

TOR-BAY, a picturesque and commodious bay on the coast of Devon, 5 m. NE of Dartmouth. It is about 12 m. round, and is formed by two capes, about 4 m. apart. An almost semicircular recess, it is a secure and general rendezvous for vessels in westerly winds. Berry-head, the W part of the bay, is in N lat. 50° 44', W long. 28° 14'. Along the W and SW coast, the shores are composed of red sandstone, the strata of which have been much disturbed and hollowed out by the action of the sea. On the E side this rock is covered by argillaceous shale, containing fragments of encrinites. The hills immediately around the N shores are composed of transition limestone, varying in colour, and containing numerous remains of shells and madrepores. From the variety of its tints, and the high polish of which it is susceptible, this marble is much used for ornamental purposes. In the bay the Prince of Orange and Marshal Schomberg landed in 1688.

TORBEC, a town of Hayti, in the dep.-du-Sud, 5 m. WSW of Cayes.

TORBOCK, or Tarbock, a township in Huyton p., co.-palatine of Lancaster, 3½ m. S by W of Prescot. Area 2,447 acres. Pop. in 1851, 681.

TORBOLE, a village of Austrian Lombardy, in the prov. and 6 m. SW of Brescia.—Also a village of the Tyrol, 7 m. SW of Roveredo.

TOR-BRYAN, a parish of Devonshire, 4 m. SW by S of Abbot's Newton. Area 2,010 acres. Pop. in 1831, 257; in 1851, 229.

TORCAZ, a village of Spain, in the prov. and 24 m. E of Madrid. Pop. 700.

TORCE, a village of France, in the dep. of Mayenne, 4 m. E of Sainte-Suzanne.—Also a village in the dep. of Sarthe, 6 m. N of Montfort-le-Rotrou.

TORCE, a town of Hindostan, in the prov. of Bahar, the capital of a small district of the same name, in N lat. 23° 42′.

TORCELLO, a town of Austrian Lombardy, in the prov. of Venice, situated on an island of the same name, in the marshy district called the Lagunes, 6 m. NE of Venice. Pop. 9,000. It is extremely unhealthy, and many of the inhabitants leave the town during the summer season. It is the see of a bishop, whose cathedral is the only building of consequence in the place.

TORCHIARA, a village of Naples, in Principato-Citra, 12 m. NW of Il Vallo. Pop. 580.

TORCOLA, an island of the Adriatic, near the S coast of Lesina, in N lat. 43° 4′. It is about 3 m. in length, low in surface, and occupied by a few shepherds whose flocks find scanty herbage upon it.

TORCY-LE-GRAND, a village of France, in the dep. of Seine-Inferieure, 3 m. E of Longueville, on the Arques. Pop. 700.

TORCZYN, a town of Russia, in the gov. of Volhynia, 41 m. N by W of Brody in Galicia.

TORDEHUMOS, a town of Spain, in the prov. and 21 m. WNW of Valladolid. Pop. 1,200.

TORDERA, a town of Spain, in the prov. and 27 m. S of Gerona. Pop. 1,440.

TORDESILLAS, a town of Spain, in the prov. and 18 m. SW of Valladolid, on the r. bank of the Douro, over which there is a fine bridge of 10 arches. Pop. 3,500. It is pleasantly situated, and for a Spanish town, well built. Its chief articles of manufacture are leather and coarse woollens.

TORDOMAR, a village of Spain, in the prov. and 27 m. SSW of Burgos, on the r. bank of the Arlanza. Pop. 570.

TORDOUET, a village of France, in the dep. of Calvados, 4 m. NW of Orbec. Pop. 1,200.

TORELLA, a town of Naples, in the Principato-Ultra, 3 m. WNW of San-Angelo. Pop. 3,300.—Also a town in the prov. of Sannio, 9 m. NW of Campobasso.

TORELLO (San-Feliu-de), a town of Spain, in the prov. and 39 m. NNW of Barcelona, at the confluence of the Ges and the Ter. Pop. 2,000. It has manufactories of woollens and cottons, paper, turnery-ware, and combs.

TORENO, a village of Spain, in the prov. and 21 m. NE of Villafranca, on the Sil. Pop. 400.

TORGAU, a town of Prussian Saxony, in the reg. of Merseburg, on the l. bank of the Elbe, 46 m. NW of Dresden, and 65 m. S by W of Berlin. Pop. 9,000. Its fortifications, formerly limited to a castle called Hastenfels, have been extended of late years, and now embrace the town itself. It has manufactories of woollens, leather, and soap; and a considerable traffic in wood and grain. T. has been the scene of several sanguinary conflicts, the most remarkable of which was the victory obtained here on the 11th November, 1760, over the Austrians, by Frederick II of Prussia.

TORGEISKOI, a village of Asiatic Russia, in the gov. of Irkutsk, 24 m. NNW of Merchinsk.

TORGELOW, a town of Prussian Pomerania, on the river Ucker, 9 m. N of Passewalk. Pop. 1,000.

TORGET, a small island of Norway, in N lat. 65° 30′. Its cliffs attain an alt. of 2,000 ft.

TORGOVITZA, a town of Russia, in the gov. of Kiev, 45 m. N of Olviopol.—Also a town in the gov. of Volhynia, 18 m. SSW of Doubno, on the r. bank of the Ikva.

TOR-HEAD, a headland of co. Antrim, on the S side of Murlough bay, 5 m. N of Cushendun. It is the point of the Irish mainland nearest to Scotland, being directly opposite the Mull of Kintyre. It consists of mica slate, with subordinate beds of limestone, sienite, and felspar porphyry, and may be regarded as a projection of the cliffs of Cushleak, being immediately overhung by the hill of Carnlea, whose summit has an alt. above sea-level of 1,250 ft. The scenery of the headland itself and of the bay which it screens, is singularly grand.

TORIESDALE HEAD, a cape on the N coast of Scotland, in N lat. 58° 30′.

TORIJA, a village of Spain, in the prov. and 15 m. NE of Guadalaxara. Pop. 760.

TORITTO, a town of Naples, in the Terra-di-Bari, 15 m. NNE of Altamura. Pop. 2,550.

TORJOK, a considerable town of Russia, in the gov. and 42 m. WNW of Twer, on the r. bank of the Tvertza. Pop. 15,000. It is a neat and regularly built place, with active manufactories chiefly of leather and leather articles, and a considerable trade in the agricultural produce of the surrounding district. A recent traveller describes the appearance of the country between T. and the city of St. Petersburg, as being dreary and monotonous in the extreme. After the first 10 or 12 versts beyond the capital, he says, "we entered a tract of forest which stretched with few intervals for more than 100 m. The whole distance indeed exhibits little but a succession of bleak open country, and thick forest. The road runs generally in a straight line, and one proceeds for miles together along a dead flat, without a human habitation; on each side, a boggy space of 50 or 100 yds. wide is kept clear of trees, beyond that lies an impenetrable mass of birch and fir wood growing up so thickly that the production of fine timber is impossible. The only part of the country through which we passed where the view is at all attractive, is in the immediate neighbourhood of Valdai, a small town about 200 versts from T."

TORKALI, a town of Russian Georgia, in Daghistan, 15 m. NW of Tarki, on a stream of the same name.

TORKINGTON, a township in Stockport p., co.-palatine of Cheshire, 3½ m. SE of Stockport. Area 820 acres. Pop. in 1831, 284; in 1851, 358.

TORKSEY, a parish in Lincolnshire, 7 m. S by E of Gainsborough, on the E bank of the Trent, including the township of Brampton. Area 3,170 acres. Pop. in 1831, 484; in 1851, 438.

TORLA, a village of Spain, in the prov. and 39 m. NNE of Huesca.

TORMANTOS, a village of Spain, in the prov. and 36 m. WNW of Logrono. Pop. 600.

TORMARTON, a parish in Gloucestershire, 4 m. SE by E of Chipping-Sodbury. Area 2,645 acres. Pop. in 1831, 402; in 1851, 463.

TORMENTOS, a volcanic mountain of Guatimala, in the district of Amatitlan, the highest of a group of three volcanoes. It is about 2 leagues distant from the village of Apacaga.

TORMES, a river of Spain, which rises among the mountains in the S of the prov. of Salamanca; crosses Leon; and falls into the Douro on the borders of Portugal, after a course of 105 m. The first half of its course is from S to N; the latter, from E to W Its banks were the scene of the battle of Salamanca, fought on the 22d of July, 1812.

TOR-MOHAM, a parish of Devonshire, 2½ m. W of Torbay. Area 1,560 acres. Pop. in 1851, 11,474.

TORMORE, an island of co. Donegal, situated within 3 furl. of the nearest part of the mainland, 1 m. NE of Toralaydan. It lifts its summit to the alt. above sea-level of about 500 ft.

TORNA, a palatinate in the NW of Hungary, situated between the com. of Zips and Aba-Ujvar.

and watered by the Bodva. It is hilly almost throughout, so that its wealth is nearly limited to its iron mines and pasturage. It has an area of 172 geog. sq. m., and about 30,000 inhabitants.—Its capital, of the same name, is a small place situated on the river Torna, 19 m. SW of Kaschau. Pop. 2,200.

TORNAVACAS, a town of Spain, in the prov. and 22 m. NNE of Caceres. Pop. 1,400. It gives name to a mountain-range, which extends N, from the Tagus, along the borders of Portugal, till it joins the Sierra-de-Guadarrama.

TORNEA, a small but remarkable port of Russian Finland, at the N extremity of the gulf of Bothnia, in N lat. 65° 50′ 50″, E long. 24° 6′ 15″, 1,735 versts from St. Petersburg. It stands on a small island in the river Tornea, and having been built by order of government, is regular in its streets; but the few wooden houses are widely separated. The pop. does not exceed 700. It is a central spot for the exports and imports into a wild and thinly peopled country. Its exports consist of timber, fish, rein-deer, hides, furs, and tar; the imports of salt, snuff, tobacco, groceries, and spirits. On the opposite side of the river is the thriving little Swedish town of Happaranda. In June the sun is here visible above the horizon at midnight. About the 23d of this month it is seen to decline to the NW, and verge more and more to the exact N, until midnight, after which it commences an ascending course to the NE. In 1736 and 1737, T. was honoured with a visit from Maupertius and other French *savans*, who, in company with the Swedish astronomer Celsius, made many observations to ascertain the exact figure of the earth. Remote as it is, this place has not always escaped the ravages of war, having been taken by the Russians in 1715, and in 1809; and by the English in 1854.

TORNEA, a river of Sweden, which issues from Lake Kiopis, in N lat. 69°; flows SE; forms the boundary between Russian and Swedish Lapland, and falls into the gulf of Bothnia, below Tornea, after a rapid course of 250 m. Its principal affluents are the Lainio and the Muonio.

TORNEA-LAPPMARK, the most northern of the six provinces of Swedish Lapland, lying between the river Tornea, Lulea Lappmark, and Norwegian Lapland. The part of it to the E of Tornea was ceded to Russia in 1809.

TORNESE (CAPE), a headland of Greece, at the NW extremity of the Morea, opposite the island of Zante, in N lat. 37° 56′.

TORNIMPARTE, a town of Naples, in Abruzzo-Ultra 2da, 7 m. SW of Aquila. Pop. 1,500.

TORNYA, a village of Hungary, in the com. of Czanad, 12 m. NW of Arad, on the Vena-Szaraz.

TORO, a town of Naples, in the prov. of Sannio, 15 m. SE of Molise. Pop. 2,400.—Also a city of Spain, in the prov. and 15 m. E of Zamora, situated on the Douro, over which it has a bridge of 22 arches. Its bishopric, one of the most ancient in the kingdom, has been for some time transferred to Zamora. It is regularly built, with wide but dirty streets, and contains several churches and convents, a theatre, and barracks. Pop. 9,500, whose chief employment is cultivating wine and corn, and linen-weaving.—Also a small island in the Mediterranean, near the SW coast of Sardinia, in N lat. 38° 52′.

TORO (MONT), a mountain in the centre of the island of Minorca, 9 m. NW of Mahon.

TORO (SAN-ANTONIO-DEL), a city of New Granada, in the prov. of Popayan, 30 m. N of Cartago.

TOROCZKO, a town of Transylvania, in the com. and 14 m. SW of Thorenburg. Pop. 1,900.

TOROGAY, an Hebridean islet of probably 2 m. in circumf., between N. Uist and Bernera.

TOROK-BECSE, a town of Hungary, in the com. of Torontal, 25 m. NNE of Peterwardein, on the l. bank of the Theiss. It has a quay extending along the bank of the river, and affording facilities for shipping corn, timber, tobacco, and other produce.

TOROK-SANKT-MIKLOS, a town of Hungary, in the com. and 30 m. S of Heves. Pop. 8,600.

TORON, a town of European Turkey, situated on a neck of land which projects into the Archipelago, between the gulfs of Monte-Santo and Cassandro, 70 m. SE of Saloniki.

TORONTAL, a comitat or county in the S of Hungary, including the western part of the Banat of Temesvar, lying along the E bank of the Theiss, and intersected by the Bega. It has an area of 2,800 sq. m., and a pop. of 343,000. About 660 sq. m. are occupied by marshes; the rest of the surface is also level, and of great fertility. Maize, rice, wine, melons, fruit, flax, and silk are raised. —The cap. is Nagy-Becskerek.

TORONTO, formerly YORK, the capital of Upper Canada, or Canada-West, in the township and co. of the same name, on the NW coast of Lake Ontario, 315 m. WSW of Montreal, in N lat. 43° 49′, W long. 79° 21′, at an alt. of 342 ft. above sea-level. It is regularly laid out, with streets running at right angles, and generally spacious. It has been much improved within the last few years, yet it still possesses only one really handsome street lined with well-filled shops and stores. The site extends about 1½ m. along the harbour. The public edifices are an Episcopalian and a Roman Catholic cathedral, a college, an assemblage of red-brick buildings called the government-house, the legislative chambers, a sessions-house, two gaols, numerous stores and buildings for the purposes of government, normal schools, a university, an observatory, and public markets. The barracks are situated to the W of the town. A battery and two blockhouses protect the entrance of the harbour. The harbour is nearly circular, and formed by a narrow peninsula, stretching W in an oblique direction for about 6 m., and terminating in a curved point nearly opposite the garrison, thus enclosing a beautiful basin about 1½ m. in diameter, capable of containing a great number of vessels, and at the entrance of which ships may lie in safety during the winter. The peninsula is in several places not more than 60 yds. in breadth, but widens towards its extremity to nearly 1 m. It is principally a bank of sand, with little grass upon it; and is so low in surface, that the expanse of Lake Ontario is seen over it. The termination of the peninsula is called Gibraltar-point, where a block-house has been erected. The E part of the harbour is bounded by an extensive marsh, through part of which the river Don runs, before it discharges itself into the basin.—In 1793, the spot on which T. stands presented only one solitary Indian wigwam. In the ensuing spring, the ground for the future metropolis of Upper Canada was fixed upon, and the buildings commenced under the superintendence of General Simcoe, then lieutenant-governor. In the space of five or six years it became a respectable place, and rapidly increased in importance. In 1817 the pop. was 1,200: in 1830, 2,860; in 1842, 15,336; in 1852, 30,763. The adjacent country for several miles round is in a high state of cultivation. T. was captured by the American army on the 27th April 1813. They held it, however, but a few days; but in that time the government-house, and all the public buildings and stores, were burnt. On the union of the two provinces, the seat of government was removed to Kingston. It suffered severely from fire in 1849.—A railway is now in progress between T. and Montreal, 345 m. in length, at an estimated cost of £3,000,000; and between T. and Tarnia, 172 m. in length, at an estimated cost of £1,376,000. These great lines are component parts of the Grand Trunk railway of Canada.—The mean temp. of the year at T. is 44° 4′; of winter, 26° 4′; of summer, 63° 8′. Recent investigations have led to the inference that opposite conditions of weather prevail simultaneously in the same parallels of latitude under different meridians; that Europe and America, in particular, usually present such an opposition, so that a severe winter here corresponds to a mild one there, and *vice versâ;* and recent theories of the distribution of heat on the surface of the globe profess to furnish the explanation. From a series of twelve years' observations Colonel Sabine has shown that, after allowance has been made for the elevation of T. above the sea, every month of the year there is colder than the normal temperature of the same month in the same parallels; that the cold reaches its extreme in February, when it exceeds 10° of Fahr.; and that on the average of the whole year it is little less than 6°.

TOROPA, a river of Russia, in the gov. of Pskov. which rises in Lake Losno, 18 m. NE of Toropetz; runs NW, and then SSW, passing through Lake Solomino; and flows into the Duno, on the r. bank, 30 m. S of Toropetz, after a course of 80 m.

TOROPALCA, a settlement of Bolivia, 56 m. SSW of Potosi.

TOROPETZ, a town of Russia, in the gov. of Pskov, situated on the river Toropa, at the point where it issues from Lake Solomino, 245 m. S of St. Petersburg. It has a cathedral and several churches and monasteries. Pop. 10,000. The principal employment of the inhabitants is the tanning of leather, and the forwarding of merchandise partly to Poland, partly to the interior of the empire, and to Riga by the Toropa, and the Dwina or Duna.

TOR-ORSAIA, a town of Naples, in the prov. of Principato-Citra, 15 m. ESE of Il Vallo. Pop. 2,500.

TOROS, a town, or group of hamlets, in Tartary, on a stream of the same name, an affluent of the Syr-derra, in N lat. 66° 22′.

TOROSAY, a parish on the E coast of the island of Mull, Argyleshire. Pop. in 1851, 1,361.

TOROX, a town of Spain, on the coast of Granada, 25 m. E of Malaga.

TORPENHOW, a parish in Cumberland, 8 m. NE by N of Cockermouth, comprising the townships of Bewaldeth-with-Snittlegarth, Blennerhasset-with-Kirkland, Bothel-with-Threapland, and Torpenhow-with-Whitrigg. Area 9,670 acres. Pop. in 1831, 1,032; in 1851, 1,175.

TORPES, a village of France, in the dep. of Doubs, 9 m. SW of Besançon, on the r. bank of the Doubs.—Also a village of the dep. of Saone-et-Loire, 4 m. SE of Pierre.

TORPHICHEN, a parish of Linlithgowshire, 4¼ m. SSW of Linlithgow. Pop. in 1851, 1,356.

TORPOINT, a chapelry in St. Anthony p., co. Cornwall, 3 m. W of Devonport.

TOR-QUAY, a picturesque watering-place and chapelry, in Tor-Moham p., Devon, 21 m. S of Exeter. It is situated in the most northerly cove of Torbay, and is sheltered on three sides by lofty hills on the declivity of which are numerous detached villas and terraces of houses. The greater part of the town, however, is built on the strand, and consists chiefly of lodging-houses for the accommodation of visitors in the bathing season. The town is open to the S, and is at the same time protected from the N and NE winds by the hills encompassing it. From the extreme salubrity and mildness of the climate, together with the picturesque scenery around it, this place is fast rising into estimation even as a winter residence. The mean temp. of the six winter months, from November to April inclusive, during the seasons 1829-30, 1830-31, and 1831-32, was 46° 43′. The mean monthly range of temp. 26° 1′; the mean daily range about 4°. The variation between the minimum of the night and the temp. of the following morning at 8 o'clock, during the entire season, is 3° 15′; and during the three winter months only 2° 29′. The prevailing winds are W and SW. Among the tender exotics which are hardy in the gardens at Tor-quay are the *Agave Americana, Cassia Capensis, Citrus Medicus, Laurus Camphora, Yucca Aloifolia, Y. gloriosa,* &c. The citron has produced fruit in the open air at Tor-quay measuring 24¾ inch. in circumference. The myrtle flourishes luxuriantly throughout the entire district: it attains a large size, and is proverbial for its longevity. T. was formerly but an insignificant fishing-town: the first great improvement was the erection of a pier, commenced in 1804 and finished in 1807, after which another was constructed, forming a basin 500 ft. long and 300 ft. broad. T. maintains coasting-vessels between the great ports; and has a regular communication with Portsmouth and Plymouth, by steam-boats, and with the southern watering-places by daily stages and by railway. There are here two Episcopalian chapels, a Roman Catholic, Wesleyan, Baptist, and two Independent chapels.

TORQUEMADA, a town of Spain, in the prov. and 15 m. NE of Palencia, on the r. bank of the river Pisuerga. Pop. 2,300. It was the birth-place of the infamous inquisitor, Torquemada.

TORQUILLA a river of South America, in the province of Darien, which enters Tarena just before the latter enters the sea.

TORRA, a village of Zanguebar, 120 m. SW of Magadoxo, on the coast.

TORRALBA, a town of Spain, in the prov. and partido and 15 m. NW of Cuenca, in low ground, commanded by a mountain, on which are the remains of a fort. Pop. 800.—Also a town in the prov. of Navarra, and partido of Estillo, 39 m. SW of Pamplona, between two streams. Pop. 395.—Also a town in the prov. of Soria, partido and 9 m. NE of Burgo-de-Osma, in a fine valley, watered by the Abion. Pop. 300. It has carding-mills.

TORRALBA-DE-CALATRAVA, a town of Spain, in the prov. and partido and 11 m. NE of Ciudad-Real. Pop. 3,016. It has a custom-house, a public granary, and a parish church, and is generally well built. Blond and lace are its chief articles of manufacture.

TORRALBA-DE-OROPESA, a town of Spain, in the prov. and 69 m. W of Toledo, and partido of Puente-del-Arzobispo, on a plateau commanding a magnificent view of the Oropesa mountains. Pop. 300. It has an hospital.

TORRANCE, a village in the p. of Campsie, Stirlingshire. Pop. 672.

TORRAO, a town of Portugal, in the prov. of Estremadura, comarca and 48 m. ESE of Setubal, in a flat and fertile country, watered by an affluent of the Caldão. Pop. 1,800. It has two hospitals, and two convents.

TORRE, a river, or rather torrent, which has its source in Austrian Lombardy, in the prov. of Usino, on the S side of Monte Musi; runs first W to the hamlet of Musi, where it receives the Rio Branco, then SSE to the junction of the Cornapo, Mattina, and Natisone; enters Illyria; and after a total course of about 42 m., throws itself into the Isonzo on the r. bank, a little above the hamlet of Turriacco, and 6 m. SSW of Gradisca.—Also a town of Naples, in the prov. of Calabria-Ultra, district and 22 m. SSW of Catanzaro, at the foot of a mountain, near the l. bank of the Ancinale, in an unhealthy locality. Pop. 1,600. It suffered severely from an earthquake in 1783.

TORRE-ALHAQUIME, a town of Spain, in the prov. and 69 m. NE of Cadiz, and partido of Olvera, on the r. bank of the Guadalete. Pop. 373.

TORRE-BLANCA, a town of Spain, in the prov. and 24 m. NE of Castellon-de-la-Plana, and partido of Albocacer, near the Mediterranean. Pop. 1,556.

TORRE-BLASCO-PEDRO, a town of Spain, in the prov. and 15 m. NNE of Jaen, and partido of Baeza, in a delightful plain, at the confluence of the Guadalquivir and Guadalimar. Pop. 480. It has several oil-mills, a soap manufactory, and a tile-work. Its oil is noted as the best in Andalusia. About ¾ of a mile to the N on the r. bank of the Guadalimar, are the ruins of the ancient *Castulum*.

TORRE-BRUNA, a town of Naples, in the prov. of Abruzzo-Citra, district and 20 m. SSW of Il Vasto, and cant. of Gelenza, on a high hill. Pop. 720. The environs are fertile.

TORRE-CAMPO, a town of Spain, in the prov. and 30 m. N of Cordoba, and partido of Pozoblanco, in a plain, near the l. bank of the Guadalmez. Pop. 2,518. It has a parish church, a custom-house, a public granary, and several convents.—Also a town in the prov. and partido and 5 m. SW of Jaen, in a plain. Pop. 3,748. It has a large square in which is an ancient fortress, and contains a parish church, a custom-house, two hospitals, and several elementary schools.

TORRE-CHICA, a fortress of Algiers, on a peninsula projecting into the bay of Sidi-Ferush, and to the W of Algiers.

TORRECILLA-DE-ALCANIZ, a town of Spain, in the prov. of Teruel, and partido of Alcaniz. Pop. 1,296. It has a parish church, a custom-house, and a public granary.

TORRECILLA-DE-LA-ABADESA, a town of Spain, in the prov. and 24 m. SW of Valladolid, and partido of Mola-del-Margues, near the r. bank of the Duero. Pop. 350. It has a tile-kiln.

TORRECILLA - DE - LA - ORDEN, a town of Spain, in the prov. and 39 m. SSW of Valladolid, and partido of Nava-del-Rey, on an elevated and arid tract. Pop. 1,315. It has manufactories of blankets, and other woollen fabrics.

TORRECILLA-DE-LOS-ANGELES, a town of Spain, in the prov. and 54 m. NNW of Caceres, on the S side of the Sierra-de-Gata. Pop. 428.

TORRECILLA-DE-LOS-CAMEROS, a judicial partido and town of Spain, in the prov. of Logrono. The partido comprises 51 pueblos. The town is 18 m. SSW of Logrono, on the slope of Monte Sarrada, on the Iruega, which is here crossed by a fine bridge. Pop. 2,027. It is well-built, and has four abundant fountains, three parish churches, one of which is surrounded by a lofty tower, a convent, and an alms-house. It possesses manufactories of cloth. In its vicinity are several thermal springs. The Iruega abounds with fish. The adjacent mountains are said to contain silver and Armenian boll.

TORRECILLAS - DE - LA - TIESA, a town of Spain, in the prov. and 36 m. E of Caceres, and partido of Trajillo, in a level locality. Pop. 583. The climate in summer is dry, and the soil arid.

TORRECILLA-SOBRE-ALESANCO, a town of Spain, in the prov. and 24 m. WSW of Logrono, and partido of Najera. Pop. 279.

TORRECUSO, a town of Naples, in the prov. of the Principato-Ultra, district and 23 m. NNW of Avellino, and cant. of Vitolano, on a hill. Pop 1,700. It has two churches, an hospital, and an alms-house.

TORRE-DE-DONA-CHAMA, a town of Portugal, in the prov. of Tras-os-Montes, comarca and 42 m. N of Torre-de-Moncorvo, on a plateau, near the base of a hill on which are the remains of an old tower.

TORRE-DE-DON-MIGUEL (LA), a town of Spain, in the prov. and 54 m. NNW of Caceres, at the foot of the Sierra-de-Gata, on a height. Pop. 1,616. It has a healthy climate, and is well supplied with good water. Wine and oil are cultivated in the environs.

TORRE-DE-EMBESORA, a town of Spain, in the prov. and 24 m. NNW of Castellon-de-la-Plana, and partido of Albocacer, in a mountainous locality.

TORRE-DE-ERGAZ, or TORRE ORGAZ, a town of Spain, in the prov. and partido and 9 m. S of Caceres, near the r. bank of the Solor. Pop. 880.

TORRE-DE-ESTEBAN-AMBRAN (LA), a town of Spain, in the prov. and 30 m. NW of Toledo, and partido of Escalona, on the r. bank of a small affluent of the Alberche. Pop. 1,257. It has a fine church, five chapels, and a convent. The environs are noted for their grapes.

TORRE-DE-JUAN-ABAD, a town of Spain, in the prov. and 45 m. SE of Ciudad-Real, and partido of Villanueva-de-los-Infantes, a little to the N of the Sierra-Morena. Pop. 1,377. In the environs are mines of lead and antimony.

TORRE-DE-LA-META, a village of Spain, in the prov. and 30 m. SSW of Alicante, and partido of Orchuela, on the Mediterranean. Pop. 176. It has a roadstead defended by a tower. A little to the W is an extensive salt lake.

TORRE-DE-LAS-GUARDAS, a fortress of Spain, in the prov. and 24 m. SE of Almeria, on the Mediterranean, near Cape Gata. It has a jasper quarry.

TORRE-DEL-COMPTE, a town of Spain, in the prov. and 75 m. SE of Zaragoza, and partido of Valderrobres, on a height, the base of which is washed by the Matarranã. Pop. 566. It has several oil and silk-mills.

TORRE-DE-LES-MANSANES, or TORREMANZANES, a town of Spain, in the prov. and 24 m. N of Alicante, and partido of Jijona. Pop. 1,029.

TORRE-DEL-GRECO, a town of Naples, in the prov. and district and 8 m. SE of Naples, on the E coast of the gulf of Naples, and near the SW base of Vesuvius. Pop. 17,502. It has two churches, both of modern structure, three convents, and a large hospital, and numerous houses of elegant architecture. The inhabitants employ themselves in fishing. The environs are noted for their fruit and wine. This town derives its name from an ancient tower in its vicinity, and from the quality of the wines of the locality, resembling those grown in the Greek islands. It was almost entirely destroyed by an eruption of Vesuvius in 1794.

TORRE-DELL'ANUNZIATA, a town of Naples, in the prov. and 13 m. SE of Naples, district and 5 m. NNW of Castel-a-Mare, near the S base of Vesuvius, in a plain, on the E side of the gulf of Naples. Pop. 12,045. It has two churches, a convent, and a port. It possesses a royal factory of arms, and a paper-mill. The trade consists chiefly in corn. This town derives its name from a tower, erected for the defence of the coast by Alphonso I.

TORRE-DELLA-PADULA, a town of Naples, in the prov. of Torre-Otranto, district and 19 m. ESE of Gallipoli, and cant. of Ruffano, at the foot of a hill. Pop. 570. It has a convent.

TORRE-DELLE-NOCELLE, a town of Naples,

in the prov. of the Principato-Ultra, and district of Avellino, 5 m. from Montefusco, on a hill. Pop. 1,275. It has a convent, and an alms-house.

TORRE-DEL'-MANGANO, a town of Austrian Lombardy, in the prov. and district and 5 m. NNW of Pavia, in a fertile locality. Pop. 450.

TORRE-DEL-VULGO (LA), a village of Spain, in the prov. and 15 m. NNE of Guadalajara, near an affluent of the Henares. Pop. 259.

TORRE-DEMBARRA, a town of Spain, in the prov. and 9 m. ENE of Tarragona, and partido of Vendrell, 3 m. from the shore of the Mediterranean. Pop. 1,868. It has manufactories of fishing-tackle, cordage, and soap, and distilleries of brandy.

TORRE-DE-MIGUEL-SESMERO, a town of Spain, in the prov. of Badajoz, and partido of Olivenza, in a flat locality. Pop. 1,378.

TORRE-DE-MONCORVO, a town of Portugal, in the prov. of Tras-os-Montes, 15 m. E of S. Joao-da-Pesqueira, on the slope of Mont-Roberedo, between the Sabor and Duero. Pop. 1,629. It has bastioned ramparts, and is defended by a square fortress. Its principal public buildings are a fine church, a convent, an alms-house, and a Latin school. Silk fabrics are woven here; and the environs are noted for their flax and hemp. This town was peopled in 1216 by Sancho II.

TORRE-DE-MORMOJON (LA), a town of Spain, in the prov. and partido of Palencia, at the foot of a mountain on which is a ruinous fortress. Pop. 708. It has an active trade in wine and in wool.

TORRE-DE-PASSERI, a town of Naples, in the prov. of Abruzzo-Ultra, district and 14 m. S of Civita-di-Penne, on the l. bank of the Pescara. Pop. 1,150. It has three churches, and contains manufactories of pottery and of dye-stuffs.

TORRE-DE-PENAFIEL, a village of Spain, in the prov. of Valladolid, partido and 8 m. SE of Penafiel, near the l. bank of the Duraton, in a woody and insalubrious locality.

TORRE-DE-VALANA (LA), a town of Spain, in the prov. and 24 m. N of Guadalajara, and partido of Tamajon, at the entrance to a valley, 1½ m. from the Rio-Sorbel. Pop. 508.

TORRE-DI-ANNIBALE, a small port of Naples, on the E coast of the prov. of Calabria-Citra. It is supposed to mark the spot at which Hannibal embarked on his return to Africa.

TORRE-DI-BELVICINO, a village of Austrian Lombardy, in the prov. of Vicenza, and district of Schio. Pop. 1,940. It has a tile-kiln.

TORRE-DI-LUSERNA, a village of Sardinia, in the div. of Turin, prov. and 12 m. SW of Pignerol, at the foot of a hill crowned by fort St. Maria, in the angle formed by the confluence of the Angrogna and Felice. Pop. 2,100. It has manufactories of silk fabrics.

TORRE-DI-MARE, a village of Naples, in the prov. of the Basilicata, district and 31 m. SSE of Matera, and cant. of Pisticoio, in an insalubrious plain, near the entrance of the Basento into the gulf of Tarento. In its vicinity are the remains of the ancient *Metaponte*, founded by Nestor, in 1280 B.C., sacked by Spartacus, 76 years before the Christian era; rebuilt afterwards by the Romans; and ultimately destroyed by the Moors. It is noted as the last residence of Pythagoras.

TORRE-DI-MONDOVI, a town of Sardinia, in the div. of Coni, prov. and 5 m. SE of Mondovi, and mand. of Vico-di-Mondovi, on the slope of a hill, between the Casotto and Consaglia, and near the confluence. Pop. 1,600.

TORRE-DI-TAGLIA, a town of Naples, in the prov. of Abruzzo-Ultra, district and 18 m. SE of Civita-Ducale, and cant. of Borgo-Colle-Fegato, on a hill near the r. bank of the Imele. Pop. 500.

TORRE-DON-JIMEINO, a town of Spain, in the prov. and 8 m. W of Jaen. Pop. 7,000. It is partly enclosed by walls, and is bathed on the W by the Salado-de-Arjona. In the centre is a large square, with a fountain and a castle, the ramparts of which form a fine promenade. It contains two parish-churches, two convents, and an hospital; and possesses extensive manufactories of linen, oil, and wine. In the vicinity is an extensive salt-work.

TORRE-FRANCA, a town of Spain, in the prov. and 33 m. NNE of Cordova, near the r. bank of the Milano, a little above its junction with the Ciguenuela, in a valley at the foot of the Sierra-Morena. Pop. 2,428. It has a church, two chapels, a convent, and an hospital. On the SSE, and separated by only the breadth of a street, is the town of Torremilano. The trade consists chiefly in pigs, sheep, and wool.—Torre-Franca is the capital of a large seignorial territory which comprises three other towns. To the S, extending to the vicinity of Cordova, are fine fertile plains. The N part of the T. comprises a portion of the Sierra-Morena.

TORRE-FUENBILLIDA, a village of Spain, in the prov. and 15 m. SE of Palencia, near the r. bank of the Esgueva. Pop. 330.

TORRE-GALINDO, a village of Spain, in the prov. and 45 m. NNE of Segovia, at the foot of a hill on the r. bank of the Riaza. Pop. 182. It has a castle in ruins, manufactories of woollen fabrics, and a fulling mill.

TORRE-GENTILE, a village of Naples, in the prov. of Abruzzo-Citra, and district of Chieti, in a fertile plain. Pop. 725.

TORRE-HERMOSA, a village of Spain, in the prov. and 24 m. SW of Calatayud. Pop. 292.

TORREILLES, a commune of France, in the dep. of the Pyrenees-Orientales, and cant. of Rivesaltes, 9 m. N of Perpignan. Pop. 1,314.

TORREJONCILLO-DEL-REY, a town of Spain, in the prov. and 24 m. SW of Cuenca. Pop. 1,889. It has a convent and an hospital; and manufactories of woollen fabrics. The trade consists chiefly in articles of agricultural produce.

TORREJON-DE-ARDOZ, a town of Spain, in the prov. and 11 m. ENE of Madrid, in a fine plain near the l. bank of the Jarama. Pop. 1,500. It has two hospitals. The manufacture of soap, and rearing of sheep form the chief objects of local industry.

TORREJON-DEL-REY, a village of Spain, in the prov. and 14 m. WNW of Guadalajara, near two streams, in a fertile locality. Pop. 484.

TORREJON DE VALASCO, a town of Spain, in the prov. and 18 m. S of Madrid, in a flat but fertile locality. Pop. 1,285.

TORREJON-EL-RUBIO, a village of Spain, in the prov. and 26 m. NE of Caceres, in a fertile locality. Pop. 400.

TORRELAGUNA, a judicial partido and town of Spain, in the prov. of Madrid. The town is 33 m. NNE of Madrid, in a fertile plain, at the foot of a chain of hills near the r. bank of the Jarama. Pop. 2,300. It has a parish-church of fine Gothic architecture, containing several marble mausolea, and a fine Magdalene by Carmona. It has also two convents and an hospital; and possesses a pottery manufactory and three oil mills. The environs afford considerable quantities of wine. In the vicinity is an aqueduct now much dilapidated, erected by Cardinal Ximenes.

TORRE-LAS-ARCAS, a village of Spain, in the prov. of Teruel, and partido of Segura, 21 m. S of Alcaniz. Pop. 444.

TORRE-LA-VEGA, a judicial partido and village of Spain, in the prov. and 15 m. SW of Santander, in a delicious valley. Pop. 700. It is well-built, and has a ducal palace and a convent. Its industry consists chiefly in the manufacture of cotton.

TORRELLAS, a village of Spain, in the prov. and 57 m. SE of Logrono, on the l. bank of the Gueilles. Pop. 722.

TORRELOBATON, a town of Spain, in the prov. and 18 m. WSW of Valladolid, and partido of Mota-del-Marques. Pop. 1,268.

TORRELODONES (LA), a village of Spain, in the prov. and 18 m. NW of Madrid, in a rugged locality, at the terminus of a canal of the same name, a branch of the canal of Castile. Pop. 175.

TORREMAGGIORE, a town of Naples, in the prov. of Capitanata, district and 5 m. W of San-Severo, and cant. of S. Paoli. Pop. 4,500. It stands on a hill, and contains a ducal palace, two churches, and two convents.

TORRE-MALAMBERTI, a village of Austrian Lombardy, in the prov. and 14 m. E of Cremona, and district of Pescarolo. Pop. 435. It consists of two parts, distinguished as Torre-d'Angiolini and Torre-de-Piccinardi, and contains a castle, a massive structure named the Torre-de-Piccinardi, enclosed by a wide ditch with a draw-bridge, and enclosing numerous relics of antiquity.

TORREMAYOR, a village of Spain, in the prov. and 18 m. E of Badajoz, and partido of Merida, on the r. bank of the Cara, a little above its confluence with the Guadiana. Pop. 659. It has a palace belonging to the Duke de-la Rosa.

TORREMENGA, a village of Spain, in the prov. and 57 m. NNE of Caceres, and partido of Jarandilla. Pop. 180.

TORREMILANO, a town of Spain, in the prov. and 33 m. NNW of Cordova, and partido of Pozoblanco, in a locality possessing little fertility, and near a small stream named the Milano. Pop. 2,754. It has an hospital; and contains manufactories of common woollen fabrics, and is noted for its hams and cheese. Separated from this town by the breadth of a street only, is the town of Torrefranca.

TORREMOCHA, a town of Spain, in the prov. and 12 m. SSE of Caceres, in a flat but fertile locality, on the r. bank of the Solor. Pop. 1,690. It is tolerably well-built, and has a parish-church, a custom-house, several convents, and a public granary. It has manufactories of coarse woollen and linen fabrics.

TORREMOCHA-DEL-CAMPO, a village of Spain, in the prov. and 39 m. NE of Guadalajara, and partido of Siguenza. Pop. 244.

TORRE-MUNA, a village of Spain, in the prov. and 18 m. SSE of Logrono, on the r. bank of the Vadillo, in a mountainous but fertile locality. Pop. 460. It has several yarn mills.

TORRENS (LAKE), a lake of South Australia, to the N of Spencer's gulf, with which it appears to be connected. It stretches northward, and is supposed to sweep round in a horse-shoe form, enclosing a peninsula intersected by Flinder's range and its ramifications. It has been coasted along a distance of nearly 400 m. Its waters are salt, and are in some parts confined to the centre of the wady.—Also a mountain of New South Wales, in the district of Wellington, near the Yammerinna river.—Also a mountain of South Australia, in Kangaroo island.

TORRENTE, a town of Spain, in the prov. and 5 m. SW of Valencia, on a height commanding an extensive district covered with gardens, orchards, and vineyards. Pop. 5,000. It has numerous villas, a parish-church, a convent, and a custom-house. The environs are noted for their wine, grapes melons, legumes, &c.

TORRENUEVA, a town of Spain, in the prov. and 33 m. SE of Ciudad-Real. Pop. 2,140. It has a church and a fine chapel. The environs are noted for their oil, and a mine of antimony exists here.

TORREPADRE, a village of Spain, in the prov. and 30 m. SSW of Burgos, in a mountainous and woody locality, near the l. bank of the Arlanza. Pop. 230.

TORRE-PERO-GIL (LA), a town of Spain, in the prov. and 27 m. NE of Jaen, in a flat but fertile locality, near the r. bank of the Guadalquivir. Pop. 5,787. It has a parish-church, four chapels, an hospital, and a fine promenade planted with trees.

TORREQUADRADA-DE-VALLES, a village of Spain, in the prov. and 45 m. NE of Guadalajara, in a mountainous locality, on a small affluent of the Tajuna. Pop. 189.

TORRES, a town of Spain, in the prov. and 18 m. ESE of Jaen, partly on the slope and partly at the foot of a mountain, in a fertile locality. Pop. 2,292. It has a parish-church, a custom-house, a public granary, and a fulling-mill.—Also a village in the prov. and 17 m. E of Madrid, in a plain, at the foot of a mountain. Pop. 700. It has manufactories of charcoal.—Also a village in the prov. of Navarra, 42 m. SW of Pamplona. Pop. 338.—Also a village in the prov. and 24 m. WNW of Teruel, on the l. bank of the Guadalaviar. Pop. 522. It has productive iron-mines and forges.

TORRES ISLANDS, a group in the Mergui archipelago, in the Indian ocean, in N lat. 11° 40′, and E long. 96° 50′.

TORRES STRAITS, a channel which divides Australia from New Guinea. It is about 80 m. in width, but is obstructed by numerous islands and reefs. Of the former the principal are the Prince of Wales islands. Towards the narrowest part the depth never exceeds 14 fathoms, and it is often not more than half as much. In clear weather the bottom is every where seen. The navigation, at all times difficult, is with a contrary wind impracticable. This is advantageous only as a route from New South Wales to China and India, and in the months of April, May, and June when the E monsoon is in its vigour. It was discovered in 1605 by the Spanish navigator Torres.

TORRE-SANDIDO, a village of Spain, in the prov. and 42 m. SSW of Burgos, on the r. bank of the Esgueva. Pop. 562. It has a Carmelite convent, and possesses manufactories of linen and of cheese.

TORRE-SANTA-SUSANNA, a town of Naples, in the prov. of the Torre-d'Otranto, district and 20 m. SW of Brindisi, in a fertile plain. Pop. 1,500. It has a convent, an hospital, and an alms-house.

TORRES-DE-ALBANCHES, a town of Spain, in the prov. and 75 m. NE of Jaen, in a valley, at the foot of a steep mountain. Pop. 396. On a height to the N is a small fort.

TORRES-DE-SEGUR, a town of Spain, in the prov. and 9 m. SSW of Lerida, in a flat locality, near the l. bank of the Segur. Pop. 902. It has a church, a custom-house, and a public granary.

TORRES-NOVAS, a town of Portugal, in the prov. of Estremadura, comarca and 21 m. E of Santarem, in a flat locality, on a small affluent of the Tagus. Pop. 4,300. It is enclosed by walls and defended by a fortress, and has four parish churches, three convents, an alms-house, and a classical school. It has manufactories of calico. This town was taken from the Moors by Alphonso in 1148, but was regained by the former in 1190 and destroyed.

TORRES-TORRES, a town of Spain, in the prov. and 21 m. N of Valencia, in a fertile plain, near a lofty mountain, on which are the ruins of a Moorish fortress. Pop. 797.

TORRES-VEDRAS, a town of Portugal, capital of a comarca of the same name, and 30 m. N of Lisbon, on the l. bank of the Sizandro, which is here crossed by five bridges. Pop. 5,500. It has a portion of its ancient walls, and is commanded by a fortress. It contains four parish-churches, three convents, two hospitals, and a classical school. The environs produce considerable quantities of wine. In the vicinity are an aqueduct of Gothic architecture and a military hospital. This town was taken from the Moors in 1148. It is remarkable for the lines of defence which Wellington erected on the adjacent heights in 1810, and behind which his army lay secure from attack by the French.

TORRETTA, a mountain of Sicily, in the prov. and district and near the town of Palermo. It has an alt. of 2,750 Parisian ft. above sea-level.

TORREVECCHIA, a village of Naples, in the prov. of Abruzzo-Citra, 3 m. NE of Chieti, in a fine plain, Pop. 450.

TORRE-VIEJA-Y-LA-MATA, a town of Spain, in the prov. and 33 m. SSW of Alicante, on the Mediterranean, on which it has a port defended by a tower and battery. Pop. 1,940. It is of modern foundation, and is well and tastefully built. It has a custom-house and extensive military and commercial magazines Fishing, the manufacture of linen, and the adjacent salt-works, form the chief branches of local industry.

TORREZELLO, a town of Portugal, in the prov. of Beira, comarca and 36 m. W of Laguarda, near the source of the Rio-das-Nabeiras, an affluent of the Mondego.

TORRIDON (Loch), a large inlet of the sea, between Gairloch and Applecross, on the W coast of Ross-shire. It consists of three compartments, connected by narrow straits. The outer loch extends 6¼ m. SE, with a mean breadth of about 2½ m.; the middle loch is identical with Shieldag; and the inner loch extends nearly 5 m. E, with a mean breadth of about 1½ m. The united loch is, as a whole, the most striking in the magnificent and frequently indented coast which it intersects; but though impressive for its grandeur, and imposing for its extent, it is not remarkable for beauty. The herring-fishery in this loch is often highly productive.

TORRINGTON, a township of Litchfield co., Connecticut, U. S., 7 m. N of Litchfield.

TORRINGTON (Black), a parish in Devon, 5 m. W by N of Hatherleigh. Area 7,280 acres. Pop. in 1831, 1,080; in 1851, 1,115.

TORRINGTON (East), a parish in Lincolnshire, 4 m. NE by N of Wragby. Area 1,498 acres. Pop. in 1831, 87; in 1851, 113.

TORRINGTON (Great), a parish, borough, and market-town in Devon, 36 m. NW by W of Exeter, on the NE bank of the river Torridge, the navigation of which is extended to T. by a navigable cut or canal joining the river about 2 m. above the town of Bideford. The parish, including the chapelry of St. Giles-in-the-wood, has an area of 3,456 acres. Pop. in 1831, 3,093; in 1851, 3,308. The town is beautifully situated on the slope and brow of a fine eminence forming the E bank of the Torridge, across which there is here a stone-bridge, uniting the hamlet of Taddy-port with the town. Some trade is carried on with Ireland through the Torridge navigation; but the glove manufacture is the chief employment of the inhabitants. At the Restoration, General Monk was made Earl Torrington. The town gives the title of Viscount to the Byng family.

TORRINGTON (Little), a parish in Devon, 1¾ m. S by W of Great Torrington. Area 2,880 acres. Pop. in 1831, 572; in 1851, 623.

TORRINGTON (West), a parish in Lincoln, 3 m. N of Wragby. Area 1,109 acres. Pop. 133.

TORRISDALE. See Tongue.

TORRY, a fishing-village in the p. of Nigg, Kincardineshire, on the r. bank of the Dee, opposite the city of Aberdeen, and 1½ m. W of the point of Girdleness. Pop. 370.

TORRYBURN, a parish in the W extremity of Fifeshire, upon the coast of the frith of Forth. Pop. in 1831, 1,436; in 1851, 1,341.—The village of T., 9 m. W of North Ferry, is situated on the coast, and was once the port of Dunfermline.

TORSA, an island off the coast of Nether-Lorn, Argyleshire, separated by only narrow straits, from Seil on the N, the continent on the W, and Luing on the E. It measures about 3 m. by 1 m. Its surface exhibits a smooth green hill, 200 ft. high. The whole of its E side, excepting a few trap rocks, and a little greywacke, consists of the same clay-slate as that for which Luing and Seil are celebrated.

TORSAAS, a town of Sweden, in the prov. of Smaaland, and 22 m. SSW of Calmar.

TORSAC, a village of France, in the dep. of Charente, 7 m. SSE of Angouleme, on the Charrau. Pop. 850.

TORSAKER, a town of Sweden, in Angermannland, 27 m. N of Hernosand, on the r. bank of the Argermann-elv.

TORSARI, an island in the gulf of Finland, 18 m. S of Viborg, in N lat. 60° 21′.

TORSHELLA, a village of Sweden, on a river which rises in Lake Hielmar, and runs into the Malär lake, 46 m. W of Stockholm.

TORSHOK. See Torjok.

TORTHORWALD, a parish in Nithsdale and Annandale, Dumfries-shire, having a superficial extent of about 5,600 acres. The sluggish Lochar traces the whole W boundary. The amount of the area under cultivation is about 2,600 acres. It is an irregular cluster of cottages, with scarcely 200 inhabitants. Collin is a larger village, on the margin of Lochar-moss, 3½ m. E of Dumfries. Pop. of p. in 1831, 1,320; in 1851, 1,313.

TORTOLA, a village of Spain, on the river Henares, in the prov. and 7 m. above Guadalaxara.

TORTOLA, one of the group of the Virgin islands, in the W. Indies, in N lat. 18° 24′, W long. 64° 32′. It is about 12 m. in length from E to W, and 3½ m. broad at its widest part. Its surface is rocky and mountainous, rising to an alt. of 1,600 ft. above sea-level in one or two points, and its shores are deeply indented with bays. It has been in the possession of England since 1666. Its pop. is about 10,000.

TORTOLES, a village of Spain, in the prov. and 12 m. SSW of Burgos, on the r. bank of the Esgueva.

TORTOLI, a town of the island of Sardinia, 9 m. ENE of Lanusei. It has a small port, and some trade in wine and cheese.

TORTONA, a province of the Sardinian states, skirted by the Po on the N, and bounded on the E by the provs. of Voghera and Bobbio; on the S by that of Novi; and on the W by Alessandria. Area 665 sq. kilom. Pop. 53,570. About one-half of the surface is covered with mountains; and 200,000 hectares are cultivated. Its agricultural productions are wheat, barley, rye, maize, and oats, flax, hemp, and silk.—Its capital, of the same name, a bishop's see, and one of the most ancient cities in the N of Italy, is seated at the foot of a hill, an extremity of one of the Apennine branches, on the Roman road from Placentia to Genoa. Pop. 10,000. Its ancient name was *Dertona*. After the death of Otho II., in 1002, it became a republic, and was besieged by Barbarossa, in 1155, who took and sacked the city. With the aid and under the protection of the Visconti, dukes of Milan, it was rendered a place of great strength. By the Emperor Charles V. and Philip II. its fortifications were further increased. In 1665, it was again placed in the highest state of defence; and finally in 1735, on being ceded to Emanuel III., king of Sardinia, he made it a place of such strength that it was considered impregnable. In later times, however, it surrendered to the French; and Napoleon having ordered the demolition of all its fortifications, scarcely a sign now remains of its ancient military strength, the lines of its fortifications presenting the more pleasing aspect of public walks shaded by beautiful trees. The present cathedral is a plain brick structure, consisting of a nave and two aisles. There are four other parochial churches, two oratories, two convents, an hospital, an asylum for exposed children, and another for orphans, two public schools, and a large seminary for clergy and youths of the higher classes. The streets are generally of very irregular construction. Important fairs are held here yearly for stock and general merchandise.

TORTORA, a town of Naples, in the prov. of Calabria-Citra, 9 m. N of Scalea, near the gulf of Policastro. Pop. 1,600.

TORTORELLA, a town of Naples, in the prov. of Principato-Citra, 6 m. NNE of Bonati, near the l. bank of the Bussento. Pop. 1,100.

TORTORETTO, a village of Naples, in Abruzzo-Ultra 1ma, 15 m. NE of Teramo.

TORTORICI, a town of Sicily, in the Val-di-Demona, 20 m. SW of Melazzo, on the Fitalia. Pop. 2,800.

TORTOSA, a city of Spain, in the prov. and 48 m. SW of Tarragona, on the l. bank of the Ebro, about 25 m. from its mouth. Pop. of town and district, 20,500. It is divided into an old and a new town, both surrounded with walls of considerable antiquity. It is the see of a very wide bishopric. The Ebro, approaching here to its influx into the sea, is wide and deep, and navigable by vessels of 100 tons; and the trade is not inconsiderable in wheat, timber, charcoal, pitch, and wine. The principal manufactures are soap, leather, cordage, glass, pottery, and coarse linens. Fine marble is quarried in the vicinity. This town, the *Dordosa* of the Romans, was more than once the scene of conflicts between the Spaniards and the Moors. In 1649 and in 1798 this town was taken by the French; in 1810, it experienced the same fate.

TORTOSA, a port of Syria, in the pash. and 32 m. N of Tripoli, opposite the island of Ruad, in N lat. 34° 50′. It is no longer a place of any importance, but it represents the ancient *Orthosia* supposed to have been built about the 5th cent., and frequently mentioned by the historians of the Crusades, as a place of great strength.

TORTOSA (Cape), a promontory of Spain, on the coast of Catalonia, in N lat. 40° 40′.

TORTSHIN, a town of Russia, in the gov. of Volhynia, 12 m. W of Lutzk.

TORTUE, an island off the N coast of Hayti, in N lat. 20° 3′. It is about 15 m. in length, and 4 m. in breadth. Its surface is mountainous.

TORTUE (La), a river of Lower Canada, which, running S, falls into the St. Lawrence, about 4 m. above Montreal.

TORTUERA, a village of Spain, in the prov. and 27 m. SSW of Calatayud.

TORTUERO, a village of Spain, in the prov. and 24 m. NNW of Guadalaxara.

TORTUGA, an island about 6 m. distant from the NE coast of the island of Cuba, in N lat. 20° 4′. It is about 60 m. in circumf.; it is fertile, and productive of tobacco, palms, sandal-wood, aloes, sugar, indigo, cotton, and fruits. It abounds in wild boars.—Also a small island in the North Atlantic, near the coast of Honduras.—Also a small island in the gulf of California, lying at about an equal distance from the coast of California and that of Mexico, 52 m. SSE of Tiburon.

TORTUGA (Punta-de), a cape on the coast of Brazil, in S lat. 3°.

TORTUGAS (Das), a group of 10 small keys or islands, off the S point of Florida, extending ENE and WSW 10 or 11 m., under the parallel of N lat. 24° 32′.

TORTUGAS (Point), a headland on the coast of Chili, in S lat. 29° 37′.

TORTUGA-SALADA, an island in the Caribbean sea, off the coast of Venezuela, in N lat. 11°. It is about 36 m. in circumf. At its SE extremity is an indifferent roadstead frequented by merchantmen which come hither to lade salt, from May to August. Near the SW extremity of the island, is a small harbour, and some fresh water.

TORTURA, a port of Syria, 15 m. S of Acre, supposed to represent the Dor or Nephath of Scripture. It was erected into a bishopric, under the archbishop of Cæsarea.

TORVESTAD, a village of Norway, in the bail. and 36 m. NNW of Stavanger, on a small island to the N of Karmoe.

TORVISCON, a town of Spain, in the prov. and 36 m. SE of Granada, near the l. bank of the Cachar.

TORY, Torry, or Troree, an island in co. Donegal, situated 4 m. N by W of Innisboffin, and 7 m. WNW of Horn-head. It extends 2¼ m. NW, with a maximum breadth of ⅜ m., and comprises an area of 785 acres. Its SE end is called Port-Doon; its NW end is surmounted by a lighthouse. The inhabitants are a semi-barbarous race, about 500 or 700 in number, who support themselves by husbandry and by fishing.

TOSA. See Toccia.

TOSA, a town on the S coast of Xicoco, in Japan, in about N lat. 33° 40′.

TOSAGUA, a river of Ecuador, in the prov. of Esmeraldas, which runs SSW, and unites itself with the Chones, in S lat. 32° 30′.

TOSANLU, a river of Asia Minor, the ancient *Lycus*, which rises in the mountains of Armenia, and after running about 200 m. almost due W, falls into the Jekil-Irmak, about 30 m. N of Amasia.

TOSCANELLA, a town of the States-of-the Church, 20 m. N of Civita-Vecchia, on the r. bank of the Marta. Pop. 3,000.

TOSCOLANO, a town of Austrian Italy, in the deleg. of Brescia, near the W shore of Lake Garda. Pop. 2,500. It is a place of considerable antiquity.

TOSELAND, a parish of Huntingdonshire, 4 m. NE by E of St. Neots. Area 1,320 acres. Pop. in 1851, 230.

TOSIA, or Tossia, a town of Asiatic Turkey, in the sanj. and 30 m. NNE of Kiangari. It is well-built, and consists of about 1,000 houses.

TOSNA, a town of Russia, in the gov. and 36 m. SSE of St. Petersburg, on a river of the same name, which joins the Neva at Pella, in the gov. of Petersburg, after a WNW course of about 60 m.

TOSS, a village and parish of Switzerland, in the cant., and 12 m. NE of Zurich, on a small affluent of the Rhine. Pop. of district, 1,732.

TOSSA, a town of Spain, in the prov. and 20 m.

S of Gerona, near the coast. Pop. 1,600. It has some coasting trade.

TOSSIA. See TOSIA.

TOSSIAT, a village of France, in the dep. of Ain, cant. and 7 m. NNW of Pont-d'Ain. Pop. 620.

TOST, or TOSCHEK, a town of Prussian Silesia, 28 m. SE of Oppeln. Pop. 964.

TOSTA, a river of Nicaragua, which enters the Pacific, in S lat. 12° 30′, after a S course in which it passes near the SW extremity of Lake Leon.

TOSTAK, a river of Asiatic Russia, which flows NNW to the Jana, in the prov. of Yakutsk, which it joins after a course of nearly 400 m.

TOTA, a lake of New Granada, in the prov. and 21 m. E of Tunja, on the top of a mountain. It is of a circular figure, and nearly 6 leagues in circumf.

TOTANA, a town of Spain, in the prov. and 27 m. SW of Murcia, on the great road through Murcia, leading from Andalusia into Valencia. Pop. 8,000. It is a poor inland place, situated in a part of the country little visited by travellers, and the inhabitants are proverbially unenterprising and indolent. The chief article of manufacture is coarse pottery.

TOTES, a village of France, in the dep. of Seine-Inferieure, 18 m. N of Rouen. Pop. 700.

TOTHAM (GREAT and LITTLE), two adjoining parishes of Essex, 3 m. NNE of Maldon. Area of Great T. 5,363 acres. Pop. in 1851, 840. Area of Little T. 1,283 acres. Pop. in 1851, 388.

TOTHILL, a parish in Lincolnshire, 4 m. NW by N of Alford. Area 854 acres. Pop. in 1851, 59.

TOTLEY, a township in Dronfield p., Derbyshire, 8 m. NW by N of Chesterfield. Pop. 403.

TOTMA, a town of Russia, in the gov. and 110 m. NE of Vologda, on the left bank of the Sukhona. Pop. 2,800. There are extensive salt-works here.

TOTNESS, a parish and parliamentary borough in Devonshire, 22¼ m. SSW of Exeter, on the river Dart, which is navigable and tidal to the town, a distance of 9 m. Area of p. 1,043 acres. Pop. in 1801, 2,503; in 1831, 3,442; in 1851, 3,828. The borough returns 2 members to parliament. The number of electors registered in 1837 was 297. The town, which has a situation eminent in point of beauty, climbs the steep declivity of a hill, and stretches itself along its brow, commanding a view of the country in its vicinity, but sheltered at the same time by higher grounds on every side. It is one of the most ancient towns in England; the piazzas in front of the houses in some parts of the upper town, and the higher stories projecting over the lower ones, are proofs of its antiquity. A number of the inhabitants are employed in agriculture, and some in fishing. The Dart is navigable for small vessels as high as the bridge. The trade in coal and culm, grain, and cider, forms the principal commerce of the place.

TOTOMINA, a large bay of Japan, on the S coast of the island of Nifon. It is 45 m. in length from N to S, and about 30 m. in breadth.

TOTON, or TOUETON, a hamlet in Attenborough p., Nottinghamshire, 5¾ m. SW of Nottingham. Pop. in 1831, 202.

TOTONICAPAN, a town of Guatimala, the cap. of a department of the same name, 90 m. N of Guatimala. Pop. 10,000. There are salt springs in the vicinity.—The dep. is watered by the Chiartan and the Sumasinta.

TOTORAL, a town of the La Plata province of Cordova, 54 m. N of Cordova.—Also a fort of Chili, 18 m. SSW of Copiapo, in S lat. 27° 30′.

TOTORKOW, or TUTURKAI, a small town of European Turkey, in Romania, with a small castle on a hill, on the Danube, 24 m. W of Silistria.

TOTSKAIA, a fortified village of Russia, in the gov. of Orenburg, 27 m. SE of Buzuluk, on the l. bank of the Samara, at the confluence of the Soroka. Pop. 600.

TOTTENHAM, a parish and village in Middlesex, 5½ m. N by E of St. Paul's, London, and intersected by the Northern and Eastern Counties railway. Area 8,403 acres. Pop. in 1801, 3,629; in 1831, 6,937; in 1851, 9,120.—The village consists of one long street on the road to Ware and Hertford, and is almost united with the metropolis by other villages, hamlets, and houses, scattered along the road between them.

TOTTENHILL, a parish in Norfolk, 6 m. NNE of Market-Downham. Area 1,590 acres. Pop. 412.

TOTTERIDGE, a parish in Hertfordshire, 1¾ m. S of Chipping-Barnet. Area 1,597 acres. Pop. in 1831, 595; in 1851, 595.

TOTTINGTON, a parish in Norfolk, 4 m. SSW of Watton. Area 3,213 acres. Pop. in 1851, 370.

TOTTINGTON (HIGHER), a township in Bury p., co.-palatine of Lancaster, 6 m. NNE of Bury. Area 3,686 acres. Pop. in 1851, 2,958.

TOTTINGTON (LOWER), a chapelry in Bury p., co.-palatine of Lancaster, 3½ m. NNW of Bury. Pop. in 1831, 9,280; in 1851, 10,691.

TOU (CAPE), a promontory on the Caucasian coast of the Black sea, 9 m. S of Chapsougo, in N lat. 44° 10′, E long. 38° 53′.

TOUAT. See TUAT.

TOUBANG, a town on the N coast of the island of Java, 470 m. E of Batavia.

TOUCHEA, a village of France, in the dep. of Saone-et-Loire, cant. and 5 m. NNW of Givry. Pop. 1,100.

TOUCHEZ (LES), a village of France, in the dep. of Loire-Inferieure, 3 m. E of Nort.

TOUCQUEZ, a river of France, which rises a little to the NE of Merlerault, in the dep. of Orne; runs N; enters the dep. of Calvados; and falls into the Channel after a course of 60 m.—There is a village of the same name on this stream, about 3 m. above its embouchure.

TOUCY, a town of France, in the dep. of Yonne, 15 m. W by S of Auxerre, on the l. bank of the Ouanne.

TOUGET, a town of France, in the dep. of Gers, 17 m. NE of Auch.

TOUGH, a parish of Aberdeenshire, 22 m. W by N of Aberdeen. Pop. in 1831, 828; in 1851, 891.

TOUGH, TUOGH, or TUORAGH, a parish in co. Limerick, 3¾ m. N by E of Pallasgreen, containing the village of Cappamore. Area 6,519 acres. Pop. in 1831, 3,594; in 1851, 2,272.

TOUGHCLUGGIN, or CLUGGIN, a parish in co. Limerick, 2¼ m. ENE of Pallasgreen. Area 2,094 acres. Pop. in 1831, 217; in 1851, 426.

TOUILLE, a village of France, in the dep. of Haute-Garonne, cant. and 2 m. SSE of Salies, near the r. bank of the Salat. Pop. 450.

TOUILLON, a village of France, in the dep. of Cote-d'Or, 10 m. WNW of Baigneux-les-Juif. Pop. 600.—Also a village in the dep. of Doubs, 7 m. S of Pontarlier, near the source of the Doubs.

TOUIVROV, a town of Russia, in the gov. of Podolia, 24 m. NW of Batslav, near the r. bank of the Bug.

TOUL, a town of France, in the dep. of Meurthe, situated on the Moselle, 12 m. W of Nancy, in a fertile valley, surrounded by a chain of hills covered with vineyards. Pop. 7,400. Its old walls were destroyed in 1700, and the place considerably enlarged and improved by the construction of a new rampart flanked with bastions. The cathedral is a fine edifice of the 17th cent. T. has an hospital, an arsenal, and barracks, and a handsome stone bridge over the Moselle. The manufactures chiefly consist of pottery and stockings. This town was a free city of the German empire until 1552.

TOULA. See TULA.

TOULON, a town of France, in the dep. of Saone-et-Loire, on the river Arroux, 20 m. W by N of Charolles. Pop. 1,600. It is joined to a village of the same name on the other side of the river, by a bridge of 13 arches.

TOULON-SUR-MER, a seaport and great naval arsenal in the SE of France, situated in the dep. of Var, on a deep double bay of the Mediterranean, in N lat. 43° 7′, E long. 5° 56′. It is built on gently rising ground at the foot of a ridge of lofty mountains, which shelter it from the N, and sweeping round on each side, enclose the bay. It is surrounded with ramparts, ditches, and bastions, of great strength; and is further defended by a fine citadel, and a number of forts and batteries distributed on the neighbouring eminences, and along a tongue of land projecting into the bay. The moles which separate the port from the roadstead are hollow, bomb-proof, and lined by batteries. The town itself is divided into an old and a new town. The former is ill built, but contains one long straight street, called the Rue-aux-Arbres from the trees with which it is shaded. The new town is better built, containing several public structures erected by Louis XIV., a few straight streets, and a Place-d'-Armes (*a*), used for exercising the garrison. The hotel-de-ville (*b*) is a handsome structure facing the harbour. Toulon has no river; but several streams descending from the neighbouring mountains, supply the fountains constructed in different parts of the town. Woollen cloth, hosiery, soap, candles, and leather, form the chief articles of manufacture. There are also ship-building docks, dye-works, and iron-foundries here; but the importance of the place is derived from its having long been one of the chief stations of the French navy, being on the Mediterranean what Brest is on the Atlantic. It has two ports, known as the old and the new ports, but which might more properly be called the commercial and the military ports. The old port or Port Marchand (A), is a basin surrounded with a handsome quay, and lined with good edifices. The new port or Port Militaire (B), immediately to the W of

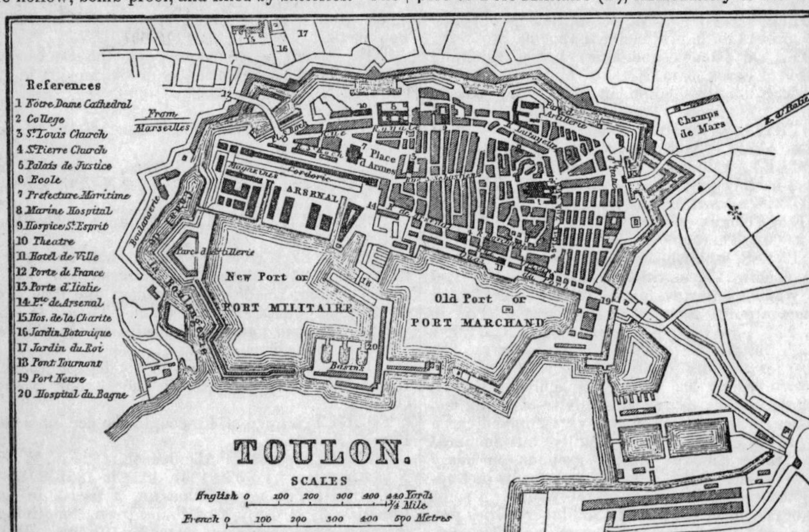

the old port, has an area of 35 acres, is one of the finest of Europe, and is said to be capable of containing 200 sail-of-the-line. It is surrounded by slips, arsenals, storehouses, and foundries of government. The *corderie* or rope-house (*d*) is nearly 1,200 ft. in length, and fire-proof. The roadstead is spacious: the sea has here no visible flux or reflux, and is tranquil in almost every wind. The arsenal (*c*) is a very large edifice, filled with every requisite for the equipment of vessels. The trade of T. is not extensive, being limited to the products of the vicinity, such as wine, oil, silk, and fruit of different kinds; but has increased considerably since the occupation of Algiers. Regular steam communication is maintained from this port with Algiers and Corsica. The tunny fishery is extensive. The pop. is about 29,000, exclusive of the workmen employed in the arsenal, amounting to about 4,000, and of the *forçats* or galley slaves, between 3,000 and 4,000 in number. T., which was made a naval station by Henry IV. and Louis XIV., has long been the scene of the equipment of naval expeditions; but the most remarkable event in its history is the occupation of the town and harbour by the British in the autumn of 1793; its subsequent siege by the republican troops of France; and the precipitate abandonment of the place by the British troops, on the 19th of December 1793, after burning 27 vessels, and carrying off 15 of the squadron contained in the port. The republicans, on obtaining possession of the place, which had been surrendered to the English by the royalist portion of the inhabitants, exercised great cruelties towards those who had or were suspected to have participated in delivering it to the English. It was from this port that the military expeditions sailed to Egypt in 1798, to the Morea in 1827, to Algiers in 1830, and to Ancona in 1832.—The arrond. of T. comprises 7 cants. with an area of 129,064 hectares. Pop. in 1846, 127,854.

TOULOS, a lake of Russia, in the gov. of Olonetz, about 120 m. WNW of Poviatnetz, under the paral lel of 63° 30′ N. It is 24 m. in length from NW to SE, and about 6 m. in breadth.

TOULOUBRE, a river of France, which rises 5 m. NNE of Aix; runs first WNW, and then SW; passes Salon; and discharges itself into the Etang-de-Berre, after a course of 24 m.

TOULOUSE, a city of France, formerly the capital of Upper Languedoc, now of the dep. of Haute-Garonne, situated on the r. bank of the Garonne, 132 m. SE of Bordeaux, 473 ft. above sea-level. Pop. in 1789, 55,068; in 1821, 52,328; in 1846, 94,236, of whom 10,747 were seamen. The Garonne is here navigable, and as wide as the Seine at Paris; the situation of T. on its banks, and near the junction of the great canal of Languedoc or Canal-du-Midi, is consequently highly advantageous for trade. It is built chiefly on the r. bank of the river, but part of it stands on an island in the river; and on the l. bank is the suburb of St. Cyprien. The buildings are almost entirely of red brick; even the town walls are of that material; and its streets are narrow, winding, and irregular. The chief embellishments of the place consist in the public promenades, the river, the quays, and the bridge over the Garonne, the last a fine structure, 810 ft. in length, and 72 ft. in breadth, built in the middle of the 17th cent. The cathedral is a handsome though somewhat irregular building. The hotel-de-ville is large, and its façade forms the side of a square called Place-du-Capitole or Place-Royale. The Place-la-Fayette is a handsome circus lined with regular mansions. The building, once the residence of the counts of Toulouse, is now used as a court of justice, though its massy construction conveys to the spectator the idea of a fort. The other buildings most worthy of notice are the old church of St. Sernin, the palace of the archbishop, the hospital, the mint, the exchange, and the theatre. There are also a museum, a public library, a picture gallery, a botanical garden, and an observatory. The manufactures are various, consisting of silks, woollens, linens, leather, vermicelli, porcelain, wax candles, cutlery, and pottery; there are also copper works, tanneries, dye-works, and distilleries, and a cannon foundry.—T. acquired an unfortunate title to notice, by an obstinate battle fought in the vicinity, on 10th April, 1814, between the British under Wellington, and the French under Soult; neither commander having been apprised of the abdication of Napoleon.—The arrond. of T., comprising 9 cants. with an area of 158,570 hectares, had a pop. of 177,329 in 1846.

TOUNG-HU, or Tongo, a town of Pegu, on the r. bank of the Sitang, in N lat. 10°. It is surrounded by a high brick wall, with a solid rampart, and a broad wet ditch, each face of which is considerably upwards of a mile in length. The town occupies but a small portion of the enclosure; temples, *kyooms* or monasteries, groves of lofty trees, and extensive fruit-gardens, are interspersed among large open spaces. There is also a large natural tank within the walls. The wall is of very ancient date, and the place is untenable as a fortification, independently of its vast size, from the extensive dilapidation of the wall in many places.

TOUPO, Towpo, or Taupo, a lake of New Zealand, in S lat. 39° 35′. It is described as "one of the most superb lakes in the world,—not from its size, although that is considerable, but from the extreme magnificence of the scenery surrounding it." Mr. Chapman considers it to be 35 m. long, and 20 m. broad. Its outline is a sort of irregular triangle, with the two most distant angles forming the N and S ends. The W shore is apparently nearly straight, and the third point of the triangle will be about the E boundary of the lake. At this E angle is a deep bay about 6 m. long, running SE, which is invisible except almost immediately opposite the entrance. "The most peculiar feature in the appearance of this lake is the immense height of the surrounding cliffs; they are always perpendicular, although in some instances rising in terraces one behind the other, and vary from 500 to 1,000 ft. high at several parts of the lake, particularly at the NNW and NE sides; these rise perpendicularly from the water to such a height, that, says Mr. Bidwill, "I never saw their tops through the clouds for above five minutes together during the eight days I was on the lake. There are but few places where a canoe can land, and at those the beaches are very short and narrow; they are covered with pumice and black sand, and always indicate the entrance of a small stream of water. There are a number of small waterfalls round the lake, but none of any consequence; the only river or stream of any size which runs into it being the Waikato, which runs in at the only part of the lake (the S end) where the banks for any distance are level and the water shallow. At the N end is a very peculiar mountain, with an outline as regular as if it had been the work of art. At the two extremes of the range are two peaks just alike, and each about one-third the height of the mountain. At about the distance of another third rise two other equal peaks, and in the centre rises a fifth, I suppose about 5,000 ft. high." This romantic region is extremely populous, although comparatively sterile.

TOUR (La), a village of Piedmont, in the prov. of Pinerolo, on the Peltise. There is here a large cotton-spinning factory.

TOURATEA, a district and town in the S part of the island of Celebes. The district is bounded by Macassar, Boutain, and the sea. The town is 36 m. S of Macassar. It has manufactories of cotton fabrics, white and coloured, for the Dutch trade in the Molucca and Philippine islands and in India, and has also several silk factories.

TOURAVAKAIRA, a town of Hindostan, in the prov. of Mysore, 60 m. N of Seringapatam. It is well defended by an earthen rampart and a ditch, and has two forts. It contains two Hindu temples of great antiquity, and about 700 houses.

TOURBA (Cape), a headland of Northern Africa, on the coast of Barca, 15 m. W of Derna, in N lat. 32° 54′ 10″, E long. 22° 2′ 45″.

TOURBALI, or Tourali, a town of Turkey in Asia, in Anatolia, in the sanj. of Aidin, 23 m. SSE of Smyrna, at the E base of a mountain, of which the ancient *Metropolis* occupied the summit.

TOURBET, or Toorbut-Hyderee, a fortified town of Persia, in the prov. of Khorassan, 75 m. SSW of Meshed. Pop. 18,000.

TOUR-BLANCHE (La), a commune of France, in the dep. of the Dordogne, cant. and 4 m. ENE of Verteillac, at the foot of a lofty mountain. Pop. 589.

TOURCOING, a canton, commune, and town of France, in the dep. of the Nord, and arrond. of Lille. The cant. comprises 9 coms. Pop. in 1831, 40,073; in 1846, 52,308. The town is 7 m. NE of Lille. Pop. in 1846, 26,834. The streets are well and regularly built, and it contains a large and handsome square in which is the town-house, two churches, the remains of the castle of the dukes of Havre, a college, and an alms-house. The industry of the place consists chiefly in the manufacture of varieties of cotton and woollen fabrics, carpets, yarn, soap, leather, bricks, sugar from beet-root, &c. It has also a salt-refinery, several breweries, and dye-works. The origin of this town is unknown; but so early as the 12th cent. it was noted for its manufactures. It has repeatedly suffered from fires.

TOUR-D'AIGUES (LA), a commune of France, in the dep. of the Vaucluse, cant. and 4 m. NE of Pertuis, in a valley. Pop. 2,300. It has a manufactory of parchment and several silk-mills.

TOUR-D'AUVERGNE (LA), a town of France, in the dep. of the Puy-de-Dome, 29 m. W of Issoire, on a lofty basaltic plateau. Pop. 1,830. In the vicinity are the remains of the ancient castle of the Auvergne family.

TOUR-DE-FRANCE (LA), a canton, commune, and town of France, in the dep. of the Pyrenees-Orientales, and arrond. of Perpignan. The cant. comprises 11 coms. Pop. in 1831, 6,205; in 1846, 6,802. The town is 14 m. WNW of Perpignan, near the r. bank of the Gly. Pop. 1,217. It has a tannery and several distilleries of brandy.

TOUR-DE-PEIL (LA), a town of Switzerland, in the cant. of Vaud, district and 1 m. SE of Vevay, from which it is separated by a charming promenade, and on the NE bank of the lake of Geneva. It is enclosed by ancient walls and ditches, and commands, from a terrace which rises from the edge of the lake, one of the finest views in the locality.

TOUR-DE-SALVAGNY (LA), a village of France, in the dep. of the Rhone, cant. and 5 m. ESE of Arbrele. Pop. 459.

TOUR-DE-TREME (LA), a village of Switzerland, in the cant. and 15 m. SSW of Friburg, near the l. bank of the Sarine. Pop. 450.

TOUR-DU-MEIX (LA), a village of France, in the dep. of the Jura, cant. and 2 m. E of Orgelet, between two mountains, near the r. bank of the Ain, here crossed by a bridge. Pop. 360.

TOUR-DU-PIN (LA), an arrondissement, canton, commune, and town of France, in the dep. of the Isere. The arrond. comprises an area of 132,306 hect., and contains 8 cant. Pop. in 1831, 126,146; in 1841, 132,960; in 1846, 136,627. The cant. comprises 16 com. Pop. in 1831, 18,084; in 1846, 19,429.—The town is 31 m. NNW of Grenoble, on the l. bank of the Bourbre, in a fertile locality. Pop. in 1789, 1,014; in 1821, 2,020; in 1831, 2,334; in 1846, 2,665. It is small and ill-built. With the exception of a sugar factory, an oil-mill, and a tannery, it possesses no industry.

TOURETZ, a town of Russia in Europe, in the gov. of Grodno and district of Novogrodek.

TOURFAN, a volcano of the Thian-shan chain, on the confines of Chinese Turkestan and the Chinese prov. of Kan-su, in N lat. 43° 30′, E long. 89° 31′. It constantly emits volumes of flame and smoke, and is noted for its sal ammoniac, which the Kalmuks transport hence to all parts of the country.

TOURFAN, or TOURPAN, a district and town of Chinese Turkestan, near the Thian-shan mountains, about 180 m. W of Khamil, between two rivers. The temperature of the locality is in summer very high, and the air is loaded with sand from the adjacent arid mountains. It produces corn, millet, lint, melons, and in considerable quantities cotton and pulse. Some distance to the W are the ruins of Old T. Goes represents T. as a strong well-fortified city. In the Jesuits' map it is placed in 43° 30′ N lat., and 89° 36′ E long.; but in Sir George Staunton's map, where it is called Tuloofan, the Chinese being unable to sound the consonant *r*, it is placed 24° 30′ W of Pekin, or 91° 58′ E of Greenwich, and in 43° N lat. By the Jesuits it is stated to be 6 days' journey WNW of Hami, over a branch of the Cobi, but 10 days' by the hills to the N, which is reckoned the safer road. The inhabitants of the country of T., says Haitho, the Armenian, in his Oriental history, are called Jegoors. They abstain rigidly from drinking wine, and eating animal food. They raise much wheat, but have no vines. Their towns are very pleasant and contain many temples sacred to the worship of idols. They cultivate the arts and sciences, but are not at all addicted to war; they have a peculiar mode of writing, [the Oigoorian character,] which has been adopted by all their neighbours [the Mongols]. The inhabitants of T., says Shadi-Khuaja, who visited this district in 1420, are idolaters [Budhists] who perform the ceremonies of their religion in spacious temples. To the W of T. 8 or 9 leagues, is Hochew, the ancient capital of the Oigoors, still called Peting-tu-hu-fu by the Chinese.

TOURIHEY, or LAHEISEY, a town of Caffraria, in the Bichuana territory, 120 m. W of Lattaku. It contains about 100 houses. The inhabitants paint their bodies red, and powder the hair with a blue shining dust. Around are extensive pasturages, on which large herds of sheep are reared.

TOURIIA, a river of Russia in Europe, in the gov. of Volhynia, which has its source in the district and 24 m. ESE of Vladimir; runs first NW, then E, into the district of Kovel; bends NNE; and after a total course of about 114 m., joins the Prepet on the r. bank.

TOURIN, a mountain of Turkey in Europe, in Rumelia, in the sanj. of Gallipoli, 42 m. SSW of Philippopoli. It connects itself on the N with the Despoto-dagh or ancient *Rhodope*.

TOURINNE, a department and commune of Belgium, in the prov. of Liege and arrond. of Huy. Pop. 191.

TOURINNES, a commune of Belgium, in the prov. of Brabant, and dep. of Tourinnes-Beauvechain. Pop. 1,107.—Also a com. in the sanj. and dep. of Tourinnes-Saint-Lambert. Pop. 522.

TOURINNES-LA-GROSSE, a department and commune of Belgium, in the prov. of Brabant and arrond. of Nivelles. Pop. 1,174.

TOURINNES-SAINT-LAMBERT, a department and commune of Belgium, in the prov. of Brabant and arrond. of Nivelles. Pop. 1,655.

TOURINSK, a town of Russia in Asia, in the gov. and 330 m. W of Tobolsk, on the Tura. Pop. 4,000, exclusive of the Tartars who reside in the suburbs. It has a suburb, six churches, a convent, and a seminary. In the vicinity are mines of iron. T. occupies the site of an ancient Tartar town named Epantchin.

TOURINSK (NIJNII), a town of Russia in Asia, in the gov. of Perm, on the r. bank of the Tura. It has extensive iron-works.

TOURINSK (VERKHNE), a town of Russia in Europe, in the gov. of Perm, district and 53 m. SSW of Verkhotourie, on the NE bank of a lagune formed by the Tura. It contains about 315 houses, and has extensive iron-works.

TOURINSKAIA, a town of Russia in Asia, in the gov. of Tobolsk, district and 47 m. ESE of Tourinsk, on the r. bank of the Tura.—Also a town in the gov. and district and 150 m. SW of Irkutsk.

TOURISK, a town of Russia in Europe, in the gov. of Volhynia, district and 11 m. SW of Kovel, on the l. bank of the Touriia.

TOURJAN, a lofty mountain of Turkey in Europe, in Rumelia, in the sanj. of Sophia, 20 m. NNW of Tatar-Bazardjik, in N lat. 42° 21′, E long. 24° 20′. It belongs to the Balkan chain.

TOURJELE, a town of Russia in Europe, in the gov. and district and 18 m. SSE of Vilna.

TOURKA, a small river of Russia in Asia, in the gov. of Irkutsk and district of Verkhne-Oudinsk, an affluent of Lake Baikal.

TOURLA, a col of the Pyrenees, between the

French dep. of the Hautes-Pyrenees and the Spanish prov. of Aragon, near the source of the Gaub, and to the SSW of Canterels.

TOUR-LANDRY (La), a commune and town of France, in the dep. of the Maine-et-Loire, cant. and 5 m. SSE of Chemille. Pop. 1,781.

TOURLAVILLE, a commune and town of France, in the dep. of the Manche, cant. and 4 m. ENE of Octeville, at the foot of a fine range of hills, on which are the remains of an ancient Roman camp. Pop. in 1846, 4,549. It has extensive manufactories of glass, pottery, and bricks. In the vicinity are quarries of slate and paving-stones. This village is supposed to occupy the site of the ancient *Coriallum*, mentioned by Ptolemy in the itinerary of Antoninus, and the Theodosian map.

TOURLY, a village of France, in the dep. of the Oise, cant. and 4 m. SE of Chaumont, on the slope of a hill, near the l. bank of the Troene. Pop. 220. It has an ancient castle.

TOURMENTE (Cape), a steep and elevated headland of Lower Canada, in the district of Quebec and county of Northumberland, on the NW coast of the estuary of the St. Lawrence, 8 m. below the island of Orleans. It has an alt. of 1,800 Parisian ft. above sea-level.

TOURMIGNIES, a village of France, in the dep. of the Nord, cant. and 2 m. SW of Pont-a-Marcq, on the Marcq. Pop. 1,015.

TOURNA, a fortress of Turkey in Europe, in Bulgaria, in the sanj. of Rustchuk, 2 m. WNW of Nikopol, on the l. bank of the Danube, a little below the confluence of the Aluta.

TOURNAGHAUT, a celebrated pass of Hindostan, in the prov. of Bujapur, in the Western Ghauts, between the state of the rajah of Sattara and the English district of Coucan, in N lat. 17° 47′.

TOURNAI, Tournay, or Doornik, an arrondissement, department, commune, and town of Belgium, in the prov. of Hainault. The town is 15 m. ESE of Lille, and 47 m. SW of Brussels, on the railway from Mons to Brussels, and on the Scheldt, by which it is divided into two parts connected by several bridges. It is enclosed by walls, bastioned and defended on the S by a citadel and other fortifications. The streets are straight and well kept, and the houses well-built. The quays, which are spacious, stretch along the banks of the river, and are in some parts lined with trees. Amongst its other edifices may be named the cathedral, a fine Gothic structure with five towers, the church of St. Martin, the Episcopal palace, the town-house, the guardhouse, the belfry, an ancient hospital, and some fine flour-mills. It has besides several churches, commercial chambers, a bank, a theatre, an athenæum, several academies, an orphan's asylum, five hospitals, and fine promenades. The pop. of T. is about 33,000. This town, one of the most commercial in Belgium, possesses manufactories of hats, printed muslin, cotton and woollen fabrics, blankets, hosiery, china and earthenware, paper, tobacco, and soap, several tanneries, dye-works, spinning-mills, limekilns, &c. It is noted for its liqueurs, especially curacao, and for its fruit. Its trade consists chiefly in articles of local manufacture, bronze, marble, chocolate, and fruit. This town returns 5 deputies to the provincial states. T., the *Tornæum* of the middle ages, one of the most ancient and important towns of Belgian Gaul, was, at the period of the conquest by Julius Cæsar, capital of the Nerviens, and bore the name of *Civitas-Nerviorum*. It has suffered much from military operations. It was formerly strongly fortified, and had one of the finest citadels in Europe, which was levelled by the French towards the middle of the 18th cent. In 1792, an action took place here between the Austrians and the French; and in May 1794, a more serious conflict between the latter and the allied troops under the duke of York. This was soon followed by the evacuation of Flanders, and T. remained in possession of the French until Feb. 1814.

TOURNAISIS, an ancient district of Belgium, in Flanders, now comprised partly in the prov. of Hainault, and partly in that of Western Flanders. Its cap. was Tournai.

TOURNAN, a town of France, dep. of Seine-et-Marne, 14 m. N by E of Melun. Pop. 1,765.

TOURNAY, a town of France, dep. of Hautes-Pyrenees, 12 m. SE of Tarbes, on the Arios. Pop. 1,258.

TOURNEPPE, a commune and village of Belgium, in the prov. of Brabant, 3 m. E of Halle, on a small affluent of the Senne. Pop. of com. 2,746.

TOURNETTE, a mountain of the Alps, in Savoy, near Annecy, having an alt. of nearly 7,200 ft. above the level of the sea.

TOURNEUR (Le), a commune and village of France, in the dep. of Calvados, 2 m. NNE of Beny-Bocage. Pop. 1,700.

TOURNON, a town of France, in the dep. of Ardeche, on a declivity near the Rhone, and opposite to Tain with which it communicates by a suspension bridge. Pop. in 1821, 3,706; in 1846, 4,807. It has a traffic in wine, charcoal, wood, silk, and corn; and some manufactures of silk and cotton stuffs, tiles, bricks, potash, and beet-root sugar. On the opposite bank of the Rhone is cultivated the well-known wine called Hermitage.—The arrond. of T. comprises 11 cantons, and an area of 184,094 hectares. Pop. in 1846, 143,365.

TOURNON, a town of France, dep. of Lot-et-Garonne, 14 m. E of Villeneuve, and 22 m. NE of Agen. Pop. of com. in 1846, 5,160.

TOURNON-SAINT-MARTIN, a canton and village of France, in the dep. of Indre, arrond. and 9 m. NW of Le Blanc. Pop. of v. in 1846, 1,299; of cant., 8,470.

TOURNUS, a town of France, dep. of Saone-et-Loire, on the r. bank of the Saone, 20 m. NE of Maçon. Pop. 4,216. It carries on a traffic in corn, wine, and wood; and was formerly the seat of a richly endowed abbey.

TOURNY, a village of France, in the dep. of Eure, cant. and 3 m. NW of Ecos. Pop. 1,050.

TOURO, a village of Portugal, in the prov. of Beira, comarca and 30 m. NE of Castello-Branco.

TOUROGEN, a town of Russia, in the gov. of Wilna, 21 m. NE of Tilset.

TOUROUVRE, a town of France, dep. of Orne, 7 m. NE of Mortagne. Pop. 1,700. Iron is wrought in the vicinity.

TOURPES, a commune and village of Belgium, in the prov. of Hainault, on the Dendre, 18 m. NW of Mons. Pop. 1,350.

TOURRETTE-DE-LEVENZO (La), a village of the Sardinian states, in the prov. and 6 m. N of Nice, in a valley watered by an affluent of the Paglion. Pop. 1,200.

TOURRETTES (Les), a village of France, in the dep. of Var, cant. and 3 m. NNE of Bar. Pop. 1,146.

TOURS, a city of France, the capital of the dep. of Indre-et-Loire, situated in a delightful plain, on the S or l. bank of the Loire, a little above the spot where that river is joined by the Cher, in N lat. 47° 23′ 47″, 120 m. SSW of Paris. The city is of an oblong form, and lies parallel to the course of the Loire, across which is here thrown a magnificent bridge of 15 arches, each of 26½ yds. space, 1,400 ft. in total length, and 75 ft. in width. Including its five suburbs, it had a pop. of 31,772 in 1789; 26,669 in 1836; and 30,766 in 1846, a number by no means proportioned to the large extent of ground enclosed by the walls; but many of the houses are small and low in structure. The Rue-Neuve or Rue-Royale, is a street of great elegance, built on a uniform plan, and traversing the city throughout its whole length in a line with the bridge over the Loire. In the same line is the Avenue-de-Grammont, leading to the bridge over the Cher; and the great walk called the Mail extends in a line with the bridges and the Rue-Neuve. The length of the whole avenues, street, and bridges, is not short of 5 m. The smaller streets to the r. and l. of the Rue Neuve, are as narrow and gloomy as in other French towns. T. is, and has long been, the seat of an archbishop. Its cathedral is remarked for its lofty spires, its huge western window, its ingenious clock, and its library. The interior is 256 ft. in length, and 85 ft. in height. The hotel-de-ville, which contains a public library of 40,000 vols., and the theatre are good provincial buildings; and the public residence of the archbishop is a very fine structure; T. has a race-course, a botanical garden, and a museum.—Of the manufactures of T. the principal is silk, which was introduced here earlier than in any other town of France, by workmen brought from Italy in the 15th cent. Their number increased in the course of the 16th to 20,000; but the competition of Lyons, situated in a climate more favourable to the silk worm, has reduced this manufacture in T. and its neighbourhood. The other manufactures of the place are woollens, woollen yarn, carpets, starch, wax-candles, and leather; some wine and brandy is made from the vines of the neighbourhood, but altogether the trade of the town, notwithstanding the vicinity of two navigable rivers, is inconsiderable. It is now connected by railway with Paris, Bordeaux, and Nantes. The great advantage of T. is its attraction to travellers, as well from the elegance of the town as the beauty of the surrounding country. It has consequently been a favourite place of resort for our countrymen since 1814. The climate is considerably warmer than in Britain during summer; and the winters are not long. T., the ancient *Cæsaromagus*, has been a place of note since the early ages of the French monarchy. It was the scene of the repulse of the Saracens by Charles Martel in 732; and in subsequent ages, its castle, built on a rock, served more than once as a place of refuge for the royal family in times of commotion.—The arrond. of T., comprising an area of 265,712 hectares, and 9 cantons, had a pop. of 157,062 in 1846; the cant., comprising 23 coms., had a pop. of 55,432.

TOURS, a commune and town of France, in the dep. of the Puy-de-Dome, cant. and 5 m. E of St. Dier, at the foot of a high hill. Pop. in 1846, 2,620.—Also a com. in the dep. of the Somme, cant. and 4 m. SW of Moyenneville. Pop. 1,158.—Also a village of Sardinia, in the div. of Savoy and prov. of Upper Savoy, 2 m. SE of Conflans, on the slope of a hill, near the r. bank of the Isere. Pop. 500.

TOUR-SAINT-GELIN (SAINT), a village of France, in the dep. of the Indre-et-Loire, cant. and 6 m. from Richelieu. Pop. 820.

TOURS-SUR-MARNE, a village of France, in the dep. of the Marne, cant. and 6 m. E of Ay, on the r. bank of the Marne. Pop. 796. It has a considerable trade in wine.

TOURTASKA, a town of Russia in Europe, in the gov. and district and 90 m. NE of Tobolsk, on the r. bank of the Oby.

TOURTENAY, a village of France, in the dep. of the Deux Sevres, cant. and 7 m. NNE of Thouars, near the l. bank of the Dive. Pop. 600. It has a distillery of brandy, and in the vicinity are quarries of free-stone.

TOURTERON, a canton, commune, and town of France, in the dep. of the Ardennes, and arrond. of Vouziers. The cant. comprises 10 coms. Pop. in 1831, 5,551; in 1846, 5,643.—The village is 11 m. NNW of Vouziers, on a small affluent of the Aisne. It has a distillery and a tannery.

TOURTOIRAC, a commune of France, in the dep. of the Dordogne, cant. and 5 m. WNW of Haute fort, on the l. bank of the Haute-Vezere. Pop. 1,406.

TOURTOUSE, a village of France, in the dep. of the Ariege, cant. and 4 m. SW of St.-Croix-de-Volvestre, on the r. bank of the Lons. Pop. 1,275.

TOURVES, a commune and town of France, in the dep. of the Var, cant. and 7 m. W of Brignolles, in a fertile plain. Pop. in 1846, 2,548. It has distilleries of brandy, manufactories of soap, and tanneries. Marble is quarried in the vicinity.

TOURVILLE, a commune of France, in the dep. of the Manche, cant. and 2 m. S of St. Malo-de-la-Lande, on a hill, near the shore of the English channel. Pop. 1,026.

TOURVILLE-LA-CAMPAGNE, a town of France, in the dep. of the Eure, 13 m. W of Louviers, on a plateau. Pop. 1,060.

TOURY, a town of France, in the dep. of the Eure-et-Loire and cant. of Janville. Pop. 1,35 It has manufactories of beet-root sugar.

TOUS, a village of Spain, in the prov. of Valencia, 21 m. NNW of San-Felipe-de-Jativa, near the l. bank of the Jucar, on a hilly locality. Pop. 1,025. It has two public schools, and manufactories of basket-ware and of charcoal.

TOUSKAR, a river of Russia in Europe, in the gov. of Kursk, which has its source on the confines of the gov. of Orel, 21 m. NE of Fatej; flows through the W part of the district of Chtchigri into that of Kursk; passes the town of that name; and a little below joins the Seim on the r. bank, after a sinuous course in a generally S direction of about 90 m.

TOUS-LES-SAINTS, or TODOS-OS-SANTOS, an island of the Japan archipelago, in the N. Pacific, to the SE of Nifon, in N lat. 30° 10′, E long. 141° 20′.

TOUSSON, a village of France, in the dep. of the Seine-et-Marne, cant. and 6 m. WNW of La Chapelle-la-Reine. Pop. 456.

TOUSSOUL, a small lake of Independent Tartary, in Badakshan, to the SE of Vushgird. It receives on the SE a considerable river, but has no apparent outlet.

TOUSTE, a town of Galicia, in the circle and 33 m. SE of Tarnopol, on the l. bank of the Gnila, which here forms a lagune.

TOUTENCOURT, a village of France, in the dep. of the Somme, 5 m. SW of Acheux. Pop. 1,140.

TOUTES-VOYES-SUR-OISE, a hamlet of France, in the dep. of the Oise and cant. of Creil, 9 m. W of Senlis, on the l. bank of the Oise. It has manufactories of iron and of tin wares.

TOUTSCHKOV, a town of Russia in Europe, in the gov. of Bessarabia and district of Ismail, on the Danube. Pop. 12,000. It is of modern foundation.

TOUVANT, a town of Algiers, in the prov. of Mascara, a little to the SW of Cape Hone, on the Mediterranean.

TOUVET (LA), a canton, commune, and town of France, in the dep. of the Isere and arrond. of Grenoble. The cant. comprises 15 coms. Pop. in 1831, 13,263; in 1846, 13,471. The town is 18 m. NE of Grenoble, in the valley of Gresivaudan, near the r. bank of the Isere. Pop. 1,796.

TOUVOIS, a village of France, in the dep. of the Loire-Inferieure, 5 m. NW of Legé. Pop. 700.

TOUVRE, a river of France, in the dep. of the Charente, which has its source 5 m. ENE of Angouleme, in a narrow basin enclosed by steep banks. Issuing thence with a breadth of 260 ft., it runs W past Touvre, Magnac, and Ruelle; and after a course of about 6 m., throws itself into the Charente on the l. bank, 1½ m. N of Angouleme. Its waters abound with fish.—The village of T. is 5 m. E of Angouleme. Pop. 230.

TOUZAC, a village of France, in the dep. of the Charente, cant. and 6 m. SW of Chateauneuf, on the slope of a hill, near the r. bank of the Né. Pop. 796. It has distilleries of brandy.

TOUZ-CASSRI, or RUSO-CASTRO, a town of Turkey in Europe, in Bulgaria, in the sanj. of Silistria, 15 m. W of Burgas, on the l. bank of the Ruso-Castro, an affluent of the gulf of Burgas.

TOUZER, or TOZER, a town of the beylik of Tunis, 150 m. WSW of Cabas, on the r. bank of Lake Loudeah. It has an active trade in dates.

TOUZ-GHIEUL, TUZLA, or SALATO, a lake of Asia Minor, in Caramania, partly in the sanj. of Ak-Serai, and partly in that of Konieh. It is about 42 m. in length from NE to SW, and 6 m. in breadth, and appears to be several lakes conjoined. Its waters are salt, and it has no apparent outlet, except in the rainy season, when they flow off on the NE to the Kizil-Irmak. This lake is the *Tattæa Palus* of antiquity.

TOUZ-KOUL, ISSI-KOUL, TEMOURTON-NOR, or JE-HAI, a lake of the Chinese empire, in the SW part of Sungaria, and div. of Eli, between the Khaisaglim-daban on the NW, and the Mussur-dagh on the SE, and to the S of Balkash-Nor. It is 120 m. in length from E to W, and 45 m. in breadth. It receives numerous tributaries from the surrounding mountains, and discharges itself on the NW into the Tzin, an affluent of the Sir.

TOUZKOURMATI, a large village of Turkey in Asia, in the pash. and 120 m. N of Bagdad, at the foot of a chain of mountains, in a plain watered by a large stream. It comprises about 400 houses, and is surrounded by numerous gardens. It derives its name from the dates and salt which abound in the locality.

TOUZLA (CAPE), a low sandy point on the coast of Anatolia, in the sanj. of Kodja-Ili, on the E side of the sea of Marmara, 3½ m. N by W of Cape Panaghia, and 21 m. SE of Constantinople. From this point the coast bends round to the E.—Also a river of Armenia, in the pash. of Erzerum, which has its source in the Ak-dagh; runs SE; and joins the Mourad-chai, between Melez-Ghird and Lysa, and after a total course of about 60 m. On its banks is a town of the same name.

TOUZLA, or TUZLA, a town of Turkey in Asia, in Caramania, in the sanj. and 60 m. N of Koniyeh, in an elevated plain, at the SW extremity of Touz-Ghieul or Lake Touzla.

TOUZLA (LOWER), a town of Turkey in Europe, in Bosnia, in the sanj. and 60 m. ENE of Trawnik, on the Idalla, an affluent of the Spritza. It has a wooden enclosure, an abundant salt-spring, and a remarkable grotto. The inhabitants are chiefly Christians. The environs abound with grain and fruit, and afford excellent pasturage.

TOUZLA (UPPER), a town of Turkey in Europe, in Bosnia, in the sanj. of Trawnik, 3 m. ESE of Lower Touzla, on the r. bank of the Idalla.

TOUZLA-CHAI, a river of Turkey in Asia, in Anatolia, in the sanj. of Bija, which descends from the W side of the Kaz-dagh, a summit of the Ida chain; runs first SW, then NW; and after a course of about 51 m., throws itself by two embouchures into the Archipelago, 12 m. N of Baba. The ancient *Pedasus* and *Smynthium* were situated on the banks of this stream.

TOVA, a small island of Patagonia, in the N part of the gulf of St. George, on the E coast, in S lat. 45° 12′, W long. 66° 10′.

TOVARRA, a town of Spain, in the prov. of Albacete, 27 m. SSE of Chinchilla, between two mountains, on each of which are a chapel, a convent, and an hospital. Pop. 6,490. It has manufactories of linen and coarse woollen fabrics.

TOVIANY, a town of Russia in Europe, in the gov. of Vilna, district and 12 m. NNW of Vilkomir, near the r. bank of the Moucha.

TOW, or TOVE, a small stream which crosses the S part of Northamptonshire, and falls into the Ouse at Cosgrove, near Stony-Stratford, Bucks.

TOWAGHTY, or TONAGHTY, a parish in co. Mayo, 6 m. NNW of Hollymount. Area 3,067 acres. Pop. in 1831, 1,258; in 1851, 884.

TOWAHNAHIOOKS, a river of N. America, which falls into the Columbia, about 275 m. above its mouth. It is 200 yds. wide at its confluence, has a rapid current, and contributes about one-fourth as much water as the Columbia possesses before the junction. Its banks are chiefly inhabited by Snake Indians, who live principally on the salmon with which this river abounds.

TOWANDA, a township and village, the cap. of Bradford co., in Pennsylvania, U. S., 100 m. N by E of Harrisburg. Pop. 1,171.

TOWARD-POINT, a cape at the S end of the co. of Argyle, on the Clyde, 24 m. S of Inverary, in N lat. 55° 51′ 45″, W long. 4° 59′ 42″.

TOWCESTER, a parish and market-town in the co. of Northampton, 8 m. SW by S of Northampton, on the small river Tow. The parish comprises the hamlets of Caldicott, Handley, and Wood-Burcott. Area 2,790 acres. Pop. in 1831, 2,671; in 1851, 2,625.—The town, which consists chiefly of one long street of well-built houses, is a thoroughfare on the road from London to Holyhead. The inhabitants are chiefly employed in the manufacture of boots, shoes, silks, and bobbin-lace.

TOWEDNACK, a parish in Cornwall, 2½ m. SW by S of St. Ives. Area 2,794 acres. Pop. 1,057.

TOWER-ISLAND, or TOREN, a small island in the Eastern seas, near the SW coast of the island of Flores, in S lat. 8° 52′ 30″, E long. 120° 10′ 00″. It has a sugar-loaf shaped peak near its W extremity.

TOWERSEY, a parish of Buckinghamshire, 9 m. SW of Aylesbury. Area 1,200 acres. Pop. 446.

TOWIE, or TOWIE-KINBATTOCK, a parish in the NW of Aberdeenshire, intersected by the Don. Pop. in 1831, 728; in 1851, 756.

TOWMORE, TOOMORE, or TUOMORE, a parish in co. Mayo, containing the town of Foxford. Area 6,787 acres. Pop. in 1831, 3,576; in 1851, 2,498.

TOWNSBURY, a village of Warren co., in New Jersey, U. S., 44 m. NW of Trenton.

TOWNSEND, a township of Middlesex co., 40 m. NW of Boston. Pop. in 1840, 1,892; in 1850, 1,947.—Also a township of Windham co., Vermont, 40 m. SSW of Windsor, intersected by West river. Pop. in 1840, 1,315; in 1850, 1,354.—Also a township of Sandusky co., in Ohio, 96 m. N of Columbus. Pop. in 1840, 692; in 1850, 969.

TOWNSEND HARBOUR, a village of Middlesex co., in Massachusetts, U. S., 37 m. NW by W of Boston.

TOWNSHEND, an island off the E coast of Australia, in S lat. 22° 13′, E long. 159° 29′.

TOWNSHEND (CAPE), a cape on the NE coast of Australia, in S lat. 22° 15′.

TOWNSHEND (Point), a cape on the SE coast of Admiralty island, in the N. Pacific, in N lat. 57° 7′.

TOWNSTALL, a parish of Devonshire, near Dartmouth. Area 1,758 acres. Pop. in 1851, 1,303.

TOWN-YETHOLM, a village in the p. of Yetholm, Roxburghshire, on the l. bank of Beaumont-water, directly opposite to Kirk-Yetholm. Jointly with the latter village it has about 1,000 inhabitants. A neat bridge connects the two villages.

TOWRIDGE, or Torridge, a river of Devon, which rises about 4 m. from Hartland-point, near the source of the Tamar; thence runs in a SE course to Iddlesly, where it is joined by the Oke; and passing by Torrington and Bidford, enters the Severn at Barnstaple bay, about 2 m. below Appledore, where it is joined by the Taw.

TOWTON, a village of the W. R. of Yorkshire, 3 m. SE of Tadcaster, famous for a sanguinary battle fought here between the forces of the houses of York and Lancaster, on Palm Sunday 1461. Pop. in 1851, 122.

TOWY. See Teify.

TOWYN, a parish of Merionethshire, situated on the river Towynmy, 11 m. S by W of Barmouth. Area 26,372 acres. Pop. in 1831, 2,694; in 1851, 2,769. The village commands a fine view of the sea, and has of late years been much frequented as a bathing-place.

TOWYNMY, a river of Montgomeryshire, which runs into the Dowy, 5 m. NE of Machynleth.

TOXTETH-PARK, a township in the p. of Walton, Lancashire, 3 m. SE of Liverpool. Area 3,768 acres. Pop. in 1801, 2,069; in 1831, 24,067; in 1851, 61,334. It forms a suburb of Liverpool.

TOYNTON (High and Low), two adjoining parishes of Lincolnshire, 2 m. E of Horncastle. Area of High T. 1,210 acres. Pop. in 1851, 212. Area of Low T. 950 acres. Pop. 133.

TOYNTON-ALL-SAINTS and ST.-PETER'S, two adjoining parishes in Lincolnshire, 2½ m. S of Spilsby. Area of T.-All-Saints, 3,120 acres. Pop. in 1851, 515. Area of T.-St.-Peter's, 2,530 acres. Pop. 486.

TOZER, a village of the Bilad-el-Jerid, 250 m. S of Tunis. It is a place of some manufacturing industry, and is the principal market for dates in the neighbourhood, and is said to carry on a considerable trade with the Niger. It is supposed to represent the ancient *Tisurus*.

TOZO, a river of Spain, which rises near Truxillo; runs S, and then WNW; and after a tortuous course of 30 m., joins the Rio-del-Monte on the l. bank.

TRABEN, a town of Prussia, in the gov. and 39 m. SW of Coblenz, on the l. bank of the Moselle, opposite to Trarbach. Pop. 1,050.

TRABI, a town of Russia, in the gov. of Wilna, 21 m. SSW of Ochmiana.

TRACHENBERG, a town of Prussian Silesia, 24 m. N of Breslau, on the Bartsh, the chief place of a principality belonging to the family of Hatzfeld. Pop. 2,450.

TRACHSELWALD, a village of Switzerland, in the cant. and 16 m. ENE of Berne. Pop. 1,717.

TRACHT, a village of Moravia, 21 m. S of Brunn, on the l. bank of the Thaya. Pop. 600.

TRACKEHNEN, a village of Prussian Lithuania, 15 m. SSW of Gumbinnen.

TRACTON, a parish in co. Cork, 3 m. S of Carrigaline, containing the villages of Minane and Tracton. Area 5,862 acres. Pop. in 1831, 2,838; in 1851, 1,796. A vein of galena intermixed with quartz and pyrites was at one time worked here; and the remarkable mineral substance called wavellite or hydragillite has been found in the vicinity of Minane-bridge. It occurs in a hill of flinty slate, both adhering to that rock and in detached nodules. The nodules vary in size from the diameter of a pea to nearly 2 inches. The lustre is splendent; the colour, bluish green, passing into yellow.

TRACUNHAEN, a town of Brazil, in the prov. of Pernambuco, 40 m. W of Goyanna, on a river of the same name, which flows into the Atlantic near Point Pedras.

TRACY, a village of France, in the dep. of Nievre, cant. and 4 m. NW of Pouilly, on the r. bank of the Loire. Pop. 1,000.

TRACY-LE-MONT, a village of France, in the dep. of the Oise, cant. and 5 m. NW of Attichy. Pop. 1,500.

TRACY'S LANDING, a village of Ann Arundel co., in Maryland, U. S., 13 m. S by W of Annapolis.

TRADATE, a town of Austrian Italy, in the Milanese, 20 m. NW of Milan, near the l. bank of the Olona. Pop. 2,146. It has some trade in the wines of the district.

TRADE-TOWN, a village of Liberia, on the Grain coast of Western Africa, 15 m. SE of Great Bassam.

TRADEWATER, a river of Ohio, U. S., which has its rise in Christian co., and, running NW, falls into the Ohio, 200 m. below the mouth of Green river. It is about 70 yds. wide at its mouth, and has a course of 80 m.

TRADOK, a village on the W coast of Borneo, in the state and 21 m. SW of Sambus.

TRAETH-BICHAN, a bay of the Irish sea, on the W coast of the co. of Merioneth, 2 m. N of Harlech.

TRAETH-MAWR, a bay of the Irish sea, on the W coast of Wales, between the cos. of Merioneth and Caernarvon, 4 m. N of Harlech.

TRAFALGAR (Cape), a low sandy point on the SW coast of Spain, on the coast of Andalusia, at the NW entrance of the straits of Gibraltar, opposite to Cape Esparte on the coast of Africa, in N lat. 36° 10′, W long. 6° 1′. On the 21st October 1805, the British fleet, commanded by Nelson, obtained a complete victory over the combined fleets of France and Spain under Villeneuve and Gravina, off this cape.

TRAFAYACH, a village of Styria, 12 m. W of Bruck, on an affluent of the Muhr. Pop. 610.

TRAFFORD-BRIDGE, a township of Plemondstall p., Cheshire, 4½ m. NE by E of Chester. Area 401 acres. Pop. in 1831, 58.

TRAFFORD-WIMBOLDS, a township of Thornton-in-the-Moors p., in Cheshire, 6 m. NE by N of Chester. Area 574 acres. Pop. in 1851, 106.

TRAGACETE, a village of Spain, in the prov. and 27 m. NE of Cuenca. Pop. 1,000.

TRAGARTH, a river of Brecknockshire, which runs into the Melta, above Istradwelthy.

TRAGEIN, or Tragwein, a village of the archd. of Austria, 12 m. SE of Freystadt, near the r. bank of the Zellerbach.

TRAGHAN, a town of Fezzan, 30 m. E of Murzuk, once considerable, but now reduced to 500 or 600 inhabitants. The vicinity is well cultivated.

TRAGONISI, or Dragonisi, a small uninhabited island in the Grecian archipelago, 2 m. E of Myconi. It affords secure anchorage to vessels, but produces nothing beyond a little pasturage.

TRAHIGUERA, a town of Spain, in the prov. of Castellon-de-la-Plana, 16 m. NW of Peniscola, near the l. bank of the Senol. Pop. 1,900. It has manufactories of earthenware.

TRAHIRAS, a town of Brazil, in the prov. and 210 m. NNE of Goyaz, on a river of the same name, an affluent of the Rio-das-Almas.

TRAHONA, a town of Austrian Italy, in the

Valteline, in a valley, to which it gives name, 12 m. S of Chiavenna.

TRAIAN, a town of Turkey, in the sanj. of Rustchuk, 30 m. W of Kabrova, on the l. bank of the Osma.

TRAILINSKAIA, a town of Russia, in the territory of the Don-Cossacks, 24 m. SE of Kundimtchevskaia, on the r. bank of the Don.

TRAINA, a town of Sicily, in the Val-di-Demona, on the river Traina, 30 m. NW of Catania. It contains a pop. of 7,000; but, like most of the inland towns of Sicily, it offers little of interest to travellers.

TRAINEL, a town of France, in the dep. of Aube, 6 m. SSW of Nogent-sur-Seine, on the Orvin.

TRAISKIRCHEN, or DRASKIRCHEN, a village of the archd. of Austria, 14 m. SW of Vienna.

TRAITORS' COVE, a harbour on the W side of the island of Revilla-Gigedo, in the N. Pacific, in N lat. 55° 40′, so called by Vancouver, from his having here been exposed to an attack of the natives, who resisted all his efforts to conciliate them.

TRAITORS' HEAD, the NE point of Erromango, one of the New Hebrides, in S lat. 18° 43′; so called by Cook in 1774, from the treacherous conduct of its inhabitants.

TRAITORS' ISLANDS, a cluster of small islands in a bay off the coast of New Guinea, in S lat. 1° 12′.

TRAITORS' ISLAND, an island in the Pacific, in S lat. 15° 55′, W long. 173° 48′, discovered by Le Maire and Schouten, in 1616, and so called from an attempt made by the natives to seize the vessel. It is low, with a hill in the centre, and divided by a channel 300 yds. wide, from the island of Kootahe. Both these islands were visited by Wallis in 1767, who called them Keppel's and Boscawen's islands; and afterwards by Perouse in 1787.

TRAJANOPOLI, or ORIKHOVA, a town of European Turkey, in Romelia, situated at the foot of the Despoto-dagh, on the r. bank of the Maritza, the ancient *Hebrus*, about 30 m. from its mouth, in N lat. 41° 7′ 30″. It is the see of a Greek archbishop, and contains about 15,000 inhabitants; but being at some distance from the most frequented roads, is little visited by travellers. It is a place of considerable antiquity, and took its name from Trajan, to whom it was indebted for considerable improvements.

TRAJETTO, a town of Naples, in the Terra-di-Lavoro, 9 m. NE of Gaeta, on the r. bank of the Ansento, on the site of the ancient *Minturnæ*.

TRAKEMBURG, a town of Styria, 19 m. SE of Cilly.

TRAKOSKIN, a small town of Austrian Croatia, 19 m. W by S of Warasdin.

TRAKTOMEROV, a town of Russia, 52 m. SE of Kiev, on the r. bank of the Dnieper.

TRALEE, a parish and town in co. Kerry. Area of p. 4,605 acres. Pop. in 1831, 11,021; in 1841, 12,534. The parish comprises a small part of the Stacks mountains, and a prominent part of the strath or valley of the little river Lee.—The town, which is also a sea-port and a parl. borough, and the assize town and capital of co. Kerry, stands on the rivulet Lee, at the intersection of the road from Ardfert to Milltown with that from Dingle to Limerick, 1¾ m. ENE of the head of Tralee-harbour, and 143¾ m. SW by W of Dublin. The site and the immediate environs of the town are low, flat, and subject to inundations from freshets of the Lee; but the middle grounds and the back grounds of the scenery around it are so composed as to form a panorama at once brilliant, beautiful, and comparatively unique. In front of the town, the vast expanse of waters and accompanying range of mountains which rise boldly from the shores of the bay, and run W to Brandon, where they dip into the ocean; and on the other side, the broad fertile valley stretching E to Castle-Island, which is bounded on the N by the Stack-hills, and on the S by the higher range of Slieve-Mish. Several of the streets are spacious, well-built, regularly laid out, and aggregately neat or even handsome; the southern outskirts embrace a large expanse of demesne-grounds disposed in public promenades. The rivulet Lee runs through the middle of the town; and is spanned in its progress by several small bridges. T. is one of the most thriving towns in the S of Ireland; and has, for a number of years past, experienced great increase of trade, both as a market-town and as a sea-port. The borough sent two members to the Irish parliament; and continues to send one to the imperial parliament. Constituency in 1841, 258; in 1851, 228. Pop. of the town and borough in 1831, 9,568; in 1841, 11,363; in 1851, 12,649.

TRALEE-BAY, a bay on the coast of co. Kerry. It consists of an outer portion, called Tralee-bay, and an inner portion, usually called Tralee-harbour. The bay, or outer portion, is screened on the W by the low and flat peninsula which runs N as a partition between it and Brandon bay. Tralee-harbour or the inner bay, opens at the Samphire islands with a width of rather more than 1½ m., contracts to a width of ¾ m., re-expands to a width of 1½ m., thence gradually contracts to the embouchure of the river Lee, and altogether penetrates the land eastward to the extent of 4⅓ m.

TRALEE-SPA, a village, and a celebrated medicinal well, in co. Kerry, on the N shore of Tralee-harbour, 2¾ W by N of Tralee.

TRALLEBORG, or TRELLEBORG, a small port of Sweden, on the Baltic, 12 m. E of Falsterbo. Its roadstead is open, excepting to N and NW.

TRALLONG, a parish of Breconshire, 5 m. W by N of Brecon. Area 3,384 acres. Pop. 319.

TRAMACASTIEL, a village of Spain, in the prov. and 24 m. WNW of Teruel. Pop. 300.

TRAMAYE, or TRAMEYES, a town of France, dep. of Saone-et-Loire, 9 m. W of Macon, at the source of the Valouze. Pop. 1,600.

TRAMAZZA, a village of Sardinia, in the prov. and 15 m. WSW of Busachi.

TRAMBLY, a village of France, in the dep. of Saone-et-Loire, 15 m. W of Macon. Pop. 750.

TRAMELAN, a parish and village of Switzerland, in the cant. and 27 m. W of Berne, on the Birse. Pop. 551.

TRAMIN, a village of the Tyrol, 9 m. SSW of Botzen. Pop. 600.

TRAMONTI, a town of Naples, in the Principato-Citra, 7 m. W of Salerno. Pop. 3,300.

TRAMORE, an open and dangerous bay, on the coast of co. Waterford. It opens between Great Newtown-head on the W, and Brownstown-head on the E,—the latter situated 6 m. W of Hook Head, or the E side of the entrance of Waterford. The open part of the bay has an almost uniform width of nearly 2⅓ m., and penetrates the land to the extent of 1¾ m. The lagune or closed part of the bay called the Back-strand, communicates with the open part only by a very narrow channel in the extreme E, is separated from all the remainder of the head of the open bay by a narrow and prolonged spit of sand-hills, alternately receives and empties through the connecting channel all the waters which occupy its area at full tide, and expands with a length of about 2 m. from E to W, and a varying breadth of from 6 furls. to 1½ m.—Also a small town, and a sea-bathing resort, in the p. of Drumcannon, co. Waterford, at the NW extremity of the open bay of Tramore, 6 m. S by W of Waterford. It is protected from the prevailing winds, commands a magnificent view of the sea, enjoys the advantages of a great extent of hard and level strand, and possesses all the other appliances which can recommend it as a sea-bathing resort. It has good markets, and commodious lodging-houses and hotels. Its strand is about 3 m. in length, and everywhere quite consolidated and smooth. Pop. in 1831, 2,224; in 1851, 1,882

TRAMUTOLA, a town of Naples, in the Basilicata, 28 m. NE of Policastro. Pop. 4,000.

TRANA, a village of the Sardinian states, in the prov. and 20 m. ESE of Susa, on the l. bank of the Sangone.

TRANBY, a village of Norway, in the bail. of Baskerud, 4 m. NE of Bragernaes.

TRANCOSO, a town of Portugal, in the prov. of Beira, 9 m. W of Pinhel. Pop. 2,000. It is surrounded by old walls, and has a castle.

TRANCOSO, a town of Brazil, in the prov. and 12 m. S of Porto-Seguro, in S lat. 16° 44′ 40″.

TRANEKIAR, an ancient fortress of Denmark, on the E side of the island of Langeland, 9 m. from Rudkioping.

TRANENT, a parish in the NW of Haddingtonshire; bounded on the S by Edinburghshire, and on the NW by the p. of Prestonpans. Area 5,415 acres. The coal-formation, with its attendant strata, but dislocated and intersected by trap-dikes, lies beneath a large portion of the parish. Coal has been worked in it in 5 seams, aggregately from 23 to 26 ft. thick, and possibly exists in other and lower seams which have never yet been explored. The battle of PRESTONPANS [see that article] was partly fought within this p. Pop. in 1801, 3,046; in 1831, 3,620; in 1851, 4,158.—The village of T. is situated on the mail-road between Edinburgh and London, 1½ m. SE of Prestonpans, and 9½ m. E of Edinburgh. It consists principally of a street which extends about ½ m., from E to W, along the public road; and of a cross street which goes off from the former near its middle, and runs about a ¼ m. down the declivity to the N; but it is built on no regular plan, and has a large aggregate of decayed and mean-looking houses.

TRANFRANT, a small port of Algiers, 30 m. SW of Oran.

TRANI, a town of Naples, situated on the Adriatic, in the prov. and 24 m. W by N of Bari. Pop. 14,000. It is better built than most towns in this part of Italy; its streets being broad, and its houses constructed of stone of a yellow tint, which gives a lively and pleasant appearance to the buildings. Its walls, in circuit about 2½ m., are built of the same materials, and most of the public edifices. The cathedral is adorned with marble pillars, and has a magnificent tower. There are 20 other churches. The principal square is handsome, and the castle has a claim to the notice of the traveller, on account both of its strength and antiquity. The harbour is tolerably secure; and fruit, corn, and oil are exported from it.

TRANMORE, a township of Bebbington p., in Cheshire, 7 m. N by E of Great Neston. Area 730 acres. Pop. in 1831, 314; in 1851, 6,519, an increase occasioned by the extension of Birkenhead.

TRANQUE, an island off the coast of Chili, separated from Quelan-point in the island of Chiloe by a channel about 1 m. wide. Its N shore is well-wooded, and it rises to 300 ft. above sea-level.

TRANQUEBAR, a port of Hindostan, in the Carnatic, and district of Tanjore, on one of the mouths of the Caveri, 20 m. N of Negapatam, in N lat. 11° 1′. It was formerly a village belonging to the rajah of Tanjore, and was purchased from him by the Danes in 1616. Its trade is chiefly carried on with Denmark, the Isle-of-France, Ceylon, Bengal, and Prince of Wales island. Coarse cotton, oil, soap, and leather are manufactured here. The pop. of the town and of the ancient territory belonging to Denmark, and which has an area of about 15 sq. m., is about 25,000. The fort, which is called Dansborg, is kept in neat order. It was taken by the British in 1787, but restored at the peace.

TRANS-CAUCASIA. See CAUCASUS.

TRANSYLVANIA, a principality in the SE of Austria, comprehending part of ancient *Dacia*, and called by the natives, ERDELY ORSZAG, that is, 'the Woody and mountainous country;' by the Germans, SIEBENBURGEN, from seven celebrated forts or castles which it contains; and deriving its name *Transylvania*, from its lying beyond the forests of the Carpathian mountains which divide it from Hungary. It is bounded on the N by Upper Hungary; on the NE by the Bukowina; on the E by Moldavia; on the SE and S by Wallachia; and on the W by Hungary. It lies between the parallels of 45° 25′, and 48° N; and the meridians of 22° 30′ and 27° E. Its greatest extent from N to S, or from the mountains of Hagymas on the N, to the point where the confluent stream of the Old and Maciar-Syl forces its way through the gorge of Mount-Mararat in the Lesser Wallachia, is 180 m. Its greatest extent from E to W, or from the frontier of Moldavia, to the western range of mountains dividing it from Upper Hungary, where the river Koros rushes through a narrow defile into the Hungarian plains, is 150 m. Blumenbach estimates the superficial extent of this principality at 1,118.70 German square miles; Liechtenstern assigns to it only 1,047.8; and Rohrer, 1,100.8. Deducting the Military frontiers included in these admeasurements, and which may be estimated at 253.30 German sq. m., these admeasurements respectively offer us 865.40, 794.5, and 856.5 German sq. m., as the probable surface of T. Stein assigns it 842.3 German, or nearly 18,103 British sq. m. The pop. in 1840 was 2,079,000; in 1850, 2,073,737, exclusive of the Military frontiers.

Physical features, &c.] In general outline T. resembles Bohemia, being nearly oval, and like it surrounded on all sides by ranges of lofty mountains, some of which are covered with perpetual snow. But though resembling Bohemia in these respects, it differs in this, that while the latter resembles a vast basin, or concavity,—the former is woody and mountainous throughout, the surrounding ranges sending out lateral ridges or branches from different points of their immense circumference, which terminate towards the centre of the country in hills covered with vineyards and rich in minerals. Owing to its superior elevation, its climate is more temperate and wholesome than that of Hungary, but the water in many places is strongly impregnated with minerals, and apt to produce colics and other dangerous distempers; even the very wine produced in the country is affected in a similar way. Environed and intersected by mountains, it contains many delightful valleys, watered by innumerable streams, which, descending from the mountains on the NE and E, disembogue themselves into the Maros and the Alauta, the two main rivers of the country; the former of which—the *Marisos* of Strabo—running SW through the whole of T., enters Hungary and joins the Theiss; the latter, after watering the E division of the country, runs S through Wallachia to the Danube. N of the Carpathians the country is drained by the Szamos, the Krasna, and the Koros. There are several lakes and marshes.

Productions.] T. produces abundance of excellent wheat, and its rich pastures feed vast numbers of black cattle. Many extensive forests occur, in which are found buffaloes, bears, lynxes, elks, wild boars, chamois, ermines, and beavers. Agriculture, in all its branches, is the principal occupation of the inhabitants of this country; but it is still carried on in a very primitive manner. Nevertheless, the fertile soil yields more than supplies the home-consumption. The vine, too, is grown on a very extensive scale, and, under good management, a considerable quantity of wine might be produced for exportation. The quantity of salt in this country is enormous; the whole of Europe might be provided with salt from hence. There are no manufactures of any importance in T. The principal articles are woollen cloths, linens, paper, glass, and iron.—The commerce of the country is in the hands of the Greeks and Armenians. The transit from the Austrian provinces to Hungary and Germany is considerable.—The mines of T. and the Bannat of Temeswar, are very numerous and valuable. Those of Najiag, 12 m. to the NE of Deva, produce gray gold ore, or that precious metal mixed with antimony, arsenic, lead, iron, and sometimes with manganese and zinc. They are the richest in all T., and are conducted with great care. At Ofenbanya, 25 m. to the N of Carlsburg, is found white gold ore, which also occurs in the hills of Fatzebay in the same quarter. To the W of Carlsburg, the country presents numerous gold mines near Zalathna. In the N are the gold mines of Kapnick, Rodna, Felsobanya, and others. Oraviza, in the Bannat of Temeswar, is the chief mining town in that quarter. Towards the S of Oraviza are found mines of copper; and gold and silver mines exist at Dognaska. At Ohlapan, near Zalathna, the finest gold in T. is found, mixed with gravel and sand.

Population, &c.] According to Hassel, there were in 1816, about 380,000 Catholics and United Greeks, 917,000 Greeks or Eastern Church, 210,000 Calvinists, 168,000 Lutherans, 45,000 Unitarians, 5,500 Armenians, and 2,500 Jews, in T.; besides about 70,000 Gipsies, who cannot with propriety be placed under any religious denomination. In 1850, it was estimated that the pop. of T. comprised 725,700 Schismatic Greeks; 605,300 Greek Catholics; 223,400 Roman Catholics; 358,300 Reformed Protestants; 220,400 Protestants of the Augsburg Confession; 44,600 Unitarians, and 7,000 Jews. The Magyars and Saxons have the best schools; those of the Vlaki are utterly miserable. The modern inhabitants of T. are a very mixed race, the descendants of various and distinct tribes. But the principal nations are the Szeckhelyi, or Scythuli, corruptly termed Siculi; the Madshars, Magyars, or modern Hungarians; the Vlakes, and the Saxons. Liechstenstern gives the following enumeration of the different races in T.;—Magyars and Szeckhely, 460,000; Saxons, 420,000; Vlakes or Wallachians, 800,000; Zigeunes or Gypsies, 70,000; Slavi or Slavonians, 7,000; Armenians, 5,500; Jews, 2,500; Italians, 2,000. The Szeckhelyi, or descendants of the ancient Huns, use a dialect of the Hunno-Scythic language; and it is believed, that several manuscripts written in the Hunno-Scythian character, and of great antiquity, are still preserved in T. The Hungarians use the same language with their brethren to the W and N, which is also spoken by the Bulgarians and Armenians. The Saxons speak the same language with the inhabitants of Lower Saxony, but so intermixed with Hungarian as makes it seem quite different from the German or High Dutch; but in the district of Burckland, on the frontiers of Wallachia, where the Saxons live unmixed with any other races, their peculiar language is spoken with the greatest purity. The language of the Vlaki—which is also spoken by the Greeks—is a dialect of the Slavonic, mixed with a great deal of bad Latin.

Government.] The constitution of T. is, or was until recently, a limited monarchy, in many respects resembling that of Hungary; but the prince enjoys here more extensive rights, and his influence in the diet is greater. The executive power belongs entirely to him, and no law can be enacted without his sanction. The deputies to the assembly of the states are elected by the three principal nations, among whom the Magyars hold the first place. The magnates consist of the great officers of state, the counts, and barons. The gentry are Magyars and Szeckhelyi, but the royal towns Saxons only. The principality is governed in the name of the prince and nobility, by the diet, the officers of state, the royal government, the exchequer, the assembly of the nobles, the tribunals of justice, and the magistrates of the Szeckhelyi and Saxons. The diets meet upon summons from the prince, now the statthalter, at Hermannstadt. The Vienna bureaucracy wish to introduce the Austrian code of laws. The Saxons wish to preserve their municipal institutions as intact as possible. The Daco-Romans wish the Austrian administration introduced, and the Saxon privileges abolished, but maintain that they ought to enjoy emoluments and offices in T. in the proportion of their pop., their numbers being nearly 1,100,000 in a population of 2,000,000; but their intelligence and capacity for employment, instead of being on a par with their numbers, are in a miserable minority. In order to render the Daco-Roman fit for employment, the government intends to carry out a system of elementary instruction, the secular part of which will be the same as that in the hereditary provs., and only the spiritual part of it left in the hands of the Daco clergy, who are of the Greek church; but the clergy are not likely to take to it more heartily than our own Catholic clergy in Ireland to the mixed system in that kingdom. The Daco clergy have also an antipathy to the use of Roman letters instead of the Cyrillian alphabet, which, with some slight variations, is the same as that of the Daco-Romans.—The revenues arise from the contributions, customs, metals, minerals, rock-salt, royal demesnes, escheats, and confiscations, and are levied by the treasury. They amount to from four to five million of florins.—The present administrative division is into the 10 circles of Hermannstadt, Kronstadt, Udvarhely, Maros-Vasarhely, Bistritz, Dees, Szilagy, Klausenburg, Karlsburg, and Broosch or Broos.

History.] The ancient inhabitants of T. in the days of Herodotus, were the *Agathyrsi*, a people of effeminate manners, "abounding in gold," and bordering on Western Scythia. About the commencement of the Christian era, T. was occupied by the warlike Daci, a Scythian tribe, of Gothic origin, according to Jornandes. The Jazyges, a Sarmatic tribe, migrated from the neighbourhood of the Palus Mæotis, or sea of Azof, and settled in T., in the neighbourhood of the Dacians, and in the country between the Theiss and Danube, a short time prior to the Roman conquest of Dacia by Trajan. Other tribes, as the Anarti and Taurisci, were dispersed round the sources of the Theiss, Marosch, and Alauta; the Carpi settled at the S foot of the Carpathian mountains, and gave their name to that range. Of these various tribes, the Daci were the most powerful; for besides T., they possessed Wallachia and part of Moldavia; but their last monarch, Decebalus, being compelled to bend before the superior power of Trajan, after a hard contested war of five years, the whole nation was transplanted into the Roman provinces S of the Danube, and their own country was taken possession of by Roman colonists from Italy, and other parts of the empire. The country thus planted with the conquered Dacians, was called *Nova Dacia*, and subdivided into the districts of *Dacia Ripensis*, *Dacia Mediterranea*, and *Dacia Prævalitana*, comprehending the present provinces of Servia and Bulgaria. The Jazyges and Carpi, notwithstanding the destruction of the Dacians, still preserved their independence, and proved very troublesome neighbours to the Roman colonists settled on the Danube. At last the whole nation of the Carpi were transplanted by

Dioclesian, into the Roman territories and Pannonia. Some remnants of the Jazyges still remain in Hungary. In the middle of the 4th cent., the Goths seized T., after defeating the Sarmatians on the Marosch, where king Visimar, with the flower of the Sarmatian nobility, perished. The Goths remained in possession of Dacia, and all the countries from the Baltic to the Black sea, till the year 375, when they were alarmed by the approach of an innumerable host of Scythians who seemed to issue from the frozen regions of Sarmatia. Re-enforced by the Alans, the Roxolani, and the rest of the numerous Sarmatian tribes whom they had subdued, the Huns fell upon the Goths with irresistible fury. The warlike Ermanaric, a conqueror of many nations, unable to survive the awful calamities that now befell his nation, laid violent hands upon himself. His son, Vithimmir, a more resolute prince, for some time made a vigorous resistance, but was finally slain, with the greatest part of the Ostrogothic nation, who retreated to the plains of Podolia. The Huns next attacked the Visigoths, who possessed Dacia, and completely defeated them, though they had constructed a strong wall extending from the Pruth to the Danube, to protect themselves from these dreadful barbarians. At last the whole nation of the Goths, unable to withstand the Huns, abandoned T. and the whole country to the N of the Danube, to their victorious invaders, who thus became masters of all the country from the Don to the Danube. Transylvania was seized by the *Gepidæ*, after the vast empire of the Huns had terminated with the death of Attila; but some tribes of the Huns still remained in the country, whose descendants are supposed to have been the *Siculi*, in the NE. The *Gepidæ* were almost wholly extirpated by the Lombards, aided by the *Avari*, another Scythian tribe, in the year 566; but some of their descendants still remain in Hungary and Transylvania. The Madshars, who conquered Hungary in the 9th cent., added T. to their dominions, in 997. In the Middle ages, T. was peopled by the Hungarians, or Madshars, in the western quarter; the Szeckhelyi, erroneously named Siculi, and believed to be the real descendants of the once formidable Huns, on the NE; a numerous colony of Saxons who penetrated into Transylvania in the 12th cent., and settled on the frontier of Wallachia, between the rivers Alauta and Maros; and the Vlaki who possessed the S, on the borders of Wallachia. T., from its conquest by Stephen, king of Hungary, in 1004, continued subject to that kingdom, and was governed by Waivodes, or governor of provinces. After the fatal battle of 1526, John Scepus, Waivode of Transylvania, married the widow of the unfortunate Louis, and became king of Hungary, under the protection of Solyman the magnificent, holding it and T. as fiefs of the Porte. Upon the death of John, Hungary was converted into a separate kingdom, by Sultan Solyman, and T. was given to his widow and her infant son; but after his death, T. was divided between two rival factions, one of which was supported by the house of Austria, and the other by the Porte, in consequence of which, the country was a scene of intestine warfare for many years. The papal party were supported by the power of Austria; and the Protestant party, headed by the celebrated Botskay, were aided by the Turks; when the successes of the latter obliged the court of Vienna, in 1606, to acknowledge the independence of T. Botskay was succeeded by Bethlem Gabor, a determined enemy of the Catholics and house of Austria, and a firm friend of the Porte, whose vassal he became in order to secure himself against the Catholic faction. He was succeeded by his son Stephen, between whom, and the family of Ragotski, a contest arose for the principality, in which the Ragotski family finally prevailed. Upon the death of Ragotski, a civil war again commenced; but the balance turned in favour of Michael Abaffi, who, aided by the Turks, under the famous vizir Cuprogli, completely defeated the imperialists, at the battle of Clausenburg, and drove them out of T. Abaffi continued in possession of T., as a vassal of the Porte, till his death, in 1690. At his death, the family of Ragotski contended for the principality with the victorious Austrians—who had by this time reconquered Turkish Hungary—but without success. Tekeli, who had married Abaffi's widow and inherited his claims, was equally unsuccessful, and was compelled to live as an exile for the remainder of his life in the Turkish dominions, where he died at Nicomedia. T. was ceded to the Austrians, by the treaty of Carlowitz, in 1699, and has ever since continued subject to the house of Austria, though an attempt was made, in 1738, by the Ragotski family, to recover their influence under the protection of the Porte, but the matter was settled by the peace of Belgrade, in 1740.

TRAONA, a village of Austrian Lombardy, in the prov. of Valteline, 14 m. W of Sondrio, in a valley of the same name, near the r. bank of the Adda. Pop. 1,015.

TRAPANI, an administrative province, district, and town of Sicily. The valle or prov. is bounded on the E by the prov. of Palermo; on the SE by that of Girgenti; and on all other sides by the sea. Pop. in 1845, 182,809. It is 51 m. in length from N to S, and 42 m. in breadth. It presents great irregularity of coast line. On the N is the long peninsula, enclosing the gulf of Castel-a-Mare on the W, and terminating in the N, in Cape S. Vito, and on the W are the peninsulas of Trapani and Marsala, the extremity of the former of which, Cape Boco, forms the most westerly point of Sicily. Stretching along the W coast are several islands, the principal of which are Favignana, Levanza, and Maretimo. The chain of mountains which traverses the island, in its entire length, terminates in this prov., and gives rise to numerous streams, of which the chief are the S. Bartolomeo, the Vitaloca, and the Fogia on the N; the Brigi and Marsala on the W; the Mazzara, Arena, and Modima on the S; and the Belici, which forms the continuous boundary with the prov. of Girgenti. The prov. of T. comprises three districts and 21 communes. The town is 53 m. W of Palermo, at the W extremity of the island. Pop. 24,000. It is built on a tongue of land, which advances into the Mediterranean, and which, with a low promontory to the S, forms a good port. It is enclosed by bastioned walls, and defended by other works. The streets are regular, and are paved with blocks of a shining species of stone. Many of the houses are handsome, and the churches and convents, which are numerous, and other public buildings, are of fine architecture. It has also a college, and several hospitals. The harbour is capable of giving accommodation to vessels of large size, and at the entrance is an island on which a lighthouse has been erected. The trade, which is extremely active, consists chiefly in salt from the adjacent works, alabaster and coral, in the natural state and wrought, soda, wine, fish, &c. At the distance of about 5 m. ENE is Mount S. Giuliano, the ancient *Eryx*, whence water is conveyed by an aqueduct to the town. T. occupies the site of the ancient *Drepanum*, noted as the theatre of frequent combats between the Romans and Carthaginians. With the exception of a single temple it contains no antiquities.

TRAPANO, Drapano, or Mussa, a headland of Turkey in Europe, on the N coast of the island of Candia, 17 m. E of Canea, in N lat. 35° 38', and E long. 24° 16'. It forms the extremity of the elevated promontory which advances between the gulfs of Sudah and Armyro.

TRAPE (La), a village of France, in the dep. of the Dordogne, cant. and 6 m. NNW of Villefranche on an elevated plateau. Pop. 100.

TRAPOHNEN, a village of Prussia, in the regency of Gumbinnen, on the Memel. Pop. 450.

TRAPPA, a town of Portugal, in the prov. of Beira, comarca and 18 m. NNW of Viseu, in a bleak mountainous locality. Pop. 300.

TRAPPE (La). See Soligny.

TRAPPES, a village of France, in the dep. of the Seine-et-Oise, cant. and 6 m. WSW of Versailles, near a lagune. Pop. 640. It has a sugar factory.

TRAPSTADT, a village of Bavaria, in the circle of Lower Franconia, 5 m. ENE of Konigshofen. Pop. 750.

TRAQUAIR, a parish in the SE border of Peebles-shire, bounded on the N by the Tweed. Area 17,200 acres. Except at the indentations, the boundary all round with Selkirkshire is high mountain water-shed, among whose summits are Gumscleuch and Minchmoor, as well as several others of an alt. exceeding 2,000 ft. above sea-level. Quair-water is the principal interior stream. The village of T. stands in the vale of the Quair, ¾ m. S of the Tweed, and 1¾ m. S of Innerleithen. Pop. 643.

TRARBACH, a town of Prussia, in the prov. of the Rhine, regency and 37 m. SW of Coblenz, on a mountain, near the r. bank of the Moselle, opposite Traben, with which it is connected by a bridge. Pop. in 1843, 1,494. It has a saw-mill. In the vicinity are the remains of a Roman way. T. was taken by the French in 1734.

TRARZAHS, a Moorish tribe who inhabit the W part of the Sahara and NW of Senegambia, comprising a district extending inland a distance of about 140 m. They are extensively engaged in the gum trade.

TRASACCO, a town of Naples, in the prov. of Abruzzo-Ultra, 81 m. SE of Avezzano, near the S bank of Lake Fucino, in a fertile plain. Pop. 760. It has a collegiate church. This town is said to have been founded by the emperor Nero.

TRASEN, a river of Austria, in the prov. of Lower Austria, which descends from the N side of the Trasenberg, near the Styrian frontier; runs NNE past Durrnitz, Wilhelmsburg, St.-Polten, Herzogenburg, and Trasmauer; and a little below the latter place throws itself into an arm of the Danube, on the r. bank, and after a sinuous but rapid course of about 57 m. In the lower part it divides into numerous branches, and from the mountains which enclose its bed it receives numerous affluents. Its banks and islands are well wooded.

TRASIMENE. See PERUGIA.

TRASMAUER, a town of Austria, in the prov. of Lower Austria, circle and 13 m. NNE of St. Polten, on the r. bank of the Trasen, near its confluence with the Danube. Pop. 545. It has a fortified castle, and manufactories of fine woollen fabrics.

TRASOBARES, a town of Spain, in the prov. and 45 m. W of Zaragoza, in a fertile locality. Pop. 1,147. It has a convent, and a manufactory of common cloth.

TRAS-OS-MONTES, a province of Portugal, bounded on the N and E by Spain, on the S by the prov. of Beira, and on the W by that of Minho, extending between 40° 48′ and 41° 50′ N lat., and between 6° 10′ and 8° W long., and comprising an area 84 m. in length from E to W, and 60 m. in breadth, and a superficies of 1,020 sq. m. Pop. in 1841, 300,840, or 77,079 families. This prov., although generally mountainous and arid, has several fine plains, watered by the Tamaga, Corgo, Sabor, and Tua, affluents of the Duero, and producing in great abundance corn, rye, maize, and fruit of all kinds; and in localities sheltered from the N, olives and silk. On the rivers, and especially the Duero, the species of grape is cultivated which produces the celebrated port wine. Cattle, horses, sheep, pigs, and bees are reared in great numbers, and game is abundant. Silk factories are numerous, and there are also numerous flax mills. The prov. comprises 2 districts, 11 comarcas, 43 concelhos, and 535 parishes. Its cap. is Braganza.

TRASPADERNE, a town of Spain, in the prov. and 36 m. NNE of Burgos, on the l. bank of the Nela, a little above its confluence with the Ebro. Pop. 356. It has two churches and a convent.

TRASPINEDO, a town of Spain, in the prov. and 15 m. SE of Valladolid, in a rugged and ill-cultivated valley, near the l. bank of the Duero. Pop. 535. It has an hospital.

TRASTE, a port of Dalmatia, in the circle and 8 m. W of Cattaro, formed by the Adriatic, in N lat. 41° 46′, E long. 18° 35′. Its entrance does not exceed 1½ m. in breadth, but it has within an extent of 5 m. from N to S, and a depth of about 3 m.

TRAU (CHANNEL OF), a strait of Dalmatia, in the circle of Spalatro, formed by the Adriatic between the island of Bua and the mainland. It runs between the channel of Solta on the NW, and the gulf of Salona on the SE; and has a total length of 20 m., with an extreme breadth of 5 m. Towards the N, in one of its narrowest parts, is a small island on which the town from which it takes its name is situated.

TRAU, or TRAGHU, a town of Austrian Dalmatia, with a small harbour on the coast of the Adriatic, in the circle and 20 m. NW of Spalatro, in N lat. 43° 40′. Pop. 3,500. It stands on an islet, having on the N side the mainland, and on the S the island of Bua. With the main it is connected by a bridge, and with the island by a mole having a drawbridge to admit the passage of vessels, which are numerous, coasters preferring this inland channel to the open sea. The vicinity is pleasant, abounding with vines and olives. The trade of T. is considerable in wine, oil, and fruit, although its harbour is small; but the island of Bua offers a very commodious harbour.

TRAUCHBURG, or TRAUENBURG, a town of Bavaria, 12 m. WSW of Kempen.

TRAUN, a navigable river of Upper Austria, which rises in the NW extremity of Styria; runs W, and then NNW; and after flowing through the lakes of Hallstadt and Traun, joins the Danube on the r. bank, 3 m. below Linz, after a course of 100 m.—The lake of T., sometimes also called the lake of Gemunden, is 9 m. long, and 4 m. broad. It lies under the parallel of 47° 48′ N, and at an alt. of 495 metres = 541 yds. above sea-level.—There is another river called Traun, in Bavaria, near the borders of the circle of Salzburg.

TRAUNKIRCHEN, a small town of Upper Austria, on the W side of the lake of Traun, 11 m. SSE of Vocklabruck.

TRAUNSTEIN, a walled town of Bavaria, 50 m. ESE of Munich, on the l. bank of the Traun, an affluent of the Alza. Pop. 2,200.—Also a village of the archd. of Austria, 12 m. SSW of Zwettel. It has extensive salt-works, which serve to evaporate the brine of springs at this place, and a part of those of Reichenhall, which is conveyed hither.—Also a mountain of the archd. of Austria, on the E shore of the lake of Traun, having an alt. of 1,578 metres = 1,725 yds. above sea-level.

TRAUNVIERTEL, a district of Upper Austria, lying along the river Traun, from the Danube on the N, to the borders of Styria on the S. It has an area of 1,955 sq. m. The N part is level and fertile, but the S is full of lofty mountains, branches of the Alps. In this part, especially along the Ens, there is a number of iron works and hardware manufactories. This quarter also abounds with salt-mines. The richest salt-mines, which form the wealth of the country, lie in the romantic district called the Salzkammergut, where 500,000 quintals of salt are produced every year, and above 1,000,000 could be produced if there was a market for it. The crown-lands belonging to the emperor, and containing the salt-mines, lie in the SW angle of the archd. of Upper Austria, which is wedged in between the frontier of Salzburg and Styria, and is traversed through its whole extent by the river Traun. The vales here have that pastoral beauty and constant verdure which is so familiar in England, with similar enclosures and hedgerows, and fruit and forest trees; but the chief picturesque beauty of the Salzkammergut lies in its numerous lakes, forming a chain strung together as it were by the Traun, the main artery of the district, passing through them in succession. They are commonly bordered with lofty mountains, whose precipices, rising abruptly from the very water's edge, without the smallest ledge or foreland, impart an extremely sublime character to these Alp-locked reservoirs. The falls of the Traun resemble those of the Rhine at Schaffhausen, though they are much inferior to them. The chief town of the T. is Steyer. In 1809, the western part of this district was ceded to Bavaria, but it was restored to Austria in 1815.

TRAVAGLIATO, an inland town of Austrian Italy, 4 m. WSW of Brescia. Pop. 3,000.

TRAVALLA, a small port on the W coast of the island of Celebes, situated on a small creek at the head of a little inlet, 9 m. S of Dungally, in S lat. 1° 10′. The neighbouring country abounds with cocoa-nuts and Indian corn.

TRAVANCORE, a province in the SW extremity of Hindostan, between the 8th and 10th parallels of N lat., and the meridians of 76° 20′ and 77° 30′ E. On the N it is bounded by the territories of the Cochin rajah; on the E by a range of woody mountains which divide it from the district of Tinnevelly; on the S and the W by the sea. Its length from N to S may be estimated at 140 m., by an average breadth of 36 m. Its area is 4,722 sq. m. Towards the E side rises a range of lofty mountains. The principal rivers are the Peraur and the Pullayur. The soil is fertile, producing grain, sago, sugar, salt, pepper, cardomums, cocoa-nuts, cassia, frankincense, and other aromatic drugs, plantains, coffee, tobacco, and silk. The wild buffalo, tiger, cheetah, and elephant are found in the forests, which are extensive, and form a valuable source of revenue. There are no fortresses of any importance in this prov.; but against hostility from Mysore and Tinnevelly a double line of works has been raised, consisting of a thick plantation, supported by a rampart with bastions and a deep ditch; while the range of mountains extending to Cape Comorin forms a natural defence on the SE. Strong currents prevail along the coast. The pop. was last returned at 1,011,824, of whom about 60,000 are Mahommedan Moplays of Arab descent, and 170,000 are Roman Catholics and Syrian Christians. This prov. having been originally one of the subdivisions of Malabar, retains many of its customs. It appears that the Tamburetties of Attingara possessed the sovereignty in this region at an early period, and that the government was always in the hands of a female, till the early part of the 18th cent., when one of these ladies not only resigned the power to her son, but enacted that in future the sovereignty should descend to the son of the senior Tamburetty, as in Malabar. The rajah thus chosen proved an ambitious and able chief. He employed an European officer to discipline his troops; and having turned his arms against the neighbouring states, subdued six of the petty rajahs, and annexed their territories to his own. He also conquered part of Cochin, and compelled the *rani* or queen of that country to adopt him as her successor. From this period the rajahs of T., by intrigue or force, gradually acquired possession of the estates of all the remaining petty chiefs; and in 1789, entered into treaty with the Dutch, for the purchase of the town and fortress of Cranganore. This circumstance, however, drew on him the hostility of Tippu Sultan, who, in April 1790, attacked the lines before described, and having succeeded in storming them, advanced as far as Virapelly, when the interference of the British alone prevented T. from becoming a prov. of Mysore. In November 1795, a treaty of alliance was concluded between the rajah and the British, by which the former agreed to subsidize 3 battalions of native infantry, to be maintained for the defence of his dominions against all enemies, and, in the event of war, to assist the British with his own forces. In 1809, the British troops took possession of the country with little opposition; and the resident, Colonel Munro, assumed the office of prime minister to the rajah, and retained it with great advantage to the state till 1813, when the office was again delivered over to the rajah upon his attaining his 16th year. The revenue of the country was at this juncture estimated at 30 lacks of rupees, or about £45,000; and £15,000 was paid to the British for their protection and the expenses of the resident. At present the revenue is 41,58,075 r., and the subsidy or tribute amounts to 7,96,430 r. The principal towns are Travancore, Trivanderam, Anjengo, Coulan or Quilon, Aibecca, and Coleshy.—Travancore, the ancient capital of the prov., formerly an extensive place, but much decayed, is situated in N lat. 8° 25′, E long. 77° 22′, 21 m. E of Trivanderam.

TRAVE, a river of Denmark, in the duchy of Holstein, which rises near Sarau, between Eutin and Ahrensbock; flows SW, and expands into Lake Warder; passes Segeberg, Oldesloe, and Lubeck; and falls into the Baltic near the small town of Travemünde, after a course of about 60 m. It is narrow, but up to the city of Lubeck is navigable for vessels drawing not above 7½ ft. A passage of 11 or 12 ft. in depth is usually kept open across its flat sandy bar. Near its mouth stands a lighthouse, in N lat. 53° 57′ 39″, E long. 10° 52′ 52″; and a lively trade is kept up from it with all the ports on the Baltic, and steam communication with Copenhagen, Stockholm, Riga, and Cronstadt.

TRAVECY, a village of France, in the dep. of Aisne, cant. and 2 m. N of La-Fere, on the r. bank of the Oise.

TRAVEMUNDE, a small town at the mouth of the Trave, 8 m. NE of Lubeck, and belonging to that city, to which it serves as a port. It is fortified towards the land, and towards the sea is defended by a small fort. The harbour is capable of containing 60 vessels, and is deep enough for those of 200 tons; but its prosperity is controlled by a monopolized corporation in Lubeck. Men-of-war lie at anchor in the road.

TRAVENDAHL, a castle and bailiwick of Denmark, in the duchy of Holstein, 15 m. W of Lubeck, and 2 m. SW of Segeberg, near the l. bank of the Trave, noted in Danish history for the treaty negociated here in 1700, between the king of Denmark and duke of Holstein.

TRAVENSALZA, a hamlet of Denmark, near the r. bank of the Trave, 1 m. SW of Oldesloe, famous for its salt works.

TRAVERS (VAL-DE), a district in the SW of the Swiss cant. of Neufchatel, consisting of a romantic valley running SW and NE among the Jura mountains, and watered by the Reuse. It contains six villages, with about 1,500 inhabitants, many of whom are lace-knitters, watchmakers, or stocking weavers. The chief place, the village of Travers, is 11 m. W of Neufchatel.

TRAVERSE, a lake in the Minnesota territory, U. S., under the parallel of 46°, occupying a depression in an elevated plateau, and giving rise to a head-branch of Red river.

TRAVERSE BAY (GRAND), a bay on the E side of Lake Michigan, in N lat. 44° 15′.

TRAVERSE ISLANDS, a chain of islands at the E end of Noquet's bay, in Lake Michigan, in N lat. 45° 15′.

TRAVNICK, a town of European Turkey, in Bosnia, at the foot of a chain of mountains, on the r. bank of the Lashva, 70 m. W of Isvornik. It has a pop. of 8,000, and is a place of considerable importance, being the point where several roads meet.

TRAWDEN-FOREST, a township and village in Whalley p., co.-palatine of Lancaster, 1¾ m. SE of Colne. Area 2,510 acres. Pop. in 1831, 2,853; in 1851, 2,601.

TRAWEENAGH, or TRYENACH, a bay or lagune on the coast of co. Donegal. It opens off the NE corner of Guibarra-bay, penetrates the land 3 m eastward, has an entrance-width of only 3 furls., and expands in the interior to a width of 1½ m.

TRAWS-FYNYDD, a parish in Merionethshire, on the post-road between Dolgelly and Maentwrog. Area 21,950 acres. Pop. in 1851, 1,498.

TRAYAN-GLAS, a hamlet in Llywell p., co. Brecon, 7 m. W of Brecon. Pop. in 1851, 648.

TRAYAN-MAWR, a hamlet in Llywell p., co. Brecon, 6¾ m. W of Brecon. Pop. in 1851, 378.

TRAZEGNIES, a commune and village of Belgium, in the prov. of Hainault, 20 m. E by N of Mons.

TRAZ-OZ-MONTES. TRAS-OS-MONTES.

TREADINGSTOWN, a parish in co. Kilkenny, 4 m. SSE of Kilkenny. Pop. in 1851, 393.

TREAL, a village of France, in the dep. of Morbihan, cant. and 3 m. NNW of Carentoir. Pop. 1,000.

TREALS-AND-WHARLES, a township in Kirkham p., co.-palatine of Lancaster, 1½ m. NE of Kirkham. Area 4,015 acres. Pop. in 1851, 696.

TREASURY ISLANDS, a group, forming part of Solomon's islands in the S. Pacific. They are low, and covered with trees. The most northern is in S lat. 7° 23′, E long. 155° 29′. The group occupies a circuit of about 10 leagues.

TREAUVILLE, a village of France, in the dep. of La Manche, cant. and 2 m. NNW of Pieux. Pop. 1,100.

TREBBIN, a town of the Prussian prov. of Brandenburg, on the river Nuthe, 22 m. S by W of Berlin. Pop. 1,200.

TREBEL, a river of Prussia on the confines of Mecklenburg and Pomerania, which runs W, and falls into the Peene, 1 m. NW of Demmin, after a course of 45 m.

TREBES, a town of France, dep. of Aude, near the canal of Languedoc, 6 m. SE of Carcassonne. Pop. 1,200.—Also a village of Bavaria, in the presidial and 3 m. SSE of Tresswitz.

TREBIA, or TREBBIA, a river of Italy, which rises among the Apennines, 10 m. NE of Genoa; runs NE; and after a course of 60 m. falls into the Po, 3 m. above Piacenza. Though not large, it is an impetuous stream. It gave name to the second victory which signalised Hannibal's invasion of Italy, and which is supposed to have been fought between the embouchure of the Trebia and Piacenza. Its banks were the scene also of sanguinary fighting in June 1799, between the French under Macdonald, and the Russians under Suwarrow, in which the latter were victorious.

TREBIGNE, a town of Turkey in Bosnia, 15 m. NE of Ragusa, on the r. bank of the Trebignitza. Pop. 1,000.

TREBISACCIA, a town of Naples, in Calabria-Citra, 18 m. ENE of Castrovillari near the coast. Pop. 1,200.

TREBITSCH, or TRZEBICZKE, a town of Moravia, on the Iglawa, 20 m. ESE of Iglau. Pop. 5,000, of whom a number are Jews. It has considerable woollen manufactures.

TREBIZOND, a considerable city of Asiatic Turkey, on the S coast of the Black sea, in N lat. 41° 1′ 0″, E long. 39° 44′ 36″, 136 m. SE of Erzerum. The city, of an oblong shape, occupies a slope gently rising from the sea. On the E and W it is defended by two deep ravines filled with trees and gardens, and crossed by long bridges. These ravines are connected by a ditch cut in the rock behind the castle, and along the skirts of which run the ancient ramparts of the city, which are built of stone of remarkable thickness, and in general lofty. The trapezoidal form, from which the city is said to have derived its name, may it is said be still traced in the arrangement of the walls. The houses for the most part are built of stone and lime, and roofed with small red tiles; like the common Turkish dwellings, they are mean in their outward appearance, and comfortless within; but all having a garden with a few fruit-trees, scarcely a house is visible from the sea. The most curious edifice in the city is the *besestein*, a huge square structure, with two small windows on each face, probably erected by the Genoese as a powder magazine. At the S extremity of the town is the citadel, which commands a full view of the city and environs. A battery of heavy guns is erected on an eminence to the left of the city. Mountains rise behind T., but of less elevation than along the rest of the neighbouring coast of Asia Minor. They are in a high state of cultivation, producing maize, barley, flax, tobacco, honey, wax, and wine. The anchorage is bad, and only available in summer, being a small open bay at the E extremity of the town. After the autumnal equinox, vessels resort to Platena, a roadstead about 7 m. to the W. T. has a pop. of 68,000, of whom about one-third are Christians; the rest are a heterogeneous mixture of Turks, Jews, Georgians, Mingrelians, Circassians, and Tartars. Its trade is very considerable with Armenia and the N of Persia. Formerly the goods manufactured in Europe and intended for Georgia and Persia were transported in Turkish vessels from Constantinople to Redut-Kaleh and to Pati, small forts in Mingrelia, near the mouth of the Riou. Armenian and Georgian caravans transported them thence to Tiflis, the cap. of Georgia, then the great commercial storehouse for a great part of Western Asia. Numerous caravans arrived from Tabriz, Astrakhan, and Baku, at Tiflis, to purchase European merchandise, by which means the Armenian merchants of that town gained immense profits, and the peasants of the Trans-Caucasus received liberal payment for the hire of their beasts of burden engaged in the conveyance of that merchandise. This flourishing state of trade in the Russian provs. beyond the Caucasus ceased suddenly when M. de Cancrin, Russian minister of finance, unfortunately proposed to the Emperor to introduce the Russian tariff into those provinces, and to extend his system of prohibition to the extreme frontier of this vast empire. The motive of this measure was to exclude British manufactures from the Trans-Caucasian market, in the hope of creating there an extensive market for the manufactures of Moscow. But the ukase of the Emperor, which placed Mingrelia and Georgia under the system of the Russian tariff, did not produce the advantage to the manufacturers of Moscow which they expected from it: the importation of British and German articles continued, but it took another direction. The ships of Constantinople conveyed them to T., where a great commerce of caravans was organized to convey the merchandise to Tabriz through Erzerum. A smuggling trade was commenced all along the Russo-Persian frontier, and assumed colossal dimensions. British cloths and German hardware entered Georgia as hitherto; but, in place of the merchants of Tiflis, it was the Persian smugglers who, passing the Araxes at night with their well-laden mules, obtained the profit. In vain did the Russian government establish on that frontier a strict military guard. The smugglers arranged with the Cossacks by means of trifling presents, so that they now pass the Araxes in full day without being molested either by the Cossacks or by the Russian custom-house officers, who are all bribed. Steam-navigation and the establishment of four wealthy commercial houses at Tabriz, by Greeks from Constantinople, principally contributed to the increase of this commercial movement. England supplies about 6,000 bales of cotton per annum, with sugar, indigo, and spices; Germany contributes hardware and Bohemian glass, which is much esteemed in Persia; France sends wine; Southern Russia, wheat and barley; Belgium, fire-arms. The exports are valued at from £300,000 to £400,000 annually in merchandise, and the remainder in Russian silver. The principal articles of export are Persian silk, Cashmere shawls, saffron, carpets, camels' skins, and cotton. All these articles come from Persia, and timber for building, fruits, wax, fish oil, and tobacco, from Anatolia. In 1845 the number of ships which entered the port amounted to 181: viz. 27 British, 90 Turkish, 38 Austrian, 19 Russian, 4 Greek, 2 Sardinian, and 1 Ionian, with a freight of 41,878 tons. The number of sailing vessels engaged in the commerce of the port during 1849 was 87, of which 10 were English, 48 Turkish, and 22 Greek; but a large portion of the trade was carried on by means of steamers, of which class 30 English, 30 Turkish, and 24 Austrian touched at the port in the course of that year. The enire value of the imports in 1849 was estimated at 53,409,215 francs, or about £2,136,000; and the exports were calculated to have amounted to 14,251,406 frs., or about £570,000. The former showed a decrease as compared with 1848; but the export trade, on the other hand, had largely increased, the total amount in 1848 being 6,239,790 frs. This large augmentation was principally owing to the greatly increased exports of copper and other produce of the Turkish territories, and also to the increased degree of favour that articles of Persian produce, such as silks, gall nuts, &c., have enjoyed in the European markets. In 1851 the value of merchandise imported into T. amounted to 182,000,000 piastres, or about £1,500,000. Although a project has been proposed by the Turkish government to construct a road from T. to Erzerum, a town of importance near the frontier of Persia, it is stated that the influence of Russia has always prevented it from being carried into execution, that power being desirous to secure the transit-trade of Persia through Georgia. The amount of shipping engaged in the trade of T. in 1851 consisted of—

Ottoman sailing vessels,	71	58,580 tons.
„ steamers,	30	
Austrian steamers (7,800-horse power),		23,300 „
British sailing vessels,	6	15,742 „
„ steamers,	17	
Greek,	9	1,698 „
Danubian,	7	1,004 „
Ionian,	3	396 „
Russian,	2	382 „
French,	1	350 „

The following were the amount of imports and exports:

	Imports.	Exports.
In British steamers, &c.,	£331,558	£114,556
Ditto, Austrian,	414,416	101,055
Native steamers and other craft,	1,396,168	424,024
Russian ditto,	6,123	8,077
Total,	£2,148,265	£647,712
Value of specie exported (chiefly in Russian coin, and for Persian account),		£221,503
Ditto for local account,		85,997
Number of passengers,	18,180	

The transit of merchandise and passengers from Constantinople to T. is now entirely in the hands of the English and Austrian steam-packet companies. The freights are almost entirely composed of English manufactured goods, part of which remain here, and are smuggled into the Russian province of Georgia; the remainder, consisting chiefly of white calicoes, are sent into Persia, where they are printed, and then introduced into Russia as of Persian manufacture. This accounts for the exportation of Russian coin for Persian account. Large quantities of silks, raw silk, tobacco, and valuable shawls, are exported by the Persian merchants. T. is a very ancient place, and is mentioned by Xenophon, under the appellation of *Trapezus*, as forming the termination of the retreat of the 10,000 Greeks after the defeat of Cyrus at Cunaxa in Mesopotamia. It is then stated to have been a colony of the Sinopians, populous, and situated in the country of the Colchians. It continued a free and independent city, till it fell under the dominion of the kings of Pontus. At the period of the Roman dominion over Asia Minor, their trade with India is supposed to have passed through this city. After the capture of Constantinople by the Latins in 1203, T. became, under Alexis Comnenus, the seat of an empire extending from the mouth of the Phasis to that of the Halys. On its surrender to Mahomet II., its commercial relations with Europe ceased entirely, and the Euxine became closed to the navigation of the Christian powers.

TREBNITZ, a town of Prussian Silesia, in the principality of Oels, 14 m. N of Breslau. Pop. 4,000. It has woollen factories, and a trade in agricultural produce and fruit.

TREBŒUF, a village of France, in the dep. of Ille-et-Vilaine, 3 m. ESE of Sel. Pop. 1,500.

TREBOROUGH, a parish of Somersetshire, 5 m. S by E of Dunster. Area 1,798 acres. Pop. in 1931, 135; in 1851, 142.

TREBSEN, a village of Saxony, on the Mulde, 18 m. ESE of Leipsic. Pop. 750.

TREBSHEN, a village of Prussia, in the prov. of Brandenburg, reg. and 60 m. SE of Frankfort.

TREBUR, a town of Hesse-Darmstadt, 11 m. WNW of Darmstadt, near the r. bank of the Rhine. Pop. 1,200.

TREBUXENA, a town of Spain, in the prov. and 23 m. NNE of Cadiz. Pop. 2,800.

TRECASTAGNI, a town of Sicily, 10 m. NNE of Catania, on the S flank of Ætna. Pop. 2,500.

TRECATE, a town of the Sardinian states, 6 m. ESE of Novara, in a fine plain, between the Ticino and Terdoppo rivers.

TRECCHINA, a town of Naples, in the prov. of Basilicata, 8 m. S of Lagonegro. Pop. 1,800.

TRECENO, a village of Spain, in the prov. and 30 m. WSW of Santander, on the l. bank of the Escudo.

TRECENTA, a town of Austrian Lombardy, 6 m. S of Badia, near the Tartaro. Pop. 2,000.

TRECENTO, a town of the States-of-the-Church, 12 m. WNW of Ferrara.

TREDARZAC, a town of France, in the dep. of Cotes-du-Nord, cant. and 4 m. W of Lezardrieux. Pop. 1,200.

TREDEGAR, a town and populous district in Bedwelty p., co. of Monmouth, 16 m. NW of Newport. Pop. in 1851, 8,305. T., about 60 years ago an insignificant hamlet, has risen to the importance of a market-town, in consequence of the extensive coal-mines and iron-works which it possesses. The Sirhowey railroad, which communicates with Newport, and is almost entirely confined to the conveyance, for shipment on the Monmouthshire canal, of the produce of the T. iron-works, is 28 m. in length.

TREDINGTON, a parish in Gloucestershire, 2½ m. SE by S of Tewkesbury. Area 870 acres. Pop. in 1851, 143.—Also a parish in Worcestershire, 2¼ m. N of Shipston-on-Stour, including the hamlets of Armscott, Blackwell, Darlingscott, and Newbold. Area 5,285 acres. Pop. in 1831, 1,036; in 1851, 1,117.

TREDONOCK, or TREDUNNOCK, a parish in Monmouthshire, 4 m. S of Usk. Area 1,393 acres. Pop. in 1831, 158; in 1851, 157.

TREDOZIO, a village of Tuscany, in the prov. and 42 m. N of Florence.

TREENE, a river of Denmark, which rises 3 m. S of Flensburg; flows SSW, and then W; and joins the Eyder, on the r. bank, at Friederichstadt, after a course of 45 m.

TREETON, a parish in the W. R. of Yorkshire, 4 m. S by E of Rotherham, comprising the townships of Brampton-en-le-Morthen, T., and Ulley. Area 3,503 acres. Pop. in 1831, 680; in 1851, 663.

TREFDRAETH, a parish in Anglesey, 5 m. NW of Llangefni. Area 3,135 acres. Pop. in 1851, 994.

TREFECCA, a hamlet in Talgarth p., Brecon, 8 m. SW of Hay. Pop. in 1831, 282; in 1851, 216.

TREFEGLWYS, a parish in Montgomeryshire, 9 m. W by S of Newtown, including the townships of Bod-aioch, Dol-Gwden, Maes-tref-gomer, and Eskirieth with Glyn-tref-nant. Area 18,166 acres. Pop. in 1831, 1,820; in 1851, 1,793.

TREFFANDEL, a village of France, in the dep. of Ille-et-Vilaine, 4 m. NE of Prelan-le-Grand. Pop. 1,100.

TREFFURTH, a town of Prussian Saxony, situated on a hill near the river Werra, 29 m. SSE of Gottingen. Pop. 2,100. It was formerly an independent principality; afterwards it belonged to the princes of Mentz, Saxony, and Hesse conjunctly; and was given in 1815 to Prussia by the congress of Vienna.

TREF-GARN, a parish in Pembroke, 5 m. N of Haverford-West. Little-Trefgarn is on the E side of the Cleddau; Great Trefgarn on the W side. Area 1,205 acres. Pop. in 1831, 112; in 1851, 99.

TREF-GARON. See CARON.

TREFILAN, a parish in Cardiganshire, 7 m. SW of Lampeter, including the hamlet of Tal-y-sarn. Area 2,201 acres. Pop. in 1831, 313; in 1851, 308.

TREFLLIS, a hamlet in Llangammarch parish, co. of Brecon, South Wales, 8 m. W of Buallt.

TREFLLYS, a parish in Carnarvonshire, 2 m. E of Criccieth. Area 999 acres. Pop. in 1851, 103.

TREFORT, a commune and village of France, dep. of the Ain, situated on a mountain called Revermont, 11 m. NE of Bourg. Pop. 2,200.

TREFRHIW, a parish and village in Carnarvonshire, 2 m. NW of Llanrwst, on the W bank of the Conway. Pop. in 1831, 413; in 1851, 428. T. may be styled the port of Llanrwst. A quay has been constructed here, and it has a brisk trade in slates, lead, bark, and oak.

TREGARE, a parish in Monmouth, 6 m. W by S of Monmouth. Area 2,387 acres. Pop. 325.

TREGAYAN, a parish of Anglesey, 2½ m. NW by N of Llangefni. Area 2,066 acres. Pop. 241.

TREGONEY, a disfranchised borough in the p. of Cuby, Cornwall, 40 m. SW of Launceston, and 11 m. NNE of Falmouth, on the river Fal. Pop. in 1831, 1,127; in 1851, 844. It occupies the summit of a hill, and consists chiefly of one long street. By means of the Fal, which is navigable for large boats, some trade is carried on, arising chiefly from the exportation of copper and tin, and the importation of coals and groceries. The borough was formerly of some consequence, but fell into decay when Truro began to flourish. It was incorporated by James I., and returned 2 members to parliament, until disfranchised by the reform act.

TREGROM, a commune of France, in the dep. of the Cotes-du-Nord, cant. and 3 m. ESE of Plouaret, on the r. bank of the Guer. Pop. 1,385.

TREGUENNEC, a village of France, in the dep. of Finistere, cant. and 5 m. NW of Pont-l'Abbe, near the bay of Audierne. Pop. 340.

TREGUIER, a canton, commune, and town of France, in the dep. of the Cotes-du-Nord, and arrond. of Lannion. The cant. comprises 10 com. Pop. in 1831, 16,770; in 1846, 19,145.—The town is 12 m. ENE of Lannion, at the confluence of the Jaudy and Guindy, which unite to form the Treguier, and 6 m. from the shore of the English channel. Pop. in 1846, 3,798. It has a safe and commodious port, and a good roadstead, and carries on an active trade in horses, grain, seed, flour, flax, hemp, honey, butter, and oysters. This town was the seat of a bishopric erected in the 9th century, and was a place of flourishing trade until 1592, when it was burnt by the Spaniards.

TREGUNC, a commune of France, in the dep. of Finistere, cant. and 3 m. SE of Concarneau. Pop. in 1846, 3,462.

TREGYNON, a parish in Montgomeryshire, 5 m. N of Newton. Area 6,760 acres. Pop. 718.

TREHEGNIER, a hamlet of France, in the dep. of Morbihan, cant. of La Roche-Bernard, and com. of Penistin, on the l. bank of the Vilaine, near its mouth. Pop. 100.

TREHOU (LE), a village of France, in the dep. of Finistere, cant. and 5 m. S of Ploudery. Pop. 1,080.

TREIDER-AA. See AA.

TREIGNAC, a canton, commune, and town of France, in the dep. of the Correze, and arrond. of Tulle. The cant. comprises 11 com. Pop. in 1831, 12,276; in 1846, 13,937.—The town is 20 m. N of Tulle, on the l. bank of the Vezere. Pop. in 1846, 3,246. It has a communal college, and possesses manufactories of fire-arms. The trade consists chiefly in sheep, wool, and fine wax.

TREIGNES, a department and commune of Belgium, in the prov. of Namur. Pop. 640.

TREIGNY, a commune of France, in the dep. of the Yonne, cant. and 5 m. N of St. Sauveur. Pop. in 1846, 2,517.

TREILLIERES, a commune of France, in the dep. of the Loire-Inferieure, cant. and 5 m. NW of La Chapelle-sur-Erdre, on a height. Pop. 1,479. Here are the castle of Gevres, built in 1653, and the chapel of Dons, founded by the dukes of Bretagne.

TREIS, a market-town of Prussia, in the prov. of the Rhine, circle and 8 m. ENE of Kochheim, on the Mozelle. Pop. 1,200. It has a castle.

TREISA, a town of Hesse-Cassel, in the prov. of Ober-Hessen, circle and 3 m. from Ziegenhain, on the l. bank of the Schwalm, at the mouth of the Wiera. Pop. in 1840, 2,482. It has manufactories of hosiery, and woollen and linen fabrics, and several dyeworks and tanneries.

TREISAM, or TREYSAM, a river of the grand duchy of Baden, in the circle of the Upper Rhine, which descends from the mountains of the Schwarzwald, 11 m. SSE of Waldkirch; runs first S, then NW through a narrow valley; passes Friburg; divides into two branches, which, at a distance of 9 m., again unite; thence it takes a N course, waters Eichstetten; and, after a course of about 36 m., joins the Elz, on the l. bank, 4 m. S of Kenzingen. Its principal affluent is the Glotter, which it receives on the r. bank.

TREIS-AN-DER-LUMBDE, a town of Hesse-Cassel, in the prov. of Ober-Hesse, circle and 12 m. from Marburg, on the Lumbde. Pop. 1,107.

TREJUNCOS, or TRESJUNCOS, a town of Spain, in the prov. and 42 m. SW of Cuenca. Pop. 929.

TREKHIBENSKAIA, a town of Russia in Europe, in the gov. of Voronej, district and 36 m. S of Starobiatlsk, on the l. bank of the Sievernoi-Donetz.

TRELAWNY. See CORNWALL and LOOE.

TRELAWNY, a parish of Jamaica, in the co. of

Cornwall, bounded on the N by the sea, on the E by the co. of Middlesex, on the S by that co. from which it is separated by Hector river and the p. of St. Elizabeth, and on the W by that of St. James. It is intersected in the S by a branch of the Blue mountains. Its principal river is the Martha Beea, which, at its mouth, forms a large harbour, on the W side of which Falmouth the chief town is situated.

TRELAZE', a commune of France, in the dep. of the Maine-et-Loire, cant. and 5 m. ESE of Angers, near the r. bank of the Authion. Pop. in 1846, 3,385. It has extensive slate quarries.

TRELECK, or TRILLECH, a parish in Monmouthshire, 4½ m. S by W of Monmouth. Area 7,061 acres. Pop. in 1831, 940; in 1851, 1,136.

TRELECK, or TRILLECH-GRANGE, a chapelry in Treleck p., Monmouthshire, 7 m. SW by S of Chepstow. Area 1,774 acres. Pop. in 1851, 148.

TRELECK-AR-BETTWS, a parish in Carmarthenshire, 12 m. NW of Carmarthen. Area 11,492 acres. Pop. in 1831, 1,565; in 1851, 1,532.

TRELLY, a commune of France, in the dep. of the English channel, cant. and 5 m. SE of Montmartin, on a hill. Pop. 1,295.

TRELODY (SAINT), a commune of France, in the dep. of the Gironde, cant. and 1 m. ESE of Lesparre. Pop. 1,589.

TRELON, a canton, commune, and town of France, in the dep. of the Nord and arrond. of Avesnes. The cant. comprises 13 coms. Pop. in 1831, 13,948; in 1846, 16,216.—The town is 9 m. SE of Avesnes, in a woody and marshy locality. It has manufactories of hosiery, crystal, potash, a tannery, several breweries, forges, and blast-furnaces.

TRELOUP, a commune of France, in the dep. of the Aisne, cant. and 5 m. NNE of Conde-en-Brie, on the r. bank of the Marne. Pop. 1,362.

TRELOVONNO, a mountain of Greece, in Livadia, in the nom. of Attica, 6 m. ESE of Athens. This is the *Hymettus* of the ancients, noted for its honey.

TREMADOC, a market-town in Carnarvonshire, 4½ m. E by N of Criccieth, near the sea-coast. It is seated on a level, 3 ft. lower than the surface of the sea, from which it is protected by an embankment. It is regularly built, and many of the erections exhibit a chaste style of architecture. The harbour, or Port Madoc, is well sheltered, and admits vessels of 300 tons.

TREMAEN, or TREMEIN, a parish in Cardiganshire, 4 m. ENE of Cardigan. Area 1,658 acres. Pop. in 1831, 241; in 1851, 276.

TREMAYNE, a parish in Cornwall, 6¾ m. WNW of Launceston. Area 1,045 acres. Pop. in 1851, 95.

TREMBLADE (LA), a canton, commune, and town of France, in the dep. of the Charente-Inferieure and arrond. of Marennes. The cant. comprises 6 coms. Pop. in 1831, 7,980; in 1846, 8,094.—The town is 5 m. SSW of Marennes, pleasantly situated on the l. bank of the Seudre, 6 m. from the sea. Pop. in 1846, 2,640. It has a Calvinist consistorial church, a custom-house, and a small port, and carries on a trade in salt, wine, brandy, and oysters.

TREMBLAY, a commune of France, in the dep. of the Ille-et-Vilaine, cant. and 3 m. S of Antrain, on a plateau, near the r. bank of the Couesnon. Pop. in 1846, 2,411. It has several tanneries and paper-mills.

TREMBLAY (LE), a town of France, in the dep. of the Eure-et-Loire, cant. and 8 m. E of Chateauneuf. Pop. 500.—Also a village in the dep. of the Seine-et-Oise, cant. and 5 m. E of Gonnesse, in a pleasant and fertile plain, watered by the Rideau. Pop. 1,200. It consists of two parts, distinguished as the Grand and Petit Tremblay. In the former is the parish-church, and in the latter a fine Gothic castle and several villas.

TREMBLECOURT, a village of France, in the dep. of the Meurthe, cant. and 1½ m. SE of Domevre, on a height. Pop. 300. It has the remains of an ancient fortress.

TREMBLEUR, a department and commune of Belgium, in the prov. of Liege. Pop. 1,729. It has a manufactory of fire-arms.

TREMBLEVIF, a village of France, in the dep. of the Loir-et-Cher, cant. and 9 m. NW of Salbris, in a mountainous locality. Pop. 1,100.

TREMBOWLA, or TREBOWLA, a town of Austrian Poland, 18 m. S of Tarnopol, on the Huizna, an affluent of the Sered. Pop. 3,000. It is celebrated for the bravery and firmness with which its inhabitants held out when besieged by the Turks in 1675.

TREM-D'USTON, a village of France, in the dep. of the Ariege, cant. of Oust and com. of Oston. Pop. 1,434.

TREMECEN. See TLEMSEN.

TREMENTINE, a town of France, in the dep. of Maine-et-Loire, cant. and 4 m. NE of Chollet. Pop. 1,100. It has manufactures of linen and other stuffs.

TREMERY, a village of France, in the dep. of Moselle, cant. and 4 m. NW of Vigy, near the Moselle. Pop. 400.

TREMEZZO, a village of Austrian Lombardy, 15 m. NNE of Como.

TREMISTIERI, a village of Sicily, in the prov. of Catania, on the S flank of Ætna. Pop. 1,200.

TREMITI, the *Diomedis Insulæ* of the ancients, five small islands in the Adriatic, about 14 m. of the Capitanata coast of Naples, in N lat. 42° 10'. They are rugged and mountainous, but present some small patches of cultivation, and have a few inhabitants. The largest of the group, San-Domenico, is 4 m. in circuit; the others are Caprara, San-Nicola, Cretaccio, and La Vecchia.

TREMLES, or STRIMILOW, a town of Bohemia, 71 m. SSE of Prague. Pop. 3,000. There are considerable linen manufactures here.

TREMOLAI, a village of France, in the dep. of Dordogne, cant. and 5 m. S of St. Alvere, near the r. bank of the Dordogne. Pop. 1,000.

TREMOLETO, a town of Tuscany, in the district of Leghorn, 15 m. SSE of Pisa.

TREMONT, a village of Tazewell co., in Illinois, U. S., 50 m. N by E of Springfield.—Also a village of Schuylkill co., in Pennsylvania, 32 m. NE of Harrisburg.

TREMONTI, a village of Naples, in Abruzzo-Ultra 2da, 18 m. SSW of Aquila. Pop. 500.

TREMOREL, a village of France, in the dep. of Cotes-du-Nord, 7 miles E of Merdrignac.

TREMOSINE, a village and district of Lombardy, in the prov. and 30 m. NE of Brescia, on the NW coast of the Lago-di-Garda. Pop. 2,000. There are iron works here.

TREMOUILLE, a town of France, in the dep. of La-Vendee, 32 m. E by S of Poitiers. Pop. 1,300.

TREMOUTHA, a haven in the p. of St. Gennys, Cornwall. It was proposed a few years ago to construct a safe and commodious harbour with breakwaters here, to which vessels might run at all times of the tide, and, in connexion with the harbour, to form a railway to Launceston, to be called the Launceston and Victoria railway.

TREMOUZEY, a village of France, in the dep. of Vosges, cant. and 2 m. SSW of Bains. Pop. 800.

TREMP, a town of Spain, in the prov. and 42 m. NNE of Lerida, on the r. bank of the Noguera-Palleresa. Pop. 1,800. Leather, brandy, and coarse linens are made here.

TREMSBUTTEL, a village of Denmark, in the duchy of Holstein, 18 m. NE of Hamburg.

TRENDELBURG, a village of Hessen, 5 m. SSW of Carlshafen. Pop. 700.

TRENEGLOS, a parish in Cornwall, 7½ m. NE by E of Camelford. Area 2,730 acres. Pop. 193.

TRENEWYDD. See NEWTOWN.

TRENT, a river of England, generally accounted the third in respect of length of course, which has its rise in Staffordshire, in three springs to the W of Leek. It soon becomes a pretty large stream, coming down from the hills with a rapid current; and having been augmented by the accession of other rivers, it flows past Trentham, to which it gives name, and thence to Burton in Derbyshire, where it becomes navigable, at a point 117 m. above its junction with the Humber. It soon after enters Nottinghamshire near Radcliffe-upon-Soar, flowing NNE, in a clear stream, and with a bold rapid current; winding round the town of Nottingham, it gives fertility to an immense range of meadows bounded by villas, villages, and luxuriant farms. It then proceeds with rather a tortuous course, through a highly cultivated country towards Newark, where it suddenly bends towards the N and pursues that direction as far as Clifton-upon-Trent, where it becomes the boundary between Nottinghamshire and Lincolnshire, and passes Gainsborough, but does not leave the co. of Nottingham until it reaches Heck-dyke, whence it proceeds to the Humber, which it joins at Trent-falls. At Gainsborough, it loses the influence of the tide, and is no longer navigable for vessels of any great burden; but vessels of a flatter construction sail up as high as Burton. Its navigation is of great importance to the country at large, and every means have been taken to afford it all the facilities possible; among other works a side-cut of 10 m. in length has been formed in order to avoid 21 shoals, which occur in little more than 13 m. of its course between Trent-bridge at the commencement of the Nottingham canal, and Sawley-ferry at the commencement of the Trent and Mersey canal. It has communication by canals with the Mersey, the Severn, and the Thames. Its chief affluents are the Sow, the Tame, the Soar, and the Devon, on the r.; and the Blyth, Dove, and Derwent, on the l. Its length of course is nearly 200 m.; and it drains an area of 4,000 sq. m.

TRENT, a parish of Somersetshire, 3 m. NE by E of Yeovil. Area 1,590 acres. Pop. in 1851, 530.

TRENT, in German TRIENT, in Italian TRIENTA, a circle of the Austrian states, in the S part of the Tyrol, comprising the eastern part of the Italian confines, and the chief part of the ancient bishopric of Trent. It is skirted by the Adige on the N, and is covered by ramifications of the Rhetian Alps.—Its cap., of the same name, is situated on the l. bank of the Adige, 72 m. NW of Venice, and 84 m. S of Insbruck. It stands in a delightful valley among the Alps, in N lat. 46° 6′ 26″. Though surrounded with walls, it is not capable of sustaining a siege. The streets are wide, well paved, and supplied with running water from a small stream which flows through the city. The chief public buildings are the palace of the archbishop; the cathedral, a structure in the Byzantine style, and the church of Sta Maria-Maggiore. The pop., about 13,000 in number, are employed partly in the manufacture of silk, partly in the culture of vines and tobacco. T. is the seat of a provincial court of justice, of a gymnasium, a lyceum, and several schools. The Romans called it *Tridentum.* Wenceslaus, emperor of Germany, made it over to the church of Rome in 1377. Its bishops became princes of the empire, and members of the diet of Tyrol. Their bishopric was included among the secularizations of 1802, given at first to the grand duke of Tuscany, and afterwards to Bavaria, but restored after 1815 to Austria. The great feature in the history of T. is its council or assemblage of Catholic prelates from all parts of Europe, opened in 1545, after the Reformation had acquired a considerable degree of consistency, and continued with more or less of interruption during 18 years. The sittings were held partly in the cathedral, more in the church of St. Maria Maggiore, where there still exists a painting representing the council in full assemblage.

TRENT, a river of Upper Canada, which rises out of Rice lake, and is the channel by which a chain of shallow lakes connected with Lake Huron are brought into Lake Ontario.

TRENTENARA, a village of Naples, in the prov. of Principato-Citra, 18 m. S of Campagna. Pop. 900.

TRENTHAM, a parish and village in Stafford, 3½ m. SSE of Newcastle-under-Lyne, watered by the Trent, and in the line of the Grand Trunk canal, comprising the chapelries of Blurton with Lightwood-Forest and Handford, and the townships of Butterton, Clayton-Griffith, Hanchurch, and T. Area 6,900 acres. Pop. in 1831, 2,344; in 1851, 2,747. The church, originally part of a priory refounded here early in the 12th cent., is a fine old building.—The village stands on the E bank of the Trent, from which it derives its name. The inhabitants are principally employed in the manufacture of bricks and tiles. Trentham-hall, the magnificent seat of the duke of Sutherland, originally erected about a century ago, but which has undergone very considerable improvement, is in this p.

TRENTISHOE, a parish in Devon, 7½ m. E by N of Ilfordcombe, on the coast of the Bristol channel. Area 1,571 acres. Pop. in 1851, 129.

TRENTOLA, a town of Naples, in the Terra-di-Lavoro, 12 m. SW of Caserta.

TRENTON, a township of Oneida co., New York, U. S., 12 m. N of Utica. Pop. in 1840, 3,178; in 1850, 3,540. The celebrated T.-falls are in this co. They are formed by Canada creek, which here falls 312 ft. within a distance of about 2 m.—Also the capital of the state of New Jersey, U. S., in Hunterdon co., on the E bank of the Delaware, opposite the falls, in N lat. 40° 14′, W long. 74° 46′, 58 m. SW of New York. Pop. in 1810, 3,003, in 1840, 4,035; in 1850, 6,766. It is a handsome and flourishing town. The river is navigable as far as this place for sloops; but above the falls it is not navigable, except for boats. This town has extensive railroad and steamboat connections.—Also the capital of Jones co., N. Carolina, on the Trent, 82 m. SE of Raleigh.—Also a township of Butler co., Ohio, 82 m. WSW of Columbus.—Also a village of Gibson co., in Tennessee, 120 m. W of Nashville.

TRENTON, a town of Upper Canada, in Northumberland co., 62 m. W of Kingston, at the entrance of the Trent into the bay of Quinté. Pop. 1,500.

TRENT-PORT, a village of Upper Canada, on a stream of the same name, which falls into the bay of Quinte. Pop. 950.

TRENTSCHIN, or TRENTSIN, a palatinate occupying the NW corner of Hungary, and adjacent to Galicia, Silesia, and Moravia. Its area is 1,860 sq. m., and its pop. 312,000, partly of Slawac, partly of Magyar descent. It is traversed by ridges of the Carpathians, and abounds in timber. The largest river is the Waag. The surface is well cultivated. —The capital, of the same name, is situated 54 m. SE of Olmutz, near the Waag. It consists of one long street and a suburb. Pop. 3,000, partly Catholics, and partly Lutherans. It has manufactories of woollens, oils, and brandy.

TRE'N-TWR, or TRETOWER, a hamlet in the p. of Llanfihangel-cwm-dû, Brecon, 3 m. NW by W of Crickhowel. Pop. in 1831, 318; in 1851, 291.

TRE-OWEN, or TREWERN, a township in Buttington p., Montgomeryshire, 4½ m. NE of Welshpool. Area 1,860 acres. Pop. in 1851, 389.

TREPORT, a town of France, dep. of Seine-Inferieure, at the mouth of the small river Bresle, 17 m. NE of Dieppe. Pop. 2,914, employed partly in yarn spinning, and partly in fishing.

TREPTOW-AM-TOLLENSE, or ALT TREPTOW, a town of Prussia, in the prov. of Pomerania, regency of Stettin, circle and 17 m. SSE of Demmin, on the Tollense. Pop. in 1843, 3,852. It is enclosed by walls with three gates, and has two hospitals. It possesses manufactories of cloth and linen, several tanneries and distilleries of brandy.

TREPTOW-AU-DER-RIGA, or NEU-TREPTOW, a town of Prussia, in the prov. of Pomerania, regency and 54 m. NE of Stettin, and circle of Greiffenberg-Osten, on the l. bank of the Riga. Pop. in 1843, 5,899. It has manufactories of cloth, serge, hosiery, and hats. The harbour at the mouth of the river is now choked with sand.

TREPUZZE, a town of Naples, in the prov. of the Terra-d'Otranto, district and 9 m. NW of Lecce, in a flat but fertile locality. Pop. 2,068. It has a convent.

TREQUANDA, a market-town of the grand-duchy of Tuscany, to the N of Montepulciano. Pop. 2,300.

TRE'R-COED, or TRECOED, a township in Diserth p., Radnor, 5 m. NE of Buallt. Pop. 241.

TRESA, or TRESSA, a river of Austrian Lombardy, which issues from the W extremity of Lake-de-Lugano; runs W; forms for some distance the line of division between the prov. of Como and the Swiss prov. of Ticino; enters the former prov.; and after a total course of about 11 m., throws itself into Lake Maggiore on the E bank, and 1 m. SW of Luvino.

TRES-AMERICANOS, a river of Brazil, in the prov. of Minas-Geraes, which has its sources in the Serra - das - Esmeraldas, and falls into the Rio-Breto.

TRES-BARRAS, a town of Brazil, in the prov. of Minas-Geraes, on the r. bank of the Velhas, at the confluence of the Sepo or Paranna.—Also an auriferous river in the prov. of Matto-Grosso, which joins the Tapajos on the r. bank.—Also a river of the prov. of Santa-Catharina, formed by the confluence of three streams, and falling into the bay of Babitonga.

TRESCLEOUX, a village of France, in the dep. of the Hautes-Alps, cant. and 3 m. N of Orpierre, on the slope of a hill, near the l. bank of the Blaisance. Pop. 480.

TRESCO, or TRESCAW, one of the Scilly islands, to the W of Annet, comprising nearly 800 acres of fertile land. The principal village upon it is Dolphin. Pop. in 1831, 470; in 1851, 430.

TRESCORRE, or TRESCORIO, a town of Austrian Lombardy, in the deleg. and 8 m. E of Bergamo, at the entrance to the Val-Cavalina. Pop. 1,630. It is noted for its thermal springs and baths, and has several forges.—Also a village in the prov. of Lodi and Crema, district and 5 m. NW of Crema. Pop. 900. It has a saltpetre refinery.

TRESEBURG, a hamlet of the duchy of Brunswick, district and circle and 6 m. S of Blankenburg, on the Bode. It has a powder-mill and a quarry of asbestos.

TRESENTA, a small district of Sardinia, in the div. of Cabo-Cagliari, prov. and to the N of Cagliari. Pop. 8,800. It is noted for its fertility, especially in corn, wine, and pasturage, and is considered the granary of Cagliari. It contains 12 villages, of which the chief is Guasile.

TRES-FORCAS, a headland of Marocco, in the prov. of Fez, to the N of Mellila, in N lat. 35° 27′ 55″, W long. 2° 56′ 25″.

TRESHINISH ISLES, a group of Hebridean islets, 5 m. NNW of Staffa, and 5½ m. SE of Coll. They are disposed in a chain of between 4 and 5 m. in length from NE to SW; and consist of five principal isles, and some intervening rocks. Their whole coasts, with little exception, present perpendicular cliffs of from 40 to upwards of 60 ft. in height; and as seen from a little distance, possess a singularly interesting appearance. They are mere uninhabited pasture-grounds. They are composed throughout of amygdaloid and basalt,—the latter of perpendicular fracture, but not columnar.

TRESILICO, a town of Naples, in the prov. of Calabria-Ultra, district and 10 m. SE of Palim, in a flat but fertile locality. Pop. 600. It has a collegiate church and a convent. It suffered severely from an earthquake in 1783.

TRESILLIAN, a village in the p. of Merthor, co. of Cornwall, on a stream which falls into one of t e creeks of Falmouth harbour, and over which there is a bridge here.

TRES-IRMAOS, a mountain of Brazil, in the prov. of Sergipe, on the r. bank of the Rio-Irapirang or Vaza-Barris, between São-Chistovão and the sea.—Also a headland on the N coast of the prov. of Rio-Grande-do-Norte, 12 m. NW of Natal, in S lat. 5° 2′, W long. 35° 58′.—Also a group of islets, three in number, to the SSE of the S extremity of Santa Catharina.

TRES-ISLAS, a group of islands, three in number, near the coast of English Guayana, opposite the mouth of the Essequibo, in N lat. 6° 5′. They serve to cover the bay, and by a fort and batteries erected by the Dutch form its defence.

TRESKOWITZ. See DRESKOWITZ.

TRESLAOUENAN, a village of France, in the dep. of Finistere, cant. and 2 m. NNE of Plowzevede. Pop. 2,660.

TRESMERE, a parish in Cornwall, 6 m. W by N of Launceston. Area 1,344 acres. Pop. in 1831, 171; in 1841, 175.

TRESNURAGHES, a village of Sardinia, in the div. of Capo Sassari, prov. and 9 m. WNW of Cagliari, in a fertile plain. Pop. 1,800.

TRESORERIE, or TREASURY ISLANDS, a group of islands in the Salomon archipelago, South Pacific. It occupies a circumference of about 30 m., and consists of 6 or 7 islands in close proximity, and generally low and woody. The central island is in S lat. 7° 23′ 30″, E long. 155° 29′ 15″.

TRES-PIES, a group of islands, 3 in number, on the coast of Peru, and dep. of Arequipa, in S lat. 20° 47′. They are sterile and desert, but abound with seals.

TRES-PRATAS, a town and parish of Brazil, in the prov. of Minas Geraes, between the rivers Grande and des Mortes, about 30 m. E of Lavras-do-Fimil. Pop. 3,000.

TRES-PUNTAS, a headland of Central America, in the state of Honduras, which advances to the NE of the Amatic gulf, in the SW part of Honduras bay.

TRESQUES, a commune of France, in the dep. of the Gard, and cant. of Bagnals. Pop. 1,095.

TRES-RIOS-DO-NORTE, a river of Brazil, in the prov. of Rio-de-Janeiro, which has its source in the Serra-de-San-Salvador, at Tres-Picos, traverses Lake Cima, and joins the Imbe to form the Ururahy.

TRESSANGE, a village of France, in the dep. of the Moselle, cant. and 5 m. NE of Audun-le-Roman. Pop. 196. It has an oil-mill.

TRESSIN, a village of France, in the dep. of the Nord, cant. and 4 m. SSW of Lannoy, on the Marcq. Pop. 409. It has a nail-work.

TRESSON, a commune of France, in the dep. of the Sarthe, cant. and 5 m. S of Boulloire. Pop. 1,504. It has a manufactory of linen, and in the vicinity are quarries of limestone.

TRESSWITZ, a village of Bavaria, in the circle of the Regen, 29 m. ENE of Amberg, on the r. bank of the Pfreimt. It has an ancient castle.

TRESTENBURG. See TASNAD.

TRESVISO, a village of Spain, in the prov. and 51 m. WSW of Santander, on the l. bank of a small affluent of the Deva. Pop. 124. In the vicinity are mines of lead and antimony.

TRESWELL, or TRUSWELL, a parish in Notts, 5¼ m. E by S of East Retford. Area 1,561 acres. Pop. in 1831, 224; in 1851, 254.

TRESZTYIA, TRESZTIE, or ROHRBACH, a village of Transylvania, in the comitat of Zarand, 19 m. N of Vajda-Hunyad, in the midst of mountains, on a small affluent of the Maros.

TRETEAU, a village of France, in the dep. of the Allier, cant. and 4 m. WSW of Jaligny, in a marshy locality. Pop. 720.

TRETIRE, a parish with Michael-Church, in Herefordshire, 5½ m. W of Ross. Area 1,356 acres. Pop. in 1831, 120; in 1851, 138.

TRETS, a canton, commune, and town of France, in the dep. of the Bouches-du-Rhône, and arrond. of Aix. The cant. comprises 8 com. Pop. in 1831, 8,695; in 1846, 8,841.—The town is 14 m. ESE of Aix, on the slope of a mountain named Regaignas, and commanding an extensive plain watered by the Arc. Pop. in 1846, 3,028. It has old ramparts now to a great extent demolished, a small suburb and several promenades. The body of the town has narrow, irregular streets, and the houses are old and many of them dilapidated. It contains four fountains, a large town-house comprising two prisons, and an hospital. It possesses manufactories of soap, salt, wax, and bricks, distilleries of brandy, coal-mines, and quarries of marble. This town is of great antiquity, and appears to have been ceded to the Marseillais by the *Albicoi* and *Salluvii* who inhabited the valley of the Arc.

TRETTO, a village of Austrian Lombardy, in the prov. and 19. m NW of Vicenza, and district of Istrio, on a mountain abounding with porcelain-clay which forms its chief article of traffic.

TREUCHTLINGEN, a market-town of Bavaria, in the circle of Middle Franconia, 10 m. SE of Heidenheim, near the r. bank of the Altmuhl. Pop. 1,590. It has two castles, and manufactories of galloon and parchment.

TREUEN, THURAN, DREWEN, or TROGEN, a town of Saxony, in the circle of Vigotland, bail. and 7 NE of Planen. Pop. 4,443. It has manufactories of cloth and linen, several tanneries, breweries, and paper-mills.

TREUENBRIEZEN, a town of Prussia, in the prov. of Brandenberg, regency and 23 m. SSW of Potsdam, and circle of Zancha-Belzig, on the Nieplitz. Pop. in 1843, 4,796. It is enclosed by walls with three gates, and has two churches and an hospital. It has manufactories of linen, cloth, and other woollen fabrics, several tanneries, and paper-mills, breweries and distilleries.

TREURENBERG BAY, a bay of the N coast of Spitzbergen, in N lat. 79° 55′. It was visited by Captain Parry in 1827.

TREVALGA, a parish in Cornwall, 6 m. N by W of Camelford. Area 1,299 acres. Pop. 126.

TREVALLYN, a town of New South Wales, in the co. of Durham, on Paterson river, 3 m. from Gresford.

TREVANDRUM, a town of Hindostan, in the principality of Travancore, in N lat. 8° 30′, E long. 77° 12′. It consists chiefly of a large quadrangular fort with extensive suburbs. Pop. 12,000.

TREVANION, or LAGUERTA, an island of the South Pacific, in the archipelego of Santa Cruz, a little to the NW of the island of that name, at the entrance to Graciosa or Trevanion bay, in S lat. 40° 30′, and E long. 185° 45′ 35″. It has a deep bay on the S side, and is very populous. It was discovered in 1767 by Captain Carteret.

TREVÈ, a village of France, in the dep. of the Cotes-du-Nord, cant. and 3 m. NW of Loudeac, on a small affluent of the Oust. Pop. 2,570.

TREVEJANO, a town of Spain, in the prov. and 15 m. S of Lagrono, amid mountains, in an arid locality, near the l. bank of the Leza. Pop. 350. Wool-spinning is the chief branch of local industry.

TREVEJO (SAN-MARTIN-DE), a town of Spain, in the prov. and 51 m. NNW of Caceres, on a mountain. It is commanded by a fortress now unoccupied. Pop. 330.

TREVENNEN. See UNAPOA.

TREVENZUOLO, a village of Austrian Lombardy, in the prov. and 13 m. SSW of Verona, and district of Isola-della-Scala, near the l. bank of the Tartaro. Pop. 900.

TREVEREY, a village of France, in the dep. of the Meuse, cant. and 9 m. NW of Gondrecourt, on the l. bank of the Ornain. Pop. 625. It has several forges and blast-furnaces. In the vicinity is a quarry of free stone.

TREVES, a canton and commune of France, in the dep. of the Gard, and arrond. of Vigan. The village is 14 m. NW of Vigan, on the Trevezet, a small affluent of the Dourbie. Pop. 477. In the vicinity are quarries of fine grained marble.—Also a village in the dep. of the Maine-et-Loire, cant. and 2 m. SE of Gennes, on the l. bank of the Loire. Pop. 280. It was formerly a town with the title of barony.

TREVES, or TRIERS, a government of Prussia, comprising the SW part of the old prov. of the Lower Rhine. It consists of part of the old electorate of Treves, a small portion of the duchy of Luxemburg, some petty districts in the co. of Sponheim, the principality of Saarbruck, and the bishopric of Metz. The area of the whole is 2,480 sq. m. Pop. in 1843, 470,330, of whom 374,002 were Catholics, and 58,870 Protestants; in 1849, 492,182. It is divided into the 13 circles of Daun, Bernkastel, Merzig, Saarlouis, Pruym, Treves, Saarbruck, Bittburg, Country district of Treves, Ottweiler, Wittlich, Saarburg, and Sanct-Wendel. It is watered by the Moselle, which is here a considerable river running in a WSW course. Its chief tributaries here are the Saar and the Nahe. The surface is hilly, and presents a considerable proportion of forest land. Iron, calamine, copper, and coal are wrought. Manufactures are conducted on a small scale, but comprise leather, linen, glass, porcelain, pottery, and hardware. In sheltered situations on the banks of the Moselle, vines are occasionally reared, and the scenery of that river is here romantic.

TREVES, or TRIERS, one of the most celebrated cities in Germany, the capital formerly of an electorate and archbishopric, and now of the above government, is situated in N lat. 49° 47′, 57 m. SW of Coblenz, and 22 m. ENE of Luxemburg, in the centre of a large valley lying along the Moselle, and open to the NW and SE, but confined on the other sides by two gentle eminences covered with vines. Pop. in 1843, 19,211; in 1846, 19,639. The chief buildings are the elector's palace, now turned into barracks; the church of Notre Dame, built about 1240, and affording a fine specimen of Gothic architecture; another church, that of St. Simeon, said to occupy the site of a building used by the Gauls for their public meetings, and by the Romans for a capitol or town-house. The cathedral is an irregular structure remarkable only for its altars, its marble gallery, and the size of the stones with which it is built. There are a public library of 60,000 vols., a museum, theatre, and several public schools. The university of T. was founded in 1454, and greatly extended in 1722. After 1794, it was converted by the French into a central school, to which the Prussians have given the name of gymnasium. Its classes are held in a building of great size, in one of the wings of which is a library. The manufactures consist of woollens and linens, porcelain, soap, leather, paper, wax, tobacco, and hats. The inhabitants, when subject to the electors, derived their support chiefly from the great ecclesiastical establishments; but the French abolished the convents and monasteries, and secularised four Benedictine abbeys, with good revenues, situated outside the walls. There is some export of Moselle wine. Few towns are richer in Roman antiquities: coins, medals, and inscriptions, are frequently dug up. The remains of the ancient Roman baths are extensive. The piers of the bridge over the Moselle are the work of either the Romans or Gauls. The Romans found a town on the site of Treves, and the inhabitants, whom they called *Treviri*, somewhat more improved than their rude neighbours. Under the name of *Augusta Trevirorum*, it became one of their chief stations, and the capital of *Gallia Belgica*. After the era of Constantine, T. was the residence of the prefect of Gaul, until the repeated inroads of the Germans necessitated the removal of the seat of administration to Arles. It was frequently a royal residence under the Franks, and was subsequently received into the German empire, and continued during many centuries under an ecclesiastical government. It was taken from them by the French between the years 1681 and 1794; and remained in their hands from 1794 to 1815.

TREVETHIN, a parish in Monmouthshire, 6½ m. W by N of Usk, including the township of Pontypool. Area 11,329 acres. Pop. in 1831, 10,280; in 1851, 16,864. The inhabitants are mostly employed in the iron and coal mines which have rapidly increased in this neighbourhood. The woollen manufacture is also carried on.

TREVI, a town of the Papal states, in the comarca, and 41 m. E of Rome, on the S slope of Monte Piaggio, near the r. bank of the Teverone. Pop. 1,025.—Also a town in the deleg. and 11 m. N of Spoleto, on the W side of Monte Sereno. Pop. 3,020. It has a considerable trade in corn.

TREVIANA, a town of Spain, in the prov. and 33 m. WNW of Logrono, on a hill, in a mountainous and infertile locality. Pop. 1,156. It has a parish-church, an hospital, and a customhouse, and several spinning-mills.

TREVICO, a town of Naples, in the prov. of the Principato-Ultra, district and 11 m. SSE of Ariano, in the midst of the Apennines. Pop. 2,000. It has a cathedral and two other churches. The environs produce valuable medicinal plants.

TREVIERES, a canton, commune, and town of France, in the dep. of Calvados and arrond. of Bayeux. The cant. comprises 28 com. Pop. in 1831, 12,082; in 1846, 12,156. The town is on the slope of a hill, 10 m. WNW of Bayeux. Pop. in 1846, 1,019. It carries on an active trade in tallow, cattle, and butter.

TREVIGLIO, or TRIVIGLIO, a town of Austrian Lombardy, in the gov. of Milan, deleg. and 14 m. S of Bergamo, in a fine plain, between the Adda and Sorio. Pop. 6,200. It has an hospital and other charitable institutions. Its industry consists chiefly in the culture of silk.

TREVIGNO. See ROVIGNO.

TREVILLERS, a village of France, in the dep. of the Doubs, cant. and 4 m. NE of Meiche. Pop. 640. It has manufactories of pendulums, and has a considerable trade in horses.

TREVINO, a town of Spain, in the prov. of Burgos, 9 m. SSW of Vitoria, on a hill, near the r. bank of the Aynda. Pop. 586. It has an old and now ruinous fortress, and an hospital. This town is the capital of a small county.

TREVISO, a town of Austrian Italy, the capital of a deleg. of the same name, situated on the rivers Sile and Piavesella, at their confluence, 15 m. NW of Venice. It is the see of a bishop, and contains 20,000 inhabitants. It is surrounded with a strong rampart and a wet ditch, and is tolerably well-built, the streets, though irregularly laid out, being wide and well-paved, and most of the houses having colonades or piazzas in front. The city contains an unfinished cathedral, and a great number of churches, monasteries, and convents, several hospitals, two theatres, a library of 30,000 vols. and a botanic garden. The industry consists in manufactures of silk, cotton stuffs, and cutlery; a traffic is carried on in these articles, and in the cattle, corn, fruit, and other productions of the adjacent country. In 1797, T. was taken by the French, who made it the capital of the department of the Tagliamento. The title of Duke of Treviso was conferred by Bonaparte on Marshal Mortier.—The deleg. of T. has an area of 45 German sq. m., with a pop., in 1843, of 279,064. It is a rich and fertile province abounding in corn, wine, wax, and timber.

TREVOIS, a village of Portugal, in the prov. of Beira-Alta, 16 m. SE of Lamego. Pop. 800.

TREVOUX, a town of France, in the dep. of Ain, 14 m. N of Lyons. It is built in the form of an amphitheatre, on the declivity of a small hill on the l. of the Saone. Pop. in 1789, 2,704; in 1846, 2,538. It has manufactures of gold and silver leaf, and jewellery, and several printing-offices. Having been the residence of Jesuits, it was remarkable for a literary journal, entitled *Memoires de Trevoux*, published here, as well as for a well-known Encyclopædia, or *Dictionnaire Universel*, which issued from the same press.—The arrond. comprises an area of 151,760 hectares, and is divided into 7 cantons. Pop. in 1846, 84,423.

TREWEN, a parish in Cornwall, 4½ m. W by S of Launceston. Area 988 acres. Pop. in 1851, 188.

TREWHET, or TREWHITT (HIGH and LOW), a township in Rothbury p., Northumberland, 13 m. WSW of Alnwick. Area 1,653 acres. Pop. 116.

TREYM, or TOUR-DE-TREME, a village of the Swiss cant. of Freyberg, 2 m. N of Gruyeres.

TREYSA, a town of Hesse-Cassel, on the l. bank of the Schwalm, the capital of the co. of Ziegenhayn. Pop. 2,600. The inhabitants are chiefly employed in the manufacture of stockings.

TREZZANO, a village of Austrian Italy, in the prov. of Milan, 2 m. SW of Corsico. Pop. 1,057.

TREZZO, a town of Austrian Italy, in the prov. and 19 m. NE of Milan. Pop. of com. 3,285.

TRIACASTELLA (SANTIAGO-DE), a village of Spain, in the prov. and 18 m. SE of Lugo. Pop. 500.

TRIADELPHIA, a township of Montgomery co., Maryland, U. S., 38 m. WNW of Annapolis.—Also a village of Ohio co., in Virginia.

TRIANA (SANTA-CRUZ-DE), a village of Chili, 25 m. S of Santiago, on the r. bank of the Rupel, in S lat. 34° 19'.

TRIANGLE ISLAND, one of the smaller Bahama islands, in N lat. 20° 51'.

TRIANGLES (NORTHERN), a reef of rocks in the bay of Honduras, in N lat. 18° 50'.

TRIANGLES (SOUTHERN), a reef of rocks and islets in the bay of Honduras, in N lat. 17° 5'.

TRIAUCOURT, a village of France, dep. of the Meuse, 17 m. NNW of Bar-sur-Ornain. Pop. 900.

TRIBALDOS, a village of Spain, in the prov. and 39 m. WSW of Cuenca. Pop. 600.

TRIBAU (BOHMISCHE), a town of Bohemia, 90 m. E by S of Prague, and 6 m. W of Landskron. Pop. 900.

TRIBAU (MAHRISCH), a town of Moravia, on the Trzebowka, 28 m. NW of Olmutz, on the railway to Prague. Pop. 3,100. It has some woollen manufactures.

TRIBESEES, or TRIBSEES, a town of Prussian Pomerania, on the r. bank of the Trebel, 12 m. E by S of Rostock. Pop. 1,700.

TRIBOLI, or TIREBOLI, a town of Asiatic Turkey, on the coast of the Black sea, 5½ m. W by S of Cape Karaburmu. The houses are scattered along the edges of the precipices and sides of three small projecting headlands. Pop. 2,000. The headlands form two small coves in which small vessels may ride. The exports consist of red wine, copper, wax, silk, and dried fruits. The surrounding hills are uncultivated, but afford good pasturage for sheep and goats.

TRIBULATION (CAPE), a cape on the NE coast of Australia, in S lat. 16° 6'.

TRIBUR. See TREBUR.

TRICALA, or TRIKALA, anciently called *Tricca*, a town of European Turkey, the capital of a sanjak comprising a considerable part of Thessaly. It stands on the declivity of a mountain-ridge, about 3 m. N of the Salembria, the ancient *Peneus*, in N lat. 39° 31', 30 m. W of Larissa, and 33 m. E by S of Janina. Its pop. has been estimated at 10,000; but being an inland place, its trade is inconsiderable, and much of it is in ruins.

TRICALORE, a town of Hindostan, in the Carnatic, in N lat. 11° 30', near the r. bank of the Panaur.

TRICARICO, a walled town of Naples, in the prov. of Basilicata, on the river Basiento, between the Basiento and the Bradano, 19 m. E of Potenza. Pop. 4,800.

TRICASE, a town of Naples, in the prov. of Terra-d'Otranto, 24 m. SE of Gallipoli. Pop. 2,800. It has manufactories of coarse cottons and of leather.

TRICERO, a walled town of Piedmont, 10 m. WNW of Casale. Pop. 1,200.

TRICESIMO, a town of Austrian Italy, in the deleg. and 7 m. N of Udina.

TRICHINDORE, a town, and a taluk or revenue-district, of Hindostan, on the SE coast of the peninsula. The whole district is flat, sandy, and saline, but has recently attracted attention from its presumed capabilities for cotton-cultivation. To the S this flat and sandy tract extends nearly to Vissiavethee, within 20 m. of Cape Comorin; and to the N it extends much further, as far as Ramnad, and perhaps nearly to Madras. The breadth of this flat sandy tract, to the NW of Trichindore, is more than 10 m.; and to the northwards it greatly increases. The country around T. may be described as uncultivated. Immense herds of antelopes roam over it, and flocks of sheep and goats, as well as herds of buffaloes and cows, are led along it for pasturage. Near the villages are some fields of dry grains, as cotton sown with cambu, rapeseed, sesamum, and castor-oil plant. There are also many topes of Palmyra trees near the villages, and likewise near the coast; but even amongst these topes, cotton could be grown, as the trees give little or no shade, and are often far apart. The soil of this sandy district contains a large amount of salt, and in consequence most plants serviceable to man will not flourish in it, and some will not even grow in it.

TRICHINOPOLY, a city and fortress of Hindostan, the cap. of a district of the same name which lies between the parallels of 10° 40' and 11° 25' N. It is advantageously situated on the S bank of the river Cavery, opposite the island of Seringam, in N lat. 10° 50', 30 m. W of Tanjore. Its pop. is about 35,000, exclusive of the troops usually numbering from 3,000 to 4,000. It has manufactories of cotton cloth, saddlery, and jewellery. The fort crowns the summit of an isolated rock 350 ft. in height. It contains the arsenal and military hospital. This city was the capital of a principality, the chief of which was entitled the Naic of Madura, until 1736, when the prime minister of the Nabob of the Carnatic got possession of it by treachery. It was taken by the Mahrattas in 1741, but recovered by the Mahommedans in 1743. From 1751 to 1755, T. being then the principal strong place remaining in possession of the nabob Mahomet Ali, was several times besieged by the French and their allies, but was rescued by the assistance of the British under Lawrence and other gallant officers. With the cession of the prov. it came into possession of the British.—The district of T. has an area of 3,000 sq. m. It is level and fertile where irrigated; but the prevailing soil is sandy and arid.

TRICHKI, a town of Russia, in the gov. of Wilna, 15 m. ENE of Telsch.

TRICIO, a village of Spain, in the prov. and 15 m. WSW of Logrono, near the r. bank of the Neijerilla.

TRICKUR, a town of Hindostan, 50 m. N by W of Cochin. It consists of a group of villages enclosed within a common rampart.

TRICOT, a town of France, in the dep. of the Oise, 15 m. NE of Clermont. Pop. 1,400. Woollen stuffs and various knitted articles in general use in France, and called from this place tricots, are manufactured here.

TRIE, a village of France, dep. of Hautes-Pyrenees, 13 m. ENE of Tarbes.

TRIEBEL, a town of Prussia, in Lower Lusatia, near the river Neisse, 47 m. S of Frankfort-on-the-Oder. Pop. 1,120.

TRIEBSWITTER, a village of Hungary, in the com. of Torontal, 6 m. from Komlol.

TRIEL, a town of France, dep. of Seine-et-Oise, 14 m. NNW of Versailles. Pop. 1,900. There are large stone quarries in the neighbourhood.

TRIE-LE-CHATEAU, a village of France, in the dep. of Oise, 16 m. SW of Beauvais, on the r. bank of the Troene.

TRIENGEN, a parish and village of Switzerland, in the cant. and 16 m. NW of Luzern. Pop. 1,917, Catholics.

TRIENT. See TRENT.

TRIER. See TREVES.

TRIESCH, or TZESZTE, a town of Moravia, 6 m. SSW of Iglau. Pop. 3,200. It has woollen manufactories.

TRIESDORF, a village of Bavaria, in the presidial and 9 m. ESE of Herrieden.

TRIESTE, a large and thriving port of the Austrian dominions, the capital of the circle of Adels berg, in the Illyrian territory, situated near the NW extremity of the gulf of Venice, in N lat. 45° 38′ 36″, E long. 13° 46′ 30″, 214 m. SSW of Vienna. Pop. in 1826, 44,234; in 1850, 63,901. It is divided into an old and a new town. The former stands on a hill which is surmounted by a castle; the new town, called also Theresienstadt, is on level ground, intersected by a broad canal, and is built with neatness and regularity. The new city presents some good streets and squares, and a number of commodious buildings, but few that are large or striking, except the cathedral, a church that formerly belonged to the Jesuits, the exchange, and the theatre. Among the public institutions of the place are a gymnasium, a society of arts and sciences, a public library, a large hospital, and a large and well-arranged lazaretto for performing quarantine. Here are also several courts for the administration of justice. The dreary masses of curiously stratified and contorted rock which form the background of T., are relieved at points by the villas of the wealthy citizens, surrounded by gardens in which the Marasca cherry, fig, grape, olive, and melon are cultivated. T. is almost the only seaport for a large tract of country, comprising the S of Germany, the Illyrian part of the Slavonian provinces, and in general the territory extending from Tyrol to Transylvania. Venice does not from its distance materially interfere with the trade of T.; while Fiume is a small place and much less advantageously situated. Among the exports from T. are the produce of the mines of Idria, and even of Hungary; linen, tobacco, and woollens from different parts of the Austrian dominions, and printed cottons from Switzerland. The amount of exports by sea in 1846 was £4,846,800; in 1851, £5,860,000. The amount of imports by sea in 1846 was £6,789,500; in 1851, £9,700,000. The imports consist of cotton, hides, raisins, silks, rice, oil from the Levant, wheat from Odessa, sugar, coffee, and other tropical products from the West Indies and Brazil. In 1844, T. exported to Malta in 22 vessels of 3,286 tons, a value of £25,100, and imported in 15 vessels of 1,846 tons, a value of £33,900. The exports to Malta consisted of flour, potatoes, glasses, tobacco, nails and steel, timber, hemp, Indian corn, sugar, and manufactured goods. The imports from Malta consisted chiefly of cotton, sandal wood, seeds, pitch, tar, &c. In the same year T. sent to Gibraltar 4 vessels of 999 tons, with cargo of £3,500 value in timber, vegetables, glass wares, and tobacco, and received from that emporium in return, £7,000 in value of cottons, lead, &c. in 2 vessels having a tonnage of 290 tons. The trade of the Adriatic is conducted in barks of from 20, 30, or 40 tons. These and vessels of 300 tons enter with ease the canal which leads from the harbour into the new town, and has on each side quays for vessels to load and unload. The number of vessels which arrived in 1850 was 2,057; in 1852, 2,858. The quantity of goods conveyed by land to and from T. is very considerable. The amount of exports by land in 1851 was £2,500,000; of imports £3,740,000.—Ship-building is carried on with activity here. The manufactures of the place are sugar-refining, white lead, soap, leather, paper, and wax. At some distance from the town there are salt-works; and coal is obtained in the vicinity. The environs of T. are stony; the land is unfit for the culture of corn. The fisheries along the coast are abundant.—T. is built on or near the site of the Roman colony of *Tergeste;* and there are some remains of the aqueduct, partly subterranean, which brought water to it from a distance of 6 m. It fell into the possession of Austria in 1382. In the 15th cent. it was a small place, without trade. In 1719 it was made a free port by the Austrian government. In 1753 the harbour was enlarged, and a mole formed to shelter it from the S; it is open, however, to the *Bora*, an impetuous NE wind, which, did it not blow off the land, would be dangerous to the shipping.

TRIESTEVITZ, a village of Prussia, in the prov. of Saxony, to the E of Torgau. Pop. 300.

TRIESTING, a river of Austria, in the prov. of Lower Austria, which descends from the mountains of the Wienerwald, in the E part of the upper circle of that name; passes Kaumberg; runs through a narrow valley; waters Altenmarkt, Poltenstein, Leobersdorf; and, after a course first in a S, and afterwards in a NE direction, joins the Schwachat, on the r. bank, 9 m. S of Vienna.

TRIEUX, a river of France, in the dep. of the Cotes-du-Nord, which issues from the etang Neuf, in the NE extremity of the cant. of Bothoa; runs N past Guingamp, Pontrieux, and Lezardrieux; and, after a course of about 42 m., throws itself into the British channel, opposite the Isles-à-Bois and Brehat. Its bed below Pontrieux is generally broad and deep, and the harbour which it forms below Lezardrieux is capable of receiving 12 vessels of the line. The Leff is the only affluent of importance.—Also a village in the dep. of the Moselle, cant. and 4 m. SSE of Audun-le-Roman. Pop. 400. It has an oil-mill.

TRIFOGLIETTI, or VALLE-DEL-BOVE, a valley of Sicily, in the prov. and district and to the N of Catania, on the E side of Mount Ætna, in the upper region, or decouverte of the mountain. It was inundated in 1755 by an aqueous eruption of the volcano.

TRIFTLARN, a market-town of Bavaria, in the circle of Lower Bavaria, to the SW of Passau. Pop. 880. It has manufactories of cloth and of violin strings.

TRIGARDON, or TRICARDON, a town of Greece, in the dep. of Acarnania and district of Xeromeros, 18 m. W of Missolonghi, in the midst of a marsh, on the r. bank of the embouchure of the Aspropotamus. It was formerly a flourishing place, but its port is now choked with sand, and its commerce consequently destroyed. This place occupies the site of the ancient *Œnidæ*, of which some remains still exist.

TRIGG, a county in the SW part of the state of Kentucky, U. S., drained by Little river and its branches, and comprising an area of 356 sq. m. The surface is undulating, and the soil fertile. Pop. in 1840, 7,716; in 1850, 10,129. Its cap. is Cadiz.

TRIGIANO, a town of Naples, in the prov. of the Terra-di-Bari, district and 5 m. SSE of Bari, and cant. of Capurso, in a plain near the shore of the Adriatic. Pop. 2,950.

TRIGL, or TRIGA, a town of Austria, in Dalmatia, in the circle and 19 m. NE of Spalatro, on the slope of a mountain, near the r. bank of the Cettina.

TRIGLAFF, or TRIEGLAFF, a village of Prussia, in the prov. of Pomerania, regency and 39 m. NE of Stettin and circle of Greiffenberg-Osten. Pop. 230.

TRIGLIA, a town of Turkey in Asia, in Anatolia, in the sanj. of Khodavendkiar, 8 m. W of Mondania, on the S side and near the entrance to the gulf of that name. It is noted for its olives.

TRIGNO, a river of Naples, in the prov. of Sannio, formed by the confluence of two streams which unite near Vasto-Girardo; runs first SE, then NE; bathes the districts of Campobasso and Larino, which it separates from the prov. of Abruzzo-Citra; enters afterwards into that prov.; and after a total course of about 66 m., throws itself by two branches into the Adriatic, 6 m. SE of Il-Vasto.

TRIGUEROS, a town of Spain, in the prov. and

11 m. NNE of Huelva, on an inclined plain stretching to the r. bank of the Tinto, and remarkable for its fertility. Pop. 3,534. It has a parish-church, a convent, two hospitals, a Jesuits' college, and a custom-house, and possesses manufactories of pottery and of tiles. The adjacent mountains contain mines which have been wrought successively by the Phœnicians, Carthaginians, Romans, and Arabs.—Also a town in the prov. and 9 m. NNE of Valladolid, in a fertile locality. Pop. 734.

TRIGUIERES, a village of France, in the dep. of the Loiret, cant. and 3 m. E of Chateau-Renard, on the Onanne. Pop. 1,360.

TRIJUEGUE, a town of Spain, in the prov. and 18 m. NE of Guadalaxara, on an elevated rock, commanding an extensive prospect. Pop. 825. It has the remains of old walls, a fine church, an hospital, and two oil-mills. In the vicinity are quarries of marble, and a mineral spring.

TRIKALA. See Tricala.

TRIKERA, a small island of the Archipelago, near the E coast of the Morea, 3 m. SE of Bear cape, and between the islands of Spezzia and Hydra.

TRIKERI, a peninsula, mountain, and town of Turkey in Europe, in the SE part of the sanj. of Tricala. The peninsula extends between the channel of the same name on the S, and the gulf of Volo on the N. The town is at the SW extremity of the peninsula, on a mountain which terminates the Pelion chain, at the entrance to the gulf of Volo, and near Cape Trikeri, 19 m. S of Zagora, and 32 m. ENE of Zeitun. Pop., chiefly Greek, 6,000. The harbour is one of the most frequented on the coast, affords good anchorage, and has several building-docks. About 5 m. ENE of the town is Mount Trikeri.—The channel of T., which washes on the S the island of Negropont, is 24 m. in length from E to W, and 8 m. in average breadth. Its W extremity divides into two parts, the more northerly opening into the gulf of Volo and Port Fetio, the southern into the gulf of Zeitun and the channel of Negropont.

TRIKERI, or Kikynthus, an island of Turkey in Europe, in the sanj. of Trikala, in the S part of the gulf of Volo, and near the N coast of the peninsula of Trikeri, from which it is separated by a channel only ¾ of a mile in width, in N lat. 39° 9′ 42″, and E long. 23° 5′ 59″. It is 2 m. in length from NE to SW, and 1 m. in breadth.

TRILBARDON, a village of France, in the dep. of the Seine-et-Marne, cant. and 5 m. E of Claye, pleasantly situated on the r. bank of the Marne, which is here crossed by a ferry, and near the canal d'Ourcq. Pop. 450. It has a fine castle. Sheep, of the merino breed, are reared in the environs.

TRILLICK, a village in the p. of Kilskeery, co. Tyrone, 4 m. S of Dromore. Pop. in 1851, 449.

TRILLO, a village of Spain, in the prov. of Guadalaxara, on the Tagus, 60 m. ENE of Madrid. Pop. 700. It was once remarkable for the number and construction of its water-mills for sawing timber brought down by the stream. It is celebrated for its medicinal waters.

TRIM, a parish and town in co. Meath. Area of p. 13,426 acres. Pop. in 1831, 5,926; in 1851, 6,226. The town of T., the county-town of Meath, and formerly a parl. borough, stands upon the river Boyne, on the road from Dublin to Enniskillen, 5 m. NNW of Summerhill. The town is very irregular, both in the alignment of its streets and the plan of its buildings, yet contains many good houses. Its castle, though partially dilapidated and destroyed, is still very extensive; and in consequence of the elevation of its site, and the grandeur of its outline, it forms a very imposing object. Pop. in 1851, 1,905.

TRIMBUCK, a fortress of Hindostan, in the prov. of Aurungabad, near the source of the river Godavery, in N lat. 20° 1′.

TRIMDON, a parish in the co. and 8 m. SE of Durham. Area 2,280 acres. Pop. in 1851, 1,598.

TRIMINGHAM, a parish of Norfolk, 5 m. SE by E of Cromer. Area 680 acres. Pop. in 1851, 243.

TRIMION, a town of Hindostan, in the Carnatic, in N lat. 10° 11′.

TRIMLEY (St. Martins), a parish of Suffolk, 8½ m. SE by E of Ipswich. Area 2,338 acres. Pop. 574.

TRIMLEY (St. Mary), a parish adjoining the foregoing. Area 2,208 acres. Pop. in 1851, 304.

TRINCOMALEE, a town, fortress, and harbour of Ceylon, situated on the NE side of the island, in N lat. 8° 33′, E long. 81° 14′. The town is of greater extent than Colombo, but contains fewer houses and less pop. The fortifications are strong, and command the principal bays, particularly the entrance to the harbour. The few native residences which compose the pettah or town, are situated on a narrow neck of land or isthmus separating what is called the inner from two outer harbours. These harbours are defended by two strong forts, called Fort-Frederick and Fort-Ostenburg, where the white troops composing the garrison are quartered. Fort-Frederick, more commonly designated Trincomalee, is the larger and more important of the two, and occupies a rocky promontory, which on the land side, rises gradually to the height of 300 ft. from the extremity of the narrow isthmus before described, and terminates in abrupt cliffs towards the sea, by which it is nearly surrounded. That portion of the isthmus nearest the fort is well cleared, covered with grass, and forms an exercising ground for the troops; beyond it is a small plain partly occupied by the houses and gardens of the pettah, but in every other direction, as far as the eye can reach, the whole country is covered with dense jungle or forest trees. The rock on which Fort-Frederick stands is composed of granite or quartz, in many parts naked, but in others slightly covered with red sand or gravel, so that the rain is either quickly absorbed, or carried off to the sea. Fort-Ostenburg, which is about 2 m. from Fort-Frederick, stands on an elevated peninsula or promontory, at the extremity of a range of hills which bounds the plain on the S. This peninsula forms the E side of the inner harbour, within which lie several small islands covered with dense vegetation. The ground on the W, and also to the S of the harbour, is flat, and for the most part overspread with jungle. There is little cultivation of any kind in the vicinity, and a salt-water lake covers a considerable portion of the surface to the westward. The soil around T. is chiefly composed of disintegrated granite and quartz, with a portion of vegetable mould; but in the low grounds the surface is in many parts covered with a thin layer of sand, under which deep beds of clay are frequently met with. About 5 m. to the S of the forts is the embouchure of the main channel of the Mahawella-Ganga, as well as of several small branches which diverge from it, and intersect the district with tortuous windings. As the country is level, these are generally sluggish in their course, and their embouchures surrounded by much swampy and muddy soil covered with rank vegetation.—The climate of T. differs materially from that of the S or W coast of Ceylon. The SW monsoon, which at Colombo is characterized by extreme moisture, is here experienced as a hot dry land wind, during the continuance of which, from April to September, there are only a few slight showers, and the therm. in the shade is sometimes as high as 95°; while the NE monsoon, so partial on the W coast, frequently prevails from October to March, with heavy rains, or occasional showers, and a range of therm. varying from 76° to 84°. Owing to their situation, the forts are much exposed to sudden gusts of wind, from which all the barracks cannot be adequately protected. The reflection of the sun's rays, too, from the buildings and rocky surface of the promontory, sometimes causes an intense heat and glare during the day, which is succeeded by chill damp winds at night.—The harbour is of great consequence; for when the violent monsoons or stormy weather sets in, all vessels on the coast of Coromandel are obliged to put to sea; but in two days ships can reach this harbour from Madras, enter it with ease, and remain in perfect security. Within the bay, the shores are so bold, and the water so deep, that vessels may moor alongside; but unfortunately the tides do not rise sufficiently high to admit of wet docks, and men-of-war are obliged to take out all their guns in order to careen. The expectations that were formed of this place by the British government have not been realized; yet, despite its capacious, sheltered, and safe anchorage, its advantageous and commanding position, overlooking as it were the whole bay of Bengal, and the healthiness of the town and neighbourhood, its capabilities have been neglected. Its situation is such that every vessel sailing to Madras, Calcutta, and all the intermediate ports, might readily find shelter; and as the straits of Malacca lie on the opposite side of the bay, a brisk and advantageous trade might be carried on with the Eastern archipelago, Siam, China, and the Austrian colonies. All down the Coromandel coast there is not a safe harbour; and Madras, as is well known, has but an open roadstead. No delay would occur, as at Colombo, in the shipment of Ceylon produce; and as the country between it and within 20 m. of Kandy is for the most part flat, good roads could be made with great facility, and even a railway would be very feasible.—The first Europeans who possessed themselves of this place were the Portuguese. It was taken from them by the Dutch, with whom it remained till 1782, when it was captured by the British; but a very inadequate garrison having been left to defend it, it was taken by the French under Admiral Suffrein, who restored it to the Dutch. In 1795, it was again taken by the British, after a siege of three weeks; and has ever since remained in their possession.

TRINDELEN, a rock in the Skagerack, almost equally distant from the coasts of Jutland and Norway, 7 m. NNE of the island of Lessoe.

TRINESIA, or **Trinassa**, a town of Greece, in the Morea, at the bottom of the gulf of Kolokythia, 8 m. SW of Barduma. Near it are the ruins of *Gythium*.

TRING, a parish and market-town of Herts, on the borders of Buckinghamshire, 30 m. NW by W of London. The town is neat, and contains several handsome houses. The church is a venerable Gothic structure, with a square tower. At Little Tring, a neighbouring village, rises one of the heads or branches of the Thames, which, leaving the county at Puttenham, runs by Aylesbury to Thame. T. was anciently a considerable town. Area of p. 7,390 acres. Pop. in 1831, 3,488; in 1851, 4,746.

TRINGANU, a state on the E coast of the peninsula of Malacca, between the parallels of 4° 35′ and 6°, between the rajahship of Calantan on the NW, and Pahang on the SE. Its cultivated surface is generally low. Its chief productions are tea, pepper, and gambier. The pop. which is almost purely Malay, has been estimated at 30,000, who are chiefly collected in a town which gives name to the state, situated in N lat. 5° 25′.

TRINIDAD, an island of the Atlantic ocean, lying between the 10th and 11th parallels of S lat. and the 61st and 62d meridians of W long., opposite the coast of Venezuela, from which it is separated by the gulf of Paria on the W. It is the most southern island of the British W. Indies; and at its SW and NW extremities, it approaches, in the straits called the Bocas, to within 7 and 13 m. of the American continent. Its general outline is that of an irregular square, having two elongated points stretching to the W from its N and S angles. Its longest lines of dimension are from Cape-Galere on the NE, to Point-Ycaque or Icacque on the SW, 79 m., and from Cape-Galeota on the SE to Cape-Blanca on the N, 56 m. Its square area is nearly 2,012 sq. m. The N side of the island presents a continued ridge of mountains, rising to an alt. of 3,000 ft., which end at Toco or Point-Galere, and probably in some remote age formed a continuation of the Parian mountains, before the great convulsion, which insulated Trinidad, and forced the passages of the Bocas. From the ridges or summits of these hills, numerous streams issue on both sides, and on the S side contribute to form the Caroni, a river which can be navigated by flats and canoes a distance of 6 leagues into the interior of the island. The S side is also bordered by a ridge of mountains of inferior height. Another ridge of hills, commencing at L'Ebranche, on the E side, and running WSW, is called the Monserrat hills. The streams which descend from this ridge are distinguished by a yellow clayish colour. The N, E, and S coasts are nearly destitute of harbours; but several good ports present themselves in the gulf of Paria, on the W side of the island. The greater part of the interior is still unexplored.

Climate.] The climate is less moist than that of Guayana, and not so dry as that of Cumana. The winter or rainy season begins here in June, and ends in October, as in all the islands of the Caribbean sea; but there is little rain. With November begins the spring, during which E and NE winds blow from the cold regions of North America. During this season the temp. in the daytime is usually 80°; during the night it falls to 60°, and sometimes even to 50° in elevated spots. The average annual fall of rain is about 65 in. Fevers and dysenteries are the prevailing diseases. T. is happily situated out of the parallel of hurricanes, which have never as yet shifted so far to the S; although, on the 12th of August, 1810, a violent indication happened. Shocks of earthquakes have been felt, but slight, and not of such consequence as to cause alarm.

Productions.] Trinidad is the largest, most fertile, and most beautiful of all the Leeward islands: yet not above a thirtieth part of the surface is under cultivation. The forests are extensive, and abound in the finest timber of large dimensions and fit for ship-building. The red cedar is the prevailing tree, and every variety of the palm class, from the imperial cabbage-tree to the *cavaat*, the leaves of which furnish covering for the roofs of houses, and will last for a length of time in such use, though exceedingly light. T. is capable of producing every article for the West India market that is under cultivation in any other of the Windward islands. Its sugar is excellent, and there is here more land of a deep, stiff, tenacious quality applicable for its culture than in Jamaica itself. Cocoa is a principal staple. The greater part of the island is excellently adapted for this tree; and Nature has here furnished a tree, called the *Bois immortelle*, or, by the Spaniards, *La madre del cacao*, which, planted at equal distances amongst the cocoa, in the driest weather collects the dews, moistens, shades, and nourishes the cocoa tree. In 1838, 21,710 acres were under sugar; 6,910 under cocoa; and 1,095 under coffee. Among the other articles of agricultural produce are rice, maize, arrow-root, and assava. The coffee also is excellent in its quality; and the indigo is equal to that of the Guatimala plant. The grape vines which have been brought from the S of France, or from Spain, are said to equal in flavour their parent-stocks, and all other fruits, or vegetables, congenial to the tropics, and even some European ones grow luxuriantly here.—In many quarters of the island limestone, resembling in some degree marble, abounds. Bricks and articles of pottery are made, clay of a superior quality having been found. On the extensive savannahs, immense herds of cattle, horses, and mules, are reared. The woods abound with game of different sorts, and opossums, armadilloes, peccaries, bears, and monkeys. The pitch or asphalt lake, called Brea, is a singular natural phenomenon on this island. See article BREA. The gulf and coasts of the island abound with fish of various kinds. Oysters are abundant, particularly the class called mangroves, which fasten and breed on the branches of trees so named. The quantity of produce shipped from T. in 1849 was as follows:—28,080 hogsheads, 2,374 tierces, and 6,429 barrels of sugar; 12,932 puncheons and 229 tierces molasses; 318 puncheons rum, 4,728,186 lbs. cocoa, 28,405 lbs. coffee, 279 bales and 275 serons cotton, and 7 serons indigo. The exports from 1845 to 1849 were as follows:

	To Great Britain.	Total.
1845	£409.219	£430.144
1846	474,859	497.317
1847	488,378	512.288
1848	254,905	208.369
1849	361,984	388,270

Government and Population.] This island is governed by a lieutenant-governor, assisted by an executive and a legislative committee of 12 members.—The laws are partly Spanish, partly English.—The public revenue in 1841 was £104,300. The pop. of T. in 1842 was 60,319; in 1852, 68,600; viz., 35,631 males, and 32,969 females. This total of 68,600 was made up of the pop. of the 8 counties into which T. is divided, in the following proportions:—

County of St. George.	38,630
Victoria,	15.940
Caroni,	7,107
St. Patrick,	4.574
Mayaro (Ward of Mayaro).	985
St. David (Wards Toco and Toururo),	913
St. Andrew (Ward of Manzanilla),	257
Nariva (Ward of Nariva),	194
	68,600

As regards the proportions of pop. in town and country, Port-of-Spain had in 1852, 17,563; and San-Fernando, 2,877; making in these two towns a total pop. of 20,440 souls,—an excess of urban to rural pop., in a colony which is entirely an agricultural one, and without manufactures of any description beyond that of baskets, sieves, carved calabashes, and coarse pottery. The total number of the native pop. of the colony in 1852 was 29,913; of Africans 8,000; the remainder are natives of every clime and country under the sun. The chief religious denominations were in 1852 returned at 43,605 Roman Catholics; 16,246 Episcopalians; 2,694 Gentoos; 2,508 Wesleyans, 1,017 Presbyterians; and 1,016 Mahommedans. As regards education, 8,710 persons were returned in 1852 as able to 'read and write,' and 5,019 who could 'read only.'

History.] T. was discovered by Columbus on the 31st July 1498. It was not, however, taken possession of by the Spaniards till 1588. Their establishment on it was preceded by the almost total destruction of the Indian pop. Sir Walter Raleigh, who visited T. in 1593, says that the inhabitants then cultivated excellent tobacco and the sugar-cane. In 1783 an edict was issued, permitting all foreigners professing the Roman Catholic religion to establish themselves in this colony; protecting at the same time, for a period of five years, those new inhabitants from all debts contracted in the countries they had quitted. In consequence of this policy, crowds of colonists crowded to it from Europe, and from the British and French possessions, bringing with them their industry and capitals; and pop. increased so rapidly, that, though in 1783 the whole amounted only to 2,763, it was estimated six years afterwards at 18.918. In 1797 the island capitulated to a British force under the command of Sir Ralph Abercrombie. From this period till the peace of Amiens in 1802, the pop. increased from 18,918 to 24,239; the produce of sugar also greatly increased, being almost doubled. The emigration which took place from St. Domingo and the British colonies, after the peace of Amiens, increased its pop. in 1807 to 31,000, of whom 21,000 were slaves. T. has since this period remained in the hands of the British.

TRINIDAD, a river of Texas, which flows into Galveston bay, after a prevailingly SSE course of 300 m.—Also a river of New Granada, in the dep. of Ystmo, which joins the Chagres, 25 m. above its mouth.—Also a town of Costa-Rica, on the river Belen, 3 leagues distant from the sea, and 82 m. ESE of Guatimala.—Also a missionary settlement in Paraguay, on the W bank of Paraná, in S lat. 27° 7.′—Also a town of Bolivia, the cap. of the dep. of Moxos, on a small affluent of the Mamore, 200 m. N of Cochabamba, and 35 m. WNW of Loretto.

TRINITA, a town of Naples, in the prov. of Naples, 4 m. from Sorrento. Pop. 1,250.—Also a town of Piedmont, in the prov. and 12 m. NW of Mondovi. Pop. 2,500.

TRINITA-VITTEREO, a village of the Sardinian states, 5 m. NE of Nice, on the Pagliona.

TRINITE′ (LA), a port of the island of Martinique, 16 m. NE of Port-Royal, in N lat. 14° 53′. The harbour is formed on the SE side by Point-Caravelle; and on the other by a high hill about 350 or 400 paces in length, which is joined to the mainland by an isthmus not above 200 ft. broad. The town is a thriving place, being the residence of several merchants, and exporting sugar and rum in large quantities. It has, however, been repeatedly devastated by hurricanes.

TRINITY BAY, a large bay on the E coast of Newfoundland, in N lat. 48°, on the N coast of the peninsula of Avalon.—Also a large but not deep bay, on the NE coast of Australia, between Cape Grafton and Cape Tribulation, in S lat. 16° 30′.

TRINITY (CAPE), a low point on the S coast of the islands of Kodiak, in the N. Pacific, in N lat. 56° 45′.

TRINITY-GASK, a parish in Strathearn, Perthshire. It lies on both sides of the Earn, but chiefly on the N. Pop. in 1831, 620; in 1851, 597.

TRINITY INLET, a bay of the N. Pacific, on the W coast of N. America, 30 m. S of Queen Charlotte's sound.

TRINITY ISLAND, an island in the N. Pacific, discovered by Cook. Vancouver, who passed this coast in 1794, says it appeared to be divided into two islands, with several others of inferior size lying to the N, about Cape Trinity. The E point of the easternmost is, according to his observation, situated in N lat. 56° 33′.

TRINITY ISLAND, an islet in Lough Key, co. Roscommon. An abbey on this islet was founded in 1215, and consumed by an accidental fire in 1466.

TRINO, a town of Sardinia, capital of a mandemento in the div. of Navaro, prov. and 13 m. SW of Vercelle, in a fertile plain near the l. bank of the Po. Pop. 5,500. It is well-built, and is divided into two unequal parts by a spacious street bordered with arcades. It has a communal college. The trade, which is considerable, consists chiefly in cattle and in hams. This town was taken in 1704 by the French under Turenne.

TRINOMALEE, a town and fortress of Hindostan, in the presidency of Madras, in the Carnatic, in N lat. 12° 11′, on a mountain, 52 m. NW of Pondicherry. It is noted for a pagoda situated at the base of a craggy mountain, the summit of which is crowned by a small chapel, held also in great veneration by the Hindus. Upwards of 200 Brahmins are constantly employed in the service of the pagoda. This town was the scene of numerous conflicts during the wars of the last century, and the combined forces of the Nizam and of Hyder-Ali were here successfully resisted by the English.

TRIOGE, a rivulet of Queen's co., which rises on the N side of Cullenagh mountain, and pursues a N course of about 12 m. to the river Barrow.

TRIONTO, a headland of Naples, in the prov. of Calabria-Citra, district and 7 m. ENE of Rossano, on the S coast of the gulf of Tarento, between the two embouchures of a river of the same name.—The river or torrent of T. descends from the E side of the Apennines, in the cant. and 5 m. W of Longobuca; runs first E, past that town, then N; and after a course of about 24 m., throws itself by two arms into the gulf of Tarento. On its banks are mines of argentiferous lead.

TRIORA, a village of Sardinia, in the div. of Nice, prov. and 13 m. N of S. Remo, in the midst of the Apennines, on a mountain near the l. bank of the Taggia.

TRIPALDA. See ATRIPALDA.

TRIPARNI, a town of Naples, in the prov. of Calabria-Ultra, district and 3 m. WNW of Monteleone, in a fertile plain. Pop. 455. It suffered severely from an earthquake in 1783.

TRIPASSUR, a town of Hindostan, in the presidency and 33 m. WNW of Madras, in the Carnatic. It was taken in 1780 by Hyder-Ali.

TRIPATUR, a town of Hindostan, in the presidency of Madras, in the prov. and district of Barramahe, and 24 m. NE of Caverypatam. It lies in the midst of plantations of fruit trees, and is well-peopled. In the vicinity is an extensive reservoir. —Also a town of the Carnatic, in the district of Tanjore, 48 m. S of Trichinopoly.—Also a town in the district of Chingleput, 84 m. NW of Madras. It is noted for its temple.

TRIPERGOLA, formerly a village of Naples, in the prov. of Naples, district and cant. of Pozzuoli. It was destroyed by an eruption in 1538, and its site is now covered by a hill consisting of lava and bitumen, and bearing the name of Montenuovo.

TRIPETTY, a town of Hindostan, in the presidency and 84 m. NW of Madras, in the Carnatic. It contains a Hindu temple,—the most celebrated S of the Krishna, and enclosing an image of one of the numerous incarnations of Vishnu.

TRIPHON (SAINT), a village of Switzerland, in the cant. of Vaud, district and 1½ m. SSW of Aigle, near the r. bank of the Rhone. It has a quarry of marble and some Roman antiquities.

TRIPOLE, a town of Russia in Europe, in the gov. and district and 32 m. SSE of Kiev, on the r. bank of the Dnieper.

TRIPOLI, a country of Northern Africa, the most easterly of the Barbary states, and politically a pashalik of the Turkish empire. As comprising Barca and Fezzan, it has a line of coast extending from Cape Razatina to Port-Bomba, a distance of nearly 800 m. Its extent inland is very irregular, owing to the frequent interruption of the Desert. Tripoli Proper lies between the parallels of 28° and 33° 30′ N, and the meridians of 10° and 20° E; and is bounded on the E by the desert of Barca; by Fezzan on the S; and on the W by Tunis and the Country of dates. This region, as a distinct district or province of the Western empire, appears to have acquired the name of *Tripolis*, from its containing the three cities of *Leptis*, *Œa*, and *Sabrata*, probably about the middle of the 1st cent. Similar appellations were given to districts of country by the ancients: thus we have *Pentapolis* applied to Cyrenaica, and *Decapolis*, a district of Palestine. This portion of Barbary is generally considered as politically comprehending not only Tripoli Proper, but the country or desert of Barca, the kingdom of Fezzan, and the oasis of Ghadames; but the bey's authority in these regions is so weak that they might almost be regarded as independent states. The gulf of Sidra, or the Greater Syrtis, occupies the eastern extremity of the coast-line of T.; that of Gabes, or the Smaller Syrtis, forms the western. The effect of the Greater Syrtis, from Mesurata, is that of a dreary moor,—a wide tract of level waste land,—without any thing to distinguish one part of it from another, but the windings of a marsh, which threads its dark surface, and is lost in different parts of the unbroken horizon. It extends in length, from Mesurata to Suleb, little less than 40 m., and in breadth from 9 to 15 m. At Suleb, its southern limit, some tolerable pasturage occurs, but the marsh, soon after passing this place, expands again and extends southward along the coast as far as Giraff, thus occupying altogether a space of 101 m. The whole of the country between Mesurata and Alexandria is described by Leo Africanus under the title of Barca. The appellation Barca is founded on the existence of this country as the independent kingdom of *Cyrene* governed by a branch of the Ptolemies. The mountains of Gerdobah intersect the southern part of this district. The western parts are the most fertile; the interior is an expanse of sandy desert. Couriers travel from T. to Cairo, across this country, in 25 or 30 days. Along the whole coast, and in many parts of the interior, are found fine specimens of classic architecture,—splendid but melancholy monuments of a civilization, a prosperity, and a glory that have completely passed away. Among the towns on the coast are Bengazi, Cyrene, Teuchera, Barca, and Bonandria, —the five cities which conferred on this district in ancient times the name of *Libya Pentapolis*.—We have sketched the geographical features of that portion of the district of Barca which extends from Alexandria to the gulf of Bomba, in our account of Lower Egypt.—The oasis of Aujela or Audjelah, the *Augila* of Herodotus, is situated in the route from the oasis of Siwah to Fezzan. Its inhabitants cultivate the soil, and escort the caravans which pass through their territory. The Gerdobah chain, which separates the desert of Barca from that of Libya, terminates at this oasis.—Fezzan, the ancient *Phazania*, *Garamantes*, or *Gamphasantes*, is bounded by Tripoli Proper on the N; by the desert of Barca on the E; and by the great desert of Sahara on the W and S. The greatest length of the cultivated country from N to S is about 255 m.; and its greatest breadth from E to W, 200 m. But the mountainous eastern region of *Harudjeh* is comprehended in its territory. In the N this country is intersected by three chains of mountains.

Climate and Productions.] The coast-line of this pashalik is indented by the great gulf of Sidra, but presents few bays or headlands. At a distance of from 12 to 20 m. from the coast, the range of the Gharian mountains extends from W to E in a direction nearly parallel to the coast, and with a breadth of from 8 to 12 m. Further inland, the Suara range runs paralled to the Gharian range; and between them lies an elevated fertile plateau. The soil along the coast is, for a few miles inland, of exuberant fertility, producing every article peculiar to tropical climates in the utmost luxuriancy, and the greater part of European vegetables. Indian wheat grows here from 5 to 6 ft. high, and barley yields twice as much as it does in Europe. But for the policy of the government, grain might be exported from T. to a large amount. The export of horses and mules is also prohibited; but the breed of both is cultivated with great care. Bullocks, sheep, and poultry are reared in immense quantities, and, owing to the small quantity of animal food consumed by the natives, afford an ample article of exportation. The fruits of T. generally possess an exquisite flavour. In the interior, which is sandy and barren, dates—which are here of a quality superior to any other to be found in Barbary—form the staple. They are of a yellow, brown, green, black, and red colour. The latter are termed horse-dates, and are given as food to that animal. The date-tree here attains a height of nearly 100 ft., and bears clusters weighing from 20 to 30 pounds. Here also is found the *lotus*, celebrated in antiquity as the food of the *Lotophagi*, a race of savages. The lotus-tree is lofty and umbrageous. The fruit is contained in a pod not unlike that of a tamarind; and when ripe, it is sweet and nutritious, tasting somewhat like gingerbread. An important article of popular diet is formed of the nutritious flour of the *cassob*, a plant unknown in Europe. It rises in the form of a reed, with a spike on the top containing the seed, which is small and of a lead colour. Cotton has been cultivated successfully, and the mulberry and castor-tree.—Gold-dust is found in the sand on the sea-shore; and veins of this precious metal occur inland as we approach to Fezzan.

Trade and Commerce.] The trade of T. is very considerable, though that with Europe is confined chiefly to Malta and the Levant. The exports are: wool of excellent quality, senna and other drugs, madder-root, barilla, hides, goat and sheep-skins dressed, salt, natron, ostrich-feathers, gold-dust, ivory, gum, dried fruit, lotus-berries, cassob, saffron, bullocks, sheep, and poultry. The imports are: cloths of every quality and colour, Manchester stuffs, sugar, tea, coffee, spices, spirits, wines, gold and silver-tissues, laces and threads, cochineal, indigo, iron, hardware of all kinds, muskets, pistols, sword-blades, &c. with naval stores of every description. The annual caravan from Marocco to Mecca, combining commercial with religious purposes, passes through the city of T.; and caravans from the interior generally arrive twice a year with slaves, gold-dust, &c. which are exchanged for coarse European cloths, silks, baracans or cloaks of the country, powder, muskets, hardware, glass-beads, toys, looking-glasses, &c. It is this latter branch of commerce with Central Africa, conducted by means of caravans from Fezzan and Ghadames, which forms the foundation of Tripolitan commerce. Murzuk, the capital of the dependent state of Fezzan, is the great depot of merchandise from the interior. In the months of December and January, caravans from Vadai, Bornu, Cashna, Soccatu, Haussa, and Timbuctu, arrive at Murzuk, where they exchange their commodities for Genoese paper, real and false corals, imitation pearls, printed cottons, silk stuffs, small mirrors, pistols, fire-arms, needles, razors, turbans, amber, porcelain vases, coffee-cups, copper vessels, caftans, embroidered muslins, white handkerchiefs, striped cotton or white calicot shirts, fine white calicots which are highly prized at Bornu and throughout Nigritia, essence of roses, and spiceries. The merchandise which the caravans bring to Murzuk from the central countries of Africa chiefly consists of slaves, gold-dust, trona, senna, ostrich-feathers, red alum, and ivory. The total number of slaves thus annually sold was estimated by M. Graaberg, the Swedish consul-general at Tripoli—to whose notice of Tripolitan commerce published in the *Antologia* of September 1827, we are indebted for these details—at 2,500. The quantity of gold-dust, exclusive of the tribute, amounts to 10,000 *metacalli*, or about 1,500 ounces, of which about one-third is used in T. for the purposes of money, and as ornaments for females; and the rest is exported to the Levant and Europe. *Trona*, a kind of natural mineral alkali, used in the manufacture of glass and the dyeing of linen, and for several domestic purposes, is annually brought to Tripoli to the amount of 7,000 cantaros. The senna of Fezzan is esteemed next in quality to that from Sidon in Syria; about 3,000 cantaros of this article are annually brought to T. The ostrich-plumes annually imported into T. are valued at from 15,000 to 18,000 piastres. They are usually sold by the skin; that of a male is reckoned worth double that of a female-ostrich. They pay a duty of 10 per cent. on their estimated value to the government. About 4,000 cantaros of alum are annually consumed in T. The quantity of ivory annually brought to Tripoli seldom exceeds 15 or 20 cantaros. All this merchandise is conveyed to T. on camels, each of which carries from 4 to 5 cantaros weight, and is valued at from 35 to 45 piastres. These animals, especially the m'herri or dromedary, are admirably fitted for the long journeys thus performed. As so much time is necessarily spent on these mercantile expeditions, a great deal is intrusted to the good faith of correspondents and agents, who are seldom found to betray the trust reposed in them. The town of Aujelah, in the oasis of that name, has long conducted a brisk trade of transit with Fezzan, especially since 1814, when, after many unsuccessful attempts to traverse the Libyan desert, a direct communication was opened with Borgu, Vadai, and Baghermi, without passing through Fezzan. The negligence of the government allows the profits of this commercial intercourse to flow into the pockets of the speculators in Grand Cairo; while the same supine indifference or extreme ignorance, has permitted a new commercial route by the way of Gorbi to be opened between Tunis and Murzuk. Although the Fezzan and Ghadames caravans are the most important in respect of commercial wealth, the most considerable in point of numbers are those which pass through this country, from Marocco and Western Barbary to Mecca. These latter caravans are chiefly composed of such persons as are performing that pilgrimage to Mecca, which all true Mahommedans are enjoined to make, at least, once in their lifetime; but mercantile speculations are conjoined by many with their religious motives. The numbers composing these caravans have greatly diminished of late years, as the greater part of the pilgrims from Western Barbary now embark in Christian vessels for the port of Alexandria, and the pasha of T. allows only a few of his own subjects to make this pilgrimage. M. Graaberg says, that the only caravan of this kind which he witnessed during four years' residence in T. was one in 1824, which consisted of about 3,000 men, a few hundreds of women and children, and 2,000 camels. It had set out from the city of Fez in Marocco; and journeying by Telonsan, and along the coasts of Algiers and Tunis, had arrived with considerably diminished numbers at T. It was commanded by an Emir-al-hoggias, a native of Kairwan in Tunis. Having rested about a month under the walls of T., the pilgrims resumed their route, part by sea, and the greater part at land; of the latter, one party travelled by way of Alexandria and Cairo, while another took the route of Aujelah, and were joined at that oasis by other pilgrims from Fezzan and the interior of Africa. These Mecca caravans carry with them gold dust, wax, ostrich plumes, silk and cotton *buracans* or haiques, muslins, morocco leather, perfumes, *khol* a kind of eye-pinguent, henna, *suak* a vermilion pigment, and a variety of medicinal drugs. On their arrival at Tripoli the pilgrims exchange a part of these articles of merchandise for European goods. On their return from the E, in about one year after, they bring Indian stuffs, pearls, Mecca balsam, musk, wood of aloes, incense, myrrh, civet, Cashmir shawls, precious stones, coffee, pistachios, naphtha, opium, and in general all the productions of Asia and Europe; but they are in general unwilling to open their packages at T., as their profits are much more considerable in their own countries.

Population.] The population of this state and its dependencies is estimated by Cortambert at 2,500,000; by Stein at only 1,350,000, by Colonel Warrington, in 1842, at 1,500,000. The rural pop. is chiefly Arab; the urban Moors. "There is probably no country on earth," says Mr. Blaquiere, "where the inhabitants are more inclined to be vicious; yet such is the promptitude with which justice is administered, that crimes bear but a small proportion to those of European countries. I have been unable to discover any of those good qualities that can be put in contrast with their well-known attributes of revenge, avarice, treachery, and deceit, which predominate alike in the prince and the peasant." Yet, in no part of Barbary, we are told, has civilization made so great progress; and in no place is so much respect paid to Christians, even their religious ceremonies being here treated with the utmost respect. Nay, Mr. Blaquiere adds, that their sensual passions "are infinitely better regulated than those of any other people on the coast of Barbary;" and upon the whole, it would seem, that T. presents by no means an unfavourable specimen either of the Moorish character or of the Barbary governments.—The pasha of T. is usually appointed by the Porte for four years. He is unfortunately left to raise his revenue by monopolies and taxes imposed at pleasure.

Tripoli, the capital, by the natives called Tarables, to the ancients known as *Æa*, is built upon a neck of land projecting into the sea, in N lat. 32° 34′, E long. 13° 11′. It is surrounded by a high wall flanked by six bastions, has a formidable line of batteries on the sea-front; and is considered impregnable to Moors and Arabs. It has a number of mosques, caravanseras and bazars. One of the latter is appropriated to the sale of woollens, Levant produce, and slaves from the interior. The police is extremely well-regulated; in point of cleanliness, T. unlike most other Moorish cities, might be a model for the best of those of Europe; acts of violence are never committed in the streets, though wine-houses are public, and intoxication by no means uncommon. The harbour is not very spacious, but is safe throughout the year, and admits frigates not drawing above 18 ft. of water. "Previously to entering the bay," says the author of *Tully's Memoirs*, "the country is rendered picturesque by various tints of beautiful verdure. No object whatever seems to interrupt the evenness of the soil, which is of a light colour, almost white, and interspersed with long avenues of trees; for such is the appearance of the numerous palms, planted in regular rows, and kept in the finest order. Their immense branches, coarse when near, are neat and distinct at a distance. The land lying low and very level, the naked stems of these trees are scarcely seen, and the plantations of dates seem to extend for many miles in luxuriant woods and groves. On a nearer view, they present a more straggling appearance, and afford neither shelter nor shade from the burning atmosphere which surrounds them. The whole town appears in a semicircle, some time before reaching the harbour's mouth. The extreme whiteness of the square, flat buildings, covered with lime, which in this climate encounters the sun's fiercest rays, is very striking. The baths form clusters of cupolas, very large, to the number of 8 or 10, crowded together in different parts of the town. The mosques have in general a small plantation of Indian figs and date-trees growing close to them, which, at a distance, appearing to be so many rich gardens, give the whole city, in the eyes of a European, an aspect truly novel and pleasing. On entering the harbour, the town begins to discover dilapidations from the destructive hand of time, large hills of rubbish appearing in various parts of it. The castle or royal palace in which the pasha resides, is at the E end, within the walls, with a dockyard adjoining. This castle is very ancient, and is enclosed by a high strong wall: it has lost all symmetry on the inside, from the innumerable additions made to contain the different branches of the royal family; for there is scarcely an instance of any of the blood royal, as far as the basha's great grandchildren, living without the castle-walls. These buildings have increased it by degrees, to a little irregular town." The town is smaller than either Algiers or Tunis. Its greatest length, Captain Beechey says, may be about 1,360 yds.; its extreme breadth about 1,000 yds. In 1805 the pop. was estimated by Ali Bey at about 15,000; in 1811 Mr. Blaquiere supposed it to amount to 25,000; in 1842, Colonel Warrington estimated it at 12,000 Turks and Moors, 1,500 Christians, and 2,000 Jews. This city has often been ravaged by the plague. The Arabic spoken here is said to be the purest known on the whole coast.

History.] After the destruction of Carthage, T. became a Roman province; and as the Roman power declined in Africa, the boundary of civilized manners and cultivated land in this region was insensibly contracted. In the reign of Valentinian, the Getulians appeared at the gates of the Tripolitan cities; and the inhabitants unprotected by their venal prefect, were compelled to join the rebellious standard of a Moor. In an evil hour Count Boniface invited the assistance of the Vandal king against his rebellious subjects. "On a sudden," says Gibbon, "the seven fruitful provinces, from Tangiers to Tripoli, were overwhelmed by the invasion of the Vandals." A series of desolating wars in the reign of Justinian completed the ruin of Roman Africa. Under the caliph Omar, the Mahommedans, crossing the desert of Barca, first invaded T. Omar's successor, Akbah, marched from Damascus at the head of 10,000 Arabs, traversed the wilderness in which his successors erected the splendid capitals of Fez and Marocco, and at length penetrated to the verge of the Atlantic and the Great Desert. On the dissolution of the caliphate, T. became an independent state. In the 16th cent. it fell into the hands of the emperor Charles V., who had landed in Africa as the ally of the Moorish king of Tunis, and who gave it to the knights of Rhodes. The Turkish corsair, Dragut-Rais, expelled the knights from T. in 1551; and after the death of Dragut, the Porte continued to send governors to T., under the titles of sanjak and pasha, till 1713, when it was erected into an independent state by Hamet Pasha, a native of Caramania, who reduced to subjection the turbulent mountaineers of Garian and Mesurata, and rendered Fezzan a tributary province. He also invited intelligent foreigners to settle under his sway, and greatly exerted himself to improve industry and manufactures. Yussuf Pasha, the late governor of T., who raised himself to that dignity by the murder of his brother, though he administered public affairs with no very gentle hand, seemed to have more liberal views than most of his predecessors. He studiously cultivated the alliance of Britain, and afforded facilities for exploring the interior of his dominions. Captain Beechy, who visited this country in 1822, says that this state was then secure under the protection of an established government; that property was respected and commerce improving; that its markets were well-supplied, its manufactures encouraged, and its population increasing.

TRIPOLI, or Tarabolus, a pashalic of Syria, comprehending all the coast from Latakia S to the Nahr-el-Kelb, which separates it from the pash. of Acre. It is bounded on the E by the chain of Lebanon which separates it from the valley of Baalbec and the valley of the Orontes; on the N by Cape-Ziarat, which divides it from the district of Antakia; and on the W by the sea. By far the greater portion of this district is mountainous, the only plain being a narrow stripe of land stretching along the coast at the foot of the mountains. It is watered by numberless mountain-streams, but is less highly cultivated than the terraces of Lebanon. The chief productions are wheat, barley, cotton, and tobacco. A great part of Mount Lebanon is included in this pash., which is divided into 14 districts: viz. 1. Jebel-Beshirai, a considerable tract to the E of Tripoli, which contains 12 villages, of which Antura and Canobin are the chief. 2. El-Zawye, a small district in Lower Lebanon. 3. Batrun, a village with a small district on the coast. 4. El-Kura in the lower part of Lebanon divided from El-Zawye by the Nahr-Kadisha. 5. El-Fetuh to the E of Jebail, and bordering on the Kesrawan. 6. Akura, a small district with a village of the same name, the seat of a Maronite bishop. 7. El-Dennye, to the N and NW of Beshirai. 8. Jebail, a district on the coast belonging to the town of that name. 9. Jebel-el-Minitra, a Mutuali district. 10. El-Hermel, on the E declivity of Lebanon towards Baalbec. 11. El-Kataa, E of Batrun. 12. El-Kella. 13. El-Shara. 14. Tortosa, on the coast. To these may be added the mountainous territories of the Ansaries and Ismaelians, extending from the Nahr-el-Kebir to the N of Latakia. The Mutualis formerly possessed several districts in this pash., but they were expelled by Yussif, emir of the Druses, and they in their turn seized the Bilad-Baalbec, and drove out the inhabitants. The pash. of Tripoli during the last century included all the line of coast from the Orontes southward to the Dog river, that is to say, not only from Suadia to the town of Tripoli, but also the Billad-Jebal, Botrun, and Kesrowan. Circumscribed as this pash. is, it presents a motley pop. of Moslems, Mutualis, Christians, and Ansaries. The stronghold of the Moslems is the town of Tripoli itself and its environs. The Mutualis are principally located in the district of Botrun and Jebail. At the period of the Druse war they assumed the superiority over the Maronites. The Ansaries are now tranquil, and in reality tolerably independent. Before the depreciation of the currency, the rent paid by this pash. to the Porte was 750 purses, which, with the Spanish dollar at 5 piastres, made 75,000 dollars a-year; but out of this the pasha was, by ancient custom, obliged to defray the expenses of the Haji to Mecca. During the Egyptian occupation, the income was about 6,000,000 of piastres, which with the dollar of 22½ piastres made 270,000 dollars. The income of the pash. is now very nearly what it was in the time of the Egyptians, the increase in the customs having made up for the diminution of the ferdeh and the arrears of the miri.—Entering this pash. from the N, Latakia is the first place of consequence we encounter. The district formed part of the ancient *Casiotis*, so called from a range of hills extending N to the Orontes. The city, the ancient *Laodicea ad mare*, built by Seleucus Nicator, who named it in honour of his mother, stands on the NW side of Cape Ziarat, an elevated promontory which advances half-a-league into the sea. To the S of this is Jebilee, the ancient *Gabala*, now a place of no consequence. From Latakia to Tortosa, an almost continuous succession of ruins is encountered scattered along a vast rich plain at the foot of the Ansairy mountains.—Tartus or Tortosa, the ancient *Orthosia*, a place of great consequence during the crusades, has nothing remaining but its castle, which is large and still inhabited. S of this is the isle of *Ruad*, the *Arvad* of Scripture, and the *Aradus* of the Greeks.

Tripoli, or Tirehboli, the capital of this pashalik, is composed of three cities, a furlong distant from each other, but which have been gradually joined together by their respective suburbs. It is built on the declivity of the lowest hills of the Libanus, near the shore, in N lat. 34° 26′, E long. 35° 46′, about 3 m. to the W of the embouchure of a river which descends from Zogana. It is the neatest town in all Syria, the houses being well-built of stone, and neatly constructed. It is surrounded with luxuriant gardens, producing abundance of oranges and lemons, and extending over the whole triangular space between the town and the sea. The city is divided into two parts by the Wady-Kadesha, which enters the plain through a beautiful narrow valley, and after traversing the town falls into the sea on the N side of the triangle. On the S side, just where the Kadesha enters the town, is the citadel, which commands both the town and the whole plain below, but which is itself commanded by a height on the opposite side of the river, only 150 yds. distant. The vicinity of T. is the most favoured spot in all Syria, as the maritime plain and neighbouring mountains place every variety of clime within a short distance of the inhabitants, and the Wady-Kadesha is the most picturesque of valleys: yet the situation, however beautiful, is not healthy, and from July to September epidemic fevers prevail here, as at Scanderun and Cyprus. The town, moreover, being open only to the west, the air has little circulation, and a constant feeling of lassitude is experienced by strangers visiting it. The pop. is estimated at 16,000, one-third of whom are Greek Christians under a bishop. Of fifteen European commercial establishments that were formerly here, only one French house remains. In the palmy days of Syrian trade, when Aleppo had its seventy English houses, and was the greatest emporium in the East, T. had also its British consul and its British factory: not a single Englishman resides here now. The principal export is long reel silk, for the Marseilles and Leghorn market; the average quantity exported is 45,000 okes, at £1 per oke. The staple manufacture used to be that of silk girdles, in which there were formerly above 1,000 looms engaged; now there are only 200. These girdles were exported in large quantities to Constantinople, but change of costume has reduced the demand. The leading manufacture at present is soap, and the largest building in the town is the Khan-el-Saboon, or Soap-khan. T. has 12 soap boilers, who manufacture on an average 3,300 cantars of this commodity, the most of which goes to Aleppo. The oil requisite is all procured from the olive-grounds to the east of the town. Candian soap, which contains very little alkali, is imported, but one-fourth of its weight of alkali being here added to it, it is resold to great advantage. Galls from the Ansairie mountains, yellow wax from Lebanon, and madder, form other articles of exportation.

TRIPOLITZA, Tripolizza, or Tripolis, a town of Greece, in the Morea, capital of the department of Arcadia, at the foot of Mount Mænalus, 22 m. SSW of Argos, and 30 m. NNW of the ruins of Sparta. It is said to have been built of the remains of three towns, *Tegea*, *Mantinœa*, and *Pallantium*, without, however, occupying the site of any of these places, which were at a considerable distance from each other. Its situation is bleak, the ground which it occupies is rugged and uneven, and the general plan of the town extremely irregular. Under Turkish regime, it was the capital of the Morea; yet the streets, with the exception of the principal one, were dirty, and paved only in the middle; and the houses reminded an English traveller of those of our poorest villages. The best buildings were the residence of the pasha, a wooden structure in the form of a square with walls and gates, and the khan, or place of accommodation for travellers. There were several mosques for the Turks, and also several churches for the Greeks. The population was computed at 12,000; it does not now exceed 2,600. The trade of the place is inconsiderable, the corn, wool, or other articles exported from it having a land carriage of 24 m., before reaching the port of Lerna in the vicinity of Argos. The fortifications of the place consisted of stone walls erected by the Albanians in the latter part of the 18th cent.; and of a small square fort built by them on an eminence to the SE. It has several schools, and a powder-mill. This town was sacked by the Greeks in 1821, and in 1825 by Mehemet Ali; and has not recovered from these disasters.

TRIPONTARY, a town of Hindostan, in the prov. and 9 m. E of Cochin.

TRIPPSTADT, a village of Bavaria, in the circle of the Pfalz, district and 19 m. ENE of Deux-Ponts, on a hill, in the midst of the Vosges.

TRIPTIS, a town of the grand-duchy of Saxe Weimar, circle and bail. and 5 m. E of Neustadt, on the Orla. Pop. 1,411. It has manufactories of hosiery, cloth, linen, and hats.

TRISOBBIO, a village of Sardinia, in the div. of Alessandria, prov. and 7 m. SE of Acqui, on a mountain. Pop. 1,050.

TRISSINO, a village of Austrian Lombardy, in the prov. of Vicenza, and district of Valdagno. Pop. 2,888.

TRISTAN D'ACUNHA, the largest and most fertile of a group of three islands in the S. Atlantic, situated between the cape of Good Hope and the coast of Brazil. It is a mass of rock 12 m. in circumf., and rising to 5,000 ft. above sea-level, in S lat. 37° 6′, W long. 12° 2′. The aspect of the NE side of the island is very striking. At the foot of an almost perpendicular mountain, thickly covered with brushwood, a fertile plain of considerable extent stretches along the shore. Wild boars, wild goats, and a species of black cock, abound on the island; while in the deep waters among the rocks are fish of almost every description. The clifts are literally covered with sea-hens, petrels, albatrosses, and other feathered tribes. Captain Denham of her Majesty's ship Herald, obtained deep sea soundings about midway between this island and Cape Horn, of 7,706 faths. or 15,412 yards. This sounding was obtained in S lat. 36° 49′, W long. 37° 6′, on a calm day, Oct. 30, 1852. The sounding line was 1–10th of an inch in diam., laid into one length, and weighing, when dry, 1 lb. for every hundred faths. The plummet weighed 9 lbs., and was 12 inches long and 1–7 inch in diam. The velocity with which the line ran out was as follows:—

	Hr.	Min.	Sec.
The first 1,000 faths. in . .	0	27	15
1,000 to 2,000 „ . .	0	39	40
2,000 to 3,000 „ . .	0	48	10
3,000 to 4,000 „ . .	1	13	39
4,000 to 5,000 „ . .	1	27	06
5,000 to 6,000 „ . .	1	45	25
6,000 to 7,000 „ . .	1	49	15
7,000 to 7,706 „ . .	1	14	15
	9	24	45

The whole time, therefore, taken by the plummet in descending to this amazing depth of 7,706 faths., or 7·7 geog. m. of 60 to a degree, was 9 hours 24 minutes and 45 seconds. The highest summits of the Himalaya are little more than 28,000 ft., or 4 7 geog. m. above the sea. The sea-bottom has therefore depths greatly exceeding the elevation of the highest pinnacle above its surface. The strength of the line, tried before the sounding, was found to be equal to bear 72 lbs. in air. The 7,706 fathoms which ran out weighed when dry 77 lbs. exclusive of the plummet. This sounding is the deepest that has ever been made.—The shores of the other islands are so wild and dangerous that it is impossible to approach them, except in the calmest weather: one is called Nightingale island, and the other Inaccessible.

TRISTE, an island of the state of Yucatan, Central America, in the SW part of the bay of Terminos, to the W of Port Royal, from which it is separated by only a narrow channel, in N lat. 18° 20′. It is about 18 m. in circumference, abounds with good water, and produces varieties of valuable wood. It is uninhabited.—Also a gulf of Venezuela, on the N coast, formed by the Caribbean sea, in N lat. 10° 30′ and W long. 68° 10′. It opens between Point Tucacas on the NNW, and Point Cabello on the SSW, and is enclosed by arid mountains. It contains 3 small islands.

TRISTE, Mego, or Ille-du-Recif, a small uninhabited island of the Indian ocean, near the SW coast of Sumatra, in N lat. 4°, E long. 100° 59′. It is surrounded by a coral reef, and in the centre is a lake. It abounds with cocoa palms, and has large numbers of rats and squirrels.

TRISTOMATO, or Skomako, a port of Turkey in Asia, in the sanj. of Rhodes, at the N extremity of the island of Scarpanto. Its entrance, formed by Cape Bonandrea on the W, and the ancient promontory of *Ephialtium* on the E, is 6 m. in width, and is divided by the island of Scarpathon into two channels. The depth of the bay is about 4 m., and it encloses several considerable branches.

TRISTOMOS, a port of Turkey in Asia, in Anatolia, sanj. of Mentesha, and 21 m. S of Almalu, formed by the Mediterranean, with which it communicates by only a narrow passage, and opposite the island of Kakava. It is nearly 3 m. in length from ESE to WSW, but is very narrow and opens to the S. The ancient *Andriace* was situated on its N shore.

TRITANI, a town of Naples, in the prov. of Calabria-Ultra, district and 18 m. NE of Palmi, on a hill, in a fertile locality. It suffered severely from an earthquake in 1783.

TRITH-SAINT-LEGER, a commune of France, in the dep. of the Nord, cant. and 4 m. SW of Valenciennes, on the l. bank of the Schelde. Pop. 1,661. It has extensive iron and copper works, manufactories of beet-root sugar, vinegar, &c.

TRITON'S ISLAND, a small island of the Paracels archipelago, in the China sea, in N lat. 15° 48′, E long. 111° 12′. It is nearly surrounded by reefs.

TRITTAU, a village of Denmark, in the duchy of Holstein, 20 m. ENE of Hamburg, on the r. bank of the Bille.

TRIUMFO-DE-LA-CRUZ, a headland, bay, and island of Honduras, Central America, to the E of the Sal. The bay affords good anchorage. On its banks are the ruins of a town of the same name.

TRIUMPHO, a town of Brazil, in the prov. of São-Pedro-do-Rio-Norte, on the l. bank of the Taguari, near the confluence of the Jacuhi. Pop. 3,462, of whom 1,766 are whites.

TRIVANDERAM, or Trivandapatam, a town of Hindostan, capital of the prov. of Travancore, on a small river, which 6 m. below falls into the Indian ocean, 1,135 m. SE of Cochin. It is the usual summer residence of the rajah, and has a spacious and substantial palace, built in European fashion, and a castle. The town is populous, and has a large garrison.

TRIVENTO, a town of Naples, in the prov. of Sannio, district and 16 m. NNW of Campobasso, on a height near the r. bank of the Trigno. Pop. 3,500. It is enclosed by walls, and has some fine edifices, amongst which are the episcopal palace and cathedral. It has besides four churches, two convents, and a seminary. This town, which is of great antiquity, was one of the most important belonging to the Samnites. It was the *Triventum* or *Treventum* of the Romans.

TRIVERO, a town of Sardinia, in the div. of Turin, prov. and 9 m. NE of Biella, at the foot of a mountain, on a small affluent of the Sessera. Pop. 3,300.

TRIVIEARY, a village of Hindostan, in the prov. of the Carnatic, on the l. bank of the Villenore or Ariancoopan river, 15 m. WNW of Pondicherry. It now consists of only a few scattered huts; but from the number and extent of its ruins it appears to have formerly been a place of considerable importance. Of these remains, the principal are a pagoda, and a tower over the gateway. The environs are noted for their petrifactions.

TRIVIER-DE-COURTES, or Trivier-de-Courtoux (Saint), a canton, commune, and town of France, in the dep. of the Ain and arrond. of Bourg-en-Busse. The cant. comprises 12 coms. Pop. in 1831, 11,679; in 1846, 12,272.—The town is 18 m. NNW of Bourg, in a pleasant and salubrious situation. Pop. 1,453. It is small, but tolerably well-built, and has a church, built it is said in the 6th cent. by St. Trivier, and an hospital. This town derives from a neighbouring village its surname, which distinguishes it from St. Trivier-en-Dombes, an adjacent village.

TRIVIER-EN-DOMBES, or Trivier-sur-Moignans (Saint), a canton, commune, and town of France, in the dep. of the Ain, and arrond. of Trevoux. The cant. comprises 19 coms. Pop. in 1831, 10,539; in 1841, 10,939. The town is 11 m. NNE of Trevoux, in a humid plain. Pop. 1,536.

TRIVIERES, a commune of Belgium, in the prov. of Hainault, arrond. and 10 m. E of Mons. Pop. 450. In the vicinity is a coal mine.

TRIVIGNO, a village of Naples, in the prov. of Basilicata, district and 11 m. SE of Potenza, amid lofty mountains, near the r. bank of the Basento. Pop. 2,500. It has a church and four chapels.

TRIZAC, a village of France, in the dep. of the Cantal, cant. and 6 m. SW of Biom-les-Montagnes, on a mountain. Pop. 1,579.

TURNAWA. See TURNAU. See also TYRNAU.

TROAD, an ancient district of Asia Minor, now comprised in the sanj. of Bigha, in Anatolia. It contained the ancient city of Troy. See MENDERE-SU.

TROARN, a canton, commune, and town of France, in the dep. of Calvados and arrond. of Caen. The cant. comprises 32 coms. Pop. in 1831, 13,083; in 1846, 13,292. The town is 9 m. E of Caen, on the Dives. Pop. 892. It has a small hospital, and contains manufactories of lace, linen, sugar from beet-root, bricks, and tiles. The trade, which is considerable, consists chiefly in butter, cheese, cider, fish, and fowls.

TROBRIAND, a group of islands in the S. Pacific, in the Louisiade archipelago. Cape Denis, the N extremity of the group, is in S lat. 8° 24′, E long. 151° 8′.

TROCCHIA, a village of Naples, in the prov. and district and 8 m. ENE of Naples, and cant. of Santa-Anastasia, at the NW base of Vesuvius. Pop. 740.

TROCHE, a village of France, in the dep. of the Correze, cant. and 4 m. W of Vigeois, near the r. bank of the Loire. Pop. 1,200.

TROCHTELFINGEN, a town of the principality of Hohenzollern-Sigmaringen, 18 m. N of Sigmaringen. Pop. 800.

TROEDYRAUR, a parish in Cardiganshire, 4 m. N by E of Newcastle-in-Emlyn. Area 4,666 acres. Pop. in 1831, 1,064; in 1851, 1,029.

TROENE, a river of France, in the dep. of the Oise, which has its source a little to the E of Ivry-le-Temple, in the cant. of Meru; flows through the marshes of Tourcy; passes Chaumont and Trie-le-Chateau; and after a course in a generally WNW direction of 17 m., joins the Epte on the l. bank, opposite Gisors.

TROENSES. See THROUSE.

TROFA, a town of Portugal, in the prov. of Beira, comarca and 11 m. ENE of Aveiro, on the l. bank of the Vouga.

TROGEN, a town of Switzerland, capital of the cant. of Appenzell-Aussen-Rhoden, 7 m. NE of Appenzell, on the Goldbach, at the foot of the Habris. Pop. 2,395. It is handsomely built,—the dwellings of the more opulent inhabitants being ornamented on the exterior with the most precious marbles,—has a handsome church, mineral baths, a public library, several schools, and an arsenal; and possesses manufactories of linen and cotton fabrics, and two printing presses.

TROGSTAD, or TRYGSTAD, a parish of Norway, in the dio. of Aggershuus and bail. of Smaalehnen, 30 m. SW of Christiania, near the SE bank of Lake Ojeren. Pop. 2,900.

TROHBERG, a mountain of Thuringerwald, in the duchy of Saxe-Coburg-Gotha, and principality of Gotha. Alt. above sea-level, 2,150 ft.

TROIANOVKA, a town of Russia in Europe, in the gov. of Volhynia, district and 28 m. ENE of Kovel.

TROIANOVO, a town of Russia in Europe, in the gov. of Volhynia, district and 12 m. SSW of Jitomir, near the r. bank of an affluent of the Teterev.

TROIS-FONTAINES, a village of France, in the dep. of the Meurthe, cant. and 7 m. SSE of Sarreburg, on the r. bank of the Bievre. Pop. 425. It has several saw-mills, and a quarry of free-stone.

TROIS-FONTAINES-LA-GRANGE, a village of France, in the dep. of the Marne, cant. and 11 m. ENE of Thieblemont, in a woody marshy valley on the Brussonelle. Pop. 200. It had a Cistercian abbey founded in 1200. Iron is wrought in the environs.

TROIS-FRERES, a group of islets, three in number, in the Indian ocean, in S lat. 6° 10′, E long. 73° 38′. Pop. 50. It depends on the English government of the Mauritius. It abounds with cocoa palms, and the coasts with fish.

TROIS-ILETS (LES), a parish and village of Martinique, arrond. and 5 m. SE of Fort Royal, on the S coast of the Royal Cul-de-Sac. It has a church, but consists entirely of wooden houses inhabited by negroes.

TROIS-MOUTIERS (LES), a canton, commune, and town of France, in the dep. of the Vienne and arrond. of Loudun. The cant. comprises 14 coms. Pop. in 1831, 8,710; in 1846, 8,675. The town is 5 m. NW of Loudun. Pop. in 1846, 1,538. The parish, which derives its name from three small islands near the coast, has a good harbour, and at the S extremity, on the islet Aux-Ramiers, is a fort, erected for the defence of the entrance to the Royal Cul-de-Sac. The coasts abound with fish, but the locality, especially near the shore, is unhealthy. The soil, which is argillaceous and marshy, possesses little fertility, but produces in some parts coffee, and in small quantities, grain. Its principal productions are sugar and pottery. Pop. 1,690, of whom 66 are whites.

TROIS-POINTES (CAPE DES), a headland of Upper Guinea, on the Gold coast, in the kingdom of Ashantee, in N lat. 4° 40′ 30″, and W long. 2° 43′ 47″. It is so named from its conformation.

TROIS-RIVIERES. See THREE RIVERS.

TROIS-RIVIERES (LES), or SAINT SAUVEUR, a town of Guadeloupe, on the SE coast of the Cabes-Terre, and 8 m. ENE of the town of Basse-Terre, at the mouth of a small river of the same name which descends from the Terre-Neuve hills, and throws itself into the bay of St. Sauveur. It has a church. It suffered severely from a Negro revolt in 1802. The quarter of Trois-Rivieres is covered with rising grounds terminating abruptly on the coast, and crowned with plateaux, the soil of which is in many parts excellent. Its principal productions are sugar, coffee, cotton, cacao, and manioc. Pop. 2,978, of whom 291 are whites.—Also a river of Hayti, in the dep. of the Nord, which has its source 5 m. NW of Marmelade, and after a sinuous course, in a generally NW direction of about 54 m., throws itself into the Tortuga channel, a little to the W of Port-de-Paix.

TROIS-ROIS. See THREE-KINGS.

TROISSEREUX, a village of France, in the dep. of Oise, cant. and 5 m. WNW of Nivillers, near the l. bank of the Therain. Pop. 680.

TROISSY, a village of France, in the dep. of Marne, cant. and 4 m. E of Dormans. Pop. 950.

TROIS VILLES, a village of France, in the dep. of Nord, cant. and 4 m. W of Chateau-Cambresis. Pop. 1,500. There are cotton-factories here.

TROITSK, a town of Asiatic Russia, in the gov. of Orenburg, 222 m. ESE of Oufa, on the l. bank of the Oui, which here receives the Uvelka. It is surrounded with wooden fortifications forming a square flanked with towers and encompassed by a ditch and glacis. The streets are straight and regular, and the town contains about 1,000 houses. It has large iron-works, and manufactories of fire-arms; but the importance of this place is derived from its being an emporium for the trade with the Asiatic tribes, particularly the Kirghiz of the Lesser Horde, who are rich in cattle, and with Bokhara and Tashkend. The trade is carried on in a large square, built on the opposite side of the Oui or Ouk. One part of this exchange is allotted to the merchants of the country, another to the Kirghiz, and a third to the Bukharians.—Also a town in the gov. and 79 m. NNW of Penza. Pop. 4,000.

TROITSKO-SASKAI See KIAKHTA.

TROITZKOI-SERGHIEVA, a town of Russia, in the gov. and 32 m. SE of Moscow, on the river Kutchura. It contains 6,000 inhabitants, who are chiefly employed in making toys of wood and stone. It is famous for its Greek monastery, commonly called the convent of the Holy Trinity, founded in the 14th cent. This monastery, which is the richest in the empire after that of Petchersk at Kiev, contains in general above 1,000 monks, and is regularly fortified. It has 7 churches within its walls, a palace, and an ecclesiastical seminary. In times of commotion or danger it has afforded a refuge to the czars, among others to Peter I. when he put an end to the administration of his sister Sophia.

TROJA, a town of Naples, in the Capitanata, 7 m. NNW of Bovino, on the small river Chilare, 33 m. SW of Manfredonia. Pop. 4,000. It forms, along with the town of Biccari, the see of a bishop.—Also a small island of the Mediterranean, off the coast of Tuscany, 13 m. E of Elba, in N lat. 42° 44′.

TROKI, a town of European Russia, in the gov. and 15 m. SW of Wilna, on a small lake. Pop. 4,000. It was built in 1321 by the duke of Lithuania, and was the residence of his successors till they removed to Wilna.

TROKI (STARO), a town of Russian Lithuania, 14 m. WSW of Wilna.

TROLLHATTA, or TROLLHETTA, a village of Sweden, in West Gothland, 15 m. ESE of Uddevalla, and 41 m. N by E of Gottenburg. The river Gotha-elf here forms a series of remarkable falls, not far from the place where it issues from Lake Wener. The river, which is very wide before reaching the falls, is here confined by the rocks within a narrow channel, and its course is further impeded by several rocky islands in the middle of the stream. The falls are three in number, and their whole descent is estimated at 100 ft. Their rapidity is very great, and their noise is heard at the distance of a league. Boats avoid the falls by a canal a quarter of a Swedish mile in length, 22 ft. wide, and 7 ft. deep, with 12 locks. See GOTHA-ELF.

TROMAUN, a village in the p. of Athleague, co. Roscommon, 3 m. S of Roscommon. Pop. 374.

TROMBETAS, or ORIXIMENA, a river of Brazil, which rises on the S flank of the Serra-Tumacumaque; runs S; and enters the Amazon on the l. bank, in S lat. 1° 50′, after a course of nearly 400 m.

TROMOE, an island off the S coast of Norway, at the entrance of Arendal bay, in N lat. 58° 14′.

TROMPERWICK (GULF OF), a bay on the E coast of the island of Usedom, in the Baltic, in N lat. 54° 40′.

TROMPIA (VAL-), a valley of Austrian Italy, in the N part of the deleg. of Brescia, traversed by the Mella. It is about 21 m. in length; runs SSW; and has good pasturage, and extensive iron-works. It is divided into 19 communes, and contains above 16,000 inhabitants.

TROMSOE, an island off the NW coast of Norway, in N lat. 69° 39′, E long. 18° 57′. It is 4 m. in length from NNE to SSW. The interior is hilly and rocky, and several of its elevations rise above the line of perpetual snow, which in this high lat. is so low as 1,200 ft. above sea-level. Corn does not succeed here; even for pasturage the only suitable tract is a narrow strip of land along the coast. There is a small town with a good harbour on the E coast, which the Danish government endeavoured, in the course of last century, to render a place of trade, from which vessels proceeding to the whale-fishery, or to the Loffoden isles, might be enabled to sail with their cargoes to the south of Europe, but the attempt did not succeed.—The province of T., comprising the N part of the Norwegian kingdom, and administratively divided into the bailiwics of Nordland and Finland, has an area of 31,560 geog. sq. m.

TRONA (WADI), a valley of N. Africa, in Fezzan, to the NE of Gherma.

TRONÆS, a village of Norway, 90 m. SW of Tromsöe, on the N coast of the island of Hindœn. Pop. 1,200.

TRONCHON, a village of Spain, in the prov. and 36 m. NE of Teruel.

TROND (SAINT), or TROIJEN, a town of Belgium, in the prov. of Limburg, on the railway from Mechlin to Liege, 10 m. SW of Hasselt. Pop. 9,000. It has a manufactory of fire-arms, and beet-root sugar-works, distilleries, and soap-works. It was formerly a place of some strength, but its walls were razed in 1675. A sanguinary action took place in the neighbourhood, between the French and Austrians, in March 1793.

TRONDA, or TRONDRAY, an island 3½ m. long, and 2 m. broad, off the W coast of Quarf, in the mainland of Shetland.

TRONDHJEIN. See DRONTHEIM.

TRONE, or TRON, a river of Prussia, in the reg. of Treves, which flows into the Moselle, after a NNW course of 30 m.

TRONEKEN, a village of Prussia, in the Hundsruck, 12 m. SSE of Traarbach.

TRONFIELD, a mountain of Norway, on the l. bank of the Glommen, in N lat. 62° 20′, having an alt. of 883 toises = 1,882 yds. above sea-level.

TRONGET, a village of France, in the dep. of Allier, cant. and 1 m. N of Mont-aux-Moines. Pop. 1,000. Coal is wrought in the vicinity.

TRONQUAY (LE), a village of France, dep. of Eure, cant. and 2 m. N of Lions-la-Foret. Pop. 1,400.

TRONS, or TRON, a village of the Swiss cant. of the Grisons, in a beautiful situation, near the l. bank of the Rhine, 18 m. SE of Altorf. Pop. 1,047, Catholics. There are iron-works in the vicinity.

TRONTANO, a village of the Sardinian states, in the div. of Novara, 2 m. ENE of Domo-d'Ossola.

TRONTO, a river of the Papal states, which rises in the Apennines, 6 m. NE of Montereale; runs N; and flowing along the border of the Neapolitan dominions, discharges itself into the Adriatic, in N lat. 42° 54′, after a course of 60 m.

TRONZZANO, a town of Piedmont, in the prov. and 12 m. W of Vercelli. Pop. 3,500.

TROO, a town of France, dep. of Loire-et-Cher, 10 m. SW of Vendome. Pop. 1,000, partly employed in the manufacture of woollen stuffs.

TROON, a promontory and a thriving port at the W end of the parish of Dundonald, in Ayrshire. The town, 6 m. N of Ayr, and 9 m. SW of Kilmarnock, is built on a belt of rock, extending 1¼ m. into the sea, and so curved as to form a large segment of a circle with the concave side facing the N. The embayed marine space embraced by the bold curvature is by far the best natural harbour in Ayrshire, affording safe anchorage-ground from every quarter except the NW. The merchants of Glasgow, aware of its advantages, made a vain effort, about the year 1700, to purchase the circumjacent property for the erection of a sea-port; and were obliged to select the very inferior site of Port-Glasgow as the station next in eligibility. The duke of Portland, about the year 1817, constructed a new pier 800 ft. long, nearly at right angles with the rock, where the depth is 19 ft. at low water, and he afterward constructed a wet dock with floodgates, a dry dock for repairing vessels, a lighthouse, and large store-houses. A railway to T. from Kilmarnock was the first public work of its class in Scotland, and brings down vast quantities of coals for shipment to Ireland, Galloway, and other destinations. A communication is maintained likewise with the Glasgow and Ayr railway, the main line of which passes in the vicinity. Many of the inhabitants are employed in a large ship-building yard, a rope and sail manufactory, and some other works; and not a few of them draw an entire or partial maintenance from letting lodgings to families of sea-bathers. In 1836, the pop. of the town was 1,088, and of the town and a district extending 2 m. from it, 1,600; in 1851, it was 2,404.

TROPEA, a town in Calabria-Ultra 2da., 12 m. W of Monteleone, on a rock overhanging the gulf of Santa-Eufemia. Pop. 4,300. It is surrounded by walls and towers, and from many points has a very picturesque effect. The rock on which it stands rises perpendicularly from the sea, so that the owners of the houses which are built on the edge of the precipice, often fish with considerable success from their windows. It has manufactories of linen and cotton. Porcelain earth is found in the vicinity. Two little streams, the Alafito and the Lamia, fall into the sea between T. and Parghelia. The neighbouring country produces a considerable quantity of cotton, which is worked upon the spot: the plants, compared with those of Egypt, have, however, a very dwarfish and miserable appearance. About 1,000 rotoli of silk are yearly produced in the environs.

TROPEZ (Saint), a small port of France, in the dep. of Var, 39 m. ENE of Toulon, in N lat. 43° 16′. It has 3,700 inhabitants, who are employed for the most part in the tunny and pilchard fishery, and in the coasting trade.

TROPIC-KEYS, a group of shoals among the Virgin islands, between Great Passage island and the E coast of Porto Rico. They are so named from their abounding in birds called *tropicos*.

TROPPAU, a principality of Silesia, bounded on the N by the principality of Oppeln, on the E by Ratibor and Teschen, and on the S and W by Moravia. It was erected into a principality in 1254. At the peace of Berlin in 1742, the part to the north of the Oppa was ceded to Prussia, but the remainder is still subject to Austria. The whole belongs in property to the princes of Lichtenstein.—Also a circle of Austrian Silesia, containing several lordships, and the parts of the principalities of T., Jagerndorf, and Neisse, that were retained by Austria, when the rest of Silesia was ceded to Frederick II. Its area is 1,080 sq. m.—Its cap. of the same name, and till lately the residence of the regency, stands at the confluence of the Oppa and Mohra, in N lat. 49° 50′. It is still surrounded with a wall, and has two public squares. It contains the ancient palace of the princes, three churches, several convents, a college, and a museum. Pop. in 1846, 12,827, who are partly employed in the manufacture of woollens, paper, rosoglio, and soap. In 1758 the greater part of this place was burnt down.

TROPPLIWODE, a village of Prussian Silesia, in the reg. and 32 m. SSW of Breslau.

TROQUEER, a parish on the E border of Kirkcudbright, containing the suburban burgh-of-barony of Maxwelltown. Pop. in 1851, 4,925.

TRORY, or St. Michael's Trory, a parish in co. Fermanagh, 3 m. N of Enniskillen. Area 6,173 acres. Pop. in 1831, 2,012; in 1851, 1,186.

TROSA, a port of Sweden, on the Baltic, in N lat. 58° 53′. Pop. 500. Its harbour is good.

TROSACHS, a surpassingly romantic mountain-vale, between lochs Achray and Katrine, in the district of Monteith, Perthshire. The opening which affords ingress is flanked on the l. by Benvenue, towering 2,800 ft. above sea-level, and on the r. by Benawn. The defile of Beal-an-Duine is in the heart of the great gorge; a little W of it is a narrow inlet; and, at a few paces farther, Loch-Katrine bursts upon the view, the Alps of Arroquhar mingling with the clouds in the distance.

TROSSIN, a village of Prussian Saxony, 9 m. NW of Torgau.

TROSTAN, a mountain in the p. of Layde, 3¾ m. SW of Cushendall, co. Antrim. Its summit has an alt. of 1,810 ft. above the level of the sea.

TROSTBERG, a village of Bavaria, on the l. bank of the Alza, 42 m. E of Munich.

TROSTON, a parish in Suffolk, 5½ m. NNE of St.-Edmunds-Bury. Area 1,764 acres. Pop. 427.

TROSTREY, a parish in Monmouthshire, 2¾ m. N by W of Usk. Area 1,255 acres. Pop. 176.

TROTTERSCLIFFE, or Trosley, a parish in Kent, 7½ m. WNW of Maidstone. Area 1,163 acres. Pop. in 1831, 310; in 1851, 283.

TROTTON, a parish in Sussex, 4 m. WNW of Midhurst. Area 3,877 acres. Pop. in 1851, 484.

TROUGHEND, a township in Elsdon p., Northumberland, 20 m. N by W of Hexham. Area 26,010 acres. Pop. in 1831, 327; in 1851, 264.

TROUGHTON, a small, low, sandy island off the NW coast of Australia, in S lat. 13° 44′.

TROUILLE, a river of Belgium, in the prov. of Hainault, which rises 3 m. NW of Merbes; runs W, turns N, and then NNW; passes Mons, and below the village of Nimy, flows into the Haine after a course of 20 m.

TROUN (Cape), a cape on the N coast of Staten island, in the N. Pacific, in N lat. 45° 30′.

TROUP, a county in the W part of Georgia, U. S., watered by the Chattahoochie. Area 417 sq. m. Pop. in 1850, 16,879. Its cap. is La Grange.

TROUP-HEAD, a cape on the N coast of the county of Banff, 10 m. W of Kinnaird's point, in N lat. 57° 39′.

TROUPSBURG, a township of Steuben co., in New York, U. S., 199 m. SW by S of Albany. Pop. 1,754.

TROUTBECK, a river of Cumberland, which runs into the Irthing, below Horsehead.—Also a river in Westmoreland, which runs into the Eden, 3 m. below Appleby.—Also a chapelry and village in Windermere p., Westmoreland, on the river of the same name, 3 m. SE by E of Ambleside. Area 5,322 acres. The village stands in a fertile and lovely valley at the foot of Wansfell. Pop. 369.

TROWBRIDGE, a parish and market-town in Wilts, 90 m. W by S of London, and 11 m. S by W of Chippenham, on the river Were. The Kennet and Avon canal passes about a mile north of the town, and a branch line of railway connects it with the Great Western. Area of p., 2,442 acres. Pop. in 1801, 5,799; in 1831, 10,863; in 1851, 11,148. The church is a spacious edifice, consisting of a nave, chancel, and two aisles, with chapels, and a large tower surmounted by a spire. The town stands on the side of a rocky hill, near the Were which runs into the Avon near Bradford. The houses are in general irregularly built, and most of the streets are narrow. The manufacture of woollen cloths and kerseymeres has long been the staple trade of T., as well as of most of the neighbouring towns and villages. It is said to have been introduced here in the reign of Henry VIII., and has probably continued to be the principal source of employment ever since its introduction.

TROWELL, a parish of Nottinghamshire, 5½ m. W of Nottingham. Area 1,570 acres. Pop. 392.

TROWERS ISLAND, or PULO-TINJIL, a small island in the Eastern seas, near the S coast of Java, in S lat. 7° 2′, 12 m. E by N of Klapper island.

TROWSE, or TROWSE-NEWTON, a parish of Norfolk, 2 m. SE by S of Norwich. Area 1,000 acres. Pop. in 1831, 583; in 1851, 1,363.

TROY, a township of Pike co., Alabama, U. S., 46 m. SSE of Montgomery. — Also a township of Waldo co., in Maine, 36 m. NE of Augusta. Pop. 1,484.—Also a township of Bristol co., Massachusetts, on the E side of Taunton river, 16 m. S of Taunton.—Also a township of Oakland co., in Michigan, 72 m. E by S of Lansing.—Also the capital of Rensselaer co., New York, on the E bank of the Hudson, 6 m. N by E of Albany. It is regularly laid out, and is a well built and flourishing town. It contains a fine court-house, and several handsome churches; and has several educational institutions of considerable repute. Its manufactures are important, embracing machinery, hardware, cutlery, cotton goods, and paper. Pop. in 1810, 3,895; in 1830, 11,405; in 1850, 28,785. The village of West Troy, on the opposite side of the Hudson, may be regarded as a suburb of this thriving town. The streams descending from the hills in the vicinity afford some of the finest mill seats in the country. —Also a township of Miami co., Ohio, on the Great Miami, 63 m. W of Columbus. Pop. 1,956.—Also a township of Walworth co., Wisconsin, 50 m. ESE of Madison.—Also a township of Orleans co., in Vermont, 40 m. N by E of Montpellier. Pop. 1,008.

TROYES, a town of France, the capital of the dep. of the Aube, situated on the l. bank of the Seine, which is here comparatively small and divided into several channels, in N lat. 48° 18′ 5″, 90 m. ESE of Paris. It is of considerable extent, and is surrounded with a wall; but is ill built, the chief material being wood, for the stone found in the neighbourhood is too soft for building. Many of the houses are curiously antique. The promenades round the town are very fine and of great extent. The cathedral is a splendid edifice; and the hotel-de-ville and palais-de-justice are both fine buildings. Pop. in 1789, 30,706; in 1831, 23,749; in 1846, 24,702. The manufactures of T. are considerable, particularly in cotton and cotton stockings; woollens, linens, leather, lace, soap, starch, and thread are also made here, but on a small scale. The corn, wine, and fruit of the adjacent country also form objects of trade. The town is connected by railway with Paris.—T. was noted in the Middle ages for its great fairs, where merchandise was sold by Troyes or Troy weight. In the campaign of 1814, this town was entered by the Allies on the 7th of February; retaken by the French on the 23d of that month; but finally re-occupied by the Allies on the 4th of March. It is the seat of an archbishop, and of the different branches of the administration of the dep. Pope Urban IV. was a native of Troyes.—The arrond. of T. comprising 7 cants. has an area of 153,000 hectares. Pop. in 1846, 93,725.

TRSCHEEMESNO, a town of Prussia, in the reg. of Bromberg, 10 m. SW of Mogillno. Pop. 3,200.

TRSTENNA, or TRSZIENNA, a town of Hungary, in the com. of Arvad, 48 m. S by W of Cracow. Pop. 3,000.

TRUB, a parish and village of Switzerland, in the cant. and 21 m. E of Berne. Pop. 2,536.

TRUBAU. See TRIBAU.

TRUBI. See AA.

TRUBIA, a town of Spain, in the prov. and 7 m. WSW of Oviedo, on the r. bank of a river of the same name. There is here a cannon foundry.

TRUBLY, or TUBBERVILLE, a parish in co. Meath, 3 m. ENE of Trim. Area 854 acres. Pop. 88.

TRUBSCHEVSK, a town of European Russia, in the gov. of Orel, on the Desna, 92 m. WSW of Orel. Pop. 4,000.

TRUCHSES. See WALDBURG.

TRUCHTERSHEIM, a village of France, in the dep. of Bas-Rhin, 9 m. NW of Strasburg. Pop. 700.

TRUCY, a village of France, in the dep. of Yonne, cant. and 8 m. NE of Coulanges. Pop. 400.

TRUEYRE, a river of France, which rises in the Cevennes, in the dep. of Lozere, 6 m. NW of St. Amand; runs SW, then NNW, and then WSW; and joins the Lot on the r. bank, at Entraigues, after a course of 90 m.

TRUFFARELLO, a village of the Sardinian states, in the prov. and 7 m. SSE of Turin, near the r. bank of the Bonna. Pop. 1,000.

TRUIM, a small river of Badenoch, Inverness-shire, which rises among the alpine recesses of the central Grampians, and runs 12½ m. NNE to the Spey, near Invernahavon.

TRULL, a parish of Somersetshire, 2 m. SSW of Taunton. Area 2,233 acres. Pop. in 1851, 716.

TRULLIKON, a village of Switzerland, in the cant. and 21 m. NNE of Zurich. Pop. 1,385.

TRUMBULL, a county in the NE of the state of Ohio, U. S., watered by branches of the Big Beaver and Grand river. Area 892 sq. m. Pop. in 1850, 30,490. Warren is the chief town.—Also a township of Fairfield co., Connecticut, 6 m. NW of Stratford. Pop. 1,313.—Also a township of Ashtabula co., in Ohio, 153 m. NE of Columbus.

TRUMIAN, a town of Hindostan, in the prov. of the Carnatic, district of Tanjore, in N lat. 10° 11′.

TRUMPINGTON, a parish of Cambridgeshire, 2 m. S of Cambridge. Area 2,200 acres. Pop. 770.

TRUN, a village of France, dep. of the Orne, 6 m. NNE of Argentan. Pop. 500.

TRUNCH, a parish of Norfolk, 3 m. N by E of North Walsham. Area 1,353 acres. Pop. 451.

TRUNS. See TRONS.

TRURO, a town of Nova Scotia, in Halifax co., at the head of the basin of Minas, 3 m. S of Onslow, and 4 m. N by W of Halifax.

TRURO, a township of Barnstable co., Massachusetts, U. S., extending across the peninsula of Cape Cod, between Provincetown and Wellfleet. Pop. in 1850, 2,051. There is a village of the same name on the NE coast of this township, in N lat. 42° 02′.

TRURO, a parliamentary borough and market-town, considered the chief town in the co. of Cornwall, 44 m. SW of Launceston, and 9½ m. N of Falmouth, on a small river called the Allen, which joins another called the Kenwyn at the quay, and forms the creek or river of Truro, opening into Falmouth harbour. Area of p. of St. Mary, 190 acres. Pop. in 1801, 2,358; in 1831, 2,925; in 1851, 3,169. The town, however, extends into the parishes of St. Clement and Kenwyn, and in 1841, contained 9,991 inhabitants; in 1851, 10,733. The scenery in the immediate vicinity is particularly fine; and the town has a picturesque appearance, situated, as it is, in a hollow surrounded by hills, indented with narrow wooded valleys. The principal streets are wide and handsome, and the houses are faced with granite or free-stone, and well-built. The principal public edifices are the church, a fine structure of the 15th cent.; the dissenting meeting-houses; the town-hall, a large substantial building, situated over the main entrance to the market-place; the coinage-hall, a heavy ancient building; the county infirmary, a spacious edifice; the theatre and ball-room, schools, &c. The museum of the Royal institution was erected at a cost of £2,100, and is supported by a proprietary besides annual subscriptions. There is also a public library containing about 6,000 vols. In the harbour or creek of Truro, the water, at spring-tide, covers a space of nearly 2 m. in length, and of sufficient depth to be navigable for vessels of 150 tons. At the recession of the tide, vessels of more than 150 tons burden must unload at Mopas. From the position of T., however, at the head of the branches of Falmouth harbour, and standing, nevertheless, about midland between the seas on the east and west, it must always constitute a great central point. The quantity of tin and copper, the chief produce of Cornwall, exported from T. has therefore been greater than from any other part of the county. The copper ore is principally shipped to South Wales to be converted into metal; and coal is brought back for the use of the mines. In other respects the commerce is not extensive, but it has been rapidly increasing in wealth as well as population, owing to the success of mining speculations carried on in the vicinity. Here is a manufactory for converting block-tin into bars and ingots, a blowing-house for refining tin, a foundry for casting iron tubes employed in the mines, two potteries, and a woollen mill. The old limits of the borough comprise the whole of the p. of St. Mary, and a portion of the p. of St. Kenwyn, but did not include the whole town. The limits therefore were extended, and the borough now includes the whole town, together with an additional portion of land along each side of the river to Calenick creek. T. returns two members to parliament. The number of electors registered in 1837, was 609; in 1852, 607.

TRUSHAM, a parish in Devon, 2 m. NNW of Chudleigh. Area 749 acres. Pop. in 1851, 205.

TRUSKMORE, a mountain in the p. of Rossinver, co. Sligo, forming the culminating part of the range of mountains which extends from Benbulben to Glenade. Its summit, on the E verge of co. Sligo, 2 m. N of the beautiful lake of Glencar, has an alt. of 2,113 ft. above the level of the sea.

TRUSLEY, a parish in the co. of Derby, 6 m. W of Derby. Area 1,078 acres. Pop. in 1851, 90.

TRUSTHORPE, a parish in Lincolnshire, 6 m. NE of Alford. Area 1,455 acres. Pop. 289.

TRUXILLO, a town of Spain, in the prov. of Caceres, 130 m. WSW of Madrid. It is situated on a hill, of which it occupies the top and the E side. The N and W sides are rugged and precipitous. On the summit of the hill is the castle, formerly a place of strength; and near it is the old part of the town, surrounded by high walls, and containing houses fortified in the feudal style; the rest of the town is modern, and extends along the slope of the hill to the plain. It contains a small but elegant square, its sides having porticoes supported by columns. The pop. in 1852 was 6,100. The environs are fertile; but the inhabitants are in general poor. Large cattle fairs are annually held here. Francis Pizarro, the conqueror of Peru, was a native of this town.

TRUXILLO, a city of Venezuela, in the prov. of the same name, 60 m. NE of Merida. The spot occupied by the city is shut in by two mountains. It is poorly built, and presents no edifices of any importance. The vicinity is pretty well cultivated, and produces sugar, cacao, indigo, coffee, and in general all the productions of the torrid and some few of the temperate zones. The pop. is about 4,000. T. was founded in 1570, and made rapid progress in industry and wealth; but in 1678 the buccaneer, Francis Gramont, with a handful of men, attacked and completely sacked and destroyed this wealthy city, killing or putting to flight its inhabitants, and reducing to ashes its superb edifices. The ruins that still remain sufficiently attest the past grandeur of this once flourishing town.—The province of T. lies between the parallels of 8° 28′ and 9° 42′, and contained in 1841, 44,788 inhabitants.

TRUXILLO, a town of Honduras, situated between two rivers, on a hill near the sea, in N lat. 15° 55′. It is at present a place of little account, but has a pretty good harbour.

TRUXILLO, a province of Peru, bounded on the N by Ecuador, and on the S by Tarma. It is a mountainous district, but fertile, and produces abundance of wheat, maize, wine, olives, sugar-canes, and fruits.—The cap. of the same name was founded in 1535 by Pizarro, at the distance of half-a-league from the sea, on the banks of a small river, in S lat. 8° 7′. The houses, which are chiefly of brick, have a neat appearance. The inhabitants amount to 8,000. The town contains a cathedral, several convents, a college, hospital, and two nunneries. It is surrounded with a low brick wall flanked by fifteen bastions; and carries on its commerce by means of its port of Guanchaco, which s about 8 m. to the northward, and is the only good harbour on the coast from Callao to Tumbez.

TRUXTON, a township of Courtland co., New York, U. S., 118 m. W of Albany. Pop. 3,629.

TRYBANE, a small fishing-harbour in the bay of Killybegs, co. Donegal. It is dry at low water, and has a very indifferent quay.

TRYBANE-MUCKRAS, a small fishing-harbour at the SW extremity of the p. of Glencollumbkill, co. Donegal. It has no quay and a difficult landing-place.

TRYSULL-AND-SEISDON, a parish of Staffordshire. Area 3,110 acres. Pop. in 1851, 559.

TRYVETT, or Travet, a parish in co. Meath, 2 m. N of Dunshaughlin. Area 4,239 acres. Pop. 285.

TSAD-SO, a river of Tibet, which rises in Khamjul, in N lat. 34° 50′; runs SE, passing Surman and Konkudsong; then turns S, and under the parallel of 27° 10′, enters Yun-nan, where it is known as the Lan-tsiang.

TSANA. See Dembea.

TSANG, or Ulterior-Tibet, called also by European geographers, Teshu, Lumbu, and the Undes, a division of Tibet, comprising the W part of that region, and bordered on the S by the Himalaya mountains, by which it is separated from Nepal. It is intersected in its entire extent by the Yaroutsangbo, and contains also numerous lakes. It comprises 6 cantons, and contains several towns, the principal of which are Teshu-Lumbu, the capital, Lingghie, Founghia, Nielam, Dsilung, Dsungar, and Ari or Ngari. See Tibet.

TSANG-BO. See Dihong.

TSANG-CHU, a district and town of China, in the prov. of Chih-le, and div. of Teen-tsin-fu, 45 m. ESE of Ho-keen-fu, on the r. bank of the Wei-ho, in N lat. 38° 22′ 20″, E long. 117° 0′ 30″.

TSANG-KE, a district and town of China, in the prov. of Sze-chuen, dep. and 11 m. N of Paou-hing-fu, on the l. bank of the Kea-ling, in N lat. 31° 40′, E long. 105° 56′.

TSANG-KEA-KEOU, a town of China, in the prov. of Chih-le, on the Thsing-shwy-ho, contiguous to the Great Wall. It is fortified.

TSANG-SHIN, a town of the Corea, in the prov. of Hwanghai, 51 m. SSW of Han-chu, on the Yellow sea.

TSANG-WU-HEEN, a district and town of China, in the prov. of Kwan-se, and div. of Wu-chu-fu.

TSAN-HWANG, a district and town of China, in the prov. of Chih-le, div. and 36 m. SSW of Ching-tih-fu, in N lat. 37° 43′, E long. 114° 32′.

TSAOU, a district and town of China, in the prov. of Gan-hwuy, div. and 39 m. ESE of Leu-chu-fu, near the E bank of Lake Tsiao-hou, and on the l. bank of the river, which forms the issue of that lake.—Also a district and town in the prov. of Shan-tung, div. and 30 m. S of Tsaou-chu-fu, in N lat. 34° 56′, and E long. 115° 38′.

TSAOU-CHU-FU, a division and town of China, in the prov. of Shan-tung. The div. comprises 11 districts. The town is 90 m. WSW of Yen-chu-fu, near the frontier of the prov. of Chih-le, in N lat. 35° 20′, and E long. 115° 36′.

TSAOU-HEA, an island in the Yang-tse-kiang river of China, commencing 2 m. above Cornwallis-Bluff, 220 m. above the mouth of the river, in N lat. 32° 10′, and stretching 7 m. in a NW direction. It is separated from the S shore by a channel 200 yds. wide.

TSAOU-HOU, a lake of China, in the prov. of Gan-hwuy, 12 m. SSE of Leu-chu-fu. It is nearly square in form, and is 30 m. in length from E to W, and 18 m. in breadth. It contains a small island, and discharges itself on the E into the Yang-tze-keang.

TSAOU-TUNG, an island of China, in the Yellow sea, off the W coast of the Corea, and prov. of Hwang-hai, in N lat. 37° 44′, and E long. 124° 55′. It contains a town of the same name.

TSAOU-YANG-HEEN, a district and town of China, in the prov. of Hu-pih, and div. of Fang-yang-fu, in N lat. 32° 10′, and E long. 112° 41′.

TSATA, Tsara, or Tchera, a small town of Chinese Turkestan, in the prov. of Thian-shan-Nanlo, on a small river 80 m. NW of Tak, and 100 m. ESE of Khoten.

TSATCHU. See Ya-lung-keang.

TSATSAK, a town of Turkey in Europe, in Servia, in the sanj. and 83 m. SSW of Semendria, on the r. bank of the Western Morava.

TSATSORKENG, a town of Tibet, in the prov. of Wei, between the supposed course of the Om-tchou or Nou-keang, and La-tchou or Lan-tsang-keang, and 90 m. NW of Bathang.

TSCH. For names commencing with these letters not found amongst the following, see *ch*, *sch*, *cs*, or *cz*.

TSCHAD. See Chad.

TSCHAEPRING. See Csepregh.

TSCHAJKISTEN, or Tsajkasok-Kerulete, a regimentary district of Slavonia, at the NE extremity of Military Slavonia, comprising an area of 134 sq. m., extending between 45° 9′ and 45° 33′ N lat., and between 19° 53′ and 20° 20′ E long., and bounded on the E by the Theiss; on the S by the Danube; and on the W and N by the Hungarian comitat of Bacs. It consists of a vast plain, intersected by canals, and possessing in some parts tolerable cultivation. Pop. in 1837, 29,027. Its capital is Tittel.

TSCHAKAKOTE, a town of Tibet, in the div. of Tsang. Pop. 5,000. It possesses an active transit trade.

TSCHERNEMBEL, or Zernamel, a town of Illyria, in the gov. and 42 m. SE of Laybach, circle and 13 m. S of Neustadtl, on the Zhermomalschiza, a small affluent of the Kulpa. Fruit and wine are extensively cultivated in the locality.

TSCHERNOWITZ. See Bukowine.

TSCHESCHEN, or Teschen, a village of Prussia, in the prov. of Silesia, regency and 33 m. NE of Breslau. Pop. 445. It has a glass-work.

TSCHINGLEN, a mountain of Switzerland, on the confines of the cantons of the Grisons and Glaris, and 14 m. SE of Glaris. It consists of a rocky mass, 8,870 ft. in height, and has on its summit a deep cavern, named St. Martin's loch.

TSCHIRNAU, or Gross-Tchirne, a town of Prussia, in the prov. of Silesia, regency and 48 m. NNW of Breslau, and circle of Guhrau. Pop. 960. It has two churches, a Catholic and a Lutheran.

TSCHIRNDORF, a village of Prussia, in the prov. of Silesia, regency and 20 m. NW of Liegnitz, and circle of Sagan, on the l. bank of the Oels-Wasser. Pop. 275. It has several iron-works.

TSCHUSSOWAJA, a river of Russia, which rises about 50 m. S of Yekaterinburg; skirts the E declivity of the Ural some distance; and then, passing through the range, runs N along the W declivity, until near 58° N lat., where it turns W, and falls into the Kama. It is navigable in spring.

TSEANG-LO, a district and town of China, in the prov. of Fo-keen, div. and 45 m. W of Yen-ping-fu, on the l. bank of a wide river, in N lat. 26° 43′, and E long. 117° 38′.

TSEEN-GAN, a district and town of China, in the prov. of Chih-le, div. and 17 m. NW of Yung-ping, in a valley, on the l. bank of the Lanho, in N lat. 40° 5′, and E long. 118° 44′.

TSEEN-KEANG, a river of China, in the prov. of Sze-chuen, which descends from the mountains running along the confines of the prov. of Shan-se; passes the towns of Tha and Kwang-yang; and after a total course in a SW direction of about 240 m. joins the Kealing, 38 m. NNW of Chung-king.

TSEEN-KEANG-HEEN, a district and town of China, in the prov. of Hu-pih, and div. of Gan-luh-fu, 50 m. E of King-chu, on a small affluent of the Han-keang, in N lat. 30° 28′, and E long. 112° 48′. —Also a district and town in the prov. of Kwan-se, and div. of Sze-gan-fu, 27 m. NNE of Ping-chu-fu, and near the l. bank of the Hung-keang, in N lat. 23° 40′, and E long. 108° 40′.

TSEEN-SHAN, a district and town of China, in the prov. of Gan-hwuy, div. and 29 m. WNW of Gan-hing-fu, in a well-watered valley, in N lat. 30° 43′, and E long. 116° 38′. It is enclosed by ramparts built of brick, 25 ft. in height, flanked with bastions. It has several gates, one of which has two triumphal arches.

TSEEN-SHAN-CHAE-TING, a district and town of China, in the prov. of Kwang-tung, in an island of the bay of Kwang-chu-fu, and to the S of Heang-shan.

TSEEN-SSUH-HEEN, a district and town of China, in the prov. of Gan-hwuy, and div. of Chu-chu.

TSEEN-TANG-HEEN, a district and town of China, in the prov. of Che-keang, and div. of Hang-chu-fu.

TSEEN-TANG-KEANG, a river of China, which has its source in the S part of the prov. of Keang-su; traverses the N part of that of Che-keang; and, after a course, in a NE direction of about 240 m., throws itself, by a wide embouchure, into the Eastern sea, a little to the E of Hang-chu.

TSEEN-WHA, an island of the Yellow sea, near the E coast of the Corea, to the S of the island of Tsiao-tung. It contains a town of the same name.

TSE-HEA, a district and town of China, in the prov. of Shan-tung, div. and 39 m. SSE of Ting-chu-fu, near the source of the Tsin-yang.

TSEIH-MIH, a district and town of China, in the prov. of Shan-tung, and div. of Lae-chu-fu, on a small river, which 9 m. below throws itself into the Yellow sea, in N lat. 36° 15′, and E long. 120° 44′.

TSEIH-YEN. See Yingkeihshar.

TSE-KE, a district and town of China, in the prov. of Che-keang, div. and 9 m. NNE of Ning-po-fu, near the sea, in N lat. 30° 1′ 24″, and E long. 121° 17′ 20″.

TSELARI, a river of Abyssinia, which rises near Gurri in Angot, and after being joined by the Zamra and the Sassella, joins the Upper Takkazie.

TSE-NAN-FU, a division and town of China, in the prov. of Shan-tung. The div. comprises 16 districts. The town, which is also the capital of the prov., is 192 m. S of Pe-king, at the foot of mountains, near the r. bank of the Ta-tsin, in N lat. 36° 44′ 24″, and E long. 117° 7′ 30″. It contains several lakes, and is intersected with canals. It is a place of considerable extent, and possesses several fine streets and buildings. It possesses an active trade in a grey-coloured silk called keen-chu, and in different kinds of lieoli or Chinese glass, the manufacture of Yen-ching, a town in the vicinity. It is large and populous, and is held in high veneration by the Chinese as the residence of a long line of kings, whose tombs are on the adjacent mountains.

TSE-NING-FU, a division and town of China, in the prov. of Shan-tung. The div. comprises three districts. The town is in N lat. 35° 33′, and E long. 116° 44′.

TSEUEN-CHU, a district and town of China, in the prov. of Kwan-se, and div. of Kwei-lin-fu, in N lat. 25° 42′ 12″, and E long. 111° 5′ 50″.

TSEUEN-CHU-FU, a division and town of China, in the prov. of Fo-keen. The div. comprises 6 districts. The town is in N lat. 24° 56′ 12″, and E long. 118° 51′ 10″, on the Tsin. Notwithstanding the disadvantages of its harbour, which is lined with rocks and sand-banks, it carries on an active trade with Formosa, the N part of the empire and Manilla. But it is infested with pirates, and the inhabitants generally are in the most wretched condition. The adjacent estuaries are crossed by bridges, one of which is nearly ¾ of a mile in length.

TSE-YANG, a district and town of China, in the prov. of Shan-tung, div. and 21 m. NNE of Tse-nan-fu, on the l. bank of the Ta-tsin, in N lat. 37° 3′, and E long. 117° 20′.

TSE-YUEN, a district and town of China, in the prov. of Ho-nan, div. and 23 m. W of Hwae-hing-fu, in N lat. 35° 7′, and E long. 112° 39′.

TSHILA-TAGH, or CHILA-TAGH, a forest-clad range of mountains in Asiatic Turkey, extending from Uskub eastwards to the Elkas-Tagh, and having an alt. of 1,350 ft.

TSIAMDO, a town of Tibet, in the prov. of Kham, in N lat. 31° 10′, in a valley surrounded by mountains between two head streams of the Ya-long-kiang. It is a poor place, with a Chinese garrison.

TSIAMPA, or BIN CHUON. See CHAMPA.

TSIBBA, a town of Japan, in the island of Nifon, prov. of Simosa, and district of the same name, near the NE coast of the bay of Yedo, and 30 m. E of the town of that name.

TSIBON-BO-SIMA, an island of Lake Biwanooumi, in Japan, in the island of Nifon. It was formed by a volcanic eruption in 82 B. C.

TSIGAAR, or TASAGOURA, a chain of mountains in Japan, in Nifon, in the prov. of Mouts, running from S to N, and terminating at Sangar strait.

TSIH, a district and town of China, in the prov. of Shan-tung, and div. of Gen-chu-fu.

TSIH-CHU-FU, a division and town of China, in the prov. of Shan-se. The div. comprises five districts. The town is 48 m. SSW of Lu-gan-fu, at the foot of mountains, in N lat. 35° 30′, and E long. 112° 49′ 30″.

TSIKA, a district of Japan, in the island of Sikokf and prov. of Syo.

TSIKONGO, a prov. of Japan, in the central part of the island of Kiu-siu, between the bay of Simabara on the W, the prov. of Tsikousen on the N, that of Boungo on the E, and the prov. of Figo on the S. It contains several lofty mountains, abounding in sulphur, silver and gold. The soil is tolerably fertile, especially in legumes and fruit. The prov. comprises 10 districts.

TSIKOUMMA, a town of Japan, in the island of Nifon, prov. of Sinano and district of the same name, 99 m. W of Yedo.

TSIN-CHU, a district and town of China, in the prov. of China, div. and 29 m. ESE of Ching-ting-fu, near the l. bank of the Hou-to, in N lat. 38° 6′, and E long. 115° 12′.—Also a division and town in the prov. of Shan-se. The div. comprises 2 districts. The town is in N lat. 36° 40′, and E long. 112° 46′.—Also a town of China, in the Corea, and prov. of Tseuen-lo, 210 m. SSE of King-ke-tao, in N lat. 35°.

TSIN-CHU-FU, a district and town of China, in the prov. of Kwan-se. The div. comprises 4 districts. The town is in N lat. 23° 26′, and E long. 109° 51′.—Also a division and town of China, in the prov. of Kan-suh. The div. comprises 9 districts. The town is in N lat. 34° 36′, and E long. 105° 46′. This div. was formerly comprised in that of Kung-chang-fu.

TSING, a district and town of China, in the prov. of Chih-le, and div. of Teen-tsin-fu, 45 m. ENE of Ho-keen-fu, in N lat. 38° 37′, and E long. 116° 54′.

TSIN-GAN, a district and town of China, in the prov. of Kan-suh, div. and 24 m. N of Tsin-chu-fu, on the l. bank of the Hwei-ho, in N lat. 34° 52′, and E long. 105° 50′.

TSIN-GAN-HEEN, a district and town of China, in the prov. of Kwei-chu, and dep. of Gan-chu-fu.

TSING-CHIN, a district and town of China, in the prov. of Kwei-chu, and div. of Gan-shun-fu, 18 m. W of Kwei-gang-fu, on an affluent of the Teou-hwei.

TSING-CHING, a district and town of China, in the prov. of Shan-tung, and div. of Wu-ting-fu, in N lat. 37° 13′, and E long. 117° 51′.

TSING-CHU, a division and town of China, in the prov. of Hu-nan. The div. comprises 3 districts. The town is 62 m. S of Yung-chu-fu, in a mountainous locality, in N lat. 26° 35′, and E long. 109° 28′.

TSING-CHU-FU, a division and town of China, in the prov. of Shan-tung. The div. comprises 11 districts. The town is on a small affluent of the Me-ho, 96 m. E of Tse-nan-fu, in N lat. 36° 44′ 22″, and E long. 118° 43′ 30″. It is a flourishing place, and carries on an extensive trade in fish.

TSING-FUNG-HEEN, a district and town of China, in the prov. of Chih-le, div. and 29 m. S of Ta-ming-fu, in N lat. 35° 58′, and E long. 115° 21′.

TSING-GAN, a district and town of China, in the prov. of Keang-se, div. and 39 m. W of Nan-chang-fu, in a well-watered valley, in N lat. 28° 48′, and E long. 115° 21′.

TSING-HAE, a district and town of China, in the prov. of Chih-le, and div. of Teen-tsin-fu.

TSING-HEEN, a district and town of China, in the prov. of Shan-se, and div. of Hin-chu.

TSING-HO, a district and town of China, in the prov. of Chih-le, div. of Kwang-ping-fu, and 69 m. E of Shun-tih-fu, in N lat. 37° 9′, and E long. 115° 46′.

TSING-HO-HEEN, a district and town of China, in the prov. of Keang-su, div. and 14 m. WNW of Hwae-gan-fu, on the l. bank of Hwang-ho, in N lat. 33° 35′, and E long. 119° 2′.

TSING-HWA-HEEN, a district and town of China, in the prov. of Chih-le, and div. of Paou-ting-fu.

TSING-KE, a district and town of China, in the prov. of Sze-chuen, and div. of Ya-chu-fu.

TSING-KEANG-HEEN, a district and town of

China, in the prov. of Keang-se, and div. of Lin-keang-fu.

TSING-KEANG-TING, a district and town of China, in the prov. of Kwei-chu, and div. of Chin-yuen-fu.

TSING-KEEN, a district and town of China, in the prov. of Shen-se, and div. of Suy-tih-chu.—Also a district and town of China, in the prov. of Shen-se, and div. of Yen-gan-fu, in N lat. 37° 12′, and E long. 110° 3′.

TSING-LAN-WEI, a town of China, in the prov. of Hu-nan, near the source of the Yuen, in N lat. 27° 4′ 48″, and E long. 108° 32′ 50″.

TSING-LEW-HEEN, a district and town of China, in the prov. of Fo-keen, div. and 42 m. NE of Ting-chu-fu, in N lat. 26° 7′, and E long. 116° 59′, on the r. bank of a river by which it is encircled on the W, N, and E.

TSING-LO, a district and town of China, in the prov. of Shan-se, div. and 54 m. NW of Tae-yuen-fu, on the l. bank of the Fuen-ho, in N lat. 38° 31′ 12″, and E long. 111° 57′.

TSING-LOU-WEI, a district and town of China, in the prov. of Kan-suh, in N lat. 37° 13′, and E long. 103° 58′.

TSING-MING-CHU, a district and town of China, in the prov. of Yun-nan, and div. of Yun-nan-fu, in N lat. 24° 47′, and E long. 102° 51′.—Also a district and town in the prov. of Kan-suh, and div. of Ping-leang-fu, 56 m. W of Ping-leang-fu, on the l. bank of the Hwei-ho, at the confluence of a small river, in N lat. 35° 35′, and E long. 105° 45′.

TSING-PING, a district and town of China, in the prov. of Shan-tung, and div. of Tung-chang-fu, in N lat. 36° 52′, and 116° 16′.—Also a district and town in the prov. of China, and Ton-yun-fu, in N lat. 26° 38′, and E long. 107° 38′.—Also a district and town of China, in the prov. of Kwei-chu, and div. of Gan-shun-fu, on the r. bank of the Cham-ke, in N lat. 26° 40′, and E long. 107° 38′.

TSING-PU, a district and town of China, in the prov. of Keang-su, div. and 12 m. NNW of Sung-keang-fu, in N lat. 31° 10′, and E long. 120° 53′, on the E of a lake. It is of considerable extent, and contains some fine houses and several temples.

TSING-SHIN, a district and town of China, in the prov. of Sze-chuen, and div. of Mei-chu, in N lat. 29° 48′, and E long. 103° 51′.

TSING-SHWUY, a district and town of China, in the prov. of Kan-suh, and div. of Kung-chang-fu, in N lat. 34° 42′, and E long. 106° 12′.—Also a district, town, and port on the S coast of the prov. of Shan-tung, and div. of Ting-chu-fu, in N lat. 36° 57′, and E long. 122° 20′.

TSING-SHWUY-HO-TING, a district and town of China, in the prov. of Shan-se, and div. of Leaon-chu.

TSING-SHWUY-PO, a lake of China, in the prov. of Shan-tung, 32 m. NNE of Tsin-chu, near the r. bank of the Siao-tsin, and near the gulf of Chih-le. It is 17 m. in length from NE to SW, 8 m. in breadth, and about 39 m. in circumference.

TSING-TEEN, a district and town of China, in the prov. of Che-keang, div. and 36 m. SE of Chu-chu-fu, on the l. bank of a large river, in N lat. 28° 10′, and E long. 120° 21′.

TSING-TEEN-CHU, a district and town of China, in the prov. of Yun-nan, and div. of Keuh-tsing-fu, in N lat. 25° 38′, and E long. 103° 19′.

TSING-TIH, a district and town of China, in the prov. of Gan-hwuy, div. and 51 m. SSW of Ning-kwo-fu, in N lat. 30° 24′ 37″, and E long. 118° 34′ 13″.

TSING-TING, a district and town of China, in the prov. of Chih-le, and div. of Ching-ting-fu, in N lat. 38° 2′ and E long. 113° 10′.

TSING-TSEUEN, a district and town of China, in the prov. of Hu-nan, and div. of Hang-chu-fu.

TSING-Y, a lake of China, in the prov. of Keang-su, 17 m. SSW of Hae-chu. It is nearly circular, with a diameter of about 9 m., and discharges itself on the S into the Tang-yang.

TSING-YANG, a district and town of China, in the prov. of Keang-su, and div. of Che-chu-fu, in N lat. 30° 45′, and E long. 117° 46′.—Also a district and town in the prov. of Gan-hwuy, div. and 23 m. E of Che-chu-fu, in a valley.—Also a river in the prov. of Shan-tung, an affluent of the Eastern sea.

TSING-YEN-HEEN, a district and town of China, in the prov. of Sze-chuen, and div. of Sze-chu, in N lat. 29° 32′, and E long. 104° 8′.

TSING-YUEN, a district, town, and fortress of China, in the prov. of Kan-suh, div. and 42 m. NNE of Lan-chu-fu, on the r. bank of the Hwang-ho, near the great wall, in N lat. 36° 38′, and E long. 104° 8′.—Also a district and town in the prov. of Kwang-tung, and div. of Kwang-chu-fu, on the r. bank of the Pe-keang, 45 m. from Canton, in N lat. 23° 44′ 24″, and E long. 112° 41′ 50″. The town is 105 m. from Canton, on the Ta. It has a fine pagoda, and a tower 5 stories in height.—Also a district and town in the prov. of Shen-se, div. and 24 m. SW of Tae-yuen-fu, in N lat. 37° 37′, and E long. 112° 16′.

TSIN-HEEN-HEEN, a district and town of China, in the prov. of Keang-se, and div. of Nan-chang-fu, in N lat. 28° 29′, and E long. 116° 17′.

TSIN-KE, or Kin-ke-heen, a district and town of China, in the prov. of Kwan-se, and div. of Wu-chu-fu, in N lat. 23°, and E long. 110° 27′.

TSIN-KEANG, a district and town of China, in the prov. of Fo-keen, and div. of Tseuen-chu-fu.—Also a district and town in the prov. of Keang-su, div. and 29 m. NE of Chang-chu-fu, on the l. bank of the Yang-tze-keang, near its mouth, in N lat. 32° 5′, and E long. 120° 14′.

TSIN-KEEN, or Tsoukota, an island of the Chinese empire, in the Lieu-kieu group, a little to the E of the Great Lieu-kieu island.

TSIN-SHWUY, a district and town of China, in the prov. of Shan-se, and div. of Tsih-chu-fu, in N lat. 35° 40′, and E long. 112° 15′.

TSIN-TING-HEEN, a district and town of China, in the prov. of Kwei-chu, and div. of Gan-shun-fu.

TSIN-UNH-FU, or Pu-urh-fu, a division and town of China, in the prov. of Yun-nan. The div. comprises 4 districts. The town is in N lat. 23° 2′, and E long. 101° 10′.

TSIN-YAN, a district and town of China, in the prov. of Ho-nan, and div. of Nan-gan-fu.

TSIN-YUEN-HEEN, a district and town of China, in the prov. of Shan-se, and div. of Tsin-chu, in N lat. 36° 35′, and E long. 112° 23′.

TSIN-YUN, a district and town of China, in the prov. of Che-keang, div. and 27 m. NE of Chu-chu-fu, on a river, in N lat. 28° 42′, and E long. 119° 53′.

TSO, a river of S. Africa, which, coming from the NW, joins the Mababe river, a feeder of Lake N'gami, on the r. bank.

TSO-CHU, a district and town of China, in the prov. of Kwan-se, div. and 23 m. NNE of Tae-ping-fu, in N lat. 22° 42′, E long. 107° 11′.

TSO-CHU, or Su-chu, a district and town of China, in the prov. of Chih-le and div. of Shun-teen-fu, 36 m. SW of Peking, in N lat. 39° 32′, E long. 116° 3′, on the r. bank of a river which is here crossed by a bridge 600 ft. in length, with parapets of white marble ornamented with sculptures of animals. The town is badly built.

TSO-KSTI, a village of Tibet, at the confluence

of the Zangskar with the Ladak Indus, at an alt. of 10,400 ft. above sea-level.

TSO-MAPHAM. See MANASAROWAR.

TSO-MORIRI, a lake of Western Tibet, in N lat. 32° 55′, E long. 78° 20′. It is about 14 m. in length from N to S, with a breadth of from 3 to 4 m.; and has an alt. of 15,200 ft. above sea-level. It receives the Gyang, the Tsakshang, and the Karzok, three small mountain-streams; but has no effluence.

TSOW, a district and town of China, in the prov. of Shan-tung, div. and 15 m. SSE of Yen-chu-fu, in N lat. 35° 30′, E long. 117° 10′, at the foot of mountains.

TSO-WAR, a lake of Western Tibet, in N lat. 33° 24′, E long. 78°, at an alt. of about 15,000 ft. above sea-level. It has an area of a few square miles, but no apparent effluence.

TSO-YUN-HEEN, a district and town of China, in the prov. of Shan-se and div. of Suh-ping-fu, to the NE of Suh-chu, in N lat. 39° 55′, E long. 112° 33′.

TSU-CHU, a district and town of China, in the prov. of Chih-le and div. of Kwang-ping-fu, in N lat. 36° 25′ 15″, E long. 114° 33′.

TSU-HEUNG-FU, a division and town of China, in the prov. of Yun-nan. The div. comprises 7 districts. The town is in N lat. 26° 6′, E long. 101° 43′ 10″, on an affluent of the Kin-cha-keang.

TSU-HEUNG-HEEN, a district and town of China, in the prov. of Yun-nan and div. of Tsu-heung-fu.

TSUH-KEANG, a district and town of China, in the prov. of Chih-le, and div. of He-chu.

TSUH-LUH, a district and town of China, in the prov. of Chih-le and div. of Paou-ting-fu.

TSU-KE, a district and town of China, in the prov. of Che-Keang and div. of Ning-po-fu, in N lat. 30° 1′ 24″, E long. 121° 17′ 20″.

TSUN, or SIUN, a district and town of China, in the prov. of Chih-le, div. and 59 m. SSW of Ta-ming-fu, on the r. bank of the Hwae-ho, in N lat. 55° 43′ 50″, E long. 114° 47′.

TSUN-E-FU, a division and town of China, in the prov. of Kwei-chu. The div. comprises 5 districts. The town is in N lat. 27° 38′, E long. 106° 58′.

TSUN-E-HEEN, a district and town of China, in the prov. of Kwei-chu, and div. of Tsun-e-fu.

TSUNG-CHIH, a district and town of China, in the prov. of Hu-nan and div. of Yung-shun-fu.

TSUNG-E, a district and town of China, in the prov. of Keang-se, div. and 21 m. N of Nan-gan-fu, in N lat. 25° 46′, E long. 114° 2′.

TSUNG-GAN, a district and town of China, in the prov. of Fo-keen, and div. of Keen-ning-fu, 45 m. NE of Shaou-wu-fu, in N lat. 27° 45′ 36″, E long. 118° 7′ 50″.

TSUNG-HWA, a district and town of China, in the prov. of Kwang-tung and div. of Kwang-chu-fu, 39 m. NE of Canton, in N lat. 23° 33′, E long. 113° 27′.

TSUNG-JIN, a district and town of China, in the prov. of Keang-se and div. of Fu-chu-fu, in N lat. 27° 44′, E long. 116°.

TSUNG-KING-CHU, a district and town of China, in the prov. of Sze-chuen, div. and 27 m. WSW of Ching-tu-fu, in an island of the Hwang-chin, in N lat. 30° 36′, E long. 103° 43′.

TSUNG-MING, a district and town of China, in the prov. of Keang-su, on the S coast of an island of the same name, 51 m. ENE of Su-chu-fu, in N lat. 31° 36′, E long. 121° 18′ 30″. The houses, excepting those of the wealthier inhabitants, which are built of brick, are constructed of reeds and thatched. The island lies in the China sea, in the embouchure of the Yang-tse-kiang, which is here about 60 m. wide, and is separated from the continent by a channel not exceeding 5 m. in width, and much obstructed by sand-banks. It is about 33 m. in length from ESE to WNW, and 9 m. in extreme width, and is intersected by numerous canals. Its SE point is in N lat. 31° 29′. The surface is level, and it possesses in some parts considerable fertility. Buffaloes, pigs, fowls, geese, wild ducks, and other aquatic birds are numerous, but the wealth of the island consists chiefly in its salt mines, which form a lucrative object of trade with the adjacent provinces. It contains several villages, and is governed by a mandarin of the third class. This island was originally a penal settlement.

TSUNG-NING, a district and town of China, in the prov. of Sze-chuen and div. of Ching-tu-fu, in N lat. 33° 35′, E long. 103° 53′.

TSUNG-PING-CHU, a district and town of China, in the prov. of Shan-tung and div. of Tae-gan-fu.

TSUNG-SHEN-HEEN, a district and town of China, in the prov. of Kwan-se and div. of Tai-ping-fu.

TSUNG-SIN-HEEN, a district and town of China, in the prov. of Kan-suh and div. of King-chu, 27 m. SE of Ping-leang, on the r. bank of an affluent of the Kiu-ho.

TSUNG-YANG, a district and town of China, in the prov. of Hu-pih, div. and 75 m. SSW of Wu-chang-fu, on the l. bank of a small affluent of the Yang-tsen-keang, in N lat. 29° 33′ 38″, E long. 113° 59′ 42″.

TSUN-HWA-CHU, a division and town of China, in the prov. of Chih-le. The div. comprises two districts. The town is 90 m. ENE of Peking, between two rivers, affluents of the San-ho, and near the great wall, in N lat. 40° 11′, E long. 117° 59′. This town is 4 m. in circumf., and is very populous. It is noted for its tobacco.

TSU-PING, a district and town of China, in the prov. of Shan-tung and div. of Tse-nan-fu, in N lat. 36° 56′, E long. 117° 50′.

TSZE-YANG, a district and town of China, in the prov. of Shen-se and div. of Hing-gan-fu.

*** TU. For many names of which the first syllable is sometimes written with this orthography, see TO, TOO, and TOU.

TUA, a river of Portugal, which rises on the borders of Galicia, in the Sierra-Segundera; flows SW through the province of Tras-os-Montes; passes Mirandela; and falls into the Douro on the r. bank, 15 m. NW of St. Joao-de-Pesqueira, after a course of 70 m.

TUABO, a village of Western Africa, in the kingdom of Kaen, on the Senegal, in N lat. 14° 56′.

TUAK, a small island in the Red sea, about 9 m. from the coast of Arabia, in N lat. 5° 58′.

TUAM, a parish and city in co. Galway. Area of p. 25,014 acres. Pop. in 1831, 14,219; in 1851, 13,060.—The episcopal city of T., the ecclesiastical metropolis of Connaught, and formerly a parl. borough, stands on the river Clare, 7¼ m. SW by S of Dunmore, and 99½ m. W of Dublin. Its site possesses one or two undulations, which relieve it from absolute monotony. Of 1,056 inhabited houses which it contained in 1841, no fewer than 825 were mere mud cabins, and each of so many of these mud cabins as 511 had only one apartment! The stone cross, now a neglected and dilapidated monument, is one of the finest specimens of the structures of its class any where to be found in Ireland. The present cathedral of the Protestant diocese is only part of the original structure, designed on a comparatively extensive scale. The present Roman Catholic cathedral is one of the costliest, best designed, best executed, and most magnificent modern Gothic ecclesiastical edifices in the three kingdoms. It exhibits, both externally and internally, all the principal and many of the minute or accessory features of the old pointed or Saracenic style of architecture; and it displays to the surrounding landscape an imposing cruciform mass, surmounted from the intersection of its nave and transept with a parapeted, pinnacled, and arch-pierced quadrangular tower. The palace or diocesan house of the bishop of T. is situated in the NE vicinity of the compact portion of the town, and is a commodious but quite plain structure. The residence of the Roman Catholic archbishop, on the opposite side of the street, is a neat and sufficiently capacious, yet comparatively unpretending house. The college of St. Jarlath, or Roman Catholic boarding and day school, principally for the training of young men for official situations in the Roman Catholic church, is a large plain building, adjacent to the archbishop's residence. The other public buildings are a Presentation nunnery, a court-house, a bridewell, a barrack, a bridge, several schools, and a work-house. T., as a seat of trade, has made more improvement during the last few years than any other inland town of Connaught. Its weekly markets are the scene of very extensive traffic, in the disposal and purchase of agricultural produce; and its daily retail business is extensive in the supply of a great and populous circumjacent country. The town has an extensive brewery, several tan-yards, several flour-mills, and some manufactories of linen and coarse canvass. Pop. of the town in 1831, 6,883; in 1851, 7,819. The bishopric of T. is little if at all known to authentic record till the middle of the 12th cent. The Church Reform act dissolved the connection between T. and Ardagh, annihilated the archiepiscopal dignity of T. and annexed to the dio. of T. the dioceses of Killalla and Achonry. The nett episcopal income of T. upon the average of 3 years ending in 1831, was £4,018; the nett episcopal income of Ardagh, held in commendam with T., was £2,977; and the episcopal income of the united sees of T., Killalla, and Achonry, under the new arrangement, is £5,000. The dio. of T. is the largest in Ireland; and comprises a great portion of co. Galway, a great portion of co. Mayo, and a considerable portion of co. Roscommon. Its area is 974,879 acres; and its pop. in 1831 was 446,703. In 1834, the pop. consisted of 44,599 Churchmen, 800 Presbyterians, 369 other Protestant dissenters, and 1,188,568 Roman Catholics.—The Roman Catholic province of T. continues to be fully recognised in the Roman Catholic hierarchical arrangements; its dioceses of Tuam, Killalla, Achonry, Elphin, and Clonfert, are severally unannexed, or have each a bishop for itself.

TUARICKS, a numerous people of Central Africa, occupying all the territory to the W and S of Sudan, as far as the Niger. They are described by Lyon as the finest race of men he ever saw; tall, erect, and handsome, with an imposing air of pride and independence. Their skin is generally white, except in those parts that are exposed to the climate, which are of a dark brown. The most remarkable feature in their costume consists in a piece of glazed cotton cloth, generally blue, with which they cover their faces as high as the eyes, and which hangs down on the breast below the chin: not to interfere with this, the beard is generally clipped very close. They wear turbans, generally blue, surmounted by high red caps; while the body is generally covered with a loose shirt of blue cotton. Their swords are straight, and of great length, and are wielded with much ease and dexterity. From the left wrist is suspended a dagger, without which no T. is ever seen: he also usually carries a light elegant spear about 6 ft. in length; but in active warfare they have three longer and heavier spears, and a strong lance, all fastened behind their saddles. They sometimes also carry a long gun, and are considered sure marksmen. They speak the Berber language. Their purifications are performed with sand. Among the most powerful tribes of T. are the Ghraat, who inhabit the neighbourhood of Ghadames; the Kolluvi, who possess the kingdom of Asben, of which Aghades is the capital; and the Tagama, who immediately border on Sudan.

TUAT, or TWAT, an oasis of Northern Africa, in the Sahara, forming in fact a large tract of country, inhabited chiefly by Tuaricks, between the parallels of 26° and 29° N. Its inhabitants consist of two distinct races, one calling themselves Zenates, of a Negro complexion; the other consisting of Arabs. They trade in *henna*, dates, butter, cheese, slaves, gunpowder, and antimony, with Timbuctu, Ghadames, and Fezzan, but seldom go so far E as Bornu. The principal town, called Ain-el-Salah, is situated on the N frontier, 40 days' journey from Murzuk, and 50 from Timbuctu.

TUBA, a small river of Asiatic Russia, which falls into the Yenisei, 16 m. SW of Abakansk.

TUBAGE, a large and deep river of New Granada, in the prov. of San-Juan-de-los-Llanos, which enters the Orinoco a little lower than the Meta.

TUBAI, or MOTUITI, a small low island in the Society group, 30 m. NE of Maurua. It is the most northerly island of the group.

TUBAL, a river of Chili, which flows through a fine district into Arauco bay, in S lat. 37° 14′.

TUBARAM (CAPE), a headland of the Brazilian coast, in S lat. 20° 16′.

TUBARAO (CAPE), a headland of the Brazilian coast, in S lat. 5° 10′.

TUBERGEN, a town of Holland, in the prov. of Overyssel, 5 m. WNW of Ootmarsum.

TÜBINGEN, a town of Würtemberg, situated on the l. bank of the Neckar, where it receives the Ammer, 18 m. S by W of Stutgard, and 59 m. E of Strasburg. Pop. 7,500, who are partly employed in woollen and linen manufactures; but the chief income of the place is derived from its university, founded in 1477. It was originally a Protestant institute, but after the numerous accessions of territory inhabited by Catholics, the separate university erected for that profession at Elwangen was suppressed, and replaced by an admittance to T., which is now the only university in the Würtemberg dominions. The number of students in 1846 was 863. There are here also a theological foundation for educating 150 persons as clergymen or schoolmasters, and a separate seminary for the sons of families of rank. The university has a library of 150,000 vols., and a botanical garden.

TUBISE, a small town of Belgium, 16 m. SSW of Brussels, on the river Senne, and the railway to Mons. Pop. 2,360.

TUBNA, a town of Algiers, 110 m. SSW of Constantina, the representative of the ancient *Tubuna*.

TUBNEY, a parish of Berkshire, 4 m. W by N of Abington. Area 1,220 acres. Pop. in 1831, 167.

TUBŒUF, a town of France, dep. of La Mayenne, 17 m. NE of Mayenne.

TUBRID, a parish in co. Kilkenny, 3½ m. ESE of Pilltown. Area 1,004 acres. Pop. in 1851, 216.—Also a parish in co. Tipperary, 3¾ m. SSW of Cahir. Area 12,572 acres. Pop. in 1851, 2,928.

TUBBRIDBRITTAIN, a parish in co. Kilkenny, 3½ m. SE of Urlingford. Area 4,140 acres. Pop. in 1831, 1,103; in 1851, 929.

TUBURBO, a town of Tunis, 16 m. WNW of the city of Tunis, supposed to be built on the site of the ancient *Tuburbum*. It was a favourite spot of the late bey, who was at great pains in establishing plantations of fruit-trees here.

TUCACAS-POINT, a low swampy point of land on the coast of Venezuela, forming the N extremity of the gulf of Triste, and 25 m. from the mouth of Porto-Cabello.

TUCAPEL-POINT, a low projecting rocky point on the coast of Chili, 8 m. NW of Molguilla-point, and 7 m. SSE of a high bold hill known as Tucapel-head.

TUCCABER, a small town of Tunis, situated on the river Mejerdeh, 24 m. W of Tunis.

TUCHAN, a town of France, dep. of the Aude, 16 m. N by W of Perpignan.

TU-CHANG, a district and town of China, in the prov. of Keang-se, and div. of Nan-kang-fu, 39 m. NW of Jao-chu-fu, on the E bank of Lake Po-jang, in N lat. 29° 20′ 24″, and E long. 116° 13′ 12″.

TUCHEL, a town of West Prussia, on the Brahe, 25 m. W of Culm. Pop. 1,800. It has a trade in horses and cattle.

TUCHEN, a village of Prussian Saxony, in the duchy of Magdeburg, near Jerichow. Pop. 1,000.

TUCKER'S ISLAND, an island of the United States, near the coast of S. Carolina, in N lat. 32° 36′.—Also a small island in the N. Pacific, in N lat. 7° 22′. It is from 2 to 3 m. in circumf. The natives, about 300 in number, are of a dark copper colour.

TUCKEREAH, a town of Algiers, 106 m. S of the city of Algiers, the representative of the ancient *Figava*.

TUCKERSVILLE, a village of Wayne co., Georgia, U. S.

TUCKERTON, a port of Burlington co., New Jersey, U. S., near the S end of Little Egg harbour bay, 45 m. SE by S of Trenton.

TUCKUM, a town of European Russia, in Courland, near the Baltic, 23 m. NNE of Mittau.

TUCOMBIRA, a river of Brazil, which runs S, and changing its course to the E, enters the sea, 21 m. N of Los Ilheos.

TUCOPIA, an island in the S. Pacific, in S lat. 12° 21′, E long. 168° 43′, discovered by Quiros in 1606. It is about 7 m. in circuit, and has a pop. of about 400.

TUCUMAN, a prov. of the Plata confederation, bounded on the NE by Salta; on the E by the Gran-Chaco; on the S by Santiago; and on the SW by the Pampas or territories inhabited by the Aucaes, Huarcas, or Pampas, Pihuenches, Puelches, Uncos, and other tribes which wander over the plains and mountains adjacent to Chili. Its surface is mountainous towards the W. The great chain of the Andes, which borders and sends forth branches into Tucuman, is so high in some parts as to reach the region of perpetual snow. In it there are several mines, formerly worked by the Spaniards, and also many vestiges of the mining operations carried on by the Peruvians. The rivers of T. are numerous; the principal are the Vermejo, the Salado, the Xuxuy, the Dulce, and the Quarto. Numerous shallow lakes occur in this province, produced by the overflowing of the rivers; but they have the singular quality of being mostly saline, particularly those in the neighbourhood of the Rio-Vermejo. Vast plains also occur which are covered with fossil salt; and there is scarcely a river in the whole country of which the waters are not brackish. The climate is hot in those parts farthest from the main chain and branches of the Andes; but in general the seasons are regular, and the soil prolific, producing all sorts of grain, and esculent plants and fruits in abundance, besides immense tracts of excellent pasture for the innumerable herds of cattle, mules, and horses which are reared in this prov. Cotton, vines, tobacco, cacao, cochineal, indigo, and flax, are cultivated; but none of these articles enter into the export trade, which chiefly consists of cattle and timber. Honey and wax are also plentifully produced. The forests, which overspread a great part of it, contain good timber, which forms one great article of its trade, supplying Santa-Fe, Buenos Ayres, and other towns, with boards, which are conveyed in carts drawn by oxen. It also supplies the timber necessary in the mining operations at Potosi. The box, laurel, pine, dragon-tree, walnut, palm, and cedar are found in the forests of T. The deserts and forests abound with all kinds of game and wild animals, as pumas, jaguars, ant-bears, bears, wild hogs, elks, deer, hares, rabbits, armadillos, guanacos, and vicunas. The American ostrich or cassowary frequents the plains. The boa, called in this country *ampolaba*, is often seen in the forests, and appears, when lying among the grass, like the huge trunk of an old tree: besides the boa, there are also rattle-snakes, vipers, and other reptiles common to warm climates. Mules constitute the great article of commerce. These animals are bought in Cordova, Santa-Fe, and Buenos Ayres; and being fattened during the winter in the valleys and plains, are driven to Peru. The amount of the pop. including the converted Indians, is about 50,000. The Jesuits formed about 24,000 of the natives into a militia, to repel the invasion of the Chaco Indians; but they are now employed in cultivating the land.—The capital and chief towns of this prov. are T., Cordova, Rioxa, Jujuy, Santiago, Londres, and Salta, with 38 other towns and villages, and 10 missions.—The Spaniards conquered this district after finishing the subjugation of Peru; and Diego-de-Roxas was its first discoverer in 1543.

TUCUMAN, or SAN MIGUEL-DE-TUCUMAN, the capital of the above prov., is situated in S lat. 27° 10′, in a pleasant plain, having a mild climate and producing abundance of fruits and grain. It contains a cathedral, a convent of Franciscans, and a college which was built by the Jesuits. It has a trade in mules; but its principal traffic consists in oxen for travelling waggons, and in the waggons themselves. There are silver mines in its neighbourhood. San-Miguel is the see of a bishop who resides at Cordova. The pop. is about 8,000.

TUCUMANILLA, a settlement in the state of Tucuman, 10 m. W of the capital.

TUCUMARE, a river which has its rise in the extensive plains which bound the Amazon, and falls into the Madeira, between the rivers Tamari and Macacipe.

TUCUPIO, a river of Venezuela, in the prov. of Cumana, which runs S, and unites with Curuma, an affluent of the Cuyuri.—Also another river of Venezuela, which takes its rise in a plain W of Guanare, and joining the Guanarito, falls into the Portugueza.

TUCURAI, a large river of Ecuador, in the prov. of Mainas, which falls into the Huallaga.

TUDDENHAM, a parish of Suffolk, 3 m. NE by N of Ipswich. Area 1,239 acres. Pop. in 1831, 369; in 1851, 425.—Also another parish in the same county, 3 m. SE by S of Mildenhall. Area 2,646 acres. Pop. in 1831, 388; in 1851, 479.

TUDDENHAM (EAST), a parish of Norfolk, 6 m. ESE of East Dereham. Area 2,065 acres. Pop. 589.

TUDDENHAM (NORTH), a parish in the above co., 4 m. E by S of East Dereham. Area 2,270 acres. Pop. in 1831, 399; in 1851, 430.

TUDDEREN, a village of Prussia, in the prov. of the Rhine, and 18 m. NNW of Aix-la-Chapelle, on the r. bank of the Suster. Pop. 400.

TUDELA, a city of Spain, in the prov. and 48 m. NW of Saragossa, at the confluence of the Queiles and the Ebro. Pop. 7,400. It is the second city in Navarre, and a bishop's see. A fine bridge of 17 arches across the Ebro, beautiful walks, and an advantageous situation, give it a picturesque appearance; but the interior is spoiled by the narrowness of the streets. The houses are lofty and well-built, and there is a good plaza. The adjacent country is fertile in corn, fruit, hemp, having also good pasturage; and the greater part of the pop. is employed in agriculture. T. was in 1314 the scene of a defeat of the inhabitants of Navarre by the Castilians, and in 1808, of a more serious overthrow of the Spaniards under Castanos, by the French under Lefebvre and Mathieu.

TUDELA-DE-DUERO, a town of Spain, on the Douro, 5 m. E of Valladolid. Pop. 1,500.

TUDELEY, a parish of Kent, 2 m. E by S of Tunbridge. Area 1,605 acres. Pop. in 1851, 642.

TUDELILLA, a village of Spain, in the prov. and 24 m. SE of Logrono. Pop. 1,000.

TUDERGA, a village of Asiatic Turkey, 32 m. NNE of Eski-Shehr.

TUDHOE, a township in Brancepeth p., co.-palatine of Durham, 4½ m. SSW of Durham. Area 1,699 acres. Pop. in 1851, 400.

TUDUAL (SAINT), a commune and village of France, in the dep. of Morbihan, cant. and 6 m. WNW of Guemene. Pop. 2,000.

TUDURA, a town of Hindostan, in the prov. of Mysore, on the W bank of the Tunga, in N lat. 13° 40′.

TUDY (SAINT), a parish of Cornwall, 5½ m. N by W of Bodmin. Area 3,257 acres. Pop. in 1851, 652.

TUEJAR, a town of Spain, in the prov. and 45 m. NW of Valencia, on the l. bank of the Chelva. Pop. 1,700.

TUERTO, a river of Spain, which rises near La-Espina-de-Tremor, in Leon, and flows, in a S and SE course of 32 m., passing Astorga and Celada, to the Orvigo, which it joins on the r.

TUFARA, a town of Naples, in the prov. of Sannio, 6 m. E of Reccia. Pop. 2,400.

TUFFE, a town of France, dep. of Sarthe, on the river Vimet, an affluent of the Huine, 20 m. NE of Mans. Pop. 1,590. It has pottery-manufactories.

TUFFER, or LASHKU, a town of Austria, in Styria, on the Save, 4 m. S of Cilly. In the neighbourhood are found coal and chalk.

TUFFLEY, a hamlet of St. Mary-de-Lode p., in Gloucestershire, 2 m. SSW of Gloucester. Pop. in 1851, 131.

TUFILLO, a town of Naples, in Abruzzo-Citra, 15 m. SSW of Il-Vasto, near the l. bank of the Trigno. Pop. 900.

TUFINO, a village of Naples, in the Terra-di-Lavoro, 3 m. NE of Nola. Pop. 1,000.

TUFO, a village of Naples, in Abruzzo-Ultra 2da, 5 m. N of Carsoli. Pop. 720.—Also a village of Principato-Ultra, 6 m. N of Avellino. Pop. 950.

TUFTON, a parish of Southamptonshire, 1 m. SW of Whitchurch. Area 1,552 acres. Pop. in 1831, 197; in 1851, 163.

TUFTONBOROUGH, a township of Strafford co., New Hampshire, U. S., on the E side of Lake Winnipiseogee, 34 m. NNE of Concord. Pop. 1,305.

TUFUA, an island in the S. Pacific, one of the Friendly group. It is volcanic, and visible from Annamuka.

TUGBY, a parish of Leicestershire, 12 m. E by S of Leicester. Area 1,830 acres. Pop. 365.

TUGELA, or TUKELLA, a division of the Natal colony in S. Africa, bounded on the N and E by the Klip river; on the S by Impafane; and on the W by the Kathlamba mountains.

TUGELOO, a river of the United States, in Georgia, one of the branches of the river Savannah, which joins the Keowee, 48 m. NW of Petersburg.

TUGFORD, a parish of Salop, 11 m. NNE of Ludlow. Area 1,320 acres. Pop. in 1851, 157.

TUGGEN, a village of Switzerland, in the cant. and 18 m. NE of Schwytz, near the river Linch.

TUGGURT, the principal village in the Rouara district, on the borders of the Sahara, in N lat. 33° 24′, E long. 5° 30′. It consists of about 500 houses, and is inhabited by an industrious and trading community. The environs are well cultivated.—Also a village of Tunis, 60 m. SW of the capital.

TUGHLIKABAD, a ruined town of Hindostan, 19 m. S of Delhi.

TUGLIE, a town of Naples, in the prov. of Terra-d'Otranto, 7 m. E of Gallipoli. Pop. 1,000.

TUH-SHAN-CHU, a district and town of China, in the prov. of Kwei-chu, div. and 33 m. S of Tu-yun-fu, on the l. bank of the U-kea, in N lat. 25° 45′, and E long. 107° 20′.

TUI, a small river of Venezuela, in the prov. of Cumana, a head-stream of the Arui.

TUICHI, a river of Peru, in the prov. of Apolabamba, which runs NE, and enters the Beni.

TUILE (LA), a village of Savoy, 7 m. WSW of Albigny. Pop. 900.

TUIRA. See SANTA-MARIA.

TUIRAN. See TOIRAN.

TUK, a village of Upper Egypt, on the l. bank of the Nile, 6 m. N of Negade.

TU-KANG-CHU, a town of China, in the prov. of Kwan-se, in N lat. 23° 4′, and E long. 106° 45′.

TU-KEANG-TING, a district and town of China, in the prov. of Kwei-chu, and div. of Tu-yun-fu.

TU-KE-CHU, a town of China, in the prov. of Kwan-se, and div. of Tae-ping-fu, in N lat. 23° 10′, and E long. 107° 8′.

TULA, a government of Russia, bounded on the N by that of Moscow; on the E by Riazan; on the S and SW by Orel; on the W by Kaluga. It extends from 52° 30′ to 55° N lat., and has a surface of 8,900 geog. sq. m., with a pop. in 1850 of 1,251,500. Its surface presents a vast undulating plain, intersected by the Oka on the NW, and centrally by the Upa. The soil is in general of good quality; and there is here more of tillage and less of pasturage than in most of the other provinces of the empire. Rye, barley, oats, hemp, and flax are largely grown. Iron is manufactured; and woollens and linen are made for home-consumption. Exports of agricultural produce take place by the rivers Oka, Upa, Don, and Plava, and the Ivanovska canal which unites the Don and the Upa.

TULA, the capital of the above government, is situated at the confluence of the Tulpa and the Upa, in N lat. 54° 11′, 115 m. S of Moscow. This place, the pop. of which now approaches to 40,000, has been called the Sheffield of Russia. Here is a cannon-foundry, and a manufactory of arms for government of great extent, and in the town there are about 800 workshops for making firearms and cutlery for private use. The ore is obtained in abundance from the vicinity, but the best quality of ore used here is brought from Siberia. The other establishments of the place are tanneries, breweries, and soap and candle works. T. has likewise a traffic in corn, hemp, and other produce of the neighbourhood. The public buildings are the churches, the residence of the bishop, the courts of justice, the hospitals, a gymnasium, a seminary for priests, and a theatre. T. was founded in the beginning of the 16th cent., but it was not till 1717, that the government manufactory of arms was established here by Peter I. The river Don has its rise in the lake of Ivanou, at a short distance.

TULA, a town of Mexico, in the state and 40 m. NNW of Mexico, in N lat. 19° 57′, at an alt. of 2,244 yds. above sea-level, on a river of the same name, an affluent of the Tampico.

TULANZINGO, a town of Mexico, in the state and 54 m. NE of Mexico.

TULARE, a county of California, U. S., lying around a large lake of the same name [see article CALIFORNIA]. Pop. in 1852, 8,575, of whom 8,408 were Indians. There are two equally good passes through the mountains into the great Tulare valley, viz.: Walker's pass, near the head-waters of Kern river, which affords an entrance into the valley somewhat to the N of its upper or southern extremity, and the Tejon or Tayhone pass, which lies at the point of junction of the Sierra-Nevado and the coast-range. This is the pass through which the valley is usually entered, either from the Great Salt lake desert or from the valley of Los-Angeles. It has frequently been called Walker's pass. Lake Buenavista, or Kern lake, is about 15 m. to the N of this pass, and it is emphatically at the head of the Tulare valley. Walker's pass is little known.

TULCZYN, a town of European Russia, in the gov. of Wilna, 14 m. S of Braslav.

TULGOM, or TULGONG, a town of Hindostan, in the prov. of Aurungabad, in N lat. 18° 46′, E long. 74° 40. This place was the scene of the defeat of the Bombay army in 1778, and the subsequent disgraceful convention.

TULJAPUR, a town of Hindostan, in the prov. of Aurungabad, in N lat. 18° 17′. It stands upon a hill, and contains a number of Hindu temples, the principal of which is dedicated to Bhavany.

TULLA, or TULLAGH, a parish and town in co. Clare. Area of p. 24,531 acres. Pop. in 1831, 7,514; in 1851, 6,760. Lough Cullaunyheeda on the S boundary, has a surface-elevation above sea-level of 97 ft.; Loughs Roslara and Cloondanagh, in the interior, have surface-elevations of respectively 141 and 177 ft.—The town is 7½ m. E by N of Ennis. Pop. in 1831, 874; in 1851, 1,215.

TULLAGH, a parish on the coast of co. Cork, containing the village of Abbeystrand, the small port of Baltimore, and the island of Innisherkin. Area 5,349 acres. Pop. in 1851, 2,374.

TULLAGHAN, a village in the p. of Rossinver, co. Leitrim, 2½ m. WSW of Bundoran.—T. bay opens immediately E of the headland which screens the E side of Blacksod bay, has an entrance width of 2¼ m., and penetrates the land NNE 5½ m.

TULLAGHANBROGUE, or GROVE, a parish in the co. and 4 m. SW of Kilkenny. Area 3,487 acres. Pop. in 1831, 627; in 1851, 750.

TULLAGHANOGE, a parish in co. Meath, 2½ m. SE of Athboy. Area 1,414 acres. Pop. 165.

TULLAGHER, or TULLOGHER, a village in the p. of Dysertmoon, co. Kilkenny, 3½ m. S of Innistiogue. Pop. in 1831, 170; in 1851, 113.

TULLAGHERIN, or TULLOWHERIN, a parish in co. Kilkenny, 3¾ m. SW by S of Gowran. Area 5,052 acres. Pop. in 1831, 895; in 1851, 847.

TULLAGHLEASE, or TULLILEASE, a parish in co. Cork, 7 m. NNE of Newmarket. Area 8,292 acres. Pop. in 1831, 2,368; in 1851, 1,000.

TULLAGHMAIN. See TULLAMAIN.

TULLAGHMELAN, or TULLOGHMELAN, a parish in co. Tipperary, 4¾ m. SW by W of Clonmel. Area 2,695 acres. Pop. in 1831, 1,100; in 1851, 730.

TULLAGHNISKEN, TULLANISKEN, or TULLYNISKEN, a parish in co. Tyrone, 3 m. NNE of the town of Dungannon. Area 4,461 acres. Pop. 3,474.

TULLAGHOBIGLY, or RYETULLAGHOBIGLY, a parish in co. Donegal, 6 m. SW of Duntanaghy. Area 68,608 acres. Pop. in 1831, 8,466; in 1851, 8,982. About one-fourth of the surface is arable land; all the vast remainder is very mountainous, moorish, or boggy. The summit of Errigal, situated in the E, has an alt. above sea-level of 2,462 ft.; the summit-ground of the peninsula which terminates in Bloody-Foreland, has an alt. of 1,035 ft.

TULLAGHOG, or TULLYHOG, a village in the p. of Desertcreight, co. Tyrone, 2 m. SSE of Cookstown. Pop. in 1831, 137; in 1851, 72.

TULLAGHORTON, a parish in co. Tipperary, 2 m. E by N of Clogheen. Area 6,889 acres. Pop. 1,504.

TULLAGHOUGHT, or TULLOGAHUGHT, a parish in co. Kilkenny, 2¾ m. SSW of Kilmaganny. Area 4,601 acres. Pop. in 1831, 1,563; in 1851, 1,364.

TULLAMAIN, a parish in co. Kilkenny, 2 m. NE of Callan. Area 1,171 acres. Pop. in 1831, 416; in 1851, 181.—Also a parish in co. Tipperary, 3 m. W of Fethard. Area 2,217 acres. Pop. 473.

TULLAMORE, the capital of King's co., in the p. of Kilbride-Tullamore, on the Grand canal, and a small affluent of the river Clodagh, 5¼ m. S of Kilbeggan, and 46½ m. W by S of Dublin. The site and the immediate environs of T. are a sort of oasis in the vast tract of morass which bears the name of the bog of Allen. The streets are regular and wide. The gaol and the court-house—usually the most conspicuous objects in an Irish county town—form a pleasing feature in T., standing adjacent to each other on a raised platform at the W end of the town, and displaying with effect their respective styles of architecture, the one castellated and the other Grecian. T., in consequence both of its position on the Grand canal, and its central situation in reference to the surrounding country, is a place of considerable business. It is the principal town and chief shipping-station on the entire line of the canal, and besides being touched by all the boats in transit to Dublin from both Ballinasloe and the Shannon, it has swift iron passage-boats of its own in communication with the metropolis. Pop. 4,630.

TULLAROAN, a parish in co. Kilkenny, 4½ m. SSW of Freshford. Area 12,359 acres. Pop. in 1831, 3,962; in 1851, 3,373. The village of T. at present consists of a few cabins, and the neighbourhood entirely depends for interest on the ruins of ancient buildings, and the tales of other times.

TULLE, a town of France, the capital of the dep. of Corrèze, situated at the confluence of the Corrèze and the Solane, partly on the declivity, partly at the foot of a mountain, 42 m. SE of Limoges, in N lat. 45° 16′. Pop. in 1789, 7,774; in 1836, 9,700; in 1846, 11,646. It is badly built, having crooked and angular streets, and no public buildings except its churches and hospital. Its pop. is employed partly in the manufacture of woollen stuffs and paper, partly in preparing brandy, chocolate, and liquors; and to a considerable extent, in a large government manufactory of fire arms. The arrond. has an area of 255,864 hectares, with a pop. in 1846 of 136,999; and comprises 11 cants.

TULLIALLAN, a parish in the detached district of Perthshire, bounded on the SW by the frith of Forth. Area 2,760 acres. Pop. in 1851, 3,043.

TULLIG, a village in the p. of Kilballyowen, co. Clare, situated in a bleak district of country, 3 m. WNW of Carrigaholt. Pop. in 1851, 193.

TULLIG-POINT, a small headland on the N coast of co. Clare, 1½ m. W by S of the village of Tullig.

TULLILEASE. See TULLAGHLEASE.

TULLINS, a town of France, dep. of Isère, situated in a rich and beautiful country, 14 m. NE of St. Marcellin. Its inhabitants, amounting to 2,400, are employed partly in the manufacture of hemp and thread, partly in iron and steel-works.

TULLIVIN. See TULLYVIN.

TULLN, a walled town of the archd. of Austria, on the r. bank of the Danube, 14 m. NW of Vienna. Pop. 1,700.

TULLOCH, a village in the parish and 2 m. NW of the town of Perth, notable as the site of the first Scottish bleachfield, and the place where the first potatoes produced in Scotland were grown. Pop. 200.

TULLOE. See TULLA.

TULLOGHER. See TULLAGHER.

TULLOW, or TULLOPHELIM, a parish in co. Carlow, containing on its W margin, a part of the town of Tullow. Area 7,989 acres. Pop. in 1831, 3,688; in 1851, 3,497.—The town stands on the river Slaney, 4¾ m. SSW of Rathvilly. Pop. of the town in 1831, 1,929; in 1851, 2,963.

TULLOWCREEN, or TULLOWCRINE, a parish in co. Carlow, 2 m. N of Leighlin-bridge. Area 5,899 acres. Pop. in 1831, 1,344; in 1851, 1,110.

TULLOWMAGRIMAH, or TULLOWMAGIMMA, a parish in co. Carlow, 3½ m. SSE of Carlow. Area 4,063 acres. Pop. in 1841, 1,125; in 1851, 984.

TULLOWMOY, a parish in Queen's co., 3¼ m. SSE of Stradbally. Area 6,007 acres. Pop. 1,015.

TULLY, or TULLYFERNE, a parish in co. Donegal, containing the villages of Millford and Ramelton. Area 16,612 acres. Pop. in 1851, 5,116.

TULLY, TULLOW, or BULLOCK, a parish in co. Dublin, 3 m. S of Kingstown, containing part of the village of Cabinteely. Area 3,285 acres. Pop. in 1831, 1,385; in 1841, 1,207.

TULLY, or COGLANSTOWN, a parish in co. Kildare, consisting of four mutually detached districts in the baronies of Kilcullen, East Ophaly. Area 5,164 acres. Pop. in 1831, 1,065; in 1851, 1,087.

TULLY, a hamlet in the p. of Ballinakill, co. Galway, 1½ m. SE of Renvyle-house.

TULLY, a township of Onondaga co, in New York, U. S., 123 m. W by N of Albany. Pop. 1,559. —Also a village of Lewis co., in Missouri, 112 m. N by E of Jefferson.

TULLYALLEN, a parish, partly in co. Meath, but chiefly in co. Louth, containing part of the town of Drogheda. Area 8,293 acres. Pop. 4,142.

TULLYBRACKEY, or TULLABRACCA, a parish in co. Limerick, 2½ m. N by E of Bruff. Area 3,209 acres. Pop. in 1831, 1,437; in 1851, 836.

TULLYCANNA, a village in the p. of Ambrosetown, co. Wexford, 3 m. N by W of Duncormuck. Pop. in 1841, 123; in 1851, 118.

TULLYCORBET, a parish in co. Monaghan, 3 m. N of Ballybay. Area 7,923 acres. Pop. 3,293.

TULLYLISH, a parish in co. Down, containing the village of Gilford. Area 11,707 acres. Pop. in 1831, 10,501; in 1851, 13,400. The river Bann passes westward through nearly the centre, in a sinuous course, along a valley of charming features. Three principal localities of manufacture along the course of the river, are the hamlet of Hall's Mills, noted so far back as a century ago, for a very extensive bleaching-green; the village of Gilford, in the centre, whose linen, flax-spinning, and bleaching trade may be regarded as a department or branch of the prosperous and interesting trade of Banbridge; and the little village of Moyallen in the west, whose environs are noted for their bleachfields and their vitriol-works.

TULLYNAKILL, a parish of co. Down, 3¼ m. SE by S of Comber, containing the village of Ardmillan. Area 2,923 acres. Pop. in 1851, 1,221.

TULLYNESSLE-AND-FORBES, a united parish nearly in the centre of Aberdeenshire. Pop. 930.

TULLYRUSK, a parish in co. Antrim, 3½ m. ENE of Glenavy. Area 4,779 acres. Pop. 1,049.

TULLYVERY, a village in the p. of Killyleagh, co. Down. Pop. in 1851, 597.

TULLYVIN, a village in the p. of Kildrumsherdin, co. Cavan, 2¾ m. SW by W of Cootehill. Pop. in 1851, 12.

TULMERO, or TURMERO, a town of Venezuela, situated in the valley of Arogoa, 36 m. E of Valencia. Pop. 8,000. It is well built, and the residence of a number of planters; but was, under Spanish regime, peculiarly the abode of all the officers, factors, and persons employed in the administration of the monopoly of tobacco, cultivated in its vicinage, on account of the king.

TULPEHOCKEN, a township of Berks co., Pennsylvania, U. S., 42 m. N by E of Harrisburg, on the N side of the Tulpehocken, a stream which runs E into the Schuylkill, above Reading. Pop. 1,800.

TULSK, a post village, and formerly a parl. borough, in the p. of Ogulla, co. Roscommon, 4¾ m. SSW of Elphin.

TU-LU-FAN-TING, a district and town of China, in the prov. of Kan-suh, and div. of Chin-se-fu.

TULUMBAH, a town of Hindostan, in the prov. of Multan, on the SW bank of the river Ravi, in N lat. 30° 58′.

TULUMBO, a village of Central Africa, in Bambarra, on the N bank of the Niger, 50 m. SE of Bambarra.

TULUN, a village of Asiatic Russia, in the gov. of Irkutsk, on the Ija, 44 m. E of Nizni-Udinsk.

TULWUNDY, a town of Hindostan, in the prov. of Lahore, in N lat. 31° 15′. This place is celebrated for having been the birth-place of Baba Nanoc Shah, the founder of the religion of the Sikhs.

TUMA, a village of the archd. of Austria, 9 m. SSE of Zlabings, on an affluent of the Theya.

TUMACO, an island in the N. Pacific, lying off the coast of Ecuador, in N lat. 1° 46′. It is 2 m. long from SE to NW, and 1 m. broad. It has a port.

TUMAS MOUNTAINS, a range of mountains in S. Africa, in the Damara territory, a little to the N of the parallel of 23° S, and intersected by the meridian of 16° E. They run NW and SE, and attain an alt. of 2,000 ft. above sea-level.

TUMAT, a river of Abyssinia, which appears to have its origin between Singe and Fazocl, and flows N to the Bahr-el-Azrek.

TUMBADO, a small island in the Bahamas, in N lat. 26° 24′, 48 m. NNW of Grand Bahama.

TUMBELAN ISLANDS, a cluster of small islands in the Eastern seas, in N lat. 1°.

TUMBEZ, a village of Peru, in the prov. of Piura, situated on the Rio-Tumbez, 280 m. N of Truxillo. It is seated in a sandy plain, and consists of about 50 houses, chiefly inhabited by Mestizoes and Indians, who supply whalers with fresh water, fruit, and vegetables. T. was the place where the Spaniards first landed in 1526.

TUMBEZ, or SARAGURA, a river of Peru, which rises in the Andes, and empties itself into the gulf of Guayaquil, opposite the island of Santa-Clara, or Amortajado. By it launches, flat-bottomed boats, and other small vessels, pass as far as the above settlement on its banks; but in winter it cannot be navigated on account of the great current arising from the extraordinary increase of its waters. The river traces the boundary between the states of Peru and Ecuador.

TUMBIA, a settlement of New Granada, in the prov. of Pasca, 24 leagues from Santa Fe.

TUMBUDRA, a river of Hindostan, formed by the junction of the Tungha and the Bhudra, both descending from the Western Ghauts, in Mysore; and flowing into the Kistna, 18 m. below Kurmel, after a course of 350 m.

TUMBUK, a town of Kordofan, 180 m. SW of Sennaar.

TUMBY, a township of Kirkby-upon-Bain p., Lincolnshire, 7 m. S by W of Horncastle. Pop. in 1831, 322; in 1851, 339.

TUMCURU, a town of Hindostan, in the prov. of Mysore, in N lat. 13° 15′. It is a place of some consequence, and is defended by a good fort.

TUMEN, a small town of Ghilan, in Persia, 18 m. W of Reshd.

TUMENE, a district of Tobolsk, in Asiatic Russia, with a town of the same name, situated on the S bank of the Tura, at its confluence with the Tumenka, in N lat. 57°. It was built in 1586, upon the site of a Tartar town, of which traces still remain. There is a suburb on the opposite side of the river, chiefly inhabited by Tartars and Bukharians. The place contains 800 houses, with several churches, a convent, and nunnery. Its wealth arises chiefly from a foundry of bells, manufactures of soap, and considerable tanneries. It was the first town built by the Russians in Siberia.

TUMIRIQUIRI, a mountain of Venezuela, in the prov. and 30 m. SSE of Cumana, near Cumanacoa, on the road to Caripe, the chief mission of the Chaymas, which passes over the summit of a lower part of the chain, which bears the general name of the Cocollar. From the summit of this last chain, at more than 2,000 ft. in height above the plain, the eye wanders over the immense plains which reach towards the banks of the Orinoco, in the ravines alone of which can be distinguished any trees, and these but thinly scattered; the remainder of the surface is covered with an uniform coat of long waving grass, intermixed with flowering shrubs. From this point the traveller ascends towards the Tumiriquiri, by a road which is partly traversed on horseback, but soon becomes too steep and slippery for these animals. The round summit of the T. is covered with turf, and is elevated more than 4,400 ft. above the ocean. This elevation gradually diminishes towards the W by a ridge of steep rocks, and is interrupted at the distance of a mile by an immense crevice, which descends towards the gulf of Cariaco. Beyond this two enormous peaks arise, the northernmost of which, named the Cucurucho of T., is more than 6,500 ft. in height, surpassing that of the Brigantin, with which it is connected. These peaks are covered with mahogany, javillo, and cedar trees, of an enormous size. The view from their summit is very fine. The chain, which extends from W to E, is seen in all its forms; its ridges running parallel to each other at short distances, forming longitudinal valleys, intersected by crevices worn by the waters in their passage to the Orinoco or the sea; the sea bounds the prospect on the N; and an immeasurable plain forms its horizon on the S.

TUMLUK, a town of Bengal, situated on the SW side of the Rupnarain river, in N lat. 22° 17′. It is the station of the superintendent of the salt manufacture on the banks of the Hugli. The country in the vicinity is very low, and requires strong embankments to preserve it from being overflowed by the rivers during the rainy season.

TUMMEL, the lower part of the northern great head-water of the Tay, in Perthshire. It is generally described by topographists as issuing from Loch-Rannoch, and expanding, near the middle of its course, into Loch-Tummel; but the stream between the two lakes really bears the name of the Rannoch, and only that between Loch-Tummel and the confluence with the Tay strictly claims to be the Tummel. Its length of run, thus limited, is only about 12 m., or, including sinuosities, 15 or 16 m.

TUMNA, Toomna, or Toemonia, a parish in co. Roscommon, extending along the r. bank of the Shannon, containing the village of Battle-Bridge. Area 9,188 acres. Pop. in 1851, 2,849.

TUMORE, a parish in co. Sligo, 3½ m. SE of Ballymote. Area 10,834 acres. Pop. in 1851, 2,351.

TUMUN, a town of Hindostan, in the prov. of Malwah, on the banks of the Batmah, in N lat. 25° 8′.

TUNA, a small island of Lower Egypt, in Lake Menzaleh, 8 m. SE of Tennis.—Also a town of the island of Porto-Rico, on the l. bank of the Quaxalaca. Pop. 1,400.—Also a small river of Brazil, in the prov. of Pernambuco, which runs E, and falls into the sea between the rivers Ilheos and Piratininga. —Also a village of Sweden, in the prefecture of Stara-Kopparberg, 15 m. SSW of Fahlun.

TUNBRIDGE, or Tonbridge, a parish and market-town in Kent, 13 m. WSW of Maidstone, on the river Medway, here divided into five branches, over which it has as many bridges; and intersected by the South-Eastern railway, which has a principal station here at the distance of 40 m. from London. Area 15,235 acres. Pop. in 1831, 10,380; in 1851, 16,548.—The town consists principally of one long and in some parts wide street. Pop. in 1851, 4,539. The Medway crosses it near the S end. The town-hall is a plain building. The principal architectural object in the town is the free grammar-school. There are some very ancient houses. The ruins of the castle, though not of great magnitude, are interesting, and the entrance tower is said to be the finest specimen of Norman architecture now existing.

TUNBRIDGE, a post township of Orange co., Vermont, U. S., 32 m. S of Montpellier. Pop. 1,786.

TUNBRIDGE-WELLS, a celebrated watering-place, partly in the p. of Tunbridge, partly in that of Speldhurst, Kent, and partly in the p. of Frant, Sussex, 18 m. SW of Maidstone, and 36 m. SE by S of London, on the Hastings branch of the South-Eastern counties railway. Pop. in 1851, 10,587. The neighbourhood abounds with mineral springs. The water is a light and pure chalybeate. A century ago, this place was only an inconsiderable village, it may now be regarded as a town of villas, consisting of four principal districts, Mount Ephraim, Mount Pleasant, Mount Sion, and the Wells. The latter stands in a valley formed by gently sloping hills, and contains, besides the springs, the public parade, ball-rooms, libraries, the theatre, market-house, and various shops for the sale of jewellery, toys, and Tunbridge-ware.

TUN-CHI, a town of China, in the prov. of Gantiwy, in N lat. 29° 48′. It has an active trade in tea.

TUNDERGARTH, a parish in Annandale, Dumfries-shire. Area 10,630 imperial acres. Pop. 577.

TUNG-CHANG-FU, a division and town of China, in the prov. of Shan-tung. The div. comprises eight districts. The town is 60 m. WSW of Tse-n-n-fu, on the Great canal, in N lat. 36° 32′ 24″, and E long. 116° 10′. The houses are better and more regularly built than in the generality of Chinese towns; and it has a pagoda, which ranks next to the Paou-gan-tsze of Nan-king, and is adorned in an equal manner. It is an octagon tower, 8 stories high, porcelain outside, incrusted in the inside with marbles of different colours, and covered with little bells. On the highest story is a statue of the idol in copper gilt. The port possesses an active and lucrative trade.

TUNG-CHING, a district and town of China, in the prov. of Gan-hwuy, div. and 36 m. N of Gan-king-fu, in N lat. 31° 7′, and E long. 116° 56′. It has bastioned brick walls, and an old tower of three stories. A small river, by which it is separated from a suburb, is crossed by a bridge. The surrounding country is well-cultivated, and has several lime-kilns.—Also a district and town in the prov. of Hu-pih, and div. of Wu-chang-fu, in N lat. 29° 15′ 36″, and E long. 113° 46′ 55″, in a mountainous locality.

TUNG-CHU, a division and town of China, in the prov. of Keang-su. The div. comprises 2 districts. —Also a district and town of China, in the prov. of

Chih-le and div. of Shun-teen-fu, on the r. bank of the Pay-ho, about 120 m. from the sea, and 15 m. E of Peking, of which it may be considered the port, in N lat. 39° 55′ 30″, E long. 116° 42′. The principal streets are straight, and paved with large square stones, with foot-paths on either side, and awnings above. It has large granaries for the market of Peking, shops well stocked with tea, porcelain, silk, furs, and European merchandise; and in the streets are numerous victual stalls. The trade with the interior, of which an important branch consists in the spawn of fish bottled, is very active. The environs possess little fertility.

TUNG-CHU-FU, a division and town of China, in the prov. of Shen-se. The div. comprises 10 districts. The town is 75 m. NE of Segan-fu, in N lat. 34° 50′ 24″, E long. 109° 50′ 55″, on the l. bank of the Lo-ho.

TUNG-CHUEN, a town of the Corea, in the prov. of Keang-yuen, on the sea of Japan, 105 m. ENE of King-ki-tao.

TUNG-CHUEN-FU, a division and town of China, in the prov. of Yun-nan. The div. comprises 5 districts. The town is in N lat. 26° 20′ 56″, E long. 103° 25′ 39″.—Also a division and town of China, in the prov. of Sze-chuen. The div. comprises 8 districts. The town is 75 m. NE of Ching-tu-fu, on the frontier of Tibet, at the confluence of the Lo-keang and Mon-koua, in N lat. 31° 9′, E long. 105° 11′.

TUNG-GAN, a district and town of China, in the prov. of Chih-le and div. of Shun-teen-fu, in N lat. 39° 22′, E long. 116° 46′.—Also a district and town in the prov. of Hu-nan, div. and 24 m. WNW of Yung-chu-fu, on the Sang-keang, in N lat. 26° 13′ 12″, E long. 111° 13′ 30″.—Also a district and town in the prov. of Fo-keen and div. of Tseuen-chu-fu, 120 m. SW of Fu-chu, in N lat. 24° 44′ 24″, E long. 118° 19′ 20″.

TUNG-GAN-HEEN, a district and town of China, in the prov. of Kwang-tung and div. of Lo-tsing-chu, 36 m. W of Chaou-king-fu, in the midst of mountains, in N lat. 23° 3′, E long. 111° 44′.

TUNG-HAE-HEEN, a district and town of China, in the prov. of Yun-nan and div. of Lin-gan-fu, in N lat. 24° 12′, E long. 102° 56′.

TUNG-HEANG, a district and town of China, in the prov. of Sze-chuen and div. of Suy-ting-fu, in N lat. 31° 27′, E long. 107° 51′.—Also a district and town in the prov. of Che-keang, div. and 12 m. SW of Kea-tung-fu, in N lat. 30° 45′, E long. 120° 23′.—Also a district and town in the prov. of Keang-se and div. of Fu-chu-fu, in N lat. 28° 15′, E long. 116° 35′.

TUNG-HEU-HEEN, a district and town of China, in the prov. of Ho-nan and div. of Kae-fung-fu, in N lat. 34° 35′, E long. 114° 35′.

TUNG-HU-HEEN, a district and town of China, in the prov. of Hu-pih and div. of E-Chang-fu.

TUNG-JIN-FU, a division and town of China, in the prov. of Kwei-chu. The div. comprises one district. The town is in N lat. 27° 38′, E long. 109°.

TUNG-KEANG, a district and town of China, in the prov. of Sze-chuen, div. and 90 m. ENE of Paou-ting-fu, in a mountainous locality, in N lat. 31° 55′, E long. 107° 16′.

TUNG-KWAN, a district and town of China, in the prov. of Shen-se, div. and 66 m. N of Se-gan-fu, on the r. bank of the Tse-ho, in N lat. 35° 6′, E long. 109° 3′.

TUNG-KWANG, a district and town of China, in the prov. of Chih-le, div. and 48 m. SE of Ho-keen-fu, on the r. bank of the Hwai-ho, in N lat. 37° 58′, E long. 116° 36′.

TUNG-KWAN-TING, a district and town of China, in the prov. of Shen-se and div. of Tung-chu-fu.

TUNG-LAN-CHU, a district and town of China, in the prov. of Kwan-se and div. of King-yuen-fu, in N lat. 24° 28′, E long. 106° 45′, 38 m. ENE of Sze-ching-fu, in a mountainous locality, on an affluent of the Po-soe-keang.

TUNG-LEANG, a district and town of China, in the prov. of Sze-chuen, div. and 39 m. NNE of Chun-hing-fu, in N lat. 29° 59′, E long. 106° 11′.

TUNG-LEW, a district and town of China, in the prov. of Gan-hwuy and div. of Che-chu-fu, on the r. bank of the Yang-tse-keang, in N lat. 30° 22′, E long. 116° 54′. It is enclosed by walls, and has two towers, one of 7 stories.

TUNG-LING, a district and town of China, in the prov. of Gan-hwuy and div. of Che-chu-fu, in N lat. 31° 4′, E long. 117° 50′.

TUNG-LU, a district and town of China, in the prov. of Che-keang, div. and 21 m. NNE of Yen-chu-fu, at the confluence of two rivers.

TUNG-LU-TING, one of the four subdivisions of the div. of Shun-teen-fu, in the Chinese prov. of Chih-le.

TUNG-MING, a district and town of China, in the prov. of Chih-le and div. of Ta-ming-fu, in N lat. 35° 29′, E long. 115° 18′.

TUNG-O-HEEN, a district and town of China, in the prov. of Shan-tung and div. of Tae-gan-fu, 26 m. SE of Tung-chang-fu, near the r. bank of the Ta-tsin, in N lat. 36° 15′, E long. 116° 20′. It is enclosed by good walls, but the houses are ill-built. It is noted for its glue.

TUNG-PING-CHU, a district and town of China, in the prov. of Shan-tung, 45 m. NW of Yen-chu-fu, on the r. bank of the Tse-nan, near the Great canal, in N lat. 36° 6′, E long. 118° 16′.

TUNG-SEANG-FU, a division and town of China, in the prov. of Shen-se. The div. comprises 8 districts.

TUNG-SHAN, a district and town of China, in the prov. of Hu-pih, div. and 69 m. S of Wie-chang-fu, near the l. bank of an affluent of the Yang-tsen-keang, in N lat. 29° 40′, E long. 114° 22′.

TUNG-SZE, a district and town of China, in the prov. of Kwei-chu and div. of Tsun-e-fu, in N lat. 28°, E long. 106° 45′.

TUNG-TAE, a district and town of China, in the prov. of Keang-su and div. of Yang-chu-fu.

TUNG-TSEE-HEEN, a district and town of China, in the prov. of Sze-chuen and div. of Kew-chu-fu, in the midst of mountains, on an affluent of the Chi-chui.

TUNG-UNG, a town of China, in the prov. of Kwang-tung and island of Hai-nan. It is enclosed by walls, and has streets regularly built and paved. It contains about 12,000 houses. The surrounding country is highly cultivated, and contains numerous triumphal arches and cemeteries.

TUNGURAGUA, a high mountain of Ecuador, in the prov. and 20 m. N of Riobamba. It is of a conical figure, regularly sloped on all sides, and has an alt. of 16,500 ft. above sea-level, rising far above the line of perpetual snows. It is volcanic, and often bursts forth into violent eruptions, by one of which the town of Riobamba was destroyed. The lower parts of the mountain are covered with thick forests.—Also a river of S. America, now identified with the Amazon, of which it forms the head-stream. See AMAZON.

TUNGUSES, a native race of Asiatic Russia, who wander over the whole SE portion of that vast territory. They are first found on the banks of the Yenisei, whence they extend eastward to the sea of Okhotsk. In the south, however, they are mixed with the Mongols and Buriats; and on the borders of the northern ocean, they give place to the Yakuts and Samoyedes.

To the W of the Yenisei, a few are mixed with the Tartars and Ostiaks; but in general the prov. of Irkutsk-proper is that of which they may be considered the denizens. They are of a middle size, of a robust constitution, and endowed with great agility. Their countenance bears a considerable resemblance to that of the Mongols, though it is larger, and still more flattened. They have small eyes, a smiling physiognomy, and long black hair, which they allow to flow freely over their shoulders. They employ themselves solely in hunting and fishing. They have no permanent abodes, but range through the woods and along the rivers, seldom remaining more than six days at a time in one spot. Their tents are formed of a few spars, put together in a conical shape, and covered with skins or with willow bark, which after being exposed for some time to the vapour of boiling water becomes equally flexible. Their arms are bows and arrows, with which they will attack the fiercest animals. The delicacy of their sight and smell is almost incredible. There are three classes of T. who take a peculiar character from the regions which they inhabit, and the tribes on which they border: the Rein-deer T., the nomadic races of the north, called also the wood or hunting T.; the Dog T., who live upon the sea of Okhotsk, and in the neighbourhood of Kamtchatka, called also Lamuts; and the Riding T., who being mixed with the Buriats and Mongols, have to a certain extent adopted their habits, and even employ themselves a little in agriculture. Their religion, like that of all the other nomadic tribes in this part of the world, consists in a species of Shamanism, formed by a mixture of the creed of the Lamas with their own original superstitions.

TUNGUSKA, three large rivers of Asiatic Russia, all tributaries to the Yenisei. The Lower T., the most northerly of the three, has the longest course. It rises in the NW part of the gov. of Irkutsk, and after running for a considerable space parallel to the Lena, turns gradually to the W, and falls into the Yenisei, near Turuchansk. Its entire course cannot be estimated at less than 1,000 m.—The Middle T. rises also in the district of Irkutsk, not far from the source of the Lower T., and after a course of between 500 and 600 m. through the district of Yeniseisk, falls into the Yenisei in N lat. 62°.—The Upper T. has a greater mass of water than either of the above streams, and might even rival the Lower T. in length of course, if counted from the commencement of its stream. It bursts from the SW part of the lake or sea of Baikal, through a broken and rocky channel; and flowing W, joins the Yenisei after a course of 750 m. It bears the name of Angara in the first part of its course, and till after being joined by the Ilim, when it receives the name of T.

TUNG-UUN-HEEN, a district and town of China, in the prov. of Chih-le and div. of Tsun-hwa-chu.

TUNG-WEI-HEEN, a district and town of China, in the prov. of Kan-suh and div. of Kung-chang-fu, in N lat. 35° 6′, E long. 105° 13′.

TUNG-YANG, a district and town of China, in the prov. of Che-keang, div. and 38 m. E of Kin-hwa-fu, in N lat. 29° 17′, E long. 120°.

TUNG-YUEN, a district and town of China, in the prov. of Kwang-tung and div. of Kwang-chu-fu.

TUN-HWANG-HEEN, a district and town of China, in the prov. of Kan-suh.

TUNIS, a considerable territory of Northern Africa, forming one of the Barbary states, lying between the parallels of 31° and 37° N, and the meridians of 8° and 11° E; and bounded on the N by the Mediterranean; on the E by Tripoli; on the S and SW by the Sahara; and on the W by Algeria. Its area has been estimated at 70,000 sq. m. It consists chiefly of a large peninsula, stretching into the Mediterranean in a NE direction, and approaching, in Ras-Adder or Cape Bon, within less than 100 m. of the coast of Sicily. Beginning at Cape Shabbu, the frontier-point of Tripoli, the coast, which is deeply indented, extends northerly with a slight declination to the E; but after turning Cape Bon, its general direction is E. The N coast line terminates at Al-Kalah, and its whole extent is about 500 m. On the E it presents the deep gulfs of Cabes and Hammamet; on the N the gulf of Tunis. The interior is as yet little known. The cultivated part reaches from 200 to 250 miles into the interior, till it terminates with the chain of Atlas, and the vast dry plains of the Biled-el-Jerid.—The ancient subdivision of this region still claims a place in geography. The fertile country in the neighbourhood of Keff and Baijah, the *Regio Zeugitania* of Pliny, is now distinguished by the name of Frigeah. The districts between the Kairwan and the Jerid correspond to the *Bozacium* of the ancients.—This regency is watered by various streams of minor importance, but the only rivers of any consequence are the Mejerdah and the Wad-el-Quibir. The Mejerdah, the *Bagrada* of antiquity, empties itself into the sea between Cape Carthage and Porto-Farma. Its sources are yet unknown; but it is formed by the confluence of numerous streams descending from various branches of the Atlas, and distributes plenty over a fine level country in which are many towns and villages containing from 5,000 to 15,000 inhabitants. The banks of this river and the country to the eastward are indeed the best cultivated parts of the regency. That on the W side having been long exposed to the inroads of the Algerines, is thinly inhabited, and in many places desert. The Wad-el-Quibir rises in the neighbourhood of El Keif, and falls into the sea near Tabarca.

Productions and climate.] To the advantage of a large sea-coast Tunis adds, in comparison with the other states of Barbary, which are mostly mountainous and intersected with the higher chains of the Atlas, or covered with wide desert wastes, that of being an open country of fertile and salubrious plains and mountains of moderate height, many of them crowned with plantations of the finest olives. The only ramification of the Atlas chain of any extent divides it from the Biled-el-Jerid, or country of dates, so that the whole regency is capable of high cultivation, and might be made to produce 500 per cent. more than its present average crops. The principal agricultural productions are grain, cattle, and flocks, olives, dates, and a variety of the same fruits which grow in the south of Europe; but oil, corn, wool, and cattle have always been the great staple exports of this country, and of nearly all Northern Africa. The mountains near the city of T. contain silver, copper, and lead ores; there is also a lead mine which was worked by the Romans; but the beys of Tunis, afraid of exciting the cupidity of the Christians, have always discountenanced the working of mines.—The heat becomes insupportable in July and August, when the S wind brings the heated air from the interior of Africa. Some branches of the Atlas contain elevated and temperate regions. The tracts to the south, forming 'the Date country,' though far from presenting the same rich and verdant aspect as those on the coast, yield not only dates, but grain of different kinds, and contain many large villages. All kinds of fruit common to warm climates are plentiful, with olive-trees, roses, and other odoriferous plants. Hammam Lif is a noted hot-bath very much resorted to by the Tunisians. Its temp. is generally 118°. The taste of the mineral is not unlike that of Glauber's salts, but by no means nauseating.

Population.] The population, which is calculated at 2,500,000 by some authors, and by Blaquiere at 4,500,000, consists, like that of the other Barbary states, of Moors, Arabs, Jews, and Christian slaves. The number of the Arabs exceeds that of the Moors, who are the agriculturists and merchants. Reading, writing, and the knowledge of the Koran, are acquirements common to the people; and they are celebrated in this quarter of the globe for superior refinement and courtesy. There are about 5,000 Turks, and Kuroghs or sons of Turks by

Moorish women; 50,000 Jews, and 12,000 Christians; nearly all of whom are Roman Catholics, and principally Maltese, Sardinians, and Neapolitans. There are also a few Levantines and Greeks; the rest are Moors, Negroes—about half free—and Arabs, but the Arabs preponderate in numbers. There are no Berebers or Kabyles, in the regency. These aborigines of North Africa are confined to Algeria and Marocco, not extending further east than the Atlas chain of the country anciently called *Numidia.*

Commerce.] The territory of T., from the manner in which it projects into the Mediterranean, coming into closer contact with Europe than any other of the Barbary states, is peculiarly favourable for carrying on trade in that sea. By these advantages, Carthage became the first commercial state of antiquity, and in the struggle for universal empire the rival of Rome itself; and though, under the relentless hatred of that iron-hearted people, the foundations of Carthage were ploughed up, yet its territories still formed the centre of the Roman African dominions, and were speedily covered with magnificent monuments of taste and wealth. In our own times the rulers of T. have been zealous in their encouragement of trade; and, accordingly, availing itself of the advantages of its situation, T. carries on a much more extensive trade than any of the other towns of Barbary. Grain was formerly a chief article of export, but the famine of 1805 induced the bey to prohibit its exportation. When it was exported, a license from the bey was obtained, and a duty paid on wheat to the amount of £1 10s. on the *caffis*, [equal to 2 English quarters,] and half that amount on barley. The principal port for shipping grain is Biserta. In a plentiful year, the state of T. is computed to produce about 900,000 quarters of wheat, and an equal quantity of barley. Olive oil pays a duty of 3s. 4d. per *metal* of 40 lbs. The principal ports for shipping oil, are Tunis, Soliman, and Susa. Wool, which is here nearly equal to the best Spanish, and having a longer staple, is much better adapted to the manufacture of shawls, is shipped in large quantities, at Jerbi, Sfax, and other ports on the eastern frontier. About 20,000 *cantars* of this article were, in time of peace, exported to France and Italy. Soap, of superior quality, is also largely exported. Sponge is collected on the shore between Sfax and Jerbi, and brings from 30s. to 42s. per hundred weight. Amid the ruins of Carthage, and on the rocky mountains to the eastward of Tunis, Orchilla weed is collected in considerable quantities. An extensive and lucrative commerce was formerly carried on with the interior, particularly with Constantina and Ghadames; but the bey's impolitic mode of governing the Arabs is said to have interrupted it. The caravans from Timbuctu, which used to arrive annually in June, furnished the merchants with gold-dust, ivory, and ostrich feathers. The imports—supplied almost exclusively, till of late years, by Marseilles and Leghorn—consist of all kinds of European manufactures, colonial produce, and East India cottons. The species of British goods chiefly demanded, is that kind of woollens called scarlet long-ells, which the caravans carry in large quantities into Central Africa. The merchant-vessels amount to about 100 of from 300 to 100 tons. The coasting trade is all carried on by *sandals*, which never exceed 40 tons.

Government.] The ruler of Tunis, under the Porte, bears the title of bey, and enjoys the same power as the other Barbary chiefs. The forces amount to about 10,000 men; his revenues have been estimated at £1,000,000. The Barbary powers up to the occupation of Algiers by France, and the usurpation of the hereditary government of Tripoli by the Porte, were *quasi* independent, and concluded treaties of amity and commerce with all the governments of Europe, without the intervention of the Turkish divan. The Porte, however, was accustomed to send the *caftan* of investment for the vice-regal authority, on the hereditary beys of T. and Tripoli succeeding to their hereditary rights; whilst the military piratical chief of Algiers, elected by that piratical republic, received in due form, and at the usual time, the confirmation of his election, without hesitation on the part of the paramount chief or sultan at Constantinople. On the other hand, the sultan exercised his rights of *paramount* sovereignty in Barbary. His money passed current in those states, his firmans of investment were received by princes and people, as the approbation of the chief of Islamism, and at various periods he received tribute in the shape of specie and magnificent presents. The amount of tribute, however, was never fixed, but was always left to the caprice or liberality of the various bashaws. The present bey of T. has at different periods despatched presents to the sultan, the last of which was, among other things, a large sum of specie, and a splendid corvette.

History.] Of the ancient Carthage, the site of which was in the close vicinity of the present capital of this state, few traces remain. "The dreadful imprecations of its eternal enemy the Romans," says a celebrated journal, "have been strictly fulfilled against this devoted city. In vain does the inquisitive traveller seek in the neighbourhood of T. for the triple wall, with its lofty towers, whose capacious chambers contained stalls for 300 elephants, and stables for 4,000 horses, with lodgings for a numerous army,—in vain does he look for those safe harbours and sheltered receptacles, for those 2,000 ships of war, and 3,000 transports, which carried Hamilcar and his warriors against Syracuse: a few remains of the public cisterns, and the common sewers, are all that is left to point out the spot where Carthage, with its 700,000 inhabitants, once stood." Though the city of Carthage was razed to the ground, the Romans made its territory the centre of their African dominion, and covered it with magnificent monuments of their taste and wealth. In the irruption of the barbarous nations upon Rome, the Vandals were the first who seized upon Africa, where Genseric erected a powerful piratical dominion, whose fleets on one occasion took and sacked the capital of the Roman empire. All other occupants, however, were destined to give way before the torrent of Saracen conquest. In the 7th cent., Sidi Occuba, a lieutenant of Caliph Othman, conquered all this part of Africa, and erected it into a kingdom, of which Kairwan, or Cairoan, was the capital. After many revolutions, towards the end of the 12th cent., the dynasty of the Almohades was established in Marocco, and T. was governed by a viceroy. Under Charles V. the city of T. was the theatre of contention between the troops of that emperor and Barbarossa, the celebrated pirate. The Spanish expedition was successful; and a treaty was concluded, stipulating for a general liberation of the captives, and for a free intercourse between Tunis and the European states. This continued till 1574, when an expedition sent by Sultan Selim drove the Spanish garrison out of the Goletta, and annexed T. to the Turkish empire. It was governed for some time by its viceroys, called deys; but the people, or rather the soldiery, soon acquired the privilege of electing their own dey. The Algerines, in the course of the last century, wrested from it the fine prov. of Constantina.

TUNIS, the capital of the above state, is situated at the SW side of a large lagune, about 10 m. SW of the site of the ancient Carthage, of which it may properly be considered as the successor, in N lat. 36° 48′, E long. 10° 24′. It is surrounded on all sides, except the east, by considerable heights, but is so encircled by lakes and marshes that in such a climate it might be expected to be much more unhealthy than it appears to be. The city is large, being supposed to contain 12,000 houses, and 130,000 inhabitants, of whom 30,000 are Jews, and about 1,500 Christians. It is built in the most irregular manner, and the streets are so narrow and filthy that they can with difficulty be passed through. Large sums have been spent in the construction of forts, and in surrounding the city with a high and double wall, yet it is by no means a strong place. The citadel, called El-Gaspa, begun by Charles V., and finished by John of Austria, is commanded by the neighbouring heights; there is also a rising ground on the north of the town, which commands both it and El-Bardo, the fortified palace of the bey, 2 m. W of Tunis. There is one great mosque, and a number of smaller ones; a Roman Catholic and a Greek church, a theatre, several bazaars, and near the centre of the city a piazza of vast extent said to have formerly contained 3,000 shops for the sale of woollen and linen manufactures. The finest structure is the new palace. The Moorish houses, according to the custom of Barbary, are only one story high, with flat roofs, and cisterns for the purpose of collecting the rain water. The city is moreover well supplied from a neighbouring spring, which is conveyed into it by a fine aqueduct built in the time of Charles V. About 6 m. to the W of the city is the Goletta, the great naval and commercial depot of the state. A basin has been formed here, sufficiently spacious to receive all the vessels of war and merchant ships belonging to Tunis; and considerable quantities of timber are brought from Tabarca, for the purpose of ship-building. A large lake, separated from the sea by a very narrow isthmus, extends between T. and the Goletta. It is not more than 6 or 7 ft. deep, and appears to have been formerly dry. At present it affords an ample supply of salt, and serves for the conveyance of passengers and goods between the harbour and the capital. See article CARTHAGE.

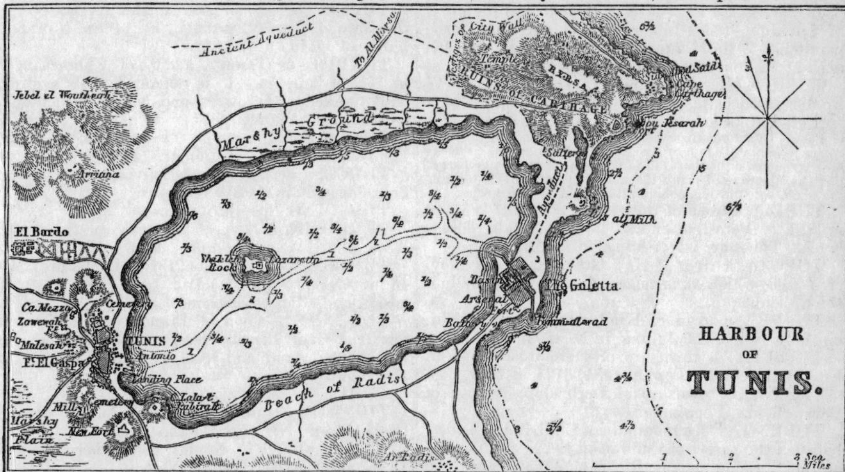

TUNICA, a county in the NW part of Mississippi, U. S. Area 600 sq. m. Pop. in 1840, 821; in 1850, 1,314.—Its cap. is Austin.

TUNJA, a town of New Granada, the capital of a district of the same name, in the prov. and 62 m. NE of Santa-Fe. It was formerly a very opulent place; and has several large churches and convents; but the present pop. of Tunja does not exceed 1,000. Its manufactures consist of coarse woollens and cottons, and formerly it had a great trade in tobacco, the quality of which was much esteemed, until it was suppressed in favour of the royal monopoly.

TUNKHANNOCK, a township of Wyoming co., Pennsylvania, U. S., 104 m. NE of Harrisburg. Pop. in 1850, 2,400. It is near the mouth of a stream of the same name, which runs SW into the E branch of the Susquehanna, about 35 m. above Wilkesbarre.

TUN-LEW, a district and town of China, in the prov. of Shan-se, and div. of Lu-gan-fu, in N lat. 36° 15′, and E long. 112° 46′.

TUNNUDTIORBIK, an island near the coast of E. Greenland, in N lat. 60° 45′.

TUNQUIN. See TONQUIN.

TUNSTAL, a parish in the co.-palatine of Lancaster, 3 m. S of Kirkby-Lonsdale, comprising the chapelry of Leck, and the townships of Burrow with Burrow, Cantsfield, and T. Area 9,224 acres. Pop. in 1831, 862; in 1851, 814.—Also a parish in Kent, 10 m. ENE of Maidstone. Area 1,196 acres. Pop. in 1831, 171; in 1851, 105.—Also a parish in Norfolk, 8 m. W by N of Yarmouth, crossed by the Norwich and Yarmouth railway. Area 1,612 acres. Pop. in 1831, 101; in 1851, 139.—Also a parish in Suffolk, 5½ m. S by W of Saxmundham. Area 2,863 acres. Pop. in 1831, 733; in 1851, 676.—Also a parish in the E. R. of Yorkshire, 14 m. E by N of Kingston-upon-Hull. Area 1,607 acres. Pop. in 1831, 172; in 1851, 159.—Also a township in Catterick p., Yorkshire, 4 m. SE of Richmond. Area 1,262 acres. Pop. in 1831, 312; in 1851, 328.

TUNSTALL-COURT, a chapelry and market-town in Wolstanton p., Stafford, 4 m. N by E of Newcastle-under-Lyne. The Grand Trunk canal passes within a ½ m. of the town, and the double tunnel, which runs for 2 m. under Hare-castle hill, is in the vicinity. The town is seated on the declivity of a considerable eminence, and contains some good houses. In the vicinity there are extensive potteries and large chemical works. Fine veins of coal, clay, and iron-ore, are found here. Pop. in 1801, 1,677; in 1831, 3,673; in 1851, 9,566.

TUNSTEAD, a parish in Norfolk, 9¼ m. NE by N of Norwich. Area 2,291 acres. Pop. in 1831; 498; in 1851, 449.—Also a small village in the p. of Tideswell, co. of Derby, 3¾ m. W by S of Tideswell.

TUNWORTH, a parish of Southamptonshire, 3½ m. SE of Basingstoke. Area 1,104 acres. Pop. in 1831, 122; in 1851, 133.

TUOLOMNE, a county of California, U. S., situated between the Stanislaus and Merced rivers, and intersected on the W by the San-Joaquin. Pop. in 1852, 17,657. Its cap. is Sonora.—Also a village in the same co., on the N side of the Tuolomne river, 30 m. above its junction with the San-Joaquin, and 78 m. SE of Vallejo.—Also a river of California, which rises in two head-branches, in the Sierra Nevada; and flows W into San Joaquin.

TUOSIST, or KILMICHAELOGUE, a parish in co.

Kerry, 11 m. SW by W of Kenmare. Area 39,341 acres. Pop. in 1831, 6,208; in 1851, 4,034. The summit of Hungry-Hill at the S extremity of the SE boundary, has an alt. above sea-level of 2,249 ft.

TUPADL, a village of Bohemia, 3 m. S of Czaslau, with manufactures of muslin and cotton stuffs.

TUPARRO, a river of New Granada, in the prov. of San Juan-de-los-Llanos, which rises under the parallel of 5° 10′ N, runs E, and falls into the Orinoco at Maypures, after a course of 200 m.

TUPHOLME, a parish of Lincolnshire, 6 m. W by S of Horncastle. Area 1,795 acres. Pop. 73.

TUPINAMBAS, a nation of Indians, in the Brazilian province of Para. They formerly dwelt opposite the islands of the river Amazon, from its mouth to the strait of Pauxis; and travellers have recounted a thousand fables of them. The greater part of them have now retired into the mountains and woods towards the south, very few remaining in the Portuguese territory.

TUPIZA, a town of Bolivia, in S lat. 21° 28′, 140 m. SSE of Potosi, at an alt. of 9,000 ft. above sea-level. There are silver mines in the vicinity.

TUPOZA, a river of New Granada, in the prov. of Ystmo, which runs nearly due W, and falls into the Chucunaqui.

TUPPEL, a town of Hindostan, prov. of Delhi, on the E side of the Jumna, in N lat. 28° 25′.

TUPSLEY, a township of Herefordshire, 2 m. ESE of Hereford. Pop. in 1831, 512; in 1851, 669.

TUPTON, a township of Derbyshire, 4 m. S of Chesterfield. Pop. in 1851, 270.

TUPUNGATO, a mountain of Chili, in the Andes, between the passes of Cumbre on the N, and Portillo on the S, rising to an alt. of 15,500 ft. above the level of the sea.

TUQUAQUE ISLANDS, some small islands off the coast of Venezuela, forming a small chain from Point-Seca to Point-Cabello.

TUQUEQUE, a river of Venezuela, a large and abundant arm of the Apure, from which it communicates with the Portuguesa, at a point 9 m. NNW of San-Fernando-de-Apure.

TUR. See THUR.

TUR, a town of Hungary, 24 m. SE of Szolnok, on the r. bank of the Beretyo.

TURA, or O'TURA, a town of Hungary, on the Waag, 23 m. N by W of Leopoldstadt.

TURA, a river of Asiatic Russia, which rises in the gov. of Perm, among the Urals, flows SE; traverses the districts of Turinsk, Teumen, and Yalutorosk; and after a course of about 260 m., falls into the Tobol below Jevlevo.—Also a small river of Irkutsk, in Asiatic Russia, which rises in the district of Nertchinsk, and falls into the Ingoda.

TURAH, a village of Kordofan, in the Habshadeh desert, 6½ hours SW of Kajebi.

TURANY, a town of Hungary, on the r. bank of the Waag, 26 m. NNW of Neusohl.

TURATE, a town of Austrian Italy, 6 m. S of Appiano, in the prov. of Cuno. Pop. of com., 2,200.

TURATTE, a town of the island of Celebes, 180 m. N of Macassar.

TURBACO, a village of New Granada, in the prov. and 12 m. SE of Cartagena. It stands at the entrance of a majestic forest, which extends toward the S and the E as far as the canal of Mahates and the river Magdalena. To avoid the excessive heats and diseases which prevail during the summer at Cartagena, and on the coast, it is the practice for those Europeans who are not seasoned to the climate, to retire inland to Turbaco during that season.

TURBAH, a town of Hindostan, in the prov. of Bahar, district of Chuta-Nagpore, in N lat. 22° 32′.

TURBENTHAL, a parish and village of Switzerland, in the cant. and 16 m. ENE of Zurich, near the r. bank of the Toss. Pop. 2,200.

TURBIGO, a town of Austrian Italy, in the Milanese, on the Naviglio-Grande, 18 m. W of Milan. Pop. 1,200.

TURBIO, a small river of Brazil, in the prov. of Rio-Janeiro, which runs W, and enters the Parana between the rivers Paranapana and Tiete.

TURBUT, a township of Northumberland co., in the state of Pennsylvania, U. S., 14 m. N of Sunbury. Pop. in 1840, 3,872.

TURBUT, or TARBUT, a city of Khorassan, 50 m. NE of Turshiz. It is defended with a strong wall flanked with towers, and is said to contain a pop. of about 18,000.

TURCKHEIM, a town of France, in the dep. of Haut-Rhin, 4 m. W of Colmar. Pop. 2,843.

TURCO, a settlement of Bolivia, in the prov. of Caranges, 120 m. WSW of Potosi.

TURCOING. See TOURCOING.

TURDOSIN, a town of Hungary, in the com. of Arva, 3 m. SW of Tistenna. Pop. 1,600.

TUREGANO, a village of Spain, in the prov. and 15 m. NNE of Segovia. Pop. 1,100. It has an old castle, and a large warehouse.

TURENNE, a town of France, dep. of Correze, near the river Tourmente, 9 m. S by E of Brives. Pop. 1,876. Iron and lead mines are wrought in the vicinity.

TURFAN. See TOURFAN.

TURGHE, a town of Hungary, in the com. of Czalad, 15 m. NE of Szala-Egaszeg.

TURI, a village of Naples, in the Terra-de-Bari, 18 m. SE of Bari. Pop. 3,500.

TURIAMO, a bay on the N coast of Venezuela, in the province of Caraccas, 3 leagues to the E of Porto-Cabello. It extends one league from N to S. It has no shelter from the N wind; and partly on this account, and partly on account of the sterility of the surrounding country, it is not much frequented by trading vessels.

TURIN, or TORINA, an administrative division of the Sardinian states, forming the greater part of Piedmont, and bounded on the N by Novara and Alessandria; on the E by Coni; on the S and W by France and Savoy. It comprises the 5 provs. of Biella, Ivrea, Pignerol, Susa, and Turin.—The prov. of T. is bounded on the N by Ivrea; on the NE by Novara; on the E by Alessandria; on the SE by Coni; on the SW by Pignerol; and on the W by Susa and Savoy. The territorial superficies of T. is 3,892 sq. kilom. Pop. 379,677. Ramifications of the Cottian Alps cover the N and W parts of the prov. The rivers are the Po and its affluents the Sangone, Dora-Riparia, Stora, Arca, and Dora-Baltea. This province is about equally divided by the Po. That portion which lies on the l. bank forms part of the plain of Lombardy; that on the r. is hilly and in part mountainous, with a soil less fertile, but in many parts fruitful in vineyards. One-ninth part of the prov. is covered with woods or forest, about four parts are employed in tillage and vineyards, three in meadows and grass-lands, which chiefly lie on the N of the Po; the remaining part is occupied by roads, buildings, gardens, &c. The mountains NW of the city of T. are rich in iron, copper, cobalt, and manganese.

TURIN, the capital and seat of the Sardinian monarchy, stands in N lat. 45° 4′, E long. 7° 40′, 75 m. WSW of Milan, and 320 m. NW of Rome, in a beautiful plain on the l. bank of the Po, which here receives the waters of the Dora-Riparia, and flows past with a copious stream, at a short distance from the walls. Its elevation above sea-level is 240 metres. It commands towards the S, W, and N views of unrivalled grandeur, embracing the whole chain of Alpine heights which bound the plains of Piedmont

On the E and SE it is sheltered by a cluster of richly wooded vine-clad hills, adorned with villas, temples, and palaces; and on the most prominent of these, about 5 m. from the city, stands the Superga, a royal mausoleum, whose splendour attracts the eye from distant points of the country. The city is of an oblong form, and its circumf., including the ramparts, is about 4 m. Its citadel and other fortifications placed it at one time in the rank of the strongest places in Europe; but its walls and bastions were demolished by the French after the battle of Marengo. The entrance of the city from the W is by a noble avenue, presenting an air of elegance which announces the seat of royalty; and the interior is worthy of the beauty of its situation. The modern streets are in general wide and straight, intersecting each other at right angles, and running in direct lines from one extremity of the city to the other: several of them have arcades or piazzas running along their sides, and affording a convenient walk for foot-passengers. The principal square, near the centre of the city, bears the name of Piazza-Reale, or Piazza-del-Castello, and ranks, both for size and beauty, among the most elegant squares of Europe. On one of its sides stands the royal palace; in the centre is a structure erected by the dukes of Savoy, commonly called the Castello. On three of the sides of the square are arcades, as in the palais royal at Paris. The Piazza-di-San-Carlo, though smaller, is entitled to notice, its façades being uniform, and its two longer sides having arcades supported by pillars. Of the streets of Turin, perhaps the finest is the Strada-di-Po, which stretches from the central square called the Piazza Reale, to the banks of the river. It is straight, broad, and bordered on each side with rows of arcades. The street called Contrada-di-Dora-Grande extends from the opposite side of the central square: it is equally straight as the Stradi-di-Po, and considerably longer, but neither so spacious nor so handsome. The houses in Turin are in general built of brick, and even the best are plastered in front with stucco. Like most towns which have been rebuilt, Turin has an old quarter, but it is very inconsiderable, hardly amounting to the sixth part of the whole. It is called Torino-Vecchio; and its streets, though less wide and handsome than those of the new town, are in general straight and regular. Of the public walks of Turin, the most frequented are the royal gardens: they form the morning rendezvous of the fashionable world. The terrace on the other side of the river, though commanding a view of great extent, is less frequented. The Rondo, extending between the city walls and the banks of the Po, is resorted to as an evening walk; while the Valentina, another promenade along the Po, about a mile from the town, is little visited on account of its distance. The materials of the public buildings are rich, consisting of marble of every vein and colour; but ornaments are perhaps too profusely scattered, and the city contains hardly one chaste model of architecture. The duomo or cathedral is an old Gothic edifice, remarkable for nothing but its marble cupola. The church of Corpus-Christi is very richly ornamented. Other churches claim attention only from their size, their pillars, or the variety of marble employed in their construction. A very handsome church was opened in Dec. 1853 for the worship of the Protestant party known as the Waldenses. The royal palace, situated in the central square, consists of three wings, surrounded by a court. Its extent is great, but in other respects it resembles the mansion of a rich individual, being built of brick, and covered with tiles. Its interior, however, is not without magnificence; and its galleries contain a number of good paintings of the Italian and Flemish schools. The Castello-Reale, situated in the midst of the square, has an elegant façade of the Corinthian order; but its other three sides are Gothic. The Palazzo-Carignano, a large building of some interest, is now occupied with government-offices. The university, a large and magnificent building, contains a court surrounded with arcades, the whole of which are covered with inscriptions and antique bas-reliefs. Its library contains 160,000 vols. The Palazzo-degli-Archivi-Reale is a vast edifice. The arsenal has a large room for arms, and workshops of some extent for the manufacture of fire-arms. The opera, or principal theatre, is of great size. In hospitals T. is richly endowed. The principal one, the Spedal-Maggiore, is on the plan of a celebrated hospital at Rome, affording both employment and support to its inmates, with education to the children. The Ospezio-de-Carita receives about 1,600 paupers. T. contains in all 14 parishes, and 31 parochial churches; besides which there are very many other churches and chapels, 13 religious houses and convents for men, and 9 for females; 16 pious and charitable institutions, and many hospitals for the sick, the lame, the blind, &c. There are fifteen establishments of public instruction, amongst which are the royal university of study, the royal academy of sciences, with the Egyptian museum, that of antiquities, and that of natural history annexed; the royal military academy, the college of the Jesuits (now suppressed), the archiepiscopal seminary, the ecclesiastic academy of the Superga, the museum, and theatre of anatomy, &c.; the cabinet of literature; the royal academy of painting, sculpture, and ornament; La Filarmonica, La Filodrammatica, the halls of trades and of agriculture and commerce, and the botanical and horticultural societies.—The pop. of Turin amounted in 1816 to 88,500; in 1820 to 90,000; in 1852, to 143,157. The character of the inhabitants is, like their dialect, Italian, with a mixture of French. Their dress differs little from that of the French, and their manners have received a polish from the long residence of the court; from which, and the consequent residence of the gentry, a number of inhabitants derive their support. "The people who pass you on foot are mostly of the lower order; the men dressed very much like French mechanics; whom they resemble, only that they are more powerful and muscular looking men—indeed, there is an annual migration of Piedmontese to Paris, where they act as porters—some of them indulging, however, in full suits of blue velvet; the women with gaudily-coloured handkerchiefs tied round their heads, and with red, yellow, and pea-green umbrellas under their arms, and gowns and cloaks of equally dazzling and forcible hues. The priests, conjoined with padres, make up a most formidable body. At every ten paces you meet, without the smallest exaggeration, at least two priests. The clerical costume is here in its greatest splendour. Nowhere else is the three-cornered shovel hat of such tremendous dimensions. In no other city are cassocks so flowing and eccentric in their cut; or coats so severe in architecture; or silver shoe-buckles and black silk stockings in such size and perfection. Gliding along, less frequently, come friars, Cordeliers, and Dominicans, with their shaven crowns and sandalled feet." The national religion of Piedmont is strictly Roman Catholic, but religious liberty is now tolerated, and there is in T. a large and respectable Piedmontese congregation of Primitive Christians and of the Swiss Protestant church, chiefly natives of the valleys of the Vaudois and of Savoy. All classes of the pop. are observant of their religious ceremonies. "They are," says a recent and most observant visitor, "an intelligent, orderly, and, generally speaking, a fine race, industrious, and attentive to their social and domestic duties, simple and courteous in their manners, sober, temperate, and frugal in their habits; and there is not, perhaps, a principal city in Europe which exhibits less appearance of immorality, or, in proportion to the pop., so small an amount of crime. I have not seen any capital so free from low, dirty streets, or from mean and miserable habitations; and among the lower orders of the people, drunkenness, abusive language, and disorderly conduct are rarely witnessed. There are so many sources whence the infirm and destitute receive assistance, and public begging is so properly discountenanced, that one sees in T. perhaps as little of the appearance of destitution and wretchedness as is to be found in any large and important city."

Silk forms the chief staple commodity of T., both in its raw and its prepared state; and the manufacture of that article is carried on to a large extent, particularly in damasks for furniture and drapery of all kinds, which are nowhere excelled in richness and quality. There are some inconsiderable fabrics of woollen, cotton and linen, plate and jewellery, iron and other hardwares, umbrellas and parasols; and the making of hats has here attained a degree of excellence vying even with Paris. The markets are well supplied with corn and grain, meat, poultry, cheese, butter, vegetables, fruit, flowers, &c. There is a special weekly mart for wine, to which from 100 to 150 hogsheads are usually brought, and there are also markets for hay, straw, wood, &c. The Po and other rivers produce good fish of various sorts, but fresh fish is generally an expensive article in Turin. Mutton is only supplied in the winter, when the flocks are brought down from the mountains. With vegetables and fruits, the markets are supplied in great profusion and fine quality. The price of labour in T. is, for the higher class of mechanics and artisans, from 3 to 4 francs per day, but the number is not considerable at the latter rate; for the second class, from 2 to 3 fr.; labourers from 1 to 1½ fr. per day. "The climate of T. is very healthy: there is a fine bracing clearness and elasticity in the air. It is often exceedingly hot during the day in summer, but towards evening a refreshing coolness is wafted from the mountains, and the nights are but seldom excessively hot. The Alps shut out the rays of the evening-sun from the plains, and close the daylight an hour before they leave the western side of those barriers. The vicinity of the capital to such a vast unbroken chain of mountains, most of which are shrouded in snow during six months of the year, occasions great coldness in winter."—A railway to connect T. with Novara is now in progress of execution; and it is in contemplation to connect Novara by railway with Milan or with Pavia.

History.] In early ages the province of T. was inhabited by one of the Ligurian tribes, called *Taurini*. They occupied both sides of the Po, and the country between that river and the Alps, as far eastward as the Orca; on the W their limit extended to Avigliana. After the conquest of Liguria by Augustus, a Roman colony was settled here, and the capital of the Taurini was honoured with the title of *Augusta Taurinorum*, whence it derived its present name—Torino or Turin. Its modern history is closely connected with the wars of Italy, of which its situation rendered it always the first theatre. It passed successively from the Romans to the Lombards, to Charlemagne, to the marquisses of Susa, and lastly to the princes of Savoy, who made it the capital of their states. It has frequently suffered from the rival pretensions of France, Spain, and Austria. In 1706 it was besieged by the French, but delivered by Prince Eugene. In 1796 it escaped, by a timely pacification, the attack of the French under Bonaparte. In 1798 it was occupied by a French army. In 1799 it was entered by the Austro-Russians, but was re-occupied by the French after the battle of Marengo in 1800, after which it remained in their possession until 1814.

TURIN, a township of Lewis co., New York, U. S., on Black river, 20 m. N of Rome. Pop. in 1840, 1,704; in 1850, 1,826.

TURINSK. See TOURINSK.

TURIRANA, a river of Brazil, in the prov. of Para, which, springing from a branch of the Tocantins, and running almost due N, serves as a line of division between the provs. of Para and Maranham, and falls into the sea opposite San-Juan.

TURIS, a village of Spain, in the prov. and 21 m. WNW of Valentia, near the l. bank of the Magro. Pop. 2,600.

TURKA, a town of Austrian Galicia, 28 m. SW of Sambor.

TURKAL, a town of Asia Minor, the ancient *Sebastopolis*, situated near a high rock crowned by an ancient fortress, 24 m. NW of Tokat.

TURK-DEAN, a parish in Gloucestershire, 2¼ m. N by W of Northleach. Area 1,890 acres. Pop. in 1831, 237; in 1851, 278.

TURKESTAN, a name often given to an extensive region of Central Asia, the original or acquired seat of the great Tartar race called Turks or Toorks, who now occupy the greater part of what is called Independent Tartary. See article TARTARY. The region north of the Oxus, corresponding to the *Sogdiana* of Ptolemy, is described in our columns under the title of GREAT BOKHARA.

TURKESTAN PROPER extends along the r. bank of the Syr, below Syram, and between it and the Aral-Nor, and stretches considerably to the N along the banks of some small rivers which come from the E and from the N. It was a rich and flourishing country previously to the invasion of Jenghiz-Khan, and included at that time the whole basin of the Syr; but the name is now confined to the NW angle of that once extensive region, now subject to Khokan. In the time of Sultan Baber, it was the seat of the Usbecs, whence, under Sheibani Khan, they came and drove Baber, and all the princes of the house of Timur, across the Syr, the Amu, and the Hindukhush. The chief rivers are the Sarasu, the Arj, the Bulat, the Talash, and the Turugay. All of these have been made to fall into the Syr; but the Sarasu is now found to lose itself in a small lake, after running a long SW course through the Kirghisian steppe. The waters of this lake sometimes communicate with those of the Tele-kol, in 45° N lat. and 67° 25′ E long. The Turugay seems to have been mistaken for the *Turgai*, which runs S from the same range as the Sarasu, and falls into the lake of Ak-su-kol or Bialawoda, to the N of the Aral-Nor. The Talash or Taraz, which originates on the NW slope of the Kinder-Tau, after running for some space NW and then SW, falls into the lake of Sakshum. Otrar, Jund, Saganak, Sabran, Yassi, Jenghikant, Osbanikhat, Esfijab, Jekel, Shalj, Taras, and Balasagun, all famous cities in this region, have now and for centuries disappeared. The only city mentioned by Nazaroff is Souzak, the frontier city on the side of the Kirghisian desert. It is said to consist of 500 stone houses, built close to each other, situated on an eminence, and surrounded with a stone wall. The inhabitants are stated to be industrious, and are employed either in agriculture or in commerce with the Kirghis, who inhabit the banks of the Sarasu and the Tchoui. Taraz, on the Arj river, 70 m. to the NE of Otrar, seems to be the same with the city called Toorkistaun by De Lisle, Strahlenberg, and others. It is situated 15 m. from the Syr, and contains about 1,000 brick houses.—NORTHERN TURKESTAN is described in this work under the title KIRGHIZ KAZAKS.—EASTERN TURKESTAN is synonymous with the LOWER BOKHARA of modern geographers. See that article.

TURKEY,

Or the OTTOMAN EMPIRE, nominally includes a large portion of the continents of Europe and Africa, whilst a most important part of the sultan's dominions is situated in Asia. Hassel calculated the superficial extent of European Turkey, according to the maps of Reichard and Riedt, at 189,925 sq. m., which is nearly in accordance with Arrowsmith and Lapie's maps. Pinkerton estimated the total extent of European Turkey, including Greece and the Morea, at 182,560 sq. m. Balbi distributed it thus:

The government of Romelia,	107,575 sq. m.
——— ——— - Bosnia,	18,890
——— ——— Kapudan Pacha,	16,589
——— ——— Kirid,	4,235
Wallachia,	25,231
Moldavia,	17,400
Total,	189,920

In the supplement to the *Encyclopædia Britannica* the total extent of European Turkey was estimated at 180,074 sq. m. Dieterici, in 1853, assigned 153,721 geog. sq. m. to the Turkish possessions in Europe; but of these, 49,021 sq. m. were only mediate possessions, or placed by their organic relations more or less under foreign influence.

Graberg estimated the superficial extent of the whole empire, previous to its recent losses, as follows:

European dominions,	206,707 sq. m.
Asiatic,	542,700
African,	291,600
Total,	1,041,007

Stein and Lichtenstern calculated the total superficies thus:

The Ottoman empire in Europe, excluding Moldavia and Bessarabia,	207,562 sq. m.
Asiatic dominions,	545,895
African,	197,887
Total,	951,344

Dieterici's estimate of the area of the Turkish empire, in 1853, is as follows:

European dominions,	153,721 geog. sq. m.
Asiatic,	505,308
African,	719,328
Total,	1,377,357

Previous to the revolt of the Greeks, European Turkey formed nearly the fourth part of the Ottoman empire; and among European states was the sixth in rank with regard to territory, and the ninth in respect of population,—being generally estimated to contain about 100,000 sq. m., with a pop. of above 11,000,000. The loss of the Morea, Livadia, and the Cyclades,—an extent of territory about as great as Portugal, or Denmark with Holstein,—reduced the Turkish dominions in Europe about one-seventh, and their population one-eighth. The emancipation of Macedonia and Epirus—which has been proposed by some—with the loss of Southern Greece, would diminish the possessions of the sultan in Europe nearly one-third; and the often-threatened defection of Bosnia, Servia, and Turkish Illyria, with the loss of Wallachia and Moldavia, would confine the Ottoman empire in Europe to the two provinces of Bulgaria and Romelia. These last do not contain above 75,000 sq. m. of surface, and 7,000,000 of inhabitants, including the pop. of the three cities of Constantinople, Adrianople, and Sophia; and of this pop. at least one-fourth of the people are in heart enemies to the religion of Mahomet. In Africa and Asia the recent losses of the Turkish sultan have been still greater than in Europe: Algiers is gone, and the remaining Barbary states yield a scant obedience Mehemet Ali, the sultan's viceroy in Egypt, not only wrested the virtual dominion of that country from his master, but actually brought under his sway the four pashalics of Syria, and the province of Adana, and would have effectually dismembered the empire but for foreign interference. Yet this feeble empire "is constituted of territories the most opulent, extensive, and imperial ever known since the days of Rome. Three quarters of the globe contribute their fairest regions to its dominions. Its possessions include the richest provinces of Asia, Africa, and Europe. The superficial extent of its territory is more than double that of Great Britain and France united, and this not constituted of swamps or solitudes, but of the most renowned and productive countries under the sun. Under the sway of the government is comprised half the great basin of the Mediterranean. All the most famous cities of Asia, all the seats of ancient power and greatness —Tyre and Sidon, Antioch and Jerusalem, Damascus and Bagdad, Babylon and Palmyra, together with all the islands and seaports of that magnificent region, form but one half of the Turkish empire. In Africa it possesses the land of Egypt, itself equivalent to a kingdom. In Europe it pushes its dependencies from the waters of the archipelago up to the latitude of Paris, and as a centre of this prodigious dominion possesses a capital so wonderfully endowed with every conceivable advantage of strength and situation that conquerors of all ages have been lost in admiration of it, and Napoleon in the height of his glory spoke of Constantinople as equivalent to it all. Judged by resources, population, and means, the Turkish empire ought to be the strongest Power on the face of the globe."

Administrative divisions.] The territories of the Ottoman empire are divided into *ejalets* or *eyalets*, the governors of which have the title of *vali*, equivalent to 'viceroy.' Each ejalet is subdivided into *sanjaks* or *livas*, which are each placed under the superintendence of a *kaimakam* or lieutenant-governor. The livas are divided into *cazas* or districts; and the cazas into *nahiyas* or villages. These jurisdictions disregard geographical boundaries. For example, the government of the Djizairs, or 'the Isles,' which belong to the Capitan-Pasha, not only embraced the Turkish isles of Europe and of Asia, but the Morea, the province of Gallipoli, and the coast of Smyrna: the sultan having thought proper to place under the immediate authority of his grand-admiral all the countries which were not accessible but by sea, or which, by their position could contribute to the armament and supply of the fleet. At the beginning of the 17th cent., when the Ottoman empire embraced in its limits the greater part of Hungary, Transylvania, Circassia, and Azerdbijan, there were 44 ejalets and 220 livas: at present not more than half of those governments belong to T.; and the limits of each province have been greatly changed.

I. TURKEY IN EUROPE comprises 15 ejalets, viz.:

1.	Ejalet of Edirne,	in Thrace.
2.	Silistré,	Bulgaria.
3.	Boghdan,	Moldavia.
4.	Eftak,	Wallachia.
5.	Vidin,	Bulgaria.
6.	Nich or Nissa,	Bulgaria.
7.	Uskup,	Albania.
8.	Syrp,	Servia.
9.	The fortress of Belgrade.	
10.	Ejalet of Bosna,	Bosnia and Croatia.
11.	Rumili,	Albania and Macedonia.
12.	Yania,	Epirus.
13.	Selanik,	Macedonia and Thessaly
14.	Djizair,	The archipelago.
15.	Cryt,	Crete

The area of European Turkey, according to Dieterici, is as follows:

1. Rumili or Rumelia,	46.119·04 geog. sq. m.	
2. Bosna or Bosnia,	20.281·44	
3. Silistré or Silistria,	27,469·28	
4. Djezair,	9,289·92	
5. Wallachia,	21.283·52	
6. Servia,	15,967·36	
7. Mŏldavia,	11,770·88	
	152,721·44	

II. TURKEY IN ASIA comprises 18 ejalets, viz.:

1. Kastamuni,	in *Paphlagonia.*
2. Khudavendiguiar,	*Bithynia.*
3. Aydin,	*Lydia.*
4. Karaman,	*Phrygia* and *Pamphylia.*
5. Adana,	*Cilicia.*
6. Bozoq,	*Cappadocia.*
7. Sivas,	do.
8. Tharabezun,	*Pontus* and *Celebes.*
9. Erzerum,	*Armenia.*
10. Mossul,	*Assyria.*
11. Kurdistan,	
12. Kharbrut,	*Comagene.*
13. Haleb,	*Syria* and *Orsoene*
14. Saida,	*Phœnicia* and *Palestina.*
15. Cham,	*Syria.*
16. Bagdad,	*Babylonia*
17. Habesh,	*Arabia* and *Ethiopia.*
18. Haremi-Nabevi,	*Arabia.*

The area of the Turkish provinces in Asia is thus estimated by Dieterici:

1. Asia Minor,	158,456·16 geog. sq. m.
2. Armenia and Kurdistan,	36.821·76
3. Syria,	33,918·88
4. Irak and Mesopotamia,	124,311·52
5. Arabia,	145,800
	505,308·32

III. TURKEY IN AFRICA is divided into 3 ejalets, viz.:

1. Missr	or Egypt,	434,672 geog. sq. m.
2. Tarablusi-Gharb	Tripoli,	225,296
3. Tunis	Tunis,	59,360
		719,328

The Turkish possessions in Europe generally pass by the name of RUMILI, and those in Asia by that of ANADOLU; though, properly speaking, Rumili is but an ejalet of Albania and Macedonia, while Anadolu means only that part of the Asiatic provinces in which the Turkish and Arabic are spoken. The African possessions are called GARB. Michelson estimates the area of the Ottoman empire at only 913,000 sq. m. The pop. was returned in 1844 as follows:

I. EUROPEAN TURKEY OR RUMILI.

Thrace,	1,800.000	
Bulgaria,	4.000,000	
Moldavia,	1,400.000	
Wallachia,	2,600.000	
Bosnia and the Herzegowina,	1.400,000	
Rumelia,	2,600,000	
Servia,	1,000,000	
Islands of the Archipelago,	700,000	
		15,500,000

II. ASIATIC TURKEY OR ANADOLU.

Asia Minor,	10,700.000	
Syria, Mesopotamia, and Kurdistan,	4,450,000	
Arabia, (Mecca, Medina, Habesh),	900.000	
		16,050,000

III. AFRICAN TURKEY OR GARB.

Egypt,	2,000.000	
Tripoli, Fezzan, Tunis,	1,800,000	
		3,800,000
		35,350,000

Dividing the above population into races and tribes, Michelson arranges them thus:

Race or Tribe.	In Europe.	Asia.	Africa.	Total.
Ottomans,	1,100,000	10,700,000	..	11,800.000
Slavonians,	7,200,000	...	...	7,200,000
Rumanians,	4.000,000	...	...	4.000.000
Arnauts,	1,500,000	...	...	1,500,000
Greeks,	1,000.000	1.000,000	...	2,000,000
Armenians,	400,000	2,000,000	...	2,400.000
Jews,	70,000	100,000	...	170.000
Tartars,	230,000	...	...	230,000
Arabs,	...	900,000	3,800,000	4,700,000
Syrians and Chaldeans.	...	235,000	...	235,000
Druses,	...	25,000	...	25.000
Kurds,	...	1,000,000	...	1,000.000
Turkomans,	...	90,000	...	90,000
	15,500,000	16,050,000	3,800,000	35,350,000

Taking the pop. according to religious creeds, the result according to Michelson is:—

Creeds.	In Europe.	Asia.	Africa.	Total.
Mahommedan,	3,800,000	12,950,000	3,800,000	20,550,000
Greeks and Armenians,	11,370,000	2,360,000	...	13,730,000
Roman Catholics,	260,000	640.000	...	900.000
Jews,	70,000	100,000	...	170,000
	15,500,000	16,050,000	3,800.000	35,350,000

These details may be compared with those given in a subsequent paragraph of this article.

Physical features.] European Turkey forms, with Greece, a large triangular peninsula, which in ancient times may have been connected with the continent of Asia: but if ever so, has been disjoined from it by some tremendous convulsion. Excluding the two principalities of Moldavia and Wallachia, the Danube, the Save, and the Unna, form the base of the triangle, and the northern boundary of the country; while Cape Matapan, in the Morea, is the apex. Its sides, excepting a small portion where the narrow strip of Dalmatia intervenes, are washed by the Euxine, the Ægean, and the Adriatic seas. The coasts, from the Cataro on the W, to the mouths of the Danube on the E, present a highly irregular configuration, being deeply indented with gulfs, between which the salient points of mountain ridges extend into the sea. The interior is everywhere intersected by long chains of mountains generally running parallel to the base, or from E to W. On the N side of the Danube, one continuous plain extends to the Carpathian mountains and the river Pruth. The declination of the whole country between the Balkan and the Carpathian mountains is towards the Black sea: the peninsular portion of Turkey declines mainly towards the Ægean. From an Alpine central plateau, a region of wild and romantic beauty, the great mountain-chains of T. run off in different directions towards the surrounding seas. Of these, the range proceeding eastwards forks into two main branches, the Balkan, and the Despoto-Dagh. The Balkan, Emineh-Dagh, or Hæmus, forms the first great mountain barrier of Turkey on the N,—and runs ENE between the 42d and 43d parallels, from the termination of the Dinaric Alps, near Perserin, and the sources of the Ibar and Vardar, to the Black sea, a distance of nearly 500 m. Its numerous ramifications intersect in various directions the whole of European Turkey S of the Danube. Except in a few places, the whole extent of the ridge is impassable: steep precipices, rugged and abrupt ascents, lofty rocks, and impending crags, render the general face of the mountains so difficult, as to repel all attempts to climb them. The chain may be said to consist of three branches: two lower ridges rising at each side parallel to the great one. The intervening valleys are exceedingly beautiful; they form extensive sequestered tracts, shut out, as it were, from the rest of the world, and abounding in every production that the fecundity of nature could supply, or the most elaborate industry produce. See article BALKAN. The Despoto-

Dagh, the *Rhodope* of the ancients, strikes off from the main chain at Dubnitza; runs SE towards the sources of the Arda, around which it takes a semicircular sweep; and then runs nearly E to Cape Makri. From Mount Scardus or Argentaro, one of the highest summits of the Balkan, and which may be regarded as the great culminating point of the whole mountain system of Turkey, a chain runs S and SE, which bore the ancient names of *Pindus*, *Cithæron*, and *Parnes*, and terminates in the *Sunium promontorium*, now Cape Colonna. This great chain divides the northern continent of Greece into two divisions of nearly equal breadth, and gives birth to all the most considerable rivers of that country which flow off, on its opposite sides, but in no instance cross it. The Dinaric Alps, which connect the mountains of T. with the vast system of the Alps, lie to the NW of the Balkan, and run through the Turkish provinces of Albania and Bosnia into the Austrian states. They divide the tributary streams of the Danube, or more properly of its ally the Save, from those rivers which flow towards the NE coast of the Adriatic. The range detaches itself from the Julian Alps at Mount Klek near the sources of the Kulpa, in N lat. 45° 28′, and running SE to the sources of the Kerka, intersects Military Croatia, and touches the frontiers of Dalmatia. From the head of the Kerka to that of the Verbas, the chain runs E through Bosnia, where it receives the denomination of Chator and Salliava. Turning SE towards the Bosnia, it forms in this part of its course Mount Ivan, and, running on towards the Drin, forms Mount Zamora. It then separates Bosnia from the sanjak of Scutari, and joins the Balkan near Perserin. This chain is known also by the names of Baba, Rachka, Bori, and Djamus-Dagh.—The Carpathian mountains run betwixt the frontiers of Austria and Turkey. Near the sources of the Alauta, the Eastern Carpathian ridge divides into two branches; and that branch which runs along the l. bank of the river forms the boundaries betwixt Transylvania and the Turkish principalities of Moldavia and Wallachia. The principal summits which rise here are the Nagy-Hagymas, the Kaszony, the Piatra-Laptuie, the Tatara, and the Magura. From Pojana-Mujeri to their termination on the Danube, the Carpathians run W by SW through Transylvania and the Banat, and along the Turkish frontiers. The Mali-Stirbacz belonging to this section of the chain, which reaches the Danube and causes that river to form a cataract, seems to be the connecting link betwixt the Balkan and Carpathian systems. See CARPATHIANS. — The Hellespont or Straits of the Dardanelles, connects the Ægean sea with the sea of Marmora, and separates Anatolia, or Asiatic Turkey, from Aktchai-Ovassi, or the Thracian Chersonnesus in European Turkey. See article DARDANELLES. The sea of Marmora, which is about 120 m. long, and in some places 40 m. broad, is connected by the Thracian Bosphorus with the Euxine, or Black sea, the *Pontus Euxinus* of the ancients, which washes the shores of Romelia, Bulgaria, and Anatolia; and the Black sea is connected with the sea of Azof by the Cimmerian Bosphorus. See articles AZOF, BOSPHORUS, BLACK SEA, and MARMORA.

The basin of the majestic Danube includes more than a third part of Turkey in Europe. It receives the Aluta or Alt in Wallachia, the Dumbovitza, the Jalonitza, the Sereth, the Pruth, the Save with its tributaries, and the Morava. The possession of Bessarabia has given to Russia the command of the mouths of the Danube; but a short and by no means difficult cut of some 30 m. from Risovata to Kustendje, would remove the mouth of this important river far from Russian obstruction, and at the same time shorten the navigation of that outlet for the industry of Central Europe, by more than 250 m., and bring its mouth, and consequently all its tributary streams, 150 m. nearer to Constantinople. The free navigation of the Danube is a point of great importance to the leading European powers. Austria, Russia, and the Porte, are the Powers chiefly concerned in the navigation of the Danube, but it is likewise connected, directly or indirectly, with the interests of so many other countries, particularly Germany and England, that the strongest co-operation of these Powers is unquestionably demanded to secure the unfettered navigation of this stream against every possible encroachment. By different early treaties connected with the Porte a perfect freedom of navigation and trade, whether the latter is carried on with domestic or foreign vessels, on the portion of the Danube which is comprised within the limits of the Turkish territory, is secured to the Austrian empire. All the subjects of Austria can within the Danubian provinces of Turkey pursue their commerce free and undisturbed, besides enjoying many special privileges. The consequence is, that in Galatz, one of the most important havens in the lower part of the Danube, one Austrian settlement has followed another, till at last it almost appears to be an Austrian place. While Austria thus took care to secure a point near the mouth of the Danube which might serve as a medium for connecting the trade of her eastern provinces with the Black sea, the Russians chiefly directed their attention to prevent the progress of navigation on the Danube from taking a direction injurious to their interests, and, indeed, to prevent the foreign trade of the principalities from being entirely absorbed by the Austrians, English, and others. To this end they wisely enough had the mouths and delta of the stream ceded to them by the peace of Adrianople. The same treaty also provides that the r. bank within the delta, so far as the same is under the Ottoman dominion, shall not be inhabited for a mile within land. In earlier times the mouths of the Danube were far more numerous than at present. Pliny mentions six: now there are but three, and even these are beginning to be more or less choked up with sand. The northern mouth has always been the least used; the middle or Sulina mouth, is the only one deep enough to admit freighted vessels of large size; while the southern arm, or that of St. George, was stopped up by the Russians. See article SULINA. The Danube forms the boundary between the body of the Turkish empire in Europe, and the principalities of Wallachia and Moldavia, which during four centuries have acknowledged the sovereign dominion of some strong power on the east or west of them. Upon the last extension of the Turkish empire in Europe, when the banner of the prophet waved in sight of the spires of Vienna, they accepted Turkish protection. Towards the close of the last cent., the sultan claimed the right of appointing the prince, called the *hospodar*, and thereafter the subjection of these parts of the empire to the Porte became almost absolute.—The basin of the ancient *Hæmus*, now the Maritza, occupies the greater part of Romelia. It rises in the Mount Rulla belonging to the Balkan chain, and runs SW to Edrene, where it turns S and falls into the gulf of Enos. Its principal tributaries are the Stanimak, the Usundscha, the Arda, the Raska, the Tundscha, and the Erkene. The White Drin and the Black Drin uniting their waters in the sanjak of Scutari, form the Drin, Drilo, or *Drinus*.—The Pruth divides the Russian prov. of Bessarabia from the principality of Moldavia, and constitutes the SW boundary of the Russian empire, from its source in the Carpathian mountains to the point where it empties into the Danube, about 75 m. from the embouchure of that river into the Black sea.—The European part of the Turkish empire enjoys a climate superior to that of almost every other European region. The seasons succeed each other with the greatest regularity; and the atmosphere is salubrious and friendly to the human constitution. In several of the large cities, indeed, the plague is a frequent visitant; but this dreadful distemper—if it originate in the nature of the corrupted atmosphere in large cities—owes its propagation and its augmented virulence to the notions of the Turks, who imagine that every precaution to avoid it is needless, and who consequently expose themselves without scruple to infection. In some quarters, the NE winds bring with them the atmosphere of Siberia; winter is colder than in any part of Great Britain, and sudden depression of temp. is frequent at all seasons. In Wallachia and Moldavia, the winter resembles that of Russia, and for some months the sledge takes the place of the wheeled carriage. Those parts of Albania which are protected by high mountains from the chilling NE winds, enjoy a delicious climate. The naked and rocky valleys of the interior, indeed, are often rendered almost uninhabitable by excessive heat; but the maritime districts, while reclining under sunny skies, are fanned by the Adriatic. The amenity of these shores is, however, counterbalanced by their liability to earthquakes.

Soil and productions.] The soil is for the most part remarkable for its fertility,—producing spontaneously most valuable grains and fruits, and being capable of producing crops sufficient for the sustenance of a much greater number of inhabitants than it at present supports. Agriculture, however, is yet in a rude state, and industry finds a thousand obstacles thrown in her way. The Oriental is slow to admit of change, especially when the change affects his daily life and daily occupation: he obstinately refuses to adopt any method of improving the soil. To marl or manure it, to grow crops in rotation, and to substitute new for effete kinds of fruits or vegetables, would never enter his thoughts. There is also a deficiency of hands to perform the operations of farming; and the farmer, narrow-minded and bigoted, occupying an exhausted soil, and often at a loss for labourers on whose efficiency he can rely, sinks into a state of poverty which becomes a fatal bar to any improvement of his condition. When he wishes to sow his land, when his store of produce is exhausted or taxes are to be paid, he is compelled to borrow money at 20 or even 30 per cent., on the security of the coming harvest; and even if he were to keep clear of usurers, and to secure the assistance of sufficient hands, he would probably gain nothing more than the power of raising an ampler maintenance for himself and his neighbours. T. has no roads. The old roads are impassable, and the means of transporting goods to and from the coast are confined to the camel and the mule, a mode of conveyance so expensive that only the more valuable goods can bear the cost. The roads are generally mere beaten pathways, made by one horseman and followed by another, and every man may make one for himself if he pleases. The only carriages are wooden planks laid upon rough wheels called *arubas*, drawn with cords by buffaloes which are seldom used except for burden. "Property in Turkey," says Mr. Macgregor in his report on the Turkish tariff, "was held after its conquest by the military chiefs and soldiery, among whom it was distributed. The *spahis* are still the chief feudal lords who hold the *zaimets* or estates. Of these there were at the time of Solyman the Magnificent, 3,192 large zaimets, and 50,150 small zaimets, less in extent than about 500 acres; in 1816 the number of feudal estates in European Turkey was stated to be under 1,000, and in Asia Minor under 1,500. Exclusive of these, there are large estates which either belong positively, or pay a heavy quit-rent to the sultan or to the church. The pashas and their fiscal officers, who are the scourges of the feudal proprietors, extort from them and from the cultivators of the soil, the taxes either directly or through the municipalities. The consequence has been that agriculture and agricultural instruments in T. are generally in the most rude condition. In Thrace, cultivation is in a better state than elsewhere south of the Balkan. The valley of Maritza is fertile, producing rice, maize in plenty, wheat, millet, onions, artichokes, vines, various fruits and vegetables. Wheat, oats, barley, buckwheat, beans, turnips, flax, madder, tobacco, cotton, saffron, olives in the southerly parts, and various other crops, are grown generally, of various qualities and quantities of production, in all European Turkey S of the Balkan. The red wines of Macedonia and Albania would be good, or at least rich in quality, if prepared with care; but as made at present, they are either sweet or sour, never dry, and to increase the quantity, often mixed with water. Notwithstanding the rude state of agriculture in European T., there is less poverty than, or not so much destitution as to the means of sustaining life as, in many other European states. The absence of beggary, and the actual means of enjoying their daily bread, is conspicuous in Roumelia, Bulgaria, Macedonia, and Albania. The latter prov. is perhaps the most independent in this, as well as in other respects, of all European T., although Bulgaria is by far the best cultivated." In the northern provinces the pasture is luxuriant, and wheat might be raised in almost any quantity.—One of the principal articles of export from T. to Great Britain is grain, but it was not until 1842 that the Turkish government permitted the shipment. Between that year and 1848, the increase in the exports of Indian corn from Galatz was from 597,062 qrs. to 1,270,745 qrs. or 110 per cent. The quantity of wheat exported from Ibraila during the same period increased from 667,909 qrs. to 1,862,909 qrs. or 180 per cent. The increase in the exports of Indian corn from the same port was from 224,310 qrs. to 1,448,619 qrs. or 545 per cent. Some opinion may be formed of the extent of the agricultural resources of T. when such results have been accomplished within the last few years. The growth of maize in the principality of Moldavia was at first a mere incidental circumstance, and the wealth derived from this source has been annually increasing for several years. The exports of grain from T. to this country have increased considerably, as the following official statement of the exports of grain from T. to the United Kingdom shows:

	Wheat.	Other grain.
1846,	33,725 qrs.	45,480 qrs.
1847,	142,899	773,582
1848,	23,169	393,749
1849,	163,387	585,717
1850,	135,558	358,475
1851,	339,939.	759,240
1852,	127,470	787,117
1853,	478,486	930,704

We doubt whether any other state in Europe can show a more rapid advancement in its agricultural resources than T. in the same period of time. The same may be said of Egypt, well known in ancient history as the most productive corn-growing country in the world. In Romelia the soil is of the richest description, especially from Philippoli to Adrianople, and thence to the Black sea; but the population is not a tenth of what would be necessary for its efficient culture, and it is only in the vicinity of towns and large villages that the land is manured. The operations of husbandry are for the most part performed by Bulgarians, who descend from the neighbouring mountains for a few weeks in the spring and autumn. From a want of the means of transport, no grain is brought to Constantinople from those fertile regions extending round its gates, although 300,000 qrs. of Russian grain are in some years imported into the capital. Next in extent to the plains of Romelia are the arable flats of Bulgaria, on a peninsula formed by the Lower Danube and the Black sea. Here the prospect is more encouraging, for the peasantry are better tillers of the soil, and the district lies open to European enterprise. Hence this division of T. has the most benefited by the treaty of 1838. During that year and the first six months of 1839, only 6,000 kilos were exported from Varna, Beljik, Kavarna, and Kostendji; but when a stop had been put to the extortions of Turkish *employés*, who monopolized trade, and were empowered to fix the price at which the peasant should furnish grain, many European agents became established at the Bulgarian ports, and within three years the exports augmented more than a hundredfold. The export of 1843 was 390 cargoes; that of 1845 about 800 cargoes. The grain of this district is now equal in hardness to any from Taganrog. In the mountain-frontiers of Bulgaria, Bosnia, and Albania, T. actually possesses but few resources.—Opium is one of the most important productions of T. It is the juice of the black poppy which is sown in November and December; and in June forms a ball which contains the seed. In these balls incisions are made, from which oozes out a milky substance, which is collected gradually, and formed either into cakes about the size of a biscuit, or balls as large as a four-pound shot, and is sent to Smyrna about the end of July. A good crop will yield 1,500 baskets of from 85 to 90 cakes each and an ordinary one from 1,000 to 1,200. Of this quantity it is known with certainty that no more than 200 baskets are used in the Turkish empire, so that the practice of chewing opium, though still considered general here, is less universal than would be imagined. The best qualities are exported by the English and Americans for their separate speculations to China, and various parts of the East Indies.—The olive crop is of great importance in T. The average supply of oil in T. may be estimated as follows, viz.: 1st, Mitylene and the Main northerly, 450,000 kintals; 2d, Smyrna and Anatolia southerly, 250,000 k.; 3d, Salonica, Volo, &c., 170,000 k.; 4th, Candia and the other Turkish islands, 200,000 k.; total kintals, 1,070,000, or about 56,000 tuns. Of this quantity the annual export to Europe and the Black sea does not exceed 250,000 to 300,000 k.—Madder is largely grown for export, and raw silk, and cotton.

Animals.] The Turkish dominions are not deficient in those kinds of cattle and beasts of burden which are common in other parts of Europe. The horses are improved by the Arabian breed; and are consequently elegant of form, spirited, and active

Mules and asses are much used; and the breed of both is more perfect than anywhere else in Europe. Cattle are not scarce; but the beef is said to be inferior to that of countries farther to the W. Sheep of various kinds are reared, and the mutton is said to be excellent. The camel is common in the southern provinces. European T. has few wild animals: the jackal is said to be the most common. The species of birds and fishes are numerous; but so little is known of the natural history of these provinces, that little can be said concerning them. The Bosphorus swarms with myriads of the finny tribe, the most ordinary of which are *scombri*, a species of mackarel, which are dried without salt by the Greeks; *palamedes* and *stavidria*, two species of dolphins; and anchovies and *nilufir*, which latter are caught by torch-light, on their migration from the Black into the White sea, during the autumn. The Bosphorus is, at times, enlivened by the gambols of shoals of dolphins, whose effigies are extant on the ancient Byzantine coins.

Minerals.] We are informed by ancients writers, that several parts of that territory which is now under the dominion of the Turks formerly had mines of gold and silver; and these metals might, perhaps, still be discovered, were the industry of the Turks equal to the search, but at present the only metals wrought appear to be copper and lead. Mineral deposits of immense value are found in Rumelia and also in Thessaly, consisting, for the most part, of galenas very rich in silver, and also in copper, iron, and alumstone. Gold intermixed with water-worn rubies occurs in the ferruginous sands of the valley of the Ardagh, which unites with the Maritza, about 2 m. NW of Adrianople. The coal formations in the neighbourhood of Adrianople are very extensive, and also near Haskoe, about midway between Adrianople and Philippoli. In the immediate neighbourhood of Adrianople the coal is of an intermediate quality, between lignite and true coal, but near Haskoe it is reported to be fully equal to good English. The mineral deposits in Thessaly are unbounded, particularly throughout that part which forms the peninsula between the gulf of Volo and the Grecian archipelago. This peninsula is an elongation of the southern arm of Mount Olympus, and embraces Mount Ossa and Mount Pelion. Mount Pelion and its slopes are also reported to present a mass of mineral wealth composed of alternating strata of crystalline limestone and schist, intersected by countless beds and veins of galena, rich in silver, copper, arsenic, antimony, iron, and a little gold.

Manufactures.] "The manufactures of T.," says Mr. Macgregor, writing in 1836, "are more important for domestic use than for exportation. Among the most distinguished are the carpets of Smyrna; crapes and gauzes of Salonica,—silks of Brussa,—printed muslins and cottons of Constantinople,—tanneries and dressed leather of Adrianople,—coarse cottons and other coarse cloths of Roumelia, Thessaly, Albania, and even of Bosnia,—various works in metal, and gold and silver lace,—some works in hair and linen,—the camlets and serges of Angora and the works of smiths and ordinary handicraft, comprise most of the manufactured products of the empire. The iron-mines of Bosnia are worked to some extent for the use of that province, where arms and some iron and copper works are made. The mines of Samakoff, near Philippoli, have recently produced cheap common iron. It would appear that not only the manufactures of cotton and silk, which are stated to have been formerly of valuable importance, have greatly declined, but that the Turks have neither taste nor inclination for the arts nor for manufactures. As the population of Armenian, Greek, and Frank races are increasing rapidly, while those of the Turkish race appear to have been diminishing in numbers; and should there be security, hereafter, for investments in manufacturing industry, we see no obstacle to the growth and prosperity of various fabrics in Turkey, excepting, that it will long be more profitable for the several Ottoman provinces in Europe and in Asia to exchange their agricultural products, and raw materials for the manufactured goods of more populous and more advanced countries." The project of home-competition with foreign manufactures has been recently followed out with some degree of success in the neighbourhood of Constantinople. "I have been informed," says a correspondent of the *Times*, writing from Constantinople under date August 9, 1845, "that, as regards the proper government works, including the Ismidt mills, the working of a coal-mine on the border of the Black sea, and of an iron mine at the Prince's islands, from which it is hoped machine iron may be obtained, they have been undertaken and are carried on solely at the expense of the sultan, who, convinced that a sure means of laying the foundation of national prosperity is by developing the productive and manufacturing resources of the country, is determined to make any pecuniary sacrifice in the attempt. The woollen trade with T. is now two-thirds in the hands of other nations; manufactured silks almost entirely so; and as to cotton, Manchester will discover that she can have but a weak competitor in her new rival; one whose success must at best be confined to a limited and inferior market. The cloth-factory near Ismidt has been built on a low tract of land, two leagues from the termination of the gulf of Nicomedia, on one of those luxuriant levels common in the Pontus, where the streams of the Olympian range produce a degree of vegetation unknown in the rest of T., but also render the undrained soil a habitat for noxious epidemics. The capability for perhaps 300 horse-power of water led to the selection of this otherwise unfavourable locality. I found the centre of the factory to comprise a single apartment or flat, 280 feet by 140; roof 35 feet, with cupolas for additional ventilation. This apartment was filled with working English machinery upon the latest principles; and here the whole of the manufacture was carried on, from the first breaking the staple of the wool to the finishing of the cloth. When the factory commenced working, 11 months back, it was the intention only to make cloths of a fine quality; but difficulties were soon experienced from the inaptitude of the workmen, particularly in the burling of the cloth, and it was changed to the lowest qualities, for the army and subaltern officers: it has since then been gradually extending its efforts to produce finer qualities. They are making at present about 1,000 yds. a-week of coarse goods. The wool for the army-cloth was ordinary Turkish, with Turkish crossed Merino and Merino of the Mohallitch flock for the superior qualities. There are 138 hands employed in the factory, 58 of whom are adults. There are 10 Englishmen as overlookers, and an English director and assistant. The Eyoub cloth factory is a remunerative speculation. The Fezhane, as this establishment is commonly called, is the property of the sultan's mother. It was commenced about 13 years ago for the manufacture of fezes, but until the last 18 months its operations were very limited: since that time additional capital has been invested, an operative director (an Englishman) and a German foreman set over the works. About nine months since the factory was further enlarged for the manufacture of cloths. This machinery was made in Germany. The average work for many weeks past has been 25 pieces, or 600 yds. a-week. All this cloth is of one quality, from 10s. to 11s. 3d. the yard, the difference being in the colours. The wool used is Merino, from Odessa and Vienna, of the middling quality. The cloth is sent into the bazaars, and finds a ready sale in competition with the foreign cloth of the same price. The simple system followed in this factory seems best adapted for an opening enterprise among a population unaccustomed to machinery. The employment of hand-looms and single cards, easier to work and keep in order, the producing one quality of cloth, and giving the workmen a high and equal monthly gage, are all calculated to insure success. The cloth manufacture promises soon a fair profit. They have made on an average at Eyoub for the last twelvemonth 1,000 fezes a-day. The first, or knitting process, is done at home by women, who take the thread from the factory. The shearing and finishing processes are done by hand. There are about 400 hands; wages from 90 to 160 piastres. The fezes are sold at 9s. and 11s. The wool is coarse Turkish with middling Odessa.—The cotton-mill is set up at Heraclea, a healthy locality, 20 m. from Ismidt. It is the property of the vizier, Riza Pasha. Although I believe more ambitious hopes are entertained for this factory than any other, I found it by far the worst conducted. The mixing is half Egyptian cotton, one-fourth Turkish, and one-fourth New Orleans. They had been trying to manufacture from Egyptian cotton singly, but the staple was so long and stubborn as to baffle all their efforts. They proposed making fresh trials with heavier machines and larger rollers." The result of these, and other similarly directed efforts has been more successful than this writer appears to have anticipated. "Amongst the contributions received from Constantinople," says the official report on the crystal palace exhibition, "will be found specimens of broad cloth equal to some of our west of England manufacture, but this has been produced at a cost which precludes all ideas of competition with other countries on the important feature of cheap production; and the same observation will apply to their cotton fabrics, and silk piece-goods, in imitation of French stuffs, all of which can be manufactured in abundance, but at too great an expense at present to meet the means of the people at large. The improvement, however, in the article of raw silks may be mentioned especially, to prove that where encouragement is given to the energies of the people, in the way of remuneration for their skill, the result is manifest. Not fifteen years since the silk imported into England was of low quality, badly reeled, hastily and dishonestly packed, and averaged in value about 9s. per lb., whilst within the last few years a new feature in the trade has been introduced, by the adoption of the improved French and Italian method of reeling, by several of the most influential firms; and the merits of the Broosa raw silk are now found to be equal to any silks brought to this market, and the consignments realize for fine ordinary qualities 14s. to 16s. 6d. per lb., whilst some of the finest have been sold within the last few months as high as 27s. per lb. Should the manufacturers still be encouraged by a liberal policy on the part of the Turkish government, there is every hope that T. may become a self-supplying country to a large extent of the useful manufactures required. In embroidery and articles of gorgeous work common in Oriental states, T. has long stood pre-eminent; but the period has arrived when a display of barbaric magnificence, although characteristic, is no longer accounted the test of a wealthy power, but the statistics of imports and exports of commodities suited to the wants of the million, are alone taken as the sure indications of the prosperity or weakness of a nation. With regard to the facilities of communication with other countries for commercial purposes, T. stands in the position of one of the 'most favoured of nations,' and her vast pop. spread over islands and on a continent margined with ports of safe and easy access to trading vessels, gives her the advantage of the ready supply to her own people of every required necessary from foreign powers, and admits of the greatest facility of exchanging her own natural productions so much required by others, and the value of which will be appreciated by a commercial people, in the list of articles enumerated in the present catalogue."

Commerce.] No country possesses greater physical advantages for carrying on an extensive commerce than T., and few have ever employed them to less beneficial purposes. Rich in native productions, and bordering upon three continents, T. might enjoy a trade with the whole world, were the industry of its inhabitants and the spirit of its merchants at all commensurate with its natural advantages. The internal trade is chiefly in the hands of Armenians, Greeks, and Jews; and the intercourse betwixt the different provinces is checked by the want of union betwixt the different component parts of the population, the absence of good roads and regular posts, the rapacity of the pashas, and the high rate of interest one per cent. per month paid for money. Jewish Armenians and Greek brokers also transact all business with the foreign market. The principal articles of exportation are: cotton from Thessaly; silk from Romelia; corn, flax, tallow, and hides from Bulgaria; tobacco, currants, raisins, wine, from the coast; and copper from Trebizond. Those of importation are: cotton-cloth, coffee, sugar, spices, glass, hardware, jewellery, paper, and slaves from Georgia, the Caucasus, and Africa. The commerce of T. is in a livelier condition than its industry. Taking the imports and the exports together, it may be estimated at somewhat under £20,000,000 per annum, exclusive of about £1,000,000 for the commerce with the European tributary provinces, and of about £5,250,000, the commerce of Egypt, as shown in the exports and imports of Alexandria, those to and from Turkish ports not being included in the calculation. The trade between Turkey and England has increased to an extraordinary extent, from about £500,000 in 1827, to upwards of £4,000,000 in 1852. The chief articles of import from T., exclusive of Syria and Egypt, to the United Kingdom, in 1849 and 1853, were as follows:

	1849.	1853.
Yellow berries, .	7,515 cwts.	6,893 cwts.
Wheat, .	103,716 qrs.	228,417 qrs.
Maize, . .	280,730 —	334,864 —
Figs, .	29,210 cwts.	28,569 cwts.
Madder, . .	99,701 —	111,920 —
Opium, .	94,844 lbs.	145,063 lbs.
Raisins, . .	31,066 cwts.	177,745 cwts.
Raw silk, .	278,238 lbs.	621,461 lbs.
Lamb skins, .	83,763 no	237,563 no.
Sponge, .	151,581 lbs.	120,192 lbs.
Tobacco, .	25,913 —	533,283 —
Wool or Goats' hair,	2,476,092 —	2,916,509 —
Sheep and Lambs' wool,	422,265 —	2,605,546 —
Shawls, . .	£12,364	£8,361

The value of the principal articles of export to T. from the United Kingdom, in the same years, was as follows:

	1849.	1850.
Cotton,	£1,658,826	£1,510,622
Cotton yarn,	179,398	158,368
Iron,	103,078	33,392
Tin,	35,499	20,214

The total exports of the produce and manufactures of the United Kingdom to T. in 1849, were £2,515,821; in 1853, £2,029,395. It should be observed, however, that not much more than half these English imports are consumed in T., the remainder passing by way of Trebizond to Persia. T. receives chiefly our cotton manufactures, linen manufactures, hardware, iron, coal, and colonial produce. Our chief imports from T. are flax, raw silk to the extent of about 220,000 lbs., cotton-wool to the extent of nearly 20,000,000 lbs., grain of various kinds, opium, &c. France is not in so favourable a position as regards her commerce with T.; the gross total of imports into T. from France being about £1,000,000, while the exports of T. to France exceed £2,000,000. The total commerce of T. with Russia (exports and imports) is somewhat more than £1,500,000, the balance of exports and imports being slightly in favour of Russia; with Austria, nearly £3,000,000; with Holland and Belgium about £360,000; with Persia about £1,000,000; with Switzerland, the United States, and other countries, about £1,500,000. These are all more or less on the increase, and do not include the commerce of Egypt by way of Alexandria.

The navigation of T., which is chiefly carried on by foreigners, is on the increase; but the most remarkable evidence of progress is to be found in the rapid and vast extension of steam navigation for mercantile purposes. Between 1841 and 1849 the number of steamers entering Constantinople had increased from 274 to 486. There is a direct and constant communication by first class steamers between Southampton and Constantinople, and recently there has been established a similar line from Liverpool. The communication between the capital and the main parts of the empire, in the Levant, the Archipelago, and the Black sea, is singularly well organized and regular. The returns for T. for 1851 do not give either the number or tonnage of the Turkish vessels which entered and cleared the port of Constantinople, neither do they afford any information of the same kind with respect to the port of Alexandretta. The returns for the Danubian principalities only give the number of Turkish and foreign vessels cleared from the ports of Galatz and Ibraila, and the nature of their cargoes, but not the tonnage employed. In 1851 there were 365 Turkish and 1,303 foreign vessels cleared from these ports. The total number and tonnage of vessels which entered and cleared with cargoes and in ballast at the various ports of the Turkish empire, exclusive of Egypt, in 1851, were as follows:

Ports.	Turkish.		Foreign.	
	Vessels.	Tonnage.	Vessels.	Tonnage.
Constantinople,	No return.		13,099	2,459,232
Salonica,	304	41,202	727	79,728
Samsun,	98	113,220	116	68,502
Trebizond,	181	114,500	142	86,267
Smyrna,	168	15,407	1,718	236,699
Rhodes,	1,274	102,408	305	44,134
Beyrout,	181	28,500	488	110,814
Alexandretta,	No return.		170	37,500
Galatz,	365	—	1,303	No return.
Total,	2,072	415,237	18,068	3,122,876

The number and tonnage of vessels which cleared from the ports of the United Kingdom to T., exclusive of Syria and Egypt, in 1849, was 434 = 100,573 tons; in 1851, 648 = 160,695 tons; in 1853, 876 = 237,860 tons.

The origin of our trade with T. is of but recent date, as previous to the last war between England and the Porte, in 1807, but two or three of our vessels annually proceeded to Constantinople with assorted cargoes. Of these cotton goods formed but a small proportion, and very few also were sent to Smyrna. An English merchant, who had conceived a just opinion of its capabilities, established a house at Constantinople in 1812, and for several years it had the entire sway of the direct trade with England. From 1813, Constantinople became a rival mart with Smyrna, which it has at length completely eclipsed for British manufactures. Goods for Persia form of late years the most valuable parts of the shipments to Constantinople, which is not only in the direct road, but offers nearly as great advantages from being an intermediate market. The shipments to T. between 1827 and 1834, advanced to £467,000 in wove cottons, £70,000 in twist, and £692,000 in the aggregate of all articles. The following quinquennial averages will show that, while our exports to T. have much more than doubled in the course of twenty years, those to Russia have remained nearly stationary:

Average.	Turkey.	Russia.
1831-5	£1,072,637	£1,488,978
1836-40	1,541,075	1,765,887
1841-5	2,301,838	1,934,213
1846-50	2,872,967	1,703,173
1851 (single year)	2,581,230	1,289,704

We find that in 1831 T. took about 24,500,000 yds. of 'cotton manufactures entered by the yard'—in 1850, 141,000,000 yds. Of 'cotton twist and yarn' she took 1,730,000 lbs. in the former year, and 6,330,000 in the latter. On the other hand, our exports of the first-named article to Russia have fallen from 2,000,000 yds. to 1,500,000; and those of the second, from 14,000,000 lbs. to less than 3,500,000 lbs. Taking values instead of quantities, our whole export dealings with the two countries in cotton and cotton manufactures would stand thus at three different periods:

	Exports to Russia.	Exports to Turkey
1831	£866,035	£687,823
1841	1,139,901	1,402,366
1851	225,117	2,077,766

During 1843 and the four following years, the amount of Austrian manufactures taken by T. had risen from £600,000 to more than £1,000,000—which was pretty exactly one-third of the total value of the manufactures exported from Austria—while the amount sent to Russia scarcely exceeded one-twentieth in any year out of the five. Commercially speaking, Russia is, indeed, wholly valueless as a neighbour to Austria. On emergencies, she supplies cereal produce to provinces which adjoin her frontier, but she contributes neither raw materials nor manufactures to any extent worth mentioning. With T., on the other hand, Austria has carried on a commercial intercourse annually increasing in value, and supplying not only food, but one-seventh of the whole raw material which enters the empire and its dependencies both by land and by sea. By a treaty between the Porte and Great Britain, all goods and merchandise, the growth of the soil, or the produce of the manufactures of Great Britain and Ireland, as well as of the countries belonging to it, and the property of British merchants, as also the goods and merchandise arriving in English vessels, or from any other country by land or by sea, is admitted into all parts of the Ottoman empire, on paying a duty of 3 per cent. *ad valorem*. In lieu of the duties which are now levied in the interior of said goods and merchandise, the merchants by whom they are imported shall be liable to an additional duty of 2 per cent., whether they sell them at the place of their arrival, or convey them for sale into the interior. British merchants are at liberty to purchase in all parts of T. all goods and merchandise imported from foreign countries. If these articles have only paid the import duty of 3 per cent., they are liable to the additional duty of 2 per cent., previous to their being conveyed for sale into the interior; but when afterwards sold in the interior, or sent to another country, they are not subjected to the payment of any other duty. No special monopoly, with the exception of that of opium, exists in T.; but there are restrictions on particular articles which call for removal. Owing to these valonea is seldom to be bought but from the authorities at the place of growth, or by special permission. With us this acorn is extensively used in tanning and dyeing, in which 7,000 to 8,000 tons are annually consumed. Cheapness, in comparison with bark and other materials used for the same purpose, is the great requisite to increase its employment without limit, and the supply in T. is now only adequate to a high-priced demand. Wool is also engrossed in the same way in some parts of Romelia, as at Adrianople and Salonica by the pashas, who re-sell to the merchants, or derive a profit from their other contracts. From the immense tracts fit for sheep-pasturage throughout Turkey it might still be raised in far greater abundance, and the quality improved by new breeds. A cargo of wheat, grown in the neighbourhood, was imported from Smyrna into England in 1831 with the permission of the governor. Oil is chiefly furnished by the aga of Mitylene, who is at the same time the contractor with the export merchants, and pays for the perquisites of his place. This island, with the country on the opposite main to the entrance of the gulf of Smyrna, and many other places around, are olive-districts, as are the southern shore of the Marmora, Scio, Samos, and the vicinity on the Turkish coast. The forests of Albania afford inexhaustible stocks of the finest oak, besides the more distant coasts of the Black sea between Samsoun and Trebizond and the banks of the Danube. The copper of Tokat is chiefly sent to France. With one or two exceptions of private property, the mines belong to the Porte, and extend from near Trebizond to Mesopotamia, from the Euphrates to the confines of Persia: their produce, if they were properly worked, would be beyond all comparison with their present amount.

Railroads.] The first want of this empire is the want of roads: instead of roads, then, to give railroads would be making a giant stride in advance. Three lines of road have already been spoken of—one from Constantinople to Adrianople, another to Smyrna, and a third to Varna. Adrianople has very extensive commercial relations with Constantinople, and yet the communication between these two first cities of the Ottoman empire is a most fatiguing, tedious, and expensive one. By land it requires four or five days' travelling on horseback on bad roads—the distance is 150 m.,—and by water there are, in every season, difficulties of another kind. The port of Enos is nearly blocked up with the accumulated deposits of the river Maritza, and is therefore a very unsafe harbour; and the river itself, full of rocks, in winter has its banks generally overflooded, and in summer has often so scanty a stream in its channels, that large boats cannot sail up to Adrianople. A railroad, consequently, on this line would offer the best prospects of success. The localities are favourable, the whole distance being plains or slightly undulated, whilst the traffic would be immense, as through the river Maritza, Arda, and Toondja, whose conflux is at Adrianople, you get into the most productive provinces of Rumelia. Engineers have recently completed the levels of the line of country from Constantinople through Adrianople, Philippoli, Sofia, and Nissa, to Belgrade, with the following branch lines:—1. From Adrianople to Shumla, Rusjuk, and Bukarest, and from Shumla to Varna. 2. From Sofia to Viddin and Orsova. 3. From Nissa, by Gorguctiervatz to Viddin. 4. From Adrianople, by the Valley of the Maritza to Enos. 5. From Sofia to Salonica, in Macedonia. 6. From Constantinople to Bugukdéré; and lastly, in Asia a line has been surveyed from Scutari on the Bosphorus, to Brussa. None of these projects, however, are likely to be realized for many years to come. "The only works which can pay in T.," it has been well observed, "are those of the simplest construction and the most common use. Whenever the subject is mentioned in conversation, some one is sure to talk about the railway from Constantinople, through Adrianople, to Belgrade. Now, this railway would cost about £7.000.000, more than the whole revenue of the country; and if the money were raised by Turkey, nearly 7 per cent. interest must be paid. To the resources of the empire the sum would be as great a burden as £120.000,000 to us. The country is a desert for hundreds of miles, and the undertaking would probably never pay 1 per cent. In fact, railways are for the present out of the question. What are wanted are roads to bring produce to market, and to facilitate the present long and dangerous journey. Where nothing more than a horse track has existed, it is impossible that a railway can be successful all at once." These remarks are unquestionably sound, and many years must elapse before a railway system can be profitably organized in T. The facilities, however, which T. offers for the prolongation of the great lines of European railway towards the regions of the East, have been ably investigated by Dr. Platé, who has pointed out the error which formerly prevailed respecting the S part of Servia, and the adjoining tracts, which were believed to be rendered inaccessible by Alpine chains. History alone showed that those parts could not but be accessible, since the great military roads of the Romans from Constantinople and Thessalonica led right through them, namely, through *Dardania* and *Mœsia*. In the middle ages, *Dardania* was the theatre of bloody wars between the Turks and the Servians; and we know that most of their battles were fought on level plains with much cavalry on either side. Dr. Boné's recent work, *La Turquii d'Europe*, informs us that this country, which is a level upland between the high Chárdágh in the W, and the Great Balkan in the E, is intersected by a few low and isolated ridges which stretch from W to E, and are between 1,000 and 1,400 feet above the level of the sea; but the most remarkable feature of them is, that they are intersected by deep gaps, some of them very wide, in the direction from N to S, forming as many easy passages from Servia into Bulgaria, Rumelia, and Macedonia. The ascent of these gaps is so imperceptible that the traveller is not at all aware of his crossing one of the most important mountain-systems, or more correctly plateaux, of T. The most important gap is the long and wide valley of Nissa and Sophia, through which the high-road from Belgrade to Constantinople winds, and which is like a trough encompassed by two parallel chains of mountains, those in the W being low and in their turn intersected by some of those gaps, so as to afford an easy communication with the valleys of the Strymon and the Vardar in Macedonia. This country opposes no physical difficulties to the construction of railways. A railway from Saloniki, across Dardania, to Belgrade, and thence to Vienna, either through Hungary by Pesth, or through Slavonia towards Mahrburg in Styria, where it would join the Vienna and Trieste railway, would reduce the journey from Alexandria to London to one-half of the time now required *viâ* Trieste—that is, supposing the projected and partly-executed lines between Vienna and Cologne, Bremen, Hamburg, and Berlin be finished. The straightest line would be from Cologne along the r. bank of the Rhine, by Frankfort, Würzburg, Nurnberg, Ratisbon, Passan, and Ling—all large and highly commercial towns—to Vienna, and thence, as mentioned above, to Belgrade and Saloniki. In Servia this railway would run along the r. bank of the great Morava, which flows through a wide and level plain as far as Jagodin where it would leave the valley, on account of the defiles near that town, and go over low upland, either towards Nissa and Sophia, through the great gap mentioned above, or descending again into the valley of the Morava, would go as far as Vrana. Between Vrana and the junction of the Bulgarska-Morava with the great Morava, there is only one obstacle, viz. a ridge of calcareous rocks, round the base of which flows the Bulgarska-Morava, leaving only a narrow space between the river and the rock. It takes ten minutes to walk round that rocky projection, and it would be easy to make a tunnel across. From Vrana the line would pass, through one of the gaps aforesaid, across the level upland to Uskub on the Vardar, whence it would go S in the valley of that river as far as Gradiska. There are rocky defiles called Demir Kápú, or the iron gate, and it would be necessary to lead the road over the uplands, which are not high, and of very gentle ascent. Once past Demar Kápú, the road would follow the Vardar in a level, wide, and most beautiful valley as far as the junction of the Vardar with the Ægean sea, a few miles W of Saloniki. The line by Nissa and Sophia could be continued to Constantinople by Adrianople, there being no serious obstacle but the ancient *Porta Trajana*, a narrow defile south of Ikhtimán in the Great Balkan. From Nissa or Sophia it would be easy to lead a line across some of the gaps west of those towns, which would run south of Kostendil—but not by Kostendil—and, following the valley of the Egerb Su, join the Vardar a little below Uskub. Dr. Platé makes the following calculations:—

From London to Ostend, 9 hours.

From Ostend to Saloniki, direct distance, 1,280 miles; add ⅛ for rounds and deviations from the straight line, or 160 miles, gives 1,440 miles. Suppose 17 miles per hour, which is the average speed on the Vienna and Olmutz line, would give nearly 85 hours, or 3 days and 13 hours; but as, no doubt, a special train would be allowed for the despatch of the India mail, the distance would be performed in 72 hours, at the rate of 20 miles per hour only; and of 57½ hours at the rate of 25 miles: say 70 hours.

From Saloniki to Alexandria, 770 miles steaming, at the average of 11 miles per hour, 70 hours: say 3 days.

Whole distance from London to Alexandria performed in either 6 days 22 hours, 6 days 9 hours, or 5 days 18½ hours.

Monies, weights, and measures.] Accounts are usually kept in piastres, which are divided into *paras* and *aspers;* 3 aspers make 1 para; 40 paras, or 120 aspers, 1 piastre; but for the convenience of reckoning accounts are often kept in piastres and centimes. Money is also reckoned by 'silver purses' of 1 *kefer* = 500 piastres, and gold purses of 1 *kitze* = 30,000 piastres. The gold coins recently in circulation were the *chequin* or *sequin* of 5 piastres, with halves and quarters of the same; their value, like that of the piastre, varies according to their purity. In 1852, the Porte published a memorandum addressed to the representatives of the different Powers, which had for its object to prohibit the circulation of foreign money. According to this note, the money coined with the name of the sultan, the paper-money issued in the capital, the silver pieces of 5, 2½, 6, 3, and 1½ piastres; those of gold of 20, 10, and 5 piastres; and the Ottoman money coined in the preceding reign, will alone be allowed in commercial transactions. All other old Ottoman coins and foreign coins are prohibited. All persons who have the prohibited coins in their possession are ordered to bring them to the mint and exchange them for legal ones. The *cantar* of cotton, in yarn or wool, is 45 okes. Silks from Persia are sold by the *catman* of 6 *okes;* and from Brussa, by the *taffee* of 610 drachms. The *chequee* of opium is 250 drachms; and of goat's wool, 800 drachms.—The *oke*, 2·272 *rotoli*, is the principal weight, and is equal to 2 lbs. 13 oz. 5 drachms avoirdupois. There is also the chequee, equal to 4,967 grains; and the *cantaro*, or quintal of 100 rotoli, is equal to 124 lbs. 9 oz. 12 drachms avoirdupois.—The *almud* is equal to 1 gallon and 3 pints English wine measure, or about 6 litres. It is generally used for wine, oil, and other liquids. But the almud of oil is estimated to weigh 8 okes.—The *pic*, or pike, is generally estimated in commerce as equivalent to ¾ of a yard, English. There are, however, the long pic, called *archim* or *kalebi*, with which silks and woollen cloths are measured; and the shorter pic, called endrezeh, for cottons and carpets. Of the *archim*, 100 are equal to nearly 77½ English yards; and of the *endrezeh*, 100 are equal to 74½ yards.—Four kisloz = 1 forteri, or fortin, and the forteri contains about 3¾ Winchester bushels, and weighs 240 lbs. The *kisloz* of rice weighs 10 okes.

Population.] It is difficult to estimate even approximatively the pop. of such a country as T., whose moral and political institutions differ so widely from our own, where no accurate census has been taken, no registers of deaths and births kept, and the number of houses is unknown. Two general censuses of the whole empire were indeed made by order of the government in the 16th and the beginning of the 17th cent., but their results are not now known; and though they were, they would not greatly assist us in estimating the probable amount of the existing pop. In the *Staatshandbuche* for 1816, the pop. of the whole Ottoman empire was calculated at 24,072,000, and this estimate is adopted by Hassel. Leichstenstern estimated it in his *Statistical Tables* of 1819 at 24,880,000, and Gräberg the same year at 24,000,000. Supposing the houses to be as thinly scattered as in the less populous parts of Spain, Malte-Brun, writing in 1820, says the pop. of the empire may amount to between 25 and 30 millions. Baron Humboldt, about the same period, estimated it at 25,330,000. But in the elements of these different calculations we find great discrepancies. One geographer, for example, assigns 6,000,000 to Asia Minor; another estimates the pop. of that part of the empire at 12,000,000; equally wide are the discrepancies in the different calculations of the pop. of Egypt. Again, with regard to the European portions of the Turkish empire, the *Staatshandbuche* for 1816 reckoned it, inclusive of Greece, at 9,482,000; while Crome, in his survey of the political strength of Europe, estimated the pop. of the same territory at only 6,700,000; and Lindner at 5,390,000. Hassel's approximation of 11,600,000, as the pop. of the European part of the empire in 1816—though considerably above Balbi's estimate in 1826—was probably pretty near the truth. This pop. Hassel distributed in the following manner:—

The whole of Ancient Greece, . .	3,000,000
The Islands, Crete, and Gallipoli, but exclusive of the small Asiatic islands, . .	600,000
Bosnia, ...	1,000,000
Wallachia and Moldavia, . .	1,500,000
Servia, ...	1,800,000
Thracia, with the two capitals, . .	2,200,000
Bulgaria, ...	1,500,000
Total,	11.000,000

This estimate is considerably under that of M. Boué, who arranges and numbers the pop. of T. thus:

Turks,	. . . . . .	1.100.000
Albanians,	(Arnaoots) (mostly Mohammedan),	1,600,000
Slaves or Slavonians,	Bosniacs (Mohammedan), .	900,000
	Hertzegovines and Croats, Christian,	400,000
	Servians, proper, . .	900,000
	Montenegrines, .	100,000
	Bulgarians, . . .	4,500,000
Romani,	Wallachians and Moldavians,	3,850.000
	Zinzars, in Epirus and Thessaly,	300,000
	Greeks, . . .	1,000,000
	Armenians, . . .	200,000
	Jews, . . .	200,000
	Gipsies and various,	250,000
		15,300,000

In an opening paragraph of this article we have given an estimate of the pop. of the Ottoman empire, founded to a certain extent on official documents, and adopted by Michelson: but as at the present crisis all information, derived from personal research and experience, with respect to the real condition of Turkey, has become doubly interesting, we here subjoin the substance of an elaborate calculation of the number and distribution of the inhabitants of European Turkey, which appears in the appendix to Mr. G. F. Bowen's recent work, entitled *Mount Athos, Thessaly, and Epirus.* "The real amount and distribution of the pop. in the heterogeneous provinces known by the general name of Turkey in Europe," Mr. Bowen remarks, "are problems which have never yet been fully solved. The European subjects of the Porte have been rated as high as thirteen or fourteen millions by Mr. Urquhart and other writers; but this number undoubtedly far exceeds the truth. All statistics of this nature are more or less guess-work in the East; for Orientals still retain the ancient prejudice against any accurate 'numbering of the people;' a process which is connected in their minds with conscriptions, poll-taxes, and other disagreeable associations. I will, however, subjoin a translation of a very able paper on this subject, which was kindly drawn up for me by my learned friend Dr. Philetas, Greek professor in the University of Corfu. He is a native of Epirus himself, and has long and deeply studied the ethnology of Turkey in Europe. The diligent inquiries which I have myself instituted in various well-informed quarters at Constantinople and elsewhere, make me feel pretty certain that his calculation is not far from the truth. If it errs at all, the mistake lies in rather underrating the number of the Mahommedans. Perhaps they approach more nearly to a third than to a fourth part of the whole population."

I. WALLACHIA and MOLDAVIA, inhabited by Wallachs, or Daco-Romans, calling themselves in their own language *Romouni*, and all of the Eastern, or Greek church. [N. B. a few villages in Epirus are inhabited by Wallachs, some of whom are also found scattered throughout most parts of T., chiefly as shepherds and carriers.] 3,000,000

II. Bosnia, Turkish Croatia, Herzegovina, Montenegro:—

Christian Slavonians (Greek church),	280,000	
Ditto ditto (Latin church),	200,000	
Mahommedan Slavonians,	400,000	
		880,000

III. Servia is entirely inhabited by Christian Slavonians of the Greek church, with the exception of Belgrade and a few other garrisons of Ottoman troops.

Christian Slavonians,	900,000	
Mahommedan Ottomans,	10,000	
		910,000

IV. Bulgaria has—

Christian Slavonians,	420,000	
Mahommedan Ottomans,	140,000	
		560,000

N. B. The Christian Slavonians in Thrace and Macedonia also speak the Bulgarian dialect of Slavonic, and are usually called Bulgarians.

V. Thrace and Constantinople have—

Christian Slavonians,	200,000	
Greeks,	200,000	
Ottomans,	500,000	
Armenians,	60,000	
Jews,	60,000	
		1,020,000

VI. Macedonia has—

Christian Slavonians,	350,000	
Greeks,	200,000	
Mahommedan Ottomans,	150,000	
Ditto Slavonians,	30,000	
Jews,	30,000	
		760,000

VII. Albania is inhabited by Christian and Mahommedan Albanians, calling themselves Skipetar in their own language. They are probably the descendants of the ancient Illyrians.

Christian Albanians (Greek church),	80,000	
Ditto ditto (Latin church),	60,000	
Mahommedan Albanians,	380,000	
		520,000

VIII. Epirus (*i. e.*, Albania south of the Acroceraunian promontory) is inhabited chiefly by Greeks, partly also by Mahommedan Albanians, and by Mahommedan Greeks, *i. e.*, the descendants of renegades of Greek race.

Greeks,	300,000	
Mahommedan Albanians,	50,000	
Mahommedan Greeks,	10,000	
		360,000

IX. Thessaly has—

Greeks,	250,000	
Ottomans,	30,000	
Jews,	10,000	

X. Crete is inhabited entirely by a race of Greek descent, but of which a portion has apostatized to Mahommedanism.

Christian Greeks,	150,000	
Mahommedan Greeks,	50,000	
		200,000

So that the total in round numbers is about 8,500,000

According to this calculation there are in the European provinces of the Ottoman empire:—

I. Christians, in all, 6,650,000, viz.:—

a. Christians of the Greek or Eastern church:—

Wallachs,	3,000,000
Slavonians,	2,150,000
Greeks,	1,180,000
	6,330,000

b. Christians of the Latin or Western church:—

Slavonians,	200,000
Albanians,	60,000
	260,000

c. Armenians,	60,000
	6,650,000

II. Mahommedans, in all, 1,750,000, viz:—

Ottomans,	830,000
Slavonians,	430,000
Albanians,	430,000
Greeks,	60,000
	1,750,000

III. Jews, in all, 100,000.

"We must add," says Mr. Bowen, "to the total a few thousand Zingaris or Gipsies. The Frank residents in European Turkey certainly do not exceed 100,000. By this term are meant the subjects of the various European powers; Ionians, who are considered as British subjects; and such Greeks as are subjects of the kingdom of Greece. So that the grand total of the inhabitants of Turkey in Europe probably somewhat exceeds 8,500,000." The country might certainly maintain a population four times more numerous without being overstocked; and indeed may have done so in the brilliant epoch of the Byzantine empire. M. Jomard—who in his *Statistique de la Turquie*, estimates the whole pop. of European T. at 11,240,000 souls—remarks that the mean number of inhabitants of this empire is but at the rate of 294 individuals per square league, whilst the number in France is 1,200, in England 1,600, and in the Netherlands and Lombardy 2,000. In order to make a *levy-en-masse* of 30,000 or 40,000 men, the whole pop. fit to bear arms of a country much larger than Sicily or Belgium would be required in this empire.

Turks.] The pop. of European Turkey, it appears from the above tables, is of very mixed descent. The Turks properly so called—although they form but a very small proportion of the pop.,—having the dominion of the country, take the precedency of all the other tribes. They are the descendants of a people who still inhabit the shores of the Caspian, and the steppes of Upper Asia. They are a fine-looking race of men, seldom below the middle size, with lofty foreheads, dark eyes, finely cut features, and limbs cast in the Grecian mould. The full form of their limbs may perhaps in some measure be attributed to their loose mode of clothing themselves, which leaves the body free from those ligatures so common among their western neighbours; but the personal elegance of the Turk is chiefly owing to the mixture of different blood in his constitution: Arab, Grecian, and Persian blood all flow in the veins of an Ottoman, and conspire to make him the handsomest of human beings. Of their women, the countenances are more striking than the form or the gait: both the latter are probably hurt by their confined mode of life which deprives them of that kind of exercise which alone can bring the human shape to perfection, and by their manner of sitting which is said almost invariably to induce a habit of stooping. The Turk still retains the same character which the histories of the 15th century attribute to him. Indolent in time of peace, when war rouses him, he becomes infuriated; unrefined and sensual in his ideas of pleasure, yet moderate withal in his enjoyments; often starting from the lap of luxuriousness to bear the severest privations without grumbling; inhuman in his vindictiveness, yet frequently carrying an exalted friendship to a point of heroism; his courage manifests itself sometimes in a chivalrous heroism, on other occasions in a stoical indifference. Lord Carlisle's estimate of Turkish character, as that of an accurate and trustworthy observer, may be here quoted: "Among the lower orders of the people," says his lordship, "there is considerable simplicity and loyalty of character, and a fair disposition to be obliging and friendly. Among those who emerge from the mass, and have the opportunities of helping themselves to the good things of the world, the exceptions from thorough-paced corruption and extortion are most rare; and in the whole conduct of public business and routine of official life, under much apparent courtesy and undeviating good breeding, a spirit of servility, detraction, and vindictiveness appears constantly at work. The bulk of the people is incredibly uninformed and ignorant. Upon the state of morals I debar myself from entering. Perhaps the most fatal, if not the most faulty, bar to national progress, is the incurable indolence which pervades every class alike, from the pasha, puffing his perfumed *narghilé* in his latticed kiosk on the Bosphorus, to the man in the ragged turban who sits cross-legged with his unadorned tchiboque in front of a mouldy coffeeshop in the meanest village. In fact, the conversation of every man whom I meet, who is well-informed on the state of the population, with very few exceptions, might be taken down as an illustration, often very unconsciously on their part, of the sense usually assigned to the prediction in the Apocalypse of the waters of the Euphrates being dried up. On the continent, in the islands, it is the Greek peasant who works, and rises; the Turk reclines, smokes his pipe, and decays. The Greek village increases its pop., and teems with children; in the Turkish village you find roofless walls and crumbling mosques." Yet, after all this has been said and admitted, it must still be allowed that the Turks are the noblest and most honourable nation, amongst that multifarious aggregation of nations who inhabit the broad realm of the house of Othman. The rayahs of all nationalities hold throughout an inferior position in point of intelligence or character. The Sclaves are more uncivilised, the Greeks and Wallachians are on a far lower standard of morality, than the Turks. Greeks, and Armenians in particular, both the clergy and the laity, will extort the last para from the Christian pop

The great majority of the Turks, on the other hand, are honest, charitable, and tolerant, though it cannot be denied that they incline to pride, indolence, and fanaticism. The Osmanli is *par excellence* called to exercise dominion over an uncultivable conglomeration of races; and as he represents the more opulent part of the pop. it is his province, even according to the spirit of modern political institutions, to be a public functionary. The immense majority of the rayahs are poor, dependent, and tillers of that soil which almost exclusively belongs to the Turkish lord of the manor. It would be tantamount to a destruction of the empire if the dominion of the Ottoman race were suddenly to cease, or to be shared by the rayah. The Turkish Sclave has no spark of sympathy for the Turkish Greek; the Arnaout has as little for the Wallachian; but they are all held in awe by the Turk. The Turks, until very recent times, were accustomed to cover the head with a turban, or a kind of cap surrounded by numerous folds of cloth. Of the turban the sizes and forms were extremely numerous: each size and form indicating something in the rank or condition of the wearer. The Turkish garments, unlike those of the western European nations, were also loose and flowing. The lower parts of the body were covered with wide drawers or rather trousers. The shirt had wide sleeves without wristbands; and over this was worn a kind of wide vest bound with a sash, and above all, a long flowing gown. The head was shaven, except a small lock upon the crown; but the beard was worn long, and frequently artificially dyed. Different professions, particularly the military and learned classes, were distinguished by peculiar habits; and different nations were known by wearing clothes, if not of a particular form, at least of a particular colour. Sumptuary laws were in force, regulating the colour of the clothes, the colour of the slippers or boots, and the size and colour of the turban. To infringe these laws was extremely dangerous for a Turk, and almost certain destruction to a Christian. But the day of the turbaned Turk has passed. "Sultan Mahmud, in his desire to imitate Peter the Great, completely metamorphosed the appearance of all persons employed under government,—that is to say, of nearly every one who was formerly possessed of wealth sufficient to wear the old costume with proper amplitude. His example has been followed by his successor; and every functionary, high or low, has been compelled to assume a kind of Frankish costume, topped by the fez or red cap, which formerly formed the skeleton as it were of the turban. I am disposed to think," says Mr. B. St. John, "that this change has gone a great way towards destroying the nationality of the Turks. It was thought that with the European dress these barbarians would assume the activity and energy of the Giaours—perhaps, also, their instruction and their civilisation. Contrary effects were produced. In the first place they do not yet know how to wear a frock-coat, and are often to be seen with one arm only in the sleeve, whilst at other times they button the garment at the throat, and allow it to hang round them like a cloak. They have not yet given up the habit of sitting cross-legged, and at their ease, so that their backs are almost always rounded, and they walk in a slouching, shambling manner, totally inappropriate to their tight dresses. Few of the Turks have really become reconciled to their new costume; they feel like winged birds, or the fox without a tail: many of them on returning to the privacy of their harems, hasten to throw aside their infidel frock-coats, and to dress themselves as their fathers dressed before them, and after the fatigues of the day, to undergo the wearisome operation of binding round their abdomens with a sextuple shawl! The Osmanli of the present era wears coat, waistcoat, and trousers; and a red skull-cap is the only mark that distinguishes him from the European. Amongst the lower orders the Oriental dress is still retained, and the priesthood still cling to the elegant Oriental robe and turban. In the interior of the country—that is to say in Asia—Turkish dress and habits still flourish in their primitive state. The female dress consists of long loose flowing robes, and differs not materially in form from that now abandoned by the male portion of the community. The head, instead of a turban, is covered with a stiff kind of cap; but with regard to this part of the body less uniformity prevails among the women than among the men. When a female appears abroad, the veil is indispensable, and so disguises her that she cannot from her appearance be known even to her most intimate friends. Cleanliness is enforced upon the Turks by their religion, which declares frequent ablutions to be absolutely necessary. It is somewhat singular, therefore, that some have attributed the prevalence of the plague in the large Turkish cities, to the uncleanliness of the Turkish mode of living. The truth is, that not only the Turks, but all other Mahommedans are, in their mode of life, much more cleanly than the generality of Christians. In eating, no people can be more moderate. Their food is simple; and is seldom taken to excess. Rice forms the principal article, and is dressed in various ways, sometimes with mutton and fowl, but oftener without either of them. The general abstemiousness of the Turks is nominally enforced by the Ramadhân or yearly fast; but is rather destroyed than aided by the way in which that fast is observed. The Ramadhân continues a month, and revolves through every month of the year. During the whole period, the most perfect abstinence is enjoined from sunrise to sunset; to taste even cold water would be to transgress the sacred law This injunction is said to be rigorously observed; but as the law which enjoins it mentions the time of sunrising and sunsetting, as containing the period during which the fast is to be observed, the Turks ingeniously suppose that no indulgence can be wrong that takes place between sunsetting and sunrising. Smoking is a species of luxury in which the Turks almost universally indulge themselves; the chief exception being in the case of young women. "In his pipe," says Dallaway, "an opulent man is extremely sumptuous; the head must be of pale amber, the stick of jasmine wood with the bark preserved, and the bowl of a delicate red clay, manufactured at Burgas in Romelia, and highly ornamented. According to the dignity of the smoker is the length of his pipe, often six or seven feet, when it is carried by two of his servants from place to place with much ceremony; and the bowl is supported by wheels as an aid to supreme indolence. In the summer, for greater coolness, the stem of the pipe is covered with cotton or muslin, and moistened with water." To procure the pleasures of inebriation without having recourse to forbidden liquors, the Turks sometimes make use of opium. In all the large cities, many coffee-houses are constantly open where coffee and opium may at any time be procured. The use of this drug, however, is said to have been formerly much more extensive than at present. According to the institutions of the Koran, every Mussulman may have four wives, and as many concubines as he can afford to maintain: reasoning from this article of their creed, and impressed with an idea of eastern voluptuousness, many Europeans are of opinion that every Turk has a seraglio, in which are immured for the pleasure of their master, four wives and an indefinite number of female slaves. But such a conclusion is not quite correct. The lady of a wealthy Turk has numerous female attendants; and as each wife occupies separate apartments and has her separate household, very few Turks treat themselves to the luxury of more than one or two wives. Polygamy exists, but by no means in the exaggerated form frequently represented. Slaves are bought, but always treated as part of the household. A slave who has become the mother of her master's child is free. The interiors of the seraglios are in general simple—a long ottoman, a bare and matted floor. Amber-mouthed pipes, inlaid with diamonds, are general; and in the establishments of some Turkish gentlemen very considerable capital is invested in this article of Turkish furniture. Turkish ladies play a great part in Eastern politics, in spite of eunuch guards and veiled faces. The life of the Turkish female is not so dreary as might be supposed; for under the cover of the *yashmak* or veil, they wander at liberty through the bazaars and busy streets of Stambul, the European quarters not excepted. They spend weeks in mutual visits at friends' houses; and notwithstanding their utter want of intellectual education, do not become the victims of *ennui*.

The Tartars.] At the mouth of the Danube, in the Dobrudsha, in the valleys of the Balkan, and around the capital, considerable numbers of Tartars are located. They are of middle stature, with a fine shape, small black and expressive eyes, fresh complexions, and dark auburn hair. They are a frank, hospitable, and peaceful race. Their dress is oriental; their dwellings are clean, and their occupation is husbandry and trade.

The Armenians.] The Armenians in European T. came from the sources of the Euphrates and Tigris in very ancient times. They have preserved their own language, but also speak that of the country. They have likewise a peculiar sacred language which is only understood by their priests. They are distinguished by their olive complexion, lively eyes, aquiline nose, regular features, and slender shape. They lead in general a sober and retired life, and are of avaricious habits.

Albanians.] The Albanians constitute the next class, and are very numerous. During the 11th century, they bore a part in some of the wars of the Greek empire. In the times of the separate principality, which, under the name of Acarnania or Ætolia, was erected by Michael Angelus in the commencement of the 13th cent., we find them extending themselves by a predatory warfare; and spreading themselves at intervals over the whole of Epirus and Thessaly. They ennobled themselves by the powerful resistance which they made to the Turkish sultans, Morad and Mohammed, under the auspices of the celebrated George Castriota, who defeated the Turks in 22 battles. He ruled over the territory extending along the coast, from the river Bojano and the *Palus Labeates* or lake of Scutari, to the mouth of the Thyamis, now Kalama, opposite Corfu. It was in his days that the designation of Albania prevailed, as the Albani of Ptolemy were properly his subjects, and at that time inhabited the districts of Kroja, Tyrano, and Dukagini and to whom he was chiefly indebted for his victories. The Albani, strictly so called, are at present denominated Merediti, and live in the pashalik of Scutari. The appellations Albani and Albania are of Roman and Greek origin; for in their own language they call themselves Arnauts, and their country Arnautlich. They differ in language, manners, and dress, from both Turks and Greeks, and as a people are greatly superior to both.

Christian population.] Three principal races constitute the Christian pop. of European T.: the Wallach, the Greek, and the Slavonic which comprises the Bulgarian, Serbian, Bosniak, and Croat. But each of these three differ as much from each other in language, origin, and character as from the Turks. The Wallachs, a race of Italian origin, speaking a bastard Latin, amount in Moldavia and Wallachia to some 4,000,000 or 4,500,000; they have offshoots in the adjacent Bessarabia, in Transylvania, and in Hungary. In the interior of T., principally about the Pindus, from a quarter to half a million of this race are located, but entirely cut off from all communication with their brethren of the Moldo-Wallachian provs. by strange populations and a wide extent of territory. The Wallach hates both the Slavonian and the Greek, though of the same religion. Moldo-Wallachia having made a seasonable capitulation with the Porte, secured to itself great privileges. The Rumane or Wallachian race, though enslaved and degraded, is intelligent, and is highly susceptible of political enthusiasm. The next important pop., numerically, is the Bulgarian. The Slavonic origin of this people is disputed, but they are identified with their Slavonic neighbours in language, feeling, and religion, though not in character.—The Bulgarians, led by their clergy, who are in the Russian interest, are Russian partisans, but they are the most industrious, and at the same time unwarlike people in European T.—The Greeks in European T., computed at from 900,000 to 1,200,000, are in their present state unimportant. "We must not confound the Turkish Greek with the independent Greek of the Morea or of the islands. Independent Greece contains locally highly courageous and energetic populations, who were formerly semi-robbers, pirates, and merchants. It is true that on examination they seem mostly if not all of Albanian origin. But the effect of diplomatic intrigues, of the ceaseless endeavours of the court to establish absolute authority, and the predominance of factious spirit and corruptibility over patriotism, have been and continue such in Greece, as to deprive her policy of all initiative power. But the Greeks of T., with trifling exceptions (such as the inhabitants of Samos or Ipsara), are unwilling and unable to fight their own battles. They are too much attached to their individual interests, too treacherous, too suspicious of each other, and for the most part too timid in this generation ever to do more than follow the initiative which the independent Hellenes might assume. The Greeks of T. are for the most part imbued with a Russian spirit, the independent Greeks on the contrary are anti-Russian, even the Russian party in Greece receiving Russian gold and making use of Russian interests to further factious views, but always with the idea of eventually deceiving her. The Greek pop. of T. is without exception the most faithless, dishonest, and superstitious the writer ever met with, but like all the Greeks, gifted with much native intelligence. They are industrious, enterprising, and above all other people distinguished by so intense a curiosity and thirst of knowledge, that probably they are destined to play in the East a part wholly disproportionate to their present insignificance." The Christian Albanians, outnumbered by their more warlike Mahommedan brethren, are divided into Greeks and Roman Catholics, who hate each other with a more fervent hatred than they bear towards the Mussulman. Montenegro, with its 100,000 independent inhabitants, represents a vast hostile fortress in the Turkish territory. The Montenegrins are directed in everything by the bishop, whose revenue is a pension from St. Petersburg, and who acts in abject obedience to its orders. Lastly, we have the Serbians who, though only numbering 1,000,000, are politically the most important of all the Christian populations because the most warlike, because universally armed, because material prosperity and prosperous indolence allows every individual to engage in warfare, and because of the influence which Serbia exercises over Bulgaria. It is an erroneous prejudice to suppose that the Christians in T. are without influence in the state. The whole of the Levant trade is in the hands of the Greeks and Armenians. They hold the most important offices in the state. The pashas choose from amongst them their secretaries and officials. They have at all times played a conspicuous part in diplomacy, as the Turks would not condescend to learn foreign languages—not even such as are more generally spoken in those parts of the Ottoman dominions where the rayah population preponderates. The Christian communities administer their own affairs; and this body, as well as the clergy, have more than once succeeded even in inducing the sultan to depose mighty pashas. A firman has recently been issued by the sultan respecting the Protestants of his empire, considering Christians, who shall so declare themselves, as a separate and united body, and as such entitling them to claim protection of the officers of the Turkish government, to demand passports, register births, and make such representations as they require through the means of a common agent to the Sublime Porte. A remarkable religious movement has declared itself amongst the Christians of Asia Minor and Syria, and more especially amongst the Armenians. These, immersed in profound ignorance, with a clergy as ignorant as themselves, have been greatly stirred by the efforts of English and American missionaries, rather to educate than proselytise them. The result has been a large dissemination of the Bible, the universal study and knowledge of it by a great number of Orientals, and the formation of several evangelical congregations and churches in the East.

Turkish religion.] The religion of the Turks is the Mahommedan, so named from its founder Mahomet. Its doctrines and rights are promulgated in the Koran, which though a wild and incongruous mass of truth and fable, like the books of the Mosaic law is at once the civil and ecclesiastical code of its votaries. The Mahommedans are divided into two great sects, viz. the followers of Omar and the followers of Ali. The former acknowledge the four Khalifs Abu-beker, Omar, Othman, and Ali, to have been the lawful successors of their prophet, and believe in the Souna or Book of Oral Traditions collected by Al-bokharee; whereas the latter pronounce Abu-beker, Omar, and Othman, to have been usurpers, and esteem Ali alone of all these to have been the lawful successor of Mahomet. They dignify Ali with the name of the 'Vicar of God,' and curse Omar, whose name in their theology is used for the name of the devil. The Persians alone are of the sect of Ali, and are regarded by the Turks and all other Mahommedans as heretics By the sect of Omar the Turkish sultan is regarded as the *khalif* or supreme head of the Mussulman faith. This high dignity was transferred by a solemn deed of the last of the Abassides in Egypt to Sultan Selim in 1516, who thus succeeded to all the prerogatives of the khalifat as the legitimate successor of Mahomet. All Mussulmans believe in the unity of God, and in Mahomet as his prophet. The commandments of the Mahommedan law are five in number: viz. 1st, The observance of prayer (*namaz*) five times a-day; 2d, The observance of the feast of Ramadhan or Ramazan; 3d, The giving of alms and other deeds of charity; 4th, The performing of a pilgrimage to Mecca (*hajilique*) at least once in the individual's lifetime; and 5th, The observing of external purity of body (*abdechte*). To the observance of these leading commandments are added some other practical injunctions which are deemed of minor importance. The principal of these are: the rigorous observance of Friday; the rite of circumcision; and abstinence from wine and every fermented liquor, and also from the flesh of pork, and of any animal which has died by suffocation. All these injunctions are scrupulously observed by the Turks, with the exception of that which prohibits the use of wine. The supreme priesthood, the administration of justice, and the sovereignty, were united in the person of Mahomet: the caliphs also who succeeded the prophet, united the temporal to the spiritual power; but from the impossibility of exercising in their own persons these prerogatives over all the extent of their vast empire, they established vicars whose knowledge and piety might recommend them to the people. This order took the name of 'The body or college of *ulemas*,' and was divided into three classes—the *imams*, or ministers of worship,—the *muftis*, or doctors of law, and the *cadis*, or judges. The mufti, in the long run, obtained the first place among the ulemas. The principal function of the mufti in the Mahommedan religion would seem to be—as the etymology of his name indicates—the pronouncing decisions on obscure or controverted points of law; but his power is not confined within limits so narrow. The mufti can be the minister of his doctrine; he can perform the duty of the imam; he is also capable of performing the functions of cadi. This affinity, or rather this association of powers among a people whose only code is religious law, explains the unalterable respect in which the administration of justice is held. The admission to the rank of the ulemas can only be obtained after long studies and frequent trial. Of the various branches of knowledge, the number of those with which some acquaintance is required, are, grammar, logic, ethics, rhetoric, theology, philosophy, jurisprudence, the Koran with its commentaries, and the *Sunna* or tradition. The ulemas thus form a respected and a powerful body—a species of nobility, in the state; they engross the most lucrative employments; the judicial power in all matters of religion is vested in them alone; their persons are sacred; and their goods are exempted from liability to confiscation. To their moral influence they also unite the authority of physical force, their numbers with that of their various dependents amounting to 30,000 in the city of Constantinople alone. Among the various religious orders or *dervishes* of the Turks, the *mehlevis* and the *ectachis* form the principal classes. These species of recluses reside in the *tecké* or monasteries, under the direction of a superior, and take upon themselves the three vows of poverty, celibacy, and obedience. The magnificence of Islamism appears in its principal mosques or *djamis*, and establishments for religious instruction. The djamis—which are mosques *par excellence*—ought only to be built in the midst of large bodies of people, in the interior or neighbourhood of cities; in the country there are none. They are ordinarily consecrated by the sovereigns. Of religious edifices of an inferior order, called *mesdjid* (chapels or oratories), the number is considerable. In the mosques are seen neither pictures, statues, nor images—all which are condemned by the law—but solely sentences in verse, extracted from the Koran, inscribed on the walls. The floor is covered by mats or carpets. On one side of the temple is the *kibla*, a kind of niche, indicating the relative situation of Mecca, towards which the Mussulman must turn during his prayers. To the right of the Kibla is a pulpit, whereon are placed the *moeddins* (or chanters), who call the people to prayer; and to the right is a chair, from which the khotba is recited. The mosques are supported from the *moucateas* or proper domains of the church—a third part of all the conquered lands having been originally set apart for this purpose by the Turkish chiefs—and the *vacoufs* which are a species of reversion heirships.

Education.] There is no country in the world which offers so many facilities for national education as Turkey. Schools are attached to every mosque throughout the empire, with lands and funded property for their support; but the number of pupils in these establishments has hitherto been very limited, and entirely out of proportion with the revenues at their disposal, and it is unnecessary to say that the system of education pursued in these schools is of the most defective kind. The studies of the pupils are entirely confined to the Koran, or to some distorted remnants of the astronomical and medical sciences, which flourished amongst the Arabs. The Turkish government having determined on the gradual reform of these schools, an ordinance was issued by a committee of public instruction, containing a plan of the studies to be pursued, with rules for the admission of the pupils and the discipline to which they are to be subjected. The system of public instruction was remodelled in 1847. Previously it had been very deficient, both as furnishing little or no practical education, and also as providing no intermediate institution as a link between the lowest and highest schools. By the new system, education is brought under three heads. That of elementary instruction comprises the subjects of reading, writing, cyphering, and religion. Parents are obliged to send their children, of either sex, to school as soon as they attain their sixth year, but the instruction is gratuitous. The next grade is occupied by the middle schools, of which there are now six at Constantinople, numbering 870 pupils. The subjects of instruction in them are—Arabic, orthography, composition, religious history, Turkish and universal history, geography, arithmetic, and geometry. Finally, there are 10 colleges, at which an education is received fitting students for the several branches of civil and military life. There are in the neighbourhood of Constantinople, a medical, a military, and a naval school, all under the direction of European professors. These schools have had considerable success. French is the only European language taught in them, and is exclusively employed as the medium of instruction.

Turkish language and literature.] The language of the Osmanlis or Turks is a dialect of the Turki, which is the common tongue of all the Nomadic tribes, from the Oxus to the Ural-Tau, and from the frontier of Tibet to the Caspian sea. It has been erroneously supposed that the language of the Turks is still more barbarous than their manners: this is a great mistake, as it is inferior to none whether ancient or modern in softness, flexibility of sound, and harmony, and its rules are so admirably simple that it seems rather to have been formed by an academy of learned men than by a number of erratic and illiterate tribes. It must be remarked, however, that the Osmanli dialect is the most corrupted of all the dialects of the Turki, because it is most mixed with foreign words and phrases, though at the same time it is the most polished and elegant. It has every title to be esteemed as one of the most refined and perfect of the Oriental languages. It is not to be considered, however, as the very same which was brought from the wilds of Tartary in the 9th cent. by the Turkomans, or by the Turki tribes who accompanied Jenghis Khan in the 13th. It has received a strong infusion of Arabic and Persian words, not merely in the terms of science and art, but in its ordinary tissue and familiar phrases. These are all connected by the regular grammar of the language, which is remarkable for clearness, simplicity, and force. Its style is far less adorned than the bombastic Persian, and as free from metaphor or hyperbole as that of a good English or French historian. On the whole, the Turki bears more resemblance to the good sense of Europe than to the rhetorical parade of Asia, and is remarkable for its downright and picturesque naivete of expression. It is written in the Arabic character, which was adopted by the Turks soon after their entrance into Persia, for they had no alphabet of their own. The Turkish is, in a great part of Asia, and on the northern coast of Africa, the common medium of communication between nations speaking different languages.

After the Turks had received a written character with the Koran of Mohammed, and at the commencement of the 14th cent., under one of their emirs, Osman, had founded an independent empire upon the ruins of the Greek, they began gradually to feel the necessity of more literary cultivation. Even Orkan, the successor of Osman, although devoted to war and conquest, founded in 1336, at Brussa, in Natolia, a literary institution, which became so celebrated for the learning of its teachers that even Arabs and Persians were not ashamed to become scholars of the Turks. Their own historians remark that the monarchs of this house until the time of Achmet I. (1603), though not all equally distinguished by glorious undertakings and princely virtues, yet all gave lustre to their reigns, by their love and encouragement of learning. The golden age of Turkish literature was in the second half of the 15th cent., during the government of Soliman, called 'the lawgiver,' the great grandson of Mahommed II., whose victories put an end to the Roman empire. One of their oldest and most esteemed annalists is Saad-ed-din, who, after having been the instructor and tutor of two sultans, died in the office of mufti at Constantinople, in 1599. His chronicle, entitled *Tadsch-et tawarich* extends from the origin of the Turks to the death of Selim I. in 1520, and is regarded by the Turks as a classical work. In the works of Naima, Raschid and Tchelebisade, the annals of the Turkish empire, from 1592 to 1727, are continued in unbroken succession. Hadschi Chalfa, surnamed Tchelebisade, who died at Constantinople in 1657, was distinguished for his historical and literary attainments. He composed a work of a cyclopædic and bibliographical character, in which the names of all the branches of science cultivated by the Arabians, Persians, and Turks, are given, and the titles of all the works written in these three languages, from the 1st to the 1050th year of the Hegira (A. D. 1640). This work served as the foundation of the encyclopædic view of Oriental science, by Joseph von Hammer (Leipsic, 1804). Their poems are chiefly of a mystical or moral cast, or devoted to love. Riddles, logogryphs, chronograms, and similar poetical trifles, are very popular among them. All their poetical productions are in rhyme. Their prosody and the technical part of their poetry are the same as those of the Arabians and Persians. The *Divan* of Baki, the principal lyric poet of Turkey, who died in 1600, has been translated into German, by Joseph von Hammer (Vienna, 1825.) In 1721, during the reign of Achmet III., Ibrahim, a native of Buda, with the assistance of a Jew of Prague, Chacham Jonas, and under the patronage of the grand vizier, Ibrahim Pacha, established a Turkish printing-press at Constantinople, at which, in 1742, seventeen works, comprising twenty-three volumes, and 13,000 copies had been printed. After a long interval of inaction, the press was again used, in 1783, during the reign of the sultan Abdolhamid; but its activity continued only about two years. In 1793 it was again brought into action by Abdorrhaman Effendi, a Turkish mathematician, who had rendered valuable services to the Porte, as commissioner for settling boundaries after the peace of Sistow. Abdorrhaman connected it with a school for teaching engineering at Chasskoi, a suburb of Constantinople. It remained there some years, and at the beginning of the present century, was removed to the barracks of the new militia at Scutari. In 1806, twenty-six works had issued from this press. The imperial historiographer, Jerome Megiser, published, at Vienna, the first Turkish grammar, in 1612; afterwards the study of the Turkish language was much attended to, particularly at Vienna, as the intercourse between Austria and the Porte had become very considerable. Meniski, court interpreter, published the best Turkish grammar (Vienna, 1680.) in Latin, folio, and likewise the best dictionary of the Turkish language. The first, with the addition of exercises in analysis and reading, was republished by Kollar (Vienna, 1756, 4to.): the last, likewise, much enlarged, by Jenisch, was republished at the expense of the emperor (Vienna, 1780—1803) There are Turkish grammars in the French language, in which the Turkish words are printed in Roman letters, by Preindl (Berlin, 1789, with a dictionary,) and by Viguier (Constantinople (1790—1794). The last grammar of the Turkish language was published by Jaubert (Paris, 1823). Many scholars, of much learning in Turkish literature have been formed at the Oriental academy in Vienna, established by Maria Theresa, in 1753, for the education of young diplomatists to conduct the intercourse with the Porte; particularly Von Jenisch, Von Sturmer, Von Chabert, and Von Hammer.—The origin of the first daily paper in T. is ascribed to the Frenchman M. Blacque, who in the year 1825 founded the *Spectateur de l'Orient* in Smyrna. The idea was, however, not new, for Verninac, ambassador of the French republic to Selim III., published in 1795, a French journal, which was printed at Pera, in the palace of the ambassador. In 1811, during the war with Russia, extracts from the bulletins of the great army were likewise printed and distributed. With exception of these beginnings, the *Spectateur de l'Orient*, which soon changed its name to *Courrier de Smyrne*, was the first real political paper published in T., and it exercised a considerable influence on the events which occurred during the latter part of the Greek rebellion from 1825 to 1828. In 1831, M. Blacque was called to Constantinople by Sultan Mahmud, and published there the *Moniteur Ottoman*, an official paper, in the French language. The following year the *Takwimi Wadai* was begun in Turkish, containing sketches of the events of the time, in fact nothing but a translation of the *Moniteur Ottoman*, yet with this difference, that the Turkish paper confined itself to publish the acts of government, while the French paper, besides the official, had an unofficial part, which had for its object to sift and defend the interests of the Porte. Smyrna, which might boast of having had the first journal in the empire, soon afterwards obtained two, and at length even four. The *Impartial* is the only one of the French journals which has remained in Smyrna; the others having been removed to Constantinople. The number of papers and periodicals has now increased to 13, which are published in the various languages of the Turkish provinces; two in the Turkish language, also translated into Armenian; four French, four Italian, one Greek, one Armenian, and one Bulgarian. The *Takwimi Wadai* is entirely supported by the government, while the other papers have but an annual allowance of about 30,000 piastres, except the *Journal de Constantinople*, which has the double amount. In Belgrade, in the Principalities, on the Danube, at Beyrut, Alexandria, &c., journals in French and in the dialect of the country are likewise published—making in all upwards of 30 periodicals at present published in the Ottoman empire.

Government.] The sultan of T. was until recently a despotic authority, uniting in himself, like the first caliphs, the whole power of the state. He not only made the laws but also executed them, and he is at the same time the supreme head of the church. It is to be observed, however, that the civil and religious law of T. having only one source, the Koran, it necessarily follows that the mufti, who is the supreme authority with regard to the legality of religious, civil, and political acts, has great influence in controlling the sultan, who can neither declare war nor conclude peace, nor undertake anything of importance, without previously asking the mufti and his ulemas if it is conformable to the law, and the mufti decides the matter by a *fetwa*, or ecclesiastical decree. The nation also sometimes applies to the mufti for similar purposes, especially in cases of rebellion, which becomes a legal opposition as soon as the motive of it is declared to be just and legal by a *fetwa*. Thus, although the sultan of T. is absolute, inasmuch as there is no political body in the empire which has any recognised power to check his will, he is obliged to reign conformably to the religious, political, and civil principles contained in the Koran. And so great, indeed, is the enthusiasm of the Mussulmans for the words of their Prophet that it would be a hopeless task to endeavour to preserve peace in the empire itself, should any insult, perpetrated against what they consider to be the true faith, be passed over unavenged. The sultan's title is that of *padishah* or 'prince.' The succession is hereditary in the male line of Osman's family, but is not always allowed to descend in regular order from father to son; the people, and the military—as is well-known—frequently exercise a violent right of election, and elevate another than the heir direct to the throne. The sultan is not crowned at his inauguration, but is solemnly girded with the sabre of Osman, and swears to support Islamism. The mother of the reigning sultan is called *sultana valide;* she enjoys distinguished prerogatives, and a fixed revenue, and is usually possessed of great influence in the state. The first of the sultan's wives who brings him a son is styled *chasseki sultana.* The sultan's titles run in the highest style of Eastern hyperbole. In 1839, a most important step in the career of civil and political reform in T. was taken in the promulgation of the constitution or declaration termed the *hatti-sheriff of Gulhani*, which was formally announced by the sultan on the 2d of November, 1839. Its primary author, Reshid Pasha, who had just returned from Paris, had been intrusted with the foreign department, and had become, by his long residence in Europe, peculiarly partial to European institutions. He now determined to try whether it was not possible to adapt the constitutional and parliamentary forms of Europe to the laws of T., and had, to that end, removed all high officers of the state resident in Constantinople who, he knew, would be adverse to the innovations he wished to introduce. On the 2d Nov., 1839, he summoned all the high dignitaries and officers of the state, together with all the more notable members of the Turkish empire, all the deputies of the various rayah nations—that is, of the Greeks, Armenians, and lastly of the Jews—to appear at the kiosk of Gulhani in the Serail. A similar invitation was given to the diplomatic body, for whom the Sultan's carriages were sent. A spacious place, extending before the kiosk, was surrounded by the guards of the sultan, who was himself seated in one of the apartments of the kiosk. To the diplomatic corps, and also to the Prince de Joinville, who was then by chance at Constantinople, other suitable apartments were allotted in the same kiosk. All the chief officers and deputies of the various nations, the chief ulemas, patriarchs, and high clergy of the Oriental church, and lastly the chief rabbi, were then assembled in the square, in the midst of which was erected a tribune. After a pause of solemn silence and expectation, a hatti-sheriff was handed to the grand vizier from the sultan. The whole assembly then saluted the document, after which it was handed over to Reschid Pasha, to be read aloud from the tribune. The sultan's court-astrologer stood by with a watch in his hand, waiting for the auspicious moment on the dial. When this arrived, he gave a signal, and the minister of foreign affairs delivered the contents of the document in a clear loud voice. When he had concluded, the ulema approached, and in a long prayer invoked the blessing of Heaven on the new work. The ceremony then concluded with a salute of 101 guns, fired from all the batteries of the Bosphorus, announcing to the people the grant of a constitution, to which, by way of adding to the unparalleled nature of the event, the sultan and high dignitaries of the church and state took the oath of fidelity. The chief provisions of this constitution were a promise on the part of the sultan to afford all his subjects, without distinction, full protection of life, honour, and property; to arrange an equal distribution of taxes; and to limit the military service to 5 years. The levies of the Mussulmans were to be made under certain restrictions, and the rayahs were entirely exempted from military service on payment of a small poll-tax. The proceedings of the courts, and especially of the religious courts, were to be subjected to strict supervision; and, lastly, the sultan surrendered the right he enjoyed to dispose of the lives and property of his functionaries without any legal process whatever. This hatti-sheriff was favourably received by the nation at large, but not so by what is emphatically styled "the old Turkish party," that is, the fanatical and retrograde adherents of the old system of administration, who have almost uniformly added their efforts to various external causes to impede and paralyze the exertions of the wiser ministers of the sultan in the path of domestic improvement.

The principal officer of state in T. is the *vizier* or *sadri-azem*, who, if he does not enjoy the whole power of his master, is made at least responsible for all his acts, and is not unfrequently sacrificed to his caprice or the resentment of the people. He has the care of the whole empire; manages the revenue, administers justice both in civil and criminal affairs, and commands the armies. Upon his appointment the sultan puts into his hands the seals of the empire, which are the badge of his office, and which he always wears upon his breast. His income amounts to 600,000 dollars a year, exclusive of presents and other perquisites.

The *divan, riquiabi-houmayoune*, or state-council, is composed of three principal ministers: the minister of the interior or *kehaya-bey;* the minister of the exterior or *reis-effendi;* and the minister of the executive or *chaush-vashi.* The ministry assembles twice a-week in ordinary council, on which occasion the mufti or the vizier presides. The ulemas, 30,000 in number, are represented in this council by two *cadi-askieres;* and its deliberations are occasionally assisted by ex-ministers who have received the sultan's commands to that effect. The extraordinary divan or *ajak-divani* is at present composed of 121 members, but is summoned only on great occasions. It consists of all the officers and chief personages of the state; but its resolutions only become important when they are sanctioned by the cabinet and the imperial council. The official interpreter is called *dragoman.*

The provinces are governed by *pashas*, *ayans*, and *waivodes*, who in their administrative form imitate those of the divan. The pashas unite the administrative and military power. The ensign of their authority is a standard to which are attached so many horse-tails, the number of which indicates the rank of the pasha. A pasha of three tails, or *miri-alem* is the highest in rank, and commands the troops of his own pashalik when ordered by the sultan. A *miri-mirani*, or pasha of two tails, must march when required under a miri-alem. A *miri-liva*, or pasha of one tail, is still further restricted in authority. The pashas of Romelia and Anatolia are the highest in rank; the former is at the head of all the European pashas; and the latter presides over the Asiatic pashas. The *ayans* are superior officers named by the Porte for particular services. They have occasionally endeavoured to make themselves independent, but their power and number have been greatly reduced by the present sultan. The *waivode* is the governor of a city or district detached from a pashalik. The seraglio has officers peculiar to itself. The superintendent of the women is called *kislar-aga;* he has the command of the black eunuchs, and is himself one. The commandant of the white eunuchs is called *capi-aga;* he receives all the petitions which are to be delivered to the emperor. The number of inferior officers is very great, but an enumeration of them is not necessary.

Laws.] The laws of the Turks are founded upon the Koran; but as Mahomet was not acquainted with life in its most civilized form, his regulations were, in the advancing stages of society, soon found to be defective. The Turkish divines have endeavoured to remedy this defect by commentaries on the Koran; and their opinions have in many cases been sanctioned as laws. The commentator who is chiefly respected in T., is said to be Abu Hanifa. The laws of the Turks appear frequently to be founded on correct notions of natural justice; but the happiness of the subject depends not only on the excellence of the laws, but on the rectitude with which they are executed, and an incorruptible judge is somewhat uncommon in this country. Turkish law is simple and summary. Civil matters are decided by the ulemas, who have the case laid before them upon a sheet of paper, space being left at the bottom for their fiat. Two or three witnesses are examined, and a decision given *instanter*. Public offences are submitted to the pashas, who have the power of inflicting capital punishment, and from whose decision there was formerly no appeal; and none daring to impugn the decision, the matter ended with the sentence and execution. The sultan, in full council, abrogated the power of the pashas to inflict capital punishment, which can now only take place on a judicial sentence having been obtained and signed by a cadi, which also allows time for appeal. Their power of administering corporal punishment, however, still exists.

Revenue.] The sources of the Turkish revenue are: the *karatsha* or capitation tax on all who are not Mussulmans,—a tax on incomes,—the customs, a kind of land-tax, amounting it is said to no less than six shillings an acre,—the confiscated estates of wealthy pashas, who are often accused of crimes for the sake of their riches,—and the yearly tribute received from Egypt, Wallachia, Moldavia, and Servia. The sum generally arising from all these sources is estimated by Balbi to amount to £10,310,000. The sum transmitted from the great pashaliks, and the duties payable both at the custom-houses and on certain articles of consumption, continually vary greatly. The annual capitation tax falls on the *rayas* or Christians alone. It is proportioned to the means of the payer: the rayas of each district being divided into three classes, who pay 16, 12, and 8 piastres respectively. From this compulsory tribute, the clergy, the women, children, and all who through age are incapable of supporting themselves, are exempted; so that the number of persons subjected to it, after deducting those on whom it does not operate, do not probably exceed 800,000, of whom one-fourth may belong to the first and second classes. The budget of the empire for 1852 showed a revenue of 750,000,000 piastres, or about £7,000,000; and a nearly equal expenditure. The ordinary income and expenditure may be stated thus:

Income.		Piastres.
Tenths or tax on produce of land,		220,000,000
Income tax,		220,000,000
Personal tax on subjects not Mussulmans,		40,000,000
Customs,		85,000,000
Tribute from Egypt,		30,000,000
Tribute from the principalities of the Danube, viz.: Wallachia,	2,000,000	
Moldavia,	1,000,000	
Servia,	2,000,000	
		5,000,000
Different indirect taxes, patents, stamps, excise, tolls, post-office, mines, &c.		150,000,000
Approximative amount,		750,000,000

Expenditure.	
The sultan's civil list,	75,000,000
Civil list of the dowager sultana and of the sultan's married sisters,	8,400,000
Army,	300,000,000
Navy,	37,500,000
War stock, ammunition, artillery, engineering fortifications,	30,000,000
Maintenance of the different offices of administration,	195,000,000
Foreign affairs, ambassadors, consulates, &c.	10,000,000
Dotation of the treasury for works of public service, roads and ways, encouragement of agriculture, &c.	10,000,000
Expenses of different kinds, such as a grant to the administration of the mosques for the maintenance of the establishments dependent thereon, annuities, and pensions,	84,100,000
	750,000,000

The tribute of Egypt has been increased, by the late arrangement made by the sultan with Abbas Pasha, from 30,000,000 to 40,000,000 Turkish piastres per annum. From this state of its budget, which may be regarded as approximatively correct, the revenue of T., previous to the breaking out of the present war, appears to have been balanced by its expenditure. It had no public debt, with the exception of its paper-circulation, amounting to 170,000,000 Turkish piastres, or about £1,600,000, bearing interest; and it was decreed that this paper-money should be withdrawn from circulation by means of a special tax, to exist only till this object was attained. A first step was made in the withdrawal of the paper-money, calling in about 4,000,000 f. worth, which was publicly burnt in the presence of the sultan. Besides this, T. had a floating debt of about 10,000,000 f. or 12,000,000 f., which it owed to the contractors from whom it buys for delivery the European products necessary for its army and navy, &c. The preceding data are drawn from the best sources; but it is, as may be easily understood, impossible to guarantee their being rigorously correct. Although the state of the Turkish people differs from that of the Christian and Jewish subjects of the Porte, by the political rights which they enjoy, and by their exemption from the poll-tax and other horribly vexatious imposts reserved for the subjects who are not Mussulmans, they nevertheless pay quite enough. The agriculturists, besides tithes,—the merchants, besides enormous custom-house duties,—the artists and the workmen, besides the expenses which they voluntarily contribute to their respective corporations, besides the innumerable vexations inflicted upon them by the local authorities, also pay a mass of taxes: such are those under the name of presents for the feast of the bairam; of expenses for the fast of the ramazan; of presents for the pasha, and for his highness's stirrup; of expenses for the repair of his seraglio and his country-houses; of expenses for his stables, his kitchen, and pantry; after which come the taxes called *equivalents;* as for example, the equivalent for the building of men-of-war; that for the sheep for provisioning the seraglio and the capital, &c. All these taxes, unequally distributed, and collected with violence, would be less ruinous were they not tripled by the rapacity of the functionaries charged with their collection; and as if the sultan were moderate in his demands, as if he left to his functionaries something to glean, his absolute lieutenant and his pashas in the provinces, who are frequently changed every six months, compel the people, for their own account, to pay an imposition which is called *kudoomiye*, or congratulation on their happy arrival! The *vergu* or income tax varies according to locality from 10 to 25 per cent., and is raised upon the presumed fortune, real or personal, of all the subjects of the sultan. The municipalities are charged with its apportionment. The tenths are collected upon all the productions of the earth, and even in some quarters, extend to sheep. At the conclusion of the war with Russia in 1776, the public debt amounted to 36,333,520 piastres, or about £3,628,350. It is now said to exceed £12,000,000.

Army.] For two hundred years, namely, from the end of the 14th to that of the 16th century, the Janissaries composed the flower of the Turkish armies, and so long as the first sultans ruled from the heart of camps, and declared their decrees from the imperial stirrup, the obedience of these warriors was secured, and there never was a fitter instrument of war and conquest. In this period, all the great successes of the Turkish army were gained; but, when the sultans began to prefer the pleasures of indolence to the visions of ambition, and exchanged the toils of the camp for the debaucheries of the harem, the discipline of this corps relaxed, and its arms became more dangerous to the ministers than to the enemies of the government. Many attempts were made to suppress it, but in vain; until the late sultan, Mahmud II., acting with more caution as well as energy, in 1826 contrived to annihilate at one blow this formidable body of warriors. The number destroyed is variously reported. All the officers, with the exception of a few of high rank who had joined the sultan's party, were known to have perished; and the general opinion is, that upwards of 20,000 were massacred during the struggle. In the recent and present struggles with Russia, and while co-operating with the British troops in Syria, the regular soldiers have exhibited the valour and discipline of European troops, and even the contingent levies fight with considerable skill. Eton calculated the Turkish army, in 1802, at 207,400 infantry of all descriptions, and 181,000 cavalry; making a total of 368,400 men, of whom 206,000 were employed as guards and in garrisons. The disposable force of T., exclusive of the contingents from Bosnia, Servia, and Egypt, was estimated in 1852 at 340,000 men, but of this number not above 170,000 could be readily mustered. At present the military force of T., as re-organized by Riza Pasha, a Prussian officer, in 1839-40, is divided into the regular active army called *nizam*, the reserve or *redif*, the contingents of auxiliaries, and the irregular troops. The regular active army under the system arranged, but not definitely organized by Riza Pasha, will consist of six *ordou* corps, or camps, under the command of a *mushir* or field-marshal, with their head-quarters at Scutari, Constantinople, Monastir, Karbront, Damascus, and Bagdad. Each *ordou* is to consist of two divisions, commanded by a *ferik* or general of division. Each division is to be divided into three brigades, commanded by *livas* or generals of brigade. The ordou consists of 11 regiments: viz. six regiments of foot, four of horse, and one regiment of artillery, amounting in all to 25,000 troops of the line. Besides the 6 ordous, there are to be three detached corps: viz. one brigade on the island of Candia, numbering 4,000 regulars, 3,500 irregulars, and 600 native cannoniers; a second brigade in Tripoli, consisting of one regiment of foot, and one of horse, to the strength of 4,000 men; and a third brigade, of 4,000 men, at Tunis. The three detached corps should thus muster 16,000 foot and horse. The special corps, under the command of the grand master of the artillery, are likewise divided into ordous. They consist of—1. The central corps of artillery, comprising 4 regiments, a reserve regiment, and 3 regiments which are quartered in the various fortresses of the empire, in the forts of the Dardanelles and of the Danube, on the coasts of Asia Minor and the Black sea; 2. The engineer brigade of 2 regiments of 800 men each. The *redif* or reserve forms a second army, with the same organization as that of the *nizam*, and consisting of the same number of regiments of the various arms. These regiments are to be divided into battalions, escadrons, or companies, and have their standing staffs of officers and corporals on active service and receiving full pay. It is the duty of the latter to live in the towns and villages among the soldiers, who, though on leave, are, nevertheless, not discharged from the service, and these soldiers they must collect and drill once a-week. The *redifs* are to gather every year for four weeks at the head-quarters of their respective ordous, and take part in the field manœuvres. While thus on service the *redifs* are to have the pay and the usual allowance of provisions. This system is admirably arranged; but is not, we fear, likely to be speedily carried out in T. Of the six ordous only the first three were completely organized in 1852, but not more than half the cavalry was mounted. The fourth and fifth ordous were very deficient, the reserve being without arms. The sixth, the army of Bagdad, had neither cavalry, artillery, nor reserve. This ordou might be looked upon as quite cut off from the rest, by its great distance, defective communications, and more particularly by the Arab tribes of the desert, always more or less in open rebellion: entire battalions have gone over to the Persians from time to time with bag and baggage. In addition to the force of T. proper, there is nominally a large auxiliary force consisting of the contingents of the tributary provinces of Servia, Bosnia, and the Herzegowina, Upper Albania, and Egypt; but it is difficult to give even an approximate idea of the numbers of the auxiliaries; it much depends on the politics of the time or the enemy against whom they are expected to take the field. In a war with Russia the number of auxiliaries may possibly amount to 105,000 or 110,000 men: viz. 6,000 from Wallachia, 40,000 from Bosnia and the Herzegowina, 20,000 from Upper Albania, 30,000 from Servia, and 25,000 from Egypt. When the Danubian principalities are occupied by the Russians, the auxiliaries from Wallachia are not permitted to join the Turkish army; but it is likely that free corps of Pandurs will assemble in Lesser Wallachia, and these free corps may possibly muster to the strength of from 6,000 to 8,000 men. The irregular troops consist of:—

1. *Kavas* or gendarmes on foot, *seymens* or mounted gendarmes, and *subishis* or county militia,	30,000
2. Tartars of the Dobrudja and Asia Minor,	5,000
3. Hungarian and Polish volunteers, .	2,000
4. Moslemim volunteers, . .	50,000
Total of irregulars,	87,000

The total of the military forces of T. are summed up in the Berlin *Zeit* as follows:—

		Regiments.	Registered.	War footing.
Infantry, . .		36	117,360	100,800
Cavalry, . .		24	22,416	17,280
Artillery, . .		6	7,800	7,800
Artillery in fortresses,		4	5,200	5,200
Engineers, . .		2	1.600	1,600
Detached corps	In Candia,	4	8,000	8.000
	Tripoli,	2	4.000	4.000
	Tunis,	2	4,000	·4,000
		80	170,376	148,680
Reserve, . .				148,680
Auxiliaries, . .				121,000
Irregulars, . .				87,009
			Total,	505,360

These 500,000 troops are not, of course, prepared to take the field in the first weeks of a campaign—first, because the distances are too great, and the various provinces so thinly populated, all of which interferes with the immediate concentration of the contingents; and secondly, because the expense of arming and keeping up all these troops would be too heavy for the imperial treasury. It ought, also, to be remembered that the numbers and services of the auxiliary contingents are affected by a variety of circumstances and events, and that the number of 121,000 may, according to circumstances, be increased or lessened by one-half. Thus, in the present war, Abbas Pasha, the viceroy of Egypt, hastened to place at the disposal of the Porte, 45,000 troops, a line-of-battle ship, 2 first-class frigates, 4 corvettes, 2 brigs, 2 steam-frigates (one of 850 and the other of 450 horse power), and a steam corvette of 300 horse power. On the other hand, Mustapha pasha, an Albanian, recently appointed governor of Herzegovine, offered to furnish the Porte with a levy of 200,000 Albanians, and to march them on any point which the government of the sultan might point out; and Rendev-han Bey also demanded from the Porte to be permitted to organise in Kurdistan a considerable body of troops. According to the Koran, infidels cannot be admitted into the army; recruits, therefore, can be found only in the Moslem pop. A recruiting commission, consisting of several staff and other officers, mollahs, doctors, and subalterns, provided with an imperial firman proceeds to the appointed district where lots are to be drawn. The period of service is 5 years. Those who escape in the lot-drawing one year must draw again every year for four more, and if they escape till the end of the five years they then enter the *redif*, which is thus composed of raw militia and only exercised one month in a year. The appointment of the recruits to the different arms of the force is not made with regard to the personal fitness of individuals, but at random or by caprice: thus the Albanians, who are good riflemen, are taken for the artillery; fishermen, sailors, and peasants, who never sat on a horse, are sometimes drawn off to the cavalry. The drilling and exercising is partly Prussian, grafted on the remains of the former French system.

Navy.] The Turkish navy consists of a number of small vessels of war and several ships-of-the-line. Before the battle of Navarino it was estimated at 21 sail-of-the-line, 31 frigates, 8 corvettes, and 30 gun-boats, carrying altogether 2,990 cannon, and 5,300 seamen. At present it consists of 2 three-deckers of 130 and 120 guns; 4 two-deckers of 74 to 90 guns; 10 sailing frigates of 40 to 60 guns; 6 corvettes of 22 to 26 guns; 14 brigs of 12 to 20 guns; 16 cutter-schooners from 4 to 12 guns; 6 steam-frigates of 400 to 800-horse power; and 12 steam-corvettes and schooners, making a total of 70 armed vessels, manned by 32,000 sailors and gunners. There is also a regiment of marines under the command of a general of brigade. The staff of the navy consists of the *kapudan pasha*, first lord of the admiralty and secretary-at-war; 5 admirals, of whom 3 are in active service; the commander of the fleet, the kapudan or commander of a squadron, the *liman-kaissi* or port-admiral, 3 vice-admirals, the vice-admiral of the fleet, the director of the dock-yard, the director of the naval academy, 7 rear-admirals, 3 of whom command on the stations of the Danube and the Black sea, the archipelago and the Persian gulfs, while 4 belong to the council of the admiralty. These commanders have the same rank and pay as the generals of the army; the admirals are on a footing with the *feriks;* the vice-admirals rank with the *livas*, and the rear-admirals with the *mir-alaïs* or colonels. The commander or flag-captain of an admiral's ship, too, has the rank of a colonel.

History of the Ottoman empire.] The finest countries of the old world—Thrace, Greece, Asia Minor, Colchis, Armenia, Mesopotamia, Syria, and Egypt, together with the islands of the Archipelago and Arabia, whose commerce connects Asia and Africa with Europe, and unites the East with the West—have been ruled for five hundred years by the Turks or Ottomans, a race of mixed blood, whose name begins to appear in history about the middle of the 6th cent., when their Scythian ancestors settled at the foot of the Altai, in the steppes of Upper Asia, between Siberia and China.

In the middle of the 7th cent. a new religion appeared in the deserts of Arabia, and rapidly propagated itself throughout Asia and Africa. Mahomet's successors, the khalifs, in 638, conquered Syria and Palestine, and in 762, Bagdad the seat of a new monarchy was founded under their auspices. The khalifs of Bagdad, admiring the reported bravery of the Turks, had long chosen their body-guard from this tribe. The chiefs of this guard gradually assumed great authority in the affairs of state. Hence arose the *Emir-al-Omrahs* of the khalifs, who, like the *Majores Domus* of the Franconian kings, under the name of prime-ministers, gradually engrossed the real power of the state, and finally elevated themselves to the throne of their masters. In the 9th and 10th centuries the Turkish dynasties of the Tulunides and Akshidides reigned in Palestine, Syria, and Egypt; and from the end of the 10th to the close of the 12th cent. the dynasty of the Gasnevides reigned in Persia and India. The Seldshucks or Seljuks, another Turkish tribe, nearly about the same period that their brethren won the dominion in these countries, successively crossed the Siphun and Oxus, attacking nations immersed in luxury, and rendered effeminate by long-continued peace, and had overrun a great part of Asia before the close of the 11th cent. and founded a mighty empire in the East with which the Crusaders contended for the possession of Palestine. In the 13th cent. the Mongols—a nation differing in language and manners from the Tartar-stem to which the Turks belonged—uniting with several other hordes, overthrew the power of the Seldshucks in Asia Minor, and founded several small Mongolian states in that country; but the chiefs and emirs of the Seldshucks and Turkomans soon led back their tribes from the valleys of the Taurus and divided Asia Minor amongst their warlike followers. One of these emirs was Osman or Othman, the founder of the Ottoman empire. This chief was of the tribe of the Oguzian Turkomans. In 1299 he took possession of the narrow passes of Olympus, and established his camp of Caucasian-Tartars in the plains of Bithynia, under the protection of the Seldshuck sultan of Iconium. Reinforced by the accession of a number of fugitive slaves, robbers, and captives, he laid waste the surrounding countries, and conquered some provinces of Asia Minor belonging to the Roman empire of the East. Upon the death of the sultan of Iconium in 1300, (Hejira 700) Othman assumed the title of sultan himself: thus was founded by a predatory chief, upon the ruins of the power of the Arabs, Seldshucks, and Mongols, the empire of the Ottoman Turks in Asia. Between 1300 and 1566, the gallantry and activity of ten great princes aided by the heroic fanaticism with which Islamism inspires its devotees, raised the Ottoman state to the rank of the first military power in Europe. Orcan, Othman's son, established the seat of his empire at Brussa, the capital of Bithynia, which he conquered in 1328. He then subdued Asia Minor to the Hellespont: and having assumed the title of *padishah* or 'prince,' obtained the hand of a daughter of the Grecian emperor Kantakuzenos in marriage. Orcan's son, the gallant Solyman I. entered Europe in 1355, and having seized Gallipoli and Sestos, made himself master of the straits which divide Europe from Asia. Amurath I. established the seat of his empire at Adrianople, and profiting by the feebleness of the Greeks, expelled them from Thrace, and defeated the Bulgarians and Servians in a dreadful battle on the plains of Cossova.

Bajazet, who succeeded Amurath, by an uninterrupted train of victories conquered the greater portion of that territory which the Turks still possess in Europe. Palæologus the Greek emperor solicited the assistance of the western princes of Christendom, and an army was speedily raised which, under Sigismund king of Hungary, succeeded for some time in counteracting the efforts of the sultan. Their hopes of ultimate success, however, were soon terminated; Bajazet's skill and bravery forced them to retreat towards the west and leave him at liberty to direct his whole strength against the city of Constantinople, which was on the eve of falling into his hands, when the unexpected tidings of an invasion by the Tartarian prince Demir, better known under the appellation of Timur or Tamerlane, forced him to hurry into his Asiatic provinces for the purpose of checking the career of this new enemy. The two hosts met each other with the intrepidity which was natural to them, and with an ardour inspired on both sides by a career of uninterrupted success. The contest was furious and sanguinary, but unfortunate for the Turks; the sultan was made prisoner, and the greater part of his army destroyed in a dreadful battle which lasted from daybreak to nightfall. The defeat of Bajazet for some time retarded the increase of the Turkish power; but Amurath II. led his army into Macedon and plundered the principal cities of that country, as well as those of Bœotia, Ætolia, and Phocis. Servia next yielded to his arms; and entering Hungary, he invested the city of Belgrade. The intrepidity and skill of John Hunniades enabled the Hungarians not only to defend themselves, but to carry on offensive operations against the Turkish forces. In the celebrated siege of Belgrade the sultan lost many of his men; no fewer than 150,000 are said to have fallen in one attack made upon them from the town, and the siege was at length raised. Irritated by discomfiture to which he had been little accustomed, the sultan sent a great army into Transylvania; but Hunniades met them, and with the loss of only 3,000 of his own men, is said to have destroyed upwards of 30,000 Turks. Another army more numerous was despatched against the Christians, and its efforts were equally unsuccessful. But these repeated defeats, far from humbling the Turks, only stimulated them to still greater efforts; another army was immediately despatched towards the west, and the Christians on the 10th of November 1444, experienced a complete defeat before Varna. In the eastern parts of Europe there now remained no power which could with any prospect of success oppose itself to the Turkish arms. The Greek emperor's power was confined within the walls of Constantinople; and for the narrow surrounding territory he was forced to pay the sultan an annual tribute of 300,000 aspers. Cannon were now introduced into the Turkish army, and the whole Peloponnesus was subjugated. Mohammed II. who, in 1449, succeeded his father Amurath, desirous of terminating a war in which he was engaged against the prince of Caramania, made peace with the emperor of the Greeks; but he had no sooner brought this contest to a close, than he provoked a fresh dispute with the Greek emperor, and laid siege to the imperial city with an army amounting it is said to 400,000 men. The defence was desperate, but crowds of Turks continually pressing forward to supply the places of those who had fallen, the strength of the heroic garrison was at length exhausted; the emperor Constantine, with all his warriors, was overwhelmed in the assault, and for three days the imperial city continued to be a scene of the most inhuman carnage. Thus terminated the existence of the western empire in the year of the Hejira 857, 1,123 years after Constantine had removed the seat of the empire from Rome to Byzantium, and had given his name to that celebrated city which was now declared to be the seat of the Turkish court, and capital of the Turkish dominions.

Mohammed's success stimulated him to new undertakings. He entered Servia with 20,000 men and made it tributary; and next with an enormous force laid siege to Belgrade, which he was only prevented from taking by the extraordinary talents of Hunniades, who defeated the Turkish fleet and army, and forced the sultan to retire into Servia, but did not long survive his glorious victory. Disappointed in his hopes of success against Hungary, Mohammed turned his attention to the Morea and islands of the Ægean sea. The former region had been seized by the Albanians,—a warlike race who offered to deliver the cities to the Turks on condition of being permitted to retain in their power the open country, for the purpose of following their ordinary occupation of shepherds; but the Turks, more willing that this territory should be possessed by the feeble Greeks than by the fierce Albanians, attacked the latter and drove them from the country. The Greeks submitted with reluctance to the Turkish yoke, and made a few efforts to regain their liberty: but in 1459, the whole country, except a few places on the coast in the hands of the Venetians, was added to the Turkish empire. The conquest of the Morea was in a short time followed by that of Trebizond, the retreat of the Comneni, who had dignified their petty monarchy with the vain title of the empire of Trebizond. After the death of Hunniades, Bosnia had fallen into Mohammed's hands, and he was meditating new enterprises, when the most formidable adversary whom he had ever been called to encounter, appeared in the person of George Castriota, an Epirote by nation, generally called Scanderbeg. Three large Turkish armies were successively overthrown by the intrepid and skilful Epirote; and tired of a contest in which he had experienced uninterrupted disasters, the sultan contracted a truce with Scanderbeg. The western Europeans now once more began to dread the Turkish encroachments; and the Venetians formed an alliance with the Hungarians for the purpose of checking the progress of their eastern adversaries. In 1466 the sultan again entered Epirus with an immense army, and compelled Scanderbeg to take refuge at Lyssa in the Venetian states, where he died in 1466. The Venetians defeated the Turkish army in 1469, but this was almost the only advantage of which they could boast,—they were driven from Negropont, and the reduction of Epirus and Albania deprived of their intrepid defendant was almost instantaneous. The Turks in an attack upon Rhodes were repulsed; but from the Venetians they took the city of Cephalonia, and invading Italy, made themselves masters of Otranto.

Bajazet II. subdued Croatia and Caramania. The Venetians recovered Cephalonia; but this slight advantage did not counterbalance the success of the Turks, who took the cities of Lepanto, Durazzo, and Modon, and plundered Syria on the one hand, and Moldavia on the other. Bajazet now formed the design of resigning his power to his eldest son Achmid. But Selim, his second son, more enterprising and ambitious than his brother, suddenly arrived from Trebizond, and having put his father to death, ascended the throne in 1512, and by the murder of his brother secured his power. Selim attacked the Persians, and along with some other places took the city of Tauris. He next turned his arms against Egypt, and finished the conquest of that country which had been commenced by his father. The shortness of his reign did not permit him to extend his conquests farther; he died in 1519.

Solyman, the tenth sultan of his race, conquered Belgrade, and drove the knights of Malta from Rhodes, after a defence which has excited the admiration of all succeeding ages. The fatal field of Mohatz witnessed the overthrow of Corvineu, who, with an army amounting only to 25,000 men had rashly but gallantly attacked 200,000 Turks. Buda immediately surrendered; and though it was retaken in 1528, it almost immediately fell again into the hands of the Turks. Moldavia was next forced to submit to the victorious Mohammedans, who then proceeded to the siege of Vienna; but the autumnal rains retarded the operations of the besiegers and forced the sultan to abandon his undertaking. The Turks had no sooner raised the siege of Vienna than their enemies acquiring new courage, assailed and drove them from every part of the German territories. Solyman now directed his arms against the nations on his eastern frontiers; Georgia was reduced, and Bagdad also fell into his hands, while Barbarossa, his celebrated admiral, caused himself to be dreaded in every part of the Mediterranean. Hungary had long been harassed by the Turkish forces: it was once more attacked and reduced to the state of a Turkish province. Selim II. who succeeded his father, for some time conducted the wars of the empire with equal vigour and prosperity; but his fleets failed to win that success which attended his army. The commercial powers in the Mediterranean combined their forces, and effected the total destruction of the Turkish fleet. The power of the Turks, and that of their European neighbours, was now nearly balanced. In the reign of Amurath III. who succeeded Selim, the advantage became more evidently in favour of the Christians: and since that time, though the Turks have sometimes enjoyed a transitory success, the real stability of their affairs has declined.

During a long period little alteration took place in the mutual relations between the Turks and western Europeans. Wars indeed were frequent; but few advantages were gained on either side, and these few were not of great importance. In 1684 the Ottoman empire in Europe seemed to be verging towards destruction, when by the wisdom and bravery of one man its affairs were once more retrieved. The vizier, Achmet Kiuperli, had the skill to call to his aid that religious enthusiasm which when properly managed conducts men to the most daring undertakings,—the merit of opposing the armies of infidels was placed in the strongest light,—and the predestinating creed of the prophet was forcibly inculcated. The effects of the vizier's policy were quickly visible, the Christians were defeated in almost every quarter, and in 1676 the Turks had retaken Candia, Podolia, and the Ukraine, and might, perhaps, have extended their conquests still farther, had they not lost their vizier at this juncture in an engagement with the Germans. After the death of Kiuperli, the Turkish affairs fell into confusion; but negotiations for a peace were commenced; and in 1698, a treaty was concluded at Carlowitz between the contending parties. The Germans retained possession of Transylvania; the Banat of Temiswar and part of Sclavonia was yielded to the Turks; the navigation of the Theisse and the Maros was declared free to both parties; a proper boundary was fixed between the German and Turkish parts of Hungary; and the truce was to continue 25 years. The peace with the Russians was to continue only for two years, but they were to retain Azof. With the Poles, the duration of the treaty was to be the same: Moldavia was yielded to the Turks, while Kaminieck, Podolia, and the Ukraine were to be restored to the Poles. The Venetians retained possession of the Morea, with several other parts of the Turkish territories, and several ports were declared free to both nations. The cession of the Morea was the provision to which the Turks most unwillingly acceded; they seized the opportunity of some commotions which prevailed among the European powers, and drove the Venetians from that part of their territories.

Under Peter the Great, the Russians continued to view with envy these fine provinces which owned subjection to the Turkish dominion, and seized the first opportunity of quarrel. They had for some time taken an active hand in the disturbances of Poland, and had effected what they were pleased to call a settlement of the affairs of that distracted kingdom: a discontented Pole retired to some part of the Turkish territories, and probably without the knowledge of the Turkish government, invited his countrymen to join him, promising them the support of the Ottoman forces,—the Russians, without inquiry, seized this opportunity of attacking the Turks, and in 1769 declared a new war. The success which attended the Russians too plainly betrayed the feeble state of their adversaries The fortress of Azof, at the mouth of the Don, was seized; and Gallitzin, passing the Dniester, endeavoured to seize the fort of Choczim, but was forced to repass the river. Under Romanzof, the Russian arms were more successful; Choczim and Yassi were successively taken, and Romanzof invested Bender, but finding the season too far advanced for carrying on the siege, he withdrew his forces. In May 1770, Romanzof again passed the Dniester; and marched from Choczim towards the Pruth, while Panin with another army arrived before Bender, which city was invested by the latter, while the former covered his operations. Romanzof attacked the Turks at Cahoul, and obtained over them a complete victory, which compelled the grand vizier to repass the Danube. In the meantime, the fortresses of Killanova and Akerman or Bialogorod, surrendered by capitulation. The strong fortress of Bender resisted until reduced to rubbish: the fortress of Brailow also was abandoned by its garrison, and immediately fell into the hands of the Russians. A Russian fleet had likewise entered the Mediterranean, and landing a body of forces upon the Morea, induced the Greeks to revolt from the Turkish government; Russian intrigues disunited the Tartar chiefs; Palestine revolted from Turkish sway, and Ali Bey chased the representative of the sultan from Cairo. In 1771, Dolgorouki, in spite of an army of 50,000 men by whom it was defended, made himself master of Crim Tartary, but the Russians durst not attack the grand vizier amid the recesses of the Balkans. The Turks had seized the fort of Giurgevo; and in Wallachia they were so powerful that Repnin refused to attack them, and the Turks resolved to winter on the north of the Danube; but ultimately found it impossible to maintain their ground on the south. The Turkish army, divided into two bodies, were attacked almost at the same instant, and both were defeated. The fort of Babadagh thereupon surrendered; the vizier with his remaining forces retired behind the Balkan; Giurgevo was retaken; and the Turks were forced to evacuate Wallachia, while the Russian fleet ravaged all their maritime possessions, and threatened Constantinople itself. The Turks now earnestly desired peace; but the Russians demanded either that they should enjoy the free navigation of the Black sea, retain possession of the city of Azof, and be allowed a certain sum as an indemnification for their expenses; or that the Crimea, Budziac Tartary, and the extensive territory bounded by the Cimmerian Bosphorus and the N bank of the Danube, should be yielded to them in perpetual possession. The negotiations which had been carried on during a whole year, terminated fruitlessly in March 1773, and each party prepared for a renewal of hostilities. After several unimportant skirmishes, Romanzof passed the Danube at the head of 87,000 men, and marched towards Silistria, which was defended by 24,000 men. To oppose the progress of the Russians, and to relieve the place, 50,000 men were despatched by the vizier, who compelled Romanzof to retire hastily, abandoning his ammunition and other military stores. At the moment of this return of prosperity the sultan Mustapha closed his days. Romanzof's army, powerfully re-enforced in 1774, repassed the Danube. Many engagements took place, in which the Russians were uniformly victorious, while from the Caucasus to the Danube the Ottoman power was pressed upon by their forces. The Turks were now entirely dispirited; multitudes laid down tneir arms; and comparatively few of them remained obedient to their officers. In such a situation, it became necessary to accede to whatever terms the Russians chose to dictate, and Romanzof, in the treaty concluded at Kutshuk-Kainarji, in July 1774, was careful that they should be sufficiently advantageous for his country. The Crimea was declared to be independent; Kilburn, Kertch, Jenikale, and the country between the Bog and the Dnieper, were ceded to the Russians, who were also to enjoy the free navigation of the Turkish seas, and to be permitted to pass the strait called the Dardanelles; some stipulations were made in favour of the inhabitants of Moldavia, Wallachia, and the Greek islands, which had been in the power of the Russians, and were now to be restored; and the Russians were to retain possession of Azof and Taganrog; but, except the places which have been mentioned, agreed to relinquish all their other conquests. Austria gained the Buckovine by the same treaty. The claim of Russia to a right of surveillance in reference to the Christian religion and its churches in T. rests on this treaty "of perpetual peace and friendship,' in which "the two empires agreed to annihilate and leave in eternal oblivion all the previous treaties between them." By art. 7, "the Sublime Porte promised to protect constantly the Christian religion and its churches; and it also allowed the ministers of the imperial court of Russia to make upon all occasions representations, as well in favour of the new church at Constantinople, of which mention was to be made in art. 14, as on behalf of its officiating ministers, promising to take such representations into due consideration, as being made by a confidential functionary of a neighbouring and sincerely friendly power." By art. 14, "after the manner of the other Powers, permission was given to the high court of Russia, in addition to the chapel built in the minister's residence, to erect in one of the quarters of Galata, in the street called Bey-Oglu, a public church of the Greek ritual, which should always be under the protection of the ministers of that empire, and secure from all coercion and outrage." By art. 8, Russian subjects were to have full liberty to visit Jerusalem, and no contribution or tax was to be exacted from these pilgrims and travellers; "but they were to be provided with such passports and firmans as were given to the subjects of the other friendly Powers." By art. 17, Russia restored the islands of the Archipelago to the Sublime Porte, "and the Sublime Porte on its part promised to observe religiously, with respect to the inhabitants of these islands, the amnesty stipulated in art. 1; and that the Christian religion should not be exposed to the least oppression any more than its churches, and that no obstacle should be opposed to the erection or repair of them, and also that the officiating ministers should neither be oppressed nor insulted." In 1787, Mansur, a sheik of the Kuban Tartars, led the hordes over whom he had established his influence, against the Russians; but found his disorderly rabble unfit to withstand the regular forces of Potemkin who vanquished him in several engagements. The island of Taman, and the Crimea, were likewise attacked by the Turks; but in this quarter too they were repelled by the Russians. Joseph emperor of Germany, though he had no legitimate cause of quarrel, assisted the Russians with an army. The opposition of the Turks, though vigorous, was not successful; Czerenitz and Bucharest in Wallachia, and Cladova in Servia, yielded to the Austrians; while a Russian army of 150,000 men, under the command of Potemkin, Romanzof, Repnin, and Soltikow, approached from the banks of the Bog. Oczakow was besieged by Potemkin, and after an obstinate defence of four months, was taken in December 1788; and the Russians made themselves masters of Akerman and Bender. The death of Abdul Hamid occurred in the opening of 1789, when Selim III., the only son of the sultan Mustapha, mounted the imperial throne. In June 1790, a conference was held at Reichenbach, at which the ministers of Britain, Holland, and Prussia, with those of Austria and Poland, were present, and peace was established between the Austrians and the Turks, upon a treaty one article of which was that all conquests made by the former should be relinquished. The Russians were thus left to manage the contest by themselves. Suwarrow, in 1790, took the fortress of Ismail. Such had been the determined valour of the defendants, that the Russians lost 10,000 men in the attack: and such was the brutal and ungenerous cruelty of the assailants, that the garrison, amounting to 30,000 men, were massacred in cold blood. Varna, the bulwark of Constantinople towards the Balkan, was now threatened, and the war on the side of the Turks had become a war of defence, and on the side of the Russians a war for the extension of their territories, when Great Britain interfered to curb that spirit of conquest which the Russians had so openly displayed. The Russians at first resented the interference; but gradually became less exorbitant in their demands, and at length consented to peace on condition of retaining Oczakow and the territory extending from the Bog to the Dniester. These were ceded to Russia, by the treaty of Yassi, in 1792, together with an extraordinary right of interfering in the provinces of Wallachia and Moldavia, that their respective hospodars should be continued in office seven years, and not removable but by the consent of Russia. The reigning hospodars were deposed before their time; and when the Russians remonstrated, the Bosphorus was closed against their ships. Taking umbrage at these causes of complaint, in 1806, Michelson was despatched with an army of 60,000 men, who crossed the Dniester, took Bender and Chotzim with little resistance, and entered Yassi, the capital of Moldavia. He then entered Bucharest, and took entire possession of the provinces of Bessarabia, Moldavia, and Wallachia. The Russians remained in almost undisturbed possession of these provinces till the year 1810; when the armies on both sides were augmented to 200,000 men, and a fierce and sanguinary contest ensued, which, perhaps, never was surpassed. The Russians passed the Danube in three places. Their direct progress would have been from Giurgevo to Rustchuk; but at this latter place the passage was impracticable. They therefore crossed over above it, at Ostrova near Widdin, and below it, at Hirsova and Turtukai, and laid siege to Rustchuk. The town was vigorously defended; and the Russians were repulsed in a desperate attack, in which they lost 6,000 men. Kaminsky made also a similar assault on the intrenched camp at Shumla; but here, too, he was driven back with great carnage. The Turks, though unacquainted with regular discipline in the field, make a fierce resistance when attacked behind their ramparts. In the month of September, Kaminsky, leaving Langeron before Rustchuk, with his disposable force suddenly attacked and defeated the Turks at Bayne; whereupon Rustchuk was compelled to surrender, with all the Turkish flotilla lying before it, and Giurgevo on the other side. In order to create a diversion, the Turks now sent a fleet into the Black sea, and threatened an attack on the Crimea. Notwithstanding this, the Russians concentrated their forces in Bulgaria, and the grand vizier was obliged to retreat before them, recross the Balkan, and take up a position at Adrianople; leaving, however, the strong and impregnable fortresses of Varna on the sea-coast, and Shumla on the ascent of the mountains, well-secured at the other side. The feeble Selim, and his successor Mustapha, had both been strangled; and Mahommed was called to the throne, who soon displayed the vigour which ever afterwards distinguished him. The new vizier, Ahmed Aga, a man of the same energy as the sultan, and who had distinguished himself by his defence of Ibrail, descended from the mountains, forced the detached corps of Russians in Bulgaria to recross the Danube, and made a fierce attack upon Rustchuk, now defended by the Russian general Kutusof. The Russians, hard pressed, transported the inhabitants to the other side of the river, set fire to the town in four quarters, and then retreated themselves. The Turks rushed into the burning town, put a stop to the conflagration, and took up their position there. The grand vizier having thus driven the Russians to the opposite bank, now determined to follow them, and made the attempt in three places, Widdin, Rustchuk, and Silistria. He succeeded at Widdin, and established 30,000 men in Wallachia. He also succeeded at Rustchuk, took possession of a large island in the river, and, in perfect confidence, passed the greater part of his army to the other side, and established them in an entrenched camp. But Kutusof immediately availed himself of the vizier's crossing over, and detached 8,000 men, under Markof, to attack the camp he had left behind, which fell into the hands of the Russians. The fugitive Turks crowded into Rustchuk, where they were cannonaded by the artillery of their own abandoned camp, while Langeron from the other side directed 100 pieces of cannon upon them. In this state they endured the severest privations, and were compelled to surrender, after having lost 10,000 men in the different assaults made on them. The exhausted state of the Turks, and the critical state of the Russians, now invaded by the French, induced an accommodation; and the peace of Bucharest, concluded in 1812, gave another accession of territory to the Russians, extending their frontier from the Dniester to the Pruth, and assigning to them all the country that lay between the two rivers, comprising Bessarabia, and a considerable part of Moldavia. The Russians upon this withdrew from the provinces of Wallachia and Moldavia, which they had occupied for seven years.

Greek revolution.] During the sitting of the congress of Laybach, at the moment when the Neapolitan states were invaded by an Austrian army, a revolt of the Greeks against their Turkish masters burst forth in Moldavia, Wallachia, the Morea, and the Grecian islands. These news were received with disapprobation by the congress of Laybach, and in 1822 the deputies were refused their demands at the congress of Verona. In the beginning of the struggle the Greeks suffered several severe defeats from the Turks; but the Greeks entered on the road to freedom with better success in the Morea and in the islands. On the 23d of September 1821, the Greeks took Tripolizza, the capital of the Morea, and on the 1st of January 1822 the political existence and independence of aii Greece was proclaimed. The Greeks gained signal naval victories over the Turks at Mitylene and in the gulf of Patras, and on the 16th of December 1822, Napoli-di-Romania was taken by capitulation, and the seat of the government transferred to that city from Tripolizza. The campaigns of 1823 and 1824 were alike inglorious and disastrous to the Turks. In March 1825, Ibrahim, son of the viceroy of

Egypt, took possession of Navarin. With this fierce and warlike chief the Greeks maintained a bloody and desolating struggle for the peninsula, until the combined fleets of Britain, Russia, and France, swept the Grecian seas of the Turkish fleet, and relieved the Morea of the presence of its ruthless invaders. Previous to this interference of arms, a treaty had been signed by these powers requiring from the belligerents an immediate armistice, as a preliminary step to a reconciliation being effected between them, on the basis, that the Greeks should hold of the sultan as of a superior lord; and should pay to the Ottoman empire an annual tribute, the amount of which should be fixed once for all by a common agreement; that they should be governed by the authorities they should themselves choose and nominate, but in the nomination of whom the Porte should have a determinate voice; and that to bring about a complete separation between the individuals of the two nations, and to prevent the collisions which are the inevitable consequences of so long a struggle, the Greeks should enter upon possession of the Turkish property situated either on the continent or in the isles of Greece, on the condition of indemnifying the former proprietors, either by the payment of an annual sum, to be added to the tribute which is to be paid to the Porte, or by some other transaction of the same nature. It was the violation of the demanded armistice in the presence of the combined fleets sent to enforce it, which led to the collision between the combined and Turkish fleets in the bay of Navarin. In March 1829, the French and British ambassadors intimated to the sultan the intention of their respective governments to acknowledge and maintain the independence of the Greeks. A treaty was finally negotiated by the three Powers, the principal articles of which will be found in our historical chapter on Greece. By this treaty, Greece is effectually separated from the sovereignty of the Porte, and pronounced a monarchical and independent state.

Recent struggles with Russia.] It has been alleged by some politicians that the real source of the Greek revolution might be found within the Russian dominions: certain it is that as soon as that movement commenced, Russia began to augment her armies on the Pruth and Dniester. On the 26th of April, 1828, the Russians passed the Pruth, and advanced into the principalities of Moldavia and Wallachia, and on the 9th of June, the emperor Nicholas crossed the Danube and entered Bulgaria. The first operations of the war were preceded or rather accompanied by a manifesto and declaration on the part of the Russians. In the first of these documents, the emperor complained of the violation of the treaty of Bucharest in 1812, and of the subsequent treaty of Ackermann,—of the illegal seizure of Russian vessels and confiscation of their cargoes,—of the closing the passage of the Bosphorus, to the great injury of the Russian trade in the Black sea, —and of the intrigues of the Porte with Persia to prevent her making peace with Russia. The declaration further charged T. with violating her pledge to the Servians, and her guarantee to the provinces of Moldavia and Wallachia: an amnesty was to be granted to the Servians; instead of which the Turks had invaded Servia, and devastated the country. The privileges of the principalities were to be guaranteed; but instead of this, a system had been established of the most sweeping plunder, while the incursions of the Turks inhabiting the l. bank of the Kuban were secretly encouraged. Finally, after enumerating his grounds of complaint, the emperor declared war against the Porte, and stated the objects of the war to be: to enforce the due and effectual observance of those treaties which Turkey had violated; and to secure the inviolable liberty of the Black sea, and the free navigation of the Bosphorus. War having commenced, the sultan did not dispute the Russian advance into the principalities of Moldavia and Wallachia, but concentrated his strength on Shumla, and determined to defend the defiles of the Balkan mountains. In the course of the month of July, 1828, the Russian main force invested the strong fortresses of Silistria on the Danube, and Varna on the coast of the Euxine. The latter place was got possession of chiefly through the treachery of Jussuff Pasha; but the Russians were compelled to raise the siege of Silistria on the 10th of November, and to retire into winter-cantonments in Moldavia and Wallachia. The campaign of 1829 was preceded by the taking of the important fortress of Kali, the *tete-de-pont* of Nicopolis, on the 24th of January, and of Turnul, three weeks afterwards. The campaign was opened in May by the Russian forces marching upon the Danube, under Diebitsch, who immediately invested Silistria and Giurgevo; at the same time Paskewitch's division advanced along the Asiatic shore of the Black sea. The Turks endeavoured to impede the Russian armies in their principal operations by constant attacks on their positions, which rendered the contest very sanguinary on both sides. On the 11th of June the grand vizier's army sustained a severe defeat near the village of Kulertsha, in the neighbourhood of Shumla; and on the 30th of the same month Silistria surrendered to the Russians. Diebitsch immediately pushed forward across the Balkan into Thrace, after having closely invested Shumla, while Erzerum, the capital of Armenia, and the most considerable place in the eastern part of Asiatic Turkey—Bagdad excepted—surrendered to Paskewitch on the 27th of June. Mesembria, Achiolon, and Bourgas, were taken possession of by the former general with little opposition; and on the 21st of August, the Russian army entered Adrianople. The affairs of the Turks being now evidently desperate, conferences on the subject of peace were entered into, and a preliminary treaty adjusted; and on the 14th of September a treaty of peace between the two belligerent powers was signed at Adrianople. By this treaty not a foot of territory was gained or lost in Europe by either party, as far as the question of positive cession was concerned; but in Asia, Russia obtained a considerable district,—the ancient *Colchis*, at the east end of the Black sea. Turkey bound herself to pay £5,000,000 to Russia as an indemnity for the expense of the war, and £750,000 as an indemnity to Russian merchants for acts of spoliation committed upon them; the Servians were put again in possession of certain rights of which the Porte had recently deprived them; the Dardanelles and Bosphorus were declared free to the ships of all nations at peace with the Porte; and the fortifications of Giurgevo were to be demolished. Russia was not to evacuate Moldavia and Wallachia until the indemnities were paid; and the Porte formally recognised the political existence of Greece, as determined by Russia, in concert with France and England. This treaty "confirmed all the treaties which had been concluded at different epochs between the two empires," except as thereby modified. It contains no new provision on the subject of religion, or of Russian interference, except in relation to Moldavia and Wallachia, which provinces stood on a peculiar footing, and formed a special case. The Porte had received them back by the treaty of Kutchuk-Kainarji upon conditions set forth in art. 16, one of which ran thus: —"The Porte likewise permits that, according as the circumstances of these two principalities may require, the ministers of the imperial court of Russia resident at Constantinople may remonstrate in their favour, and promises to listen to them with all the attention which is due to friendly and respected Powers" These principalities had been the subject of further treaties at Jassy in 1792, and at Bucharest in 1812; and this treaty of Adrianople in 1829, speaking of these provinces, states, that "Russia had guaranteed their prosperity." The treaty of Balta-Liman, in 1849, recognized Russia as a protecting power in relation to these principalities. Immediately after the peace of Adrianople, in 1830, a most formidable insurrection burst forth in Albania. The grand vizier was sent to quell it; and, by dint of prudent and conciliating conduct, effectually succeeded in his mission.

Dispute with Mehemet Ali.] Though nominally professing allegiance to the head of the Ottoman empire and the Mussulman faith, the two pashas of Acre and of Egypt had been for many years nearly independent of his authority. Mehemet Ali of Egypt, though by far the more powerful of the two, persevered in a show of reverence and obedience to the Sublime Porte: not so Abdallah of Acre. In 1822, he boldly threw aside his mask of submission, and at the head of an army composed of Arabs, Druses, and mercenaries from all parts of the Turkish empire, he endeavoured to seize on the pashalic of Damascus. He was foiled, however, in his attempt, and speedily obliged to retreat. He shut himself up in his strong fortress of Acre, where, by command of the Porte, he was besieged by five neighbouring pashas. After a ten months' siege, Mehemet Ali, pasha of Egypt, having become mediator with the Porte, the daring Abdallah was pardoned by the sultan, and restored to his former honours, on condition of paying a heavy sum of money as a fine. In this transaction Mehemet Ali conceived he had laid both Sultan Mahmud and Abdallah Pasha under obligations to himself. From the latter, it is certain, he thenceforward exacted more deference than that proud chieftain was inclined to pay. At last, in November 1831, the quarrel between Mehemet Ali and Abdallah Pasha of Acre, broke out. Proceeding now without waiting for orders from the Porte, an Egyptian army laid siege to Acre, and took it on the 27th of May, 1832, after a bold resistance of six months. The sultan however, naturally averse to see his pashas carrying on war on their own account, had meanwhile espoused the cause of Abdallah. The successful Egyptian, therefore, sent forward his son Ibrahim with an army which overran Syria, penetrated into Asia Minor, and advanced towards the capital of the Turkish empire. The sultan, conscious of his want of strength, called in the aid of the Russians; but even after the arrival of a Russian fleet and army at his capital, he had to make peace with Ibrahim through the medium of a French diplomatist, and to grant the pasha of Egypt the very terms he demanded before the arrival of the sultan's Muscovite allies. Sultan Mahmud died on June 30, 1839. Though he had seen some of the largest and fairest provinces of his immense empire wrested from him by conquest, or partly alienated by treaty and usurpation of his sovereign rights, Moldavia and Wallachia on the N, and Greece, Egypt, and Syria, on the S and SE, Mahmud, one of the greatest reformers of his age, freed the Porte from the military domination of the janissaries, and had gone far towards effecting a most salutary social reorganization of the whole Turkish nation before his death.

Abd ul Mejid.] Abd ul Mejid succeeded his father on the 1st of July, 1839, in the midst of the most formidable crisis which had ever threatened the existence of the Ottoman empire. The energy and courage of Mahmud had given way after the rout of his army at Nezib, which seemed destined to bring Ibrahim to the promontory of Scutari, and to place the Porte at the mercy of its imperious vassal. At such a moment it seemed like a bitter jest of fortune to place a youth who had not yet completed his seventeenth year at the head of the tottering empire. The most urgent danger which threatened the very existence of the house of Osmanlis at the moment of his accession, was terminated in 1840 by the active intervention of the Great powers of Europe; and an arrangement concluded between the pasha of Egypt and his imperial master which has fortunately proved the basis of a lasting peace. The assistance which had been rendered to the Porte at the moment of its utmost need by the Christian powers was an indelible memorial of the abject condition to which the empire had been reduced. In the personal character of the young sultan, however, there was ground of hope for his country. He assumed and has since exercised the supreme power in a way which exhibited no want of activity, courage, or resolution. In 1839, he gave his people what may be regarded as their Magna Charta or bill of rights; and in 1846 and 1853, by numerous *hatti-sheriffs* he established perfect toleration in his empire. The treaties existing till the late declaration of war between Russia and the Porte are of no great antiquity. The first concession of liberty to trade was made in the treaty of the Pruth in 1711, and in 1739 Russian merchants were first allowed to navigate the Black sea, though still in Turkish vessels. It is from the treaty of Kainarji, in 1774, of which we have given an outline, that the diplomatic relations of the two countries really commence. Her treaties—in as much as the stronger Power always has the means of interpreting treaties to the disadvantage of the weaker—form in fact, the machinery by which Russia has gradually increased her pressure on T., and endeavoured to draw into her own hands that protectorate of the Christian subjects of the Porte, which the Western Powers may claim to share in an equal degree. The first disturbance of the *status quo* originated in a dispute between the Greek and Latin churches respecting the Holy places. The affair of the sanctuaries at Jerusalem having been disposed of, and from the testimony even of Count Nesselrode himself, satisfactorily arranged, another difficulty sprung up. Menchikoff demanded guarantees for the maintenance of the privileges of the Greek church. The cabinet of St. Petersburg did not attempt to show, by alleging any particular instance. that these privileges had been violated, and the Porte, on the contrary, solemnly confirmed the religious immunities of his Christian subjects. The court of St. Petersburg seemed indeed to regard itself as charged with the exclusive right of protection over the Christian subjects of the Porte; but the other states of Europe felt that their interest in the Christians of the East was in no degree secondary to that of Russia, and that it was a perversion both of treaties and of facts for Russia to claim an influence over that body which ought to be shared, as far as it exists at all, by the other Christian powers; for practically every one of the important measures adopted of late years by the Porte for the relief of its Christian subjects has had the active and enlightened support of the British or French Ambassador, and has been opposed by Russian influence. The demands of Russia were felt to be incompatible with the preservation of Turkish independence, and were firmly rejected by the Porte: whereupon the czar declared his intention of invading the Danubian principalities, and holding them as "a material guarantee" for the concessions he demanded. A Russian army which had already been collected for the emergency, accordingly crossed the Pruth, and, in the midst of profound peace, invaded two provinces of the Ottoman empire, on the 2d of July 1853. The progress of the struggle since that date has hitherto reflected little honour either on Russian arms or Russian diplomacy; and the attitude which France and England have assumed, will, it may well be hoped, not only restore peace to Europe, but also exact important guarantees from Russia against the early renewal of fresh trouble from her ambitious designs. The treaties between Russia and the Porte, the fruit of eighty years of encroachment and intrigue on the part of Russia, are now happily abrogated, never again we trust to be re-enacted; and the two great Western powers, France and England, will not we hope and believe desist from the enterprise in which they are engaged until they have, in the phrase taught them by the czar himself in his opening of the present struggle, secured "material guarantees" against the aggressive views of Russia. These guarantees, says an able despatch of M. Drouyn de Lhuys, the French minister, "result even from the present position which manifests the danger arising from the want of them. Thus Russia has profited by the exclusive right of *surveillance* which treaties conferred upon her in regard to Moldavia and Wallachia, conferring sovereign powers upon her to enter those provinces, as if they formed part of her own territory. The privileged position of Russia upon the Black sea has enabled her to found establishments there and to employ a naval force which without a counterpoise forms a perpetual menace against the Ottoman empire. The possession without control of the principal mouth of the Danube by Russia has created in the navigation of that great river moral and material obstacles which affect the commerce of every nation. In short, the articles of the treaty of Kutchuk-Kainardji relating to religious protection become, by an abuse of their interpretation, the original cause of the contest now existing in T. On all these points there are new regulations to establish, and important modifications to introduce into the *status quo ante bellum*. It may be said, I think, that the common interest of Europe demands 1st. That the protectorate exercised up to this moment by the imperial court of Russia over the principalities of Wallachia, Moldavia, and Servia cease for the future, and that the privileges granted by the sultans to these provinces, as dependencies of their empire, by virtue of an agreement concluded with the Sublime Porte, be placed under the collective guarantee of the Powers. 2d. That the navigation of the Danube, to its mouths, be freed from all restraint, and subjected to the application of the principles established by the acts of the congress of Vienna. 3d. That the treaty of July 13, 1844, be revised in concert by the high contracting Powers, with a view to establishing an European equilibrium, and to limiting the power of Russia in the Black sea. 4th. That no power claim a right to exercise an official protectorate over the subjects of the Sublime Porte, to whatever sect they may belong, but that France, Austria, Great Britain, Prussia, and Russia shall lend their mutual aid to obtain in the initiative from the Ottoman government a respect for and observance of the religious privileges of the different Christian communities; and to turn to advantage, in the reciprocal interests of their co-religionists, the generous intentions manifested by his majesty the Sultan, without there resulting any attempt to control the dignity and independence of his crown."

TURKEY-IN-ASIA, a great division of the Turkish empire, bounded by Persia on the E; by the Mediterranean sea, the isthmus of Suez, the Syrian and Arabian deserts, and the Persian gulf on the S; by the Black sea, the Bosporus, the sea of Marmora, the Hellespont, and the Archipelago on the W; and by the Black sea and the dominions of Russia on the N. On the S and W these boundaries are clear and distinct, following those of Nature; on the side of Persia they have continued much the same since 1638. A line drawn southwards from the Luristan mountains, in 34° N lat. and 46° E long., till it strike the l. bank of the Tigris above Kut-al-Hamara, in 32° N lat. forms the boundary of Khuzistan and Irak-Arabi. Thence it is continued along the l. bank of the Tigris, to its junction with the Euphrates at Kornah, and from thence to the sea. From the point in the Luristan mountains above specified, the boundary runs nearly due N along the extensive range of mountains called Ayagha-Dagh —the ancient *Zagros*,—till it strike the Kurdistan mountains in 36° N lat., passing in its progress the sources of the Dialla, the Tahite, and Little Zab. From thence it runs NW along that range, passing to the NE of the sources of the Great Zab, as far as N lat. 39°. Thence it runs due N, passing to the E of Bayazid, and across the summit of Ararat in 45° E long., and from thence NW to the junction of the Aras and Harpasu, in 40° N lat. and 44° E long. Since the cession to Russia of all Persian Armenia N of the Araxes, what was once the Persian boundary on that side is now the Russian, and therefore in this part, Asiatic T. is separated from the Russian dominions on the E by the Harpasu, as far as its source in the Tchildir-Dagh, which divides the basin of the Kur from that of the Araxes. The greatest extent of Asiatic T. from W to E, or from Cape Baba in 25° 51′ E long., and 39° 40′ N lat., to the W base of Ararat in nearly the same lat., and 45° E long., is upwards of 1,000 m.; and the greatest breadth, or from the head of the Persian gulf, to the mouth of the Natonabi, 920 m. in a NW direction. But as this extensive region consists of two great integral divisions—Asia Minor or the peninsular division, and the eastern or continental division —the length and breadth are exceedingly various. The western division, as far as the Euphrates, contains a superficies of more than 200,000 sq. m.; the eastern, upwards of 300,000 sq. m. The peninsular division of Asiatic T. has been described in our article ASIA MINOR. The continental division comprises the upper basin of the Euphrates and Tigris, comprehending ARMENIA and KURDISTAN,—the region of the Orontes and Mount Lebanon, comprehending SYRIA and PALESTINE,—and the lower basin of the Euphrates and Tigris, containing MESOPOTAMIA or Al-Jezirah, together with BABYLONIA and CHALDÆA, now denominated IRAK-ARABI, or the Arabian Irak, in contradistinction to Irak-Ajemi, the ancient *Media*. Each of these divisions is described under its respective head in our columns.

TURKEY, a township in Essex co., New Jersey, U. S., on the Passaic, 14 m. NW of Amboy.

TURKEY-FOOT, a township of Somerset co., in the state of Pennsylvania, U. S., 15 m. SW of Bedford, drained by Castleman's river and Laurel-Hill creek. Pop. in 1840, 1,422.

TURKEY-POINT, a cape on the coast of Maryland, U. S., at the mouth of the Susquehanna, where it takes the name of Chesapeak. Here the British army landed in August 1777, in their advance upon Philadelphia.

TURKEY-TOWN, a village of Cherokee co., in the state of Alabama, U. S., 128 m. NE of Tuscaloosa, on the Alabama and Tennessee railway.

TURKEYEN, a commune and village of Belgium, in the prov. of E. Flanders and dep. of Bassevelde. Pop. 292.

TURKHAL, a town of Turkey-in-Asia, in the pash. and sanj. of Sivas, 21 m. NW of Tokat, on the l. bank of the Tosanlu. It contains about 800 houses, and behind the town is an immense rock crowned by a fortress. It is the *Sebastopolis* of antiquity. Its environs are well cultivated.

TURKHEIM, a market-town of Bavaria, in the circle of Swabia, 25 m. SSW of Augsburg, on the l. bank of the Wertach, at the confluence of the Flossach. Pop. 1,500. It has a castle.

TURKHEIM (Ober and Unter), two adjacent villages of Würtemberg, in the circle of the Neckar, and both on the river of that name, containing respectively 900 and 2,100 inhabitants. The environs produce good wine, and contain several gypsum quarries.

TURKHEIM, or Thuringheim, a commune and town of France, in the dep. of the Haut-Rhin, and cant. of Wintzenheim, on the Fecht. Pop. in 1846, 2,858. It has manufactories of calico, and several spinning and paper-mills. The environs are noted for their wine. A victory was here gained by Turenne over the Imperialists in 1675.

TURKOCHORIO, or Esed-Abad, a town of Greece, in Rumelia, in the diocese of Violia, 18 m. NW of Livadia, to the N of the Gavrios.

TURKOMON, a town of Persia, in Azerdbijan 21 m. NW of Miana. Rice, cotton, and lint are cultivated in the environs.

TURK'S ISLAND, one of the Bahama group, having its N extremity in 21° 31′ N lat., and 71° 7′ 40″ W long. A reef projects 3 m. from the N point, and continues along the whole E side of this island.

TURK'S ISLANDS, or Turques, a group of islets in the SE extremity of the Bahama archipelago, and to the N of Hayti. Pop. 1,200. These islands are the most SE of all the Bahama islands. The principal is Grand Key, where there is established a port-of-entry. Many of the inhabitants migrate yearly from the Bermudas for the purpose of salt-raking, and return to Bermuda after the season is over for that purpose, so that the number is very fluctuating. There is no other produce exported from the Turk's islands.

TURLEGUE, a town of Spain, in the prov. and 24 m. SE of Toledo, in a fertile but insalubrious locality. Pop. 1,963. It has an hospital.

TURLER-SEE, a lake of Switzerland, in the cant. and 8 m. SSW of Zurich, and bail. of Knonau, at the W base of Mount Albis. It is about 2 m. in length, and ¾ m. in breadth, and abounds with fish.—Also an island at the head of Cook's inlet, in N lat. 61° 8′.

TURLOUGH, a parish in co. Mayo, 3¾ m. NE by E of Castlebar. Area 24,566 acres, of which 368 acres are in Lough Conn and 1,483 acres in Lough Cullen. Pop. in 1831, 6,929; in 1851, 4,416.

TURN-AGAIN, an island of Torres strait, in S lat. 9° 34′, and E long. 140° 55′. It is about 3 m. in length and 1½ m. in breadth. It is low, level, and marshy, and surrounded by reefs and sand-banks. Its shores are covered with mangroves, and swarm with mosquitoes.—Also a headland of New Zealand, on the SE coast of Eaheino-Mauwe, in S lat. 40° 34′, and E long. 177° 5′.

TURN-AGAIN (Cape), a headland of N. Australia, on the E side of the gulf of Carpentaria, in S lat. 13° 53′.

TURNAK, a river of Afghanistan, which has its source on the S side of the mountains which run between the Huzareh and Suliman ranges, at an alt. of 7,500 ft. above sea-level, and about 5 m. N of Oba, and joins the Urgundab in N lat. 31° 23′, and E long. 65° 8′. In the upper part of its course it is a considerable stream, 8 yds. wide and 2½ ft. deep; but in consequence of its waters being drawn off for irrigation, it becomes in the lower part of its course in summer completely dry.

TURNASTON, a parish in Herefordshire, 10 m. W by S of Hereford, watered by the Dore. Area 530 acres. Pop. in 1831, 54; in 1851, 70.

TURNAU, or Trnawa, a town of Austria, in Bohemia, in the circle of Bunzlau, 18 m. NE of Jung-Bunzlau, on the l. bank of the Iser. Pop. 3,505. It has manufactories of calico, printed muslins, and bijouterie. Garnets, and other precious stones, and sharpening and polishing stones, are found in the environs. In the vicinity are the ruins of the ancient castle of Trossky.—Also a town of Moravia, in the circle and 27 m. WNW of Olmutz, on the E bank of a small lake at the foot of a mountain, the summit of which is crowned by the old castle of Cunbury. Pop. 400.

TURNBY, an island of S. Australia, in Spencer's gulf, in Sir Joseph Bank's group, to the NW of Beevesly island.

TURNDITCH, a chapelry in Duffield p., Derbyshire, 4¾ m. S by E of Wirksworth. Pop. 380.

TURNEFFE, a group of islands in Honduras bay, 45 m. from the coast of Yucatan, in N lat. 17° 20′, and W long. 87° 40′. It is about 45 m. in length from N to S, and 18 m. in breadth, and is surrounded by sand-banks. It is frequented by fishermen, but is uninhabited.

TURNER, a river of W. Australia, in the SW part of the co. of Sussex, and which flows into the sea to the SE of Cape Hamelin.

TURNER, a township of Oxford co., in the state of Maine, U. S., on the W side of Androscoggin river, 25 m. WSW of Augusta. The surface is pleasantly diversified, and the soil fertile. Pop. in 1850, 2,537.

TURNER'S-PUDDLE, a parish in Dorsetshire, 7½ m. NW by W of Wareham. Area 1,983 acres. Pop. in 1831, 82; in 1851, 109.

TURNHAM-GREEN, a hamlet in Chiswick p., Middlesex, 8 m. W by S of St. Paul's, London.

TURNHOUT, an arrondissement, department, commune, and town of Belgium, in the prov. of Antwerp. Pop. of arrond. 80,000; of dep. 13,298. —The town is 26 m. ENE of Antwerp, in the midst of waste lands. Pop. 8,836. It is well-built, has a college, and possesses manufactories of linen and woollen fabrics, carpets, printed muslins, lace, numerous bleacheries, and dye-works, several breweries, tanneries, tile, brick, and lime-kilns. It returns three deputies to the provincial states. This town, which occupies a portion of a forest belonging to the dukes of Brabant, was founded in 1209. Its castle was built by Marie, duchess of Gueldres. After the peace of Munster, Philip IV., king of Spain, made a gift of T. to the widow of Frederick Henry of Nassau, through whom it descended to the house of Orange. The locality is noted for the victory of Prince Maurice of Nassau over the Spaniards in 1597, and for the defeat of the Austrians by the Brabançons in 1789.

TURNICH, a village of Prussia, in the prov. of

the Rhine, regency and 11 m. WSW of Cologne, and circle of Bergheim, on the r. bank of the Erfft. Pop. 1,784. It has several spinning-mills.

TURNISCHA, or TURNISCHE, a town of Hungary, in the comitat of Szalad, 20 m. SW of Lövö, near the r. bank of the Lendoa. Pop. 650.

TURNITZ, TWRDONICE, or TWORDONICZE, a town of Austria, in Moravia, in the circle and 36 m. SSE of Brunn, near the r. bank of the March. Pop. in 1834, 1,254.

TURNWORTH, a parish in Dorsetshire, 4¾ m. W by N of Blandford-Forum. Area 1,560 acres. Pop. in 1831, 78; in 1851, 103.

TURNY, a commune of France, in the dep. of the Yonne, cant. and 7 m. NE of Brienon, in a valley, on a small affluent of the Armançon. Pop. 1,308.

TUROBIN, a town of Poland, in the gov. of Lublin, obwod and 23 m. SW of Krasnistaw, in the midst of a marsh, on the Por. Pop. 1,447. It has two churches.

TURON, a town of Spain, in the prov. and 51 m. SE of Granada, in a valley enclosed by mountains, in a dry but fertile locality, at some distance from the r. bank of the Adra. Pop. 2,655. It has a parish church, a customhouse, and a public granary; and in the vicinity are mines of lead and antimony.

TURON, a river of New South Wales, in the co. of Roxburgh, which flows into the Macquarie. The district through which this river flows has been one of the most productive fields of gold-digging enterprises. Both on the river itself, and on the table-land in the vicinity, whence its tributaries take their rise, the diggings have been successfully pursued. Many of these tributaries, Big Oakey and Little Oakey creek especially, have yielded a large amount of gold. The bed of the river is auriferous throughout its entire extent.

TURON, or NORTH CAPE, a headland of Annam, in the China sea, in N lat. 16° 7′, and E long. 108° 13′, at the NE extremity of an island of the same name, which encloses Turon harbour on the SE. It forms a lofty projection terminating in a rock, bearing in configuration a striking resemblance to that of a lion couchant. The harbour, locally named Han-san, forms one of the finest in existence. It is spacious and well-defended, and receives on the SE a small navigable river communicating by a canal with the town of Fai-fo. On the river, about a mile above its mouth, and 45 m. SE of Hue, is the town of Huron, formerly a flourishing and populous place. The houses are mostly built of bamboos covered with straw, and the market is plentifully supplied with poultry, especially ducks, fish, and vegetables. An extensive trade in cotton was formerly carried on with China, but the culture of the tree is now neglected. Marble now forms the chief article of export. In 1847 this place was visited by Sir John Davis on a commercial mission from the British government. Very little communication could be obtained with the shore for about six days, the alarm at first being evidently great; in addition to which it rained incessantly, and there was besides a severe typhoon. Two commissioners ultimately came on board, and invited the plenipotentiary to an entertainment and conference on shore the next day, when a very handsome reception was prepared —the commissioners appearing in their dresses of ceremony, the ancient costume of the Chinese. The first commissioner was assistant member of the board of revenue, the second treasurer of Kwangnan prov. in which T. lies. The rains continued incessantly, impeding communication, and the country was reported to be inundated and nearly impassable. A large present to the crews of the two ships, of bullocks, fowls, hogs, fruit, &c., was offered and accepted, on condition of a return being made, which was done accordingly. A party went on shore to view some very singular marble rocks about 5 m. distant; a guide was given them, and orders transmitted to afford every facility. These rocks rise at once out of the sandy flat between the river and the sea, and are covered in most parts with trees and luxuriant vegetation, peopled by numerous monkeys. The principal masses are 5 in number, of which the one nearest the sea contains some splendid caverns and galleries, which have been improved by art, and the caverns converted into temples of Buddha. The finest of these is quite a natural pantheon, being a dome of 80 ft. high, lit from the top by three openings. The floor is about 70 ft. in every direction, and has been paved. Several idols and shrines of Buddha and his disciples adorn the inside, and the whole is approached by an arched gallery with descending steps at intervals. The entire rock is laid out in a corresponding manner, with small dwellings and gardens in the open and level spaces. It would appear that our envoy, though respectfully and even kindly treated, was unable to obtain an audience of the king; and this has been attributed, "chiefly, if not altogether, to the plain and simple manner in which the mission was got up." But this failure may be regarded as of little moment. Our settlement on the NW coast of Borneo will inevitably bring us more closely into contact with the Hindu-Chinese nations, and furnish us with a point on which to place our commercial lever for moving them effectually. They will be unable to resist our civilization, when it shall have been brought into their immediate neighbourhood; and experience will demonstrate to them the futility of standing aloof, when our steamers are found to swarm in all the seas which wash their coasts, and our political power has become paramount.

TUROPOLYA, or TUROVOPOLJE, a district of Civil Croatia, in the comitat and to the SSE of Agram, between the r. bank of the Save, on the N, and a branch of the Alps on the S. It is 14 m. in length, and 12 m. in breadth, and presents the appearance of a fine well-watered plain. Its principal rivers are the Lomnicza, Odra, and Runa. It enjoys some peculiar political privileges.

TURPAUGE, a commune of Belgium, in the prov. of Luxemburg, and dep. of Messancy. Pop. 249.

TURQUEL, a town of Portugal, in the prov. of Estremadura, comarca and 4 m. ESE of Alcobaça, on the r. bank of the Alcoa.

TURQUESTAIN, a village of France, in the dep. of the Meurthe, cant. and 8 m. SSE of Lorquin, on a steep mountain to the l. of the Sarre Blanche. Pop. 112. It has the remains of an ancient castle.

TURRABEILE, a river of New South Wales, in the district of Liverpool plains, which flows into Lammoy river.

TURRACH, a village of Styria, in the circle and 41 m. WSW of Judenburg, near the frontier of Illyria, on the Schwarzbach, in the midst of mountains which are here crossed by a good road. It has mines of iron and copper, and several forges.

TURRAH, or THERAH, a town of Hindostan, in the prov. of Gujerat, about midway between Theraud and Rahdunpur. Pop. in 1809, 12,500, chiefly Coolies.

TURREEHEY, a town of Caffraria, in the Betchuana country, 120 m. W of Littaku. It consists of about 100 houses. The inhabitants paint their bodies red, and use on the hair a blue shining powder. They have numerous herds of cattle.

TURRET, a small lake and a rivulet in Strathearn, Perthshire. The lake is about 1 m. long, and a ½ m. broad. The rivulet issues from the end of it, and has a run of about 5 m. SE to the Earn.

TURRI, a town of Naples, in the prov. of Abruzzo-Citra, district and 11 m. SW of Chieti, on a sandy rock, near the r. bank of the Pescara. Pop. 500. It has three churches.—Also a village of Sardinia, in the prov. and 8 m. WSW of Isili, in a fertile but insalubrious plain. Pop. 2,800.

TURRIERS, a canton, commune, and town of France, in the dep. of the Basses-Alpes, and arrond. of Sisteron. The cant. comprises 11 com. Pop. in 1831, 3,548; in 1846, 3,817. The town is 19 m. NE of Sisteron, in a valley. Pop. 613.

TURRIFF, a parish on the NW border of Aberdeenshire, divided chiefly by the Deveron, and partly by a small brook from Banffshire. Area 16,896 Scottish acres. Pop. in 1801, 2,090; in 1831, 2,807; in 1851, 3,584.—The town of T. near the centre of the p., is 11 m. SSE of Banff, and 34½ m. NNW of Aberdeen. An extensive bleachfield has existed here for about 75 years.

TURRITANO, a river of Sardinia, in the div. of Capo-Sassari, which has its source in the prov. of Alghero, flows thence into that of Sassari, and after a total course in a generally NNW direction of about 36 m., throws itself into the Mediterranean, 13 m. NNW of Sassari. Its principal affluents are the Campo-Lazzaro and the Mascaris.

TURRUNCUN, a town of Spain, in the prov. and 30 m. SE of Logrono, amid mountains. Pop. 305.

TURRYVENE, a commune of Belgium, in the prov. of Limburg, and dep. of Oostham. Pop. 126.

TURSAC, a village of France, in the dep. of the Dordogne, cant. and 7 m. N of St. Cyprien, near the l. bank of the Vezere. Pop. 740.

TURSAN, an ancient district of France, in the provinces of Guyenne and Gascogny, now comprised in the departments of the Landes and Gers. Its capital was Aire.

TURSI, a town of Naples, in the prov. of Basilicata, district and 50 m. S of Matera, on a hill, between the Sinno and Agri. Pop. 4,600. It has a cathedral, a collegiate and a parish church, several convents, an hospital, a diocesan seminary, and several handsome dwellings. This town is said to be of Saracenic foundation.

TURTEE, a village of Sind, on the r. bank of the Western Narra.

TURTLE, a small island of the South Pacific, in the Fiji archipelago, in S lat. 19° 50', and E long. 182°.—Also an island off the NW coast of Australia, in S lat. 19° 53' 48".—Also a group of islands, 6 in number, off the NE coast of Australia, to the N of Point Look-out. They are all low, covered with wood, and encircled by coral reefs.

TURTON, a chapelry in Bolton p., co.-palatine of Lancaster, 4 m. N of Great Bolton. Area 4,110 acres. Pop. in 1831, 2,563; in 1851, 4,158. Many of the inhabitants are employed in cotton-spinning, bleaching, and calico-printing.

TURTUKAI, a town of European Turkey, in Bulgaria, on the r. bank of the Danube, opposite Oltenitza, 27 m. W by S of Silistria, situated on the sloping bank of the river, which here flows from W to E, and swells to a breadth of 1,200 yds. a little below. The guns of its fortifications command the plain on the l. or Wallachian bank, which is here quite flat for a long distance northwards. Opposite to the town there is an island in the Danube, which divides the stream into two channels, that on the S being from 600 to 700 yds. broad; that on the N 200 yds.

TURTZ, or TURCZ, a village of Hungary, in the comitat of Ugocs, 9 m. E of Halmi, at the foot of mountains, on a small river of the same name, an affluent of the Thur. It has silver mines.

TURVEY, a parish in Bedfordshire, 8 m. WNW of Bedford, on the E bank of the Ouse. Area 3,944 acres. Pop. in 1831, 988; in 1851, 1,028.

TURVEY-BRIDGE, a hamlet in the p. of Lusk, co. Dublin, 2¼ m. SSW of Lusk.

TURVILLE, or TURFIELD, a parish in Bucks, 6 m. NW by W of Great Marlow. Area 2,315 acres. Pop in 1831, 442; in 1851, 436.

TURVO, a parish of Brazil, in the prov. of Minas-Geraes, on a river of the same name.—Also a river in the SW part of the prov. of Goyaz, an affluent of the Cururuhi.—Also a river of the prov. of Rio-de-Janeiro, in the district of Parati.

TURVO (GRANDE E PEQUENO), two rivers of Brazil, in the prov. of Minas-Geraes, affluents of the Chopoto.

TURWESTON, a parish in Bucks, 6 m. WNW of Buckingham. Area 1,240 acres. Pop. 322.

TURWICK, or TERWICK, a parish in Sussex, 5 m. W by N of Midhurst. Area 718 acres. Pop. 95.

TURY, or TURI, a small town of Brazil, in the prov. and 99 m. NNW of Maranhao, and comarca of Guimaraens, on the SE coast of the bay of Tury-Açu, in S lat. 1° 20', in a fertile locality. Pop. 5,000.

TURY-AÇU, a river of Brazil, which has its source in the W extremity of the prov. of Maranhao, near the confluence of the Araguaia, with the Tocantins; runs along the conterminous confines of the prov. of Para, traversing vast forests; and, after a total course of upwards of 375 m., in a generally NNE direction, throws itself into a bay of the same name. Boavista is the chief place on its banks.—The bay is formed by the Atlantic, between the provinces of Para and Maranhão, in S lat. 1° 22', and W long. 45° 1'. It is 36 m. wide at its entrance from NW to SE, and 21 m. in depth, and receives on the S a river of the same name, opposite Turi. The chief place in the bay is the island of Joao.—Also a headland at the NW extremity of the bay, which it separates from that of Motuoca, and 33 m. NNW of Turi.

TURYNANA, or TUROROMA, a river of Brazil, in the prov. of Maranhão, and district of Guimaraens, which throws itself into the Atlantic, in S lat. 2° 18', and W long. 44° 8'. Its embouchure forms a good port for small vessels.

TUSA, a town of Sicily, in the prov. of Messina, district and 8 m. NW of Mistretta, on a mountain, near the shore of the Tyrrhenian sea.

TUSBY, a town of Russia in Europe, in Finland, in the gov. and 48 m. SSE of Tavastehus and district of Borgo, on the S side of a small lake.

TUSCALOOSA, a central county of the state of Alabama, U. S., comprising an area of 1,507 sq. m., intersected by Black Warrior and its affluents, and by Sipsey river. The surface is undulating, and the soil highly productive. Pop. in 1840, 16,583; in 1850, 18,056. The capital, which bears the same name, is on the SE bank of Black Warrior or Tuscaloosa river, and 94 m. NW by W of Montgomery, in N lat. 32° 12', and W long. 87° 42'. Pop. in 1840, 1,949; in 1850, 2,961.

TUSCANY, in Italian, TOSCANA, a grand-duchy of Northern Italy, bounded on the NW by Lucca and Modena; on the NE, the E, and SE by the Papal dominions; and on the SW and W by the Mediterranean, here called the Tuscan or Tyrrhenian sea. Without these boundaries, there lies, within the circumference of Lucca, the Vicariato-di-Pietra-Santa, belonging to T.; and between Modena, Parma, Genoa, Massa, and Lucca, what is called the Lunigiana of T. Besides those insulated districts, the islands of Elba, Pianosa, Palmajola, Gorgona,

Giglio, Montechristo, Melora, La Formiche, and Gianuti belong to this country. Liechtenstern calculates the Tuscan dominions at 8,415 sq. m.; Mayer's maps exhibit it at 8,460; and Balbi states it at 8,430 sq. m.

Physical features.] T. embraces three distinct regions. In the bottom of a beautiful valley, the famous Arno forms a basin of which Florence is the centre, and which on the S stretches to Crotona, and on the E to Pisa. This basin terminates near the sea in a dead level country resembling land which has been abandoned by the sea. The r. banks of the Arno are surrounded by the high chain of the Apennines; and the l. bounded by the sea and the Papal dominions, present the appearance of a desert, barren, flat country, where the soil is sterile, the air unhealthy, and the hills covered with ruins of all ages. The region of the Apennines covers one-third of the surface of Tuscany; the rich valley of the Arno only one-sixth, and the Maremna of Sienna, the remaining three-sixths. The coast-line is about 150 m. in length. The soil in the delightful valley of the Arno is rich and well-cultivated; that on the Apennines is stony. The soil near the coast is low, sandy, and at some parts swampy. The Maremna consists of a white clay, impregnated with sulphur; everywhere in this desolate country sulphureous springs and mephitic flames are seen issuing from the soil. The principal river is the Arno which is navigable for barges from Florence to the sea, and receives a great number of small rivers, as the Chiana, the Sieve, and the Evola. The Arno is of great use to the country by feeding upwards of ,000 small canals of irrigation. The rivers were formerly very destructive, but are now all well-built up with dykes which protect the adjacent lands. The Ombrone is the largest river next to the Arno, but it is not navigable; it flows through the Maremna. The Tiber has its source here in the Apennines. The canal of San-Giovanni, and the Fosso-di-Navicelli are navigable; all the other numerous canals are merely for irrigation. The largest lake, the Lago-di-Castiglione-di-Piscaja, about 25 m. in circumference, is in communication with the sea by a canal. The climate is as much diversified as the country itself. The numerous mountains and the sea-breezes diminish the heat; though it sometimes becomes exceedingly oppressive. The mean temp. at Florence is 10° of R.; at Leghorn, 11°. On the high mountains, snow remains for weeks in winter, and the nights are very cold even in August; in the valleys, winter is so mild that snow scarcely remains for one day. Rain is not frequent; but copious dews refresh the plants. In summer the *sirocco* and the *libecchio* winds are very oppressive, and exert a pernicious influence on the vegetation. On the Apennines, and in the valley of the Arno, the climate is everywhere good; in the swamps near Pisa fevers and agues are common; the sulphureous soil of the Maremna sends forth pestilential exhalations, and it is only in winter that the herdsmen dare to visit these desolate regions long since forsaken by their inhabitants.

Productions.] The valley of the Arno is finely cultivated, and is divided into numerous small farms, separated by rows of trees or small canals: in no country of Europe perhaps is the soil so minutely subdivided as here. In the Maremna, the vegetation is so plentiful in summer that the pastures furnish supply for additional flocks of 400,000 sheep, 30,000 horses, and a great number of cattle and goats. The corn raised supplies the consumption of the numerous population only in ordinary years; in bad years supplies must be got from Leghorn. Straw has great value in Tuscany. The finest—that of the *marzolano*, a species of wheat—being used for making bonnets. One of the most important productions is chestnuts, which on the Apennines, and the hills of the Maremna, are used for bread. The produce of chestnuts is calculated at 800,000 bushels per annum. Wine is made in considerable quantity; the best qualities are found at Monte-Pulciano, Chianti, and Pomino; most of it is consumed in the country. Oil is an object of great importance; the exportation being great, and the home-consumption amounting to about 11,050,000 lbs. per annum. Wood is abundant. Horses are numerous, but small and feeble. One of the most famous studs is at San-Rossore, where also a herd of about 200 camels has been kept from the time of the crusades. At San-Rossore there is also a herd of 1,800 wild cows, and a flock of merinoes. The asses of T. are thought the finest and strongest in Europe. Goats are numerous, and good cheese is prepared from their milk. The iron-mines of the isle of Elba are very extensive. Sulphur is exported in great quantities, and saltpetre, alum, and vitriol. The Tuscan marble is much valued. In statuary and coloured marbles, granites, chalcedony, alabaster, and serpentine, it is very rich.

Manufactures and commerce.] T. is one of the most industrious countries in Italy. In the valley of the Arno, in the neighbourhood of Florence, the manufacture of straw-hats is carried on by women, and annually produces 3,000,000 *liri.* Silk manufactures are still a principal branch of industry in the Florentine cities, and there are also some manufactures of linen, broad cloth, coral, soap, perfumeries, liquors, leather, paper, china, marble, alabaster, and mosaics. Leghorn has been declared a free port, and maintains a considerable commerce with the East. The balance of commerce is in favour of this country, which also derives considerable advantages from the numerous strangers who are constantly present in its fine capital, or visiting the baths at Pisa.

Inhabitants.] Balbi estimated the pop. in 1826, at 1,275,000 souls. The Tuscans are of Etrurian descent, but are now quite amalgamated with the Italians. The purest Italian is spoken here; but the pronunciation is guttural and harsh. There are about 15,000 Jews, and a good many French, English, German, and even Turkish settlers, particularly in Leghorn. The clergy were formerly very numerous, but Leopold abolished most of the convents, and the French secularised the rest; and only those orders which are devoted to the care of the sick, and the objects of education, were allowed. Of late years, however, T. has earned a disgraceful reputation for religious intolerance; and in 1851, a concordat was concluded between Rome and T., and which does away with many of the guarantees decreed in the last cent. against the authority of the church by the grand duke Peter Leopold. The leading points of this concordat are: that the ecclesiastical authorities are free in the exercise of their sacred office, and that the lay authorities are to aid them in the protection of morality and religion, and in the maintenance of the episcopal authority; bishops are free to publish whatever documents relate to their functions; and art. 3d provides that the bishops alone shall have the right of censorship over works treating *ex-professo* of religion, and shall moreover have the power of warning their flock or, in other words, of compelling them to avoid reading any book they may consider contrary to religion and morality. All cases relating to the faith, the sacraments, and other matters belonging to the spiritual jurisdiction by the sacred canons, must be deferred to the ecclesiastical authorities;

lay tribunals shall only be competent to inflict fines upon ecclesiastics detected in the perpetration of minor offences against the revenues, the game laws, &c. Church property is under the control of the bishops and rectors.—There was a time when the arts and sciences rendered Florence 'the brightest star of star-bright Italy;' but that age is gone; the Medici, Dante, Michael, Angelo, and Leonardo da Vinci are no more; nevertheless Florence, on account of her glorious monuments of art, and the remembrances of former times, is still one of the most interesting places in Europe. There are three universities at Florence, Pisa, and Sienna; that of Pisa has the most reputation.—The elementary schools are mostly in the hands of the clergy. More, however, is done here for popular education than in any other part of Italy, with the exception of the Sardinian dominions. There are academies and literary societies in all the important towns. The most remarkable are the Academia delle belle arti, and the Academia Fiorentina, which latter consists of the united societies Del-Cimento, Della-Crusca, and Del-Disegno.

Government.] The government is an absolute monarchy. The state of justice here, as in every country where the sovereign's will is law, is extremely low. There are courts of law, and there is a Florentine code,—but it avails little when inimical to the views of the duke; and into the former it is impossible to procure admission, as all the proceedings are private, and the accused, with the horrible spirit of the inquisition, are not permitted to see their accusers. In 1852, a decree of the grand duke constituted a council-of-state, independent of the council of ministers. The council-of-state is composed of a president, 8 councillors in ordinary service, councillors extraordinary in unspecified number, and a first and second secretary. The ministers of state cannot at the same time be members of the active council-of-state; and they will only attend at the express desire of his imperial highness, but even then they will have no vote. The salary of the president of this council-of-state is fixed at 9,800 *lire* per annum; that of the councillors at 7,000 *lire*. In 1853 another decree was published, constituting on new bases the representation of the communes. Every commune is to be represented by a council-general, a tribunal, and *gonfalonier*, the latter being nominated by the grand-duke. The council-general is to be composed of the *gonfalonier*, priors residing within the jurisdiction of the tribunal, and of the number of members determined by law. These members are to be drawn by lot from amongst the landed proprietors paying taxes in each commune. In towns where the pop. exceeds 2,000, the priors are to be 3, and the other councillors 6; over 2,000 and less than 5,000, the priors 4, and the others 8; over 5,000 and less than 10,000, 5 and 10; over 10,000 and under 15,000, 6 and 12; over 15,000 and under 20,000, 7 and 14; over 20,000, 8 and 16. In the town of Lucca there are to be 9 and 18, and in Florence and Leghorn 10 and 20. The *gonfalonier* is appointed for four years.—There are about 4,000 regular troops, besides militia. The few frigates and vessels which formerly protected the coast were lost during the French domination.—T. is administratively divided into 3 provinces, containing 36 towns, 135 boroughs, and 2,570 parishes.

Revenue.] The *Monitore Toscano* published the budget of the revenue and expenditure of T. for 1853, as follows:

Revenue,	36,376,400 lire.
Expenditure,	36,308,800
Surplus,	67,600

Consequently, leaving aside the extraordinary expenses occasioned by the auxiliary force of 1,000 Austrians, the ordinary administration of the state seems to balance itself. The revenue for 1853 was 576,600 lire more than in 1852. The indirect taxes amounted to 25,699,100 lire. The expenditure for 1853 was thus estimated:—

Ministry of Interior,	2,780,000 lire.
„ Grace and Justice,	3,494,600
„ War,	7,000,000
„ Foreign affairs,	267,400
„ Public Instruction,	866,500
„ Ecclesiastical affairs,	789,200
„ Finance,	21,111,100
Total,	36,308,800

In the expenditure for 1853 the responsibility taken upon itself by the government for guaranteeing a minimum of interest to the contractors of the Siennese railway, and of the Lucca, Pistoja, and Pescia line is noted, the sum of 250,000 *lire* put down to that end. The conversion of the old loans and debts of T. into a consolidated fund at 3 per cent. was projected, as follows:—

1. Divers debts, bearing interest,	13,025,900 lire.
2. Lucca loan and loan of 1849,	31,306,000
3. Loan of 30th Sept., for 12,000,000 with mortgage on mines for thirty years,	8,811,000
4. Different debts to be converted,	21,096,000
Actual total debt,	74,238,900
Add to this proposed new loan,	21,000,000
	95,238,900
or in sterling money,	£3,174,630

The budget for 1855 is as follows:—

Receipts,	37,608,400 lire.
Expenses,	37,546,700
Surplus,	61,700

This balance in favour of the revenue proceeds in part from the re-establishment of an estate tax at its original rate, and from a decree that creates throughout all the communes, from and after the 1st of February next, a tax on butchers' meat from which the cities liable to the salt-tax are alone to be exempted. The credits opened to the seven ministerial departments are distributed in the manner following:—

Ministry of the Interior,	2,741,100 lire.
„ War,	7,883,900
„ Grace and Justice,	3,464,000
„ Foreign affairs,	240,600
„ Public instruction,	867,600
„ Ecclesiastical affairs,	819,500
„ Finance,	21,530,000
Total,	37,546,700

History.] T., anciently called *Etruria*, was inhabited in early times by the *Etrusci*, whose antiquity is such that it is lost in the obscurity of ages. Some maintain that they were of Egyptian descent; many regard them as of Lydian origin; a still greater number will have them to have been the ancient *Pelasgi*, and of late they have been supposed to be the offspring of the Phœnicians or Canaanites. Nothing certain has yet resulted from all these learned investigations. We however know that they occupied a territory much more extensive than modern T., to which they were gradually circumscribed by the Gauls, the Samnites, and the Umbrians. They equalled the Egyptians in the solidity, and surpassed them in the beauty of their edifices, while they excelled them in the arts, and rivalled them in the luxuries of life, while the Greeks were still barbarians, and Rome was yet without a name; and though obliged in after times to bend under the Roman power, they could boast that they communicated to their conquering mistress the skill that erected her temples, the ceremonies that graced her religion, the robes that invested her magistrates, the pomp that accompanied her triumphs, and even the military music that animated her legions. Etruria was a confederacy of twelve republics, each governed by a chief called *lucomon*. These chiefs—who were priests at the same time—used to assemble near the temple of Volturna to consult on the affairs of their country. After the downfall of the Roman empire, this country came consecutively into the power of the West-Goths, Longobards, Byzantinians, and Franks, and was named Tuscany. From the time of Charlemagne it had its own margraves, who soon succeeded in becoming independent. That great revolution in the Italian cities which infused new life into the whole of Upper Italy, and effected a rapid transformation on the face of society there, was slowly developed in T. where the power of the margraves was more firmly established. Pisa was the first town which began to rise, and it was not till its power was destroyed by internal dissensions that Florence became powerful. About the middle of the 15th cent., Sienna was formed into a flourishing republic. These three cities divided the power among themselves; but Florence soon became preponderant; and the House of Medicis, especially distinguished by its great influence and patriotism, at length acquired the sovereignty of Pisa in 1407. In 1538, Florence 'with the Tuscan fields and hills' was erected into a duchy by the emperor Charles V. in favour of Alexander-de-Medicis. He was of the elder branch of the Medician family; which derived its descent from Philip-de-Medicis, who lived in the 13th cent., and was a chief opposer of the Ghibeline faction. The Medici were noted for their taste in literature, poetry, and the fine arts; and learning and genius flourished under their fostering care. Cosmo I., in 1557, received Sienna from Philip II. of Spain; and in 1569 the title of grand-duke, which was confirmed by Maximilian II. in 1575. In 1737, the family of Medicis became extinct; and Francis duke of Lorrain obtained the duchy as a compensation for the loss of his hereditary dominions, which had been seized by France, and given to Stanislaus, the father-in-law of Louis XV. By the marriage of Francis with Maria Theresa, queen of Hungary, and daughter of Charles VI. he became emperor of Germany, and T. in 1745 became an appanage of the Austrian family. Francis was succeeded in the duchy by his second son Leopold, who became emperor in 1790, upon the decease of his brother Joseph. In 1792, Ferdinand, second son of Leopold, was named grand-duke of Tuscany. He was driven out of his territories in 1799 by the French; who erected them into a republic under the name of the Etruscan commonwealth; afterwards made a kingdom of it under the name of Etruria; and finally incorporated it with France as an integral part of the French empire, Napoleon's sister, the duchess of Lucca, being named grand-duchess of Tuscany. In 1814, Ferdinand was restored to his duchy, which now comprehends the imperial fiefs, the principality of Piombino, the whole island of Elba, and the duchy of LUCCA: see that article. Ferdinand died in 1824, and was succeeded by his son Leopold II.

TUSCARAWAS, a county in the E part of the state of Ohio, U. S., comprising an area of 704 sq. m., drained by a river of the same name, and by Sandy, Conotten, and Sugar creeks, and intersected by the Springfield, Mount-Vernon, and Pittsburg railway, and by the Ohio canal. Pop. in 1840, 25,631; in 1850, 31,761. It contains a village of the same name, 84 m. ENE of Columbus. Pop. 100.—Also a township of Coshocton co., in the same state. Pop. in 1840, 455.—Also a township of Stark co., in the same state. Pop. 1,792.—Also a river of Ohio, U. S., which rises in Portage co., and flows in a W course of 80 m. to its junction with the Walhonding, at Coshocton, with which it forms the Muskingum.

TUSCARORA, a village of Lewiston township, Niagara co., in the state of New York, U. S. Pop. in 1840, 300.—Also a township of Bradford co., in the state of Pennsylvania, watered by a creek of the same name. Pop. 1,020.—Also a township of Juniata co., in the same state. Pop. 1,018.—Also a creek of the same co., which has its source in Tell township, Huntingdon co., and joins the Juniata.—Also a range of mountains in the same state, in Schuylkill co., between Rush and West Penn townships, and forming a continuation of Mauch Chunk mountains. They abound in anthracite coal.

TUSCHKAU-OB-DER-MIES, TAUSSKOW, or TUSSKOW-NAD-MZY, a town of Austria, in Bohemia, in the circle and 6 m. WNW of Pilsen, on the l. bank of the Mies. Pop. 954.

TUSCOLA, a county in the E part of the peninsula of Michigan, comprising an area of 800 sq. m., bounded on the NW by Saginaw bay of Lake Huron, and drained by Cass river and its branches.—Also a township of the same state, in Livingston co. Pop. in 1840, 247; in 1850, 544.—Also a township of the same state, in Saginaw co. Pop. 55.

TUSCUMBIA, a village of Franklin co., in the state of Alabama, U. S., 2 m. S of Tennessee river, and intersected by the Memphis and Charleston railway. Pop. in 1840, 2,000; in 1850, 2,300.—Also a village of Millar co., in the state of Missouri, on the r. bank of Osage river, 28 m. SW by S of Jefferson city. Pop. in 1840, about 50.

TUSHINGHAM-WITH-GRINDLEY, a township in Malpas p., co.-palatine of Chester, 3 m. NW by N of Whitechurch. Area 1,292 acres. Pop. in 1831, 328; in 1851, 315.

TUSKAR, or TUSCAR ROCKS, a group of rocks off the coast of co. Wexford, 4¼ m. SE by E of Greenore-point. It consists of four masses of rock, arranged in a line from N to S, bearing the names of Tuscar *par excellence*, the North Hen and Chickens, and the South Hen and Chickens, and all lying within a space of about 2 furlongs from extremity to extremity. A lighthouse on the Tuscar rocks guides the mariner throughout the W side of the southern entrance of St. George's channel.

TUSKEVAR, a town of Hungary, in the com. and 31 m. W of Veszprim, on the Marczal.

TUSKI. See TCHUKTSHES.

TUSMORE, a parish in Oxfordshire, 6 m. N by W of Bicester. Area with that of Hardwick, 900 acres. Pop. in 1831, 16; in 1851, 52.

TUSPAN, a port of Mexico, in the dep. and 136 m. WNW of Vera Cruz, on the l. bank of a small stream of the same name, at its influx into the gulf of Mexico.

TUSSI, a town of Naples, in the prov. of Abruzzo-Ultra, district and 17 m. SE of Aquila, and cant. of Santo Demetrio, on a hill. Pop. 550.

TUSSON, a village of France, in the dep. of the Charente, cant. and 4 m. NE of Aigre. Pop. 1,050. It has distilleries of brandy.

TUSTEREN, an island off the W coast of Norway, in the prov. of Drontheim, in N lat. 63° 11′, E long. 8°, between the Drontheimleed on the N, by which it is separated from the island of Smoeelen; the Solsund on the E; and the Tuster-fiord on the S. It is 12 m. in length from NE to SW, and 5 m. in medium breadth, and has deeply indented coasts, especially on the W. It is generally mountainous, and contains several villages.

TUSTLA, TUXTLA, or TOCHLAN, a volcano of Mexico, in the state and 90 m. SE of Vera-Cruz. The last eruption, which took place in 1793, threw ashes as far as Perote, at the distance of 171 m. In its vicinity is an Indian village of the same name.

TUSZYN, a town of Poland, in the gov. and 71 m. ESE of Kalisch, and obwod of Petrikau. Pop. 700.

TUTA, a town of New Granada, in the dep. of Boyaca and prov. of Tunga. Pop., chiefly Indian, 200.

TUTAM-DARA, a village of Afghanistan, at the E extremity of the fertile valley of the Ghorbund. The locality is noted for the defeat of Ali-khan, by a British force under Sir Robert Sale, in Sept., 1840.

TUTBURY, a parish and town in Staffordshire, 4½ m. NW by N of Burton-upon-Trent, on the S bank of the Dove. Area of p. 4,001 acres. Pop. in 1801, 844; in 1831, 1,553; in 1851, 1,798. The ancient trade of the town, that of woolstapling, is extinct; but there are here a cotton factory, a corn mill, and a glass work. T. is celebrated for its ruined castle, once the seat of the Mercian kings, and afterwards of the earls and dukes of Lancaster. In the civil wars of the 17th cent., the castle, after a long siege, surrendered to the parliamentarians, and was soon afterwards dismantled.

TUTHI, a village of the Punjab, to the N of Swan river, an affluent of the Sinde.

TUTICORIN, a town of Hindostan, in the presidency of Madras, and prov. of the Carnatic, district and 42 m. ENE of Tinnevelly, near the shore of the gulf of Manaar. It has a pearl fishery, but the pearls are inferior in purity of colour to those of Ceylon. The first European settlement here was formed by the Dutch. It was finally ceded to the English in 1824.

TUTINO, a town of Naples, in the prov. of the Terra-d'Otranto, district and 25 m. SE of Gallipoli, and cant. of Tricase, at the foot of a hill. Pop. 330.

TUTNAL-WITH-COBLEY, a hamlet in the p. of Tardebigg, Worcestershire, 2 m. ESE of Bromsgrove. Area 3,450 acres. Pop. in 1851, 492.

TUTOYA, a town of Brazil, in the prov. and 165 m. E of Maranhao, on the l. bank of a river of the same name, the most westerly of the six arms by which the Paranahiba discharges itself into the ocean. The inhabitants are chiefly of European extraction, and carry on an active trade.

TUTT, a village of Sind, on the E bank of the Indus.

TUTTEROEEN, an island of the North sea, off the W coast of Norway and dio. of Drontheim. It is noted for its apples and cherries.

TUTTINGTON, a parish in Norfolkshire, $2\frac{1}{4}$ m. E of Aylsham. Area 830 acres. Pop. in 1851, 213.

TUTTLINGEN, or DUTLINGEN, a bailiwick and town of Würtemberg, in the circle of the Schwarzwald, 22 m. WSW of Sigmaringen, on the r. bank of the Danube. Pop. in 1840, 5,302. It has manufactories of shawls, gloves, hosiery, bonnets, woollen and cotton fabrics, cutlery, nails, paper, &c.; and has also a considerable trade in corn. In the vicinity are the iron forges of Ludwigsthal, and the ruins of the ancient castle of Honberg. T., which is of great antiquity, was in 1643 the theatre of a battle between the French and Imperialists. Pop. of bail. 21,580.

TUTUILA, an island of the S. Pacific, in the Samoan group, in S lat. 14° 30′, W long. 170° 41′. It is 17 m. in length, and 5 m. in breadth; and attains an elevation, in its loftiest peak, of 2,327 ft. It is inhabited by a considerable population, and has numerous villages upon it. On its S side is a good port called Pago-Pago. According to the testimony of Captain Wilkes, "the men of T. are a remarkably tall fine-looking set, with intelligent and pleasing countenances. The women are far from being good-looking, but are remarkably domestic and virtuous," exhibiting a strange contrast to those of Tahiti. "Travelling is believed to be safe throughout the island, and the natives," says Captain Wilkes, "as far as our experience goes, are not the bloodthirsty race they have been reported to be. The unfavourable estimate of their character has, I presume, been derived from those who first knew them, and particularly from their attack upon the expedition of La Perouse." Of this conflict, Captain Wilkes obtained the following particulars. On the morning of the massacre, the vessel stood in towards the land, and about noon the boats went ashore. While on shore, a number of canoes, belonging to the island of Upolu (to which T. was at the time subject), put off to the vessels. When alongside, a young man in one of them having laid his hand on a bolt, with the intention, as it was supposed, of stealing it, he was fired upon by the French, and mortally wounded. The natives immediately left the vessels, and hastened to the shore, where the boats had gone to get water; and on reaching them they began the attack, which resulted in the massacre of M. De Langle, and of those who were with him on shore. When the natives began this attack, the great body of the French were absent from their boats; but on the commencement of the disturbance, they all rushed towards the boats, and the confusion became general. The Tutuilians maintain that they endeavoured to save the lives of the French; and on the following day, as soon as they dared to venture from the mountains, whither they had fled during the massacre, they collected the bodies, dressed them in native cloth, and buried them in the beach, as they were accustomed to bury their own chiefs. The actors in the massacre proceeded at once to Upolu, which will account for their having been afterwards seen there, and recognized by the French.

TUTWANU, a village of the Punjab, on the r. bank of the Ravi.

TUTWERLU, a commune of Belgium, in the prov. of Antwerp and dep. of Baelen. Pop. 116.

TUXFORD, a parish and market-town in Nottinghamshire, 24 m. NE by N of Nottingham. Area of p. 3,000 acres. The town, situated on an eminence, near the banks of a small stream which flows into the Trent, has a modern appearance; being built on the site of the old town, the greater part of which was destroyed by fire in 1702. The chief trade is in hops, considerable quantities of which are grown in the neighbourhood. Pop. in 1851, 1,211.

TUXTLA. See TEXTLA.

TUY, a judicial partido and town of Spain, in the prov. of Pontevedra. The partido comprises 51 parishes. The town is 18 m. SSE of Vigo, on a height surrounded by three streams, affluents of the Minho, by which it is separated from Portugal. Pop. 4,212. It has good ramparts, several forts, and a citadel furnished with numerous cannon, and stands opposite Valencia, a strong Portuguese fortress. It is well-built, and has regular streets paved and well-kept, several squares, and promenades, three fountains, a fine episcopal palace, a cathedral, several chapels, three convents, two hospitals, a foundling asylum, a college, a seminary, and barracks. The industry of the town consists chiefly in the manufacture of table-linen, hats, and leather. The climate is mild, but in consequence of adjacent marshes, insalubrious. Grain, wine, fruit of all kinds, &c. are extensively cultivated in the environs. T. is of great antiquity, and is supposed to be the *Castellum* of Pliny.

TUY, a river of Venezuela, which has its source in the mountains of San Pedro, 30 m. SW of Caraccas; flows in a generally E direction through the valleys of Aragoa, Tacata, Ocumare, and Santa-Teresa; and after a total course of upwards of 105 m., throws itself into the Caribbean sea, 27 m. SSE of Cape Codera. Its principal affluent is the Guayana.

TUYLK, a commune of Belgium, in the prov. of Limburg, and dep. of Curange. Pop. 238.

TUYRA, TUIRA, or SANTA-MARIA, a river of New Granada, which rises in the Spiritu-Santo range, in 7° 40′ N lat.; flows N, and then NW, to its junction with the Chuamaque, coming from the N, in N lat. 8° 10′, W long. 77° 45′; and then pursues a NW course to Darien harbour at the head of the gulf of San-Miguel. It is joined by the Nique, the Tapanacu, the Paya, the Puero, and the Yape on the r. bank; and by the Lemon, Bico, Grandi, Pirre, San-Josef, and Tavina, on the l. bank.

TUYTENBERG, a commune of Belgium, in the prov. of Brabant and dep. of Lennick-Saint-Quentin. Pop. 252.

TU-YUN-FU, a division and town of China, in the prov. of Kwi-chu. The div. comprises 8 districts. The town is 30 m. S of Ping-yue-chu, in a valley on the l. bank of the Cham-ke, in N lat. 26° 12′, E long. 107° 22′. It is exposed to the incursions of hordes who maintain themselves in independence in the adjacent mountains, and is consequently ill-built, and without any edifice worthy of notice. In this division is also Tu-yun-heen.

TVAREN, a gulf formed by the Baltic, on the coast of Sweden, in the pref. of Nykoeping.

TVEDESTRAND, a port of Norway, in the dio. of Christiansand, and bail. of Nedenaes, formed by the Skager-Rack, in N lat. 58° 37′, E long. 9°, and 12 m. S of Œster-Risoer. It is 12 m. in depth, and very narrow, and is entered by the Oxe-fiord, in which are the islands of Boroe and Tverdalsoe.

TVER, a central government of European Russia, lying between the parallels of 56° and 58° 40′ N lat. and from 32° to 38° 20′ E long. It is bounded by the gov. of Novgorod on the NW and N; by Jaroslav on the E; by Vladimir on the SE; by Moscow and Smolensk on the S; and by Pskov on the W. It has a superficial extent of 19,650 geog. sq. m., with 1,359,920 inhabitants, in 1851, partly of Russian, partly of Finnish descent. The surface presents a wide plain, with numerous undulations, but nothing that can be called hills, and immense pine-forests. The climate is temperate, but changeable. The rivers are frozen over from the beginning of December to the end of March: the principal are the Volga, the Dwina, the Msta, the Tverza, the Mologa, and the Medveditza. The chief lakes are the Seliger and the Dvinez. The soil is of moderate fertility; and though fruit seldom succeeds, corn, hemp, cattle, and wood are raised in sufficient quantity to enable the inhabitants to export these productions. There are also iron mines. Manufactures are not numerous or important; but the transit trade is considerable. It is administratively divided into 12 circles.

TVER, the capital of the above government, and an archbishop's see, stands on the great road from Petersburg to Moscow; at the confluence of the Tvertza, the Volga, and the Tmaka, which three rivers divide the town into four parts, united by bridges, that over the Volga being of boats that may be removed during winter. Pop. 20,000. T. has been repeatedly ravaged by fire, particularly in 1763, when only a few houses were saved. The empress Catherine, after its last destruction, ordered a plan of a new town to be made out, and gave orders that the houses should be rebuilt on the prescribed model, erecting the public edifices at her own expense, and assisting individuals by loan. In consequence, T. is the most regularly built city in the empire, after St. Petersburg and Moscow. The streets are broad, and perfectly straight; and they all proceed from an open space in the centre of the town. The houses which surround this space, as well as those in the principal streets, are of brick, stuccoed white; there are also several handsome squares; but the city has upon the whole an inactive desolate appearance. The chief public buildings are the cathedral, which is old and in the Gothic style, the palace, the courts of justice, the government-offices, and the hospital. The building called the palace was fitted up for the archduchess Catherine, and her husband, a prince of Oldenburg, to whom, in his capacity of governor of the provs. of Tver, Novgorod, and Vladimir, the town was indebted for various improvements. The seminaries consist of one for the clergy; a gymnasium for education in classics, history, and philosophy; and a school for youths of rank.—The manufactures comprise linen, cordage, wax, leather, candles, bells, and hardware. Having the advantage of water-communication, T. has a considerable traffic in the supplying of the adjacent country, and the exporting of its produce, viz.: corn, hemp, and iron. Steamboats ply on the Volga between T. and the sea.

TVER, a town in the gov. of Vilna, district and 17 m. S of Telch.

TVERDOSSIN. See TURDOSIN.

TVERETCH, a town of Russia in Europe, in the gov. of Vilna, district and 20 m. NE of Sventziany.

TVERTZA, a river of Russia in Europe, in the gov. of Tver, which issues from a small lake, in the district and 8 m. NNE of Vichni-Volotchok; traverses the district of Torjok; passes the town of that name; and after a total course in a generally SE direction of about 126 m., joins the Volga on the l. bank, in Tver. This river communicates by the canal of Vichni-Volotchok with the Tzna, and thus connects the basins of the Volga and Neva. Upwards of 6,000 vessels annually pass to the Neva by this route.

TVIBERG, an island of the North sea, off the W coast of Norway, in the dio. of Bergen, in N lat. 61° 21′, E long. 4° 52′.

TWEED, a river which, directly or by tributaries, drains nearly the whole area of the central and eastern border counties of Scotland, and detached districts of Durham and the north corner of Northumberland in England. It usually ranks in importance as the fourth of the Scottish rivers; but, estimated according to the extent of country which it drains, it surpasses every Scottish river except the Tay; for, while the Spey, which this rule places next to it, drains only 1,300 sq. m., and the Clyde, which is next to the Spey, drains only 1,200 sq. m., the Tweed drains 1,870 sq. m. In two head-waters, the one of which bears the name of Tweed from the outset, while the other is throughout called the Lyne, it rises respectively in the SW and the NW extremities of Peebles-shire; by the head-streams of the Ettrick, the Yarrow, and the Teviot, it drains Selkirkshire and Roxburghshire from the furthest W; by the Gala, the Leader, the Whitadder, and other streams, it draws off the waters from the S acclivities of the Moorfoot and the Lammermoor hills, even from a line but 11 or 12 m. S of Edinburgh; and from its remotest source to the sea at Berwick, it performs, irrespectively of windings, a run of about 100 m., about one-third of which is in Peebles-shire, and about another third through or in contact with Roxburghshire. The really remotest source of the T. is a paltry fountain called Tweed's-well, at the foot of a hill named Tweed's-cross, on the SW boundary of the p. of Tweedsmuir, in Peebles-shire, at an alt. of 1,500 ft. above the level of the sea. Over 22 m. from its origin, the T. runs prevailingly NE; and over the remaining 14 m. of the Peebles-shire part of its course, in the direction of E by S. The chief of the larger Peebles-shire tributaries, on the l. bank, are the Biggar, between Glenholm and Stobo; the Lyne, between Stobo and Peebles; the Eddlestone, at the town of Peebles; and the Leithen, at the village of Innerleithen; on the r. bank, the Fruid and the Talla, in Tweedsmuir; the Manor, between Manor and Peebles; and the Quair, nearly opposite the Leithen. During its connexion with Selkirkshire and Roxburghshire the T. continues, as in the latter part of its course in Peebles-shire, to run prevailingly E, till the last 9 or 10 m.; and over these it runs in the direction of NE by E. Its larger tributaries here, from the N, are the Cadon, between Stow and Galashiels; the Gala, between Galashiels and Melrose; the Allan, in Melrose; the Leader, between Melrose and Merton; and the Eden, in Ednam; from the S, the Ettrick, between Selkirk and Galashiels; and the Teviot, at the town of Kelso, the former previously augmented by the Yarrow, and the latter by far the largest of the Tweed's tributaries, and almost a rival of its upper stream in importance. After leaving Roxburghshire the T. is but partially a Scottish river; it divides Berwickshire from England till within 4½ m. of the sea, and then bids adieu to Scotland, and runs between England and the Liberties of Berwick. Its course for 4 or 5 m. after leaving Roxburghshire, and again for about 5 m. before entering the sea, is E; and over the intermediate distance it is in the direction of NE by N. The chief tributaries which enter it below Roxburghshire are, on the l. bank, the Leet, in Coldstream; and the Whitadder, in the liberties

RUINS OF TYRE.

county, to the sea at Tynninghame, 2¾ m. NW of Dunbar, performing a run of 16 m., or, if measured from its remotest source, 28 m.

TYNEHAM (West, or Little), a parish in Dorsetshire, 6¾ m. W by S of Corfe-Castle. Area 2,915 acres. Pop. in 1831, 247; in 1851, 274.

TYNEMOUTH, a borough and seaport in Northumberland, at the mouth of the river Tyne, on its N bank, 8 m. ENE of Newcastle by the Newcastle and North Shields railway, which terminates here. The parish, including the townships and chapelries of North Shields, T., Chirton, Cullercoats, Monk-Seaton, Murton or Moortown, Preston, and Whitley, has an area of 7,222 acres. Pop. in 1801, 14,345; in 1831, 24,778; in 1851, 30,524.—The town of North Shields partly occupies the townships of T., Preston, and Chirton. The mouth of the Tyne forms a haven of sufficient depth for vessels of any tonnage except on the bar of sand which crosses it, where there is little more than 7 ft. depth of water at ebb. The principal manufactures carried on in the p. are those connected with the supply of the shipping in the port and in the ship-building yards; but there are considerable manufactories of chemical substances, tobacco, hats, and gloves. T. village, consisting principally of one good street and two smaller ones, is situated about 1 m. to the E of the market-place of North Shields. The houses are for the most part well-built; and during the bathing season, considerable numbers resort hither for bathing. The ruins of the ancient priory of T., situated on a peninsula of rocks at the mouth of the Tyne, have been appropriated as a magazine for military stores. T. and North Shields were enfranchised by the Reform act, and now return one member to parliament.

TYNGSBORO, a township and village of Middleton co., in the state of Massachusetts, U. S., on both sides of the Merrimac, and intersected by the Nashua and Lowell railroad. Pop. in 1840, 670; in 1850, 799.

TYNIEC, a village of Galicia, in the circle and 20 m. ENE of Wadowice, on the r. bank of the Vistula. It has a Benedictine convent.

TYNISH, an inhabited island in the p. of Moyrus, co. Galway, 3 furls. SSW of the nearest part of the mainland, and 3¾ m. NW by W of Garomna. Its length, south-south-westward, is 1¼ m.; and its extreme breadth is 3 furls. Its inhabitants, amounting to about 150, are engaged chiefly in the fisheries.

TYNRON, a parish in Nithsdale, Dumfries-shire; bounded on the W by Kirkcudbrightshire. Area 11,990 acres. The surface consists principally of the glen or strath of the Shinnel, and two ranges of hills which form its screens. The highest hills, Lamgarroch and Coremilligan, are each about 1,800 ft. above sea-level. The v. of Tynron-kirk, with scarcely 100 inhabitants, stands on the Minnyhive road, 2½ m. from Minnyhive. Pop. of the p. in 1831, 493; in 1851, 482.

TY-PIN-SHAN, or Tae-ping-shan, an island of the Eastern sea, in the group sometimes known as the Majicosima islands, of which it is the largest and most easterly, in N lat. 24° 50′, and E long 125°. It is about 20 m. in length, by 10 m. in breadth, and enclosed on the N and NW by an extensive reef. Captain Sir Edward Belcher has given the first account of the Ty-pin-san group which consists of 5 islands, namely, T., Ku-ri-măh, Y-drah-bu, Y-ki-măh, and U-ga-mi: "In describing the principal features which seem to mark the character of the people of this hitherto unknown, or rather unvisited group, termed Meia-co-shimahs by the authorities, it will be immediately apparent," says Captain Belcher, "how very near they approximate, in general, to the inhabitants of Loo-Choo, so well detailed by Mac Leod and Basil Hall in the voyages of her majesty's ships *Alceste* and *Lyra*. Like those mild and inoffensive islanders, their physical appearance much resembles that of the Koreans and Japanese; their dress also consists of a similar loose robe, of varied pattern, having large wide sleeves, and which is secured about the waist by a long sash. Like them, also, they strain back their long black hair, which is secured in an elegant top-knot, through which they pass their ornamental hair-pins. On state occasions the grandees wear red, yellow, or blue caps of office; and in pursuance of the customs of the Chinese their women are strictly secluded. It would be an easy task to designate this people as a set of tea-drinking old women, imbecile and apathetic; void of energy and enterprise, living in contentment on a group of islands the value and facilities of which they are almost entirely ignorant of, and of whose position and resources they are unable to take advantage. But on contrasting them with the insidious, fawning, and deceitful Chinese, or the savage and vindictive bloodythirsty Malays, I cannot but fancy their character amiable and their condition one to be envied. Their food is extremely simple, consisting chiefly of the batata, rice, and other vegetables, varied with the produce of the deep, including molluscous animals, such as the cuttle-fish, the large clam (*Tridacna gigas*), and others. They appear to indulge constantly in smoking tobacco, and seem as passionately fond of tea as their continental friends the Chinese.

TY-PO-HOY, a bay formed by a series of small islands in the bay of Canton, on the S coast of China, and prov. of Kwang-tung, in N lat. 22° 32′, and E long. 114° 25′. It is about 12 m. in length, and nearly equal in breadth.

TYRAWA-SOHIA, a village of Galicia, in the circle and 6 m. NE of Sanok, on the Tyrawa, a small affluent of the San. It has a salt-refinery, several silk mills, and petroleum springs.

TYRAWA-WOLOSKA, a town of Galicia, in the circle and 9 m. ENE of Sanok, on the Tyrawa.

TYRCONNEL, an ancient Irish toparchy, nearly identical with the present county of Donegal, in the NW of Ulster.—Also a small island in the mouth of Murchison strait, at the E coast of Baffin bay, in N lat. 77° 36′.

TYRE, or Sur, a small port of Syria, 25 m. SW of Saida or Sidon, the representative of the ancient Tyre, called by Joshua, 'the strong city of Tzoor,' and the most celebrated of all the Phœnician cities. It seems to have been originally founded by a colony from Sidon, and at a subsequent period to have eclipsed that city itself in commercial wealth and political importance, being the greatest mart in the ancient world, and possessed of all the trade of the Mediterranean sea. So great was its naval power that it baffled the arms of the Assyrian conqueror Shalmanazar, and withstood the arms of Nebuchadnezzar for 13 years. When at length their city was taken by that powerful prince, the inhabitants retired into the insular Tyre where they built a new town, which soon rose to equal importance with the continental city. The renovated Tyre stood a siege of seven months from Alexander the Great, who treated the captive inhabitants with unrelenting cruelty; but it was again rebuilt and repeopled by the conqueror who had destroyed it, and rose again to importance, though not to its pristine grandeur, as the monopoly of commerce which the former city enjoyed was now annihilated for ever. After the Saracenic conquest of Syria, T. gradually declined, and at the end of the last century it was a miserable village with hardly ten poor fishermen in habiting it. But within this century T. has again begun to lift up its head. In 1816, according to Buckingham, it contained 800 substantial stone built buildings, with suitable appendages, besides other smaller houses for the poorer classes, a mosque, three Christian churches, a public bath, and three bazaars. He calculated the increasing pop. at from 5,000 to 8,000 souls, three-fourths of whom were Arab Catholics; the rest Mussulmans, Arabs, and Turks. Mr. Connor, however, in 1820, gives a much lower estimate of the pop., making it only 1,900, and of these 100 only were Turks; and still later, M. de Bertou states the pop. at only 1,200. The chief staple is tobacco, which is sent to Cairo and Damietta, where it is sold at from £20 to £24 a cantar. To the same places are sent charcoal, dried figs, and faggots of wood. The walls of the ancient city may still be traced, forming an irregular square of about 1 m. in circuit. The harbour, which will now admit only boats, stretches 80 ft. along the shore, and 150 ft. along banks that run into the sea. Near the walls at the E end, are the ruins of two square towers, which Pocock supposes may have served as reservoirs of water from the aqueduct, for its distribution over the city. Without the walls are the ruins of a large church, built in the Syrian style.

TYRE, a township of Seneca co., in the state of New York, U. S., 150 m. W by N of Albany, drained by Clyde river.

TYRELLA, a parish of co. Down, 3 m. E of Dundrum. Area 1,999 acres. Pop. in 1851, 762.

TYRENASCRAGH, or TIRANASCRAGH, a parish in co. Galway, 3¾ m. SSW of Eyrecourt. Area 3,629 acres. Pop. in 1831, 1,093; in 1851, 719.

TYRESIOE, an island of Sweden, in the prefecture of Stockholm, and haerad of Sotholm. It contains the town of Dalard.

TYRGHETOU, a mountain of Mongolai, in the Khalkas country, between Urga and Kiakhta. It gives rise to the Chara.

TYRIE, a parish near the NE extremity of Aberdeenshire. Area about 15 sq. m. Pop. 2,830.

TYRI-FIORD, a lake of Norway, in the dio. of Aggershuus and E part of the bail. of Buskerud, in N lat. 60°, E long. 10°. It is 18 m. in length from NE to SW, but is very unequal in breadth, branching into several extensive arms. On the NW it receives the Beina-elv, and on the SW it discharges itself, by the Drammen-elv, into the gulf of Christiania.

TYRINGHAM, a township and village of Berkshire co., in the state of Massachusetts, U. S., 112 m. W of Boston, watered by branches of Housatonic river. Pop. in 1840, 1,477; in 1851, 821.

TYRNAU, or NAGY-SZOMBAT, a free royal town of Hungary, in the comitat and 28 m. NE of Presburg, in a fertile plain, and on a small river of the same name, an affluent of the Dudvag. Pop., chiefly Slovacks, Hungarians, and Germans, and nearly all Catholics, 7,000. It contains 9 Catholic churches—of which one is surmounted by two fine towers,—two convents, two seminaries, a Catholic gymnasium and observatory, a public library, an infirmary, an alms-house, and a lunatic-asylum. It has manufactories of linen and cloth, and carries on an extensive trade. It was formerly the seat of a university.

TYROL, a province of the Austrian empire, comprising an extensive and very mountainous country, on the SW frontiers of Germany, extending from Italy to Suabia, and from the duchy of Carinthia to the Grisons. Under the appellation are included, not only Tyrol properly so called, but the bishoprics of Trent and Brixen, the seven lordships of the Vorarlberg, and the counties of Feldkirch, Schellenberg, Monfort, Pludenz, Hohenems, and Sonneberg: the whole—excepting the seigniory of Weiler—comprehending, according to Lichtenstern, 514 German sq. m., or 10,280 British sq. m. Its greatest extent from N to S, or from the southern frontiers of the bishopric of Trent to the northern frontier of Suabia, is 140 m.; its greatest length from E to W, or from the western boundary of Carinthia, to the lake of Constanz, is 160 m.; but its figure is extremely irregular. In the S its breadth is not above 70 m.; in the NW not above 40 m.; in the NE, where it forms a narrow angle between Upper Bavaria and Salzburg, not more than 35 m. and decreasing from that to 15 m., being merely a long valley, 60 m. in length, watered by the Inn. Another angle, reaching from the moor of Sterzing to the frontiers of Carinthia, interposes between Salzburg and Italy; extending 50 m. in an E and W direction, and from 30 to 20 m. in breadth from N to S.

Physical features.] Tyrol, and the whole SW part of the country above the Ens, may justly be denominated German Switzerland,—being, in respect of physical features, a continuation of that country. Of this extensive and highly diversified region, the bishopric of Trent forms the southern division; that of Brixen, the NE; the Vorarlberg, the NW; while Tyrol Proper occupies the centre. It is bounded on the N by Bavaria and Suabia; on the E by Salzburg and Illyria; on the S by Austrian Italy; and on the W by Austrian Italy, the country of Bormio, the Grisons, and the lake of Constanz. The Vorarlberg, separated on the E by a chain of mountains denominated 'the Mountains of Eagles,' from Tyrol, is sometimes known as the county of Bregenz; having the counties of Pludenz, Sonnenberg, Feldkirch, and Schellenberg, on the S. This region was the ancient *Rhetia;* and was by the Romans divided into Upper and Lower, the former answering to the Grisons and the Italian Tyrol, and the latter to the German Tyrol and Salzburg, the heights of the Brenner being the boundary between the two. It is everywhere covered with mountains; but the principal chain stretches from the Valteline on the SW, to the duchy of Salzburg on the NE, which, as Saussure remarks, is the general course of the Alpine chains. The Brenner, or 'burning hill,' as it is called in German, the modern name of this chain, rivals the grand Alps of Switzerland in its glaciers; and like other grand chains, presents exterior barriers, that on the N being distinguished by the name of Spitz, while that on the S is termed Vedretta. The breadth of the Tyrolian chain from Trent to Innspruck is 70 m. directly across from S to N. The primitive or highest elevations are to the N of Sterzing, whence precipitous streams descend to the river Inn on the N. The naked and rugged peaks of the mounts of Lorinzen, Fartschel, and Tschafateh, raise their towering heads towards the NW; on the SE are those of Glander, Schloss, Pragels, and Pallanser; whose summits are entirely bare, and seem to be composed of granite. The glacier most easy of access is that of Stuben, 4,692 ft. above the level of the sea. The Brenner is, according to Beaumont, only 5,109 ft. above sea-level. The Bok-kogel, another vast peak, in the opinion of Mr. Eustace equal in height if not superior to Mont Blanc, presents a most sublime and majestic appearance. Towards the W and N of Innspruck are several detached mountains covered with constant snow, amongst which those of Verner are the most remarkable. The Great Glockner, between Tyrol and Salzburg, is said to be 12,780 ft. in height; and the Orteles has been computed at 12,864 ft. From Schellenberg and Feldkirch, a chain of moun tains runs NW and SE as far as the Ober-Innthal, separating the Vorarlberg and Western Tyrol from the Grisons. It then runs S, separating the Lower Engadine from Tyrol; and stretching eastward, separates the canton of Bormio and the sources of the Adda on the S, from the Munsterthal on the N. This chain, which is the highest of the Julian Alps, is called the Wurmser-Joch, and is noted for the daring exploit of the French army, under Dessoles, in 1799, when, in order to reach the Munsterthal, and make themselves masters of the head-valley of the Adige, notwithstanding the ices and snows which covered the Wurmser-Joch, they climbed the mountain, and by this manœuvre turned the intrenched defiles; attacked Laudohn, in his intrenchments at Glurns and Tauffers; and compelled his army to surrender with its cannon and baggage, Laudohn himself escaping with a few of his men into the Val-de-Venosta.—As might be expected in a country of mountains, rivers descend here in every direction, and separate to different seas: the Inn, the Drave, and the Lech, to the Danube; the Ill and the Bregenz, to the Rhine; the Adda, to the Po; the Adige, the Eisach, the Rienz, the Brenta, the Piave, and the Tagliamento, to the Adriatic. The Inn, the principal river of T., rises in the canton of the Grisons, and waters the Upper and Lower Engadine, before it enters T. at St. Jacob and Finstermunz. During this part of its course, amounting to 50 m., as its descent is more gradual, so its course is less precipitous than other Alpine streams; but after its entrance into T., where it is joined by numberless rapid torrents issuing from the mountains bordering on both sides of a long, steep, and narrow valley of 100 m., it runs with greatly increased velocity, till after a comparative course of 270 m., during which it receives the Zill, the Aicha, the Salza, and the Rot, it enters the Danube, at Passau, with a volume of water equalling, if not surpassing that of its rival stream. The Adige rises in the valley of the Malsheide, which is separated from the Ober-Innthal on the N by a chain of heights. After running 20 m. S it is joined by another branch, coming from the N side of the Wurmser-Joch, through the Munsterthal. The confluent stream at Glurns obtains the name of the Etsch, which it retains till joined by the Eisach at Botzen, when it receives that of the Adige. After receiving the Non and the Lavis, and passing to the south by Trent and Verona, it runs E and falls into the Adriatic, 10 m. to the N of the Po, after a comparative course of 200 m. The Eisach is larger and more rapid than the Adige, its sister-stream. To the E of the source of the Rienz, on the opposite side of the mountain, and above the village of Innichin, rises the Drave. In descending the banks of this river, through the steep and winding defiles of the Tyrolian mountains, till it arrives at the town of Lienz, only 15 leagues from its source, this stream must be crossed no less than fifteen times. From Lienz it pursues an E and SE course, through Carinthia, Styria, and Lower Hungary, and falls into the Danube, 16 m. below Esseck. Its comparative course may be estimated at 400 m.; and it is frequently impassable on account of its rapidity. The Lech rises in the mountains of the Vorarlberg, and after a course, through that elevated district, of 50 m., enters Bavaria, separating it from the circle of Suabia, and flows into the Danube, 10 m. E of Donauwert. Its comparative course is 150 m. The Iser, a river of equal length and magnitude, rises in T., and runs along the foot of the mountains that skirt the valley of Innspruck, till it enters Bavaria. After visiting Munich, the capital of Bavaria, it runs a NE course of 80 m., and enters the Danube at Dickendorff. The Iller rises from the W side of the Mountains of Eagles; and after running through part of the Vorarlberg and Suabia, and passing Kempten, enters the Danube, almost opposite Ulm, after a course of 100 m.

Productions and industry.] The Tyrolese mountains present every aspect, from the ever-blooming verdure of perpetual spring, to the dreary sterility of the frigid zone. Though covered with eternal snow, their sides are clothed with fine forests, abounding in every variety of forest tree, and sheltering numerous species of game. The valleys, though rocky in soil, have rich and extensive fields of corn, flax, and tobacco. On the eminences which crown these fertile vales, various sorts of fruit are grown; the vine is reared as far as Brixen, but the wine is not skilfully managed. The rugged aspect of this elevated country, contrasted with the beauty and fertility of its vales, gave rise to a saying of the emperor Maximilian, that T. was like a peasant's frock,—coarse indeed, but right warm. But the chief wealth of T. lies in its mineral productions and precious stones, such as agates, cornelians, rubies, a species of diamonds, amethysts, emeralds, chalcedonies, and granites. Mines of silver and lead, which have been long celebrated, occur near Lermos. The mines of Nasereit, in the same quarter, among the Verner mountains, 32 m. NW of Innspruck, are opulent in silver, lead, copper, and iron. Nor is the S region of T. deficient in mines; but the principal mines are at Schwatz, where one containing silver and copper, discovered in 1448, still employs 1,000 miners. The salt mines at Halle, 6 m. NE of Innspruck, produce vast quantities of that commodity, and yield a revenue of 300,000 florins, or £35,000 annually. The Tyrolese are a very industrious people. They are often seen climbing the steepest rocks with a basket of manure on their head, or letting themselves down by a rope to some little insulated spot of garden-ground which they have discovered amid the cliffs of their mountains. They work, both men and women, from the earliest dawn to the last moment of twilight. In autumn, they rear silk-worms, and cultivate hemp, flax, and tobacco. Many of them, like the Irish, in summer, emigrate for the season, —leaving their homes with a hurdy-gurdy, a knapsack, and a stock of oaten cakes, and returning at the end of autumn, after a tour in the surrounding countries, with the profits of their summer's toil. These annual emigrants are said to amount to 30,000. During winter, while the snow and the torrents block up their villages, every house exhibits a scene of industry,—the women are employed at their spinning-wheels, or in knitting stockings, or embroidering cloth,—while the men and boys make boxes, cases of instruments, and toys, which find their way not only to most parts of Europe, but even to America. A particular branch of industry in this country is the breeding of Canary birds, which are reared in great quantities for sale.

Population.] The population, according to Lichtenstern, excluding the lordship of Wieler in the N of the Vorarlberg, amounted in 1812 to 692,000; according to Stein, the pop. in 1831 was 755,401; in 1850, it was 859,706. The Tyrolese are chiefly of German origin; but there are about 160,000 Italians in the southern part of the country. A very small part of this pop. is collected in towns, as Innspruck the capital contains little more than 9,000 inhabitants, and Trent 11,000.—Tyrol and Vorarlberg are divided into 7 circles, containing 22 towns, 36 boroughs, and 3,150 villages, and bearing the names of the principal valleys and districts.

History.] Tyrol or 'the Tyrol,' as it is most frequently called, is one of the most ancient possessions of the House of Austria, to which it fell by inheritance in 1363. Although from the commencement of their connexion with Austria, the Tyrolese had on various occasions given proofs of attachment and loyalty, it was not till the war of Succession that the Austrian family were made duly sensible of the worth of such subjects. Even when Frederic, the founder of the Tyrol line of Austrian princes, yielded himself and his possessions to the emperor Sigismund, these mountaineers continued faithful to him, fortified their passes, set the imperial troops at defiance, and preserved for him a country of which he was unworthy. The ungrateful Frederic rewarded them by the imposition of heavy taxes; but at the same time secured to them their rights and liberties. Villanage and servitude continued unknown; and the land was tilled by a free peasantry, whose representatives formed one of the branches of the legislature. Amidst the mountains of Tyrol, and their hardy mountaineers, an asylum was found in the worst ages of persecution, by many of the Waldenses. But in the latter part of the 17th cent., the bishop of Brixen, and the archbishop of Salzburg, having discovered that the posterity of these good men continued in the faith of their fathers—which was neither, strictly speaking, the system of Luther nor Calvin, and therefore not within the letter of the law which had compelled them to tolerate these two main sects of the reformers—ordered them to go to mass; whereupon, in 1681, 20,000 Tyrolese left their mountains and valleys, and went forth to seek for liberty of conscience in the Protestant states of Germany and Switzerland. Happily, however, the house of Austria soon perceived it to be its interest to pursue a milder policy; freedom of conscience was allowed, and the Tyrolese peasantry were more gently taxed than any other of the emperor's subjects. The Tyrol, therefore, is almost the only corner of the dominions of Austria whose inhabitants, though inspired with the true love of liberty, have remained attached to the dynasty of Hapsburg. During the war of Succession, the French entered T. on one side, and the Bavarians on another, at a time when there were no troops to defend the country; but the peasants blocked up the passes, broke down the bridges, and prevented their junction, nor was a single man found who could be bribed to carry intelligence from one army to the other. In 1744, when the French, at that time in alliance with Prussia and Bavaria, invaded T., the people rose in a mass and compelled the French to retire. In 1796, they again drove the French, under Vaubois, out of their country; and in 1797, when Bonaparte was adding conquest to conquest, they rose *en masse*, under Laudohn, and descending like a torrent from their native mountains, drove the French out of their country. Had not the Austrian emperor been terrified into the preliminaries of Leoben, by the menaces of Bonaparte, who well knew the danger of his own situation, the French army, entangled in the defiles of the Styrian mountains, might have been destroyed, for the Tyrolese had made themselves masters of Verona, and were joined by all the neighbouring mountaineers, to the number of 50,000. In 1799, they drove Massena out of the Vorarlberg with great slaughter. In 1801 and again in 1805, they were also successful, particularly in 1805, when they thrice defeated Ney and the Bavarians; but by the treaty of Presburg, in 1806, their country was delivered up to Napoleon's Bavarian ally, with a futile stipulation that their ancient privileges should be preserved to them. By the constitution of Tyrol, the Austrian sovereign did not acquire a right to the allegiance of the Tyrolese until the oath of fealty had been taken, in the name of the community, by the four estates convened in full assembly at Innspruck: the Bavarian government neglected this ceremony, and took possession of the country by a set of French commissioners; the estates remonstrated, and the new monarch answered them with gracious promises. At last the constitution was abolished by a royal ordonnance, and the country deprived of its very name by its subdivision into the circles of the Inn, the Eisach, and the Etsch, under which denominations it was incorporated in the Bavarian monarchy, then newly remodelled into a dwarfish resemblance of its great foster-mother the French empire. When war again commenced, in 1809, between Austria and France, the Tyrolese, under the command of the gallant Hofer, took arms to emancipate themselves from the Bavarian yoke, and were for a time eminently successful, till, deprived of Austrian aid by the defeat of Wagram and the armistice that followed, they were left to maintain the contest alone, and after a brave, but ineffectual defence, were compelled to yield to the power of Bavaria, aided by the legions of France. In the last action, near Brixen, where the wife fought by her husband, and the maiden by the side of her father or betrothed husband, 350 Tyrolese women fell under the sabres of the enemy's cavalry. All the patriots who survived were delivered over to the military tribunals. The gallant Hofer was seized in his hut, and conducted barefooted through the snow, to Botzen, and thence to Mantua, where he was shot, pursuant to sentence of a military tribunal, on the 24th of February, 1810. The congress of Vienna, in 1815, released the Tyrolese and Vorarlbergers from their hated subjection to the Bavarian yoke, and restored their country to its former sovereign, the emperor of Austria.

TYROL, a town of the Tyrol, in the circle of Botzen, 1½ m. N of Meran, on the l. bank of the Adige. On an adjacent mountain is an ancient fortress of the same name. In the vicinity is a quarry of fine white marble.

TYRONE, an inland county of the Irish province of Ulster, bounded on the N and NE by co. Londonderry; on the E by Lough Neagh; on the SE by co. Armagh; on the S by cos. Monaghan and Fermanagh; on the SW by Fermanagh; and on the W and NW by co. Donegal. The outline of the co. in a loose or general sense, exhibits a broad parallelogram, extending in the direction of SE by E. Its greatest length in the direction of SE by E, and along the S border, from the summit of Croagh mountain, 2½ m. E of the Gap of Barnesmore, to the river Blackwater at Caledon, is 38½ m.; its greatest breadth, in the opposite direction, and over Strabane and Lough Derg, is 30 m. Its area comprises 450,286 acres of arable land, 311,867 of uncultivated land, 11,981 of continuous plantations, 710 of towns, and 31,796 of water,—in all, 806,640 acres. The extreme SW district consists of a portion of the great alpine region which occupies the greater portion of co. Donegal, steeply dissevered into sections here by the glens of the rivulets Derg and Mournebeg. The river Blackwater rises near the S boundary of the co. and flows along the S border, to the SW corner of Lough Neagh. The Torrent, the only considerable tributary of the Blackwater within Tyrone, rises on the SW side of Bartack mountain, and flows sinuously eastward. The Ballinderry rises in the moorland region, N of Pomeroy; and flows eastward along the N border, to Lough Neagh. The Tullyclea or Ballinamallard drains the district on the SW border around Trillick, into co. Fermanagh, toward Lower Lough Erne. All the other streams of the county consist of the head-waters and forming-tributaries of the river Foyle; they comprise all that great river's affluents of any consequence except the Finn; and they are noticed, with sufficient detail, in the article FOYLE. The county of Tyrone comprehends a small district within the basin of the Erne, a large district within the basin of Lough Neagh, and a still larger district within the basin of the Foyle.

Minerals.] Granitic rocks form a district of country about 19 m. in length, and 2½ or 3 m. in mean breadth, extending WSW from the NE extremity of the co. Rocks of old red sandstone and sandstone-conglomerate constitute a great central district of the co., extending 36 m. W, with a mean breadth of about 10 m., and continued into the co. of Fermanagh down to the E side of Lower Lough Erne. Rocks of carboniferous limestone, exhibiting some diversity of character, constitute a great district in the S, 30 m. in length, and from 4 to 11 m. in breadth. The coal formation, situated to the N of Dungannon, though small, contains a greater number of valuable beds of coal, than any other in Ireland. At present nine workable seams are known, which vary from 3 to 9 ft. in thickness.

Agriculture.] In consequence of the characteristic configuration of the co., the soils of the low grounds consist, in a large degree, of the detritus of the rocks of the hills and mountains; and in consequence of the great diversity of the geognostic formation of the uplands, the soils are exceedingly various. Yet, in a general view, reclaimed moor or bog may be exhibited as the prevailing soil of the arable lands in the great central district of the co., a gravelly or sandy soil on the skirts of the hill and mountain flanks of many of the glens and vales, and a loam of diversified character, but, to a large extent, light and friable, over the rich fertile grounds of the S and E. Few farmers hold a larger quantity of land than 40 or 50 Irish acres. The average size of tillage farms is about 12 acres. The dairy and sheep produce bear a very small proportion to the tillage, there being but little dairying and few sheep. In 1841, the number of farms, within the rural districts of the county, of from 1 acre to 5 acres in extent, was 14,555; of from 5 to 15 acres, 14,671; of from 15 to 30 acres, 3,776; and of upwards of 30 acres, 1,139. In 1841, the live stock in the rural districts was valued at a total of £745,496. In 1851, of 273,125 acres under crop, 149,763 acres were in oats, and only 7,994 under wheat. Potatoes were cultivated to the extent of 37,937 acres; turnips, 16,150 acres; and flax, 18,908 acres.

Divisions and Towns.] T. is divided into the 4 baronies of Dungannon on the E, Clogher on the S, Omagh on the SW, and Strabane on the NW. The towns and principal villages are Dungannon, Caledon, Carnteel, Aughnacloy, Benburb, Clonfeacle, Moy, Cookstown, Grange, Tullyhog, Coal-Island, Stewartstown, Castle-Caulfield, Donaghmore, Pomeroy, and Coagh, in the barony of Dungannon; Clogher, Fintona, Ballygawley, Five-mile-town, and Augher, in the barony of Clogher; Omagh, Beragh, Seskinore, Dromore, Trillick, Drumquin, Castle-Derg, Six-mile-cross, and Termonrock, in the barony of Omagh; and Strabane, Ardstraw, Douglas-Bridge, Newtown-Stewart, Gortin, Dunnamanna, Ballymagorry, and Clady, in the barony of Strabane.—The county is ecclesiastically distributed among the dioceses of Clogher, Armagh, and Derry. The county sent 10 members to the Irish parliament, or two from the county at large, and two from each of the boroughs of Dungannon, Strabane, Clogher, and Augher; but it sends to the imperial parliament only two from the county at large, and one from the borough of Dungannon. Constituency of the county at large in 1842, 2,521. Pop. of the county in 1831, 304,468; in 1841, 312,958; in 1851, 255,819. Houses, in 1831, 54,663; in 1841, 54,919. First-class inhabited houses in 1841, 664; second-class, 11,705; third-class, 24,443; fourth-class, 18,107. Families employed chiefly in agriculture in 1841, 37,276; in manufactures and trade, 16,923; in other pursuits, 3,138. Families dependent chiefly on property and professions, 1,034; on the directing of labour, 19,354; on their own manual labour, 36,392; on means not specified, 557.

History.] The county of T. is conspicuous in history; it was the principal arena of the contests incident to the rebellion, recognised as 'the Tyrone rebellion,' of Hugh O'Neill in 1597. Previous to the flight and attainder of O'Neill, T. was divided into the districts of North T. and South T.; but about the period of 'the plantation of Ulster,' North T. was transferred to the county of Londonderry.

TYRONE, a township of Livingston co., in the state of Michigan, U. S., 40 m. E of Lansing, drained by On creek, a branch of Shiawassee river. Pop. in 1840, 394; in 1850, 867.—Also a township and village of Steuben co., in the state of New York, 172 m. W by S of Albany. It has a hilly surface, and is drained by Mud creek. The soil is clay and calcareous loam. Pop. in 1840, 2,122; in 1850, 1,894; and of village, 250.—Also a township of Fayette co., in the state of Pennsylvania. Pop. in 1840, 1,189.—Also a township of Adams co., in the same state, 10 m. NE of Gettysburg, bordered by Bermudian creek, on the E, and by Conewago creek on the S. The surface is level, and the soil red shale and gravel. Pop. 756.—Also a township of Perry co., in the same state, drained by Sherman's creek and its branches. The soil in the valleys is calcareous loam. Pop. 2,391.—Also a township of Huntingdon co., in the same state. Pop. 1,226.—Also a village of Blair co., in the same state, 76 m. WNW of Harrisburg, on the Pennsylvania railway.

TYRREL, a county in the E part of the state of North Carolina, U. S., comprising an area of 727 sq. m., drained by Alligator river and its branches. The surface is low and wet, but produces, in some parts, rice, cotton, grain, &c. Pop. in 1840, 4,675; in 1850, 5,133. Its capital is Columbia.

TYRREL'S BAY, a bay and town in the SW part of the island of St. Vincent, in the parish of St. George. It has a fortress, the best in the island, and carries on an active trade.

TYRREL'S-PASS, a market-town in the parishes of Newtown and Clonfad, co. Westmeath, 7½ m. NNE of Tullamore. It presents a remarkably clean, orderly, neat, and respectable appearance.

TYRRINGHAM-WITH-FILGROVE, a parish in Bucks, 2¼ m. NNW of Newport-Pagnell, nearly encircled by the Ouse. Area 1,736 acres. Pop. in 1831, 227; in 1851, 188.

TYSMIENICA, a town of Austria, in the prov. of Galicia, circle and 8 m. SSE of Stanislawow, near the l. bank of the Worona, called also Tysmienica, a little above its confluence with the Bistrica. Pop. 2,800. It has several Catholic, United Greek, and Armenian churches, and contains several tanneries. It has an active trade in horses and in wax.

TYSNOES, a parish of Norway, in the diocese and 27 m. SSE of Bergen, and bail. of South Bergenshuus, on the N coast of the island of Tysnoes-Oe. This island is situated in the North sea, between the Strand-fiord on the N, and the Hardanger-fiord on the S, a little to the NE of the island of Storoe, in N lat. 60°, and E long. 9° 35′. It is 15 m. in length from N to S, and 6 m. in medium breadth, and has a deeply indented coast.

TYSOE-WITH-WESTCOTE, a parish in Warwickshire, 5 m. S by E of Kington. Area 4,710 acres. Pop. in 1831, 1,007; in 1851, 1,049.

TYSSNAR, a lake of Sweden, on the confines of the prefectures of Linkoeping and Nykoeping.

TYSSOWICE, a town of Poland, in the gov. of Lublin, obwod and 17 m. SW of Hrubieszow, on the Hulzwa. Pop. 2,531.

TYSY, or TISI, a town of Senegambia, in the kingdom of Kasson, 15 m. from the r. bank of the Senegal, and 36 m. WNW of Kuriakary.

TYTERS (LILLE), a small island of Russia in Europe, in the gov. of Esthonia, and district of Vezenberk, in the gulf of Finland, about 20 m. from the continent, in N lat. 59° 50′.

TYTERS (STORE), a small island of Russia in Europe, in the gov. of Esthonia, and district of Vezenberk, in the gulf of Finland, 27 m. from the continent, and 9 m. ENE of Lille Tyters.

TYTHBY, a parish in Nottinghamshire, 9 m. E by S of Nottingham, including the township of Cropwell-Butler. Area 3,010 acres. Pop. 811.

TYTHEGSTON, a parish in Glamorganshire, 3 m. W by S of Bridgend. Area 2,871 acres. Pop. in 1831, 404; in 1851, 1,152.

TYTHERINGTON, a township in Prestbury p., co.-palatine of Chester, 1½ m. N by W of Macclesfield. Area 993 acres. Pop. in 1831, 427; in 1851, 374.—Also a parish in Gloucestershire, 3 m. SE by E of Thornbury, including the tithing of Itchington. Area 2,100 acres. Pop. in 1831, 476; in 1851, 465.—Also a parish in Wilts, 4½ m. SE of Warminster. Area 1,650 acres. Pop. 95.

TYTHERLY (EAST), a parish in Hants, 6 m. SW by S of Stockbridge. Area 1,560 acres. Pop. in 1831, 294; in 1851, 398.

TYTHERLY (WEST), a parish in Hants, 6½ m. SW of Stockbridge, including Buckholt-Farm. Area 2,270 acres. The 'Rationalists,' under the superintendence of the celebrated Robert Owen, have or had an establishment here, conducted on a system of Home Colonization, and named 'Harmony.' Pop. in 1831, 497; in 1851, 447.

TYVERA, a village of Afghanistan, between Subzawur and Giriskh.

TYWARDRETH, a parish in Cornwall, 3 m. WNW of Fowey. Area 3,252 acres. Pop. 3,287.

TZABA, a mountain of Turkey in Europe, in Bosnia, in the sanj. of Herzegovina, on the Dalmatian frontier, and 20 m. W of Ghoubigne.

TZABATZ, a fortress of Turkey in Europe, in Servia, in the sanj. of Semendria, 41 m. W of Belgrade, on the r. bank of the Save.

TZAGLAIK, a town of Turkey in Europe, in Rumelia, in the sanj. of Gallipoli, 17 m. ENE of La Cavale, between two mountains, on the r. bank of the Kara-su.

TZANA, Dek, or Daga, the largest of the islands in Lake Dembea, in Abyssinia. See Dembea.

TZAPOTECA, a mountainous district of Mexico, in the state of Oaxaca, inhabited by the Tzapotecs, between the Oaxaca and Tehuantepec.

TZAREVOKCHAISK, a district and town of Russia in Europe, in the gov. and 75 m. NW of Kazan, on the Maloia-Kokchaga.

TZAREVOSANTCHOURSK, a town of Russia in Europe, in the gov. and 147 m. SW of Viatka, and district of Jaransk, on the r. bank of the Bolchaia-Kokchaga.

TZARITZIN, a fortress of Russia in Europe, capital of a district of the same name, in the gov. of Saratov, and 93 m. NW of Tchernoilar, on the r. bank of the Volga, at the confluence of the Tzaritza. Pop. 5,000.

TZARITZINO, a village of Russia in Europe, in the gov. and 8 m. S of Moscow, on the road from that town to Kachira, near the SE bank of a small lake. It possesses an imperial palace, erected by Catherine II., the architecture of which is a strange mixture of the Greek and Gothic. Surrounding it are fine gardens, and a noble avenue of trees extends from it to the village of Kolomenskoe, a distance of nearly 4 m. T. was purchased by Catherine II. in 1774 from Prince Dmitri Kantimar, hospodar of Moldavia, to whom it was given by Peter the Great.

TZATZIN, a town and fort of Turkey in Europe, in Bosnia, in Croatia, 11 m. NNE of Bihach, on a mountain.

TZAVO, a river of Eastern Africa, which rises on the NW flank of Kilimanjaro in Jagga, and runs E towards the Adi, or the bay of Melinda. At the point where Dr. Krapf crossed it, he says, he "found it about 20 or 25 ft. in breadth, with banks 15 to 18 ft. in height. The water, which was very cool, run with great speed over a fine reddish sand, and was about 2½ ft. deep. Rocks I could not observe in the river's channel, nor did I hear the least noise —quite in opposition to the rivers which I crossed on my journey to Usambára, where the river's noise was heard at a long distance. Indeed, nobody could suppose a river to be in this country, so gently glides the T. along its sandy bed. Only large mikóma trees indicate to the traveller the river's course. I saw these trees in great abundance on the coast of Malinde, into the bay of which the T. seems to empty itself. The T. runs near a range of red hills into the Galla country, where the Wói joins it. All these rivers seem to form the river Sabáki, which has probably its outlet in the bay of Malinde. However, I wish not to be too positive in this matter, as the Sabáki must be more examined at some distance from Malinde. No doubt, the T. takes up all, or at least most part, of the water running from the E of the Kilimanjaro, whilst the rivers Góna and Lúmi carry their contents to the Pangany river. After crossing the river T., we had a fine view of Mount Théuka, which rises from the W bank of the river to a great height. There is no doubt that the whole country around the river T. has in former ages undergone great changes by volcanic action."

TZEKHANOVETZ, a town of Russia in Europe, in the prov. and 48 m. SW of Bialistok, on the l. bank of the Nourtchek, by which it is separated from Poland.

TZEKINOVKA, a town of Russia in Europe, and gov. of Podolia, 3 m. SE of Soroka, in a peninsula formed by the Dniester.

TZERKOUNOUI, a town of Russia in Europe, in the gov. and district and 11 m. NNE of Kharkov, on the r. bank of a river of the same name.

TZERNA, a mountain of Turkey in Europe, in Bosnia, in the sanj. of Herzegovina, 23 m. ENE of Macarsca. It attaches itself on the SE of Mount Tertla.

TZERNA-GORA, a mountain of Turkey in Europe, in Bosnia, in Croatia, sanj. and 42 m. SSW of Bagna Luka, which attaches itself on the NE to the Tzervlievitza, and in the N side of which the Sanna has its source.

TZERVLIEVITZA, a chain of mountains in Turkey in Europe, in Bosnia, running between the Lopata mountains on the NW, and the Tzerna-Gora on the SE, a distance of about 21 m.

TZIETZIERJIN, a town of Russia in Europe, in the gov. and district and 36 m. NW of Mohilev, on the r. bank of the Droutz.

TZIMNITZA. See Simnitza.

TZINATZIGOS, a mountain of Turkey in Europe, in Albania, in the central part of the sanj. of Monastir. It runs from E to W between the mountains of Mouritchi and Sarakini.

TZINTZOUTZAN, a town of Mexico, and ancient capital of the kingdom of Mechoacan. Pop. 2,500.

TZIOUMLIANSKAIA, a town of Russia in Europe, in the gov. of the Don Cossacks, 39 m. SW of Potemkinskaia, on the r. bank of the Don.

TZIRIN, a town of Russia in Europe, in the gov. of Grodno, district and 23 m. SE of Novogrodek.

TZITON-HAVA, a mountain of Turkey in Europe, in Lower Wallachia, on the frontier of Transylvania, and 27 m. ESE of Kronstadt.

TZITOVIANY, a town of Russia in Europe, in the gov. of Vilna, district and 25 m. SSW of Chavli.

TZKHET. See Mtskhetka.

TZOUROUKHAITOU, a small town of Russia in Asia, in the gov. of Irkutsk, district and 150 m. SSE of Nertchinsk, near the Chinese frontier.

TZYPA, a river of Russia in Asia, in the gov. of Irkutsk and district of Nertchinsk, which runs NE, and, after a course of about 300 m., throws itself into the Vitim.

TZYVILSK, a district and town of Russia in Europe, in the gov. and 130 m. W of Kazan, on the r. bank of the Tzyvil.

U

UADANHAN, a lake of Brazilian Guayana, which discharges itself into the Rio-Branco.

UAIHUNANA, a small river of Brazilian Guayana, an affluent of the Rio-Negro.

UANAPU, Anapu, or Guanapu, a river of Brazil, in the prov. of Para, which has its source in the country of the Mamayanases, and, after a course in a generally N direction of about 330 m., throws itself into the Tajipuru, opposite the island of Marajo. The clove tree abounds on its banks.

UARACA, a river of Brazilian Guayana, an affluent of the Rio-Negro, on the r. bank.

UARANACUA, a small river of Brazilian Guayana, an affluent of the Rio-Negro, on the l. bank.

UARANAPU, one of the numerous arms by which the Hyapura discharges itself into the Amazon, to the W of the principal embouchure of that river.

UARAPIRANGA, an island of Brazil, in the prov. of Para, at the mouth of the Tocantins, to the E of Belem.

UARBRY, a village of New South Wales, in the co. of Bligh, or Talbragen river, 202 m. from Sydney.

UATUMA, a river of Brazilian Guayana, in the prov. of Para, which flows first S, then SE, and joins the Amazon, on the l. bank, nearly opposite the confluence of the Furo or Abacaxis.

UAUPES, or Uacaiari, a river of Brazil, which has its sources in New Granada; flows ESE; and, after a total course of nearly 350 m., throws itself into the Rio-Negro, on the r. bank, at San-Joaquim-de-Coanne, nearly on the equatorial line, and the intersection of the 68th meridian. It is nearly 1 m. wide for 130 m. above its confluence with the Rio-Negro. Its banks are inhabited by an Indian tribe of the same name.

UAUTAS, or Uhautas, a lake of Brazil, in the prov. of Para, and territory of Solimoens. It is of considerable extent, and contains several woody islands. It receives several rivers, and discharges itself by a river of the same name, which, after a course of 36 m., throws itself into the Madeira, 15 m. above Borba, and by several other outlets, affluents of the Amazon.

UBA, a town of Brazil, in the prov. of Minas-Geraes, near the Parahibuna.

UBACH, a village of Prussia, 12 m. N of Aix-la-Chapelle. Pop. of p. 1,400.

UBAHI, a river of Bolivia, which issues from a lake of the same name, in the S part of the dep. of Santa-Cruz-de-la-Sierra, in the Pampas-de-Huanacos; flows NNW; and joins the Guapore, on the l. bank, a little to the NW of Fort Principe-de-Beira, and after a course, to a great extent through marshes, of about 150 m. It is also called the Magdalena, Branco or San-Miguel.

UBATUBA, a maritime town of Brazil, in the prov. of São-Paulo, 30 m. NE of San Sebastian, at the mouth of a river, and at the edge of a fertile plain, in N lat. 33° 26′, and W long. 45° 7′. Pop. 3,000. It has a church and a chapel. Fishing, and the culture of manioc, rice, and coffee, form the chief objects of local industry.

UBAYE, a river of France, in the dep. of the Basses-Alpes, which has its source in the Alps, at the NE extremity of the dep.; and, after a course in a generally W direction, throws itself into the Durance, on the l. bank, at Saulze, on the confines of the dep. of the Hautes-Alpes. Barcelonnette is the chief place on its banks.

UBBERGEN, a village of Holland, prov. of Gueldres, and cant. of Wychen, 3 m. E of Nimeguen. Pop. 1,450.

UBBERSAL, a commune of Belgium, in the prov. of Limburg, and dep. of Heusden. Pop. 175.

UBBESTON, a parish in Suffolk, 5¾ m. SW by W of Halesworth. Area 1,212 acres. Pop. 208.

UBEDA, a judicial partido and town of Spain, in the prov. of Jaen. The partido comprises 6 pueblos. The town is 20 m. NE of Jaen, at the foot of a hill of the same name, in a luxuriant plain, between the Guadalquivir and Guadalimar. Pop. 13,089. It has the remains of ancient walls, and contains some well-built houses, spacious and well-paved streets, three public squares and fountains, a fine promenade, well-shaded and watered by a fountain and several streams, eleven parish churches, numerous convents, a Latin school, three hospitals, an asylum, a public granary, a prison, &c. It has manufactories of varieties of woollen fabrics, and of pottery, and carries on an active trade in wine, oil, fruit, and horses. In the environs are some fine country houses, and several salt-works, wrought by government. U. was taken by Alphonso the Good from the Moors in 1212, and finally by Ferdinand I. in 1239.

UBERAVA, a town of Brazil, in the prov. of Minas-Geraes, and comarca of Patrocinio. Pop. 4,000.—Also a lake of the prov. of Matto-Grosso, at the foot of the Serra-Ponta-dos-Limites of Chamez. It is about 9 m. in circumference, and communicates on the S with Lake Gahiba, and on the E with the Rio Paraguay. Its banks are inhabited by Indians of the Guatos tribe.

UBERAUA-FALSA, a river of Brazil, in the prov. of Goyaz, and district of Rio-das-Velhas, which has its source in the Serra-da-Canastra; runs first E, then SW; and, after a course of about 60 m., throws itself into the Parana.

UBERKINGEN, a village of Würtemberg, in the circle of the Danube, W of Geislingen. Pop. 500.

UBERLINGEN, a town of the grand-duchy of Baden, capital of a bail. of the same name, in the circle of the See, 9 m. N of Constanz, and on the NE bank of the Uberlinger-see, an arm of the Lake of Constanz. Pop. 2,800. It is pleasantly situated in the midst of vineyards and gardens, and has a suburb, a small port, five churches, one of which, named the Munster, has a tower 300 ft. in height, a customhouse, an hospital, and a gymnasium. The ramparts are now falling into ruins. It contains several tanneries, manufactories of linen, hosiery, and pottery, and oil, tobacco, and saw-mills, and carries on trade chiefly with Switzerland.

UBERLINGER-SEE, or Botmer-See, an arm of Lake Constanz, of which it forms the most northerly part, in the grand-duchy of Baden. It is about 14 m. in length from NW to SE, and 1½ m. in medium breadth. On its NE bank is Uberlingen, and in the SE part is the island of Meinan.

UBERSDORF, a village of Switzerland, in the cant. and NE of Freiburg. Pop. 1,033.

UBIGAU, a town of Prussia, in the prov. of Saxony, regency of Merseburg and circle of Liebenwerda, on the Elster, 12 m. ENE of Torgau. Pop. 1,400.—Also a village in the circle of the Meissen, on the Elbe, to the N of Dresden. Pop. 160. It has a castle.

UBLEY, a parish in Somersetshire, 8 m. N by W of Wells. Area 1,811 acres. Pop. in 1851, 294.

UBRIQUE, a town of Spain, in the prov. and 51 m. E of Cadiz, amid sterile mountains, near the Mayacente. Pop. 5,439. It has a poor appearance, and, excepting a parish-church, a convent, a public granary, and a customhouse, contains nothing worthy of note. It has some manufactories of coarse woollen fabrics, and in the vicinity are some iron-mines.

UBSTADT, a village of the grand-duchy of Baden, in the circle of the Middle Rhine, NE of Bruchsal. Pop. 1,200. It has a salt-spring.

UBU, a river of Brazil, in the prov. of Bahia, and comarca of Porto-Seguro, which joins the Jequitinhonha, on the r., near the confluence of the Grapiuna with that river.

UCAIARI, or UAHUPE, a river of Brazilian Guayana, which joins the Rio-Negro above fort São-Gabriel.

UCANHA, a town of Portugal, in the prov. of Beira, comarca and 4 m. E of Lamego, on the r. bank of the Barosa, over which is a bridge defended by a tower. Pop. 356. In the vicinity is a Franciscan convent.

UCAYALE. See PARO.

UCCLE, a department and commune of Belgium, in the prov. of Brabant, and arrond. of Brussels. Pop. of dep. 4,972. The town is 2 m. S of Brussels, in a fine valley. Pop. 212.

UCHANDURRI, a town of Spain, in the prov. of Logrono, 6 m. NW of S. Domingo-de-la-Calzada, in a level country on the Tiron. Pop. 250.

UCHANIE, a town of Poland, in the gov. of Lublin, obwod and 14 m. NW of Hrubieszow. Pop. 920, of whom 400 are Jews.

UCHAUD, a commune of France, in the dep. of the Gard, and cant. of Vauvert. Pop. 847.

UCHDA, a town and cultivated oasis of Marocco, 16 hours from Tlemcen, in the desert of Angad. The town is enclosed by the walls of its gardens, and protected by a large fortress. Its pop. is estimated at 1,000, Moors and Arabs. It has a mosque and three chapels dedicated to ancient santons. The houses, built of clay, are low and of a wretched appearance, the streets winding, and covered with flints. The fortress, where the *kaid* of the district resides, is guarded in ordinary times by a dozen of soldiers, but could not be defended in consequence of its wretched condition. A spring of excellent water, at a little distance from U., keeps up, the whole year round, verdure and freshness in the gardens, through which it is distributed by the system of irrigation that the Arabs practise so well. The aridity of the surrounding country, in summer particularly, makes those gardens appear delicious, as melons, olives, figs, and other fruits, are produced there in abundance. The desert of Angad commences beyond the river Malouia, at more than 40 leagues from Tlemcen. Like the Algerine Angad, which extends to the S of Tlemcen, it is of frightful aridity, particularly in summer, though here and there may be seen spots where irrigation is practised, and where tillage is carried on.

UCHEE, a village of Walton co., in the state of Florida, U. S., on the W side of a branch of Choctawhatchee river, 115 m. W by N of Tallahassee. Pop, in 1850, 300.

UCHISY, a village of France, in the dep. of the Saone-et-Loire, cant. and 5 m. SW of Tournus. Pop. 1,100.

UCHOS, a town of Peru, in the dep. of Truxillo, district and 45 m. SSE of Pataz, on the r. bank of the Tunguragua. It has a good port.

UCHPAS, a tribe of Indians of Ecuador, who inhabit the district in the dep. of Assuay, extending between the Pastaza and Tigre, in S lat. 3°.

UCHTE, a bailiwick and town of Hanover, in the lndr. of Hanover, 18 m. SW of Nienburg, in the midst of a marsh, on the Mulbach. Pop. 1,168.

UCIMONT, a commune of Belgium, in the prov. of Luxemburg and dep. of Sausanrutz. Pop. 273.

UCKANGE, a commune of France, in the dep. of the Moselle, cant. and 4 m. SSW of Thionville, on the Moselle. Pop. 783. It has a custom-house, manufactories of mechanical instruments, and several tanneries.

UCKER. See UKER.

UCKERATH. See SIEGBURG-UCKERATH.

UCKERMUNDE. See UKERMUNDE.

UCKERSDORF, a village of the duchy of Nassau, in the bail. and 4 m. NW of Herborn. Pop. 326. It has a paper-mill.

UCKFIELD, a parish and village in Sussex, 8 m. NNE of Lewes. Area of p. 1,717 acres. Pop. 1,590.

UCKINGTON, a chapelry in the p. of Elmstone-Hardwicke, Gloucestershire, 2¾ m. NW of Cheltenham. Pop. in 1831, 175.

UCLES, a judicial partido and town of Spain, in the prov. of Cuenca. The partido comprises 18 pueblos. The town is 30 m. WSW of Cuenca. Pop. 1,602. It encircles a hill, of which the summit is crowned by a fine monastery of the order of Santiago, and is well-fortified. It has a church, two convents, an hospital, and a public granary. This town is of considerable antiquity. It was long an object of conflict between the Spaniards and Moors, and was the scene of the signal defeat of the latter in 1108.

UCRIA, a town of Sicily, in the prov. of Messina, district and 8 m. SW of Patte.

UDALA, or UDALACH, a mountain of Spain, on the confines of the provinces of Vizcaya and Guipuzcoa, near Mondragon. It contains iron and Armenian bol.

UDANGE, a commune of Belgium, in the prov. of Luxemburg and dep. of Toernich. Pop. 581.

UDDEN, a town of Arabia, in Yemen, 54 m. NE of Mocha, on the Wady Zebid. It contains about 300 houses. The governor's palace is on an adjacent mountain. The coffee grown in the environs is deemed the best in Yemen.

UDDEVALLA, a town of Sweden, in the län and 45 m. N of Goteborg, on a bay of the Skager-Rack. Pop. 3,917. It has a small port. The streets are spacious, and the houses, although of wood, are substantially built. It has manufactories of cordage and of iron ware. The trade consists chiefly in fish, especially salt-herring, wood, and iron. In 1806, this town was almost entirely destroyed by fire; and again, in 1824, it suffered severely from inundation.

UDDINGSTON, a village of Lanarkshire, in the p. of Bothwell, 4 m. NW of Hamilton.

UDDUREE, a village of Sind, near the Phurraun river, to the SE of Hydrabad.

UDEN, a parish of Holland, in the prov. of N. Brabant, to the E of the Bosch. Pop. 4,500.

UDENHOUT, a village of Holland, in the prov. of N. Brabant, 17 m. E of Breda. Pop. 2,000.

UDERWANGEN, a village of Prussia, in the regency of Konigsberg, to the SW of Wehlau. Pop. 600.

UDESTADT, a village of the duchy of Saxe-Weimar, on the Gramme. Pop. 895.

UDHURANA, a town of the Punjab, on the Kasu.

UDIMORE, a parish in Sussex, 3½ m. WNW of Winchelsea, on the road from Rye to Battle, included in the new borough of Rye. Area 2,221 acres. Pop. in 1831, 454; in 1851, 435.

UDINE, a town in the Austrian prov. of Friuli, situated in an extensive plain, on the Roja, at an alt. of 452 ft. above sea-level, 51 m. E of Belluno, and 66 m. NE of Venice. Pop. in 1840, 22,179; in 1850, 26,700. It is surrounded by walls, crenulated, and in some parts terraced, and is well fortified. In its centre is a hill surmounted by a castle. It contains a cathedral, several churches and convents, some handsome private dwellings, a lyceum, several gymnasia, and an episcopal seminary. Its cemetery is considered one of the finest in Europe. The culture of silk, manufacture of woollen fabrics, and of liqueurs, form the chief objects of local industry. The adjacent mountains contain quarries of marble and minerals. In the vicinity are the village and castle of Campo-Formio. See Friuli.

UDINGEN, a village of Prussia, in the prov. of the Rhine and regency of Aachen, on the r. bank of the Roer. In its vicinity are an iron-foundry and a paper-mill.

UDLIGENSCHWEIL, a parish of Switzerland, in the cant. and bail. of Lucerne. Pop. 430.

UDNY, a parish in the district of Formartin, Aberdeenshire. Area 8,380 acres. Pop. 1,513.

UDONG, a city of Cambodia, about 200 m. NE of Kampot, the present residence of the sovereign of the country. According to the report of a recent European visitor, it has little in its appearance which indicates it to be the residence of a prince who formerly ranked amongst the most powerful rulers of the East. It has suffered greatly during the last fifteen or twenty years from almost uninterrupted ruthless incursions by the Siamese and Cochin-Chinese. The houses, or rather huts, are almost without exception built of bamboo and attap. The inhabitants have so often seen their homes consumed by the enemy's torch, that they no longer erect permanent buildings: even their cocoa-nut and other fruit-trees have been destroyed by their merciless enemies, the Cochin-Chinese. The pop. may amount to about 10,000 souls, principally Cambojans, with a few Siamese, Cochin-Chinese, and Chinese. In the centre of the town is a spacious square surrounded by a wall, with a gate on each side defended by a kind of tower: within this the king's palace is situated, surrounded by another wall, not however like the first one, calculated for defence. The buildings occupied by the king are without any architectural ornaments or spires, and principally constructed of wood. See article Kampot.

UDORF (Ober and Unter), two villages of Switzerland, in the cant. and bail. of Zurich and parish of Dietikou, in a fine valley. It has mineral baths, and extensive manufactories of hosiery.

UDSIRE, an island off the W coast of Norway, in the dio. of Christiansand and bail. of Stavanger, in N lat. 59° 18′, E long. 4° 54′.

UDSTERN, an island of Norway, in the dio. of Christiansand and bail. of Stavanger, in the gulf of Bukke. It formerly had a monastery.

UDVARHELY, or Udvarhely-Szek, a sthul or administrative province of Transylvania, in the country of the Szeklers, bounded on the N by the comitat of Thorenburg; on the NE and E by the sthul of Csek; on the S by those of Haromszek and Reps, and the comitat of Upper Weissenburg; and on the W by that of Kockelburg. It is 51 m. in length from WNW to ESE, and 33 m. in breadth; and comprises an area of 381 sq. m. Pop. in 1837, 81,422. It is almost entirely covered by the Carpathian mountains, the loftiest summits of which are on the NE and E. The principal river is the Gross-Kockel, by which the prov. is intersected from NE to SW. The chief productions are corn, cattle, tobacco, wood, and iron. The prov. is divided into 2 circles, Upper and Lower, and comprises 8 districts, 2 towns, of which the chief is also named Udvarhely, Udvarhely-Szekely or Oberhellyen, and 128 villages.—The town of U. is 26 m. ENE of Segesvar, on the l. bank of the Gross Kockel. Pop. 6,150, chiefly Szeklers, of whom this town is considered the capital, and whose national archives and seal it possesses. It has an ancient fortress of the counts of Giulay, a fine town-house, three churches, a Reformed and two Catholic, a convent, a Reformed college, a gymnasium, and a Catholic seminary. It contains numerous tanneries and manufactories of cordage, and carries on an active trade in tobacco and honey, the produce of the environs.

UEBELBACH, a town of Styria, in the circle and 17 m. NW of Gratz, on the Haselbach, in the midst of mountains. Pop. 500.

UEBERKINGEN, a village of Würtemberg, in the circle of the Danube, bail. and 2 m. SW of Geislingen. It has a mineral spring. Pop. 400.

UEBERLINGEN. See Uberlingen.

UEDEM, a town of Prussia, in the prov. of the Rhine, regency and 41 m. NW of Dusseldorf, and circle of Cleves, in a wide plain. Pop. 1,800. It is enclosed by walls, and has four gates. Its manufacture of linen and distillation of brandy are the chief objects of local industry.

UELITZ, a parish of the grand-duchy of Mecklenburg-Schwerin and bail. of Hagenow. Pop. 1,360.

UELMEN, a market-town of Prussia, in the prov. of the Lower Rhine, regency and 30 m. WSW of Coblenz, and circle of Kochem, on the Uelmenermeer.

UELTZEN, Uelzen, Ultza, or Ulyssea, a town of Hanover, in the gov. and 24 m. SSE of Lunneburg, on the l. bank of the Ilmenau. Pop. 2,900. It is enclosed by walls, and has three gates. It has a gymnasium and two hospitals, and possesses manufactories of woollen and linen fabrics, paper-mills, tile-kilns, and distilleries. Its trade consists chiefly in linen, yarn, wax, and honey.

UERDINGEN, a town of Prussia, in the prov. of the Rhine, regency and 11 m. NW of Dusseldorf, and circle of Crevelt, on the l. bank of the Rhine. Pop. 2,300. It has manufactories of cloth, hats, iron-ware, soap, comfits, pottery, and nails, several breweries, and oil-mills. Its trade consists chiefly in articles of local produce, and in coal. This town is of Roman origin. It was enclosed by walls in 1330.

UERERE, or Batac, a small river of Brazilian Guayana, which runs S, and joins the Rio-Negro on the l. bank, between Thomar and Moreira.

UESSLINGEN, a village of Switzerland, in the cant. of Thurgau and bail. of Frauenfeld, on the r. bank of the Thur, here crossed by a bridge.

UETERSEN, a market-town of Denmark, in the duchy of Holstein and seignory of Pinneberg, on the Pinnan, 18 m. NW of Hamburg. Pop. 2,600. It has a convent, a burgess school, a poor's-house, and manufactories of sugar, starch, and pottery.

UETIKON, a parish of Switzerland, in the cant. of Zurich, bail. and 2 m. from Meilen, a little to the E of Lake Zurich. Pop. 1,100.

UETLIBERG, a mountain of Switzerland, in the cant. and a little to the NW of Lake Zurich. It makes part of the Albis chain, and has an alt. of 2,918 ft. above sea-level.

UFA, a considerable city of Asiatic Russia, the capital of the gov. of Orenburg, situated on the banks of the Bielaïa, near its junction with the Ufa, about 730 m. E by N of Moscow. Its site is intersected by numerous torrents and ravines. It was once fortified; but the works have been allowed to fall into decay. This place contains about 10,000 inhabitants. It has a cathedral, 7 churches, 2 convents, an academy, and some mosques. The neighbourhood is inhabited by a race of Tartars who appear to be diligent and industrious cultivators, and have acquired considerable wealth. It was built in 1573, by Ivan Vassilivitch, in order to collect the tribute from the Bashkirs, as also to serve as a barrier against the inroads of the Kirghiz. It is asserted that there was anciently upon this spot a great Tartar city, the capital residence of the khans of the Nogais.

UFA, or OUFA, a river of Asiatic Russia, which rises in the Ural mountains; flows WNW and then SSW through a mountainous and fertile country, till it falls into the Bielaïa, near the city of its own name, after a course of 350 m. Several of its tributary streams have on their banks rich mines of iron.

UF-DER-SCHYNEW, a mountain of Switzerland, in the cant. of Berne, and bail. of Signau, 6 m. from Langnau. It is noted for its cheese.

UFFCULME, or UFFCOLUMB, a parish and village in Devon, 4 m. NE by N of Collumpton, on the river Culm. The village was at one period a considerable manufacturing place. Area of p. 6,122 acres. Pop. in 1831, 2,082; in 1851, 2,098.

UFFENHEIM, a presidial and town of Bavaria, in the circle of Middle Franconia, and 27 m. NW of Anspach, on the Gollach. Pop. 1,650. It is enclosed by walls, with two gates, and has a castle, three churches, an hospital, and a Latin school. It has manufactories of cloth, and several tanneries, and carries on an active trade in cattle and wool. Pop. of presidial, 10,000.

UFFHOLTZ, or USHOLTZ, a commune of France, in the dep. of the Haut-Rhin, cant. and ¾ of a mile N of Cernay. Pop. in 1841, 1,852.

UFFHOVEN, a village of Prussia, in the prov. of Saxony, and regency of Erfurt, to the W of Langensalza. Pop. 760.

UFFINGTON, a parish of New South Wales, in the co. of Durham, bounded on the E by William's river.

UFFINGTON, a parish and village in Berks, 6½ m. W by N of Wantage, intersected by the Wilts and Berks canal, and the Great Western railway. On the hill above the village, which rises to the height of 893 ft., is a British encampment. Area of p., 6,650 acres. Pop. in 1801, 432; in 1831, with the chapelry of Baulking and township of Woolstone, 1,019; in 1851, 1,170.—Also a parish in Lincolnshire, 2½ m. E by N of Stamford, on the N bank of the Welland. Area 3,996 acres. Pop. in 1831, 481; in 1851, 573.—Also a parish in Salop, 2¼ m. ENE of Shrewsbury. Area 2,110 acres. Pop. 209.

UFFORD, a parish in Northamptonshire, 7½ m. NW by W of Peterborough. Area 2,320 acres. Pop. in 1831, 309; in 1851, 297.—Also a parish in Suffolk, 2½ NE by N of Woodbridge, on the river Deben. Area 1,156 acres. Pop. in 1851, 729.

UFLEN. See SALZUFLEN.

UFNAU, UFFNAU, or AUFNAU, a small island of Switzerland, in the cant. of Zurich, and lake of that name, below the Rapperschwyl bridge, and opposite the castle of Pfäffikon. It lies in the wildest part of the lake, and is remarkable for the beauty of its situation and fertility. It belongs to the convent of Einsiedlen, and contains a church built in 973, in which are the tombs of the German heroes Ulrich and Hutten.

UFTON, or UFTON-NERVET, a parish in Berks, 6¼ m. WSW of Reading. Area 2,080 acres. Pop. 421.

UFTON, or OLOUGHTON, a parish and village in Warwickshire. The v. is beautifully situated on a rocky eminence 3½ m. W by N of Southam. Area of p., 1,920 acres. Pop. in 1851, 195.

UFTRUNGEN, a village of Prussia, in the prov. of Saxony, and circle of Sangerhausen, near Rossla, and 6 m. SE of Holtberg. Pop. 860.

UGARTE-DE-MUJICA, or MUGICA, a town of Spain, in the prov. of Vizcaya, 12 m. from Bilbao, at the foot of a mountain, and bathed by two streams. Pop. 996. It contains a residence belonging to the Mujica family, and has several mills.

UGBOROUGH, a parish in Devon, 2½ m. NNE of Modbury. Area 8,659 acres. Pop. in 1851, 1,863.

UGENA, a town of Spain, in the prov. and 39 m. NW of Toledo, in an elevated situation. Pop. 450. It has a parish church, two chapels, and a palace with fine gardens belonging to the counts of Saceda.

UGENTO, a town of Naples, in the prov. of the Terra d'Otranto, district and 15 m. SE of Gallipoli, on a hill, in a pleasant and salubrious situation near the gulf of Tarento. Pop. 1,500. It has a cathedral, two convents and a seminary. This town, the *Uxenthum* of the Romans, was one of the ancient cities of Greece. It was destroyed by the Saracens in the 8th century, and again in 1527.

UGGESHALL, a parish in Suffolk, 12 m. NNE of Saxmundham. Area 1,473 acres. Pop. 293.

UGGIANO-DELLA-CHIESA, a town of Naples, in the prov. of the Terra d'Otranto, district and 26 m. SE of Lecce. Pop. 1,517.

UGGIANO-MONTEFUSCOLI, a town of Naples, in the prov. of the Terra d'Otranto, 26 m. ESE of Tarento, in a fertile and salubrious plain. Pop. 500.

UGGLEBARNBY, a chapelry in the p. of Whitby, Yorkshire, 3½ m. SSW of Whitby. Pop. 426.

UGIE, a river in the NE of Aberdeenshire, rising 2⅔ m. from the N coast at the village of Aberdour, and entering the ocean on the E coast, 1 m. NW of the town of Peterhead. It has a run of about 21 m., generally in a SE direction.

UGIE, a parish of Ross-shire, comprising the W part of the island of Lewis and four inhabited islands, the chief of which is Bernera, eight pasture islands, and various waste islets, in and near Loch-Roag, and the group of Flannan or Flannel islands. The pop. is segregated in hamlets and villages; and in each of several of them amounts to 200 or 250. Pop. in 1831, 3,041; in 1851, 3,209.

UGINE, a town of Sardinia, capital of a mandemento, in the prov. of Upper Savoy, 8 m. N of Conflans, near the r. bank of the Arly, a little above the confluence of the Cheze. Pop. 2,500. An active transit trade is here carried on in the wines of the valley of the Isere. To the N of U. is a lofty square tower, the remains of an ancient castle destroyed in the 13th century.

UGLEY, a parish in Essex, 5 m. N by E of Bishop-Stortford, intersected by the road from London to Newmarket. Area 2,038 acres. Pop. 450.

UGLIAN, an island of Austria, in Dalmatia, in the circle and opposite the town of Zara, and a little to the NW of the island of Pasman, with which it forms the W side of the Zara channel. It is 11 m. in length, and 3 m. in breadth, and possesses considerable fertility. Fresh water is scarce. Its inhabitants, who are numerous, are said to be milder in character than the Dalmatians in general.

UGLITSCH, a town of Russia in Europe, in the

gov. of Jaroslav, on the Wolga. Pop. 8,000. It has a cathedral, numerous churches, and 3 schools.

UGOTS, Ugocsa, or Ugotscha, a gespanschaft or comitat of Hungary, in the circle beyond the Theiss, bounded on the NW by the comitat of Beregh, on the NE by that of Marmaros, and on the S by the comitat of Szathmar. It is 32 m. in length from N to S, and 27 m. in breadth, and comprises a superficies of 186 sq. m. Pop., chiefly Sclaves and Catholics, 46,003. The surface is generally mountainous, and possesses little fertility. The principal rivers are the Theiss, by which it is intersected from E to W, the Borsova, the Batar, and the Tur. It has some silver mines, but its forests and fisheries form its chief resources. Grain, lint, hemp, and tobacco, and in small quantities wine are the chief productions of the soil. The comitat comprises 6 towns, of which the principal is Nagy-Szöllös, and 65 villages.

UGTHORPE, a township in the p. of Lythe, Yorkshire, 17 m. W by N of Whitby. Pop. 260.

UHAUTAS, Uautos, or Catauixis, a lake of Brazil, in the prov. of Para, to the W of the Madeira, about 80 m. above the confluence of that river with the Amazon. It receives several streams, and discharges itself into by Paratari, which issues from its W side, and flows NW to the Amazon, and by a river of the same name, which flows into the Madeira and Amazon, biforking at the confluence of those rivers. The lake is studded with islets.

UHERSKY-BROD. See Ungarisch-Brod.

UHLINGEN, a village of the grand-duchy of Baden, in the circle of the Lee, and to the NE of Berau. Pop. 700.—Also a village of Prussia, in the prov. of Pomerania, regency and 75 m. ENE of Koslin, and circle of Lauenburg-Bütow, near the shore of the Baltic. It has beds of fine potter's clay.

UHLSTADT, a town of Bavaria, in the circle of Middle Franconia, 24 m. N of Anspach.—Also a village in the duchy of Saxe-Altenberg, on the Saali, to the SW of Arlamünde. Pop. 850.

UHNOW, a town of Gallicia, in the circle and 24 m. NNW of Zolkiew, near Zolokia.

UHRICKSVILLE, a village of Tuscarawas co., in the state of Ohio, U. S., 95 m. ENE of Columbus, and intersected by the Steubenville and Indiana railway. Pop. in 1850, 577.

UIG, a parish in the SW of Lewis, Ross-shire. Its greatest length, from NE to SW, is 22½ m.; its greatest breadth is 17¼ m.; and its area, exclusive of water, is probably between 220 and 225 sq. m. The bay of Uig penetrates the interior to the extent of 3¼ m., and has a mean breadth of about 1 m. Pop. in 1831, 3,041; in 1851, 3,209.

UINTAH, a river of Utah Territory, an affluent of the Rio Colorado, which it joins in Utah co., in N lat. 40°.

UIST (North), a large Hebridean island, nearly in the centre of the Long Island chain, and politically belonging to Inverness-shire. It has on the NE the sound of Harris; on the SW a narrow, complicated, and shallow strait, separates it from Benbecula. Its greatest length, from E to W, is 17 m.; its breadth varies between 3¼ and 13¼ m. The E division is cut into indescribable labyrinths by intersections of sea and lake. The W division is, comparatively speaking, continuous land; and sends up, in lines from SE to NW, three distinct groups or ranges of heights. One of these ranges bounds the sound of Harris; and, although lifting its chief summits of Ben-Breach and Ben-More, to nearly 1,000 ft. of alt., is of tame appearance. The second range extends almost from end to end of the district along very nearly its middle, and sends up its principal eminence, Ben-Croghan, to a height of 1,500 ft. The third range is a prolonged and irregular group of less elevation than the others, of a smooth and undulating surface, and with declivities which fall off in gentle slopes to the SW. A belt of uneven low land between this last group and the sea forms both the chief and the most profitable area of arable ground in the island. The island has nearly 50 m. of statute-labour roads, and about 30 m. additional of good roads; but its nearest mart, or that at which any suitable interchange of commodity can be effected, is Greenock or Glasgow, distant about 200 m. The chief islands comprised in the p. are Kirkibost, Illery, Balishare, Gramesay, Kaillen, Valley, and Oronsay, all connected with the mainland by dry sands at low water, and thickly inhabited; Borreray, about 2 m. to the N; and Heiskar, about 10 m. to the W. Pop. 3,918.

UIST (South), an island of Inverness-shire, in the section of the Long Island Hebrides. Its greatest length, from NNW to SSE, is 22 m.; its greatest breadth, 7⅝ m.; its area, including interior and intersecting waters, is about 110 sq. m. Loch-Skiport, penetrating the E coast, nearly 7 m. from the N extremity, runs NW across the island, assumes at the further end the name of Loch-Gamoslechan, and cuts off, from the main body, a low flat island which bears the designation of Iachdar. Loch-Eynort, penetrating the same coast 7½ miles farther to the S, runs WNW to within 3 furl. of the W coast, and sends off, in its progress, a profusion of raggedly outlined and forking bays. Loch-Boisdale, penetrating also the E coast, 3½ m. from the SE extremity of the island, extends 4½ m. W with several ramifications, and a mean breadth of about 1 m.; and at its head it has communication with Loch-Allan,—a narrow fresh-water lake of 3 m. in length, which ranges along the W coast at the mean distance of 3 or 4 furl. from the beach. The universal prevalence of hard gneiss rock, or of gneiss nearly granitic, presents neither subterranean receptacles for water, nor fissures to transmit it, and occasions, throughout the island, an almost total absence of springs. The W side presents an uniform alluvial plain of peat, interspersed with numerous lakes. The E division is upland; but, in consequence of its dissseverment by sea-lochs its heights have not the character of a range, and may, in a general view, be regarded as consisting of three separate groups. The group N of Loch-Eynort occupies apparently between a fifth and a fourth of the island, and forms a conspicuous ridge rising to an alt. of between 2,500 and 3,000 ft. The post communication is maintained by way of N. Uist, Skye, and Inverness,—the last nearly 200 m. distant; and market communication is maintained chiefly with the Clyde. S. Uist, along with Benbecula, Eriskay, and Rona, and some smaller low islands, now forms a parochial district, 40 m. in extreme length, and 8 m. in extreme breadth. Pop. in 1851, 6,173.

UITENHAGE, a district and village of S. Africa, in the colony of the Cape-of-Good-Hope. The district, which is chiefly pastoral and agricultural, is bounded on the NW and N by that of Graaf Reynet; on the NE by the districts of Somerset and Albany; on the S by the ocean; and on the W by the district of George. Pop. 12,000. The Sunday or Zondag river, which flows into Algoa bay, and the Camtoos or Great river, with its affluent the Konga, are the chief rivers by which it is intersected.—The village is 435 m. E of the Cape, on the l. bank of the Zwartkops, a small river which 17 m. below flows into Algoa bay. To the SE of Uitenhage are several salt lakes.

UJ, or VI, a Hungarian word meaning 'new.' Names of places beginning with Uj or Vi not found in this part of the Gazetteer, may be sought for under the name that follows: thus UJARAD, see ARAD (*New*), UJPALANKA, see PALANKA.

UJ-BANYA. See KONIGSBERG.

UJ-EGYHAZ. See LESCHKIRCH.

UJEIN, OUJAIN, or UGIN, a large city of Hindostan, the capital of a district of the same name, in the prov. of Malwah, on the Sipperah river, in N lat. 23° 12′, E long. 75° 50′. The modern city is of an oblong form, about 6 m. in circuit, and surrounded by a stone wall with round towers. The houses are of brick, covered with tiles. The principal bazaar forms a spacious and regular street paved with stone, having houses of two stories in height on each side. The principal buildings are the mosques and temples. The S quarter of the city, called Jeysingpur, contains an observatory erected by rajah Jysing of Jyenagur, in the early part of the last cent. At the distance of 2 m. from the city, Mahdaji Sindia laid the foundations of an extensive fort and citadel, containing a palace; but his death put a stop to the works, his successor wisely giving the preference to Gwalior as his strong hold and place of refuge. It is one of the most ancient cities of Hindostan, and was known to the Greeks under the name of *Ozene*. Its Sanscrit name is *Ujayini*, and it is stated to have been the capital of Bickermajít, a short time after the commencement of the Christian era. At some distance from the town, on an island of the river, there is a subterranean palace said to have been built about 1500, by Nasir-Addin-Khilije, king of Malebah, who used to spend the hot season of the year in this place. An immense fair, or rather religious festival, is held here every year.

UJEST, a town of Prussia, in the prov. of Silesia, regency and 29 m. SE of Oppeln, and circle of Gross-Strelitz, on the r. bank of the Klodnitz. Pop. in 1843, 2,312. It has an hospital and mineral baths, and possesses manufactories of linen and hosiery.

UJFALU, or BERETTYO-UJFALU, a town of Hungary, in the comitat of Bihar, 24 m. SSE of Debreczin.

UJHELY, or SATOR-ALLYA-UGHELY, a town of Hungary, in the gov. and 10 m. SW of Zemplin. Pop. 6,548. It has a college and a gymnasium. The environs afford good wine.

UJHELY (VAG). See NEUSTADTL-AU-DER-WAAG.

UJHELY-KISZUCZA. See NEUSTADTL (OBER).

UJIJAR, a judicial partido and town of Spain, in the prov. of Granada. The partido comprises 18 pueblos. The town is 51 m. SE of Granada, on the Adra, in the midst of the Alpujarras mountains. Pop. 3,016. It has a collegiate church, a convent, an hospital, a public granary, and a glass manufactory.

UJ-LAK, a town of Hungary, in the comitat and 7 m. W of Neutra. It contains several distilleries. See also ILLOK.

UJ-PETS, a village of Hungary, in the comitat of Torontal, 14 m. SW of Temeschwar. It has extensive rice plantations.

UJSZIGET. See SZIGETVAR.

UJVAR (MAROS). See MAROS-UJVAR.

UJVAROS, a town of Hungary, in the comitat of Szabolcs, 17 m. NW of Debreczin, on the Kadarz. It has two churches, a Lutheran and a Reformed. Cattle are extensively reared in the environs.

UJVAROSKA, a town of Hungary, in the comitat and 15 m. WSW of Neutra.—Also a town in the same comitat, to the NW of Leopoldstadt.

UJVIDEK. See NEUSATZ.

UKER, a river of Prussia, which issues from a lake of the same name, in the prov. of Brandenburg and circle of Prenzlow; passes the town of that name; runs W into the prov. of Pomerania; and, after a course, in a generally N direction, of about 45 m., throws itself into the Kleine-haff, an arm of the Pommersche-haff, at Ukermunde.

UKERMUNDE, a town of Prussia, capital of a circle of the same name, in the prov. of Pomerania, regency and 33 m. NW of Stettin, at the confluence of the Uker with the Kleine-haff. Pop. in 1843, 3,575. It is enclosed by walls now in a ruinous condition, and has two gates, two suburbs, a poors' house, and the remains of a castle. Boat-building forms the chief branch of local industry. Pop. of circle, 25,229.

UKER-SEE, a lake of Prussia, in the prov. of Brandenburg, and circle of Prenzlow. It is 6 m. in length from N to S, and 1½ m. in breadth. It discharges itself from the N by a river of the same name. Prenzlow is the chief place on its banks.

UKRAINE, an extensive region in the SE of Russian Poland, which, under the new administrative division of the Russian empire, forms the four governments of Kiev, Podolia, Poltava, and Charkov. It lies between the 48th and 52d parallels of N lat., corresponding to the north of France and the central part of England; but has the characteristics of the climate of those of the S of Russia generally,—in summer, great heat, in winter, intense cold. In natural fertility no part of Poland, and few districts in Europe, surpasses the Ukraine. Wheat, oats, and barley are here raised with comparatively little labour, and the pastures are of great luxuriance. Fruit is abundant, and the kermes, or Polish cochineal, forms one of the leading products of the country. The forests, consisting of oak, larch, and other trees, are turned to little account, horses, goats, and other animals in an unconfined state being allowed to occupy them. The inhabitants of the Ukraine are called Malo-Russians. The chief town is Kiev, once the capital of the Russian dominions, in their circumscribed state and previous to the founding of Moscow. This province, situated between Russia and Poland, was the scene of repeated invasions, of which that by Charles XII. of Sweden, in 1709, terminated in the fatal battle of Poltava. The great natural feature of the country is the river Dnieper, which intersects it in a winding direction from N to S, and affords a channel for the conveyance of products to the Black sea, in which Odessa, situated between the mouths of the Dnieper and Dniester, forms the principal outlet. The state of productive industry is that of a country wholly agricultural, possessing little trade, and few manufacturing establishments.

UKRAINE (SLOBODSK). See CHARKOV.

UKRIE, a village of Sind, 8 m. from Sehwan, and on the road thence to Kurachee.

UKUMBANI, a district of Eastern Africa, recently visited by the missionary Dr. Krapf, who describes its N boundary as formed by the Dana, and its SW boundary by the river Adi, which, at the point where he crossed it in November 1849, was flowing with a gentle current, and a width of 170 yds. In the rainy season, the Adi is a mighty stream which the natives cannot cross. It is said to have its principal source in the mountains of Kikuyu, and to receive additional contributions from the mountains Kilungo, Iwéti, Muka, Kú, and Nsáo Wi. The river runs in U. along the mountain-wall which stretches from Ndungúni, along the Galla country, to U., and even to Kikuyu. "My guide told me," says Dr. Krapf, "that a branch of the Adi comes from Kilimanjaro; and it is highly probable that the N part of the snow-mountain issues a portion of its water to the region of Ukambáni. The Adi joins the Tzávo, and has its probable outlet in the bay of Malinde. Having rested for a while on the E bank of the Adi, we commenced slowly to ascend the mountain-range which stretches from Kikuyu to Ndungúni. We ascended to a height of about 1,800 ft., when we arrived on the plain of Yata, where we had a majestic view of the whole region around. We viewed the serpentine course of the Adi toward the W and NW; we saw the hills and plains of the wild Wakuafi; we noticed the mountains Noka, Julu, Engolia, Théuka, in whose vicinity lay the road we had taken to Kikúmbuliu. Eastward we saw the mountains of Munumóni, which separate the Galla country from U. To the N, Ukambáni-Proper lay before our view. Had I been a mere traveller, pursuing only geographical objects, I would, standing on the plain of Yata, have considered myself amply compensated for the troubles I had sustained on the road; for a great many geographical problems were solved in an instant on the height of Yata."

ULANOW, a town of Austria, in Gallicia, in the circle and 36 m. N of Breszow, on the San.

ULASH, a village of Asiatic Turkey, in the pash. and 18 m. SW of Sivas. It contains about 60 Armenian families. There are salt-springs near it.

ULASKOWA, a town of Austria, in Gallicia, in the circle of Czortkow, on the Sered.

ULATTS VALLEY, a valley of Solano co., in the state of California, U. S., to the NE of Suisan valley, which runs NW, and opens into the valley of the Sacramento. It is a great thoroughfare.

ULBEEK, a department and commune of Belgium, in the prov. of Limburg and arrond. of Tongres. Pop. of dep. 538; of com. 394.

ULBO, a small island of Dalmatia, in the circle of Zara, and at the NW entrance to the channel of that name. It is destitute of fertility, and inhabited chiefly by fishermen.

ULCEBY, a parish in Lincolnshire, 6½ m. SE of Barton-on-Humber. Area 2,220 acres. Pop. in 1831, 694; in 1851, 191.—Also a parish in Lincolnshire, 3½ m. SW of Alford. Area 3,790 acres. Pop. with that of Forthington, in 1831, 218; in 1851, 959.

ULCOMBE, a parish in Kent, 7¼ m. SE by E of Maidstone. Area 3,529 acres. Pop. in 1851, 438.

ULDALE, a parish in Cumberland, 9 m. S by W of Wigton, on the E bank of the river Ellen. There are two small lakes in the p., which are well stocked with fish, and from which the river Ulne or Ellen takes its rise. Area 5,500 acres. Pop. 388.

ULEA, or ULBO, a river of Russia in Europe, in the grand-duchy of Finland, and Uleaborg. It issues from the NW extremity of the Ulea-trask; runs NW; and after a course of about 90 m., in which it forms several falls, throws itself by four embouchures into the gulf of Bothnia.

ULEABORG, a läen or gov. and town of Russia, in the N part of the grand-duchy of Finland. The gov. is bounded on the NW by the Tana, by which it is separated from Norway; on the W by the Tornea, by which it is divided from Sweden, and by the gulf of Bothnia; on the S by the governments of Vasa and Kuopio; and on the E by those of Olonetz and Arkhangel. Pop. in 1840, 125,114. It comprises an area, extending between 63° 30′ and 70° N lat., and between 20° 30′ and 30° 20′ E long., of 13,080 sq. m., being 495 m. in length from NNE to SSW, and 180 m. in breadth. The soil produces rye and barley, but not in quantity adequate to the local consumption. It is to a great extent covered with wood, and in many parts marshy. The inhabitants of the coasts are Swedes, but those of the interior are chiefly Fins. They employ themselves chiefly in agriculture, and make considerable quantities of butter and cheese. The gov. comprises 5 haerads or districts. It was annexed to Russia in 1809.—The town is 345 m. N of Helsingfors, on a peninsula, at the confluence of the Ulea with the gulf of Bothnia. Pop. 5,000. Its streets are long and straight. The trade consists chiefly in pitch, pulse, fish, and salt butter. It was built in 1610, and taken by the Russians in 1714. In 1773 it suffered severely from fire. On a small adjacent island is the castle of Uleaborg, which was built and fortified in 1590.

ULEASTRE. See HONIMA.

ULEA-TRASK, a lake of Russia in Europe, in the grand-duchy of Finland, and gov. of Uleaborg. It is about 45 m. in length from ESE to WNW, and upwards of 30 m. in breadth; and may be viewed as a series of lakes connected together by narrow channels. It gives rise on the NW to the Ulea.

ULEILA-DEL-CAMPO, a town of Spain, in the prov. of Almeria, 15 m. SSE of Purchena, in a mountainous and infertile locality. Pop. 1,690. It has a parish church, and a public granary.

ULEY, a parish in Gloucestershire, 2¼ m. E by N of Dursley. Area 1,492 acres. Pop. 1,327.

ULFON, an island of the gulf of Bothnia, near the E coast of Sweden, in N lat. 63° 2′, and E long. 18° 29′.

ULFSTEIN, a parish of Norway, in the diocese of Drontheim, and bail. of Romsdal, in the island of Hareid. Pop. 1,700.

ULGHAM, a parochial chapelry in Northumberland, situated on the Line-water, 5 m. NE by N of Morpeth. Area 3,615 acres. Pop. 329.

ULGUZ-DAGH, a mountain of Turkey in Asia, in Anatolia, and sanj. of Kankeri, to the NW of Tosia.

ULITEA. See RAIATEA.

ULLA. See OOLLA.

ULLA, a rivulet of co. Limerick, which rises between the mountains Roscagh and Menyeen, and runs about 6 m. SW to a confluence with the Feale.

ULLA, a river of Spain, in Galicia, which has its source 3 m. W of Puerto-Marin, in the prov. of Lugo; runs WSW between the provinces of Corunna and Vigo; and after a total course of about 90 m. throws itself into Arosa bay, near Carril. Its principal affluent is the Sar, which it receives on the r.

ULLADULLA, a maritime town of New South Wales, in the co. of St. Vincent, 136 m. SSW of Sydney. It has a boat harbour.

ULLAHAN, a rivulet of co. Limerick, which rises on the SE side of the mountain Menyeen, and runs about 7 m. SW to the Feale.

ULLAH BUND, a low earthern ridge on the confines of Sind and Cutch, extending it is said a distance of 50 m. from E to W, and in breadth about 16 m. It was upheaved by an earthquake in 1819, and was subsequently cut through by an inundation of the Phurran branch of the Indus, which formed a channel 35 yards wide, and about 30 deep, and immediately below spread into a lake named by Burnes the Lake of Sindree, and covering an area of 2,000 sq. m.

ULLAPOOL, a village on the W coast of Ross-shire, 30 m. NE of Poll-Ewe. It has a small but safe, excellent, and well-kept harbour. It was built, in 1788, by the British Fishery society; and large sums were afterwards expended upon both it and kindred establishments on the isles Martin and Tanera, at the mouth of Loch-Broom; but the failure, for a long series of years, of the herring-fishery has reduced it to insignificance when compared with its pristine condition. Pop. of the village, in 1836, 730.

ULLARD, a parish partly in co. Kilkenny, and partly in co. Carlow. It lies along the Barrow, the Kilkenny section on the W bank, and the Carlow section on the E bank, 2½ m. NNE of Graiguenamanagh. Area 5,846 acres. Pop. in 1851, 1,681.

ULLDECONA, a town of Spain, in the prov. of Tarragona, 18 m. S of Tortosa, in a fertile valley, 1½ m. from the l. bank of the Cenia. Pop. 3,364. It has a parish church, two convents, and a custom-house. The streets are straight and well-built. It possesses manufactories of calico, and distilleries of brandy.

ULLENHALL, a chapelry in the p. of Wootton-Wawen, Warwickshire, 1¾ m. NW of Henley-in-Arden. Pop., including the hamlet of Fordhall, in 1831, 465; in 1851, 459.

ULLENSAGER, a parish of Norway, in the diocese and bail. of Aggershuus, 18 m. NE of Christiania. Pop. 4,000.

ULLENSDORF, a village of Moravia, in the circle of Olmutz, in the midst of mountains. It has a paper-mill, a glass-work, and a mineral spring.—Also a village of Prussia, in the prov. of Silesia, circle and 9 m. S of Lowenberg. Pop. 930. It has a paper-mill.

ULLESKELFE, a township in the p. of Kirkby-Wharfe, Yorkshire, 2 m. SW of Tadcaster, intersected by the York and North Midland railway, which has a station here, 8 m. 54 chains from York. The railroad between York and its junction with the Leeds railway, after being carried across the Wharfe by a viaduct, 274 ft. in length, runs across the village-green and common. Area 1,299 acres. Pop. in 1831, 339; in 1851, 483.

ULLESTHORPE, a hamlet in the p. of Claybrooke, Leicestershire, 3½ m. NW of Lutterworth. The village is close upon the line of the Midland Counties railway, which has a station here, 7¾ m. from Rugby. Pop. in 1831, 599; in 1851, 592.

ULLEY, a township in the ps. of Aston and Treeton, Yorkshire, 8 m. E of Sheffield. Area 850 acres. Pop. in 1831, 193; in 1851, 182.

ULLFORS, a parish of Sweden, in the prefecture of Upsal, and haerad of Œrbyhus. It has extensive forges.

ULLIBARRI, a village of Spain, in the prov. of Alava, 5 m. SW of Orduna, in a valley. Pop. 153. It has quarries of fine stone.

ULLID, or ILLUD, a parish in co. Kilkenny, 2¾ m. NE of Mountcoin. Area 2,248 acres. Pop. in 1851, 577.

ULLINGSWICK, a parish in Herefordshire, 5 m. SW by W of Bromyard. Area 1,245 acres. Pop. in 1831, 293; in 1851, 356.

ULLOA (SAN VICENTE-DE), a town of Spain, in the prov. and 15 m. from Lugo. Pop. 200. It is the centre of extensive linen manufactures.

ULLOCH, with PARDSEY and DEANSCALES, a township in the p. of Dean, Cumberland, 5½ m. SW by S of Cockermouth. Pop. in 1851, 321.

ULLSWATER, a lake situated partly in the mountainous SW region of Cumberland, and partly in Westmoreland, at an elevation of 460 ft. above sea-level. Its length is 9 m.; average breadth 1 m.; greatest depth 210 ft. Forced by the bold hills which environ it to assume a zigzag direction, its form is nearly that of a Z. The lower reach of the lake—about 3 m. in length—is tame, when compared with the middle and upper reaches.

ULLUA, a river of the state of Honduras, which descends from the mountains in the interior; runs first NW, then N, and after a total course of about 180 m., throws itself into the gulf of Honduras. It is navigable for vessels of 200 tons burthen.

ULLY-SAINT-GEORGES, a commune of France, in the dep. of the Oise, and cant. of Neuilly-en-Thelle. Pop. 1,210.

ULM, a village of the grand-duchy of Baden, in the circle of the Middle Rhine, 17 m. NNE of Offenburg. Pop. 700. In its vicinity are the ruins of a castle.—Also an oberamt and town of Würtemberg, in the circle of the Danube. The town is on the l. bank of the Danube, which is here crossed by a fine bridge, at the confluence of the Blau, 51 m. SE of Stuttgard. Pop. in 1840, 16,231. Many of the houses are in the old German style, the streets are tortuous, some of them only are paved and drained. It contains several fine specimens of the architecture of the middle age. Of these the principal is the minster erected between 1377 and 1500, and one of the finest, largest churches in Germany. The church of Saint Michael and the town-house are also ancient edifices. It has besides a commandry of the Teutonic order, an arsenal, a theatre, an hospital, a gymnasium, a house-of-detention, and numerous benevolent institutions. Its industry, formerly very active, consists chiefly in the manufacture of linen, silk and woollen fabrics, tobacco, pipes, copper and tin ware, clocks, surgical instruments, chemical substances, sugar from beetroot, and floor-cloths. It has also numerous breweries and tanneries. The trade consists chiefly in grain, horses, and snails. The snails are bred for the supply of various markets in Austria and Germany, especially for that of Vienna, where they are esteemed a great delicacy after having been fed on strawberries. Ulm, which is of great antiquity, was constituted a free imperial town in 1802. It was formerly a place of great great commercial activity and opulence, but has suffered severely from monopolies. It occupies an important military position, and has repeatedly taken part in the wars of Germany. Its fortifications were destroyed after its capitulation to Napoleon in 1805. In 1810, it was united to Würtemberg.

ULME, a town of Portugal, in the prov. of Estremadura and comarca of Alemquer, in a fertile valley, 14 m. ENE of Santarem. Pop. 450.

ULMERFELD, or UNGERFELD, a town of Austria, in the prov. of Lower Austria, and upper circle of the Wienerwald, on the r. bank of the Ips, 10 m. NNE of Waidhofen. Pop. 800.

ULMHAUSEN, a large village of Würtemberg, in the circle of the Danube, bail. and 1 m. from Goppingen. It is noted for its mineral waters.

ULNES-WALTON, a township in the p. of Croston, co.-palatine of Lancaster, 5¼ m. W by N of Chorley. Area 2,087 acres. Pop. in 1851, 586.

ULOG, or ILLEUK, a town of Turkey in Europe, in the prov. of Bosnia, 63 m. NNW of Trawnik.

ULPHA, a chapelry in the p. of Millom, Cumberland, 8 m. E by S of Ravenglass, stretching along the Duddon. Pop. in 1831, 405; in 1851, 370.—Also a joint township with Methop, in the chapelry of Witherslach, Westmoreland, 9 m. SSW of Kendal.

ULRAKA-KHAN, a summit of Hindostan, in the prov. of Delhi, 17 m. SW of Almora, near a small lake. It has an alt. of 7,366 ft. above sea-level.

ULRICEHAMN, or BOGESUND, a village of Sweden, in the prefecture of Elfsborg, 60 m. SE of Wenersborg. Pop. 800. It has tobacco manufactories.

ULRICH, a village of the grand-duchy of Baden, in the circle of the Upper Rhine, to the S of Freyburg. Pop. 150.

ULRICH (SAINT), a village of the Tyrol, in the circle of the Lower Inn-thal, on the Piller-see. It has a blast furnace and a powder-mill.

ULRICHEN, a village of Switzerland, in the cant. of the Valais, 60 m. NE of Sion, pleasantly situated in a plain, at the foot of the Grimsel. Pop. 240. It has a sulphureous spring.

ULRICHSHALBEN, a village of the grand-duchy of Saxe-Weimar, to the NE of Weimar, on the Ilm. Pop. 440.

ULRICHSKIRCHEN, a town of Austria, in the prov. of Lower Austria, circle and 12 m. ENE of Korneuberg, on the Russbach. Pop. 940.

ULRICHSTEIN, a town of the grand-duchy of Hesse-Darmstadt, in the prov. of Upper Hesse and bail. of Scholten, 24 m. E of Giessen. Pop. 891. It has linen factories, and an active trade in cattle.

ULROME, a township and chapelry in the p. of Barmston and Skipsea, Yorkshire, 7½ m. S by W of Bridlington, bounded on the E by the North sea. Area 1,651 acres. Pop. in 1831, 166; in 1851, 221.

ULSTER, the northern of the four provinces of Ireland; bounded on the W and the N by the Atlantic ocean; on the E by the North channel and the Irish sea; on the S by the prov. of Leinster; and on the SW by the prov. of Connaught. The longest line which can be drawn westward through the prov., extends from Donaghadee in co. Down, to Feelin-head in co. Donegal, and measures 103 m.; the longest line which can be drawn southward, extends from Malin-head in co. Donegal to the village of Finea in co. Cavan, and measures 89 m. The area of the prov. comprises 3,407,539 acres of arable land, 1,764,370 of uncultivated land, 79,783 of continuous plantations, 8,790 of towns, and 214,956 of water,—in all 5,475,438 acres. The surface is much more diversified, more beautiful in the aggregate, and greatly freer from bogs and level grounds, than any of the other three provinces of Ireland. One vast congeries of mountains, exceedingly various in outline, in grouping, and in the character of their intersecting glens and vales, occupies most of the W section of the province, or the greater portion of co. Donegal, a large portion of co. Londonderry, a large portion of co. Tyrone, a considerable portion of co. Fermanagh, and a small portion of co. Cavan; and, excepting in the range of the Sperrin mountains, along the mutual border of Londonderry and Tyrone,—this enormous congeries—exceedingly the largest in Ireland—is nowhere drawn out into ranges or systems of any considerable extent, but consists of mere groups, assemblages, and even isolated masses of individual mountains. Very many of the summits have altitudes of between 1,500 and 2,000 ft. above sea-level; a few have altitudes even upwards of 2,000 ft.; and by far the majority are too lofty and conspicuous to be designated mere hills. The S portion of the W district, or that which lies S of the Sperrin mountains, and occupies all the central regions of the county of Tyrone, fuses out into a tabular expanse of moorland, prevailingly russet in dress, little diversified in surface, tame, bleak, and dismal; and a rich low country, of various breadth and various character, but comprising a large proportion both of flat meadowy land and fertile undulated arable land, flanks all the E side of the great mountain congeries, and passes around Lough Neagh and across a portion of the Upper Bann and the whole of the Lower Bann, so as to constitute all the valley of the Lower Bann, all the shores and immediate basin of Lough Neagh, and the terminating portion of the valley of the Upper Bann. Most of the region S of the Antrim trap-mountain system, and of the eastern and central sections of the great mountain congeries of Ulster,—or the region which constitutes most of the county of Down, about two-thirds of the county of Armagh, the south-eastern district of the county of Tyrone, more than one-half of the county of Monaghan, a small part of the county of Cavan, and a considerable part of the eastern half of the county of Fermanagh,—may be summarily described as a constant, intricate, rich, and beautiful intermixture of hills and hollows, swells and dells, totally unlike any other great district in either Ireland or Great Britain. The largest lake in Ireland, and one of the largest in Europe, is Lough Neagh on the mutual border of the counties of Londonderry, Antrim, Down, Armagh, and Tyrone. Two large and very beautiful lakes, are Upper Lough Erne and Lower Lough Erne, both principally within co. Fermanagh, but the former also slightly within co. Cavan, and the latter slightly within co. Donegal. The other principal lakes of the prov. are Loughs Melvin, Upper-Macnean, Gowna, Kinnail, Sheelin, Oughter, Ramor, the Cootehill lakes, and Loughs Derg, Esk, and Veagh. The only great rivers are the Erne, from the S boundary of the prov., through cos. Cavan and Fermanagh, and across the SW wing of co. Donegal, to Donegal bay at Ballyshannon harbour; the Foyle, from many and voluminous head-waters in cos. of Donegal and Tyrone, along the N part of the boundary between these cos., and across the W wing of co. Londonderry, to the head of Loch Foyle; the Bann, from the Mourne mountains through cos. Down and Armagh, through Lough Neagh, along the boundary between cos. Londonderry and Antrim, and across the NE wing of co. of Londonderry, to the Atlantic ocean; and the Lagan, from the Slievecroob mountains, through co. Down, and along the boundary between cos. Down and Antrim, to the head of Belfast-lough.

The manufacture of linen, which has for so long a period made a chief figure among the industrial occupations of the Irish, belongs chiefly to Ulster. At present the annual value of the linen cloth manufactured in U. cannot be less than £4,000,000 stg. The number of persons employed in all branches of the manufacture is about 170,000; we may safely assert, that 500,000 derive their subsistence from it. The roads of U., in spite of the prevailing hilliness of the surface, are, in a general view, well constructed, minutely ramified, and excellently maintained.

Divisions, &c.] U. is divided into the 9 counties of Donegal in the NW, Londonderry in the N, Antrim in the NE, Down in the SE, Armagh in the E of the S, Monaghan in the W of the S, Cavan in the SW, Fermanagh in the W, and Tyrone in the centre. The pop. in 1831 was 2,286,622. Families employed chiefly in agriculture, 268,864; in manufactures and trade, 88,421; in other pursuits, 68,029. The pop. in 1841 was 2,386,373. Families employed chiefly in agriculture, 267,799; in manufactures and trade, 141,801; in other pursuits, 30,205. Families dependent chiefly on property and professions, 9,443; on the directing of labour, 152,081; on their own manual labour, 271,509. Males at and above 5 years of age who could read and write, 412,697; who could read but not write, 237,687; who could neither read nor write, 358,659. Females at and above 5 years of age who could read and write, 205,945; who could read but not write, 382,127; who could neither read nor write, 480,058. Pop. in 1851, 2,100,289.

History.] The tribes whom the Roman geographer represents as inhabitants of Ulster in the second century were the *Voluntii*, the *Venicnii*, the *Robagdii*, the *Darnii*, and the *Erdini*. Some portion of the ancient history of the province is glanced at in the articles DALRIADA and AILEACH; and the principal passages of its history during the middle ages down to the overthrow and expulsion of its native septs and chieftains are noticed in the articles TYRONE, DONEGAL, LONDONDERRY, DOWN, CLANEBOY, and BELFAST: see these articles. At the accession of James I. measures were adopted for pacificating and civilizing Ireland: and, during his reign, the project was formed and executed of what is known in history as the Plantation of Ulster—com pletely breaking the power of native chiefs throughout this prov., colonizing with British Protestants the vast extent of lands forfeited by the recent rebellions, expelling to Connaught, or at least from the north of Ireland, the native Irish resident on these lands, the creation of a large number of parliamentary boroughs, with an exclusively Protestant franchise, and the reduction of the civil government of the several counties into a form accordant with that of the counties of England. In consequence of a very large proportion of the colonists at 'the Plantation' being immigrants from Scotland, the Presbyterian form of church government became extensively diffused in the province, an extensive engraftment of the Scottish dialect was made upon the native Hibernico-English, and so marked a Scottish tone was given to the entire social character that in other parts of Ireland, particularly in Connaught, the modern inhabitants of U. continue till the present day to be popularly called Scotsmen. In 1784, His Royal Highness, Frederic, Duke of York, was created Earl of U; and this title, so long in use, and so various in possession, is now restricted to princes of the royal family.

ULSTER, a river which has its source in Bavaria, in the Rhöne-gebirge, to the N of Bischofsheim; flows thence into the grand-duchy of Saxe-Weimar; thence into Electoral Hesse, and after a course in a generally N direction of about 36 m., joins the Werra, on the l. bank, 1½ m. below Vach.

ULSTER, a county in the SE part of the state of New York, U. S., comprising an area of 1,076 sq. m., drained by Walkill, Esopus, Rondout, and Shawangunk creeks. It is intersected by the Shawangunk and Blue mountains, and possesses considerable diversity of soil. It is crossed by the Albany and Hoboken railway, and by the Delaware and Hudson canal. Pop. in 1850, 59,384. Its capital is Kingston.—Also a township of Bradford co., in the state of Pennsylvania, 107 m. N of Harrisburg, bordered on the E by the Susquehanna.

ULTING, a parish in Essex, 4 m. SSW of Witham, on the N bank of the river Chelmer. Area 1,147 acres. Pop. in 1831, 158; in 1851, 164.

ULVA, one of the Argyleshire Hebrides, lying between Loch-Tua and the entrance of Loch-na-Keal, on the W coast of Mull. Its length from E to W, is 4½ m.; its breadth is nearly 2 m. The island is distinguished for grand basaltic colonnades and picturesque combinations of these with amorphous masses of trap. Its surface rises from the shore in successive ranges of terraces to an extreme alt. of 1,300 or 1,400 ft.—Also a parish and hamlet of Van Diemen's Land, in the co. of Monmouth.

ULVERSTON, a parish in the co.-palatine of Lancaster. It contains the market-town and township of the same name, the chapelries of Blawith, Church-Coniston, Egton with Torver, Lowick, and Satterthwaite; also the townships of Newland and Osmotherley, and Mansriggs. Area 24,586 acres. Pop. in 1831, 7,741; in 1851, 10,623.—The township of U. has an area of 2,900 acres. In 1801, the pop. was returned at 2,937; in 1831, at 4,876; in 1851, 6,742. The village of U., which has risen with the decadence of Dalton, and may now be considered the capital of the Furness district of the co.-palatine, is 21 m. distant from Lancaster, over the sands, and 33¼ m. by way of Cartmel. It is a neat, clean, cheerful-looking town, consisting principally of modern stone-houses, not very regularly arranged, but substantial and well-built. It stands upon uneven ground, sheltered on the N and W by gentle eminences, and in the distance by the fells. The fells produce great quantities of blue slate; the lower parts abound in the richest iron-ore found in Great Britain, and the mineral treasures of the earth do not here, as in many places, detract from the fertility of the surface. U. is not now a sea-port, for the estuary of the Leven has receded so greatly as to place the town above 1 m. inland; but a canal, 1¼ m. in length, connects a capacious basin at the town with Morecambe-bay, and is so wide and deep as to admit vessels of 200 tons burthen. The principal commerce of U. is in iron-ore, pig, and bar iron, limestone, slates, gunpowder, and grain. The manufactures are coarse cottons, and linens, ropes, hats, and woollen-yarn; shipbuilding is also carried on to a small extent. On the NE of the town rises the Hoad-Hill, on which a handsome monument 100 ft. high has been erected by private subscription, to perpetuate the memory of Sir John Barrow, Bart., a native of the adjoining village of Dingleybeck, a great promoter of geographical science, and eminent in literature, also a distinguished traveller. Sir John was 40 years secretary of the admiralty.

ULVESUND, a strait of Denmark, between the SE coast of the island of Sieland and the island of Moeen. It is about 1½ m. in breadth.

ULVSFIORD, a gulf of the Arctic ocean, in Norway, in the bail. of Finmark, to the SE of the island of Ringvadsoe, and to the W of the Lyngen-fiord. It is 36 m. in length from N to S, and about 6 m. in medium breadth.

ULYSSES, a township of Tompkins co., in the state of New York, U. S., on the W side of Cayuga lake, 8 m. NW of Ithaca, drained by several small streams. Pop. in 1840, 2,979; and in 1850, 3,122.—Also a township of Potter co., in the state of Pennsylvania, 121 m. N by W of Harrisburg, drained by the head waters of Pine creek. Pop. in 1840, 371; in 1850, 542.

ULZEN, a town of Hanover, in the principality of Luneburg, on the Ilmenau. Pop. 2,900. It has manufactories of tobacco, chicory, linen, &c.

UMA, a town of Prussia, in the prov. of Westphalia, circle and 15 m. SSW of Hamm, on the Kottelbecke. Pop. 8,247. It has manufactories of linen and of pottery, several distilleries and breweries, and an extensive salt-work.

UMAGO, or OMAGO, a town of Illyria, in the gov. and 26 m. SW of Trieste, and circle of Istria, on the Adriatic, in N lat. 45° 25′ 20″, and E long. 13° 21′ 50″. The situation is insalubrious.

UMARI, a town of Brazil, in the prov. of Ceara, in the mountains, on the confines of the prov. of Parahiba.

UMARINAUHI, a river of Brazilian Guayana, an affluent of the Cauhaburi or Caburi.

UMBAGOG, a lake partly in the state of Maine, U. S., and partly in that of New Hampshire. It is about 15 m. in length, and in some parts 10 m. in width. It discharges itself into the Androscoggin, which it enters in Essex township.

UMBECTA, a town of Hindostan, in the prov. of Delhi, 15 m. SW of Saharunpur.

UMBER, a town of Hindostan, in the prov. and 35 m. SE of Aurungabad.

UMBRAIL, or WORMSER-JOCH, a mountain on the confines of the Swiss canton of the Grisons and of Lombardy, between the Munster-thal and Bormio, 48 m. SE of Coire.

UMBRETE, a town of Spain, in the prov. and 8 m. WSW of Seville, in a pleasant locality. Pop. 1,215. It has a fine church of modern structure, a palace and gardens belonging to the archbishop of Seville, an hospital, and several distilleries of brandy. Oil and wine of the best quality are cultivated in the environs.

UMBRIATICO, a town of Naples, in the prov. of Calabria-Ultra, district and 24 m. NNW of Cotrone, on a steep mountain, and enclosed on all sides by inaccessible precipices. Pop. 2,500. It has a cathedral, but is a small and ill-built place. It has a considerable trade in grain, wax, and wine. Gypsum and alabaster are found in the environs.

UMCAMAS, a river of Caffraria, in the district of Victoria, which descends from the Giant's Cup, a summit of the Quathlamba mountains; runs in a generally SE direction; and falls into the ocean at the port of Natal. It receives the Umpalani and Umagorda river on the l.

UMEA, or UHMA, a town and port of Sweden, 495 m. NNE of Stockholm, capital of the län of the same name, and at the confluence of the Umea-elv with the gulf of Bothnia. Pop. 1,465. It is built of wood, and has been twice destroyed by the Russians. Its trade consists chiefly in timber, hides, fish, butter, and pitch. "This, and all the towns I have passed," says Mr. Laing, "are, in sober reality, very like our own coast-side towns of the same population. The people earn their living in the same way, by the fisheries, the trade of ship-building, and the supplying the neighbouring country with wares. The people are clad in the same way —the peasantry very like our own Scotch country people: in some respects the difference appears to me in favour of the little towns here. They are more open and airy, the streets better paved and cleaner, the houses more roomy and nice, the meanest with window-curtains or blinds and flower-pots in the windows, and much better washed and scoured. The inns are better. I am here in a more comfortable, cleaner house, than any of our smaller towns in the north of Scotland, excepting perhaps Inverness, can boast of. In this little town of 1,100 inhabitants, at the distance of 470 m. from the capital, there are two booksellers' shops; in which I found a good stock of modern books. All the comforts, conveniences, and, to judge by the appearance of the ladies and gentlemen, the elegancies of a refined life are to be found in as great abundance as in our small towns, and perhaps even extending lower in society, from the daily mode of living being less costly. In the appearance or habits of the people, there is nothing to give you the idea of ignorance, rawness, or a low state of manners. There is nothing of Lapland here, except perhaps in the food."

UMEA, or WASTER-BOTTEN, a län or province of Sweden, on the gulf of Bothnia, watered by the Umea-elv. Pop. in 1840, 57,164. It produces rye and potatoes, and pastures considerable numbers of cattle. It is divided into 3 haerads, 3 fogderi, and 25 parishes.—The river U. has its source in a lake, on the E side of the Dofrines; runs SE; expands into lake Stor-Uman, and after a total course of about 300 m., throws itself into the gulf of Bothnia.

UMEA-LAPPMARK, a division of Sweden, in the W part of the län of Umea. Lyckscle is its chief place.

UMENAK, or OMENAK, a district of Greenland, on the W coast, to the N of Jacob bay, in N lat. 72°. It has extensive seal fisheries, and contains mines of coal.

UMGANI, a river of Caffraria, which has its source in the Quathlamba mountains, runs SE, and falls into the ocean to the NE of Port Natal.

UMLYALLA, a town of Hindostan, in the prov. of Guzerat, 30 m. NNE of Ahmedabad.

UMMANZ, an island of Prussia, in the prov. of Pomerania, and circle of Bergen, a little to the W of the island of Rugen. It is 5 m. long, and 2 m. in breadth, and contains a village of the same name.

UMMELN, a village of Hanover, in the gov. and principality of Hildesheim. Pop. 145.

UMMENDORF, a village of Prussia, in the prov. of Saxony, regency and 20 m. W of Magdeburg, and circle of Neu-Haldensleben. Pop. 1,000. It has a quarry of freestone.—Also a village of Würtemberg, in the circle of the Danube, between the Riss and Umlach. Pop. 500. It has a castle.

UMMERSTADT, a town of the duchy of Saxe-Meiningen, principality and 15 m. SSE of Hildburghausen, and bail. of Hildburg, on the Rodach. Pop. 900. It has cotton spinning-mills, and manufactories of pottery.

UMPFERSTADT, a village of the duchy of Saxe-Weimar, to the E of Weimar. Pop. 385.

UMPQUA, a county, village, and river of Oregon Territory. The village is on the shore of the Pacific, at the mouth of the river of the same name, 92 m. SSW of Salem. It has a good harbour, and promises to become a place of importance.

UMRITZIR. See AMRITSIR.

UMSEMLABAA, or CHASE'S RIVER, a river of Caffraria, which flows into the ocean about 30 m SE of the 2d port of Natal.

UMSLANYAN, a river of Caffraria, which flows into the ocean about 70 m. NE of the 2d port of Natal.

UMSTADT, a town of the grand-duchy of Hesse-Darmstadt, in the prov. of Starkenburg, 12 m. E of Darmstadt. Pop. 3,140. It has manufactories of linen, and of leather. In the vicinity is the ancient castle of Otzberg.

UMTATA, or NOSTRIL RIVER, a river of Caffraria, which has its sources in the Umtate mountains, runs SE, and falls into the sea about 20 m. NE of the 1st port of Natal.

UM-TAVOONE, a river of Caffraria, which flows SE, and falls into the 2d port of Natal.

UM-TOOMBIC, a river of Caffraria, in the district of Victoria, which flows into the ocean about 20 m. SW of the port of Natal.

UMUR-FAKHI, a town of Turkey in Europe, in Rumelia, in the sanj. and 40 m. N of Kirk Kilissia, on the Serpa.

UMZIMCULU, or ANT'S RIVER, a river of Caffraria, which has its sources in the Quathlamba mountains, in the district of Victoria, flows first S to the confines of the district, and falls into the ocean about 20 m. SW of the 3d port of Natal.

UM-ZIMVOOBO, or ST. JOHN'S RIVER, a river of Caffraria, which has its source in the Quathlamba mountains in Cavern Glen, flows in a generally SSE direction, and falls into the sea between the 1st and 2d ports of Natal.

UMZINYATI, a river of Natal, which rises at the base of the Draakberg, in S lat. 27° 46′, and E long. 29° 25′; from this its course is about ESE, until it falls into the Utukela, below the confluence of the Mooi, and not far from the mouth of the Utukela, which appears to be in S lat. 29° 16′, and E long. 31° 30′. The country below the rise of the Umzinyati, and for the distance of about 20 m., is for the most part flat and undulating, with little or no wood, but covered with sweet grass, and a good tract of country for sheep.

UNA, a town of Spain, in the prov. and 14 m. NE of Cuenca, in a valley enclosed by steep mountains. Pop. 330. In its vicinity is a lake containing a floating island on which sheep are pastured.

UNA, a parish of Brazil, in the prov. of Pernambuco, on the coast, watered by a river of the same name, which has its source in the Serra-Garanhuus; separates the prov. of Pernambuco from that of Alagoas; and flows into the ocean 9 m. N of Barra Grande. Its principal affluent is the Jacuhipe.—Also a river of the prov. of São Paulo, which flows into the ocean by a wide embouchure, 30 m. N of the embouchure of the Iguape.—Also a river of the prov. of Bahia, in the comarca of Ilheos, which has its source in the Serra d'Itarca, and flows into the ocean to the S of the bay dos Ilheos.—Also a river of the same prov., in the comarca de-Valenca, which has its source in the Serra-de-Pedra-Branca, and throws itself into the ocean, between the island of Tinhare and the continent.—Also a river of the same prov., in the comarca of Rio-de-Contas, an affluent of the Paraguacu.—Also a river of the prov. of Santa Catharina, which flows into the Laguna, near the S extremity.—Also a river of the prov. of Rio-de-Janeiro, in the district of Cabo-Frio.—Also a river of the prov. of Espirito Santo, in the district of Guarapari, and which flows into the ocean 6 m. S of the town of that name.

UNA-DE-QUINTANA-DEL-MARQUESADO (LA), a town of Spain, in the prov. and 39 m. from Leon, at the foot of woody mountains, in a humid locality. Pop. 416. It has manufactories of linen and of coarse cloth.

UNADILLA, a township of Livingston co., in the state of Michigan, U. S., 31 m. SE of Lansing, drained by Portage river. Pop. in 1840, 643; in 1850, 1,027.—Also a township of Otsego co., in the state of New York, 30 m. SW of Cooperstown. It has a hilly surface, and is drained by Unadilla and Susquehanna rivers. Pop. in 1840, 2,272; in 1850, 2,463. The river U. rises in Oneida co., and flows S to the Susquehanna, which it joins in the NE corner of Brainbridge.

UNADILLA FORKS, a village of Plainfield township, Otsego co., in the state of New York, U. S., at the junction of the E and W branches of the Unadilla. Pop. in 1840, 250; in 1850, 300.

UNADINGEN, a village of the grand-duchy of Baden, in the circle of the See, bail. and 3 m. E of Löffingen. Pop. 650. It has a gypsum quarry and coal-mines.

UNALASHKA. See OONALASKA.

UNARE, a river of Venezuela, which has its source near a town of the same name, in the prov. of Barcelona, and after a rapid course in a generally N direction of about 135 m., throws itself into the Antilles sea, a little below Tocayo. It receives the Guere on the r., and is navigable for about 18 m.

UNCASTILLO, a town of Spain, in the prov. and 57 m. NNW of Zaragoza, encircling a mountain crowned with an ancient castle. Pop. 2,113. It has a square in which there are two fine towers with casemates, two churches, and a convent. In the vicinity are the remains of an aqueduct.

UNDAL, a parish of Sweden, in the diocese and 30 m. W of Christiansand, and bail. of Mandal. Pop. 4,000.

UNDANA, a town of Hindostan, in the prov. of Ajmir, consisting in 1820 of about 400 houses.

UNDERBARROW, a chapelry in the p. of Kendal, Westmoreland, 3 m. W of Kendal. Pop. 503.

UNDERCLIFF, a celebrated district of the isle of Wight, forming a narrow belt of land of about 5 m. in length, running between the cliffs on the SE side of the island and the sea, and not less celebrated for its salubrity than for its beauty. With regard to temperature, the mean range for the year at U. is, compared with Devon and London, as 70 to 83 and 101. In respect to the daily range of temp., it is even more equable than Madeira, as appears by the following table drawn up by Dr. Martin:

Mean daily range of temperature.

	Undercliff.	Madeira.
January,	7·46	9·15
February,	8·20	10·17
March,	10·42	9·79
April,	12·84	9·39
May,	12·82	9·05
June,	11·90	8·73
July,	10·62	9·82
August,	11·39	10·05
September,	11·09	9·83
October,	9·53	10·56
November,	7·78	10·76
December,	6·96	10·48

If we compare the difference of the mean diurnal range of U. with that of London for a similar series of years, we shall have, says Dr. Martin, another example of the greater equability of temp. which the U. presents, and we will add Newport to show the difference which exists at so short a distance as 10 m. from the spot we are describing.

Mean daily range of temperature.

	Undercliff.	Newport.	London,
January,	7·46	9·03	8·92
February,	8·20	10·11	10·93
March,	10·42	14·03	12·77
April,	12·84	19·56	15·95
May,	12·82	19·00	17·52
June,	11·90	21·25	18·61
July,	10·62	17·41	17·66
August,	11·39	18·33	17·29
September,	11·09	15·91	16·99
October,	9·53	14·28	13·54
November,	7·78	10·91	10·73
December,	6·06	8·50	8·76

The following table exhibits the average direction and duration of the winds at U.

SW,	96·97 days.	NW,	30·95 days.
E.	60·34 „	S,	26·72 „
NE,	54·61 „	N,	24·46 „
W,	52·24 „	SE,	18·85 „

UNDERMILLBECK, a township partly in the p. of Windermere, partly in that of Kendal, Westmoreland, 6½ m. W by N of Kendal. Area 3,925 acres. Pop. in 1801, with the chapelry of Winster, 500; in 1831, 854; in 1851, 1,421.

UNDER-SKIDDAW, a township in the parish of Crossthwaite, Cumberland, 6 m. NNW of Keswick. Pop. in 1831, 477; in 1851, 508.

UNDERVILIER, or UNDERWEILER, a village of Switzerland, in the cant. of Berne, bail. and 8 m. SW of Delemont. Pop. 650. It has several iron-works.

UNDERWALD. See UNTERWALD.

UNDES. See URNA-DESA.

UNDY, a parish in Monmouth, 8½ m. E by S of Newport. Area 3,717 acres. Pop. in 1851, 373.

UNGARISCH-BROD, or UNERSKY-BROD, a town of Moravia, in the circle and 11 m. SE of Hradisch, near the r. bank of the Olschawa. Pop. 2,900, of whom a large portion are Jews.

UNGARISCH-HRADISCH. See HRADISCH.

UNGH, a river of Hungary, in the comitat of Unghvar, which has its source in the Carpathian chain; runs SW, and joins the Laborcza, on the l. bank, after a course of about 90 m. Unghvar and Nagy-Berczna are the chief places on its banks.

UNGHVAR, a comitat and town of Hungary, in the circle beyond the Theiss. The comitat is bounded on the N by Galicia; on the E by the comitat of Beregh; on the SE by that of Szabolcs; and on the SW and W by the comitat of Zemplin. It is 60 m. in length from NE to SW, and 48 m. in breadth, and comprises an area of 498 sq. m. Pop. in 1837, 136,908. It is intersected by the Carpathian chain, and is generally mountainous. The principal rivers are the Ungh and Laborcza. The climate in the lower grounds is mild, and the soil favourable to the culture of the vine and of silk. Its principal productions are grain, especially oats, hemp, wine, fruit, wood, honey, iron, and marble. The comitat comprises 5 towns and 206 villages.—The town is 186 m. NE of Buda, on the r. bank of the Ungh. Pop. 6,224. It occupies a fine situation, and has two suburbs, three churches, two Catholic, and a united Greek, a Catholic gymnasium, and a theological seminary. It has an active entrepot trade in salt; but their cattle and the produce of the vineyards form the chief resources of the inhabitants. It has a castle on a lofty eminence.

UNGUNATUA, a small village of Bornu, in Central Africa, 6 days distant from Kuka. Richardson, the enterprising traveller, died here in March 1851. Early in January of that year, he and the companions of his mission, Drs. Barth and Overweg, arrived at the immense plain of Damergu, when, after remaining a few days, they separated, Barth proceeding to Kanu, Overweg to Guber, and Richardson taking the direct route to Kuka, by Zinder. There it would seem his strength began to give way, and before he had arrived 12 days' distant from Kuka, he became seriously ill, suffering much from the oppressive heat of the sun. Having reached a large town called Kangarrua, he halted for three days, and feeling himself rather refreshed renewed his journey. After two days more travelling, during which his weakness greatly increased, he arrived at the Wady Mellaha. Leaving this place on the 3d of March, he reached in two hours the village of U., when he became so weak that he was unable to proceed. In the evening he took a little food and tried to sleep—but became very restless, and expired about two hours after midnight. In the morning, the body, wrapped in linen and covered with a carpet, was borne to a grave which was dug four feet deep under the shade of a large tree close to the village, followed by all the principal sheiks and people of the district.

UNHOSCHT, a small town of Bohemia, in the circle and 17 m. E of Rakonitz.

UNIA, an island of Illyria, in the gov. of Friuli, a little to the W of the island of Osero, in N lat. 44° 38′, and E long. 14° 15′. It is 12 m. in length from N to S, and 1½ m. in breadth. Pop. about 220.

UNIACK POINT, a headland of New South Wales, in the county of Stanley, at the N side of the embouchure of the Brisbane river.

UNICZOW. See NEUSTADT (MAHRISCH).

UNIEGOW, a town of Poland, in the gov. and obwod and 39 m. ENE of Kalisch, on the Wartha. Pop. 800. It has manufactories of cloth, linen, and gloves, and several tanneries.

UNIEH. See EUNIEH.

UNIGUESAS, a tribe of Indians who inhabit the N part of Peru, to the E of the Ugayle, and near the sources of the Mugua and Capanachuas.

UNINI, a natural canal of Brazil, in the prov. of Para, connecting Lake Cadaya with the Rio-Negro. It extends from SW to NE a distance of about 120 m.

UNION, a central county in the state of Pennsylvania, U. S., comprising an area of 576 sq. m., drained by Penn's, Middle, Buffalo, and Mahantanga creeks, affluents of the Susquehanna, by the W branch of which it is bounded on the E. It is hilly in some parts, but generally fertile. Pop. in 1840, 22,787; in 1850, 26,038. Its capital is New Berlin.—Also a county in the S part of the state of Arkansas, comprising an area of 1,230 sq. m., drained by Wachita river and its affluents. It is level and fertile. Cotton is its chief production. Pop. in 1840, 2,889; in 1850, 10,298. Its capital is El-Dorado.—Also a county in the N part of the state of Georgia, comprising an area of 706 sq. m., drained by head branches of Hiawassee river, and other small streams. It is bordered on the S and E by mountains; but possesses considerable fertility. Pop. in 1840, 3,152; in 1850, 7,234. Its capital is Blairsville.—Also a county in the SW part of the state of Illinois, comprising an area of 390 sq. m., drained by Clear creek, an affluent of the Mississippi, by which it is bordered on the W, and by Cash river, an affluent of the Ohio, and intersected by the Illinois Central railway. It has a hilly surface, but is in some parts very fertile. Pop. in 1840, 5,524; in 1550, 7,615. Its capital is Jonesboro'.—Also a county in the E part of the state of Indiana, containing a surface of 168 sq. m., drained by the E fork of White river, and by Hannal, Rockland, and Silver creeks, and intersected by the Indiana Junction railway. Pop. in 1840, 8,017; in 1850, 7,144. Its capital is Liberty.—Also a county in the SW part of the state of Iowa, containing a superficies, generally level and fertile, of 432 sq. m.—Also a county in the NW part of the state of Kentucky, comprising an area of 385 sq. m., drained by Highland and Tradeswater creeks and their branches, and bordered on the W by the Ohio. It has an undulating surface. Pop. in 1840, 6,673; in 1850, 9,012. Its capital is Morganfield.—Also a county in the S part of the state of North Carolina, comprising an area of 475 sq. m., drained by branches of Yadkin river. It is hilly, but generally fertile. Pop. in 1850, 10,051. Its capital is Mourse.—Also a central county of the state of Ohio, comprising an area of 384 sq. m., drained by Big Darby and Bosques, and by other affluents of the Scioto river, and intersected by the Springfield, Mount Vernon, and Pittsburg, the Columbus, and Bellefontaine railways, &c. It is nearly quite level, and is very fertile. Its capital is Marysville.—Also a district in the N part of the state of South Carolina, comprising an area of 749 sq. m., drained by Pacolet and Tyger rivers and their branches, and intersected by the Union and Spartanburg railway. It has an undulating surface, and possesses considerable fertility. Pop. in 1840, 18,936; in 1850, 19,852. Its capital is Unionville.—Also a parish in the N part of the state of Louisiana, comprising an area of 1,027 sq. m., drained by D'Arbonne river and its branches, and by branches of Wachita river. It has a level surface, and is in some parts very fertile. Pop. in 1850, 8,203. Its capital is Farmersville.—Also a township of Tolland co., in the state of Maine, 31 m. NE by E of Hartford, drained by head branches of Willimantic river, and by other small streams. Pop. in 1850, 729.—Also a township of Lincoln co., in the state of Maine, 27 m. ESE of Augusta, drained by the George and Muscungus rivers. Pop. in 1850, 1,974.—Also a township of Branch co., in the state of Michigan, 52 m. SW by S of Lansing, intersected by St. Josephs. Pop. in 1840, 507; in 1850, 1,271.—Also a township of Essex co., in the state of New Jersey, 42 m. NE of Trenton, drained by Elizabeth and Rahway rivers. Pop. in 1840, 1,482; in 1850, 1,662.—Also a township of Broome co., in the state of New York, on the N side of the Susquehanna, 128 m. WSW of Albany, and drained by Nanticoke river, and intersected by the New York and Erie railroad. Pop. in 1840, 3,165; in 1850, 2,143.—Also a township of Rock co., in the state of Wisconsin, 20 m. S of Madison, drained by Sugar creek. Pop. in 1850, 1,050.—Also a township of Gloucester co., in the state of New Jersey, 35 m. NE of Hartford. Pop. in 1840, 1,074.—Also a township of Berks co., in the state of Pennsylvania, drained by Mill, French, and Sixpence creeks. Pop. 1,272.—Also a township of Erie co., in the same state, 22 m. SE of Erie, drained by the S branch of French creek. Pop. 593.—Also a township of Huntingdon co., in the same state, 19 m. S of Huntingdon. Pop. 817.—Also a township of Bedford co., in the same state. Pop. 1,029.—Also a township of Luzerne co., in the same state, drained by Hemlock Shinkshinny, and Huntingdon creeks. Pop. 1,242. Also a township of Washington co., in the same state. Pop. 858.—Also a township of Mifflin co., in the same state. Pop. 1,221.—Also a township of Schuylkill co., in the same state. Pop. 906.—Also a township of Fayette co., in the same state. Pop. 2,723.—Also a township of Union co., in the same state. Pop. 1,630.—Also a township of Allen co.

in the state of Ohio. Pop. 669.—Also a township of Belmont co., in the same state, drained by Stillwater creek. Pop. 2,126.—Also a township of Brown co., in the same state, bounded on the SW by the Ohio, and watered by Red Oak and Eagle creeks. Pop. 3,316.—Also a township of Butler co., in the same state. Pop. 1,828.—Also a township of Carrol co., in the same state. Pop. 889.—Also a township of Champaigne co., in the same state. Pop. 1,249.—Also a township of Clermont co., in the same state. Pop. 1,423.—Also a township of Clinton co., in the same state. Pop. 3,284.—Also a township of Hancock co., in the same state. Pop. 637.—Also a township of Highland co., in the same state. Pop. 1,089.—Also a township of Lawrence co., in the same state. Pop. 1,034.—Also a township of Knox co., in the same state, watered by Mohiccan river. Pop. 1,098.—Also a township of La Fayette co., in the same state. Pop. 1,945.—Also a township of Licking co., in the same state, on the Central Ohio railroad, and bordered on the SE by the Ohio canal. Pop. in 1840, 1,637; in 1850, 2,368.—Also a township of Logan co., in the same state. Pop. in 1840, 832.—Also a township of Madison co., in the same state. Pop. 1,350.—Also a township of Mercer co., in the same state. Pop. 566.—Also a township of Miami co., in the same state. Pop. 1,967.—Also a township of Monroe co., in the same state. Pop. 1,356.—Also a township of Morgan co., in the same state. Pop. 1,328.—Also a township of Muskingum co., in the same state. Pop. 1,625.—Also a township of Putnam co., in the same state. Pop. 401. —Also a township of Scioto co., in the same state, drained by Little Scioto river. Pop. 575.—Also a township of Tuscarawas co., in the same state. Pop. 945.—Also a township of Ross co., in the same state. Pop. 2,632.—Also a township of Union co., in the same state. Pop. 694.—Also a township of Warren co., in the same state. Pop. 1,617.—Also a township of Grant co., in the state of Indiana. Pop. 593.—Also a township of Hancock co., in the same state. Pop. 476.—Also a township of Miami co., in the same state. Pop. 584.—Also a township of Rush co., in the same state. Pop. 1,480.—Also a township of Shelby co., in the same state.—Also a township of Montgomery co., in the same state. Pop. 2,943.—Also a township of Johnson co., in the state of Arkansas. Pop. 114.—Also a township of Greene co., in the same state. Pop. 526.—Also a township of Lawrence co., in the same state. Pop. 259.—Also a township of Marion co., in the same state. Pop. 287.—Also a township of Pope co., in the same state. Pop. 291.—Also a township of St. Francis co., in the same state. Pop. 613.—Also a village of M'Henry co., in the state of Illinois, on the Galena and Chicago Union railway, 62 m. NW of Chicago.—Also a village of Randolph co., in the state of Indiana, at the junction of the Indianapolis and Bellefontaine and the Ohio lines to Cleveland, Pittsburg, and Cincinnati.—Also a village of Franklin co., in the state of Missouri, 73 m. E of Jefferson City, on Bourbeuse creek. Pop. 200. —Also a river of Hancock co., in the state of Maine, which flows S, and discharges itself into a bay of the same name.

UNION-BRIDGE, a village of Belknap co., in the state of New Hampshire, U. S., on the Boston, Concord, and Montreal railroad.

UNION-HALL, a village in the p. of Myross, co. Cork, on an isthmus between the head of Castlehaven, and a creek of Glandore-harbour. Pop. 463.

UNION-SPRINGS, a village of Cayuga co., in the state of New York, U. S., 168 m. W by N of Albany, on the W side of Cayuga lake, and intersected by the Cayuga and Susquehanna railroad. Pop. 600.

UNIONTOWN, a township of Union township Fayette co., in the state of Pennsylvania, U. S., on Red-Stone creek. Pop. in 1840, 1,710; in 1850, 2,333 —Also a village of Lake township, Stark co., in the state of Ohio, 12 m. N of Canton. Pop. in 1840, 300.—Also a village of Carroll co., in the state of Maryland, 51 m. NW by N of Annapolis. Pop. in 1840, 100; in 1850, 339.—Also a village of Perry co., in the state of Alabama, 60 m. SE of Tuscaloosa. Pop. in 1840, 450; in 1850, 500.—Also a village of Wheeling township, Belmont co., in the state of Ohio.

UNION-VALE, a township of Duchess co., in the state of New York, U. S., drained by Fishkill creek. Pop. in 1840, 1,498; in 1850, 1,552.

UNION-VILLAGE, a village of Warren co., in the state of Ohio, U. S., 4 m. W of Lebanon. Pop. in 1840, 500.—Also a village of Greenwich township, Washington co., in the state of New York, on Battenkill. Pop. 1,300.

UNIONVILLE, a village of Hartford co., in the state of Connecticut, U. S., on the E side of Farmington river, and on the Collinsville branch of the New Haven and Northampton or Canal railroad.—Also a village of Orange co., in the state of New York, 98 m. SSE of Albany. Pop. in 1850, 180.—Also a village of Lake co., in the state of Ohio, on the Cleveland and Erie railroad, 53 m. from Erie.—Also a village of Union district, in the state of South Carolina, on a small branch of Tiger river, and intersected by the Union and Spartanburg railway.

UNITED STATES
OF NORTH AMERICA,

A federal republic, comprising the great central division of North America, between the parallels of 49° and 24° N; and bounded on the N by New Britain, and Upper and Lower Canada; on the E by New Brunswick and the Atlantic; on the S by the gulf of Mexico, and Mexico; and on the SW and W by Mexico and the Pacific.

Boundaries.] With respect to the boundary line of the republic, the following details may find a fitting place here. On the 7th of April 1682, the Sieur-de-la-Salle having sailed down the Mississippi, and reached the ocean, took possession of the whole country watered by that river, in the name of Louis XIV., and bestowed upon it the title of Louisiana. It remained a French colony until 1763; but in that year Canada was ceded to England, and the French settlements were fixed by a line drawn along the middle of the Mississippi from its source to the Iberville, and thence, by a line drawn along the middle of that river, and of lakes Maurepas and Pontchartrain, to the sea. France afterwards ceded to Spain all the country known as Louisiana, and also New Orleans, that is, so much of Louisiana as was not transferred by France to Great Britain. By treaty in 1783 between Great Britain and the United States, the independence of these states was recognised, and their NW, W, and S boundaries were thus described: "By a line through the middle of Lake Erie until it arrives at the water-communication between that lake and Lake Huron; thence along the middle of the said water-communication into Lake Huron; thence through the middle of the said lake to the water-communication between that lake and Lake Superior; thence, through Lake Superior, northward of the isles Royal and Philipeaux, to the Long Lake; thence, through the middle of Long Lake and the water between it and the Lake-of-the-Woods, to the Lake-of-the-Woods; thence, through the said lake, to the most NW point thereof; and thence, in a due W course, to the river Mississippi; thence, by a line drawn along the middle of the said river Mississippi, until it shall intersect the northernmost part of the 31st parallel of N lat.; thence, by a line drawn due E from the determination of the line last-mentioned in the lat. of 31° N, to the middle of the river Apalachiola or Catahouche; thence along the middle thereof to its junction with the Flint river; thence straight to the head of the St. Mary's river; and thence along the middle of St. Mary's river to the Atlantic ocean." There was only one error in this otherwise clearly defined boundary. The headwaters of the Mississippi are to the S of the Lake-of-the-Woods. On the 1st of October, 1800, Louisiana was retroceded by Spain to France "with the same extent that it now has in the hands of Spain, and that it had when France possessed it, and such as it should be after the treaties subsequently entered into between

PRINCIPAL CITIES OF THE UNITED STATES

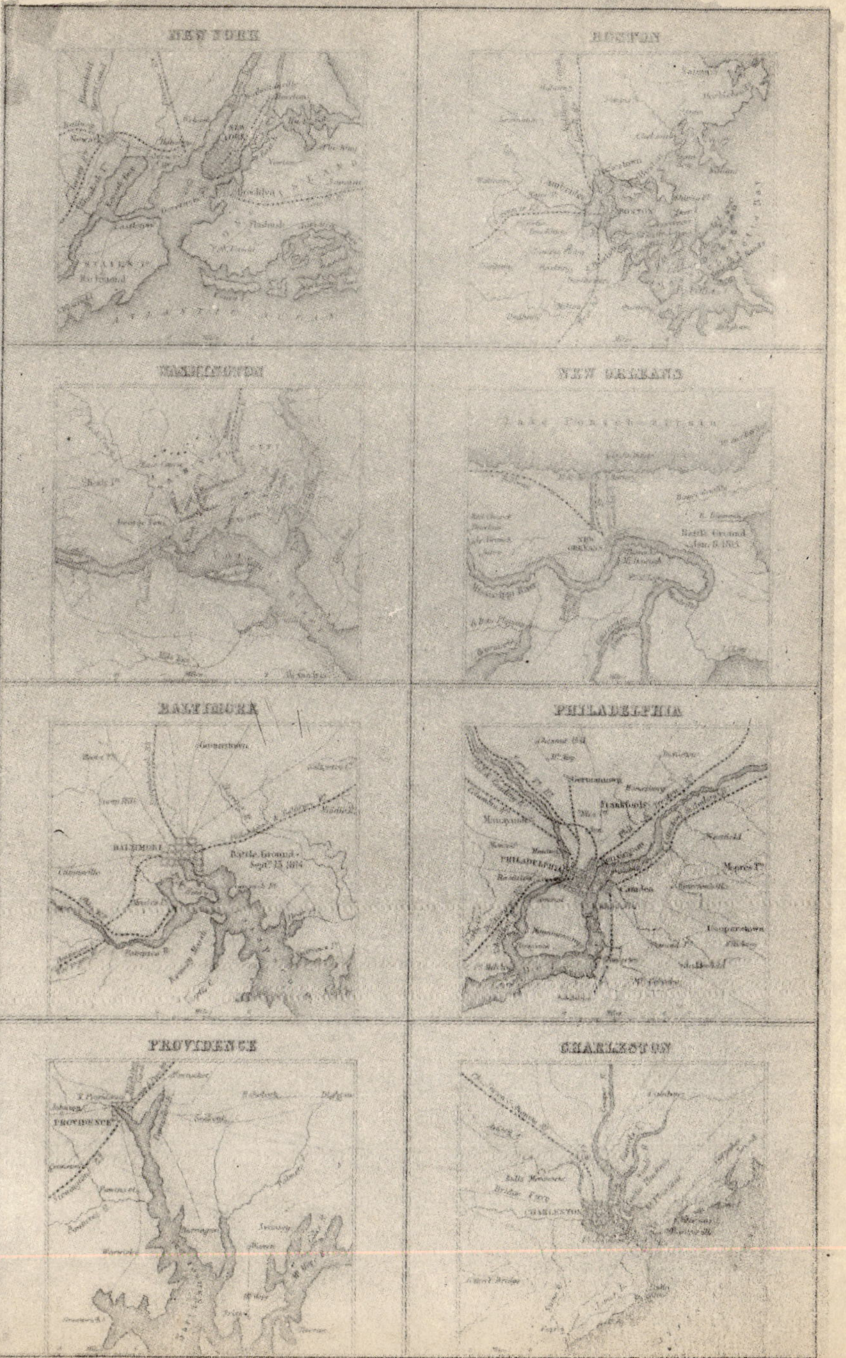

Spain and other states." In 1803, France sold Louisiana to the United States. The purchase included all lands "on the *east* [west?] side of the Mississippi, as far as the great chain of mountains [the Rocky mountains] which divide the waters running into the Pacific and those falling into the Atlantic; and from the said chain of mountains to the Pacific, between the territory claimed by Great Britain on the one side, and by Spain on the other;" but no point was mentioned where the line in the chain of mountains was to commence, nor where the last mentioned tract of land lay. France, however, had actually nothing to sell but what constituted Louisiana after the cession made to Great Britain in 1763. In defining the limits of the United States, therefore, no small difficulty has been experienced, and various discussions have taken place; first, as to the Mexican or SW boundary; second, as to those affecting the NW or Canadian boundary; and, lastly, as to those respecting the Oregon or W boundary. After the United States obtained possession of Louisiana, the boundaries on that side were, till 1821—when they were settled by a convention between the two powers—matter of dispute with the Spanish Americans. The United States claimed as part of Louisiana all the country to the SW as far as the mouth of the Rio-del-Norte. They alleged that all that region, called by the Mexicans the prov. of Texas, anciently belonged to Louisiana; and therefore that America ought to possess all Louisiana, as it was possessed by France before its cession to Spain. But the maps of Humboldt and Pike show that the bay of St. Bernard is at least 200 m. N of the mouth of the Rio-del-Norte; besides, the Spanish pop. of Texas was brought, in the early stages of the conquest, from the interior of Mexico, or New Spain; moreover the claim which the line proposed to be drawn along the Rio-del-Norte included, was founded on the discoveries of La Salle; but the landing of La Salle in Texas was purely accidental, he having gone by mistake beyond the mouth of the Mississippi, while the Spaniards had previously explored and taken possession of both banks of the Rio-Norte; and the measures of La Salle and of those Frenchmen who followed him in Texas were always regarded as hostile by the Spaniards. The negotiations thus raised were brought to a termination by the convention of 1821, under which the Americans, in return for the cession of the Floridas, abandoned to the Mexicans all claim to the territory SW of the Sabine river, which on that side was taken as the settled boundary of Louisiana, to 32° N lat.; whence the boundary-line now turns due N to Red river, and up that river to 100° W long. These boundaries were again disturbed by the revolution in Texas, its recognition as an independent state in 1837, and subsequent reception into the North American federacy. See article TEXAS.—By recent treaties the boundary of the United States on the side of Mexico has been considerably advanced. See article MEXICO.—The extent of Louisiana to the N was, in the first instance, to be only to the Illinois: this is distinctly noted in the grant made, in 1712, by Louis XIV. to Crozat. Subsequently the Illinois was added to it; but all to the N of the Illinois remained part of Canada. On the treaty for the cession of Canada to Great Britain in 1763, this subject was fully discussed, and the result was, not merely an unconditional cession of Canada "in full right with all its dependencies," but the Illinois was also ceded. As a subordinate province of Canada, then, Louisiana extended no further than the distinct boundaries of it could be defined; it never extended further N than the Illinois; Canada was ceded to Great Britain in its fullest extent; and the official map used by France in her negotiations with Great Britain, proves that the country N and NW of the Mississippi was ceded with the prov. of Canada. At the period of the peace, in 1783, the line of demarcation between the British possessions and those of the United States, in this quarter, as far as the NW extremity of the Lake-of-the-Woods, in 49° 37′ N lat., and 94° 31′ W long., was indeed pretty clearly defined; but beyond this limit owing to the geographical ignorance of both parties—it became a matter of uncertainty how or where to fix it. Both parties believed the source of the Mississippi to be in the same parallel of latitude with the NW extremity of the Lake-of-the-Woods, but at least 6° of long. W of that position; and the line of boundary was made to run from the NW extremity of the lake, due west, till it should strike the supposed source of the Mississippi, and thence in a SE direction, down the middle of that stream as far as the 31st parallel. But as the head of the Mississippi had been found, by the successive observations of Thomson, astronomer to the North-west company, and Pike, who, in 1806, by orders of the American government, ascended the river to its source, to lie in 47° 42′ N lat., and 95° 8′ W long., or only 37′ W, and 1° 55′ S of the NW extremity of the Lake-of-the-Woods, a difference of opinion arose of course between the North-west company and the American government, respecting this boundary. The Americans maintained that if the pretensions of the North-west company were to be admitted in making the line of boundary run from the NW extremity of the Lake-of-the-Woods, almost due S instead of due W, till it touch the source of the Mississippi, it would deprive them not only of the upper course of the Red river, but of nearly two-fifths of Louisiana. Commissioners had been appointed to settle this boundary, but their powers, according to the provisions of the treaty of 1814, were limited to the determining what is the middle of the river St. Lawrence, and of the lake and water communications as far as the Lake-of-the-Woods. Delay in settling the boundary westward, by gradually unfolding the value of the country, only increased the difficulty of affixing the limits, and afforded matter of serious disputes between both countries. By convention in 1827 it was agreed that the country between the Rocky mountains and the Pacific should remain open to both countries. Under the purchase from France, the United States had no well-founded claim to any territory beyond the Rocky mountains; but that government contended, that their claim dated, not from that purchase, but from the treaty with Spain in 1819. By that treaty the northern boundary of Mexico was defined by a line running along the 42d parallel from the mountains to the Pacific, and it was accompanied by a cession of all Spanish rights to the N. It was therefore contended, on the part of the United States, that Great Britain had no title to any territory N of that parallel, on the ground that no other country but Spain had a right to such territory. These various grounds were ultimately adjusted by treaty in 1846, the tenor of which has been set forth in our article on OREGON.—With respect to the line of demarcation between our Canadian frontier and that of the United States, the point of departure on the NE was confidently laid down in the treaty of 1783 as at the source of the St. Croix; but the river which was best entitled to be regarded as the St. Croix of the treaty was not ascertained until 1798. The defective description of the boundary in that treaty was acknowledged by both parties. The north-eastern boundary of the United States depended upon ascertaining the position of high lands dividing rivers which empty themselves into the St. Lawrence from rivers which fall into the Atlantic ocean: upon such high lands the NW angle of Nova Scotia, according to the treaty, was to be found. Commissioners of boundary, under the 5th article of the treaty of Ghent, in September 1816, could not agree upon the position of the high lands, and the king of the Netherlands declared it to be his opinion, after due consideration of the statements submitted to him, that the real NW angle of Nova Scotia was unknown when the treaty of 1783 was concluded; and his final award was a compromise of the question, which the United States government refused to abide by. In 1840, Colonel Mudge and Mr. Featherstonhaugh, having examined the physical character of the territory in dispute, reported that the line claimed by America was inconsistent with both the physical geography of the country and the intentions of the language of the treaty of 1783; that they had discovered that a line of high lands did exist to the S of the one claimed by America, and to the S of the St. John, which was in accordance both with the language of the treaty and the physical geography of the country. The disputed territory now became resolved into an area of country consisting of about 7,000,000 acres of land lying between two distinct lines claimed adversely to be the high lands of the treaty, and through the centre of which the western part of the course of the river St. John flowed. Lord Ashburton was now sent out, and the treaty of Washington, as set forth in our article CANADA, was the result of his mission.

Extent and area.] In 1790 the territory of the United States extended from the frontier of Canada to that of Florida, and from the Atlantic to the Mississippi; and the population of the republic was only 3,929,827. In 1850, the territory of the United States extended from the Atlantic to the Pacific, and from the Canadian frontier to the gulf of Mexico and San Francisco; while the pop. had increased to 23,267,498. The greatest longitudinal extent of territory possessed by the United States, from E to W, namely, from the E coast of Maine to the Oregon shores of the Pacific, is 2,900 m. Its greatest extent from N to S is 1,700 sq. m. In general terms it may be regarded as a parallelogram of 2,400 m. in mean length from E to W, and 1,300 m. in average breadth. The territory of the United States when confined to the E side of the Mississippi, was computed to contain 1,000,000 sq. m., or 640,000,000 acres, of which the water-surface amounted to 79,698¼ sq. m. In this computation, however, Hutchins committed two errors; taking for granted that the lakes were accurately delineated in existing maps; and extending the territory of the United States too far to the west by laying down the course of the Mississippi, the grand inland boundary, several degrees farther W than the truth. By these two material errors, he made the superficial extent of the United States, east of the Mississippi, 100,000 sq. m., or 64,000,000 of acres too much. By official returns received in 1820, it appears that the superficial area of the United States was estimated at that period at 2,364,400 sq. m. In this estimate, the two Floridas, and the territory watered by the Columbia and its numerous branches W of the Rocky mountains, were included. The area of each of these divisions was thus estimated: 1st division, or eastern territory, including the Floridas, 965,100 sq. m.; 2d division, W of the Mississippi, 1,111,300 sq. m.; 3d division, W of the Rocky mountains, 288,000 sq. m.: total, 2,364,400 sq. m. The annexation of Texas and California and the acquisitions from Mexico in 1848 and 1854, have added considerably to this area. The following tables and computations are taken from the 7th census report of the United States:

Area of the Pacific slope, or of the region watered by rivers falling into the Pacific,		778.266 sq. m.
Area of the Mississippi valley, or of the region watered by the Mississippi, Missouri, and their tributaries, . . .		1.237,311
Area of the Atlantic slope proper,	637,100	
Area of the Atlantic slope including only the waters falling into the gulf of Mexico W of the Mississippi, ...	183,646	
Area of the Atlantic slope including only the waters falling into the gulf of Mexico E of the Mississippi, ...	146,830	
Total of the Atlantic slope, or of the regions whose waters fall into the Atlantic, . . .		967,576
Total area of the United States and their territories in 1853, . . .		2,981,153

This estimate is by Colonel Abert, and has some claims to authenticity, which cannot be urged for those more commonly used, but in a subsequent part of the introduction to the report on the census, the aggregate area of the union, given by states and territories is stated at 3,306,000 sq. m. The latter amount is the result of an examination of various official reports from the land-office and the state department. A statement given in the census report, published in 1852, raises the territory of the United States to a still larger area. De Bow remarks upon the estimate of 3,306,865 sq. m. as follows:—"The territorial extent of the republic is nearly ten times as large as that of Great Britain and France combined; three times as large as France, Great Britain, Austria, Prussia, Spain, Portugal, Belgium, Holland, and Denmark together; one and one-half times as large as the Russian Empire in Europe; one-sixth less only than the area covered by the 59 or 60 empires, states, and republics of Europe; of equal extent with the Roman empire, or that of Alexander, neither of which is said to have exceeded 3,000,000 sq. m. The area of all the states of Europe is given in the American census report at 3,684,832 sq. m. The areas of the different countries on the American continent are given as follows:

United States, by detailed estimate, .	3,306,865 sq. m.
British America, . . .	3,050,398
Mexico, ...	1,038,834
Central America, . . .	203,551
Russian America, . . .	394,000
Danish America (Greenland), .	380,000
Total area of North America, .	8,373,648

The shore line of the United States, as furnished by the coast-survey office, is as follows:

	Main shore, including bays, sounds, &c.	Islands.	Rivers to head of tide.	Total.
Atlantic coast,	6,861	6,328	6,655	19.844
Pacific coast,	2,281	702	712	3.695
Gulf coast,	3,467	2,217	3,846	9,530
	12,609	9,247	11,213	33,069

Table of distances.] The following table will show the distances between some of the leading points of the United States by the nearest mail-routes:

Pittsburg to Boston, . . .	616 m.
New York to Mobile, . . .	1,476
Philadelphia to Pensacola, . . .	1,443
Boston to Nashville, . . .	1,590
Albany to Richmond, . . .	506
New York to Charleston, . . .	790
New York to Cleveland (Ohio), . . .	671
Boston to Galveston (Texas), . .	2.256
New York to Astoria (land route), . .	3.523
New York to Astoria (*via* Cape Horn), .	17,500
New York to Astoria (*via* Panama), .	6.260
New York to San Diego, Cal. (land route), .	3,732
Charleston to Hartford, . .	900
New York to New Orleans, . . .	1.640
Falls of St. Anthony to mouths of Mississippi,	2.200
Sources of Mississippi to mouths of Mississippi,	2,986
Pittsburg to New Orleans *via* river, .	2,175

The citizen of the United States arriving at New Orleans from New York has passed over a distance more than equal to that separating London from Constantinople, or Paris from St. Petersburg. If he has taken the land route to Astoria his travel will be nearly as great as from New York to Bremen; if the water route, he will have made a voyage nearly equal to one from London to Canton.

The following table is taken from the census report of 1850:

POPULATION, ETC., OF THE UNITED STATES.

	Total pop. in 1850.	Area in sq. m.	Inhabitants to sq. m.
1 Alabama, .	771,623	50,722	15.21
2 Arkansas, .	209,897	52,198	4.02
3 California, .	92.597	188,982	0.49
4 Carolina (North),	369,039	45,550	19.1
5 Carolina (South),	668,507	28,000	23.87
6 Columbia District,	51,687	50	1033.74
7 Connecticut, .	370,792	4,750	78.06
8 Delaware, .	91,532	2.120	43.17
9 Florida, .	87,445	59,268	1 48
10 Georgia, .	906,185	58,400	15.62
11 Illinois, .	851,470	55,409	15.37
12 Indiana, .	988,416	33,809	29.24
13 Indian Territory,	—	187,171	—
14 Iowa, .	192,214	50,914	3.77
15 Kentucky, .	982,405	37,680	26.07
16 Louisiana, .	517,762	41,346	12.52
17 Maine, .	583,169	35 000	16.66
18 Maryland, .	583,034	11,000	53.00
19 Massachusetts,	994,514	7,250	137.17
20 Michigan, .	397,654	56,243	7.07
21 Minnesota Territory,	6,077	141,838	0.04
22 Mississippi, .	606,526	47,151	12.86
23 Missouri, .	682,044	65,037	10.49
24 Nebraski Territory,	—	136,700	—
25 New Hampshire,	317,976	8,030	39.6
26 New Mexico Territory.	61.547	210,774	0.29
27 New York, .	3,097,394	46.000	67.44
28 New Jersey,	489,555	6,851	71.46
29 Northwest Territory,	—	528,725	—
30 Ohio, . .	1,980,329	39.964	49.55
31 Oregon Territory,	13,294	341,463	0 04
32 Pennsylvania, .	2,311,786	47,000	49.29
33 Rhode Island, .	147,545	1,200	122.95
34 Tennessee, .	1,002,717	44,000	22.79
35 Texas, .	212,592	325,520	0.65
36 Utah Territory,	11.380	187,923	0.06
37 Virginia, .	1,421.661	61.352	23.17
38 Vermont, .	314,120	8,000	39.26
39 Wisconsin, .	305,391	53,294	5.66
Total United States,	23,191,876	3,306,865	7.01

1 ALABAMA, formed out of territory ceded to United States by S. Carolina and Georgia, was admitted into the Union Dec. 14, 1819.

2 ARKANSAS, formed from territory ceded to United States by France, was admitted into the Union June 15, 1836.

3 CALIFORNIA, formed of territory ceded by Mexico, was admitted into the Union Sept. 9, 1850.

4 CAROLINA (NORTH), one of the thirteen original states, ratified the constitution of the United States Nov. 21, 1789.

5 CAROLINA (SOUTH), one of the thirteen original states, ratified the constitution of the United States May 23, 1788.

6 COLUMBIA (DISTRICT OF), formed from territory ceded by Maryland and Virginia, was established as seat of government July 16, 1790. Alexandria retroceded July, 1846.

7 CONNECTICUT, one of the thirteen original states,

ratified the constitution of the United States Jan. 9, 1788.

8 DELAWARE, one of the thirteen original states, ratified the constitution of the United States Dec. 7, 1787.

9 FLORIDA, formed from territory ceded to United States by Spain, was admitted into the Union March 3, 1845.

10 GEORGIA, one of the thirteen original states, ratified the constitution of the United States January 2, 1788.

11 ILLINOIS, formed out of territory ceded to the United States by Virginia, was admitted into the Union December 3, 1818.

12 INDIANA, formed from territory ceded to the United States by Virginia, was admitted into the Union December 11, 1816.

14 IOWA, formed from part of the territory of Wisconsin, was admitted into the Union December 28, 1846.

15 KENTUCKY, formed from the territory of Virginia, was admitted into the Union June 1, 1792.

16 LOUISIANA, formed from territory ceded to United States by France, was admitted into the Union April 1, 1812.

17 MAINE, formed out of part of the territory of Massachusetts, was admitted into the Union March 15, 1820.

18 MARYLAND, one of the thirteen original states, ratified the constitution of the United States April 28, 1788.

19 MASSACHUSETTS, one of the thirteen original states, ratified the constitution of the United States February 6, 1788.

20 MICHIGAN, formed from territory ceded to United States by Virginia, was admitted into the Union January 28, 1837.

21 MINNESOTA TERRITORY had its territorial government established March 3, 1849.

22 MISSISSIPPI, formed from territory ceded to United States by S. Carolina, was admitted into the Union December 10, 1817.

23 MISSOURI, formed from territory ceded to United States by France, was admitted into the Union August 10, 1821.

25 NEW HAMPSHIRE, one of the thirteen original states, ratified the constitution of the United States June 21, 1788.

26 NEW MEXICO TERRITORY, formed from territory ceded by Mexico and Texas, had its territorial government established September 9, 1850.

27 NEW YORK, one of the thirteen original states, ratified the constitution of the United States July 26, 1788.

28 NEW JERSEY, one of the thirteen original states, ratified the constitution of the United States December 18, 1787.

30 OHIO, formed out of territory ceded to United States by Virginia, was admitted into the Union November 29, 1802.

31 OREGON TERRITORY had its territorial government established August 14, 1848.

32 PENNSYLVANIA, one of the thirteen original states, ratified the constitution of the United States December 12, 1787.

33 RHODE ISLAND, one of the thirteen original states, ratified the constitution of the United States May 29, 1790.

34 TENNESSEE, formed of territory ceded to United States by North Carolina, was admitted into the Union June 1, 1796.

35 TEXAS, an independent republic, was admitted into the Union Dec. 29, 1845.

36 UTAH TERRITORY had its territorial government established Sept. 9, 1850.

37 VIRGINIA, one of the thirteen original states, ratified the constitution of the United States June 26, 1788.

38 VERMONT, formed from part of the territory of New York, was admitted into the Union March 4, 1791.

39 WISCONSIN, formed from part of the territory of Michigan, was admitted into the Union May 29, 1848.

Comparative density of population.] The density of population is shown in the following table exhibiting the population and number of inhabitants to the sq. m. in various American and European countries:

Countries.	Population.	Density.
United States,	23,191,876	7.90
Canada,	1,842,265	5.31
Mexico,	7,661,919	7.37
Central America,	2,049,950	10.07
Brazil,	6,065,000	2.19
Peru,	2,106,492	3.63
Russia in Europe,	60,315,350	28.44
Austria,	36,514,466	141.88
France,	35,783,170	172.74
England,	16,921,888	332.00
Great Britain and Ireland,	27,475,271	225.19
Prussia,	16,331,187	151.32
Spain,	14,216,219	78.03
Turkey in Europe,	15,500,000	73.60
Sweden and Norway,	4,645,007	15 83
Belgium,	4,426,202	388.60
Portugal,	3,473,758	95 14
Holland,	3,267,638	259.31
Denmark,	2,296,597	101.92
Switzerland,	2,392,740	160.05
Greece,	998,266	55.70

General physical features.] "The general physical character of the country," says Professor Wilson, "is very distinctly marked—a range of mountainous land traverses the extent of the Union, from its NE limits in the state of Maine to its SW extremity in Texas. The New England states are principally occupied by disturbed portions of this range, which in some instances, as in Vermont, rises to an alt. of 6,000 ft. The area lying E of this mountain-chain, between it and the Atlantic sea-board, is occupied by formations of the secondary and tertiary series, while the older formations are developed to a vast extent in the area on the W side. The general characteristics of the vegetation of this portion of the American continent are thus sketched by Malte Brun: "From the shores of the Atlantic to the Mississippi, the United States present an immense natural forest, interspersed with open and naked plains called *prairies*, which are numerous on the W side of the Alleghanies, but rare on the E. In the country W of the Mississippi, wood is comparatively scarce; and in the arid and desert plains occupying a breadth of three or four hundred miles to the E of the Rocky mountains, only a few trees are seen on the banks of the rivers. Some naturalists have distinguished the vegetation of the United States into five regions. 1*st.* The region of the NE, bounded by the Mohawk and Connecticut rivers, where the firs, pines, and other evergreens of Canada prevail. 2*d.* The region of the Alleghanies, where the red and black oak, the beech, the balsam-poplar, the black and red birch, overshadow the plants and shrubs of Canada, at least as far as North Carolina. The valleys among these mountains are remarkably fertile in corn. 3*d.* The upland country, extending from the foot of the mountains to the falls of the rivers. Here the prevailing trees are the red maple, the red and black ash, the walnut, the sycamore, the acacia, and the chestnut. In the S, the magnolia, the laurel, and the orange are interspersed through the forest. Tobacco, with the indigo and cotton plants, succeed as far N as the Susquehanna, beyond which pastures prevail. 4*th.* The region of maritime pines, which extends along the Atlantic coast, from the sea to the first elevations. Here the long-leafed pine, the yellow pine, and the red cedar, occupy the dry grounds; and the acacia-leafed cypress, the low and moist soils, as far as the Roanoke, or even the Chesapeake. Further to the N we find the white pine, the black and Canadian fir, and the *Thuya occidentalis.* The rice-grounds commence where the tide-water becomes fresh, and terminates where it ceases to be felt. 5*th.* The western region, which, no doubt, admits of sub-division, but in which, generally speaking, the forest trees are, the white oak, the black and scaly walnut, the walnut hickory, the cherry, the tulip-tree, the white and gray ash, the sugar-maple, the white elm, the linden, and the western plane, which all grow to a great size upon the Atlantic coast. The United States, generally speaking, do not present the beautiful verdure of Europe; but among the larger herbs which cover the soil are the *Collinsonia*, which affords the Indians a remedy for the bite of the rattlesnake; several species of *phlox;* the golden lily; the biennial *œnothera*, with several species of star-flower, of *monarda*, and of *Rudbeckia.* It is in Virginia, and in the S and SW states, that the American flora displays its wonders, and the savannas wear their perpetual verdure. There the magnificence of the primitive forests, and the exuberant vegetation of the marshes, captivate the senses by the charms of form, of colour, and of perfume. The lands to which the tide reaches on these shores, are distinguished from the lands which remain dry by the moving and compressed stalks of the *Arundo gigantea*, the light foliage of the *Nyssa aquatica*, the *taccamahaca*, and the white cedar, which perhaps, of all the trees of America, presents the most singular aspect: its trunk being composed, where it issues from the ground, of four or five enormous bearers, which uniting at the height of 7 or 8 ft., form a sort of open vault, from the summit of which rises a single straight stem of 18 or 20 ft. in height, without a branch, but terminating in a flat canopy shaped like a parasol, garnished with leaves curiously figured, and of the most delicate green. The name of *savannahs* is given to those vast prairies of the western region, which display a boundless ocean of verdure, and deceive the sight by seeming to rise towards the sky, and of which the only inhabitants are immense herds of bisons or buffaloes: the name is also given to those plains which skirt the rivers, and are generally inundated in the rainy season. The *Magnolia glauca*, the American olive, the *Gordonia argentea* with its odorous flowers, are seen here, isolated, or in groups, while the general surface exhibits a long and succulent herbage, mixed with plants and shrubs. Here the wax-myrtle appears conspicuous among many species of *azalia, kalmia, andromeda*, and *rhododendron;* the margins of the pools and the low and moist spots are adorned with the brilliant azure flowers of the *ixia*, the golden petals of the *canna lutea*, and the tufted roses of the *hydrangea;* while an infinite variety of species of the pleasing *phlox*, the retiring and sensitive *dionea*, and the flame-coloured *Amaryllis atamasco*, in those places where the tide reaches the impenetrable ranks of the royal palms, form a fanciful girdle to the woods, and mark the doubtful limits where the savannah rises into the forest. The calcareous districts, which form the great portion of the region W of the Alleghanies, present certain tracts entirely divested of trees, which are called *barrens*, though capable of being rendered productive. The cause of this peculiarity has not yet been accurately examined." "Tameness is the great fault of American scenery. It is the last thing one expects. A traveller comes from the old world filled with ideas of the sublimity and majesty which the boundless forests and vast lakes and mighty rivers of the American continent must possess and display. Nothing can be more unfounded. In imagination, indeed, he may revel in the thought of the immense solitudes that stretch on all sides around him, and may moralise on the littleness of man in the presence of his Creator's works, and the trifling part he plays upon the mighty scene; but as far as regards the direct effect of landscape upon the eye, all this, of course, does not apply. Vision has but a limited range; and if all within that range be tame and monotonous, it is but a poor consolation to reflect upon the geographical extent of country similarly characterised. Now, in the northern part of America there are few striking features; you hardly ever see a bold and rugged outline of mountain, or a naked precipitous rock. The hills are generally round and low, and covered with vegetation to the top; the colouring too, is monotonous, except for a brief season in autumn; you have not the variety produced either by sterility or by cultivation; for the purple heather and the cold grey stone of European mountain scenery are wanting (heath, I believe, being absolutely unknown), and so, of course, is the smiling richness of a fully peopled country." [*Godley's Letters.*]

Geology.] The territory of the United States lying between the Atlantic and the Missouri, the gulf of Mexico and the great lakes, may be divided, according to Volney, into the five following regions, namely, the granitic, sandstone, the calcareous, the sea-sand, and river alluvions.—The first, or granitic region, extends from the entrance of the St. Lawrence, down to Long island. In New Hampshire and Maine it is mixed with some sandstone and lime; but the White mountains, in New Hampshire, are granitic. The granitic region in the state of New York seems to be divided from that of sandstone by the Mohawk river. There is some granite, however, on the banks of the Susquehannah, and many blocks of the same substance at the foot of the south-west chain in Virginia.—The region of sandstone comprises the western branch of the Green mountains, in Vermont (the eastern declivity being granitic),—the Katskill mountains,—all the mountainous country of Blue ridge, Alleghany, Laurelhill, —the Gauley mountains, at the sources of the Great Kanhaway, —the two great knots, or concentrated transversal ranges of the Alleghanies,—and in general, all their chain, as far as the angle of Georgia, and the Apalachies, to the S. Towards the NW, this sandstone region terminates on the southern sides of the Genessee, Ontario, and Erie lakes, in a district of slaty schist and blue marl, which even seems to form the beds of these watery reservoirs, as is evident from the soundings taken in them, and the stones in their bottoms and on their banks.—The third region of calcareous or limy earth includes all the western or back country, extending from the western foot of the Alleghanies, as far as the alluvion of the Missouri, and NW, across the rivers and lakes, to the sources of the Saskatchewine, and the Chipewyan, or Rocky mountains, according to Mackenzie. All this country, from the Tennessee on the S, to the head-waters of the St Lawrence on the N, has for its basis an immense stratum, or layer of limestone, disposed nearly in a horizontal direction, in laminæ or plates of one or more inches thick, of a smooth, close grain, and of a gray colour. This stratum, or layer, in its turn, rests in some places on a bed of clay, in others on gravel; and covered on the surface of the ground with a layer of excellent black mould, deepest in the river bottoms, where it is sometimes 15 ft. thick; and shallowest on the rising grounds and heights, where it occasionally does not exceed six or eight inches. Besides this immense calcareous region, there are two calcareous districts worthy of notice, as forming an exception in what was denominated the sandstone region, as being embosomed in the mountainous country, namely: the fertile valley of limestone, bounded by the lateral chains of the North mountain and the Blue ridge—and another valley contiguous to that of Limestone, and extending along the back of Blue ridge on the east, from the gap made by the Potomac, to the vicinity of the Schuylkill, in the county of Lancaster. The region of sea-sand comprises the whole shore, from Sandy Hook, facing the granitic shore of Long island, as far as Florida. The whole of this region is bounded inland, by a stratum or layer of talcky granite, called foliated stone, or Muscovy glass, or isinglass, which runs constantly in the direction of the coast, that is from NE to SW. This talcky ridge, everywhere—as Evans, an old American geographer, observes—marks its course by the falls which it occasions in the rivers before they reach the sea, and these falls are the extreme limits of the tide. This seam of mica, or rather talc, is from two to six m. broad. The land between it and the sea, varying in breadth from 30 to 100 m., is evidently sand brought by the ocean, originally bounded by the talcky ridge. The alluvial region rises in gentle undulations from the talcky ridge to the foot of the mountains of sandstone or granite. This tract is marked by its undulations, consisting in some places of isolated risings, in others, of ridges of little hills, as in N. Carolina; and by its soil being composed of different kinds of earth and stones, sometimes jumbled together, and sometimes regularly stratified, which are interrupted, or succeed each other, several times, from the coast to the mountains; but constantly exhibiting the marks of matter rolled down from the declivities above by the waters, and this is in fact the origin of all this country. This is not at all surprising, if we consider the number and magnitude of the rivers that descend from the Alleghanies to the sea; and reflect that most of them are from 1,200 to 4,000 yds. broad, and from 20 to 50 ft. deep, long before they reach their mouths; and that, in their annual inundations, they sometimes overflow the flat country, to the depth of 20 ft. It is easy to conceive, that prodigious changes must, in the lapse of ages, have been made in the interval between the mountains and the sea, by such masses of water; and that the changes must have been also much more rapid than now, when the interjacent space is much more elevated, by the depositions brought down from the high grounds, thus rendering the action of the waters less forcible. This tract of river alluvions extends from the Hudson to the Mississippi; gradually enlarging in its progress, as the mountains recede from the coast, or region of talcky granite; covers all the delta of the Missouri; and slowly contracts its dimensions as it ascends this river, till it finally disappears above the confluence with the Illinois. It varies in breadth from 10 to 200 m. "The tertiary beds," says Professor Wilson, in his recent report, "cover an extensive area in the northern and eastern parts of the States, occupying the valleys of the St. Lawrence and of Lake Champlain, and spreading over a great portion of the surface between the great mountain-chain and the coast-line. They then stretch southward and westward, spreading over the south part of Georgia, Alabama, Mississippi, and including a large portion of Louisiana and of Texas. On the western side of the mountain-chain the alluvium and drift cover up a large portion of the older formations, of whose characters they largely partake. The cretaceous system is first seen occurring to a considerable extent in New Jersey and Delaware. It also occupies a considerable area in Virginia and in North and South Carolina, and may be followed stretching across the northern part of Georgia, and the centre portion of the states of Alabama and Missouri. It then appears to a smaller extent in Arkansas, and expands to a considerable width in Texas. The triassic is first met occurring in Massachusetts, and then is seen occupying the valley of the Connecticut river and a portion of the valley of the Hudson;—thence it passes southward through the states of New Jersey and Pennsylvania, occurring to a small extent in Maryland, and occupying a more extended area in Virginia and North Carolina. The great range of elevated rocks, chiefly of the metamorphic age, which, commencing at the NE limits of the States, extends southward in an uninterrupted line down to Alabama, forms the principal depository of the mineral wealth of the country. In the upper part of New York state, on the S shores of Lake Superior, and in the Southern and Western states, Texas, Missouri, Arkansas, and Illinois, other distinct metalliferous districts exist: but it is believed that, beyond the limits of these well-defined ranges, no regular mineral deposits of any importance occur. The enormous development of the coal-formation secure everywhere an illimitable supply of that all-important material. Distributed throughout the strata of the coal series, and equally along the range of the metamorphic rocks, are found beds of iron ore, some of them of enormous extent, and many of them extremely rich in quality. Thus, in these formations, the carboniferous and the metamorphic, are stored away the two great sources of a country's wealth—coals and iron—the backbones of its industry and the basis of its commercial prosperity. Happily, these, as well as the produce of the other mineral districts, are now within the range of transport communication with the great centres of consumption; ready markets absorb even their largest supplies; the various obstacles attendant upon the establishment of new industries have been gradually overcome, and the only element now wanting for their full and successful development—is time.".

Mountains.] The territory of the United States, though embracing in its wide extent several elevated ranges of great length and breadth, cannot on the whole be considered as a mountainous country. The land along the whole line of sea-coast is level for a considerable distance into the interior. The breadth of this level tract expands from 50 m. in the NE extremity, gradually as we advance to the SW, till, in the state of Georgia, it has attained an extent of nigh 200 m.; beyond this the land gradually rises into mountains, which are much more remarkable for their length and breadth than their height. They sometimes consist of numerous parallel ridges, rising successively behind each other; at other times they run into knots; and sometimes they recede from their parallel direction into what are called spurs. These ranges, or belts of mountainous country, though receiving a vast number of different appellations, are most usually known by the name of the Alleghanies. The French and Spaniards, who first became acquainted with it in Florida, applied to this range or long plateau, throughout its whole extent, the name of Appalachian, which is still retained by a considerable river of that country. Geographers and geologists are not yet agreed as to the precise commencement and termination of the Alleghany chain. E of the Hudson, this chain is chiefly granitic in its geological structure, and is distributed in irregular groups of mountains with rounded summits. In Pennsylvania and Virginia, they assume the form of long parallel ridges, occupying a breadth of 100 m., and rising from 2,500 to 4,000 feet. In the N part of Georgia, they again break into groups of isolated mountains. See ALLEGHANIES. According to Michaux, the Alleghanies attain their highest elevation, where they separate N. Carolina from the state of Tennessee. From the sea to the base of Tyron and Hogback mountains, on the frontier of N. Carolina, the ascent is 800 ft.; and these mountains are 3,840 ft. from this base, or 4,640 ft. above the level of the sea. The Alleghany mountains were, with peculiar felicity of expression, denominated by Jefferson the spine of the United States; for, in his time, they divided the eastern from the western waters, and the whole territory, from the Mississippi to the Atlantic, into three natural divisions, materially differing in climate, configuration, soil, and produce, namely: the coast, the mountains, and the western territory. The last is denominated, by the inhabitants of the coast, the *back country*. No other mountains worthy of notice occur in the United States, between the Atlantic and the Mississippi. The whole country N of the Ohio as far as the Illinois, with few exceptions, has neither hills nor mountains, especially the tract lying between the Ohio and Lake Erie, which is one large flat, or high upland level, with its opposite slopes scarcely perceptible. The highest hill in the vicinity of Pittsburg is only 460 ft. above the bed of the Ohio, and 1,200 feet above the sea.

In extent, in elevation, and in breadth, the Rocky mountains far exceed the Alleghanies of the eastern states. Their mean breadth is 200 m., and where broadest, 300 m.; and their height in several instances exceeds 12,000, and even 13,000 ft. Many of the ranges are covered with snow in the middle of June. The longitudinal extent of this chain is immense, running as far NW as 60° N lat., and perhaps to the Frozen ocean itself. It may even be considered as a continuation of the cordilleras of Mexico. See article ROCKY MOUNTAINS. The immense valley between the Alleghanies and the Rocky mountains forms the basin of the Mississippi. The country W of the Mississippi presents, in the Washington and Oregon territories, the Cascade Range, with its continuation, the Sierra Nevada in California. In this chain are found the greatest ascertained heights within the United States territory: viz. Mount Shasta, Mount St. Helen's, Mount Hood, and Mount Rainier, summits which range in elevation from 12,000 to 14,000 ft. A low plain, from 60 to 100 m. wide, extends along the Atlantic coast; beyond this the land rises toward the interior till it terminates in the Alleghanies. The rest of the country E of the Mississippi is agreeably diversified with hills and valleys, plains and mountains. The soil of the low country, except on the banks of creeks and rivers, is sandy, and comparatively unproductive; but the remainder has a strong, fertile soil, capable of supporting a dense population. The Pacific section is a highly varied country, abounding in wild and majestic scenery, and traversed by magnificent rivers. Much, however, of its interior is desert, and will probably never be inhabited by civilized man. The great California desert, indeed, is one of the dreariest regions of the world, the solitude being relieved only by a few oases in the neighbourhood of streams.

Coasts.] The shores of the United States are washed by three seas: the Atlantic ocean on the E, the gulf of Mexico on the S, and the Pacific ocean on the W. The shore-line on the Atlantic, exclusive of bays, sounds, islands, and lakes, measures 6,861 m.; on the Gulf, 3,467 m.; on the Pacific, 2,281 m. The principal bays and sounds on the Atlantic are Passamaquoddy bay, between Maine and the British prov. of New Brunswick; Massachusetts bay, between Cape Ann and Cape Cod; Long Island sound, between Long Island and the coast of Connecticut; Delaware bay, which sets up between Cape May and Cape Henlopen, separating the states of New Jersey and Delaware; Chesapeake bay, which communicates with the ocean between Cape Charles and Cape Henry, extending in a northern direction for 200 m., through the states of Virginia and Maryland; Albemarle sound, and Pamlico sound, on the coast of N. Carolina. There are no large bays or sounds on the coast of the gulf of Mexico. On the Pacific coast, however, there are several excellent bays, but the principal and only one necessary to mention is the bay of San Francisco, in the state of California. It is one of the finest bays in the world.

Lakes.] With the exception of Michigan and Champlain, none of the great lakes of North America lie wholly within the territory of the Unites States; the rest are on the N boundary, where they form a connected chain, extending through a distance of more than 1,200 miles. The first in the chain is Lake Superior, the largest body of fresh water on the globe. Few persons are really aware of the magnitude of these great lakes; they are truly inland seas, and navigation is as dangerous, and subjected to all the vicissitudes which are connected with the navigation of the Baltic, the Black sea, or the Mediterranean. The following statement of the extent of these fresh-water seas, with the mean depth of their waters, and their elevation above the sea, is adopted by some recent authorities:—

	Mean length.	Mean breadth.
Lake Superior, .	400 m.	80 m.
Michigan, .	320	70
Huron, . .	240	80
Green bay, .	100	20
Erie, . .	240	40
Ontario, . .	180	35
St Clair, . .	20	14

	Area.	Mean depth.	Elevation.
Lake Superior,	32,000 sq. m.	900 ft.	596 ft.
Michigan,	22,400	1,000	578
Huron,	20,400	1,000	578
Green bay,	2,000	1,000	578
Erie, .	9,600	84	565
Ontario,	6,300	500	232
St. Clair,	360	20	570

Lake Champlain, lying between Vermont and New York, is 128 m. long, and from 1 to 16 m. wide, and discharges its waters through the Sorel into the St. Lawrence. It is computed that the lakes contain above 14,000 cubic miles of water—a quantity more than five-sevenths of all the fresh water on the earth. The extent of country drained by the lakes, from the NW angle of Lake Superior to the St. Lawrence, including also the area of the lakes themselves, is estimated at 335,515 sq. m.

Rivers.] The principal rivers of the United States may be divided into four classes. 1st, The Mississippi and its wide-spread branches, which drain the waters of the whole country included between the Alleghany and Rocky mountains; 2d, The rivers E of the Alleghany mountains, which, rising from their eastern declivity, water the Atlantic plain, and hence flow into the ocean; 3d, The system of rivers flowing into the gulf of Mexico, which may be subdivided into those flowing from the S slope of the Alleghanies, and those having their source in the NW highlands of Texas; and 4th, Those streams on the W of the Rocky mountains, which flow into the Pacific. The Mississippi—now the great central stream of the United States—rises on the great tableland of Minnesota, in N lat. 47° 47′, amid lakes and swamps; and after a SE course of about 500 m. reaches the falls of the St. Anthony, where it descends perpendicularly 16 ft., and where are numerous rapids. From these falls it pursues at first a SE, and then a S direction; and, after forming the boundary between Iowa, Missouri, and Arkansas, on the W, and Wisconsin, Illinois, Kentucky, Tennessee, and Mississippi, on the E, passes through Louisiana, and discharges itself through a delta of many mouths into the gulf of Mexico. It is nearly 3,200 m. in length, and is navigable, with few obstructions, to the falls of St. Anthony. Its principal tributaries from the E are—1. The Wisconsin, which joins it between the parallels of 42° and 43° N lat;—2. The Illinois, a navigable river, which joins it near lat. 38° 40′ N;—3. The Ohio, which is itself formed by the junction of the Alleghany and Monongahela rivers at Pittsburg. It flows in a SW direction for 945 m., separating the NW States from Virginia and Kentucky, and falls into the Mississippi in 37° N lat. The chief tributaries of the Ohio are the Wabash, the Cumberland, and the Tennessee, which last is formed of several streams from the W parts of Virginia and the Carolinas, which unite a little W of Knoxville, in the state of Tennessee, and runs at first SW into Alabama, where it turns and runs NW, through Tennessee into Kentucky, and joins the Ohio, 10 m. below the mouth of the Cumberland;—and 4. The Yazoo, which rises in the N part of the State of Mississippi, and, running SW, joins the Mississippi, 100 m. above Natchez. The tributaries from the West are—1. The Minnesota or St. Peter's, which joins it about 9 miles below the falls of St. Anthony, after a SE course of several hundred miles;—2. The Des Moines, which joins it near the parallel of 40° N lat., after a SE course of more than 800 m.;—3. The Missouri, which is formed by three branches, called Jefferson's, Madison's, and Gallatin's rivers, all of which rise and unite in the Rocky mountains. The whole length, from the highest point of Jefferson's river, to the confluence with the Mississippi, is, by actual course, about 2,500 m., and to the gulf of Mexico nearly 4,000 m.; during the whole of which distance there is no cataract or considerable impediment to the navigation, except at Great Falls, which are above 2,000 m. from the Mississippi. At these falls, the river descends, in the distance of 18 m., 362 ft. The principal tributaries of the Missouri are the Yellow Stone, which rises in the Rocky mountains, and joins it after a NE course of 600 m.; the Nebraska or Platte, which rises also in those mountains, and, after an E course of 800 m., joins the Missouri, in lat. 41° N; and the Kansas, which joins it near lat. 39° N, after an easterly course of more than 600 m.;—4. The Arkansas, which rises in the Rocky mountains, and pursuing a SE course, forms, for some distance, the boundary between the Indian Territory and Texas; after which its course lies principally in the state of Arkansas, till it joins the Mississippi in 34° N lat. Its length is more than 1,300 m.;—5. The Red River, which also rises in the Rocky mountains, below Sante Fé, and, after a SE course of more than 1,000 m., falls into the Mississippi in lat. 31° N.

The principal rivers E of the Alleghanies are—1. The Connecticut, which rises in the highlands separating the United States from Canada, and running S, divides New Hampshire from Vermont, and passing through Massachusetts and Connecticut, falls into Long Island sound;—2. The Hudson, which rises W of Lake Champlain, and pursuing a S course of more than 300 m., falls into the bay of New York, after receiving numerous affluents. It is connected with Lakes Champlain, Erie, and Ontario, by means of canals from Albany, and with the Delaware by a canal from Rondout;—3. The Delaware, which rises in New York, and flowing S, separates Pennsylvania from New York and New Jersey, and falls into Delaware bay, after a course of 300 m.;—4. The Susquehanna, which also rises in New York, and, pursuing a S course through Pennsylvania, falls into the head of Chesapeake bay, near the NE corner of Maryland;—5. The Potomac, which rises in the Alleghanies, and, after forming, during its whole course, the boundary between Maryland and Virginia, falls into Chesapeake bay. It is navigable for ships of the largest dimensions to Washington, the federal capital, about 200 m. from the ocean;—6. James river, which rises in the mountains, and falls into the southern part of Chesapeake bay; and—7. The Savannah, which forms the dividing line between S. Carolina and Georgia, and falls into the Atlantic in N. lat. 32°.

The principal rivers which rise S of the Alleghanies, and fall into the gulf of Mexico, are—1. The Appalachicola, which discharges itself into Apalachee bay, in Florida. It is formed by the union of the Chattahoochee and Flint rivers, the former of which rises in the N part of Georgia, and flowing S, receives the Flint at the SW extremity of the State. During the latter part of its course, the Chattahoochee forms the boundary between Georgia and Alabama;—2. The Mobile, which discharges itself into Mobile bay. It is formed by two large rivers, the Alabama and Tombigbee, which unite near N lat. 31°, after having pursued each a separate course of many hundred miles. There is another system of rivers flowing into the gulf from the highlands of northern Texas, consisting of the Sabine, Trinidad, and Brazos. The rivers flowing from the Rocky mountains to the Pacific, consist of—1. The Columbia, which rises near N lat. 55°, and running SW, falls into the ocean in N lat. 46° 15′, after a course of 1,500 m. Its principal tributaries are Clark's river, Lewis' river, and the Multnomah or Willamette, all of which join it on its l. bank. This river was discovered in 1792, and settlements were made in the neighbourhood by Americans in 1810. The mouth of the river is obstructed by flats, but vessels of 300 tons can ascend to the distance of 125 m., and large sloops farther;—2. The Sacramento and San Joaquin, emptying into the bay of San Francisco;—3. The Buenaventura, rising in the coast range of the California mountains, empties into Monterey bay;—4. The Colorado, and River Gila (which separates Mexico from the United States), flow from the mountains near Santa Fé, and would, if not received by the gulf of California, empty into the Pacific; they belong, however, to the same system of rivers. On all these several physical features now enumerated, distinct articles will be found in our pages. The snows and fountains of this range, from the 38th to the 48th parallel, feed, with never-failing supplies, the Missouri and its powerful auxiliary streams.

Climate.] In a country so extensive as that of the United States, there must be a great variety of climate. Were latitude alone to determine the temperature of the North American climate, the southern parts would have the same temp. as Marocco, Barbary; and Egypt: the mouths of the Missouri corresponding with those of the Nile, while the New England states correspond with the S of France, the centre of Italy, and European Turkey. America, however, affords a striking instance of the fallacy of analogical reasoning, and proves that the temp. of any region is not determined by the circumstance of latitude alone, but that a number of other circumstances must be taken into account, as modifying it, namely, the high or low situation of the country,—its being bare or woody,—its aspect,—and more particularly, the quantum and direction of its currents of air. The general characteristic of the American climate is its sudden transitions from extreme heat to extreme cold.

Northern climate.] In the northern states, situated between 42° and 45°, and corresponding to the S of France and N of Spain, the ground is sufficiently covered with snow for three or four months in winter, to make the use of sledges general during that period. In New Hampshire, they reckon upon having eight cold months in the year. The therm., at that season, varies from 8° or 10° to 18° below zero. The climate of the Genessee and Oneida tracts, which lie to the S of Lake Ontario, though half a degree farther N than Albany, is much milder in winter than that of the latter on the E of the mountains, where no month of the year is exempt from frost, and where neither peaches nor cherries will ripen. The climate of the territory S of Lake Ontario is milder than that of Philadelphia, 3° farther south; for the winter there does not generally exceed three months, nor do their snows last more than two months. Carver's assertion, that the climate to the W of the Mississippi is much milder than that of the Atlantic coast in the same parallel, is to be taken with great restrictions. Beyond the parallel of 45° it is equally cold with its opposite parallels on the Atlantic coast. There winter commences early in October with great severity; and beyond this parallel, as we advance westward, the cold increases rapidly in intensity. The reason of this phenomenon can only be found in the very high elevation of this place, and the prevalence of the NW wind, which increases in intensity of cold as we advance westward, and blows from the frozen summits of the Rocky mountains. No softening winds from the gulf of California on the SW, or the Pacific on the W, exist to ameliorate or mitigate the cold; these being either chilled during their progress over these frozen summits, or arrested in their passage by the Mexican Alps or on the ranges which form their NW continuation. Near the Rocky mountains, indeed, the SW wind is more cold and violent than that which comes from the NW. The clouds rise suddenly near these mountains, and distribute their contents partially over the neighbouring plains. The same cloud will discharge hail alone in one part, hail and rain in another, and rain only in a third, and all within the space of a few miles; while at the same time it is snowing on the mountains. The intensity of the summer heat in North America is equal to that of the cold in winter, reckoning from the summer solstice, or longest day. There are few years in which the heat does not rise at Salem as high as 30° and 31° R, which is the temp. of the Persian gulf and the Arabian coast.

Central climate] In the central states, the winters are shorter; but the cold is neither less rigorous nor less piercing. It commonly sets in about the shortest day, and continues very severe for six or seven weeks; but it begins to be felt as early as the end of October. At Philadelphia, in 39° 55′ N lat. or the same parallel with Madrid, Naples, and Valencia, and 1,100 m. S of the parallel of Edinburgh, the therm. is several days every winter at 8° and 10° below 0°. Notwithstanding the motion of a tide which rises and falls 6 ft., the Delaware—here a mile broad—has been frozen over in 24 hours, and continued in that state for 20, 30, and sometimes 40 days. The summer heats are equally intense as the winter cold. From the time of the summer solstice, and even for three weeks before, the heats are so oppressive at Philadelphia, that the streets are deserted from noon to 5 o'clock, and most of the inhabitants go to rest after dinner. At this season, Reaumur's therm. will frequently rise to 25°, and even sometimes to 28° and 30°. But what renders the heat most insupportable is the almost total want of wind, particularly from three o'clock in the afternoon, and the moisture with which the air is loaded on all the coast; so that Volney says, he would infinitely prefer the heat of Grand Cairo, in Egypt, where a constant brisk wind and very dry air contribute to render the heat supportable, to that of Philadelphia. In New York, the therm. never rises higher than 86° of Fahrenheit for 24 hours together, and the mean heat is 65°.

Southern climate.] In the southern states, as Virginia, the Carolinas, and Georgia, the intensity of the cold, and its duration, diminish in pretty regular proportion to the latitude. The course of the Potomac, or more precisely that of the Patapsco, forms a striking line of demarcation in this respect. Here the domain of snow terminates; and the traveller from the north, who has hitherto seen at every house a sledge, sees not another after he has descended the steep hill which overlooks that river. This coast has, however, pretty sharp attacks of frost during the 40 days succeeding the winter solstice. On the other hand, on all the coast beyond the Potomac, the heat, from a month before the longest day, is so great, that the therm., for the space of four months, commonly rises in the afternoon to 22° or 24° R., or 82° and 86° F., notwithstanding a gentle sea-breeze. It rises even to 32° and 33° at Savannah, which is much greater than in Egypt, where 25° R, or 88° F., is the common term in the shade. From the above facts, we have a scale of variation from heat to cold, in the northern states, of 60° or 66° R.; in the middle states, the variation is 46° or 48°; and in the southern states, 32° or 34° of the same scale. But it is not merely the great intensity of cold in winter and heat in summer that characterises the climate of the American states, in the northern, middle, and southern parts: another peculiarity, and one which is very destructive of health, are the great and sudden variations along the whole tract E of the Alleghanies, and between them and the sea, but more particularly in the middle states. After a day in which the therm. has stood at 90°, it will sometimes fall in a single night to 65°, or even 60°. There are generally few nights in which a fire could be dispensed with, except in the months of July and August.

Western climate.] The climate of the western states, after making every reasonable deduction from the inflated statements of some geographers, is much more mild and genial, and much less subjected to the extremes of cold and heat, and those frequent and sudden variations which characterise the climates of the northern, middle, and southern states. As we advance westwardly from the sea to the mountains, the climate becomes gradually colder, on the same parallel of latitude, just as if we were proceeding northward. Having arrived at the summit of the Alleghanies, where the cold has attained its maximum, a change of temp. commences as we begin to descend, and continues in an opposite ratio, till it becomes warmer on the shores of the Mississippi than in the same lat. on the sea coast. In 1799, when the therm. was at Monticello at 90°, and 96° at Williamsburg, it was at 110° at Kaskaskias. Winter does not commence in the western states and territories till the approach of the solstice; and the cold weather is felt only in the six or seven weeks following; though even then it is not fixed and constant, as there are intervals of temperate and warm days. The therm. does not in general fall lower than 19° or 21° F.; the frosts which at first show themselves a few days in October disappear, but return again toward the end of November, though they do not become settled till towards January, when the brooks, small rivers, and standing waters, freeze, but seldom continue frozen over more than from 3 to 15 days. The Wabash is frozen almost every winter, but only from 3 to 15 days. Throughout Kentucky and the basin of the Ohio, snow commonly remains only from 3 to 8 or 10 days, and seldom falls deep. Spring brings on rain and showers, with NE and NW winds; but in about six weeks after the vernal equinox, the heats commence, and exist in all their intensity, during the 60 or 70 days that succeed the summer solstice. The therm. is then between 90° and 93°. At the autumnal equinox rains come on, with winds from the NE, SE, and even NW. The whole of the autumn is more serene, pleasant, and temperate, than the other two seasons; for, strictly speaking, throughout the continent of North America there is no spring. Such is the climate of Kentucky, and all the basin of the Ohio; and we must proceed as high as the Upper Mississippi, in N lat. 45°, before we perceive any sensible difference in it. Even as high up as Niagara, it is still so temperate, that the cold does not continue with any intensity more than two months. In Ohio, according to Fearon's account, the therm. ranges from 85° to 105°, in the months of July and August. On the S this temperature extends no farther than the 35th degree of N lat.; beyond this it is subjected to the same common law as that of the Atlantic states. The annual and mean quantity of rain that falls in the United States is much greater than in most countries of Europe, certain mountainous regions and heads of gulfs excepted. No observations have as yet been made, to determine the annual and mean quantity of rain in the western states; the following table, therefore, is restricted to the eastern:

At Charleston, according to Dr Ramsay, in 1795,	71.8
At Charleston, at a medium from 1750 to 1759,	41.75
At Charleston, from 1797 to 1807,	49.3
The greatest quantity in one of the years immediately above,	83.4
The least quantity in one of the years immediately above,	38.6
The greatest monthly quantity in any one of the 10 years immediately above,	12.9
At Williamsburg, Virginia, (Jefferson),	47
At Cambridge. near Boston, (Williams),	47.5
At Andover. in Massachusetts, (ditto),	51
At Salem, (ditto),	35
At Rutland, in Vermont, (ditto),	41
At Philadelphia, (Rush),	30

The habitual dryness of the American climate increases as we advance W and NW from the Missouri, where there frequently is not a drop of rain for six months. This is owing to the greater distance from any sea,—the superior elevation,—and the comparative want of timber. combined with the greater intensity and longer duration of the NW wind, which sweeps with unobstructed force over the naked plains. Upon the whole, it appears that more rain falls in fewer days, in America, than in Europe; and that there are fewer cloudy days, more fair days, and quicker evaporation.

There are, strictly speaking, only three seasons throughout the whole of North America, summer, autumn or fall, and winter. Winter does not really commence till the middle of December, or a little before the shortest day; and it is preceded by several grand crises in the atmosphere, before a general change of temperature is accomplished by the NE or NW winds driving back those of the S. In February the greatest snows fall, and the most piercing cold is experienced. In March, or the approach of the vernal equinox, it is cold and tempestuous, with snowy blasts from the NW and NE. The predominance of these winds at this period, combined with the coldness imparted to the earth by the snows, retard vegetation so much that April passes away and leaves the ground as bare as March. It is not till the beginning of May,—even in Virginia, in the latitude of 36° and 37°, —that the forests are re-clothed with foliage, though the solar rays are insufferably hot from the middle of April. Hence it follows that there is no spring in America, but a sudden transition from severe cold to severe heat, with the incongruous circumstances of a freezing wind and a scorching sun, a winter landscape and a summer-sky. When vegetation at last bursts forth, its progress is so rapid, that the flowers are quickly succeeded by the fruit, which latter also ripens more speedily than in Europe. Very perceptible changes have taken place in the American climate, and have displayed themselves in proportion as the land has been cleared. These changes were long ago observed by different writers and travellers, from La Hontan, down to those of the present day: they are to be ascribed to the clearing of the land, and the extensive openings made through the woods. The settler has scarcely cleared a few acres of the forest, before the ground, exposed to all the action of the solar rays, acquires, at the depth of near a foot, a heat of 10° or 12° beyond that of the land still covered with wood. But though a change has taken place in the climate, it is questionable still whether it has improved; because, though the winters are shorter, the summers longer, and the autumns later, yet these advantages seem to be fully overbalanced by the increasing variations of the temperature, which are more injurious to the human constitution than long-continued cold or heat.

Soil and produce.] In a country so extensive as that of the United States, every diversity of soil, not less than of climate, is to be found—from that which mocks the industry of man, up to the highest degree of fertility. Taken on a large scale, the soil of the old states may be divided into two regions distinct from each other, namely: the lands between the sea and the Alleghanies, and those beyond them. The lands bordering on the Atlantic are seldom good; whilst those W of the Alleghanies, in Lower and part of Upper Louisiana, are in general not exceeded by the best in Europe, and have besides the advantage of being nearly all virgin soil, and easily cultivated. In the northern states, or those of New England, including Vermont, one description of trees predominating in each soil has originated the descriptive names of 'oak-lands,' 'chestnut-lands,' 'pine-barrens,' 'cedar-swamps,' &c., &c. The best lands produce walnut and chestnut; the next, beech and oak; the third, fir and pitch pine; the fourth, whortleberry and barberry bushes; and the last or poorest produces nothing but marshy shrubs. The soil of the New England states is, however, much more adapted for grazing than for grain; and the former is pursued with great industry and success. The soil of the central states does not materially differ from that of the northern states, unless in the back parts of the state of New York, where it is much more productive, as in the Genessee, Onondaga, and Oneida tracts. The soil of the southern states is in general better than that of the central and northern states, except on the coast, where in many places almost absolute barrenness prevails. Wheat and tobacco are the chief articles of agricultural produce; but the labour is chiefly managed by slaves, and the plantations are too extensive to be well-cultivated.

Nothing in all the western territory equals in fertility the banks of the Ohio, where the soil is a true vegetable earth, produced by the thick bed of leaves which are annually collected on the ground, and are soon converted into mould by the humidity prevalent in these sequestered forests. The best lands in Kentucky and Western Tennessee yield abundant harvests; but there the trees do not attain a bulk or an elevation comparable to those on the Ohio. Along the banks of the Missouri, as far as the confluence of the river Platte—a space of above 600 miles by the windings of the river—the land is well timbered, and esteemed among the best in America, in point of fertility. The soil is a deep black vegetable mould. Along the Platte, it is still richer, and has the advantage of better and more abundant streams of water. Between the Osage and the Kanzas rivers, the soil is perhaps preferable to any in the United States, if taken in connection with the climate, for the purpose of settling. The watershed of the Pacific presents but a comparatively small portion of cultivable land. A barren lava covers the western slopes of the Rocky Mountains. To the W of the Cascade Range, we meet with some rich tracts of land lying along the Willamette, the Sacramento, and the San Joaquin.

Agriculture.] The science of agriculture is less understood in the central states than in the New England states. Wheat is there the staple article. Pennsylvania is a good grazing country, where large dairies are kept. The richness of the staple compensating for the small quantity of the productions, the southern planters prefer their own agricultural system to that of the northern states, which is more conformable to the European method. The cultivation of the tobacco-plant is much more troublesome than that of wheat, and very prejudicial to the soil. More attention, indeed, has of late years been paid in the south to rearing wheat; but still, as the wheat is to be raised from an exhausted soil, the produce cannot be equal to that even of the central and northern states. The system of slave-labour is vastly prejudicial both to the population and agriculture of the southern states; and it may be remarked that, in the southern Atlantic states, the kind of agricultural produce being almost entirely regulated by foreign demand, at one time the soil is devoted almost exclusively to the cultivation of tobacco; at another, to the cultivation of rice; at another, to the cultivation of wheat and Indian corn; at another, to that of indigo; and finally, to that of sugar and cotton. The subjoined statement exhibits the number of bushels of wheat, Indian corn, and potatoes raised in each of the southern or slave states, in the year 1848, according to the estimate of the commissioner of patents:—

	Wheat.	Indian corn.	Potatoes.
Delaware, .	450.000	3,850,000	200,000
Maryland, .	5,150,000	8,800,000	1,000,000
Virginia, .	12,250,000	38,000,000	3,500,000
North Carolina,	2,450,000	26,000,000	3,200,000
South Carolina,	1,400,000	13,500,000	4,200,000
Georgia, .	2,100,000	27,000,000	2,000,000
Alabama, .	1,300,000	28,000,000	2,500,000
Mississippi, .	550,000	17,000,000	2,600,000
Louisiana, .	...	10,600,000	1,800,000
Tennessee, .	9,000,000	76,600,000	3,000,000
Kentucky, .	6,500,000	65,000,000	2,200,000
Missouri, .	2,000,000	28,000,000	1,200,000
Arkansas, .	500,000	8,000,000	800,000
Florida, .	...	1,250,000	500,000
Texas, . .	1,300,000	1,800,000	800,000
Total Southern states, .	44,950,000	353,400,000	29,000,000
Total Northern states, .	81,414,000	234,750,000	85,475,000

It will be seen by this table that in these three articles the North is not so far in advance of the South. The three leading articles of food—those upon which man most depends—are much more equally divided in the cultivation between the North and the South than is generally imagined. Taking the value of these articles, the North shows a preponderance.

SOUTHERN STATES.

	Bushels.	Price.	Value.
Wheat, .	44,950,000	1 00 dol.	44,950,000 dol.
Indian corn,	353,400,000	50	176,700,000
Potatoes, .	29,000,000	1 00	29,000,000
Total,	.	. .	250,650,000

NORTHERN STATES.

	Bushels.	Price.	Value.
Wheat, .	81,414,600	1 00 dol.	81,414,600 dol.
Indian corn,	234,750,000	50	117,375,000
Potatoes, .	85,475,000	1 00	85,475,000
Total,	.	. .	284,264,600

This shows an excess in favour of the North of 33,614,600 dollars in the three leading articles.

The following table shows the preponderance of the agricultural over the other interests of America, if reckoned according to the number of persons employed, according to the census of 1840:

Geographical Divisions.	Pop.	No. of Persons employed in Agricul.	Com.	Manu.
New England states,	2,234,822	414,138	17,157	187,258
Middle states, .	5,118,076	808,633	50,077	333,947
Southern states, .	3,333,483	955,729	12,962	87,955
South-western states,	2,245,602	650,546	14,496	37,899
North-western states,	4,131,370	890,905	22,315	144,690
Total,	17,069,453	3,719,951	117,607	791,749

It appears from official tables, that the several states of the American union produced in 1840 the following quantities of agricultural produce:—Wheat, 84,823,272 bushels; oats, 123,071,341 bushels; Indian corn, 377,531,875 bushels (between four and five times the quantity of wheat produced in the same year). Of "neat cattle", there were 14,971,586; also 26,301,293 swine; 790,479,275 lbs. of cotton; and 219,163,319 lbs. of tobacco. Virginia and Kentucky were the two chief tobacco-growing states, producing upwards of 128,800,000 lbs. out of the above-named quantity. Ohio was the greatest wheat-growing state, producing in the one year named 16,571,661 bushels; and Pennsylvania the next, producing 13,213,077 bushels in the like period. New York state, however, nearly equalled the latter, its amount of wheat being 12,286,418 bushels. It is remarkable how little wheat is grown, however, in some of the states. The area of the states, estimated in 1840 at 690,747,040 acres, gave an average of between 40 and 41 acres to each person living in those states, while the average in Great Britain was little more than 3 acres to each person. According to the census of 1850, the quantity of improved land in the United States, by which is meant such only as produces crops, or adds in some way to the productions of the farmer, is reported at 118,457,622, or about 7⅓ acres to each inhabitant. But, if it be assumed that two-fifths of the pop. are engaged in other pursuits than that of agriculture, the average extent of improved land to be assigned to each person working and occupying it is not less than 12 acres. In the New England states the average for the whole pop. was in 1850 a little more than 4 acres to each person; in New York and Pennsylvania, 3·9 acres; in the other middle states, the same. In Virginia the proportion was about 7 acres; in S. Carolina, 6 acres; in Kentucky, 12 acres; and in Tennessee, 5 acres. The value of the farms in the United States was returned at 3,270,733,093 dollars in 1850. The average value of land attached to farms and plantations, both improved and unimproved, is given at 10 dols. 79c. an acre.

The agricultural products of the United States, in 1844 were as follows:

Wheat,	. .	95,607,000 bushels.
Barley,	. .	3,672,000 —

Oats, . .	172.247.000 —
Rye, . . .	26.450,000 —
Buck wheat, . .	9.071,000 —
Indian corn, . .	421.953,000 —
Potatoes, . . .	99.493.000 —
Hay, . . .	17.715.000 tons.
Flax and hemp, . .	22.800.000 lbs.
Tobacco, . . .	161.705.000 —
Cotton, . . .	872,107.000 —
Rice, . . .	311,759,000 —
Silk, . . .	396.790 —
Sugar, . . .	210.107,000 —

From the report of the census commissioners for 1850, we find that, in the years from 1840 to 1850, the number of cattle had increased from 14,971,586, to 18,355,287, rather more than 20 per cent. During those 10 years the production of Indian corn rose from 377,531,875 bushels in 1846, to 592,326.612 in 1850, an increase of nearly 60 per cent.; while the pop. of the Union had increased only 35 per cent. As regards wheat the progress has not been so remarkable. It rose from 84,823,272 bushels in 1840, to 100,503,899 in 1850, but the crop of 1849, to which the return of 1850 relates, was far below the average in some of the largest wheat-growing states; so that we must not look upon the last decennial increase as a fair measure of the progress of agriculture. Compared with the production of Indian corn, the quantity of wheat seems so small as hardly to afford enough for home consumption; but we must remember, that throughout the greater part of the Union Indian corn is the staple article of food. The increase in the production of wheat between 1840 and 1850 was confined mainly to the Western states. In 1840, Illinois, Indiana, Michigan, and Wisconsin, contributed only 9,800,000 bushels to the general aggregate; in 1850, those four states produced upwards of 25,000,000 bushels. The total produce of all kinds of grain in the United States, in 1850, was in round numbers 108,000,000 quarters; or rather more than double the estimated annual produce of the United Kingdom, for a pop. nearly five millions larger. One remarkable item in the agricultural statistics of the United States is the large number of horses and horned cattle. In that respect, America presents a great contrast to this country. So far as can be ascertained from Parliamentary returns and other sources of information, it appears that there are altogether only about 1,500,000 horses in Great Britain. In the United States there are 4,335,358; nearly three to one. This great disparity is mainly owing, no doubt, to the wide extent of country over which the pop. is dispersed, and the consequent distance of markets. In horned cattle there is a still greater superiority. In Great Britain the total number is estimated at about 5,220,000; in the United States there are, according to the last census, no fewer than 18,355,287. Assuming that the same proportion is annually slaughtered in America as in this country, and that the carcasses are about the same weight, the consumption of animal food must be about three or four times greater than it is in England. As regards mutton the comparison is in our favour. The number of sheep in the United States was 21,721,514 in 1850; in Great Britain, taking the commonly received estimates, the number of sheep is about 40,000,000; so that the larger quantity of mutton which this indicates will partly make up for the deficient consumption of beef.

Wheat.] The most important grain crops of the United States, next to maize or Indian corn, is wheat. The entire amount produced in 1849, the year to which the last census returns apply, was 100.503,899 bushels, being an increase over that of 1839 of 15,645,378 b. In the six New England states the culture of this grain appears to be declining, as the crop has fallen off, in 10 years, 50 per cent. In the six old wheat producing states, from New York to Virginia, including both, there was an increase of 6,140,900 b., equal to 15·27 per. cent. This portion of the Union seems better adapted, from the character of its soil and climate, to the permanent culture of the grain. In the four principal states of the NW the wheat crop was augmented, during the 10 years, from about 10,000,000 to 25,000,000 b. It is from this district that the largest supplies for export are derived. The crop of 1849 was a short one, particularly in the NW states. The exports of flour, wheat, and breadstuffs from the United States, it will be seen from the tables in our paragraphs on the commerce of the United States, have declined since 1846, but it appears that the cause of this has been rather the cessation of an extraordinary demand than any diminution in the ability of the country to supply. The aggregate amount of grain convertible into breadstuff or its substitutes annually produced throughout the Union is estimated at 900.000,000 bushels, of which nearly one-half is Indian corn, while the total of all kinds required for home-consumption does not exceed 300.000.000 b. The quantity of wheat is supposed to be about 100,000,000 b. With regard to the amount of water contained in the various sorts, the results obtained by Professor Beck give 16 to 20 per cent. for Alsatian, 14 to 17 per cent for English, 12 to 14 per cent. for American, and 9 to 11 per cent. for African and Sicilian. In relation to the amount of gluten in various samples of flour from different parts of the United States and Europe, the preference is awarded to the Kubanka variety.

Rye.] The production of rye had decreased in 1849, 4,457,000 b. in the aggregate; but in New York it was greater than in 1840 by about 40 per cent. Pennsylvania, which is the largest producer, had fallen off from 6,613.373 to 4,805,160 b. During the year ending June 1, 1850, there were consumed, of rye, about 2,144.000 b. in the manufacture of malt and spirituous liquors. According to the census returns of 1840 the product of the country was 18.645.567 b.; in 1850, 14,188.637 b.

Indian corn.] The most universally cultivated crop of the United States is maize or Indian corn. It is a principal product of the farm in nearly every state of the Union. It is a favourite article of food with numerous classes of the people, and is used in all parts of the country for the subsistence and fattening of stock. According to the census of 1840, the maize crop of the United States was 377,531,875 b.; in 1849 it was 592,326,612 b., an increase equal to 56 per cent. The production of New England had advanced from 6,993,000 to 10,377,000 b., showing an increase of nearly 50 per cent. In the production of this crop no state has retrograded. Ohio, which in 1840 occupied the fourth place as a corn-producing state, now ranks as the first, Kentucky second, Illinois third, Tennessee fourth. The crop of Illinois had increased from 22,000,000 to 57,500,000 b., or at the rate of 60 per cent. in 10 years. The rate of production is from 30 to 140 b. per acre. More than 11.000,000 b. of Indian corn were consumed in 1850 in the manufacture of malt and spirituous liquors.

Oats.] The crop of oats increased from 123,071,344 b. in 1840 to 146,678.000 in 1850. It is a crop which does not enter into the foreign commerce of the country.

Barley.] In 1840 there were produced 4.161,504 bushels of barley; in 1850, 5.167,016 b. The consumption of barley in the manufacture of malt and spirituous liquors amounted in 1849 to 3,780.000 b.

Potatoes.] The production of the commonest species of this root, called 'the Irish potato,' though indigenous to the American continent, has fallen off since 1840. The crop of 1849 was 65,796,000 b, supposed to be about 18,000,000 b. less than that of 1839, although the exact truth cannot be ascertained, as, in the tables of 1840, the two varieties of the common and the sweet potato were included in the same statement. The crop of sweet or Carolina potatoes increased from an estimated quantity of 25.000,000 b. in 1839, to 38,000,000 b. in 1849.

Flax and Hemp.] The census returns of 1840 show that there were raised in the United States 95,251⅘ tons of flax and hemp; in 1849, 35,093 tons of hemp, and 7,715,961 lbs. of flax. The correctness of the returns as to hemp, in the seventh census, has not yet been perfectly verified. If the returns are allowed to stand without reduction, it would appear that the cultivation of hemp and flax has not materially changed since 1840. The capacity of the United States to produce flax and hemp is almost without limit. Efforts are now making to introduce Mr. Claussen's method for the separation of the fibre of flax from the woody and useless parts of the plant into the flax-growing districts of the country, and, should they succeed, the returns of the next census will prove the cultivation and manufacture of flax to be one of the most important interests in the United States.

Rice.] The production of rice increased from 80,841,422 lbs. in 1840 to 215,312,710 lbs. in 1849. The present culture of rice in the United States is chiefly confined to S Carolina, Georgia, Florida, Alabama, Mississippi, Louisiana, and Texas. The yield per acre varies from 20 to 60 b, weighing from 45 lbs. to 48 lbs. when cleaned. Under favourable circumstances as many as 90 bushels to an acre have been raised. The consumption of rice is increasing in the United States, and the exportation is rather decreasing.

Tobacco.] According to the census returns of 1840, the amount of tobacco raised in the United States was 219,163,319 lbs.; in 1849, 199,752,646 lbs., showing a decrease in its culture of 19,410,673 lbs. It seems probable that this crop, hitherto a most important staple, will gradually decline until it ceases to be a leading article in the agricultural and commercial statistics of the United States.

Cotton.] Cotton was raised in gardens in the U. S. before 1784, which was the first year in which it was cultivated by planters as a crop. In 1791, the whole export of this article was under 200,000 lbs. According to the census returns, the cotton crop of 1839 was 790,479,275 lbs.; of 1849, 987,449,600 lbs.; showing an increase of 196,970,325 lbs. It appears that the culture of cotton is rapidly diminishing in Virginia and N. Carolina. In those states it is doubtless giving place to other productions of the soil. There has been a very heavy falling off also in Louisiana, and no appreciable increase in Mississippi; but the diminution in the former state, and the failure of any advance in the latter, are accounted for by the terrible inundations of the Mississippi and its tributaries. The cotton zone of the United States, Professor Wilson informs us in his special report, may be considered as extending from the Atlantic to the Rio-del-Norte, and includes the states of Georgia, S. Carolina, Alabama, Mississippi, Louisiana, those portions of the states of North Carolina, Tennessee, and Arkansas that lie below 35° N lat., all the state of Florida above the 27th parallel, and all the state of Texas between the gulf of Mexico and the 34th parallel. This region occupies an area of upwards of 450,000 sq. m., but it is estimated that, from various causes, more than two-thirds of the territory in each state mentioned is not adapted to the advantageous growth of cotton. In 1852 there were 6,300,000 acres of it under cultivation, and the total cotton crop raised amounted to 3,150,000 bales of 400 lbs. each; of which Alabama produced 750,000 bales, and Georgia 740,000 bales. The number of labourers employed to produce this crop was 787,000. There was, besides, an unemployed area of 39,200,000 acres, suitable for the growth of cotton, on which might have been grown an additional crop of 19,600,000 bales of 400 lbs. each, giving employment to 4,900,000 additional labourers. The labour is estimated at an average of one man to each 4 bales; and the land is estimated at 200 lbs. produce per acre, or 8 acres to each labourer. The cost of producing cotton necessarily varies according to the soil, climate, season, and skill employed. It appears from the most reliable estimates that it averages from 4·7 cents to 6 cents per lb.; and that the net profit to the producer may be taken at from 10 to 20 per cent. upon his capital. Professor Wilson states that the per cent. increase of the production of cotton in the United States has been regularly diminishing since the commencement of the present century. The total increase per cent. for the first 20 years of the century may be taken at 177, or 8·85 per annum; for the next 15 years it was 119, or 7·66 per annum; for the next 10 it declined to 58, or 5·95 per annum; for the five years ending 1850 it reached only 15, or 3 per cent. per annum. In the same time the rate of home consumption has been increasing until it has reached 9 per cent. per annum. The per cent. increase in the production of the world has fallen in the last five years as low as 1 8 per cent. per annum, while the gross percentage of increase in consumption has been just 3·8 per cent. per annum. Commercially speaking, two distinct kinds of cotton are cultivated in the states, the long staple and the short staple. The first generally commands a considerably higher price than the latter, and the finest quality, 'Sea Island,' often brings a price very much exceeding the top market price. The long staple yields from 75 lbs. to 150 lbs. of cleaned fibre to the acre; and the short staple averages from 150 lbs. to 250 lbs. to the acre. The cultivation of the long staple is confined to Georgia, S. Carolina, and Florida. Twenty years ago there were retained for home consumption in the United Kingdom 208,987,000 lbs. of cotton wool. Ten years ago there were retained 672,000,000 lbs. During the last ten years the increase has been less rapid, being not more than that which the following table indicates:

1839,	1,054,000 bales.
1840,	1,293,000
1841,	1,150,000
1842,	1,193,000
1843,	1,388,000
1844,	1,426,000
1845,	1,566,000
1846,	1,599,000
1847,	1,105,000
1848,	1,505,000

Yet even this rate of progression, it is clear, far outstrips the rate of extended cultivation which the capital and labour of America afford. And were it not for the supplementary supplies drawn from other quarters, there can be little doubt that ere this the evils of an inadequate market for the raw material would have been much more severely felt than they have yet been. The consumption of America herself, also, begins to form an important item in the general calculation. In 1832 but 173,000 bales were retained for domestic consumption in the United States; whereas there were last year retained 500,000 bales, or one fifth of the entire crop of 1848. Supposing the American demand to proceed during the next few years at anything like this rate, it is manifest that either our cotton manufacture must become factitiously contracted within dimensions wholly irrespective of its natural capabilities, or the wants of the world; or new and additional cotton-fields must be found whence the raw material may in future be derived. It may, perhaps, not be out of place here to show the great fluctuations that have heretofore marked the growth of cotton in the United States:—

Growth of Cotton in the United States.

1823—24,	509,158 bales.
1833—34,	1,205,394
1839—40,	2,177,835
1840—41,	1,634,945
1841—42,	1,683,574
1842—43,	2,378,875
1843—44,	2,030,409
1844—45,	3,394,503
1845—46,	2,100,537
1846—47,	1,778,651
1847—48,	2,347,634
1848—49,	2,728,596
1849—50,	2,096,706
1850—51,	2,355,257
1851—52,	3,015,257
1853—54,	3,262,882

Silk.] The attempt to introduce the silk culture into the United States has been a failure. In 1840 there were produced 61,552 lbs. of cocoons; in 1849 but 14,763 lbs. The decrease dates from an earlier period than 1840.

Sugar.] The cultivation of sugar is carried on in Louisiana, Texas, Florida, Alabama, and to some extent in Georgia and California. It is now a very important interest, and is rapidly increasing. Hitherto the amount of sugar and molasses consumed in the United States has exceeded the quantity produced; consequently there has been no direct occasion for their exportation. In 1815, it was estimated that the sugar made on the banks of the Mississippi alone amounted to 10,000,000 lbs. In 1818, the entire crop of Louisiana was only 25,000,000 lbs.; in 1849, it had reached the enormous quantity of 226,001,000 lbs. besides about 12,000,000 gallons of molasses. According to the census of 1840, the amount of cane and maple sugar raised in 1839 was 155,100,809 lbs., of which 119,947,720 lbs. were raised in Louisiana. By the census of 1850 the cane-sugar made in the United States in 1849 was 247,581,000 lbs., besides 9,700,606 gallons of molasses; maple sugar, 34,249,888 lbs., showing an increase, in 10 years, of 126,730,077 lbs.

Domestic animals.] According to the census returns of 1840, there were in the United States in 1839, 4,335,669 horses and mules, 14,971,586 neat cattle, 19,311,374 sheep, and 26,301,293 swine; in 1849, the numbers were 4,335,358 horses, 559,229 asses and mules, 28,360,141 horned cattle (including 6,392,044 milch-cows, and 1,699,241 working), 21,721,814 sheep, and 30,316,608 swine. The number of horses has not increased with the same rapidity as other kinds of stock, for the reason that the great extension of railroads in the northern states, between 1840 and 1850, had superseded the necessity for a large proportion of those employed for travelling purposes. In the new states of the NW, into which at the date of the census railroads had not been generally introduced, the increase of horses had kept pace with the pop. There is one horse to each 5 persons in the United States. The 500,000 asses and mules returned are almost confined to the southern states, where the climate is better adapted to this animal than to the horse. In 1850 there were returned 18,355,287 milch-cows, working oxen, and other kine, being an increase of 3,383,701 for the 10 years since 1840, or about 20 per cent. The amount of butter produced in 1850 was 313,266,962 lbs., and of cheese, 105,535,219 lbs. The average value of the exports of these two articles from the United States during 10 years has been 1,000,000 d.; during the last five years of the period, it has been 1,400,000 d.

Sheep and wool.] There was between 1840 and 1850 an increase of 2,309,108 in the number of sheep in the United States. In New England there has occurred a remarkable decrease in their number. There were in that division of the Union, in 1840, 3,811,307; in 1850 the number had declined to 2,164,452; being a decrease of 1,646,855, or 45 per cent. In the five Atlantic middle states—New York, New Jersey, Pennsylvania, Delaware, Maryland—there was a decrease from 7,402,851 to 5,641,391, equal to 1,761,460, or about 22½ per cent. In Pennsylvania there was a gain, however, during this period of 155,000. While there has been a positive diminution of 3,408,000 in the states above-named, there has been an augmentation of 5,717,608 in those S of Maryland and W of New York. Ohio has gained most largely, having been returned as pasturing, in 1839, 2,028,401; in 1849, 3,942,929; an increase of 1,914,528, or nearly 100 per cent. In each of the states S and W of the lines above indicated there has been a large proportional increase in this kind of stock; and there is reasonable ground for the opinion, that the hilly lands of Virginia, N. and S. Carolina, Tennessee, and the prairies of Illinois, Iowa, and Texas will prove highly favourable for the rearing of sheep for their wool and pelts. New Mexico has the extraordinary number of 377,271 sheep, or more than 6 to each inhabitant; proving the soil and climate of that new territory to be well adapted to this description of stock, and giving promise of a large addition from that quarter to the supply of wool. But there is no reason to expect that the American Union will produce wool for export for many years to come. The average amount imported annually, from 1840 to 1850, was 14,000,000 lbs. mostly from South America, and the quantity is rapidly increasing. In 1851, the importation amounted to 32,548,693 lbs. The growth of wool in 1849 is returned at 52,789,174 lbs., being an increase on the production of 1840 of nearly 17,000,000 lbs., or about 46 per cent. In 1840 the average weight of the fleece yielded by each sheep was 1·84 lb. In 1850 it was 2·43 lb., exhibiting an increase in the average product of each sheep of 32 per cent., and thus affording an indication of a great improvement in the breeds of sheep in the United States. This improvement is chiefly manifested in the returns relative to Vermont, Massachusetts, and New York. The value of the live stock in all the states and territories is returned at the considerable amount of 543,969,420 dollars.

The aggregate of the value of the leading products of the United States in 1840 and 1850 is thus stated:—

	1840.	Value.
Wheat, bushels, . .	84,823.272	69.840,000 d.
Corn, „ . .	377,531,875	94,333,000
Cotton, bales, . .	1,976,198	59,285,940
Sugar, lbs., . .	155,100,809	10,500,000
Hay, tons, . .	20,248,108	11,240,546
Dairy, . . .	—	33,787,008
Wool, . . .	35,862.114	11,740,634
Coals, tons, . .	359,686	3,680,000
Total, ...		293,497,162 d.

	Value.	1850.
Wheat, bushels, . .	104,799,230	83,839,384 d.
Corn, „ . .	591,586,053	147,896,513
Cotton, bales, . .	2,474,214	97,968,540
Sugar, lbs., . .	350,403,263	17,897,755
Hay, tons, . .	13,605,384	68,026,920
Dairy, . . .	—	37,412,303
Wool, . . .	52,422,797	15,726,839
Coals, tons, . .	4,000,000	16,000,000
Total, ...		384,768,254 d.

The following table exhibits the number of acres of improved and unimproved land in each state of the Union in 1849; the average cash value per acre; the produce of wheat and Indian corn in bushels of each state; the number of sheep, and the value of the live stock in each state, and the total value of the home-made manufactures, all founded on the returns made to the census commissioners in 1850:—

States and Territories.	Acres of improved land.	Acres of unimprov. land in farms.	Average cash value per acre.	Wheat, bushels.	Ind. Corn, bushels.	Sheep.	Value of live-stock.	Value of home-made manuf.
Maine,	2,039,596	2.515.797	12 04 d.	296.259	1.750,056	451.577	9,705,726 d.	513,599 d.
New Hampshire,	2,251,488	1,140,926	16 28	185.658	1.573.670	384,756	8.871.901	393,455
Vermont,	2.601,409	1,524.413	15 36	535.955	2.032.396	1.014,122	12,643.228	267,710
Massachusetts,	2,133.436	1,222.576	32 50	31,211	2,345,490	188.651	9,647,710	205,333
Rhode Island,	356.487	197.451	30 82	49	539,201	44.296	1.532,637	26.495
Connecticut,	1.768.178	615,701	30 50	41.762	1,935,043	174.181	7.467.490	192,252
New York,	12,408,968	6.710,120	29 00	13.121.498	17,858.400	3,453.241	73,570,499	1,230,333
New Jersey,	1,767,991	984.955	43 67	1.601,190	8,759.704	160.488	10,679,291	112.781
Pennsylvania,	8,628,619	6.294,728	27 33	15,367.691	19,835.214	1,822,357	41,500,053	749,132
Delaware,	580,862	375.282	19 75	482.511	3,145.542	27,503	1,849,281	38,121
Maryland,	2,797,905	1,836,445	18 81	4,494.680	11,104,631	177,902	7,997,634	111,828
District of Columbia	16.267	11,187	63 03	17.370	65.230	150	71,643	2.075
Virginia,	10,360.135	15.792,176	8 27	11.232,616	35.254.319	1,310,004	33.656,659	2,156,312
N. Carolina,	5.453,977	15.543,010	3 23	2.130.102	27.941.051	595.249	17.717,647	2,086,522
S. Carolina,	4,072,651	12.145.049	5 08	1.066.277	16.271.454	285.551	15,060.015	909.525
Georgia,	6,378,479	16,442,900	4 19	1,088.534	30.080.099	560,435	25.728,416	1,838,968
Florida,	349,049	1.236.240	3 99	1,027	1,996,809	23,311	2.880,058	75,582
Alabama,	4,435,614	7,702,067	5 30	294.044	28.754.048	371.880	21,690,112	1.934,120
Mississippi,	3,444,358	7,046,061	5 22	137,990	22.446.552	304.929	19,403,662	1,164,020
Louisiana,	1,590,025	3,939,018	13 71	417	10,266.373	110.333	11.152,275	139.232
Texas,	639,107	14,454,669	1 09	41,689	5.926,611	99,098	10.266,880	255,719
Arkansas,	781,531	1,816,684	5 88	199.639	8,893,939	91.256	·6.647,969	638.217
Tennessee,	5,175,173	13.808.849	5 16	1,619,381	52,276,223	811.591	29.978.016	3,137,810
Kentucky,	11,368,270	10.972.478	6 91	2,140,822	58.675,591	1,102,121	29,591.387	2,458,128
Ohio,	9,851,493	8,146,000	19 93	14.487.351	59,078,695	3.943,929	44,121.741	1,712.196
Michigan,	1,929.110	2.454,780	11 83	4.925,889	5,641.420	746,435	8,008,734	340.947
Indiana,	5,046.543	7,746,879	10 66	6,214.458	52.964.363	1,122,493	22.478,555	1.631.039
Illinois,	5.039.545	6,997.867	7 99	9,414,575	57.646,984	894.043	24.209.258	1,155,902
Missouri,	2,938,425	6,794.245	6 50	2,981.652	36,214,537	762.511	19,892.580	1.674,705
Iowa,	824,682	1,911,382	6 09	1,530,581	8,656,799	149.960	3,689,275	221,292
Wisconsin,	1,045,499	1,931,159	9 58	4,286,131	1,988.979	124,892	4,897,385	43,624
California,	62,324	3,831.571	99	17,328	12,236	17,574	3,351.058	7,000
Minnesota Territory,	5,035	23,846	5 61	1.401	16,725	80	92,859	—
Oregon do.	132,857	299.951	6 58	211.493	2,918	15,382	1,876.189	—
Utah do.	16,333	30.516	6 65	107.702	9,899	3,262	546.968	1.392
New Mexico do.	166,201	124,370	5 69	196,516	365,411	377,271	1,494,629	6,033
Aggregate,	118,457,622	184,621,348	av. 10 79 d	100,503,899	592,326,612	21,721,814	543,960,420 d.	27,481,399 d.

Zoology.] The wild animals of the United States nearly correspond with those of Canada. The class mammalia, in North America, as far as known, includes 118 quadrupeds, and 28 cetacea. In the countries between the Missouri and the Atlantic, their numbers are constantly decreasing, in proportion as the population is increasing. The Moose-deer is becoming very rare, and will probably be soon extirpated, as the wolf and wild boar have been in Britain. The buffaloes have almost all crossed the Mississippi and Missouri, and taken refuge in the vast plains of Louisiana. Among the animals common to North America, are six kinds of squirrels, which greatly injure the maize-plantations and the fields of wheat. The species called the Carolinian squirrel is grayish, and a little larger than the European one. Vast numbers of antelopes frequent the plains of the Missouri. Beyond the Mandan villages, and from thence to the Rocky mountains, are found large herds of those animals which, by the French and natives, are called wild or mountain-sheep; but they little resemble sheep, except in the head, horns, and feet. They are of a dun colour, except on the belly and round the rump, where they are white. In size, they are somewhat larger than a deer; they have fine soft hair, and very large horns, especially the males. The European animals which have been naturalized in America are the cow, horse, ass, hog, sheep, goat, and dog; and these have multiplied to such a degree, as to exceed the native quadrupeds greatly in number. The finest horses in the United States are those of Pennsylvania, Virginia, and Kentucky, which, by their mien and the elegance of their form, may be compared with the best in Europe. Their beauty arises from crossing the English and Spanish breeds. Vast numbers of wild horses are to be found in Louisiana and Mexico. Merino, or Spanish sheep, have been lately introduced into the United States. Of all the domestic animals, pigs are the most numerous, in the western states: every inhabitant keeps some, and several have as many as 150 or 200. These animals never quit the forests, where they always find food, particularly in autumn and summer. Sometimes they penetrate into the recesses of the forests, and do not re-appear for several months; they are, however, accustomed to return to the plantations occasionally, by feeding them with maize once or twice a-week. The American hare is not above 18 inches long, and seldom weighs above 2 lbs. Of amphibious animals, many different species are to be found in the lakes and rivers of the United States; the Missouri, in the lower part of its course swarms with alligators; as also a number of its tributary streams, and all the rivers of the southern states. Of serpents, Dr. Morse enumerates near 40 kinds in the United States. Virginia, in particular, produces great numbers. The rattle-snake is the most dreadful. Among the fish are most of those esteemed in Europe; and among those that are peculiar may be mentioned the large species of white trout found in the lakes, and the cat-fish in the Ohio, Mississippi, and Missouri, some of which weigh above 100 lbs. The quantity of fish in the Missouri and its tributary streams is astonishing. Charles Lucien Bonaparte has arranged the birds of the United States into 28 families, 81 genera, and 362 species, viz.: 209 land and 153 water-birds. Of these 81 genera, 63 are common to Europe and America, while 18 have no representatives in Europe. Pheasants and partridges do not exist on the American continent. Vast numbers of wild turkeys are to be met with in the western territory and Louisiana; but as a bird of sport, the turkey is no longer found on the E side of the Appalachian chain.

Mineralogy.] The gold region of the Atlantic states is chiefly confined to a narrow belt or strip of schistose rocks, extending from the Rappahannock in Virginia, to the Coosa river in Alabama, varying in width from a few yards to several miles, with its continuity often interrupted by intruding rocks of a more ancient date. The general course of this belt from Virginia to Georgia is NE and SW. After entering the latter state, it bends somewhat more to the W, until all traces of it are lost in the state of Alabama. It may be said to extend, however, to Canada on the N, gold having been found at Canaan, N.H., and in the state of Maine; also at Middle Haddam, on the Connecticut river. Up to the present time in the states of Virginia, N. and S. Carolina, and Georgia only have gold mines been worked. The gold-bearing rocks *par excellence* are the talcose and talco-micaceous slates, rocks belonging to the metamorphic series of geologists, ranking next in age to the granite, greasy to the feel, and of all degrees of hardness and shades of colouring. Gold is also found in gneiss, sienite, hornblende, mica slate and granite. Occasionally it is disseminated throughout the rock, but most generally occurs in veins of quartz, oxide of iron, copper or iron pyrites. It is usually in a state of minute division, some of the best ores giving no external indications of the richness, the gold only becoming visible on pulverising the specimen, and carefully washing off the sand. The yield of the mines in the Atlantic states became important first in 1824, and the first supply of California gold was in 1848. The deposits from all these sources at the mines, down to July 31st, 1852, were as follow:

	Atlantic mines.	California.	Total.
1824 to 1847,	12,808,575	—	12,808,575
1848,	851,374	45,301	896,675
1849,	927,784	6,151,360	7,079,144
1850,	665,217	36,273,097	36,938,314
1851,	602,380	56,938,232	56,540,612
1852,	702,263	53,794,700	54,506,963
1853,	508,564	55,113,487	55,622,051

Silver ore has not been extensively wrought in the older states. There is however every reason to expect it in New Mexico and on the Rio Gila; and very rich quicksilver mines occur at New Almaden in California. Indications of silver are also, said to occur on the Little Missouri, a small stream that runs into the Washita.—Copper is found in the state of Massachusetts. Within a few miles of New York is a rich copper mine, yielding from 60 to 70 lbs. of fine copper per cwt. Rich copper mines are found in New Jersey along the course of the rivers Passaik and Rariton. Other mines of the same mineral occur in Virginia, and near Lake Superior; at Copper-harbour, in the state of Michigan, on the shores of Lake Superior, a company in Boston has secured what they consider the richest copper mines in the world, extending over 250,000 acres. The metallic copper is found in laminæ of varying thickness in the rocks on the S shore of Lake Superior, in the Keeweenaw district, where considerable mining operations are carried on. The entire produce of copper within the United States, in 1854, was nearly 4,000,000 lbs. The Michigan mines produced 2,000 tons. Associated with the metallic copper, silver has been found, sometimes in the copper or attached to the exterior of the mass, and at others in small distinct nodules or irregular masses. Lumps, or nuggets of larger size, have been met with in the debris of the rocks, accompanied by lumps of metallic copper. Zinc has been discovered in Pennsylvania. Lead appears on the banks of the Connecticut, in the state of New York, New Hampshire, Virginia, Kentucky and Tennessee. A rich and extensive vein of lead retches along the E bank of the Missouri, from the mouth of Rocky river, more than 100 m. upwards; but the most valuable lead-mines are found in Upper Louisiana, near the banks of the Marrammeg, which falls into the Missouri between St. Louis and St. Genevieve. They cover an area of more than 3,000 sq. m., and are said to be the most extensive on the globe. The number of mines is 165, in which 1,100 men are employed, producing annually 3,000,000 lbs., valued at 120,000 d. The entire yield in 1853 was 33,000,000 lbs.

There is abundance of iron ore, of various qualities, in the United States. In the state of Maine, the founderies are supplied with bog-ore. There are iron mines in Rhode island, and in S. Carolina. Pennsylvania also abounds in this mineral, and immense quantities of iron ore are found in various parts of Massachusetts, Maryland, Tennessee, and Kentucky. The number of furnaces, forges, and bloomeries, in 1810, was 530, and the value of the iron manufactured annually was estimated at from 12,000,000 to 15,000,000 dollars. The production of iron in 1844 was 700,000 tons; in 1854, 1,000,000 tons. The census of 1850 represents the capital invested in the iron manufacture of the United States as follows:

I. Capital invested in manufacture,	17,348,000 d.
Raw metal consumed,	7,000,000
Cost of labour,	5,066,000
Value of products,	12,740,000
Hands employed,	20,458
II. In the manufacture of cast iron:	
Capital invested in manufacture,	17,456,000 d.
Raw material,	10,346,000
Labour,	7,000,000
Value of product,	25,000,000
Number of hands,	20,507

In 1821 the value of imported iron into the United States was 1,213,000 d.—in 1850, 9,150,000 d.—in 1851, 9,000,000 d. The subjoined statements show the value of iron and articles which are wholly or in part composed of iron imported into the United States in 1851 and 1852:—

	1851.	1852.
Pig,	950,660 d	1.142,717 d.
Railroad,	7,397,166	10,650,194
Manufacture of cast iron,	1.437,133	1.219,202
" wrought,	2,278,089	3,469,112
" not specified,	3.427.180	2,319,280
Cutlery,	1,274,838	1,607,158

Other manufactures composed of steel in part, . . .	668,531	916,889
Old, ...	202,099	311,444
	17,635,507 d.	21,626,993 d.

30,000 men are employed in the United States in iron castings; 25,000 in the manufacture of pig-iron; and 14,000 in wrought iron.

According to Professor H. D. Rogers, the Eastern or Great Appalachian coal-field extends from the N limits of Pennsylvania to the central part of Alabama, having a length of about 750 m., and widening out in some places to a breadth of 180 m. He estimates the entire area at about 63,000 sq. m. The Great Western coal-field, so far as it has been well identified, occupies the largest portion of the state of Illinois, a considerable portion of Indiana, and a small area of the state of Kentucky. The late surveys of Dr. Owen have, however, shown that in Iowa and Missouri this same coal-field has a much greater extension than had been previously supposed, which would give reasonable grounds for believing that it occupies even a greater area than the Eastern or Great Appalachian field. The portion lying E of the Mississippi is assumed to be equal to the entire Appalachian field, while that lying west is fully half as much more. Following Dr. Owen's report, and taking its extreme width from Green river in Kentucky to Des Moines river in Iowa, its range will be seen, in that direction, to be between 500 and 600 m.; while in an E and W direction, across the states of Illinois and Missouri, its greatest breadth will be fully 400 m.; and from its N termination in Iowa to its present fixed S limits on the Osage river, its range is more than 300 m. Still further S in Arkansas, a coal-field of considerable extent is known to exist. In Michigan the existence of a large coal deposit has been for some time well known, though hitherto it has been allowed to remain unproductive. In Rhode island also, irregular detached masses of a highly disturbed coal formation are met with and worked economically, though only to a limited extent. As may be supposed, these enormous areas are intersected and traversed by several of the large rivers, such as the Mississippi, the Missouri, the Ohio, Illinois, Wabash, Des Moines, and others, which offer great facilities for transportation to their produce, and open up communication with even the most distant markets." [*Professor Wilson's Report.*] The coal product of the world for 1853 is estimated by able statisticians at 75,000,000 tons, of which amount 40,000,000 tons are produced by Great Britain, 9,000,000 by the United States, France 4,500,000, Prussia 3,500,000, Austria 700,000 tons, and Belgium 5,000,000. It will thus appear that the United States ranks second among the coal producing countries, although her mines, especially the anthracite, are but the discovery of yesterday, comparatively speaking. It is generally believed that because Pennsylvania furnishes by far the largest quantity of coal for consumption and export, that her coal fields are the most extensive in the Union: this, however, is far from being the case. Out of twelve coal-producing states, she only ranks eighth on the list, having the smallest coal area of all except four: yet by diligence, skill, and enterprise she mines at least two-thirds of the whole that is dug in the United States. The following is a table of all the coal-fields of the United States:—

States.	Area of State.	Coal areas.	Proportion of coal.
1. Alabama,	50,875 sq. m.	3,400 sq. m.	1-14th.
2. Georgia,	58,200	150	1-386th.
3 Tennessee,	44,720	4,300	1-40th.
4. Kentucky,	39,015	13,500	1-31st.
5. Virginia,	61,080	21,195	1-3d.
6. Maryland,	10,829	550	1-20th.
7. Ohio,	38,850	11,900	1-3d.
8. Indiana,	34,800	7,700	1-5th.
9. Illinois,	59,130	44,000	3-4ths,
10. Pennsylvania,	43,969	15,437	1-3d.
11. Michigan,	60,820	5,000	1-20th.
12. Missouri,	60,324	6,000	1-10th.

It will be seen that the aggregate area in the 12 states is about 565,000 sq. m., of which 133,000 m., or nearly one-fourth, are composed of coal-beds. More of these beds are bituminous than anthracite; for example, all in Virginia are bituminous, and all in Pennsylvania W of the Alleghany mountains; while nearly all E of the mountains is anthracite; and this latter is the mineral that supplies the cities of Philadelphia and New York, and finds its way to tide-water for export. Some idea of the advantages of this vast mineral treasure to the country and its future importance, will be seen by the rapid increase of production, for several periods since 1840, including the imports of foreign coal:—

	Anthracite.	Cumberland.	Imports.
1840,	865,414 tons.		162,877 tons.
1845,	2,023,052	24,653 tons.	85,776
1848,	3,089,238	79,571	196,188
1850,	3,556,899	196,848	198,213
1851,	4,395,209	257,769	180,439
1852,	4,996,130	534,178	183,015
1853,	5,195,151	533,980	147,000
1854, (Estimat.)	5,600,000	650,000	175,000

The above figures only represent what was actually brought to tide-water, not taking into view the consumption of the interior. An increase of ten per cent. per annum will show a demand for anthracite coal in 1870 of 27,573,065 tons. The great scarcity in the eastern markets, and consequently high price of bituminous coal, has been the means of opening several coal fields W of the Alleghanies. Hitherto nearly all the gas coal consumed in the eastern markets have been imported from England and the British North American provinces.

Limestone is rare to the E of the Blue ridge. From Milton, near the foot of South mountain, a vein of limestone, formed like schist, and placed between layers of perfect slate never more than 10 ft. thick, but frequently less, stretches 220 m. in a NE and SW direction. This limestone, when calcined, makes excellent lime. Abundance of marbles are found in the western states; the beautiful red serpentine, of which the pipes of the Indian chiefs are made, comes from the W of the Mississippi. Plumbago or black lead, aluminous slate, and asbestos, are found in Massachusetts. Zinc occurs in Connecticut and New York. Fine basalt, talc, white and yellow quartz, gray and green *lapis ollaris*, and *petrosilex*, occur in various parts of the United States. The Indian hatchets were formerly made of fine basalt; their knives of quartz and *petrosilex;* their kettles of *lapis ollaris*, as also their tobacco-pipes; but the calumets, or pipes of peace, were made of a species of fine serpentine. Of precious stones, properly so called, there are few or none in the United States, with the exception perhaps of topazes, which are said to have been found in Connecticut. Georgia contains no minerals whatever, except a bank of oyster-shells, 90 miles from the sea. Gypsum, or plaster of Paris, is found in the state of New York; and it is supposed that immense beds of it exist on the high grounds to the west of the Washita; it is also found in several places in the vicinity of the Missouri. White pipe-stone clay, marle, red and yellow ochre, and fuller's earth, are found in the northern states.—That useful mineral, salt, is found in various parts of the United States, but not yet in such quantities as fully to supply the home-consumption. The eastern states are chiefly supplied with this necessary article from the salt springs of Onondago and Cayuga. The salt works of Onondago are in the state of New York, about 5 or 6 m. N of Onondago village, in the village of Salina, which is situated on a bank 50 ft. above the creek and marsh in front. The country in the immediate vicinity appears to be a stiff barren clay; and wood, from the vast consumption of the furnaces, has already become scarce. The salt springs are found on the margin of an extensive marsh not unlike in appearance to the salt marshes of Hoboken. The total produce of the district in 1797 was 25,474 bush.; in 1853, 5,404,524 bush. Salt springs are much more abundant in the western states than in those to the E of the Alleghanies. There are 12 in the single state of Kentucky, between the Great Sandy and Cumberland rivers. These springs are called *licks*, from the earth about them being furrowed out in a curious manner by the buffalo and deer, which lick the earth on account of the saline particles with which it is impregnated. According to all accounts, the water of these licks or salt-springs, is not so strong as sea-water; as it requires 400 gallons to make one bushel of salt, which is more by one-half than would be wanted of sea-water to produce that quantity. The springs at the Great Kanhaway are far superior to any other in the western country, both with respect to the strength of the water, and the quantity and quality of the salt, as it requires only from 90 to 130 gallons of salt water to produce a bushel of that useful mineral. There are salt springs on the Wabash; and more salines to the W of the Missouri than would be sufficient to supply the wants of a pop. double that of the United States. About 4 m. below St. Genevieve is a copious salt spring, where all the salt used in the Illinois and Indiana territories is made. On the banks of the Washita, under the parallel of 34° N, are two remarkably strong salines, where the water oozes from a bed of quicksand, as saline and bitter as the water of the ocean. The country on the S of the Arkansas abounds in salines to such a degree as to induce an almost complete sterility of soil for some hundred miles in the upper part of its course. Saltpetre is also very abundant in this quarter.

The aggregate value of the whole mineral produce of the United States, in 1839-40, was as follows:—

	Quantity	Value
Cast iron, . .	286,903 tons.	22,952,240 d.
Bar iron, . .	107,200 „	10,904,803
Lead, . .	31,239,453 lbs.	917,183
Anthracite coal, .	863,489 tons.	6,406,167
Bituminous do., .	27,603,191 bushels.	6,900,797
Salt, . .	6,170,174 „	1,599,793
Granite and other stone, . .		3,695,884
Gold, . . .		529,605
Other metals, ...		370,614
Aggregate value in dollars, .		59,768,088

Manufactures and commerce.] The territories now composing the American confederation, were, when originally discovered, in their physical aspect and productions, vastly different from those which tempted the cupidity of the Spanish adventurers. They possessed neither gold nor silver; and the appearance of the country convinced the first settlers that they had exchanged their native home for no paradise. The difficulties of the colonists impelled them to habits of active and laborious exertion. The country, though wild and wooded, was gradually cleared; and provisions were at length procured in such abundance as not only to supply their own necessities, but even enable them to procure some of the conveniences and luxuries of their neighbours. The parent-state, at that time rising into commercial eminence, soon began to perceive that her once-despised colonies would become to her a source of extensive trade; and to secure this market for herself, prohibition succeeded prohibition; and the interests both of Britain and the colonies were often sacrificed to short-sighted maxims of policy, and the narrow views of a few selfish traders. A report laid before the house of commons in 1732, by the board-of-trade, represents New Hampshire as manufacturing a small quantity of linen-goods for its own use, and trading in naval stores, lumber, and fish. Massachusetts, it was stated, manufactured a small quantity of woollen and linen-goods, but not nearly sufficient to supply its own consumption. Some iron-works had been established in that state, which did not supply one-twentieth of the demand, and ships were sometimes built there, and exchanged with the French and Spaniards for molasses, wines, rum, and silks. No manufacture deserving of notice existed in the state of New York in 1720; its commerce consisted in furs, whalebone, oil, pitch, tar, and provisions; and it received from the West Indies rum, sugar, molasses, cocoa, indigo, and cotton. The commerce of New Jersey, in 1720, chiefly consisted in the export of provisions to New York and Pennsylvania. Pennsylvania exported provisions and lumber, but at that time possessed no manufactures. In 1731 some coarse woollen and linen-goods were made in the last-mentioned state for home-consumption, and a few small vessels were built for the West India market. Rhode island had no manufactures, but possessed a few iron-works: one-fourth of the metal, however, necessary for its own consumption, was not produced. Connecticut seems to have equalled, if not surpassed, any of the other provinces in trade and industry; its inhabitants exported all kinds of lumber, grain, hemp, flax, tobacco, cattle, and horses. The West Indies was their chief market, whence they received, in return, sugar, molasses, and rum. The Carolinas were in their infancy at the period referred to; but S. Carolina had begun to export rice. Such were the humble beginnings of North American commerce. England, however, had now become fully alive to its importance, and spared no pains to monopolize the whole; but in spite of restrictions imposed for this purpose, it continued rapidly to increase. The trade of the Spanish and Portuguese colonies was at this period confined, by their respective parent-countries, to particular ports, and allowed to be carried on only by particular ships; the British colonies, on the contrary, were not only allowed to trade with every part of the mother-country indiscriminately, but—what was to them of vast importance—they were not obliged to bring the whole of their exported produce to Britain. A few articles—on this account denominated 'enumerated commodities'—could be carried only to Britain; all other 'non-enumerated commodities' might be carried any where at first, and afterwards to any port S of Finisterre. This privilege was the more valuable, because, among the non-enumerated commodities were several which constituted the staple of the commercial states, as grain, lumber, salt provisions, and fish. The intercourse of the colonies with each other however was always entirely free; and this single advantage of free mutual intercourse amongst a people whose numerical population was rapidly increasing, even though their foreign commerce had been more shackled than it was, could not fail of insuring future wealth and power to the Americans. The trade of the colonies consequently continued to increase annually, till their political separation from Great Britain. That event flattered them with a degree of prosperity much greater than they had ever enjoyed; but a little experience convinced them that their hopes were too sanguine. The first congress had not power to mark out such rules of national trade as were necessary to establish the general interest of the states. Each province consulted its own emolument; and none of them evinced so much patriotism as to be above enriching themselves at the expense of their neighbours. The French revolution, which desolated every continental state, and burst asunder every tie which had hitherto bound together the various members of the European commonwealth, communicated a mighty impulse to American commerce. The annihilation of the French commerce, and eventually that of all the other European states, with the exception of Britain herself, combined with the awful catastrophe of St. Domingo, which drove the remaining planters to America, was to the United States a source of wealth and prosperity. By the destructon of the Dutch commerce, she obtained the carrying-trade of Europe; and the inability of France, either to support her own colonies, or to import their produce, transferred to the United States all her colonial commerce. These advantages America enjoyed till the abuses which the Americans made of the neutral flag, led to disputes with Great Britain, which terminated in a war that deprived them of the carrying-trade and foreign commerce. America, deprived of her foreign commerce, was now necessitated to pay more than usual attention to domestic manufactures, in order to supply the absence of European but especially British manufactures. The following branches of manufacture were among the earliest established, and now supply, in several instances, the greater, and in all a considerable part of the consumption of the United States, namely: iron and ironware, manufactures of cotton, wool, and flax, hats, paper, printing-types, printed books, spirituous and malt-liquors, hemp-manufactures, gunpowder, window-glass, jewellery, clocks, lead, straw-bonnets, candles, leather, soap, and sugar. The great obstacles to the introduction and improvement of manufactures, in the present state of things, may be reduced to the three following, namely: high wages,—deficiency of capital,—and the want of the subdivision of labour. The first is occasioned by the scarcity of hands, and the great number requisite for agricultural purposes; America has not yet a surplus population for manufacturing purposes, nor can she have for a long time to come. While good land remains to be cleared and settled, people will continue to emigrate. While agriculture and clearing new lands continue to be still the most profitable way of employing capital, little of it will receive any other direction.

Cotton manufactures.] The suspension and non-intercourse acts of the American government, depriving the Americans of the usual market for their cotton wool, and of the accustomed importations of cotton-goods, induced them to pay redoubled attention to the cotton-manufacture, both to supply their own consumption, and to avail themselves as much as possible of the superabundance of the raw material. But this attempt was successful at first only for a short interval, being forced upon them by an unnatural state of circumstances. The return of peace crushed their cotton manufactures, and ruined nearly all the individuals concerned in these branches. The first cotton mill was erected in the state of Rhode Island, in 1791; another succeeded in the same state, in 1795; and two more, in Massachusetts, in 1803 and 1804. During the three succeeding years, ten more were erected in the state of Rhode Island, and one in Connecticut; making 15 cotton mills in all, erected before the year 1808, working at that time about 8,000 spindles, and producing 360,000 ℔s. of yarn annually. In 1808, the cotton mills were increased to 87 in number, before the end of that year; 62 of which (48 water and 14 horse-mills) were in operation, and worked at that time 34,000 spindles: 25 more were in operation, in 1811, making a total of 80,000 spindles. The capital vested in these mills was estimated at 4,800,000 d.: the cotton wool employed at 3,600,000 ℔s.; value, 720,000 d.: yarn spun, 2,880,000 ℔s.; value, 3,240,000 d.: and persons employed, 4,000; namely, 500 men, and 3,500 women and children. It must be observed, however, that the yarn made was very coarse, and the goods manufactured of the same description. The number of bales of cotton manufactured from 1800 to 1815 were as follows: 1800, 500 bales; 1805, 1,000 bales; 1810, 10,000 bales; 1815, 90,000 bales. The whole yearly value of the manufactures of cotton, in 1815, was 24,000,000 d., and of the capital employed in manufacturing cotton by machinery, 40,000,000 d. At that period the value of the manufactures of cotton in England might be estimated at 95,000,000 d. In 1835, the cotton manufactures in America amounted to from 45,000,000 to 50,000,000 d.; while those of England might be about 170,000,000 d. The capital embarked in manufacturing cotton by machinery within the States, in 1835, was estimated at 80,000,000 d. The following table exhibits the statistics of this manufacture in 1849:

States.	No. of estab. in operation.	Capital invested.	Value of raw material.	No. of hands employed.	Value of entire products.
Maine, . .	12	3,329,700 d.	1,573,110 d.	3,739	2,596,356 d.
New Hampshire, .	44	10,950,500	4,839,429	12,122	8,830,619
Vermont, . .	9	202,500	114,415	241	196,100
Massachusetts, .	213	28,455,630	11,289,309	28,730	19,712,461
Rhode Island, .	158	6,675,000	3,484,579	10,875	6,447,120
Connecticut, . .	128	4,219,100	2,500,062	6,186	4,257,522
New York, . .	86	4,176,920	1,985,973	6,320	3,591,989
New Jersey, . .	21	1,483,500	666,645	1,712	1,109,524
Pennsylvania, .	208	4,528,925	3,152,530	7,663	5,322,262
Delaware, . .	12	460,100	312,068	838	538,439
Maryland, . .	24	2,236,000	1,165,579	3,022	2,120,504
Virginia, . .	27	1,908,900	828,375	2,963	1,486,384
North Carolina, .	28	1,058,800	531,903	1,619	831,342
South Carolina, .	18	857,200	295,971	1,019	748,338
Georgia, . .	35	1,736,156	900,419	2,272	2,135,044
Florida, . .	—	80,000	30,000	95	49,920
Alabama, . .	12	651,900	237,081	715	382,260
Mississippi, . .	2	38,000	21,500	36	30,500
Louisiana, . .	—	—	—	—	—
Texas, . .	—	—	—	—	—
Arkansas, . .	3	16,500	8,975	31	16,637
Tennessee, . .	33	669,600	297,500	891	510,624
Kentucky, . .	8	239,000	180,907	402	273,439
Ohio, . . .	8	297,000	237,060	401	394,700
Michigan, . .	—	—	—	—	—
Indiana, . .	2	43,000	28,220	95	44,000
Illinois, . .	—	—	—	—	—
Missouri, . .	2	102,000	86,446	155	142,200
Iowa, . .	—	—	—	—	—
Wisconsin, . .	—	—	—	—	—
California, . .	—	—	—	—	—
District of Columbia, .	1	85,000	67,000	144	100,000
Total, . .	1,094	74,501,031	34,835,056	92,286	61,869,184

The following list shows the amount of dividends paid by the largest cotton-spinning and weaving establishments in New Hampshire, for a series of years:

	Capital.	Years.	Dividends.
Nashua company,	1,000,000 d.	10	85 per cent.
Jackson company,	480,000	10	120 ...
Stark company,	1,250,000	8	73 ...
Merrimac mills,	1,200,000	$2\frac{1}{8}$	— ...
Cocheco company,	1,300,000	10	40 ...
Great Falls company,	1,500,000	9	90 ...
Exeter company,	162,000	10	55 ...
Salmon Falls company,	1,000,000	4	41 ...
Newmarket company,	600,000	10	44 ...
Portsmouth steam mills,	537,000	$2\frac{1}{8}$	— ...
Columbian company,	180,000	21	133 ...
Pittsfield company,	150,000		2 div. in 13 years.
Portsmouth company South Berwick,	183,000	15	40 per cent.
Total, .	8,612,000 d.		

The average dividend is less than 6 50-100 per cent. per annum. If the difference between the cost of the stocks and their present market value be taken from the amount of the dividends, the rate will be less than 5 per cent. per annum. To show the great reduction that has taken place in the prices of cotton fabrics manufactured in the United States, and the vigilance required by our manufacturers to keep pace with their Transatlantic brethren, it may be stated that in 1828 Amoskeag tickings were sold at 27½ c. and 23 c. a yard, the same as are now sold at 11 c. and 9¼ c. Merrimac prints fell from 17 c., in 1837, to 9-28 c. in 1849. Cocheco prints, in 1837, were 14 14-100 c., and now 8 92-100 c. Printing-cloths, which in 1835 were sold at 10 c., are now at 5 c. to 6 c. The prices of Newmarket cottons in 1837 were 12, 13, and 11½ cents; they are now 7½, 6½, and 6 cents. The Boott and Suffolk drillings in 1836 were sold at 14 cents, and now at 7 cents. Neither has this reduction been sudden, but has gone on steadily to the present time. The following table indicates the price of cotton to the manufacturer, for the under-noted years; and also the cost of the cloth per pound during the same time. It embraces two qualities of cotton as well as of cloth:

	Fine cotton.	Coarse.	Fine cloth	Coarse.
1835,	20.5	18.36	35.39	28.12
1836,	21.6	19.36	43.9	29.22
1837,	19.8	17.44	43.86	26.92
1838,	13.28	11.55	36.19	19.67
1839,	16.5	14.53	34.08	23.66
1840,	11.	9.78	33.33	18.12
1841,	12.5	10.58	29.55	18.31
1842,	11.	9.2	28 21	16.90
1843,	8.	6.88	25.52	13.57
1844,	9.5	8.71	21.48	14.91
1845,	7.25	6.03	22.49	12.43
1846,	8.12	7.61	22.37	15.42
1847,	10.4	10.4	21.69	18.26
1848,	9.34	8.	23.62	15.12
1849,	8 to 11¾	7 to 11.6 mos.	20.36 mos.	14.65

Showing that the reduction in the price of cotton has been 10 and 9 c., while in the cost of cloth it has been 15.03, and 13.47 c. a pound, or rather less, if carried out for the whole of 1849.

Woollen manufactures.] In addition to the high prices of labour, a principal obstacle to the extension of woollen manufactures in America was for a length of time the want of wool, which was also very deficient in quality, little attention having been paid to the rearing of sheep. The woollen imported for consumption in 1826 was estimated at 8,000,000 d.; and it was then calculated that about two-thirds of the clothing, including hosiery, and of the house and table-linen worn and used by the inhabitants of the United States, who did not reside in cities, was the produce of family-manufacture. Mallory estimated the capital invested in the manufacturing of woollen, in 1827, at 40,000,000 d.; and the agricultural capital invested in raising the supply of wool, at 40,000,000 d. more. The following table shows the statistics of the woollen manufactures of the United States in 1849:—

States.	No. of estab. in operation.	Cap. invested.	Value of raw material.	No. of hands employed.	Value of entire products.
Maine,	36	467,600 d.	495,940 d.	624	753,300 d.
New Hampshire,	61	2,437,700	1,267,329	2,127	2,127,745
Vermont,	72	886,300	830,684	1,393	1,579,161
Massachusetts,	119	9,089,342	8,671,671	11,130	12,770,565
Rhode Island,	45	1,013,000	1,463,900	1,758	2,381,825
Connecticut,	149	3,773,950	3,325,709	5,488	6,465,216
New York,	249	4,459,370	3,838,292	6,674	7,030,604
New Jersey,	41	494,274	548,367	898	1,164,446
Pennsylvania,	380	3,005,064	3,282,718	5,726	5,321,866
Delaware,	8	148,500	204,172	140	251,010
Maryland,	38	244,000	165,568	362	295,140
Virginia,	121	392,640	488,899	668	841,013
N. Carolina,	1	18,000	13,950	30	23,750
S. Carolina,	—	—	—	—	—
Georgia,	3	68,000	30,392	78	88,750
Florida,	—	—	—	—	—
Alabama,	—	—	—	—	—
Mississippi,	—	—	—	—	—
Louisiana,	—	—	—	—	—
Texas,	1	8,000	10,000	8	15,000
Arkansas,	—	—	—	—	—
Tennessee,	4	10,900	1,675	17	6,310
Kentucky,	25	249,820	205,287	318	318,819
Ohio,	130	870,220	578,423	1,201	1,111,027
Michigan,	15	94,000	43,402	129	90,242
Indiana,	33	171,545	120,486	246	205,802
Illinois,	16	154,500	115,367	178	206,572
Missouri,	1	20,000	16,000	25	56,000
Iowa,	1	10,000	3,500	7	13,000
Wisconsin,	9	31,225	32,630	25	87,992
California,	—	—	—	—	—
District of Columbia,	1	700	1,630	2	2,400
Total,	1,559	28,118,650 d.	25,755,988 d.	39,252	43,207,555 d.

Iron manufacture.] The iron manufacture of this country is one of the most important branches of domestic industry. The census returns of the manufacture of iron castings exhibit the following results:—

	Establishments.	Capital.	Tons pig iron.	Products.
Alabama,	10	216,625 d.	2,348	271,126 d.
California,	1	5,000	75	20,740
Columbia Dist.,	2	14,000	545	41,696
Connecticut,	60	580,800	11,396	981,400
Delaware,	13	373,500	4,440	267,462
Georgia,	4	35,000	440	46,200
Illinois,	29	260,400	4,818	441,185
Indiana,	14	82,900	1,968	149,430
Iowa,	3	5,500	81	8,500
Kentucky,	20	502,200	9,781	744,316
Louisiana,	8	205,000	1,660	312,500
Maine,	25	150,100	3,591	265,000
Maryland,	16	359,100	7,220	685,000
Massachusetts,	68	1,499,050	31,134	2,335,635
Michigan,	63	195,450	2,494	279,697
Mississippi,	8	100,000	1,197	117,400
Missouri,	6	187,000	5,100	336,495
N. Hampshire,	26	232,700	5,673	371,710
New Jersey,	45	593,250	10,866	686,430
New York,	323	4,622,482	108,945	5,921,980
N. Carolina,	5	11,500	192	12,867
Ohio,	183	2,063,650	37,555	3,069,350
Pennsylvania,	320	3,422,924	69,501	5,354,881
Rhode Island,	20	428,800	8,918	728,705
S. Carolina,	6	185,700	169	87,683

Tennessee,	16	139,500	1,682	264,325
Texas,	2	16,000	250	55,000
Vermont,	26	290,720	5,279	460,831
Virginia,	54	471,160	7,114	674,416
Wisconsin,	15	116,350	1,371	216,195
Total,	1,391	17,416,361 d.	345,553	25,108,155 d.

Whale fishery.] The importance of this fishery to the United States will be shown by the annexed statement of the vessels employed in the trade, and the quantities of oil brought by them to each port in 1850; as also for the previous nine years. The towns now engaged in the trade are, with the exception of New York, all in the New England states, and include those ports from whence whaling was carried on in the last century, which, though first commenced by the colonists in boats and in sight of their own homes, now employ upwards of 140,000 tons of shipping in every quarter of the globe:—

	Ships.	Brigs.	Schs.	Barrels Sperm.	Barrels Whale.
New Bedford,	65	0	0	38,059	92,144
Nantucket,	10	1	0	16,095	980
Fairhaven,	9	0	0	8,561	9,979
New London,	14	0	0	2,149	28,862
Sag Harbour,	11	1	0	635	25,205
Greenport,	1	0	0	500	800
Westport,	6	1	0	3,697	333
Boston,	4	2	0	2,680	200
Providence,	2	0	0	107	3,353
Edgartown,	1	0	0	2,000	150
Warren,	1	0	0	1,035	0
New York,	1	1	1	2,054	1,310
Mattapoisett,	3	1	0	2,702	81
Mystic,	3	0	0	500	2,500
Stonington,	7	0	0	766	14,927
Provincetown,	0	5	15	2,751	165
Holmes' Hole,	3	0	0	982	4,890
Sippican,	1	0	0	43	1,453
Cold Spring,	1	0	0	0	1,660
Wareham,	1	0	0	200	2,800
Yarmouth,	0	1	0	68	13
Dartmouth,	0	1	0	33	7
Truro,	0	1	0	130	0
Beverly,	0	2	0	410	0
Total,	144	17	16	86,157	191,752
Receipts in 1849,	165	9	14	99,433	256,183
1848,	178	12	6	108,531	243,876
1847,	214	28	11	121,410	320,545
1846,	158	30	11	92,877	219,763
1845,	207	19	16	158,484	274,843
1844,	—	—	—	138,585	267,082
1843,	104	28	13	167,134	205,861
1842,	162	38	7	163,697	163,816
1841,	171	42	9	157,643	205,164
1840,	165	42	6	156,445	203,441

Whalebone also forms an important item in the products of the fisheries. The quantities imported into the United States during ten years have been as follows:—

1850,	2,242,012 lbs.
1849,	1,990,640
1848,	1,755,892
1847,	3,450,124
1846,	2,723,939
1845,	3,116,100
1844,	3,015,145
1843,	1,933,321
1842,	1,329,234
1841,	1,942,885

Domestic exports.[The following table shows the value of the exports of domestic produce, &c., for the year ending 30th June, 1852:—

Produce of the sea,	2,282,342 d.
Produce of the forest,	7,864,220
Produce of agriculture and of animals,	6,323,439
Vegetable food,	19,886,588
Cotton,	87,965,732
Tobacco,	10,031,283
Hemp,	18,649
Other agricultural products,	780,732
Miscellaneous,	6,957,280
Cotton piece goods,	7,672,151
Miscellaneous,	46,815,811
Articles not enumerated,	4,073,434 d.
Total amount,	192,368,984

The subjoined table shows the value of the exports of domestic produce each year since 1840:—

Year ending on	Specie and bullion.	Produce and manfs.	Total.
1840, 30th Sept.	2,235,073	111,660,561	113,895,634
1841,	2,746,486	103,636,236	106,382,722
1842,	1,170,754	91,799,242	92,969,996
1843, 9 months, June 30,	107,429	77,686,354	77,793,783
1844, year,	183,405	99,531,744	99,715,179
1845,	844,446	98,455,330	99,299,776
1846,	423,851	101,718,042	102,141,893
1847,	62,620	150,574,844	150,637,464
1848,	2,700,412	130,203,709	132,904,121
1849,	956,874	131,710,081	132,666,955
1850,	2,046,679	134,900,233	136,946,912
1851,	18,069,580	178,620,138	196,689,718
1852,	37,437,837	154,931,147	192,368,984

The exports of the U. S. for the fiscal year from 30th June 1853 to 30th June 1854 were in detail as follows:—

I. PRODUCT OF THE SEA.

Fisheries:—			
Oil, spermaceti,	1,418,845 d.		
Oil, whale and other fish,	223,247		
Whalebone,	1,063,705		
Spermaceti candles,	112,600		
Dried or smoked fish,	371,607		
Pickled fish,	89,409		
		3,279,413 d	

II. PRODUCT OF THE FOREST.

Wood:—			
Staves, shingles, hewn timber, &c.,	2,578,149 d.		
Other lumber,	123,743		
Masts and spars,	129,628		
Oak bark and other dye,	118,894		
Manufactures of wood	2,294,122		
Naval stores, tar, pitch, rosin, and turpentine,	1,406,488		
Ashes, pot and pearl,	334,311		
		6,985,345	
Ginseng,		133,813	
Skins and furs,		796,101	
			7,915,219

III. PRODUCT OF AGRICULTURE.

Of Animals:—			
Beef, tallow, hides, and horned cattle,	2,214,554 d.		
Butter and cheese,	862,343		
Pork (pickled), bacon, lard, and live hogs,	6,202,324		
Horses and mules,	246,731		
Sheep,	17,808		
Wool,	26,567		
		9,570,327	
Vegetable food:—			
Wheat,	4,354,403		
Flour,	14,783,394		
Indian corn,	1,374,077		
Indian meal,	709,974		
Rye meal,	34,186		
Rye, oats, and other small grains, and pulse,	165,824		
Biscuit or shipbread,	454,020		
Potatoes,	152,569		
Apples,	107,283		
Rice,	1,657,658		
		23,793,388	
			33,363,715
Cotton,			109,456,404
Tobacco,			11,319,319
Hemp,			18,195
Other agricultural product:—			
Flax-seed,		7,719	
Hops,		40,054	
Brown sugar,		33,856	
Indigo,		36	
			81,663

Manufactures:—	
Wax,	113,602
Refined sugar,	375,770
Chocolate,	10,230
Spirits from grain,	141,173
... molasses,	329,381
Molasses,	17,582
Vinegar,	20,443
Beer, ale, porter, and cider,	64,677
Linseed oil,	15,468
Spirits of turpentine,	347,492
Household furniture,	714,556

Hats,		91,261	
Saddlery,		48,229	
Tallow candles and soap,		681,342	
Snuff and tobacco,		1,671,500	
Leather, boots and shoes,		673,708	
Cables and cordage,		103,216	
Gunpowder,		180,048	
Salt,		119,729	
Lead,		5,540	
Coaches and other carriages,		184,497	
Iron—pig, bar, and nails,		181,998	
... castings,		220,420	
... all manufactures of,		2,097,234	
Copper and brass, all manufactures of,		108,205	
Medical drugs,		327,073	
			8,844,404
Cotton piece goods:—			
Painted or coloured,	1,086,167		
Uncoloured,	6,926,485		
Cotton thread and yarn,	22,594		
All manufactures of,	733,648		
		8,768,894	
Flax and hemp:—			
Cloth and thread,		2,924	
Bags and all manufactures of,		13,860	
Wearing apparel,		239,733	
Earthen and stoneware,		53,685	
Combs and buttons,		31,395	
Brushes of all kinds,		6,612	
Billiard tables and apparatus,		1,673	
Umbrellas, parasols, and sun-shades,		6,183	
Morocco and other leather, not sold by the pound,		6,448	
Fire engines and apparatus,		9,662	
Printing presses and types,		32,250	
Musical instruments,		52,397	
Books and maps,		142,604	
Paper and stationery,		122,212	
Paints and varnish,		83,020	
Glass,		170,561	
Tin,		22,988	
Pewter and lead,		14,064	
Marble and stone,		47,628	
Gold and silver, and gold leaf,		11,873	
Gold and silver coin,		23,548,535	
Artificial flowers and jewellery,		66 397	
Trunks,		27,148	
Brick and lime,		32,625	
			33,515,361
Coal,			336,003
Ice,			175,056
Articles not enumerated:—			
Manufactured,			3.788,700
Raw produce,			1,324,205
Total,			213,417,697 d.

About two-thirds of the total value, it will be perceived, is composed of the four great staple products—cotton, wheat, tobacco, and gold; but nearly one-half of the whole is cotton. The exports in 1858 amounted to 293,758,279 d., in which cotton alone amounted to 131,386,661 d.—The subjoined table gives the export of domestic manufactured articles, and of wheat, for a period of 21 years:—

	Manufactures.	Wheat.
1831,	5,804,199 d.	10,461,715 d.
1832,	5.42[illegible],014	4,974,123
1833,	6,888.229	5,642.802
1834,	6,013,385	4,560,797
1835,	6,937.999	4,446,182
1836,	6,915,748	3,574.561
1837,	7,811,848	3,014,415
1838,	9,010,358	3,617.724
1839,	8,019,271	7,069,361
1840,	10,613,767	11,779,098
1841,	10,775,500	0,502,527
1842,	9,769,851	8,292,308
1843,	7,354,726	4,027,182
1844,	9,680,534	7,232,898
1845,	11,007,121	5,735,372
1846,	11,482,854	13,360,644
1847,	11,847,288	32,183,161
1848,	14,711,175	15,863,284
1849,	12,947,705	73,037,430
1850,	17.145,203	7,742,315
1851,	22,209,262	11,543,063
Total,	212,376,329 d.	186,740,962 d.

This statement shows an excess of manufactured articles over the exports of wheat of nearly 26,000,000 d. in the space of 21 years, being an annual average of more than 1,200,000 d. The annexed statement of the amount, value, and average price per pound of cotton exported from the United States, from 1821 to 1858 inclusive, is valuable as showing the fluctuations in the price of that important staple, also the increase both in the quantity exported and its aggregate value. This vast increase is to be mainly attributed to the rapid growth of the cotton manufactories of Great Britain, as nearly two-thirds of the whole quantity exported has been brought to this country.

	Sea Island. Lbs.	Other. Lbs.	Total. Lbs.	Value. Dollars.	Aver. per lb. Cts.
1821,	11,344,066	113,549,339	124,893,404	20,257,048	16·2
1822,	11,250,635	133,424.460	144,675,095	24,035,058	16·6
1823,	12,136,688	161,586.582	173,723,270	20,445,520	11·8
1824,	9,525,722	132.843.941	142.369,663	21,947,401	15·4
1825,	9,665,278	166,784,629	176,449.907	36,846,649	20·9
1826,	5.972.852	198,562,563	204,535.415	25,025,214	12·2
1827,	15,140,798	299,169,317	294,310,115	29,359,545	10·
1828,	11,288,419	179,302,044	210.590,463	22,487,229	10·7
1829,	12,833,307	252.003,879	264.837,186	26,575,311	10
1830,	8.147,165	290,311.937	298,459.102	29,674.883	9·2
1831,	8.311.762	268,668,022	216,979,784	25,289,492	9·1
1832,	8,743.373	313,471,749	322,215,122	31,724,682	9·8
1833,	11,142,987	313.565,617	324,698,604	36,191,105	11·1
1834,	8.085.937	376,631,970	384,717.907	49,448,402	12·8
1835,	7,752,736	379.606,256	387.358,992	64,961,302	16·8
1836,	7,849.597	415,781.710	423,631,307	71,284,925	16·8
1837,	5.286,971	438.924,566	444.211,537	63.240,102	14·2
1838,	7,286.340	588,665,957	595.952,297	61,556,811	10·3
1839,	5,107,404	408,516,808	413.621,313	61,238,982	14 8
1840,	8,779,609	735,161,392	743,941.061	69,870.307	8·5
1841,	6,257.424	523,966,676	530.204,100	54,330,341	10·2
1842,	7,254.099	577.462.918	584,711,017	47,593,464	8·1
1843,	7,515,079	784,782,027	792,297,105	49.119,805	6·2
1844,	6.099.076	657,534.379	663,633,455	54,063.501	8·1
1845,	9.389,625	863,515,371	872,905,996	51,739,643	5·92
1846,	9.388,533	538,169,522	547,558.055	42,767,341	7·81
1847,	6,293.973	520,925,985	527.219,958	53.415.848	10·34
1848,	7,724,148	806,556.283	814,274,431	61,998,294	7·61
1849,	11.969,259	1,014,633,010	1,026,603,269	66.396,967	6·4
1850,	8.236,403	627.145,141	635,381,604	71,984,616	11·3
1851,	8,299,655	918.937,433	927,237,089	112,315,317	12·11
1852,			1,093,230,639		8·0
1853,			1,111,570,370		9·8
1854,			987,833.106		9·5
1855,			1,008,424,601		8·74
1856,			1.351,431,701		9·49
1857,			1.048,282,475		12·55
1858,			1,118,624,012		11·70

General Imports and Exports.] In 1774, the whole of the American commerce, both exports and imports, amounted only to 13,000.000 d. In 1784, it had risen to 21,500,000 d.; in 1794, to 67,043,725 d. In 1806, American commerce reached its acme during the period of the great war: the exports being 103,000,000 d., of which 60,283,236 d. were for foreign produce, exclusive of freight and charges. Her imports, in the same year, amounted to 100,000,000 d., making a total of £17,500,000. This extensive foreign commerce employed upwards of 1,500,000 tons of shipping, and 100,000 sailors. The whole exports of America, in 1826, amounted to 78,000,000 d., of which 6,000,000 were domestic manufactures, being more than double the exports of domestic manufactures in 1821. The exports of domestic produce, in 1828, amounted to £11,050,000; and of manufactures to £1,600,000; the exports of foreign produce and manufactures to £5,398,750. The total value of the exports of the United States, for the year ending the 30th September, 1828, was £18,048,750. The total value of the exports of domestic produce, in the year ending 30th June, 1846, was 102,141,893 d.; of which 23,507,483 d. was carried in foreign, and the remainder in American vessels. On the other hand, the value of the imports of foreign merchandise was 121,691,797 d., of which 11,346,623 d. was afterwards exported; leaving the balance 110,345,174 d., and showing an excess of imports over exports for the year of 8,203,281 d. Of the total imports, 15,683,624 d. was brought in foreign vessels. The proportions in which the trade of the country was carried on with the various parts of the world, in 1846, appears from the following summary of the principal items of exports to, and imports from, different countries:—

	Exports.	Imports.
Great Britain and Ireland,	45,501,957 d.	45,150,025 d.
Great Britain, dependencies of,	12,691,358	4,516,397
France and her dependencies,	14,277,286	24,330,882
South American states,	6,622,032	13,408,502
Spain and her dependencies,	5,931,310	12,376,482
Hanse towns,	4,008,315	3,149,864
Holland and her dependencies,	2,469,883	1,971,680
Belgium,	1,632,607	836,372
China,	1,178,188	6,593,881
Hayti,	1,114,013	1,542,962
Austria,	1,104,468	379,719
Denmark,	1,057,198	753,927
Italy,	942,263	1,189,786
Russia,	535,388	1,570,054
Sweden and her dependencies,	500,913	730,150
Prussia,	396,210	31,584
Sicily,	319,441	513,235
Sardinia,	283,283	—
Portugal and her dependencies,	192,581	547,474
Turkey,	126,193	760,998
Africa, generally,	544,467	475,045
Asia, generally,	202,232	361,988
South Seas and Pacific ocean,	278,765	153,029
Sandwich islands,	—	243,034
Marocco,	—	4,554

In the year ending 30th June 1858, the total exports to Great Britain and Ireland amounted to 168,095,848 d. The total exports to all countries, in that year, amounted in value to 321,644,421 d., whereof 293,758,279 d. consisted in domestic produce. The item of grain amounting to 35,924,948 d.; of tobacco, 17,009,767 d.; of cotton, 131,386,681 d. The amount of goods and merchandise imported into and exported from the states during 1851 with foreign countries, and of the relative value of each article of merchandise was as follows:

	Imports.	Exports.
United Kingdom,	£19,551,643	£24,572,086
France,	6,607,407	5,885,864
Cuba,	3,551,444	1,359,192
British North America,	1,394,400	2,503,109
Hanse Towns,	2,085,076	1,259,885
Brazil,	2,401,105	781,857
China,	1,471,905	517,768
Holland and Belgium,	922,987	1,051,496
Spain,	450,536	1,157,206
British West Indies and Guayana,	218,351	968,291
Italy, with Sardinia and Sicily,	600,130	467,597
Chili,	569,739	394,855
Argentine republic,	680,288	223,910
British East Indies,	695,070	143,414
New Granada,	144,918	633,504
Hayti,	393,743	378,602
Porto Rico,	516,735	212,212
Venezuela,	495,895	217,609
Mexico,	375,995	329,538
Austria,	152,248	520,097
Russia,	290,163	335,769
Other countries,	1,477,081	1,583,640
Total,	£45,046,859	£45,497,501

Imports in 1851.	Value.
Silk manufactures,	£5,838,806
Cotton do.,	4,775,228
Woollen do.,	4,064,023
Iron and steel, raw and manufactured,	3,932,658
Sugar,	2,884,571
Coffee,	2,677,306
Linen manufactures,	1,878,928
Hides and skins, raw,	1,242,675
Specie and bullion,	1,136,165
Tea,	999,584
Tin and manufactures thereof,	827,301
Wool,	798,574
Molasses,	772,413
China, earthenware, &c.,	695,963
Tobacco, raw and manufactured,	643,609
Leather,	586,598
Watches and parts thereof,	585,281
Copper and manufactures thereof,	562,241
Spirits,	547,471
Other goods,	9,597,464
Total,	£45,046,859

Exports in 1851.

Home Produce and Manufactures:—	Value.
Cotton, raw,	£23,399,024
Gold and silver coin,	3,764,496
Grain and flour,	3,032,549
Tobacco, raw and manufactured,	2,158,916
Cotton manufactures,	1,508,584
Wood, manufactured and unmanufactured,	979,229
Bacon, pork, &c.,	910,003
Fish, fish oil, bone, &c.,	686,394
Iron manufactures,	469,937
Rice,	452,276
Beef, tallow, hides, and cattle,	352,075
Other goods,	3,263,540
	£40,977,023
Foreign Produce, &c.:—	
Bullion and specie,	£2,375,661
Grain and flour,	366,104
Tea,	276,794
Other goods,	1,501,919
	£4,520,478
Total,	£45,497,501

The following table exhibits the imports, exports, custom-house duties, average percentage of duties, and proportion of imports per head during each 10 years since the commencement of the government, and for the last five years, ending June 30, 1854. It shows the increased rate of consumption in 1850-1854, as compared with the previous 10 years, when the business of the country was managed with more discretion than since the gold discoveries in California:—

	Imports.	Exports.	Excess Imports.
1789 to 1799,	500,592,686 d.	410,997,338 d.	89,595,348 d.
1800 — 1809,	933,516,278	751,150,299	182,365,979
1810 — 1819,	819,069,274	586,958,905	232,110,369
1820 — 1829,	802,206,480	761,590,913	40,615,567
1830 — 1839,	1,266,411,483	1,034,105,465	232,306,018
1840 — 1849,	1,196,786,983	1,240,747,504	43,960,521
1850 — 1854,	1,184,071,983	810,380,606	69,128,996
Total,	6,712,655,167 d.	5,595,931,030 d.	802,161,756 d.

	Customs.	Average per cent.	Pop.	Proportion of imports to each inhabitant each year.
1789 to 1799,	50,321,525 d.	10½	3,929,827	12·74 d.
1800 — 1809,	120,957,109	13	7,239,814	12·90
1810 — 1819,	157,411,864	19¼	9,638,131	8·50
1820 — 1829,	191,606,399	23 7-8	12,866,920	6·23
1830 — 1839,	213,126,802	16 5-6	17,069,453	7·42
1840 — 1849,	217,497,496	18	23,267,498	5·14
1850 — 1854,	259,162,591	22	26,511,384	8·93
Total,	1,210,083,786 d.	18		

The subjoined table shows the exports of the produce and manufactures of the U. S. in 1851, not only as a total, but also to each particular country. It will be seen that nearly two-thirds of the whole amount are sold to the British empire:—

	In American vessels.	In Foreign vessels.	To each country.
Russia,	1,187,116 d.	278,588 d.	1,465,704 d.
Prussia,	5,152	75,317	80,469
Sweden and Norway,	198,269	562,531	760,800
Swedian W. Indies,	58,924	2,233	61,157
Denmark,	2,913	89,344	92,257
Danish W. Indies,	804,909	97,778	902,687
Hanse Towns,	550,542	4,855,414	5,405,956
Holland,	711,724	1,199,391	1,911,115
Dutch E. Indies,	168,226	36,204	204,430
Do. W. Indies,	34,397	25,501	366,898
Do. Guayana,	85,491	—	85,491
Belgium,	2,335,077	374,316	2,709,393
England,	2,200,571	32,921,350	105,121,921
Scotland,	2,004,306	1,806,697	3,811,003
Ireland,	203,335	395,353	598,688
Gibraltar,	91,606	86,288	177,904
Malta,	60,261	3,800	64,061
British E. Indies,	454,670	58,236	512,906
Cape of Good Hope,	158,666	3,225	161,891
Mauritius,	—	16,882	16,882

Honduras,	190,507	23,299	213,806
British Guayana,	384,266	156,288	540,554
Do. West Indies,	2,292,923	1,650,637	3,943,560
Canada,	3,585,571	2,250,263	5,835,834
Brit Amer. Colonies,	492,627	2,731,926	3,224,553
France on Atlantic,	23,864,292	702,775	24,567,057
Do. Mediterranean,	588,172	146,846	735,018
French W. Indies,	217,319	72,260	289,579
Miquelon and other French fisheries,	3,715	—	3,715
French Guayana,	45,693	—	45,693
Bourbon,	16,607	3,246	19,853
Spain on Atlantic,	759,853	198,860	958,713
Do. Mediterranean,	87,638	4,369,693	4,457,331
Teneriffe and other Canaries,	8,765	4,775	13,540
Manilla and Philippine Islands,	125,544	—	125,544
Cuba,	5,039,718	199,558	5,239,726
Other Span. W. Indies,	861,286	100,124	961,410
Portugal,	83,945	83,397	167,342
Madeira,	68,474	26,115	94,589
Fayal and other Azores,	15,411	4,829	20,240
Cape de Verdes,	57,476	—	57,476
Italy generally,	906,791	830,043	1,736,834
Sicily,	3,305	38,438	41,743
Sardinia,	136,361	174,527	310,888
Trieste & other Austrian Adriatic ports,	1,465,822	793,751	2,265,573
Turkey, Levant, &c.,	162,204	—	162,204
Hayti,	1,380,447	298,925	1,679,372
Mexico,	916,173	98,517	1,014,690
Central Republic of America,	217,691	5,611	223,202
New Grenada,	2,413,568	94,133	2,507,701
Venezuela,	757,003	97,776	854,779
Brazil,	2,841,983	286,973	3,128,956
Cisplatine Republic,	25,804	6,907	32,711
Argentine do.,	463,535	196,317	659,852
Chili,	1,581,798	27,079	1,608,877
Peru,	186,320	63,440	249,760
China,	2,111,029	44,916	2,155,945
West Indies generally,	68,761	8,175	76,936
S. America do.,	36,119	—	36,119
Asia do.,	70,586	—	70,586
Africa do.,	1,175,049	70,312	1,245,361
South Seas and Pacific Ocean,	601,146	—	601,146
Total,	137,934,539 d.	58,755,179 d.	196,689,718 d.

The *New York Herald* contains a very interesting table of the movements of bank-operations in the United States, and the progress of foreign trade in each of the last 26 years, during which four different tariffs have been in operation,—two protective, one compromise, and one revenue tariff. The bank-returns are made up to the 1st January, and the trade-returns up to the 30th of June each year. The table is as follows:—

	Loans and Discounts.	Imports.	Exports.
1830,	200,451,214 d.	70,070,020 d.	73,810,508 d.
1831,	215,631,784	103,191,124	81,310,583
1832,	238,165,587	101,029,266	87,176,943
1833,	267,951,684	108,118,311	90,140,433
1834,	324,119,499	120,521,099	104,336,073
1835,	365,163,834	149,895,742	121,693,577
1836,	457,506,080	189,980,035	128,663,040
1837,	525,115,702	140,989,217	117,419,376
1838,	485,631,687	113,717,404	108,486,616
1839,	492,278,015	162,092,132	121,028,416
1840,	462,896,523	107,141,519	132,085,946
1841,	386,487,662	127,946,177	121,851,803
1842,	323,957,569	100,162,087	104,691,534
1843,	254,544,937	64,753,799	84,346,480
1844,	264,905,814	108,435,035	111,200,046
1845,	288,617,131	117,254,564	114,646,606
1846,	312,114,404	121,691,797	113,488,516
1847,	310,282,945	146,545,638	158,648,622
1848,	344,476,582	154,998,928	154,032,131
1849,	332,323,195	147,857,439	145,755,820
1850,	364,204,078	178,138,318	151,898,720
1851,	413,756,799	220,779,355	218,388,011
1852,	475,500,000	212,613,282	209,641,625
1853,	550,500,000	267,978,647	230,452,250
1854,	607,287,428	304,562,381	278,241,064

The following table exhibits the value of the exports of domestic produce from each state and territory in 1840 and in 1850; and the total exports of domestic and foreign produce, and of imports in 1850:—

	Exports of domestic produce. 1840.	1850.	Tot. exports of American and foreign produce.	Total imports in 1850.
Maine,	1,009,910 d.	1,536,818 d.	1,556,912 d.	856,411 d.
N. Hampshire,	20,761	8,722	8,927	49,079
Vermont,	305,150	404,749	43,906	463,092
Massachusetts,	6,268,158	8,253,473	10,680,763	30,374,684
Rhode Island,	203,006	206,299	215,265	258,303
Connecticut,	518,210	241,262	241,930	372,390
New York,	22,676,609	41,502,800	52,712,789	111,123,524
New Jersey,	14,883	1,655	1,655	1,494
Pennsylvania,	5,736,456	4,049,464	4,501,606	12,066,154
Delaware,	37,001	—	—	—
Maryland,	5,495,020	6,589,481	6,967,353	6,124,201
Dist. Columbia,	751,429	80,388	80,588	59,819
Virginia,	4,769,937	3,413,158	3,415,646	426,599
N. Carolina,	387,484	416,501	416,501	323,692
S. Carolina,	9,981,016	11,446,892	11,447,800	1,933,785
Georgia,	6,862,959	7,551,943	7,551,943	636,964
Alabama,	12,856,694	10,544,858	10,544,858	865,362
Mississippi,	—	—	—	—
Louisiana,	32,998,059	37,698,277	38,105,350	10,760,499
Ohio,	991,954	217,532	217,632	582,504
Kentucky,	—	—	—	190,987
Tennessee,	—	—	—	27,966
Michigan,	162,229	132,045	132,045	144,102
Florida,	1,895,634	2,607,968	2,623,624	95,709
Missouri,	—	—	—	359,643
Illinois,	—	17,649	17,699	15,705
Texas,	—	24,958	24,958	25,650
Total,	113,895,634	136,946,912	151,898,720	178,138,310

Inspection of these tables shows that in most of the states of the union the domestic exports have steadily increased. In the western states of Ohio and Michigan we find a great increase in the exports to the Canadas. The greatest increase of exports has been in New York. In 1858, of the total value of exports from the states, amounting to 324,644,421 d., 108,340,924 d. were from New York, and 88,875,995 d. from Louisiana.—A Washington statistician, while estimating the comparative resources of the free and the slave-holding states of the Union, or of the western and the southern states, calculated the value of such articles of agriculture produced in the south as will always command a foreign market, for the three years, from 1847 to 1849, as follows:—

	1849.	1848.	1847.
Cotton,	95,250,000 d.	74,620,000 d.	72,905,000 d.
Tobacco, .	6,616,741	8,756,360	11,008,200
Rice, .	3,841,964	3,575,895	3,091,215
Naval stores, .	1,624,190	1,864,319	1,798,612
	107,332,895	88,816,574	88,803,027
To the above may be added sugar and molasses, .	18,417,500	16,486,000	22,746,430
Total agricultural productions of the slave states, .	125,756,395	105,302,574	111,549,457

Of which there were exported to foreign countries, during the same period, according to official returns, viz.:—

	1849.	1848.	1847.
Cotton, .	66,396,967 d.	61,998,294 d.	53,415,878 d.
Tobacco, .	5,804,207	7,551,122	7,242,086
Rice, .	2,569,362	2,331,824	3,605,896
Naval stores, .	845,161	752,303	759,221
	75,615,700 d.	72,633,543 d.	65,023,051 d.

The value of imports into the U. S. for three years, distinguishing the proportions received at the north and at the south, was on the same authority, as follows:

	1849.	1848.	1847.
New York, .	92,737,497 d.	94,525,141 d.	84,167,352 d.
Boston, .	26,327,874	28,647,707	34,477,008
Other Northern ports,	14,716,030	14,200,043	11,161,667
Total North,	133,780,361 d.	137,372,891 d.	129,806,027 d.

New Orleans, .	8.077.910	9,380,439	9,222.969
Charleston, .	1.310.591	1,485,299	1,580.658
Other Southern ports,	4,688,577	6,760,298	5,934,978
Total South,	14,077,078 d.	17,626,036 d.	16,738,605 d.

The following is a summary of the value of exports of such articles as were produced by the free states, or from abroad by the capital of such as are identified with the interests of the free states, viz:—

	1849.	1848.	1847.
Fisheries. .	512,177 d.	718.797 d.	795.850 d.
Oil and Whalebone,	1,876,074	1,075.327	2,480,716
Candles, .	159,403	186,839	191.467
Skins, furs, and ginseng, .	839,194	770,427	811,612
Lumber and articles manufactured from wood, .	3,718,033	5,066.877	3,806.341
Ashes, .	515,603	466.477	618.000
Provisions—estimate,	10.000.000	8,800,000	7,300,000
Breadstuffs. —	19.000.000	18,000.000	42,000.000
Miscellaneous, —	1,800,000	1,500,000	1,200,000
	38,420,484	36,584,744	59,203,986
Add manufactured goods, — estimate,	12,000,000	11,000,000	9,000,000
	50,420,484 d.	48,584,744 d.	68,203,984 d.

Another authority in Washington gives the following as the value of southern agricultural products exported from the U, S. for the under-named years:—

	1845.	1847.	1849.
Naval stores, .	1,085,712 d.	759,221 d.	845,164 d.
Rice, . .	2.564,991	3,605,896	2,569,362
Tobacco, .	8.478,270	7.242,086	5,804.207
Cotton, .	42,767,341	53,415,848	66,396.967
Sugar, . .	7,235	25,483	24,906
Southern states,	54.903,549	65.048.534	75,640,606
Total exports,	102,141,893	150,637,464	132,666,955
Northern states,	47,238,344 d.	85,588,930 d.	55,026,349 d.

An able political pamphlet known to be written by a distinguished citizen of Virginia, and designed to exhibit the superiority in amount and value of the commercial products of the southern states over those of the north, and the superior commercial advantages of the south, states the exports of southern and northern domestic produce as follows:

I. SOUTHERN EXPORTS.

	Cotton.	Tobacco.	Rice.
1790-1800, .	19.105,755 d.	49,311,260 d.	27,554,109 d.
1801-1810, .	88.073,500	54,773,730	21.592,660
1811-1820, .	156,726,604	50,051,999	23.021,144
1821-1830, .	256,554,294	57,788.204	20,070,244
1831-1840, .	528,506,110	74,457,223	22.257.791
1841-1845 (June 30),	256,846,765	42,605.511	9,886,144
1846, . .	42,767,341	8,478.270	2,564,991
1847, . .	53,415,848	7,242,086	3,605,896
1848, . .	61.998,294	7,551,122	2,331,824
1849, . .	66,396,967	5,804,207	—

II. TOTALS OF SOUTHERN EXPORTS FOR THE SAME PERIODS.

	Total of cotton, tobacco, and rice.	Southern share of the other domestic exports in ratio of pop.
1790-1800, .	95,971,124 d.	104.071,190 d.
1810, .	164,439.890	126,860,890
1820, .	229,799.747	111,706,852
1830, .	334,412,742	97,488,201
1840, .	625,191,124	118,215.517
1845, .	309,338,410	71.041,660
1846, .	53,810,603	21,000,000
1847, .	64,263,830	37,736,170
1848, .	71,881,240	26.203.943
1849, .	75,000,000	24,500,000

III. TOTAL SOUTHERN AND NORTHERN EXPORTS.

	Total exports of southern produce.	Total exports of northern produce.
1790-1800, .	200,042.314 d.	111,248,511 d.
1801-1810, .	291,300,780	133,071,866
1820, .	341,506,599	121,194,689
1830, .	431,900,943	104,204,080
1840, .	743.407,467	149,482,842
1845, .	380,380,070	95,781,386
1846, .	74.810.603	27,831,290
1847, .	102.000,000	48,637.464
1848, .	98,085,183	34,878.938
1849, .	99,500,000	32,210,081

IV. IMPORTS.

	Actual imports into the South.	North.
1790-1800, .	244,000.000 d.	370,845,454 d.
1801-1810, .	232,000,000	695,663.510
1811-1820, .	202,000,000	606.119,274
1821-1830, .	135.475.117	663,158.283
1831-1840, .	220.382,975	1.082,295,107
1846, .	13,435.111	108,256,686
1848, .	17,628,137	137,370,791

The following is an extract from a table showing the gain of the northern imports and exports of foreign produce and exports of domestic produce, and the consequent loss upon southern capital employed in product and consumption:—

	On exports of domestic produce.	On imports.	On exports of foreign produce.
1790-1800, .	29,500.660 d.	153.000,000 d.	64,078,520 d.
1801-1810, .	29,225,666	399,680.000	100,444,140
1811-1820, .	65,645,323	389,119,274	68,484.313
1821-1830, .	121,434,846	508,158,283	158.303,657
1831-1840, .	66.976,981	864,293,107	128.062.554
1846, .	17,382.459	75,956,686	7,631.940
1848, .	22,150,613	97,446,825	13,875,389

Trade with Great Britain.] In 1740 the imports of New York from Great Britain were £72,390, and the exports £171,000; but in 1836 the value of the imports in that city was estimated at £23,000,000, of which probably above £11,000,000 consisted of British manufactures and commodities. The 16,000,000 of our American brethren were better customers for our manufactures than France and Germany with a pop. of 70,000,000, and as the latter countries are approaching or have reached a state in which the progress of manufactures is more strikingly displayed than that of agriculture, they are daily becoming our rivals in the manufacturing markets, while in the United States industry is still most profitably employed in developing the resources of agriculture, and we, by our advancement in non-agricultural industry and arts, materially assist them in the rapid creation of wealth from the cultivation of the soil. The following facts show the proportions in which the commercial interests of the two countries were blended: In 1821 the proportion which the trade with England bore to the foreign trade of the United States was 35 per cent.; in 1835 it was 41 per cent. The proportion which the trade with the United States bore to the foreign trade of England was 17 per cent. in 1821, and 22 per cent. in 1835. In 1805 the proportion was 28 per cent., but in the interval our aggregate trade with all other countries had increased in a greater ratio than that with the United States. The proportion of British to American shipping which entered the ports of the United States averaged 9½ per cent. annually from 1822 to 1830, but from 1831 to 1836 the average was 35½ per cent. Professor Tucker, of Virginia, estimated, on the basis of the census of 1840, that the annual productions of his country amounted to 965,413,650 d. The commissioner of patents made the amount in 1849, 2,330,000,000 d. Making use of the estimate of Professor Tucker as the smallest, and referring to official returns of exports, the proportion taken by Great Britain, as compared with the whole, is as follows:—

	Annual value.	Exports to Great Britain.	Total Exports.
Agriculture,	654.387.597 d.	73,495,549 d.	111.059.378 d.
Manufactures,	239.836,224	541,540	11,809,501
Mining,	42.358.761	194	875,154
Forests,	16,835,060	1,649,869	5.917.994
Fisheries,	11,996,008	843,053	2,547,654
Total,	965,413,650	76,534,205	131,710,081

According to the estimate of the professor, one-seventh of the annual production in 1840 was exported, according to that of the commissioner one-twentieth. The imports of manufactured goods, according to the first, are one-fifth of those manufactured at home; and, according to the commissioner, one-tenth. The British returns give £247,000,000 as the annual value of the production of great Britain in 1840, and £9,564,902 as the export thence to the United States. Reducing these sums to dollars, the following comparison has been made of the mutual interchange of each other's industry:—

	Great Britain.	United States
Annual production, .	1,185,600,000 d	965,413,650 d.
Annual total export, .	187,200,000	131,710,081
British export to U. S., .	45,949,926	
U. S. export to Great Britain, . . .		76,530,205 d.

Assuming the pop. of Great Britain at 31,000,000, and that of the United States at 20,000,000, the consumption in each of domestic and imported goods per head was as follows:—

British consumption.		United States consumption.	
Home goods,	30.13	Home goods,	41.73
United States' goods,	2.50	British goods,	2.25
Total,	32.63	Total,	43.98

By which it appears that of the British consumption 8 per cent. only was produced in the United States; and of the consumption of the United States 5 per cent. was the produce of British industry.

The following table shows the value of British and Irish produce and manufactures exported to the United States; and of the amount of raw cotton imported into the United Kingdom from the United States, according to British official returns, between the years 1820 and 1853:—

	Declared value.	Cotton imported.
1820, . .	—	89,999,174 lbs.
1821, . .	—	93,470,745
1822, . .	—	101,031,766
1823, . .	—	142,532,112
1824, . .	—	92,187,662
1825, . .	—	139,908,699
1826, . .	—	130,858,203
1827, . .	£7,018,272	216,924,812
1828, . .	5,810,315	151,752,289
1829, . .	4,823,415	157,187,396
1830, . .	6,132,346	210,885,358
1831, . .	9,050,500	210,333,698
1832, . .	5,468,272	219,756,753
1833, . .	7,579,699	237,506,758
1834, . .	6,844,989	269,203,075
1835, . .	10,568,455	281,155,819
1836, . .	12,425,605	289,615,692
1837, . .	4,695,225	320,651,716
1838, . .	7,585,760	431,437,888
1839, . .	8,839,294	311,597,798
1840, . .	5,283,020	487,856,504
1841, . .	7,098,642	358,240,964
1842, . .	3,528,027	414,036,779
1843, . .	5,013,514	374,738,520
1844, . .	7,938,079	517,218,022
1845, . .	7,142,839	626,650,412
1846, . .	6,830,460	401,949,393
1847, . .	10,974,161	364,599,291
1848, . .	9,564,909	600,247,488
1849, . .	11,971,028	634,504,050
1850, . .	14,891,961	493,153,112
1851, . .	14,362,976	596,638,962
1852, . .	16,567,737	765,630,544
1853, . .	—	658,451,794

Currency and Banks.] All the banks now established in the United States are joint-stock companies incorporated by law, with a fixed capital, a portion of which is sometimes vested in public stocks, but this is not obligatory. The business of these banks consists in receiving money on deposit, in issuing bank-notes, and in discounting bills of exchange. The capital of the state-banks existing in 1830, amounted to nearly 110,500,000 d. The total amount of the paper currency on 1st January 1830, was estimated at 63,500,000 d. There are no means of ascertaining correctly what portion of this sum consisted of the precious metals. In 1850, the whole number of banks and branches in operation in the Union was 824. The total capital paid was 217,317,211 d.

RESOURCES.

Loans and discounts, . . .	364,204,078 d.
Stocks, ...	20,606,759
Real estate, ...	20,582,160
Other investments, . . .	11,949,568
Due by other banks, . . .	41,631,853
Notes of other banks, . . .	16,303,245
Specie funds, ...	11,603,289
Specie, ...	45,379,345

LIABILITIES.

Circulation, ...	121,366,520 d.
Deposits, ...	109,586,585
Due other banks, . . .	36,717,450
Other liabilities, . . .	8,835,309
The circulation and deposits, . .	240,953,121
Aggregate immediate means, . .	114,917,234

The Eastern states had	297 banks with	61,882,154 d. capital.
Middle states	293	75,933,881
Southern states	87	38,873,251
South-western states	73	31,923,250
Western states	74	9,604,667

The circulation and deposits in the several sections were as follows:—

	Circulation.	Deposits.
Eastern, . . .	29,000,000 d.	14,923,000 d
Middle, . . .	41,862,000	64,764,000
Southern, . . .	25,220,000	9,678,000
South-western, . .	20,819,000	16,334,000
Western, . . .	14,635,000	5,586,000

A report of the secretary of the treasury has lately been published, giving an account of the condition of the banks in the United States for a series of years. The following table shows some curious facts:—

I. LIABILITIES.

	Banks.	Capital.	Circulation.	Deposits.	Total.
		Dollars.	Dollars.	Dollars.	Dollars.
1834,	506	200,005,944	71,957,299	75,666,986	869,448,084
1835,	704	231,250,337	100,815,209	83,081,865	437,156,148
1836,	713	251,875,292	132,710,548	115,104,440	538,204,670
1837,	783	290,772,091	151,969,860	127,397,185	610,292,735
1838,	829	317,636,778	153,990,871	84,691,184	613,731,130
1839,	840	327,132,512	170,981,428	90,240,146	622,111,799
1840,	907	358,442,692	132,465,294	75,696,857	595,011,801
1841,	784	313,608,959	114,631,651	69,786,327	534,622,516
1842,	692	260,171,797	72,187,704	62,408,870	421,577,164
1843,	691	228,821,948	53,404,283	56,168,628	359,189,574
1844,	696	210,872,056	65,474,277	84,550,785	379,067,189
1845,	707	206,045,969	80,140,021	88,029,646	392,250,205
1846,	707	196,894,309	94,498,168	96,913,070	411,012,402
1847,	715	203,070,622	93,864,620	91,702,533	407,789,669
1848,	751	204,838,175	118,089,622	103,226,177	456,594,860
1849,	782	207,309,361	106,108,715	91,178,628	439,731,659
1850,	824	217,317,211	118,984,142	109,586,595	474,325,141
1851,	879	227,807,553	140,106,408	128,957,712	529,313,728
1852,	992	237,164,895	156,137,654	162,273,856	602,287,433
1853,	1008	249,726,816	172,885,924	178,807,535	680,257,530
1854,	1208	301,376,071	190,274,494	188,188,744	765,836,165

II. RESOURCES.

	Banks.	Notes, Bills of exchange, &c.	Specie.	Real estate.	Total.
		Dollars.	Dollars.	Dollars.	Dollars.
1834,	506	331,956,241	26,641,753	10,850,090	369,448,084
1835,	704	382,078,356	43,937,625	11,140,167	437,156,148
1836,	713	483,990,701	40,019,594	14,194,375	538,204,670
1837,	783	553,312,944	37,915,340	19,064,451	610,292,735
1838,	829	544,638,414	35,184,112	38,908,604	613,731,130
1839,	840	560,371,294	45,132,673	16,607,832	622,111,799
1840,	907	533,524,727	33,105,155	29,181,919	595,811,301
1841,	784	465,487,662	34,813,958	34,320,896	534,622,516
1842,	692	359,794,753	28,440,423	33,341,988	421,577,164
1843,	691	302,846,961	33,515,806	22,826,807	359,189,574
1844,	696	306,648,057	49,898,269	22,520,863	379,067,189
1845,	707	325,831,693	44,241,242	22,177,270	392,250,205
1846,	707	349,901,307	42,012,095	19,099,000	411,012,402
1847,	715	351,437,188	35,132,516	21,219,865	407,789,669
1848,	751	389,094,140	46,369,765	20,530,955	456,594,860
1849,	782	372,540,716	43,619,368	23,571,575	439,731,659
1850,	824	408,363,630	45,379,345	20,582,166	474,325,141
1851,	879	460,422,956	48,671,048	20,219,724	529,313,728
1852,	992	527,863,521	53,167,429	21,256,483	602,287,433
1853,	1,093	604,129,867	54,011,288	21,716,375	680,257,530
1854,	1,208	684,058,440	59,410,253	22,367,472	765,836,165

It appears by this that the number of banks has nearly doubled since 1844; their capital has increased 40 per cent., their circulation nearly 200 per cent., and their deposits over 100 per cent., while their specie has increased only 18 per cent., and their real estate has actually diminished. Their notes, &c., in 1854, were 684,000,000 against 306,000,000 in 1844.

For the ten years ending with 1850, as between Europe and the United States, the exports of specie were far below the imports, as the subjoined table will demonstrate:—

	Imported.	Exported.
1841, . . .	4,988,633 d.	10,034,382 d.
1842, . . .	4,087.016	4,813,539
1843, . . .	22.320,335	1,520,791
1844, . . .	5.830,429	5,454,214
1845, . . .	4,070,242	8.606,495
1846, . . .	3,777,732	3,905.268
1847, . . .	24,121,280	1,907.739
1848, . . .	6.360,224	15,841.620
1849, . . .	6,651.240	5,404.648
1850, . . .	4,628,792	7,522.994
	86,835,932 d.	65,011,640 d.

Coins.] The metallic currency of the United States forms scarcely one-tenth part of the amount of the general circulation. Its copper coinage consists of a *half-cent*—the circulation of which is studiously prevented—and a *cent*, or *penny*, as it is termed, in value about the same as the English halfpenny, 100 of which go to the dollar. The silver is composed of 5 cent and 10 cent pieces, and 25 and 50 cent pieces, the *quarter* and the *half-dollar;* the latter of which are comparatively numerous. The gold coinage is the *quarter eagle* = 250 cents; the *half eagle* = 500 cents; and the *eagle* = 1,000 cents. The American dollar contains 371¼ grains of pure silver, or 416 grains of standard silver. The American eagle of the old coinage, previous to the 31st of July, 1834, contained 247½ grains of pure gold. The American eagle of the new coinage contains 232 grains of pure gold. The British sovereign, when coined, contains 113 grains, and 18-1214th parts of a grain of pure gold, worth, according to our present mint valuation, 4·87 7-120th dollars. The Spanish doubloons should, according to the regulations which have nominally prevailed since the year 1772, contain 376 grains of pure gold, which would, at the American mint valuation, be worth 16 dollars 20 cents; but, according to assays made at the London and Philadelphia mints, previous to the year 1829, Spanish doubloons contained only from 360 to 362 4-10th grains of pure gold. This would make their average value 15·56¾ ds.

The following are the respective values, in dollars and cents, of the foreign coins circulating in the U. S. as fixed by law in assessing duties:—

	d.	c.
Franc of France and Belgium, . .	0	18 6-10
Florin of the Netherlands, . .		40
Florin of Southern states of Germany, .		40
Guilder of Netherlands, . .		40
Livre Tournois of France, . .		13½
Lira of the Lombardo-Venetian kingdom,		16
Lira of Tuscany, . . .		16
Lira of Sardinia, . . .		18 6-10
Milrea of Portugal, . . .	1	12
Milrea of Azores, . . .		82½
Marc Banco of Hamburgh, . .		35
Pound sterling of Great Britain, .	4	84
Pound sterling of British provinces of Nova Scotia, New Brunswick, Newfoundland, and Canada, . . .	4	00
Dollars of Mexico, Peru, Chili, and Central America, ...	1	00
Pagoda of India, . . .	1	84
Real Vellon of Spain, . . .		5
Real Plate of Spain, . . .		10
Rupee (Company's), . . .		44¼
Rupee of British India, . . .		44½
Specie dollar of Denmark, . .	1	05
Rixdollar or thaler of Prussia and the Northern states of Germany, . .		69
Rixdollar of Bremen, . . .		78¾
Rouble, silver, of Russia, . .		75
Specie dollar of Sweden and Norway, .	1	06
Florin of Austria, . . .		48½
Ducat of Naples, . . .		80
Ounce of Sicily, . . .	2	40
Tael of China, . . .	1	48
Leghorn livres. . . .		16

The amount of the U. S. coinage up to August 1852 was—

	Gold.	Silver.	Total.
1793 to 1820,	7,431.545 d.	10,898.431 d.	18,420,976 d.
1821 to 1851,	172.752,745	66,972,977	289,725,722

In the 5 years from 1850 to 1854 inclusive, the total average of gold was 258,750,915 d.; of silver, 21,646,893 d.; and of copper, 304,432 d. And the total coinage of the republic up to Jan. 1, 1855, was 442,884.650 d.; of which 80 per cent. was gold; and 75 per cent. of this gold coinage was the result of the last 5 years.

Commercial navy.] The total merchant tonnage of the United States, corresponding to the British registered tonnage, amounted, in 1826, to 1,534,000 tons, being an increase of 11,070 tons within the year, and more than double the increase in any one of the twelve preceding years. That of the United Kingdom, in 1827, was 2,105,605 tons. The aggregate tonnage of the Union for 1829 was 1,260,798 tons, of which 650,143 was employed in foreign trade, and 610,655 in coasting trade and fishery. In 1831 the amount of tonnage was stated at 1,191,776, of which the registered tonnage was 576,475 tons, and the enrolled 615,301. The amount of the mercantile tonnage in 1845 was 2,417,002 tons, of which the enrolled and licensed tonnage amounted to 1,321,829 tons. The latter tonnage is almost exclusively engaged in the coasting-trade, or in the cod or mackeral fisheries; it also includes the steamers on the rivers. In 1848, the sum total of the commercial tonnage was 3,154,051 tons, being an increase of 56 per cent. in 10 years. In 1852-3, the tonnage had reached 4,407,010 tons. The total tonnage in 1854 was 4,802,903 tons; in 1858 it had increased to 5,049,808 tons. The return of clearances shows that foreigners have a large share in the carrying trade. The total clearances for the fiscal year ending 30th June 1850 was 4,361,002. The total number of vessels was 18,195; of which 8,379 were American, and 9.816 foreign. The tonnage of those cleared in New York was 2,149,096; the number of American vessels being 3,610; of foreign, 3,693. In the year ending 30th June 1858, 4,490,033 tons of American shipping, and 2,312,759 tons of foreign shipping, cleared for foreign countries. The returns for 1850, compared with those of 1849, were as follows:—

	1850.	1849.
Tonnage—American, . .	2.632,788	2,753.724
Foreign, . .	1,728,214	1,675,709
Vessels—American, . .	8,379	11.466
Foreign, . .	9,816	8,847
Deduct tonnage of California, 180.128 tons, from the aggregate Foreign and American this year, and we have a balance of . .		4.180,874
Compare with last year, . . .		4,429,433
Showing a falling off of . . .		248,559
Also a decrease of American tonnage of .		120.936
An increase of foreign tonnage, . .		52,505

The number of men and boys clearing as crews in 1850 and 1858 was as follows:—

Crews.	1850.	Total.	1858.	Total.
Men (American vessels),	102,888		144,657	
Boys, . .	3,865		405	
		106,753		145,082
Men (foreign vessels),	86,886		102.476	
Boys, . .	2,232		1,047	
		89,118		103,824
		195,871		248,906

Steam navigation] The first steam-boat successfully used in the United States was constructed by Robert Fulton in 1807. It plied on the Hudson river, with an engine of only 18 horse-power, and made the passage between Albany and New York (from 150 to 155 m.) in 33 hours. Her engine was manufactured in England by Boulton and Watt. In March 1811, the first steam-boat built by Fulton, in Pittsburg, called the New Orleans, was launched on the Ohio, and commenced, in December of the same year, to make regular trips between Natchez and New Orleans. The time required to make the trip of 300 m. between the two places was 3 days down stream, and 7 to 8 days up. The boat performed in a year only thirteen trips up and down, or 7,800 m. On the 6th of May 1817, the first steam-boat went up the Mississippi and Ohio, from New Orleans to Louisville, and arrived there on the 30th of May, or in 25 days. In 1818, the first steam-boat was launched on the great North-western lakes; in 1835, they were navigated by 25 steam-boats; in 1838, the number of steam-boats was 70. The largest ship-yards for building steam-boats, are at New York, Philadelphia, Baltimore; at Louisville, New Albany, Cincinnati, Pittsburg, and St. Louis. In 1848, there were 8 lines of steam-packets with 28 steam-ships, and 58 regular lines of sailing vessels, going to almost every part of the world, comprising 188 vessels, besides 50 line or 179 vessels whose times of sailing were not regular. In addition to these, there were the Cunard line, the Collins line, the New York and Liverpool, New York and Bremen, New York and Havre, two Havana, Aspinwall's Panama and Oregon, and the Charleston and Havana lines, comprising some of the largest and finest steam-ships. The total steam-tonnage in 1853 amounted to 514,094 tons; and in 1858 to 729,390 tons. The number of steamers built in 1853 was 271; in 1858, 226, of a total tonnage of 242,286 tons.

Canals.] It appears from official statistics that, on the 1st of January 1843, the extent of canals in actual operation amounted to 4,333 m., and that there were then in progress 2,359 m., a considerable portion of which has since been completed, so that it is probable that the actual extent of artificial water communication now in use in the United States considerably exceeds 5,000 m. The average cost of executing this prodigious system of artificial water-communication was at the rate of £6,432 per mile, so that 5,000 m. would have absorbed a capital of above £32,000,000. This extent of canal-transport, compared with the pop., exhibits in a striking point of view the activity and enterprise which characterize the American people. In the United States there is a mile of canal-navigation for every 5,000 inhabitants, while in England the proportion is 1 to every 9,000 inhabitants, and in France, 1 to every 13,000. The ratio, therefore, of this instrument of intercommunication in the U. S. is greater than in the United Kingdom, in proportion to the pop. as 9 to 5; and greater than in France in the ratio of 3 to 5.

Railways.] In 1840, railways had been introduced into 24 states of the Union: and the number of railways, either completed or in progress, were 179, with a total length of 9,378 m., of which however, only 3,430 m. was in operation or nearly completed. An official statement concerning the railroads of the United States gives the number of miles of railroad in operation on the 1st of January 1852, as 10,814 m., and on 1st January 1859 at 27,857 m. Many of the new lines projected have been substituted for plans for the construction of canals and turnpike roads. The cost of the railroads in the six New England states is about £9,000 per mile; in New York, Pennsylvania, and Maryland it is about £8,000. In the southern states, and in the valley of the Mississippi, it is about £4,000 per mile. The railroad system in the U. S. was commenced in 1830. The first road in operation was built for the transportation of ice from a small lake to the sea in the state of Massachusetts. The length of this road was 4 m. In the same year, the state of S. Carolina caused to be commenced a railroad from Charleston, its principal port, to Augusta, in the state of Georgia. The distance is 135 m., and the road was completed in 1833, at the remarkably small cost of about £267,000. The longest continuous line in the country, and that in the building of which the greatest natural obstacles have been overcome, is that which extends from the Hudson river, through the southern counties of New York, to Lake Erie. Its length is 469 m., and its branches are in the aggregate about 68 m. The aggregate cost of this important road was about £4,716,000, or about £8,666 per mile. Grading operations were commenced in 1833, and it was completed in May 1851. The state of New York advanced £1,200,000 towards the work, and afterwards released the company from obligation to pay the loan. In the infancy of the American railroad system, and for ten years thereafter, it was the rule to extend to every important enterprise of the kind the assistance of the state through which the road was to pass. In 1850, congress passed an act, after a protracted discussion, granting to the state of Illinois about 2,700,000 acres of public lands to aid in the construction of the central railroad above mentioned; but the precedent is not likely to be followed.—The following table presents, in a convenient form, some of the principal facts connected with railroads in the U. S., drawn from official sources in May 1852:—

	Miles of railroad complete.	Miles of railroad in progress.	Area of sq. m. in whole state.	Inhabitants to sq. m.
Maine,	315	127	30,000	19.44
New Hampshire,	489	47	9,280	34.26
Vermont,	380	59	10,212	30.76
Massachusetts,	1,009	07	7,000	127.10
Rhode Island,	50	32	1,306	112.97
Connecticut,	547	261	4,674	79.33
New York,	1,826	745	46,000	67.33
New Jersey,	226	111	8,320	58.84
Pennsylvania,	1,146	774	46,000	50.25
Delaware,	16	11	2,120	43.17
Maryland,	376	125	9,356	63.31
Virginia,	478	818	61,352	23.17
N. Carolina,	249	385	45,000	19.30
S. Carolina,	340	298	24,500	27.28
Georgia,	754	229	58,000	15.62
Alabama,	121	190	50,722	15.21
Mississippi,	93	273	47,156	12.86
Louisiana,	63	—	46,431	11.15
Texas,	—	32	237,321	0.89
Tennessee,	112	748	45,600	21.98
Kentucky,	93	414	37,680	26.07
Ohio,	828	1,892	39,964	49.55
Michigan,	427	—	56,243	7.07
Indiana,	600	915	33,809	29.23
Illinois,	176	1,409	55,405	15.36
Missouri,	—	515	67,380	10.12
Wisconsin,	20	421	53,924	5.65
	10,814	10,898		

The management of the American, like that of the English railroads, is entirely in private hands. Their concerns are managed by corporations, chartered by the respective states, and having for officers a president, secretary, and directors. Each of the directors must be a stock-holder. The president and secretary have liberal salaries, but the services of directors are gratuitous. The rate of railroad speed in America is not so great as in England. The ordinary velocity of a passenger train is 20 m. an hour, though on some lines it is as great as 28 and 30 m. Special trains frequently keep up a speed of 45 m. an hour for a long distance, and on one road, that from New York city to Albany, 40 m. an hour is the average rate for all passenger trains. In New England the average price per mile in the first class is about a penny; the second class throughout the country is only used by railroad labourers, emigrants, negroes, and other persons of the same class. From New York to Boston the fare is about one penny farthing per mile; from New York to Philadelphia about a penny three farthings; and from Philadelphia to Baltimore, three halfpence. From New York to Cincinnati the distance is 857 m., of which 143 m. are travelled by steam-boat. The price of passage for the whole distance is about £3 6s. 8d. The entire paid in capital of the railroads, as on 8th Dec. 1857, was 491,435,661 d., of the debt, 417,243,664 d. The available income was 24,290,326 d.

Electric telegraph.] Several of the states of the Union are completely interlaced by electric telegraphs, which are more numerous than rivers or railroads. Every leading city or town has a telegraph office or agency, and even the children at the public schools are taught the principle of this remarkable discovery and its various applications. The first telegraph line established in the United States was Morse's, in May 1844, between Baltimore and Washington, and an appropriation was made by congress to test the practical operation of the invention. There are now 17,500 m. of telegraph working under the Morse patent, and between 9,000 and 10,000 m. otherwise. A telegraph to the Pacific is contemplated, to start from Natchez, Mississippi, through Texas, crossing the gulf of California to San-Diego, and thence up the coast to Monterey and San Francisco. When a steamer arrives from Europe in the morning, whether at Halifax, Boston, New York, or Philadelphia, the news she brings is published on the same day in every principal city in the Union.

Population of North America.] If the aboriginal population be rapidly declining, the superinduced European pop. is proportionally increasing. It was provided by the constitution that the first census of the U. S. should be made "within three years after the first meeting of congress, and within every subsequent term of ten years, in such manner as they shall by law direct." The first census was accordingly taken in 1790, and the seventh in 1850. In the first census the whole pop. of the U. S. was divided into only five classes, in which the total amount of the several classes was as follows:—

1. Free white males under 16 years,	802,127
2. Do do. of 16 years and upwards,	813,365
3. Free white females,	1,556,626
4. All other free persons except Indians not taxed,	59,511
5. Slaves,	697,697
Total,	3,929,328

In the first three enumerations "all other free persons except Indians not taxed" were thrown into one mass, without distinction of age or sex, and the same course was adopted respecting the slaves; but in the fourth census each sex of both these descriptions of persons was distinguished, according to age, into four classes, and each sex of the free white inhabitants was divided, as in the second and third censuses, into five classes; and in addition, the number of free white males between 16 and 18 years was exhibited in a distinct column. Persons engaged in agriculture, commerce, and manufactures, were also distinguished into three several classes; and "foreigners not naturalized" formed an additional class. Since the census of 1840 there have been added to the territory of the republic, by annexation, conquest, and purchase, 833,970 sq. m.; and its title to a region covering 341,463 sq. m., previously claimed and occupied by a foreign power, has been established by negotiation, and it has been brought within the acknowledged boundaries of the Union. By such means the area of the U. S. has been extended during the past ten years from 2,055,163 to 3,230,572, sq. m., without including the great lakes or the bays which indent the Atlantic and Pacific coasts. It is estimated that these new territories brought into the Union, at the time of their incorporation with it, 172,000 inhabitants. The total number of inhabitants in the U. S. was on the 1st of June, 1850, 23,263,488. The absolute increase from the 1st of June, 1840, was 6,194,035; and the actual increase per cent. 36·23; but the probable amount of population acquired by additions of territory should be deducted in making a comparison between the results of the present and the last preceding census. These deductions diminish the total pop. of the country, as a basis of comparison, to 23,091,488, and the increase to 6,022,035. The relative increase, after this allowance, is found to be 35·27 per cent. The aggregate number of whites in 1850 was 19,630,738, exhibiting a gain upon the number of the same class in 1840 of 5,434,933, and a relative increase of 38 28 per cent.; but, excluding the 153,000 free pop. supposed to have been acquired by the addition of territory since 1840, the gain was 5,281,933, and the increase per cent. 37·20. The number of slaves, by the census of 1850, was 3,204,089, which shows an increase of 716,733, equal to 28·81 per cent. If we deduct 19,000 for the probable slave pop. of Texas in 1840, the result of the comparison will be slightly different. The absolute increase will be 697,733, and the rate per cent. 28·05. The number of free-coloured in 1850 was 428,661; in 1840, 386,292. The increase of this class has been 42,369, or 10·96 per cent. From 1830 to 1840 the increase of the whole pop. was at the rate of 32·67 per cent. At the same rate of advancement the absolute gain for the ten years from 1840 to 1850, would have been 5,576,590, or 445,445 less than it has been, without including the increase consequent upon additions of territory. The aggregate increase of pop. from all sources shows a relative advance greater than that of any other decennial term, except that from 1800 to 1810, during which time the country received an accession of inhabitants by the purchase of Louisiana, amounting to something more than 1 per cent. of the whole number. Rejecting from the census of 1810 1·45 per cent. for the pop. of Louisiana, and from the census of 1850, 1 per cent. for that of Texas, California, &c., the result is in favour of the last ten years by about one-fourth of one per cent., the gain from 1800 to 1810 being 35·00 per cent., and from 1840 to 1850, 35·28. The decennial increase of the most favourite portions of Europe is less than 1½ per cent. per ann.; with the U. S. it is at the rate of 3½ per cent. According to past progress, viewed in connexion with that of European nations, the pop. of the U. S. in 40 years will exceed that of England, France, Spain, Portugal, Sweden, and Switzerland combined. The relative progress of the several races and classes of the population is shown in the following statement:—

INCREASE PER CENT. OF EACH CLASS OF INHABITANTS IN THE UNITED STATES FOR 60 YEARS.

Classes.	1790 to 1800.	1800 to 1810.	1810 to 1820.	1820 to 1830.	1830 to 1840.	1840 to 1850.
Whites,	35·7	36·2	34·19	33·95	34·7	38·28
Free coloured,	82·2	72·2	25·25	36·85	20·9	10·96
Slaves,	27·9	33·4	29·1	30·61	23·8	28·81
Total coloured,	32·2	37·6	28·58	31·44	23·4	26·41
Total,	35·01	36·45	33·12	33·48	32·67	36·28

This is a view officially put forth by an officer of the American government of the past and probable future progress of the country. Mr. Darby, a writer of some reputation upon statistics in the United States, assumes the average annual increase of the pop. as equal to 3 per cent. If the ratio of increase be taken at 3 per cent. per ann., the pop. doubles in about 24 years; therefore, if no serious disturbing influence should interfere with the natural order of things, the aggregate pop. of the U. S., at the close of this century, must be over 100,000,000.

The following table shows the increase of pop. from 1790 to 1850, without reference to intervening periods:—

	1790.	1850.	Absolute Increase in 60 years.	Increase per cent. in 60 years.
Whites,	3,172,464	19,630,738	16,458,274	518·78
Free coloured,	59,466	428,661	369,195	620 85
Slaves,	697,897	3,204,089	2,506,192	359·10
Total free, coloured, and slaves,	757,363	3,632,750	2,875,387	379·65
Total,	3,929,827	23,263,488	19,333,661	491·97

Sixty years since, the proportion between the whites and blacks, bond and free, was 4·18 to 1. In 1850, it was 5·4 to 1, and the ratio in favour of the former race is increasing.

The following remarks upon the comparative advance of population in the U. S. and Great Britain accompany the census report of 1852:—"During ten years the pop. of the entire kingdom of Great Britain and Ireland increased from 26,833,496 to 27,452,262, or at the rate of a little more than half-a-million in 10 years. In the last 50 years England and Wales increased 102 per cent. (males 105, females 97·50), and Scotland 78 per cent. (males 84, females 73). The pop. of the U. S., during the past 50 years, has increased at the rate of 337 per cent., and in 10 years intervening between the last two censuses it increased from 17 and a fraction millions to over 23,000,000, or 36 per cent. During the same period (leaving Ireland out of view) the pop. of Great Britain increased at the rate of 12 per cent. during 10 years, or $1\frac{2}{10}$ per cent. per annum." The comparative house-accommodation of the people in the two countries are described in the subjoined statements: "By the last census it appears that in the U. S. the number of houses occupied by free persons amounted to 3,363,427. It would seem from the British reports that the pop. of that country is supplied with houses almost in the precise proportion as in our own country. In comparing the pop. of Great Britain and Ireland with the inhabited houses, it appears that the whole number of houses in Great Britain amounts to 3,669,437, being nearly one house to each 6 persons; and that in Ireland the number of inhabited houses amounts to 1,047,735, being in the proportion of two houses to each 13 persons."

The density of the pop. of the country at large and of its several parts is thus given: "Taking the 31 states together, their area is 1,486,917 sq. m., and the average number of their inhabitants is 15·54 to the sq. m. The total area of the U. S. is 3,230,572 sq. m., and the average density of pop. is 7·2 to the square mile."

From the most authentic *data* in possession of the public departments, the following table has been prepared:—

TABLE OF THE AREA AND THE NUMBER OF INHABITANTS TO THE SQUARE MILE IN EACH STATE AND TERRITORY IN THE UNION:—

States.	Area in sq. m.	Population in 1850.	Number of inhabitants to the sq. m.
Maine,	30,000	583,188	19·44
New Hampshire,	9,280	317,964	34·26
Vermont,	10,212	314,120	30·07
Massachusetts,	7,800	994,499	126·11
Rhode Island,	1,306	147,544	108·05
Connecticut,	4,674	370,791	79·83
New York,	46,000	4,097,394	67·66
New Jersey,	6,320	489,333	60·04
Pennsylvania,	46,000	2,311,786	50·25
Ohio,	39,946	1,980,408	49·55
Indiana,	33,809	988,416	29 23
Illinois,	55,405	851,470	15 37
Iowa,	50,914	192,214	3·77

Wisconsin, .	53,924	305,191	5·45
Michigan, .	56,243	397,654	7·07
California, .	188,982	165,000	·87
Minesota Territory,	83,000	6,077	·07
Oregon Territory,	341,463	13,293	·04
New Mexico Territory,	219,774	61,547	·28
Utah Territory,	187,923	11,380	·06
Total Free states and Territories,	1,474,993	13,419,190	

States.	Area in sq. m.	Population in 1850.	Number of inhabitants to the sq. m.
Delaware, .	2,120	91,535	43·64
Maryland, .	9,356	583,036	62 31
District of Columbia,	60	51,687	861·45
Virginia, .	61,351	1,421,661	23·17
North Carolina,	45,000	868,903	19·30
South Carolina,	24,500	668,507	27·28
Georgia, .	58,000	905,999	15·68
Florida, .	59,268	87,401	1·47
Alabama, .	50,723	771,671	15·21
Mississippi, .	47,126	606,555	12·86
Louisiana, .	46,437	511,974	11·02
Texas, .	237,321	212,592	·89
Arkansas, .	52,198	209,639	4·01
Tennessee, .	45,600	1,002,625	21·98
Kentucky, .	37,680	982,405	26 07
Missouri, .	67,380	682,043	10·12
Total Slaveholding states, .	844,115	9,638,223	

From the location, climate, and productions, and the habits and pursuits of their inhabitants, the states of the Union may be properly arranged into the following groups.

	Area in sq. m.	Pop.	No of inhabitants to sq. m.
I. New England states, (6)	63,272	2,728,106	43·11
II. Middle states—including Maryland, Delaware, and Ohio, (6) . .	151,760	8,553,713	56·36
III. Coast Planting States—including South Carolina, Georgia, Florida, Alabama, Mississippi, and Louisiana, (6) . .	286,077	3,557,872	12·43
IV. Central Slave states—Virginia, North Carolina, Tennessee, Kentucky, Missouri, Arkansas, (6)	309,210	5,167,276	16 71
V. North-Western states—Indiana, Illinois, Michigan, Wisconsin, and Iowa, (5)	250,295	2,734,745	10·92
VI. Texas, . .	237,321	212,592	·89
VII. California, .	188,982	165,000	·87

The ratios of increase in which the several classes of the pop. of the U. S. have advanced is shown in the following statement of population by classes decennially, from 1790 to 1850 inclusive.

	1700.	1800.	Inc. p. ct.	1810.	Inc. p. ct.
Whites, .	3,172,464	4,304,489	35·7	5,862,004	36 2
Free coloured,	59,400	100,805	69·0	186,446	72·0
Slaves, .	697,897	893,057	27·9	1,191,364	33·4
	3,929,827	5,305,941	—	7,239,814	—
Total free,	3,231,930	4,412,884	36·4	6,048,450	37·
Total coloured pop., free and slaves, .	757,363	1,001,452	32·2	1,377,810	37·6

	1820.	Inc. p. ct.	1830.	Inc. p. ct.
Whites,	7,866,569	34·19	10,532,060	33·95
Free coloured,	233,524	25·25	319,599	30·85
Slaves, .	1,538,098	29·1	2,009,043	30·61
Seamen in the United States' service,	—	—	5,318	
	9,638,191	—	12,866,020	—
Total free,	8,100,093	33·92	10,856,977	34·03
Total coloured pop., free and slaves, .	1,771,622	28.58	2,328,642	31·44

	1840.	Inc. p. ct.	1850.	Inc. p. ct.
Whites,	14,189,705	34·71	19,630,738	38·28
Free coloured,	386,292	20·86	428,661	10·96
Slaves, .	2,487,356	23·8	3,204,089	28·81
Seamen in the United States' service,	6,100	—	—	—
	17,069,453	—	23,263,488	—
Total free,	14,582,097	34·31	20,059,399	37·56
Total coloured pop., free and slaves,	2,873,648	23·4	3,632,750	26·41

Immigrants.] The reports of the registrar-general show the departure from Liverpool and other British ports of 630,661 persons for the U. S. during the years 1848, 1849, and 1850; but the American returns indicate the arrival of only 541,000, exhibiting a disparity in the accounts of 89,000 during those three years, of 11 to 14 per cent. After a careful investigation of authentic records, and making due allowances for the imperfections of official returns, the following statement of immigration into the U. S. from all the rest of the world, since the date of the first census up to that of the last, a period of 60 years, has been prepared:—

Number of foreigners arriving from 1790 to 1810,	120,000
Natural increase, reckoned in periods of 10 years,	47,560
Number of foreigners arriving from 1810 to 1820,	114,000
Increase of the above to 1820, ...	19,000
Increase from 1810 to 1820 of those arriving previous to 1810, ...	58,450
Total number of immigrants and descendants of immigrants in 1820, ...	359,010
Number of immigrants arriving from 1820 to 1850,	203,979
Increase of the above, ...	35,728
Increase from 1820 to 1830 of immigrants and descendants of immigrants in the country in 1820,	134,130
Total number of immigrants and descendants of immigrants in the U. S. in 1830, . .	732,847
Number of immigrants arriving from 1830 to 1840.	778,500
Increase of the above, ...	135,150
Increase from 1830 to 1840 of immigrants and descendants of immigrants in the U. S. in 1830,	254,445
Total number of immigrants and descendants of immigrants in the U. S. in 1840, . .	1,900,942
Number of immigrants arriving from 1840 to 1850,	1,542,850
Increase of the above at 12 per cent., . .	185,142
Increase from 1840 to 1850 of immigrants and descendants of immigrants in the U. S. in 1840,	722,000
Total number of immigrants into the U. S. since 1790, living in 1850, together with their descendants, ...	4,350,934

The inter-migration of the native citizens among the states is a branch of the general inquiry embracing the nativity of the inhabitants, which is curious and interesting. It is found that out of 17,736,792 free native inhabitants, 4,112,433 have migrated and settled beyond the states of their birth. Of white natives of Virginia, 335,000, equal to 26 per cent. of the whole, have found homes beyond her borders. S. Carolina has sent forth 163,000, which is 36 per cent. of all native white citizens of that state, and is equal to 59 per cent. of all remaining in the state of their nativity. This is the highest proportion observable in the statistics of any of the states. N. Carolina has lost by emigration 261,575 free inhabitants, equal to 31 per cent. Among the Northern states Vermont and Connecticut have contributed most largely, in proportion to their total pop., to the settlement of other parts of the country. This roving tendency of the American people is incident to the peculiar condition of their country. When the fertile plains of the West shall have been filled up, and men of scanty means cannot by a mere change of location possess themselves of homesteads, the inhabitants of each state will become comparatively stationary.

Nativity of the Population.] The British colonies in America, forming the original 13 states, were settled by emigrants, a large proportion of whom were natives of Great Britain. No considerable emigration of Celtic Irish, or other people of Celtic origin, took place until after the commencement of the present cent. The New England states, New Jersey, Pennsylvania, Delaware, Maryland, Virginia, North Carolina, South Carolina, and Georgia, were mainly settled by Englishmen. New York, the only Dutch colony, passed under British dominion, with a small pop., partly Dutch and partly English, in 1674. A few Dutch, Germans, and Poles settled in New Jersey; a few Swedes in Delaware, many Germans in Pennsylvania, where they afterwards became one-third of the pop.; and some French Protestants, called Huguenots, in New York, New Jersey, and South Carolina. Besides the small Polish colony in New Jersey, another branch of the Slavonic race was represented by a colony of a few Moravians and Bohemians in Pennsylvania. With the exception of a few Scotch Highlanders who settled in North and South Carolina and Georgia, no Celtic colony is to be found among the settlements in the British North American colonies of either the 17th or 18th centuries. Settlements of Lowland Scotch and Scotch Irish from the north of Ireland were made in Pennsylvania and the Carolinas, and a small number of Irish Protestants settled in the town of Londonderry in New Hampshire. The considerable numbers of Irish Protestants from Ulster and other parts of Ireland, of Lowland Scotch and not of Celtic origin, who have, from time to time, emigrated to the U. S., have led to error in investigating the elements of American pop.; but no considerable Celtic element existed in the pop. of America previous to the census of 1790. The Welsh, considered by some as Celts, furnished a small proportion of the early emigrants to British America. There were a few Welsh colonies in the U. S., in the last century, where the emigrants retained their language, manners, and customs. Such is the county of Cambria, in Pennsylvania, and some smaller settlements in New York and other states. It is probably fair to estimate the Welch element in the present pop. of the United States at 500,000. The greater proportion of the Jewish race now found in the U. S. is of recent emigration. Of the free inhabitants of the U. S., 17,737,505 were, according to the census of 1850, natives of its soil, and 2,210,828 were born in foreign countries, while the nativity of 39,014 could not be determined. It is shown that 1,965,518 of the whole number of foreign-born inhabitants were residents of the free states, and 245,310 of the slave states. The persons of foreign birth formed 11·06 per cent. of the whole free pop. The countries whence have been derived the largest portions of these additions to its population appear in the following statement:—

Natives of Ireland in the U. S. in 1850,	961,719
... Germany, ...	573,225
... England, ...	278,675
... British America, ...	147,700
... Scotland, ...	70,550
... France, ...	54,069
... Wales, ...	29,868
... All other countries, ...	95,022
	2,210,828

The proportion in which the several countries above-named had contributed to the aggregate immigrant pop. was as follows:

Ireland,	43·04 per cent.
Germany,	25·09 ...
England,	12·06 ...
British America,	6·68 ...
Scotland,	3·17 ...
France,	2·44 ...
Wales,	1·34 ...
Miscellaneous,	4·47 ...

By scanning the foregoing statement respecting the nativity of the foreign pop. of the Union, it will be seen that 1,488,507, or 67 per cent. of its whole amount, are natives of the British islands and the British American provinces. The following classification of the inhabitants of the United States, according to their origin, was made by an intelligent commentator upon the census returns:—

Descendants of the European colonists previous to 1776,	14,280,865
Ditto of people of Louisiana, Texas, and other acquired territories,	1,000,000
Immigrants since 1790 and their descendants,	4,350,934
Descendants of Africans,	3,626,961
Total,	23,258,700

Rate of mortality.] For general estimates, adopting the current classification of the states, the American census exhibits the following ratios of mortality, disregarding the ages at death:—

	Annual deaths per cent.	Ratio to the number living.
New England states,	1·55	1 to 64
Middle states, with Ohio,	1·39	1 72
Central slave states,	1·38	1 73
Coast Planting states,	1·37	1 73
North-Western states,	1·24	1 80
Total,	1·38	1 73

The values for the three middle divisions agree with the average for the U. S. as a whole, representing one death to 73 living. It is stated in the report for 1851, that the statistics of mortality for the census year represent the number of deaths occurring within the year as 320,433, the ratio being as one to 72·6 of the living, or as ten to each 726 of the population. The rate of mortality in this statement, taken as a whole, seems so much less than that of any portion of Europe, that it must at present be received with some degree of allowance. With respect to the longevity and vital characteristics of the slaves and free-coloured, the following epitome of life-tables is given on the expectation of life for coloured persons:—

Completed age.	New England. Coloured, Male.	New England. Coloured, Female.	Louisiana. Coloured, Male.	Louisiana. Coloured, Female.
	Years.	Years.	Years.	Years.
0	39·75	42·20	28·89	34·09
10	42·92	45·75	35·92	40·69
20	35·87	39·92	30·48	35·36
30	29·77	34·96	26·87	30·86
40	22·83	28·75	23·25	25·85
50	18·27	22·11	19·13	21·07
60	13·89	17·31	14·75	15·27
70	9·42	13·06	11·33	10·93
80	6·44	7·87	5·38	6·16
90	3·69	4·61	3·43	3·34

Deaf, Dumb, Blind, Idiotic, and Insane.] It appears from the census of 1852, that there is one insane person for every 2,464 of the whole free pop.; one idiot in every 2,088 persons; one deaf and dumb person for every 3,353; and one blind person for every 4,705. These calculations are made only for the free persons born in the states in which they resided when the census was taken. To include the whole number would be unfair to those states whose benevolent and well-regulated institutions for the afflicted have attracted many from other states not possessing similar advantages, and a large number are foreigners. Rhode Island has the greatest number of insane persons, New Hampshire the greatest number of idiots, Connecticut the largest number of deaf and dumb, and Delaware the most blind persons in proportion to their respective population, while the western states and territories are almost exempt from these objects of sympathy and philanthropy. It also appears that one slave is deaf and dumb out of every 6,553 of the slave pop.; one blind in 2,646; one insane in 11,011; one idiotic in 3,081. Delaware has the greatest proportion of deaf and dumb; Virginia the greatest proportion of blind; district of Columbia the greatest proportion of insane; and Delaware the greatest proportion of idiots. A comparison of the proportion between the afflicted free and slave pop. shows a much greater number of deaf and dumb, insane and idiotic of the former, while the relative number of blind slaves far exceeds those of the free. The following table shows the comparative ratios:—

	Free pop.	Slaves.
Insane,	1 in every 2,463	1 in every 11,011
Idiotic,	1 in every 2,087	1 in every 3,082
Deaf and dumb,	1 in every 3,352	1 in every 6,553
Blind,	1 in every 4,705	1 in every 2,646

DEAF AND DUMB.		
Whole No. in U. S.,	Whites,	9,469, or 1 in 2,073
... ...	Free coloured,	145, or 1 in 2,956
... ...	Slaves,	489, or 1 in 6,553
BLIND.		
Whole No. in U. S.,	Whites,	7,997, or 1 in 2,454
... ...	Free coloured,	494, or 1 in 867
... ...	Slaves,	1,211, or 1 in 2,646
INSANE.		
Whole No. in U. S.,	Whites,	15,156, or 1 in 1,295
... ...	Free coloured,	321, or 1 in 1,335
... ...	Slaves,	291, or 1 in 11,011
IDIOTIC.		
Whole No. in U. S.,	Whites,	14,230, or 1 in 1,309
... ...	Free coloured,	436, or 1 in 983
... ...	Slaves,	1,040, or 1 in 3,081

The above aggregates and proportions show that the deaf and dumb and insane are the most numerous in the white pop.; the blind and idiotic among the free coloured; while the slaves are almost exempt from insanity and idiocy.

Indian tribes within the States.] The most numerous and most important of the native tribes which still continue in the immediate vicinity of the whites, are those which occupy reservations in Georgia, the Floridas, Alabama, Mississippi, and Tennessee. The only tribes lately residing E of the Mississippi, who could aspire to the names of nations, were the Creeks, the Choctaws, the Chickasaws, the Cherokees, and the Seminoles. As a rule, the red man disappears before the superior moral and physical influence of the white. In Malte Brun's statement—furnished by his translator, and taken from *Nile's American Register* for June, 1822—the whole number of Indians then dwelling on the E side of the Mississippi was estimated at 120,346; those to the W of the Mississippi, as far as the Rocky mountains, and S to the Red river, were stated at 134,220. To the W of the Rocky mountains, the number was estimated at 171,200; and between the Red river and the Rio-del-Norte, at 45,370. Grand total, 470,000. On the first of December, 1837, Mr. C. A. Harris, commissioner of Indian affairs, laid the following calculations before congress, showing the number of Indians then E of the Mississippi; of those that had emigrated from E to W of that river; and of those within striking distance of the western frontier; also the estimated number of warriors they might collectively be able to bring into the field:—

I. TRIBES EAST OF THE MISSISSIPPI.

1st. *Under treaty stipulations to remove W of the Mississippi.*

Winnebagoes,	4,500
Ottawas of Ohio,	100
Potawatamies of Indiana,	2,950
Chippewas, Ottawas, and Potawatamies,	1,500
Cherokees,	14,000
Creeks,	1,000
Chickasaws,	1,000
Seminoles,	5,000

Apalachicolas,		400
Ottawas, and Chippewas in the peninsula of Michigan,		6,500
		36,950

2d. Not under treaty stipulations to remove.

New York Indians,	4,176	
Wyandots,	575	
Miamies,	1,100	
Menomonies,	4,000	
Ottawas, and Chippewas of the lakes,	2,564	
		12,415
Total,		49,365

II. NUMBER WHO HAVE EMIGRATED FROM THE E TO THE W OF THE MISSISSIPPI.

Chickasaws,	549
Chippewas, Ottawas, and Potawatamies,	2,191
Choctaws,	15,000
Quapaws,	476
Creeks,	20,437
Seminoles,	407
Apalachicolas,	265
Cherokees,	7,911
Kickapoos,	588
Delawares,	826
Shawnees,	1,272
Ottawas,	374
Weas,	222
Piankeshaws,	162
Peorias and Kaskaskias,	132
Potawatamies of Indiana,	53
Senecas,	251
Senecas and Shawnees,	211
Total,	51,327

III. NUMBER OF INDIGENOUS TRIBES WITHIN STRIKING DISTANCE OF THE W FRONTIER.

Sioux,	21,600
Iowas,	1,500
Sacs,	4,800
Foxes,	1,600
Sacs of the Missouri,	500
Osages,	5,120
Kansas,	1,606
Omahas,	1,600
Ottoes and Missourias,	1,000
Pawnees,	12,500
Camanches,	19,200
Kioways,	1,800
Mandans,	3,200
Quapaws,	450
Minatarees,	2,000
Pagans,	30,000
Assinaboins,	15,000
Appaches,	20,280
Crees,	3,000
Arrepahas,	3,000
Gros Ventres,	16,800
Eutaws,	19,200
Crows,	7,200
Caddoes,	2,000
Poncas,	900
Arickarees,	2,750
Cheyennes,	3,200
Blackfeet,	30,000
Total,	231,806

In 1854, the commissioner of Indian affairs reported that "the whole number of Indians within the U. S. limits was estimated at 400,000. About 18,000 yet linger in some of the states E of the Mississippi, principally in New York, Michigan, and Wisconsin; the remainder, consisting of Cherokees, Choctaws, and Seminoles, being in North Carolina, Mississippi, and Florida. The number in Minnesota, and along the frontiers of the Western states of Texas, comprising mainly emigrated tribes, is estimated at 110,000; those of the Plains and Rocky mountains, and not within any of our organized territories, at 63,000; those in Texas at 29,000; those in New Mexico at 45,000; those in California at 100,000; those in Utah at 12,000; and those in the territories of Oregon and Washington at 23,000." The aboriginal pop. does not exceed one person for every 20 sq. m.; it is constantly declining, and probably in the lapse of two centuries at most, the native races of this part of the continent will be extinct. From time to time they are bought out—to roam westward, 'like bees whose hive has been destroyed.' Besides this helpless state of things, the continual feuds between savages, their inordinate propensity for spirituous liquors, and the ravages of small-pox, are causes powerfully and constantly operating to diminish their numbers. The Cherokees alone have made considerable progress in the cultivation of their lands, and even in literature and the arts of civilized life. The Creeks have abundance of tame cattle, swine, and poultry; and cultivate tobacco, rice, maize, potatoes, beans, cabbage, melons, grapes, strawberries, and other fruits.

Slaves and Free Blacks.] The taint of black slavery "is the plague-spot upon the social condition of the Americans;" for "be the blight of slavery more or less pernicious to its victim, it reaches the master caste as well as that of the slave." In the Northern and Middle states—where the numbers never were very great—a system of gradual emancipation was many years since adopted, which, by giving freedom, at a certain age, to the children born after a given date, has put an end to slavery in these districts. The slaves are chiefly to be found in the southern and south-western states, where the culture of the soil is similar to that of our West India islands. It is unlawful to import slaves from abroad into any part of the United States; but it is legal to transport them from one slave-state to another; and the northern plantations—which offer very few of the ordinary excuses for the employment of Negro labour, and where in many districts the land is so exhausted that the slaves cannot reproduce as much as they consume—are becoming a vast breeding-ground, on which blacks are reared for the express purpose of supplying the terrible demands of an internal slave-trade which is carrying on both by sea and land. "The Americans," says Captain Basil Hall, "are perpetually taunting England with having entailed slavery upon their country." But, as regards the validity of the plea of necessity, the whole case has completely altered its character since the acquisition of Florida and Louisiana, countries cultivated wholly by slaves; and still more, by the introduction of states into the Union, where no similar necessity for slave-cultivation, on account of climate, was ever pretended to exist, but where, nevertheless, slavery has been introduced by the solemn act of the legislature. A recent observer, Sir Charles Lyell, while he gives a uniformly favourable account of the condition of the slave pop. of the U. S., is favourable to Negro emancipation, and takes every opportunity to stand up for the improvableness of their mental faculties. "It is useless," he says, "now to look back and wish, for the sake of civilization, that no Africans had ever crossed the Atlantic. Their number in the Union now exceeds 3,000,000, and, as they have doubled in the last twenty-five years, we must expect, unless some plan can be devised to check their increase, that they will amount, before the close of this century, to 12,000,000, by which time the white pop. will have augmented to 80,000,000. Notwithstanding this increase of Negroes, were it not for disturbing causes, I should cherish the most sanguine hopes of their future improvement and emancipation, and even their ultimate amalgamation and fusion with the whites, so highly has my estimate of their moral and intellectual capabilities been raised by what I have lately seen in Georgia and Alabama." It had previously been imagined that an impassable gulf separated the two races; but now it is proved that more than half that space can, in a few generations, be successfully passed over, and the humble Negro of the coast of Guinea has shown himself to be one of the most imitative and improvable of human beings. Yet the experiment may still be defeated, not so much by the fanaticism of abolitionists, or the prejudices of those slave-owners who are called perpetualists, who maintain that slavery should be permanent, and that it is a blessing in itself to the Negro, but by the jealousy of an unscrupulous democracy invested with political power. The abolitionists of the North endeavour to bring the powerful lever of public opinion to act upon their brethren of the South; and at every meeting of congress they also try to make slavery a national affair in that body, though the spirit of the constitution guarantees to each state the integrity of its separate laws and institutions. The first Abolition society was established in Philadelphia in 1833. The balance of political power between the slave and the free states in congress is still too closely adjusted to afford any prospect of a speedy termination to Negro slavery by any measures that would not endanger the permanence of the union. In the senate every state has an equality of representation with every other; and the slave states are 15 in number, the free states 16. In the house-of representatives the altered relations of pop. have increased the representatives of the free states by 3, and diminished those of the slave states by the same number since 1840. But there are other considerations to be taken into account besides the mere aggregate numbers of the representatives of slave states and free states. In the 32d congress, 8 of the old Atlantic free states will have fewer members by 20 than they had in the 27th, and 6 of the old Atlantic slave states by 25. On the other hand 6 of the new western free states will have more members by 20 in the 32d than they had in the 27th congress, and 6 of the new western slave states by 14. The old Atlantic states are becoming less powerful in the house-of-representatives; the more reckless pioneers of settlement in the far west are becoming more powerful there; and the ratio in which the representatives of the western free states are increasing exceeds that at which those of the western slave states are increasing even more than the ratio at which the representatives of the Atlantic slave states is decreasing exceeds that at which those of the Atlantic free states are decreasing. In a recent American journal, we find the following estimate of the number of the coloured race now believed to exist on this continent in slavery and in freedom:—

United States,	3,650,000
Brazil,	2,050,000
Spanish colonies,	1,470,000
South American republics,	1,130,000
British colonies,	750,000
Hayti,	850,000
French colonies,	270,000
Dutch colonies,	50,000

Danish colonies,	45,000
Mexico,	70,000
Canada,	35,000
Total,	16,370,000

Of these, 7,500,000 are in slavery in the United States, Brazil, and the Spanish and Dutch colonies; 250,000 are in progress of emancipation in the South American republics; and the remainder, 3,620,000, are free.

Physical characteristics.] The Americans are generally tall, with strong and well-proportioned frames, and fresh and ruddy complexions; but their features are said to want delicacy, and their physiognomy has little expression; the women have more of that delicate beauty which belongs to their sex, and generally have finer features and more expression in their countenances. They are mostly tall, with light and airy shapes, fine heads, and complexion of a dazzling whiteness. "Without participating," says Ward, "in all Mr Waterton's enthusiasm with regard to the American women, it is difficult for the most casual observer not to be struck with their appearance; nor do I know any part of the world, not even excepting England, where the display of female beauty is more striking than at New York and Philadelphia, where from the warmth of the climate, the light dress of the south of Europe is seen in conjunction with the freshness and bright colouring of a northern complexion." Sir Charles Lyell expresses his opinion that the principal cause of the different aspect of the Anglo-Saxon race in England and America will be found to be the climate. "Many who have been born in America, of families settled there for several generations, find their health improved by a visit to England, just as if they had returned to their native air; and it may require several centuries before a race becomes thoroughly acclimatized. The great difference of the species of indigenous animals and plants in North America, those of the middle and southern states being almost all distinct from the European, points to a wide diversity of climate, the atmosphere being drier, and there being a much greater annual range of the thermometer than in corresponding latitudes on the E side of the Atlantic. Even so cosmopolite a being as man may demand more than two centuries and a quarter before he can entirely accommodate his constitution to such altered circumstances, and before the successive generations of parents can acquire themselves, and transmit to their offspring, the new and requisite physiological peculiarities. English travellers often ascribe the more delicate health of the inhabitants here, to their in-door habits and want of exercise. But it is natural that they should shrink from exposing themselves to the severe frosts and long-continued snows of winter, and to the intense heat of the summer's sun. An Englishman is usually recognised at once in a party by a more robust look, and greater clearness and ruddiness of complexion, and it is surprising how distinguishable he is even from persons born of English parents in the United States. It is also a curious fact, which seems generally admitted, that the native Anglo-Australians bear a considerable resemblance to the Anglo-Americans in look and manner of speaking, which is a mystery, for there is certainly in that case no analogy between the climates of the two countries." There is undoubtedly considerable danger of a degeneracy in the physical characteristics of the white man in America. So long as the pop. is recruited by an immense influx from Europe, the change will be less perceptible; but that immigration once checked, it may become a very serious question.

Languages.] The languages spoken in the United States, exclusive of those of the indigenous tribes, are as numerous and diversified as the nations to which its emigrant population belong, namely: English, French, German, Dutch, Swedish, Spanish, and Gaelic. All these emigrants retain, in a greater or less degree, their native language, in which they perform their respective religious services, converse, and transact mutual business; the English language, however, is by far the most prevalent, and must in process of time become the universal instrument of oral communication. In it, all the public business of the different states is transacted, laws are promulgated, records kept, and diplomatic correspondence carried on in English. In the northern states, English is spoken in great purity, and pronounced with propriety, by persons of education. In the middle states, where the pop. is more mixed, the language is proportionally corrupted, especially in pronunciation. There are no provincial dialects in America. Emigrants from all parts of Great Britain have met and intermixed with each other, and with the natives of the country, and the peculiarities of dialect, therefore, have necessarily melted down into the general speech, which is common English. The low-bred Londoner does not there transmit his vulgar *shibboleth;* and the son of the Northumbrian is free from the *burr* which sticks in the throat of his father. Dialects can only be preserved by collective bodies speaking the language which they acquired in their youth, and cannot therefore continue in promiscuous colonies. The French and Spanish languages are confined chiefly to the states of Upper and Lower Louisiana, and must at no very distant epoch be completely superseded by the English, which will in process of time be the vernacular language from the Atlantic to the Pacific, and from the gulf of Mexico to the Frozen sea.

Literature and Public press.] "With their due proportion of genius, in common with all other nations, and with the advantage of a more general diffusion of popular intelligence than is to be found in any other community, too many of our citizens in *all* the learned professions," says an American author, "begin, continue, and end their career, on much narrower ground than their native capacity, properly unfolded by previous general information, would enable them to cover. Another obstacle to the growth of literature in the U. S., arises from the great propensity to consume the talent of the country in the effusion of newspaper essays and political pamphlets, instead of concentrating it in the production of some regular, consecutive work. Add to this, the universal vice of the U. S., a perpetual craving after novelty. The charge which Demosthenes brought against his own countrymen, that they were continually running about, and asking, 'Is there anything new?' is equally applicable to the Americans. This eternal restlessness and desire of change, pervade the whole structure of our society: the same man will start into life as a clergyman, then turn lawyer, next convert himself into a farmer and land-jobber, and, taking a seat in congress, or some state legislature, by the way, end his days as a merchant and money-broker. The people are incessantly shifting their habitations, employments, views, and schemes: the residence of a servant does not average two months in each place; the abode of a whole household is generally changed once a-year, and sometimes oftener; numerous families, that have been previously settled in the elder states of New York, Connecticut, and Massachusetts, are continually migrating into Ohio, or the territories of Alabama, Illinois, and Mississippi; the executive, the legislators, the magistrates, and officers of all kinds, are changed biennially, or annually, or half-yearly, according to the greater or less infusion of the restless spirit of democracy into our various forms of government." These causes are all hostile to the cultivation of literature in this country; at the same time there is a wide diffusion of knowledge throughout the states. The cheap publication of books encourages a demand and supply in the U. S. almost unexampled in any European state. There have been eight editions, comprising 7,500 copies, of Stewart's *Moral Philosophy* published during the last twenty years,—a greater number, we suspect, than has been sold in our own country. Periodical works and newspapers abound in extraordinary and countless quantity. In most cases the editions of one and the same work are larger and more frequent in the U. S. than in any other country. Many reprinted English works have here passed through three or four editions, while the publishers of the original in England have but one. In one instance, the sale of a book in America amounted to 100,000 copies; whereas in England, only four editions, of 1,000 copies each, were disposed of. The sale of five book-selling establishments amounted, in 1836, to 1,350,000 dollars. A single publisher paid, in the five years preceding 1834, 135,000 d. for copyrights, out of which 30,000 d. were for two works only. The number of volumes issued in the U. S., from the 1st of July, 1850, to the same date in 1851, was 1,298. The number of pages in these vols. amounts to 213,049. The distinct works composing the vols. are 1,176. Of these 817 were published in New York, 223 in Philadelphia, and 203 in Boston. The department of fiction, including every class of novels and tales in prose, comprised 249 distinct works. —The first newspaper published in America was *The Boston News Letter*, which appeared in 1704. There were but 7 papers in the United States in 1750; in 1810 there were 359 (including 25 published daily), which circulated 22,200,000 copies in the year. In 1823 they had increased to 588, and in 1828 they amounted to 892, of which Pennsylvania alone had 185, and New York 161. The number of copies circulated in the year by these journals exceeded 30,000,000; the circulation in 1842 did not fall short of 100,000,000. The whole number of journals in the U. S. on the 1st of June, 1850, amounted to 2,800. The aggregate circulation of these papers and periodicals is about 5,000,000, and the entire number of copies printed annually in the U. S. amounts to 422,600,000. The following table shows the number of daily, weekly, monthly, and other issues, with the aggregate circulation of each:

	Number.	Circulation.
Dailies,	350	750,000
Tri-weeklies,	150	75,000
Semi-weeklies,	125	80,000
Weeklies,	2,000	2,875,000
Semi-monthlies,	50	300,000
Monthlies,	100	900,000
Quarterlies,	25	20,000
Total,	2,800	5,000,000

Since the census was taken a large number of new journals have been established, chiefly in the state of California and in the new territories of Oregon, New Mexico, Utah, and Minesota. There is a journal, edited with spirit and ability, among the Cherokee Indians, which is printed in the Cherokee language and characters. Copies of a great many daily papers are sold at 1 cent each; but nearly two-thirds are sold at from 2 to 4 cents. Of the above number of publications, 424 are issued in the eastern states, 876 in the middle states, 716 in the southern states, and 784 in the western states. There is one journal for every 7,184 free inhabitants. In the great majority of cases the editor is both the printer and publisher, or proprietor, of his paper.

Education.] The northern states have made the most liberal provisions for public education; but it has been uniformly a part of the Land-system of the U. S. to provide for public schools. The land thus appropriated, up to the end of 1853, amounted in all to 48,909,505 acres; besides special state appropriations. But the chief dependency for the costs of popular education is direct taxation. The total expenditure on public schools in 1853 was 9,549,542 d., of which 4,653,006 d. were raised by direct taxation. The great body of the New England pop. can read and write, and many are acquainted with the classical languages. The school-fund of Connecticut appears to be 1,858,074 d. = £371,000. The amount of division to the common schools annually, in 1822, was 67,791 d., which enables the state to expend 60,000 d. annually on schools. There is a rule to enforce a distribution of this fund to the different district-schools, according to the number of children in each, between the ages of 4 and 16. The mode pursued is this: the inhabitants establish school-districts, and appoint a committee in each district, whose duty it is to manage the concerns of the district and provide an instructor for the school with the consent of the district and the approbation of the visitors. There were 1,580 schools in the state of Connecticut alone in 1824. The same principle and regulations apply to all the western states in the Union. The new states have made immense appropriations of land (from the sale of which the fund is raised) for the purpose of general education; thus is laid the foundation to support schools and colleges to the fullest extent wanted. In the southern states, the state of education is deplorable. Before the revolution, the wealthy Carolinians sent their children to Europe for their education; since that period, they have sent them to the northern states,—an unequivocal testimony of their own inferiority.

Colleges.] There were in the United States 43 incorporated colleges or universities, in 1839; and the statistics of these were as follows:

The number of instructors in 32 colleges was	217
The number of graduates in 30, in 1828, was	652
The number of under-graduates in 33 colleges, in 1828 and 1829,	2,809
Number of volumes in 30 college libraries,	128,118
Number of volumes in 25 students' libraries,	66,730

From calculations made in the year 1830, it appears that the number and proportion of college students in each state of the Union, compared with the pop., was at that date as follows:—In Maine, 128 students, being 1 student in 3,300 persons; New Hampshire 139, or 1 in 2,200; Vermont 138, or 1 in 2,000; Massachusetts 543, or 1 in 1,070; Rhode Island 34, or 1 in 2,700; Connecticut 205, or 1 in 1,400; New York 543, or 1 in 3,700; New Jersey 115, or 1 in 2,900; Pennsylvania 436, or 1 in 3,200; Delaware 10, or 1 in 8,000; Maryland 171, or 1 in 2,600; District of Columbia 38, or 1 in 1,400; Virginia 370, or 1 in 3,200; North Carolina 96, or 1 in 7,500; South Carolina 214, or 1 in 2,800; Georgia 107, or 1 in 3,800; Alabama 33, or 1 in 12,000; Mississippi 28, or 1 in 4,700; Louisiana 18, or 1 in 17,000; Kentucky 250, or 1 in 2,600; Ohio 169, or 1 in 5,800; Tennessee 144, or 1 in 4,000; Indiana 4, or 1 in 75,000. It is estimated, that, in the New England states, there is now, on the average, one student in college for 1,650 inhabitants; in the middle states, 1 for 3,400; and in the states south and west of Pennsylvania, 1 for 4,400. The chief literary institution in the U. S. is Harvard university in the state of Massachusetts, at Cambridge, 3 m. W of Boston. Its professorships now amount to upwards of 20, including the medical department, which is an integral part of the institution, although for the greater convenience of medical students, it occupies buildings in Boston. The university comprises five colleges and a new hall. The academical course is completed in four years, at an expense of about 100 guineas a- year.

The whole number of incorporated colleges in the U. S., in 1840, was 100. All the states in the Union have one or more of these institutions. The whole number of students, according to the catalogue for 1840, was 9,142, giving on an average about 90 students to each college. Most of the colleges in New England and the middle states, with a very few of those in the south and the west, have able faculties. The collegiate system of the U. S. has been derived from the collegiate system of England. There are, indeed, many points of dissimilarity, but these have grown out of the peculiar circumstances which have attended the carrying out of the English system. In all the older colleges, and in very many of the more recent ones, the students reside together in the collegiate buildings, under the immediate eye of the tutors or of the unmarried professors, and they eat at a common table. They are divided into four classes, called *freshmen*, *sophomores*, *juniors*, and *seniors*, the whole course thus extending through four years. There are, however, comparatively but few students who spend so long a time in college; inasmuch as they are allowed to enter any class if they can bear the requisite examination. The collegiate year, like that of the English, embraces about nine months, and is divided into three terms of scarcely three months each, with two short and one long vacation between them. The whole property of an American college is held by a corporation, usually known as a board of trustees. There are in the U. S. at least 81 libraries of 5,000 volumes and upwards, to which the public, more or less restrictedly, have access, and of these 49 are immediately connected with colleges or public schools. The aggregate number of volumes in these libraries is about 980,413, 200,000 of which are in the state of Massachusetts, 159,000 in the state of Pennsylvania, 158,000 in the state of New York, 81,000 in the state of Connecticut, 53,000 in the district of Columbia, 41,000 in the state of Virginia, 38,000 in the state of Maine, 37,000 in the state of Rhode Island, 30,000 in the state of Ohio, 30,000 in the state of South Carolina, 28,000 in the state of New Jersey, 22,000 in the state of New Hampshire. The number of volumes thus provided in these states respectively, excluding from notice those contained in libraries under 5,000 volumes, appears to be: In Rhode Island about 34 to every 100 of the pop., in Massachusetts about 27, in Connecticut 26, in Pennsylvania 9, in New Jersey 7, in New Hampshire 7, in Maine 7, in New York 6, in South Carolina 5, in the district of Columbia, containing the capital of the Union, it is 121. The city of Boston possesses about 73 volumes to every 100 of its inhabitants; Richmond, the capital of Virginia, 59; Philadelphia 53, New York 35, Washington 119. The library of oldest foundation, and also of greatest extent in the U. S., is that of Harvard University, which dates from 1638. Its present number of books, including the "society libraries" of the students, is stated at 68,500. The library of Yale college, founded in 1700, contains 34,500 vols. The library of the American Antiquarian society at Worcester is particularly rich in American history, and contains many valuable MSS.; its present number of volumes exceeds 13,000. The Mercantile Library of New York contains about 25,000.

Religion.] Since the American revolution, no religious establishment is recognised by the federal government. As this is a matter of some importance, it shall be given in Dr. Morse's own words: "The constitution of the U. S. provides against the making any law respecting an establishment of religion or prohibiting the free exercise of it: and in the constitution of the respective states, religious liberty is a fundamental article. Religion is here placed on its proper basis, without the feeble and unwarranted aid of the civil power: it is left to be supported by its own evidence, and the lives of its professors, and the Almighty care of its divine author." It would be an erroneous conclusion, were it inferred from this, as has been attempted by some, that the people of the U. S. are inattentive to religious observances: the fact is directly the reverse; in no other country are there so many churches in proportion to the numbers of the pop., and in no other country are the working clergy so well paid. Religion, although deriving no pillars from the state, is possessed of 8,000 places of public worship, and "intolerance is disarmed by being let alone," and the various Christian sects agree to differ. There are no rich livings, indeed, but, on the other hand, none of the ministers of religion are subjected to the miserable penury on which many of their brethren in the gospel are allowed to starve in our own country. There are, in certain parts of the Union, lands which were given to the Episcopal church anterior to the Revolution, and which in these particular cases render any further contribution unnecessary; but the clergy of the U. S. are almost uniformly supported by assessments, voluntary so far as the law is concerned, but which public opinion renders in no small degree obligatory. The Americans are distributed into all the different sects which are to be found on this side of the Atlantic. The most numerous are Baptists, Methodists, Congregationalists or Independents, and Presbyterians. The churches or edifices for public divine worship in the U. S. in 1850, numbered 36,000 (36,011), of which the Methodists owned one-third, or 12,467; the Baptists nearly one-fourth, or 8,791; the Presbyterians, the next number or 4,584; and if we count the Dutch Reformed, Congregational, Lutheran, and German Reformed with the Presbyterian (and the differences between all these seem slight and unessential), the total was 8,112. But the estimated capacity of the Presbyterian and allied churches was greater in the average than that of the Baptist and Methodist churches, so that, while all the Methodist churches would accommodate but 4,209,333 worshippers, and all the Baptist but 3,130,878, the Presbyterians and related churches aforesaid had room for 3,705,211 worshippers. The Catholics had but 1,112 churches accommodating 620,950 worshippers. The Episcopalians had 1,422 churches, accommodating 625,213 worshippers. The average number that each church edifice in the Union will accommodate is 384; the total value of church property, in 1850, was 86,416,639 d. If all the churches should be filled at one time they would hold 13,849,896 persons—probably something near the total pop. that could at one time attend church.

Federal constitution.] The U. S. of North America are a political aggregation of independent but confederated republics, each of which has a particular constitution of its own, and a distinct and separately organized government. The form of constitution adopted by the particular states is nearly the same; but it is proper here to premise, that the colonies now forming the U. S. were settled at different times, and by different classes of emigrants, and that the degree of power retained by the crown over them was extremely various. In some instances, the governor and council were named in England; in others, the governor alone was appointed there; while, in the case of Rhode Island and Connecticut the people enjoyed by charter so uncontrolled an independence in the regulations of their local concerns, that the revolution did not render necessary the slightest alteration in the forms of internal administration. One of these colonies, indeed, retains even to this day the charter of Charles II., as its system of state-government; another only parted with the royal charter for a constitution of its own making in 1818. With distinct legislatures, and with such discrepancies in the sources from which their respective executives derived their authority, it need hardly be observed, that each colony was wholly independent of every other. The idea of a congress was first partially acted on in 1722; and one composed of deputies from all the eastern and middle colonies was held at Albany in 1754 with the approbation of the mother-country, for the purpose of considering the best means of defending the provinces against the French, then rulers of Canada, and their Indian allies. On this occasion a plan was proposed of a federal government; but the scheme was not relished either by the ministry at home or by the colonies, and was never put into execution. The first congress held on account of the alleged grievances sustained from England, and which met at New York in 1765, maintained that the power of taxation resided solely in the colonial legislatures. The language adopted at the meeting of 1774 was still more decided; but it was in the ensuing year that the congress assembled, which, on the 4th of July, 1776, declared the independence of the United States. On the separation from England, the little power which had been either claimed or exercised by the British government, was at once assumed by the individual states. The authority of congress was at first wholly of a recommendatory character, and rested on the general conviction of the necessity of co-operation, rather than on any defined principles. But the importance of some understood compact was early felt; and even before the declaration of independence was formally made, a committee was appointed to prepare an act of confederation, which having been agreed to in congress, and proposed to the several states for their ratification in 1777, was generally assented to by them in the course of the ensuing year. This act was essentially between 13 independent powers for specific purposes. It was indeed resolved that the union of the states was to be perpetual; but it was also declared that "each state retains its sovereignty, freedom, and independence, and every power, jurisdiction, and right, which is not by this confederation expressly delegated to the United States in Congress assembled." The idea of an association of equal sovereignties was so fully persevered in, that each colony, without regard to its extent or population, had but one vote in the congress: thus recognizing a very pernicious principle, and giving, it might be, to the opinions of an actual minority, a power as great as to those of the majority. The principal object of the league was to act with energy for the common defence; and for this purpose a general treasury was established for the payment of the troops, and such national matters. Congress had no power to raise taxes; even the land-forces for the common defence were raised by the several states, the quotas only being fixed by congress. The limitations of the powers of the superintending authority occasioned a degree of inaction which was very far from according with the pressing nature of the demands made by the general or the local governments; and when fear from abroad had ceased, the whole confederacy was found inefficient for the promotion of the general interests of the several states. The present general constitution was adopted in 1787, and carried into effect on the 4th of March, 1789. The "constitution," the "supreme law," as it is called, begins thus: "We, the people of the United States, in order to form a more perfect union, establish justice, ensure domestic tranquillity, provide for the common defence, promote the general welfare, and secure the blessings of liberty to ourselves and our posterity, do ordain and establish this constitution for the United States of America." The whole constitution, therefore, proceeds from the people of the states; it is their act, and it is passed for the especial and admirable purposes enumerated. Not only the living people, therefore, of the states have a right to alter, as their predecessors had to annul this constitution—they are bound, in order to attain the objects and ends described, to alter and amend it, if it fail to attain these ends. The great distinction which exists between the present system, and that for which it was substituted is, that the federal government, instead of being a confederation of the states, now acts, in those cases in which it possesses jurisdiction directly as individuals, without the intervention of any other authority, while at the same time the powers which are not transferred by the constitution to the general government, remain in the individual states as independent republics. Thus the American people are under a double government, each of which is independent of the other. If a subject fall within the cognizance of the state, the law respecting it must be passed by the local legislature, the judiciary authorities who interpret it are appointed under the sanction of the provincial constitution, as also the executive or ministerial officers who give effect to the decision. So, on the other hand, should the matter appertain to the general government, congress provides the law in the case, the United States Judiciary construe it, and it is enforced by persons deriving their authority from the federal power. Piracy and murder might be adduced as familiar instances of this distinction: these crimes being cognizable by different jurisdictions, could in no part of the investigation come before the same tribunals. Many imperfections have, by politicians, been pointed out in this constitution. It wants, say they, a declaration of rights. The people are not secured in the benefits of even the common law. The number of representatives is too small for that of the people represented. The senate has too great preponderance in the constitution, and is in such a situation as to be daily acquiring more. The judiciary department is so constructed, that, by continued appeals, a process may be extended to an extraordinary length, and it possesses the anomalous power of annulling a law by declaring it unconstitutional. The president has no constitutional council; and will be naturally led to choose one for himself from the great officers of the state. Hence, too, arises the power of the senate, in the appointment of public officers; and a connexion between that branch of the legislature and the executive branch by far too close for the safety of the state. Treaties are declared to be supreme laws of the land, although, in negotiations concerning them, the opinion of the representatives be not asked. The general clause at the end of the enumerated powers of congress may be construed into a sense very dangerous to the liberty of the country. The power of congress is superior to that of the governments of the particular states, so that the declarations of rights made by these states are of no consequence. No provision is made for the preservation of the liberty of the press; trial by jury in civil causes is not secured; nor is the danger of standing armies in time of peace sufficiently obviated. Among the unavoidable evils of a double government are increased expense and a jealousy among the component parts, whether directed against the whole, or against each other: 'a sectional feeling,' as it is named, is called forth on all questions that suppose a divided and opposite interest. Much has been said by the advocates for American wisdom, towards the extenuation of these blemishes; but, after all that their ingenuity has been able to suggest, they are obliged to confess, that many things exist in the constitution of the U. S. loudly requiring amendment.

Congress.] In the federal democratic republic of the U. S., all legislative powers are vested in Congress, which consists of a senate and a house-of-representatives. The house-of-representatives is composed of members chosen every second year by the people of the several states, and in number in accordance with the pop. of each; and in order to ascertain the number each state is entitled to, a census is taken every 10 years, excluding from the enumeration for this object two-thirds of the slaves, and all Indians not taxed. Each state is entitled to at least one representative. Vacancies are filled by intermediate elections. The house chooses its speaker and other officers. No person under 25 years of age, who has been less than seven years a citizen of the U. S., and who is not a resident of the state electing him, is qualified for representative. The constitution provided for a specific number of representatives from each state to compose the House until the ascertainment of the pop. under the census of 1790; but since then legislation has decennially fixed the number to be elected. From the 3d March, 1793, the apportionment was one representative to every 33,000 of the representative pop.; after 1803, one to every 33,000 also; after 1813, one to every 35,000; after 1823, one to every 40,000; after 1833, one to every 47,000; after 1843, one to every 70,680; and after 3d March, 1853, 233 representatives to be divided *pro rata* to the several states. The following table shows the number of representatives to which each state has been entitled since the establishment of the government:—

States.	1787.	1793.	1803.	1813.	1823.	1833.	1843.	1853.
Maine,	—	—	—	—	7	8	7	6
New Hampshire,	3	4	5	6	6	5	4	3
Massachusetts,	8	14	17	20	13	12	10	11
Rhode Island,	1	2	2	2	2	2	2	2
Connecticut,	5	7	7	7	6	6	4	4
Vermont,	—	2	4	6	5	5	4	3
New York,	6	10	17	27	34	40	34	33
New Jersey,	4	5	6	6	6	6	5	5
Pennsylvania,	8	13	18	23	26	28	24	25
Delaware,	1	1	1	2	1	1	1	1
Maryland,	6	8	9	9	9	8	6	6
Virginia,	10	19	22	23	22	21	15	13
North Carolina,	5	10	12	13	13	13	9	8
South Carolina,	5	6	8	9	9	9	7	6
Georgia,	3	2	4	6	7	9	8	8
Alabama,	—	—	—	—	3	5	7	7
Mississippi,	—	—	—	—	1	2	4	5
Louisiana,	—	—	—	—	3	3	4	4
Tennessee,	—	—	3	6	9	13	11	10
Kentucky,	—	2	6	10	12	13	10	10
Ohio,	—	—	—	6	14	19	21	21

Indiana,	—	—	—	—	3	7	10	11
Illinois,	—	—	—	—	1	3	7	9
Missouri,	—	—	--	—	1	2	5	7
Arkansas,	—	—	—	—	—	—	1	2
Michigan,	—	—	—	—	—	—	3	4
Florida,	—	—	—	—	—	—	1	1
Texas,	—	—	—	—	—	—	2	2
Iowa,	—	—	—	—	—	—	2	2
Wisconsin,	—	—	—	—	—	—	3	3
California,	—	—	—	—	—	—	2	2

In addition to these representatives from states, the house admits a delegate from each organized territory, who has the right to debate on subjects in which his territory is interested, but cannot vote. California has 2 members by a special act. The senate consists of 2 members from each state, elected by the legislatures thereof respectively for 6 years. One-third the whole body is renewed biennially, and if vacancies happen, by resignation or otherwise, during the recess of the legislature of any state, the executive of such state makes a temporary appointment until the next meeting of the legislature, which fills such vacancy. Senators must be at least 30 years old, must have been citizens of the U. S. for 9 years, and be residents of the state by which chosen. Each senator has one vote. The vice-president of the U. S. is *ex officio* president of the senate, but a president *pro tempore* is elected by and from among the senators, who, in the absence of the president, acts in his stead. Besides its ordinary legislative capacity, the senate is vested with certain judicial functions, and its members constitute a high court of impeachment. No person can be convicted by this court unless on the finding of a majority of senators, nor does judgment extend further than to removal from office and disqualification. Representatives have the sole power of impeachment.

President.] The executive power is vested in a president, who is elected by an electoral college, chosen by popular vote, or by the legislature of the state, the number of electors being equal to the number of senators and representatives from the states to congress. His term of office is 4 years, but he is eligible for re-election indefinitely. The electors forming the college are themselves chosen in the manner prescribed by the laws of the several states. A majority of the aggregate number of votes given is necessary to the election of president and vice president, and if none of the candidates has such a majority, then the election of president is determined by the house-of-representatives, and that of the vice-president by the senate, from among the three candidates having the highest number of electoral votes, and in doing so, the vote is taken by states, the representatives of each state having only one vote, which must, of course, be determined by a majority of their number. No person can be president or vice-president who is not a native-born citizen, of the age of 35 years, and who has been a resident of the U. S. for 14 years. The president is commander-in-chief of the army and navy, and of the militia when in the service of the Union. With the concurrence of two-thirds of the senate, he has the power to make treaties, appoint civil and military officers, levy war, conclude peace, and do all that rightly belongs to the executive power. He has a *veto* on all laws passed by congress, but so qualified, that notwithstanding his disapproval, any bill becomes a law on its being afterward approved of by two-thirds of both houses of congress. The president has a salary of 25,000 d. per annum, and a house at Washington for a residence during his official term. The vice-president is *ex-officio* president of the senate; and in case of the death, resignation, or other disability of the president, the powers and duties of that office devolve upon him for the remainder of the term for which the president had been elected. This provision of the constitution, for the first time since the foundation of the government, came into operation in 1841, on the demise of General Harrison, who died just one month after his inauguration, when John Tyler, the vice-president, succeeded. In case of the disability of the vice-president, the president of the senate *pro tempore* takes his place.—The offices of president and vice-president have been occupied by the following gentlemen since the adoption of the constitution:

	Presidents.	Vice-presidents.	Terms of Office.	
1.	George Washington,	John Adams,	30th April, 1789,	to 4th March, 1793
2.	Do. do.	Do. do.	4th March, 1793,	— 1797
3.	John Adams,	Thomas Jefferson,	— 1797,	— 1801
4.	Thomas Jefferson,	Aaron Burr,	— 1801,	— 1805
5.	Do. do.	George Clinton,	— 1805,	— 1809
6.	James Madison,	Do. do. (d. 20th Apr., 1812),	— 1809,	— 1813
7.	Do. do.	Elbr. Gerry (d. 23d November, 1814),	— 1813,	— 1817
8.	James Monroe,	Daniel D. Tompkins,	— 1817,	— 1821
9.	Do. do.	Do. do.	— 1821,	— 1825
10.	John Quincy Adams,	John C. Calhoun,	— 1825,	— 1829
11.	Andrew Jackson,	Do. do.	— 1829,	— 1833
12.	Do. do.	Martin Van Buren,	— 1833,	— 1837
13.	Martin Van Buren,	Richard M. Johnson,	— 1837,	— 1841
14.	William Henry Harrison,	John Tyler,	— 1841,	4th April, 1841
	John Tyler (on the death of General Harrison),		4th April, 1841,	4th March, 1845
15.	James K. Polk,	George M. Dallas,	4th March, 1845,	— 1849
16.	Zachary Taylor,	Millard Fillmore,	— 1849,	9th July, 1850
	Millard Fillmore (on the death of General Taylor),		9th July, 1850,	4th March, 1853
17.	Franklin Pierce,	Wm. Rufus King (d. 18 Ap., 1853),	4th March, 1853,	— 1857
18.	James Buchanan,	James C. Breckinridge,	4th March, 1857	

Cabinet.] The administrative business of the nation is conducted by several officers, with the title of secretaries, etc., who form what is termed 'the Cabinet.' These are the secretary-of-state, the secretary-of-war, the secretary-of-the-navy, the secretary-of-the-treasury, the postmaster-general, the secretary-of-the-interior, and the attorney-general—the last being the official law authority for advisement in administrative affairs. Each of these presides over a separate department.

The Judiciary.] The judicial power of the U. S. is vested in one supreme court, and in such inferior courts as congress may from time to time establish. The present judicial establishment consists of a supreme court, district courts, and circuit courts. The supreme court is composed of one chief justice, and six associate justices, who hold a court in the city of Washington annually; besides which, each of these justices attends in a certain circuit, comprising two or more districts appropriated to each, and, together with the judge of the district, composes a circuit court, which is held in each district of the circuit. The district courts are held respectively by the district judge alone. Appeals are allowed from the district to the circuit courts in cases where the matter in dispute, exclusive of costs, ex ceeds the sum or value of fifty dollars, and from the circuit courts to the supreme court, in cases where the matter in dispute, exclusive of costs, exceeds the sum or value of 2,000 d.; and in some cases, where the inconvenience of attending a court by a justice of the supreme court is very great, the district courts are invested with circuit court powers. Each state is one district, for the purpose of holding district and circuit courts therein, with the exception of New York, Pennsylvania, Virginia, Tennessee, Louisiana, and Alabama, each of which is divided into two districts. There are, besides, territorial courts, which are temporary, and lose that character whenever a territory becomes a state. In addition to the district court for the district of Columbia, there is a circuit court for that district, which exercises, under the authority of congress, common law and equity jurisdiction similar to that of the county courts of Maryland and Virginia. Each court has a clerk, a public attorney, or prosecutor, and a marshal, all of whom are appointed by the president of the United States, with the exception of the clerks, who are appointed by the courts. The compensation of the judges is fixed by law; that of the clerks, attorneys, and marshals, consists of fees. The judges of the U. S. have adopted a dictum of Lord Hale's, and declared that the judiciary is not a subordinate but co-ordinate power of the state. They have declared too that the "Constitution is the supreme power," that the judicial power extends to all cases of law and equity under it; and on these premises they have assumed to be the interpreters of the supreme power, and have gone so far as to declare that any "act of the Congress of the U. S. which shall be, in the opinion of the supreme court of the U. S., contrary to the written words of the constitution, is *ipso facto* void and of no effect." Justly alarmed at such inordinate pretensions, President Jefferson proposed to appoint the judges for a term of years. President Jackson renewed the proposition, and though it has never been carried into effect as to the supreme court, in no less than 22 states of the Union, the judges are now elected by the people for terms of years.

State governments] The subdivisions into which the several states of the Union are formed, are styled *counties*, with the exception of the states of S. Carolina and Louisiana. In S. Carolina the subdivisions are termed *districts*, and in Louisiana, *parishes*. In the six New England states, and also in the states of New York, New Jersey, Pennsylvania, and Ohio, the counties are subdivided into *townships*, and in Delaware into *hundreds*; but in the rest of the states no such subdivision as that of township is known. In the New England states these townships are commonly styled *towns*. They differ considerably in size, generally varying from about 5 to 6 m. square. They are incorporated by the legislature of the respective states with certain rights and privileges, and have a distinct police, which is conducted by officers elected annually by the citizens. Some of the principal of these officers are a town-clerk, selectmen, assessors, school committee, overseers of the poor, surveyors of highways, &c. The townships in the New England states, and in the state of New York, are subdivided into school districts of convenient size; and in these districts common schools are maintained, at least a part of the year, to which all the inhabitants, both poor and rich, have an equal right to send their children for instruction. The money necessary for the support of the schools and of the poor, for the repair of roads, &c., in the several towns, is derived from a tax upon the inhabitants, or from funds appropriated to the object. In those states in which the subdivision of township is unknown, the word *town* is used in a more restricted and appropriate sense for a compact collection of houses. The differences which exist between the general government and the state governments are unimportant. They consist each of an executive and legislative department; but each state has prescribed the manner in which it pleases to have its own internal administration regulated, as well as the extent of authority which it deems necessary to confide to its local rulers. The only restraint on the unlimited freedom of the inhabitants to decide for themselves on this matter is to be found in the provision in the federal constitution, that the U. S. shall guarantee to every state a republican form of government. But though there is no concurrence in the particular details of the systems established by the several states, they agree with one another, and with the constitution of the Union, in the general distribution of powers. The governors are elected more frequently than the president, in most cases annually or biennially. They are in some states chosen by the legislature, in others by the people at large. The lower house of the legislature—except in one instance—is elected annually, and the term of service of the councillors or senators differs from one to four years. The qualification of the electors is various. In the eastern and middle states it is so small as to amount almost to universal suffrage, paupers and criminals being alone excluded. Voting by ballot very generally prevails, and is easily managed.

Revenue of the States.] On the subject of American finance and taxation, it ought always to be kept in mind that each of the states in the union has a distinct government and legislature of its own, which levies taxes for defraying the expenses of its own militia, and its own civil administration, including the salaries of the governors, judges, and all other public officers, and the expense of the state legislature, &c. These state-taxes are levied differently in different states. The revenues of the state of New Jersey, for instance, are raised by taxes on land, iron-works, mills, distilleries, breweries, ferries, fisheries, carriages, stages, taverns, horses, cattle, &c. But the most common way of raising the state-taxes is by a poll-tax and a property-tax. These state-taxes are never paid into the treasury of the government of the U. S.; nor do they make any part of the ways and means in the budget of the federal government presented yearly to congress. It is necessary to remark this in order to prevent our readers from falling into the common mistake of those whose whole knowledge of American revenue and expenditure is derived from the newspaper reports of the president's speech, and the American budget as presented to congress, and who consequently suppose that the receipts and disbursements of the federal government comprise all that the people of the U. S. pay in taxes.

State debts.] The following table exhibits the debts of the states, and the annual interest. It may not be strictly accurate, but it is as nearly so as possible. It will be seen that Pennsylvania exhibits a very prominent attitude in this table, the gross amount of her indebtedness being nearly 40,000,000 d.:—

	Debt in 1839.	Debt in 1848.	Debt in 1858.
Maine,	554,976 d.	979.000 d.	1,034,277 d.
N. Hampshire,	—	None.	—
Vermont,	—	None.	—
Massachusetts,	4,290,000	6,091,047	6,313,456
Rhode Island,	—	—	386,311
Connecticut,	—	33,212	—
New York,	16,015,631	23,937,249	32,441,944
New Jersey,	—	62,596	95,000
Pennsylvania,	30,784,003	40,424,737	39,268,111
Delaware,	—	None.	—
Maryland,	15,000,000	15,900,000	14,854,204
Virginia,	4,129,700	14,400,507	33,605,159
N. Carolina,	—	977,000	7,181,923
S. Carolina,	5,753,770	3,622,039	6,192,743
Georgia,	500,000	1,903,472	3,354,750
Florida,	3,950,000	—	158,000
Alabama,	13,882,000	10,385,938	5,098,000
Mississippi,	12,000,000	7,271,707	7,271,707
Louisiana,	19,735,000	16,238,131	10,701,642
Texas,	—	11,050,201	
Arkansas,	3,100,000	3,862,172	2,981,133
Tennessee,	2,600,000	3,337,856	16,643,607
Kentucky,	3,765,000	4,531,913	5,574,244
Ohio,	10,592,498	19,173,223	17,131,919
Michigan,	5,340,000	2,849,939	2,337,630
Indiana,	10,064,000	6,556,437	7,357,874
Illinois,	10,880,000	16,612,795	11,138,454
Missouri,	2,500,000	956,261	19,938,000
Iowa,	—	55,000	128,010
Wisconsin,	—	None.	100,000
Total,	175,466,578 d.	211,252,432 d.	254,080,083 d.

Of the above sum of 254,080,083 d., being the amount of state debt on 1st January 1859, 60,940,580 d. was contingent debt; the rest absolute. The interest on the absolute debt amounted at that date to 11,672,110 d.; in 1849 it was 8,684,035 d.

The stocks and bonds issued for banking purposes, and to be lent to railroad and canal companies, are treated in this table as state debts. So far as the general interests of the state are affected by state debts, there is no difference between bonds issued and lent to corporations, and those issued to procure money for constructing public works by the state. In both cases, the stock is disposed of in the foreign market, and the specie must be sent from the U. S. to pay the interest; and it is material to the credit of the stocks, that all loans of credit certificates to incorporations should be at once recognised and acknowledged as state debts. There are a great number of cities in the Union, besides those enumerated in the present table, which have contracted debts, the amounts of which are unknown. The bank of the U. S. issued bonds, post-notes, and other evidences of debt, and put them afloat in Europe, to the amount, probably, of 20,000,000 d. There are also a class of monied incorporations, which possess, or have assumed, the right of issuing bonds, and these have been sold or hypothecated in Europe, to the amount of several millions of dollars, which are an addition to the debt against this country. Of the states which now compose the Union, the following nine have no public debt:—New Hampshire, Vermont, Rhode Island, Delaware, Florida, and Wisconsin. The following nine states have regularly paid their interest without interruption: Massachusetts, New York, Ohio, Virginia, S. Carolina, Georgia, Kentucky, Tennessee, and Alabama. Pennsylvania suspended payment in August 1842, and resumed in February 1845. The following eight have failed to pay their debts: Mississippi failed to pay in May 1841; Indiana, July 1841; Illinois, January 1842; Florida, January 1842; Michigan, January 1842; Maryland, January 1842; Louisiana (on part of her debt), December 1842; Arkansas, 1841. By the constitution of 1787, the judicial power of the supreme court of the U. S. was extended to all controversies "between a state and the citizens thereof and foreign states, citizens, or subjects." Had this article remained in force, the foreign holders of state-bonds could have enforced the payment of their debts through the supreme court. This provision of the constitution was vindicated in the *Federalist* in these terms: "The Union will undoubtedly be answerable to foreign powers for the conduct of its members, and the responsibility for an injury ought ever to be accompanied with the faculty of preventing it. As the denial or perversion of justice by the sentences of courts is with reason classed among the just causes of war, it will follow that the federal judiciary ought to have cognizance of all causes in which the citizens of other countries are concerned. This is not less essential to the preservation of public faith than to the security of public tranquillity." To these reasons Judge Story adds this one: "It is of the greatest national importance to advance public as well as private credit in our intercourse with foreign nations and their subjects. Nothing can be more beneficial in this respect than to create an impartial tribunal to which they may have resort upon all occasions when it may be necessary to ascertain or enforce their rights." The constitution, however, was amended, and the 11th amendment is, that "The judicial power of the U. S. shall not be construed to extend to any suit in law or equity commenced or prosecuted against one of the U. S. by citizens of another state, or by citizens or subjects of a foreign state." The object of this amendment is thus related by Judge Story in his *Commentary on the Constitution:* "It is a part of our history that, at the adoption of the constitution all the states were greatly indebted, and the apprehension that these debts might be prosecuted in the federal courts, formed a very serious objection to that instrument. Suits were instituted, and the supreme court maintained its jurisdiction. The alarm was general, and to quiet the apprehensions that were so extensively entertained, this amendment was proposed in congress, and adopted by the state legislatures." The constitution of the U. S. was thus distinctly altered in order to deprive foreign creditors of the means which the law gave them to recover their debts. It was changed to relieve "the apprehension" of certain states, that the general government would be compelled to enforce the payment of their foreign debt.

General Expenditure and Revenue.] The following is a statement of the annual expenditure, exclusive of the public debt, from the commencement of the U. S. constitutional government to the 31st of December, 1837:—

For the year	1791,	1,919,590 d.	For the year	1815,	26,953,571 d.
...	1792,	1,877,904	...	1816,	23,373,433
...	1793,	1,710,070	...	1817,	15,454,610
...	1794,	3,500,546	...	1818,	13,808,674
...	1795,	4,350,658	...	1819,	16,300,273
...	1796,	2,531,930	...	1820,	13,134,530
...	1797,	2,833,591	...	1821,	10,723,479
...	1798,	4,623,223	...	1822,	9,827,643
...	1799,	6,480,167	...	1823,	9,784,154
...	1800,	7,411,370	...	1824,	15,330,144
...	1801,	4,981,669	...	1825,	11,490,459
...	1802,	3,737,080	...	1826,	13,062,316
...	1803,	4,002,824	...	1827,	12,653,095
...	1804,	4,452,859	...	1828,	13,296,041
...	1805,	6,357,225	...	1829,	12,660,460
...	1806,	6,081,109	...	1830,	13,229,533
...	1807,	4,984,572	...	1831,	13,864,067
...	1808,	6,504,339	...	1832,	16,516,388
...	1809,	7,414,672	...	1833,	22,713,755
...	1810,	5,311,082	...	1834,	18,425,417
...	1811,	5,592,604	...	1835,	17,514,950
...	1812,	17,829,499	...	1836,	30,868,164
...	1813,	28,082,397	...	1837,	39,164,745
...	1814,	30,127,687			

The following table for the year ending June 30, gives the U. S. revenue and expenditure for the fiscal years ending June 30, 1848 and 1859:—

	1848.	1859.
Customs,	31,671,443 d.	49,565,824 d.
Lands,	3,228,521	1,756,687
Miscellaneous,	275,745	2,082,559
Loans,	17,299,270	28,287,400
Total revenue,	52,477,979 d.	81,692,470 d.

	1848.	1859.
Civil,	5,598,064 d.	5,963,795 d.
Army,	25,198,108	23,243,822
Indian, &c.	2,631,035	4,753,972
Navy,	9,394,391	14,7[illegible],640
Interest,	2,037,527	
Debt,	7,808,892	17,405,285
Foreign intercourse and Miscellaneous,		18,272,025
Total expenditure,	52,668,017 d.	83,751,511 d.

By way of comparison we subjoin the tables of revenue and expenditure of the general government for nine years, from 1845 to 1853 inclusive:—

REVENUE.

Year ending June 30.	Customs.	Miscellaneous.	Total.
1845,	27,528,113 d.	2,241,021 d.	29,769,134 d.
1846,	26,712,668	2,786,579	29,499,247
1847,	23,747,864	2,598,926	26,346,790
1848,	31,757,070	3,679,679	35,436,750
1849,	28,346,738	2,727,608	31,074,347
1850,	39,668,686	3,707,112	43,375,798
1851,	49,017,568	3,295,412	52,312,979
1852,	47,320,326	2,425,273	49,745,598
1853,	58,931,865	2,405,707	61,337,572

EXPENDITURE

Year ending June 30.	Civil List and Miscellaneous.	War.	Navy.	Total.
1845,	5,608,207 d.	9,533,203 d.	6,228,639 d.	21,370,049 d.
1846,	6,783,000	13,579,428	6,450,862	26,813,290
1847,	6,715,854	41,281,606	7,931,633	55,929,093
1848,	5,585,070	27,820,163	9,406,737	42,811,970
1849,	14,017,640	17,290,936	9,869,818	57,631,667
1850,	14,839,725	12,801,764	7,923,313	43,002,168
1851,	17,872,967	11,814,793	8,987,798	48,005,879
1852,	17,361,164	13,424,075	8,928,236	39,713,475
1853,	17,174,955	15,476,825	10,891,639	43,543,419

Public debt.] The public debt of the U. S. at different times was as follows:—

Before the adoption of the constitution,	75,416,476 d.
1790,	60,219,022
1800,	82,976,294
1810,	53,173,217
1816,	127,334,933
1820,	91,015,566
1830,	48,565,406
1844,	24,748,188

On the 1st day of July, 1844, the debts of the U. S. consisted of the following particulars:—

	Dollars.	Cts.
Old funded debt,	178,034	84
Outstanding certificates of do.,	24,214	29
Treasury notes of 1812,	4,317	44
Certificates of Mississippi stock of 1814 and 1815,	4,320	9
Debts assumed by Congress, district of Columbia, in 1836, now reduced to	1,260,000	0
Outstanding Treasury notes of two acts,	2,256,207	31
Loan due 1845, 5½ per cent.,	5,672,976	88
Do. 1863, 6 per cent.,	8,343,886	88
Do. 1853, 5 per cent.,	7,004,231	35
Total debt,	24,748,188	23

We find from official documents that the principal and interest, including the stock of Texas, payable annually from the 1st of January, 1851, to its final redemption, will be as follows:—

Years when the principal will become due.		Principal. Dol.	Cts.	Interest. Dol.	Cts.	Total. Dol.	Cts.
1851,	10th Aug.,	303,573	92	4,209,735	68	4,513,309	60
1852,				4,194,556	94	4,194,556	94
1853,	1st July,	6,468,231	35	4,032,851	16	10,501,082	51
1854,				3,871,145	38	3,871,145	38
1855,				3,871,145	38	3,871,145	38
1856,	12th Nov.	4,999,149	48	3,829,485	72	8,828,635	20
1857,				3,573,688	16	3,573,688	16
1858,				3,573,688	16	3,573,688	16
1859,				3,573,688	16	3,573,688	16
1860,				3,573,688	16	3,573,688	16
1861,				3,573,688	16	3,573,688	16
1862,				3,573,688	16	3,573,688	16
1863,	1st Jan.	8,198,686	03	3,081,767	00	11,280,453	03
1864,				3,081,767	00	3,081,767	00
1865,	1st Jan.	10,000,000	00	2,581,767	00	12,581,767	00
1866,				2,581,767	00	2,581,767	00

1867,		2 581,767 00	2,581,767 00
1868,*	43,029,450 60	372,200 00	43,401,650 00
	72,999,090 78	59,732.084 22	132,731,175 00
Add debt of the district cities assumed by congress, whereof there is payable annually, 60,000 d. and interest, . . .			990,600 00
The old funded and unfunded debt, a great proportion whereof will never be called, .			119,585 98
And outstanding treasury notes, which, when issued in stock, will be payable in 1868, .			209,561 61
			74,288.238 37
Deduct Texas stock, .			10,000,000 00
Debt 1st Dec. 1850, as per statement,			64,288,238 37

* 1868, {1st Jan., 27,289.450 00 d.
{1st July, 15,740,000 00

43,029,450 00 d.

The total federal debt on 1st July 1858 was stated at 44,910,777 d.; and on 1st July 1859, 58,851,777 d.

The following table purports to show the amount of the federal, state, municipal and corporate stocks of the U. S., held in Europe, with the increase thereof during four years from 1848 to 1852:—

	July 1, 1848.	July 1, 1852.	Increase.
Federal, .	11.000,000 d.	45.000,000 d.	34.000,000 d.
New York, .	16,000,000	40,000,000	24,000,000
Pennsylvania,	30,000,000	47,000.000	17,000.000
Ohio, .	15,000,000	30.000,000	15.000,000
Michigan, .	2,000,000	6.000.000	4,000,000
Massachusetts,	2.000,000	7.000,000	5,000.000
Maryland, .	6,000.000	10,000,000	4,000,000
Virginia, .	6,000,000	12.000,000	6,000.000
S. Carolina,	2,000,000	2,500,000	500,000
Georgia, .	700,000	1,200.000	500,000
Alabama, .	7,000,000	8,000,000	1,000,000
Mississippi, .	7,000,000	7,000,000	—
Texas, .	2,000,000	2,000,000	—
Arkansas, .	2.500.000	2,500,000	—
Tennessee, .	2,500.000	4,000,000	1,500,000
Kentucky, .	2,000,000	3,000,000	1.000,000
Indiana, .	5,000,000	10,000,000	5,000,000
Illinois, .	10,000,000	13.000.000	3,000,000
Missouri, .	500,000	1.000,000	500,000
Louisiana, .	12,000,000	10,000,000	—
Total, .	141,200,000 d.	261,200,000 d.	122,000,000 d.
Less decrease in Louisiana, . .			2,000,000
Net increase in four years, . . .			120,000,000 d.
Dividend as follows,—			
Transfers of Federal stock, . . .			34,000,000
Transfers of State stock, . . .			16.000,000
Sales of New State stock, . . .			10.000.000
Transfers city and Railroad bonds, . .			10,000.000
Sales of New Railroad, City and County bonds,			50,000,000
Total as above, . . .			120,000,000 d.

Post-office.] The postmaster-general reported the receipts, 3,487,199 d. during the nine years ending 30th June, 1845, as having averaged 4,364,625 d. a year. At the close of the fiscal year ending June 30, 1851, there were in operation within the U. S., 6,170 mail routes; their aggregate length was 196,290 m., and 5,544 contractors were employed thereon. The annual transportation of the mails on those routes was 53,272,252 m.; the annual cost thereof, 3,421,754 d.; being about 6c. 4 mills per m. There were on the 30th day of June six foreign mail routes of the estimated aggregate length of 18,349 m. The number of miles of annual transportation thereon was estimated at 615,206. The whole number of post-offices in the United States was 19,796. In the year ending 30th June 1858, the number of mail-routes had increased to 8,296; of a total length of 260,603 m. The amount of annual transportation was 78,765,491 m., at a cost of 7,795,418 d. Of these 19,555,734 m. were by coach; 25,763,452 m. by railroad; 4,569,540 by steam-boat. The number of post-offices was 28,573, of which 400 were managed by parties appointed by the president. The gross receipts of the postal department were 8,186 792 d. The expenditure amounted to 12,722,470 d:—The earliest reliable statistics of the post-office are those for the year 1790, when the number of post-offices was 75, the extent of post routes 1,875 m., and the revenues of the department 37,935 d. In the United Kingdom there are 15 letters written in the year for every one of the pop.; in the U. S. the number is only 4. In 1854 the letters which passed through the American post-offices were 119,000,000, while the number that passed through the post-offices in this country was nearly 450,000,000. In London, in 1852, there were 41 letters written for every one of the inhabitants; in New York, the number was only 23. The cause of this discrepancy may be the want of postal facilities in America. There is no such thing as prompt and daily deliveries of letters to every house, cottage, and hovel in that country such as in England, and although there are 24,000 post-offices in America, and only 10,000 in the United Kingdom, the former do not afford the accommodation which the latter do on account of the vast area over which they are spread, and are a poor substitute for the letter carrier and rural postman. The pop. of the following 8 places, viz., London, Liverpool, Manchester, Birmingham, Bristol, Preston, Leicester, and Limerick, in 1854, was 3,500,000; the postal receipts were £1,500,000; and the postal local expenditure was £260,000. The pop. of the 8 following places, viz., New York, Philadelphia, Baltimore, Boston, New Orleans, Cincinnati, St. Louis, and San Francisco, was in the same year 1,500,000; the postal receipts were £300,000; and the local expenditure, £80,000. The local cost of working a million of letters in England is £550, and in America £2,400. This difference arises from the cheapness with which letters charged an uniform postal rate can be worked as compared with the cost of working letters charged a variable postal rate. In America there are three inland rates, and in England there is only one. In 1839, before the introduction of an uniform rate of postage in England, the gross expense of working a million of letters was £10,000; at the present time it is only £3,000, or less than one-third.

Military force.] It is evident that the warlike establishment of a great nation, removed by distance from the immediate danger of collision with powerful neighbours, can give no approximate idea of the resources of the country. There has always been a jealousy in the republic against any unnecessary standing army. The principal reliance of the federal union for defence is on the militia of the several states, amounting, by the returns of 1859, to 2,727,486 men. Of this number, about one-fifth, say 300,000 men, are uniformly equipped, and are mustered and drilled several times a-year. The *Army Register* for 1843 contains the appalling announcement, that this multitudinous host was commanded by 627 generals, 2,670 general staff officers, 13,813 field officers, 44,938 company officers, being in all 62,205 officers. Service in the militia entitles to the elective franchise; and all persons able to bear arms are called on to perform military duty thrice a year. But, according to Captain Hall, the training does not exceed four or six days altogether in the year, and the militia-exercise is so loose, that they would require to be untrained whenever they were called out to actual service. The regular army in 1841 was 12,539; in 1842, 9,012; in 1844, 8,616; in 1853, 11,000. This force was employed chiefly to garrison the fortifications on the sea-coast and the frontiers. In 1846 it was augmented to 30,000 men, chiefly volunteer forces, to meet the demands of the war with Mexico. The officers of the regular army are chiefly educated at the government military academy at West Point. The army expenses in 1844 were 6,174,485 ds.

Marine force.] A navy department was not created till 1798, nor a permanent naval peace-establishment till 1801. The force was then reduced to 9 frigates; and, excepting an attack on Tripoli, was employed only on coasting-duties, until the war with England in 1812. The military marine in 1828 consisted of 7 ships of 74 guns; 6 frigates of 44 guns; 4 of 36 guns; 2 corvettes of 24; 12 sloops of 18; and 7 schooners; besides 5 ships of the line, with 6 frigates, in course of building. The expenses of the naval establishment for 1828, were 3,786,649 d., or about £750,000; in 1844, 4,703,950 d. The force of the American marine in 1845 was as follows: 11 ships-of-the-line, viz. one of 120 guns, 7 of 80, and 3 of 74; 17 frigates, mostly of 44 guns, with a total of 738 guns; 34 smaller vessels of 30 to 6 guns; 2 steamboats of 500 horse-power, one of 220 horse-power, and one of 120 horse-power—total 1,360 horse-power. "The ratings of these vessels will, however," says Captain Marryat, "very much mislead people as to the real strength of the armament. The 74's and 80's are in weight of broadside equal to most three-decked ships; the first-class frigates are double banked of the scantling, and carrying the complement of men of our 74's. The sloops are equally powerful in proportion to their ratings, most of them carrying long guns. Although flush vessels, they are little inferior to a 38-gun frigate in scantling, and are much too powerful for any that we have in our service, under the same denomination of rating. It is impossible," Captain Marryat continues, "not to be struck with the beautiful architecture of most of these vessels. The Pennsylvania rated 120 guns, of four decks, carrying 146, is not by any means so perfect as some of the line-of-battle ships. The Ohio is, as far as I am a judge, the perfection of a ship of the line. But in every class you cannot but admire the superiority of the models and workmanship. The dockyards in America are small, and not equal to what they may eventually require. They certainly do not require such establishments or such store-houses as we have, as their timber and hemp are at hand when required; but they are very deficient both in dry and wet docks. Properly speaking, they have no great naval depot. This arises from the jealous feeling existing between the several states. A bill brought into congress to expend so many thousand dollars upon the dockyard at Boston in Massachusetts, would be immediately opposed by the state of New York, and an amendment proposed to transfer the works intended to their dockyard at Brooklyn. The other states who possess dockyards would also assert their right, and thus they will all fight for their respective establishments until the bill is lost, and the bone of contention falls to the ground. It is remarkable that along the whole of the eastern coast of America, from Halifax in Nova Scotia down to Pensacola in the gulf of Mexico, there is not one good open harbour. The majority of the American harbours are barred at the entrance, so as to preclude a fleet running out and in to manœuvre at pleasure; indeed, if the tide does not serve, there are few of them in which a line-of-battle ship, hard pressed, could take refuge. A good spacious harbour, easy of access, like that of Halifax in Nova Scotia, is one of the few, perhaps the only, natural advantage wanting in the U. S." The *New York Herald*, in 1853, thus describes the condition and organization of the federal navy: "The Mississippi is so badly armed, and her guns so light, that an English war-steamer, with only one large gun, could destroy her before her fire could take the slightest effect. Then look at the miserable sailing qualities of the ships and war-steamers of our navy. Notwithstanding the improvements of the age in naval construction, the fleet—or rather the apology for a fleet, which we possess—stands just where it did 40 years ago. Our war-steamers are the worst of all. In the present state of things two or three British steamers, with guns of long range, could sink every ship and steamer in possession of the government. Yet this country has resources which, if properly wielded, could soon turn the tables against the combined fleets of the world. Our steam-tonnage, including the river-navigation, doubles that of Great Britain, and exceeds that of all nations put together. Here, then, we have scientific, practical, and mechanical skill sufficient to raise a steam fleet that would outnumber, outrun, and defeat all others. But before this could be accomplished the amount of disasters that might be sustained in encounters by our navy and merchant marine would be terrible, to say nothing of the exposure of our badly fortified ports. The vast extent of our seaboard demands protection far beyond what a navy four times as large as our present one could afford. Then look at the list of our old foggy captains in the navy, promoted, not by merit, but seniority—men unfit to take a command. The low wages paid to seamen—being less by one-third than they can procure in merchant-vessels—together with the barbarous system of flogging that prevailed until lately, have driven or kept away almost every good man from the service, and left the dregs behind. In short, the face of things in the navy is changed since the days of Bainbridge, Lawrence, and Decatur; and, had we suddenly a repetition of the war of 1812, we should have a very different tale to tell." The U. S. *Navy Register* for 1852 represents the navy as thus composed in that year:—

	Vessels.	Guns.
Ships of the Line,	12	914
Frigates,	14	600
Sloops of War,	21	402
Brigs,	4	40
Schooners,	3	4
Steamers.	16	63
Store Ships,	5	24
Total,	75	2,047

No estimate can be formed of the real strength of the crews of the American navy, because it is well known that a considerable proportion of this fleet is manned by British seamen, who are attracted by the high pay of the American navy in time of peace. In 1845, of the 6,100 seamen employed in the federal fleet, only 960 were native Americans. But that very circumstance shows the extreme difficulty of manning an American fleet in war, and especially in a war with England. Their ships are, in fact, worked to a great extent by English mercenaries, a large proportion of whom would return to their own colours on the outbreak of hostilities.

Comparative cost of government.] The following remarks on the comparative cost of government in the U. S., and in the United Kingdom, from Mackay's *Western World*, apply to the state of things in 1846: The American people being now about 20,000,000, their expenditure for army, navy, ordnance, pensions, civil contingencies, and foreign intercourse—in short, for everything but the debt, amounts to about 5s. sterling per head. We, being about 30,000.000 of people, have to pay 14s. 8d. per head to defray the expenses of the state, exclusive of the debt; that is to say. for the mere expenses of government, we pay absolutely between four and five times as much as the American; and individually, nearly three times as much. When the debt of both countries is thrown into the scale, this difference is, of course, greatly increased, inasmuch as we pay yearly, as interest upon our debt, more than the whole principal of the debt of America, even after the war, amounts to. The expense of all the state governments taken together does not exceed 5,500,000 of dollars, which is but a trifle over £1,250,000. This added to the £5,000,000 for the support of the general government, gives £6,125,000, as the aggregate cost of government, both general and local, in America. This makes the cost per head of government in America, exclusive of the debt, 6s. 3d., to contrast with 14s. 8d., the cost per head of the government of England, exclusive of her debt.—The following curious comparative estimates are from another pen. "Beneath the despotism of the Czar, the Russian lives, trades, and contributes to the revenue as follows: his staple food is rye and cabbage; his average consumption of colonial produce is 1 lb. 4 oz. to 1 lb. 9 oz. of sugar, and 2½ oz. of tea. He trades to the extent of 8 shillings, and contributes to the state 5 shillings. The Austrian, under an absolutism locally modified, and with a rigorous censorship, lives on inferior grains, consumes 2lb. 11 oz. of sugar, and contributes to the state about 7s. 7d. The Prussian and German, under illusory constitutions, and with a preventive censorship, live—the Prussian on potatoes and rye—the German on inferior grains. They average a consumption of 4 lb. 11 oz. of sugar, and 1½ lb. coffee. The estimate of animal food in Prussia is averaged from 15 to 30 lb. and the Prussian contributes rather less that 10s. to the royal treasury. If we now take France, when under a representative form of government, in which at least one adult male in every 44 shared, and with a press which the law may persecute, but whose voice no censorship could silence; the Frenchman's staple food is wheat; he consumes from 6 lb. 10 oz. to 7 lb. 4 oz. of sugar, he trades at the rate of 54s., and contributes upwards of 39s. to the state. The Belgian, with a more liberal constitution, which gives to one adult male in every 21 a vote, lives chiefly on wheat, consumes 14 lb. of sugar, contributes 18s. to the state, and trades to the extent of 70s. The inhabitant of the United Kingdom (whose electors are one in 7) lives on wheat and animal food, consuming, at the lowest calculation, 77 lb. of meat, 19 lb. of sugar, 1¼ lb. of coffee, and 1 lb. 9 oz. of tea. He trades to the extent of 107s., contributes 35s. to the state, and in its need has contributed upwards of 100s. with far more facility than the Russian pays his 5s. The United States-man, thoroughly self-governed, and whose vote is his birthright, has secured a share of material comfort proportioned to the freedom of his institutions. A choice of grains, 19 lb. of sugar, near 6½ lb. of coffee, nearly 1 lb. of tea, and 200 lb. of meat, constitute his average food. His trade averaged, in 1840, 57s., and the cost of his government was 5s. or 6s.

History.] The history of the United States naturally divides itself into two periods: the first embracing the annals of the British North American colonies, which separated from the mother-country in 1776; and the second, the history of the republic established by the colonists.

I. *The Colonies from* 1607 *to* 1776.] "In the course of the 17th century," says Everett, "the French and English took possession of all this part of North America, which was not pre-occupied by the Spaniards. The French entered by the St. Lawrence; followed that noble artery to the heart of the continent; traced the great lakes to their parent rivulets and weeping fountains; descended the Mississippi. Miracles of humble and unavailing heroism were performed by their gallant adventurers and pious missionaries in the depths of our western wilderness. The English stretched along the coast. The geographer would have pronounced that the French, in appropriating to themselves the mighty basins of the Mississippi and the St. Lawrence, had got possession of the better part of the continent. But it was an attempt to compose the second volume of the Fortunes of America, in advance of the first. This it was ordained should be written at Jamestown and Plymouth. The French, though excelling all other nations of the world in the art of communicating for temporary purposes with savage tribes, seem, still more than Spaniards, to be destitute of the august skill required to found new states. I do not know that there is such a thing in the world as a colony of France growing up into a prosperous commonwealth. A half a million of French peasants in Lower Canada, tenaciously adhering to the manners and customs which their fathers brought from Normandy two centuries ago; and a third part of that number of planters of French descent in Louisiana are all that is left to bear living witness to the amazing fact, that not a century ago France was the mistress of the better half of North America. It was on the Atlantic coast and in the colonies originally planted or soon acquired by England, that the great work of the 17th century was performed:—slowly, toilsomely, effectively. A mighty work for America and mankind, of which even we, fond and proud of it as we are, do but faintly guess the magnitude! It could hardly be said, at the time, to prosper in any of its parts. It yielded no return to the pecuniary capital invested. The political relations of the colonies from the first were those of encroachment and resistance; and even the moral principle as far as there was one, on which they were founded, was not consistently carried out. There was conflict with the savages,—war with the French and Spaniards,—jarring and feuds between neighbouring colonies,—persecution of dissenting individuals and sects,—perpetual discord with the crown and the proprietaries. Yet in the main and on the whole, the work went on. Things that did not work singly worked together; or if they did not work together they worked by reaction and collision. Feeble germs of settlement grew to the consistency of powerful colonies; habits of civil government rooted themselves in a soil that was continually stirred; the frame of future republics knit itself as it were in embryo, under a monarchical system of colonial rule; and in the middle of the 18th century the approach of mighty changes began to be dimly foreseen by gifted spirits. A faint streak of purple light blushed along the eastern sky." Of the thirteen colonies whose delegates signed the declaration of independence, twelve were settled in the 17th cent., and the colonists, with a few trifling exceptions, were Englishmen. In 1630 the number of English colonists in North America did not exceed 4,000; in 1660 it was not less than 80,000, or had increased twenty-fold in the space of 30 years: in 1701 the pop. of the colonies is estimated to have been about 262,000. The period of colonization was one of great intellectual and political excitement in the mother-country. The principles of liberty, the rights of Englishmen, the nature, use, and objects of government, were topics of general interest and discussion in England at this period. It is an observation of Fox, that from 1588 to 1640, the general improvement in all the arts of civil life, and, above all, the astonishing progress of literature, were the most striking among the general features of that period, and were in themselves causes sufficient to produce effects of the utmost importance. Another remarkable element in the society from which swarmed the American colonists, was the state of religion. An imperfect reformation, favoured by the government, and amounting to little more than a secession from the Catholic church, was accompanied by a popular reformation ready to follow out its principles to their results. The ***Puritans***, as they were reproachfully called, or Non-conformists, were compelled to attack the temporal power, and to defend civil liberty, while defending freedom of conscience. This mixture of religious faith in the contest for political rights, gave the English Puritans the zeal, firmness, and boldness of religious reformers. It is further to be considered, that while the English colonists brought with them to America the broadest and most generous principles of liberty, and those free institutions which convert general maxims into practical truths and make them a part of the daily life of men, they, at the same time, left behind them those restraints which checked their free action in England. They brought with them, across the Atlantic, trial by jury and the right of representation, but left behind them feudal services, privileged orders, corporations and guilds, with other similar burdens upon industry. "Two things worth mentioning," says Everett, "contributed to the successful result of North American colonization. One was the absence of the precious metals. The British colonies were rich in the want of gold. As the abundance of gold and silver in Mexico and Peru contributed, in various ways, to obstruct the prosperity of the Spanish colonies, the want of them acted not less favourably here. In the first settlement of a savage wilderness, the golden attraction is too powerful for the ordinary routine of life. It produces a feverish excitement unfavourable to the healthy growth and calm action of the body politic. Although California from the first has had the advantage of being incorporated into a stable political system, of which as a sister State she forms an integral part, it is quite doubtful whether, looking to her permanent well-being, the gold is to be a blessing to her. It will hasten her settlement; but that would at any rate have advanced with great rapidity. The other circumstance which operated in the most favourable manner upon the growth of the Anglo-American colonies was the fact, that they were called into existence less by the government than by the people; that they were settled not by bodies of colonists, but by individual emigrants. The Crown gave charters of government and grants of lands; and a considerable expenditure was made by some of the companies and proprietors who received these grants; but upon the whole, the U. S. are settled by individuals,—the adventurous, resolute, high-spirited, and many cases, persecuted men and women, who sought a home and refuge beyond the sea; and such was the state of Europe in the 17th and 18th centuries, that it furnished a succession of victims of a long series of political and religious disasters, who found, one after another, a safe and a congenial retreat in some one of the American colonies." In 1606 two companies of merchants and others were incorporated, under the names of the London company and the Plymouth company, with the exclusive right of settling and trading within their respective limits. The former began the colonization of British America in 1607, by sending to Virginia a colony of 100 men, which, before the end of the year, was reduced by suffering and the badness and scarcity of food, to 38 souls. In October, 1609, the number had been increased by new colonists to 500; but famine again reduced them in six months to 60 persons. In 1619 the first colonial assembly was convoked, consisting of representatives elected by the boroughs, the concerns of the colony having been previously managed by the company in England. As the colonists were mostly adventurers without families, 90 young girls were sent over by the company in 1620, and sold to the young planters, at the rate per head of from 100 to 150 pounds of tobacco. In 1621 the company passed an ordinance vesting the government of the colony in a governor, council, and general assembly, the latter chosen by the inhabitants, with power to enact laws. In 1622, 347 men, women, and children, were massacred by the Indians; a general Indian war followed, and the number of settlements were reduced from 80 to 8. In 1624 the company was dissolved by the crown, and the colony taken into the hands of the king. Such are a few incidents from the humble annals of the first colonists, presenting a picture of suffering too often renewed in other parts of the country. The Plymouth company, to which was granted the exclusive right to trade and settle in N. Virginia, did nothing effectual towards the colonization of their jurisdiction. But, in 1620, a number of Puritans (Brownists), who had set sail for Virginia, were landed either by accident or treachery within the limits of the Plymouth company. Ten years afterwards they obtained from the company a grant of the land to which they had previously no title but occupancy; but they were never incorporated as a body politic by royal charter, and they therefore remained a mere voluntary association, under laws and magistrates formed and chosen by themselves, until their union with Massachusetts in 1692. The germ of the New England colonies was the Massachusetts colony, settled in 1628 by a company incorporated that year by royal charter, the land having been previously purchased from the Plymouth company. The government of the colony was transferred to Massachusetts in 1630, by vote of the company; and a few years later, the freemen adopted the plan of acting by delegates or representatives; courts were also established, and the charter of a trading company was thus tacitly converted into the constitution of a commonwealth. The Massachusetts colonists were Puritans, and rendered themselves not less obnoxious to the court-party at home by their religious principles, than by this unwarranted assumption of political power. This, with other circumstances, led the Plymouth company to resign their charter to the king in 1635. Such, however, was the distracted state of England at the time, that these remote colonies attracted little attention, and were left to grow up in habits of self-government, while their numbers and wealth were increased by successive emigrations of parties worsted in the civil strife at home. The persecuted Puritans fled to New England; the oppressed Catholics to Maryland; the defeated cavaliers or royalists to Virginia. The first settlement, that of Virginia, says Everett, "was commenced in the spirit of worldly enterprise, with no slight dash however of chivalry and romance on the part of its leader. In the next generation this colony became the favourite resort of the loyal cavaliers and gentlemen, who were disgusted by the austerities of the English commonwealth or fell under its suspicion. In the meantime New England was founded by those who suffered the penalties of nonconformity. The mighty change of 1640 stopped the tide of emigration to New England, but recruited Virginia with those who were disaffected to Cromwell. In 1624 the island of Manhattan, of which you have perhaps heard, and if not, you will find its history related with learning, judgment, and good taste, by a loyal descendant of its early settlers, was purchased of the Indians for 24 dollars, a sum of money by the way, which seems rather low for 22,000 acres of land including the site of this great metropolis, but which would, if put out at compound interest at 7 per cent. in 1624, not perhaps fall so very much short even of its present value; though I admit that a dollar for a thousand acre is quite cheap for choice spots on the fifth avenue. Maryland next attracted those who adhered to the ancient faith of the Christian world. New Jersey and Pennsylvania were mainly settled by persecuted Quakers; but the latter offered an asylum to the Germans whom the sword of Louis XIV. drove from the Palatinate. The French Huguenots, driven out by the revocation of the edict of Nantes, scattered themselves from Massachusetts to Carolina. The Dutch and Swedish settlements on the Hudson and the Delaware provided a kindred home for such of their countrymen as desired to try the fortune of the new world. The whigs of England, who rebelled against James II. in 1685 and were sent to the transatlantic colonies, lived long enough to meet in exile the adherents of his son, who rebelled against George I. in 1715. The oppressed protestants of Salzburg came with Oglethorpe to Georgia; and the Highlanders, who fought for Charles Edward in 1745, were deported by hundreds to North Carolina. They were punished by being sent from their bleak hills and sterile moors to a land of abundance and liberty; they were banished from oatmeal porridge to meat twice a-day. The Gaelic language is still spoken by their descendants, and thousands of their kindred at the present day would no doubt gladly share their exile." It would lead us greatly beyond our necessary limits to attempt even a sketch of the annals of the colonies. We must satisfy ourselves with a hasty view of the forms of government which prevailed in them. Of these forms there were three: the royal, the charter, and the proprietary governments. 1. The charter governments were confined to New England. The people of the New England colonies, by their charters, were entitled to the privileges of natural born subjects, and invested with the powers of government, legislative, executive, and judicial; they chose their own governors, elected legislative assemblies, established courts of justice, and in many points exceeded the powers conferred by the charters. The only limitation to their legislative power was, that their laws should not be contrary to those of England. The crown claimed indeed the right of revoking the charters; but the colonists maintained that they were solemn compacts, irrevocable unless for cause. 2. The royal governments were those of Virginia, New York, and at a later period, the Carolinas (1728) and the Jerseys (1702). In these colonies the governor and council were appointed by the crown, and the colonists chose representatives to the colonial assemblies. The governors had a negative on the proceedings of the legislatures, which were composed of the councils and the popular assemblies: the judges and most of the officers were also appointed by the king, although, in many cases, paid by the colony. 3. The proprietary governments were those of Maryland, Pennsylvania, and, at first, the Carolinas and the Jerseys,—the two former remained such till the revolution. These colonies were in the hands of proprietors or individuals to whom grants of land had been made by the crown, with authority to establish civil governments, and make laws under certain restrictions. The history of the proprietary governments is little more than that of a perpetual quarrel between the people and the proprietors, chiefly owing to the manner in which the latter exercised their prerogative of repealing or negativing the acts of the colonial assemblies; for even in these colonies representative bodies, partly elected by the freemen, and partly summoned by the proprietors, were soon introduced. The common English right of assent to taxes, as the colonists themselves called it, was asserted in the fullest manner by nearly all the colonies as early as the middle of the 17th cent.; and it was a commonly received opinion in America, that the authority of parliament in the colonies did not extend to taxation or internal legislation, but was confined to the regulation of trade. For a long time the colonists enjoyed almost unlimited freedom of commerce; it was not until they had surmounted the hardships and difficulties incident to their first establishment, and had begun to increase rapidly in wealth, that their commerce began to be submitted to restrictions intended to secure its benefits to the mother-country. In the two French wars, 1744—1748, and 1755—1763, the colonists took an active part, and favourably distinguished themselves on several occasions. The various Indian wars which prevailed at intervals all along the extensive and exposed line of frontiers, were likewise in general conducted by the colonists themselves. Primary schools were first established in New England, and it early became a favourite object, in that part of the country, to enable every individual to read and write. Colleges were also founded for the higher education of youth; and, at the beginning of the revolutionary war, there were eight institutions of this kind in the country. The peace of Paris, in 1763, by annihilating the French power in North America, relieved the colonies from the pressure of foreign hostilities. But a course of policy was now adopted by the British ministry, which finally led to the separation of the American colonies from the British empire. In 1764 an act of parliament was passed, continuing the duties on certain articles imported into the colonies, towards raising a revenue; and it was provided that all penalties for breach of the acts of trade and revenue committed in America might be recovered in any vice-admiralty court in the colonies. In 1765 (March 22) the stamp act was passed; and soon after, a bill was brought in authorizing the quartering of troops in the colonies. These measures were received in America with universal opposition; and a general congress of the colonies was held at New York, which adopted a declaration of rights and grievances, asserting taxation by themselves, and trial by jury, to be the inherent rights of the British subjects in the colonies. In this state of affairs the stamp-act was repealed, March 18, 1766; at the same time, a declaratory act was passed, asserting the right of parliament to bind the colonies in all cases whatsoever. The Rockingham ministry, by which these measures were carried, was soon displaced; and in June, 1767, the new chancellor of the exchequer brought in a bill imposing duties on glass, paper, pasteboard, white and red lead, painters' colours, and tea, imported into the colonies. This act, after giving rise to combinations among the Americans against the importation of the articles specified, was also repealed, March 5, 1770, the duty on tea being alone continued. The colonists accordingly renounced the use of that article, or obtained it from foreign countries. In 1773 the ministry procured an act, permitting to East India a drawback on teas exported to America, for the purpose of rendering it cheaper in this country, and inducing the Americans to submit to the small duty imposed upon it. Large shipments were made; but in Philadelphia and New York the vessels were not allowed to land their cargoes; in Charleston it was put into stores, but not permitted to be offered for sale; and in Boston, where the British authorities refused to allow the vessels to return without having been entered, the tea was thrown overboard. This act of violence was followed by the Boston port bill, interdicting all commercial intercourse with the town of Boston, and by a bill for entirely subverting the government of Massachusetts. In this crisis the other colonies made common cause with Massachusetts; and a general congress met at Philadelphia on the 5th of September, which adopted a declaration of rights, asserting the exclusive power of legislation, in all cases of taxation and internal policy, to be in the provincial legislatures. They next proceeded to interrupt all commercial intercourse with Great Britain, by the non-importation, non-consumption, and non-exportation agreement, binding themselves not to import, consume, or export certain articles, until the acts complained of should be repealed. These proceedings had no effect on the policy of the British government, and general preparations now began to be made in the colonies for armed resistance. A detachment of troops sent from Boston to seize some provincial stores collected at Concord, fired upon the citizens who assembled to oppose them, and actual hostilities were commenced. The second congress, which met in May, 1775, determined to organize an army; and Washington was appointed commander-in-chief of the colonial forces. "We have counted the cost of this contest," said congress, "and are with one mind resolved to die freemen rather than to live slaves." On the 4th of July 1776, congress adopted a declaration of independence.

II. *War of independence.*] 1. The declaration of independence, after recapitulating those acts of the British government which were viewed as subversive of the liberties of the British subjects in America, declares the united colonies to be free and independent states absolved from all allegiance to the British crown, and all political connection between them and the state of Great Britain to be totally dissolved. The affairs of Lexington and Bunker hill, the capture of several posts, and an unsuccessful expedition against Canada, were among the acts of hostility which had preceded the declaration of independence. By the beginning of March, 1776, 14.000 regular troops had been enlisted, and the British were obliged to evacuate Boston on the 17th of March. The military operations which closed the year were the capture of New York by the British forces under Lord Howe and Sir W. Howe, after the defeat of the Americans on Long Island; the battle of White Plains, and the retreat of the American forces through the Jerseys and across the Delaware; and the subsequent affairs at the battles of Trenton and Princeton. The campaign of 1777 was more favourable to the American cause. General Howe had indeed taken possession of Philadelphia, and the Americans had been defeated on the Brandywine, and at Germantown; but the northern army, under the command of General Gates, after some partial success, compelled General Burgoyne to surrender his army at Saratoga on the 17th of October, and thus cut off the communication between New York and Canada. On the reception of this intelligence in Europe, the French government, on 6th Feb. 1778, entered into treaties of amity and commerce with the U. S., and thus recognised their existence as an independent nation. The British commander now came to the resolution to evacuate Philadelphia, and concentrate his forces at New York. About the same time a French fleet arrived on the coast, having on board about 4,000 French troops. Congress had hitherto consisted of an assembly of delegates from 13 independent states, with little more authority than that of advising the states to adopt certain measures; money was raised only by consent of the states; congress had no power to enforce obedience, and the states were held together only by external pressure. After long and embarrassing discussions, articles of confederation were agreed upon by congress, Nov. 15, 1777, and submitted to the state legislatures for ratification. By these articles, the exclusive cognizance of foreign relations, the rights of war and peace, and the right to make requisitions of men and money, were confided to congress. But, like all former confederations, the decrees of the federal government operated upon the states in their sovereign capacity, and not upon the individual citizens; and as soon as danger from abroad was past, this confederacy was found to be utterly incompetent to govern the country. In 1779 the south became the principal theatre of warlike operations. In October 1781, the contest was finally closed by the surrender of Cornwallis at Yorktown, to the combined French and American forces under Washington, and Rochambeau. In the following year a treaty was concluded between Holland and the U. S.; and after long-protracted negotiations, a definitive treaty of peace with Great Britain was signed, Sept. 23, 1783. Denmark, Spain, Sweden, and Russia, had previously recognised the U. S. as a sovereign power. The war had been attended with great sacrifices on the part of the Americans. Without arms, ammunition, or pecuniary resources, congress was obliged to have recourse to a paper medium. Bills of credit to the amount of 300.000,000 dollars had been emitted during the first five years of the war, and no provisions were made for redeeming them. In 1780 these bills had depreciated to such a degree as to cease to circulate: the treasury was empty, and the army unpaid, without clothing, and often without food. At this crisis the French government made a grant of 6,000,000 livres by way of subsidy, and a further sum by way of loan; and finally a loan of 10,000,000 livres was raised in Holland. These supplies, with a new organization of the finance department, and the establishment of a national bank, contributed to relieve the pressure. At the peace the public debt was found to amount to 42,000,000 d., on which congress was unable to pay even the interest; but the requisitions and regulations of that body were little regarded by the states, and the country was becoming a prey to anarchy, when a convention, composed of delegates from the several states, met at Philadelphia in May, 1787, for the purpose of revising articles of confederation. On the 17th of September, this convention, under the presidency of Washington, agreed on a federal constitution to be proposed to the people in state conventions.

From the establishment of a federal government] The thirteen united colonies, which issued the declaration of independence, entered into the confederation of 1778, and formed the constitution of 1789, were New Hampshire, Massachusetts, Rhode Island, Connecticut, New York, New Jersey, Pennsylvania, Delaware, Maryland, Virginia, N. Carolina, S. Carolina, and Georgia. At an early period of the struggle, congress had recommended to the assemblies and conventions of the several colonies, to establish governments suited to the new exigencies of the country: not a difficult task to a people accustomed to manage their own affairs, and whose legislative and judicial institutions in some cases needed only a slight reform. The constitutions of the states were in general modelled on the same plan, having, in imitation of the British constitution, a legislative authority vested in two houses, one or both of which were chosen directly by the people; and an executive, with definite power, chosen by the people or by their representatives. These documents are the first written constitutions of government with which history makes us acquainted. These governments were all highly popular in their principles, simple in their machinery, and well-calculated for the management of home affairs; but the general regulation of commerce on uniform principles, the management of the intercourse with foreign states, the support of a military peace establishment, the settlement of disputed questions between thirteen sovereign powers, required a general authority competent to decide these and similar questions. The convention of Philadelphia, after a session of about four months, finally adopted on 17th September, 1787, the present federal constitution of the U. S., with the provision, that the ratification of it by 9 states should be sufficient for the final establishment of the new system. The first president was Washington, who was chosen by a unanimous vote of his fellow-citizens. The first congress assembled at New York, March 4, 1789, and immediately proceeded to raise a revenue by imposing duties; to establish a federal judiciary consisting of a supreme court, with circuit and district courts; to organize the executive administration, by creating the departments of war, of foreign affairs, and the treasury; to fund the debt of the United States, and assume the state debts; and to create a national bank. The jealousy of state rights, which had made the adoption of a federal constitution a matter of great delicacy and difficulty, soon began to manifest itself in opposition to the federal government, and the two parties of federalists and anti-federalists were distinctly marked from an early period. Indian hostilities required an increase of the military establishment; and a duty imposed on domestic spirits met with open resistance from the people. Meanwhile the progress of the French revolution excited a lively interest in America. The president indeed attempted to maintain a strict neutrality in the contest between the new republic and the European cabinets; but the minister of the French convention, offended with this neutral policy, actually undertook to raise troops against the Spanish possessions, to fit out naval expeditions in the ports of the U. S., in direct defiance of the orders of the executive; and was supported in this conduct by a strong party in different parts of the country. A treaty settling the boundaries between the U. S. and the Spanish colonies, and conceding to the former the right to navigate the Mississippi, was concluded Oct. 27, 1795; and a treaty with Great Britain was ratified Aug. 14, 1795. This latter treaty—commonly known as Jay's treaty—created such divisions in the U. S. as to put in danger the existence of the government. Petitions against it were sent in from all parts of the country, and it was the subject of a violent debate in the next session of congress. "The objections of those opposed to carrying the treaty into effect, were, generally, that it wanted reciprocity; that it gave up all claim of compensation for negroes carried away contrary to the treaty of peace, and for the detention of the western posts; that it contravened the French treaty, and sacrificed the interest of an ally to that of Great Britain; that it gave up, in several important instances, the law of nations, particularly in relation to free ships making free goods, cases of blockade, and contraband of war; that it improperly interfered with the legislative powers of congress, especially by prohibiting the sequestration of debts; and that the commercial part gave few advantages to the U. S."—[*Pitkin's History of the United States.*] But public opinion gradually settled in favour of the treaty, as being the only possible means of saving the country from becoming involved in the wars of the French revolution. The depredations of French cruisers upon the commerce of the U. S. were such, that, on the 7th of July, 1797, the treaties with France were declared to be no longer obligatory. But the revolution of 18th Brumaire changed the policy of the French republic, and a new treaty was concluded between the two powers, Sept. 30, 1800. On the retirement of Washington, John Adams was elected to succeed him; and, in 1801, Thomas Jefferson was chosen third president of the U. S. One of the most important events of this period was the purchase of Louisiana from the French government, in 1803, for the sum of 15,000,000 dollars, 2,500,000 of which were retained as a compensation for illegal captures made by France. After the rupture of the peace of Amiens the Americans became possessed of an extensive and lucrative carrying-trade; but the continental system of Napoleon, and the counter measures of the British government, directed to the interruption of all commercial intercourse between the respective belligerents and neutrals, threatened the entire destruction of American commerce. Great Britain also claimed the right of searching American vessels, and impressing from them British seamen. To counteract the imperial decrees and the British orders-in-council, the general government of the U. S. in 1807 prohibited the exportation of any article from the U. S. territories; but this measure was received with so much opposition that in 1809 the embargo was repealed; at the same time, all trade and intercourse with France and England was prohibited by act of congress. In June, 1812, war was declared against Great Britain, and was continued, with varied fortune, for three years, during which the Americans attempted, without success, the conquest of Canada, and the British were repulsed in several attacks upon the maritime cities of the U. S. Peace was finally concluded at Ghent, December 24, 1814, by a treaty which settled nothing, and made no allusion to the causes of the war.

The change in European affairs produced by the peace of Paris in 1815, contributed to introduce a considerable change in the policy of the American government, and the relations of domestic parties, which still continues in operation. Previous to this time, party-divisions were founded, in a great measure, upon points of foreign policy, and on foreign predilections or aversions. The facilities for commerce had diverted a great portion of the national capital into commercial channels, and manufactures had received little attention. Since that period European politics have exercised less influence in the country; manufacturing industry has been developed to an astonishing degree; internal improvements have been pushed with vigour; several new states have been admitted into the Union; and many years of peace and prosperity have well developed the resources of the country The following extract from the message of President Monroe to congress, in 1823, develops what may be regarded as the dominant idea in the Union with relation to foreign politics. "In the wars of the European powers, in matters relating to themselves, we have never taken any part, nor does it comport with our policy so to do. We owe it to candour, and to the amicable relations existing between the U. S. and those Powers, to declare, that we should consider any attempt on their part to extend their system to any portion of this hemisphere, as dangerous to our peace and safety. With the existing colonies and dependencies of any European power, we have not interfered, and shall not interfere. But with the governments who have declared their independence, and maintained it, and whose independence we have, on great consideration and on just principles, acknowledged, we could not view any interposition for the purpose of oppressing them, or controlling in any other manner their destiny by a European power, in any other light than as the manifestation of an unfriendly disposition towards the U. S. It is impossible that the Allied powers should extend their political system to any portion of either continent, without endangering our peace and happiness, nor can any one believe that our southern brethren, if left to themselves, would adopt it of their own accord. It is equally impossible, therefore, that we should behold such interposition in any form with indifference." The maxim that no European nation has a right to interfere in the policy or affairs of the American continent is a sufficiently arrogant one of itself; but appears more singularly preposterous as put forth by a government which has increased its powers by extending territory and jurisdiction on every possible occasion and on the slightest pretexts.

Immediately after the establishment of Mexican independence, the country lying between the Sabine and the Rio-Grande was overrun by emigrants from the U. S., whose influence prevailed on Texas to declare for the federal constitution of the Mexican republic, when Santa Anna, in 1834, had succeeded in establishing a military despotism. The contest between Texas and Santa Anna terminated in the capture of the Mexican general, and the recognition by him, as the price of his liberty, of the independence of Texas. The same influence which had overthrown Santa Anna's power in this quarter of the Mexican territory, ultimately succeeded in annexing it to the federal union of the North American states in 1845; and in 1848 New Mexico and California were added to the union. The history of these latter transactions is sketched in the respective articles CALIFORNIA, MEXICO, and TEXAS.

UNITY, a township of Waldo co., in the state of Maine, U. S., 31 m. NE of Augusta, drained by a branch of Sebasticooti river. Pop. in 1840, 1,457; in 1850, 1,557.—Also a township of Sullivan co., in the state of New Hampshire, 45 m. W by N of Concord. It has a diversified surface, and is drained by Little Sugar and Beaver Meadow rivers. Pop. in 1840, 1,238.—Also a township of Westmoreland co., in the state of Pennsylvania, 130 m. W of Harrisburg, drained by Crabtree and Big Sewickly creeks, and by Nine Mile river. Pop. in 1840, 3,003; in 1850, 3,100.—Also a township of Columbiana co., in the state of Ohio, 149 m. NE by E of Columbus, drained by tributaries of Little Beaver river, and intersected by the Ohio and Pennsylvania railroad. Pop. in 1840, 1,896; in 1850, 2,095.—Also a township of Sullivan co., in the state of New Hampshire.—Also a village of Alexander co., in the state of Illinois, 5 m. N of the Mississippi.

UNKEL, a village of Prussia, in the prov. of the Rhine, regency and 26 m. NW of Coblenz, and circle of Neuwied, on the r. bank of the Rhine. Pop. 680. The environs are noted for their wine.

UNKELBACH, a village of Prussia, in the reg. of Coblenz, circle of Ahrweiler. Pop. 380.

UNKENBACH, a village of Bavaria, in the circle of Pfalz, district of Overmoschel. Pop. 364.

UNKIAR-SKELESSI, a village of Turkey, on the Bosphorus, 8 m. NNE of Constantinople, celebrated as the spot where a treaty of defensive alliance, between Russia and Turkey, was concluded on the 8th of July 1833, on "the principle of conservative and mutual defence." By the 2d article of this treaty preceding treaties were confirmed; and by the 3d art. Russia engages to furnish by land and by sea as many troops as the Sublime Porte might call for. A separate and secret art. appended to this treaty, recites that, "in virtue of one of the clauses of the first article of the patent treaty of offensive alliance concluded between the imperial court of Russia and the Sublime Porte, the two high contracting parties are bound to afford to each other mutually, substantial aid, and the most efficacious assistance for the safety of their respective dominions. Nevertheless, as his Majesty the emperor of all the Russias, wishing to spare the Sublime Ottoman Porte the expense and inconvenience which might be occasioned to it, by affording substantial aid, will not ask for that aid if circumstances should place the Sublime Porte under the obligation of furnishing it, the Sublime Ottoman Porte, in the place of the aid which it is bound to furnish in case of need, according to the principle of reciprocity of the patent treaty, shall confine its action in favour of the imperial court of Russia to closing the strait of the Dardanelles—that is to say, to not allowing any foreign vessel of war to enter therein under any pretext whatever."

UNKIE-TUNKI, a hill-fort of Hindostan, in the prov. of Aurungabad, on a rock which rises precipitously to the height of 200 ft.

UNNA, a town of Prussia, in the prov. of the Rhine, regency and 18 m. NW of Arnsberg, and circle of Hamm. Pop. 5,663. It is enclosed by walls and ditches, and has an hospital. It has manufactories of calico, hats, distilleries of brandy, and breweries.—Also a river of Turkey in Europe, in Croatia, which has its source in Mount Sabor, the Dinaric Alps, on the confines of Dalmatia; runs first NW, then bends NE, and afterwards E, forming for some distance the line of separation from Austria, and joins the Save on the confines of Slavonia. Its principal affluents are the Unnatz and Sanna, both of which it receives on the r. The chief towns on its banks are Ostrovitz, Bihach, Novi, and Turk-Dubitza.

UNNATZ, a town of Turkey in Europe, in Croatia, on the l. bank of a river of the same name, an affluent of the Unna, 20 m. SE of Ostrovitz.

UNNERSDORF, a village of Bavaria, in the circle of Upper Franconia, lordship of Banz. Pop. 240.

UNSANG, a district in the NE part of Borneo, forming the E extremity of a peninsula which projects between the Mindoro and Celebes seas, and presents a narrow isthmus between Sandakan and Darvel bays.

UNSEBURG, a village of Prussian Saxony, in the reg. of Magdeburg, circle of Wanzleben, on the Bode. Pop. 850.

UNSHA, a town of Russia, in the gov. of Kostroma, circle of Makariev, on a stream of the same name which flows into the Volga.

UNST, an island and a parish in the extreme N of Shetland. The island is separated on the SW from Yell by Blumol-sound, a rapid tideway about 1 m. broad, and from Fetlar on the S by a sound between 3 and 5 m. broad. It forms a slender, though indented oblong, extending from NNE to SSW; and measures 11½ m. in length, about 3½ m. in mean breadth, and upwards of 36 sq. m. in area. The coast, over much the larger part of its extent, is a constant alternation of headlands, and of indenting bays and creeks. Balta-sound on the E, and Uyea-sound on the S, are so covered by isles of their own name, and screened by projecting headlands at their entrances, as to afford good shelter to shipping. The isles Balta and Uyea, the islets Hafgruna and Hunie, and the holms Sha, Burrafrith, Woodwick, Newgord, Hogoland, and Weatherholm, skirt the coast of U., and are comprehended in its parish. The surface, compared with that of the other Shetland islands, is reckoned level. About

2,000 acres are arable, and nearly an equal number of acres are meadow and grass-lands, which might easily be brought into tillage. Five-sixths of the whole area are in commonage; and might, to a considerable extent, be improved. Gneiss, serpentine, chlorine, schist, and diallage rock, are the principal rocks of the island; and talcose and micaceous schists, primitive limestone, and quartz and hornblende rocks, also occur. Fishing here, as throughout Shetland, forms the prime employment of the inhabitants. Articles of Shetland hosiery form a considerable article of manufacture. A village has recently been built at Uyea-sound Pop. of the p. of U., in 1831, 2,909; in 1851, 2,977.

UNST (North), an isolated rocky islet lying to the N of the Shetland isles, on which a lighthouse has recently been erected.

UNSTONE, a township in the p. of Dronfield, Derbyshire, 4½ m. N by W of Chesterfield. Pop. in 1831, 586; in 1851, 776.

UNSTRUT, a river of Prussia, which has its source in Saxony, in the Dun-gebirge, a little to the W of Dingelstadt; runs first SE; enters the duchy of Saxe-Coburg-Gotha; flows thence NE to the confines of the principality of Schwarzburg-Rudolstadt, and afterwards SE to the Saxon-Saale, which it joins on the l. bank, a little to the N of Naumburg, and after a total course of about 120 m. Its principal affluents are the Wipper and the Holme on the l., and on the r. the Gera-Muhlhausen. Thamsbruck, Artern, Laucha, and Freyburg, are the chief towns on its banks.

UNTER, a German word signifying Lower. Names with which it is compounded, not found amongst the following, are to be sought for in the subsequent word.

UNTERACH, a village of Austria, in the gov. of Ems, circle of Kogel. Pop. 600.

UNTERAGERI, a commune in the Swiss cant. of Zug. Pop. in 1852, 2,243.

UNTERBACH, a village in the Swiss cant. of Valais, bail. of Raron. Pop. 228.

UNTERFRANKEN. See Franconia (Lower).

UNTERHALLAU, a village and district in the Swiss cant. of Schaffhausen, bail. of Unterklettgau. Pop. 2,607.

UNTER-INNTHAL. See Inn.

UNTERJESINGEN, a town of Würtemberg, in the circle of the Schwarzwald, on the l. bank of the Ammer. Pop. 1,400.

UNTERLAND, a district in the W part of the duchy of Saxe-Meiningen. It contains the town of Meiningen.

UNTERSCHACHEN, a village of Switzerland, in the cant. of Uri, 9 m. E of Altorf. Pop 475. It has several mineral springs.

UNTERSCHUTZEN, a village of Prussia, in the prov. of the Rhine, and circle of Altenkirchen. Pop. 200. It has a powder-mill.

UNTER-SEE. See Zeller-See.

UNTERSEEN, a small town of Switzerland, in the cant. and 30 m. SE of Bern, and bail. of Interlachen, on the Aar, between lakes Thun and Brienz. Pop. 1,360. The houses are generally of grotesque structure, and black in colour. It has several saw-mills and box-manufactories, and in the vicinity are mines of silver and copper.

UNTER-THURGAU, a village of Bavaria, in the circle of Swabia, SE of Kaufbeuren. Pop. 600.

UNTER-TURKHEIM, a town of Würtemberg, in the circle of the Neckar, bail. and 2 m. SE of Canstadt. Pop. 1,700. It has a gypsum quarry. Wine is cultivated in the environs.

UNTERVAZ, a district and v. of Switzerland, in the cant. of the Grisons, in the Unterland. Pop. 1,097.

UNTERWALDEN, a canton situated nearly in the centre of Switzerland, between the parallels of 46° 44′ and 46° 57′ N; bounded on the N by the lake of Luzern; on the E by the canton of Uri; on the S by Bern; and on the W by the canton of Luzern. It is about 25 m. in length from NE to SW, and 20 m. in its greatest breadth; with an area of only 260 sq. m. Pop. in 1837, 22,571; in 1850, 25,188, of whom only 28 were Protestants. Of this pop. 13,799 belonged to the Upper U., and 11,339 to the Lower U.; and only 52 in the whole number were foreigners. It consists of four principal valleys, covered with meadows and pasture lands, and surrounded by lofty Alps, of from 3,000 to 10,000 ft. elevation above sea-level, and some of which, as the Titlis and Surene, are covered with glaciers and perpetual snow. The general slope of the surface is from the S towards the N, or the shores of Lake Luzern, into which the two principal streams of the canton, the Aa and the Melch, discharge themselves. The eastern part of the canton has a favourable exposure, and a mild temperature; but the vine is reared with difficulty. The exports consist of cattle, hides, cheese, butter, honey, tallow, timber, and fuel; the imports, of corn, wine, and various manufactures. In point of manners, this is altogether a primitive district, a spot where, in conjunction with the neighbouring cantons of Schwytz and Uri, the independence of the republic had its origin, and the simple habits of the assertors of Swiss liberty are in a great degree preserved. The constitution is democratic. The chief towns are Stanz, the capital of Bas-Unterwald; Sarnen, the capital of Haut-Unterwald; and Engelberg. The contingent is 677 men.

UNTERWYL, a hamlet of Switzerland, in the cant. of Argau and parish of Gebistorf. Marble and coal are found in the environs.

UNTHANK, a township in the p. of Skelton, Cumberland, 5½ m. NW of Penrith. Pop. 255.

UNTSCHEN, a village of the duchy of Saxe-Altenburg, and bail. of Altenburg. Pop. 150.

UNVERRE, a commune of France, in the dep. of the Eure-et-Loir, cant. and 5 m. ESE of Brou, near the Ozanne. Pop. in 1846, 2,478.

UNWATTEN, a central county of the state of Michigan, U. S., comprising an area of 576 sq. m., drained by Maskegon river and other small streams.

UNZMARKT, a market-town of Styria, in the circle and 9 m. W of Judenburg, on the l. bank of the Muhr. It has nitre works and a forge.

UOMO-MONTO (Punta dell'), a headland of the N coast of Sicily, in the prov. and district of Palermo, in N lat. 38° 11′ 20″, and E long. 13° 7′ 30″.

UPA, a river of Russia, in the gov. of Tula, which flows first NNW, and then WSW to the Oka. It has a total length of about 150 m., and is navigable to the town of Tula.

UPADRANG, a town of Nepal, 40 m. WSW of Catmandu, on the l. side of the Tirsul-Ganga.

UPAIX, a town of France, in the dep. of the Hautes-Alpes, and cant. of Laragne, 11 m. NW of Sisteron. Pop. 720.

UPANEMA, a river of Brazil, in the prov. of Rio-Grande-do-Norte, an affluent of the Appodi.

UPAR, or Upari, a valley of New Granada, in the dep. of the Magdalena, intersected by the Cesare. It runs from N to S, and is 45 m. long and 30 in breadth. It produces sugar, and has mines of silver, copper, iron, and lead. It is well-peopled.

UPATA, a town and canton of Venezuela, in the prov. of Cartagena. The town is situated in N lat. 7° 49′, and W long. of Caracas 4° 31′, at an alt. of 351 yds. above sea-level, 31 leagues from Angostura. It contains a handsome church, and a pop.

of about 700.—The canton, containing 6,000 sq. leagues, and a pop. of 12,000, three-fourths of whom are independent Indians, Guayanas, Caribisce, Guaycas, Warrows, Pariagotos, and Aruaaks, is the most beautiful district in the whole course of the Orinoco. It consists of undulating elevated savannahs, traversed by ranges of wooded hills, and presenting varied scenery. The climate is more temperate than in other parts of the Lower Orinoco. Vast herds of wild cattle and horses graze over these immense plains. The Capuchins of Catalonia had formerly thirty missions in the neighbourhood of the capital of the canton; but at the breaking out of the war of independence, the civilized Indians dispersed, and the missions were mostly abandoned. A great deal of coffee, cotton, sugar, and excellent tobacco is exported from the *labranzas* in the neighbourhood of the capital; and it has recently been reported that gold dust has been observed in large quantities in the bed of the Yurukei river in this canton. The nearest port on the Orinoco to Upata is Puerto-de-las-Tablas, which is reached by a ride of 17 leagues over a beautiful country. Puerto-de-las-Tablas is near the mouth of the Caroni, and a few hours' pull above Guayana-Vieja, which Keymis took, and where Sir Walter Raleigh's son was killed. The steamer that now runs between Trinidad and Angostura calls at Puerto-de-las-Tablas.

UPATOIE, a village of Muscogee co., in the state of Georgia, U. S., on the W branch of a creek of the same name, an affluent of the Chattahoochee river, and intersected by the Muscogee railway. Pop. in 1840, 100.

UP-AVON, or UPHAVEN, a parish in Wilts, 9 m. ESE of Devizes. Area 3,329 acres. Pop. in 1831, 498; in 1851, 510.

UPCHURCH, a parish in Kent, 5 m. E of Chatham. Area, with the hamlets of Ham, Haliwell, and Otterham, 5,138 acres. Pop. in 1851, 497.

UPERNAWICK, a district of Greenland, on the shore of Baffin's bay.

UPHALL, a parish in Linlithgowshire, bounded on the SE and S by Edinburghshire. Area 3,922 acres. The rivulet Broxburn passes eastward through it, and gives its name to a village, with 500 inhabitants, a little NE of the centre of the p. The village of U., with 160 inhabitants, is situated in the W side of the p., 13 m. from Edinburgh. Pop. in 1831, 1,254; in 1851, 1,331.

UPHAM, a parish in Hants, 2¾ m. NW by N of Bishop's-Waltham. Area 2,852 acres. Young, author of *The Night Thoughts*, was born here in 1681. Pop. in 1831, 511; in 1851, 550.

UPHILL, a parish in Somersetshire, 8 m. NW by W of Axbridge, at the mouth of the Axe. Area 1,697 acres. Pop. in 1831, 306; in 1851, 422.

UPIE, a commune of France, in the dep. of the Drome, cant. and 8 m. SSW of Chabeuil. at the foot of a mountain. Pop. 1,098.

UPIGNY, a department and commune of Belgium, in the prov. of Namur. Pop. 268.

UPITANGA, a river of Brazil, in the prov. of Santa Catharina and district of São-Francisco.

UPLAND, an ancient prov. of Sweden, now comprised in the prefectures of Upsal and Stockholm.

UPLEADON, a parish in Gloucestershire, 3 m. E by N of Newent, on the river Leadon. Area 1,207 acres. Pop. in 1831, 241; in 1851, 275.

UPLEATHAM, a parish in the N. R. of Yorkshire, 3 m. NNE of Guisborough. Area 1,100 acres. Pop. in 1831, 265; in 1851, 447.

UPLOWMAN, or UPLOMAN, a parish in Devon, 4½ m. ENE of Tiverton. Area 2,912 acres. Pop. in 1831, 335; in 1851, 386.

UPLYME, a parish in Devon, 1 m. NNW of Lyme-Regis, on the borders of Devonshire. Area 3,199 acres. Pop. in 1831, 975; in 1851. 1,032.

UPMINSTER, a parish in Essex, 3[illegible] ESE of Romford. Area 3,373 acres. Pop. in 18[illegible] [illegible],228.

UPOLU, or OATUAH, one of the Samoan group, in S lat. 13° 49′. It is about 40 m. in length, and 10 m. in breadth; and is richly covered with timber. On its N coast is Apia harbour, in the above lat., and W long. 171° 40′.

UP-OTTERY, a parish in Devon, 5 m. NNE of Honiton. Area 5,830 acres. Pop. in 1851, 1,042.

UPPER, a township of Cape May co., in the state of New Jersey, U. S., 13 m. NE of Cape May, bounded on the SE by the Atlantic, and intersected by the Tuckahoe and Cedar Swamp creeks. Pop. in 1840, 1,217.—Also a township of Lawrence co., in the state of Ohio. Pop. 1,180.—Also a township of Carroll co., in the state of Arkansas. Pop. 478.—Also a township of Crawford co., in the same state. Pop. 1,144.

UPPER ALLOWAY'S CREEK, a township of Salem co., in the state of New Jersey, U. S., 7 m. SE of Salem, drained by Alloway's and Stow creeks. Pop. in 1840, 2,235.

UPPER ALTON, a village of Madison co., in the state of Illinois, U. S., 2½ m. E of the Mississippi. Pop. in 1840, 1,002; in 1850, 2,000.

UPPER BERN, a township of Berks co., in the state of Pennsylvania, U. S., bounded on the N by the Blue mountains, and on the E by Schuylkill river. Pop. in 1840, 2,906.

UPPER CHICHESTER, a township of Delaware co., in the state of Pennsylvania, U. S., watered by Hook and Naaman's creeks. Pop. in 1840, 475.

UPPER DARBY, a township of Delaware co., in the state of Pennsylvania, U. S., drained by Cobb's and Darby's creeks. Pop. in 1840, 1,489.

UPPER DUBLIN, a township of Montgomery co., in the state of Pennsylvania, U. S., 99 m. E of Harrisburg, drained by Darby and Cobb's creeks. Pop. in 1840, 1,322; and in 1850, 1,550.

UPPER FREEHOLD, a township of Monmouth co., in the state of New Jersey, U. S., 12 m. E by S of Trenton, bordered on the NE by Assunpink creek, and on the S by Crosswicks creek, and drained by affluents of these rivers. Pop. in 1850, 2,556.

UPPER HANOVER, a township of Montgomery co., in the state of Pennsylvania, U. S., 81 m. E by N of Harrisburg, drained by Perkioman creek and its branches. Pop. in 1840, 1,467; in 1850, 1,570.

UPPER MACUNGY, a township of Lehigh co., in the state of Pennsylvania, U. S., 74 m. ENE of Harrisburg, drained by Lehigh creek. Pop. in 1840, 1,769; in 1850, 1,900.

UPPER MAHANTANGO, a township of Schuylkill co., in the state of Pennsylvania, U. S., 37 m. NE by N of Harrisburg, drained by two branches of Mahantango creek. Pop. in 1840, 1,291.

UPPER MAHANY, a township of Northumberland co., in the state of Pennsylvania, U. S. Pop. in 1840, 1,131.

UPPER MAKEFIELD, a township of Bucks co., in the state of Pennsylvania, U. S., 25 m. NNE of Philadelphia, bordered on the N by Pidcock's creek. Pop. in 1840, 1,490.

UPPER MARIETTA, a village of Lancaster co., in the state of Pennsylvania, U. S., on the E bank of the Susquehanna, and on the Columbia branch railroad.

UPPER MARLBORO', a village of Prince George co., in the state of Maryland, U. S., 7 m. SW of Annapolis, on the S side of the W branch of Patapsco river. Pop. in 1850, 800.

UPPER MIDDLETOWN, a village of Middletown township, Middlesex co., in the state and on the W side of the Connecticut, U. S.—Also a village of Manallen township, Fayette co., in the state of Pennsylvania, on Redstone creek, 155 m. W by S of Harrisburg.

UPPER MILFORD, a township of Lehigh co., in the state of Pennsylvania, U. S., drained by the N branch of Perkiomen creek, and by the head waters of Saucon creek. Pop. in 1840, 3,081.

UPPER MOUNT BETHEL, a township of Northampton co., in the state of Pennsylvania, U. S., on the W side of the Delaware. Pop. in 1840, 2,643.

UPPER NAZERETH, a township of Northampton co., in the state of Pennsylvania, U. S., drained by two branches of Manookisy creek. Pop. in 1840, 1,118.

UPPER OXFORD, a township of Chester co., in the state of Pennsylvania, U. S. Pop. in 1840, 1,277.

UPPER PAXTON, a township of Dauphin co., in the state of Pennsylvania, U. S., 22 m. N of Harrisburg, drained by Mahantango and Great and Little Wiconisco creeks, and bordered on the W by the Susquehanna. Pop. in 1840, 1,814.

UPPER PENN'S NECK, a township of Salem co., in the state of New Jersey, U. S., bounded on the W by Delaware river. Pop. in 1840, 1,854.

UPPER PROVIDENCE, a township of Delaware co., in the state of Pennsylvania, U. S., 13 m. W of Philadelphia, bordered on the E by Crum creek, and on the W by Ridley creek. Pop. in 1840, 780.—Also a township of Montgomery co., in the same state, bordered on the SW by Schuylkill river, and drained by Perkiomen and Mingo creeks. Pop. 2,244.

UPPER RED HOOK, a village of Redhook township, Dutchess co., in the state of New York, U. S., 46 m. S of Albany, on the E bank of Hudson river. Pop. in 1840, about 300.

UPPER SAINT CLAIR, a township of Alleghany co., in the state of Pennsylvania, U. S., drained by Chartier's creek. Pop. in 1850, 2,700.

UPPER SALFORD, a township of Montgomery co., in the state of Pennsylvania, U. S., 10 m. NNW of Norristown, drained by Perkiomen and Rich Valley creeks. Pop. in 1840, 1,301.

UPPER SANDUSKY, a village of Wyandott co., in the state of Ohio, U. S., on the W side of Sandusky river, 60 m. NW of Columbus. Pop. in 1850, 783.

UPPER SAUCON, a township of Lehigh co., in the state of Pennsylvania, U. S., 6 m. SE of Northampton, drained by Saucon creek, an affluent of Lehigh river. Pop. in 1840, 2,072.

UPPER STILLWATER, a village of Penobscot co., in the state of Maine, U. S., on the W side of Penobscot river, opposite Marshe's island, and intersected by the Bangor and Piscataquis railroad.

UPPER SWATARA, a township of Dauphin co., in the state of Pennsylvania, U. S., drained by Swatara creek. Pop. in 1840, 1,205.

UPPER TULPEHOCKEN, a township of Berks co., in the state of Pennsylvania, U. S., drained by Northkill and Little Northkill creeks, and intersected by the Union canal. Pop. in 1840, 2,941.

UPPERVILLE, a village of Tanquier co. in the state of Virginia, U. S., 129 m. N by W of Richmond. Pop. in 1850, 350.

UPPINGHAM, a market-town and parish in the co. of Rutland, 12 m. WSW of Stamford. The town stands on an elevated situation, and contains several good houses chiefly disposed in the form of a square. Area of p. 1,210 acres. Pop. in 1851, 2,068.

UPPINGTON, a parish in Shropshire, 7 m. ESE of Shrewsbury. Area 755 acres. Pop. in 1831, 117; in 1851, 98.

UPRIGHT POINT, a headland of New South Wales, in the co. of St. Vincent, in S lat. 35° 50′.

UPSALA, or UPSAL, an ancient town in Sweden, in a laen or district of the same name, situated in a flat country, 39 m. NNW of Stockholm, on the small river Sala or Fyris, which divides it into two parts, and opens a communication with Lake Mälar. It is the see of an archb., and had a pop. in 1845 of 5,680. Most of the private houses are of wood, but the public buildings are of stone or brick. The cathedral is a large structure of the 13th cent., on the model of Notre-Dame at Paris, but built in great part of brick. The university of U., founded in the 15th cent., has 47 professors; and is usually attended by from 800 to 1,000 students. It can boast of Linnæus, Celsius, Bergmann, and a number of distinguished professors, and possesses a valuable library founded by Gustavus Adolphus, and now numbering 150,000 vols. The other objects worthy of attention are the observatory, the botanical garden, cabinets of natural history and mineralogy, the royal castle and gardens, and the residence of the archb. This town was long the residence of the Swedish monarchs, who bore the title of kings of U., till the beginning of the 10th cent. The coronation of the kings of Sweden is still performed here. About 3 m. to the N of the more recent town is Old U., a decayed place with a very curious old church.—The laen of U., bounded on the N by the gulf of Bothnia, on the E by the laen of Stockholm, on the S by lake Mälar, on the W by the laen of Westeraas, and on the NW by Gefle, has an area of 1,556 geog. sq. m., with a pop. in 1845 of 87,700. Its chief river is the Sala.

UPSHIRE, a hamlet in the p. of Waltham Abbey, Essex; 12 m. N by E of London. Pop. in 1831, 745; in 1851, 845.

UPSHUR, a county in the NE part of the state of Texas, U. S., comprising an area of 1,053 sq. m., drained by Little Cypress river and its branches, and by affluents of Sabine and Big Cypress rivers. Pop. in 1850, 3,394. Its capital is Gilmer.

UPSON, a central county of the state of Georgia, U. S., comprising an area of 208 sq. m., drained by branches of Flint river, by which it is bounded on the SW and W. Pop. in 1840, 9,408; in 1850, 9,424. Its capital is Thomaston.

UPTON, a township of Worcester co., in the state of Massachusetts, U. S., 35 m. WSW of Boston, on West river. Pop. in 1840, 1,466; in 1850, 2,023.

UPTON, a chapelry in the parish of Blewberry, Berks, 4 m. NNE of East Ilsley. Pop. in 1851, 337. Also a parish in Bucks, 2 m. NNE of Windsor, intersected by the Western railway. Area 1,950 acres. Pop. in 1801, 1,018; in 1831, 1,502; in 1851, 3,573.—Also a township in the p. of St. Mary, co.-palatine of Chester, 2 m. N of Chester. Area 1,106 acres. Pop. in 1831, 289; in 1851, 555.—Also a parish in the co.-palatine of Chester, 7¾ m. N by W of Great Neston. Area 929 acres. Pop. in 1831, 191; in 1851, 227.—Also a parish in Huntingdonshire, 6 m. NW of Huntingdon. Area 970 acres. Pop. in 1831, 150; in 1851, 169.—Also a township in the p. of Sibson, Leicestershire, 3½ m. SW of Market-Bosworth. Pop. in 1831, 148; in 1851, 152.—Also a parish in the co. of Lincoln, 5 m. SE by E of Gainsborough. Area 3,180 acres. Pop. in 1831, 460; in 1851, 577.—Also a parish in Norfolkshire. Area 1,693 acres. Pop. in 1831, 510; in 1851, 564.—Also a parish in Northamptonshire, 2 m. W of Northampton. Area 979 acres. Pop. in 1831, 48; in 1851, 42.—Also a parish in Nottinghamshire, 2½ m. E of Southwell. Area 1,860 acres. Pop. in 1831, 533; in 1851, 629.—Also a parish in Somersetshire, 7 m. W by N of Wiveliscombe. Area

3,779 acres. Pop. in 1831, 344; in 1851, 344.—Also a township in the p. of Badsworth, Yorkshire, 6 m. S by E of Pontefract. Area 1,026 acres. Pop. 255.

UPTON, a township of Worcester co., in the state of Massachusetts, U. S., 35 m. WSW of Boston, drained by West river. Pop. in 1850, 2,023.

UPTON (BISHOP'S), a parish in Herefordshire, 4 m. NE by E of Ross. Area 3,391 acres. Pop. 693.

UPTON-CRESSETT, a parish in Salop, 4½ m. W by S of Bridgenorth. Area 1,603 acres. Pop. 58.

UPTON-GRAY, a parish in Hants, 3 m. SW by W of Odiham. Area 2,234 acres. Pop. 422.

UPTON-HELIONS, or HILION, a parish in Devon, 8 m. NW of Exeter. Area 819 acres. Pop. in 1831, 152; in 1851, 137.

UPTON-ST.-LEONARD, a parish in Gloucestershire, 3¼ m. SE of Gloucester. Area 2,975 acres. Pop. in 1831, 898; in 1851, 1,124.

UPTON-LOVEL, a parish in Wilts, 2 m. SE of Heytesbury. Area 1,399 acres. Pop. in 1851, 233.

UPTON-MAGNA, a parish in Salop, 4 m. E of Shrewsbury. Area 3,260 acres. Pop. in 1851, 488.

UPTON-NOBLE, or LOVEL, a parish in Somersetshire, 3½ m. NNE of Bruton. Area 677 acres. Pop. in 1831, 282; in 1851, 238.

UPTON-PYNE, a parish in Devon, 3 m. N by W of Exeter. Area 1,852 acres. Pop. in 1851, 491.

UPTON-SCUDAMORE, a parish in Wilts, 2 m. N by W of Warminster. Area 2,500 acres. Pop. in 1831, 392; in 1851, 407.

UPTON-UPON-SEVERN, a market-town and parish in Worcestershire, 9 m. SE of Worcester. Area of p. 3,170 acres. Pop. in 1831, 2,343; in 1851, 2,290. The town, supposed to have been erected on the site of a Roman station called *Upoessa*, stands on a cliff on the r. bank of the Severn, and consists of several well-paved streets containing good houses. The river is navigable here for vessels of 110 tons, and is crossed by a stone-bridge of 6 arches. It has some trade in cider.

UPTON-SNODSBURY, a parish in Worcestershire, 6 m. E of Worcester. Area 1,661 acres. Pop. in 1831, 316; in 1851, 341.

UPTON-WARIN, a parish in Worcestershire, 3½ m. NNE of Droitwich. Area 2,600 acres. Pop. 422.

UPTON-WATERS, or PARVA, a parish in Salop, 5½ m. N by W of Wellington. Area 732 acres. Pop. in 1831, 193; in 1851, 202.

UPWAY, a parish in Dorset, 4 m. S by W of Dorchester, including the tythings of Elwell and Stottingway. Area 1,385 acres. Pop. in 1851, 637.

UPWELL, a village and parish partly in Cambridgeshire, and partly in Norfolk, comprising the chapelry of Welney. The village, sometimes called Well, is situated on both sides of the navigable river Welney, 6 m. SE of Wisbeach. Area of p., 21,746 acres. Pop. in 1831, 3,471; in 1851, 5,377.

UPWOOD, a parish in Huntingdonshire, 2½ m. SW of Ramsey. Area 1,809 acres. Pop. 416.

UR, a village of France, in the dep. of the Pyrenees Orientales, and cant. of Saillagouse. Pop. 250. It has an iron mine.

URABA. See DARIEN.

URACH, a town of Würtemberg, capital of a bail. of the same name, in the circle of the Schwarzwald, 24 m. SE of Stuttgard, on the Ems and Elsach, in a deep valley of the Alps. Pop. in 1852, 5,952. It has a castle belonging to the counts of Würtemberg, built in 1443, two churches, and a theological seminary, and possesses extensive manufactories of linen, cutlery, and printed goods, several yarn and paper mills, bleacheries, &c. The trade consists chiefly in sheep, and in articles of local manufacture. This town is of great antiquity, and was frequently the residence of the sovereigns of Würtemberg. In its vicinity are the remains of the ancient fortress of Hohen-Urach, and of the convent of Güterstein.—Also a village of Baden, in the circle of the See. Pop. 700.

URAGA, a town and port of Japan, on the W coast of the bay of Jeddo. It was visited by the American expedition under Captain Perry in 1854. The squadron, consisting of the steam-frigates Susquehannah and Mississippi, and the sloops of war Plymouth and Saratoga, sailed from the harbour of Napa-kiang, in Lieu-kieu on the 2d of July. On the morning of the 8th they made Cape Idzu, near the S entrance of the bay of Jeddo, and sailing directly up the bay, anchored in the afternoon off the town of U. The appearance of the steamers—the first ever seen in Japanese waters—with the other vessels in tow, moving with all sails furled, at the rate of nine or ten knots an hour, appeared to produce considerable sensation among the Japanese, and all the trading junks, with which the bay was crowded, carefully kept out of their way. As the vessels were coming to anchor, two shells or rockets were fired into the air from a battery about a mile distant, but apparently as a signal, and not as a token of hostility. The next morning, the governor of U., and a nobleman of the third rank, came off, and, after ascertaining the object of the visit, asked for time to despatch an express to Jeddo, in order to obtain instructions how to act. During the three days which elapsed before the answer arrived, the Mississippi made a trip of about 10 miles further up the bay, finding everywhere deep soundings. Beyond the promontory of U., a point which no foreign vessels had passed before, she discovered a large and beautiful bight, which was perfectly land-locked, and offered secure and commodious anchorage. She was followed at a distance by a number of government boats, but none of them attempted to interfere with her, or with the cutters of the different vessels which were sounding in advance of her. The presence of the squadron appeared to cause no interruption to the inland commerce, for the bay was at all times studded with large junks and hundreds of small craft, passing up and down.

URAGO D'OGLIO, a village of Austria, in Lombardy, in the prov. and 19 m. W of Brescia, and district of Chiari, on the r. bank of the Oglio. Pop. 1,358. It has a castle.

URAHI, a small river of Brazil, in the prov. of Rio-de-Janeiro, and district of Campos, an affluent of the Imbe.

URAL MOUNTAINS, an extensive chain stretching from the gulf of Kara, in the Arctic ocean, southwards to about the parallel of 50°, forming a natural physical line of demarcation between Asia and Europe, throughout an extent of 900 m.; and traceable perhaps, on a meridian elevation, to an extent of 30°. Throughout a great part of its course the chain is made up, in breadth, of numerous embankments and lateral ridges, the aggregate of which is never under 30 m. in breadth, and, in some points, reaches 120 m. In several places the range does not constitute the actual watershed between the rivers of Europe and Asia, for some streams, which rise on the E declivity of the chain, after skirting it for some distance, break through the range and mingle their waters with the rivers of Europe. The N division of the chain, beyond the parallel of 65°, is covered with forests and morasses. The average elevation between the parallels of 58° and 56° is only about 1,350 ft., although the base on which the range rests has an average alt. of 900 ft. above sea-level. The highest part of the range is to the N of the parallel of 59°. The highest summit N of the parallel of 66° is not more than 3,000 ft. high. About the parallel of Obdorsk, the chain turns suddenly to the E for nearly 25 m., after which it recovers its direction of N and S, and rapidly declines towards the marshes. The culminating point of the whole chain is supposed to be the Konshakofskoi-Kamen, in N lat. 59° 30′, which attains an alt. of 5,720 ft. above sea-level; but later observations appear to indicate that the D'eneshkin-Kamen, which lies to the N of the parallel of 60°, has a greater alt. than this. Both these points, however, are to the eastward of the principal range. The S division, within that of the N, is composed of several ridges, arranged in a fan-shape and bending SSE, S, and SW. It is a diversified and picturesque region, and rises in the Iremel, to an alt. of 5,200 ft. above sea-level. The highest part of the range is traversed by two roads. One of these begins at Werkoturie; passes to the S of the Tawdinskoi-Kamen; and crosses the chain in about 59° 15′, to the town of Solikamsk. The other, and more northern route, passes from Bogosslavsk in Siberia, to Tcherdin in Anope; crossing the principal range in 60° N lat., and on the N side of the Kakwinskoi-Kamen.

Geological structure.] For our knowledge of the geology of this great mountain-barrier, we are mainly indebted to the recent researches of our countryman, Sir Roderick J. Murchison. "Considering the short space which has elapsed since the conquest of Siberia," says that distinguished geologist, "and up to how recent a time these mountain-tracts remained in a state of impenetrable forest inhabited by idolatrous Voguls and Ostiaks upon the N, and Mahomedan Bashkirs on the S, we ought rather to feel astonishment at the rate with which the region has been cleared and civilized through the introduction of European manners and mining industry. When Peter the Great, with a keen perception of the surest methods of advancing his empire, selected the first Demidoff, to explore the iron ores of these mountains, he laid the foundation of the great native mineral wealth, which now so conspicuously distinguishes Russia from all the surrounding nations. The earliest mining-establishments or *zavods* planted by that great sovereign are still the centres of activity, and have served as models after which numerous other works have been formed, both by the government and private speculators. In the days of Pallas, geology was so little understood (a few gold-mines only being known, and a great portion of the country unreclaimed), that the descriptions of the great naturalist are chiefly to be viewed as vivid portraits of living nature: as such, indeed, his observations have well stood the test of time, and small gleanings only have remained for those who followed him. Since that time, the Russian miners, learning their first lesson from foreigners, have become a well-informed class, independent of extraneous aid, and their directors (officers of the imperial school-of-mines) have described the lithological and mineral characters of the country, around their respective posts, with great fidelity." The axis, or central portion of the chain, consists to a great extent of talcose schists, or chlorite and quartzite, ancient sedimentary strata, for the most part in a highly metamorphic condition, in consequence of numerous syenitic and trap rocks; but on account of the presence of certain organic remains, traceable at intervals in limestones to the silurian series. Along the E flank, the most accessible by reason of the mining establishments, the strata are also greatly altered by the prevalence of igneous rocks; it was only along the W flank that Sir Roderick and his companions were enabled to establish a clear succession of carboniferous, devonian, and silurian deposits. True granite is of very rare occurrence along the axis of the chain, and has at a comparatively recent period burst through the eastern dislocations. The periods of dislocation, the change of relative level of land and water, and of the protrusion of igneous rocks, appear to have been phenomena repeated at different geological epochs. From the occurrence of cupriferous minerals diffused throughout the Permian strata, Sir Roderick infers that anterior to the deposition of those beds, metallic veins must have existed in the U.; and from the abundance of the remains of terrestrial plants in the same deposits, that the chain must have been raised to a certain extent above the level of the then existing ocean. Subsequent periods of disruption are proved by the lines of disturbance in the Permian series on the immediate flank of the U., and connected with dislocations which have affected them. The patches of Jurassic rocks at the N and S extremities of the range, are considered to have been subsequently desiccated, and the absence of strata of that age throughout the great mass of the chain, or for 12° of lat., to prove that it was constantly above the level of the sea during the Jurassic epoch. Between that period and the accumulation of the gold alluvia, there are no signs of any great changes in the physical structure of the U., and the only deposits assignable to that interval are certain trachytic grits and beds of lignite, which, it is conceived, may have been formed in lakes. On the W side of the chain, the order of the metalliferous rocks is best developed; but it is on the E igneous slopes that the miner is best repaid by ores. Sir Roderick concludes that "the widely-spread cupriferous deposits of Permia, which occupy all the low country to the W of these mountains, have been derived from pre-existing eastern lands, upon which the plants and vegetables enclosed in the Permian conglomerates must have grown. Judging from its composition—it is entirely made up of fragments of ancient Uralian rocks—the great Permian deposit must have been accumulated, not only after the completion of the silurian, devonian, and carboniferous systems, but after their consolidation, and either after or during their mineralization with copper-ores. This is a clear and undeniable conclusion, at which the field-geologist who has examined this region arrives; for, in whatever parallel of latitude he may trace this ancient detritus, he invariably finds it to be more coarse and metalliferous as it approaches the mountains from which its materials have been derived, whilst in receding from them, such mineral matter (always in the form of deposit, and never in the condition of veins) as regularly dies away and is lost in marine marls, sand, and limestone. But if the Ural mountains were, as we contend they must have been, the source whence all these cupriferous sediments, as well as detritus and fossil vegetables were supplied, very different indeed must have been their former outline from that which now prevails; for on the W slope of the axis down which the waters now flow into Permia, there are no great veinstones and original sources from which such debris could have been derived. All the spots where the largest veins, masses, and original centres of copper-ore occur, whether at Bogoslofsk, Nijny-Tagilsk, Gumeshefsk, and Polofsk, south of Miask, or other and intermediate places, are on *the eastern* [illegible]*e of the chief ridge.* Supposing that these mines were in the process of forming, or having been formed, were undergoing destruction, during an era in which the land had assumed its present outline, almost every cupriferous particle and drop of water impregnated with or transporting such mineral matter must have descended into the adjacent low country of Siberia. By no natural agency could any considerable quantity of such coarse materials be now carried to the low countries on the west, between which and all the great copper sources which are known lies the ridge of the U. Now, as all the cupriferous detritus has been carried to the western flank of the mountains, and not a particle of it into the low country of Siberia, it follows that by far the greatest variation in physical outline which the region has undergone—one by which a lofty wall was thrown up between Permia and the original copper-sites of the U.—took place at a period posterior to the formation of the Permian deposits."

Gold deposits.] When the region of Permia was submerged beneath the sea, and the Permian deposits were in process of formation, the U. mountains formed the rocky shore of a low continent, from which powerful streams poured into a western sea. That old continent contained iron and copper, but neither gold nor platinum; for traces of those metals have never been found in the Permian debris. In rocks still older—such as the carboniferous conglomerates—there is no trace of gold; nor in rocks far younger, such as certain tertiary grits. From these, and other reasons equally strong, Sir Roderick concludes that the U. chain became auriferous during the most recent disturbances by which it was affected, when its highest peaks were thrown up, the present water-shed established, and the greenstone, porphyries, syenitic granites, and other comparatively recent igneous rocks, intruded through the palæozoic rocks, along its E slopes; in short, that gold was one of the most recent mineral productions anterior to the historic era, and coeval with mammoths and rhinoceroses. Their bones are seldom detected out of the line of the gold-works; and the Bashkirs regard them with superstitious respect, saying to the Russian miners, "Take from us our gold if you will; but, for God's sake, leave us the bones of our ancestors." Along with these gigantic quadrupeds are found the remains of the *Bos Urus*, now the only survivor of this ancient *fauna.* A question of interest arises from the total absence, on both flanks of the U., of erratic blocks, and of any traces of those scratches, grooves, and polishings, which are considered, by the advocates of the glacial theory, to be proofs of the former existence of glaciers. Its drift is all local and not transported; and in the northern portion of the chain, between 60° and 65° of lat., no glaciers are found on peaks constantly covered with snow, and attaining an alt. exceeding that of the highest mountains of the British Isles. This absence of all the phenomena of glacial action seems to exclude the possibility of the lower or flat regions of Russia having been once invested in a cerement of ice. The problem connected with the entombment of mammalian remains in the gold alluvia, as well as in alluvium generally, is difficult of solution, but by whatever means the universal destruction of those great mammals, during one particular period, may be attempted to be explained, Sir Roderick answers, that it was owing, in the district under consideration, including the low regions extending from each flank of the U., to an elevation whereby a change to a colder climate was effected. To the general view of Baron Humboldt, that the richest gold-deposits are those which are derived from ridges having a meridian direction, M. Erman is opposed; but Sir Roderick is of opinion, that it is a fact that the greatest quantity of gold ore has been obtained from chains having a nearer relation to N and S than to equatorial or E and W directions, due perhaps to the general form of the chief masses of land, and the prevai'ˊg strike of the palæozoic rocks. In a paper reaa before the British Association, Sir Roderick points out the erroı into which some persons had fallen, that the Uralian mines were worked underground; the only small subterraneous work being one near Ekaterinburg, which affords a very slight profit. All the other mines along the Ural chain, throughout 8° of lat. are simply diggings and washings which are made in the detritus or shingle accumulated on the slopes of the ridges and in the adjacent valleys, and, with one exception, are all upon the E side of the range. This phenomenon is a necessary result of the structure of the chain; the older deposits through which the eruptive rocks have risen constituting chiefly the crest and E slopes of the chain, whilst the W slopes are occupied by deposits of younger or Permian age. Comparing California with the U., Sir Roderick shows that there is a very great coincidence of mineralogical structure, and that with these *constants* the same results obtain; the chief distinction consisting in the apparently larger proportion of gold in the detritus of the newly-discovered deposits in California than in those of the U. More recently, Sir Roderick, and other geologists, have observed a striking resemblance between the geology of the Ural chain and that of the Blue mountain ranges in Australia, which run, in a general line, not far from the meridian, through 30 degrees of lat. in Tasmania and New Holland, and New Guinea, having their highest points at an alt. of from 5,000 to 6,000 ft. above the sea; with an axis of chloritic and talcose schist, and quartzites, with occasional limestones of Silurian age, in which occur metallic ores and gold in veins of quartz. Moreover, granitic rocks of a comparatively younger age break through them. As on the flanks of the U., the carboniferous formation repose, so in Australia occurs the very same order of deposits. Sir Roderick, therefore, anticipated the extraordinary discovery of gold in the Australian continent, thinking it most highly probable, that besides the lead and copper which exist in the Blue mountain ranges, auriferous sands would be found in the rivers flowing from them. In respect of other phenomena, the dividing ranges of New Holland are similar to the U., with the exception that the slope in this hemisphere is to the W, and the escapement edges of the deposits to the E. So far as has yet been ascertained, the drift of the U. mountains seems to have its parallel in that of Australia, which is all local, and much of it fluviatile, deposited at the ancient mouths of the present rivers when the country was at a lower level. The phenomena pointed out by Capt. Sturt render it far from improbable that the interior depressed area of Australia is of similar character to the great Aralo-Caspian country. From these and other facts communicated to the Geological society, Sir Roderick predicted the probable mineral wealth of Australia, and in a letter addressed to Sir C. Lemon, and published in the *Philosophical Magazine*, was the first to advise that a person well acquainted with the washing of mineral sands should be sent out to Australia, speculating on the probability of auriferous alluvia being abundant in that region.

In the paper read before the British Association, already referred to, Sir Roderick adverts to the distinctions between such surface mining-operations as those of Siberia, California, and the Brazils, and those works in which besides the ores of silver, copper, &c. gold also had been extracted from the veins in the solid or parent rock, as in Mexico. Sir Roderick also traces the history of gold and its development as known to the ancients and our ancestors of the middle ages; and shows that in all regions where rocks similar to those he had described occurred, there gold has been found in more or less quantities; and that just in proportion to the time that a country has been civilized has the extraction and produce of the precious metal diminished; so that in many tracts where it formerly prevailed to some extent, it has been either worked out or the mines have been almost forgotten. The continental tracts formerly so rich, as cited by Strabo, with the exception of the N. Ural or country of the Arimaspes, whence the Scythian ores came, are no longer gold-bearing districts. The Scythian oı Uralian tract had, in fact, remained unknown and unattended to from the classical age until this century, and so completely ignorant were the modern Russians of the existence of gold in the U. mountains, or that they had in their hands the country which supplied so much gold to Greece and Rome, that excellent German miners had long worked the iron and copper mines of that chain before any gold veins were discovered. These also were worked as solid veins in the rock for some time before the accidental discovery of a small per-centage of gold ore in the ancient alluvium or drift led to the superficial diggings, which produced at an infinitely less expense ten times the amount of produce of the mines in the solid rock near Ekaterinburg. All the energy displayed by the Russian miners having failed to augment the amount of Uralian gold, and as it has never much exceeded half-a-million sterling, the period is gradually arriving when the local depressions or basins of auriferous detritus of that region will be successively dug and washed out, and the U. will then resemble many other countries in possessing actual mines of iron and copper, but merely a history of its gold. Russia, however, has also the golden key of all Eastern Siberia, in which various offsets from the Altai chain, chiefly those which, separating the rivers Lena, Jenisei, &c., stretch along the shores of the Baikal lake, have proved so very productive, that for some years they have afforded three millions sterling average, exclusive of the U.

URAL, or JAIK, a river of Russia, which rises in the above chain, under the parallel of 55°; flows S to Orsk; then W through the mountain-chain to Orenburg; and from the latter place SW and S to the Caspian, into which it discharges itself by several mouths below Gurjev. The waters of the U., which have a breadth of from 150 to 200 fath. at Glinyoi, divide into two branches, about 4 m. below Gurjev. The first branch flows into the Beloi-Ilmen, 8 m. SE of Gurjev; the other, the Poltavai, flows westwards, and again separates into 4 branches before it reaches the Caspian. All the mouths of the U. appear to be shallow and filled with reeds.

URALSK, a town of Russia, in the gov. of Orenburg, on the r. bank of the Ural, in N lat. 51° 11′ 23″, 150 m. WSW of Orenburg. Pop. 15,000.—There is another small place of the same name on the Ural, in N lat. 53° 42′.

URARIEVERA, a river of Brazil, in the prov. of Para, which runs NE, receives the Uraricapara on the l., and after a total course of about 135 m., joins the Idome.

URARINHAS, a town of Ecuador, in the dep. of Assuay, on the N bank of the Amazon, at the confluence of the Xucaray, 30 m. NE of La Laguna.

URARIRA, a river of Brazilian Guayana, an affluent of the Rio Negro, which it joins on the r. bank, between Moreira and Thomar.

URARTE, a village of Spain, in the prov. of Alava, 9 m. SSE of Vitoria. Pop. 170.

URAS, a village of the island of Sardinia, in the div. of Cape Cagliari, and prov. of Busachi, in a marshy but fertile locality. Pop. 1,500.

URATUBA, a small town of Independent Tartary, in the khanat of Kokan, on the r. bank of the Sehun, 30 m. NE of Khojend.

URBACH (OBER), a town of Würtemberg, in the circle of the Jaxt, bail. and 3 m. E of Schorndorf. Pop. in 1840, 2,073. It has a tobacco factory.—Also a parish and village of Hanover, in the reg. of Hildesheim, near Buckenem. Pop. 780.—Also a village of Prussia, in the reg. of Coblenz, circle of Neuwied. Pop. 462.

URBAIN (SAINT), a town of France, in the dep. of the Haute Marne, and cant. of Donjeux-Sur-Marne, 16 m. SE of Vassy. Pop. 928.

URBAN, a village of Austria, in Moravia, in the circle of Znaim. Pop. 1,100.

URBANA, a township of Steuben co., in the state of New York, U. S., 183 m. W of Albany. Pop. in 1840, 1,884; in 1850, 2,079.—Also a township of Champaign co., in the state of Ohio, 39 m. W of Columbus, and intersected by the Columbus, Piqua, and Indiana, and the Mad river and Lake Erie railroads. Pop. in 1840, 1,386; in 1850, 3,414.—Also a village of Champaign co., in the state of Illinois, on the S side of the Salt Fork of Vermillion river, and intersected by the Chicago branch of the Illinois Central railroad. Pop. in 1840, 150; in 1850, 400.—Also a village of Jackson co., in the same state, 155 m. S by E of Springfield, and near the Illinois Central railroad.

URBANIA, a town of the Papal States, in the deleg. of Urbino and Pesaro, on the Metauro, 8 m. SW of Urbino. Pop. 3,716. It has a china manufactory.

URBANNA, a village of Middlesex co., in the state of Virginia, U. S., on the SW side of Rappahannock river, near the mouth of a river of the same name, and 46 m. E by N of Richmond.

URBANO, a fortress of the Papal States, in the leg. and 17 m. WNW of Bologna, on the frontier of the duchy of Modena. It has 5 bastions, and a small garrison.

URBAR, a village of Prussia, in the prov. of the Rhine, regency and circle of Coblenz. Pop. 508.

URBINO, a town of the Papal States, capital of the leg. of Urbino and Pesaro, and of the district of the same name, 45 m. W of Ancona, and 138 m. N of Rome, on a mountain. Pop. 4,418. It is the see of an archb., a college, and a ducal palace. Needles are its chief article of manufacture. It is noted as the birthplace of Raphael.

URBINO-AND-PESARO, a legation or province of the Papal States, bounded on the N by the legation of Forli and the republic of San Marino; on the NE by the Adriatic; on the SE by the deleg. of Ancona; on the S and SW by that of Perugia; and on the W by Tuscany; and comprising an area of 540 sq. m. Pop. in 1843, 237,966. It is to a great extent covered by ramifications of the Apennines, and is watered by the Metauro, and its affluents, the Cantiano, Cesano, Foglia, and Marecchia. The mountains are, for the most part, naked and sterile, but the hills are covered with olive and mulberry trees and vines, and in the valleys, corn, legumes, lint, hemp, and fruit, are grown in great abundance. It has numerous herds of cattle, sheep, and pigs, and the silk of the Fossombrone is noted as the finest in Europe. The leg. comprises 5 districts, 28 towns, and 410 villages.

URBIONE, a lake of Spain, in the prov. and 42 m. NW of Soria, at a considerable elevation, amid lofty mountains. It is, according to Garibay, 4,800 ft. in depth, and appears to have been the crater of a volcano. It discharges itself into the Najerilla.

URÇAY, a village of France, in the dep. of the Allier, cant. and 12 m. W of Cerilly. Pop. 500. It has an iron-mine.

URCEL, a village of France, in the dep. of the Aisne, and cant. of Anizy-le-Chateau. Pop. 400. It has manufactories of sulphate of iron and of alum.

URCHFONT, or ERCHFONT, a parish in Wilts, 3½ m. SSE of Devizes, containing the tythings of Eastcott and Wedhampton. Area 6,873 acres. Pop. in 1831, 1,389; in 1851, 1,604.

URCISSE (SAINT), a village of France, in the dep. of the Tarn, cant. and 5 m. NW of Salvagnac. Pop. 340. In the environs are quarries of marble.

URCIZE (SAINT), a commune of France, in the dep. of Cantal, cant. and 12 m. S of Chaudesaigues. Pop. 1,878.

URDA, a town of Spain, in the prov. and 33 m. SSE of Toledo, on two hills enclosed in a valley, watered by the Amarguillo, which is liable to extensive inundations. Pop. 4,100. It has a church, a handsome chapel, and an hospital. The manufacture of charcoal and of hosiery form the chief objects of industry. In the vicinity are quarries of jasper, and mines of silver and antimony.

URDAX, a town of Spain, in the prov. of Navarra, 36 m. NNE of Pamplona, in the valley of Bastan, near the French frontier. Pop. 600. It has a church, and a convent, and an iron-work, and carries on an active trade in faggots.

URDEL, a village of France, in the dep. of the Basses-Pyrenees, and valley of Aspe. In its vicinity is an alabaster quarry.

URDEN, a mountain of Switzerland, in the cant. of Grisons, enclosing a small deep lake, liable by vicissitudes of weather to violent agitation.

URDENBACH, a village of Prussia, in the prov. of the Rhine and circle of Dusseldorf, on the Rhine. Pop. 1,000. It has manufactories of woollen fabrics.

URDINGEN. See UERDINGEN.

URDOS, a commune of France, in the dep. of the Basses-Pyrenees, and cant. of Accons, 30 m. S of Oloron. Pop. 406. It has a custom-house.

URE, or YORE, a river of Yorkshire, which, in conjunction with the Swale, with which it unites at Myton-hall, forms the Ouse. From the elevated moorland between Yorkshire and Westmoreland, in which it rises, its general course is E, passing Hawes and Askrigg. From this point to Ripon, it runs SE, passing Middlesham and Mesham; 3 m. below which latter place it becomes the boundary between the N and W ridings. From Ripon its course is S and then E, past Boroughbridge and Aldborough till its junction with the Swale.

UREI, or OURI, a river of Sumatra, which has its

source in the E part of the district of Redjang; runs along the confines of that of Anak-Sungei; and which, after a total course of about 60 m., throws itself into the Indian ocean, to the SE of the embouchure of the Caytun. Near the sources of this river are the grottoes of Goha, from which saltpetre in great quantities is extracted.

URENHAL, a village of Kashmir, to the NW of Islamabad.

URFAH. See ORFA.

URFAHR, or UFER LINZ, a market-town of Austria, in the circle of the Muhl, on the Danube, opposite Linz. Pop. 2,660.

URFT, a village of Prussia, in the prov. of the Rhine, reg. of Aix-la-Chapelle, and circle of Gemünd. It has several forges.

URGUB, a town of Asiatic Turkey, in the pash. of Caramania, 30 m. SW of Kaisariyeh, on an affluent of the Kizil-Irmak. It is singularly built on a number of conical eminences.

URGUNDAB, a river of Afghanistan, which has its source in the Huzareh country, about 100 m. NE of Candahar, flows past that town, and 25 m. below is joined by the Turnak; thence it takes a W direction, and after a total course of about 250 m., falls into the Helmund in N lat. 31° 30′. Its banks are fertile, and its waters are extensively employed in irrigation. On its bank, about 25 m. W of Kelah-i-Ghiljie, is a village of the same name.

URGEL. See SEU-D'URGEL.

URGLIN, or RUTLAND, a parish in co. Carlow, 2¾ m. ENE of the town of Carlow. Area 3,149 acres. Pop. in 1831, 970; in 1851, 873.

URGNANO, a village of Austria, in Lombardy, in the prov. and 8 m. S of Bergamo, and dist. of Verdello, on the road by Canonica from Brescia to Milan. Pop. 2,596. It was formerly enclosed by walls, defended by towers, of which only a small portion now exists, and has an ancient castle.

URI, a central canton of Switzerland, bounded on the N by Schwyz; on the NE by Glaris; on the E by the Grisons; on the S by Tessin; on the SW by Valais; and on the W by Unterwalden and Bern. Schöpf estimates the superficial surface of this canton at 24 German sq. m.; but according to Meyer's chart, its area is nearer 34 such sq. m. The whole canton consists of the two valleys of the Reuss and the Urseren. The most remarkable alp is the St. Gotthard, over which one of the main roads to Italy is here led, with a summit-level of 6,790 ft. above the Mediterranean. The chain of mountains which immediately surround this place, takes the general appellation of St. Gotthard, but each summit has its own particular name. No spot in Europe gives birth to so many noble rivers as this: from it the Reuss flows towards the N; the three sources of the Rhine towards the E; the Ticino towards the S; and the Rhone towards the W. No mountain in this cant. is under 5,000 ft. in alt. above sea-level; and the greater number of the summits range from 8,000 to 9,000 ft. in alt. The pop. of the canton in 1850 was 14,505, in 2,855 families, being an increase since 1837 of 986, or a mean annual increase of 1 in 184, while 1 in 144 was the average for all Switzerland. Of the pop. in 1850, only 40 were returned as foreigners, and only 12 as of the protestant religion. Every male above 20 years of age is a member of the general assembly, which is held once a year to choose the officers of the executive department. The revenue is small, there being almost no taxes, and little trade; it is surprising how three such poor cantons as Luzern, Uri, and Unterwalden, could raise sums sufficient for the execution of their share of the superb carriage-road over the St. Gotthard. The chief town is Altdorf, with 3,000 inhabitants.

URIBE, a district of Spain, in the prov. of Vizcaya. It extends N to the gulf of Gascogny; is watered by the Salcedon and Nervion; and is one of the most extensive and most fertile in the prov. It contains 32 communes, and has several iron-works.

URIE, a river of Aberdeenshire, which rises in the district of Strathbogie, and flows successively E, S, and SE. Its entire length of course is about 20 m.

URIMARAS, a tribe of Indians of Central America, who inhabit the SE part of the state of Costa-Rica.

URITANA. See ORIA.

URIUPINSKAIA, a town of Russia, in the gov. of the Don Cossacks, 25 m. SW of Novo-Cherkask, on the l. bank of the Choper.

URK, an island of the Zuider-see, in Holland, in the prov. of that name, 24 m. E of Hoorn, in N lat. 52° 39′ 47″. It does not exceed 3 m. in circumference, but has a village with 575 inhabitants employed chiefly in fishing.

URLAND, a parish and village in Norway, in the diocese and 75 m. NE of Bergen, and bail. of Northern Bergenshuus. Pop. 2,000.

URLINGFORD, a parish, containing a small town of the same name, co. Kilkenny. Area 3,497 acres. Pop. in 1831, 2,492; in 1851, 2,010. The town stands on the mail-road from Dublin to Cork, 8¾ m. NE by E of Thurles. Pop. in 1851, 1,168.

URLOFFEN, a village of the grand-duchy of Baden, in the circle of the Middle Rhine, bail. and 8 m. N of Offenburg, on the railway from Carlsruhe to Freiburg. Pop. 2,250.

URMENY, a market-town of Hungary, in the comitat and 9 m. S of Neutra.

URMITZ, a village of Prussia, in the prov. of the Rhine, on the Rhine, SE of Neuwied. Pop. 598.

URMSTON, a township in the p. of Flixton, co.-palatine of Lancaster, 5½ m. SW by W of Manchester, on the N bank of the Mersey. Pop. 730.

URNA-DESA, or UNDES, a district of Little Tibet, extending between the Himalaya and Cailas mountains, and watered by the Sutlej. It is noted for its wool, which is said to be the finest in Tibet, and has several iron mines. The mountaineers cultivate the vine. Deba is the chief town.

URNASCH, a village of Switzerland, in the cant. and 8 m. WSW of Appenzell, on a small river of the same name. Pop. 2,464.

URNEN (NIEDER), a village of Switzerland, in the cant. and 9 m. N of Glaris, on the Linth, at its exit from lake Wallenstadt. Pop. 1,505. It has mineral baths, and manufactories of silk and leather.—In the vicinity is another village distinguished as Ober U., with a pop. of 691.

URNEY, a parish in co. Cavan, containing the town of Cavan. Area 7,934 acres. Pop. in 1831, 6,050; in 1851, 5,239.—Also a parish, partly in co. Donegal, but chiefly in co. Tyrone, containing the village of Clady, and a small part of the town of Strabane. Area 14,488 acres. Pop. in 1851, 6,317.

URNIETA, a town of Spain, in the prov. of Guipuzcoa, 6 m. S of San-Sebastian. Pop. 1,625. It has a parish church, two chapels, and an hospital, and possesses several iron-works. The environs are fertile, producing maize, legumes, and fruit.

URNIYARAH, a village of Sind, in the Little Desert.

URNSHAUSEN, a village of the duchy of Sachsen-Weimar, in the circle of Eisenach. Pop. 700.

UROLA, a river of Spain, in the prov. of Guipuzcoa, which has its source in Mount Araya, in the S part of the prov.; waters Legazpia, Zumarraga, Villareal, Azcoytia, and Azpeytia; and after a course in a generally N direction of about 30 m., throws itself into the gulf of Gascogny at Zumaya.

URPETH, a township in the parish of Chester-le-Street, co-palatine of Durham, 2 m. SW of Birtley, intersected by the Great North of England railway. In the vicinity are extensive coal-mines. Area 1,671 acres. Pop. in 1831, 716; in 1851, 952.

URQ, a village of France, in the dep. of the Ariege, cant. and 3 m. ESE of Cabanes. Pop. 188.

URQUHART, a parish on the coast of Morayshire. Pop. in 1831, 1,019; in 1851, 1,331.

URQUHART-AND-GLENMORISTON, a united parish in Inverness-shire, having a length from E to W of 30 m., and in general from 8 to 12 m. in breadth. The peopled districts are the narrow slopes along Loch-Ness, and the glens of Urquhart and Moriston, the only transverse valleys which branch off from the NW side of the Ness part of the Glenmore-nan-Albin. The largest of several hamlets, Milntown, has about 150 inhabitants. Good roads go up the two principal glens toward Kyle-Rhea, the ferry to Skye; and two roads connect the glens respectively along Loch-Ness, and 8 or 9 m. to the W. Pop. in 1831, 2,942; in 1851, 2,380.

URQUHART-AND-LOGGIE-WESTER, a parish chiefly in the SE of Ross-shire, but comprehending also the well-known detached district of Nairnshire called Ferrintosh. It extends from NNE to SSW 10 m. About 4,860 acres of the whole area are in tillage. The village of Conan Bridge has upwards of 300 inhabitants; and at the hamlet of Culbokie are held four annual fairs. Pop. in 1831, 2,864; in 1851, 3,153.

URR, a parish in the SE of Kirkcudbrightshire. Area about 12,000 acres. The surface, compared to that of most Galloway parishes, is level, few of the hills being of considerable height. On the W bank of Urr-water is the celebrated Moat of Urr, an artificial mount or table-land, rising from concentric and successive terraces, and anciently used as a seat of judicature by the petty chiefs of the feudal times. This is probably the largest work of its kind in Scotland. The villages of the parish, with their pop., in 1836, are Dalbeattie with 1,393; Haugh, with about 220; and Springholm, with about 300. Pop. of the p. in 1831, 3,098; in 1851, 3,362.—Also a river of Kirkcudbrightshire, which issues from the lake below described, and runs 26 m. to the Solway frith, midway between the Nith and the Dee. Its prevailing direction is toward the S. About 4½ m. before losing itself in the Solway, it begins to expand into an estuary; and at its embouchure, between Heston island and the headland on the E, it has a breadth of very nearly 2 m. It is navigable for considerable craft 7¼ m. above its embouchure.

URR (LOCH), a small lake on the boundary between Dumfries-shire and Kirkcudbrightshire, about 3 m. in circumf., and 5 fath. in extreme depth. Its water appears extremely black, the ground under and around it being in general a heathy moss. Pike, and a few large trout, are the only fish which it produces. The surrounding scenery is a bleak and dreary expanse of tame, naked, heathy hill.

URRACAL, a town of Spain, in the prov. of Almeria, on the Almanzora, 5 m. ENE of Purchena. Pop. 850. It has a soap manufactory, and several linen bleacheries.

URRAN, a rivulet of co. Wexford, which rises among the SE skirts of Mount Leinster; receives various early affluents from the Blackstairs; and flows about 11 m. SE to a confluence with the Slaney at a point about 5 furl. below Enniscorthy.

URRAO, a town of New Grenada, in the dep. of the Cauca, and prov. of Choco, 24 m. SW of Santa Fe-de-Antioquia. Pop. 620.

URRAY, a parish partly on the N border of Inverness-shire, but chiefly in the S end of Wester Ross, Ross-shire. Pop. in 1851, 2,621.

URREA-DE-GAEN, a town of Spain, in the prov. of Teruel. Pop. 1,426. It has a parish church and a custom-house. The manufacture of linen and rearing of cattle form the chief branches of local industry.

URREA-DE-XALON, a town of Spain, in the prov. and 18 m. W of Zaragoza, near the Xalon, in a fertile locality. Pop. 740.

URRIBARI, a hamlet of Spain, in the prov. of Vitoria, and district of Aramayona. It has several mineral springs.

URROZ, a town of Spain, in the prov. of Navarra, 14 m. E of Pamplona, in a plain, at the foot of a mountain. Pop. 680. It has manufactories of coarse cloth.

URRUGNE, a commune and town of France, in the dep. of the Basses-Pyrenees, cant. and 3 m. SW of St. Jean-de-Luz. Pop. in 1846, 3,633.

URSA (PUNTA-DELL'), a headland of Sicily, on the N coast, in the prov. and district of Palermo, in N lat. 38° 18′, and E long. 13° 11′.

URSANNE (SAINT), or SANKT URSITZ, a small town of Switzerland, in the cant. of Berne, bail. and 9 m. SE of Porentruy, in a wild and rocky locality, on the Doubs, at an alt. of 471 yds. above sea-level. Pop. 726. It has several tanneries, and in the vicinity are extensive iron-mines.

URSBERG, or URSPERG, a hamlet of Bavaria, in the circle of Schwaben, to the NE of Memmingen. Pop. 240. It has a castle and a fine church.

URSEKAU, a village of Prussia, in the reg. of Breslau, circle of Steinau. Pop. 660.

URSEL, a commune and village of Belgium, in the prov. of E. Flanders, 12 m. NW of Ghent. Pop. 2,500.

URSEL (OBER), a town of the duchy of Nassau, bail. and 5 m. NE of Königstein, on a river of the same name. Pop. 2,400. It has several paper and tobacco mills, and copper works.

URSERNTHAL, or URSERENTHAL, a valley of Switzerland, in the cant. of Uri. It is about 9 m. long, and 2 broad, and is watered by the Reuss. The mountains by which it is enclosed afford excellent pasturage. Cheese forms the chief article of produce. See ANDERMATT.

URSETSH, a town of European Russia, in the gov. of Minsk, circle of Sluzk. Pop. 1,500.

URSINS, a village of Switzerland, in the cant. of Vaud, and district of Yverdun, 18 m. N of Lausanne. Pop. 252. It has some Roman ruins.

URSLINGEN, a village of Würtemberg, in the Schwarzwald circle, bail. of Rottweil. Pop. 509.

URSPITZ, a village of Moravia, in the circle of Brunn, on the Iglawa. Pop. 640.

URSPRING, a village of Bavaria, in the prov. of Upper Bavaria, district of Schongau. Pop 200.—Also a village of Würtemberg, in the circle of the Danube, near Ulm. Pop. 293.

URSWICK (GREAT), a village, township, and parish in the co.-palatine of Lancaster, 3 m. SSW of Ulverstone, containing the townships of Adgarley, Bardsea, Bolton, Stainton, and Little U. Area 4,100 acres. Pop. in 1831, 752; in 1851, 891.

URT, a commune of France, in the dep. of the Basses-Pyrenees, cant. and 5 m. NW of La Bastide-Clerence, on the l. bank of the Joyeuse. Pop. 1,657.

URTEBIZE, a commune of Belgium, in the prov. of Hainault, and dep. of Mainvault. Pop. 318.

URU'A, a village of Brazil, in the prov. of Rio-Grande-do-Norte, and district of Villa-Flor.—Also an island of the Madeira, in the prov. of Para, below the confluence of the Mataura.

URTSCHITZ, a town of Moravia, in the circle of Olmutz, lordship of Plumenau. Pop. 1,050.

URUBAMBA, a district and town of Peru, in the dep. of Cuzco. The district produces cacao, and varieties of fruit. Pop. 9,250. The town is 39 m. N of Cuzco, on the r. bank of the Vilcabamba.

URUBAQUIS, a tribe of Indians of Brazil, who inhabit the district between the Uatuma and Anavilhana, to the N of the Amazon.

URUBU, a comarca and town of Brazil, in the prov. of Bahia. The town is finely situated on the r. bank of the Rio-de-São-Francisco, 84 m. NW of Rio-de-Contas. Pop. 1,000. It has a church and a chapel. The environs are fertile, and produce sugar in considerable quantities.—Also a town of the prov. of Maranhão, on the l. bank of the Itapicurie, and comarca of Caxias. Its inhabitants, who are Indians, cultivate cotton and rice.—Also a river of the prov. of Para, in Brazilian Guayana, which throws itself into the Amazon, on the left bank, after a course of about 150 m., and nearly opposite the confluence of the Madeira. See also PROPIHA.

URUBUQUARA, a river of Brazil, in the prov. of Para, which has its source in the Serra-Velha; runs S; traverses the lake of the same name, and after a total course of 195 m., throws itself by the l. bank into the Amazon.

URUBURETAMA, a mountain-range in Brazil, in the prov. of Ceara, and district of Villa-da-Imperatriz, running from N to S between the rivers Curu, and Acaracu.

URUCAIA, a village of Brazil, in the prov. of Minas-Geraes, and district of São-Romão, on the l. bank of a river of the same name, and about 90 m. above its confluence with the São-Francisco.—The river rises in the Serra-Tabatinga, and flows E to the São-Francisco, which it joins on the l. bank after a course of 200 m. It has a narrow channel.

URUÇANGA, a river of Brazil, in the prov. of Santa-Catharina, which runs E to the ocean. Its mouth is obstructed by sand-banks.

URUÇUHI, a river of Brazil, in the prov. of Piauhy, which has its source on the E side of the Serra-Gurguea; runs N, and joins the Parnahiba, on the r. bank, nearly opposite the confluence of the Rio-das-Ballas.

URUCUPARANA, a river of Brazil, in the prov. of Para, which runs first NE, then N, and after a total course of about 135 m., joins the Amazon, on the l. bank, a little below the confluence of the Cuari.

URUCURICAIA, or IGUARAPE, a natural canal of Brazil, in the prov. of Para, which connects the rivers Guajara and Xingu.

URUCURUNI, a tribe of Indians of Brazil, who inhabit the Cordilheira-dos-Parecis, in the prov. of Matto-Grosso, and N part of the comarca of that name.

URUCUYA. See URUCAIA.

URUENA, a town of Spain, in the prov. of Segovia, and partido of Sepulveda, on Mount Almarez. Pop. 706. It has a parish church, a customhouse, and a public granary.

URUGUAY. See BANDA-ORIENTAL.

URUGUAY, a great navigable river of S. America, which rises in Brazil, under the 28th parallel, on the W flank of the Sierra-de-Santa-Catharina, and collecting various streams, traverses a vast extent of country in a W, NW, and finally SSW course; and, under the 34th parallel, unites with the Parana to form the Rio-de-la-Plata. The U., though not equal either to the Paraguay or the Parana for length of course, surpasses both in the accession of waters it receives, and near its confluence with the latter, it is its equal if not its superior in breadth. Its entire length of course exceeds 800 m. The quick accumulation of waters from the mountain-torrents renders it an extremely rapid stream in the upper part of its course; and when it leaves the hilly country, it attains so great a breadth 680 m. above its junction with the Parana, that a ten-oared boat, it is said, requires half-an-hour to cross it. The most remarkable tributaries of the U. are the Pelotas, the Uruguay-Pitu, the Piratiny, the Ibicui, the Guarey, the Arapey, and the Rio-Negro. On the l., the U. communicates by the river Merim or Mirinai, with the lake that bears the latter name. Upon the whole, the U. is a rocky stream, of difficult navigation, except by the *balzas*, a species of craft peculiar to South America. It abounds in fish, and the country through which it passes is romantic, beautiful, and fertile in pasturage. Hides and tallow form as yet the chief articles of export from the regions through which it flows. From the junction of the Uruguay-Pitu, a little below the parallel of 27°, to the junction of the Uary in S lat. 30° 20′, it forms the boundary between Brazil and the La-Plata federacy; from the latter point to its estuary, it separates the republic of Uruguay or the Banda-Oriental from the same federacy. The chief towns on or near its banks are Concepcion in the Missiones prov., and Belem and Purificacion in Uruguay. By special treaty with Brazil, the navigation of this river is open to all countries.

URUKSTADT. See KARGE.

URUMJIK, a town of Turkey in Europe, in Rumelia, in the sanj. and 50 m. NW of Gallipoli, and near the r. bank of the Maritza.

URUMEA, a river of Spain, which has its source in the mountains near the village of Goyzueta, in the W part of the prov. of Navarra; traverses the valley of Basaburna-Menor; flows afterwards into the prov. of Guipuzcoa; passes near Ernani and Astigarraga, and under the walls of St. Sebastian; and after a total course, in a generally NW direction, of about 15 m., throws itself into the gulf of Gascogny.

URUMIYAH, or OORMIAH, a town and lake of Persia, in the prov. of Azerdbijan. The town is situated in N lat. 37° 35′, E long. 45° 10′, about 10 m. W of the lake, at an alt. of 4,300 ft. above sea-level. It is walled, and contains about 25,000 inhabitants, of whom 900 are Nestorians, 2,000 Jews, and the rest Mahommedans of the Shyite sect. The streets are irregular and dirty, and have a gloomy appearance, all the larger houses being shut in by lofty mud walls. A very efficient educational agency amongst the Nestorians has existed here since 1833 under the American board of foreign missions. A printing-press has been set up, and upwards of 1,300 scholars, throughout the various Nestorian districts, are now receiving instruction at the schools established by the missionaries. U. was the birthplace of Zoroaster, and for several subsequent ages the sacred city of the fire-worshippers, of whom a few yet survive in the remote valleys of Khorassan. Of the author of the Zendavesta and his followers, however, U. preserves but few traces. Several lofty mounds, composed entirely of ashes, and varying in height from 60 to 100 ft., remain to point out the site of the altars whereon their rites were celebrated; but except these and some temple-caves in the neighbouring mountains, there is little to remind the traveller of the sacred fires and their worshippers. The lake, which takes its name from the town, is little less than an inland sea, being 80 m. in length from N to S, and 20 m. in breadth, with a circuit of rather over 300 m. Its waters are so salt that fish cannot live in them; and from them and the broad girdle of pure salt washed up all round its banks, the greater part of Kurdistan is supplied with this article. Of late years this vast sheet of water has sensibly diminished, and on many parts of the shore whole square miles of white sparkling salt now spread out to where once was the water-mark. The greatest depth of the lake does not exceed 4 fath.; the average is only 2 fath.; but the shores shelve so gradually, that even this depth is rarely attained within 2 miles of the land. From the lake to the commencement of the Kurdistan mountains on the W stretches out a splendid and fertile plain of some 30 m. long, by half the number broad, and scattered over this magnificent area are not less than 300 villages, with varying populations of from 100 to 1,200. Indeed, as seen from the summit of one of the neighbouring mountains, it appears like one vast garden, richly studded with every kind of fruit tree—peach, nectarine, apricot, cherry, plum, pear, quince, and apple-trees clustering forest-like wherever the ground is not occupied with the vine or other crop.

URUNUELA, a town of Spain, in the prov. of Logrono, 48 m. E of Burgos, on the r. bank of the Najerilla. Pop. 530. It has manufactories of household linen.

URUOCA, a defile of the Serra-Hibiapaba, in Brazil, in the prov. of Ceara and district of Granja.

URUP, or COMPANY'S ISLAND, an island of the N. Pacific, in the group of the Kurile islands, to the N of Hurup or Staaten island, from which it is separated by Vries strait.

URUPES, a group of islets in the Madeira, in the Brazilian prov. of Para.

URURAHI, a parish of Brazil, in the prov. of Rio-de-Janeiro, and comarca of Campos, near a river of the same name. Pop. 2,000. The culture of sugar and mandioc and distillation of rum form the chief objects of local industry. The river U. connects lakes Aina and Feia.

URURI, a town of Naples, in the prov. of Sannio, district and cant. and 6 m. E of Larino, pleasantly situated on a hill. Pop. 1,228. It is enclosed by old walls, and contains a fine episcopal palace, a church, and two chapels.

URVILLE, a village of France, in the dep. of the Calvados, cant. and 2 m. SSE of Breteville-sur-l'Aize. Pop. 625. In the vicinity is a productive iron-mine.

URVOLGY. See HERRENGRUND.

URWITZ, a village of Prussia, in the prov. of the Rhine, and circle of Coblenz. Pop. 200.

URZAINQUI, a town of Spain, in the prov. of Navarra, 30 m. ENE of Pamplona, in the valley of Roncal, on the Ezca, which is here crossed by a bridge. Pop. 360 It has a fulling-mill.

URZEDOW, a town of Poland, in the gov. and 26 m. SW of Lublin. Pop. 1,800.

URZULIN, a village of Poland, in the gov. of Lublin, obwod and 39 m. SE of Radzyn, in a marshy locality. Pop. 180.

URZY, a village of France, in the dep. of the Nievre, cant. and 6 m. SE of Pougues, on the Nievre. Pop. 520. It has several forges and nail-works.

USACA. See OSAKA.

USAGRE, a town of Spain, in the prov. and 54 m. SE of Badajos, partly on a hill, and partly on a river of the same name, which is here crossed by a bridge. Pop. 2,260. It has an oil-mill. Woad and madder are cultivated in the vicinity.

USAN, a fishing-village on the coast of the p. of Craig, 3 m. SSE of Montrose, in Forfarshire. Its inhabitants are about 200; and, in employments and character, are akin to their neighbours of FERRYDEN: which see.

USANAS, a town of Spain, in the prov. and 12 m. NW of Guadalajara. Pop. 932. It has a parish-church, a custom-house, and a public granary.

USBECS, or UZBEKS, a predominating race in Turkestan, and especially in the three khanates of Khiva, Bokhara, and Khokan. The founder of the Usbec power was Sheibani Khan, a remote descendant of Tushi Khan, son of Jenghiz Khan. Abulghazi, their historian, tells us their name originated with a khan named Usbec, who reigned in Kiptchak in the 14th cent. Yet it seems strange, that, if the hordes of Kiptchak took the name of Usbecs from Usbec Khan, none but the hordes of Khiva, Bokhara, and Khokan, should retain it. Abulghazi tells us further, that the U. are composed of four tribes: Vighurs or Oigurs, Naimans, Durmans, and Kongorats. But little dependence can be placed on an author who confounds the Mongolian with the Turkish language, and derives all the tribes of Central Asia from Turk, the son of Japhet. The ancestors of Sheibani Khan possessed the principality of Tura, to the E of the Ural-Tau, and their subjects dwelt in summer on the banks of the Jaik, in winter on those of the Syr. At the commencement of Sheibani's reign, his dominions were confined to the district of Turkistan, to the NW of Tashkend. His followers were then an agglomeration of tribes of Turki, Mongol, and probably of Finnic race, moulded down into one people, with a preponderance of Turks. His army was latterly swelled by predatory volunteers, from all the nomadic tribes from Khashgar to the Volga. The whole region possessed by the race of Timur Bek, was at this juncture split into a confused mass of petty principalities, whose chiefs, continually at war with each other, could never be brought to act in concert against their formidable enemy, Sheibani Khan, who subjugated them all in succession, till not one foot of ground was left of all that Timur had conquered, between the Kynder-Tau and the Hindu-kush, and from the Bellur to the Caspian. Sheibani wished also to subjugate Persia; but in this attempt he was defeated and slain by Shah Ismail, the founder of the Suffi dynasty, in 1510. This decisive defeat saved Persia, but the successors of Sheibani retained their conquests between the Amu and the Syr. The male line of Sheibani ended in Abul Fyze Khan, who was stripped of all his possessions S of the Amu, by the celebrated Nadir Shah. His misfortunes drew upon him the contempt of his chiefs; and he was assassinated by a Mungut, named Rahim, who succeeded him, and who was in his turn succeeded, after a reign of 12 years, by Dauniar Khan, a descendant of Sheibani, whose son, Shah Murad, celebrated under the appellation of *Beggi Jan*, an active, ambitious prince, recovered almost all the territories which had been lost by his predecessors since the time of Nadir; nothing but the singular abilities of Killich-Ali preserved Balkh from falling into his hands. This royal priest reigned sixteen years, and was succeeded by Mir Hyder who died in 1826. See articles KHOKAN, KHIVA, and BALKH.

USCLITZ, or USCZ, a town of Prussia, in the prov. of Posen, circle and 11 m. NW of Chodziesen, on the Netze, opposite the confluence of the Kuddow. Pop. 1,784. It has extensive glass-works.

USCIE, a town of Galicia, in the circle and 15 m. NE of Stanislawow, at the confluence of the Dniester and Knechenin.

USCIE-BISKUPIE, a town of Galicia, in the circle of Czortkow, on the Dniester, 15 m. E of Taleszczyk.

USCIECZKO, a town of Galicia, in the circle and 30 m. SSE of Czortkow, on the l. bank of the Dniester, at the confluence of the Siclawa.

USCOQUES, a Slavonic tribe who inhabit Illyria, Croatia, and Dalmatia, and were formerly noted for their piracies. They speak the Illyrian and Croat languages, and profess Catholicism.

USDUM, a long range of low hills in Syria, near the S end of the Dead sea. It runs from NNW to SSE, giving the same direction to the shore of the sea. At the S extremity of the Dead sea, it bends SSW. It is in general nearly 150 ft. high, and is a mass of solid rock salt, covered with a thin stratum of limestone and marl. See article DEAD SEA.

USE MOUNTAINS, a small mountainous range in co. Cork, extending about 4½ m. in length. It flanks the W side of the vale of the Allua.

USED, a town of Spain, in the parish of Zaragoza, 12 m. SW of Daroca, at the foot of a chain of sandy hills. Pop. 1,150. It has a parish-church, a custom house, and a public granary. A little to the S is the salt lake of Gallocanta.

USEDOM, or UESEDOM, an island, circle, and town of Prussia, in the prov. of Pomerania, and regency of Stettin. The island lies between the Baltic on the N, the Pommersche-haff on the S, the Peene on the W, and the Swiene, by which it is separated from the island of Wollin, on the E. It is 33 m. in length from NW to SE, and 15 m. in extreme breadth. It contains several lakes, of which the principal is the Achterwasser, an expansion of the Peene. It has extensive level tracts, and is in some parts covered with wood. The soil possesses little fertility. The inhabitants, about 12,000 in number, employ themselves chiefly in fishing and navigation. It contains two towns, viz. Swienemunde and Usedom.—The town is 11 m. NW of Stettin, in N lat. 53° 47′ 24″, and E long. 14° 9′ 25″, on the Usedom-see, an arm of the Kleine-haff, and on an isthmus of the S coast of the island of the same name. Pop. in 1853, 1,504. It is enclosed by walls, with three gates, and has two suburbs. It has distilleries of brandy, but fishing and agriculture form the chief objects of industry.

USELDINGEN, a village of Holland, in the prov. of Luxemburg, circle of Diekirch. Pop. 560.

USERAS, a town of Spain, in the prov. and 12 m. NNW of Castellon-de-la-Plana, and 32 m. SW of Peniscola, in an irregular but fertile locality. Pop. 1,900. It has a church, several chapels, and an hospital.

USERIN, a village of the grand-duchy of Mecklenburg-Strelitz, bail. and 5 m. W of Alt-Strelitz. It has tan and saw-mills. Pop. 300.

USH, or OUJ, a river of Russia, which rises in Volhynia, and flows in an ESE course of 120 m. to the Pripet.

USHAK, a town of Asiatic Turkey, 70 m. SW of Kutahiyah. It has a considerable manufacture of carpets.

USHANT, or OUESSANT, an island of France, off the coast of Finistere. It is 5 m. in length, and 3 m. in breadth; and had a pop. in 1841 of 2,194, chiefly fishermen. It has a lighthouse on its NE point, in N lat. 48° 25′ 31″, E long. 5° 3′ 19″.

USHLAWRCOED, a hamlet in the p. of Bedwelty, Monmouth. Area 6,150 acres. Pop. 5,359.

USINGEN, a town of the duchy of Nassau, capital of a bail. of the same name, and 24 m. NE of Wiesbaden, on the Usbach. Pop. 200. It is enclosed by walls, and has a suburb, a fine castle, two churches, an hospital, and a polytechnic school. It possesses manufactories of hosiery, flannel, tanneries, and several tan and fulling-mills. Pop. of bail., which has an area of 26 sq. m., 22,000.

USITZA, a district and town of Turkey in Europe, in Servia, 100 m. SSW of Belgrade, on one of the head streams of the Morava.

USITENNA. See VOUSITRIN.

USK, a market-town and parish in the co. of Monmouth, 12½ m. SW of Monmouth. The p., which includes the hamlets of Glascoed and Gwchellog, has an area of 4,956 acres, of which 410 belong to the township of U. Pop. in 1801, 734; in 1831, 1,775; in 1851, 2,038. The town, which receives its name from the river on the banks of which it is situated, is supposed to stand on the site of the Roman station *Burrium* or *Bullæum*. On an abrupt eminence to the E are the remains of an ancient castle, destroyed in the Parliamentary war, when the town also suffered considerably. The town chiefly consists of several streets, disposed generally in the form of an oblong square. The high-road to Abergavenny passes the principal street; the houses in which are much scattered and separated by gardens and orchards. There is a bridge of 5 arches over the Usk. The Mamhilad railway runs from this bridge to the Brecknock and Abergavenny canal, in Mamhilad p., a distance of about 5 m. The town-hall is a neat structure, erected by the Duke of Beaufort. The inhabitants are employed in the salmon-fishery on the Usk, in husbandry, and the manufacture of japanned tin ware. The quarter-sessions for the county are held alternately here and at Monmouth. The county house-of-correction is in the neighbourhood of the town. This town returns a member to Parliament, in conjunction with Newport and Monmouth. The number of voters registered in 1832 was 107.

USK, a river of South Wales, which rises in a lake on the N side of the Bannau-Sir-Gaer in Caermarthenshire; runs first N and then E as far as Brecknock, where it is joined by the Honddu; it then runs SE to Abergavenny, and in this part of its course is joined by the Grwyneu-fawr. About 3 m. below this it enters Monmouth, and its course from this point is described under the head MONMOUTHSHIRE. Its total course is about 60 m.

USKE, a parish, 3¾ m. NE by N of Ballytore, in co. Kildare. Area 1,763 acres. Pop. in 1851, 722.

USKEANE, a parish in co. Tipperary, 2 m. NE of Borris-o'-kane. Area 7,656 acres. Pop. in 1831, 1,459; in 1851, 1,076.

USKEVAGH (LOCH), an inlet of the sea on the E side of Benbecula, in the Outer Hebrides, the largest of the singularly ramified sea-lochs which cut that island into a labyrinth of land and water.

USKONG, a parish of Norway, in the diocese and bail. of Aggershuus. Pop. 2,500.

USKUP, a sanjak and town of Turkey in Europe, in the W part of Rumelia, and forming a portion of the ancient Macedonia. The sanj. lies between 41° 20′ and 42° 15′ N lat., and between 21° 15′ and 22° 20′ E long.; and is bounded on the W and NW by the sanj. of Scutari, from which it is separated by the Tchar-Dagh and the mountains of Gliubotin; on the NE and E by the sanj. of Ghiustendil, from which, in the first of these directions, it is separated by the mountains of Argentaro or Egrisu; and on the SW by the sanj. of Monastir. It is 69 m. in length from E to W, and 39 m. in breadth, and has generally a mountainous surface. The principal rivers are the Vardar, and its affluents the Uskup and Lepentz and the Psigna. The sanj. comprises 4 districts or jurisdictions. The inhabitants are chiefly Greeks.—The town is situated on the l. bank of the Vardar, at the confluence of the river of the same name, and 120 m. NW of Salonica. Pop. 12,000. From the beauty of its situation and buildings, it has been named the fiancee of Greece. It contains several mosques and Greek churches, and fine promenades. This town, from the number and extent of the ruins by which it is surrounded, appears to have formerly been a place of great size. It is built on irregular ground, but is filthy in the extreme. In the W part, on a hill near the Vardar, are the ruins of an old castle.

USLAR, or UESLAR, a town of Hanover, capital of an amt of the same name, in the gov. of Hildesheim, 23 m. N of Gottingen, at the foot of the Solling, in a valley watered by the Ahle. Pop. 2,780. It is enclosed by walls, and has an hospital. It has manufactories of linen and pipes, several bleacheries, a copper-work, forges, and paper-mills.

USMAN, a town of Russia, in the gov. and 90 m. SW of Tambof. Pop. 3,800.

USMASTON, or UZMASTON, a parish in Pembrokeshire, 1¾ m. SE of Haverford-West, on the river Cleddy. Area 2,070 acres. Pop. in 1851, 683.

USPALLATA, a town of Chili, in the prov. and 100 m. NE of Santiago, in a valley of the same name, formed by a bifurcation of the Andes, and traversed by the Mendoza and San-Juan rivers, which flow respectively S and N, and escape on the E by two narrow gorges. The pass of U. forms the most frequented road between Chili and La Plata. In the vicinity is an extensive silver-mine.

USSAC, a commune of France, in the dep. of the Correze, cant. and 3 m. NNE of Brives. Pop. in 1846, 2,175.

USSANA, a village of Sardinia, in the div. of

Cape Cagliari, prov. and 12 m. N of Cagliari, in a fertile but unhealthy plain. Pop. 1,000.

USSAT, a village of France, in the dep. of the Ariège, cant. and 2 m. SE of Tarascon, near Mount Segur. It has several mineral springs and baths, and in its vicinity are beds of fine alabaster.

USSEAU, a commune of France, in the dep. of the Deux Sevres, and cant. of Mauze. Pop. 1,230.

USSEL, an arrondissement, canton, commune, and town of France, in the dep. of the Correze. The arrond. comprises an area of 89,586 hect., and contains 7 cant. Pop. in 1831, 57,278; and in 1846, 64,836. The cant. comprises 13 com. Pop. in 1831, 10,910; in 1846, 12,036. The town is 36 m. NE of Tulle, on the Sarsonne, which is here crossed by a fine bridge, near the Diege, and at an alt. of 792 yds. above sea-level. Pop. in 1789, 2,652; in 1821, 3,497; in 1831, 3,963; and in 1846, 4,350. It has manufactories of woollen fabrics, sail-cloth, and saddlery, a tannery, and a nail-work. Granite is found in the environs.

USSELBY, a parish in Lincolnshire, 3 m. N by W of Market-Raisen. Area 1,110 acres. Pop. in 1831, 84; in 1851, 69.

USSES, a river of Sardinia, in the div. of Savoy, and prov. of Genevois, which has its source in the mountain Des Saleves; runs first W, and after a total course of about 30 m., flows into the Rhine on the l. bank, a little above Seyssel. It is joined by the Petites Usses, a mountain torrent.

USSON, a commune of France, in the dep. of the Puy-de-Dôme, cant. and 2 m. NE of Sauxillanges. Pop. 881. It contains the ruins of an ancient fortress of the same name, formerly one of the strongest in France. It was noted as the residence of Marguerite, first wife of Henry IV. It was destroyed by Louis XVIII. in 1634.

USSON-EN-FOREZ, a commune and town of France, in the dep. of the Loire, cant. and 8 m. SW of Bonnet-le-Chateau. Pop. in 1846, 3,780.

USSON-DE-POITOU, a commune and town of France, in the dep. of the Vienne, and cant. of Gençay, 15 m. NE of Civray, on the Cluere. Pop. in 1846, 2,183.

USSORA, a river of Turkey in Europe, in the prov. of Bosnia, which, after a course in a generally NE direction of about 45 m., joins the Bosna.

USTAD, a village of Cutch Gundava, near a lake on the road from Sind to Candahar.

USTARITS, or USTARITZ, a canton, commune, and town of France, in the dep. of the Basses-Pyrenees, and arrond. of Bayonne. The cant. comprises 8 com. Pop. in 1831, 8,433; and in 1846, 9,491. The town is 8 m. S of Bayonne, on the l. bank of the Nive. Pop. in 1846, 2,465. It has manufactories of common woollen fabrics, glass, earthenware, and nails, and several tanneries.

USTARROZ, a town of Spain, in the prov. of Navarra, 30 m. ENE of Pamplona, in the valley of Boncal, partly at the foot, and partly on the slope of a mountain, in a fertile locality and near the frontier of France. Pop. 619. It is divided into two parts by a river of the same name.

USTELLEGNY, a mountain of France, in the dep. of the Basses-Pyrenees, near the valley of Bargorry. It contains extensive seams of iron.

USTER, a village of Switzerland, in the cant. and 13 m. ESE of Zurich, and bail. of Greiffensee on the Aabach. It contains the ruins of a castle of the same name. Pop. of parish, 4,496. It has extensive cotton factories. Wine is cultivated in the locality.

USTICA, an island of Sicily, in the prov. of Palermo, in the Tyrrhenian sea, in N lat. 38° 45′, and E long. 13° 11′. Pop. 1,300. It is 3 m. in length, and 1½ in breadth, and is nearly oval in form. It contains several mountains, of which the principal are Guardia, in the centre; Larso, in the S; and in the NE, the hill on which is fort Falconara. The coasts are in most parts steep. The soil possesses considerable fertility, and produces wine, corn, olives, cotton, &c. Fresh water is scarce, and the inhabitants depend chiefly on rain water, which they collect in cisterns. In the vicinity are several coral banks. The chief town is Santa Maria.

USTON, a commune of France, in the dep. of the Ariège, and cant. of Oust, on the Aleth. Pop. in 1846, 3,391. In the vicinity are extensive silver-mines and a forge.

USTRON, a village of Moravia, in the circle and 9 m. ESE of Teschen, on the l. bank of the Vistula. It has extensive iron-works.

USTRZGKY, a town of Galicia, in the circle and 21 m. ESE of Sanok.

USUGE (SAINT), a commune of France, in the dep. of the Saone-et-Loire, cant. and 4 m. N of Louhans, near the r. bank of the Seille. Pop. 2,391.

USUMASINTA, a river of Guatimala, which rises in the district of Peten, in Vera-Paz—on the opposite side of the ridge on which the Belize of Honduras takes its rise under the name of the Rio-de-la-Passion—and flowing from E to W, it enters the country of the Mayas Indians, in crossing which, its navigation is interrupted by a considerable cataract, near the ruins of Palenque. Below this, it is joined by the San-Pedro, the Chacamas, and the Tulija; and near the sea the Tabasco also joins it. It discharges itself by numerous channels, the principal one of which flows into the bay of Campeachy, to the W of the Lago-dos-Terminos, at the port of Victoria.

USUPAMA, a mountain of Venezuela, in the dep. of Caracas, which runs from NW to SE, and terminates on the r. bank of the Coroni.

USURBIL, a town of Spain, in the prov. of Guipuzcoa, 6 m. SW of San-Sebastian, on a hill which is washed by the Orrio. It consists of five parts, some of which are well built. Pop. 1,400. It contains a parish church, several chapels, and an hospital; and has manufactories of iron-ware and anchor-forges. This place formerly bore the name of Belmonte-de-Usurbil.

USWORTH (GREAT and LITTLE), a united township in the p. of Washington, co. of Durham, 4 m. SE of Gateshead. Pop. with North Biddick in 1801 was 1,236; in 1831, 1,477.

USZIKA, a district and town of Russia in Europe, in the gov. of Podolia, 25 m. ESE of Kamieniec, on the Dniester. Pop. 1,250.

UTA, a village of Sardinia, in the div. of Cape Cagliari, prov. and 12 m. W of Cagliari, between the Mannu and Sixerris, in a fertile plain. Pop. 1,200. It has some trade in corn.

UTAH, the name now assumed by the United States territory which was originally organized under the title of the state of DESERET, and is physically described under that name, and in the article GREAT SALT LAKE, in our columns. The territory, as now remanded back by congress, in the session of 1849–50, to a territorial rank, under its present title, and which previously constituted a portion of the Mexican province of California Alta, lies between the parallels of 37° and 42° N, and the meridians of 106° and 120° W; and is bounded on the N by Oregon; on the E by Nebraska and New Mexico; on the S by New Mexico; and on the SW and W by California. Its extreme length from E to W is 740 m.; its breadth 340 m. Its area has been recently computed at 187,923 sq. m. The great basin, which the larger portion of it constitutes, is about 500 m. in length by 265 m. in breadth; and contains the Great Salt Lake, about 70 m. in length by 30 to 35 m. in breadth. None of its streams or lakes have any visible outlet either to the Atlantic or the Pacific. In 1850 it was divided into the 7 cos. of Davis, Great Salt Lake, Iron, San-Peto, Tule, Utah, and Weber, having an aggregate pop. of 11,380.—Also a county of Utah territory, extending from the mountains to the W of Utah lake to the Rocky mountains, a distance of about 25 m. It is generally hilly, and is drained by Green river and its affluents. Pop. in 1840, 2,026. Its capital is Provo city.—Also a lake of the same territory, to the SE of the Great Salt Lake, with which it is connected by the Jordan. It receives Timpanogoo river and Spanish Fork on the E, and Utah river on the S. The latter river has its source in the W part of San-Peto co.

UTANDE, a town of Spain, in the prov. and 15 m. NE of Guadalajara, in a valley watered by the Vadiel. Pop. 292. In the vicinity is a Benedictine abbey, the oldest in the prov.

UTATLAN. See QUICHE.

UTAWA. See OTTAWA.

UTCH, a village of Sind, in the delta of the Indus, near the Joa mouth.

UTCUBAMBA, or HUAHUA, a river of Peru, in the dep. of Truxillo, which has its source in the Andes passes, to the W of Chachapoyas; enters Ecuador, and joins the Tunguragua on the r. bank, about 20 m. SSE of Jaen-de-Bracamoros, and after a course, in a generally NW direction, of about 210 m.

UTELLE, a town of Sardinia, capital of a mandemento, in the prov. and 21 m. N of Nice. Pop. 1,600. The culture of the vine, olive, and chestnut, and manufacture of cheese, form the chief objects of local industry.

UTENDORF, a village of Saxe-Meiningen, at the foot of the Dolmars. Pop. 230.

UTERSEN, or YTERSON, a town of Denmark, in the duchy of Holstein, 12 m. SE of Glückstadt, on the Pinnau. Pop. 2,200. It has manufactories of pottery and sugar refineries, and carries on an active trade in cattle.

UTHLEBEN, a village of Prussia, in the reg. of Merseburg, circle of Sangerhausen. Pop. 595.

UTICA, a city of Oneida co., in the state of New York, U. S., on the S side of the Mohawk river, 84 m. distant by railroad from Albany, and in N lat. 43° 6′ 49″, and W long. 75° 13′. It stands on a beautiful inclined plane, sloping towards the river, and is surrounded by one of the best agricultural and manufacturing districts in the Union. It is intersected by the Albany and Buffalo railroad, and the Erie and Chenango canals. Pop. in 1820, 2,972; in 1830, 8,323; in 1840, 12,782; and in 1850, 17,565. —Also a village of Licking co., in the state of Ohio, on the E side of the N fork of Licking river, and 47 m. by the Central Ohio and the Sandusky, Mansfield, and Newark railway from Columbus. Pop. in 1840, 300; and in 1850, 420.—Also a village of Shelby township, Macomb co., in the state of Michigan, 22 m. NNE of Detroit, on the E bank of Clinton river.—Also a village of Clarke co., in the state of Indiana, 109 m. S by E of Indianapolis, on the NW bank of the Ohio. Pop. about 300.—Also a village of Livingstone co., in the state of Missouri, 167 m. of Jefferson, on the W side of Grand river. Pop. about 100.

UTIEL, a town of Spain, in the prov. and 54 m. W of Valencia, near the gorge of the mountain of Bunol. Pop. 5,700. It is built without regularity. It possesses manufactories of silk and linen fabrics, paper, and soap, and brandy distilleries and flour-mills.

UTILA, a small island of the Caribbean sea, near the coast of Honduras, to the NW of the island of Roatan, in N lat. 16° 14′, and W long. 87° 10′. It is 7 m. in length, rises to a considerable height, and is covered with wood. It was formerly inhabited.

UTKINTON, a township in the p. of Tarporley, co.-palatine of Chester, 1½ m. N by W of Tarporley. Pop. in 1801, 458; in 1831, 564.

UTLAENGE, or UTLANGEN, an island of Sweden, in the prefecture of Bleking, in the Carlskrona channel. It has a village of the same name, and possesses considerable cultivation.

UTOE, a small island of Russia in Europe, off the SW coast of Finland, in N lat. 59° 47′, and E long. 21° 25′. It has a light-house.—Also a small island of Sweden, in the archipelago of Stockholm, in the Baltic. It has productive iron-mines.

UTOROP, a village of Galicia, in the circle and 15 m. S. of Kolomea.

UTRECHT, a province of Holland; bounded on the N by N. Holland and the Zuyder-Zee; on the E and S by Gelderland, and on the W by N. Holland. Its surface is level, except toward the S and the E, where the monotonous aspect of the scenery begins to be exchanged for a slight variety of vale and eminence. The soil in some parts is sandy, and fit for little but raising wood; in general, however, it affords good pasture. The extent of the province is about 490 sq. m. It is traversed by the Vecht, the Amstel, and other branches of the Rhine, and has a better supply of springs and streams than the maritime provinces, as well as a purer atmosphere. Agriculture is here active and prosperous. Wheat, rye, barley, oats, buckwheat, beans, and potatoes, are extensively cultivated. Cattle and horse-breeding is also pursued on a very extensive scale. The exportation, chiefly to England, from this province in 1849, was 72,522 head; in 1850 it had risen to 75,966. Of horses the exportation in 1849 was 11,214; in 1850, 11,801. Of sheep, the exportation in 1849 was 27,671; in 1850, 30,749. Of swine, the exportation in 1849 was 19,921; in 1850, 22,549. Cheese is also manufactured on a large scale; and as a consequence of the extensive reclamation of waste lands, turf is raised in large quantities, and the exportation of that article is considerable. The pop. in 1851 numbered 150,279—a small amount by comparison with the other provinces of Holland. Since 1830, however, the augmentation of pop. has been 17,253. The trade of this district is inconsiderable. There is in progress a very extensive reclamation and tillage of waste land, assisted by a regulation of the government which exempts from the land-tax all land taken for the purpose of reclamation. Education is well provided for. Besides the university of Utrecht, which now contains 351 students, there are 149 elementary schools in the district, attended by 18,911 pupils; of whom 10,427 are boys, and 8,484 girls. This prov. is likewise well supplied with charitable establishments for the poor. There are altogether 148, of which 131 are entirely maintained and conducted by ecclesiastics, and with ecclesiastical funds; 14 are civil establishments; and 3 mixed, civil and ecclesiastic. In addition to the elementary schools before mentioned, the writer whose recent details we have transcribed in this article states that there are 16 for pauper children, 67 communal schools, 10 infant schools, and 6 Sunday-schools.

UTRECHT, the capital of the preceding province, is situated 23 m. SE of Amsterdam, on the Vecht, where it branches off from the Old Rhine, by which it is divided into two parts. The position is healthy, and exempt from the disadvantages of damp so common in Dutch towns, the soil being dry, and having a degree of elevation considerably above the level of the canals; and nothing can surpass the beauty of the approaches to the town, particularly that from Amsterdam, which consists of a broad avenue bordered with rows of trees. U. is surrounded with an earthen mound which has been levelled and planted. It is of a form nearly square, and, exclusive of the suburbs, about 3 m. in circuit. Its aspect is antique, many of the houses being built in the Gothic style. The streets are of considerable width, and intersected by canals, the level of which is in some instances 20 ft. below that of the pavement. Of the public edifices, the most remarkable is the cathedral, a considerable part of which is now in ruins, but the detached tower, which still remains entire, is a very remarkable object. Its height is said to be 388 ft., and from a point, at the height of 300 ft., the eye commands, towards the W, the cities of Amsterdam and Rotterdam, and the whole coast between them. There are 11 other Protestant, and 8 Roman Catholic churches. The stadhuis or town-house is a good structure: the other objects worthy of notice are the charitable establishments, and hospitals. A beautiful public walk, called the Maliebaan, outside the walls, is upwards of a ½ m. in length, and bordered with several rows of lime-trees. The university of U., though not so celebrated as that of Leyden, is of considerable note. It was founded in 1636; is attended by about 300 students; and has professors in the classical languages, mathematics, medicine, divinity, and law. Its buildings are plain; but attached to it are a library, an anatomical theatre, a botanical garden, a cabinet of natural history, and an observatory. The town likewise possesses a hall-of-paintings, schools for the fine arts, and several valuable private libraries and other collections. The pop. in 1849 was 49,176. The manufactures are on a small scale, and consist chiefly of cotton and woollen spinning, silk-twisting, and carpet weaving, velvet, soap, leather, and cordage; but the trade in grain and in cattle is not inconsiderable, having the advantage of inland navigation. U. is the first town in the Dutch provinces where the traveller coming from the westward perceives an uneven surface, and begins to exchange the monotony of Holland for the diversified scenery of Gelderland. "Here the flat, dead level of Holland seems to end. The neighbourhood, it is true, can still be flooded for purposes of defence, by means of the works erected as recently as 1820; and by this contrivance the prov. of Holland is capable of being separated from the land to the east, like a fortress within a moat, and defended by its pop. when the rest of the country is in the hands of an enemy. It is because U., as it were, commands the outworks of the fortress, that so much pains have been taken at various times with the works constructed for its defence. But as you travel E from this city, you find that, not many miles off, defence by means of inundations becomes no longer possible: there is no longer the flat, open country level as a bowling-green, the line around the horizon being as uniform as if drawn by a ruler; canals become less frequent; the rows of trees and windmills are no longer to be seen; the rich level pastures studded with cattle, have been left behind; and in their stead, are waving fields of grain and buckwheat, spreading over a gently undulating country, well timbered, full of glade, and copse, and woody avenues. Here, in the primeval age of Holland, the sea intruded not; the mud-flats out of which the prov. of Holland proper was afterwards scraped, could not have extended eastward of U which must then have lain upon the verge of the peopled continent."—U. was the *Trajectum ad Rhenum* of the Romans; and *Ultra Trajectum* of the Middle ages. It was the birthplace of pope Adrian VI.; and is memorable as the place where, in 1579, was concluded the act-of-confederation of the seven provinces, and in 1713, the well-known treaty of peace between the allies and French.

UTRERA, a judicial partido and town of Spain, in the prov. of Seville. The town is 20 m. SSE of Seville, in a luxuriant valley. Pop. 12,000. It is surrounded by old walls, and has the remains of an ancient castle. The streets are spacious and well-paved, and the houses generally well-built. It has a large square adorned with two fine fountains. It contains two parish churches, the principal of which, of Gothic architecture, is remarkable for its fine naves and lofty clock tower—seven convents, numerous hospitals, a fine town-house, cavalry barracks, and a prison. It possesses manufactories of hats, pottery, soap, and starch, a tannery, and numerous oil-mills. Cattle and horses are among its most important objects of trade. In the vicinity are several salt works. U. is the ancient *Iliturge*.

UTRILLAS, a village of Spain, in the prov. and 51 m. from Saragoza. Pop. 100. In the vicinity are several coal-mines.

UTTENDORF, a town of Austria, in the gov. of Linz, and circle of the Inn, 29 m. N of Salzburg, on the Mattig. It has manufactories of hosiery.

UTTENHAGE. See Uitenhagen.

UTTERBY, a parish in Lincolnshire, 4¼ m. N by W of Louth. Area 1,564 acres. Pop. in 1851, 246.

UTTOXETER, a market-town and parish in the co. of Stafford, 14 m. NE by E of Stafford, containing the townships of Crakemarsh, Creighton, Stramshall, and Woodlands, and the v. of Woodgate, which last is pleasantly situated about a ½ m. S of Uttoxeter. Area 8,973 acres, of which the township of U. contains 2,440 acres. Pop. of p. in 1801, 2,779; in 1831, 4,864; in 1851, 4,990. The town is a place of great antiquity. Leland calls it Uttok-Cester. It is situated upon a gentle eminence on the W bank of the Dove, about ¾ m. from the river, in the heart of a rich pasture district. It consists of several streets with a market-place in the centre. An ancient bridge over the Dove here unites the counties of Stafford and Derby. The town has a considerable traffic in corn and cheese, which has been facilitated by the formation of a branch-canal from the Caldon canal near Froghall, a branch of the Trent and Mersey navigation, which comes up to a wharf in the W end of the town.

UTTWEIL, or Utwyl, a village of Switzerland, in the cant. of Thurgovia, and bail. of Arbon, 26 m. ENE of Frauenfeld, on lake Constance. Pop. 600.

UTZE, a village of Hanover, in the gov. of Lüneburg, 15 m. SSE of Celle.

UUA, or Oa, a river of New Granada, which issues from a lake of the same name, and after a course in a generally SE direction of about 100 m., joins the Guaviare on the l. bank, in W long. 70°.

UVALDE, a central county of the state of Texas, U. S., comprising an area of about 1,300 sq. m., drained by Rio Frio and its branches, and bordered on the W by Nueces river. It is mountainous in the N, but is in other parts fertile and well-cultivated.

UWCHLAND, a township of Chester co., in the state of Pennsylvania, U. S., 85 m. ESE of Harrisburg. It is generally level, and is drained by Brandywine creek. Pop. in 1850, 1,600.

UWUR, a village of the Punjab, on the l. bank of the Chenab.

UXBRIDGE, a market-town and chapelry in the p. of Hillington, Middlesex, 15 m. W by N of London. The town is situated on a gentle declivity on the banks of the river Colne, over which there are here two bridges. The Grand Junction canal, on which are wharfs, warehouses, and other conveniences for trade, passes through it. The town consists chiefly of a single well-paved street, nearly 1 m. in length, and having a commodious market-house in the centre. It has a considerable trade in flour, for grinding which there are numerous mills on the river, while the conveyance of flour to the metropolis is facilitated by the Paddington canal. Pop. in 1801, 2,111; in 1831, 3,043; in 1851, 3,236.

UXBRIDGE, a parish of Van-Diemen's Land, in the co. of Buckingham, bounded on the E by the Plenty, and on the W by the Styx.

UXBRIDGE, a township of Upper Canada, in the Home district. Pop. in 1842, 810.

UXBRIDGE, a township of Worcester co., in the state of Massachusetts, U. S., 36 m. SW of Boston, watered by Blackstone river and its tributaries, and intersected by the Providence and Worcester railway. Pop. in 1840, 2,004; in 1850, 2,457.

UXELOUP, or UXLOUP, a village of France, in the dep. of the Nievre, cant. and 10 m. NE of St. Pierre-le-Moutier, in a plain, on the l. bank of the Loire. Pop. 125.

UXMAL, a ruined city of Yucatan, 45 m. SSW of Merida, and 10 m. from the ruins of Zayi. Mr. Norman describes a vast pile of ruins here which he calls the governor's house, as standing upon three ranges of terraces: the great platform, or terrace, measuring upwards of 500 ft. long, and 415 ft. broad. It is encompassed by a wall of fine hewn stone 30 ft. high, with angles rounded, still in good preservation. Along the S edge of this platform are the remains of a range of small pillars. Upon the NW corner of the platform is an edifice, the smallest of all the ruins, with few and plain ornaments, the most remarkable of which is a continuous line of turtles, cut from stone of about a foot square, arranged under the cornices. Upon the main terrace stands another of smaller dimensions, constituting the foundation of the governor's house. The measurement of this terrace is 338 ft. long, 82 ft. broad, and 30 ft. high, having a majestic flight of stone steps in front of the entrance. It is divided into double ranges of rooms, from front to rear. Two of the principal, situated in the centre, are 55 ft. long, 10 ft. broad, and about 19 ft. high, with an angular ceiling occupying one half of the whole. The walls, of a durable kind of limestone, are upwards of 3 ft. thick, and of fine hewn stone laid with the greatest care. The ornaments are composed of small square pieces of stone, shaped with skill, and inserted between the mortar and stone with the greatest care and precision. An extensive pile of ruins designated as the nuns' house, is situated a few rods distant, in a N direction from the governor's house. It comprises four great ranges of edifices, placed on the sides of a quadrangular terrace, measuring about 1,100 ft. around, and varying in height from 15 to 24 ft., its sides corresponding to the cardinal points. The principal entrance is through an acute-angled arched doorway into a spacious court. This range is upwards of 200 ft. long, 25 ft. broad, and 16 ft. high, containing eight rooms on either side of the principal entrance. The opposite, or northern range, by its superior elevation, and more elaborate work, was evidently the principal portion of this immense structure. The walls of the rooms are finished with a clean, white, hard substance. On the western façade are the remains of two great serpents; their heads turned back, and entwining each other. They extend the whole length of the façade, through a chaste ground-work of ornamental lines, interspersed with various rosettes; and are put together by small blocks of stone, exquisitely worked, and arranged with skill and precision. The pyramid, situated about two rods E from the ruins of the nuns' house, with which it appears in some way to have been connected, presents a fine exterior of hewn stones, large at the base, and diminishing in size as they approach the platform. Its base measures 500 ft.; from the base to the summit or platform it is 100 ft. The summit is reached, on the E side, by a flight of 100 stone steps, each one foot high, and about 6 inches deep. The area of the platform measures 72 ft. in length and 21 ft. in width, and is occupied by an edifice 60 ft. long, 12 ft. wide, and 20 ft. high. A few rods distant, in a SW direction from the governor's house, are the remains of an extensive range, or succession of ruins composed of terraces, walls, rooms and corridors, and court-yards.

UXUE, a town of Spain, in the prov. of Navarra, and 27 m. SE of Pamplona, on the summit of a mountain of the same name, and one of the highest in the prov. Pop. 1,440. It was formerly surrounded by walls, of which a small portion only now remains, and has a cistern of extraordinary dimensions. Its trade consists chiefly in charcoal, faggots, and oil.

UYCKHOVEN, a department and town of Belgium, in the prov. of Limburg, watered by the Meuse. Pop. 435.

UYEA, an isle, harbour, and village, in the p. of Unst, in Shetland. It is the grand rendezvous of shipping for the deep sea-fishing, and the entrepot for goods to the fishing-stations in Unst and neighbouring islands.—Also a pasture-island, about 2½ m. in circumf., on the W coast of Northmaven, 4½ m. from the N extremity of the mainland of Shetland. It covers a safe harbour on the coast of the mainland.

UYLEBROCK, a commune of Belgium, in the prov. of E. Flanders, dep. of Vlierzele. Pop. 323.

UZBEKS. See USBECS.

UZE, a commune of France, in the dep. of the Drôme, cant. and 2 m. from St. Vallier, on the Galaure, near its confluence with the Rhone. Pop. 600. It has manufactories of pottery and bricks.

UZECH-DES-OULE'S, a commune of France, in the dep. of Lot, cant. of St. Germain. Pop. 706. It has extensive manufactories of pottery, and several oil-mills.

UZEDA, a town of Spain, in the prov. and 24 m. NW of Guadalajara, on a rising ground, the base of which is washed by the Xarama. Pop. 700. It has the remains of towers, a church, and a convent. This town has the title of a county, and was formerly fortified. At the distance of about 5 m. amid the mountains is the famous village of Patones.

UZEGEOIS, an ancient district of France, in Languedoc, now comprised in the departments of the Gard and Lozere. It derived its name from Uzès, the capital.

UZEL, a canton, commune, and town of France, in the dep. of the Côtes-du-Nord. The cant. comprises 7 com. Pop. in 1831, 14,205; in 1846, 12,390. The town is 9 m. NNW of Loudeac, near the l. bank of the Oust. Pop. in 1841, 2,090. It has three fine halls, and five different market-places, and is the centre of an extensive trade in linen and thread, and in cattle, fruit, and shoes.

UZEN (BOLCHOI), a river of Russia, which rises in the gov. of Saratov; and flows SW, and then SE, to the salt lake of Kamych, into which it flows after a course of 260 m.

UZERCHE, a canton, commune, and town of France, in the dep. of the Correze, and arrond. of Tulle. The cant. comprises 9 com. Pop. in 1831, 13,340; in 1846, 14,213.—The town is 15 m. NW of Tulle, on a steep hill, near the l. bank of the Vezere, which is here crossed by a substantial bridge, and near the confluence of the Bradascon. Pop. in 1846, 3,435. It is ill-laid out and badly built, but has a generally picturesque aspect. It has a communal college. The locality is noted for its horses. U. formerly bore the name of Usreca.

UZERO, a village of Spain, in the prov. and 30 m. W of Soria, on a river of the same name, which is here crossed by a bridge, and a little above its confluence with the Duero. Pop. 200. It has an old castle, a fulling-mill, and a dye-work.

UZE'S, an arrondissement, canton, commune, and town of France, in the dep. of the Gard. The arrond. comprises an area of 148,010 hect., and contains 8 cant. Pop. in 1831, 83,732; in 1846, 89,536.—The cant. comprises 15 com.—Pop. in 1831, 14,489; in 1846, 16,204. The town is 14 m. NNE of Nimes, on a height near the r. bank of the Auzon. Pop. in 1789, 7,566; in 1821, 5,622; in 1831, 7,274; in 1846, 7,215. It is generally ill-laid out, and ill-built, and with the exception of the episcopal palace, and a fine terrace near the church, contains few objects of interest. It has a consistorial church, a communal college, and possesses manufactories of hosiery, in silk and cotton and common cloth, and earthenware, several tanneries, oil, and silk spinning-mills; and a trade in grain, wine, brandy, cattle, oil, and silk. This town, which is of great antiquity, was formerly the seat of a bishop. In the 16th century, it was fortified by its inhabitants, who had embraced the Calvinist faith, and who maintained their independence within its walls till 1629, when it was taken by Louis XIII, and its fortifications razed to the ground.

UZESTE, a town of France, in the dep. of the Gironde, cant. and 3 m. ESE of Villandrau. Pop. 915.

UZNACH, a district and town of Switzerland, in the cant. and 30 m. SW of St. Gall, on a height, near the E extremity of Lake Zurich. Pop., Cath., 1,536. It is well-built, and is surrounded by a finely wooded country. In the vicinity is a quarry of lignites in which naphtha is found.

UZZANO, a commune of the grand-duchy of Tuscany, in the prov. of Florence, and vicariat of Pescia, in the Val-di-Nievole. Pop. 3,280.

V

*** For articles not found under V, look under B, F, or W.

VAACKE, a village of Electoral Hesse, in the prov. of Lower Hesse and circle of Hof-Geismar, on the Weser. Pop. 792.

VAAGE, a parish of Norway, in the diocese of Aggershuus and bail. of Christian, and 156 m. NNW of Christiania. Pop. 3,400.

VAAGOE (Ost), an island of the Lofoden group, to the W of Hindoen, in N lat. 68° 50', and E long. 114° 10'. It is 36 m. in length from NE to SW, and 18 m. in breadth. It contains the parish and port of Vaage. Pop. 1,600.

VAAGOE (Vest), an island of the Lofoden group, to the WSW of Ost-Vaagoe, in N lat. 68° 25', and E long. 13° 10'. It is 36 m. in length from NNE to SSW, and 15 m. in breadth, and contains several parishes. Both these islands are chiefly inhabited by fishermen.

VAAGS-OE, an island off the W coast of Norway, in the diocese of Bergen and bail. of Northern Bergen, in N lat. 61° 58', to the SW of Stadland.

VAAL. See Gariep and Orange River.

VAALE, a parish of Norway, in the diocese of Aggershuus and county of Jarlsberg. Pop. 1,700.

VAALER, a parish of Norway, in the diocese of Aggershuus and bail. of Smaalehen, and 30 m. S of Christiania. Pop. 1,800.

VAAS, a commune and town of France, in the dep. of the Sarthe, cant. and 8 m. S of Mayet, on the r. bank of the Loir, between the Brault and Guichorat. Pop. 1,998. It has several tanneries. This town was formerly fortified, and was for some time held by the English.

VAASSEN, a village of Holland, in the prov. of Guelders, 23 m. N of Arnheim. Pop. 1,300.

VAAST (Saint), a village and port of Belgium, in the prov. of Hainault, 10 m. E of Mons. Pop. 1,615.—Also a town of France, in the cant. and 3 m. E of Quettehou, a little to the NE of La Hogue. Pop. 880. It is protected by a battery on the E.

VAAST. See Hougue (La).

VAAST-DIEPPEDAL (Saint), a village of France, in the dep. of the Seine-Inferieure, cant. and 7 m. ENE of Ourville. Pop. 1,026.—Also a commune in the dep. of the Aveyron, and cant. of Rieupeyroux. Pop. 2,156.

VABRE, or Vabres-de-Senegats, a canton, commune, and town of France, in the dep. of the Tarn, and arrond. of Castres. The cant. comprises 6 com. Pop. in 1831, 9,027; in 1846, 9,579.—The town is 12 m. N of Castres, on the Berlon. Pop. in 1846, 2,583. It has a Calvinist consistorial church, and possesses manufactories of cotton and woollen fabrics, and of leather.

VABRES, a commune and town of France, in the dep. of the Aveyron, cant. and 3 m. WSW of St. Affrique, on the r. bank of the Dourdan, and near its confluence with the Sorgues. Pop. 730. It was formerly a bishop's see.

VACAHI, a river of Brazil, in the prov. of São-Pedro-do-Rio-Grande, and district of Cacapaba, which has its source in the mountains to the E of the Rio-Ibicui; runs ENE, and joins the Jacuhi, on the r. bank, some leagues above the confluence of the Santa-Barbara.

VACALLO, a village of Switzerland, in the cant. of Tesino and district of Mendrisio. Pop. 550.

VACARAPI, a river of Brazilian Guayana, in the district of Macapa, which joins the Para, 18 m. above its confluence with the Amazon.

VACARIA, a river of Brazil, in the prov. of Minas-Geraes, and comarca of Rio-de-Jequitinhonha, which has its source in the Serra of Grão-Mogor, and joins the Jequitinhonha on the l. bank, above the confluence of the Aracuahi.

VACAS (Rio-de-Las), a river of Guatimala, which has its source in the E part of the territory of Sacatopeq; runs SW; passes Guatimala-la-Nueva; and after a course of about 90 m., throws itself into the port of Guatimala.

VACCA, a small island of the Mediterranean, near the coast of Sardinia, in the gulf of Palmas, and to the NE of the island of Toro, with which it was comprised by the ancients under the name of *Plumbeæ*.

VACCARIZZO, a town of Naples, in the prov. of Calabria-Citra, district and 14 m. W of Rossano, on a hill. Pop. 978. It has two churches, a Latin and a Greek.

VACH, a village of Bavaria, in the circle of the Rezat and presidial of Kadolzburg, 8 m. NW of Nuremberg. Pop. 883. It has glass-works.

VACH, or Vacha, a town of Saxe-Weimar-Eisenach, capital of an amt of the same name, in the principality and 20 m. WSW of Eisenach, on the l. bank of the Werra. Pop. in 1852, 2,458. It is enclosed by walls, with three gates, and has a school and an hospital. It possesses several paper and spinning-mills, and tanneries.

VACHE, an island near the S coast of Hayti, 8 m. ENE of Point Abacou, in N lat. 18° 4', and W long. 73° 40'. It is 12 m. long from E to W, and 3 m. in breadth, and has two good ports. It is fertile and well supplied with good water.

VACHE (LA), a village of France, in the dep. of the Drome, cant. and 6 m. S of Valence, on the Viore. Pop. 260.

VACHERESSE, a village of Sardinia, in the div. of Savoy and prov. of Chablais, on the r. bank of the Drance, 7 m. SE of Evian. Pop. 1,000.

VACHKA, a river of Russia in Europe, which has its source in the gov. of Vologda, in the district and to the NW of Jarensk; flows through the N part of the gov. into that of Arkhangel, and after a course, in a generally NNW direction, of 225 m., joins the Mezen on the l. near Oust-Vachka.

VACIA-MADRID, a village of Spain, in the prov. and 9 m. SE of Madrid, at the confluence of the Manzanares and Jarama. It contains an ancient royal palace, several mineral springs, and some Roman antiquities.

VACQUERIE (LA), a village of France, in the dep. of the Herault, cant. and 8 m. NE of Lodeve. Pop. 500.

VACQUIES, a village of France, in the dep. of the Haute-Garonne, cant. and 7 m. SE of Fronton. Pop. 550.

VACRI, a town of Naples, in the prov. of Abruzzo-Citra, district and 23 m. SE of Chieti, and cant. of Bucchianico, on a hill. Pop. 1,180.

VACZ. See WAITZEN.

VADA, or BURDJ-KADIJA, a headland of Tunis, to the N of the gulf of Cabes, in N lat. 35° 11′ 52″. On its summit is a lofty tower, visible from a distance at sea of 18 m., and forming an important land-mark to navigators.

VADAGARRY, a town of Hindostan, in the presidency of Madras, in the Carnatic, in the district and 42 m. NW of Tinnevelly, at the foot of the Western Ghauts.

VADAGHERRY, a town of Hindostan, in the prov. of Malabar, 24 m. N by W of Calicut.

VADANS, a town of France, in the dep. of the Jura, cant. and 4 m. NW of Arbois, near the Cuisance. Pop. 672.

VADENROD, a village of Hesse, in the prov. of Ober Hesse. Pop. 540.

VADI, or VADO, a market-town of Sardinia, in the prov. and 4 m. SSW of Savona, on the gulf of Genoa. Pop. 2,000. It has productive fisheries.

VADILLO, a town of Spain, in the prov. of Zamora, on the Guarena. Pop. 500.

VADKERT, a market-town of Hungary, in the comitat of Neograd, to the SW of Balassa-Gyarmath. It has mineral baths. 2,400.

VADO, a market-town of Spain, in the prov. of Guadalajara. Pop. 300.

VADOCONDIS, a market-town of Spain, in the prov. of Burgos and partido of Aranda-de-Duero, on the Duero. Pop. 720.

VADONVILLE, a village of France, in the dep. of the Meuse, on the Maas. Pop. 248.

VADRUP, a village of Prussia, in the circle of Munster. Pop. 420.

VADSOE, a parish of Norway, in Finmark, on the Waranger-fiord. It has a harbour and productive fisheries. Pop. 1,200.

VADUTZ, a parish and market-town of the principality of Lichtenstein, 48 m. SSW of Bregenz, near the r. bank of the Rhine. Pop. 1,000.

VAELS, or VAALS, a town of Holland, in the prov. of Limburg, 16 m. SE of Maestricht, with 3,000 inhabitants, employed chiefly in the manufacture of woollens, needles, and wax-candles.

VAERDAL, a town of Norway, in the prov. and 32 m. NE of Trondhjein, at the mouth of the Vaerdals-elv. Pop. 1,900.

VAETHEN, a village of Prussia, in the reg. of Magdeburg, circle of Stendal. Pop. 320.

VAGA, a river of Russia, which rises in the gov. of Vologda; flows NE, and joins the N. Dwina, near Ust-Vaga.

VAGAI, a river of Asiatic Russia, which, after a course of upwards of 150 m., falls into the Irtish, 34 m. SE of Tobolsk.

VAGHA, a town of Hungary, in the com. of Presburg, on the Vagh. Pop. 1,400.

VAGLI, a town of the duchy of Modena, in the district of Canara. Pop. 1,300.

VAGLIANO, a town of Tuscany, in the prov. of Florence, on the river Chiana.

VAGLIO, a town of Naples, in the Basilicata, 7 m. E of Potenza. Pop. 2,680.—Also a village in the Swiss cant. of Ticino, near Lugano. Pop. 236.

VAGNEY, a town of France, dep. of Vosges, on the small river Vagney, 6 m. E of Remiremont. Pop. 2,962.

VAGOS, a town of Portugal, in the prov. of Douro, 6 m. S of Aveiro.

VAGUARE, a river of New Granada, in the prov. of Neiva, which runs E, and enters the Magdalena.

VAHLBRUCH, a village of Hanover, in the reg. of Calenberg, near Hameln. Pop. 500.

VAHR, a village in the territory of Bremen, on the r. bank of the Elbe. Pop. 430.

VAHRENHOLZ, a village of the principality of Lippe-Detmold, on the Weser. Pop. 730. It has a castle.

VAHRN, a village of Austria, in Tyrol, in the circle of Brixen. Pop. 770.

VAIGATZ, an island, or assemblage of islands, belonging to European Russia, in the Frozen ocean, between Nova-Zembla and the continent, having an entire length of 95 m., with a breadth of 32 m. Their surface is low, rocky, and incapable of cultivation equally from the soil and climate, so that they are only visited occasionally by hunters from the adjacent parts of Russia on the main land. The fixed inhabitants are Samoyedes. The strait, which lies between the gov. of Arkhangel and the island of V., and between the Frozen ocean and the gulf of Cara, called the straits of Karsky, is 60 m. in length from NW to SE. This passage was discovered by the Dutch in 1594, and carefully explored by the Russians between 1735 and 1739.

VAIHINGEN, a town of Würtemberg, on the river Enz, 11 m. NW of Stuttgard. Pop. 3,324.—Also another large village of Würtemberg, near Esslingen, on the Fildem. Pop. 1,464.—The district of V., in the Neckar circle, has an area of 3·3 German sq. m., with a pop. in 1852 of 23,151.

VAILLAC, a town of France, in the dep. of the Lot, 16 m. N of Cahors.

VAILLY, a town of France, in the dep. of the Aisne, 9 m. N of Soissons. Pop. 1,594.

VAIMIRGA, a river of Russia, which joins the N. Dwina, after an E course of 70 m.

VAIRANO, a town of Naples, in the prov. of Lavoro, 7 m. NE of Teano.

VAIRO, a town of the duchy of Parma, 32 m. S of Parma, on the Enza.

VAISON, a town of France, dep. of Vaucluse, on an eminence near the small river Ouvèze, 9 m. E by N of Orange. Pop. 2,879. It has a castle, and was the birthplace of the Roman historian Trogus Pompeias. It still exhibits several Roman ruins.

VAJASD, a small town of Transylvania, in the co. of Lower Weissenburg, near the Marosch.

VAJDA-HUNYAD. See EISENMARKT.

VAKH, a river of Siberia, in the gov. of Yeniseisk, which joins the Obi, after a W course of 300 m.

VAKUP. See AK-HISSAR.

VAL (Le), a commune of France, in the dep. of the Var, cant. and 4 m. N of Brignoles. Pop. 1,752.

VALACHIA. See Wallachia.

VALADARES, a village of Portugal, in the prov. of Minho, comarca and 3 m. E of Valença-de-Minho. Pop. 500.

VALADY, a village of France, in the dep. of Aveyron, cant. and 3 m. SW of Marcillac. Pop. 500.

VALAIS, or Wallis, a canton in the south of Switzerland; bounded on the N by the cantons of Bern and Vaud; on the NE by Uri and Ticino; on the E and S by Piedmont; and on the W by Savoy. It has an area of 192 sq. leagues of 25 to a degree, or 1,106 geog. sq. m. It is watered in its whole extent by the Rhone, and is bordered on the N by the Bernese Alps, and on the S by the Pennine Alps, ranges embracing some of the loftiest mountains in Europe. No country exhibits a greater diversity in its productions, the low ground being hot, the ascent of the mountains temperate, while the high grounds approach to the climate of Norway and Iceland. The vine is cultivated to an alt. of 2,000 ft. above sea-level; and in the valley of the Rhine the silk-worm is reared. Harvest in the valley is finished by the end of May, while in other districts its activity is greatest in September and October. The lower slopes of the mountains are covered with magnificent forests of chestnuts, walnuts, and pines. The pop. in 1852 was 81,559, of whom 1,654 were foreigners; and 1,204 Swiss from other cantons. The increase of the pop. since 1837 has been 4,969, or a mean annual increase of 1 in 207. The religion of the inhabitants is the Catholic; the number of Protestants in 1852 being only 463. The language of the Upper Valais, or mountainous parts, is German; that of the low ground, or Lower Valais, a dialect of French, the inhabitants being a mixed race descended from the aborigines of France and Italy. The rearing of cattle is the only employment of consequence; traffic is confined to the sale of cattle, and of a little corn and wine. This petty community was formerly an independent republic in alliance with Switzerland; it was declared one of the cantons of Switzerland by the congress of Vienna. The capital is the town of Sion, with a pop. in 1852 of 3,516. The next town in importance is St. Maurice. Pop. 1,327. The cant. is administratively divided into 13 *dizaines* or districts, each of which has its own council, elected by the male citizens, and sends 4 deputies to the legislative diet. The executive is vested in a council of 5.

VALANIDI, a town of Naples, in Calabria-Ultra 1ma, 6 m. SSE of Reggio. Pop. 1,200.

VALATIA, a village of Columbia co., in the state of New York, U. S., at the confluence of the Valatia and Kinderhook creeks, 2 m. NE of Kinderhook. Pop. in 1850, 1,650.

VALBASES (Los), a village of Spain, in the prov. and 18 m. SW of Burgos. It is walled, and has two churches, and two convents. Coarse woollens are woven here.

VALBENOITE, a commune of France, in the dep. of the Loire, cant. and 6 m. SE of St. Etienne, on the Furens. Pop. in 1846, 5,504. It has manufactories of instrument strings and vegetable dyes, and several forges.

VALBERT, a village of Prussia, in the reg. of Arnsberg and circle of Olpe. Pop. 468. It has several iron-works.

VALBONA, a village of Spain, in the prov. of Teruel. Pop. 545. It has manufactories of cloth.

VALBONNAIS, a canton, commune, and town of France, in the dep. of the Isere, arrond. of Grenoble. The cant. comprises 11 com. Pop. in 1831, 6,003; in 1846, 6,277. The town is 27 m. SE of Grenoble, on the r. of the Marsanne. Pop. 1,386.

VALBONNE, a village of France, in the dep. of the Gard. Pop. 1,060.—Also a village in the dep. of the Var. Pop. 1,010.

VALBUENA, a village and port of the Argentine prov. of Tucuman, on the Salado, 120 m. NE of San-Miguel-de-Tucuman.

VALCANVILLE, a commune of France, in the dep. of the Manche, and cant. of Quettehou, 15 m. NE of Valognes. Pop. 1,176.

VALCARES, a salt lagune of France, in the dep. of Bouches-du-Rhone, and in the delta of that river. It is 9 m. in length and 7 m. in breadth.

VALCARLOS, a village of Spain, in the prov. of Navarra, 35 m. NE of Pamplona. Pop. 500. Iron is wrought in the vicinity.

VALCIMARA, a market-town of the Papal States, in the prov. of Macerata, 5 m. NE of Camarino. Pop. 850.

VALCIVIERES, a village of France, in the dep. of Puy-de-Dome, cant. and 4 m. NE of Ambert. Pop. 1,200.

VALCONETE, a market-town of Spain, in the prov. of Guadalajara. Pop. 340.

VALCOUR, an island of the state of New York, U. S., in Lake Champlain, 5 m. SE of Plattsburg. It is 3 m. long and 1 broad.

VALDAGNO, a market-town of Austria, in Lombardy, 15 m. NW of Vicenza, on the r. bank of the Agno. Pop. 3,300. It has manufactories of cloth.

VALDAHON (Le), a commune of France, in the dep. of the Doubs, and cant. of Vercel, 18 m. S of Baume-les-Dames. Pop. 985.

VALDAI, a district and town of Russia in Europe, in the gov. and 99 m. SE of Novgorod, on a lake of the same name. Pop. 3,000. It is enclosed by earthen walls, and contains a palace and several churches. The lake, which is about 6 m. long and 3 m. broad, is fed by a small stream called the Valdaika, and discharges itself into the Msta.

VALDAI HILLS, a range of hills in the same gov., the culminating point of which does not exceed 1,100 ft., and the general elevation of which is only 300 ft., but which form the line of separation between the basins of the Black sea and the Baltic. The Volga rises on their S slopes, at an alt. of not less than 800 ft. It has long been supposed that these hills form the highest ground in the interior of European Russia; but near Kremenetz in Podolia, there is a point of 1,328 ft. alt. above sea-level, whence the Dniester and Dnieper flow to the Black sea; and the Bug and the Vistula to the Baltic. See article Russia.

VAL-D'AJOL (Le), or Laitre, a commune and town of France, in the dep. of the Vosges, and cant. of Plombières, 9 m. SW of Remiremont on the Combant. Pop. in 1846, 7,040. It has manufactories of kirchenwasser, and of straw-hats, calico, and cordage, and spinning and saw-mills and wax-works. Coal is wrought in the vicinity.

VALDARACHAS, a village of Spain, in the prov. and 6 m. ESE of Guadalaxara.

VALDARNO. See Arno.

VAL-DE-AVELLANO, a town of Spain, in the prov. of Soria. Pop. 881. It has a parish church, a custom-house, and a public granary.

VAL-DE-CANAS, a market-town of Spain, in the prov. of Estremadura, on the Tajo. Pop. 1,700.

VAL-DE-FUENTES, a town of Spain, in the prov. of Salamanca, on the Sangusin. Pop. 680. It has a parish church, a custom-house, and a public granary.—Also a town in the prov. of Caceres, to the r. of the Sierra Montanchez. Pop. 1,134. It has a parish church and a custom-house.—Also a market-town in the prov. of Estremadura, near Montanches. Pop. 1,768.

VALDEGANGA-DE-JORQUERA, a town of Spain, in the prov. of Albacete. Pop. 1,470. It has a parish church and a custom-house.

VALDEGATE, a market-town of Spain, in the prov. of Burgos. Pop. 378.

VAL-DEL-ALGORFA, a village of Spain, in the prov. of Teruel, near Alcaniz. Pop. 1,430.

VAL-DEL-ANDE, a market-town of Spain, in the prov. of Burgos. Pop. 300.

VAL-DEL-ARENAS, a market-town of Spain, in the prov. of Guadalajara. Pop. 525.

VAL-DEL-AVERO, a market-town of Spain, in the prov. of Guadalajara. Pop. 498.

VAL-DEL-COLMENAS, a market-town of Spain, in the prov. of Cuenca, near Huete. Pop. 768.

VAL-DEL-CONCHA, a market-town of Spain, in the prov. of Guadalajara. Pop. 566.

VAL-DEL-LAGUNA, a market-town of Spain, in the prov. of Madrid. Pop. 478.

VAL-DEL-LOSA, a market-town of Spain, in the prov. of Salamanca. Pop. 655.

VAL-DEL-MORILLO, a market-town of Spain, in the prov. of Madrid. Pop. 1,335.

VAL-DEL-PIELAGO, a market-town of Spain, in the prov. of Leon. Pop. 400.

VAL-DEL-RIBELLO, a market-town of Spain, in the prov. of Guadalajara. Pop. 210.

VAL-DEL-ROBLES, a market-town of Spain, in the prov. of Zaragoza, near Alcaniz. Pop. 2,255. It has several paper-mills.

VAL-DEL-SAZ, a market-town of Spain, in the prov. of Guadalajara. Pop. 445.

VAL-DEL-UNCAR, a market-town of Spain, in the prov. of Caceres, near Plasencia. Pop. 412.

VAL-DEL-VIMBRE, a market-town of Spain, in the prov. of Leon. Pop. 308.

VALDEMORO, a market-town of Spain, in the prov. of Madrid. Pop. 1,880. It has a parish church, two hospitals and a custom-house.

VAL-DE-MORO-DEL-REY, a market-town of Spain, in the prov. of Cuenca, near Huete. Pop. 408.

VAL-DE-NUZ, a valley of Switzerland, in the cant. of Neufchatel, extending between the lake of that name on the S, and the Chasseral on the N. It contains 24 villages.

VAL-DE-OLIVAS, a market-town of Spain, in the prov. of Cuenca. Pop. 2,700. It has a parish church, several schools, and a convent. Its industry consists chiefly in the manufacture of linen and woollen fabrics, and the rearing of cattle.

VAL-DE-OMILLAS, a market-town of Spain, in the prov. of Valencia. Pop. 510.

VAL-DE-PENAS, a judicial partido and town of Spain, in the prov. of Ciudad-Real. The partido comprises 6 pueblos. The town is 30 m. SE of Ciudad-Real. Pop. 10,200. It has a parish-church, several convents, an hospital, and a public granary, and possesses manufactories of woollen and linen fabrics, and several mills.—Also a market-town in the prov. of Jaen. Pop. 4,000.

VAL-DE-PENAS-DE-LA-SIERRA, a market-town of Spain, in the prov. of Madrid, near Alcala-de-Henares, on the Jarama. Pop. 355.

VAL-DE-PIELAGOS, a town of Spain, in the prov. of Guadalajara. Pop. 355.

VALDERADUEY, a river of Spain, which has its source in the prov. of Leon, enters that of Zamora, and joins the Duero on the r. bank.

VALDERCE'S, a canton, commune, and town of France, in the dep. of the Tarn, and arrond. of Albi. The cant. comprises 6 com. Pop. in 1846, 6,016. The town is 9 m. NE of Albi. Pop. 1,544.

VALDERROBRES, or VALDERROBLES, a town of Spain, in the prov. of Teruel. Pop. 2,250. It has a parish-church, a custom-house, and a public granary, and possesses manufactories of paper and cloth.

VAL-DE-SANTO-DOMINGO, a market-town of Spain, in the prov. of Toledo. Pop. 1,825. It has a parish church, a custom-house, and a public granary, and possesses manufactories of common cloth.

VAL-D'ESNE, a market-town of France, in the dep. of the Haute-Marne. Pop. 815.

VALDESOTO (SAN FELIX), a town of Spain, in the prov. of Oviedo. Pop. 1,253.—It has a parish church and a custom-house.

VAL-D'ESPINA, a market-town of Spain, in the prov. of Palencia. Pop. 377.

VALDIERI, a village of Sardinia, in the prov. and 12 m. SW of Coni. Pop. 2,600. It has iron works.

VAL-D'ILLIEZ, a village of Switzerland, in the cant. of Valais, and circle of Montheli. Pop. 835.

VALDIVIA, a province of Chili, situated in the midst of the country mainly occupied by the Araucanian Indians. It lies upon the coast, on both sides of the river Valdivia, and is bounded on the N by the river Imperial; on the S by the Maypul or Rio-Negro. The Guinchi or Cunchi Indians are in possession of its S part. It is about 180 m. long, and 70 m. broad. The climate is colder than in the N provs. of Chili, and snow covers the peaks of the Andes chain which bounds this prov. on the W, though of considerably lower elevation than those towards the N. From the coast to the Andes, the surface is generally level, and the soil good; while intercourse is facilitated by a number of navigable rivers. The principal towns are Valdivia and Osorno. The pop. is about 25,000.

VALDIVIA, the capital of the above province, is situated on the S shore of the river of its name, at 9 m. distance from the sea, in S lat. 39° 49′, W long. 73° 15′. The harbour is situated in a beautiful bay formed by the river, and has been represented as the safest, the strongest from natural position, and the most capacious of any ports in the South sea; but the anchorage for large ships is limited, and a bank with shallow water upon it detracts very materially from its value. It is known that the coasts of Chili have been elevated several feet, apparently by internal pressure, of late years; and Dr. Wayman reports that in 1820 he had observed that there was at this point of the coast, only 2 ft. water where 6 Dutch line-of-battle ships had anchored about 70 years before. The island of Mancera, situated in the mouth of the river, forms two passages, which are bordered by steep mountains and strongly fortified. The town contains a college built by the Jesuits, several convents, a parochial church, and a royal hospital. The private houses are chiefly of wood. The adjacent country is hilly, rising to an alt. of 1,000 ft. Water is plentiful, and provisions are cheap. This city was founded in 1551, by Pedro-de-Valdivia, who gave it his name, and obtained immense sums of gold from its vicinity. Its wealth allured many, and it soon became one of the most populous cities in the kingdom. It was twice besieged ineffectually by the Toqui Caupolican, but it was not so fortunate in resisting the celebrated Paillamachu. In 1590 it was surprised at night by that leader, who killed the greater part of the garrison consisting of 800 soldiers, and, having burned the city, carried off immense booty, together with a number of prisoners. The Spaniards rebuilt it anew, and fortified it so strongly that it has resisted all the attempts of the Araucanians to take it. It was, however, taken in 1640 by the Dutch, who, frustrated in their attempts to form an alliance with the Araucanians and the Cunchi, were compelled to abandon it. In 1820 it was captured by Lord Cochrane with a patriot force.

VALDIVIA, a river of Chili, which has its rise in the lake of Huanahué, in the Andes, and runs W into the Pacific in a course of about 120 m. Its stream is clear, and it is so deep that ships of the greatest burden come close up to the city of Valdivia, and lie close to the quays. The banks of this river are fertile, and yield wheat, pulse, and fruit. The territory through which it flows abounds in cattle and in timber.

VALDOBIADENE, a town of Austrian Lombardy, in the prov. and 20 m. NW of Treviso. Pop. 2,480. It has several silk-mills.

VALDROME, a commune and village of France, in the dep. of Drome, arrond. and 20 m. SSE of Die. Pop. 1,288.

VALDUGGIA, a commune and village of the Sardinian states, in the div. of Novara, 3 m. E of Borgosesia. Pop. of com., 3,158.

VALDUNQUILLO, a village of Spain, in the prov. and near the city of Leon. Pop. 910.

VALEDIA, a small port of Marocco, 27 m. SSW of Mazagan. The coast is here very rocky, and though there is a spacious natural harbour, it is little frequented, on account of its difficult and dangerous entrance.

VALEGGIO, a town of Austrian Lombardy, on the Mincio, SW of Verona. Pop. of com., 4,558.

VALENÇA, a comarca and town of Brazil, in the prov. of Bahia. The town is on the r. bank of the Una, and near a canal by which that river is connected with the Jiquie. Pop. 1,200. It has a parish church and two schools. Coffee is cultivated to a considerable extent in the locality.—Also a town of Rio-de-Janeiro, between the Parahiba and Preto. Pop. 5,000. It has a parish church and an hospital. An active trade is here carried on in coffee, millet, mandioc, and other productions of the locality.—Also a town of the prov. of Piauhi, about 60 m. N of Ociras, on the Catinguinha. Pop. 3,000. The annexed district is extensive, but thinly peopled. Its chief productions are rice, millet, mandioc, tobacco, saltpetre, and cattle.

VALENÇA-DO-DOURO, a town of Portugal, in the prov. of Beira, 9 m. W of St. João-de-Pesqueira, near the l. bank of the Douro. Pop. 635.

VALENÇA-DO-MINHO, a town and fortress of Portugal, in the prov. of Minho, on the l. bank of the Minho, opposite the fortress of Tuy in Spain, 56 m. N of Oporto, and 72 m. WNW of Braganza. Pop. 1,020. It is supposed to have been founded by the soldiers of Viriatus.

VALENÇAY, a town of France, dep. of the Indre, 25 m. NNW of Chateauroux, on the Nahon. Pop. of com., 3,229. Hosiery and woollen goods are manufactured here. It has a fine castle, where Ferdinand VII. of Spain resided from 1808 to 1813. Pop. of cant. comprising 10 coms., 12,315.

VALENCE, a town of France, the capital of the dep. of the Drome, agreeably situated on a declivity on the l. bank of the Rhone, 42 m. SW of Grenoble, in N lat. 44° 55′ 50″, E long. of Paris, 2° 33′ 9″, at an alt. of 332 ft. above the Rhone. It is surrounded with walls, which give it at a distance rather a gloomy appearance; and it is ill built and irregular, with narrow and winding streets. The cathedral is a building of considerable antiquity, but has little claim to attention in point of architecture. The Gothic façade of the old castle is said to be one of the finest specimens of that style in France. In front of the Place d'armes stands the citadel, in which Pius VI. died in 1799. V. is the see of a bishop, and the seat of a diocesan college, and a communal school. Its pop. in 1841 was 13,407; in 1846, 13,901. Its manufactures comprise silk, cotton, and leather, but each on a small scale. It has also dye-works, tanneries, breweries, and marble-works. Olives grow in the neighbourhood, and the town contains a number of oil mills.—The cant. comprises 7 coms. Pop. in 1846, 23,011.—The arrond. has an area of 188,368 hectares, and comprises 10 cants. Pop. in 1846, 149,278.

VALENCE D'AGEN, a commune and town of France, dep. of Tarn-et-Garonne, 14 m. SE of Agen. Pop. 3,088. The preparation of quills and feathers forms a chief branch of trade here.

VALENCE-EN-ALBIGEOIS, a village of France, dep. of the Tarn, 14 m. NE of Albi. Pop. 1,278.

VALENCE-SUR-BAISE, a town of France, dep. of the Gers, on the small river Baise, 6 m. S of Condom. Pop. 1,205.

VALENCIA, a province of Spain, bounded on the N by the provs. of Castellon-de-la-Plana and Teruel; on the E by the Mediterranean; on the S by the provs. of Alicant and Albacete; and on the W by Albacete and Cuenca. Area 2,754 geog. sq. m Pop. in 1834, 388,759; in 1849, 500,000. It is administratively divided into 15 partidos, and 245 pueblos.—The kingdom of V., an ancient division of Spain, is bounded on the N by Aragon and Catalonia; on the E and S by the Mediterranean; on the SW by Murcia; and on the NW by New Castile. It comprises the three western provs. of Castellon, V., and Alicant; and has an area of 7,193 geog. sq. m. In 1787 the pop. of this province was 825,059; in 1808, Antillon and Rehfues state it to have been about 1,200,000; in 1849, it was 1,110,950. The greater part of the surface is mountainous. The principal plains stretch along the coast, which, in the S is bordered with cliffs and rocks, and in the N and beyond Cape San Antonio, is flat and full of sandbanks, among which the large laguna of Albufera is situated. The plain adjacent to the capital is above 80 m. in length; those to the southward, which adjoin the towns of Alicant and Orihuela, if inferior in extent, may challenge a comparison in beauty and fertility. This prov. is watered by three great rivers, the Xucar, the Segura, and the Guadalaviar; also by the Murviedro, the Palencia, the Millares, and others of less size, all flowing from the mountains of the interior to the Mediterranean. The soil consists partly of clay mixed with sand, and partly of lime and chalk; the first is predominant in the plains, which are of a luxurious fertility; the latter is chiefly found in the mountainous parts. The mountains of this region belong to the Iberian chain, which here runs into the sea in the promontories of Oropesa, Antonio, Martin, Nao, and Cervera. The principal ridges are the Sierra-de-Espadan in the N, the Monte-Penaglosa on the boundaries of Aragon; the Sierra-Picochera, the Sierra-de-Meriola, and the Bujeron. All these mountains are steep and of a wild appearance; but there is seldom much snow upon them, and a few are covered with wood. The climate is delicious, and the temp. is refreshed by sea-breezes. The heat in summer is seldom above 17° or 18° R.; in winter the temp. is from 7° to 13° R. Hoar frost and mists are uncommon, and there are scarcely 18 or 20 rainy days in the whole year. The *solano* and earthquakes, however, sometimes visit this paradise; and myriads of insects annoy the inhabitants during the summer. Agriculture is well-conducted. The lands are divided into *huertas* and *serranas*. Irrigation is here, as in the S of France, an indispensable requisite to productiveness. The fields are divided into tracts supplied with water, and eminences which do not admit of irrigation, and are planted with vines, olives, figs, and whatever requires comparatively little humidity. Wheat, Indian corn, and barley, are grown; but the kingdom does not produce sufficient grain for its own consumption, notwithstanding that chestnuts, rice, and algarrobas, are extensively used as substitutes for bread. The finest vegetables are grown in great quantities. Among the wines of this province, are the white and red Alicant, the strong Benicarlo, the Vino-de-la-Torre, and the Cartuxo. Brandy is exported into France. The sugar-cane is grown around Gandia, but only the fresh juice of the plant is made use of. Silk is grown in considerable quantity; but the management of this article is not well understood. Flax, hemp, and rice, are likewise reared here; yet, notwithstanding all these products, the peasantry live in general in great poverty. In manufactures the Valencians have made very little progress. The Carthaginians, and after them the Romans, worked innumerable mines in the mountains of the south of Spain, and procured from them great quantities of gold, silver, lead, and other metals. The sides of mountains in V, are still seen perforated with the mouths of caverns, whence came the silver and gold that glittered round the throne of the Cæsars; yet it did not until very recent times occur to any of the modern occupiers of the soil, to try if some of the precious metals were not still left within. Several foundries have been established along the coast, for the purpose of smelting the ores, and separating the silver from the lead, the argentiferous lead-mines only being those which are thus explored. Near Pinnosa there is a stratum of mineral salt, but it is not made use of, as there are many rich salt-springs, three of which annually yield 300,000 cwt., and could be made to produce a great deal more. The language is a peculiar dialect, kindred to that of Catalonia. There are 9 cividades, 161 villas, 378 villages, 562 parishes, 171 monasteries, 54 nunneries, and 39 establishments of charity in this prov. This part of Spain, early invaded by the Cathaginians, was conquered by the Romans, and at a subsequent date by the Goths, from whose hands it fell into those of the Moors. The latter established the kingdom of V. in 713, and retained it under several vicissitudes of fortune, until 1238, when it was definitively taken from them, and united to Aragon. It afterwards formed a component part of the Spanish monarchy, but continued to preserve its representative body and its privileges, till the early part of the 18th cent., when having, in the war for the succession to the crown, taken part against the Bourbon dynasty, it was, on their establishment on the throne, deprived of its old constitution, and obliged to conform to the laws of Castile.

VALENCIA, the chief town of the above province and kingdom, is situated upon the Guadalaviar, here about 100 yds. broad, at the distance of 2 m. from the sea, in N lat. 39° 28′ 45″, W long. 0° 24′ 23″. It is surrounded with ramparts whose walls are entire and well-preserved; but the citadel is useless, being poorly fortified, and not even commanding the town. It has 8 gates, and 5 faubourgs or suburbs. The interior, far from meriting the flattering epithet of 'Valencia la Bella,' consists of narrow and winding streets, crossed by a multiplicity of lanes in many of which there is no thoroughfare. Of the different public walks, the chief is that which extends along the banks of the river. Few towns contain a greater number of public buildings, but they are less remarkable for elegance than for antiquity and profuse decoration. The ancient palace, called El-Real, is now the residence of the captain-general. The other remarkable structures are a Moorish mosque; a church built for Christian worship in the time of the Goths; and several modern edifices, such as the college of Pio Quinto, the convent of the Carmelites, the custom-house, and the college of the patriarch. V. is the see of an archbishop, whose revenue amounts to £34,375 annually. The number of regular and secular clergy is very great, being no less than 2,610 persons. The cathedral is of great antiquity, and has undergone many changes, having been a pagan temple under the Romans,—a church under the Goths,—a mosque under the Moors,—being reconverted into a Christian temple dedicated to St. Paul after Valencia was conquered by the Cid in 1094; again turned into a Mahommedan mosque upon its re-capture by the Moors in 1100; and again converted into a Romish cathedral by Jayme, king of Aragon, in 1230, the era of its final conquest by the Christians. The university of this city was founded in 1470, and is on an extensive scale, but the course of study is antiquated. The Valencians have also an academy of painting, and reckon several eminent artists among their number. There are two public libraries.—Of manufactures, the most extensive one is of silk; but there are, on a small scale, fabrics of leather, glass, woollens, cordage, tobacco, and lace. The maritime trade of the place is carried on by lighters, which load and unload vessels at the village of Grao, near the mouth of the river.—The commerce of this city was recently very extensive; the value of its foreign and interior commerce amounting annually to £773,416, besides its exportation of silk, wrought and raw, amounting to £1,875,000, at a moderate calculation, and employing 22,000 hands. The chief articles of export are rice, saffron, almonds, and oranges.—The environs form a delicious garden, the air of which is continually loaded with perfumes. The temp. of Valencia is remarkably mild and pleasant, the intensity of the summer heat being lessened by the vicinity of the sea, and in some measure by the moisture of the adjacent plains. The spring is sometimes rainy, but the autumn is a very pleasant season, the trees and the fields not losing their verdure until December. As a resort for invalids from the north of Europe, V. has a claim to rank with Lisbon, Nice, Pisa, and other well-frequented towns. V. was the birthplace of the lawyer Mayans, of the celebrated Ant. Jos. Cavanilles, and of several renowned painters, as Espinosa, and Francisco Bibalta.—Valencia is a place of antiquity, and was probably the *Valentia Edetanorum* of the Romans. On the invasion of the Goths, it fell, with the rest of the peninsula, into their hands, and continued so until 715, when it was occupied by the Moors. From these invaders it was taken by the Cid in 1094, and given to that warrior to govern as a dependency of the king of Castile. It was consequently called 'Valencia of the Cid,' and devolved on his death to his widow Ximene, a heroine almost equally noted in Spanish history as her husband, and who sustained in it a long siege against the Moors in the year 1100. At last, however, it surrendered, and was occupied by the Moors until 1238, when it was finally retaken by the king of Aragon. In the war of the succession, the inhabitants adhered to the Austrian candidate for the crown, until the unfortunate battle of Almanza (in 1707) obliged them to submit to the Bourbons. A number of them were famished, the city was deprived of its privileges, and obliged to adopt the customs of Castile. V. declared against the French, and baffled the first attempts made on it in 1808 by Moncey; but in the end of 1811 it was attacked by

Suchet, and, after a vigorous siege and bombardment, surrendered in January 1812.

VALENCIA, a city of Venezuela, in the department of Caracas, and province of Carabobo, situated in a fine plain, half-a-league W of the lake of the same name, in N lat. 10° 10′ 11″, 77 m. WSW of Caracas. Pop. about 16,000. The houses are in general low and irregular, though some of the streets are broad and well-built. The parish-church, and a handsome square in which it stands, form the principal ornaments of the city. The inhabitants, formerly noted for their indolence, have lately become active and industrious, and the situation of the place is favourable for trade, being separated from Puerto-Cabello by only 10 leagues of good road. The adjacent country produces provisions and fruits in great abundance, and the plains feed immense herds of cattle, sheep, horses, and mules. V., with the towns of Victoria and Barquisimeto, suffered much from the earthquake which overthrew Caracas, La-Guayra, Merida, and several villages, in March, 1812. It remained firm to the cause of Ferdinand up to the year 1810, when it was compelled to submit to Miranda, and shortly after became the seat of the new congress of the united states of Venezuela. It was soon afterwards occupied by the royalists; but in 1812 its possession was disputed by Miranda; and in 1813, it was entered by Bolivar, without opposition.

VALENCIA (Lake of), a fine inland sheet of water, in the dep. of Caracas, in Venezuela. It stretches 13½ leagues from ENE to WSW; its greatest breadth is 4 leagues. It lies at the distance of one league from the city of the same name, and is situated in a valley surrounded with mountains excepting on the W, where it extends into the interior. The waters of twenty rivers are discharged into it; but it has no visible outlet; and the space which separates it from the sea is 6 leagues in width, and is filled with inaccessible mountains. It is reported to have been diminishing for several years; and its waters are still receding, leaving behind them a rich and productive soil. This lake has about 27 islands scattered over its surface. The largest of them, called Caratapona, is highly fertile, and contains a pop. sufficient to raise provisions, fruits, and vegetables, for the market. The produce which grows on the borders of the lake is transported across its waters, or by the rivers which fall into it. The navigation, however, is not very easy, owing to the numerous islands with which the lake is interspersed.

VALENCIA-DE-ALCANTARA, a strong town of Spain, in the prov. of Caceres, on the frontiers of Portugal, 25 m. SW of Alcantaro, on the l. bank of the Aird. Pop. 4,798. Some coarse linens are made here.

VALENCIA-DE-DON-JUAN, a town of Spain, in the prov. and 18 m. S of Leon, on the Esla, on the site of the ancient *Coiaca*. Pop. 1,710.

VALENCIA-DE-LAS-TORRES, a village of Spain, in the prov. and 40 m. S of Badajoz. Pop. 695.

VALENCIA-DEL-VENTOSO, a town of Spain, in the prov. and 48 m. SSE of Badajoz. Pop. 3,880.

VALENCIANA, a town and a celebrated mine of Mexico, in the dep. and 10 m. N of Guanaxuato. In the mine the great vein was 22 ft. in breadth; and as the chasm is dry, it was easier worked than almost any other American mine. The pits extended, in 1790, to the breadth of 4,900 ft., and the lowest was 1,640 ft. in depth. The undertaking employed upwards of 900 men in carrying the ores to the surface; 1,800 workmen in procuring and sorting the ore; and 400 women and children carrying the minerals to the sorters. The expenses of materials, workmen, overseers, &c. was £187,500 per ann.; the net profit, during the same period, to the proprietors, after deductions of the king's fifth, and all expenses, was from £82,500 to £123,759 per ann. This mine is no longer wrought.

VALENCIENNES, a fortified town of France, in the dep. of Nord, 28 m. SE of Lille, situated on the Scheldt, which traverses the town, and here becomes a navigable river, though small in its volume of water, and sluggish in its course. The form of the town is circular. The streets are narrow and crooked, and the houses are in general ill-built, many of them being of wood. The principal edifices are, the public square, the church of Notre-Dame, the town-hall, the artillery-house, the theatre. The public library has 18,000 vols. Pop. in 1852, 24,343. The chief manufactures are lace of great fineness, cambric, gauze, linen-stuffs, hosiery, wire-cloth, pottery ware, and Prussian blue. This town belonged to the Netherlands until 1677, when it was taken by the French, and confirmed to them by treaty. In 1793 it formed one of the first objects of attack by the allies, after the defeat of Dumourier. The siege, conducted under the command of the duke of York, was long and obstinate, and part of the town was laid in ashes before the capitulation. It was retaken by the French in 1794; escaped attack in the invasion by the allied powers in 1814 and 1815, and was definitively confirmed to France by the treaties of these years.

VALENDAS, a village of the Swiss cant. of Ungusan, in the Glenner circle, NW of Ilanz. Pop. 555.

VALENS, a commune and village of Switzerland, in the cant. and 32 m. SE of St. Gall. Pop. 1,315.

VALENSOLE, a town of France, in the dep. of Basses-Alpes, 14 m. SE of Forcalquier. Pop. 3,135.

VALENTANO, a town of the States-of-the-Church, deleg. of Viterbo, 14 m. SW of Orvieto. Pop. 1,030.

VALENTIA, an island and village off the W coast of co. Kerry, near the S entrance of Dingle bay. The island is about 5 m. in lengh, and 2½ m. in breadth, with an area of 6,371 acres. Bray-head, forming the SW extremity, projects about 1 m. beyond the S screen of the southern entrance of the harbour, flings sheer down to the ocean cliffs of 628 ft. in alt., and is one of the boldest and most striking of the many sublime promontories of Kerry. The cliffs of Fohilly, which screen the NW extremity of the island, are separated from the entrance of Lough Kay by Rinnadrolan point, and soar aloft to the elevation of 886 ft. The N or principal entrance to Valentia harbour, between Beg-Innis and Cromwell's fort, is very fine in scenery. A great part of the surface is under tillage, and there is a considerable range of pasture. A slate quarry on the island is worked for export. It is used for flagging, for fish-slabs, and for many purposes to which marble has been usually applied, and finds a ready market in England. At present about 120 hands are employed in this quarry. The hamlet of Valentia is situated near the NE extremity of the island, in the vicinity of the pier. The chief trade, additional to the quarrying and export of its slates and flags, consists in the conducting of somewhat extensive fisheries, in the exporting of corn and butter, and in the importing, chiefly from Liverpool, of iron, coals, and timber. The Spaniards occupied the island and harbour of V. up to the period of the English commonwealth; but the lieutenants of Oliver Cromwell expelled them, and erected two forts at respectively the north entrance and the south entrance of the harbour, to put an end to the privateering enterprises for which it had been used.

VALENTIA, or Dissee, an island off the coast of Abyssinia, about 2½ m. long and ¾ m. broad. It forms a bay, affording good anchorage for large vessels, and is situated a little to the SE of Massuah. It is supposed to be the ancient *Orine*.

VALENTINE, a town of France, on the Garonne, 3 m. S by W of St. Gaudens. Pop. 1,000.

VALENTINE'S BAY, a bay on the SE coast of Tierra-del-Fuego, to the W of Cape Success, in S lat. 52° 55′, W long. 74° 15′.

VALENTINO (San), a town of Naples, in Abruzzo-Citra, 12 m. SW of Chieti. Pop. 2,460.

VALENTINSHAUSER, a village of Austria, in Tyrol, in the bail. of Nanders. Pop. 840.

VALENZA, a town of the Sardinian states, in the prov. of Alessandria, situated on an eminence near the Po, 12 m. SE of Casale. It is surrounded with a wall and defended by a castle which formerly rendered it a place of strength; but its fortifications have been neglected, and it did not sustain a siege in the campaigns of 1796 or 1799. It contains several churches, a civic palace, a college, 7 hospitals, and some vestiges of antiquity. Its pop., employed partly in cultivating the vicinity, partly in manufacture, amount to 7,400. The district of V. comprises 4,364 hectares of surface; about 3,500 of which are *colline*, with a soil composed in some parts of cretaceous clay, in others pure argil, or calcareous. More than two-thirds of this land are appropriated to vineyards, calculated to produce upwards of 40,000 hectolitres of wine annually. It has some brick and tile kilns, which yield about 96,000 lire; linen and canvas are woven to the value of about 30,000 lire, and thrown silk realizes yearly about 130,000 lire.

VALENZA, a town of Spain, in the prov. of Leon. 18 m. SSW of Leon.

VALENZUELA, a town of Spain, in the prov. and 26 m. E of Cordova. Pop. 2,710.—Also a town in the prov. and near the town of Ciudad-Real. Pop. 1,280.

VALERA-DE-ARRIBA and VALERA-DE-ABAXO, two villages of Spain, in the prov. and about 18 m. S of Cuença. The former has a pop. of 1,250; the latter of 1,500.

VALERIEN (Saint), a town of France, dep. of the Yonne, 9 m. W of Sens. Pop. 1,030.

VALE-ROYAL, a beautiful valley in the p. of Whitegate, co.-palatine of Chester, intersected throughout its length by the river Weaver, which is crossed here by the Grand Junction railway, on a viaduct of 5 arches of 63 ft. span each, at the height of 66 ft. above the level of the water.

VALERY-EN-CAUX (Saint), a town of France, dep. of Seine-Inferieure, 33 m. N by W of Rouen. Pop. 5,370. It has a small but good harbour, and considerable fisheries particularly of herring. The other branches of its industry are ship-building and the manufacture of soda.

VALERY-SUR-SOMME (Saint), a small port of France, situated on the r. bank of the Somme, near its mouth, 12 m. NW of Abbeville. Pop. 3,241. It carries on a considerable coasting and transit trade, serving as port to Amiens and Abbeville. Its harbour is capacious, but the entrance is difficult on account of sand-banks. It was here that, in 1066, William the Conqueror embarked for the invasion of England.

VALET, a town of France, dep. of the Loire-Inferieure, 6 m. N of Clisson.

VALETTA (La). See Malta.

VALETTE (La), a town of France, dep. of Charente, situated on a mountain, 14 m. SE of Angouleme. Pop. 955.—Also a town and fort in the dep. of Var, 4 m. NE of Toulon.

VALEZO. See Valeggio.

VALFF, a village of France, in the dep. of Bas-Rhin, cant. and 4 m. SE of Obernay.

VALGARA, a village of Italy, 22 m. NW of Como. Pop. 1,000. Marble is wrought in the vicinity.

VALGORGE, a town of France, dep. of Ardeche, on the small river Baune, 25 m. SE of Privas. Pop. 1,436.

VALGRANA, a town of Piedmont, prov. of Coni, situated on the river Grana, 8 m. W of Coni. Pop. 1,625.

VALGREGHENTINO, a village of Austrian Italy, in the prov. and 18 m. NE of Como. Pop. 1,100.

VALIDIA. See Valedia.

VALINCOURT, a town of France, dep. of Nord, 9 m. SE of Cambray, and 23 SE of Douay.

VALKI, a town of European Russia, on the river Mscha, in the Slobodsk-Ukraine, 27 m. W by S of Charkov. Pop. 5,000. The industry of its inhabitants is confined to cultivating the environs, distilling spirituous liquors, and trafficking in cattle.

VALLADA, a town of Spain, in the prov. and 40 m. SW of Valencia, on the Canolas. Pop. 1,500.

VALLADARES, a town of Spain, in the prov. of Galicia, near Tuy. Pop. 1,200.

VALLADOLID, an inland prov. of Spain, forming part of the kingdom of Leon, and bounded on the N by the provs. of Leon and Palencia; on the E by Burgos; on the S by Avila; on the SW by Segovia; and on the W by Salamanca and Zamora. Area 2,237 geog. sq. m. Pop. in 1848, 210,000. The surface lies in general high, and its soil is sandy and barren; yet there is no want of water, the prov. being traversed by the Duero, which is joined by the Cea, the Valderaduey, the Pisuerga, the Esla, the Arlançon, and other rivers. The climate is cold. Some corn is exported in good years; but the chief dependence of the inhabitants is on pasturage. Flax and hemp are cultivated to a considerable extent. The manufactures are trifling; and this is, on the whole, one of the poorest regions of Spain. It is divided into 22 districts.

Valladolid, the capital of the above prov., is situated on the banks of the Esgueva, which divides it into two, and of the larger stream of the Pisuerga which bathes its walls, in N lat. 41° 39′ 14″, 98 m. NNW of Madrid. The town stands in the midst of an extensive plain, which might be rendered more productive were the streams that traverse it made to serve effectually for the purpose of irrigation. The streets are dirty, and many of the houses in decay; there are, however, two squares, one of which, the Plaza Mayor, contains some good buildings. The cathedral, begun by Philip II., has never been finished. The town has six gates, one large bridge, and a number of small ones, 16 churches, and 24 convents. It is the seat of a bishop, of a provincial court of justice, and of an university, which is at present the most frequented in Spain. The manufactures are confined to silks, lace, paper, coarse woollens, and earthenware. The produce of the environs consists of corn, wine, and silk.

VALLADOLID, a town of Yucatan, 80 m. ESE of Merida, near the source of the Bullina.

VALLADOLID-DE-MECHOACAN, a town of Mexico, in the state of Mechoacan, on the Olid, in N lat. 19° 42′, 120 m. W of Mexico, at an alt. of 69,396 ft. above sea-level. It has a fine cathedral, and several handsome houses. Pop. 25,000.

VALLANCA, a village of Spain, in the prov. and NE of Valencia. Pop. 1,025. It has some trade in wine, honey, and wax.

VALLATA, a town of Naples, in the prov. of Principato-Ultra, and district of Ariano. Pop. 4,400.

VALLAURIS, a commune of France, in the dep. of the Var, and cant. of Antibes, 12 m. SE of Grasse. Pop. in 1846, 2,482. It has manufactories of pottery. Manganese is wrought in the locality.

VALLAY, a village of France, in the dep. of the Haute-Saone. Pop. 485.

VALL-DE-MOSA, a market-town of Spain, in the prov. of Baleares and island of Mallorca. Pop. 1,580.

VALL-DE-UXO, a market-town of Spain, in the prov. of Castellon-de-la-Plana and partido of Nulez. Pop. 5,500. It has a parish-church, an hospital, and a custom-house.

VALLDONA, a village of Spain, in the prov. of Castellon-de-la-Plana, and to the NW of Peniscola.

VALLDORF, a village of Prussia, in the reg. of Minden and circle of Herford. Pop. 1,000. It has mineral-springs.

VALLE, a market-town of Austria, in Illyria, in the gov. of Trieste and circle of Istria, near Rovigno. Pop. 1,050.—Also a market-town of Naples, in the prov. of Terra-di-Lavoro, near Caserta. Pop. 810. —Also a market-town of Spain, in the prov. of Badajoz, near Villa-Nueva-de-la-Serena. Pop. 878. —Also a parish of Norway, in the district of Nedenas, on the Odderen. Pop. 1,600.

VALLE, or VALLE-D'ALESANI, a canton, commune, and town of the island of Corsica, and arrond. of Corte. The cant. comprises 9 com. Pop. in 1831, 3,057; in 1846, 3,243. The town is 15 m. E of Corte. Pop. 616.

VALLECAS, or BALLECAS, a market-town of Spain, in the prov. and 5 m. SE of Madrid. Pop. 1,542. It has a parish-church, two hospitals, and a custom-house. Agriculture, and the manufacture of linen, form the chief objects of local industry.

VALLECORSA, a market-town of the Papal States, in the delegation and 18 m. S of Frosinone, on the Rio-di-Valle-Frata. Pop. 3,000.

VALLE-D'ALESANI. See VALLE.

VALLE-DE-ADDALAGIS, a market-town of Spain, in the prov. and NNW of Malaga. Pop. 3,300.

VALLE-DE-CABUERNIGA, a judicial partido and town of Spain, in the prov. of Santander. The partido comprises 36 pueblos. The town is situated in a valley of the same name. Pop. 287. It has a parish-church and a custom-house. The manufacture of linen and of agricultural implements form the chief objects of local industry.

VALLE-DE-CERRATO, a market-town of Spain, in the prov. of and near Palencia. Pop. 425.

VALLE-DEL-ORO (SANTA CECILIA), a parish of Spain, in the prov. of Lugo, and partido of Monforte, watered by a river of the same name. Pop. 780. It has a parish-church and a custom-house.

VALLE-DEL-ORO (SANTA-CRUZ-DEL), a parish of Spain, in the prov. of Lugo and partido of Mondonedo. Pop. 962. The manufacture of linen and rearing of cattle form the chief employment.

VALLE-DE-MATAMOROS, a market-town of Spain, in the prov. of Badajoz, near Xerez-de-los-Caballeros. Pop. 1,250.

VALLE-DE-SANTA ANNA-LA-REAL, a market-town of Spain, in the prov. of Teruel, to the W of Henares. Pop. 474.

VALLE-DE-TABLADILLO, a town of Spain, in the prov. of Segovia, between two hills, in a valley of the same name. Pop. 517. It has a parish-church, a custom-house, and a public granary.

VALLE-D'IGUAPE, a valley of Brazil, in the prov. of Bahia, watered by an affluent of the Paraguacu. It contains extensive sugar plantations, and is considered one of the most fertile tracts in the Brazilian territory.

VALLEES (PAYS-DES-QUATRE), a district of France, in the ancient prov. of Guyenne, containing an area of 160,000 hect., and now comprised in the departments of the Haute-Garonne and Hautes-Pyrenees. Its cap. was Castelnau-de-Magnoac.

VALLEJO, a village of Solano co , in the state of California, U. S., on the Napa straits, 25 m. NNE of San-Francisco It is at present the capital of the state and the seat of government.

VALLE-LUNGA, a market-town of Naples, in the prov. of Calabria-Ultra, and near Monteleone. Pop. 1,000.—Also a town of Sicily, in the intendancy and to the NW of Caltanisetta. Pop. 3,500.

VALLE-MAROBBIA-IN-PIANO, a village of Switzerland, in the cant. of Tessina and bail. of Bellinzona. Pop. 794.

VALLEN, an island of France, off the coast of the dep. of Finistere.

VALLENDAR, a town of Prussia, in the prov. of the Rhine, regency and circle and 3 m. NNE of Coblenz, on the Rhine. Pop. in 1843, 3,135. It has a castle, and possesses manufactories of cloth, and other woollen fabrics, leather, tobacco, and dye-works. The culture of the vine and navigation afford also important employment to the inhabitants.

VALLENDAS, a village of Switzerland, in the cant. of the Grisons and bail. of Glenner. Pop. 555.

VALLENGIN, or VALANGIN, a market-town of Switzerland, in the cant. and to the NW of Neuchatel. Pop. 6,250. It has a castle.

VALLERANGUE, a canton, commune, and town of France, in the dep. of the Gard, and arrond. of Le-Vigan. The cant. comprises 3 com. Pop. in 1831, 6,521; in 1846, 6,972. The town is on the Herault, 15 m. N of Le-Vigan. Pop. 3,853. It has a consistorial Calvinist church, and possesses extensive manufactories of silk and cotton hosiery, and numerous silk-mills. The trade consists chiefly in timber. Chestnuts are extensively cultivated in the vicinity of this town.

VALLE-ROTONDA, a town of Naples, in the prov. of the Terra-di-Lavoro, and to the NW of Benafro, in a valley. Pop. 2,210.

VALLERUELA-DE-PEDRAZA, a town of Segovia, in the partido of Sepulveda. Pop. 760.

VALLERY, a village of France, in the dep. of the Yonne. Pop. 681.

VALLES, a market-town of Central America, in the state of San Louis Potosi, U. S., on the l. bank of the Montezuma. Pop. 3,500. It has a trade in sugar, and other articles of local production.

VALLESESIA. See VALSESIA.

VALLET, a canton, commune, and town of France, in the dep. of the Loire-Inferieure, and arrond. of Nantes. The cant. comprises 4 com. Pop. in 1831, 10,304; in 1846, 9,875. The town is 15 m. E of Nantes. Pop. 5,583.

VALLEY, a village of Bavaria, in the circle of Upper Bavaria, and presidial of Wiesbach. Pop. 200. It has a castle.

VALLEY-FORGE, a village of Chester co., in the state of Pennsylvania, U. S., on the S side of Schuylkill river, near the mouth of Valley creek, 70 m. E of Harrisburg, and intersected by the Philadelphia railroad.

VALLEYRES, a village of Switzerland, in the cant. of Vaud and bail. of Orbe. Pop. 498.

VALLEYRES-SOUS-URSIUS, a village of Switzerland, in the cant. of Vaud, and bail. of Isserten. Pop. 227.

VALLEYRES-SUR-MONTAGNY, a village of Switzerland, in the cant. of Vaud and bail. of Isserten. Pop. 215.

VALLFONGA, a market-town of Spain, in the prov. of Gerona, near Campredon. Pop. 748.

VALLIBONA, a market-town of Spain, in the

prov. of Castellon-de-la-Plana, near Morolla. Pop. 790. Wool-spinning forms an important branch of local industry.

VALLIER (Saint), a commune of France, in the dep. of the Saone-et-Loire, cant. of Mont-Saint-Vincent, 33 m. SW of Chalon-sur-Saone, on the Saone. Pop. in 1852, 1,145.—Also a canton and commune of France, in the dep. of the Var and arrond. of Grasse. The cant. comprises 5 com. Pop. in 1831, 4,282; in 1846, 4,346. The village is 8 m. NW of Grasse. Pop. in 1852, 856.—Also a canton, commune, and town of France, in the dep. of the Drome and arrond. of Valence. The cant. comprises 16 com. Pop. in 1831, 15,086; in 1846, 17,878. The town is 24 m. N of Valence, on the r. bank of the Rhone, at the confluence of the Galaure. Pop. in 1852, 1,145.

VALLIERES, a commune and market-town of France, in the dep. of the Creuse, and cant. of Felletin, 9 m. NE of Aubusson. Pop. in 1852, 2,400. Cattle are extensively reared in the locality.

VALLIEVO, or Valliova, a town of Turkey in Europe, in Servia, 60 m. SW of Belgrade, on the l. bank of the Kolubara. Pop. 4,100.

VALLIGUIERES, a commune of France, in the dep. of the Gard, and cant. of Remoulins, 20 m. E of Uzès. Pop. 492.

VALLIQUERVILLE, a commune of France, in the dep. of the Seine-Inferieure, cant. and 4 m. W of Yvetot. Pop. 1,615. It has manufactories of woollen fabrics.

VALLMOLL, a market-town of Spain, in the prov. and near Tarragona. Pop. 874. It has distilleries of brandy.

VALLO (Il), a town of Naples, in the prov. of the Principato-Citra, 15 m. SE of Salerno. Pop. 2,480. Leather is its chief article of manufacture.

VALLOE, a small island of Norway, in the prov. of Jarlsberg, in the Christiana-fiord.—Also a stift and town of Denmark, in the island of Sieland, 20 m. NNE of Praestoe.

VALLOIRES, a market-town of Sardinia, Savoy, and in the prov. of Maurienne. Pop. 1,915.

VALLON, a canton, commune, and town of France, in the dep. of the Ardeche, and arrond. of Largentiere. The cant. comprises 11 com. Pop. in 1831, 8,579; in 1846, 10,422. The town is 15 m. SSE of Largentiere. Pop. 2,627. It has a Calvinist consistorial church. The culture of wine and silk, and manufacture of leather, form the chief branches of local industry.—Also a commune and market-town of France, in the dep. of the Sarthe, and cant. of Loue, 18 m. W of Le Mans, on the Geay. Pop. 1,550. It has manufactories of common woollen and linen fabrics.

VALLON-EN-SALLY, a market-town of France, in the dep. of the Allier, and arrond. of Montlucon, on the Cher. Pop. 980.

VALLONISE, a commune of France, in the dep. of the Hautes-Alpes, and cant. of Largentiere, 9 m. WSW of Briancon. Pop. in 1841, 1,193.

VALLS, a judicial partido and town of Spain, in the prov. of Tarragona. The partido comprises 30 pueblos. The town is 12 m. NNW of Tarragona, in a level tract watered by the Francoli. Pop. 9,000. It has a parish-church, 2 convents, and an hospital, and contains several distilleries of brandy, tanneries, wool-spinning and paper-mills.

VALLUHN, a village of the grand-duchy of Mecklenburg-Schwerin, on the Boize. Pop. 290.

VALM (Alt and Neu), two villages of Prussia, in the regency of Kostin, and circle of Neustettin. Pop. 960.

VALMALEDA, a town of Spain, in the prov. of Viscayo, to the SW of Bilboa, on a hill, near the l. bank of the Salcedon, which is here crossed by two bridges. Pop. 2,000. It has a parish-church, a convent, and an hospital; and carries on an active trade in copper and iron ware, and other articles of local produce, and in cattle.

VALMARTEN, a hamlet of France, in the dep. of the Seine-Inferieure, cant. of Clères, and com. of Le Bocasse.

VALMONT, a canton and commune of France, in the dep. of the Seine-Inferieure, and arrond. of Yvetot. The cant. comprises 23 com. Pop. in 1831, 17,082, and in 1846, 16,999. The town is 18 m. NW of Yvetot. Pop. 1,112. It has manufactories of calico, and several bleacheries.

VALMONTONE, a town of the Papal States, in the leg. of Velletri, 5 m. SSE of Pelestrina. Pop. 1,088.

VALMUNSTER, a village of France, in the dep. of the Moselle, and arrond. of Metz. Pop. 421. It has manufactories of magnesia and vitriol.

VALMY, a commune of France, in the dep. of the Marne, cant. and 7 m. from Sainte Menehould. Pop. 466. A victory was here gained by the French over the Prussians on the 20th September 1792.

VALNAY, a village of France, in the dep. of the Cher, and arrond. of St. Amand. Pop. 500.

VALOGNES, an arrondissement, canton, commune, and town of France, in the dep. of the Manche.—The arrond. comprises an area of 103,995 hect., and contains 7 cant. Pop. in 1831, 95,660; in 1841, 95,370; and in 1846, 93,857.—The cant. comprises 9 com. Pop. in 1831, 17,032; and in 1846, 15,838.—The town is 39 m. NNW of Saint Lo, on the Merderet. Pop. in 1789, 4,732; in 1821, 6,858; in 1831, 6,940; and in 1846, 6,579. It has a communal college, a public library and a savings' bank. It possesses manufactories of lace and blond, several dye-works and tanneries; and carries on a trade in grain, butter, eggs, fowls, game, honey, wax, feathers, fresh-fish, shell-fish, flax, and linen.

VALOIRES, a town of Sardinia, in the prov. of Savoy, 13 m. SSE of St.-Jean-de-Maurienne. Pop. 2,000.

VALOIS, a district of France, in the ancient prov. of the Ile-de-France, now comprised in the departments of the Aisne and Oise. It had an area of 181,877 hect., and contained several towns, of which Crespy was the capital.

VALONA. See Avlona.

VALONGO, a market-town of Portugal, in the prov. of Beira, and to the NE of Oporto.

VALOR, a town of Spain, in the prov. of, and 55 m. from Granada, at the foot of the Sierra Nevada. Pop. 1,600. It has a parish-church, a custom-house, a public granary, and several mineral springs, and carries on a considerable trade in grain, wine, fruit, and wool.

VALORBE, a market-town and parish of Switzerland, in the cant. of Vaud, and bail. of Orbe, 20 m. NW of Lausanne, near the source of the Orbe. Pop. 1,491. It has extensive iron-works.

VALORIA-DEL-ALCOR, a market-town of Spain, in the prov. of and near Palencia. Pop. 234.

VALORIA-LA-BUENA, a judicial partido and market-town of Spain, in the prov. of Valladolid. The partido comprises 41 pueblos. The town is 15 m. ENE of Valladolid, on the l. bank of the Arrayo-Maderazo. Pop. 795. It has a parish-church and a custom-house. Linen is manufactured here.

VALORICKE, a district and town of Russia in Europe, in the gov. and 156 m. SSW of Voronej, on the Oskol.

VALPARAISO, a province of Chili, bounded on the N by the lake of Catapillo; on the E by the Cuesta-de-Calavera; on the S by the dep. of Melipilla; and on the W by the Pacific. Pop. in 1847, 75,962. The soil is generally stony and arid, and rain seldom falls; but when irrigated, the ground is highly productive; and some of the valleys are very beautiful, and richly clothed with vegetation. The only considerable stream is the Quillota. The principal places are the city which gives name to the prov., and the towns of Quillota and Casa-Blanca.

Valparaiso, the capital of the above prov., and the principal emporium of Chili, whence all the commerce with Peru is carried on, is situated in S lat. 33° 1′ 56″, W long. 71° 41′ 45″, about 95 m. direct distance WNW of Santiago, of which it is the port, and at the foot of a range of steep hills about 1,600 ft. high. The city itself lies about 3 m. from the harbour; but the suburb called Almendral lies close upon the harbour, and is inhabited by those employed about the shipping. The harbour is defended by a fort. V., though the chief commercial entrepôt of Chili, and indeed one of the most flourishing sea-ports of South America, is nevertheless peculiarly unfortunate in its local situation. The bay is open to the N, and affords no shelter to shipping during the winter months; the country which surrounds it is of the most barren description, and the want of space at the base of the cliffs prevents the improvement of the town. But in spite of these disadvantages, the place has increased rapidly of late years, and it is supposed to contain at present above 50,000 inhabitants. Of this number 6,000 are strangers. The main body consists chiefly of one long, straggling street, mostly composed of low white-washed houses, which runs parallel to the beach, and wherever a ravine comes down, the houses are piled up on each side of it. Of late years many handsome houses of four and five stories have been erected; and there are several fine public buildings, amongst which are the custom-house, the Merced and Matriz churches, and the Plaza-de-la-Victoria is a fine open space. The surrounding hills, being only partially protected by a very scanty vegetation, are worn into numberless little gullies which expose a singularly bright red soil. In a NE direction there are some fine glimpses of the Andes; the volcano of Aconcagua is particularly magnificent; but the immediate neighbourhood is not very attractive to a European eye. During the long summer the wind blows steadily from the S, and a little off-shore, so that rain never falls; during the three winter-months, however, it is sufficiently abundant. The vegetation in consequence is very scanty; except in some deep valleys, there are no trees; and only a little grass and a few low bushes are scattered over the less steep part of the hills. When we reflect that at the distance of 350 m. to the S, this side of the Andes is completely hidden by one impenetrable forest, the contrast is very remarkable. "We," says Dr Poeppig, "who during a long voyage had seen nothing but black and dreary rocky islands and a cheerless sky, now—in the bay of Valparaiso—looked round us in vain to discover some green tree or grassy slope; in whatever direction the eye turned, it was sure to encounter monotonous steep cliffs of grey syenite, which are all connected with the same semicircular mountain ridge. The sides and summit of this ridge are painted in the colours of an old brick, brown and red, and not a single group of trees breaks this dull uniformity, inasmuch as the parched ground is capable only of nourishing a few bushes with woody boughs and greyish leaves. Many broad slopes are quite destitute even of these, and are either covered with heaps of stones, or split into reddish gaps and ravines. Narrow and dangerous paths wind over the summit of the ridge, and along these, involved in ruddy clouds of dust, long trains of mules laden with the productions of the country, descend to the port. The deep and dark gullies which sink down from the mountain-top to the sea-side, have no signs of a living brook; their perpendicular rocky sides are still barer than the rest. It is only at the bottom of these gullies where some soil has been heaped together, that one remarks a faint tint of green, a trace of the careful garden-culture of the poor peasant, during a short period of the year."

V. is, as it were, a city of yesterday. Under the rule of the Spaniards it was a place of no importance, and never could have been, for they vigorously repressed the trading spirit, and wished to confine its entire operations to an exchange of the manufactures of Catalonia for the agricultural produce of this dependency. Even so late as 1819, V. was only a village, with ten or a dozen huts. A little more than a century ago the whole township was sold for 1,500 d.; now, we are told, many separate houses fetch from 4,000 to 5,000 d. of rent; and building is still going on very extensively. Wherever the tide recedes a few yards, the spot is appropriated and covered with bricks and mortar. Much if not most of the property of the town is said to belong to Englishmen and other foreigners. The merchants are nearly all English, American and German; but nearly all the shops are kept by Frenchmen. The posts are regular; the police well-organized; the streets are kept clean, and crime is repressed. The number of vessels which entered the port of V. in 1845 was 859; in 1849, 968; in 1850, 1,465 = 412,240 tons; in 1851, 1,561. The customs' revenue in 1849 was £337,096; in 1851, £487,400. Of the total imports into the republic of Chili during the first six months of 1852, amounting to 8,404,539 d. in value, 7,790,241 d. were imported at V. See article Chili. The project of a railway to Santiago has been entertained for some time.

In 1822, V. was greatly injured by a severe earthquake; and on the 2d of April 1851 it was again visited by a very severe shock. Many geologists are of opinion that the bay of V. is being gradually but very sensibly affected by the general elevation of the coast in this quarter of America. In an article which appeared in the South American journal, *El Arancano*, Don Mariano Rivero dissents entirely from the opinion that earthquakes have produced changes of level in the bay of Valparaiso; and Colonel Walpole says, "I have heard, from persons long resident on the spot, both previous to the earthquake of 1822, and still residing there,—accurate observers of events,—statements in direct corroboration of M. Rivero, and whom he could not have consulted: and it is difficult to conceive why those whose opinions he considers as inconsiderately put forth, should have drawn their conclusions from the mere appearance of a piece of ground not more than 200 yds. of the whole circumf. of a bay of considerable extent." The observations made by Mr. Darwin during the surveying voyage of the Beagle, extended over part of the coast from 60 m. S of Valparaiso to 80 m. N of it; and throughout the whole of this line extensive beds of shells were found, elevated at various heights, from the level of the present sea-beach to 250 ft. above it. The most unequivocal proof of a recent rise is drawn from *balanidæ* found adhering to the rock at the height of 14 ft. above high-water mark. Mr. Darwin also observed *balanidæ* similarly situated at the Rapel, 60 m. S of Valparaiso, and at Quintero, a few miles to the N of it. The present position of an ancient sea-wall, which was built in 1680 at V., cannot be otherwise explained than by a change of level. The appearance of the granitic rocks, both to the N and

S of the bay, also bear testimony, in Mr. Darwin's opinion, to the fact of an elevation to the amount of about 14 ft. The church of San Augustin was found to stand 19 ft. 6 inches above high-water mark; allowing for its probable position when built in 1614, the greatest amount of possible change cannot have exceeded 15 ft. in the period of 220 years. Mr. Darwin considers it certain that the land was elevated during the earthquake of 1820. The rise of the land, even in the bay of Valparaiso, was not equal; for a part of a fort which was not formerly visible from a certain spot, subsequently to the earthquake fell within the line of vision. There is good reason to believe that part of the most recent rise of the land, attested by the *balanidæ* and position of the sea-wall, has been due to changes acting previously to 1822, as well as to that earthquake itself, and likewise to an elevation produced by insensible degrees since that period.

VALPARAISO-DE-ABAJO, a market-town of Spain, in the prov. of Cuenca, and to the W of Huete. Pop. 540.

VALPARAISO-DE-ARRIBA, a market-town of Spain, in the prov. of Cuenca, near Huete. Pop. 210.

VALPELLINA, a village of Sardinia, in the prov. of Aosta. Pop. 535.

VALPERPA, a market-town of Sardinia, in the prov. and 11 m. SSW of Ivrea. Pop. 3,500.

VALPO, a town of Austria, in Slavonia, in the comitat of Vezocze, on an affluent and near the S bank of the Drave, and 20 m. WNW of Esseg.

VALREAS, a canton, commune, and town of France, in the dep. of the Vaucluse, and arrond. of Orange. The cant. comprises 4 com. Pop. in 1831, 8,425; in 1846, 9,060. The town is 24 m. NNE of Orange, on the Coronne. Pop. in 1846, 4,690. It has manufactories of silk and of leather. Oil, silk, madder and chestnuts are cultivated in the environs. V. was the birth-place of Cardinal Maury.

VALS, a commune and town of France, in the dep. of the Ardeche, and cant. of Aubenas, 21 m. SW of Privas, on the l. bank of the Ardeche, near the confluence of the Volane. Pop. in 1846, 2,834. It has mineral baths, silk and paper-mills. The vine and mulberry, olive and almond trees are cultivated in the locality. In the vicinity on the Volane are basaltic rocks called the Chaussée des Geans.—Also a village of Austria, in the Tyrol, in the circle of Bruneck. Pop. 360.

VAL-SASSINA, a district of Austrian Italy, adjoining Lake Como, between the Grisons and the Valteline. Its length is about 18 m. Though very mountainous, it has some fruitful vales. The German family of Tour-and-Taxis take the title of dukes of Valsassina.

VALSEQUILLO, a town of the island of Gran Canaria, in the Canary group. Pop. 2,700.

VALSESIA, or VALLESESIA, a mountainous territory of Piedmont, in the prov. of Nevara, extending from the source of the Sesia on the NW, to the Lago-d'Orta on the SE, about 36 m. Of its whole surface only 17,700 hectares are cultivated. Pop. 35,145. It is administratively divided into the 3 mandamenti of Borgosesia, Varallo, and Scopio. It physically comprehends four principal valleys, respectively denominated the Valle-Grande-della-Sesia; Valle-Piccola, which is watered by the torrent Sermenta; Valle-Mastallone, bordering the torrent of that name; and Valle-d'Uggia, commonly called Valduggia, irrigated by the torrents Strona and Uggia. The first of these—the great valley of the Sesia—commences above Alagna, its highest village at the foot of the Alps, and extends down the river to Ponte-di-San Quirico, below Borgosesia; the torrent Sermenta and the Valle-Piccola descend southward, or nearly so, from the chain of mountains which divides this prov. from the valley of Anzasca, the torrent joining the Sesia at Balmuccia; and the Mastallone, which has its source in the same direction, but more eastward, runs parallel with it and falls into that river at Varallo. The Strona, bordered by the Valle-d'Uggia, rises in Monte-Navigno, on the W of the Lago-d'Orta, and issues in the Sesia at Borgosesia. Besides these there are about twenty other valleys or inlets of smaller extent and isolated among the folds of the mountains, difficult of access, but generally watered by little tributary streams. There is a good deal of manufacturing industry in this district, employing 8 paper-mills, 10 cotton factories, 5 hat-works, and 5 establishments for dressing and dyeing skins. At Valduggia there is a bell-foundry; and Varallo is famous for its brass-castings.

VALSYIO, a town of Hungary, in the com. of Baranya, 5 m. W of Siklos.

VALTELINE, a lordship of Austrian Italy, in the gov. of Milan, now forming the greater part of the delegation of Sondrio. Its superficial extent is about 1,270 sq. m. It consists of a long valley traversed by the Adda, and surrounded by the Alps in all directions, except where it opens to the lake of Como on the W. Cattle form an object of export; silk is cultivated to a considerable amount. The inhabitants are of Italian origin; but were politically alienated to the Grisons until 1797, when their country was incorporated with the dependances of France in Italy, as afterwards, in 1815, with the dominions of Austria. The chief town is Sondrio. See SONDRIO.

VALTIERRA, a town of Spain, in Navarre, 8 m. NW of Tudela. Pop. 1,500.

VALTOUMANCHE, a village of the Sardinian states, 18 m. NE of Aosta. Pop. 1,500.

VALTRI, a town of the Sardinian states, in the duchy of Genoa, among the Apennines.

VALTRIE (LA), a small stream in Lower Canada, which falls into the St. Lawrence from the north, 40 m. below Montreal.

VALUIKI, a town of Russia, in the gov. of Voronetz, in N lat. 50° 12′ 34″. Pop. of circle 4,000.

VALVERDE, a town of Spain, in the prov. and 14 m. S of Badajos. Pop. 909.—Also a town in the prov. and 25 m. S. of Cuença. Pop. 1,700.—Also a town in the prov. of Cuença, on the Guadalquivir, 18 m. NNE of Seville.—Also a town of the prov. of Huelva, in the Sierra Morena, 21 m. N of Moguer, and 50 m. W by N of Seville. Pop. 5,200.

VAMKAOSE, a small island in the Chinese archipelago, 62 m. SW of Macao, where the celebrated St. Francis Xavier was buried.

VAN, a large and fortified city of Turkish Armenia, situated about 1 hour's distance from the E shore of a lake of the same name, 145 m. SE of Erzerum. It is surrounded with a wall of mud and stone, with large round and small square bastions, and a deep ditch, and has four gates. Cotton and corn are imported from Persia. To the N of the city, on a high, narrow, inland rock rising abruptly from the plain, and about 600 yds. in its entire length, stands the castle, which can be approached only by one passage admitting no more than two persons abreast, and, unlike most Turkish strongholds, is not commanded by neighbouring heights. The city is well built; the houses are of stone and tile; and the streets though narrow are well paved and comparatively clean. Most of the private dwellings have each their own inner court, and are entered through an outer gate. The town contains two large churches, four mosques, two baths, and two caravanserais. The great charm of V. are its gardens which stretch over a level area of 7 m. by 4 m. between the city and the mountains to the E. It is abundantly supplied with water. Its pop. has been recently stated at 15,000, two-thirds of which are Turks, and the remainder Kurds and Armenians. The chief occupation of the inhabitants consists in the manufacture of coarse cotton chintzes. V. is said to have been founded by Semiramis. In 1533 it was taken by Solyman.

VAN (LAKE OF), the Dzow-WANAÏ of the Armenians, and VAN-GOL of the Turks, a large inland sheet of water in Turkish Armenia, between the parallels of 38° and 39°. In the maps of D'Anville it is represented as occupying a space of 90 m. long by more than half that in breadth; and in that of Rennel it is nigh 60 m. in length by 30 m. in breadth; Kinneir prodigiously reduced its dimensions, making its greatest length only 30 m., and its breadth only from 9 to 12 m. The latter authority says that a vessel with a fair wind can sail the whole length of the lake in 4 hours, while he states, at the same time, that its bays and inlets are so numerous that to make the circumference of the lake would require a journey of 90 hours or 300 m. Recent observations report its length at about 70 m. from NE to SW; its extreme breadth at 28 m., and its alt. above sea-level 5,467 ft. It contains several islands. A few crazy boats navigate the lake, which abounds in fish and water-fowl. The water is brackish. The Bendi-Mahi-Su flows into the NE arm of this lake; the Kara-Chai into its E side; the Choseh-Ab into its SE extremity. But all these streams are of small extent. The Nimrud-Dagh and the Sapan-Dagh divide its basin, on the NW and N, from the basin of the Murad-Chai or Upper Euphrates. This lake is the *Arsissa Palus* of Ptolemy, probably so named from the city of Arzis or Arjish, on its N shore. It is called *Beznunius* by Moses of Chorene, and the district round it *Beznunia.* It is not mentioned by either Pliny or Strabo.

VAN-BUREN, a central county of the state of Arkansas, U. S., comprising an area of 1,456 sq. m., drained by Little Red river and its branches. Pop. in 1840, 1,518; in 1850, 2,864. Its capital is Clinton.—Also a county in the SE part of the state of Iowa, comprising an area of 450 sq. m., drained by Des Moines river and its branches. It is undulating but generally fertile. Pop. in 1840, 6,146; in 1850, 12,270. Its capital is Keosanque.—Also a county in the SW part of the S peninsula of the state of Michigan, comprising an area of 620 sq. m., drained by Pawpaw river and its branches; by the S. branch of Black river; and by Brush and Dowagiake creeks, and intersected by the Michigan Central railway. The surface is level, and the soil fertile. Pop. in 1840, 1,910; in 1850, 5,800. Its capital is Pawpaw.—Also a central county of the state of Tennessee, comprising an area of 312 sq. m., drained by Laurel creek, and by other branches of Caney fork of Cumberland river, which flows on the N border. The surface is undulating, and the soil fertile. Pop. in 1850, 2,674. Its capital is Spencer.—Also a county in the W part of the state of Missouri, comprising an area of 648 sq. m., drained by Grand river and Big creek, and consisting chiefly of prairie. Pop. in 1850, 4,693. Its capital is Harrisonville.—Also a township of Onondaga co., in the state of New York, on the S side of Seneca river, and 144 m. N by W of Albany. It has an undulating surface, drained by Camp brook and other affluents of Seneca river, and intersected by the Oswego and Syracuse railway. Pop. in 1840, 3,021; in 1850, 3,873.—Also a township of Hancock co., in the state of Ohio, 96 m. NNW of Columbus. Pop. in 1840, 402.—Also a township of Drake co., in the same state Pop. 421.—Also a township of Shelby co., in the same state. Pop. 596.—Also a township of Wayne co., in the state of Michigan. Pop. 940.—Also a township of Brown co., in the state of Indiana. Pop. 417.—Also a township of Lagrange co., in the same state. Pop. 284.—Also a village of Crawford co., in the state of Arkansas, on the N side of Arkansas river, and 124 m. WNW of Little Roch. Pop. in 1840, 500; in 1850, 547.—Also a village of Clay co., in the state of Indiana, 55 m. WSW of Indianapolis, and near the Terre-Haute and Richmond railway. Pop. in 1840, 538.—Also a village of Itawamba co., in the state of Mississippi, on the W side of Tombigbee river, and 165 m. NE by N of Jackson. Pop. in 1850, 100.—Also a village of Ripling co., in the state of Missouri, on the E bank of Currant river, and 123 m. SE by S of Jefferson.

VAN-BUREN HARBOUR, a village of Chautauque co., in the state of New York, U. S., on the E side of Lake Erie, and near the Buffalo and State Line railway.

VANCOUVER'S ISLAND, or QUADRA, a large island off the coast of North-West America, extending from N lat. 48° 17′ to 50° 55′, and from W long. 123° 10′ to 128° 30′. Its extreme length from NW to SE is 276 m.; its breadth varies from 50 m. to 65 m. On the E and NE it is separated from the mainland of British America by the gulf of Georgia and Queen Charlotte's sound; on the S, the strait of San-Juan-de-Fuca separates it from the Oregon territory of the United States. Little is yet known of the interior of this extensive island, but its natural resources and capabilities are reported to be very considerable; and it occupies a highly important position both with regard to the adjoining territories, the Japan and China trade, and the islands of the Pacific, and may probably at some future period command a great portion of the trade between the archipelago of the Pacific and the continent of America. It is only 4,078 m. distant from England, and but a few days sail from California and the Sandwich islands. It possesses numerous excellent harbours, the sea and rivers abound with fish, and whales are found in great numbers near the coast. Coal and limestone are abundant, and it produces the finest timber of several kinds, including the oak, pine, cedar, beech, and ash. The climate resembles that of England, but is milder, and the soil, so far as cultivation has yet extended, proves to be well adapted for the production of wheat and other crops grown in England. Stock of all kinds is easily reared, and thrives well. Capt. Wilkes, of the United States' navy, speaking of the country generally, estimates the produce, whether from farm or garden, of the finest character. He tells us, that the wheat produced weighs 63 lbs. to the bushel, and 600 acres produced 7,000 bushels; that barley yields 20 bushels to the acre, though oats do not thrive well; but pease, beans, and potatoes yield abundantly; strawberries and gooseberries, the former nearly ripe, and salad gone to seed, were seen by him at Nisqually on the 15th May. Upon the whole, he considers V. three times more fertile than the United States; cattle, he says, find natural hay all the year round, and multiply with astonishing rapidity. The climate, even in the northern part, is considerably milder than that of England; and for settlers the facilities, both as regards the raising of produce and the erection of works and dwellings, appear to be greater than had been anticipated. Game is everywhere abundant, and is calculated to prove an important item in economising domestic expenditure, while, in addition to wild fowl on the waters, there are many valuable fisheries, besides oyster-beds. Deer are often met with, and there are also some black bears, which are considered harmless. One of the most remarkable peculiarities of the island is the great number of inlets, or arms of the sea, which, being deep and narrow, and penetrating a considerable distance inland, furnish in many instances, from the great rush of water in and out, an important and permanent motive power. The natural features of the country are described as being very beautiful, owing to its lakes, hills, woodlands, and occasional patches of prairie. The oak is abundant, frequently growing in glades, with park-like regularity. There is also a cypress which grows to a large size, and is particularly valuable for building purposes. The pines are of three sorts, and are found 160 ft. in height, and from 12 to 16 ft. in circumf. One description of pine is extremely plentiful, and possesses superior qualities for ship-building purposes. Varieties of maple and other trees are likewise common, and the valuable hemp known as the *urtica canabina* grows wild and luxuriantly in the woods. Artificial grasses, it is believed, might be cultivated with great advantage, and seeds of several descriptions have been sent out. The soil generally is a dark vegetable mould, averaging 18 inches in depth, and unusually fertile. Contrary to what is seen in England, the best lands are mostly those where pine trees grow. The *flora* of the island is much the same as that of Great Britain. With regard to its mineral resources, the principal rocks are stated to be of an excellent kind for building materials, both as respects appearance and durability. It is also considered that the geological features of the districts surveyed are all such as strongly to favour the anticipation that gold may be found throughout an extensive range. There is an abundance of felspar and quartz. Ironstone is likewise found in several places, but its value has not yet been determined. The quantity of iron is sufficient to affect the compass considerably, and captains of vessels who have visited the island are of opinion that this disturbance is unusually great along its shores. The native Indians behave well, and a small party of settlers penetrating into the interior with a canoe met a number of them, by whom they were warmly welcomed. They described a large lake, which no white man had ever yet seen, but which has since been visited, and found to be between 20 and 30 m. in length, and from 2 to 3 m. in breadth. A considerable quantity of wheat is raised, which is chiefly bought by the Hudson's Bay company, and exported in flour to the Sandwich islands, California, and to the Russian settlement of Sitka. The representatives of the Hudson's Bay company at Fort Vancouver, which stands about 100 m. from the mouth of the Columbia, and is the company's chief establishment on the W side of the Rocky mountains, have ingratiated themselves with the natives, with whom they carry on an extensive trade in skins. The port of Camosack, or Victoria, with its excellent harbour, promises to become of great importance as a port of refuge to ships trading along the W coast of America. The point of scientific interest, however, is the coal-field of the NE district, which is worked so near the surface, that the Cormorant steam-sloop was supplied by the natives with 62 tons of coal within three days. Specimens of this coal have been examined for the Admiralty, and although it yields a considerable per-centage of ash, it is reported to be not much inferior to the coal of South Wales.

The native pop. has been estimated at about 12,000 Indians of various tribes. The number of settlers in 1852, most of whom were sent out by the Hudson's Bay company and by the Puget Sound company, was about 400. The Hudson's Bay company having obtained from the Crown a grant of this island, in 1849, under the condition of colonizing it, but revocable at the end of 11 years, it is of importance that emigrants should know upon what terms and conditions they can settle and obtain subgrants of land on the island. They are 1. That no grant of land shall contain less than 20 acres. 2. That purchasers of land shall pay to the Hudson's Bay company the sum of £1 per acre for the land sold to them, to be held in free and common soccage. 3. That purchasers shall provide a passage to Vancouver's island for themselves and their families; or be provided with a passage—if they prefer it—on paying for the same at a reasonable rate. 4. That purchasers of larger quantities of land shall pay the same price per acre, namely, £1, and shall take out with them 5 single men, or 3 married couples, for every 100 acres. 5. That all minerals, wherever found, shall belong to the company, who shall have the right of digging for the same, compensation being made to the owner of the soil for any injury done to the surface; but that the owner shall have the privilege of working for his own benefit any coal mine that may be on his land, on payment of a royalty of 2s. 6d. per ton. 6. That every freeholder shall enjoy the right of fishing all sorts of fish in the seas, bays, and inlets of, or surrounding, the said island; and that all the ports and harbours shall be open and free to them, and to all nations, either trading or seeking shelter therein. In 1852, the Hudson's Bay company reported that, up to the 27th of April in that year, there had been sold 1,478½ acres, to 11 persons, and that the fur-trade branch of the Hudson's bay company was in possession, under the sanction of her Majesty's government, of 3,084 acres, which were occupied by them previous to the date of the boundary treaty with the United States, of which they had sold portions to some of their retired servants, who have settled themselves upon it. The company further reported that to obviate the difficulty of providing the means of subsistence for settlers on their first arrival, the Puget Sound company had made arrangements with four competent and experienced agriculturists to cultivate four farms. The Puget Sound company supply the capital, and give an interest in the profits of cultivations to the persons employed to conduct the farms with the option of purchasing them at their cost price. There had been sold and exported by the Hudson's Bay company 1,315 tons of coals, which were collected from the surface seams by the Indians; the company had incurred considerable expense in searching for coal, by boring or otherwise, without success until lately, when promising appearances had been discovered about 80 m. N of Fort Victoria, on the E coast of the island, nearly opposite the mouth of Frazer's river. The company stated that the high rate of wages in Oregon and California, and the attraction of the gold districts, in the latter country, had not only operated to prevent persons of capital settling in Vancouver's island, but had also obstructed the Hudson's Bay company and the Puget Sound company in their endeavours to bring land into cultivation, and provide means of subsistence for settlers: hitherto it had been necessary to import flour for that purpose, and for the use of the company's establishments on the island.

The government of the island has been arranged on the following constitutional basis:—The governor is appointed by the Crown, with a council of 7 members, likewise so appointed. The governor is authorised to call assemblies, to be elected by inhabitants holding 20 acres of freehold land; but it is left to the discretion of the governor to fix the number of representatives; and to divide the island into electoral districts, if he shall think such division necessary. The governor has the usual powers of proroguing, or dissolving such assembly. Laws will be passed by the governor, council, and assembly. The legislature, thus constituted, will have full power to impose taxes, regulate the affairs of the island, and to modify its institutions, subject to the usual control of the Crown.

VANCOUVER (Cape), a headland of Western Australia, in Plantagenet co., between King George sound on the W, and Port Free People on the N, in S lat. 32° 2′.

VANDALIA, a village of Fayette co., in the state of Illinois, U. S., on the W bank of Kaskaskia river, and 66 m. SE by S of Springfield, in N lat. 38° 50′, and W long. 89° 2′. It is intersected by the Illinois Central, the Massac, and Sangamon, and the Mississippi and Atlantic railroads. Pop. in 1840, 800; in 1850, 1,100.

VANDANS, a village of Austria, in the Tyrol, district of Vorarlberg and bail. of Montafon. Pop. 900.

VANDEL, a parish of Norway, in the bail. of Romsdal, and on the Vandelv-fiord. Pop. 1,480.

VANDENESSE, a commune of France, in the dep. of the Nievre, and cant. of Moulins-Engibert, 18 m. SW of Château-Chinon, on a river of the same name. Pop. 1,718. It has a blast-furnace, forges, and foundry.

VANDENESSE-LES-CHAROLLES, a village of France, in the dep. of the Saone-et-Loire, cant. and 4 m. E of Charolles, near the Semeuse. Pop. 1,630.

VANDERBURG, a county in the SW part of the state of Indiana, U. S., comprising an area of 240 sq. m., drained by Big Pigeon, Blue Grass, Little, and Locust creeks, and intersected by the Wabash and Erie canal, and by the Evansville and Illinois railway. Pop. in 1840, 6,250; in 1850, 11,415.

VANDERLIN, an island of the gulf of Carpentaria, of which the N extremity, Cape Vanderlin, is in S lat. 15° 34′ 30″, and E. long. 137° 8′.

VANDERSCHELLING, a group of islands in the N. Pacific, near the N coast of New Guinea, in the bay of Geelvink, in N lat. 3° 32′, E long. 136° 17′.

VAN DIEMEN'S LAND, or Tasmania, a large island which lies to the southward of the eastern part of Australia, between 40° 45′ and 44° S lat., and 145° and 148° 30′ E long. Its form is rudely triangular, and its superficial extent is estimated at 23,437 sq. m. Its average length is 165 m., its average breadth 155 m. Approached from the E, it has a picturesque and beautiful appearance from the sea, presenting an endless succession of lofty mountains covered to their summits with wood; while tall rocks and precipices, glens and hills, give additional interest to the scenery of this romantic island; but the W side is wild and barren. On traversing the island, it is found to present a constant alternation of hill and dale, with occasional flats or plains; but these are comparatively few in number, though some of them are of great extent, consisting in several instances of not less than from 3,000 to 10,000 acres. "The interior," says Mr Dixon, "is characterised by its uncommon features, being in many respects unlike any other part of the world. The surface heaves up into irregular ranges of mountain-scenery, crowded over and intersected in all fashions and directions. We never see one mountain alone; but where one is, there are chains of others running up, round, and about, in open wildness and disorder; towering here and sinking there, in bewildering yet striking confusion. From April to October, the highest are capped with snow. The whole face of the country is covered thickly with trees of immense height and circumference, growing close together, and reaching to a great loftiness before they shoot out their branches. Their leaves are unfading, but dusky and mournful, and seem in the distance nearly black, throwing an air of heavy gloom over the face of nature. The valleys are circumscribed within narrow limits, and, like the mountains, spread over with high sturdy forests. Those tracts which have never been disturbed in their primitive wildness, are called, not the forest, but 'the bush.' The native tree and shrub are not subject to the annual decay and fall of leaf which are common in other regions; but here the leaf is as lasting as the stem, and both live and die together. Nature, too, clothes herself with a constant verdure, and even in the midst of winter retains her greenness. While the earth is covered often ankle-deep in snow, the grass, the flowers, the shrubs, and the trees remain unchanged, and are as green and as healthy as in the middle of summer. But notwithstanding its exterior garment, no country can be imagined more dreary than 'the bush' of Van Diemen's Land. Travel through it hundreds of miles and your prospect is the same. Tiers of mountains still environ you, and you see nothing but a wearisome uniformity. Nothing in the shape of diversity presents itself, either to attract the eye or captivate the imagination. The stately gum-tree, with its dark and heavy foliage, shoots up before you wherever you turn your head; birds flutter in the branches, and at intervals emit a wild chirp, a dolorous scream, or a dull caw; but the note is without melody—harsh and displeasing to the ear. The plumage of many of the species, however, is pretty, and that of the cockatoo very handsome. Tracts of land are at times seen, bursting with fertility, and many more rocky, sandy, and sterile. Some have a park-like appearance, free from underwood and obstruction, and others are so overgrown with these as to be almost impervious. A few of the native shrubs are pretty, but so sparingly scattered that they have to be sought out. The wild-flowers are in small variety; but some of those that exist are very handsome when closely examined, although, being all diminutive, they are often trampled over. The indigenous trees grow up to an enormous height. The gum is the highest and greatest, and in height and circumference is reckoned to be the biggest in the world. It grows straight upward and shoots out its branches above. The root works deep in the earth, and spreads itself about in all ways and to some distance. The variety of trees is not great, but the gum is that which predominates. The foliage of all is scanty, the leaves small and far apart, and such as to afford no shade in hot weather."

Shores, Bays, &c.] The shores of Van Diemen's Land are bolder in character than those of New Holland, and advantageously broken up by deep bays and inlets. The NE shore, near Cape Portland, consists of a sandy beach occasionally broken by rocky headlands; behind it are gently rising hills; and over their green slopes the mountains of the interior rear their lofty and naked summits. As the entrance of the river Tamar is approached, the country becomes fertile, and continues so for a considerable distance to the westward. To the west of Circular Head—a remarkable promontory running 12 m. out into the sea—the shore is low and sandy, and the interior covered with heath. Cape Grim, a high, bold, dark-coloured cliff, forms the NW extremity of the island. About 70 m. to the S of this, is a narrow inlet which has not yet been explored, but which is thought to be the estuary of a considerable river. About the middle of the W shore is situated Macquarie harbour, opening from the NW, and deeply intersecting the land in a SSE direction. To the S of this the coast is wild and rocky. The southern coast, taking its general outline, is the most elevated of any. The rocks at South-West cape are remarkably bold. About 16 m. to the E of this cape lies D'Entrecasteaux's channel, which, passing between the north end of Bruni island and a peninsula from the main island, reaches a considerable way above Hobart town, under the name of the Derwent river, and presents a succession of fine harbours. Cape Pillar is a striking basaltic point on the SE corner of the island, which has to be doubled by vessels to and from Sidney. To the N of Pirate's bay, the land is barren, low, and sandy.

Mountains and Rivers.] Van Diemen's Land consists of two ridges, with a comparatively low country in the middle. It is diversified in features, but upon the whole is mountainous in character. According to Count Strzelecki, the space occupied by granite, porphyry, quartz, and other crystalline rocks, is to that of the limestone, clay, slate, and other sedimentary rocks, as 7 to 1. One ridge appears to stretch in the direction of Cape Grim and South cape above 230 m., dividing the streams that run W, from those of the Derwent and the Tamar. Snow lies upon one of the peaks of this ridge, in S lat. 43° 25′, called Mount Humboldt, which has an alt. of 5,520 ft., for some months of the year. On Table-Mountain, or Mount-Wellington, which rises to 4195 ft., about the height of Ben-Nevis in Scotland, snow lies for three-fourths of the year. The southern mountains near Port Davey are very lofty; but this SW portion of the island has yet been but imperfectly explored.—The three principal rivers of Van Diemen's Land are the Derwent, running to the S; the Tamar, running to the N; and the Huon, which runs W, and joins the sea not many miles from Hobart-Town. The remotest source of the Derwent is a lake, in S lat. 52°, within 50 m. of the N coast. This river drains an extent of country amounting to nearly 12,000 sq. m. The fresh water which gets the name of the Derwent is not the principal branch; the main stem of the river is the Ouse. The Clyde and the Jordan are the other principal branches of the Derwent on the E side. The Macquarie is the largest branch of the Tamar, and passes through fine rich grounds. The North Esk and South Esk also flow into or rather form the Tamar, which flows NNW to Port Dalrymple. The Gordon and the King flow into Macquarie harbour on the W coast. About the parallel of 43° there are numerous lakes, one of which, Lake Clarence, is 15 m. long and 5 m. broad.

Climate.] Van Diemen's Land enjoys a salubrious climate, well adapted to European constitutions. If the summer heat in New South Wales be taken at 90°, that of Van Diemen's Land cannot be rated much above 70°; although there are times when the therm. exhibits sudden elevations, even to 100° or 110°. The mean annual heat of Hobart-Town is 52°; of summer 63°, of winter 42°. Thunder-storms are seldom experienced. September, October, and November form the spring, March, April, and May the autumn. The average of the therm. for these months is from 50° to 60°. The average number of days on which rain falls throughout the year, is from 50 to 60. Frosts are sometimes severe. The farmer sows his grain in July, August, and September, which are the spring months; in October he prepares the land for Swedish turnips; in November he gets in his potato and turnip crops; December is the height of his hay-harvest. At about the middle of January his wheat-harvest commences, and continues through February; in March he pays attention to his fallowing and husbandry; in April he gathers his second crop of potatoes; in May he lays down his English grasses; and in June he continues his ploughing and harrowing.

Vegetation.] The natural trees of this island are nearly the same with those of New Holland; the gigantic blue gum is the prevailing tree; but neither mahogany nor rose-wood has been found here. There is, however, a species of oak called black-wood, which, with the Huon pine, the black and silver mimosas, and the pencil-cedar, serve as good substitutes for these useful trees. All the trees are evergreens; but the foliage is generally of a dark and sombre green. They often attain a vast size and height; but there is little variety in their forms. The dog-wood, pink-wood, and musk-wood, are fine-grained trees. The variety of shrubs is great, and some of them are extremely beautiful, particularly the native cherry and the fern. The myrtle often forms dense forests, and will occasionally attain a huge girth. The indigenous botany is, like that of New Holland, exceedingly scanty in articles fit for human sustenance. The sands produce a species of plantago which forms a good salad, and is one of the most useful plants that this island furnishes. In the woods of the interior, a new species of *ficoides* is found, the fruit of which is eaten by the natives; but the delicious fruits that abound in latitudes nearer the equator, and attain such perfection even in New South Wales, are unknown here. A large species of truffle occurs, which in a roasted state forms a good substitute for bread.

Animals.] The native animals are few and diminutive; and, like those of Australia, chiefly of the marsupial genus. The three different species of kangaroo, the opossum, the squirrel, and the wild cat, are the most numerous. The native dog, so well known in New South Wales, is not found here. The hyæna-opossum, though it flies from man with the timidity of a hare, is extremely destructive to the flocks, amongst which it frequently commits the most dreadful havoc. This animal attains considerable size, having been found in many instances to measure six feet from the snout to the extremity of the tail. The *Dasyurus ursinus* is found in the interior.—The birds have fine plumage. We here find roselle parrots,—black, white, and satin cockatoos,—quails, widgeons, mountain and wild ducks, eagles, black swans, emus, and a bird resembling the thrush called the wattle-bird. Pheasants, partridges, snipes, and wood-cocks, have been imported from England for the amusement of the colonists; and the hunting of the emu also affords excellent sport.—Some though not all of the poisonous serpents found in New Holland are also seen here; in the reptile-family may also be mentioned guanas and lizards, centipedes, scorpions, and tarantulas. The wasp of Van Diemen's Land is a smaller but much more splendid insect than the English wasp; it has four orange-coloured wings, and horns and legs of the same colour, a hard body, and a formidable sting. Bees have been recently introduced, and promise to do well. The rivers abound with mullet, perch, craw-fish, and oysters; and the surrounding seas with whales, dolphins, and seals.

Soil.] From the hilly nature of the country, there is but a comparatively small proportion of it adapted for the plough, though presenting abundance of excellent pasturage. The extent of really available land throughout the explored part of the island has been estimated at one-third of the whole, and this again divided into four parts, giving one for the plough and the other three for pasture; thus, out of 1,000 acres of land, about 100 will be found fit for cultivation, and from 300 to 400 for grazing. Of the whole island, which comprises 16,000,000 acres, there will be about 3,000,000 acres located, or private property; the remaining 13,000,000 acres are Crown land, about 2,000,000 acres of which are occupied under grazing-licenses. The staple crops are wheat and potatoes. Wheat may be considered the chief export of the colony; it is sent to Sydney, the Mauritius, and Rio Janeiro. The principal corn-district is at Port Dalrymple, on the NE side of the island, 150 miles from Hobart-Town. Potatoes and hops are cultivated with success; and all the choicest vegetables of an English garden are raised with great facility, and pay uncommonly well for the expense and labour of planting. The seed-time is a month or six weeks later than in New South Wales, though wheat may be sown in November and reaped in March. The farms in the neighbourhood of Hobart-Town average little more than 50 acres each; in the district of New Norfolk they often exceed 2,000 acres. The average returns per acre in this latter district are,—wheat, 15 bushels; barley, 18; oats, 30; pease, 18; potatoes, 4¼ tons; turnips, 4½ tons. The whole quantity of wheat grown upon the island in 1831, was estimated at 383,000 bushels, with an excess over the consumption of the preceding year of 70,000 bushels, making in all a stock of 453,000 bushels; while the consumption was not reckoned at more than 250,000, thus leaving a surplus of 253,000 bushels. In 1836, 485,969 bushels were raised. In 1845, the land in cultivation was 140,949 acres.—The live stock, in 1850, was estimated at 2,000,000 sheep, 85,000 cattle, and 16,000 horses.

Minerals.] This island produces copper, iron, alum, coal, slate, limestone, asbestos, and basaltes. It also affords cornelian, rock crystal, chrysolite, jasper, marble, and many petrifactions. In the neighbourhood of Launceston there are mountains of iron-ore, which must prove a source of great wealth to the island when once it is sufficiently populous. Coal also is met with in extensive beds, particularly near Port-Arthur, and over the E division of the island. Limestone is common, and of excellent quality. In 1836, 47,600 bushels of lime were manufactured. In the district of Oatlands are three small lakes, or ponds, which are so strongly impregnated with salt that they yield, by a natural process, many tons of that essential article of life annually. This salt is not equal in quality to English salt, but nevertheless brings 10s. a hundred-weight in the colony.

Population.] The colonial pop. of Van Diemen's Land in 1821 was 7,185; in 1830, 10,390; in 1835 it was 41,512, of which 18,268 were convicts. In 1840 the free inhabitants were 28,294. In March 1851, the pop. was returned at 70,130 in the proportion of 43,127 males to 25,482 females, excluding the military, showing an excess of males to the number of 17,645. The military numbered 953 men, women, and children. Allowing for 2,813 convicts taken twice in the census of 1847, the total increase in the pop. of both sexes since that date, amounted to 2,779. The free pop. had been increased by 7,055 souls, but the decrease of convicts amounted to 4,276. Of the general total of the male pop. 10,649 were free and born in the colony. The bond male population were numbered in the following proportion—

Holding tickets of leave,	7,295
In government employment,	1,723
In private assignment,	1,816
In private employment,	1,660
In public works,	568
Total,	13,052

The total of free females was 21,445; of bond as follows:

Holding tickets of leave,	1,254
In government employment,	1,312
In private assignment,	604
In private employment,	867
Total,	4,037

Of the total female pop. 9,771 were married, and 15,709 were single. Excess of single males, 17,638. The professed religions of the military and convicts on public works are not stated, but the remainder of the population are classed as follows:—

Church of England,	45,073
Church of Scotland,	4,485
Wesleyan Methodists,	3,772
Other Protestant dissenters,	2,379
Roman Catholics,	12,444
Jews,	435
Mahommedans and Pagans,	21

We find 39,611 persons whose occupations are not included in the classes enumerated as under:—

Landed proprietors, merchants, bankers, and professional persons,	1,577
Shopkeepers and other retail dealers,	1,415
Mechanics and artificers,	5,687
Farmers,	2,533
Market gardeners,	281
Shepherds and others in care of sheep,	1,445
Gardeners, stockmen, and farm servants,	9,731
Domestic servants,	5,600
Licensed boatmen,	346
Seamen employed in coasting and river craft,	341
Whalers,	42

The total number of houses in the colony was 11,844; of which 5,723 were built of stone or brick, and 6,121 of wood. The total number of unfinished houses was 561, and 599 were uninhabited. From another return it appears that the number of free males of the ages of 21 years and upwards in the 15 electoral districts were 18,130, and the number of inhabited houses 11,392. In each district each number is as follows:—

	Males.	Houses.
Hobart-Town,	4,097	3,127
Launceston,	2,381	6,644
Cornwall,	800	596
Westbury,	1,391	755
Longford,	957	424
Morven,	949	516
Campbell-Town,	872	471
Oatlands,	898	452
Cumberland,	822	462
Richmond,	992	417
Brighton,	702	409
Sorell,	940	476
Buckingham,	967	714
New Norfolk,	603	362
Huon,	805	567

In the remoter settlements, the aborigines proved themselves very destructive and implacable. They are in complexion perfectly black; their features are flat, and hair woolly. All attempts to induce them to leave their native woods, and mingle in a friendly manner with the colonists, were yet more ineffectual than in the case of the natives of New South Wales, and in 1854 it was calculated that not above 28 of the aboriginal race were in existence.

In October 1848 there were receiving from the public funds, and under the immediate direction or general surveillance of the government, 65 schools, containing 3,147 children, of whom 1,733 were boys, and 1,414 were girls; and 2,323 pupils were in attendance at 100 private schools in various parts of the island. The total number of children from 2 to 14 years of age, in the various districts of the island, according to the census, appeared to be 12,994; while the number of children of these ages attending at public day-schools was only 2,917.

Government.] Up to the year 1825, Van Diemen's Land was merely a dependency on New South Wales. In that year, on the petition of the colonists, the colony was declared amenable to the mother-country only. Until 1850, the lieutenant-governor of this colony was assisted by two councils, an executive and legislative: the latter consisting of 15 members named by the king. Its duty was to make laws and ordinances for the well-being of the colony, such laws being always submitted to parliament within six weeks after the opening of each session. By 13th and 14th Vict., c. 59, the government is vested in a governor and executive council nominated by the Crown, and a legislative council of at least 24 members, of whom one-third is nominated by the Crown, and two-thirds are elected. There are also here a chief-justice, attorney-general, and all the other appendages of a supreme court of judicature,—courts-of-requests, attorneys, barristers, solicitors, proctors, sheriffs, justices of the peace, and the whole of the paraphernalia of civil and criminal jurisprudence known in this country. There are besides, as in New South Wales, a number of stipendiary police-magistrates, each having a separate and distinct district under his judicial authority. There is generally one entire regiment of the line on this island.

Commerce.] Van Diemen's Land produces for exportation corn, wool, oil, timber, mimosa bark, cattle, and sheep, and salt meat and hides. Grain is exported to New South Wales, the Swan river, the Mauritius, and recently to England. An article of more regular export is wool; next in order of value is the produce of the whale-fishery. The progressive increase of exports to the mother-country may be seen in the following table:—

	In 1827.	1831.	1836.
Bark for tanning,	9.122 cwts.	39.264 cwts.	1.542 tons.
Train oil and sperm,	179 tons.	848 tons.	598,224 galls.
Whale fins,	168 cwts.	818 cwts.	2.842 cwts.
Sheep's wool,	192,075 lbs.	1,359,203 lbs.	1,727,258 lbs.

The total value of the imports in 1836 was £332,548; in 1848, £490,281. The bulk of the imports is from the mother-country, comprehending spirits, malt-liquors, ironmongery, woollens, linens, stationery, and soap. Sugar is imported from the Isle of France; tea from China; silks, rice, wine, and spices, are imported from the East Indies; tobacco from Rio; cheese, oranges, horses, butter, coals, and many other articles which Van Diemen's Land itself produces, from New South Wales. The ports of the island of Van Diemen's Land and of New South Wales are free ports, and placed on the same footing, as respects trade, with those of Malta and Gibraltar. Launceston, the nearest port to the continent of Australia, is 300 m. sailing distance from Port-Phillip. Hobart-Town is 800 m. sailing distance from Sydney.

Revenue.] The ordinary revenue of Van Diemen's Land arising from import duties, law fees, licenses, and fines, was, in 1831, £67,300; while the total expense of government amounted to £60,738. It is to be remarked, however, that a parliamentary grant of about £100,000 was annually made on account of the criminal population of this colony. The revenue in 1847 was £148,763; in 1849, £152,668. The expenditure in 1847 was £140,157; in 1849, £154,634. The expense of the judicial department amounts to about £13,000 per annum, and the ecclesiastical to about £6,500. The governor's situation is reckoned worth £5,000 per annum, although his nett salary is only £2,500. The salary of the governor's private secretary is £500 per annum. British silver is used, besides dollars, rupees, and several other foreign coins.

Divisions.] Originally Van Diemen's Land was divided into two counties only, viz., Buckingham and Cornwall. In 1826 these were subdivided into several police districts, viz., Hobart-Town, comprising about 400 sq. m.; Richmond, containing 1,050 sq. m., with the towns of Richmond, Sorell, and Brighton; New Norfolk, containing 1,500 sq. m., with Elizabeth-Town, or, as it is commonly called, New Norfolk, and Hamilton; Clyde, comprising 1,700 sq. m., and the town of Bothwell; Oatlands, containing 900 sq. m., with the towns of Oatlands and Jericho; Campbell-Town, containing about 1,260 sq. m., with the towns of Campbell-Town, Ross, Lincoln, and Fingall; Norfolk Plains, a district of great extent, comprising above 2,250 sq. m., but of which the greater proportion is rugged, inaccessible land, with the townships of Latour and Westburg; Launceston, an extensive but unproductive district of 3,800 sq. m., with the town of Launceston, the second town in the colony, and also the towns of Perth and George-Town; and Oyster bay, a small district of about 900 sq. m. To these divisions may be added the penal settlements of Macquarie island and Port Arthur, with numerous islands in the straits that separate the colony from New South Wales; Maria island, formerly a penal settlement; and the territories of the Van Diemen's Land Company, comprehending nearly 500,000 acres in the NW corner of the island, and presently inhabited by about 400 souls.—The capital is HOBART-TOWN on the estuary of the Derwent: see that article. Next to Hobart-Town in importance is Launceston, situate at the confluence of the North and South Esk rivers, or head of the Tamar navigation, 121 m. from the seat of government, and 40 m. within the N line of coast. Its present pop. is about 2,000. From the favourable nature of its situation for commercial purposes—the river being navigable for vessels of 400 tons up to the town—its maritime trade is very considerable, and is every day increasing. The chief exports from Launceston are wheat, bark, wool, and whale-oil. Of these £60,000 are shipped annually. The custom-house revenue of the port, for the quarter preceding July, 1832, amounted to about £12,000.—Elizabeth-Town is a very agreeable township, 22 m. from Hobart-Town, on the highest navigable point of the Derwent. It is a place of considerable traffic between the capital and the rich pasturage and village districts of the interior. Spring-bay on the E, and Port-Davey, and Macquarie harbour, on the W, afford shelter for vessels.

History.] More than 120 years elapsed before the discovery of Van Diemen's Land by Tasman was followed by any event of the slightest importance to its annals. Cook, as well as his companion Furneaux, in the course of their voyage of circumnavigation in 1773, and again in 1777, visited the shores of Van Diemen's Land without discovering its insularity. In 1803 it was formally taken possession of by the English; a small detachment under the command of Lieutenant Bowen, having arrived from Sydney, with a view of forming a penal settlement for convicts transported from that colony. Risdon or Restdown, as it is sometimes called, on the E bank of the Derwent, a few miles up the river, was the spot selected for the settlement; but beyond this, little was effected at that time. Early in 1804, Lieutenant-governor Collins, who had recently left England with a considerable expedition, having in view the formation of a settlement at Port Phillip, on the southern coast of New Holland, altered his destination by reason of the insurmountable difficulties which then appeared to attend the establishment of a colony at that place, and arrived in the Derwent, when the island was formally taken possession of in the name of his Britannic majesty; and after various surveys of the Derwent, the present site of Hobart-Town was decided upon for head-quarters. Governor Collins was accompanied by several respectable gentlemen, to fill the various situations of his infant government, and had about 400 prisoners under him, with about 50 marines. In the course of the same year, a settlement was formed on the other side of the island, under the command of Colonel Paterson, of the 102d, who arrived from Sydney, and in the first instance, made choice of a spot beyond George-Town, calling it York-Town, but which was afterwards abandoned. After the island had been settled about three years, the first sheep and cattle were imported. Fresh arrivals of prisoners were constantly taking place from Sydney, and the colony continued to increase, although still preserving its original character of being a place of punishment for the convicted felons of New South Wales. In 1813, Lieutenant-Colonel Davey arrived as lieutenant-governor, and about this time the embryo importance and value of the colony began to be developed. Until this period all communication between Van Diemen's Land and other places, excepting England and New South Wales, had been interdicted by certain penalties upon merchant vessels that might attempt to enter the ports; but these were now done away with, and the colony placed precisely on the same footing with respect to commerce as New South Wales. The consequence of this and of other measures that were adopted about the same time, soon became obvious. The colony began to wear the appearance of an abode of Englishmen; and although emigrants from the mother-country had not yet directed their steps hither, what with the officers of different regiments who remained in the colony—with the number of individuals who had been brought here by government upon the evacuation of Norfolk Island, with occasional arrivals from New South Wales—and the crown prisoners who had become free either by servitude or indulgence, the pop. increased rapidly. Land was also more and more cultivated, houses were erected, farms enclosed; every thing, in short, assumed an improving aspect. About 1821, the tide of emigration set in from England towards Tasmania; and the natural consequence of the capital introduced was an extension of the colony within itself in every shape. Trade began to assume regularity; distilleries and breweries were erected; the Van Diemen's Land bank was established; and many other steps taken equally indicative of the progress the place was making. Originally a dependency of New South Wales, it now holds the rank of an independent colony.

VAN DIEMEN, a river of N. Australia, which flows into the gulf of Carpentaria, on the SE side, in S lat. 17°.

VAN DIEMEN GULF, a gulf formed by Melville island and Coburg peninsula and the continent of Australia, and entered by Dundas strait on the N, and by Clarence strait on the W. On the S it receives South Alligator river. It contains several islands.

VANDOEVRE-EN-BRENNE, a village of France, in the dep. of the Indre, cant. and 8 m. SW of Buzançois. Pop. 1,050. It has an iron-work.

VANDOEVRES, a commune of Switzerland, in the cant. of Geneva. It has important coal-mines.

VANDRILLE-RANCON (Saint), a commune of France, in the dep. of the Seine-Inferieure, and cant. of Candebec, 9 m. S of Yvetot. Pop. 1,008. It has a cotton-spinning mill.

VANDSBURG, a town of Prussia, in the prov. of West Prussia, circle and 20 m. E of Flatow, on a lake of the same name, in which is an island with a castle. Pop. 1,000.

VANDSOCE, a parish of Norway, in the diocese of Christiansand and bail. of Stavanger, on a peninsula. Pop. 5,000.

VANDY, a river of France, in the dep. of the Oise, formed by the confluence of two streams which have their sources in the dep. of the Aisne, and which, after a course of about 8 m., joins the Aisne, on the l. bank, at the foot of La Motte.

VAN DYKE (Jost and Little), two islands of the Antilles, in the group of the Virgin islands, to the NW of Tortola. The larger, which is 8 m. NW of Tortola, is 7 m. in length, and 2 m. in breadth.

VANG, a parish of Norway, in the diocese of Aggershuus and bail. of Christian, on the Reina. Pop. 2,100.—Also a parish in the bail. of Hedemarken, 68 m. NNE of Christiania, on the NE bank of the Mjosen-see. Pop. 5,110.

VANGE, a parish in Essex, 13 m. S of Chelmsford. Area 2,250 acres. Pop. in 1851, 164.

VANGEN, a parish of Norway, in the bail. of Jarlsberg, on the Christiania-fiord, to the NE of Tonsberg.

VANGEROW, a village of Prussia, in the reg. of Koslin, and circle of Neustettin. Pop. 250.

VANIAMBADDY, a town of Hindostan, in the presidency and 120 m. WSW of Madras, and prov. of Salem and Barahmahl, pleasantly situated in an island of the Palar. It is enclosed by an earthen wall, and has two celebrated Hindu temples.

VANIKORO, two islands in the South Pacific, in the archipelago of Santa Cruz, in S lat. 11° 40′, and E long. 165° 50′. They are unequal in extent, are close together, and surrounded by immense reefs—penetrable by only a few intricate passages—upwards of 30 m. in circuit, containing some small islands, and extending nearly continuously to the archipelago of the New Hebrides on the S. Pop. 1,500. These islands rise to the height of about 3,114 ft. above sea-level, and are densely covered with wood. On the E side of the larger island is a bay, dividing the lesser V. into two parts, distinguished as Manevai and Tevai bays, within which is a basin sheltered on all sides from the storms which frequently break with violence on the reefs without. On the N is also another passage in which the remains of the vessels of La Perouse, wrecked forty years previously, were discovered by Dillon in 1827. The islands of V. are extremely insalubrious to Europeans. The aborigines belong to the black Oceanic race, are diminutive, squalid in appearance, and hostile in disposition to strangers. They wear in the ears strings of rings made of the tortoise-shell, and a feather passed through the cartilage of the nose. The bread-fruit, cocoa-nut, and fish form their chief articles of subsistence. These islands were named by Captain Edwards, by whom they were visited in 1791, Pitt islands; and subsequently from Dillon they received the appellation of La Perouse; by D'Entrecasteaux, in 1793, Recherche. The smaller islands of the group are named Manevai and Nanounha. Paion and Vanon are districts of the larger Vanikoro.

VANIKUM, a village of Prussia, in the regency of Dusseldorf, and circle of Reuss. Pop. 480.

VANITCHES, a town of Turkey in Europe, in the sanj. and 78 m. SE of Monastir, near the l. bank of the Indje-Carasu.

VANKLEEK-HILL, a village of Upper Canada, in the W part of the township of West Hawkesbury, 8 m. S of the Ottawa. Pop. 250.

VANNEN, an island of the Lofoden group, to the NE of Ringvadsoe, in N lat. 70° 10′, and E long. 20° 5′. It is 18 m. in length from N to S, and 12 m. in breadth. Pop. 500.

VANNES, a river of France, which has its source a little to the E of Fontaines, in the dep. of the Aube, flows thence into that of the Yonne, and after a course of about 39 m., in a generally W direction, joins the river of that name, on the r. bank, a little above Sens. Its principal affluents are the Lancre, St. Mards, Pouy, and Vandeurs. Villeneuve-sur-Vannes is the chief place on its banks.

VANNES, an arrondissement, canton, commune, and town of France, in the dep. of the Morbihan. The arrond. comprises an area of 163,419 hect., and contains 10 cant. Pop. in 1831, 119,774; and in 1846, 129,816.—The cant. comprises 13 com. Pop. in 1831, 27,274; and in 1846, 31,280.—The town, which is also capital of the dep., is 72 m. NW of Nantes, and 336 m. WSW of Paris, near the N extremity of the small gulf of Morbihan, in N lat. 47° 39′ 31″, and W long. 7° 25′ 41″. Pop. in 1789, 8,814; in 1821, 11,289; in 1831, 10,395; and in 1846, 12,974. It is enclosed by two small rivers, one of which bears the same name, and forms at its mouth a small harbour. It has two suburbs, each larger in extent than the town. Of these, one named the Marche is separated by walls, defended by strong towers and by a ditch. The other, distinguished as St. Paterne, contains a handsome church, a fine mall, and a general hospital. The town possesses a custom-house, a public library, a communal college and museum, a free school of navigation, a theatre, a house of correction, and two printing-establishments. It has also an agricultural society, a polymathic society for the promotion of science, literature, and the arts. It has manufactories of varieties of woollen, linen, and cotton fabrics, hats, and lace, several tan-mills, tanneries, and iron and copper-foundries. The trade consists chiefly in grain, salt, hemp, honey, wax, butter, tallow, cider, iron, wines of Bordeaux, &c. Vannes, from the number of its ruins, appears to have been the capital of the ancient *Venetes*. It was erected by its sovereigns, the dukes of Bretagne, into a county. The name signifies in the low vernacular Guenet, handsome. In the adjacent peninsula of Rhuys is the ancient abbey of St. Gildas.

VANNES (Les), a group of islets of France, in the Atlantic, opposite the mouth of the Vilaine, in the dep. of the Morbihan.

VANOSC, or Vanosc-en-Vancance, a commune of France, in the dep. of the Ardeche, cant. and 7 m. SW of Annonay. Pop. 1,660.

VANS (Les), a canton, commune, and town of France, in the dep. of the Ardeche. The cant. comprises 21 com. Pop. in 1831, 17,643; and in 1846, 20,165. The town is 15 m. SW of Largentiere, near the Chassezac. Pop. 2,240. It has manufactories of silk and several spinning-mills, and carries on an active trade in silk, woollen, and linen fabrics, grain, cattle, and chestnuts. Wine, mulberries, olives, and chestnuts are cultivated in the vicinity.

VANSITTART-BAY, an indentation of the NW coast of Australia, to the E of Cape Bougainville, by which it is separated from Admiralty gulf. It is 30 m. in depth from N to S, and 27 m. in extreme breadth. Its shores present extreme irregularity of outline, and the mouth is studded with islands.

VANSITTART-HILLS, a range of hills in New South Wales, in the district of Liverpool Plains, running from E to W along the S side of Gorke river, to its confluence with Bowen river.

VANTSIOEGUTSROER, a lake of Sweden, in the prefecture of Stora-Kopparberg. It gives rise to the Eastern Dal-elv.

VANVES, or Vanvres, a commune of France, in the dep. of the Seine, cant. and 4 m. N of Sceaux, and 2 m. SW of the walls of Paris. Pop. 2,506. It has a lunatic asylum, and possesses manufactories of locks, extensive bleacheries, and quarries of freestone. In the vicinity is an ancient castle belonging to the Prince-de-Conde.

VANVEY, a town of France, in the dep. of the Cote-d'Or, cant. and 8 m. ESE of Châtillon-sur-Seine, on the Ource. Pop. 795.

VAN-WERT, a county in the NW part of the state of Ohio, U. S., drained by St. Mary's and Little Anglaize rivers, and other small streams, and intersected by the Ohio and Indiana railway, the Miami canal, &c. It has a level surface, but is generally fertile. Pop. in 1840, 1,580; in 1850, 4,794.—The chief village, which bears the same name, is 105 m. NW by W of Columbus. Pop. 270.

VANXAINS, a commune of France, in the dep. of the Dordogne and cant. of Riberac. Pop. 2,010.

VANZAY, a village of France, in the dep. of the Deux-Sevres, 5 m. SE of Chenay. Pop. 350.

VANZE, a town of Naples, in the prov. and 9 m. E of Lecce, and cant. of Vernole, in a low and unhealthy situation. Pop. 200.

VAOUR, a canton and commune of France, in the dep. of the Tarn, and arrond. of Gaillac. The cant. comprises 10 com. Pop. in 1831, 5,919; in 1846, 5,672.—The town is 14 m. NNW of Gaillac, on the most elevated point of the ridge which separates the valleys of the Tarn and Aveyron. Pop. 600. It has a fine castle.

VAPIXANA, a tribe of Indians, who inhabit the N part of Brazilian Guayana, near the confines of the British possessions.

VAPNA-FIORD, a harbour of Iceland, on the E coast, in N lat. 65° 40'. It receives the river Hof on the W side.

VAPRIO, a market-town of Sardinia, in the div. and prov. and 12 m. NNW of Novara, and mande. of Momo. Pop. 6,000. It has extensive paper-mills. A sanguinary engagement took place here in 1324 between the Guelphs and Ghibelines, in which the former were defeated.—Also a market-town of Austrian Lombardy, in the gov. and delegation of Milan, district and 7 m. NE of Gorgonzola, on the Martesana canal, and r. bank of the Adda. Pop. 1,880. It has a large paper-mill.

VAR, or Varo, a river which has its source in the Alps, in France, in the dep. of the Basses-Alpes, 9 m. E of Colmar. After running a few leagues, it enters the kingdom of Sardinia; forms afterwards for some distance the line of separation between the div. of Nice and the French dep. of the Var; and after a total course of about 78 m., throws itself into the Mediterranean near St. Laurent-du-Var. It flows first S to Entrevaux, then E to the confluence of the Tinea, and afterwards S. Timber is its chief article of transit. Its principal affluents are the Colon and the Esteron in France, and the Tinea and the Vesoubia in Sardinia. The V. in the lower part of its course forms numerous islands. It is liable in spring to extensive floods, and is remarkable for the frequency with which it changes its bed. In summer it is reduced to an insignificant stream.

VAR, a department of France, forming the SE extremity of the kingdom; and bounded on the N by the dep. of Basses-Alpes; on the NE by Piedmont; on the E and S by the Mediterranean; and on the W by the dep. of Bouches-du-Rhone. It has an area of 2,800 sq. m. Branches of the Alps extend into the N and NE of this dep.; so that it is in general rugged and uneven, the only extensive plains being along the coast, which is penetrated by several deep bays. The rivers descending from the mountains are the Var, Verdon, Esteron, and Argens, with a number of smaller streams. The climate varies according to elevation, being in some parts, especially under a NW wind, bleak even in this southern latitude; while in others it is mild and warm; but in several districts of the lower grounds, and in the vicinity of the lagunes on the coast, the air is infected by stagnant water. The soil is in many parts dry and stony, in others, such as on the banks of the Var, and the vicinity of the town of Grasse, it is more productive. Not above a sixth of the surface is arable, and the corn raised is not equal to the consumption. Wine, brandy, and vinegar are made in considerable quantities from the grape, the culture of which occupies a twelfth of the surface. Dried fruits, cork, and silk, are likewise exported. The pastures are good in particular spots, where the command of water makes up for the infrequency of rain. The animals chiefly reared are sheep, goats, and asses; the sheep are sent in summer to mountain-pasture. Honey and wax are exported. Of minerals, in whatever variety they may exist, hardly any have been worked except coals and a little iron and lead. The manufactures are of insignificant amount; the chief articles are silk, paper, leather, and coarse woollens. The fisheries along the coast are considerable.—The pop. in 1846 was 349,859; in 1852, 357,967. The common language is a dialect composed of French, with a mixture of Italian.—This department forms part of the ancient Provence. It contains no town of consequence except Toulon. It is subject to the royal court of Aix, and is divided into the 4 arrondissements of Toulon, Brignolles, Grasse, and Draguignan. The last, though a small place, is, from its central situation, the capital of the dep. It is subdivided into 35 cantons, and 202 communes.

VARA (Pico), a mountain of the Azores, near the E extremity of the island of San-Michael. It has an alt. of 5,000 ft. above sea-level, and is occasionally capped for a short period with snow. This peak terminates the chain of craters which extends to the centre of the island.

VARAD, a market-town of Hungary, in the comitat of Schemegh, on the l. bank of the Oker, 6 m. SE of Istvándi. It has a saltpetre factory.

VARA-DE-REY, a market-town of Spain, in the prov. of Cuenca, 6 m. E of San-Clemente, in a fertile plain. Pop. 1,690.

VARADES, a canton, commune, and town of France, in the dep. of the Loire-Inferieure, and arrond. of Ancenis. The cant. comprises 5 com. Pop. in 1831, 10,766; in 1846, 10,247.—The town is 8 m. E of Ancenis, near the r. bank of the Loire. Pop. 3,483. It has a glass-work, and carries on an active trade in wine and timber, both the produce of the locality. About 3 m. N of the town are the coal-mines of Montrelais.

VARADIN. See Wardein (Gross).

VARADOURO, a port of Brazil, in the prov. of Parahiba, on the river and to the E of the town of that name.—Also a river of the prov. of São-Paulo, which flows through the districts of Cananea and Paranagua, and joins the Ararapira.

VARAGES, a commune of France, in the dep. of the Var, and cant. of Barjols, 21 m. NW of Brignols. Pop. 1,478. It has manufactories of china. Oil of fine quality is cultivated in the locality.

VARAGGIO, or Varazze, a town of Sardinia, in the prov. and 6 m. NE of Savona, on the gulf of Genoa, on which it has a small port. Pop. 1,800. The manufacture of anchors and boatbuilding form the chief objects of local industry.

VARAIGNES, a village of France, in the dep. of the Dordogne, and cant. of Bussiere-Badil, on an affluent of the Bandiat. Pop. 1,000. In the vicinity are mines of iron and antimony.

VARAIZE, a village of France, in the dep. of the Charente-Inferieure, cant. and 1 m. ESE of St. Jean d'Angely. Pop. 710.

VARALLO, a town of Sardinia, capital of the prov. of Val-Sesia, on the l. bank of the Sesia, and 36 m. NW of Novara. Pop. 3,300. It has an hospital, a gymnasium, and a communal college, and contains brass-foundries, and manufactories of iron and copper from the mines in the adjacent mountains. On a mountain in the vicinity is a church, containing 52 chapels, and an image of the virgin which attracts numerous votaries.

VARALLYA (Znio), or Kloster-Kuhhorn, a town of Hungary, in the com. of Thurotz, near the r. bank of the Vricza, and 20 m. NNW of Körmöcz-Banya. To the W is the castle of Znio.

VARAMBON, a village of France, in the dep. of the Ain, cant. and 2 m. from Pont-d'Ain. Pop. 500. It was formerly a town, and possessed a fortress which was destroyed in 1451. It was erected into a marquisate in 1576.

VARANACO, a small river of New Granada, in the prov. of San-Juan-de-los-Llanos, which rises near the source of the Paucana, and runs in an easterly course to the Orinoco.

VARANGER-FIORD, a gulf of Norway, in the diocese of Nordland, and E part of Finmark, in N lat. 70°, and E long. 30° 20′. It is 6 m. in length from E to W, and about 15 m. in medium breadth. It receives the Palsjoki, which falls into one of its small bays distinguished as the Pasvig-fiord.

VARANGEVILLE-LES-DEUX-EGLISES, a village of France, in the dep. of the Seine-Inferieure, cant. and 3 m. NE of Duclair. Pop. 300. It has paper-mills, and in the vicinity are mines of coal.

VARANGEVILLE-SUR-MER, a village of France, in the dep. of the Seine-Inferieure, cant. and 5 m. NW of Offranville, on the English channel. Pop. 1,200.

VARANGUEBEC, a village of France, in the dep. of Manche, cant. and 4 m. NE of La-Haye-du-Puits.

VARANO, a small lagune in the E of the kingdom of Naples, in the Capitanata, near Monte-Gargano. It is about 20 m. in circuit, and discharges itself into the Mediterranean by a small stream on the NE.

VARANO, or Wrano, a town of Hungary, 53 m. N of Tokay. Pop. 1,500.

VARAPODIO, a market-town of Naples, in the prov. of Calabria-Ultra, 10 m. SE of Gioja. Pop. 1,200. It was devastated by an earthquake in 1783.

VARAYRE, a village of France, in the dep. of Lot, cant. and 3 m. SW of Limogne.

VARAZZE. See Varaggio.

VARBERG, a port of Sweden, in the laen and 36 m. NW of Helmstadt, in N lat. 57° 6′. Pop. 1,700.

VARDA, or Kis-Varda, a town of Hungary, on the Theiss, 53 m. N by E of Debreczin.

VARDAC, a town of France, dep. of Lot-et-Garonne, on the small river Bayse, with 1,400 inhabitants. The cork forests of the neighbourhood supply an object of manufacture to the inhabitants.

VARDAR, a large river of European Turkey, the *Axius* of the ancients, which rises on the E flank of the Tchar-dagh; flows from NW to S through Macedonia, passing Uskup and Koprili; and empties itself into the gulf of Salonika, 10 m. W by S of Salonika, after a course of 170 m.

VARDE, a village of Denmark, in the prov. of Jutland, on the r. bank of a stream of the same name, near its embouchure in the gulf of Hierting.

VARDOC, an island of Norway, in N lat. 70° 21′, E long. 31° 10′.

VAREL, a town of Germany, in the grand-duchy of Oldenburg, near the mouth of the Hase or Jahde, in N lat. 53° 24′, 17 m. N of Oldenburg, and 35 m. NW of Bremen. Pop. 3,200.

VARELA (Point), or Cape Muinai, the easternmost point of Annam, in N lat. 12° 53′, E long. 109° 24′. It forms the extremity of a narrow sandy peninsula.

VARELLA (Pulo), an island off the NE coast of Sumatra, in N lat. 3° 47′, about 10 m. from Jaboeng-point.—Also an islet off the E coast of the peninsula of Malacca, in N lat. 3° 16′.

VARENHOLZ, a village of the German principality of Lippe-Detmold. Pop. 690.

VARENNA, a town of Austrian Italy, situated on the E side of the lake of Como, 30 m. N by E of Milan.

VARENNES-SUR-ALLIER, a town of France, on the river Allier, 14 m. WNW of La Palisse, on the Valaçon, an affluent of the Allier. Pop. 2,296.

VARENNES-EN-ARGONNE, a town of France, dep. of the Meuse, on the river Aire, 18 m. NW of Verdun. Pop. 1,600. It has manufactures of leather and paper; but is chiefly remarkable as the place where Louis XVI. was stopped in his flight from

Paris in June 1791.—Also a small town in the dep. of Haute-Marne, 14 m. NE of Langres.

VARENNES-SOUS-DUN, a commune and village of France, in the dep. of Saone-et-Loir, 1 m. E of La Clayette.

VARENSELL, a village of Prussia, in the reg. of Minden, circle of Weidenbruck. Pop. 1,200.

VARENT (SAINT), a town of France, dep. of Deux-Sevres, on the small river Thouaret, 14 m. E of Bressuire. Pop. 735.

VARESE, a town of Austrian Italy, in the gov. of Milan, situated on the small river Verbano, 13 m. W of Como, near a lake. It has about 8,500 inhabitants, who cultivate and manufacture silk. In its vicinity are a number of villas belonging to Milan families, and an elegant modern palace situated on an eminence, and surrounded by pleasant gardens and fountains.—The lake of Varese is a fine expanse of water, of an oblong form, about 6 m. in length from NW to SE, having its banks covered with luxuriant vegetation. Its elevation above the level of the Adriatic is 236 metres—774 ft.

VARGEL, or GREAT VARGULA, a town of Prussian Saxony, on the Unstrut, 14 m. NW of Erfurt. Pop. 1,000.

VARGEM-FORMOSA, a village of Brazil, in the prov. of Mato-Grosso, 28 m. W of Cuiaba.

VARGO, an island in the gulf of Bothnia, in N lat. 62° 59′, E long. 21° 3′.

VARHELY, a village of Hungary, in the com. of Hunyad, 8 m. WSW of Hatzig.

VARICK, a township and village of Seville co., in the state of New York, U. S., 160 m. W of Albany. Pop. 1,872.

VARIETY, a village of Nelson co., Virginia, U. S., 85 m. W by N of Richmond.

VARIGNANO, a town of the States of the Church, deleg. and 12 m. SE of Bologna. Pop. 3,100.

VARILHES, a town of France, situated on the Arriege, 6 m. S of Pamiers, and 6 m. N of Foix. Pop. 1,700.

VARIN, a town of Hungary, in the com. and 45 m. NE of Trentsin, near the Waag.

VARINAS, a prov. of Venezuela, bounded on the N by the prov. of Maracaybo; on the E by the plains of Caracas and the Orinoco; on the S by Arhaguas; and on the W by Merida. Pop. 109,497. Its chief products are tobacco, well known in the European markets; cattle, sheep, mules, sugar, coffee, cotton, indigo, and all the fruits of the torrid zone. The produce is exported chiefly by water to Guayana; the place of embarkation being at a spot called Tocunos, 5 leagues below the cap. The most remarkable features of this country are the extensive plains of which it is composed, and which are covered with a luxuriant herbage, feeding innumerable herds of cattle, flocks of sheep, and droves of mules and horses.—The capital of the prov., of the same name, is situated about 800 m. SE of Caracas. It is a little place; but the public edifices are reduced to one parish-church and an hospital. Pop. 12,000.

VARIRIN, a river of Brazil, in the prov. of Seara, which rises from the mountains in the interior, and enters the Parana.

VARKENY, a village of Hungary, in the com. of Borsod. Pop. 1,100.

VARLOSEN, a village of Hanover, in the reg. of Minden, near Dransfeld. Pop. 425.

VARNA, a town of European Turkey, in the sanj. of Silistria, in Bulgaria, situated on the N side of a bay on the SW coast of the Black sea, at the mouth of the river Varna, opposite Galata, in N lat. 43° 12′ 17″, E long. 27° 56′ 31″, 14 hours' sailing distance from Constantinople. The semicircular bay of V. which is about 1½ m. in length, and 2 m. across, shoals up to the beach at the apex of the semicircle formed by its shores, and the land is so low at that point that the fresh waters from the neighbouring hills form the Devna lakes, which extend in a SW direction for 7 or 8 m., with a breadth of from 2 to 3 m. towards Shumla. V. is built on a slightly elevated bank of sand, of such varying height that, in some places, the base of the walls around it is on the level of the water, and at others 20 or 30 ft. above it. Below this bank a series of plains spread all round the landward side of the town, till they are lost in the hills which, dipping into the sea in an abrupt promontory on the NE side, rise in terraces to the height of 700 or 800 ft., at the distance of 3 m. from the walls, and trend westwards to meet the corresponding chain of hills on the S extremity of the bay, thus enclosing the lake and plains in a sort of natural wall, which is, like all the rest of the country, covered with brushwood and small trees. A stone wall 10 ft. high, painted white and loop-holed, is built all around the place, so as to enclose a large irregular four-sided figure, each side of which averages 1¼ m. in length, and some detached batteries, well provided with heavy guns, chiefly 68-pounders, are erected in advance of the walls on the land side. On the sea-face are four batteries, provided with heavy guns also, and at every favourable spot bastions have been thrown out. A ditch, 18 ft. wide by 12 ft. deep, further protects the *enceinte*. There are 5 gates in the exterior wall. Peering above these walls, in an irregular jumble of red-tiled roofs, are seen the houses of the place, with a few minarets towering from the mosques above them. The angles of the works are irregular, but in most instances the walls are so constructed as to admit of a fair amount of flanking fire on an assailing force. Nevertheless, the bank on which the town is built is so uneven, that a portion of the inner side of the walls could be swept from a fleet in the bay, and other parts are equally accessible to the fire of batteries on the trifling hillocks around the town. The houses of the town, about 3,000 in number, are chiefly built of wood, painted white, blue, brown, saffron, and yellow; but, for the most part, they exhibit little of the brightness of the original colouring, and the roofings of broken red tile, combined with the general dilapidated look of the plank sheathings of the side-walls, give the whole place an appearance of decay, which is not much belied by the interiors of the habitations. The houses, with few exceptions, present nothing but a door and a low gable-wall towards the street. Generally, they look out on a court-yard provided with some sheds, a well, and as many flowers and trees as can force their way through the hardened earthen floor of the enclosure. A high stone wall shuts out each mansion from its neighbour, and the doors, which perforce must open on the street, are shut the instant the occasion for their opening ceases. "Thus," says the graphic pen which has furnished the greater part of the present article, "as you wander through hot lanes presenting one eternal blank of stone walls and unpainted doorways, topped by tiled gables, and hear no sound within, except the wailing of an infant, the sharp cry of the kite soaring overhead, or the growl of the dog half-awakened by your step as he basks in the middle of the narrow path, you might think the place stricken by the plague and destitute of life, but that now and then a door opens and a pile of red or yellow or bright blue cloth, surmounted by the white folds of the *yashmak*, and terminating in light yellow boots, emerges, and at your presence rushes in again, or takes a precipitate motion across the street and dashes in at an opposite door, or a ragged porter with a water-jar, shuffles along and eyes the infidel sullenly as he goes by. In the heat of the mid-day, when the sun blazes down into those straitened streets, the silence is absolutely oppressive, and it is a relief to hear the twittering song of the swallow as he clings to the roof. V., nevertheless, contains 13,000 or 14,000 inhabitants, chiefly Bulgarians. The garrison is generally 5,000. There is more bustle, animation, and life in the smallest hamlet in Dorsetshire than here, unless you go down to the landing-place, or visit the bazaar, where the inhabitants flock for work or business. Towards evening there are more people moving about, and the *muezzin* breaks the silence with his shrill and boylike voice." The town contains four mosques, three Greek churches, and one Armenian church. There are numerous and abundant fountains within the walls. There are three small jetties of wood opposite the principal gate of the town; and a beach of a few yards broad between the sea and the foot of the walls serves as a landing-place for lighter boats. The roadstead of V. is spacious, with good holding-ground, and well-protected from S, W, and N winds, but is open and exposed to the E. In spite of this, from the growing importance of V. as the principal port of Bulgaria and the readiest point of connexion with central Europe, it has been surveyed for its capacities both as a commercial port and as a naval arsenal. Not only is it the natural point for the concentration of the defensive resources for the protection of the eastern shore of Turkey in Europe, but, if a railway be carried out to Shumla and Rustchuk, V. must become a city of the first importance both for goods and passenger traffic. Two projects are, therefore, now on the *tapis* for improving this port: one is the construction of a couple of breakwaters on the sunk ledges of rocks in prolongation of the bluff cape to the south, and the shallows to the north of the roadstead; but this would involve a vast expense, and after all leave too large a space open to the easterly gales. The other project is to form the lower lake of Devna into a harbour. This lake is 60 ft. deep in the middle; and the isthmus between it and the sea is a flat of land a little raised above the level of either, but not exceeding a ½ m. in breadth. Thus a ship-canal of a few hundred yards would open a capacious basin for the largest warlike or mercantile navy, and at the same time a secure arsenal for the construction, repair, and armament of a fleet. The direct high road to Silistria strikes off close to V. through the mountains which run parallel to the lakes of Devna or Devne; and the lower or summer-road to Paravati skirts the N shore of the lakes. Austrian steamers generally put in twice a-week at this place on their passage to and from Constantinople and Galatz. In 1847, the value of exports from V., chiefly corn, timber, charcoal, eggs, and fowls, was £600,000.—In 1828, this place was captured by the Russians, who directed their main attack upon it from the N. In August 1854, it was devastated by fire, the work probably of a Russian incendiary. In the same year, the great military expedition of the Allies sailed from this place to disembark on the Crimea.

VARNA, a village of Tompkins co., in the state of New York, U. S., 140 m. W by S of Albany, on Fall creek. Pop. in 1850, about 200.

VARNAVIN, a town of Russia in Europe, in the gov. and 168 m. ESE of Kostroma, in a district of the same name, on the Vetluga. Pop. 668. The houses are built of wood. The trade consists chiefly in pitch, woodware, and mats.

VARNEVILLE-AUX-GRIS, a village of France, in the dep. of the Seine-Inferieure, cant. and 2 m. S of Totes. Pop. 339. In the vicinity are the extensive freestone quarries from which the village takes its name.

VARNHAGEN, a village of the principality of Waldeck and bail. of Eder, on a river of that name, a little to the SW of Waldeck.

VARNHALT, a village of Baden, in the circle of the Middle Rhine, and bail. of Buhl. Pop. 660.

VARNHOVEL, a parish of Prussia, in the reg. of Munster, circle of Ludinghausen. Pop. 280.

VAROCHI, a town of Greece, in the Morea, near the E coast of the gulf of Coron, and 12 m. SE of Calamata. Pop. 450. It is the residence of a bishop.

VAROE, an island of Norway, in the Loffoden archipelago, to the N of the Maelstrom.

VAROKH, a village of Turkey in Europe, in Albania, in the pash. and 30 m. NW of Uskup, and near the S base of the Tcha-Dagh.

VARONNE, a village of Switzerland, in the cant. of Valais, bail. and 1½ m. NW of Leuk, about the same distance NNW of the confluence of the Dola with the Rhnie. Pop. 413.

VAROSCH, a market-town of Turkey in Europe, in Bosnia, in the sanj. of Trawnik, 15 m. N of Bosna-Seraï, on the r. bank of the Sondica. In the vicinity are several mines and forges.

VAROUNDAS, a people of Africa, who inhabit the territory of Moro-Pooa, to the E of Lower Guinea. They cultivate maize and manioc.

VARRAIGNES. See VARAIGNES.

VARREL, a village of Hanover, in the bail. of Ehrenburg. Pop. 770.—Also a village of the duchy of Oldenburg and bail. of Delmenhorst. Pop. 340.

VARRIGSEN, a village of Brunswick, in the circle of Wolfenbuttel. Pop. 200.

VARRONE, a mountain of Austria, in Lombardy, on the confines of the provinces of Como and Bergamo, in N lat. 46° 0′ 25″, and E long. 9° 31′ 33″.

VARS, a commune and town of France, in the dep. of the Charente, and cant. of Saint-Amant-de-Brixe, on the l. bank of the Charente. Pop. 1,987. The locality affords good wine.

VARTOFTA, a haerad of Sweden, in the prefecture of Skaraborg, bathed by Lake Vetter.

VARTREY, a rivulet of co. Wicklow, which rises on the S side of Douce mountain, at an elevation of upwards of 1,030 ft. above sea-level; and flows to the head of the long estuarial belt of water which bounds the W side of the Murrough.

VARTSALA, an island of the Baltic, off the SW coast of Finland, in the archipelago of Abo, in N lat. 60° 30′, and E long. 21° 14′.

VARU. See BARU.

VARYSBURG, a village of Wyoming co., in the state of New York, U. S., on the E side of the head branch of Tonawanda creek, 236 m. W of Albany. Pop. in 1850, about 250.

VARZEA, a valley of Brazil, in the prov. of Rio-Grande-do-Norte, and district of Villa-da-Princeza.

VARZEA-DAS-RAIZES, a village and fertile valley of Brazil, in the prov. of Santa-Catherina, watered by the Tubarão.

VARZEA-DE-VACA, a village of Brazil, in the prov. of Ceara, and district of Villa-do-Principe.

VARZI, a market-town of the duchy and 30 m. SW of Parma, in the district of Borgotaro, near the r. bank of the Zeno, and at the foot of a hill. Pop. 1,488. —Also a town of Sardinia, capital of a mandemento, in the prov. and 11 m. NW of Bobbio, on the r. bank of the Staffora. Pop. 1,598.

VARZY, a canton, commune, and town of France, in the dep. of the Nievre, and arrond. of Clamecy. The cant. comprises 12 com. Pop. in 1831, 12,048; in 1846, 13,472.—The town is 11 m. SW of Clamecy, at the foot of a mountain covered with vines, and watered by a stream of the same name, an affluent of the Sozay. Pop. 3,132. It has a college and an hospital, and possesses extensive manufactories of linen and of leather. This town was formerly defended by a fortress, and sustained several sieges. It was taken by the Protestants in 1590.

VAS, a village of Brazil, in the prov. of Rio-Janeiro, a short distance N of Villa-do-Principe.

VASA, or WASA, a port of Russian Finland, in the district of Sodra-Korsholm, on a small bay in the gulf of Bothnia, in N lat. 63° 4′ 20″, 350 m. NW of St. Petersburg. Pop. 2,700. It carries on an active trade in timber, fish, pitch, and leather. In August 1852, this town, founded in 1611, was burnt down, and a new town has been begun at a short distance from the original site, which is in future to bear the name of Nicolaistad.

VASARHELY (HOD-MEZÖ), a town of Hungary, in the com. of Csongrad, on the lake of Hold. Pop. 25,600, a number not frequently found collected in this agricultural country. They are for the most part Calvinists, and their chief employment is the culture of tobacco and wine.

VASARHELY (KEZDI), a town of Austria, in the Szekler-Land, NW of Kronstadt, in N lat. 46° 30′. Pop. 4,850.

VASARHELY (SOMYLO), a town of Hungary, on the river Torna, 72 m. S by E of Presburg.

VASARUT, a town of Hungary, in the island of Schütt, formed by the Danube. Pop. 1,000.

VASBECK, a village of the German principality of Waldeck, bail. of Arolsen. Pop. 470.

VASCONAS, or PROVINCIAS VASCONGADAS, a general term for the three Spanish provinces of Biscay, Guipuscoa, and Alava. See BISCAY.

VASE, a river of North America which empties into the Mississippi, 3 m. below the Great Rock, and about 55 m. NW by N of the mouth of the Ohio. It is navigable about 60 m., through a rich country abounding in extensive natural meadows.

VASEUX, a river of the state of Illinois, U. S., which rises between the Kaskaskia and the Little Water rivers, and, flowing nearly parallel to the former, falls into the Mississippi some distance above Cape Girardeau.

VASHON'S ISLAND, an island near the W coast of America, at the bottom of Admiralty inlet, in N lat. 47° 10′.

VASIETTA, a river of America, which runs into Lake Michigan, in N lat. 44° 38.

VASIL, or VASIL-SURSK, a town of European Russia, near the confluence of the Sura and the Volga, 87 m. ESE of Nijni-Novgorod. Pop. 1,600.

VASILIKA, or BASILICA, a village of the Morea, about 9 m. WNW of Corinth, situated on the angle of a rocky ascent, on the site of the ancient *Sicyon*, in N lat. 37° 59′.

VASILKOV, a town of European Russia, in the gov. of Kief, on the banks of the Stugma, 24 m. SSW of Kief.

VASINGAN, a river of Asiatic Russia, which joins the Obi, 30 m. below Narym, after an ENE course of 170 m.

VASKUT, a village of Austrian Galicia, in the Bukowino, near the river Czeremos. Here is kept a large stud of Tartar horses for the Austrian army.

VASLES, a town of France, in the dep. of Deux-Sevres, cant. and 6 m. NNE of Menigoutte. Pop. 2,200.

VAS-MARTIN, an island of the Atlantic, in S lat. 20°, to the SW of Picos. It is desert, and covered with thick woods.

VASONY, a town of Hungary, in the com. and near the town of Veszprim. Pop. 1,800.

VASPINGE, a town of Persia, in Azerdbijan, about 9 m. from Tauris.

VASQUEZ, a settlement of the island of Cuba, 66 m. NW of Villa-del-Principe.—Also an island in the S. Pacific, in S lat. 24° 44′.—Also a river of Nicaragua, which flows into the Caribbean sea, a little to the N of Cartago.

VASSALBOROUGH, a township of Kennebec, co. Maine, U. S., on the E side of the Kennebec, 8 m. N by E of Augusta. Pop. 3,100.

VASSELAY, a town of France, in the dep. of Cher, cant. and 7 m. SW of Meneton. Pop. 1,000.

VASSEROLLE (COL-DE-LA), a pass between the French deps. of Ain and Jura, having an alt. of 3,480 French ft. above the lake of Geneva.

VASSOUVAR, a district and village of Brazil, in the prov. of Rio-de-Janeiro, intersected by the Pato-do-Alferes, an affluent of the Parahiba.

VASSY, a town of France, dep. of Haute-Marne, 12 m. NW of Joinville. Pop. 2,600, employed partly in the manufacture of woollens. It is a well-built place; and has a considerable trade.—Also a town of France, dep. of Calvados, 12 m. NE of Vire. Pop. 3,150.

VAST (LE), a town and commune of France, in the dep. of Manche, cant. and 4 m. SE of St. Pierre-Eglise, near the Seine. Pop. 2,500. There are cotton and spinning-mills here.

VASTAN, a town of Armenia, on the SE coast of the lake of Van, 20 m. SW of Van.

VASTO (IL), or VASTO d'AMMONE, a town of Naples, in Abruzzo-Citra, on the coast of the Adriatic, in N lat. 47° 7′, 18 m. WNW of Termoli. Pop. 9,000. It is walled, and well-built. It was severely damaged by an earthquake in 1706; and in 1816, the ground giving way, many of the buildings sunk suddenly into the earth and nearly disappeared. Its pop. before this dreadful calamity was about 5,000. An active fishery is conducted from this place.

VASTO-GIRARDO, a town of Naples, in the prov. of Sannio, 15 m. NNE of Isernia.

VATAN, a town of France, dep. of the Indre, 14 m. NW of Issoudun. Pop. 2,000.

VATHI, a small port and the chief place of the island of Theaki or Ithaca, in the Ionian group. It is situated at the extremity of a spacious bay, and contains 1,800 inhabitants.—Also a port on the N coast of the Greek island of Samos, containing about 500 houses, the inhabitants of which support themselves by fishing, and by the export of a good wine produced in the neighbourhood.

VATICANO (CAPE), a cape on the W coast of Naples, in the prov. of Calabria-Ultra 2da, in N lat. 38° 37′.

VATISA. See FATSA.

VATTETOT-SUR-MER, a village of France, in the dep. of Seine-Inferieure, 18 m. N of Havre. Pop. 800.

VATTEVILLE-LA-RUE, a village of France, in the dep. of Seine-Inferieure, 10 m. NE of Yvetot.

VATU-LELO, one of the Fiji islands, in S lat. 18° 30′.

VATU-MANDRE′, a town on the E coast of Madagascar, in S lat. 19° 40′, at the mouth of a small river of the same name.

VAUBE′COURT, a town of France, dep. f Meuse, on the l. bank of the Aisne, 12 m. N of Bar. Pop. 1,270.

VAUCE′, a village of France, in the dep. and 12 m. NNE of Mayenne. Pop. 1,700.

VAUCHRETIEN, a village of France, in the dep. of Maine-et-Loire, cant. and 5 m. NE of Thouarce. Pop. 1,200.

VAUCLUSE, a department of France, between the parallels of 43° 35′ and 44° 25′ N; bounded on the N by the dep. of Drome; on the E by that of Basses-Alpes; on the S by that of the Bouches-du-Rhone; and on the W by the dep. of Gard. Area 1,328 sq. m. Pop. in 1841, 251,080; in 1852, 264,618. Its surface in the NE is mountainous, being traversed by branches of the Alps, some of which, such as Monts Ventoux, Lure, and Lebéron, rise to a considerable elevation. The W section consists of a broad valley extending along the Rhone. The principal rivers are the Rhone and the Durance. The former receives the Aigues and the Sorgues; the latter, the Cavallon. The soil is rich in the low grounds; in the mountains it is stony and unproductive. About one-half of the surface is arable. The climate is warm and suitable to the culture of fruit, silk, olives, vines, maize, and wheat. To pasturage, from deficiency of moisture, it is less favourable, and the cattle are in general of a diminutive breed. The manufactures consist of silk-stuffs, velvet, woollen cloths, and, on a small scale, linen, leather, paper, and vegetable essences. The dep. belongs to the diocese of Avignon, and to the jurisdiction of the court of Nimes. It is divided into the 4 arrondissements of Avignon, the capital, Orange, Carpentras, and Apt, which are subdivided into 22 cantons, and 149 communes.

VAUCLUSE, a village and remarkable fountain in the SE of France, which gives name to the above dep. The village is romantically situated on the r. bank of the Sorgues, 15 m. E of Avignon, at an alt. of 107 metres = 351 ft. above sea-level. Pop. 455. The fountain, which issues from an immense cavern overhung and surrounded by huge rocks and mountains, 1 m. E of the village, is remarkable chiefly for the quantity of water discharged, which forms at once a river, the Sorgues, an affluent of the Nesque, capable of driving mills and bearing boats. In summer, and during dry seasons, its waters issue tranquilly from a vast basin of a semicircular form; but in spring, and after heavy falls of rain, they overflow the basin, and precipitate themselves over huge fragments of rock in a number of cascades. The valley of the V. is extremely narrow, bounded by high rocks of a brownish grey tint, relieved by a sprinkling of olive and fig trees. This fountain, the VALCHIUSA of the Italians, is celebrated for the ill-fated loves of Petrarch and Laura, whose residence was in the vicinity.

VAUCOULEURS, a town of France, dep. of the Meuse, situated on the declivity of a hill on the l. bank of the river Meuse, 14 m. SE of Commercy. Pop. 2,600. It has considerable cotton factories.

VAUD (PAYS-DE), WAADT, or WAADTLAND, a canton of Switzerland, bounded on the N by the cant. and lake of Neufchatel; on the E by the cant. of Friburg; on the S by the lake of Geneva; on the W by France. Its superficial extent is nearly 1,180 sq. m. It is in general less mountainous than most other parts of Switzerland, consisting of beautiful valleys and plains intersected by small cultivated hills. The ridge of the Jorat forms the dividing line in this cant. between the basin of the Rhine and that of the Rhone. The principal rivers are the Rhone, the Orbe, and the Broye. The tract bordering on the lake of Geneva, in particular, may be ranked among the most beautiful districts in Europe. The valleys and plains are appropriated to the culture of corn; the eminences and hills, to that of vines. The climate, comparatively mild in the W of the canton, becomes colder towards the E, from its vicinity to mountains. Part of the arable ground is applied to the culture of hemp and flax; other districts to pasturage, plantations, and orchard grounds. The mineral products are iron, salt, coal, lead, sulphur, and gypsum. The chief exports are wine, cattle, leather, filberts, walnuts, chestnuts, and cheese.—The pop. in 1852 was 199,585, being an increase of 15,968 since 1837. Of the pop. in 1852, 5,293 were returned as foreigners; 6,962 as Roman Catholics; and 388 as Jews. The inhabitants are mostly strict Calvinists, and remarkable for the careful education given to their youth. Geneva is in the vicinity; and the chief town of the cant. Lausanne has a seminary for theological study. At another of its towns, Yverdun, is the well-known institution founded by Pestalozzi. The current language is French, spoken with purity in the towns, but only as a dialect in the country. The chief towns are Lausanne, with a pop. of 14,500; Vevey, with 5,201; Yverdun, with 3,619; Morges, with 3,941.—After the decline of the Roman empire, the Pays-de-Vaud formed a part of the kingdom of Burgundy, and was afterwards annexed to Savoy. It was conquered by the Swiss, and was annexed to that country as a dependency of the canton of Berne; but was acknowledged as a separate canton in 1803 and 1814. Its government is less democratic than that of most of the other cantons.

VAUDEMONT, a town of France, in the dep. of the Meurthe, situated in a fertile country, 22 m. SSW of Nancy. It was formerly the chief place of a county, from which a branch of the Lorraine family, now extinct, took the title of prince of Vaudemont.

VAUDOIS, an interesting race of people inhabiting the valleys which descend from the sides of the snowy Alps into the rich plains of Piedmont. The best known of these valleys is that of Susa, due W of Turin, through which passes the great route of the Mont Cenis, the principal highway between Italy and France, and down which flows the Dora to unite its waters with those of the Po at Turin. Just S of the valley of Susa, and betwixt the Mont Cenis and Monte-Viso, are the valleys of the Vaudois. The largest and most northerly of these valleys is that of Perouse, extending from the town of Pignerol to the town of Perouse, and there branching off in a NW direction into the valley of Pragelas, and W into the valley of St. Martin; which latter again divides into two, namely, that of the Balsille and that of Prali. More southerly is the valley of Luzern, or Lucerna, which runs W from the town of Luzern to the roots of the Col-de-la-Croix and Monte-Viso; and this valley is joined at its embouchure upon the plain by the smaller valley of Angrogna. The principal valleys occupied by the Vaudois are that of Perouse, watered by the Clusone; that of St. Martin, watered by the Germanasea; that of Angrogna, with a streamlet of the same name; and that of Luzern, watered by the Pelice. All these streams flow at no great distance into the Po, which, issuing from the foot of Monte-Viso, fertilizes the great plains of Piedmont and Lombardy, and at length pours its noble flood into the Adriatic. The Italian name of this remarkable community of Protestants is VALDESI, or VALLESI, that is, 'inhabitants of the valli or valleys;' their French name is Vaudois; and they are not unfrequently designated by English writers WALDENSES, but they are not to be confounded with the Waldenses of the south of France, who occupy the valleys on the French side of the Alps near Briançon, from whom they are separated only by the breadth of the Alpine range: still less are they to be indentified with the Vaudois of Switzerland, so called from inhabiting the Pays-de-Vaud, on the N side of the lake of Geneva. "It is somewhat remarkable," says a recent most intelligent tourist, "however, that the V. of Piedmont, though historically distinct from the Waldenses of the S of France, should speak the French language, or a *patois* which is more French than Italian, whilst their ecclesiastical language is also French; but they are now teaching the Italian in all their schools; they consider themselves as belonging to Italy, not only geographically and politically, but by feeling; and they regard Italy as their future field of usefulness." The V. trace their history with good documentary evidence to the 11th cent.; but they confidently believe that their ancestors through every age from the Apostolic time to the present have been the depositaries of the simple faith of the Gospel, and protesters against the corruptions of the church of Rome. In 1476, their first persecution was commenced by Zolande, sister of Louis XI., and widow of a duke of Savoy. In 1486, Pope Innocent VIII. issued a bull of extermination against them. England repeatedly interfered to shelter them from persecution; but on the revocation of the edict of Nantes they were nearly exterminated. On the 17th of February 1848, full civil and religious rights were awarded to the V by a special edict of the Piedmontese government.

VAUDREVANGE, or WALLENFANGEN, a village of Prussia, in the prov. of the Rhine, and regency of Treves, circle and 2 m. NW of Sarrelouis, on the l. bank of the Sarre. Pop. 1,320. It has manufac tories of china. Copper and azure stone are found in the vicinity.

VAUDREY, a village of France, in the dep. of the Jura, cant. and 4 m. S of Montbarey. Pop. 742.

VAUFFELIN, a village of Switzerland, in the cant. of Berne, and bail. of Courlelarn. Pop. 255.

VAUFREY, a village of France, in the cant. of the Doubs, cant. and 6 m. ENE of St. Hippolyte-sur-le-Doubs. Pop. 300.

VAUGHAN, a township of Upper Canada, in the Home District, watered by branches of the Humber. Pop. in 1842, 4.300.

VAUGIRARD, a commune of France, in the dep. of the Seine, cant. and 6 m. N of Sceaux, and contiguous to Paris on the SW. Pop. 13,817. It contains a large number of handsome dwellings, and possesses numerous manufactories of chemical substances, Prussian blue, chimney tiles, candles, glue, musical instrument strings, paper, card-board, pottery, pipes, Roman cement, bricks, articles in caoutchouc, sealing-wax, starch, hardware, machinery, and leather. Freestone is wrought in the environs, and large numbers of milk-cows are kept for the supply of the metropolis. Fire-wood and charcoal form also important objects of trade. V. was formerly a seignory belonging to the abbey of St. Germain-des-Pres at Paris.

VAUGNERAY, a canton and commune of France, in the dep. of the Rhône, and arrond. of Lyon. The cant. comprises 15 com. Pop. in 1831, 13,949; in 1846, 16,609. The village is 9 m. W of Lyon. Pop. 1,970. Coal and lead are found in the environs.

VAUGUEUX (Le), a hamlet of France, dep. of Calvados, cant. and com. of Caen. Pop. 1,500.

VAUJANY-EN-OYSANS, a village of France, in the dep. of the Isere, cant. and 9 m. NNE of Bourg-d'Oysans. It has quarries of marble.

VAUJOURS, a village of France, in the dep. of the Seine-et-Oise, cant. and 7 m. SE of Gonesse. Pop. 600. It has a fine castle.

VAULH, a village of France, in the dep. of the Yonne, cant. and 3 m. W of Avallon. On an adjacent hill are the ruins of a small temple of Mercury.

VAULION, a village of Switzerland, in the cant. of Vaud, and circle of Orbe, at the foot of Dent-de-Vaulion. Pop. 1,000. The Dent-de-Vaulion is in the Jura chain, between the valleys of Romainmotiers and Joux, and at an alt. of 3,420 Parisian ft. above the level of the lake of Geneva. The Col-de-Vaulion, which is a carriage way, is 2,430 Parisian ft. above the level of the lake.

VAULRUZ, a village of Switzerland, in the cant. of Friburg, and bail. of Greierz. Pop. 496.

VAULVANEYS-LE-HAUT, a commune of France, in the dep. of the Isere, and cant. of Vizille. Pop. 1,664.

VAULX, a department and commune of Belgium, in the prov. of Hainault, and arrond. of Tournai, watered by the Scheldt. Pop. 996.—Also a department and commune of the same prov., in the arrond. of Charleroi, watered by the Blanche. Pop. 152. It has several marble-works.

VAULX-VRAUCOURT, a commune of France, in the dep. of the Pas-de-Calais, cant. and 5 m. SE of Croisille. Pop. 1,640.

VAUMARCUS, or Famergu, a village of Switzerland, in the cant. and 11 m. SW of Neufchatel, and bail. of Boudry, on the lake of Neufchatel. Pop. 200. It has a castle and fine gardens.

VAUMAS, a village of France, in the dep. of the Allier, cant. and 6 m. SW of Dompierre. Pop. 815. It has several iron-works.

VAUMEIL, a village of France, in the dep. of the Basses-Alps, cant. and 6 m. SW of La Motte-du-Caire. Pop. 600.

VAUNAVES, a village of France, in the dep. of the Drome, cant. and 3 m. N of Brest. Pop. 400.

VAUNAVEY-LE-HAUT, a village of France, in the dep. of the Isere, cant. and 5 m. NNE of Vizille. Pop. 1,185.

VAUPE'. See Uaupes.

VAUQUELIN, or Wiles (Cape), a headland of South Australia, in Eyre Land, on the W side of Sleaford bay, in S lat. 36° 57', and E long. 135° 38'.

VAUREAS. See Valreas.

VAUREILLES, a village of France, in the dep. of the Aveyron, cant. and 6 m. N. of Monbazens. Pop. 350.

VAURENARD, a town of France, in the dep. of the Rhone, cant. and 5 m. from Beaujeu. Pop. 1,074. It is noted for its cheese.

VAUTEBIS, a market-town of France, in the dep. of the Deux-Sevres, cant. and 6 m. NW of Menigoutte, in a woody locality. Pop. 489.

VAUTORTES, a commune of France, in the dep. and 12 m. W of Mayenne. Pop. 1,632.

VAUVENARGUES, a town of France, in the dep. of the Bouches-du-Rhone, and cant. of Trets, 8 m. ENE of Aix, on a plateau commanding a valley. Pop. 400. It has a castle, said to have been built in the 14th cent.

VAUVERT, a canton, commune, and town of France, in the dep. of the Gard, and arrond. of Nismes. The cant. comprises 12 com. Pop. in 1831, 15,710; and in 1846, 16,668. The town is 12 m. SW of Nismes. Pop. 4,175. It has a Calvinist consistorial church, manufactories of silk, hosiery, oil-mills, distilleries of brandy and liqueurs, and saltpetre works. Wine is cultivated in the locality.

VAUVILLE, a village of France, in the dep. of the Manche, on the N coast. Pop. 720.

VAUVILLERS, a canton, commune, and market-town of France, in the dep. of the Haute-Saone, and arrond. of Lure. The cant. comprises 23 com. Pop. in 1831, 11,150; and in 1846, 11,666. The town is 28 m. NW of Lure. Pop. 1,246. It has manufactories of candles, tanneries, glass-works, lime and tile-kilns, &c.

VAUVRIER, a village of Switzerland, in the cant. of Valais, on the Rhine. Pop. 950.

VAUX, a commune of France, in the dep. of the Rhone, cant. and 10 m. NW of Villefranche-sur-Saone. Pop. 1,934.—Also a commune in the dep. of the Seine-et-Oise, cant. and 2 m. E of Meulan, at the foot of hills, on the r. bank of the Seine. Pop. 1,025. It contains an ancient castle belonging to the Dukes de Praslin, and many other fine residences. Gypsum is wrought in the environs.—Also a commune of Belgium, in the prov. of Liege, and dep. of Vaux-et-Borset. Pop. 735.

VAUX (La), a district of Switzerland, in the cant. of Vaud, to the E of Lausanne. It is noted for the excellence of its wine, and for the number of the terraces which here rise amphitheatrally from the banks of the Geneva. The district comprises the circles of Saint Saphorin, Cully, and Lutry, with a pop. of 6,810 souls.

VAUX (Saint-Laurent-de), a village of France, in the dep. of the Rhone, and cant. of Vangueray, 11 m. from Lyon. Pop. 1,130.

VAUXAINS, a village of France, dep. of the Dordogne, cant. and 4 m. SW of Riberac. Pop. 1,820.

VAUXCHAVANNE, a department and commune of Belgium, in the prov. of Luxemburg, and arrond. of Marche. Pop. 385.

VAUX-CHIVREMONT, a department and commune of Belgium, in the prov. and arrond. of Liege, watered by the Vesdre. Pop. 2,262.

VAUX-EN-ARROUAISE, a commune of France, in the dep. of the Aisne, and cant. of Wassigny, 2?

m. NW of Veronas. Pop. 1,621. It has manufactories of shawls.

VAUX-ET-BORSET, a department and commune of Belgium, in the prov. of Liege, and arrond. of Huy. Pop. 735.

VAUX-SUR-BLAISE, a village of France, in the dep. of the Haute-Marne, cant. and 3 m. SE of Vassy, near the Blaise. Pop. 350. It has manufactories of cotton fabrics.

VAUX-SUR-MER, a village of France, in the dep. of the Charente-Inferieure, cant. and 3 m. NW of Royan. Pop. 400.

VAUX-DE-VIRE (Les), a valley of France, in the dep. of Calvados, near Vire. It has a large paper-mill, and a manufactory of Prussian-blue.

VAVAO, a small archipelago of the Pacific, in the group of the Friendly Islands, in S lat. 18° 30′ 2″, and W long. 173° 55′. The principal island of the group, which also bears the same name, is 12 m. in length and 6 m. in medium breadth. Pop. 6,000. The S coasts are deeply indented and form several narrow peninsulas. On the E side are port Valdes and the roadstead of Del Refugio.

VAVINCOURT, a canton and commune of France, in the dep. of the Meuse, and arrond. of Bar-le-Duc. The cant. comprises 15 com. Pop. in 1831, 8,260; in 1846, 7,452. The village is 6 m. NE of Bar-le-Duc. Pop. 772.

VAVIORKA, a town of Russia in Europe, in the gov. of Grodno, district and 15 m. SW of Lida.

VAVITAO, or Raivavae, an island of the South Pacific, in the group of the Austral islands, in S lat. 23° 42′, and W long. 147° 22′. It rises to a considerable height, but is generally fertile. Its inhabitants speak the dialect of Otaheite. It was discovered in 1791 by Broughton, and received from him the name of High island.

VAXHOLM, a town of Sweden, in the prefecture of Stockholm, haerad of Danderyd, and parish of Rid, in the island of Vaxoe. It has 4 ports opposite Vaxholm. On a rocky islet is a fort erected for the defence of the Stockholm canal.

VAXOE, an island of Sweden, in the prefecture of Stockholm, haerad of Danderyd, and parish of Rid. It contains the town of Vaxholm.

VAY, a commune and village of France, in the dep. of Loire Inferieure, cant. and 5 m. W of Nozay. Pop. 2,000.

VAYA, an island of the Orinoco, one of those which, with the point of Galera, off the island of Trinidad, form the entrance of the bay of Charaguanas.

VAYGARU, a river of Hindostan, in the Carnatic, which rises on the E flank of the Western Ghauts; runs E and then SE, passing Madura, Patinur, and Peringary; and falls into the gulf of Manaar, after a course of 180 m.

VAYNOR, a parish of Breconshire, near Brecon. Pop. in 1841, 2,286.

VAYRES, a commune and town of France, in the dep. of Gironde. Pop. 1,600.

VAZABARRIS, a river of Brazil, in the prov. of Sergipe-del-Rey, which rises near the coast; runs SSE; and enters the bay of Sergipe, after a course of 120 m.

VAZIHABAD, a town of Hindostan, on the Punjab, 62 m. NW of Lahore, near the l. bank of the Chenab. It is a handsomely built place.

VEAGH, or Beagh, a glen and a lake in the p. of Gartan, 10 m. NW by W of Letterkenny, co. Donegal. The glen is usually called Glenveagh, and forms a continuation of Glendowan on the S, while itself is continued by the mountain-vale called emphatically the Glen on the north. The lake, the wildest and the most picturesque of the lakes of Donegal, extends 3 m. NE, with a mean width of between 2 and 3 fur.; it has a surface-elevation of 149 ft. above the level of the sea; and, notwithstanding its limited width, occupies the entire breadth of the lower part of the glen.

VEARN, a village of Herefordshire, 6 m. NNE of Hereford.

VEBRET, a village of France, in the dep. of Cantal, cant. and 2 m. ENE of Saignes. Pop. 1,200.

VECHEL, a large village of Holland, in the prov. of N. Brabant, 12 m. ESE of Bois-le-Duc. It is traversed by the Aa.

VECHELDE, a village of the duchy and 6 m. W of Brunswick, on the Aue. Pop. 300.

VECHENSKAIA, a town of Russia, in the Don Cossack territory, on the l. bank of the Don, 60 m. SE of Bogutshar.

VECHERA, a village of Turkey, on the l. bank of the Upper Kamchik, in N lat. 42° 50′, E long. 26° 30′, 9 m. NNE of Selimné. It consists of about 80 houses.

VECHINGEN, a parish and village of Switzerland, in the cant. of Berne, on the r. bank of the Worblen. Pop. 2,597.

VECHT, an arm of the Rhine, in the Netherlands, which separates from that river near Utrecht, flows northward, and falls into the Zuyder-Zee at Muiden, after a course of 18 m.

VECHTA, a town of Germany, in that part of the bishopric of Munster given to Oldenburg in 1802, on a river of the same name, 28 m. S of Oldenburg. Pop. 1,976. It gives name to a circle and bailiwic.

VECHTE, a river of the Netherlands, which rises in the Prussian gov. of Munster; runs NW and then SW through Hanover; enters Holland; and falls into the Zuyder-Zee above Swarte-Sluys, after a NW course of 90 m.

VECKERHAGEN, a village of Hesse-Cassel, on the Weser, 13 m. N by E of Cassel. Pop. 1,900. A saltpetre manufactory was established here in 1815.

VEDDAHS. See article Ceylon.

VEDANO, a village of Austrian Lombardy, in the prov. and 12 m. N of Milan. Pop. 1,200.

VEDEM, a town of Prussian Westphalia, in the duchy and 9 m. SSE of Cleves.

VEDEN, a village of the Netherlands, in N. Brabant. Though at a distance from the sea, it has the benefit of water communication by its vicinity to a navigable river.

VEDRA (Cape), a cape in the NW of Spain, on the coast of Galicia, in N lat. 42° 19′.

VEDRENE, or Vedrin, a village of Belgium, in the prov. and 3 m. N of Namur. Pop. 1,400.

VEEN, a village of Holland, in N. Brabant, 10 m. NW of Hertogenbosch.

VEENDAM, a town of Holland, in the prov. and 15 m. SE of Groningen. Pop. 5,200. It has saw-mills, tanneries, rope-works, and boat-building yards.

VEENENDAAL, a village of Holland, in the prov. of Utrecht and cant. of Rheenen, 12 m. SE of Amersfoort. Pop. 2,960. It has a yarn-mill.

VEENHUFEN, a village of Hanover, in the bail. of Leer. Pop. 260.

VEENHUESEN, a parish of Holland, in the prov. of North Holland, to the NE of Hoorn. Pop. 300.—Also a parish in the prov. of Drontheim. Pop. 5,000.

VEERAWOW, a town of the Rajput territory of Parkur, on the NW frontier of India, situated on a lake about 3 m. in circumf., in N lat. 24° 31′ 6″, near the ruins of the ancient city of Pareenuggur.

VEERE. See Ter-Vere.

VEERLE, a department and commune of Belgium, in the prov. and 29 m. ESE of Antwerp, and arrond. of Turnhout. Pop. of dep. 1,395; of com. 790.

VEERSSEN, a village of Hanover, in the prov. of Luneburg and bail. of Oldenstadt, near Uelzen. Pop. 385.

VEERST, a parish of Denmark, in Jutland, in the stift of Ripen, to the W of Fredericia.

VEERT, a village of Prussia, in the regency of Dusseldorf and circle of Geldern. Pop. 684.

VEFSEN-ELV, a river which has its source in Sweden, in the prefecture of West Bothnia, in a lake in the midst of the Dofrines, flows thence into Norway, traverses the S part of the bail of Nordland, and after a course in a generally NW direction of about 90 m. throws itself into the Vefsen-fiord.

VEFSEN-FIORD, a parish of Norway, in the diocese and bail. of Nordland, on a bay of the Atlantic of the same name, and at the mouth of the Vefsen-elv, 195 m. NNE of Drontheim. Pop. 3,500.

VEGA, a village of Spain, in the prov. and 45 m. NW of Oviedo. Pop. 525.

VEGA (LA), a village of the island of Porto-Rico, in the jurisdiction of San Juan, on a river of the same name, 6 m. S of Toa-Alta. Pop. 1,230.

VEGA (LA), or LA CONCEPCION-DE-LA-VEGA, a town of the island of Hayti, in the dep. of the Cibao, on the Camus, in a fine plain 75 m. NNW of Sante Domingo. About 6 m. W are the ruins of a town of the same name, founded by order of Columbus, and destroyed by an earthquake in 1564.

VEGA-DE-ESPINAREDA, a market-town of Spain, in the prov. of Leon, and 9 m. from Pontferrada, in a locality watered by the Cua, and liable to extreme variations of temperature. Pop. 810. It has a Benedictine monastery. The manufacture of linen is its chief branch of industry.

VEGA (NUESTRA-SENORA-DE-LA), a market-town of Spain, in the prov. and 24 m. SSE of Santander, in a cold mountainous locality. Pop. 4,055. It has a parish-church and a custom-house.

VEGA-DEL-RIO-TIRON. See HERRAMELLURI.

VEGA-DE-RIO-PONCE, a market-town of Spain, in the prov. and 33 m. from Leon, near Valderaduey. Pop. 500. It has two parish churches.

VEGA-DE-VILLALOBOS, a village of Spain, in the prov. and 33 m. from Leon. Pop. 290.

VEGANZONES, a market-town of Spain, in the prov. and 18 m. NE of Segovia, in a fine plain, near the Cega. Pop. 615.

VEGAS-DEL-CONDADO, a village of Spain, in the prov. and 12 m. from Leon. Pop. 217. It has a palace belonging to the dukes of Frias, and possesses manufactories of linen, a tile-foundry, and productive trout fisheries.

VEGAS-DE-MATUTE, a market-town of Spain, in the prov. and 12 m. from Segovia, in a level tract near the Sierra-de-Guadarrama. Pop. 635.

VEGAS-DE-MONTE, a village of Spain, in the prov. of Segovia, in a level tract encircled by the Sierra-de-Guadarrama. Pop. 552. It is of modern foundation, and has a parish church, a custom-house, and a public granary. Lime burning and cattle-rearing form the chief objects of local industry.

VEGEN-OE, an island of Norway, near the coast of Nordland, with a small village on its NW side.

VEGESACK, a bailiwick and market-town of the territory of Bremen, on the Weser, 12 m. NW of Bremen, in N lat. 53° 10′ 31″. It has building docks, an iron foundry, and several breweries and brandy distilleries. Pop. 2,460.

VEGIA, or BEJA, a town of Tunis, near the frontier of Algeria, on the Wady-Zain, in N lat. 36° 42′. It is noted for its horses, and carries on an active trade in corn.

VEGLIA, an island of Dalmatia, in the gulf of Quarnero, to the E of the island of Cherso, and separated from the continent on the E by the Morlacca channel. Its N point is in N lat. 45° 14′ 20″, and E long. 14° 31′ 30″. Pop. 17,000. It is about 24 m. in length from NW to SE, and 15 m. in extreme breadth. The surface is mountainous and rocky, sterile in the N and E, but in other parts fertile. It abounds in wood, pastures large numbers of horses, sheep, and goats, and yields considerable quantities of fruit, wine, silk, and salt. It has quarries of fine marble, and fish abound on its coasts. The inhabitants are chiefly of Morlachian descent, and are notorious for their superstition and credulity. There are on the coast several spacious harbours, but none sufficiently protected from the gales which prevail in the Adriatic.—The capital, which bears the same name, is 81 m. SE of Trieste, on the SW coast of the island. Pop. 3,500. It is built partly on a hill, and is commanded by two mountains. It is a bishop's see, and has a cathedral, and several monasteries. The harbour is capable of receiving vessels of medium size, and is defended by a castle.

VEGLIE, a market-town of Naples, in the prov. of the Terra d'Otranto, district and 24 m. S of Brindisi, in a plain. Pop. 1,355. It has a church, a convent, and two charitable institutions.

VEGUILLAS, a town of Spain, in the prov. and 21 m. from Guadalaxara, at the foot of the Sierra-de-Zarzuela, in a low and bleak locality. Pop. 200.

VEHAM, an island of the S. Pacific, in the N part of the group of the Arru islands, to the SW of New Guinea. It has a village named Vorkey. Near it is a pearl bank.

VEHLAGE, a village of Prussia, in the regency of Minden and circle of Lubbeke. Pop. 758.

VEHLEFANZ, a village of Prussia, in the regency of Potsdam and circle of Osthavelland, near Kremmen. Pop. 680.

VEHLEN, a village of the principality of Schauenburg-Lippe, in the co. of Schauenburg and bail. of Bückeburg, on the Aue. Pop. 410. It has a coal-mine.—Also a village in the regency of Magdeburg and circle of Jerichow. Pop. 200.

VEHLGAST, a village of Prussia, in the regency of Potsdam and circle of West Priegnitz. Pop. 260.

VEHLIN, a village of Prussia, in the regency of Potsdam and circle of Ost-Priegnitz. Pop. 390.

VEHLINGEN, a village of Prussia, in the regency of Dusseldorf and circle of Rees. Pop. 540.

VEHLINGSDORF, a village of Prussia, in the regency of Stettin and circle of Saatsig. Pop. 240.

VEHLITZ, a village of Prussia, in the regency of Magdeburg and circle of Jerichow, on the Elbe. Pop. 360. It has several mills.

VEHLOW, a village of Prussia, in the regency of Potsdam and circle of Ost-Priegnitz. Pop. 280.

VEHRA, a village of Prussia, in the regency of Erfurt and circle of Weissenser. Pop. 220.

VEHRAB-JAGOTE, a village of Beluchistan, in the district of Lus, on the mountains to the SE of Beila.

VEHRINGEN, a village of the principality of Hohenzollern-Sigmaringen. Pop. 880.

VEHRINGENDORF, a village of the principality of Hohenzollern-Sigmaringen, on the Lauchart. Pop. 440.

VEHRNIKA. See LAYBACH (OBER).

VEHRTE, a village of Hanover, in the prov. and to the N of Osnabruck. Pop. 650.

VEHS, a village of Hanover, in the prov. of Osnabruck, to the N of Bersenbruck. Pop. 550.

VEHY, a district of Hindostan, in the prov. of Cashmere, traversed by the Jelum. It is noted for its saffron. Pamper is its chief place.

VEI, or VAI, a nation, or collection of tribes inhabiting the country near Cape Mount on the W coast of Africa, and some distance into the interior, about 5 days' sail SE of Sierra Leone

They are in possession of a written character of recent invention. The writing itself is syllabic. About 200 symbols—though about half of them appear superfluous, as they are never used—represent the sounds of all the syllables occurring in the language, which is of simple construction. The inventor, Doalu Bukara, a man of about 40 years of age, of great intelligence, and much religious feeling, lives about 20 m. in the interior. When a child he had for a few weeks learnt the ordinary phonetic alphabet from an American missionary. He states that after he was grown up, he received in a dream the first impulse to express his language by written syllabic symbols. He imparted the dream to a few companions, who assisted him to invent the characters, and procure through the favour of the king of the tribe the means of establishing schools and of teaching the people generally. War soon broke out, Jondu, the town of the Veis, was destroyed by fire, and the tribe depressed and dispersed. They have not had schools since; but it seems very probable that they may be re-established in the new town of Bandakoro—literally 'cotton-tree ground'—so named from the abundance of cotton trees growing near. There are as yet no *data* for stating with precision the geographical extension of the Vei language; but several closely allied to it—the Mandingo, Bambarra, Mendi, Susu, &c., are spoken over a very large area, stretching further than from the Gambia to the Gallinas rivers, over a seaboard of 650 m., and extending probably 300 m. into the interior. It has been generally asserted and believed that among the 150 supposed dialects of Africa, not one has till now been raised by the natives themselves to a written language

VEIGNE, a village of France, in the dep. of the Indre-et-Loire, 1½ m. E of Montazon. Pop. 890.

VEILE, a town of Denmark, in Jutland, in the stift and 45 m. NE of Ripen, on a river of the same name, which is here crossed by a substantial bridge, and near its confluence with the Veile-fiord. Pop. 1,600. It has three gates, a market-place, an hospital, a house of detention, and a small port admitting vessels drawing under 10 ft. water; and possesses manufactories of cloth, leather, soap, starch, hair-powder, and earthenware, and productive salmon fisheries in the Veile-fiord. The trade consists chiefly in wood, grain, and tallow.—The V.-fiord is formed by the Cattegat, and opens between Cape Biörus-knude on the NE, and the Trelde-naes on the SW. It is 9 m. wide at the entrance, and about 6 m. in length, with a depth of 8 fath., and gradually shoaling water. The town is situated at the head of the bay.

VEILLY, a village of France, in the dep. of the Côte-d'Or, cant. and 5 m. WNW of Bligny-sur-Ouche. It has a mineral spring.

VEILSDORF, a village of Saxe-Meiningen, near Hildburghausen. Pop. 840.

VEIROS, a town of Portugal, in the prov. of Alemtejo. It has a castle.—Also a town of Brazil, in the prov. of Para, on the r. bank of the Xingu, about 60 m. above its confluence with the Amazon.

VEISCHEDE, a village of Prussia, in the regency of Arnsberg, and circle of Olpe. Pop. 200.

VEISENCHTEIN. See WISSENSTEIN.

VEISSEMBURG. See WISSEMBURG.

VEIT, a village of Bavaria, circle of Upper Bavaria, presidial of Neumarkt, on the Rott. Pop. 240.

VEIT, or VEIT-AN-DER-WEIN (SANKT), a village of Austria, in the circle of the Lower Weinerwald, 5 m. W of Vienna, on the Wein. Pop. 1,220. It consists of an upper and a lower town. In the former is a castle formerly belonging to the archbishop of Vienna. The latter contains manufactories of white lead and Prussian blue.

VEIT (SANCT), a town of Austria, in Illyria, in the circle and 15 m. N of Klagenfurt, on the Glan, in N lat. 46° 44′ 52″, and E long. 15° 37′ 30″. Pop. 1,525. It has manufactories of iron-ware and of paper. Until 1518 this town was the capital of Carinthia.—Also a market-town in the circle of the Upper Weinerwald, 12 m. S of St. Polten. Pop. 1,020. It has an iron-work, and carries on an active trade in wood. This town occupies the site of the ancient castle of Hohenstaufen destroyed in 1299.—Also a village in the circle of the Muhl, and bail. of Wachsenberg. Pop. 360.—Also a village near Salzburg Pop. 310.—Also two villages of Styria, in the bail. of Rothenthurm and Buchenstein, containing respectively 250 and 380 inhabitants.—Also a village in the circle of Judenburg. Pop. 330.—Also a village of Tyrol, circle of Bruneck. Pop. 420.

VEIT (SANKT), or VEIT-AN-DER-TRIESTING, a village of Austria, in the circle of the Lower Weinerwald, 21 m. SSW of Vienna, on the Triesting. Pop. 918. It has copper-works, and extensive manufactories of machinery.

VEIT-AN-FLAUM (SANKT). See FIUME.

VEITLAHN, a village of Bavaria, circle of Upper Franconia, presidial of Culmbach. Pop. 180.

VEITSAURACH, a village of Bavaria, circle of Middle Franconia, presidial of Heilsbronn. Pop. 198.

VEITSBRONN, a village of Bavaria, in the circle of Middle Franconia, and presidial of Radolzburg, near Muhlen. Pop. 260.

VEITSCH (GROSS and KLEIN), two villages of Styria, in the circle of Bruck. Pop. 1,060. It has mines of iron and copper.

VEITSHOCHHEIM, a town of Bavaria, in the circle of Lower Franconia, presidial and 5 m. NW of Wurzburg, on the r. bank of the Main. Pop. 1,355. It has a fine castle.

VEITSRODE, a village of Oldenburg, in the bail. of Birkenfeld, near Oberstein. Pop. 300.

VEIVODE. See VOÏAVAT.

VEJA, a village of Austria, in Lombardy, in the del. of Verona.

VEJENOE, an island of Norway, near the W coast, in the bail. of Nordland.

VEKCHNY, a town of Russia in Europe, in the gov. of Vilna, district and 39 m. NW of Chavli.

VELA (CAPE), a promontory on the coast of the prov. of Santa-Marta, New Granada, in N lat. 12° 13′, W long. 72° 12′. It is lofty, and was discovered by Alonso de Ojeda, who gave it this name, in 1499.

VELA (LA), a port of Venezuela, on the SE coast of the gulf of Coro, in N lat. 11° 25′, 1½ m. E of the mouth of the river Coro.

VELAAR, or VELAUR, a river of Hindostan, which rises in the district of Salem, and flows across the Carnatic into the gulf of Bengal, in N lat. 10° 6′, in an E course of 120 m.—There is another stream of the same name, which rises in the Eastern Ghauts, and flows into the bay of Bengal, in N lat. 11° 29′.

VELADA, a village of Spain, prov. and 6 m. N of Talavera. Pop. 800.

VELAINE, a village of Belgium, prov. and 1 m. S of Namur, on the r. bank of the Sambre. Pop. 1,500.

VELAINES, a commune and village of Belgium, prov. of Hainault, 6 m. NE of Tournay. Pop. 2,150.

VELAIS-SUR-OUCHE, a village of France, in the dep. of Cote-d'Or, cant. and 7 m. W of Dijon. Pop. 500.

VELAS, a port of Costa-Rica, between Cape Santa Catalina and the Morro-Hermoso, in N. lat. 10° 30′.

VELATE, a village of Austrian Italy, in the prov. and 16 m. NE of Milan.

VELAUX, a town of France, dep. of Bouches-du-Rhone, 12 m. W of Aix, in the valley of the Arc. Pop. 1,180.

VELAY, a small and mountainous district of France, in the Cevennes, lying between the Vivarais on the E, Gevaudan on the S, Auvergne on the W, and Forez on the N, and now forming part of the department of Haute-Loire. Its cap. is Le Puy.

VELAZGHIRD, a town of Kerman, in Persia, on the river Karun, 54 m. NE of Gombrun.

VELBERT, a village of Prussia, in the duchy of Berg, 14 m. ENE of Dusseldorf, and 7 m. NW of Elberfeld. It is chiefly inhabited by artisans, who manufacture various articles of steel, iron, and brass.—

VELBURG, or VILDBURG, a town of Bavarian Franconia, on the Laber-Noir, 24 m. NW of Ratisbon.

VELDEN, a town of Austrian Illyria, in Carinthia, on the lake of Wordt, 13 m. W of Klagenfurt.—Also a village of Bavaria, in the presidial of Vilsbiburg, 10 m. SSE of Landshut.—Also a village of Bavaria, on the Regnitz, 21 m. NE of Nuremburg.

VELDENZ, or THAL-VELDENZ, a town of Prussia, prov. of the Rhine, 19 m. ENE of Treves.

VELDES, a village of Austrian Illyria, on the lake of Frauen, 20 m. S by W of Klagenfurt.

VELDHOVEN, a village of Holland, in N. Brabant, 4 m. W of Eindhoven.

VELDSCHTERIN. See USITERNA.

VELE, a river of France, which joins the Aisne on the l. bank, between Soissons and Vailly, after a NW course of 75 m.

VELEIA, a village of the duchy of Parma, 24 m. S of Piacenza, on the site of the ancient *Velleia.*

VELEN, a village of Prussia, in the regency of Muster, circle of Borken. Pop. 728.

VELENCZE, or VARAD-VELENCZE, a town of Hungary, 2 m. E of Varadein. Pop. 1,200.

VELE-RETE, a cluster of rocks in the Eastern seas, situated to the S of the island of Formosa, in N lat. 21° 55′, E long. 121° 30′. The largest of these rocks is about the height of a small ship's hull out of the water, and in clear weather may be discerned at the distance of 8 m. It is surrounded by several smaller ones.

VELESTINA, a town of Turkey, in the sanj. of Trikala, 24 m. SE of Larissa.

VELETCHICO, a mountain on the N frontier of Greece, under the parallel of 39° 13′, 9 m. N of Mount Chelona, and 12 m. NE of the town of Arta. It presents a bleak and barren surface.

VELEZ, a city of New Granada, in the prov. of Tunja, on the river Suarez, at the foot of a ridge of mountains, in N lat. 6° 10′. Pop. 2,500. It is built on swampy soil, and its streets are at times impassable in consequence. The temperature is moderated by the frequency of tempests in the surrounding regions. The city contains a very handsome church and two convents.

VELEZ-DE-BENAUDALLA, a town of Spain, in the prov. and 30 m. SE of Granada. Pop. 3,000. Lead is wrought in the vicinity.

VELEZ-EL-BLANCO, a town of Spain, in the prov. and 57 m. NE of Almeria, on the borders of Murcia, 4 m. NNE of Velez-el-Rubio. The pop. is considerable, but chiefly agricultural.

VELEZ-EL-RUBIO, a town of Spain, in the prov. and 53 m. NNE of Almeria, on the Guadalentin river. Its pop. amounts to 11,000; but notwithstanding this number, and its having been a place of strength under the Moors, it is neither remarkable for its buildings nor the industry of its inhabitants. Its manufactures consist chiefly of coarse woollen and hemp fabrics, oil, and flour.

VELEZ-MALAGA, a town of Spain, in the prov. and 14 m. NE of Malaga, in N lat. 36° 45′. Pop. 13,000. It stands on the slope of a hill, amidst vineyards and plantations, the produce of which, consisting of raisins, olive oil, lemons, oranges, citrons, and almonds, forms the chief articles of its trade. The town is 2 m. distant from the sea, and takes its name from the river Velez, which passes its walls, flowing S from the chain of mountains separating Granada from Andalusia. The pop. of the town amounted to 16,000 till 1804, when nearly half that number were carried off by a dreadful fever which ravaged Cadiz, Malaga, and other parts of the S of Spain. In the 15th cent. this district formed the last retreat of the Moors, who built small towns on the tops of the hills, and carried on from these fastnesses an inveterate warfare against their enemies.

VELHAS (RIO DAS), a river of Brazil, in the prov. of Minas-Geraes, which runs N and enters the Sao-Francisco, near Bana-das-Velhas.—There is another river of the same name, which runs WNW, and enters the Paranaiba, after a course of 240 m.

VELICA, a town of Austrian Slavonia, between Gradiska and Zagrab.

VELICALA, a town of the peninsula of California, near the coast, in N lat. 20° 35′.

VELICSNA, NAGY-FALU, or WELKA WES, a village of Hungary, on the Arva, 32 m. N of Neusohl.

VELIJ, a town of Russia, in the gov. and 52 m. NE of Vitebsk, on an affluent of the Southern Dwina.

VELIKAIA, a river of Russia, which rises in the gov. of Pskov, and flows into the SE extremity of Lake Pskov, after a course of 100 m.

VELIKA-GUBAVIZA, a village of Austrian Dalmatia, on the river Cettina, which forms here a cataract nearly 150 ft. in height.

VELIKI-USTJUG, a town of Russia in Europe, in the gov. and 249 m. NE of Vologda, on the Soukhona, a little above its confluence with the Jug Pop. 10,000. It is nearly 5 m. in circumference, and although containing about 1,450 houses, not more than twenty are built of stone, and of the entire number not more than four-score are fit for habitation. It possesses a cathedral, 26 churches, 2 convents, a foundling hospital formerly a bank, a town-hospital, and a post-office, all built of stone. There are besides 3 salt-magazines, several establishments connected with the town-house, numerous taverns, several tanneries, manufactories of bricks, soap, tallow, and leaven. The era of the foundation of this town is unknown. It stood previous to 1212 on the r. bank of, and considerably lower down, the Sukhona, and nearly opposite the confluence of Jug, at a place now occupied by a convent and hermitage. In 1226, a Tartar chief, having embraced Christianity, founded a convent in the locality, and subsequently the existing cathedral, and to defend their new settlement, which they named Ustjug, surrounded it with an earthen rampart and ditch. From the advantages of its situation it rapidly rose in importance. In 1324 and again in 1398, it was taken and pillaged by the Novgorodians. In 1635 it was ravaged by pestilence, and again in 1761, by a disastrous inundation of the Soukhona.

VELIKIJA-LUKI, a town of Russia, in the gov. of Pskov, at the junction of the Kolomenka with the Lovat, 130 m. SE of Pskov. Pop. 4,000. It has a considerable trade in flax and hemp.

VELILLA-DE-EBRO, a market-town of Spain, in the prov. and 36 m. SE of Zaragoza, in a fertile locality, on the l. bank of the Ebro. Pop. 1,138.

VELILLA-DE-SAN-ANTONIO, a market-town of Spain, in the prov. and 12 m. ESE of Madrid, on the l. bank of the Jarama. Pop. 228.

VELILLE, a town of Peru, in the dep. and 75 m. SSE of Cuzco, and prov. of Chumbivilcas.

VELINES, a canton and commune of France, in the dep. of the Dordogne, and arrond. of Bergerac. The cant. comprises 16 com. Pop. in 1831, 9,548; and in 1846, 8,791. The village is 20 m. W of Bergerac, on an affluent of the Dordogne. Pop. 787.

VELINO, a mountain of Naples, in the prov. of Abruzzo-Ultra, and near the NW side of Lake Fucino. It has an alt. of 7,668 Parisian ft. above sea-level, and is one of the principal summits of the Apennines.—Also a river which has its source on the W side of the Apennines, in the NW part of the Neapoline prov. of Abruzzo-Ultra; flows thence into the Papal States; traverses the N part of the delegation of Rieti, and the S of that of Spoleto, and after a course first SW and afterwards NW of about 60 m., joins the Neva, on the l. bank, 6 m. above Terin. Its principal affluents are the Salto and the Turano, both of which it receives on the l. The V. forms fine cascades.

VELITZ-DAGH, a mountain of Turkey in Europe, in Rumelia, between the sanjaks of Ghiustendil and Salonica. It makes part of the chain separating the basins of the Vardar and Radovitz.

VELITZNA. See VELICSNA.

VELKOPOLYA. See HOCHWIESEN.

VELLACH, a town of Austrian Illyria, in Carinthia, on the river Moll, 11 m. NNW of Sachsenburg. It is divided into an upper and a lower town, and has in its vicinity mines and ironworks.

VELLAHN, a village of Mecklenburg-Schwerin, in the circle of Mecklenburg. Pop. 540.

VELLBERG, a village of Würtemberg, in the circle of Jaxt, on the Bühler. Pop. 596.

VELLETRI, a town of Italy, in the States of the Church, in the deleg. of Rome, on the declivity of Monte-Artimisio, 21 m. SE of Rome. Its pop. amounts to about 12,000. Though pleasantly situated, it is an ill-built and irregular town, the streets being narrow and dirty, while the houses bear in general the appearance of decay. It contains, however, several detached buildings of considerable beauty, such as the Palazzo-Ginetti, with its elegant front, and the Palazzo-Borgio, with its fine collection of paintings and antiques. The town-house, also, is a good building, and several of the fountains are handsome. The principal square contains a bronze statue of Pope Urban VIII. by Bernini. Velletri, originally a town of the Volsci, became at an early period a Roman colony, and being the seat of the Octavian family, had the honour of giving birth to Augustus.

VELLEXON, a village of France, dep. of Haute Saone, cant. and 4 m. NW of Frene-St.-Mametz.

VELLO, a town of Austrian Italy, in the gov. of Milan, deleg. of Sondrio, in the district of the Sette-Communi, 18 m. NW of Vicenza.

VELLON, a town of Spain, in the prov. and 24 m. E of Madrid. Pop. 700.

VELMERSTOOT, a mountain of Lippe-Detmold, forming the highest summit in the Teutoburgerwald. Alt. 1,460 Parisian ft.

VELSEN, a town of Holland, in the prov. of N. Holland, 5 m. N of Haarlem.

VELSHIN, a village of Bulgaria, 24 m. ESE of Shumla, near the N bank of the Paravati, and 5 m. NNW of the town of that name.

VELLORE, a fort of Hindostan, in the Carnatic, on the r. bank of the Palaar, 15 m. NW of Ascot.

VELP, a village of Holland, in the prov. of Gelderland, 4 m. ENE of Arnhem.

VELPKE, a village of the duchy and 21 m. NE of Brunswick. Pop. 500.

VELSIGNE, a town of Belgium, in E. Flanders, 7 m. ENE of Audenarde.

VELTHERIA, a village of Switzerland, in the cant. of Aargau, district of Brugg. Pop. 900.

VENADILLO, a settlement of New Granada, in the prov. of Mariquita, 14 leagues SW of Santa-Fe.

VENAFRO, a town of Naples, in the Terra-di-Lavoro, 30 m. NE of Gaeta. Pop. 2,800.

VENANGO, a county in the NW of Pennsylvania, U. S. Area 1,120 sq. m. It is watered by the Alleghany river, and various other smaller streams. Pop. in 1850, 17,900.—Also a township of Crawford co., Pennsylvania. Pop. 1,200.—Also a township of Butler co., in the same state. Pop. 822.

VENANT (SAINT), a town of France, dep. of Pas-de-Calais, situated in a marshy district on the Lys, 6 m. NE of Lillers. Pop. 1,000. It contains several oil-mills and breweries, rape-seed and barley being the products of the neighbourhood.

VENASCA, a town of Piedmont, situated on the river Vraita, 9 m. S of Saluzzo. Pop. 2,264.

VENASQUE, a town of France, dep. of the Vaucluse, near the river Nasque, 18 m. NE of Avignon. Pop. 1,100.

VENASQUE, or BENASCA, a small town and fortress of the NE of Spain, in Aragon, among the Pyrenees, near the source of the river Esuera, 50 m. N by E of Balbastro. In the neighbouring mountains are found silver, copper, and lead.

VENAUS, a village of the Sardinian states, in the div. of Turin, 2 m. WNW of Susa. Pop. 1,200.

VENCE, a town of France, dep. of Var, situated on the borders of Piedmont, 12 m. NE of Grasse.

VENDA (MONTE), the highest summit of the Euganean hills, in the Lombardian prov. of Padua, in N lat. 45° 18', E long. 11° 41'. It has an alt. of 300 toises = 1,918 ft. above sea-level.

VENDAIS, a village of France, in the dep. of Gironde, cant. and 7 m. NW of Lesparre. Pop. 1,600.

VENDE'E (LA), a department of France, comprising a part of Poitou; and bounded on the N by the deps. of Loire-Inferieure and Maine-et-Loire; on the E by Deux-Sevres; on the S by Charente-Inferieure; and on the SW and W by the bay of Biscay. Its area, 2,600 sq. m., is equal to two of the largest counties in Britain; its pop., thinly scattered, was 376,184 in 1846; in 1852, 383,734. Its surface is almost entirely level, presenting no eminence whose elevation exceeds 450 ft. It is divided into three parts; the *bocage* or wood, the *marais* or marsh, and the plain. The first, taking its name from the tangled coppice-woods and woody thickets with which it is covered, is fertile in corn, wine, and pasture. The marsh, comprising the stripe contiguous to the coast on the S and W, is a succession of sand-hills, heaths, bogs, and salt-pools; but the pasturages are rich, and are covered with cattle and sheep of a larger size than is usual in France. The plain, formed of the tongue of land comprised between the woody part and the N limit of the department, is fertile, and fit for various kinds of culture: here hedges, timber-clumps, coppice, heath, and furze, alternate with rich corn fields, and smiling vineyards. About two-thirds of the surface is arable. The principal rivers are the two Sevres, the Vendée, the Lay, the Vie, and the Autise; but the general intersections of the dep. are by small canals and ditches. The exports consist of corn, cattle, bay salt, coal, flax, and, in a smaller degree, of iron, wool, and hides. The dep. is divided into the three arrondissements of Napoleon-Vendée, Sables-d'Olonne, and Fontenay. These are subdivided into 30 cantons, and 294 communes. It is memorable for the resistance here made to the republican army in 1793, 1794, and 1795, a resistance singularly favoured by the woods, thickets, and ditches of the country. It was attended for a time with great success, though commenced without any concert with the other royalists of France, and carried on for a season with very limited support from England. This region was also the scene of some sharp fighting in 1815. The inhabitants are a credulous and simple race, attached to old usages, and easily led by their clergy.

VENDE'E, a river of France, which rises in the dep. of Deux-Sevres; traverses the department of La-Vendee; and discharges itself into the Sevre-Niortaise, above Marans, after a SW course of 45 m.

VENDEN, a town of European Russia, in Livonia, near the Aa, 33 m. E by N of Riga. Pop. 1,300.

VENDENHEIM, a town of France, in Alsace, with 1,100 inhabitants.

VENDEUVRE, a commune and town of France, dep. of Aube, 13 m. W of Bar-sur-Aube. Pop. 1,880.

VENDHUILE, a commune and village of France, in the dep. of Aisne, arrond. of St. Quentin. Pop. 1,262.

VENDOLA, one of the Admiralty islands, of which it is the most eastern, in S lat. 2° 14', E long. 148° 9'. It is about 3 m. in circuit, is covered with cocoa trees, and populous. Being surrounded by a reef of rocks, it is inaccessible almost to boats.

VENDOME, a town of France, the capital of the dep. of Loir-et-Cher, situated on the r. bank of the Loir, 30 m. NE of Tours. For an old town it is tolerably well-built, containing several churches, an hospital, and the remains of a once magnificent castle. It has also pleasant walks, but is chiefly remarked for its public school or college, an establishment with considerable endowments, and good buildings. V. has a population of 8,000, and manufactures of coarse cottons, paper, leather, and gloves. Its environs are fertile, and picturesque.

VENDOTENA. See PANTELARIA.

VENDRANGE, a village of France, in the dep. of the Loire, cant. and 5 m. W of St. Symphorien-de-Lay. Pop. 500.

VENDRELL, a judicial partido and market-town of Spain, in the prov. of Tarragona. The partido comprises 48 pueblos. The town is 42 m. W of Barcelona, on a height, 1½ m. from the shore of the Mediterranean, on which it has a small port, which is defended by two towers. Pop. 3,700. It has a parish-church and an hospital. Of its walls only portions now exist. The trade consists chiefly in wine and brandy.

VENDRES, a village of France, in the dep. of the Herault, cant. and 5 m. S of Beziers, on the bank of a lagune of the same name. Pop. 750. It has mineral baths.—The etang or lagune is 5 m. in length from N to S, 3 m. in breadth, and communicates on the SSE with the Mediterranean. It was formerly traversed by the Aude.

VENDRESSE, a village of France, in the dep. of Ardennes, 12 m. SSE of Mezières. Pop. 745. It has iron-forges.

VENEO, a district and town of Russia in Europe, in the gov. and 36 m. ENE of Toula, on the Venevka. Pop. 3,400.

VENER. See WENER.

VENERATA (MONTE), a mountain of Sicily, in the prov. of Messina, near Taormina, which rises to the height of 2,900 Parisian ft. above sea-level.

VENERIA-REALE, a market-town of Sardinia, in the div. and prov. and 6 m. NW of Turin, near the Stura. Pop. 2,780. It has a castle, with fine gardens, and a superb orangery. The culture of silk-worms forms the chief object of local industry.

VENERQUE, a town of France, in the dep. of the Haute-Garonne, cant. and 8 m. N of Auterive, on the r. bank of the Ariege, at the confluence of the Hesse. Pop. 1,100.

VENERSBORG. See ELFSBORG.

VENES, a commune of France, in the dep. of the Tarn, and cant. of Lautrec, 10 m. NNW of Castres. Pop. 1,210.

VENESME, a commune of France, in the dep. of the Cher, and cant. of Chateau-Neuf, near the Cher, 15 m. NW of Saint-Amand-Mont-Rond. Pop. 1,013.

VENETICO, a small island of the Ionian sea, Cape Gallo, near the SW extremity of the Morea, in N lat. 36° 41' 40", and E long. 21° 55' 20".

VENEUCO, a town of Sicily, in the prov. and district and 14 m. W of Messina.

VENEV, a town of Russia in Europe, in the gov. and 33 m. ENE of Tula, on the Venevka, an affluent of the Osetr. It contains, besides the cathedral, seven parish churches, and about 600 houses, and has manufactories of silk fabrics and sail-cloth.

VENEZIA. See VENICE.

VENEZUELA, a republic of South America, comprising the territories of the old Spanish captain-generalship of Venezuela or Caracas; and extending between the parallels of 2° and 12° S, from New Grenada on the W, and Brazil on the S, to the Caribbean sea on the N, and the Atlantic on the W. The boundaries with the coterminous states are somewhat undetermined. Its area has been estimated at 410,000 sq. m.

Physical features.] The great eastern chain of the Andes enters this republic from New Grenada, and runs in a direction nearly parallel with the coast, across the N part to the gulf of Paria. It is sometimes known under the name of 'the Maritime cordilleras of Caracas.' The most remarkable feature of the country is the vast plain which, extending from the Caqueta in New Grenada to the mouths of the Orinoco, constitutes the greater part of the surface. It chiefly consists of *llanos* or open plains, intersected by numerous rivers. The *mesas* or elevated table-lands, also constitute a peculiar feature of this country. The Orinoco is the great river of the country, and has its whole course within the frontiers of V. See article ORINOCO. The rivers are numerous: those belonging to the basin of the Orinoco alone exceed 400. Every valley has its stream: and though many of them are not of sufficient size to be navigable, yet all afford ample supplies of water to irrigate the plantations on their banks. The principal of these, which run from the mountains of Caracas and Coro into the Caribbean sea, are the Zulia, the Metatan, the Guiges, the Tocuyo, Aroa, Yaracuy, and the Tuy. The rivers which rise on the S side of the chain, and flow to the Orinoco, are the Guarico, which receives some of the branches of the Apure, and then following a course parallel to that river, enters the Orinoco a short distance E of it. The Guarico, a very fine river, is joined, near its confluence with the Orinoco, by the Rio-Maneapra which flows through the plains of Calabozo. The Iguane, the Cachivamo, and several others, which fertilise the vast uninhabited plains of the Orinoco, flow into that river to the W of the junction of the Apure. Most of these rivers swell in the month of April, and continue to overflow their banks during three or four months. They abound in alligators and fish. The Portuguesa, formed by the union of the Pao and the Barquisimeto, flows through the greater part of V., and joins the Apure, 40 m. NW of its mouth. The lakes are not numerous, for we can hardly give that appellation to the sheets of water produced by the periodical swell of the Orinoco, or the rains, and which are generally without any depth. The lake of Valencia or Tacarigua is, however, a beautiful sheet of water, and of great extent. The lake of Maracaybo is rather a gulf of the Caribbean sea.

Climate and soil.] The climate of V. is modified according to the situation. On the coast, and in the plains, a scorching heat prevails, accompanied in the latter with deluges of rain. In the mountain-valleys the air is in general pure and mild, and in some elevated parts even cold. On these mountains, which form a part of the great branch extending from the W to the gulf of Paria, divide the lands of the coast from the plains of the valley of the Orinoco, the climate is so singularly altered, that a traveller may observe the fruits of the tropics luxuriating at a short distance from those of Europe. To the S of this chain, lie the llanos or plains, which stretch to the Orinoco, and are inhabited solely by herds of cattle, tended by mulattoes nearly as rude in nature and habits as the beasts they guard. On these plains the rainy season commences in April, and continues till November. The rain falls oftener in the morning than in the evening, and, on an average, generally occupy three hours of each day, during which period the plains nearest the rivers are converted into lakes of immense extent. From August to November, and the months of February and March, form the dry season. Earthquakes are of frequent occurrence.—The soil is fertile, and yields in abundance all the products of the West Indies, besides others which those islands do not possess. Its most noted commercial article is cacao, which is inferior to none grown in America; vanilla, maize, indigo, cotton, sugar, tobacco, and coffee are among the principal objects of cultivation; wild cochineal, dye-woods, medicinal drugs, gums, resins, balsams, sarsaparilla, sassafras, liquorice, squills, storax, cassia, and aloes, here find that climate the most favourable to their growth; the immense plains in the interior feed multitudes of cattle, horses, and mules; in the valleys and on the mountains, sheep and deer are numerous. The jaguar and puma occur in the forests. All kinds of game are found in this country, the rivers of which also abound with fish.—Silver, tin, copper, coal, and salt are wrought.

Commerce.] The official tables of returns of trade by sea and navigation for the year ending 30th of June, 1844, show in their aggregate, as in their separate amounts of imports and exports, a considerable decrease as compared with the returns for the fiscal year preceding of 1842-3, and a larger decrease still, as compared with 1841-2:—

1843-4,	Imports and exports together, about		£1,660,100
1842-3,	ditto	ditto	1,961,100
1841-2,	ditto	ditto	2,225,200

The decrease in 1843-4, as measured by 1841-2, was equal, therefore, to about 25 per cent. and of about 12 per cent. by 1842-3. The causes of this relaxation in the progress of national industry, which for several years down to 1841-2, in the exportation of indigenous products above all, had been regularly on an ascending scale, was attributable, in the first place, to unfavourable seasons, by which production was sensibly diminished, and to the extent the means of exchange and internal consumption were checked. V. is necessarily an agricultural country, whose resources principally consist in the cultivation and export of coffee, cacao, tobacco, *dibi divi*, dry hides, and the breeding of cattle, the latter being exported chiefly to the British and French West Indies. Although so small a state, it is already a considerable consumer of British wares. The value of British produce exported to V. in 1840 was £379,743; in 1845, £390,149; in 1850, £301,094; in 1852, £273,738. The duties on the traffic of the port of La Guayra, the port of the capital, Caracas, for four years was as follows:—

From 1842	to 1843,	831,848 d.
1843	1844,	722,115
1844	1845,	795,651
1845	1846,	891,502

The trade of the port of La Guayra, for the years stated, compiled up to the 5th of October in each year, was as follows:—

	Coffee. Quintals.	Cacao. Fanegas.	Cotton. Quintals.	Sugar. Quintals.	Indigo. Quintals.	Hides. Number.
1843,	147,474	28,624	510	3,268	1,074	31,684
1844,	132,086	38,739	235	2,508	702	45,242
1845,	122,912	29,418	767	5,153	489	44,873
1846,	151,975	32,476	240	4,562	517	35,560
1847,	126,812	37,876	1,128	7,040	621	49,773

The total value of exports in 1845 was £894,745; of imports, £793,877.

Population.] The pop. of the republic was returned at 780,000 in 1800; in 1839 at 945,247; in 1851, at 1,356,000, whites, Indians, and Negroes. Upwards of one-half are people of colour. The Negroes, 49,000 in number in 1839, were emancipated in 1854. The independent Indian tribes may have an aggregate pop. of 50,000.

Government and revenue.] The government is vested in a senate and a house-of-representatives. The revenue,—of which almost the whole is derived from customs' duties,—suffered a material diminution from the reverses of industry and commerce in 1843-4. In 1840-1, the customs rendered nearly £360,000 to the treasury; in 1843-4, the return was only £264,400. By executive decree of the 16th of September, 1840, the debt was acknowledged to be 28½ per cent. of £6,625,950, equal to £1,888,395 15s., or 11,302,473 d. 44 c. of active debt, bearing interest payable half-yearly from April 1, 1840. Of deferred debt, the same amount, bearing interest from October 1, 1852. No interest has been paid on the active debt since the 1st of April, 1847. The arrears of interest to the 1st of April, 1851, were £192,146. The first year's interest on the deferred debt, payable on the 1st of April and the 1st of October, 1853, at 1 per cent. per ann., is £15,239, increasing yearly ¼ per cent.: that is £3,809, until it attains the *maximum* rate of 5 per cent. per ann.

Divisions.] The republic is divided into the 5 departments of Maturin, Caracas, Zulia, Apure, and Orinoco, which are subdivided into 13 provs. Its chief harbours are those of Cumana in the gulf of Cariaco and Puerto-Cabello. Codazyi gives the following as the pop. of these provinces in 1841:—

Carácas,	242,888	85·4	to the sq. league.
Carabobo,	96,967	1·45	"
Barquisimeto,	112,755	144·	"
Coro,	40,476	43·	"
Maracaibo,	27,800	15·	"
Truxillo,	44,788	123·7	"
Mérida,	62,116	68·4	"
Bainas,	109,497	54·9	"
Apure,	15,479	8·5	"
Barcelona,	52,103	45·1	"
Cumana,	50,671	54·6	"
Margarita,	18,305	494·7	"
Guayana,	20,149	2·2	"

VENGEONS, a commune of France, in the dep. of the Manche, cant. and 2 m. NE of Sourdeval. Pop. 1,700. It has manufactories of iron ware and of paper.

VENGOVA, a town of Russia in Europe, in the gov. of Vilna, district and 33 m. SE of Telch.

VENIA, a town of Spain, in the prov. and 36 m. SW of Oviedo. Pop. 420. In the vicinity are copper-mines, now unwrought.

VENICE (PROVINCE OF). See article LOMBARDO-VENETIAN KINGDOM.

VENICE, in Italian VENEZIA, in German VENEDIG, the capital of Austrian Italy, or of the Lombardo-Venetian kingdom, situated in the dogado of the same name, on 72 little islands, among the shallows which occur near the head of the Adriatic gulf, in N lat. 45°25′9″, E long. 12°20′2″, 70 m. NE of Mantua, 115 m. NE of Florence, 140 m. E of Milan, and 226 m. N of Rome. A railway viaduct of 222 arches (*a a a*) connects the city with the railway to Padua on the mainland. The lagune, on the N division of which the city stands, is about 5 m. distant from the mainland, being a kind of small inner gulf separated from the larger one by some islands, or tongues of land, which in a great measure break the force of the Adriatic storms before they reach the *laguna* or lake; yet, in high winds, the navigation of the lake is dangerous to gondolas, and sometimes the gondoliers will not even trust themselves in the canals that intersect the city. This lagune is from 25 to 30 m. long, and about 5 m. broad. It is protected from the fury of the Adriatic, by three tongues of land, which are hung in a disunited chain from N to S, the most northern, which is immediately before the city at 1 m. distance, being called Lido, the second Palestrina, and the last Salto-Marina. Between those tongues of land are *portos* or channels from the Adriatic, by which ships not exceeding in the main passage 17 ft., and in the other 10 ft. draught of water, enter; but these *portos* are de-

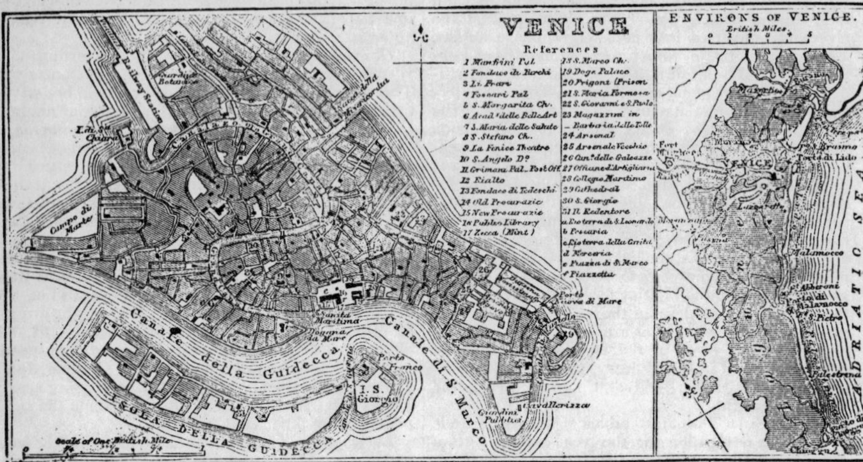

fended by forts on both banks, so that each passage is considered to be hermetically closed to a hostile squadron. The most southern entrance, where exists the greatest facility for landing, is protected by an additional fort, called the Brondolo, built on the canal leading to the lagune, and this place, though 20 m. distant from the city, is considered to be the most important point, and great care has been taken in rendering it very strong. The passages are called, beginning at the N, Porto-di-Lido, Porto-di-Malinocco, Porto-di-Chioggu, and Porto-di-Brondolo. The whole of these *portos*, and the entrance to the lagune, is again secured by a long bar or sand bank, which breasts them from N to S, and where the deepest water, at nearly 2 m. distance, is 26 ft., shoaling on the bank itself to 5 or 6 ft., and near the shore to 3 or 4 ft. Nature and art have thus rendered V., in common opinion, inaccessible to hostile operations on the side of the Adriatic. The N extremity of the lagune is broken up into marshes and little specks of land where no footing for man or beast exists, and therefore on that side no precaution need be taken. The accessible part may be said to be the W shore of the lagune, from the said marshes on the N, to the fort of Brondolo on the S, making a line to be defended of nearly 40 m. in entire extent. The high road from Padua to Venice touches on the nearest point, about 4 m. distant from the city, and at that point an immense fort has been built, which is called Malghera, on the possession of which the safety of the city depends. [*Times.*] Most of the houses of the city have a door opening upon a canal, and another communicating with the street, by means of which, and of the bridges, a person may visit any part of the city either by land or water. The streets and canals are generally very narrow, but the great canal is 100 ft. broad, and meanders through the middle of the city. Here "the whole scene, under the brilliant light of a noon-day sun, is full of movement and colour. As soon as the steamer has dropped anchor at the entrance of the Grand canal, a little fleet of gondolas crowds round her, and, as they receive the passengers, dart off in the most easy and graceful manner possible, their steel prows flashing in the sun, and their keels tracing a line of pearl upon the bright green water. The long wave which the prow turns over is dashed against a wall of marble-fronted palaces, the names of which, carelessly mentioned by the gondolier, awaken trails of golden memories in the mind. The breadth of the 'silent highway' allows the sun to lie in broad rich masses upon this imposing gallery of architectural pictures, and to produce those happy accidents of light and shade which the artist loves. High in the air arise the domes and spires of the numerous churches with which wealth and devotion have crowded the islands of V., the bells of which are ever filling the air with their undulating streams of music. Everything is dreamlike and unsubstantial; a fairy pageant floating upon the waters, a city of cloud-land rather than of the earth. The gondola itself in which the traveller reclines, contributes to weave the spell in which his thoughts and senses are involved. No form of locomotion ever gratified so well the two warring tendencies of the human soul, the love of movement and the love of repose. There is no noise, no fatigue, no danger, no dust. It is managed with such skill and so little apparent effort, that it really seems to glide and turn by its own will." This position of V. in the midst of waters, gives it indeed a singular appearance from a distance. Its domes and spires, its churches and public buildings, appear to the spectator, particularly when approaching by sea, to float, as our quotation expresses it, on the surface of the waves. The marble domes of the churches are very white; the palaces are of a pleasing grey colour; and the varied mass of buildings is chequered with intermingling red and white, as bricks or stucco prevail in their external construction. The streets, or rather lanes, are paved with flags or marble slabs, having small sewers for carrying off the filth. The ordinary dwellings are built of brick, and in general covered with wood. Without having arcades, as is the case in many towns in the N of Italy, they are in general provided with balconies. From the extreme narrowness of the streets, their houses are in general gloomy, and are miserably deficient in that commodious distribution of parts which marks the dwellings of Britain, the Netherlands, and the improved parts of France and Germany. The rooms too are often wretchedly small. The general height is three or four stories. The larger houses are commonly of a square form, with an inside court containing a cistern, into which water flows from the roof, which, after being filtrated, serves for domestic purposes.

Quarters and Buildings.] The length of the city is somewhat more than 2 m.; its breadth 1½ m.; its circuit 6 m.; so that its form, without being either square or circular, is compact. It is divided into two parts nearly equal, by the Canalazo or Grand canal (*b b b*), which winds through its whole length in a serpentine form. Exclusive of the general division into north and south by this canal, V. is separated for the purposes of police into six wards, of which the most eastern adjoins the castle, and bears the name of Sestieri-di-Castello; the Sestieri-di-San-Marco lies more towards the centre of the city; that of Canareggio comprises the NW division. These three are to the N of the great canal; the remaining quarters are situated to the S of it. That of San-Paolo is in the SE; that of Sante-Croce in the W, including several small gardens; that of Dorso-Duro forms the most southern division of the city, bordering on the Canale-della-Guidecca (*c c*). The aspect of V. is stately, and even magnificent, whether we look to its public or its private edifices; for though few of the buildings are in a pure style of architecture, the general effect is grand and imposing. In regard to the streets, as they are termed, their breadth is in general only 5 or 6 feet: in many places still less. The only exception is the street called the Merceria, situated near the centre of the city, and containing shops of all kinds; but even of that the breadth is insignificant, varying only from 12 to 20 ft. The only open place entitled to the name of square, is the Piazza-di-San-Marco (A), an oblong of 280 ft. in length, by nearly 100 ft. in breadth, enclosed by handsome and even splendid buildings, singularly contrasted in their outward decorations. Of these the principal are the churches of San-Marco and Geminiano; the palace formerly occupied by the doge, and the buildings called the Procurazie. This small but elegant square, a miniature of the Palais-Royal of Paris, is bordered by arcades containing elegant shops and coffee-rooms, which, when lighted at night, have a splendid appearance. It forms the central point of the gaiety and amusements of V., the resort of foreigners, and of loungers of every description. The Piazetta (B) is a smaller opening leading, at right angles, from the square of San-Marco to the sea, and having on the one side the palace of the doge, on the other the public library, with its pillars of granite. This spot presents, from the concourse of people, an animated and interesting scene. The only other open spaces in the city are in the front of some of the churches, and at each end of the Rialto.

V. contains 360 bridges; but the most of them are only single arches thrown over the canals. The Rialto (*d*), a single arch thrown over the narrowest part of the Grand canal, is of marble, 90 ft. in span, and 24 ft. in height; but its beauty is impaired by two rows of booths or shops which divide its upper surface into three narrow streets. The prospect from the Rialto is lively and magnificent; but it is almost the only one in Venice; for, except the Canal-Grande and the Canaletto-dello-Canaregio (*n n n*), all the other streets are narrow: some of them have no quays, and the water literally washes the houses. In rowing along these canals, there is scarcely an agreeable object to cheer the view; and the stench which at certain seasons exhales from the waters is most offensive.—Some parts of the city exhibit fine specimens of the architecture of Palladio. The chief buildings are the ancient ducal palace, the churches of San-Marco and San-Geminiano, the old and new Procurazie, which contain the museum, the library of St. Mark, consisting of 65,000 volumes and 5,000 manuscripts, and nine large apartments belonging to the procurators of St. Mark. All these buildings are of marble, and are situated in and around the Piazza-di-San-Marco. The patriarchal church of San-Marco, though one of the richest and most expensive in the world, does not immediately attract the eye. Its architecture is of a mixed kind, mostly Gothic, yet many of the pillars are Grecian. The outside is incrusted with marble, and the inside ceiling and floor are of the finest marble; the whole is crowned with five domes. The front which looks to the palace has five brazen gates with historical bas-relievos; over the principal gates are placed the four famous bronze horses of incomparable workmanship, originally brought to V. from the hippodrome of Constantinople, and said to have been executed by the famous Lysippus. The principal entrance of the ducal palace, 'a vast and sumptuous pile,' is called the Giant's Stair on account of two colossal statues of Mars and Neptune placed at the top. Under the portico, were the gaping mouths of two brazen lions. In an opening from the ducal palace to the sea, stood two granite pillars, lofty and massive, between which criminals were publicly executed. There is a communication between the ducal palaces and the state-prisons by a gloomy bridge,—Il Ponte-de'-Sospiri or 'The Bridge of Sighs.' The cells of this prison, made of massy marble, the architecture of the celebrated Sansovino, are not only dark and black as ink, but being surrounded and confined with huge walls, the smallest breath of air can scarcely find circulation in them. They are about 9 ft. square on the floor, arched at the top, and between 6 and 7 ft. high in the highest part. There is to each cell a round hole of eight inches diameter, through which the prisoner's daily allowance of twelve ounces of bread and a pot of water is delivered. The furniture of each cell is a little straw, and a small tub; nothing else. These Publiche-Prigioni can contain about 500 prisoners. They range along the narrow canal which skirts the palace of San-Marco.

The churches built or designed by Palladio, bear witness to the taste and genius of that distinguished architect. Others, and indeed the greater number, are in the Saracenic, or, as it is currently termed, the Gothic style, and are less distinguished by elegance of structure, than by richness of interior decoration. The church of Santa-Maria-della-Salute, the work of Palladio, is an elegant structure, open to the great canal, and built outside and in, of marble, with a moderate share of ornament. The church Il-Redemptore is also an elegant building. The cathedral of Venice, dedicated to St. Peter, on an island at the E end of the city, is built of Istrian marble, and adjoins the former residence of the patriarch of Venice. The church of San-Georgio is remarkable for its front of marble and its cupola; that of San-Giovanni-e-San-Paolo is a large Gothic edifice surmounted by a cupola, and is the Westminster abbey of Venice, containing the tombs of many of its defenders and doges.—Of the palaces or mansions of the great families, the most conspicuous are those of Grimani, now the post-office, Grassi, now an hotel, Pisani, Corra-della-Regina, Foscari, Manfrini, and Cornaro, all more remarkable for their size than for elegance or symmetry. The arsenal is a commodious and even magnificent building, situated on an island near the E end of the city. It is defended by a rampart of nearly 2 m. in extent, as well as by the surrounding water; and has before its gates two great pillars, with the four gigantic lions in granite, which stood formerly on the Piræus at Athens. Its halls are lofty and commodious; and it contains four basins or wet docks (*o o o o*), and several dry docks and slips. On the Grand canal, not far from the Rialto, stands the Fondaco-di-Tedeschi (*m*), long a depot for the goods of German merchants, now the council-house of the city. Of theatres, V. has no less than 6 great and small; but several of them are open only during the carnival. The lyceum has a rich cabinet of natural history; and among the educational institutions are two royal gymnasia, a normal high school, a marine college, and the seminary of the Saluta. The Armenian convent of San-Lazaro is a fine establishment situated upon a small island to the NW of the Lazaretto.

Population.] The pop. of V. at the end of the 17th cent., amounted to nearly 200,000. The census of 1825 returned it at 109,927; that of 1846, at 127,925. Venice has a fine and salubrious climate. While Paris has one person of 60 years of age to every 200 inhabitants; in V. the proportion is one to every 116. In Paris you have one octogenarian to every 900 inhabitants; in V. the proportion is one to every 491. The characteristics of the climate of V. are a summer-heat much greater than is experienced in England; a winter not of great length, but sharp, particularly during the prevalence of a NW wind which blows across the interior of Switzerland and the Alps. Rains are frequent, particularly in spring; and there being no springs or wells, the inhabitants are supplied—as in many towns of Holland—with water collected in cisterns, from the tops of the houses: this frequently wants the freshness of running water. But the chief privation to a Venetian is an exclusion from fields and gardens, the inconvenience of walking in narrow lanes, and the ascending and descending steps at every bridge which crosses a canal.—The artisans of this city form several corporations, and each corporation maintains a school, yet the Venetians are little more informed generally speaking than other Italians. Their costume has not materially changed during four centuries. As for the wealthy classes, Simond informs us they pass their time in the following manner: people of fashion rise at 11 or 12 o'clock,—pay a few visits and idle away their time till three, when they dine; in summer they take an hour's siesta after dinner; they dine and go to the coffee-house or casino till nine,—then to the opera, which is another casino,—then to the coffee-house again for another hour or two,—and seldom go to bed in summer before sunrise. There remain only two or three families of the old patrician caste in V. The *gondolieri* form a distinct class of the pop. Venice was the birthplace of Algarotti, Gazzi, Goldoni, the Paoli, and Bembo.—The civil administration of V. does little to benefit the city. The water, collected, as described, in cisterns, is very bad—in fact, all the potable water is brought in boats from the land; and though a French company has, with the permission of the government, made a number of Artesian wells, the water obtained from them has a sulphurous taste, which renders it quite unfit for drinking. It would be easy to lay pipes along the railway bridge that connects V. with the mainland, but nothing is done, and this necessary element continues to be brought in boats, as it was centuries ago. During the late siege, upwards of 5,000 persons died of cholera within the city. There are numerous benevolent institutions within the city, the total annual endowments of which have been stated at 50,000 florins = £58,000; and it has been calculated that out of a pop. of 130,000, not fewer than 52,000 are in receipt of charitable relief in one form or another.

Manufactures and commerce.] The manufactures of V., if not extensive in any one branch, are of considerable diversity: they consist of woollens, serges, canvas and ropes, gold and silver stuffs, velvet, silk stockings, and lace, which is made chiefly on the adjacent island of Murano, artificial gems, glass, glass beads, soap, and paper. The goldsmiths of V. have ever been famed for the skill and beauty of their productions. The glass-works are said to employ 4,500 men. Printing is carried on here more extensively than in any other town in Italy; and books are supplied by wholesale to the Grecian islands, Constantinople, Spain, and Portugal. The price is as low as that of books in France; the type is usually good, but the quality of the paper inferior.—Napoleon is said to have made the following memorable remark with regard to this city of the seas:—" V. is the best-situated port for commerce in Italy. The goods from Constantinople and the Levant reach it by the shortest way in traversing the Adriatic. From there they can be spread to Upper Italy and Turin by the Po, and to Germany by mounting the Adige to Batzen, and from thence on to Ulm, Augsburg, Munich, and Nuremberg. V. is the port of the higher Danube, the Po, and of the Adige; nature intended it to be the depot for the Levant, Italy, and southern Germany." Napoleon at that time knew nothing either of railways or steamers; but V. certainly possesses the elements of a prosperous commercial place. The canals, that come to the very doors of the houses, render the discharging of vessels easy; indeed, almost all circumstances combine to create a hope that it may yet revive from neglect and decay. In the time of the city's prosperity the Venetian ships of the largest class, denominated *galeasses* were fitted up for the double purpose of war and commerce. Some of them carried 50 pieces of cannon and crews of 600 men. These vessels, sometimes also called *argosas* or *argosies*, had an early intercourse with England. Sir William Monson mentions that the last argosie that sailed from V. for England was lost with a rich cargo and many passengers on the coast of the isle of Wight, in 1487. Some idea of the commerce and maritime power of V., in the 15th century, may be formed from the following account given by Daru of her trade at that epoch:—" In the beginning of the 15th cent., the annual value of the goods exported from V. by sea, exclusive of those exported to the states adjoining her provinces in Lombardy, was estimated, by contemporary writers, at 10,000,000 ducats; the profits of the out and home voyage, including freight, being estimated at 4,000,000 ducats. At that period the Venetian shipping consisted of 3,000 vessels of from 100 to 200 tons burden, carrying 17,000 sailors; 300 ships with 8,000 sailors, and 45 galleys of various size kept afloat by the republic for the protection of her trade, having 11,000 men on board. In the dockyard 16,000 labourers were usually employed. The trade to Syria and Egypt seems to have been conducted principally by ready money, for 500,000 ducats are said to have been annually exported to those countries; 100,000 were sent to England." Subsequent to the cession of V. to Austria, it seems to have been the policy of the government to encourage Trieste in preference to V.; and the circumstance that Trieste was a free port whilst Venice was not, placed the latter at a disadvantage. Since 1830 a more equitable course has been pursued; for in that year V. was also constituted a free port, and has continuously shared almost the same privileges as were conferred on Trieste. The limits of the free port are declared to extend from the port of Malamocco to the dyke of Garzina near Sant' Erasmo. Within these limits, trade is perfectly free, and no customs' duties are levied. There are custom-houses at Treporti, Mazzorbo, Campalto, Fusina, and San Pietro in Volta, forming a second line around the first. The intermediate space is the *circondario doganale*, or 'territory of the customs.' Vessels of all nations are free to enter the ports of Lido and Malamocco without paying customs' duties. Certain canals are pointed out, by which merchandise may be shipped to the main land. Fishermen's boats may go to any spot of the main land, if licensed, and not laden with goods. Salt, tobacco, nitre, and gunpowder, being monopolies of the state, are not allowed to be unshipped within the precincts of the free port, except in the custom-house. "Owing to the circumstance of V. being a free port, and the fact of her adjacency to the Papal states, a comparatively large contraband trade in various goods imported to the city, or manufactured there, is well known to be carried on with those states and the small duchies of Northern Italy. It has been estimated that full two-thirds of the coffee consumed in Lombardy is smuggled. Some trade, in vessels of insignificant size, is also carried on with the opposite coasts of Dalmatia, Albania, and the Morea. The imports of V. consist of wheat and various kinds of grains, from the adjoining provinces and from the Black sea; olive oil, principally from the Ionian islands; cotton stuffs and hardware from England; colonial produce—through England, the United States, and Brazils—with dried fish and dyeing materials. Her exports comprise grain, raw and wrought silk, paper, fruit, cheese, some oil, and wine. The trade with the interior is carried on by means of flat-bottomed vessels, which, passing through the lagunes, enter the numerous rivers and canals, freely traversing the river Po and its tributaries, and penetrating the country to a great distance. These convey salt, sugar, coffee, and other imported or manufactured necessaries, returning with an exchange of grain, silk, flax, and hemp, cheese, wine, and fire-wood. I was particularly struck," says a recent writer in the *Morning Chronicle*, "with the insignificant size and shallow draught of the trading vessels in the port of V. Many scores, if not hundreds, of two-masted nondescript craft—more like our pilot boats than any kind of vessel we use—whose burthen at most cannot be more than 20 tons, lie moored in tiers before the Molo, and off the Canale-della-Giudecca. The merchandise principally transported by these vessels which I personally noted, were corn and rice, wine and oil, timber and fire-wood—a large trade in the latter—fruits, including vast quantities of melons, and pumpkins of every variety of form, size, and colour, bricks, bottles, and matting. A few brigs and schooners engaged in the foreign trade, amongst which I recognised but one English vessel, lay at anchor before the town, outside the Austrian frigate before-mentioned, a few hundred yards from the shore; in short, the shipping business appeared to go forward languidly. The soul of trade seemed to have departed. The sailors, sleeping in rows on the marble-paved quays, or shading themselves under the boats slung on the deck, and fanned by idly flapping sails, seemed as sluggish as the sea, which, so completely landlocked in the harbour by its outlying islands, has here no swell, nor any further motion than a gentle ripple. Such as I have described was the flourishing commerce of Venice at the zenith of her prosperity—such is its degenerate and languishing condition at the present day. It would seem as though physical as well as institutional causes are operating against her; for her harbour is choking, her trade transferred to Trieste, and though restored to the dignity of what is there termed a free port, she has still fiscal regulations that impede her advance by incumbering trade. The physical position of V. precludes her from emulating, much less rivalling, Trieste as a port for commerce; a wise policy, therefore, should suggest to the Venetians that the only hope they have of recovering their ancient wealth and splendour is by application to manufactures, for which the abundance of cheap houses, warehouses, and factories, united with the low value of labour, offer them peculiar advantages. The commerce of Trieste is more than fourfold that of V. It has a port always approachable—so deep are the soundings—whereas the entrances to V. by the most favoured channels—those of the Lido and Malamocco—are insufficient for ships of heavy draught, having at most but 16½ ft. of water, and the port is gradually though not rapidly filling."

History.] The history of V. is sketched in the historical paragraph to our article on the LOMBARDO-VENETIAN KINGDOM During the revolutionary movement of 1848, V. endured four successive weeks' storm of shot and shell from the Austrian batteries, but no mischief of any serious extent was done to the buildings, and only four non-combatants lost their lives, though the men in arms were slaughtered by wholesale in the lines, and want of food, fever, and the cholera made unheard-of ravages among the pop. This extraordinary fact is thus accounted for. In the first place the imperial batteries were 3 m. removed from the centre of the town—a distance at which no shells reached, and at which cannon shot fell half-spent and nearly powerless. The bombs, fortunately for the city, only demolished the poorest quarter bordering on the Laguna; the balls knocked in a hole when they struck a palace, and then their vital force expired. The nearest point on which a battery could be placed was a French league, and a third removed from the centre of the city; and the cannon were all fired at an angle of 45, so that the effect was what is called that of 'curved artillery.' Had V. been well provisioned, or the blockade by sea and land not been strictly enforced, the city would not have been captured; but hunger and the cholera did the work of the besiegers effectually, and it surrendered only when hope of relief became extinct.

VENICE, a township of Cayuga co., in the state of New York, U. S., 144 m. W of Albany. The surface is undulating, and is drained by Salmon river. The soil consists chiefly of gravel and clay loam. Pop. in 1840, 2,105; and in 1850, 2,028.—Also a village of Erie co., on the S side of Sandusky bay, at the confluence of Cold creek, and 99 m. N by E of Columbus. Pop. in 1840, 400; and in 1850, 450. —Also a township of Seneca co., in the same state. Pop. 1,222. It contains two villages named Attica and Caroline.

VENICE (GULF OF). See ADRIATIC.

VENISSIEUX, a commune of France, in the dep. of the Isere, cant. and 6 m. N of St. Symphorien-d'Ozon. Pop. 2,415.

VENIZY, a commune of France, in the dep. of the Yonne and cant. of Brienon. Pop. 1,573.

VENLOO, a fortified town of Holland, capital of a cant. in the prov. of Limburg, 14 m. NNE of Ruremonde, on the r. bank of the Maas, in N lat. 51° 22′ 16″, and E long. 6° 10′ 15″. Pop. 7,179. It has numerous distilleries, oil-mills, bleacheries, tanneries, breweries, manufactories of spangles and needles and of snuff, and salt-refineries, and carries on an active trade. The environs are unhealthy, consisting chiefly of marshy plains. V. is of considerable antiquity. It made part of the Hanseatic league. In 1702 it was taken by the allies under the Duke of Marlborough. Opposite on the l. bank of the river is the fort of St. Michel.

VENN, a village of Prussia, in the regency of

Dusseldorf, and circle of Gladbach. Pop. 352.—Also a village of the circle of Dusseldorf. Pop. 200.

VENNEBECK, a village of Prussia, in the regency and circle of Minden. Pop. 448.

VENNEBERG, a haerad of Denmark, in Jutland, in the stift and to the NW of Aalborg. Pop. 6,600.

VENNHAUSEN, a village of Prussia, in the regency and circle of Dusseldorf. Pop. 320.

VENNIKEL, a village of Prussia, in the regency of Dusseldorf, and circle of Geldern. Pop. 260.—Also a village in the circle of Krefeld. Pop. 480.

VENNINGEN, a village of Bavaria, in the circle of the Pfalz, and cant. of Edenfoben. Pop. 1,036.

VENN-OTTERY, a parish of Devonshire, 3 m. SW by S of Ottery-St.-Mary. Area 918 acres. Pop. in 1831, 133; in 1841, 105.

VENOE, an island of Denmark, in Jutland, in the stift of Aalborg, bail. of Ringkioebing, in the Lymfiord. It is 6 m. in length from N to S, and 1½ m. in breadth. Its coasts abound with fish.

VENOGE, a river of Switzerland, in the cant. of Vaud, which issues from the heights commanding the village de L'Isle, runs first E, then S, passing near Cossonay, and after a course of about 15 m., discharges itself into the lake of Geneva, between Lausanne and Morges.

VENOISE (LA), a mountain of Sardinia, in Savoy, between the provinces of Maurienne and Tarantaise, in N lat. 45° 24′ 15″, and E long. 6° 49′ 12″. It has an alt. of 3,863 metres, or 12,674 ft. above sea-level, and is one of the chief summits of the Alps between the Arc and upper course of the Isere.

VENOSA, a town of Naples, in the prov. of the Basilicata, district and 11 m. E of Melfi, on a plateau near the Ofanto. Pop. 6,000. It is well built and has a fine square, many handsome private houses, a magnificent cathedral, 5 parish-churches, and an abbey, in the church of which are several fine tombs in marble, amongst others, that of William surnamed Bras-de-fer; there are besides several convents, an hospital, two charitable institutions, a bank and some Roman antiquities. In the vicinity is a fine aqueduct, by which the town is supplied with water. V. is noted as the birthplace of Horace. It was the *Venusia* of the Samnites, and was taken by the Romans at an early period of the Republic.

VENOY, a village of France, in the dep. of the Yonne, cant. and 3 m. E of Auxerre. Pop. 1,000.

VENRAY, a town of Holland, in the prov. of Limburg, district and 21 m. N of Ruremonde. Pop. of district, 4,340. It has extensive manufactories of linen, shoes, leather, and vinegar, several dye-works and breweries.

VENTABREN, a commune of France, in the dep. of the Bouches-du-Rhone and cant. of Barre, on a height commanding a fine prospect. Pop. 1,714. It has an oil-mill, a bleachery, and a copper-foundry.

VENTANA (SIERRA-), a mountain-ridge of Buenos-Ayres, in S lat. 38° 5′. It attains an alt. of at least 3,500 ft.

VENTAS-CON-PENA-AGUILERA, a village of Spain, in the prov. and 18 m. SW of Toledo, amid the mountains of that name. Pop. 1,150. The making of charcoal, and the granite quarries of the adjacent mountains, form the chief employments of the inhabitants.

VENTAS-DE-RETAMOSA, a town of Spain, in the prov. and 21 m. N of Toledo. Pop. 515.

VENTAVON, a commune of France, in the dep. of the Hautes-Alpes and cant. of Laragne, 21 m. SW of Gap. Pop. 1,100.

VENTEROL, a village of France, in the dep. of the Drome, cant. and 3 m. NW of Nyons, on the slope of a hill. Pop. 1,458. It has an active trade in oil and in truffles.

VENTEUIL, a commune of France, in the dep. of the Marne, cant. and 7 m. NW of Epernay, near the Marne. Pop. 1,102. It has distilleries of brandy.

VENTIMIGLIA, a town of Sardinia, capital of a mandemento, in the prov. of San-Remo, 24 m. NNE of Nice, on the Roya, at its entrance into the gulf of Genoa. Pop. 5,000. It is the see of a bishop, and has a college, a theological seminary, an arsenal, and a small port. The fortifications, which were destroyed by the French, were renewed in 1832.

VENTIPUR, a village of Cashmere, on the r. bank of the Jelum, 22 m. SE of Sirinagur.

VENTLUE (LA), a market-town of France, in the dep. of the Pas-de-Calais, 11 m. NE of Bethune. Pop. 3,950.

VENTNOR, a village on the SE coast of the Isle of Wight, in the Undercliff district, between Bonchurch and St. Lawrence. It occupies a fine commanding terrace on the cliff; and has rapidly risen into estimation as a watering-place. Pop. in 1851, 2,569.

VENTOLIENE, an island of Naples, in the prov. of the Terra-di-Lavoro, in the gulf of Gaeta, to the N of Ischia. Pop. 400. It has a harbour and extensive fisheries.

VENTOLO, an island of the South Pacific, in the group of the Admiralty islands, in S lat. 2° 14′, and E long. 148° 10′. It is surrounded by coral reefs, and from a distance appears a barren rock. It contains, however, some fertile tracts, and maintains a numerous population.

VENTOSA (LA), a town of Spain, in the prov. and 15 m. NW of Cuenca, in a hilly but fertile locality. Pop. 686. It has an hospital and a fortress now in ruins. The spinning of yarn for the cotton manufactures of Cuenca forms the chief object of local industry.

VENTOSA-DE-LA-CUESTA, a town of Spain, in the prov. and 18 m. S of Valladolid, on an arid hill. Pop. 320. It has a distillery of brandy.

VENTOSA-DEL-RIO-PISUERGA, a town of Spain, in the prov. and 27 m. from Palencia, in a fine valley, near the Pisuerga. Pop. 300. It has several spinning-mills.

VENTOUX, a mountain of France, in the dep. of the Vaucluse, and cant. of Malaucene, to the NE of Carpentras. It forms a ramification of the Alps, running from E to W, and attains an alt. of 1,960 metres = 6,430½ ft. above sea-level. It derives its name from the violence of the winds which rage around its summit.

VENTRON, a commune of France, in the dep. of the Vosges, cant. and 5 m. E of Saussure. Pop. 1,118. It has manufactories of calico, and a considerable trade in cheese.

VENTROSA, a town of Spain, in the prov. and 33 m. from Soria, at the foot of the Sierra d'Orbion, near the Najerilla. Pop. 694.

VENTRUP, a village of Prussia, in the regency of Munster, and circle of Narendorf. Pop. 200.

VENTRY, a parish, containing a village of the same name, 3½ m. WSW of Dingle, in co. Kerry. Area 4,439 acres. Pop. in 1851, 1,339.

VENTURA, an island of Brazil, in the bay of Angra-dos-Reis, in the prov. of Rio-de-Janeiro.

VENTURADA, a village of Spain, in the prov. and 27 m. of Madrid, in a sterile locality. Pop. 168. It suffered much from the French in 1808.

VENUS (POINT), a headland forming the N point of the island of Tahiti or Otaheite, in the S Pacific, near the bay of Matavae, in S lat. 17° 29′ 21″, and W long. 149° 29′ 18″.

VENUSBERG, a village of Saxony, in the gov. of Zwickau and bail. of Wolfenstein. Pop. 932.

VENWEGEN, a village of Prussia, in the regency and circle of Aachen. Pop. 448.

VENZOLASCA, a commune of France, in Corsica, capital of the cant. of Cavinca, in the arrond. and 17 m. S of Bastia. Pop. 728. It has a small port.

VENZONE, a town of Austria, in Lombardy, in the gov. of Venice, in the deleg. and 21 m. NNW of Udino, on the road thence to Tarvis, and on the Tagliamento. Pop. 3,380.

VEOVETZ, a town of Russia in Europe, in the gov. and 66 m. ESE of Grodno, and district of Honim.

VEP. See WEPPENDORF.

VEPRIK, a town of Russia in Europe, in the gov. of Poltava, district and 9 m. E of Gadratch.

VEPRINITZ, a market-town of Austria, in Illyria, in the gov. of Trieste, and circle of Istria, 3 m. SW of Castua.

VER, a commune of France, in the dep. of the Manche and cant. of Gavray, 15 m. S of Coutances, at the confluence of the Sienne and Airon. Pop. 1,188.—Also a village in the dep. of the Calvados, and cant. of Ryes, 10 m. NE of Bayeux. Pop. 1,400.—Also a village in the dep. of the Oise, cant. and 7 m. SW of Nanteuil-le-Haudouin. Pop. 650. It has a quarry of freestone. This village appears to be the *Palatium Vernum* of ancient chronicles.

VERA, a judicial partido and town of Spain, in the prov. of Almeria. The partido comprises 10 pueblos. The town is 42 m. NE of Almeria, in a level tract near the sea, and enjoying, it is said, the mildest climate in Andalusia. Pop. 8,000. It has a parish-church, a convent, an hospital, and possesses numerous saltpetre works, manufactories of ropes and of pottery, and extensive fisheries. It has a small port, and carries on an active trade in corn. In the vicinity are the ruins of the ancient *Urci*. The adjacent Sierra-Cabrera contains mines of iron, lead, and ochre.—Also a market-town in the prov. of Navarra, 35 m. N of Pamplona, in the valley of Santisteban-de-Lerin. Pop. 1,888. It has a church, a convent, and an hospital. This town was taken and for a short time held by a party of Spanish constitutionals, under General Mina, in 1830.—Also a market-town in the prov. of Barcelona, and 9 m. S of Tarazona, on the road thence to Calatayud. Pop. 648. It has a church and a convent.

VERACIEUX, or VARACIEUX, a village of France, in the dep. of the Isere, cant. and 5 m. WNW of Vinay. Pop. 1,000.

VERA-CRUZ (LA), a federal prov. of Mexico, extending along the Mexican gulf, from the Rio-Baraderas—or de los-Lagartos—to the river of Panuco, which rises in the metalliferous mountains of San-Luis-Potosi. Its length, from the bay of Terminos, near the island of Carmen, to the port of Tampico, is 210 leagues; its breadth is in general from 25 to 28 leagues. It is bounded on the N by New Santander; on the E by the peninsula of Merida; on the S by the gulf of Mexico; and on the W by Oaxaca, Puebla, and Mexico. Its area has been estimated at 27,660 sq. m. Pop. in 1841, 250,380. The only rivers are those which form its limits on the E and W. All the W part of the prov. is covered with the eastern range of the Cordillera d'Anahuac. The E part presents a surface of wide and arid plains. All the W part of the prov. forms the declivity of the cordilleras of Anahuac. Here, in the space of a day, the inhabitants descend from the regions of eternal snow to the plains in the vicinity of the sea, where the most suffocating heat prevails. The order with which different tribes of vegetables rise above one another, in strata as it were, is nowhere more perceptible than in ascending from the port of Vera-Cruz to the table-land of Perote. The inferior limit of oaks warns the colonist who inhabits the central table-land, how far he may descend towards the coast without dread of yellow fever. Forests of liquid amber, near Xalapa, announce by the freshness of their verdure that this is the elevation at which the clouds suspended over the ocean come in contact with the basaltic summits of the cordillera. A little higher, near La-Banderilla, the nutritive fruit of the banana comes no longer to maturity. At the height of San-Miguel, pines begin to mingle with the oaks, which are found by the traveller as high as the elevated plains of Perote, where he beholds fields sown with wheat. At 2,600 ft. higher, the coldness of the climate will no longer admit of the vegetation of oaks; pines alone there cover the rocks, whose summits enter the limit of eternal snow. Thus within the compass of not many miles, the naturalist in this prov. ranges through the whole scale of vegetation. The prov. of Vera-Cruz is enriched by nature with the most precious productions. In the ever-green forests, at the foot of the cordillera, grows the tree of which the odoriferous fruit is employed for perfuming chocolate. The myrtle is produced in the forests which extend towards the Rio-Baraderas, in the E part of the prov. The cocoa of Acayucan would be in request if the natives were to apply themselves more assiduously to the cultivation of cocoa trees. On the E and S declivities of the Pic-d'Orizaba, in the valleys which extend towards the town of Cordoba, tobacco of an excellent quality is cultivated, which yielded an annual revenue to the crown of Spain of more than 18,000,000 francs. The smilax, of which the root is the true sarsaparilla, grows in the humid and umbrageous ravines of the cordillera. The cotton of the coast is celebrated for its fineness and whiteness. The sugar-cane here yields nearly as much sugar as in the island of Cuba, and more than in the plantations of St. Domingo; but the effect of the bounty of nature, and the facility of providing without effort for the most urgent wants of life, impedes the progress of industry. Although on the first arrival of the Spaniards, the coast of Vera-Cruz, and all the country from the river Alvarada to Huaxtecupan, was no doubt better inhabited and better cultivated than it now is, the conquerors found the old pop. of Mexico chiefly concentrated in the interior of the country, on the table-land. The Spaniards, who generally followed the traces of the civilization which they found established in the country, had very powerful motives for settling on the table-land of Anahuac. They dreaded the heat and the diseases which prevail in the plains. The search after the precious metals, the cultivation of European grain and fruit, and the analogy of the climate with that of the Castilles, all concurred to fix them on the ridge of the cordilleras. To this region, therefore, they transported great numbers of the Indians, either to work in the mines, or that they might be nearer the habitation of their masters. For two centuries also the trade in indigo, sugar, and cotton being next to nothing, Whites could by no means be induced to settle in the plains, where the true Indian climate prevails; and it will require centuries to re-people those deserts, in which spaces of many square leagues are now only occupied by two or three huts around which stray herds of half wild cattle. A small number of powerful families, who live on the central table-land, possess the greater part of the sea coasts of Vera-Cruz and San-Luis-Potosi. No agrarian law forces these rich proprietors to sell their estates, if they persist in refusing to bring the immense territories which belong to them under cultivation. To this evil other causes of depopulation are to be added. The militia of Vera-Cruz was long too numerous for a country so thinly inhabited: the service oppressed the labourer, who fled from the coast to avoid the militia conscription. The marine conscriptions were also too frequent, and executed in too arbitrary a manner. The prov. of Vera-Cruz contains within its limits two colossal summits, of which one, the volcano of Orizaba, is, after the Popocatepetl, the most elevated mountain of New Spain. The other summit, the Coffre-de-Perote, which bears 58 m. N 72° 55′ W from the town of Vera-Cruz, has an alt. of 2,186 toises = 4,669·5 yards, and according to the measurement of Humboldt, is nearly 1,312 ft. higher than the Peak of Teneriffe. The small volcano of Tuxtla is situated 4 leagues from the coast, SE of the port of Vera-Cruz.

VERA-CRUZ, the grand sea-port of Mexico, or New Spain, and formerly the residence of the intendant of the province, is situated in 19° 12′ 34″ N lat., 96° 10′ 30″ W long., 190 m. ESE of Mexico, 205 m. sailing distance from Tampico; 810 m. from the Havannah; 4,964 m. from Southampton *via* Nassau and Bermuda; and 5,899 m. *via* Mobile and Bermuda. It fronts the sea in a semicircle, and is enclosed with a wall or parapet, 6 ft. high and 3 ft. broad. On the shore to the SE and NW, are two redoubts, to defend the port, which is not commodious, being merely a bad anchorage among shallows, formed by the walls of the town on the S side, by the walls of the castle on the N, and three or four small islands on the SE and E. The principal entrance is on the NW side (*a a*). From 40 to 60 ships of war, or 100 merchantmen, might anchor here, in from 4 to 10 fath.; but the N winds are

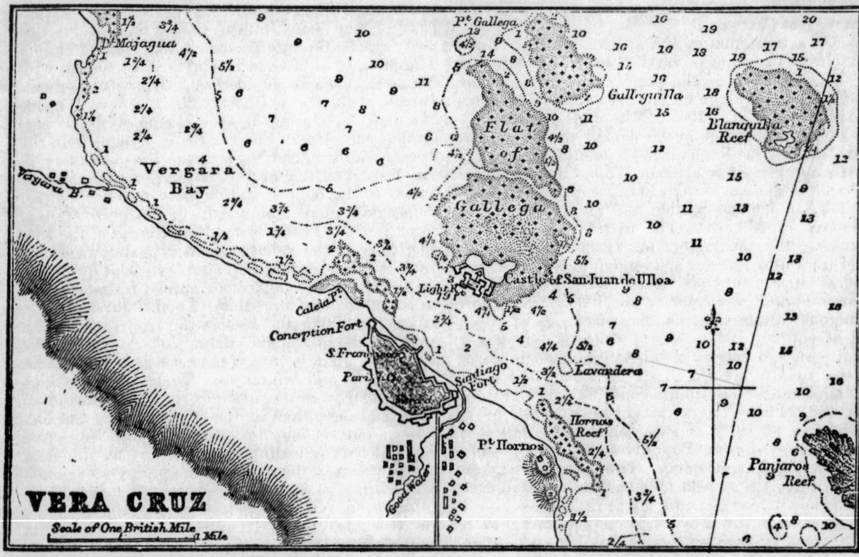

VERA CRUZ

Scale of One British Mile — 1 Mile

terrible, and often drive vessels on shore. In the rainy season, the marshes on the N are haunted by alligators. Vera-Cruz was long one of the most considerable places for trade in the Spanish dominions, being the centre of European and West Indian commerce, and receiving also great quantities of East Indian produce, by way of Acapulco, from the Philippine islands. The city is situated in an arid plain, without running water, and on which the N winds, which blow with dreadful impetuosity from October to April, have formed hills of moving sand. These downs, which change their form and situation every year, are from 26 to 38 feet in height, and contribute much, by the reverberation of the sun's rays, and by the high temperature which they acquire during the summer-months, to increase the suffocating heat of the atmosphere. The breeze, which blows feebly during summer from the SE, is only felt on the terraces of the houses; and the inhabitants, whom the N winds frequently prevent in winter from crossing the streets, in the hot season breathe only a stagnant and burning air. Between the city and the Aroya Gavilan, in the midst of the sandy downs, are marshy grounds covered with mangles and other brushwood; while the stagnant water of several small lakes occasions intermittent fevers among the natives, and is one of the fatal causes of the yellow fever, which in this country is so dreadful a scourge. Many of the houses are large, some three stories high, built in the old Spanish or Moorish style, and generally enclosing a square court with covered galleries. They have flat roofs, glass windows, and are well adapted to the climate: most of them have balconies of wood in front, and the interior arrangement is the same as in old Spain. Most of the town, as well as the castle, is built of coral (the *madrepora meandrities*), drawn from the bottom of the sea, and the lime that forms the cement is of the same material: it is used for the roofs and foot-pavement, and is so hard that in some places it receives from friction a polish like marble. There is one tolerably good square, of which the government-house forms one side, and the principal church the other. The foot-paths are frequently under piazzas, which protect the passengers from the sultry heat of the sun, and the heavy rains which descend in torrents in the wet season. Sixteen cupolas or domes are counted from the sea, but only six churches are now in use. Nothing is more repulsive to strangers accustomed to the bustle of European cities, than the gloomy death-like appearance of the place. Of any other city it is considered a disgrace to say that grass grows in the streets, but here it would be a compliment, for no vegetation is to be observed even for miles around; and fish is the only article of provision not brought from a distance. The only water fit to drink is what falls from the clouds, and is preserved in tanks; that from the castle and the convent of Franciscans being the best. Though the markets are tolerably well supplied by the Indians, living at the hotels is expensive and very uncomfortable. Provisions are dear, with the exception of fish, which is in abundance and good. Milk is scarcely to be had, as not a cow is kept within many miles, and what is perhaps peculiar to Vera-Cruz, there is not a garden even near it. The absence of vegetation attests at once the poverty of the soil and the insalubrity of the climate. Water is found, on digging the sandy soil at the depth of 9½ ft.; but this water proceeds from the filtration of the marshes formed in the downs; it is rain water, which has been in contact with the roots of vegetables, and is of a very bad quality, being only used for washing. The want of good water has for centuries past been regarded as one principal cause of the unhealthiness of this place; and in 1764, a project was formed for conducting part of the river of Xamapa to the port. On this undertaking large sums have been expended. A stone aqueduct capable of furnishing a section of water of about 18 sq. inches, was constructed, for a length of about 2,000 ft.; yet notwithstanding all those exertions and expenses, the waters of the Xamapa are still more than 12 m. distant from the town of Vera-Cruz. There is about a mile's breadth of sea between the town of Vera-Cruz and the castle of San-Juan-de-Ulloa. The outer walls of the fortress are of immense thickness—upwards of 12 ft.; and in positions most exposed, the walls are 17 or 18 ft. in depth. They are formed of white stone, very porous, and rather soft, so that balls do not split or crack it so much as quietly imbed themselves. These outer walls have batteries all round. The inner walls are so constructed, that if the outer walls are gained it would still be at a slaughterous expense to the besiegers, if the garrison were at all competent to avail themselves of their position. "We entered the fortress," says a recent writer, "from below, at the principal gate, which was of great strength, and very skilfully contrived, and then went along a stone passage, which had several gateways, and cunningly devised narrow passes, with high stone walls on each side. This was terminated by a canal, or moat, with a drawbridge over it. We next arrived at flights of stairs, and passing up several vault-like ascents, we gained the top of the grand batteries. Their general characteristic is that of great strength, and plenty of room to work in. They mounted 120 long 24-pounders, all of brass, and for the most part in excellent condition. The mortars were of large calibre, though not in such good order as the guns. The powder-magazines were each literally a dry stone well, plugged at the top with blankets, and having a round metal lid over the mouth that opened upon the batteries. We next descended to the inner works, and gained the secondary walls by a circuitous route. Besides the necessity to the besiegers of having guides, who well knew every turn of the works, the excitement and smoke are almost certain to produce a confusion, in which the voice or presence of the guides would be lost, and the party dashing onward might only arrive at a dead wall, a gap looking out upon the sea, or the mouth of a 24-pounder. The circuitous route of descent from the upper to the lower range of walls, is entirely exposed to the batteries, the guns grinning at us all the way as we traversed stone causeways and narrow passes. Whole regiments might here be raked down, after they had conquered the outer walls. A wide and steep flight of stone stairs led us down into the grand castle square or little town, as one might almost call it. We entered at the bottom, through stone gateways, and found ourselves in a large open square, enclosed on all sides by lofty walls, the lower part of which displayed doors and entrances into barracks, guard-houses, and shops of various kinds for the sale of such articles as a garrison would need. The governor's house is at the further end. It was a genuine soldier's lodgment, and very bare of all ornament, except those of war, for it was riddled all over with the marks of shot and shell. Its strong covered balcony, intended to serve both as a protection from the broiling sun, and from the fall of missiles, was in many places torn in long gaps. All the towers and buildings of any elevation had also been knocked about and defaced by the shot and shells from Vera-Cruz, previous to the surrender of the castle. But the mutilations and destruction did not materially affect the strength of the place. Very few of the guns had been dislodged; even the outer batteries were not injured so as to render them ineffective, with the exception of a gap of ruins in one or two places. How strongly and skilfully this fortress is protected by art the reader has now some idea; but San-Juan-de-Ulloa is equally protected by nature; for, while the defences of art which I have briefly described are chiefly devoted to the side and angle facing the town, those angles which face the main ocean on the opposite side, or back of the castle, are protected by long successions of rocky reefs, utterly defying the approach of any vessels of war."

The habitual pop. of Vera-Cruz, without including the militia and seafaring people, was, under Spanish regime, 16,000; but upon the arrival of the flota from Old Spain, a fair was opened which lasted many weeks, and during this period there was a great resort of strangers to the place. The chief disadvantage of Vera-Cruz is its unhealthy situation and climate. The yellow fever generally begins its ravages when the mean temp. of the air rises to 24° of the centigrade therm., or to 75° of Fahrenheit. In December, January, and February, when the heat remains below this limit, the distemper generally disappears. The Mexicans who inhabit the high table land in the interior, where the climate is more temperate, and Europeans newly landed, are peculiarly liable to the infection. Every imaginable precaution has been taken for the safety of the inland inhabitants of the country, when they visit the coast, but hitherto without success. It has never been found that they were attacked by the distemper with less violence and rapidity than the soldiers, for whom none of these precautions were taken. In the season when the yellow fever rages, the shortest stay at Vera-Cruz is sufficient to communicate the contagion; insomuch that the inhabitants of Mexico, who have merely passed through the town in a litter, and instantly embarked, have been attacked by the disease after they had sailed for Europe. The quantity of rain which falls annually at this port amounts to 73 inches, which, joined to the heat of the climate, is extremely favourable to the progress of disease. The ascent from the city into the interior of the country, which is an elevated plain more than 8,000 ft. above the level of the sea, being through difficult and narrow roads, the merchants of Vera-Cruz have constructed a magnificent causeway, which is continued into the interior as far as Mexico, a distance of more than 200 m. W. In 1825 the castle of San-Juan-de-Ulloa surrendered to the patriots after a long siege, during which its fire greatly devastated the city. In 1829, it was taken by the French. It 1847, it was bombarded, along with the city, by the American forces, and capitulated within five days.

VERA-CRUZ, a small island in the Atlantic, on the coast of Brazil, at the entrance of the bay of Todos-Santos.

VERA-CRUZ (OLD), a port of Mexico, in the province of Tlascala, 15 m. N of the present city of Vera Cruz. This is the port where Cortez landed in 1518. It stands in a very unhealthy situation, in a marsh, and on a river which is full of alligators.

VERAGUA, a province of New Granada, bounded on the N by the Caribbean sea; on the E by the prov. of Panama, which is separated from it by the ridge of Canatagua; on the S by the Pacific; and on the W by Costa-Rica. It is a mountainous, rugged country, covered with vast forests, interspersed with luxuriant and fertile valleys. V. was first discovered by Columbus in 1503; and was granted to him, as a reward for his services, with the title of duke. He gave the river which he ascended the appellation of Verdes-Aguas, on account of the colour of its waves; and this has gradually been corrupted into Veragua, and from this river the prov. takes its name. It is administratively divided into the 4 cantons of Santiago, La-Mesa, Remedios, and Alange. Pop. in 1840, 35,367.

VERAGUA (SAN JAGO DE), the capital of the above prov., is a handsome town, situated 20 m. N of Montijo bay, and 120 m. WSW of Panama. It has some trade in copper, cotton, and dye-woods. The Indians in the vicinity dye their cottons, manufactured by themselves, with the juice of shell fish found in the bay of Salinas and on the coast, affording a rich and delicate purple.

VERAMALLAY, a town of Hindostan, in the presidency of Madras, in the Carnatic, district and 27 m. WSW of Trichinopoly.

VERAN (SAINT), a village of France, in the dep. of the Côtes-du-Nord, cant. and 3 m. NNW of Merdrignac. Pop. 1,130.

VERAN-LA-VILLE (SAINT), a village of France, in the dep. of the Hautes-Alpes, and cant. of Aiguilles-en-Guerras. Pop. 625.

VERAND (SAINT), a commune of France, in the dep. of the Isere, cant. and ¾ of a mile from Saint-Marcellin. Pop. 1,041. It affords good wine.

VERANO, a village of Austria, in Lombardy, in the gov. and 17 m. NNE of Milan.

VERANS, a mountain of Sardinia, in the div. of Savoy, and prov. of Faucigny, near the r. bank of the Arve. It commands the village of St. Martin, opposite Sallanches, and has an alt. of 2,705 metres above the level of the Mediterranean. It consists chiefly of limestone and slate.

VERA-PAZ, a province of Guatimala, bounded on the N by Yucatan; on the E by Honduras, and the bay or gulf of Honduras; on the S by the prov. of Guatimala; and on the W by Guatimala and Totonicapan. It is about 120 m. in length, and 74 in breadth. The surface, generally rough and broken, full of deep ravines, and covered with thick and impenetrable woods, is divided into two physical regions, the *Alta* or high country, and the *Vaxa* or low. The elevated country is cold and humid; the low, warm and dry. It is watered by the Motagua, the Usumasinta, and the Polochic. The chief productions are sugar, dye-woods, cacao, cotton, and cedar. The gulf of Dolce or Dulce, a sort of large lake, which communicates with the sea by means of the gulf of Amatique, might greatly facilitate the commerce of this prov. The capital, now called Coban, is situated on the Rio-Coban, which falls into the gulf of Dulce, 600 m. SE of Mexico.

VERAPELLY, or VERAPOLI, a town of Hindostan, in the pres. of Madras, prov. and 9 m. NE of Cochin. It stands on an island in a shallow lagune; and has a cathedral, a seminary, and a Carmelite convent.

VERAT, a small island of the South Pacific, in the group of the Viti islands, in S lat. 17° 50′, and E long. 180° 23′.

VERAUX, a river of France, in the dep. of the Creuse, which has its source to the SE of Cressat, and after a course in a generally N direction of 15 m. joins the Petite-Creuse, on the l. bank at Batisse.

VERAWOW, a village of Sind, 15 m. N of Nuggur-Parker.

VERBA, a town of Russia in Europe, in the gov. of Volhynia, district and 12 m. SSW of Dubno.

VERBACZ, or VERBAS. See VERBITZA.

VERBENICO, a village of Illyria, in the gov. of Trieste, in the island of Veglia, 6 m. ENE of the town of that name. Pop. 1,200. It has an castle.

VERBERG, a village of Prussia, in the regency of Dusseldorf and circle of Krefeld. Pop. 360.

VERBERIE, a commune and town of France, in the dep. of the Oise, and cant. of Pont-Sainte-Maxence, 11 m. NE of Senlis, on the l. bank of the Oise, delightfully situated at the foot of a mountain. Pop. 1,321. It contains some handsome dwellings, and near the house named St. Corneille, from the abbey of that name to which it belonged, is a ferruginous spring. It has manufactories of chemical substances, oil, flour, and paper-mills, and tileworks. Onions and hemp are cultivated in the vicinity. There was formerly a castle belonging to the 1st dynasty of the kings of France, and in which the councils of 853 and 859 were held.

VERBICERO, a market-town of Naples, in the prov. of Calabria-Citra, district and 29 m. N of Paola, and 6 m. from the shore of the Tyrrhenian sea, in a fertile valley. Pop. 3,300.

VERBICZ, or WRIBICA, a town of Hungary, in the comitat of Lyptau, on the r. bank of the Waag, a little to the SE of St. Miklos. Pop. 1,100.

VERBITZA, VERBACZ, or VERBAS, a river of Turkey in Europe, in Bosnia, which has its source in the mountain of Vranja, on the N side of the Dinaric Alps; separates the sanj. of Croatian Turkey from that of Trawnik; passes Bagnaluka; and after a total course, in a N direction, of 120 m., joins the Save on the r. bank, 15 m. E of Gradiska. It is fordable at all seasons. The country through which it flows is fertile, well-cultivated, and abounding with fruit.

VERBO, WERBAU, or WERBOWE, a town of Hungary, in the gsp. and 27 m. NW of Neutra, on the Holleska. Pop. 2,100. It has factories of cloth, and an extensive grain-market.

VERBOSCA, or VERBOVSZKO, a market-town of Austria, in Croatia, in the gsp. of Agram, on the Dobra, 29 m. WSW of Carlstadt. Pop. 1,100. It has a castle.

VERBOVETZ, a small town of Russia in Europe, in the gov. of Podolia, 18 m. NE of Ouchitza.

VERCEL, a canton, commune, and town of France, in the dep. of the Doubs, and arrond. of Baume-les-Dames. The cant. comprises 30 com. Pop. in 1831, 10,061; in 1846, 10,579. The village is 14 m. S of Baume-les-Dames. Pop. 1,210. It has manufactories of hats and hosiery, and several tanneries.

VERCELLI, a province of Piedmont, stretching between the river Sesia on the E, which divides it from the prov. of Novara, and the Dora-Baltea, which divides it from the prov. of Biella on the W. It has an area of 1,421 sq. kilom., with a pop. in 1848 of 127,955. Its N portion is mountainous; its

S, a dead unbroken plain, generally covered with rich vegetation. Its productions are wheat, rice, barley, flax, hemp, fruit, walnuts, and wine. The cultivated land amounts to 112,000 hectares; 15,000 h. are under wood. The rearing of the silk-worm forms an important branch of rural industry. The prov. is divided into 12 mandamenti or districts, which are subdivided into 65 communes.—Its cap. of the same name is situated on the l. bank of the Sesia, at an alt. of 78 metres above sea-level. It is an archiepiscopal see, and contains a noble duomo or cathedral, and 9 other churches. Among the public buildings are the palace-of-justice, and the tribunal-of-prefecture. The city is well-paved, and has large markets. It contains 862 houses, and 4,205 families, with an aggregate pop. of 18,353.

VERCHENY, a village of France, in the dep. of the Drome, cant. and 4 m. NE of Saillians.

VERCHERES, a town of Lower Canada, on the r. bank of the St. Lawrence, 23 m. below Montreal.

VERCHERS (Les), a commune of France, in the dep. of the Maine-et-Loire, cant. and 4 m. SW of Doué, near Layon, on the r. bank of the Loire. Pop. 1,501.

VERCIEN, a village of France, in the dep. of the Isere, cant. and 11 m. NNW of Mirestel. Pop. 500.

VERDACHELLUM, a town of Hindostan, in the Carnatic, 44 m. SW of Pondicherry, on the r. bank of the Tripalapur.

VERDALL, a village of France, in the dep. of Tarn, cant. of Dourgne, on the Soult, 7 m. SW of Castres.

VERDAS-NOVAS, a village of Portugal, 4 m. NE of Oporto, remarkable only for an action between the van of the British and the rear of the French troops, in May 1809, at the commencement of Soult's retreat from Oporto.

VERDE (Cape), a headland on the coast of the territory of Genoa, in N lat. 43° 50'.—Also a cape on the E coast of the straits of Magalhaens.

VERDE, a river of Ecuador, in the prov. of Esmeraldas, which runs into the Pacific in the bay of Tola.—Also a river of Peru, in the prov. of Tacunga, which runs S, and enters the Pastaza, near its source.—Also a river of Mexico, which runs S, and enters the Salado.—Also a river of Brazil, in the prov. of Minas-Geraes, which rises in the mountains of Gram Mogol, near the coast, and flowing W, and then N, enters the Rio-Francisco.—Also a river of Brazil, in the prov. of Goyaz, which runs SE to the Parana.—Also a river of St. Domingo, which rises near the N coast, between the La Vega and Santiago, and running W, enters the Yaque.—Also a river of Paraguay, which runs SE, and enters the Amambay.

VERDE, a small island of New Granada, on the coast of the prov. of Cartagena, at the mouth of the Magdalena.—Also a small island of the Atlantic ocean, near the coast of Vera-Cruz, and not far from the island of Sacrifices.—Also a small island of the Atlantic, near the coast of Cumana, between the islands of Testigos and Frayles.—Also an island near the S coast of Luçon, in N lat. 13° 30', E long. 121° 2'.

VERDE (Cape). See Cape Verde.

VERDE ISLANDS (Cape). See Cape-Verde Islands.

VERDELLO-MAGGIORE, a village of Austrian Lombardy, 7 m. SW of Bergamo.

VERDELOT, a village of France, in the dep. of Seine-et-Marne, cant. and 8 m. NE of Rebuis, on the Petit-Morin.

VERDELPINO, a village of Spain, in the prov. of Cuenca, 3 m. from Hueta. Pop. 580.

VERDEN, a district of Hanover, bearing the title of duchy, lying between Luneburg and Bremen. Its area is about 520 sq. m.; its population 28,000. Its surface is a continued level, and covered to a great extent with heath. It produces a little corn, but is better adapted to pasture. After being long a Catholic bishopric, it was secularized in 1648. and given to Sweden; but has belonged to Hanover since 1715. It is divided into the two bailiwicks of V. and Rotenburg.

Verden, the chief town of the above district, is situated on the r. bank of the Aller, 53 m. SW of Hamburgh. Pop. 3,600.

VERDIER (Le), a village of France, in the dep. of Tarn, 7 m. NNW of Gaillac. Pop. 630.

VERDON, a river of France, dep. of Basses-Alpes, which rises in the Alps, to the SW of Barcelonetta; runs S; and falls into the Durance, on the l. bank, after a course of 110 m.

VERDOUBLE, a river of France, which rises in the dep. of Aude, near Monthoumet; runs SSE; and joins the Gly, near Estagel.

VERDU, a town of Spain, in the prov. and 22 m. E of Lerida. Pop. 1,500. Linen is made here.

VERDUN, a village of Spain, in the prov. and 48 m. NNW of Huesca.

VERDUN, a town of France, dep. of the Meuse, 35 m. W of Metz, in N lat. 49° 9' 31", E long. 5° 22' 3". It is traversed by the Meuse, here in a comparatively early part of its course, and is divided into three parts, an upper, lower, and new town. The first forms the larger portion, and stands on an eminence sloping towards the Meuse. It is strongly fortified, and has a citadel situated on an eminence. It suffered severely in 1755, from the blowing up of a powder-magazine. It is better known to English readers as the place of confinement of their countrymen detained in France after the rupture of 1803. Its manufactures consist of woollen's, leather, articles of confectionary, liquors, and jewellery. It is the see of a bishop.—Also a small town of France, situated at the confluence of the rivers Saone and Doubs, 12 m. N by E of Chalons-sur-Saone. Pop. 1,800.—Also a village in the dep. of Aveyron, cant. and 4 m. NE of Nancelle.

VERDUN-SUR-GARONNE, a town of France, in the dep. of Tarn-et-Garonne, on the Garonne, 20 m. NW of Toulouse. Pop. 3,500.

VERE. See Tervere.

VERE, a parish of Jamaica, in the county of Middlesex, and forming the most southerly portion of the island. The E coast is flat and marshy, and presents numerous indentations, of which the principal forming West harbour is enclosed on the S by Portland Bridge—an extensive peninsula terminating in the SE in Portland point. The entrance to this harbour, in common with all this portion of the coast, is studded with reefs and islands. In the W part of the p. is the range of the Braziletto hills, and in the N Harrison's plain. The principal river is the Mintro or Dry river, which flows into the sea to the NW of Carlisle bay. The soil is fertile and well-cultivated.

VEREBELY, or Wrablany, a town of Hungary, in the gsp. or comitat of Bars, 12 m. SE of Neutra, on the l. bank of the Tsitva, in a richly agricultural district.

VERECZKE (Also), a market-town of Hungary, in the comitat of Beregh, 30 m. NE of Munkacs, at the foot of the Carpathian chain.

VEREIA, a town of Russia in Europe, cap. of a district of the same name, in the gov. and 64 m. WSW of Moscow, on both banks of the Protva. Pop. 6,800. The era of the foundation of this town, as well as of the name of its founder, is unknown; but it is of great antiquity, and has always been known to have followed the fate of Moscow. It has been ravaged both by the Tartars and Poles, but the castle, which still stands, and the fortifications of which consist chiefly in a high earthen rampart and a ditch, has never been taken. V. contains four churches inclusive of the cathedral, which is a handsome edifice, a school, magazines of salt and brandy, and about sixty shops. The trade in hemp, tallow, wood, and corn, is considerable. Large numbers of fishing-nets are here made by the women, for export to the Ukraine, Don Cossacks, and Astrakhan. Onions are extensively grown in the environs.

VERENAY, a hamlet of France, in the dep. of the Rhone, cant. of Sainte-Colombe, near the r. bank of the Rhone. There is a lead mine in the vicinity.

VERESPATEK, a village of Transylvania, in the com. of Lower Weissenburg, 26 m. NW of Carlsburg. Gold is found in the vicinity.

VERESVAGAS, or OZERWENICZA, a village of Hungary, in the com. of Saros, 14 m. SE of Epiries. In the vicinity is the only mine of the true opal in Europe.

VERETZ-SUR-CHER, a village of France, in the dep. of the Indre-et-Loire, cant. and 6 m. SE of Tours, on the l. bank of the Cher. Pop. 900. It has an ancient castle.

VERFEIL, a canton and commune of France, in the dep. of the Haute-Garonne, and arrond. of Toulouse. The cant. comprises 8 com. Pop. in 1831, 5,218; in 1846, 5,632. The town is 14 m. ENE of Toulouse. Pop. 2,420.—Also a commune in the dep. of the Tarn-et-Garonne, cant. and 8 m. ENE of St. Antonin. Pop. 1,103.

VERGA (CAPE), a headland of Upper Guinea, on the Sierra Leone coast, in N lat. 10° 18′ 52″.

VERGADA, a small island of Austria, in Dalmatia, in the circle of Zara, and channel of that name, and to the E of the island of Pasman. It contains a village of the same name, and about 200 inhabitants.

VERGARA, or BERGARA, a judicial partido and market-town of Spain, in the prov. of Guipuzcoa. The partido comprises 15 pueblos. The town is 33 m. SW of San-Sebastian, in a fertile valley on the r. bank of the Deva. Pop. 4,000. The streets are spacious and well-paved, and the houses generally well-built. It has two squares, in one of which is the town-house, a handsome edifice, supported by a range of 12 arcades, and contains also three parish-churches, several convents, a Jesuits' college, and a seminary. It has extensive iron and steel-works. This town is noted as the seat of the first patriotic society for the improvement of education in Spain. In the vicinity are several mineral springs.

VERGATO, a market-town of the Papal States, in the leg. and 14 m. SW of Bologna. Pop. 2,850.

VERGAVILLE, a commune and town of France, in the dep. of the Meurthe, cant. and 2 m. NE of Dieuze, on the Spin. Pop. 1,145.

VERGENNES, a town of Addison co., in the state of Vermont, U. S., 33 m. W by S of Montpelier, on Otter creek, which is here divided by two islands into three channels, and intersected by the Rutland and Burlington railway. Pop. 1,378.

VERGES, a market-town of Spain, in the prov. and 15 m. E of Gerona, on the road from that town to the port of the Escala. Pop. 1,500. It has several oil-mills.

VERGEZE, a commune of France, in the dep. of the Gard, cant. and 5 m. NW of Vauvert. Pop. 1,316. It has several mineral springs.

VERGNE (LA), a village of France, in the dep. of the Charente-Inferieure, cant. and 3 m. NW of St. Jean d'Angely. Pop. 300.—Also a village in the Lot, cant. and 2 m. NE of Gramat, on the r. bank of the Alzon. Pop. 700.

VERGNIES, a department and commune of Belgium, in the prov. of Hainault, and arrond. of Charleroi. Pop. 309.

VERGNIOZ, a village of France, in the dep. of the Isere, cant. and 6 m. NE of Roussillon. Pop. 400.

VERGORAZ, or VERGORATZ, a village of Austria, in Dalmatia, in the circle and 18 m. SE of Macarsca, between lakes Rastok and Jesero. Pop. 800. It has a productive pitch-mine.

VERGT, or ST. JEAN-DE-VERGT, a canton and commune of France, in the dep. of the Dordogne, and arrond. of Perigueux. The cant. comprises 16 com. Pop. in 1831, 11,089; in 1846, 11,456.

VERIA, or KARA-VERIA, a town of European Turkey, in Macedonia, a few miles from the coast of the gulf of Salonika, and 38 m. W by S of that town. It contains about 8,000 inhabitants, mostly Greeks; and has extensive manufactories of cotton and cotton thread. The neighbouring country produces rice and fruit in abundance.

VERIN (SANTA-MARIA-DE), a judicial partido and town of Spain, in the prov. of Orense. The partido comprises 86 parishes. The town is 30 m. SE of Orense and 5 m. from the Portuguese frontier, on the Tamaga, which is here crossed by a substantial bridge, and at the foot of a mountain. Pop. 1,000. It has a church, two convents, a Bernardine priory, and a custom-house. The manufacture of linen, and the adjacent tin-mines, form the chief objects of local industry.

VERINA, a town of Venezuela, in the dep. of Maturin and prov. of Cumana, on the Atlantic. It is famous for its tobacco.

VERINES, a commune of France, in the dep. of the Charente-Inferieure, and cant. of La Garrie, 7 m. ENE of La Rochelle. Pop. 1,264.

VERISSIMO, or VIRISSIMO, two rivers of Brazil, in the prov. of Goyaz, both of which rise on the S side of the Serra-dos-Pirenneos, and which, after a considerable course, throw their united waters into the Paranaiva.

VERJON, a village of France, in the dep. of the Ain, and cant. of Coligny, 13 m. NNE of Bourg. Pop. 500.

*** VERKHNE, VERKHNII, or VERKHNAÏA, a Russian adjective signifying 'upper.' For names to which it is prefixed not found amongst the subjoined, see the following word.

VERKHNE'DNIEPROVSK, a town of Russia in Europe, in the gov. and 36 m. WNW of Yekaterinoslav, on the r. bank of the Dnieper. Pop. 1,000.

VERKHNE'-KOLYMSKI, a town of Russia in Asia, in the prov. of Yakutsk, on the Kolyma, 210 m. SSW of Sredne-Kolymsk.

VERKHNE'-VILIOUISK, a town of Russia in Asia, in the prov. and 330 m. WNW of Yakutsk, on the r. bank of the Vilionï.

VERKHNII (SALLOV), a town of Russia in Europe, in the gov. of Kharkov, district and 18 m. SSW of Voltchansk. Pop. 1,500.

VERKHOIANSK, a town of Russia in Europe, in the prov. and 330 m. NNE of Yakutsk, on the l. bank of the Jana.

VERKHOLENSK, a town of Russia in Asia, in the gov. and 135 m. NNE of Irkutsk, on the Lena, near its source. Pop. 510.

VERKHO-SOSENSK, a town of Russia in Europe, in the gov. and 12 m. NW of Biriutsk, and near the sources of the Tikhaïa-Sosna. Pop. 5,000. It contains four churches. It was founded in the reign of Alexis-Mikhaïlovitch, when it made part of the line of defence between Oskol and Korotoïak.

VERKHOTURIE, a district and town of Russia in Asia, in the gov. of Perm. The district comprises several copper-mines and iron-works.—The town is situated on the l. bank of the Tura, 210 m. ENE of Perm. It is walled, and conducts considerable transit-commerce with Siberia.

VERKHOVKA, a town of Russia in Europe, in the gov. of Podolia, 36 m. SE of Bratslav.

VERKHOWASHKOI, a town of Russia in Europe, in the gov. of Vologda. Pop. 1,540.

VERL, a parish and v. of Prussia, in the reg. of Minden, and circle of Wiedenbruck. Pop. 1,250.

VERLAINE, a department and commune of Belgium, in the prov. of Liege and arrond. of Huy. Pop. of dep. 1,260; of com. 819.

VERLAR, a village of Prussia, in the regency of Minden, and circle of Buren. Pop. 360.

VERLAUTENHEIDE, a village of Prussia, in the regency of Aachen. Pop. 630.

VERLEE, a department and commune of Belgium, in the prov. of Namur and arrond. of Dinant. Pop. 108.

VERLIEHAUSEN, a village of Hanover, in the prov. of Gottingen, near Uslar. Pop. 318.

VERLIKA, or VERLICCA, a market-town of Austria, in Dalmatia, in the circle and 48 m. NW of Spalatro, near the Cettina. Pop. 300.

VERLINGHEM, a commune of France, in the dep. of the Nord, cant. and 3 m. S of Quesnoy-sur-Deule. Pop. 1,915.

VERLORENWASSER, a village of Austria, in Silesia, in the circle of Troppau. Pop. 350.—Also a village of Prussia, in the regency of Breslau, and circle of Habelschwerdt. Pop. 520.

VERMAND, a canton, commune, and town of France, in the dep. of the Aisne, and arrond. of Saint-Quentin. The cant. comprises 24 com. Pop. in 1831, 13,380; in 1841, 14,270. The town is 7 m. NW of St. Quentin, near the Avignon. Pop. 1,100.

VERMANDOIS, formerly a district of France, in the arrond. and prov. of Picardy, comprising an area of 124,170 hect. and now divided between the deps. of the Aisne and Somme. Its chief towns were Saint-Quentin, the capital, Vermand, and Ham. It derived its name from the *Veromanduentes*, a Gallic people, by whom it was inhabited. It was united to France under Philip Augustus.

VERMANTON, a canton and commune of France, in the dep. of the Yonne, and arrond. of Auxerre. The cant. comprises 14 com. Pop. in 1831, 11,216; in 1846, 11,160. The village is 14 m. SE of Auxerre, on the r. bank of the Aire. Pop. 2,616. It has an active trade in wine, casks, and faggots.

VERMDOE, an island of Sweden, in the Baltic, a little to the E of Stockholm, in N lat. 59° 20′. and E long. 18° 25′.

VERMEJA (SIERRA), a range of mountains in Spain, on the confines of the provinces of Malega and Cadiz, in the Ronda chain. It consists of two parallel ranges, united at the base, one of which is reddish, from the colour of its soil, whence the name Vermeja; the other is white. The first, which is the loftier, is covered with pines, and the other is snow-capped throughout the greater part of the year. It has some oak plantations, and contains mines of iron and mineral springs.

VERMEJO, or RIO-GRANDE, a river which has its source in Bolivia, to the NW of San-Bernardo-de-Tarija, forms for some distance the boundary line of La Plata; traverses the N part of that territory; and after a course, in a generally SE direction, of about 1,200 m., joins the Paraguay, on the r. bank, 90 m. SSW of L'Assompcion.

VERMELEN, a lake of Sweden, in the prefecture of Carlstadt, to the N of Lake Wener, into which it discharges its waters.

VERMELHA (SIERRA), a mountain-range of Brazil, in the prov. of Piauhy, which runs from NE to SW, near the confines of the prov. of Pernambuco, and of the Rio-Itatimi.

VERMELHO, a river of Brazil, which rises near Villaboa, in the prov. of Goyaz; runs NW; and flows into the Araguay in S lat. 13° 35′, after a course of 220 m.

VERMILLION, a county in the E part of Illinois, U. S. Area 1,208 sq. m. Pop. in 1840, 9,303; in 1850, 11,492. Its cap. is Dunville.—Also a county in the W part of Indiana. Area 280 sq. m. Pop. in 1840, 8,274; in 1850, 8,661. Its cap. is Newport.—Also a river of Louisiana, which rises in the district of the Opelousas, and after running a S course, discharges itself into Vermillion bay, in the gulf of Mexico, about 200 m. W of the mouth of the Mississippi. The bay is 20 m. in length, and has 10 or 12 ft. water.—Also a river of Louisiana, which runs into the tributary branches of the Missouri.—Also a river of Ohio, which falls into Lake Erie, 9 m. E of Huron river.—Also a river of Illinois, which runs into the Illinois river, 150 m. from the Mississippi.—Also a township of Huron co., Ohio, on Lake Erie, at the mouth of the Vermillion, about 40 m. W of Cleveland.

VERMO, a town of Austrian Illyria, in the prov. of Istria, 21 m. WSW of Fiume.

VERMONT, one of the northern United States, situated between 42° 44′ and 45° N lat., and 71° 38′ and 73° 26′ W long. The boundary line that separates it from Lower Canada on the N is 90 m. long; on the E it is bounded by New Hampshire; on the S by Massachusetts; and on the W by Lake Champlain and New York. Its mean length from N to S is 157 m. The distance from the ocean to the nearest point of this state is about 80 m. Its area is stated at 10,212 sq. m., or 6,535,680 acres.

Physical features.] The general character of this state is that of a hilly region. The Green mountains, from 10 to 15 m. in breadth, traverse its whole length. These mountains begin in Canada, whence they extend through the states of Vermont, Massachusetts, and Connecticut, and terminate within a few miles of the coast. Their general direction is from NNE to SSW; and their extent is not less than 400 m. in length. They run nearly parallel with the course of the Connecticut river, and are intersected by numerous valleys, the soil of which is deep, rich, and loamy. Towards the centre of the state, the range divides into two, the western of which continues to run N, and the eastern runs NE, near the base of Camel's Rump [alt. 4,188 ft.]. On the S side of Onion river, a line of road crosses the Green mountains from Burlington to Hartford. The most level tract occurs on the borders of Canada. Between the banks of Lake Champlain and the mountains there is a valuable tract of arable land, extending 100 m. in length and 30 m. in breadth. The surface in its natural state is thickly wooded. Along the banks of the rivers, white oak, beech, and elm are abundant. The higher parts are covered with white oak, sugar-maple, butternut, ash, and birch; and the mountains are clothed with evergreens to their summit. The highest summits of the Green mountains in Vermont, are Killington-peak, alt 3,675 ft.; Camel's Rump, and Mansfield mountain, on the N side of Onion river, alt. 4,279 ft. above sea-level.—All the streams and rivers of V. have their origin among the Green mountains. About 30 of them have an easterly direction, and fall into Connecticut river; about 25 run westerly, and discharge themselves into Lake Champlain; and two or three, running in the same direction, fall into Hudson's river. In the NE part of the state are four or five streams which have a northerly direction tc lake Memphremagog, on the frontiers of Canada. The most considerable streams on the W side of the Green mountains are Otter-creek, 85 m. in length, Onion river, 80 m. in length, the Lamoille, and Michiscoui or Missisque. On the E side of the Green mountains, the rivers are not so large as those on the west, but they are more numerous: the largest are Wantastitquek or West river, White river, and Poousoomsuck. The river Connecticut, into which these rivers fall, forms the E boundary of the state. Above two-thirds of Lake Champlain lie in this state. The islands in this lake, belonging to the state of V., constitute Grand Isle co. The principal territories on the lake are St. Alban's, Burlington, and Vergennes. A canal connects the S side of the lake with the Hudson river, and the New York and Erie canal. The N line of Vermont passes over the S part of Lake Memphremagog. This lake is about 40 m. in length, and 6 or 7 m. wide. The river St. Francis forms a communication between it and the the St. Lawrence river. Round this lake there is a rich soil and a fine level country.

Climate.] The climate is subject to great extremes of heat and cold. The mean annual range of the therm. is 120°; the temp. of winter sinking sometimes to 25° below zero, while that of summer rises to 100°. Snow lies from the middle of December to the middle of March, during which period it is customary to travel in sledges. It disappears about the middle of April, except on the highest parts of the mountains, where it lies till May. It is generally permanent from the 10th or 12th of December to the beginning of April, when it suddenly dissolves by the influence of a warm sun. In the low grounds it remains till about the 20th of March. Wheat and oats are sown about the middle of April and are reaped about the middle of August. The frosts commence in the middle of September. Notwithstanding the severity of winter, which is 10° or 11° colder than in the same latitude in Europe, young trees are seldom killed by the frost. Lake Champlain is seldom totally frozen over before the beginning of February. The ice of lakes and stagnant waters, in the severest winter, seldom exceeds 30 inches in thickness. It generally dissolves in the last days of March. The most agreeable season is from the beginning of September to the middle of October, after which, to the close of November, there are frequent rains, winds, and snow. The N, NW, and W winds, which are the most prevalent, are dry, elastic, and invigorating; those from the S and SW are warm and relaxing.

Productions.] Iron ore exists in great abundance on the W side of the Green mountains, and near Lake Champlain. Lead, copper, and ochre are wrought. Porcelain clay is found, which retains its white colour in the fire; also soapstone, slate, granite, marble, clay for bricks, pipe-clay, mill-stones, and marl. There are lead, zinc, and copperas mines. Among the forest-trees are the pine, maple, buttonwood, elm, hemlock, oak, basswood, ash, and birch. The most common are the hemlock, elm, spruce, sugar-maple, and beech. On the summit of the highest mountains, the trees, which are chiefly spruce, hemlock, and pine, are very low, and their branches are interwoven so as to form an impenetrable thicket. There is a much greater proportion of evergreen trees on the W than on the E side of the mountains. Williams enumerates thirty-six species of quadrupeds in this state, of which the most remarkable are the bear, black-cat, wild-cat, catamount, deer, fox, hare, martin, ermine, mole, mouse, porcupine, rabbit, racoon, skunk, squirrel, weasel, wolf, wood-chuck, beaver, mink, musk-rat, and otter. In lakes Champlain and Memphremagog, and also in several rivers, are found bass, bream, bret, chub, dace, eel, minow, red-perch, white-perch, pickerel or pike, salmon, salmon-trout, shad, skinner, sturgeon, sucker, and trout.

Agriculture.] The soil of V. is rich and loamy, and well adapted to wheat, rye, barley, corn, oats, pease, flax, hemp, and culinary plants. The wheat crop in 1840 amounted to 495,800 bushels; in 1850, to 525,925 b. The oats raised in 1840 amounted to 2,222,584 b.; in 1850 to 2,307,714 b. The Indian corn crop in 1840 was 1,119,678 b.; in 1850, 2,032,016 b. In 1850, 176,297 b. of rye, 42,147 b. of barley, and 208,699 b. of buckwheat were raised. Spring-wheat, barley, oats, and pease, are sown from the 16th to the 20th of April. The first is ripe about the middle of August; barley on the 1st, and oats about the 20th of that month. Indian corn is sown about the middle of May, and is ripe towards the 1st of October. The natural pasture is excellent. In winter, the beeves when grown are fed with hay, clover, turnips, pumpkins, &c., and the milk cows with wheat, bran, oats, and Indian corn. The sheep weigh 120 pounds, and yield three or four pounds of wool. When the acorn, beech-mast, and other nuts begin to fall, the swine are driven to the woods, where they fatten to an enormous size. In 1850, the extent of improved farm lands in this state was returned at 2,591,379 acres; of unimproved, 1,525,368 acres; and the number of farms at 29,687. The live stock in 1850 comprised 61,057 horses, 146,146 milk cows, 48,497 working oxen, 919,992 sheep, 66,278 swine.

Manufactures and commerce.] The manufactures in this state are not considerable. In 1850 the number of manufacturing establishments was 1,835, of which 9 were cotton factories, 72 woollen factories, 37 iron factories, and 152 tanneries. Pot and pearl ashes, and yellow ochre, are manufactured.—The exports consist of grain, flour, bar iron, nails, pot and pearl ashes, live cattle, horses, beef, pork, cheese and butter, lumber, peltry, and flax, which are sent to Montreal in Canada, and, by river and canal communication, to the cities and towns of New York, Portland, Hartford, and Boston. The value of the exports from V., which are chiefly made from Burlington on Lake Champlain, amounted, in 1799, to 20,480 dollars; in 1803 to 117,450 d.; in 1810 to 432,631 d.; in 1816 to 892,594 d.; in 1850 to 430,906 d., of which 404,749 d. were domestic produce.

Population.] The population of this state in 1790 was 85,416, including 272 Blacks; in 1800, 154,465, including 557 Blacks; in 1810, 217,895, Blacks 750; in 1820, 280,652, of whom 881 were Blacks; in 1830, 280,652; in 1840, 291,948; in 1850, 313,611 of whom 709 were free Blacks.—There are three colleges in V., one at Burlington founded in 1791; another at Middlebury founded in 1800; and a third at Norwich founded in 1834. Common schools are supported throughout the state, and in 1850 amounted to 2,594 with 90,110 scholars.—The principal denominations of Christians in V. are Congregationalists, who had, in 1850, 168 churches. Methodists, who had in 1850, 128 schools; and Baptists, who had 88 churches. There were 34 Universalist churches in the same year, 25 Episcopal, and 8 Roman Catholic.

Government.] The legislative power is vested in a house of representatives at present numbering 230 members, one from each town, and a senate of 30 members apportioned in the ratio of the pop., but each county is entitled to send one. Every town has the right to send one representative. The executive power is vested in a governor, lieutenant-governor and 12 councillors. All these officers, together with the representatives, are chosen annually. The state sends 4 members to congress. Judges, sheriffs, and other officers are appointed for one year by the general assembly in ballot. The constitution prohibits slavery. Montpelier is the political capital.—The revenue of the state arises from taxes on persons, lands, and cattle, assessed in proportion to their value. The revenue in 1850 was 125,375 d.; the disbursements only 122,253 d. V. has no public debt.—The number of paupers in 1850 were 3,654, of whom 2,043 were natives. The expense of their support was 120,462 d.

History.] The first settlements in this state began about the year 1724; in 1760 a number of towns were already formed by emigrants from New Hampshire, Massachusetts, and Connecticut. V. was considered as under the jurisdiction of New Hampshire till 1764, when, by act of parliament, it was annexed to New York, under the name of Cumberland and Gloucester counties; but the inhabitants finding, that by this union they would be obliged to purchase the lands they already occupied, and preferring to be incorporated with New Hampshire, remonstrated against this project. In the mean time, the war commenced against England, and the inhabitants seized the opportunity to declare themselves free and independent; a circumstance which inspired the English with the hope of detaching them from the American cause. Various means were employed to effect this, which were at last defeated by an arrangement with the state of New York, in 1790, as to their mutual claims, and by her admission into the federal union the year following. The new constitution of this state, which had been prepared and sanctioned in 1778, and was afterwards revised and improved in 1786 and 1792, was finally adopted in 1793; but has been modified at various subsequent periods.

VERMONTVILLE, a township and village of Eaton co., in Michigan, 24 m. W by S of Lansing. Pop. 324.

VERN, a commune and village of France, in the dep. of Ille-et-Vilaine, 5 m. SE of Rennes. Pop. 1,700.—Also a com. and v. in the dep. of the Maine-et-Loire, 15 m. NW of Angers.

VERNAISON, a village of France, in the dep. of Rhone, 9 m. S of Lyons, on the railway to St. Etienne. Pop. 950.

VERNAL, an island near the coast of Mexico, in S lat. 16° 35′.

VERNAMO, a village of Sweden, in the laen and 42 m. S of Jönkioping.

VERNANTE, a town of Piedmont, on the great road from Nice to Turin, 11 m. S of Coni. Pop. of com., 3,171.

VERNANTES, a town of France, dep. of Maine-et-Loire, 13 m. SE of Baugé. Pop. 2,050.

VERNANTOIS, a village of France, dep. of Jura, cant. and 4 m. SSE of Lons-le-Saunier. Pop. 950.

VERNAZZO, a town of the duchy of Genoa, 5 m. SW of Spezzia.

VERNEIL, a village of France, in the dep. of Sarthe, cant. and 3 m. SE of Mayet. Pop. 1,200.

VERNET, a town of France, dep. of Orientales-Pyrenees, 4 m. S of Prades.

VERNET (LE), a village of France, in the dep. of Puy-de-Dome, cant. and 10 m. SW of St. Amand-Talende. Pop. 1,000.

VERNET-SUR-SIOULE (LE), a village of France, in the dep. of Allier, cant of Escurolles, on the r. bank of the Sioule.

VERNEUIL, a town of France, dep. of Eure, on the small river Arve, 20 m. SW of Evreux. It is still surrounded with a wall, but contains no buildings worthy of notice, with the exception of an old tower. Pop. 3,500. It has manufactures of woollens, cottons, druggets, leather, and paper.—Also a com. and v. of France, in the dep. of Haut-Vienne, 6 m. WNW of Limoges. Pop. 2,000.

VERNEUIL-SUR-OISE, a village of France, in the dep. of Oise, 6 m. NW of Senlis. Pop. 1,180.

VERNHAM-DEAN, a village of Hampshire, 8 m. W by N of Andover.

VERNON, a county of New South Wales, skirted by the Macleay river on the NE, and intersected by the Apsley river.

VERNON, a town of France, dep. of Eure, on the l. bank of the Seine, and on the Paris and Havre railway, 28 m. S by E of Rouen. Pop. 3,950. It has an old castle, two churches, and an hospital. It has some manufactures of leather, cotton-caps, and velvet, and carries on a traffic with Paris in corn and meal.—Also a com. and v. in the dep. of Indre-et-Loire, 9 m. E of Tours. Pop. 1,800.

VERNON, a township of Tolland co., Connecticut, U. S., 6 m. WSW of Tolland. Pop. 2,900.—Also a county in the W of Missouri. Area 612 sq. m., drained by the Marais-des-Cygnes river.—Also a township of Oneida co., New York, 17 m. W of Utica. Pop. 3,098.—Also a township of Sussex co., New Jersey, 68 m. N by E of Trenton. Pop. 2,619.—Also a township of Trumbull co., Ohio, 20 m. NE of Warren. Pop. 828.—Also a township of Crawford co., Ohio.—Also a village of Windham co., Vermont, 105 m. S by E of Montpelier. Pop. 821.—Also a township of Jennings co., Indiana, 25 m. NW of Madison. Pop. 600.—Also a village of Suttee co., in California, on the E bank of Feather river, at its confluence with the Sacramento.

VERNON ISLES, a group of 3 small islands at the entrance of Clarence-strait, on the NW coast of Australia, in S lat. 12° 5′.

VERNOU, a village of France, in the dep. of Indre-et-Loire, cant. of Vouvray, on the Brenne, 6 m. E of Tours. Pop. 1,500.—Also a village of France, in the dep. of Deux-Sevres, cant. and 2 m. NW of Briou. Pop. 400.

VERNOUX, a town of France, dep. of Ardeche, 12 m. NE of Privas. Pop. 1,546. It has a trade in woollens and agricultural produce.

VERNOUX-EN-GATIVE, a commune of France, dep. of Deux-Sevres, 12 m. W of Parthenay. Pop. 1,400.

VEROCZA, or VEROVITZ, a county of Slavonia, comprising the NW of the province, and bounded by the Drave on the N, and by the Danube on the E. It has a territorial extent of nearly 1,800 sq. m., and is divided into 6 marches. Its surface is in general level, and is covered in many parts with extensive marshes; in general, however, the soil is fertile. The export of cattle forms a principal branch of trade. The capital is Esseck or Esseg.

VEROCZA, or VEROVITICZA, a small town of Austrian Slavonia, formerly the capital of the preceding palatinate, situated on the Drave, 62 m. WNW of Esseck, in N lat. 45° 51′ 14″. It was taken in 1626 by the Turks, and in 1684 by the Imperialists.

VEROLA-ALGHISE, a town of Austrian Italy, in the gov. of Milan, deleg. and 18 m. SW of Brescia, on the river Savorola.—About 1 m. to the E of this town is Verola-Vecchia.

VEROLENGO, a town of Piedmont, in the prov. and 18 m. NE of Turin. Pop. 4,000.

VEROLI, a town of the Papal states, 55 m. E by S of Rome. Though small, it is the see of a bishop, and stands at the foot of the Apennines, on the river Cosa, 22 m. N by E of Terracina.

VEROMILES (SAN), a village of the island of Sardinia, in the prov. of Busachi, 9 m. N of Oristano. Pop. 1,500. Lead, copper, and iron are wrought in the vicinity.

VERON, a town of France, dep. of the Yonne, 6 m. S of Sens. Pop. 1,160.

VERONA, a province of Austrian Italy, in the W part of the gov. of Venice, with a superficial extent of 1,330 sq. m. It is bounded on the N by Tyrol; on the E by the provs. of Vicenza and Padua; on the SE by the Polesina; on the S by Mantua; and on the W by the Lago-di-Garda. It is watered from N to S by the Adige, and on the W by the Mincio; and though partly mountainous, it has an agreeable climate and fertile soil. The principal productions are corn, wine, oil, flax, silk and beautiful marble. Pop. in 1846, 302.902.

Verona, the capital of the above province, and now the metropolis and seat of government of the Lombardo-Venetian kingdom, stands in N lat. 45° 26′, 22 m. NE of Mantua, and 60 m. W of Venice. It occupies a pleasant and picturesque situation, partly on a declivity, partly on the border of a large plain which stretches far to the southward, and extends along the banks of the Mincio and the Po. To the N are the Tyrolese Alps, the first step to the ascent of which may be said to take place in this city. The Adige, descending from Tyrol, flows with a full and rapid stream and serpentine course through the city, dividing it into two unequal parts. The form of the city is irregular, the modern buildings extending considerably beyond the old walls. Its circuit is about 6 m. It retains its former fortification of a moat and earthen mound, and has also two castles on high ground, with a third on the plain. Its fortifications have been greatly strengthened of late years. The garrison of V. amounts at present to 16,000 or 18,000 men. At the side of the city, leading to Santa-Lucia, an immense plain has been laid out for military evolutions, and several important points have been strengthened by the erection of forts capable of containing 2,000 men. "V. is a complete military stronghold, and will certainly be the last place relinquished by Austria, in the event of its dominion in Italy being again disputed. It is surrounded on all sides by fortifications; and several new barracks are now being constructed in the interior of the town. It is the first place, leading towards the Tyrol, which bears decided marks of its vicinity to Germany. Though Italian is the predominant language, what with the military and the great number of German officials, you might fancy yourself in 'Faderland.' The town is externally quite Italian; but the signs over the shops are a true type of its present pop., for they are inscribed in both the Italian and German tongues." The interior of V. does not correspond with the beauty of its position, several of the streets being narrow and dirty; others, in particular that which leads to the Mantua gate, and the Corso, or street where horse-races are held, are spacious. The houses, though built in general in an antique style, are of good appearance, from the quantity of marble employed in their construction. The best buildings are in the principal square. V. contains a cathedral and a number of churches, some noted for their paintings, others for their architecture. The latter forms the characteristic of the church of San-Zeno, remarkable for its façade, its vast portal, and the rows of columns, each of a single piece of marble, which support its roof. The building called the royal palace has never been completed. The Palazzo-Bevilacqua, said to be the oldest building in V., is a stately edifice, but is going fast to decay. Opposite to it is the Palazzo-Cannossia, admired for its front, and the prospect it commands. The museum of the academy is admired for its portico, with its peristyle of large Ionic columns. Of all the monuments of V., the most interesting is the Roman amphitheatre, situated in the spacious square called the Piazza-del-Bra, one of the most magnificent remains of Roman architecture that has descended to modern times. The arena, in the centre and of an oval form, is 220 ft. in length, in breadth 130 ft. The seats rise in successive ranges to a height corresponding to the top of the second row of outward arches. These seats, as well as the different passages, the stair-cases and galleries of communication, remain entire. The number of ranges of seats is 45; that of spectators which might be contained within them, about 22,000; the outward circumference of the amphitheatre 1,290 ft. The whole consists of vast blocks of marble, forming a solid mass, resting on a double row of massy vaults, which, in former ages, were appropriated to the custody of the lions, tigers, and other wild beasts. V. is the seat of one of the five sections of the imperial and royal institute of Austrian Italy. It contains, besides, a lyceum or great school, a gymnasium or classical school, an academy of painting, a public library, and several private collections. The pop. is about 45,000. Its principal manufacture is that of silk; the lesser manufactures are woollens, leather, gloves, and shoes.

History.] The date of the foundation of V. is not known; but Julius Cæsar established a colony here. On the decline of the empire, V. experienced the fate of the other towns in the north of Italy. It was taken by Charlemagne in 774; became subsequently a free town; fell, in the course of time, under the sway of leading families; and in 1405, was united to the territorial possessions of Venice. With these it enjoyed many ages of peace and tranquillity, until the year 1796, when Italy was invaded by the French. It was then added to the kingdom of Italy. In 1814 it again fell into the hands of Austria. A congress was held here in 1822. V. is the place which has profited most by the late revolutionary war: it has despoiled Milan and Venice of their official advantages, and reduced them to the humble rank of provincial cities. "It is the residence of the lieutenant-general—the office of viceroy being for the present suspended—and from it the mandates of the imperial cabinet flow, from the Tessin to the Po, and from Alp to Alp. Politically speaking, nothing can be wiser than this arrangement; for, as long as the forts on the hills which command the city are held by Austrian troops, the fidelity of the place is insured, and, no matter what event takes place in any other part of the kingdom, the seat of government cannot be disturbed. The campaign of 1848 shows that the line of the Adige is the true strategical position, and that, whilst it is preserved, even though the Adda, the Oglio, and the Mincio be lost, the imperial government need not despair."

VERONA, a township of Oneida co., New York, U. S., on Wood-creek, 20 m. W of Utica. Pop. in 1840, 4,504; in 1850, 5,568.

VERPELETH, a town of Hungary, in the com. of Heves, 9 m. SW of Erlau. Pop. 1,600.

VERPILLIERE (La), a canton, commune, and town of France, in the dep. of the Isere, and arrond. of Vienne. The cant. comprises 16 com. Pop. in 1841, 13,477; in 1846, 14,658.—The town is 17 m. NE of Vienne, on the Alliat. Pop. 1,186. The manufacture of sugar from beet-root forms the chief branch of local industry.

VERPLANK, a village of Westchester co., in the state of New York, U. S., on Hudson river, 42 m. from New York.

VERRE, a river of France, which has its source in the N part of the dep. of the Tarn, near Taix; runs first W, then NW; enters the dep. of the Tarn-et-Garonne; and after a course of about 30 m., joins the Aveyron, on the l. bank, at Bruniquel.

VERREBROCK, a department and commune of Belgium, in the prov. of E. Flanders, and arrond. of Termonde, watered by the Scheldt. Pop. of dep., 1,272; of com., 476.

VERRERIE, a rivulet of France, in the dep. of the Côte-d'Or, which has its source near that of the Tustie and the Junie, and after a course of about 3 m., joins the Seine, on the r. bank, between Billy and Oigny.

VERRES, a market-town of Sardinia, capital of a mandemento, in the div. and prov. of Aosta, on the Lovinson, near its confluence with the Dora-Baltea, and 21 m. NW of Ivrea. Pop. 3,000. It has iron manufactories. Goitre and cretinism are here extremely prevalent.

VERRICO, a market-town of Naples, in the prov. of Abruzzo-Ultra, district and 20 m. NW of Aquila

VERONA

VERRIE, a village of France, in the dep. of the Vendée, cant. of Mortagne-sur-Sevre. Pop. 1,906.

VERRIE'RES, a commune of France, in the dep. of the Seine-et-Oise, cant. and 3 m. N of Palaiseau, on the Bievre. Pop. 1,011.—Also a com. in the dep. of the Loire, cant. and 5 m. SW of Montbrison. Pop. 1,175. It contains many handsome dwellings. The locality is noted for its bricks and tiles.—Also a com. in the dep. of the Marne, cant. and 3 m. S of Sainte-Menehould, near the l. bank of the Aisne. Pop. 1,051.—Also a com. in the dep. of the Vienne, cant. and 7 m. W of Lussac-les-Chateaux. Pop. 788. It has blast furnaces and manufactories of agricultural implements.—Also a village in the dep. of the Orne, cant. and 5 m. E of Nocé. Pop. 1,050. —Also a village of Switzerland, in the cant. and 23 m. WSW of Neufchatel, near the French frontier. Pop. 1,635.

VERRIE'RES (Les), a village of France, in the dep. of the Charente, cant. and 5 m. SW of Segonzac. Pop. 740. It is noted for its brandy.

VERRIE'RES-SUR-ECOTAY, a village of France, in the dep. of the Loire, cant. and 5 m. SW of Montbrison. Pop. 1,900.

VERRIE'RES-DE-JOUX, a commune of France, in the dep. of the Doubs, cant. and 5 m. E of Pontarlier. Pop. 616. It has a custom-house, and possesses an active trade in horses. In the vicinity is a spring, called the Fontaine-Ronde, remarkable for its periodical flux and reflux.

VERRIE'RES-SOUS-CELLES, a commune of France, in the dep. of the Deux-Sevres, and cant. of Celles, 3 m. SE of Melle, on the Belle. Pop. 1,133.

VERRIZ, a parish and v. of Spain, in the prov. of Vizcaya, near Durango, in a mountainous locality. Pop. 1,200. It has a sulphureous spring.

VERRO, a small town of Russia in Europe, in the gov. of Livonia, in the district and 45 m. SSE of Derpt, 135 m. NE of Riga, on the r. bank of the Tonda, and enclosed on three sides by a lake and its outlet, by which it is connected with the Tonda. In the vicinity are the remains of a fortress.—A canal of the same name connects the Tonda, an affluent of Lake Pskov, and thus of the gulf of Finland, with the Aa, a tributary of the gulf of Livonia.

VERRUA, a fortified town of Sardinia, in the div. and prov. and 23 m. NE of Turin. Pop. 2,600. In 1610, this place successfully defended itself against a large Spanish force under Feria.

VERRUCHIO, a market-town of the Papal states, in the deleg. and 27 m. SE of Forli.

VERRUYE, a commune of France, in the dep. of the Deux-Sevres, cant. and 2 m. SE of Mazières. Pop. 1,555.

VERS, a town of France, in the dep. of Lot, cant. and 1 m. NW of St. Gery, on the river Lot. Pop. 1,000.

VERSAILLES, a city of France, in N lat. 48° 47' 56", at an alt. of 174 metres = 571 ft. above sea-level, in the dep. of Seine-et-Oise, 12 m. W by S of Paris. Pop. in 1789, 44,200; in 1846, 34,901. Its streets cross at right angles, and though less spacious than might have been expected in a town built so recently, and with such great advantages, they are wider than those of Paris. The market-places and squares are also larger than is common in France. The avenues leading to the palace divide the town into two parts, of which the one situated to the left is called the old, the other the new town. The latter contains the principal church, and the greater proportion of elegant buildings. The removal of the royal residence in 1789, was a great blow to the prosperity of V. A manufacture of fire-arms has been established here. Clocks and watches are also made; and the spinning and weaving of cotton, as well as the bleaching of linen, are carried on on a small scale. V. is the see of a bishop, and has a cathedral, with eight churches. It has also a high school, several private seminaries of education, a library of 40,000 vols., a cabinet of natural history, and a botanical garden, and a royal college.—The arrond. of V., comprising 8 cants., has an area of 84,751 hectares. Pop. in 1846, 159,779.—In 1666, V. was little more than a village, with a hunting-lodge for the royal family, when Louis XIV., pleased with the situation, and desirous of residing out of Paris, began to erect a splendid palace, which required twelve years to build. Its front and wings are of polished stone, ornamented with statues, and a central colonnade of the Doric order. A broad central avenue, and two lateral ones, named respectively the Avenues de-Paris, de-St.-Cloud, and de-Sçeaux, converge in a large open space, 800 ft. broad, called the Place-d'Armes, into which project the outer gates of the palace. Passing these we stand in an open area called the Cour-des-Statues or the Cour-d'Honneur. This grand court, 380 ft. broad, is flanked by a plain range of buildings on each side, erected by Louis XIV. for the ministers of state. In front of these stand 16 marble statues, including those of Richelieu, Bayard, Colbert, Jourdan, Massena, Turenne, Du Guesclin, Sully, Lannes, Mortier, Condé, &c. In the midst, at the upper part of this outer court, is a fine colossal equestrian statue of Louis XIV. Beyond this and formerly separated from the great court by a railing, is the court proper of the palace, formerly named the Cour-Royale. This, and the inner court, stretch far into the body of the palace, whose wings clasp and enclose the courts. On the N side are a wing and pavilion in the Corinthian style, built under Louis XV.; on the S side, are those completed under Louis XVIII. Next comes the Cour-de-Marble, with its marble pavement, encircled by the old palace of Louis XIII. which is all of red brick coped with stone; crowned with balustrades and sculpture, and ornamented with vases, trophies, busts, and statues. The busts, nearly all of white marble and antique, number 80, and are placed on brackets between the windows. In the centre is a balcony of white marble, supported by eight Doric columns of beautifully coloured marble. The N wing is divided from the chapel by the Cour-de-la-Chapelle, its architecture being in the best style of the age of Louis Quatorze. It has Corinthian pilasters between the windows, with sculpture, and a balustrade crowned by statues. Its external dimensions are 148 by 75 ft., and 90 ft. high; and the height of its roof, richly ridged with iron work, enables it to be seen over the palace from almost every side. The western or garden-façade of the palace, the grandest specimen of the style in France, presents a large projecting central mass of building, with two immense wings; and consists of a ground floor and first floor in the Ionic style, and an attic. The wings each exceed 500 ft. in length; the central front is 320 ft. long, and each of its retiring sides 260 ft. The number of windows and doors is 375. The palace is open from eleven to four every Sunday, and on the first Sunday in the month, it is usual for the *petits eaux*—the smaller fountains—to play; the great fountains only playing on great occasions, costing from 8,000 to 10,000 francs every time they play. From the vestibule of the chapel a fine view of the interior is had. Restored by Louis Philippe to its original splendour, when the grandeur and taste of the age of Louis XIV. were concentrated here, it still exhibits one of the most magnificent spectacles at Versailles. It is 114 ft. long to the altar, 60 ft. broad, and 86 ft. high, consisting of a nave and aisles, formed by arches with finely sculptured spandrils, supporting side galleries fronted with Corinthian columns. The Grand Galerie-des-Batailles, 393 ft. long, 42 ft. broad, and 42 ft. high, is filled with battle pictures from that of Tolbiac, gained by Clovis, in 496 to that of Wagram, in July 1809. One room is devoted to celebrated academicians, inventors, and discoverers, and men of learning, in science and literature. The theatre or Salle de-l'Opéra is 144 ft. long, divided into equal parts by the curtain; its breadth, 60 feet.; height, 50 ft. Ionic columns, fluted and gilt, separate the upper boxes. The decorations are crimson and gold, with a profusion of mirrors and chandeliers. It required 10,000 wax candles to illuminate this theatre, and the cost of performing here a grand opera is said to be 100,000 francs. A gallery 300 ft. long contains the busts, statues, and monumental effigies of the kings, queens, and illustrious personages of France to the reign of Louis XV. In the middle of this gallery to the left, is the entrance to the Salles-des-Croisades, five splendid Gothic rooms, full of pictures of the crusades. V. having, from its situation, little water, it was necessary to bring a supply from a distance, by machinery. This was executed on a grand scale; and no city or royal residence can boast a greater display of reservoirs, fountains and canals. The spacious park and gardens situated behind the palace, contain parterres, jets-d'eau, cascades, and thickets in pleasant variety; also a magnificent orangery. At some distance from the palace, and in a retired spot, are the two Trianons, which may be termed royal residences in miniature,—seats to which the king and queen retired, when desirous to lay aside their rank and state.

VERSAILLES, the capital of Woodford co., Kentucky, U. S., situated on a creek running into Kentucky river, 12 m. SW of Lexington.—Also a village of Ripley co., in Indiana, 65 m. SE of Indianapolis. Pop. 800.—Also a village of Morgan co., in Missouri, 37 m. WSW of Jefferson. Pop. about 600.

VERSAM, a village of Switzerland, in the cant. of the Grisons and bail. of Glenner, to the S of Palendas. Pop. 529.

VERSBACH, a village of Bavaria, circle of Lower Franconia, presidial of Wurtzburg. Pop. 1,160.

VERSBUHEL, a village of Austria, in Styria, in the circle of Bruck, near Rottenmann. Pop. 360.

VERSCIO, a village of Switzerland, in the cant. of Ticino, and bail. of Locarno. Pop. 347.

VERSHIRE, a township of Orange co., in the state of Vermont, U. S., 24 m. SE by S of Montpelier, and drained by Ompomponoosuc river. Pop. in 1850, 1,071.

VERSIGNY, a village of France, dep. of Oise, cant. and 3 m. NW of Nanteuille-Haudouin. Pop. 412.

VERSIN, a village of Prussia, in the regency of Koslin and circle of Rummelsburg. Pop. 310.

VERSITZ, or VERSCHITZ, a free royal town of Hungary, in the gsp. of Temes, 48 m. S of Temeswar, at the foot of a mountain of the same name. Pop. 16,550, of whom 10,190 are Greeks, and 533 Cath. It has a Greek gymnasium, an hospital, and large cavalry barracks. Wine, silk, and rice, are cultivated in the environs.

VERSMOLD, a town of Prussia, in the prov. of Westphalia, regency of Minden, circle and 9 m. W of Halle. Pop. 1,425. It has manufactories of linen and tobacco, and is noted for its butter.

VERSOIX, or VERSOY, a town of Switzerland, in the cant. and 7 m. N of Geneva, on the W bank of the lake of that name, and at the mouth of a small river also named Versoix. Pop. 937. It has a large square from which the streets, some of which are still unfinished, run in straight lines. It has extensive manufactories of lamps and lackered ware. This town previous to 1816 belonged to France.

VERSON, a commune of France, in the dep. of the Calvados, cant. and 5 m. NE of Evrecy, on the Odon. Pop. 1,190.

VERSOUD (LE), a village of France, dep. of the Isere, cant. and 4 m. NE of Domène. Pop. 600.

VERSOVAH, a town of Hindostan, in the presidency of Bombay and prov. of Aurungabad, on the W coast of the island of Salsette, and 15 m. N of Bombay. It has a small port.

VERT, a river of France, in the dep. of the Basses-Pyrenees, which has its source in the cant. and 11 m. S of Aramitz, runs NNE, waters the valley of Baretons, and after a course of 19 m. throws itself into the Gave d'Oleron, 3 m. NW of Oleron.

VERT (CAPE). See VERDE (CAPE).

VERTAISON, a canton and commune of France, in the dep. of the Puy-de-Dome, and arrond. of Clermont-Ferrand. The cant. comprises 11 com. Pop. in 1831, 12,753; in 1846, 12,449. The village is 11 m. E of Clermont-Ferrand. Pop. in 1846, 2,470.

VERTAMBAZ, a village of France, in the dep. of the Jura and cant. of Clairvaux-les-Vaux-Dain, 11 m. SE of Lons-le-Saunier. Pop. 300. It has several iron-works.

VERTAVILLO, a town of Spain, in the prov. and 15 m. SE of Palencia.

VERTE (ISLAND), an island of Russian America, in Prince William's sound, in N lat. 60° 18', and W long. 146° 50'.

VERTEILLAC, a canton, commune, and town of France, in the dep. of the Dordogne, and arrond. of Riberac. The cant. comprises 16 com. Pop. in 1831, 12,961; in 1846, 12,657. The town is 11 m. N of Riberac. Pop. 1,076.

VERTENTIS, a river of Brazil, in the prov. of Matto-Grosso, an affluent of the Araguaya, which it joins on the l. bank, 60 m. below the confluence of the São-João, opposite the island of Bananal.

VERTEUIL, a commune of France, in the dep. of the Charente, cant. and 4 m. S of Ruffec, on the r. bank of the Charente. Pop. 1,448. It has several tanneries, and carries on an active trade in grain and saffron. It is commanded by a castle erected in 1459, and had formerly a cordelier's convent, which, for upwards of 500 years, was the place of sepulture of the Rochefaucauld.—Also a commune in the dep. of the Lot-et-Garonne, cant. and 7 m. NW of Castelmoron. Pop. 1,818.

VERTEUIL, or VERTHEUIL, a commune of France, in the dep. of the Gironde, cant. and 6 m. NW of Pauillac. Pop. 1,000.

VERTH, a village of Prussia, in the regency and circle of Munster. Pop. 310.

VERTLIANSKAIA, a town of Russia in Europe, in the gov. of Astrakhan, on the r. bank of the Volga, between Tchernoiar and Enotaevsk. It is surrounded by an earthen rampart and ditches.

VERTOLAYE, a village of France, in the dep. of the Puy-de-Dome, cant. and 5 m. SE of Olliergue. Pop. 790.

VERTON, a village of France, in the dep. of the Pas-de-Calais, cant. and 7 m. SW of Montreuil-sur-Mer. Pop. 1,050.

VERTOU, a canton and commune of France, in the dep. of the Loire-Inferieure, and arrond. of Nantes, on the r. bank of the Sevre-Nantaise. Pop. in 1846, 5,635. It has a brick-kiln.

VERTOVA, a village of Austrian Lombardy, in the prov. and 11 m. NE of Bergamo, and district of Gandino, on the r. bank of the Serio, at the confluence of the Vertova. Pop. 1,544. It has manufactories of coarse cloth, and other woollen fabrics, and iron-works, and numerous flour and fulling-mills.

VERTRYCK, a department and commune of France, in the dep. of the Brabant, watered by the Velpe. Pop. of dep. 498. It has factories of beet-root sugar.

VERTUS (Les), a canton and commune of France, in the dep. of the Marne, and arrond. of Chalons-sur-Marne. The cant. comprises 29 com. Pop. in 1831, 8,519; in 1846, 8,576. The village is 18 m. WSW of Chalons-sur-Marne, in a plain at the foot of a hill. Pop. 2,200. The trade consists chiefly in wine, oil, and timber. This village is a place of considerable antiquity.

VERULAM, a township of Upper Canada, in the Colborne district. Towards its centre, stretching from E to W, is Sturgeon lake.

VERULAM, a village and settlement of the colony of Natal, in South Africa. The village is on the S side of the Umhloti, on the main road from the coast to the Zulu country.

VERVIERS, a town of Belgium, in the prov. of Liege, on the small river Wesdre, and on the railway to Cologne, 14 m. E of Liege, in a deep valley. Pop. in 1850, 20,000. It is a clean, cheerful town, with extensive manufactories of woollens. It has likewise, on a small scale, dye-works, and establishments for making soap, vitriol, and aquafortis. A visitor, in 1851, reports of V. that it is "the great seat of the woollen trade of Belgium. Its manufacture of cloths treads closely on the heels of that of England; and no inconsiderable portion of the industry of the place is devoted to the fabrication of cloths which are afterwards sold in America as the produce of English looms. It is extremely difficult," says this well-informed visitor, "to determine whether, in the main, the manufactures of V. are advancing or declining. Together with the greater portion of Belgian industry, the woollen manufacture has suffered from the active competition of England, and the negative opposition of continental rivals. V. was injured by the separation of Belgium and Holland, in the loss of the Dutch market, while the Zollverein also closed some very important channels for Belgian cloths; still, the present condition of affairs cannot be considered as unfavourable, if it be compared with the past. The exports of woollen cloths, which, during the whole of 1849 had amounted to 675,269 kilogs., very greatly exceeded that amount in 1850, but although there has been of late years a steady increase, the case is very different from what it was in 1831 and 1832, when Germany alone took 1,000,000 kilogs. of Belgian cloth. This, in fact, like so many other branches of Belgian industry, suffers from the want of an extended market. Deprived of the demand afforded by Dutch commerce, and excluded, or nearly excluded, from Germany, the Belgian manufacturer maintains an unequal competition with England and France in the general markets of the world."

VERVINS, a town of France, in the dep. of Aisne, arrond. and 23 m. NNE of Laon. Pop. 2,500. It has manufactories of linen and cambric.

VERWIG, a parish of Cardiganshire, 3 m. from Cardigan. Area 3,062 acres. Pop. 378.

VERYAN, a parish of Cornwall, 3½ m. SW of Tregony. Area 5,627 acres. Pop. 1,488.

VERZINO, a town of Naples, in Calabria-Ultra 2 da, 21 m. NW of Cotrone. Pop. 850.

VERZUOLO, or Verzolo, a town of Piedmont, 2 m. S of Saluzzo. Pop. of com., 3,900.

VERZY, a town of France, dep. of Marne, 9 m. SE of Rheims. Pop. 1,200. Champagne wine is largely made here.

VESAY (Cape), a headland on Lake Ontario, Upper Canada, the N point of Prince Edward's bay.

VESCOVANA, a town of Austrian Lombardy, in the prov. and 21 m. SSW of Padua. Pop. 3,200.

VESCOVATO, a town of Austrian Lombardy, in the Milanese, 8 m. NNE of Cremona. Pop. 2,400.

VESCOVIO, a town of Italy, in the States-of-the-Church, 12 m. S of Narni.

VESDRE, a river of Prussia, which rises near Roetgen, and flows into the Ourthe, 3 m. SE of Liege, after a course of 45 m.

VESDUN, a village of France, in the dep. of Cher, 12 m. E of Chateau-Meillant. Pop. 1,000.

VESE'GINSK, a town of Russia, in the gov. and 130 m. NE of Twer, on the river Mologa. Pop. 2,800. It has a traffic in salt and timber.

VESIN, a hamlet of France, in the dep. of Moselle, cant. and 4 m. NW of Longuyon.

VESLY, a village of France, in the dep. of Manche, 12 m. W of Caventan. Pop. 1,300.

VESOUL, a town of France, dep. of the Haute-Saone, situated near the small river Durgeon, 26 m. N by E of Besancon. Pop. 7,100. The town stands in a fertile district, corn and vines being cultivated around it; and has an active trade in iron, wine, corn, cattle, and skins.—At the adjacent village of Frotte there is a spring of the extraordinary width of 90 ft., and depth of 120 ft., from which water sometimes issues in such quantities as to overflow the surrounding country.

VESPOLATO, a town of Piedmont, 6 m. SSE of Novara. Pop. 1,800.

VESSJESSGONSK. See Veseginsk.

VESTENE, a town of Austrian Lombardy, in the prov. and 20 m. NE of Brescia. Pop. 1,300.

VESUVIUS, a celebrated volcanic mountain in the S of Italy, about 9 m. SSE of Naples, in N lat. 40° 49′, E long. 14° 26′. It rises in a gentle swell from the bay of Naples, to an elevation of 3,642 ft. The view from its summit is very beautiful, including Naples, with its bay, its islands, and its promontories, as well as the scenery of the Campagna-Felice. "The larger crater encircles you entirely for a mile, cutting off the view of the sides of the mountain, and, from the elevation of the new cone, you look over the rising edge of this black field of smoke and cinders, and drop the eye at once upon Naples, lying asleep in the sun, with its lazy sails upon the water, and the green hills enclosing it clad in the indescribable beauty of an Italian atmosphere." [*Willis.*] To the W the prospect loses itself in the immensity of the sea; to the E it extends far into the interior, until arrested by the Apennines. The upper part of the mountain, torn by a series of convulsions, is strewn with its own fragments; the part next in the descent, is mixed with dried lava extending in wide black lines over its surface; while the lower part is covered with villages and country-seats, fields of maize, groves of fruit-trees, and other luxuriant productions, displaying the great fertility given by the ashes to the soil. The summit has the form of a cone, and consists of masses of burned earth, ashes, and sand, thrown out in the course of ages by the volcano. It is steep, and difficult of ascent, from the looseness of the materials. The crater is nearly ½ m. in circumf., but has not above 350 ft. of depth or descent from the ridge. Its sides or interior surface have been progressively formed of ashes and cinders, intermixed with some rocks and dried lava. The lower part of the crater is a level spot, nearly ¾ m. in circumf., composed of a sort of crust of brown burned earth, and containing several orifices emitting a thin vapour. Professor Silliman thus describes the great crater: "after passing through thick clouds of steam charged with the suffocating sulphurous acid gas which greatly annoyed our lungs, when the wind cleared our vision, we found ourselves on the narrow rim of the great crater. It was so narrow that only two persons could walk upon it abreast. Our present position was on the side of V. contiguous to Somma on the N. At this place the descent of the exterior surface of the mountain was very steep,—a rapid slope of cinders and loose lava, down to the deep and narrow valley—the Atrio-del-Cavallo—between V. and Somma—a descent on the slant of probably 1,800 ft. at an angle of 50 or 60 degrees. On the other side was the crater, whose angle of descent was not less. On this narrow rim we walked on a surface not more than 6 or 8 ft. wide, with the terrific crater, 1,000 ft. deep on one side, and the abrupt descent of 1,800 ft. on the other side, into the valley of Somma. It was a spectacle truly sublime, awfully grand and appalling. The heat below sent up, in throes and spasms, dense clouds of steam and sulphurous acid gas, which at short intervals, filled the crater, and all that appeared was a pillar of a cloud, in which we were often involved, and half suffocated; we stood with arms locked, for then it was unsafe to move until a whirlwind swept the crater clear, and we could see distinctly into its profound abyss. Nothing could be more perfectly formed than this crater. It was a magnificent hollow cone, whose nether apex opened into the great world of fire below; still the fire we did not see, although we had the most decisive proof of its existence and continued action in the violent ejection of steam and gas, filling every few minutes this vast funnel, whose diam. across from side to side was probably 1,000 ft.; but the dense cloud of steam and noxious gas which hovered to leeward over a large portion of the circular orifice, rendered it impossible to walk around it, or even to ascertain whether a continuity of surface, free from cross fractures and chasms, would render it practicable without the most imminent danger. The interior of the great crater appeared to be composed of stratified volcanic beds, resembling solid lava rocks, arranged in a rude stratification. There were tempting deposits of yellow sulphur, green chloride of copper, sal-ammoniac and other substances on the walls of the crater. By throwing large stones into the crater, we made several rough measurements of its probable depth inferred from the time consumed in the returning sound, and we judged in this way that the depth of the crater was from 1,000 to 1,100 ft." V., says Shelley, "is, after the glaciers, the most impressive exhibition of the energies of nature I ever saw. It has not the immeasurable greatness, the overpowering magnificence, nor, above all, the radiant beauty of the glaciers; but it has all their character of tremendous and irresistible strength. From Resina to the hermitage you wind up the mountain, and cross a vast stream of hardened lava, which is an actual image of the waves of the sea, changed into hard black stone by enchantment. The lines of the boiling flood seem to hang in the air, and it is difficult to believe that the billows which seem hurrying down upon you are not actually in motion. This plain was once a sea of liquid fire. From the hermitage we crossed another vast stream of lava, and then went on foot up the cone—this is the only part of the ascent in which there is any difficulty, and that difficulty has been much exaggerated. It is composed of rocks of lava, and declivities of ashes; by ascending the former, and descending the latter, there is very little fatigue. On the summit is a kind of irregular plain, the most horrible chaos that can be imagined; riven into ghastly chasms, and heaped up with tumuli of great stones and cinders, and enormous rocks blackened and calcined, which had been thrown from the volcano upon one another in terrible confusion. In the midst stands the conical hill from which volumes of smoke, and the fountains of liquid fire, are rolled forth for ever. The mountain is at present in a slight state of eruption; and a thick heavy white smoke is perpetually rolled out, interrupted by enormous columns of an impenetrable black bituminous vapour, which is hurled up, fold after fold, into the sky with a deep hollow sound, and fiery stones are rained down from its darkness, and a black shower of ashes fell even where we sat. The lava, like the glacier, creeps on perpetually, with a crackling sound, as of suppressed fire. There are several springs of lava; and in one place it gushes precipitously over a high crag, rolling down the half-molten rocks and its own overhanging waves: a cataract of quivering fire. We approached the extremity of one of the rivers of lava; it is about twenty feet in breadth, and ten in height; and as the inclined plane was not rapid, its motion was very slow. We saw the masses of its dark exterior surface detach themselves as it moved, and betray the depth of the liquid flame. In the day the fire is but slightly seen; you only observe a tremulous motion in the air, and streams and fountains of white sulphurous smoke. At length we saw the sun sink between Capreæ and Inarime, and, as the darkness increased, the effect of the fire became more beautiful. We were, as it were, surrounded by streams and cataracts of the red and radiant fire; and in the midst, from the column of bituminous smoke shot up into the air, fell the vast masses of rock, white with the light of their intense heat, leaving behind them through the dark vapour trains of splendour." After the eruption of 1794, the cone lost much of its elevation, sinking down into the crater, and almost filling up the cavity. The fire having thus lost its upward vent, burst through the side, and poured out lava which rolled down the declivity all the way to the sea, burning up the cultivated ground, and covering with a fluid, which afterwards became solid and hard, the chief part of the town of Torre-del-Greco. "Both in ascending and descending this mountain," says Professor Silliman, "we are impressed beyond the power of language to describe, with the incalculable amount of lava which has been ejected from V. during the ages in which it has existed. It is mantled all over, with robe upon robe, of its own congealed overflowings, and the robes have flowed down its flanks, and covered its feet. Immense fields of bristling, ragged, black and billowy lava, now extend down the sides of V., quite to the plains, and the lava has often flowed across the towns at its base, and invaded the sea. In no position so well as from the summit can these vast tracts of lava be seen all in one view. Their thickness is enormous; probably several thousand feet near the mountain, which stills rears far above them its venerable and beautiful form, and will probably continue for ages to come, as it has done for ages past, to afford a vent for the internal fires of this part of the globe until it shall be sealed up like Rocca-Monfina on the road to Rome, Mount-Vultur in Apuleia, or Mount-Epimeo in Ischia; and until a new orifice shall be formed wherever the fire shall force a passage. The height of V. varies with its eruptions, as the cone is elevated or depressed." The total number of great eruptions of V. on record is above 30, reckoning from the celebrated one of A. D. 79, which proved destructive to *Herculaneum*. One of the latest, though not most formidable eruptions, took place October 24, 1822, when the ashes emitted darkened the sky at Naples, and spread as far as Cassano, 105 m. distant. In 1855, a considerable eruption of V. took place. Professor Palmieri reporting on this latest eruption states that some days before the commencement of that phenomenon, the most singular irregularity was observed in the dipping-needle, the variations of which became so strong and frequent during the last two days before the eruption as to amount to what the professor calls a magnetic storm. The magnetic vibrations continued with increased intensity during the eruption; and the electrical state of the atmosphere was equally remarkable, being greater than the maximum in ordinary times; its diurnal period was disturbed, greater electricity having often been observed during the night than during the day; and, what was still more singular, during the eruption of ashes the fixed conductors gave but slight indications of negative electricity, while the moveable ones gave the strongest possible signs of positive tension. In general the electrical current appeared to follow the course of the smoke. The barometer remained constantly low, the therm. on the side where the lava flowed rose eight degrees, and the wind kept veering round to all the points of the compass.

VESZPRIM, a palatinate in the W of Hungary, bounded on the N by Oldenburg and Raab; on the NE by Komorn; on the E by Stuhl-Weissenburg; on the S by Tolna, Schemegh, and Zarad; and on the W by Eisenburg. It has a superficial extent of nearly 1,600 sq. m. The forest of Bakony occupies the centre of this com. In the S is the lake of Balaton. The soil is fertile. The pop. is 152,000, mostly of Magyar descent, but with a few Slavonians and Germans.

VESZPRIM, the chief place of the preceding palatinate, is situated in N lat. 47° 5′ 8″, on the r. bank of the Sed, 60 m. SW of Buda. Pop. 9,000. It has a trade in salt, corn, wine, and timber. Being a bishop's see, it has a cathedral.

VETA, a village of New Granada, in the prov. and 18 m. ESE of Pamplona. It has productive gold-mines.

VETA-NEGRA-DE-SOMBRERETE, a village of Mexico, in the dep. of Zacatecas, near Sombrerete. It has silver mines, considered by Humboldt the richest known. They are now wrought by an English company.

VETERAN, a township of Chemung co., in the state of New York, U. S., 160 m. W by S of Albany. Pop. in 1850, 2,698.

VETERANISCHE-HOHLE, a grotto of Hungary, in the Banat, 8 m. SW of Alt-Orschova. It receives its name from the famous general Veterani, who with a small body of men defended himself here against the Turks in 1694. The grotto is in a lofty rock, on the l. bank of the river, above the village of Agradina, inaccessible on all sides except one. The entrance is only 4 ft. high, and 2 ft. broad, but the interior can afford accommodation to a thousand men, and is supplied by wells with fresh water.

VETERES, a Negro tribe who inhabit the Issini territory, in Upper Guinea, on the Ivory coast.

VETERNICO, a town of Turkey in Europe, in the sanj. and 21 m. W of Trikala, in the midst of mountains.

VETKA, a village of Russia in Europe, in the gov. of Mohilev, district and 15 m. NE of Novo-Bielitza, on the Soj. It was originally a colony formed previous to the reduction of this territory by Russia, by a Russian religious sect named *rascolnike* or heretics, and numbering 40,000 individuals. The colony was broken up by order of the government in 1764, and 20,000 of the sectaries sent to Siberia.

VETLUGA, a river of Russia in Europe, which has its source in the W part of the gov. of Viatka, flows thence into that of Vologda, forms for some distance its dividing line from the gov. of Kostroma, and also a portion of the E boundary of that of Nijni-Novgorod, and after a course of upwards of 300 m., first W, and latterly S, flows into the Volga a little below Kozmodemiansk. It takes its course to a great extent through vast forests. Vetluga and Varnavin are the chief places on its banks.—Also a town in the gov. and 186 m. E of Kostroma, on the river of the same name, at the confluence of the Neia and Kikschina. Pop. 1,200. Matting is its chief article of manufacture.

VETRALTA, a town of the Papal states, in the delegation and 6 m. SW of Viterbo. Pop. 1,020.

VETRANA. See AVETRANA.

VETSCHAU, a town of Prussia, in the regency of Frankfurt and circle of Kalau, 45 m. SW of Frankfurt on the Oder. Pop. 1,505.—Also a village in the regency and circle of Aachen. Pop. 220.

VETTELHOFEN, a village of Prussia, in the regency of Coblenz and circle of Ahrweiler. Pop. 250.

VETTELSCHLOSS, a parish of Prussia, in the regency of Coblenz, and circle of Neuwied. Pop. 400.

VETTER. See WETTER.

VETTERNIG, a village of Austria, in Styria, in the circle of Cille. Pop. 450.

VETTIN, a village of Prussia, in the regency of Potsdam and circle of Ost-Priegnitz. Pop. 240.

VETTLAX, an island of Russia in Europe, in the gulf of Bothnia, off the W coast of Finland, in N lat. 59° 58′, and E long. 40° 52′.

VETTWUSS, a village of Prussia, in the regency of Aachen and circle of Düren. Pop. 600.

VETZBERG, a village of Prussia, in the regency and circle of Wetzlar. Pop. 240. It has the ruins of a castle.

VEUIL, or VOEUIL, a village of France, in the dep. of the Indre, cant. and 4 m. SW of Valençay. Pop. 650.

VEULES, a commune of France, in the dep. of the Seine-Inferieure, cant. and 5 m. E of St. Valery-en-Caux, on the English channel. Pop. 1,552.

VEURDRE (LE), a commune and town of France, in the dep. of the Allier, and cant. of Leury-Levy, on the Allier, at the confluence of the Bieudre. Pop. 1,049. It has extensive iron-works, and carries on also an active trade in wood.

VEURNE. See FURNES.

VEUXAULLES, a village of France, in the dep. of the Côte-d'Or, cant. and 1½ m. E of Montigny-sur-Aube. Pop. 630. It has blast furnaces and forges.

VEUZAC, a village of France, in the dep. of the Aveyron, cant. and 2 m. NNE of Villefranche. Pop. 100. It has an iron-mine.

VEVAY, a village of Switzerland co., in the state of Indiana, U. S., on the Ohio, 88 m. SE by S of Indianapolis. Pop. in 1850, 1,500.

VEVAY, or VIVIS, a district and town of Switzerland, in the cant. of Vaud, on the NE bank of the lake of Geneva, at the mouth of the Vevayse, and 12 m. ESE of Lausanne, in N lat. 38° 46′. Pop. 5,201. This town, which is esteemed the prettiest in the cant., and next to Lausanne the most considerable, is built with great regularity, in the form of a triangle, with the longest side towards the lake. The streets are spacious and well-kept, and the houses, though generally low, are to some extent handsome. Its principal public edifices are the church of Saint-Claire, the town-house, the corn-hall with its range of marble columns, the bridge over the Vevayse, erected in 1808, the hospital, containing a public library, and the fountains, of which the most remarkable, surmounted by an obelisk, is near the market-place, a parallelogram 600 Parisian ft. in length, and 400 ft. in breadth, 3 sides of which are lined with handsome buildings, and the fourth extending to the lake by ranges of trees. A delightful promenade runs along the lake, and another above the town round St. Martin's church. Its industry consists in the manufacture of chocolate and tobacco. It possesses an active transit trade. It is supposed to be the *Vibiscum* or *Viviscum* of the Romans. Under the house of Savoy it formed part of Chablais.

VEVAYSE, or VEVAISE, a river of Switzerland, in the cant. of Vaud, which has its source at the foot of the Molesson, in the cant. of Friburg; flows thence into the cant. of Vaud; and after a course through a picturesque valley, in a generally SW direction of about 15 m., throws itself into the cant. of Vaud at Vevay.

VEX, a village of Switzerland, in the cant. of Valais, 2 m. SE of Zion. Pop. 800.

VEXAINCOURT, a village of France, in the dep. of the Vosges, cant. and 13 m. NE of Raon.

VEXALA, an island of Russia in Europe, in the gulf of Bothnia, near the W coast of Finland, in N lat. 63° 30′, and E long. 22° 20′.

VEXERA, a town of Spain, in the prov. and 33 m. SE of Cadiz, and 6 m. from the straits of Gibraltar.

VEXIN, formerly VEULGSIN, an ancient district of France, situated on the banks of the Epte, by which it was divided into two parts, distinguished as Vexin-Francais and Vexin-Normand, and lying respectively in the provs. of the Isle of France and Normandy. The former, which was on the l. bank of the river, contained three towns, of which Pontoise was the capital. The latter, which was on the r. bank of the river, contained five towns, of which Rouen was the chief.

VEXIOE, a town of Sweden, capital of the stift of Kronoberg, and of the haered of Kinneveld, on the Gulds-Meltsbacke, by which it is divided into two parts, and on Lake Bergunda, 225 m. SW of Stockholm. Pop. 1,200. It is well-built, and has a gymnasium, carpet-factories, and in the environs several paper-mills.

VEYLATS, or VAYLAS, a village of France, in the dep. of the Lot, cant. and 6 m. E of Lalbenque. Pop. 660.

VEYLE, a river of France, in the dep. of the Ain, which has its source near Charlemont; traverses the centre and W of the dep.; runs first NNW, then W; and after a course of about 69 m., throws itself into the Saone, on the l. bank, a little below Macon. Pont-de-Veyle is the chief place on its banks.

VEYNES, a canton, commune, and town of France, in the dep. of the Hautes-Alpes, and arrond. of Gap. The cant. comprises 10 com. Pop. in 1831, 4,172; in 1846, 4,074.—The town is 14 m. W of Gap, on the r. bank of the Buech. Pop. 1,859. It has several tanneries, and carries on an active trade in wool. In the vicinity are the remains of the ancient *Mons-Seleucus*, and a small village named Mont-Saleon.

VEYPAR, or BIPAR, a town of Hindostan, in the presidency of Madras, in the Carnatic district, and 54 m. NE of Tinnevilly, near the mouth of the Veyp-Aru.

VEYP-ARU, a river of Hindostan, in the presidency of Madras, in the Carnatic, which descends from the Western Ghauts; runs ESE; and after a course of 90 m., throws itself into the gulf of Maanar, near Veypar.

VEYRAC, a commune of France, in the dep. of the Haute-Vienne, cant. and 5 m. SW of Nieuil. Pop. 1,518.

VEYRE, a river of France, which has its source in the dep. of the Cantal, to the N of Parlan; flows thence into the dep. of the Lot, and after a course, in a generally SSW direction, of about 18 m., throws itself into the Selle, opposite Bouluech.

VEYRE, or VEYRE-MONTON, a canton, commune, and town of France, in the dep. of the Puy-de-Dome, and arrond. of Clermont-Ferrand. The cant. comprises 7 com. Pop. in 1831, 13,594; in 1846, 12,706. The town is 9 m. SE of Clermont-Ferrand. Pop. 2,987.

VEYRE. See VERE.

VEYROS, or VEIROS, a town of Brazil, in the prov. and 285 m. WSW of Para, and district of Xingutania, 48 m. above Porto-de-Moz, on the E bank of the Xingu, near the confluence of a small river of the same name.

VEYUT. See JELUM.

VEZELAY, a canton, commune, and town of France, in the dep. of the Yonne, and arrond. of Avallon. The cant. comprises 17 com. Pop. in 1831, 11,989; in 1846, 12,325.—The town is 8 m. W of Avallon, on a mountain, near the Cure. Pop. 1,193. The houses are old, and only tolerable in structure, and with the exception of a church, it has no edifice worthy of notice. It has some hardware factories and rope-works, and carries on an active trade in wine, cattle, and wood. V. is noted as the birth-place of Theodore-de-Beza, and its vicinity as that of Vauban. It was formerly fortified. In 1146, a council was held here in which St. Bernard preached the crusade. It was taken, and for some time held by the Protestants, in the reign of Charles IX.

VEZELISE, a canton, commune, and town of France, in the dep. of the Meurthe, and arrond. of Nancy. The cant. comprises 33 com. Pop. in 1831, 14,022; in 1846, 13,766.—The town is 18 m. SSW of Nancy, at the confluence of the Brenon and Uvry. Pop. 1,603. It has a church, the spire of which is remarkable for its height and delicacy of structure, and an hospital, and possesses manufactories of calico, muslin, and other cotton fabrics, and brocades, several oil and tan-mills, tanneries and bleacheries. Fine porcelain clay occurs in the vicinity.

VEZENBERK. See WESENBERG.

VEZENOBRES, a canton, commune, and town of France, in the dep. of the Gard and arrond. of Alais. The cant. comprises 17 com. Pop. in 1831, 6,372; in 1846, 6,730. The town is 6 m. SSE of Alais. Pop. 1,066. It has a consistorial church.

VEZERE, a river of France, which has its sources in the dep. of the Correze: waters the N and W part of the dep.; runs afterwards SE into that of Dordogne; and after a course, in a generally SW direction, of about 12 m., joins the Dordogne on the r. bank, at Limeuil. Its principal affluent is the Correze, which it receives on the l., and thence becomes navigable.

VEZERE (HAUTE), a river of France, which has its source in the NW part of the dep. of the Correze, in the cant. and to the NW of Uzerche; runs N into the dep. of the Dordogne; and after a course, in a generally SW direction, of 54 m., joins the Isle, on the l. bank, 6 m. E of Perigueux.

VEZERONCE, a village of France, in the dep. of the Isere, cant. and 2 m. S of Morestel. Pop. 800.

VEZEZON, a village of France, in the dep. of the Haute-Loire, cant. and 1 m. N of Auzon, on the r. bank of the Allier. About 1,600 boats are annually constructed here for the trade of the river.

VEZIA, a village of Switzerland, in the cant. of Tessino and district of Lugano. Pop. 302.

VEZIN, a department and com. of Belgium, prov. of Namur. Pop. of dep., 1,161; of com., 455.

VEZINS, or VESINS, a canton and commune of

France, in the dep. of Aveyron and arrond. of Millau. The cant. comprises three com. Pop. in 1831, 4,493; in 1846, 5,005. The village is 15 m. NNW of Millau. Pop. 648.—Also a commune and town in the dep. of the Maine-et-Loire, and cant. of Chollet, 18 m. SSE of Beaupreau. Pop. 1,825. It has a tannery and a dye-work.

VEZIR-KEUPRI, formerly GEDA-KARA, a town of Turkey in Asia, in the pash. of Sivas and sanj. of Djanik, on a small affluent of the Kizil-Ermak, 45 m. NW of Amasia. It contains 2,000 families, 13 mosques, two public baths, and a well-furnished bazaar. In the vicinity is Gouza, famous for its mineral wells and extensive ruins.

VEZOUZE, a river of France, in the dep. of the Meurthe, which has its source in the Vosges, in the cant. of Lorquin, where it is formed by the junction of two principal streams. Thence it flows W past Cirey, Blamont, and Luneville; and after a course of about 45 m., throws itself into the Meurthe, on the r. bank, a little below Luneville. Its principal affluents are the Chatillon, Blette, and Verdierette. On its banks are numerous mills.

VEZZANI, a canton and commune of Corsica, in the arrond. of Corte. The cant. comprises 8 com. Pop. in 1831, 5,174; in 1846, 5,252. The village is 12 m. ESE of Corte, on the Tagnone. Pop. 984.

VEZZANO, a market-town of Sardinia, in the div. of Genoa, prov. and 21 m. E of Levante, near Monte-Trebbiano. Pop. 1,038.

VIADANA, a town of Austrian Italy, 20 m. SSW of Mantua, on the r. bank of the Po. It has a small fort and several churches. Its manufactures of linen and leather are considerable. Pop. 6,200.

VIADEIROS (SERRA-DOS-), a mountain-chain of Brazil, in the prov. of Goyaz, between the rivers Maranhão and Paranan. It gives rise to the Tocantins-Pequeno, an affluent of the Maranhão.

VIA-GRANDE, a town of Sicily, on the SE slope of Mount Ætna, 7 m. NE of Catania. Pop. 3,100.

VIAISON, a river of Savoy, which rises near Vernet; runs from SW to NE, and joins the Arve, opposite Pont-Notre-Dame.

VIALA (LE), a village of France, dep. of Aveyron, cant. and 8 m. W of St. Affrique. Pop. 500.

VIALA-DU-TARN (LE), a village of France, dep. of Aveyron, cant. and 8 m. SW of Levezow. Pop. 700.

VIALAS, a village of France, in the dep. of Lozère, cant. and 9 m. E of Pont-de-Montvert.

VIA-MALA, a celebrated defile in the Swiss cant. of the Grisons, extending along the Rhine from about 1 m. beyond Thusis, to the opening of the valley of Schams, a distance of about 4 m. It is a narrow gorge cleft in the mountain, the nearly perpendicular walls of which are in some places 1,600 ft. in height, while they are not above 30 ft. apart at the level of the road.

VIAMAO, a parish and village of Brazil, in the prov. of São-Pedro-do-Rio-Grande, near the E bank of a lake of the same name, and 7 m. ESE of Porto-Alegre. Pop. of district, 2,816.—The lake of V. extends between the mouth of the Rio-Jacuhi and the Lagoa-dos-Patos, and is chiefly fed by the Sino, a navigable stream.

VIANA, a town of Portugal, in the prov. of Minho, on the N side of the river Lima, 42 m. N by W of Oporto. Pop. 8,000, whose chief employments are navigation, fishing, and trading in wine, corn, oil, and fruit. The river Lima is here of considerable depth, but its channel is so much obstructed by sand-banks, that ships do not come up to the town except at high water. The entrance is defended by two strong forts.—Also a town in the prov. of Alemtejo, 15 m. SSW of Evora. Pop. 1,500.

VIANA, a town of Spain, prov. and 4 m. NE of Logrono, near the l. bank of the Ebro. Pop. 3,150.

VIANA-DEL-BOLLO, a village of Spain, in the prov. and 45 m. SE of Orense.

VIANAZ, a village of Switzerland, cant. of Schwytz, on the l. bank of the Rhone, NW of St. Maurice.

VIANCE (SAINT), a village of France, in the dep. of La Correze, cant. and 4 m. WSW of Dordogne. Pop. 1,100.

VIANDEN, a town of Holland, in the prov. and 21 m. N of Luxemburg, on the Our. Pop. 800, employed in the manufacture of leather and woollens.

VIANE, a village and canton of Belgium, in the prov. of E. Flanders, arrond. of Alost. Pop. 1,556.

VIANEN, a town of Holland, in the prov. of S. Holland, on the branch of the Rhine called the Leck, in N lat. 51° 59′ 35″, 7 m. S by W of Utrecht. Pop. 2,500. In the vicinity is the Amaliabei castle and park, and the ruined Vakestein.

VIANNA, a parish and village of Brazil, in the prov. of Maranhão, 60 m. SSE of São-Luiz, near the Macacu, a tributary of the Pindaré.—Also a recently founded town in the prov. of Espirito-Santo, on the Cordilheira-dos-Aimores, 28 m. NE of Victoria.

VIANNE, a town of France, dep. of Tarn, 7 m. ENE of Castres. Pop. 1,600.

VIANOS, a village of Spain, in the prov. of Chinchilla, 3 m. N of Alcaraz, in a picturesque valley.

VIAREGGIO, a port of Italy, in the duchy of Lucca, 16 m. N by W of Leghorn. Pop. 5,850. It has several churches, one of which is of great size. The chief export is statuary marble.

VIARIGGI, a village of the Sardinian states, in the prov. and 9 m. S of Casale. Pop. 1,300.

VIAS, a village of France, in the dep. of Herault, cant. and 4 m. W of Agde. Pop. 1,400.

VIAST, a town of Prussian Silesia, in the gov. of Oppeln, 8 m. E by N of Kosel. Pop. 1,100.

VIATKA, a gov. in the E of European Russia, between the parallels of 55° 25′ and 60°, bounded on the N by the gov. of Vologda; on the E by Perm; on the S by Nijni-Novgorod, Orenburg, and Kasan; and on the W by Kostroma. It has an area of 120,958 sq. versts, or nearly 54,000 sq. m. Pop. in 1846, 1,662,800. Of the surface, several thousand sq. m. are occupied by ramifications of the Ural mountains, and by marshes, or by immense forests. The soil is in general clayey, the only large tract of rich fertile mould is on the banks of the Kama. The climate is cold, particularly in the N; but a considerable part of the natives being of Finnish origin, are diligent cultivators, and a surplus of corn is annually exported. The other articles of export are wood, timber, tar, game, peltry, copper, iron, honey, and wax. Manufactures are altogether in their infancy; and though there are several navigable rivers, such as the Kama, the Viatka, the Tcheptza, and the Kilmes, the export trade is insignificant. This gov. is divided into 10 circles.

VIATKA, the capital of the above government, seated at the confluence of the rivers Viatka and Klinovitza, in N lat. 58° 24′, 420 m. ENE of Moscow, contains 8,000 inhabitants, and is the see of a bishop of the Greek church. It has a seminary for priests. The inhabitants are in possession of the larger share of the traffic of the prov., and send to Archangel country produce, corn, flax, linseed oil, tallow, honey, and wax.

VIATKA, a river of Russia, which rises in the E of the gov. to which it gives name; runs N, then SW, and then SE; and joins the Kama on the r. bank, near Mamaditch, after a course of 500 m., in which it passes V., Orlov, Kotelnitch, Malmisch, and Mamadich. Its principal affluents are the Kobra the Letka, the Velikaia, the Maloma, on the r.; and the Tcheptza and Kilmes on the l.

VIATLSK, or VELSK, a town of Russia, in the gov. and 150 m. NNE of Vologda, on the l. bank of the Vaga.

VIAU (SAINT), a village of France, dep. of Loire-Inferieure, 2 m. SE of Paimbœuf.

VIAUR, a river of France, which rises near La Clau, in the dep. of Aveyron, and flows into the Aveyron, near Saint-Amans, after a course of 50 m.

VIAZMA, a town of Russia, in the prov. of Smolensk, at the junction of the rivers Viazma and Bebri, 110 m. E by N of Smolensk. Pop. 10,000. It has a number of tanneries, and exports leather, flax, flax-seed, and hemp to Riga and St. Petersburg. V. was on the road of the French, both in their advance and retreat in 1812, and was the scene of partial engagements on both occasions, in the course of which the town was burnt.

VIAZNIKI, a town of Russia, in the gov. and 84 m. ENE of Vladimir, on the Kliazma. Pop. 1,900.

VIBORG. See WIBORG.

VIBRAE, a village of France, dep. of Charente, cant. and 4 m. N of Chateauneuf, on the r. bank of the Charente. Pop. 500.

VIBRAYE, a town of France, dep. of the Sarthe, 25 m. E of Le Mans. Pop. 1,000. Iron is wrought in the vicinity.

VIC, a town of France, dep. of Meurthe, 4 m. S of Chateau-Salins, 21 m. ENE of Nancy. Pop. 3,500. It has considerable manufactures of stockings and needles. A mine of salt was discovered here in 1820. There are here the remains of a castle founded in 1190.

VIC, or VICJ, a commune and village of France, in the dep. of Haute-Vienne, cant. and 6 m. NW of Saint-Germain-les-Belles-Filles. Pop. of cant. 2,600.

VIC-EN-BIGORRE, a town of France, dep. of the Hautes-Pyrenees, 10 m. N of Tarbes. Pop. 3,500.

VICALBARO, a village of Spain, in the prov. and 4 m. from Madrid. Pop. 1,500.

VICARELLO, a town of the States-of-the-Church, 17 m. S of Viterbo. In the vicinity are hot-springs, the *Thermæ Aureliæ* of the Romans.

VICCHIO, a village of Tuscany, in the prov. and 21 m. NE of Florence.

VIC-DESSOS, a town of France, dep. of Ariége, 6 m. SW of Tarascon, on a stream of the same name, an affluent of the Ariége. Pop. 1,000.

VICENTE (SAN), an island of Brazil, in the prov. of São-Paulo, in the bay of Santos. It is about 15 m. in circuit; and has a town of the same name on its S coast, in S lat. 24° 1′.

VICENTE (SAN), a town of Spain, in the prov. and 27 m. from Ciudad-Real, near the frontier of Portugal. Pop. 8,000.

VICENTE (SAN), a settlement of New-Granada, in the dep. of Aneropica, in N lat. 6° 19.

VICENTE (SAN), or LORENZANA, a town of San-Salvador, 40 m. E of San-Salvador. Pop. 8,000, chiefly employed in cultivating indigo and tobacco.

VICENTE (SAN), a settlement of Chili, a little to the N of the embouchure of the Biobio.

VICENTE (SAN), a city of Brazil. It formerly enjoyed a great commerce, and it has a commodious and secure port, with a good bottom. It is also defended by a castle. The cathedral church is handsome.

VICENTE-DE-LA-BARQUERA (SAN), a coast-town of Spain, in the prov. and 32 m. WSW of Santander. Pop. 3,000. It has certain fisheries.

VICENTE-DE-LA-SONSIERRA (SAN), a town of Spain, in the prov. and 15 m. NW of Logrono, near the l. bank of the Ebro. Pop. 2,200.

VICENZA, a prov. of Austrian Italy, in the gov. of Venice; bounded on the NW by Tyrol; on the NE by the prov. of Belluno; on the E by Treviso; on the SE by Padua; and on the SW by Verona. It is 90 m. in length from N to S, and 34 m. in greatest breadth. Its surface is in general flat, though diversified in the N and W by a few spurs of the Alps, and in the S by the hills of Berici. The river Bacchiglione intersects the prov. and passes its cap.; the Gua waters its W quarter; the Brenta, its E. It is a fertile and well-cultivated country, containing on a superficial extent of 51·12 German sq. m. 315,785 inhabitants in 1834; and 340,694 in 1850.

VICENZA, the capital of the above prov., is pleasantly situated between two mountains, at the confluence of the Recone with the Bacchiglione, 38 m. W by N of Venice, in N lat. 45° 32′ 24″. Pop. 30,000. Its circumf. is about 4 Italian miles; and though not a place of strength, it is surrounded by a double wall and dry moat, and is defended by a fort. It is in general regularly built, and contains several good streets, and has a market-place of an oblong form, of which the chief ornament is the front of the town-house. V. has several modern structures of elegance, in particular those of the celebrated Palladio, who was a native of this place. The principal are the town-house, which is equally remarkable for the beauty of its architecture, and for the paintings contained in it; the theatre, built in the ancient Roman taste, the seats forming a semicircle, and rising one above the other; the bridge of San-Michele consisting of one lofty arch, resembling the Rialto at Venice. V. contains a number of churches and religious houses. The cathedral is a large edifice, encumbered by surrounding houses; the church Della Gratie, the work of Palladio, inferior in size, is far superior in design and execution. V. is the see of a bishop, and the seat of a lyceum or high school, a clerical seminary, and several hospitals. Of its manufactures, the chief are silks and linen, porcelain, and paper; woollens and leather are made to a small extent. V. was a Roman station, and suffered greatly on the irruption of the northern tribes. During the middle ages, it had at times an independent government, but passed, in the beginning of the 15th cent., into the hands of the Venetians, remaining in the enjoyment of peace until the invasion of Italy by Buonaparte in 1796. In November of that year, its vicinity was the scene of sanguinary conflicts between the French under that commander, and the Austrians under Alvinzi.

VICH, or VIQUE, a town of Spain, in the prov. and 37 m. N of Barcelona. Pop. 10,667. It has large wide streets, a few houses of a superior class, and a cathedral. Its manufactures consist of linens, cottons, and coarse woollens.

VICHADA, a large and rapid river of New Granada, which flows down the mountains of Bogota; runs E through the plains of Cazanare; and enters the Orinoco, on the l. bank, in N lat. 5° 6′.

VICHEGRAD, a town of Bosnia, in the sanj. of Novibazar, on the r. bank of the Drin, 48 m. ESE of Bosna-Serai.

VICHERA, a river of Russia, in the gov. of Perm, which rises on the W flank of the Urals; runs S, and then SW; and joins the Kama on the l., after a course of 240 m.

VICHI, a river of Bolivia, in the prov. of Atacames, which runs N, and enters the Guaillabamba.

VICHNI-VOLOTSHOK, a town of Russia, in the gov. and 73 m. NNW of Tver, near the Tzna, and on the canal which unites the Tvertza with the Tzna, or the Caspian with the Baltic. Pop. 4,000.

VICHOTNA, a village of Hungary, 25 m. W by S of Kesmark. Pop. 1,300.

VICHTACH, a town of Bavaria, 36 m. E of Ratisbon. Pop. 1,300.

VICHY, a town of France, near the r. bank of the Allier, 14 m. SW of La-Palisse. Pop. 1,800. It is greatly resorted to for its hot medicinal waters, which were known to the Romans.

VIC-LE-COMTE, a town of France, in Auvergne, near the river Allier, 17 m. SE of Clermont.

VICIOSAS, a cluster of small islands near the coast of Honduras, in N lat. 15° 12'.

VICKSBURG, a town of the state of Mississippi, U. S., 40 m. WNW of Jackson, on the l. bank of the Mississippi, 400 m. above New Orleans. Pop. in 1840, 3,104; in 1850, 3,678. It has an excellent harbour for steamers. Its chief exports are cotton and tobacco.

VICO, a small town in Corsica, 28 m. NE of Ajaccio. Pop. 2,000. It has a trade in wine and oil.

VICO, a town of the States-of-the-Church, in the deleg. of Frosinone, 3 m. N of Alatri. Pop. 3,000. It stands on the E bank of a small lake of the same name, formed by the Rio-Ricano.

VICO, a considerable town of Naples, in the prov. of Capitanata, 10 m. W of Viesti, on Monte-Gargano. Pop. 9,000.—Also a village of Naples, in the prov. of Terra-di-Lavoro, 6 m. from Nola.

VICO, a village of Austrian Dalmatia, on the river Norin, near the town of Narenze, on the site of the ancient *Narona*. It contains a variety of antiquities, such as remains of Roman roads, aqueducts, baths, temples, walls, and private houses.

VICO-DELLA-BARONICA. See TREVICO.

VICO-DI-MONDOVI, a village of the Sardinian states, prov. and 2 m. SE of Mondovi. Pop. 2,000.

VICO-EQUANA, or VICO-DI-SORRENTO, a town of Naples, 15 m. SSE of Naples. Pop. 2,000. It is the see of a bishop, and the neighbouring country produces good wine. It suffered greatly by an earthquake in 1694.

VICO-MOSCANO, a village of Austrian Italy, prov. of Cremona, 1 m. E of Casale-Maggiore. Pop. 750.

VICO-PISANO, a town of Tuscany, in the prov. and 5 m. E by S of Pisa.

VICOSA, a town of Brazil, on the coast of the prov. of Bahia, in the comarca of Caravillas, on the S bank of the Rio-Peruhipe.—Also a town in the prov. of Ceara, on the Serra-Hibiapaba, 24 m. SE of Granja. Pop. of district 5,000, chiefly Indians.

VICO-SOPRANO, or VESPRAN, a village of Switzerland, in the cant. of the Grisons, near the l. bank of the Maira, 30 m. SSE of Coire. Pop. 287.

VICO-VARO, a town of the Papal States, 6 m. NE of Tivoli, with the title of a principality. Horace's villa stood in the neighbourhood of this place.

VICQ, a village of France, in the dep. of Allier, arrond. and 7 m. NW of Gannat. Pop. 1,000.—Also a village of Haute-Marne, 2 m. NW of Varennes.

VICQ (SAINT-ETIENNE-DE), a village of France, in the dep. of Allier, cant. and 7 m. SW of La Palisse, on the Mourgon. Pop. 650.

VIC-SUR-AISNE, a village of France, in the dep. of Aisne, arrond. and 9 m. W of Soissons. Pop. 500.

VIC-SUR-ALLIER, a town of France, in the dep. of Puy-de-Dome, arrond. and 13 m. SE of Clermont-Ferrand, near the Allier. Pop. 2,000.

VIC-SUR-CERE, or VIC-EN-CARLADES, a village of France, in the dep. of Cantal, arrond. and 10 m. ENE of Aurillac, on the Cere. Pop. 2,000.

VIC-SUR-CHARTIER, a village of France, in the dep. of Indre, cant. and 4 m. NNW of La-Chatre.

VIC-SUR-GARTEMPE, a village [illegible]rance, in the dep. of Vienne, cant. of Saint-Sa[illegible]n the l. bank of the Gartempe.

VIC-SUR-LOSSE, or VIC-FEZENZAC, a town of France, in the dep. of Lers, arrond. and 18 m. NW of Auch, on the l. bank of the Losse. Pop. 2,700.

VICTOIRE ISLE, a small island in the Eastern seas, in N lat. 1° 39', E long. 106° 30'.

VICTOR, a village of Ontario co., New York, U. S., 9 m. NW of Canandaigua. Pop. 350.—Also a village of Clinton co., in Michigan, 13 m. NE of Lansing.

VICTOR (SAINT), a commune and village of France, in the dep. of Ardeche, arrond. and 9 m. WNW of Tournon. Pop. 1,500.—Also a com. and v. in the dep. of Aveyron, 8 m. NNW of Saint-Affrique.—Also a com. and v. in the dep. of Loire, cant. and 6 m. NE of Saint-Symphorien-de-Luy.

VICTOR-DE-BUTHON (SAINT), a village of France, in the dep. of Eure-et-Loir, cant. and 3 m SW of La Louppe. Pop. 1,000.

VICTOR-DE-LA-COTE (SAINT), a village of France, in the dep. of Gard, cant. and 7 m. W of Roquemaure. Pop. 1,000.

VICTOR-DE-RENO (SAINT), a commune and village of France, in the dep. of Orne, cant. and 4 m. SE of Longny. Pop. 1,400.

VICTOR-L'ABBAYE (SAINT), a village of France, in the dep. of Seine-Inferieure, 18 m. S of Dieppe. Pop. 600.

VICTOR-SUR-LOIRE (SAINT), a village of France, in the dep. of Loire, cant. and 4 m. NW of Chambon. Pop. 1,100.

VICTOR-SUR-THIERS (SAINT), a commune and village of France, in the dep. of Puy-de-Dome, cant. and 3 m. NNE of Saint-Remy. Pop. 1,500.

VICTORIA, formerly PORT PHILLIP, or AUSTRALIA FELIX, a British colony on the SE coast of Australia, the youngest of all the Australian provinces, but the progressive prosperity of which has been unparalleled by any colony in ancient or modern times. It lies between the parallels of 34° and 39° S, and the meridians of 141° and 150°, and has an area of 98,000 sq. m. In our general article on AUSTRALIA, we have sketched the features of this district, and inserted some statistical details; but the rapid and astonishing development of the resources of this colony, since its separation in 1850, from the colony of New South Wales, demands that it form the subject of a separate article in this work.

Physical features.] The coast-line from Cape Howe to Cape Wilson, extends SW 200 m.; from Cape Wilson to the mouth of the Glenelg, on the 141st meridian, it runs NW and then W, 500 m. To the N of Cape Wilson is the settlement of Port-Albert, on Comer inlet. About 82 m. to the W of Cape Wilson lies Cape Western-port; and a line beyond it is the entrance to the noble bay of PORT PHILLIP and the harbour of GEELONG: see these articles. Cape Otway lies 60 m. to the SW of Port Phillip; and between this basin and Cape Wilson, the W coast of Portland bay, are numerous flourishing settlements on the coast.—The Australian Alps spread their ramifications over the E part of the prov. The lower elevations, known as the Pyrenees and the Grampians, fill up the Central and the W division with a series of gentle undulations; and divide the streams flowing N to the Murray, from those which flow directly towards the coast. It is amongst these hills that MOUNT ALEXANDER, the head-quarters of the gold-diggings in this part of Australia, is situated [see that article].—The great river of this prov. is the Murray which skirts its northern frontier, and receives the Nutta-Nutta, the Lodden, and the Ovens, from the hills to the S. See article MURRAY. The Yarra-Yarra flows into the head of Port-Phillip and is navigable to the city of Melbourne, the cap. of the prov., 8 m. inland.

The Glenelg waters the W extremity of the prov. The site of the capital was chosen that the future inhabitants might enjoy the supply of fresh water afforded by the Yarra-Yarra. But the convenience seems to have been dearly purchased. The town is situated 9 m. by the river's course from the anchorage in the nearest point of Port-Phillip, and between Melbourne and the river's mouth there are two bars or shallows, having in general only about 9 ft. of water at full tide. The river, too, is so circuitous that from no direction will any one wind send a vessel up to the town. The light river craft have been found quite inadequate to satisfy the requirements of a commerce increasing with a rapidity beyond all calculation. The freight for goods has actually been doubled by the unfortunate necessity of ascending the petty stream of the Yarra after a long ocean-voyage, and the rate for the last 9 m. of the voyage from London to Melbourne has been equal to that of the first 14,000.

Climate, &c.] The climate is mild and agreeable, the annual range of the therm. on the coast not exceeding 19°. The one great drawback to the otherwise genial climate of the SE of Australia is the prevalence at certain periods of parching winds from the N, heated by traversing the vast desert of the interior. Victoria derived its old name of Felix, or 'the Happy,' from the boundless expanse of clear pasture well shaded with trees, but not overgrown with brushwood, which it presented to the eyes of its first explorers. Mr. Westgarth says: "V., and indeed Australia generally, has been by the natural features of the country more favoured than most others of our colonial settlements, as regards that preliminary physical drudgery of every new country, by which colonial society is long withheld, more or less, from the higher aspects of refinement and civilisation. The beauties of cultivation presented almost everywhere throughout England, as we rapidly traverse, by modern modes of travel, the varied surface of our noble country—the alternate park and garden, the every aspect of the arts and appliances of man, represent the accumulated achievements of human toil, graduated over centuries since the age when the forest, the bog, and the morass formed the conspicuous features. If we transfer our view to the backwoods of America, we perceive the recommencement, as it were, of the same labour, and a society depressed in its outward aspects by the daily sweat and toil in which it is continually immersed. But the free grassy surface of a great part of Australia, ready-made for the plough, has overstepped a long age of such customary colonial toil, and saved her fortunate sons from a century of physical warfare."

State of agriculture, &c.] An import trade in live stock in 1835 had become an export trade in 1838. The gold discoveries in 1851 gave a great impulse to the prosperity of this colony, and so enhanced the value of land that an allotment of the value of £50 in 1837 had risen in 1851 to the value of £4,000, and in 1853 had actually realized £15,000! The quantity of crown land sold in the colony of V. in 1852 was 258,144 acres, and the purchase-money upwards of £701,000. Previous to the discovery of the gold-fields, the staple commodity of the colony was wool. In 1852 the quantity of wool exported was upwards of 20,000,000 lbs. weight, valued at a little over £1,000,000. The live stock in the colony in 1852 was 6,551,501 sheep, 531,380 horned cattle, 34,021 horses. Another important article of export is tallow. The great increase in the flocks depreciated the price of a sheep to 2s. 6d., 1s., and eventually to 7d. a head; hence the origin of the practice of boiling down the carcase for the sake of the tallow. In 1850 the quantity of tallow exported was 4,489 tons; but the rapid increase of pop. since the gold discoveries had led to a greater demand for animal food, and consequently to a gradual falling off in the export of this commodity. The number of acres in crop in 1853, was 16,823 under wheat; 411 under barley; 2,947 under oats; 1,978 under potatoes; 14,107 under sown grasses; and 401 in green food for cattle. The produce of grain in the same year was 498,704 bushels of wheat; 61 b. of maize, 9,431 b. barley, 96,980 b. oats, 4,512 tons of potatoes, and 21,286 tons of hay. The number of acres planted with vines in 1853 was 107; the produce, 4,500 gallons of wine, and 500 gallons of brandy.

Geology and gold-diggings.] The conformity of the geological structure of the Australian mountain-ranges to those of the Ural mountains [see article URAL MOUNTAINS], led our distinguished countryman, Sir R. Murchison, to predict that extraordinary discovery which, since 1851, has enabled the squatters of V. to pass from the state of wandering nomades to that of a landed and wealthy interest. The rock structure, from the Glenelg on the W to beyond the Australian Alps on the E, exhibits great parallel bands of schistose and granitic rock, having a N and S strike, and a vertical position. Along the line of section, Mount-William, in the Grampians, consists of granite capped by sandstone formed of decomposed granite. Mount-Cole, in the Pyrenees, next in order, consists of hornblend schists. The valley of the Loddon and Forest-creek are on the edges of a band of argillaceous schists interlaminated with auriferous quartz. These schists constitute 'the great auriferous band,' extending N and S from Ballarat to Bendigo. The granite of Mount-Alexander succeeds, then the argillaceous band of the Campache valley, and the M'Ivor. Granite, 'the Goulburn auriferous band,' the granite of Violet-Town, 'the Ovens river auriferous band,' succeed in order to the eastward, and then the great granitic bands of the Australian Alps occasionally interlaminated with clay-slate, &c. The E flanks of the Alps are covered by thick unconformable beds of sandstone with thin seams of coal. The auriferous deposits are derived by decomposition from the edges of the quartz-rock, &c., and exist under various conditions of depth and material, according to the conformation of the surface and the structure of the rock. The richest diggings are carried along the N and S extension of the quartz ore bands. The quantity of gold exported from Victoria from August 1851, to April 1852, was 563,471 oz.; and it has been estimated that the quantity produced up to the end of 1852 was 4,891,000 oz. The total produce for 1853 was 3,090,342 oz., which, valued at 75s. per ounce, would give £11,588,782. Of this £9,365,488 was exported to the mother country. The annual report of the Melbourne Chamber of Commerce for 1853, has the following passage:—"The greater proportion of our colonial export produce now consists of gold, and it may readily be apprehended that a large quantity of this commodity is constantly being exported from the colony without any official record. The customs' returns gave 1,975,000 oz., as the quantity exported for the year 1852; but an additional quantity of 1,600,000 oz. had been traced into the adjacent colonies, or otherwise exported without official record. It may also be remarked that the quantity of gold as officially recorded, was valued at the very low rates that were then temporarily current. A careful estimate on this subject, exhibited in the minute alluded to, gave us the value of export produce raised in Victoria during the year 1852, as no less a sum than eighteen and a half millions sterling. But deducting the quantity of gold assumed to have been on hand throughout the colony at that date, this amount is reduced to £14,880,000, or about twice the amount set forth in the official records of the custom-house."

Commerce.] A return of the imports and exports each year since 1843, shows that the commercial resources of the colony have rapidly increased, but emigration up to the year 1851 did not advance sufficiently to develop its abundant resources. Since then the increase has been enormous, but it has been almost entirely absorbed by the attractions of the gold-fields. The following is a table of the imports and exports from 1844 to 1851, inclusive:—

	Imports.	Exports.	Total.
1844,	£151,000	£257,000	£408,000
1845,	248,000	464,000	712,000
1846,	316,000	425,000	741,000
1847,	438,000	669,000	1,107,000
1848,	374,000	675,000	1,049,000
1849,	480,000	755,000	1,235,000
1850,	745,000	1,042,000	1,787,000
1851,	1,056,000	1,423,000	2,479,000

The principal production of the colony, prior to June 1851, was wool, and the exports of this staple rapidly increased during the years above specified, as appears by the subjoined statement:—

QUANTITY OF WOOL EXPORTED.

1844,	4,326,229 lbs.
1845,	6,841,813
1846,	6,496,950
1847,	10,210,038
1848,	10,524,663
1849,	14,567,005
1850,	18,091,207
1851,	16,345,468

This return would appear to indicate that this important branch of commerce, after rapidly increasing up to 1850, received a check in the following year, caused by the discovery of gold, and the abstraction of labour from its accustomed pursuits to the lucrative attraction of the gold-fields. The total imports for the year 1852 were—from Great Britain, £1,752,216; from New Zealand, £25,499; other British colonies, £2,102,297; South Sea islands, £266; United States of America, £60,363; foreign states, £129,101. Total, £4,069,742. The total exports were:—Produce or manufacture of Victoria, £7,337,925; of the United Kingdom, £54,898; of British possessions, £6,434; of foreign states, £52,292. In all, £7,451,549. The exports were to the following places:—To Great Britain, £6,198,433; New Zealand, £1,775; other British colonies, £1,232,236; to foreign states, £19,105.

The quantity and value of grain, &c., imported into V. during 1852, was as follows:—Wheat, 87,570 bushels; maize, 81,182 bls.; barley, oats, and pease, 254,803 bls.; flour and bread, 23,101 tons; rice, 969,920 lbs.; potatoes, 3,475 tons; malt 2,621 bushels. Total value, £564,881. The quantity and value exported in 1852 was—wheat 25,249 bushels; barley, oats, and pease, 627 bls.; rice, 31,696 lbs.; potatoes, 34 tons. Total value, £12,304.

The live-stock imported was 3,481 horses, 824 sheep, 430 cattle; total value, £34,305. The live-stock exported was—horses 29, horned cattle 4,645, sheep 24,876, pigs 10; total value, £36,129.

The value imported of manufactured leather was, £103,489; unmanufactured leather, £6,469; total value, £109,958. The value exported—hides, £7,523; manufactured leather, £2,256; unmanufactured leather, £663; total value, £10,442. The quantity of wool exported was 20,047,453 lbs.; value, £1,062,787. The quantity of tallow exported was 1,995 tons, 4 cwt.; value, £60,261. The native gold exported in 1852 was 1,988,526 oz. 10 dwts. 13 grs.; value, £6,135,728. The oil exported was—black, 160 gallons, £40; salad, 2,210 gals., £517. Total value, £557. The vessels built and registered in 1852 were—2 schooners, 182 tons; 1 sloop, 21 tons. Vessels registered: 3 barques, 628 tons; 10 brigs, 1,474 tons; 9 brigantines, 1,171 tons; 1 hermaphrodite, 139 tons; 31 schooners, 2,360 tons; 7 sloops and cutters, 172 tons; 1 steamer, 44 tons. Totals—built, 3 vessels, 203 tons; registered, 62 vessels, 5,988 tons, 399 men.

The imports for 1853 amounted to £15,842,637, being at the rate of nearly £80 per head for a pop. averaging about 200,000. Of these imports, the principal articles were as follows:—

IMPORTS INTO VICTORIA IN 1853.

Beer and cider, 2,988,549 gallons,	£614,692
Bread and flour, 37,147 tons,	851,261
Building materials and houses,	2,004,798
Cotton, linen, silk, and woollen goods,	2,869,542
Furniture,	308,752
Horsefeed,	751,258
Iron, and iron manufactures,	1,059,261
Leather and leather manufactures,	461,994
Provisions and groceries,	1,460,245
Specie,	1,163,344
Spirits, 1,973,393 gals.,	1,045,053
Sugar and molasses, 17,455 tons,	486,614
Tea, 3,195,188 lbs.,	205,364
Tobacco, 1,836,179 lbs.,	357,396
Wine, 815,615 gals.,	373,529

THE COUNTRIES FROM WHICH EXPORTS WERE MADE WERE AS FOLLOWS:

Great Britain,	£8,288,226
West Indies (British),	14,973
North America (British),	13,563
Other British colonies,	5,036,311
United States of America,	1,668,606
Foreign States,	820,961
	£15,842,637

THE EXPORTS IN 1853 AMOUNTED TO £11,061,543, OF WHICH THE PRINCIPAL ITEMS WERE:

Bread and flour, 6,891 tons,	£155,753
Cottons, silks, linens, woollens,	78,795
Gold, 2,497,723 oz. 15 dwt. 16 grs.,	8,644,529
Provisions and groceries,	69,411
Spirits, 115,357 gallons,	44,999
Tobacco, 311,907 lbs.,	40,022
Wool, 20,842,591 lbs.,	1,051,071

THE COUNTRIES TO WHICH IMPORTS WERE MADE WERE AS FOLLOWS:

Great Britain,	£9,875,624
British colonies,	942,741
United States of America,	19,646
Foreign States,	223,532
	£11,061,543

Banking.] The astonishing progress of this colony cannot be shown better, or in a more striking light, than by its banking statistics. The following were the aggregate amounts, under the respective heads, of the banks existing here in the periods mentioned:—

	1851.	1852.
Notes in circulation,	£102,415	£942,545
Deposits (not bearing interest),	703,167	2,731,797
Total liabilities,	819,303	3,746,228
Coin,	273,185	952,923
Gold, valued at cost,	Nil.	420,110
Bills discounted, &c.,	615,311	1,046,126
Total assets,	908,807	2,561,675

Mr. Westgarth, the compiler of the annual report of the Melbourne chamber of commerce for 1852, enters into the following striking comparative estimates of the progress of a colony, in which "eighteen years ago there was not a civilized human being residing. The value of the produce of this colony actually exported in 1852, was," he says, "in round numbers £15,000,000. With so encouraging a fact, we may venture upon an interesting research, and compare the results of the commerce of V. with those in several other instances that are naturally suggested to the mind by our own present position and most recent history. 1. California is our great competitor in the production of gold. The latest statements that have come into my hands on the subject of the gold produce of that country are up to March of last year. California had then entered the sixth year of her golden harvest, while V. had attained the second. At that time the produce of gold, the sole export produce of California, amounted in value, by official record, to £12,000,000 annually, and to this quantity it was estimated that one-fourth should be added for the unrecorded export: making a total of £15,000,000, the amount which we have just ascertained to be the annual value of the export of this colony, about the same period. 2. Among British colonies those of India have hitherto stood first, as far exceeding all others in the magnificent scale of their wealth and commerce. The exports for the year 1851 from Calcutta, the capital and seaport of Bengal, the greatest of the Indian presidencies, amounted in value to £11,040,000 or rather less than three-fourths of the amount of the exports of this colony for the year succeeding. 3. To proceed to still higher standards of comparison, let us take the export commerce of Britain itself. The average annual value of the exports for the four years, 1848 to 1851, amount to £65,565,000; so that the value of the export produce of this colony already approaches to one-fourth of that of the parent state." When the railway to connect Melbourne with its harbour is opened, the perils, the waste of money and time to which the voyager to the land of gold was exposed in the last 9 m. of his course will exist no more. The navigation of the Yarra will be in a great measure superseded by the possibility of unloading at a sea-harbour, and then 10 minutes of a railway will take the eager traveller into the midst of the ugliest, worst-built, richest, and dearest town in the globe. Another railway, again, will soon connect Melbourne with Geelong. V. has now two capitals, each possessing some conveniences and advantages, and both labouring under some counterbalancing deficiencies; but when a railway has bound them together, each will share the advantages of the other. And these undertakings seem small and insignificant by the side of the great scheme which proposes to connect Melbourne with Sydney. A line is in contemplation which will first carry the digger to the gold rocks of Mount-Alexander, and then sweep away to the NE. The line will find an opening through the great chain of mountains by means of one of the passes that admit the waters of the Yass, and then, entering on the long valley which gives a channel to the Shoalhaven and the Hawkesbury, will terminate at the harbour of Port-Jackson. It will probably be some years before the London and Liverpool of Australia, as they have been termed, will be thus united; but colonial energy will not suffer the time to be a day longer than is absolutely necessary. Of the 500 m. over which the line is to extend, one half is said to be on a dead flat, and the gradients of the remaining half are by no means great.

Population.] In 1846, the district of Port Phillip had a pop. of 32,879. In March 1851, the pop. was 77,345; on 26th April, 1854, it amounted to 236,798, of whom only 80,912 were females. The discovery of the gold fields in June 1851 attracted a vast number of persons to the colony. A return of immigrants who arrived in V. in 1852, distinguishing those who arrived at the public expense from those who arrived at their own expense, showed that the immigrants were 15,477; the unassisted, 79,187. Grand total: 94,664.

The countries from which the immigrants last came were as under:

England,	28,159
Scotland,	11,814
Ireland,	4.477
Other countries,	1.961
Australian colonies,	48.253
	94,664

The total number of schools in the province in 1852 was 115, at which attended 4,322 boys, 3,519 girls. Total, 7,841. Aid from Government, £7,763; amount paid from school fees, £5,731.—The number of churches and chapels in the colony on the 31st of December, 1852, was 49. Estimated to accommodate 16,060 persons, and attended generally by 14,520 persons. The churches and chapels belonged to the following bodies:—Church of England, 13; Presbyterians, 7; Wesleyans, 17; Independents, 3; Baptists, 2; Roman Catholics, 6; Jews' Synagogue, 1.—The council of the Melbourne University includes the bishop of Melbourne, and of the Roman Catholic church, two Presbyterian ministers, and lay members of different denominations. The university is incorporated by an act of council. Forty acres of valuable land to the N of the city have been appropriated to the use of the council; £30,000 have been granted for the building, and an endowment of £9,000 per annum. The four professors, for the Greek and Latin languages, Natural Science, Mathematics, and Logic, Modern Literature, and Political Economy, are to receive a salary of £1,000 per annum with chambers.

Government.] The administration of this colony is now conducted by a governor, an executive council, and a legislative assembly, the latter of which is about to be divided into an upper and a lower house. The schedule for the upper house, which is to consist of 30 members, has been agreed to in committee; they are distributed according to the following arrangement, which also shows the administrative subdivisions of the province:

CENTRAL PROVINCE, including the city of Melbourne and the country extending three miles round the corporate limits of the city,	5
SOUTH PROVINCE, including the counties of Bourke, Evelyn, and Mornington,	5
SOUTH-WESTERN PROVINCE, including the counties of Grant, Grenville, and Polwarth,	5
WESTERN PROVINCE, including the counties of Ripon, Hampden, Heytesbury, Villiers, Normanby, Dundas, and Follett,	5
NORTH-WESTERN PROVINCE, including the counties of Talbot and Dalhousie, and the Wimmera and Loddon districts,	5
EASTERN PROVINCE, including the county of Anglesey, the proposed county of Rodney, and the Murray and Gipps' Land districts,	5

The revenue in 1852 was £1,123,825, of which £558,500 was territorial. In 1853, the revenue was £1,577,181, of which £846,214 was territorial. The following is a return of the revenue of the colony in each of the four quarters of 1854 ending 30th September:—

	Quarters ending			
	Dec. 31, 1853.	March 31, 1854.	June 30, 1854.	Sept 30, 1854.
Gold licenses,	£130,727	£92,062	£77,835	£89,629
Escort and Treasury fees,	18,122	14.659	16,456	15.570
Customs,	325,015	196,092	218,247	186,666
Port & harbour dues,	4,922	5.208	5,996	8.576
Postage,	8,421	9,628	12.465	24,909
Licenses,	5,675	7,814	67,049	47,420
Fines and forfeitures,	17,404	14,044	14,576	17,112
Fees of office,	7,098	7,744	11,306	12,748
Miscellaneous,	7,101	14,173	11.961	13,940
Land revenue,	548,854	231,075	377,201	511,642

The expenditure in 1853 was £734,961, of which the expences of the civil establishment amounted to £93,273; of the judicial, to £18,990; of the police, £51,726; of the gaol and penal establishment.

£24,782; of the ecclesiastical establishment, £4,219; and of education, £7,591. The public works and buildings amounted to £96,977; and the military establishment to £2,873; while the gold-workings were debited with £146,335.

VICTORIA, a large river of Northern Australia, discovered by Stokes in 1839. Its estuary between Turtle-point and Pearce-point, in Queen's channel, is 26 m. wide, and it is navigable for large vessels at least 60 m. from its mouth. It has been traced to 135 m. from its mouth, where it was found flowing from the SE. It has recently been proposed to trace this important river to its source, and thence strike across the country, in a S direction, to Adelaide, if practicable; or if not, in an E direction to the SE corner of the gulf of Carpentaria. The advocates of this line of exploration argue that the mouth of this river is about the nearest point of Australia to England in a direct line. London (the straits of Dover excepted) will shortly, say they, be connected with Constantinople by means of railways. The passage is short and easy across the Dardanelles to Scutari. From Aleppo, either by the N or the S side of the range of the Kurin mountains, there are probably no difficulties which engineers will not in time overcome. Turning again to Mercator's chart, it will be found that from Aleppo to the great river Euphrates the distance does not exceed 60 m. Along the banks of that once famous and still noble river will, no doubt, be traced a comparatively easy road for railway communication to Bussora at the head of the Persian gulf. And along the NE shores of that gulf there do not appear to be any great engineering obstacles to the formation of a railway thence to Choubar, on the coast of Beluchistan. Thence the course would be across the Indus to Hyderabad, and through the important commercial districts of Cambay and Surat to Bombay. When at Bombay, a day and a-half or so, at any ordinary speed, would carry us by railway through Punah to the ancient city of Goa, and through Seringapatam to Cochin or to Cape Comorin, touching at Ceylon and at Sumatra, and Java if desirable, a steamer would reach the river V. in about 8 days, at the present rate of speed. From that part of the V. river we could, by means of railroads if the country is traversable, reach Adelaide, Melbourne, or Sydney in 2 or 3 days at the most. Thus our goldfields in Australia may be reached from England, at only moderate rates of speed, by steam, in 22 days; and with that object in view, the V. river is likely to become the great hall door of Australia. Independent, however, of all this, and presuming that the interior of Australia is perfectly inaccessible—which we have no reason to suppose—and independent of the deep scientific interest attached to the proposed expedition, they believe there is every probability that such an expedition will open up new and valuable lands; and while the proposed starting point is the best for penetrating into the unknown parts of the interior, it is in itself a site of which it is desirable to have a better knowledge, since it promises to be an important station for the rapidly developing commerce between India, China, and the Indian archipelago. Further, the expedition could usefully survey a probable station for the steam communications which must yet be established between India, China, and New South Wales. The lands also upon the northern rivers of Australia promise to be suitable for many valuable purposes besides the ordinary pastoral business of that country, and especially for the growth of cotton. Again, the station which would there be formed would, in all probability, concentrate the trepang fishery and other existing trades, at present chiefly in the hands of Malay adventurers. To these considerations it is replied by the advocates of another line of exploration that the merits of the V. river, as a central point from which our knowledge of the interior could most readily be extended, appear by no means equal to those of the gulf of Carpentaria. The only territory to which the V. offers peculiar facilities of access, is on the NW water-shed; and it is not unlikely that mountainous country may render access from its head-waters to the E and S difficult and perhaps impracticable. On the other hand, the S shores of the gulf of Carpentaria run inland to the extent of 500 m.; and the numerous rivers which flow into it command as close an approach to the NW water-parting as the V. In addition to this, the following immense sweep of country is most accessible from the gulf, viz. from the drainage area of the E coast of the continent, extending from Leichhardt's route to the gulf, W to Mitchell's furthest; thence to the central point attained by the expedition of Sturt; and so on, completing the unknown extent of the N water-shed and its relations to the vast centre, up to the comparatively limited range naturally approachable by the NW rivers. A series of movements radiating from the gulf along its numerous rivers, and connecting it with the line thus indicated, would, it is argued, intercept, in every case, the course intended to be traversed by the missing explorer, Dr. Leichhardt; and if this were the only advantage, it ought not to be overlooked. Captain Stokes, while recording the exploration of the Victoria, as well as that of the gulf and some of its rivers, often makes reference to the remarkable advantages offered by the gulf, both for exploration and settlement. This evidence in favour of the gulf comes from the experienced discoverer who explored the V. as well as the gulf, and it cannot therefore be considered *ex parte*. "It is of importance in the establishment of a settlement on the N coast of Australia," say the advocates of the gulf route of exploration, "that it should be accessible and attractive to the native navigators and traders of the adjacent Indian archipelago: all along that remarkable maritime channel which penetrates between two continuous and parallel chains of important islands, from Singapore to the gulf of Carpentaria, there is now no depot for British merchandise, eastward of Singapore. The smooth seas, the periodical winds, and land within sight, to which the native traders are accustomed, would render the boisterous ocean, within the region of hurricanes, through which the V. must be approached, an impediment to that important traffic. We must endeavour to take timely warning from the failure of Port-Essington, in proceeding to bring about any new settlement on the N coast. Recurring to the gulf of Carpentaria, it is but a prolongation of the frequented line of traffic through the archipelago. By way of the coast of New Guinea the line of land is scarcely interrupted between the archipelago and the shores of the gulf; and the native traders will derive from the land which surrounds the gulf that protection from violent winds to which they are accustomed amongst their own contiguous islands. Safe harbours, numerous rivers, rich soil, fine climate, and proximity to the eastern trade, must cause the territory around the gulf to be rapidly settled as soon as its merits are better known.

VICTORIA, a recently founded town, on the N side of the island of Hong-Kong, on Victoria bay. See article Hong-Kong.

VICTORIA, a district of Upper Canada, skirted on the S by the bay of Quinti, and watered by the Trent, the Mocra, and the Salinas. Its district town is Belleville. Pop. in 1842, 15,842.

VICTORIA, a county in the SE of the state of Texas, U. S. Area 804 sq. m. It is intersected by the Guadalupe, and by the Garatas, the Zanilla, and the Casa-Blanca. Pop. 289.—Its cap. of the same name is situated on the E bank of the Guadalupe, 113 m. SSE of Austin, in N lat. 28° 56′ 47″, W long. 97° 7′ 30″.—Also a village in Knox co., Illinois, 87 m. N by W of Springfield.—Also a village of Davies co., in Missouri, 138 m. NW of Jefferson.—Also a village of Bolivar co., in Mississippi, 100 m. NNW of Jackson.

VICTORIA, a settlement of Brazil, in the prov. of Bahia, comarca of Urubu.—Also an island in the prov. of São-Paulo, to the E of the island of São-Sebastião, in S lat. 23° 47′ 42″.

VICTORIA, a town of Mexico, in the prov. of Tabasco, on the coast of the gulf of Campeachy. It has a small fort for its defence.

VICTORIA, a Dutch fort on the SW coast of the island of Amboyna, in S lat. 3° 41′ 41″.

VICTORIA (Cape), a cape on the W coast of Patagonia, in S lat. 52° 35′.

VICTORIA (Fort), a factory of the Hudson's Bay company, in Vancouver's island, on Esquimalt-bay on the SE coast of the island, in N lat. 48° 25′. The fort is an oblong stockade, with a blockhouse at the opposite corners, and a few guns in each. There are also some storehouses, and two wharfs, one of which has 18 ft. water alongside.

VICTORIA-HARBOUR, a large roomy inlet, on the S side of Choiseul sound, in the Falkland archipelago. It is formed by a long peninsula running parallel to the shore for about 7 m.

VICTORIA (Lake). See Alexandrina.

VICTORIA-LAND, a recently discovered region in the southern Atlantic ocean, nearly upon the same meridian with New Zealand, and 2,000 m. further S, discovered by Sir James Ross. "On the morning of the 11th January 1841, when in S lat. 70° 41′, and E long. 172° 36′, land was discovered," says that gallant officer, "at the distance, as it afterwards proved, of nearly 100 m. directly in the course we were steering, and therefore directly between us and the pole. Although the circumstance was viewed at the time with considerable regret, as being likely to defeat one of the more important objects of the expedition, yet it restored to England the honour of the discovery of the southernmost known land, which had been nobly won, and for more than twenty years possessed by Russia. Continuing our course towards this land, for many hours we seemed scarcely to approach it. It rose in lofty mountain-peaks of from 2,000 to 12,000 ft. in height, perfectly covered with eternal snow; the glaciers that descended from the mountain summits projected many miles into the ocean, and presented a perpendicular face of lofty cliffs. As we neared the land, some exposed patches of rock appeared; and steering towards a small bay for the purpose of effecting a landing, we found the shore so thickly lined for some miles with bergs and pack-ice, and with a heavy swell dashing against it, we were obliged to abandon our purpose, and steer towards a more promising looking point to the S, off which we observed several small islands. On the morning of the 12th I landed, accompanied by Commander Crozier, and a number of the officers of each ship, and took possession of the country in the name of her most gracious majesty. The island on which we landed is composed wholly of igneous rocks, numerous specimens of which, with other imbedded minerals, were procured; it is in S lat. 71° 56′, and E long. 171° 7′. Observing that the E coast of the main land trended to the southwards, whilst the N shore took a north-westerly direction, I was led to hope, that by penetrating to the south as far as practicable, it might be possible to pass beyond the magnetic pole, which our combined observations placed in 76° (?) nearly; and thence by steering W, complete its circumnavigation. We accordingly pursued our course along this magnificent land, and on the 23d of January, we reached 74° 15′ S., the highest southern lat. that had ever been attained by any preceding navigators. Although greatly impeded by strong southerly gales, thick fogs, constant snow-storms, we continued the examination of the coast to the southward, and on the 27th we again landed on an island in S lat. 76° 8′, and E long. 168° 12′, composed, as on the former occasion, entirely of igneous rocks. Still steering to the southward, early the next morning, a mountain of 12,400 ft. above the level of the sea, was seen emitting flame and smoke in splendid profusion: this magnificent volcano, which received the name of Mount-Erebus, is in S lat. 77° 32′, and E long. 167°. An extinct crater to the E of Mount-Erebus, of somewhat less elevation, was called Mount-Terror. The mainland preserved its southerly trending, and we continued to follow it until, in the afternoon, when close in with the land, our further progress in that direction was prevented by a barrier of ice, stretching away from a projecting cape of the coast, directly to the ESE. This extraordinary barrier presented a perpendicular face of at least 150 ft., rising, of course, far above the mast-heads of our ships, and completely concealing from our view everything beyond it, except only the tops of a range of very lofty mountains in a SSE direction, and in S lat. 79°. Pursuing the examination of this splendid barrier to the eastward, we reached the latitude of 78° 4′ S. the highest we were at any time able to attain, on the 2d of February; and having, on the 9th, traced its continuity to the long. of 191° 23′, in S lat. 78°, a distance of more than 300 m., our further progress was prevented by a heavy pack, pressed closely against the barrier; and the narrow lane of water, by means of which we had penetrated thus far, became so completely covered by rapidly-forming ice, that nothing but the strong breeze with which we were favoured enabled us to retrace our steps. When at a distance of less than half a mile from its lofty icy cliffs, we had soundings with 318 fath. on a bed of soft blue mud. With a temperature of 20° below the freezing point, we found the ice to form so rapidly on the surface, that any further examination of the barrier in so extremely severe a period of the season being impracticable, we stood away to the westward for the purpose of making another attempt to approach the Magnetic pole, and again reached its lat. 76° S (?) on the 15th of February, and although we found that much of the heavy ice had drifted away since our former attempt, and its place, in a great measure, supplied by recent ice, yet we made some way through it, and got a few miles nearer to that pole than we had before been able to accomplish, when the heavy pack again frustrated all our efforts, completely filling the space of 15 or 16 m. between us and the shore. For several days we continued to examine the coast to the westward, tracing the pack-edge along, until, on the 25th of February, we found the land abruptly to terminate in S lat. 70° 40′, and E long. 165°, trending considerably to the S of W, and presenting to our view an immense space occupied by the newly-formed ice, so covered by recent snow as to present the appearance of one unbroken mass, and defying every attempt to penetrate it. The great southern land we have discovered, and whose continuity we have traced from nearly the 70th to the 79th degree of latitude, I am desirous to distinguish by the name of our most gracious sovereign Queen Victoria."

VICTORIA-LAND, a recently discovered tract of Arctic America, lying between the parallels of 70° and 72°, along the W side of Victoria strait, which separates it from Boothia Felix, and continuous with Wollaston-Land on the SW, and Prince Albert-Land on the NE. It was discovered by Mr. T. Simpson in 1838. Having passed the farthest point reached by Franklin in 1821, Simpson saw from a low cape, which he named Cape Franklin, a new land or crowded chain of islands of great extent, in many places high and covered with snow, stretching from W to NE. The mainland here turned off to ENE, which continued to be its general bearing for the three following days. It is flat in its outline, the path leading alternately over soft sand, sharp stones, and swampy ground. At the distance of from 1 to 2 miles the coast is skirted by a range of low stony hills partially clothed with dull verdure, which send down to the sea numberless brooks and small streams: none of the latter at that season (August) reached above the waist, though the deep and rugged channels of many of them showed that in the spring they must be powerful torrents. Two leagues inland, a hill—which was named Mount George, after Governor Simpson—rises to the height of 600 ft., and forms a conspicuous object for a day's journey on either side. The ice all along lay immoveably aground upon the shallow beach, and extended in every direction as far as the eye could reach. The great northern land still stretched out before the explorers, and kept alive doubts of their being engaged in exploring an immense bay, which even the increase in the tides, the quantity of seaweed and shells, and the discovery of the remains of a large whale and of a Polar bear could not altogether dispel. These doubts seemed almost converted into certainty as they drew near, on the fourth evening, an elevated cape, and saw land apparently all around. With feelings of disappointment Simpson ascended the height from whose summit a splendid and unlooked-for view suddenly burst upon him. "The ocean, as if transformed by enchantment, rolled its free waves beneath and beyond the reach of vision to the eastward, islands of various shape and size overspread its surface, and the northern land terminated in a bold and lofty cape, bearing NE at least 40 m. distant, while the coast of the continent trended away to the south-east. I stood, in fact," continues Mr. Simpson, "on a remarkable headland at the E entrance of an ice-obstructed strait. The extensive land to the northward I have called 'Victoria Land,' in honour of our youthful sovereign; and its eastern extremity Cape Pelly, after the governor of the honourable company. To the promontory, where we encamped, I have attached the name of Cape Alexander, after an only brother, who would give his right hand to be the sharer of my journeys. The rise and fall of the tide there was about 3 ft., being the greatest observed by us in the Arctic seas. The coast here changes its character, the water becomes deep, the approach easy, and I have little doubt that the islands contain secure harbours for shipping. Next morning, at the distance of 8 or 9 m., we crossed another high cape, formed of trap rocks, in N lat. 68° 52′ 18″·5, the variation of the compass being 63° E. The travelling had become more and more toilsome; our road new passing over some miles of round loose stones, and then through wet mossy tracks, sown with large boulders, and tangled with dwarf willows. At our usual camping hour, we opened a large bay studded with islands, which ran in for 5 m. to the SSW, and then turned off in a bold sweep of rounded granitic hills dipping to the sight in the ESE, at the distance of 30 m. To walk round even the portion of the bay in view would have consumed three days: the time allotted for outgoing was already expired, and two or three of my men were severely lame from the fatigue of their burdens, the inequalities of the ground, and the constant immersion in icy cold water. The morning of the 25th of August was devoted to the determination of our position, and the erection of a pillar of stones on the most elevated part of the point. After which, I took possession of the country, with the usual ceremonial, in the name of the honourable company, and for the Queen of Great Britain. In the pillar I deposited a brief sketch of our proceedings, for the information of whoever might find it. Its situation is in N lat. 68° 43′, W long. 106° 3′. The compass grew sluggish and uncertain in its movements as we advanced eastward, and frequently had to be shaken before it would traverse at all. Two miles to the S of our encampment, a rapid river of some magnitude discharges itself into the bay, the shores of which seemed more broken and indented than those along which we had travelled."

VICTORIA (La), a village of Venezuela, on the road leading from Caracas to Puerto-Cavello, in N lat. 10° 13′ 35″. The principal ornament of the place is a handsome church, so large that it might well be termed a cathedral.

VICTORIA (North and South), two districts of British Caffraria, in S. Africa.

VICTORIA (Nossa-Senhora-da), a town of Brazil, on an island off the coast of Espirito-Santo, in S lat. 20° 18′, W long. 40° 0′. It is well-built.

VICTORIA STRAIT, a channel of the Arctic ocean, separating Victoria-Land on the W, from Boothia-Felix on the E; and communicating on the N, by Ommaney inlet, with Barrow's strait.

VID, or Urak, a river of Turkey in Europe, in Bulgaria, which has its source in the sanj. of Sophia, on the N side of the Balkan, and to the NW of Ihleman; flows thence into the sanj. of Rustchuk, and after a course in a generally NNE direction of about 150 m., joins the Danube on the r. bank, about 5 m. above the confluence on the opposite side of the Aluta.

VIDALIA, a village of Concordia parish, in the state of Louisiana, U. S., on the W bank of the Mississippi, opposite Natchez, and 77 m. N by W of Baton-Rouge.

VIDAUBAN, a commune of France, in the dep. of the Var, and cant. of Le Luc, 8 m. SSW of Draguinan, on the r. bank of the Argens. Pop. 2,203.

VIDDIN. See Widdin.

VIDEM, a village of Austria, in Styria, in the circle of Cilli. Pop. 260 It has a school and an Armenian institute.

VIDEROE, an island in the NE part of the group of the Faröe islands, between the islands of Fuglöe and Bordöe. It is 9 m. in length, and nearly equal in breadth, and contains a village of the same name, with 600 inhabitants.

VIDEY, an island near the W coast of Iceland, a little to the NE of Reikiavick. It has a royal farm, formerly a monastery. The eider duck is found here in great numbers.

VIDIGUEIRA, a town of Portugal, in the prov. of Alemtejo, and comarca of Beja, delightfully situated. Pop. 2,400.

VIDIGULFO, a market-town of Austria, in Lombardy, in the prov. and 9 m. NNE of Pavia, and district of Landriano, in a fertile locality. Pop. 1,200. A victory was here gained over the Pavesians by the Milanese in 1146.

VIDIO (Cape), a headland of Spain, in the prov. of Oviedo, WSW of Cape Penas.

VIDOURLE, a river of France, which has its source in the W part of the dep. of the Gard, near Saint Hippolyte; forms for some distance the dividing line between this dep. and that of the Herault; divides into two branches, of which one joins the Canal-de-la-Radelle, and the other entering the dep. of the Herault, flows into the Etang-de-Mauguio. This river flows in a generally SSE direction, and has a total course of 51 m. Saint-Hippolyte, Sauve, Quissac, and Sommieres, are the chief places on its banks.

VIDRERAS, a market-town of Spain, in the prov. and 15 m. from Gerona, between that town and Barcelona, and near the coast. Pop. 1,600.

VIDZICHKI, a town of Russia in Europe, in the gov. of Vilna, district and 6 m. N of Vilkomir.

VIDZY, a town of Russia in Europe, in the gov. and 75 m. NE of Vilna, on a small affluent of the Dwina.

VIE, a river of France, which has its source in the dep. of the Orne, a little to the W of the village of Grand-Val; flows thence into the dep. of the Calvados; waters Vimoutiers, Livarot, and St. Julien-le-Faucon; and after a course of about 33 m., throws itself into the Dives, on the r. bank. It runs in a generally NW direction.—Also a river in the dep. of the Vendee, which has its source a little to the S of Belleville, in the cant. of Poire-sous-la-Roche; runs W; and after a course of 33 m., throws itself, with the Jauney, into the Atlantic, at the port of Saint-Gilles.

VIECHT (Gross and Klein), two villages of Bavaria, in the circle of Upper Bavaria. Pop. 260.

VIECHTACH, a village of Austria, in Upper Austria, in the circle of the Traun, on the Alm. It has several saw-mills, and a trade in timber.

VIECHTACH (Ober), a market-town of Bavaria, in the circle of the Upper Pfalz, and to the E of Raabburg. Pop. 1,225.

VIECHTACH (Unter), a market-town of Bavaria, in the circle of the Regen, on the river of that name, and 47 m. NW of Passau. Pop. 1,600. It has a poor-house and an hospital, and contains manufactories of linen and of tobacco, dye-works, and breweries.

VIECHTENSTEIN, a village of Austria, in the circle of the Inn, and on the river of that name. Pop. 460. It has a castle.

VIECHTWANG, a village of Austria, in the circle of Traun, on the Inn. Pop. 1,400.

VIEHAN, a village of Prussia, in the regency of Breslau, and circle of Neumarkt. Pop. 360. It has a castle.

VIEHBACH (Ober and Unter), two villages of Bavaria, in the circle of Lower Bavaria, and containing respectively 260 and 300 inhabitants. They contain a convent and several mills.

VIEHHAUSEN, a village of Bavaria, in the circle of Upper Bavaria. Pop. 360. It has a castle.

VIEHHOFEN, a village of Austria, in the circle of the Upper Wienerwalde, and near St. Polten. It has a glass manufactory.

VIEHLAND, a bailiwick of Hanover, in the gov. of Stade and duchy of Bremen. Pop. 2,600. Schiffdorf is its chief place.

VIEIL-HESDIN. See Hesdin-le-Vieux.

VIEIL-SAINT-REING, a commune of France, in the dep. of the Ardennes, and cant. of Novron. Pop. 1,291.

VIEIL-SALM, a department and commune of Belgium, in the prov. of Luxemburg and arrond. of Marche, watered by the Glain. Pop. 2,961.

VIEJO, a market-town of Central America, in the state of Nicaragua, near a volcano of the same name, in the dep. and 30 m. NNE of Realejo. Pop. 3,000.

VIELANT, a village of the grand-duchy of Mecklenburg-Schwerin, in the prov. of Mecklenburg. Pop. 440.

VIELAU, a village of Saxony, in the circle of Zwickau. Pop. 750.

VIELBAUM, a village of Prussia, in the regency of Magdeburg and circle of Osterburg. Pop. 350

VIELBRUNN, a market-town of Hesse, in the prov. of Starkenburg. Pop. 780.

VIELGOROD. See BELGOROD.

VIELGUT, a village of Prussia, in the regency of Breslau and circle of Oels. Pop. 724.

VIELIST, a village of the grand-duchy of Mecklenburg-Schwerin, and prov. of Mecklenburg. Pop. 240.

VIELITZ, a village of Prussia, in the regency of Potsdam and circle of Ruppin. Pop. 210.

VIELLA, a commune and town of France, in the dep. of the Gers, cant. and 5 m. SW of Riscle. Pop. 1,600. The locality affords good wine.

VIELLA, a judicial partido and market-town of Spain, in the prov. of Lerida. The partido comprises 34 pueblos. The town is 84 m. NNE of Lerida, in the valley of Aran, in the Pyrenees, on the l. bank of the Garonne. Pop. 700. It has a parish church and a custom-house. The manufacture of cotton and linen fabrics, and rearing of cattle form the chief objects of local industry.

VIELLA (COL DE), a pass of the Pyrenees, in Spain, on the S side of the valley of Aran, between Viella and Senet. The hospital of V. is near the S entrance to this pass.

VIELLE, or VIELLE-AURE, a canton and commune of France, in the dep. of the Hautes-Pyrenees, and arrond. of Bagneres-de-Bigorre. The cant. comprises 15 com. Pop. in 1831, 3,665; in 1846, 3,805. The village is 21 m. SE of Bagneres-de-Bigorre, on the Neste, in a valley of that name. Pop. 481.

VIELLE-LOYE, a commune of France, in the dep. of the Jura, and cant. of Montbarrey, 9 m. from Dôle. Pop. 730. It has a bottle-work.

VIELLE-LYRE, a commune of France, in the dep. of the Eure, and cant. of Rugles, on the Rille, 27 m. SW of Evreux. Pop. 779. It has a foundry and blast furnace.

VIELLEVIGNE, a commune and town of France, in the dep. of the Loire-Inferieure, cant. of Aigrefeuille, on the Ognon. Pop. in 1846, 5,375.

VIELLY, a village of France, in the dep. of the Nord, cant. and 3 m. SW of Solemas. Pop. 1,670.

VIELMANSDORF, a village of Austria, in Styria, in the circle of Bruch and bail. of Rottenmann. Pop. 210.

VIELMUR, a canton and commune of France, in the dep. of the Tarn and arrond. of Castres. The cant. comprises 9 com. Pop. in 1831, 5,508; in 1846, 6,279. The village is 8 m. W of Castres, on the r. bank of the Agout. Pop. 1,178. It has manufactories of hosiery.

VIELSK. See VIATLSK.

VIELS-MAISONS, a commune of France, in the dep. of the Aisne, and cant. of Charly. Pop. 884. It has manufactories of articles in caoutchouc, of pottery, and of baskets, and a tannery.

VIELSTEDE, a village of Oldenburg, in the bail. of Delmenhorst. Pop. 550.

VIELVERGE, a village of France, in the dep. of the Côte-d'Or, cant. and 5 m. SE of Pontarlier.

VIENAN, a village of Prussia, in the regency of Magdeburg and circle of Salzwedel. Pop. 200. It has salt-works and several mills.

VIENENBURG, a village of Hanover, in the gov. and 33 m. SE of Hildesheim, on the Rradan. Pop. 924.

VIENNA, [in German, WIEN; in French, VIENNE,] the capital of the Austrian empire, and the largest city in Germany, situated in the midst of a plain diversified by a number of picturesque eminences and hills, near the r. bank of the Danube, where it receives a small stream called in German, the Wien—whence the city has its name—which passes through the city and suburbs, near the site of the ancient *Vindobona* in *Pannonia Superior*, in N lat. 48° 12′, E long. 16° 23′, at an alt. of 480 ft. above sea-level, 380 m. SSE of Berlin, and 650 m. E by S of Paris. To the E and N, the surrounding country is entirely level; but to the W and S is seen, at the distance of from 9 to 12 m., a range of mountains thickly planted with trees and vines. The Danube, which is here very wide, divides itself in the W part of the city into several arms forming islands covered with wood. The neighbourhood of this vast river, its agreeable islands, its variety of prospect, and the fertility of the soil along its banks, all concur to beautify the appearance of V., and would make it a very pleasant abode, if a variable climate and foggy atmosphere did not frequently overcast the cheerfulness of the scene. Few spots are so intersected with water as the vicinity of V. The Danube, broken into a variety of channels, loses its usual rapidity, and seems as if disposed to linger in this beautiful scenery; but scarcely has it left the neighbourhood of the capital, and advanced into the great level of Hungary, than it rolls along, in all its former impetuosity. One of the arms of this river, flowing between the city and the suburb of Leopoldstadt, serves for the purposes of navigation, and is crossed in four different places by wooden bridges. Its streets are unfortunately as narrow as those of any town in the south of Europe. It is divided into two great parts: viz. the Innerestädt or Vienna Proper, and the Vorstädt suburbs; and nothing can exhibit a greater contrast than these component parts of the same capital,—the suburbs surprising us by their extent and beauty, while the city disappoints us by its mean and irregular buildings. A stranger is perpetually impressed with the belief that the inhabitants of the Innerestadt are, as it were, imprisoned in their crowded dwellings; and this impression is confirmed by the impatience of the citizens to exchange them in the spring for the free air of the suburbs, where wide streets, extensive gardens, and large edifices, unite to enable the inhabitants to pass the summer to their satisfaction. In 1683, when V. was besieged by the Turks, the suburbs were only three or four in number; and a century ago, several of them were only villages or country-seats. Their increase has been chiefly owing to the abrogation by Joseph II., of the feudal rights possessed by the landed proprietors of the spot, after which the district became entitled to the same privileges as the rest of the capital. Fifty years ago, Vienna was considered to be well-fortified; having a rampart, twelve strong bastions, ten ravelins, deep and wide square ditches, and outworks of proportionable strength. The old works were said to be built with the money extorted by Leopold of Austria from the people of England, as the ransom of the gallant Richard-Cœur-de-Lion. With this sum, amounting to 140,000 marks of silver, Cologne weight, Leopold not only walled and fortified Vienna, but likewise the cities of Ens, Hainburg, and Neustadt. The entire extent of the lines enclosing both the city and suburbs is 16 m. Howitt, in his *Rural Life in Germany*, thus describes the city generally: "The city is great and compact, that is, so far as it is included within the walls, while far around there is an immense circle built upon, called the Vorstädte or suburbs, formed in segments radiating from the centre of the city, six-and-thirty in number. The city itself is still surrounded by its lofty walls and broad moat. Without this moat lies a broad open space, [600 yds. wide] called the glacis, consisting of plots of grass divided by walks and roads, and by lines of trees; without this green open circle commences the Vorstädte. These are interspersed with gardens, public walks, churches, palaces, and theatres, so that as you walk round the ramparts, now converted into a public promenade surrounding the whole city, you behold within the city a dense mass of noble, though narrow streets, immense piles of princely buildings, and a crowding, bustling population. On the other hand, that is, outwardly, you overlook, wherever you are, a more scattered but wide-spread scene, as of an eastern city, with towers and domes, gardens and masses of trees: the suburbs, in fact, form the much greater part of V. "V.," says a lively American writer, "is an immense city, but its heart only is walled in. You may walk from gate to gate in twenty minutes. In leaving the walls you come upon a feature of the city which distinguishes it from every other in Europe: its rampart is encircled by an open park, called the glacis, a ¼ m. in width, and perhaps 3 m. in circuit, which is, in fact, in the centre of V. The streets commence again on the other side of it, and, on going from one part of the city to another, you constantly cross this lovely belt of verdure which girds her heart like a cestus of health. The top of the rampart itself is planted with trees, and, commanding beautiful views in every direction, it is generally thronged with people. Between this and the glacis

lies a deep trench, crossed by draw-bridges at every gate, the bottom of which is cultivated prettily as a flower-garden. Altogether V. is a beautiful city. Paris may have single views about the Tuileries that are finer than any thing of the same kind here, but this capital of western Europe, as a whole, is quite the most imposing city I have seen." [*Willis.*] "As you enter the streets of the city," says Mr. Howitt, "you are surprised at the life and stir. Streams of well-dressed people are pouring along them; handsome carriages and equipages are seen driving as rapidly as in London; the shops present brilliant fronts; cafés in open places project their cool awnings, and set out their scores of chairs for luxurious smokers; all is motion, life, splendour, and crowds; and you feel for the first time since you left London, as if you were once more in a great capital. You are made sensible, too, how far east you have got, and in the chief city of what a variously compounded empire you are. Picturesque groups of foreigners are seated at the doors of various coffee-houses and hotels, and the throng in the streets is brightly variegated with the costumes of Turks, Albanians, Tyrolese, Jews, Wallachians, Hungarians, Armenians, and Italians. There is no capital in Europe of the same extent in which so much of what you want to visit, so many of the resorts of business or amusement, of literary or scientific institutions, are set down so near together. Palaces, theatres, houses of the nobility, libraries, collections of subjects of natural history, of arms, trophies, and jewellery, institutions for the education and assistance of its citizens, stand thickly, all within a very moderate space. The finest collections of works of art and of armour, with some palaces of the nobility, it is true, lie in the suburbs; but the imperial palace, the university, the arsenal, the treasure-chamber, the principal theatres and churches, lie within the walls. Within the walls, too, reside the highest classes chiefly; and wherever you go you obtain views of vast hotels of the nobles, built round courts, the splendour of which in a great measure is lost in the general view of the city, but which surprise you wherever you come upon them." Of the whole space occupied by the city, the Innerestadt does not comprise a tenth part; but its pop. is of course denser than that of any other quarter. It is subdivided into 4 quarters: the Schotten viertel on the W and NW; the Stuben-viertel on the NE; the Wimmer-viertel on the SE; and the Kärnter-viertel on the S. Its wall is pierced by 7 large and 5 small gates.

As a city, V. is a remarkable contrast to Berlin; the streets of the latter are very filthy, while those of V. are clean, thoroughly drained, and paved with great exactness, from the walls of the houses on one side to the walls on the other. There are no defined foot-paths for pedestrians, and in the narrow streets, especially, the vehicles frequently drive to within a few inches of the doors and windows on either side. As an imperial city, and the seat of government, the streets are pretty well occupied with private and public carriages, but of carts or waggons, very few are there to disturb either the comfort or safety of the multitudes who traverse the thoroughfares. The pavement of the street slopes from both sides to the centre, and hence, every falling shower carries every particle of dust and dirt to the gratings and drains in the centre. The city is crowded with public buildings, and some magnificent churches. V. contains 50 churches, 21 convents, several nunneries, 70 coffee-houses, 300 taverns, 5 theatres, and 6,518 houses, of which 1,387 are within the walls. Most of the houses are well-built of freestone, six stories high, with flat roofs; those of a different description are covered with pieces of timber shaped like tiles. Many of them have four cellars, one under another, with an open space in the middle of each arched roof for the purposes of ventilation; and from the lowermost there is a tube to the top, to let in air from the streets.

The principal places or squares of the city proper are the Josephsplatz, the Burgplatz, the Stephensplatz, the Neumarkt, and the Hohemarkt, the Herrengrasse, and the Wallnergrasse. The dom-kirche or cathedral of St. Stephen, the protomartyr, in the centre of the Stephensplatz, is a fair and stately Gothic fabric, but somewhat gloomy, owing to the painted glass in the windows. It was founded by Henry I. of Austria, and finished by Henry II. The building, which is of freestone, is 342 ft. long, by 144 ft. broad. The steeple is 447 ft. high, and is one of the finest in Germany, and much stronger than that of Strasburg, though not so elegant.—The university of Vienna was founded in 1365, and is divided into four faculties, and four nations,—Austrian, Saxon, Hungarian, and Rhenish. Upwards of 5,000 students attend this university, which has a library of 90,000 volumes.—There are several other excellent libraries in Vienna, as the Theresianum, and the imperial and archducal libraries. Of these, the imperial library in the Josephsplatz is by far the largest, whether for printed books or manuscripts. This building contains, according to the details of Balbi, about 270,000 volumes printed since the year 1500; 12,000 *incunabula* (books printed previous to the year 1500); 16,016 manuscripts, and 11,240 portfolios, containing one of the richest collections of engravings in Europe. These treasures are principally contained in one grand room, 240 (Vienna) ft. long, by 45 ft. wide, and 62 ft. high; having an oval dome of 30 ft. elevation above the general ceiling, and in five subsidiary rooms of smaller dimensions; but as the annual increase is from 3,500 to 3,800 volumes, the want of additional space is severely felt. This increase arises partly from the deposit of one copy of every work published in the Austrian territories, and partly from the purchase of foreign books—for which latter object, together with the cost of binding, and the purchase of engravings and manuscripts (the salaries of officers being paid separately), there is a fixed annual donation of 19,000 florins, or £1,900 sterling; besides such further funds as are required, and are readily granted by the government, for the purchase of any specific works of expense. For five hours in every day the library is open to the public. No introduction is requisite. The imperial palace, on the SW side of the Innerestadt, is an ancient and irregularly built structure. The imperial and the civil arsenals are of great extent. The imperial printing-office was founded by the late emperor Francis, for the purpose of supplying the printed returns, bulletins, and orders of the various branches of the administration; but it was also intended, under the auspices of the government, to give an impetus to, and become a kind of training school of the typographic and graphic arts in the empire. Its extent grew apace with its importance, and at the present moment, besides vast numbers of artists, men of science and mechanics, there are not less than 700 persons regularly employed in its various offices. It has, moreover, extended its influence to and established branch-offices in Lemberg, Temesvar, Salzburg, and other provincial capitals of the empire. The machinery department of the imperial printing-office is extremely well-supplied. An engine of 20-horse power moves not less than 48 printing and 24 copperplate presses, and 10 glazing machines. There are, moreover, 36 large and 12 small iron handpresses, 12 numbering and embossing machines, and 30 lithographic presses. A fresh supply of types is constantly supplied by 12 casting-machines and nine ovens, and 3,000 cwt. of type is kept on the premises. 500,000 sheets, or 1,000 reams of paper *per diem* are required for the consumption of the establishment. The botanic garden is on the SE of the city, a short distance beyond the open zone; and opposite the entrance gate is the chief residence of Prince Metternich. The garden is entered by an open iron gateway, from a court-yard, a short distance from the street or road, and on either side of the entrance are, apparently, well-worn public buildings, of not very imposing aspect. The form of the garden is a parallelogram, rather more than twice as long as broad, sloping to the north; and in extent, it may contain about 7 or 8 acres.—The faubourgs present, among their more remarkable buildings, the Belvidere palace, the Schwarzenberg palace, the Hotel-des-Invalides, the beautiful church of San-Borromeo, the Kaunitz palace, the Auersperg, the grand hospital, and the imperial porcelain manufactory.

Population, &c.] According to the official census, the pop. amounted, in 1829 to 289,785; in 1840, to 357,927, of whom 204,298 were Austrians, and the rest foreigners; the increase since 1837 being 23,427, chiefly foreigners, residing in the suburbs: the city, properly so called, having only 52,593 inhabitants. Dr. Ungewitter states the pop. in 1844 at 375,000. The garrison generally amounts to 8,000 or 12,000 men. The climate of V. is very variable, and can by no means be called healthy: the annual bills of mortality exhibiting a list of deaths, in the proportion of one in fifteen, according to De Serres; and one in nineteen, according to Nicolai. Various hypotheses have been framed to account for this extraordinary mortality. Nicolai attributes it to hard eating and drinking, as if all the people of Vienna were drunkards and gluttons. De Serres supposes it to arise chiefly from the resort of country invalids to the numerous hospitals and infirmaries of the capital; and affirms, in perfect contradiction to Nicolai, that the inhabitants are sober and temperate. The charitable institutions at Vienna are numerous and richly endowed. The Allgemeine-Krankenhaus, or great hospital, equal in extent to any in Paris or London, has 2,000 beds, and receives often 16,000 patients in the course of a year. There are separate hospitals for the soldiers, for Jews, for foundlings, orphans, and aged persons. Several of these charitable establishments are served by nuns. The lying-in hospitals are also on a liberal plan, and under good management. V. is the seat of an archbishop.

Manufactures, &c.] Manufactures in V., if not very extensive in any particular department, are of great variety, and comprise silks, ribbons, cottons, gloves, lace, saddlery, paper, earthenware, philosophical and musical instruments, cutlery, bronze goods, jewellery, liqueurs, chemical substances, maps, engravings, coaches, and carriages in general. The transactions of merchants, and the business conducted on commission and speculation, are also extensive, the consumption of the city being large, and the Danube giving it the benefit of navigation both to E and W. V. is the central place of commerce in the Austrian states; the place for exchange operations, for extensive sales and purchases, for loans and contracts.

History.] Vienna owes its first aggrandizement to Henry I. in 1142, who then made it the place of his residence, and in 1158 surrounded it with a wall. In 1198, it obtained its municipal privileges. In 1241 it was captured by Frederick II. In 1477 it was unsuccessfully besieged by the Hungarians, who took it, however, in 1485, under the command of Matthias Corvinus, their king. It sustained two sieges from the Turks, in 1529 and 1683, the last of which was infinitely more terrible than the first, as it lasted upwards of ten weeks. The road to Vienna was laid open to the Turks by Tekeli, whom Leopold would not subdue by clemency, and could not reduce by force. The garrison only consisted of 16,000 men, commanded by Count Stahremberg. Never, till the days of Napoleon, was such an army as that of the Turks seen at Vienna It consisted of 140,000 regular troops; 18,000 Wallachians, Moldavians, and Transylvanians, led by their respective princes; 15,000 Hungarians, under Tekeli; 50,000 Tartars under the famous Selim Gyeray Khan: the whole, including volunteers, and the usual cumbrous appendages of an oriental army, amounting to 300,000 men, 31 pashas, and 5 sovereign princes, with 300 pieces of cannon; all under the command of Kara Mustapha, nephew of the great Cuprogli, but destitute of his abilities. The whole plain that encircles V., comprehending a space of three leagues, was filled in all its extent with the Turkish camp, which abounded in every thing necessary for so vast a multitude. The different quarters were commanded by pashas, who displayed the magnificence of kings; but all this magnificence was eclipsed by the pomp of the vizier, who combined in his character, the opposite extremes of avarice and prodigality. His retinue consisted of 4,000 officers and servants; and the park that enclosed his tents was as extensive as the city he besieged. The counterscarp was lost by the besieged; after a series of successive combats of 23 days' continuance; and on the 22d of August, the engineer, Caplieri, who had made an exact computation of his means of defence, judged that he could not hold out three days longer, if the Turks should make a general assault. No army had yet appeared to help the place: and Stahremberg wrote these emphatic words to the duke of Lorrain: "No more time to lose, my lord; no more time to lose." But the avarice of the vizier proved the salvation of Vienna; for it is certain, that if he had given the orders for a general assault, the city must have fallen. Entertaining the notion that the capital of the German emperors must contain immense treasures, he was afraid of losing this imaginary wealth, if the city should be taken by storm; as in that case it would inevitably be subjected to universal and indiscriminate pillage. He therefore forbade the army to assault the place, but rather to wait patiently till it should surrender. This enraged the janizaries, and damped the ardour of the besiegers, who plainly saw that the vizier meant to be the sole gainer by the business. John Sobieski, with 25,000 men, at last arrived, on the 7th of September. On the 9th, the whole Christian army, 74,000 strong, was in motion. On the 12th, they approached the last heights of the Kalemberg range. It was yet in the power of the vizier to repair his faults, for he had nothing to do but occupy the hill, and mask the defiles, in order to prevent the advance of the Christian army; but he neglected the opportunity. John Sobieski was appointed generalissimo of the Christian force; and in company with the duke of Lorrain, the electors of Saxony and Bavaria, and the other commanders, took the sacrament two hours before day-break of the 13th of September,—a day that was to decide whether the cross or the crescent should prevail,—whether Vienna, under Mahommed IV. should share the fate of Constantinople, under Mahommed II.,—whether the star of the house of Hapsburg, or of Othman, should be lord of the ascendant, and whether the western empire should be re-united to that of the east. The eventful problem was settled ere night, by the utter defeat of an army three times more than that of the assailants. V., though dreadfully damaged, was delivered; and Sobieski made a splendid entry, through smoking ruins, into the city, amidst the rapturous acclamations of a saved people. It was indeed a great political deliverance; for had Vienna been taken, the Christian churches, as was the case at Constantinople, would have been converted into mosques; the tide of conquest would have rolled to the Rhine, and the power of Louis XIV. been the only opposing mound to the swell of Mussulman ambition. In 1797, V. was menaced by the French, but the treaty of Leoben saved it from the impending peril. In 1806, and again in 1809, it was taken by the French. In 1848, it was the scene of a revolutionary rising, which was for the moment successful. It was ravaged by pestilence in 1679 and 1713; and by cholera in 1831-2. See for further details the historical paragraphs of AUSTRIAN EMPIRE.

VIENNA, a village of Pickens co., in the state of Alabama, U. S., on the Tombigbee river, near the junction of the Sipsey, and 120 m. WNW of Montgomery. Pop. in 1850, 140.—Also a village of Dooly co., in the state of Georgia, on a branch of Flint river, and 78 m. SSW of Milledgeville.—Also a village of Johnson co., in the state of Illinois, 167 m. S by E of Springfield, on Cash creek. Pop. 142.—Also a village of Scott co., in the state of Indiana, 68 m. S by E of Indianapolis, and near the Jeffersonville railway.—Also a township of Kennebec co., in the state of Maine, 15 m. NW of Augusta, and watered by a branch of Sandy river. Pop. in 1840, 891.—Also a village of Dorchester co., in the state of Maryland, on the W side of Nanticoke river, and 49 m. SE of Annapolis.—Also a township of Oneida co., in the state of New York, 112 m. WNW of Albany, bounded on the SW by Oneida lake. It has an undulating surface, and is drained by Fish creek. Pop. in 1840, 2,530; in 1850, 3,393.—Also a township of Trumbull co., in the state of Ohio, 144 m. NE by E of Columbus. It has a good soil, and is drained by several streams.—Also a village of Philps township, Ontario co., in the state of New York, on the outlet of Canandaigua lake, at the junction of Flint creek. Pop. in 1840, 1,400.

VIENNA, a village of Upper Canada, in the township of Bayham, on Otter creek, 3 m. N of Port Burwell, in a hilly country. Pop. 300.

VIENNE, a river of France, which rises on the plateau of Millevache, in the Limousin; flows northward, passing Limoges, Confolens, and Chatellerault; and joins the Loire, on the l. bank, in the dep. of the Indre-et-Loire, 2 m. above Saumur, after a course of 200 m. It gives name to two departments, and is navigable at some distance above its influx into the Loire.

VIENNE, a department in the NW of France, formed out of the ancient prov. of Upper Poitou, and bounded on the N by the deps. of Indre-et-Loire, and Maine-et-Loir; on the E by the dep. of Indre; on the S by Haute-Vienne and Charente; and on the W by Deux-Sevres. It has a superficial extent of 2,600 sq. m., and a pop. in 1846 of 308,391; in 1852, 317,305. The surface is for the most part level, or varied only by low hills. The principal rivers are the Vienne, the Charente, the Dive, the Clain, and the Creuse. The soil is somewhat light and sandy, partly fertile, partly covered with heath. About one-third of the surface is arable. The principal productions are corn, pulse, potatoes, hemp, flax, and wine. Agriculture is backward; and the wine, though made in large quantities, is of inferior quality, and is generally manufactured into brandy. Chestnuts form a great part of the food of the lower orders. The dep. contains mines of iron and marble, but its manufactures are inconsiderable. The dep. is divided into the arrondissements of Chatellerault, Civray, Loudun, Montmorillon, and Poitiers. The cap. is Poitiers.

VIENNE, a town of France, on the r. bank of the Rhone, in the dep. of Isère, 45 m. WNW of Grenoble, on the railway from Lyons to Avignon. It is pleasantly situated on the declivity and base of cultivated hills; but stands on uneven ground, and has streets in general narrow and steep. It contains several public buildings, a cathedral, erected on an eminence, two good churches, a small theatre, a high school, and a museum. The pop. in 1846 was 18,610. It has manufactures of woollen, linen, hard-ware, leather, and coloured paper. There are here numerous Roman remains, several traces of aqueducts, and the piles of an ancient Roman bridge. V. was the seat of a church-council in the 14th cent., and has long been the see of an archbishop. Vines are cultivated in the neighbourhood, which produce excellent wine known as *Cote rotie*.

VIENNE (HAUTE), a department in the W of France, including the greater part of the Limousin; bounded on the N by the dep. of Vienne and Indre; on the E by Creuse; on the S by Correze and Dordogne; and on the W by Charente. It is traversed by the river Vienne flowing northward to the Loire. It has a superficial extent of 2,260 sq. m., and a pop. in 1846, of 314,739; in 1852, of 319,379. This dep. is mountainous, and produces comparatively little corn, but has extensive pasturages. Its highest summit, Mont-Jargeau, has an alt. of 3,116 ft. above sea-level. Its forests also are extensive, and its game abundant. Its mineral products consist of marble, coal, iron, lead, and antimony. The manufactures, necessarily limited in extent, consist of coarse woollens, nails, paper, pottery, and wax. The dep. is divided into the four arrondissements of Limoges (the capital), Bellac, Rochechouart, and St. Yrieix.

VIENNE-LE-CHATEAU, a commune and town of France, in the dep. of the Marne, and arrond. of Sainte-Menehould and cant. of Ville-sur-Tourbe, on the Biesme. Pop. 1,766. It has extensive manufactories of hosiery, bottle-works, and bell founderies.

VIENNOIS, formerly a district of France, in the ancient prov. of Dauphiny, and now comprised in the departments of the Drôme and Isère. Its capital was Vienne.

VIENS, a commune of France, in the dep. of the Vaucluse, cant. and 13 m. E of Apt, on the r. bank of the Calavon. Pop. 1,207.

VIERA, a tribe of Samoïdes, of Russia in Europe, who inhabit the gov. of Arkhangel, along the Timani coast, to the W of Petchora and near Cape Sviatoï.

VIERBAUM, a village of Prussia, in the reg. of Dusseldorf and circle of Geldern. Pop. 350.

VIERET, or VIHRET, a village of Bavaria, in the circle of Upper Franconia, and bail. of Bamberg. Pop. 558. It has two synagogues.

VIERHOF, a village of Austria, in Bohemia, in the reg. and circle of Budweis. Pop. 210.

VIERHOFN, a village of Prussia, in the regency of Breslau, and circle of Glatz. Pop. 350.

VIERINGHAUSEN, a village of Prussia, in the reg. of Dusseldorf, and circle of Lennep. Pop. 220.

VIERITZ, a village of Prussia, in the regency of Magdeburg, and circle of Jerichow. Pop. 240.

VIERKIRCHEN, a village of Bavaria, in the circle of Upper Bavaria, and bail. of Dachan. Pop. 430.

VIERLANDE, a small district lying between the Elbe and the Bille, in the territories of Hamburg and Lubeck, and comprising the villages of Curslack, Alten-Gamm, Neuen-Gamm, and Kirchwarder, with a pop. of 10,000, who are chiefly engaged in rearing cattle and in market-gardening.

VIERLINGSBEEK, a parish of Holland, in the prov. of North Brabant. Pop. 1,150.

VIERMUNDEN, a village of Hesse-Darmstadt, in the prov. of Upper Hesse, and circle of Frankenberg. Pop. 550. It has a castle and two mills.

VIERNAU, a village of Prussia, in the regency of Erfurt, and circle of Schleusingen. Pop. 898.

VIERNHEIM, a town of Hesse, in the prov. of Starkenburg, and bail. of Lorch. Pop. 2,000.

VIERQUARTIREN, a district of Prussia, in the reg. of Dusseldorf, circle of Geldern. Pop. 1,670.

VIERRADEN, a town of Prussia, in the prov. of Brandenburg, circle and 14 m. ENE of Angermünde, on the Welse, near its junction with the Oder. Pop. 1,544. Tobacco is extensively cultivated in the environs.

VIERSEL, a department and commune of Belgium, in the prov. and arrond. of Antwerp. Pop. 500.

VIERSELDYKE, a commune of Belgium, in the prov. of Antwerp, and dep. of Viersel. Pop. 250.

VIERSEN, a town of Prussia, in the prov. of the Rhine, regency of Dusseldorf, circle and 4 m. NNW of Gladbach, near the Niers. Pop. 2,645. It has several churches belonging to different denominations, and possesses manufactories of silk, linen and cotton fabrics, printed muslin, velvet-gilt-ware, cotton-spinning-mills, bleacheries, tile-kilns, distilleries of brandy and vinegar.

VIERSET, a commune of Belgium, in the prov. of Liege, and dep. of Vierset-Barse. Pop. 320.

VIERSET-BARSE, a department and commune of Belgium, in the prov. of Liege, and arrond. of Huy, watered by the Hoyaux. Pop. 669.

VIERTEL, a village of Austria, in Bohemia, in the reg. of Pilsen, and circle of Taus. Pop. 245.

VIERVES, a commune and village of Belgium, in the prov. of Namur, arrond. of Philippeville. Pop. 540.

VIERWALDSTADTER-SEE. See LUCERNE (LAKE OF).

VIERZEHNHEILIGEN, a village of Saxe-Meiningen, in the bail. of Camburg, near the battle-field of Jena. Pop. 160.

VIERZEHNHOFEN, a village of Austria, in Bohemia, in the regency of Eger, and circle of Saatz. Pop. 240.

VIERZIGHUBEN, a village of Austria, in Moravia, in the gov. of Olmutz, and district of Zwittau. Pop. 1,100. It has a castle.

VIERZIGHUBEN (ALT and NEU), two villages of Prussia, in the reg. of Konigsberg, circle of Allenstein, containing respectively 240 and 220 inhabitants.

VIERZON, or VIERZON-VILLE, a canton, commune, and town of France, in the dep. of the Cher, and arrond. of Bourges. The cant. comprises 10 com. Pop. in 1831, 14,813; in 1846, 18,843.—The town is 21 m. NW of Bourges, on the r. bank of the Evre, near its confluence with the Cher, in a fertile plain. Pop. 5,679. It is tolerably well-built, and the greater number of the houses are roofed with slate. It has an almshouse, several tanneries, manufactories of parchment, pottery and porcelain, extensive iron and steel forges, and founderies, breweries, tile-works, &c., and carries on an active trade in wool and timber.

VIERZON-VILLAGE, a commune of France, in the dep. of the Cher, and cant. of Vierzon. Pop. in 1846, 4,547. It has manufactories of earthenware, porcelain, and of bricks, several flour-mills, and extensive iron-works.

VIESCAS, a town of Spain, prov. of Huesca, to the NW of Jaca, on the Gallego. Pop. 1,000.

VIESCH, or VIEGE, a market-town of Switzerland, in the cant. of the Valais, near the r. bank of the Rhone, on a torrent which descends from the glacier of Viesch, in the Bernese Alps, and 42 m. ENE of Sion. Pop. 400.

VIESCHERHORNER, a summit of the Alps, between the cantons of Berne and Valais, and extending between the Grindelwald and Aletschgletscher. It has an alt. of 12,500 Parisian ft. above sea-level.

VIESEBECK, a village of Hesse-Darmstadt, prov. of Lower Hesse, and circle of Wolfhagen. Pop. 400.

VIESECKE, a village of Prussia, in the regency of Potsdam, and circle of Westpriegnitz. Pop. 250

VIESELBACH, a village of the grand-duchy of Saxe-Weimar, and circle of Weimar-Jena, 90 m. WNW of Weimar. Pop. 540.

VIESLY, a commune of France, in the dep. of the Nord, and cant. of Solesmes, 12 m. from Cambrai. Pop. 2,761.

VIESSOIX, a commune of France, in the dep. of the Calvados, and cant. of Vassy, 5 m. E of Vire. Pop. 1,302.

VIESTI, a town of Naples, in the prov. of Capitanata, district and 47 m. NE of Foggia, on the Adriatic, by which it is bathed on three sides, and at the NE extremity of the promontory of Monte-Gargano, in N lat. 41° 54′ 10″, E long. 16° 10′ 10″. It is enclosed by walls, and has a fortress, a cathedral, and two convents. It has a small fishing port. This town, the *Apenestæ* of the ancients, had formerly a temple dedicated to Vesta. It has suffered much from earthquakes.

VIESVES, a department and com. of Belgium, in the prov. of Namur, arrond. of Dinant. Pop. 540.

VIESVILLE, a department and commune of Belgium, in the prov. of Hainault, and arrond. of Charleroi, and watered by the Pieton and Teinsean. Pop. 972. It has manufactories of cutlery and of nails, and several lime-kilns.

VIETEGEST, a village of the grand-duchy of Mecklenburg-Schwerin, circle of Wendischer, and bail. of Gustrow. Pop. 250.

VIETLOW, a village of Prussia, in the reg. of Koslin, and circle of Stolpe. Pop. 290.

VIETLUBBE, a village of the grand-duchy of Mecklenburg-Schwerin, circle of Wendischer, and bail. of Plau. Pop. 280.

VIETMANSDORF, a village of Prussia, in the regency of Potsdam, and circle of Zemplin. Pop. 240.

VIETNITZ, a village of Prussia, in the regency of Frankfurt, and circle of Konigsberg. Pop. 320.

VIETRI, a market-town of Naples, in the prov. of the Principato-Citra, district and 1½ m. W of Salerno, and about ¾ of a mile from the shore of the Tyrrhenian sea, on a hill. Pop. 2,110. It has a convent, with an hospital, and several fine private houses, and contains several paper-mills, a glass-work, and iron-forges. The harbour is small, but its trade, especially in fish, is considerable. V. probably occupies the site of the ancient *Marcina*.

VIETRI-DI-POTENZA, a market-town of Naples, in the prov. of Basilicata, district and 17 m. WSW of Potenza. Pop. 2,625. It has a convent, an hospital, and two charitable institutions. This town, which is of considerable antiquity, formerly bore the name of *Campi-Veteres*.

VIETZE, a village of Prussia, in the reg. of Frankfurt, circle and 17 m. WSW of Landsberg. Pop. 1,528. It has an iron-foundry.

VIETZIG, a village of Prussia, in the regency of Koslin, and circle of Lauenburg-Buton. Pop. 410.

VIETZNITZ, a village of Prussia, in the reg. of Potsdam, and circle of Westhavelland. Pop. 270.

VIETZOW, a village of Prussia, in the reg. of Koslin, and circle of Belgard, near the Persante. Pop. 260.

VIEURE, a village of France, in the dep. of the Allier, cant. and 11 m. SW of Bourbon-l'Archambault, on the Morgon. Pop. 830.

VIEUSSAN, a village of France, in the dep. of the Herault, cant. and 4 m. ESE of Olargues, near the l. bank of the Orbe, at the foot of a mountain, in which is an intermitting spring. Pop. 750. In the vicinity is a copper-mine.

*** VIEUX, VIELLE. Names with this prefix, not to be found amongst the following, are to be sought for under the second word.

VIEUX, a village of France, in the dep. of the Tarn and cant. of Castelnau-de-Montmirail, 8 m. N of Gaillac, near the Verre. Pop. 400. It was formerly noted for its monastery.

VIEUX-BERGUIN, a commune of France, in the dep. of the Nord and cant. of Bailleul. Pop. 3,642. It has an oil-mill and a brewery.

VIEUX-BOUCAN, a town of France, in the dep. of the Landes, and cant. of Soustans. Pop. 272. It has a small fishing-port. Wine of good quality is grown in the locality.

VIEUX-BOURG (LA), a commune of France, in the dep. of the Cotes-du-Nord, cant. and 16 m. W of Quintin. Pop. 1,215.

VIEUX-BRISACH. See BRISACH (VIEUX).

VIEUX-CHAMPAGNE, a village of France, in the dep. of the Ain, cant. and a little to the SE of Champagne. Pop. 528.

VIEUX-D'IZENAVE, a village of France, in the dep. of Ain, cant. of Brenod, 6 m. SSW of Nantua. Pop. 560.

VIEUX-GENAPPE, a commune and v. of Belgium, in the prov. of Brabant, arrond. of Nivelles. Pop. 1,153.

VIEUX-MAISON-LE-VIDAME, a village of France, in the dep. of Aisne, cant. of Charly, 9 m. S of Chateau-Thierry. Pop. 850.

VIEUX-MARCHE' (LE), a commune and v. of France, in the dep. of Cotes-du-Nord, cant. and 1 m. ESE of Plouaret. Pop. 3,500.

VIEUX-SAINT-REMY, a village of France, in the dep. of Ardennes, cant. of Nouvion-en-Porcien, 9 m. NNE of Rethel. Pop. 1,200.

VIEUXVY-SUR-COUESNON, a commune and village of France, in the dep. of Ille-et-Vilaine, cant. and 8 m. NE of Saint-Aubin-d'Aubigne. Pop. 1,900.

VIEVY, a village of France, in the dep. of Cote-d'Or, cant. and 6 m. S of Arnay-le-Duc. Pop. 1,300.

VIEVY-LE-RAYE', a village of France, in the dep. of Loir-et-Cher, cant. of Ouzouer-le-Marche, 21 m. N of Blois. Pop. 480.

VIF, a town of France, in the dep. of Isère, 10 m. S of Grenoble. Pop. 1,200. It has pottery-works and silk-mills.

VIG, a river of Russia, which rises in the gov. of Olonetz, in a lake of the same name; issuing from the NNW side of which, it flows in a NW and broken course of 170 m. to the White sea.

VIGAN, a port on the NW coast of the island of Luçon, in the Philippine group, at the mouth of the Avra. Pop. 18,000. It is an active trading-place.

VIGAN (LE), a town of France, in the dep. of Gard, arrond. and 42 m. WNW of Nismes, on the l. bank of the Arre. Pop. 4,600. It has manufactories of silks and cottons; and coal is wrought in the vicinity.—Also a village in the dep. of Lot, cant. and 3 m. E of Gourdon, on the Bleu.

VIGATTO, a commune and village of the duchy of Parma, 6 m. S of Parma. Pop. 4,500.

VIGEAN (LE), a village of France, in the dep. of Cantal, cant. and 2 m. ENE of Mauriac.—Also a village in the dep. of Vienne, cant. and 3 m. WSW of L'Ile-Jourdain. Pop. 1,200.

VIGEANS (SAINT), a parish of Forfarshire, on the coast, comprising part of the town of Arbroath. Area 9,385 Scotch acres. Pop. 8,793 in 1841; in 1851, 9,799.

VIGEDAL, a parish and village of Norway, in the laen and 36 m. NNE of Stavanger, on the N shore of the Bukkefiord.

VIGEIR (LE), a commune and village of France, in the dep. of Haute-Vienne, cant. and 6 m. S of Limoges, near the Briance. Pop. 1,800.

VIGENTINO, a village of Austrian Lombardy, 1 m. SE of Milan. Pop. 600.

VIGEOIS, a commune and village of France, in

the dep. of Correze, 15 m. NW of Tulle, near the Vezere. Pop. 2,400.

VIGEREN, an island off the W coast of Norway, in N lat. 62° 34′.

VIGEVANO, a town of the Sardinian states, in the div. of Novara, prov. of Lomellina, 10 m. NE of Mortara, in N lat. 45° 18′ 54″, on the Mora, an affluent of the Ticino. It is a comparatively well-built place, with fine cavalry barracks, a college, two hospitals, and several churches. It has manufactories of silks, hats, macaroni, and soap. Pop. 15,000.

VIGGIANELLO, a town of Naples, in the prov. of Basilicata, 3 m. NE of Rotonda. Pop. 3,200.

VIGGIANO, a commune and town of Naples, in the prov. of Basilicata, 24 m. S of Potenza. Pop. 5,700.

VIGGIN, a small village of Austrian Lombardy, in the prov. and 20 m. W of Cano. Pop. 2,100. Marble is wrought in the vicinity.

VIGIA, or SAO-JORGE-DOS-ALAMOS, a town of Brazil, in the prov. and 50 m. ENE of Para, on the r. bank of the Para. It has a considerable trade in cotton, coffee, cocoa, the produce of the environs.

VIGILIO (SANTO), a village of Austrian Lombardy, in the prov. and 7 m. NNW of Brescia, near the Mella. Pop. 600.—Also a village in the prov. and 18 m. NW of Verona, on a headland projecting into the Lago-di-Garda, in which it has a good port.

VIGNAC (LE), a village of France, in the dep. of Landes, cant. and 7 m. N of Castels. Pop. 900.

VIGNALE, a town of the Sardinian states, in the prov. and 6 m. S of Casale. Pop. 2,100.

VIGNANELLO, a town of the Papal States, in the deleg. and 9 m. ESE of Viterbo, near the Maggiore. Pop. 1,000

VIGNARA, a village of Naples, in Calabria-Ultra 2da, on the Alli.

VIGNEMALE, the most elevated summit of the French Pyrenees, in the dep. of Hautes-Pyrenees, on the Spanish frontier, between the Pic-de-Taillou and the Pic-de-Fontry, in N lat. 42° 26′ 29″. It has an alt. of 1,722 toises above sea-level, and is covered with glaciers.

VIGNETTE, a village of France, in the dep. of Marne, cant. of Sainte-Menehould. Pop. of com. 1,600. There are glass-works here.

VIGNEUELLES-LES-HATTONS-CHATEL, a village of France, in the dep. of Meuse, arrond. and 18 m. NNE of Commercy. Pop. 780.

VIGNEUX, a village of France, in the dep. of Loire-Inferieure, arrond. and 9 m. E of Savenay. Pop. 2,500.

VIGNOC, a village of France, in the dep. of Ille-et-Vilaine, cant. and 4 m. S. of Hedée. Pop. 1,000.

VIGNOLA, a village of the duchy of Modena, on the l. bank of the Panaro, 12 m. SSE of Modena.—Also a town of Naples, in the prov. of Basilicata, 6 m. S of Potenza. A large annual fair is held here. Pop. 4,000.

VIGNOLO, a commune and village of the Sardinian states, 5 m. SW of Coni. Pop. 1,200.

VIGNONE, a village of Tuscany, in the prov. of Sienna, 4 m. S of Pienza, near the Orcia.

VIGNORY, a village of France, in the dep. of Haute-Marne, arrond. and 12 m. N of Chaumont. Pop. 900.

VIGNOT, a village of France, in the dep. of Meuse, cant. and 1 m. NNE of Commercy. Pop. 750.

VIGNOUX - SUR - BARANJON, a village of France, in the dep. of Cher, cant. and 6 m. E of Vierzon. Pop. 900.

VIGNY, a village of France, in the dep. of Seine-et-Oise, arrond. and 9 m. WNW of Pontoise, on the l. bank of the Guiry. Pop. 500.

VIGO, a town of Spain, in the prov. and 15 m. S of Pontevedra, in N lat. 12° 15′, on the S coast of a bay of the same name, from the shore of which it rises in a series of terraced streets. It is an ill-built place, but has considerable commercial activity, being always touched-at by the peninsular steamers; its fisheries are also a source of considerable traffic. Its port is small but good; and it is well-defended by several forts. Pop. 5,500.—The bay of V., on the NW coast of Galicia, to the S of the bay of Pontevedra, is 20 m. in length from NE to SW, and 6 m. in its greatest breadth.

VIGO, a county in the W part of the state of Indiana, U. S. Area 408 sq. m. It is intersected by the Wabash, and has a level surface and fine soil. Pop. in 1840, 12,076; in 1850, 15,289. Its cap. is Terre-Haute.

VIGO (LAKE), a lake of Russia, on the NW of the gov. of Olonetz, to the NW of Lake Onega, and NE of Lake Sego. It is fed by the Vig, which enters it on the SE, and flows from it on the NNW.

VIGOLENO, a commune and v. of the duchy of Parma, 20 m. SE of Placentia. Pop. of com., 4,267.

VIGOLZONE, a commune and v. of the duchy of Parma, 10 m. S of Placentia. Pop. of com., 3,038. There are paper-mills here.

VIGONA, a town of the Sardinian states, in the prov. and 7 m. ESE of Pignerol, near the source of the Langiate. Pop. 6,000.

VIGTEN ISLANDS, a group of three small islands off the W coast of Norway, in N lat. 64° 46′.

VIGUERA, a village of Spain, in the prov. and 15 m. SW of Logrono. Pop. 1,300. It has some trade in charcoal, and in agricultural produce.

VIGUZZOLO, a town of the Sardinian states, in the prov. and 3 m. NE of Tortona. Pop. 2,100.

VIGY, a village of France, in the dep. of Moselle, arrond. and 8 m. NE of Metz. Pop. 850.

VIHIERS, a town of France, in the dep. of Maine-et-Loire, arrond. and 24 m. WSW of Saumur. Pop. 1,292.

VIJAYAPUR. See BEJAPUR.

VIKEN, a lake of Sweden, which is united with Lake Wener by a canal 35,860 metres in length. Its elevation above Lake Wener is 40 metres.

VIKKUR, a town of Sind, on the r. bank of the Hujamri branch of the Indus, in N lat. 24° 14′. It is a poor place, having been ruined by a change in the depth of the arm of the river on which it stands.

VIKTRING, a village of Austria, gov. of Laybach, circle and 3 m. SW of Klagenfurt. Pop. 250.

VILABERTRAU, a village of Spain, in the prov of Gerona, near Figueras. Pop. 900.

VILACH, a village of Spain, in the prov. of Lerida, in the valley of Aran, 2 m. from Viella. Pop. 265.

VILAFAFELA, a village of Spain, in the prov. and 18 m. NE of Zamora.

VILAFAMES, a town of Spain, in the prov. and 6 m. NNE of Castellon-de-la-Plana. Pop. 1,910. It has a fine church.

VILAGOS, or SIRI, a town of Hungary, in the com. and 15 m. NE of Arad. Pop. of district, 6,200. Wine is extensively made in the vicinity.

VILAGRASA, a village of Spain, in the prov and 24 m. ENE of Lerida. Pop. 545.

VILAINE, a river of France, which rises in the dep. of Mayenne, to the W of Ernée; flows W through the dep. of Ille-et-Vilaine, and enters that of Loire-Inferieure; and enters the Atlantic at Penerf after a course of about 126 m. Its principal affluents are the Canlache, the Chevre, the Ille, the Meu, and the Oust, on the r.; and the Seiche, the Semnon, the Cher, the Don, and the Isac on the l.

VILALONGA, a town of Spain, in the prov. and 7

m. from Taragona. Pop. 1,100. It has manufactories of cotton and woollen goods.

VILAMITJANA, a village of Spain, in the prov. of Lerida, 3 m. from Talarn. Pop. 420.

VILAMOS, a village of Spain, in the prov. of Lerida, in the Val-d'Aran, 6 m. N of Viella.

VILARODONA, a town of Spain, in the prov. and 12 m. NNE of Taragona, on the Rio-Gaya. Pop. 1,400. It has manufactories of brandy and of paper.

VILASA-DE-MAR, a village of Spain, in the prov. and 20 m. E of Barcelona, on the coast. It has manufactories of woollens and cottons, and conducts an active fishing-trade. Pop. 1,900.

VILASECA, a town of Spain, in the prov. and 6 m. SW of Taragona, on the coast. Pop. 4,000.

VILBEL, a town of Hesse-Darmstadt, on the l. bank of the Nidda, 6 m. N of Frankfort-sur-Main. Pop. 2,400.

VILCABAMBA, a river of Peru, which rises in the dep. of Cuzco, and flows E to the Paucartambo, which it joins on the l. bank, after a course from SE to NW of 300 m.

VILCHEZ, a town of Spain, in the prov. and 30 m. NNW of Jaen, on the S flank of the Sierra-Morena. Pop. 2,000. The vicinity is understood to be rich in minerals.

VILEIKA, a village of Russia, in the gov. and 52 m. NNW of Minsk, on the r. bank of the Viliia.

VILENA, a village of Spain, in the prov. and 30 m. from Burgos, on the r. bank of the Oca. Pop. 280.

VILHAC, a village of France, dep. of Dordogne, cant. and 6 m. NW of Terrasson. Pop. 1,000.

VILIIA, or VILIA, a river of Russia, which rises in the gov. of Minsk, runs E through that of Vilna; and flows into the Niemen, on the r. bank, near Kovno, after a generally W course of 240 m.

VILIKUI, a river of Asiatic Russia, which rises in the W part of the district of Jakutsk; runs E, passing Verkhne-Viliusk; and flows into the Lena.

VILKOMIR, a town of Russia, in the gov. and 50 m. NNW of Vilna, on the r. bank of the Svizta, in N lat. 55° 15'.

VILLA, a village of Switzerland, in the cant. of the Grisons, and bail. of Glenner. Pop. 229.—Also a village in the cant. of Ticino, and bail. of Lugano. Pop. 204.

VILLA, a market-town and harbour of New Granada, in the dep. of Istmo, and prov. of Veragua, at the mouth of a river of that name.—Also a town in the dep. of Cundinamarca, in the prov. and 15 m. N of Neiva.

VILLA-ALFONSINA, a market-town of Naples, in the prov. of Abruzzo-Citra, district and 9 m. NW of Il Vasto, in a fertile plain. Pop. 1,000. This town was founded by a colony of Sclaves towards the end of the 15th century.

VILLA-ALONSO, a market-town of Spain, in the prov. of Zamora, 8 m. NW of Toro, in a plain, on the Jaraniello. Pop. 690. It has an ancient fortress and an hospital. In the vicinity are quarries of freestone.

VILLA-ALPANDO, a town of Spain, in the prov. of Zamora. Pop. 2,600.

VILLA-ALVA, a market-town of Portugal, in the prov. of Alemtejo. Pop. 1,500.

VILLA-ANTONIA, a town of Brazil, in the prov. of São-Paulo, on the bay of Paranagua. Pop. about 4,000.

VILLA-BANEZ, a market-town of Spain, in the prov. and 12 m. ESE of Valladolid, in the valley of Cerrato, on the Jaraniello. Pop. 695.

VILLA-BARTOLOMEA, a village of Austrian Lombardy, in the del. of Verona, district and 5 m. SE of Legnago, on the r. bank of the Adige. Pop. 2,300. The trade consists chiefly in rice, of which large quantities are grown in the locality.

VILLABARUZ, a market-town of Spain, in the prov. and 30 m. from Valladolid, on the slope of a mountain, the base of which is washed by the Sequillo. Pop. 458. It has manufactories of hosiery.

VILLA-BASILICA, a town of the duchy of Lucca, in the district of Borgo, on the r. bank of the Pescia-di-Colladi, and 11 m. NE of Lucca.

VILLA-BELLA. See MATO-GROSSO.

VILLA-BELLA-DA-PRINCEZA, a small town of Brazil, in the prov. of São-Paulo, on the W side of the island of São-Sebastião, 96 m. E of São-Paulo. The district contains 3,000 inhabitants, who employ themselves chiefly in the culture of sugar and tobacco, and in the distillation of rum.

VILLABIANCA, a town of Sicily, in the prov. and district and 18 m. WSW of Palermo, near the Tyrrhenian sea. Pop. 2,500.—Also a town in the prov. of Girgenti, near Vivona. Pop. 3,000.

VILLA-BLANCA, a market-town of Spain, in the prov. and 78 m. W of Seville, in a mountainous locality. Pop. 1,100.

VILLA-BOA. See GOYAZ.

VILLABOIM, a market-town of Portugal, in the prov. of Alemtejo, comarca and 17 m. NE of Villa-Vicosa.—Also a market-town of Brazil, in the prov. of Para, and comarca of Mundrucania, on the l. bank of the Tapajoz, 45 m. from the Amazon, and at an equal distance SW of Santarem, in a fertile locality. It is inhabited chiefly by Indians.

VILLABONA, a market-town of Spain, in the prov. of Guipuzcoa, in a plain near the Oria, between Tolosa and Andoain. Pop. 358. It has two churches, an hospital, and several iron founderies and anchor-forges.

VILLA-BRAGIMA, a market-town of Spain, in the prov. and 18 m. from Valladolid. Pop. 1,510.

VILLABUENA, a village of Spain, in the prov. of Logrono, 6 m. SW of La-Guardia. Pop. 500.

VILLAC, a commune and town of France, in the dep. of Dordogne, cant. of Terrasson, on an affluent of the Vezere, 27 m. N of Sarlat. Pop. 1,170. Lignite and slate are wrought in the environs.

VILLACANAS, a town of Spain, prov. of Toledo, 15 m. NW of Alcazar-de-St.-Juan. Pop. 5,720. There are salt-marshes in the vicinity.

VILLACARI, a town of Peru, in the dep. of Lima, 30 m. E of Pisco.

VILLACARLOS, a market-town of Spain, in the island of Minorca, about 1 m. from Mahon, in a fertile locality. Pop. 4,200. It has a battery for the defence of the entrance to the port of Mahon. Soap is its chief article of manufacture. The environs are noted for their wine.

VILLACARRILLO, a town of Spain, in the prov. of Jaen, 18 m. NNE of Ubeda, near the Guadalquivir. Pop. 6,000.

VILLACASTIN, a market-town of Spain, prov. and 21 m. SW of Segovia, at the foot of the Sierra-de-Guadarama. Pop. 1,225. It has two convents.

VILLACH, a circle and town of Austria, in Illyria, in the NW part of the gov. of Laybach. The circle, formed from the W part of Carinthia, is bounded on the N by the archduchy of Austria and Styria; on the E by the circle of Klagenfurt; on the S by those of Laybach and Görz and Lombardy; and on the W by the Tyrol, comprising an area of 99·3 German sq. m. Pop. in 1837, 122,860; in 1843, 124,800.—The town, also called Belak, is 63 m. NW of Laybach, on the Drave, a little above the confluence of the Gail, in N lat. 46° 36' 50", and E long. 33° 50' 41". It is enclosed by walls, defended by a castle, and is well-built. It has two churches, and possesses manufactures of ceruse, litharge, vermillion, minium, and of iron-ware, and carries on an active transit-trade in iron, lead, and calamine. In the environs are several iron-forges, copper and lead-mines, marble quarries and mineral springs. V. was extensively damaged by an earthquake in 1348, and has since repeatedly suffered from fires.

VILLA-CIDRO, a market-town of Sardinia, in the prov. and 18 m. NE of Iglesias, at the foot of a high mountain. Pop. 6,800. Lead is wrought, and there are hot-springs, in the environs.

VILLA-CINTOR, a market-town of Spain, in the prov. and 18 m. E of Leon, on the r. bank of the Cea. Pop. 325. It has manufactories of coarse woollen fabrics.

VILLA-CLARA, a town in the central part of the island of Cuba, 162 m. ESE of La-Havana.

VILLA-COJA. See Coja.

VILLA-CONANCIO, a market-town of Spain, in the prov. and 18 m. ESE of Palencia, in one of the two valleys which form that of the Cerrato. Pop. 530. The locality produces in abundance, grain, honey, and medicinal plants.

VILLA-CONEJOS, a market-town of Spain, in the prov. and 24 m. N of Cuenca. Pop. 900. It has manufactories of cloth, and other woollen fabrics.—Also a market-town in the prov. of Toledo, 6 m. from Aranjuez, in a fertile plain. Pop. 746.

VILLACOURT, a village of France, in the dep. of the Meurthe, cant. and 3 m. SE of Bayon, on the l. bank of the Euron. Pop. 1,125.

VILLA-COVA, a market-town of Portugal, in the prov. of Beira, at the foot of the Serra-Estrella, and E of Coimbra.—Also a market-town in the prov. of Entre-Minho-e-Douro, near the coast. Pop. 3,200. It has a harbour and fisheries.

VILLA-CUPELLA, a market-town of Naples, in the prov. of Abruzzo-Citra, district and 3 m. SW of Il Vasto. Pop. 1,515. It was founded in the 15th cent. by a colony of Slaves.

VILLADA, a market-town of Spain, in the prov. and 27 m. NW of Palencia, on the r. bank of the Sequillo, which is here crossed by a bridge. Pop. 3,325. It has two churches, one convent, and a palace. Coarse woollen fabrics are manufactured here.

VILLA-DA-HORTA, a town of Portugal, in the Azores, capital of the island of Fayal, on a spacious bay on the SE coast. Pop. 4,000. It has a good port and an active trade. See Fayal.

VILLA-DA-IMPERATRIZ, a town of Brazil, in the prov. of Ceara, comarca of Baturite, 52 m. W of Fortaleza.

VILLA-DA-INDEPENDENCIA, a town of Brazil, in the prov. of Parahiba, comarca of Brejo-d'Area.

VILLA-DA-LAGUNA, a town of Portugal, in the Azores, on the SE coast of the island of Pico. It has a good port.

VILLA-DA-PRAYA. See Praya.

VILLA-DA-PRINCEZA, a district and town of Brazil, in the prov. of Rio-Grande-do-Norte, in a vast plain, on the l. bank of the Rio-das-Peranhas, 80 m. NE of Natal. Pop. of dist. 5,000.

VILLA-DEATI, a village of Sardinia, div. of Alessandria, prov. and 12 m. W of Casal. Pop. 2,600.

VILLA-DE-FRADES, a town of Portugal, prov. of Alemtejo, comarca and 18 m. NNE of Beja.

VILLA - DE - LA - PURIFICACION, a town of Mexico, in the state of Xalisco, 159 m. SW of Guadalaxara, and 24 m. from the shore of the Pacific.

VILLA-DE-LEON, a town of Mexico, in the state and 36 m. WNW of Guanaxuato. Pop. 1,200. It has a college, an hospital, and three convents. It has considerable manufacturing industry, and an active trade in corn.

VILLA-DEL-FUERTE, or Montes-Claros, a town of Mexico, in the state of Cinaloa, on the l. bank of the Rio-del-Fuerte, and 180 m. NW of Culiacan. Pop. 5,000.

VILLA-DEL-PRINCIPE, a town of Cuba, 30 m. SSE of Puerto-Principe, on the S flank of the central chain of mountains. Pop. 30,000.

VILLA-DEL-REY, a market-town of Spain, in the prov. and 27 m. WNW of Caceres. Pop. 710.—Also a town of Portugal, in the prov. of Estremadura, comarca and 15 m. E of Thomar.

VILLA-DEL-RIO, a town of Spain, in the prov. and 27 m. from Cordova, near the Guadalquivir. Pop. 3,500.

VILLA-DE-MONTELEONE, a town of Sardinia, prov. of Sassari, and SE of Alghero. Pop. 3,000.

VILLA-DE-MOR, a town of Spain, in the prov. and 18 m. from Leon. Pop. 880. It has an hospital.

VILLA-DE-NORDESTE, a town of Portugal, in the Azores, on the NE coast of the island of Miguel.

VILLA-DE-SAN-JUAN-BAPTISTA. See Tabasco.

VILLA-DE-VALLE-FERTIL, a town of La Plata, in the prov. of Cuyo, 165 m. NE of San Juan-de-la-Frontera.

VILLA-DI-CASTELLO, a town of Tuscany, in the prov. of Florence, vicariat and 1 m. SE of Petra-Santa. In the adjacent mountain of Argentiera are mines of silver and copper.

VILLA-DIEGO, a judicial partido and market-town of Spain, in the prov. of Burgos. The town is 27 m. NW of Burgos on the Brulls. Pop. 1,700. It has two churches, five chapels, a convent, an hospital, and a Latin school, but is ill-built, and contains little worthy of note. The trade, which is considerable, consists chiefly in grain.

VILLADIEZMA, a market-town of Spain, in the prov. of Palencia, 11 m. from Carrion-de-los-Condes. Pop. 515. It has distilleries of brandy and manufactories of linen and coarse woollen fabrics.

VILLADIN, a village of France, in the dep. of the Aube, cant. of Marcilly-le-Hayer, 18 m. W of Troyes. Pop. 400.

VILLA-DI-ROVIGNO, a village of Illyria, in the gov. of Trieste, circle of Istria, 4 m. NE of Rovigno.

VILLA-DI-SAN-GERMANO, a town of Naples, in the prov. of Terra-di-Lavoro, cant. of San-Germano. Pop. 1,200.

VILLA-DI-VILLA, a town of Austrian Lombardy, in the prov. of Padua, 4 m. SE of Este. Pop. 2,200.

VILLA-DO-BISPO, a town of Portugal, in the prov. of Algarva, comarca and 17 m. W of Lagos.

VILLA-DO-CONDE, a market-town of Portugal, in the prov. of Entre-Douro-e-Minho, and comarca of Barcellos, 14 m. NNW of Oporto, in a fertile and salubrious plain, about a mile above the embouchure of the Ave, on the Atlantic. Pop. 4,000. It has a collegiate church, two convents, and two hospitals. The harbour, accessible only to small vessels, is defended by a fort. In the environs are numerous gardens, and an extensive aqueduct. This town is said to have existed in the time of the Romans. It derives its present name from one of its ancient possessors Count Mendo-Paez-Rofino.

VILLA-DO-FANADO. See Bom-Successo.

VILLA-D'OGNA, a village of Austrian Lombardy, in the prov. of Bergamo, and district of Clusone, in the Val-Seriana, on the l. bank of the Serio, 24 m. NNE of Bergamo. Pop. 240. It has a saw and paper-mills, forges, and nailworks.

VILLA-DO-IMPERADOR, a town of Brazil, in the prov. and comarca of Parahiba. Sugar and cotton are the chief productions of the locality.

VILLA-DO-PRINCIPE, a town of Brazil, in the prov. of São-Pedro, 75 m. SSW of Curitiba, near the

l. bank of the Curitiba river. Pop. of district, 5,000. The environs are extremely fertile, producing varieties of European fruits, flax, millet, and pasturing large numbers of cattle, horses, and mules.

VILLA-DOSE, a village of Austrian Lombardy, in the prov. of Polesina, on the r. bank of the Adigetto, and 8 m. E of Rovigo. Pop. 1,200. The environs are noted for their wine.

VILLAESCUSA, a town of Spain, in the prov. of Zamora, and 18 m. from Toro, on a hill, in a mountainous locality. Pop. 2,030.

VILLAESCUSA-DE-HARO, a town of Spain, in the prov. of Cuenca, and 21 m. from S. Clemente. Pop. 2,000. It has two convents, and manufactories of soap and linen, and in the vicinity is a silver-mine. This town is supposed to occupy the site of one destroyed by Hannibal.

VILLAESTAR, a town of Spain, in the prov. and 5 m. SSW of Teruel, on the Guadalaviar. Pop. 400.

VILLAFAFILA, a town of Spain, in the prov. and 24 m. NNE of Zamora, in a marshy locality. Pop. 1,103. It has four parish churches, and in the vicinity is a salt-mine.

VILLA-FALETTO, a town of Sardinia, capital of a mandemento, in the div. and prov. and 6 m. N of Coni, near the r. bank of the Maira. Pop. 3,000.

VILLAFELICHE, a market-town of Spain, in the prov. of Calataqua, and 6 m. NW of Daroca, on r. bank of the Xiloca. Pop. 1,450. It has a large powder-mill, and manufactories of earthenware.—Also a market-town in the prov. of Castellon-de-la-Plana, to the NW of Morella. Pop. 2,000.—Also a market-town in the prov. of Leon. Pop. 3,000.—Also a market-town in the prov. of Pamplona. Pop. 3,000.

VILLA-FERNANDO, a town of Portugal, in the prov. of Alemtejo, comarca and 15 m. NE of Villa-Vicosa.

VILLA-FLOR, a town of Portugal, in the prov. of Tras-os-Montes, comarca and 12 m. NW of Torre-de-Moncorvo. Pop. 3,200. It is enclosed by walls, with four gates, and has a parish church, 12 chapels, an hospital, and numerous fountains. The trade consists chiefly in leather. Also a town in the prov. of Alemtejo, comarca and 27 m. NW of Portalegre, on the l. bank of the Tagus.—Also a town of the Canary Islands, in the island of Teneriffe, and partido of Orotava.

VILLA-FLOR, a town of Brazil, in the prov. of Rio-Grande-do-Norte, 36 m. SSE of Natal, near the Cunhau. Pop. of district, 2,500. It is built of lath, and with considerable regularity.

VILLA-FLORES, a town of Spain, in the prov. and 21 m. from Salamanca, in a fertile valley. Pop. 500.

VILLA-FORMOSA. See SERENHEN.

VILLA-FORTE. See ASSUMCAO (N. SENHORA D').

VILLAFRADES, a town of Spain, in the prov. and 21 NW of Palencia, on the Sequillo. Pop. 250.

VILLA-FRANCA, a town of Brazil, in the prov. of Para, on the bank of a lake which communicates with the Amazon on the r., and with the Tapajoz on the l., 18 m. SW of Santarem. Pop. of district in 1842, 4,000.—Also a town on the S coast of the island of St. Michael, in the Azores group, in N lat. 37° 42′, W long. 25° 29′. Pop. of town and district, 7,800.

VILLA-FRANCA, a market-town of Spain, in the prov. of Navarra, and 39 m. S of Pamplona, at the base of a mountain. Pop. 2,762. It has a parish-church, an hospital, and a custom-house. The environs are extremely fertile, and are noted for their wine.—Also a town of Tuscany, in the prov. of Florence, vicariat and 5 m. SW of Bagnone, near the l. bank of the Bagra. Pop. 3,000.—Also a town of Austria, in the gov. of Venice, deleg. and 9 m. SW of Verona. Pop. 5,000.—Also a town of Sicily, in the prov. and 30 m. NW of Girgenti, and district of Borona. Pop. 3,000. The locality is noted for its agates.

VILLA-FRANCA-DE-LA-MARISMA, a market-town of Spain, in the prov. and 15 m. from Seville. Pop. 3,527. On an adjacent height is a fine ducal palace.

VILLA-FRANCA-DE-LAS-ABUJAS, a market-town of Spain, in the prov. and 12 m. ENE of Cordova, on the r. bank of the Guadalquivir. Pop. 3,596. It has a parish-church, two custom-houses, a convent, and an hospital.

VILLA-FRANCA-DE-LA-SIERRA, a market-town of Spain, in the prov. and 33 m. N of Avila. Pop. 1,119. It has a parish-church, a custom-house, and a public granary, and possesses manufactories of coarse woollen fabrics.

VILLA-FRANCA-DE-LOS-BARROS, a market-town of Spain, in the prov. of Badajoz, and 21 m. S of Merida. Pop. 5,877. It has a parish-church, several convents, an hospital, a custom-house, and a public granary, and possesses numerous oil-mills.

VILLA-FRANCA-DE-LOS-CABELLEROS, a market-town of Spain, in the prov. and 45 m. SE of Toledo. Pop. 2,800. It has a church and a chapel, and manufactories of saltpetre and gunpowder.

VILLA-FRANCA-DEL-VIERZO, a judicial partido and town of Spain, in the prov. of Leon. The partido comprises 143 pueblos. The town is on the SW of Leon, on the road from Corunna to Madrid, and at the confluence of the Valcarce and Burbia. Pop. 2,996. It is of considerable antiquity, and contains a palace belonging to the Marquisate of the same name, several churches, three convents, and an hospital. Its industry consists chiefly in the manufacture of linen and woollen fabrics, and rearing of cattle.

VILLA-FRANCA-DE-MONTES-DE-OCA, a market-town of Spain, in the prov. and 21 m. E of Burgos, at the foot of the Oca mountains. Pop. 692. It has a fine parish-church, and a large hospital, and manufactories of linen and woollen.

VILLA-FRANCA-DE-PANADES, a judicial partido and town of Spain, in the prov. of Barcelona. The partido comprises 46 pueblos, 36 m. W of Barcelona, on the Tet, which is here crossed by a stone bridge. Pop. 5,516. It is enclosed by lofty walls, and contains a parish-church, several convents, an hospital, and barracks. The streets are narrow and ill-built. The distillation of brandy and manufacture of calico and of leather form the chief objects of local industry. This town derives its name from the franchises granted to it by one of the Counts of Toulouse, by whom it was subjected in 1000.

VILLA-FRANCA-DE-XIRA, a town of Portugal, in the prov. of Estremadura, and comarca of Ribatejo, on the r. bank of the Tagus, 32 m. NE of Lisbon. Pop. 4,700. It is the residence of the military of Ribatejo, and contains a parish-church, an almshouse, and an hospital. It has extensive salt-works on the Tagus, and a considerable trade. This town was founded in 1160 by the English, to whom it was ceded by D. Alphonso Euriquez, in return for their services in the conquest of Lisbon, and by whom it was named Cornuella, its present appellation being derived from the peculiar privileges it enjoyed.

VILLA-FRANCA-DI-PIEMONTE, a town of Sardinia, capital of a mandemento, in the div. of Turin, prov. and 17 m. SE of Pignerol, on the l. bank of the Po, on which it has the first port. Pop. 6,700. It is well built, and has a communal college and several convents.

VILLA-FRANCA-DO-CAMPO, a town of the Azores, in the island of St. Michael. Pop. 3,800. It has warm springs.

VILLAFRANQUEZA, a town of Spain, in the prov. of Alicante. Pop. 1,300.

VILLAFRECHOS, a market-town of Spain, in the prov. and 36 m. NW of Valladolid. Pop. 1,325.

VILLAFRUELA, a market-town of Spain, in the prov. and 36 m. SSW of Burgos. Pop. 400.

VILLAGARCIA, or SANTA-EULALIA-DE-AREA-LONGA, a market-town of Spain, in the prov. of Pontevedra, 30 m. NNE of Vigo, on the SE bank of the bay of Arosa, a little to the S of the embouchure of the Ulla. Pop. 1,443. It has a parish-church, a custom-house, a convent, and a palace belonging to the counts of Villagarcia, and possesses manufactories of cotton and linen fabrics, and a tannery. The port is good, and well-sheltered, and is capable of receiving vessels of war of small size. The adjacent fisheries are extremely productive.

VILLAGARGIA, a market-town of Spain, in the prov. of Badajoz, 3 m. from Llerena. Pop. 2,000. It contains a parish-church, a custom-house, and a public granary, and has manufactories of linen.—Also a market-town in the prov. and 18 m. from Valladolid. Pop. 1,145.—Also a market-town in the prov. of Cuenca, 24 m. ESE of S. Clemente. Pop. 1,480.

VILLA-GONZALO, a town of Spain, in the prov. of Badajoz, 9 m. from Merida, in a fine plain, bathed on the NW by the Guadiana. Pop. 1,360. It has distilleries of brandy.

VILLAGORDA-DE-XUCAR, a market-town of Spain, in the prov. of Albacete. Pop. 1,422.

VILLAGRA, a market-town of Spain, in the prov. and 42 m. NW of Valladolid, near the Valderaduey. Pop. 798. It has two parish-churches.

VILLA-HAMETE, a town of Spain, in the prov. and 39 m. from Leon, on the Valderaduey. Pop. 432.

VILLAHARTA. See VILLARTA.

VILLA-HERMOSA, a market-town of Spain, in the prov. and 30 m. NW of Castellon-de-la-Plana, on a mountain, the base of which is washed by a river of the same name, an affluent of the Mijares. Pop. 1,760. It has a parish-church, a custom-house, and a public granary. Its industry consists chiefly in the manufacture of bolting-cloth.—Also a town in the prov. of Ciudad-Real, 6 m. E of Villaneuva-de-los-Infantes. Pop. 2,600. It has a parish-church, an hospital, and a public granary.

VILLA-HERMOSA, a market-town of Mexico, in the state of Tabasco, to the N of Chiapas, on the Grijalva.—Also a town of Central America, in the state of Costa-Rica, about 9 m. W of Villa-Vieja. Pop. 4,000.

VILLA-HERREROA, a town of Spain, in the prov. of Palencia, 9 m. from Carrion-de-los-Condes, near the Cieza. Pop. 500. It has manufactories of linen.

VILLA-HORNATE, a town of Spain, in the prov. and 27 m. S of Leon, on the Esla, in a plain. Pop. 610. Its industry consists chiefly in the manufacture of linen and coarse woollen fabrics, and in the rearing of cattle.

VILLA-HOZ, a town of Spain, in the prov. and 24 m. SSW of Burgos. Pop. 1,090.

VILLAHUMBRALIS, a market-town of Spain, in the prov. and 6 m. from Palencia, on the W bank of the Canal-de-Campos, which is here crossed by a stone-bridge. Pop. 1,210. It has a parish-church, a convent, and a custom-house.

VILLAINE-EN-DUEMOIS, a town of France, in the dep. of the Côte-d'Or, and cant. of Baigneux, 14 m. S of Chatillon-sur-Seine. Pop. 784.

VILLAINES, a commune of France, in the dep. of the Indre-et-Loire, and cant. of Aizey-le-Indian. Pop. 1,102. Its industry consists chiefly in the culture of the willow, and in the manufacture of basket-ware.

VILLAINES-LA-JUHEL, a canton, commune, and town of France, in the dep. and arrond. of the Mayenne. The cant. comprises 10 com. Pop. in 1831, 12,256; in 1846, 13,134. The town is 18 m. E of Mayenne. Pop. in 1846, 2,508. It has a considerable trade in grain and cattle.

VILLAINE-SOUS-LUCE', a village of France, in the dep. of the Sarthe, cant. and 1 m. E of Grand Luce. Pop. 1,260.

VILLAJOYOSA, a judicial partido and town of Spain, in the prov. of Alicante. The partido comprises 6 pueblos. The town is 18 m. NE of Alicante, on the slope of a hill, which descends to the Mediterranean. Pop. 8,097. It has a small port, defended by a massive tower, a parish-church, an hospital, a convent, a public granary, building-docks, and an active coasting-trade. The ramparts of this town were destroyed during the war of succession.

VILLAJUAN, a town of Spain, in the prov. of Pontevedra, on the Arosa. Pop. 1,967. It has a parish-church and a custom-house. Its industry consists chiefly in sardine fishing, and in the manufacture of linen.

VILLALACO, a town of Spain, in the prov. and 18 m. NE of Palencia, on the Pisuerga. Pop. 600.

VILLALBA, a village of Spain, in the prov. and 18 m. NNW of Lugo, on the Magdalena, over which a bridge is here thrown. Pop. 500.—Also a v. in the prov. of Taragona, 30 m. NNW of Tortosa.

VILLALBA-DE-ADAJA, a village of Spain, in the prov. and 18 m. S. of Valladolid, on the Adaja.

VILLALBA-DE-DUERO, a village of Spain, in the prov. and 48 m. S of Burgos, near the r. bank of the Duero. Pop. 500.

VILLALBA-DE-LA-LAMPREANA, a village of Spain, in the prov. and 15 m. from Zamora, near the Salado. Pop. 500.

VILLALBA-DEL-ALCOR, a town of Spain, in the prov. and 27 m. WSW of Seville. Pop. 2,400. Numerous Roman remains occur in the vicinity.—Also a walled town in the prov. and 21 m. SE of Valladolid. Pop. 1,400, chiefly charcoal-burners.

VILLALBA-DE-LA-SIERRA, a village of Spain, in the prov. and 9 m. from Cuenca, near the Jacar. Pop. 200. There are mineral baths here.

VILLALBA-DE-LOS-LLANOS, a village of Spain, in the prov. and 18 m. SW of Salamanca. Pop. 300.

VILLALBA-DEL-REY, a town of Spain, in the prov. of Cuenca, 15 m. NE of Huete, on the Guadamejuz. Pop. 1,200.

VILLALBA-DE-RIOJA, a village of Spain, in the prov. and 48 m. N of Burgos. Pop. 300.

VILLALBARO, a village of Spain, in the prov. and 60 m. N of Burgos. Pop. 300.

VILLALBILLA, a village of Spain, in the prov. and 18 m. ENE of Madrid. Pop. 400.

VILLALCAMPO, a village of Spain, in the prov. and 18 m. W of Zamora, near the confluence of the Esla and the Duero. Pop. 800.

VILLALCAZAR-DE-SIRGA, a village of Spain, in the prov. and 18 m. N of Palencia. Pop. 500.

VILLALCON, a village of Spain, prov. of Palencia, 12 m. from Carrion-de-los-Condes. Pop. 240.

VILLALIER, a village of France, dep. of Aude, arrond. and 7 m. NE of Carcassonne. Pop. 450.

VILLALOBON, a village of Spain, in the prov. and 2 m. from Palencia. Pop. 500.

VILLALOBOS, a town of Spain, in the prov. and 45 m. from Leon. Pop. 1,480.

VILLALON, a town of Spain, in the prov. and

36 m. NW of Valladolid. It has several handsome churches and public buildings, and is a place of considerable manufacturing activity, having manufactories of chocolate, cheese, and oil; also linen and woollen factories. Pop. 4,700.

VILLALPANDO, a town of Spain, in the prov. and 36 m. WNW of Valladolid. Pop. 2,700. It is situated on an extensive but ill-cultivated plain; and has a trifling trade in nuts, wool, and wine.

VILLALPARDO, a village of Spain, in the prov. and 72 m. SE of Cuenca. Pop. 800.

VILLALTA, a village of Mexico, in the state and 72 m. NE of Saxaca.

VILLALUENGA, a village of Spain, in the prov. and 15 m. NW of Calatayud, on the l. bank of the Monubles. Pop. 650.—Also a village in the prov. and 15 m. E of Toledo.—Also a village in the prov. of Malaga, 3 m. from Grazalema.

VILLALUMBROSO, a village of Spain, in the prov. of Palencia, 15 m. from Carrion-de-los-Condes, on the Val-d'Eginate. Pop. 500.

VILLALVA, a village of Portugal, in the prov. of Alemtejo, comarca of Beja, 21 m. S of Evora.

VILLALVERNIA, a village of the Sardinian states, in the prov. and 4 m. SSW of Tortona, on the l. bank of the Scrivia.

VILLA-MAGNA, a town of Naples, in Abruzzo-Citra, cant. and 4 m. SE of Chieti. Pop. 2,000. It has a fine church.—Also a village of Naples, in Principato-Ultra, 7 m. NW of San-Angelo-de-Lombardi. Pop. 870.

VILLAMALEA, a town of Spain, in the prov. of Cuenca, 30 m. S of San-Clemente. Pop. 2,000.

VILLAMANAN, a town of Spain, in the prov. and 18 m. S of Leon, near the Esla. Pop. 1,860.

VILLAMANRIQUE, a town of Spain, in the prov. and 21 m. SW of Seville. Pop. 2,000.

VILLAMANRIQUE-DE-MONTIEL, a town of Spain, in the prov. and 60 m. ESE of Ciudad-Real, on the S flank of the Sierra-Morena. Pop. 1,100. Copper is wrought in the vicinity.

VILLAMANRIQUE-DE-TAJO, a village of Spain, in the prov. and 36 m. SE of Madrid, on the l. bank of the Tagus. Pop. 400.

VILLAMANTA, a village of Spain, in the prov. and 21 m. SW of Madrid. Pop. 300.

VILLAMANTELLA, a village of Spain, in the prov. and 21 m. SW of Madrid. Pop. 400.

VILLA-MARIA, a town of Brazil, in the prov. of Mato-Grosso, comarca of Cuiaba, in S lat. 16° 3′ 33″, on the l. bank of the Paraguay, 60 m. NE of Cuiaba. Pop. 1,000. The surrounding territory is fertile and well-cultivated.

VILLA MARTIN, a town of Spain, in the prov. of Cadiz, 42 m. S of Seville, at the confluence of the Guadalete and the Serrecin. Pop. 2,800. The largest cattle-fairs in Andalusia are held here.—Also a v. in the prov. and 6 m. W of Palencia. Pop. 300.

VILLA-MARTIN-DE-SAN-SANCHO, a village of Spain, in the prov. and 27 m. E of Leon, on the r. bank of the Cea. Pop. 320.

VILLA-MASSARGIA, a commune and v. of the island of Sardinia, 6 m. SE of Iglesias, in a marshy plain. Pop. of com. 2,000.

VILLA-MAYOR-DE-CALATRAVA, a town of Spain, in the prov. and 18 m. SSW of Ciudad-Real. Pop. 1,200.

VILLA-MAYOR-DE-CAMPOS, a town of Spain, in the prov. and 40 m. NW of Valladolid. Pop. 2,500, chiefly muleteers.

VILLA-MAYOR-DE-LOS-MONTES, a village of Spain, in the prov. and 18 m. from Burgos. Pop. 780.

VILLA-MAYOR-DE-SANTIAGO, a town of Spain, in the prov. and 45 m. ENE of Toledo. Pop. 2,600.

VILLAMEDIANA, a town of Spain, in the prov. and 3 m. from Logrono, in a plain watered by the Iregua. Pop. 1,200.—Also a town in the prov. and 9 m. S of Palencia. Pop. 1,300.

VILLAME'E, a commune and v. of France, in the dep. of Ille-et-Vilaine, cant. and 7 m. W of Louvigne-du-Desert, on the Beuvron. Pop. 2,400.

VILLAMENIL, a village of Spain, in the prov. and 39 m. N of Palencia. Pop. 200.

VILLA-MESIA, a village of Spain, in the prov. of Badajoz; 12 m. SW of Truxillo. Pop. 800.

VILLAMIEL, a market-town of Spain, in the prov. of Salamanca, 27 m. from Ciudad Rodrigo, at the foot of a mountain.—Also a market-town in the prov. and 9 m. N of Toledo. Pop. 622. It has an hospital.

VILLAMIZAR, a market-town of Spain, in the prov. and 24 m. E of Leon, on the r. bank of the Cea. Pop. 410. It has manufactories of linen and woollen fabrics.

VILLAMOR-DE-LOS-ESCUDEROS, a market-town of Spain, in the prov. and 21 m. SE of Zamora, in a picturesque but insalubrious locality. Pop. 1,055.

VILLAMUELAS, a market-town of Spain, in the prov. of Toledo, and 15 m. from Olana. Pop. 540. The manufacture of mats forms the chief branch of local industry.

VILLAMUERA, a market-town of Spain, in the prov. of Palencia, and 6 m. from Carrion-de-los-Condes. Pop. 432.

VILLAMUNIO, a town of Spain, in the prov. and 21 m. SE of Leon, in a fertile plain. Pop. 200. It has manufactories of linen and other fabrics.

VILLAMURIEL, a market-town of Spain, in the prov. and 3 m. S of Palencia, near the Castile canal. Pop. 755.

VILLAN-DE-PALENZUELA, a market-town of Spain, in the prov. and 18 m. SE of Palencia. Pop. 774. Hosiery is extensively manufactured here, and in the environs are quarries of gypsum and free-stone. In the vicinity is the fine convent of San-Salvador-del-Moral.

VILLANDRAUD, or VILLANDRADE, a canton, commune, and town of France, in the dep. of the Gironde, and arrond. of Bazas. The cant. comprises 7 com. Pop. in 1831, 8,702; in 1846, 9,085. The town is 9 m. NW of Bazas, on the r. bank of the Ciron. Pop. 885. It has a glass-work, and is noted as the birthplace of Pope Clement V.

VILLANDRY, a commune of France, in the dep. of the Loire, cant. and 10 m. W of Tours, on the l. bank of the Cher. Pop. 1,026.

VILLANOVA, a town of Sardinia, in the prov. and 24 m. from Alghero. Pop. 3,200.—Also a town, capital of a mandemento, in the prov. of Coni, 7 m. SW of Mondovi, on the Loriggia. Pop. 3,000.

VILLA-NOVA, a town of Portugal, in the prov. of Beira, and comarca of Guarda.

VILLA-NOVA, a township of Chautauque co., in the state of New York, U. S., 280 m. W of Albany, drained by Conewango creek. Pop. in 1850, 1,536.

VILLA-NOVA, a town of Brazil, in the prov. of Rio-de-Janeiro, on the l. bank of the Macacu, 9 m. above its confluence with the bay of Nitherohi. Pop. 2,000. Hats are here extensively manufactured.—Also a town in the prov. of Santa-Catherina, 45 m. S of Nostra-Santa-do-Desterro, to the S of Mount Tapiruva, and N of Lake Panema, with two ports. It has extensive fisheries, and the surrounding district, which is very fertile, produces in great abundance flax, millet, and mandioc.—Also a town of Brazil, in the prov. of Para, in Guayana, on the

l. bank of the Anarapucu, 21 m. above its confluence with the Amazon.

VILLA-NOVA-DA-ASSEMBLEA, a district and town of Brazil, in the prov. of Alagoas. Pop. of district, 1,200.

VILLA-NOVA-D'ASTI, a town of Sardinia, capital of a mandemento, in the prov. of Asti, 14 m. ESE of Turin. Pop. 2,800.

VILLA-NOVA-DA-BARONIA, a town of Portugal, in the prov. of Beira, and comarca of Evora.

VILLA-NOVA-DA-CERVEIRA, a town of Portugal, in the prov. of Entre-Douro-e-Minho, and comarca of Viana, 8 m. SW of Valentia, on the l. bank of the Minho, in a fertile country, near the frontier of Galicia. Pop. 1,148. It lies at the base of a height crowned by a castle, and is fortified. It contains a convent and an hospital, and has extensive fisheries.

VILLA-NOVA-DA-FORMIGA. See FORMIGA.

VILLA-NOVA-DA-PRINCEZA, a town of Brazil, in the prov. of Rio-Grande-do-Norte, 135 m. WNW of Natal, on the l. bank of the Rio-dos-Paranhes, and 21 m. from its entrance into the Atlantic. It has considerable trade in cattle and in salt.

VILLA-NOVA-DA-RAINHA, a district and town of Brazil, in the prov. of Bahia, comarca and 60 m. N of Jacobina, and near the Itapicuru-Mirim. The houses are small, and constructed chiefly of wood. The inhabitants of the district, about 2,000 in number, employ themselves chiefly in rearing cattle, in the culture of cotton, millet, rice, mandioc, and sugar, and manufacture of rum. Rock-crystal and saltpetre are found in the locality.—Also a small town in the prov. of Para, at the confluence of the Mauhé or Canoma with the Amazon. It is inhabited by Indians of the Mauhé tribe, who employ themselves in agriculture.

VILLA-NOVA-DE-ANCOS, a market-town of Portugal, in the prov. of Beira, comarca and 12 m. SW of Coimbra, on the r. bank of the Anzos. Pop. 900. It has a church, two chapels, and two hospitals.

VILLA-NOVA-DE-FAMELICAO, a town of Portugal, in the prov. of Entre-Douro-e-Minho, and comarca of Barcellos, 9 m. S of Braga, in a low, woody locality. Pop. 700. It has a palace belonging to the dukes of Braganza.

VILLA-NOVA-DE-FOSCOA, a market-town of Portugal, in the prov. of Beira, and comarca of Trancoso, 9 m. SSW of Torre-de-Moncorvo, near the confluence of the Coa and Douro.

VILLA-NOVA-DE-GAYA, or VILLA-NOVA-DE-PORTO, a town of Portugal, in the prov. of Entre-Douro-e-Minho, and comarca of Porto, on the l. bank of the Douro, opposite Porto, of which it is considered a suburb. Pop. 5,300. The streets, with the exception of one parallel to the river, are steep and tortuous, and the houses are generally mean-looking. It contains two convents, and two hospitals, and in the vicinity is a fine Augustine convent. Fishing and boat-building form the chief branches of local industry.

VILLA-NOVA-DE-GOA. See GOA.

VILLA-NOVA-DE-LAS-AVELLANAS, a town of Spain, in the prov. and 27 m. from Lerida, in a fertile plain. Pop. 352. It has saline springs.

VILLA-NOVA-DEL-REY, a town of Brazil, in the prov. of Ceara, 60 m. SE of Villavicosa, on the Monte-dos-Cocos, a summit of the Serra-Hybiappaba, and near the source of the Macambyra. It contains about 50 dwellings earth-built, and covered with palm-leaves. The culture of mandioc and millet, and rearing of cattle, form the chief objects of local industry.—Also a market-town in the prov. and 36 m. NE of Para, and district of Xingutania, 12 m. SSE of Cape Tigioca, at the head of a bay of the same name, in S lat. 0° 42′ 30″, and W long. 47° 51′ 52″.

VILLA-NOVA-DE-MIL-FONTES, a market-town of Portugal, in the prov. of Alemtejo, comarca and 39 m. W of Ourique, on the Atlantic, at the mouth of the Odemira, which here forms a small port. Pop. 1,858. It has two hospitals.

VILLA-NOVA-DE-MONTELEONE, a village of Sardinia, in the prov. and 8 m. SE of Alghero. Pop. 3,000.

VILLA-NOVA-DE-PORTIMAO, a market-town of Portugal, in the prov. of Algarve, comarca and 15 m. NE of Lagos, on the Portimão, near the Atlantic. Pop. 3,200. It is well-built, and has two hospitals, a college, and a convent. The harbour, which is defended by two forts, is spacious, but wanting in depth.

VILLA-NOVA-DE-PORTO. See VILLA-NOVA-DE-GAYA.

VILLA-NOVA-DE-SAN-JOSE-D'EL-REY, a town of Brazil, in the prov. and district of Rio-de-Janeiro, 6 m. SSW of Macacu, 1½ m. from the l. bank of the Macacu, and about 3 m. from the sea, in N lat. 22° 45′, and W long. 42° 43′.

VILLA-NOVA-DE-SANTE-CRUZ, a town of Brazil, in the prov. and 510 m. WSW of Para, on the l. bank of the Tapajoz, about 12 m. above Pinhel, and almost opposite Aveyro. It is small and ill-built. The inhabitants are Indians of the Mundurucus. Cacao is the chief production of the locality.

VILLA-NOVA-DE-SANTO-ANTONIO, a district and town of Brazil, in the prov. and 90 m. NE of Sergipe-d'El-Rey, on a height commanding the r. bank of the São Francisco, and about 30 m. from the Atlantic. It has three churches, and a Latin school. The district contains about 4,000 inhabitants, chiefly Indians.

VILLA-NOVA-DE-SAO-JOAO-DE-CACHOEIRA. See CACHOEIRA.

VILLA-NOVA-DE-SAO-LUIZ. See GUARATUBA.

VILLA-NOVA-DE-SOUZA, a town of Brazil, in the prov. and 255 m. WNW of Parahiba, on the l. bank of the Peixe, in a plain, near the foot of the Serra-do-Commissario. The houses are earth-built. It has a church and a school, and is advantageously situated for trade. The district contains 5,000 inhabitants.

VILLA-NOVA-DO-PRINCIPE, a town of Brazil, in the prov. of Rio-Grande-do-Norte, 188 m. WSW of Natal, on the Rio-Serido.

VILLA-NOVA-DOS-CARDIGOS, a town of Portugal, in the prov. of Alemtejo, comarca and 26 m. NW of Crato. Pop. 1,100.

VILLANTERIO, a town of Austrian Lombardy, in the prov. and 12 m. ENE of Pavia, on the Lambro. Pop. 2,000.

VILLANUBLA, a town of Spain, in the prov and 6 m. from Valladolid. Pop. 1,200.

VILLANUEVA-DE-ALCOLEA, a town of Spain, in the prov. of Castellon-de-la-Plana, 30 m. SE of Morella. Pop. 1,200.

VILLANUEVA-DE-ALCORON, a village of Spain, in the prov. and 36 m. N of Cuenca. Pop. 600.

VILLANUEVA-DE-ALGECILLA, a village of Spain, in the prov. and 18 m. S of Guadalaxara, near the Rio-Henares. Pop. 250.

VILLANUEVA-DE-AROSA, a village and port of Spain, in the prov. and 18 m. N of Vigo, in the Val-de-Salnes, on the S coast of the bay of Arosa. Pop. 780.

VILLANUEVA-DE-BOGOS, a village of Spain, in the prov. of Toledo, 9 m. from Ocana. Pop. 350

VILLANUEVA-DE-CAMEROS, a village of Spain, in the prov. and 30 m. from Soria, on the r. bank of the Iragua. Pop. 700.

VILLANUEVA-DE-CANEDO, a village of Spain, in the prov. and 10 m. from Salamanca. Pop. 300.

VILLANUEVA-DE-CASTELLON, a town of Spain, in the prov. and 6 m. S of San-Felipe. Pop. 2,000.

VILLANUEVA-DE-CORDOVA, a town of Spain, in the prov. and 42 m. NNE of Cordova, at the foot of the Sierra-Morena. Pop. 5,300. It has a trade in agricultural produce, and pigs.

VILLANUEVA-DE-CUBELLAS-Y-LA-GEL-TRU, a town and small port of Spain, in the prov. of Taragona, 26 m. SW of Barcelona. Pop. 9,000. It has manufactories of soap, paper, hats, and brandy, and some boat-building yards.

VILLANUEVA-DE-DUERO, a village of Spain, in the prov. and 15 m. SSW of Valladolid, on the l. bank of the Duero. Pop. 300.

VILLANUEVA-DE-GOMEZ, a village of Spain, in the prov. and 18 m. N of Aorla. Pop. 790.

VILLANUEVA-DE-GUADAMEJUZ, a village of Spain, in the prov. of Cuenca, 9 m. from Huete. Pop. 400.

VILLANUEVA-DE-LA-FUENTE, a town of Spain, in the prov. of Cuidad-Real, 15 m. NW of Alcaraz. Pop. 2,300. Iron is mined in the vicinity.

VILLANUEVA-DE-LA-HUERVA, a village of Spain, prov. and 24 m. SSW of Saragossa. Pop. 900.

VILLANUEVA-DE-LA-REYNA, a town of Spain, in the prov. and 18 m. NW of Jaen, on the Guadalquivir. Pop. 1,800.

VILLANUEVA-DEL-ARISCAL, a town of Spain, prov. and 12 m. from Seville. Pop. 1,100.

VILLANUEVA-DEL-ARZOBISPO, a town of Spain, in the prov. and 45 m. NE of Jaen, on the Guadalquivir. Pop. 3,500. It has oil and flour-mills.

VILLANUEVA-DE-LA-SAGRA, a town of Spain, prov. and 15 m. N of Toledo. Pop. 1,100.

VILLANUEVA-DE-LA-SERENA, a town of Spain, in the prov. and 50 m. E of Badajoz, near the l. bank of the Guadiana. Pop. 7,000. It has a trade in wine and linens.

VILLANUEVA-DE-LA-SIERRA, a village of Spain, prov. of Caceres, on the W side of the Sierra-de-Dios-Padre, 15 m. from Cura. Pop. 1,400. It has a trade in oil and wine.

VILLANUEVA-DE-LAS-MANZANAS, a village of Spain, in the prov. and 9 m. from Leon. Pop. 300.

VILLANUEVA-DE-LA-VERA, a town of Spain, in the prov. of Caceres, 42 m. E of Plasencia, near the Laguna-de-Gredoz. Pop. 2,200.

VILLANUEVA-DE-LA-XARA, a town of Spain, in the prov. and 52 m. SSE of Cuenca, on the Valdemembra. Pop. 2,200, chiefly employed in agriculture.

VILLANUEVA-DEL-CAMPELLO, a village of Spain, prov. and 24 m. from Aville.

VILLANUEVA-DEL-CAMPO, a town of Spain, in the prov. and 9 m. W of Leon. Pop. 2,400. It has well-frequented agricultural fairs.

VILLANUEVA-DEL-CARDETE, a town of Spain, in the prov. and 52 m. ESE of Toledo, near the Giguales. Pop. 1,500. It is a comparatively well-built place, and has a traffic in agricultural produce.

VILLANUEVA-DEL-CONDE, a town of Spain, prov. and 48 m. S of Salamanca. Pop. 1,400.

VILLANUEVA-DEL-DUQUE, a town of Spain, in the prov. and 30 m. NNW of Cordova. Pop. 1,700.

VILLANUEVA-DEL-FRESNO, a town of Spain, in the prov. and 50 m. SSW of Badajoz, near the Portuguese frontier. Pop. 1,800. It has a trade in iron and cattle.

VILLANUEVA-DE-LOS-CABALLEROS, a village of Spain, in the prov. and 24 m. WNW of Valladolid, on the Sequillo. Pop. 500.

VILLANUEVA-DE-LOS-CASTILLEJOS, a town of Spain, in the prov. and 70 m. W of Seville. Pop. 3,000. It has manufactories of metal buttons, hats, and leather.

VILLANUEVA-DE-LOS-INFANTES, a town of Spain, in the prov. and 48 m. ESE of Ciudad-Real, in a fine and fertile plain. Pop. 17,000. It is a well-built place, with regular streets, and several handsome public buildings. Its manufactures consist of oil, pottery, soap, woollens, leather, and brandy.—Also a village of Spain, in the prov. and 12 m. E of Orense, in the valley of the Celanova.

VILLANUEVA-DEL-REY, a village of Spain, in the prov. and 36 m. NW of Cordova. There are important coal-mines here.—Also a village in the prov. and 18 m. NW of Murcia.

VILLANUEVA-DE-MESIA, a village of Spain, prov. and 18 m. W of Grenada, on the r. bank of the Genil. Pop. 800.

VILLANUEVA-DE-SAN-CARLOS, a village of Spain, prov. and 24 m. from Ciudad-Real. Pop. 600.

VILLANUEVA-DE-SAN-MARCIO, a village of Spain, prov. and 30 m. from Valladolid. Pop. 400.

VILLANUEVA-DE-SAN-MARCOS, a town of Spain, in the prov. of Cordova, 18 m. N of Antequera, near the Genil. Pop. 4,000.

VILLANUEVA-DE-TAPIA, a village of Spain, prov. and 36 m. NNE of Malaga, on the Cereyo. Pop. 980.

VILLANUEVA-DE-VALDEJAMUZ, a village of Spain, in the prov. and 30 m. from Leon.

VILLA-NUOVA, a village of Austrian Italy, in the prov. of Brescia, 4 m. WSW of Salo. Pop. 400.

VILLA-NUOVA, a strong fort on the coast of Naples, in the cant. and 6 m. N of Ostuni, in N lat. 40° 47′ 36″.

VILLA-NUOVA, a village of Tuscany, prov. and 18 m. NE of Florence.

VILLA-NUOVA, a commune and v. in the duchy of Parma, 30 m. NW of Parma, on the l. bank of the Arda. Pop. 3,000.

VILLA-NUOVA-D'ASTI, a walled town of the Sardinian states, in the prov. and 15 m. WNW of Asti. Pop. 3,100.

VILLA-NUOVA-DI-MONDOVI, a town of the Sardinian states, in the prov. and 6 m. SW of Mondovi, on the Loriggia. It has two fine churches. Pop. 3,600.

VILLA-NUOVA-SOLARO, a village of the Sardinian states, in the prov. and 8 m. NNE of Saluces, on the r. bank of the Vraita.

VILLA-PALACIOS, a town of Spain, in the prov. of Chinchilla, 10 m. SW of Alcaraz. Pop. 1,200.

VILLA-PIEDMONTE (LA), a town of Naples, in the prov. of Terra-di-Lavoro, cant. and 4 m. WNW of Santo-Germano. Pop. 1,400.

VILLAPUZZU, a town of Sardinia, 34 m. NE of Cagliari, on the l. bank of the Flumendosa. Pop. 2,200.

VILLAR (EL), a village of Spain, in the prov. of Caceres, 6 m. N of Plasencia. Pop. 700.—Also a village in the prov. of Logrono, 3 m. E of La-Guardia.

VILLAR (LE), a village of France, dep. of Lozere, cant. and 3 m. ENE of Chanac.

VILLARABOUD, a village in the Swiss cant. of Friburg, bail. of Glane. Pop. 247.

VILLARALTO, a town of Spain, in the prov. and 45 m. NW of Cordova. Pop 1.220.

VILLARAMIEL, a town of Spain, in the prov.

and 21 m. W of Palencia, near the Sequillo. Pop. 3,500. It has some woollen-factories, tanneries, and oil-mills.

VILLARCAYO, a village of Spain, prov. and 39 m. N of Burgos. Pop. 500.

VILLARD-BONNOT, a village of France, in the dep. of Isere, cant. and 4 m. NE of Domène. Pop. 700.

VILLARD-D'ARENNE, a village of France, in the dep. of Hautes-Alpes, arrond. and 18 m. NW of Briancon.

VILLARD-DE-LANS, a canton, commune, and town of France, in the dep. of the Isere, and arrond. of Grenoble. The cant. comprises 4 com. Pop. in 1831, 5,217; in 1846, 6,184. The town is 15 m. SW of Grenoble, near the Bourne. Pop. 2,298. It has a coal-mine.

VILLAR-DE-CANAS, a town of Spain, in the prov. and 24 m. from Cuenca, on the Zancara. Pop. 1,500. It has dye-works, and a trade in corn and agricultural produce.

VILLAR-DE-CIERVOS, a village of Spain, in the prov. and 45 m. NW of Zamora. Pop. 1,000.

VILLAR-DE-DOMINGO-GARCIA, a town of Spain, in the prov. and 24 m. from Cuenca. Pop. 686. It has manufactories of coarse cloth.

VILLAR-DE-FRADES, a town of Spain, in the prov. and 27 m. WNW of Valladolid. Pop. 800. It has two churches.

VILLAR-DE-LA-ENCINA, a town of Spain, in the prov. and 39 m. SSW of Cuenca, on the Zancara. Pop. 480.

VILLAR-DE-LA-GIGUA, a town of Spain, in the prov. of Salamanca, and 15 m. W of Ciudad-Rodrigo, on a mountain, in a fertile, wild, and healthy locality.

VILLAR-DEL-AGUILA, a town of Spain, in the prov. of Cuenca, and 12 m. from Huete. Pop. 400.

VILLAR-DEL-HORNO, a town of Spain, in the prov. and 18 m. SW of Cuenca, and 18 m. SE of Huete, on a height. Pop. 440. It has manufactories of coarse linen and woollen fabrics.

VILLAR-DEL-OLMO, a town of Spain, in the prov. and 20 m. from Madrid. Pop. 430.

VILLAR-DEL-REY, a town of Spain, in the prov. of Estremadura, 24 m. N of Badajoz, and 6 m. from the Portuguese frontier, amid mountains. Pop. 1,950.

VILLAR-DEL-SAZ-DE-DON-GUILLEN, a town of Spain, in the prov. and 21 m. from Cuenca. Pop. 850. It consists of two parts distinguished as Abajo and Arriba, and has two churches. The manufacture of coarse linen and woollen fabrics forms its chief object of local industry.

VILLAR-DEL-SAZ-DE-NAVALON, a town of Spain, in the prov. and 9 m. from Cuenca, near the sources of the Huete and Guadamejuz. Pop. 300.

VILLAR-DEL-VARO, a village of Sardinia, in the div. and prov. and 20 m. NNW of Nice, on the l. bank of the Var.

VILLAR-DE-RENA, a town of Spain, in the prov. and 45 m. ENE of Badajoz. Pop. 280.

VILLAR-DE-SAN-CONSTANZO, a market-town of Sardinia, in Piedmont, in the prov. of Coni.

VILLARDON-PARDO, a town of Spain, in the prov. and 12 m. from Jaen, on the summit of a hill, near the Salado-de-Arjona and Salado-de-Porcuna. Pop. 660.

VILLARD-REYMOND, a village of France, in the dep. of the Isere, cant. and 1½ m. S of Bourg-d'Oysans. Pop. 270.

VILLARDS, a commune of France, in the dep. of Dordogne, and cant. of Champagnac, 9 m. S of Nontron. Pop. 1,710.

VILLAREAL, a judicial partido and town of Spain, in the prov. of Castellon-de-la-Plana, 6 m. SW of Castellon-de-la-Plana, in a fertile plain, on the Mijares, which is here crossed by a bridge of 39 arches. Pop. 7,900. It is enclosed by old walls now much dilapidated, and has four gates and two suburbs, one of which is larger than the town. The principal street is straight and well-built. With the exception of the convents, one of which has a fine statue by Vergara, it contains no edifices worthy of note. It has several distilleries of brandy, and manufactures of tape. This town derived its origin from a royal palace erected by one of the kings of Aragon. In 1706 it was taken and to a great extent destroyed by Philip V.—Also a market-town of the prov. of Vizcaya, 24 m. SW of San-Sebastian, on the road from Irun to Vittoria, and a little to the S of the Urola. Pop. 800. It is of considerable extent, and contains some well-built streets, several fine residences belonging to the nobility, two large squares lined with handsome houses, and a fountain of excellent water. It has manufactories of hats, linen, coarse woollen fabrics, and cord, and several tile-kilns. In the vicinity are ferruginous springs, and quarries of marble.—Also a market-town in the prov. of Toledo, 6 m. from Ocana, on the plateau of that name. Pop. 600. It has a church, in which is the fine tomb of St. Raymond, founder of the order of Calatrava. Wool-spinning is its chief branch of industry.—Also a market-town of Portugal, capital of a comarca of the same name, in the prov. of Tras-os-Montes, 15 m. N of Lamego, on the Corgo, which is here traversed by a bridge. Pop. 4,250. It is the largest and finest town in the prov., and contains two churches, three convents, and an hospital. Pop. of comarca, 90,300.

VILLA-REAL-DA-CONCEIÇAO, a town of Brazil, in the prov. of Mato-Grosso, near the l. bank of the Paraguay, 141 m. N of L'Assompcion.

VILLAREAL-DE-ALAVA, a town of Spain, in the prov. of Alava, and 9 m. N of Vittoria, in a mountainous locality. Pop. 645. It has the remains of walls and of a fortress. In the environs are quarries of marble, freestone, and sharping-stone.

VILLAREAL-DE-EBRO, a town of Spain, in the prov. and 30 m. from Logrono, on a hill, 1½ m. from the Ebro. Pop. 2,000. It has two oil-mills, and manufactories of shoes and of linen.

VILLAREAL-DE-SAN-ANTONIO, or VILLAREAL-DE-ARAUILHA, a town of Portugal, in the prov. of Algarva, comarca and 6 m. E of Tavora, at the mouth of the Guadiana, which here forms a port. Pop. 2,000. It is the residence of a military governor, and is fortified on the sea-ward side. The streets, which are spacious, straight and well-paved, cross at right angles, and form 40 equal squares. In the centre is a spacious square, adorned with a fine marble fountain, and with the custom-house, a magnificent edifice. It has little trade, and the harbour although spacious is wanting in depth, but has extensive fisheries. Wine is extensively cultivated in the environs. This town was founded in 1774 by the famous marquis of Pombal.

VILLAREAL-DE-SANTA-LUZIA, a town of Brazil, in the E part of the prov. of Sergipe d'El Rey, and pleasantly situated near the Guararema, an affluent of the Rio Real.

VILLAREAL-DO-BREJO-D'AREA, a town of Brazil, in the prov. and 135 m. WNW of Parahiaba. Cotton is extensively cultivated in the environs.

VILLA-REAL-DO-SABARA. See SABARA.

VILLAREJO-DE-FUENTES, a market-town of Spain, in the prov. and 27 m. SW of Cuenca. Pop. 2,600. It has a convent, and manufactories of coarse woollen and linen fabrics.

VILLAREJO-DE-LA-PENUELA, a town of Spain, in the prov. and 12 m. from Cuenca. Pop. 250. It has manufactories of coarse linen and woollen fabrics.

VILLAREJO-DEL-VALLE, a market-town of Spain, in the prov. and 36 m. WSW of Toledo. Pop. 1,158.

VILLAREJO-DE-SALVANES, a market-town of Spain, in the prov. and 30 m. SE of Madrid, between the Tagus and Tajuna. Pop. 2,800. It has a parish church, a fine Gothic structure, a convent, and an hospital, and possesses manufactories of oil of excellent quality, and of mats, and a pottery. The trade consists chiefly in wine.

VILLAREJO-SOBRE-HUERTA, a market-town of Spain, prov. and 15 m. SW of Cuenca. Pop. 555.

VILLARELHO-DA-CASTANHEIRA, a town of Portugal, in the prov. of Tras-os-Montes, comarca and 8 m. NW of Torre-de-Moncorvo, on a hill, crowned with the ruins of a castle. Pop. 600. It has a convent.

VILLAREPOS, a village of Switzerland, in the cant. of Freiburg, and bail. of Murten. Pop. 310.

VILLARES (Los), a market-town of Spain, in the prov. and 5 m. from Jaen, in a fine valley on the Guadalquivir. Pop. 1,825. It has manufactories of soap, distilleries of brandy, and numerous oil-mills, and lime-kilns. In the vicinity are coal-mines, in which are found yellow amber and jet.

VILLARET (Cape), a headland of the North West coast of Australia, in De Witts land, enclosing Roebuck bay on the S, in S lat. 18° 18' 50", and E long. 121° 56'.

VILLAR-FOCCHIARDO, a village of Sardinia, in the div. of Coni, prov. and 11 E of Susa, and mande. of Bussolino, near the r. bank of the Dora-Ripaira. Pop. 1,400.

VILLARGORDO, a village of Spain, in the prov. and 9 m. from Jaen, on the Guadalquivir. Pop. 1,425.

VILLARGORDO-DE-CABRIEL, a market-town of Spain, in the prov. and 45 m. SE of Cuenca, near the l. bank of the Cabriel. Pop. 815. It has manufactories of coarse woollen fabrics.

VILLARGORDO-DEL-MARQUESADO, a town of Spain, in the prov. of Cuenca, and 9 m. from Belmonte. Pop. 380.

VILLARGORDO-DE-XUCAR, a town of Spain, in the prov. of Cuenca, and 6 m. WSW of Taragona, on the r. bank of the Xucar. Pop. 1,280.

VILLARIAZ, a village of Switzerland, in the cant. of Freiburg, bail of Glane. Pop. 201.

VILLARICA, a town of Chili, 90 m. NE of Valdivia, on the Tolten, a little to the W of a volcanic peak of the Andes to which it gives name.—Also a town of Paraguay, 60 m. ESE of Assumpcion. Pop. 3,000.

VILLARICA. See Ouro-Preto.

VILLARICA (Serra de). See Espinhaço.

VILLARINO, a town of Spain, in the prov. and 45 m. WNW of Salamanca, near the confluence of the Tormes and the Duero. Pop. 2,000.

VILLARLUENGO, a town of Spain, in the prov. and 36 m. NE of Teruel. Pop. 1,500. It has paper-mills, and some trade in agricultural produce.

VILLARO, a village of Spain, in the prov. and 15 m. SE of Bilbao. Pop. 750.

VILLA-ROBLEDO, a town of Spain, in the prov. and 82 m. ENE of Ciudad-Real. It is a comparatively well-built town, with two or three large churches and other public buildings; and has manufactories of linen and woollen goods, bricks, and pottery ware. Pop. 6,500.

VILLA-RODONA, a town of Spain, in the prov. and 16 m. NE of Taragona.

VILLA-RODRIGO, a village of Spain, in the prov. and 72 m. SE of Ciudad-Real.

VILLA-ROYA, a town of Spain, in the prov. and 9 m. NNW of Calatayud. Pop. 1,600. It has woollen manufactories.—Also a town of Spain, in the prov. and 24 m. NE of Teruel. Pop. 1,800.

VILLARRIN-DE-CAMPOS, a village of Spain, in the prov. and 24 m. NNE of Zamora. Pop. 640. Saltpetre is produced in the environs.

VILLAR-RUBIA-DE-OCANA, a town of Spain, in the prov. and 30 ENE of Toledo. Pop. 2,000. It is pretty well built, and has manufactories of coarse woollens.

VILLARS, a village of France, in the dep. of Ain, cant. of Saint-Trevier-en-Dombes. Pop. 570.—Also a village of France, in the dep. of Dordogne, cant. of Champagnac-de-Bel-Air, on the Tricon. Pop. 1,600.—Also a village of France, in the dep. of Vaucluse, 6 m. N of Apt.

VILLAR-SAN-COSTANZO, a village of the Sardinian states, in the prov. and 6 m. WNW of Coni. It has a fine abbey and church.

VILLAR-SECCO-DA-LOMBA, a town of Portugal, in the prov. of Tras-os-Montes, comarca and 15 m. W of Vinhaes. Pop. 2,000.

VILLARTA, a village of Spain, in the prov. and 36 m. NE of Ciudad-Real, on the l. bank of the Guadiana. Pop. 900.—Also a village in the prov. and 18 m. NNW of Cordova, on the Guardabarbo.—Also a village in the prov. and 48 m. W of Ciudad-Real.

VILLA-RUBIA-DE-LOS-OJOS-DE-GUADIANA, a town of Spain, in the prov. and 18 m. ENE of Ciudad-Real, at the foot of a range of hills to which it gives name, in a fertile district of the plain of La Mancha watered by the Giguela and the Guadiana. It contains several good houses, and a palace of the dukes of Hijar.

VILLA-RUBRAS, a village of Spain, in the prov. of Salamanca, 15 m. from Ciudad-Rodrigo.

VILLASALTA, a town of the island of Sardinia, in the prov. of Isili, 30 m. NE of Cagliari. Pop. 1,300.

VILLASANDINO, a town of Spain, in the prov. and 18 m. W of Burgos. Pop. 1,600. It has manufactories of linens and woollens.

VILLA-SAN-FIORANO, a village of Austrian Lombardy, in the prov. of Milan, 3 m. NE of Monza.

VILLA-SAN-GIOVANNI, a village of Naples, in the prov. of Calabria-Ultra 1ma, 9 m. N of Reggio, on the strait of Messina. Pop. 1,250. It has silk-manufactories.

VILLA-SANTA-MARIA, a town of Naples, in Abruzzo-Citra, 21 m. SSW of Lanciano, on the l. bank of the Sangro. Pop. 1,700.

VILLASARRACINO, a village of Spain, in the prov. and 27 m. N of Palencia. Pop. 1,000.

VILLASAVARY, a town of France, in the dep. of Aude, arrond. and 9 m. SSE of Castelnaudary, on the Mezuran. Pop. 1,700.

VILLASAYAS, a village of Spain, in the prov. and 27 m. SSW of Soria. Pop. 650.

VILLASECA-DE-LA-SAGRA, a town of Spain, in the prov. and 12 m. NE of Toledo. Pop. 1,600. It is celebrated for its coarse earthenware.

VILLASMUNDO, a village of Sicily, in the prov. and 21 m. NW of Syracuse, on the San-Giuliano.

VILLASOR, a village of the island of Sardinia, in the prov. and 15 m. NW of Cagliari, in a fertile plain. Pop. 1,700.

VILLATOBAS, a town of Spain, in the prov. of Toledo, 9 m. SE of Ocana, in a fertile plain. Pop. 2,300.

VILLATORO, a village of Spain, in the prov. and 21 m. WNW of Avila. Pop. 300.

VILLAUDRIC, a village of France, in the dep. of Haute-Garonne, cant. and 2 m. ESE of Fronton. Pop. 1,600.

VILLAUIS, a town of Spain, in the prov. and 1 m. NNE of Pampeluna, on the r. bank of the Ulzama, near its confluence with the Arga. Pop. 400.

VILLA-VELHA. See ESPIRITO-SANTO.

VILLA-VELHA-DE-RODAO, a town of Portugal, in the prov. of Beira, comarca and 12 m. SW of Castello-Branco. Pop. 1,800.

VILLAVERDE, a town of Brazil, in the prov. of Bahia, and 15 m. WSW of Porto-Seguro, on the S bank of the Rio-da-Cachoeira, in S lat. 16° 10'. It has some trade in wood and in cotton.

VILLAVERDE, a town of Spain, in the prov. of Chinchilla, 24 m. S of Alcaraz. Pop. 900.

VILLAVERDE-DOS-FRANCOS, a town of Portugal, in the prov. of Estremadura, comarca of Torres-Vedras, 10 m. NE of Lisbon.

VILLAVERDE-MOJINA, a village of Spain, in the prov. and 21 m. S of Burgos, near the Arlanzon. Pop. 500.

VILLAVICENCIO-DE-LOS-CABALLEROS, a town of Spain, in the prov. and 39 m. from Leon, on the l. bank of the Valderaduey, in a fertile valley.

VILLAVICIOSA, a town of Spain, in the prov. and 18 m. WNW of Cordova, near the r. bank of the Guadiatu. Pop. 1,600.—Also a village in the prov. and 15 m. E of Guadalaxara. Pop. 300. An important battle was gained in the vicinity of this place by Philip V. in 1710.—Also a village of Spain, 9 m. from Madrid.—Also a town of Spain, in the prov. and 24 m. NE of Oviedo, at the embouchure of the Lonares. It has a small but good port.

VILLAVICIOSA-DE-LA-RIBERA, a village of Spain, in the prov. and 15 m. from Leon.

VILLA-VICOSA, a town of Brazil, in the prov. and 150 m. W of Ceara, on the Serra-Hibiapaba.—Also a town of Brazil, in the prov. and 70 SW of Para, on the l. bank of the Tocantins.

VILLA-VIÇOSA, a town of Portugal, in the prov. of Alemtejo, 15 m. SW of Elvas, in a fine valley. It is a place of considerable strength, is well-built, and possesses some fine buildings, amongst which is a palace of the dukes of Braganza, and a handsome collegiate church. Pop. 3,000. Wine and oil are extensively grown in the vicinity.

VILLAVIEJA, a town of Spain, in the prov. and 7 m. SSW of Castellon-de-la-Plana, on the Rio-Uxo, near the coast. Pop. 1,600. It is frequented for its baths.—Also a town in the prov. of Salamanca, 18 m. N of Ciudad-Rodrigo. Pop. 1,000.—Also a village in the prov. and 45 m. SE of Orense. Pop. 300.

VILLA-VISTOZA-DA-MADRE, a town of Brazil, in the prov. and 240 m. NW of Para, on the Ananirapucu.

VILLAVIUDAS, a village of Spain, in the prov. and 12 m. E of Palencia. Pop. 720.

VILLAZOPEQUE, a village of Spain, in the prov. and 18 m. SW of Burgos.

VILLE', a town of France, in the dep. of Bas-Rhin, arrond. and 10 m. NW of Schelestadt. Pop. 2,000. It has bleach-works and tanneries.

VILLE'-A-POURÇON, a village of France, in the dep. of Nievre, cant. and 8 m. SE of Moulens-Engelbert. Pop. 2,000.

VILLE-AUX-CLERCS (LA), a village of France, in the dep. of Loir-et-Cher, arrond. and 11 m. N of Vendome. Pop. 650.

VILLEBAROU, a village of France, in the dep. of Loir-et-Cher, cant. and 2 m. NNW of Blois. Pop. 1,400.

VILLEBERNIER, a commune and v. of France, in the dep. of Maine-et-Loire, cant. and 2 m. E of Saumur. Pop. 1,200.

VILLEBOIS, a village of France, in the dep. of Ain, cant. of Lagnieu, 15 m. NW of Belley.

VILLEBRUMIER, a village of France, in the dep. of Tarn-et-Garonne, on the r. bank of the Tarn, 4 m. NW of Villeneuve. Pop. 700.

VILLECHENE'VE, a village of France, in the dep. of Rhone, cant. and 7 m. NNW of Saint-Laurent-de-Chamousset. Pop. 1,200.

VILLE-COMTAL, a town of France, in the dep. of Aveyron, cant. and 6 m. WSW of Estaing, on the l. bank of the Dourdon. Pop. 1,000.

VILLECOMTE, a village of France, in the dep. of Cote-d'Or, cant. and 4 m. WSW of Issar-Tille. Pop. 1,000. Iron is wrought in the vicinity.

VILLECROZE, a village of France, in the dep. of Var, cant. and 1 m. NE of Salernes. Pop. 1,100.

VILLE-D'AVRAY, a village of France, in the dep. of Seine-et-Oise, cant. and 1 m. WNW of Sevres. Pop. 400.

VILLEDIEU, a market-town and commune of France, in the dep. of the Indre, cant. and 6 m. SE of Buzancois, on the Tregouze, near its confluence with the Indre. Pop. 810. It has a porcelain manufactory and several fulling-mills.—Also a village in the dep. of the Vaucluse, and cant. of Vaison, 18 m. NE of Orange. Pop. 1,100.—Also a village in the dep. of Loir-et-Cher, 12 m. E of Chateau-du-Loir. Pop. 1,180.

VILLEDIEU, or VILLEDIEU-EN-BEAUCE, a commune and town of France, in the dep. of the Loir-et-Cher, and cant. of Montoire, 22 m. WSW of Vendôme. Pop. 1,303.

VILLEDIEU (LA), a village of France, dep. of Tarn-et-Garonne, and cant. of Montech, 6 m. E of Castel-Sarassin. Pop. 640.—Also a canton and com. in the dep. of the Vienne, and arrond. of Poitiers. The cant. comprises 10 com. Pop. in 1831, 5,292; in 1846, 6,074. The village is 9 m. S of Poitiers. Pop. 320.

VILLEDIEU-LE-BLOUE'RE, a village of France, in the dep. of the Maine-et-Loire, cant. and 36 m. SW of Beaupreau. Pop. 1,050.

VILLEDIEU-LES-POE'LES, a canton, commune, and town of France, in the dep. of the Manche. The cant. comprises 10 com. Pop. in 1831, 5,292; in 1846, 6,074. The town is 14 m. NE of Abranches, on the r. bank of the Sienne. Pop. 3,763. It has extensive copper foundries, manufactories of copper and of plated ware, parchment, hair-cloth, linen, and several tanneries.

VILLEDOMER, a commune of France, in the dep. of the Indre-et-Loire, cant. and 4 m. SSW of Chateau-Regnault, near the Brienne. Pop. 1,060. It has several flour-mills.

VILLEDOUX, a village of France, in the dep. of the Charente-Inferieure, cant. and 7 m. SW of Marans. Pop. 1,210.

VILLE-DU-PONT, a market-town of France, in the dep. of the Doubs, cant. and 1½ m. NE of Montbenoit, on the r. bank of the Doubs. Pop. 628. It has a gypsum quarry.

VILLEFAGNAN, a canton, commune, and town of France, in the dep. of the Charente, and arrond. of Ruffec. The cant. comprises 21 com. Pop. in 1831, 13,595; in 1846, 13,562. The town is 6 m. W of Ruffec. Pop. 1,668.

VILLEFORT, a canton, commune, and town of France, in the dep. of the Lozere, and arrond. of Mende. The cant. comprises 10 com. Pop. in 1831, 5,872; in 1846, 6,027.—The town is 24 m. ESE of Mende, in a narrow valley, at the junction of the dep. of the Ardeche and Gard. Pop. 1,625. It has manufactories of caddis and a dye-work, and carries on an active transit trade in wine, silk, salt, coal, corn, flour, and chestnuts. In the vicinity are productive lead and silver mines.

VILLEFRANCHE, a town of France, in the dep. of the Allier, cant. and 8 m. NW of Montmarault. Pop. 700. In the environs are coal-mines and a glass-work.—Also a village in the dep. of the Drome, cant. and 2 m. NW of Sederon. Pop. 90. In the environs are considerable copper-mines.—Also a canton, commune, and town in the dep. of the Tarn. The cant. comprises 8 com. Pop. in 1831, 8,299; in 1846, 8,715. This town owes its foundation to Philippe-de-Montfort, descendant of Simon, count of Montfort.

VILLEFRANCHE-DE-CONFLENT, a town of France, in the dep. of Pyrenees-Orientales, cant. of Prades, 30 m. WSW of Perpignan, on the r. bank of the Tet. It is walled, and defended by a strong fort on the NW. Its principal buildings are the barracks and military hospital.

VILLEFRANCHE-DE-LAURAGAIS, a town of France, in the dep. of Haute-Garonne, 21 m. SE of Toulouse, near the Canal-du-Midi. Pop. 2,350. It is a well-built place, with manufactures of sail-cloth and earthenware, and a trade in agricultural produce.—The arrond. has an area of 93,586 hectares, and comprises 6 cants. Pop. in 1846, 65,040.

VILLEFRANCHE-DE-LONCHAPT, a village of France, in the dep. of Dordogne, 24 m. WNW of Bergerac. Pop. 680.

VILLEFRANCHE-DE-PERIGORD, a town of France, in the dep. of Dordogne, 12 m. SSE of Belva. Pop. 1,400.

VILLEFRANCHE-DU-QUEYRAN, a town of France, in the dep. of Lot-et-Garonne, cant. and 6 m. E of Castel-Jaleux. Pop. 2,500.

VILLEFRANCHE-SAINT-PHAL, a village of France, in the dep. of Yonne, cant. of Charny, 24 m. NW of Auxerre. Pop. 800.

VILLEFRANCHE-SUR-AVEYRON, a town of France, in the dep. of Aveyron, on the r. bank of the Aveyron, 24 m. W of Rhodez. Pop. 9,500. It has a considerable trade in corn, cattle, and wine; manufactures of paper, hats, and muslin, and copper foundries.

VILLEFRANCHE-SUR-SAONE, a town of France, in the dep. of Rhone, cant. and 18 m. NNW of Lyons, on the Morgon, an affluent of the Saone. Pop. 7,000. It is a well built place, with considerable manufactures of linens, cottons, and cotton-yarn.—The arrond. comprises an area of 149,796 hectares, subdivided into 9 cants. Pop. in 1846, 161,451; in 1836, 151,980.

VILLEFRANQUE, a village of France, in the dep. of Basses-Pyrenees, cant. and 4 m. N of Ustarits, on the r. bank of the Nive. Pop. 290.

VILLEGAILHENE, a village of France, in the dep. of Aude, 4 m. N of Carcassonne, on the Trapel. Pop. 1,000.

VILLEGAS, a village of Spain, in the prov. and 18 m. NW of Burgos. Pop. 610.

VILLEGAUDEN, a village of France, in the dep. of Seine-et-Loire, cant. and 2 m. SE of Saint-Martin. Pop. 350.

VILLE-HOUDLEMONT, a village of France, in the dep. of Moselle, cant. and 4 m. E of Longwy. Pop. 691.

VILLE-JESUS, a village of France, in the dep. of Charente, cant. and 1 m. SE of Aigre. Pop. 1,000.

VILLEJUIF, a town of France, in the dep. of Seine, 3 m. NE of Sceaux, on a rising ground, 4 m. S of Paris. Pop. 1,400. It contains numerous villas and country-seats of Parisian citizens, and has manufactures of soap, wax, and straw-plait.

VILLEL, a village of Spain, in the prov. and 60 m. E of Guadalaxara, near the Tagus.—Also a town in the prov. and 15 m. SSW of Teruel, on the r. bank of the Guadalaviar. Pop. 1,900. There are hot springs in the vicinity.

VILLELOIN, a village of France, in the dep. of Indre-et-Loire, 12 m. E of Loches.

VILLELONGUE, a village of France, in the dep. of Hautes-Pyrenees, cant. and 4 m. SE of Argeles. Pop. 570.

VILLELOUP, a village of France, in the dep. of Aube, cant. and 12 m. WNW of Troyes. Pop. 360.

VILLEMAGUE, a village of France, in the dep. of Aude, cant. and 9 m. NE of Castelnaudary.—Also a village in the dep. of Herault, cant. and 5 m. SE of Saint-Gervais-la-Ville. Pop. 590.

VILLEMBLARD, a village of France, in the dep. of Dordogne, 15 m. SW of Perigueux. Pop. 1,200.

VILLEMEUX, a village of France, in the dep. of Eure-et-Loire, cant. and 5 m. NW of Nogent-le Roi. Pop. 1,200.

VILLEMOISAN, a commune and v. of France, in the dep. of Maine-et-Loire, cant. and 4 m. S of Louroux-Beconnais. Pop. 1,500.

VILLEMONSTAUSSON, a village of France, in the dep. of Aude, cant. of Conques, near Trapel. Pop. 800.

VILLEMONTOIS, a village of France, in the dep. of Loire, cant. and 7 m. SW of Raonne. Pop. 990. Coal and lead are wrought in the vicinity.

VILLEMORT-SUR-VANNES, a village of France, in the dep. of Aube, cant. of Estissac, near the r. bank of the Vannes, 18 m. WSW of Troyes. Pop. 500.

VILLEMUR, a canton, commune, and town of France, in the dep. of the Haute-Garonne, and arrond. of Toulouse. The cant. comprises 4 com. Pop. in 1831, 7,404; in 1846, 6,783. The town is 20 m. N of Toulouse, on the r. bank of the Tarn. Pop. 5,472.

VILLENA, a judicial partido and town of Spain, in the prov. of Alicante. The town is 33 m. NW of Alicante, in a fine and fertile plain, at the foot of the mountain of São-Christoval, which is crowned with an old castle. Pop. 9,800. It contains 14 principal streets, all tolerably well-built, four public squares, several fountains, fine promenades, a palace belonging to the Marquis of Vallena, a town-house, the front of which is much admired, two parish-churches, three convents, numerous chapels, an hospital and barracks. Beyond the walls, of which the remains only now exist, is an extensive suburb, larger and more modern than the town itself, and nearly completely encircling it. Soap, brandy, salt, cloth, and linen are the chief productions of the locality. This town occupies the site of the *Arbacula* or *Turbula* of the Romans.

VILLENAUXE, a canton, commune, and town of France, in the dep. of the Aube, and arrond. of Nogent-sur-Seine. The cant. comprises 7 com. Pop. in 1831, 5,000; in 1846, 5,266. The town is 9 m. NNE of Nogent-sur-Seine, in a valley, on a small river of the same name. Pop. 2,669. It has manufactories of basket and of hard-ware, leather, cordage, and vinegar.

VILLENAVE, a village of France, in the dep. of the Landes, cant. and 11 m. N of Tartas. Pop. 550.

VILLENAVE-D'ORNON, a commune of France, in the dep. of the Gironde, cant. and 5 m. SE of Pessac. Pop. 1,377. It is noted for its wine.

VILLENEUVE, a town of France, in the dep. of the Ain, cant. of Coligny and com. of Domsure, 19 m. N of Bourg. Pop. 1,100.—Also a canton, com. and town in the dep. of the Aveyron, and arrond. of Villefranche-de-Rouergue. The cant. comprises 12 com. Pop. in 1831, 9,546; in 1846, 10,115. The town is 6 m. N of Villefranche. Pop. 3,251.—Also a river in the dep. of the Creuse, which has its source in the hamlet of Lez-Maison; runs NNW; and after a course of about 12 m., throws itself into the Taurion, on the r. bank, below Banise.—Also a village in the dep. of Lot-et-Garonne, cant. and 3 m. S of Mezin. Pop. 400.

VILLENEUVE, or NEUSTADT, a town and port of Switzerland, in the cant. of Vaud, and district of Aigle, 18 m. SE of Lausanne, in a marshy plain, towards the E extremity of the lake of Geneva, and 3 m. NE of the embouchure of the Rhone. Pop. 1,160. It is enclosed by a wall in a ruinous condition; is ill-built, badly-paved, and insalubrious. Near the entrance to the harbour is a small island, the only one in the lake. This town originally bore the name of *Pennilucus*, and is of considerable antiquity.—Also a village of Switzerland, in the cant. of Freiburg and bail. of Broya. Pop. 295.

VILLENEUVE (LA), a village of France, in the dep. of the Allier, cant. and 9 m. NW of Moulins, near the r. bank of the Allier. Pop. 685.—Also a hamlet in the dep. of the Creuse, cant. of Crocy and com. of Basville. Pop.

VILLENEUVE-D'AGEN, or VILLENEUVE-SUR-LOT, an arrondissement, canton, commune, and town of France, in the dep. of the Lot-et-Garonne. The arrond. comprises an area of 153,587 hect., and contains 10 cant. Pop. in 1831, 97,587; in 1841, 97,324; in 1846, 96,992. The cant. comprises 7 com. Pop. in 1831, 14,957; in 1846, 17,096.—The town is 15 m. NNE of Agen, on the Lot, which is here crossed by a fine old bridge. Pop. in 1789, 6,162; in 1821, 9,211; in 1831, 10,652; in 1841, 12,337; in 1846, 13,088; in 1851, 13,337. It is regularly built, the streets are straight, but the houses generally are old, and there are no buildings worthy of note. It has a communal college, an agricultural society, a savings' bank, a printing establishment, and a house-of-correction, and possesses manufactories of linen, hats, china, and nails, marble-works, copper-founderies, tanneries, dye-works, and extensive tile, brick, and lime-kilns. The trade consists chiefly in grain, wine, prunes, and brandy. V. was founded in the 13th cent. by Duke Alphonso. It suffered much during the religious wars.

VILLENEUVE-DE-BERG, a canton, commune, and town of France, in the dep. of the Ardeche, and arrond. of Ardeche. The cant. comprises 16 com. Pop. in 1831, 10,781; in 1846, 12,832. The town is 15 m. SSW of Privas. Pop. 2,600. Silk is extensively cultivated in the environs.

VILLENEUVE-DE-DURAS, a village of France, in the dep. of the Lot-et-Garonne, cant. and 6 m. NE of Duras. Pop.

VILLENEUVE-DE-MARC, a village of France, in the dep. of the Isere, cant. and 4 m. SW of St. Jean-de-Bournay, between the Valaize and Girand. Pop. 1,200.

VILLENEUVE-DE-MARSAN, a canton, commune, and town of France, in the dep. of the Landes, and arrond. of Mont-de-Marsan. The cant. comprises 12 com. Pop. in 1831, 8,781; in 1846, 9,667. The town is 11 m. E of Mont-de-Marsan, on the l. bank of the Midon. Pop. 1,568. It has manufactures of coarse woollen fabrics, and a considerable trade in feathers.

VILLENEUVE-DE-RIVIERE, a commune of France, in the dep. of the Haute-Garonne, cant. and 4 m. WNW of St. Gaudens, near the confluence of the Fayet and Garonne. Pop. 1,594.

VILLENEUVE-EN-MONTAGNE, a village of France, in the dep. of the Saone-et-Loire, cant. and 7 m. NW of Brixy. Pop. 250.

VILLENEUVE-LA-GUYARD, a commune and town of Belgium, in the dep. of the Yonne, and cant. of Pont-sur-Yonne, 15 m. NW of Sens, on the Yonne, Pop. 1,794.

VILLENEUVE-L'ARCHEVEQUE, or SUR-VANNE, a canton, commune, and town of France, in the dep. of the Yonne, and arrond. of Sens. The cant. comprises 16 com. Pop. in 1831, 9,237; in 1846, 10,077. The town is 14 m. E of Sens, on the Vanne. Pop. 1,925. It has extensive manufactories of cloth and of lace, tan, fulling, and wool-spinning-mills, and carries on an extensive trade in wool, hemp, blankets, and other woollen fabrics.

VILLENEUVE-LE-COMTE, a village of France, in the dep. of the Seine-et-Marne, cant. and 5 m. SW of Crecy.

VILLENEUVE-LE-ROI, or VILLENEUVE-SUR-YONNE, a canton, commune, and town of France, in the dep. of the Yonne, and arrond. of Joigny. The cant. comprises 8 com. Pop. in 1831, 10,394; in 1846, 11,474. The town is 15 m. NNW of Joigny, on the r. bank of the Yonne, which is here crossed by an old stone-bridge. Pop. 4,525. The principal street is spacious and handsome, and contains a fine church. It has manufactories of cloth, enamelled-ware, and casks, wax-works, tanneries, and dye-works, and carries on an active trade in wood, charcoal, coarse cloth, leather, tiles, and in the wines of the locality.

VILLENEUVE-LES-AVIGNON, a town of France, in the dep. of Gard, 24 m. NE of Nismes, on the r. bank of the Rhone, opposite to Avignon. Pop. 3,600. It has an old abbey and monastery, and a considerable public library. Silks, linens, and cordage are manufactured here.

VILLENEUVE-LES-CHANOINES, a village of France, in the dep. of Aude, arrond. and 9 m. NE of Carcassonne. Pop. 600.

VILLENEUVE-LES-GENETS, a village of France, in the dep. of Yonne, arrond. and 24 m. SW of Joigny. Pop. 600.

VILLENEUVE-SAINT-GEORGES, a town of France, in the dep. of Seine-et-Oise, cant. and 4 m. SW of Boissy-St.-Leger, on the r. bank of the Seine, at the confluence of the Yeres. Pop. 1,035. There is a fine chateau here, and several pleasant villas adorn the vicinity.

VILLENEUVE-SUR-BELLOT, a village of France, dep. of Seine-et-Marne, cant. and 6 m. E of Rebais, on the Petit-Morin. Pop. 1,250. It has pottery-works.

VILLENEUVE-SUR-VERRE, a village of France, in the dep. of Tarn, cant. and 8 m. NW of Alby, on the Verre. Pop. 700.

VILLENOUVELLE, a village of France, in the dep. of Haute-Garonne, cant. and 4 m. NW of Villefranche. Pop. 1,000.—Also a v. in the dep. of Herault, cant. and 3 m. SE of Beziers, near the Orbe. Pop. 500.

VILLEPINTE, a town of France, in the dep. of Aude, cant. and 7 m. ESE of Castelnaudary, at the confluence of the Trebone and the Fresquel. Pop. 1,230.

VILLEPOT, a village of France, dep. of Loire-Inferieure, 6 m. NE of Chateaubriant. Pop. 1,300.

VILLEPREUX, a village of France, in the dep. of Seine-et-Oise, cant. and 5 m. SW of Marly-le-Roi. Pop. 954. Fine merino cloths are made here, and there are paper-mills in the vicinity.

VILLER, a village of France, in the dep. of Haut-Rhin, 21 m. NNE of Belfort, on the l. bank of the Thure. Pop. 1,848. It has woollen-yarn mills.

VILLE-REAL, a commune and v. of France, in the dep. of Lot-et-Garonne, 32 m. NNE of Agen. Pop. 1,687.

VILLERET, a village of Switzerland, in the cant. of Bern, bail. of Courtelary. Pop. 970.

VILLE-REVERSURE, a village of France, in the dep. of Ain, cant. of Ceyseriat. Pop. 1,120.

VILLERIAS, a village of Spain, in the prov. and 15 m. from Palencia. Pop. 500.

VILLEROUGE-DE-TERMENES, a village of France, in the dep. of Aude, cant. of Monthoumet, 6 m. S of Lagrasse. Pop. 350.

VILLERS-AU-TERTRE, a village of France, in the dep. of Nord, and 4 m. NE of Arleux. Pop. 420.

VILLERS-AUX-TOURS, a commune and v. of Belgium, in the prov. of Liege, arrond. of Huy. Pop. 276.

VILLERS-BOCAGE, a village of France, in the dep. of Somme, 7 m. N of Amiens. Pop. 1,498.

VILLERS-BRETONNEUX, a village of France, in the dep. of Somme, cant. and 3 m. S of Corbie. Pop. 1,880. It has woollen-yarn mills.

VILLERS-COTERETS, a town of France, in the dep. of Aisne, 30 m. SW of Laon, 18 m. SE of Compeigne. Pop. 3,481. It has manufactories of steel-wares, honey, and turnery-ware.

VILLERS-DEUX-EGLISES, a commune and v. of Belgium, in the prov. of Namur, arrond. of Philippeville. Pop. 387.

VILLERS-DEVANT-ORVAL, a commune and v. of Belgium, in the prov. of Luxemburg, arrond. of Verton, 6 m. N of Malmedy. Pop. 668.

VILLERS-EN-CAUCHIE, a village of France, in the dep. of Nord, cant. and 6 m. NNE of Carnieres. Pop. 1,815.

VILLERS-EN-FAGNE, a commune and v. of Belgium, in the prov. of Namur, arrond. of Philippeville. Pop. 195.

VILLERS-FARLAY, a village of France, dep. of Jura, 27 m. NNE of Lons-le-Saunier. Pop. 983.

VILLERS-FAUCON, a village of France, in the dep. of Somme, cant. and 2 m. N of Roisel. Pop. 1,225, partly muslin-weavers.

VILLERS-GUILAIN, a village of France, dep. of Nord, cant. and 6 m. SW of Marcoing. Pop. 1,280.

VILLERS-LA-BONNE-EAU, a commune and v. of Belgium, in the prov. of Luxemburg, arrond. of Bastogne. Pop. 608.

VILLERS-LA-LOUE, a commune and v. of Belgium, in the prov. of Luxemburg, arrond of Verton. Pop. 983.

VILLERS-LA-MONTAGNE, a village of France, in the dep. of Moselle, cant. and 4 m. SE of Longwy. Pop. 900. Iron is wrought in the vicinity.

VILLERS-LA-TOUR, a commune and v. of Belgium, in the prov. of Hainault, arrond. of Thuin. Pop. 570.

VILLERS-LA-VILLE, a commune and v. of Belgium, in the prov. of Brabant, arrond. of Nivelles. Pop. 827.

VILLERS-LE-BOUILLET, a canton and v. of Belgium, in the prov. of Liege, arrond. of Huy. Pop. 1,667.

VILLERS-LE-GAMBON, a commune and v. of Belgium, in the prov. of Namur, cant. of Florennes. Pop. 396.

VILLERS-LE-PEUPLIER, a commune and v. of Belgium, in the prov. of Liege, cant. of Hannut. Pop. 638.

VILLERS-LE-TEMPLE, a commune and v. of Belgium, in the prov. of Liege, cant. of Seny. Pop. 755.

VILLERS-L'EVEQUE, a commune and v. of Belgium, in the prov. of Liege, cant. of Alleur. Pop. 941.

VILLERS-NOTRE-DAME, a commune and v. of Belgium, in the prov. of Hainault, cant. of Chievres. Pop. 225.

VILLERS-OUTREAU, a commune and v. of France, in the dep. of Nord, cant. and 6 m. SW of Clary. Pop. 2,248.

VILLERS-PENVIN, a commune and v. of Belgium, in the prov. of Hainault, cant. of Gossebis. Pop. 617.

VILLERS-POL, a village of France, in the dep. of Nord, cant. and 4 m. NW of Quesney. Pop. 980.

VILLERS-POTTERIE, a village of Belgium, in the prov. of Hainault, cant. of Charleroi. Pop. 432.

VILLERS-SAINT-ARNAUD, a village of Belgium, in the prov. of Hainault, cant. of Chievres. Pop. 654.

VILLERS-SAINTE-GERTRUDE, a village of Belgium, in the prov. of Luxemburg, cant. of Barvaux. Pop. 306.

VILLERS-SAINT-GHISLAIN, a village of Belgium, in the prov. of Hainault, cant. of Rœulx. Pop. 660.

VILLERS-SAINT-SIMEON, a village of Belgium, in the prov. of Liege, cant. of Alleur. Pop. 300.

VILLERS-SEXEL, a canton and v. of France, in the dep. of Haute-Saone, arrond. and 9 m. S of Lure. Pop. 1,400.

VILLERS-SIRE-NICOLE, a village of France, in the dep. of Nord, cant. and 4 m. NNE of Maubeuge. Pop. 1,324. There are cotton-spinning works here.

VILLERS-SUR-LESSE, a commune and v. of Belgium, in the prov. of Namur, cant. of Rochefort. Pop. 495.

VILLERS-SUR-SEMOIS, a commune and v. of Belgium, in the prov. of Luxemburg, cant. of Etaille. Pop. 680.

VILLERUPT, a village of France, in the dep. of Moselle, cant. and 9 m. ESE of Longwy. Pop. 748.

VILLES, a village of France, in the dep. of Vaucluse, cant. of Mourmoiron, on the Auzon. Pop. 1,400.

VILLE-SOUS-ANJOU, a commune and v. of France, in the dep. of Isere, cant. and 2 m. E of Roussillon, on the l. bank of the Sonne. Pop. 1,060.

VILLE-SUR-AUJON. See CHATEAU-VILAIN.

VILLE-SUR-HAINE, a commune and v. of Belgium, in the prov. of Hainault, cant. of Rœulx. Pop. 500.

VILLE-SUR-ILLON, a commune and v. of France, in the dep. of Vosges, cant. and 4 m. S of Dompaire. Pop. 1,000.

VILLE-SUR-JARNIOST, a commune and v. of France, in the dep. of Rhone, cant. of Bois-d'Oingt. Pop. 1,384.

VILLE-SUR-TOURBE, a canton and v. of France, in the dep. of Marne, arrond. and 7 m. NNW of Sainte-Menehould. Pop. 511.

VILLETANEUSE, a village of France, in the dep. of Seine, cant. and 2 m. NNW of Saint-Denis. Pop. 837.

VILLETERTRE (LA), a village of France, in the dep. of Oise, cant. and 6 m. SSE of Chaumont, on the Viosne. Pop. 480.

VILLETOUREIX, a village of France, in the dep. of Dordogne, cant. and 2 m. NE of Riberac, near the r. bank of the Dronne. Pop. 1,200.

VILLETTA, a village of Naples, in the prov. of Abruzzo-Citra 2da, 18 m. S of Sulmona, on the l. bank of the Sangro. Pop. 700.

VILLETTA (LA), a town of France, in the dep. of Seine, arrond. and 4 m. SSE of Saint-Denis, forming a suburb of Paris on the N. Pop. 13,485.

VILLETTE, a village of the Sardinian states, in the prov. of Savoy, 4 m. NE of Moutiers, on the r. bank of the Isere. Pop. 450.—Also a v. of the Swiss cant. of Vaud, in the bail. of Lavaux. Pop. 270.

VILLETTE-LES-ROLLOT (LA), a commune and v. of France, in the dep. of Somme, cant. and 5 m. SE of Montdidier. Pop. 1,800.

VILLEURBANNE, a commune and v. of France, in the dep. of Isere, cant. and 4 m. W of Meyzieux. Pop. 1,698. It has chemical works and yarn-spinning-mills.

VILLEVALLIER, a village of France, in the dep. of Yonne, arrond. of Joigny. Pop. 518.

VILLEVAYRAC, a commune and v. of France, in the dep. of Herault, cant. and 7 m. N of Meze. Pop. 1,125.

VILLEVEQUE, a commune and v. of France, in the dep. of Maine-et-Loire, cant. and 9 m. NE of Angers, on the l. bank of the Loir. Pop. 1,735.

VILLEVIEILLE, a village of France, dep. of Gard, cant. and 1 m. NE of Sommieres. Pop. 500.

VILLEVIEUX (LA), a commune and v. of France, in the dep. of Jura, cant. of Bletterand, 1 m. NW of Lons-le-Saunier. Pop. 1,100.

VILLIE', a commune and v. of France, dep. of Rhone, cant. and 4 m. E of Beaujeu. Pop. 1,754.

VILLIERS, a commune and v. of France, in the dep. of Loir-et-Cher, cant. and 4 m. W of Vendome. Pop. 1,175.—Also a village of the Swiss cant. of Neufchatel, in the Val-de-Ruz. Pop. 288.

VILLIERS-CHARLEMAGNE, a commune and v. of France, in the dep. of Mayenne, cant. and 9 m. NW of Grez-en-Bouere. Pop. 1,400.

VILLIERS-EN-DESEVRE, a village of France, in the dep. of Eure, cant. and 7 m. SE of Pacy-sur-Evreux. Pop. 450.

VILLIERS-EN-PLAINE, a village of France, in the dep. of Deux-Sevres, cant. and 6 m. SE of Coulonges-les-Royaux. Pop. 1,008.

VILLIERS-LE-BEL, a commune and v. of France, in the dep. of Seine-et-Oise, cant. and 1 m. SE of Ecouen. Pop. 1,328. It contains a number of pleasant villas.

VILLIERS-LE-DUC, a village of France, in the dep. of Cote-d'Or, cant. and 7 m. SE of Chatillon-sur-Seine. Pop. 380.

VILLIERS-SAINT-BENOIT, a village of France, in the dep. of Yonne, 18 m. SSW of Joigny, on the r. bank of the Ouanne. Pop. 800.

VILLIERS-SAINT-GEORGES, a commune and v. of France, in the dep. of Seine-et-Marne, arrond. and 8 m. NE of Provins. Pop. 838.

VILLIERS-SAINT-PAUL, a village of France, dep. of Oise, cant. and 2 m. NE of Creil. Pop. 500.

VILLIERSTOWN, a chapelry and v. in co. Waterford, 3½ m. S of Cappoquin. Pop. 159.

VILLIGAS, a village of the Swiss cant. of Aargau, in the bail. of Brugg. Pop. 733.

VILLINGEN, a town of the grand-duchy of Baden, on the Bug, 40 m. NW of Constanz, in N lat. 48° 3′ 12″. Pop. 3,700. It is surrounded by walls and ditches, and is finely supplied with fountains of water. Its manufactures consist of woollen and hosiery goods, leather, articles of cutlery and linen.—Also a village in the duchy of Hesse, in the landg. of Hangen. Pop. 900.—Also a parish and v. of Würtemberg, in the Schwarzwald circle, bail. of Rottweil. Pop. 572.

VILLMAR, a town of the duchy of Nassau, in the bail. and 1 m. SE of Runkel, on the l. bank of the Lahn. Pop. 1,770. It has marble and granite works.

VILLMERGEN, a village of Switzerland, in the cant. of Aargau, circle of Bremgarten, 10 m. ESE of Aarau. Pop. 1,594.

VILLNACHEM, a village of the Swiss cant. of Aargau, bail. of Brugg. Pop. 492.

VILLOGNON, a village of France, in the dep. of Charente, cant. and 4 m. WSW of Mansle, on the l. bank of the Charente. Pop. 500.

VILLOLDO, a village of Spain, in the prov. of Palencia, 9 m. from Carrion-de-los-Condes, on the l. bank of the Carrion. Pop. 300.

VILLONNE, a village of France, in the dep. of Meuse, cant. and 5 m. SSE of Dun-sur-Meuse, on the Meuse. Pop. 450.

VILLORA, a village of Spain, prov. and 48 m. SE of Cuenca. Pop. 300.—Also a village in the duchy and 4 m. SE of Lucca.

VILLORIA, a village of Spain, in the prov. and 30 m. SW of Leon, on the r. bank of the Orbigo. Pop. 768.—Also a village in the prov. and 18 m. E of Salamanca. Pop. 665.

VILLORUELA, a village of Spain, in the prov. and 12 m. from Salamanca. Pop. 600.

VILLOSANGES, a commune and v. of France, in the dep. of Puy-de-Dome, cant. and 4 m. NW of Pont-au-Mur. Pop. 1,100.

VILLOSLADA, a town of Spain, in the prov. and 24 m. from Logrono. Pop. 1,900. It has woollen factories.

VILLOTTE, a village of France, in the dep. of Vosges, cant. and 2 m. N of La Marche. Pop. 380.—Also a village in the dep. of Melise, arrond. of Bar-le-Duc. Pop. 600.—Also a village in the dep. of Cote-d'Or, cant. and 7 m. NE of Vitteaux, on the Ource. Pop. 380.

VILLOUXEL, a village of France, in the dep. of Vosges, cant. and 6 m. W of Neufchatel.

VILLOVELA, a village of Spain, in the prov. and 32 m. S of Burgos, on the Esguera. Pop. 410.

VILLUSTO, a village of Spain, in the prov. and 21 NW of Burgos. Pop. 315.

VILLY-EN-AUXOIS, a village of France, in the dep. of Cote-d'Or, cant. and 5 m. ENE of Vitteaux. Pop. 798.

VILLY-EN-TRODE, a village of France, in the dep. of Aube, cant. and 7 m. N of Bar-sur-Seine, on the Bodronne. Pop. 550.

VILM, a small island off the coast of Prussian Pomerania, to the SE of the island of Rugen. Pop. 10.

VILMANSTRAND, a fortified village of Russian Finland, in the gov. and 32 m. NW of Viborg, on the S bank of the Saïma lake. Pop. 320. On the 23d of Aug. 1741, the Russians won an important victory over the Swedes in the vicinity of this place, which was followed by the treaty of Abo.

VILNA, or WILNA, a government in the W of European Russia, comprising a portion of the ancient Lithuania, and extending from the parallels of 53° 35′ to 56° 25′ N. It is bounded on the N by the gov. of Courland; on the E by that of Minsk; on the S by the gov. of Grodno, and by Poland; and on the W by Prussia, and for a few miles by the Baltic. Its area is estimated at 12,325 geog. sq. m. Pop. in 1846, 863,700; in 1851, 898,000. Its surface is generally flat, and in some places marshy; in the E are a number of small lakes. All the rivers belong to the basin of the Baltic. The principal are the Niemen which separates this gov. from Poland; the Viliia, the Mertshanka, the Western Berezina, the Svizta, the Pevieja, the Dubiza, the Vindau, and the Southern Aa. Much of the surface is under wood. The soil is light, sandy, and fertile. Grain, hemp, and flax are largely grown. Iron ore occurs in several places. Manufacturing industry has received little development in this gov. hitherto. In 1833, there were 12 towns in this gov. with a united pop. of 64,213; and 191 villages, with a united pop. of 82,911. The gov. is administratively subdivided into 11 circles. This section of Poland was appropriated by Russia in the infamous partition of 1793.

VILNA, the capital of the above prov., is situated upon the Viliia, or Vileika, in N lat. 54° 41′ 2″, E long. 23° 16′ 16″, 170 m. SSE of Riga. Pop. about 36,000.

It contains two or three squares or open places, and a few good houses, and large public buildings, but a great proportion of the inhabitants are still lodged in old wooden houses. Among the public buildings are an observatory of some repute, a museum, a library of 30,000 vols., a synagogue, and a Tartar mosque. The Roman Catholic educational institute has 11 professors, the medico-chirurgical academy, 23 professors. There were 37 public schools within the place in 1833, with 1,497 pupils. The academy of V. founded in 1578 was suppressed in 1832. The chief trade is connected with the transmission of agricultural and raw produce to the ports of Memel, Konigsburg, and Riga. This town was founded in 1322, and rapidly rose into eminence. Its pop. about the middle of the 16th cent. was 100,000. It has suffered from fire in 1737, 1748, and 1749.

VILS, a river of Bavaria, which rises to the E of Erding; runs ENE; and falls into the Danube, on the right bank, at Vilshofen, after a course of 72 m. —Also a river of Bavaria, which joins the Raab, on the r. bank, after a course of 60 m. from N to S, at Kallmunz.

VILS, a village of Tyrol, on the Bavarian frontier, 4 m. SW of Fussen. Pop. 588.

VILSBIBURG, a town of Bavaria, on the Vils, 12 m. SE of Landshut. Pop. 1,250. The district of V. has an area of 12·5 German sq. m., with a pop. of 25,000.

VILSECH, a village of Bavaria, in the presidial and 12 m. NNW of Arnberg. Pop. 1,300.

VILSEN, a village of Hanover, in the county of Hoya, bail. and 2 m. SW of Alten-Bruchhausen, on the Eycher. Pop. 972, chiefly linen weavers.

VILSENDORF, a walled town of Bavaria, at the confluence of the Vils and the Danube, 15 m. WNW of Passau, in N lat. 48° 38′ 20″. Pop. 2,180.

VILTERS, a parish and village of the Swiss cant. of St. Gall, in the circle of Sargans. Pop. 1,659.

VILUENA (La), a town of Spain, in the prov. and 6 m. SW of Calatayud. Pop. 615.

VILVESTRE, a town of Spain, in the prov. and 52 m. WSW of Salamanca. Pop. 1,904.

VILVIESTRE DEL PINAR, a village of Spain, in the prov. and 45 m. S of Burgos. Pop. 554.

VILVIESTRE-DEL-MUNO, a village of Spain, in the prov. and 12 m. WSW of Burgos. Pop. 270.

VILVORDE, a town of Belgium, in the prov. of S. Brabant, 7 m. NE of Brussels, at the influx of the Wolawe into the Senne, and on the canal to Brussels, and railway to Malines. Pop. 2,800, employed in manufactories of calicoes, glue, vermicelli, chemical substances, breweries, bleachfields, and oil and corn mills. The old castle of this town has been converted into a very extensive house-of-correction composed of 8 distinct buildings, each walled, and the whole surrounded by one large and strong wall. Tindal, the first English translator of the Bible, suffered martyrdom here in 1536.

VIMBUCH, a village of Baden, in the Middle Rhine circle, bail. of Buhl. Pop. 440.

VIMENET, a village of France, in the dep. of Aveyron, cant. and 6 m. ENE of Laissac. Pop. 1,108.

VIMERCATO, a town of Austrian Italy, in the prov. and 14 m. NE of Milan, on the Molgora. Pop. 3,800. There is here a fine church. The place has a trade in corn, wine, and silk.

VIMIEIRO, or Vimeiro, a village of Portugal, in the prov. of Estremadura, comarca and 21 m. N of Evora, in a valley watered by the Maceira. Pop. 300. Wellington here defeated Junot on the 21st August 1808.

VIMINES, a village of the Sardinian states, in Savoy, 3 m. SW of Chambery. Pop. 1,140.

VIMIOSO, a village of Portugal, in the prov. of Tras-os-Montes, 15 m. NW of Miranda, on the Sabor. Pop. 955.

VIMMERBY, a town of Sweden, in the prefecture and 75 m. NNW of Calmar, and haerad of Seevedco. Pop. 860.

VIMONT, a commune of France, in the dep. of Calvados, and cant. of Troarn, 9 m. ESE of Caen.

VIMOUTIERS, a canton, commune, and town of France, in the dep. of the Orne, and arrond. of Argentan. The cant. comprises 20 com. Pop. in 1831, 14,218; in 1846, 13,974.—The town is 18 m. NE of Argentan, on the Vire, in a valley. Pop. 4,200. It has manufactories of linen fabrics, numerous tanneries and bleacheries, and an active trade.

VIMY, a canton, commune, and town of France, in the dep. of the Pas-de-Calais, and arrond. of Arras. The cant. comprises 28 com. Pop. in 1831, 17,417; in 1846, 17,881. The town is 7 m. NNE of Arras. Pop. 1,075.

VINADIO, a town of Sardinia, capital of a mandemento, in the prov. of Coni, div. and prov. and 21 m. WSW of Coni, on the l. bank of the Stura. Pop. 2,600. It has thermal baths, and in the environs are mines of argentiferous lead.

VINAGRE (Rio), a river of New Granada, in the dep. of the Cauca, which descends from the volcano of Purace, and after a short course, in which it forms a cascade 369 Parisian ft. in depth, joins the Cauca.

VINAIGRE, a mountain of France, in the dep. of the Var, and arrond. of Draguignan, between Frejus and Cannes. It runs from NE to SW, and is crossed by a road, from the summit of which the snowy crest of the Alps is visible at the distance of about 20 leagues.

VINALMONT, a department and commune of Belgium, in the prov. of Liege and arrond of Huy, watered by the Mehaigne. Pop. 248.

VINAMARCA, a lake of Peru, in the dep. of Cazco, on the confines of Bolivia. It forms the SE part of Lake Titicaca, from which it is separated by a long narrow peninsula. It is 75 m. in length from NW to SE, and 30 m. in medium breadth.

VINARA, a settlement in the Argentine prov. of Tucuman, 56 m. NNW of San-Jago-del-Estero.

VINAROZ, a judicial partido and town of Spain, in the prov. of Castellon-de-la-Plana, 9 m. NNE of Peniscola, to the SW of the embouchure of the Ebro. Pop. 10,000. It has no regularly formed port, but admits of the approach of small vessels, and has an active fishing and coasting trade, and a traffic in brandy and in salt.

VINASPRE, a town of Spain, in the prov. of Logrono, 9 m. SE of La-Guardia. Pop. 2,000.

VINAY, a canton and town of France, dep. of Isere, on the Trery, 17 m. W of Grenoble. Pop. in 1846, 3,386. It has steel manufactories and ironworks.—The canton had a pop. of 10,314 in 1846.

VINCA, a canton and town of France, dep. of Pyrenees-Orientales, 4 m. ENE of Prades, near the r. bank of the Tet. Pop. of town, 2,020; of cant., 12,160.

VINCENNES, a town of France, situated near the confluence of the Seine and Marne, about 3 m. E of Paris. Pop. in 1846, 7,697. It has distilleries, and saltpetre works, and vinegar manufactories. It is remarkable for its castle, built in a remote age as a country-residence of the royal family, and after successive enlargements, finished in the 14th cent.; but since Louis XIV. removed the court to Versailles, it has been used as a state prison. It is still of considerable strength, being surrounded by a deep ditch, and having nine towers of great height and solidity. The largest is surrounded by a separate ditch of 40 ft. in depth. It was here that the unfortunate Duc d'Enghien was shot on 21st March,

1804. Adjoining to the castle is a fine park, of 732 hectares, and a forest called the Bois-de-Vincennes, a favourite resort of the Parisians. Among the public institutions at V. is the *Ecole Normale de Tir*, founded in 1845. In 1842, a school for rifle practice was established here, the object being to instruct the 10 battalions of chasseurs who were formed at that time, and whose name is now so well-known throughout Europe. In 1845, this school was made a normal school, with a view of instructing professors and soldiers, who should teach the various *corps d'armée* in the different parts of France. Subordinate establishments were also formed at Grenoble, St. Omer, and Toulouse. One uniform system of instruction in rifle-practice has been adopted, and a school formed in every regiment in the service, where this system was taught to the soldiers by persons educated at V. The *Ecole Normale de Gymnastique* established at La Faisanderie, near V., on the 11th of January, 1853, is presided over by a General officer; and soldiers from the different regiments are regularly sent there, who, after completing their gymnastic education, return to their corps, and impart to their comrades the instruction they have received. Branch-schools have also been established at Arras, Metz, Strasburg, and Montpelier. Each of these receives 200 pupils, who must be soldiers of good character and constitution, active and energetic, and not less than three years in the army.

VINCENNES, a town of Knox co., Indiana, U. S., on the E bank of the Wabash, here 230 yds. wide, 120 m. W of Louisville, in N lat. 38° 43′, W long. 87° 25′, 100 m. above its junction with the Ohio, in a direct line, and nearly 200 m. by the course of the river. Pop. in 1840, 2,000; in 1850, 2,070. The streets are wide, and cross each other at right angles. Among the public edifices are a Roman Catholic cathedral, and architectural seminary. It has considerable trade in furs and peltry. V. was first settled by the French from Lower Canada in 1730.

VINCENT, a township of Chester co., Pennsylvania, U. S., on the Schuylkill. Pop. 1,630.

VINCENT, a parish and hamlet in Monmouth co., Van Diemen's Land, skirted on the W and NW by the Clyde river; on the S by the cos. of Stradbroke and Henry.

VINCENT (SAINT), a very beautiful island in the British West India group, of which the capital, Kingston, on its S coast, lies in 13° 5′ N lat., and 61° 15′ W long. The island is about 17 m. long and 10 broad, and lies 95 m. W of Barbadoes. St. Vincent is in many parts very fertile; and is extremely well calculated to produce sugar of the best quality, and every other article cultivated in the neighbouring islands. It is watered by above 20 small rivers; the country is, however, in many parts mountainous, and has suffered by volcanic eruptions. A bold ridge of hills runs through it from N to S, at nearly equal distances, on both sides, from the coast. In its NW quarter is the celebrated volcanic mountain La-Souffriere. The climate is humid. The principal productions are sugar, rum, molasses, arrow-root, and dye-woods. The island contained, in 1783, 61 sugar-estates, 500 acres in coffee, 200 acres in cocoa, 400 in cotton, 50 in indigo, and 500 in tobacco, besides land appropriated to the raising of provisions, such as plantains, yams, and maize. In 1800, the quantity of sugar made amounted to 16,518 hds. In 1812, the quantity and value of the productions were stated as follows: sugar, 17,491 hds.; rum, 7,358 punchs.; molasses, 2,718 casks; coffee, 8,740 lbs.; cotton, 208,690 lbs. Estimated value of the preceding articles, £541,220. Estimated value of miscellaneous articles, including cattle, esculents, and fruits, £270,861. Estimated value of exports to the United Kingdom, £472,815; to all other parts, £43,186. Estimated value of imports from the United Kingdom, £154,995; from all other parts, £42,093. In 1845, the exports were £189,241; in 1851, £218,521 in value; the imports in 1851 were £198,679. Among other articles of recent export is pozzolano, a description of earth with which the island abounds, and of a quality far superior to what may be had from any other island. Nearly 2,000 tons have been exported hence to Bermuda during the last three years, where it is used by government upon the erection of an extensive breakwater at Ireland island, and is found to be superior to anything hitherto used for similar purposes. It has also been recently analysed by eminent chemists in England, and pronounced by them an invaluable manure. The pop. in 1812 was 1,280 whites, 1,172 free coloured, and 27,156 slaves. The slave-pop. in 1825 amounted to 24,252. The pop. in 1844 was 27,248; in 1851, 30,128.—The gov. is vested in a governor, a council of 12, and a house-of-assembly of 19. The gov. of St. V. has attached to it 8 small islands: viz. Union, containing 2,150 acres; Bequia, 3,700; Canouane, 1,777; Mustique, 1,200; besides the smaller islands of Petit St. Vincent, Petit Martinique, Maillereau, and Balleseau, each of which produces some cotton. Among these, Bequia is valuable, from the commodiousness of its fine harbour, called Admiralty bay.

History] St. V. was discovered by Columbus; but it does not appear that the Spaniards ever formally took possession of it. When discovered it was inhabited by the Yellow Caribbs, a warlike people, who were numerous on this island, attracted perhaps partly by its natural fertility, and partly by its affording them a point of rendezvous in their expeditions to the continent. About 1680, a ship from Guinea, with a large cargo of slaves, was either wrecked or run on shore upon the island of St. Vincent, when great numbers of the Negroes escaped into the woods and mountains. These people, in contradistinction to the aborigines, received the name of Black Caribbs, and, by intermarriages with the natives, and the accession of runaway Negroes from Barbadoes, became at length so very powerful, that in the early part of the last cent. they compelled the native Caribbs to evacuate in their favour the NW part of the island. In 1672, Charles II. attempted to assume the sovereignty of this and the neighbouring islands of St. Lucia, Dominica, and Tobago, and until 1748 great contentions prevailed between England and France respecting them; by the treaty of Aix-la-Chapelle, they were declared neutral, and such of the European proprietors as remained were left in undisturbed possession. The peace, however, which terminated the Seven years' war in 1763, gave to Great Britain in perpetuity the islands of St. V., Dominica, and Tobago, while the sovereignty of St. Lucia was allotted to France. In this treaty no mention was made of the Caribbs. In 1779, during the French American war, the island of St. V. was captured by a body of French troops sent from Martinique. The terms of capitulation, however, were favourable, and the island was restored to the dominion of Great Britain by the general pacification of 1783. The Caribbs continued to harass and distress the British from the time the island was first ceded in 1763, until their final removal in 1797. Instigated by the French residing on the neighbouring islands, who landed troops on St. V. in 1795, they waged a new war with the British settlers, which continued for more than two years. The ravages and devastations committed by them during this sanguinary war, cost the proprietors of the lands fully one-third of the value of their estates; but they were at length subdued, and removed by order of government to the island of Rattan, in the bay of Honduras, with provisions sufficient to subsist them for a time, and implements of husbandry to enable them to cultivate the land, and to raise that species of food to which they had been accustomed.

VINCENT, a small island in the N. Pacific ocean, at the entrance into Portlock's harbour, in N lat. 57° 48.

VINCENT (SAINT), or SAO-VICENTE, one of the Cape Verd islands, in N lat. 16° 54′ 42″, W long 25° 1′ 15″ [*Raper*], 2,340 m. from Plymouth. It is about 12 m. in length, and 9 m. in breadth. It is an old volcano, and nearly the whole island is rocky, or covered with beds of lava, ashes, or sand; there are no trees left, and the little grass is generally withered up: sometimes no rain falls for a year and upwards. The water in the few wells is brackish. The harbour, Porto-Grande, however, is the best of the whole group; and being a coaling-place for the Brazilian, Cape, and Australian steamers, has become of some importance. There are a Portuguese governor, a few black soldiers, some prisoners, two or three Englishmen, including a consul, and a population of about 500 Negroes and their descendants, of all shades between jet black and whitey-brown. The inhabitants are of good stature, the complexion copper-colour, the hair dark and straight. A coarse woven material of peculiar manufacture is used as a dress. Their food consists of bananas, oranges, several kinds of vegetables, besides Indian corn, which they raise in small quantities. Their huts are formed of pieces of iron-stone, roofed with bamboo and thatched with grass. The language spoken is Portuguese.

VINCENT (Saint), a village and fort of France, in the dep. of Basses-Alpes, cant. and 3 m. WNW of Barcelonette, on the r. bank of the Ubaye. Pop. 700.—Also a commune and v. of France, in the dep. of Cantal, cant. and 6 m. NE of Salers, near the r. bank of the Mars. Pop. 1,400.—Also a commune and v. in the dep. of Herault, cant. and 2 m. W of Olargues. Pop. 1,500.—Also a commune and v. in the dep. of Haute-Loire, 7 m. N of Puy. Pop. 1,400. —Also a commune and v. in the dep. of Lot, cant. and 2 m. SE of Saint-Ceré. Pop. 1,050.

VINCENT (Cape Saint), or Cabo-de-Sao-Vicente, a rocky peninsula, forming the extreme W point of Portugal, and also of Europe, in N lat. 37° 2′ 54″, W long. 9° 0′ 0″. It is noted for the naval victory over a Spanish naval force gained off it on 14th February 1797, by Sir John Jarvis, afterwards Earl St. Vincent.—Also a cape on the W coast of Madagascar, in S lat. 21° 54′.

VINCENT (Saint), a village of the Sardinian states, 2 m. E of Chatillon, at the foot of a high rock of the same name, in the prov. of Aosta.

VINCENT (Point Saint), a point of land on the W coast of Van Diemen's Land, in S lat. 43° 25′.

VINCENT BAY (Saint), a bay on the N coast of Tierra-del-Fuego, a little to the E of Cape St. Vincent. The country about the bay is in general flat.

VINCENT (Gulf of Saint), a large inlet on the S coast of Australia, extending between the parallels of 34° and 35° 40′ eastward, about 45 m., and from thence in a N direction, about 60 m. further. It is separated from Spencer gulf by York peninsula on the W. At its mouth it is not more than 9 m. broad, but it afterwards widens to 40 m. See article Adelaide.

VINCENT (Point), a cape on the coast of New Albion, in the N. Pacific, in N lat. 33° 44′.

VINCENT D'ARDENTES (Saint), a town of France, dep. of Indre, 7 m. E of Chateauroux. Pop. 1,800.

VINCENT-DE-CONOYAC (Saint), a village of France, in the dep. of Dordogne, 7 m. SSE of Riberac. Pop. 920.

VINCENT-DE-RHINE (Saint), a commune and v. of France, in the dep. of Rhone, cant. and 4 m. WSW of Saint-Mezier. Pop. 2,516 in 1846.

VINCENT-DE-RIVEDOLT (Saint), a commune and v. of France, in the dep. of Lot, 7 m. WNW of Cahors. Pop. 1,100.

VINCENT-DES-BARRES (Saint); a village of France, in the dep. of Ardeche, cant. of Rochemaure, 7 m. SE of Privas. Pop. 500.

VINCENT-DES-LANDES (Saint), a commune of France, in the dep. of the Loire-Inferieure, and cant. of Derval, 11 m. SW of Chateaubriant. Pop. 1,125.

VINCENT-DES-PRES (Saint), a commune of France, in the dep. of the Sarthe, cant. and 6 m. SSE of Mamers. Pop. 1,236.

VINCENT-DE-TYROSSE (Saint), a canton and commune of France, in the dep. of the Landes, and arrond. of Dax. The cant. comprises 12 com. Pop. in 1831, 10,105; in 1846, 10,723. The village is 14 m. WSW of Dax. Pop. 754.

VINCENT-DU-LOROUER (Saint), a commune of France, in the dep. of the Sarthe, cant. and 4 m. S of Grand Luce, on the r. bank of the Veuve. Pop. 1,728.

VINCENT-SUR-GRAON (Saint), a village of France, in the dep. of the Vendée and cant. of Moutiers-les-Maufaits. Pop. 4,100.

VINCENT-SUR-OUST (Saint), a village of France, in the dep. of the Morbihan, cant. and 5 m. NE of Allaire. Pop. 1,000.

VINCENTON, a village of Burlington co., in the state of New Jersey, U. S., on the S branch of Rancocus river, at the junction of Stop-the-Jade creek and 19 m. S of Trenton. Pop. in 1840, 200.

VINCENZO (San), a market-town of Naples, in the prov. of Terra-di-Lavoro and district of Piedimonte, cant. and to the W of Castellone. Pop. 800.

VINCHIATURO, a market-town of Naples, in the prov. of Sannio, district and 5 m. SSW of Campobasso, and cant. of Baranello, on a height. Pop. 3,100. It has two churches and a convent. In the environs are the abbey of Monteverde, and several mineral springs.

VINCHUTA, a village of Bolivia, in the prov. of Moxos, near the r. bank of the Paracti, an affluent of the Chapare, and 16 m. SSW of Chimore.

VINCI, a market-town of Tuscany, in the prov. and 15 m. W of Florence, and vicariat of Empoli, in the Val-di-Nievoli. Pop. 980. It is noted as the birth-place of Leonardo-di-Vinci.

VINCON, a river of France, in the dep. of the Haute-Vienne, which has its source 5 m. from Compreignac; runs first W, then NW; receives the Basine at Bellac, and, after a course of 22 m., joins the Gartempe on the l. bank, 3 m. W of Bellac.

VINCZ (Fel), or Ober-Wincz, a town of Transylvania, capital of the stuhl and 4 m. S of Karlsburg, on the l. bank of the Maros. Pop. 3,500. Its industry is chiefly agricultural.

VINDAU, a river of Russia in Europe, which has its source in the NW part of the gov. of Vilna, to the SE of Telch, runs NW, and after traversing the W part of Courland, throws itself into the Baltic at Vindau, and after a course of about 180 m. It is joined at Niomon by a canal.—The town of V. is in the district and 36 m. NW of Goldingen, on the Baltic, on the l. bank of the embouchure of the river of the same name. Pop. 3,000. It has a fortress, and carries on an active trade in corn, timber, flax, hemp, and salted meat. This town is of considerable antiquity, and was formerly of importance.

VINDEFONTAINE, a commune of France, in the dep. of the Manche, and cant. of Lahaye-du-Puits, 36 m. N of Coutances. Pop. 1,050. It has numerous manufactories of pottery.

VINDEL, a town of Spain, in the prov. and 36 m. of Cuenca. Pop. 360.

VINDEL-ELV, a river of Sweden, in the prefecture of Western Bothnia, which has its source on the E side of the Dofrines, runs SE and joins the Umea-Elf, 9 m. NW of the town of that name, and after a course of 240 m. It forms, in the upper part of its course, an extensive lake.

VINDEROD, a parish of Denmark, in the island of Seeland and bail. of Fredericksburg, on the N bank of Lake Arre, which is connected by a canal with the gulf of Roeskilde, and 30 m. NW of Copenhagen. It contains the village of Frederikswoerk.

VINDHYA MOUNTAINS, a chain of mountains in Hindostan, extending from E to W, between the parallels of 22° and 24° N, from Benares to the gulf of Cambay. It separates the basin of the Ganges from that of the Nerbudda, parallel to the l. bank of which it runs in nearly its entire length. Its elevation is from 2,000 to 2,600 feet. Its highest points are the Chaizgur, alt. 800 metres, and the Ambawara, alt. 585 metres.

VINDICARI, a town and fort of Sicily, in the prov. of Syracuse, district and 9 m. S of Noto, on the Ionian sea. It has a small fort. In the vicinity is a group of islands of the same name.

VINDO, an island of the gulf of Bothnia, near the E coast of Sweden, in N lat. 59° 20', and E long. 38° 36'. It is comprised in the prefecture of Stockholm.

VINDUR, a river of Beluchistan, an affluent of Sonmeani bay. It forms, in rainy weather, an impetuous torrent, but in summer becomes dry.

VINEUF, a village of France, in the dep. of the Yonne, cant. and 7 m. W of Sergines. Pop. 1,120.

VINEUIL, a commune of France, in the dep. of the Loir-et-Cher, cant. and 2 m. E of Blois, near the r. bank of the Cosson. Pop. in 1846, 2,255.

VINEYARD-MILLS, a village of Hundingdon co., in the state of Pennsylvania, U. S., 65 m. W of Harrisburg.

VINGA. See Theresiopel.

VINGEANNE, a river of France, which has its source in the dep. of the Haute-Marne, flows thence into that of the Cote-d'Or, and after a course in a generally S direction of about 27 m., joins the Saone on the r. bank, 5 m. NE of St. Jean-de-Losne.

VINGER, a parish of Norway, in the diocese of Aggershuus and bail. of Hedemarten. Pop. 6,200.

VINGOADOMPY, a country of Upper Guinea, on the gold coast, in the empire of Ashantee, bounded on the S by the gulf of Guinea, and on the N by the kingdom of Assine.

VINGORLA, a town of Hindostan, in the presidency of Bombay, prov. of Bejapur and district of Concan, 33 m. NNW of Goa, on the sea of Oman. It has a fort. Opposite are Burnt islands, a small rocky group surrounded by reefs.

VINHAES, a small town of Brazil, in the centre of the island of Maranhão, on a river of the same name. The houses are constructed of wood, thatched, and built with some regularity. Its inhabitants, who are chiefly Indians, manufacture mats and cordage.

VINHAES, a town of Portugal, in the prov. of Tras-os-Montes, comarca and 45 m. NW of Miranda. Pop. 525. It is enclosed by walls, with two gates, an old fort flanked with two towers, two parish-churches, a convent, and two hospitals, and carries on an active trade in wine and corn.

VINICZA, a market-town of Civil Croatia, in the comitat and 5 m. NW of Warasdin. It has a castle.

VINIEGRA-DE-ABAXO, a market-town of Spain, in the prov. and 39 m. from Soria, near the Najerilla. Pop. 855.

VINIEGRA-DE-ARRIBA, a market-town of Spain, in the prov. and 33 m. from Soria, in a cold and elevated situation. Pop. 508.

VINKOVCZE, or Vinkovicz, a market-town of Austria, in Hungary, capital of the regimentary district of Brod, and 24 m. SSE of Essek, in a wild and picturesque situation on the Bossuth, in N lat. 45° 17' 33", and E long. 18° 51' 20". Pop. 2,288. It has a Greek and two catholic churches, and a German Normal school.

VINKOVTZY, a town of Russia in Europe, in the gov. of Podolia, district and 33 m. N of Ouchitza.

VINKRATH, a village of Prussia, in the regency of Dusseldorf and circle of Kempen. Pop. 580.

VINN, a village of Prussia, in the regency of Aachen, and circle of Heinsberg. Pop. 310.—Also a village in the regency of Dusseldorf and circle of Geldern. Pop. 246.

VINNEMER (Saint), a village of France, in the dep. of the Yonne, cant. and 8 m. W of Crucy-le-Chatel, near the Armancon. Pop. 800.

VINNEN, a village of Hanover, in the principality of Osnabruck, and bail. of Ahremberg-Meppen. Pop. 350.

VINNESAS, a market-town of Spain, in the prov. and to the NW of Soria, on the Duero. Pop. 1,328. It has yarn and paper-mills.

VINNINGEN, a parish and village of Bavaria, in the Pfalz, and cant. of Pirmasens. Pop. 1,076. It has a paper-mill.

VINNITZA, a town of Russia in Europe, capital of a district in the gov. of Podolia, 84 m. NE of Kamenetz, on the r. bank of the Bug, in N lat. 49° 14' 16", and E long. 28° 27' 20". It is enclosed by a ditch, and has a castle, several Catholic churches, and, in the old Jesuits' college, a gymnasium. Pop. 2,000, of whom a large proportion are Jews.

VINNEUF, a commune of France, in the dep. of the Yonne, and cant. of Sergines, 18 m. N of Sens. Pop. 1,401. It has manufactures of hardware, and of cordage, and lime-kilns, and a considerable trade in cattle and in hemp.

VINNOVO, a village of Sardinia, in Piedmont, in the prov. of Turin. Pop. 3,007.

VINNUM, a village of Prussia, in the regency of Munster, and circle of Ludinghausen. Pop. 500.

VINSEBECK, a village of Prussia, in the regency of Minden, and circle of Horter. Pop. 650.

VINSOBRES, a commune of France, in the dep. of Drome, cant. and 5 m. SW of Nions, in a fertile locality. Pop. 1,576. It has a silk spinning-mill, and a considerable trade in olive-oil.

VINTAM, a town of Senegambia, in the kingdom of Firia, on a river of the same name, which a little below throws itself into the Gambia, near the mouth of that river, and nearly opposite fort James.

VIN-TAM-PHU, a trading town of Lower Cambodia, on the Pong-som river.

VINTIMIGLIA, or Ventimiglia, a town of Sardinia, capital of a mandemento, in the div. and 20 m. ESE of Nice, and prov. of St. Remo, at the confluence of the Roya with the Mediterranean. Pop. 5,000. It has the remains of a fortress. The environs are noted for their oil, wine, and fruit.

VINTON, a county in the SE part of the state of Ohio, U. S., comprising an area of 416 sq. m. drained by Racoon creek and its affluents, and by branches of Scioto river, and intersected by the Scioto and Hocking valley, and by the Cincinnati and Marietta railways. It is hilly, but generally fertile, and has some coal. Pop. in 1850, 9,353.—Also a village of Benton co., in the state of Iowa, on a branch of Cedar creek.—Also a village of Gallia co., in the state of Ohio, on the W side of Racoon river, and 75 m. SE by E of Columbus.

VINUDAUA, a mountain of British Guayana, in S lat. 1° 50'. W long. 0° 14', near the sources of the Takutu.

VINUELAS, a town of Spain, in the prov. and 15 m. NW of Guadalaxara, in a fertile plateau. Pop. 340. It has manufactories of linen.

VINUESA, a town of Spain, in the prov. and 21 m. NW of Soria, on the l. bank of the Duero. Pop. 1,310.

VINXEL, a village of Prussia, in the regency of Cologne and circle of Sieg. Pop. 270.

VINZAGLIO, a market-town of Sardinia, in the div. of Novara, and prov. of Lomellina, 5 m. E of Vercelli, near the Sesia. Pop. 1,128.

VINZELBERG, a village of Prussia, in the reg. of Magdeburg, and circle of Gardelegen. Pop. 240.

VINZELLES, a village of France, dep. of the Puy-de-Dome, cant. and 8 m. N of Lezoux. Pop. 560.

VINZENSENBRONN, a village of Bavaria, in the circle of Middle Franconia, and presidial of Radolzburg. Pop. 210.

VINZIER, a village of Denmark, in Holstein. Pop. 200.

VIOL, a parish of Denmark, in the duchy of Heswig, to the N of Husum. Pop. 1,400.

VIOLET, a township of Fairfield co., in the state of Ohio, U. S. Pop. in 1840, 2,378.

VIOLEY, a commune of France, in the dep. of the Loire, cant. and 7 m. E of Neronde, 27 m. from Roanne. Pop. in 1841, 1,585.

VIOMENIL, a village of France, in the dep. of the Vosges, cant. and 8 m. NW of Bains, near the source of the Saone. Pop. 598.

VION, a village of France, in the dep. of the Ardeche, cant. and 5 m. NNW of Tournon, near the r. bank of the Rhone. Pop. 650.

VIONNAZ, a village of Switzerland, in the cant. of Valais, and bail. of Mouthey. Pop. 776.

VIQUE. See VICH.

VIQUE. See CRAB ISLAND.

VIRA, a hamlet of Sweden, in the prefecture and 30 m. NE of Stockholm. There is here a government manufactory of fire-arms.

VIRANDEVILLE, a village of France, in the dep. of La-Manche, cant. and 6 m. SW of Octeville. Pop. 600.

VIRANTSHEPURA, a town of Hindostan, in the Carnatic, on the r. bank of the Palaar, 30 m. W of Arcot.

VIRAZEIL, a commune and v. of France, in the dep. of Lot-et-Garonne, cant. and 3 m. E of Marmande. Pop. 1,200.

VIRE, a river of France, which rises in the SW of the dep. of Calvados, and flows NNW, past Saint-Lö, to the English channel, into which it flows, near Isigny, after a course of 70 m. It is navigable to Coquet. Its principal affluent is the Aure.

VIRE, a town of France, dep. of Calvados, situated near the river Vire, 23 m. SE of Saint-Lö. Pop. in 1789, 6,786; in 1821, 8,453; in 1846, 7,658. It has extensive manufactories of woollens, chiefly of the coarser qualities, paper-mills, and cotton-yarn spinning-mills. In time of war, it supplies to government a large proportion of the army-clothing.—The arrond. of V., comprising 6 cants., has an area of 95,637 hectares. Pop. in 1846, 84,048.

VIREY, a commune and v. of France, in the dep. of Manche, cant. of Saint-Hilaire-du-Harcouet. Pop. 1,500.

VIRGIL, a township of Courtland co., U. S., 135 m. W of Albany. Pop. 2,410.—Also a township of Fulton co., in Illinois, 64 m. NW of Springfield.

VIRGIN, a river of Utah territory, U. S., which rises in the Wahatsch mountains, and flows S to the Rio-Colorado.

VIRGIN (CAPE), a steep white cliff not unlike the North Foreland, which forms the N point of the entrance of the strait of Magalhaens, in S lat. 52° 24′.

VIRGIN-GORDA, one of the Virgin islands, in the West Indies, in N lat. 18° 30′. It is of irregular figure, 8 m. in length from NE to SW. It lies 11⅜ nautical m. S of Anegada, and forms the easternmost islet of the main group. Primitive rocks predominate in its geological formation.

VIRGIN ISLANDS, a cluster of islands in the West Indies, situated to the E of Porto-Rico, between the parallels of 18° 5′ and 18° 50′ N. They are upwards of 90 in number, but for the most part desert and barren. The group extends 90 m. in length, and upwards of 30 m. in breadth. The English and Danes divide most of them; but the Spaniards claim those near Porto-Rico. The island of Virgin-Gorda, on which depend Anegada, Nicker, Prickley Pear, and Mosquito islands, Cammanoes, Dog-islands, the Fallen City, with the Round Rock, Ginger, Cooper's, Salt island, Peter's island, and Dead-Chest, belong to the English; they possess also Tortola, on which depend Jost, Van-Dyke's, Little Van-Dyke's, Guano, Beef, and Thatch islands. To the Danes belong St. Thomas's island, on which Brass, Little Saba, Buck island, Great and Little St. James, and Bird island, are dependent; with St. John, to which belong Lavango, Cam, and Witch islands; they have also Santa or St. Croix.—The first possessors of such of these islands as now belong to the British government were a party of Dutch Buccaneers, who settled at Tortola about the year 1648, and built a fort for their protection. In 1666, they were driven out by a stronger party of the same class of adventurers, who, calling themselves English, pretended to take possession for the crown of England; and the English monarch, if he did not commission the enterprise, made no scruple to claim the benefit of it, for Tortola and its dependencies were soon afterwards annexed to the Leeward island government, in a commission granted by Charles II. to Sir William Stapleton. The Dutch had made but little progress in cultivating the country when they were expelled from Tortola; and the chief merit of its subsequent improvements was reserved for some British settlers from the little island of Anguilla, who, about the year 1694, embarked with their families, and settled in the V. islands. In 1756, the inhabitants, whose number amounted to 1,263 whites, and 6,121 blacks, wished to be put on the same footing as the other islands, by the establishment of a civil government and courts of justice; but in this expectation they were not gratified until the year 1773. The chief and almost the only productions of these islands are sugar and cotton. The cultivated acres of the English islands, in 1812, amounted to 12,000. The pop. at the same date was 300 whites, 400 free persons of colour, and 10,000 slaves. In 1825 the slave pop. was returned at 5,436; and in 1844 the total pop. at only 6,689.

VIRGIN-ROCKS, a cluster of rocks in the Atlantic, 60 m. SE of Cape Race, off the coast of Newfoundland, in N lat. 46° 20′, W long. 50° 0′.

VIRGINAL-SAMME, a commune and v. of Belgium, in the prov. of Brabant, on the Samme, 16 m. S of Brussels. Pop. 1,200.

VIRGINIA, one of the United States of North America, situated between the parallels of 36° 30′ and 39° 43′ N, and between 75° 40′ and 83° 33′ W long. It is bounded on the N by Pennsylvania and Maryland; on the E by Maryland and Delaware bay; on the S by N. Carolina and Tennessee; and on the W by Kentucky and Ohio. Its length, from the Atlantic on the E, to the Cumberland mountains on the W, is 410 m. Its greatest breadth, from N to S, is 290 m.; its area is 64,000 sq. m., according to one computation; or 61,352 sq. m. according to another.

Physical features.] The Blue-ridge mountains divide V. from E to W; but the State-constitution recognises four physical districts, viz.: the Tide-water, or district below the lower falls of the larger rivers; the Piedmont, or district between these falls and the Blue ridge; the Valley, or district between the Blue ridge mountains and the Alleghany; and the Trans-Alleghany, comprising all the country W of the mountain-ranges. The first of these districts, extending 100 m. inland, from the coast to the termination of tide at Fredericksburg, Richmond, &c., is low and flat, sometimes fenny, sometimes sandy, and on the margin of the rivers is composed of a rich loam covered with a luxuriant and even rank vegetation. This part is very unhealthy in the months of August, September, and October. The next division extends from the head of tide-water to the Blue ridge, 150 m. The surface of this division near tide-water is level; higher up the rivers it becomes swelling; near the mountains it is often abrupt and broken. The soil is here divided into sections of unequal quality, running parallel to each other, and extending throughout the state. The pop. of this section, especially near the mountains, is more robust and healthy than that of any other part of the state. The scenery of the upper part of this division is highly picturesque and romantic. The third division, or valley between the Blue ridge and North and Alleghany mountains, extends, with little interruption, from the Potomac, across the state, to N. Carolina and Tennessee. Its soil is a mould formed on a bed of limestone. The surface of the valley is sometimes broken by sharp and solitary mountains, the sides of which are nearly bare, or thinly covered with blasted pines. The bed of the valley is fertile, producing good crops of Indian corn, wheat, rye, oats, buckwheat, hemp, flax, and clover. This valley has inexhaustible mines of excellent iron ore. The fourth division, extending from the Alleghany mountains to the river Ohio, is wild and broken, in some parts fertile, but generally barren, but having mines of iron, lead, coal, and salt. The Blue ridge, although pierced by the Potomac, the James, and the Staunton rivers, constitutes a continuous chain of 250 m. in length, rising in the peaks of Otter to an alt. of 4,260 ft. The Kittatiny, or Blue mountain, forms the centre of the great table-land of V. To the W of this lie several detached masses; and still further W is the Alleghany chain, of which Powell's mountain [alt. 4,500 ft.] is an outlier or spur. To the W of the Alleghany, the surface slopes to the Ohio.

Rivers.] The principal rivers of this state are the Potomac, which after a course of 350 m. reaches the Chesapeake; the Shenandoah, a principal tributary of the Potomac; the Rappahannock, which reaches its tidal point at Fredericksburg; the James river, which is the principal river wholly belonging to this state, and is navigable to Richmond, 100 m. from its mouth, by vessels of 140 tons; the Nottoway and the Meherrin, which unite to form the Chowan; the great rivers of Western V. flowing to the Ohio are the Monongahela, the Kanawha, the Greenbrier, the Covington, and the Guyandotte.—The bay of Chesapeake extends inland 200 m., to its termination in Maryland. Between the capes, its width is 12 m.; a little above, it increases to 30 m.; and gradually diminishes to 5 m., at its northern extremity.

Climate.] The states of V. and Maryland lie between those parallels which include the finest climate in the old continent, viz.: Marocco, Fez, Syria, Asia Minor, Greece, Sicily, Naples, and the southern provinces of Spain. Jefferson, in his 'Notes on Virginia,' observes, that, proceeding on the same parallel of lat. westwardly, the climate becomes colder, till you reach the summit of the Alleghany ridge. Thence descending to the Mississippi, the temperature again increases, and to such an extent, that the climate is several degrees warmer than in the same lat. on the shores of the Atlantic. The total range of the therm. is from 6° below zero to 98°. Of late years, snow does not lie below the mountains more than a few days; the rivers seldom freeze, and the heat of summer is also more moderate. The temp. is much influenced by the winds; those from the N and NW bringing cold and clear weather; those from the SE haziness, moisture, and warmth. The pleasantest months are May and June: July and August are intensely hot: September and October are generally rainy.

Productions.] The principal forest-trees in the state of V. are the wild or sweet-scented crab, ash, aspen, beech, black and white birch, catalpa, cherry, chestnut, horse chestnut, cucumber-tree, cypress, dogwood, elder, elm, fir-hemlock, spruce, fringe or snow-drop tree, sweet gum, hawthorn, hickory, Indus red-bud; juniper or Virginia cedar, laurel-swamp, linden or American lime, the locust, the sugar and red-flowering maple, red mulberry, black chestnut, red and white oak, *pacan* or Illinois nut, *persimon*, the black spruce, white and yellow pine, plane-tree, poplar, sassafras, spindle tree, and black and white walnut. The forests have little underwood; and it is easy to travel through them on foot or on horseback, except on the lowlands, in the eastern parts, which are covered with cedars, pines, and cypresses. A few wild animals are still found in the western parts, namely, the wolf, the bear, the deer, the racoon, the squirrel, and the oppossum. At the approach of winter, the bear descends from the mountains in search of the fruit of the persimon tree, when it is pursued and taken by dogs. On the E side of the mountains, animals have become rare, and peltries are no longer an article of exportation.—Among the bird kind is the wild turkey, which is yet found on the branches of the Kanawha, and other streams, where they weigh, when full grown, from 12 to 30 lbs. Partridges are numerous. The shell-drake, or canvas back duck, is found on James river. The *sora*, or American ortolan, appears with the first white frost, early in September, and disappears with the first black or hard frost,—an interval which varies from one to nine weeks. The turkey-buzzard, so called from its red gills resembling those of a turkey, is nearly of the size of the eagle. The Virginia nightingale, or mocking-bird, derives its name from its extraordinary imitation of all other songsters. The red bird and the humming-bird are admired for their beautiful plumage.—The rivers contain sturgeon, cat-fish, herring, perch, drum, carp, bass, oysters, cod, sun-fish, and crabs, the sea-dog, gar, ray-fish, sword-fish, and frog-fish. Some of the largest sturgeon weigh from 100 to 200 lbs. The shell-fish are oysters, lobsters, crabs, land-turtle, sea-turtle, loggerhead, and terrepin.

Soil and agriculture.] The soil of V. though often highly fertile, has been exhausted in some of the older settlements by an injudicious system of tillage. The E and Piedmont sections are chiefly engaged in the production of Indian corn and tobacco. The produce of Indian corn is from 12 to 50 bushels an acre, according to the nature of the soil. Of wheat, which is much cultivated, the greatest produce is about 50 b. an acre; but the average crop does not exceed 15 b., owing to the previous exhaustion of the soil by tobacco and Indian corn. White buckwheat, or French wheat, is raised in considerable quantities; and rice on the borders of the Dismal swamp, where it is very productive. Hemp is cultivated to a considerable extent, and has become a great article of export to the northern states: on the borders of rivers, and between the ridges of mountains, it is raised of such a quality as to bring from 150 to 300 dollars a ton. Almost every planter cultivates cotton for his own use; and along the Roanoke it is found to be more profitable than any other crop. From 5,000 to 10,000 bags, averaging each 300 lbs., are yearly brought to market, chiefly at Petersburg, and fetch as good a price as any short staple cotton. The culture of indigo is now abandoned. *Palma-Christi* is cultivated for the oil which it affords; and *benne*, from the seed of which, a fine oil is extracted in the proportion of 3 gallons to a bushel. The grasses are the white and red clover, which grow luxuriantly; the former natural to the country. Hay and oats are given for fodder; but not many years ago, leaves of Indian corn were chiefly used for this purpose. The following were the returns of the grain crops raised in the state in the years 1840 and 1850:

	1840.	1850.
Wheat, . .	10.104.716 bush.	11.232.616 bush.
Indian corn, .	74.577,591	35.254.319
Oats, . .	13.451.062	10.179,045
Rye, . .	1,482.799	458.980
Barley, . .	87.430	25.437
Buckwheat, . .	243.822	214.898
Rice, . .	2.596 lbs.	17,154 lbs.

Of miscellaneous crops, the returns in 1840 and 1850 were as follows:—

	1840.	1850.
Tobacco, . .	75.347.106 lbs.	56.803.218 lbs.
Cotton, . .	3.494.483	1,578.800
Hay, . .	364,708 tons.	369.048 tons.
Hops, . .	10.597 lbs.	11.506 lbs.
Silk cocoons, .	3.191	517
Maple-sugar, .	1,541.833	1,227.665
Wine, . .	13,911 galls.	5.498 galls.

The wool raised in 1850 was 2,860.765 lbs.; butter, 11.089,359 lbs.; cheese, 436,298 lbs.; value of slaughtered animals, 7,503,006 dols. The wool crop of 1840 was 2,538,374 lbs. The average crop per fleece in 1840 was 31·4 oz.; in 1850, 34·9 oz.

Minerals.] V. presents an immense field of mineral wealth. Iron-ore is in great abundance on the banks of James river. Brown scaly iron ore, or the brown oxide of iron, occurs on the Shenandoah. Plumbago, or carburet-of-iron, is in great abundance between the Blue ridge and the extremity of tide-water. Copper, in a native state, has been found in Orange co.; and the ore of this metal on both sides of James river, in Amherst co. Gold ore is found on both the North and Rapid-Ann rivers, the North and the South Anna, the Rivanna, and James river. Sulphuret-of-antimony is said to exist near Richmond. Manganese is found in Albemarle co., and also in Shenandoah, on the North mountain. Lead ore abounds on the banks of the Kanawha, and opposite the mouth of Cripple creek. Limestone is procured everywhere W of the Blue ridge. Talc, or soapstone, which is used for chimneys and tobacco pipes, is found; ochre occurs in different places, and when calcined, forms a valuable red paint. Coal is found in great abundance above Richmond, and on the Appomatox-branch of James river, where it extends in veins of 20 m. in length, and 10 m. in breadth, on an elevation of nearly 200 ft. above the level of the river. The coal of the Richmond basin is extensively wrought; that of the Wheeling basin, on the Ohio, is highly bituminous, and upwards of 16 ft. thick. A very rich coal bed occurs in the Great Kanawha. Saltpetre is found in considerable quantity. The salines of V. are almost wholly in the W part of the state, and important salt-works are situated on the Great and the Little Kanawha rivers. There are many mineral springs in V. The hot and warm springs of Bath co., the sweet springs of Monroe co., the sulphur-springs of Greenbrier and of Montgomery cos., and the baths of Berkeley co. are much frequented.

Manufactures.] In 1850, the state of V. contained 27 cotton-factories, 121 woollen factories, 122 iron manufactories, and 341 tanneries. Maple sugar is generally made, and distilleries and breweries are numerous. Ship-building is carried on to a considerable extent. There are rope-walks at Norfolk, Petersburg, Richmond, Alexandria, Fredericksburg, and Lynchburg. Sugars are boiled, baked, and refined, at Norfolk and Alexandria. The total amount of manufactures in 1810, was estimated at 12,263,473 d., besides those of a doubtful kind, amounting to 5,715,252 d., and consisting of flour, meal, maple, sugar, and saltpetre. The total capital invested in manufactures in 1840 was 11,360,861 d. The amount of cotton sheeting produced in 1850 was 15.640,107 yards; of cotton-yarns, 1,755,915 lbs. The amount of woollen cloth produced in 1850 was 2.037,025 yards; and of woollen yarn, 398,705 lbs. The capital invested in these manufactures in 1850 was 1,776,171 d.; the value of all the raw material used, 1,046,769 d.; the produce, 22,163 tons of pig iron, 5,577 of wrought iron, and 15,928 of cast iron; and the total value of the year's produce 2,451,335 d.

Commerce and Exports.] In respect of foreign commerce, V. ranks seventh or eighth among the states of the union. The chief exports are tobacco, wheat, Indian corn, lumber, tar, pitch, turpentine, beef, and pork. From the south are sent to Europe tobacco, wheat, flour, Indian corn, cotton, pease, white oak, staves, tar, pitch, turpentine, pork, bacon, ginseng, rattle and black snake root, indigo, oak bark, charcoal, lamp-black, panther, wolf, and squirrel skins. From the northern parts, hemp, saltpetre, gunpowder, lead, coals, cypress, and pine shingles, to the N of Europe and West India islands. The following is the amount of the value of the exports and imports for different years:—

	Exports.	Imports.
1791, . .	3,131,865 d.	...
1799, . .	6.292.986	...
1802, . .	3.978.363	...
1811, . .	4.822.307	...
1816, . .	8.212.860	...
1820, . .	4.557.957	...
1824, . .	3.277.564	639.787 d.
1830, . .	4.791.644	405,739
1835, . .	6.064.063	691.255
1840, . .	4.778.220	545.685
1845, . .	2.104.581	267.658
1850, . .	3,415,646	426,599

The shipping employed in the foreign commerce which entered in 1850 was 157 vessels = 30,96[illegible] tons, of which 88 = 18.775 tons were foreign vessels. The shipping cleared in 1850 amounted to 74,971 tons.

Railroads and canals.] The railroads and canals of this state have their central points at Alexandria, Richmond, and Norfolk, on the Atlantic, whence they extend westwards towards Ohio, Kentucky, and Tennessee. In the NW several lines connect the railway-systems of Pennsylvania and Maryland with those of Ohio and the West. In the SE, the Roanoke railroad furnishes an important route into N. Carolina. The great northern lines of railroad pass through Fredericksburg, Richmond, and Petersburg, and are continued to Wilmington and Columbia.—The canals are the James river and Kanawha canal, the Dismal Swamp canal, and the Alexandria canal, in continuation of the Baltimore and Ohio canal from Georgetown to Alexandria. The Dismal Swamp canal, 28 m. in length, connects Chesapeake bay with Albemarle sound.

Towns, &c.] V. is divided into 137 counties. Richmond is the metropolis of the state. The other most considerable towns are Norfolk, Petersburg, Lynchburg, Fredericksburg, Winchester, Portsmouth, Williamsburg, Staunton and Wheeling. There are 10 colleges, and 8 professional schools in this state. William and Mary college, at Williamsburg, founded in 1692, is the oldest college; but that of Central college, at Charlottesville, or the university of Virginia, as it is now called, is the principal institute. There were 3,904 schools in different parts of the state in 1850-1. The Methodists had, in 1850, 1,002 churches with 315,763 members; Baptists, 639 congregations with 241,689 members; the Episcopalians, 167 churches with 79,884 members; the Presbyterians, 296 churches with 101,625 members.

Government, &c.] The present constitution was ratified in 1851. It gives the right of voting to every white male citizen 21 years old, and resident in the state 2 years. The legislature is composed of a senate and house-of-delegates. The house-of-delegates consists of 152 members, chosen annually by districts, one-fourth being chosen every year. The senate consists of 50 members, one-half of whom are chosen biennially, two from each county, and one from several cities and boroughs. The governor is chosen annually, and holds office 4 years. There is also a lieutenant-governor, a secretary, a treasurer, and an auditor. At present V. elects 13 representatives to congress.—The judiciary is vested in a supreme court of appeal, district courts, and circuit courts.—The militia force in 1851 was returned at 125,128 men. All whites between 18 and 45 are subject to military duty.—The aggregate debt on 1st April 1851 was 11,248,962 dollars, exclusive of guaranteed bonds to the extent of 3,947,894 d.

Population.] The following table shows the decennial progress of the pop. as ascertained by census:

	Pop.		Slaves.	Increase.
1790,	748.308	including	298.427	
1800,	880.200		345.796	17·6 per cent.
1810,	974.622		392.518	10·7
1820,	1,065.379		425.158	9·3
1830,	1,211.405		469.757	13·7
1840,	1.299.797		448,987	2·3
1850,	1,421,661		472,528	14·6

The number of foreign births in the pop. of 1850 was 22,394, of whom 2,998 were Englishmen, 11.643 Irish, 947 Scotch, 5,511 German, 321 French. The whole number of dwellings in 1850 was 165,815; of families, 167,530. The number of paupers in 1850 was 5,118, of whom 4,933 were natives of the state. The expense of their relief during the year was 151,722 d.

History.] The English attribute the discovery of this part of the American continent to John Cabot; the French to Verrazam, who took possession of it in the name of Francis I.; but the first settlements were made in 1587, by a company of English merchants, under the auspices of Sir Walter Raleigh, who, in 1584, had obtained a grant from Queen Elizabeth, for all remote barbarous and heathen lands he should discover and settle. In 1606, a company of adventurers, London merchants, having obtained a patent from James I., sent two vessels under the command of Captain Newport, with a number of men, who laid the foundation of James town on the peninsula which advances into the river of the same name. In 1609, two little colonies, of 120 persons each, proceeding from James town, established themselves, the one at Nansemond, the other at Powhattan, an Indian town, which had been purchased from the king of the country, whose name it bore. The Indians, being afterwards disturbed in their possessions, waged an active war against the new colonists, who, reduced by famine from 500 to 60 persons, were on the point of departing for Newfoundland, when the arrival of Lord Delaware, with 3 vessels, containing provisions and stores, induced them to remain. In 1611, Lord Delaware returned to Europe, leaving about 200 men, and was replaced, as governor, by Sir Thomas Dale, who brought with him 3 vessels containing provisions and new settlers. In 1619, the company of Virginia sent thither a fleet laden with cattle, provisions, and about 1.300 men. Under the governorship of George Hardby, the culture of tobacco was encouraged, and a council and general assembly were instituted. Some years afterwards, the concession made by the company at London was revoked, and the province was placed under the immediate government of the Crown. The track S of 36° 30′ was separated from V. in 1630, and called Carolina; Maryland was taken from it two years afterwards. In 1661, the laws of England were adopted as the provincial laws of V. The colonists suffered great injury in 1673, from a Dutch squadron which ravaged the coast, and also from insurrections, which broke out in 1675 and 1676. In 1754, Washington surprised and took Fort-du-Quesne; but was afterwards obliged to yield to superior force. V. showed great opposition to the arbitrary measures of the British government, in 1765 and 1769. In 1781, and again in 1812, it became the theatre of war. V. has given no fewer than 7 presidents to the union.

VIRGINIA, a village in co. Cavan, 9 m. NW of Kells. Pop. in 1851, 859.

VIRGINIA, a village of Cass co., in the state of Illinois, U. S., 32 m. WNW of Springfield. Pop. in 1850, about 300.

VIRGINSTOW, a parish and hamlet of Devonshire, 6 m. NNE of Launceston. Area of p., 1,274 acres. Pop. in 1831, 136; in 1851, 173.

VIRIEU, a canton, commune, and town of France, in the dep. of the Isere, and arrond. of Tour-du-Pin. The cant. comprises 13 com. Pop. in 1831, 11,105; in 1846, 11,014. The town is 6 m. SSE of La-Tour-du-Pin, on the Bourbre. Pop. 1,825.

VIRIEUX-LE-GRAND, a canton, commune, and town of France, in the dep. of the Ain, and arrond. of Belley. The cant. comprises 13 com. Pop. in 1831, 7,748; in 1846, 7,953. The town is 8 m. NNW of Belley. Pop. 775. The locality affords good wine.

VIRIEUX-LE-PETIT, a village of France, in the dep. of the Ain, and cant. of Champagne, near the Seran and Arvierè. Pop. 618. In the vicinity is the fine cascade of Cerveysien.

VIRIVILLE, a commune of France, in the dep. of the Isere, cant. and 5 m. NNW of Roybon, on the Perouse Pop. in 1846, 2,092.

VIRLE, a village of Austrian Lombardy, in the prov. and district and 8 m. E of Brescia. Pop. 918. In the adjacent mountain of Rizzato are quarries of freestone and marble.

VIRNAUGH, a village of Hindostan, in the prov. and 37 m. SE of Cashmere, in N lat. 34°, E long. 74° 13′.

VIRNEBURG, a market-town of Prussia, in the reg. and 24 m. W of Coblenz, and circle of Adenau, on the Eifel. Pop. 300. It has a stone-quarry.

VIRNHEIM, a town of Hesse-Darmstadt, in the prov. of Starkenburg, and bail. of Bensheim, 8 m. NE of Manheim. Pop. 2,483.

VIROFLAY, a commune of France, in the dep. of the Seine-et-Oise, cant. and 3 m. E of Versailles. Pop. 1,025. It is surrounded by hills covered with coppice and wood, and numerous villas are scattered over it.

VIRTON, a department, commune, and town of Belgium, in the prov. of Luxemburg, and arrond. of Arlon. Pop. of dep. 1,701. The town is 13 m. SW of Arlon, on the Ton. It has several breweries, a cotton spinning-mill and potteries, and carries on an active trade in wood and iron.

VIRTY, or VIRTZERV, a lake of Russia in Europe, in the gov. of Livonia, between Derpt and Fellin. It is 30 m. in length from N to S, and 9 m. in extreme breadth, and is formed by the Embach, which issues from its NE extremity, and falls into Lake Peïpus. On the NW it discharges itself by the Fellin into the Navast, the latter of which streams flows into the gulf of Livonia.

VIRUA, a river of Brazil, in the N part of Gauyana, near the Serra-Pacarima.

VIRY, a village of France, in the dep. of the Saone-et-Loire, cant. and 5 m. NE of Charolles. Pop. 800.—Also a village of Sardinia, in the div. of Savoy and prov. of Carouge, mand. and 3 m. SW of St. Julien. Pop. 1,100.

VIRY-CHATILLON, a commune of France, in the dep. of the Seine-et-Oise, cant. and 5 m. ESE of Longjumeau, pleasantly situated on the slope of a hill. Pop. 404. It contains a fine castle, and many handsome villas. It has manufactories of white arms, and lime and gypsum-kilns, and is noted for its cream cheese. It possesses also considerable trade in coal and mill-stones.

VIRY-NOUREUIL, a commune of France, in the

dep. of the Aisne and cant. of Chauny, 27 m. W of Laon. Pop. 1,448.

VIRZON, an imperial chateau of Russia in Europe, in the gov. of Courland, and district of Mitau, on a river of the same name.

VISA. See VIZA.

VIS-AKNA. See SALZBURG.

VISAN, a commune of France, in the dep. of Vaucluse, cant. and 6 m. SSW of Valreas. Pop. in 1846, 2,280.

VISBECK, a parish and village of the duchy of Oldenburg, in the circle and bail. and 8 m. NE of Vechte. Pop. 2,400.—Also a village of Prussia, in the regency and circle of Arnsberg. Pop. 200.

VISCARDO (CAPE), a headland of the island of Cephalonia, one of the Ionian islands, at the N extremity of the island, opposite Cape Ducato, the SW extremity of Santa Maura, and to the NW of the channel of the same name, in N lat. 38° 27′ 10″, and E long. 20° 33′ 10″.

VISCARDO, or CEPHALONIA (CHANNEL OF), a channel of the Ionian islands, between Cephalonia and Theaki. It is 15 m. in length from N to S, and 3 m. in medium breadth.

VISCARI. See BISCARI.

VISCH. See FISH RIVER.

VISCHBECK, a village of Hanover, in the principality of Osnabruck. Pop. 500.—Also a village of Hesse, in the prov. of Lower Hesse. Pop. 700.

VISCHNAVETZ, a village of Austria, in Styria, in the circle of Cilly. Pop. 240.

VISCHNEI. See VOLOTSCHOK.

VISCIANO, a market-town of Naples, in the prov. of the Terra-di-Lavoro, district and 5 m. E of Nola, on a mountain. Pop. 1,415. It has four churches, two convents, and a marble-quarry.

VISCOS, a village of France, in the dep. of the Hautes-Pyrenees, cant. and 2 m. NW of Luz, near the Gave-de-Barreges. Pop. 114. In the adjacent mountains are quarries of marble and a lead-mine.

VISE, or WESET, a department, commune, and town of Belgium, in the prov. and arrond. of Liege, watered by the Maese. Pop. of dep. 2,051. The town is 9 m. NE of Liege, on the r. bank of the Maese. Pop. 1,333. It has a fine town-house, and manufactories of beet-root sugar, tiles and hosiery, and several breweries.

VISERNY, a village of France, in the dep. of the Cote-d'Or, cant. and 6 m. SW of Montbard.

VISEU, or VIZEU, a town of Portugal, capital of a comarca, in the prov. of Beira, 51 m. NE of Coimbra, and 168 m. NNE of Lisbon, in an elevated but fertile locality, covered with vines, olive plantations, and fruit-trees. Pop. 10,000. It is tolerably well-built, and is the residence of the governor of the prov. The cathedral, situated in the highest part of the town, is surmounted by two towers of Roman structure, one of which serves as a belfry. It contains also two parish-churches, two convents, two hospitals, and a college. The squares and promenades are spacious and handsome. The trade consists chiefly in cattle, and in articles of local manufacture. Pop. of comarca, 150,000.

VISEU, or VIZEU, a village of Brazil, in the prov. of Para, and comarca of Bragança, near the shore of the Atlantic, and to the E of the embouchure of the Tocantins. It is inhabited by Indians, who employ themselves in fishing, and in the culture of the soil. —Also a village in the prov. of Mato-Grosso, on the l. bank of the Rio-Corumbiara, an affluent of the Guapore. It is inhabited by Indians of the Gurrajus tribe.

VISGE, a river of France, which rises near La-Villate, in the dep. of Orense, and joins the Taurion, near St. Martin, after a course of 20 m.

VISINADA, a village of Illyria, in the gov. and 30 m. S of Trieste.

VISINGEN, a parish and v. of Sweden, in the laen and 18 m. NNE of Jönkœping, occupying the island of Visingsœ in Lake Wetter.

VISK, a town of Hungary, in the com. of Marmaroz, 9 m. SE of Honth, on the Theiss. Pop. 2,000. It has a trade in cattle and flax.

VISNIO, a town of Hungary, in the com. of Borsod, 7 m. NE of Apatfalva.

VISO-DEL-ALUR, a town of Spain, in the prov. and 15 m. E of Seville. Pop. 4,400.

VISO-DE-LOS-PEDROCHES, a town of Spain, in the prov. and 36 m. NNW of Cordova. Pop. 2,600.

VISO (EL), a neatly-built town of Spain, in the prov. and 40 m. N of Cordova. Pop. 2,700.

VISO (MONTE), a lofty mountain of the Cottian Alps, between Piedmont and France, in N lat. 44° 39′ 44″, E long. 7° 5′ 11″. It has an alt. of 1,945 toises = 4,150 yds. above sea-level.

VISONE, a town of the Sardinian states, situated at the confluence of the Carmagna and the Bormida, 3 m. E of Acqui. Pop. 500.

VISP, or VISPACH, a town of Switzerland, in the Valais, on the Visp or Visbach, 25 m. E of Sion.

VISTAVILLE-DEL-MAESTRAZGO, a town of Spain, in the prov. and 27 m. NW of Castellon-de-la-Plana. Pop. 1,300.

VISTRITZA, a river of European Turkey, the *Astræus* of the ancients, which issues from Lake Ostrovo in Romelia, and falls into the Vardar, a little above the influx of the latter stream into the gulf of Salonika, after an ESE course of 50 m.

VISTULA, or WEICHSEL, the great river of Poland, which rises in Austrian Silesia, at the foot of the Carpathians, in the outlying spur of the Schafberg, at an alt. of about 2,000 ft. above sea-level; and flowing eastward, enters Poland on the S frontier; passes Cracow; and after bathing the walls of Sandomir, receives, in the San, a great addition to its waters. Turning NNW, it traverses a considerable tract of country to Warsaw, at some distance from which it receives the Bug, a river almost equal to itself in magnitude, and bringing with it the waters of the SE and the N of Poland. The V., now become one of the great rivers of Europe, continues to hold a N course inclining to W; enters Prussia; passes the towns of Thorn and Culm; and after flowing with a wide channel and undiminished volume, 10 m. below Marienwerder, divides into two branches, of which one, called the Nogat, and another the Old Vistula, flow NE to the Frischehaff, while the largest stream preserves the name of V., and turning W, falls into the Baltic at Dantzic, after a course of 680 m. The V. flowing generally through a level country, is navigable as far up as Cracow. It is the great channel for the conveyance of corn and other products from the interior of Poland. It is connected by canal with the Netze, an affluent of the Oder. See articles DANTZIC, POLAND, and PRUSSIA.

VIT (SAINT), a village of France, in the dep. of Doubs, cant. and 4 m. WNW of Boussiere. Pop. 780.

VITA, a town of Sicily, to the SW of Palermo. Pop. 2,800.

VITAGLIANO (SAN), a village of Naples, in the prov. of Terra-di-Lavoro, 3 m. W of Nola.

VITCHEGDA, a river of Russia, which, rising near the NE frontier of the prov. of Vologda, flows W to the Dwina, into which it falls at Solvitschegodsk, after a course of 400 m.

VITCHUGA, a village of Russia, in the gov. of Kostroma, 18 m. SW of Kineshma. It wholly belongs to the Tatitchev family, and is handsomely planned and laid out.

VITENZ, or CHTELNITZA, a town of Hungary, 35 m. NNE of Presburg.

VITEPSK, a government of European Russia, bounded on the N by Pskov; on the NW by Livonia; on the E by Smolensk; on the S by Minsk and Mohilev; and on the W by Livonia, lying between 55° Y' and 57° 20' N lat. Its territorial extent is about 18,000 sq. m. Pop. in 1851, 742,811, partly Poles, Lithuanians, and Letts, partly Russians, Germans, and Jews. The surface is generally flat, and, though in the NE, the elevation is considerable. There are no mountains, and hardly any hills. Its chief rivers are the Dwina, which traverses its S part, the Ula, the Drissa, the Svoiana, the Lovat, and the Viteba, all belonging to the basin of the Baltic. The forests are very extensive; and potash is made in large quantities for exportation; hemp and flax are also raised for exportation. The pastures are good: cattle are reared and exported. There is also some traffic in honey and bees' wax. The manufactures consist of iron, brandy, woollen cloth, leather, pottery-ware, and glass.

VITEPSK, the capital of the above government, stands on the Dwina, at the influx of the Viteba, which divides it into two parts, in N lat. 55° 11' 43", 330 m. S of St. Petersburg. Pop. 17,000, of whom a considerable proportion are Jews. The town is surrounded by a wall, but made no regular defence in the campaign of 1812, having been alternately occupied by French and Russians, according to their superiority in the field. Though not well-built, it contains a proportion of neat wooden houses. Its principal manufactures consist of woollen and linen cloth.

VITERBO, a town of the States-of-the-Church, the capital of a deleg. of the same name, situated at the foot of Monte-Cimone, on the Arciona, in N lat. 42° 25', 39 m. NNW of Rome. Pop. 14,000. It is a place of antiquity, and is supposed by some to be the representative of the ancient *Fanum Voltumnæ.* It is still surrounded with a wall, and a number of round towers. Its streets are broad and well-paved, its market-place neat, and several of the principal buildings are constructed with taste. It is the see of a bishop, and has some manufactures on a small scale, and a trade in sulphur and in iron. Its chief attractions to strangers are the warm mineral springs in the neighbourhood, and the number of villas belonging to families of rank in Rome. About a ½ m. from the town is a small lake, called Bulicame, the waters of which emit a sulphureous smell, and appear to be in a state of continual agitation.—The deleg. has an area of 818 geog. sq. m., with a pop. of 121,000. The surface is mountainous towards the N. The lake of Bolsena lies near the centre of the deleg., and discharges itself by the Marta into the Tyrrhenian sea.

VITERSEN, a town of Denmark, in the duchy of Holstein, 5 m. W of Pinneberg.

VITEZ, a town of Bosnia, in the sanj. and 7 m. ESE of Traunick, on the Lashva.

VITH (SAINT), a village of Prussia, in the reg. and 34 m. S of Aix-la-Chapelle. Pop. 868.

VITICUSO, a village of Naples, in the prov. of Terra-di-Lavoro, 4 m. NW of Venafro.

VITIGUDINO, a village of Spain, in the prov. and 30 m. W of Salamanca. Pop. 1,000. Coarse linens are made here.

VITI-LEVU, the largest of the Fiji islands, in the S. Pacific, in S lat. 18° 11', W long. 178° 31'. It is about 80 m. in length by 50 m. in breadth, and appears to have a considerable population.

VITIM, a river of Asiatic Russia, which rises in the mountains skirting Lake Baikal on the E; runs NE and then NNW; and joins the Lena on the r. bank, after a course of 650 m. Its principal affluents are the Karengha, Kazatchia, Jaugha, Tzipa, Nerpa, and Nijniaia-Mama.

VITIMSK, a town of Asiatic Russia, on the Lena, at the confluence of the Vitim, 580 m. NNE of Irkutsk.

VITO (SAN), a town of Austrian Lombardy, in the prov. and 22 m. SW of Udino, on the Lemone. Pop. 4,000.—Also a town of Naples, in Abruzzo-Citra, 6 m. NE of Lanciano. Pop. 2,000.—Also a town of Naples, in Calabria-Ultra 2da, 15 m. SSW of Catanzaro.—Also a village of Sicily, 18 m. NE of Trapani.

VITO-DEGLI-SCHIAVI (SAN), a town of Naples, in the prov. of Terra-d'Otranto, 15 m. W of Brindisi. Pop. 3,600.

VITOLANO, a town of Naples, in the Principato-Ultra, 21 m. NW of Avellino. It has considerable manufactures of leather and woollen goods.

VITORCHIANO, a village of the Papal States, in the deleg. and 7 m. NNE of Viterbo. Pop. 1,000.

VITORIA, a town of Spain, the capital of the prov. of Alava, situated 30 m. SSE of Bilbao, partly on the slope of a hill, partly at the entrance of a beautiful valley watered by the Zadora. Pop. 10,000. It is surrounded by a wall, but is, with some exceptions, a scattered and ill-built place. It has various manufactories, amongst which are paper, pottery, leather, cabinet-ware, hardware, combs, and mirrors. Its trade consists chiefly in the transit between Castile and the northern provinces. The vicinity of this town was the scene of a general engagement on 21st June 1813, in which the French, under Jourdan, were defeated by Wellington.

VITRAC, a village of France, in the dep. of Cantal, cant. and 3 m. S of Mamet. Pop. 700.—Also a village in the dep. of Charente, cant. and 4 m. NW of Montembœuf.

VITRE', a walled town of France, in the dep. of Ville-et-Vilaine, situated on the Vilaine, which is here a small stream, 20 m. E of Rennes. Pop. 6,800. Though an ill-built place, it has considerable manufactures of woollens, serge, flannel, canvas, casks, wooden-ware, and linen; also tanneries and manufactures of leather. In January 1817, it suffered severely by fire.

VITREY, a town of France, dep. of Haute-Saone, 6 m. W of Jussey. Pop. 1,200.

VITROLLES, a village of France, in the dep. of Bouches-du-Rhone, 12 m. SW of Aix.

VITRY, a town of France, dep. of Pas-de-Calais, situated on the Scarpe, 4 m. SW of Douay.

VITRY-AUX-LOGES, a village of France, in the dep. of Loiret, cant. and 6 m. N of Chateauneuf.

VITRY-LE-BRULE, a village of France, dep. of Marne, 3 m. NE of Vitry-le-François, on the Ornaine.

VITRY-LE-FRANÇOIS, a town of France, in the dep. of Marne, cant. and 21 m. SE of Chalons, on the Marne, which here becomes navigable for barges. Pop. 7,389. It is still surrounded by a wall, and though built chiefly of wood, has a pleasant appearance; its largest square being neat and regular, and its principal church built in the Italian style. The addition of Le François to its name is derived from Francis I., in whose reign it was built. It was the scene of important military operations in 1814. It has a trade in corn, wool, wood, and charcoal.

VITRY-SUR-SEINE, a town of France, in the dep. of Seine, 2 m. SE of Paris. Pop. 2,500.

VITTEAUX, a town of France, dep. of Cote-d'Or, on the Brenne, 31 m. W of Dijon. Pop. 2,000. It has woollen-factories.

VITTEFLEUR, a village of France, in the dep. of Seine-Inferieure, cant. and 2 m. N of Cany, on the l. bank of the Durdan.

VITTEL, a town of France, dep. of Vosges, 14 m. SW of Mirecourt.

VITTORE (MONTE), a summit of the Apennines, in the Papal States, deleg. of Spoleto, to the N of the Monte-della-Sibilla. Alt. 2,478 metres = 8,293 ft.

VITTORE (SAN), a village of Naples, in the prov. of Terra-di-Lavoro, 6 m. SE of San-Germano.

VITTORIA, a town of Sicily, in the Val-di-Noto, 40 m. SW of Syracuse. Its population has been stated at 11,000, a number rendered probable by the fertility of the surrounding country, and the special privileges enjoyed by the whole of the adjacent district, called the Contado-di-Modica.

VITTORIA, a village of Upper Canada, in Norfolk co., 7 m. N of Simcoe. Pop. 600.

VITTORINO (SAN), a village of Naples, in Abruzzo-Ultra 2da, 6 m. NW of Aquila.

VITZELS, a town of Asiatic Turkey, in the pash. and 60 m. ENE of Trebizond.

VITZENBURG, a village of Prussian Saxony, reg. of Merseburg, 6 m. SSW of Querfort. Pop. 600.

VIU, a town of Piedmont, on the river Stura, 4 m. NW of Turin, at an alt. of 782 metres above sea-level. Pop. 3,745. It occupies an elevated position, commanding splendid views, and is much resorted to in summer by families from Turin.

VIVARAIS, an ancient district of France, in the NE of Languedoc, now included in the dep. of Ardeche. The cap. was Viviers.

VIVEL, a town of Spain, in the prov. of Valentia, district of Morella, 27 m. W of Castellon-de-la-Plana, near the Palencia river. Pop. 2,000.

VIVERO, a town of Spain, in the prov. and 45 m. N of Lugo, on the coast, near the embouchure of the Landrova, which forms a capacious harbour. Pop. 4,000. It is a walled place, and has some public buildings. The trade is chiefly coasting.

VIVEROLS, a town of France, dep. of Puy-de-Dome, 9 m. SSW of Ambert.

VIVIERS, a town of France, dep. of Ardeche, cant. and 12 m. SE of Privas, situated on the Rhone. Pop. 1,700. Though a small and ill-built place, it is the see of a bishop, and has manufactures of woollen, linen, and silk.

VIVIERS-LES-MONTAGNES, a village of France, in the dep. of Tarn, 4 m. SW of Castres. Pop. 1,200.

VIVIERS-SUR-CHIERS, a village of France, dep. of Moselle, cant. and 4 m. N of Longuyon. Pop. 600.

VIVOIN, a village of France, dep. of Sarthe, cant. and 1 m. NE of Beaumont-le-Vicomte, on the r. bank of the Sarthe.

VIVONNE, a town of France, dep. of Vienne, on the Clain, 12 m. S by W of Poitiers. Pop. 1,470.

VIX, a town of France, dep. of La Vendée, 9 m. S. of Fontenay.

VIZA, a town and sanjak of Turkey, in Romelia. The town is 72 m. NW of Constantinople, on a small stream which flows into the Black sea. It is walled and slightly fortified.—The sanj. is bounded on the SE by the straits of the Dardanelles; on the NE by the Black sea; and is watered by the Erkene, an affluent of the Maritza.

VIZAGAPATAM, a town of Hindostan, the capital of a district of the same name, on the coast, in N lat. 17° 41′, E long. 83° 24′, 40 m. NE of Madras. There are a fort, an arsenal, and cantonments here.—The district has an area of 7,650 sq. m., with a pop. of 1,254,272. It stretches between the Eastern Ghauts and the Orissa coast, from the parallel of 17° 15′ to 19° 3′.

VIZIADRUG, a port on the Malabar coast of Hindostan, 160 m. S of Bombay.

VIZIANAGROM, a town of Hindostan, 35 m. NNE of Vizagapatam.

VIZILLE, a town of France, dep. of the Isere, 9 m. SSE of Grenoble. Pop. 2,500. It has manufactures of linen and paper, and in the neighbourhood are extensive quarries.

VIZZINI, a town of Sicily, in the Val-di-Noto, in a mountainous district, on the road from Calatagirone to the coast, 28 m. SW of Catania. Pop. 9,000.

VLAARDINGEN, a town of Holland, in the prov. of South Holland, on the l. bank of the Maese, 6 m. W by S of Rotterdam. Pop. 7,600. It has extensive fisheries, and is the principal place of rendezvous of the Dutch herring-vessels.

VLADIKAUKAS, a Russian fortress in Circassia, on the r. bank of the Terek, 120 m. N of Tiflis, commanding the route from Caucasus into Georgia.

VLADIMIR, a government of European Russia, bounded on the N by the govs. of Jaroslav and Kostroma; on the E by Nijni-Novgorod; on the S by Riazan; on the W by Moscow; and on the NW by Tver. Area 864·6 German sq. m. Pop. in 1851, 1,168,303. It presents in general a flat surface, fertile, and sloping towards the E, but woody in the W. The chief rivers are the Volga, the Kliasma, and the Oka. Rye, barley, and oats are cultivated, and fruit is abundant. Iron, linen, and woollen goods are manufactured.—Also a town of the gov. of Volhynia, on the r. bank of the Louïa, an affluent of the Bug. Pop. 2,000.

VLADIMIR, on the river Kliasma, the capital of the above government, is 112 m. E by N of Moscow, and 500 m. SE of St. Petersburg, in N lat. 56° 8′, E long. 38° 1′. Pop. 4,000. It has manufactories of silk goods, earthenware, leather, and soap.

VLADIMIRETZ, a town of Russia, in the gov. of Volhynia, 60 m. NE of Lutsk.

VLAEMERTINGHE, a town of Belgium, in the prov. of E. Flanders, 3 m. W of Ypres. Pop. 2,500.

VLASKA, a district of Upper Wallachia, bounded on the NE by the district of Imbrovitza; on the SE by the Danube; on the SW by the district of Teliorma; and on the NW by that of Mountchel. It is a fertile and chiefly pastoral region intersected by the Koman.

VLASSANITZA, a town of Bosnia, 30 m. NE of Bosna-Serai.

VLETEREN (OOST), a town of Belgium, in the prov. of W. Flanders, 9 m. NW of Ypres.

VLIAN, a town of Austrian Dalmatia, on the island of St. Michael.

VLIE, or FLIE, the passage from the Zuyder-zee into the German ocean, between the islands of Schelling and Vlieland.

VLIELAND, a small island of Holland, at the entrance of the Zuyder-zee, about 5 m. NE of the Texel. It is about 10 m. long, and 3 m. broad. It had formerly two villages, East and West V., but the latter is now covered by the sea. Pop. 800.

VLODZIMIRZ. See VLADIMIR.

VLODZIMIRZETZ, a town of Russia, in the gov. of Volhynia, on the river Styr, 155 m. WNW of Zytomiers.

VLOTHO, a town of Prussian Westphalia, in the co. of Ravensburg, 6 m. S of Minden, on the l. bank of the Weser. Pop. 1,600.

VOBARNO, a town of Austrian Italy, in the gov. of Milan, near the river Chiese, 13 m. NNE of Brescia.

VOBURG, or VOHBURG, a town of Bavaria, on the Danube, 10 m. E of Ingolstadt. Pop. 900.

VOCKLABRUCK, a town of Upper Austria, on the Vockla, 36 m. SW of Lintz, near the confluence of the Vöckla and the Agger. It has manufactures of ticking and coarse linen. Pop. 1,200.

VODABLE, a village of France, in the dep. of Puy-de-Dome, 6 m. WSW of Issoire. Pop. 750.

VODE, or VUDE, a river of Upper Wallachia, which rises to the NW of Pitesti; runs SE; and flows into the Danube near Rustchuk, after a course of 120 m.

VODINA, or VODENA, a town of Turkey, in the sanj. of Salonika, on the Vistritza, 50 m. WNW of Salonika. It has manufactories of cotton and woollen fabrics. Pop. 12,000.

VODLA, a river of Russia, in the gov. of Olonetz, which flows from a lake of the same name, in a SSE and then in a SSW direction, to Lake Onega, into which it falls after a course of 100 m.—The lake lies to the NNE of Pudoj, and is 26 m. in length from N to S, and 14 m. in breadth.

VODOLAGA, a village of Russia, in the gov. and 27 m. SW of Charkov.

VODONIKO, a village of European Turkey, in Macedon, near the Venetiko, 40 m. N by W of Joannina.

VOEL (LOCH), a lake of Perthshire, in the p. of Balquhidder, about 3 m. long, and 1 m. broad; the source of the river Balvaig, one of the principal branches of the Teith.

VOEUIL, a village of France, in the dep. of Charente, cant. and 4 m. S of Angouleme. Pop. 300.

VOGEL (LA), a small island in the Eastern seas, to the E of Hoorn island, in S lat. 5° 12′.

VOGEL ISLANDS, a cluster of small islands near the W coast of Siam, in N lat. 7° 38′.

VOGELSBERG, a lofty mountain of the Swiss cant. of the Grisons, 35 m. ESE of the Saint Gothard, having an alt. of about 10,280 ft. above the level of the sea.

VOGELSBERG, a range, or rather group of mountains in the W of Germany, between Upper Hesse-Fulda and the Wetterau, on the basin of the Main and that of the Weser. It is connected on the SE with the Rhone-Gebirge; on the SW with the Hoheberg; and on the NW with the Rothhaargebirge. It is productive in flax, and has good pasturage. The Oberwald, its highest point has an alt. of 2,500 ft.

VOGHERA, a province of the Sardinian states, bounded on the N and NE by the Po; on the E by the duchy of Piacenza; on the S by the prov. of Bobbio; and on the W by that of Tortona. Area 797·35 sq. chilom. Pop. 97,162. About five-sevenths of the surface are plain; the remainder is covered with hills; but all is under cultivation. The agricultural produce consists of wheat, Indian corn, barley, and oats; wine, flax, hemp, and silk, are also cultivated. Sulphur, chalk, and limestone, are wrought. This prov. was formerly annexed to Pavia, and was denominated Oltrepo-Pavise; by the treaty of Worms, it was transferred to Sardinia. It comprises 12 mandementi, which are subdivided into 77 communes.

VOGHERA, the chief place of the preceding prov., is situated on the river Staffora or Iria, at an alt. of 426 ft. above sea-level. Pop. 10,766, in 1,045 houses, and 2,241 families. It is tolerably built, having a good market-place, an hospital, and six churches and convents, a civic palace, and extensive barracks. It is the see of a bishop, and its cathedral, rebuilt in the 17th cent., is a noble structure. The principal branches of industry are the spinning and weaving of silk. V. was formerly a fortified place of considerable strength, but public walks now occupy the place of the fortifications.

VOGOGNA, a town of the Sardinian states, in the prov. of Novara, 12 m. NE of Casale, near the confluence of the Anza with the Toccia. It is a place of high antiquity.

VOGOULIIA, a river of Asiatic Russia, in the gov. of Tobolsk, which rises on the E flank of the Ural; runs E; and flows into the Sosva on the l.

VOGTLAND, or VOIGTLAND, a circle or district of the kingdom of Saxony, occupying an area of 700 sq. m., in the SW corner of the kingdom, between the parallels of 50° 10′ and 50° 42′. It is a mountainous territory, and bears a great resemblance to the adjoining circle of the Erzgebirge, except that it has fewer mines, and that the great branch of employment is the weaving of muslin. The streams belong to the basin of the Elbe. The principal river is the Elster. The chief town is Plauen.

VOGUE′, a village of France, in the dep. of Ardeche, cant. and 5 m. WSW of Villeneuve-de-Berg.

VOGULS, a Finnish race inhabiting the govs. of Perm and Tobolsk in Asiatic Russia. They are supposed by some to be a branch of the Ostiaks. They are of small stature, but of a robust constitution, and frequently live to an advanced age. Their chief subsistence is derived from the herds of cattle which they rear, and from fishing.

VOHBURG, a town of Bavaria, in the circle of Regen, 9 m. E of Ingolstadt, on the r. bank of the Danube. Pop. 1,300.

VOHEMARO (CAPE), a cape on the E coast of the island of Madagascar, in S lat. 13° 20′.

VOHENSTRAUS, a town of Bavaria, 45 m. N by E of Ratisbon. Pop. 1,100.

VOHL, a town of Hesse-Darmstadt, on the Asel, 30 m. NE of Marburg. Pop. 1,200.

VOHRENBUCH, a village of Baden, in the bail. of Neustadt, on the Bregenz, 50 m. N of Constanz.

VOHRINGEN, a village of Hohenzollern-Sigmaringen, 7 m. N of Sigmaringen, on the Lauchart. Pop. 700.

VOIAVAT, a town of Asiatic Turkey, in the sanj. and 50 m. ENE of Kastamuni, on an affluent of the Kizel-Irmak. Pop. 2,000.

VOID, a town of France, in the dep. of the Meuse, 6 m. NW of Commercy, on the Meholle.

VOIDE (LE), a commune and village of France, in the dep. of Maine-et-Loire, 1 m. NNW of Vihiers.

VOIGTSBERG, a village of Saxony, in the Voigtland, 6 m. SSE of Plauen, on the r. bank of the Elster.

VOIKER, a river of Asiatic Russia, which takes its rise in the Ural chain, and falls into the Obi in N lat. 65° 50′.

VOIRON, a town of France, dep. of Isere, on the river Morges, 14 m. NW of Grenoble. Pop. 5,630. It is the staple place for linen manufactured here and in the surrounding villages. It has likewise bleaching-grounds, paper-mills, tanneries, and steel-works, where arms are manufactured.

VOIRONS (MONT), a mountain of Switzerland, on the borders of the lake of Geneva, having an alt. of about 3,400 ft. above the level of the lake.

VOISENES, a village of France, in the dep. of Yonne, cant. of Villenueve-L'Archeveque. Pop. 700.

VOISEY, a commune of France, in the dep. of Haute-Marne, cant. and 4 m. NE of La Ferte.

VOISSAU, a village of France, dep. of Isere, cant. and 4 m. NE of Saint-Geoire.

VOISSENON, a village of France, in the dep. of Seine-et-Marne, cant. and 3 m. NNE of Melun. Pop. 340.

VOITEUR, a town of France, dep. of the Jura, 6 m. N of Lons-le-Saulnier. Pop. 900.

VOITSBERG, a village of Styria, in the circle and 12 m. W of Gratz. Pop. 800.

VOIVRES (LA), a village of France, in the dep. of Vosges, cant. and 3 m. NE of Bains. Pop. 620.

VOIUSSA, a river of Albania, which rises in Mont Zigoz; runs NW; passes Konitza; and flows into the Adriatic a little to the N of the gulf of Avlona, after a course of 120 m.

VOIVE, a river of France, in the dep. of Haute-Marne, which flows into the Aube, on the r. bank, opposite Molins, after a W course of 24 m.

VOJE, a lake of Russia, in the E part of the gov. of Novgorod, to the NE of Lake Bielo. It is 28 m. in length from N to S, and 12 m. broad.

VOLA (CAPE), a cape of South America, on the N coast of Venezuela, in N lat. 12°, W long. 72°.

VOLANON, a small town of the States-of-the-Church, on the Adriatic, 28 m. E of Ferrara.

VOLARNO, a village of Venetian Lombardy, in the prov. of Brescia, 2 m. WNW of Salo.

VOLCANELLO, or VULCANO, a small islet of the Mediterranean, belonging to the Lipari group. It is of volcanic origin, and though formerly separated from the larger island of Volcano by a narrow channel, it is now connected with it by a neck of land which was formed during a violent eruption. It is of a triangular form, and uninhabited. It continues to emit smoke from different parts of its surface.

VOLCANO, or VULCANO, an island of the Mediterranean belonging to Sicily, in N lat. 28° 21′, the most southern of the Lipari group, situated between the island of Lipari and the N coast of Sicily, and separated from the former by a channel 2 m. in breadth. It consists of a single volcanic mountain of rather more than a ½ m. in height, which descends by successive gradations till it loses itself in the sea, where the circumf. of its base, and consequently of the island, is about 12 m. In all parts of the island, the traces of fire are distinctly visible. The portion of it opposite to Lipari is without the smallest trace of vegetation; towards the S and E, however, it is covered with trees and other vegetable productions. The mountain called Monte-Aria, presents the appearance of a cone, and is about 3,000 ft. in height. Its crater is oval, and in circuit about 1 m. It has no opening of consequence, but a smooth surface throughout, emitting smoke and vapour.

VOLCANO, an island in the Pacific, about 24 m. N of Egmont island, one of those called Queen-Charlotte's islands, in S lat. 10° 17′, E long. 165° 4′. Its summit rises above 2,000 ft. above the level of the sea, and sometimes emits a bright flame.—Also an island in Dampier's straits, off the NE coast of Papua, in S lat. 5° 30′, so named from its being the seat of a volcano, which was in an active state in 1792.—Also an island in the N. Pacific, in N lat. 34° 5′, E long. 139° 35′.

VOLCANO-BAY, an extensive bay in the SE extremity of the island of Jesso, so called by Captain Broughton from the volcanoes on its shore. This bay is very capacious; its entrance is between the parallels of 42° and 42° 30′, and it has 50 fath. water in the centre. The N side of the bay presents an agreeable diversity of rising grounds.

VOLCHOV. See VOLKHOV.

VOLCKACH, a town of Bavarian Franconia, on the Maine, 15 m. E by N of Wurtzburg. Pop. 1,500.

VOLCONDA, a town of Hindostan, in the Carnatic, 50 m. NNE of Trinchinopoly.

VOLCZYSK, a town of European Russia, in the gov. of Podolia, district of Kaminiec.

VOLENDAM, a village of Holland, in the prov. of N. Holland. Pop. 800.

VOLGA, or WOLGA, a river of Russia, which has the longest course, and, with the exception of the Danube, the largest volume of water of any river in Europe. It rises among the Valdai mountains, in Lake Selingher, in N lat. 57°, E long. 33° 10′, at an alt. of 550 ft. above sea-level, and takes a direction in general to the eastward, but with many windings, until it reaches the city of Kazan. The rivers that flow into it above Kazan are, if we except the Oka, of second rate magnitude; but below Kazan it receives the Kama, which brings to it the tribute of a great extent of country. It now flows southward, with a great volume of water, and forms the boundary between Europe and Asia for several hundred miles, till reaching Tzarystyn, it turns E, approaches the Caspian, and after separating into a number of branches, estimated at from 60 to 70, discharges itself into that sea, 85 versts = 56⅔ m. below Astrakhan. Its course is computed at 2,700 m.; and the area of its drainings at 400,000 sq. m. The vast tract of country through which it flows being in general level, it is navigable, after passing Tver, in the early part of its course. In May and June its waters receive great increase from the melting of the snow and ice; and the boatmen take advantage of the increased depth, to descend its current, and avoid those shallows and islands which at other seasons obstruct their course. From the vicinity of Tver northward, a communication is opened to the Msta, a river flowing N to the Nieva; and steam-boats ply between Tver and the Caspian. The principal rivers which join the V. are the Tvertza, the Mologa, the Sestra, the Soscha, the Oka, the Sura, the Kasanka, the Kama, the Sok, and the Samara. Its waters abound in fish, particularly sturgeon. See article RUSSIA.

VOLHYNIA, or VOLYNIEN, a government of Russia, lying between the parallels of 49° 25′ and 52° 10′ N; and bounded on the N by the govs. of Minsk and Grodno; on the E and SE by Kiev; on the S by Podolsk; on the SW by Austrian Galicia; and on the W by Poland. Area 1300·4 German sq. m. Pop. in 1851, 1,469,442. The surface is low towards the N; but diversified by a few hills towards the S. The river Bug skirts it on the W; the other principal rivers are the Styr and the Gorin. While Poland was independent, V. formed a prov. of that kingdom, which bordered with the Ukraine on the SE. The soil is generally chalky, but in some places marshy, and in some a rich vegetable mould. The climate is temperate. The produce consists in wheat, millet, and rye; hops and fruit are also grown. Pasturages are extensive, but a great part of the surface is still forest land. Hemp, potash, pitch, tar, iron, hides, cattle, tallow, honey, and wax are exported. From its situation on the frontier of Poland, V. has often been exposed to the evils of invasion. Since 1793, it has been in the possession of Russia. Its chief town is Jitomir.

VOLIANO, a town of Tuscany, in the prov. of Florence, vicariat and 10 m. E of Montepulciano, on the l. bank of the Chiana-Pontifica, which is here crossed by a bridge.

VOLINIA, a township of Cass co., in the state of Michigan, U. S., 85 m. SW of Lansing. Pop. in 1840, 411; in 1850, 607.

VOLISSO, a village of Turkey in Asia, on the W coast of the island of Chio, and a little to the N of a bay of the same name.

VOLKACH, a town of Bavaria, capital of a presidial, in the circle of Lower Franconia, on the l. bank of the Main, 15 m. NE of Wurzburg. Pop. 1,984.

VOLKARDINGHAUSEN, a village of the principality of Waldeck, district of Twiste. Pop. 240.

VOLKARLSHAIN, a village of Hesse, in the prov. of Upper Hesse, and circle of Nidda. Pop. 270.

VOLKEGEM, a department and commune of Belgium, in the prov. of E. Flanders, and arrond. of Audenarde. Pop. of dep. 747. It has linen factories.

VOLKEN, a village of Switzerland, in the cant. of Zurich, and bail. of Andelfingen. Pop. 385.

VOLKENODA, a bail. and village of the duchy

of Saxe-Coburg-Gotha, and principality of Gotha. Pop. 100.

VOLKENSHAGEN, a village of the grand-duchy of Mecklenburg-Schwerin, in the circle and bail. of Rostock. Pop. 220.

VOLKERMARKT, a town of Illyria, in the gov. of Laybach, circle and 20 m. E of Klagenfurt, on the l. bank of the Drave. Pop. 1,000.

VOLKERODE, a village of Hanover, in the bail. of Gottingen. Pop. 300.—Also a village of Prussia, in the regency of Erfurt, and circle of Heiligenstadt. Pop. 364.

VOLKERSDORF, a village of Prussia, in the regency of Liegnitz and circle of Lauban. Pop. 705. It was formerly famous for its manufacture of garnets, and has several linen factories.—Also a village of Saxony, in the regency of Dresden, and circle of Moritzburg. Pop. 330.

VOLKERSHAUSEN, a village of the grand-duchy of Saxe-Weimar, and circle of Eisenach, 3 m. SE of Vach. Pop. 1,000.

VOLKERSHEIM, a village of the duchy of Brunswick, in the circle of Wolfenbuttel. Pop. 760. —Also a village of Würtemberg, in the circle of the Danube. Pop. 329.

VOLKERTSHAUSEN, a village of the grand-duchy of Baden, in the circle of the See, bail. and 6 m. WSW of Stockach, on the Aach. Pop. 328. It has a forge, a paper-mill, and a bleachery.

VOLKERTSHOVEN, a village of Bavaria, in the circle of Swabia. Pop. 200.

VOLKESFELD, a village of Prussia, in the regency of Coblenz, and circle of Mahen. Pop. 280.

VOLKETSWEIL, a village of Switzerland, in the cant. of Zurich, and bail. of Uster. Pop. 2,028.

VOLKHOV, a river of Russia in Europe, which issues from Lake Ilmen, in the gov. of Novgorod, a little above the town of that name, which it intersects; thence it runs NNE; and after traversing the E part of the gov. of St. Petersburg, throws itself into the S part of Lake Ladoga, near Novaïa-Ladogo, and after a total course of about 150 m. This river is deep and rapid, but except when its waters are low, when it forms cascades, it is navigable.

VOLKHOUSKI, a chain of low woody mountains, in Russia in Europe, in the gov. of Tver, to the SW of the Valdai chain. It makes part of the ridge which forms the great water-shed of Europe.

VOLKMANNSDORF, a village of Prussia, in the regency of Appeln, and circle of Neisse. Pop. 961.—Also a village of Saxe-Weimar, in the circle and bail. of Neustadt.

VOLKMANSGRUN, a village of Bavaria, in the circle of Upper Franconia, and presidial of Naila. Pop. 310.

VOLKMARITZ, a village of Prussia, in the regency of Merseburg, and circle of the See. Pop. 200.

VOLKMARSDORF, a village of the duchy of Brunswick, and circle of Helenstadt. Pop. 210.—Also a village of Saxony, in the bail. and near Leipzig. Pop. 1,751.

VOLKMARSEN, a town of Hesse-Cassel, capital of an amt in the prov. of Lower Hesse, circle and 9 m. NW of Wolfagen, on the Twiste. Pop. 2,970. It is enclosed by walls, with four gates, and has two churches and an hospital, and the ruins of the ancient fortress of Kugelburg. It has manufactories of woollen fabrics and hosiery, and several tanneries. Pop. of bail., 6,750.

VOLKOVISK, a town of Russia in Europe, capital of a district in the gov. and 48 m. SSE of Grodno, on a small affluent of the Niemen.

VOLKSHAGEN, a village of the grand-duchy of Mecklenburg-Schwerin, circle of Wendisch, and bail. of Nibrutz. Pop. 350.—Also a village of Prussia, reg. of Stralsund, and circle of Rugen. Pop. 200.

VOLKSTEDT, a village of the principality of Schwarzburg-Rudolstadt, bail. and 2 m. SSW of Rudolstadt, on the l. bank of the Saale. Pop. 300. It has a manufactory of porcelain.

VOLLEGE, a village of Switzerland, in the cant. of Valais, and bail of Entremont. Pop. 869.

VOLLEKIRCHEN, a village of Prussia, in the regency of Coblenz, and circle of Metzlar. Pop. 230.

VOLLENBORN, a village of Prussia, in the regency of Erfurt, and circle of Worbis. Pop. 330.

VOLLENHOVEN, a town of Holland, in the prov. of Ober-Yssel, 14 m. NNW of Zwolle, on a headland of the Zuider-zee, in N lat. 52° 40′ 51″. Pop. 2,200. It has an active trade. In the vicinity are the ruins of the castle of Tontelburg.

VOLLORE, a town of France, dep. of Puy-de-Dome, arrond. and 6 m. SE of Thiers. Pop. 4,300.

VOLMA, a town of Russia, in the gov. and 30 m. W of Minsk.

VOLMAR, a town of Russia, in the gov. of Livonia, 65 m. NE of Riga, on the r. bank of the Aa. It has suffered repeatedly from fire. Pop. 1,700.

VOLME, a small river of Prussian Westphalia, which falls into the Roer.

VOLMERSTEIN, a village of Prussian Westphalia, in the co. of Mark, 8 m. SW of Schwiert, at the confluence of the Volme and the Ruhr. Pop. 400.

VOLMUNSTER, a village of France, in the dep. of Moselle, 56 m. E of Metz.

VOLNAY, a town of France, in the dep. of Cote-d'Or, cant. and 4 m. SW of Beaune. Pop. 650.—Also a commune and v. in the dep. of Sarthe, cant. and 4 m. SE of Bouloire, on the Hanes. Pop. 1,200.

VOLNEY, a township of Oswego co., New York, U. S., 50 m. W of Rome. Pop. in 1850, 5,310.

VOLO, a town of European Turkey, in Thessaly, situated on the gulf of Volo, 32 m. SE of Larissa, in N lat. 39° 24′. Pop. 5,000. Its harbour is large, and is resorted to by vessels from Egypt, Candia, and Smyrna. In 1685, it was taken and partly destroyed by the Venetians. It occupies the site of the ancient *Demetrias*.

VOLO (GULF OF), a gulf of the Archipelago, on the coast of Thessaly, between the parallels of 39° and 39° 25′ N, and intersected by the meridian of 23° E, to the NW of Negropont, from which it is separated by the channel of Trikeri, having the peninsula of Zagora on the E, and that of Trikeri on the SE. It is the *Sinus Pelasgicus* of the ancients.

VOLOCSA, a town of Austria, in the territory of Istria, 7 m. NNW of Fiume. It has two harbours, and a traffic in fruit and in wine.

VOLOGDA, a vast province of European Russia, lying between the parallels of 58° 20′ and 64° 30′ N; and bounded on the N by the gov. of Arkhangel; on the E by the Uralian chain; on the S by the govs. of Perm, Viatka, Kostroma, and Yaroslav; and on the W by those of Novgorod and Olonetz. Area estimated at 6,967 German sq. m., or 337,111 sq. versts. Pop. in 1853, 893,000. The surface is generally flat. Mountains are rare, but a succession of hill and dale is very common; and in many places these inequalities produce scenery which is not deficient in beauty. Nearly all the rivers belong to the basin of the Arctic ocean. The principal are the Northern Dwina, the Sukhona, the Louza, Vega, Vitchegda, Mezen, Pisega, and Petchora. A small portion of the S part of the gov. is drained by affluents of the Volga. To the N the trees lose their leaves in August, and the rivers are frozen over from the end of October to the middle of April. In the S there are large tracts occupied by forests, lakes, and morasses. Agriculture is followed to a certain extent, but the severity and changeable state of the weather render it precarious, though a considerable quantity of wheat and barley is grown. The produce of the pasture-grounds, of the chase and fishing, tend to make up for this uncertainty, and the woods supply potash, tar, and other materials for export, as well as domestic use. The gov. is administratively divided into 10 circles.

VOLOGDA, the capital of the above province, stands on the river Vologda, in N lat. 59° 13′ 0″, E long. 40° 11′, 350 m. E by S of St. Petersburg. Pop. 12,000. It is a bishop's see, and has a theological seminary, and a gymnasium. The inhabitants appear to be enterprising, having established a number of manufactures, such as silk, linen, canvas, dye-stuffs, white-lead, candles, and leather; and they carry on a considerable trade with St. Petersburg and Arkhangel, China, and the Aleutian islands.

VOLOGDA, a river of Russia, which rises in a marsh in the gov. of the same name; and flows into the Sukhona, on the r. bank, after a course of 90 m.

VOLOGEN, a town of Russia, in the gov. of Vilna, 36 m. SE of Otchmiana.

VOLOGUE, a river of France, in the dep. of Vosges, which rises in a series of small lakes, and flows into the Moselle, on the r. bank, at Jarmeuil, after a course of 36 m.

VOLOKOLAMSK, a town of European Russia, 73 m. WNW of Moscow, on the Lama. Pop. 2,600.

VOLONNE, a town of France, dep. of Basses-Alpes, 6 m. SSE of Sisteron, near the Durance.

VOLOTSCHINSK, a town of European Russia, in the gov. of Volhynia.

VOLPE (CAPE), a cape on the N coast of Sardinia, in N lat. 40° 4′.

VOLPEDO, a town of the Sardinian states, in the prov. of Tortona, on the small river Curone, 6 m. E of Tortona. Pop. 1,070.

VOLPERSDORF, a village of Prussian Silesia, 14 m. S of Glatz. Pop. 1,100.

VOLPIANO, a town of Piedmont, in the prov. and 12 m. N by E of Turin. Pop. 1,000; of district, comprising 4 communes, 3,663, in 437 houses. Corn, wine, fruit, and silk are produced in this district.

VOLSK, a town of European Russia, in the gov. and 70 m. NE of Saratov, on the Volga. Pop. of district, 11,000; of town, 3,500. It has manufactures of earthenware and tile-works; but the principal employments are agriculture, and the sale of its produce.

VOLTA, a town of Milan, near the Mincio, 12 m. N by W of Mantua. Pop. 4,000.

VOLTA (RIO), or ASWADA, a considerable river of Western Africa, rising in the mountains of Kong; flowing SE to Addah; and forming the boundary between the Gold and Slave coasts. It overflows its banks in the rainy season. The channel is obstructed with rocks and sand-banks, and does not admit ships of burden. Its chief affluents are the Senne and the Loka.

VOLTAGGIO, a town of the Sardinian states, in the prov. and 15 m. N of Genoa, at the foot of the Borchetta. Pop. 2,100.

VOLTAIRE (CAPE), a headland of N. Australia, in S lat. 14° 15′, E long. 125° 43′.

VOLTAS (CAPE), a headland on the coast of S. Africa, at the mouth of the Gariep or Orange river, in S lat. 28° 44′, E long. 16° 32′.

VOLTERRA, a town of Tuscany, 32 m. SW of Florence. Pop. 4,600. It is built on a lofty sandstone rock, which is touched by the Era on the N, and by the Cecina on the S, at an alt. of 1,870 ft. above sea-level. It contains a cathedral, an hospital, a palazzo-pubblico containing a museum, and several churches. It is the see of a bishop, has a public seminary of education, and appears from its monuments, particularly its walls of Etruscan construction, to have been a place of great antiquity. It is generally identified with the *Velittus* or *Voliterra* of the Etruscans. At present its citadel serves as a state prison; and the chief employment of the inhabitants is derived from the sale of the produce, corn, wine, and olive oil, of the adjacent country. Here are also quarries of alabaster, marble, and gypsum, and considerable salt-works.

VOLTORINO, a town of Naples, in the prov. of Capitanata, cant. and 3 m. E of Volturava.

VOLTOYA, a river of Spain, which rises to the W of Epinar, on the Sierra-de-Guadarama; flows W, and then N; and joins the Eresma, near Coca, after a course of 60 m.

VOLTRI, a small town of the Sardinian states, 9 m. W of Genoa, at the entrance of the Ceruso. Pop. 3,000.

VOLTSCHANSK, a town of European Russia, in the Slobolsk-Ukraine, 50 m. NE of Charkov. Pop. 2,500.

VOLTURARA, a town of Naples, in the Capitanata, 27 m. W of Foggia, on a spur of the Apennines. Pop. 2,200.—Also a town in the prov. of Principato-Ultra, 15 m. WNW of San-Angelo-di-Lombardi. Pop. 4,000.

VOLTURNO, a river of Naples, which takes its rise in the Apennines; flows SSE, and then W; and falls into the gulf of Gaeta, 20 m. NW of Naples, after a course of 90 m.

VOLUNTOWN, a township of Windham co., Connecticut, U. S. 16 m. N by E of Norwich. Pop. 1,064.

VOLUSKA, a small port of Austrian Illyria, in Istria, 6 m. W of Fiume, at the bottom of the gulf of Quarnero.

VOLUTZA, a mountain of Turkey, on the frontier of the sanjaks of Monastro and Trikala, between the bases of the Salembria and the Jnje-Karasu, and connected on the E with Mount Olympus.

VOLVIC, a town of France, in the dep. of Puy-de-Dome, cant. and 4 m. WSW of Riom. Pop. 2,400. Lava is largely quarried in the vicinity.

VOMANO, a river of Naples, in Abruzzo-Ultra 1ma, rising to the NE of Aquila; running NW, and then E, and flowing into the Adriatic, 6 m. NE of Attri.

VOMO, one of the Fiji group, in S lat. 17° 29′, E long. 177° 13′.

VONC, a village of France, in the dep. of Ardennes, cant. and 1 m. E of Attigny, near the r. bank of the Aisne. Pop. 1,100.

VONIZZA, a town of European Turkey, in Epirus, at the foot of Mount Zoromeros, on the S coast of the gulf of Arta, in N lat. 39° 15′. It is supposed to be the ancient *Anactorium*, and is built at the bottom of a bay between two promontories. It has a good roadstead for vessels, and carries on a little trade in oil and grain, but is important chiefly for the quantities of wood for ship-building produced in its territory.

VOORBURG, a village of Holland, in S. Holland, 2 m. E of the Hague, and on the railway from Delft. Pop. 1,700.

VOORDE, a village of Belgium, in E. Flanders, 22 m. SSE of Ghent. Pop. 1,100.

VOORN, an island of Holland, situated between the Old Maese and the Haringvliet. It is about 18 m. in length, and 5 m. in breadth, and is divided into East and West Voorn. It is tolerably fertile in corn. The chief place is Brielle.

VOORSCHOTEN, a village of Holland, 3 m.

SSW of Leyden, on the railway to Rotterdam. Pop. 1,000.

VOORST, a town of Holland, in the prov. of Gelderland, 12 m. E of Arnheim.

VORARLBERG, a mountainous district of the Austrian states, bounded on the NNE by Bavaria; on the NE by the Ober-Innthal; on the S by the Swiss cant. of the Grisons; on the W by the principality of Lichtenstein, and the cant. of St. Gall; and on the NW by the lake of Constanz. It has an area of 746 geog. sq. m. Pop. in 1849, 98,531. It takes its name from a mountain-ridge called Arlberg, which separates it from Tyrol and forms a branch of the Alps. It is watered by the Aach, the Ill, the Fussach, the Lech, and the Iller. The Rhine skirts it on the NW. The V. consisted of a number of petty lordships, which were ceded to Bavaria at the peace of Presburg in 1806, but were restored after the fall of Buonaparte. It now forms a circle of Tyrol. The V. sends out masons and housebuilders to nearly the whole of Switzerland, and the neighbouring provinces of France. They leave early in spring, and live sparingly during the summer; and they return home in autumn, where they have little to do during winter, excepting to fell wood in the forests and other chance work. The children leave the country at the same time in thousands to herd cattle in Suabia and Bavaria, and return with nearly the whole of their earnings. The women who remain, and the elder men, cultivate the land, and the girls and many of the young men weave, and are employed in manufactures. The chief town is Bregenz.

VORAU, a town of the Austrian states, in Styria, 61 m. S by W of Vienna. Pop. 600.

VORAY, a village of France, in the dep. of Haute-Saone, cant. and 7 m. S of Rioz. Pop. 600.

VORDATE, an island of the Eastern seas, in S lat. 7° 50′; in the Timor-Laut group.

VORDE, a village of Prussian Westphalia, 21 m. S of Arensberg.

VORDEN, a village of Holland, in the prov. of Gelderland, 15 m. E by S of Zutphen.—Also a town of Hanover, 15 m. N by E of Osnabruck. Pop. 700.—Also a village of Prussian Westphalia, circle and 7 m. NW of Hoxter, on the Brucht. Pop. 750.

VORDENBERG, a town of the Austrian states, Upper Styria, 4 m. N of Leoben. It has iron-works.

VORDINGBORG, a port of Denmark, in the island of Zieland, situated on a bay opposite to Falster, 52 m. SSW of Copenhagen.

VORDONI, a small town of the Morea, situated on the river Vasilipotamo, 7 m. S by E of Misitra.

VOREPPE, a town of France, dep. of Isere, on the river Roise, 9 m. NW of Grenoble. Pop. 1,300. It has manufactories of hats, leather, oil, and glass.

VOREY, a town of France, dep. of Haute-Loire, 11 m. NE of Le-Puy.

VORIA, a river of Russia, which rises near Jar, in the gov. of Smolensk, and flowing SSW joins the Ugra, after a course of 60 m.

VORMS, an island at the entrance of the gulf of Finland, in N. lat 59°, E long. 23° 12′, to the E of Dago. It is 9 m. in length from E to W, and has a pop. of 1,000, chiefly of Swedish descent.

VORNHAGEN, a village in the principality and 1 m. S of Waldeck.

VORNY, a village of Russia, in the gov. and 150 m. NW of Vilna.

VORONA, a river of European Russia, which rises in the gov. of Penza, and joins the Khoper on the r. after a course of 200 m. It is navigable through the chief part of its course.

VORONEJ, or Voronez, a river of European Russia, which rises on the frontier of the gov. of Tambov; runs W, past Lipetzk; and flows into the Don on the l., after a course of 250 m. It is connected by a canal with the Riaza, a tributary of the Oka, and runs into the Volga.

VORONEZ, or Woronesk, a prov. or gov. in the interior of European Russia, bounded on the N by the gov. of Tambov; on the NE by Saratov; on the E and SE by the country of the Don Cossacks; on the S by Ekaterinoslav; and on the W by Charkov and Kursk. Its area, equal to that of Scotland, is nearly 20,000 geog. sq. m. Pop. in 1851, 1,691,000. It has in general a good soil and climate. It is traversed by three navigable rivers, the Don, the Voronez, and the Donetz, besides a number of smaller streams. The chief objects of tillage are corn, flax, hemp, and, in a small degree, tobacco. Immense herds of cattle and sheep are also reared. Various kinds of fruit are raised, and water-melons. Manufactures are confined to woollens, linen, brandy, and hard-ware for home consumption. The gov. is administratively divided into 12 circles.

Voronez, the capital of the above government, situated in N lat. 51° 40′ 30″, 130 m. E of Kursk, on the r. bank of the river Voronez. Pop. 20,000. It is the see of an archbishop, and its principal buildings are the cathedral, the residence of the governor, and the public offices of the province. Here are a number of manufacturing establishments, for woollens, leather, soap, vitriol, and hardware, but each on a small scale. It also exports iron, sailcloth, oil, flour, wine, and lime. The trade of the place is carried on by the Voronej, on the banks of which a quay, with docks for barges, was built by Peter I.—Also a town in the gov. of Chernigov, 21 m. S of Novgorod-Sieverskoi.

VORONKI, a town of Russia, in the gov. of Poltava, 18 m. SW of Lokhovitza.

VORONOV, a town of Russia in Europe, in the gov. of Grodno, district and 21 m. N of Lida.

VORONOVITZA, a town of Russia in Europe, in the gov. of Podolia, district and 27 m. NW of Bratzlav.

VOROSPATAK. See Verespatak.

VOROSVAGAS. See Cservenitza.

VOROSVAR, or Eisenthurm, a town of Hungary, in the comitat of Eisenburg, 18 m. SW of Guns, on the Pinka. Pop. 600.

VOROTYNSK, a town of Russiain Europe, in the gov. of Kalouga, district and 12 m. from Peremyschl, on the Vissa. It is enclosed by an earthen rampart. Pop. 1,200.

VOROUX-GOREUX, a department and commune of Belgium, prov. and arrond. of Liege. Pop. 260.

VOROUX-LEZ-LIERS, a department and commune of Belgium, in the prov. and arrond. of Liege. Pop. 259. It has several oil and flour-mills.

VORRA, a village of Bavaria, in the circle of Upper Franconia, and presidial of Burgebrach, on the Ebrach. Pop. 214. It has a brewery.—Also a village in the circle of Middle Franconia, and presidial of Pegnitz. Pop. 450.

VORSBACH, a village of Prussia, in the regency of Cologne, and circle of Mühlheim. Pop. 524.

VORSCHEIDT, a village of Prussia, in the regency and circle of Aachen. Pop. 932.

VORSEC, a village of Würtemberg, in the circle of the Danube, and bail. of Ravensburg. Pop. 120.

VORSFELDE, a market-town of the duchy and 21 m. NE of Brunswick, district of Shoningen, and circle of Helmstadt, on the Aller. Pop. 1,448. It has several distilleries and tanneries. Tobacco, hops, and flax are cultivated in the vicinity.

VORSKLA, a river of Russia in Europe, which has its source in the gov. of Kharkov, to the NW of the town of that name, traverses the SE part of the gov. of Poltova, passing Poltova and Kobiliaki, and after a course in a generally SW direction of about 150 m., joins the Dnieper on the l. bank, 36 m. below Krementchong.

VORSSELAER, a department and commune of Belgium, in the prov. of Antwerp and arrond. of Turnhout, watered by the Aa.

VORST, a department and commune of Belgium, in the prov. of Antwerp, and arrond. of Turnhout, watered by the Beke and Laek. Pop. 1,813.—Also a village of Prussia, in the regency of Dusseldorf, and circle of Kempen. Pop. 1,208.—Also a village in the circle of Gladbach. Pop. 370.

VORST-HARDT, a village of Prussia, in the reg. of Dusseldorf, and circle of Gladbach. Pop. 370.

VORWALDE, a village of Hanover, in the principality of Osnabruck, and bail. of Wittlage-Hunteburg. Pop. 520.

VORWEIDEN, a village of Prussia, in the regency and circle of Aachen. Pop. 448.

VORWERK, a village of Austria, in Bohemia, in the regency of Bohmisch-Leipa, and bail. of Rumburg. Pop. 500.—Also a village of Prussia, in the regency of Konigsberg, and circle of Mohrungen. Pop. 300.

VORWERK (Neu), a village of Prussia, in the regency of Breslau and circle of Ohlau. Pop. 350.—Also a village in the regency of Stettin, and circle of Demmin. Pop. 300.

VORWERK (Gross and Klein), two villages of Prussia, in the regency of Leignitz and circle of Glogau. Pop. 520.

VORWOHLDE, a village of Brunswick, in the circle of Holzminden. Pop. 290.

VOSGES, or Wasgau, a chain of mountains in the E of France, extending from N to S, in a line nearly parallel to the course of the Rhine, from Bale to Spires. It may be regarded as a continuation of the Jura mountains, for it begins on the SE, in the Ballon-d'Alsace, nearly where the latter end, and is separated from them only by a valley. The length of the main chain of the V., running from SSW to NNE, is about 130 m.; but a farther chain or range, called the Faucilles, extends westwards, traversing the dep. of the Vosges, and ending, near Sedan, in the hills and forest land of the Ardennes. The V. have in various parts an elevation of 4,000 or 5,000 ft., and are covered with snow during eight or nine months in the year. They form a natural barrier between France and Germany; and although Alsace, situated on the German side, has been a century and a half in the possession of France, the language of the country is still German. Great part of the V. mountains are covered with forests; and they contain copper, lead, salt, and coal. The Thure, the Techt, the Liepvrette, the Brucht, the Wissig, the Zorn, the two Zentzels, the Moselle, the Surbach, the Lauter, the Queich, the Speyerbach, the Eiss, and the Selz, all rise on the E flank of the V. The W flank gives birth to the Meurthe, the Sarre, the Erbach, and the Lautern. The great rivers leading from the V. to the sea are the Meuse and the Moselle. The Ill forms the principal channel from them to the Rhine.

VOSGES, a department in the NE of France, formed of part of the old prov. of Lorraine, and bounded on the N by the deps. of Meuse and Moselle; on the E by Haut-Rhin and Bas-Rhin; on the S by Haute-Saone; and on the W by Haute-Marne. Its area is about 2,400 sq. m. Pop. in 1836, 397,987; in 1846, 427,894; in 1852, 427,409. The surface is rugged, consisting of mountains and forest-land, interspersed with valleys of greater or less extent. Several large rivers, amongst which are the Meuse, the Moselle, the Meurthe, and the Saone, have their source in this dep. The smaller streams and mountain-torrents are numerous, as are the mineral waters, of which the best known are those of Plombieres. The soil on the mountains is often stony; in the plains chalky and sandy. Oats, barley, rye, potatoes, flax, and hemp are raised; in the more fertile tracts, wheat, and in situations of favourable exposure, vines. The rearing of cattle and sheep forms the principal employment. The mineral products are iron, lead, copper, and, in a few situations, silver, marble, and potter's-earth. The cap. is the small town of Epinal. The pop., chiefly agriculturists, is far from dense, and would be still less, did not the iron-mines afford work to a number of hands. The cheapness of provisions, and consequently of labour, has led to the introduction of linen manufactures, and cotton cloth, the spinning of yarn, the making of lace, and likewise the manufacture of steel and iron goods. These articles furnish, along with cattle, butter, cheese, paper, leather, glass, earthen-ware, and timber, the chief exports of the dep.

VOSKOPOLIS, a town of Turkey in Europe, in the sanj. and 36 m. SSW of Okhrida.

VOSKRESENSK, a town of Russia in Europe, in the gov. and 36 m. WNW of Moscow, and district of Zvenigorod, on the l. bank of the Istra. Pop. 1,000. It was originally a village which was purchased by the patriarch Nikon, for the monastery of New Jerusalem, founded by him in the locality in 1656, on the plan of the Holy Sepulchre in Jerusalem.

VOSKRESENSKAIA, a town of Russia in Europe, in the gov. of Orenburg, district and 105 m. E of Tchéliabinsk, on the l. bank of the Miias. Pop. 1,000.

VOSNACKEN, a village of Prussia, in the reg. of Dusseldorf, and circle of Elberfeld. Pop. 360.

VOSNES, a commune of France, in the dep. of the Côte-d'Or, and cant. of Nuits. Pop. 501. It is noted for its wine.

VOSPORSKOI-PROLIO. See Yenikale.

VOSSELAER, a department and commune of Belgium, in the prov. of Antwerp, and arrond. of Turnhout. Pop. 750.

VOSSELAERE, a department and commune of Belgium, in the prov. of East Flanders. Pop. 944. It has manufactures of linen.

VOSSEM, a department and commune of Belgium, in the prov. of Brabant, and arrond. of Louvain, watered by the Voer. Pop. 878.

VOSSENACK, a village of Prussia, in the regency of Aachen, and circle of Montjoie. Pop. 654.

VOSSEVANGEN, a parish of Norway, in the diocese of Bergen, and bail. of Southern Bergenshuus. Pop. 6,000.

VOSTITZA, or Bostitza, a town and port of Greece, capital of the deme of the same name, N coast of the diocese of Achaia, on the gulf of Lepanto, and 20 m. E of Patras. Pop. 2,500. It has a custom-house, and carries on an active trade in cheese, grapes, silk, brandy, wine, cochineal, and gum. This town occupies the site of the ancient *Ægium*, of which some traces still remain.

VOSTZBERG, a village of Prussia, in the regency of Stettin, and circle of Saatrig. Pop. 530.

VOSTZWINKEL, a village of Prussia, in the regency and circle of Arnsberg. Pop. 430.

VOSTROI-KAMEN, one of the principal summits of the Ural chain in Russia, gov. of Perm.

VOTIAKS, a half-civilized tribe of Finnish descent, who are scattered through the Russian provs. of Viatka, Orenburg, and Kazan, or that part of Russia in Europe which adjoins Russia in Asia. They take the names of Udi and Mordi, and are said to have preserved in a great measure their original language, as well as their former distinction into tribes. By the Slaves, they are called Votes; and by the Tartars, Ars. Their number is considerable, being computed at 30,000 in the gov. of Orenburg, and 60,000 in that of Viatka. They are cultivators of the soil.

VOTKA, a town of Russia in Europe, in the gov. of Viatka, and district of Sarapoul, on the Ij. Pop. 10,000. It has an imperial manufactory of arms of great extent, an arsenal, provision magazines, an hospital, a prison, and a school.

VOTTAN, or BITOW, a town of Moravia, in the circle and 17 m. WSW of Znaym, on the Zelitawa, near its confluence with the Thaya. Pop. 300.

VOTTEM, a department and commune of Belgium, in the prov. and arrond. of Liege. Pop. 1,675. It has a forge, a manufactory of arms, and several coal-mines.

VOUGA, a river of Portugal, in the prov. of Beira, which has its source in the Serra-de-Arad; runs E; joins the Ovar, and forms a small lake, and after a total course of about 99 m. throws itself into the Atlantic below Aveiro, of which it forms the port, and 48 m. from the embouchure of the Douro.

VOUGAY (SAINT), a commune of France, in the dep. of the Finistere, and cant. of Plouzevede. Pop. 1,255.

VOUGEOT, a commune of France, in the dep. of the Côte-d'Or, and cant. of Nuits, 11 m. SSW of Dijon. Pop. 263. It is noted for its wine.

VOUGY, a commune of France, in the dep. of the Loire, and cant. of Charlieu, near the r. bank of the Loire. Pop. 1,911.

VOUIGA. See VIG.

VOUIHA, a district in the SW part of the island of Paou, in the Viti archipelago. It is noted for its sandal-wood.

VOUILLE, a commune of France, in the dep. of the Deux-Sevres, cant. and 5 m. N of Prahecq. Pop. 1,395.—Also a canton and com. in the dep. of the Vienne, and arrond. of Poitiers. The cant. comprises 11 com. Pop. in 1831, 11,657; in 1846, 3,110. The village is 11 m. WNW of Poitiers, near the r. bank of the Auzance. Pop. 1,435. The locality is noted for a battle between Clovis and Alaric in 507.

VOUILLE-LES-MARAIS, a commune of France, dep. of the Vendee, and cant. of Chaille-les-Marais, 11 m. SW of Fontenay-le-Comte. Pop. 1,517.

VOUILLON, a village of France, in the dep. of the Indre, cant. and 11 m. SW of Issondun. Pop. 450. In its vicinity is an iron-mine.

VOUIMARA, a town of Madagascar, in the Betsiminsares territory, to the S of Foulpoint.

VOUISE, a river of France, in the dep. of the Creuse, which has its source a little to the NW of Chenerailles, passes Gouzon, and after a course of 21 m. throws itself into the Tardes, on the l. bank, at Chambon.

VOUJAUCOURT, a commune of France, in the dep. of the Doubs, and cant. of Audincourt, 4 m. SW of Montbeliard, near the l. bank of the Doubs. Pop. 525. It has a large harbour, communicating with the junction canal of the Rhône and Rhine, and carries on an active trade in wood and other mountain productions.

VOULAINE, a village of France, in the dep. of the Côte-d'Or, and cant. of Recey, 11 m. SE of Châtillon-sur-Seine, near the r. bank of the Ource. Pop. 715. It has two forges and a blast-furnace.

VOULF, a small island of the gulf of Finland, near the N coast of Esthonia, in Russia in Europe, near Revel, in N lat. 59° 34', and E long. 24° 43'. It is inhabited by Swedes, who employ themselves in agriculture.

VOULLEDJ, a town of Arabia, in Yemen, 60 m. NE of Hodeida.

VOULOUN-VILON, a lofty summit of Madagascar, in the central chain of the island.

VOULPAIX, a village of France, in the dep. of the Ain, cant. and 4 m. W of Vervins. Pop. 1,050.

VOULTE (LA), a canton, commune, and town of France, in the dep. of the Ardeche, and arrond. of Privas. The cant. comprises 11 com. Pop. in 1831, 11,670; in 1846, 12,353. The town is 11 m. ENE of Privas, on the r. bank of the Rhone, at an alt. of 393 yds. above sea-level. Pop. in 1846, 3,155. It is commanded by an old castle. It has a consistorial Calvinist church, and possesses several forges, blast-furnaces and founderies. The vine and mulberry are cultivated, and large numbers of pigs are reared in the environs.

VOULX, a commune of France, in the dep. of the Seine-et-Marne, cant. and 5 m. NE of Lorrez-le-Bocage. Pop. 1,170.

VOUNEUIL-SOUS-BIARD, a village of France, in the dep. of the Vienne, cant. and 4 m. SW of Poitiers. Pop. 700.

VOUNEUIL-SUR-VIENNE, a canton and commune of France, in the dep. of the Vienne, and arrond. of Poitiers. The cant. comprises 8 com. Pop. in 1831, 8,276; in 1846, 8,737. The village is 15 m. NE of Poitiers, on the l. bank of the Vienne. Pop. 1,326.

VOURCANO, a mountain of Greece, in the Morea, in Messenia, near the r. bank of the Pirnatza, connected on the WSW with Mount Elias.

VOUREY, a village of France, in the dep. of the Isere, cant. and 3 m. S of Rives. Pop. 1,050.

VOURLA, or DURLAK, a town of Turkey in Asia, in Anatolia, in the sanj. of Saghala, 24 m. WSW of Smyrna, and about 3 m. S of the gulf of Smyrna, on which it has a port. A little to the N of V. is supposed to be the site of the ancient *Clazomene*.

VOURLES, a village of France, in the dep. of the Rhône, and cant. of St. Genis-Laval. Pop. 930.

VOUSINSK, a mountain of Turkey in Europe, in the N part of Bosnia, between Bosna on the E, and Okrina on the W, and between Dobor and Derbent.

VOUSSAC, a village of France, in the dep. of the Allier, and cant. of Chantel-le-Chateau. Pop. 1,100.

VOUTE (LA), a canton and v. of France, in the dep. of Ardeche, arrond. and 9 m. ENE of Privas, on the r. bank of the Rhine. Pop. 1,500.—Also a v. of France, in the dep. of Herault, cant. and 3 m. NE of Olargues. Pop. 1,200.

VOUTENAY, a village of France, in the dep. of Yonne, cant. of Vezelay, 8 m. NW of Avallon, near the Cure. A bloody engagement was fought here in 842, in which Lothaire, son of Louis-le-Debonnaire, was defeated by his two brothers.

VOUTE-SUR-ALLIER (LA), a village of France, in the dep. of Haute-Loire, 9 m. S of Brioude.

VOUTEZAC, a town of France, in the dep. of Correze, cant. of Jaillac, 10 m. NNW of Brives-la-Gaillarde. Pop. 2,200.

VOUTOIRY, a small river of France, which rises to the SE of Saint-Michel-de-Vesse, and flows into the Creuse, on the l. bank, at Chave, after a course of 9 m.

VOUTRE', a village of France, in the dep. of Mayenne, cant. and 6 m. SE of Evron.

VOUVANT, a village of France, in the dep. of La Vendee, cant. of La Chataigneraie, 9 m. SSW of Fontenay-le-Comte. Pop. 500. Coal is wrought in the vicinity.

VOUVRAY-SUR-LOIRE, a canton and town of France, in the dep. of Indre-et-Loire, 9 m. W of Amboise, at the confluence of the Cize and the Loire. Pop. 2,500.

VOUZAILLES, a village of France, in the dep. of Vienne, cant. and 7 m. SW of Mirebeau.

VOUZIERS, an arrondissement, canton, and town

of France, in the dep. of Ardennes.—The arrond., comprising 8 cants., has an area of 140,101 hectares, with a pop. in 1846 of 42,374.—The cant. had a pop. of 9,744.—The town is situated on the l. bank of the Aisne, 27 m. S of Mezieres. It has a traffic in grain, wine, and iron. Pop. in 1831, 2,003; in 1846, 2,771.

VOUZON, a commune and v. of France, dep. of Loir-et-Cher, 27 m. NE of Romorantin. Pop. 1,300.

VOVES, a canton and v. of France, in the dep. of Eure-et-Loire, 15 m. SSE of Chartres. Pop. 1,200.

VOWCHURCH, a parish of Herefordshire, 10 m. WSW of Hereford. Area 2,690 acres. Pop. in 1841, 355; in 1851, 323.

VOY-DE-PONAS (SAINT), a canton and v. of France, in the dep. of Haute-Loire, cant. and 6 m. SSW of Tence. Pop. 2,400.

VRACAS, an island of the Pacific, in the N part of the Marianne archipelago, in N lat. 20° 45′.

VRACHIE, a town of Belgium, in the prov. of E. Flanders, arrond. and 12 m. N of Dendermonde. Pop. 5,600. It is a well-built place, with dye-works, and active factories, and a considerable trade in agricultural produce.

VRACHORI, town of Greece, 24 m. NW of Lepanto, 5 m. NW of a lake of the same name, about 5 m. in length, and 4 m. in breadth, the *Trichonis* of the ancients.

VRAIN (SAINT), a commune and v. of France, in the dep. of Nievre, cant. and 4 m. SSW of Saint Amand. Pop. 800.

VRAITA, a river of the Sardinian states, which rises on the E flank of the Maritime Alps, and flows E, and then NE to the Po, into which it falls, on the r. bank, after a course of 60 m.

VRANA, a town of Austria, in Dalmatia, in the circle and 15 m. SE of Zara, on the N bank of the lake of that name. It has the ruins of a fortress destroyed by the Venetians, and was the residence of the Grand-master of the Templars.—Also a village of Greece, in Livadia, near the plain of Marathon, and 21 m. NE of Athens. It appears to occupy the place by which the l. flank of the Athenians was covered in the battle of Marathon.

VRANA, VIVARINA, or IVARINA, a town of Turkey in Europe, in Rumelia, in the sanj. and 48 m. WNW of Ghiustendil, on a small river of the same name, an affluent of the Morava. Pop. 3,000. It has extensive manufactories of arms, and in the vicinity are the celebrated iron mines of Olassina.

VRANCOURT, a village of France, in the dep. of the Pas-de-Calais, and cant. of Croisilles, and com. of Vaulx-Vrancourt. Pop. 1,715.

VRANDOUK, a small town of Turkey in Europe, in Bosnia, in the sanj. and 17 m. NE of Trawnik, on the l. bank of the Bosna. It is circular in outline, and enclosed by a wall, which still exhibits the breaches made by the Turks in taking the town. It is commanded by a fortified tower. The environs are populous, and well-cultivated.

VRANGEL, or VRAANG, two islands of the gulf of Finland, distinguished as Great and Little, belonging to the Russian gov. of Esthonia, and to the NE of Revel. The larger is in 59° 37′ N lat., and 24° 57′ E long. They are both inhabited by Swedes, who employ themselves in rearing cattle, and in fishing.

VRANJA, a mountain of Turkey in Europe, in Bosnia, 12 m. S of Trawnik. It belongs to one of the N branches of the Dinaric Alps.

VRANOGRAKH, a fortress of Turkey in Europe, in Bosnia, in the sanj. of Croatia, 72 m. WNW of Bagnaluka, and near the Austrian frontier.

VRASSELT, a village of Prussia, in the regency of Dusseldorf and circle of Rees. Pop. 408.

VRATZA, a market-town of Turkey in Europe, in Bulgaria, in the sanj. of Sophia, and near the confines of that of Widden, on the r. bank of the Insika, and 21 m. ESE of Berkovatz.

VRECOURT, a commune of France, in the dep of the Vosges, and cant. of Bulgueville, 14 m. from Neufchateau, on the Mouzon, which is here crossed by a fine bridge. Pop. 931. It has several cotton factories, tanneries, and blast furnaces.

VREDEN, a town of Prussia, in the regency of Munster, circle and 9 m. SW of Ahaus, on the Berkel. Pop. 2,920. It has manufactories of linen, and cloth, and of chicory.

VREES, a village of Hanover, principality of Osnabruck, bail. of Ahrenberg-Meppen. Pop. 420.

VREESWIK, a village of Holland, in the prov. and to the S of Utrecht, on the Leck. Pop. 800.

VREM, a market-town of Austria, in Illyria, in the gov. of Laibach, circle and 14 m. S of Adelsberg, on the Reka.

VRENDE, a department and commune of Belgium, in the prov. of Brabant and arrond. of Antwerp. Pop. 750.

VRESSE, a department and commune of Belgium, in the prov. of Namur and arrond. of Dinant, watered by the Semoi. Pop. 126.

VRETA, a parish of Sweden, in the prefecture of Linkœping and haerad of Gullberg.

VRETOT (LE), a commune of France, in the dep. of the Manche, and cant. of Bricquebec, on the l. bank of the Scie. Pop. 1,216.

VREXEN, a village of the principality of Waldeck, and bail. of Diemel, on a river of that name, at the confluence of the Urbe, 3 m. N of Rhoden. It has a paper-mill.

VRIES, a village of Holland, in the prov. of Drenthe, to the N of Assen. Pop. 1,759.

VRIES STRAIT, a channel of the Kurile islands, between those of Iturup and Urup.—On the NE coast of the former island is Cape Vries.

VRIESEVEEN, a village of Holland, prov. of Ober-Yssel, cant. and 5 m. N of Almelo. Pop. 2,893.

VRIESLAND. See FRIESLAND.

VRIGNE AUX BOIS, a commune of France, in the dep. of the Ardennes, cant. and 5 m. NW of Sedan, on the Vrigne. Pop. 1,115. It has extensive iron-works, and manufactories of iron-ware.

VRIN, a river of France, in the dep. of the Yonne, which has its source to the SE of Fourolles; runs N; and after a course of 21 m., throws itself into the Yonne, on the l. bank, at Cezy, and 4 m. WNW of Joigny.—Also a village of Switzerland, in the cant. of the Grisons, and bail. of Glenner, in the Lugnetz-thal. Pop. 520.

VRITZ, a village of France, in the dep. of the Loire-Inferieure, and cant. of St. Mars-de-la-Jaille. Pop. 1,900.

VROENHOVEN, a department and com. of Belgium, in the prov. of Limburg, and arrond. of Tongres. Pop. 1,314.

VROI, an island of Denmark, near the N coast of that of Seeland.

VRON, a commune of France, in the dep. of the Somme, and cant. of Rue, 21 m. NW of Abbeville. Pop. 1,370. It has manufactories of porcelain.

VSELIOUB, a town of Russia in Europe, in the gov. of Grodno, district and 9 m. N of Novogrodek.

VUACHE (MONTAGNE DA), or DES VACHES, a mountain of Sardinia, in the W part of the div. of Savoy, and to the SE of Fort-de-l'Ecluse, which, from the similarity of geological structure, appears to have formed part of the Jura chain previous to its separation by the Rhone. It runs from NW to SE, a distance of 8 m., and rises at the central point to the height of 800 metres above the level of the Rhone.

It is almost entirely covered with wood, in which bears and wild boars are frequently to be met with.

VUADENS, a village of Switzerland, in the cant. of Freiburg, and bail. of Greierz. Pop. 948.

VUARRENS, a village of Switzerland, in the cant. of Vaud, district and 5 m. N of Echallens, and 12 m. N of Lausanne. Pop. 520.

VUCHERENS, a village of Switzerland, in the cant. of Vaud, and bail. of Milden. Pop. 458.

VUCHIN, or MESIN, a town of Civil Slavonia, in the comitat and 17 m. SSE of Verovitz, on the Vuchinszka, an affluent of the Karaschitza.

VUCHT, a department and commune of Belgium, in the prov. of Limburg, and arrond. of Tongres, on the Maas. Pop. 391.

VUE, a commune of France, in the dep. of the Loire-Inferieure, and cant. of Le Pellerin, 11 m. SE of Paimbœuf. Pop. 1,235.

VUERSTEIN, a village of Austria, in Styria, and circle of Celli. Pop. 400.

VUFFLENS-LA-VILLE, a village of Switzerland, in the cant. of Vaud, and bail. of Cossonay. Pop. 315.

VUFFLENS-LE-CHATEAU, a village of Switzerland, in the cant. of Vaud, and bail. of Mordec. Pop. 247.

VUFLLENS, a village of Switzerland, in the cant. of Vaud, bail. of Sron. Pop. 461.

VUILLAFANS, a village of France, in the dep. of the Doubs, cant. and 5 m. SE of Ornans. Pop. 1,195. It has a paper-mill, several tanneries, forges, and manufactories of kirschwasser.

VUILLERENS, a village of Switzerland, in the cant. of Vaud, and bail. of Mordec. Pop. 512.

VUIPPENS, a village of Switzerland, in the cant. of Freiburg, and bail. of Greierz. Pop. 207.

VUISSENS, a village of Switzerland, in the cant. of Freiburg, and bail. of Broye. Pop. 284.

VUISTERNENS, a village of Switzerland, in the cant. of Freiburg, and bail. of Glane. Pop. 314.

VUISTERNENS-EN-OGAZ, a village of Switzerland, in the cant. of Freiburg, and bail. of Saane. Pop. 537.

VUITTEBOEUF, a village of Switzerland, in the cant. of Vaud and bail. of Orbe. Pop. 437.

VUKOVAR, or ALT-VUKOVAR, a town of Hungary, in Civil Slavonia, one of the capitals of the gov. of Syrmia, on the Danube, at the confluence of the Vuka, and 23 m. SE of Eszek, in N lat. 45° 21′ 9″, and E long. 9° 1′ 10″. It is divided by the V. into two parts, distinguished as the Old and New towns, and has two churches, and a Franciscan convent. Wine and silk are cultivated in the environs.

VULCAIN, one of the Banana islands, in the Atlantic, off the coast of Sierra-Leone, in Upper Guinea.

VULCANO. See VOLCANO.

VULKAN PASS, a pass of the Carpathian chain, Transylvania, in the comitat of Hunyad, leading thence into Wallachia.

VULLIENS, a village of Switzerland, in the cant. of Vaud and bail. of Oron. Pop. 458.

VULPMES, a village of Austria, in Tyrol. Pop. 800.

VULSI. See ZARACA.

VUNG, a village of the Punjab, 10 m. N of Julalpur, on the road thence to Rawul-Pindee.

VUOXEN, a river of Russia in Europe, in the grand-duchy of Finland, which has its source in N lat. 64° 4′, in the N part of the gov. of Kuopio, and which, after a total course, in a generally SSE direction, throws itself into Lake Ladoga, near Kexholm. In the upper part of its course it expands into numerous lakes, and near the fortress of Nyslot enters Lake Saïma, which soon after leaving it forms a cataract named Imatra, one of the grandest in Europe.

VUPABUCA, a lake of Brazil, in the prov. of Minas-Geraes, in the Serra-de-São-Sunao, near the comarca of Porto-Seguro.

VURGURU, or SAINT-NICOLAS (PORT), a harbour of Greece, in the sanj. of Salonica, formed by the gulf of Monte-Santo, on the E coast of the isthmus of the peninsula of Coron. A little to the S of its entrance is the island of Diapori.

VURSTE, a department and commune of Belgium, in the prov. of East Flanders and arrond. of Ghent, watered by the Schelde.

VUSSEM, a village of Prussia, in the regency of Aachen and circle of Schleiden. Pop. 238. It has several iron-works.

VUSTRIN, or VELTCHISTERIN, a town of Turkey in Europe, in Rumelia, in the sanj. of Scutari, district and 15 m. NW of Pristma, on the Sidnitza, an affluent of the Ibar. Pop. 3,000. This town is the *Viciterum* of the ancients.

VUSUVA, a village of the Punjab, at some distance from the r. bank of the Chenab.

VYBORG. See WIBORG.

VYCHGORODOK, a town of Russia in Europe, in the gov. of Volhynia, district and 36 m. SSE of Kremenetz, near the Galician frontier.

VYCZDNAIA, a town of Russia in Europe, in the gov. of Nijni-Novgorod, district and 3 m. W of Arzamas.

VYJVA, a town of Russia in Europe, in the gov. of Volhynia, district and 21 m. NW of Kovel, on the r. bank of the Vyjovka, an affluent of the Touriia.

VYLE-ET-THAROUL, a department and com. of Belgium, in the prov. of Liege, and arrond. of Huy, watered by the Hoyaux. Pop. 323.

VY-LE-FEROUX, a village of France, in the dep. of the Haute-Saone, cant. and 6 m. S of Scey-sur-Sayone. Pop. 280.

VY-LES-LURE, a commune of France, in the dep. of the Haute-Saone, cant. and 4 m. SW of Lure. Pop. 1,217. It has manufactories of muslin and of brocades.

VYNET, a department and commune of Belgium, in the prov. of East Flanders, and arrond of Ghent. Pop. 2,186. It has manufactures of linen.

VYTCHEGDA, a river of Russia in Europe, in the gov. of Vologda, which issues from a marsh in the district of Oust-Sisolsk, in the E part of the gov.; runs first S, then W, passing near Oust-Sisolsk and Yarensk, and watering Solvitchegodsk; and which, after a total course of 450 m., joins the Dwina of the north, on the r. bank, a little below Solvitchegodsk. Its principal affluents are the Yulva and Yarenga on the r., and the Sisola on the l. It is at all times navigable.

VYTEGRA, a river of Russia in Europe, in the gov. of Olonetz, which has its source in the SE of the gov. of that name; runs NW; passes a town of the same name; and, after a course of about 75 m., falls into Lake Onega, on the SE. It is navigable, and is connected by the Mariienskoi canal with the Kovja, and by that river with lake Belo and the Volga.—Also a district and town in the same gov. The town is on the river of the same name, 15 m. SE of Lake Onega, and 90 m. SE of Petrozavodsk, in N lat. 61° 0′ 16″, and E long. 36° 15′ 15″. Pop. 2,500. Its situation in the line of communication by the Onega with Lake Ladoga and St. Petersburg on one side, and by the Mariienskoi canal with the Volga on the other, is one of great commercial importance. It has two churches, and possesses manufactories of linen and of candles, and extensive building-docks. In the environs are extensive glass-works.

W

*** Articles beginning with the letter W, but not found under it in this work, may be sought under V or Ou.

WAABE, a river of the duchy of Brunswick, which falls into the Ocker.

WAACKE, a village of Hanover, in the principality and 6 m. ENE of Gottingen. Pop. 700.

WAACKHAUSEN, a village of the duchy of Bremen, near the river Hamme.

WAADT. See VAUD.

WAAG, or VAG, a river of the west of Hungary, which rises at the foot of the lofty Krivan, in the palatinate of Lyptau; flows from E to W through the palatinates of Thurotz, Trentschin, and Neutra; and falls into the Danube 6 m. below Comorn, after a course of 220 m. In the upper part of its course, it has a number of rapids and falls; in other parts it is easily navigated. It receives the Arva and the Kiszucza on the r.; and the Thurocz, Zilincza, and Neutra on the l.

WAAG, a bay on the W coast of Norway, on which is the town of Bergen.

WAAGOE, one of the Faroe islands, to the W of Stromoe, from which it is separated by Westmannshaven. It is 9 m. in length, and 5 m. in breadth.

WAAL, the name borne by the bifurcation of the Rhine, which branches off from a point 12 m. below Auchan, and joins the Maas at Garinshem. See article RHINE.

WAALIA, a collection of villages on the top of a hill in Abyssinia, between Gondar and Tcherkin.

WAALWYCK, or WAALWIJK, a town of Holland, in N. Brabant, 10 m. W of Bois-le-Duc. Pop. 2,600. It has a trade in grain and cattle.

WAARSCHOOT, or WAERSCHOOT, a town of Belgium, in E. Flanders, 9 m. NW of Ghent. Pop. 5,500. It has manufactures of linen, besides breweries and distilleries.

WAASMUNSTER, a village of Belgium, in E. Flanders, on the l. bank of the Durne.

WAATSCH, or VAZHE, a town of Austrian Illyria, in Carniola, on a hill, 10 m. ENE of Laybach.

WABASH, a river of the United States of North America, which rises in the NW of Ohio, and flows NW to its junction with the St. Joseph; then flows W and then S through Indiana, watering the middle and western parts of that state, and flows into the Ohio, on the N side, 68 m. above Cumberland river, after a course of 600 m. It is navigable for keel-boats 400 m. to Ouiatan, and for steam-boats to La Fayette, 450 m. It meanders in most of its course through an extensive plain of fertile soil, diversified by lofty groves and beautiful prairies. By the Wabash and Erie canal, it connects Lake Erie with the Ohio.

WABASH, a county in the S of Illinois, U. S. Area 198 sq. m. It is watered by the Jordan, Coffee, and Crawfish creeks. Pop. in 1840, 4,240; in 1850, 4,690. The cap. is Mount Carmel.—Also a county in the N of Indiana, watered by the Wabash and the Eel. Area 420 sq. m. Pop. in 1840, 2,756; in 1850, 12,138. Its cap., Wabash, 72 m. N by E of Indianapolis, on the Wabash and Erie canal, has a pop. of about 1,100.—Also a township in Fountain co., Iowa. Pop. 1,200.—Also a township in Tippecanoe co., in Iowa. Pop. 1,200.—Also a township in Wayne co., Illinois, 155 m. SE of Springfield.

WABASH (LITTLE), a river in the United States, in Indiana, which runs SE into the Wabash, a few miles above the Ohio.

WABASHAW, a county or district of Minnesota territory, stretching from the Mississippi to the Missouri, between the parallels of 44° and 40° N; and drained on the E by the Minnesota, and on the W by the Riviere-à-Jacqua, Sioux, and other affluents of the Missouri. Its pop. in 1850 was only 243.

WABBERTHWAITE, a parish of Cumberland, $1\frac{1}{2}$ m. SE of Ravenglass. Area 1,901 acres. Pop. in 1841, 146; in 1851, 212.

WABERN, a village of Hesse-Cassel, 4 m. SE of Fritzlar. Pop. 1,000.

WABI, a river of NE. Africa, which is reported to rise in a district of the same name, and to flow SE through the territory of the Galla tribe of Maroko, in the Somali territory, to the sea-coast.

WABISINEKAN, a river of Missouri, U. S., which runs into the Mississippi above the Missouri.

WABUSKAGAMA, a river of Canada, which flows into the Saguenay, in N lat. 48° 20′.

WACAHATCHA, a river of Louisiana, in the district of Opelousas, which falls into the Sabine.

WACCAMAW, a river of Carolina, U. S., which rises in a lake of the same name, flows into the Great Pedee, after a WSW course of 100 m. It is navigable 40 m.

WACHAS, a lake of Louisiana, U. S., to the W of the Mississippi, and 22 m. from New Orleans. It is 23 m. long, and 6 m. wide; and communicates with the gulf by several outlets.

WACHBACH, a village of Würtemberg, circle of Jaxt, 3 m. S of Mergentheim. Pop. 1,000.

WACHENHEIM, a town of the Bavarian circle of the Rhine, at the foot of the Hart mountain, 12 m. W of Manheim. Pop. 1,800.

WACHITA. See WASHITA.

WACHOKO, a town of Poland, on the Kamilena, an affluent of the Vistula, 24 m. SW of Radom.

WACHSHOLM. See WAXHOLM.

WACHSHURST, or WAGSHURST, a village of Baden, 10 m. E by N of Strasburg. Pop. 800.

WACHTENDONK, a town of Prussian Westphalia, on the river Niers, 20 m. NW of Dusseldorf. Pop. 1,100, many of whom are silk or velvet-weavers.

WACHTERSBACH, a town of Hesse-Cassel, on the Kinzig, 20 m. E by N of Hanau. Pop. 1,200.

WACHUSETT, a mountain of Worcester co., in Massachusetts, U. S., 42 m. W by N of Boston, having an alt. of 2,021 ft. above the level of the sea.

WACKEN, or WAKKEN, a town of Belgium, in the prov. of W. Flanders, near the river Lys, 16 m. SW of Ghent. Pop. 2,500.

WACKERFIELD, a township of the co. of Durham, NE of Staindrop.

WACTON, a parish of Hereford, $4\frac{1}{2}$ m. NW by W of Bromyard. Area 1,002 acres. Pop. in 1841, 109; in 1851, 129.—Also a parish of Norfolk, 4 m

SW by S of St. Mary-Stratton. Area 1,044 acres. Pop. in 1841, 267; in 1851, 261.

*** WAD, WADY, or WADI, a frequent prefix in Arabian names, signifying 'a valley.' For localities with this prefix, see the second word.

WADAN, a town of Fezzan, on the route from Tripoli to Murzuk, eight days' journey S of Mesurata.

WADBOROUGH, a hamlet of Worcester, 3 m. W of Pershore, 5 m. SE of Worcester.

WADDAM'S GROVE, a territory and v. of Stephenson co., Illinois, U. S., 174 m. N of Springfield. Pop. 450.

WADDESDON, a parish of Buckinghamshire, 5½ m. NW by W of Aylesbury. Area 6,010 acres. Pop. in 1841, 1,750; in 1851, 1,743.

WADDINGHAM, a parish of Lincolnshire, 8 m. S by W of Glanford Bridge. Area 3,720 acres. Pop. in 1841, 678; in 1851, 834.

WADDINGSVEEN (NORTH and SOUTH), two adjoining villages of Holland, in the prov. of S. Holland, 10 m. SE of Leyden.

WADDINGTON, a parish of Lincolnshire, 4 m. S. of Lincoln. Pop. in 1841, 814; in 1851, 962.—Also a township in the W. R. of Yorkshire, situated on the river Ribble, 19 m. WSW of Skipton.

WADDINGTON, a village of St. Lawrence co., in the state of New York, U. S., on the St. Lawrence, 168 m. NNW of Albany. Pop. 650.

WADDINGWORTH, a parish of Lincolnshire, 5 m. W by N of Horncastle. Area 928 acres. Pop. in 1841, 64; in 1851, 71.

WADDON (EAST and WEST), two small hamlets in the p. of Abbotsbury, Dorsetshire.

WADE'S POINT, a cape of the United States, on the coast of N. Carolina, in N lat. 36° 7′.

WADEIJ, a town of Yemen, in Arabia, 80 m. SSW of Saad.

WADELS, a river which rises in Radnorshire, and falls into the Lug, near Combe, in Herefordshire, about 3 m. E of Presteign.

WADENHOE, a parish of Northamptonshire, on the river Nen, 3½ m. N by E of Thrapston. Area 1,150 acres. Pop. in 1841, 287; in 1851, 290.

WADENSCHWEIL, a town of Switzerland, in the cant. and 12 m. SE of Zurich, on the S side of the lake of Zurich. Pop. of com. in 1851, 5,841, of whom 178 were Catholics. It is well-built, and has manufactures of cotton and silk, soap, and soda.

WADERN, a town of the Prussian prov. of the Rhine, 18 m. ESE of Treves. Pop. 700.

WADERO, a small island on the W coast of Sweden, in N lat. 56° 24′.

WADESBOROUGH, a township of Anson co., N. Carolina, U. S., 70 m. SSE of Salisbury. Pop. 650.—Also a village of Callaway co., Kentucky, 2 m. W of Clark's river.

WADEY, or WADAI. See BORGU.

WADHAM ISLANDS, a cluster of small islands near the NE coast of Newfoundland, in N lat. 49° 57′.

WADHURST, a parish of Sussex, 5 m. SE of Tunbridge Wells. Area 10,147 acres. Pop. in 1841, 2,491; in 1851, 2,802.

WADING-RIVER, a village of Suffolk co., New York, U. S., 130 m. SSE of Albany.

WADJO, or WAJU, a state or confederacy in the island of Celebes, situated to the N of the Bugis territories. In 1775, this country was governed by 40 regents, among whom women were admitted as well as men.

WADOR, a town of Afghanistan, in the Derajat, 17 m. W of Dera-Ghazi-Khan.

WADOWICE, a town of Austrian Poland, in the circle of Myslenice, on the l. bank of the Skawa, 22 m. SW of Cracow. Pop. 3,000.

WADSLEY, a chapelry and village of Yorkshire, in the p. and 3 m. NW of Sheffield. Pop. 4,333.

WADSTENA, or WADSTEIN, a town of Sweden, in E. Gothland, on Lake Wetter, 20 m. W of Linkoping. It has manufactures of lace and woollens.

WADSWORTH, a township in the W. R. of Yorkshire, 7½ m. WNW of Halifax. Area 10,080 acres. Pop. in 1841, 5,583; in 1851, 4,491.

WADWORTH, a parish in the W. R. of Yorkshire, 5½ m. S of Doncaster. Area 2,930 acres. Pop. in 1841, 681; in 1851, 724.

WADY. See explanation of this prefix under WAD.

WAELHELM, a commune and v. of Belgium, in the prov. and 12 m. S of Antwerp, on the Nethe. Pop. 1,075.

WAE-PING-FU, a town of China, in the prov. of Che-kiang, 100 m. SW of Hang-chew. It is said to have a pop. of 150,000.

WAEREGHEM, a commune and town of Belgium, in the prov. of W. Flanders, 9 m. NE of Courtray. Pop. 700. The manufacture of linens forms the chief article of local industry.

WAFTIB, a small island in the Eastern seas, on the W coast of the island of Waygiou.

WAGARFELD, a village of Hanover, in the co. and 9 m. SE of Diepholz.

WAGENDRUSSEL, a village of Hungary, in the com. of Zips, on the Golnitz, 5 m. S of Kapsdorf. Pop. 2,300. Iron is wrought in the vicinity.

WAGENINGEN, a town of Holland, in the prov. of Gelderland, situated in a marshy district on the N side of the Leck, a branch of the Rhine, 10 m. W by S of Arnheim, on the railway to Utrecht. Pop. 2,100. It is surrounded with a wall, and has considerable traffic in corn, cattle, and tobacco. It is an ancient place, and supposed to be the *Vadu* of Tacitus.

WAGER'S STRAITS, or RIVER, a large inlet of Hudson's bay, in N lat. 65° 8′, W long. 87°.

WAGHEN, or WAWN, a parish in the E. R. of Yorkshire, 5 m. SE by E of Beverley. Area 5,065 acres. Pop. in 1841, 362; in 1851, 347.

WAGING, a village of Bavaria, near the W bank of Lake Tachen, 56 m. SE of Munich.

WAGNAGUR, a town of Hindostan, in the prov. of Gujerat, situated on the coast of the peninsula, in N lat. 21° 3′.

WAGONTOWN, a village of Chester co., in Pennsylvania, U. S., 60 m. ESE of Harrisburg. There are iron works here.

WAGRAM, or DEUTSCH-WAGRAM, a village of Lower Austria, on the l. bank of the Rusbach, 10 m. NE of Vienna, celebrated for the battle which took place in the vicinity between the Austrians and the French, on 6th July 1809, and in which the latter were victorious.

WAGRAM, or WAGELIN, a village of Austria, in the circle and 36 m. SSE of Salzburg, on the Little Arlbach.

WAGSTADT, or BILOWES, a town of Austrian Silesia, 17 m. S by E of Troppau, on the Waybach. Pop. 2,400.

WAGUR, a district of Hindostan, prov. of Cutch, of which it forms the eastern boundary. It is elevated and woody, and intersected by a number of small streams which fall into the Runn.

WAHAL. See WAAL.

WAHKOU, a county in the W part of the state of Indiana, U. S., comprising an area of 770 sq. m., drained by Little Sioux river and its branches. It is hilly but generally fertile.

WAHLBACH, a village of Prussia, in the regency of Arnsberg, and circle of Siegen. Pop. 540.—Also a village in the regency of Coblenz, and circle of Simmern. Pop. 240.

WAHLE, a village of the duchy of Brunswick, in the bail. of Bechelde. Pop. 340.

WAHLEN, a village of Prussia, in the regency of Treves, and circle of Merzig. Pop. 910.—Also a village of the regency of Aachen, and circle of Schleiden. Pop. 260.—Also a village of Hesse-Homburg, in the circle of Upper Hesse, and presidial of Homberg. Pop. 420.—Also a village of the prov. of Starkenburg, and presidial of Furth. Pop. 260.—Also a village of Austria, in Tyrol, in the circle of Bruneck. Pop. 260.—Also a village of Saxony, in the circle of Zwickau, and bail. of Werdau. Pop. 466.—Also a village of Switzerland, in the cant. of Bern, and bail. of Laufen. Pop. 328.

WAHLENAU, a village of Prussia, in the regency of Coblenz, and circle of Zele. Pop. 200.

WAHLENHORST, a parish of Hanover, in the gov., prin. and bail. and 6 m. NNW of Osnabruck.

WAHLERN, a village of Switzerland, in the cant. of Bern, and bail of Schwarzenburg. Pop. 5,414.

WAHLEROD, a village of the duchy of Nassau, bail. and 8 m. WSW of Hachenburg. Pop. 320.

WAHLERSHAUSEN, a village of Hesse-Cassel, in the prov. of Lower Hesse, and circle of Cassel. Pop. 1,800.

WAHLERT, a village of Hesse-Cassel, in the circle of Schlüchtern. Pop. 200. It has two mills.

WAHLHAUPTEN, a village of Bavaria, in the circle of Swabia, and presidial of Buchloe. Pop. 220.

WAHLHEIM, a village of Hesse, in the circle of Alzey. Pop. 360.—Also two villages of Würtemberg, in the circle of the Neckar, and bail. of Besigheim, on the Neckar. Pop. 1,196.

WAHLSCHEID, a village of Prussia, in the regency of Treves, and circle of Saarbruck. Pop. 250.

WAHLERSHAUSEN, a village of Hesse-Cassel, in the prov. of Lower Hesse, circle and 42 m. W of Cassel, on the Drusel. Pop. 1,800. In its vicinity is the magnificent castle of Wilhelmshöhe, commenced in 1606 by the landgrave Maurice, on the site of the ancient convent of Weissenstein, and completed in 1787. This castle, the finest sovereign residence in Germany, and in grandeur of site unequalled in Europe, stands at the foot of the Karlsberg, upwards on which its castle and dependencies extend. In the environs are also the coal-mines of Habichtswald.

WAHLSHAUSEN, or VERNA-WAHLSHAUSEN, a village of Hesse-Cassel, in the prov. of Lower Hesse, and circle of Hofgeismar, on the Schwülme. Pop. 600. It has an extensive bleachery, and a coal-mine.

WAHLSDORF, a village of Prussia, in the regency of Potsdam, and circle of Juterbogk. Pop. 206.

WAHLSTATT, a village of Prussia, in the regency and circle and 5 m. SE of Liegnitz. Pop. 378. An engagement took place here in 1241, between Duke Henry of Liegnitz and the Tartars, and another in 1813 between Blucher and the French.

WAHLSTETT, a village of Denmark, in the duchy of Holstein, to the NW of Segeberg. Pop. 362.

WAHLWIES, a village of Baden, in the circle of the See, and bail. of Stockach. Pop. 340.

WAHLWINKEL, a village of Saxe-Coburg-Gotha, in the bail. of Tenneberg. Pop. 250.

WAHMBECK, a village of Hanover, in the bail. of Nienover. Pop. 590.

WAHN, a village of Hanover, in the circle of Meppen. Pop. 610.—Also a village of Prussia, in the regency of Cologne, and circle of Muhlheim. Pop. 538.

WAHNS, a village of Saxe-Meiningen, in the bail. of Wasungen. Pop. 360.

WAHNSDORF, a village of Saxony, in the bail. of Dresden. Pop. 320.

WAHNWEGEN, a village of Bavaria, in the circle of the Pfalz, and cant. of Rusel. Pop. 400.

WAHRBURG, a village of Prussia, in the regency of Magdeburg, and circle of Stendal, on the Uchte. Pop. 250.

WAHREN. See WAREN.

WAHRENBERG, a village of Prussia, in the regency of Magdeburg, and circle of Osterburg, on the Elbe. Pop. 810.

WAHRENBRUCK, a town of Prussia, in the prov. of Saxony, regency of Merseburg, and circle of Liebenwerda, on the l. bank of the Black Elster, and 27 m. E of Torgau. Pop. 608. It has a paper-mill.

WAHRENHOLZ, a village of Hanover, in the prov. of Luneburg, near Gifhorn. Pop. 448.

WAHRLANG, a village of Prussia, in the regency of Koslin, and circle of Neustettin. Pop. 240. —Also a village in the regency of Stettin, and circle of Uckermunde. Pop. 320.

WAHRSOM, a village of Mecklenburg-Strelitz and circle of Ratzeburg. Pop. 200.

WAHSALCH, a range of mountains in Utah Territory, U. S., extending in a SSW direction from the Green river range of the Rocky mountains, to the Sierra Nevada range of California.

WAIBLINGEN, a town of Würtemberg, capital of a bail. of the circle of the Neckar, 8 m. NE of Stuttgart, on the l. bank of the Rems. Pop. 3,131. It has manufactories of cloth, tanneries, and extensive tile-kilns. This town is of great antiquity, and is said to have been the birthplace of the emperor Frederick Barberousse. From this town also the Hohenstaufen party received the name of Webelingen or Ghibelins. Pop. of bail., 27,947.

WAIBSTADT, a town of Baden, in the circle of the Lower Rhine, and bail. of Neckar-Bischofsheim, on the Rodenbache. Pop. 1,874.

WAIDERSFELDEN, a town of Upper Austria, 12 m. E of Freystadt.

WAIDHOFEN (BAYERISCH), a town of Lower Austria, on the l. bank of the Ips, 72 m. W by S of Vienna. Pop. 3,500. It is noted for its manufactures of hardware.

WAIDHOFEN (BOHMISCH), a town of Lower Austria, on the river Thaya, 65 m. WNW of Vienna. Pop. 1,400. It has manufactures of linen and tape.

WAIGIOU, or WAYGEOU, an island of the S. Pacific, to the NW of New Guinea, the SE point of which is placed by Raper in S lat. 0° 21′, E long. 131° 18′. It is 70 m. in length from E to W, and 20 m. in breadth. It is inhabited by Malays, Alforas, and Papuans.

WAIHOU, a river of the North island of New Zealand, which runs SSE, and flows into the ocean in S lat. 42° 41′.

WAIKATO, a river of the N. island of New Zealand, which rises in Lake Taupo; flows N, NW, and W; and enters a narrow channel on the W coast, 30 m. S of Manukas, after a total course of 140 m., of which 100 m. are navigable by vessels of 30 tons.

WAIMATE, a settlement on the W coast of the North island of New Zealand, 15 m. W of the Bay-of-Islands.

WAINFLEET, a town of Lincolnshire, situated on a creek at the E part of the county, 4 m. above its entrance into the Wash, 14 m. NNE of Boston. The creek is navigable for vessels of small burden. The town is neat and well-built. Pop. in 1841, 2,117; in 1851, 2,255.

WAINGAROA, a harbour on the W coast of the North island of New Zealand, in S lat. 35°, consisting of a long inlet into which two rivers empty themselves.

WAIPAPA POINT, a headland on the E coast of the Middle island of New Zealand, in 42° 9′ S lat.

WAIPAPAPA POINT, a headland on the S coast of the Middle island of New Zealand, in N lat. 46° 40′, E long. 168° 54′.

WAIRAU, a river of the Middle island of New Zealand, which rises in the Snowy Alps, and flows ENE into Cloudy bay, in S lat. 42° 30′. Its chief tributary is the Waiopai, which joins it on the r. bank a little above its embouchure.

WAIROA, a river of the North island of New Zealand, which flows into Kaipara harbour.

WAITANGI, a river of the Middle island of New Zealand, which flows into a considerable estuary on the E coast, in S lat. 44° 38′.

WAITANGI BAY, a deep bay on the W coast of Chatham island, in S lat. 43° 58′, W long. 176° 38′. It presents good anchorage in 5 faths., and on firm sand.

WAITHE, a parish of Lincolnshire, 7 m. SSE of Great Grimsby. Area 780 acres. Pop. in 1841, 49.

WAITSFIELD, a township of Washington co., Vermont, U. S., 18 m. SW of Montpelier. Pop. in 1840, 1,048.

WAIT'S RIVER, a river of Vermont, U. S., which runs into the Connecticut, 12 m. below Well's river.

WAITZEN, or VACZ, a town of Hungary, beautifully situated on a bend of the Danube, 19 m. N of Pest, and 95 m. E of Presburg, in N lat. 47° 47′. It is a place of antiquity, having been a bishop's see in 1074. It is tolerably built, and contains a square of some extent. The public structures are a cathedral, several Catholic, and a Protestant church, a government-house, an episcopal seminary, and a Piarist college, and a military school.

WAITZENKIRCHEN, a town of Upper Austria, 19 m. W of Lintz.

WAJIDA, a village of Tlemcen, in Algiers, 25 m. SW of Tlemcen.

WAJORA, a village of Kaarta, in Central Africa, in N lat. 14° 48′.

WAKAMBA, a territory of Central Africa, lying between the Kiskambutia mountains and the Kilimandjaro chain, and intersected by the river Adi. It is inhabited by a people of comparatively fair complexion, who have commercial intercourse with Kaffa and Enarea.

WAKASASA BAY, a bay on the SW coast of Florida, which receives several rivers, and contains a number of islands.

WAKASP, a river of the Middle island of New Zealand, which flows into False bay, on the W coast, in S lat. 44° 20′, E long. 168° 24′.

WAKATOMIKA, a river of Ohio, U. S., which runs SE, and joins the Muskingum, 13 m. above Zanesville.

WAKE, a central co. of the state of N. Carolina, U. S., comprising an area of 1,018 sq. m., drained by Neuse river and its branches, and intersected by North Carolina Central, and Gaston and Raleigh railways. Pop. in 1840, 21,118; in 1850, 24,887. Its capital is Raleigh.

WAKEFIELD, a parish and a municipal and parl. borough, in the wapentake of Agbrigg, W. R. of Yorkshire. The p. contains the town of W., the chapelries of Horbury and Stanley, and the townships of Alverthorpe, Thornes, Wakefield, and Wrenthorpe; and is intersected by the Manchester and Leeds railway, which is here carried over the Calder at Broadreach by a viaduct of three arches, each 63 ft. 7 ins. on the skew. The North Midland railway, from Derby to Leeds, joins the Manchester and Leeds line a little to the east of Wakefield. Area of p. 9,311 acres. Pop. of p. in 1801, 16,597; in 1831, 24,538; in 1851, 33,117. The church, originally erected in 1329, is in a mixed style of architecture, 156 ft. in length by 69 ft. in breadth; with a square embattled tower, and an octagonal spire 237 ft. in height. The interior consists of a nave and aisles, divided by pointed arches. The township of W. comprises the greater part of the town; on the SW side of the township, the buildings of the town advance, in a continuous street, into the neighbouring agricultural township of Alverthorpe. The town—which in many respects may be considered the capital of the West Riding—is situated on the N or l. bank of the Calder, on a sloping eminence, nearly in the centre of the parish. It is built with considerable regularity, and contains many large and lofty houses chiefly of brick. In Wood-street is a handsome structure, containing a library and news-room, with rooms for concerts and assemblies. The market-place is not of great extent; but a handsome corn-exchange, with a saloon over it, was erected in Westgate in 1838. The river is here crossed by a stone-bridge of eight arches, erected in the reign of Edward III. In the centre of the bridge, projecting over the eastern side, and partly resting on the starlings, is a chapel, which is commonly supposed to have been erected by Edward IV. on the site of a more ancient building. It is a beautiful structure, about 30 ft. long by 18 ft. broad. The court-house is a very elegant edifice, with a Doric portico. The house-of-correction for the West riding, is an extensive structure at the bottom of Westgate. The new gaol, a grand massive imposing building, is the largest prison in England, accommodating on the separate system, 732 prisoners. It stands on about 14 acres of ground, and cost £120,000. The Reform act conferred the privilege on W. of returning one member to parliament. The East moor, the village of Thornes, and parts of the townships of Alverthorpe and Stanley, are included within the boundaries of the parl. borough, of which the pop. in 1851 was returned at 22,057. The number of electors registered in 1837 was 733; in 1852, 850. W. is also a polling-station in the election of members for the West riding.—W. was formerly celebrated for its manufactures of woollen cloth and worsted yarn, but both these, though still considerable, have greatly declined. There are dyeing works, breweries, foundries, and starch works in the town. The chief trade now consists in the export of corn, coal, and wool. The wool is sent to factors in this town for sale from various parts of England. The coals are sent down the Calder, and thence by the Ouse to York, and by the Humber to Hull. The corn market is the most important in the north of England. The river was rendered navigable to W. in 1698; and in 1760 the navigation was extended to Salter-Hebble near Halifax. The town is, by railway, distant from Leeds 12½ m.; from Manchester 57½ m. It is supposed that a town existed on the site of W. in the time of the Romans. A battle was fought here on the 24th of December, 1460, between the princess Margaret of Anjou, wife of Henry VI., and Richard, duke of York, in which the latter was defeated and slain, with upwards of 3,000 Yorkists. During the Parliamentary wars W. suffered severely, having been alternately in the hands of the royalist and parliamentary parties.

WAKEFIELD, a township in Carroll co., New Hampshire, U. S., 46 m. NE of Concord. Pop. in 1840, 1,396; in 1850, 1,405.—Also a village of Kingston township, in the state of Washington, Rhode island, 11 m. WSW of Newport.

WAKEFIELD (PORT), a village and port of S. Australia, on the gulf of St. Vincent. It is a shipping-place for the adjacent northern districts, and especially for the export of the London copper-smelting company. The tonnage at Port W., in 1851, was in excess of 20,000 tons.

WAKEMAN, a township of Huron co., in the state of Ohio, U. S., 89 m. NNE of Columbus, and intersected by the Toledo, Norwalk, and Cleveland railroad. Pop. in 1840, 702; in 1850, 704.

WAKENDORF, a village of Denmark, in the duchy of Holstein, and bail. of Georgenthal. Pop. 200.—Also a village of the bail. of Segeberg. Pop. 480.—Also a village of the bail. of Traventhal. Pop. 240.

WAKENETZ, a river which issues from the N extremity of Lake Ratzeburg, on the confines of the Dutch territory of Lauenburg, and territory of Lubeck; enters the latter; runs NNW; and after a course of 12 m., throws itself into the Trave, on the r. bank, at Lubeck.

WAKERING (GREAT), a parish in Essex, stretching along the coast, at the mouth of the Thames. The v. is 5 m. SE of Rochford. Area of p., 5,484 acres. Pop. in 1831, 834; in 1851, 905.

WAKERING (LITTLE), a parish in Essex. 4 m. E by N of Prittlewell. Area 5,862 acres. Pop. in 1831, 297; in 1851, 292.

WAKERLEY, a parish in Northamptonshire, 8 m. NE of Rockingham, Area 2,130 acres. Pop. in 1831, 218; in 1851, 232.

WAKRU, a village of the district of Martaban, on a small river of the same name, about 10 m. E of Amherst. It is said to have formerly been a place of importance, and is still a valuable outpost.

WAKULLA, a county in the N part of the state of Florida, U. S., comprising an area of 492 sq. m., drained by branches of Ockatockany river, and generally fertile. Pop. in 1850, 1,955. Its capital is Newport.

WALACHIA, or WALLACHIA [GERM. *Walachei;* FRENCH, *Valachie*], a northern principality of European Turkey, which derives its name from its inhabitants, the *Vlaques* or *Wallachians.* The Turks call it EFLAKE; the Walachians, ZARA-RUMANASKA, and themselves RUMUNI or Romans. It lies between the parallels of 43° 40' and 45° 40' N; and is bounded on the NW by the Austrian principality of Transylvania; on the NE by Moldavia; on the SE by the sanjak of Silistria; on the S by the sanjak of Nicopoli; on the SW by the sanjak of Widin; and on the W by the military frontiers of Hungary. Hassel and Balbi estimate the superficial extent at 25,231 sq. m.; Ritter at 25,310, and Sulzar at 29,767 sq. m. It is surrounded on the W and NW by high mountains belonging to the Carpathian system, which separate the country from Austria. The advanced ridges of these heights extend into the country on the W, as far as to the banks of the Danube; on the E side they form a narrow ridge. The rest of the surface is an extensive plain, through which large rivers, descending from the Carpathian range, flow slowly towards the Danube. This plain is diversified by some forests; but is scarcely broken by a single eminence. Thornton describes the appearance of both principalities in rapturous terms: the fertility of the soil,—the torrents rushing down the precipices and winding through the valleys,—the fragrance of the lime-flower, and of the herbs crushed by the browsing flocks,—the solitary huts of the shepherd on the brow of the mountains,—the mountains themselves rising far above the clouds with their snow-clad summits, and beneath adorned with lofty and majestic trees,—all constitute ever-present elements of beauty in the landscape. Many of the mountain summits are composed of gneiss and mica-schist. There are five important passes between W. and Transylvania. The Danube is the principal river; its main tributaries are the Schyl, the Aluta, the Arjish, the Dumbrovitza, and the Jalomnitza. "These rivers," says Mr. O'Brien, "could, at a small comparative expense, be made navigable for the passage of rafts. The mountains in which they have their source are covered to their summits with magnificent forests. At the base of the mountains are oaks, midway up are beech-trees, and above these are pines of extraordinary height and girth. All this wealth of wood lies utterly useless, and trees and branches blown down by the storm rot where they fall for want of the means of transport. In the Moldavian parts of the Carpathians, the case, however, is not exactly the same. The river Sereth, which separates the two principalities, is larger and deeper than the rivers flowing through W., and its waters are consequently covered during the season with immense rafts of wood, consisting chiefly of oak and pine, both of which are good in building and masting ships." There are no lakes of any considerable size, but numerous ponds and marshes.—The climate is temperate, though not so mild as on the S side of the Balkan; the Danube occasionally freezes, and winter, commencing in November, lasts till March. The spring is rainy. During three months the heat of summer is very great, and thunder-storms are frequent. Autumn is the finest season of the year. Earthquakes sometimes occur; in 1802 the city of Bucharest was nearly destroyed by one. The nights are cold and attended with copious dews. In general the air is healthy; but in the mountain districts goitre is occasionally seen. Bilious fevers and agues are common but not dangerous.

Soil and Productions.] The soil is productive in the valleys between the mountains; and of luxurious fertility in the plains. It is, however, very indifferently cultivated, the inhabitants generally preferring the life of herdsmen to the occupations of the agriculturist. The possessors of land all reside in Bucharest, and care nothing about their property or the people who cultivate it, so long as the rents are regularly paid. The farmers usually occupy much smaller pieces of ground than in England, and seldom get rich; indeed, they are rather serfs than farmers. The arrangement between them and the landlords is made at stated periods of five years, fixed by government. If the farmer does not like his landlord, he must give one year's notice that he is going to leave his village, and under no circumstances can he leave it except at the expiration of one of these quinquennial periods. He is forced to cultivate ground somewhere, and consequently the cultivators are nearly tied to the soil, as every obstacle is thrown in the way of their changing landlords. The rent is paid partly in kind, partly in money, and partly in labour. The landlord gets the tenth part of all corn which is grown, and the corn may not be removed until he has taken his share. The people appear very poor, much more so than the Bulgarian pop. on the S of the Danube, under the Mussulman government. The latter grow rich frequently in spite of the oppression they may be sometimes individually subjected to, and their cottages and the furniture and cooking utensils are always much better than those of the corresponding class in W. Mr. O'Brien, in his recent account of this country, says of its resources: "There is certainly not in Europe a soil more rich, and scarcely a climate more favoured than that of W. The country is literally overflowing with grain of every sort, and out of France I have never drank a *vin ordinaire* as good as in this country. The common white wine of W., when kept for two or three years in bottle, is equal to anything of the kind produced on the banks of the Rhine. The rich meadow-lands afford pasturage to numberless herds of cattle and flocks of sheep, whilst in the neighbourhood of the forests there are immense herds of swine. It is on account of this great abundance of meat of all kinds, that Mr. Goldner's successors have established themselves in the neighbourhood of Galatz, and that another English establishment largely engaged in the pork trade has been formed at Kalafat. The proprietors of both these establishments find, that notwithstanding the high wages they are forced to give their English workmen, and the expenses of transport, they can sell their merchandise cheaper in England than if they purchased the materials for their trade in any of the markets of Great Britain or of Ireland. Game of almost all kinds abound in W. Wild turkeys are met with in hundreds in the steppes or great open plains. Hares were sold until lately in the market of Bucharest at 4d. each, and a brace of blackcock at about the same price. There is also an abundance of fish in the inland rivers, and some of it of very exquisite flavour." Turkish corn is the grain most extensively cultivated; but a good deal of excellent wheat also is raised and exported; melons form a principal article of food, and fruit is excellent and abundant. Wine might be extensively manufactured in this country if the people were industrious. The principal wine districts are Dragishan, Cemez, Greca, and Lacuisni. There are extensive forests, which are peopled with flocks of singing-birds, particularly nightingales. The pastures are good, and support a number of cattle. The native oxen are all white, small, thin, and long-haired. The horses are small, but swift and strong. The wool is excellent, the number of sheep in the country exceeds 3,000,000. Upwards of 100,000 are yearly exported to Constantinople. There are three species of sheep fed in this country: the *zurkan* with long and coarse wool; the *zigey* with short and fine wool; and the Tartarian breed with enormous tails and a middling fleece. Game, as already observed, is abundant; but the *bojards* or lords of the soil—as in more civilized countries—reserve to themselves the exclusive privilege of hunting. All the forests and fastnesses abound in wolves, bears, wild-boars, and deer. Bees are kept in great quantities, and leeches are exported. There are three mines of rock-salt in W. which yield 20,000,000 okas, of 2 ℔ 14 oz., annually. Gold occurs in the sands of several rivers. Mercury abounds in the vicinity of Piteshti. Iron occurs at Bara, Annua, and elsewhere. At Gesseni there are coal-mines. Good building-stone is scarce; chalk is abundant. There is little manufacturing industry. The common people make their own clothes and tools; the rich are supplied with articles of luxury from abroad.

Commerce.] The two principalities export annually nearly 4,000,000 hectolitres of grain. Notwithstanding the impediments at the mouth of the Sulina, and the difficulties laid in the way of trade at Braila, 1,128 ships were laden in 1851. British ships alone carried 345,000 hectolitres of grain, whilst 350,000 hectolitres were conveyed to England in foreign vessels. In 1852, upwards of 700 ships were laden at Galatz. The following is a recent estimate of the importations: Viennese and Transylvanian wares, 3,200,000f.; manufactures of Leipzic, 6,000,000f.; Russia, 350,000f.; the Levant, 4,600,000f.; British manufactured iron wares, 2,000,000f. From Vienna the principalities import cloth, shoes, gloves, carriages, musical instruments, and glass; from Transylvania, coarse linens and leather; from the Levant, colonial wares, coffee, sugar, tobacco, cotton twist, British and French manufactures, and British iron; from Russia, salt-fish and furs; from England, iron, steel, machinery, and utensils; from France, silk cloths, perfumery, fashionable articles, and books. The commerce of the principalities is free from prohibition. The custom-house taxes hitherto amounted to 3 per cent. for both importation and exportation: they have lately been raised to 5 per cent. In 1847 the custom-houses were let out for the annual sum of 1,200,000f. To the Danube W. is indebted for the greater part of its commerce, and for the means of communicating both with the E and W. Vienna is not more than six days' distant from Giurgevo, and the Black sea not more than three. It is at once the channel through which W. disposes of the raw materials and the immense quantities of corn which have procured for her and Moldavia the designation of the granary of Europe, and through which W. receives the products of European art and industry. As yet, however, the Walachians take hardly any part in its navigation, which remains wholly in the hands of Austria, if we except the wretched flat-bottomed boats which bring a few necessaries from Galatz to other points on the river, or transport the immense masses of salt, which form one of the principal items in Walachian commerce to Bulgaria, Hungary, and Servia. The steamboats of the Austrian Lloyd keep up a regular communication, in time of peace, between Vienna and Constantinople. Three stations in W. are established—Turnu-Severin, Giurgevo, and Ibraila, where they have agents, who are mostly Austrian consuls as well. On issuing from W. the Danube is divided into numerous branches, separated by islands of greater or less dimensions. The most remarkable of these mouths are those of Kilia, St. George, and Sulina. In the old and golden days of the East, the Genoese had a factory at Kilia; but at this point the river has now dwindled into insignificance. The St. George mouth is accessible only to fishing-boats, and as regards Sulina, everybody knows how successfully Russia has managed to destroy it also, so as to drive the trade of this quarter into Odessa. In the best of times, this mouth required considerable care to keep it open, and a project was at one time set on foot to run a canal across, between Rassova and Kustendje, the expenses of which Austria offered to bear, so as to render the navigation independent of the shifting sands of the river.

Population, religion, &c.] Mr. Wilkinson, the English consul, estimated the total pop. of W. at only 1,000,000. Stein in 1826 estimated the pop. at 950,000. A census taken in 1853, returned the pop. at 2,324,484. The latter census returned the number of married priests at 5,454; deacons married, at 840; monks, 2,860; nuns, 2,050; titled boyards, whose nobility is not hereditary, 3,807; hereditary nobility (cultivators), living in the villages, 5,510; persons exempt from capitation for services rendered to the state, 509; persons holding privileges of various sorts for the same reason, 6,041; merchants holding patents or licenses to trade, 17,634; tax-paying cultivators or peasants, 339,907; *Csigany* or gipsies belonging to the government, but emancipated and paying taxes, 18,273; ditto, slaves of private individuals, 20,500; foreigners, 4,297; maimed, deformed, or infirm, 30,000; persons without any fixed residence, 741; Jews living only in the towns, 4,700; artizans, 5,600.—The Walachians are descendants of the ancient Romans, Dacians, Mœsians, and Bulgarians. They are an indolent and superstitious race of men, knowing no better amusement than to dance to the sound of the pipe under the shade of their woods. Their dress consists of a large loose shirt fastened by a belt, wide trousers, a sheep-skin cloak, and sandals of untanned leather

The Greek portion of the pop. are wealthier and more active. Turks are not allowed to settle in the principality. The Gypsies amount to about 50,000. The Walachians and Greeks generally belong to the Greek church; the head of which in this country is the archb. of Bucharest. The Roman Catholics are under the bishop of Nicopolis. The clergy possess a third part of the landed property in the country; and the annual income of their archbishop is said to be equal to 400,000 piastres. The abbeys and dioceses are exposed to sale, and the price is put into the coffers of the hospodar. The church property in W. is very large, and might be diminished with great advantage. Reformers complain that the highest and most valuable offices in the church are, through the medium of Russian influence, filled by Greeks, men who do not understand one word of the language, and are therefore utterly unfitted for a pastoral charge among the people where they are sent. Nay, more, in some instances they are not qualified for their holy office even by the correctness of their lives and deportment, and, after debauchery in private during the eight or ten years they reside in the country, return with large fortunes to their native land. The Walachian is not the written language: the only books used are printed in Greek, and Greek is also taught in the schools. There are no country schools; the only seminaries of education are to be found in the towns; and such is the prevailing ignorance, that even Greek bishops are occasionally found who can neither read nor write. It is, or very recently was customary in this country, to open the sepulchres of the dead every seven years, and if the body had not returned to its kindred dust, the relations judged that the soul that once animated it was in a state of condemnation, or that the deceased had become a vampire. The rich bojards either employ foreign tutors, or send their sons to Germany or Italy for their education. Marriages are easily dissolved.

Government, revenue, &c.] "The territories of the old Byzantine monarchy, which the Turks appropriated, had been bounded on the N by the Danube, but, as the power of the Ottoman sovereigns increased, they carried their conquests beyond this frontier, and established themselves for a considerable period even in the capital of Hungary. Among the states in these parts which they reduced to obedience were two principalities of some antiquity termed W. and Moldavia, governed by rulers of their own and comprising very nearly the same districts which they include at the present day. These were brought by the great Solyman into a species of subjection to the Ottoman crown. That the conquest was in itself incomplete is not very probable, considering the strength of the Turks at that period, but, for some reason or other, the provinces in question were never actually incorporated in the Ottoman empire, but were left under their own princes, with a reservation only of sovereignty and tribute to the Sublime Porte." W. and Moldavia were governed by native princes, under the authority of the Sublime Porte, until 1731, when the divan deposed the indigenous princes and clothed the Fanariotes in their spoils. The divan did not deprive the natives of all influence in the government; and various posts were reserved for the native boyards, such as those of chief-justice, mayor, secretary-general of the districts and cantons. But the situation of minister of the interior and for foreign affairs, of the police, the sheriffs, the military officers, and a multitude of other posts, were usually given to the Fanariotes in the suite of the hospodar, who from the moment of their appointment took the title of boyar or *boyard*. The only allegiance which W. owns to the sultan, under latter arrangements, is an annual payment of £50,000 as tribute, and the necessity of having his approbation of the person who may be elected as prince. The prince is elected by a council of about 60 persons, half of whom are elected by the landed proprietors of the various districts, and the other half consists of the most influential boyards in Bucharest and the ministers of state. When they have made choice of any one, his name is sent to the emperor of Russia and the sultan for approbation, without which he cannot assume office. The Russians exert every means of intrigue to secure the election of a person in their interest, and they had obtained such influence in the country of late years that they were always sure of gaining their point. The W. noblesse dates its origin from the year 1241. Previous to that time the country had been desolated by the continued incursions of the Tartars; the pop. was terribly reduced in numbers,—the slaves only ventured to make head against the enemy; and W. and Moldavia put themselves under the protection of Hungary, then great and flourishing, under Louis I. In the year above mentioned the Walachs of Transylvania, no longer able to endure the inroads of the Tartars, emigrated into their old fatherland under the command of Roddoulo Negro, or Black Rodolph, as he is called in W. annals. He was accompanied by a considerable number of warriors of greater or less renown, to whom he confided the principal charges and dignities of his new state, and in these personages, says tradition, you will find the origin of the boyards or nobles. The word *boyard*, or *boier*, is, according to some, Slavonic, and means 'soldier,' or 'armiger;' others derive it from the Latin *bos*, 'an ox,' and affirm it was the designation in the old Roman times of the large farmers, distinguished beyond their neighbours by the possession of great herds of cattle. The prince, and the boyards, are alike distinguished from their inferiors by the length of their beards; the latter are also easily discernible from a common inhabitant by the enormity of their *kalpaks*, or head-dress, which is composed of black lamb-skins in the form of a balloon. By the porte, the hospodar is allowed a limited revenue arising from a tithe upon sheep, bees, altogether valued at only £800 a year: in a short time, however, he generally contrives to amass immense treasures. The hospodar has not the military force of the prov. at his disposal like the pashas; his divan or court is modelled after that of the Byzantine emperors.—There is no written code of laws; sometimes indeed reference is made to the Basilian and Roman law; but the hospodar is always supreme judge, and all legal proceedings are very summary. The native militia used to consist of about 6,000 men; occasionally reinforced by a body of Albanians—The revenue must be considerable to supply the enormous drain occasioned by the rapacity of the hospodars, and the demands of the porte. The revenue nominally paid to Turkey used to be 600 purses or 300,000 piastres; but seldom fell short of 500,000 piastres. The inhabitants were likewise obliged to supply the Turkish capital every year with 8,000 horses, 28,000 sheep, and 187,500 quarters of wheat. The tributary peasants in 1817 paid 1,800,000 piastres, or £360,000. A complete system of bribery and corruption pervades all the departments of government, and especially that which has charge of the administration of justice. In a lawsuit the poorer is always sure to have the worst of it, and even the lower class of boyards say that it is useless for them to carry a dispute with one of the higher nobility into court. All the officers of state are poorly paid—a minister only gets £600 per annum

—but they make it up by other means. There have been very large sums laid aside from the state revenues, and subscribed by private individuals, for keeping the streets of Bucharest well paved, but their present state is ruination to carriage springs, and most uncomfortable for those inside.

Topography.] W. is divided into W. Proper, comprising Upper and Lower W. on the E of the Aluta; and Little W., sometimes called the Banat of Crayova, on the W of the same river. Bucharest is the capital of W. Proper. Fokshani, or Foktchan, situated on the r bank of the Milkov, was once a considerable commercial town: but was burned in 1822. In the neighbourhood of Waleni are the extensive salt-mines of Stanikul, which recently furnished 900,000 cwts. of that article annually. The Transylvanian trade is chiefly conducted by way of Kimpina, a town on the Braowa. Brahilow or Ibrail, situated on the Danube, near the mouth of the Sereth, is the capital of a district, and has a strong citadel. This city capitulated to the Russians in 1828, after a gallant defence.—Giurgevo, Djordjova, or Jerköki, which politically belongs to the sanjak of Rutshuk, is situated on the l. bank of the Danube nearly opposite Rutshuk,—forming the *tete-de-pont* of that town. The citadel is situated on the island of Hobodyie.—Tergovisto, the ancient residence of the waivodes, is situated upon the Jalonitza.—Little W. was ceded to Austria in the peace of Passowitz; but given back to the Turks by the treaty of Belgrade in 1739. Crajova, the capital, is situated at a little distance from the l. bank of the Schiul.

History.] This prov. was unknown in authentic history, until its invasion and conquest by the Romans in the reign of Trajan. That prince sent hither several colonies, who cultivated tracts of land, and built, in prescribed situations, towns and villages. On the decline of the empire, W. shared the fate of other frontier provinces, being alternately in possession of the Greek emperors and the barbarians. The 9th cent. is said to have been the era of their embracing the doctrines of the Greek church; while the early part of the 12th is given as the date of the foundation of Bucharest, the capital, and of some smaller towns. In the 13th and 14th cents. W. was in some degree subject to Hungary. In the beginning of the 15th, the Turks penetrating in this direction, long before they accomplished the conquest of Constantinople, laid waste the country, and subjected it to a tribute. The conditions imposed upon the principalities were mild. The prince of W. occupied the *suzerainty* of the Porte, and agreed to pay an annual tribute; but at the same time preserved the right of declaring peace or war, and the other prerogatives of an independent ruler. "The other conquests of the Sultan to the N of the Danube were all ultimately lost, but 'the Danubian principalities,' as they were denominated, remained under relations with the Turkish monarchy until its fortunes began to be affected by the ascendency of Russia. Peter the Great was contented with obtaining the support or connivance of their princes in his contests with the sultan; but, as the ambition of the czars expanded, it became a leading object of Russian policy to detach these semi-independent states from the Turkish dominion altogether, and to transfer to St. Petersburg, by gradual encroachments, the allegiance due to Constantinople. These designs were aided by a concurrence of favourable circumstances. As the Ottoman power declined, the independence of W. and Moldavia became somewhat more substantial, and, as their pop. was exclusively Christian, the pretext of a protectorate was brought plausibly into requisition. The Russian government affected, therefore, to defend against the Porte, the rights and immunities constitutionally reserved to the principalities, and so successfully were the operations of Russia conducted that, in the end, the authority of the sultan was practically divided for the benefit of the czar: every victorious war was made to contribute its impulse to this consummation. As long ago as 1792, it was stipulated and agreed that the governors, or, as they are styled, 'hospodars' of the Danubian principalities, should neither be appointed nor removed without license obtained from the Russian court, and so rapidly since that time has the work of encroachment proceeded, that the sultan now stands debarred by his own surrenders from most of the privileges of a supreme lord. The 'sovereignty' of the provinces, it is true, still nominally pertains to him, but no Turk can settle in them, nor can any Turkish troops be quartered there, except under specified circumstances." The court of St. Petersburg had, however, hitherto contented itself with simply detaching the principalities from their old allegiance without openly claiming any sovereignty of its own. It pretended merely to secure their "independence." The invasion of the principalities could easily enough be effected by Russia without any military difficulties, and, practically speaking, without opposition. To invade Turkey proper, Russia would have to cross the Danube—always an arduous operation—or to effect a landing at some point of the Black Sea coast, and in either of these cases she would be brought into inevitable collision with the Ottoman forces. But by simply occupying Walachia and Moldavia, she confines her expedition to an unresisted march, encounters no enemy, and can reckon upon retiring without damage when her desires have been satisfied for the time. Hence the Russians have repeatedly invaded the principalities: "They have been in and out of them, in fact, for the last 50 years. As long ago as 1806, they were there, and found means to prolong their occupation in spite of opposition. They were upon the point of entering in 1821; they actually entered in 1828; and on the 30th September 1848, they entered again. It was on the last occasion that a special treaty was signed between Russia and the Porte, stipulating the particular circumstances under which *either* power might occupy the provinces in question for the future, and it is this treaty which the czar violated." The convention of Balta-Liman, on 1st May 1849, recognised the right which had always been denied to Russia of military occupation of the two principalities and of interference in their internal administration. The privileges which the Moldo-Walachian principalities acquired from their capitulations with the Turks, and which were recognised by subsequent treaties, are (1) The right of electing their hospodars; (2) Administrative and legislative independence; (3) Inviolability of their territories. The rights of sovereignty appertaining to the Porte: (1) The purely honorary right of confirming the hospodars elected by the nation. (2) An annual tribute of 3,000,000 of Turkish piastres from the two principalities. The right of Russia is confined to preventing the Porte from trespassing on the privileges accorded to the Moldo-Walachians, and which the latter had guaranteed by the treaty of Adrianople. The convention of Balti-Liman maintains the following principles: "two committees will be appointed to revise the organic laws, whose labours are to be submitted to the examination of the Ottoman government, which, after coming to an understanding on the subject with the court of Russia, will promulgate them in the form of a *hatti-scheriff.* (Art. 2.)—Each of the two governments will maintain in the principalities armies numbering from 20,000 to 30,000 men, until the entire pacification of the country; subsequently to the pacification, and until the completion of the constitution, for the principalities, each of the two powers is to reduce its army to 10,000 men; immediately afterwards the two powers will completely and simultaneously evacuate the principalities. (Art. 4.)—It is evident then that the convention of Balta-Liman not only violated all the privileges of the Moldo-Walachians, but also gave to Russia equal rights of sovereignty with those possessed by the Porte. See articles MOLDAVIA and TURKEY.

WALBECK, a town of Prussia, in the prov. of Saxony, regency and 27 m. WNW of Magdeburg, and circle of Gardelegen, on the Allier. Pop. 1,200. It has manufactories of pipes and of potash.

WALBERSWICK, or WALDERSWICK, a parish in Suffolk, at the mouth of the river Blyth, 2 m. SW of Southwold. Area 1,960 acres. Pop. in 1851, 357.

WALBERTON, a parish in Sussex, 3½ m. WSW of Arundel. Area 1,722 acres. Pop. in 1851, 578.

WALCHBERG, a mountain of Austria, in Styria, in the circle and 36 m. NW of Judenburg. It has mines of sulphur and of argentiferous copper.

WALCHEN-SEE, or WALLER-SEE, a lake of Bavaria, in the circle of Upper Bavaria, and presidial of Weilheim. It is 5 m. in length from NE to SW, and 2 m. in breadth, and discharges itself by the Jaishna into the Iser. It abounds in fish.

WALCHEREN, or WALCHERN, an island of Holland, in the prov. of Zealand, situated in the German ocean, at the mouth of the Scheldt, and separated from the islands of Beveland by a narrow channel called the Sloe. Pop. 43,000. If not the largest, it is the most populous and best-cultivated of the different islands composing the prov. of Zealand. It is of an oblong form; in length from NW to SE about 12 m.; in breadth from NE to SW, 8. m. It lies low, and would be subject to inundations from the sea, were it not protected by strong dykes. The dyke of West-Cappel, in particular, is of great size and strength. The greater part of the surface is rich meadow. This island contains the towns of Middleburg, the capital, Flushing, and Veere. The villages are numerous. Agues and bilious complaints prevail in spring and autumn, in consequence probably of the quantity of fresh water in the canals with which the island is intersected: hence the sickness prevalent among the British troops during their occupation of it in 1809.

WALCHFING, a village of Bavaria, in the circle of Lower Bavaria, and presidial of Vilshofen. Pop. 410.

WALCHOW, a village of Prussia, in the regency of Potsdam, and circle of Ruppin. Pop. 240.

WALCHUM, a village of Hanover, in the principality of Osnabruck, and bail. of Ahremberg-Mippen. Pop. 230.

WALCHWYL, a village of Switzerland, in the cant. and bail. of Zug. Pop. 1,039.

WALCOT, or WALCOT-FALKINGHAM, a parish in Lincolnshire, $1\frac{1}{2}$ m. NW of Falkingham. Area 1,747 acres. Pop. in 1831, 183; in 1851, 152.

WALCOT, or WALCOT-BILLINGHAY, a hamlet and chapelry in the p. of Billinghay, Lincolnshire, 2 m. N of Billinghay. Pop. in 1831, 514; in 1851, 617.

WALCOT, a parish partly within the city of Bath, Somersetshire. Area 1,023 acres. Pop. in 1831, 26,023; in 1851, 25,457.

WALCOT-CUM-MEMBRIS, a hamlet in the p. of Holy-Cross, Worcestershire, 2 m. NE of Pershore. Pop. in 1831, 375; in 1851, 506.

WALCOTT, a parish in Norfolk, 5 m. E by N of North Walsham. Area 735 acres. Pop. 167.

WALCOURT, a department, commune, and town of Belgium, in the prov. of Namur, and arrond. of Dinant. Pop. of dep., 910. The town is 29 m. SSW of Namur, on the Heure. Pop. 461. It is enclosed by walls, and has an ancient church, noted as a resort of religious devotees. It possesses extensive iron-works. It is a place of great antiquity. In 1615 it was reduced to ashes.

WALD, a village of Bavaria, in the circle of Upper Bavaria, and presidial of Alt-Setting. Pop. 200. —Also a village in the presidial and 12 m. SW of Burghausen, on the r. bank of the Alz. Pop. 260. It has a royal castle.—Also a village of the circle of Middle Franconia, and presidial of Gunzenhausen. Pop. 200.—Also a market-town of the circle of Swabia, and presidial of Furtheim. Pop. 640. It has a castle in ruins.—Also a village of Austria, in the regency of Salzburg. Pop. 240.—Also a village in Styria, in the circle of Bruch. Pop. 640.—Also a village in the circle of Judenburg. Pop. 420.—Also a village of Tyrol, in the circle of the Imst. Pop. 410.—Also a village of Prussia, in the regency and 14 m. ESE of Dusseldorf, and circle of Solingen. Pop. 700. It has woollen, linen, and cotton factories, iron founderies, manufactories of copper, steel and iron-ware, of ribbon, and tobacco.—Also a village in the circle of Geldern. Pop. 790.—Also a village of Saxony, in the circle of Bautzen, and bail. of Lobau. Pop. 396.—Also a village of Switzerland, in the cant. of Appenzell, on the Sitter. Pop. 1,432.—Also a town in the cant. of Zurich, and bail. of Hinweil. Pop. 3,808. It has cotton and silk factories.

WALD (OBER and NIEDER), two villages of Austria, in Styria, in the circle of Gratz. Pop. 700.

WALDAI. See VALDAI.

WALDALGESHEIM, a village of Prussia, in the regency and 30 m. SSE of Coblenz, and circle of Kreuznach. Pop. 840.

WALDAMORBACH, a village of Hesse, in the prov. of Starkenburg, and bail. of Hochst. Pop. 300.

WALDANBACH, a village of Nassau, in the bail. of Herborn. Pop. 240.

WALDANGELLOCH, a village of the grand-duchy of Baden, in the circle of the Upper Rhine, and bail. of Hoffenheim. Pop. 1,000.

WALDASCHAFF, a village of Bavaria, in the circle of Lower Franconia, and presidial of Rothenbuch, at the source of the Aschaff. Pop. 1,324. It has several iron-works.

WALDAU, a village of Anhalt-Bernburg, in the circle of Bernburg. Pop. 600.—Also a village of Baden, in the circle of the Upper Rhine, and bail. of Freiburg. Pop. 250.—Also a village of Bavaria, in the circle of the Upper Pfalz, and presidial of Vohenstrauss. Pop. 379.—Also a village of the circle of Upper Franconia, and presidial of Culmbach. Pop. 240.—Also a village of Electoral Hesse, and prov. of Lower Hesse, circle and $1\frac{1}{2}$ m. S of Cassel, on the Wahlebach. Pop. 810.—Also a village of Prussia, in the regency of Erfurt, and circle of Schleusingen. Pop. 401.—Also a village of the regency of Liegnitz, and circle of Bunzlau. Pop. 2,560. It has a castle, and contains several mills, factories, and bleacheries.—Also a village in the regency and circle of Liegnitz. Pop. 625.—Also a village in the regency and circle of Konigsberg, between two lakes. Pop. 280.—Also a village in the regency of Marienwerder, and circle of Flatow. Pop. 300.

WALDAU (NEU), a village of Prussia, in the regency of Liegnitz, and circle of Sagan. Pop. 688.

WALDBACH, a village of Würtemberg, in the circle of the Neckar, and bail. of Weinsberg. Pop. 693.—Also a village of Hungary, in the district of Kovar.

WALDBACH (OBER and UNTER), two villages of Bavaria, in the circle of Swabia, and presidial of Burgau, containing respectively 608 and 31 inhabitants.

WALDBILLIG, a town of Belgium, prov. of Luxemburg, and com. of Grevenmacher. Pop. 560.

WALDBOCKELHEIM, a village of Prussia, in the circle of the Rhine, regency of Coblenz, circle and 8 m. WSW of Kreutznach. Pop. 1,168.

WALDBREDIMUS, a village of Belgium, in the prov. of Luxemburg, and com. of Grevenmacher. Pop. 380.

WALDBROL, a circle and village of Prussia, in the prov. of the Rhine, regency and 32 m. ESE of Cologne. Pop. of circle, 19,609; of village, 350. It has lead and iron-mines.

WALDBRUNN, a village of Bavaria, in the circle of Lower Franconia, and presidial of Wurzburg. Pop. 550.

WALDBURG, a village of Würtemberg, in the circle of the Danube, and bail. of Ravensburg. Pop. in 1840, 241.—It has a castle which belongs to the ancient and formerly powerful family of the same name.

WALDBUTTELBRUNN, a village of Bavaria, in the circle of Lower Franconia, and presidial of Wurzburg. Pop. 550.

WALDCAPPEL, a town of Electoral Hesse, in the bail. of Bischhausen, circle and 8 m. WSW of Eschwege, and 20 m. SE of Cassel. Pop. 1,000.

WALDDORF, a market-town of Baden, in the circle of the Lower Rhine, and bail. of Heidelburg, 17 m. SE of Manheim. Pop. 1,800.—Also a village of Würtemberg, in the circle of the Schwarzwald, and bail. of Nagold. Pop. 1,140. It has cloth factories.—Also a village in the bail. and 81 m. NE of Tubingen. Pop. 1,404. It has manufactories of linen, and of cutlery.—Also a village of Prussia, in the regency of Oppeln, circle of Neisse. Pop. 700.

WALDDORF (GROSS and KLEINE), two villages of Prussia, in the regency and circle of Dantzig, on the Mottlau. Pop. 400.

WALDECK, a principality in the W of Germany, consisting of the two counties of Waldeck and Pyrmont, having a united area of 460 sq. m., with a pop. in 1843 of 58,753; in 1852, 59,697.—The county of W., forming the chief part of the principality, has an area of 424 sq. m., and is bounded on the W and N by Rhenish Prussia; and on the E and S by Hesse-Cassel. It is one of the most elevated districts in Germany. The climate is cold. The mountainous tracts contain iron, copper, and lead. The rivers are the Eder on the N, and the Diemel on the S.—Pyrmont lies wholly detached from W., and about 30 m. to the N of its frontier. It had a pop. of 6,623 in 1852.

WALDECK, a market-town of Bavaria, in the circle of the Upper Pfalz, and presidial of Kemnath, on the Hor, 20 m. ESE of Bayreuth. Pop. 450.—Also a village of Silesia, in the circle of Troppau, near Johannisberg. Pop. 620.—Also a village of Prussia, in the regency of Konigsberg, and circle of Preuss-Eilau. Pop. 200.—Also a village of the grand-duchy of Saxe-Weimar, in the principality of Weimar, and circle of Weimar-Jena, 11 m. E of Jena. Pop. 270.—Also a town of the principality of Waldeck, 20 m. W by S of Cassel. Pop. 1,000. —Also a village of Bohemia, in the circle of Leitomischt. Pop. 300.

WALDEGRAVE'S ISLANDS, two small rocky islands, lying off the S coast of Australia, in S lat. 33° 35′. They are included by Flinders under the general title of the Investigator's Group.

WALDEN, a township of Caledonia co., Vermont, U. S., 22 m. NE of Montpelier. Pop. 910.—Also a village of Orange co., in New York, 80 m. S by W of Albany.

WALDEN (KING'S). See KING'S WALDEN.

WALDENBACH, a village of Bavaria, in the Upper Palatinate, on the Regen. Pop. 800.

WALDENBUCK, a town of Würtemberg, 10 m. S by W of Stuttgard, on the Aach. Pop. 2,000.

WALDENBURG, a town of Würtemberg, principality of Hohenlohe, 7 m. NW of Hall.—Also a town of Saxony, 49 m. WSW of Dresden, on the l. bank of the Mulde, in N lat. 50° 52′. Pop. 2,553. It is divided by the river into a new and old town. It has manufactories of woollens, linen, cotton, and earthen-ware.—Also a town of Prussian Silesia, 10 m. WSW of Schweidnitz. Pop. 3,500. It has a trade in linen and linen yarn.—Also a town of Switzerland, in the cant. of Bale, at the foot of the Hauenstein, 15 m. NE of Soleure. Pop. 756.

WALDENRATH, a village of Prussian Westphalia, in the duchy of Juliers, circle of Hemsberg. Pop. 630.

WALDEN-SAFFRON, or SAFFRON-WALDEN, a market-town and parish in Essex, 40 m. NNE of London. Area of p., 7,416 acres. Pop. in 1801, 3,181; in 1831, 4,762; in 1851, 5,911. The church is a beautiful and conspicuous edifice, 200 ft. in length and 82 ft. in breadth, with a tower and spire 193 ft. high. It consists of a nave, chancel, and side-aisles of mixed decorated and perpendicular Gothic architecture.—The town is supposed to have derived its name from the Saxon *weald*, 'a wood,' and *den*, 'a valley.' The adjunct *Saffron* is derived from the once extensive culture of that plant in the neighbourhood. It was formerly called Waldenburg. The situation of the town is a narrow tongue of land encompassed with a valley in the form of a horse-shoe, enclosed by distant hills. On the bottom of the tongue stand the ruins of a castle, and on the top the church, round which, on the side of the hill and in the valley, the town is built. The houses are principally of lath and plastered brick; but some of them are very ancient. The market-place has been rendered spacious, and contains a handsome town-hall. There is a considerable trade carried on here in malt and barley, and an extensive general retail trade.—The borough is co-extensive with the p., and includes the hamlets of Little Walden, Sewer's End, North End, and Audley End, a magnificent structure, situated about 1 m. W of the town. It was erected in the time of James I., and was long regarded as the largest mansion in the kingdom. Only a small part of the original building is now remaining, and this portion forms the splendid mansion of Audley-End, the seat of Lord Braybrooke.

WALDENSBERG, a village of Hesse-Cassel, in the co. of Isenburg, built by Protestant refugees from Piedmont, towards the close of the 17th cent. Pop. 800.

WALDENSES. See PIEDMONT.

WALDEN-ST.-PAUL'S, a parish in Hertfordshire, 5 m. NW of Welwin. Area 3,678 acres. Pop. in 1831, 1,058; in 1851, 1,175.

WALDEN-STUBBS, a township in the p. of Womersley, Yorkshire, 7 m. SE of Pontefract. Area 1,350 acres. Pop. in 1831, 139; in 1851, 185.

WALDERNBUCH, a village of Nassau, in the bail. of Hadamar. Pop. 530.

WALDERSHARE, a parish in Kent, 5 m. N by W of Dover. Area 1,242 acres. Pop. in 1851, 105.

WALDERSHOP, a town of Bavaria, 62 m. N of Ratisbon. Pop. 1,000.

WALDFEUCHT, a village of Prussia, in the reg. of Aachen, circle of Heinsberg. Pop. 925.

WALDFISCHBUCH, a village of Bavaria, on the Pfalz, circle of Pirmasens. Pop. 723.

WALDGREBWEILER, a village of Bavaria, on the Pfalz, circle of Rockenhausen. Pop. 540.

WALDHAUSEN, a village of Upper Austria, 4 m. ESE of Zwetl.—Also a village of Baden, in the circle of the Lower Rhine, bail. of Adelsheim. Pop. 420.—Also a v. of Nassau, in the bail. of Weilburg. Pop. 470.—Also a v. of Prussia, in the reg. and circle of Arnsberg. Pop. 300.—Also a village of Würtemberg, in the Danube circle, bail. of Geislingen, on the Alp. Pop. 1,240.

WALDHEIM, a town of Saxony, on the river Zschoppau, 31 m. W of Dresden. Pop. 4,000. It has manufactories of linens and woollens.—Also a v. of Bohemia, 44 m. W of Pilsen, in the bail. of Tachau. Pop. 1,490.

WALDHUTTE, a village of Lower Austria, near Burkersdorf. Pop. 1,200.

WALD-HWOZD, a mountainous district of Bohemia, in the circle of Prachin, on the confines of the Upper Palatinate.

WALDINGFIELD (GREAT), a parish in Suffolk, 3 m. NE of Sudbury. Area 2,423 acres. Pop. 659.

WALDINGFIELD (LITTLE), a parish in Suffolk, 5 m. NE of Sudbury. Area 1,574 acres. Pop. 404.

WALDITCH. See WALLDITCH.

WALDITZ (GROSS and KLEIN), villages of Prussian Silesia, in the circle of Lowenberg. Pop. 1,223.

WALDKAPPEL, a parish and v. of Kur-Hesse, in the circle of Eschwege. Pop. 1,304.

WALDKIRCH, a parish and v. of Switzerland, in the cant. of St. Gall, on the l. bank of the Sitter. Pop. 2,681.

WALDKIRCHEN, a town of Baden, on the l. bank of the Elz, 8 m. NE of Freiburg. Pop. 2,500. —Also a town of Bavaria, 14 m. NNE of Passau. —Also a v. of Saxony, in the circle of Zwickau, 7 m. SE of Chemnitz. Pop. 1,200.

WALD-LEININGEN, a town of the Bavarian circle of the Rhine, in the district of Spire.

WALDMICHELBACH, a village of Hesse Darmstadt, 15 m. NNE of Heidelberg. Pop. 1,600.

WALDMOHR, a town of Bavaria, in the duchy and 11 m. N of Deux-Ponts on the Glau. Pop. 1,200.

WALDMUNCHEN, a town of Bavaria, in the Upper Palatinate, 31 m. NE of Ratisbon. Pop. 2,132. It has manufactories of linen and glass.

WALDNIEL, a town of Prussian Westphalia, in the duchy of Juliers, 2 m. E of Ruremond. Pop. 1,200.

WALDO, a village of Prussia, in the reg. of Frankfort, cant. of Luckau. Pop. 362.

WALDO, a county in the S of Maine, U. S., drained by the Marsh and St. George river. Area 981 sq. m. Pop. in 1850, 47,230. Its cap. is Belfast.—Also a township and v. in Waldo co., 37 m. E by N of Augusta. Pop. 812.

WALDOBOROUGH, a port of Lincoln co., Maine, U. S., 22 m. ENE of Wiscasset. Pop. in 1840, 3,661; in 1850, 4,199. The shipping belonging to this port, in 1816, amounted to 19,743 tons; in 1850, to 96,380 tons. In 1851, 28,632 tons of shipping were built here.

WALDORF, a village of Prussia, in the reg. of Coblenz, circle of Ahrweiler. Pop. 788.—Also a v. in the reg. of the Rhine, circle of Bonn. Pop. 939.

WALDRIDGE, a township in the p. of Chester-le-Street, co.-palatine of Durham, 5½ m. N by W of Durham. Area 795 acres. Pop. in 1851, 747.

WALDRINGFIELD, a parish in Suffolk, 3½ m. S by E of Woodbridge. Area 1,156 acres. Pop. in 1831, 166; in 1851, 169.

WALDRON, a parish situated on the Forestridge, Sussex, 6 m. ESE of Uckfield. Area 6,218 acres. Pop. in 1831, 997; in 1851, 1,106.

WALDSASSEN, a town of Bavaria, in the Upper Palatinate, 69 m. N by E of Ratisbon. Pop. 1,522. It has manufactories of paper, and of woollen and cotton stuffs.

WALDSEE, a town of Würtemberg, 32 m. SW of Ulm. Pop. 1,570.

WALDSHOT, a town of Baden, on the Rhine, at the entrance into the Black Forest, 20 m. W by S of Schaffhausen. Pop. 1,500.

WALDSTADT, a parish and v. of Switzerland, in the cant. and 4 m. WNW of Appenzell. Pop. 983. There are mineral baths here.

WALDSTADTER-SEE. See Luzern (Lake of).

WALDSTETTEN, a town of Bavaria, 7 m. SW of Burgau. Pop. 758.—Also a village of Würtemberg, 29 m. E of Stutgard. Pop. 1,500.

WALDTHARM, a village of Würtemberg, in the Jaxt circle, near Grailsheim.

WALDTHURN, a village of Bavaria, near the borders of Bohemia, 24 m. NE of Amberg. Pop. 874.

WALDUDDA, a tract of hot and level country, extending along the N frontier of Abyssinia, upon the banks of the Tacazze. It is about 80 m. long, and 30 m. in breadth. Its marshy districts are occupied by bands of savage Shangalla.

WALDURN, a town of Baden, 30 m. WSW of Wurzburg. Pop. 3,200.

WALDWIMMERSBACH, a parish and v. of Baden, circle of the Lower Rhine, bail. of Neckargemund. Pop. 554.

WALDZELL, a village of Bavaria, in the Lower Rhine circle, bail. of Rothenfels. Pop. 260.

WALEN (El), a town of the Sahara, in Central Africa, in the district of Twat, 115 m. W of Ghadames.

WALES, a broad and sweepingly indented peninsular principality, projecting from the W side of England, and bounded on the N by the Irish sea and the estuary of the Dee; on the E by Cheshire, Shropshire, Herefordshire, and Monmouthshire; on the S by the Bristol channel; and on the W by St. George's channel. Its E boundary is traced to a considerable extent by the river Dee toward the N,—still more by the river Rumney toward the S,—and over various distances by the Vrynwy, the Wye, and other intermediate streams. The principality, while uninterruptedly washed on three sides by the sea, or by broad estuaries, is so formed that its own territorial length constitutes the breadth of its peninsula; the country extending in length from N to S, and the peninsula projecting towards the W. It lies between the parallels of 51° 40′ and 53° 26′ N, and the meridians of 2° 41′ and 4° 56′ W. Its length, in a straight line from N to S, is 115 m.; its breadth averages about 60 m., and varies between 37 m. and a little upwards of 90 m.; but, in a general view, it may be regarded as a slender oblong of about 46 m. in breadth, subtended westward at the two ends by two large projections. The oblong is contiguous along all its E side with England, and is washed along the central part of its W side by the long sweep of Cardigan bay; the S projection is a semi-ellipsoidal peninsula, about 55 m. in length, and 27 m. in mean breadth, indented at its extremity by St. Bride's bay, and comprehending all Pembrokeshire, and about the half of Cardiganshire and Caermarthenshire; the N projection forks into two parts, the one extending NW, and consisting of the county of Anglesea, or islands of Anglesea and Holyhead, and the other extending SW, and consisting of the elongated and intricately outlined peninsular portion of Caernarvonshire.—The area of the country is usually stated at 5,206,900 acres, or 8,125 sq. m.; but it is estimated in the reports to the Board of agriculture at 5,100,000 acres, and is alleged in the third report of the Emigration committee of the House of Commons, to comprise only 7,425 sq. m., or 4,752,000 acres. According to the reports to the Board of agriculture, 900,000 acres are in tillage, 2,500,000 are under pasturage, 700,000 are uncultivated, but capable of cultivation, and 1,000,000 are wholly unprofitable; and according to the report of the Emigration committee, 3,117,000 are arably and pastorally cultivated, 530,000 are uncultivated, but capable of cultivation, and 1,105,000 are wholly unprofitable.

Surface.] North W. is imposingly mountainous, South W. grandly hilly, and both are a land of soaring heights and magnificent acclivities,—of limited plains, narrow vales, and gorgy ravines,—of uplands endlessly varied in contour, and intersecting stripes and patches of lowland, rich in the dresses of river, lake, and woodland scenery. The mountains, however distributed, dissevered, or occasionally isolated among themselves, are so closely connected as to constitute one system, and are divided from the high grounds of England by the champaign territory of the plains of Cheshire and Salop, and the valley of the Lower Severn; yet the valley of the Upper Severn, from the boundary with England SW to Llanidloes, and the valley of the Dyfi, from Machynlleth, or from a point 15 m. NW by W of Llanidloes to Cardigan bay, cut the system into two sections very nearly coincident with the political divisions of North and South W., or leave them to be connected by an intermediate mountain-band of only 15 m. in breadth. The mountains of the northern section, or of North Wales—which includes Anglesey, Carnarvon, Denbigh, Flint, Merioneth, and Montgomeryshires—rise from all sides of the country towards a great central five-sided tabular upland; or bear a relation to it whose character may be fancifully described as intermediate between that of radii and buttresses. The five sides of the central plateau are, in a general view, continuous mountain-ridges; and apart from their elongations beyond the limits of contact with the plateau, they measure respectively 10, 10, 20, 25, and 25 m. in length. The NW side is the Snowdon range, sending up the triple summit which constitutes the loftiest ground in South Britain, and extending from a point a few miles NW of the N angle of Cardigan bay to the mouth of the river Conway; the N side is formed from the point of intersection with the Snowdon range to the town of Llanrwst on the Conway, by a range which comes in from Caernarvon on the Menai Strait, and is prolonged eastward from Llanrwst to the Hiraethog mountains; the NE side is formed by a chain of hills which flanks the Conway river, is cloven through by the river Dee at its confluence with the Alwen, 1 m. above Corwen, and falls by prolongation upon the Berwyn range at the SW ascent of the summit of Moel-Ferna; the SE side is formed by the Berwyn range which begins at Moel-Ferna, a mile or two S of the Dee, extends to the head of the valley of the Dyfi, and is thence prolonged down the r. bank of that stream; and the W side is an irregular line of heights commencing at the summit of Cader-Idris off the Berwyn range, proceeding along the numerous lofty summits which extend between the rivers Maw and Dwyryd, at a distance of about 5 m. from Cardigan bay, and falling upon the Snowdon range not far from the village of Beddgelart. The country included within these five sides, comprehends about one-fourth of the whole area of North Wales; lies, with very slight exceptions, upon a basis of probably not less than 600 ft. of alt. above sea-level; and consists of a congeries, a segregation, a sea of hills, generally rapid in acclivity, yet seldom equal in elevation to the summits of either the Snowdon, the Berwyn, or the west boundary ranges; while the country lying exterior to the plateau, among the numerous spurs and elongations of the ranges which form it, or between the seranges and the sea, or at the subsession of the country's upland upon the margin of the Cheshire and Shropshire plains, presents ramifications of hill and intersections of valley and ravine far too multitudinous and intricate to be succinctly indicated.—South W., which comprehends Brecon, Cardigan, Caermarthen, Glamorgan, Pembroke, and Radnor shires—though decidedly inferior to North W., in both the amassedness and the alt. of its mountains, possesses not a few conspicuous summits, and several lofty and far-stretching chains of upland. A range of high ground, so broad as to be a kind of tableau, extends from the vicinity of Llanfair, in the E division of Montgomeryshire; sends up, on the N frontier of South W., the proud and lofty crest of the celebrated Plynlimmon; subsides in the course of a long stretch to the W, and stoops precipitously down in the abrupt cliffs which bound Cardigan bay in the vicinity of Aberystwith. An extensive range, whose most conspicuous eminences are the group called 'the yellow mountains,' Cwn Rhysglog, Pen-y-cader, Mynydd-castel, Newydd 'Carreg Wen, and Llanvernach, commences at Bleddva-forest, to the NE of Llandinrod-wells in Radnorshire, crosses the N part of Brecknockshire, extends SW through Caermarthenshire, and terminates in the bold and prominent mass of Procelly, near the centre of the N division of Pembrokeshire. Another line or band of heights commences in the Fothoc hills on the E side of Brecknockshire, and is prolonged westward, with various degrees of ridgy regularity, but not without intersections of water-course to the W extremity of the Mynydd-Dhu or Black mountains, and the high table-land closed by the isolated mountain of Penbre-hill, in the county of Caermarthen. Of the principal summits in both sections of the principality, Snowdon has an alt. above sea-level of 3,571 feet, Cader Idris 2,914, the Beacons 2,862, Cader-Ferwin 2,563, Plynlimmon 2,463, and many from 1,500 to 2,300 ft. The lengths of single or twin mountains, and the elongations of minor ridges, and of great ranges, are prevailingly in the direction of NW to SE; and their escarpment, or more water-worn declivitous abrupt descent, generally faces the NW.—Green and luxuriant vales, while the seat of pop., are the boast of the inhabitants, and often present a delightful contrast to the wood-clad declivities, the rugged ascents, the frowning cliffs, and the soaring peaks, by which many of them are enclosed and sheltered. "In these," says a well-informed and judicious contemporary, "a fine deep soil is not unfrequently found, and abundant harvests often reaped. The vale of Clwyd in North W. is not less remarkable for the noble panorama of natural graces displayed in a length of 30 m., than for the verdure of its pastures and the value of its grain harvests. This, however, is the elysium of Wales. The vale of Eideirnion is also fruitful, but less extensive and less beautiful; and the vale of the Conway is a grassy mead of 20 m. in length, where some of the best cattle brought into England annually are reared. Though less interesting from the tameness of the impending mountains, the vales of South Wales are still to be classed among the characteristic features of the country. There is more breadth between the hills, less accession from the accompaniment of waving wood, and less grandeur of height and variety of form in the mountain chains. The celebrated vale of Glamorgan, extending along the shores of the Bristol channel, does not belong to the class of natural enclosures here alluded to, but is merely a vast sloping bank, falling gradually from the mountain base to the water's edge, and basking in a southern sun. There is yet another species of valley which appears almost peculiar to North Wales: it is called in the native language *cwm*, a hollow, and is a perfect amphitheatre or basin-formed glen, accessible only by descent, and sometimes through a small gap, occurring, as it were, in the imperfect or broken rim of the great crater."—[*Gorton.*]

Climate.] The climate of W. aggregately possesses a much closer resemblance to that of the W division of Scotland, than to that of by far the greater portion of England; and, in detail, it is not a little various,—bleak and severe on the mountain table-lands, sharp on the hills, moderate in the smaller inland vales, mild in the great interior valleys, and genial on the southern seaboard or vale of Glamorgan. Snow is almost everywhere more frequent than in England, accumulates to a greater depth, and lies during a much longer period,—often sheeting for days together the sides of the E frontier hills, when the plains of Cheshire and Salop and the Lower Severn are all green, and usually covering the tops of the loftiest mountains during a succession of many months in the year. Rain is frequent, not in winter only, but in the three other seasons, and is believed to fall to the average depth of 34 in., while the fall in England averages only 22. Yet the air, though humid, is in general highly salubrious; and so powerfully does it operate, in conjunction with the simple and frugal dietetics of the country, to prolong life, that scarcely a cemetery in the principality is destitute of some monumental testimony to the longevity of the inhabitants.

Islands.] Anglesey, with its adjunct of Holy Island, differs so much in contour from the other five cos., and, at the same time, is so strictly insular, and constitutes so large an area, that it has not been included in our general view of the principality's surface; yet, in its character of a county, it is so fully noticed in its own place in our alphabetical arrangement, that even a sentence respecting it here would be idle repetition. Excepting Anglesey, the islands of Wales are few and tiny; in no instances entitled to be called more than islets, and in most constituting mere *skerries* or rocks. The islets and rocks called skerries, off the NW extremity of Anglesey, are noticeable chiefly for their lighthouse, and for their hazards to navigation. Bardsey island, off the W point of the Carnarvonshire peninsula, supports a few fishermen, feeds some cattle, has a lighthouse, and gives name to the sound between it and the mainland; St. Tudwall's islands, off the E side of the eastern point of the same peninsula, are also grazing ground, and have a few inhabitants. Of the numerous islets which straggle seaward from the N side of the entrance of St. Bride's bay, Ramsey is inhabited; the Bishop and his Clerks are the most menacing group of rocks in St. George's channel; and all possess a melancholy interest in the annals of navigation. The Skomar and the Skoham islands, seaward of the S side of the entrance of St. Bride's bay and the Caldy islands, off the SE extremity of Pembrokeshire, are partly grazing grounds, and partly perilous rocks. Barry island, near the Glamorgan shore, SSW of Llandaff, possesses some antiquarian interest; Sully island, two or three miles farther E. is little known; and nearly all other islets than we have noticed, however dignified in local or topographical nomenclature, are little else than dots of land upon the sea.

Bays and Harbours] Natural harbours are in W. fewer and aggregately less commodious than in an equal extent of coast in most other northern countries of mountainous character; yet such as occur are in some instances important and excellent, and in others artificially improveable. The estuary of the Dee is dotted up to the head of its Welsh side by a series of shipping quays. Various creeks, small bays, and the embouchures of streams between the Dee and Great Orme-head, are shipping-places of collieries, lime-works, and the general agricultural market. The narrow but somewhat prolonged estuary or tidal space of the Conway river is believed to be rapidly undergoing natural improvement by the diminution of the bar across its entrance, and might easily be improved into the best post of communication, on the North W. coast, with the metropolis of Ireland. Beaumaris bay, while capacious, is well-sheltered, and contains a good roadstead; and its interior expansion of Port-Penrhyn, at Bangor, though small, is so covered and screened as to be a retreat from nearly all winds. Numerous creeks serrate the outline of Anglesey; yet possess even inferior importance only, in the aggregate, and in a few instances offer safe berthage for coasting-vessels employed in the export of mining produce and cattle. Holyhead bay, between Anglesey and Holy island, forms on its W side the small but sheltered and well-improved harbour which has the peculiar distinction of being the station of the Irish packets. The S end of the Menai strait is obstructed at its entrance by a shifting bar; but, when this has been passed, it presents, in the ramified and generally landlocked expansion S of Carnarvon, a thoroughly sheltered and safe anchoring-ground. The Carnarvonshire creeks of Nevin and Portdinlleyn admit small craft; its bay of Aberdaron is larger but not so important; its still more capacious and almost semicircular bay of Hell's mouth, appals by its very name; its bay of St. Tudwall's road is commodious, safe, and a good outlet; and its creeks at Pwllheli and Port-Madoc are of note, the former as a place from which ships hail, and the latter as the place of shipment for the Festiniog slates. Four small estuaries are the only noticeable marine inlets on the coast of Merionethshire; and but two of these, the estuaries of the Maw and the Dyfi, are ports of any consequence. The embouchure of the Ystwyth and the small estuary of the Tyfi form tiny harbours, and the chief scenes of the Cardiganshire commerce. Newport and Fishguard bays, on the N of the Pembroke coast, are comparatively expansive, and witness considerable traffic. Pwllcrochen bay, on the same coast, is of no note. St. Bride's bay, farther S in Pembroke, is a miniature gulf in area, totally unsheltered from the wild winds of the west, and replete with appalling historical associations. Milford haven, the next great opening, flaunts from 4 or 5 to 15 or 16 m. into the interior, in a series of ramifications; presents a landlocked, natural harbour, sufficiently capacious to hold all the navy of Britain; and, but for the sinuosity of its entrance, and the remoteness of its local position, would have been adopted as a principal naval station. Carmarthen bay is a miniature sea, open, expansive, and destitute of shelter; but its offshoots, the small estuaries of the Tawe and the Towy, the Gwendraeth-Fawr, and the fine estuary of the Burri, or Llwchwr river, are safe retreats for shipping, and the scenes of much commercial stir. All the inlets hence, to the E boundary of the principality, are, with the single exception of Swansea bay, small bays, or mere creeks, possessing little character as either landscapes or natural harbours, yet in some instances improved by art to subserve, on a small scale, the purposes of commerce. "Swansea bay," remarks the writer from whom we quoted before, "is a scene of matchless beauty; it is frequently compared to the scenery of the bays of Dublin and Naples. Pennarth harbour, the most easterly on the coast, is also artificial: from this the manufactures of Merthyr-Tydvil are sent forth to the public market, and the commercial affairs of Cardiff are transacted."

Rivers and Lakes.] The Dee rises in two head-streams on the S frontier of Merionethshire; receives, on its l. bank, the tributary of the Alwen; and on its r. bank the Ceirog; and, debouching from among the mountains, turns N, passes into England, expands into estuary, and thence continues on the boundary to the sea.—The Clwdd rises on the confines of Denbighshire and Merionethshire; glides past St. Asaph; receives the Elwy; washes the castle-walls of Rhyddlan; and 3 m. below, falls into the Irish sea.—The Conway issues from Llyn-Conway near the junction of Denbigh, Merioneth, and Carnarvon cos.; descends into the beautiful vale of Nant-Conway; receives about 20 affluents, the chief of which are the Machno, the Lledder, and the Llugwy; washes the town of Llanwryst; flows in beautiful curvatures to Trefrhiw; and, thence to the mouth of its little estuary at Aberconway, is tidal and navigable.—The Ogwen, entirely a rivulet of Carnarvonshire, forms the cove of Penrhyn at Bangor, on Beaumaris bay.—The Sciont, cradled among the loftiest uplands of Snowdon, traverses the two beautiful lakes of Llanberis; laves the ancient *Segontium* and the modern Carnarvon; and forms a safe and commodious little harbour on the Menai strait.—The rivulets of Anglesey, though many in number and profuse in irrigation, are all so small and prosaic as to challenge only collective notice.—Criccieth and Pwllheli rivers, and other streams of the Carnarvonshire peninsula, are also inconsiderable in both character and volume.—The Gwynedd tumbles from the SE side of the Snowdonian alps; and forms the commodious estuary at the NE corner of Cardigan-bay, called the Traeth-Mawr.—The Drwydd rises near Bwlch-Carreg-y-fran, in the mountainous district of Merionethshire; flows along the picturesque vale of Festiniog; and forms the estuary of Traeth-Vach, a twin expansion with the Traeth-Mawr.—The Maw or Avon rises SW of the source of the Dee; receives from the N the Mowddoc; and expands into the prolonged and slender estuary whose mouth forms the harbour of Barmouth or Barmaw.—The Dyfi springs from the foot of Arran-Fowddy; traverses a fine vale from Dinas-y-mowddu to Machynlleth; is joined by the Ceiriog; and forms an elongated estuary to the middle of the Cardigan bay coast at Aberdyfi.—The Rheidiol rises on the Cardigan side of Plynlimmon, within 4 m. of the sources of the Severn and the Wye; and seeks the sea in the port of Aberystwith.—The Ystwith has a common embouchure with the Rheidiol, and runs nearly parallel to it in direction.—The Eiron or Aeron traverses a lofty hill-screened valley to the sea at Abereiron.—The Tyfi or Tivy rises among the mountains of the NE corner of Cardiganshire; traverses the plain of Llanbeder; and passes Newcastle-Emlyn; and finally moves across low ground, to the sea below Cardigan. This stream is navigable from the sea to Lampeter, a distance of nearly 30 m.—The Nevern and the Gwyn form, at their respective embouchures, the harbours of Newport and Fishguard.—The Western Cleddau, the Eastern Cleddau, and various other streams, enter the inner ends of various ramifications of Milford haven; and the two Cleddaus are navigable respectively to Haverford-West and to Caniston-bridge.—The Taff or Tave rises on the E side of the Precelly mountain, and runs to Carmarthen bay, a little below the town of Llaugharne. The Towy issues from a lake near Cwn-Berwyn, in the SE of Cardiganshire; becomes tidal and navigable at Carmarthen; and forms a partially conjoint estuary with the Taff at the head of Carmarthen bay.—The Llwchwr, Lwghor, or Loughor, divides the counties of Carmarthen and Glamorgan; and loses itself in the capacious eastward projection of Carmarthen bay, called the Burry river.—The Tawy and the Neath run SSW from the base of the Black mountains in Brecknockshire; traverse a rich and various mining district; and fall into the bay of Swansea.—The Avon rises in Glen-Corwg; runs SW to the coal and copper-mining field of Aberavon; and 3 m. thence glides over a bar into the east side of Swansea-bay.—The Ogmore runs S from among the central uplands of Glamorganshire; is joined by the Lanfy and the Ewenny; and falls into the British channel below Merthyr-Mawr.—The Taffe exceeds most of the Welsh rivers in the wildness of its features, and the vagaries of its course; it runs prevailingly SE from the S base of Tre Beddw mountain; enters the vicinity of Merthyr-Tydvil; debouches into the plain of Llandaff; and rolls past Cardiff to the Bristol channel near the point of Penarth.—The

Rumney or Rhymny rises in Glyn-Colwyn, near the junction of the counties of Glamorgan, Brecknock, and Monmouth, runs S, and enters the Bristol channel about 3 m. N of Penarth point.—Most of the streams we have named, from the Taffe and the Towy to the Rumney, though very limitedly or not at all navigable as rivers, are valuable feeders, or even ultimate outlets of different lines of canal navigation, formed, for the accommodation of the extensive mining and manufacturing districts of Glamorgan, Carmarthen, Pembroke, and Brecon. The Usk rises on the N side of the Black mountain; runs E to a confluence with the Honddy at Brecknock; and hence into Monmouthshire, and becomes a river of England.—The Wye rises on the S side of Plynlimmon about 1 m. from the source of the Severn; flows past the town of Rhayadar, commences at Hay to trace for 3 or 4 m. the boundary with England, and passes away into the spacious plain of Herefordshire.—The Severn gushes from a spring on the NW side of Plynlimmon; passes the towns of Llanidloes and Newtown; receives the Rhiw, and various other tributaries; and, at its confluence with the Vyrnwy, leaves the principality, enters the great plain of Salop, and begins to wear the lofty distinction of the second river of England.—To describe, or even to enumerate, all the lakes of Wales, would be both a thankless and a very difficult task. The most distinguished for extent or beauty, or a combination of both, are, in North Wales, Llyniaw - Nantle, Llyn - Cywellin, Llyniaw-Llanberris, Llyn-Conway, Pemble - meer, and Tulyllyn; and, in South W., Llyn-Bychlyn in Radnorshire, and Llyn-Savathan in Brecknockshire. The lake of Bala, the most extensive in the principality, is only 8 m. long and 1 m. broad. The lakes of Llanberris lie amidst the sublime scenery of the Snowdon mountains; and, jointly with a few other lakes and pools of Carnarvonshire, monopolize probably four-fifths or more of the entire beauty and fame of Welsh lake scenery.

Minerals.] Anglesey, on its W side, exhibits granite, mica-slate, and serpentine: in its central district, it consists prevailingly of slate; and on its SE side it has a small coal-field, apparently incumbent on mountain-limestone. A remarkable trap dyke intersects the rocks of the island. The Snowdon mountains, with their offshoots to the Rivals, at the S extremity of Carnarvon bay, consist principally of clay-slate, chlorite-slate, other schists, and felspar porphyry. Many of the mountains which range from Snowdon into Merionethshire are composed of porphyritic trap. On some of the Carnarvon mountains, at an alt. of 1,400 ft. above sea-level, beds of sand and gravel have been observed, intermixed with marine shells, said to resemble broken shells found on the sea-beach. A great part of the counties of Denbigh, Montgomery, Radnor, Cardigan, Carmarthen, and Pembroke, and the N division of Brecknock, display great sameness of geognostic and mineral character, and consist chiefly of the formations which convenience, till modern theories became better adjusted, requires still to be called greywacke or transition rocks. The formations conterminous with the great transition district, and extending from Pembrokeshire to Shropshire, are various beds of old red sandstone, alternating in the upper part with beds of greenish limestone and mountain limestone, and occasionally passing into greywacke, imperfect porphyry, and schistose sandstone. A range of beds of the lower transition limestone runs along the northern margin of the old red sandstone from Carmarthen to the boundary with Shropshire, in the direction of Wenlock edge. A considerable extent of mountain limestone, often resting on a thick bed of coarse conglomerate, and containing many valuable lead-mines, occurs in the Denbigh and Flint sections of the E side of North Wales. A coal-field, nearly 100 m. in length from east to west, and 5 to 20 m. in breadth, the most extensive coal-field in Great Britain, covers the transition rocks of the southern border of South W., from the boundary with England to the interior of Pembrokeshire, and, except in Swansea and Carmarthen bays, is almost everywhere surrounded by mountain limestone and old red sandstone. If a view, full, minute, and in the phraseology of the most recent theories, be desired of the geognosy of Wales, particularly of its great transition district, the ordinary reader will find quite as ample a one as he can use in the work of R. J. Murchison, Esq. Silver ore is obtained in considerable quantities, in combination seriously with lead ore, spar, quartz, and copper. Copper occurs diffusedly and in considerable plenty. Parys mountain in Anglesey, discovered in 1768 to be richly metallic, was long supposed to consist wholly of copper, in a state of either native copper, sulphate, black ore, or malachite; but, though for many years singularly productive, it has lost its high fame, and is in a great degree abandoned. A richer ore than that of Parys, such as yields from 8 to 10 per cent. of the pure metal, is found at Llanberris, in Carnarvonshire; and ores of various value and extent occur near the extremity of the E screen of Beaumaris bay, at Evcaer Vraith in Cardiganshire, at Llanymynich in Montgomeryshire, and in several other localities. The sales of copper in W., in the year ending 30th June 1853, amounted to 29,244 tons; but a very large proportion of this was foreign copper. Lead is extensively diffused, particularly in the cos. of Flint, Denbigh, Cardigan, Carnarvon, Carmarthen, and Montgomery. A mineral tract in Cardiganshire was not long ago esteemed the richest in lead of any in Britain; but it has suffered greater declension in fame than even the Parys copper mountain, and has been wholly abandoned. In 1828, the lead mines of Flintshire were the most productive in the principality, those of Denbighshire the second in productiveness; and, in common with all other lead mines in the country, they yielded a total of 12,000 tons of pig lead. The lead ore raised in W. in 1852 amounted to 18,379 tons of ore, yielding 13,708 tons of lead. Iron-ore, in the various forms of hematites, brown ironstone, black ironstone, bog-ore, and pyrites, is partially diffused through Denbighshire, occurs to a noticeable amount in other parts of North W., and is abundant in the geognostic secondary district of South W. Though the ironstone is poorer than the ores of Staffordshire, and of other parts of England, its inferior richness is more than compensated by the contiguity of coal, and the comparatively high facility of mining. Iron-works were commenced at Merthyr-Tydvil, Aberdare, and Cyfartha in Glamorganshire, and at various places in Brecknockshire, not long after the middle of last century, others about 35 years ago at Rhiwabon in Denbighshire, and some, at various other places in the intervals; and so steadily and greatly have they succeeded that, in 1830, they were estimated to produce 2.000 tons more than the iron mines of all England,—the produce of England being 300,000 tons, that of South Wales 277.000 tons, and that of North Wales 25.000 tons. The iron made in W. in 1848 was 16.120 tons in North W., and 706.680 tons in South W.; or 722,800 tons in all. The total for England, Wales, and Scotland being 1.998,568 tons. —The great coalfield of the S division of South W. is so rich and extensive as to have been pronounced inexhaustible. Coal is raised in vast quantities, from this field, in the counties of Glamorgan, Brecknock, and Carmarthen; and occurs, in the vicinity of the upper ridge of limestone, in many parts of Pembrokeshire. The coal of North W. is inferior in quality, very much less in extent, and aggregately much more impracticable in position than that of South W. Roofing-slates are a very plentiful mineral production of North W.; and the quarrying and dressing of them constitute a chief support of the operative classes of the community. The largest quarries are in the hills over Dolawen or Llandegai, and export their enormous produce from the town of Bangor. Quarries at Llanberris are connected by railway with Moel-y-Don, on the Menai strait: quarries Llanllyffni, Cylgwyn, and adjacent places, make their shipments at Carnarvon; the Festiniog quarries, at Port Madoc and Maentwrog; the quarries of Llanrhwychwyn, at Trefrhiw; and even the far inland and inconveniently situated quarries of the parish of Penmachno are able, from the superior quality of their produce, to hold free and brisk connection with the general market.

Botany and Zoology.] So numerous are the rare plants of W., and so profuse the variety of the more common kinds, that to name the whole would be to compose a botanical nomenclature of probably two-thirds or more of the whole British Flora. Almost every mountain and hill possesses natural memorials of having formerly been covered with wood; and during two generations past, many square miles of upland have been annually laid under plantation. Though large tracts of waste or feebly pastoral land are still bare, wood is aggregately abundant, and, at the same time, so rapidly increases as to exhibit promise of soon acquiring as much of its ancient dominion as cannot be profitably subjected to the plough, or disposed in pasture. Ash, beech, Welsh elm, walnut, sycamore, mountain-ash, holly, willow, and some other trees and shrubs are indigenous; oak acquires a good size in half-a-century; and the larch, though unknown in the country at the close of the last century, is the tree most approved, quickest in growth, and most extensively planted.—Wolves anciently infested the principality, but were attacked by system and premium, and, about a century and a-half ago, were exterminated. The wild goat, larger in size, and finer in hair than in most other mountainous countries, continued to bound from crag to crag during years after the commencement of the present century; but is now completely banished. Roebucks were anciently numerous; but, so far back as 50 years ago, they were confined to the most intricate parts of the country, and could very rarely be seen. The martin, an old inhabitant, still haunts the woods of Merioneth and Caernarvon. The beaver, described by Giraldus as a native of Wales, and distinctly named in the Welsh laws, has long ceased to be traceable. The otter is a constant attendant on the banks of fishy secluded lakes and streams. Wild cats, weasels, badgers, squirrels, hedgehogs, hares, and rabbits are very numerous. Foxes are comparatively few. Seals are noticed by Pennant as natives of the coasts, and as seen most frequently between Llyn in Carnarvonshire, and the northern parts of Anglesey. Nearly the whole of the feathered tribe of W. are common to it with England; and the eagle, the falcon, the kite, and the hawk, are both more numerous and of rarer species. The golden eagle has his eyry among the mountains of Snowdon; the peregrine falcon breeds prolifically among the rocks of Llanidno; and the smerlus or merlin autumnally migrates from Wales to England. The long-eared owl, if not peculiar to the country, is characteristic. The nutcracker was once seen in Flintshire; and the shag has occasionally appeared on the shore near Holyhead. Game is aggregately abundant; grouse least plentiful, pheasants more so, and woodcocks and partridges very numerous. Reptiles and insects are of the kinds known in England. The chief fish which have been claimed as peculiar to Wales, but some of which are found in either England or Scotland, are the Beaumaris shark, the trifurcated hake, the gattorugine, and the trimaculated with the striped and gibbous wrape, the crooked perch, the salmo erwn or sewin, the deformed trout of the river Eynion, the samlet, the torgoch, or red char, the gilt or silver char, and the salmo lavaretus, or gwiniad. The domestic animals aggregately present little peculiarity in either species or breed.

Agriculture.] Forty years ago, when agricultural enterprise was careering athwart England, and wearing the laurels of a successful revolution, a well-informed writer asserted that "Wales, in a general view, may be considered a century at least behind England as to its state of agriculture." Improvements in Welsh agriculture have been fewer, slower, and less expansive, than in the agriculture of almost any other section of the empire, and have hitherto worked out such limited results as, in the words of Macculloch, to leave "the agriculture altogether bad." "Wales in general," say Messrs. Kennedy and Grainger, "does not produce half what it is capable of doing under proper management." A proper rotation of crops continues to be limitedly known, or but seldom introduced. Black oats are the prevailing cereal crop on high arable grounds; and oats of several varieties are elsewhere cultivated far more extensively than other grains. The vales of Glamorgan and the Clwyd, and some other low grounds of rich argillaceous or otherwise alluvial soil, are the main or only grounds on which wheat is raised. The cultivation of the potato and the turnip is extending,—of the potato, rapidly,—and of the turnip, considerably. The farms are in some instances large, in the great majority small; and average, in South W., from 50 to 60 acres. Leases, though not uncommon, are usually so devoid of restrictions as to management, that, before they expire, the lands held under them are scourged almost into sterility. Horses, cattle, and sheep are larger in size, and more intermixed with introduced breeds in the vales than in the uplands. Merlins—the small, hardy, sure-footed, native Welsh ponies—are still reared in considerable numbers in Merionethshire, and in less considerable numbers in Montgomeryshire, but are elsewhere generally superseded or neglected. A superior breed of horses has long been reared in the vales of Montgomeryshire. Black cattle are a main object of Welsh husbandry, and are principally depended on for the means of paying rent and other current demands. The upland sheep are in high repute for their mutton; the total stock of sheep in the principality is estimated at 1,250,000; and the total annual produce in wool is supposed to be upwards of 10,000 packs. The total rental of land in W. in 1810-11 was £1,622,985 or an average of 6s. 10d. per acre; in 1841, it was £2,206,146, or an average of 9s. 3d. per acre; the average for England being 19s. 2d. per acre.

Manufactures and commerce.] Metallic, lapidose, and earthy subjects of manufacture, figure largely in the principality's productive industry; but in the great majority of instances, or in their connection with native mining, they have been already noticed in the section on minerals. The Welsh are equal to the best, and superior to the most, of the other divisions of British population, in all the arts of smelting and refining ores; hence, though tin is not a native metal, Wales is a great seat of the manufacture of tin-plate or white iron; and almost all the copper raised in Cornwall is carried to Swansea and other places in the South Wales coalfield to be smelted. A manufacture of writing-slates, slate chimney-pieces, slate inkstands, and other articles formed of slate, makes a prominent figure in Bangor.—The woollen manufacture, in the articles of hosiery, strong cloth, small cloth, but principally the well-known Welsh flannel, is diffused, in one or more of its departments, over the whole principality; yet, while everywhere visible, is most prominent in Montgomeryshire. Thirty years ago strong cloth was made principally in the neighbourhood of Dolgelly and Machynnlleth in Merionethshire; small cloth, in the large parish of Glynn in Denbighshire; flannel, in Montgomeryshire, and within a circle of about 20 m. round Welshpool; and stockings, wigs, socks, gloves, and other small knit articles, in the town and neighbourhood of Bala. Considerable manufactures of cotton and cotton twist were established 50 years ago or upwards, in the counties of Flint and Denbigh; and principal factories continue to be in operation at Northop, Greenfield, Scaviog, Denbigh, Holywell, and Caerwys. The silk manufacture was several years ago introduced. Notwithstanding its great extent of sea-coast, the number of its harbours, the excellence of many of its roadsteads and landing-places, the facility of its communication all round with the open sea, and the great aggregate amount of its interior productive industry, W. has surprisingly little commerce. Its exports are principally the produce of its mines, its black cattle, and its woollen fabrics; and its seaward trade, with a small total of exceptions, chiefly at Swansea and Carnarvon, all terminates in British ports. Few of the Welsh ports possess vessels of very considerable tonnage.

Divisions and Towns.] W. was anciently bounded, nearly all the way on the E, by the Dee and the Severn; and is now minus its ancient extent, the whole of Monmouthshire, and part of Gloucestershire, Herefordshire, Salop, and Cheshire. In the time of Llewellyn ap Gruffydr, the last prince of North Wales, it was divided into the three dynasties of Gwynedd, Deheubarth, and Powysland. Gwynedd, or North W., comprehended Anglesey, Carnarvon, and part of Merioneth, Denbigh, and Flint. Deheubarth, or South W., comprehended Cardigan,

Pembroke, Carmarthen, Glamorgan, Monmouth, and part of Brecknock, Radnor, Gloucester, and Hereford; and Powysland, or East W., comprehended part of Merioneth, Denbigh, Flint, Radnor, Brecknock, Salop, and Chester. In 940, the whole tripartite country was consolidated under one great government; but, previous to its falling under the power of the Anglo-Normans, it was partitioned into the divisions which it still retains of North and South W.; at some period not well ascertained, but supposed to have been about the epoch of the Anglo-Norman possession, it was subdivided into counties; and it then retained, and continues still to do so, a much earlier distribution into *cantrefs* or hundreds, and *cwmwds*, *comots*, or communities. North W., according to the present civil division, comprehends the six cos. of Anglesey, Carnarvon, Merioneth, Denbigh, Flint, and Montgomery; and South W., the 6 cos. of Cardigan, Radnor, Brecknock, Glamorgan, Carmarthen, and Pembroke. The parishes do not correspond with the civil divisions; they are variously stated in number at 751 and 869; and they are included in the dio. of St. Asaph on the NE, that of Bangor on the NW, that of St. David on the SW, and that of Llandaff on the SE.—The number of principal or returning boroughs is 14; and of market-towns is 58; the largest towns in the order of pop. are Merthyr-Tydvil, Swansea, Pembroke, and Cardiff; and the principal coast-towns are Bagillt, Rhyl, Aberconway, Bangor, Beaumaris, Amlwych, Holyhead, Carnarvon, Nevin, Pwllheli, Port-Madoc, Barmouth, Aberfydi, Aberystwith, Cardigan, Newport, Fishguard, Milford, Haverford-West, Tenby, Carmarthen, Kidwelly, Oxwich, Swansea, and Cardiff.

Statistics.] Two members of parliament are returned by each of the cos. of Carmarthen, Denbigh, and Glamorgan; one by each of the other nine counties; one by each of the boroughs of Brecknock and Merthyr-Tydvil; and one by each of the districts of boroughs whose principal or returning individuals are Beaumaris, Carmarthen, Cardigan, Carnarvon, Denbigh, Flint, Cardiff, Swansea, Montgomery, Pembroke, Haverford-West, and Radnor. The total number of members returned is thus 29.—The sum expended for the relief of the poor, in 1836, was £255,549; in the year ended March 25, 1841, it was £266,000. The rate of expenditure for relief of the poor, per head of pop., in the latter year, was 5s. 10d. The total amount of money levied by assessment, in name of poor rates, in this last year, was £342,264. The total number of in-door and out-door paupers relieved throughout Wales during the quarter ending Lady-day, 1841, was 73,405, or a proportion of 9 per cent. to pop. in 1851, it was 68,224.—The annual value of real property, in 1815, was £2,323,970.—Houses assessed to hearth-tax, in 1690, 77,291; inhabited houses in 1841, 188,196; in 1851, 200,087. Pop. in 1700, 366,500; in 1750, 450,200; in 1801, 541,546; in 1821, 717,438; in 1831, 805,236; in 1841, 911,321; in 1851, 1,005,721.—The total number of persons committed for trial or bailed, throughout Wales, in 1848 was 570; in 1841, 628; in 1838, 441; in 1834; 442.

History] Ancient Wales, or all the country west of the Dee and the Severn, was inhabited, at the period of the Roman invasion, by the *Ordovices*, the *Silures*, and the *Dimetæ*; and owing to these tribes having been Celtic in origin, and known under the comprehensive name of *Cymri* or *Cimbri*, the country received from the Romans, and long continued to wear the appellation of *Cambria*. The Romans encountered fierce and prolonged resistance from the Cambrian tribes; but eventually subdued them, took possession of their country, dignified it with the name of *Britannia Secunda*, and proceeded, as they had previously done in Britannia Prima, to build towns, appoint and garrison stations, construct iters, and establish jurisdictions. After the Roman abdication, Cambria was for many years convulsed with intestine feuds, yet it acquired consolidation and strength from the necessity of checking the predatory incursions of the Saxons, and was long enabled to maintain a rude but sturdy independence. From the date of the arrival of the Saxons, till 703, a succession of pendragons, or sovereign princes, held their court at Diganway on the river Conway, and at Caer Segont near Carnarvon; and aided by the *reguli* or toparchs of six principalities, into which Cambria was divided, waged a various but generally successful warfare of resistance against the Saxons. The Cambrians, in the course of the prolonged war seemed so repugnant to an union with their would-be masters, that they were now designated by them Welsh,—a name which has been variously supposed to mean strangers, wanderers, runaways, Celts. In 703, the Cambrian *pendragonate*, or imperial sovereignty, terminated at the death of Cadwallader; and about 20 years after, Offa, king of Mercia, conquered a belt of the country along the E, erected the celebrated artificial boundary from the Wye to the mouth of the Dee, which is still called Clawdd-Offa, or Offal's Dyke, and thus compressed the principality within nearly its present limits. Yet, in spite of the great frequency of their inroads, and the apparent security of their eventual partial conquest, the Saxons appeared to have acquired no permanent footing, and, with the exception of the work of Offa, have left to modern times scarcely a vestige of their presence. The tripartite division of Wales terminated in the accession to their several thrones of the reputedly benevolent prince and wise legislator Edwin Dda, or the Good, who died in 948. After Edwin's death, the Danes and the Saxons alternately, or at different points, scattered anarchy and murder athwart the country; and, in 1091, Robert Fitzhamon, a Norman baron, commenced a course of invasion which terminated in the subjugation of Glamorganshire, the conquest of all other low grounds of Wales, and the establishment of an absolute jurisdiction and a grinding despotism, administered by nobles who were called Lords Marchers. The resolute Cambrians still and long continued to struggle for independence; they performed deeds of the most heroic courage, and made a persevering and lion-hearted stand for liberty unsurpassed in the annals of almost any nation in the world; yet, in spite of oft renewed efforts, and a sort of dropping fire maintained for ages, they virtually lost all independence, in 1282, at the death of their last reigning prince, Llewellyn ap Gryffdh. Edward I., who completed their subjugation, allowed them to retain possession of their estates, and to enjoy as much liberty as comported with the laws of England. But the Lords Marchers practically annulled his benign provisions, and the Welsh did not experience relief till the enactment, in Henry VIII.'s reign, of the statute which enjoined the appointment of justices-of-peace, justices-of-the-quorum, and justices-of-gaol-delivery, invested with the same powers as those of England, and which originated the system of judicature called the great sessions of Wales, and superseded only in the year 1831.

WALES, a parish partly within the E. R., and partly in the W. R. of Yorkshire, 7 m. SE of Rotherham. Area 1,950 acres. Pop. in 1851, 268.

WALES, a township of Kennebec co., in the state of Maine, U. S., 18 m. SW of Augusta. Pop. in 1850, 612.—Also a township of Hampden co., in the state of Massachusetts, 63 m. WSW of Boston, drained by a head stream of Quinnebaug river. Pop. in 1840, 686; in 1850, 711.—Also a township of Erie co., in the state of New York, 248 m. W of Albany. It has an undulating surface, drained by Seneca creek, and is generally fertile. Pop. in 1840, 1,987; in 1850, 2,124.—Also a township of St. Clair co., in the state of Michigan, 98 m. E by N of Lansing. Pop. in 1850, 189.

WALES (New South). See New South Wales.

WALESBY, a parish in Lincolnshire, 2½ m. E of Market-Raisen. Area 2,580 acres. Pop. in 1831, 247; in 1851, 331.—Also a parish in Nottinghamshire, 3 m. NE of Ollerton. Area 1,260 acres. Pop. in 1831, 340; in 1851, 362.

WALESVILLE, a village of Whitestone township, Oneida co., in the state of New York, U. S., 90 m. WNW of Albany.

WALFISH. See Walvisch.

WALFORD, a parish in Herefordshire, 2¾ m. SW of Ross. Area 4,241 acres. Pop. in 1831, 1,158; in 1851, 1,217.—Also a joint township with Letton and Newton, in Leintwardine p., Hereford. Pop. in 1831, 212.

WALGAU, a village of Bavaria, in the circle of Upper Bavaria, and presidial of Werdenfels. Pop. 220.

WALGHERTON, a township in the p. of Wybunbury, Cheshire, 3¾ m. SE by E of Nantwich. Area 838 acres. Pop. in 1831, 231; in 1851, 213.

WALGRAVE, a parish in Northamptonshire, 7 m. NW of Wellingborough. Area 2,040 acres. Pop. in 1831, 575; in 1851, 613.

WALHAM-GREEN, a chapelry in the p. of Fulham, Middlesex, 6 m. SW by W of St. Paul's. Pop. in 1851, 4,380.

WALHAM-SAINT-PAUL-SART-LEZ-WALHAM, a department and commune of Belgium, in the prov. of Brabant, and arrond. of Nivelles, watered by the Hain. Pop. of dep. 1,705; of com. 972.

WALHAUSEN, a village of Prussia, in the reg. of Erfurt, and circle of Heiligenstadt. Pop. 528.

WALHONDING, a river of the state of Ohio, U. S., formed in the NW part of Holmes co., by the confluence of the Muddy and Black forks; flows first S, then E; joins the Tuscarawas at Coshocton, and thence takes the name of Muskingum.

WALINCOURT, a commune of France, in the dep. of the Nord, and cant. of Clary, 9 m. SE of Cambray. Pop. in 1846, 2,602. It has a silk factory.

WALJIERS, a lake of New South Wales, near the l. bank of the Lachlan, about 100 m. above the confluence of that river with the Murrumbidgee, and connected on the NW with Lake Boyongo.

WALKENAAM, or WAKENAAM, an island of British Guayana, at the mouth of the Essequibo, to the NW of Leguan island, in S lat. 7° 5′. It is well-cultivated, and produces sugar and coffee in great abundance. Near the N extremity is a village named Zelandia.

WALKENRIED, a town of Brunswick, in the circle and 36 m. SW of Blankenburg, at the foot of the Hartz, on the Wieda. Pop. 507. It is enclosed by walls, and has an ancient abbey and an hospital. In the vicinity are coal and agate mines.

WALKER, a county in the N part of the state of Alabama, U. S., comprising an area of 1,687 sq. m., drained by Mulberry fork of Black Warrior and its branches, Sipsey river, and Blackwater and Lost creeks. It is hilly, but is in some parts fertile, producing cotton in considerable quantities. Pop. in 1840, 4,032; in 1850, 5,124. Its capital is Jasper. Also a county in the NW part of the state of Georgia, comprising an area of 700 sq. m., drained by Chattooga and Chickamanga rivers, and intersected by the Western and Atlantic rivers. It has a diversified surface, but is generally fertile. Pop. in 1840, 6,572, in 1850, 13,109. Its capital is Lafayette.—Also a county in the S part of the state of Texas, comprising an area of 1,042 sq. m., drained by head branches of San Jacinto river, and by affluents of Trinity river, by which it is bordered on the NE. The surface is hilly, but affords fine pasturage. Pop. in 1850, 3,964. Its capital is Huntsville.—Also a township of Centre co., in the state of Pennsylvania, 60 m. NW of Harrisburg, drained by Little Fishing river. Pop. in 1840, 11,810.—Also a lake of Mallard co., Utah Territory, 358 m. W of Fillmore, at the base of the Sierra-Nevada range. It is 30 m. in length, and 7 m. wide, and receives a river of the same name.

WALKER, a township in the p. of Long-Benton, Northumberland, on the N bank of the Tyne, 3 m. E of Newcastle-upon-Tyne. There are extensive chemical works and iron foundries here. Pop. in 1851, 3,963.

WALKER (CAPE), a headland forming the NE extremity of Russell island, at the western entrance of Barrow strait, in the Arctic ocean, in N lat. 74° 10′, W long. 98° 10′.

WALKERIA, a settlement of Western Africa, on the Rio Nunez, 35 m. from the mouth of the river, which 2 m. above this settlement is much interrupted by a series of basaltic rocks.

WALKERINGHAM, a parish in Notts, 4 m. NW of Gainsborough. Area 3,000 acres. Pop. 608

WALKERN, a parish in Hertfordshire, 4 m. E by N of Stevenage. Area 2,924 acres. Pop. 738.

WALKERVILLE, a town of South Australia, in the co. of and 3 m. from Adelaide.

WALKHAMPTON, a parish in Devon, 4 m. SE of Tavistock. Area 10,540 acres. Pop. in 1851, 751.

WALKI. See VALKI.

WALKINGTON, a parish in the E. R. of Yorkshire, 2½ m. SW of Beverley. Area 5,460 acres. Pop., with the township of Provost's-Fee, in 1801, 403; in 1831, 558; in 1851, 699.

WALKREN. See WALCHEREN.

WALL, a chapelry in the p. of St. John Lee, Northumberland, 3 m. N of Hexham. Area 1,623 acres. Pop. in 1831, 495; in 1851, 474.

WALLABOUT BAY, an indentation of the coast of New York, U. S., between Brooklyn and Williamsburg.

WALLABY, a river of Tasmania, in the co. of Monmouth, an affluent of Coal river.—Also a group of islands, two in number, in the Abrolhos, distinguished as East and West Wallaby.

WALLACE, a parish of Van Diemen's Land, in the co. of Monmouth, bounded on the E by the Jordan.

WALLACEBURGH, a settlement of Upper Canada, in the township of Sombra, at the forks of Bear creek, 9 m. from the river St. Clair. Pop. 60.

WALLACETOWN, a modern but populous suburb of Ayr, in Ayrshire, situated on the E side of Newton-upon-Ayr, forming with it one compact town, and separated from the royal burgh only by Ayr river. See AYR.

WALLAGORING, a lake of New South Wales, in the co. of Argyle.

WALLAU, a village of Nassau, in the bail. of Hochheim. Pop. 770.—Also a village of Hesse-Darmstadt, in the prov. of Upper Hesse, and 27 m. NNW of Geissen. Pop. 790. It has a paper-mill.

WALLAROBBA, a village of New South Wales, in the co. of Durham, on a creek of the same name.

WALLASEY, WALLASEA, or WALL-FLEET ISLAND, a peninsula about 4 m. in length and 1 m. in breadth, between Paglesham and the river Crouch, in the hundred of Rochford, Essex.

WALLAWALLAU, a river of Oregon territory, which has its source in the Blue mountains; flows first N, then W, and joins the Columbia river near Fort Wallawalla.

WALLAZEY, anciently called KIRKBY-IN-WALLEY, a parish in the co.-palatine of Chester, 20 m. NE of Chester, containing the townships of Liscard, Poulton-with-Seacomb, and W., and forming the NW peninsular corner of the county. Area 17,775 acres. Pop. in 1831, 3,247; in 1851, 8,339.—A creek of the Mersey, known as Wallazey-pool, runs inland a considerable way in this p. It has been projected to erect a bridge across this pool,—an undertaking of no slight importance to the inhabitants on the other side of the water, as it will have the effect of shortening the distance between Birkenhead, Seacombe, and New Brighton about 7 or 8 m. An embankment at Wallazey-Leasowes was commenced about 25 years ago, to resist the encroachments of the sea. It is situated on the Cheshire coast, about half-way between New Brighton and Hoylake, and equi-distant between the Dee and the Mersey, where a tract of land, running for the most part parallel with the coast, and comprising 1,400 Cheshire or nearly 3,000 statute acres, is below the level of a tide 25 ft. above the old dock-sill at Liverpool. The owners of this area of land pay one moiety of the expenses of constructing this embank ment, and the corporation of Liverpool the other. The latter do this in order to preserve the port, since, had it not been for the embankment, the sea would, ere this, have forced a passage over the flats, in the direction of Wallazey-pool, and the channels of navigation would have been choked by accumulations of silt, and the formation of banks.

WALLBACH, a village of Baden, in the circle of the Upper Rhine, and bail. of Sackingen. Pop. 418. —Also a village of Hesse-Darmstadt, in the prov. of Starkenburg and presidial of Hochst. Pop. 226. —Also a village of Saxony, in the circle of Leipsic, and bail. of Leisnig. Pop. 298.—Also a village of Saxe-Meiningen, and bail. of Wasungen. Pop. 270. —Also a village of Switzerland, cant. of Aargau, and bail. of Rheinfelden. Pop. 651.

WALLBERG, a village of Austria, in Moravia, in the circle of Olmutz. Pop. 570.

WALL-BOTTLE, a township in the p. of Newburn, Northumberland, 4 m. W by N of Newcastle-upon-Tyne. Pop. in 1831, 688; in 1851,

WALLBURG, a village of Baden, in the circle of the Upper Rhine, and bail. of Ettenheim. Pop. 55[illegible].

WALLDITCH, a parish in Dorsetshire, 2 m. E by S of Bridport. Pop. in 1831, 164.

WALLDORF, a village of Baden, in the circle of the Lower Rhine, and bail. of Wiesloch. Pop. 2,140.—Also a village of Hesse-Darmstadt, prov. of Starkenburg, and presidial of Langen. Pop. 560.— Also a village of Austria, in Moravia, in the circle of Iglau. Pop. 240.—Also a village of Prussia, in the regency of Dantzig, and circle of Elbing. Pop. 240.—Also a village of Saxe-Meiningen, and bail. of Meiningen. Pop. 1,570.

WALLDURN, a cant. and town of Baden, in the circle of the Lower Rhine, in the Odenwalde. Pop. of bail., 11,000. The town is 15 m. SW of Wertheim. Pop. 2,980. It has a castle, a church, and a Capuchin convent.

WALLE, a village of Hanover, in the district of Ostfriesland. Pop. 290.—Also a village in the prov. of Slade, and bail. of Verden. Pop. 270.—Also a village in the prov. of Bremen. Pop. 615.

WALLEFELD, a village of Prussia, in the reg. of Cologne, and circle of Gummersbach. Pop. 9[illegible].

WALLEGA, an extensive district to the S of Abyssinia, stretching from the Dedhesa river westwards to the river Baro. Its surface is described as level and chiefly desert.

WALLEN, a village of Prussia, in the regency of Konigsberg, and circle of Ortelsburg. Pop. 220.— Also a village in the regency of Arnsberg, and circle of Meschede. Pop. 315.

WALLENBERG, a mountain of Switzerland, in the cant. of Glaris, near Mollis. It is covered with wood, affords fine pasture, and has a mineral spring.

WALLENBON, a valley of Prussia, in the regency of Treves, and circle of Daun. Pop. 290.

WALLENBRUCK, a village of Prussia, in the regency of Minden, and circle of Herford. Pop. 1,100.

WALLENDORF, a village of Prussia, in the regency of Breslau, and circle of Ranislau. Pop. 400. —Also a village in the regency of Konys, and circle of Niedenburg. Pop. 272.—Also a village in the regency and circle of Merseburg. Pop. 180.— Also a village in the regency of Treves, and circle of Bitteburg. Pop. 490.—Also a village of Saxe-Meiningen, in the principality of Saalfeld, on the Lichte. Pop. 778. It has manufactories of porcelain and of colours.—Also a village of Nassau, in the bail. of Herborn, near Beilstein.

WALLENDORF, or Olaszi, a town of Hungary, in the comitat of Zips, near the l. bank of the Hermad, and 12 m. SE of Lutschau. Pop. 3,000. It has a Catholic and a Lutheran church.

WALLENFELS, a market-town of Bavaria, in the circle of Upper Franconia, and presidial of Kronach, on the Rodach, 26 m. N of Bayreuth. Pop. 1,121. It has several saw-mills, and quarries of sharping stones. Agates are found in the vicinity.

WALLENHAUSEN, a village of Bavaria, circle of Swabia, and presidial of Roggenburg. Pop. 396.

WALLENHORST. See Wahlenhorst.

WALLENRIED, a village of Switzerland, in the cant. of Freiburg, and circle of the See. Pop. 220.

WALLENROD, a village of Hesse-Darmstadt, in the circle of Upper Hesse, and presidial of Lauterbach. Pop. 760.

WALLENSEN, a market-town of Hanover, in the principality of Kalenburg, and bail. of Lauenstein, on the Saale, 27 m. SSW of Hanover. Pop. 580.

WALLENSTADT (Lake of), a lake of Switzerland, in the cants. of Glarus and St. Gall, 9 m. SE of the lake of Zurich, at an alt. of 1,385 ft. above sea-level. It is 11 m. long, and 2 m. wide, and communicates with that of Zurich by the river Limmat. It has a depth of 500 ft., and is surrounded on all sides, except at the two extremities, with lofty rocks and mountains, interspersed with forests, meadows, and villages. Its chief feeder is the Seez. Its navigation is occasionally attended with hazard from sudden squalls.

WALLENSTADT, a town of Switzerland, in the cant. of St. Gall, situated near the E extremity of the lake of Wallenstadt, 40 m. ESE of Zurich. Pop. 1,800. The trade between Zurich and Italy is carried on chiefly through this town.

WALLERN, a town of Bohemia, 85 m. SSW of Prague. Pop. 2,100.

WALLER-SEE, a small lake of Upper Austria, in the circle of Salzburg.

WALLERS, a village of France, in the dep. of Nord, 6 m. NW of Valenciennes. Pop. of com., 2,700.

WALLERSTEIN, a town of Bavarian Franconia, 4 m. N of Nordlingen. Pop. 1,950.

WALLHAUSEN, a town of Prussian Saxony, on the river Helme, 4 m. W of Sangarshausen.

WALLI, a small kingdom of Western Africa, extending along the N bank of the Gambia, having Yani on the W, and Woolli on the E.

WALLINGA, a village of New South Wales, in the co. of St. Vincent, on the coast.

WALLINGFORD, a parl. borough in the hund. of Moreton, Berks, containing four parishes, with the liberty of the Castle, and the liberty of Clapcott. Area 821 acres. Pop. in 1801, 1,744; in 1831, 2,467; in 1851, 2,819.—The town is situated on the W bank of the Thames, about 3 m. from the high road between London and Oxford and on the road between Reading and Oxford, 46 m. from London. The river is here crossed by a stone-bridge about 300 yds. in length. Having no manufactures, and little thoroughfare, its chief importance is derived from being the market-town for the surrounding county. There is some small trade in malt, and in corn and flour. It has canal-communication through the Thames, with Bath, Bristol, and Birmingham. This town was a borough by prescription, and returned 2 members to parl. since the 23d Edward I. By the Reform act its representation was reduced to one; and the several ps. of Brightwell and Sotewell, with part of the p. of Cholsey and Clapcott liberty, all adjoining; with the ps. of North and South Moreton, and parts of Aston-Tirrell, and Aston-up-Thorpe, all in Berkshire; and the ps. of Bensington, Crowmarsh, and Newnham-Murren, in Oxfordshire, were united with it. The pop. within the boundaries of the parl borough, in 1831, was 7,159; in 1851, 8,064. The number of electors in 1852 was 428.—The name of this town is supposed to be derived from the British word *Guallen*, or the Roman *Vallum*, each signifying 'a fort,' combined with the circumstance of there having anciently been a ford over the Thames at this place.

WALLINGFORD, a township of Rutland co., Vermont, U. S., 32 m. W of Windsor. Pop. 1,600.—Also a township of New Haven co., Connecticut, 12 m. NNE of New Haven. Pop. 2,200.

WALLINGROVE, a village of New South Wales, in the district of New England, on a branch of the Severn.

WALLINGTON, a parish in Hertfordshire, 3 m. E by S of Baldock. Area 1,950 acres. Pop. in 1831, 213; in 1851, 254.—Also a parish in Norfolk, 3 m. N by E of Downham. Area 1,460 acres. Pop. with Thorpland in 1831, 47; in 1851, 58.—Also a hamlet in the p. of Beddington, Surrey, 3 m. WNW of Croydon, on the banks of the Wandle. Pop. in 1831, 933; in 1851, 881.—Also a township in the p. of Hartburn, Northumberland, 3 m. SW of Hartburn. Area 1,781 acres. Pop. in 1831, 183; in 1851, 204.

WALLIS. See VALAIS.

WALLIS, a lake of New South Wales, in the co. of Gloucester, near Cape Hawke.—Also a creek in the co. of Northumberland, an affluent of the Hunter.—Also a plain in the same co., to the S of the Hunter, and W of Wallis creek.—Also an island of Port Hunter, at the entrance to Fullerton Cove.—Also an island near the NW coast of Australia, in S lat. 10° 51′ 25″.

WALLISCHBICHEN, a town of Bohemia, in the circle of Prachin. Pop. 2,100. It has tile works, distilleries, and woollen and linen factories.

WALLIS'S BAY, a bay in the straits of Magalhaen, 12 m. NE of Cape Forward.

WALLIS'S ISLAND, a small island near the SE coast of New Ireland, at the entrance of Gower's harbour, 9 m NW of Cape St. George.

WALLIS'S ISLANDS, a group of 9 islands, in the S. Pacific, in S lat. 13° 24′, discovered by Wallis in 1767. The inhabitants are robust and active.

WALLKILL, a township of Orange co., New York, U. S., 20 m. W of Newburgh. Pop. 4,200.

WALLKILL, a river of North America, which rises in Sussex co., in New Jersey, U. S., and flows NNE into the Rondoutkill, 7 m. above the entrance of the latter into the Hudson.

WALLOE, or WOTO, a small port on the Ivory coast of Africa, in N lat. 5° 20′.

WALLOOSTOOK, a river of Maine, U. S., which rises in Somerset co., and flows NNE to the St. Francis river.

WALLOP-NETHER, a parish in Hants, 4 m. W by N of Stockbridge. Area 7,201 acres. Pop. in 1831, 900; in 1851, 962.

WALLOP-OVER, a parish in Hants, 5½ m. WNW of Stockbridge. Pop. in 1841, 481.

WALLRABONSTEIN, a village in the duchy of Nassau, 4 m. NW of Idstein. Pop. 300.

WALLS-AND-FLOTA, a united parish at the SW extremity of Orkney. W. consists of the S half of the island of HOY [which see]; and is about 8 m. long, and 7 m. broad. Flota consists of the islands of Flota, Fara, Cava, and Gransey [see FLOTA]; and, including intervening marine straits, is about 5½ m. long, and 4 m. broad. The inhabitants are small farmers and fishermen, and are dispersed over the parish. Pop. in 1831, 1,436; in 1851, 1,667.

WALLS-AND-SANDNESS, a united parish in Shetland, consisting of the districts of Walls and Sandness on the mainland; and of the islands of FOWLA, PAPA-STOUR, and VAILA: which see. The mainland districts lie in the most westerly part of the mainland. They are jointly about 7½ m. long from N to S; and about 5 m. broad. The inhabitants, as elsewhere in Shetland, all unite the vocations of farming and fishing. The rocks are gneiss, granitic porphyry, quartz rock, and old red sandstone. Pop. in 1831, 2,143; in 1851, 2,442.

WALLSDORF, a village in the duchy of Nassau, 4 m. NNE of Idstein. Pop. 600.

WALLSEL, a village of Austria, on the r. bank of the Danube, 45 m. W of Saint-Polten. Pop. 500.

WALLSEND, a parish of Northumberland, 4 m. ENE of Newcastle, stretching along the Tyne, and intersected by the Newcastle and N. Shields railway. It is about 1½ m. in breadth from N to S; and 3 m. in length from E to W; and contains the townships of Howden-Pans, W., and Willington. Area 2,787 acres. Pop. in 1801, 3,120; in 1831, 5,510; in 1851, 5,721. The v. situated near the Shields road, is large and well-built; contains several good houses, and has a spacious green in the centre. Extensive coal-mines exist in the neighbourhood; and the place gives its name to an excellent variety of coal, immense quantities of which are shipped for London. Near W. church, a four-inch pipe connected at the pit bottom with an insulated portion of coal strata of about 4 acres, is carried up as high as the head-gear; and from the orifice of this tube there constantly issues an ignited stream of gas forming a flag of flame 8 or 9 ft. in length. Mr. Lyell calculates that ten times as much gas is evolved annually by this pipe as is used in illuminating the town of Sheffield.

WALLSTOWN, a parish in co. Cork, 3 m. E by S of Doneraile. Area 3,056 acres. Pop. 531.

WALL-TOWN, a township in the p. of Haltwhistle, Northumberland, 2½ m. NW of Haltwhistle. Here, according to Hedley, was the site of the Roman station *Vindolano*. Area 2,956 acres. Pop. in 1831, 96; in 1851, 81.

WALLUBGHUR, a hill-fortress of Hindostan, in the prov. of Bejapur, district of Darwar.

WALMER, a parish and town in Kent, on the high road from Deal to Dover. Area of p. 1,079 acres. Pop. in 1831, 1,779, in 1851, 2,105.—The town, which is much resorted to for sea-bathing, and contains several good houses and marine villas, s situated on a rising ground, about 1 m. S of Deal, and ½ m. from the sea-shore. Close to the shore stands Walmer castle, erected by Henry VIII. at the same time with those of Deal and Sandown, for the defence of the coast. It consists of 4 round lunettes of very thick arched stone-work, with large port-holes; and a round tower in the centre; the whole being compassed by a fosse. This fortress, which is the residence of the lord-warden of the cinque-ports, and was fitted up in modern times for that functionary's use, commands a fine view of the Downs, the adjoining channel, and, in a clear day, of the cliffs of Boulogne and Calais. The duke of Wellington, who held the office of warden, died in this castle on the 14th September, 1852.

WALMERSLEY, a township and chapelry in the p. of Bury, co.-palatine of Lancaster, 2½ m. N of Bury. There are coal-mines and extensive spinning-mills here. Area, 5,056 acres. Pop. in 1831, 3,456; in 1851, 4,560.

WALMESGATE, or WALMSGATE, a parish in Lincolnshire, 8 m. NNW of Spilsby. Area 920 acres. Pop. in 1831, 72; in 1851, 79.

WALMESLEY, or WALMSLEY, a chapelry in the township of Turton, p. of Bolton-le-Moors, co.-palatine of Lancaster, 4½ m. N of Bolton.

WALNEY (ISLE OF), an island and chapelry in the p. of Dalton-in-Furness, co.-palatine of Lancaster. Pop. in 1831, 848; in 1851, 921. The island is about 10 m. in length from NW to SE, and almost touches the mainland at its N extremity. There is a road to it, across the sands at low-water, from Hawcoat. It lies on a bed of moss, and all round the island moss is found by digging through a layer of sand and clay which covers it. It has the appearance of a bank or wall in the sea; hence it was called by the Saxons Waghney, Woney, or Walney, 'a walled island.' It contains two hamlets, Biggar and Scale. The castle, known as the 'Pile of Fowdrey,' stands on the N extremity of the island, and is also insulated at high-water. There are, indeed, several insulated portions of land or rock around it, which appear connected with it at low-water. On the S end of W. is a light-house, in N lat. 54° 2′ 54″, W long. 3° 10′ 30″. There are some curious intermitting springs in this island, which, though perfectly fresh, ebb and flow with the tide.

WALNUT, a township of Juniata co., in Pennsylvania, U. S., 35 m. NW of Harrisburg.—Also a township of Jefferson co., Iowa, 48 m. SW of Iowa.

WALNUT (BIG), a river of Ohio, which rises in Delaware co., and joins the Scioto, about 10 m. below Columbus.

WALNUT-COVE, a village of Stoke co., in N. Carolina, 100 m. WNW of Raleigh.

WALNUT-CREEK, a river of Ohio, which runs into the Scioto, 6 m. above Circleville.—Also a township of Holmes co., in Ohio, 74 m. NE of Columbus. Pop. 1,077.—Also a v. in Buncombe co., N. Carolina.

WALNUT-GROVE, a village of Knox co., Illinois, 91 m. N by W of Springfield.—Also a village of Mercer co., New Jersey, 41 m. N of Trenton.—Also a village of Kanawha co., in Virginia.

WALNUT-HILL, a village of La Fayette co., Arkansas, U. S., 130 m. SW of Little Rock.—Also a v. of Marion co., Illinois, 95 m. S by E of Springfield.

WALNUT-VALLEY, a township of Warren co., in New Jersey, 52 m. N by W of Trenton.—Also a village of Rock-Island co., in Illinois, 120 m. NNW of Springfield.

WALPOLE, a township of Talbot district, in Upper Canada, skirting on the S on Lake Erie, and on the NE on Tuscarora township.—Also an island in the NE part of Lake St. Clair, in Upper Canada. It is 10 m. in length by 3½ m. in breadth, and is inhabited chiefly by Indians, whose number in 1842 amounted to 1,140.—Also a township and village of Norfolk co., in Massachusetts, U. S., 18 m. SW of Boston. Pop. 1,929.—Also a township and v. in Cheshire co., New Hampshire, U. S., 41 m. WSW of Concord. Pop. 2,034.

WALPOLE, a parish in Suffolk, 2 m. SW of Halesworth. Area 1,750 acres. Pop. in 1851, 563.

WALPOLE-ST.-ANDREW, a parish in Norfolk, 7 m. W by S of Lynn-Regis. Area 3,494 acres. Pop. in 1831, 514; in 1851, 737.

WALPOLE-ST.-PETER, a parish in Norfolk, 8 m. WSW of Lynn-Regis. Pop. in 1851, 1,361.

WALSALL, a parish and parl. borough in the S division of the hund. of Offlow, co. of Stafford, comprising the chapelry of Bloxwich, and the township of Walsall-Foreign, as distinguished from Walsall-Borough. Area of p. 8,182 acres. Pop. in 1821, 11,914; in 1831, 15,066; in 1851, 26,822.—The town, situated on a bold eminence, at the foot of which flows a small stream, falling into the river Tame, is distant by coach-road 8 m. N by W of Birmingham; 6 m. E of Wolverhampton; 87 m. SE of Liverpool; and 119 m. NW of London. It consists of several handsome streets, and presents the general appearance of a compact and flourishing town. It ranks as the second manufacturing town in the county. Among its staple manufactures are bridle-bits, stirrups, spurs, saddle-trees, spoons, snuffers, buckles, plated ware, locks, coach-harness, and every description of saddlers' and coachmakers' ironmongery; there are also several brass and iron foundries; and the p. contains extensive iron, coal, and lime-works. The trade and commerce of the town are greatly facilitated by an extension of the old Birmingham canal, which, passing the W side of the town, joins the Wyrley and Essington canal, and sends off several ramifications to different eligible points in the town and parish. The Grand Junction railway passes within 1½ m. on the SW. The nearest station, that of Bescott-bridge, is 9¾ m. from Birmingham, and 88 from Liverpool or Manchester by railway. W. under the reform act returns one member to parliament. The boundary of the parl. and of the municipal borough corresponds with that of the ancient borough, with the exception of not including Walsall-wood. Pop. of the parl. borough in 1831, 15,066; of whom 6,401 resided in the Borough, and 8,665 in the Foreign; in 1851, 25,680. Parl. constituency in 1835-6, 679; in 1839-40, 837; in 1852, 1,026. W. is a polling-station for the southern division of the county.

WALSALL-FOREIGN, a township in the above p., including the W and S parts of the town itself, and a number of villages and hamlets stretching 3 m. northwards. Area 7,820 acres. Pop. in 1831, 6,665; in 1851, 12,442.

WALSHAM (NORTH), a parish and town in Norfolk. Area of p. 4,252 acres. Pop. in 1831, 2,615; in 1851, 2,911. The church is a spacious structure, 156 ft. in length, by 68 ft. in breadth, situated in the central area of the town. The town is pleasantly built on a declivity sloping to the river Ant, which flows about 1 m. N of the town, and is navigable from it to Yarmouth. On the 25th of June, 1600, it was terribly devastated by fire.

WALSHAM (SOUTH), a considerable village and district in Norfolk, containing two parishes: that of St. Mary and that of St. Lawrence. Area of both ps. 3,149 acres. Pop. in 1831, 575; in 1851, 689. The village is 11 m. NW by W of Yarmouth.

WALSHAM-IN-THE-WILLOWS, a parish in Suffolk, 5 m. E of Ixworth. Area 2,800 acres. Pop. in 1831, 1,167; in 1851, 1,297.

WALSHTOWN, a village in the p. of Templenacarriga, 5 m. NNW of Middleton, co. Cork. Pop. in 1851, about 70.

WALSINGHAM (GREAT), or OLD WALSINGHAM, a parish in Norfolk, containing the united ps. of All-Saints and St. Peter. United area 2,170 acres. Pop. in 1831, 434; in 1851, 476. The v. of Old Walsingham is about 1 m. N of the town of New Walsingham.

WALSINGHAM (LITTLE), or NEW WALSINGHAM, a parish and market-town in Norfolk, 28 m. NW of Norwich. Area of p. 860 acres. Pop. in 1831, 1,067; in 1851, 1,207. A famous chapel here, founded in 1061, appears to have been in as great repute as that of "our Lady of Loretto," or St. Thomas à Becket at Canterbury. Some interesting ruins of this once magnificent edifice still exist in the pleasure-grounds of W. abbey.—The market-town is situated in a picturesque and rugged valley, on the banks of the Stiffkey, 5 m. S by E of Wells, and 27 m. NW of Norwich. Here is a house-of-correction, which has recently been enlarged. The family of De Grey takes the title of baron from this place.

WALSINGHAM (CAPE), a headland of British North America, on the E coast of Cumberland Island, to the S of Exeter-sound, in N lat. 66°, and W long. 60° 50′ 10″.

WALSLEBEN, a village of Prussia, in the regency of Magdeburg and circle of Osterburg, on the

Uchte. Pop. 320.—Also a village in the regency of Potsdam, and circle of Ruppin. Pop. 460.—Also a village in the regency of Stettin, and circle of Naugard. Pop. 320.

WALSMUHLEN, a village of Mecklenburg-Schwerin, in the prov. of Mecklenburg, and bail. of Schwerin. Pop. 210.

WALSOKEN, a parish and village in Norfolk. The v. is about 1 m. E of Wisbeach, and is almost joined to that town by a pleasant suburb called New Walsoken. Area of p. 4,658 acres. Pop. in 1831, 1,856; in 1851, 2,740.

WALSRODE, a town of Hanover, in the gov. and 47 m. SW of Luneburg, and bail. of Rethem, on the Bohme. Pop. 2,100. It has a suburb, and contains a convent. It possesses manufactories of cloth, articles in straw, thread, cordage, and saltpetre, paper and flour-mills, and carries on an active trade in wool, gin, hats, and articles of local manufacture.

WALSTON, a small parish in the upper ward of Lanarkshire. It lies high; Walton Black mount is said to stand fully 1,500 ft. above the level of the sea. There are two small villages, viz., Walston and Elsridgehill, or Elsrickel. Pop. in 1831, 429; in 1841, 493; in 1851, 497.

WALTENBERG, Zillenmarkt, or Zilah, a town of Transylvania, capital of the gov. of Kozep-Szolnok, 39 m. NW of Klausenburg, near the source of the Zilah, an affluent of the Kraszna, and at the foot of the Meszes. Pop. 8,450. It has a Calvinist gymnasium.

WALTENDORF, a village of Austria, in Styria, in the circle of Alli. Pop. 260.

WALTENHAUSEN, a village of Bavaria, in the circle of Swabia, and presidial of Krumbach. Pop. 380.

WALTENSBURG, a village of Switzerland, in the cant. of the Grisons and bail. of Glenner, on the l. bank of the Vorder-Rhein. Pop. 443.

WALTENSWEIL, a village of Switzerland, in the cant. of Aargau, and bail. of Muri. Pop. 684.

WALTERBORO, a village of Colleton district, in the state of South Carolina, U. S., 94 m. S by E of Columbia. Pop. in 1850, 300.

WALTERKEHMEN, a village of Prussia, in the regency and circle of Gumbinnen. Pop. 350.

WALTERMENBURG, a village of Prussia, in the regency of Magdeburg, and circle of Jerichow. Pop. 598. It has a castle.

WALTERSBERG, a village of Bavaria, in the circle of Middle Franconia, and presidial of Beilngries. Pop. 200.

WALTERSBRUCH, a village of Hesse, in the prov. of Lower Hesse, and circle of Fritzlar, on the Schwalin. Pop. 500.

WALTERSCHLAG, a village of Austria, in Moravia, in the circle of Iglau. Pop. 450.

WALTERSDORF, a village of Austria, in the country below the Ens, and lower circle of the Weinerwald, 18 m. S of Vienna, on the Riesenbach. Pop. 500. It has a paper-mill.—Also a village in the lower circle of the Manhartsberge, on the March. Pop. 300.—Also a village in the circle of Staatz. Pop. 280.—Also a village of Bohemia, in the regency of Pardubitz, and bail. of Landsfron. Pop. 1,700.—Also a village of Moravia, in the circle of Olmutz, lordship of Goldenstein. Pop. 365.—Also a village in the lordship of Karsberg. Pop. 500.—Also a village in the lordship of Fulnek. Pop. 515.—Also a village of the archb. of Lebengut. Pop. 1,500. It has mills and bleacheries.—Also a village of Styria, in the circle of Gratz. Pop. 440. It has an Armenian institute.—Also a village of Prussia, in the regency of Konigsberg, and circle of Mohrungen. Pop. 410.—Also a village of the circle of Heiligenbeil. Pop. 315.—Also a village of the regency of Liegnitz, and circle of Schonau. Pop. 540.—Also a village of the circle of Sprottau. Pop. 1,810.—Also a village of the circle of Lowenburg. Pop. 320. It has a castle, mills, and salt-works.—Also a village of the regency of Frankfort, and circle of Coran. Pop. 440.—Also a village of the circle of Luckau. Pop. 345. It has three mills.—Also a village of the regency of Potsdam, and circle of Telton. Pop. 260.—Also a village of the regency of Erfurt, and circle of Weissensee. Pop. 220.—Also a village of Reuss-Schleiz, in the circle of Jera. Pop. 250.—Also a village of Saxony, in the circle of Zwickau, and bail. of Annaberg, on the Zschopau. Pop. 266. It has two mills.—Also a village in the circle of Dresden, and bail. of Pirna, near Barenstein. Pop. 242.—Also a village near Prossen. Pop. 275.—Also a village of Saxe-Altenburg, and bail. of Altenburg. Pop. 250.—Also a village of Saxe-Weimar, and bail. of Neustadt. Pop. 550.

WALTERSDORF (Alt), a village of Prussia, in the regency of Breslau, and circle of Habelschwert. Pop. 1,070. It has several mills and salt-works.—The v. of New W. in the vicinity has a pop. of 900.

WALTERSDORF (Alt and Neu), two adjoining villages of Saxony, in the circle of Bautzen, and bail. of Lobau, near Zittau. Pop. 2,000. They have manufactories of linen and woollen fabrics.

WALTERSDORF (Lang), a village of Prussia, in the regency of Breslau, and circle of Waldenburg. Pop. 830. It has two churches, and a work house, and possesses manufactories of woollen and linen fabrics, and bleacheries.

WALTERSDORF (Nieder), a village of Prussia, in the regency of Breslau, and circle of Walden burg. Pop. 200. It has several mills.

WALTERSDORF (Roth), a village of Prussia, in the regency of Breslau, and circle of Glatz. Pop. 660. It has several mills and salt-works.

WALTERSDORF (Wuste), a village of Prussia, in the regency of Breslau, and circle of Waldenburg. Pop. 1,500. It has a castle, a workhouse, several mills, and bleacheries.

WALTERSHAUSEN, a village of Bavaria, in the circle of Lower Franconia, and presidial of Konigshofen. Pop. 600. It has a castle and several mills.—Also a town of Saxe-Coburg-Gotha, in the principality and 8 m. WSW of Gotha, on the Bodewasser, at the foot of the Burgberg. Pop. 3,400. It is enclosed by walls with four gates, and has four suburbs, two churches, two hospitals, and several schools. It possesses manufactories of linen and woollen fabrics, tanneries, breweries, paper mills, and tile kilns.

WALTERSHOF, a market-town of Bavaria, in the circle of the Upper Pfalz, and presidial of Waldsassen, 26 m. E of Bayreuth. Pop. 1,200. It has manufactories of woollen and linen fabrics, breweries, and iron-works.

WALTERSHOFEN, a village of Baden, in the circle of the Middle Rhine, and bail. of Freiburg. Pop. 938.

WALTERSKIRCHEN, a village of Austria, in the country below the Ens, circle of the lower Manhartsberge, and presidial of Feldsberge. Pop. 630.

WALTERSLEBEN, a village of Prussia, in the regency and circle of Erfurt. Pop. 260.

WALTERSMUHL, a village of Prussia, in the regency of Konigsberg, and circle of Heilsberg. Pop. 370.

WALTERSTONE, a parish in Herefordshire, 15 m. SW of Hereford. Area 1,241 acres. Pop. 147.

WALTERSTOWN, a parish in co. Kildare, 3¼ m. SSW of Kildare. Area 1,501 acres. Pop. 376.

WALTERSWEIER, a village of Baden, in the circle of the Middle Rhine, and bail. of Offenburg. Pop. 458.

WALTERSWEIL, a village of Switzerland, cant. of Berne, and bail. of Trachselwald. Pop. 850.

WALTHAM, or TEMPLE-WALTHAM, a parish and v. of Kent. Area of p., 3,215 acres. Pop. 576.

WALTHAM, a parish in Lincolnshire, 3½ m. S by W of Great Grimsby. Area 2,350 acres. Pop. in 1831, 545; in 1851, 782.

WALTHAM, a township and village of Middlesex co., in Massachusetts, U. S., 8 m. WNW of Boston. Pop. in 1840, 2,504; in 1850, 4,464.

WALTHAM-ABBEY, or HOLY-CROSS, a parish and market-town in Essex, 12 m. S by E of London. Area of p., including the hamlets of Hallifield or Holy-field, Sewardstone, and Upshire, 10,876 acres. Pop. in 1831, 4,104; in 1851, 4,303. The town, which is large and irregularly built, is 12 m. S by E of London. It is situated near the river Lea, which here divides into diverse streams, and is skirted by low meadows partly occupied by government for gunpowder-mills and other works which have been erected here, and which give employment to many of the inhabitants. Here are several malt-kilns, a pin-manufactory, and a silk-mill.—The convent of W. appears to have been originally founded in the reign of Canute. Harold is commonly stated by historians to have been killed at the battle of Hastings, and interred in W. abbey, where, during a long period, a tomb was shown as the sepulchral monument of the last of our Saxon kings; but Palgrave seems to consider this tomb as merely a cenotaph.

WALTHAM (BISHOP'S), a parish and market-town in Hants, 10 m. ENE of Southampton. Area 7,388 acres. Pop. in 1831, 2,181; in 1851, 2,265. The town is situated on the high road from Winchester to Portsmouth, in the midst of a fertile country. The ancient palace of the bishops of Winchester stands on the SW side of Waltham, and still exhibits traces of its former magnificence.

WALTHAM-BRIGHT, a parish in Berks, 5 m. WSW of East Ilsley. Area 2,038 acres. Pop. in 1831, 442; in 1851, 465.

WALTHAM (GREAT), a parish in Essex, 4½ m. N by W of Chelmsford, including the hamlet of North-End. Area 7,335 acres. Pop. in 1851, 2,335.

WALTHAM-ST.-LAWRENCE, a parish in Berks, 5½ m. SW of Maidenhead, in the line of the Great Western railway. Area 3,468 acres. Pop. in 1831, 739; in 1851, 783.

WALTHAM (LITTLE), a parish in Essex, 4¼ m. N by E of Chelmsford, on the river Chelmer. Area 2,227 acres. Pop. in 1831, 674; in 1851, 651.

WALTHAM (NORTH), a parish in Hants, 6 m. SW by W of Basingstoke. Area 1,938 acres. Pop. in 1831, 458; in 1851, 506.

WALTHAM (UP), a parish in Sussex, 6½ m. SSW of Petworth. Area 1,245 acres. Pop. 67.

WALTHAM-ON-THE-WOLDS, a parish in Leicestershire, 5 m. NE of Melton-Mowbray. Area 2,870 acres. Pop. in 1831, 653; in 1851, 732.

WALTHAMSTOW, a parish in Essex, 6 m. NE by N of London, on the river Lea, here navigable, and in the line of the Eastern Counties railway. Area 4,436 acres. Pop. in 1831, 4,258; in 1851, 4,959. The village, which is chiefly inhabited by wealthy merchants of the metropolis, stands in a pleasant situation on the border of Epping-Forest, and consists of numerous detached houses, encompassed with trees and woodland.

WALTON, a county in the central part of the state of Georgia, U. S., comprising an area of 354 sq. m., drained by the head branches of Oconee and Ocmulgee rivers, and intersected by the Georgia railroad. It has an undulating surface, and is extremely fertile. Pop. in 1840, 10,209; in 1850, 10,821. Its capital is Monroe.—Also a county in the W part of the state of Florida, comprising an area of 1,157 sq. m., drained by Yellow and Shoal rivers, and by White and Alequa creeks. It has an undulating surface, and is generally fertile. Pop. in 1840, 1,401; in 1850, 1,379. Its capital is Uchee Anna.—Also a township of Delaware co., in the state of New York, 78 m. SW by W of Albany, drained by the W branch of Delaware river and its affluents. Pop. in 1840, 1,840; in 1850, 2,271.

WALTON, a parish of Bucks, 2 m. NE of Fenny-Stratford. Area 757 acres. Pop. in 1831, 114; in 1851, 95.—Also a parish in Cumberland, 3 m. N by W of Brampton, containing the townships of High and Low W. Area 3,592 acres. Pop. in 1831, 481; in 1851, 455.—Also a chapelry in the p. of Chesterfield, Derbyshire, 3 m. SW by W of Chesterfield. Pop. in 1831, 935; in 1851, 1,114.—Also a hamlet in the p. of Knaptoft, Leicestershire, 4 m. NE by E of Lutterworth. Pop. in 1831, 234; in 1851, 223.—Also a hamlet in the p. of Paston, Northamptonshire, 2¾ m. NNW of Peterborough. Pop. in 1831, 160; in 1851, 216.—Also a township in the p. of Old Radnor, co. of Radnor. Pop. in 1831, 186; in 1851, 207.—Also a parish in Somersetshire, 3½ m. SW by W of Glastonbury. Area 2,502 acres. Pop. in 1831, 732; in 1851, 768. —Also a parish in Suffolk, 10 m. SE by E of Ipswich, on the coast of the North sea. Area 1,988 acres. Pop. in 1831, 887; in 1851, 897.—Also a parish in the E. R. of Yorkshire, 2 m. E of Wetherby. Area 1,670 acres. Pop. in 1831, 237; in 1851, 245.

WALTON-CARDIFF, a parish in Gloucestershire, 1 m. SE by E of Tewkesbury. Area 650 acres. Pop. in 1831, 57; in 1851, 60.

WALTON-LE-DALE, a chapelry in the p. of Blackburn, co.-palatine of Lancaster, 2 m. SE of Preston, on the banks of the Derwent, near the Wyre and Preston railway. Area 4,630 acres. The inhabitants are principally employed in the cotton manufacture. Pop. in 1831, 5,767; in 1851, 6,855.

WALTON (EAST), a parish in Norfolk, 7½ m. NW of Swaffham. Area 2,459 acres. Pop. in 1851, 202.

WALTON (EAST), a parish in Pembrokeshire, 6½ m. NE of Haverford-West. Area 1,893 acres. Pop. in 1831, 229; in 1851, 279.

WALTON-IN-GORDANO, a parish in Somersetshire, 11½ m. W of Bristol. Area 1,315 acres. Pop. in 1831, 297; in 1851, 225.

WALTON-ON-THE-HILL, a parish in the co.-palatine of Lancaster, 3 m. N by E of Liverpool, in the line of the Leeds and Liverpool canal, comprising the chapelries of Everton, Formby, Kirkby, West Derby, Kirkdale, and Bootle, and the townships of Fazakerley, Simonswood, and W. In consequence of its proximity to Liverpool, it contains many elegant mansions—the residence of merchants, manufacturers, and retired tradesmen. Area of p. 29,233 acres. Pop. in 1831, 22,575; in 1851, 46,302.—Also a parish in Surrey, 4½ m. S by E of Epsom. Area 2,592 acres. Pop. in 1831, 352; in 1851, 426.

WALTON (INFERIOR and SUPERIOR), townships in the p. of Runcorn, co.-palatine of Chester, 2 m. S of Warrington, in the line of the Grand Junction railway, and of the Mersey and Irwell canal. Area of both 961 acres. Pop. in 1831, 578; in 1851, 575.

WALTON-LE-SOKEN, a parish in Essex, 12 m. SE by E of Manningtree, on the sea-coast. The village, which is romantically situated on the cliffs next the sea, has obtained considerable celebrity as a watering-place. Area 3,260 acres. Pop. in 1831, 469; in 1851, 729.

WALTON-ON-THAMES, a parish in Surrey, 14 m. NE by N of Guildford, in the line of the South-Western railway. The p. contains the divisions of Commonside, Hersham, Burnwood; and Town. Area 6,834 acres. Pop. in 1851, 2,861.

WALTON-ON-TRENT, a parish in Derbyshire, 4 m. SW of Burton-upon-Trent, near the Birmingham and Derby railway. Area, 2,309 acres. Pop. in 1831, 408; in 1851, 465.

WALTON (WEST), a parish in Norfolk, 3 m. NE by N of Wisbeach. Area 5,219 acres. Pop. in 1831, 905; in 1851, 999.

WALTON (WEST), a parish in Pembrokeshire, 5½ m. SW of Haverford-West. Area 1,408 acres. Pop. in 1831, 411; in 1851, 518.

WALTON-ON-THE-WOLDS, a parish in Leicestershire, 4 m. E of Loughborough, on the river Soar. Area 1,720 acres. Pop. in 1851, 260.

WALTON-WOOD, a parish in Huntingdonshire, 6 m. N by W of Huntingdon. Area 3,830 acres. Pop. in 1831, 305; in 1841, 273.

WALTONHAM, a village of St. Louis co., in the state of Missouri, U. S., 9 m. NW of St. Louis. Pop. in 1840, about 50.

WALTRA, a village of Austria, in Styria, in the circle of Gratz. Pop. 260.

WALTRACH, a village of Prussia, in the regency and circle of Treves. Pop. 688.

WALTRINGEN, a village of Prussia, in the regency of Arnsberg, and circle of Goest. Pop. 300.

WALTRINGHAUSEN, a village of Hesse, prov. of Lower Hesse, and circle of Rinteln. Pop. 420.

WALTROP, a village of Prussia, regency of Munster, and circle of Recklinghausen. Pop. 1,108.

WALTROWITZ, a village of Austria, in Moravia, in the circle of Znaim. Pop. 410.

WALTRUP, a village of Prussia, in the regency of Munster, and circle of Steinfurt. Pop. 300.

WALTWELDER, a department and commune of Belgium, in the prov. of Limburg, and arrond. of Tongres. Pop. 604.

WALVISCH, or WALFISCH BAY, a large bay on the SW coast of Africa, the S point of which is in S lat. 22° 52′. It forms the estuary of the Kiusip, but that river is only a periodical one, having its waters frequently dried up before it reaches the sea. From the coast of this bay, which presents an excellent harbour, a journey of 16 hours across sand, takes the traveller to Oosop, on the Swakop river, in S lat. 22° 45′ 38″. Communication has recently been opened between this bay and Lake N'gami in the interior.

WALWORTH, a township in the p. of Heighington, co.-palatine of Durham, 4 m. NW of Darlington. Area 2,132 acres. Pop in 1831, 155; in 1851, 142.

WALWORTH, a chapelry in the p. of Newington-Butts, Surrey, 2¼ m. S of St. Paul's. Pop. 29,861.

WALWORTH, a county in the SE part of the state of Wisconsin, drained by Turtle creek and Fox river and their affluents, and intersected by the Chicago and Wisconsin, the Milwaukie and Mississippi railways, and the Racine and Janesville plank-road. It has a diversified surface and contains several lakes. Pop. in 1840, 2,611; in 1850, 17,861.—Also a township of the same co., 46 m. SE of Madison, bordered on the E by Geneva lake. It has a diversified surface, and is highly cultivated. Pop. in 1840, 226; in 1850, 987.—Also a township of Wayne co., in the state of New York, 168 m. W by N of Albany. It has an undulating surface, and is drained by affluents of Lake Ontario. Pop. in 1840, 1,734; in 1850, 1,981.

WALWYN'S-CASTLE, a parish in Pembrokeshire, 6 m. SW of Haverford-west. Area 2,904 acres. Pop. in 1831, 311; in 1851, 353.

WALXHEIM, a village of Würtemberg, circle of the Danube, and bail. of Ellwangen. Pop. 215.

WALZEN (GRAFLICH and SCHLOTZ), two villages of Prussia, in the reg. of Oppeln, circle of Neustadt, containing respectively 599 and 328 inhabitants.

WALZEN (UNTER), a village of Prussia, in the regency of Oppeln, and circle of Neustadt. Pop. 880.

WALZENHAUSEN, a village of Switzerland, in the cant. of Appenzell, bail. and NE of Trogen. Pop. 1,794.

WAMBACH, a village of Nassau, in the bail. of Langenschwalbach. Pop. 299.—Also a village of Prussia, in the regency of Coblenz, and circle of Altenkirchen. Pop. 275.

WAMBEKE, a department and commune of Belgium, in the prov. of Brabant, and arrond. of Brussels. Pop. of dep. 1,514; of com., 746.

WAMBELN, a village of Prussia, in the regency of Arnsberg, and circle of Dortmund. Pop. 380.—Also a village in the circle of Hamm. Pop. 300.

WAMBERG, a town of Austria, in Bohemia, in the regency of Gotschin, and bail. of Reichenau. Pop. 2,500.

WAMBRECHIES, a commune and town of France, in the dep. of the Nord, cant. and 4 m. N of Lille. Pop. in 1846, 3,542. It has manufactories of thread, of beet-root sugar, and oil; a distillery of gin, a tannery, and building-docks.

WAMBROOK, a parish in Dorset, 2 m. SW of Chard. Area 1,857 acres. Pop. in 1851, 245.

WAME, a river of Belgium, which takes its rise in the prov. of Luxemburg, flows thence into that of Namur, and after a course, first in a N direction and afterwards to the W, joins the Homme, on the r. bank, at Gamelle.

WAMEKON, a village of Mecklenburg-Schwerin, in the prov. of Mecklenburg, and bail. of Enwitz. Pop. 200.

WAMEL, a village of Prussia, in the regency of Arnsberg, and circle of Goest. Pop. 260.

WAMLITZ, a village of Prussia, in the regency of Stettin, and circle of Randow. Pop. 200.

WAMMER, an island of the Austral ocean, in the group of the Aru islands, to the SW of Wofan.

WAMMONT, a department and commune of Belgium, in the prov. of Liege and arrond. of Huy. Pop. 435. It has manufactories of beet-root sugar.

WAMPACH (OBER), a commune and town of Belgium, in the prov. of Luxemburg, arrond. and 15 m. W of Diekirch. Pop. 818.

WAMPHRAY, a parish in Upper Annandale, Dumfries-shire, skirted on the W by the river Annan. It has an area of about 12,000 acres. Pop. in 1831, 580; in 1851, 523.

WAMPU, an island of China, in the bay of Canton.

WAN, a district and town of China, in the prov. of Chih-le, and div. of Paou-ting-fu, in N lat. 38° 50′, and E long. 115° 13′.—Also a district and town in the prov. of Ho-nan, and div. of Hwae-ting-fu, in N lat. 30° 57′, and W long. 108° 32′.

WANBOROUGH, a liberty in Surrey, 4 m. W of Guildford. Area 156 acres. Pop. in 1831, 111; in 1851, 208.—Also a parish in Wilts, 3½ m. E by S of Swindon. Area 4,440 acres. Pop. in 1831, 1,016; in 1851, 964.

WAN-CHANG, and district and town of China, in the prov. of Kwang-ting, and div. of Keun-chu-fu, or island of Hae-nan, of which it occupies the eastern part. The town is a little to the N of the Wan-chang-keang, which is here crossed by a bridge; and, although small, is fortified, in N lat.

19° 36′, and E long. 110° 43′. To the N of the town is the temple of Ling-wang, and to the SE is the fine monument of Chao-ting, said to be the place of sepulture of a king of Cochin-China. Pop. in 1819, 116,784.—Also a district and town in the prov. of Che-keang, and div. of Gen-chu-fu, in N lat. 35° 50′, and E long. 114° 20′.

WAN-CHU, a district and town of China, in the prov. of Kwang-ting, and div. of Keun-chu-fu, or island of Hae-nan, in N lat. 18° 49′, and E long. 110° 20′. It is enclosed by brick walls from 18 to 20 ft. in height, and of great thickness; and the parapets, about 4 ft. in height, are pierced with numerous embrasures, separated by loop-holes. The gates, 4 in number, open to the four cardinal points. The streets, which are paved, are narrow, and the houses, although built of brick and stone, are only a single story in height, and have a mean appearance. There are some temples of handsome structure, and several triumphal arches, and there are various indications of the town having formerly been one of great importance. The environs are fertile and well cultivated. They abound in game, and contain several villages.

WAN-CHU-FU, a div. and town of China, in the prov. of Che-keang. The div. comprises 6 districts. The town is on a small bay of the China sea, at the mouth of a river, 180 m. SSE of Hang-chu-fu, in N lat. 28° 2′ 15″, and E long. 120° 49′ 37″.

WANDERSLEBEN, a town of Prussian Saxony, in the gov. and 10 m. SW of Erfurt. Pop. 914.

WANDIA, a town of Hindostan, in the prov. of Cutch, on the N side of the Runn, opposite the fortress of Mallia, in N lat. 23° 3′.

WANDIPORE, a town and fortress of Bhotan, in N lat. 27° 51′, E long. 89° 57′. It stands on the narrow extremity of a rock, between three rivers whose streams, uniting at its point, form a river of considerable magnitude called the Chantchieu, which, continuing its course to the southward, enters the district of Rungipur, where it takes the name of Gudhadhar, and at last falls into the Brahmaputra. The town is principally inhabited by *gylongs* or priests of the Grand Lama.

WANDIWASH, a town and fortress of Hindostan, in the Carnatic, in N lat. 12° 30′, E long. 79° 38′. In September, 1759, the British troops, in an attack on this place, were repulsed with great slaughter, but in November following, it was taken by Sir Eyre Coote, with scarcely any loss. In January 1760, a decisive battle was fought in the vicinity of this town, between the British and the French, with their respective allies, in which the latter were defeated. The fortress was demolished by orders of General Stuart, in 1783.

WANDLE, a small river, which, rising near Carshalton, in Surrey, flows N till it falls into the Thames at Wandsworth.

WANDORF, a village of Hungary, on the borders of Lower Austria, 2 m. from Oedenburg. Coal is extensively wrought in the vicinity.

WANDRÉ, a canton and village of Belgium, in the prov. and 4 m. NE of Liege. Pop. 2,700. It has iron-works.

WANDSBECK, a town of the duchy of Holstein, 3 m. NE of Hamburgh, on the Wanne, an affluent of the Alster. Pop. 4,000. It has manufactures of cotton and linen stuffs.

WANDSWORTH, a parish in Surrey, 6 m. SW of London, on the river Wandle. The South Western railway passes across Wandsworth common. Area 2,478 acres. Pop. in 1831, 6,879; in 1851, 9,611.—The village is finely situated on the declivities of two hills, and contains many handsome houses. Here are manufactories of hats, and bolting cloths for dressing flour. There are also distilleries, vinegar-works, mills for the preparation of iron, white-lead, and linseed oil, besides flour-mills, and establishments for calico-printing. The environs are adorned with elegant mansions, commanding fine views of the Thames, the metropolis, and a great part of Middlesex.—About 1½ m. from the Clapham station of the South-Western railway, a fine astronomical tower has been erected. The building consists of a central tower of brick, the walls of which are about 18 inches in thickness, the height 64 ft., and the diam. 15 ft. It is erected upon a solid bed of concrete. By the side of this tower hangs a gigantic telescope, the immense tube of which, measuring 85 ft. in length, is shaped something like a cigar. At one end is the eye piece, at the height of about 5 ft. from the ground, and at the other the dew-cap, or covering, the object of which is to prevent the absorption and condensation of moisture, which takes place during the night when the instrument is most in use. The exterior is of bright metal, and the interior painted black to absorb the divergent rays of light. The instrument has a focal distance which varies from 76 to 85 ft.; its greatest circumf. is 13 ft., the widest portion of the tube being about 24 ft. from the object-glass. The lower end of the tube is supported on a wooden frame work, to which wheels are attached, and these wheels traverse a circular iron railway at a distance of 52 ft. from the base of the tower. The other end is suspended by a strong iron chain, capable of sustaining a weight of 15 tons, from the top of the tower, and this chain, which passes across to the other side through the top over pullies, can be elevated or depressed by means of a windlass, so as to raise or lower the tube from azimuth to an angle of 80° of elevation with as much ease and rapidity as an ordinary telescope. The instrument has an object-glass of 2 ft. aperture.

WANFERCE′E-BAULET, a commune and v. of Belgium, in the prov. of Hainault, 8 m. NE of Charleroi. Pop. 2,540. Coal is wrought in the vicinity.

WANFRIED, a town of Hesse-Cassel, on the r. bank of the Werra, 30 m. E by S of Cassel. Pop. 2,400.

WANGANIN, a river of the North island of New Zealand, which flows into the sea, on the W coast, 60 m. SE of New Plymouth.

WANGARA, a country of Central Africa, much celebrated in the early descriptions of that continent, and represented by Arabian travellers in the 12th cent. as the grand source of African wealth. Edrisi describes it as traversed and intersected by branches of the Nile of the Negroes, which form it into a species of island, 300 m. in length, and 150 m. in breadth. Modern geographers identify this region with

WANGARI-BAY, a bay on the E coast of the North island of New Zealand, 65 m. NNW of Auckland.

WANGEN, a town of France, in the dep. of Bas-Rhin, 12 m. WNW of Strasburg.—Also a town of Switzerland, in the cant. of Berne, on the r. bank of the river Aar, 5 m. NE of Soleure.—Also a village of Switzerland, in the cant. and 6 m. NE of Zurich.—Also a town of Würtemberg, 11 m. NNE of Lindau. Pop. 1,680. It has manufactories of linen, paper, and fire-arms.—Also a village of Baden, 6 m. SSW of Radolfzell. Pop. 500.

WANGENHEIM, a village of Saxe-Coburg, in the principality and 6 m. NNW of Gotha, on the Nesse. Pop. 500.

WANGERIN, a town of Prussian Pomerania, 22 m. NE of Stargard, between two small lakes. Pop. 1,638.

WANGEROOG, an island on the coast of the duchy of Oldenburg, about 4 m. from the mainland. Pop. 356, chiefly Friesian fishermen. It is 12 m. in circumf. It has a light-house, which is in N lat. 53° 49′ 36″, E long. 7° 51′ 30″.

WANGFORD, a parish in Suffolk, 3 m. NW of Southwold, containing the hamlet of Henham. Area 2,600 acres. Pop. in 1831, 792; in 1851, 814.—Also a parish in Suffolk, 3½ m. SW by W of Brandon-Ferry. Area 3,252 acres. Pop. in 1851, 33.

WANGWELL, a small island in the Pacific, near the S coast of Waygiu, in S lat. 0° 23′.

WANJI, a commune and village of the Swiss cant. of Thurgau, on the Murg, 5 m. SE of Frauenfeld. Pop. 1,586.

WANKANIER, a fortified town of Hindostan, prov. of Gujerat, on an angle formed by the conflux of the rivers Muchu and Pataliah, in N lat. 22° 57′. It contains about 500 houses, and a very handsome mosque, but is commanded by a range of lofty mountains.

WANKUM, a village of Prussia, prov. of Cleves, 7 m. S by W of Guelders. Pop. 1,000.

WANLIP, a parish in Leicestershire, 3½ m. SE by S of Mount Sorrel, on the river Soar, which is here crossed by a bridge. Area 952 acres. Pop. 137.

WANLIN, a village of Belgium, in the prov. of Namur, cant. of Beauraing. Pop. 229.

WANLOCK-HEAD, a large mining-village at the head of the Wanlock, a head-branch of the Crawick, 1¼ m. S of Leadhills, and 8½ m. ENE of Sanquhar, Dumfries-shire, situated close on the wild boundary with Lanarkshire, at an alt. of about 1,380 ft. above sea-level. The mines, which alone maintain the v., are continuous with the mines of Leadhills, on the Crawford or Lanarkshire side of the frontier, and jointly with these, shoot out to a circumf. whose diam. is about 2½ m. The mines are said to have been discovered by a German in the reign of James V. The veins, five in number, have been worked to the depth of from 60 to 136 faths., and measure from a few inches to 4 ft. in thickness. Along with the lead ore or lead glance, are small quantities of manganese, ochre, blende, brown hematite, copper pyrites, green lead ore, white lead ore, and lead-vitriol. Pop. in 1851, 872.

WANNE, a commune and v. of Belgium, in the prov. of Liege, cant. of Chevron. Pop. 854.

WANNEBECQ, a village of Belgium, in the prov. of Hainault, 24 m. NE of Nems. Pop. 1,052.

WANNEGEM-LEDE, a canton and v. of Belgium, in the prov. of E. Flanders, cant. of Cruyshantem. Pop. 1,493.

WANSBECK. See NORTHUMBERLAND.

WANSEN, a town of Prussian Silesia, in the reg. of Oppeln, 22 m. S by E of Breslau. Pop. 1,500.

WANSFORD, or WALNESFORD, a parish in Northamptonshire, 89 m. N by W of London. Area 469 acres. Pop. in 1831, 179; in 1851, 184.—Also a township in the p. of Nafferton, Yorkshire, 3 m. ESE of Great Driffield, on the river Hull. There is a considerable cotton and carpet manufactory at this place. Pop. in 1831, 378; in 1841, 242.

WANSTEAD, a parish in Essex, 7 m. NE of London, on the river Rodon, or Roding. The p. contains many handsome houses,—the abodes of opulent merchants and tradesmen of London, and a large orphan asylum. Area 2,004 acres. Pop. in 1831, 1,403; in 1851, 2,207.

WANSTROW, a parish in Somersetshire, 5 m. N by E of Bruton, including the hamlet of Weston. Area 2,054 acres. Pop. in 1831, 410; in 1851, 472.

WANTAGE, a parish and market-town of Berks, 60 m. W of London, on a branch of the river Ock. A branch of the Wilts and Berks canal, affording a communication with Bath, Bristol, and London, comes up to the town; and the Great Western railway passes about 3 m. to the N of it, and has an intermediate station here 63¾ m. distant from the London terminus, and 7½ m. from Shrivenham. The p. comprises the hamlets of Charlton and Grove. Area 7,530 acres. Pop. in 1831, 3,282; in 1851, 3,860. The town, which is situated on the margin of the Vale of White-horse, is irregularly built. The inhabitants are principally employed in the manufacture of coarse cloth and twine, and in the flour and malt trade. W. was a royal seat in the Saxon times, and appears to have been then of some consequence. King Alfred was born here in 849.

WANTESDEN, a parish of Suffolk, 4½ m. NW by W of Orford. Area 2,126 acres. Pop. 107.

WANTSUM, a branch of the river Stour, in Kent. See KENT.

WANTZENAU, a town of France, in the dep. of Bas-Rhin, 6 m. NE of Strasburg, at the influx of the Ill into the Rhine. Pop. 2,343.

WANZER (GROSS and KLEIN), two villages of Prussia, in the reg. of Magdeburg, circle of Oesterburg. United pop. 584.

WANZLEBEN, a town of Prussian Saxony, in the reg. and 10 m. WSW of Magdeburg. Pop. 3,200. It has breweries and flour-mills.—In the immediate vicinity is the v. of Klein-W., with a pop. of 508.

WAPAKONETTA, or WAPAGHKANETTA, a town of Allen co., in Ohio, U. S., on the Auglaise river, at the head of navigation, 7 m. SE of Taway. Pop. 504.

WAPELLO, a county in the SE of Iowa, U. S. Area 432 sq. m. Pop. in 1850, 8,471. Its cap. is Ottamwa.—Also a village of Louisa co., in Iowa, 30 m. SE by S of Iowa. Pop. 1,000.

WAPENBURY, a parish of Warwickshire, 5 m. NNW of Southam. Area 1,550 acres. Pop. 271.

WAPESSAGA, a lake of Canada, in N lat. 48° 10′, W long. 71° 40′.

WAPITWAGO ISLANDS, a cluster of islands near the S coast of Labrador, in N lat. 50° 4′.

WAPLEY, or WAPELEY, a parish of Gloucestershire, 2 m. SW by S of Chipping Sodbury. Area 2,418 acres. Pop. with Codrington, in 1851, 305.

WAPNO, a village of Bohemia, in the reg. of Gitschin, circle of Chlumetz. Pop. 260.

WAPPENHAM, a parish of Northamptonshire, 5 m. WSW of Towcester. Area 2,980 acres. Pop., with Astwell, in 1831, 458; in 1851, 599.

WAPPING, a parish in the Tower division, Middlesex, 2 m. ESE of St. Paul's cathedral, forming an extensive portion of the metropolis, and sometimes called the port of London. Area 80 acres. Pop. in 1831, 3,564; in 1851, 4,477. It was originally a low district overflowed by the waters of the Thames, and was not properly recovered till the reign of Queen Elizabeth, when it was drained, enclosed with walls, and converted into a kind of meadow-ground under the name of Wapping-Wash. Besides numerous streets, it now contains some of the most extensive and important of the metropolitan docks and warehouses: see article LONDON. The inhabitants are chiefly employed in various manufactures connected with shipping.

WAPPINGER'S CREEK, a river of the state of New York, U. S., which runs into the Hudson, 8 m. S of Poughkeepsie. Length 33 m.

WAPPO, a small port on the Grain coast of Africa.

WAPPOCOMO, a river of Virginia, U. S., which runs into the Potomac, 9 m. ESE of Fort Cumberland.

WAPUWAGAN ISLANDS, a cluster of islands near the coast of Labrador, in N lat. 50° 2′

WARA, or HARA, a city of Central Africa, the capital of the kingdom of Borgu or Wadai, in about N lat. 13° 30′. It is a well-built town.

WARADIN. See GROSSWARDEIN.

WARANG. See FORMOSA.

WARANKUL, or WARANGOL, a city of Hindostan, in the prov. of Hyderabad, belonging to the Nizam, in N lat. 17° 54′. It was formerly the capital of the kingdom of Telingana, and was first taken by the Mahommedans in about the year 1323.

WARASDIN, a county of the Austrian states, in Croatia, having Military Hungary on the NE; the com. of Agram on the E; and Syria and Illyria on the SW and W. Its area is about 720 sq. m.. Its population, about 158,000 in number, are partly Catholics, partly of the Greek church. The river Drave forms the N boundary of the prov. The other rivers are the Socta, Kropina, Behnya, and Plitvitza.

WARASDIN, the capital of the above county, is situated on the N bank of the Drave, in N lat. 46° 18′ 20″, 38 m. NNE of Agram. The town proper is small, but the suburbs, being more extensive, the total population is about 9,000. The chief public structure is the court-house, or place of meeting of the county diets. Its traffic in silk, wine, and vinegar is pretty active. In the neighbourhood there are warm mineral springs.

WARASDIN (GENERALATE OF), a district of Croatia, adjoining to Slavonia, and separated from Hungary by the Drave. More extensive, but less populous, than the county of the same name, this district contains 1,440 sq. m., with only 70,240 inhabitants. Its physical aspect is chiefly that of a great valley, surrounded by mountains sloping towards the Drave. Its cap. is Belovar.

WARBERG. See OURBERG.

WARBERG, a village of Brunswick, in the circle of Konigslutter. Pop. 560.

WARBLETON, a parish of Sussex, 6½ m. N by E of Haylsham. Area 5,763 acres. Pop. 1,509.

WARBLINGTON, a parish of Southamptonshire, 1 m. ESE of Havant. Area 3,848 acres. Pop. in 1831, 2,118; in 1851, 2,302.

WARBOROUGH, a parish of Oxfordshire, 10½ m. SE by S of Oxford. Area 1,673 acres. Pop. 729.

WARBOYS, a parish of Huntingdonshire, 4 m. SSE of Ramsay. Area 8,300 acres. Pop. 1,996.

WARBSEN, a village of Brunswick, in the circle of Holzminden, bail. of Zorstbuche. Pop. 320.

WARBSTOW, a parish of Cornwall, 8½ m. NE of Camelford. Area 4,104 acres. Pop. in 1851, 470.

WARBURG, a town of Prussian Westphalia, on the l. bank of the Dymel, 22 m. SE of Paderborn. Pop. 3,925. It has linen manufactories.—The circle of W. has an area of 9·36 German sq. m. Pop. 34,668.

WARBURTON, a parish of the co. of Chester, 9½ m. NNW of Nether Knutsford. Area 1,747 acres. Pop. in 1831, 510; in 1851, 489.

WARCOING, a commune and village of Belgium, in the prov. of Hainault, on the Scheldt, 32 m. WNW of Meux. Pop. 1,100.

WARCOP, a parish of Westmoreland, 3 m. W by N of Brough. Area 10,020 acres. Pop. in 1851, 740.

WARD, a rivulet of co. Dublin, an affluent of the Swords or Broadmeadow.—Also a parish in co. Dublin, 4½ m. NNW of Finglass. Area 1,349 acres. Pop. in 1831, 251; in 1851, 128.

WARD, a township of Worcester co., Massachusetts, U. S., 45 m. WSW of Boston.

WARD'S CREEK, a river of Virginia, U. S., which runs into James river, in N lat. 37° 10′.—Also a river of Maryland, which runs into the Chesapeake, in N lat. 38° 8′.

WARDE, a town of Denmark, near the W coast of Jutland, on a river of the same name, 20 m. NNW of Ribe. Pop. 1,600. It has manufactories of pottery.

WARDE (POINT), a cape on the W coast of North America, at the upper end of Prince Ernest's sound, in N lat. 56° 9′.

WARDEIN. See GROSSWARDEIN.

WARDEN, a parish of Kent, 6 m. E of Queenborough. Area 796 acres. Pop. in 1831, 27; in 1851, 58.—Also a township in Northumberland, 3 m. NNW of Hexham. Area 3,122 acres. Pop. 646.

WARDEN. See CHIPPING-WARDEN.

WARDEN (OLD), a parish in Bedfordshire, 3½ m. W by S of Biggleswade. Area 3,330 acres. Pop. in 1831, 660; in 1851, 627.

WARDER, a village of Denmark, in the duchy of Holstein, on a small lake of the same name, to the NW of Segeberg, in N lat. 53° 58′. Pop. 350.

WARDINGTON, a chapelry in the p. of Cropredy, Oxfordshire, 4½ m. NE by N of Banbury. Area 2,600 acres. Pop. in 1851, 862.

WARDLAW, a hill of Selkirkshire, in the p. of Etterick, elevated 1,986 ft. above the level of the sea.

WARDLE, a township in the p. of Banbury, in Lancashire, 3 m. NNE of Rochdale. Pop. with Waerdle, in 1831, 6,754; in 1851, 7,855.—Also a township in the p. of Nantwich, co. Chester, 5 m. NW of Nantwich. Area 1,036 acres. Pop. 178.

WARDLEWORTH, a township in the p. of Rochdale, Lancashire. Pop. in 1851, 14,103.

WARDLEY, a parish of Rutlandshire, 3 m. WSW of Uppingham. Area 1,550 acres. Pop. 59.

WARDLOW, a township in the p. of Hope, Derbyshire, 2 m. E by S of Tideswell. Area 629 acres. Pop. in 1831, 149; in 1851, 191.

WARDOE, an island on the W coast of Norway, in the bishopric of Drontheim, 70 m. E by S of Hammerfest. Pop. 200.

WARDOEHUUS, the chief town of the above, in N lat. 70° 23′, E long. 31° 7′. It is inhabited by fishermen.

WARDSBOROUGH, a township of Windham co., Vermont, U. S., 20 m. NE of Bennington. Pop. 1,125.

WARE, a parish and town in Hertfordshire, 3 m. ENE of Hertford, and 21 m. N of London, on the river Lea, which is here navigable. Area of p., 4,700 acres. Pop. in 1801, 2,950; in 1831, 4,214; in 1851, 5,088.—The town, situated on the river Lea, which flows here with considerable rapidity, consists of several streets, one of which is about a mile in length. The principal trade consists in malt and corn, which is carried on to a great extent. In the neighbourhood are springs, which, under the superintendence of the New river company, supply the metropolis with a portion of its water.

WARE, or WEAR-GIFFORD, a parish in Devonshire, 2½ m. NNW of Great Torrington. Area 1,587 acres. Pop. in 1831, 547; in 1851, 551.

WARE, a county in the SE of Georgia, U. S., intersected by the Santilla river. Area 1,652 acres. Pop. in 1850, 3,888. Its cap. is Waresboro'—Also a township of Hampshire co., Massachusetts, 60 m. W of Boston. Pop. in 1840, 1,890; in 1850, 3,785. Also a river of Worcester co., Massachusetts, which runs SW, and unites with the Chiccapee.—Also a river of Virginia, which runs into the Chesapeake, in N lat. 37° 25′.

WAREBRIDGE. See WADEBRIDGE.

WAREHAM, a borough and market-town, containing the three parishes of the Holy Trinity, Lady-St.-Mary and Out-parish, and St. Martins, with the liberty of Stoborough, in the S division of Dorset, 18½ m. ESE of Dorchester. Area of p., 8,366 acres. Pop. in 1801, 1,627; in 1831, 2,325; in 1851, 3,078.—The town, situated between the mouths of the rivers Frome and Piddle, is laid out with considerable regularity. It consists chiefly of four principal streets diverging at right angles, and several smaller ones. The S entrance to the town is over a bridge of 5 arches which crosses the Frome. The N entrance, in like manner, crosses a bridge of 3 arches over the Piddle. A great proportion of the females of the town are employed in the manufacture of straw-plait, stockings, and shirt-buttons. Pipe-clay is found in the neighbourhood in great abundance; about 10,000 tons of it are annually exported to London, Hull, Liverpool, and Glasgow. W. was formerly a port of some repute, but its harbour has been destroyed by the retreat of the sea.—The parl. borough, which returns one member to parliament, had a pop. of 7,218 in 1851. Electors in 1852, 418.

WAREHAM, a township of Plymouth co., Massachusetts, U. S., at the head of Buzzard's bay, 17 m. S of Plymouth, and 45 m. SSE of Boston. Pop. in 1840, 2,002; in 1850, 3,186.

WAREHORNE, a parish of Kent, 7½ m. S by W of Ashford. Area 2,870 acres. Pop. 507.

WAREMME, a town of Belgium, in the prov. of Liege, on the Jaar or Geer, 13 m. W by N of Liege. Pop. 1,488.

WAREN, a town of Mecklenburg-Schwerin, in the principality of Schwerin, to the N of the Muritz-see. Pop. 4,624. It has woollen and linen manufactures.

WARENDORF, a town of Prussian Westphalia, on the Ems, 15 m. E of Munster. Pop. 4,480. It is a place of active manufacturing industry.

WARESLEY, a parish of Huntingdonshire, 5½ m. SE of St. Neot's. Area 1,979 acres. Pop. 295.

WARFFUM, a village of Holland, in the prov. and 12 m. N of Groningen. Pop. 1,200.

WARFIELD, a parish of Berkshire, 4½ m. ENE of Workingham. Area 3,239 acres. Pop. 1,374.

WARFORD (GREAT), a township in Alderley p., Cheshire, 5 m. E by S of Nether Knutsford. Area 1,272 acres. Pop. in 1831, 349; in 1851, 348.

WARGITTEN, a village of Prussia, in the reg. of Konigsberg, circle of Heilsberg. Pop. 370.

WARGOLSHAUSEN, a village of Bavaria, in the circle of Lower Franconia, landgraviate of Neustadt. Pop. 364.

WARGOWO, a village of Prussia, in the reg. of Posen, circle of Obornik. Pop. 210.

WARGRAVE, a parish and village of Berkshire, 6½ m. NE by E of Reading. Area of p., 4,314 acres. Pop. in 1831, 1,423; in 1851, 1,773. The v. was formerly a market-town, and has a ferry over the river Thames.

WARHAM (ALL-SAINTS), a parish of Norfolk, 3 m. N of Walsingham. Area 1,774 acres. Pop. 343.

WARHAM (ST. MARY), a parish adjoining the above. Area 3,066 acres. Pop. 61.

WARHEM, a commune and v. of France, in the dep. of Nord, 6 m. SE of Dunkirk. Pop. 2,500.

WARI, a country of Western Africa, situated to the SE of Benin, near the river Formosa. The surface is covered to a great extent with an impenetrable forest growing upon land which forms a complete marsh, being covered with water, even in the dry season, to nearly the depth of a foot. The capital is situated on an island in the river.

WARIN, a town of Mecklenburg-Schwerin, in the principality and to the SW of Schwerin. Pop. 1,038.

WARING, a village of Austria, a little to the W of Vienna, with a number of villas.

WARINGSTOWN, a town in the p. of Donaghcloney, co. Down, 2½ m. SSW of Magheralin. Pop. in 1851, 956. There is a considerable manufactory of cambrics here.

WARISOULX, a canton and town of Belgium, in the prov. and arrond. of Namur. Pop. 2,024.

WARK, a parish and village in Northumberland, situated on the North Tyne, 12 m. NW of Hexham. Area of p., 22,986 acres. Pop. in 1851, 865.

WARKA, a town of Poland, 30 m. S of Warsaw, on the river Pilica. Pop. 2,000.

WARKEN, a village of Holland, in the bishopric of Diekirch, in Luxemburg. Pop. 200.

WARKLEIGH, a parish of Devonshire, 5½ m. WSW of South Molton. Area 2,451 acres. Pop. in 1831, 283; in 1841, 337.

WARKOTSCH, a village of Prussia, in the reg. of Breslau, circle of Strehlen. Pop. 200.

WARKSBURN, a township of Northumberland, 13 m. NW of Hexham. Area 9,929 acres. Pop. in 1831, 278; in 1851, 217.

WARKTON, a parish of Northamptonshire, 2 m. ENE of Kettering. Area 1,810 acres. Pop. 309.

WARKULLEN, a village of Prussia, in the reg. and circle of Gumbinnen. Pop. 280.

WARKWORTH, a parish in Northumberland, 7 m. SE of Alnwick, on the river Coquet, comprising the townships of Amble, Acklington, Acklington-Park, Birling, Botherick, Bullockshall, High and Low Buston, East and West Chivington, Gloster-hill, Hauxley, Hadston, Morrick, Starton-Grange, Walk-Mill, W., and the liberty of Togston. Area 19,365 acres. Pop. in 1831, 2,478; in 1851, 4,439. There are quarries within the parish of coal and limestone, and two woollen mills. The v., near the sea, and almost surrounded by the river Coquet, which is here crossed by an ancient stone-bridge, consists principally of one street. Breakwater piers are erected at the mouth of the Coquet. The castle of W. stands on the S side of the river, and about 1 m. from its mouth, and presents a more perfect representation of the stronghold of a feudal baron of the 14th and 15th centuries, than any other castle on the borders. The greater part of the outer walls, enclosing the castle-yard, are yet standing; and the walls of the keep and its adjoining towers are, for the most part in comparatively good repair.

WARLEGGAN, a parish of Cumberland, 5½ m. ENE of Bodmin. Area 2,055 acres. Pop. 295.

WARLEY, a township in the p. and 3 m. W of Halifax, Yorkshire. Area 3,980 acres. Pop. in 1831, 5,685; in 1851, 6,408. There are woollen mills and iron foundries here.

WARLEY (GREAT and LITTLE), adjoining parishes of Essex, 3½ m. S of Brentwood. Area of Great W., 2,793 acres. Pop. 952. Area of Little W., 1,051 acres. Pop. 344.

WARLEY-WIGORN, a township of the p. of Hales-Owen, Worcestershire, 3 m. N by E of Hales-Owen. Pop. in 1851, 989.

WARLIN, a village of Mecklenburg-Strelitz, in the circle of Stargard. Pop. 240.

WARLINGHAM, a parish in Surrey, 5 m. SSE of Croydon. Area 1,777 acres. Pop. 505.

WARLOW, a village of Mecklenburg-Schwerin, in the bail. of Neustadt. Pop. 490.—Also a village of Prussia, in the reg. of Oppeln, circle of Lublinitz. Pop. 270.

WARLOY-BAILLON, a canton and v. of France, in the dep. of Somme, arrond. and 19 m. NE of Amiens. Pop. 2,742.

WARMAERDE, a canton and v. of Belgium, in W. Flanders, arrond. of Courtray. Pop. 1,075.

WARMBRUNN, a town of Prussian Silesia, in the reg. of Liegnitz, among the Riesengebirge mountains, 60 m. WSW of Breslau. Pop. 1,930. It has warm baths.

WARMENSTEINACH, a village of Bavaria, in Upper Franconia, in the landgraviate of Weidenberg. Pop. 760.

WARMFIELD, a parish in the W. R. of Yorkshire, 3 m. E of Wakefield. Area 2,618 acres. Pop. in 1831, 995; in 1851, 960.

WARMINGHAM, or WARMICHAM, a parish of Cheshire, 3½ m. W of Sandbach. Area 4,730 acres. Pop. in 1831, 1,167; in 1851, 1,271.

WARMINGHURST, a parish of Sussex, 5½ m. NW of Steyning. Area 1,051 acres. Pop. 116.

WARMINGTON, a parish of Warwickshire, 5½ m. ESE of Kineton. Area 1,750 acres. Pop. in 1831, 470; in 1851, 523.—Also a parish of Northamptonshire, on the river Nen, 3 m. NE of Oundle. Area 3,150 acres. Pop. in 1831, 617; in 1851, 671.

WARMINSTER, a parish and market-town in Wilts, 20 m. WNW of Salisbury. Area 6,370 acres. Pop. in 1831, 6,115; in 1851, 6,285. The town, situated on the river Willey, consists principally of one street about 1 mile in length. The principal trade is that of malting. The clothing manufacture, at one time carried on here extensively, has declined of late years. Camden supposes this town to have been the *Verlucio* of the Romans.

WARMINSTER, a township of Bucks co., Pennsylvania, U. S.—Also a township of Nelson co., Virginia, on James river, 70 m. above Richmond.

WARMISRIED, a village of Bavaria, in the prov. of Suabia, landg. of Mindelheim. Pop. 340.

WARMOND, a town of Holland, in the prov. of S. Holland, 4 m. N of Leyden. Pop. 1,100.

WARMSDORF, a bailiwick and village of Anhalt-Dessau, on the Wipper. Pop. 1,100.

WARMSROTH, a village of Prussia, in the reg. of Coblenz, circle of Kreuznach. Pop. 220.

WARM-SPRING MOUNTAINS, a ridge of the Alleghany mountains, in Bath and Pendleton cos., Virginia, U. S., remarkable for warm springs. Warm Spring, in Bath co., issues in a very bold stream, sufficient to keep its basin, which is 30 ft. in diam., at the warmth of 96°.

WARM-SPRINGS, a village of Buncombe co., N. Carolina, U. S., 237 m. W of Raleigh.

WARMSWORTH, a parish of the W. R. of Yorkshire, 3 m. SW of Doncaster. Area 1,042 acres. Pop. in 1831, 362; in 1851, 389.

WARMWELL, a parish of Dorsetshire, 5½ m. SE by S of Dorchester. Area 1,531 acres. Pop. 149.

WARMUNKOWITZ, a village of Prussia, in the regency of Oppeln, circle of Gross-Strelitz. Pop. 250.

WARNA, a river of Mecklenburg, which flows W, NW, and N, and falls into the sea near Warnemunde.

WARNANT, a village of Belgium, in the prov. of Namur, arrond. of Dinant, on the Maaz. Pop. 308.

WARNBOROUGH (SOUTH), a parish of Southamptonshire, 3 m. SW by S of Odiham. Area 2,569 acres. Pop. in 1831, 374; in 1851, 410.

WARNDON, or WARMEDON, a parish of Worcestershire, 3 m. NE by E of Worcester. Area 988 acres. Pop. in 1831, 171; in 1851, 193.

WARNE, a river in Northumberland, which falls into the German ocean, about 4 m. S of Holy island.

WARNEMUNDE, a town and fort of Germany, in the duchy of Mecklenburg-Schwerin, district and 6 m. NW of Rostock, in N lat. 54° 10′, at the mouth of the Warna. It has a small harbour, at which, in 1849, 435 vessels cleared; in 1853, 613.

WARNER, a township of Hillsborough co., New Hampshire, U. S., 17 m. WNW of Concord. Pop. 2,038.—Also a river of New Hampshire, which runs into the Contoocook.

WARNETON, a town of Belgium, in the prov. of W. Flanders, on the river Lys, 10 m. SE of Ypres. Pop. 6,000. It has considerable manufactures of linen and lace, and extensive distilleries.

WARNFORD, a parish of Southamptonshire, 6 m. NE of Bishop's Waltham. Area 3,057 acres. Pop. in 1831, 418; in 1851, 414.—Also a hamlet in the p. of Bamborough, Northumberland.

WARNHAM, a parish of Sussex, 3 m. NNW of Horsham. Area, 4,920 acres. Pop. in 1851, 1,016.

WARNING (MOUNT), a mountain of New South Wales, in S lat. 28° 24′. Alt. 3,300 ft.

WARNITZ, a village of Prussia, in the regency of Frankfort, circle of Konigsberg. Pop. 417.—Also a v. in the reg. of Stettin, circle of Pyritz. Pop. 240.

WARNITZA, a village of Austria, in Steiermark, in the circle of Marburg. Pop. 370.

WARNOW, a village of Mecklenburg-Schwerin, in the bail. of Grevismuhlen, on the Sartover-see. Pop. 341.—Also a v. of Prussia, in the reg. of Stettin, circle of Wollin. Pop. 270.

WARNSDORF (OLD), a town of Bohemia, in the circle of Letmeritz, 4 m. SE of Romberg. Pop. 4,800.—Also a town of Moravia, in the regency of Olmutz, division of New-Titschein. Pop. 1,180.

WARNSVELD, a town of Holland, in the prov. of Gelderland, 2 m. E of Zutphen. Pop. 1,948.

WARP (NEU), a town of Prussian Pomerania, 24 m. NW of Stettin. Pop. 1,800.

WARPSGROVE, a parish of Oxfordshire, 7½ m. E of Stadhampton. Area 460 acres. Pop. 30.

WARREN, a parish of Pembrokeshire, 4 m. from Pembroke. Area 1,169 acres. Pop. in 1851, 124.

WARREN, a county in the NE of Georgia, U. S. Area 436 sq. m. Pop. in 1840, 9,789; in 1850, 12,425. Its cap. is Warrenton.—Also a co. in the NW of Pennsylvania. Area 832 sq. m. Pop. in 1840, 9,278; in 1850, 13,671. Its cap. of the same name is situated on the Alleghany, 158 m. NW of Harrisburg. Pop. in 1840, 797; in 1850, 1,013. The Alleghany is here 150 yds. wide, and the Connewango 100 yds. There is a safe and easy harbour abreast the town, at which almost all boats and rafts descending the Alleghany from Olean, or the Connewango from Chataque lake stop.—Also a co. in the NW of Illinois. Area 540 sq. m. Pop. in 1840, 6,739; in 1850, 8,176. Its cap. is Monmouth.—Also a co. in the SW of Ohio, traversed by the Little Miami. Area 400 sq. m. Pop. in 1840, 23,141. Its cap. is Lebanon.—Also a central co. of Iowa. Area 432 sq. m. Pop. in 1850, 961. Its cap. is Indianola.—Also a co. of Kentucky. Area 546 sq. m. Pop. in 1840, 15,446; in 1850, 15,128. Its cap. is Bowling-Green.—Also a co. in the W of the state of Mississippi. Area 570 sq. m. Pop. in 1840, 15,820; in 1851, 18,121. Its cap. is Vicksburg.—Also a co. in the NW of New Jersey. Area 474 sq. m. Pop. in 1840, 20,336; in 1850, 22,358. Its cap. is Belvidere.—Also a co. in the NE of North Carolina. Area 439 sq. m. Pop. in 1840, 12,919; in 1850, 13,912. Its cap. is Warrenton.—Also a central co. of Tennessee. Area 372 sq. m. Pop. in 1840, 10,803; in 1850, 10,179. Its cap. is M'Mennsville.—Also a township of Lincoln co., Maine, on St. George's river, 30 m. E by N of Wiscasset. Pop. in 1850, 2,428.—Also a township of Addison co., Vermont, 20 m. SW of Montpelier.—Also a township of Grafton co., New Hampshire, 11 m. SE of Haverhill. Pop. 872.—Also a town of Bristol co., Rhode Island, on the NE part of Narraganset bay, 10 m. S of Providence. Pop. in 1840, 2,437; in 1850, 3,103. It carries on considerable trade, and is remarkable for ship-building.—Also a township of Litchfield co., Connecticut, 9 m. W of Litchfield. Pop. in 1850, 831.—Also a township of Herkimer co., New York, 10 m. S of Herkimer. Pop. 1,756.—Also a township of Franklin co., Pennsylvania.—Also a township of Albemarle co., Virginia, on the river James, 10 m. NE of Warminster.—Also a township of Cuyahoga co., Ohio, 8 m. ESE of Cleveland. Pop. 1,408.

WARRENBURG, a township of Warren co., New York, U. S., on Scroon river, 7 m. NW of Caldwell. Pop. 1,874.—Also a township of Green co., Tennessee. Pop. 100.

WARREN'S ISLAND, a high island in the N. Pacific ocean, in N lat. 55° 56′, near the middle of the entrance into the Duke of Clarence's straits, so named by Vancouver, in honour of Sir John Borlase Warren.

WARRENPOINT, a parish containing a small town of the same name, in co. Down. Area 1,178 acres. Pop. in 1831, 1,573; in 1851, 2 152.—The sea-port town stands at the embouchure of the Narrow or Newry into the head of Lough Carlingford, 2 m. W of Rostrevor. It has been almost wholly built since 1780; and owes its prosperity, partly to the beauty of its situation, partly to its numerous attractions for sea-bathers and occasional visitors, and partly to its facilities for serving as a sub-port to Newry. Pop. in 1831, 1,856; in 1851, 1,769.

WARRENSTOWN, a village in the p. of Knockmark, co. Meath, 2¾ m. WNW of Dunshaughlin.

WARRENSVILLE, a township of Du-Page co., in Illinois, U. S., 158 m. NE by N of Springfield.

WARRENTON, the capital of Fauquier co., Virginia, U. S., 40 m. NNW of Fredericsburg. Pop. 1,300.—Also the capital of Warren co., N. Carolina, 56 m. NNE of Raleigh. Pop. 750.—Also the cap. of Marshall co., in Alabama, 137 m. N of Montgomery.—Also the cap. of Warren co., Mississippi, on the E bank of the Mississippi, 18 m. below Walnut hills.—Also a township of Warren co., Georgia, 55 m. NNE of Milledgeville.—Also a township of Jefferson co., Ohio, on the Ohio, 13 m. below Steubenville. Pop. 300.

WARRI, or SAWUNT-WARRI, a district of Hindostan, in the prov. of Bejapur, situated between the sea and the Western Ghauts. It is about 40 m. in length, by 25 m. in breadth, forming what in old maps is designated 'the pirate coast.' For a number of years, the inhabitants of this rocky sterile stripe of coast land were excessively troublesome to British commerce, till at length, in 1818, an expedition entered the country, took the capital and the fortresses of Rairi and Newti, and put an end to the government.

WARRICK, a county in the SW of Indiana, U. S., watered by the Ohio. Area 370 sq. m. Pop. 8,911. Its cap. is Bonneville.

WARRINGTON, a parish and parliamentary borough in the hund. of West Derby, co.-palatine of Lancaster, on the river Mersey, which is here crossed by a stone-bridge, and in the line of the Grand Junction section of the Great North-Western railway, equidistant, 16 m., from Manchester and Liverpool, by railway. Besides the township of W., the parish contains the chapelry of Burtonwood, and the townships of Poulton-with-Fearnhead, Rixton, and Woolston-with-Martinscroft. Area of p., 12,168 acres. Pop. in 1831, 19,155; in 1851, 23,561. Area of township, 2,600 acres.—The town consists chiefly of several streets, some of which are open, and contain handsome buildings; while others are long and narrow. Amongst the public buildings, besides the places of worship, are a town-hall, market-hall, cloth-hall, assembly-rooms, and theatre. The position of the town is highly favourable for trade. Besides its water communication through the Mersey from Liverpool in vessels of 80 to 100 tons burthen; and through the Manchester and Irwell navigation and the Sankey canals, the formation first of the Manchester and Liverpool, and then of the Grand Junction railways, have opened up unlimited communication in every direction. In the 17th cent. it was noted for its coarse linens and checks. These manufactures were afterwards superseded by that of sail-cloth; but latterly fustian weaving has superseded all others here as the staple trade. W. also possesses manufactures of hardware goods, and the files made here are celebrated throughout Europe. There are also extensive tanneries and soap manufactories; and ale is made both for consumption and export. Under the reform act W. returns one member to parliament. The borough boundaries include the respective townships of Warrington and Latchford, and also those two detached portions of the township of Thelwall which lie between the boundary of the township of Latchford and the river Mersey. The number of electors registered in 1837 was 635; in 1852, 701. W. is supposed by some to have been a British town, even before the Roman invasion, and to have been converted by Agricola into a Roman station. In 1648 it was the scene of an obstinately fought contest between General Lambert, commanding the forces of the parliament, and a body of Scottish troops, in retreat after the battle of Ribbledale, under the duke of Hamilton. The battle, in which 1,000 men were slain, terminated in favour of the parliamentarians, who again, in 1651, met the royalists here under Charles II., and again defeated them with great loss.

WARRINGTON, a township of Bucks co., Pennsylvania, U. S., 90 m. E of Harrisburg.—Also a township of York co., Pennsylvania.—Also a township of Escambia co., Florida, 182 m. W by N of Tallahassee.

WARRIOR'S MARK, a township of Huntingdon co., Pennsylvania, U. S., 60 m. WNW of Harrisburg. Pop. 2,000. There are iron-works here.

WARRUNA, a town of Hindostan, prov. of Berar, in N lat. 19° 37′.

WARRUS, a village of Prussia, in the reg. of Gumbinnen, circle of Heidekrug. Pop. 210.

WARSAW, the WARZAWA of the Poles, and WARSCHAU of the Germans, a large city, formerly the capital of the whole of Poland, and now of the Russian kingdom of that name, situated on the l. bank of the Vistula, in N lat. 52° 13′ 5″, E long. 21° 2′ 14″, 320 m. E of Berlin. It covers a great extent of ground, the length of the town and suburbs being between 3 and 4 m.; its breadth between 2 and 3 m.; but large spaces are occupied by gardens. A rampart and fosse surround the city, and the Russians have constructed a large citadel for its defence. The pop., it is said, before it lost—in 1795—its character of capital of the whole of Poland, exceeded 90,000. In subsequent years, when the Polish court was feebly replaced by a Prussian or Saxon viceroy, it fell to 70,000; when again, after 1815, rendered the resort of a legislative body, the pop. increased; and in 1851, it was returned at 164,115, of whom one-fifth were Jews. The garrison usually numbers 20,000.

The town is divided into an old and a new quarter, exclusive of four suburbs, of which one, Praga, lies on the r. bank of the river. The old town, consisting of one main street, with some smaller streets joining it on either side, is miserably built, with the exception of a few public edifices, such as the council-house, the cathedral of St. John, and a collegiate church, belonging formerly to the Jesuits. The new town extends along the banks of the Vistula to the extent of nearly 3 m. It contains several churches, public buildings, and barracks, an hotel-de-ville, and the Jardin-de-Saxe. The castle stands near the river, and is a large quadrangle, with halls where the diet and senate held their sittings. "W. is, upon the whole," says a recent writer, "a splendid city. Viewed from Praga, as it slopes upwards on the l. bank of the Vistula, the general effect is very imposing; and a nearer examination of its numerous churches and fine public buildings, which there is plenty of room to see to advantage, as well as of the palaces of the nobility, confirms the first favourable impressions of its architectural magnificence. A closer acquaintance, however, will dispel much of the pleasure with which a stranger will be disposed to regard the external appearance of the streets and buildings. On passing the arched gateway which pierces many a noble-looking front, he will be surprised to see the mean and irregular erections which compose the rest of the quadrangle, and his sight will not be the only sense which will receive an unpleasant shock. Though admirably situated for drainage, there being hardly a street in the city in which there is not a good natural fall, W. has no sewers. The streets are tolerably well paved, but seldom swept, and gas has not yet been introduced." The suburbs are less badly built than the old town. Praga is memorable in history for the assaults made on it in the autumn of 1794, by the Russian army under Suwarrow. Of the castles or mansions in the vicinity, one of the most remarkable is that which was once the residence of Sobieski, and is still remarked for its beautiful gardens and statues. About 2 m. W of the town is the village and field of Wola, the scene, in former ages, of the assemblage of the national diet. The course of the Vistula is here from S to N; its depth is less than that of the Thames at London; but its breadth is fully equal. The bridge across it from the suburb of Praga is a floating bridge, 3,000 ft. in length.

Of the public establishments of Warsaw, the principal are the offices of government, especially the government palace, an immense building of recent construction. Towards the end of 1816, there was established here a university, consisting, like those of Germany, of classes in theology, law, philosophy, and several of the sciences, including political economy; but this institution has been suppressed. There are schools for surgery and drawing, a lyceum or high school, a college for the sons of the Catholic nobility, a military academy, and a Piarist college.—The situation of W. is favourable to trade; and the improvements in the river navigation, and the opening up a communication by railroad with Berlin and Leipzig, bid fair to increase its commercial activity. The Vistula, here near the middle of its course, is navigable to a great extent upwards as well as downwards: at some seasons, however, great inconvenience is experienced from the extent of its inundations, and at others from the shifting of sand-banks. The middle of summer, and during the interval that the channel is full without overflow, it is computed that nearly 100 boats or barges, laden with the produce of the country, corn, spirits, and wine, are daily sent down its stream. The manufactures consist of cotton and woollen stuffs, jewellery, paper, chemical substances, carriages, harness, and carpeting. Since 1817, two annual fairs have been established here, on the plan of those of Frankfort, and other towns of Germany.

History.] W. is a town of old date, but was a place of insignificance till the annexation of Lithuania to Poland; after which the territory of the republic being extended to the west, Cracow was no longer sufficiently central to be the capital. In 1566, the diet was transferred from the old to the new capital of Poland. In the war with Sweden, in the middle of the 17th cent., W. was occupied by the invaders. When Charles XII. advanced in 1703 to W., it surrendered to him without opposition. In 1793, the Russian garrison that occupied it were expelled by the Poles, on receiving intelligence of the success of Kosciusko near Cracow. That gallant leader, when obliged to change the scene of contest, retreated on W., and defended it with success against the Prussians, during the summer of 1794, obliging them eventually to raise the siege. A different fate awaited Warsaw on the arrival of Suwarrow and the Russians. Praga being taken by assault, and delivered to pillage, the capital submitted without opposition. On the final partition of Poland in 1795, this part of the country fell to the share of Prussia, and W. had no other rank than that of capital of a province, until the end of 1806, when the overthrow of Prussia led to the formation of an independent state, called the duchy of Warsaw. Of this state it continued the capital, until the evacuation of Poland by the French in January 1813. Since 1815, it has, in a manner, retained its character of a capital, being the residence of a viceroy representing the emperor of Russia.

WARSAW, a township of Hancock co., in Illinois, U. S., 100 m. WNW of Springfield.—Also the cap. of Kosciusko co., in Indiana, 100 m. N by E of Indianapolis. Pop. 300.—Also the cap. of Gallatin co., in Kentucky, on the Ohio, 36 m. N of Frankfort. Pop. 700.—Also the cap. of Wyoming co., in New York, 200 m. W of Albany. Pop. 2,624.

WARSAW, a village of Mecklenburg-Schwerin, in the circle of Mendisch, bail. of Neukaiden. Pop. 210.—Also a v. of Mecklenburg, in the bail. of Hagenow, on the Tude. Pop. 280.—Also a v. of Prussia, reg. of Stettin, circle of Randow. Pop. 400.

WARSLEBEN, a village of Prussia, in the reg. of Gumbinnen, circle of Niederung. Pop. 260.

WARSLOW, a township in the p. of Alstonefield, in Staffordshire, 6 m. ENE of Leeke.

WARSOP-CHURCH, a parish of Nottinghamshire, 5 m. NNE of Mansfield. Area 6,710 acres. Pop. in 1831, 1,203; in 1851, 1,398.

WARSTEIN, a town of Prussian Westphalia, in the gov. and 14 m. E of Arnsberg. Pop. 1,720.

WARTA, a river of Poland, which rises about 30 m. NW of the city of Cracow; flows N through Kalisch; then taking a direction W, traverses the grand-duchy of Posen, and part of Brandenburg, and joins the Oder at Custrin, on the r. bank, after a course of 400 m.—Also a town of Poland, on the Warta, 104 m. W by S of Warsaw. Pop. 2,650.

WARTAU, a town of Switzerland, in the cant. of St. Gall, 20 m. N of Sargans. Pop. 2,087.

WARTBERG, a village of Styria, in the circle of Bruck, on the l. bank of the Murz, in N lat. 47° 31′.

WARTBERG, or SZENCZ, a town of Hungary, 13 m. ENE of Presburg. Pop. 1,700.

WARTENBERG (POLNISCH), a town of Prussian Silesia, 32 m. ENE of Breslau. Pop. 2,600.

WARTENBURG, a town of East Prussia, 60 m. S of Konigsberg, on the Pisch. Pop. 3,500.

WARTER, a parish of Yorkshire, 5 m. E by N of Pocklington. Area 7,000 acres. Pop. 488.

WARTHAUSEN, a village of Würtemberg, in the Danube circle, on the Ritz. Pop. 451.

WARTHE, a village of Prussia, in the reg. of Stettin, circle of Usedom-Wollin. Pop. 220.—Also a v. in the reg. of Potsdam, circle of Templin. Pop. 200.

WARTHILL, a parish of the N. R. of Yorkshire, $5\frac{1}{2}$ m. NE by E of York. Area 860 acres. Pop. 169.

WARTHOLM, a small island in Orkney, near S. Ronaldshay.

WARTLING, a parish of Sussex, $4\frac{1}{2}$ m. E by S of Haylsham. Area 4,736 acres. Pop. 1,039.

WARTON, a parish of Lancashire, 8 m. N by E of Lancaster. Area 11,141 acres. Pop. in 1851, 2,099.—Also a township in the parish and 3 m. from Kirkham. Area 3,939 acres. Pop. 473.—Also a hamlet in the p. of Rothbury in Northumberland, 14 m. SW by W of Alnwick. Area 624 acres Pop. in 1851, 59.

WARTSENSTADT, a village of Prussia, in the bail. of Wohldenburg. Pop. 206.

WARUPOHNEN, a village of Prussia, in the reg. of Gumbinnen, circle of Pillkullen. Pop. 300.

WARWASCHAU, a village of Bohemia, in the reg. of Pilsen, circle of Mirowice. Pop. 880.

WARWICK, a parliamentary borough, the capital and locally situated in the S division of the co. of Warwick, 90 m. NW of London, 20 m. SE of Birmingham, on the banks of the Avon, and in the line of the Birmingham and Warwick, and Warwick and Napton canal. The Warwick and Leamington railway, running from Warwick and Leamington to Coventry, 8½ m. in length, forms a junction with the London and Birmingham section of the Great North-Western line. Area of the borough, 5,360 acres. Pop. in 1801, 5,592; in 1831, 9,109; in 1851, 10,973. The town is finely situated, in the midst of a richly diversified country, on a rocky eminence, which rises abruptly from the N side of the Avon. About the close of the 17th cent. it was nearly destroyed by fire, and that part which has been rebuilt is regular and well-arranged, and contrasts strikingly with the small part of the ancient town which still remains. The principal street, which runs parallel with the river, and at a short distance from it, is of considerable length, and contains many handsome houses. It is intersected by a wide street and several smaller ones. At each end of the main street is a gateway; that on the W being surmounted by an ancient chapel. The court-house, erected shortly after the fire, is a respectable building. The county-hall is an elegant structure in the Corinthian order of Grecian architecture. The county-jail is a substantial modern fabric, surrounded by a brick wall 23 ft. high. The environs of the town are adorned with several elegant modern villas. The castle, exhibiting, as it does, the beau ideal of a feudal fortress, is in point of grandeur and completeness second to none in the kingdom; and its accompaniments are in true keeping with the character of the building. It is situated on the SE side of the town, on the solid rock, at an elevation of nearly 100 ft. higher than the level of the river, which washes the foot of the rock, although on the N side it is even with the town. The church is a magnificent pile, in a mixed style of architecture, erected in the 14th cent. Among those parts of the original edifice which escaped the conflagration, is the beautiful Beauchamp chapel. W. returns 2 members to parliament. Electors registered in 1837, 1,013; in 1852, 723. The town is not noted for its manufactures. An iron foundry has been established, and the manufacture of hats is carried on to a considerable extent. The general trade, which is good, is greatly augmented by the proximity of the town to Leamington. In 1312, Warwick castle was repaired by the celebrated Guy, earl of Warwick, who here beheaded Piers Gaveston. During the great civil war the castle was garrisoned for the parliament. A large portion of the town was burnt, as already noticed, in 1694.

WARWICK, a parish in Cumberland, 5 m. E by N of Carlisle, on the river Eden, which is here spanned by a bridge of four arches, and in the line of the Carlisle and Newcastle railway. Area 1,845 acres. Pop. in 1831, 686; in 1851, 363.

WARWICK, a township of Upper Canada, in the Western district, intersected by the N branch of Bear creek, and watered also by branches of River Aux Salles. Pop. in 1845, 1,235.

WARWICK, a county in the SE part of the state of Virginia, U. S., comprising an area of 55 sq. m., drained by affluents of James river. It has a level surface, and is tolerably fertile. Pop. in 1840, 1,456; in 1850, 1,546. Its capital, which bears the same name, is 58 m. SE by E of Richmond.—Also a township of Franklin co., in the state of Massachusetts, 68 m. WNW of Boston. It is hilly, but affords good pasturage. Pop. in 1840, 1,071; in 1850, 1,021.—Also a township of Kent co., in the state of Rhode Island, on the Stonington and Providence railroad, and 8 m. S of Providence, bounded on the E by Narraganset bay, and drained by Pawtuxet river. Pop. in 1840, 6,726; in 1850, 7,740.—Also a township of Orange co., in the state of New York, 24 m. SW of Newburg, drained by Wallkill river, and by tributaries of Passaic river. Pop. in 1840, 5,113; in 1850, 4,902.—Also a township of Bucks co., in the state of Pennsylvania, 5 m. SE of Doylestown. It has a hilly surface, and is drained by Nishaming creek. The soil is sandy loam and gravel. Pop. in 1840, 1,259.—Also a township of Lancaster co., in the same state, 9 m. N of Lancaster. It has a rolling surface, and is drained by Great Chiques, Hanmer, Cocalico, and Marovia creeks. Pop. 3,725. —Also a township of Tuscarawas co., in the state of Ohio, 7 m. S of New Pennsylvania. Pop. 846.

WARWICK-BRIDGE, a township in the p. of Wetheral, Cumberland, 5 m. E of Carlisle. Area, with Great Corby, 2,747 acres. Pop. 1,180.

WARWICK-PLAINS, a level tract of New South Wales, in the co. of Bathurst, on the river Lachlan, and 170 m. from Sydney.

WARWICKSHIRE, an inland county of England, bounded on the NE by Leicestershire; on the E by Northamptonshire; on the SE by Oxfordshire; on the SW by Gloucestershire; on the W by Worcestershire; and on the NW by Staffordshire. An insulated portion of it lies in Worcestershire, and another in Gloucestershire; while it includes within its limits a portion of Worcestershire, a few miles S of Stratford-on-Avon. The extreme length of this co. from N to S is about 50 m.; breadth 35 m. The area of the whole is 902 sq. m., or 577,280 square acres. It is subdivided into 209 parishes, containing, exclusive of the borough of Coventry, 1 county-town, Warwick; 2 parl. boroughs, Warwick, and Birmingham; and 10 market-towns, Atherstone, Alcester, Coleshill, Henley, Kineton, Nuneaton, Rugby, Southam, Stratford-on-Avon, and Sutton-Coldfield. The pop. in 1801, was 208,109; in 1831, 336,610; in 1841, 402,121; in 1851, 475,013.

Rivers.] The co. is watered by numerous streams, the principal of which are the Avon and the Tame. The Avon, which rises in Northamptonshire, enters this county near Clifton-upon-Dunsmore, runs with a tortuous course towards the W and SW, and passes into Worcestershire, a short distance below Salford. Its current is gentle, and vessels of 40 tons burden can sail up to Stratford. Its tributary streams are the Dove, the Sow, and the Leam. The Stour rises to the E of Long-Compton, and runs in a NW direction, receiving many small tributary streams, and joins the Avon a few miles below Stratford-upon-Avon. The Alne rises near Baddesley-Clinton, and joins the Avon near the SW angle of the county. The Tame rises in Staffordshire, enters the county a few miles to the N of Birmingham, runs nearly due N, and leaves the co. at Tamworth. The Blythe runs with a devious course towards the N, and falls into the Tame near Lea-Marston. W. possesses an extensive artificial navigation.

Climate, Soil, and Aspect.] The insulated position of this co., and its freedom from any great inequalities of surface, render the climate mild and the vegetation early. The soil is fertile and of great variety; indeed, nearly every description of soil, except that incorporated with chalk and flint, is to be met with. The only extensive commons are those of Sutton-Coldfield and Sutton-park. The highest points of land are near Corley, and in the vicinity of Pakington. A ridge on the SE, comprising the Brailes and Edge-hills, is also much elevated, and commands a variety of picturesque views. The whole county is in general well cultivated. The meadow and pasture-lands are computed at 235,000 acres, and 60,000 acres are covered with sown grasses. The tracts of permanent pasture abound chiefly on the borders of Leicestershire.

Minerals and Manufactures.] The mineral productions of the co. consist chiefly of coal, limestone, iron-stone, free-stone, and a kind of blue flag-stone. Manufactures of various descriptions are extensively carried on, particularly at Birmingham, which has long been celebrated for the variety and excellence of its hardware goods, especially muskets, swords, army accoutrements, and machinery. The manufacture of watches at Coventry has so greatly increased within the last 40 or 50 years, that the number of watches annually made there far exceeds that furnished by the metropolis. In the city of Coventry and suburbs, the ribbon trade gives employment to between 17,000 and 18,000 persons. There are considerable flax-mills in different parts of the county. Fish-hooks and needles are extensively manufactured at Alcester, and horn combs at Kenilworth. The principal exports consist of manufactured articles. See articles BIRMINGHAM and COVENTRY.

Roads and Railways.] Of the principal lines of road crossing this co., one enters near Long Compton, and, passing Shipston-on-Stour, divides into two great branches, one of which passes Warwick, Kenilworth, Coventry, and Nuneaton, leaving the county near Hinckley; the other, passing Stratford-on-Avon, Henley-in-Arden, and Birmingham, enters Staffordshire at Hackley-Brook. The London and Birmingham section of the Great North-Western railway enters the co. near Hill-Moreton, and passes through Rugby, Coventry, Hampton-in-Arden, to Birmingham. At Rugby a junction is formed with the Midland Counties line, which leaves this co. at Willy. The Birmingham and Derby Junction railway starts from the Hampton station on the London and Birmingham line, and after passing through Coleshill, Kingsbury, Tamworth, Walton-on-Trent, and Burton-on-Trent, forms a junction with the Midland and North Midland Counties line at Derby. The Birmingham and Gloucester railway leaves the county near Edgbaston. The Warwick and Leamington Union railway joins the Great North Western line at Coventry.

Franchise, &c.] Since the passing of the reform act, 4 members are returned for the co.; 2 for the N division, who are polled for at Coleshill, Nuneaton, Coventry, Birmingham, and Dunchurch, the principal place of election being Coleshill; and 2 for the southern, who are polled for at Warwick, Kineton, Stratford, Henley, and Southam, the principal place of election being at Warwick. The number of electors registered for the county in 1837, was 10,935; in 1853, 11,182. The city of Coventry and the boroughs of Birmingham and Warwick also return 2 members each. The county is included in the Midland circuit. The assizes and quarter-sessions are held at Warwick. The greater part of Warwickshire, comprising the deaneries of Arden, Coventry, Marton, and Stonely, in the archd. of Coventry, formerly lay within the diocese of Lichfield and Coventry. By order in council, however, of date 22d December 1836, the archd. of Coventry was transferred to the dio. of Worcester, within which the co. is now wholly included. The poor-rate returns for 3 years to Easter, 1750, show an average expenditure of £10,445 on the poor of this co.; in 1840 the expenditure was £109,522.

History.] The N part of Warwickshire, at the period of the Roman invasion, was in the territory of the *Cornavii*, or *Carnabii*, and the southern in that of the *Wigantes*, or *Wiccii*. On the establishment of the Saxon octarchy W. formed an integral part of the kingdom of Mercia, or of the middle Angles. In the civil wars in the reign of Charles I, the inhabitants of this co. almost unanimously sided with the parliament.

WARY. See CARLSBAD.

WARZA, a village of Saxe-Coburg, in the bail. of Gotha. Pop. 480.

WARZENBACH, a village of Hesse-Darmstadt, in the circle of Marburg. Pop. 420.

WARZENOW, a village of Prussia, in the reg. of Danzig, and circle of Karthaus. Pop. 200.

WARZENRIED, a village of Bavaria, in the circle of Lower Bavaria, and presidial of Kotzting. Pop. 580.

WASA. See VASA.

WASAGY, a small low island of the South Pacific, near the W coast of the island of Waïgiou, to the NW of New Guinea. It is covered with trees.

WASBECK, a village of the duchy of Holstein, in the bail. of Neumunster. Pop. 280.

WASCHELWITZ, a village of Prussia, in the regency of Oppeln, and circle of Neustadt. Pop. 260.

WASCHENBACH, a village of Hesse, in the prov. of Starkenburg, and bail. of Lichtenberg. Pop. 275.

WASCHENBEUREN, a town of Würtemberg, in the circle of the Jaxt, and bail. of Lorch, 20 m. E of Stuttgart. Pop. 1,000.

WASCHGRUN, a village of Austria, in Bohemia, in the regency of Eger, and bail. of Plau. Pop. 200.

WASCHKE, a village of Prussia, in the regency of Posen, and circle of Kroben. Pop. 215.

WASCHLEITHEN, a village of Saxony, in the circle of Zwickau, and bail. of Grunhain. Pop. 316.

WASCHON. See VASONY (NAGY).

WASCHULKEN, a village of Prussia, in the regency of Konigsberg, and circle of Neidenburg. Pop. 215.

WASDALE, or NETHER WASDALE, a chapelry in the p. of St. Bees, 7 m. NNE of Ravenglass. Pop. 200.

WASDALE-HEAD, or UPPER WASDALE, a chapelry in the p. of St. Bees, Cumberland, 11 m. SW by S of Keswick, at the head of Wast-Water. Area 7,000 acres. Pop. 47.

WASEN, a village of Switzerland, in the cant. of Bern, and bail. of Trachselwald, E of Summiswald. Pop. 5,564.—Also a village in the cant. and bail. of Uri, on the Reuss. Pop. 1,394.

WASEN-HANSCHAG. See HANSAG.

WASENBACH, a village of Nassau, in the bail. of Dietz. Pop. 280.

WASENBERG, a village of Hesse-Darmstadt, in the prov. of Upper Hesse, and bail. of Ziegenhain. Pop. 700.

WASENWEILER, a village of Baden, in the circle of the Upper Rhine. Pop. 724.

WASH (THE). See NORFOLK.

WASHBOURNE (GREAT), a parish in Gloucestershire, 4½ m. NNW of Winchcomb. Area 470 acres. Pop. in 1831, 87; in 1851, 117.

WASHBROOK, or GREAT BELSTEAD, a parish in Suffolk, 4 m. W by S of Ipswich. Area 1,443 acres. Pop. in 1831, 418; in 1851, 514.

WASHFIELD, a parish in Devonshire, 1¾ m. NNW of Tiverton. Area 3,319 acres. Pop. 452.

WASHFORD-PYNE, a parish in Devon, 8½ m. N by W of Crediton. Area 1,140 acres. Pop. 192.

WASHINGBOROUGH, a parish in Lincolnshire, 3½ ESE of Lincoln, near the river Witham. Area 5,190 acres. Pop., with the township of Heighington, in 1831, 1,124; in 1851, 1,180.

WASHINGLEY, a parish in Huntingdonshire, 1½ m. NW of Stilton. Area 1,260 acres. Pop. 58.

WASHINGTON, the federal capital and seat of the supreme government of the United States, situated on the E bank of the Potomac, in N lat. 38° 55′ 48″, and W long. 77° 1′ 30″. Pop. in 1800, 3,210; in 1840, 23,364; in 1850, 40,001. "The situation of the city, chosen by Washington himself at the junction of the Potomac and of its eastern bend, the Anacosta, is really worthy the metropolis of a great empire. The finely-wooded hills of Virginia, seen across the river; the forest-clad ranges indented with valleys, which appear on the Maryland side; and the beautiful heights above Georgetown encircle the smooth plain upon which the city is placed, and which is relieved from dulness, and lighted up with life, by the passage through it of a noble river, fringed by belts of trees, and lively traces of high cultivation." The ground on which W. is built is elevated about 40 ft. above the river, from which it has a gradual ascent. It is regularly laid out in streets running due N and S, intersected by others at right angles. Besides these streets, which are from 80 to 110 ft. wide, there are 15 avenues from 130 to 160 ft. wide, which diverge from centres in various parts of the city, crossing the other streets transversely, and named after the several earlier states of the Union. The streets which run E and W, are named after the letters of the alphabet, as A street east, B street west, &c.; those which run N and S are numbered 1, 2, 3, &c., as First street north, and Second street south, &c. At the points from which the avenues diverge, are spacious squares. The ground embraced in the plan of the city is very extensive, but only a small portion of it is yet occupied by buildings. Five of the avenues radiate from the president's house, and five from the capitol, which afford ready communication from all parts to these central and important points. Pennsylvania avenue, extending between these edifices, is the most compactly built and the handsomest thoroughfare of the city. "W. is well entitled the 'city of magnificent distances;' for it has been laid out, on the most extensive scale, with streets of miles in length, and magnificent in width. But alas! as there is but little to be 'got out of it,' except in congress time, and then by hungry politicians only—for there are no manufactures, and no large back country to feed it—the city may boast of its future rather than of its present. Seen from the Capitol, it presents long lines of broad streets, diverging from the Capitol like rays, with smaller streets crossing and encircling them again almost to the limit of vision. But though around the Capitol, and along the Pennsylvania and other main avenues, both sides of the streets present fine blocks of building, houses, shops, hotels—as the eye wanders further, it sees a gradually decreasing quantity of stone and brick; first, a street with one side only, then blank spaces of increasing width, till at last the actual city shades off in whit dots of houses, standing in proud individuality, as if waiting for coming companions; while beyond the utmost limit marked by any residence, the streets provided for posterity stretch out into the distance. The whole looks very much like a spider's web half filled with flies." The whole area of the city is about 3,016 acres; its circumf., about 14 m.

The Capitol, which is finely situated on an eminence, commands a view of every part of the city, and is a grand and imposing structure. It stands on an elevation of 72 ft. above tide-water, and is the first object that attracts the eye on approaching the city. The building is of white free-stone, and occupies an area of 1½ acre. Including the wings, which are each 100 ft. front, it is 362 ft. long, and 121 ft. in depth. The projection on the E or main front, inclusive of the steps, is 65 ft., and on the W, 83 ft. The first is ornamented with a portico of 22 Corinthian columns; the latter, with a like portico supported by 10 columns. The building is surmounted by a dome 120 ft. high, under which, in the middle of the building, is the rotunda, 95 ft. in diameter and 95 ft. in height. The room is also adorned with various alto-relievo groups, each representing some great national event. A colossal statue of Washington, by Greenough, is placed in the rotunda. The congressional library-room, on the W of the rotunda, is 92 ft. long, 34 ft. broad, and 36 ft. high, and previous to the recent fire contained from 50,000 to 60,000 volumes, and a collection of medals, paintings, and statuary. The Senate-chamber, in the N wing of the building, is 78 ft. long and of a semi-circular form. Under this is the room in which the supreme court sits. The house-of-representatives is in the opposite wing, and is somewhat larger than the senate chamber, and semi-circular. These several apartments are ornamented with statuary and paintings. An extension of the building is now making by the addition of two wings at the ends of the present building, with which they will be connected by corridors 44 ft. long and 50 ft. wide. The wings are to be each 143 ft. from N to S, by 238 ft. from E to W, exclusive of the porticos and steps. The entire length of the whole building, when completed, will be 751 ft., including porticos and steps; its area on the ground 153,112 superficial ft., or over 3½ acres,—the additions covering nearly 2 acres. These additions are to be constructed of white marble. The general style of architecture will be consistent with that of the present building, with such projections and porticos upon all the disconnected sides as tend to produce an agreeable composition with the present structure. They will contain nearly 100 additional rooms, for the accommodation of committees and other purposes connected with legislation. The new representative hall, which will be located in the second story of the south wing, will have 300 seats within the bar.

The White House, the residence of the President of the United States, is situated at the junction of Pennsylvania, New York, Connecticut and Vermont avenues. It is a splendid free-stone edifice, 170 ft. long, 85 ft. deep, and two stories high. It is ornamented on the front, facing on La-Fayette square, with an Ionic portico, and the garden front is embellished by a circular colonnade of 6 Ionic columns. Adjoining the president's mansion are extensive buildings appropriated by the department of war, navy, state, and treasury. The new treasury building is 300 ft. long, and in the rear is a wing 100 ft. long. Along the front is a colonnade supported by 32 columns of massive dimensions. The General Post-office is a large marble building 204 ft. in length, with two wings 116 ft. deep, highly adorned by large fluted columns of marble. The Patent-office is a large and splendid building, and exhibits great architectural skill. The Smithsonian Institute is a splendid edifice of reddish free-stone, 450 ft. in length, and 140 ft. in width.

The Navy-yard, situated on the Anacosta branch of the Potomac, about three-quarters of a mile from the capitol, has an area of 27 acres, and is enclosed by a substantial brick wall. Within this enclosure are the officers' quarters, shops, warehouses, two large ship houses, and an armory. These are all fine establishments of their kind, and are kept in the best of order. The navy magazine is an extensive building of brick, in which are employed a large number of artisans for the manufacture of combustibles for warlike purposes.—Within the limits of the city there are numerous churches belonging to the different denominations. There are two orphan asylums, the Washington and St. Vincent's, which are supported, one by the ladies of the city, and the other by the Sisters of Charity. Among the public buildings of Washington, the theatres and places of amusement are not the least important. The Washington theatre, the National theatre, the Assembly Rooms, are conspicuous in this class. Columbia college is situated on an elevation, commanding a splendid view of the surrounding country. The Theological seminary, the Columbian institute, the American historical society, the Columbian horticultural society, the City library, the Athenæum, the National institution for the promotion of science, and some others, afford to the citizens every advantage in the pursuit of knowledge, and the edifices are highly ornamented and rich in taste and design. The City-hall, intended for the use of the corporate authorities of Washington, is a magnificent structure. The penitentiary and city jail are large buildings.

The Congressional cemetery occupies an area of 10 acres, near the E branch of the Potomac, about 1½ m. from the capitol. The grounds are surrounded by a high brick wall, and have a considerable elevation above the river, commanding an extensive prospect of the beautiful scenery which surrounds the city. They are tastefully laid out with ornamental trees and shrubbery. Washington is connected with the N and S by railway, and is accessible from the Atlantic for ships of the largest class.

WASHINGTON, a county in the SW part of the state of Alabama, U. S., comprising an area of 1,049 sq. m., bounded on the E by Tombigbee river, and watered by its branches, and by Escatawpa creek. It has an undulating surface, and possesses considerable fertility. It is intersected by the Mobile and Ohio railway. Pop. in 1840, 5,300; in 1850, 2,713. Its capital is Washington.—Also a county in the NW part of the state of Arkansas, comprising an area of 929 sq. m., drained by the Illinois, and by a head branch of White river. The surface is mountainous, but the soil fertile. Pop. in 1840, 7,148; in 1850, 9,849. Its cap. is Fayetteville.—Also a county in the state of Columbia. See District of Columbia.—Also a county in the W part of the state of Florida, comprising an area of 1,500 sq. m., bounded on the SW by the gulf of Mexico, and on the W by Choctawhatchee, and drained by affluents of that river, and by the Econfina. The surface is level, and the soil generally poor. Pop. in 1840, 859; in 1850, 1,950. Its cap. is Roche's Bluff.—Also a central co. in the state of Georgia, comprising an area of 962 sq. m., drained by branches of Oconee and Ogeechee rivers, by which it is bounded on the NE and W. The surface is diversified, but the soil is chiefly of a light sandy nature. It is intersected by the Georgia Central railroad. Pop. in 1840, 10,565; in 1850, 11,766. Its cap. is Sandersville.—Also a county in the SW part of the state of Illinois, comprising an area of 583 sq. m., drained by Elkhorn and Muddy creeks, and by branches of Kaskaskia, by which it is bounded on the NW, and intersected by the Illinois Central and Sangamon and Massac railroad. Pop. in 1840, 4,810; in 1850, 6,953. Its cap. is Nashville.—Also a county in the S part of the state of Indiana, comprising an area of 540 sq. m., drained by Muscatatuck, the E fork of White river, Lost river, and the head streams of Blue river, and intersected by the New Albany and Salem railway. It has a finely diversified surface. Pop. in 1840, 15,269; in 1850, 17,040. Its cap. is Salem.—Also a county in the SE part of the state of Iowa, comprising an area of 569 sq. m., intersected by Iowa river and its branches, by Skunk river, and by Long and Crooked creeks. The surface is undulating, and the soil generally fertile. Pop. in 1840, 1,594; in 1850, 4,957. Its cap., which bears the same name, is on the E side of Adam's creek, 25 m. S by W of Iowa city, and intersected by the Dubuque and Keokuk railway. Pop. in 1840, 100; in 1850, 742.—Also a central county of the state of Kentucky, comprising an area of 307 sq. m., drained by Chaplin's fork of Salt river and its branches. The surface is level, and the soil generally fertile. Pop. in 1840, 10,596; in 1850, 12,194. Its cap. is Springfield.—Also a parish in the NE part of the state of Louisiana, comprising an area of 858 sq. m., drained chiefly by Bogue Chitto and Chifuncte rivers. It has a hilly surface, and possesses little fertility. Pop. in 1840, 2,649; in 1850, 3,408. Its cap. is Franklintown.—Also a county in the SE part of the state of Maine, comprising an area of 2,470 sq. m., drained by Machias, Pleasant, and Naraquagas, and intersected by the Franklin, the European, and North American, and the Calais and Baring railroads. It has a hilly surface diversified with lakes, and is in some parts very fertile. Pop. in 1840, 28,327; in 1850, 38,810.—Also a county in the W part of the state of Maryland, comprising an area of 518 sq. m., bounded on the E and NE by the Patuxent, and on the SW by the Potomac rivers, and drained by the Piscataway, Broad, Conecocheague, and other creeks, and intersected by the Baltimore and Ohio, and the Franklin railways, and by the Chesapeake and Ohio canal. It has a diversified surface, and is in some parts extremely productive. Pop. in 1840, 28,850; in 1850, 30,848. Its cap. is Hagerstown.—Also a county of Minnesota, separated by the St. Croix from the Wisconsin territory, and bounded on the S by the Mississippi. It has a level surface, drained by affluents of the St. Croix, and well-cultivated. Pop. in 1850, 1,056. Its cap. is Hillwater.—Also a county in the W part of the state of Mississippi, comprising an area of 1,219 sq. m., drained by Williams, Bayon, and Sunflower river, and other affluents of Yazoo river. It contains numerous lakes, is generally diversified, and very fertile. Pop. in 1840, 7,287; in 1850, 8,389. Its cap. is Greenville.—Also a county in the SE part of the state of Missouri, comprising an area of 937 sq. m., drained by Big river and its branches, and by the headwaters of St. Francois and Big Blackwater rivers. It contains the celebrated Iron Mountain, which has an alt. of 300 ft. above the plain, and yields from 70 to 80 per cent. of good metal. It is in some parts level and very fertile, and is intersected by Iron Mountain railway. Pop. in 1840, 7,231; in 1850, 8,811. Its cap. is Potosi.—Also a county in the NE of the state of New York, comprising an area of 807 sq. m., drained by Hoosick and Baltenkill rivers, affluents of the Hudson, and by Pawlet and Poultney rivers, Wood creek, and Moseskill, and intersected by the Rutland and Washington and the Saratoga and Washington railways, and by the Champlain canal. It is finely diversified, and generally fertile. Pop. in 1840, 41,080; in 1850, 5,666. Its cap. is Salem and Sandy Hill.—Also a county in the SE part of the state of Ohio, comprising an area of 612 sq. m., intersected by Muskingum and Muskingum rivers, and by Duck, Wolf, Bear, and Cat creeks, and by the Cincinnati and Marietta, and the Cincinnati and Belpré railroads. The banks of the rivers are level and fertile. Pop. in 1840, 20,823; in 1850, 29,540. Its cap. is Marietta.—Also a county of Oregon territory, bounded on the NE and E by the Columbia and Willamette rivers, and comprising an area of 5,000 sq. m. Pop. in 1850, 2,651. Its cap. is Hillsboro.—Also a county in the E part of the state of North Carolina, drained by affluents of Lake Phelps in the E, and Albemarle in the N. It is in many parts low and swampy. Pop. in 1840, 4,525; in 1850, 5,666.—Also a county in the SW of Pennsylvania, com prising an area of 795 sq. m., bounded on the E by the Monongahela river, and drained by Buffalo, Raccoon, and Cross creeks, and intersected by Pittsburg, and Steubenville, and Hempfield railways. Its capital, which bears the same name, is 170 m. W of Harrisburg. Pop. in 1840, 2,062; in 1850, 2,662.—Also a county in the S part of the state of Rhode Island, comprising an area of 367 sq. m., drained by Pawcatuck, and Charles, and Wood rivers, and intersected by the Providence and Stonington railroad. Pop. in 1840, 14,342; in 1850, 16,430. Its cap. is Kingston.—Also a county in the E part of the state of Tennessee, comprising an area of 512 sq. m., drained by Nolichucky river and its affluents, and by branches of Watanga river. It is in some parts hilly, but generally fertile. Pop. in 1840, 11,751; in 1850, 13,821. Its cap. is Jonesboro.—Also a central county of the Texas, drained by Gegua, Jackson, New Year's, and Caney creeks. It is generally fertile. Pop. in 1850, 5,983. Its cap. is Brenham.—Also a central county in the state of Vermont, comprising an area of 550 sq. m., drained by Onion river and its branches, and intersected by the Vermont Central railroad. Pop. in 1840, 23,506; in 1850, 24,654. Its cap. is Montpelier.—Also a county in the SW part of the state of Virginia, comprising an area of 547 sq. m., bounded on the NW by Blue Ridge, and intersected by the Virginia and Tennessee railroad, drained by New river, and by branches of Great Kanawha and Clinch rivers. Pop. in 1840, 13,001; in 1850, 14,612. Its cap. is Abingdon.—Also a county of Utah territory, extending 35 m. in width along the S confines of the Ter between the Rocky mountains, to the Californian frontier, and drained by the Colorado.—Also a county in the E part of the state of Wisconsin, comprising an area of 663 sq. m., drained by Milwaukie and Menomonee rivers. Pop. in 1840, 343; in 1850, 19,484. Its cap. is Port Washington.—Also a township of Lincoln co., in the state of Maine, 22 m. ESE of Augusta, drained by branches of Damariscotta and Muscongus rivers. Pop. in 1840, 1,600; in 1850, 1,756.—Also a township of Sullivan co., in the state of New Hampshire, 27 m. W by S of Concord, drained by Ashuelot and Contoocook rivers. Pop. in 1840, 1,103; in 1850, 1,054.—Also a township of Orange co., in the state of Vermont, 15 m. SE of Montpelier, watered by Jail branch of Winooski or Onion river, and by branches of Waits and White rivers. Pop. in 1840, 1,359; in 1850, 1,348.—Also a township of Berkshire co., in the state of Massachusetts, 128 m. W of Boston, watered by branches of Westfield and Housatonic rivers, and intersected by the Western railway. Pop. in 1840, 991; in 1850, 953.—Also a township of Litchfield co., in the state of Connecticut, 32 m. W by S of Hartford, drained by Shepaug river. Pop. in 1840, 1,622; in 1850, 1,802.—Also a township of Dutchess co., in the state of New York, 60 m. S of Albany, drained by several small creeks. Pop. in 1840, 2,833; in 1850, 2,805.—Also a township of Bergen co., in the state of New Jersey, 37 m. N by W of Trenton, drained by Muskonetong river, and by Pohatcong creek. Pop. in 1850, 1,567.—Also a township of Gloucester co., in the same state, 12 m. W of Woodbury, drained by Pensauken and Cooper's creeks, and several affluents of Atsion river. Pop. in 1840, 1,545.—Also a township of Morris co., in the same state, drained by the S branch of Raritan river. Pop. 2,451.—Also a township of Luzerne co., in the state of Pennsylvania. Pop. in 1840, 1,255.—Also a township of Berks co., in the same state. Pop. 1,210.—Also a township of Franklin co., in the same state, 11 m. SW of Chambersburg, bounded on the E by South Mountain, and drained by two branches of Anticlam creek. Pop. 2,404.—Also a township of York co., in the same state. Pop. 1,226.—Also a township of Fayette co., in the same state. Pop. 1,515.—Also a township of Cambria co., in the same state, drained by a branch of Kiskiminetas. Pop. 1,266.—Also a township of Erie co., in the same state. Pop. 1,551.—Also a township of Greene co., in the same state, drained by branches of Ten Mile creek. Pop. 936.—Also a township of Indiana co., in the same state, drained by Crooked and Plum creeks. Pop. 1,893.—Also a township of Jefferson co., in the same state. Pop. 367.—Also a township of Westmoreland co., in the same state, 16 m. N of Greensburg, drained by White Deer Hole and Black Hole creeks. Pop. 2,004.—Also a township of Allen co., in the state of Ohio. Pop. 457.—Also a township of Belmont co., in the same state, drained by Captina creek. Pop. 1,387.—Also a township of Brown co., in the same state, 12 m. NE of Georgetown. Pop. 848.—Also a township of Carroll co., N of Carrollton. Pop. 1,024.—Also a township of Clermont co., in the same state, bounded on the S by the Ohio, and containing two villages. Pop. 2,100.—Also a township of Clinton co., in the same state. Pop. 1,170.—Also a township of Coshocton co., in the same state. Pop. 1,029.—Also a township of Columbiana co., in the same state. Pop. 8,112.—Also a township of Darke co., in the same state. Pop. 889.—Also a township of Franklin co., in the same state, bounded on the E by Scioto river. Pop. 843.—Also a township of Guernsey co., in the same state. Pop. in 1840, 1,353; in 1850, 1,729.—Also a township of Hancock co., in the same state. Pop. in 1840, 809.—Also a township of Hardie co., in the same state. Pop. 203.—Also a township of Harrison co., in the same state. Pop. 1,022.—Also a township of Hocking co., in the same state. Pop. 1,124.—Also a township of Holmes co., in the same state. Pop. 1,461.—Also a township of Jackson co., in the same state. Pop. 481.—Also a township of Licking co., in the same state. Pop. 1,348.—Also a township of Marion co., in the same state. Pop. 880.—Also a township of Mercer co., in the same state. Pop. 214.—Also a township of Montgomery co., in the same state. Pop. 2,210.—Also a township of Monroe co., in the same state. Pop. 535.—Also a township of Pickaway co., in the same state. Pop. 1,194.—Also a township of Sandusky co., in the same state. Pop. 1,074.—Also a township of Shelby co., in the same state, watered by a branch of Miami river. Pop. 1,688.—Also a township of Muskingum co., in the same state. Pop. 1,488.—Also a township. of Preble co., in the same state. Pop. 2,459.—Also a township of Richland co., in the same state, to the S of Mansfield. Pop. 1,914.—Also a township of Scioto co., in the same state, bounded on the S by the Ohio, and on the E by Scioto river, and intersected by the Ohio canal. Pop. 653.—Also a township of Tuscarawas co., in the same state. Pop. 978.—Also a township of Miami co., in the same state. Pop. 1,161.—Also a township of Starke co., in the same state. Pop. 1,389.—Also a township of Union co., in the same state. Pop. 151.—Also a township of Van Wert co., in the same state. Pop. 47.—Also a township of Wood co., in the same state. Pop. 196.—Also a township of Williams co., in the same state. Pop. 98.—Also a township of Adams co., in the state of Indiana. Pop. 262.—Also a township of Allen co., in the same state. Pop. 595.—Also a township of Blackford co., in the same state. Pop. 143.—Also a township of Brown co., in the same state. Pop. 584.—Also a township of Clay co., in the same state. Pop. 1,046.—Also a township of

Gibson co., in the same state. Pop. 640.—Also a township of Grant co., in the same state. Pop. 539. —Also a township of Hamilton co., in the same state. Pop. 1,285.—Also a township of Harrison co., in the same state. Pop. 721.—Also a township of Marion co., in the same state. Pop. 1,859.—Also a township of Noble co., in the same state. Pop. 259.—Also a township of Putnam co., in the same state. Pop. 1,872.—Also a township of Randolph co., in the same state. Pop. 1,219.—Also a township of Rush co., in the same state. Pop. 1,163. —Also a township of Tippecanoe co., in the same state. Pop. 809.—Also a township of Wayne co., in the same state. Pop. 1,935.—Also a township of Washington co., in the same state. Pop. 1,992.—Also a township of Davies co., in the same state, 4 m. E of the White fork of White river. Pop. in 1840, 650; in 1850, 2,578.—Also a township of Macomb co., in the state of Michigan, 76 m. E of Lansing, drained by branches of Clinton river. Pop. in 1840, 1,314; in 1850. 1,541.—Also a township of Buchanan co., in the state of Missouri. Pop. in 1840, 484.—Also a township of Clark co., in the same state. Pop. 314.—Also a township of Johnson co., in the same state. Pop. 476.—Also a township of Lafayette co., in the same state. Pop. 458.—Also a township of Livingston co., in the same state. Pop. 167.—Also a township of Monroe co., in the same state. Pop. 1,367.—Also a township of Polk co., in the same state. Pop. 417.—Also a township of Taney co., in the same state. Pop. 275.—Also a township of Sevier co., in the state of Arkansas. Pop. 516.—Also a village of Rappahannock co., in the state of Virginia, 87 m. NNW of Richmond. Pop. in 1840, 375; in 1850, 400.—Also a village of Beauford co., in the state of North Carolina, on the N side of Tar river, and 91 m. E by S of Raleigh. Pop. in 1840, 1,200; in 1850, 1,300.—Also a village of Wilkes co., in the state of Georgia, 52 m. NE by N of Milledgeville. Pop. in 1840, 800; in 1850, 850.—Also a village of Antauga co., in the state of Alabama, on the N side of the Alabama river, 10 m. below Montgomery. Pop. in 1840, 200.—Also a village of Adams co., in the state of Mississippi, 6 m. NE of Natchez. Pop. in 1840, 400.—Also a village of Landry co., in the state of Louisiana, on the SW of Bayou Beche. Pop. in 1840, 125; in 1850, 250.—Also a village of Rhea co., in the state of Tennessee, on the W side of Tennessee river, and 134 m. ESE of Nashville. Pop. in 1840, 110; in 1850, 150. —Also a village of Wills township, Guernsey co., in the state of Ohio, 85 m. E of Columbus. Pop. in 1840, 500.—Also a village of Tazewell co., in the state of Illinois, 71 m. N of Springfield. Pop. in 1840, 350.—Also a village of Golo co., in the state of California, on the W bank of the Sacramento.—Also a village of Mason co., in the state of Kentucky, 60 m. ENE of Frankfort, and intersected by the Lexington and Maysville railroad. Pop. in 1850, 650.—Also a village of Washington co., in the state of Texas, on the W bank of Branos river, opposite the mouth of Navasote river, and 94 m. E of Austin city.

WASHINGTON (Territory of), a district of the United States, which, until 1853, formed a portion of Oregon territory, occupying all the country lying between the parallels of 46° and 49° N, and extending from the Rocky mountains on the E, to the Pacific ocean on the W. The Columbia river forms the S boundary below the line of 46° to the sea. It comprises somewhat more than one-third of the original territory, or about 130,000 sq. m. It includes the waters of Puget's sound. Vancouver's island and the gulf of Georgia indent its north-western extremity. The Columbia river is its great interior water-course, and there are numerous other important streams which empty either into Puget's sound or into the Pacific. Fort Nesqually and Olympia, at the S extremity of Puget's sound, are already important places. The interior has been as yet but imperfectly explored.

WASHINGTON, a parish in the co.-palatine of Durham, 5 m. SE by S of Gateshead, comprising the townships of Barmston, Great and Little Usworth, and W. Area 5,335 acres. Pop. in 1831, 2,673; in 1851, 3,485. The inhabitants are chiefly employed in collieries.—Also a parish in Sussex, 4 m. W by N of Steyning. Area 3,162 acres. Pop. in 1851, 884.

WASHINGTON (Cape), a headland of the Antarctic continent, in S lat. 74° 37′, E long. 165° 10′.

WASHINGTON (Mount), a summit of the White mountains, in New Hampshire, U. S., having an alt. of 6,265 ft. above tide-level.

WASHINGTON-POINT, a point of land formed by the junction of the S and E branches of Elizabeth river, Virginia, 1 m. S of Norfolk.

WASHINGTONVILLE, a village of Sandy Creek township, in Oswego co., in the state of New York, U. S. Pop. 250.—Also a v. of Orange co., New York, 10 m. W of Newburgh. Pop. 200.

WASHITA, or Ouachitta, a large river of Arkansas and Louisiana, U. S., which rises between the Arkansas and Red rivers, in N lat. 34° 30′, about 70 m. to the W of the Mississippi, and falls into Red river, 30 m. above its junction with the Mississippi, after a winding course of more than 500 m. It is sometimes called the Black river, from its junction near the Tenzas; and it only receives the name of the Washita, at the point where its waters unite with those of the Catahoola and the Tenzas. Its current is gentle, and in many places the river does not exceed 80 yds. in breadth. The river is navigable for large boats at all times of the year, except in very dry seasons.—Also a parish of the state of Louisiana, U. S., situated in a valley of the Washita, in the NE part of the state. Area 2,090 sq. m. Pop. in 1840, 4,640; in 1850, 5,008. Its cap. is Monroe.

WASHMINSKER ISLANDS, a cluster of islands, near the S coast of Labrador, in N lat. 50°.

WASHTENAVO, a county in the SE of Michigan, U. S. Area 920 sq. m. Pop. in 1840, 23,571; in 1850, 25,567. Its cap. is Ann-Arbor.

WASING, a parish in Berks, 7½ m. SE by E of Newbury. Area 682 acres. Pop. in 1851, 58.

WASKEMASHIN, an island in the gulf of St. Lawrence, near the coast of Labrador, in N lat. 50° 3′.

WASKUACHAOUIPIOU, a river of Canada, which runs into the Saguenay, in N lat. 48° 20′.

WASLING, a parish of Berkshire, 7½ m. ESE of Newbury. Area 682 acres. Pop. 88.

WASLUI, a town of Moldavia, on the Birlad, 32 m. S of Jassy.

WASMES, a canton and town of Belgium, in the prov. of Hainault, 6 m. W of Mons. Pop. 6,000.

WASMES-BRIFFOEIL, a town of Belgium, in the prov. of Hainault, 7 m. ESE of Tournay. Pop. 1,100.

WASMUNSTER, a town of Belgium, in East Flanders, on the Durme, 16 m. W by N of Ghent.

WASNIOW, a village of Poland, in the woivode of Sandomir, 36 m. S of Radom. Pop. 200.

WASPERTON, a parish of Warwickshire, 4 m SSW of Warwick. Area 1,619 acres. Pop. 292.

WASPIK, a village of Belgium, in the prov. of N. Brabant, 10 m. NW of Breda.

WASS, a township of Kilbrun p., Yorkshire, 6½ m. SW of Helmesley. Pop. 132.

WASS ISLAND, an island of the Atlantic, near the coast of America, in N lat. 44° 28′.

WASSAH, a town of Hindostan, in the prov. of Gujerat, district of Cambay, in N lat. 22° 39′.

WASSAW (Great), an island in the Atlantic, near the coast of Georgia, in N lat. 32° 52′. It is 16 m. in circumf.—Little W. lies to the SW.

WASSAWA LAKE, a lake in the SW of La Pointe co., in Wisconsin, U. S. It is 7 m. in length and 4 m. in breadth, and discharges itself into the St. Croix.

WASSEIGES, a village of Belgium, in the prov. of Liege, arrond. of Huh. Pop. 832.

WASSELA, a mountainous country of Senegambia, in Central Africa, situated to the E of Kong, and having to the N and W Manding and Bambarra.

WASSELONNE, a commune and town of France, in the dep. of Bas-Rhin, on the Mussig, 14 m. NW of Strasburg. Pop. 4,400. It has woollen and cotton-yarn mills.

WASSEN, a village of Switzerland, in the cant. of Uri, on the Reuss. Pop. 1,349.

WASSENAER, a village of Holland, in the prov. of S. Holland, 5 m. W by S of Leyden. Pop. 1,768.

WASSENBERG, a town of Prussia, in the prov. of Aachen, on the Roer, 9 m. ESE of Ruremond. Pop. 1,100.

WASSER (Alt), a village of Prussia, in the reg. of Breslau, circle of Waldenburg. Pop. 1,068.—Neu W. is a village in the reg. of Koslin, circle of Schlawe. Pop. 270.

WASSERBURG, a town of Bavaria, on the Inn, 28 m. ESE of Munich. Pop. 2,238. It has a trade in hops and hemp.—Also a v. of Bavaria, on a point of land projecting out into the lake of Constanz. Pop. 800.

WASSERLEBON, a village of Prussian Saxony, in the reg. and 12 m. SW of Magdeburg, on the Ilse. Pop. 1,100.

WASSERMUNGENAU, a village of Bavaria, in Middle Franconia, landg. of Heilbronn. Pop. 429.

WASSER-TRUDINGEN, a town of Bavarian Franconia, on the Wernitz, 19 m. S of Anspach. Pop. 2,017.

WASSIBU, a small town of Bambarra, in Central Africa, 75 m. ESE of Benowm.

WASSIGNY, a town of France, dep. of Aisne, on the small river Vaux. Pop. 1,100.

WASSOTAH, a celebrated fortress of Hindostan, in the prov. of Bejapur, 30 m. SSW of Saturah. There are two forts, about 1,000 yds. from each other, both situated on rocks nearly 3,000 ft. high. The adjacent scenery is of the grandest description.

WASTWATER, a lake of Cumberland, at the foot of Scawfell. It is nearly 3 m. in length, and about a ½ m. broad, and discharges itself by the Irt. It is environed by lofty and very precipitous mountains, and has a stern and gloomy aspect.

WASUNGEN, a town of the duchy of Saxe-Meiningen, on the Werra, 6 m. N of Meiningen. Pop. 2,500.

WASZILKOW, a town of Russian Poland, 5 m. N of Bialystock. Pop. 900.

WATAB, a village of Benton co., in Tennessee, U. S., on the E bank of the Mississippi, 72 m. NW of St. Paul.

WATAGUAKI, a river of Labrador, which runs into the gulf of St. Lawrence, in N. lat. 50° 12′.

WATANGA, a county in the NW of North Carolina, U. S. Area 548 sq. m. Pop. in 1850, 8,400.

WATCH-POINT, a cape on the E coast of Rhode Island, U. S., in N lat. 41° 13′.

WATCHER ISLANDS, a cluster of small islands in the straits of Macassar, near the W coast of Celebes, in S lat. 0° 27′.

WATCHER (South), a small island in the straits of Macassar, near the W coast of Celebes, in S lat. 0° 3′.

WATCHET, a seaport in the co. of Somerset, situated on the Bristol channel, 15 m. NW of Taunton. Pop. in 1841, 916.

WATCHFIELD, a hamlet in the p. of Great Farringdon.

WATERBEACH, a parish of Cambridge, 5½ m. NNE of Cambridge. Area 5,556 acres. Pop. 1,440.

WATERBOROUGH, a township of York co., Maine, U. S., 25 m. N of York. Pop. 1,989.

WATERBURY, a township of Washington co., Vermont, U. S., on Onion river, 12 m. NW of Montpelier. Pop. 2,352.—Also a township of Newhaven co., Connecticut, 20 m. NNW of Newhaven. Pop. in 1840, 3,608; in 1850, 5,137. There are pin manufactories and corn and rolling-mills here.

WATERDEN, a parish of Norfolk, 4 m. W of Little Walsingham. Area 763 acres. Pop. 39.

WATER-EATON, a hamlet of Bucks, adjoining to Fenny-Stratford, in Bletchley p. Area 1,040 acres. Pop. 241.—Also a hamlet of Oxfordshire, 3½ m. N of Oxford, in the p. of Kidlington. Area 1,630 acres. Pop. 119.

WATEREE. See Catabaw.

WATERFALL, a parish of Staffordshire, 7 m. SE of Leek. Area 1,530 acres. Pop. 521.

WATERFORD, a maritime county of Ireland, in the SE of the prov. of Munster, bounded on the N by the cos. of Tipperary and Kilkenny; on the E by Wexford; on the S by the Atlantic; and on the W by the co. of Cork. The greatest length of the co., in a line due W from Croden-head in Waterford-harbour, to a point near that at which the river Blackwater enters Waterford from Cork, is 39½ m.; its greatest breadth from a point on the Suir 2¾ m. above Clonmel, southward to Ram-head, is 22¾ m. Its area comprises 325,345 acres of arable land, 105,496 of uncultivated land, 23,408 of continuous woods, 1,525 of towns, and 5,779 of water—in all, 461,553 acres. A broad, boldly-featured range of mountains, called the Cummeragh or Monavoulagh mountains, rises suddenly up from the vicinity of Clonmel, and extends nearly quite across the co., to within about 2 m. of Dungarvan. The river Suir flows in a NE and latterly S direction to the ocean. The Nier runs W to the Suir at nearly the point of that river's first contact with the county. The Clodagh runs E to the Suir. The Mahon flows SSE to the ocean. The Tay also flows SSE to the ocean. The Blackwater comes in from co. Cork with majestic volume, and flows E past Lismore to Cappoquin, and thence S to the head of Youghal harbour. The Bride also comes in from co. Cork, and flows E past Tallow. The Goish flows WNW to the Blackwater, at a point a little above the influx of the Bride. The only navigations within the county, or upon its boundaries, are the Suir to Clonmel, the Blackwater, for small craft, to a point above Cappoquin, and the Bride for small craft to the vicinity of Tallow. Transition rocks, consisting principally of clay slate, greywacke, and greywacke slate, constitute the greater portion of the eastern half of the co.; rocks of old conglomerate, and of purple, red, green, and grey clay slate, constitute the greater portion of the western half. Slate quarries occur at Whitfield. Silver ores are found in the vicinity of Waterford, at Don-Isle, and at Bonmahon. Lead ores occur in the vicinity of Waterford, at Annestown, at Bonmahon, at Kilkeamy, and in the vicinity of Lismore. Iron ores are found in the Cummeragh mountains, at Ardmore, and in the vicinity of Lismore. Copper ores are found at Bonmahon, at Annestown, at Dromana, and in the vicinity of Lismore. The mineral district at Bonmahon and its vicinity, round the mouth of the rivulet Mahon, ranks as the most valuable in Ireland.

Agriculture.] A large proportion of the land capable of cultivation is in tillage. The common Irish cow is the general dairy stock. Most of the sheep are of a large coarse breed. The horses are light-boned, active animals. In 1841, there were within the rural districts of the co. 15,902 horses and mules, 2,407 asses, 42,980 cattle, 46,511 sheep, 67,547 pigs, and 185,487 poultry. Estimated value of live stock in the rural districts of the co., £549,226. The plantations within the co., in 1841, amounted to 26,536 acres.

Manufactures and Trade.] The principal seats of manufacture and productive industry are the city of Waterford and the towns of Portlaw, Dungarvan, Bonmahon, Lismore, and Cappoquin; and the leading facts respecting the most prominent departments of manufacture and productive industry are stated in the articles on these towns and those of Clonmel and Carrick-on-Suir.

Towns, &c.] The towns and principal villages are Dunmore, Cheekpoint, and Passage, in Gualtier barony; Tramore and Annestown, in Middlethird; Portlaw, Carrickbeg, Rathgormack, Scronthea, and part of Clonmel, in Upperthird; Bonmahon, Dungarvan, Kilmacthomas, Kill, Knockmahon, and Stradbally, in Decies-without-Drum; Aglish, Villierstown, Ardmore, Clashmore, and Ringville, in Decies-within-Drum; and Lismore, Cappoquin, Tallow, Tallow-bridge, and Ballyduff, in Coshmore and Coshbride.—The co. is ecclesiastically divided into the dio. of Waterford and the larger part of the dio. of Lismore. The assizes are held at Waterford; courts of quarter-sessions at Dungarvan, Lismore, Waterford, and Carrickbeg. The county sent 10 members to the Irish parliament, or 2 from the county at large, 2 from the city of Waterford, and 2 from each of the boroughs of Dungarvan, Lismore, and Tallow; but—irrespective of its small portion of the borough of Clonmel—it sends only 5 to the imperial parliament, or 2 from the county at large, 2 from the city of Waterford, and 1 from the borough of Dungarvan. Constituency of the county at large, in 1844, 880; in 1851, 3,248. Pop. in 1831, exclusive of the co. of the city, 148,233. Houses, 21,234. Families employed chiefly in agriculture, 15,202; in manufactures and trade, 3,535; in other pursuits, 5,627. Pop. in 1841, exclusive of the municipal district of the city, 172,971. First-class inhabited houses, 610; second-class, 6,225; third-class, 11,116; fourth-class, 7,416. Pop. in 1851, 138,574.

History.] In the time of the geographer Strabo, near the middle of the second century, a people called the *Menappii*, whose origin and character are matter of much dispute among antiquaries, inhabited the territories which now constitute the counties of Waterford and Wexford. The next people we meet with in this country were a powerful clan called the *Desii*. In 1169, Melaghlin O'Feolain, prince of the Desii, was taken prisoner by Earl Strongbow when the city of Waterford was stormed. In him ended the chieftainry of the Desii, and no traces of consequence remain of this territory, except in the large extended barony of Desies in this county, which was soon after established.

WATERFORD, a city and parl. borough, the capital and near the SE extremity of co. Waterford, on the river Suir, 6 m. NNE of Tramore, and 75¾ m. SSW of Dublin. The river Suir, while passing the town, has a breadth of from 230 to 350 yds. The northern façade and chief thoroughfare of the city, which assumes the name of the Quay, and which has a noble appearance, measures one statute mile in length; but over the 320 yards of its extent above the bridge, it consists of a series of buildings pressing down to the water's edge, and bisected by three or four narrow lanes. The Mall extends 300 yds. SW from the Quay, and is the most spacious street in the interior of the city. The principal streets in the vicinity of the Quay and the Mall, and particularly the Quay and the Mall themselves, possess the best houses, and are the scene of the chief trade and the principal attractions of the city. The cathedral is a light and beautiful building, 170 ft. in length. The largest of the Roman Catholic chapels, and that which serves as the cathedral of the dio., stands in Barron-strand-street, is capable of accommodating 11,000 persons, and is said to be one of the largest buildings in Ireland. It was erected at the cost of £20,000. It displays a beautiful architectural front of hewn stone in the Ionic order. The present great timber-bridge, so prominent and singular a feature in all the river views of the city, was built in 1793-4. It measures 832 ft. in length, and 42 ft. in breadth; has 40 sets of piers, each of seven pieces, besides cut-waters. A draw-bridge exists at the S end of the bridge, to permit the transit of river-craft to Carrick-on-Suir and Clonmel. The depth of the Suir directly in front of the quay, varies from 20 to 65 ft. at low water, and from 37 to 82 ft. at high water, of the greatest spring-tides. A brisk trade in cattle and butter was, at an early period, carried on with the English colonies and with Spain; but it was eventually destroyed, in consequence chiefly of the establishment of close commercial relations between the West Indies and America. The manufactures consist of glass, starch, and beer. A large and excellent ship-yard, with a patent slip, possesses considerable celebrity. The old borough is co-extensive with the quondam county of the city, and continues also to be the parl. borough. It has an area of 10,059 acres. The municipal borough, or existing district of the liberties of the city, includes little more than the actual town, and comprises an area of only 668 acres. The parl. borough sends two members to the imperial parliament. Constituency, in 1841, 1,499; in 1852, 1,135.

The Diocese.] The dio. of W. is alleged to have been founded in the 11th cent. The see of W. and Lismore has been united, in terms of the church reform bill, to the sees of Cashel and Emly. The episcopal income amounts to £3,933. The diocese of W. is, in territorial extent, the smallest diocese in Ireland, and lies wholly in the eastern portion of the county of Waterford. Its length is 13 statute miles; its breadth is 9 statute miles; and its area is 513,239 acres. Pop. in 1831, 45,730. Number of parishes, 29. The Roman Catholic dioceses of Waterford and Lismore remain mutually consolidated as before the Reformation; but, in other respects, continue unannexed, and are regarded as constituting one diocese. This united diocese is distributed into 37 parishes. In 1649, Cromwell proceeded to invest W., which was surrendered to him on the 10th of August 1650. W. gives the title of Earl, in the peerage of Ireland, to the noble family of Talbot; and the title of Marquis, in the peerage of Ireland, to the noble family of De La Poer Beresford.

WATERFORD, a village of Upper Canada, township of Townsend, 7 m. from Simcoe. Pop. 200.

WATERFORD, a township of Oxford co., Maine, U. S., 12 m. SW of Paris. Pop. 1,448.—Also a township of Caledonia co., Vermont, on the Connecticut, 14 m. E of Danville. Pop. 1,412.—Also a township of New London co., Connecticut, 4 m. NW of New London. Pop. 2,262.—Also a township and village of Saratoga co., New York, on the W bank of the Hudson, 10 m. N of Albany. Pop. in 1840, 1,824; in 1850, 2,683.—Also a township of Erie co., Pennsylvania, on French creek, 15 m.

PORTS IN IRELAND EAST COAST St GEORGE'S CHANNEL

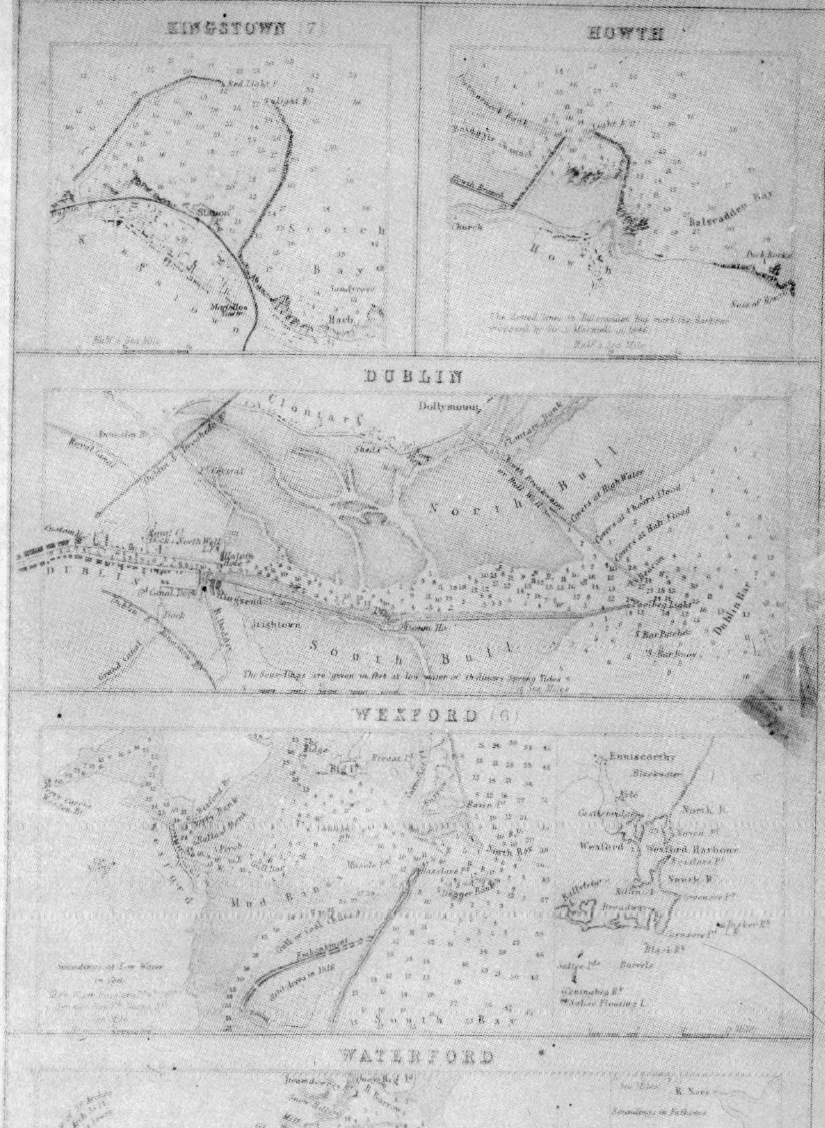

Constructed & Engd. from the Admiralty Survey by G. H. Swanston Edinburgh

A. Fullarton & Co. London & Edinburgh

SSE of Erie. Pop. 1,300.—Also a township of London co., Virginia. Pop. 500.—Also a township of Washington co., Ohio, on the Muskingum, 16 m. S of Erie. Pop. in 1840, 1,166; in 1850, 1,698.

WATERFORD-HARBOUR, either the joint estuary of the rivers Suir and Barrow, between co. Waterford, Munster, and co. Wexford, Leinster, or likewise the portion of the river Suir downward from the city of Waterford, partly through the county of Waterford, Munster, and partly between that county and the county of Kilkenny, Leinster. The estuary is entered from the ocean between Hook-head on the E and Swiney-head on the W. The entrance to the harbour is open, sufficiently facile, and comparatively very safe; but Tramore bay, situated a short distance to the W of it, has frequently been mistaken for it. The E shore trends 2¾ m. NNE to the head of Oldtown bay; 3⅓ m. N by W to Duncannon-fort; 2½ m. in the direction of NW by N, to a point about 1 m. above Ballyhack; and 2¼ m. N to a point about 1¼ m. N of Cheek Point, and at the extremity of a small ramified lagoon, formed to the N and the NE of Cheek Point. Most of the shore is rocky, yet interspersed with sandy beach. and belted at low water with a portion of strand.

WATERGRASSHILL, a village of co. Cork, 9 m. NE of Cork. Pop. 651.

WATERHEAD, a hamlet of Cumberland, 24 m. NE of Carlisle.—Also a village of Lancashire, 2 m. E by N of Oldham.

WATERHOUSE, a small island lying off the N coast of Van Diemen's Land, in S lat. 40° 48′. It is about 4 m. in length, and rises abruptly to a moderate elevation.

WATERHOUSE (Point), a cape on the N coast of Van Diemen's Land, forming the W point of Ringarooma bay.

WATERINGBURY, a parish of Kent, 5 m. W by S of Maidstone. Area 1,470 acres. Pop. 1,448.

WATERINGEN, a village of Holland, in the prov. of South Holland, 12 m. NW of Rotterdam. Pop. 1,450.

WATER-KEY, a small island in the bay of Honduras, near the coast of Mexico, in N lat. 17° 30′.—Also a small island in the Caribbean sea, near the Mosquito shore, in N lat. 12° 15′ N.

WATERLAND, a district of Holland, in the prov. of N. Holland, to the N of Amsterdam, and SE of Alkmaar.

WATERLAND, an island in the S. Pacific, in S lat. 14° 46′, discovered by Le Maire and Schouten, in 1616. It is low, sandy, and full of rocks.

WATERLAND-OUDENAU, a commune and village of Belgium, in the prov. of E. Flanders, arrond. of Eecloo. Pop. 681.

WATERLOO, a commune and village of Belgium, 10 m. S of Brussels, on the road from Brussels to Namur. Pop. 2,756. The village is now nearly joined to that of Mont St. Jean, memorable for the greatest, and, let us hope, the last battle in which the French and British measured their strength against each other, fought on the 18th of June 1815. The forces engaged were, until late in the evening, nearly equal, the French reckoning 71,000 men, while the troops under Wellington were about 54,000, of whom 32,000 were British, and those under Bulow, who came up early in the afternoon, were about 15,000. The ground on which the British were drawn up was a gentle ascent from a shallow valley, the corresponding ridge, on the opposite side, forming the French position. The contest opened by Napoleon ordering an assault towards the right of the British, on the Chateau-de-Goumont, or Hougoumont, a post which, as the key of the British position, was defended with such firmness, that the French could gain possession only of the plantation, and failed in their attempts to drive the British from the chateau. This encounter, which took place between 11 A. M. and 1 P. M., was followed by a more serious onset on the British left and centre. The British plan of battle, as regarded the infantry, was defensive. Their battalions drawn up in squares, and protected by a number of field-pieces, awaited attack, as at Vimiera and Talavera; but their cavalry stood ready to seize any favourable opening to attack the advancing enemy. This occurred on more than one occasion, in the early part of the battle; and conflicts took place with that varied result which will always prevail when, in armies of equal, or nearly equal discipline, there remain regiments in reserve, to fall on a body of their opponents disordered even by success. A charge made by a body of British horse, on the flank of a French column, when marching from left to right, was attended with great success; but this was soon found to be dearly purchased, when fresh regiments of the enemy galloped up against the disordered line of the British; the field was covered with a number, apparently equal, of French and British uniforms; and the French cavalry were repulsed only by the arrival of a fresh body of British dragoons. On the part of the British infantry, the defensive plan being strictly followed, the resistance was almost uniformly successful. The French generals witnessed a dreadful carnage; but observing that the British never advanced, and being unable to see clearly the whole of their battalions, they ventured between 4 and 5 o'clock P. M., to do what should never be hazarded against highly disciplined infantry until fatigued and disordered, they brought forward to the charge their cavalry-reserve of the imperial guards. Twice did this "fiery mass of living valour" rush on the British field-pieces and battalions; though partially successful against the former, they failed against the latter; and were obliged each time to retreat with heavy loss. "This charge," said Bonaparte, who stood on high ground at some distance, "is too early by an hour!" "Ney," rejoined Soult, "commits us as he did at Jena!" After two hours more of firing and partial attacks, Napoleon thought it time to bring forward his final reserve, the imperial foot-guard. This took place at 7 o'clock, and brings us to the most remarkable juncture of the battle, that juncture in which, on almost every former occasion, whether against Austrians, Prussians, or Russians, the attack of a corps fresh and high-spirited, caused the rout of the opposing line. Here, however, the case was different; the British troops had been well supported; and though fatigued were not shaken. The blanks in their ranks had been successively filled by drafts from the reserve; and, if few expected victory, all were determined rather to fall than yield. Wellington alone had higher hopes; he knew that the Prussian main body was approaching; and making his squares dissolve order, and form into a continuous line, obtained, in their musketry fire, a great advantage over an enemy formed chiefly in close column. Affairs were now drawing to a crisis. Lord Wellington, observing the march of Blucher, ordered a forward movement; and the French, seeing on one side the British advance, on the other that the high road in their rear was on the point of being forced by the Prussians, relinquished the field of battle, and sought safety in retreat. Napoleon is stated to have criticised the position of W., and to have observed that it had a very material defect—namely, that of only having one communication to the rear across the forest of Soignes, and consequently the total ruin of Wellington's army must have taken place had his grace been found to retreat. From the position of W., however, so far from there being only one road towards Brussels,—there are several. From Brain-la-leud—to begin with the r.—there is a direct road, and almost the whole of it a good *chaussée;* then comes the Waterloo *chaussée.* Close to W. are the small villages of Raupart and Rambuck, from both of which there are good cart roads across the forest. From La-Roche, a little more than 2 m. to the l. of Rambuck, and only about 3 m. from W., there is a *chaussée* to Brussels; from Wavre a *chaussée* leads to the same city. The forest of Soignes is the place whence the inhabitants of Brussels draw their fuel. There is no under-wood: it is composed entirely of beech, and intersected in every direction by vistas and alleys in which the nobility and gentry of Brussels ride and drive. The forest is practicable for cavalry, artillery, and infantry, in almost every direction. "If the allied army," says Sir James Carmichael Smyth, "had been obliged to fall back from W., what steps the duke of Wellington might have thought proper to take, it is impossible to say; but the loss of the army does by no means appear to have necessarily followed as the consequences of a retrograde movement, as has been insinuated. The roads from Brain-la-leud, from W. and from Wavre, meet upon very strong ground, about 1½ m. in front of Brussels. There exists a very excellent position, extending across these three *chaussées*, the right of which may be said to be at Weill, upon the Senne, its centre at Vleugat, and its left thrown back to Exilles, and on which the Netherland troops from Brain-la-leud, the British from W., and the Prussians from Wavre, might have been assembled without any great difficulty. As the division from Tubice, and the Netherland corps under Prince Frederick of Orange, would in that case have fallen back by the *mons chaussée* towards Brussels, they might have occupied the right of the new position, and in which, supposing the French to have been successful at W., they might have found an army more numerous than that with which they had been engaged, drawn up, and in readiness to receive them."

WATERLOO, a village of Seneca co., New York, U. S., on Seneca river, 14 m. W of Auburn. Pop. of t. in 1840, 3,036; in 1850, 3,795.—Also a v. of Fayette co., in Indiana, 52 m. E by S of Indianapolis. Pop. 1,000.—Also a township of Jackson co., in Michigan, 29 m. SE by S of Lansing. Pop. 1,090.—Also a township of Jefferson co., in Wisconsin, 22 m. E by N of Madison.

WATERLOO, a township in Wellington district, Upper Canada, intersected by Grand river.—Also a village in the township and 3 m. W of Kingston, Upper Canada.—Also a v. in the township of Bertie, on Niagara river.

WATERLOO, a village of Lancashire, 5 m. NW of Liverpool, at the mouth of the Mersey. Pop. 750.

WATERMAEL-BOITSFORT, a commune and village of Belgium, in Brabant, 4 m SE of Brussels.

WATER-MILLOCK, a chapelry in the p. of Greystock, Cumberland, 7 m. SW of Penrith, on the NW bank of Ulleswater. Pop. in 1841, 524.

WATER-NEWTON, a parish of Southamptonshire, near Peterborough.

WATER-OVERTON, a hamlet of Warwickshire, 2½ m. NW of Coleshill.

WATER-PERRY, a parish of Oxfordshire, 5½ m. W of Tame. Area 2,620 acres. Pop. in 1851, 258.

WATERS (POINT), a cape on the W coast of North America, in N lat. 60° 5′.

WATERSAY, one of the Hebrides, lying to the S of the island of Barray, from which it is distant about 1 m. It is about 3 m. long, and in some places tolerably fertile. It possesses an excellent harbour for vessels of any burden, sheltered from all storms by the island of Sanderay, and Muldonich or the Deer island.

WATERSTOCK, a parish of Oxfordshire, 5 m. W of Tame. Area 653 acres. Pop. 141.

WATERSTREET, a village of Huntingdon co., Pennsylvania, U. S., on the Juniatta river, 70 m. WNW of Harrisburg.

WATERTOWN, a township of Middlesex co., Massachusetts, U. S., on Charles river, 7 m. W of Boston. Pop. in 1840, 1,810; in 1850, 2,837.—Also a township of Litchfield co., Connecticut, 12 m. SSE of Litchfield. Pop. 1,533.—Also a capital of Jefferson co., New York, at the mouth of Black river, 145 m. NW by W of Albany. Pop. in 1840, 5,027; in 1850, 7,201. It has large woollen and cotton factories, and is a place of deposit for the military stores of New York.

WATERVILLE, a township and village of Kennebec co., Maine, U. S., on the W side of the Kennebec, opposite Winslow, and 17 m. NNE of Augusta. Pop. in 1840, 2,971; in 1850, 3,965. It is finely situated at the head of boat-navigation, opposite Ticonic falls, which present a majestic and beautiful cascade of 18 ft. in height, extending across the river and affording immense water-power.—Also a village of Haven co., in Connecticut, 22 m. WNW of New Haven.—Also a v. of Oneida co. in New York, 86 m. W by N of Albany. Pop. 1,000.—Also a v. of Louisville co., in Vermont, 13 m. W by N of Montpelier.

WATERVLIET, a town of Belgium, in the prov. of E. Flanders, 12 m. NW of Ghent.

WATERVLIET, a town of Albany co., New York, U. S., on the W side of the Hudson, 6 m. N of Albany. Pop. in 1840, 10,141; in 1850, 16,675.

WATFORD, a parish and market-town in the county of Hertford, in the line of the London and Birmingham railway, on which there is here a principal station, 17½ m. from London. The Birmingham line here passes through the Watford tunnel, which is 1 mile and 70 yds. in length, 25 ft. in height, and 24 ft. in width, and is cut through an earth composed of sand and gravel. The parish includes the hamlets of Cashio, Levesden, Oxhey, and Town. Area 10,792 acres. Pop. in 1801, 3,530; in 1831, 5,293; in 1851, 6,546.—The town consists of a well-built and paved street nearly 1 m. in length, rising on a hill with a gentle ascent and nearly surrounded by the Colne, on the banks of which are several mills. The principal manufacture is the throwing of silk. The Grand Junction canal, which passes within 1 m. to the W of the town, and the London and Birmingham line of railway, open up an extensive communication with the northern parts of the kingdom as well as with the metropolis.—Also a parish in Northamptonshire, 5 m. NNE of Daventry, in the line of the London and Birmingham railway, and the Union canal. Area 3,080 acres. Pop. in 1831, 353; in 1851, 503.

WATH, a parish, comprising the chapelries of Norton-Conyers, Melmerby and Middleton-Quernhow, with the township of W., in the wapentake of Hallikeld, all in the N. R. of the county of York, on the river Ouse, and at an average distance of 4 m. N of Ripon. Area 3,568 acres. Pop. in 1831, 730; in 1851, 747.—Also a township in the p. of Hovingham, N. R. of the co. of York.

WATHLINGEN, a village of Hanover, in the gov. of Lüneberg, near the l. bank of the Fuse, 8 m. SSE of Celle. Pop. 599.

WATH-UPON-DEARNE, a parish in the W. R. of Yorkshire, 6 m. N of Rotherham, in the line of the North Midland railway, and the Dearne and Dove canal. It comprises the chapelries of Nether-Hoyland, Swinton, and Wentworth, and the townships of Brompton-Bierlow, and W. Area 10,709 acres. Pop. of the parish, in 1851, 9,521.

WATIETITZ, a village of Austria, in Bohemia, in the regency of Pilsen, and bail. of Schultenhofen. Pop. 840. It has a castle, and manufactories of glass and of paper.

WATIGNIES, or WATTIGNIES, a commune of France, in the dep. of the Nord, and cant. of Seclin, 4 m. SW of Lille. Pop. in 1846, 2,226.

WATIGNY, a village of France, in the dep. of the Aisne, and cant. of Herson, 5 m. N of Aubenton. Pop. 650. It has iron-works.

WATIHO, an island in the S. Pacific ocean, one of the Harvey group, in S lat. 19° 58′, discovered by Captain Cook in 1777. It is about 8 m. in length, finely varied by hills and plains, and covered with verdure.

WATLING'S ISLAND, an island of the West Indies, in the Bahama archipelago, in N. lat. 24°.

WATLINGTON, a parish and town of Oxfordshire, situated on a small brook among the Chiltern hills, 24 m. SE of Oxford. Area 3,440 acres. Pop. in 1851, 1,884.—Also a parish of Norfolk, 5½ m. N of Market-Downham. Area 1,709 acres. Pop. 557.

WATLISLAW, a village of Austria, in Bohemia, in the regency of Böhmisch-Leipa, and bail. of Leitmeritz. Pop. 250.

WATOU, a department and commune of Belgium, in the prov. of West Flanders, and arrond. of Furnes. Pop. of dep., 2,939; of com., 1,039. It has a considerable trade in tobacco and sugar.

WATRA, a town of Galicia, in the circle and 69 m. SSW of Tschernowitz.

WATRELOS, a village of France, in the dep. of the Nord, and cant. of Roubaix, 9 m. ENE of Lille. Pop. 4,485. It has extensive cotton factories, and several brick-kilns.

WATSCHENOWITZ, a village of Austria, in Moravia, in the circle of Hradisch. Pop. 690.

WATSON, a township of Lewis co., in the state of New York, U. S., 10 m. NE of Martinsburgh. Pop. 1,707.

WATTAWA, a river of Bohemia, which rises near the frontier of Bavaria, and falls into the Muldau, near Prague.

WATTEN, a parish in the centre of the co. of Caithness, about 14 m. long, and 10 or 11 broad, watered by the river Wick. Pop. in 1851, 1,351.

WATTENBUCH, a village of Kircherren, in the prov. of Lower Hesse, circle of Kassel. Pop. 416.

WATTENHEIM, a village of Bavaria, on the Pfalz, cant. of Grunstadt. Pop. 1,215.

WATTENS, a village of Austria, in Tyrol, landg. of Hall. Pop. 1,710.

WATTENSCHEIDT, a village of Prussia, in the reg. of Arnsberg, circle of Bochum. Pop. 1,035.

WATTENWEIL, a village and district of the Swiss cant. of Bern, circle of Septigen. Pop. 2,300.

WATTENWEILER, or WATWILLER, a town of France, dep. of Haut-Rhin, com. of Belfort, at the foot of the Bogesen. Pop. 1,854.

WATTENWEILER, a town of Bavaria, in the circle of Pfalz, cant. of Zweibruchen. Pop. 462.

WATTESFIELD, a parish of Suffolk, 12½ m. NE by E of St. Edmund's Bury. Area 1,517 acres. Pop. 603.

WATTISHAM, a parish of Suffolk, 2 m. NE of Bildeston. Area 1,298 acres. Pop. 220.

WATTON, a parish and town of Norfolk, 18 m. SW of Norwich. Area of p. 1,807 acres. Pop. in 1851, 1,353. Great quantities of butter are sent from this place to London.—Also a parish of Hertfordshire, 5½ m. NNW of Hertford. Area 3,499 acres. Pop. in 1851, 976.—Also a parish of Yorkshire, 5½ m. S by W of Great Driffield. Area 3,720 acres. Pop. in 1831, 345; in 1851, 315.

WATTSBOROUGH, a village of Lunenburg co., Virginia, U. S., 116 m. SW of Richmond.

WATTSBURG, a township of Erie co., in Pennsylvania, U. S., 263 m. NW of Harrisburg. Pop. 131.

WATTS-NESS, a cape on the W coast of the island of Shetland, in N lat. 60° 14′, W long. 2° 6′.

WATTWEIL, a parish and village of Switzerland, in the cant. of St. Gall, on the r. bank of the Thur, 16 m. SW of St. Gall. Pop. 4,540. There are cotton-mills here.

WATZENHORN, a village of Hesse-Darmstadt, circle of Giessen. Pop. 640.

WATZETSDORF, a village of Moravia, in the circle of Olmutz, lordship of Hohenstadt. Pop. 450.

WATZLAWITZ, a village of Moravia, in the circle of Olmutz, lordship of Bosowitz. Pop. 330.

WATZMANN, a mountain of Bavaria, belonging to the Noric Alps, in N lat. 47° 33′ 34″, rising to the height of 9,650 ft.

WAUCHOPE, a river of Dumfries-shire, which, augmented by the Laggan, after a course of some miles falls into the Esk at Langholm.

WAUFGONG, a town of Hindostan, in the prov. of Aurungabad, 24 m. N of Punah, whence the family of the celebrated Mahratta chief, Holkar, originated.

WAUJPORE, a town of Hindostan, in Gujerat, 45 m. from Surat, in N lat. 21° 24′.

WAURICHEN, a village of Prussia, in the reg. of Aachen, circle of Geilenkirchen. Pop. 330.

WAUSSIM, a town of Hindostan, prov. of Berar, cap. of a small district of the same name intersected by the Paya-Ganga.

WAVENDON, a parish of Bucks, 3½ m. NE of Fenny-Stratford. Area 2,665 acres. Pop. 935.

WAVENEY, a river of Suffolk, which rises near the source of the Little Ouse; runs past Bellingford, Harleston, Bungay, and Beccles; and joins the Yare, 5 m. above Yarmouth, after a course of 50 m. It is navigable from Bungay.

WAVERLEY, a hamlet of Surrey, near the river Way, in the p. of Farnham. Here are the ruins of the first Cistercian abbey founded in England.

WAVERLY, a village of Pike co., in Ohio, U. S., 60 m. S of Columbus. Pop. 250.

WAVERTON, a parish of Cheshire, 4½ m. SE by E of Chester. Area 4,027 acres. Pop. in 1851, 788.

WAVERTON (HIGH and LOW), adjoining hamlets in Cumberland, in the p. and 3 m. W by S of Wigton. Pop. 600.

WAVERTREE, a township in the p. of Childwall, Lancashire, 3½ m. ESE of Liverpool. Pop. 2,660.

WAVRE, a town of Belgium, in South Brabant, 15 m. SE of Brussels. Pop. in 1842, 5,250. It has manufactories of leather, cotton-yarn, and hats. On 18th June 1815, the French marshal Grouchy, at the head of 34,000 men, was engaged here by a body of Prussians under Thielmann, and prevented marching to join the French army at the battle of Waterloo.

WAVRE-NOTRE-DAME, a commune and v. of Belgium, in the prov. and 13 m. SE of Antwerp. Pop. 2,172.

WAVRE-ST.-CATHERINE, a commune and v. of Belgium, adjacent to the above. Pop. 3,174.

WAVRIN, a commune and v. of France, in the dep. of Nord, 7 m. SW of Lille. Pop. 1,768.

WAWARSING, a township of Ulster co., in New York, U. S., 22 m. SW of Kingston. Pop. 4,050.

WAWAY, a small island in the eastern seas, near

the E coast of the island of Celebes, in S lat. 4° 8′. It is about 40 m. in circumference.

WAWRA, a small town of Kaarta, in Central Africa, 60 m. ESE of Benowm. It is surrounded by high walls, and is inhabited by Negroes.

WAWUL, a town of Hindostan, in Gujerat, situated on the banks of the Sereswati river, SE of Rahdunpore.

WAXHAM, a parish of Norfolk, 12 m. ESE of North-Walsham. Area 2,087 acres. Pop. 76.

WAXHOLM, a town of Sweden, on the island of Vaxo, in the Baltic, at the entrance of a narrow passage leading to Stockholm, 16 m. E of that city. It has a strong castle, built in 1649, to guard the entrance to the capital. Pop. 1,000.

WAXWAY, an island in the Euxine sea, off the W coast of Celebes, in S lat. 3° 33′.

WAY, a hamlet of England, in the p. of Throwley, Devonshire.

WAY, or Pulo-Way, an island off the NW coast of Sumatra, in N lat. 5° 40′. It is 12 m. in length.—Also an island in the gulf of Siam, in N lat. 9° 58′.

WAYBARI, a river of Guayana, which runs into the Atlantic, in N lat. 6° 25′.

WAYFORD, a parish of Somersetshire, 2 m. SW of Crewkerne. Area 1,618 acres. Pop. in 1851, 238.

WAYGIOU. See Waigiou.

WAYLAND, a township of Middlesex co., in Massachusetts, 17 m. W of Boston. Pop. 998.

WAYNE, a county of New York, U. S., in the N part of the state. Area 572 sq. m. Pop. in 1840, 42,057; in 1850, 44,953. Its cap. is Lyons.—Also a co. in the NE of Pennsylvania. Area 762 sq. m. Pop. in 1840, 11,848; in 1850, 21,890. Its cap. is Bethany.—Also a co. on the E side of Indiana. Area 420 sq. m. Pop. in 1840, 23,292; in 1850, 25,320.—Also a county in the central part of N. Carolina. Pop. in 1840, 10,981; in 1850, 13,486. Its cap. is Waynesboro'.—Also a county in the interior of Ohio. Area 660 sq. m. Pop. in 1840, 35,908; in 1850, 32,981. Its cap. is Wooster.—Also a county on the S side of Kentucky. Area 637 sq. m. Pop. in 1840, 7,399; in 1850, 8,692. Monticello is the chief town.—Also a county in the SE of Georgia. Area 716 sq. m. Pop. in 1840, 1,258; in 1850, 1,499.—Also a county in the E of Virginia. Area 412 sq. m. Pop. in 1850, 4,760. Its cap. is Wayne.—Also a township of Kennebec co., Maine, 20 m. W of Augusta. Pop. 1,367.—Also a township of Steuben co., New York, 15 m. E of Bath. Pop. 1,347.—Also a township of Lycoming co., Pennsylvania. Pop. 1,000.—Also a township of Mifflin co., Pennsylvania, on the Juniatta, 15 m. ESE of Huntingdon.—Also a township of Stark co., Ohio. Pop. 500.—Also a township of Allen co., Indiana, 100 m. NNE of Indianapolis. Pop. in 1850, 5,283.—Also a township of Erie co., in Pennsylvania. Pop. 1,000.

WAYNESBOROUGH, the cap. of Greene co., Pennsylvania, U. S., 22 m. S of Washington. Pop. 852.—Also a township of Augusta co., Virginia, 12 m. ESE of Staunton.—Also the cap. of Wayne co., N. Carolina, on the Neuse, 50 m. SE of Raleigh. Pop. 400.—Also the cap. of Burke co., Georgia, 24 m. ENE of Louisville. Pop. 200. Also a village of Franklin co., in Pennsylvania. Pop. 850.

WAYNESVILLE, a town of Warren co., Ohio, on the Little Miami, 40 m. NE of Cincinnati. Pop. 600, chiefly Quakers.—Also a township of Pulaski co., in Missouri, 50 m. S of Jefferson. Pop 200.—Also the cap. of Haywood co., in N. Carolina, 240 m. W by S of Raleigh.

WEALD, a chapelry in the p. of Seven-Oaks, Kent, 2½ m. S of Seven-Oaks. Pop. in 1851, 963.

WEALD (North), a parish in Essex, 3¼ m. NE of Epping, containing the hamlets of Haslingwood and Thornwood. Area 3,377 acres. Pop. 842.

WEALD (South), a parish in Essex, 1½ m. W of Brentwood, near the Eastern Counties railway, containing the chapelry of Brentwood, the township of South W., and the hamlet of Brookstreet. Area 5,037 acres. Pop. in 1831, 1,183; in 1851, 3,588.

WEAR, a river in the co. of Durham, which flows E and NE by Auckland, Binchester, and Durham, and falls into the German ocean at Sunderland, after a course of 70 m.

WEARDALE-ST.-JOHN, a chapelry in the p. of Stanhope, Durham, 6¼ m. WNW of Stanhope, in the vale of the river Wear.

WEARDLEY, a township in the p. of Harewood, Yorkshire, 6¼ m. E of Otley. Area 1,080 acres. Pop. in 1831, 894; in 1851, 144.

WEARE, a parish in Somerset, 1½ m. S by W of Axbridge. Area 2,146 acres. Pop. in 1851, 715.

WEARMOUTH (Bishop or South), a parish in the co.-palatine of Durham, forming the SW part of what is commonly called Sunderland.

WEARMOUTH (Monk or North), a parish in the co.-palatine of Durham, on the N bank of the Wear, united with Sunderland by its celebrated bridge, and by ferry-boats forming a constant thoroughfare. The p. comprises the townships of Monk-Wearmouth, Monk-Wearmouth-Shore, Fulwell, Hylton, and Southwick. Area of p., 5,419 acres. Pop. in 1831, 9,428; in 1851, 16,911.

WEARMOUTH-SHORE (Monk), a township in the p. of Monk-Wearmouth, co.-palatine of Durham, on the N bank of the river Wear, and adjacent to, and built on continuous with, the town of Sunderland: which see. Area 308 acres. Pop. in 1831, 6,051; in 1851, 10,109.

WEASENHAM (All-Saints), a parish in Norfolk, 7½ m. SW of Fakenham. Area 1,988 acres. Pop. in 1831, 313; in 1851, 363.

WEASENHAM (St. Peter), a parish in Norfolk, 7 m. SW of Fakenham. Area 1,423 acres. Pop. 326.

WEATHERSFIELD, or Wethersfield, a parish in Essex, 6¾ miles NNW of Braintree. Area 4,228 acres. Pop. in 1831, 1,698; in 1851, 1,770.

WEAVER, a river of Cheshire: which see.

WEAVERHAM, a parish in the co.-palatine of Chester, 3¼ m. W by N of Northwich, on the line of the Grand Junction railway. Area 7,634 acres. Pop. in 1831, 2,321; in 1851, 2,745.

WEAVERTHORPE, a parish in the E. R. of the co. of York, 10½ m. NNW of Great-Driffield. Area 5,100 acres. Pop. in 1831, 753; in 1851, 1,066.

WECHERSDORF, a village of Bohemia, in the reg. of Getschin, bishopric of Brannau. Pop. 1,700.

WECHSELBURG, a village of Saxony, in the circle of Leipzig, bail. of Rochlitz. Pop. 1,192.

WEDDINGTON, a parish in Warwick, 1½ m. N of Nuneaton. Area 911 acres. Pop. in 1851, 54.

WEDMORE, a parish in Somerset, 4¾ m. S by E of Axbridge. Area 9,986 acres. Pop. in 1851, 3,905.

WEDNESBURY, or Wodensbury, a parish and market-town in the co. of Stafford, 19 m. SSE of Stafford, near the source of the river Thame, and in the line of the London and Birmingham section of the Great North-Western railway, and the Birmingham and Walsall canal. Area 2,175 acres. Pop. in 1831, 8,437; in 1851, 14,281. The district is celebrated for its mines of coal and ironstone, and for the manufacture of guns, gun-locks, gun-barrels, and various other articles of iron and steel, hinges, nails, screws, files, edge-tools, gas tubes, stove-grates, and cast-iron articles. The coal is considered the best in England for the smith's forge, on account of its peculiar intensity of heat. It is found in separate beds from 3 to 14 ft. thick. A reddish earth, called *kipp*, found in the vicinity, is used in glazing vessels of various kinds.

WEDNESFIELD, a chapelry and township in the p. of Wolverhampton, 2 m. NE of Wolverhampton, on the Wyreby and Essington canal, and 1½ m. N of the London and Birmingham railway. Area 3,700 acres. Pop. in 1831, 1,879; in 1851, 4,858. Immense quantities of locks, keys, and traps of every description, are manufactured here for the Birmingham and Wolverhampton merchants. It is memorable as the scene of a decisive victory which Edward the Elder obtained over the Danes in 910.

WEEDON, a hamlet in the p. of Hardwicke, Bucks, 3 m. N of Aylesbury. Area 1,860 acres. Pop. in 1831, 405; in 1851, 447.

WEEDON-BECK, or WEEDON-ON-THE-STREET, a parish in Northamptonshire, 4 m. SE of Daventry, in the line of the London and Birmingham section of the Great North-Western railway, which is here carried through a tunnel 400 yds. in length, and on which there is here a principal station 69¼ m. from London. The p. is also crossed by the Grand Junction canal, which is carried here across a valley by means of a noble embankment. Area of p., 1,710 acres. Pop. in 1831, 1,439; in 1851, 1,996. Here are spacious barracks, containing an hospital, parade, &c., and forming an important military establishment, and a depot capable of receiving 200,000 stand of arms, and a great quantity of warlike stores. The Grand Junction canal communicates with the storehouses, and the railway passes close to them.

WEEDON-LOYS, or PINKENEY, a parish in Northamptonshire, 6 m. W by S of Towcester. Area 1,050 acres. Pop. with the hamlet of Weston, in 1831, 526; in 1851, 545.

WEEFORD, a parish in Staffordshire, 4 m. SSE of Lichfield, containing the hamlets of W., Thickbroom, and Swinfen, and the liberty of Packington. Area 4,556 acres. Pop. in 1831, 306; in 1851, 425.

WEEK, a parish in Hants, 1 m. NW by W of Winchester. Area 1,050 acres. Pop. in 1851, 446.

WEEKLEY, a parish in Northamptonshire, 2 m. NE by N of Kettering. Area 1,800 acres. Pop. in 1831, 273; in 1851, 265.

WEEK-ST.-LAWRENCE, or WICK, a parish in Somersetshire, 8½ m. NNW of Axbridge, near the mouth of the river Yeo, and 3 m. NW of the Bristol and Exeter railway. Area 1,000 acres. Pop. in 1831, 281; in 1851, 300.

WEEK-ST.-MARY, a parish in Cornwall, 6 m. S of Stratton. Area 5,824 acres. Pop. in 1851, 641.

WEELEY, or WYLEY, a parish in Essex, 8 m. SSE of Manningtree. Area 2,087 acres. Pop. 617.

WEEM, a dissevered and fragmentary parish in Perthshire, dispersed in separate and far-distant portions, over nearly a fourth of that county from near the head of Glenlochy on the W, to the vicinity of Strathbran on the E, and from 3 m. S of Loch Tummel on the N, to the vicinity of Loch-Earn on the S. W. Proper, or the district in which the church is situated, lies in Strath-Tay, along the l. bank of the river, opposite the village of Aberfeldy. Pop. of p. in 1831, 1,209; in 1851, 740.

WEENDAM, a village of Holland, in the prov. and 17 m. SE of Groningen, and arrond. of Winschoten. Pop. 6,000.

WEENDE, or WEHNDE, a village of Hanover, in the gov. of Hildesheim, 6 m. N of Gottingen, on the Leine. Pop. 1,080. It has manufactories of cloth, paper, and tobacco.

WEENER, a market-town and bailiage of Hanover, in the gov. and 24 m. SSW of Aurich, on the l. bank of the Ems, with which it communicates by a canal. Pop. of town, 2,658; and of bail., 12,000. It has two churches, a Catholic and a Reformed, and carries on an active trade in horses, and in oats, barley, butter, and cheese.

WEENERMOOR, a village of Hanover, in the gov. of Aurich. Pop. 380.

WEER, a village of Austria, in the Tyrol, in the reg. of Innspruck, bail. of Schwatz. Pop. 450.

WEERBURG, a village of Austria, in the Tyrol, in the regency of Innspruck, and bail. of Schwatz. Pop. 1,160.

WEERDE, a department and commune of Belgium, in the prov. of Brabant, and arrond. of Brussels, watered by the Jenne. Pop. 580.

WEERDT, a town of Belgium, in the prov. of Limburg, arrond. and 14 m. WNW of Ruremonde, near the Bree. Pop. in 1838, 6,095. It has a fine church, and a college, and contains numerous distilleries of brandy, manufactories of cloth, hats, and hosiery, and carries on an active trade in grain, linen, yarn, and cattle. It was taken by the French in 1792.

WEERDT (NEDER), a village of Belgium, in the prov. of Limburg, arrond. and 14 m. WNW of Ruremonde, and cant. of Weerdt. Pop. 4,173. It has several breweries, distilleries of brandy, tanneries, oil and saw-mills, and manufactories of printed calico.

WEERE, a town of Holland, in the prov. of Zeeland, and island of Walcheren, near the mouth of the East Scheldt. Pop. 1,500.

WEERE'S CREEK, a river of New South Wales, in the district of Liverpool Plains, an affluent of the Condailly river.

WEERSELO, a market-town of Holland, in the prov. of Over-Yssel, to the SE of Almelov. Pop. 4,300.

WEESENSTEIN, a village of Saxony, in the bail. and SW of Pirna, on the Muglitz. Pop. 251. It has a castle belonging to Prince Maximilian of Saxony, with extensive domains.

WEESHANA, a village of Afghanistan, 35 m. S Cabul.

WEESOW, a village of Prussia, in the regency of Potsdam, and circle of Oberbarmin. Pop. 220.

WEESP, a town of Holland, in the prov. of N. Holland, arrond. and 8 m. SE of Amsterdam, on the l. bank of the Vecht, in N lat. 52° 18′ 29″. Pop. 895. It has distilleries of brandy.

WEETFELD, a village of Prussia, in the regency of Arnsberg, and circle of Hamm. Pop. 300.

WEETHLEY, or WETHELE, a parish in Warwickshire, 3 m. SW by W of Alcester. Area 638 acres. Pop. in 1831, 62; in 1851, 48.

WEETING-WITH-BRUMHILL, a parish in Norfolk, 1½ m. N of Brandon-ferry. Area 6,187 acres. Pop. in 1801, including the hamlet of Brumhill, 368; in 1831, 357; in 1851, 429.

WEETON, a township in the p. of Kirkham, co.-palatine of Lancaster, 3½ m. NW by W of Kirkham, in the line of the Preston and Wyre railway. Area 2,876 acres. Pop. in 1831, 477; in 1851, 465.—Also a township in the p. of Harewood, W. R. of the co. of York, 6 m. ENE of Otley. Area 1,230 acres. Pop. in 1831, 322; in 1851, 300.

WEEVER, a township in the p. of Middlewich, co.-palatine of Chester, 4 m. WSW of Middlewich, near the river Weaver. Area 1,200 acres. Pop. in 1831, 196; in 1851, 140.

WEEZE, a market-town of Prussia, in the regency of Dusseldorf, circle and 11 m. NNW of Geldern, on the l. bank of the Nierse. Pop. 451. It has two churches, a Catholic and a reformed, and possesses manufactories of linen.

WEFENSLEBEN, a village of Prussia, in the regency and 21 m. W of Magdeburg, and circle of

Neu Haldensleben. Pop. 598. It has manufactories of alum and vitriol. Coal and free-stone are wrought in the environs.

WEFERLINGEN, a market-town of Prussia, in the regency of Magdeburg, and circle of Gardelegen, on the Aller. Pop. 1,755. It has a castle, and possesses salt-works.

WEGBERG, a village of Prussia, in the regency of Aachen, and circle of Erkelenz. Pop. 780. It has manufactories of linen, ribbon, velvet, and woollen fabrics, and several distilleries of brandy.

WEGE, a village of Waldeck, on the Eder. Pop. 310.

WEGEFAHRT, a village of Saxony, circle and prov. of Dresden. Pop. 748.

WEGELEDEN, a town of Prussia, in the regency of Magdeburg, and circle of Oschersleben, 2 m. ESE of Halberstadt, on the l. bank of the Bode. Pop. 2,400. It is enclosed by walls with two gates, and has a castle and an hospital, and possesses manufactories of linen and earthen-ware. In the vicinity, on the r. bank of the river, is an ancient abbey, founded in 1282, by Albert I., bishop of Anhalt.

WEGENSTADT, a village of Prussia, in the regency of Magdeburg, and circle of Gardelegen. Pop. 340.

WEGENSTETTEN, a village of Switzerland, in the cant. of Aargau, district and 8 m. ESE of Rheinfelden, and circle of Stein. Pop. 755.

WEGFURT, a village of Bavaria, in the circle of Lower Franconia, and presidial of Bischofsheim. Pop. 636.

WEGGIS, a village of Switzerland, in the cant. and 6 m. E of Lucerne, on the N bank of lake Waldstettes. Pop. 1,176.

WEGGUM, a village of Prussia, in the regency of Potsdam, and circle of Zemplin. Pop. 290.

WEGNEZ, a department and commune of Belgium, in the prov. of Liege, and arrond. of Verviers, watered by the Vesdre. Pop. of dep., 884.

WEGROW, a town of Poland, in the gov. of Podlachie, obwod and 21 m. NW of Siedlec, on the r. bank of the Liwiec. Pop. 2,000. It has a school.

WEGSCHEID, a village of Bavaria, in the circle of Upper Bavaria, 12 m. NE of Passau. Pop. 245.

WEGSCHEIDT, a village of Bavaria, in the circle of Lower Bavaria, NW of Passau. Pop. 245.

WEGSTADT, a town of Bohemia, in the circle of Melnik, on the Elbe, 22 m. W of Jung-Bunzlau.

WEHEN, a village of Nassau, 14 m. NNW of Mentz. Pop. 684.

WEHINGEN, a village of Prussia, in the regency of Treves, circle of Saarburg. Pop. 340.

WEHLAU, a town of Prussia, in the reg. of Konigsberg, at the junction of the Alle and the Pregel, with a bridge over the latter river, 28 m. E by S of Konigsberg, in N lat. 54° 36′ 35″. Pop. 4,194. It has a considerable trade in agricultural produce.

WEHLEN, a village of Prussia, in the regency of Treves, circle of Bernkastel. Pop. 983.—Also a v. of Saxony, in the bail. of Hohnstein, on the r. bank of the Elbe. Pop. 932.

WEHR, a village of Prussia, in the regency of Aachen, 9 m. S of Bonn. Pop. 358.—Also a v. in the regency of Coblenz, circle of Maren. Pop. 539.—Also a bailiwick and village of Baden, 6 m. N by W of Sackingen. Pop. 2,000.—Also a river of Westphalia, which joins the Aa at Hervorden.

WEHRAU, a village of Prussia, in the reg. of Liegnitz, on the Queiss. Pop. 524.

WEHRDA, a village of Kurhessen, in the prov. of Fulda, circle of Hunfeld. Pop. 995.—Also a village of Upper Hesse, in the circle of Marburg. Pop. 478.

WEHRDEN, a village of Prussia, in the regency of Minden, on the Waer. Pop. 480.

WEHRENDORF, a village of Hanover, in the circle of Osnabruck, bail. of Mittlage-Hauteburg. Pop. 808.—Also a v. of Prussia, in the reg. of Minden, circle of Herford. Pop. 425.

WEHRINGEN, a village of Bavaria, in the circle of Swabia, landg. of Schwabmünchen. Pop. 706.

WEHRSTADT, a village of Prussia, in the regency of Magdeburg, circle of Halberstadt. Pop. 375.

WEHRSTEDT, a village of Hanover, in the bail. of Waldenberg, near Hildersheim. Pop. 315.

WEI, a town of China, in the prov. of Chih-le, and div. of Ta-ming-fu, in N lat. 36° 25′, and E long. 115° 6′.—Also a district and town in the prov. of Chih-le, and div. of Kwang-ping-fu, in N lat. 37° 5′, and E long. 115° 25′.—Also a district and town in the prov. of Shan-tung, and div. of Lae-chu-fu, in N lat. 36° 46′, and E long. 119° 20′.

WEIBERG, a village of Prussia, in the regency of Minden, circle of Buren. Pop. 380.

WEIBERSBRUNN, a village of Bavaria, in the circle of Lower Franconia, landg. of Rothenbuch. Pop. 864. It has glass-works.

WEICHSELBURG, a town of Austrian Illyria, 9 m. ESE of Laybach. Pop. 4,000. It is the centre of the wool-trade of Carniola.

WEICHTURGEN, a village of Bavaria, in Lower Franconia, landg. of Munnerstadt. Pop. 330.

WEI-CHU, a town of China, in the prov. of Szechuen, and div. of Ching-tu-fu, in N lat. 31° 25′ 12″, and E long. 103° 40′ 30″.

WEIDA, a town in the grand-duchy of Saxe-Weimar, on the river Weida, 34 m. ESE of Weimar. Pop. 3,800.—Also a small river of Saxony, in the Voigtland, which joins the Elster near Veitsberg.—Also a river of Prussian Silesia, which rises on the confines of Poland, and falls into the Oder, 9 m. below Breslau.

WEIDACH, a village of Bavaria, in Upper Bavaria, landg. of Wolfraeteshausen. Pop. 240.

WEIDELBOCH, a village of Bavaria, in Middle Franconia, landg. of Dinkelsbuhl. Pop. 260.

WEIDEN, a town of Bavaria, 42 m. N of Ratisbon. Pop. 2,200.—Also a town of Bavarian Franconia, 4 m. E of Weiss-Mayn. Pop. 2,280.—Also a village of Würtemberg, in the Schwarzwald circle, near Salz. Pop. 300.

WEIDENAU, a town of Austrian Silesia, 11 m. SW of Neisse. Pop. 1,700.—Also a v. of Kur-Hesse, in the circle of Fulda. Pop. 370.

WEIDENBACH, a town of Bavaria, in Middle Franconia, 5 m. SSE of Anspach. Pop. 820.—Also a v. of Prussia, in the regency of Breslau, circle of Oelz. Pop. 230.—Also a v. of Hungary, in the com. of Wieselburg, on the Neusiedler-see.

WEIDENBERG, a market-town of Bavaria, in Upper Franconia, 7 m. ESE of Bayreuth. Pop. 1,400.

WEIDENECK, a village of Austria, in the circle of the Upper Manhartsberg, on the l. bank of the Danube, 20 m. SW of Krems.

WEIDENHAIN, a village of Prussia, in the reg. of Merseburg, circle of Torgau. Pop. 630.

WEIDENSTETTEN, a village of Würtemberg, in the circle of the Danube, near Ulm, on the Alp. Pop. 684.

WEIFA, a village of Saxony, in the circle of Bautzen, bail. of Stolpen, near Bautzen. Pop. 753.

WEIGELSDORF, a village of Prussian Silesia, 6 m. S of Reichenbach. Pop. 1,400.—Also a v. of Moravia, in the lordship of Hertsek. Pop. 240.—Also a v. of Prussia, in the reg. of Breslau, circle of Munsterberg. Pop. 730.

WEIGELSHOFEN, a village of Bavarian Franconia, 5 m. SSW of Schweinfurt. Pop. 320.

WEIGENHEIM, a village of Bavaria, in Middle Franconia, lordship of Hohenlandeberg. Pop. 538.

WEIGHTON-MARKET, a parish and market-town in the E. R. of the co. of York, 19 m. ESE of York, near a branch of the river Foulness, at the foot of the Wolds, and communicating by a canal with the Humber. Area of p., 7,248 acres. Pop. in 1801, 1,508; in 1831, 2,169; in 1851, 2,427. The town consists of one principal street, with several small ones intersecting it.

WEIGSDORF, a village of Prussia, in Upper Lusatia, 8 m. NNE of Krottau.—Also a v. of Bohemia, in the reg. of Bohm-Leipa. Pop. 340.

WEI-HAI, a port of China, on the N coast of the prov. of Shan-tung, and div. of Ting-chu-fu, in N lat. 37° 30′, and E long. 122° 10′.

WEIHER, a village of Bavaria, on the Pfalz, cant. of Edenkyben. Pop. 799.—Also a v. of Baden, circle of the Middle Rhine, bail. of Burchsal. Pop. 739.

WEIHERS, a village of Bavaria, in Lower Franconia. Pop. 824.

WEI-HWUY-FU, a division and town of China, in the prov. of Ho-nan. The div. comprises 10 districts. The town is in N lat. 35° 27′ 40″, and E long. 114° 16′.

WEIKENDORF, a town of Lower Austria, 16 m. NE of Vienna. Pop. 1,700.

WEIKERSDORF, a town of Lower Austria, 22 m. WNW of Vienna. Pop. 1,250.

WEIKERSHEIM, a town of Würtemberg, on the Tauber, in the bail. of Mergentheim. Pop. 1,923.

WEIL, a town of Würtemberg, on the river Wurm, 11 m. WSW of Stutgard. Pop. 1,900. The celebrated Kepler was born here.—Also a town of Switzerland, in the cant. of Aargau, circle of Brangarten. Pop. 485.—Also a village in the cant. of Basle, 2 m. N of Basle. Pop. 779.

WEILBUCH, a village of Bavaria, in Lower Franconia, lordship of Amorbach, on the Mudau. Pop. 926.—Also a v. of Nassau, in the bail. of Hochheim. Pop. 735.

WEILBURG, a town of the duchy of Nassau, in the circle of Hadamar, on an eminence on the Lahn, 35 m. N by E of Mentz. Pop. 2,680.

WEILER, a village of Baden, in the Lake circle, bail. of Radolphzell. Pop. 223.—Also a v. in the bail of Pforzheim, on the Pfenz. Pop. 510.—Also a v. in the circle of Kreuznach. Pop. 650.

WEILERSBUCH, a village of Baden, in the Lake circle, bail. of Bellingen. Pop. 480.—Also a v. of Bavaria, in Upper Franconia. Pop. 680.

WEILHEIM, a village of Baden, in the Upper Rhine circle, bail. of Waldshut. Pop. 732.—Also a town of Würtemberg, 19 m. ESE of Stutgard. Pop. 710.—Also a village of Upper Bavaria, on the Amper, 26 m. SW of Munich. Pop. 2,500.

WEILMUNSTER, a village of the duchy of Nassau, 29 m. N by E of Mentz. Pop. 1,321.

WEILNAU, a village in the duchy of Nassau, 18 m. NE of Mentz. Pop. 260.

WEILTINGEN, a village of Bavaria, 42 m. SW of Nuremberg. Pop. 821.

WEIMAR (Saxe). See Saxe-Weimar.

WEIMAR, the capital of the grand-duchy of Sachsen-Weimar-Eisenach, situated on the l. bank of the Ilm, 50 m. WSW of Leipzic, and 13 m. W of Erfurt, in N lat. 50° 59′ 12″. Pop. 12,000. It stands in a valley, with a wooded mountain on the N, and hills of some elevation to the S and E, while the river winds along the S side of the town, which is built in a plain and somewhat antique style, but the number of walks and gardens which adorn and separate its different quarters, justify Madame de Stael's description of it as "not so much a town, as the country with houses in it." Its public buildings are a work-house, an hospital, two Lutheran churches, of which one, the Stadtkerche, is deserving of notice as the burying-place of the reigning family. The grand ducal residence is finely situated to the E of the town, with a park extending along the banks of the Ilm, and open to the public. Its imposing architecture has been completed by a wing recently built. The four principal halls are consecrated to Wieland, Herder, Schiller, and Goethe. The first is the most remarkable for its paintings. On a red ground detached medallions represent in fresco the principal scenes of 'Oberon,' the great work of the German Voltaire. The hall of Herder was executed by Jolger. On an escutcheon in the centre of the porquet a winged lyre is represented, the arms granted to Herder by Charles Augustus. The hall of Goethe is illustrated by scenes from his principal works. A mythological scene from the second part of 'Faust' covers a considerable portion of the walls. The Belvedere, another residence of the reigning family, is situated on an eminence to the S. W. has long held the same rank in Germany for literature, as Dresden for the fine arts; and in consequence of the liberal patronage of the court, a number of the best writers of Germany have either been educated or residents here. In the early years of the present cent., W. reckoned among its residents Schiller, Goethe, Herder, and Wieland, to all of whom statues and monuments have been erected here. Goethe's house is a large building, whose exterior presents nothing to strike the attention. Arrived in the vestibule, however, we discover a staircase adorned with antique statues and bas-reliefs; while marbles, frescoes and mouldings, strike the eye, and form an imposing entrance to the gallery containing the collections. A vast number of statues and busts outline all the apartments. Around the first hall, glass cases are arranged, which contain antique gems, bas-reliefs, Etruscan vases and a collection of David's medallions. Between the windows of the next gallery a rich collection of ancient engravings bound in immense folios is disposed. Between the massive bookcases, which contain these, glasses containing a collection of the medals of all nations are arranged. The gallery is painted in fresco, after the manner of Pompei. In the same room are some of Canova's figures, and a bust of Goethe himself, which although inferior to that of David is more like the original. The garden is tolerably large, but laid out rather for utility than refined enjoyment. A wooden summerhouse, which stands before the door with the air of a Swiss chalet, gives a peculiar character to the whole. The public institutions of W., if not numerous or extensive, are well managed. The inhabitants derive their support chiefly from the residence of the court, their manufactures being insignificant.

WEIMAR (Ober), a village of the grand-duchy of Saxe-Weimar, bail. and 1 m. SE of Weimar, on the Ilm. Pop. 550. It has an agricultural and educational establishment.—Also a village of Hesse, in the prov. of Ober Hesse, and circle of Marburg. Pop. 302.

WEIMARSCHMIEDEN, a village of Bavaria, in the circle of Lower Franconia, and presidial of Mellrichstadt. Pop. 240.

WEIMARSHEIM, a village of Bavaria, in the circle of Middle Franconia, and presidial of Ellingen. Pop. 440.

WEIMARSKIRCHEN, a village of Holland, n.

the prov. of Luxemburg, and cant. of Esch, on the Alzette. Pop. 670.

WEIMISLITZ, a market-town of Austria, in Moravia, in the circle and 15 m. NE of Znaim, on the Rokitma. Pop. 560.

WEINAHR, a village of the duchy, in the bail. and 4 m. NE of Nassau. Pop. 398.

WEI-NAN, a district and town of China, in the prov. of Shen-se, and div. of Se-gan-fu, in N lat. 34° 29′, and E long. 109° 27′.

WEINBACH, a village of the duchy of Nassau, in the bail. of Weilburg. Pop. 500.

WEINBERG, a village of Bavaria, in the circle of Middle Franconia, and presidial of Feuchtwang. Pop. 485.—Also a colony of Austria, in Moravia, in the circle of Brünn, near Königsfeld. Pop. 260.—Also a colony in the circle of Znaim, near Mahrisch-Kruman. Pop. 300.—Also a village of Styria, in the circle of Gratz, and bail. of Hohenbruck. Pop. 380.—Also a village of the same circle, in the bail. of Harlberg. Pop. 270.—Also a village of the same circle, in the bail. of Marburg. Pop. 210.—Also a village of Transylvania, in the comitat of Unter-Weissenburg, to the SW of Carlsburg.—Also a village of Prussia, in the regency and circle of Liegnitz. Pop. 410.

WEINBERGE, two villages of Prussia, in the regency of Merseburg, and circle of Liebenwerda. Pop. 360.

WEINBOHLA, a village of Saxony, in the gov. of Dresden, and bail of Meissen, in N lat. 51° 9′ 47″. Pop. 1,059.

WEINBURG, a village of Austria, in Styria, in the circle of Gratz. Pop. 260. It has a castle.

WEINDORF, or Borband, a market-town of Austria, in Transylvania, in the comitat of Unter-Weissenburg, and 2 m. NE of Carlsburg.

WEINE, a village of Prussia, in the regency of Minden, and circle of Büren. Pop. 320.—Also a village in the regency of Posen, and circle of Fraustadt. Pop. 600.

WEINER, a village of Prussia, in the regency of Munster, and circle of Steinfurt. Pop. 690.

WEINFELDEN, a market-town and amt of Switzerland, in the cant. of Thurgau, 9 m. E of Frauenfeld, on the r. bank of the Thur, which is here crossed by a fine bridge. Pop. 2,256. This town is considered the second capital of the cant., and is the place of assembly of the great council. The environs are noted for their wine.

WEINGARTEN, a market-town of the grand-duchy of Baden, in the circle of the Middle Rhine, and bail. of Durlach, 6 m. ENE of Carlsruhe, on the Schussen. Pop. 3,150. Wine and madder are extensively cultivated in the environs.—Also a hamlet of Würtemberg, in the circle of the Danube, bail. and 2 m. NE of Ravensburg, near Altorf. It is noted for its castle, formerly an imperial abbey, and now converted into an orphan's asylum. It was founded in 920 by Duke Guelph or Welf, and has a fine church erected in 1715.—Also a village of Bavaria, in the circle of the Pfalz, and cant. of Germansheim. Pop. 1,126.—Also a village of Switzerland, in the cant. of Thurgau, and bail. of Tobel. Pop. 215.

WEINGARTEN (Gross), a village of Bavaria, in the circle of Middle Franconia, and presidial of Pleinfeld. Pop. 415.

WEINGARTENHAUSER, a village of Austria, in Bohemia, in the regency of Prague, and bail. of Melnik. Pop. 770.

WEINGARTS, a village of Bavaria, in the circle of Upper Franconia, and presidial of Grafenburg. Pop. 340.

WEINGARTSGEREUTH, a village of Bavaria, in the circle of Upper Franconia, and presidial of Hochstadt. Pop. 270. It has a castle.

WEINGASSE, or Winari, a village of Prussia, in the regency of Oppeln, and circle of Neustadt. Pop. 240.

WEINHAUS, a village of Austria, in the district below the Enns, and bail. of Hietzing. Pop. 400.

WEINHEIM, a bail. and town of the grand-duchy of Baden, in the circle of the Lower Rhine, 11 m. NE of Manheim, on the Weschnitz, in a fertile and picturesque locality. Pop. 5,000. It is enclosed by walls with four gates, and has a suburb and six churches, a synagogue, and mineral baths. It has manufactories of woollen fabrics, leather, and chocolate, several oil-mills.—Also a village of Hesse, prov. of the Rhine, and circle of Alzes. Pop. 950.

WEI-NING-CHU, a district and town of China, in the prov. of Kwei-chu, and div. of Ta-ting-fu, in N lat. 26° 43′, and E long. 104° 15′.

WEININGEN, a village of Switzerland, in the cant. of Zurich, and bail. of Steckborn, on the Limmat. Pop. 832.

WEINITZEN, a village of Austria, in Styria, in the circle of Gratz. Pop. 440.

WEINOLDSHEIN, a village of Hesse, in the prov. of the Rhine, and circle of Mainz. Pop. 620.

WEINOTHEN (Alt and Neu), two villages of Prussia, in the regency of Gumbinnen, and circle of Tilsit. Pop. respectively 800 and 200.

WEINRIED, a village of Bavaria, in the circle of Swabia, and presidial of Babenhausen. Pop. 400.

WEINSBERG, a bail. and town of Würtemberg, in the Neckar circle. The town is situated on the Sulm, 3 m. ENE of Heilbronn. Pop. 2,180.

WEINSDORF, a village of Prussia, in the reg. of Konigsberg, circle of Mohrungen. Pop. 525.

WEINSHEIM, a village of Prussia, in the reg. of Coblenz, circle of Kreuznach. Pop. 800.

WEISBACH, a village of Baden, in the circle of the Lower Rhine, bail. of Rendenau, on the Musbach. Pop. 300.—Also a village of Bavaria, in the circle of Lower Franconia, landg. of Bischofsheim. Pop. 650.

WEISCHLITZ, a village of Saxony, in the circle of Zwickau, bail. of Plauen. Pop. 660.

WEISEL, a village of Nassau, in the bail. of St. Goarshausen. Pop. 800.

WEISENBACH, a village of Baden, in the Upper Rhine circle, bail. of Fryberg. Pop. 675.

WEISENBORN, a village of Saxony, in the circle of Dresden, bail. and near Freiberg. Pop. 546.—Also a village of Kurhesse, in Lower Hesse, circle of Eschwege. Pop. 514.

WEISENHEIM-AM-SAND, a town of Bavaria, in the Pfalz, near Frankenthal. Pop. 1,760.

WEI-SHE, a district and town of China, in the prov. of Ho-nan, and div. of Kae-fung-fu, in N lat. 34° 18′, and E long. 114° 5′.

WEISSACH, a village of Würtemberg, in the Neckar circle, on the Strudelbach. Pop. 1,190.

WEISSBACH (Upper), a village in the principality of Schwartzburg-Rudolstadt. Pop. 230.

WEISSENBERG, a town of Saxony, in Upper Lusatia, near Bautzen, bail. of Lobau. Pop. 990.

WEISSENBURG (Lower), or Also-fejer-Varmegye, a county of Transylvania, bounded on the N by the cos. of Thorenburg and Klausenburg; on the E by that of Kokelburg; on the S by Zarand and the Hunyad; and on the W by Hungary. It has an area of 89 German sq. m., with a pop. in 1839 of 204,300. Its surface is mountainous. The principal town is Carlsberg.

WEISSENBURG (Upper), or Felso-fejer-Varmegye, a county in the SE part of Transylvania.

watered by the Aluta. Its area is 29 German sq. m. Pop. in 1839, 48,800. Its cap. is Furstenburg.

WEISSENBURG, a walled town of Bavaria, in Middle Franconia, on the r. bank of the Rezat, 27 m. SE of Anspach. Pop. 4,194. It has manufactories of jewellery, pins, and needles.

WEISSENBURG, a township of Lehigh co., Pennsylvania, U. S., 72 m. E by N of Harrisburg. Pop. 1,427.

WEISSENFELS, a town of Prussian Saxony, on the Saale, in N lat. 51° 12′ 7″, 20 m. WSW of Leipsic. Pop. 7,800. It has woollen and cotton factories, tobacco-mills, and porcelain works.—Also a town of Austrian Illyria, in Carniola, 4 m. WSW of Tarvis.

WEISSENHORN, a town of Bavaria, on the river Roth, 8 m. SE of Ulm. Pop. 1,605.

WEISSENKIRCHEN, a village of Austria, on the l. bank of the Danube, 6 m. W of Krems.

WEISSENSEE, a town of Prussian Saxony, 11 m. N of Erfurt, on the Helbe. Pop. 1,800.

WEISSEN-SEE, a lake of Austrian Illyria, in Carinthia, between the Drave and the Dail, 30 m. W of Klagenfurt. It is about 4 m. in length, and discharges itself into the Drave.

WEISSENSTADT, a town of Bavaria, in Upper Franconia, on the Eger, 17 m. NE of Bayreuth. Pop. 1,400.

WEISSENSTEIN, a mountain of Switzerland, in the cant. and 6 m. N of Soleure, having an alt. of 3,966 ft. above sea-level.—Also a v. of Würtemberg, 24 m. NNW of Ulm.

WEISSENTHURN, a town of Prussia, in the prov. of the Rhine, situated on the Rhine, opposite to Neuwied.

WEISSERITZ, a river of Germany, which rises in Bohemia, and joins the Elbe, near Dresden.

WEISSKIRCHEN, a town of Hungary, in the bannat of Temesvar, 58 m. S by E of Temesvar. Pop. 4,808, chiefly employed in agriculture.—Also a town of Moravia, 20 m. E of Olmutz, on the Betshwa. Pop. 5,800, many of whom are woollen and silk weavers.—Also a village of Styria, in the circle and 6 m. SSE of Judenburg.

WEISSMAIN, a town of Bavaria, in Upper Franconia, on the White Main, 14 m. NNE of Bamberg. Pop. 1,050.

WEISSPODOL, a village of Bohemia, in the reg. of Pardubitz, circle of Czaslau. Pop. 485.

WEISSWASSER, a town of Austrian Silesia, 4 m. SW of Patschkau. Pop. 1,100.—Also a town of Bohemia, 6 m. NW of Jung-Bunzlau, on the Bila. Pop. 1,100.—Also a town of Moravia, 60 m. NW of Troppau.

WEISWEIL, a village of Baden, 17 m. NW of Freyburg, near the r. bank of the Rhine. Pop. 1,035.

WEISSWEILER, a village of Prussian Westphalia, in the circle of Duren, on the Jude. Pop. 110.

WEISTRITZ, a river of Prussian Silesia, which rises in the Riesengebirge, passes Schweidnitz, joins the Polsnitz, and flows into the Oder, on the l. bank.

WEIT, a village of Illyria, in the gov. of Laybach, 9 m. N of Klagenfurt.

WEITEN, a town of Lower Austria, in the circle of Upper Mannhartsberg, 18 m. WSW of Krems.

WEITENAU, a village of Bavaria, 8 m. NW of Immerstadt.

WEITERODE, a village of Kur-Hesse, near the river Fulda, in the circle of Rotenburg. Pop. 820.

WEITERSTADT, a village of Hesse, in the prov. of Starkenburg, landg. of Langen. Pop. 600.

WEITRA, or WEITRACH, a town of Lower Austria, 75 m. WNW of Vienna. Pop. 1,150.—Also a town of Austria, in the circle of the Upper Wieverwald, on the Leinsitz. Pop. 1,800.

WEITZ, a town of Austria, in Styria, 11 m. NE of Gratz. Pop. 1,000.

WEITZENKIRCHEN, a village of Austria, on the Hansruck, near Linz. Pop. 600.

WEI-YUEN, a district and town of China, in the prov. of Kan-suh, and div. of Lan-chu-fu, in N lat. 35° 8′, and E long. 104° 16′.—Also a district and town in the prov. of Szechuen, and div. of Keatsing-fu, in N lat. 29° 39′, and E long. 104° 40′.

WEI-YUEN-TING, a district and town of China, in the prov. of Yunnan, and div. of Tsin-unh-fu, or Pu-urh-fu, in N lat. 23° 29′, and E long. 101° 5′.

WELBORN, a township of Conway co., in Arkansas, U. S. Pop. 683.

WELBORNE, a parish of Norfolk, 6½ m. NNW of Wymondham. Area 3,270 acres. Pop. 265.

WELBOURNE, a parish of Lincolnshire, 9½ m. NW of Sleaford. Area, 3,270 acres. Pop. 592.

WELBURN, a township in the p. of Balmer, N. R. of Yorkshire, 5 m. SW by W of New Malton. Area 750 acres. Pop. in 1851, 520.—Also a township in the parish and 4½ m. E of Helmesley.

WELBURY, a parish in the N. R. of Yorkshire, 6½ m. NNE of Northallerton. Area 2,569 acres. Pop. in 1841, 266; in 1851, 249.

WELBY, a township in the parish of Melton-Mowbray, Leicestershire, 2½ m. NW by W of Melton-Mowbray. Area, 1,582 acres. Pop. in 1851, 141.—Also a parish of Lincolnshire, 4½ m. E by N of Grantham. Area 2,740 acres. Pop. 481.

WELD, a township of Franklin co., Maine, U. S., 47 m. WNW of Augusta. Pop. 1,045.

WELDON (GREAT), a market-town and parish of Northampton, near the river Willy, 4½ m. ESE of Rockingham. Area 2,350 acres. Pop. 321.

WELDON (LITTLE), a hamlet of Northamptonshire, adjoining the village of Great Weldon. Area 1,330 acres. Pop. in 1851, 537.

WELDON, a village of Halifax co., in N. Carolina, U. S., 95 m. NE of Raleigh, on the SW side of the Roanoke, at the great falls, which are passed by a canal 12 m. in length, with a lockage of 100 ft. Pop. 600.

WELFORD, a parish of Berkshire, 5 m. NW of Speenhamland. Area 5,173 acres. Pop. 1,115.—Also a parish of Gloucester, 9 m. N of Chipping Campden. Area 3,550 acres. Pop. 659.—Also a hamlet in the parish of Kempford, Gloucestershire.

WELFORD, or WELLESFORD, a parish of Northamptonshire, 15 m. NNW of Northampton. Area 3,650 acres. Pop. in 1851, 1,153.

WELHAM, or WELLANDHAM, a parish of Leicestershire, 4 m. NE by N of Market-Harborough. Area 1,109 acres. Pop. in 1851, 68.

WELHAM, a hamlet of Yorkshire, 1½ m. S of New Malton.

WELIKOWA, a village of Moravia, in the circle of Hradisch, lordship of Lukow. Pop. 330.

WELITSCHNA, a town of Hungary, 32 m. N of Neusohl.

WELKA, a town of Moravia, in the circle of Hradisch. Pop. 1,600.

WELKERSDORF, a town of Prussia, in the reg. of Liegnitz, circle of Lowenberg. Pop. 1,488.

WELL, a parish of Lincolnshire, 1½ m. SSW of Alford. Area 2,110 acres. Pop. in 1851, 80.—Also a parish of the N. R. of Yorkshire, 5 m. S of Bedale. Area 6,451 acres. Pop. in 1851, 1,044.

WELLAND, a parish of Worcestershire, 3 m. WSW of Upton-upon-Severn. Area 2,027 acres. Pop. 582.

WELLAND, a river of England, which takes its rise in the co. of Northampton; flows NE, passing

Rockingham and Stamford; turns N, and passes Market-Harborough, Market-Deeping, and Spalding; and below the latter place enters the Wash, after a total course of 70 m. It is navigable by locks from Stamford downwards.

WELLAND, or CHIPPEWA, a river of Upper Canada, which flows through a fertile country in an E course of 60 m., to the Niagara, and forms a part of the canal navigation known as the Welland canal, 28 m. in length, formed to connect Lakes Erie and Ontario, and avoid the falls of Niagara. Up to January 1851 this canal had cost £954,824. Its revenue in 1852 was £58,272.

WELLCOMBE, a parish of Devonshire, 17 m. SW by W of Bideford. Area, 1,751 acres. Pop. in 1851, 234.

WELLESBOURNE-HASTINGS, a village and parish of Warwickshire, 5½ m. NW of Kineton. Pop. 797. Area, with W. Mountford, 4,740 acres.

WELLESLEY'S ISLANDS, a group so called by Flinders, at the head of the gulf of Carpentaria, on the N coast of Australia. They consist of Mornington isle, Pisonia, Bentinck, Sweer's island, and several other small and rocky islets.

WELLFLEET, a town and port of Barnstable co., Massachusetts, U. S., on a bay of the same name, 31 m. ENE of Barnstable, and 97 m. SE of Boston. Pop. in 1840, 2,377; in 1850, 2,411.

WELLHAUGH, a township in the parish of Falstone. Area 29,475 acres. Pop. in 1831, 272; in 1851, 333.

WELLIN, a department, commune, and town of Belgium, in the prov. of Luxemburg, and arrond. of Neufchateau. Pop 494.

WELLING, a village of Prussia, in the regency of Coblenz, and circle of Mayen. Pop. 300.

WELLINGBOROUGH, a parish and town of Northampton, 10 m. NE of Northampton. Area of p., 4,490 acres. Pop in 1831, 4,688; in 1851, 5,297.—The town consists of four good streets, diverging from the market-place. The inhabitants were formerly occupied in the manufacture of worsted stuffs; but this employment declined, and of late years they have been chiefly engaged in the manufacture of boots and shoes, and bobbin-lace.

WELLINGDORF, a village of Denmark, in the duchy of Holstein, on the Schwentine. Pop. 366.

WELLINGEN, a village of Prussia, in the regency of Treves, and circle of Merzig. Pop. 250. —Also a village of Würtemberg, in the circle of the Danube, and bail. of Kirchheim. Pop. 320.

WELLINGERODE, a village of Hesse-Darmstadt, in the prov. of Lower Hesse, bail. of Abterode, on the Kupferbache. Pop. 353.

WELLINGHAM, a parish in Norfolk, 6 m. SW by S of Fakenham. Area 1,066 acres. Pop. 163.

WELLINGHOFEN, a village of Prussia, in the reg. of Arnsberg, circle of Dortmunde. Pop. 462.

WELLINGHOLZHAUSEN, a village of Hanover, in the principality of Osnabruck, and bail. of Groningen. Pop. 934.

WELLINGORE, a parish in the co. of Lincoln, 9½ m. NW by N of Sleaford. Area 2,400 acres. Pop. in 1831, 752; in 1851, 914.

WELLINGSBUTTEL, a village of Denmark, in the duchy of Holstein, near Homberg. Pop. 320.

WELLINGTON, a county of New South Wales, bounded on the NW and NE by the Cudgeegong, by which it is separated from the counties of Bligh and Phillip; on the SE by the co. of Roxburgh, from which it is divided by the Turon; on the S by the co. of Bathurst; and on the W by the Bell, an affluent of the Macquarie, by which the co. is intersected from S to N. It is traversed by several mountain ranges. In the W., near the junction of the Bell and Macquarie rivers, is a fine valley of the same name. The pop. in 1851 was returned at 1,609. The chief town, Mudgee, is on the Cudgeegong, 150 m. from Sydney.—Also a district comprising an area of about 10,000,000 acres, extending between the Lachlan and Macquarie, and adjacent to the counties of Wellington and Bathurst. Pop. in 1851, 1,512.—Also a co. of Western Australia, bounded on the N by the co. of Murray; on the E by that of Wicklow; on the S by the co. of Nelson; and on the W by the Pacific. It is intersected from N to S by the Roe Range, a portion of the Darling chain, and is watered by several streams, the principal of which are Murray river in the NE corner, and Preston and Brunswick rivers, affluents of Leschenault inlet, a considerable sheet of water separated from the sea by a narrow tongue of land, and opening to the E of Point Casuarina into Koombanah bay. The shore is generally low, and is studded with lagunes.—Also a parish of Van Diemen's Land, in the co. of Buckingham, bounded on the N by the Derwent.—Also a mountain which rises above Hobart-Town to an alt. of 4,000 ft. above sea-level.

WELLINGTON, a parish in Hereford, 5½ m. N of Hereford. Area 2,538 acres. Pop. in 1831, 630; in 1851, 689.—Also a township and village in the p. of Walls-End, Northumberland, 3 m. W by S of North Shields, on the river Tyne, and in the line of the Newcastle and North Shields railway. There are extensive collieries in the vicinity.—Also a parish and market-town in Salop, 11 m. E of Shrewsbury, containing the townships of Aston, Hodley, Horton, Ketley, Lawley, and Walcott. Area 8,757 acres. Pop. in 1831, 9,671; in 1851, 11,544. The town is about 2 m. S from the Wrekin. The parish is well supplied with coal, ironstone, and limestone; and the town has long been the seat of extensive operations in the manufacture of iron and iron articles, particularly of nails.—Also a parish and market-town in Somersetshire, 150 m. WSW of London. Area of p., 5,195 acres. Pop. in 1831, 4,762; in 1851, 6,415. It consists of several streets, the principal of which is upwards of 1 m. in length. Druggets and serges were formerly manufactured here to a considerable extent; but these branches have of late years considerably fallen off. Arthur Wellesley was created Viscount Wellington on the 4th September, 1809; Earl on the 28th February, 1812; Marquis on the succeeding 18th of August; and Duke on the 3d May, 1814.

WELLINGTON, a settlement on the SW coast of the North island of New Zealand, extending along the crescent-shaped beach of Port Nicholson. The town, which is a long straggling village of wooden houses often far apart, but chiefly near the beach on the W and S sides of the splendid harbour or basin, had a pop. of 2,649 in 1848. See articles PORT NICHOLSON and NEW ZEALAND.

WELLINGTON, a district of Upper Canada, consisting of the co. of Waterloo, and comprising 26 townships. It is watered by the Grand River and some of its affluents, and contains some of the best farms in the prov. Pop. in 1841, 13,851; in 1848, 36,865.—Also a village partly in the township of Hillier and partly in that of Hallowell.

WELLINGTON, a township of Piscataquis co., in the state of Maine, U. S., 53 m. NE of Augusta. Pop. in 1840, 722; in 1850, 600.—Also a township of Lorain co., in the state of Ohio, 89 m. NNE of Columbus, drained by the W branch of Black river, and intersected by the Cincinnati, Cleveland, and Columbus railroad. Pop. in 1840, 781; in 1850, 1,556.

WELLINGTON-CHANNEL, an arm of the Arctic ocean leading from Barrow strait north

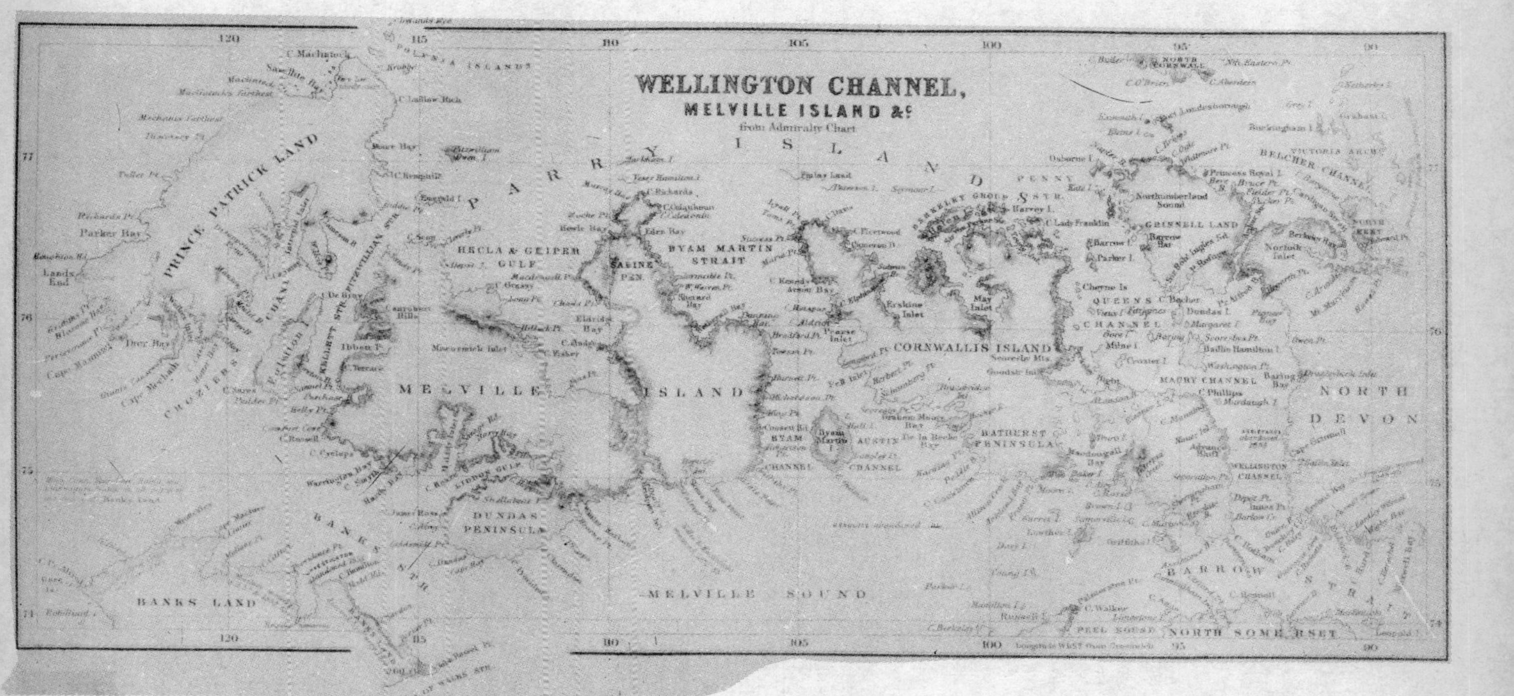
WELLINGTON CHANNEL,
MELVILLE ISLAND &c.
from Admiralty Chart
PARRY ISLANDS
PRINCE PATRICK LAND
MELVILLE ISLAND
CORNWALLIS ISLAND
HECLA & GRIPER GULF
BYAM MARTIN STRAIT
SABINE PEN
QUEENS CHANNEL
GRINNELL LAND
BELCHER CHANNEL
NORTH DEVON
WELLINGTON CHANNEL
MAURY CHANNEL
BATHURST PENINSULA
BYAM MARTIN CHANNEL
AUSTIN CHANNEL
DUNDAS PENINSULA
MELVILLE SOUND
BARROW STRAIT
NORTH SOMERSET
BANKS LAND
PEEL SOUND
Northumberland Sound
Erskine Inlet
May Inlet
Parker Bay
Lands End
Longitude West from Greenwich

ward along the W side of North Devon, between the parallels of 74° 30′ and 77°. Beechey island, the scene of Franklin's first winter encampment, in the last and ill-fated expedition, lies at the SE extremity of this channel. It has been supposed that Franklin may have gone up this channel, if he could not get to the westward, and found it comparatively free from ice. Sir Edward Parry, the discoverer of this channel or strait, says: "When I was going up westward from Melville island we saw Wellington strait perfectly free from ice, and so I marked it on my chart. It was not my business to go N as long as I could get W; and therefore we ran past, and did not examine it; but it had always been a favourite idea of those who imagined that the NW passage was to be easily made by going N. That we know was the favourite idea of Franklin, and we know he did intend, if he could not get westward, to go up Wellington channel. We have it from his own lips. My belief is still that after the first winter he did go up that channel, and that, having steam power—which I had not in my time—it is possible he may have gone up in a favourable season. For anything more different than a favourable and an unfavourable season in those regions cannot be imagined, or the changes that take place in the ice there. I have been sometimes beset for two or three days together by the ice in such a way that from the masthead I could not see sufficient water to float a bottle in; and in 24 hours there was not a bit of ice to be seen. Therefore in a favourable season he may have gone up that inlet, and may, by the power of steam and favourable circumstances, have got so far to the NW that in an ordinary season he could not get back again. And those who knew Franklin know this—that he would push on year after year so long as his provisions lasted. Nothing could stop him. He was not a man to look back if he believed the thing was still possible." The last we know of the expedition is that it passed the winter of 1845-6 in a small cove between Cape Riley and Beechey-island, facing Lancaster-sound; and in the neighbourhood of Wellington-channel traces of a travelling party were subsequently discovered, which seems to have been despatched from the same expedition. From these facts, it has been inferred by some geographers, "that Franklin's expedition never got far up Wellington strait; that they had, near its mouth, or in attempting to proceed westward, been embedded in the ice; that they had clung to their ships till their provisions began to fail, and then commenced their dreary journey to the southward, for they were in want of provisions when met by the Esquimaux. The date of these events cannot be later than 1850, when the solitary traces of the party were found at Cape Riley and Beechey-island. They may have happened as late as 1849, for to that time, according to letters from Franklin, he thought his provisions might be stretched out."

WELLINGTON-ISLAND, an island of the Patagonian archipelago, extending between the gulf of Penas, by which it is separated from the peninsula of Tres-Montes on the N, and the gulf of Trinidad on the S, and divided from the continent on the E by the Messier and Wide channels. It lies between 47° 30′ and 50° 20′, and is 138 m. in length, but of small comparative breadth.

WELLINGTON-SQUARE, a village of Upper Canada, in the township of Nelson, on Lake Ontario, 81 m. from Hamilton. Pop. 400.

WELLMICH, a market-town of Nassau, in the bail. of St. Goarshausen, on the Rhine. Pop. 492.

WELLOW, a parish in Notts, 1½ m. SE of Ollerton. Area 991 acres. Pop. in 1831, 473; in 1851, 597.—Also a parish in Somersetshire, 4 m. S by W of Bath, containing the hamlets of Beggeridge, Hassage, Peglinch, Shascombe, Stony-Littleton, Twinney, or Twyniho, Whiteoxmead, and Woodborough. Area 5,292 acres. Pop. in 1851, 1,142.

WELLOW (East), a parish in Hants, 4 m. W of Romsey. Area 2,373 acres. Pop. in 1851, 289.

WELLOW (West), a parish in Wilts, 4¾ m. W by N of Romsey. Area 1,344 acres. Pop. 407.

WELLS, a parish and seaport town in Norfolk, 5 m. N by W of Walsingham, and 32 m. NW by N of Norwich. Area of p., 4,510 acres. Pop. in 1831, 3,624; in 1851, 3,675. The town, situated on a small creek which has a winding course of about 2 m. in length, through salt marshes, to the German ocean, consists chiefly of two main streets. The entrance to the harbour has been rendered difficult by the gradual accumulation of silt and other obstructions; but it has of late been considerably improved under the inspection of harbour-commissioners. Spring tides rise 11 ft. at the quay, and vessels of 150 tons can come up at high water. The exports consist chiefly of grain, malt, and oysters.

WELLS, a city and borough in Somersetshire, 19 m. S of Burton, and 19 m. SW of Bath, on the south side of the Mendip hills. The town, which is small, has an interesting appearance, from the number of ancient ecclesiastical edifices which it contains, particularly its magnificent cathedral. It consists chiefly of four principal streets. There is a large stocking manufactory which sometimes employs as many as 1,500 persons; the market for corn is no longer considerable; but the cheese market is the greatest in the west of England. The pop. in 1831 was 4,603; in 1851, 4,734. The burgh returns 2 members to parliament. Electors in 1852, 325. Bishop Robert, about the year 1139, decided that the diocesan should be styled bishop of Bath and Wells, and be enthroned, on his admission, in both churches. At the Reformation, the chapter of Bath was dissolved; and the election of the bishops has ever since been vested in the dean and chapter of Wells. The cathedral church, a magnificent cruciform structure, chiefly in the earliest style of English architecture, stands at the E extremity of the city. Its form is the usual one of a cross, extending from E to W, in extreme length, 415 ft.; the transept measuring 155 ft. From the intersection springs a quadrangular tower, 178 ft. in height; and two towers, each 126 ft. in height, rise from the N and S sides of the W entrance. The most remarkable portion of the building is the west front, which is covered with elaborate sculpture of light and airy design. The choir is 108 ft. in length. The Lady-chapel is an architectural gem 52 ft. in length. The episcopal palace is an ancient castellated mansion to the S of the cathedral, surrounded by a wall and moat, enclosing a space of 7 acres, and communicating with the cathedral through a venerable gateway leading over a bridge.

WELLS, a parish in cos. Carlow and Kilkenny, partly in the barony of Gouran, but chiefly in that of West Idrone. Area 2,726 acres. Pop. 1,298.

WELLS, a parish of Tasmania, in the co. of Devon, bounded on the E by the Tamar.—Also a town of South Australia, in the co. of Flinders, on an arm of Coffin bay.—Also a town in the co. of Hindmarsh, on Encounter bay.

WELLS, a port on the N coast of Russian America, in the NW part of Prince William's strait, in N lat. 61°, and W long. 147° 27′.

WELLS, a central county of the state of Indiana, U. S., comprising an area of 372 sq. m. drained by Upper Wabash and Salamonie rivers, and intersected by the Fort Wayne and Southern railway. Pop. in

1840, 1,822; in 1850, 6,152.—Also a township of York co., in the state of Maine, 81 m. SSW of Augusta. It is low and marshy, drained by affluents of the Atlantic, and intersected by the Portland, Saco, and Portsmouth railway. Pop. in 1840, 2,978; in 1850, 2,945.—Also a township of Hamilton co., in the state of New York, 69 m. NNW of Albany, drained by the Sacandaga river. Pop. in 1840, 365; in 1850, 400.—Also a township of Rutland co., in the state of Vermont, 67 m. SSW of Montpelier. It has a diversified surface, and contains a large lake. Pop. in 1850, 804.—Also a township of Bradford co., in the state of Pennsylvania, drained by South creek and by a branch of Seely creek. Pop. in 1840, 1,492.—Also a township of Macon co., in the state of Missouri. Pop. 680.—Also a river of the state of Vermont, which has its source in Long Pond, in Groton, and joins the Connecticut in the NE of part of Newbury.

WELLSBORO', a village of Tiogo co., in the state of Pennsylvania, U. S., 102 m. N by W of Harrisburg. Pop. in 1850, 620.

WELLSBURG, a village of Chenning co., in the state of New York, U. S., on the Erie railway, 276 m. from New York.—Also a village of Brook co., in the state of Virginia, on the E side of the Ohio, at the mouth of Buffalo creek. Pop. in 1840, 2,000; in 1850, 2,200.

WELLSEE, a village of Denmark, in the duchy of Holstein, near Kiel, on a small lake of the same name. Pop. 278.

WELL'S ISLAND, an island of Jefferson co., in the state of New York, U. S., and one of the largest of the Thousand Islands, in the St. Lawrence.

WELLSPITZ, a village of Austria, in Moravia, in the circle of Brunn. Pop. 660.

WELL'S RIVER, a village of Orange co., in the state of Vermont, U. S., on the W side of the Connecticut, at the mouth of a river of the same name, 26 m. ESE of Montpelier, and on the Boston, Concord, and Montreal railroad.

WELLSTRASSE, a village of Prussia, in the regency and circle of Munster. Pop. 200.

WELLSVILLE, a village of Scioto township, Alleghany co., in the state of New York, U. S., 227 m. W by S of Albany, and on the E side of Genesee river. Pop. in 1840, 240.—Also a village of Yellow Creek township, Columbiana co., in the state of Ohio, on the W bank of the Ohio, 126 m. ENE of Columbus, and on the Cleveland and Pittsburg railroad. Pop. in 1850, 1,549.

WELLWITZ, a village of Prussia, in the regency of Marienwerder, and circle of Flatow. Pop. 270.

WELLWOOD, a town of New South Wales, in the co. of and near Bathurst.

WELMLINGEN, a village of Baden, in the circle of the Middle Rhine, and bail. of Lorrach. Pop. 345.

WELMSCHLOSS, a village of Austria, in Bohemia, in the regency of Eger, and bail. of Saatz. Pop. 390. It has a castle.

WELNA, a village of Prussia, in the regency and 24 m. N of Posen, and circle of Obernik, on a small river of the same name, an affluent of the Warthe. Pop. 428.

WELNETHAM (GREAT), a parish in Suffolk, 4 m. SSE of Bury-St.-Edmunds. Area 1,493 acres. Pop. in 1831, 422; in 1851, 552.

WELNETHAM (LITTLE), a parish in Suffolk, 4 m. SE of Bury-St.-Edmunds. Area 592 acres. Pop. in 1831, 180; in 1851, 178.

WELNEY, a chapelry in the p. of Upwell, partly in Cambridgeshire, and partly in Norfolk, 6 m. ENE of March. Area 5,292 acres. Pop. in 1851, 1,206.

WELNITZ, a village of Austria, in Bohemia, in the regency of Bohm-Leipa, and bail. of Haida. Pop. 800.

WELPER, a village of Prussia, in the regency of Arnsberg, and circle of Bochum. Pop. 290.

WELPLAGE, a village of Hanover, in the principality of Osnabruck, and bail. of Wittlage-Huntebury, near Lemforde. Pop. 824.

WELS, a town of Austria, capital of the circle of Hausruck, on the l. bank of the Traun, 18 m. SW of Lintz, and 120 m. W of Vienna, in N lat. 48° 9′ 32″, and E long. 14° 1′ 13″. Pop. 5,550. It is regularly built, and contains two castles belonging to the princes of Auersberg and Polhaim, several handsome churches, a Protestant chapel, a large hospital, and several schools. It has manufactories of cotton fabrics, tanneries, and copper and powder-mills, and carries on an active trade in wood, corn, and cattle.

WELSBACH (GROSS and KLEIN), two villages of Prussia, in the regency of Erfurt, and circle of Langenfalza. Pop. 430.—Also a village in the circle of Treves, and circle of Saarbruck. Pop. 280.

WELSBERG, a village of Austria, in Tyrol, in the prov. of Brixen, and circle of Bruneck, on the Rienz. Pop. 660.

WELSCHBILLIG, a town of Prussia, in the reg. and circle and 11 m. NNW of Treves. Pop. 698.

WELSCH-BIRKEN, a market-town of Austria, in Bohemia, in the regency of Pilsen, and circle of Prachin, to the NW of Huffenez. Pop. 458.

WELSCHEID, a village of Belgium, in the prov. of Luxemburg, and bail. of Diekirch. Pop. 260.

WELSCHE-KONFINIEN. See ROVEREDO.

WELSCHENNEXT, a village of Prussia, in the regency of Arnsberg, and circle of Olpe. Pop. 330.

WELSCHENROHR, a village of Switzerland, in the cant. and to the NE of Soleure, and bail. of Balstale. Pop. 721. It has an iron-mine.

WELSCHNEUREUTH, a village of Baden, in the circle of the Middle Rhine, and bail. of Karlsruhe. Pop. 700.

WELSCH-STENIACH, a village of Baden, in the circle of the Middle Rhine, and bail. of Haslach. Pop. 460.

WELSE, a river of Prussia, in the regency of Potsdam, and circle of Angermunde, an affluent of the Oder.

WELSEDE. a village of Hesse-Darmstadt, in the circle of Kinteln, and bail. of Oldendorf. Pop. 468.

WELSHEIM, a market-town of Bavaria, in the circle of Middle Franconia, and presidial of Eichstadt. Pop. 500. It has a castle.

WELSHOFEN, a village of Bavaria, in the circle of Upper Bavaria, and presidial of Dachau. Pop. 220.

WELSHPOOL, or POOL, anciently TRALLWNG, a borough and parish in Montgomeryshire, North Wales, 170 m. NW of London. Area 6,801 acres. Pop. in 1831, 4,538; in 1851, 4,391.—The town, situated in a valley near the banks of the Severn, in the vicinity of the Montgomeryshire canal, consists in fact of two towns, Pool town and Welsh town; which are however so united as to form a continuous town. The Ellesmere canal, which passes close on the E side of the town, communicates with Chester and Liverpool; and the Severn is navigable to a place called the Pool stake, 200 m. from its mouth, and within a short distance of the town, during several months of the year. The inhabitants are partly employed in the manufacture of flannels. W. was anciently a borough, and contributed with Montgomery in returning a member to parliament, but was disfranchised in 1728. The parl. borough, which comprises the whole of the p. of Pool, and the township of Gunrog-Fechan, in the parish of Guilsfield, is now contributary with Montgomery, Llanfyllin, Llanidloes, Machynlleth, and Newtown, in returning one member to parliament.

WELSIGKENDORF, a village of Prussia, in the reg. of Potsdam, and circle of Juterbogk. Pop. 240.

WELSLEBEN, a village of Prussia, in the reg. of Magdeburg, and circle of Wanzleben. Pop. 1,152.

WELST, a town of Russia in Europe, in the gov. of Vologda, on the Waga. Pop. 900.

WELSTORF, a village of Lippe-Detmold, in the bail. of Barenholz. Pop. 440.

WELT (NEU) a village of Bohemia, in the circle of Bidschow, 38 m. N of Neu Bidschow, in the midst of the Riesen-Gebirge. It is noted for its glass-works.

WELTDER, a valley of New South Wales, in the district of Liverpool Plains.

WELTE, a village of Prussia, in the regency of Munster, and circle of Koesfeld. Pop. 450.

WELTEM, a town of Holland, in the prov. of Limburg, 12 m. E of Maestricht. Pop. 3,080.

WELTENBURG, a village of Bavaria, in the prov. of Lower Bavaria, and presidial of Kelheim. Pop. 245.

WELTENSCHWANN, a village of Würtemberg, circle of the Schwarzwald, on the Calw. Pop. 220.

WELTEROD, a village of Nassau, in the bail. of St. Goarshausen. Pop. 370.

WELTERSBACH, a village of Bavaria, in the circle of Pfalz. Pop. 260.

WELTESBURG, a village of Nassau, in the bail. of Walmerod. Pop. 200.

WELTEWITZ, a village of Saxe-Weimar, in the bail. of Neustadt, on the Arla. Pop. 230.

WELTIESCH, a village of Austria, in Bohemia, in the regency of Bohmisch-Leipa, and bail. of Leitmeritz. Pop. 250.

WELTON, a parish in Lincolnshire, 6 m. NE of Lincoln. Area 3,690 acres. Pop. in 1831, 516; in 1851, 604.—Also a parish in Northamptonshire, 3 m. NE of Daventry.—Area 1,690 acres. Pop. in 1831, 600; in 1851, 663.—Also a township in the p. of Ovingham, Northumberland, 9 m. ENE of Hexham.—Also a parish, including the chapelry of Melton, in the E. R. of Yorkshire, 3½ m. SE of South Cave. Area 3,553 acres. Pop. in 1851, 856.

WELTON-IN-THE-MARSH, a parish in Lincolnshire, 5 m. ENE of Spilsby. Area 2,600 acres. Pop. in 1831, 363; in 1851, 431.

WELTON-LE-WOLD, a parish in Lincolnshire, 4 m. W of Louth. Area 2,520 acres. Pop. 368.

WELTRUB, a village of Austria, in Bohemia, in the regency of Pardubitz, and bail. of Neu-Kolin. Pop. 880.

WELTRUSS, a village of Austria, in Bohemia, in the regency of Prague, and bail. of Welwarn, in an island of the Moldau. It has a castle.

WELTSCHOWITZ, a village of Austria, in Moravia, in the circle of Prerau. Pop. 260.

WELTSIN, a village of Prussia, in the regency of Stettin, and circle of Demmin. Pop. 400.

WELVER, a village of Prussia, in the regency of Arnsberg, and circle of Goest. Pop. 238.

WELWARN, or BELWAR, a town of Austria, in Bohemia, in the regency of Prague, and bail. of Schlan, on the Rothenbach, 18 m. NW of Prague. Pop. 1,758. It is enclosed by walls, with four gates. It has some linen manufactories.

WELWICK, a parish in the E. R. of Yorkshire, 2 m. SE of Patrington. Area 6,674 acres. Pop. in 1831, 401; in 1851, 468.

WELWYN, a parish of Hertfordshire, on the line of the Great Northern railroad, which is here carried al[illegible] viaduct of 42 arches, each 30 ft. wide, and [illegible]h. The village consists of one street of [illegible] houses, thro[illegible]h which runs the great road from Londo[illegible] Yor[illegible]. Area of p. 2,987 acres. Pop. in 1831, 1,369; in 1851, 1,557.

WELZHEIM, a town of Würtemberg, in the circle of Jaxt, 20 m. NE of Stutgard. Pop. 1,680. It has a trade in flax and timber.

WEM, a parish in Salop, comprising the market-town of W.; the chapelries of Edstaston and Newtown; and the townships of Aston, Cotton, Horton, Lacon, Lowe with Ditches, Northwood, Sleap, Soulton, Tilley, Wem, and Woolverley. Area of p., 13,841 acres. Pop. in 1841, 4,119; in 1851, 3,747. —The town of W., 11 m. N by E of Shrewsbury, consists of one large street, with several cross streets and lanes; and stands on a level ground near the N bank of the Roden. The inhabitants are partly employed in the manufacture of leather, and in malting.

WEMBDON, a parish in Somersetshire, 1 m. NE of Bridgewater. Area 2,471 acres. Pop. in 1851, 819.

WEMBURY, a parish in Devon, on the W bank of the Yealm, near the point at which it falls into the English channel, 5 m. S by W of Earl's Plympton. Area 3,205 acres. Pop. in 1851, 577.

WEMBWORTHY, a parish in Devon, 4 m. SSW of Chumleigh. Area 2,411 acres. Pop. 444.

WEMDAL, a village of Sweden, in the district of Gefleborg, 120 m. WNW of Sundswal.

WEMDING, a town of Bavaria, 10 m. N of Donauworth, on the Schwald. Pop. 2,000.

WEMELDINGE, a village of Holland, in the prov. of Zieland, 6 m. E of Goa. Pop. 800.

WEMYSS, a parish in Fifeshire, on the shore of the Forth, 5¾ m. in length, with a breadth of from 1 to 2 m. Pop. in 1831, 5,001; in 1851, 5,647. There are seven villages in the parish: West W., the pop. of which, in 1831, was 858; in 1851, 1,013; East W., in which is the parish-church, pop. 802; Buckhaven, a fishing-town, pop. in 1831, 1,363; in 1851, 1,769; Methil; Kirkland, where there are extensive spinning-mills; West-Coaltown, and East-Coaltown, both inhabited by colliers.

WEN, a town of China, in the prov. of Kan-suh, and div. of Kung-chang-fu, in N lat. 32° 55′, and E long. 105° 8′.

WEN-CHING-CHU, a town of China, in the prov. of Kwan-se, and div. of Tae-ping-fu, in N lat. 23° 2′, and E long. 107° 8′.

WEN-CHU, a town of China, in the prov. of Kwang-tung, and div. of Kwang-chu-fu, in N lat. 18° 10′, and E long. 110° 20′.

WEN-CHUEN-HEEN, a district and town of China, in the prov. of Sze-chuen, and div. of Mowchu, in N lat. 31° 22′, and E long. 103° 36′.

WENDEL (SAINT), a town of Prussia, in that part of the old department of the Sarre assigned to Saxe-Coburg, on the Blies, 29 m. SE of Treves. Pop. 2,400.

WENDELL, formerly SAVILLE, a township of Sullivan co., New Hampshire, U. S., 22 m. NE of Charlestown. Pop. 795.—Also a township of Franklin co., Massachusetts, 13 m. E of Greenfield. Pop. 875.

WENDELSTEIN, a market-town of Bavaria, in Middle Franconia, on the Schwarzach, 9 m. S of Nuremberg. Pop. 1,200. It has some manufactures of cutlery and distilleries.—Also a village of Prussia, in the reg. of Merseburg, circle of Querfurt. Pop. 200.

WENDEN, a principality of Germany, in the duchy of Mecklenburg-Gastrow, lying contiguous to Brandenburg and Pomerania.—Also a village of Prussia, in the reg. of Konigsberg, circle of Rastenburg. Pop. 400.—Also a town of Russia, in the gov. of Livonia, in N lat. 57° 18′. Pop. 1,200.

WENDINGEN, a town in Würtemberg, at the confluence of the Lauter and Neckar, 12 m. ESE of

Stutgard. Pop. 1,100.—Also a village of Baden, in the Upper Rhine circle, bail. of Freiburg. Pop. 535.

WENDISCHHIPPA, a village of Saxony, in the circle of Leipzig. bail. of Oschutz. Pop. 443.

WENDLEBURY, a parish in Oxfordshire, 2½ m. SW of Bicester. Area 1,050 acres. Pop. 242.

WENDLING, a parish in Norfolk, 4¼ m. W of East Dereham. Area 1,436 acres. Pop. 385.

WENDON-LOUGHTS, or LOFTS, a parish in Essex, 5 m. W by N of Saffron-Walden. Area 1,520 acres. Pop. in 1831, 54; in 1851, 59.

WENDOVER, a parish and town in Bucks, 23 m. SE of Buckingham. Area of p., 5,719 acres. Pop. in 1801, 1,397; in 1831, 2,008; in 1851, 1,937. The town is situated at the foot of the Chiltern-hills. Lace-making is carried on to a considerable extent by the female inhabitants. Until the passing of the reform act, W. returned 2 members to parliament.

WENDRON (SAINT), a parish in Cornwall, including the town of HELSTON: which see. Area 13,320 acres. The pop., including the borough of Helston, in 1801, was 5,254; in 1851, 8,675.

WENDY, a parish in Cambridgeshire, 6½ m. NW of Royston. Area 947 acres. Pop. in 1851, 154.

WENDZIN, a village of Prussia, in the reg. of Oppeln, circle of Lublinitz. Pop. 660.

WENER, a lake of Sweden, bounded by the old provs. of Wermeland, Dalecarlia, and West Gothland. It is 100 m. long, from NE to SW, and 50 m. broad at the widest part. Its surface, amounting to 780 geog. sq. m., is about 150 ft. above the level of the Cattegat. It is divided into two parts by two necks of land which approach each other within 15 m. Though the navigation is not exempt from hazard, it is of importance for the transport of iron and other bulky commodities. It receives the waters of a number of streams which descend from the mountains of Norway; the largest of which is the Klar. The only river that flows out of it is the Gotha-elv, which issues from its SW extremity. The coasts are lined with islands, which are for the most part inhabited and cultivated.

WENERSBORG, a town of Sweden, in the laen of Elfsborg or prov. of West Gothland, situated on a bay, near the efflux of the Gotha-elv from Lake Wener, 52 m. N of Gottenburg. Pop. 2,500. It is the principal staple for the iron and timber sent from Wermeland to Gottenburg.

WENEW, a town of Russia, in the gov. and 30 m. NE of Tula, on the Wenewka. Pop. 3,700.

WENFOE, or WENVOE, a parish in Glamorganshire, 5 m. SW of Cardiff. Area 2,955 acres. Pop. in 1831, 432; in 1851, 475.

WENG, a village of Bavaria, in Lower Bavaria, landb. of Griesbach. Pop. 240.—Also a village of Austria, on the Upper Em circle, in the district of Braunau. Pop. 410.—Also a village of Syria, in the circle of Judenburg. Pop. 440.

WEN-GAN, a district and town of China, in the prov. of Chih-le, and div. of Shun-teen-fu, in N lat. 38° 53′, and E long. 116° 34′.—Also a district and town in the prov. of Keang-se, and div. of Keih-gan-fu.

WENGEN, a village of Bavaria, in the circle of Suabia, district of Kempten. Pop. 666.

WENGENS, a village of Prussia, in the reg. and circle of Oppeln. Pop. 318.

WENGI, a village of Switzerland, in the cant. of Berne, circle of Buren. Pop. 744.

WENG-KEANG, a district and town of China, in the prov. of Jar-moug, and div. of Gan-king-fu, in N lat. 30° 15′, and E long. 116° 40′.

WENG-TU-HEEN, a district and town of China, in the prov. of Chih-le, and div. of Paou-ting-fu, in N lat. 38° 45′. and E long. 115° 14′.

WENHAM, a township of Essex co., in Massachusetts, U. S., 20 m. N by E of Boston. Pop. in 1840, 689; in 1850, 977. From a lake or pond of about 1 m. sq. on its S border, many thousand tons of ice are annually exported.

WENHAM (GREAT or BURNT), a parish in Suffolk, 4½ m. SE by S of Hadleigh. Area 1,123 acres. Pop. in 1831, 181; in 1851, 269.

WENHAM (LITTLE), a parish in Suffolk, 5 m. SE by E of Hadleigh. Area 931 acres. Pop. 72.

WENHASTON, a parish in Suffolk, 3 m. SE of Halesworth. Area 2,326 acres. Pop., including the hamlet of Mells, 578; in 1851, 1,008.

WEN-HE, a district and town of China, in the prov. of Shan-se, and div. of Keang-chu, in N lat. 35° 25′, and E long. 111° 3′.

WEN-HEANG, a town of China, in the prov. of Ho-nan, and div. of Ho-nan-fu, in N lat. 34° 38′, and E long. 110° 30′.

WEN-HEEN, or WAN-HEEN, a district and town of China, in the prov. of Kan-suh, and div. of Keae-chu, in N lat. 35° 6′, and E long. 113° 13′.—Also a district and town in the prov. of Sze-chuen, and div. of Kew-chu-fu, in N lat. 30° 57′, and E long. 108° 32′.

WENIGERN, a village of Prussia, in the reg. of Arnsberg, circle of Hagen. Pop. 477.

WEN-KEANG, a district and town of China, in the prov. of Sze-chuen, and div. of Ching-tu-fu, in N lat. 30° 40′, and E long. 103° 55′.

WEN-LEEN-CHU, a town of China, in the prov. of Yun-nan, in N lat. 24° 28′, and E long. 99° 33′.

WENLOCK (LITTLE), a parish in Salop, 3¼ m. S of Wellington. Area 2,745 acres. Pop. 1,030.

WENLOCK (MUCH), a parish and parliamentary borough of Salop, 12 m. SE of Shrewsbury. Area 8,846 acres. Pop. in 1831, 2,487; in 1851, 2,398. The town consists principally of two streets at right angles to one another. The parl. borough comprises 17 parishes, with a pop. in 1851 of 20,588, chiefly engaged in potteries, mines, iron-works, and coal-pits. Electors in 1852, 905.

WENNE (SAINT), a parish in Cornwall, 4 m. NE by E of St. Columb-Major. Area 4,546 acres. Pop. in 1831, 649; in 1851, 650.

WEN-NEEN, a district and town of China, in the prov. of Keang-se, and div. of Faou-chu-fu, in N lat. 28° 40′, and E long. 116° 51′.

WENNINGTON, a parish in Essex, 2 m. N by W of Purfleet. Area 1,570 acres. Pop. in 1831, 127; in 1851, 177.—Also a township in the p. of Melling, co.-palatine of Lancaster, 6½ m. S by E of Kirkby-Lonsdale. Area 830 acres. Pop. 189.

WEN-SHAN. See LADRONES.

WEN-SHANG, a district and town of China, in the prov. of Shan-tung, and div. of Yu-chu-fu, in N lat. 35° 50′, and E long. 116° 40′.

WEN-SHWUY, a district and town of China, in the prov. of Shan-se, and div. of Tae-yuen-fu, in N lat. 37° 29′, and E long. 111° 58′.

WENSLEY, a parish and township in the N. R. of Yorkshire, 3 m. NW of Middleham, including the chapelry of Bolton-castle; the townships of Preston-under-Scar, Redmire, and W.; and the market-town of Leyburn. Area of p., 14,280 acres. Pop. in 1801, 1,505; in 1831, 2,266; in 1851, 2,105.—Wensley-dale is an extensive tract of country, banked on each side by lofty moors, in two straight unbroken ridges, leaving open, as far as the eye can reach, the long dale or valley of the river Ure.—Also a township and village in the p. of Darley, Derbyshire, 3½ m. WNW of Matlock. Pop. including the hamlet of Snitterton, in 1851, 557.

WEN-TING, a district and town of China, in the

prov. of Shan-ting, and div. of Ting-chu-fu, in N lat. 37° 12', and E long. 122° 20'.

WENTNOR, a parish in Salop, 5½ m. NE of Bishop's castle. Area 6,698 acres. Pop. 646.

WEN-TSAE, a district and town of China, in the prov. of Keang-se, and div. of Yuen-chu-fu, in N lat. 28° 5', and E long. 114° 20'.

WEN-TSEUEN, a district and town of China, in the prov. of Chih-le, and div. of Seuen-hwa-fu, in N lat. 40° 45', and E long. 114° 38'.—Also a district and town in the prov. of Shan-se, and div. of Pu-chu-fu, in N lat. 35° 25', and E long. 110° 46'.

WENTWORTH, or Wingford, a parish in Cambridgeshire, 4½ m. WSW of Ely. Area 1,437 acres. Pop. in 1831, 144; in 1851, 189.—Also a chapelry and township in the p. of Wath-upon-Dearn, Yorkshire, 5½ m. NW by N of Rotherham. Area 2,830 acres. Pop. in 1831, 1,394; in 1851, 1,556. Wentworth-house, the seat of Earl Fitzwilliam, is situated in this chapelry between Barnsley and Rotherham. The principal facade of this noble mansion is of 600 ft. in front, and the interior corresponds in magnificence. The park contains 1,500 acres.

WENTWORTH, a township of Grafton co., New Hampshire, U. S., 48 m. NW of Concord. Pop. 1,197.—Also a village in Rockingham co., N. Carolina, 76 m. NW of Raleigh.

WEOBLEY, a parish and town of Hereford, 12 m. NW of Hereford. Area 3,309 acres. Pop. in 1831, 919; in 1851, 908. The town, situated on the road from Hereford to Knighton, consists of one principal street. Under the reform act it was disfranchised.

WEONARD'S (Saint), a parish in Herefordshire, 7 m. W by N of Ross. Area 4,536 acres. Pop. in 1831, 564; in 1851, 648.

WEPION, a department and commune of Belgium, in the prov. and arrond. of Namur, watered by the Meuse. Pop. of dep., 748; of com., 340.

WEPPENDORF, or Vep, a market-town of Hungary, in the comitat of Eisenburg, 6 m. E of Steinamanger. Pop. 1,000. It has a castle belonging to the Counts of Erdöly.

WEPPERATH, a village of Prussia, in the prov. of the Lower Rhine, regency of Treves, and circle of Bernkastel.

WEPRIKAU, a village of Austria, in Bohemia, in the regency of Pardubitz, and bail. of Choticor. Pop. 440.—Also a village of the same reg., in the bail. of Polna. Pop. 400.

WEPRITZ, a village of Prussia, in the regency of Frankfort, and circle of Landsberg. Pop. 625.

WERANG-GHATI, a pass of the Himalaya, leading across the ridge which separates the valley E of Pangi from that of Lipa, in Kunawur. The crest of the pass, which has an alt. of 13,145 ft. above sea-level, is a vast mass of loose rock. The view from it is only remarkable for its barrenness. The descent on the NE side to Lipa, a distance of 5 m., is long and steep.

WERATSCHE, a village of Austria, in Styria, in the circle of Cilli. Pop. 260.

WERBACH, a village of Baden, in the circle of the Lower Rhine, and bail. of Tauber-Bischofsheim. Pop. 1,190.

WERBACHHAUSEN, a village of Baden, in the circle of the Lower Rhine, and bail. of Tauber-Bischofsheim, on the Tauber. Pop. 360.

WERBAN. See Verbo.

WERBELLIN, a village of Prussia, in the regency of Potsdam, and circle of Angermunde. Pop. 250.—A canal extends from a lake of the same name in the vicinity of this v. to the Finow canal.

WERBELOW, a village of Prussia, in the regency of Potsdam, and circle of Prenzlaw. Pop. 230.

WERBEN, a town of Prussia, in the regency and 57 m. NNE of Magdeburg, and circle of Osterburg, near the l. bank of the Elbe, opposite the confluence of the Havel, in N lat. 52° 51'. Pop. 924. It has a fine church erected by the knights of Malta, an hospital, and a tannery.—Also a village in the regency of Frankfort, and circle of Kottbus. Pop. 924.—Also a village in the regency of Merseburg, and circle of Weissenfels. Pop. 352.—Also a market-town in the regency of Stettin, and circle of Pyritz, on the E bank of Lake Madüe. Pop. 615. It has an hospital and extensive fisheries.

WERBERG, a village of Bavaria, in the circle of Upper Franconia, and presidial of Bruckenau. Pop. 492.

WERBIETZE. See Werbicz.

WERBIG, a village of Prussia, in the regency of Frankfort, and circle of Kustrin. Pop. 320.—Also a village in the regency of Magdeburg, and circle of Jerichow. Pop. 210.—Also two villages, distinguished as Hohen and Neider, in the regency of Potsdam, and circle of Zauch-Belzig. Pop. 260.

WERBITSK, a village of Russia in Europe, in the gov. of Moskow, and circle of Dmitrow, on the Jachroma. It has a large porcelain-manufactory.

WERBLIN, a village of Prussia, in the regency of Danzig, and circle of Neustadt. Pop. 340.

WERBLITZ, a village of Prussia, in the regency of Frankfort, and circle of Soldin. Pop. 348.

WERBOMONT, a department and commune of Belgium, in the prov. of Liege, and arrond. of Huy. Pop. of dep., 276; of com., 118.

WERCHAU, a village of Prussia, in the regency of Magdeburg, and circle of Schweinitz. Pop. 200.

WERCHHABEN, a village of Austria, in Bohemia, in the circle of Bunzlau, near Perstein. Pop. 270.

WERCHLUGAU, a village of Prussia, in the regency of Merseburg, and circle of Schweinitz. Pop. 260.

WERCHNE-UDINSK, a town of Russia in Asia, in the gov. of Irkutsk, at the mouth of the Uda, on the E side of Lake Baikal. Pop. 3,500.

WERCHNE-URALSK, a town of Russia, in the gov. of Orenburg, on the Ural, to the S of Ufa. Pop. 1,500. See also Uralsk.

WERCHNIJ-LAMOW, a town of Russia in Europe, in the gov. of Penza, and circle of Nisny-Lamow, on the Lamovn. Pop. 1,000.

WERCHNOI-KAMTCHATKA, a village of Russia, in Kamtchatka, to the N of Petropaulowsky. Pop. 300. It has a military hospital.

WERCHO, a village of Prussia, in the regency of Frankfort, and circle of Kalau. Pop. 260.

WERCHOWASHK, or Werchowaschkoi-Posad, a village of Russia in Europe, in the gov. of Vologda, and circle of Welsk, to the NW of Zolma, on the Waga.

WERCHTER, a department and commune of Belgium, in the prov. of Brabant, and arrond. of Louvain, watered by the Dyer, the Demer, and the Laeke. Pop. of dep., 1,195; of com., 551.

WERCKEN, a department and commune of Belgium, in the prov. of W. Flanders, and arrond. of Furnes, watered by the Handzaem-Vaerdeken. Pop. of dep., 1,373; of com., 170.

WERDA, a village of Saxony, in the circle of Zwickau, and bail. of Voightsberg, near Falkenstein. Pop. 731.

WERDAU, a town of Saxony, in the circle, bail., and 6 m. WNW of Zwickau, on the Pleisse. Pop. in 1837, 5,432; in 1843, 6,218; in 1849, 6,966. It has two churches, and possesses manufactories of woollen and cotton fabrics, and of shoes, several dye-works, and calico-printing-mills.

WERDE, a village of Austria, in Styria, in the circle of Cilli. Pop. 290.

WERDEK, a village of Austria, in Bohemia, in the regency of Gitschin, on the Elbe. Pop. 400.

WERDEN, a town of Prussia, in the regency and 15 m. NE of Dusseldorf, and circle of Duisberg, on the l. bank of the Ruhr. Pop. 4,688. It has an ancient abbey, now used as a house of correction, and possesses manufactories of silk and woollen fabrics, and of chemical substances, a paper-mill, and copper and iron-forges. Coal is found in the environs.

WERDENBERG, a town of Switzerland, in the cant. and 20 m. SSE of St. Gall, and district of Sargans, near the l. bank of the Rhine. Pop. 855.

WERDENFELS, a presidial of Bavaria, in the circle of Upper Bavaria. Pop. 9,800.

WERDER, a town of Prussia, in the regency and 6 m. W of Potsdam, and circle of Zauch-Belzig, on an island of the Havel. Pop. 2,000. It has manufactories of linen, breweries, and distilleries of brandy, building-docks, and extensive fisheries. Wine and varieties of fruit are extensively cultivated in the environs.

WERDERHAUSEN, a village of Anhalt-Dessau, and bail. of Grobzig. Pop. 240.

WERDOHL, a village of Prussia, in the prov. of Westphalia, regency of Arensberg, circle and 5 m. SE of Altena, on the Lenne. Pop. 606. It has salt-works.

WERDOM, a village of Prussia, in the regency of Posen, and circle of Obornik. Pop. 240.

WERDORF, a village of Prussia, in the regency of Coblenz, and circle of Metzlar, on the Dill. Pop. 734.

WERDUM, a village of Hanover, in the gov. of Aurich, and bail. of Esens. Pop. 400.

WERECHOV, a village of Austria, in Bohemia, in the regency of Pilsen, and bail. of Strakonitz. Pop. 260.

WEREHAM, a parish in Norfolk, 2 m. NW of Stoke-Ferry. Area 2,231 acres. Pop. 609.

WEREJA, a town of Russia in Europe, in the gov. of Moscow, on the Protwa. Pop. 6,000. It has a trade in hemp, corn, and other agricultural produce.

WERENZHAIN, a village of Prussia, in the regency of Frankfort, and circle of Luckau. Pop. 340.

WERESZYCA, a river of Galicia, which has its source in the NW part of the circle of Lemberg, traverses the W part of that circle, and the NE of that of Sambor, and after a course in a generally S direction of about 45 m., in which it forms several lakes, joins the Dniester, on the l. bank. Grudek and Komarno are the chief places on its banks.

WERFEN, a town of Upper Austria, in the circle of Salzburg, on the river Salza, 26 m. SSW of Salzburg. Pop. 1,510.—Also a village of Prussia, in the reg. of Minden, circle of Herford. Pop. 400.

WERGELA, or WURGLA, a town of the Beled-el-Jerid, 300 m. S of Algiers.

WERHENDAM, a village of Holland, in the prov. of N. Brabant, 24 m. NW of Hertogenbosch. Pop. 1,500.

WERKEL, a village of Kur-Hesse, in Lower Hesse, circle of Fritzlar. Pop. 640.

WERKLEIZ, a village of Prussia, in the reg. of Magdeburg, circle of Kalbe, on the Saal. Pop. 320.

WERL, a town of Prussian Westphalia, 19 m. WSW of Lippstadt. Pop. 4,000. There are large salt-works in the vicinity.

WERLTE, a village of Hanover, in the prov. of Osnabruck, near Meppen. Pop. 1,100.

WERM, a village of Belgian Limburg, in the cant. of Cortessem, arrond. of Tongres. Pop. 208.

WERMELSKIRCHEN, a village of Prussian Westphalia, in the duchy of Berg. Pop. 1,100. It has manufactures of coarse woollens.

WERMSDORF, a village of Moravia, in the circle of Olmutz, near Weisenberg. Pop. 1,200.—Also a town of Saxony, 22 m. ESE of Leipzig. Pop. 1,600.

WERNE, a river of Prussian Westphalia, in the principality of Paderborn, which falls into the Weser.—Also a town of Prussian Westphalia, on the Lippe, 19 m. S of Munster. Pop. 1,790, chiefly linen weavers.

WERNECK, a village of Bavarian Franconia, 18 m. NNE of Wurzburg. Pop. 460.

WERNERSDORF, a village of Prussia, in the reg. of Danzig, circle of Marienburg. Pop. 410.—Also a village of Bohemia, circle of Gitschin. Pop. 1,100.

WERNERSREUTH, a village of Bohemia, in the circle of Elbogen, on the Elster. Pop. 1,050.

WERNETH, a township and village in the parish of Stockport, co.-palatine of Chester, 4 m. ENE of Stockport. Area 1,560 acres. Pop. in 1831, 3,462; in 1851, 3,635. The manufacture of cotton goods and calico-printing is carried on here.

WERNIGERODE, a district, with the title of county, in Prussian Saxony, lying between the principality of Halberstadt and the states of Brunswick and Hanover. Its area is about 100 sq. m.; its population 13,000. It lies in the Harz forest, and is consequently covered with mountains. Its principal streams are the Ilse and the Ecker. The capital, of the same name, stands on a small stream called the Zillicherbach or Holzemme, at the N extremity of the Harz, 12 m. WSW of Halberstadt, in N lat. 51° 50′ 34″. It is divided into an old and a new town, and the suburb of Nessenrode. Pop. 5,600. It has a considerable trade in corn, chicory, and spirituous liquors; also in woollens, which are manufactured in the town. Adjoining to the town is a castle, the residence of the prince.

WERNITZ, a river of the Bavarian states, which rises near Rothenburg, 14 m. W of Anspach; flows SSE; and runs into the Danube, near Donauwert, after a course of 60 m.

WERNSDORF, a small town of Saxony, in the circle of Zwickau, 23 m. E of Leipsic. Pop. 400.—Also a village of Bohemia, in the circle of Saatz, 3 m. NW of Kadan. Pop. 806.

WERNSTADTL, a town of Bohemia, 38 m. N of Prague. Pop. 1,700. It has extensive cotton manufactures.

WERRA, a river of Germany, which has its source in the forest of Thuringia, 16 m. N of Coburg, at an alt. of 2,179 ft. above sea-level; flows WSW, and then NNW, traversing most of the principalities of the house of Saxe, and part of the electorate of Hesse; at Allendorf it becomes navigable for small boats; and at Münden, after a course of 140 m., joins the Fulda, after which the united stream takes the name of the Weser. Its chief affluents are the Nessa and the Ulster.

WERRAGEBIRGE, a name sometimes given to the northern portion of the Rhangebirge, in Kur-Hesse, the highest point of which is the Meisner, 2,184 ft. above sea-level.

WERRAY, a river of Cardiganshire, which runs into the Irish channel, near Arth, about 7 m. S of Aberystwith.

WERRE, a small river of Germany, which rises in Lippe-Detmold; flows NW and WNW; and falls into the Weser, on the l. bank, after a course of 40 m. The Bega and the Else are its chief affluents.

WERRIBI, a river of S. Australia, which flow

into Port Phillip, 16 m. SW of Melbourne, after a SE course of 60 m.

WERRINGTON, a parish of Devonshire, 2 m. N of Launceston. Area 5,000 acres. Pop. in 1831, 661; in 1851, 657.—Also a chapelry in Northamptonshire, 3½ m. NW of Peterborough. Pop. 669.

WERRUFELD (GROSS and KLEIN), two villages of Bavaria, in Lower Franconia. United pop. 720.

WERSE, a small river of Prussian Westphalia, in the principality of Munster, which falls into the Ems near Telligt.

WERSMENINGKEN, a village of Prussia, in the reg. of Gumbinnen, circle of Tilsit. Pop. 420.—Also a v. in the same reg., circle of Pillkallen. Pop. 540.

WERSTADT, a town of Hesse-Darmstadt, district of the Rhine. It is neatly built. Pop. 1,200.

WERTACH, a river of Bavaria, which rises on the frontiers of Tyrol; flows NNW; and falls into the Lech below Augsburg.

WERTH, a village of Prussia, in the reg. of Munster, circle of Borken, on the Yssel. Pop. 575.—Also a village in the reg. and circle of Coblenz. Pop. 728.—Also a town in the reg. of Dusseldorf, circle of Elberfeld. Pop. 2,020.—Also a lake of Carinthia, in the circle of Klagenfurt.

WERTHEIM, a town of Baden, the capital of the circle of the Maine and Tauber, and of the county of the same name, situated in a narrow valley at the confluence of the Maine and the Tauber, 20 m. W of Wurzburg. Pop. 4,900. It is surrounded with a wall, and is divided into four quarters. The ancient residence of the counts of Wertheim, situated on a hill outside the town, is now in a decayed state; but the counts have two modern buildings for their residence.

WERTHER, a town of Prussian Westphalia, 5 m. NNW of Bielefeld. Pop. 1,90 '.

WERTINGEN, a town of Bavaria, on the Zuzam, 14 m. NNW of Ulm. Pop. 1,500.

WERVICK, a canton and town of Belgium, in the prov. of W. Flanders, 8 m. SE of Ypres, on the Lys. Pop. 5,700. It is a place of considerable trade, and has manufactories of starch, leather, ropes, and oil.

WESCHNITZ, a small river of Germany, which rises in the Odenwald; passes Weinheim; and falls into the Rhine near Stein.

WESEL, or LOWER WESEL, a town of Prussia, in the duchy of Cleves, at the confluence of the Lippe and the Rhine, 28 m. ESE of Cleves, in N lat. 51° 39'. It is a place of old date, having formerly belonged to the Hanseatic confederacy; but in consequence of wars and other calamities, its population was reduced in 1740 to 5,700; and in 1770 it did not exceed 4,500; in 1849 it amounted to 16,228. Its manufactures are considerable, particularly those of woollens, hosiery, and spirituous liquors. It is strongly fortified, and has a good citadel.

WESEL (OBER), a town of Rhenish Prussia, in the gov. and 22 m. SE of Coblenz, on the l. bank of the Rhine. Pop. in 1852, 3,686.

WESELA, a village of Bohemia, in the reg. of Budweis, 62 m. S by E of Prague. Pop. 470.

WESELBERG, a village of Bavaria, in the Pfalz, canton of Waldstschbach. Pop. 350.

WESELERWALD, a village of Prussia, in the reg. of Dusseldorf, circle of Rees. Pop. 240.

WESELICHE, a town of Bohemia, in the reg. of Budweis, district of Muhlhausen. Pop. 1,460.

WESELY, a town of Moravia, on an island in the March, 40 m. S of Olmutz. Pop. 2,530.—Also a town of Bohemia, in the reg. of Budweis, at the confluence of the Luscheritz and the Nescharka, in N lat. 49° 11' 3". Pop. 1,800.

WESEN, a town of Switzerland, in the cant. of St. Gall, at the W extremity of the lake of Wallenstadt, 7 m. S of Uznach. Pop. 642. The inhabitants are employed partly in fishing, partly in the transit-trade over the lake to Zurich. The town has of late received injury from the inundations of the lake.

WESENBERG, a village of Mecklenburg-Strelitz, circle of Stargard, bail. of Alt-Strelitz, on the Havre. Pop. 1,310.—Also a village of Holstein, in the bail. of Reinfeld, on the Trave. Pop. 254.

WESENCARL, a village of Belgium, in the prov. of S. Brabant, near Lowen. Pop. 1,100.

WESENITZ, a small river of Saxony, which falls into the Elbe near Pirna.

WESENSTEDT, a village of Hanover, in the prov. of Oberhoya, bail. of Ehrenburg. Pop. 400.

WESENSTEIN, a village of Saxony, on the Elbe, 8 m. SSE of Dresden.

WESER, one of the principal rivers of Germany, in Hanover, formed by the junction of the Werra and the Fulda, at Munden. From the point of confluence, it flows in a NW and WNW course through the territories of Hanover, Brunswick, Prussia, Bremen, and Oldenburg; and reaches the German ocean, by an estuary 24 m. wide, between the last mentioned principality and the prov. of Bremen, after a course of 280 m. Ships go up the stream as far as Vegesack, and boats navigate its whole course. At Munden it has a breadth of 300 ft.; at Minden, of 420 ft.; at Bremen, of 700 ft.; and at Vegesack, of 2,000 ft. Its affluents on the l. are the Diemel, the Werre, the Bastan, the Aue, the Ochte, and the Hunte; and on the r. the Aller, Wumme, Dreyte, Luhne, Geeste, and Lessum. This river flows through a rich and fertile region, with a dense pop.; but the bordering states have so greatly neglected the navigation, that it is, says a correspondent of the *Daily News*, "no longer navigable for the larger descriptions of inland vessels, and scarcely even for small ones; for the sand and detritus of rocks washed down from the mountains have been allowed to accumulate so shamefully in the more level part of the country, where, of course, the current is not so strong as in the mountainous regions, that barges and steamers drawing only 18 inches of water are frequently unable to navigate the W., particularly in the summer. It is only below Bremen that the river becomes really important: above that city it is nowhere wider than the Isis at Oxford, and the navigation is so impeded by low bridges, and locks and weirs, to say nothing of the fiscal regulations and duties, that the only wonder is that commercial people should think it worth while at all to use it as a highway of communication for the transit of merchandise. These fiscal enactments are enforced by virtue of the convention of the 10th September, 1823, entered into between the seven states bordering on the W., together with the supplementary treaties of 1825 and 1839. But the stipulations of these conventions have exclusive reference to the amount of toll which each state is empowered to levy on the unfortunate skippers, and not at all to the general regulations affecting the navigation of the river. The consequence is that each state acts with such sovereign independence of the rest, that between Hanoverian Münden and Bremerhafen the navigation is subject to no less than 35 different codes of regulations. The inhabitants of Bremen have found out that, for the modern requirements of commerce, they are located some 50 m. too far from the sea, as the river will not admit of vessels drawing more than 10 or 12 ft. water going up to the town. Fortunately for them they some years ago induced the Hanoverian government—when Hanover was united to the British Crown—to cede to them a few hundred acres of marsh land near the mouth of the Weser, where they have founded a port and town called Bremerhafen, and made extensive docks and other works, capable of affording ample accommodation to the largest ships, and where the greater part of the Bremen trade is carried on."

WESERAM, a village of Prussia, in the reg. of Potsdam, circle of Westhausleand. Pop. 300.

WESERITZ, a village of Bohemia, in the reg. of Eger, district of Plau. Pop. 960.

WESLEY, a township of Washington co., Maine, U. S., 169 m. E of Augusta.—Also a village of Washington co., Ohio, 94 m. ESE of Columbus.

WESSELHUREN, a town of Holstein, in North Ditmarschen, to the W of Herde. Pop. 1,300.

WESSEL'S ISLANDS, a chain of islands which extend from the NW entrance of the gulf of Carpentaria, in a NE direction about 50 m. The N extremity of the chain is in S lat. 10° 59′, E long. 176° 45′.

WESSEM, a town of Dutch Limburg, on the Maese, 22 m. N by E of Maestricht. Pop. 2,500.

WESSINGTON, or WASHINGTON, a township in the p. of Crick, in Derbyshire, 3½ m. NW by W of Alfreton. Area 958 acres. Pop. in 1851, 575.

WEST, a township of Huntingdon co., in Pennsylvania, U. S., 8 m. N of Huntingdon. Pop. 1,629. —Also a township of Columbiana co., in Ohio. Pop. 1,915.—Also a river of Maryland, which flows, by a large estuary, into Chesapeake bay, in the S part of Anne-Arundel co.

WEST (CAPE), a headland of New Zealand, in S lat. 45° 54′, E long. 166° 32′.

WEST-ACRE, a parish in Norfolk, 5¼ m. NW by N of Swaffham. Area 3,400 acres. Pop. 475.

WESTALL (POINT), a headland of South Australia, in S lat. 32° 52′, E long. 133° 59′.

WEST-ALMOND, a township of Alleghany co., in New York, U. S., 6 m. E of Angelica. Pop. 976.

WEST-ALTON, a village of Belknap co., in New Hampshire, U. S., 28 m. NE of Concord.

WEST BEND, a township and village of Washington co., in Wisconsin, U. S., 68 m. ENE of Madison. Pop. 672.

WESTBERE, a parish in Kent, 3½ m. NE by E of Canterbury. Area 1,185 acres. Pop. 197.

WEST BERGEN, a village of Genesee co, in New York, U. S., 22 m. from Rochester.

WEST BETHEL, a village of Oxford co., in Maine, U. S., 54 m. W of Augusta.

WEST BLOOMFIELD, a village of Ontario co., in New York, U. S., 198 m. W of Albany. Pop. 1,698.—Also a v. of Essex co., in New Jersey, 48 m. NE of Trenton.

WESTBORO', a village of Worcester co., in Massachusetts, U. S., 28 m. W by S of Boston. Pop. 2,376.

WESTBOROUGH, a parish in Lincolnshire, 8 m. NW of Grantham. Area 890 acres. Pop. 236.

WESTBOURNE. See BOURNE (WEST).

WEST BOYLSTON, a village of Worcester co., in Massachusetts, U. S., 36 m. W of Boston. Pop. 1,749.

WESTBROMWICH. See BROMWICH (WEST).

WEST BROOK, a village of Middlesex co., in Connecticut, U. S., 36 m. S by E of Hartford. Pop. 1,298.

WEST BROOKFIELD, a village of Worcester co., in Massachusetts, U. S., 57 m. W by S of Boston. Pop. 1,344.

WESTBURTHHAUSEN, a village of Prussia, in the regency of Minden, circle of Halle. Pop. 321.

WESTBURY, a parish in Bucks. Area 2,547 acres. Pop. in 1831, 391; in 1851, 458.—Also a parish in Salop, 8¾ m. W by S of Shrewsbury, containing the chapelry of Minsterley, and the township of W. Area, 11,274 acres. Pop. in 1831, 2,228; in 1851, 2,485.—Also a parish in Somersetshire, on the river Ax, which separates it from Wedmore, 4 m. NW by W of Wells. Area 2,968 acres. Pop. in 1831, 681; in 1851, 625.—Also a hamlet joined to that of Peake, in the p. of East Meon, Hants, 6¼ m. W of Petersfield.—Also a borough and parish in Wiltshire, comprising the chapelries of Braton and Dilton, and the townships of Leigh, Hawkeridge, and Haywood. Area 11,901 acres. Pop. of the p. in 1821, 6,846; in 1831, 7,324; in 1851, 7,029.—The town, which lies in the centre of the parish, consists of three streets irregularly built, and is 34 m. NW by W of Salisbury. Broad cloths and kerseymeres are manufactured here to a considerable extent. The whole p. is included in the borough for parliamentary purposes. Pop. of the borough in 1801, 1,837; in 1831, 2,495; in 1851, 7,029. Electors in 1852, 314.

WESTBURY-ON-SEVERN, a parish in Gloucestershire, 3 m. NE of Newnham. Area 8,695 acres. Pop. in 1831, 2,032; in 1851, 2,498.

WESTBURY-ON-TRIM, a parish in Gloucestershire, 3 m. NW of Bristol, containing the chapelry of Shirchampton, and the tything of Bishop's-Stoke. Area 5,456 acres. Pop. in 1831, 4,263; in 1851, 6,728.

WESTBY-WITH-PLUMPTONS, a joint township in the p. of Kirkham, co.-palatine of Lancaster. Area 3,426 acres. Pop. in 1851, 707.

WEST CALDER. See CALDER (WEST).

WEST CAMBRIDGE, a village of Middlesex co., in Massachusetts, U. S., 6 m. NW of Boston. Pop. in 1840, 1,363; in 1850, 2,202.

WEST CANADA CREEK, a river of New York, U. S., which flows by a SW and S course into the Mohawk, 1 m. below Herkimer, forming Trenton falls, 20 m. above its mouth.

WESTCAPPELLE, a village of Belgium, in W. Flanders, 8 m. NE of Bruges.

WEST CHESTER, a county in the SE of New York, U. S., containing 470 sq. m. Pop. in 1840, 48,686; in 1850, 58,263. Its cap. is Bedford.—Also a township and v. in the same co., 123 m. SW of Albany. Pop. in 1850, 2,492.—Also the cap. of Chester co., in Pennsylvania, 70 m. E by S of Harrisburg. Pop. 2,500.

WEST-CLIFFE, a parish in Kent, 3 m. NE of Dover. Area 1,194 acres. Pop. in 1851, 129.

WEST CONCORD, a village of New Hampshire, U. S., 3 m. WNW of Concord.—Also a v. of Essex co., in Vermont.

WEST CORNWALL, a village of Litchfield co., in Connecticut, U. S., 36 m. W by N of Hartford.

WESTCOTE, a parish in Gloucestershire, 4 m. SE by S of Stow-on-the-Wold. Area 1,503 acres. Pop. in 1831, 188; in 1851, 242.—Also a hamlet in the p. of Waddesdon, Bucks, 7 m. WNW of Aylesbury. Pop. 273.

WESTCOVE, a hamlet in the p. of Dunkerrin, co. Kerry, on the N side of the Kenmare estuary, 7 m. SW of Sneem.

WEST DEDHAM, a village of Norfolk co., in Massachusetts, U. S., 12 m. SW of Boston.

WESTERAAS, or VESTERAS, a laen of Sweden, bounded on the N by Fahlun and Gefle; on the E by Upsala; on the S by the Mälar lake; and on the W by the laen of Oerebro. It has an area of 2,000 geog. sq. m. Pop. in 1833, 91,300; in 1845, 94,850. —The cap., of the same name, is situated on the N side of Lake Mälar, in N lat. 59° 37′, 54 m. NW of Stockholm. Pop. 3,800. It has a trade in timber and iron.

WESTERBECK, a village of Prussia, in the regency of Munster, circle of Warendorf. Pop. 380.

WESTERBOKUM, a village of Oldenburg, in the circle and bail. of Vechta. Pop. 260.

WESTERBOTTEN. See UMEA.

WESTERBURG, a town of Nassau, in the bail. of Rennerode. Pop. 1,538.

WESTERBUTTEL, a village of Holstein, in South Dithmarsch, bail. of Meldorf. Pop. 350.

WESTERDALE, a parish in the N. R. of Yorkshire, 7½ m. SE of Guisborough. Area 15,930 acres. Pop. in 1831, 281; in 1851, 286.

WESTERFELDE, a village of Bavaria, in the regency of Arnsberg, circle of Dortmund. Pop. 200.

WESTERFIELD, a parish in Suffolk, 2½ m. NNE of Ipswich. Area 1,071 acres. Pop. in 1851, 324.

WESTERHAM, a parish and market-town in Kent, 22 m. W of Maidstone. Area of p. 5,676 acres. Pop. in 1831, 1,985; in 1851, 2,113. The town is situated on a gentle declivity, on both sides of the river Darent, which rises in the vicinity.

WESTERHAUSEN, a town of Prussia, in the regency of Magdeburg, circle of Aschersleben. Pop. 1,645.

WESTERHOFEN, a village of Würtemberg, in the Jaxt circle, bail. of Ellwangen. Pop. 370.

WESTERHUSEN, a village of Prussia, in the regency of Magdeburg, circle of Wanzleben, on the Elbe. Pop. 710.

WESTERKIRK, a parish of Eskdale, in Dumfries-shire. Area 27,307 acres. Pop. in 1851, 658.

WESTERLEIGH, a parish in Gloucestershire, 3 m. SW of Chipping-Sodbury. Area 4,009 acres. Pop. in 1831, 1,709; in 1851, 1,679. There is a railway 9 m. in length between the floating-dock in Bristol, and Coal-pit-heath in this parish.

WESTERLOO, a township and village of Albany co., in New York, U. S., 17 m. SW of Albany. Pop. 2,860.

WESTERLY, a township and village of Washington co., in Rhode island, 26 m. W by S of Newport. Pop. in 1850, 2,766.

WESTERN AUSTRALIA. See AUSTRALIA.

WESTERN PORT, a natural harbour of South Australia, 35 m. SE of Melbourne, 12 m. SE of Port-Phillip. It is about 20 m. in length, and 5 m. in width at its mouth.

WESTERNPORT, a village of Alleghany co., in Maryland, U. S., on the N side of the Potomac, 135 m. W by N of Annapolis.

WEST FAIRFIELD, a village of Westmoreland co., in Pennsylvania, 115 m. W of Harrisburg. There are iron-works here.

WESTFIELD, a parish of Norfolk, 3 m. S of Dereham. Area 569 acres. Pop. in 1831, 127; in 1851, 143.—Also a parish in Sussex, 4½ m. E by S of Battle. Area 4,272 acres. Pop. in 1851, 900.

WESTFIELD, a township and village of Hampden co., in Massachusetts, 74 m. W by S of Boston. Pop. in 1840, 3,526; in 1850, 4,186.—Also a township of Essex co., in New Jersey, 36 m. NE of Trenton. Pop. 1,575.—Also a township of Chautauque co., in New York, 57 m. from Buffalo, by railway. Pop. 3,100.—Also a township and village of Monroe co., in Ohio, 31 m. N of Columbus. Pop. 1,414.—Also a township of Tioga co., in Pennsylvania, 117 m. W by N of Harrisburg. Pop. 1,000.—Also a township of Orleans co., in Vermont, 41 m. N of Montpelier. Pop. 502.

WEST FINLEY, a township of Washington co., in Pennsylvania, U. S., on Wheeling creek. Pop. 1,200.

WESTFORD, a township and village of Middlesex co., in Massachusetts, U. S., 23 m. NW of Boston. Pop. 1,473.—Also a township of Otsego co., in New York, 53 m. W of Albany. Pop. 1,428.—Also a township of Chittenden co., in Vermont, 31 m. NW of Montpelier. Pop. 1,458.

WEST GREENVILLE, a village of Mercer co., in Pennsylvania, U. S., 190 m. NW of Harrisburg.

WEST GREENWICH, a township and v. of Kent co., in Rhode island, U. S, 23 m. NW of Newport. Pop. 1,350.

WESTHALL, a parish in Suffolk, 4 m. NE of Halesworth. Area, 2,316 acres. Pop. in 1851, 496

WESTHAM, a parish in Sussex, 5¾ m. SE of Hailsham. Area 4,718 acres. Pop. in 1851, 761.

WESTHAMPNETT. See HAMPNETT (WEST).

WESTHARPTREE. See HARPTREE (WEST).

WEST HARTFORD, a township and village of Windsor co., in Vermont, U. S., 41 m. S by E of Montpelier.

WEST HAVEN, a township and v. of Rutland co., in Vermont, U. S., 58 m. SW of Montpelier. Pop. 718.

WESTHEIM, a village of Würtemberg, in the Jaxt circle. Pop. 459.—Also a village of Bavaria, in Middle Franconia, bail. of Heidelheim. Pop. 524.

WEST HOBOKEN, a village of Hudson co., in New Jersey, U. S., 3½ m. NW of New York.

WEST HOFEN, a town of Hesse-Darmstadt, 20 m. S of Mainz, on the Seebach. Pop. 1,850.—Also a town of France, in the dep. of Bas-Rhin, 5 m. W of Strasburg. Pop. 2,500.

WESTHORPE, a parish in Suffolk, 7 m. N of Stow-Market. Area 1,322 acres. Pop. 240.

WEST INDIES. See INDIES (WEST).

WESTINGTON, a hamlet of Chipping-Camden p., in Warwickshire. Pop., with Combe, 146.

WEST JAMES, a township and village of Westchester co, in New York, U. S., 126 m. S of Albany. Pop. 4,436.

WESTKAPELLE, a town of Holland, in the prov. of Sieland, 7 m. NW of Middleburg. Pop. 1,820.

WESTLEIGH. See LEIGH (WEST).

WESTLETON, a parish in Suffolk, 2¾ m. E of Yoxford. Area 6,103 acres. Pop. in 1851, 903.

WESTLEY, a parish in Suffolk, 2 m. W by N of Bury-St.-Edmunds. Area 1,216 acres. Pop. 575.

WESTLEY-WATERLESS, a parish in Cambridgeshire, 5 m. SW of Newmarket. Area 1,102 acres. Pop. in 1831, 158; in 1851, 214.

WEST LINTON, or LEVINGTON, a township in the p. of Kirk-Linton, Cumberland, 4 m. SE of Longtown. Pop. in 1831, 629; in 1851, 575.

WESTMALLE, a commune and village of Belgium, in the prov. and 12 m. NE of Antwerp. Pop. 1,000.

WESTMANN, or WESTMANNA ISLANDS, a group of islets off the coast of Iceland, in N lat. 63° 20', W long. 20° 30'. They are 14 in number, but only 4 of them produce any vegetation or pasturage, and of these only one, called Heimaey or Home island, is inhabited. It is 15 m. from the coast, and has a harbour partly encircled by a high perpendicular rock. A precipitous path leads to the top of the island, where the people, with their habitations, a few sheds, and their little church, remain 2,000 ft. above the ocean. The islands are basaltic, like Fingal's cave and the Giant's causeway; but instead of being 100 or 200 ft. in height, rise like immense columns, nearly a ½ m. above the sea. The inhabitants draw their entire subsistence from the ocean and the cliffs, catching codfish and killing sea-birds, myriads of which haunt the rocks of their sea-girt shores. The sea-fowls furnish them feathers; some sorts are used for food, and some for fuel. They split them open, dry them, and then burn them, feathers and all. These islands, called in Icelandic VESTMANNŒYJAR, were settled by a colony of Irish in 875. A Norwegian pirate cruising in the Atlantic came upon the coast of Ireland, landed, and captured 40 or 50 persons, men, women, and children, whom he carried off as slaves. Before he got home they rose on their captors, slew them, and went ashore on the first land they met. This was the largest of the Westmann islands, and that name was given them by the Icelanders as these people came from the West.

WESTMEATH, an inland county in the W part of the Irish province of Leinster; bounded on the NW by co. Longford; on the N by co. Cavan; on the NE and E by co. Meath; on the S by King's co.; and on the W by co. Roscommon. The greatest length of the co., in a line drawn SW, is 33 m.; its greatest breadth, SE, is 16½ m. Its area comprises 365,218 acres of arable land, 56,392 of uncultivated land, 8,803 of continuous woods, 628 of towns, and 22,427 of water,—in all, 453,468 acres. W. was pronounced by Wakefield the most beautiful co. in Ireland next to Kerry, Fermanagh, Wicklow, and Waterford. Its surface may, in a general view, be characterized as a luxuriant plain, diversified with swells, gravelly ridges, and a few considerable hills, —sombrely patched, in various large districts, particularly in the south and east, with bog,—and largely and beautifully gemmed in the N, on the W, and S through the centre, with handsome and expressive lakes. Mountains do not exist; and hills are, for the most part, mutually isolated, and considerably apart. The basis elevation above sea-level varies from 125 ft. in the W, to about 328 ft. in the centre, and probably averages about 265 ft.—The principal lakes are Lough Ree, on the W boundary; Lough Glen or Deerin, on the NW; Loughs Sheelin and Kinnail on the N; and Loughs Dereveragh, Ennel, Owhel, Lane, and Iron, in the interior. The river Inny, one of the chief tributaries of the Shannon, traverses the NW district; several tiny affluents of the Shannon, drain the remainder of the W. The Brosna, another chief tributary of the Shannon, rises in the interior, and drains a large proportion of the centre and all the S into King's co.; and the rivulets Stonyford and Deel drain the E district into Meath toward a confluence with the Boyne. The eastern district of the county thus belongs to the basin of the Boyne; and all the other districts belong to the basin of the Shannon.

Soil and agriculture.] This co. presents much less variety of surface-rock than any other co. of Ireland, and, at the same time, excels every other in the aggregate wealth of the prevailing substrata of its soils. Mr. Wakefield, who wrote in 1812, and distributed all Ireland as to its agricultural condition into nine districts, classified W. with Meath, Louth, Dublin, Kildare, and Kilkenny. Considerable improvements have been effected in the practices of husbandry since Mr. Wakefield wrote; but the agricultural condition of the co. is still much inferior to that of the best districts of England and Scotland, and unworthy of the opulent character of the soil. In 1841, there were within the rural districts of the co., 4,266 farms of from 1 acre to 5 acres, 4,076 of from 5 to 15 acres, 1,648 of from 15 to 30 acres, and 1,385 of upwards of 30 acres. The black cattle of W. are among the best in Ireland, and are held in high repute. The horses are, in general, of very superior breeds. The sheep include some excellent specimens of the long-woolled breed. The estimated value of live stock in the rural districts in 1841 was £505,105.

Population, &c.] Pop., in 1831, 136,872; in 1841, 141,300. Inhabited houses, 24,002. First-class inhabited houses, 529; second-class, 4,796; third-class, 10,841. Families employed chiefly in agriculture, 18,090; in manufactures and trade, 4,860; in other pursuits, 2,743. Families dependent chiefly on property and professions, 804. Males who could neither read nor write, 28,000; females who could neither read nor write, 36,938. Pop. in 1851, 111,409. The towns and principal villages are Moate, in Clonlonan barony; part of Athlone, in Brawney barony; Glassen and Auburn, in Kilkenny-West; Ballymore, Rathconrath, and Moyvore, in Rathconrath; Ballinacarrig and Rathowen, in Moygoish; Finea, Castle-Pollard, Coole, and Fore, in Demifore; Ballinalack and Multifarnham, in Corkaree; Castletown, Kilbeggan, Ballinagore, and Killavally, in Moycashel; Tyrrel's-Pass, Rochfort Bridge, and Milltown, in Fartullagh; Mullingar, in Moyashel and Magheradernon; Castletown-Delvin, Clonmellan, and Drumcree, in Delvin; and Kinnegad, Killucan, Raharney, and Rathwire, in Farbill.—In the ecclesiastical divisions, W. lies partly in the diocese of Ardagh, but chiefly in that of Meath. The head-quarters of the constabulary are at Mullingar. The assizes are held at Mullingar, and quarter sessions at Mullingar and Moate. —The county sent 10 members to the Irish parliament, or two from the county at large, and two from each of the boroughs of Athlone, Mullingar, Kilbeggan, and Fore; and though it continues to send 2 members to the imperial parliament from the county at large, it now contains no parl. borough except part of Athlone. County constituency in 1841, 1,125; in 1851, 3,132.

History.] Westmeath afforded one of the earliest settlements to the Anglo-Normans in the 12th cent., and constituted a portion of the palatinate of Hugh de Lacy. The O'Melaghlins were ancient sovereigns of the kingdom of Meath, which territory comprised the district comprising in modern times East and West Meath, with part of the King's co. and co. Dublin. The Anglo-Norman settlers uniformly derived under Hugh de Lacy, lord and earl of Meath, who partitioned his prov. amongst his relations and followers, many of whose descendants remain in high consideration at the present time.—W. gives the title of Marquis, in the peerage of Ireland, to the family of Nugent.

WEST-MESTON, a parish in Sussex, 7 m. NW of Lewes. Area 4,074 acres. Pop., exclusive of that of the chapelry of East-Chiltington, in 1801, 205; in 1831, 236; in 1851, 617.

WEST MILFORD, a township and village of Passaic co., in New Jersey, U. S., 60 m. NE of Trenton. Pop. 2,624.

WESTMILL, a parish in Hertfordshire, 1¼ m. S by E of Buntingford. Area 2,137 acres. Pop. 380.

WESTMINSTER. See LONDON.

WESTMINSTER, the capital of Carroll co., in Maryland, U. S., 47 m. NW of Annapolis. Pop. 835. —Also a township and village of Worcester co., in Massachusetts, 45 m. W by N of Boston. Pop. 1,916.—Also a village of Windham co., in Vermont, 84 m. S by E of Montpelier. Pop. 1,721.

WEST MONROE, a township of Oswego co., in New York, U. S., 140 m. NW of Albany. Pop. 1,197.

WESTMORE, a township of Orleans co., in Vermont, U. S. Pop. 130.

WESTMORELAND, an inland county in the north of England; bounded on the N by Cumberland; on the NE by Durham; on the E and SE by Yorkshire; on the S and SW by Lancashire; and on the W by Cumberland. It extends from 54° 11′ to 54° 42′ N lat., and from 2° 20′ to 3° 12′ W long.; and contains 763 sq. m., or 488,320 square acres, of which 40,000 acres are computed to be in tillage, and 140,000 in pasturage. It is divided into the two baronies of Kendal and Westmoreland; that of Kendal being divided into the wards of Kendal and Lonsdale; and that of Westmoreland, which is occasionally called the barony of Appleby, or 'the Bottom' of Westmoreland, into the East and the West ward. It contains 32 parishes, one parliamentary borough, Kendal; and 8 market-towns, Appleby, Ambleside, Brough, Burton, Kendal, Kirkby-Lonsdale, Kirkby-Stephen, and Orton.—The pop. in 1801, was 41,617; in 1831, 55,041; in 1841, 56,469; in 1851, 58,287.

Climate, Soil, &c.] The climate of this co. is exceedingly humid, owing to its contiguity to the western ocean, from which the winds blow during two-thirds of the year, carrying along with them exhalations which fall in rain on the mountains. On an average of five years the quantity of rain which fell amounted to 64 ins. The air, however, is remarkably pure and healthy, though, in the mountainous districts, it is cold and piercing. The general aspect of the county is marked with immense tracts of mountains,—beautiful but contracted valleys,—extensive lakes,—and large rocky districts presenting high, steep, and projecting crags on every side. Farm-houses, and small villages covered with blue slate, and whitened with lime-wash, are scattered about the base of the hills, and their small irregular fields spreading up the sides of the mountains, are almost always divided by stone walls,—a circumstance which gives the country a naked appearance. Several low heathy commons are seen towards the E side of the county; the W is characterized by high rugged prominences, and large tracts of low flat peat-moss. The occurrence of numerous trees in the mosses proves that W. was originally a wooded district; and several extensive tracts, especially on the NE boundary, still retain the name of forests. Wood, however, at present exists chiefly in plantations, and in detached groves round the country-seats and farm-houses. In some districts, considerable portions of land are covered with coppice which chiefly consist of oak, alder, ash, birch, and hazel. These underwoods are cut down usually every 16th year, and are either converted into hoops or into charcoal for the iron furnaces in the vicinity.—Helvellyn, a mountain rising, according to the Ordnance survey, to the height of 3,055 ft., is more particularly noted as a Cumberland height. Carterfell, High-street, and Kidsey-pike, are stupendous heights, within a few miles of the southern end of Haws-water. Langdale-pikes, in the W corner of the county, are conical hills of great height, with pyramidal rocky tops. The chain on the E, which is continued N and S through the adjoining counties, is regular, with mossy and heathy tops, except two or three conical green hills opposite Appleby. In other districts the hills are partly heathy, but more generally green, dry, and rocky. The soils of this co. are in general dry and gravelly: in some parts, particularly in the E and N, sand and hazel mould appear. The soil that rests on limestone is esteemed the most profitable.

Rivers.] The principal rivers are the Eden, Eamont, Lowther, Lune, and Kent. The Eden rises in one of the hills at the top of Mallerstang, near the SE extremity of the county; and runs at first in a NE direction, passing Kirkby-Stephen; it then turns NW, in which direction it proceeds by Appleby, to Brougham, where it enters Cumberland.—The Eamont takes its rise in Ulleswater lake, and flows in a NE direction along the Cumberland border to the Eden, into which it falls at the point where the co. is quitted by that river.—The Lowther rises at the foot of the Shap-fells in the middle of the co., and flowing at first NE, and gradually afterwards due N, joins the Eamont at a point about 2 m. SE of Penrith.—The Lune rises in Ravenstone-dale; flows for a few miles towards the N, and then turning suddenly towards the W reaches Tebay, where it again turns, and enters Lancashire near Kirkby-Lonsdale.—The Kent rises on the S side of High-street, and proceeds in a direction gradually becoming SE to the town of Kirkby-Kendal, whence it flows nearly due S to the spacious bay of Morecambe. No portion of the Kent is navigable until it enters the sands of Morecambe bay. Besides those already noticed, there are numerous minor streams, such as the Brathay and the Rothay, which fall into Windermere near its head, and the Troutbeck, which falls into the same lake near Culgarth.—The principal lakes are Windermere, situated partly in Lancashire; Ulleswater, extending a little into Cumberland; Hawswater, between Ulleswater and Shap; and Grasmere, a few miles north of Ambleside. There are various smaller lakes, the principal of which are Rydal-water, below Grasmere; Elter-water, 2 m. W of Ambleside; Broadwater, at the head of Patterdale; Kentmere-tarn, in Kentmere dale; Skeggles-water, about 3 m. NE of Kentmere tarn; Sunbiggin tarn, near Orton; and Whinfell tarn, near Kendal.—The only line of inland navigation in this co. is the Lancaster and Kendal canal, which enters the co. from Lancaster, near Burton-in-Kendal, and runs N to Kirkby-Kendal. A ship canal was projected by the Port-Fleetwood company, to run from Bardsea, through Ulverstone to the foot of Windermere lake, at Newby-bridge, so as to permit steamers to ply direct from Fleetwood to Ambleside Waterhead, and thus open up a new route to the lakes of Westmoreland and Cumberland, with the aid of a good road over the steep ascent of Kirkstone to Patterdale. The road from London to Glasgow and Edinburgh, by Carlisle, enters the co. from Yorkshire, a little to the SE of Brough, passes through that town and Appleby, and leaves the county near Penrith. The London road to Carlisle, through Lancashire, enters at Burton-in-Kendal, and runs N by Milnthorpe, Kendal, Shap, and Clifton into Cumberland, by Penrith. A railway runs from Lancaster, by Penrith to Carlisle; and a branch railway connects Kendal and Windermere with this line.

Agriculture.] Turnip and clover husbandry have made some progress in this co., and considerable quantities of wheat are now annually grown, though oats are principally cultivated. Fold-yard dung, and peat-ashes, are almost the only manures; for though the limestone of the county is inexhaustible, the want of coal for burning it prevents the general use of this manure. The farms are in general small, and the rents are chiefly made up by the sale of cattle, sheep, wool, butter, eggs, and hams. The wool is used in the manufactures of Kendal, and of Bradford in Yorkshire; part of it is wrought into knitted stockings about Kirkby-Stephen, Orton, and Ravenstonedale. The cattle bred in this county are long-horned, much resembling the Lancashire breed; and, when kept to a proper age, they grow to a great size. The breed of sheep on the mountains and commons is either native or a cross with Scottish rams. They are horned, dark or grey-faced, and thickly pitted with coarse, strong, hairy wool. Geese are bred on the mountains in great numbers, and afford an article of exportation. Grouse and other game are also abundant on the mountains. Veins of metallic ore, chiefly copper and lead, have been occasionally discovered; and some attempts have been made to work them, but many of these lie so deep, or are so minutely situated, that not having in general paid the expense, the greater part have been abandoned. The strata of coal in this co. are too compact in their nature to afford much prospect of a large supply of pit-coal being raised, and, in general the quality is inferior. The mountainous parts abound with various sorts of slate, the best of which are either carried by sea to London, Liverpool, Lynne, or Hull, or by land, into the adjoining counties of Durham, Cumberland, Lancaster, and Northumberland. The general colour of the slate is blue, varying in tint from a pale to a very deep blue.

Commerce and manufactures.] W. can scarcely be said to be either a commercial or a manufacturing county. Its manufactures are principally confined to Kendal and its vicinity, and mainly consist of a kind of coarse woollen cloth, and worsted stockings. See article KENDAL. The exports of the co. are principally cloths and other articles of home-manufacture, with numbers of cattle, sheep, geese, gunpowder, hoops, charcoal, and hams.

Franchise, &c.] Previous to the passing of the reform act, 2 members were returned for the shire, and 2 for Appleby. Under the reform act Appleby has been disfranchised, and Kendal returns one member. The county still returns 2 members, who are polled for at Appleby, Kirkby-Stephen, Shap, Ambleside, Kendal, and Kirkby-Lonsdale. The number of electors registered for the county in 1837 was 4,683; in 1852, 4,062. W. is included in the northern circuit. The assizes are held at Appleby. The barony of Westmoreland, constituting the deanery of Westmoreland, is in the dio. of Carlisle; that of Kendal, which is divided into the two deaneries of Kendal and Kirkby-Lonsdale, is in the dio. of Chester.—Prior to the arrival of the Romans, this county was occupied by the tribes of the Brigantes, the eastern parts being possessed by the *Voluntii*, and the western by the *Sistuntii*. During the Roman government, it was included in their province of *Maxima Cæsariensis*, which reached from the Humber to the Tyne. When Alfred made the division of his kingdom into counties, this co., with those of Northumberland and Cumberland, were not mentioned, and are therefore supposed to have then belonged to the Scots.

WESTMORELAND, a county in the SW of Pennsylvania, U. S. Area 989 sq. m. Pop. in 1840, 42,699; in 1850, 51,726. Its cap. is Greensburg.—Also a co. in the NE of Virginia. Area 162 sq. m. Pop. in 1840, 8,019; in 1850, 8,080. Its cap., of the same name, is 50 m. NE of Richmond.—Also a township of Cheshire co., in New Hampshire, 45 m. SW of Concord. Pop. 1,677.—Also a township of Oneida co., in New York, 98 m. NW of Albany. Pop. 3,291.

WESTOE, a township and chapelry in the p. of Jarrow, co.-palatine of Durham, forming a populous suburb to South Shields. Pop. in 1841, 13,990; in 1851, 19,349.

WESTON, a township in the p. of Runcorn, co.-palatine of Chester, 2 m. NNW of Frodsham. Area 1,282 acres. Pop. in 1831, 532; in 1851, 933.—Also a chapelry and township in the p. of Wybunbury, co-palatine of Chester, 6 m. E by N of Nantwich, near the Grand Junction railway. Area 1,851 acres. Pop. in 1831, 401; in 1851, 514.—Also a tything in the p. of Stalbridge, Dorset, 7½ m. E by N of Sherborne. Pop. in 1831, 225; in 1851, 250.—Also a parish in Hertfordshire, 4 m. NE by N of Stevenage. Area 4,530 acres. Pop. in 1831, 1,046; in 1851, 1,186.—Also a parish in Lincolnshire, 4 m. NE of Spalding. Area 5,386 acres. Pop. in 1831, 567; in 1851, 759.—Also a parish in Somersetshire, 1¾ m. NW by W of Bath, on the N bank of the river Avon, at the point at which a small tributary stream rising on Lansdown hill, joins that river. Area 2,650 acres. Pop. in 1831, 2,560; in 1851, 3,088.—Also a parish in Suffolk, 3 m. S of Beccles. Area 1,550 acres. Pop. in 1831, 233; in 1851, 243.—Also a parish in the W. R. of Yorkshire, 2 m. NW by W of Otley, on the river Wharfe. Area 4,552 acres. Pop. in 1851, 492.

WESTON, a village of Upper Canada, 10 m. NW of Toronto. Pop. 550.

WESTON, a township and village of Missouri, U. S., 160 m. WNW of Jefferson.—Also a township and v. of Fairfield co., in Connecticut, 60 m. SW of Hartford. Pop. 1,968.—Also a v. of Middlesex co., in Massachusetts, 11 m. W of Boston. Pop. 1,200.—Also a v. of Lewis co., in Virginia, 188 m. NW of Richmond.

WESTON-WITH-ALCONBURY, a parish in Hunts, 6 m. NW of Huntingdon. Area 1,540 acres. Pop. in 1831, 441; in 1851, 516.

WESTON-UPON-AVON, a parish partly in Warwickshire, and partly in Gloucestershire, 9 m. N by E of Chipping-Campden. Area 1,540 acres. Pop., in 1831, 108; in 1851, 115.

WESTON-BEGGARD, a parish in Herefordshire, 5 m. E of Hereford. Area 934 acres. Pop. in 1831, 281; in 1851, 267.

WESTON-BAMPFYLDE, a parish in Somersetshire, 5 m. SSW of Castle-carey. Area 631 acres. Pop. in 1831, 123; in 1851, 119.

WESTON-BIRT, a parish in Gloucestershire, 4 m. SW by W of Tetbury. Area, with Lashborough, 1,904 acres. Pop. in 1831, 138; in 1851, 234.

WESTON (COLD), a parish in Salop, 7½ m. NE by N of Ludlow. Area 670 acres. Pop. 28.

WESTON-COLVILLE, a parish in Cambridgeshire, 6 m. NNE of Linton. Area 2,943 acres. Pop. in 1831, 444; in 1851, 574.

WESTON-CONEY, a parish in Suffolk, 7 m. SW of East Harling. Area 1,741 acres. Pop. 266.

WESTON-COYNEY, a joint township with Hulme, in the p. of Caverswall, Staffordshire, 5 m. W of Cheadle. Pop., with Hulme, in 1851, 778.

WESTON-SUB-EDGE, a parish in Gloucestershire, 2 m. W by N of Chipping-Camden. Area 2,632 acres. Pop. in 1831, 367; in 1851, 358.

WESTON-FAVELL, a parish in Northamptonshire, 2½ m. ENE of Northampton. Area 1,050 acres. Pop. in 1831, 443; in 1851, 508.

WESTON-IN-GORDANO, a parish in Somersetshire, 10 m. W by N of Bristol. Area 733 acres. Pop. in 1831, 124; in 1851, 203.

WESTON-ON-THE-GREEN, a parish in Oxfordshire, 4½ m. SW by W of Bicester. Area 2,466 acres. Pop. in 1831, 494; in 1851, 517.

WESTONING, or WESTON-INGE, a parish in Bedfordshire, 4 m. S by W of Ampthill. Area 1,715 acres. Pop. in 1831, 627; in 1851, 782.

WESTON-LAURENCE, a tything in the p. of Henbury, Gloucestershire, 5¼ NW by N of Bristol. Pop. in 1831, 329; in 1851, 334.

WESTON-UNDER-LIZARD, a parish in Staffordshire, 5¾ m. NE by E of Shiffnal. Area 2,398 acres. Pop. in 1831, 257; in 1851, 248.

WESTON-MADOC, a township in the p. of Church-Stoke, Montgomeryshire, 3½ m. SE of Montgomery. Pop. 273.

WESTON-MARKET, a parish in Suffolk, 7 m. S of East Harling. Area 1,083 acres. Pop. 317.

WESTON-SUPER-MARE, a parish and watering-place in the co. of Somerset, 9¾ m. NW by W of Axbridge, and 20 m. SW of Bristol, and within 1½ m. of the Bristol and Exeter railway, with which it is connected by a branch here. Area of p., with the hamlets of Ashcombe and Milton, 2,770 acres. Pop. in 1831, 1,310; in 1851, 4,034. The purity of its air, combined with a smooth and extensive sand-beach, of nearly 3 m. in extent, added to its vicinity to Bath and Bristol, have within these few years raised W. from a small fishing-village to the rank of a handsome watering-place containing nearly 600 houses, and supplied with accommodation for numerous visitors, and possessing two churches and five dissenting chapels.

WESTON (Old), a parish in Hunts, 8 m. N of Kimbolton. Area 2,012 acres. Pop. in 1851, 432.

WESTON-PATRICK, a parish in Hants, 4½ m. SW by W of Odiham. Area 1,402 acres. Pop. 215.

WESTON-UNDER-PENYARD, a parish in Herefordshire, 2 m. ESE of Ross. Area 3,142 acres. Pop. in 1831, 639; in 1851, 733.

WESTON-PEVERELL, or Penny-Cross, a chapelry in the p. of St. Andrew, Devon, 3 m. N by W of Plymouth. Area 1,281 acres. Pop. 321.

WESTON-UNDER-RED-CASTLE, a chapelry in the parish of Hodnet, Salop, 4 m. E of Wem. Area 1,289 acres. Pop. in 1851, 302.

WESTON-RHYN, a joint township with Bron-y-gath, in the p. of St. Martin, Salop, 3 m. N of Oswestry. Pop., with Bron-y-gath, in 1851, 1,165.

WESTON (South), a parish in Oxford, 3 m. S by E of Tetsworth. Area 560 acres. Pop. in 1851, 198.

WESTON-UPON-TRENT, a parish in Derby, on the river Trent, 7 m. SE by S of Derby. Area 1,820 acres. Pop. in 1831, 387; in 1851, 399.—Also a parish in Stafford, on the river Trent, 4½ m. NE of Stafford. Area 825 acres. Pop. in 1831, 608; in 1851, 570. This p. is the seat of extensive operations in the preparation of salt. The brine is conveyed to the works by means of pipes across the Trent from the p. of Ingestre, where it is raised by machinery. About 1,400 hogsheads of brine are daily consumed, and the weekly quantity of salt produced is about 250 tons.

WESTON-TURVILLE, a parish in Bucks, 2¼ m. N by W of Wendover. Area 2,450 acres. Pop. in 1831, 637; in 1851, 748.

WESTON-UNDERWOOD, a parish in Bucks, 1¾ m. WSW of Olney, on the N bank of the river Ouse. Area 1,300 acres. Pop. in 1851, 405.—Also a township in the p. of Stanton-by-Dale, Derby, 5¾ m. NW by N of Derby. Area 1,300 acres. Pop. in 1831, 272; in 1851, 405.

WESTON-ZOYLAND, a parish in Somersetshire, 4 m. ESE of Bridgewater, on the N bank of the river Parret, which separates it from North Petherton. Area, with the hamlets of Bussex and Liney, 2,729 acres. Pop. in 1851, 1,007.

WESTOUTRE, a village of Belgium, in W. Flanders, 32 m. W of Bruges. Pop. 1,222.

WESTOW, a parish in the E. R. of Yorkshire, 5½ m. SSW of New Malton, near the river Derwent, containing the townships of Eddlethorpe, Firby, Menethorpe, and W. Area 2,917 acres. Pop. 592.

WESTPALSTOWN, a parish in co. Dublin, 3½ m. S of Naul. Area 1,595 acres. Pop. in 1851, 167.

WEST-PANS, a village on the coast of the p. of Inveresk, Edinburghshire, 1¼ m. ENE of Musselburgh. Pop. 300.

WESTPHALIA, a province belonging to the western division of the Prussian monarchy; bounded on the NW by Holland; on the N by Hanover; on the E by Hanover, Schauenburg, Lippe-Detmold, Brunswick, and Hesse-Cassel; on the SE by Hesse-Cassel, Waldeck, and Hesse-Darmstadt; and on the SW and W by the province of the Rhine. Its total area is 5,888 geog. sq. m. The northern part of this prov. belongs to the great plains of the north of Germany, and has only a few small elevations running from the Egge to the Weser. The southern part from the S side of the Lippe, consists in numerous ridges of small mountains and hills. The soil in the N is mostly sandy, with some large moors, heaths, and bogs. In the S, the soil, though not sandy, is frequently stony, and not very productive. The mountains of the prov. belong to the lower chains of the Harz. The most remarkable are the Weser mountains; the Teutoburgerwald, a remarkable chain, which runs like a wall from the Rhine to almost the middle of the Ems; and the Porto-Westphalien, a beautiful picturesque chain, which has been abruptly cleft by water, about 1 m. to the S of Minden. This opening forms the celebrated Porto-Westphalien, or pass of Westphalia. The largest river is the Weser, but it only flows on the W side of the prov., where it forms the boundary for about 60 m. It then breaks through the Porto-Westphalien, and runs for 28 m. through the district of Minden. It receives several small rivers as the Diemel and the Emmer. The next river is the Ems, which is not navigable, and receives no river of considerable size. The Lippe is partly navigable. The Röhr, which falls into the Rhine, was made navigable by Frederic II; the Vechta runs through the Netherlands into the Zuydersee. There are no lakes or ponds, but a number of bogs, and some mineral springs. The climate is temperate and healthy; the air is colder but more pure in the S than in the N, where there are many bogs. The winter is everywhere cold and damp; in summer the heat is sometimes very great. Horses, and the common domestic animals, swine, in great number, fowls, game, corn, potatoes, vegetables, fruit, hemp, flax, copper, lead, iron, marble, lime, salt, peat, and coals, are the productions of this prov. Agriculture is everywhere the principal branch of industry. The richest corn-districts are the vicinity of Paderborn, the N part of the duchy of W., and some districts of Minden and Münster. There are other parts where the soil is so stony that nothing but oats can be cultivated. Rye is the most common corn; from it is prepared the brown bread called *pumpernickel*, celebrated throughout Germany. The rearing of cattle is considerable, particularly herds of swine, which furnish the celebrated Westphalian hams. There are extensive salt-works; iron and linen are the most important articles of manufacture. The inhabitants, returned at 1,464,921 in 1849, and 1,516,046 in 1853, are all of the Low-German race. In a great part of the country the inhabitants live dispersed in small isolated farms. The Catholics have two universities, one at Münster, and one at Paderborn; they have also seminaries for the clergy, gymnasia, and elementary schools. The Protestants have gymnasia, and city and parochial schools; but the state of general information, and school establishments, is here decidedly inferior to those of other Prussian provinces. Westphalia is divided into the three governments of Münster, Minden, and Arnsburg, which are subdivided into circles. It contains 128 towns, 60 boroughs, and 4,045 villages. The chief town of the prov. is Münster, on the Aa.

WEST POINT, a village of Troup co., in Georgia, U. S., 120 m. W by S of Milledgeville. Pop. 150.

WEST POINT, a village of Orange co., in the state of New York, U. S., on the W bank of the Hudson, 53 m. from New York, in N lat. 41° 23′ 31″, W long. 73° 57′ 03″. The military academy of the Union was founded here in 1802. The 'Point' consists of an irregular angle or point, elbowing the Hudson into the sharpest curve of its entire navigable course. A plain of about 160 acres, elevated over 150 ft. above the river, crowns this point, while the limiting bluffs and slopes coming down to the water's edge offer many beautiful clusterings of foliage and out-looking granite spurs. On the plain is the academic hall, the chapel, hospital, library, cadet-barracks, and mess hall, the houses of the professors and officers, and an open area for military evolutions. Under a hill to the NW lies the quarter known as Camptown, which consists of the soldiers' barracks, and various small tenements demanded by the motley academic retainers of all minor degrees. In the rear or to the W, the plain is shut in by a range of hills, Mount Independence being just abreast. About 1 m. W, Redoubt hill rises still higher, and between this and lordly Crow's Nest winds the valley threaded by the Canterbury road. Across the river is Constitution island, crowned with ruins. Fort Montgomery is about 6 m. below the

Point. The new cadet barracks compose a noble edifice in the Elizabethan style, with towers, battlements, and embrasures. The academic hall contains the recitation, and drawing-rooms, the picture gallery, the cabinets of engineering and geology, the laboratory, the fencing-rooms, and the riding-hall. The library building has an imposing appearance. The library-room, which is used for the examinations and for winter concerts, contains a valuable selection of 15,000 volumes. The observatory, which occupies three towers, possesses a fine transit instrument, and a large mural circle and an indifferent equatorial. The hospital, hotel, professors' houses, laboratory, and storehouse, the dragoon stables, the commissaries' store, the band and engineer barracks, and various minor tenements complete the buildings of the establishment. For each congressional district one cadet is allowed, whose appointment is practically in the gift of the representative in Congress from that district. Contingencies considered, a vacancy occurs about once in three years for each district. The corps of cadet usually numbers about 250, and is organised into a battalion of 4 companies, all officered by cadets. Over these is the commandant of cadets, an army captain, who is the immediate military head of this battalion; also four army lieutenants, commanding the four companies, as assistant instructors of tactics. The cadet first class furnishes the requisite cadet captains and lieutenants; the second class, the sergeants; and the third class, the corporals; all other cadets serve indiscriminately as privates. The usual staff of teachers consists of the superintendent, 2 officers in the engineering department, 3 in philosophy, 6 in mathematics, 2 in chemistry, 3 in ethics, 2 in drawing, 4 in practical engineering, 3 in French, 5 in infantry tactics, 4 in artillery and cavalry, an adjutant, a surgeon, and an assistant-surgeon; being 38 in all. These officers are all detached from the regular army for the duty at West Point, and are in turn greatly instructed by the necessities of teaching. The system is excellent in always keeping the academy young and vigorous, while the army is leavened by the higher progress in science thus wrought out among its officers, who are in time returned to their field-duties.

WESTPORT, a market-town, and nominally a seaport, in the p. of Aughavel, co. Mayo, 6½ m. S of Newport. It is one of the most regular small towns of Ireland. The commerce, which belongs rather to Westport-Quay than to the town itself, consists in the export of corn, meal, and flour, provisions, spirits, wool, linen, and flax and tow. Pop. in 1831, 4,448; in 1851, 4,121.

WESTPORT, a village of Fairfield co., in Connecticut, U. S., 55 m. SW of Hartford, on the Sangatuck river. Pop. in 1840, 1,808; in 1850, 2,649.—Also a township and village of Bristol co., in Massachusetts, 51 m. S of Boston. Pop. 2,795.—Also a township and v. of Essex co., in New York, 106 m. N by E of Albany. Pop. 2,352.

WESTPORT BAY, the south-eastern arm or offset of Clew bay, co. Mayo. The S side of the small island of Innisgort, which is surmounted by a lighthouse, and is situated 5¼ m. WNW of Westport-Quay, marks the entrance. The channel, after passing about 2 m. up from Innisgort, expands into a lagune nearly 2 m. wide, and almost perfectly landlocked; after this lagune is passed, the bay gradually contracts, up to its termination at the village of Westport-Quay.

WESTPORT-QUAY, a seaport village in the p. of Aughavel, co. Mayo, at the head of Westport bay, 1 m. W of the town of Westport, and the port of that town. Pop. in 1851, 694.

WESTPORT-ST.-MARY, a parish in Wilts, adjacent to Malmesbury. Area 2,036 acres. Pop. 1,661.

WEST RANDOLPH, a village of Orange co., in the state of Vermont, U. S., 25 m. S of Montpelier, and on the Vermont Central railway.

WESTRAY, an island at the NW extremity of the Orkney group, 9½ m. NW of Costa-head, the nearest point of Pomona, and 4½ m. N of Rousay, from which it is separated by the Westray frith. Within distances of from ½ to 2½ m. from its E and SE coasts lie Papa-Westray, Eday, Faray, and Farayholm. Its extreme length, from NNW to SSE, is 9 m.; and its extreme breadth is 4 m. Pop. in 1811, 1,396; in 1831, 1,712; in 1851, 2,083.

WESTRAY AND PAPA-WESTRAY, a united parish in the NW of Orkney, comprehending the parish and island of Papa-Westray, and North and East, or Mary and Lady parishes, which divide between them the island of Westray. Pop. 2,459.

WEST ROCKHILL, a township of Bucks co., in the state of Pennsylvania, U. S., 33 m. NNW of Philadelphia, drained by Perkiomen creek and a branch of Tohickon creek. Pop. in 1840, 1,631.

WESTRUTHER, a parish lying debateably between Lauderdale and Lammermoor in Berwickshire. Its superficial extent is about 22½ sq. m. The village of W. stands in the centre of the parish, 7 m. E of Lauder. Pop. in 1831, 830; in 1851, 791.

WEST RUTLAND, a village of Rutland co., in the state of Vermont, U. S., on the W side of Otter creek, 51 SSW of Montpelier, and intersected by the Rutland and Washington railroad.

WEST SPRINGFIELD, a township of Hampden co., in the state of Massachusetts, U. S., on the W side of Connecticut river, and 82 m. W by S of Boston, intersected by the Western railway. Pop. of township in 1840, 3,626; in 1850, 2,979.

WEST STOCKBRIDGE, a township of Berkshire co., in the state of Massachusetts, U. S., drained by Williams river, and intersected by the Western and Stockbridge railway. Pop. in 1850, 1,713.

WEST TOWNSEND, a village of Middlesex co., in the state of Massachusetts, U. S., on a branch of Nashua river, 41 m. NW by W of Boston, and intersected by the Peterboro and Shirley railroad.

WEST TROY, a village of Albany co., in the state of New York, U. S., on the W bank of Hudson river, and 6 m. N by E of Albany. Pop. 7,564.

WEST UNION, a village of Adams co., in the state of Ohio, U. S., 79 m. S by W of Columbus. Pop. in 1840, 500; in 1850, 462.

WESTWARD, a parish in Cumberland, 3 m. SE by S of Wigton, on the W bank of the river Wampool, containing the townships of Brocklebank, with Stoneraise, Rosley, and Woodside. Area 13,120 acres. Pop. in 1831, 1,253; in 1851, 1,283.

WEST WAREHAM, a village of Wareham township, Plymouth co., in the state of Massachusetts, U. S., on Cape Cod branch railroad, and 46 m. SSE of Boston.

WEST WATERVILLE, a village of Kennebeck co., in the state of Maine, U. S., on the Androscoggin and Kennebeck railroad.

WESTWELL, a parish in Kent, 2 m. ESE of Charing. Area 5,199 acres. Pop. in 1831, 861; in 1851, 1,030.—Also a parish in Oxfordshire, 2 m. WSW of Burford. Area 890 acres. Pop. 186.

WEST WHITELAND, a township of Chester co., in the state of Pennsylvania, U. S., 72 m. from Harrisburg, and drained by Southern Valley creek. Pop. in 1840, 1,085.

WESTWICK, a parish in Norfolk, 2¾ m. S. of North Walsham. Area 1,043 acres. Pop. 204.

WEST WINSTED, a village of Litchfield co., in the state of Connecticut, U. S., on Mad river, and at the terminus of the Nangatuck railroad.

WEST WOODVILLE, a village of Claremont co., in the state of Ohio, U. S., on the W side of Stonelick creek, and near the Cincinnati, Hillsboro, and Parkersburg railroad.

WESTWOOD-WITH-IFORD, a parish in Wilts, 2 m. S of Bradford, containing the hamlet of Iford. Area 813 acres. Pop. in 1831, 390; in 1851, 605.

WESUWE, a village of Hanover, in the principality of Osnabruck, near Mippen. Pop. 300.

WESZE, a village of Austria, in Moravia, in the circle of Iglau. Pop. 220.

WESZENINGKEN, a village of Prussia, in the reg. of Gumbinnen, circle of Ragnit. Pop. 220.

WESZMAR, a village of Prussia, in the regency and circle of Merseburg. Pop. 230.

WESZPRIM. See VESPRIM.

WESTPORT.

WETERE, a town of New South Wales, in the district of Murrumhdjee, on the N bank of the river Murray, in E long. 148°.

WETERSCHAN, a village of Austria, in Moravia, in the circle of Brunn. Pop. 670.

WETHAU, a village of Prussia, in the regency of Merseburg, and circle of Naumburg. Pop. 332.

WETHEN, a village of Waldeck, in the bail. of Diemel. Pop. 543.

WETHERAL, a parish in Cumberland, 5 m. ESE of Carlisle, on the river Eden, including the townships of Cumwhinton with Coathill, Scotby, Wetheral, Great Corby, and Warwick-bridge. Area 11,778 acres. Pop. in 1831, 4,149; in 1851, 3,165.

WETHERBY, a market-town and chapelry in the p. of Spofforth, W. R. of York, 13 m. W of York, on the river Wharfe, over which is a handsome stone-bridge of 6 arches. Area 1,570 acres. Pop. in 1831, 1,321; in 1851, 1,494. The town consists chiefly of one street, and is situated in a fine agricultural district.

WETHERDEN, a parish in Suffolk, 4¼ m. NW of Stow-market. Area 1,830 acres. Pop. 541.

WETHEREDVILLE, a village of Baltimore co., in the state of Maryland, U. S., on Gwinn's Falls, 4½ m. WNW of Baltimore. Pop. in 1850, 750.

WETHERINGSET, a parish in Suffolk, 5 m. SSW of Eye, in the projected line of the Eastern Counties railway. Area 3,783 acres. Pop. in 1841, 1,065; in 1851, 1,101.

WETHERSFIELD. See Weathersfield.

WETHERSFIELD, a township of Hartford co., in the state of Connecticut, U. S., on the W side of Connecticut river, 3½ m. S of Hartford. Pop. in 1840, 3,824; in 1850, 2,523.—Also a township of Wyoming co., in the state of New York, 235 m. W of Albany. Pop. in 1840, 1,728; in 1850, 1,489.—Also a township of Trumbull co., in the state of Ohio, 169 m. NE of Columbus, drained by Mahoning river, and intersected by the Pennsylvania and Ohio canal. Pop. in 1840, 1,447.

WETHERSFIELD SPRINGS, a village of Wethersfield township, Wyoming co., in the state of New York, U. S. Pop. in 1850, 160.

WETLER, a town of Prussia, in the prov. of Westphalia, regency of Arensberg, and circle of Hagen, 11 m. S of Dortmund. Pop. 540. It has a china manufactory, and coal mines.

WETLIN, a town of Prussia, in the prov. of Saxony, in the regency and 20 m. NNW of Merseburg, on the r. bank of the Saale. Pop. 2,540. It is enclosed by walls, with 3 gates, and has 4 suburbs, a castle, the ancient residence of the Counts of Wetlin, 2 churches, a Lutheran and a reformed, and an hospital. It contains several breweries, and coal mines are wrought in the vicinity.

WETLUGA, a town of Russia in Europe, in the gov. of Kostroma, on a river of the same name, an affluent of the Volga. Pop. 1,800.

WETSCHEN, a village of Hanover, in the bail. of Diepholz. Pop. 461.

WETSCHEWELL, a village of Prussia, in the reg. of Dusseldorf, circle of Gladbach. Pop. 280.

WETTAU, or Wetter, a large island of the Eastern seas, lying to the N of Timor, from which it is separated by a channel 35 m. wide. Its E point is in S lat. 7° 40′, E long. 126° 45′. It is 57 m. in length from ENE to WSW, and 25 m. broad; and is generally high and bold.

WETTBERGEN, a village of Hanover, in the bail. of Wenningsen, near Hanover. Pop. 378.

WETTE (Deutsch and Polnisch), two villages of Prussia, in the regency of Oppeln, and circle of Niesse, and containing respectively 532 and 426 inhabitants.

WETTEBORN, a village of Hanover, in the bail. of Bilderlake, near Lamspringin. Pop. 398.

WETTEL, a village of Austria, in Bohemia, in the circle of Prague, bail. of Raudnitz. Pop. 270.

WETTELBRUNN, a village of Baden, circle of the Upper Rhine, bail. of Staufen. Pop. 378.

WETTELDORF, a village of Prussia, in the regency of Treves, and circle of Prum. Pop. 280.

WETTELRODE, a village of Prussia, in the regency of Merseburg, and circle of Sangershausen. Pop. 380.

WETTELSHEIM, a village of Bavaria, in the circle of Middle Franconia. Pop. 731.

WETTEN, a village of Prussia, in the regency of Dusseldorf, and circle of Geldern. Pop. 609.

WETTENHALL, a township and chapelry in the parish of Over, co.-palatine of Chester, 7 m. NNW of Nantwich. Area 1,954 acres. Pop. 294.

WETTENHAUSEN, a village of Bavaria, in the circle of Swabia, on the Kamlach. Pop. 542.

WETTER, a village of Prussia, in the regency of Arnsberg, and circle of Hagen, on the Ruhr. Pop. 550. It has manufactories of china, and several iron-works.—Also a town of Hesse-Cassel, capital of a bail., in the prov. of Upper Hesse, circle and 7 m. NNW of Marburg, on the Wettschaft. Pop. 1,516. It contains a fine abbey, and an hospital, and has several paper-mills.—Also a river of the grand-duchy of Hesse-Darmstadt, which has its source near Laubach, in the prov. of Upper Hesse; runs first W, then S, and after a course of about 36 m. joins the Nidda, a little to the SW of Assenheim. It gave its name to the district of Wetterau.

WETTER. See Wettau.

WETTER (Lake), a lake of Sweden, between the parallels of 57° 47′ and 58° 53′ N, and stretching nearly 85 m. from NNE to SSW, with a medium breadth of 15 m., at an alt. of 290 ft. above sea-level. It is united, through the small lake of Viken, and by the West Gotha canal, with Lake Wever; and by the Motala canal with the Baltic.

WETTERAU, or Wettergau, a district of Germany, watered by the Wetter, which gives it its name. It comprises a great part of Upper-Hesse, Hesse-Darmstadt, and of the territory of Frankfort-on-the-Main. Its chief towns are Friedberg, Vilbel, and Staaden.

WETTERBURG, a village of Waldeck, in the bail. of Twiste, and circle of Arolsen. Pop. 540.

WETTEREN, a department, commune, and town of Belgium, in the prov. of West Flanders, arrond. and 11 m. W of Dendermond or Termond, on the r. bank of the Scheldt. Pop. of dep., 8,705; and of com., 3,802. It has manufactories of linen and woollen fabrics, and of lace.

WETTERFELD, a village of Bavaria, in the circle of the Upper Pfalz, and presidial of Roding. Pop. 337.—Also a village of Hesse-Darmstadt, in the prov. of Upper-Hesse, 14 m. ESE of Giessen, on the Wetter. Pop. 496.

WETTERHORN, a mountain of Switzerland, in the cant. of Berne, between the valleys of Hasle and Grindelwald, and in the N spur of the Bernese Alps, in N lat. 46° 38′ 22″. It is pyramidal in form, and has an alt. of 12,154 ft. above sea-level. The summit of this mountain was attained, for the first time on the 28th of Aug. 1844, after a toilsome ascent of three days from the valley of the Grindelwald. On the N, says Mr. Spiers, we perceived from the summit of the peak, "the Faulhorn and the chain of mountains which borders the lake of Brienne, and behind them the pass of Brunig; then the lakes of Lungerne and Lucerne, on the banks of which rose the Righi and Mount Pilate, whose summits appeared to us like molehills. On the E, the eyes wandered over an indefinite chain of mountain-peaks. On looking towards the S, we perceived on that side the two monarchs of the Bernese Alps, raising their heads side by side—viz., the Rosenlani and the Berglistock. Separated from them by the neck of Lauteraar, we descried the red Schreckhorn, very properly called the 'Peak of Terror,' whilst the last of the groups, the Finsteraarhorn, appeared to overtop its companions. On the right of these two peaks was seen the brilliant Vischerhorner, the nearest of all, beyond which we discovered the three rival peaks, the Eiger, the Mönch, and the Jungfrau, rising together to a height of more than 12,000 ft. At the base of these gigantic mountains lay the Wengern of the Alps, which looked like a mere undulation, whilst we could scarcely distinguish, far beneath us, the village of Grinderwald and the river of Lutchinen, winding like a silver thread in the midst of the valley. On all sides of the peak upon which we were stationed (and on the summit of which 12 persons could hardly find room) we had before our eyes vast and profound precipices, at the foot of which were stretched the plains of snow, and the numerous glaciers, situated still lower down. On our left were observed the higher glaciers of Grinderwald, and those of Lauteraar; and on the right, those of Gauti, Reufen, and Rosenlani, above which rose the peaks of Wellhorn, Losenhorn, and Engelhornet."

WETTERINGEN, a village of Prussia, in the regency of Munster, and circle of Steinfurt. Pop. 583.

WETTERSTEIN, a mountain of Bavaria, in the circle of Upper Bavaria, and presidial of Werdenfels, which has an alt. of 9,619 ft. above sea-level.

WETTERSWALDA, a village of Saxony, in the circle of Meissen, bail. and 4 m. N of Oschatz.

WETTERWITZ, a village of Saxony, in the circle of Dresden, and bail. of Meissen. Pop. 350.

WETTESINGEN, a village of Hesse-Cassel, in the prov. of Lower-Hesse, circle of Wolfhagen, and bail. of Wolfmarsen. Pop. 776.

WETTIN, a town of Prussia, in the regency and 27 m. NNW of Merseburg, and circle of the Saal, on the river of that name. Pop. 3,200. It has manufactories of chicory, tobacco, and oil.

WETTINGEN, a village of Switzerland, in the cant. of Aargau, bail. and 3 m. SE of Baden, on the Limmat. Pop. 1,612. It has a Benedictine abbey.

WETTLENSTADT, a village of Brunswick, in the circle of Brunswick. Pop. 220.

WETTMANNSTETTIN, a village of Austria, in Styria, in the circle of Marburg. Pop. 210.

WETTMAR, a village of Hanover, in the principality of Luneburg, and bail. of Burgweddel, near Hanover. Pop. 516.

WETTOLSHEIM, a commune of France, in the dep. of the Haut-Rhin, and cant. of Wintzerheim, 3 m. SW of Colmar. Pop. 1,392.

WETTON, a parish in Stafford, 7½ m. NW by N of Ashbourne. Area 2,600 acres. Pop. 466.

WETTRINGEN, a village of Bavaria, in the circle of Middle Franconia, and presidial of Rothenburg, near a branch of the Tauber. Pop. 553.—Also a village of Lower Franconia, in the presidial of Hofheim. Pop. 260.

WETTRUP, a village of Hanover, in the principality of Osnabruck, and bail. of Freeren. Pop. 553.

WETTSCHUTZ, a village of Prussia, in the regency of Liegnitz, and circle of Glogau. Pop. 260.

WETTSLATTEN, a village of Bavaria, in the circle of Upper Bavaria, and presidial of Ingolstadt. Pop. 500.

WETTSWEIL, a village of Switzerland, in the cant. of Zurich, and bail. of Affoltern. Pop. 317.

WETTZEL, a village of Bavaria, in the circle of Upper Bavaria, and presidial of Oberviechtach. Pop. 290.

WETUMPKA, a town of Montgomery co., in the state of Alabama, U. S., 110 m. SE of Tuscaloosa, on the E side of Corsa river. Pop. in 1840, 2,600.

WETWANG, a parish in the E. R. of Yorkshire, 5¾ m. W by N of Great Driffield, including the township of W., and the chapelry of Fimber. Area 5,740 acres. Pop. in 1831, 621; in 1851, 750.

WETZ, a village of Prussia, in the regency of Coblenz, and circle of Wetzlar. Pop. 400.

WETZAWINKEL, a village of Austria, in Styria, circle of Gratz, and bail. of Freiburg. Pop. 310.

WETZBACH, a village of Prussia, in the regency of Coblenz, and circle of Wetzlar. Pop. 230.

WETZDORF, a village of Austria, in Bohemia, in the regency of Gitschin, and bail. of Seftenberg. Pop. 610.—Also a village of Saxe-Weimar, in the district of Weimar, and bail. of Burgel. Pop. 200.

WETZEL, a county in the NW part of the state of Virginia, U. S., comprising an area of 558 sq. m., drained by branches of the Ohio, and intersected by the Baltimore and Ohio railroad. Pop. 4,284.

WETZELSDORF, three villages of Austria, in Styria, in the circle of Gratz. United pop. 1,200.

WETZENDORF, a village of Bavaria, in the circle of Middle Franconia, and bail. of Nurnberg. Pop. 250. It has tobacco-mills.—Also a village of Prussia, in the regency of Merseburg, and circle of Ouerfurt. Pop. 260.

WETZHAUSEN, a village of Bavaria, in the circle of Lower Franconia, and presidial of Holstein. Pop. 240. It has a castle, an hospital, and hot baths.

WETZISREUTE, a village of Würtemberg, in the circle of the Danube, and bail. of Ravensburg. Pop. 200.

WETZLAR, a circle and town of Prussia, in the prov. of the Rhine, and regency of Coblenz. Pop. of circle, 41,666. The town is 45 m. NE of Coblenz, on a height, and near the r. bank of the Lahn, at the confluence of the Dille, and at an alt. of 417 ft. above sea-level. Pop. in 1831, 4,677; in 1840, 5,146; in 1849, 5,250. It is enclosed by walls, with five gates, and has two suburbs. It contains several churches, both Catholic and Protestant, an hospital, a Lutheran and a Catholic gymnasium, a school of industry, and an institute for the poor. In the vicinity are the ruins of the castle of Kalschmitt or Carlsmund. Leather is its chief article of manufacture. An engagement between the French and Austrians took place under its walls in 1796.

WETZLAS, a village of Austria, in the prov. of Lower Austria, and upper circle of the Mannhartsberg, 3 m. W of Krumau. It has an observatory situated in N lat. 48° 36′ 30″, E long. 15° 19′ 46″.

WETZLOS, a village of Hesse-Cassel, principality of Fulda, and circle of Hunfeld. Pop. 200.

WEVELGHEM, a department and commune of Belgium, in the prov. of West Flanders, arrond. and 4 m. WSW of Courtrai, watered by the Lys. Pop. of dep., 4,045. It has manufactories of linen and cotton fabrics, and a trade in colza and flax.

WEVELINGHOVEN, a town of Prussia, in the regency and 11 m. SW of Dusseldorf, and circle of Grevenbroich, near the r. bank of the Erft. Pop. 2,400. It contains several breweries, distilleries, vinegar-works, and manufactories of linen and woollen fabrics.

WEVERTON, a village of Frederick co., in the state of Maryland, U. S., on the W side of the Potomac, at the confluence of Israel creek, 66 m. WNW of Annapolis, intersected by the Chesapeak and Ohio canal, and the Baltimore and Ohio railroad.

WEVELSFLETH, a village of Denmark, in the

duchy of Holstein, and bail. of Steinburg, on the Storen, near its confluence with the Elbe, and to the N of Gluckstadt. Pop. 1,100. It has a customhouse, building docks, numerous breweries and distilleries, and manufactories of casks.

WEWELSBURG, a village of Prussia, in the regency of Minden, and circle of Buren. Pop. 899.

WEWER, a village of Prussia, in the regency of Minden, and circle of Paderborn. Pop. 720.

WEXFORD, a maritime county in the SE of the Irish province of Leinster, bounded, on the N, by co. Wicklow; on the E, by the Irish sea and St. George's channel; on the S, by the Atlantic; and on the W, by cos. Waterford, Kilkenny, and Carlow. The area comprises 510,702 acres of arable land, 45,501 of uncultivated land, 14,325 of continuous woods, 2,392 of towns, and 3,668 of water, —in all, 576,588 acres.—The coast, along all the E and all the S, is low, and, for the most part, beachy. A lofty, magnificent, and boldly charactered range of mountains extends along most of the boundary with the county of Carlow, and forms an imposing background to a large proportion of the larger landscapes within the co. of Wexford. Tara-hill makes a conspicuous figure on the N portion of the E coast. A series of lofty hills, both too broad and too straggling to be called either range, chain, or group, commences 3½ m. S by W of New Ross, and extends about 10½ m. to the NE. The district S of the Forth mountain is almost a dead level. The whole of the E and S maritime frontier presents a district in many parts consisting of a deep alluvial soil well adapted for tillage,—abounding in various species of marl, calcareous sand, and, in some situations, limestone. In the more internal parts of the co., the river Slaney, traversing the country in a direction from NW to SE, marks out a district of great beauty and fertility on both sides of its verdant banks. The district to the eastward stretches to the sea-coast, abounds in marl, and is productive in grain.

The river Slaney flows a brief distance upon the NW boundary, and then runs SSW through the interior to the head of Wexford-harbour. It is navigable for river-craft to Enniscorthy. The Derry comes down from co. Wicklow, and runs a few miles on the NW boundary to a confluence with the Slaney. The Bann rises on the NE frontier, and drains the larger portion of the NE district to the Slaney. Three confluent rivulets, constituting the Corug, run to the head of Bannow-bay. The Barrow, large in volume, facile for navigation, fluctuating with the tide, and receiving, in its progress from co. Kilkenny, the magnificent tributary of the Nore, flows along the whole of the boundary with Kilkenny.

Minerals and Soil.] A band of lofty granitic country, forming part of the great granite district of the counties of Carlow, Wicklow, and Dublin, constitutes all the higher portion of the Mount Leinster and Blackstairs mountains within co. Wexford. Carboniferous limestone constitutes the surface rock of the Hook peninsula. The surface rocks throughout all the vast remainder of the co. are of the transition series, and consist principally of clay slate, greywacke, and greywacke slate. A lead mine is in operation at Cairme. A small vein of copper ore, of the kind called malachite, or carbonated green copper ore, occurs at Kirlogue, in the vicinity of Wexford. The soil, generally speaking, cannot boast of the superior degrees of fertility which is found in many other districts in Ireland, being mostly of a cold clayey nature.

Climate and Agriculture.] The E and S districts of the county of Wexford, in consequence of lying low and quite open to the sea, enjoy a much milder climate than the adjoining counties of Carlow and Kilkenny. The average climate of the county appears to be somewhat more genial than the average climate of the portions of Wales and England situated within the same parallels. The harvests of W. are sometimes finished when those of Haverfordwest and Carmarthen in South Wales, and the district around Barnstaple in the N of Devonshire, are only in progress. The county of W. and the S part of the county of Wicklow constitute one of the nine agricultural districts into which Mr. Wakefield distributed the whole of Ireland. The quantity of wheat used for seed in this district he states at 186 lbs. per acre, and the produce at 2,020 lbs.; the quantity of barley used for seed at 296 lbs. per acre, and the produce at 2,614 lbs.; the quantity of oats used for seed at 368 lbs. per acre, and the produce at 2,606 lbs.; the quantity of potatoes used for seed at 2,632 lbs. per acre, and the produce at 21,140 lbs.; and the produce of flax per acre at 896 lbs. "Happily," say Mr. and Mrs. Hall, in their recent work on Ireland, "W. is, in one respect, highly privileged—few of its landed proprietors are absentees. There are no huge estates, over which several agents must of necessity be placed; and as very few of its gentry have involved properties, it follows, as matter of course, that the tenants are in easy circumstances, and are neither rack-rented nor pressed for sudden payments." In 1841, there were within the rural districts of the county 5,219 farms of from 1 acre to 5 acres, 6,313 of from 5 to 15 acres, 4,151 of from 15 to 30 acres, and 2,457 of upwards of 30 acres. The totals of live stock, together with their respective estimated value, in 1841, were 27,474 horses and mules, £219,792; 2,891 asses, £2,891; 55,449 cattle, £360,419; 65,670 sheep, £72,237; 71,354 pigs, £89,192; and 251,929 poultry, £6,298. Grand total of estimated value of live stock in the rural districts, £750,829.

Manufactures and Trade.] The commerce of probably three-fourths of the county is identical with that of the port of Wexford; and the commerce of the remainder may be estimated by assigning to it a proportion of the exports and imports of New Ross and Waterford.

Communications.] The Barrow navigation is available to the southern part of the W side of the county; and the Slaney navigation is available to the districts around Enniscorthy, and thence down the valley of the Slaney to the sea. A railway has been proposed to connect the town of Wexford with the Dublin and Kilkenny railway, in the immediate vicinity of Carlow, and pass up the valley of the Slaney by way of Enniscorthy and Newtown-Barry. Another has been projected to connect Wexford on the one hand with Wicklow and Dublin, and on the other hand with New Ross, Waterford, Clonmel, Limerick, Cork, and Valentia; and this, if executed, would place the southern counties of Ireland in easy communication, by Wexford, with the Irish Wales railway,—the distance between Fishguard and Wexford being short, and favourable as to tides and prevailing winds.

Divisions, Towns, &c.] The towns and principal villages are Ballycanew, Coolcraney, and part of Gorey, in the barony of Gorey; Riverchapel, Ballaghkeen, Ballinamuddagh, Blackwater, Oilgate, Ford, Oulart, and parts of Gorey and Enniscorthy, in the barony of Ballaghkeen; Wexford and Broadway, in the barony of Forth; Tullycanna, Carrick, Duncormack, and Kilmore, in the barony of Bargie; Fethard, Churchtown, Slade, Arthurstown, Ballyhack, Duncannon, Ramsgrange, and Saltmills, in the barony of Shelburne; New Ross, Clonroche, and Maudlins, in the barony of Bantry; Ferns, Newtown-Barry, Cloghamon, Watchhouse, Camolin, and part of Enniscorthy, in the barony of Scarewalsh; Taghmon and Foxmills, in the barony of West Shelmalier; and Castle-bridge, in the barony of East Shelmalier. Pop. in 1831, 182,713. Males, 87,995; females, 94,718; families, 32,856. Inhabited houses, 29,923; uninhabited complete houses, 790; houses in the course of erection, 529. Families employed chiefly in agriculture, 21,465; in manufactures and trade, 6,155; in other pursuits, 5,236.—Pop. in 1841, 202,033. Males, 97,918; females, 104,115; families, 36,594. Families residing in first-class houses, 1,413; in second-class houses, 10,611; in third-class houses, 16,165; in fourth-class houses, 8,405. Families employed chiefly in agriculture, 24,664; in manufactures and trade, 8,147; in other pursuits, 3,783. Families dependent chiefly on property and professions, 985. The population by the last census (1851), was 179,793—viz., males, 86,770; females, 93,020. The decrease of pop., therefore, in the ten years, amounts to 22,243. The total number of inhabited houses, in 1851, was 29,479; of uninhabited, 1,776; and building, 50. The result of the return, therefore, shows a decrease of pop. of 22,243; of inhabited houses, 3,828.—The county gaol is at Wexford; and bridewells are at Gorey, Enniscorthy, and New Ross. The assizes are held at Wexford; quarter-sessions at Wexford, Gorey, Enniscorthy, and New Ross. The county sent no fewer than 18 members to the Irish parliament; or 2 from the county at large, and 2 from each of the boroughs of Wexford, Enniscorthy, Gorey, New Ross, Fethard, Bannow, Clonmines, and Taghmon; but it sends only 4 to the imperial parliament, or 2 from the county at large, one from the borough of Wexford, and one from the borough of New Ross. Constituency of the county at large, in 1841, 1,739; in 1851, 5,917.

History.] The Danes, previous to the Anglo-Norman period, held similar possession of the county of W., to that which they held of the counties of Waterford and Dublin. Wexford-harbour was one of the four principal stations which the Danes used for their fleets, and one of the fiords or havens to which they gave permanent designations. The name Wexford is thus a corruption of the Danish Wess-fiord, signifying "West-Haven.' The little adventurous advanced body of the army of the Anglo-Norman invasion, under Robert Fitzstephen, Maurice Fitzgerald, and Hugh de Montmorency, landed at Bagenbon-Head near the SW extremity of this county. Dermod MacMurrough, king of Leinster, who invited and aided the Anglo-Norman invasion, had previously the seat of his court and government at FERNS: which see. In 1210, the whole territory was erected into a county by King John. Cromwell, on his arrival in Ireland, found her sons here engaged in cutting each other's throats with the same blind fury which has so often rendered her energies of no avail. He dispossessed all the descendants of the old Anglo-Norman proprietors, long since become 'more Irish than the Irish themselves,' and divided their lands amongst his followers, in the possession of whose descendants they still remain. In 1774, the county was disturbed by the illegal associations called Whiteboys; in 1793, it was again disturbed by the illegal associations called the Defenders; and in 1798, it acquired permanent and most painful notoriety as the scene of the busiest and worst acts of the rebellion.

WEXFORD, a parliamentary borough, and the capital of the county of Wexford, barony of Forth, situated on the W shore of the upper part of Wexford harbour, 3 m. S by W of Castle-Bridge, 29 m. ENE of Waterford, and 64 m. S by W of Dublin. The site of most of the town is low, flat ground along the shore of the harbour; that of the remainder is part of the NE skirts of the Forth mountains. The harbour, at a point about ¾ m. above the town, begins to expand northward to the vicinity of the village of Castle-Bridge; and at a point about 2 m. WNW of the upper extremity of the town, makes a rapid and final contraction to mere fluviatile breadth, and is there crossed by the romantic bridge of Ferrycarrig. The W and SW environs of the town, consisting of the range, offsets, and skirts of the Forth mountain, are broken and picturesque, and command a noble prospect of the town, the harbour, and the circumjacent country. The quay and some of the modern parts of the town present an appearance, occasionally respectable, and generally pleasant; portions of even the old streets contain a number of well-built and respectably inhabited houses; and several of the scattered and isolated edifices in the outskirts appear cheerful and handsome; yet the town, as a whole, is in most of its thoroughfares orientally narrow, and multitudes of its houses are squalid and disgusting. The most interesting and conspicuous ecclesiastical ruin in Wexford, is the abbey of St. Peter and St. Paul, usually called Selsker-abbey, situated in the N district of the compact portion of the town. The present parish-church of St. Iberius is a plain edifice, surmounted by a small cupola. The parish-church of St. Selsker is connected by a small vestibule with the massive tower and other architectural remains of the old abbey. The meeting-houses for Presbyterians, Independents, and Methodists, are plain buildings; and the Roman Catholic chapel of St. John, and the chapel attached to the Franciscan friary, are unornamental structures. St. Peter's college, adjoining the parochial Roman Catholic chapel of St. Peter on Summerhill, rises conspicuously over the general elevation of the town's architecture, and forms an arresting feature in the general landscape of the town and environs. The bridge of W. is in itself an interesting and picturesque object. It consists of two raised causeways of respectively 650 and 188 ft. in length at the ends, a timber bridge of 733 ft. in length in the middle, and a drawbridge, for admitting vessels up and down the Slaney, near the town end. The county court-house faces the bridge. The custom-house is a small structure. The poor-law union workhouse occupies a conspicuous site near the N entrance of the town; and is a handsome building. The gaol is situated in the NW outskirts of the town. A magnificent monumental column on Carrig-a-Dagon, though not properly belonging to the town, is too arresting a feature to be unworthy of incidental notice. It is a copy of Pompey's pillar at Alexandria in Egypt; and was erected to commemorate the conquest of Egypt by the army under Sir Ralph Abercromby. During 1843, 653 sailing vessels, of aggregately 36,662 tons, entered the port coastwise; 469 sailing vessels, of aggregately 23,873 tons, cleared coastwise; 82 steam-vessels, of aggregately 14,542 tons, entered coastwise; 83 steam-vessels, of aggregately 14,770 tons, cleared coastwise; 10 British vessels, of aggregately 2,165 tons, entered from the colonies; 7 British vessels, of aggregately 1,582 tons, cleared for the colonies; 2 British vessels, of jointly 231 tons, entered from foreign ports; and 1 British vessel, of 280 tons, cleared for a foreign port. In 1851, 647 sailing vessels = 38,531 tons, and 45 steamers = 10,434 tons, entered coastwise. Weekly communication is maintained with Liverpool by steam navigation. The shops of the town are good; and the retail trade is extensive. The borough sent 2 members to the Irish parliament; but it sends only 1 member to the imperial parliament. Constituency of the borough in 1841, 301; in 1851, 343. Pop. in 1831, 10,673; in 1841, 11,252; in 1851, 12,819. Since the arrival of the English in the 12th century, the inhabitants of W. have been principally composed of the descendants of Frissians, Anglo-Saxons, Anglo-Normans, and a few Welsh. Cromwell, soon after his sanguinary reduction of Drogheda and his immediately subsequent successes, marched toward W., which was hastily surrendered by treachery, and became the scene of a horrible carnage. In 1798, the town fell under the power of the rebels, and while in their possession was the scene of a horrible massacre of Protestant prisoners.

WEXFORD, a county in the NW part of the southern peninsula of the state of Michigan, U. S., comprising an area of 576 sq. m., drained by the Manistie river and its branches.

WEXFORD-HARBOUR, an estuarial lagune in co. Wexford, communicating with the Irish sea by a mouth only 7 furl. wide, between Roslare point on the S, and Raven-point on the N; and elsewhere separated from the sea by two long and narrow peninsulæ of sandhills, which appear to have been formed by the conflict of the tidal currents from without, and the fluviatile currents from within. The distance from the mouth of the harbour direct to the town of Wexford is only 3½ m.; but the distance from the extreme N to the extreme S is 6½ m. Most of the expanse of this great lagune, except along a belt from Wexford to its mouth, swept by the fluviatile and the tidal currents of the Slaney, is so very shallow as to be dry at low water. So high a bar stretches across the mouth of the harbour, that vessels of more than 200 tons burden cannot cross, even at high water of spring tides, without being lightened of part of their cargo.

WEXHAM, a parish in Bucks, 1½ m. NE of the Slough station on the Great Western railway. Area 670 acres. Pop. in 1831, 181; in 1851, 201.

WEXIO, a laen of Sweden, bounded on the N by the laen of Jönköping; on the E by that of Kalmar; on the S by Carlscrona and Christianstadt; and on the W by Helmstadt. It has an area of about 2,900 geog. sq. m., and formed part of the ancient division of Smäland. Pop. in 1833, 111,500; in 1843, 129,186.—The capital, of the same name, is situated on the Sodre, 210 m. SW of Stockholm. Pop. 1,844.

WEYBOURNE, a parish in Norfolk, 3¾ m. NE of Holt, near the coast of the North sea. Area 1,680 acres. Pop. in 1831, 273; in 1851, 301.

WEYBREAD, a parish in Suffolk, 2 m. SSW of Harleston, on the southern side of the river Waveney. Area 2,476 acres. Pop. in 1851, 745.

WEYBRIDGE, a parish in Surrey, 12 m. NE by N of Guildford, on the river Wey, a small affluent of the Thames. Area 1,292 acres. Pop. 1,225.

WEYDA, or WEIDA, a town of the grand-duchy of Saxe-Weimar, in the circle and 14 m. ENE of Neustadt, on a small river of the same name, an affluent of the Auma, by which it is divided into two parts, distinguished as the old and new town. Pop. 3,800. It has an hospital, and a school, and contains manufactories of cloth, coarse woollen fabrics, and sailcloth, shoes and pottery, and several paper-mills. On an adjacent height is the castle of Osterburg. Pop. of bail., 17,000.

WEYER, a department and commune of Belgium, in the prov. of Limburg, and arrond. of Hasselt. Pop. 453. It has a brewery.—Also a village of Bavaria, in the circle of Lower Franconia, and district of Sulzheim. Pop. 240.—Also a village of Nassau, in the bail. of Runkel. Pop. 768.—Also a village in the bail. of St. Goarshausen. Pop. 370. —Also a market-town of Austria, in the circle of the Traun, 20 m. SE of Steyer, on the r. bank of the Ens. Pop. 1,350. It has extensive iron-works.—Also a village of Austria, in Styria, in the circle of Judenburg. Pop. 200.—Also a village of Prussia, in the regency of Aachen, and circle of Schleiden. Pop. 340.

WEYERIGG, a village of Austria, in the circle of Hausruck, on the Attersee. Pop. 310.

WEYERSBACH, a village of Oldenburg, in the circle of Birkenfeld. Pop. 240.

WEYERSHEIM, or WIRSCHEN, a commune of France, in the dep. of the Lower Rhine, and cant. of Brumath, 17 m. N of Strasburg. Pop. in 1846, 2,125. It has an oil-mill.

WEYERSWEIL, a village of Switzerland, in the cant. of Thurgau, and bail. of Weingelden. Pop. 316.

WEYHE, a village of Hanover, in the principality of Hoya, and bail. of Syke. Pop. 800.

WEYHILL, a parish in Hants, 3 m. W by N of Andover. Area, with Pen-Grafton, 1,888 acres. Pop. in 1831, 429; in 1851, 419. A great fair for sheep is held here on October 10th; and for horses, leather, hops, and cheese, on the five following days.

WEYLAND (CAPE), a headland of South Australia, in Eyre Land, on the N side of Anxious bay, in S lat. 33° 14′, and E long. 134° 31′ 45″.

WEYMOUTH, a parliamentary borough and port in the county of Dorset, on a small bay in the English channel, at the mouth of the river Wey, 8 m. S by W of Dorchester, and 128 m. SW by W of London. Area of chapelry of Weymouth, 52 acres; of the p. of Melcombe-Regis, 1,548 acres. Pop. of Weymouth, in 1801, 1,267; in 1831, 2,529; in 1851, 2,957. Pop. of Melcombe-Regis, in 1801, 2,350; in 1831, 5,126; in 1851, 5,273. The two portions of this fashionable and celebrated watering-place,—namely, Weymouth and Melcombe-Regis, —are separated by the narrow mouth of the Wey, and connected by a handsome stone-bridge of two arches, with a swivel in the centre, to admit of the entrance and exit of the shipping. Weymouth stands on the SW, and Melcombe on the NE, side of the river; the former at the foot of a considerable eminence; the latter on a peninsula, the isthmus of which separates the waters of the bay from those of the river estuary within. The streets in Melcombe-Regis are chiefly disposed along the margin of the sea, near which are many elegant edifices, especially on the esplanade, one of the finest promenades in England, nearly a mile in length. Weymouth still retains a good deal of the character of the fishing town, in which the modern watering-place originated. The channel to the tidal harbour is about 14 ft. deep at high water; at the quays at each side the vessels lie aground at low water. The bay is well sheltered by the surrounding hills, and there is an excellent beach of uniform depth. As the distance between the English-channel and Bridgewater bay on the Bristol-channel does not at this point exceed 60 m., W. enjoys many of the advantages of an ocean climate, the seasons being temperate and equable. Weymouth with Melcombe returns 2 members to parliament, till the passing of the Reform act, the united borough returned 4 members. The number of electors registered for the borough, in 1837, was 629; in 1852, 679. Weymouth was a place of considerable trade at an early period; and its commerce with France, Spain, and Newfoundland sustained the maritime importance of the town. About the year 1763, it became resuscitated in the form of a watering-place; and since that period its trade has revived.

WEYMOUTH, a township and village of Norfolk co., in the state of Massachusetts, U. S., 11 m. SSE of Boston, watered by affluents of Boston harbour, and intersected by the South Shore railroad. Pop. in 1840, 3,738; in 1850, 5,369.

WEYMOUTH BAY, a bay of North-east Australia, between Cape Fair on the NW, and a headland of the same name, by which it is separated on the SE from Lloyd bay, in S lat. 12° 35′, and E long. 143° 23′. The bay is about 18 m. in length, and 5 in breadth.—Also a town on the N coast of

Tasmania, in the co. of Dorset, at the mouth of Piper's river.

WEYPERT, a town in Bohemia, in the regency of Eger, circle and 26 m. NE of Ellbogen. Pop. 3,400. It has mines of silver and cobalt, 2 paper-mills and manufactories of small arms.

WEYRA, a village of Saxe-Weimar, in the circle of Neustadt. Pop. 730.

WEYRE, a town of Hindostan, in the prov. and 50 m. W of Agra, and principality of Bhurtpoor. It is enclosed by an earthen wall, flanked with bastions, and by a spacious ditch. It contains an old castle, and some other ancient buildings.

WEYROW, a village of Austria, in Bohemia, in the reg. of Pilsen, and bail. of Kralowitz. Pop. 270.

WEYSCHITZ, a village of Austria, in Bohemia, in the reg. of Pilsen, and bail. of B zeznitz. Pop. 300.

WEYWANOWITZ, a village of Austria, in Bohemia, in the regency of Pardubitz, and bail. of Chrudim. Pop. 340.

WEZEREN, a department and commune of Belgium, prov. of Liege, and arrond. of Huy. Pop. 125.

WEZIKON, a district and village of Switzerland, in the cant. and 14 m. ESE of Zurich, and bail. of Hinweil. Pop. 3,364.

WEZ-VELVAIN, a department and prov. of Belgium, in the prov. of Hainault, and arrond. of Tournai. Pop. 942.

WHAAL. See WAAL.

WHADDON, a parish in Bucks, 5 m. S by E of Stony-Stratford. Area 3,730 acres. Pop. in 1851, 987.—Also a parish in Cambridge, 4½ m. N of Royston, intersected by the river Cam. Area 1,463 acres. Pop. in 1831, 339; in 1851, 340.—Also a parish in Gloucestershire, 3 m. SW of Gloucester. Area 727 acres. Pop. in 1831, 152; in 1851, 120.—Also a parish in Wilts, 2¾ m. NE of Trowbridge, intersected by the river Avon, and the Kennet and Avon canal. Area 438 acres. Pop. in 1851, 44.

WHALAN, a creek of New South Wales, an affluent of the river Darling.

WHALE-HEAD, a headland of Tasmania, on the S coast, between South Cape and Recherche bays.

WHALE ISLANDS, a group off the coast of Greenland, in N lat. 68° 59′ 13″, W long. 53° 13′ 0″, on the S side of, and close to, Disco. They present bare masses of dark rock, with no vegetation save short grass, and coarse brown moss. A few Esquimaux inhabit the most verdant of the group, from whom the Danes collect oil and skins.

WHALEY, a joint-township with Yeardsley, in the p. of Taxall, co.-palatine of Chester, 9¼ m. SE of Stockport, in the line of the Peak-forest canal. Area 1,320 acres. Pop. in 1831, 403; in 1851, 693.

WHALLEY, a very extensive parish of Lancashire, containing the borough and market-town of Clitheroe; the market-towns of Burnley and Colney; the chapelries of Old Accrington, Altham, Bacup, Chatburn, Christ-church, Cliviger, Downham, Fence-in-Pendle, Goodshaw, Habergham, Holme, Marsden, Newchurch-in-Pendle, Newchurch-in-Rossendale, Padiham, Rawtenstall, and Worsthorn. Altogether this immense parish contains about 50 market-towns, chapelries, and townships; and is about 30 m. in length, and 15 m. in breadth, having an area of 104,689 acres; yet it is by no means so considerable in size as it formerly was, having anciently included those districts which are now the parishes of Blackburn, Chipping, Mitton, Ribchester, Rochdale, and Slaidburn. The rivers Ribble and Calder unite their waters near the W extremity of the parish, and it is intersected by the Liverpool and Leeds canal. Pop. of the entire parish in 1831, 97,868; in 1851, 134,196. Area of the township of W., 1,890 acres. Pop. in 1801, 876; in 1831, 1,151; in 1851, 945. The village, 14 m. ENE of Preston, has an ancient church in the early English style of architecture. In the interior are 18 stalls of ancient workmanship, and some fine screen-work.

WHALTON, a parish in Northumberland, 6 m. SW by W of Morpeth, intersected by the rivers Blythe and Howburn, and containing the townships of Newham, Ocle or Ogle, Riplington, and W. Area 5,918 acres. Pop. in 1831, 548; in 1851, 461.

WHAMPOA, an island and town of China, on the canton river, and 10 m. SE of the town of Canton. The foreign shipping trading with Canton have their chief anchorage here, 19 m. above Tiger island.

WHAND, a village of Lower Sind, 30 m. NE of Omercote.

WHAPLODE, a parish in Lincolnshire, 2¾ m. W of Holbeach. Area with Whaplode-Drave, 10,164 acres. Pop. in 1831, 1,998; in 1851, 2,564.

WHARFE, a river of Yorkshire, which rises in Langstrath dale; flows SE to near Ilkley, and then E to Wetherby, 13 m. above York, where it enters upon a plain of nearly a dead level, through which it flows SE to Tadcaster, and joins the Ouse, on the l. bank, 9 m. below York. It is navigable to Tadcaster.

WHARRAM-PERCY, a parish in the E. R. of Yorkshire, 8 m. SE of New Malton. Besides the township of W., it contains those of Raisthorpe-with-Birdall, Thixendale, and Towthorpe. Area 9,017 acres. Pop in 1831, 330; in 1851, 685.

WHARRAM-LE-STREET, a parish and township in the E. R. of Yorkshire, 6¾ m. SE by E of New-Malton. Area 2,024 acres. Pop. 131.

WHARTLEE, a mountain of Hindostan, 7 m. NE of Kotegur, in N lat. 31° 5′, and E long. 77° 29′. It has an alt. of 10,673 ft. above sea-level, and is wooded to the summit.

WHARTON, a county in the SE part of the state of Texas, U. S., comprising an area of 1,256 sq. m., drained by Colorado river and by branches of the San Bernardo, and is generally level. Pop. in 1850, 1,242. Its chief town, which bears the same name, is on the NE of the Colorado, 117 m. SE by E of Austin city.—Also a township of Fayette co., in the state of Pennsylvania, 11 m. SE of Union Town, bounded on the E and N by the Goughiogheny river, and watered by its branches. Pop. in 1840, 1,325.—Also a township of Potter co., in the same state, 179 m. WSW of Harrisburg, watered by a branch of Sandy creek. Pop. 171.

WHARTON, or WINSFORD, a township in the p. of Davenham, co.-palatine of Chester, 2½ m. WNW of Middlewich, in the line of the Grand Junction railway, the Middlewich canals, and the river Weaver. Area 1,224 acres. Pop. in 1851, 1,775.

WHASHTON, a township in the p. of Kirkby-Ravensworth, Yorkshire, 4 m. W of Richmond. Area 1,195 acres. Pop. in 1831, 159; in 1851, 104.

WHATCOTE, a parish in Warwickshire, 2¼ m. NE of Shipton-on-Stour. Area 1,400 acres. Pop. in 1831, 219; in 1851, 177.

WHATELEY, a township of Franklin co., in the state of Massachusetts, U. S., 80 m. W of Boston, bounded on the E by the Connecticut, and intersected by the Connecticut River railroad. Pop. 1,101.

WHATFIELD, or WHEATFIELD, a parish in Suffolk, 3¼ m. SE of Bildestone. Area 1,570 acres. Pop. in 1831, 377; in 1851, 380.

WHATLEY, a parish in Somerset, 2¾ m. W by S of Frome. Area 1,259 acres. Pop. in 1851, 361.

WHATLINGTON, a parish in Sussex, 2 m. N by E of Battle. Area 1,255 acres. Pop. in 1851, 458.

WHATTON (LONG), a parish in Leicestershire, 4¼ m. NW by W of Loughborough. Area 2,050 acres. Pop. in 1831, 855; in 1851, 838.

WHATTON-UPON-SMITE, a parish in Notts, 3 m. E of Bingham, containing the chapelry of Aslacton. Area 3,100 acres. Pop. in 1851, 764.

WHEATACRE-ALL-SAINTS, a parish in Norfolk, 4¾ m. NE by E of Beccles, between the river Waveney and the Lowestoft navigation. Area 1,163 acres. Pop. in 1831, 186; in 1851, 180.

WHEATENHURST, or WHITMINSTER, a parish in Gloucestershire, 8 m. NW by W of Stroud, on the S bank of the river Severn, and intersected by the Gloucester and Berkeley and the Thames and Severn canals. Area 1,247 acres. Pop. 380.

WHEATFIELD, a parish in Oxfordshire, 2 m. S of Tetsworth. Area 778 acres. Pop. 85.

WHEATFIELD, a township of Niagara co., in the state of New York, U. S., 12 m. W of Lockport, bounded on the W by the Niagara, and drained by Tonawanda creek. Pop. in 1840, 1,057.—Also a township of Perry co., in Pennsylvania. Pop. 617.

WHEATHAMPSTEAD, a parish in Hertfordshire, 4¼ m. WSW of Welwyn, intersected by the river Lea. Area 5,033 acres. Pop. in 1851, 1,908.

WHEATHILL, a parish in Salop, 9½ m. NE by E of Ludlow. Area 1,415 acres. Pop. in 1831, 123. in 1851, 143.—Also a parish in Somerset, 4 m. W by S of Castle-Cary. Area 314 acres. Pop. 36.

WHEATLAND, a township of Monroe co., in the state of New York, U. S., 214 m. W by N of Albany, drained by Allen's creek. Pop. in 1840, 2,871; in 1850, 2,917.—Also a township of Kenreha co., in the state of Wisconsin, 71 m. SE by E of Madison. Pop. in 1850, 1,193.—Also a township of Hillsdale co., in the state of Michigan, 88 m. SW of Detroit. Pop. in 1840, 164.

WHEATLEY, a chapelry in the p. of Cuddesden, Oxfordshire, 5½ m. E by S of Oxford. Area 970 acres. Pop. in 1831, 976; in 1851, 1,037.

WHEATLEY (NORTH), a parish in Notts, 5½ m. NE of East Retford. Area 2,181 acres. Pop. 427.

WHEATLEY (SOUTH), a parish in Notts, 5½ m. NE by E of East Retford. Area 641 acres. Pop. in 1831, 35; in 1851, 34.

WHEATON-ASTON, a chapelry in the p. of Lapley, Staffordshire, 5¼ m. W by S of Penkridge, in the line of the Birmingham and Liverpool Junction canal. Pop. in 1801, 563; in 1851, 711.

WHEELER, a township of Steuben co., in the state of New York, U. S., 187 m. W of Albany. It has a hilly surface, drained by Five Mile creek, and by the branches of Conhocton river. Pop. 1,471.

WHEELERSBURG, a village of Scioto co., in the state of Ohio, U. S., on the Ohio, 56 m. S by E of Columbus. Pop. in 1840, 200; in 1850, 300.

WHEELING, a city, port of entry, and capital of Ohio co., in the state of Virginia, U. S., 247 m. NW of Richmond, in N lat. 40° 7′, and W long. 80° 42′. Pop. in 1810, 914; in 1820, 1,567; in 1830, 5,221; in 1840, 7,885; in 1850, 11,391. It is finely situated on the E bank of the Ohio, at the confluence of the Wheeling creek, which is here crossed by a fine bridge, and at the terminus of the Baltimore and Ohio railroad. A suspension bridge of great dimensions also crosses the Ohio.—Also a township of Belmont co., in the same state, drained by Indian Wheeling creek. Pop. in 1840, 1,389.—Also a township of Guernsey co. Pop. 909.

WHEELOCK, a township of Caledonia co., in the state of Vermont, U. S., 29 m. NE of Montpelier, drained by branches of Patsampsie river. Pop. 885.

WHEELOCK, a township in the p. of Sandbach, co.-palatine of Chester, 1½ m. SSW of Sandbach, in the line of the Grand Trunk canal, and the Manchester and Birmingham railway. A small river of the same name passes through the p., and falls into the Dane below Middlewich. On both sides of this river, at the distance of about 60 yds. from the bed, large quantities of salt are procured from the brine which is found upon digging to the proper depth. Area 666 acres. Pop. in 1831, 440; in 1851, 548.

WHEEL PONDS, a group of ponds in New South Wales, in the district of Liverpool plains. They discharge themselves into the Gwyder.

WHEELTON, a township in the p. of Leyland, co.-palatine of Lancaster, 4 m. NE of Chorley. Area 1,669 acres. Pop. in 1831, 1,519; in 1851, 1,041

WHELDRAKE, a parish containing the township of Langwith, partly in the E. R. of Yorkshire, 7½ m. SE of York. Area 4,921 acres. Pop. 722.

WHENBY, a parish in the N. R. of Yorkshire, 8 m. E of Easingwold. Area 1,010 acres. Pop. 128.

WHEPSTEAD, a parish in Suffolk, 4½ m. SSW of Bury-St.-Edmunds. Area 2,670 acres. Pop. 652.

WHERRY, or FERBANE, a parish in King's co., containing the greater part of the town of FERBANE: which see. Area 18,556 acres. Pop. in 1851, 2,901.

WHERSTEAD, a parish in Suffolk, 3 m. S by W of Ipswich. Area 2,154 acres. Pop. 238.

WHERWELL, a parish in Hants, 3¾ m. SSE of Andover, intersected by the rivers Ande and Test. Area, with Westover and Fallerton and the chapelry of Bullington, 3,546 acres. Pop. in 1851, 632.

WHESSAL, or WHESSOE, a township in the p. of Houghton-le-Skerne, co.-palatine of Durham, 2½ m. NW by W of Darlington, and in the line of the Great North of England, and the Stockton and Darlington railways. Area 1,412 acres. Pop. in 1851, 110.

WHETSTONE, a parish in the co. of Leicester, on the SE bank of the river Soar, 2 m. distant from the Wigston station. Area 1,680 acres. Pop. in 1831, 903; in 1851, 986.

WHICHAM, or WHITEHAM, a parish in Cumberland, 10 m. SSE of Ravenglass. Area 7,502 acres. Pop. in 1831, 285; in 1851, 329.

WHICHBURY, a parish in Wilts, and partly in Hants, 8½ m. N by W of Fordingbridge. Area 1,850 acres. Pop. in 1831, 183; in 1851, 170.

WHICHFORD, a parish in Warwickshire, 6 m. SE of Shipston-on-Stour, containing the hamlets of Ascot and Stourton, or Stowerton. Area 3,100 acres. Pop. in 1831, 638; in 1851, 744.

WHICKHAM, a parish in the co.-palatine of Durham, 3 m. WSW of Gateshead, containing the townships of Fellside, Lowside, Swalwell, and W. Area 3,963 acres. Pop. in 1851, 5,565.

WHIDBEY, an island near the coast of New Georgia, in the SE part of the gulf of Georgia, and to the N of the entrance of Admiralty gulf, in N lat. 48° 7′, and W long. 122° 20′.

WHIDBEY, or LA PLACE POINT, a headland of South Australia, in Eyre Land, in the co. of Flinders, at the E side of Avoid bay.—Also a group of islands near the entrance of Avoid bay, and to the SSW of Whidbey Point, in S lat. 34° 40′, and E long. 135° 15′.

WHIDDY, an island near the head of Bantry bay, co. Cork, Munster. It extends SW, and lies at the distance of from a ½ m. to 1½ m. of the SE coast. Its length is 2¾ m.; and its extreme breadth is a little upwards of 1 m.

WHILTON, a parish in Northamptonshire, 5 m. ENE of Daventry, intersected by the London and Birmingham railway, and the Grand Junction canal. Area 1,270 acres. Pop. in 1831, 397; in 1851, 357.

WHIMPLE, a parish in Devon, 4½ m. WNW of Ottery St. Mary. Area 3,019 acres. Pop. 704.

WHINBURGH, or WHINBERGH, a parish in Norfolk, 3½ m. SSE of East Dereham. Area 1,241 acres. Pop. in 1831, 219; in 1851, 262.

WHINFELL, a township in the p. of Kirkby-in-

Kendal, Westmoreland, 6½ m. NE by N of Kendal. Pop. in 1831, 214; in 1851, 181.

WHIPPINGHAM, a parish in the Isle of Wight, 3½ m. N by E of Newport. Area 5,208 acres. The hamlet of East Cowes, separated from West Cowes by the river Medina, is within the parish; also the smaller hamlets of Burton, Coombly, and Fairlee. Pop. in 1831, 2,229; in 1851, 3,100.

WHIPSNADE, a parish in Bedfordshire, 3 m. S by W of Dunstable. Area 1,205 acres. Pop. 183.

WHISSENDINE, a parish in Rutlandshire, 4½ m. NW by N of Oakham. Area 3,870 acres. Pop. in 1831, 800; in 1851, 795.

WHISSONSETT, a parish in Norfolk, 5 m. S of Fakenham. Area 1,344 acres. Pop. in 1851, 666.

WHISTON, a township in the p. of Prescot, co.-palatine of Lancaster, 1 m. S of Prescot, in the line of the Liverpool and Manchester railway. Area 1,793 acres. Pop. in 1831, 1,468; in 1851, 1,825.—Also a parish in Northamptonshire, 6¾ m. E by S of Northampton. Area 809 acres. Pop. in 1831, 64; in 1851, 69.—Also a township in the p. of Kingsley, Staffordshire, 3¾ m. NE of Cheadle. Pop. in 1831, 549; in 1851, 675.—Also a parish in the W. R. of Yorkshire, 2½ m. SE of Rotherham. Area 3,190 acres. Pop. in 1831, 927; in 1851, 1,050.

WHITACRE (Nether), a parish in Warwickshire, 3¼ m. NE of Coleshill, on both sides of the river Tame. Area 2,210 acres. Pop. in 1851, 517.

WHITACRE (Over), a parish in Warwickshire, 3½ m. ENE of Coleshill. Area 1,375 acres. Pop. in 1831, 288; in 1851, 316.

WHITBECK, or Whitebeck, a parish in Cumberland, 8¾ m. SSE of Ravenglass, near the foot of Black-Comb hill. Area 5,372 acres. Pop. in 1831, 234; in 1851, 217. The sea has gained on the land in many places in this parish, and old roads and hedges have been visible a considerable way beyond water-mark.

WHITBOURNE, a parish in Herefordshire, 5 m. E by N of Bromyard. Area 3,056 acres. Pop. in 1831, 899; in 1851, 826.

WHITBURN, a parish and village in the co.-palatine of Durham, 3½ m. N of Sunderland, on the sea-coast, containing the township of Cleedon. Area 4,594 acres. Pop. in 1831, 1,001; in 1851, 1,203. The village is frequented as a sea-bathing place, and many lodgings have been erected for the accommodation of visitors.

WHITBURN, a parish and village in the SW of Linlithgowshire. Area 18 sq. m. Pop. in 1831, 2,075; in 1851, 3,792. The village is 21 m. WSW of Edinburgh, with a station on the Edinburgh and Glasgow railway.

WHITBY, a sea-port and parliamentary borough in the N. R. of the county of York, on the coast of the North-sea, at the mouth of the river Eske, in N lat. 54° 29′ 42″, W long. 0° 36′ 45″, 22 m. NNW of Scarborough, and 246 m. N of London. Besides the town of W., the parish comprises the chapelry of Eskdaleside, part of the township of Aislaby, and the townships of Hawsker-with-Stainsacre, Newholme-with-Dunsley, Ruswarp, and Ugglesbarnby. Area 15,918 acres. Pop. in 1801, 10,974; in 1831, 11,725; in 1851, 12,875. The appearance of the town is that of a substantial and wealthy sea-port town. A part of it is situated on the E bank of the river Eske; but the larger and better built portion has been erected on the western. The two parts are connected by a bridge, the middle of which is moveable, for the purpose of allowing ships to pass through. The town contains many excellent houses; but there are few public buildings worthy of notice. The port of W. must have been of considerable consequence at the period of the Conquest. The discovery of an alum mine at the close of Elizabeth's reign, was the original cause that raised Whitby from its obscurity, and, by opening a channel for commerce, enabled the town to attain a degree of maritime consequence. The trade of the port was so greatly increased by the improvements in the harbour about the beginning of last century, that, in 1734, no fewer than 130 vessels, of 80 tons burden, belonged to W., and in that year three spacious dry docks were constructed on the E side of the Eske. In 1850, 625 sailing vessels = 27,256 tons, and 47 steam-vessels = 2,115 tons entered this port coastwise; 3 = 881 tons entered from the colonies; and 22 sailing vessels = 2,400 tons, and 6 foreign vessels = 386 tons, entered from foreign ports. There are here bonded warehouses for all foreign goods, except East India goods and tobacco, unless brought coastwise for home use and ship stores. The exports to foreign parts consist principally of alum, whale-oil, and dried fish. The imports are chiefly articles of Baltic produce, comprehending timber, deals, hemp, flax, and ashes. The coasting trade is also considerable, and the shipments made hence to other parts of England consist principally of freestone from the quarries at Aislaby, alum, sail-cloth, butter, bacon, grain, and leather. The town now returns a member to parliament. Electors in 1837, 464; in 1852, 454.

WHITBY, a town of Upper Canada, on the N shore of Lake Superior, 28 m. E of Toronto. Pop. 1,700. It has a good harbour, and conducts a large lumber trade.

WHITCHURCH, a parish and village of Bucks, 4½ m. N by W of Aylesbury. Area 1,586 acres. Pop. in 1831, 928; in 1851, 915.—Also a parish in Devon, 1½ m. SE of Tavistock. Area 5,079 acres. Pop. in 1831, 791; in 1851, 1,156.—Also a parish in Glamorganshire, 4 m. NW of Cardiff, in the line of the Taffvale railway, near the Taff and the Merthyr canal. Area 3,192 acres. Pop. in 1831, 1,184; in 1851, 1,661.—Also a parish in Oxfordshire, 6½ m. NW of Reading, on the river Thames, and in the line of the Great Western railway. Area 2,070 acres. Pop. in 1831, 745; in 1851, 893.—Also a parish and town in Hants, 12 m. N by E of Winchester, on the river Teste, a noted trouting stream. Besides the village, the parish comprises the tythings of Charlecot, Cold Henley, and Freefolk with Freefolk-Prior. Area 6,142 acres. Pop. in 1831, 1,674; in 1851, 1,911. The town consists of one street only, but of late years it has been much improved. The inhabitants are chiefly employed in the manufacture of shalloons and serges, and in a large paper-mill at which most of the paper used for the Bank of England notes has been made. Before the passing of the reform act W. returned 2 members to parliament. Under the provisions of the reform act it has been disfranchised.—Also a parish in Warwickshire, 5¼ m. SSE of Stratford-on-Avon, near the Moreton and Stratford railway, comprising the hamlets of Broughton, Crimscott, and Wimpstone. Area 1,942 acres. Pop. in 1831, 261; in 1851, 257.

WHITCHURCH, or Eglwys-Wen, a parish in Pembrokeshire, 8 m. SE of Cardigan. Area 3,138 acres. Pop. in 1831, 377; in 1851, 1,252.

WHITCHURCH, or Tre-Croes, a parish in Pembrokeshire, 12½ m. WNW of Haverford-West, near St. David's, on the shore of St. Bride's bay. Area 3,138 acres. Pop. in 1831, 1,028; in 1851, 1,252.

WHITCHURCH, or Blancminster, a market-town and parish partly in Salop, and partly in the co.-palatine of Chester, 20 m. N by E of Shrewsbury. Part of the p. which is in Chester contains only part of the township of Wirswall; the other portion includes the townships of Alkington, Great

Ash, Little Ash, Black Park, Broughall, Doddington, Edgeby, Hinton, and Hollyhurst with Chimnal, the chapelry of Tilstock, the town of W., and part of Wirswall. Area 14,445 acres. Pop. in 1801, 4,618; in 1831, 5,902; in 1851, 5,976. The town is well-built, and stands on a gentle eminence with many fine points of view in the vicinity. Its trade is principally in grain and malt.

WHITCHURCH, or FELTON, a parish in Somersetshire, 3 m. N of Pensworth, 2 m. SW of the Great Western railway. Area 2,194 acres. Pop. 428.

WHITCHURCH-CANONICORUM, a parish and town of Dorset, 5 m. WNW of Bridport, containing Abbotsworton, Barn, Vale, and Wild-Quarters. Area 6,113 acres. Pop. in 1851, 1,532.

WHITCOMBE, a parish in Dorset, 2¾ m. SE of Dorchester. Area 540 acres. Pop. in 1851, 61.

WHITCOMBE-MAGNA, a parish in Gloucestershire, 3½ m. NE by N of Painswick. Area 918 acres. Pop. in 1831, 174; in 1851, 167.

WHITE, a creek of New South Wales, in Camden co., 82 m. from Sydney.

WHITE, a central co. of the state of Indiana, U. S., comprising an area of 504 sq. m., drained by Tippecanoe river and its tributaries, and intersected by the New Albany and Salem extension railway. Pop. in 1840, 1,832; in 1850, 4,761. Its cap. is Monticello.—Also a central county of the state of Tennessee, comprising an area of 346 sq. m., drained by Caney Fork of Cumberland river, and its affluents, and intersected by the Lexington and Mackmimville railway. Pop. in 1840, 10,747; in 1850, 11,444. Its cap. is Sparta.—Also a county in the SE part of the state of Illinois, comprising an area of 4,809 m., drained by Little Wabash and Skillet Fork. Pop. in 1840, 7,919; in 1850, 8,925. Its cap. is Carnie.—Also a central county of the state of Arkansas, comprising an area of 1,043 sq. m., bordered on the E by White river, and intersected by Red river, one of its affluents. Pop. in 1840, 929; in 1850, 2,619. Its cap. is Searcy.—Also a township of Cambria co., in the state of Pennsylvania. Pop. 438.—Also a township of Carrol co., in the state of Arkansas. Pop. 222.—Also a range of mountains in the state of New Hampshire. This range, the highest in New England, and among the highest in the United States, is 26 m. in length from N to S, and 22 m. in breadth. Their highest summits, the principal of which, Mount Washington, is 6,226 ft. above sea-level, are covered with snow during a great part of the year, and are visible at the distance sea-ward of upwards of 50 m.—Also a river of the state of Vermont, which has its source in Hancock township, Addison co., and after a course of 55 m. joins the Connecticut in Hartford township. —Also a river of the state of Indiana, which has its source in Johnson co., and, after a course of 225 m., joins the Wabash.—Also a river of the state of Arkansas, which has its source in the Black mountains, and after a winding course of 1,000 m., joins the Mississippi in Desha co.—Also a river which has its source in Newago co., flows SW through Oceana co., and falls into Lake Michigan.—Also a river of Utale territory, and forming one of the chief head-streams of the Colorado.

WHITECHAPEL. See LONDON.

WHITECHURCH, a parish in co. Cork, 5½ m. NNW of the city. Pop. in 1831, 2,856; in 1851, 2,547.—Also a parish in co. Dublin, 1½ m. S of Rathfarnham, containing the villages of Ballyboden and Edmonstown. Pop. in 1831, 1,705; in 1851, 1,275.—Also a parish in co. Kildare, 2½ m. N of Kill. Area 3,165 acres. Pop. in 1831, 279; in 1851, 269. —Also a parish in co. Kilkenny, 2 m. NW of Pilltown. Area 2,186 acres. Pop. in 1831, 778; in 1851, 637.—Also a parish in co. Tipperary, 3 m. SW of Cahir. Area 3,921 acres. Pop. in 1831, 1,218; in 1851, 1,310.—Also a parish in co. Waterford, 4½ m. NNW of Dungarvan. Area 9,951 acres. Pop. in 1831, 3,176; in 1851, 2,672.—Also a parish in co. Wexford, 4½ m. S by W of New Ross. Area 5,342 acres. Pop. in 1831, 1,328; in 1851, 1,186.

WHITECHURCH-GLYNN, a parish of co. Wexford, 2½ m. N by E of Tagmon. Area 7,187 acres. Pop. in 1831, 1,738; in 1851, 1,599.

WHITE-CREEK, a township of Washington co., in the state of New York, U. S., 30 m. NE of Albany, drained by Owl and Little White creeks. Pop. in 1840, 2,195; in 1850, 2,994.

WHITE-DEER, a township of Union co., in the state of Pennsylvania, U. S., bounded on the N by White Deer mountain, and drained by White Deer creek, an affluent of the Susquehanna. Pop. 1,252.

WHITE-EYES, a township of Coshocton co., in the state of Ohio, U. S., drained by affluents of the Tuscarawas. Pop. 997.

WHITEFIELD, a township of Lincoln co, in the state of Maine, U. S., 14 m. N of Wiscassel, drained by Sheepscot river. Pop. in 1840, 2,150; in 1850, 2,160.—Also a township of Coos co., in the state of New Hampshire, 9 m. S of Lancaster, drained by John's river. Pop. in 1840, 751; in 1850, 857.

WHITEFIELD, a township in the p. of Deerhurst, Gloucestershire, 4 m. SSW of Tewkesbury. Pop., with Apperley, in 1851, 548.

WHITEFORD, or WHITFORD, a parish in Flintshire, 4 m. NW of Holywell, near the estuary of the river Dee, including the townships of Bychton, Eden-Owain-Merton, Is-Glan, Merton-Uwch-Glan, Mostyn-Tre-Lan, Tre'r-Abbot, and W. Area 13,065 acres. Pop. in 1831, 3,333; in 1851, 3,884. This parish abounds with lead, zinc, and copper ore, coal, calamine, chertz, and limestone, all which are wrought to a considerable extent.

WHITEGATE, or NEW CHURCH, a parish in the co.-palatine of Chester, 3¼ m. SW by S of Northwich, on the river Weaver, which partly intersects the parish, and in the line of the Grand Junction railway. It comprises the townships of Darnhall and Marton. Area 4,384 acres. Pop. 1,529.

WHITEGATE, a fishing village in the p. of Aghada, co., Cork, on the E side of Cork harbour, 2 m. N of Roche's Point. Pop. in 1851, 1,228.

WHITE-HALE, a township of Washington co., in the state of New York, U. S., 58 m. NNE of Albany, drained by Pawlet and Poultney rivers and Wood creek, affluents of Lake Champlain, and intersected by the Saratoga and Washington rail road. Pop. in 1840, 3,813; in 1850, 4,726. It contains a village of the same name with 3,000 inhabitants.—Also a village of Baltimore co., in the state of Maryland, on the Baltimore and Susquehanna railroad, 26 m. N of Baltimore.

WHITEHAVEN, a port and parliamentary borough, in the p. of St. Bees, Cumberland, 40 m. SW of Carlisle, and 13 m. S by W of Maryport. Pop. of the township in 1831, 11,393; in 1851, 14,190. The creek, at the mouth of the Poebeck, on which W. is built, is so deeply seated that the adjacent lands overlook it on every side. The town itself is one of the most handsome in the northern counties; the streets being regular and spacious, and crossing each other at right angles. The castle, the seat of Viscount Lowther, at the SE end of the town, is a large quadrangular building. The harbour is formed by the mouth of the Poebeck, and has been protected by piers of stone, bending across from each side, with lighthouses on each. In the interior of the harbour are three piers running outwards in parallel lines from the town, and others connected with the exterior protecting piers. A narrow vale extending to the village of St. Bees, supposed to have been formerly occupied by the sea, might be cut through and made navigable for large vessels at a moderate expense. The harbour is capable of sheltering several hundred sail of vessels. The trade of this port chiefly consists in the exportation of coal, lime, iron, freestone, gypsum, and grain; and in the importation of West Indian, American, and Baltic produce; and of flax and linen from Ireland, and pig-iron from Wales. The jurisdiction of the port extends from Maryport northwards to the river Duddon, including the intermediate ports of Millern, Harrington, Ravenglass, and Workington. The shipping belonging to W., in 1828, amounted to 195 vessels = 30,960 tons; besides a number of small vessels entirely engaged in the coasting trade. In 1851, the entries of sailing vessels were 897 = 50,564; and of steamers, 191 = 56,365 tons. Steamers sail weekly between this port and Liverpool, and occasionally to Dublin, the isle of Man, and Dumfries. Extensive collieries are in the immediate vicinity of the town, and, indeed, under its very streets and harbour. Some of these collieries are worked to the extent of 2 m. under the sea. They are excavations which have in their structure a considerable resemblance to the gypsum quarries of Paris; and are of such a magnitude and extent, that, in one of them alone, a sum exceeding £500,000 was, in the course of a century, expended by the proprietors. Their principal entrance is by an opening at the bottom of a hill, through a long passage hewn in the rock, leading to the lowest vein of coal. The greater part of this descent is through spacious galleries, which continually intersect other galleries, all the coal being cut away, with the exception of large pillars, which, where the mine runs to a considerable depth, are 9 ft. in height, and about 36 ft. square at the base. The mines are sunk to the depth of 130 fath. There are three strata of coal, which lie at a considerable distance, one above the other, and are made to communicate by pits; but the vein is not always continued in the same regularly inclined plane, the miners frequently meeting with hard rock, by which their further progress is interrupted. W. was enfranchised by the reform act, and now returns one member to parliament. The borough boundaries include the whole town and suburbs, and a considerable circuit of ground in the immediate vicinity of the town. The number of electors for the borough, in 1837, was 463; in 1852, 512.

WHITEHAVEN, a port of Nova Scotia, in N lat. 45° 10′, W long. 61° 10′. It has a splendid and convenient port, as capacious as Halifax harbour, and as safe and commodious, and its approaches are safe and under any circumstances easily attainable from open sea. The haven finishes to the N at Pleasant river, which is navigable for 2 m. by vessels of any burden, and for small craft 2 m. further to its head, which is only 4 m. from the high road from Guysboro' and the port of Canso. Whitehaven-island, the outer point to seaward of the haven, is 140 ft. high, and may be considered as the NE extremity of Nova Scotia, and the nearest available point of the American continent to the British islands, although itself isolated. The Acadian settlement of Molasses-harbour is separated to the W by a very narrow isthmus of mere beach from the western part of the haven; besides which there are not now more than 8 or 10 establishments around W. The shores of this port are said to offer no impediments to railroad termini wherever convenient, and the vicinity is practicable for rail communications.

WHITEHILLS, a fishing-village in the p. of Boindie, Banffshire, 2½ m. WNW of Banff. Pop. about 600.

WHITEHORSE, a fishing-harbour in the p. of Durrus, co. Cork, on the SE side of Bantry bay, 4¾ m. SW of the town of Bantry.

WHITEHOUSE, a manufacturing locality, and a group of three villages, in the p. of Carnmoney, co. Antrim, on the NW side of Belfast-lough, 3½ m. N by E of Belfast. Pop. of the village of Whitehouse-abbey in 1831, 391; in 1851, 852. Pop. of Upper and Lower Whitehouse in 1851, 2,236.

WHITEHOUSE, a village of Hunterdown co., in the state of New Jersey, U. S., on the E side of Rockaway creek, 26 m. N of Trenton, on the New Jersey Central railroad.

WHITE-ISLAND, an island in the Scilly group, situated about 2½ m. W of St. Mary's. Its area is about 50 acres.

WHITEKIRK, a parish on the coast of Haddingtonshire. Area about 6,000 acres. About 300 acres at the mouth of the Tyne are sandy marsh. Pop. in 1831, 1,109; in 1851, 1,131. The present p. comprehends the ancient parishes of Tyninghame, Aldham, and Hamer or Whitekirk.

WHITE-LADY-ASTON, or Aston-Episcopi, a parish in Worcestershire, 4 m. SE of Worcester. Area 1,230 acres. Pop. in 1831, 381; in 1851, 356.

WHITELETS, a village in the parish of St. Quivox, 1½ m. NE of Ayr.

WHITELEY, a township of Greene co., in the state of Pennsylvania, U. S., 168 m. W by S of Harrisburg, drained by Whiteley creek, an affluent of Monongahela river. Pop. in 1850, 2,500.

WHITE-MARSH, a township of Montgomery co., in the state of Pennsylvania, U. S., 85 m. E of Harrisburg, drained by Wissahicton creek. Pop. in 1840, 2,079; in 1850, 2,400.

WHITE-MOUNTAIN, one of the eastern summits of the Blackstairs mountains, in co. Wexford, 3 m. NE of the junction-point of the counties of Wexford, Carlow, and Kilkenny. It has an alt. of 1,259 ft. above the level of the sea.—Also one of the Sperrin mountains, in co. Londonderry, overhanging the SW side of the pass of Glenshane, confronted by Carntopher on the opposite side of the pass. Its summit has an alt. of 1,996 ft. above the level of the sea.

WHITE-MOUNTAINS. See Alleghany.

WHITE-OAK, a township of Ingham co., in the state of Michigan, U. S., 522 m. SE by E of Lansing. Pop. in 1840, 270; in 1850, 508.

WHITEPARISH, a parish in Wilts, 8 m. SE by E of Salisbury. Area 6,284 acres. Pop. with Earldoms, in 1831, 1,254; in 1851, 1,391.

WHITE-PIGEON, a township of St. Joseph's co., in the state of Michigan, U. S., 80 m. SW by S of Lansing, drained by Pigeon and Fawn rivers branches of St. Joseph river, and intersected by the Michigan Southern railway. Pop. in 1850, 794.

WHITE-PLAINS, a township of Westchester co., in the state of New York, U. S., 116 m. S of Albany, bounded on the W by Broad river, and on the E by Mamaronick creek, and intersected by the Harlem railroad. Pop. in 1840, 1,087; in 1850, 1,414.

WHITEPOINT, a village in the p. of Clonmel, co. Cork, situated on Great Island, in the vicinity of Cove.

WHITE-RIVER, a township of Izard co., in the state of Arkansas, U. S. Pop. in 1840, 551.—Also a township of Independence co., in the same state. Pop. 556.—Also a township of Marion co., in the same state. Pop. 339.—Also a township of Randolph co., in the state of Indiana. Pop. 2,227.

WHITE-RIVER, a village of Windsor co., in the

state of Vermont, U. S., on the W side of the Connecticut, at the confluence of White river and at the junction of the Vermont Central, the Connecticut, and Passampsic, and the Northern railways.—Also a village of Windsor co., in the state of Vermont, U. S., on White river, and on the Vermont Central railroad, 1 m. from White River junction.

WHITESBORO, a village of Oneida co., in the state of New York, U. S., on the SW bank of Mohawk river, 8 m. WNW of Albany, and intersected by the New York Central railroad. Pop. about 2,000.

WHITESIDES, a county in the NW part of the state of Illinois, U. S., comprising an area of 729 sq. m., intersected by rock river, and by the Chicago, Albany, and Camanche railroad. Pop. 5,361.

WHITE-STANTON, or WHIT-STAUNTON, a parish in Somersetshire, 3½ m. NNW of Chard. Area 1,960 acres. Pop. in 1831, 318; in 1851, 261.

WHITESTONE, or WHITSTONE, a parish in Devon, 3¾ m. WNW of Exeter. Area 4,077 acres. Pop. in 1831, 643; in 1851, 624.

WHITESTOWN, a township and semi-capital of Oneida co., in the state of New York, U. S., 88 m. WNW of Albany, bounded on the NE by Mohawk river, and drained by Jadaguada and Oriskany creeks, and intersected by the Erie canal, and by the Central railroad. Pop. in 1850, 6,810.

WHITESTOWN, a village in the p. of Carlingford, co. Louth, 1 m. WSW of Bellagan, 3½ m. SE of Carlingford. Pop. in 1831, 370; in 1851, 223.

WHITE SULPHUR SPRINGS, a village of Greenbriar co., in the state of Virginia, U. S., on a branch of Greenbriar river, 156 m. W by N of Richmond, and on the Covington and Ohio railroad.

WHITESVILLE, a village of Columbus co., in the State of North Carolina, U. S., on the W side of Whitemarsh creek, on the Wilmington and Manchester railroad. Pop. in 1850, 350.

WHITEWATER, a township of Hamilton co., in the state of Ohio, U. S., 102 m. SW by W of Columbus, drained by Whitewater river. Pop. in 1840, 1,883; in 1850, 1,567.—Also a township of Walworth co., in the state of Wisconsin, 39 m. SE by E of Madison, watered by tributaries of Rock river, and intersected by the Milwaukie and Mississippi railroad. Pop. in 1850, 1,252.—Also a river of the state of Indiana, which is formed at Brookville, Franklin co., by the confluence of the E and W forks, and thence flows SE and S to Miami river, 6 m. above its mouth.—Also a river which has its source in Cape Guardian co., in the state of Missouri, by the junction of several streams, and flows nearly parallel with the Mississippi to St. Francis bay, Mississippi co., in the state of Arkansas.

WHITEWELL, or WHITEWELL-IN-BOWLAND, a chapelry in the p. of Whalley, W. R. of Yorkshire, 7 m. NW by W of Clitheroe.

WHITFIELD, a township of Indiana co., in the state of Pennsylvania, U. S. Pop. in 1840, 1,664.

WHITFIELD, a township in the p. of Glossop, Derby, 8½ m. N by W of Chapel-en-le-Frith. Pop. in 1831, 1,734; in 1851, 4,776, an increase attributed to the great extension of the cotton manufacture in this township.

WHITFIELD, anciently BEAUXFIELD, a parish in Kent, 3½ m. NNW of Dover. Area 893 acres. Pop. in 1851, 210.—Also a parish in Northamptonshire, 2 m. NE by N of Brackley. Area 1,210 acres. Pop. in 1831, 328; in 1851, 326.—Also a parish in Northumberland, 11½ m. WSW of Hexham. Area 12,125 acres. Pop. in 1851, 340.

WHITGIFT, a parish and village in the W. R. of Yorkshire, 6½ m. SE of Howden, on the navigable river Ouse, near its junction with the Trent at the Humber. It includes the chapelry of Swinfleet, and the townships of Ousefleet, Reedness, and W Area 8,328 acres. Pop. in 1851, 2,394.

WHITHORN, a parish in the district of Machers, Wigtonshire. Area about 10,000 acres. The coast-line extends 5½ m., runs out into the little promontories of Port-Yarrock-head and Stun-head, and admits the little bays of Isle of Whithorn, Port-Allan, and Port-Yarrock. The only village, as well as noticeable port, is Isle-of-Whithorn. Pop. in 1831, 2,415; in 1851, 3,001.—The royal burgh, situated near the centre of the parish, stands 11 m. S of Wigton. It unites with Wigton, Stranraer, and New Galloway, in sending a member to parliament. W. boasts a very high antiquity. It was the capital of the British tribe of Novantes, who possessed all Galloway west of the river Dee; and, on the influx and ascendency of the Romans, it was adopted by that people as a station. Ptolemy mentions it under the name of *Leucophibia*. Bede emphatically notices the humble edifice of Ninian as the first church which was built of stone, and says that, on that account, it was called *Candida Casa*. David I. set up at W. an episcopal see, which, under the wide name of the bishopric of Galloway, held coeval sway with that of popery and of Stuart prelacy till the final triumph of presbyterianism in 1689.

WHITING, a township of Washington co., in the state of Maine, U. S., 109 m. E by N of Augusta. Pop. in 1840, 462; in 1850, 470.—Also a township of Addison co., in the state of Vermont, 41 m. SW of Montpelier, intersected by the Rutland and Burlington railroad. Pop. in 1840, 659; in 1850, 629. —Also a township of Windham co., in the state of Vermont, 106 m. S by W of Montpelier, drained by Deerfield river. Pop. in 1840, 1,391; in 1850, 1,380.

WHITINS, a village of Worcester co., in the state of Massachusetts, U. S., on the Providence and Worcester railway, 26 m. from the latter place.

WHITKIRK, a parish in the W. R. of Yorkshire, 4 m. E of Leeds, including the townships of Austhorpe, Seacroft, Temple-Newsham, and Thorp-Stapleton. Area 5,930 acres. Pop. in 1851, 2,794.

WHITLEY, a hamlet in the p. of St. Giles, Berks, 2 m. S of Reading. Pop. in 1831, 363; in 1851, 639.—Also a chapelry and village in the p. of Tynemouth, Northumberland, 2½ m. N by E of North Shields, with which it communicates by a railroad. Area 531 acres. Pop. in 1831, 632; in 1851, 431. The inhabitants are principally occupied in mining operations,—large quantities of coal and magnesian limestone being raised in the parish, and iron-ore also in considerable quantities. The village is in general well-built, and contains several very excellent houses.—Also a township in the p. of Kellington, Yorkshire, 5 m. W by S of Snaith. Area 1,679 acres. Pop. in 1831, 310; in 1851, 320.

WHITLEY, a county in the NE part of the state of Indiana, U. S., comprising an area of 324 sq. m., and drained by Eel river and its tributaries. It is in some parts undulating. Pop. in 1840, 1,237; in 1850, 5,190. Its capital is Columbia.—Also a county in the SE part of the state of Kentucky, comprising an area of 704 sq. m., drained by Cumberland river and its branches, and intersected by the Knoxville and Lexington railroad. It is generally fertile. Pop. in 1840, 4,673; in 1850, 7,447.

WHITLEY (INFERIOR), a chapelry in the parish of Great Budworth, co.-palatine of Chester, 5 m. NW by N of Northwich. Area 1,118 acres. Pop. in 1831, 237; in 1851, 216.

WHITLEY (LOWER), a township in the p. of Thornhill, Yorkshire, 5½ m. SW of Wakefield. Area 1,011 acres. Pop. in 1831, 1,012; in 1851, 1,068.

WHITLEY (OVER), a township in the p. of Kirk

Heaton, Yorkshire, 6 m. E by N of Huddersfield. Area 1,953 acres. Pop. in 1851, 1,002.

WHITLEY (Superior), a township in the p. of Great Budworth, co.-palatine of Chester, 5¾ m. NNW of Northwich, comprehending the hamlets of Norcot, Anterbus, Middle-Walke, Seven-Oaks, and Crowley. Area 1,011 acres. Pop. in 1851, 322.

WHITLINGHAM, a parish in Norfolk, 2½ m. ESE of Norwich. Area 542 acres. Pop. 19.

WHITMORE, a parish in Staffordshire, 4 m. SW of Newcastle-under-Lyne. Area 2,023 acres. Pop. in 1831, 281; in 1851, 377.

WHITNASH, or Witnash, a parish in Warwickshire, 3 m. ESE of Warwick. Area 1,242 acres. Pop. in 1831, 260; in 1851, 346.

WHITNEY, a parish in Herefordshire, 5 m. NE of Hay. Area 1,483 acres. Pop. in 1851, 232.

WHITNEY'S VALLEY, a village of Alleghany co., in the state of New York, U. S., on the Buffalo and New York City railroad, 78 m. from Buffalo.

WHITNEYVILLE, a village of Washington co., in the state of Maine, U. S., on the SW side of Matthias river, and on the Franklin railroad, 9 m. from Machiasport.

WHITSOME AND HILTON, a united parish in the district of Merse, Berwickshire, with a superficial extent of 7¾ sq. m. Pop. in 1851, 612.

WHITSTABLE, or Whitstaple, a parish and small port in Kent, 6 m. NNW of Canterbury, near the entrance to the Swale, opposite to Sheppey isle, and connected by railroad with Canterbury. It contains the hamlet of Harwich. Area of p., 4,075 acres. Pop. in 1851, 2,746.

WHITSTON, a parish in Monmouthshire, 5¾ m. SE by E of Newport. Area 1,073 acres. Pop. 76.

WHITSTONE, a parish in Cornwall, 5½ m. SSE of Stratton, intersected by the Bude and Launceston canal. Area 3,787 acres. Pop. in 1851, 425.

WHITSUNDAY. See Pentecost.

WHITSUNDAY PASSAGE, a passage on the NE coast of Australia, between Cumberland Islands and the continent, discovered by Captain Cook in 1770.

WHITTINGHAM, a parish in Haddingtonshire. Area about 20,700 acres, of which about five-eighths is moorish upland, declining from both sides of the summit-range of the broad Lammermoors. Pop. in 1851, 671.

WHITTINGHAM, a township in the p. of Kirkham, co.-palatine of Lancaster, 6 m. NNE of Preston. Area 4,322 acres. Pop. in 1831, 710; in 1851, 677.—Also a parish in Northumberland, 8¾ m. W by S of Alnwick, intersected by the river Aln, comprising the townships of Callaley-with-Yetlington, Glanton, Lorbottle, Great Ryle, Little Ryle, Shawdon, and W. Area 17,484 acres. Pop. in 1831, 1,790; in 1851, 1,905.

WHITTINGTON, a parish in Derbyshire, 2 m. N of Chesterfield, near the North Midland railway. Area 2,640 acres. Pop. in 1831, 740; in 1851, 874. —Also a parish in Gloucestershire, 4 m. ESE of Cheltenham, intersected by the small river Colne. Area 1,422 acres. Pop. in 1831, 247; in 1851, 233. —Also a parish in the co.-palatine of Lancaster, 2 m. SSW of Kirkby-Lonsdale, on the river Lune. Area 4,322 acres. Pop. in 1831, 542; in 1851, 414. —Also a parish and village in Salop, 3 m. ENE of Oswestry, on both sides of the river Perry. Area 8,296 acres. Pop. in 1831, 1,788; in 1851, 1,927.—Also a parish in Stafford, 4 m. NW by W of Tamworth. Area 2,921 acres. Pop., with the hamlet of Tamhorn, in 1831, 773; in 1851, 809.—Also a chapelry in the parish of St. Peter, Worcestershire, 2½ m. SE of Worcester. Area 989 acres. Pop. in 1831, 279; in 1851, 289.—Also a township in the parish of Corbridge, Northumberland, 7 m. NE of Hexham.

WHITTLE, a township in the p. of Glossop, Derbyshire, 6 m. NW of Chapel-en-le-Frith. Pop. in 1831, 2,266; in 1851, 2,691.—Also a township in the parish of Shilbottle, Northumberland, 5 m. S of Alnwick. Area 555 acres. Pop. in 1851, 40.

WHITTLE (Welch), a township in the p. of Standish, co.-palatine of Lancaster, 3 m. SW of Chorley. Area 594 acres. Pop. in 1851, 140.

WHITTLE-LE-WOODS, a township and chapelry in the p. of Leyland, co.-palatine of Lancaster, 3½ m. N by E of Chorley. Area 1,357 acres. Pop. in 1831, 2,015; in 1851, 2,310.

WHITTLEBURY, a parish in Northamptonshire, 4 m. S by W of Towcester. Area 2,870 acres. Pop. in 1831, 670; in 1851, 707.

WHITTLESEY, or Whittlesea, a parish and village in the isle of Ely, 6 m. E by S of Peterborough. Area 2,531 acres. Pop. in 1851, 5,472.

WHITTLESFORD, a parish in Cambridgeshire, 6 m. W by N of Linton. Area 1,915 acres. Pop. in 1831, 524; in 1851, 719.

WHITTON, a parish in Lincolnshire, 11 m. WNW of Barton-upon-Humber. Area 2,440 acres. Pop. in 1831, 245; in 1851, 190.—Also a parish in Radnorshire, 4 m. NW of Presteign. Area 1,549 acres. Pop. in 1831, 131; in 1851, 118.—Also a parish within the liberties of the borough and 2¾ m. NNW of Ipswich, intersected by the Stow-market and Ipswich canal. Area with Thurlston, 1,459 acres. Pop. in 1831, 346; in 1851, 476.

WHITTONSTALL, a chapelry in the p. of Bywell-St.-Peter, Northumberland, 10 m. SE by E of Hexham. Area 1,968 acres. Pop. in 1851, 198.

WHITWELL, a parish in Derbyshire, 11 m. ENE of Chesterfield. Area 4,680 acres. Pop. in 1831, 1,007; in 1851, 1,355.—Also a parish in Norfolk, 1 m. SW of Reepham. Area 1,511 acres. Pop. in 1831, 483; in 1851, 477.—Also a parish in Rutland, 4½ m. E of Oakham. Area 602 acres. Pop. in 1831, 124; in 1851, 129.—Also a parish in the isle of Wight, 8 m. S by E of Newport. Area 1,963 acres. Pop. in 1831, 556; in 1851, 637.

WHITWELL, or Whitwell-on-the-Hill, a township in the p. of Crambe, Yorkshire, 6 m. SW of New-Malton. Area 1,640 acres. Pop. 205.

WHITWICK, a parish in Leicestershire, 5½ m. E by S of Ashby-de-la-Zouch, containing the chapelry of St. George and the townships of Swannington and Thringstoe. Area 6,229 acres. Pop. 4,956.

WHITWOOD, a township in the p. of Featherstone, Yorkshire, 5 m. NW by W of Pontefract, on the Calde navigation. Area 1,045 acres. Pop. 576.

WHITWORTH, a parochial chapelry in the co.-palatine of Durham, 4 m. NE by N of Bishop-Auckland. Area, with Tudhoe, 1,465 acres. Pop. in 1831, 337; in 1851, 650.—Also a chapelry in the p. of Rochdale, co.-palatine of Lancaster, 3 m. N by W of Rochdale.

WHIXLEY, a parish in the W. R. of Yorkshire, 6½ m. SSE of Aldborough, comprising the township of Thornville, Green-Hamerton, and W. Area 3,470 acres. Pop. in 1831, 968; in 1851, 1,041.

WHIXOE, or Wixoe, a parish in Suffolk, 4 m. WSW of Clare, on the river Stour. Area 600 acres. Pop. in 1831, 146; in 1851, 168.

WHOAHOU. See Oahu.

WHORLTON, a chapelry in the p. of Gainford, co.-palatine of Durham, 4 m. ESE of Barnard-Castle, on the river Tees. Area 1,760 acres. Pop. in 1831, 311; in 1851, 296.—Also a parish in the W. R. of Yorkshire, 5½ m. SW by S of Stokesley, including the chapelry of Faceby, and the townships of Potto and W. Area 9,690 acres. Pop. in 1851, 865.

WHYDAH, a district and town of Western Africa, on the Bight-of-Benin, about 80 m. S of Abomey. The town is in N lat. 6° 9′, E long. 2° 5′. The district is fertile, and is well cultivated. A considerable trade in palm-oil, ivory, and salt, is conducted from this place.

WIANBEEN, a creek of New South Wales, in the co. of St. Vincent, an affluent of Shoalhaven river near Curwary.

WIAZOWNA, a town of Poland, in the woiwodie of Masow, and obwod of Stanislawow, 15 m. SE of Warsaw. Pop. 300.

WIBLING, or WIBLINGEN, an oberamt and village of Würtemberg, in the circle of the Danube. Pop. of amt, 22,854. The village is near the confluence of the Iller and Danube, 3 m. S of Ulm. Pop. 900. It has a royal castle.

WIBLINGWERDER, a village of Prussia, in the prov. of Westphalia, regency and 23 m. WSW of Arensberg, and circle of Altena. Pop. 1,240.

WIBORG, WYBORG, or VIBORG, a district of European Russia, lying between Finland Proper and the lake of Onega. Its area is computed at 16,000 sq. m.. The pop., chiefly Finns, is 200,000. —Its cap., of the same name, stands in a valley opening on the gulf of Finland, in N lat. 60° 42′ 40″, 70 m. NW of Petersburg. Pop. 3,600. It consists of several parts, town, castle, and suburbs, and has a large military hospital, and a gymnasium. Its exports consist of wood, rosin, pitch, potash, and fish. Its trade is considerable. Its proper harbour is about 8 m. distant at Stralsund.

WIBORG, or WYBORG, an ancient town of Denmark, situated on a small lake, nearly in the centre of Jutland, and 24 m. W of Randers. It is tolerably built, having been in some degree reconstructed after a great fire in 1726. Pop. 3,400. It has few manufactures, but is of some importance as a military point.

WIBRIN, a department and commune of Belgium, in the prov. of Luxemburg, and arrond. of Marche. Pop. 1,193.

WICHAWSKA-KHOTA, a village of Austria, in Bohemia, in the reg. of Gitschin, and bail. of Hohenelbe. Pop. 940.

WICHDORF, a village of Hesse-Darmstadt, prov. of Lower Hesse, circle of Fritzlar. Pop. 587.

WICHELEN, a department and commune of Belgium, in the prov. of East Flanders, district and 6 m. WSW of Termonde, on the r. bank of the Scheldt. Pop. of dep., 4,143; of com., 459.

WICHELSDORF, a village of Prussia, in the reg. of Liegnitz, and circle of Sprottau. Pop. 440.

WICHENFORD, a parish in Worcestershire, 6 m. NW by N of Worcester. Area 2,669 acres. Pop. in 1831, 355; in 1851, 336.

WICHENSTOCK, a mountain of Switzerland, in the cant. and 9 m. S of Glaris.

WICHLINGHAUSEN, a village of Prussia, in the regency of Dusseldorf, circle and 3 m. NE of Elberfeld. Pop. 1,500.

WICHMANSDORF, a village of Prussia, in the regency of Potsdam, and circle of Zemplin. Pop. 300.

WICHMANNSHAUSEN, a village of Hesse-Darmstadt, in the prov. of Lower Hesse, and circle of Eschwege. Pop. 794.

WICHMAR, a village of Saxe-Meiningen, in the bail. of Earnburg, on the Saale. Pop. 250. It has a paper-mill.

WICHNOR, a chapelry in the parish of Tatenhill, Stafford, $5\frac{1}{2}$ m. NE of Lichfield, in the line of the Grand Trunk canal. Pop. in 1851, 131.

WICHRAU, a village of Prussia, in the reg. of Oppeln, and circle of Rosenberg. Pop. 440.

WICHROWITZ, a village of Prussia, in the reg. of Konigsberg, and circle of Neidenburg. Pop. 210.

WICHSTADLE, or WIEGSTADLE, a village of Austria, in Bohemia, in the regency of Witschin, and bail. of Seftenberg. Pop. 890.

WICHTE, a village of Hesse-Darmstadt, in the prov. of Lower Hesse, and circle of Melfungen. Pop. 334.

WICHTERICH, a village of Prussia, in the reg. of Cologne, and circle of Enskirchen. Pop. 490.

WICHTRACH, a village of Switzerland, in the cant. of Berne, and bail. of Konolfingen. Pop. 562.

WICHTRINGHAUSEN, a village of Hanover, in the prov. of Calenberg, and bail. of Wennigsen. Pop. 250.

WICHTSHAUSEN, a village of Prussia, in the regency of Erfurt, and circle of Schleusingen. Pop. 500.

WICK, a parish in Glamorganshire, 5 m. SE of Bridgend. Area 1,370 acres. Pop. 404.

WICK, a parish on the E coast of Caithness-shire, having a superficial extent of 75 sq. m. Keiss-bay, called also Sinclair-bay, penetrates the N half of the coast to the extent of $3\frac{3}{4}$ m. Wick-bay, measuring 1 m. across the entrance, and stretching inland to the extent of $1\frac{1}{2}$ m., opens 3 m. S of Keiss-bay. The hills of Camsten, Yarrow, and Bean, in the SW, are the only high ground in this p. The fishing-village of Broadhaven, situated about 1 m. E of the town of Wick, has about 170 inhabitants. The other villages, either suburban to the burgh or situated at a distance, are Pulteney-town, Louisburgh, Staxigoe, and Sarclet. Pop. in 1831, 9,850; in 1851, 11,851.

WICK, a royal burgh, the capital of Caithness, and the principal seat of the northern-herring fishery, is situated at the head of Wick-bay, $18\frac{1}{2}$ m. S of John o'Groat's, $20\frac{3}{4}$ m. SE by E of Thurso, and $119\frac{1}{4}$ or $139\frac{1}{4}$ m., according to the route, NNE of Inverness. It lies low, and is encompassed by a tame and level strath. The town consists of three parts, Wick-Proper, Louisburgh, and Pulteney-town. W. consists principally of one crooked street about 220 yds. long, winged with lanes, and situated on the north side of the river. Louisburgh, situated quite adjacent to it on the N, nearly vies with it in bulk, and far excels it in regularity and freedom from unpleasantness. Pulteney-town, situated on the S side of the river, but extending eastward till it lines part of the coast of the bay, is the division of prime importance, and contains several hundreds more than one-half of the pop. A bridge of 3 arches, flung across a narrow part of the river, a little above the head of the bay, connects Pulteney-town with the other two divisions. The present manufactures of the town are all of modern origin, and seem to have sprung up in connection with its fishing-establishment. Rope-works, the dressing pavement-flags for exportation, boat-building, a recently-erected iron-foundry, the making of herring-nets, and the spinning of material for them, employ a large number of hands. In the town are a distillery, a brewery, and some grain and sawmills. A large proportion of the trade of the port is conducted at the little harbour of Staxigoe. A steam-vessel plies once a-fortnight, during eight months of the year, between W. and Leith, calling at some intermediate ports, and extending her trips to Orkney and Shetland. In 1808, the establishment of Pulteney-town was commenced by the British fishery society; and, since that date, the annual fishings, during eight or ten weeks succeeding the middle of July, have been regularly conducted both at Pulteney-town itself, and at the subordinate stations of Broadhaven, Staxigoe, Keiss, and Sarclet. The burgh unites with Dingwall, Tain, Cromarty, Kirkwall, and Dornoch, in sending a member to parliament; and is the returning burgh. Total parl. constituency in 1852, 699. Pop. of royal burgh in 1841, 1,333; in 1851, 1,514.

WICK, a town and port of Denmark, in the duchy and 47 m. WNW of Sleswig, and bail. of Tondern, on the E coast of the island of Foehr. Pop. 700. It has a nautical school, a port capable of receiving a considerable number of large vessels, and a good roadstead. See also VICK.

WICK-NEAR-PERSHORE, a chapelry in the p. of St. Andrew Pershore, Worcestershire, 1½ m. ESE of Pershore. Pop. in 1831, 280; in 1851, 339.

WICK-CHAMPFLOWER, a chapelry in the p. of Bruton, Somersetshire, 1½ m. W of Bruton. Pop. in 1851, 107.

WICKEDE, a village of Prussia, in the regency of Arnsberg, and circle of Dortmund. Pop. 638.—Also a village in the circle of Goest. Pop. 440.

WICK-ST.-LAWRENCE, a parish in Somersetshire, 8½ m. NNW of Axbridge. Area 1,900 acres. Pop. in 1831, 1,281; in 1851, 300.

WICKEN, or WYKES, a parish in Cambridgeshire, 7 m. NW by W of Newmarket. Area 3,812 acres. Pop. in 1831, 892; in 1851, 1,054.—Also a parish in Northamptonshire, 3½ m. WSW of Stony-Stratford, intersected by the Buckingham canal. Area 2,280 acres. Pop. in 1831, 536; in 1851, 487.

WICKEN-BONANT, or BONHUNT, a parish in Essex, 5 m. SW by S of Saffron-Walden, in the line of the Northern and Eastern counties railway. Area 841 acres. Pop. in 1831, 134; in 1851, 156.

WICKENBY, a parish in Lincolnshire, 5½ m. NW of Wragby. Area 1,997 acres. Pop. 289.

WICKENDORF, a village of Bavaria, in the circle of Upper Franconia, presidial of Lugwigstadt. Pop. 300.—Also a village of Prussia, in the regency of Breslau, and circle of Schweidnitz. Pop. 360.

WICKENRODE, a village of Hesse, in the prov. of Lower Hesse, circle of Witzenhausen, and bail. of Gross-Almerode. Pop 1,076. It has manufactories of alum and vitriol.

WICKENRODT, a village of the principality of Oldenburg, and bail. of Oberstein. Pop. 260.

WICKER, a village of the duchy of Nassau, and bail. of Hockheim. Pop. 640.

WICKERADBERG. See WICKRATHBERG.

WICKERAN (GROSS and KLEIN), a village of Prussia, in the regency of Danzig, and circle of Elbing. Pop. 400.

WICKERODA, a village of Prussia, in the regency of Merseburg, circle and 6 m. WNW of Sangerhausen. Pop. 330. It has a copper-work.

WICKERSLEY, a parish in the W. R. of Yorkshire, 3½ m. E by S of Rotherham. Area 1,230 acres. Pop. in 1831, 527; in 1851, 700. This p. is famous for a bed of fine stone peculiarly adapted for the making of grindstones.

WICKFORD, a parish in Essex, 6 m. E by S of Billericay. Area 1,758 acres. Pop. in 1851, 490.

WICKFORD, a village of North Kingston township, Washington co., in the state of Rhode island, U. S., 22 m. S of Providence, on the W side of Narragansett bay, and intersected by the Stonington and Providence railroad. Pop. in 1850, 500.

WICKHAM, a parish in Hants, 4 m. S by E of Bishop's-Waltham. Area 2,433 acres. Pop. in 1851, 1,049.

WICKHAM, a river of N. W. Australia, an affluent of Limnen Bight river.

WICKHAM (BISHOP'S), a parish in Essex, 2¾ m. S by E of Witham. Area 1,534 acres. Pop. 577.

WICKHAM-BREAUX, or WICKHAM-BREUX, a parish in Kent, 5 m. E by N of Canterbury. Area 2,310 acres. Pop. in 1831, 486; in 1851, 481.

WICKHAM-BROOK, a parish in Suffolk, 6¼ m. N by W of Clare. Area 4,348 acres. Pop. 1,597.

WICKHAM (CHILD'S), a parish in Gloucestershire, 5 m. W by S of Chipping-Campden. Area 2,040 acres. Pop. in 1831, 415; in 1851, 466.

WICKHAM (EAST), a parish in Kent, 3½ m. WNW of Crayford. Area 885 acres. Pop. 666.

WICKHAMFORD, a parish in Worcestershire, 2 m. SE by E of Evesham. Area 1,242 acres. Pop. in 1831, 136; in 1851, 129.

WICKHAM-MARKET, a parish and village in Suffolk, 2½ m. NE of Ipswich, on a rising ground near the river Deben. Area 2,310 acres. Pop. in 1831, 1,202; in 1851, 1,697. It is said that no fewer than fifty other churches are visible from the eminence on which the church of this p. is built.

WICKHAM-ST.-PAUL, a parish in Essex, 3¼ m. E by N of Castle-Hedingham. Area 1,225 acres. Pop. in 1831, 388; in 1851, 425.

WICK-HAMPTON, a parish in Norfolk, 4½ m. SSE of Acle, in Halvergate Marsh. Area 1,605 acres. Pop. in 1831, 122; in 1851, 131.

WICKHAM-SKEYTH, a parish in Suffolk, 2¾ m. N of Mendlesham. Area 1,770 acres. Pop. 613.

WICKHAM (WEST), a parish in Cambridgeshire, 4½ m. NE by E of Linton. Area 2,937 acres. Pop. in 1831, 529; in 1851, 570.—Also a parish in Kent, 2¾ m. SW by S of Bromley. Area 2,645 acres. Pop. in 1831, 614; in 1851, 732.

WICKLEWOOD, a parish and village in Norfolk, 3 m. WNW of Wymondham. Area 1,564 acres. Pop. in 1831, 787; in 1851, 894.

WICKLOW, a maritime county of the Irish province of Leinster; bounded on the N, by co. Dublin; on the E, by the Irish sea; on the S, by co. Wexford, on the SW, by co. Carlow; and on the W, by co. Kildare. The longest line which can be drawn within the co. in any direction, extends SW by S from the N extremity of the coast-line to the W side of Slieveboy, and measures 34½ m.; and the longest line which can be drawn in a direction at right angles with the preceding, falls upon the S extremity of the coast-line, and measures 26 m. The area of the co. comprises 280,393 acres of arable land, 200,754 of uncultivated land, 17,600 of continuous woods, 341 of towns, and 1,090 of water,—in all, 500,178 acres. Of the unimproved land, 130,000 acres have an elevation exceeding 1,000 ft. The whole coast, generally viewed, is inhospitable to navigators; it offers no better retreat to even the smallest craft than the poor harbours of Wicklow and Arklow; and presents to the sea not one salient angle, and scarcely even a considerable projection.

Surface.] The sea-board or eastern district of the co., to the mean breadth of between 4 and 5 m., is a band of tumulated ground, decidedly mountainous in the extreme N. The central district, extending from the extreme N to near the extreme S, with a breadth of from 7½ to 13 m., is grandly mountainous, and possesses many picturesque or beautifully romantic scenes. Three of the *gaps* or alpine defiles, through which roads cross the summit-line of this great district, are Sally-gap, near the N, with an elevation of 1,631 ft. above sea-level; Wicklow-gap, 6¼ m. SW by S of Sally-gap, with an elevation of 1,569 ft.; and the gap at the S side of Table-mountain, 3 m. SW of Wicklow-gap, with an elevation of 2,266 ft. The N part of the western district of the co. has a small average breadth, and consists of the vale of the Liffey, with the convergence upon it of the vale of the King's river; the central part of the western district consists of a far-extending ramification of the great mountain-region of the co., having for its chief feature the summit of Slieve-Gadoe, and acting as a watershed between the river-system of the Liffey and the river-system of the Slaney; and the S part of the western district consists principally of the valley of the Slaney. The south-western district forms a wing or projecting district of the co., consisting of an irregular and flowingly-outlined series of pleasant vales and lofty hills. The principal summits of the co., together with their respective altitudes above sea-level, are Kippure, on the N boundary, 8½ m. W by S of Bray, 2,473 ft.; a height on the boundary, 1¼ m. WNW of Kippure, 2,364 ft.; a height 3 m. E by S of Sorrelhill, 2,364 ft.; Thonelagee, 6¾ m. SE of Blessington, 2,683 ft.; a height 1¼ m. NNE of Thonelagee, 2,307 ft.; Comaderry, 2½ m. S of Thonelagee, 2,296 ft.; Lugduff, 2 m. SE of Comaderry, 2,148 ft.; and Croghan, 4 m. SE of Lugnaquilla, 2,175 ft.

Minerals and mines.] Granitic rocks constitute a conspicuous district of the co., about 8½ m. in mean breadth, extending SSW from the middle of the northern boundary. Metamorphic rocks constitute considerable bands of country along both sides of the granitic district. Transition rocks, consisting principally of clay-slate, greywacke, and greywacke slate, constitute the platform of all the other districts of the co.—The principal mines worked within the county in 1838, were the Luganure lead mines, employing from 150 to 300 men, and annually producing from 1,200 to 3,600 tons; the Ballymurtagh copper mine, annually producing from 5,000 to 6,000 tons; the Cronbane and Tigroney copper mines, annually producing from 5,000 to 6,000 tons; the Connoree copper mine, annually producing from 1,000 to 1,500 tons; the Ballygahan copper mine, annually producing from 200 to 300 tons; and the Glenmalure lead mine, annually producing from 180 to 200 tons. The gold mines of W., situated in a vale among the declivities of Croghan-Kinshela, on the S margin of the co., have been noticed in the article CROGHAN-KINSHELA: which see.—Bisulphuret of iron occurs in vast beds in association with the copper lodes of the clay-slate district. Porcelain clay, of various degrees of fineness, formed by the decomposition of the felspar of granite, constitutes the agricultural soil of some tracts of the county; and kaolin of fine quality, occurs at Kilranelagh, in the vicinity of Baltinglass.

Rivers and lakes.] The river Liffey, and its tributary, the King's river, drain the NW district into the cos. of Kildare and Dublin; the river Slaney, and its tributaries, the Derreen and the Derry, drain the SW district into the cos. of Carlow and Wexford; the Ovoca, formed by the large indigenous rivulets, Avonmore, Avonbeg, and Aughrim, drains the central and the SE districts into the sea at Arklow; five or six small rivulets drain part of the E district into the sea between Arklow and Wicklow-head; the Vartrey river drains a chief part of the E district into the sea at Wicklow; and the Bray river drains the NE district into the sea, on the boundary with co. Dublin, at Bray. The principal lakes, together with their respective surface-elevation above sea-level, are Lower Lough Bray, 1,225 ft.; Upper Lough Bray, 1,458 ft.; Lough Tay, 897 ft.; Lough Dan, 685 ft.; Lough Ouler, 1,829 ft.; Lough Nahanagan, 1,384 ft.; the lower lake of Glendalough, 411 ft.; and Broad Lough, the estuarial expansion of the Vartrey parallel to the sea above the town of Wicklow.

Soils and agriculture.] Bog of considerable depth, and covered with heath, overlies a large part of the great granitic district. Vast numbers of the rounded summits and curving acclivities of the granitic mountains are either quite bare rock, or so thinly covered with soil as to possess the most meagre vegetation. The soil of a large proportion of the other mountainous grounds of the county, like the substratum of the bog on the granite mountains, consists simply of the disintegrations of the underlying rocks; that of the valleys of the NE district of the co., from the boundary with co. Dublin to a point 3 m. S of the town of Wicklow, consists of a diluvium, which bears evidence of having been swept hither from the breaking up of a limestone country. In 1841, there were within the rural districts of the county, 2,620 farms of from 1 acre to 5 acres, 2,922 of from 5 to 15 acres, 1,891 of from 15 to 30 acres, and 2,000 of upwards of 30 acres. The totals of live stock, together with their respective estimated value, were 12,314 horses and mules, £98,512; 1,917 asses, £1,917; 45,022 cattle, £292,643; 75,743 sheep, £83,317; 19,348 pigs, £24,185; and 120,785 poultry, £3,019.—The plantations within the county consisted of 17,600 continuous acres, and 1,015,301 detached trees.

Manufactures and Trade.] The commerce of the co. is of small extent, and is chiefly conducted through the ports of Wicklow and Arklow. The most considerable seat of manufacture is the small town of Stratford-on-Slaney. A large aggregate of productive industry is concentrated upon mining, and has been noticed in the section upon minerals. —A line of railway has been proposed to connect Bray with the Dublin and Kingstown railway. The line of railway from Wexford to the Kilkenny and Dublin railway at Carlow, passes near the SW boundary of the barony of Shillelagh.

Divisions and Towns.] The co. is divided into the 8 baronies of Rathdown, in the NE; Newcastle, in the middle-E; Arklow, in the SE; North Ballinacor, in the N-centre; South Ballinacor, in the S-centre; Lower Talbotstown, in the NW; Upper Talbotstown, in the middle-W; and Shillelagh, in the SW. The towns and principal villages are Enniskerry, Delgany, Killincarrig, and part of Bray, in the barony of Rathdown; Newtown-Mount-Kennedy, Glenealy, Downs, Kilcoole, Killisky, Newcastle, Ballinalee, Rathnew, and part of Wicklow, in the barony of Newcastle; Arklow, Redcross, and part of Wicklow, in the barony of Arklow; Rathdrum, in the barony of North Ballinacor; Tinnehely, in the barony of South Ballinacor; Dunlavin, Blessington, and Donard, in the barony of Lower Talbotstown; Baltinglass and Stratford-on-Slaney, in the barony of Upper Talbotstown; and Carnew and Shillelagh, in the barony of Shillelagh. The assizes are held at Wicklow; quarter-sessions at Wicklow, Arklow, Baltinglass, Bray, and Tinahely. The co. sent 10 members to the Irish parliament, or 2 for the county at large, and 2 for each of the boroughs of Wicklow, Blessington, Baltinglass, and Carysfort; but it sends to the imperial parliament only 2 members,—those for the county at large. Constituency in 1841, 1,503; in 1851, 3,330. Pop. in 1831, 121,557, in 19,970 families. Pop. in 1841, 126,143. Males, 63,489; females, 62,654; families, 21,182. Inhabited houses, 19,210. First-class inhabited houses, 616; second-class, 5,488; third-class, 7,709; fourth-class, 5,397. Families residing in first-class houses, 716; in second-class houses, 6,141. Families employed chiefly in agriculture, 14,032; in manufactures and trade, 4,740, in other pursuits, 2,410. Families dependent chiefly on property and professions, 662. Pop. in 1851, only 99,287.

WICKLOW, a port and county-town, and formerly a parliamentary borough, partly in the parishes of Drumkey and Kilpoole, and partly in the parish of Rathnew, co. Wicklow, situated on the estuary of the Vartrey, at the head of a small bay, 2 m. NW of Wicklow-head, and 25½ SSE of Dublin. Its streets are narrow, irregular, and badly-aligned, and the town, in an architectural view, is the poorest of the assize-towns of Ireland. The harbour, in ordinary tides, admits vessels drawing between 7 and 8 ft. of water; but it is a bar harbour, with a narrow entrance in any state of the tide, and inaccessible at low water. Pop. in 1831, 2,472; in 1851, 3,141. It was anciently called *Wickenloe* by the Danes, and *Kilmantan* by the Irish. It gives the title of Earl, in the peerage of Ireland, to the noble family of Howard.

WICKLOW, a county of Western Australia, bounded on the N by the co. of Grantham; on the E by that of Seel; on the S by the co. of Goderich; and on the W by that of Wellington and Nelson. It contains the William, Buchanan, Arthur, and Beaufort rivers, and several salt lakes. The chief towns are Bannister and Williamsburg.

WICKMERE, a parish in Norfolk, 5 m. NNW of Aylesham. Area 820 acres. Pop. in 1851, 302.

WICKRATH, or WICKERATH, a village of Prussia, in the regency and 18 m. WSW of Dusseldorf, and circle of Grevenbroich, on the l. bank of the Niers. Pop. 855. It has a fine castle, and possesses manufactories of linen and cotton fabrics, and of iron ware, tanneries, and paper-mills.

WICKRATHBERG, a village of Prussia, in the regency of Dusseldorf, and circle of Grevenbroich, on the Niers. Pop. 880. It has manufactories of linen and damask.

WICKRATHHAN, a village of Prussia, in the regency of Dusseldorf, and circle of Grevenbroich, near Wickrath. Pop. 260. It has manufactories of linen and damask.

WICKTON, a township in the p. of Stoke-Prior, Herefordshire, 4½ m. SE of Leominster. Pop. in 1831, 143; in 1851, 124.

WICKWAR, a parish and town in Gloucestershire, 24 m. SSW of Gloucester. Area of p. 2,307 acres. Pop. in 1831, 972; in 1851, 966.

WICONISCO, a township of Dauphin co., in the state of Pennsylvania, U. S., 22 m. N of Harrisburg. Pop. in 1850, 500.

WICZLIN, a village of Prussia, in the regency of Danzig, and circle of Neustadt. Pop. 430.

WICZOW, a village of Austria, in Moravia, in the circle of Olmutz. Pop. 470.

WIDACH, a village of Austria, in Bohemia, in the regency of Gitschin, and bail. of Neu Pefa. Pop. 600.

WIDA (GULF OF), a gulf in the Molucca archipelago, between the S and SE peninsulas of the island of Gilolo. It is 75 m. in depth. Near its widest part to the S of its entrance is a group of small islands of the same name.

WIDAWA, a town of Poland, in the gov. of Kalisch, obwod and 18 m. SE of Sieradz, on the Widawka, an affluent of the Wartha. Pop. 1,510. It has manufactories of cloth, linen, hosiery, and hats.

WIDCOMBE, a parish in Somersetshire, 1½ m. SE by E of Bath. Area with Lyncombe, 1,845 acres. Pop., with Lyncombe, in 1831, 8,704; in 1851, 9,974.

WIDDECOMBE-IN-THE-MOOR, a parish in Devon, 6 m. NW by N of Ashburton. Area 10,614 acres. Pop. in 1831, 959; in 1851, 974.

WIDDERN, a town under the common government of Würtemberg, and of the grand-duchy of Baden, on the l. bank of the Jaxt, 10 m. N of Oehringen. The part belonging to Würtemberg is in the circle of the Neckar, bail. and 12 m. NE of Neckarsulm. The Badois portion is in the circle of the Lower Rhine, and bail. of Adelsheim. Pop. 1,419. It has two castles and a tile-kiln.

WIDDERSBACH, a village of Bavaria, in the circle of Middle Franconia, and presidial of Leutershausen. Pop. 330.

WIDDERSDORF, a village of Prussia, in the regency and circle of Cologne. Pop. 360.

WIDDERSHAUSEN, a village of Hesse, in the prov. of Fulda, and circle of Hersfeld. Pop. 523.

WIDDERT (OBER and UNTER), two villages of Prussia, in the regency of Dusseldorf, and circle of Solingen. Pop. 460.

WIDDESHOVEN, a village of Prussia, in the regency of Dusseldorf, and circle of Grevenbroich. Pop. 580.

WIDDIG, a village of Prussia, in the regency of Cologne, and circle of Bonn. Pop. 432.

WIDDIN, or VIDIN, a town of Bulgaria, the cap of a sanjak of the same name, situated on the r. bank of the Danube, opposite Kalafat, and 130 m. SE of Belgrade. The outer town forms a semicircular outwork on the land side, the houses being all of one story, built of mud, and each having a small farmyard planted with trees and enclosed with wattling. This large agglomeration of farmyards, with the appertaining land scattered out at a distance, reminds one of Debreczin and other towns of Hungary, which cover so much ground, and are in fact but large villages. Within this semicircle, which has the Danube, here about 1 m. wide, for its base, are the inner town, or citadel, built by the Genoese, now employed as an arsenal, and in which is situate the seraglio of the pasha; the principal mosque, and a population exclusively Moslem. It is much better built than the *veras* or outer town. The total number of houses in the two towns is 1,500 Turkish, 1,250 Christian or Bulgarian, and 360 Jewish; and the total civil pop. may amount to 16,000. The outer town is only fortified towards the land side. The inner town is separated from it by a wide and deep fosse, with solid bastions and curtains. The town has some trade in corn, wine, and salt.

WIDDIN, a creek of New South Wales, which forms the line of separation between the counties of Phillip and Hunter, and flows into the Goulburn.

WIDDINGTON, a parish in Essex, 5 m. W by N of Thaxted, near the line of the Northern and Eastern Counties railway. Area 2,028 acres. Pop. in 1831, 386; in 1851, 447.

WIDFORD, a parish in Essex, 1 m. SW of Chelmsford, in the line of the Eastern Counties railway. Area 692 acres. Pop. in 1831, 157; in 1851, 184.—Also a parish in Gloucestershire, 1½ m. ESE of Burford, isolated from the rest of the county, and locally in Oxfordshire. Area 564 acres. Pop. in 1831, 51; in 1851, 43.—Also a parish in Hertfordshire, 4 m. E by N of Ware, intersected by the river Ash. Area 1,137 acres. Pop. in 1851, 519.

WIDHOSTITZ, a village of Austria, in Bohemia, in the regency of Eger, and bail. of Saatz. Pop. 240.

WIDIM, a village of Bohemia, in the regency of Bohmisch-Leipa, and bail. of Dauba. Pop. 270.

WIDIM-KOKORRIN, a village of Bohemia, in the regency of Bohmisch-Leipa, and bail. of Dauba. Pop. 322.

WIDITZ, a village of Bohemia, in the regency of Pardubitz, and bail. of Kultenberg. Pop. 200.

WIDLEY, a parish in Hants, 5½ m. E by N of Fareham. Area 1,090 acres. Pop. in 1851, 565.

WIDLITZ, a village of Bohemia, in the regency of Eger, and bail. of Tachan. Pop. 235.

WIDMERPOOLE, a parish in Notts, 9 m. SSE of Nottingham. Area 2,190 acres. Pop. 147.

WIDMMENAR, a market-town of Prussia, in the reg. of Gumbinnen, circle of Loken. Pop. 532.

WIDNAU, a parish of Switzerland, in the cant. of St. Gall, and bail. of the Lower Rheinthal. Pop. 2,586.

WIDONITZ, a village of Bohemia, in the reg. of Gitschin, and bail. of Horzitz. Pop. 255.

WIDONNER, a village of Prussia, in the reg. of Konigsberg, and circle of Kaslenberg. Pop. 406.

WIDOYE, a department and commune of Belgium, in the prov. of Limburg, watered by the Molebeck. Pop. 192.

WIDRINGTON, a parochial chapelry in Northumberland, adjacent to Morpeth, including the township of Linton. Area 4,530 acres. Pop. 429.

WIDWORTHY, a parish in Devon, 3½ m. E by S of Honiton. Area 1,437 acres. Pop. in 1851, 192.

WIDZIM, a village of Prussia, in the reg. of Posen, and circle of Bomst, near Wollstein. Pop. 500.

WIDZY, a town of Russia in Europe, in the gov. of Wilna, on the Desna. Pop. 4,000.

WIEBELITZ (GROSS and KLEIN), two villages of Prussia, in the regency of Magdeburg, and circle of Salzwedel. Pop. 220.

WIEBELSBACH, a village of Hesse, in the prov. of Starkenburg, and presidial of Hochst. Pop. 335.

WIEBELSBERG, a village of Bavaria, in the circle of Lower Franconia, and presidial of Gerolzhofen. Pop. 166.

WIEBELSHEIM, a village of Prussia, in the regency of Coblenz, and circle of St. Goar. Pop. 345. —Also a village in the regency of Treves, and circle of Saarbruck, at the confluence of the Osterbachs with the Blies. Pop. 1,320.—Also a village of Bavaria, in the circle of Lower Franconia, and presidial of Uffenheim. Pop. 200.

WIEBELSKIRCHEN, a bail. of Prussia, in the reg. of Treves, and circle of Ottweiler. Pop. 1,310.

WIEBLINGEN, a village of Baden, in the circle of the Lower Rhine, and bail. of Heidelberg, on the Neckar. Pop. 1,110. It has a tobacco factory.

WIECHNOW, a village of Austria, in Moravia, circle of Iglau, and district of Bistritz. Pop. 460.

WIECHS, a village of Baden, in the circle of the Upper Rhine, and bail. of Schopfheim. Pop. 607. —Also a village in the circle of the See, and bail. of Blumenfeld. Pop. 401.

WIECK, a village of Prussia, in the regency of Stralsund, and circle of Griefswald. Pop. 478.

WIECK-ON-THE-DARS, a village of Prussia, in the regency of Stralsund, and circle of Franzburg. Pop. 754.

WIECK AND WITTOW, a village of Prussia, in the regency of Stralsund, and circle of Bergen. Pop. 798.

WIECK (DOMLAPOTELS), a village of Prussia, in the regency of Stettin, and circle of Kammin. Pop. 497.—Also a village in the regency of Koslin, and circle of Schlawe. Pop. 441.

WIECZIN, a village of Prussia, in the regency of Posen, and circle of Pleschen. Pop. 312.

WIED, a river of the duchy of Nassau, which has its source in the Westerwald, and flows past Neuwied into the Rhine.—Also a village in the bail. and 3 m. SW of Hachenburg. Pop. 200.

WIED (ALTEN), a market-town of Prussia, in the regency and 9 m. NW of Coblenz, and circle of Neuwied, on the Wied. Pop. 314. It has copper and lead mines, a silver finery, and copper forges.

WIEDA, a village of Brunswick, in the circle of Blankenburg, and bail. of Walkenried. Pop. 1,130.

WIEDELAH, a village of Hanover, in the prov. of Hildesheim, and bail. of Vienenburg, near Salzauler. Pop. 615.

WIEDEMAR (GROSS and KLEIN), two villages of Prussia, in the regency of Merseburg, and circle of Delitzsch. Pop. 440.

WIEDEN (OBER and UNTER), two villages of Baden, in the circle of the Upper Rhine, and bail. of Schonau. Pop. 274.

WIEDENBRUCK, a circle and town of Prussia, in the prov. of Westphalia, and regency of Minden. The circle comprises an area of 180 sq. m. Pop. in 1849, 39,492. The town is 42 m. SW of Minden, on the Ems, which is here crossed by four bridges. Pop. 2,716. It contains three Catholic churches, a gymnasium, and an hospital, and has manufactories of woollen hosiery and of tobacco.

WIEDENBRUGGE, a village of Shaumburg-Lippe, and bail. of Hagenburg. Pop. 200.

WIEDENFELD, a village of Prussia, in the regency of Cologne, and circle of Bergheim. Pop. 210.

WIEDENSAHL, a market-town of Hanover, in the prov. of Kalenberg, and bail. of Loccum, near Rehburg. Pop. 745.

WIEDENSHAUSEN, a village of Bavaria, in Upper Bavaria, presidial of Dachau. Pop. 260.

WIEDERAU, a village of Prussia, in the regency of Merseburg, and circle of Liebenwerda. Pop. 345.—Also a village of Saxony, in the circle of Liepsig, and bail. of Pegau. Pop. 240.—Also a village in the bail. of Rochlitz. Pop. 944.

WIEDERGELTINGEN, a village of Bavaria, in the circle of Swabia, and presidial of Turkheim, on the Wertach. Pop. 455.

WIEDERGRUN, a village of Austria, in Silesia, in the circle of Troppau, and district of Freudenthal. Pop. 330.

WIEDERMUTH, a village of Schwarzburg-Sondershausen, and bail. of Ebeleben, on the Helbe. Pop. 260.

WIEDERSBACH, a village of Prussia, in the reg. of Erfurt, and circle of Schleusingen. Pop. 320.

WIEDERSBERG, a village of Saxony, in the circle of Zwickau, and bail. of Voigtsburg. Pop. 245

WIEDERSEE, a village of Prussia, in the reg. of Marienwerder, and circle of Graudenz. Pop. 300.

WIEDERSTADT (OBER), a village of Prussia, in the regency of Merseburg, and circle of Mannsfeld, near Hettstadt, on the Wipper. Pop. 535.

WIEDERSTEIN, a village of Prussia, in the regency of Arnsberg, and circle of Siegen. Pop. 260

WIEDIKON, a village of Switzerland, in the cant. of Zurich, bail. and near Zurich. Pop. 328.

WIEDLISBACH, a town of Switzerland, in the cant. of Berne, and bail. of Wangen, 6 m. NE of Soleure, near the l. bank of the Aar. Pop. 914. It has quarries of mill-stone.

WIEDNITZ, a village of Prussia, in the reg. of Liegnitz, and circle of Hohersweda. Pop. 340. It has a castle.

WIEFELSTEDE, a parish of the duchy and 11 m. NNW of Oldenburg, circle of Neuenburg, and bail. of Kastede. Pop. 1,925.

WIEGANDSTHAL, a market-town of Prussia, in the regency of Liegnitz, circle and 15 m. SSE of Lauban, at the confluence of the Tafelfichte and Lausitzbach. Pop. 490. It has manufactories of linen, and several furnaces.

WIEGE-FATY, a commune of France, in the dep. of the Aisne, and cant. of Sains, 9 m. NNW of Vervins. Pop. 1,647. It has manufactories of beet-root sugar.

WIEGERSDORF, a village of Hanover, in the prov. of Hildesheim, and bail. of Ilfeld. Pop. 412.

WIEGLEBEN, a village of Saxe-Coburg, in the bail. of Tonna. Pop. 270.

WIEGSTADTL, a town of Austrian Silesia, 28 m. NE of Olmutz. Pop. 1,300.

WIEHE, a town of Prussian Saxony, in the reg. of Merseburg, 26 m. NE of Erfurt. Pop. 1,700.

WIEHL, a village of Baden, in the circle of the Upper Rhine. Pop. 1,400.

WIEHRE, a village of Baden, in the Upper Rhine circle, bail. of Freiburg. Pop. 800.

WIEKEVORST, a commune and v. of Belgium, in the prov. and 20 m. ESE of Antwerp. Pop. 1,100.

WIELD, a parish in Hants, 6 m. W of Alton. Area 2,087 acres. Pop. in 1831, 248; in 1851, 307.

WIELICKZKA, a town of Austria, in Galicia, 8 m. SE of Cracow, at the N extremity of a spur of the Carpathian mountains. Pop. 4,500. The exclusive trade of this place is connected with its salt mines, which have been wrought for six centuries, and are still perhaps the richest in Europe

The works at present extend over an area of 35,000 sq. fathoms; and the entire length of their galleries and passages is about 37 m., and the greatest depth yet attained is 743 ft. They are entered by a broad flight of steps cut through the superincumbent strata of marl, clay, and sand. The first traces of salt present themselves at a depth of about 15 fath. In some places the beds are only 20 ft. thick; in others, they attain a thickness of 100 ft. Every 400 cubic fath. of rock yield 100,000 cwts. of salt; and the annual yield is 700,000 cwts. On the occasion of the partition of Poland in 1772, Russia and Prussia introduced a stipulation into the infamous treaty; according to which Austria cedes from the 900,000 cwts. of salt annually raised in the mines of Bothnia and W. 200,000 cwts. to Prussia, and 150,000 cwts. to Russia.

WIELLRANDOROV, a village of Prussia, in the reg. of Danzig, circle of Stargard. Pop. 400.

WIELONA, a small town of Russian Lithuania, in the gov. of Wilna, on the Niemen, 20 m. S of Rosienne.

WIELSBEKE, a commune and town of Belgium, in the prov. of W. Flanders, 18 m. S of Bruges, on the Lys. Pop. 1,900.

WIELUN, a town of Poland, 65 m. E of Breslau. Pop. 2,000.

WIEN. See VIENNA.

WIENERWALD, a mountain range of Lower Austria, a branch of the Styrian Alps, extending from Kahlenberg southward, across the valley of the Danube, towards St. Polten, about 70 m. It separates and gives name to the two circles of the Upper and Lower Wienerwald, otherwise known as the quarters 'above' and 'below' the forest of Vienna.

WIEPRZ, a river of Poland, which, rising from a lake to the SE of Zamosc, flows NNW, and joins the Vistula, near Stericza, after a course of 140 m.

WIERDE, a commune and village of Belgium, in the prov. and 4 m. SE of Namur. Pop. 1,200.

WIERDEN, a village of Holland, in the prov. of Overyssel, 14 m. NE of Deventer.

WIERINGEN, an island of Holland, in the Zuyder-Zee, opposite to the coast of N. Holland, and 11 m. SE of the Helder. It is about 10 m. in length, and 2 m. broad, and has a well-cultivated surface. Pop. 1,500. Its chief place is Hypolitushof.

WIERS, a commune and village of Belgium, in the prov. of Hainault, 22 m. W of Mons. Pop. 3,300.

WIERUM, a small port of Holland, in Friesland, 6 m. N of Dokkum.

WIERUSZOW, a town of Poland, on the Prosna, 31 m. S of Kalisch. Pop. 1,000.

WIESA, a river of Germany, which rises on the Feldburg, and flows WSW to the Rhine, which it joins 2 m. below Barb.

WIESBADEN, the capital of the duchy of Nassau, situated on the Salzach, 20 m. WSW of Frankfort, and on the Frankfort railway. It has a neat, clean, and animated appearance, and is yearly increasing in size by the construction of handsome streets and detached houses. Pop. 13,000, being double what it was twenty years ago. Its great attractions are its baths, and hot alkaline springs, which are annually visited by upwards of 15,000 strangers from all parts of Europe. Placed in a valley, with an open and fruitful country in the direction of the Rhine and the Maine, and bounded on the N and E by hills covered with apple, cherry, and walnut trees, behind which rises the Taunus range of mountains, W. is effectually protected from the cold winds from these quarters. On account of its position, it is at times exceedingly hot and oppressive in the summer months

WIESE, or LAUKA, a town of Moravia, on the river Iglau, 6 m. E of Iglau.

WIESELBURG, or MOSONY, a county of Hungary, bounded on the N by the Danube, and on the W by Lower Austria. Its area is about 740 sq. m. Pop. 26,000, a mixture of Germans and Croats.—Its chief town of the same name is on a branch of the Danube, 21 m. SSE of Presburg. Pop. 2,500.

WIESEN, a district of the grand-duchy of Baden, of which it forms the SW corner. It contains the south part of the Brisgau, and the district of Sausenburg.—Also a small river of Germany, which rises in the Black forest, and falls into the Rhine opposite to Huningen.

WIESENBRONN, a town of Bavarian Franconia, near Kitzingen. Pop. 100.

WIESENSTEIG, a town of Würtemberg, 18 m. NW of Ulm. Pop. 1,200.

WIESENT, a small river of Bavaria, which falls into the Regnitz at Forch-heim.

WIESENTHAL, a village of Baden, 4 m. E of Philipsburg. Pop. 900.—Also a town of Saxony, 12 m. SSE of Schwarzenburg. Pop. 1,500.

WIESENTHEID, a village of Bavaria, in Lower Franconia, 19 m. E of Wurzburg. Pop. 1,100.

WIESLOCH, a town of Baden, 8 m. S of Heidelberg. Pop. 2,700.

WIETLISBACH, a town of Switzerland in the cant. of Bern, 6 m. ENE of Soleure.

WIGAN, a parish and parl. borough in the co.-palatine of Lancaster, intersected by the Leeds and Liverpool canal, the North Union line of railway from the Liverpool and Manchester, and the Grand Junction lines. Besides the town of W, the parish comprises the chapelries of St. George, Abram, Billinge-Chapel-end, Haigh, Hindley, Holland, Pemberton, and Scholes-Church; and the townships of Billinge-higher-end, Dalton, Ince, Orrel, and Winstanly. Area of the parish, 24,433 acres. Pop. in 1801, 25,552; in 1831, 44,486; in 1851, 63,287. Area of the township, 2,161 acres. Pop. in 1801, 10,989; in 1831, 20,774; in 1851, 31,941. The borough having a pop. of 31,941 in 1851, returns 2 members to parliament. The number of electors registered, in 1837, was 551; in 1852, 718. The town, which is situated on a rising ground, is bounded on the W by the river Douglas, has increased rapidly of late years. Many of the houses are old and ill built, but the more modern streets contain several good houses. W. is well situated for manufactures by steam power, being in the midst of coal mines. The vicinity is celebrated for producing coal of the species well-known as cannal or gamal coal, which may be turned in a lathe, and gives out a bright light when burning. The chief manufactures are those of cotton and iron. Amongst the fabrics produced here have been calicoes, fustians, checks, linens, &c., besides spun cotton yarn in large quantities; there are also manufactories of edge-tools, brazen articles, and spades, and other agricultural implements.

WIGBOROUGH (GREAT), a parish in Essex, 7 m. SSW of Colchester, bounded by a creek of the Blackwater river called Verley. Area 2,585 acres. Pop. in 1831, 434; in 1851, 471.

WIGBOROUGH (LITTLE), a parish in Essex, 7½ m. SSW of Colchester, bounded on the E by the Mersea channel, and on the S by the Verley; both creeks are branches of the Blackwater. Area 1,168 acres. Pop. in 1831, 123; in 1851, 89.

WIGGENHALL-ST.-GERMAN, a parish in Norfolk, 4 m. SW by S of Lynn-Regis, on the river Ouse, near the Eaubrink cut. Area 1,219 acres. Pop. in 1831, 552; in 1851, 750.

WIGGENHALL (ST.-MARY), a parish in Norfolk, 5 m. SW of Lynn-Regis, on the river Ouse, near the Eaubrink cut. Area 2,807 acres. Pop. 325.

WIGGENHALL (St. Mary Magdalene), a parish in Norfolk, 5½ m. SSW of Lynn-Regis, on the river Ouse. Area 4,248 acres. Pop. in 1851, 845.

WIGGENHALL (St. Peter), a parish in Norfolk, 5 m. SW of Lynn-Regis, on the river Ouse. Area 944 acres. Pop. in 1831, 114; in 1851, 162.

WIGGENHOLT, a parish in Sussex, 7¾ m. NNE of Arundel, on the E bank of the river Arun. Area 841 acres. Pop. in 1831, 37; in 1851, 39.

WIGGINTON, a parish in Hertfordshire, 1¼ m. SE of Tring, near the Grand Junction canal, and the London and Birmingham railway. Area 1,662 acres. Pop. in 1831, 536; in 1851, 643.—Also a parish in Oxford, 5½ m. WNW of Deddington. Area 1,790 acres. Pop. in 1831, 327; in 1851, 314.—Also a chapelry in the parish of Tamworth, Staffordshire, 1¾ m. W of Tamworth station, on the Birmingham and Derby railroad. Pop. in 1851, 593.—Also a parish in the N. R. of Yorkshire, 5 m. N of York. Area 1,465 acres. Pop. in 1831, 359; in 1851, 374.

WIGGLESWORTH, a township in the p. of Long Preston, W. R. of Yorkshire, 6½ m. SSW of Settle. Area 4,089 acres. Pop. in 1851, 350.

WIGGONBY, a township in the p. of Aikton, Cumberland, 5 m. NE of Wigton. Pop. in 1851, 219.

WIGHILL, a parish in Yorkshire, 3 m. N by W of Tadcaster. Area 2,588 acres. Pop. in 1851, 296.

WIGHT (Isle of), a beautiful island, politically belonging to Hampshire, washed, on the NW and NE, by the Solent sea, and on the other sides by the English channel. It lies opposite the mainland part of Hampshire, from Portsmouth on the E, to a point 4 m. beyond Lymington on the W. The sound or channel called the Solent sea, varies in breadth from 2 to 5 m.; and is believed to have been formed by a gradual encroachment of the ocean upon a connecting belt of low l. nd, which in the time of the Romans may have been an isthmus or passable at low water. The island has, in a general view, a four-sided outline, whose angles face the cardinal points of the compass; but it is acutely prolonged on the W, somewhat rounded on the N and S, and gently curved and indented in its general coast-line. Its greatest length from E to W, or from the Needles to Foreland-farm, is nearly 23 m.; its greatest breadth, from its N to its S angle, is about 13 m.; and its superficial extent is estimated at from 105,000 to 125,000 acres. A steamer of ordinary size can easily circumnavigate the island in about 8 hours, in good weather. The Southampton railway has brought the isle of W. within 4 hours' distance of the metropolis. The passage from Southampton to Cowes is usually performed by the steamers in 1½ hour; that from Portsmouth to Ryde in half-an-hour. Along nearly the whole of the NW, the coast slopes gently to the water's edge, and possesses no feature of scenic interest; around and toward the W angle it becomes rocky, broken, precipitous, and romantic; and along the SW and the SE, commonly called 'the Back of the Island,' it breaks down in a richly varied series of cliffs, often sheer and abrupt, or high and very bold, and everywhere replete with interest to both the geologist and the lover of landscape. The general surface of the island lies at a considerable height above sea-level. The water-shedding line uniformly follows the trending of the S coast, and is distant from it never more than 2½ m., and generally less than 1 m.; so that all the rivers and rivulets flow leisurely to the N. Blackgang-chine comes down the seaward face of a hill 830 ft. above sea-level, the loftiest height in the island. From a point 1½ m. E of Blackgang, a singularly romantic and grand tract, called Undercliff, whose characteristic features are those of a great landslip, extends nearly 6 m. to Bonchurch. Shanklin-chine, 2¼ m. NE of Undercliff, is about ½ m. in length, and its width and depth, at the mouth, are respectively about 180 and 270 ft.—Culver cliffs, 4 m. farther to the NE, are a series of chalky precipices 500 ft. high.—Bembridge-down, which extends inland from the cliffs to the vale of the Brading, and forms the east end of the great bisecting range of the island, commands a panoramic view which, for combination of extent, variety, and magnificence, excels most others seen in the country.

Bays and Rivers.] Alum, Toland, and Colwell bays indent the NW coast, the first in the immediate vicinity of the W extremity; the third at the distance of 3½ m.; and the second in the centre. The river Yar enters the sea 2 m. NE of Colwell bay; it is between 3 and 4 m. long, and over much the greater part of its length it is tidal, and has the character of an estuary. Newton bay opens nearly 4 m. farther NE. Thorness and Gurnet bays are situated respectively 2¾ and 4 m. NE of Newton bay. The Medina—anciently called the Mede, and disputing with the Brading the honour of being the monarch-stream of the island—enters the sea at Cowes harbour, the N point of the country, 1½ m. from Gurnet bay. This river rises near Down court, within less than 1½ m. of the S angle of the island; and it performs due northward an entire run of about 12 m. It is tidal, navigable, and slenderly of an estuary character, 4¾ m. up to Newport. Brading river opens upon St. Helen's roads, 1½ m. NW of Foreland, on the E extremity of the island. This river rises near Niton, 1 m. from the source of the Medina, and flows upwards of 7 m. N and NE to Horringfield. The bays or creeks along the whole of the SE and SW coasts, though numerous, are all so small as simply to render the coast in a gentle degree curvilinear. Pure water springs abound. At Shanklin is a spring which holds in solution some proportion of alum; at Pitland is one impregnated with sulphur; and at Blackgang, under Chale cliff, is a very strong aluminous spring.

Mineralogy.] Chalky formations constitute all the higher parts of the island. Schists of various kinds, but all secondary, lie under these, and occasionally look out on the south coast. Fossils—particularly bivalve and turbinated shells, and more particularly ammonites of from an inch to 1½ foot in diameter—are very abundant in the cliffs, and may often be gathered in large quantity from fragments of rock which have fallen 200 or 300 ft., and lie upon the beach. Some of the schists are of a kind to make good whetstones. Limestone, unusually rich in fossils, is burnt for manure. Coal of good quality, about 3 ft. thick in the upper stratum, and dipping to the N, commences at Bembridge cliff, and extends across the S district of the island. A vein of red ochre, and veins of white sand and fuller's earth, occur on respectively the south and the north sides of the coal-bed. Sandstone of various kinds is plentiful; but, as a building-material, though at one time worked, it has long been abandoned in favour of the beautiful Portland stone. A siliceous and silvery-looking sand occurs in immense beds at Freshwater and Alum bays, and has, for a long period, been exported to a great annual amount for the glass and china works of London, Bristol, and Worcester. Brighter and various coloured sands, at the same places, are used in smaller quantities, for being so disposed in phials, as to represent fancy scenes behind the glass, and form mantel-piece ornaments. Pipe clay abounds in various localities. Native alum, in large quantities, and red and yellow ochres, in such strata as beautifully variegate the tinting of the cliffs, occur at Alum bay. Small masses of native sulphur are often seen beneath the cliffs; and copperas stones were, for many years, so abundant on the shores, that several small vessels were employed in conveying them to London for the extraction of their copperas.

Climate and Soil.] The Isle of Wight has obtained a reputation for salubriousness and genial warmth, and most of the amenities of a benign climate, so high and of so long standing as ill to brook dispute. The temperature in winter, owing to saline exhalations from the encincturing sea, is two or three degrees higher than on most of the adjacent mainland, and considerably higher than in the colder parts of England; but in consequence of the hilly ranges constantly attracting vapours, all the central districts are subject to heavy rains, and, in winter, fling humidity and gloom down all the valleys and the slowly descending slopes. See article UNDERCLIFF.—The soil is very various; to a large extent, a poor, yellowish, ill-drained clay; but chiefly a strong loamy earth, easy of culture, and exceedingly fertile—so much so, that the natives are in the habit of boasting that their fortunate island yields seven times as much as its own population consumes. The uplands are disposed in pasturage, and maintain about 40,000 fine-fleeced sheep. Farms are generally of a moderate size, and pay a rental of from £100 to £400; but a few are so high as £800.

Trade.] The principal manufactures are those of starch and salt. There are numerous corn-mills on the island, and several breweries. The chief exports are grain, malt, wool, salt, and siliceous sand; and the chief imports are coals, timber, iron, hemp, fruits, and wine. The harbours of the island, particularly that of Cowes, have increasingly become ports of call for outward-bound ships making the final arrangements for their voyage, and inward-bound ships landing their letters and passengers.

Divisions and General Statistics.] The island is divided, politically, into the two hundreds of East and West Medina, which are separated from each other by the river whence they have their name; and ecclesiastically into 30 parishes, which constitute an archdeaconry in the dio. of Winchester. The civil administration is conducted by a governor, a vice-admiral, and other honorary functionaries. The island, which is considered as a county of itself under the Reform act, returns a member to parliament. The polling-stations are Newport and West Cowes. Electors registered in 1837, 1,167; in 1850, 1,877. Pop. in 1377, 8,000; in 1777, 18,024; in 1801, 22,602; in 1821, 31,611; in 1831, 35,363; in 1841, 42,547; in 1851, 50,324.

NEWPORT is the capital of the island. COWES, YARMOUTH, CARISBROOKE, RYDE, BRADING, and NEWTON, are separately described. Ventnor is a little but rapidly rising and elegant town of villas on the skirt of Undercliff. See VENTNOR. St. Helen's Green is a village of neat, thatched, scattered cottages on a creek of Brading harbour, 4 m. E of Ryde. St. Laurence is celebrated for the smallest parish-church in the kingdom. The other noticeable villages are Newchurch, Niton, Chale, Shorwell, Brixton, Gatcombe, Whippingham, Clatterford, Brook, Shalfleet, Binstead, Arreton, Godshill, Yaverland, Motteston, Calbourne, Sandown, Wootton, and Bembridge, most of which will be found separately described in our pages.

History.] The name of the Isle of Wight among the Britons was *Guith* or *Guict*—a word signifying 'divorced,' or 'separated,' and supposed by some writers to have been suggested of the island's disseverment from the mainland, by the gradual formation of the Solent sea. The name known to the Romans was *Vecta*, or *Vectis*. In the year 43, according to Tacitus' Life of Agricola, Plautius, the Roman general, conquered the Isle of Wight. In 530, Cerdic, a Saxon chieftain, founder of the kingdom of Wessex, subdued the island, and colonized it with Jutes and Saxons. In 787 the Danes seized it. William Fitz-Osborne, afterwards Earl of Hereford, received from William I. the Isle of Wight as an independent lordship. Richard de Redvers, Earl of Devon, received from Henry I. a grant of the lordship of the isle. Edward I. obtained the reversion of the lordship for the sum of 6,000 marks. In the reign of Richard II., the French made a descent upon the isle, burnt Yarmouth, Newton, and Rye, and unsuccessfully besieged Carisbrooke castle. Henry Beauchamp, duke of Warwick, was made king of the Isle of Wight by Henry VI., the king placing the crown on the duke's head with his own hand. In 1545, 2,000 French, from the fleet commanded by D'Annebaut, landed on the island, but were obliged to withdraw hastily. Colonel Robert Hammond, being governor of the isle at the time when Charles I., unconditionally, surrendered to him at Titchfield-house, the fallen monarch was conveyed by him to Carisbrooke.

WIGHTON, a parish in Norfolk, $2\frac{1}{2}$ m. N by E of Little Walsingham. Area 2,982 acres. Pop. 582.

WIGLAND, a township in the p. of Malpas, co.-palatine of Chester, 2 m. SSE of Malpas. Area 562 acres. Pop. in 1831, 265; in 1851, 203.

WIGMORE, a parish in Herefordshire, 10 m. NW of Leominster. Area 3,441 acres. Pop. 494.

WIGNEHIES, a commune of France, in the dep. of the Nord, and cant. of Trelon, 9 m. SE of Avesnes, near the r. bank of the Petite-Helpe. Pop. in 1846, 2,413. It has manufactories of hosiery, wood-ware, staves, and charcoal, and a bleachery.

WIGOLDINGEN, a village of Switzerland, cant. of Thurgau, and bail. of Weinfelden. Pop. 359.

WIGRY, a lake of Russian Poland, in the woiwodie and obwod of Augustowo, 6 m. SE of Suwalki. It is upwards of 6 m. from N to S, and $1\frac{1}{2}$ m. in medium breadth. Near the N extremity is an island on which is a convent which forms the residence of the bishop of Lithuania.

WIGSTADT. See WIEGSTADT.

WIGSTADTL, a market-town of Austria, in Bohemia, in the circle and 39 m. ESE of Koniggratz. Pop. 600.

WIGSTEIN, a village of Austria, in Silesia, in the circle of Troppau. Pop. 300.

WIGSTON (MAGNA), a parish and village of Leicestershire. Area 2,780 acres. Pop. in 1831, 2,174; in 1851, 2,441. The v. is situated $\frac{1}{2}$ a m. W of the Wigton station on the Midland Counties railway, by which it is 4 m. distant from Leicester.

WIGTOFT, a parish in Lincolnshire, 7 m. SW of Bolton. Area 3,300 acres. Pop. in 1851, 741.

WIGTON, a parish in Cumberland, containing the market-town of the same name, the townships of Oulton, High and Low Waverton, and Woodside-Quarter. Area 11,800 acres. Pop. in 1831, 6,501; in 1851, 6,229. The town, situated in a tract of country which embraces little level land, yet scarcely can be called hilly, contains several well-built houses, and consists principally of two streets, the smaller one placed transversely at the extremity of the other. Some cotton goods are manufactured here, principally ginghams, muslins, and other fine fabrics.—Also a township in the p. of Harewood, W. R. of Yorkshire, $5\frac{1}{2}$ m. N by E of Leeds. Area 1,170 acres. Pop. in 1831, 168; in 1851, 147.

WIGTON, a parish, port, and parl. borough, the capital of Wigtonshire, situated on the W side of Wigton-bay, 94 m. SW of Edinburgh. Pop. of p. in 1831, 2,337; in 1851, 2,824; of burgh in 1831, 1,972; in 1851, 1,556. It has no manufactures except the ordinary artisan products for local use, and little trade beyond the transfer of commodities for the supply of the circumjacent country. The most valuable part of the export trade is the annual shipment of probably, on the average, 10,000 or 12,000 bushels of grain, with a proportionate quantity of potatoes, and a small quantity of oatmeal. It unites with Stranraer, Whithorn, and New-Galloway in sending a member to parliament. Total constituency, in 1840, 320; in 1851, 400. W. gave the title of Earl to the family of Fleming.

WIGTON-BAY, the estuary or frith of the Cree, running NNW between the cos. of Wigton and Kirkcudbright. Its extremities or headlands are Burrowhead in Wigtonshire, and the Slack of Ross in Kirkcudbrightshire. Its entrance, measured in a straight line between these points, is 12 m. wide. Its length, measured from the middle of this line to a point a little N of Creetown, where the river begins to be estuary, is 15 m. Its breadth over the upper half slowly expands from 6 furls. to 4 m.; and over the lower half averages about $7\frac{1}{2}$ m. Fleet-bay opens from about the middle of its Kirkcudbrightshire side. Of smaller bays which open from it, the chief on the Wigtonshire side are Isle of Whithorn-bay, Port-Yarrock, Rigg-bay, Garlieston-bay, and the little estuary of the Bladenoch: on the Kirkcudbrightshire side, Bridgehouse, Boreness, Kirkandrew, and Knockbrex bays.

WIGTONSHIRE, a county forming the W division of Galloway, and occupying the SW extremity, as well as containing the most southerly land of Scotland. It is bounded on the N partly by the Irish channel or frith of Clyde, principally by Ayrshire, and to a small extent by Kirkcudbrightshire; on the E by Kirkcudbrightshire; on the S by the Irish sea; and on the W by the Irish channel. The county lies between the parallels of 54° 36′ 45″ and 55° 3′ 40″ N lat., and between 4° 15′ 50″ and 5° 7′ 10″ W long. It extends from E to W between 30 and 31 m.; and from N to S between 28 and 29. Its form would be proximately square but for the enormous indentation of Luce-bay on the S, and the smaller though very considerable indentation of Loch-Ryan on the N. The superficial extent is variously stated by the New Statistical Account at 459 sq. m.; by the Rev. Samuel Smith, in his *View of the Agriculture of Galloway*, at 485·5 sq. m., or 244,498 Scottish acres; and by Chalmers, in his *Caledonia*, at 484 sq. m., or 309,760 statute acres. The district or double peninsula W of Luce-bay and Loch-Ryan, and of the isthmus of 6½ m. which lies between them, is called the Rhinns of Galloway. The district which forms the broad large peninsula between Luce-bay and Wigton-bay is called the Machers. The remaining district, or all the country N of the Machers and E of Loch-Ryan, bears the loose general name of the Moors. Ordinary reckoning restricts the Machers to little more than one-half of the large peninsula, or to about 64 sq. m.; and, estimating the Rhinns at 116, it consequently assigns to the Moors about 305 sq. m., or an area not much more than double of the joint area of the other districts. The physical aspect of W. is not striking or broadly varied, and offers few large interesting landscapes. While almost uniformly tumulated, a restless and continuous sea of knolls and hillocks and hills, it aggregately rises less above sea-level than probably any other co. in Scotland. A plain or valley, bearing marks of having at a comparatively recent period lain under the sea, extends from Luce-bay to Loch-Ryan, and measures 6½ m. in length, and about 3 m. in breadth. A considerable extent of low and level ground, also at a comparatively late date submarine, lies along the lower Cree, and the upper part of Wigton-bay. The heights, as the county recedes northward, become bolder and of a more decidedly hilly character than near the coast; and in the vicinity, or along the line of the boundary with Ayrshire, they often rise to 1,000 ft., and occasionally to not very far from 1,500 ft. above sea level. The coast, except at the head of the three very large indentations of the sea, and in a few of the minor bays and creeks, is very generally—indeed almost continuously—bold and rocky; and though it aggregately extends to something upwards of 120 m., it contains comparatively few recesses where a large ship may safely ride at anchor, or attempt to land a cargo. The streams of W. are, for the most part, small and unsuited to topographical notice; yet, in a few instances, they afford admission, for a brief way, to light coasting-vessels, and are ornamental to a country of no great scenic beauty. The principal are the Cree and the Bladenoch, southward to Wigton-bay; the Tarf, a tributary of the Bladenoch; and the Luce southward; and Piltanton-burn south-eastward, to the head of Luce-bay. The lakes are very numerous; but are all small, and aggregately cover only 7½ sq. m.

Geology, Soil, and Agriculture.] The most prevalent rocks are primary, transition, and secondary schists. Greywacke, greywacke slate, and argillaceous schist, are particularly abundant. Hillocks and hills occur of till and gravel, evidently formed by water, and singularly contrasting in the round ness and smoothness of their surface, to the roughness of the rest of the country. Slate quarries have been worked in Kirkmaiden. Sandstone occurs in the N of the Rhinns. Limestone occurs in no great plenty, and is generally of too poor a quality to be worked for the uses of the farmer. The soil of nearly all the Machers and a considerable part of the Moors is of a hazel colour, and of the species sometimes termed dry loam, though it often inclines to gravel. The carse of Baldoon, which constitutes the larger part of the low ground along the Cree, and measures about 3,000 acres, has all the qualities of carse, or strong clay soil, and justly possesses the fame of being the richest and best cultivated land in W. The lands of the Rhinns, in a general view, possess all the characters of the best soils of Galloway, and are, to a large extent, arable. The central and northern sections of the district of the Moors present a pervading aspect of extreme barrenness. The land of the county lies, for the most part, under entail; and is little subdivided among proprietors. Farms are generally of medium size, and are usually let on leases of 19 years. The proportions of the whole area which are arable and pastoral are to each other respectively as 7 to 11; or the arable grounds amount to about 35 per cent. of the entire county. The average rent of land throughout the co. is at present little more than 6s. per acre; but, in 1810, it amounted to 8s. 6d. W. is celebrated, in common with the eastern and larger section of Galloway, for the excellence of its pastures, the good points of its peculiar breed of black cattle, and the largeness of its produce in oxen and sheep for the markets of England. It is an agricultural and a grazing co., to the almost total exclusion both of manufactures and of all such commerce as is unconnected with the exchange of the produce of the soil for foreign articles of local consumpt. The only exports consist of grain, wool, sheep, and black cattle, which are sent to England and the west of Scotland; the imports are chiefly lime, coals, and all sorts of groceries, and manufactured goods from the markets to which the exports are taken, and occasionally timber from America, and timber and iron from the Baltic.

Towns, &c.] The royal burghs are Wigton, Stranraer, and Whithorn; and they, jointly with New Galloway in Kirkcudbrightshire, send a member to parliament. The existing burghs-of-barony are Portpatrick, Glenluce, and Newton-Stewart. The extinct burghs of barony are Invermessan, on the shore of Loch-Ryan; Myreton, in the parish of Penninghame; and Merton, in the parish of Mochrum. The principal villages, additional to the existing burghs-of-barony, are Garlieston, Isle-of-Whithorn, Port-Logan, and Stewarton. Wigtonshire sends a member to parliament. Constituency in 1840, 1,038; in 1851, 1,272. Pop. in 1801, 22,918; in 1811, 26,891; in 1821, 33,240; in 1831, 36,258; in 1841, 39,179; in 1851, 43,389.

History.] In ancient times, and even for ages after the abdication of the Roman government in Britain, the powerful tribe of the *Novantes* inhabited the territory which now forms W. and the western half of Kirkcudbrightshire. They had *Leucophibia* on the site of the present Whithorn for their chief town, and possessed also another town, called *Rerigonium*, on the *Rerigonius Sinus*, the modern Loch-Ryan. The civil history of the co. will be found in our article GALLOWAY.

WIGUM, a village of Prussia, in the regency of Munster and circle of Ahaus. Pop. 260.

WIHERIES, a department and commune of Belgium, in the prov. of Hainault, arrond. and 12 m. SW of Mons. Pop. 1,548.

WIHLAW, a village of Austria, in Bohemia, in the reg. and bail. of Budweis. Pop. 210.

WIHLEN, a village of Baden, in the circle of the Upper Rhine, and bail. of Lorrach. Pop. 1,105.

WIHOGNE, a department and commune of Belgium, in the prov. and arrond. of Liege. Pop. 210.

WIHORAN, a village of Austria, in Bohemia, in the reg. of Pilsen, and bail. of Neugedein. Pop. 590.

WIHR-AU-VAL, a town of France, in the dep. of the Haut-Rhin, and cant. of Wintzenheim, 8 m. WSW of Colmar. Pop. 822.

WIHURGHULLY, a village of New South Wales, in the co. of Durham, at the confluence of Myall creek with Williams river, and 158 m. from Sydney.—Also a hill of the same co. near Tatsit creek.

WIJUHEN, a village of Holland, in the prov. of Gelderland, 6 m. SW of Nijmegen. Pop. 1,000.

WIK, a village of Holstein, in the bail. of Kronshagen. Pop. 610.

WIKANAMSH, a village of Vancouver's island, on a bay of the W coast, to the S of Nootka.

WIKAU, a village of Austria, in Bohemia, in the reg. of Pardubitz, and bail. of Bohimsch-Brod.

WIKE, a township in the p. of Birstall, W. R. of Yorkshire, 3½ m. NW of Bradford. Area 920 acres. Pop. in 1831, 1,918; in 1851, 2,916.—Also a township in the p. of Harewood, Yorkshire, 6½ m. NNE of Leeds. Area 880 acres. Pop. in 1851, 131.

WIKEORNST, a village of Belgium, prov. of Antwerp and arrond. of Mecheln. Pop. 1,021.

WIKISHKEN, a village of Prussia, in the reg. of Gumbinnen, and circle of Darkehmen. Pop. 210.

WIKLANTITZ, a village of Austria, in Bohemia, in the regency of Eger, and bail. of Patzau. Pop. 210. It has a castle and a potash-factory.

WIKLEK, a village of Austria, in Bohemia, in the reg. of Gitschin, and bail of Elumetz. Pop. 500.

WIKOLINE, a village of Prussia, in the regency of Breslau, and circle of Guhrau. Pop. 240.

WILANTITZ, a village of Bohemia, reg. of Gitschin, and bail. of Konigenhof. Pop. 640.

WILATKOWO, a village of Prussia, in the reg. of Bromberg and circle of Gnesen. Pop. 370.

WILATOWO, or WILLATOWO, a town of Prussia, in the regency of Bromberg, circle and 35 m. S of Mogillno, between two lakes. Pop. 430.

WILAWITAL, a creek of New South Wales, in the co. of Gloucester, an affluent of Port Stephens.

WILBARSTON, a parish in Northamptonshire, 6 m. E of Market-Harborough. Area 2,800 acres. Pop. in 1831, 681; in 1851, 754.

WILBERFORCE, a town of Upper Guinea, on the Sierre Leone coast, 2 m. SW of Freetown.—Also a colony of New South Wales, in the co. of Cook, on the left bank of the Hawkesbury, opposite Windsor, and 33 m. NW of Sydney. This colony is also called Gabramatta.—Also a headland on the N coast of New South Wales, to the NW of Melville bay, in S lat. 11° 53′, and E long. 136° 39′.

WILBERFOSS, a parish in the E. R. of Yorkshire, containing the townships of Newton-upon-Derwent and W. Area 2,870 acres. Pop. 367.

WILBRAHAM, a township of Hampden co., in the state of Massachusetts, U. S., 72 m. W by S of Boston, drained by the head branches of Scantic river, and other affluents of the Connecticut, and intersected by the Western railway. Pop. of township in 1840, 1,864; in 1850, 2,127.

WILBRAHAM, or WILBURGHAM (GREAT,) a parish in Cambridgeshire, 6 m. E of Cambridge. Area 1,800 acres. Pop. in 1831, 510; in 1851, 644.

WILBRAHAM (LITTLE), a parish in Cambridgeshire, 7 m. E of Cambridge. Area 1,300 acres. Pop. in 1831, 315; in 1851, 397.

WILBRENNING, a village of Prussia, in the regency and circle of Munster. Pop. 270.

WILBRINGHOVEN, a village of Prussia, in the regency of Cologne, and circle of Waldbrol. Pop. 280.

WILBURGSLATTEN, a village of Bavaria, in the circle of Middle Franconia, and presidial of Dinfelsbuhl. Pop. 390.

WILBURTON, a parish in Cambridgeshire, 6½ m. SW of Ely. Area 2,233 acres. Pop. 562.

WILBY, a parish in Norfolk. Area 1,400 acres Pop. in 1831, 123; in 1851, 106.—Also a parish in Northamptonshire, 2½ m. SW by W of Wellingborough. Area 1,120 acres. Pop. in 1831, 386; in 1851, 468.—Also a parish in Suffolk, 6 m. SE of Eye. Area 1,844 acres. Pop. in 1851, 620.

WILCHENREUTH, a village of Bavaria, in the circle of the Upper Pfalz, and presidial of Weiden. Pop. 200.

WILCHINGEN, a village of Switzerland, in the cant. of Schaffhausen, and bail. of Unterklettau. Pop. 1,345.

WILCHWA, a village of Prussia, in the regency of Oppeln, and circle of Ribnitz. Pop. 330.

WILCOT, a parish in Wilts, 2 m. WNW of Pewsey, containing the chapelry of Draycott-Foliatt, and the tythings of Pare and Stowel. Area 2,668 acres. Pop. in 1831, 677; in 1851, 702.

WILCOTE, a parish in Oxfordshire, 4 m. N by E of Witney. Area 270 acres. Pop. in 1851, 10.

WILCOX, a county in the SW part of the state of Alabama, U. S., comprising an area, generally level and productive, of 906 sq. m., drained by the Alabama and its branches. Pop. in 1840, 15,278; in 1850, 17,352. Its cap. is Camden.

WILCZA, two villages of Prussia, in the regency of Posen, and circle of Pleschen. Pop. 300.

WILCZAK, a village of Prussia, in the regency and circle of Bromberg. Pop. 530.

WILCZE, a village of Prussia, in the regency and circle of Bromberg. Pop. 350.

WILCZYN, a town of Russia in Europe, in Poland, in the gov. of Warsaw, and obwod. of Konin. Pop. 400.

WILDA (OBER and UNTER), a village of Prussia, in the regency and circle of Posen. Pop. 420. It has an hospital.

WILD-ALPEN, a range of mountains on the confines of Styria and of the archduchy of Austria, to the NNE of Bruck, in the chain of the Noric Alps. They contain marble-quarries.

WILDAU, a village of Prussia, in the regency of Potsdam, and circle of Jüterbogk. Pop. 200.

WILDBACH, a village of Saxony, in the circle of Zwickau, and bail. of Stein, near Schneeberg. Pop. 502.

WILDBAD, a town of Würtemberg, in the Schwarzwald circle, 20 m. WSW of Stuttgard. Pop. 1,800. It is situated in a sombre valley of the Black forest, enclosed by steep pine-covered mountains, at an alt. of 1,335 ft. above sea-level. It is celebrated for its hot springs and baths, of a temp. ranging from 82° to 100°. Their chemical composition is the same, the amount of saline and gaseous substance being exceedingly small, and the solid residuum after an analysis of a pint of the water scarcely exceeding three grains: so that the medicinal properties of the water may be considered as principally depending upon its temp., in which it differs from most of the other German waters, which require to be either warmed or cooled down to the proper degree of heat for bathing. The water rises through a layer of sand, to a height of from 2 to 3 ft., into basins in which several persons bathe at the same time. The rapid rivulet Ens flows behind the town; and the only carriage-road through the valley is constructed on its banks, which, beyond the baths, are planted with trees for a mile, forming the public promenade. There is in fact, notwithstanding the florid descriptions of W. indulged in by some writers, little either of beauty or romance in the environs of this spa, and its poor looking village. Wetzler states, that during the greater part of the year the climate is raw and cold; snow not unfrequently remaining on the neighbouring mountains till the middle of May: whilst from the deficiency of free ventilation, to which narrow valleys between wood-covered mountains are subject, the atmosphere must be considerably charged with moisture for some time after the falling of rain or heavy dew.

WILDBAD-GASTEIN, a village of Austria, in the regency of Linz, and circle of Salzburg, in the valley of Gastein, 4 m. from Hof. It is noted for its thermal springs and baths, and has an imperial castle formerly belonging to the princes of Salzburg. An aqueduct constructed in 1829, conduct the mineral waters to Hof.

WILDBAHN, a village of Prussia, in the reg. of Breslau and circle of Militsch. Pop. 250.

WILDBANK, a village of Prussia, in the reg. of Aachen and circle of Gupen. Pop. 210.

WILDBERG, a town of Würtemberg, in the circle of the Schwarzwald, bail. and 6 m. N of Nagold, on the river of that name. Pop. 1,787. It has manufactories of woollen fabrics, and several paper-mills.—Also a village of Prussia, in the regency of Potsdam, and circle of Kuppin. Pop. 647.—Also a village in the regency of Stettin, and circle of Demmin. Pop. 268.—Also a village of Saxony, in the circle and bail. of Dresden. Pop. 212.—Also a village of Switzerland, in the cant. of Zurich, and bail. of Pfaffikon, on the Toss. Pop. 990.

WILDBOAR-CLOUGH, a township in the p. of Prestbury, co.-palatine of Chester, 6½ m. SE of Macclesfield. Area 4,980 acres. Pop. 447.

WILDDEPENRIED, a village of Bavaria, in the circle of the Upper Pfalz, and presidial of Oberbiechtach. Pop. 310.

WILDEK, a hamlet of Hesse-Cassel, in the prov. of Lower Hesse, circle and 15 m. E of Rotenburg, on the Suhla. It has an ancient castle, formerly the residence of the landgraves of Hesse-Rotenburg.

WILDEMANN, a town of Hanover, in the territory and 3 m. NW of Klausthal, on the Innereste. Pop. 1,300. It is built in the form of a horse-shoe.

WILDEN, a parish in Bedfordshire, 5½ m. NE by N of Bedford. Area 2,160 acres. Pop. 483.

WILDEN, a village of Prussia, in the regency of Arnsberg, and circle of Siegen. Pop. 374.

WILDENAU, a village of Bavaria, in the circle of the Upper Pfalz, and presidial of Terschenreuth. Pop. 386.—Also a village of Austria, in the district above the Ens, and circle of the Inn. Pop. 280.—Also a village of Prussia, in the regency of Merseburg, and circle of Schweiritz. Pop. 260.—Also a village of Saxony, in the circle of Zwickau, and bail. of Grunham, near Schwarzenberg. Pop. 300.—Also a village in the bail. of Plauen, near Auerbach. Pop. 560.

WILDENBERG, a village of Bavaria, in the circle of Lower Bavaria, and presidial of Rottenburg. Pop. 200. It has a castle.

WILDENBRUCH, a village of Prussia, in the regency of Stettin, and circle of Griefenbergen. Pop. 815.—Also a village of the regency of Potsdam, and circle of Zauch-Belzig. Pop. 210.

WILDENFELS, a seignory and town of Saxony, in the circle and bail. of Zwickau, and 17 m. SW of Chemnitz. Pop. in 1837, 2,202; in 1843, 2,374; in 1849, 2,723. It has a castle belonging to the counts of Solm, and possesses manufactories of linen and hosiery.

WILDENHAG, a village of Austria, in the district above the Ens, circle of Hausruck, and bail. of Rogel. Pop. 260.

WILDENHAGEN, a village of Prussia, in the circle of Frankfurt, and circle of Sternberg. Pop. 220.—Also a village in the regency of Stettin, and circle of Kammin. Pop. 200.

WILDENHAIN, a village of Saxony, in the circle of Dresden, and bail. of Grossenhain. Pop. 441.—Also a village of Prussia, in the regency of Merseburg, and circle of Torgau, near Gelenburg. Pop. 450.

WILDENHOLZ, a village of Bavaria, in the circle of Middle Franconia, and presidial of Feuchtwang. Pop. 473.

WILDENHORTEN, a village of the duchy of Saxe-Altenburg, and bail. of Lumpzig. Pop. 200.

WILDENOW, a village of Prussia, in the reg. of Frankfurt, and circle of Friedeberg. Pop. 260.

WILDENRATH, a village of Prussia, in the reg. of Aachen, and circle of Heinsberg. Pop. 280.

WILDENREUTH, a village of Bavaria, in the circle of the Upper Pfalz, and presidial of Neustadt. Pop. 445. It has a castle.

WILDENROTH, a village of Bavaria, in the circle of Upper Bavaria, and presidial of Bruck. Pop. 240.

WILDENSCHWERT, or Austjnad-Orlicy, a town of Austria, in Bohemia, in the circle and 29 m. E of Chrudim, on the Adler, at the confluence of the Trebowka and Stillen-Adler. Pop. 3,500. It has manufactories of varieties of linen and woollen fabrics, and of hosiery.

WILDENSEE, a village of Bavaria, in the circle of Lower Franconia, and presidial of Klingenberg. Pop. 200.

WILDENSFORG, a village of Bavaria, circle of Upper Franconia, presidial of Banberg. Pop. 290.

WILDENSPRING, a village of Schwarzburg-Rudolstadt. Pop. 460.

WILDENSTEIN, a village of Würtemberg, in the circle of the Jaxt, and bail. of Crailsheim. Pop. 564.—Also an ancient fortress of Baden, in the circle of the See, bail. and 8 m. NW of Möskirch, on a rocky island of the Danube.

WILDENTHAL, a village of Saxony, in the circle of Zwickau, and bail. of Eibenstock. Pop. 469.

WILDENTHIERBACH, a village of Würtemberg, in the circle of the Jaxt, and bail. of Gerabronn. Pop. 236. It has a castle.

WILDEREN, a department and commune of Belgium, in the dep. of Limburg, and arrond. of Hasselt. Pop. 163.

WALDERMIEMINGEN, a village of Austria, in the Tyrol, in the circle of the Inn, and bail. of Silz. Pop. 420.

WILDERSWEIL, a village of Switzerland, in the cant. of Berne, and bail. of Interlaken. Pop. 1,145.

WILDERVANK, a village of Holland, in the prov. of Groningen, arrond. and 11 m. SW of Winschoten. Pop. 2,800.

WILDESHAUSEN, a parish-town of the duchy and 18 m. SSE of Oldenburg, and circle of Delmenhorst, on the l. bank of the Hunte. Pop. 2,100. It has two churches, a Catholic and a Lutheran, a poor-house, and an asylum for the deaf and dumb, and contains manufactories of hats and pottery, and several tanneries and breweries.

WILDE TAUBE, a village of Reuss-Greiz, and bail. of Dolau. Pop. 330.

WILDFLEEKEN, a village of Bavaria, in the

circle of Lower Franconia, and presidial of Bischofsheim. Pop. 490.

WILDGRUB, a village of Austria, in Silesia, in the circle of Troppau. Pop. 1,200.

WILDHAUS, a village of Switzerland, in the cant. and 15 m. S of St. Gall, district of Ober-Toggenburg, and circle of Alt-St.-Johann. Pop. 1,163. It is noted as the birthplace of Ulrich-Zwingli.

WILDON, a market-town of Austria, in Styria, in the circle and 15 m. SSE of Gratz, on the r. bank of the Muhr. Pop. 600. It has an hospital, and contains the ruins of an hospital.

WILDPRECHTSRODE, a village of Saxe-Meiningen, in the bail. of Salzungen. Pop. 220.

WILDRA, a creek of New South Wales, which flows between the counties of Bligh and Phillip, and discharges itself into Cudgegong river.

WILDSACHSEN, a village of Nassau, in the bail. of Hochheim. Pop. 200.

WILDSCHITZ, a village of Austria, in Bohemia, in the reg. of Pilsen, bail. of Rokitzan. Pop. 380.

WILDSCHONAU, a village of Austria, in the Tyrol, in the circle of Schwartz, and bail. of Rattenberg. Pop. 2,120.

WILDSCHUTZ, a village of Austria, in Bohemia, in the regency of Gitschin, and bail. of Trantenau. Pop. 1,300. It has a castle and a paper-mill.—Also a market-town of Silesia, in the circle of Troppau, and valley of Buchberg. Pop. 1,400. It has an hospital.—Also a village of Prussia, in the regency of Merseburg, and circle of Torgau. Pop. 220.—Also a village in the regency of Breslau, and circle of Dels. Pop. 380. It has a castle.

WILDSWORTH, a hamlet in the p. of Laughtone, Lincolnshire, 7½ m. N of Gainsborough. Area 750 acres. Pop. in 1831, 132; in 1851, 144.

WILDSTEIG, a village of Bavaria, in the circle of Upper Bavaria, presidial of Schongau. Pop. 600.

WILDSTEIN, a village of Bavaria, in the circle of the Upper Pfalz, and presidial of Oberviechtach. Pop. 200.—Also a town of Austria, in Bohemia, in the regency and bail. of Eger, and circle of Elnbogen. Pop. 2,500. It has 2 castles, and possesses manufactories of woollen and linen fabrics.

WILDTHAL, a village of Baden, in the circle of the Upper Rhine, near Freiburg. Pop. 428.

WILDUNGEN (Alt and Neu), two towns of the principality of Waldeck, in the bail. of Eger, and near the river of that name, and containing respectively 450 and 1,725 inhabitants. The latter contains a gymnasium, a lyceum, an hospital, and a ducal castle.

WILENZ, a village of Austria, in Moravia, in the circle of Iglau. Pop. 280.

WILER, or Wyler, a village of Prussia, in the reg. of Arnsberg, and circle of Siegen. Pop. 474.

WILES, or Vanquilen (Cape), a headland of South Australia, in the S part of the peninsula of Eyre Land, and on the E side of Sleaford bay, in S lat. 34° 56′ 30″, and E long. 135° 38′.

WILEYKA. See Vileika.

WILFERDINGEN, a village of Baden, in the circle of the See, and bail. of Durlach. Pop. 824.

WILFERSDORF, a village of Austria, in Styria, in the circle of the Gratz. Pop. 440.

WILFLINGEN, a village of Prussia, in Hohenzollern, in the bail. of Hechingen. Pop. 560.—Also a village of Würtemberg, in the circle of the Danube, and bail. of Riedlingen. Pop. 334.

WILFLINGEN (Ober and Unter), two villages of Würtemberg, in the circle of the Jaxt, and bail. of Ellwangen, containing respectively 96 and 570 inhabitants.

WILFORD, or Wildford, a parish in Notts, 2 m. SW of Nottingham. Area 1,450 acres. Pop. 570.

WILGARTSWEISEN, a village of Bavaria, in the circle of the Pfalz, and cant. of Annweiler. Pop. 1,083.

WILGERSDORF, a village of Prussia, in the regency of Arnsberg, and circle of Siegen. Pop. 474.

WILHAMSTEAD, or Wilshampstead, a parish in Bedfordshire, 4 m. S by E of Bedford. Area 3,027 acres. Pop. in 1831, 753; in 1851, 923.

WILHELM, an island of the Polar sea, near the N extremity of Nova-Zembla.

WILHELMAN, a village of Austria, in Bohemia, in the regency of Pardubitz, and bail. of Deutschbrod. Pop. 400.

WILHELMINEN, a village of Prussia, in the regency of Koslin, and circle of Schlawe, near the Wipper. Pop. 270.

WILHELMINENORT, a village of Prussia, in the regency of Breslau, and circle of Dels. Pop. 484. It has a castle.

WILHELMINENSWALDE, a village of Prussia, in the reg. of Frankfurt, circle of Goldin. Pop. 200.

WILHELMSANE, a village of Prussia, in the reg. of Frankfort, and circle of Kustrin. Pop. 440.

WILHELMSBAD, or Wachenbuchen, a village of Hesse, in the prov., circle, presidial, and 1½ m. NW of Hanau, near a volcanic hill. Pop. 70.

WILHELMSBRUCH, a village of Prussia, in the reg. of Marienwerder, circle of Kulm. Pop 260.

WILHELMSBURG, a market-town of Austria, in the circle of the Upper Weinerwald, on the l. bank of the Trasen, at an alt. of 296 yds. above sea-level, 8 m. S of St. Polten. It has manufactories of fire-arms, money, and stone-vases.—Also an island of Hanover, in the Elbe, opposite Hamburg.—Also a village of Prussia, in the regency of Stettin, and circle of Uckermunde. Pop. 349.

WILHELMSDORF, a village of Bavaria, in the circle of Middle Franconia, and presidial of Erlbach, on the Zein, 20 m. NW of Nuremberg. Pop. 647. It has a castle, and contains manufactories of hosiery, and several breweries.—Also a village of Prussia, in the regency of Leignitz, and circle of Goldberg-Haynau. Pop. 662.—Also a village of the regency of Erfurt, and circle of the Zeigenrück. Pop. 275.—Also a village of Würtemberg, in the circle of the Danube, and bail. of Ravensburg. Pop. 259.

WILHELMSFELD, a village of Baden, in the circle of the Lower Rhine, and bail. of Heidelburg. Pop. 628.

WILHELMSGLUCK, a hamlet of Würtemberg, in the circle of the Jaxt, and bail. of Hale. Pop. 33. It has extensive salt-works.

WILHELMSHAUSEN, a village of Hesse, in the prov. of Lower Hesse, and circle of Hofgeismar, on the Fulda. Pop. 587.

WILHELMSHUTTE, a hamlet of the duchy of Brunswick, in the circle and 8 m. NNW of Seisen, on the Notte. Pop. 585. It has ironworks.

WILHELMSHULD, a village of Prussia, in the reg. of Danzig, and circle of Karthaus. Pop. 270.

WILHELMSTEIN, a small fortress of the principality of Lippe-Schauenburg, and bail. of Hagenburg, on an island of Lake Steinhude, and 20 m. NE of Bückeburg.

WILHELMSTHAL, or Neuenbach, a village of Bavaria, in the circle of Upper Franconia, and presidial of Kronach. Pop. 528.

WILHELMSTHAL, or Neustadtel, a town of Prussia, in the regency of Breslau, circle and 11 m. ESE of Habelschwert, at the foot of the Schneeberg. Pop. 530. It has manufactories of musical instruments, and several mills.

WILHERING, a village of Austria, in the circle

of Hansruck, on the r. bank of the Danube, 5 m. W of Linz. It is noted for its abbey.

WILHERMSDORF, a market-town of Bavaria, in the circle of Middle Franconia, and presidial of Radolzburg. Pop. 1,302.

WILIA. See VILIA.

WILIMOW, a village of Bohemia, in the regency of Pardubitz, and bail. of Czaslau. Pop. 1,100.

WILIMOWITZ, a village of Austria, in Bohemia, in the regency of Pardubitz, and bail. of Deutschbrod. Pop. 300.—Also a village in the bail. of Ruttenberg. Pop. 200.—Also a village of Moravia, circle of Iglau, bail. of Trebitsch. Pop. 600.

WILKA, a village of Prussia, in the regency of Liegnitz, and circle of Lauban. Pop. 200.

WILKASSEN, a village of Prussia, in the regency of Gumbinnen, and circle of Lötzen. Pop. 200.—Also a village of the circle of Oletsko. Pop. 240.

WILKAU, a village of Prussia, in the regency of Breslau, and circle of Namslau. Pop. 710. It has two castles.—Also a village in the circle of Neumarkt. Pop. 400. It has a castle.—Also a village of the circle of Schweidnitz. Pop. 490. It has a castle, and several mills.—Also a village of the regency of Liegnitz, and circle of Glogau. Pop. 520. It has a castle.—Also a village of the regency of Frankfurt, and circle of Zullichau. Pop. 290.—Also a village of Saxony, in the circle and bail. of Zwickau. Pop. 245.

WILKAU (GROSS), a village of Prussia, in the regency of Breslau, and circle of Nimptsch. Pop. 440. It has a castle and two mills.

WILKAU (OBER), a village of Prussia, regency of Oppeln, and circle of Neustadt. Pop. 290.

WILKAWE, a village of Prussia, in the regency of Breslau, and circle of Trebnitz. Pop. 310.

WILKE, a village of Prussia, in the regency of Posen, and circle of Fraustadt. Pop. 620.—Also a village in the circle of Koslin. Pop. 440.

WILKEIM, a village of Prussia, in the regency and circle of Konigsberg. Pop. 210.

WILKENBERG, a village of Prussia, in the regency of Arnsberg, and circle of Olpe. Pop. 230.

WILKENDORF, a village of Prussia, in the reg of Konigsberg, and circle of Wehlau. Pop. 440.

WILKENDORF, or WILKOVO, a village of Prussia, in the regency of Konigsberg, and circle of Kastenburg. Pop. 390.

WILKERSDORF, a village of Prussia, in the reg. of Frankfurt, and circle of Kustrin. Pop. 500.

WILKES, a county in the NE part of the state of Georgia, U. S., comprising an area generally level of 486 sq. m., drained by Fishing creek and its affluents, and by branches of Broad and Little rivers. Pop. in 1840, 10,148; in 1850, 12,107. Its capital is Washington.—Also a county in the NW of the state of North Carolina, comprising an area finely diversified of 570 sq. m., bounded on the NW by Blue Ridge, and drained by Gadkin river and its branches. Pop. in 1840, 12,577; in 1850, 12,099. Its capital is Wilkesboro.

WILKESBARRE, a township of Luzerne co., in the state of Pennsylvania, U. S., on the SE bank of Susquehanna river, and 82 m. NE by N of Harrisburg. Pop. in 1840, 1,718; in 1850, 2,723.

WILKESBORO, a village of Wilkes co., in the state of North Carolina, U. S., on the S side of Gadkin river. Pop. in 1850, 180.

WILKESVILLE, a township of Vinton co., in the state of Ohio, U. S., 68 m. SE by S of Columbus, drained by Racoon creek, and intersected by the Hillsboro and Parkersbury railroad. Pop. in 1840, 738; in 1850, 1,037.

WILKINS, a township of Alleghany co., in the state of Pennsylvania, U. S., on the Pennsylvania railroad, 7 m. from Pittsburg. Pop. in 1840, 2,165; in 1850, 2,500.

WILKINSON, a central county of the state of Georgia, U. S., comprising an area, finely undulating and generally fertile, of 388 sq. m., drained by Big-Sandy and Commissioner creeks, and affluents of Oconee river, by which it is bounded on the E, and intersected by the Georgia Central, and the Milledgeville and Gordon railroad. Pop. in 1840, 6,842; in 1850, 8,212. Its capital is Irwinton.—Also a county in the SW part of the state of Mississippi, bounded on the W by the Mississippi, and intersected by Buffalo creek. Pop. in 1840, 14,193; in 1850, 16,914. Its capital is Woodville.

WILKINSTOWN, a village in the p. of Kilberry, co. Meath, 5 m. N by W of Navan.

WILKISCHEN, a village of Bohemia, in the reg. and bail. of Pilsen. Pop. 350. It has a castle.

WILKITEN, a village of Prussia, in the regency of Konigsberg, and circle of Memel. Pop. 840.

WILKOMIERZ, a district and town of Russia in Europe, in the gov. and 60 m. from Wilna, on the Iventa. Pop. 5,000.

WILKOWITZ (GROSS and KLEIN), two villages of Prussia, in the regency of Oppeln and circle of Beuthen. Pop. 400.

WILKOWYSZKI, a town of Poland, in the gov. of Augustowo, obwod and 27 m. NNW of Kalwarya, Pop. 3,511, of whom $\frac{5}{8}$ are Jews, and the remainder Christians.

WILKSBY, a parish in Lincolnshire, 5 m. SE of Horncastle. Area 670 acres. Pop. in 1851, 66.

WILL, a county in the NW part of the state of Illinois, U. S., comprising an area of 1,152 sq. m., drained by Iroquois and the Kankakee, affluents of the Illinois, and intersected by the Chicago branch of the Central railroad, the Chicago and Mississippi railroad, and by the Illinois and Michigan canal.

WILLAMETTE, a river of Oregon, which rises at the foot of the Cascade Range in California, in N lat. 43°, and flows in a N course of about 200 m. to the Columbia, which it joins about 8 m. below Fort Vancouver, in N lat. 46° 30′, W long. 122° 30′. It winds through fertile prairies now and then skirted by forests of fir, pine, and cedar, and receives many considerable tributaries, all of which have traversed their whole length through a fine country interspersed with groves of oak. About 25 m. from its confluence with the Columbia, the W. is interrupted by a series of falls. On arriving near these, the river spreads out into a wide, deep basin, and runs slowly and smoothly until within $\frac{1}{2}$ a mile of the falls, when its velocity increases, its width diminishes, eddies are formed in which the water turns back as if loth to make the plunge but is forced forward by the water in the rear, and when still nearer, it breaks upon the volcanic rocks scattered across the channel, and then precipitates itself over an almost perpendicular ledge of 25 ft., presenting a somewhat whitened column. Below the falls, for the distance of $\frac{1}{2}$ a m., the channel of the river is confined by ledges of basaltic rocks to about 200 yds. in width. About 1 m. below the falls, the Clackamas river enters from the E, which takes its rise near Mount-Hood, one of the snow peaks of the Cascade mountains. Here the Clackamas rapids form another obstacle in the navigation of the W.; but the difficulties here presented are of minor importance — at almost any stage of water, the boats of the Hudson-bay company ascend and descend with but little difficulty. From this place to the mouth of the W., a distance of 25 m., navigation is good, as the tide runs up to the foot of the Clackamas rapids. Oregon city—as it is called—has been founded on the E side of the Willamette, at the falls, on a piece of land which a few years ago was a dense forest of fir and underbrush. The city is now incorporated with a pop. of above 500.

WILLAND, a parish in Devon, 2½ m. NE of Collumpton. Area 989 acres. Pop. in 1851, 398.

WILLANMEZ, a group of islands of the South Pacific, in the archipelago of New Britain, and to the N of the island of that name. The most northerly of the group, also called Willanmez, is in S lat. 5°, and E long. 150° 10′. It is mountainous and indented by deep bays.

WILLANPUIS, a department and commune of Belgium, in the prov. of Hainault, and arrond. of Tournai. Pop. 581.

WILLANZHEIM, a village of Bavaria, circle of Middle Franconia, presidial of Bobart. Pop. 732.

WILLASTON, a township in the p. of Wybunbury, co.-palatine of Chester, 1½ m. E by N of Nantwich. Area 967 acres. Pop. in 1851, 147.—Also a township in the p. of Nestor, co.-palatine of Chester, 3 m. E of Great Nestor. Area 1,941 acres. Pop. in 1851, 317.

WILLATOWO, or Wilatowo, a town of Prussia, in the regency of Bromberg, circle and 7 m. S of Mogilno. Pop. in 1843, 510.

WILLBICH, a village of Prussia, in the reg. of Erfurt, and circle of Heiligenstadt. Pop. 513.

WILLEBADESSEN, a town of Prussia, in the regency of Minden, and circle of Warburg, on the Nette. Pop. 1,315. It is enclosed by walls, now much dilapidated, and entered by two gates.

WILLEBRINGEN, a department and commune of Belgium, in the prov. of Brabant, and arrond. of Louvain. Pop. of dep., 607; of com., 411.

WILLEBROECK, a commune and town of Belgium, in the prov. and 13 m. S of Antwerp, and arrond. of Malines, on the Rupel at the junction of the Brussels canal. It has building docks, distilleries of gin, vinegar manufactories, breweries, and carries on an active trade in butter, chicory, and grain.

WILLEM, a river of Australia, in the SW part of De Witt Land, which throws itself into the Indian ocean, near the Cape, and to the SW of a bay of the same name. It has a narrow mouth, encumbered with rocks.—Cape Willem or Murat is in S lat. 21° 37′, and E long. 60° 20′.

WILLEM, Guillaume, or Exmouth, a bay of the W coast of Australia, in De Witt Land, in S lat. 22°, enclosed on the W by Cape North-West or Vlaming.

WILLEMEAU, a department and commune of Belgium, in the prov. of Hainault, and arrond. of Tournai. Pop. 462.

WILLEMS, a commune of France, in the dep. of the Nord, and cant. of Lannoy, 10 m. E of Lille. Pop. 1,735.

WILLEMSTAD, a town of Holland, in the prov. of North Brabant, arrond. and 18 m. WNW of Breda, on the l. bank of the Maas, called the Hollands-Diep, in N lat. 41° 33′. Pop. 1,980. It is fortified, and has a harbour. This town was founded in 1583, by the Prince of Orange.

WILLEMSTAD, a town of the Dutch Antilles, the capital of and on the S coast of the island of Curaçao, in N lat. 12° 7′, and W long. 69°. Pop. 8,000. It is large, handsome, and well-built, and the port is one of the safest of the New world. The entrance is narrow, but within are two fine basins, perfectly sheltered, and capable of containing vessels of the largest size. This town is the entrepot of an active trade.

WILLEN, or Wyllien, a parish in Bucks, 1½ m. S of Newport-Pagnell. Area 450 acres. Pop. 98.

WILLENBERG, a town of Prussia, in the reg. and 90 m. SSE of Konigsberg. Pop. 2,000.

WILLENHALL, a chapelry in the p. of Wolverhampton, Staffordshire, 3 m. W of Walsall. Area, with Wednesfield township, 5,580 acres. Pop. of chapelry in 1831, 5,834; in 1851, 11,931. Coal and ironstone are raised here to a considerable extent, and sent along the Wyrley and Easington canal. The village—which is midway between Wolverhampton and Walsall—was in a flourishing condition in the reign of Queen Elizabeth. Its iron manufactures, especially that of locks, bitts, and keys, have now arrived at a remarkable degree of perfection.—Also a hamlet in the parish of the Holy Trinity, Coventry, 2½ m. SE by E of Coventry. Area 440 acres. Pop. in 1831, 120; in 1851, 107.

WILLERBY, a parish in the E. R. of Yorkshire, 6 m. S by W. of Scarborough, containing the townships of Binnington, Staxton, and W. Area 4,180 acres. Pop. in 1831, 356; in 1851, 422.

WILLERSEY, a parish in Gloucestershire, 3 m. W of Chipping-Camden. Area 1,344 acres. Pop. in 1831, 327; in 1851, 372.

WILLERSLEY, a parish in Herefordshire, 7 m. ESE of Hay. Area 230 acres. Pop. in 1851, 5.

WILLESBOROUGH, a parish in Kent, 2 m. SE by E of Ashford. Area 1,457 acres. Pop. in 1841, 641; in 1851, 1,022.

WILLESDEN, or Wilsdon, a parish in Middlesex, 8 m. NW by W of St. Paul's. Area 4,190 acres. Pop. in 1831, 1,876; in 1851, 2,939.

WILLESLEY, a parish of Derby, locally within Leicestershire, 2½ m. SW of Ashby-de-la-Zouch. Area 910 acres. Pop. in 1831, 63; in 1851, 48.

WILLEY, a parish in Salop, 5 m. NW by N of Bridgenorth. Area 1,330 acres. Pop. in 1831, 159; in 1851, 144.—Also a parish in Warwickshire, 7 m. N by W of Rugby. Area 970 acres. Pop. 127.

WILLIAM (Fort). See Fort-William.

WILLIAM (Mount), a mountain in South Shetland, in S lat. 64° 45′, W long. 63° 51′.—Also a mountain of Victoria, in Australia, in S lat. 37° 15′, E long. 142° 25′, the loftiest summit in the Grampian range.

WILLIAM (Point), a cape on the W coast of North America, forming the S point of the entrance into Bellingham's bay, in N lat. 48° 37′.

WILLIAMS, a county in the NW of Ohio, U. S. Area 432 sq. m. Pop. 8,108. Its cap. is Bryan.

WILLIAM'S ISLAND, an island of Australia, in Spencer's gulf, in S lat. 35° 6′, E long. 136°.

WILLIAM'S RIVER, a river of Vermont, U. S., which rises in Chester, and runs into the Connecticut, 3 m. N of Bellows-falls.

WILLIAMSBOROUGH, a township of Granville co., North Carolina, on a creek which falls into the Roanoke, 42 m. N by E of Raleigh.

WILLIAMSBURG, a township of Hampshire co., Massachusetts, U. S., 90 m. W of Boston. Pop. 1,500.—Also a township of Clermont co., Ohio, 82 m. SW of Columbus. Pop. 1,900.—Also a village of James co., Virginia, 46 m. ESE of Richmond. It is situated between two rivulets, one of which flows into York, and the other into James river. Pop. 877. The college of William and Mary, founded here in 1691, has 7 professors.—Also a district in the NE part of S. Carolina. Area 1,087 sq. m. Pop. in 1850, 12,477. Its cap. is Kingston.—Also a village of Jackson co., Tennessee, on the N side of the Cumberland, about 15 m. ENE of Carthage.—Also a township of Dauphin co., Pennsylvania, at the junction of the Swatara and Little Swatara, 23 m. NNE of Harrisburg.—Also the cap. of King's co., in New York, on the East river, 2 m. E by N of New York. Pop. in 1840, 5,094; in 1850, 30,786.

It has extensive ship-yards, and varied and extensive manufactures.

WILLIAMSFIELD, a township of Ashtabula co., Ohio, U. S., 158 m. NE of Columbus.

WILLIAMSON, a central county of the state of Illinois, U. S., comprising an area of 432 sq. m., and intersected by the Sangamon and Massac railway, drained by Muddy and Saline creeks. The surface is diversified, but generally fertile. Pop. in 1840, 4,457; in 1850, 7,216. The capital is Marion.—Also a central county of the state of Tennessee, comprising an area of 536 sq. m., drained by Harpeth river, and intersected by the Nashville and New Orleans railway. It has a finely undulating surface, but is generally fertile. Pop. in 1840, 27,006; in 1850; 27,201. Its capital is Franklin.—Also a central county of the state of Texas, comprising an area of 1,187 sq. m., drained by San Gabriel river and its N and S forks, and by Brushy creek. It is finely diversified, and generally productive. Pop. in 1850, 1,568. Its capital is Georgetown.—Also a township of Wayne co., in the state of New York, 182 m. W by N of Albany, drained by affluents of Lake Ontario.—Pop. in 1850, 2,380.

WILLIAMSPORT, a village of Warren co., in the state of Indiana, U. S., on the W bank of Wabash river, 69 m. NW by W of Indianapolis, and intersected by the Chesapeake and Ohio canal, and by the Franklin railroad. Pop. in 1840, 350; in 1850, 450.—Also a village of Lycoming co., in the state of Pennsylvania, on the W branch of Susquehanna river, 66 m. N by W of Harrisburg, and on the Williamsport and Elmira, and the Sunbury and Erie railroads. Pop. in 1840, 1,353; in 1850, 1,615.

WILLIAMSTON, a village of Martin co., in he state of North Carolina, U. S., 88 m. E of Raleigh. Pop. in 1840, 250; in 1850, 300.

WILLIAMSTON, a parish in Pembrokeshire, 4 m. NW of Tenby. Area 1,431 acres. Pop. 551.

WILLIAMSTOWN, a maritime town of Australia Felix, in the county of Bourke, on Point Gellibrand, a low promontory on the S side of Hobson's bay. Pop. 322. It is badly supplied with water.

WILLIAMSTOWN, a village of Upper Canada, in the township of Charlottenburg, on the river Aux Raisins. Pop. about 200.

WILLIAMSTOWN, a township of Orange co., in the state of Vermont, U. S., 10 m. S by E of Montpelier. Pop. in 1840, 1,020; in 1850, 1,452. —Also a township of Berkshire co., in the state of Massachusetts, 19 m. N of Pittsburg, drained by Hoosic river and its branches. Pop. in 1840, 2,153; in 1850, 2,626.—Also a township of Oswego co., in the state of New York, 124 m. WNW of Albany, and intersected by the Watertown and Rome railway. Pop. in 1840, 842; in 1850, 1,121.—Also a village of Grant co., in the state of Kentucky, 49 m. NNE of Kentucky, on the side of Eagle creek.

WILLIAMSTOWN, a village in the p. of Booterstown, co. Dublin, on the S shore of Dublin bay. Pop. in 1851, 1,416.—Also a village in the p. of Templetogher, co. Galway, 7 m. NE by N of Dunmore.

WILLIAMSVILLE, a village of Amherst township, Erie co., in the state of New York, U. S., on the NE side of Ellicott's creek.—Also a settlement of Upper Canada, on the N shore of Lake Erie, between Walpole and Rainham. Pop. about 30.

WILLIAN, a parish in Hertfordshire, 3 m. E of Hitchin. Area 1,854 acres. Pop. in 1851, 322.

WILLICHOW. See WIELICHOWO.

WILLICH, a village of Prussia, in the regency of Dusseldorf and circle of Krefeld. Pop. 828.

WILLIES, a village of France, in the dep. of the Nord, and cant. of Trelon, near the Grande-Helpe. Pop. 320.

WILLIMAN, a village of Austria, in Moravia, in the circle of Olmutz, and district of Chudwein. Pop. 660.

WILLIMANSETT, a village of Hampden co., in the state of Massachusetts, U. S., on the E bank of the Connecticut river, and intersected by the Connecticut railroad.

WILLIMANTIC, a township of Windham co., in the state of Connecticut, U. S., on the N side of a river of the same name, a head-stream of the Shetucket river, and intersected by the Hartford, Providence, and Fishkill, the New London, Willimantic, and Palmer, and the New York and Boston lines of railway.

WILLIMETSCH, a village of Austria, in Moravia, circle of Iglau and district of Reish. Pop. 260.

WILLIMOW, a market-town of Austrian Bohemia, reg. of Eger, and bail. of Saatz. Pop. 800.

WILLIMOWITZ, a village of Austria, in Moravia, in the circle of Brunn, and district of Raitz. Pop. 500.

WILLIMS, a village of Prussia, in the regency of Konigsberg, and circle of Rossel. Pop. 260.

WILLINBOROUGH, a township of Burlington co., in the state of New Jersey, U. S., 7 m. NW of Mount Holly, and bounded on the NW by the Delaware. Pop. in 1840, 900.

WILLING, a village of Bavaria, in the circle of Upper Bavaria, and presidial of Aibling. Pop. 241.

WILLINGALE-DOE, or WILLINGEHALE-DE-OU, a parish in Essex, 4¾ m. NE of Chipping-Ongar. Area 1,739 acres. Pop. in 1831, 466; in 1851, 518.

WILLINGALE-SPAIN, or WILLINGEHALE-DE-ISPANIA, a parish in Essex, 4½ m. NE of Chipping-Ongar. Area 1,200 acres. Pop. in 1851, 216.

WILLINGBUSEN, a village of Holstein, in the bail. of Reinbeck. Pop. 220.

WILLINGDON, a parish in Sussex, 2½ m. N by W of East Bourne. Area 4,259 acres. Pop. 678.

WILLINGEN, a village of Nassau, in the bail. of Marienburg. Pop. 821.—Also a village of Waldeck, and district of Eisenberg. Pop. 990.

WILLINGHAM, or WIVELINGHAM, a parish in Cambridgeshire, 6¼ m. E by S of St. Ives, and 10 m. NW of Cambridge, on the S bank of the Ouse. Area 4,663 acres. Pop. in 1831, 1,403; in 1851, 1,604. Also a parish in Lincolnshire, 6 m. SE of Gainsborough. Area 2,170 acres. Pop. in 1831, 392; in 1851, 499.—Also a parish in Suffolk, 4½ m. S of Beccles. Area 1,023 acres. Pop. in 1851, 160.

WILLINGHAM-CHERRY, a parish in Lincolnshire, on the N bank of the Witham, 3½ m. E by N of Lincoln. Area 980 acres. Pop. in 1851, 148.

WILLINGHAM (NORTH), a parish in Lincolnshire, 4 m. E by S of Market-Raisen. Area 1,990 acres. Pop. in 1831, 223; in 1851, 861.

WILLINGHAM (SOUTH), a parish in Lincolnshire, 5 m. ENE of Wragby. Area 1,900 acres. Pop. in 1831, 212; in 1851, 341.

WILLINGSHAIN, a village of Hesse, in the circle of Hersfeld. Pop. 352.

WILLINGSHAUSEN, a village of Hesse, in the prov. of Lower Hesse, circle and bail. of Ziegenhain. Pop. 663.

WILLINGTON, a parish in Bedfordshire, on the N bank of the Ouse, 4 m. E of Bedford. Area 1,648 acres. Pop. in 1831, 332; in 1851, 282.—Also a township in the p. of Whalley, co.-palatine of Chester, 3 m. NW of Tarporley. Pop. in 1841, 103.—Also a parish of Derbyshire, on both sides of the river Trent, and intersected by the Birmingham and Derby junction railway, which has a station-house here, 6¼ m. from Derby, and 4¾ m. from Burton. Area 1,270 acres. Pop. in 1831, 402; in 1851, 442. —Also a township in the p. of Brancepeth, union and co.-palatine of Durham, 4 m. N of Bishop-Auckland. Area 1,485 acres. Pop. in 1831, 216; in 1851, 965.—Also a township in the p. of Hanmer, co. of Flint, 5 m. NE of Ellesmere. Area 1,874 acres. Pop. in 1831, 388; in 1851, 358.—Also a township in the p. of Wallsend, Northumberland, on the Tyne, 3 m. W by N of North Shields. Coal is raised to a great extent within the township. Pop. 2,284.

WILLINGTON, a township of Tolland co., in the state of Connecticut, U. S., 21 m. E by N of Hartford, drained by branches of Willimantic river, and intersected by the New London, Willimantic, and Palmer railroad. Pop. in 1840, 1,268; in 1850, 1,399.—Also a village of Piscataquis co., in the state of Maine. Pop. in 1840, 722.

WILLIS, a parish of New South Wales, in the co. of Brisbane. See also WILIS (CAPE).

WILLISAU, a parish and town of Switzerland, cant. and 18 m. WNW of Lucerne, and bail. of the same name, at the foot of the Williberg, near the l. bank of the Wigger. Pop. of bail. 23,250; and of town, 2,859.

WILLISHAM, a parish in Suffolk, 3 m. SSW of Needham. Area 927 acres. Pop. in 1851, 213.

WILLISTON, a township of Chittenden co., in the state of Vermont, U. S., 26 m. WNW of Montpelier, on the S side of Onion river, and intersected by the Vermont central railroad. Pop. in 1850, 1,699.—Also a village of Barnwell district, in the state of S. Carolina, on the S. Carolina railroad.

WILLISTOWN, a township of Chester co., in the state of Pennsylvania, U. S., 20 m. NW of Philadelphia, drained by Ridley and Crum creeks. Pop. in 1840, 1,460.

WILLITSCH, a village of Austria, in Styria, in the bail. of Marburg. Pop. 210.

WILLITZ, a village of Austria, in Bohemia, in the reg. of Budweis, and bail. of Tabor. Pop. 220.

WILLKISCHKEN, a village of Prussia, in the reg. of Gumbinnen, and circle of Tilset. Pop. 428.

WILLKOWE, a village of Prussia, in the regency of Breslau, and circle of Militsch. Pop. 220.

WILLKOWEN, a village of Prussia, in the regency of Gumbinnen, circle of Angerburg. Pop. 210.

WILLKUHNEN, a village of Prussia, in the regency and circle of Konigsberg. Pop. 210.

WILLMANDINGEN, a village of Würtemberg, in the circle of the Schwarzwald, and bail. of Neutlingen, on the Alp. Pop. 747.

WILLMANSDORF, a village of Prussia, in the regency of Leignitz, and circle of Jauer. Pop. 240.

WILLMANSTRAND, a town of Finland, in the laen and 50 m. N of Viborg, on Lake Saima. Pop. 308. It suffered extensively from fire in 1741.

WILMARS, a village of Bavaria, circle of Lower Franconia, and presidial of Mellrichstadt. Pop. 691.

WILLMENDINGEN, a village of Baden, circle of the Upper Rhine, bail. of Waldshut. Pop. 270.

WILLMERSBACH, a village of Bavaria, in the circle of Middle Franconia, and presidial of Neustadt. Pop. 200.

WILLMETSHOFEN, a village of Bavaria, in the circle of Swabia, and presidial of Zusmaurshausen. Pop. 220.

WILLOFS, a village of Bavaria, in the circle of Swabia, and presidial of Obergünzburg. Pop. 220.

WILLOMITZ, a town of Austria, in Bohemia, in the regency of Eger, and bail. of Kaaden. Pop. 800.

WILLOUGHBY, a parish in Lincolnshire, 3½ m. SSE of Alford. Area 4,280 acres. Pop.—including the hamlet of Sloothby—in 1831, 557; in 1851, 748. —Also a parish in Warwickshire, 4½ m. NW of Daventry. Area 2,290 acres. Pop. in 1851, 373.

WILLOUGHBY, a parish of New South Wales, in the co. of Cumberland, between Middle Harbour on the NE, and Port Jackson on the E. Pop. 987.

WILLOUGHBY, a bay of the island of Antigua, Little Antilles, on the SW coast, between the parishes of St. Phillip and St. Paul, and opening between Hudson Point on the NE, and Sandy island on the SW.

WILLOUGHBY, a township of Upper Canada, in the Niagara district, bounded on the E by the Niagara, and on the N by the Welland river, by which it is separated from the township of Stamford. Pop. in 1841, 895.

WILLOUGHBY, a township of Lake co., in the state of Ohio, U. S., 133 m. NE by N of Columbus, bounded on the E by Chagrin river, and on the N by Lake Ontario, and intersected by the Cleveland and Erie canal. Pop. in 1840, 1,943; in 1850, 2,081.—Also a lake of Orleans co., in the state of Vermont, which discharges itself by a river of the same name into Barton river.

WILLOUGHBY (CAPE), a headland of Australia, on the E extremity of Kangaroo island, having a lighthouse elevated 241 ft. above sea-level, and in S lat. 35° 49′ 20″, E long. 138° 12′ 30″.

WILLOUGHBY (SCOTT), a parish in Lincolnshire, 4 m. NW of Folkingham. Area 556 acres. Pop. in 1831, 24; in 1851, 23.

WILLOUGHBY (SILK), a parish in Lincolnshire, 2½ m. SW of Sleaford. Area 2,505 acres. Pop. 256.

WILLOUGHBY-WATERLESS, or WATER-LEYS, a parish and village in Leicestershire, intersected by the Midland Counties railway, from the Broughton-Astley station, 5½ m. NNE of Lutterworth. Area 1,151 acres. Pop. in 1831, 327; in 1851, 761.

WILLOUGHBY-ON-THE-WOLDS, a parish in Notts, 7¾ m. NE of Loughborough. Area 2,080 acres. Pop. in 1831, 465; in 1851, 600.

WILLOUGHTON, a parish in Lincolnshire, 8½ m. E by N of Gainsborough. Area 2,460 acres. Pop. in 1831, 477; in 1851, 657.

WILLOW FORK, a township of Morgan co., in the state of Missouri, U. S. Pop. in 1840, 1,146.

WILLOW SPRINGS, a township of Lafayette co., in the state of Wisconsin, 45 m. SW by W of Madison, and watered by Pekatomca river. Pop. in 1850, 606.

WILLSBACH, a village of Würtemberg, in the circle of the Neckar, and bail. of Weinsberg, on the Sulm. Pop. 1,080.

WILLSBOROUGH, a township of Essex co., in the state of New York, U. S., on the W side of Lake Champlain, 118 m. N by E of Albany. Pop. 1,932.

WILLSHIRE, a township of Van-Wert co., in the state of Ohio, U. S., 93 m. WNW of Columbus, drained by St. Mary's river. Pop. in 1840, 434; in 1850, 1,053. It contains a village of the same name.

WILLSINGEN, a village of Würtemberg, in the circle of the Danube, bail. of Munsingen. Pop. 220.

WILLSTEDT, a market-town of the grand-duchy of Baden, in the circle of the Middle Rhine, bail. of Kork, 6 m. NNW of Offenburg. Pop. 1,449.

WILLUDDEN, a village of Prussia, in the regency of Gumbinnen, circle of Angerburg. Pop. 220.

WILLUHNEN, a village of Prussia, in the regency of Gumbinnen, circle of Pillkallen. Pop. 310.

WILLUNGA, a town of South Australia, at the W base of Mount Barker, and near the gulf of St. Vincent.

WILLWILLBROOK, a parish of Australia Felix, in the co. of Bourke, bounded on the W by the Monee Monee ponds.

WILLWITZ, a village of Prussia, in the regency of Breslau, and circle of Munsterburg. Pop. 220.

WILMARSDONE, a department and commune of Belgium, prov. and arrond. of Antwerp. Pop. 775.

WILMENROTH, a village of Nassau, in the bail. of Rennerod. Pop. 410.

WILMERSDORF, a village of Prussia, in the regency of Frankfurt, and circle of Lebus. Pop. 200.—Also a village in the regency of Potsdam, and circle of Teltow. Pop. 280.—Also a village of Schwarzburg-Sondershausen, and bail. of Gehren. Pop. 355.

WILMERSHAGEN, a village of Prussia, reg. of Stralsund, and circle of Grimmen. Pop. 200.

WILMINGTON, a parish of Kent, 1 m. S of Dartford. Area 1,715 acres. Pop. in 1831, 724; in 1851, 915.—Also a parish of Sussex, 4½ m. SW of Haylsham. Area 1,744 acres. Pop. in 1851, 288.

WILMINGTON, a township of Windham co., Vermont, U. S., 21 m. E of Bennington. Pop. 1,372. —Also a township of Middlesex co., Massachusetts, 16 m. N of Boston. Pop. 876.—Also a town and port of entry, in Newcastle co., Delaware, between the Brandywine and Christiana creeks, 1 m. above their confluence, and 37 m. N of Dover, in N lat. 39° 41′, W long. 75° 28′. It is built on gently rising ground, the most elevated part of which is 112 ft. above tide-water, and is regularly laid out, the streets intersecting each other at right angles. It contains a town-hall, several county-buildings, an arsenal, and 19 churches. Pop. in 1810, 4,416; in 1830, 6,628; in 1850, 13,979. The Christiana is navigable as far as W., for vessels drawing 14 ft. of water. The shipping owned here, in 1850, amounted to 9,460 tons. The trade of the town is considerable; its exports consist chiefly in flour and leather, and gunpowder. On the Brandywine, at a little distance from the body of the town, there are extensive flour-mills. The Brandywine and the Christiana, with their branches, afford great water-power for mills and manufactories. W. was severely injured on the 31st of May 1854, by the explosion of three waggons each loaded with 150 barrels of gunpowder, while passing through from Dupont's mills.—Also the cap. of Clinton co., Ohio, 54 m. SW of Columbus. Pop. 1,238.—Also a port of entry and the cap. of New Hanover co., N. Carolina, on the E side of Cape Fear river, situated immediately below the confluence of the NE and NW branches, about 35 m. from the sea, and 120 m. SSE of Raleigh, in N lat. 34° 11′, W long. 78° 10′. It is well situated for trade; but is accounted unhealthy from the lowness of its site. The total tonnage of the district was in 1850, 15,198 tons. The exports consist of lumber, turpentine, rosin, tar, pitch, peanuts, and cotton. The harbour admits vessels of 300 tons; but the entrance is rendered dangerous and difficult by a large shoal. Opposite the town are two islands, which afford the finest rice fields in the state. W. is connected with Weldon and with Manchester by railroads.—Also an island of the United States, near the coast of Georgia, at the mouth of the Savannah, in N lat. 32°.

WILMOT, a township of Merrimac co., New Hampshire, U. S., 29 m. NW of Concord. Pop. 1,272.—Also a township of Noble co., Indiana, 114 m. NNE of Indianapolis.

WILMSLOW, or WIMBOLDSHEY, a parish and village of Cheshire, 7 m. NW of Macclesfield. Area of p., 8,028 acres. Pop. in 1851, 4,952.

WILMSTORF (OLD), a village of Prussian Silesia, in the co. of Glatz. Pop. 900.

WILNA. See VILNA.

WILNA, a township of Jefferson co., New York, U. S., 134 m. NW of Albany.—Also a village of Houston co., in Georgia, 50 m. SW of Milledgeville.

WILNE (GREAT), a parish of Derbyshire, 7½ m. SE of Derby. Area 8,500 acres. Pop. in 1851, 2,096.

WILNECOTE, a hamlet in Tamworth p., Warwickshire, 2½ m. SE of Tamworth. Pop. in 1851, 824.

WILNIS, or WILLIS, a village of Holland, in the prov. of Utrecht, 7 m. N of Woerden. Pop. 1,041.

WILPSHIRE, a township of Blackburn p., in Lancashire, 4 m. N by E of Blackburn. Area 940 acres. Pop. in 1831, 337; in 1851, 237.

WILRYK, a commune and village of Belgium, in the prov. and 4 m. S of Antwerp. Pop. 2,700.

WILSCHDORF, a village of Saxony, in the bail. of Dresden. Pop. 368.—Also a village in the circle of Bautzen, bail. of Stolpen. Pop. 740.

WILSDEN, a township of Keighley p., in Yorkshire, 4 m. SE of Keighley. Area 2,450 acres. Pop. in 1831, 2,252; in 1851, 3,454.

WILSDRUFF, a town of Saxony, 9 m. W of Dresden. Pop. in 1849, 2,342.

WILSELE, a commune and village of Belgium, prov. of Brabant, 15 m. NE of Brussels. Pop. 1,200.

WILSFORD, or WIVELSFORD, a parish in Lincolnshire, 4 m. WSW of Sleaford. Area 2,860 acres. Pop. in 1831, 393; in 1851, 484.—Also a parish of Wiltshire, 2 m. SW of Amesbury. Area 1,591 acres. Pop. in 1831, 119; in 1851, 135.

WILSFORD-DAUNTSEY, a parish in Wilts, 4 m. SW of Pewsey. Area 2,997 acres. Pop. 575.

WILSNACK, a town of Prussia, in the prov. of Brandenburg, 63 m. WNW of Berlin. Pop. 1,848.

WILSON, a township of Niagara co., New York, U. S., 206 m. W by N of Albany. Pop. in 1850, 2,955.—Also a county in West Tennessee, on the S side of the Cumberland. Area 468 sq. m. Pop. in 1850, 27,446. Its cap. is Lebanon.

WILSON'S ISLES, a group of three small low islands in the Pacific, in N lat. 7° 11′, E long. 145° 8′. Pop. about 140.

WILSON (POINT), a cape on the W coast of North America, in the gulf of Georgia, in N lat. 48° 10′.

WILSON'S PROMONTORY, a remarkable headland, about 20 m. long, and from 5 to 14 m. in breadth, forming the southernmost point of Australia, in S lat. 39° 8′, E long. 146° 23′. It is of granitic formation, and attains an alt. of 3,000 ft.

WILSONTOWN, a village of Lanarkshire, in the p. of Carnwath, 23 m. SW of Edinburgh, and 8½ m. E of Lanark.

WILSTER, a town of Holland, near the Elbe, 34 m. NW of Hamburg. Pop. 2,900. It has a trade in corn and cattle, and several distilleries. The surrounding district is called Wilstermarsch.—Also a river of Holstein, properly a branch of or separation from the Cudensee. It falls into the Stor, and is navigable for small craft.

WILSTHORPE, a township of Lincolnshire, 5 m. NW of Market-Deeping. Pop. in 1851, 66.—Also a hamlet in the p. and 3 m. N by W of Bridlington in Yorkshire. Area 712 acres. Pop. 147.

WILTHEN, a village of Saxony, on the borders of Lusatia, to the S of Bautzen. Pop. 429. The vicinity was the scene of a battle between the French and the Allies, in May 1813.

WILTON, a parish of Roxburghshire, lying on the banks of the Teviot. Pop. in 1851, 2,557.

WILTON, a parish in Somersetshire, 1 m. SW of Taunton. Area 700 acres. Pop. in 1831, 795; in 1851, 949. The p. contains the hamlets of Gaunton and Sherford, and forms an extensive suburb to the town of Taunton.—Also a borough and parish in Wilts, including the tythings of Bulbridge and Ditchampton. Area 1,791 acres. Pop. in 1801, 2,144; in 1831, 1,997; in 1851, 1,804.—The town, situated in a valley 3 m. W by N of Salisbury, in the peninsula formed by the confluence of the Madder and Wily, consists principally of one street. It is well-known for the manufacture of the kind of carpets called after its name, and which was introduced from France by one of the earls of Pembroke, who brought workmen to England for the purpose of instructing the inhabitants in the art. The borough formerly sent 2 members to parliament; under the reform act it has been deprived of one member. Number of voters in 1852, 219.—The name of this place is probably derived from its situation on the river Wily. It is a town of great antiquity, and was anciently of considerable importance, giving name to the co. In the beginning of the 10th cent., it became the seat of the diocese of Wiltshire; and was such until the two bishoprics having been united, the see was removed to Sherborne, and ultimately to Old Sarum.—Also a township and chapelry in the p. of Kirk-Leatham, N. R. of Yorkshire, 3½ m. NW of Guisborough. Area 6,928 acres. Pop. in 1831, 411; in 1851, 518.—Also a township in the p. of Ellerburn, N. R. of Yorkshire, 3½ m. E of Pickering. Area 2,060 acres. Pop. in 1851, 202.

WILTON, a township of Franklin co., Maine, U. S., 27 m. NW of Augusta. Pop. 1,908.—Also a township of Hillsborough co., New Hampshire, 40 m. SSW of Concord. Pop. 1,161.—Also a township of Fairfield co., Connecticut, 7 m. N of Norwalk. Pop. 2,066.—Also a township of Saratoga co., New York, 32 m. N of Albany. Pop. 1,458.

WILTON (BISHOP'S), a parish in the E. R. of Yorkshire, 4½ m. N of Pocklington, containing the townships of Bishop-Wilton with Belthorpe, Bolton, and Youlthorpe with Gowthorpe. Area 4,970 acres. Pop. in 1831, 831; in 1851, 652.

WILTS, or WILTSHIRE, an inland county of England; having Gloucestershire on the N; Berkshire and Hampshire on the E; Hampshire and Dorsetshire on the S; and Somersetshire and Gloucestershire on the W. The area of the county is, according to the latest admeasurements, 869,620 acres.—Pop. in 1801, 185,107; in 1811, 193,828; in 1821, 222,527; in 1831, 240,200; in 1841, 260,007; in 1851, 254,221. The form of this co. is somewhat that of an ellipse, measuring nearly 57 m. from N to S, by 34 m. from E to W. It is naturally divided into two portions, by an irregular range of hills, which extends transversely through the greater part of the co., in a direction inclining from NE to SW, or from Savernake forest, by the Black-heath, to Westbury. The districts on either side of this range are usually denominated South and North Wiltshire; and differ materially from each other, not only in appearance, but in almost every distinguishing quality. South W. forms the western division of a vast tract of chalk-hills, comprising a considerable part of Hampshire, and presents, when surveyed from a distance, the appearance of a large elevated plain. Long-Knoll, near Maiden-Bradley on the borders of Somersetshire, the highest point in this division of W., is only 973 ft. above sea-level.—North W. differs completely from the southern division of the co. in its general appearance. Instead of a constant series of 'chalky waves,' the aspect of this district—which extends from the verge of the Downs to the hills of Gloucestershire—is nearly that of a perfect flat; the few deviations from the ordinary level being, for the most part, so gradual as scarcely to be perceptible. The air in South W.—as on all open downs—is generally cold and bracing. In North W. the climate is milder than in the southern parts; but is, nevertheless, likewise cold.

Rivers.] W. abounds with rivers and streams, most of which take their rise within itself. The Isis, afterwards called the Thames, rises in the Cotswold hills, in Gloucestershire; enters W. near Ewen; forms its N boundary for a short distance; and leaves the co. at its NE extremity. It receives the small streams of the Swillbrook, the Key, and the Churn, from the N part of the co.—The Lower Avon rises on the borders of Gloucestershire; proceeds SE, and, on reaching Chippenham, flows in a winding direction W to a point, near the borders of Somersetshire, by Bradford and Melksham; and finally turning N, enters the last named co. a little to the S of Winsley.—The Kennet rises to the W of East Kennet; passes Marlborough; and flows in a direction at first SW, and afterwards S, to the West Kennet; and passing Ramsbury and Chilton, reaches Hungerford in Berks.—Several small streams uniting in the vale of Pewsey, form the river called the Wiltshire and Hampshire Avon, which, passing Amesbury and Salisbury, enters Hampshire, near Downton. It receives several streams and small rivers, of which the principal is the Wily, which rises at the foot of the Clayhills, and after passing Warminster, Heytesbury, and Wilton, is joined by the Nadder, and discharges itself into the Avon by two channels. The Nadder rises near Shaftesbury, in Dorsetshire, and joins the Wily near Wilton. The Bourne rises near Easton, passes the three Winterbournes, and joins the Avon near the city of Salisbury. The bed of this river is generally dry in summer.

Agriculture, &c.] About 800,000 acres of the superficies of W. are arable, meadow, or pasture-land. The average rent of land is 18s. 6d. per acre. With regard to agriculture, the S district may almost be regarded as one vast sheep-farm; the greater part of its extensive downs, and large portions of the vales being devoted to the rearing and feeding of that animal. The sheep stock—consisting chiefly of South Downs—is estimated at 700,000, of which nearly six-sevenths are depastured on the downs. The distinguishing feature of the agriculture of North W. is the pasturage, or grazing system, which is conducted principally with a view to the supply of the dairy with milk for the manufacture of cheese. W. had formerly abundance of wood, but the forests have now in a great measure disappeared. There are few wastes, and these are principally marshy. Owing to the scarcity of coal—the supply of which is in a great measure derived from the coal-fields of Somersetshire—fuel is scarce. The mineral productions of W. are of little importance. Chalk forms the chief part of the substructure of all that extensive assemblage of hills which constitute the downs. On the W side of the co., bordering on Somersetshire, are several freestone quarries.

Manufactures.] The manufactures of W. are of considerable importance. Woollens were formerly made to a great extent, but the manufacture has greatly fallen off within these thirty years. Salisbury carries on a considerable manufacture of cutlery and steel goods. Carpets of an excellent quality and kerseymere and linsey, are made at Wilton. Broad cloths and other descriptions of woollen goods are made at Bradford, Devizes, Heytesbury, Chippenham, Westbury, Trowbridge, and all the adjacent towns and villages. Linen is manufactured at Mere; fustians and thicksets at Aldbourn. Linen and silk are made to a considerable extent at Stourton and Maiden-Bradley. The principal exports of the county consist of its agricultural produce; besides supplying its own extensive cloth manufactories, it exports a large proportion of its wool.

Canals and roads.] The Thames and Severn canal, which commences at Stroud in Gloucestershire, and terminates near Lechlade in the same co., passes through a small portion of the N boundary of W. The Kennet and Avon canal, which connects the navigation of these two rivers, crosses the whole of the co., and nearly divides it into two equal parts.—The Wiltshire and Berkshire canal strikes off from the Kennet and Avon canal at Semington, and running in a direction almost due N, passes Melksham; and pursues nearly the same course as the Great Western railway, by Wootton-Basset, Chaddenton, and Swindon, to within a mile of Highworth, near which town it leaves this co., and passes into Berkshire, on its way to join the Thames, a short distance below Abingdon.—The North Wilts canal branches off from the Wiltshire and Berkshire canal, a little to the NW of Swindon, and running in a NW direction, joins the Thames and Severn canal at Easton.—The roads in Wiltshire are both numerous and good; indeed few counties are so well provided in this respect.—The Great Western road, which leads from London to the most western part of Cornwall, enters W. at two points, about a mile distant from one another; the one branch is from Andover, and the other from Stockbridge; they join about a mile after entering the co., and pass through Salisbury to Blandford.—The London road to Bath, by Chippenham, passes through Hungerford in Berkshire; immediately after enters W.; and passes through Marlborough, Calne, and Chippenham, to Bath.—The Great Western railroad enters W. at a point about 3 m. NE of Bath, and soon after passes Box, where it passes through a ridge of lias limestone, Bathstone, and hard dry quartz, by a tunnel 3,168 yds. in length, and 30 ft. wide. It then passes on the S of Corsham, on the N of Chippenham, S of Wootton-Basset, and N of Swindon; and enters Berkshire at a point about 4 m. to the NE of Swindon.—There is a branch railway from the Swindon station, by Purton and Minety, to Cheltenham.

Ecclesiastical and Civil Divisions, &c.] This co., although now in the dio. of Salisbury, was in the time of the West Saxons included in that of Winchester. It forms the two archdeaconries of Salisbury and of Wilts; the former comprising the deaneries of Amesbury, Chalke, Salisbury, Wilton, and Wylie; the latter those of Avebury, Marlborough, and Potterne. In 1829, W. was divided into 295 parishes; viz. 143 rectories, 104 vicarages, and 48 perpetual curacies.—The co. is divided, for the purposes of civil government, into 28 hundreds, viz. Alderbury, Amesbury, Bradford, Branch and Dole, Calne, Cawden and Cadworth, Chalke, Chippenham, Dammerham (North and South), Downton, Dunworth, Elstub and Everley, Frustfield, Heytesbury, —Highworth, Cricklade, and Staple,—Kingsbridge, Kinwardstone, Malmesbury, Melksham, Mere, Potterne and Cannings, Ramsbury, Selkley, Swanborough, Underditch, Warminster, Westbury, and Whorwelsdown. It contains the city of New Sarum, commonly called Salisbury; the borough and market towns of Calne, Chippenham, Cricklade, Devizes, Malmesbury, Marlborough, and Westbury; the borough of Wilton; and the market-towns of Amesbury, Great Bradford, Market-Lavington, Melksham, Mere, Swindon, Trowbridge, Warminster, Hindon, and Wootton-Basset. The two last named towns, with Great Bedwin, Downton, Heytesbury, Ludgershall, and the well-known Old Sarum, returned members to parliament until the passing of the Reform act. The represented towns are Chippenham, Devizes, Calne, Marlborough in the N division; and Wilton, Salisbury, and Westbury in the S. By the Reform act this co., formerly returning 2 members, returns 4; namely, 2 for the northern, and 2 for the southern division. The number of registered voters was, in 1832, in the N division, 3,614; in the S, 2,540. In 1852 the voters in the N division were 4,955; in the S, 3,256. The total number of members returned by the co. and boroughs, under the new act, is 18. W. is within the western circuit, the assizes being held at Salisbury.

History.] When Cæsar wrote his Commentaries, Wiltshire was occupied by the Belgæ throughout the greater part of its extent. The Romans included W. in their province of *Britannia Prima*. W. suffered much from the Danes; the Great Alfred was defeated at Wilton in 871. During the civil wars which raged between the Empress Maude and Stephen, the latter seized the castles of Salisbury, Devizes, and Malmesbury; but was defeated, with great slaughter, at Wilton. Shortly after the battle of Lansdown, the Parliamentary army was defeated on Roundaway-hill, and suffered great loss. Devizes was taken by Cromwell in 1645.

WILTSHIRE (Fort), a military station in South Africa, 44 m. NNE of Graham's Town, near the Keiskamma river.

WILY, or Wylye, a parish in Wilts, 7 m. ENE of Hindon. Area, with Deptford, 2,279 acres. Pop. in 1831, 476; in 1851, 510.

WILZ, a town of Holland, in the prov. of Luxemburg, bail. and NW of Diekirch, on the r. bank of a river of the same name, an affluent of the Sure. Pop. 2,500. It has manufactories of cloth and linen, and several paper-mills.

WILZENBURG, a village of Oldenburg, in the bail. of Birkenfeld. Pop. 200.

WIMBACH, a village of Prussia, in the regency of Coblenz, and circle of Adenau. Pop. 290.

WIMBERG, a village of Prussia, in the regency of Arnsberg, and circle of Iserlohn. Pop. 200.

WIMBISH, a parish in Essex, 4¼ m. ESE of Saffron-Walden. Area, with Thunderley, 4,862 acres. Pop. in 1831, 921; in 1851, 1,004.

WIMBLEDON, a parish and village in Surrey, intersected by the Southampton railway, by which the village is 6 m. distant from London. Area of p., 3,700 acres. Pop. in 1831, 2,195; in 1851, 2,693.

WIMBLINGTON, a hamlet in the p. of Doddington, Cambridgeshire, 4 m. S of March. Area 7,589 acres. Pop. in 1831, 965; in 1851, 1,158.

WIMBORNE-ST.-GILES, or Upwinborn-Malmaynes, a parish in Dorset, 2½ m. SW by W of Cranborne. Area, including Wimborne-all-Saints, 3,978 acres. Pop. in 1831, 384; in 1851, 495.

WIMBORNE-MINSTER, a parish in Dorset, containing the market-town of the same name, and the tythings of Abbot-street, Barnesley, Cowgrove, Leigh, Petersham, Stone, Thornhill, and Wimborne-Borough. Area 11,966 acres. Pop. in 1801, 3,039; in 1831, 4,009; in 1851, 4,759. The church, or minster, is a spacious cruciform structure of early Norman, with an admixture of early English architecture. In the chancel—which is raised several steps above the level of the nave— are 16 stalls, surmounted by canopies of carved oak. The town, which stands in a valley near the confluence of the Stour and Allen, consists of a few irregular and ill-constructed streets. The borough contains the N part of the town, and consists of two streets. It has been supposed by some that this is the place designated in the Saxon Chronicle under the name of *Wicganbeorche*, as the scene of a great battle fought in 851, between the Danes and the Devonians; but Bishop Gibson refers the scene of the encounter to Wenbury, in the county of Devon. In the time of Leland it was a respectable town.

WIMESBOTHAM, or Wimbotsham, a parish in Norfolk, 1½ m. NE of Downham-Market. Area 2,015 acres. Pop. in 1831, 476; in 1851, 606.

WIMESWOULD, a parish in Leicestershire, 5 m. NE by E of Loughborough. Area 4,220 acres. Pop. in 1831, 1,276; in 1851, 1,235.

WIMILLE, a commune of France, in the dep. of the Pas-de-Calais, cant. and 3 m. NNE of Boulogne-sur-Mer, on the Vimeux. Pop. 1,779. Iron and potter's clay are found in the vicinity.

WIMMELBACH, a village of Bavaria, in the circle of Upper Franconia, and presidial of Forchheim. Pop. 240.

WIMMELBURG, a village of Prussia, in the regency of Merseburg, and circle of Mannsfelder See. Pop. 458.

WIMMENTHAL, a village of Würtemberg, in the circle of the Neckar, and bail. of Weinsberg. Pop. 394.

WIMMER, a village of Hanover, in the principality of Osnabruck, and bail. of Wittlage-Huntebure, near Bohmte. Pop. 759.

WIMMERA, a district of Australia Felix, comprising an area of about 15,000,000 acres, extending S of the Murray from the South Australian frontier, to the dividing ridge of the basins of the Murray and of Portland bay, and S to a line between that ridge and Mount Cole. Pop. 876. It is watered by numerous streams, of which the principal are the Murray, Wimmera, and Avoca, and contains several lakes.—The river W. rises in Mount Cole, and flows into Lake Hindmarsh.

WIMMERBY, a town of Sweden, in the län of Calmar, 40 m. WSW of Westerwick. Pop. 1,465.

WIMMERSBACH (Neckar), a village of Baden, in the circle of the Lower Rhine, and bail. of Oberbach, on the Neckar. Pop. 448.

WIMMERSBACH (Wald), a village of Baden, in the circle of the Lower Rhine, and bail of Neckar. Pop. 561.

WIMMERTINGEN, a department and commune of Belgium, in the prov. of Limburg, and arrond. of Hasselt, on the Laak. Pop. 131.

WIMMIS, a village of Switzerland, in the cant. of Berne, and bail. of Nieder-Simmenthal. Pop. 1,369.

WIMPFEN-AM-BERG, a town of Hesse-Darmstadt, capital of a bail. of the same name, in the prov. of Starkenburg, near the l. bank of the Neckar, opposite the confluence of the Jaxt, and 8 m. NNW of Heilbronn. Pop. 2,340. In consists of two distinct parts, one on a height, and the other in a valley, and is surrounded by walls. It has five gates, two churches, a Catholic and a Lutheran, an hospital, a poor's house, and three schools. This town is of great antiquity, and was erected in 1230 into a free imperial school.

WIMPFEN-IM-THAL, a town of Hesse-Darmstadt, in the prov. of Starkenburg, and bail of Wimpfen-am-Berg. Pop. 410.

WIMPOLE, a parish in Cambridgeshire, 6 m. E of Caxton. Area 2,428 acres. Pop. in 1851, 452.

WIMSBACH, a market-town of Austria, in the county above the Ens, and circle of the Traun, on the l. bank of the Alm, and 27 m. W of Steyer. Pop. 550. It has a castle.

WIMSHEIM, a village of Würtemberg, in the circle of the Neckar, and bail. of Leonberg. Pop. 669.

WIMY, a village of France, in the dep. of the Aisne, and cant. of Hirson, near the Oise, 9 m. NNE of Vervins. Pop. 610. It has extensive bottle-works.

WINAR, a village of Austrian Bohemia, in the regency of Gitschin, and bail. of Neu Bidschow. Pop. 440.—Also a village of the regency of Pardubitz, and bail. of Czaslau. Pop. 540.—Also a village of the same regency, in the bail. of Hohenmauth. Pop. 340.—Also a village in the regency of Prague, and bail. of Karolinenthal. Pop. 670. It has a castle.—Also a village of Moravia, in the circle of Prerau. Pop. 270.

WINARITZ, a village of Austrian Bohemia, in the regency of Gitschin, and bail. of Jungbunzlau. Pop. 350.—Also a village in the regency and bail. of Pardubitz. Pop. 240.—Also a village in the reg. of Prague, and bail. of Laun. Pop. 370. It has the ruins of a castle.

WINAU, a village of Austrian Bohemia, in the regency of Budweis, and bail. of Gratzen. Pop. 440. —Also a village in the regency of Pardubitz, and bail. of Deutschbrod. Pop. 240.—Also a village of Moravia, in the circle of Znaim, and bail. of Jaispitz. Pop. 620.—Also a village of Prussia, in the regency and circle of Oppeln. Pop. 220.

WINBURNDALE, a river of New South Wales, in the co. of Roxburgh, an affluent of the Macquarie.

WINCANTON, a parish of Tasmania, in the co. of Somerset, to the W of the Macquarie.

WINCANTON, or Wincanuton, a market-town and parish in Somerset, 108 m. W by S of London. A large portion of the town was consumed by fire in 1747, but it has since been rebuilt in a more regular form. It is situated on the W declivity of a wooded hill, at the foot of which flows the Cale, and consists of four streets regularly built, and containing some good houses. The parish contains the hamlets of Barrow-Common, Marchcourt, Stavordale, Sutton, the Tything, and Wincanton-Common. The inhabitants are partly employed in the manufacture of linen and bed-ticking, also in a branch of the silk trade; but the chief trade is in corn, cattle, butter, and cheese. Area of p., 4,130 acres. Pop. in 1831, 2,123; in 1851, 2,488.

WINCEBY, an island of South Australia, in Spencer's gulf, in the group of Sir Joseph Banks' islands.

WINCEBY, a parish in Lincolnshire, 4 m. E by S of Horncastle. Area 842 acres. Pop. in 1851, 78.

WINCH (East), a parish in Norfolk, 6 m. SE by E of Lynn. Area 2,530 acres. Pop. in 1851, 469.

WINCH (West), a parish in Norfolk, 3 m. S of Lynn. Area 1,170 acres. Pop. in 1851, 477.

WINCHAM, or Winsham, a township in the p. of Great Budworth, co.-palatine of Chester, 2 m. NE by E of Northwich. Area 941 acres. Pop. 684.

WINCHCOMBE, or Winchelcombe, a market-town and parish in Gloucestershire. The parish, which comprises the chapelries of Grest and Gretton, with the hamlets of Contes, Cockbury, Corndean, Langley, Naunton with Frampton, Postlip, and Sudley-Tenements, has an area of 5,700 acres. Pop. in 1831, 2,514; in 1851, 2,824. The town is situated at the base of the Cotswold hills, on the N side of the range, and to the SE of the river Islip, in a sheltered valley. The chief streets, three in number, run from E to W, and are crossed by a few of smaller extent. The houses are in general old. The inhabitants were at one time much employed in the culture and manufacture of tobacco. Here are paper mills, a tanyard, and a silk-mill, and some small manufactories of stockings and thread.

WINCHELSEA, a market-town and parish in Sussex, 64 m. SE of London, and 2 m. SW of Rye. Area of p., 1,510 acres. Pop. in 1831, 772; in 1851, 778. The town consists of several streets running at right angles to one another, and is said to have been built after the plan of the old town which was swept away by the sea in 1287. It gives the title of Baron to the family of Finch. Until the passing of the reform act, the borough returned 2 members to parliament. It now unites with Rye in returning a member.

WINCHENDON (Nether), a parish in Bucks, 6 m. W of Aylesbury. Area 1,520 acres. Pop. in 1831, 294; in 1851, 284.

WINCHENDON (Upper), a parish in Bucks, 5 m. W by N of Aylesbury. Area 1,086 acres. Pop. in 1831, 223; in 1851, 186.

WINCHENDON, a township of Worcester co., in the state of Massachusetts, U. S., 54 m. WNW of Boston, drained by Miller's creek, and intersected by the Cheshire railroad. Pop. in 1840, 1,754; in 1850, 2,445.

WINCHESTER, a township of Cheshire co., in the state of New Hampshire, U. S., 53 m. SW by W of Concord, drained by Ashuelot river, and intersected by the Ashuelot railroad. Pop. in 1840, 2,065; in 1850, 3,296.—Also a township of Litchfield co., in the state of Connecticut, 26 m. NW of Hartford, and drained by Mad river. Pop. in 1840, 1,667; in 1850, 2,179.—Also a village of Frederick co., in the state of Virginia, 116 m. NNW of Richmond, and at the junction of the Winchester and Potomac and the Baltimore railways. Pop. in 1840

3,454; in 1850, 4,300.—Also a township of Adam's co., in the state of Ohio, drained by a branch of Brush creek. Pop. in 1840, 1,121.—Also a village Clarke co., in the state of Kentucky. Pop. in 1840, 1,047; in 1850, 1,100. -Also a village of Franklin co., in the state of Tennessee, 94 m. SSE of Nashville, on a small branch of Elk river. Pop. in 1840, 700; in 1850, 800.—Also a village of Scott co., in the state of Illinois, 15 m. SW of Jacksonville. Pop. in 1840, 720; in 1850, 700.—Also a township of Middlesex co., in the state of Massachusetts, on the Boston and Lowel railroad. Pop. in 1850, 353.—Also a village of Wayne co., in the state of Mississippi, on the E bank of Chickasawha river, and 104 m. ESE of Jackson.

WINCHESTER, an ancient and peculiarly constituted city, once the metropolis of England, situated nearly in the centre of Hampshire, 12 m., by railway, N of Southampton, and 64½ m., by railway, SW by W of London. The river Itchin approaches its E side, in such a tiny maze of channels as to form a series of islets; concentrates its waters to cut off, over the whole breadth of the town, a large eastern suburb; and again divides and subdivides its stream in wandering toward St. Cross. On the E of the vale rise high downs; on the W stretch spreading hills, which acknowledge the dominion of the plough, and permit, through a deep cut, the passage of the London and Southampton railway. The site of the town, though pleasant, cannot be called beautiful; and is so far from being advantageous that one wonders why it should ever have held a city of importance, and still more a metropolis of England. As seen from the railway station it is "a wide extended mass of old brick houses, grey church towers, and red-tiled roofs, with immediately in front, on the brow of the descending eminence, a huge square edifice, now answering as a barrack for soldiers, but formerly a royal residence." The barrack, fitted up in 1811 for the accommodation of 2,000 men, occupies the site of an ancient and demolished castle. The guild-hall, built in 1711, is a neat structure in the Doric style, and has in front a statue of Queen Anne, bearing the sceptre. The county-gaol is a large and substantial building, with an imposing front. The county house-of-correction is an old building, on a low site. The county hospital is a capacious building, partly of recent date. The city-cross, in the centre of High-street, is an elegant pyramidal structure, 43½ ft. in height, of three stages, in the later Gothic style, with arches, niches, and figures. The other noticeable buildings—additional to the cathedral and the college of Winchester, and the hospital of St. Cross—are the market-house, a theatre, and assembly-rooms.

The Cathedral.] The original cathedral of W. claims—on very apocryphal authority—to have been built, in 177, on the site of a great pagan temple. The present pile, regarded in its various parts, is of at least seven principal dates; and, if viewed in its minor features, its smaller accessories, its remodellings, and its restorations, must be pronounced to belong to upwards of ten centuries. The oldest parts of the edifice belong to a pile constructed, in the Saxon style, by the famous architect and bishop of Winchester, St. Ethelwold, in 980. Apart from minute or secondary restorations, W. cathedral may be viewed as nearly a continuous exemplification of all the successive styles of ecclesiastical architecture in England, from the Saxon to the debased variety of the Gothic. The grand western front, the most conspicuous part of the contribution made to the pile by William of Wykeham, is 118 ft. long, embellished below with tabernacle-work, and everywhere magnificent. The cathedral is cruciform. Its length, from E to W, is 545 ft.; and from N to S, within the walls, 208 ft. The nave is 78 ft. high; and, including the aisles, 86 ft. wide; and the central tower is 138 ft. in height, and 50 by 48 ft. in breadth. The altar is ornamented by West's picture of the raising of Lazarus from the dead; and behind the altar is a stone screen of wonderful elaboration and beauty.

The College.] Winchester college, founded by William of Wykeham, is the oldest institution of its class in England, and served as a model for the schools of Eton and Westminster. It was instituted about the year 1387, for a warden, 10 fellows or perpetual chaplains, and 70 scholars, besides conductors, clerks, choristers, and some other parties; and, in the 26th year of Henry VIII., it had endowments to the amount of £639 yearly value. It is simply a school of the higher order, and somewhat in the style of the middle ages, for Latin and Greek; it forms a kind of literary porch to New college, Oxford, and trains young gentlemen as candidates for that college's fellowships, scholarships, and exhibitions; it equips its masters and scholars in black gowns and caps in the manner of the Oxonians, and enacts from them the observance of similar rules of discipline; and, in addition, it admits a certain number of young men to be educated, under the name of 'commoners,' on the terms of a boarding-school, and has attached to it, as a kind of fagend, the singing boys of the cathedral. The buildings and grounds of the college are situated on the verge of the green vale of the Itchin, in the lower environs of the town, and are surrounded by a protecting wall. The original buildings are in the early style, and form a large quadrangle fronting the street, and a smaller quadrangle behind. The school-room is a finely proportioned apartment, measuring 90 ft. by 36. Adjoining the school, and constituting the extreme SE of the college buildings, are the cloisters or covered walks, very neat but more modern than the time of Wykeham, and enclosing a quadrangular area of 17,424 sq. ft.—The ruins of Wolvesey castle are at a short distance SE from the college.

The Hospital and Church of St. Cross.] The hospital of St. Cross, situated in the suburban village of its name, within a mile of the city, was founded and endowed, in 1132, by Henry de Blois, bishop of Winchester, for the residence and support of a master, steward, 4 chaplains, 13 clerks, 7 choristers, and 13 poor brethren. The buildings of the hospital consist of an outer and an inner court. The former is entered by a spacious gateway; but is small, and presents little matter of interest except 'the Hundred-men's-hall,' or the refectory for the quondam 100 outboarders. Over the centre of the range of building which separates the two courts, rises a lofty and beautiful tower. The inner or principal court presents, on three sides, the neat uniform buildings of the hospital, two stories high, and ornamented with fig-trees and vines. The brethren wear black cloaks, with a silver cross on the left breast. The church of St. Cross, built partly by Henry de Blois, and partly by William of Wykeham, like Winchester cathedral, has been so much and often modified as to exhibit a record of architectural styles from the Norman to the most highly decorated Gothic. It is cruciform; and consists of three aisles, with a transept, a cross aisle, and a low broad tower rising over the intersection, Its length, from E to W, is 160 ft.; its breadth, including the transepts, is 120 ft. The interior of the tower, which serves as a lanthorn to the chancel, is open to a considerable height above the vaulting of the nave, and is ornamented with galleries. The present income of the hospital is £1,500, but, on the expiry of certain life-leases, will amount to £8,000.

Institutions, Charities, &c.] A mechanics' institution was established in W. in 1835. A subscription library is well supported. Concerts, balls, and assemblies are frequently held in St. John's rooms.—The dissenting congregations, are a Baptist, an Independent, a Wesleyan Methodist, and a Roman Catholic. Opposite the latter—which is a handsome Gothic structure—stands a Benedictine nunnery, inhabited by a community who removed hither from Brussels. A new cemetery was formed in 1840, in a field to the SW of the King's-house. The city returns 2 members to parliament. Constituency, in 1837, 500; in 1852, 788. The making of sacks, and that to no great extent, is the only manufacture. Coal is supplied by the Itchen navigation from Southampton. Pop. of the city, in 1801, 3,555; in 1831, 5,280; in 1851, 6,446. Pop. within the parliamentary boundary, in 1831, 9,292; in 1851, 13,704.

History.] The city, no doubt, occupies one of the most ancient seats of population in Britain; but it does not become known to history till it figures as the Roman station of *Venta Belgarum*. It afterwards became the principal seat of the Saxon, the Danish, and the Norman dynasties; it was the grand scene of Alfred's and of Canute's glories; it did not lose its character, as the capital of England, till London became supereminent in magnitude, influence, and wealth; and it continued to be a chief or frequent place of residence for the English kings and princes, till the death of the husband of Queen Anne. In the reign of Henry Beauclerc, when the city reached its culminating point, the city extended a mile in every direction beyond its present limits; it had densely crowded fairs, a considerable woollen manufactory, and an extensive commerce with the continent, and was the general thoroughfare from the eastern to the western parts of the kingdom. It was the chief seat of government, and the site of the royal mint, the royal treasury, and the national archeries; and it had a great and magnificent royal palace, and possessed in its cathedral the remains of more personages of the various royal families, than all the other towns and places of the island.

WINCHFIELD, a parish in Hants, 2½ m. NE of Odiham. Area 1,543 acres. Pop. in 1851, 327.

WINCHMOREHILL, a chapelry in the p. of Edmonton, Middlesex, 8 m. N of London.

WINCKLEY, or WINKLEIGH, a parish in Devon, 5 m. SW of Chulmleigh. Area 9,118 acres. Pop. in 1831, 1,596; in 1851, 1,554.

WINCLE, a township and chapelry in the p. of Prestbury, co.-palatine of Chester, 4½ m. SE of Macclesfield. Area 1,980 acres. Pop. in 1851, 336.

WINDACH (OBER, MITTEL, and UNTER), three villages of Bavaria, in the circle of Upper Bavaria, and presidial of Landsberg. Pop. 440.

WINDANG, an island of New South Wales, in the co. of Camden, at the entrance of Lake Illawarra.

WINDAU, or WINDAWA, a town of Russia in Europe, in the gov. of Courland, and district of Goldingen, to the NW of Mitau, on the Baltic, in N lat. 57° 23′ 52″. Pop. 1,756. It has a port, and carries on an active trade in grain, flax, hemp, and timber. This town is one of considerable antiquity.

WINDBERG, a village of Bavaria, in the circle of Lower Bavaria, and presidial of Bogen. Pop. 270.—Also a village of Prussia, in the regency of Dusseldorf and circle of Gladbach. Pop. 240.

WINDBERGEN, a village of Holstein, in the bail. and 5 m. SSE of Mildorf. Pop. 520.

WINDEBERG, a village of Prussia, in the reg. of Erfurt, and circle of Muhlhausen. Pop. 360.

WINDECKEN, a town and bail. of Hesse-Cassel, in the prov. and circle and 5 m. N of Hanau, on the l. bank of the Nidda. Pop. 1,612. It has an hospital. It is enclosed by walls, with 5 gates, and contains two churches, a Lutheran and a Reformed, and an hospital.

WINDEHAUSEN, a village of Prussia, in the reg. of Merseburg, circle of Sangerhausen. Pop. 505.

WINDELLAMA, a creek of New South Wales, in the co. of Argyle, an affluent of Werriaming river.

WINDELSBACH, a village of Bavaria, in the circle of Middle Franconia, and presidial of Leutershausen. Pop. 270.

WINDEN, a village of Bavaria, in the circle of the Pfalz, and cant. of Kandel. Pop. 538.—Also a village of Baden, in the circle of the Middle Rhine, and bail. of Baden. Pop. 380.—Also a village of Nassau, in the bail. of Nassau. Pop. 700.—Also a village of Prussia, in the regency of Aachen, and circle of Duren. Pop. 370.

WINDENREUTHE, a village of Baden, in the circle of the Upper Rhine, and bail. of Emindingen. Pop. 493.

WINDERMERE, a parish in Westmoreland, 9 m. NW of Kendal, comprising the chapelry of Troutbeck, and the townships of Applethwaite and Under-Milbeck. Area 17,139 acres. Pop. in 1831, 1,632; in 1851, 3,280.—The large and beautiful lake called Windermere, in the Applethwaite division of the p., is nearly 11 m. in length, and has an average breadth of ¾ m. It runs N and S, with a very slight curve near the centre. The middle and lower parts of the lake lie between bold and steep hills. Towards its foot, where its undulating shores are adorned with forest trees, it contracts in breadth considerably. Towards the head the hills recede, leaving a broad margin of woods, meadows, and lawns, beautifully spread over a waving surface. It is a mile in width in the upper part, and presents a very noble expanse of water. It contains 14 islands, most of them forming a cluster near the middle. The largest, Belle isle, which runs obliquely across the centre of the lake, opposite Bowness, contains 28 acres. The lake has an extreme depth of 240 ft., and abounds in trout, perch, pike, and char, the last being a rare and delicate fish, found only in the deepest of the lakes. The whole of the W shore, and nearly half of the E, are in Lancashire, but the lake is considered to belong to Westmoreland. There is a ferry over the lake, a little below Bowness. Its alt. above sea-level is 116 ft.; and it discharges its waters by the Leven river, which flows, from the southern end of the lake, into the deep inlet called Leven sands. The Kendal branch railway from the Lancaster and Carlisle line, has been prolonged to the thriving and beautiful village of Birthwaite or Windermere on this lake.

WINDESHEIM, a village of Prussia, in the regency of Coblenz, circle of Kreutznach. Pop. 904.

WINDFOCHE, a village of Prussia, in the reg. of Dusseldorf, and circle of Elberfeld. Pop. 380.

WINDGALLEN, a rocky mountain of Switzerland, in the cant. of Uri, 8 m. SE of Altorf. It has an alt. of upwards of 9,000 Parisian ft. above Lake Waldstetten.

WIND-GAP, a pass of the Blue Mountains, in Northampton co., in the state of Pennsylvania, U. S., 90 m. ENE of Harrisburg.

WINDHAG, a market-town of Austria, in the country above the Ens, and circle of the Muhl, 8 m. NE of Freistadt. Pop. 400.

WINDHAGEN, a village of Prussia, regency of Cologne, and circle of Gummersbach. Pop. 260.

WINDHAM, a township of Upper Canada, in the Talbot district, watered by Big creek. Pop. 1,568.

WINDHAM, a county in the NE part of the state of Connecticut, U. S., drained by Quinnebaug and Shetucket rivers, and their branches, and intersected by the Norwich and Worcester, the New York and Boston, the Hartford, Providence, and Fishkill, and the Plainfield railways. Pop. in 1840, 28,080; in 1850, 31,079. Its capital is Brooklyn.—Also a county in the SE part of the state of Vermont, comprising an area of 765 sq. m., drained by Williams, West, and Saxton rivers, and the tributaries, and intersected by the Vermont Valley railway. It is generally hilly. Pop. in 1840, 27,442; in 1850, 29,062. Its capital is Fayetteville.—Also a township of Windham co., in the state of Connecticut, 28 m. E by S of Hartford, drained by Shetucket river and its branches. Pop. in 1840, 3,382; in 1850, 4,636.—Also a township of Cumberland co., in the state of Maine, 47 m. SW of Augusta, drained by branches of Presumpscot river. Pop. in 1840, 2,303; in 1850, 2,380.—Also a township of Rockingham co., in the state of New Hampshire, 35 m. SSE of Concord, watered by several ponds, and by branches of Merrimac river, and intersected by the Manchester and Lawrence railroad. Pop. in 1840, 926; in 1850, 818.—Also a township of Green co., in the state of New York, 32 m. SW of Albany, drained by Pataviakill and Red-kill creek. Pop. in 1840, 2,417; in 1850, 2,048.—Also a township of Portage co., in the state of Ohio, 130 m. NE of Columbus, drained by a branch of Mahoning river. Pop. in 1840, 907; in 1850, 813.—Also a township of Bradford co., in the state of Pennsylvania, 115 m. N by E of Harrisburg, drained by Wepassening creek. Pop. in 1840, 879; in 1850, 1,000.—Also a township of Lucerne co., in the same state, 25 m. NW of Wilkesbarre, drained by Big and Little Mahoopany creeks. Pop. in 1840, 1,647.—Also a township of Windham co., in the state of Vermont, 76 m. S of Montpelier, drained by branches of Williams, West, and Saxton's rivers. Pop. in 1850, 763.

WINDHAM CENTRE, a village of Greene co., in the state of New York, U. S., on Red-kill creek, 33 m. SW of Albany. Pop. in 1850, 700.

WINDHAUSEN, a village of Brunswick, in the circle of Wolfenbuttel. Pop. 490.—Also a village of Hesse, in the prov. of Upper Hesse, and circle of Grunberg. Pop. 460.

WINDHEIM, a village of Prussia, in the reg. and circle of Minden, on the Weser. Pop. 400.—Also a village of Bavaria, in the circle of Upper Franconia, and presidial of Ludwigstadt. Pop. 390.—Also a village in the circle of Lower Franconia, and presidial of Hammelburg. Pop. 424.—Also a village in the same circle, in the presidial of Munnerstadt. Pop. 182.—Also a village in the same circle, in the lordship of Rotherfels. Pop. 300.

WINDHOF, a village of Austria, in Styria, circle of Gratz, near Peckan. Pop. 800.

WINDHORING, a village of Bavaria, circle of Swabia, and presidial of Altotingen. Pop. 350.

WINDIGSTEIG, a market-town of Austria, in the circle of the Upper Mannhardsberg, on the Thaya, 30 m. NW of Krems. Pop. 335.

WINDIG-YENIKAU, a town of Bohemia, in the circle and 33 m. S of Czaslau.

WINDISCH, a parish and village of Switzerland, in the cant. of Argau, district and 1½ m. E of Brugg, on the Reuss, near its confluence with the Aar. Pop. 1,294. This village occupies the site of the *Vindonissa* of the Romans, one of the most important towns of Helvetia, destroyed by the Germans in 570, and of which some remains still exist. It was the seat of the bishopric transferred after its destruction to Constance.

WINDISCHBACH, a village of Würtemberg, circle of the Jaxt, and bail. of Oehringen. Pop. 390.

WINDISCH-BLEYBERG, a village of Illyria, in the gov. of Laybach, and circle of Klagenfurt. It has a lead mine and mineral springs.

WINDISCHBUCH, a market-town of Baden, in the circle of the Lower Rhine, bail. and 3 m. S of Boxberg. Pop. 450.

WINDISCHENLAIBACH, a village of Bavaria, in the circle of Upper Franconia, and presidial of Weidenberg. Pop. 280.

WINDISCHESCHENBACH, a market-town of Bavaria, in the circle of the Upper Pfalz, and presidial of Neustadt. Pop. 917.

WINDISCH-FEISTRITZ. See FEISTRITZ-WINDISCH.

WINDISCHGARSTEN, a market-town of Austria, in the country above the Ens, and circle of the Traun, 27 m. SSW of Steyer, in the midst of mountains, on the Teichel. It has mineral springs and baths, and in the environs are numerous forges and saw-mills.

WINDISCH-GRATZ, or SLAVENI-GRADEZ, a town of Austria, in Styria, in the circle and 20 m. ESE of Cilly, on the Missling. Pop. 700. It has a castle and an hospital.

WINDISCHHAUSEN, a village of Bavaria, in the circle of the Middle Rhine, and presidial of Heidenheim. Pop. 230.

WINDISCH-HOLZHAUSEN, a village of Prussia, in the reg. and circle of Erfurt. Pop. 200.

WINDISCHIN-STAIG, a market-town of Austria, in the country below the Ens, and circle of the Upper Mannhardsberg, on the Thaya. Pop. 400.

WINDISCH-KAPPEL, a market-town of Austria, in Illyria, in the gov. of Laibach, and circle of Klagenfurt. It has a mine of quicksilver.

WINDISCHLANDSBERG, a market-town of Austria, in Styria, in the circle and 20 m. ESE of Cilly, on the r. bank of the Sotta. Pop. 400. It has a castle.

WINDISCHLEUBA, a village of Saxe-Altenberg. Pop. 480.

WINDISCHMATREY, or WINDISCH-MATTRAY, a market-town of Austria, in the Tyrol, in the circle of the Pusterthal. Pop. 558.

WINDISCHPROBEN, a market-town of Hungary, in the comarca of Thurocz. It has a castle.

WINDISCHSCHLETTEN, a village of Bavaria, in the circle of Upper Franconia, and presidial of Schesslitz. Pop. 270.

WINDLACH, a village of Switzerland, in the cant. of Zurich, and bail. of Regensberg. Pop. 390.

WINDLE, a township in the p. of Prescot, co.-palatine of Lancaster, containing the hamlets of Hardshaw and St. Helen's. Area 2,907 acres. Pop. in 1831, 5,825; in 1851, 9,310.

WINDLESHAM, a parish in Surrey. Area 5,874 acres. Pop. in 1831, 1,912; in 1851, 1,794.

WINDLESTONE, a township in the p. of St.-Andrews-Auckland, co.-palatine of Durham, 4 m. ESE of Bishop-Auckland. Area 1,250 acres. Pop. in 1831, 201; in 1851, 153.

WINDLEY, a township in the p. of Duffield, Derbyshire, 6¾ m. NNW of Derby. Pop. 219.

WINDMAIS, a village of Bavaria, in the circle of the Upper Pfalz, and presidial of Oberviechtach. Pop. 114.

WINDMILL POINT, a promontory of Upper Canada, on the N side of St. Lawrence, 1 m. E of Prescott.

WINDO, an island of Sweden, in the laen of Stockholm, to the NE of Rimaroe.

WINDORF, a market-town of Bavaria, in the circle of Lower Bavaria, and presidial of Vilshofen, on the Danube. Pop. 524.—Also a village of Saxony, in the bail. of Leipzig. Pop. 355.

WINDRATH, a village of Prussia, in the regency of Dusseldorf and circle of Elberfeld. Pop. 440.

WINDRUSH, a parish in Gloucestershire, 5¼ m. E of Northleach. Area 1,710 acres. Pop. 332.

WINDSBACH, a town of Bavaria, in the circle of Middle Franconia, and presidial of Heilsbronn, on the Rezat. Pop. 1,196. It is surrounded with walls, with two gates, and has two churches and a Latin school. It has tobacco-mills, and carries on an active trade in wood.

WINDSBERG, a village of Bavaria, in the circle of the Pfalz, and cant. of Pirmasens. Pop. 540.

WINDSCHACHT. See SCHEMNITZ.

WINDSCHLAG, a village of Baden, in the circle of the Middle Rhine, bail. of Offenburg. Pop. 884.

WINDSFELD, a village of Bavaria, in the circle of Middle Franconia, and presidial of Heidenheim. Pop. 285.

WINDSHAUSEN, a village of Bavaria, in the circle of Lower Franconia, and presidial of Neustadt, on the Saale. Pop. 308.

WINDSHEIM, a town and presidial of Bavaria, in the circle of Middle Franconia, on the Aisch, at the confluence of the Raunach, and 18 m. NW of Anspach. Pop. 3,335. It is enclosed by walls, with 5 gates, and has a castle, two churches, an hospital, and a Latin school.

WINDSOR (NEW), a borough, market-town, and parish, in the NE corner of Berkshire. The borough is pleasantly situated on the r. bank of the Thames, 19 m. E by N of Reading, 40 m. SE by E of Oxford, and 22 m. W by S of London. Its principal thoroughfare bears the names successively of Park, High, Castle, and Thames street; and describes, through the centre of the town, from the E end of the Thames, a curvature of about ½ a m. in length. The smaller streets so ramify or intersect the greater, as to give the town no pretension to regularity of street arrangement. York-place, Gloucester-place, Brunswick-terrace, the New-road, Grove-place, and Clarence-crescent, are the chief lines of new houses erected in the S and W outskirts. The royal palace, with its sumptuous suites of accompanying edifices, sweeps eastward from the centre of the town, away into the royal demesne, and looks out from the summit of a rising-ground upon a brilliant and far-extending landscape. The infantry barracks, built in 1795, and situated in Sheet-street, accommodate nearly 1,000 men. The cavalry barracks, situated at Spital, ½ a m. to the south, have less accommodation, but a better appearance. The bridge connecting Windsor with Eton, and placed on a line with Thames street, is 200 ft. long and 29 ft. wide, and consists of 3 cast-iron arches, resting on piers of granite. The town-hall, situated in High-street. is a plain edifice, with an interesting interior. The theatre, built in 1815, at a cost of nearly £6,000, has little exterior ornament, but is said to possess some interior embellishment. The parish-church, built in 1822, is a spacious structure in the plain Gothic style, surmounted by an embattled and pinnacled square tower. The town has good public and subscription libraries. Though numerous craft are borne past by the Thames, the town profits little by the navigation. Ale is made in large quantity, and sent chiefly to London. Omnibuses ply to the Slough station of the Great Western railway. The p. has an area of 3,237 acres. The parish of New W. is all comprehended within the borough-boundaries. Windsor comprises the hamlet of Dedworth, and a part of the p. of Clewer. Pop. of parl. burgh in 1831, 7,103; in 1851, 9,596. The borough sends two members to parliament. Electors in 1852, 712.

WINDSOR-CASTLE, a royal palace—a suite of majestic structures, quaintly yet expressively pronounced, for its combination of grandeur, beauty, strength, and antiquity, "an emblem of the British constitution"—lies on the E side of the town. "Windsor," says Von Raumer, "has far surpassed my expectation, and produced a greater impression on my mind than any other castle I have ever seen. It combines the bold peculiarity of the middle ages with all the comfort and magnificence of the present day. There is not a tedious symmetrical repetition of similar apartments, not the same thing over and over again; but every staircase, every gallery, every saloon, every window is different, striking, and peculiar—in a word, it is poetical. These giant towers, bastions, balconies, chapels, and halls of chivalry—the ever-changing prospect over stream and valley, wood and pasture—the fantasy of a thousand years compressed within a single moment, all combine to produce an effect exceeding everything that scenic decoration can produce on canvass." A Saxon palace stood at old Windsor, 1½ m. from the present castle, and was occasionally the residence of the kings of England. A fortress at New Windsor was given, along with the town, by Edward the Confessor, to the abbey of St. Peter at Westminster; but was purchased back, reconstructed, and made a place of considerable strength by William the Conqueror. It was enlarged by Henry I., and, in the time of Stephen, it was esteemed the second fortress in the kingdom. It became the rendezvous of the royal party in the civil wars of Henry III.; and it was the favourite residence of Edwards I. and II., and the birth-place of Edward III., and other princes. Edward III. commenced, and extensively achieved, the construction of the whole pile on its present plan of vastness and grandeur. Edward IV. rebuilt, upon an extended plan, St. George's chapel; Henry VII. made several additions to the chapel, and to the upper ward; Henry VIII. rebuilt the principal gateway—that which still bears his name; Elizabeth built, in the peculiar architectural style of her age, a gallery still named after her, the northern terrace; and Charles II. made numerous additions and changes, some enriching, others not a little debasing. The castle passed to George III., nearly in the state in which it was left by Charles II.; but, during the former's reign, it was much renovated and embellished in its chapel, in the N front of its upper ward, and in some other of its parts. Soon after George III.'s death, a purpose was adopted, on a grand scale, to strip the entire building of excrescences, to free it from occasional features of doubtful taste, to mould it into symmetry, and to extend and beautify it into uniform magnificence; and the designs of Mr. Jeffry Wyatt, afterwards Sir Jeffry Wyattville, were adopted for the purpose. The castle crowns a hill, or rising ground, which extends from E to W, and is somewhat precipitous on the N, but, on the other three sides, slopes gently to the plain. The Thames makes so noble sweep round it, gemmed with islets and gorgeously decorated along the banks, as to be 1 furlong from the W end, 5 furls. from the middle of the N, and 3½ from the NE corner. The magnificent pile appears more or less imposing and picturesque from almost every point of view within a radius of several miles. Its N front, which presents the longest continuous range of its buildings, and comprises Queen Elizabeth's picture-gallery, the state apartments, and Winchester, George III.'s, and Brunswick towers, is seen with impressive effect from the Bath road and the Great Western railway, but especially from the play-grounds of Eton college. The E front, which, with its four grand towers seems of itself to be larger than most other great edifices—which comprises the suite of apartments at present occupied by the Queen—and which has, as a beautiful fore-ground, the new Pentagon terrace, with bastions at the angles, and an enclosed flower-garden—is strikingly seen on the road to Windsor from Datchet. The S front, and the general outline, combined with the sylvan and luxuriant frame-work of scenery in which the whole is set, produce, to a spectator at the extremity of the Long walk, and especially to one on Snow-hill, perhaps the most beautiful, though not the most striking, of the entire series of views. The W front, and the combination of it and the north one, form impressive though close views, as seen from the Thames. The Round tower, 'the proud keep of Windsor,' as seen round the whole circumference of the range of vision, rises above the whole group of edifices, and fixes the attention by the vastness of its bulk. It crowns the summit of a lofty artificial mound between the two wards of the castle, and is surrounded by a belt of low ground, once a fosse, but now filled up, or transmuted into a shrubbery and garden. The tower is 302½ ft. in circumf., and lifts the top of its flag-pole to the alt. of 295½ ft. above the level of the Little park.—The State apartments occupy the N side of the great quadrangle.—The chapel, dedicated to St. George, and designed for the special use of the order of the Garter, founded at Windsor by Edward III., but chiefly constructed by Henry VII., is remarkable for its exquisite proportions, and the rich yet solemn ornaments of the interior. It is in the purest style of the later pointed architecture.

WINDSOR, a town of New South Wales, in the p. of St. Matthew, co. of Cumberland, 20 m. from Parramatta, 35 m. from Sydney, at the confluence of South creek with Hawkesbury river, on a hill, at an alt. of 100 ft. above the level of the river. Pop. 1,679. The streets are regularly built, but the houses are chiefly of wood. It has a church, a Methodist chapel, a public school, and an hospital. It was originally named Greenhill.—Also a river of New Zealand, on the SW coast of the island of Tavaï-Poenammon, which flows into the ocean in S lat. 46° 10′.

WINDSOR, a town of Nova Scotia, in the co. of Hants, of which it is the capital, on the r bank of a river of the same name, also called Pigaquit, and near its entrance into Minas bay. It is well built, and has a university.

WINDSOR, a village of Upper Canada, in the township of Sandwich, on the Detroit river, opposite the town of that name, on a bank, at an alt. of about 50 ft. above the river, which is here about 1 m. in width. Pop. 500.—Also a village in the township of Whitby, 2 m. from Windsor bay, on Lake Ontario, and 31 m. from Toronto. Pop. 500.

WINDSOR, a county in the SE part of the state of Vermont, U. S., comprising an area of 912 sq. m., drained by Black, Queechee, Williams, and White rivers, and intersected by the Vermont Central and the Rutland and Burlington railroads. It is hilly, but is in some parts very fertile. Pop. in 1840, 40,356; in 1850, 38,320. Its cap. is Woodstock.—Also a township of the above-named co., on the W side of Connecticut river, 54 m. S by E of Montpelier, drained by Mill branch, and intersected by the Sullivan railway. It is generally hilly, and in the S is Ascutney mountain, which has an alt. of 3,320 ft. above sea-level. Pop. in 1840, 2,744; in 1850, 1,928.—Also a township of Kennebec co., in the state of Maine, 10 m. E of Augusta, drained by Sheepscot river. Pop. in 1840, 1,789; in 1850, 1,793.—Also a township of Hillsboro' co., in the state of New Hampshire, drained by branches of Contoocook river. Pop. in 1840, 177.—Also a township of Berkshire co., in the state of Massachusetts, 83 m. W by N of Boston, drained by branches of Westfield river. Pop. in 1840, 897; in 1850, 899.—Also a village of Broom co., in the state of New York, 106 m. WSW of Albany, drained by Susquehanna river, on the W bank of which the village is situated. Pop. of township in 1840, 2,368; in 1850,

2,645; and of village, 440.—Also a township of York co., in the state of Pennsylvania, 32 m. SE of Harrisburg, watered by Grist and Fish creeks, and by Cabin branch and Beaver creeks.—Also a township of Berks co., in the same state, bordered on the W by Schuylkill river and canal, on the E by Maiden creek, and on the NW by Blue mountain. Pop. in 1840, 2,882.—Also a village of Bertie co., in the state of North Carolina, on Cashie river, 93 m. E by N of Raleigh.—Also a township of Ashtabula co., in the state of Ohio, 146 m. NE of Columbus, drained by Grand river and its branches. Pop. in 1840, 876; in 1850, 1,033.—Also a township of Lawrence co., in the same state, drained by Symmes and Indian Guyandot creeks. Pop. in 1840, 815.—Also a village of Bureau co., in the state of Illinois, 119 m. N of Springfield. Pop. 150.—Also a township of Eaton co., in the state of Michigan, 8 m. SW of Lansing, drained by the S branch of Grand river. Pop. 253.—Also a township of Dane co., in the state of Wisconsin, 12 m. N by W of Madison, drained by affluents of Fourth lake. Pop. 860.

WINDSOR-LOCHS, a village of Hartford co., in the state of Connecticut, U. S., on the W bank of Connecticut river, 12 m. N of Hartford, and on the New Haven, Hartford, and Springfield railway.

WINDSOR (OLD), a parish in Berks, 2 m. SE by E of New Windsor, on the SW bank of the Thames. Area 5,401 acres. Pop. in 1851, 1,785.

WINDWARD ISLANDS. See ANTILLES.

WINEGA (EL), a village of Fezzan, 100 m. WNW of Mourzuk, visited by Hornemann in 1798.

WINENNE, a department and commune of Belgium, in the prov. of Namur, and arrond. of Dinant. Pop. 701.

WINESTEAD, a parish in the E. R. of Yorkshire, 2 m. NW by W of Patrington. Area 2,570 acres. Pop. in 1831, 145; in 1851, 131.

WINFARTHING, a parish in Norfolk, 4 m. N of Diss. Area 2,620 acres. Pop. in 1851, 691.

WINFIELD, a township of Herkimer co., in the state of New York, U. S., 74 m. W by N of Albany, drained by head branches of Unadilla river. Pop. in 1840, 1,652; in 1850, 1,481.

WINFORD, a parish in Somersetshire, 6½ m. SW by W of Bristol, containing the tythings of Felton, Reghill, and W. Area 2,991 acres. Pop. in 1831, 865; in 1851, 1,022.

WINFORTON, a parish in Herefordshire, 6 m. NE by E of Hay, N of the river Wye. Area 1,099 acres. Pop. in 1831, 158; in 1851, 159.

WINFRITH-NEWBURGH, a parish in Dorset, 8 m. W by S of Wareham, containing the hamlets of Bromhill, Burton, Foreshill, Knighton, Lancotes, Bindon, Burngate, and Hamburgh. Area 4,496 acres. Pop. in 1831, 891; in 1851, 1,101.

WING, or WENGE, a parish in Bucks, 4 m. SW by W of Leighton-Buzzard, containing the hamlets of Ascot, Burcot, Cottesloe, and Crofton. Area 5,310 acres. Pop. in 1831, 1,052; in 1851, 1,376.—Also a parish in Rutlandshire, 4 m. NE by E of Uppingham. Area 1,050 acres. Pop. in 1851, 334.

WING, a township of Lucas co., in the state of Ohio, U. S. Pop. in 1840, 145.

WINGATE, or WINDEGATE-GRANGE, a chapelry and township in the p. of Castle-Eden, co.-palatine of Durham, 6½ m. ESE of Durham, in communication with Hartlepool by railway. Area 4,154 acres. Pop. in 1801, 135; in 1831, 115; in 1851, 2,456. The great increase in pop. has been occasioned by the opening of coal mines.

WINGATES, a township in the p. of Long Horsley, Northumberland, 9 m. NNW of Morpeth, comprising the hamlets of Chirm, Garrotlee, Whelm, and Wingates-Moon. Area 2,642 acres. Pop. 186.

WINGECARRIBEE, a river of New South Wales, in the co. of Camden, which flows through the town of Berrima, and joins the Wollondilly.

WINGEN, a village of France, in the dep. of the Lower Rhine, cant. and 7 m. WSW of Wissembourg. Pop. 500. It has the remains of a fortress and ditch.

WINGEN, a volcanic mountain of New South Wales, near the river Hunter.

WINGENDORF, a village of Prussia, in the regency of Coblenz, and circle of Altenkerchen. Pop. 221.—Also a village of the regency of Liegnitz, and circle of Lauban. Pop. 221. It has a paper-mill.—Also a village of Saxony, in the circle of Dresden, and bail. of Freyburg. Pop. 325.

WINGERATH, a village of Prussia, in the reg. of Treves, and circle of Bernkastel. Pop. 200.

WINGERODE, a village of Prussia, in the reg. of Erfurt, and circle of Worbis. Pop. 970.

WINGERSHAUSEN, a village of Hesse, in the prov. of Upper Hesse, and circle of Nidda, near Satten. Pop. 260.

WINGERSHEIM, a commune of France, in the dep. of the Lower Rhine, and cant. of Hachfelden. Pop. 1,379.

WINGERWORTH, a parish in Derby, 2¾ m. S by W of Chesterfield, intersected by the North Midland railway. Area 2,906 acres. Pop. in 1831, 471; in 1851, 463. Coal, ironstone, and freestone, are found in abundance in this vicinity.

WINGESHAUSEN, a village of Prussia, reg. of Arnsberg, and circle of Wittgenstein. Pop. 750.

WINGFIELD, a parish in Suffolk, 5¾ m. E by N of Eye. The church is a handsome edifice of high antiquity, constructed of flints and stone. Area 2,441 acres. Pop. in 1831, 668; in 1851, 654.

WINGFIELD, or WINKFIELD, a parish in Wilts, 2 m. WSW of Trowbridge. Area 1,383 acres. Pop. in 1831, 288; in 1851, 305.

WINGFIELD (NORTH), a parish in Derbyshire, 4½ m. SSE of Chesterfield, near the line of the North Midland railway, comprising the townships of Claylane, Stretton, Tupton, and Woodthorpe. Area 7,603 acres. Pop. in 1801, 1,452; in 1831, 1,691; in 1851, 4,351; an increase attributable to the opening of coal mines in the township of Claylane.

WINGFIELD (SOUTH), a parish in Derby, 2 m. W of Alfreton, crossed by the North Midland railway, on which there is here a station. Area 3,308 acres. Pop. in 1831, 1,091; in 1851, 1,092.

WINGHAM, a parish and village in Kent, 6 m. E by S of Canterbury, on the high road from Canterbury to Deal. Area 2,841 acres. Pop. 1,083.

WINGHE - SAINT - GEORGES, a department and commune of Belgium, in the prov. of Brabant, and arrond. of Louvain. Pop. 1,071. It has a castle and an ancient convent, and carries on an active trade in wood.

WINGRAVE, a parish in Bucks, 5½ m. NE of Aylesbury. Area 2,600 acres. Pop. in 1851, 813.

WINHALL, a township of Bennington co., in the state of Vermont, U. S., 80 m. S by W of Montpelier, and drained by a river of the same name. Pop. in 1840, 576; in 1850, 762.

WINHORING, a village of Bavaria, in the circle of Upper Bavaria, and presidial of Altötting. Pop. 350. It has a castle.

WINIARY, a village of Prussia, in the regency and circle of Posen. Pop. 350.

WINICKY, a village of Austria, in Galicia, in the circle of Lemberg. Pop. 2,300. It has an extensive manufactory of tobacco.

WINIKON, a village of Switzerland, in the cant. of Luzern, bail. of Sursee, on the Suren. Pop. 879.

WINITZ, a village of Austria, in Bohemia, in

the regency of Gitschin, and bail. of New-Bidschow. Pop. 220.

WINKBOURNE, a parish in Nottinghamshire, 3 m. N of Southwall. Area 2,240 acres. Pop. 129.

WINKEL, a village of Bavaria, in the circle of Upper Bavaria, and presidial of Berchtesgaden. Pop. 250.—Also a village of the same circle, in the presidial of Landsberg. Pop. 250.—Also a market-town of the duchy of Nassau, in the bail. of Budesheim, on the r. bank of the Rhine. Pop. 1,700.—Also a village of Austria, in the Tyrol, in the circle of Brixen, and bail. of Lienz. Pop. 350.—Also a village of Saxe-Weimar, in the bail. of Altstadt. Pop. 433.—Also a village of Switzerland, in the cant. and 4 m. S of Luzern, on an arm of the Waldstetten-see.—Also a village in the cant. of Zurich, and bail. of Bülach. Pop. 728. See also WETTIN.

WINKELDORF, a village of Hanover, in the gov. of Stade, and bail. of Ottersberg. Pop. 200. —Also a village of Prussia, in the regency of Breslau, and circle of Habelschwerdt. Pop. 280.

WINKELHAUSEN, a village of Prussia, in the reg. of Dusseldorf, and circle of Lennep. Pop. 330.

WINKELHORST, a village of Prussia, in the reg. of Munster, and circle of Beckum. Pop. 220.

WINKELS, a village of Nassau, and bail. of Weilburg. Pop. 450.

WINKEL-SAINT-CLOI, a village of Belgium, in the prov. of West Flanders, and arrond. of Courtrai. Pop. 2,350.

WINKELSAS, a village of Bavaria, in the circle of Lower Bavaria, presidial of Rottenburg. Pop. 200.

WINKELSDORF, a village of Austria, in Moravia, in the circle of Olmutz, and lordship of Ullersdorf. Pop. 760. It has a glass-work.

WINKELSHUTTEN, a village of Prussia, in the reg. of Minden, and circle of Halle. Pop. 628.

WINKELSTETTIN, a village of Hanover, in the principality of Osnabruck, and bail. of Iburg. Pop. 350.

WINKFIELD, a parish and village in Berks, 6 m. SW of New Windsor, including the hamlets of Ascot, Bulbrook, and Hateley. Area 8,982 acres. Pop. in 1831, 2,009; in 1851, 2,185.

WINKHAUSEN, a village of Prussia, in the reg. of Dusseldorf, and circle of Duisburg. Pop. 860.

WINKLARN, a market-town of Bavaria, in the circle of the Upper Pfalz, and presidial of Oberviechtach. Pop. 862. It has a castle, and possesses important manufactories of cloth and other fabrics, and fisheries.

WINKLERN, a village of Austria, in Styria, in the circle of Judenberg. Pop. 350.

WINKLERSBERG. See SCHNEEGEBIRGE.

WINKLING, a village of Bavaria, in the circle of Lower Bavaria, and presidial of Bogen. Pop. 280.

WINKOWITZ. See VINKOVCZE.

WINKSLEY, a chapelry in the p. of Ripon, W. R. of Yorkshire, 4¾ m. W of Ripon. Area 703 acres. Pop. in 1831, 259; in 1851, 214.

WINLATON, a parochial chapelry and township in the p. of Ryton, co.-palatine of Durham, 6¼ m. W by S of Gateshead, on the river Derwent, over which there is here a bridge. A handsome chain bridge also has been thrown across the Tyne, between this place and Scotswood in Northumberland. Area 8,261 acres. Pop. in 1831, 3,951; in 1851, 6,085. Here are extensive iron-works, established in 1690, in which anchors, even of the largest size, anvils, chain-cables, cylinders, edge-tools, nail-rods, files, &c., are manufactured. On the banks of the Tyne are works for refining lead.

WINMARLEIGH, a township in the p. of Garstang, co.-palatine of Lancaster, 3 m. NW of Garstang, near the Preston and Lancaster canal and railway. Area 2,282 acres. Pop. in 1851, 262.

WINNAGORA, a village of Prussia, in the reg. of Posen, and circle of Schroda. Pop. 400.

WINNALL, a parish in Hants, 1 m. NNE of Wi chester. Area 561 acres. Pop. in 1851, 117.

WINNAMAC, a village of Pulaski co., in the state of Indiana, U. S., on the W side of Tippecanoe river, and 90 m. N by W of Indianapolis, and intersected by the Indiana Central railroad. Pop. in 1850, 240.

WINNBACH, a village of Austria, in the Tyrol, circle of Brixen, and bail. of Bruneck. Pop. 500.

WINNEBAGO, a county in the N part of the state of Illinois, U. S., comprising an area of 534 sq. m., intersected by Rock river, and by its branches Kishwaukie and Pekatomca, and by the Galena and Chicago, and the Belvoit branch railroads. Pop. in 1840, 4,609; in 1850, 11,773.—Also a county in the N part of the state of Iowa, comprising an area of 432 sq. m., drained by affluents of Mankato river. —Also a central county of the state of Wisconsin, comprising an area of 471 sq. m., drained by Neenah or Fox, and Panwaggun rivers, and containing several lakes. Pop. in 1840, 135; in 1850, 10,125. Its cap. is Ashkosh.—Also a lake in the E part of the state of Wisconsin, 30 m. W of Lake Michigan, between Winnebago and Calumet counties, and in the N part of Fond-du-Lac co. It is 32 m. in length, and 8 m. in average breadth. It receives several streams, and discharges itself into Green bay.

WINNEBAH, a town of Africa, in Upper Guinea, on the Gold coast, in the kingdom of Ashantee, to the W of Acorah. Pop. 4,000.

WINNELENDONK, a village of Prussia, in the reg. of Dusseldorf, and circle of Geldern. Pop. 845.

WINNEN, a village of Hesse, in the prov. of Upper Hesse, circle of Marburg, and bail. of Treis, on the Lumbde. Pop. 200.—Also a village of Nassau, and bail. of Kennerod. Pop. 250.

WINNENDEN, a town of Würtemberg, in the circle of the Neckar, bail. and 6 m. NE of Waiblingen. Pop. 3,200. It has a lunatic asylum, formerly the royal castle of Winnenthal, and several schools, and contains extensive tanneries, dye-works, and yarn-mills. The trade consists chiefly in grain and wood.

WINNENTHAL, a market-town of Prussia, in the regency and 32 m. NW of Dusseldorf, and circle of Geldern. Pop. 1,300.

WINNESHEIK, a county in the NE part of the state of Iowa, U. S., comprising an area of 732 sq. m., drained by Upper Iowa and Turkey rivers and their branches. Pop. in 1850, 546. Its cap. is Winnesheik.

WINNEZELLE, a village of France, in the dep. of the Nord, and cant. of Steenwoorde. Pop. 1,550.

WINNIBIGOSHISH LAKES, a group of lakes in the Minnesota Territory, 180 m. N by W of St. Paul. It is 14 m. long, and 9 m. wide.

WINNICA. See VINNITZA.

WINNIFELD, or VARUSFELD, a territory of the principality of Lippe-Detmold, and bail. of Detmold.

WINNINGEN, a market-town of Prussia, in the regency and circle and 5 m. SW of Coblenz, on the Moselle. Pop. 1,548, chiefly Protestants. The vine is assiduously cultivated in the vicinity of this place, and the wine produced here is the best in the regency.—Also a village in the regency of Magdeburg, and circle of Aschersleben. Pop. 552.

WINNINGTON, a township in the p. of Great Budworth, co.-palatine of Chester, 1 m. NW of Northwich, on the river Weever, over which there is here a bridge, and in the line of the Grand Junction canal. Area 597 acres. Pop. in 1831, 256;

in 1851, 405.—Also a township in the p. of Mucklestone, Staffordshire, 4½ m. NE of Drayton-in-Hales. Pop. in 1831, 249; in 1851, 200.

WINNIPEG, a river of North America, which rises in the Lake-of-the-Woods, and flows in a W course to Lake Winnipeg, into which it enters at Fort Alexander, in N lat. 50° 36′ 30″, after a singularly broken course of 170 m. It forms the chief water-communication between Lake Superior and Lake Winnipeg. Major Long, in his account f the route pursued by his exploring expedition between the two lakes, says that the W. at its confluence with the lake of the same name, presents a majestic and impressive stream of considerable but variable width. "It is composed of a series of deep and broad basins rising one above another, and serving as the channel of a huge volume of water, which is precipitated from one basin to another in tumultuous cataracts of the most romantic character. Of these waterfalls, there are no less than 31 in the route to the Lake-of-the-Woods, which interrupt the passage of canoes, and at all of which are carrying places. The aggregate descent of water in the W. may be estimated at 410 ft., which may be considered as the elevation of the Lake-of-the-Woods above Lake W. The route by Covert and Sturgeon-Dam rivers is probably the most direct, (the lower portion of which is the same with that above mentioned,) but the obstructions are said to be more numerous and formidable, especially in a low stage of water. Besides these there are numerous other deviations from the main route, some of which have been traversed, but the number that remains to be explored is doubtless far greater. At the distance of about 60 miles below the Lake-of-the-Woods, the W. receives a large tributary from the north, called English river, which is of a character similar to that of the principal, and nearly as large as the latter above their junction. Its head-waters interlock with those of Albany river, which empties into James' bay, and is the principal channel of intercourse between Lake W. and the trading establishments on that river. The Lake-of-the-Woods is about 75 m. long, and of irregular widths, from 10 to 35 or 40 m. Compared with other lakes, it deserves a high rank on the scale of beauty. The scenery is wild and romantic in a high degree, its shores being faced with precipices and crowned with hills and knobs of variable heights, clad with a dense foliage of shrubbery and evergreens. Its surface is beautifully studded with countless islands of various sizes and forms, disclosing between them the continued sheet of its wide-spreading waters, the extent of which enlarges upon the vision as the traveller advances upon the lake, till the main land is shut out from the view by the islands that multiply around him. The 49th parallel crosses the lake within the distance of about 12 m. from its S extremity. The region bordering upon the waters above described, is one of the most dreary imaginable. Its climate is rigorous, its surface exceedingly rugged and broken, and its products so limited and meagre, that it seems never to have been claimed as a residence either by man or beast. A solitary moose, caraboo, or bear, is occasionally to be found; and a half-starved family of savages sometimes fix a temporary residence upon some of the water-courses, and subsist miserably upon fish, but it seems as if comfort and competency were denied to both. The prevailing rocks are primitive, and are almost exclusively the ingredients of which the hills are composed, while the earthy portions of the valley are made up of the coarse and unproductive detritus afforded by their disintegration. The soil is uniformly thin, and in many places totally wanting. The stinted growth of the woodlands, and the dwarfish character of vegetation which prevail generally throughout this region, are attributable to these causes. The islands of the lakes and river are similar to the circumjacent highlands, being uniformly based upon rock and presenting rugged and broken surfaces. The growth found on the lower part of the river, comprises only the aspen, white birch, spruce, tamarack, and scrub-oak, none of which attain any considerable magnitude. As we approach the Lake-of-the-Woods, the following trees make their appearance, viz., two species of pine, called the white and red epinette, the former of which is more commonly called the larch. From the latter is extracted the gum employed for pitching canoes, which usually have their ribs and lining constructed of its timber; a small species of pitch pine called by the Canadians cypress, which also furnishes a gum inferior to that above mentioned; and the liard, a variety of the poplar, more commonly called the balm of Gilead. The undergrowth is dense in many places, and consists of stinted oak, chokecherry, hazle, pembina or bush cranberry, service-berry, arrow-wood, wild plum, raspberry, briar-bush, whortleberry, sumac, wild rose, sweet briar; sand cherry, red and black cherry, pea-vine, gooseberry, currant, bear-berry, &c. &c. Above the Lake-of-the-Woods, Rainy river becomes the channel of communication, and extends 100 m. to the lake of the same name. It has an average breadth of about 300 yds., is deep and gentle, and has no obstructions to its navigation, within 48 m. of its mouth; at this distance are situated the rapids of Rainy river, which are about one mile long, and have an aggregate descent of about 10 ft. About 10 m. further up is another inconsiderable rapid, with a fall of 3 ft. At the outlet of Rainy lake is a rapid of about 5 ft. descent, and 2½ m. below are the falls of Rainy river, down which the torrent pours with terrific grandeur through an aggregate descent of 25 ft. in the distance of but a few yards. At this place are situated an establishment of the Hudson's Bay company on the N side of the river, and one belonging to the American Fur company on the S. Twenty miles below the falls is the entrance of a considerable tributary from the SW, called the Grand fork, which affords a channel of communication between the principal and Little Winnepeek lake of the Mississippi, navigable in wet seasons. It receives several other streams of less note. The contrast between the country of Rainy river and that before described, is no less striking than that between the two water-courses themselves. Here bottoms and table-lands of considerable extent are often to be met with, wearing the aspect of a secondary region; these are, however, generally terminated, at no great distance, by tracts of a rugged and broken character. Rocks are seldom to be found in the immediate valley or bed of the river. The forests are more dense and heavy, and contain several trees not enumerated in the foregoing list, viz. white oak, ash, hickory, water-maple, white walnut, linden, elm, &c. The pine and white birch become more abundant, and attain a more stately size. From the estimates above given, making some allowance for the general descent of the river, it will appear that the surface of Rainy lake is elevated about 60 ft. above that of the Lake-of-the-Woods, or 470 ft. above that of Lake W. Pursuing our course upwards we passed through Rainy, Sturgeon, La Croix, Upper Sturgeon, Doré, Cannibal, and Thousand lakes, besides numerous others of less note and size, and also the channels by which they are connected, which, in the language of the voyagers, are denominated rivers, and known by various names.

before we reach the dividing ridge between the waters of lakes W. and Superior. Several routes are practicable through this part of the country, but the one we pursued is said to be the most frequented. The route connected with the Grand Portage, which was the only route frequented till within a few years, is united to the new route by two channels, one branching off at Lac-de-la-Croix, and the other at Upper-Sturgeon lake, both of which unite in Lake Saganaga, and enter Lake Superior at Pigeon river. The lakes of this region are of all possible shapes, exceedingly numerous, and thronged with islands. As on the route before described, the lakes here rise one above another in continual gradations, but less abruptly, giving an altitude to the uppermost on the route at least equal to 130 ft., making the entire elevation above Lake W. about 600 ft. The alt. of the dividing ridge, above the water-table of the adjacent country, is no where greater than about 150 ft., the head-waters of the streams tributary to Hudson's bay being somewhat more elevated than those of Lake Superior. The channel of communication thence to Lake Superior is through Dog river, the lower portion of which is more commonly called the Kamanatekwoya. This river is exceedingly serpentine in its course, has a regular bed, and a rapid current. About 40 m. below the point where we entered it, is a lake of the same name, about 12 m. long and from 1 to 5 m. broad. The river receives several tributaries, the most considerable of which is the Cypress, entering from the W above Dog lake. Two others (names unknown) one of which enters Dog lake from the NE, and is said to communicate with the English river before noticed, and the other into the SW part of the same lake, affording a communication with the Thousand lakes, which is sometimes travelled; besides these there are two others of considerable size. On this route the portages are equally as numerous, and more extensive than on the other, nor are its other impediments less formidable. Rapids and cataracts abound; among the latter is a magnificent cascade, denominated by the Indians the falls of Kakabikka or 'Cleft Rock,' situated about 30 m. upward from the mouth of the river, which is here contracted to the width of about 50 yards, and supplied with a volume of water unusually large for that width. Thus confined, the whole body of the river is precipitated, in a dense sheet, down a perpendicular precipice more than 130 ft. into a deep chasm, bounded by perpendicular cliffs of the height just mentioned; the banks of the river, for a distance of nearly ½ a mile below, are completely insurmountable, rising perpendicularly, and in many places overhanging their bases. Below the falls of Kakabikka, the river presents a continued rapid for the distance of about 20 m., below which it quietly passes through serpentine folds to its mouth, which is in an arm of the lake called Kamana bay. The whole descent of the water from Coldwater lake, (the first water eastward of the dividing ridge), to Lake Superior, may be estimated at about 600 ft."

WINNIPEG (Lake), a large and interesting lake nearly in the centre of the great middle plain of the North American continent, comprised within what is called the Hudson's bay territories, and lying between the parallels of 50° 30′ and 53° 45′ N lat., and the meridians of 95° 30′ and 99° 30′ W long. It is about 280 m. long, with a breadth varying from 15 to 80 m. Its area has been estimated at 9,000 sq. m.; and its alt. above sea-level by some at 853 ft., by others at only 630 ft. It appears to be the same lake which was called by early travellers the Lake of the Assinipoils. Carver gave it its present appellation, which is said to be derived from the muddy or sallow appearance of its waters: *we*, signifying 'muddy,' and *nepe*, 'water' in the Chippewa language. Its banks on the S and SW are low and marshy; on the NE the shore swells into broad hills of no great elevation, which are covered with a thin growth of pine, spruce, juniper, red-cedar, and white birch. The whole of the eastern shore appears to be of primitive formation; while the western is composed of secondary rocks. Its shores are much indented, but its general direction is about NNW. Few lakes receive so many and such large streams; and by many of them and of the rivers which flow from it, a communication is kept up between the Atlantic and Pacific. The Saskatchawan flowing to the NW, enters this lake after a course of 650 m.; and issuing from its NE extremity, under the name of the Nelson river, flows into the SW side of Hudson's bay, after a further course of 350 m. The Winnipeg river enters its SE extremity; and the Red river, which has its remotest sources in Lake Travers, pursues a nearly N course of about 600 m. to the SW extremity. The latter river—which is to be distinguished from the Red river affluent of the Mississippi—rises in about N lat. 45° 40′, W long. 96° 34′ in the Riviere-des-Sioux, which, after a course of 30 m. from Lake Travers, is joined by the Ottertail river, which rises near the head-waters of the Mississippi, 150 m. NE of Lake Travers. The general course of the confluent streams is N, inclining a little to the W, and exceedingly tortuous. Its length, following its meanders, is, according to Long, more than 500 m., while in the direction of its valley, it does not exceed 360 m. It receives numerous tributaries, of which the following are the principal, and enter it in the order here mentioned, commencing with the uppermost, viz.:—The Pse or Rice river, the Shienne, the Buffalo, Elm, Wild-rice, the Plum and Sandhill rivulets. The Goose river, the Red Fork, the Turtle, Big Saline and Park rivers, the Swamp Brook, the "Two Rivers," and Pembina river, all within the territory of the United States. The Reed Grass, Scotchman, Rat, Muddy, Assiniboin, and Death rivers, entering northwardly of the 49th parallel. Red river is navigable for canoes, and even pirogues of two tons burthen, from its mouth to its source, as also to the sources of several of its tributaries when swollen by freshets. On such occasions, canoes have been known to pass from Lake Travers, its source, and into the St. Peter, and back again without inconvenience. The voyage down the river is now seldom performed, owing to the limited nature of the trade in this direction. Formerly the Hudson's Bay company had a trading establishment on Lake Travers, between which and their establishments lower down the river considerable intercourse existed. It abounds in rapids, which, together with its numerous and extensive windings, render the passage by water very tedious. The aggregate descent from Lake Travers to Lake W., or from the source to the debouchure of Red river, a distance of about 600 m., following the meanders of that stream, amounts, Major Keating estimates, to about 200 ft. Ottertail river is navigable as above to its source, through which a water communication is continued, forming a connection with the Riviere-de-Corbeau of the Mississippi, and the Medicine-Bark of the St. Peter The Wild-rice river and the Red Fork are connected in the same manner with other sources of the Mississippi, affording navigable communications between the subsidiary lakes at their respective heads, and others tributary to that great river. Reed Grass river has a two-fold connection with other waters, serving as a channel of intercourse between Red river and the source of the Mississippi, as also that of one of the tributaries of Rainy river. Rat river also affords a communication between Red river and the Lake-of-the-Woods. These several communications are only practicable in very wet seasons, and the transition from their summits into other waters is interrupted by portages of greater or less extent, but in no instances by ridges of any considerable height. The Goose and Turtle rivers, both of which take their rise in Devil lake, are navigable to that place, and the lake itself, which is said to be made up of a multiplicity of small pools connected by navigable channels, affords an extent of navigation of about 100 m. The Assiniboin is the largest of all the tributaries of Red river, and in point of magnitude and extent, vies even with the principal. Their point of junction is in N lat. 49° 53′ 35″, and W long. 97° 00′ 50″. Its sources mingle with the waters of the Saskatchawan, NW from the point just mentioned. In its progress downwards it forms an extensive curve with a convexity to the SW, and receives numerous subsidiaries, among which is a stream of respectable size, called Mouse river, that is said to receive some of its waters from a point within one mile of the Missouri. The Assiniboin is navigable at all stages of the water to a great distance, and is the channel of continual intercourse between several British trading establishments located on its waters, and one of their principal depots situated at the mouth of the river. The immediate valley of Red river is not bounded by parallel ranges of bluffs or banks like that of the St. Peter and other tributaries of the Mississippi, but expands to a great width, terminated on the west by highlands connected with the Coteau-des-Prairies, and on the E by the ridges and swamps in which the waters of the Mississippi, St. Lawrence, and Nelson's rivers find a common origin. A broad expanse of verdant prairie, says Keating, spreading beyond the utmost extent of vision, is here presented to the view. If we except the margin of the river and those of its tributary streams, which are fringed with trees and shrubbery, there is very little to interrupt the simplicity and uniformity of the scenery; scarcely is there an undulation to variegate the prospect, save what is afforded by an optical illusion that makes the traveller fancy himself in the centre of a basin, and surrounded by an amphitheatre of rising ground at no great distance, which constantly eludes his approach. The soil is generally thin, of a light complexion, and argillaceous structure. The dwarfish appearance of the herbage which it supports, especially on the upper portion of the valley, indicates either a want of fecundity or the admixture of salts or other ingredients not congenial to vegetation. In many places, however, the soil appears rich, supporting a dense and luxuriant growth of grass, weeds, &c. As we descend along the river, the indications of fertility multiply, the soil becomes deeper and the vegetation more thrifty, woodlands become more frequent, and the trees attain a larger size. The flatness of surface that almost uniformly prevails throughout the valley of Red river, may be regarded as a defect in its natural character that cannot easily be remedied. The colony planted by the Earl of Selkirk occupies two positions on the banks of this river, one at the confluence of the Assiniboin, usually called Fort Douglas, and the other about 60 m. above, called Pembina. See Red River Settlement.

WINNIPISEOGEE (Lake), a lake of the state of New Hampshire, U. S., between Belknap and Carrol counties, and at an alt. of 472 ft. above sea-level. It is 22 m. in length, from 1 to 10 m. in width, and discharges itself on the SW, by a river of the same name, into Pemigewasset river, a head-stream of the Merrimac. It contains above 300 small islands, some of which contain from 100 to 200 cultivated acres.

WINNISBORO, a village of Fairfield district, in the state of South Carolina, U. S., on the Charlotte and South Carolina railroad. Pop. in 1850, 1,050.

WINNOWE (Saint), a parish in Cornwall, 2½ m. SE of Lostwithiel, on the river Fowey, which is here navigable, and is crossed by the high road from Bodmin to Plymouth. Area 6,137 acres. Pop. in 1831, 1,048; in 1851, 1,076.

WINNSDORF, a village of Prussia, in the regency of Oppeln, and circle of Neisse. Pop. 310.

WINNWEILER, a town and presidial of Bavaria, in the circle of the Pfalz, 15 m. NE of Kaiserslautern, on the Alsenz. Pop. 1,284. It has iron and copper-works.

WINOOSKI-FALLS, a village of Chittenden co., in the state of Vermont, U. S., on the N bank of Onion river, and intersected by a branch railroad from Burlington to Essex junction.

WINSCALES, or Winskel, a township in the p. of Workington, Cumberland, 2½ m. SE of Workington. Pop. in 1831, 100; in 1851, 200.

WINSCHOTEN, a town of Holland, in the prov. and 22 m. ESE of Groningen, on the Bensel, canalised under the name of Winschoter-Trekvaart, and 8 m. S of Dollart. Pop. 3,500. It has ancient fortifications, and is noted for a victory by the Prince of Orange over the Spaniards in 1568.

WINSCOMBE, a parish in Somerset, 2 m. N by W of Axbridge. Area 4,140. Pop. in 1851, 1,439.

WINSELDORF, a village of Holstein, in the district of Breitenburg. Pop. 279.

WINSELER, a village of Holland, in the prov. of Luxemburg, bail. of Dickirch. Pop. 240.

WINSEN-AN-DER-ALLER, a village and parish of Hanover, in the gov. and 48 m. SSW of Luneburg. Pop. 928.

WINSEN-AN-DER-LUHE, a village and parish of Hanover, in the gov. and 11 m. NW of Luneburg, on the Luhe, which here becomes navigable. Pop. 1,821. It has a castle and an hospital, and possesses extensive fisheries, manufactories of vinegar, and breweries.

WINSFORD, a parish in Somersetshire, 5 m. N by W of Dulverton crossed by the river Ax, containing the hamlets of Northern-Mill and Garleycombe, also the ancient, though now depopulated, villes of Quarum-Kitnor and Quarum-Monceaux. Area 8,656 acres. Pop. in 1831, 524; in 1851, 604.

WINSHAM, a parish in Somersetshire, 4 m. SE by E of Chard. Area 2,953 acres. Pop. 1,062.

WINSHILL, a township in the p. of Burton-upon-Trent, Derbyshire, 1½ m. ENE of Burton-upon-Trent. Area 1,150 acres. Pop. in 1851, 405.

WINSLADE, a parish in Hants, 3 m. S by E of Basingstoke, containing the tything of Kempshot. Area 1,235 acres. Pop. in 1831, 134; in 1851, 174.

WINSLEY, a hamlet in the p. of Darley, Derbyshire, 3½ m. WNW of Matlock, near the river Derwent. Area, with Snitterton, 2,200 acres. Pop. in 1831, 671; in 1851, 557.—Also a chapelry in the p. of Bradford, Wilts, 1½ m. W of Bradford, near the Kennet and Avon, and the Wilts and Berks canals. Pop. in 1831, 2,847; in 1851, 1,587.

WINSLOW, a parish and market-town in Bucks, 6½ m. NE of Buckingham. Area 1,920 acres. Pop. in 1831, 1,290; in 1851, 1,889. The town, which has existed from a remote period, and is mentioned as having been given, in 794, to the abbey of St. Albans, by Offa, consists chiefly of three streets, neat and well-built. The inhabitants are partially employed in the manufacture of lace.—Also a township in the p. of Bromyard, Herefordshire, 2½ m. SW by W of Bromyard. Pop. in 1851, 491.

WINSLOW, a township of Kennebec co., in the state of Maine, U. S., on the E bank of Kennebec river, 17 m. NNE of Augusta, and intersected by the Kennebec and Penobscot railroads. Pop. in 1840, 1,722; in 1850, 1,796.—Also a township of Camden co., in the state of New Jersey, 38 m. S of Trenton, drained by Great Egg Harbour, and by a branch of Little Egg Harbour. Pop. in 1850, 1,540.

WINSTANLEY, a township in the p. of Wigan, co.-palatine of Lancaster, 3¼ m. SW by W of Wigan. Area 1,853 acres. Pop. in 1851, 675.

WINSTER, a chapelry and market-town in the p. of Youlgrave, Derbyshire, 4 m. W by N of Matlock, and 3 m. NE of the High-Peak railway. Pop. in 1831, 962; in 1851, 928.—Also a chapelry in the parish of Kendal, Westmoreland, 7 m. W of Kendal.

WINSTON, a parish in the co.-palatine of Durham, 6½ m. E of Barnard-Castle, on the river Tees, across which there is here a bridge. It includes the hamlets of Burford, Newsham, Osmondcroft, and Stubhouse. Area 2,961 acres. Pop. in 1831, 327; in 1851, 301.—Also a parish in Gloucestershire, 6 m. NW by N of Cirencester. Area 1,437 acres. Pop. in 1831, 164; in 1851, 252.—Also a chapelry in the p. of Bibury, Gloucestershire, 4½ m. SSW of Northleach. Area 1,437 acres. Pop. in 1831, 176; in 1851, 252.—Also a parish in the co. of Pembroke, 5 m. NE of Haverford West, containing the town of the same name, which is incorporated, and contributes with Pembroke in returning a member to parliament. Pop. in 1831, 745; in 1851, 683.—Also a parish in Suffolk, 7 m. W by S of Framlingham. Area 1,476 acres. Pop. in 1851, 392.

WINSTON, a central county of the state of Mississippi, U. S., comprising an area of 718 sq. m., drained by head branches of Pearl river, and by affluents of Noxubee river. It is level, but generally fertile. Pop. in 1840, 4,650; in 1850, 7,956. Its capital is Louisville.

WINSUM, a village of Holland, in the prov. and 8 m. N of Groningen. Pop. 1,525.

WINTDORF, a village of Prussia, in the reg. of Frankfort, and circle of Rottbus. Pop. 200.

WINTER (OBER), a village of Prussia, in the regency of Coblenz, and circle of Ahrweiler. Pop. 691.

WINTERBACH, a village and parish of Bavaria, in the circle of Swabia, and presidial of Dillingen. Pop. 305.—Also a village in the circle of the Pfalz, and cant. of Zweibrucken. Pop. 260.—Also a village of Baden, in the circle of the Middle Rhine, and bail. of Oberkirch. Pop. 300.—Also a village of Prussia, in the regency of Coblenz and circle of Kreuznach. Pop. 316.—Also a village of the reg. of Treves, and circle of St. Wendel. Pop. 425.—Also a parish and village of Würtemberg, in the circle of the Jaxt, and bail. of Schorndorf. Pop. 2,007.

WINTERBERG, a town of Austria, in Bohemia, in the regency of Pilsen, on the Wolinka, 30 m. SW of Pisek. Pop. 2,238. It has a castle, a paper-mill, and glass-works.—Also a town of Prussia, in the regency of Arnsberg and circle of Brelon, on the Orke, amid steep mountains. Pop. 1,308. It has a considerable trade in iron and wood-ware.—Also a village of the principality of Lippe-Detmold, and bail. of Scheider, 14 m. E of Detmold. Pop. 350. It has a glass-work, and in the vicinity, on Mount Armenius, are the ruins of a fortress.

WINTERBERG (GROSS), a mountain of Saxony, in the circle of Dresden, near Schandau, in N lat. 50° 53′ 56″. It has an alt. of 1,706 Parisian ft. above sea-level.

WINTERBORN, a village of Bavaria, in the circle of the Ffalz, and cant. of Obermoschel. Pop. 300.

WINTERBOURN, a chapelry in the p. of Chevely, Berks, 3½ m. NNW of Speenhamland. Pop. 395.

WINTERBOURNE, a parish in Gloucestershire, 6½ m. NNE of Bristol, containing the hamlet of Hambrook. Area 3,170 acres. Pop. in 1851, 2,876.

WINTERBOURNE-ABBAS, a parish in Dorset, 5 m. W of Dorchester, intersected by a stream called the Winterbourne. Area 1,500 acres. Pop. 195.

WINTERBOURNE-BASSET, a parish in Wilts, 8 m. NW by W of Marlborough. Area 2,210 acres. Pop. in 1831, 288; in 1851, 289.

WINTERBOURNE-CAME, a parish in Dorset, 3 m. SE of Dorchester, including the district of Farringdon, and the tything of Cripton. Area 1,970 acres. Pop. in 1831, 80; in 1851, 150.

WINTERBOURNE-CLENSTONE, a parish in Dorsetshire, 4 m. SW of Blandford-Forum. Area 1,406 acres. Pop. in 1831, 84; in 1851, 97.

WINTERBOURNE-DANTSEY, a parish in Wilts, 4 m. NNE of Salisbury. Area 1,162 acres. Pop. in 1831, 161; in 1851, 187.

WINTERBOURNE-EARLS, a parish in Wilts, 3 m. NNE of Salisbury. Area 1,663 acres. Pop. in 1831, 243; in 1851, 281.

WINTERBOURNE-GUNNER, or CHERBOROUGH, a parish in Wilts, 4 m. NNE of Salisbury. Area 1,562 acres. Pop. in 1831, 166; in 1851, 167.

WINTERBOURNE-KINGSTON, a parish in Dorset, 12 m. NE of Dorchester. Area 2,508 acres. Pop. in 1831, 564; in 1851, 584.

WINTERBOURNE (ST. MARTIN), a parish in Dorset, 3 m. WSW of Dorchester. Area 3,503 acres. Pop. in 1831, 369; in 1851, 434.

WINTERBOURNE-MONKTON, or WAST, a parish in Dorset, 2 m. SW by W of Dorchester. Area 431 acres. Pop. in 1831, 101; in 1851, 87.

WINTERBOURNE-STEEPLETON, or STAPLETON, a parish in Dorset, 4 m. W by S of Dorchester. Area 1,783 acres. Pop. in 1831, 176; in 1851, 206.

WINTERBOURNE-STOKE, a parish in Wilts, 5 m. W by S of Amesbury. Area 3,419 acres. Pop. in 1831, 272; in 1851, 365.

WINTERBOURNE-STRICKLAND, a parish in Dorset, 4 m. WSW of Blandford-Forum. Area 1,340 acres. Pop. in 1831, 401; in 1851, 595.

WINTERBOURNE-WHITCHURCH, a parish in Dorset, 5½ m. SW of Blandford-Forum. Area 2,841 acres. Pop. in 1831. 513; in 1851, 595.

WINTERBOURNE-ZELSTONE, a parish in Dorset, 6 m. S by E of Blandford-Forum. Area 823 acres. Pop. in 1831, 233; in 1851, 224.

WINTERBURG, a village of Prussia, in the reg. of Coblenz, and circle of Kreuznach. Pop. 325. It has a castle.

WINTERBURN, a township in the p. of Gargrave, Yorkshire, 7 m. NW by N of Skipton. Area, with Flasby, 4,311 acres. Pop. in 1851, 124.

WINTERFELDE, a village of Prussia, in the reg. of Stettin, and circle of Greiffenhagen. Pop. 245.

WINTERFELDT, a village of Prussia, in the reg. of Magdeburg, circle of Salzwedel. Pop. 200.

WINTERFULGEN, a village of Baden, in the circle of the See, and bail. of Heiligenberg. Pop. 300.

WINTER-HARBOUR, an inlet on the SE coast of Melville island, in N lat. 74° 47′ 2″, W long. 110° 48′ 2″, in which the expedition-ships Hecla and Griper, under Parry, wintered from Sept. 1819 to 1st August 1820. Measures were adopted for fitting up the ships so as to preserve the health and promote the comfort of the people on board, during the long and inclement winter upon which they had just entered. The boats, spars, and rigging were removed ashore, and a sort of housing, formed of planks and the kind of canvass used for covering waggons, was erected over the decks, to protect the men from snow and wind, and to enable them to take exercise whenever the weather should be too inclement for walking on shore. The greatest obstacle to health and comfort that our navigators had to encounter, arose from the circumstance, that when the temp. of the atmosphere began to fall considerably below the zero of Fahrenheit, the steam from the coppers, as well as the breath and other vapour generated in the inhabited parts of the ship, was found to condense into drops upon the beams and sides, to such a degree as to keep them constantly wet. The invention of the officers, therefore, and particularly of the medical gentlemen, was employed in devising means whereby at once to create warmth and secure ventilation. About the end of October the atmosphere became much colder than it had been hitherto experienced. "On the 29th," says Captain Parry, "the weather was calm and clear; and we remarked, for the first time, that the smoke from the funnels scarcely rose at all, but skimmed nearly horizontally along the housing, the therm. having got down to 24°, and the mercury in the barom. standing at 29° 70′. It had become rather a painful experiment to touch any metallic substance in the open air with the naked hand; the feeling produced by it exactly resembling that occasioned by the opposite extreme of intense heat, and taking off the skin from the part affected." The cold at length became so great as to burst not only the beer barrels, but even the bottles containing lemon-juice. To while away the dreary hours of an Arctic winter, the officers got up a newspaper, which was published on board the Hecla, and denominated the *North Georgia Gazette and Winter Chronicle*,—a production which does no small credit to the talents and pleasantry of the several contributors. It was conducted by Captain Sabine; and was subsequently printed in London. The sun, which had ceased to rise on the 4th of November, did not reappear above the horizon till the same day in February. As, however, that luminary was at no period so much as 18° below the horizon, there was a perceptible twilight even on the 22d of December. It was possible to read the smallest print, when the page was turned to the S. On the first of September the expedition left Lancaster's sound, and proceeded down the W side of Davis' straits on their way to England. The most westerly point in the polar sea, to which discovery was actually extended in this expedition by the squadron under Captain Parry, is situated in N lat. 74° 26′ 25″, and W long. 113° 46′ 43″. The mean degree of cold and heat for twelve months, beginning with September 1819, when the expedition took up their residence in Winter-harbour, was as follows:

September,	+ 22.54
October,	— 3.46
November,	— 20 60
December,	— 21.79
January,	— 30.09
February,	— 32.19
March,	— 18.10
April,	— 8.37
May,	+ 16.66
June,	+ 36.24
July,	+ 42.41
August,	+ 32 68

The lowest temp. observed in the course of the winter, was on the 15th day of February, when a therm. fixed on a pole between the ships and the shore, fell as low as — 55°, that is 87° below the freezing-point of Fahrenheit. During the temp. now mentioned, "which," says Captain Parry, "was the most intense degree of cold marked by the spirit thermometer, during our stay in Winter-harbour, not the slightest inconvenience was experienced from the exposure to the open air by a person well clothed, as long as the weather was perfectly calm; but in walking against a very light air of wind, a smarting sensation was experienced all over the face, accompanied by a pain in the middle of the forehead, which soon became rather severe. We amused ourselves freezing some mercury during the continuance of this cold weather, and by beating it out on an anvil previously reduced to the temperature of the atmosphere; it did not appear to be very malleable when in this state, usually breaking after two or three blows from the hammer."

WINTERHAUCH, a mountain of the Odenwald, in the grand-duchy of Baden, and circle of the Lower Rhine, near Mosbach. It has an alt. of 1,680 ft. above sea-level.—Also mountains in the N part of the principality of Lichtenberg, belonging to the duchy of Saxe-Coburg-Gotha. It is one of the most northerly ramifications of the Vosges.

WINTERHAUSEN, a village of Bavaria, in the circle of Lower Franconia, and seignory of Sommershausen, on the Main, 7 m. SE of Wurzburg. Pop. 1,000.

WINTERHEVEN. See WINTER-HARBOUR.

WINTERHOEK, a mountain of South Africa, in the Tulbagh district connected with the Khamuberg, and rising to an alt. of 3,100 ft. above sea-level.

WINTERHUDE, a village belonging to Hamburg, in the parish of Eppendorf, on the Allster. Pop. 380.

WINTERICH, a village of Prussia, in the regency of Treves, and circle of Berncastell. Pop. 750. Slate is quarried in the vicinity.

WINTERINGHAM, a parish in Lincolnshire, 7½ m. W of Barton-upon-Humber, on the S bank of the Humber. Area 5,675 acres. Pop. 824.

WINTER ISLAND, an island of British North America, near Melville peninsula, in N lat. 66° 11′.

WINTERITZ, a village of Austria, in Bohemia, in the regency of Eger, and bail. of Raaden. Pop. 290. It has a castle, several flour and saw mills, and manufactories of potash.

WINTERKASTEN, a village of Hesse, in the prov. of Starkenburg, and presidial of Michelstadt. Pop. 490.

WINTERLINGEN, a village of Würtemberg, in the circle of the Schwarzwald, and bail. of Balingen. Pop. 1,817. It has muslin factories.

WINTERRAM, a village of Prussia, in the reg. of Dusseldorf, and circle of Geldern. Pop. 588.

WINTERRIEDEN, a village of Bavaria, in Swabia, in the seignory of Burheim. Pop. 360.

WINTERSBACH, a village of Bavaria, in the circle of the Lower Rhine, and presidial of Rothenbuch. Pop. 639.

WINTERSCHEID, a village of Hesse, in the prov. of Upper Hesse, and circle of Ziegenhain. Pop. 210.

WINTERSCHEIDT, a village of Prussia, in the regency of Cologne, and circle of Sieg, near Siegburg. Pop. 374.

WINTERSDORF, a village of Baden, in the circle of the Middle Rhine, and bail. of Rastadt. Pop. 610.—Also a village of Austria, in Styria, in the circle of Marburg. Pop. 200.—Also a village of Prussia, in the regency and circle of Treves. Pop. 250.—Also a village of Saxe-Altenburg. Pop. 960.

WINTERSETT, a township in the p. of Wragby, Yorkshire, 6 m. SE by E of Wakefield, near the North Midland railway.

WINTERSHAGEN, a village of Prussia, in the regency of Koslin, and circle of Stolpe, on the river of that name. Pop. 210.

WINTERSHEIM, a village of Hesse, in the prov of the Rhine, and circle of Mainz. Pop. 315.

WINTERSHOVEN, a department and commune of Belgium, in the prov. of Limburg, and arrond. of Tongres. Pop. 274.

WINTERSINGEN, a village of Switzerland, in the cant. of Balse, and bail. of Gissach. Pop. 575.

WINTERSLOW, a parish, including the tything of East W. or Easton, West W. or Weston, and Middle W. or Middleton, in the hund. and union of Alderbury, Wilts, 6 m. ENE of Salisbury. Area 4,843 acres. Pop. in 1831, 749; in 1851, 913.

WINTERSPELT, a village of Prussia, in the regency of Treves, and circle of Prum. Pop. 215.

WINTERSPUREN, a village of Baden, in the circle of the See, and bail. of Stockach. Pop. 420.

WINTERSTEIN, a village of Saxe-Coburg-Gotha, in the principality and 14 m. WSW of Gotha. Pop. 789. It has manufactories of linen fabrics, and of baskets.

WINTERSTETTENDORF, a village of Würtemberg, in the circle of the Danube, and bail. of Waldsee, on the Riss. Pop. 118. It has saw-mills and tile-kilns.

WINTERSTETTENSTADT, or Markt-Winterstetten, a village of Würtemberg, in the circle of the Danube, and bail. of Waldsee, in the valley of the Riss. Pop. 394.

WINTERSWEILER, a village of Baden, in the circle of the Upper Rhine, and bail. of Lorrach. Pop. 298.

WINTERSWYK, a market-town of Holland, in the prov. of Guelders, and arrond. of Zutphen. Pop. 5,800. It has extensive bleachfields.

WINTERTHUR, a town of Switzerland, capital of a bail., in the cant. and 14 m. NE of Zurich, on the Enlach, in a fine valley. Pop. 2,158. It consists of two principal streets, and is well-built. Its principal edifices are the parish church, the town-house, and an hospital. It has a gymnasium, a public library, and a printing-press, and possesses manufactories of woollen and cotton fabrics, and of chemical substances and tanneries. Pop. of bail., 22,000. This town is of considerable antiquity. It was erected into an imperial town by Sigismund, and was ceded by Austria to Zurich in 1552.

WINTERTHUR (Ober), a village of Switzerland, in the cant. of Zurich, circle and 1½ m. NE of Winterthur. It is the *Vitodurum* of the ancients, and contains some antiquities.

WINTERTON, a parish in Lincolnshire, 8 m. WSW of Barton-upon-Humber, at the termination of the ancient Fosseway. Area 3,628 acres. Pop. in 1831, 1,295; in 1851, 1,665.—Also a parish in Norfolk, 5½ m. N by W of Caistor, on the coast of the North sea. Area 1,515 acres. Pop. 722.

WINTERTON, a parish of Tasmania, in the co. of Monmouth, bordered on the W by the Jordan, and on the S by the Quoin rivers.

WINTERVELD, a district of South Africa, in the prov. of Graaf Reynet, to the N of the Sneeuw-Berg or Snow mountains.

WINTHORPE, a parish in Lincolnshire, 11 m. E by N of Spilsby, on the coast of the North sea. Area 2,339 acres. Pop. in 1831, 244; in 1851, 299.—Also a parish in Notts, 3½ m. NNE of Newark, on the river Trent, and intersected by the ancient Fosseway. Area 680 acres. Pop. in 1851, 243.

WINTHROP, a township and village of Kennebec co., in the state of Maine, U. S., 10 m. W by S of Augusta, drained by Cobbesconte river, and intersected by the Androscoggin and Kennebec railway. Pop. in 1840, 1,915; in 1850, 2,154.

WINTON, a township in the p. of Kirkby-Stephen, Westmoreland, 1½ m. NE of the town of Kirkby-Stephen. Area 3,383 acres. Pop. in 1831, 267; in 1851, 257.—Also a township in the p. of Kirby-Siggerton, Yorkshire, 4 m. NE by E of North-allerton. Area 1,354 acres. Pop. in 1851, 108.

WINTON, a village of Hertford co., in the state of North Carolina, U. S., on the SW bank of Chowan river, and 102 m. NE by E of Raleigh. Pop. in 1840, 100; in 1850, 140.

WINTRICH, a village of Prussia, in the reg. of Treves, and circle of Bernkastel. Pop. 862.

WINTRINGEN, a village of Holland, in the prov. of Luxemburg, and bail. of Grevenmacher. Pop. 400.

WINTRINGHAM, a parish in the E. R. of Yorkshire, 7 m. E by N of New-Malton, comprising the townships of W. with Linton and Newton, and the chapelry of Knapton. Area 8,480 acres. Pop. 588.

WINTZ, a village of Prussia, in the reg. of Arnsberg, and circle of Bochum. Pop. 400.

WINTZENHEIM, a canton, commune, and town of France, in the dep. of the Haut-Rhin, and arrond. of Colmar. The cant. comprises 11 com. Pop. in 1831, 14,779; in 1846, 15,772. The town is 4 m. W of Colmar. Pop. in 1846, 3,896. It has manufactories of cotton and woollen fabrics, and of soap, and an iron foundry.

WINWEILER, a town of Bavaria, in the circle of the Pfalz, 11 m. NE of Kaiserslausen, on the Alsenz. Pop. 1,000. It has a castle and several forges.

WINWICK, a parish, partly in Northamptonshire, but chiefly in Hunts, 6 m. SE of Oundle. Area 1,710 acres. Pop. in 1831, 326; in 1851, 431.—Also a parish and township in the co.-palatine of Lancaster, 2½ m. N of Warrington, at the junction of the Liverpool and Manchester, the Grand Junction, and the North Union section of the Great North-Western railway, and intersected by the Sankey canal. It contains the chapelries of Ashton-in-Makerfield, Croft, Lowton, Newchurch, Newton and Makerfield, and Downall-Green; with the townships of Culcheth, Golborne, Haydock, Houghton with Middleton and Arbury, Kenyon, and W.; the borough of Newton in Makerfield, and the lordship of Risley. Area 25,148 acres. Pop. in 1801, 12,290; in 1831, 17,961; in 1851, 19,954. The living of this parish, amounting in 1831 to £4,220 gross, has been regarded as the largest in England; but its division into several districts for ecclesiastical purposes, and the appropriation of its income amongst the several incumbents is proposed.—Also a parish in Northamptonshire, 9½ m. NNE of Daventry, intersected by the Grand Union canal. Area 2,038 acres. Pop. in 1831, 159; in 1851, 155.

WINXELE, a department and commune of Belgium, in the prov. of South Brabant, and arrond. of Louvain. Pop. of dep. 1,170; and of com., 567.

WINXELE-DELLE, a commune of Belgium, in the prov. of South Brabant, and dep. of Winxele. Pop. 603.

WINYAW-BAY, an indentation of the coast of Georgetown district, in the state of South Carolina, U. S. It runs in a SW by S direction, and is 14 m. long, and 2 m. wide. It receives Great Pedee, Black, and Waccamaw rivers.

WINZA, a market-town of Austria, in Transylvania, in the comitat of Carlstadt, on the Marosch. Pop. 4,000. It has manufactories of china.

WINZELDORF, a village of Denmark, in Holstein, in the bail. of Pinneberg. Pop. 210.

WINZELN, a village of Bavaria, in the circle of the Pfalz, and cant. of Pirmasens. Pop. 640.—Also a village of Würtemberg, in the circle of the Schwarzwald, and bail. of Oberndorf. Pop. 1,149.

WINZENBERG, a village of Prussia, in the reg. of Oppeln, and circle of Grottkau. Pop. 558.

WINZENBURG, a village and bail. of Hanover, in the prov. of Hildesheim, near Alfeld. Pop. of town, 1,172; and of bail., 13 500. It contains the ruins of a castle.

WINZENDORF, a village of Austria, in Styria, in the circle of Gratz. Pop. 240. See also ALVINCZ.

WINZENHEIM, a village of Prussia, in the reg. of Coblenz, and circle of Kreuznach. Pop. 646.

WINZENHOFEN, a village of Baden, in the circle of the Lower Rhine, and bail. of Krantheim. Pop. 270.

WINZENHOHL, a village of Bavaria, in the circle of Lower Franconia, and presidial of Aschaffenberg. Pop. 184. It has considerable iron-works.

WINZER, a village of Bavaria, in the prov. of Lower Bavaria, and presidial of Hengersberg. Pop. 694.—Also a village of Swabia, in the presidial of Mindelheim. Pop. 359.—Also two villages distinguished as Ober and Nieder, in the circle of the Upper Pfalz, and presidial of Regenstauf, containing respectively 167 and 177 inhabitants. They have several breweries.

WINZERHAUSEN, a village of Würtemberg, in the circle of the Neckar, and bail. of Marbach. Pop. 928. It has a castle.

WINZERLA, a village of Saxe-Weimar, in the prov. of Weimar, and bail. of Jena. Pop. 260.

WINZIG, a town of Prussia, in the regency and 32 m. NW of Breslau, and circle of Mohlau. Pop. 2,000. It is enclosed by walls with three gates, and has a suburb, a Catholic and two Lutheran churches, and an hospital. It contains manufactories of tobacco and of earthenware, and several breweries.

WINZINGEN, a village of Würtemberg, in the circle of the Jaxt, and bail. of Gmund, on the Lauter. Pop. 418. It has a castle.—Also a village of Bavaria, in the circle of the Pfalz, and cant. of Neustadt. Pop. 712.

WINZINGERODE, a village of Prussia, in the reg. of Erfurt, and circle of Worbis. Pop. 360.

WINZLAR, a village of Hanover, in the principality of Kalenberg, and bail. of Loccum, to the SW of Lake Steinhude. Pop. 468.

WINZNAU, a village of Switzerland, in the cant. of Solothurn, and bail. of Eosgen. Pop. 312.

WIOSTE, a village of Prussia, in the reg. of Posen, and circle of Bomst. Pop. 600.

WIPFELD, a market-town of Bavaria, in the circle of Lower Franconia, and presidial of Werneck, on the r. bank of the Main, 17 m. NNE of Wurzburg. Pop. 742. It has several mineral springs, and wine is cultivated in the environs. In the vicinity is the castle of Klingenberg.

WIPKENGEN, a village of Switzerland, in the cant. bail., and to the NW of Zurich, on the Limmat. Pop. 887.

WIPPACH, WIBBACH, or VIPAVA, a market-town of Austria, in Illyria, in the gov. of Laibach, circle and 20 m. WNW of Adelsburg, on a river of the same name, an affluent of the Isonzo. Pop. 1,050. It has a fine castle, and possesses manufactories of cloth, and of copper ware. The valley of the W., noted as the garden of Carniola, abounds in fruit and wine.

WIPPER, or WUPPER, a river which has its source in Prussia, in the prov. of Saxony, regency of Erfurt, and circle of Worbis. It traverses the principality of Schwarzburg-Sondershausen; re-enters the Prussian prov. of Saxony; and, after a course of about 15 m., joins the Unstrut, on the l. bank, to the SW of Heldrungen. Sondershausen is the chief place on its banks.—Also a river which has its source in the Prussian prov. of Saxony; in the reg. of Merseburg, and circle of Sangerhausen, to the E of Stollberg; passes Wippra, Leimbach, Hettstadt: enters the duchy of Anhalt-Bernburg, in which it waters Sandersleben and Gusten; and after a course, in a generally NE direction, of about 36 m., joins the Saale, on the l. bank, a little above Bernburg.—Also a river of Prussia, in the prov. of Pomerania, which flows into the Baltic a little below Rugenwalde.—Also a river which has its source in the prov. of Westphalia, in the regency of Merseburg, and circle of Hagen; flows thence into the regency of Dusseldorf; waters the circles of Elberfeld and Solingen; and, after a course, in a generally SW direction of about 30 m., joins the Rhine, on the r. bank, 8 m. N of Cologne.—Also an affluent of the Gera, in the duchy of Saxe-Coburg-Gotha.

WIPPERFURT, a circle and town of Prussia, in the prov. of the Rhine, and regency of Cologne. Pop. of circle in 1849, 26,121.—The town is 24 m. NE of Cologne, on the Wipper. Pop. 1,795. It has a gymnasium, and possesses manufactories of woollen and cotton fabrics, and of iron-ware. In the vicinity is a copper-mine.

WIPPERGEN, a village of Hanover, in the principality of Osnabruck, and bail. of Mippen. Pop. 400.—Also a village of Würtemberg, circle of the Danube, and bail. of Blaubeuren. Pop. 382.

WIPPERODA, a village of Saxe-Coburg-Gotha, in the bail. of Georgenthal. Pop. 210.—Also a village of Hesse, in the prov. of Lower Hesse, and circle of Eschwege. Pop. 300.

WIPPERSDORF, a village of Prussia, in the reg. of Potsdam, and circle of Juterbogk. Pop. 230.

WIPPERSHAIN, a village of Hesse, in the prov. and circle of Fulda. Pop. 414.

WIPPRA, a market-town of Prussia, in the reg. of Merseburg, and circle of Giberge-Mannsfeld, on the Wipper. Pop. 880. It has extensive bleacheries.

WIPS, a village of Prussia, in the regency of Konigsberg, and circle of Allenstein. Pop. 200.

WIPSHAUSEN, a village of Hanover, in the principality of Luneberg, and bail. of Meinersen, near Dhof. Pop. 334.

WIRAWA, a village of Austrian Bohemia, in the reg. of Gitschin, and bail. of Königgratz. Pop. 650.

WIRBALLEN, or WIERZBOLOW, a town of Russia, in Poland, in the gov. of Augustowo and obwod of Kalwary, 41 m. N of Suwalki. Pop. 1,800.

WIRBELAU, a village of Nassau, in the bail. of Runkel. Pop. 360.

WIRBENZ, a village of Bavaria, in the circle of the Upper Pfalz, and presidial of Remnath. Pop. 300.

WIRBLEN, a village of Prussia, in the reg. of Gumbinnen, and circle of Insterburg. Pop. 420.

WIRCHLESCHE, a village of Prussia, reg. of Oppeln, and circle of Gross Strehlitz. Pop. 210.

WIRDUM, a village of Hanover, in Ostfriesland, in the bail. of Greetsyhl, near Emden. Pop. 405.

WIRFKALLEN, a village of Prussia, in the reg. of Gumbinnen, and circle of Insterburg. Pop. 260.

WIRFUS, a village of Prussia, in the regency of Coblenz, and circle of Kochem. Pop. 361.

WIRGES, a village of Nassau, in the bail. of Montabaur. Pop. 740.

WIRINGTON, or WITHERINGTON, a hamlet in the p. of Paston, Northamptonshire, 3½ m. NNW of Peterborough. Pop. in 1831, 537; in 1841, 629.

WIRKSWORTH, a parish including the market-town of Wirksworth, the chapelry of Cromford, and the townships of Hopton and Ible, the hamlets of Callow and Middleton, in the hund. of W.; the chapelry of Alderwasley, and the townships of Ashleyhay, Biggen, and Idridgehay with Allton, in the hundred of Appletree; and the hamlet of Iron-Brook-Grange, in the hund. of High-Peak, Derbyshire. The town is distant 13 m. NNW of Derby, and the parish is intersected by the High-Peak railway and the Cromford canal, through which it communicates with the North Midland railway. Area 14,154 acres. Pop. in 1801, 5,862; in 1831, 7,754; in 1851, 7,480. The town is situated in a valley. The principal portion of the inhabitants are employed in mining operations. The manufacture of cotton goods, principally ginghams, is also carried on in this parish.

WIRMINGHAUSEN, a village of Waldeck, in the bail. of Arolsen. Pop. 250.

WIRMO, a parish of Russia in Europe, in Finland, to the NW of Abo.

WIRMSTHAL, a village of Bavaria, in the circle of Lower Franconia, and presidial of Guerdorf. Pop. 284.

WIRRING LAKE, a lake of Australia Felix, in the district of Portland bay, near the Woady Galoak river.

WIRRWITZ, a village of Prussia, in the reg. and circle of Breslau. Pop. 641.

WIRSCHAWITZ, a village of Austria, in Silesia, in the circle of Troppau. Pop. 440.

WIRSCHBEL, or WIERSPEL, a village of Prussia, reg. of Oppeln, and circle of Falkenberg. Pop. 448.

WIRSCHEN. See WEYERSHEIM.

WIRSCHKOWITZ (ALT and NEU), two villages of Prussia, in the regency of Breslau, and circle of Militsch, containing respectively 530 and 380 inhabitants.

WIRSCHWEILER, a village of Prussia, reg. of Treves, and circle of Bernkastel. Pop. 462.

WIRSCTZ, a circle and town of Prussia, in the prov. of Posen and regency of Bromberg. Pop. of circle, 47,286. The town is 33 m. WNW of Bromberg, on the Lobsonka. Pop. 908. It has Lutheran and Catholic churches, and possesses manufactories of cloth, linen, and lace, and tanneries.

WIRT, a township of Alleghany co., in the state of New York, U. S., 280 m. W by S of Albany. Pop. in 1840, 1,207; in 1850, 1,544.—Also a county in the NW of the state of Virginia, comprising an area of 296 sq. m., intersected by the Little Kanawha, on which its capital is situated. Pop. 3,353.

WIRTEMBERG. See WURTEMBERG.

WIRTHEIM, a village of Bavaria, in the circle of Lower Franconia, and presidial of Orb, on the Biber. Pop. 955.

WIRTKALLEN, a village of Prussia, in the reg. of Gumbinnen, and circle of Insterburg. Pop. 250.

WIRY, a village of Prussia, in the regency and circle of Posen. Pop. 460.

WIRZBICA. See WIERNZICA.

WIRZCHUCIN, a village of Prussia, in the reg. and circle of Bromberg. Pop. 412.

WIRZERW, a lake of Russia in Europe, in the gov. of Livonia, 16 m. in length, which discharges itself into Lake Peipus.

WIRZFELD, a village of Prussia, in the reg. of Aachen, and circle of Malmedy. Pop. 460.

WISA. See VISA.

WISBAUM, a village of Prussia, in the reg. of Treves, and circle of Daun.—Pop. 360.

WISBADEN. See WIESBADEN.

WISBEACH, or WISBECH, a borough and port in the isle of Ely, co. of Cambridge, 44 m. N of Cambridge, on the river Nene or Wisbeach, which is here crossed by a bridge of one elliptical arch 72 ft. in span. The navigation of the port is through a canal connecting the river with the Old Nene at Outwell, at the commencement of Well-creek, which connects it with the river Ouse. The borough comprises the parishes of St. Mary and St. Peter, with the chapelries of New Chapel and Gyhirn, in the archd. and dio. of Ely. Area of the whole, 16,038 acres. Pop. in 1831, 8,777; in 1851, 12,709. The town, which is the most considerable place in the isle of Ely, stands on both sides of the Nene, and is in general well-built. It possesses assembly-rooms, a theatre, spacious baths, and one or two good libraries. There are no manufactures carried on in the town excepting those of cordage, iron, and malt. There are several large malting-houses, one extensive brewery, an iron-foundry, and yards for building and repairing small vessels and boats. The principal exports are corn, of which above 100,000 quarters are annually sent by the Ouse and canals to Cambridge, Lynn, &c., and rape-seed and long wool which are sent into Yorkshire. Hemp, peppermint, and several other articles not everywhere found, are grown in the vicinity, and exported through Wisbeach. W. owes much of its present prosperity to the improved state of the fens by which it is surrounded. The drainage of these fens has much increased the produce of the land in this vicinity. Trade has also been much facilitated by improvements made in the outfall of the Nene, and in its channel from Wisbeach to the Crosskeys Wash.

WISBOROUGH-GREEN, a parish in Sussex, 6 m. NE by E of Petworth, intersected by the Arun and Wey canal. It contains the chapelry of Loxwood-End. Area 8,484 acres. Pop. in 1851, 1,740.

WISBU, a village of Prussia, in the regency of Stettin, and circle of Regenwalde. Pop. 500.

WISBUHR, a village of Prussia, in the reg. of Stettin, and circle of Furstenthum. Pop. 380.

WISBY, a town of Sweden, capital of a laen of the same name, 130 m. from Stockholm, on the W coast of the island of Gothland, in N lat. 57° 38′ 34″. Pop. 4,550. It is fortified, and has a cathedral, a gymnasium, and a good port. The laen comprises the island of Gothland only. Its chief manufactures are leather and tobacco.

WISCASSET, a port of entry, and the cap. of Lincoln co., in the state of Maine, U. S., on the W bank of Sheepscot river, 22 m. SSE of Augusta. The v. has an elevated and very beautiful site, and contains the county buildings, several churches, a bank, and many handsome dwellings. Pop. in 1850, 2,343. It is one of the most important seaports in the state, being largely engaged in the coasting trade, fisheries, ship-building, &c. Its foreign commerce is not so extensive as formerly. The total tonnage of this district on June 30, 1850, was 18,241 tons. The registered tonnage amounted to 6,024 tons, of which 4,748 tons were permanent, and 1,276 tons temporary.

WISCH, a village of Holstein, to the NE of Kiel, in the parish of Schonberg. Pop. 440.

WISCH, a commune of France, in the dep. of the Vosges, and com. of St. Die. Pop. 1,137.

WISCH (GREU), a village of Holstein, to the SW of Elmshorn. Pop. 246.

WISCHAU, or WISKOW, a town of Austria, in Moravia, in the circle, and 21 m. ENE of Brunn, on the Hanna. Pop. 3,360. It is well-built.

WISCHDORF, a village of Austria, in Styria, in the circle of Marburg. Pop. 320.

WISCHEHORSCH, a village of Austria, in Moravia, in the circle of Olmutz and seignory of Hohenstadt. Pop. 300.

WISCHEN, a village of Prussia, in the regency of Posen, and circle of Meseritz. Pop. 300.

WISCHENAU, or WISZNOW, a market-town of Austria, in Moravia, in the circle of Znaim. Pop. 510. It has a castle.—Also a village of Bohemia, in the regency of Budweis, and bail. of Neuhaus. Pop. 260.

WISCHENIOWETZ, a village of Austria, in Bohemia, in the reg. and bail. of Pardubitz. Pop. 300.

WISCHERAD, a town of Austria, in Bohemia, in

the circle of Kaurim, near Prague, of which it forms a suburb.

WISCHERWITZ, a village of Austria, in Bohemia, in the regency of Pardubitz, and bail. of Bohmisch-Brod. Pop. 350.

WISCHEZAN, a village of Austria, in Bohemia, in the circle of Pilsen and seignory of Tepl. It has an iron-mine.

WISCHHAVEN, a bailliage of Hanover, in the duchy of Bremen, Neustadt. Pop. 82. It contains the village of Neu-Wischhaven, on the Elbe, and near the island of Wischhavener-Sand.

WISCHIN, a village of Prussia, in the prov. of Posen, reg. of Bromberg, and circle of Chodziesen. Pop. 360.—Also a village in the regency of Danzig, and circle of Berent. Pop. 320.

WISCHIN-HANLAND, a village of Prussia, in the regency of Bromberg, and circle of Chodziesen. Pop. 420.

WISCHIN-NENDORFA, a village of Prussia, in the reg. of Bromberg, and circle of Chodziesen. Pop. 350.

WISCHKOWA, a village of Austria, in Bohemia, in the reg. of Eger, and bail. of Saatz. Pop. 210.

WISCHKOWITZ, a village of Austria, in Moravia, in the regency of Olmutz, and bail. of Gross-Peterswald. Pop. 300.—Also a village of Silesia, circle of Troppau. Pop. 220. It has a castle.

WISCHLAU, a village of Prussia, in the reg. of Oppeln, and circle of Neisse. Pop. 240.

WISCHNAGORA. See Weichselburg.

WISCHNEI-WOLOTSCHOK, a town of Russia in Europe, in the gov. of Twer, on the Tzna. Pop. 6,000. It has a cathedral, a school, and a glass-work.—Also a canal which connects Lake Ilmen with the Wolga.

WISCHNIT, a village of Prussia, in the regency of Oppeln, and circle of Tost. Pop. 300.

WISCHNITZA, or Wisnitz, a town of Austria, in Galicia, in the circle and 33 m. WSW of Tchernowitz, on the r. bank of the Czeremosz, opposite Kuty.

WISCHOPOL, a village of Austria, in Bohemia, in the regency and bail. of Gitschin. Pop. 260.

WISCHORU, a village of Turkey in Europe, in Great Walachia, on the Danube.

WISCHUTZ, a village of Prussia, in the regency of Breslau, and circle of Mohlau. Pop. 470.

WISCHWILL, a village of Prussia, in the regency of Gumbinnen, and circle of Ragnit, 20 m. E of Tilsit, near the r. bank of the Niemen. Pop. 525. It has a copper foundry.

WISCONSIN, one of the most flourishing and healthy states of the North American union, stretching between the parallels of 42° 30′ and 47° N lat., and between 87° and 94° 30′ W long.; and bounded on the N by the territory of Minesota, Lake Superior, and the northern peninsula of Michigan; on the E by Lake Michigan; on the S by the state of Illinois; and on the W by the Mississippi, which separates it from the state of Iowa and the Western territories. It is 300 m. in length, and 240 m. broad, with an area of 73,711 sq. m., or 47,175,292 acres, of which only 13,955,825 are surveyed. Its surface is one vast plain, varied only by river hills and the gentle swells and undulations of country usually called "rolling." This plain is elevated from 600 to 1,500 ft. above the level of the ocean. The highest lands are those dividing the waters of the lakes from those of the Mississippi. From these there is a gradual descent towards the S and W, which, however, is frequently interrupted by ridges and mounds, the latter of which, rising above the general landscape, present an anomaly in the contour of the country. The slope towards Lake Superior is abrupt, and, as a consequence, the rivers intersecting this portion are short, rapid, and broken by falls. They are unfit for navigation, but possess abundance of water-power which, at no distant period, will become useful to the settler. There is another ridge of broken land, running SW from Green bay, forming the dividing ridge between the waters of Lake Michigan and those of the bay and the Neenah. After pursuing a similar direction, this ridge passes into the state of Illinois. Besides the great lakes on the N and E, a vast number of smaller ones are scattered over the northern portion of the state. They are from 1 to 20 m. in extent, and many lie amid beautiful and picturesque scenery. In the shallow water of the bays, the *Zigania aquatica*, or wild rice, is abundant, and attracts immense flocks of water-fowl. Among the small lakes may be mentioned Lakes Winnebago, St. Croix, Cass, Pepin, Four Lakes, the Mille Lac, Ottawa, Pewaugau, Pewaukee, Geneva, Greene, and many others. The natural feature peculiar to W., says Mr. J. W. Hunt, in his gazetteer of the state, published in 1853, is the uniformity of its elevation, and shape of its surface, which is neither mountainous, hilly, nor flat, but gently undulating. The country W of Sugar-river, and S of the Wisconsin, is somewhat broken, principally by the dividing ridge upon which the road from Madison to Prairie-du-Chien passes. In this section, known as the mines, are several peculiar elevations called mounds. W of the Wisconsin river is a range of high hills, being the only elevations in the state either deserving or assuming the dignity of mountains. The SE portion of the state is marked by ravines at the streams, but little depressed below the surrounding level. Its prominent features are the prairie, destitute of tree or shrub, covered only by a luxuriant growth of grass, interspersed with flowers of every hue; the oak-opening, the lake, the woodland on the border of streams, and the natural meadow. Proceeding N, to the Fox and Wisconsin rivers, and Green bay, the timber increases, and the soil gradually changes from the vegetable mould of the prairie to a sandy loam. The surface also becomes somewhat depressed and uneven, diversified with timber, rolling prairie, large marshes, and extensive swamps, having an abundant growth of cranberries and wild rice. Still, N and W the surface becomes more uneven, and the streams rapid, affording an abundance of water-power for the manufacture of lumber from the immense forests of evergreen, scarcely surpassed on the Western continent.

Soil and Productions.] The soil of the prairie consists of a dark brown vegetable mould, from 1 to 2 ft. in depth, very mellow, and entirely destitute of stone or gravel, which, for fertility and agricultural properties, cannot be surpassed. The subsoil is a clayish loam, similar to the soil of the timbered lands, and is also suitable for cultivation. The soil of the timbered land is less rich than the prairie, not so deep, and contains less carbonate of lime, which enters into the composition of the latter in the proportion of from 20 to 40 per cent. The mining region, unlike that of any other mineral district, promises a liberal reward as well to the farmer as to the miner. The soil of the evergreen district is mostly sandy, and not so rich as in other portions of the state; it is nevertheless well adapted to agriculture and grazing. The *prairies* of W. are not so extensive as those of other states, and are so skirted and belted by timber that they are well adapted to immediate and profitable occupation. The *openings*, which comprise a large portion of the finest land of the state, owe their present condition to the action of the annual fires which have kept under all other forest growth, except those varieties of oak which can withstand the sweep of that element. The limestone underlying the coal-fields of Illinois forms the immediate basis of the alluvion of Southern W. This geological district, in addition to that portion of the state which lies S of the valley of the Wisconsin river, comprises the whole of the slope towards Lake Michigan. In many portions of this district, the lime rock disappears, and the out-cropping sandstone furnishes a fine material for building. The lead-bearing rock of the mineral region is a porous limestone, prevailing throughout Grant, Lafayette, and Iowa cos., and comprising four-fifths of the 'lead district' of the Upper Mississippi, the remaining one-fifth being in the states of Illinois and Iowa. Deposits of iron-ore, water-limestone, and beds of gypsum, together with other varieties of minerals, are found in localities more or less numerous throughout the limestone region. That section of the state which lies between Lake Superior on the N and the falls of St. Anthony on the Mississippi, and the falls of the other rivers flowing S, is primitive in its prevailing geological character; and it is within this primitive region that the copper-mines of Lake Superior are found. In all that portion of the state lying between the primitive region just described and the limestone formation of the south and east, the transition sandstone prevails interspersed with limestone, and more sparsely with rock of a primitive character. This formation comprises that section of the country drained by the Wisconsin and other rivers tributary to the Upper Mississippi and below the falls of those streams.

Rivers.] The Mississippi, which, as before observed, forms the western boundary, is augmented from this state by the waters of the Chippewa and Wisconsin, which, though themselves considerable rivers, scarcely perceptibly increase the volume of 'the father of waters.' Innumerable smaller streams and branches run through the whole extent of the state, so that no portion of it is without a plentiful supply of good and generally pure water. The Mississippi is navigable as far up as the falls of St. Anthony, and small steamboats ply on the Wisconsin and some other rivers. The rivers running into the Mississippi take their rise in the vicinity of the sources of those running into the lakes, and often originate in the same lake or swamp, so that the communication from the Mississippi to the lakes is rendered comparatively easy at various points. Some of the rivers are supplied from the Tamarack swamps, from which the water takes a dark colour.

Soil, Climate, &c.] W. is composed of timbered and prairie land, pretty equally divided, with some swamps and wet prairies having generally a soil from 1 to 10 ft. deep. All kinds of crops which are raised in northern latitudes may be here cultivated with success; and owing to the great range of pasturage on the prairies, it is a fine grazing country. The cos. of Grant and Iowa abound with lead and copper ore. Bordering the Mississippi and Wisconsin rivers the soil is rich, and the surface generally covered with a heavy growth of timber. The proximity of W. to the great lakes ensures it a softer climate than its geographical position would assign to it. The extremes, however, are great—the therm. sometimes during the summer marking 100°, and in winter receding to 40° below zero; but the salubrity of the country is not disturbed by these circumstances, as the heat and cold are seasonal, and approach not by sudden flaws but by a gradation that renders the change almost unfelt. The lowest mean monthly temp. is 13° 58′, in January; the highest, 75° 47′, in July. The average annual amount of rain is from 27.96 to 38.83 inches. The N and NW, and the S and SW winds, are those most prevalent; the former in the winter, the latter in the summer season. The salubrity of the climate, the purity of the air and of the water, the coolness and short duration of the summers, and the dryness of the winters, conspire to render W. one of the most favoured regions of the United States. Its swamps, marshes, and wet meadows, are constantly supplied with pure water from springs; and as they are seldom exposed to long continued heats, they do not send forth those noxious vapours so much dreaded in the more southern sections of the union. It is especially adapted for agricultural purposes, and for grazing or sheep-farming. Its river and lake facilities are great, and the busy strife of commercial activity has already penetrated to its very centre. Steamboats ply on its waters; and roads built by the late territorial government, greatly facilitate the development of the natural wealth of this fine country. Lead, copper, and iron are found. The SW portion of the country is one vast mine of lead, extending over a region of at least 100 m. in circumf. The ore is supposed to be inexhaustible in quantity, and yields about 70 per cent. pure metal. The lead is generally found in a light ochrous earth, free from rock, at a depth of 10 or 15 ft.; but the greatest beds have been found at about 40 ft. below the surface. The copper-mines in the N bordering on Lake Superior, are among the richest in the world, and are at the present day worked to a considerable amount. The ore yields from 60 to 75 per cent. of metal. To the lumberman, the pineries of W. present great inducements for investment and settlement. That of the Upper W. and its tributaries is the most extensive; and distinguished still more for the fine quality than the inexhaustible quantities of its timber. The other localities of the white pine and other evergreens are mainly on the Wolf, the great northern affluent of the Fox, the tributaries of Green Bay, and on the La Crosse, the Black, Chippewa, and the St. Croix, branches of the Upper Mississippi. The rapids of these streams furnish abundant water-power for the manufacture of lumber, and on the annual spring rise, and occasional freshets at other seasons of the year, the yield of the mills is floated from the Wolf into Lake Winnebago and the Lower Fox, and from most of the other streams into the Mississippi. Scarcely ten years have elapsed since the Alleghany pine of Western New York and Pennsylvania had undisputed possession of the market, not only of the Ohio Valley, but of the Mississippi and its tributaries above New Orleans, at which point it competed with the lumber of Maine and New Brunswick. The course of the lumber trade may now be considered as permanently changed. The pineries of Wisconsin now control, and will hold exclusive possession of the market of the valleys of the Mississippi and its great western affluents. The amount of pine lumber estimated to be sawed in Wisconsin annually, is as follows: Black River, 15,000,000 ft.; Chippewa, 28,500,000 ft.; Green Bay, 21,000,000 ft.; Manitouwoc, 24,500,000 ft.; St. Croix, 20,000,000 ft.; Wisconsin, 58,500,000 ft.; Wolf, 25,500,000 ft.—total number of feet, 193,000,000. Aside from the manufacture of pine lumber, reaching as it does nearly 200,000,000 feet per year, saw-mills, driven by both steam and hydraulic power, are now in operation in every section of the state where timber is found, and large quantities of oak scantling and plank, and basswood siding and lath, are yearly manufactured.

Population, &c.] In regard to pop., W. has outstripped some of the old established states. The progressive increase of its pop. has been as follows:

	Population.	Increase. Numerical.	Per cent.
1830,	3,245		
1836,	11,686	8,441	72·7
1838,	18,149	6,463	35·3
1840,	30,945	12,796	41·4
1842,	46,678	15,733	33·9
1844,	117,000	70,322	60·1
1846,	155,277	38,277	24·7
1848,	215,800	60,523	28·0
1850,	304,565	274,246	886·2

The number of free persons born in the U. S., in the pop. of 1850 was 197,912; of foreign birth 106,695; of birth unknown 784. The foreign pop. was composed of persons from—England 18,952, Ireland 21,043, Scotland 3,527, Wales 4,319, Germany 34,519, France 775, Spain 4, Portugal 4, Belgium 45, Holland 1,157, Italy 9, Austria 61, Switzerland 1,244, Russia 71, Denmark, 146, Norway 8,651, Sweden 88, Prussia 3,545, Sardinia 1, Greece 1, Asia 17, Africa 1, British America 8,277, Mexico 9, Central America 11, South America 6, West Indies 20, and other countries 191. The general statistics of the resources, productions, manufactures, wealth, and institutions of W., in accordance with the census of 1850, and other authentic reports referring to the same period of time, are as follows: The improved farm lands amounted to 1,045,499 acres; unimproved lands to 1,931,159 acres. The total number of farms in operation on the 1st June, 1850, was 20,177. The live stock of 1840 and of 1850 was as follows:

Live Stock.	1840.	1850.
Horses,	5,735 head	30,179 head
Asses and Mules,		156 „
Milch Cows,	30,269 „	64,339 „
Working Oxen,		42,801 „
Other Cattle,		76,293 „
Sheep,	34,624 „	124,892 „
Swine,	51,383 „	159,276 „

The amount of animal produce was, wool, 253,963 pounds; butter, 3,633,750 pounds; cheese, 400,283 pounds; and the value of animals slaughtered in the year 920,178 d. The wool crop of the census year of 1840 amounted to 6,777 pounds, hence that of 1850 exhibits an increase of 247,186 pounds, or of 3,499·8 per centum. The clip per fleece, in 1840, was 31·3 ounces, and in 1850, 32·5 ounces being an average increase of 1·2 ounces per fleece. The grain crops of 1850, compared with the crops of 1840, was as follows:

Crops.	1840.	1850.
Wheat,	212,116 bushs.	4,286,131 bushs.
Rye,	1,965 „	81,253 „
Indian corn,	379,359 „	1,988,979 „
Oats,	406,514 „	3,414,672 „
Barley,	11,062 „	209,692 „
Buckwheat,	10,654 „	79,878 „

Pease and beans were grown in 1850 to the extent of 20,657 bushels; potatoes, Irish, 1,402,077 b., and sweet 879 b. The potato crop of 1840 amounted to 419,608 b. The principal miscellaneous crops, compared with similar crops of 1840, exhibit the results appended:

Crops.	1840,	1850.
Hay,	30,938 tons	275,662 tons
Hops,	133 pounds	15,930 pounds
Hemp—dew-rotted,	2 tons	300 tons
„ water-rotted,		2 „
Flax,		68,393 pounds
Maple-sugar,	135,288 pounds	610,976 „
Wine,	0 gallons	113 gallons

Manufactures.] The whole number of manufacturing establishments in operation on the 1st June, 1850, and manufacturing to the value of 500 d. and upward annually, was 1,273, the great portion of them being mills of various descriptions, and other agricultural manufactories, and of the strictly national manufactures the following: 9 woollen factories, 16 iron works—one making pig iron, and 15 making castings; and 8 tanneries. The total capital invested in manufactures, in 1840, amounted to 635,926 d. In the woollen manufacture there was a capital of 31,225 d. invested; wool consumed, 134,200 pounds; value of all raw material, &c., 32,630 d.; hands employed, 25; products of the year—cloth 36,000 yards, and yarn 94,350 pounds; value of products 87,992 d. The capital invested in the manufacture of pig iron was 15,000 d.; ore used, 3,000 tons; coke and charcoal consumed, 150,000 bushels; hands employed 60; pig iron produced in the year 1,000 tons; value of entire products, 27,000 d. In the manufacture of iron castings, a capital of 116,350 d. was invested, pig iron used, 1,371 tons; old metal, 15 tons; coal, 595 tons; and coke and charcoal, 2,700 bushels; value of all raw material, &c., used in the year, 86,930 d.; hands employed, 228; products of the year—castings 1,342 tons, and, including 64,025 d., the value of other products, these were valued at 216,195 d. In 1840, there was but one furnace in the State, which produced only 3 tons cast iron, and employed 3 hands, and a capital of 800 d. In tanneries, the capital invested amounts to 78,950 d.; value of hides and skins used, 93,380 d.; hands employed, 75; monthly cost of labour, 1,710 d.; tanned during the year, 14,900 skins, and 59,600 sides of leather, valued at 175,710 d. In 1840, but one tannery was reported, employing 3 hands and 2,000 d., and producing annually 150 sides of sole leather, and 150 sides of upper leather. The capital invested in the manufacture of malt and spirituous liquors amounted to 98,700 d. in 1850; quantities and kinds of grain, &c., consumed—barley, 91,020 bushels; corn, 29,900 bushels; rye, 9,200 bushels; hops, 28 tons, hands employed, 98; quantities of liquors produced—ale, &c., 31,320 barrels; whisky, 127,000 gallons. In 1840, there were 3 breweries, producing 14,200 gallons ale, &c. and 3 distilleries, producing 8,300 gallons spirits, &c., together employing 11 hands, and a capital of 14,400 d.

Commerce and Internal improvements.] The commerce of W., chiefly carried on at its lake ports, is extensive. The statistics of its foreign trade with Canada are embraced in those of the districts of Michillimackinac and Chicago, and cannot be stated separately. The coasting trade is relatively of immense importance, both on the lake and on the Mississippi. The chief ports of the Mississippi are Prairie-du-Chien and Cassville, and the exports from those are chiefly mineral. The internal commerce of the state is carried on with great activity, radiating from the principal commercial towns to every portion, and large supplies are needed in the northern pineries, which supply so valuable a material to general commerce. The internal improvements of W. are more prospective than existing, but great progress has been made in building railroads, one of which will extend along the lake shore from Chicago to Fond-du-Lac, and other roads diverge from Chicago, Milwaukie, &c., toward the Mississippi, centering at Madison. A line will also extend from the latter point to Portage city on the N, and to Beloit and Janesville to the S, and these will be carried to a junction with the Galena and Chicago Union R. R. The only line yet complete is that from Milwaukie to Madison, which, under the name of Milwaukie and Mississippi R. R., will stretch across the State. The improvement of Fox and Wisconsin rivers is in progress, with the intention ultimately of connecting the rivers by a canal, and thus giving uninterrupted navigation from the lake to the Mississippi.

Government.] The constitution now in force was adopted in convention on the 1st February, 1848, and approved by the people at the election held in April following. All males 21 years old, residents of the state for one year next before the election, who are white citizens of the U. S., or white foreigners who have declared their intention to become such citizens, or persons of Indian blood once declared by the laws of the U. S. to be citizens, subsequent laws to the contrary notwithstanding, or "civilized persons of Indian descent" not members of a tribe, are entitled to vote at all elections; and the right of suffrage may be extended by a law approved by a majority of votes at a general election. The legislature consists of an assembly and senate. Members of the assembly (in number between 54 and 100) are chosen annually, and senators (not more than one-third nor less than one-fourth the number of assemblymen) are chosen for two years, one-half each year. Both must be qualified voters in their districts. The legislature meets at Madison on the first Monday of January annually. The governor shall be a qualified voter and citizen of the U. S., and is chosen for two years by a plurality of votes, and if there be no choice the legislature shall, by joint ballot, choose one of the persons having the highest equal number of votes. In default of the governor, his duties are discharged by the lieutenant-governor, who is chosen with the same qualification and for the same term as the governor. He is *ex-officio* president of the senate. If the lieutenant-governor defaults, then the secretary of state fills the office. The governor's *veto* may be defeated by a vote of two-thirds of the members present in each house. The judiciary is comprised in a supreme court, circuit-courts, county courts, and justices of the peace. Justices of the peace are elected in the several towns, hold office two years, and have jurisdiction throughout their counties in civil matters when the debt or damages claimed do not exceed 100 dol. A county court is established in each county, and the judge is elected by the voters of the county for the term of four years; his court has jurisdiction concurrent with the circuit courts in all civil actions arising within the county, and in civil actions originating out of the county, when the damages claimed are not more than 500 dol., and exclusive appellate jurisdiction in all cases of appeal or *certiorari* from justices of the peace, also probate powers and jurisdiction. The State is divided into six judicial circuits, the voters of each of which elect a judge for six years. The circuit courts have general original jurisdiction in actions civil and criminal at law and equity within their circuits. The supreme court is composed of the judges of the circuits, one of whom is elected by the other judges chief justice. Except the power to issue writs of *mandamus. quo warranto*, and the like, this court has appellate jurisdiction only, and is the court of last resort. There can be no trial by jury in this court. Tribunals of conciliation may be established, and in every organized county the legislature may appoint one or more persons with powers not exceeding those of a circuit judge at chambers. Sheriffs, coroners, registrars of deeds, and district attorneys are chosen by the people for two years. Among the miscellaneous provisions of the constitution are the following: no lottery or divorce can be granted by the legislature; laws shall be passed providing a way for suing the state; the credit of the state shall never be lent, nor shall any debt be contracted, nor money paid for internal improvements, unless the state hold trust property dedicated to such uses; except in cases of war, invasion, or insurrection, no debt exceeding 100,000 d. shall be contracted; a university without sectarian instruction shall be established; the legislature shall prevent towns and cities from contracting debts; no general or special law to create a bank or banks shall be passed till a majority of the votes at a general election shall have been in favour of a bank, and until such a majority have afterward approved the act as passed; any person implicated in a duel loses the right of suffrage and of holding office; no public defaulter shall hold office; in criminal prosecutions for libel the jury are judges of law and fact; all lands are allodial, and leases of agricultural lands, with rent or service reserved for more than 15 years, are void; resident aliens have all the property rights of citizens; imprisonment for debt on contract is prohibited; no religious opinions shall disqualify a witness, &c. Amendments to the constitution, agreed to by a majority of members of both houses shall be published for three months before the next election, and referred to the legislature then chosen; and if again approved by such majority, shall be submitted to the people. A convention may be called in like manner.

Federal representation, &c.] W., in accordance with the apportionment act of the 23d May, 1850, has *three* representatives in the U. S. congress. The militia force of the state, as reported to the adjutant-general of the U. S. army, consists of 32,203 men of all arms, of which 1,804 are commissioned officers, and 30,399 non-commissioned officers, musicians, artificers, and privates. Of the commissioned officers, 9 are general officers, 31 general staff officers, 212 field officers, &c., and 1,552 company officers. The principal benevolent institution of the state is the asylum for the education of the blind at Janesville, a tax of one-fifteenth of a mill on every dollar of taxable property in the state is levied for its aid. —The total receipts for the year ending 31st December, 1850, amounted to 93,889 d., and the expenditures to 62,746 d. The total receipts since the organization of the state government had been to date, 160,218 d., and the total expenditures 152,969 d. The chief sources of revenue are taxes and the sales of lands. The principal expenditures are for salaries of executive and judicial officers, 13,592 d.; legislative expenses, 14,309 d.; printing, 8,793 d.; State convicts (there being no prison), 3,093 d.; miscellaneous (including revision of statutes in 1849), 16,356 d. There is a territorial debt of 12,892 d., which would be due in June, 1851.—Madison, on the isthmus between Third and Fourth lakes, is the political capital.

Statistics of education.] The superintendent of public instruction reported the capital of the school fund as amounting on the 1st June, 1851, to 538,094 d., which, at 7 per cent. interest, yielded an annual revenue of 37,666 d. It is expected that 5,301,943 d. will be eventually realized from the sales of the school-lands. For the year ending August 1, 1850, out of 2,105 districts, 1,700, and out of 764 part districts, 650 had reported schools. In the districts reported, each school had been taught on an average 4·96 months, and all received 65,604 d. of public money. Out of 92,047 children between the ages of 4 and 20, 61,597 attended schools, and 1,020 under 4, and 688 over 20, also attended. There were also in the state 87 private or select schools, with an average of 75 pupils. The state university is located at Madison; it was founded in 1848, and is well endowed. The Beloit college was founded in 1846, and in 1850 had 6 professors, and 25 students; and its library contained about 2,000 volumes. The whole number of periodicals published in W., in 1850, was 49, of which 11 were conducted on whig principles, 29 on democratic principles, and 9 were neutral in politics, including literary, religious, and all issues, the character of which had not been ascertained.

Religious denominations.] The statistics of the several religious denominations in the state, in 1850, were as follows:—

Denominations.	No. of Churches.	Church accom.
Baptist,	28	9,505
Christian,	2	700
Congregational,	33	10,585
Dutch Reformed,	2	550
Episcopal,	19	5,140
Free,	2	275
Lutheran,	18	5,000
Methodist,	54	15,008
Presbyterian,	21	6,000
Roman Catholic,	57	23,717
Union,	1	400
Universalist,	1	300

Pauperism and Crime.] The whole number of paupers who received support within the year ending 1st June 1850, was 666, of which number 169 were native, and 497 foreign born; and the whole number of paupers at the date specified, was 338, of which 72 were natives, and 166 foreigners. Cost of support within the year, 14,743 d. A state-prison has just been erected at Waupun, in Fond-du-Lac co.

Historical sketch.] Wisconsin, says J. A. Lapham, Esq., does not fall behind the other portions of the Western country in the monuments it affords of the existence of an ancient people who once inhabited North America, but of whom nothing is known except what can be gathered from some of the results of their labours. The works at Aztalan, in Jefferson county, are most known and visited; but there are many other localities which are said to equal them in interest and importance. The substance called brick at this place is evidently burnt clay, showing marks of having been mixed with straw, but they were not moulded into regular forms. There is a class of ancient earth-works in Wisconsin, not before found in any other country, being made to represent quadrupeds, birds, reptiles, and even the human form. These representations are rather rude, and it is often difficult to decide for what species of animal they are intended; but the effects of time may have modified their appearance very much since they were originally formed. Some have a resemblance to the buffalo, the eagle or crane, or to the turtle or lizard. One representing the human form, near the Blue Mounds, is, according to R. C. Taylor, Esq., 120 ft. in length; it lies in an east and west direction, the head towards the west, with the arms and legs extended. The body or trunk is 30 ft. in breadth, the head 25, and its elevation above the general surface of the prairie is about 6 ft. Its conformation is so distinct that there can be no possibility of mistake in assigning it to the human figure. A mound at Prairieville, representing a turtle, is about 5 ft. high, the body is 56 ft. in length; it represents the animal with its legs extended and its feet turned backward. It is to be regretted that this interesting mound is now nearly destroyed. The ancient works are found in all parts of the territory, but are most abundant at Aztalan, on Rock river, near the Blue Mounds, along the Wisconsin, the Neenah, and the Pishtaka rivers, and near Lake Winnebago. The mounds are generally scattered about without any apparent order or arrangement, but are occasionally arranged in irregular rows, the animals appearing as if drawn up in a line of march. An instance of this kind is seen near the road, 7 m. E from the Blue Mounds, in Iowa co. At one place near the Four lakes, it is said that one hundred tumuli, of various shapes and dimensions, may be counted—those representing animals being among others that are round or oblong. Fragments of ancient pottery of a very rude kind are often found in various localities. They were formed by hand, or moulded, as their appearance shows evidently that these vessels were not turned on a "potter's wheel." Parts of the rim of vessels, usually ornamented with small notches or figures are most abundant. A mound is said to have been discovered near Cassville, on the Mississippi, which is supposed to represent an animal having a trunk like the elephant, or the now extinct mastodon. Should this prove true, it will show that the people who made these animal earth-works, were cotemporaries with that huge monster whose bones are still occasionally found; or that they had then but recently emigrated from Asia, and had not lost their knowledge of the elephant.—W. was first visited by French missionaries in 1660, in October of which year Mesnard reached Che-goi-me-gon, on Lake Superior. In 1672, Aloues and Dablon visited Green bay, and the country between the Fox river and the south end of Lake Michigan. In the year following, on the 13th of May, Marquette, a Jesuit missionary, and Joliet, an agent of the government of France, with five other Frenchmen, embarked from their mission near Mackinac, and arrived at Green bay, where they found an Indian village and procured guides to accompany them up Fox river to the portage with the Wisconsin. They descended this river to its mouth, where they arrived on the 17th of June 1673, and made the first discovery of the Upper Mississippi. The territory remained under the government of the French, who claimed it, until 1763, when, at the treaty of Paris, it was ceded to Great Britain, who retained it until the independence of the U. S. was acknowledged by that country in 1783, when it was claimed by Virginia as a part of the Illinois country conquered by Col. George Rogers Clark. It, however, remained in the possession of Great Britain until 1796, when it was surrendered in accordance with Jay's treaty, ratified the previous year. On the first day of March 1784, it was ceded by Virginia to the U. S. By the ordinance passed the 13th of July 1787, a governor was provided for the territory NW of the Ohio river, which territory was divided into two separate governments, the western called Indiana, by an act passed May 7th, 1800. An act dividing the Indiana territory, and organizing Illinois, was passed and approved February 3d, 1809. By the act of Congress to enable the people of Illinois to form a state government, approved April 18th, 1818, all that portion of said territory north of the parallel of latitude 42° 30′ west of the middle of Lake Michigan, was attached to the territory of Michigan, which had been set off from Indiana in 1805. In 1835, Michigan having assumed a state government, a session, convened at Green bay, from the remainder of said territory, signed several memorials to congress, among which was one asking for the organization of the territory of W., with the seat of government at Cassville, on the Mississippi. An act establishing the territorial government of W., was passed and approved April 20th, 1836, and the territory fully organized July 4th, 1836. On the 12th of June 1838, an act was passed dividing the territory of W., and establishing that portion on the W side of the Mississippi (which had been attached to Michigan in 1834) into a separate government, under the name of Iowa. A convention was held at Madison, October 5th, 1846, for the purpose of drafting a state constitution, which was adopted in convention, December 16th, 1846, but rejected by the people at the election held on the first Tuesday in April 1847. A second convention was held December 16th, 1847, and a constitution agreed to February 1st, 1848, which was approved of by the electors, at the election held April 1848, and W. was admitted into the union, on an equal footing with the other states, on the 29th day of May 1848.

WISCONSIN, a river which rises in Vieux Desert lake, on the NE boundary of the above State, and flows S in general course for 250 m. until it reaches Portage, when it turns, and flows SW and WSW 114 m. to its confluence with Mississippi river, near Prairie-du-Chien. Above Portage it has numerous rapids, which afford great water-power, used at many places by saw-mills. At Portage it is 1,200 ft. wide, and gradually increases in width to its mouth, where it is 1,800 ft. wide. Its current is usually quite rapid, but is shallow, and there are numerous islands and shifting sand-bars, rendering navigation somewhat difficult. Steam-boats of light draft can usually ascend to Portage. In Richmond co., where it is 1,350 ft. wide, it is bordered by high sandstone bluffs, from 150 to 200 ft. high, presenting grand and beautiful scenery. Proposed improvements of this river will render it a very important channel of trade and travel between the great lakes and the western rivers.

WISDORF, a village of Prussia, in the regency of Dusseldorf, and circle of Solingen. Pop. 725.

WISEGO, a village of Prussia, in the regency of Konigsberg, and circle of Ortelsburg. Pop. 240.

WISEK, a village of Austria, in Bohemia, in the regency of Pilsen, and bail. of Bokikan. Pop. 260.

WISELL, a village of Austria, in Styria, in the circle of Cilli. Pop. 200.

WISEMBARCK, a commune of France, in the dep. of the Vosges, cant. and 9 m. E of Saint Die. Pop. 1,137.

WISENT. See WIESENT.

WISERNES, a village of France, in the dep. of the Pas-de-Calais, and cant. of Tumbres, near the Aa. Pop. 610. It has a flour and three paper mills.

WISETON, or WISSTON, a township in the p. of Clayworth, Notts, 5 m. ESE of Bawtry, intersected by the Chesterfield canal. Pop. in 1851, 127.

WISGOLDINGEN, a village of Würtemberg, circle of the Jaxt, and bail. of Gemund. Pop. 699.

WISHART, an island of South Australia, in the archipelago of New Britain, 6 m. from New Ireland, in S lat. 2° 20′, and E long. 150° 49′.

WISHAW, a parish in Warwickshire, 3½ m. ESE of Sutton-Coldfield, intersected by the Birmingham and Fazeley canal, and containing the hamlet of Moxhall. Area 1,196 acres. Pop. in 1851, 282.

WISHAWTON, a village of Lanarkshire, in the p. of Cambusnethan, 14 m. E by S of Glasgow. Pop. 3,271, chiefly engaged in coal mines and iron works in the vicinity.

WISHFORD, or WICHFORD (GREAT), a parish in Wilts, 2½ m. NNW of Wilton. Area 1,610 acres. Pop. in 1831, 361; in 1851, 373.

WISINGSOE, an island of Sweden, in the Wettern-See. Pop. 900. It has a botanic garden, and a gymnasium.

WISKA, a village of Austria, in Bohemia, in the regency of Gitschin, and bail. of Weiswasser. Pop. 760.—Also a village of the regency of Pardubitz, and bail. of Neu-kolin. Pop. 260. It has manufactories of potash.

WISKITKI, a town of Russia in Europe, in Poland, in the gov. and WSW of Warsaw, on the Pisia. Pop. 1,000.

WISKITTNO, a village of Prussia, in the reg. and circle of Bromberg. Pop. 250.

WISKOW. See WISCHAU.

WISLA. See WEICHSEL.

WISLEY, or WISHLEY, a parish in Surrey, 2½ m. N by E of Ripley. Area 1,321 acres. Pop. 167.

WISLICA, a town of Poland, in the gov. and 39 m. S of Kielce, obwod of Stobinca, on the Nida, in a marshy locality. Pop. 800. It is enclosed by walls, and has two churches.

WISLOCH. See WIESLOCH.

WISLOK, or WISLOKA, a river of Austria, in Galicia, which has its source on the N side of the Carpathian chain, in the circle of Sanok; runs first NW, then NE; and, after a course of 120 m., joins the San, on the l. bank. Krasno, Fryztak, Stryszow, and Rzeszow, are the chief places on its banks.—Also a river of the same prov., which also has its source on the N side of the Carpathian chain; traverses the circles of Jaslow and Tarnow; and after a course, in a generally N direction, of 90 m., joins the Vistula, on the r. bank. Zmygrod, Osiek, Jaslo, Kolaczyce, Brzostek, Pilsno, and Dembica, are the chief places on its banks.

WISMANNSDORF, a village of Prussia, in the reg. of Treves, and circle of Bittburg. Pop. 220.

WISMAR, a town of the grand-duchy of Mecklenburg-Schwerin, capital of a seignory of the same name, 18 m. N of Schwerin, on a bay of the Baltic, named Walfisch, and at the mouth of the outlet of Lake Schwerin, in N lat. 53° 53′ 31″. Pop. 12,000. It is enclosed by walls and ditches, with four gates, and has a suburb, four Lutheran churches, three hospitals, an orphans' asylum, and a school. It possesses manufactories of tobacco, playing cards, sail-cloth, and woollen and linen fabrics, distilleries of brandy, breweries, and building-docks, and carries on an active trade in grain, wood, and other articles of local produce. The harbour is good, but wanting in depth for large vessels. The number of vessels which entered the port during 1849 was 328, and 343 cleared outwards, exclusive of steamers; of these 77 vessels were English, 29 Russian, 39 Danish, 64 Swedish, 3 Norwegian, 23 Dutch, and 2 French.

WISNA, a town of Poland, in the gov. of Augustowo, obwod and 15 m. E of Lamoa, on the Narew. Pop. 2,013.

WISNICE, or WISNICZ, a town of Poland, in the woiwode of Siedlec, obwod and 29 m. E of Radzyn.—Also a town of Austria, in Galicia, in the circle and 5 m. S of Bochnia. It has a castle and a Carmelite convent.

WISNIEWKE (GROSS, KLEIN, NEU and POLNISCH), four villages of Prussia, in the regency of Marienwerder, and circle of Flatow. Pop. 980.

WISNIEWO, a village of Prussia, in the reg. of Bromberg, and circle of Wongrowiec. Pop. 200.

WISNITZ. See WISCHNITZA.

WISOCKEN, a village of Prussia, in the reg. of Gumbinnen, and circle of Lyk. Pop. 200.

WISOKA, a village of Austria, in Moravia, in the circle of Olmutz, and seignory of Biskupitz. Pop. 220.—Also a village of the circle of Prerau, and seignory of Hustopetsch. Pop. 230.

WISOKEIN, a village of Austria, in Moravia, circle of Znaim and seignory of Bottau. Pop. 240.

WISOKEPOLE, a village of Austria, in Moravia, and seignory of Brunow. Pop. 570.

WISOKI-MAZOWIECKI, a town of Russia in Europe, in Poland, in the gov. of Augustowo, obwod and 29 m. SE of Lomza. Pop. 1,000, of whom a large proportion are Jews.

WISOKY, a village of Austria, in Moravia, circle of Iglau, and seignory of Saar. Pop. 240.

WISOTSCHAU, a village of Austria, in Bohemia, in the regency of Gitschin, and bail. of Neu-Bidschow. Pop. 240.

WISOWITZ, or WYSOWICE, a town of Austria, in Moravia, circle and 27 m. NE of Hradisch, on the Drewnitza. Pop. 2,700. It has a castle, and possesses manufactories of cloth and paper.

WISP-HILL, a mountain of Roxburghshire, near the source of the Teviot, 13 m. ESE of Ettrick-Pen, in N lat. 55° 17′, W long. 2° 57′, having an alt. of 1,830 ft. above sea-level.

WISPINGTON, a parish in Lincolnshire, 4 m. W by N of Horncastle. Area 1,570 acres. Pop. 83.

WISPITZ, a village of Anhalt-Dessau, in the bail. of Nienburg, on the Saale. Pop. 260.

WISSANT, a commune of France, in the dep. of the Pas-de-Calais, cant. and 7 m. NNW of Marquise. Pop. in 1841, 1,012. It has a port, now much choked with sand, supposed to have been the *Itius-Portus* of the Romans.

WISSCHRAD. See WISCHEHRAD.

WISSEK, or WISOKA, a town of Prussia, in the prov. of Posen, regency and 42 m. WNW of Bromberg, and circle of Wirsitz. Pop. in 1842, 915.

WISSEK, a village of Austria, in Moravia, in the circle of Brunn, and bail. of Boskowitz. Pop. 260.

WISSEL, a village of Prussia, in the regency of Dusseldorf, and circle of Kleves. Pop. 730.

WISSELSHEIM, a village of the grand-duchy of Hesse-Darmstadt, in the prov. of Upper Hesse, to the SE of Rutzbach, on the Wetter. Pop. 200. It has a salt-work.

WISSELSING, a village of Bavaria, in the circle of Lower Bavaria, presidial of Deggendorf. Pop. 400.

WISSEMBOURG, WEISSENBURG, or KRON-WEISSENBURG, an arrondissement, canton, commune, and town of France, in the dep. of the Bas-Rhin. The arrond. comprises an area of 78,450 hect., and contains 6 cant. Pop. in 1831, 96,349; in 1836, 95,873; in 1846, 94,821. The cant. comprises 13 com. Pop. in 1831, 17,398; in 1846, 16,598. The town is 36 m. NNE of Strasburg, on the r. bank of the Lauter, and on the frontier of Rhenish Bavaria. Pop. in 1789, 5,694; in 1821, 5,714; in 1831, 6,097; in 1841, 6,184; in 1846, 6,273. It stands in a gorge running between the Vosges and the plains of the Rhine, and is enclosed by a strong wall and a ditch, forming a portion of the fortifications known as the 'Lignes-de-Wissembourg.' It contains a church, a Lutheran chapel, a synagogue, a communal college, an hospital, and an alms-house, and possesses manufactories of hosiery, felt, straw-hats, china, pottery, soap, and turnery, dye-works, bleacheries, wax-works, oil-mills, brick, tile, and lime-kilns, &c. The environs are noted for their wine. It is of great antiquity, and is reputed the *Sebusium* of the ancients. It had an abbey founded by Dagobert I., the abbe of which ranked third among the abbe-princes of the empire. It became a free imperial town in 1247, and was united to France by the treaty of Ryswyk. It was dismantled by Louis XIV. In 1744, it was taken by Charles of Lorraine, and in 1793 by the Austrians.

WISSEN, a village of Prussia, in the regency of Dusseldorf, and circle of Geldern. Pop. 350.—Also a village of the regency of Coblenz, and circle of Altenkirchen, on the Sieg. Pop. 615. It has copper-works.

WISSENBACH, a village of Nassau, in the bail of Dillenburg. Pop. 470.

WISSENTRUP, a village of Lippe-Detmold, in the bail. of Lage. Pop. 260.

WISSERSHEIM, a village of Prussia, in the regency of Achen, and circle of Düren. Pop. 500.

WISSET, a village of Austria, in Bohemia, in the reg. of Eger, bail. of Saatz. Pop. 240.

WISSETT, a parish in Suffolk, 2 m. NW of Halesworth. Area 2,259 acres. Pop. in 1851, 490.

WISSING, a village of Bavaria, in the circle of Middle Franconia, presidial of Beilngries. Pop. 302.

WISSOKA, a village of Prussia, in the regency of Danzig, and circle of Stargard. Pop. 300.

WISSOWATTEN, a village of Prussia, in the regency of Gumbinnen, circle of Lotzen. Pop. 340.

WISSULKE, a village of Prussia, in the regency of Marienwerder, circle of Deutsch-Crone. Pop. 200

WISTANSTOW, a parish of Salop, 9½ m. NW by

N of Ludlow. Area 5,160 acres. Pop. in 1831, 989; in 1851, 1,044.

WISTASTON, a parish in the co.-palatine of Chester, 2½ m. NE by E of Nantwich, near the junction of the Chester and Crewe, the Grand Junction, and the Manchester and Birmingham railways.—Area 1,465 acres. Pop. in 1831, 350; in 1851, 298.—Also a chapelry in the p. of Marden, co. of Hereford, 7 m. N by E of Hereford, on the river Lug.

WISTENLACH, a village of Switzerland, in the cant. of Freiburg, and bail. of Murten. Pop. 1,037.

WISTERNITZ, a market-town of Austria, in Moravia, in the circle and 5 m. E of Olmütz, on the Fistritz. Pop. 1,250.

WISTERNITZ (Unter), or Dolny-Wisternice, a market-town of Moravia, in the circle and 23 m. S of Brunn, on the r. bank of the Thaya. Pop. 660.

WISTERZA. See Feistritz-Windisch.

WISTITTEN. See Wystyten.

WISTON, a parish in the upper ward of Lanarkshire, on the N bank of the Clyde. It is irregular in form, and about 6 m. in length, by from 3 to 4½ m. in breadth. The hill of Tinto, which stands on its N boundary, raises its beautiful crest 2,000 ft. above the level of the sea, and is green to the very summit. From this elevation may be enjoyed one of the most extensive views in Scotland; for the eye embraces within its range an immense tract of country and sweep of sea, including Hartfell and Queensberry, in Dumfries-shire; Goatfell, in the Isle of Arran; the Bass; and even the hills in the north of England and north of Ireland. There are three villages in the p., viz., Roberton, Wiston, and Newton of Wiston. Pop. in 1831, 940; in 1841, 929; in 1851, 819.

WISTON, or Wissington, a parish in Suffolk, 1 m. WSW of Nayland, on the N bank of the river Stour, over which there is here a bridge. Area 7,030 acres. Pop. in 1831, 249; in 1851, 683.—Also a parish in Sussex, 1½ m. NNW of Steyning. Area 2,865 acres. Pop. in 1831, 296; in 1851, 301.

WISTOW, a parish in Huntingdonshire, 3½ m. SSW of Ramsey. Area 2,070 acres. Pop. in 1831, 404; in 1851, 553.—Also a parish in Leicestershire, 7 m. SE by S of Leicester, intersected by the Union canal. It includes the chapelry of Newton-Harcourt. Area 2,100 acres. Pop. in 1831, 298; in 1851, 261.—Also a parish in the E and W ridings of Yorkshire, 3 m. NW by N of Selby, near the river Ouse. Area 3,870 acres. Pop. in 1851, 788.

WISTRITZ, a village of Austria, in Bohemia, in the reg. of Eger, and bail. of Teplitz. Pop. 310.

WISTRSCHENOWITZ, a village of Austria, in Moravia, in the circle of Znaim, and seignory of Neu-Reisch. Pop. 200. It has two mills.

WISWALL, or Wiswell, a township in the p. of Whalley, co-palatine of Lancaster, 3 m. S of Clitheroe, and within the borough boundaries. Area 1,410 acres. Pop. in 1831, 724; in 1851, 747.

WISZGOLDINGEN, a village of Würtemberg, circle of the Jaxt, and bail. of Gmünd. Pop. 669.

WISZLAUF, or Wiezlauf, a river of Würtemberg, an affluent of the Rems.

WISZNICZ. See Wisnicz.

WISZNIEWEN, a village of Prussia, in the reg. of Gumbinnen, and circle of Lyk. Pop. 470.

WISZNIOWIEC, a town of Russia in Europe, in the gov. of Volhynia, and circle of Kremenitz. Pop. 2,000.

WISZNOW. See Wischenau.

WISZTICKEN, a village of Prussia, in the regency and circle of Gumbinnen. Pop. 300.

WITANOW, or Wital, a village of Austria, in Bohemia, in the regency of Pardubitz, and bail. of Hohenmauth. Pop. 400.

WITASCZYCE, a village of Prussia, in the reg. of Posen, and circle of Pleschen. Pop. 640.

WITCHAM, a parish in the isle of Ely, co. of Cambridge, 5½ m. W of Ely. Area 2,671 acres. Pop. in 1831, 519; in 1851, 594.

WITCHFORD, a parish in the isle of Ely, co. of Cambridge, 3 m. WSW of Ely. Area 2,376 acres. Pop. in 1831, 461; in 1851, 620.

WITCHINGHAM (Great), a parish in Norfolk, 2 m. S of Reepham. Area 2,245 acres. Pop. 669.

WITCHINGHAM (Little), a parish in Norfolk, 2 m. SE by S of Reepham. Area 738 acres. Pop. in 1831, 62; in 1851, 42.

WITCHLING, a parish in Kent, 10 m. E of Maidstone. Area 1,309 acres. Pop. in 1851, 124.

WITCOMBE, or Whitcombe (Great), a parish in Gloucestershire, 3½ m. NE by N of Painswick. Area 918 acres. Pop. in 1831, 174; in 1851, 167.

WITHAM, a parish and market-town in Essex, 8 m. NE by E of Chelmsford, in the line of the Eastern Counties railway. Area of p. 3,633 acres. Pop. in 1831, 2,735; in 1851, 3,303. The town, which stands in an agreeable situation near the confluence of the rivulet Braine, with the river Blackwater, consists principally of one long street of a respectable aspect, through which runs the high road from London to Colchester. It is the supposed site of the Roman station *Canonium;* and there is a Danish camp in the vicinity.

WITHAM, a river in Lincolnshire, which rises near South Witham, and passing Colsterworth, Grantham, and Lincoln, unites with the sea at Boston deeps. From Lincoln it becomes navigable, and it has also a communication with the Trent by the Foss-dyke canal.

WITHAM. See Wytham.

WITHAM-ON-THE-HILL, a parish in Lincoln shire, 4½ m. SW of Bourne, containing the hamlets of Manthorpe, and Toft with Lound. Area 4,210 acres. Pop. in 1831, 540; in 1851, 635.

WITHAM (North), a parish in Lincolnshire, 1½ m. S by W of Colsterworth, intersected by the river Witham. It contains the hamlet of Lebthorpe. Area 2,373 acres. Pop. in 1831, 273; in 1851, 309.

WITHAM (South), a parish in Lincolnshire, 3½ m. S by W of Colsterworth. Area 3,230 acres. Pop. in 1831, 410; in 1851, 544.

WITHAM-FRIARY, or Charterhouse-Witham-in-Selwood, a parochial chapelry in Somersetshire, 5½ m. SSW of Frome. Area 4,970 acres. Pop. in 1831, 574; in 1851, 556.

WITHAMSVILLE, a village of Union township, Clermont co., in the state of and 3 m. N of the Ohio, U. S., 128 m. SW of Columbus. Pop. in 1840, 250.

WITHCALL, a parish in Lincolnshire, 4 m. W by S of Louth. Area 2,650 acres. Pop. 117.

WITHCOTE, a parish in Leicestershire, 9 m. SSE of Melton-Mowbray. Area 777 acres. Pop. 40.

WITHERIDGE, a parish, formerly a borough and market-town, in the co. of Devon, 8 m. E of Chumleigh, on the small river Sturcomb. Area 9,048. Pop. in 1831, 1,263; in 1851, 1,309.

WITHERLEY, a parish in Leicestershire, 1 m. E of Atherstone, comprising the hamlet of Atterton. The counties of Leicester and Warwick are here separated by the river Anker and the Watling-street, which crosses the Anker at Witherley-bridge. The tower of the church, surmounted by a spire 156 ft. high, is considered the finest in the county. Mancetter, the site of the Roman station *Manduessuedum*, is in this p. Area 1,570 acres. Pop. in 1851, 540.

WITHERNE-WITH-STAIR, a parish in Lincolnshire, 3½ m. NW by N of Alford. Area 2,669 acres. Pop. in 1831, 390; in 1851, 503.

WITHERNSEA, a chapelry in the p. of Hollym,

E. R. of Yorkshire, $2\frac{1}{4}$ m. N of Patrington. Area 1,210 acres. Pop. in 1821, 130; in 1851, 109.

WITHERNWICK, a parish in the E. R. of Yorkshire, 8 m. N by E of Hedon. Area 2,600 acres. Pop. in 1831, 443; in 1851, 513.

WITHERSDALE, a parish in Suffolk, 3 m. SE by E of Harleston. Area 880 acres. Pop. 199.

WITHERSFIELD, a parish in Suffolk, 9 m. W by N of Clare. Area 2,509 acres. Pop. 642.

WITHERSLACK, a township and chapelry in the p. of Beetham, Westmoreland, $7\frac{1}{2}$ m. WNW of Milnthorpe, on the river Belo. Area 4,689 acres. Pop. in 1831, 488; in 1851, 487.

WITHERSTONE, a parish in Dorset, 5 m. ENE of Bridport. Pop. returned with that of Poorstock.

WITHIEL, a parish in Cornwall, 5 m. W by S of Bodmin. Area 3,005 acres. Pop. in 1851, 452.

WITHIELL-FLOREY, a parish in Somerset, 5 m. NE of Dulverton. Area 2,485 acres. Pop. in 1851, 104.

WITHINGTON, a parish in Gloucestershire, 6 m. W of Northleach, intersected by the river Colne. Area 5,830 acres. Pop. in 1831, 743; in 1851, 823.—Also a parish in Herefordshire, 4 m. ENE of Hereford. Area 2,392 acres. Pop. in 1831, 723; in 1851, 881.—Also a township in the p. of Manchester, co.-palatine of Lancaster, 4 m. S of Manchester. Area 2,498 acres. Pop. in 1831, 1,048; in 1851, 1,492. There is here a well-organized college for the education of students of the Congregational body of dissenters.—Also a parish in Salop, 6 m. E of Shrewsbury, intersected by the Shrewsbury canal. Area 1,135 acres. Pop. in 1851, 266.

WITHINGTON (LOWER), a township in the parish of Prestbury, co.-palatine of Chester, 7 m. NNW of Congleton, in the line of the Manchester and Birmingham railway, and near the Macclesfield branch. Area 2,265 acres. Pop. in 1831, 584; in 1851, 570.

WITHINGTON (OLD), a township in the p. of Prestbury, co.-palatine of Chester, $7\frac{1}{2}$ m. NNW of Congleton, in the line of the Macclesfield branch of the Manchester and Birmingham railway. Area 1,840 acres. Pop. in 1831, 191; in 1851, 189.

WITHNELL, a township in the p. of Leyland, co.-palatine of Lancaster, 5 m. NE by N of Chorley, near the Liverpool and Leeds canal. Area 3,557 acres. Pop. in 1831, 1,251; in 1851, 1,975.

WITHYBROOK, a parish in Warwickshire, 8 m. NE by E of Coventry, intersected by the Oxford canal. Area 2,520 acres. Pop. in 1851, 334.

WITHYCOMBE, a parish in Somersetshire, $2\frac{1}{2}$ m. SE of Dunster. Area 1,787 acres. Pop. 329.

WITHYCOMBE-RAWLEIGH, a parish in Devon, $2\frac{1}{2}$ m. E of Exmouth, bounded on the west by the navigable river Exe. It contains a portion of the town of Exmouth. Area 2,617 acres. Pop. in 1831, 1,063; in 1851, 1,811.

WITHYHAM, a parish in Sussex, $6\frac{1}{2}$ m. ESE of East Grimstead. Area 8,086 acres. Pop. 1,692.

WITHYPOOLE, a parish in Somersetshire, 7 m. NW of Dulverton. Area 3,630 acres. Pop. 259.

WITIEGITZ, a village of Austria, in Bohemia, in the reg. of Pilsen, bail of Prachatitz. Pop. 610.

WITIM. See VITIM.

WITIN, a village of Austria, in Bohemia, in the reg. of Budweis, and bail. of Neuhaus. Pop. 370.

WITINOWES, a village of Austria, in Bohemia, in the reg. and bail. of Gitschin. Pop. 420.

WITKOW, a market-town of Austria, in Galicia, in the circle and 36 m. NNW of Zloczow.

WITKOWITZ, a village of Austria, in Bohemia, in the regency of Gitschin, and bail. of Hohenelbe. Pop. 2,200.

WITKOWKI, a village of Prussia, in the reg. of Posen, and circle of Kosten. Pop. 200.

WITKOWO, a town of Prussia, in the regency of Bromberg, circle and 12 m. SE of Gnesen. Pop. 1,958. It has several Catholic and Lutheran churches, and possesses manufactories of linen and woollen fabrics, and nails, and several tanneries.

WITKOWOWIES, a village of Prussia, in the reg. of Bromberg, and circle of Gnesen. Pop. 230.

WITLEY, a parish in Surrey, $3\frac{1}{2}$ m. S by W of Godalming. Area 6,328 acres. Pop. in 1851, 1,546.

WITLEY (GREAT), a parish in Worcestershire, 5 m. SW by S of Stourport. It contains the hamlet of Redmarley. Area 2,633 acres. Pop. 408.

WITLEY (LITTLE), a chapelry in the p. of Holt, Worcestershire, $8\frac{1}{2}$ m. NW by N of Worcester. Area 960 acres. Pop. in 1831, 287; in 1851, 206.

WITLINGHAM, or WHITLINGHAM, a parish in Norfolk, 3 m. ESE of Norwich, on the river Yare. Area 542 acres. Pop. in 1831, 45; in 1851, 19.

WITMAARSUM, a parish of Holland, in the prov. of Friesland, bail. of Leeuwarden. Pop. 800.

WITMUND. See WITTMUND.

WITNESHAM, a parish in Suffolk, $4\frac{1}{2}$ m. N by E of Ipswich, in the projected line of the Eastern Counties railway. Area 1,906 acres. Pop. 575.

WITNEY, a parish and market-town in the co. of Oxford, 11 m. W by N of Oxford. The parish contains the chapelry of Hailey, and the hamlets of Crawley, Curbridge, and Lew. Area 7,083 acres. Pop. in 1801, of the township, 2,584; of the entire parish, 4,349; in 1831, of the former, 3,190; of the latter, 5,336; in 1851, of the latter, 5,437. The town, which consists principally of two streets, well-built houses arranged with considerable regularity, is situated on the small river Windrush, over which there is here a bridge of 3 arches. The road from London to Cheltenham and Gloucester passes through it. The vicinity abounds in agreeable scenery. The inhabitants are partly employed in the manufacture of gloves and malt, and in various branches of the woollen manufacture. The chief source of employment is the manufacture of blankets, duffells, pilot cloth, waggon tilts, &c., for which W. has long been celebrated. Dr. Plott, an old historian of Oxfordshire, writing in 1675, attributes this celebrity, in part at least, to the detersive qualities of the waters of the Windrush. "The Isis, 'tis true," says he, "till it comes to Newbridge, receives not (that I find) any eminently salt or sulphureous waters; but there it admits the nitrous Windrush, so well impregnated with that abstersive salt, that no place yields blanketing so notoriously white as is made at Witney." In the reign of Queen Anne, the blanket weavers in the town, and within 20 miles around it, were incorporated by the style of "The Master, Assistants, Wardens, and Commonalty of the blanket weavers of Witney, in Oxfordshire." At that period, 150 looms and 3,000 persons were constantly employed, and the weekly consumption of wool amounted to about 1,000 packs; but the charter was found to lead to bad consequences, the rules framed by the corporation having retarded the interests of the trade, so that for many years the privilege has not been acted upon. The manufacture is by no means in such a flourishing condition as it once was; but the trade is steady. The average annual number of blanket pieces, including pilot cloth, &c., manufactured, each containing about half a peck of 120 lbs. of wool, is about 10,000. The average value of a piece is £9, but some are double that amount. There is a handsome blanket-hall in the High-street, erected in 1721. In the market-place is a neat town-hall, the lower part of which consists of piazzas for the use of the market.—W. existed long before the Conquest as a town of some importance, under the name of Whitney or Whitteney. In the reign of Edward the Confessor the manor was given by Bishop Ailwyn to the monastery of St. Swithin at Winchester, in gratitude for the deliverance of Emma, mother to the king, from the fiery ordeal which she underwent in the cathedral of that city. It was formerly a royal borough, and sent 2 members to parliament.

WITOSES, a village of Austria, in Bohemia, in the regency of Eger, and bail. of Saatz. Pop. 470.

WITOSLAW, a village of Prussia, in the prov. of Posen, regency of Bromberg, and circle of Wirsitz. Pop. 280.

WITOSLAWITZ, a village of Prussia, in the prov. of Silesia, regency of Oppeln, and circle of Kosel. Pop. 280.

WITOWITZ, a village of Austria, in Moravia, in the circle of Brunn, and seignory of Posoritz. Pop. 300.

WITOWO, a village of Prussia, in the regency of Posen, and circle of Schroda. Pop. 300.

WITRKOWITZ, a village of Austria, in Moravia, in the circle of Prerau, and seignory of Hochwald, on the Lubina. Pop. 400.

WITRY, a department and commune of Belgium, in the prov. of Luxemburg, and arrond. of Arlon, watered by the Sure. Pop. 622.

WITRY-LES-REIMS, a commune of France, in the dep. of the Marne, and cant. of Bourgogne, 5 m. NE of Reims. Pop. 1,161.

WITSCHAPP, a village of Austria, in Moravia, in the circle of Znaim. Pop. 490. It has a castle and a brewery.

WITSCHE, a village of Austria, in Moravia, in the circle of Olmutz, and bail. of Wallachisch-Meseritsch. Pop. 1,530.

WITSCHEIN, a village of Austria, in Styria, in the circle of Marburg. Pop. 1,600. It has a castle.

WITSCHIN, a village of Austria, in Bohemia, in the reg. of Eger, and bail. of Tepel. Pop. 400.

WITSCHITZ, a village of Austria, in Moravia, in the circle of Olmutz, and seignory of Moritz. Pop. 400.

WITSEMBERG, a summit of South Africa, in the district of Talbagh, near the valley of Rodezand. It has an alt. of 2,900 Parisian ft. above sea-level.

WITT'S (DE) LAND. See Australia.

WITTAU, a village of Austria, in the country below the Ens, bail. of Gross Enzersdorf. Pop. 400.

WITTBACH, a village of Bavaria, in the circle of Lower Bavaria, and presidial of Kreuzwertheim. Pop. 330.

WITTBRIETZEN, a village of Prussia, in the regency of Potsdam, and circle of Zauch-Belzig, near Belitz. Pop. 488.

WITTEKIND. See Giebichenstein.

WITTELBACH, a village of Baden, in the circle of the Lower Rhine, and bail. of Lahr. Pop. 220.

WITTELSBACH (Ober and Unter), two villages of Bavaria, in the circle of Upper Bavaria, and presidial of Aichach, 17 m. NE of Augsburg, on a mountain, containing respectively 90 and 185 inhabitants. In the vicinity are the ruins of the castle of Wittelsbach, the cradle of the royal house of Bavaria.

WITTELSBURG, a village of Hesse, in the prov. of Upper Hesse, circle and presidial of Marburg. Pop. 538.

WITTELSHEIM, or Wettelsen, a commune of France, in the dep. of the Haut-Rhin, and cant. of Cernay, 27 m. NE of Belfort. Pop. in 1841, 1,565.

WITTELSHOFEN, a village of Bavaria, in the circle of Middle Franconia, and presidial of Dinkelsbuhl. Pop. 646, of whom 200 are Jews. It has a synagogue.

WITTEM, a commune of Belgium, in the prov. of Limburg, and arrond. of Maestricht. Pop. 3,007. It has a paper-mill.

WITTEN, a town of Prussia, in the regency of Arnsberg, and circle of Bochum, on the Ruhr. Pop. 2,480. It has manufactories of silk and woollen fabrics, iron and steel ware, and several oil and paper-mills. The trade consists chiefly in corn. In the vicinity is a coal-mine.

WITTENBACH, a village of Switzerland, in the cant. of St. Gall, and bail. of Tablat. Pop. 1,301.

WITTENBERG, a circle and town of Prussia, in the prov. of Saxony, and regency of Merseburg. Pop. of circle, 48,181. The town is 48 m. NE of Merseburg, on the r. bank of the Elbe, which is here crossed by a long wooden bridge, and at an alt. of 229 ft., above sea-level, in N lat. 51° 52′ 13″, and E long. 12° 45′ 29″. Pop. in 1831, 7,084; in 1841, 8,241; in 1849, 11,052. It is fortified, and has two suburbs, a fortress, and five churches, in one of which belonging to the castle are the tombs of Luther and Melancthon. It has also a colossal statue of Luther, an orphans' asylum, a theological seminary, and gymnasium. The ancient university was joined to that of Halle in 1816. Its industry consists chiefly in the manufacture of woollen and linen fabrics, and distillation of brandy. Wood forms one of its chief articles of trade.

WITTENBERGE, a town of Prussia, in the prov. of Brandenburg, regency and 84 m. NW of Potsdam, and circle of West Priegnitz, at the confluence of the Stepnitz. Pop. 2,780. The manufacture of earthenware, fishing, and navigation form the chief objects of local industry.

WITTENBURG, a village of Hanover, in the principality and bail. of Kalenburg, near Hanover. Pop. 138.—Also a town and bail. of the grand-duchy of Mecklenburg-Schwerin, district and 17 m. SW of Schwerin. Pop. 1,760. It has distilleries of brandy, breweries, dye-works, tanneries, manufactories of linen, tobacco, pipes, and nails, and a tin-foundry.

WITTENDORF, a village of Würtemberg, in the circle of the Schwarzwald, and bail. of Freudenstadt. Pop. 536.

WITTENFELDE, a village of Prussia, in the regency of Stettin, and circle of Naugard. Pop. 220.

WITTENFORDEN, a village of Mecklenburg-Schwerin, in the prov. of Mecklenburg, and bail. of Schwerin. Pop. 630.

WITTENHAGEN, a village of Prussia, in the reg. of Stralsund, and circle of Grimmen. Pop. 270.

WITTENHAM (Little), a parish in Berks, 4 m. NW by N of Wallingford, south of the Thames. Area 870 acres. Pop. in 1831, 113; in 1851, 128.

WITTENHAM (Long), a parish in Berks, 5 m. NW of Wallingford. Area 2,280 acres. Pop. 608.

WITTENHEIM, or Wetteneu, a commune of France, in the dep. of the Haut-Rhin, and cant. of Mulhausen, 15 m. NNE of Altkirch. Pop. 1,260.

WITTENMOOR, a village of Prussia, in the reg. of Magdeburg, and circle of Slendal. Pop. 270.

WITTEN-SEE, a lake of Denmark, in the duchy of Sleswig, and bail. of Heitten, 6 m. NE of Rendsburg. It is upwards of a mile in length from NE to SW, and 1½ m. in breadth. It discharges itself in the S into the Eyder.

WITTENWEIER, a village of Baden, in the circle of the Middle Rhine, bail. of Lahr. Pop. 520.

WITTENWEIL, a village of Switzerland, in the cant of Thurgau, and bail. of Fruenfeld. Pop. 411.

WITTER, a parish, consisting of a main body and a detached district, in co. Down. The main body lies at the extremity of the Ardes peninsula, 2 m. SE of Portaferry. Area of the whole 2,529 acres. Pop. in 1831, 1,116; in 1841, 1,053.

WITTERDA, a village of Prussia, in the regency and circle of Erfurt. Pop. 781.

WITTERING, or WHITTERING, a parish in Northamptonshire, 2½ m. NW by N of Wansford. Area 2,690 acres. Pop. in 1831, 216; in 1851, 262.

WITTERING (EAST), a parish in Sussex, 6 m. SW of Chichester. Area 1,505 acres. Pop. 233.

WITTERING (WEST), a parish in Sussex, 7½ m. SW by W of Chichester. The English Channel bounds the parish on the SW, and the mouth of Chichester harbour lies on the NW side. Area 3,615 acres. Pop. in 1831, 606; in 1851, 609.

WITTERSCHLICK, a village of Prussia, in the regency of Cologne, and circle of Bonn. Pop. 384.

WITTERSEE, a village of Belgium, in the prov. of Brabant, and arrond. of Nivelles. Pop. 958.

WITTERSHAM, or WITTESHAM, a parish in Kent, 4½ m. S by E of Tenterden. Area 3,601 acres. Pop. in 1831, 919; in 1851, 987.

WITTERSHAUSEN, a village of Bavaria, in the circle of Lower Franconia, and presidial of Guerdorf. Pop. 329.—Also a village of Würtemberg, in the circle of the Schwarzwald, bail. of Sulz. Pop. 654.

WITTERSHEIM, a village of Bavaria, in the circle of Pfalz, and cant. of Blieskastel. Pop. 356.

WITTERSWEIL, a village of Switzerland, in the cant. of Soleure, and bail. of Dornach, in the Leiman-thal, near Blauen. Pop. 291.

WITTESHEIM, a village of Bavaria, in the circle of Swabia, and presidial of Monheim. Pop. 316.

WITTGENAU, a village of Prussia, in the reg. of Leignitz, and circle of Grunberg. Pop. 240.

WITTGENBORN, a village of Hesse, in the prov. of Hanau, circle of Gelnhausen, and bail. of Wachtersbach. Pop. 695.

WITTGENDORF, a village of Prussia, in the regency of Leignitz, and circle of Landshut. Pop. 915.—Also a village of the same regency, in the circle of Sprottau. Pop. 731.—Also a village of the regency of Merseburg, and circle of Zeitz. Pop. 220.—Also village of Saxony, in the circle of Bautzen, and bail. of Lobau, near Zottau. Pop. 1,046.

WITTGENDORF (NIEDER and OBER), two villages of Prussia, in the regency of Leignitz, and circle of Goldberg, containing respectively 261 and 110 inhabitants.

WITTGENSDORF, a village of Saxony, in the circle of Zwickau, and bail. of Chemnitz. Pop. 2,089. It has extensive spinning-mills and bleacheries.—Also a village in the circle of Dresden, and bail. of Pirna, near Dohna. Pop. 104.

WITTGENSTEIN, a circle of Prussia, in the prov. of Westphalia, and reg. of Arnsberg. Pop. 22,500. Its cap. is Berlenburg. It belongs to the princes of Sayn-Wittgenstein, and has a castle of the same name. It has considerable manufactures.—Also a village of the regency of Leignitz, and circle of Landshut. Pop. 1,000.—Also a village of the same regency, and circle of Sprottau. Pop. 730.—Also a village of the regency of Merseburg, and circle of Zeitz. Pop. 260.—Also a village of Saxony, in the circle of Bautzen, and bail. of Lobau. Pop. 1,120.—Also a village of Schwarzburg-Rudolstadt, and bail. of Königsee. Pop. 380.

WITTGENSTEIN, a group of islands of the South Pacific, in the Low Archipelago, to the SE of the Palliser islands, in S lat. 16° 1′, and W long. 145° 50′.

WITTGENSTEIN (OBER and UNTER), two villages of Prussia, in the regency of Leignitz, and circle of Goldburg-Hainau. Pop. 500.

WITTICHENAU, or KULOW, a town of Prussia, in the prov. of Brandenburg, regency and 75 m. SSW of Frankfort-on-the-Oder, and circle of Spremberg, on the Black Elster. It has two churches and an hospital, manufactories of linen and hosiery, and a considerable trade in cattle.

WITTICHENAU, or WITTGENAU, a town of Prussia, in the regency of Liegnitz, and circle of Hoyerswerda, on the Black Elster, at an alt. of 344 ft. above sea-level. Pop. 2,200. It has manufactories of woollen and linen fabrics, iron-works, a paper-mill, and important cattle-markets.

WITTICHOW, a village of Prussia, in the regency of Stettin, and circle of Pyritz. Pop. 440.

WITTICHSTHAL, a village of Saxony, in the circle of Zwickau, and bail. of Erbenstock. Pop. 280.

WITTICHTHAL, a village of Austria, in Bohemia, reg. of Eger and bail. of Tachau. Pop. 200.

WITTICHWALDE, a village of Prussia, in the reg. of Königsberg, circle of Osterode. Pop. 260.

WITTIG (OBER and UNTER), two villages of Austria, in Bohemia, in the regency of Bohmisch-Leipa, and bail. of Reichenberg. Pop. 1,000.

WITTIGHAUSEN (OBER and UNTER), two villages of Baden, in the circle of the Lower Rhine, and bail. of Gerlachsheim, containing respectively 220 and 590 inhabitants.

WITTINGAU, WITGOW, or TRZEBEN, a town of Austria, in Bohemia, in the regency and 15 m. ENE of Budweis, and bail. of Neuhaus, on the Goldbach. Pop. 4,800. It has a fine castle belonging to the princes of Schwarzenberg.

WITTINGEN, a town of Hanover, in the gov. and 42 m. SSE of Lüneburg, and bail. of Knesebeck. Pop. 1,100.

WITTINGREUTH, a village of Austria, in Bohemia, in the regency of Eger, and bail. of Tachau. Pop. 300.

WITTINSBURG, a village of Switzerland, in the cant. of Basle, and bail. of Giffach. Pop. 245.

WITTISHEIM, or WITTSEN, a commune of France, in the dep. of the Bas-Rhin, and cant. of Marckolsheim, 12 m. NE of Schelestat. Pop. 1,034.

WITTISLINGEN, a village of Bavaria, in the circle of Swabia, and presidial of Dillingen. Pop. 1,172.

WITTITZ, a village of Austria, in Bohemia, in the regency of Pardubitz, and bail. of Schwarz-Kosteletz. Pop. 420.

WITTKAU, a village of Prussia, in the regency of Marienwerder, and circle of Flatow. Pop. 200.

WITTKOW, a village of Prussia, in the regency of Marienwerder, circle of Deutsch-Crone. Pop. 270.

WITTKOWITZ, a village of Austria, in Moravia, in the circle of Olmutz, and seignory of Hochwald. Pop. 200.

WITTLAGE-HUNTEBERG, a village and bail. of Hanover, in the gov. and 15 m. ENE of Osnabruck, near Oster-Kappeln. Pop. 258.

WITTLENSWEILER, a village of Würtemberg, in the circle of the Schwarzwald, and bail. of Freudenstadt. Pop. 665.

WITTLICH, a circle and town of Prussia, in the prov. of the Rhine, and regency of Treves. Pop. of circle, 35,496. The town is 21 m. NNE of Treves, on the Lieser. Pop. 2,974. The vine and tobacco are cultivated in the environs. On an adjacent summit are the ruins of an ancient castle.

WITTLINGEN, a village of Würtemberg, in the circle of the Schwarzwald, and bail. of Urach, on the Alp. Pop. 584.—Also a village of Baden, in the circle of the Upper Rhine, and bail. of Lorrach. Pop. 322.

WITTMANSDORF, a village of Austria, in Styria, in the circle of Gratz. Pop. 420.—Also a village in the circle of Marburg. Pop. 260.—Also a village of Prussia, in the regency of Frankfurt, and circle of Lubben. Pop. 260.

WITTMUND, or WITMUND, a market-town and

bail. of Hanover, in the gov. and 15 m. NE of Aurich, on the Harl, a canal connecting it with the sea. Pop. 1,815. It has a school and a poor-house, and possesses manufactories of linen. The trade consists chiefly in horses, grain, and butter.

WITTNAU, a village of Switzerland, in the cant. of Aargau, and bail. of Laufenburg, on the confines of Basle. Pop. 939.—Also a village of Baden, in the circle of the Upper Rhine, and bail. of Freiburg. Pop. 318.

WITTOCHOW, a village of Austria, in Moravia, in the circle of Iglau, and seignory of Bistritz. Pop. 260.

WITTON, a parochial chapelry, locally in the p. of Great Budworth, co.-palatine of Chester, ½ m. E of Northwich, containing the township of Twambrooke and part of that of Rudheath. The village, which forms a suburb to the town of Northwich, stands on an elevated spot near the banks of the river Dane. Area 588 acres. Pop. in 1831, 2,912; in 1851, 3,493. Also a parish in Huntingdonshire, 2½ m. W by N of St. Ives, bounded on the N by the river Ouse. Area 1,690 acres. Pop. in 1831, 277; in 1851, 207.—Also a township in the parish of Blackburn, co.-palatine of Lancaster, 2 m. N by W of Blackburn, on the river Derwent. Area 650 acres. Pop. in 1831, 1,047; in 1851, 1,367. The inhabitants are chiefly employed in some cotton-spinning manufactories.—Also a parish in Norfolk, 3½ m. E of Norwich. Area 1,746 acres. Pop. in 1851, 299.

WITTON (East), a parish in the N. R. of Yorkshire, 2 m. SE of Middleham. Area 7,720 acres. Pop. in 1831, 687; in 1851, 610. In this parish are the remains of Jerveaux abbey.. The late earl of Aylesbury caused the whole to be explored and cleared out in 1806 and 1807; when the abbey, church, and choir, with the transepts and high altar, several tombs, the chapter house, abbot's house, kitchen, &c., were discovered.

WITTON (Gilbert), a parish in the co.-palatine of Durham, 3½ m. NW of Durham, N of the river Browney. Area 2,535 acres. Pop. in 1851, 1,758.

WITTON-LE-WEAR, a parish in Auckland, co.-palatine of Durham, 5 m. WNW of Bishop-Auckland, on the N bank of the Wear, and in the line of the Bishop-Auckland and Weardale railway, which is here crossed by a bridge. Area 2,955 acres. Pop. in 1831, 502; in 1851, 918.

WITTON (Long), a township in the p. of Hartburn, Northumberland, 8 m. W by N of Morpeth. Pop. in 1831, 143; in 1851, 149.

WITTON (Nether), a parish in Northumberland, 8 m. WNW of Morpeth, including the townships of Coat-Yards, Ewesley, Healy-with-Comb-hill, Nether-Witton, Nunnikirk, Ritton-Coltpark, and Ritton-Whitehouse. Area 6,850 acres. Pop. in 1851, 489.

WITTON-SHIELDS, a township in the p. of Long Horsley, Northumberland, 7 m. NW by W of Morpeth. Area 574 acres. Pop. in 1851, 25.

WITTON (West), a parish in the N. R. of Yorkshire, 5 m. W of Middleham. Area 3,715 acres. Pop. in 1831, 552; in 1851, 550.

WITTONITZ, a village of Austria, in Moravia, in the circle of Prerau, and seignory of Leipnitz. Pop. 700.

WITTORF, a village of Hanover, in Luneburg, in the bail. of Winsen-on-the-Luhe, near Luneburg. Pop. 435.—Also a village of Stade, in the bail. of Rottenberg, near Visselhövede. Pop. 4,200.

WITTSTOCK, a town of Prussia, in the regency of Potsdam, and circle of Ost-Priegnitz, on the Dosse, 63 m. NW of Berlin. Pop. 6,873. It is enclosed by walls, with four gates, and has an hospital and a poor's-house. It possesses manufactories of woollen and linen fabrics, and of tobacco, distilleries, and breweries, and several tanneries. It was founded at the beginning of the thirteenth century.—Also a village in the same circle. Pop. 260.—Also a village in the regency of Frankfurt, and circle of Kustrin. Pop. 260.

WITTUN, a village of Prussia, in the regency of Marienwerder, and circle of Flatow. Pop. 220.

WITT-WATER'S RAANDT, a high tract of land in S. Africa, on the N side of the Vaal river, running parallel to the Magalie's Berg, and giving rise to the Oori or Krokodil river, the main source of the Limpopo.

WITTWEIL, a village of Switzerland, in the cant. of Argau, and circle of Zofingen. Pop. 526.

WITUCHOWO, a village of Prussia, in the regency of Posen, and circle of Birnbaum. Pop. 200.

WITZELRODA, a village of Saxe-Meiningen, in the bail. of Salzungen. Pop. 230.

WITZEN, a village of Prussia, in the regency of Frankfurt, and circle of Sorau. Pop. 270.

WITZENHAUSEN, a town of Hesse-Cassel, capital of the circle and bail. of the same name, in the prov. of Nieder-Hesse, on the l. bank of, and at the confluence of the Gelster with the Werra, which is here crossed by a bridge, and 14 m. E of Cassel. It is enclosed by walls, with three gates, and has a synagogue, an hospital, and a school. Pop. 3,235. It has several dye-works, manufactories of linen, tobacco, and vinegar, a paper-mill and tanneries, and possesses an active trade. The environs are noted for their beans, fruit, wine, and tobacco.—The circle comprises an area of 38,500 hect., of which 10,159 hect. are arable. Pop. in 1840, 32,828. It is generally mountainous, and contains valleys of great fertility, especially in grain, legumes, cherries, and other fruit. Salt, alum, coal, and potter's clay, are its chief mineral productions. It is divided into four bailiwicks, and contains 5 towns, and 54 villages.

WITZENITZ, a village of Austria, in Moravia, in the circle of Brünn, near Znaim. Pop. 300.

WITZETZE, a village of Denmark, in the prov. of Lauenburg. Pop. 270.

WITZHAVE, a village of Holstein, in the bail. of Trottau, to the E of Hamburg. Pop. 307.

WITZHELDEN, a village of Prussia, in the regency of Dusseldorf, and circle of Solingen. Pop. 220

WITZIN, a village of Mecklenburg-Schwerin, in the prov. of Mecklenburg, and bail. of Sternberg. Pop. 390.

WITZISCHKEN, a village of Prussia, in the regency of Gumbinnen, circle of Niederung. Pop. 338.

WITZKE, a village of Prussia, in the regency of Potsdam, and circle of Westhavelland. Pop. 200.

WITZLESHOVEN, a village of Bavaria, in the circle of Ober-Franconia, and presidial of Berneck. Pop. 200.

WITZNITZ, a village of Prussia, in the regency of Stettin, and circle of Regenwalde. Pop. 388.

WITZOMIELITZ, a village of Austria, in Moravia, in the circle of Brünn. Pop. 490. It has a brewery.

WITZOMIERITZ, a village of Austria, in Moravia, circle of Olmutz. Pop. 380. It has a castle.

WITZSCHDORF, a village of Saxony, in the circle of Zwickau, bail. of Augustusburg. Pop. 776.

WITZSCHERSDORF, a village of Prussia, in the regency and circle of Merseburg. Pop. 200.

WIVELISCOMBE, a parish and market-town, situated, with three other parishes, in a detached portion of the hund. of Kingbury, co. of Somerset, 28 m. W of Somerton. Area of p. 5,984 acres. Pop. in 1801, 2,571; in 1831, 3,047; in 1851, 2,861. The town is situated on a low hill, in the midst of a valley of some extent, bounded by lofty hills interrupted by numerous ravines. It consists of several streets containing some handsome modern houses, as well as a number which bear the stamp of considerable antiquity. One part of the town is called the borough, and at an early period the inhabitants sent members to parliament, but were relieved from that burden at their own request. The inhabitants are employed to a considerable extent in blanket-weaving and other branches of the woollen manufacture.

WIVELSFIELD, a parish in Sussex, 4 m. SE of Cuckfield, in the line of the London and Brighton railway. Pop. in 1831, 559; in 1851, 608.

WIVENHOE, a parish in Essex, 4½ m. SE by E of Colchester, on the river Colne. Area 1,597 acres. Pop. in 1831, 1,714; in 1851, 1,672. This is regarded as the port to Colchester. A plan was recently formed for the construction of a ship canal hence to the Hythe, at an estimated expense of £56,000. The quay is very commodious, and vast numbers of the celebrated Colchester oysters are shipped hence for London and other markets. There is a custom-house subordinate to Colchester.

WIVETON, a parish in Norfolk, ½ m. W of Clay. Area 1,042 acres. Pop. in 1831, 218; in 1851, 245.

WIX, or WEEKS, a parish in Essex, 4½ m. ESE of Manningtree. Area 3,090 acres. Pop. in 1851, 778.

WIXFORD, or WICKLESFORD, a parish in Warwickshire, 2 m. S of Alcester. Area 530 acres. Pop. in 1831, 108; in 1851, 117.

WIXHAUSEN, a village of Hesse, in the prov. of Starkenburg, and presidial of Langen. Pop. 560.

WIXWEN, a village of Prussia, in the regency of Gumbinnen, and circle of Niederung. Pop. 220.

WIZARD-PEAK, a summit of Western Australia, in the district of Victoria, between Mount Fairfax on the N, and Greenough river on the S. It rises conically to the height of 715 ft. above sea-level, and consists of blocks of iron-stone.

WIZAYNY, a town of Russia in Europe, in Poland, in the gov. of Augustowo, near Segny. Pop. 1,400.

WIZERNES, a village of France, in the dep. of the Pas-de-Calais, and arrond. of St. Omer. It has a paper-mill and a brewery.

WIZNA, a town of Russia in Europe, in Poland, in the gov. of Augustowo, on the Narew. Pop. 2,160.

WIZZELSDORF, a market-town of Austria, in the circle of the Unter-Mannhardsberg, 29 m. SE of Korneuberg, near the l. bank of the Danube.

WKRA, a river which has its source in the S part of the prov. of East Prussia, in Prussia; flows under the name of Soldau past the town of that name; forms for some distance the dividing line between Prussia and Poland; enters the latter in the gov. of Plack, and after a total course, first in a SW and afterwards in a SE direction, of about 120 m., joins the Bug on the r. bank, a little above Modlin.

WLACHEN. See WALACHIA.

WLACHOWITZ, a market-town of Austria, in Moravia, in the circle of Hradisch. Pop. 780.

WLADISLAW, a market-town of Austria, in Moravia, in the circle and 23 m. ESE of Iglau, on the l. bank of the Iglawa. Pop. 600.

WLADISLAWOW, a town of Poland, in the gov. of Warsaw and obwod of Konin, 33 m. NE of Kalisch. Pop. 500. It has two churches, a Catholic and a Lutheran, and manufactories of linen and of hats.—Also a town in the gov. of Augustowo, and obwod of Mariampol, on the Szezupa, at its confluence with the Szyrwinta. Pop. 3,550.

WLASATICI. See WOSTITZ.

WLASCHIM, a town of Austria, in Bohemia, in the regency of Budweis, circle and 23 m. SSW of Kaurzim. Pop. 2,400. It has a castle, a Franciscan convent, and an hospital, and possesses manufactories of woollen fabrics.

WLASCHKA, a bail. of Turkey in Europe, in Upper Walachia, on the Danube.

WLASENITZ, a village of Austria, in Bohemia, reg. of Budweis, and bail. of Pilgram. Pop. 230.

WLASTETZ, a village of Austria, in Bohemia, in the reg. of Pilsen, and bail. of Pisek. Pop. 230.

WLC, a village of Austria, in Bohemia, in the reg. of Prague, and bail. of Laun. Pop. 200.

WLCETIN, a village of Austria, in Bohemia, in the regency of Gitschin, and bail. of Bomisch-Aicha. Pop. 260.—Also a village in the regency of Budweis, and bail. of Pilgram. Pop. 290.

WLCOWES, a village of Austria, in Bohemia, in the regency of Budweis, and bail. of Tabor. Pop. 400. It has manufactories of beet-root sugar, and of potash, and paper-mills.

WLECKOWITZ, a village of Austria, in Bohemia, in the reg. of Budweis, and bail. of Wotitz. Pop. 220.

WLKANCITZ, a village of Austria, in Bohemia, in the regency of Pardubitz, and bail. of Schwarzkosteletz. Pop. 260.

WLKANETZ, a village of Austria, in Bohemia, reg. of Pardubitz, and bail. of Habern. Pop. 280.

WLKANOW, a village of Austria, in Bohemia, regency of Gitschin, and bail. of Horzitz. Pop. 300.

WLKAU, a village of Austria, in Moravia, in the circle of Iglau, near Ossowa. Pop. 320.

WLKAWA, a village of Austria, in Bohemia, in the regency of Gitchin, and bail. of Nimburg. Pop. 340. It has a castle.

WLKONITZ, a village of Austria, in Bohemia, in the regency of Pilsen, and bail. of Strakonitz. Pop. 370.—Also a village in the bail of Horazdiowitz. Pop. 270.

WLKOSCH, a village of Austria, in Moravia, in the circle of Hradisch. Pop. 630.—Also a village in the circle of Brerau. Pop. 580.

WLKOW, a village of Austria, in Bohemia, in the regency of Budweis, and bail. of Neuhaus. Pop. 200.—Also a village of the regency of Gitschin, and bail. of Koniginhof. Pop. 380.

WLKSITZ, a village of Austria, in Bohemia, in the regency of Budweis, and bail. of Muhlhausen. Pop. 220.

WLOCLAWEK, a town of Poland, capital of the obwod of Kujawie, in the gov. of Warsaw, on the Vistula. Pop. 4,000.

WLODAWA, a town of Poland, in the gov. of Lublin, and obwod of Radzyn, on the l. bank of the Bug, at the confluence of the Wlodawka. Pop. 3,180.

WLODOWICE, a town of Poland, in the gov. of Lublin, and obwod of Olkusz, 60 m. WSW of Kielce. Pop. 900. It has a castle, two churches, and a synagogue, and has manufactories of cloth.

WLODZIMIERZ. See VLADIMIR.

WLOSCIEJEWKI, a village of Prussia, reg. of Posen, and circle of Schrimm. Pop. 350.

WLOSZEZANOW, a town of Russia in Europe, in Poland, in the gov. of Lublin, near Kielce, near Lake Wloszczowka. Pop. 1,210.

WLTSCHKOW, a town of Austria, in Bohemia, reg. of Pardubitz, and bail. of Landskron. Pop. 310.

WLTSCHKOWA, a village of Austria, in Moravia, in the circle of Hradisch. Pop. 410.

WLTSCHKOWITZ, a village of Austria, in Bohemia, in the regency of Pilsen, and bail. of Planitz. Pop. 220.—Also a village of the regency of Gitschin, and bail. of Koniggratz. Pop. 230.

WLTSCHNAU, a village of Austria, in Moravia, in the circle of Hradisch. Pop. 1,290.

WLTSCHY, a village of Austria, in Bohemia, reg. of Pilsen, and bail. of Przecstitz. Pop. 360.

WNISTERNING, or WELSCHWINTERLINGEN, a village of Switzerland, cant. of Freiburg. Pop. 537.

WOADY GALOAK, a river of Australia Felix, in the district of Portland bay, which has its source in the range of the Stringy-Bark, and falls into Lake Carangainote.

WOAHU. See OAHU.

WOBBEL, a village of the principality of Lippe-Detmold, and bail. of Schneider, 11 m. E of Detmold, on the Emmer. Pop. 640.

WOBBELIN, a village of the grand-duchy of Mecklenburg-Schwerin, and bail. of Neustadt-an-der-Elbe, 20 m. SSE of Schwerin. The poet Korner died here in 1813.

WOBECK, a village of Brunswick, in the circle of Helmstadt. Pop. 280.

WOBECKNITZ, a village of Bohemia, in the circle and 20 m. SSW of Beraun, in the Brdy-Wald.

WOBERMIN, a village of Prussia, in the reg. of Stettin, and circle of Pyritz. Pop. 230.

WOBERN, a village of Austria, in Bohemia, in the regency of Bohmisch-Leipa, and bail. of Danba. Pop. 330.

WOBESDE, a village of Prussia, in the regency of Koslin, and circle of Stolpe. Pop. 390.

WOBESER, a village of Prussia, in the reg. of Koslin, and circle of Rummelsburg. Pop. 210.

WOBLANSE, a village of Prussia, in the reg. of Koslin, and circle of Rummelsburg. Pop. 240.

WOBORA, a village of Austria, in Bohemia, in the regency of Gitschin, and bail. of Nimburg. Pop. 240.—Also a village in the regency of Pilsen, and bail. of Plass. Pop. 370.

WOBRAMKOSTEL. See WOLFRAMITZKIRCHEN.

WOBRATAIN, a village of Austria, in Bohemia, reg. of Budweis, and bail. of Patzau. Pop. 400.

WOBRUB (GROS and KLEIN), two villages of Austria, in Bohemia, in the regency of Gitschin, and bail. of Sobolka. Pop. 400.

WOBS, a village of Oldenburg, in the territory of Lubeck, and bail. of Gutin. Pop. 200.

WOBURN, or OLD WOBURN, a parish and market-town in the co. of Bedford, 15 m. SW by S of Bedford, on the high road from London to Leeds. Area of p. 3,200 acres. Pop. in 1831, 1,827; in 1851, 2,049. The church, which was erected by Robert Hobbs, the last abbot of Woburn, is a curious structure, nearly covered with ivy. Within the last few years it has received great architectural improvements, and now presents a very noble appearance. The town is now chiefly of modern erection, and consists of several broad and well-built streets intersecting one another at right angles, with a handsome market-house. The inhabitants are partly employed in agriculture, partly in the manufacture of straw-plait and thread-lace. Woburn abbey, the stately and elegant seat of the dukes of Bedford, was erected here on the site of an ancient abbey of that name, granted 1st Edward VI. to Lord John Russell.

WOBURN, or WOOBURN (WEST or BISHOP'S), a parish, formerly a market-town, in the co. of Buckingham, 3 m. WSW of Beaconsfield. Area 2,850 acres. Pop. in 1831, 1,927; in 1851, 2,026. The church is a spacious ancient structure containing some interesting monuments, and a curious font. The town stands in a pleasant valley, near the small river Wycombe, on which are several corn and paper mills. Lace is manufactured here to a small extent.

WOBURN, a small fishing harbour in the p. of Donaghadee, 3 m. SSE of the town of Donaghadee, co. Down.

WOBURN, a township of Middlesex co., in the state of Massachusetts, U. S., 10 m. NW by W of Boston, drained by Mystic river, and intersected by the Middlesex canal and the Boston and Lowell railroad. Pop. in 1840, 2,993; in 1850, 3,954.

WOCHEINER-SEE, a lake of Illyria, in the gov. and circle and 54 m. WNW of Laibach. It is 3 m. in length from NE to SW, and about 1 m. in breadth. It discharges itself by a small stream, which issues from it on the E, and joins the Save below Radmansdorf.

WOCHERN, a village of Prussia, in the regency of Treves, and circle of Saarburg. Pop. 270. It has three mills.

WOCHOS, a village of Austria, in Bohemia, in the reg. of Pardubitz, and bail. of Chrudim. Pop. 200.

WOCHOW, a village of Austria, in Bohemia, in the regency of Gitschin, and bail. of Reichenau. Pop. 300.

WOCKENDORF, a village of Austria, in Silesia, in the circle of Troppau. Pop. 670.

WOCKENRATH, a village of Prussia, in the reg. of Aachen, and circle of Erkelenz. Pop. 200.

WOCKLITZ, village of Prussia, in the reg. of Danzig, and circle of Elbing. Pop. 210.

WODADDY, a town of Hindostan, in the presidency of Madras, and Northern Circars, district and 24 m. WNW of Vizagapatam, on a small river which flows into the gulf of Bengal.

WODDOW, a village of Prussia, in the regency of Potsdam, and circle of Prenzlow. Pop. 338.

WODECQ, a department and commune of Belgium, in the prov. and 17 m. ENE of Hainault, and arrond. of Tournai. Pop. 2,919.

WODERAD, a village of Austria, in Bohemia, in the regency of Pardubitz, and bail. of Leitomischl. Pop. 380.—Also a village of the same regency, in the bail. of Schwarzkostelitz. Pop. 450.

WODIERAD, a village of Austria, in Moravia, in the circle of Brunn. Pop. 300.

WODISLAW. See LOSLAU.

WODITZ, a village of Austria, in Bohemia, in the regency of Budweis, and bail. of Patzau. Pop. 270.—Also a village of the regency of Pilsen, and bail. of Netolitz. Pop. 240.

WODKE, a village of Prussia, in the reg. of Stettin, and circle of Greiffenberg. Pop. 388.

WODLOCHOWITZ, a village of Austria, in Bohemia, in the reg. of Budweis, and bail. of Wotitz. Pop. 360.

WODNA, a village of Austria, in Bohemia, in the regency of Budweis, and bail. of Kamenitz. Pop. 300.

WODNIAU, a town of Austria, in Bohemia, in the regency of Pilsen, and bail. of Pisek, on the r. bank of the Blanitz. Pop. 2,800. It is enclosed by walls and ditches.

WODOCHOD, a village of Austria, in Bohemia, in the regency of Prague, bail. of Karolinenthal. Pop. 310.—Also a village of the same regency, in the bail. of Melnik. Pop. 400.

WODOKRT, a village of Austria, in Bohemia, reg. of Pilsen, and bail. of Przestitz. Pop. 220.

WODOLKA, a village of Austria, in Bohemia, in the regency of Prague, and bail. of Karolinthal. Pop. 500. It has a castle.

WODOLOW, a village of Austria, in Bohemia, reg. of Gitschin, and bail. of Trantenau. Pop. 220.

WODSCHWIENER-SEE, a lake of Prussia, in the regency of Stettin, and circle of Saatzig, 4 m. in length, and ½ a m. in breadth. It abounds with fish.

WODZEK, a village of Prussia, reg. of Bromberg, and circle of Inowraclaw. Pop. 270.

WODZICZNA, a village of Prussia, in the reg. of Oppeln, and circle of Schildberg. Pop. 400.

WODZISLAW, a town of Russia in Europe, in

Poland, in the gov. and obwod of Kielce, on the Wodzislawa. Pop. 1,795.

WOEL, a village of France, in the dep. of the Meuse, and cant. of Frêne-en-Voivre, 20 m. SE of Verdun. Pop. 520.

WOENSDRECHT, a parish of Holland, in the prov. of North Brabant, and arrond. of Eindhoven. Pop. 1,260.

WOENSEL, a village of Holland, in the prov. of North Brabant, arrond. and 2 m. N of Eindhoven. Pop. 2,200.

WOERDEN, or WARDEN, a fortress of Holland, in the prov. of Holland, and arrond. of Leyden, 11 m. W of Utrecht. Pop. 2,780. It has a gymnasium, and a harbour, and possesses several breweries, and salt refineries.

WOERDEN (OLDEN), a town of Holland, in the prov. of Holland, and arrond. of Leyden, on the Old Rhine, in N lat. 52° 5′ 11″, and E long. 4° 52′ 52″.

WOERTH-SUR-SAUER, a canton and commune of France, in the dep. of the Bas-Rhin, and arrond. of Wissembourg. The cant. comprises 21 com. Pop. in 1831, 13,017; in 1846, 12,649. The town is 14 m. SW of Wissembourg. Pop. 1,240. It has a Calvinist consistorial church. Potters' clay is found in the vicinity.

WOESTEN, a department and com. of Belgium, in the prov. of West Flanders, and arrond. of Ypres. Pop. 1,190.

WOFILA, a province of Abyssinia, in the S part of the kingdom of Tigre, near Lake Ashangi.

WOGADEEN SOMAULI, a tribe of the Somauli, who inhabit the territory of that name, bordering on Abyssinia.

WOGNUMIESTECZ, a town of Bohemia, in the circle and 32 m. SE of Czaslau. Pop. 550. It has a glass work.

WOGUNI, a village of the Punjab, on a small branch of the Chenab.

WOHL, a town of Hesse-Darmstadt, in the prov. of Ober-Hesse, on the Asel. Pop. 614.

WOHLAU, a town of Prussia, in the prov. of Silesia, regency and 23 m. NW of Breslau, between two small lakes. Pop. 1,700. It is enclosed by walls with three gates, and has a suburb, a castle, two Lutheran and the same number of Catholic churches, and an hospital. It possesses manufactories of cloth, hosiery, hats, and linen, and carries on an active trade in cattle.

WOHLE, a village of Hanover, in the prov. of Hildesheim, and bail. of Steinbruch. Pop. 360.

WOHLEN, a village of Switzerland, in the cant. of Argau, and district of Bremgarten, 12 m. ESE of Aarau, in a fine plain. Pop. 1,700.

WOHLEN (GROSS and KLEIN), villages of Austria, in Bohemia, in the regency of Bomisch-Leipa, and bail. of Bensen. Pop. 900.

WOHMAN PEAK, a summit of New South Wales, in the district of Clarence river.

WOHNFURT, a town of Bavaria, in the presidial of Hassfurt, on the Main, 30 m. NE of Würzburg. Pop. 350. It has a castle.

WOHRA, a village of Hesse-Cassel, in the prov. of Ober-Hesse, and circle of Kirchhain, on a river of the same name, an affluent of the Werra. Pop. 685.

WOHRD. See WORTH.

WOHRDEN, a market-town of Holstein, in the Sudcidithmarsch. Pop. 688.

WOHYN, a town of Poland, in the woiwodie and 39 m. SE of Siedlec, and obwod of Radzyn.

WOIEREZ. See HOYERSWERDA.

WOIMBEG, a village of France, in the dep. of the Meuse, and cant. of Pierrefitte, near the l. bank of the Meuse, 20 m. NNW of Commercy. Pop. 250. It has a paper-mill.

WOINCOURT, a commune of France, in the dep. of the Moselle, and cant. of Ault, 18 m. W of Abbeville. Pop. 704. It has manufactories of ironware, locks, and cylinders.

WOJJERAT, a province of Abyssinia, in the S part of the gov. of Tigré, and to the S of the prov. of Enderta. It is covered with forests abounding with elephants, lions, and rhinoceroses. The inhabitants, who exceed the generality of Abyssinian tribes in strength and stature, are said to be descended from Portuguese soldiers who settled in the locality about the middle of the 15th century. Satarro and Andono are the chief towns.

WOKALUP, a river of Western Australia, in the co. of Wellington, an affluent of Koombaunah bay.

WOKEFIELD, a tything in the p. of Stratfield-Mortimer, Berks. Area 600 acres. Pop 158.

WOKING, a parish, and formerly a market-town, in Surrey, 2½ m. W by N of Ripley, on the river Wey, and intersected by the Basingstoke canal. The parish is divided into the tythings of Townstreet, Goldsworth, Heathside, King-field, Mayford, Shackelford, Hale-end, Crastock, and Sutton; all of which, Crastock and Sutton excepted, are in the manor of Woking. Area 7,332 acres. Pop. in 1831, 1,975; in 1851, 2,837.

WOKINGHAM, or OAKINGHAM, a parish and a corporate market-town, partly in Berks, and partly in Wilts, 7 m. ESE of Reading. Area of p. 8,131 acres. Pop. in 1831, 2,692; in 1851, 3,752. The town, which stands in a healthy situation within the limits of Windsor forest, consists of several streets diverging irregularly from the market-place, occupying a triangular area, and containing several good houses. The inhabitants are partly employed in the making of malt and flour, and in the manufacture of shoes, gauze, and silks. In the centre of the market-place is the town-hall, an ancient building.

WOLA, a considerable river in the E part of Upper Guinea, to the NE of the Calbongas.

WOLA, a village of Poland, in the woiwodie of Masow, and obwod of Warsaw, in a fine plain.

WOLAKA, a prov. of Abyssinia, bordered on the W by the Blue Nile, and intersected by the Sainbu or Jamba.

WOLAN, or VALEINE, a parish and town of Austria, in the gov. of Gratz, circle and 18 m. NW of Cilly.

WOLANDRA, a creek of New South Wales, in the district of Wellington.

WOLANOW, a town of Poland, in the gov., obwod and 12 m. W of Radom. Pop. 250.

WOLAR, a creek of New South Wales, in the co. of Phillip, an affluent of the Goulburn.

WOLARIONZENSA, a village of Prussia, in the regency of Posen, and circle of Pleschen. Pop. 220.

WOLAU, a circle and town of Prussia, in the prov. of Silesia, in the regency and 30 m. NW of Breslau. Pop. in 1843, 2,424.

WOLBECK, a town of Prussia, in the prov. of Westphalia, regency and circle and 5 m. SE of Munster, on the Angel. Pop. 990. It has a castle.—Also a village in the vicinity. Pop. 300.

WOLBORZ, a town of Russia in Europe, in Poland, in the gov. of Warsaw, obwod and 11 m. NE of Petrikau, on the Wolborka. Pop. 1,250. It has a castle, forming the usual residence of the bishop of Kujavie, and three churches.

WOLBROM, a town of Russia in Europe, in Poland, in the gov. of Kielce, obwod and 15 m. NNE of Olkusz. Pop. 550.

WOLCOTT, a township of Lamoille co., in the state of Vermont, U. S., 23 m. N of Montpelier, watered by Lamoille river and its branches. Pop.

in 1840, 910; in 1850, 909.—Also a township of New Haven co., in the state of Connecticut, 54 m. S of Hartford, drained by Mad river. Pop. in 1840, 633; in 1850, 603.—Also a township of Wayne co., in the state of New York, 162 m. W by N of Albany. It has a generally level surface, and is drained by several small streams, affluents of Lake Ontario, by which it is bounded on the N. Pop. in 1840, 2,481; in 1850, 2,751. It contains a village with about 650 inhabitants.

WOLCOTTSVILLE, a village of Torrington township, Litchfield co., in the state of Connecticut, U. S., at the confluence of the head branches of Nangatuck river, and on the Nangatuck railroad. Pop. in 1840, 200.

WOLD. See OLD.

WOLDE, a village of the grand-duchy of Mecklenburg-Schwerin, in the circle of Wendischer, and bail. of Ivenack. Pop. 250.

WOLDECK, or WOLDEGK, a town of the grand-duchy of Mecklenburg-Strelitz, in the circle of Stargard, 23 m. NE of Neu-Strelitz. Pop. 2,000. It is enclosed by walls with three gates, and has a school, numerous mills, manufactories of linen and woollen fabrics, tanneries and breweries.

WOLDENBERG, a village and bail. of Hanover, in the principality and 12 m. SE of Hildesheim, near Bockenem. Pop. 50.—Also a village of Prussia, in the regency and 69 m. NE of Frankfurt, and circle of Friedeburg, on a small lake. Pop. 3,440. It is enclosed by walls, and has manufactories of cloth and paper, and several tanneries.

WOLDENHAUSEN, or AHRENSBURG, a village of Holstein, in the bail. of Ahrensburg. Pop. 1,000.

WOLD-NEWTON, a parish in the E. R. of Yorkshire, 8 m. NE of Bridlington. It includes the township of Forder. Area 1,960 acres. Pop. in 1831, 252; in 1851, 276.

WOLDHAM. See OLDHAM.

WOLDINGHAM, a parish in Surrey, 3 m. NE by N of Godstone, in the line of the London and Dover railway. Area 1,570 acres. Pop. 48.

WOLDUCH, a village of Austria, in Bohemia, in the reg. of Pilsen, and bail. of Rokitzan. Pop. 1,100.

WOLEIN, or MIRZIN, a market-town of Austria, in Moravia, in the circle of Iglau. Pop. 1,068. It has manufactories of cloth.

WOLENITZ, a village of Austria, in Bohemia, in the regency of Pilsen, and bail. of Brzeznitz. Pop. 250.

WOLENOW, a village of Austria, in Bohemia, in the regency of Pilsen, and bail. of Klattau. It has manufactories of woollen fabrics and paper.

WOLESCHETZ, a village of Austria, in Bohemia, in the regency of Pardubitz, and bail. of Schwarzkosteletz. Pop. 460.

WOLESCHNA, a village of Austria, in Bohemia, in the regency of Pilsen, and bail. of Pisek. Pop. 420.—Also a village of the regency of Pardubitz, and bail. of Pilna. Pop. 380.—Also a village of the regency of Prague, and bail. of Rakonitz. Pop. 720. It has a castle.—Also a village in the regency of Budweis, and bail. of Neuhaus. Pop. 420.—Also a village of the same regency, and bail. of Pilgram. Pop. 320.—Also a village of Moravia, in the circle of Iglau. Pop. 550. It has a castle.

WOLESCHNIK, a village of Austria, in Bohemia, in the reg. and bail. of Budweis. Pop. 350.

WOLESCHNITZ, a village of Austria, in Bohemia, in the regency of Gitschin, and bail. of Adlerkosteletz. Pop. 550.—Also a village of the same regency, in the bail. of Neu-Bidschow. Pop. 300.—Also a village of the same regency, in the bail. of Eisenbrod. Pop. 1,400.—Also a village of the same regency, in the bail of Nachod. Pop. 1,100.—Also a village of the same regency, in the bail. of Turnau. Pop. 350.—Also a village of the regency of Pardubitz, and bail. of Ebotiebor. Pop. 220.

WOLF, a village of Prussia, in the regency of Treves, and circle of Bernkastel, on the Mosel. Pop. 540.

WOLFACH, a town and bail. of the grand-duchy of Baden, in the circle of the Middle Rhine, on the Kinzig, at the confluence of a river of the same name, and 18 m. SE of Offenburg. Pop. of town, 1,630; of bail. 10,800. It has a castle, and several saw-mills, and carries on an active trade in wood, and in pigs. In the vicinity are extensive mines of lead, silver, copper, cobalt, and antimony.

WOLFARTSWEILER, a town of Würtemberg, in the circle of the Danube, and bail. of Saulgau. Pop. 230.

WOLFARTSWEYER, a village of Baden, in the circle of the Middle Rhine, and bail. of Durlach. Pop. 359.

WOLFBORO', a township and village of Carroll co., in the state of New Hampshire, U. S., 29 m. NE by N of Concord. Pop. in 1840, 1,918.

WOLF-CREEK, a township of Mercer co., in the state of Pennsylvania, U. S., 184 m. WNW of Harrisburg. Pop. in 1840, 1,732; in 1850, about 2,000.

WOLFECK, a village of Bavaria, in the circle of the Ober-Pfalz, and presidial of Regenstauf. Pop. 270.

WOLFEGG, a village of Würtemberg, in the circle of the Danube, near the Waldsee. Pop. 1,610. It has a castle, a museum, a library, &c.

WOLFE-ISLAND, an island of Upper Canada, in the NE extremity of Lake Ontario, near the entrance of the St. Lawrence, and nearly opposite the town of Kingston. It is long and irregular in outline, and forms a township of the Middle district. It is 20 m. in length, and about 6 m. in extreme breadth, and well clothed with oak, elm, ash, and pine timber. On the S side, a deep bay runs in so far as to leave an isthmus of no more than one mile in breadth. Up this bay is the general route from the S side of the St. Lawrence to Kingston. Pop. 1,289.—Also a small island of the United States, near the coast of Georgia, in N lat. 31° 19'.—Also an island near the E coast of Labrador, in N lat. 53° 55'.—Also an island in the gulf of St. Lawrence, near the S coast of Labrador, in N lat. 50° 2'.

WOLFELSDORF, a village of Prussia, in the prov. of Silesia, regency and 66 m. SSW of Breslau, circle and 5 m. SE of Habelschwert, on the slope of the Schneeberg, and near a small affluent of the Neiss, by which a fine cascade is formed.

WOLFEN, a village of Prussia, in the regency of Merseburg, and circle of Bitterfeld. Pop. 290. It has a paper-mill.

WOLFENBRUCK, a village of Würtemberg, in the circle of the Jaxt, and bail. of Gaildorf. Pop. 200.

WOLFENBUTTEL, a circle and town of the duchy of Brunswick. The circle, which lies to the W of that of Schöningen, and NE of that of the Harz, comprises an area of 320 sq. m. Pop. in 1852, 52,662. The town is situated on the Ocker, 8 m. S of Brunswick, in N lat. 52° 9' 29", and E long. 10° 31' 39". Pop. 9,003. It is enclosed by fortifications, now much dilapidated, and has three gates. It consists of three parts, viz., the castle, the town properly so called, or Heinrichstadt, and the suburbs of Augusstadt and Gotteslager. It is "a neat primitive little place," and contains four squares, of which the finest are those of the castle and market. The streets are straight, spacious, well-paved, and watered by canals, branching from the Ocker. The old ducal castle is now converted into a manufac tory, and the arsenal into barracks. It has a library of 220,000 volumes, which are neatly arranged in galleries in a circular building, three Lutheran churches, of which the handsomest is that of St. Mary, a Catholic church, a synagogue, an orphans' asylum, an infirmary, a poors' house, a prison, formerly the castle of the lords of Wolfenbuttel, a gymnasium, three schools, and a convent. It has manufactories of lacker, bronze, paper-hangings, leather, liqueurs, tobacco, ribbon, and linen, and carries on an active trade in corn and flax. In the environs are numerous gardens.

WOLFENHAUSEN, a village of Nassau, in the bail. of Runkel. Pop. 590.—Also a village of Würtemberg, in the circle of the Schwarzwald, and bail. of Rottenburg. Pop. 294.

WOLFENSCHIESSEN, a village of Switzerland, in the cant. of Unterwalden, and bail. of Niddem-Walde. Pop. 1,301.

WOLFENWEILER, a village of Baden, in the circle of the Upper Rhine, and bail. of Freiburg. Pop. 678.

WOLFERBORN, a village of Electoral Hesse, in the prov. of Hanau, circle of Gelnhausen, and 9 m. NW of Salmunster. Pop. 660.

WOLFERING, a village of Bavaria, in the presidial of Nabburg. Pop. 200. It has a castle.

WOLFERLOW, a parish in Hereford, 4 m. N by E of Bromyard. Area 1,535 acres. Pop. 135.

WOLFERN, a village of Austria, in the circle of the Muhl, and bail. of Ort. Pop. 370.

WOLFERODE, a village of Electoral Hesse, in the circle of Kirchhain, and bail. of Rauschenberg. Pop. 373.

WOLFERODE (Alt and Neu), villages of Prussia, in the regency of Merseburg, and circle of the See. Pop. 590.

WOLFERSDORF, a village of Bavaria, in the circle of Upper Bavaria, and presidial of Morsburg. Pop. 200.—Also a village of Austria, in Bohemia, in the regency and bail. of Bohmisch-Leipa. Pop. 2,300. It has manufactories of glass, and bleacheries.—Also a village of Prussia, in the regency of Leignitz, and circle of Sprottau. Pop. 300.—Also a village of Saxe-Weimar, in the circle of Neustadt. Pop. 400.

WOLFERSDYK, or Wolfaartsdyk, a small island of Holland, in the prov. of Zeeland, and arrond. of Goes, between the islands of North and South Bereland. It contains only one village named Oosterland.

WOLFERSGRUN, a village of Saxony, in the circle of Zwickau, and bail. of Kirchberg. Pop. 345.

WOLFERSHAUSEN, a village of Electoral Hesse, circle of Melfungen, and bail. of Felsberg. Pop. 382.

WOLFERSHEIM, a village of Bavaria, in the circle of the Pfalz, and cant. of Blieskastel. Pop. 313.—Also a town of Hesse-Darmstadt, in the prov. of Upper Hesse, 7 m. from Hungen. Pop. 709.

WOLFERSTADT, a village of Bavaria, in the circle of Swabia, presidial of Wemding. Pop. 644.

WOLFERSTEDT, a village of Saxe-Weimar, and bail. of Allstedt. Pop. 890.

WOLFERSWEILER, a village of Oldenburg, in the bail. of Nohfelden. Pop. 760. It has two mills.

WOLFERSZELL, a village ot Bavaria, in the circle of Lower Bavaria, and presidial of Straubing. Pop. 200.

WOLFERTSCHWENDEN, or Wolfartschwenden, a village of Bavaria, in the circle of Swabia, and presidial of Ottobeuren. Pop. 301.

WOLFFURT, a village of Austria, in the Tyrol, and district of Vorarlberg. Pop. 930.

WOLFGANG (Sanct), a market-town of Austria, in the country above the Ens, and circle of Hausruck, on the NE bank of a lake of the same name—also called the Aber-see, and 45 m. SW of Wels. Pop. 555. See Aber-See.

WOLFGANG-AM-BERG, a market-town of Bavaria, in the circle of Upper Bavaria, and presidial of Wasserburg, on the Schwindau. Pop. 505.

WOLFGANG-AM-WEICHSELBACH, a town of Austria, in the presidial of Linz, and circle of Salzburg. It is noted for its mineral springs and baths.

WOLFHAG, a village of Baden, in the circle of the Middle Rhine, and bail. of Oberkirch. Pop. 230.

WOLFHAGEN, a circle and bail of Electoral Hesse, in the prov. of Lower Hesse. The circle comprises an area of 170 sq. m., of which nearly one half is arable. Pop. 26,500. It produces in considerable quantities rye, oats, wheat, flax, and oleaginous grains. Its live stock averages 3,238 horses, 5.500 head of cattle, 27,700 sheep, 4,600 pigs, and 2,900 goats. The trade consists chiefly in flax, linen, grain, and poultry. It is divided into 4 bailiwicks, and contains 4 towns and 36 villages.—The town stands on a height, 17 m. W of Cassel. Pop. 3,187. It is enclosed by walls with four gates, and has a suburb, a castle, and an hospital. It possesses manufactories of linen and woollen fabrics, and of paper, and several bleacheries.

WOLFHALDEN, a village of Switzerland, in the cant. of Appenzell, and bail. of Trogen. Pop. 2,212. It has muslin manufactories.

WOLFINGHAM, a parish of New South Wales, in the co. of Durham, bounded on the S by the river Hunter.

WOLFIRSCH, a village of Austria, in Moravia, in the circle of Iglau. Pop. 580. It has two mills.

WOLFIS, a village of the duchy of Saxe-Coburg-Gotha, in the principality and 12 m. SSE of Gotha.

WOLFISBERG, a village of Switzerland, in the cant of Bern, and bail. of Wangen. Pop. 250.

WOLF-ISLANDS, a group of islands in the bay of Fundy, to the SE of the bay of Passamaquoddy, and 9 m. N of Great Manan island, in N lat. 44° 58'.

WOLFLEBEN, a village of Prussia, in the reg. of Erfurt, circle and 6 m. NW of Nordhausen, on the Zorge. Pop. 350. It has a quarry of alabaster.

WOLFLISROYL, a village of Switzerland, in the cant. of Argau, and district of Lauffenburg, 6 m. NNW of Aarau. Pop. 1,000.

WOLFMANNSHAUSEN, a village of Saxe-Meiningen, in the bail. of Romhild. Pop. 360.

WOLFORD, a township of Upper Canada, in the Johnstown district, separated from the township of Montague by the Rideau river and canal, and watered by branches of the former. Pop. 2,422.

WOLFORD. See Woolford.

WOLFPASSING, a village of Austria, in the country below the Ens, circle of the Lower Mannhartsberg, in the presidial of Niederfreuzstetten. Pop. 660.—Also a village in the same circle, and presidial of Leobendorf. Pop. 360.

WOLFRAMITZ, or Olbramowice, a market-town of Austria, in Moravia, in the circle of Znaim. Pop. 528.

WOLFRAMITZKIRCHEN, or Wobramkostel, a market-town of Austria, in Moravia, in the circle of Znaim. Pop. 690. It has several mills and breweries.

WOLFRAMS, a village of Austria, in Moravia, in the circle of Iglau. Pop. 370.

WOLFRATHSHAUSEN, or Wolfertshausen, a presidial and town of Bavaria, in the circle of Upper Bavaria. Pop. of presidial, 14,700. The town is on the Loisach, 17 m. SSW of Munich. Pop

1,200. It has three churches, and possesses manufactories of glass, soda, breweries, &c.

WOLF-RIVER, a river of Louisiana, U. S., which falls into the Missouri, 464 m. from the Mississippi. It is about 60 yds. wide at its mouth. It has its source near to that of the Kansas, and is navigable for boats some distance up.—Also a river of Tennessee, which runs into the Mississippi, N of Chickasaw-Bluff.—Also a river of Mississippi, which runs S into the gulf of Mexico, 22 m. E of Pearl river. It forms a considerable bay at its mouth called St. Louis' bay.

WOLF-ROCK, a low, flat, rocky islet in the North Pacific, about 12 m. from the S shore of Prince of Wales's archipelago.

WOLFSANGER, a village of Electoral Hesse, in the prov. of Lower Hesse, and circle and presidial of Hesse. Pop. 990.

WOLFSBACH, a village of Bavaria, in the circle of the Upper Pfalz, and presidial of Amberg. Pop. 260.—Also a village of Austria, in the country below the Ens, circle of the Upper Wienerwald, and presidial of Salaberg. Pop. 550.

WOLFSBEBRINGEN, a village of the duchy of Saxe-Coburg, and bail. of Gotha. Pop. 360.

WOLFSBERG, a village of Austria, in Bohemia, in the regency of Gitschin, and bail. of Podiebrad, Pop. 220.—Also a village in the regency of Bohmisch-Leipa, and bail. of Harnspach. Pop. 800. It has several bleacheries.—Also a town of Illyria, in the gov. of Laibach, and circle of Klagenfurt, on the Lavant. Pop. 1,470. It has a castle, and possesses iron-mines and founderies, and manufactories of white lead.—Also a village of Styria, in the circle of Gratz, and bail. of Labek. Pop. 600.—Also a village of Prussia, in the regency of Gumbinnen, and circle of Niederung. Pop. 210.—Also a village of the regency of Merseburg, and circle of Sangershausen. Pop. 240.

WOLFSBRON, a village of Bavaria, in the circle of Middle Franconia, and presidial of Heidenheim. Pop. 120. It has several paper-mills.

WOLFSBUCH, a village of Bavaria, in the circle of the Upper Pfalz, and presidial of Niedenburg. Pop. 270.—Also a village of Würtemberg, in the circle of the Jaxt, and bail. of Mergentheim. Pop. 410.

WOLFSBURG, a village of Brunswick, in the circle of Helmstadt. Pop. 340.—Also a village of Prussia, in the regency of Magdeburg, and circle of Gardelegen. Pop. 310.

WOLFSCHLUGEN, a village of Würtemberg, in the circle of the Schwarzwald, and bail. of Nurtingen. Pop. 1,205.

WOLFSDORF, a village of Austria, in Moravia, in the circle of Olmutz, and bail. of Titschein. Pop. 310.—Also a village of Silesia, in the circle of Troppau and seignory of Oderau. Pop. 210.—Also a village of Styria, in the circle of Marburg. Pop. 230.—Also a village of Prussia, in the regency of Liegnitz, and circle of Goldberg. Pop. 780.—Also a village of the regency of Konigsberg, and circle of Heilsburg. Pop. 474.

WOLFSDORF (Gross), a village of Prussia, in the regency of Konigsberg, and circle of Kastenberg. Pop. 380.

WOLFSDORF-AUF-DER-HOHE, a village of Prussia, in the reg. of Danzig and circle of Elbing. Pop. 270.

WOLFSDORF-IN-DER-NIEDERUNG, a village of Prussia, in the regency of Danzig and circle of Elbing, on the Nogat. Pop. 268.

WOLFSECK, a village of Bavaria, in the circle of the Upper Pfalz, and presidial of Regenstauf, Pop. 200. It has a castle.—Also a town of Austria, in the country above the Ens, and circle of Hausruck, 21 m. WSW of Wels. Pop. 560. It has a castle, and possesses mineral baths and coal-mines.

WOLFSEIFEN, a village of Prussia, in the prov. of the Rhine, regency of Aachen, and circle and 3 m. W of Gemund.

WOLFSGRUBEN, a village of Austria, in Styria, in the circle of Gratz, and bail. of Freiberg Pop. 480.

WOLFSHAGEN, a village of Prussia, in the reg. of Konigsberg, and circle of Kastenberg. Pop. 355.—Also a village of the reg. of Potsdam, and circle of Prenslau. Pop. 311.—Also a village of the same regency, in the circle of Westpriegnitz. Pop. 200.—Also a village of the duchy of Brunswick, and circle of Wolfenbuttel. Pop. 1,000.

WOLFSHEIM, a village of Hesse, in the prov. of the Rhine, and cant. of Alzey. Pop. 580.

WOLFSHOFIN, a village of Prussia, in the reg. of Konigsberg, and circle of Labrau. Pop. 220.

WOLFSHORST, a village of Prussia, in the reg. of Stettin, and circle of Randow. Pop. 200.

WOLFSKEHLEN, a village of Hesse, in the prov. of Starkenburg, and presidial of Gross-Gerau. Pop. 840.

WOLFSKIRCH, a village of France, in the dep. of the Bas-Rhin, and cant. of Drulingue, 20 m. NW of Saverne. Pop. 570. It has manufactories of pottery and cordage.

WOLFSMUNSTER, a village of Bavaria, in the circle of Lower Franconia, and presidial of Gmunden, on the Saale, 27 m. NNW of Wurzburg. Pop. 200. It has a castle.

WOLF'S-NEWTON, or Wolves-Newton, a parish in Monmouthshire, 5½ m. E by S of Usk. Area 2,649 acres. Pop. in 1831, 248; in 1851, 219.

WOLFSORDEN, a village of Würtemberg, circle of the Neckar, and bail. of Marbach. Pop. 290.

WOLFSTEIN, a canton and town of Bavaria, in the circle of the Pfalz. Pop. of cant. 12,500. The town is on the l. bank of the Lauter, 15 m. NW of Kayserslautern. It contains two churches, and has mines of coal and quicksilver.

WOLFSTHAL, a village of Austria, in Bohemia, in the regency of Bohmisch-Leipa, and bail. of Niemer. Pop. 300.

WOLFSWINKEL, a village of Prussia, in the prov. of Brandenburg, regency of Potsdam, and circle of Oberbarmin, near Neustadt-Eberswalde. It has a paper-mill.

WOLFTERODE, a village of Electoral-Hesse, in the prov. of Lower Hesse, and circle of Eschwege. Pop. 230.

WOLFTITZ, a village of Saxony, in the circle of Leipsig, and bail. of Borna. Pop. 210.

WOLFWYL, a village of Switzerland, in the cant. of Solothurn, and bail. of Balsthal, on the Aar. Pop. 1,033.

WOLGA. See Volga.

WOLGAN, a river of New South Wales, in the co. of Cork, an affluent of the Colo.

WOLGAST, a town of Prussia, in the prov. of Pomerania, regency and 30 m. SE of Stralsund, and circle of Greifswald, on the l. bank of the Peene, at some distance from the Baltic. Pop. in 1831, 1,514; in 1842, 1,669; in 1849, 1,830. It has three suburbs, two churches, and an hospital, and possesses the ruins of a castle, and a safe port and building docks. Tobacco and soap are its chief articles of manufacture.—Also a village of the same province, in the reg. of Frankfort, and circle of Friedeberg. Pop. 260.

WOLHOPE, or Woolhope, a parish in Herefordshire, 8 m. SE by E of Hereford. It contains the hamlets of Bartree, Bucken-hill, and Putley.

Area 4,653 acres. Pop. in 1831, 880; in 1851, 902.

WOLHYNIA. See VOLHYNIA.

WOLIKA-KOZIA, a village of Prussia, in the regency of Posen, and circle of Pleschen. Pop. 228. It has a paper-mill.

WOLIN, or ZWOLIN, a market-town of Austria, in Bohemia, in the circle of Prachin, 20 m. SW of Pisek, on the l. bank of the Wolinka, an affluent of the Wotawa. Pop. 1,900. It has manufactories of cloth.

WOLKAU, a village of Saxony, in the circle of Dresden, and bail. of Meissin. Pop. 240.

WOLKENBERG, a summit of Prussia, in the prov. of the Rhine, in the chain of the Siebengebirge, and circle of Sieg, which has an alt. of 1,480 ft. above sea-level.—Also a village of the regency of Frankfort, and circle of Steinberg.

WOLKENBURG, a village of Saxony, in the circle of Leipsig, and bail. of Borna, near Penig, on the Mulde. Pop. 588. It has a castle, with a library, and park, a church, and several spinning-mills.

WOLKENSHAGEN, a village of Mecklenburg-Schwerin, circle and bail. of Rostock. Pop. 250.

WOLKENSTEIN, a town of Saxony, capital of the bail. of Wolkenstein and Annaberg, in the circle of Zwickau, in a romantic situation on the Tschopa. Pop. in 1837, 1,870; in 1843, 1,901; in 1849, 2,010. It has an ancient castle, an hospital, an orphans' asylum, and mineral baths, and possesses manufactories of hosiery, lace, parchment, and linen.

WOLKENWEHE, a village of Holstein, in the bail. of Blumendorf. Pop. 230.

WOLKERING, a village of Bavaria, in the circle of the Ober-Pfalz, and presidial of Stadtamhof. Pop. 200.

WOLKERSDORF, a market-town of Austria, in the circle of the Lower Mannhardsberg, on the Rossbach. Pop. 1,715. It has a castle.

WOLKERSDORF (OBER and UNTER), two villages of Bavaria, in the circle of Middle Franconia, and presidial of Schwabach. Pop. 260.

WOLKERTSHOFEN, a village of Bavaria, in the circle of Middle Franconia, and presidial of Eichstadt. Pop. 210.

WOLKOW, a village of Prussia, in the regency of Stettin, and circle of Demmin. Pop. 330.

WOLKOWIST. See VOLKOWIST.

WOLKOWO, a village of Prussia, in the reg. of Posen, and circle of Kosten. Pop. 270.

WOLKRAMSHAUSEN, a village of Prussia, in the regency of Erfurt, circle and 8 m. SW of Nordhausen, on the r. bank of the Wipper. Pop. 450.

WOLKSHAUSEN, a bail. and village of Bavaria, in the circle of Lower Franconia, and presidial of Aub. Pop. 325.

WOLLABERG, a village of Bavaria, in the circle of Lower Bavaria, and presidial of Wegscheid. Pop. 400.

WOLLA-GALLE, a tribe of the Galla, in Abyssinia, who inhabit the territory to the S of the Wochale-Galla.

WOLLAND, or WOOLLAND, a parish in Dorset, 9 m. W by N of Blandford-Forum. Area 1,098 acres. Pop. in 1831, 119; in 1851, 107.

WOLLASTON, a parish in Northamptonshire, 3 m. S by E of Wellingborough, intersected by the river Nen. Area 3,640 acres. Pop. in 1851, 1,261. The church is a handsome cruciform structure, with a tower surmounted by a lofty spire.

WOLLASTON'S LAND, a region in the Arctic seas, lying along the N side of Dolphin and Union strait, between the parallels of 69° and 71° N, and intersected by the meridian of 115° W.

WOLLATON, anciently OLAVESTON, a parish in Nottinghamshire, 3 m. W of Nottingham, intersected by the Nottingham canal. Area 2,340 acres. Pop. in 1831, 537; in 1851, 581. Coal mines have been wrought here from time immemorial.

WOLLAU, a market-town of Austria, in Styria, circle and 3 m. NW of Cilly, on the Paackbach.

WOLLAWOLLAS, an Indian tribe of the Oregon territory. United States, to the W of the Rocky mountains, who inhabit the banks of the Columbia and Lewis rivers.

WOLLBACH, a village of Bavaria, in the circle of Lower Franconia, and presidial of Rissengen. Pop. 593.—Also a village of the same circle, in the presidial of Neustadt. Pop. 451.—Also a village of Swabia, in the presidial of Zusmarshausen. Pop. 220.—Also a village of the grand-duchy of Baden, in the circle of the Upper Rhine, and bail. of Lorrach. Pop. 892.

WOLLBECK. See WOLBECK.

WOLLBRANDSHAUSEN, a village of Hanover, in the principality of Grubenhagen, near Sieboldehausen. Pop. 548.

WOLLBRECHTSHAUSEN, a village of Hanover, in the principality of Gottingen, and bail. of Moringen-Hardegsen. Pop. 345.

WOLLEIN, a market-town of Austria, in Moravia, in the circle of Iglau, and seignory of Tscherna. Pop. 1,400.

WOLLEN, a mountain of New South Wales, in the co. of Durham, at the head of Rouchel and Stewart's brooks.

WOLLENBERG, a village of Baden, in the circle of the Lower Rhine, and bail. of Neckar-Bischofsheim. Pop. 432.

WOLLENDORF, a village of Prussia, in the regency of Coblenz, and circle of Neuwied. Pop. 435.

WOLLENHAGEN, a village of Prussia, in the regency of Magdeburg, and circle of Gardelegen. Pop. 200.

WOLLENTHAL, a village of Prussia, in the regency of Danzig, and circle of Stargard. Pop. 440.

WOLLENTSCHIN, a village of Prussia, in the regency of Oppeln, and circle of Rosenberg, near the source of the Prosna. Pop. 200.

WOLLERAN, a village of Switzerland, in the cant. of Zurich, and bail. of Hofe, on Lake Zurich. Pop. 1,168.

WOLLERSHAUSEN, a village of Hanover, in the principality of Grubenhagen, and bail. of Herzberg, near Sieboldehausen. Pop. 308.

WOLLERSHEIM, a village of Prussia, in the regency of Aachen, and circle of Duren. Pop. 390.

WOLLERSLEBEN, a village of Prussia, in the regency of Erfurt, circle of Nordhausen. Pop. 290.

WOLLERT, a parish of Australia Felix, in the co. of Bourke.

WOLLERTON, a hamlet in the p. of Hodnet, Salop, 5 m. SW of Drayton-in-Hales. Pop. 134.

WOLLESCHNA, a village of Austria, in Bohemia, in the regency of Pilsen, and bail. of Rok kan. Pop. 300.—Also a village of Moravia, in the circle of Iglau, and seignory of Studein. Pop. 240.

WOLLESCHNITZ, a village of Austria, in Bohemia, in the regency of Gitschin, and bail. of Gemel. Pop. 1,400.

WOLLEY, a parish in Somersetshire, 3 m. N of Bath. Area 365 acres. Pop. in 1851, 91.

WOLLHAUSEN-UN-MARKT, a village of Switzerland, in the cant. of Lucern, and bail. of Sursee, on the Emmen. Pop. 272.

WOLLH-WIGGERN, a bail. and village of Switzerland, in the cant. of Lucern, near Wollhausen-un-Markt. Pop. 1,381.

WOLLIM, a creek of New South Wales, in the co. of Cork, an affluent of the Colo river.

WOLLIN, a market-town of Prussia, in the prov. of Pomerania, regency and 32 m. N of Stettin, and circle of Usedom-Wollin, on the E coast of an island of the same name, on the Dievenow, in N lat. 53° 49′, and E long. 14° 39′. Pop. 4,400. It has manufactories of woollen fabrics and of leather, and extensive fisheries, and carries on an extensive trade in wood. This town is supposed to occupy the site of the town of Julin, destroyed in 1116.—The island of Wollin, or Wollinscher-Werder, lies in the estuary of the Oder, between the Baltic on the N, and the Grosse-haff, the E part of the Stettiner-haff, on the S, the Dievenow on the E, and the Swiene on the W, and is united by three bridges to the continent. It is 18 m. in length from E to W, and 15 m. in extreme breadth. It is noted for its pasture and its fisheries, especially of eels.—Also a village of the regency of Koslin, and circle of Stolpe. Pop. 278.—Also a village of the regency of Magdeburg, and circle of Jerichow. Pop. 392.—Also a village of the regency of Potsdam, and circle of Prenzlow. Pop. 253.—Also a village of the regency of Stettin, and circle of Randow. Pop. 195.

WOLLINGST, a village of Hanover, in the principality of Bremen, and bail. of Beverstedt. Pop. 220.

WOLLISHAUSEN, a village of Bavaria, in Swabia, in the presidial of Zusmarshausen. Pop. 300.

WOLLISHOFEN, a village of Switzerland, in the cant. and bail. of Zurich, on the W bank of the lake of that name. Pop. 1,093.

WOLLISWYL, a village of Switzerland, in the cant. of Berne, and bail. of Wangen. Pop. 204.

WOLLMADINGEN, a bail. and village of the grand-duchy of Baden, in the circle of the See, and bail. of Constanz. Pop. 798.

WOLLMAR, a village of Electoral Hesse, in the prov. of Lower Hesse, circle of Marburg, and bail. of Wetter. Pop. 600.

WOLLMERSHEIM, a village of Bavaria, in the circle of the Pfalz, and cant. of Landau. Pop. 746.

WOLLMERSLEBEN, a village of Prussia, in the regency of Magdeburg, and circle of Wanzleben, on the Bude. Pop. 788.

WOLLMETSHOFEN, a village of Bavaria, in Swabia, in the presidial of Krumbach. Pop. 230.

WOLLMIRSTADT, a circle and town of Prussia, in the prov. of Saxony, regency of Magdeburg. The circle comprises an area of 350 sq. m. Pop. 38,300. The town is on the l. bank of the Ohre, 8 m. N of Magdeburg, in N lat. 52° 14′ 58″. Pop. 3,140. It is enclosed by walls, with three gates, and has two churches and an hospital. It possesses manufactories of linen, plain and damask, and cloth; tanneries, distilleries of rum and liqueurs, breweries.—Also a village of the regency of Merseburg, and circle of Eckartsberga. Pop. 458.

WOLLNZACH, a market-town of Bavaria, in the circle of Upper Bavaria, and presidial of Pfaffenhofen. Pop. 1,090. It has a castle, and numerous breweries.

WOLLOMBA, a river of New South Wales, in the co. of Gloucester, an affluent of Wallace lake.

WOLLONDILLY, a river of New South Wales, which has its source in the Cullarin range, a little to the SW of Lake Bathurst; flows into the co. of Argyle, separates it from those of Camden and Westmoreland; and after a course in a generally NE direction, joins the Warragamba river, at its confluence with Cox's river.

WOLLONGONG, a parish and maritime town of New South Wales, in the co. of Camden. The town is 60 m. from Sydney. Pop. 515.

WOLLONGONGINE, a mountain of New South Wales, in the co. of Northumberland, near Wyong creek, and 10 m. from Gosford.

WOLLPADINGEN, a village of Baden, in the circle of the Upper Rhine, and bail. of St. Blasten. Pop. 582.

WOLLRODE, a village of Electoral Hesse, in the prov. of Lower Hesse, circle and bail. of Melfungen. Pop. 392.

WOLLSDORF, a village of Austria, in Styria, in the circle of Gratz. Pop. 330.

WOLLSERFEN, a village of Prussia, in the regency of Aachen, and circle of Schleiden. Pop. 431.

WOLLSTEIN, a town of Hesse-Darmstadt, capital of a cant. of the same name, in the prov. of Rhein-Hesse, on the Appelbach, 21 m. SW of Mayence. Pop. 1,462, of whom 560 are Catholics, and 26 Jews. It has three churches.

WOLLSTEIN, or Wolsztyn, a town of Prussia, in the prov., regency, and 42 m. SW of Posen, and circle of Bomst, between two lakes. Pop. 2,700, of whom one-fourth are Jews. It has a synagogue and a gymnasium, and possesses manufactories of woollen and linen fabrics and of leather.

WOLLUM HILLS, a range of hills in New South Wales, in the co. of Brisbane.

WOLMAR. See Volmar.

WOLMIRSTADT. See Wollmirstadt.

WOLMSDORF, a village of Prussia, in the regency of Breslau, and circle of Habelschwerdt. Pop. 152.—Also a village of the same regency, in the circle of Frankenstein. Pop. 440.

WOLMSDORF (Nieder, Ober, and Stadtisch), three villages of Prussia, in the regency of Liegnitz, and circle of Wolfenheim, containing respectively 308, 369, and 50 inhabitants.

WOLMUNSTER. See Volmunster.

WOLNZNACH, a town of Bavaria, in the circle of the Upper Rhine, presidial and 9 m. NE of Pfaffenhofen. Pop. 1,023. It has several breweries.

WOLODIMER. See Wladimir.

WOLOGDA. See Vologda.

WOLOMBI, a parish and town of New South Wales, in the co. of Northumberland. The parish is bounded on the W and S by a stream of the same name, an affluent of the Hunter river. The town is on the W, 93½ m. from Sydney. Pop. 76.

WOLOW, a village of Poland, in the wöiwodie of Sandomir, and obwod of Opoczno.

WOLOWIETZ, a village of Turkey in Europe, in Moldavia, on the Pruth.

WOLOWITZ, a village of Austria, in Bohemia, in the regency of Prague, and bail. of Rakonitz. Pop. 500.

WOLOWOLAR, a mountain of New South Wales, in the co. of Argyle, to the SW of Lake Bathurst, and 147 m. from Sydney. The outlet of this lake, the Boro creek, forms the S boundary of the co.

WOLPADINGEN, a village of the grand-duchy of Baden, in the circle of the Upper Rhine, and bail. of St. Blasten. Pop. 200.

WOLPE, a village of Hanover, in the principality of Kalenberg, 3 m. NE of Nieuberg. Pop. 70.

WOLPERNDORF, a village of Saxe-Altenberg. Pop. 210.

WOLPERTSCHWENDE, a bail. and village of Würtemberg, in the circle of the Danube and bail. of Ravensberg. Pop. 285.

WOLPHAMCOTE, a parish in Warwick, 5½ m. SE by S of Dunchurch, intersected by the Oxford canal. Area 3,470 acres. Pop. in 1851, 464.

WOLSCHAU, a village of Austria, in Bohemia, in the regency of Pilsen, and bail. of Horazoiowitz. Pop. 210.—Also a village of the regency of Prague, and bail. of Schlau. Pop. 240.—Also a village of Moravia, in the circle of Iglau, and seignory of Datschitz. Pop. 350.

WOLSCHEN, a village of Austria, in Bohemia,

in the regency of Bohmisch-Leipa, and bail. of Niemes. Pop. 220.

WOLSCHI, a village of Austria, in Moravia, in the circle of Iglau, and bail. of Gross Miseritsch. Pop. 440.

WOLSCHOW, a village of Prussia, in the regency of Potsdam, and circle of Prenzlau. Pop. 260.

WOLSCHY, a village of Austria, in Bohemia, in the regency of Budweis, and bail. of Muhlhausen. Pop. 260.

WOLSDORF, a village of Prussia, in the regency of Cologne, and circle of Sieg. Pop. 330.—Also a village of the duchy of Saxe-Coburg-Gotha, principality and bail. and 1½ m. NW of Saalfeld.

WOLSEIFEN, a village of Prussia, in the regency of Aachen, and circle of Schleiden. Pop. 420. It has iron-works and several mills.

WOLSFELD, a village of Prussia, in the regency of Treves, and circle of Bittburg. Pop. 230.

WOLSIER, a village of Prussia, regency of Potsdam, and circle of Westhavelland. Pop. 210.

WOLSINGHAM, or WALSINGHAM, a parish and market-town in the co.-palatine of Durham, 16 m. WSW of Durham, on the river Wear, and near the Bishop-Auckland and Weardale railway. Area 20,403 acres. Pop. in 1831, 2,239; in 1851, 4,585. The town, which stands on the N bank of the Wear, is irregularly built. The inhabitants are partly employed in the numerous coal, lime, and lead works in the vicinity, and partly in the manufacture of linen, woollen cloth, edge-tools, and agricultural implements.

WOLSKO, a village of Prussia, in the prov. of Posen, regency of Bromberg, and circle of Wirsitz. Pop. 220.

WOLSTAN, or WOLFUCHESTON, a parish in Warwickshire, 6½ m. W of Rugby, intersected by the London and Birmingham railway. It includes the hamlets of Brandon, Bretsford, and Marston. Area 2,770 acres. Pop. in 1831, 968; in 1851, 1,209.

WOLSTANTON, a parish and township in Staffordshire, 1½ m. N by E of Newcastle-under-Lyne. intersected by the Grand Trunk canal, which here passes under a tunnel 1½ m. in length. The parish includes the chapelries of Thurstall, Kidgrove, and New-chapel, the hamlet of Brerehurst, the liberty of Tunstall-Court, and the townships of Chatterley, Chell, Chesterton, Knutton, Oldcott, Rainscliff, Stadmerslow, Wedgwood, and W. Area 10,739 acres. Pop. in 1831, 10,853; in 1851, 22,191.

WOLSTENHOLME ISLAND, an island in Baffin's Bay, lying in N lat. 76° 43′, W long. 70° 55′, W of Cape Atholl, at the SW entrance to Wolstenholme Sound, of which Cape Abernethy and Cape Atholl form the extreme N and S points.

WOLSTON, WOLSTON-MINE, or WESTON-NEGEND, a chapelry in the p. of Worthen, Salop, 7½ m. NE of Montgomery.

WOLSTORF, a village of Brunswick, in the circle of Helmstadt, and bail. of Konigslutter. Pop. 300.

WOLSZTYN. See WOLLSTEIN.

WOLTA, a village of Austria, in Bohemia, in the reg. of Gitschin, and bail. of Trautenau. Pop. 550.

WOLTERDINGEN, a village of Hanover, in the gov. of Stade, district of Verden, and bail. of Rotenburg. Pop. 240.—Also a bail. and village of Baden, in the circle of the See, and bail. of Donaueschingen. Pop. 671.

WOLTERITZ, a village of Prussia, in the regency of Merseburg, and circle of Delitzsch. Pop. 220.

WOLTERSDORF, a village of Hanover, in the gov. of Luneburg, near Luchow. Pop. 412.

WOLTERSDORF (GROSS), a village of Prussia, in the regency of Potsdam, and circle of Ruppin. Pop. 361.—Also a village of the regency of Koslin, and circle of Dramburg. Pop. 220.—Also a village of the regency of Stettin, and circle of Randow. Pop. 300.—Also a village of the same regency, in the circle of Greiffenhagen. Pop. 340.—Also a village of the regency of Frankfurt, and circle of Soldin. Pop. 280.—Also a village of the regency of Potsdam, and circle of Jüterbogk. Pop. 240.

WOLTERSDORF (GROSS and KLEIN), two villages of Prussia, in the regency of Potsdam, and circle of Ostpriegnitz. Pop. 270.—Also a village of the regency of Magdeburg, and circle of Jerichow. Pop. 320.—Also a village of the same regency, in the circle of Jerichow. Pop. 220.

WOLTERTON, a parish in Norfolk, 4 m. NNW of Aylsham. Area 877 acres. Pop. in 1851, 44.

WOLTHUSEN, a village of Hanover, in East Friesland, near Emden. Pop. 384.

WOLTIN, a village of Prussia, in the regency of Stettin, and circle of Greiffenhagen. Pop. 698.

WOLTINGERODE, a hamlet of Hanover, in the gov. and principality of Hildesheim, and bail. of Vienenburg. It had formerly a Benedictine convent.

WOLTMERSHAUSEN, a village of Bremen, on the Ufer. Pop. 420.

WOLTORF, a village of Brunswick, in the circle of that name, and bail. of Bochelde. Pop. 250.—Also a village of Hanover, in the principality of Hildesheim, near Peine. Pop. 372.

WOLTRUP, a village of Hanover, in the principality of Osnabruck, and bail. of Bersenbruck, near Ancum. Pop. 334.

WOLTUSCH, a village of Austria, in Bohemia, in the reg. of Pilsen, and bail. of Brzeznitz. Pop. 320.

WOLTWIESCHE, a village of Brunswick, in the circle of Wolfenbüttel. Pop. 620.

WOLUWE-SAINT-ETIENNE, a department, commune, and town of Belgium, in the prov. of Brabant, and arrond. of Brussels. The town is 5 m. ENE of Brussels, on the Woluwe. Pop. 795. It has a paper-mill.

WOLUWE-SAINT-LAMBERT, a department and commune of Belgium, in the prov. of Brabant, and arrond. of Brussels. Pop. of dep. 1,070. The village is 2 m. SSW of Woluwe-Saint-Etienne, on the Woluwe. Pop. 382.

WOLUWE-SAINT-PIERRE, a department and commune of Belgium, in the prov. of Brabant, and arrond. of Brussels. Pop. of dep. 1,130. The village is 2 m. SSW of Woluwe-Saint-Etienne, on the Woluwe. Pop. 591.

WOLVELINGEN, a village of Holland, in the prov. of Luxemburg, and bail. of Diekirch. Pop. 200.

WOLVERCOTE, or WOOLVERCOTT, a parish in Oxfordshire, 3 m. NNW of Oxford, intersected by the river Isis and the Oxford canal. Area 1,600 acres. Pop. in 1831, 524; in 1851, 631.

WOLVERHAMPTON, a parish, borough, and market-town, in the co. of Stafford. The parish, comprising the chapelries of Pelsall, Wednesfield, and Willenhall, the market-towns of Bilston and Wolverhampton, and the townships of Featherstone, Hatherton, Hilton, and Kirwaston, has an area of 16,680 acres. The parish is intersected by, or in immediate communication with, the Birmingham, the Wyrley and Essington, the Stafford and Birmingham, and the Birmingham and Liverpool, canals, the Grand Junction section of the Great North-Western, and the Shrewsbury and Birmingham railways. Pop. in 1801, 24,536; in 1831, 48,080; in 1851, 92,287.—W. was enfranchised under the Reform act, and now returns 2 members to parliament. The borough boundaries comprehend the townships of Wolverhampton, Bilston, Willenhall, and Wednesfield, and the parish of Sedgley. Pop. of parl. borough in 1851, 49,985. The number of electors registered for the borough in 1837, was 2,170, in 1852, 2,600. The town consists of several streets diverging from a centre, in which stands the market-place. It is lighted with gas, paved, and supplied with water chiefly from numerous wells sunk to a great depth through the solid rock on which the town is built. The church of St. Peter's is a stately and venerable cruciform structure, consisting of a lofty nave, two side aisles, and a chancel, with a very fine embattled and richly ornamented Gothic central tower, 120 ft. in height. The old town-hall was the place in which Mrs. Siddons and John Philip Kemble first developed their extraordinary histrionic talents. The public subscription library is a handsome building, containing a news-room, and assembly and concert-rooms. In August, races are annually held on a fine course near the town. W. has been long celebrated for its lock and key smiths; and the inhabitants were formerly much employed in the manufacture of the finer kinds of steel ornaments, but this latter branch of trade has declined, or been transferred to other places. Wares of brass and tin, japanned wares, and papier-mache articles, locks, files, nails, gun-locks, screws, machinery of every kind, and ironmongery articles in general, are largely made. The smelting of iron, and the making of boiler-plates and iron-castings, are now a principal portion of the manufactures of the place. Here are also chemical works of some extent; and there are two mills for spinning worsted, and a variety of other manufacturing establishments. The situation of W., nearly in the centre of England, in the midst of exhaustless mines of coal and ironstone, and in communication with some of the most important lines of road, canal, and railway, in the kingdom, is one highly advantageous to its commercial advancement.

WOLVERLEY, a parish in Worcestershire, 2 m. NW of Kidderminster, intersected by the river Stour, and the Stafford and Worcester canal. It contains the hamlets of Blakesall, Caunsall, Cookley, Haustelev, Horseley, Kingsford, Low, Sion-hill, Wolverley, and Little Wolverley. Area 5,532 acres. Pop. 2,441.

WOLVERTHEM, a department, commune, and town of Belgium, in the prov. of Brabant, and arrond. of Brussels. Pop. of dep. 3,221. The town is 8 m. NNW of Brussels. Pop. 1,230.

WOLVERTON, or WOOLVERSTON, a parish in the co. of Buckingham, 1 m. ENE of Stony-Stratford, in the line of the Grand Junction canal, and the London and Birmingham section of the Great North-Western railway, on the latter of which the railway company have here their central station, depôt, and factory, distant 52½ m. from London, and 59¾ from Birmingham. A 'railway church,' the first of its kind, for the religious instruction of those employed by the railway company, was recently erected here, near the station. The railway directors have also established a 'railway school,' with an infant school, in addition to a reading-room and library for the artificers and others employed by the company, whose magnificent establishment here comprises a new town, with a market-place, streets, factories, &c. Area 2,260 acres. Pop. in 1851, 2,070.

WOLVERTON, or WOLFERTON, a parish in Norfolk, 3 m. N of Castle-Rising, on the coast of the Wash. Area 5,634 acres. Pop. in 1851, 165.

WOLVERTON, or WOOLVERTON, a parish in the co. of Somerset, 4½ m. N by E of Frome, on the river Frome. Area 736 acres. Pop. in 1851, 174.

WOLVERTON, or WOLVERDINGTON, a parish in the co. of Warwick, 5½ m. WSW of Warwick. Area 1,320 acres. Pop. in 1831, 166; in 1851, 174.

WOLVEY, a parish in the co. of Warwick, 5 m SE of Nuneaton, at the source of the river Anker. Area 3,790 acres. Pop. in 1831, 935; in 1851, 963.

WOLVISTON, a chapelry in the p. of Billingham, co-palatine of Durham, 4½ m. N by E of Stockton-upon-Tees, in the line of the Clarence and Hartlepool railway. Area 2,396 acres. Pop. 750.

WOLWARK, a village of Prussia, in the prov. of Posen, regency of Posen, and circle of Schubin. Pop. 240.

WOLXHEIM, or WOLIXEN, a commune of France, in the dep. of the Bas-Rhin, and cant. of Molsheim. Pop. 1,140. It is noted for its wine.

WOLZ (GROSS), a village of Prussia, in the regency of Marienwerder, and circle of Graudenz. Pop. 488.

WOLZ (OBER), a town of Austria, in Styria, in the gov. of Gratz, circle and 20 m. W of Judenburg, on the Wolzbach, at the confluence of the Schoderbach. Pop. 585. It has an old castle, and possesses several iron and steel works, and a saltpetre refinery.

WOLZHAUSEN, a village of Hesse, in the circle of Ober-Hesse, and presidial of Bredenkopt. Pop. 310.

WOMBA, a town of Africa, in Sudan, in the Fellatah territory, and prov. of Guri, to the SSE of Sakkatu. It is surrounded by an earthen wall, and has a daily well-frequented market. The surrounding country is well-cultivated, and produces a good deal of rice. It is at present the residence of an independent sultan.

WOMBACH, or WOHMBACH, a village of Bavaria, in the circle of Lower Franconia, and presidial of Lohr. Pop. 334.

WOMBLETON, or WIMBLETON, a township in the p. of Kirkdale, N. R. of Yorkshire, 5 m. E by S of Helmsley. Area 1,040 acres. Pop. in 1851, 335.

WOMBMELLORE, or OMALORE, a town of Hindostan, in the presidency of Madras, district and 9 m. NNW of Salem, on an affluent of the Cavery.

WOMBO, a mountain of New South Wales, in the co. of Hunter, between the river of that name and a small stream named Wollombe.

WOMBOURNE, a parish in the co. of Stafford, 4 m. SW by W of Wolverhampton, intersected by the Stafford and Worcester canal. Area 4,680 acres. Pop. in 1831, 1,647; in 1851, 2,007.

WOMBRIDGE, a parish in the co. of Salop, 2 m. E of Wellington, at the junction of the Shrewsbury, Shropshire, and Marquis of Stafford's canals; and intersected by the ancient Watling-street, the Holyhead parliamentary road, and various railways connected with extensive coal-mines and iron-works in the vicinity, particularly at Ketley. Area 698 acres. Pop. in 1831, 1,855; in 1851, 2,166.

WOMBRZEZNO. See BRIESEN.

WOMBWELL, a chapelry in the p. of Darfield, W. R. of Yorkshire, 5 m. SE by E of Barnesley, in the line of the Dearn and Dove canal, and the North Midland railway. Area 5,557 acres. Pop. 1,627.

WOMELSDORF, a town of Berks co., in the state of Pennsylvania, U. S., on the S side of Tulpehocken creek, and 36 m. E by N of Harrisburg. Pop. in 1840, 849; in 1850, 950. Its inhabitants are chiefly of German origin.

WOMENSWOLD, or WIMLINGSWOLD, a parish of Kent, 5 m. SW of Wingham. Area 1,010 acres. Pop. in 1831, 263; in 1851, 256.

WOMERSLEY, a parish in the W.R. of Yorkshire, 5½ m. ESE of Pontefract, containing the townships of Cridling-Stubbs, Little Smeaton, Walden-Stubbs, and W. Area 7,780 acres. Pop. in 1851, 998.

WOMITZ, a village of Austria, in Moravia, in the circle of Brunn, and seignory of Rossitz. Pop. 310 It has two flour-mills.

WOMMELGEN, a department and commune of Belgium, in the prov. and arrond. of Antwerp, watered by the Schyn. Pop. of dep. 1,800; and of com. 750.

WOMMELSHAUSEN, a village of Hesse, in the prov. of Upper Hesse, circle of Biedenkopf, and presidial of Gladenbach. Pop. 230.

WOMMEN, a village of Hesse, in the prov. of Lower Hesse, and circle of Eschwege. Pop. 333.

WOMMERSOM, a department and commune of Belgium, in the prov. of Brabant, and arrond. of Louvain, watered by the Ghète. Pop. of dep. 562; and of com. 429.

WOMPIERSK, a village of Prussia, in the regency of Marienwerder, and circle of Strasburg. Pop. 250.

WOMRATH, a village of Prussia, in the regency of Coblenz, and circle of Simmern. Pop. 350.

WOMWELNO, a village of Prussia, in the prov. of Posen, regency of Bromberg, and circle of Wirsitz. Pop. 320.

WOMZO, a village of Central Africa, in Sudan, SSW of Womba.

WONASQUATUCKET, a river of Providence co., in the state of Rhode Island, U. S., which flows SE into Providence lake.

WONASTOW, or WONASTON, a parish in Monmouthshire, 2 m. WSW of Monmouth, on the river Trothy. Area 1,599 acres. Pop. in 1851, 141.

WONCK, a department and commune of Belgium, in the prov. of Limburg, and arrond. of Tongres, watered by the Jaer. Pop. of dep. 1,480. The village is 8 m. E of Tongres, on the Jaer.

WONDA, a river of Senegambia, in the kingdom of Fuladu, which, after a course of about 150 m., joins the Kokoro, on the l. bank, 24 m. W of Keminoum. It receives the Comersay on the l.—Also a walled town of the state of Manding, 45 m. ENE of Kamalia.

WONDELGEM, a department and commune of Belgium, in the prov. of E. Flanders, and arrond. of Ghent. The dep. is intersected by the Canal-du-Sas. Pop. 1,181. The village is 2 m. N of Ghent, on the above-named canal. Pop. with Kerkdriesch, 409.

WONDOLEK, a village of Prussia, in the regency of Gumbinnen, circle and S of Johannisburg. It has extensive iron-works.

WONDREB, a bail. and village of Bavaria, in the circle of the Upper Pfalz, and presidial of Tirschenreuth. Pop. 333. It has iron-works and manufactories of porcelain.

WONERSH, or OGNERSH, a parish in Surrey, 3½ m. SSE of Guildford, in the line of the Arun and Wey canal. Ironstone is abundant, and is worked to some extent in this vicinity. Area 4,427 acres. Pop. in 1831, 1,069; in 1851, 1,280.

WONFURT, or WOHNFURT, a village of Bavaria, in the prov. of Lower Franconia, and presidial of Hassfurt. Pop. 632.

WONGAN, a range of hills in Western Australia, in the co. of Glenelg.

WONGROWITZ, or WONGROWIEC, a circle and town of Prussia, in the prov. of Posen, and regency of Bromberg. The circle comprises an area of 24·5 German sq. m. Pop. 46,519.—The town is 60 m. SW of Bromberg, on the Wilna, at the extremity of a small lake. Pop. 2,849. It has a Cistercian convent, a synagogue, and a good market.

WONISCHEN, a village of Austria, in Bohemia, in the regency of Pilsen, and bail. of Bischofteinitz. Pop. 870. It has extensive bleacheries.

WONNERUP, a town of Western Australia, in the co. of Sussex, on Vasse Inlet, Bay-du-Geographe.

WONNIBERG, a village of Prussia, in the regency of Konigsberg, and circle of Rössel. Pop. 240.—Also a village of the regency and circle of Danzig. Pop. 495.

WONNO, a village of Prussia, in the regency of Marienwerder, and circle of Lobau. Pop. 350.

WONOMISCHEL, a village of Prussia, in the regency of Pardubitz, and bail. of Neukolin. Pop. 270.

WONSCHOW, a village of Austria, in Bohemia, in the regency of Pardubitz, and bail. of Ledetsch. Pop. 600. It has a castle, and possesses manufactories of potash and several mills.

WONSCHOWITZ, a village of Austria, in Bohemia, in the regency of Pilsen, and bail. of Winterberg. Pop. 325.

WONSEES, or WOHNSEES, a market-town of Bavaria, in the circle of Upper Franconia, and presidial of Hollfeld, 17 m. W of Bayreuth, on the Kainach, near the castle of Sanspareil. Pop. 1,700. It has manufactories of woollen and linen fabrics.

WONSHEIM, a village of Hesse, in the prov. of Rhine-Hesse, and circle of Bingen. Pop. 680.

WONSOSZ, a town of Russia in Europe, in Poland, in the gov. of Augustowo. Pop. 1,255.

WONSOWO, a village of Prussia, in the regency of Posen, and circle of Buk. Pop. 458.

WONSTON, a parish in the co. of Southampton, 6 m. S of Whitchurch. Area 4,160 acres. Pop. 716.

WONTERGEM, a department and commune of Belgium, in the prov. of E. Flanders and arrond. of Ghent. Pop. of dep. 1,067.

WONVOLNICA, a town of Russia in Europe, in Poland, in the gov. of Lublin. Pop. 1,150.

WOOBURN. See WOBURN.

WOOD, a county in the NW part of the state of Virginia, U. S., comprising an area of 476 sq. m., drained by Little Kanawha river and its branches, and intersected by the North-Western railway. The surface is hilly, but generally fertile. Pop. in 1840, 7,923; in 1850, 9,450. Its capital is Parkersburg.—Also a county in the NW part of the state of Ohio, comprising an area of 612 sq. m., drained by Portage river and its branches, and by branches of Maumee river, and intersected by the Toledo and Port Wayne railroad, and the Wabash and Erie canal. It has a diversified surface, and is in some parts fertile. Pop. in 1840, 5,357; in 1850, 9,157. Its capital is Perrysburg.—Also a county in the NE of the state of Texas, comprising an area of 1,046 sq. m., drained by Sabine river and its branches. It has a finely undulating surface, and is very fertile. Its capital is Quitman.—Also a creek of Oneida co., in the state of New York, which has its source in Rome, and falls into Oneida lake. It now forms a part of Oneida Lake canal.—Also a creek of Washington co., in the same state, an affluent of Lake Champlain, and forming for some distance a part of the canal of that name.

WOOD, or WOODVILLE, a member of the port and liberty of Dover, locally situated in the isle of Thanet, co. of Kent, 3 m. SW by W of Margate. Pop. in 1831, 292; in 1851, 254.

WOOD-BASTWICK, a parish in Norfolk, 5 m. NW of Acle. Area 2,163 acres. Pop. in 1851, 293.

WOODBINE, a village of Carroll co., in the state of Maryland, U. S., on the Baltimore and Ohio railroad, and 50 m. NW of Annapolis.

WOODBOROUGH, a parish in Nottinghamshire, 7½ m. NE by N of Nottingham, intersected by the Doverbeck. Area 1,940 acres. The inhabitants are chiefly employed in the stocking manufacture on the loom or frame invented in 1528, by William Low, a mechanic of the village of Woodborough. Pop. 852.

WOODBOROUGH, a parish in Wilts, 3½ m. W of Pewsey, intersected by the Kennet and Avon canal. Area 1,015 acres. Pop. in 1851, 424.

WOODBRIDGE, a parish and market-town in Suffolk, 7½ m. ENE of Ipswich, on the NW bank of the navigable river Deben. Area of the p. 1,059 acres. Pop. in 1831, 4,769; in 1851, 5,161. The church is a spacious edifice of black flint and freestone. There are two Independent chapels, and places of worship for Baptists, Wesleyan Methodists, and the Society of Friends. The town consists of two principal streets, with a number of lanes, and a spacious square. The eminence on which it stands is surrounded by pleasant walks, and the situation is remarkably healthy. The divisional county prison, or house-of-correction, stands on an elevated spot at the extremity of the town. The mouth of the river forms an excellent haven, and is navigable as far as the town for vessels of 120 tons. There are bonded warehouses for wine and spirits, wood and corn; two quays, one for general commerce, the other for the accommodation of vessels loaded with lime; and two small docks. Vessels of war of small magnitude, and coasting-craft are occasionally built here. A regular trade is conducted with London, Newcastle, Hull, Liverpool, the continent of Europe, and the Baltic. The principal articles of import are coal, timber, wine, spirits, porter, salt, grocery, drapery, and iron-wares. The exports consist chiefly of corn, flour, and malt.

WOODBRIDGE, a township of Middlesex co., in the state of New Jersey, U. S., 33 m. NE of Trenton, and on the New Jersey railroad. Pop. in 1840, 4,821; in 1850, 5,141. It contains several flourishing villages.—Also a township of Hillsdale co., in the state of Michigan. Pop. in 1840, 226.

WOODBURN, a manufacturing locality in co. Antrim, on the Woodburn rivulet, and on the road from Carrickfergus to Ballinure, 2 m. NW of Carrickfergus. Extensive calico print-works at this place, during a number of years preceding 1823, afforded employment to about 80 or 100 persons, and annually bleached and printed about 15,000 pieces of calicoes, shawls, and muslins. The cotton factories of W., though distant 2 miles from Carrickfergus, are usually ranked among the manufactories of that town. See article CARRICKFERGUS.

WOODBURY, a parish, formerly a market-town, in the co. of Devon, 3 m. E by S of Topsham, on the navigable river Exe. It contains the hamlets of Ebford, Exton, Grindle, Gulliford, Higher Nutwell, Salterton, and Woodmanton. Area 7,804 acres. Pop. in 1831, 1,673; in 1851, 2,014.

WOODBURY, a township and village of Litchfield co., in the state of Connecticut, U. S., 31 m. WSW of Hartford, drained by Pomperang river and its branches. Pop. of township, 2,150. The village is near the junction of the head streams of the Pomperang river.—Also a township of Huntingdon co., in the state of Pennsylvania. Pop. 2,102.—Also a township of Bedford co., in the same state, 113 m. W of Harrisburg, drained by Yellow creek and its branches, and by the Frankstown branch of the Juniata. Pop. in 1840, 3,944; in 1850, 4,000.—Also a village of Oyster Bay township, Queen's co., in the state of New York.—Also a village of Gloucester co., in the state of New Jersey, 33 m. SW by S of Trenton, and watered on the N by a creek of the same name, an affluent of the Delaware. Pop. in 1840, 800; in 1850, 900.—Also a village of Cannon co., in the state of Tennessee, 42 m. SE by E of Nashville. Pop. in 1840, 150; in 1850, 200.—Also a township and village of Washington co., in the state of Vermont, 17 m. NE by N of Montpelier, watered by numerous ponds, and by branches of Lamoille and Onion rivers. Pop. in 1850, 1,070.

WOODCHESTER, a parish in Gloucestershire, 2¼ m. SW of Stroud. Area 1,293 acres. Pop. in 1831, 885; in 1851, 893.—Also a parish and township in the co.-palatine of Chester, 6½ m. N by W of Great Neston. The parish contains the townships of Arrow, Barnston, Irby, Landican, Noctorum, Oxton, Pensby, Prenton, Thingwell, and W. Area 5,792 acres. Pop. in 1801, 725; in 1831, 929; in 1851, 2,927.—Also a parish in Kent, 5 m. E of Tenterden. Area 6,948 acres. Pop. in 1831, 1,887; in 1851, 1,286.

WOODCOCK, a township of Crawford co., in the state of Pennsylvania, U. S., 187 m. NW by W of Harrisburg, drained by Woodcock creek. Pop. in 1840, 1,921; in 1850, 2,000.—Also a valley of Hopewell township, Huntingdon co., in the state of Pennsylvania, between Allegripus and Tussey's mountains.

WOODCOTE, or WOODCUT, a parish in the co. of Southampton, 5 m. NNW of Whitchurch. Area 1,350 acres. Pop. in 1831, 90; in 1851, 100.—Also a chapelry in the p. of South Stoke, Oxfordshire, 5½ m. SSE of Wallingford, on the Thames.—Also a township in the p. of Sheriff-Hales, co. of Salop, 3 m. SE by S of Newport. Area 1,248 acres. Pop. in 1831, 195; in 1851, 166.

WOOD-DALLING, a parish in Norfolk, 3 m. N by W of Reepham. Area 2,444 acres. Pop. 574.

WOOD-EATON, a parish in the co. of Oxford, 4 m. NNE of Oxford, on the river Cherwell. Area 639 acres. Pop. in 1831, 86; in 1851, 89.

WOOD-EATON. See CHURCH-EATON.

WOODEN-BRIDGE, a retreat of tourists at the confluence of the Ovoca and the Aughrim or Second Meeting of the Waters, 3¾ m. NW of the town of Arklow, in co. Wicklow.

WOODFORD, a township in the p. of Prestbury, co.-palatine of Chester, 5 m. S by W of Stockport, near the Manchester and Birmingham railway, and its Macclesfield branch. Area 1,449 acres. Pop. in 1831, 403; in 1851, 430.

WOODFORD, a parish in the co. of Northampton, 7½ m. SSW of Daventry. Area with Membris, 2,655 acres. Pop. in 1831, 827; in 1851, 800.—Also a parish in the co. of Northampton, 2½ m. SW by W of Thrapston, on the river Nen. Area 1,750 acres. Pop. in 1831, 639; in 1851, 726.—Also a parish in Wilts, 4½ m. NNW of Salisbury, on the river Avon. Area 2,780 acres. Pop. in 1851, 496.

WOODFORD, a small river of Ireland, which rises among the SE declivities of the Slieve-an-Ierin mountains, in co. Leitrim, and flows about 17 m. eastward, past Ballinamore, through Lough Garadice, across the NW wing of co. Cavan, and along the boundary between co. Cavan and co. Fermanagh, to the head of the W side of Upper Lough Erne.—Also a rivulet of the p. of Ballinakill, co. Galway, which rises among the Slieve-Baughta mountains, on the S verge of the co., and runs about 4 m. ENE to the village of Woodford, and about the same distance ESE to the head of a small bay of Lough Derg.—Also a village in the p. of Ballinakill, co. Galway, on the Woodford rivulet, 8 m. SW by W of Portumna. The portion of the vale and screens of the Woodford rivulet which immediately adjoins it, is largely enriched with wood; and the country lying immediately to the west and the south climbs speedily into the acclivities; and toward the summits, of the Slieve-Baughta mountains. Pop. in 1851, 267.

WOODFORD, an island of New South Wales, in the river Clarence and district of that name.

WOODFORD, a central county of the state of Kentucky, U. S., comprising an area of 141 sq. m., drained by branches of Kentucky river, by which it is bounded on the W, and intersected by the Lexington and Frankfort railway. Pop. in 1840, 11,740.

in 1850, 12,423. Its capital is Versailles.—Also a central co. of the state of Illinois, comprising an area of 532 m., drained by the Mackinan and other branches of the Illinois, by which it is bounded on the W, and intersected by the Central Illinois railroad. Pop. in 1850, 4,416. Its capital is Metamera.—Also a township of Bennington co., in the state of Vermont, 103 m. SSW of Montpelier, on the Green Mountain range, and watered by several streams. Pop. in 1840, 487; in 1850, 425.—Also a village of Cumberland co., in the state of Maine, on the York and Cumberland railway, 2 m. W of Portland.

WOODFORD (Cape), a headland of Albemarle island, Gallipagos islands, South Pacific, on the SE coast, in N lat. 0° 56′, and W long. 90° 52′.

WOODFORD-ST.-MARY, a parish in Essex, 8½ m. NE of St. Paul's, London, near the river Roding. Area 2,148 acres. Pop. in 1831, 2,548; in 1851, 2,774. The village, which stands on the main road from London to Newmarket, consists chiefly of detached houses, surrounded by trees and gardens, and of handsome villas, which are much resorted to by wealthy merchants of the metropolis.

WOODGATE. See Uttoxeter.

WOODHALL, a parish in the co. of Lincoln, 3 m. WSW of Horncastle. Area 2,240 acres. Pop. in 1831, 196; in 1851, 275.

WOODHAM, a township in the p. of Aycliffe, co.-palatine of Durham, 7 m. ESE of Bishop-Auckland, in the line of the Great Northern railway. Area 3,705 acres. Pop. in 1831. 204; in 1851, 209.

WOODHAM-FERRERS, or Ferris, a parish in Essex, 4½ m. SSE of Danbury. Area 4,481 acres. Pop. in 1831, 826; in 1851, 981.

WOODHAM-MORTIMER, or Little-Woodham, a parish in Essex, 2½ m. SW by W of Maldon. Area 1,380 acres. Pop. in 1831, 339; in 1851, 326.

WOODHAM (Walter), a parish in Essex, 2½ m. ENE of Danbury, near the Chelmer and Blackwater navigation. Area 2,421 acres. Pop. 585.

WOODHAY (East), a parish in the co. of Southampton, 10½ m. NNW of Whitchurch. Area 4,966 acres. Pop. in 1831, 1,269; in 1851, 1,550.

WOODHAY (West), a parish in Berks, 6 m. SW by W of Newbury. Area 1,407 acres. Pop. 115.

WOODHEAD, a chapelry in the p. of Motram-in-Longden-dale, co.-palatine of Chester, 12½ m. ENE of Stockport. Returns with the parish.

WOODHORN, a parish in Northumberland, 6½ m. ENE of Morpeth. It contains the chapelries of Newbiggen and Cresswell, the demesne of Woodhorn, and the townships of Ellington, Hurst, Linmouth, North Seaton, and W. Area 8,456 acres. Pop. in 1831, 1,416; in 1851, 1,598.

WOODHOUSE, a township and chapelry in the p. of Barrow-upon-Soar, co. of Leicester, 3½ m. W of Mountsorrel. Pop. in 1831, 1,262; in 1851, 1,201.

WOODHOUSE, a township of Upper Canada, in the Talbot district, bordered on the S by Lake Erie. Pop. in 1840, 1,694.

WOODHULL, a township of Steuben co., in the state of New York, U. S., 194 m. W by S of Albany, drained by Tuscarora creek. Pop. in 1840, 827; in 1850, 1,769.—Also a township of Shiawassee co., in the state of Michigan, 10 m. ENE of Lansing, drained by branches of Looking-glass river. Pop. in 1840, 147; in 1850, 259.

WOODHURST, a parish in the co. of Huntingdon, 4 m. N of St. Ives. Area 2,130 acres. Pop. 533.

WOOD-ISLAND, an inhabited islet in the p. of Tullynakill, co. Down, on the W side of Lough Strangford, ¾ m. WNW of Island-Magee, and 2 m. NNE of Killinchy. The tide entirely forsakes it at low water.

WOODLAND, a chapelry in the p. of Ipplepen, Devon, 2 m. E by S of Ashburton. Area 1,606 acres. Pop. in 1831, 237; in 1851, 188.—Also a township in the p. of Staindrop, co.-palatine of Durham, 4½ m. NE of Barnard-Castle. Pop. 240.

WOODLAND, a chapelry in the p. of Kirkby-Ireleth, co.-palatine of Lancaster, 7 m. NE by N of Ulverstone.

WOODLANDS, or Woodland-Eyam, a township in the p. of Eyam, co. of Derby, 2 m. N of Stoney-Middleton. Pop. in 1831, 213; in 1851, 275.

WOODLANDS, or Woodland-Hope, a hamlet in the p. of Hope, co. of Derby, 6 m. NE of Tideswell. Pop. in 1831, 273; in 1851, 256.—Also a tything in the p. of Horton, Dorset, 4 m. SSW of Cranbourne. Area 2,561 acres. Pop. in 1831, 423; in 1851, 476. After the battle of Sedgmoor, the Duke of Monmouth is said to have fled to this place, and to have been taken in a ditch, under an ash tree still in existence.

WOODLAWN, a village of Cecil co., in the state of Maryland, U. S., 51 m. NE by N of Annapolis. Pop. in 1850, 130.

WOODLE, a populous island of the North Pacific, in the Mulgrave archipelago, in N lat. (NW part) 0° 15′ 55″, and E long. 168° 45′ 58″. It is separated from the island of Enderville, by a channel 6 m. in width. It was discovered in 1809.

WOODLEIGH, or Woodley, a parish in the co. of Devon, 3 m. N of Kingsbridge. Area 2,319 acres. Pop. in 1831, 279; in 1851, 233.

WOODMANCOT, a hamlet in the p. of Bishop's-Clive, co. of Gloucester, 3½ m. W by S of Winchcombe, and about 2 m. E of the Birmingham and Gloucester railway. Pop. in 1831, 267; in 1851, 432.

WOODMANCOTE, a parish in the co. of Southampton, 8 m. SW of Basingstoke, near the South-Western railway. Area 1,396 acres. Pop. in 1831, 92; in 1851, 76.—Also a parish in Sussex, 5 m. NE by E of Stevning. Area 2,236 acres. Pop. 326.

WOODMANSEA, a township within the liberties of the town of Beverley, Yorkshire, 2 m. SE by E of Beverley. Pop. in 1831, 360; in 1851, 241.

WOODMANSTONE, a parish in Surrey. Area 1,590 acres. Pop. in 1831, 184; in 1851, 271.

WOODNESBOROUGH, or Winsborough, a parish in Kent, 2 m. WSW of Sandwich. The church is an elegant structure in the decorated style of English architecture. Area 2,944 acres. Pop. 813.

WOOD-NORTON, a parish and village in Norfolk, 7 m. NW of Reepham. Area 1,726 acres. Pop. in 1831, 315; in 1851, 308.

WOOD-RISING, a parish in Norfolk, 2 m. WNW of Hingham. Area 1,363 acres. Pop. in 1851, 127.

WOODS (Lake of the). See Lake-of-the-Woods.

WOODSETTS, a township, partly in the p. of Auston, partly in that of Loughton-in-le-Morthen, partly within the liberty of York, partly in the wapentake of Strafforth, W. R. of York, 4½ m. NW by W of Worksop. Pop. in 1831, 146; in 1851, 173.

WOODSFIELD, a village of Centre township, Monroe co., in the state of Ohio, U. S., on the S side of Sunfish creek, 96 m. E of Columbus. Pop. in 1840, 300; in 1850, 395.

WOODSFORD, a parish in the co. of Dorset, 5½ m. E of Dorchester, on the river Frome. It consists of the villes of East and West W. or Woodsford-Strangeways. Area 1,742 acres. Pop. 183.

WOOD'S HOLE, a village of Barnstable co., in the state of Massachusetts, U. S., 62 m. SSE of Boston. It has a good harbour.

WOOD'S MILLS, a village of Vigo co., in the state of Indiana, U. S., on Terre-Haute and Richmond railway, 8 m. E of Terre-Haute.

WOODSIDE, or WOODSIDE-QUARTER, a township in the p. of Wigton, Cumberland, 3 m. E by N of Wigton, in the line of the Carlisle and Maryport railway. Pop. in 1841, 415; in 1851, 404.

WOODSIDE, a joint township with Hatton, in the p. of Shiffnall, Salop, 3 m. SSE of Shiffnall. Pop. in 1831, 379; in 1851, with Hatton, 1,085.—Also a township in the p. of Elsdon, Northumberland, 1½ m. N of Elsdon. Area 6,437 acres. Pop. 112.

WOODSTOCK, a chapelry, borough, and market-town, in the co. of Oxford, 8 m. NNW of Oxford. It includes the extra-parochial precincts of Blenheim-park. Area 360 acres. Pop. in 1831, 1,380; in 1851, 1,262. The town, which consists of several well-built and regular streets, chiefly of stone, with a large proportion of handsome houses, stands on an eminence, skirted on the west by the river Glynn, which, after expanding into a sheet of water in Blenheim-park, is joined by the Evenlode, and shortly after falls into the Isis. It contains a handsome town-hall, erected in 1766. It formerly possessed a wool staple, granted by Queen Elizabeth, but there is now no trade in wool. It was also celebrated for the manufacture of articles in fine polished steel, formed of the iron nails extracted from old horse shoes; but this branch of trade has nearly disappeared since the advance of Birmingham and Sheffield. The manufacture of gloves has long been the staple of the town, and gives employment to many of the inhabitants, both of the town and the neighbouring villages. The number of pairs made weekly at W. is said to average 500 dozen; and about 100 men, and 1,500 women and girls, are said to receive constant employment from it. Under the reform act, the borough boundaries, which were previously confined within the limits of the town, even excluding part of it, were extended so as to comprehend, besides the old borough, the several parishes of Bladon, Begbrook, Shipton-on-Cherwell, Hampton-Gay, Tackley, Wooton, Stones-field, Coombe, and Handborough; the parish of Kidlington, except the respective hamlets of Gosford and Wator Eaton; the hamlet of Old Woodstock, and Blenheim-park. The borough was also deprived of one member. The number of electors registered for 1837, was 385. This parish became, at an early period, a favourite abode of royalty, and at present it contains the splendid demesne and residence of the duke of Marlborough, called Blenheim. It appears that Alfred the Great resided here while employed in the translation of Boetius, and it is said to have been the birth-place and occasional residence of Chaucer. The duke of Portland takes his title of Viscount from Woodstock.

WOODSTOCK, a rivulet of co. Wicklow, which rises on the E side of the Douce mountain, and runs about 7 m. ESE to the Irish sea at the Breaches, forming the only outlet for drainage between the Three-Trouts' Stream at Delgany, and the mouth of the Leitrim or Vartrey river at Wicklow.—Also an old castle in the p. of Churchtown, co. Kildare, on the r. bank of the Barrow, in the N environs of the town of Athy. The walls are of great thickness, and, considering the attacks they have been exposed to, in good preservation.

WOODSTOCK, a town of Upper Canada, in the Brock district, in the township of Blandford, 32 m. from London. Pop. 1,085.

WOODSTOCK, a township of Oxford co., in the state of Maine, U. S., 37 m. W of Augusta, drained by a branch of Little Androscoggin river, and intersected by the Atlantic and St. Lawrence railroad. Pop. in 1840, 819; in 1850, 1,012.—Also a township of Grafton co., in the state of New Hampshire, 54 m. N by W of Concord, drained by Penngewasset river. Pop. in 1840, 472; in 1850, 418.—Also a township of Windsor co., in the state of Vermont, 45 m. S by E of Montpelier, drained by Otta-Queechee river and its branches. Pop. in 1840, 3,315; in 1850, 3,041.—Also a township of Windham co., in the state of Connecticut, 43 m. ENE of Hartford, drained by Muddy Brook and its branches, and by other streams. Pop. in 1840, 3,053; in 1850, 4,636.—Also a township of Ulster co., in the state of New York, 45 m. SSW of Albany. It has a mountainous surface, and is drained by Saghkill creek. Pop. in 1840, 1,691; in 1850, 1,650.—Also a village of Shenandoah co., in the state of Virginia, 106 m. NW by N of Richmond. Pop. in 1840, 1,000; in 1850, 1,000.—Also a village of Cherokee co., in the state of Georgia, on a branch of Etowah river, and 108 m. NW of Milledgeville.—Also a village of Champaign co., in the state of Ohio, 30 m. NNW of Columbus. Pop. in 1850, 205.—Also a township of Lenawee co., in the state of Michigan, 48 m. S by E of Lansing, watered by the head branches of Raisin river. Pop. in 1840, 674; in 1850, 949.

WOODSTONE, a parish in the co. of Huntingdon, 1 m. S of Peterborough. Area 1,050 acres. Pop. in 1831, 243; in 1851, 320.

WOODSTOWN, a village of Salem co., in the state of New Jersey, U. S., on the N side of Salem creek, 48 m. SW by S of Trenton. Pop. in 1840, 700; in 1850, 750.

WOODSVILLE, a village of Hopewell township, Mercer co., in the state of New Jersey, U. S., 13 m. NNW of Trenton.

WOODTHORPE, a township in the p. of North Wingfield, co. of Derby, 6 m. ENE of Chesterfield, near the North Midland line of railway. Pop. in 1831, 231; in 1851, 267.

WOODTON, or WOOTTON, a parish in Norfolk, 5 m. NW of Bungay. Area 2,124 acres. Pop. 588.

WOODVILLE, a village of Ellisburg township, Jefferson co., in the state of New York, U. S., on the S side of Sandy creek, and 147 m. NW by W of Albany. Pop. in 1840, 150; in 1850, 200.—Also a village of Perquimans co., in the state of North Carolina, on the W side of Little river. Pop. in 1840, 75.—Also a township of Sandusky co., in the state of Ohio, 104 m. N by W of Columbus, drained by Portage river, on which is a village of the same name. Pop. in 1840, 486; in 1850, 1,069.—Also a village of Wilkinson co., in the state of Mississippi, 103 m. SW by S of Jackson, and intersected by the West Feliciana railroad. Pop. in 1840, 800; in 1850, 800.—Also a village of Jackson co., in the state of Alabama, and near the Memphis and Charleston railway.—Also a village of Tyler co., in the state of Texas, on the W side of Turkey creek, 200 m. E by N of Austin city.—Also a village of Greene co., in the state of Georgia, on Athens Branch railway, 48 m. N by E of Milledgeville.

WOODY, an island of the Asiatic archipelago, in the group of the Banda Islands, to the S of Timor Laut.

WOOKEY, or WOKEY, a parish in Somersetshire, 2 m. W of Wells. It contains the hamlets of Bladney, Castle, Hinton, Kalleyhill, Oure, Turlichmead, and Yardley. Area 3,420 acres. Pop. 1,158.

WOOL, a parish in Dorset, 6 m. W by S of Wareham, on the river Frome. It includes the hamlets of Great Bindon, Bovington, and Woodstreat. Area 2,550 acres. Pop. in 1851, 545.

WOOLASTON. See WOLLASTON.

WOOLASTON, a parish in Gloucestershire, 5 m. NE of Chepstow, on the river Severn. Area 3,640 acres. Pop. in 1831, 880; in 1851, 1,261.

WOOLASTON (GREAT), a township and chapelry

in the p. of Alberbury, Salop, 11 m. W of Shrewsbury. Pop. in 1831, 383; in 1851, 344.

WOOLAVINGTON, a parish in Somersetshire, 4 m. NE of Bridgewater, near the Glastonbury canal and the Bristol and Exeter railway. Area 1,725 acres. Pop. in 1831, 412; in 1851, 405.

WOOLAVINGTON (East and West), a parish divided into two tythings, in Sussex, 4½ m. SW by S of Petworth. Area 2,530 acres. Pop. 462.

WOOLBEDING, a parish in Sussex, 1¾ m. NW of Midhurst, intersected by the river Rother. Area 2,253 acres. Pop. in 1831, 307; in 1851, 320.

WOOLBOROUGH, or Wolborough, a parish in Devon, 1 m. S of Newton-Abbots. Area 1,231 acres. Pop. in 1831, 2,194; in 1851, 3,227.

WOOLDALE, a township in the p. of Kirkburton, W. R. of the county of York, 6¾ m. S of Huddersfield. Area 2,370 acres. Pop. in 1851, 5,600.

WOOLER, a parish and market-town in Northumberland, 46 m. NNW of Newcastle-upon-Tyne. Area of p., 4,852 acres. Pop. in 1831, 1,926; in 1851, 1,911. The town, which is situated in a healthy district on the eastern declivity of the Cheviots, consists of several streets diverging from a central area used as the market-place. Through the centre of it runs a good trout stream, which subsequently falls into the river Till.

WOOLFARDISWORTHY, a parish in Devonshire, 9½ m. SW by S of Bideford. Area 5,798 acres. Pop. in 1831, 840; in 1851, 824.—Also a parish in Devon, 6 m. N by W of Crediton. Area 1,815 acres. Pop. in 1831, 226; in 1851, 208.

WOOLFORD (Great), a parish in the co. of Warwick, 4 m. S by W of Shipston-on-Stour. It includes the hamlet of Little Woolford. Pop. 290.

WOOLHAMPTON, or Wolhampton, a parish in Berks, 7 m. E of Newbury, on the Kennet navigation. Area 694 acres. Pop. in 1851, 602.

WOOLIBARRY, an estuary of Clarence river, in the district of that name, New South Wales.

WOOLLEN-GRANGE, or Moll-Grange, a parish in co. Kilkenny, 3 m. NNW of Thomastown. Area 1,663 acres. Pop. in 1831, 373; in 1851, 265.

WOOLLEY, a parish in Huntingdonshire, 5 m. NE by N of Kimbolton. Area 1,420 acres. Pop. in 1831, 58; in 1851, 90.—Also a chapelry in the p. of Royston, W. R. of the co. of York, 6 m. N by W of Barnesley, near the North Midland railway. Area 2,569 acres. Pop. in 1831, 553; in 1851, 450.

WOOLLOOMOOLLOO. See Sydney, New South Wales.

WOOLOWARDALLA, a town of New South Wales, in the co. of King, on the Narrawa or Lachlan river, and 149 m. from Sydney.

WOOLPIT, a parish in Suffolk, 5¾ m. NW by W of Stow-Market. Area 1,877 acres. Pop. 1,071.

WOOLSTASTON, a parish in the co. of Salop, 10½ m. SSW of Shrewsbury. Area 843 acres. Pop. in 1831, 89; in 1851, 72.

WOOLSTHORPE, or Woolstrop, a parish in Lincolnshire, 6 m. W by S of Grantham, in the line of the Nottingham and Grantham canal. Area 2,600 acres. Pop. in 1831, 650; in 1851, 632.

WOOLSTON, a parish in Gloucestershire, 4 m. WNW of Winchcombe, near the line of the Birmingham and Gloucester railway. Area 787 acres. Pop. in 1831, 92; in 1851, 86.—Also a township with Martinscroft, in the p. of Warrington, co.-palatine of Lancaster, 2½ m. E by N of Warrington, and the Grand Junction railway. Area 1,444 acres. Pop. in 1831, 578; in 1851, 516.

WOOLSTON (Great), a parish in Buckinghamshire, 3½ m. N by W of Fenny-Stratford, on the river Ousel, and intersected by the Grand Junction canal. Area 760 acres. Pop. in 1851, 72.

WOOLSTON (Little), a parish in Buckinghamshire, 3½ m. N of Fenny-Stratford, on the river Ousel, and intersected by the Grand Junction canal. Area 613 acres. Pop. in 1831, 121; in 1851, 102.

WOOLSTONE, anciently Wolverichton, a chapelry in the p. of Uffington, Berks, 5½ m. S by E of Great-Farrington. Pop. in 1831, 270; in 1851, 288.

WOOLTON (Little), a township in the p. of Childwall, co.-palatine of Lancaster, 4½ m. SW by S of Prescot. Area 1,200 acres. Pop. in 1851, 1,016.

WOOLTON (Much), a chapelry in the p. of Childwall, co.-palatine of Lancaster, 5½ m. SW by S of Prescot. Area 930 acres. Pop. in 1851, 3,616.

WOOLVERSTONE, a parish in Suffolk, 4½ m. S by E of Ipswich, skirted on the NE by the navigable river Orwell. Area 1,256 acres. Pop. 241.

WOOLVERTON. See Wolverton.

WOOLVERTON, a parish in Southamptonshire, 7½ m. NW of Basingstoke. Area 736 acres. Pop. in 1831, 229; in 1851, 174.

WOOLWICH, a parish and market-town in Kent, on the S bank of the Thames, 2½ m. E of Greenwich, and 8½ m. E by S of St. Paul's, London. The parish extends to the Essex side of the Thames, where it comprehends the Devil's-house, and certain lands running along the N bank of the river, but included in the county of Kent. The surface soil is generally loam; the sub-soil is in many places clayey; the sub-strata is sand to a considerable depth, and then chalk. Area of p. 1,596 acres. Pop. in 1801, 9,826; in 1831, 17,661; in 1841, 26,311; in 1851, 32,367. This pop. includes the garrison, whose numbers, including artillerymen, sappers, miners, and marines, may amount to 3,600, and the convicts in the hulks usually exceeding 1,000.—The town stands on an elevated site, close to the S bank of the Thames, and gradually rises on the gravel ridge which extends from Greenwich to Erith, here known as Shooter's hill. There are marshes on the E and W of the town, which are sometimes flooded by the river. The principal street runs parallel to the river, and is crossed at right angles by others of smaller extent. It adjoins Plumstead on the E, and Charlton on the W; and stands partially within the parish of Plumstead. The most agreeable part of the town occupies the upper part of the elevated site, near the Charlton road, and the scenery of the environs is of a rich and diversified character, well-wooded, and containing a number of respectable villas. The lower and more densely peopled parts of the town, occupying a marshy site between the ridge and the river, are less healthy. The inhabitants are not engaged in any particular branch of manufacture, nor is there any trade of importance; by far the larger part of the population is supported by the government army and navy establishments. The common is open, spacious, and park-like in its general character, and there are good public walks. Off the town are moored hulks for the confinement of convicts. W. by the reform act joins with Deptford and Greenwich, under the name of the borough of Greenwich, in returning two members to parliament.

Government works.] The most interesting and important objects in Woolwich are the dockyard, the arsenal, and other government-works. The dock-yard here is supposed to be the most ancient royal dock-yard in the kingdom. It consists of a narrow strip of land on the banks of the river, about a furlong in breadth, and rather more than half a mile in length. The outer basin comprises an area of 120,000 sq. ft. The inner basin has an area of 160,000 sq. ft. Recent extensive improvements in this great naval depot are such as must render it superior to any in the kingdom. The dry dock, at the east end of the yard, is without exception the finest and most commodious that has ever been built, and is constructed of the most durable materials, being first laid in the bottom with one foot of brickwork, and over it large granite stones, about 3 ft. 6 in. in thickness, and each many tons in weight. The base of this dock is 230 ft. in length, and of a proportionate breadth; but it will contain vessels of upwards of 300 ft. in length on the upper deck. At the west end of the dock-yard, a dock of still greater dimensions than the old one was begun in 1840, and finished in July 1843. The dimensions of the dock are 300 ft. by 80 at the top of the water, and 245 ft. at the bottom. This is considerably larger than any ever yet made, but it is anticipated that steam-vessels of this gigantic size will be laid down when docks capable of containing them for fitting and repairing are ready for their reception. In 1839, two ranges of building, each 440 ft. in length by 26 in breadth, were built and fitted up with machinery for the manufacture of steam-engines and boilers. The number of workmen employed in the dock-yard in 1842 was 1,300, exclusive of about 700 convicts. The royal arsenal on the east side of the town, extends into the parish of Plumstead. It is the grand depot of artillery, &c., for the army and navy. It contains barracks and foundries, factories, &c., for the manufacture of various warlike stores; and, including a part of the Plumstead marshes, used as an artillery practising ground, it extends over about 300 acres. A practice-range of nearly 3 m. in extent has also been formed towards Erith. All the government ordnance are first proved here. The quantity of cannon, cannon-balls, bomb-shells, &c., is immense. There are four air-furnaces in the foundry, the largest of which will melt 325 cwt. of metal, a sufficient quantity to cast 14 guns, and the smallest 20 cwt. On the E side of the artillery barracks are military hospitals, and on the W side is a piece of water on which experiments are occasionally made with boats. Between the dock-yard and the arsenal is a rope-walk, an extensive building about 400 yds. in length, where, in time of war, several hundred workmen are engaged in making cables of all dimensions. On the SW side of the barrack-field is the Repository, containing an extensive collection of military models and trophies. An observatory for the officers of the royal artillery and engineers was lately in course of erection, in the barrack field between the repository and the mortar-battery on the right wing of the barracks. An institution for the better education of engineer and military officers was founded here in 1719. It is under the direction of the master-general and board-of-ordnance. The education is excellent; and the sons of officers are admitted upon a scale of payment corresponding to the father's rank, and varying from £80 to £20 per annum; while the sons of private gentlemen pay £125 per annum. The cadets receive an annual allowance of £45 12s. 6d.; and commissions according to their merit. It is understood that the establishment, which numbers from 150 to 170 pupils, now nearly maintains itself.

WOOLWICH, a township of Upper Canada, in the Wellington district, intersected by Grand river. Pop. in 1842, 1,009. It contains a village of the same name.

WOOLWICH, a township of Lincoln co., in the state of Maine, U. S., 26 m. S of Augusta, drained by small affluents of the Kennebec, by which it is bounded on the W. Pop. in 1840, 1,416; in 1850, 1,420.—Also a township of Gloucester co., in the state of New Jersey, 11 m. SW of Woodbury, bounded on the NW by Delaware river, and drained by Pepaups, Little Timber, Racoon, and Oldman's creeks. Pop. 3,676.

WOOLWORTH POINT, a promontory of Tasmania, of which it forms the NW extremity.

WOONSOCKET FALLS, a village of Smithfield township, Providence co., in the state of Rhode Island, U. S., on Blackstone river, 15 m. NNW of Providence, and intersected by the New York and Boston Direct railroad. Pop. in 1850, 3,000.

WOOPERTON, a township in the p. of Eglingham, Northumberland, 6½ m. SE by S of Wooler. Area 923 acres. Pop. in 1831, 107; in 1851, 86.

WOOPPOOSE, a small uninhabited island of Upper Canada, in Prince Edward's bay, Lake Ontario, between the forks of the township of Marysburgh.

WOORE, or WORE, a chapelry in the p. of Mucklestone, Salop, 6¾ m. NNE of Drayton-in-Hales. It includes the townships of Bearston, Dorrington, and Gravenhanger. Area 5,291 acres. Pop. in 1831, 890; in 1851, 1,860.

WOORNYALOON, a parish of Australia Felix, in the co. of Grant.

WOOSPEER, a village of the grand-duchy of Mecklenburg-Schwerin, in the prov. of Mecklenburg, and bail. of Domitz. Pop. 500.

WOOSTEN, a village of the grand-duchy of Mecklenburg-Schwerin, in the circle of Wendisch, and bail. of Goldberg. Pop. 200.

WOOSTER, a township of Wayne co., in the state of Ohio, U. S., 73 m. NE of Columbus, drained by Killbuck creek and its branches, and intersected by the Ohio and Pennsylvania railroad. Pop. 4,080.

WOOTTON, a parish in the co. of Bedford, 4½ m. SW of Bedford. Area 3,711 acres. Pop. in 1831, 1,051; in 1851, 1,204.—Also a parish in Berks, 4 m. NW by N of Abingdon. Acreage with the parish of Cumner. Pop. in 1831, 340; in 1851, 370.—Also a hamlet in the p. of St. Mary de Lode, Gloucestershire. Area 260 acres. Pop. in 1831, 804; in 1851, 1,174.—Also a parish in Kent, 9½ m. SE by S of Canterbury. It includes the hamlet of Giddings. Area 1,019 acres. Pop. in 1831, 128; in 1851, 153. —Also a parish in Lincolnshire, 5¾ m. SE of Barton-upon-Humber. Area 2,980 acres. Pop. in 1831, 459, in 1851, 606. Also a parish in Northamptonshire, 2½ m. S by E of Northampton. Area 1,420 acres. Pop. in 1831, 643; in 1851, 877.—Also a parish in Oxfordshire, 2½ m. N by W of Woodstock. Area 3,720 acres. Pop. in 1831, 1,060; in 1851, 1,250.—Also a parish in the isle of Wight, 4 m. NE of Newport, between the river Motherbank on the N, and an inlet of the sea on the E; the latter crossed by a causeway 900 feet in length, on the high road to Newport. Area 1,360 acres. Pop. in 1831, 55; in 1851, 58.—Also a township in the p. of Eccleshall, co. of Stafford. Area 680 acres. Pop. in 1831, 150; in 1851, 182.—Also a township in the p. of Ellastone, Staffordshire, 4½ m. W by S of Ashbourne. Pop. in 1831, 269; in 1851, 218.

WOOTTON-BASSET, a parish, disfranchised borough, and market-town, in Wilts, 36 m. N by W of Salisbury, in the line of the Great Western railway, and the Wilts and Berks canal, on the former of which there is here a principal station, distant 82½ m. SW of London. Area 4,778 acres. Pop. in 1801, 1,244; in 1831, 1,896; in 1851, 2,123. Two members were regularly returned to parliament for the burgh from 26° Henry VI., till disfranchised under the Reform act. The town consists of a single street, about a ½ m. in length, occupying an elevated site in the middle of a beautiful and fertile district. The inhabitants are chiefly engaged in agriculture and the ordinary handicrafts.

WOOTTON-COURTNEY. See WOTTON-COURTNEY.

WOOTTON-GLANVILLE, a parish in Dorset

7½ m. SSE of Sherborne. It contains the tything of Wootton-Newland. Area 1,665 acres. Pop. 328.

WOOTTON (St. Lawrence), a parish in the co. of Southampton, 3 m. W by N of Basingstoke. Area 3,957 acres. Pop. in 1831, 847; in 1851, 922.

WOOTTON (North), a parish in the co. of Dorset, 2 m. SE by S of Sherborne. Area 1,536 acres. Pop. in 1831, 78; in 1851, 75.—Also a parish in Norfolk, 2 m. W by S of Castle-Rising. Area 4,968 acres. Pop. in 1831, 179; in 1851, 188.—Also a parish in Somersetshire, 4 m. WSW of Shepton-Mallet. Area 1,536 acres. Pop. in 1851, 335.

WOOTTON-RIVERS, a parish in Wilts, 3 m. NE of Pewsey, in the line of the Kennet and Avon canal. Area 1,179 acres. Pop. in 1851, 427.

WOOTTON (South), a parish in Norfolk, 2½ m. SW by S of Castle-Rising. Area 1,874 acres. Pop. in 1831, 177; in 1851, 155.

WOOTTON-WAWEN, a parish in the co. of Warwick, 1½ m. S of Henley-in-Arden, intersected by the Birmingham and Stratford-on-Avon canal. It contains the chapelries of Henley-in-Arden and Ullenhall, and the hamlet of Aspley with Fordhall. Area 8,700 acres. Pop. in 1851, 2,306.

WOPERSNOW, a village of Prussia, in the reg. of Koslin, and circle of Schiefelbein. Pop. 237.

WOPLAN, a village of Austria, in Bohemia, in the regency of Pardubitz, and bail. of Schwarz-Kosteletz. Pop. 340.

WOPLANKEN, a village of Prussia, in the reg. of Konigsberg, and circle of Kastenburg. Pop. 281.

WOPORZAN, a market-town of Austria, in Bohemia, in the regency of Budweis, and bail. of Muhlhausen. Pop. 830. It has a paper-mill.

WOPPAM, a village of Austria, in Bohemia, in the regency of Bohmisch-Leipa, and bail. of Leitmeritz. Pop. 240.

WOPPEN, a village of Prussia, in the regency of Konigsberg, and circle of Braunsberg. Pop. 251.

WOPPENHOF, a village of Bavaria, in the circle of the Upper Pfalz, and presidial of Bohenstrauss. Pop. 250. It has a castle.

WOPPENRIETH, a village of Bavaria, in the circle of the Upper Pfalz, and presidial of the Bohenstrauss. Pop. 90.

WOPPENRODT, a village of Prussia, in the reg. of Treves, and circle of Bernkastel. Pop. 200.

WORADA, a state of Senegambia, to the S of Konkadu, to the W of the Bafing or Black river, and intersected by an affluent of that river.

WORASITZ, a village of Austria, in Bohemia, in the reg. of Prague, and bail. of Laun. Pop. 260.

WORATSCHEN, a village of Austria, in Bohemia, in the reg. of Eger, bail. of Saatz. Pop. 420.

WORATTA, a country which is reputed to lie to the S of Abyssinia, and E of Kaffa, and to be inhabited both by Christians and Pagans of various shades of colour.

WORB, a village of Switzerland, in the cant. of Bern, and bail. of Konolfingen, surmounted by an ancient castle. Pop. 3,185.

WORBEN, a village of Switzerland, in the cant. of Bern, and bail. of Nidau. Pop. 454.

WORBIS, a circle and town of Prussia, in the prov. of Saxony, and regency of Erfurt. The circle comprises an area of 8·26 German sq. m. Pop. 42,589. The town is on the Wipper, near the source of that river, and 45 m. NW of Erfurt. Pop. 2,078. It has manufactories of woollen and cotton fabrics.

WORBIS (Breiten), a village of Prussia, in the prov. of Saxony, and regency of Erfurt, circle and 5 m. ESE of Worbis, on the Wipper. Pop. 1,300. It has manufactories of linen fabrics, and spinning mills.

WORBLINGEN, a village of the grand-duchy of Baden, in the circle of the See, and bail. of Radolphzell. Pop. 588.

WORBSCHEIDT, a village of Prussia, in the reg. of Arnsberg, and circle of Olpe. Pop. 260.

WORCESTER, a city and county of itself, the seat of a see, and the capital of the co. of Worcester, locally situated in the division, union, and county of Worcester, on the river Severn, which is here crossed by a handsome bridge of 5 arches, and distant 3½ m. W of the Birmingham and Gloucester railway at Spetchley, from which a branch line runs to the city. In the parliamentary returns for 1831, Worcester is held to comprise the parishes of St. Albans, All Saints, St. Andrew, St. Helen, St. Nicholas, and St. Swithin, parts of the parishes of Claines, St. Clement, St. Martin, and St. Peter, and the extra-parochial parts of the Blockhouse, the jail and bridewell, and the precincts of the college. Area 220 acres. Pop. in 1801, 11,300; in 1831, 18,610; in 1851, 27,528. This city, which is one of the most ancient in the kingdom, stands on a gentle slope on the E bank of the Severn. It is sheltered from the E by a finely wooded hill: on all the other sides it is open, being in what may be called the great plain of the Severn. It was formerly surrounded by a strong wall, of which some slight vestiges remain. The streets in general are wide, well-paved, and lighted, clean, and neat; and the principal streets are regular and handsome. The principal public edifices, besides the cathedral and the parish churches, nine of which stand within the walls, and two without, are the guild-hall, a handsome modern structure; new county-courts recently erected; the new county-jail, a building on Howard's plan; the city-jail erected in 1824; the house-of-industry, the infirmary, a theatre, and various hospitals and schools. The principal manufactures are those of gloves and porcelain. There is a very extensive demand for W. gloves, both for home-consumption and for exportation, and the largest number of the working classes are employed in this branch of manufacture, which is recommended to them by the facility which it allows to those engaged in it to work in their own houses, and at their own convenience. There are several large porcelain establishments. The beauty of W. china is well known in most parts of the world, it being a considerable article of trade. The trade of the city is very considerable, not only in the articles of its more immediate produce, but also owing to the great advantages of its central situation for inland navigation, independently of its recent valuable railway communications. The assizes and general quarter-sessions for the shire are held in W. as the county-town. The city has regularly returned 2 members to parliament since 23° Edward I. The number of electors registered, in 1837, was 3,196; in 1852, 2,651. W. is a polling-place, and the principal place of election for the members representing the western division of the shire.

Diocese and See.] The diocese of W., which is said to have been founded at the request of Oshere, a prince of the Wiccii, by Ethelred, king of Mercia, about the year 680; was formerly of much greater extent than latterly, even though re-enlarged by recent orders in council. The whole of the co. of Gloucester anciently belonged to it. Latterly it contained all Worcestershire (excepting 15 parishes and 8 chapelries belonging to Hereford), besides about one-third of Warwickshire, and the parishes of Brome and Clent in Staffordshire, and of Halesowen in Shropshire. By order in council of date 22d December, 1836, the archd. of Coventry was transferred to this dio. from that of Lichfield and Coventry; and part of the deanery of Burford. The total amount of the average gross yearly income of the see of W. for three years, ending 31st December, 1831, was £6,916: nett yearly income £6,569. The average nett income of the dean and chapter or corporation of the cathedral, as a corporation aggregate, during the three years ending 1841, was £8,479; the corporation consisting of a dean and 10 prebendaries or canons.—The Cathedral is a noble specimen of Gothic simplicity: the exterior is extremely plain, and devoid of all laboured ornaments. Its beauty consists in its height, space, and the lightness of its architecture, which is greatly aided by the lofty pinnacles rising from every termination of the building. This edifice was first erected by Ethelred, king of Mercia, in 680, when it was a convent of secular priests. The greatest part of the present buildings, consisting of the hall, refectory, cloister, watergate, &c., were all built between 1320 and 1386, by Bishop Wakefield. The form is the usual one of a double cross, displaying the grand features of the Gothic style, which consist in extent and strength, and to which we may add the solemnity of the high pointed arch, and the beauty of diminutive ornament. The proportions of the exterior are on a grand scale: it is in length 514 ft.; in breadth 78 ft.; and in height 68 ft.; and the tower, which rises from the centre of the cross aisle to the altitude of 200 ft., is ornamented at the corners by four lofty pinnacles, and with elegant battlements of light open work. This tower has been reckoned heavy, particularly in the upper tier of ornamental windows. Nothing can be imagined more august, and yet more simple, than the choir, in which the best effect is produced by its clustered pillars, the exquisite open-worked mouldings of its pointed arches, and its general arrangement. The cloisters are 125 ft. by 120 ft., and 16 ft. in width. The bishop's palace stands near the cathedral, in a commanding situation, on the banks of the Severn. The charities of W. are very numerous, and in 1830 yielded in all an annual revenue of about £4,500.

History.] The city of W. boasts of an antiquity as high as any in the island. In the time of the ancient Britons, according to the list of their cities given by Nennius, it was a place of considerable consequence, and was called Caer Guorangon. Perhaps it owed its consequence to its situation, being protected, on the one hand, by an impassable forest, and, on the other, by the river Severn, besides having the advantage of an adjacent ford over the river. It was taken possession of by the Romans at an early period of their dominion in Britain, and, although some doubts exist as to whether it was a principal station of that people, yet it seems certain that a fortress was erected here to protect the passage of the Severn, and secure the conquests in the neighbourhood. After the general union of the small Saxon kingdoms, the bishop of W. obtained certain royal grants, which enabled him to fortify the cathedral and close. In the 12th and 13th centuries, the city suffered severely and repeatedly from fires, partly accidental, and partly by military and other incendiaries. During this period, and the two succeeding centuries, it was also the scene of several recorded historical events of some importance, which, however, our limits do not permit us more particularly to notice. During the civil wars in the reign of Charles I., this city embraced the royal cause, and was, in fact, the first city which openly did so. It afterwards fell into the hands of the parliamentary forces, and a fine of £5,000 imposed on the inhabitants, besides a forced loan of £3,000. These measures did not subdue the spirit of the citizens, who, after the removal of the army, continued to exert themselves in the cause of the king, and in consequence, it was again besieged in 1646, when, after an obstinate resistance of four months, the garrison capitulated on honourable terms. The city still remained favourable to the royal cause, and on the 22d August, 1651, notwithstanding the opposition of the garrison, the gates were opened to a Scottish army under Charles II. From various untoward circumstances, the king, on arriving at W., found that his forces, fatigued by a long and harassing march, were not more numerous than when he rose from his camp at Torwood. On the other hand, he found all the militia and regular forces of the country everywhere opposed to him, and in consequence his situation was one of great difficulty. With an army of about 30,000 men Cromwell fell upon the city, and attacking it on all sides and meeting with little opposition, except from the Duke of Hamilton and General Middleton, broke in upon the disordered royalists, and the whole Scottish army were either killed or taken prisoners. The country people, inflamed with national antipathy, put to death the few that escaped from the field of battle. The king left W. in the afternoon, and travelled to Boscobel in Staffordshire. After his escape, the citizens made their last stand in the town-hall, but were finally overwhelmed by superior numbers.

WORCESTER, a county and division of Cape Colony in S. Africa. The division is bounded on the N by the division of Clanwilliam; on the E by Beaufort; on the S by George and Zwillendam; and on the W by Stellenbosch and Cape. It has an area of about 20,000 sq. m., and is watered by the Breede, the Great Berg, and the Doom. The town is situated on the Breede, 60 m. ENE of Cape Town. Between Cape Town and the rich corn-growing regions of W. there interposes, at a distance of about 40 m., the Bokkeveld range of mountains stretching across the country, and nearly cutting off the capital from all communication with the interior in that direction. These mountains are only passable at a few points, where it was formerly necessary to unload a waggon at one side, carry over the packages by hand, or on the backs of horses, and reload them into another at the other side; in short, they form as complete a barrier to intercourse by wheel carriages as the isthmus of Darien to ships. At an expense of £50,000, with the labour of our colonial convicts, this mighty barrier has been cut through from one side to the other,—upwards of 18 m., and a magnificent road constructed, by which the wealth of the interior may pour down unobstructed to the shores of Table Bay.

WORCESTER, a central county, township, and city of the state of Massachusetts, U. S. The township, comprising an area of 1,450 sq. m., is generally undulating, and possessing considerable fertility, drained by Blackstone, Guinnebaug, Ware, Millar's, Nashua, and Mill rivers, and intersected by the Boston and Worcester, the Providence and Worcester, the Worcester and Fitchburg, the Western, the Nashua and Worcester, the Fitchburg, the Vermont and Massachusetts, the Norwich and Worcester, the Cheshire, &c. railroads. Pop. in 1840, 95,313; in 1850, 130,789. Its capital, which bears the same name, is 45 m. by railroad W by S of Boston, and in N lat. 42° 16′ 17″, and W long. 71° 18′ 18″. It is pleasantly situated in a valley, surrounded by hills, and drained by branches of Blackstone river. It contains some handsome streets and public buildings, and is one of the most important thoroughfares in New England, being situated in the centre of five extensive lines of intercommunication, viz. the Boston and Worcester, the Western, the Providence and Worcester, the Norwich and Worcester, the Worcester and Nashua, and the Fitchburg and Worcester.—Also a county in the SE part of the state of Maryland, comprising an area of 616 sq. m. intersected from N to S by Pocomoke, and drained by its branches. Pop. in 1840, 18,377; in 1850, 18,859. Its capital is Snowhill.—Also a township of Otswego co., in the state of New York, 57 m. SW of Albany. It has a hilly surface, and is drained by Charlotte river and its branches. Pop. in 1840, 2,390; in 1850, 2,047. It has a village of the same name, with about 200 inhabitants.—Also a township of Montgomery co., in the state of Pennsylvania, 89 m. E of Harrisburg Pop. in 1840, 1,200; in 1850, 1,400.—Also a township of Washington co., in the state of Vermont, 7 m. N of Montpelier, drained by the N branch of Onion river. Pop. in 1840, 587; in 1850, 702.

WORCESTERSHIRE, a central county of England, bounded on the N by Shropshire and Staffordshire; on the E by Warwickshire; on the S by Gloucestershire; and on the W by Herefordshire and Shropshire. In form it is a very irregular quadrangle, with detached portions on all sides surrounded by other counties, while portions of other counties are insulated within it. Its greatest length is 36 m.; breadth 26 m.; circumf. 130 m.; area 723 sq. m., or 462,720 acres. It was divided into hundreds very irregularly shaped and singularly intermixed, until by an act passed in 1831, new divisions were formed. The county is now accordingly divided into 10 portions, called divisions, politically separated into an eastern and a western parliamentary district or division; the eastern comprising the Droitwich, Northfield, Pershore, Stourbridge, Dudley, and Blockley, divisions, and the western, those of Worcester, Kidderminster, Hundred House, and Upton. There are 197 parishes in the county, comprising 1 city and county town, Worcester; 6 parliamentary boroughs, Worcester, Evesham, Droitwich, Kidderminster, Bewdley, and Dudley; and these and the following six other market-towns, namely, Stourbridge, Pershore, Shipston, Tetbury, Broomsgrove, and Upton. The number of inhabited houses in the county, in 1831, was 41,646; in 1841, 46,962; in 1851, 51,943. The pop. of this county, in 1801, was 439,333; in 1831, 211,400; in 1841, 233,484; in 1851, 276,926. The pop. of 1831 consisted of 45,512 families, 14,654 of whom were chiefly employed in agriculture, and 19,030 in trade, manufactures, and handicraft.

Climate and soil.] The climate of this co. is mild and salubrious, and in the vales vegetation is very early. On the Malvern-hills the air is extremely mild as well as pure. The same character prevails on some of the elevated parts in the NW; but on the Lickey, the Bredon, and the Broadway-hills, the climate is rather colder. The general appearance of the county is rich and beautiful, being finely diversified with hill and valley, and richly wooded. There is scarcely any tract so sterile as to be destitute of verdure, and in most parts the cultivation is good. Almost the entire central and southern parts of the district are occupied by the vales of Worcester and Evesham. The vale of Worcester is bounded on the SW by the Malvern-hills, which form part of the boundary between this county and Herefordshire: see article MALVERN-HILLS. Connected with these are the Abberley-hills which occupy the NW part of the county, and bound on the W the vale which stretches on the N into Shropshire and Staffordshire. On the E side of the vale are the Lickey or Hagley-hills, which, commencing at Hagley, continue in nearly a SE direction to Headless Cross, whence they stretch nearly S along the boundary of the co., and for a short distance enter it again, at length sloping off near Evesham. Connected with the Coteswold-hills of Gloucestershire, are the Bredon-hills, in the SE district of the county, a mile or two S of Pershore. These hills separate the vale of Gloucester from that of Evesham. A ridge of high ground beginning a little to the E of Worcester, and stretching nearly S between the Severn and the Avon, separates the vales of Worcester and Evesham, the former of which is, in fact, part of the vale of the Severn; and the second of its tributary the Avon. The soil of this co. is various. In the northern parts it consists of rich loamy sand, united with a small proportion of gravel. Towards the east, light sand and peat earth are sometimes found; but the prevailing soil there is a strong clay. In the vale the upper soil is particularly deep: it is a dark-coloured earth, resting on a substratum of clay in some parts, and of gravel in others. On some parts of the banks of the Severn and its tributary streams, there is a rich alluvial soil, extremely favourable to vegetation. Wheat is very extensively cultivated, and yields a good return, as well as barley and beans. Hops in some districts form a principal crop, and orchards of apple and pear trees, for cider and perry, are numerous. Perry is in some parts so plentiful as to form the common drink of the peasantry. Cattle and sheep are fattened on the rich pasturage of this county. The new red sandstone formation occupies the principal part of the co. Rocks of an older epoch, as usual, form the higher ranges. The Malvern-hills are composed of granite, syenite, and syenitic greenstone. The Lickey-hills are a mass of quartz, a rock which is also found in the Malvern-hills. Limestone occurs in the Abberley-hills, on the NW side of the co. At the SE corner in Cleeve Prior parish, there are quarries of very good stone; and still farther S, in the Broadway-hills, stone of a reddish colour is quarried. Freestone and calcareous flagstone are quarried in various quarters. The NW corner of the co. is for a small space occupied by part of the Bewdley coal-basin, and coal-pits are sunk there at Mamble and at Bayton. The detached division of Dudley forms part of the Dudley coal-basin, in which the beds of coal alternate with strata of ironstone; and coal is raised, and iron made in large quantities. The Dudley limestone quarries are very extensive and curious excavations. Brick earth abounds in this co. At Droitwich there are rich and long celebrated brine-springs, supposed to have been worked even by the Romans. There are singular springs and beds of rock-salt at Stoke Prior, near Bromsgrove.

Rivers.] The Severn, the Avon, the Stour, and the Teme, are the principal rivers in Worcestershire. The Severn enters the co. from Salop, near Bewdley, and taking nearly a southward course, runs through the length of the co., passing by Stourport, Worcester, and Upton, and finally quitting it for Gloucestershire, a little before it is joined by the Upper Avon. The Severn is navigable for vessels of 80 tons as far as Worcester, and for those of 60 tons as far as Bewdley-Bridge. The whole navigation of this river from its mouth, without the aid of locks, is 160 m. The mass of the co., as already seen, is within the basin of the Severn, of which it occupies one of the finest districts. The Upper Avon enters from Warwickshire on the E, and thence takes a SW course to Evesham, round which town it flows. Running then in a general western direction, with many bendings, by Pershore, and after a very winding course nearly S, it passes into Gloucestershire. The Stour enters the co. from Staffordshire, passes through Kidderminster, and joins the Severn at Stourport. The Teme, which has its source in Shropshire, enters this co. near Tenbury, on the NW, whence it takes a very irregular course to the SE, and joins the Severn about $1\frac{1}{2}$ m. below Worcester.

Roads and Railways.] This co. is intersected by several canals, and water communication is thus kept up with all parts of England. The Staffordshire and Worcestershire, or Stourport canal, which opens into the Severn at Stourport, meets the old Birmingham canal, a little to the N of Wolverhampton in Staffordshire, and thence communicates with the Grand Trunk canal. In its course of 46 m., it is carried by aqueducts over the Trent, Sow, Stour, and two or three other smaller streams, and by a tunnel under the town of Kidderminster. There are likewise two other tunnels, two large reservoirs, and a number of bridges and small aqueducts on its course. The Dudley canal proceeds from the Worcester and Birmingham canal, about 4 m. S of Birmingham, and joins the Stourport canal a little west of Stourbridge. Its whole course is 13 m., and in this short distance it passes through three tunnels of more than 4 m. in aggregate length. About 1 m. N of Stourbridge a branch runs to that town. The Leominster and Kington canal runs through a small portion of the NW part of this co. where there are coal-mines. The Droitwich canal runs from the salt-works at Droitwich to the Severn, a distance of 5 m., in which it descends $56\frac{1}{2}$ ft. by 8 locks. The Worcester and Birmingham canal connects the two towns so named, and joins the Severn at the former. About 15 m. from this junction, at Tardebig, 3 m. ESE of Bromsgrove, it runs through a tunnel 500 yds. in length, and in the remainder of its course there are three other shorter tunnels. All of these tunnels are 18 ft. high, and 18½ ft. wide, with 7 ft. 2 in. depth of water. The Birmingham and Stratford canal communicates with this one at King's Norton.—The principal roads are those from London to Welshpool, Aberystwith, Worcester, Shrewsbury and Holywell, Kidderminster and Birmingham.—The Birmingham and Gloucester railway intersects this county from its NE to its SE extremity. From its junction with the London and Birmingham line at Birmingham, it proceeds across the Birmingham and Warwick canal, and by Camphill, Bromsgrove, Stoke Prior, Droitwich, Spetchley near Worcester, Defford, Eckington, and Bredon, (at all of which places there are stations), into Gloucestershire, east of Tewkesbury. This line is 45 m. in length, with branches 8 m. The London and Birmingham railway crosses a small portion of the co. at its NE corner. A line runs from Worcester by Pershore, Evesham, and Banbury, to the London and Birmingham line.

Trade, Manufactures, &c.] Besides the general county produce of corn, cattle, fine wool, hops, fruit, cider, and perry, the trade and manufactures of W. consist chiefly of carpets, &c., at Kidderminster; glass and iron at Dudley and Stourbridge; gloves, porcelain, iron, &c., at Worcester; bar, sheet, and other iron, to a great extent, in the villages near the river Stour; salt at Droitwich and Stoke Prior; and needles at Redditch.

Franchise and Government.] Previous to the passing of the reform act, two members were returned to parliament for the co., two each for Worcester, Evesham, and Droitwich, and one for Bewdley. Under that act Droitwich was deprived of one member, and Kidderminster and Dudley were each authorized to return one; and the county four, two for the eastern division, who are polled for at Droitwich, Stourbridge, Pershore, and Shipston, the principal place of election being at Droitwich; and two for the western, who are polled for at Worcester, Upton, Tenbury, Bewdley, and Stourport, the principal place of election being at Worcester. The number of electors registered for the co. in 1837 was, for the eastern division, 5,995; for the western, 4,654. In 1852 the electors were 6,354 for the eastern division, and 4,102 for the western. This county is included in the Oxford circuit. The assizes and quarter-sessions are held at Worcester.—In ecclesiastical jurisdiction, this co. is in the archdeaconry and dio. of Worcester, and prov. of Canterbury. It includes the deaneries of Kidderminster, Droitwich, Wighorn, Powick, Pershore, and Evesham.

History and Antiquities.] The principal military transactions connected with the history of this co. occurred, the first in the vale of Evesham, where fell Simon de Montfort, with many of his party, under the arms of Prince Edward, son of Henry III., in the barons' wars against that king; and the second at Worcester, in 1651, where Cromwell obtained what he called his crowning victory over the Scotch army, which had marched into England for the purpose of placing Charles II. on the throne. In the time of the Britons, Worcestershire was inhabited by the *Cornavii;* the Romans comprised it in the province of *Flavia Cœsariensis;* and during the heptarchy it belonged to Mercia. There are traces here of the Iknield-street and the Ridge-way. Upton and Worcester are supposed to have been Roman stations.

WORCHEN. See Borgo-di-Val-Lugano.

WORDEL, a village of Prussia, in the regency of Marienwerder, and circle of Deutsch-Krone. Pop. 200.—Also a village of the same regency, in the circle of Flatow. Pop. 260.

WORDINGBORG, a town of Denmark, on the S coast of the island of Seeland. Pop. 1,500. It has manufactories of woollen fabrics.

WORDWELL, a parish in Suffolk, 6 m. W by N of Bury-St.-Edmunds. Area 2,299 acres. Pop. 56.

WORECH, a village of Austria, in Bohemia, in the regency of Prague, bail. of Simchow. Pop. 310.

WOREL, a village of Austria, in Bohemia, in the reg. of Parbubitz, and bail. of Chrudim. Pop. 590.

WORFELDEN, a village of Hesse, in the prov. of Starkenburg, presidial of Grossgerau. Pop. 460.

WORFIELD, a parish in Salop, 3½ m. NE by E of Bridgnorth. Area 10,320 acres. Pop. 1,735.

WORININGEN (Gross and Klein), two villages of Prussia, in the regency of Gumbinnen, and circle of Pilkallen, containing respectively 230 and 320 inhabitants.

WORITTEN, a village of Prussia, in the regency of Konigsberg, and circle of Allenstein. Pop. 300.

WORKHEIM, a village of Prussia, in the reg. of Konigsberg. and circle of Heilsberg. Pop. 230.

WORKINGTON, a parish, market-town, and sea-port, in Cumberland, 34 m. SW of Carlisle, at the mouth of the river Derwent, near the sea. The parish, besides the town of W., contains the chapelry of Great Clifton, and the townships of Little Clifton, Stainburn, and Winscales. Area 8,310 acres. Pop. in 1831, 7,196; in 1851, 7,159. The town is situated on the S bank of the Derwent, which is here crossed by a handsome bridge, and from which it receives a good supply of water. The older part consists of narrow and irregular streets, but a few good streets have been formed of late years. In the vicinity are extensive coal mines, and the chief portion of the trade consists in the exportation of their produce to Ireland and other places. Shipbuilding and manufactures of ropes, cables, and other articles in connection with shipping, are carried on.

WORKSALL, or Worsall (High), a township and chapelry in the p. of Northallerton, Yorkshire, 4 m. SW of Yarm. Area 1,511 acres. Pop. 130.

WORKSALL, or Worsall (Low), a township in the p. of Kirkleavington, Yorkshire, 3 m. SW of Yarm. Area 1,312 acres. Pop. in 1851, 171.

WORKSOP, a parish and market-town in the co. of Nottingham, 10 m. E of the Eckington station, on the North Midland railway, and intersected by the Chesterfield canal. It contains, besides the town of W., the chapelries of Scofton and Shireoaks, and the hamlets of Gateford, Kilton, Osberton, Radford, Ratcliffe, Ryton, and Sloswicks. Area 18,220 acres. Pop. in 1831, 5,566; in 1851, 7,215. The church, which formed the western part of the church of an ancient priory, is one of the most perfect specimens of the Anglo-Norman style in England. It has two beautiful towers, one of which was formerly the great central tower of the pile. The situation of the town is in a valley at the N extremity of Sherwood forest, near the source of the river Ryton. The principal street is of considerable length, and is crossed by another of smaller dimensions, both consisting of good houses. The principal trade is in malt, which is chiefly sent to Manchester by the Chesterfield canal. The vicinity is covered with wood, and is, in general, picturesque: there are several handsome seats belonging to the nobility, on account of which the vicinity of W. is well known under the cognomen of 'the Dukery.' Among these splendid seats are Clumber, the residence of the Duke of Newcastle; Welbeck abbey, that of the Duke of Portland; and Thoresby, that of Earl Manvers.

WORKUM, a town of Holland, in the prov. of Friesland, arrond. and 11 m. WSW of Sneek, and about ⅔ of a m. from the Zuider-Zee. Pop. 3,000. It has a harbour and considerable fisheries.

WORKUM, or Wondrichem, a fortified town of Holland, in the prov. of North Brabant, arrond. and 18 m. NW of Bois-le-Duc, on the l. bank of the Maas, opposite the confluence of the Whaal. Pop. 1,000.

WORLABY, or WORLETBY, a parish in Lincolnshire, 5½ m. N by E of Glandford-bridge. Area 3,210 acres. Pop. in 1831, 309; in 1851, 500.

WORLDHAM (EAST), a parish in the co. of Southampton, 2½ m. E by S of Alton. Area 1,587 acres. Pop. in 1831, 212; in 1851, 258.

WORLDHAM (WEST), a parish in the co. of Southampton, 2½ m. SE by E of Alton. Area 447 acres. Pop. in 1831, 96; in 1851, 98.

WORLE, a parish in the county of Somerset, 8 m. NW of Axbridge, in the line of the Bristol and Exeter railway. Area 1,810 acres. Pop. 960.

WORLESTON, a township in the p. of Acton, co.-palatine of Chester, 1⅝ m. N of Nantwich, in the line of the Chester and Crewe railway. Area 1,122 acres. Pop. in 1831, 367; in 1851, 337.

WORLINGHAM, a parish in Suffolk, 2 m. ESE of Beccles, on the navigable river Waveney. Area 1,631 acres. Pop. in 1831, 202; in 1851, 174.

WORLINGTON, a parish in Suffolk, 1 m. WSW of Milden-hall, on the navigable river Lark, over which there is here a ferry. Area 2,080 acres. Pop. in 1831, 368; in 1851, 394.

WORLINGTON (EAST), a parish in Devonshire, 6 m. E of Chumleigh. Area 2,363 acres. Pop. in 1831, 292; in 1851, 277.

WORLINGTON (WEST), a parish in Devonshire, 5½ m. E of Chumleigh. Area 2,683 acres. Pop. in 1831, 187; in 1851, 229.

WORLINGWORTH, a parish in Suffolk, 5 m. NW of Framlingham. Area 2,446 acres. Pop. 811.

WORLITSCHKA, a village of Austria, in Bohemia, in the regency of Pardubitz, and bail. of Landskron, at the confluence of the Adler and Buchberges. Pop. 1,300.

WORLITZ, a town of the duchy of Anhalt-Dessau, in the bail. of Orienbaum, 1½ m. from the l. bank of the Elbe, and on the S side of a lake of the same name. Pop. 1,895. It is well-built, and has several breweries, and in the vicinity is a ducal castle and park, reputed the finest in Germany.

WORLOW, a village of Austria, in Bohemia, in the reg. of Pardubitz, and bail. of Lietomisahl. Pop. 200.

WORM, a village of Prussia, in the regency and circle of Aachen. Pop. 200.

WORMBERG, a summit of the Harz mountains, on the confines of the duchy of Brunswick, which has an alt. of 2,880 ft. above sea-level.

WORMBRIDGE, a parish in Herefordshire, 9 m. SW of Hereford, in the line of the Abergavenny and Hereford railroad. Area 720 acres. Pop. 91.

WORMDILT, or ORNETA, a town of Prussia, in the regency and 47 m. SSW of Konigsberg, and circle of Braunsberg, on the Dreweion. Pop. 3,630. It has a castle, two suburbs, and two hospitals, and possesses manufactories of cloth and gloves, and tanneries. This town was founded in 1316.

WORMELDANGE, or WORMELDINGEN, a village of Holland, in the prov. and 10 m. E of Luxemburg, and cant. of Grevenmachern, on the l. bank of the Moselle. Pop. 1,510.

WORMELN, a village of Prussia, in the reg. of Minden, and circle of Warburg. Pop. 435.

WORMENHALL, or WORMINGHALL, a parish in Buckinghamshire, 12 m. WSW of Aylesbury. It includes the hamlet of Thumley in the county of Oxford. Area 1,469 acres. Pop in 1851, 340.

WORMER, a village of Holland, in the prov. of North Holland, and bail. of Hoorn, 9 m. N of Amsterdam. Pop. 1,210.

WORMERSDORF, a village of Prussia, in the reg. of Cologne, and circle of Rheinbach. Pop. 268.

WORMGAY, or WERMEGAY, a parish in Norfolk, 7½ m. NNE of Downham-Market. Area 2,788 acres. Pop. in 1831, 323; in 1851, 440.

WORMHILL, a chapelry in the p. of Tideswell, co. of Derby, 2¾ m. WSW of Tideswell, on the river Wye, amidst picturesque scenery. Pop. 369.

WORMHOUDT, a canton, commune, and market-town of France, in the dep. of the Nord, and arrond. of Dunkerque. The cant. comprises 10 com. Pop. in 1831, 15,342; in 1846, 15,730. The town is 16 m. SSE of Dunkerque, on the Penne. Pop. in 1846, 3,991. It has several breweries, a wax-work, an oil-mill, tanneries, and brick-kilns.

WORMINGFORD, a parish in Essex, 3⅝ m. WSW of Nayland, on the navigable river Stour. Area 2,331 acres. Pop. in 1851, 535.

WORMINGTON (LITTLE), a parish in the co. of Gloucester, 5 m. N by E of Winchcombe. Area 560 acres. Pop. in 1831, 97; in 1851, 62.

WORMLAGE, a village of Prussia, in the reg. of Frankfort, and circle of Kalau. Pop. 322.

WORM LAKE, or BAHR-EL-DUD, a small lake of Fezzan, in Central Africa, in N lat. 26° 42′ 0″, E long. 0° 54′ 57″, according to Dr. Vogel. It is one of the Trona lakes, as they are called, a small group scattered around a labyrinth of sand hills, valleys, and precipices, along the N side of the Wadi-Shergi and the Wadi-Gharbi. Its waters, which are of a dark colour, of from 18 to 24 ft. deep, abound in a species of worm called *dud*.

WORMLEIGHTON, a parish in Warwickshire, 8 m. E by N of Kingston, crossed by the Oxford canal. Area 2,320 acres. Pop. in 1851, 194.

WORMLEY, a parish in Hertfordshire, 2½ m. N by E of Cheshunt, crossed by the New river, bounded on the E by the river Lea, and intersected by the Northern and Eastern railway. Area 940 acres. Pop. in 1831, 471; in 1851, 511.

WORMS, a city of Hesse-Darmstadt, in the prov. of Rheinhessen, on the l. bank of the Rhine, a few hundred yards from the river, and 26 m. SSE of Mentz. The streets are dark and narrow, and the old massive walls, by which it is surrounded, are in a dilapidated state. The cathedral is a ponderous Gothic building, with a magnificent portal. The other public buildings are the mint, the town-house, in which Luther appeared before the diet in 1521, Trinity church, and two synagogues. The population is said to have been formerly considerable; but the city having been laid waste by the French in the general devastation of the palatinate in 1689, part of the inhabitants retired to Frankfort-on-the-Maine, or to Holland; and many of the houses were never rebuilt, the ground belonging to them being laid out in gardens. The present pop. is about 9,500, chiefly Lutherans. The manufactures are leather, tobacco, and lacquered ware. The environs are fertile. The title of the town to historic notice, arises from its having been frequently the seat of the diet of the empire. Few places have suffered more from war and other calamities. So early as 407 W. was ravaged by the Vandals; in 451, and 538, by the Huns. It has suffered also by fires; and on one or two occasions, by earthquakes. The former bishopric of this name comprised a territory in the W of Germany, of the extent of 170 sq. m., with 18,000 inhabitants. It was divided into two parts by the Rhine; and, at the peace of Luneville in 1801, the portion to the W of that river was ceded to France, while that to the E was given to Hesse-Darmstadt. At present, the whole belongs to the latter state.

WORMSDORF, a village of Prussia, in the regency of Magdeburg, and circle of Neuhadensleben. Pop. 545.

WORMSER-JOCH. See STILFSER-JOCH.

WORMSFELD, a village of Prussia, in the reg. of Frankfurt, and circle of Landsberg. Pop. 362.

WORMSHILL, or WORMSELL, a parish in Kent, 5 m. SSW of Sittingbourne. Area 1,467 acres. Pop. in 1831, 186; in 1851, 209.

WORMSLEY, a parish in Herefordshire, 3½ m. SE by S of Weobly. Area 1,293 acres. Pop. in 1831, 102; in 1851, 125.

WORMSTEDT, a village of Saxe-Weimar, in the prov. of Weimar, and bail. of Dornburg. Pop. 400.

WORMZE, a village of North West Africa, in Sudan, in the Fellatah state, to the SE of Saccatu.

WORNITZ. See WERNITZ.

WORNSDORF, a village of Baden, in the circle of the See, and bail. of Stockach. Pop. 389.

WORONETSCH. See VORONEZ.

WORONETZ, a village of Galicia, in the circle and 54 m. from Tschernowitz, near the r. bank of the Moldawa. It is noted for its manufacture of pottery.

WORPLACK, a village of Prussia, in the reg. of Konigsberg, and circle of Rossel. Pop. 200.

WORPLESDON, a parish in Surrey, 3½ m. NNW of Guildford. It contains the tythings of Bingham, Perry-hill, Westend, and Wyke. Area 7,140 acres. Pop. in 1831, 1,360; in 1851, 1,549.

WORPSWEDE, a village of Hanover, in the gov. of Stade, and principality of Bremen, near Osterholz. Pop. 461.

WORRINGEN, a town of Prussia, in the prov. of the Rhine, regency and 9 m. NNW of Cologne, on the l. bank of the Rhine. Pop. 1,415. It has several breweries, and distilleries of brandy.

WORSBOROUGH, a chapelry in the p. of Darfield, Yorkshire, 2½ m. S by E of Barnesley, and connected with the Barnesley and Dearne and Dove canals, by a branch in Worsborough-dale. Area 3,594 acres. Pop. in 1831, 2,677; in 1851, 4,250.

WORSLEY, a township in the p. of Eccles, co.-palatine of Lancaster, 5½ m. SSE of Great Bolton, intersected by the Wigan and Leigh canal, and near the Liverpool and Manchester railway. Area 6,240 acres. Pop. in 1801, 5,061; in 1831, 7,839; in 1851, 10,189. Here are extensive coal-mines, and subterranean canals formed by the late duke of Bridgewater. The Bridgewater canal was one of the earliest of our inland navigations, and was formed soon after the construction of the Sankey canal. The first portion of this great work, the Manchester and Worsley branch, was commenced in 1758, and completed in 1762; but the plan of a canal was first conceived early in the 18th century by Sir Wm. Egerton. In 1737 an act of parliament was obtained for rendering Worsley brook navigable, but was never carried into execution. The act which was destined to be worked out was passed in 1758, and the scheme included about 30 m. of underground tunnelling, connecting together the extensive collieries of Worsley. The other portions of the canal—those to Runcorn and Preston Brook—were subsequently grafted upon the Manchester and Worsley. So highly successful was the undertaking, that at the death of the duke, in 1803, it is said to have been realising a profit of £50,000 to £80,000 a-year. The mouths of the colliery tunnels at W. form a singular feature in the landscape. They are low, dark, cavernous openings, only constructed for very small boats, which, when loaded, sink almost to the water's edge. The whole length of the tunnels is about 14 m.—the canals being upon two levels. "The first level strikes into the solid rock at W., and follows the higher seam of coal: in other words, the miners, in getting the coal from this field, take out a portion in the form of a canal, so that wherever they go the water follows them, and they are enabled to load directly into the boats. The level of the water in this subterranean canal is a little higher than the level of the canal outside, and the water is held up into it during the day by sluices at the mouth of the tunnel. When the miners have completed their toil in the evening, the boats they have filled during the day, holding about 10 tons each, and chained together, are carried out with the current created by raising the sluice at the mouth of the tunnel, so as to let the surface of the subterranean canal find its level. The floating of the coal out of the mines is thus performed at little or no cost beyond the first outlay. The sluices of the subterranean canals are closed again every morning, and the water constantly filtering through the earth and accumulating in the mines is sufficient to raise the level of the water so as again to float out the coal boats. These underground canals thus drain the workings to some extent, whilst they supply cheap water carriage. Another bed of coals, lying at a depth of 60 yds. below this, has a similar subterranean canal, but it has no outlet, and merely floats the boats to a shaft, whence they are lifted by steam to the canal previously mentioned, on the upper level. The total craft employed upon the navigation (including 10 steamers on the Mersey, connecting the canal at Runcorn with Liverpool) are 497 in number. Eleven of these are passenger packets, known by the name of 'swift boats,' and are each drawn by a pair of horses at a speed of eight miles per hour. The number of workmen in the employ of the earl, including colliers, is not far short of 5,000."

WORSSTADT, a town of the grand-duchy of Hesse-Darmstadt, and prov. of the Rhine, 14 m. SSW of Mayence. Pop. 1,500. It has a castle and an hospital, and carries on an active trade in cattle.

WORSTEAD, a parish in Norfolk, 4 m. SSE of North Walsham. Area 2,603 acres. Pop. in 1831, 830; in 1851, 827. The village of W. was formerly a market-town, and had extensive manufactures of worsteds, which indeed took their name from it, [see article NORWICH,] but neither market nor manufactures now exist.

WORSTHORN, a township in the p. of Whalley, co.-palatine of Lancaster, 2 m. E of Burnley. Area 2,800 acres. Pop. in 1831, 798; in 1851, 909.

WORSTON, a township in the p. of Whalley, co.-palatine of Lancaster, 1 m. NE of Clitheroe. Area 860 acres. Pop. in 1831, 129; in 1851, 89.

WORTEGHEM, a department and commune of Belgium, in the prov. of Antwerp, and arrond. of Oudenarde. Pop. 3,142. The village is 5 m. WNW of Oudenarde. Pop. 707.

WORTEL, a department and commune of Belgium, in the prov. of Antwerp, and arrond. of Turnhout. Pop. 702. The village is 26 m. NE of Antwerp, on the r. bank of the Merck. In the vicinity is Champ Frederich, a colony founded in 1822 for the repression of mendicity.

WORTELSTETTEN, a village of Bavaria, in Swabia, in the presidial of Wertingen. Pop. 407.

WORTH, a county in the N part of the state of Iowa, U. S., comprising an area of 432 sq. m., drained by English river and Lime creek. It has a generally level surface, and contains several large lakes.—Also a township of Tuscola co., in the state of Michigan, 77 m. NE of Lansing. Pop. 600.

WORTH, a township in the p. of Prestbury, co.-palatine of Chester, 6 m. SE of Stockport. Area 450 acres. Pop. in 1831, 490; in 1851, 885.—Also a parish in Sussex. Area 13,250 acres. Pop. in 1831, 1,859; in 1851, 2,475.

WORTH, or WORD, a parish in Kent, 1½ m. S of Sandwich. Area 7,431 acres. Pop. in 1851, 471.

WORTH, or WOHRD, a town of Bavaria, in the

circle of Middle Franconia, near Nuremburg. Pop. 2,243. It has manufactories of cloth, chemical substances, and paper.

WORTH-MATRAVERS, a parish in Dorset, 3½ m. SSE of Corfe-Castle. Area 2,645 acres. Pop. in 1831, 356; in 1851, 396.

WORTHAM, a parish in Suffolk, 5½ m. NW of Eye. Area 2,736 acres. Pop. in 1851, 1,086.

WORTHEN, a parish, and formerly a market-town, chiefly in the co. of Salop, but extending into the adjoining co. of Montgomery, 9 m. NE of the town of Montgomery. The parish includes Bing-Weston, Bromblow, Upper-Heath, and Worthen, quarters, also Aston-Pigott, Aston-Rogers, Beachfield, Brockton, Habberley, Netherheath, and Walton with Rewins-Farm. Area 4,227 acres. Pop. in 1831, 2,290; in 1851, 3,325. Lead-mines were wrought here to a considerable extent in the reign of Adrian. A curious line of stones here is supposed to have been the ancient boundary between England and Wales.

WORTHENBURY, a parish in Flint, 8 m. N of Ellesmere, on the river Dee. Area 3,279 acres. Pop. in 1831, 623; in 1851, 543.

WORTHING, a parish in Norfolk, 4 m. N by E of East Dereham. Area 690 acres. Pop. in 1831, 138; in 1851, 170.—Also a sea-port and chapelry in the p. of Broadwater, 20 m. E by S of Chichester. This town has risen from a poor fishing village to be a well-frequented watering-place. The lodging-houses and inns are of the best description, and the streets are well-paved and lighted. Pop. 5,370. The inhabitants are extensively employed in the herring and mackerel fisheries.

WORTHINGTON, a township in the p. of Standish, co.-palatine of Lancaster, 3½ m. N by W of Wigan, in the line of the Leeds and Liverpool canal, and the North Union railway. Area 657 acres. Pop. in 1831, 124; in 1851, 176.—Also a chapelry in the p. of Breedon-on-the-hill, co. of Leicester, 4¼ m. NE of Ashby-de-la-Zouch, with which, and with the Swannington and Leicester railway, it is connected by a branch line of railway. Pop. with the liberty of Newbold, in 1801, 1,096; in 1831, 1,211; in 1851, 1,203.

WORTHINGTON, a township of Hampshire co., in the state of Massachusetts, U. S., 17 m. WNW of Northampton. It has a finely diversified surface, and is drained by Weetfield river and its branches. Pop. in 1840, 1,197; in 1850, 1,134.—Also a village of Sharon township, Franklin co., in the state of Ohio, on the E side of Olontangy river, 9 m. N by W of Columbus, and intersected by the Cleveland, Columbus, and Cincinnati railroad. Pop. in 1840, 440; in 1850, 501.—Also a village of Armstrong co., in the state of Pennsylvania, 145 m. WNW of Harrisburg. It has a large charcoal hot-blast furnace. —Also a township of Richland co., in the state of Ohio. Pop. in 1840, 1,842.

WORTHY (KING'S), a parish in the co. of Southampton, 2¼ m. NNE of Winchester, in the line of the Southampton railway. Area 2,216 acres. Pop., with the tything of Abbots-Worthy, in 1801, 161; in 1831, 345; in 1851, 382.

WORTHY (MARTYR), a parish in the co. of Southampton, 3 m. NE by N of Winchester. It includes the hamlet of Chilland. Area 1,800 acres. Pop. in 1831, 219; in 1851, 193.

WORTING, a parish in the co. of Southampton, 2 m. W of Basingstoke, in the line of the Southampton railway. Area 1,139 acres. Pop. in 1851, 158.

WORTLEY, a township and chapelry in the p. of St. Peter, Yorkshire, 2¼ m. WSW of Leeds. Area 1,036 acres. Pop. in 1801, 1,995; in 1831, 5,944; in 1851, 7,896. The inhabitants are extensively employed in the woollen manufacture of the Leeds district. Here also coarse earthenware and tobacco pipes are made from clay found in the neighbourhood.—Also a chapelry in the p. of Tankersley, Yorkshire, 5½ m. SSW of Barnesley, in the line of the Manchester and Sheffield railway. Area 5,984 acres. Pop. in 1831, 918; in 1851, 1,095.

WORTON, a tything in the p. of Potterne, Wilts, 3½ m. SW of Devizes. Pop. in 1851, 349.

WORTON (NETHER), a parish in the co. of Oxford, 3 m. SW of Deddington. Area 790 acres. Pop. in 1831, 94; in 1851, 62.

WORTON (OVER), a parish in Oxford, 4 m. SW by S of Deddington. Area 623 acres. Pop. 85.

WORTOWA, a village of Austria, in Bohemia, in the regency of Pardubitz, and bail. of Hohenmauth. Pop. 280.

WORTWELL, a hamlet in the p. of Reddenhall, Norfolk, 6 m. SW of Bungay, on the Waveney. Pop. in 1831, 537; in 1851, 541.

WOSANT, a village of Austria, in Bohemia, in the regency of Eger, bail. of Tachau. Pop. 280.

WOSCHITZ (JUNG), a town of Austria, in Bohemia, in the regency of Budweis, and bail. of Tabor. Pop. 1,800. It has a castle belonging to the counts of Khuenburg, an Armenian institute, an hospital, and a paper-mill. In the vicinity is the village of Alt-Woschitz.

WOSCHNICK, a market-town of Prussia, in the regency of Oppeln, and circle of Lublinitz. Pop. 1,000. It has a castle.

WOSCHTSCHUTZ, a village of Prussia, in the regency of Oppeln, and circle of Pless. Pop. 402. It has an iron-work.

WOSCZELLEN, a village of Prussia, in the regency of Gumbinnen, and circle of Lyk. Pop. 240.

WOSEHNO, a village of Austria, in Bohemia, in the regency and bail of Budweis. Pop. 260.

WOSEK, a village of Austria, in Bohemia, in the regency of Budweis, and bail. of Mühlhausen. Pop. 330.—Also a village of the regency of Gitschin, and bail. of Podiebrad. Pop. 260.—Also a village of the same regency, in the bail. of Gitschin. Pop. 260. —Also a village of the regency of Pardubitz, and bail. of Neu-Colin.—Also a village of the regency of Pilsen, and bail. of Przestitz. Pop. 200.—Also a village of the same regency, in the bail. of Rokitzan. Pop. 1,000. It has a synagogue, and possesses several iron-works. — Also a village of the same regency, in the bail. of Strakonitz. Pop. 400. It possesses a castle and a synagogue, and has manufactories of potash.

WOSELEK, a village of Austria, in Bohemia, in the regency of Pilsen, and bail. of Horazdiowitz. Pop. 500.

WOSERIN, a village of Mecklenburg-Schwerin, in the prov. of Mecklenburg, and bail. of Sternberg. Pop. 200.

WOSH, WEDGE, or WIJH, a small creek on the E coast of the Red sea, in N lat. 26° 13′, E long. 36° 32′, having a shallow natural anchorage, with an entrance 250 yds. wide. There is a village or collection of huts upon an elevated rock on the N side of the creek, distant about 5 m. from an inland fort of the same name, which has a garrison of 25 men, and is used as a depot for grain for the supply of the caravans going to Mecca.

WOSISCHEK, a village of Austria, in Bohemia, in the regency and bail. of Pardubitz. Pop. 300.

WOSITZ, a village of Austria, in Bohemia, in the regency and bail. of Pardubitz. Pop. 370.

WOSKORINEK, a village of Austria, in Bohemia, in the regency of Gitschin, and bail. of Nimburg. Pop. 800.

WOSKRESENSKOE, a village of Russia in Europe, in the gov. of Moscow. Pop. 1,100.

WOSLAWITZ, a village of Austria, in Moravia, in the circle of Iglau, and seignory of Meseritsch. Pop. 340.

WOSLOW, a village of Austria, in Bohemia, in the regency of Pilsen, and bail. of Pisek. Pop. 410.

WOSNESSENSK, a market-town of Russia in Europe, in the gov. of Kherson. Pop. 1,000.

WOSNITZ, a village of Austria, in Bohemia, in the regency of Gitschin, and bail. of Königgrätz. Pop. 240.

WOSPORK. See WEISSENBERG.

WOSSARKEN, a village of Prussia, in the regency of Marienwerder, and circle of Graudenz. Pop. 250.

WOSSIDEN, a village of Prussia, in the regency of Konigsberg, and circle of Heilsberg. Pop. 200.

WOSSIK, a village of Austria, in Bohemia, in the regency of Eger, circle of Klattau, and seignory of Lukawitz. Pop. 230.

WOSSINGEN, a town of the grand-duchy of Baden, in the circle of the Middle Rhine, and bail. of Stein, 11 m. E of Carlsruhe. Pop. 1,300. It has a castle and two churches.

WOSSITZ, a village of Prussia, in the regency and circle of Danzig. Pop. 335.

WOSTEGKOWITZ, a village of Austria, in Moravia, in the circle of Znaim, and seignory of Jamnitz. Pop. 300.

WOSTITZ, or WASATICE, a market-town of Austria, in Moravia, in the circle and 18 m. SSW of Brunn, between two small lakes. Pop. 1,100.

WOSTRATSCHIN, a village of Austria, in Bohemia, in the regency of Pilsen, and bail. of Bischofteinitz. Pop. 550. It has a castle.

WOSTREDEK, a village of Austria, in Bohemia, in the regency of Budweis, and bail. of Beneschau. Pop. 460. It has a castle.

WOSTRETIN, a village of Austria, in Bohemia, in the regency and bail. of Pardubitz. Pop. 800.

WOSTROMUR, a village of Austria, in Bohemia, in the regency of Gitschin, and bail. of Horzitz. Pop 550.

WOSTROW, a village of Austria, in Bohemia, in the regency of Pardubitz, and bail. of Hohenmauth. Pop. 300.—Also a village of the same regency, in the bail. of Leditsch. Pop. 440. It has a castle and manufactories of potash.—Also a village of the regency of Budweis, and bail. of Beneschau. Pop. 240.—Also a village of the regency of Pilsen, and bail. of Brzeznitz. Pop. 250.

WOSTROWETZ, a village of Austria, in Bohemia, in the reg. of Pilsen, bail of Pisek. Pop. 300.

WOSTRUZNO, a village of Austria, in Bohemia, in the regency and bail. of Gitschin. Pop. 300.

WOTAWA, a river of Austria, in Bohemia, in the circle of Prachin, which has its source on the NE side of the Bohmerwald; flows first NE, then E, and afterwards NE; and after a total course of 75 m., joins the Moldau on the l. bank. Schüllenhofen, Horazdiowitz, Strakonitz, and Pisek, are the principal places on its banks.

WOTENIK, a village of Prussia, in the regency of Stralsund, and circle of Grimmen, on the Tribel. Pop. 820.

WOTENITZER, a village of Mecklenburg-Schwerin, in the prov. of Mecklenburg, and bail. of Grerismuhlen. Pop. 220.

WOTERSIN, a village of Mecklenburg-Strelitz, and regency of Ratzeburg. Pop. 640.

WOTITZ, a town of Austria, in Bohemia, in the regency of Budweis. Pop. 2,000. It has a castle.

WOTRUB, a village of Austria, in Bohemia, in the reg. of Pardubitz, bail. of Ledetsch. Pop. 210.

WOTSCH, a village of Austria, in Bohemia, in the regency of Eger, and bail. of Saatz. Pop. 220.

WOTTON-COURTNEY, a parish in Somersetshire, 4 m. W of Dunster. Area 3,145 acres. Pop. in 1831, 426; in 1851, 411.

WOTTON-FITZ-PAINE, a parish in Dorset, 4½ m. NNE of Lyme-Regis. Area 1,679 acres. Pop. in 1831, 455; in 1851, 361.

WOTTON-UNDER-EDGE, a parish and market-town in the co. of Gloucester, 19 m. S by W of Gloucester. Besides the town of W., this p. contains the tythings of Huntingford, Sinwell with Bradley, Symond's-hall with Combe, and Wortley. Area 4,880 acres. Pop. in 1831, 5,482; in 1851, 4,224. The town was burned in the reign of John, and the ancient site is called 'the Brands.' The present town stands on a rising ground, and consists of two parallel streets, containing some neat houses.

WOTTON-UNDERWOOD, a parish in Bucks. Area 3,220 acres. Pop. in 1851, 253.

WOTUNDO, or ATUNDO, a river of Zanguebar, which throws itself by two embouchures into the Indian ocean, opposite the island of Zanzebar.

WOTWOWITZ, a village of Austria, in Bohemia, in the reg. of Prague, bail. of Wolwarn. Pop. 660.

WOTZLAF, a village of Prussia, in the regency and circle of Danzig. Pop. 350.

WOUBRECHTEGEM, a department and commune of Belgium, in the prov. of East Flanders, and arrond. of Audenarde. Pop. 2,800.

WOUBRUGGE, a village of Belgic Holland, in South Holland. Pop. 1,168.

WOUDSEND, a village of Holland, in the prov. of Friesland, to the N of Slooten.

WOUGHTON-ON-THE-GREEN, a parish in Bucks, 2½ m. N by W. of Fenny-Stratford, on the river Ouse, and crossed by the Grand Junction canal. Area 890 acres. Pop. in 1831, 303; in 1851, 333.

WOULDHAM, a parish in Kent, 3 m. SW of Rochester, on the river Medway. Area 612 acres. Pop. in 1831, 411; in 1851, 343.

WOUMEN, a department and com. of Belgium, in the prov. of West Flanders, and arrond. of Furnes, watered by the Canal-de-l'Yser. Pop. 2,330.

WOUTSCHUA, a village of Austria, in Styria, in the circle of Cilly. Pop. 538.

WOUW, a village of Holland, in the prov. of North Brabant, to the N of Bergen-op-Zoom. Pop. 2,600.

WOW, a fortified town of Hindostan, capital of a small state of the same name, in the prov. of Gujerat, and district of Neyer, near the E side of the Bunn, and 10 m. SW of Theraud. The inhabitants are chiefly Coolies.

WOWAMIA, a town of Hindostan, in the prov. of Gujerat, and district of Jhallavar, on the S bank of the Bunn, opposite its junction with the gulf of Kutch, and 24 m. NW of Murevy. It is inhabited chiefly by fishermen.

WOXEN, a river of Russia in Europe, in Finland.

WOXFELDE, a colony of Prussia, in the regency of Frankfurt, and circle of Sternberg. Pop. 540.

WOXNA, a village of Sweden, in Jefleborg, on the Worna-Elf. Pop. 700. It has iron-works.

WOXREHNE, or WOXHOLLANDER, a colony of Prussia, in the regency of Frankfurt, and circle of Landsberg. Pop. 360.

WOYANOW, a village of Prussia, in the regency and circle of Danzig. Pop. 370.

WOYCIECHOWO, two villages of Prussia, in the regency of Posen, and circle of Bomst. Pop. 460.

WOYCIN, a village of Prussia, in the regency of Bromberg, and circle of Inowraclaw. Pop. 330.

WOYDEHNEN, a village of Prussia, in the regency of Gumbinnen, and circle of Ragnitz. Pop. 220.

WOYNASSEN, a village of Prussia, in the reg. of Gumbinnen, and circle of Oletsko. Pop. 200.

WOYNICZ, a town of Austria, in Galicia, in the circle and 18 m. E of Bochnia, on the l. bank of the Dunajec. Pop. 1,200.

WOYNILOW, a market-town of Austria, in Galicia, in the circle of Stry, on the Woynilowka. It has manufactories of common earthenware.

WOYNITZ, a village of Prussia, in the regency of Posen, and circle of Kosten. Pop. 260.

WOYNOWICE, a village of Prussia, in the reg. of Posen, and circle of Buk. Pop. 579.

WOYNOWITZ, a village of Prussia, in the reg. of Oppeln, and circle of Ratibor. Pop. 540.

WOYNOWO, a village of Prussia, in the regency of Posen, and circle of Bomst, near Karge. Pop. 110. It has a castle.

WOYSLAWICE, a town of Russia in Europe, in Poland, in the gov. and 53 m. ESE of Lublin. Pop. 1,000.

WOYTIECHOW, a village of Austria, in Bohemia, in the regency of Pardubitz, and bail. of Richenburg. Pop. 480.

WOYWERG, a village of Russia in Europe, in Poland, in the gov. of Augustowo, to the NE of Suwalki.

WOYZISKWO, a town of Russia in Europe, in Poland, in the gov. of Lublin. Pop. 200.

WOZNIKI, a village of Prussia, in the reg. of Posen, and circle of Buk. Pop. 362.

WOZNITZEN, a village of Prussia, in the reg. of Gumbinnen, and circle of Sensburg. Pop. 550.

WRAB, a market-town of Austria, in Bohemia, in the regency of Prague, and bail. of Karolinenthal, and a little to the S of Brandeis. Pop. 260.

WRABLANG. See VEREBELY.

WRABNESS, a parish in Essex, 4¾ m. E of Manningtree, on the river Stour estuary. Area 1,491 acres. Pop. in 1831, 248; in 1851, 261.

WRACKOWITZ, a village of Austria, in Bohemia, in the regency of Budweis, and bail. of Veneschau. Pop. 660.

WRACLAW. See BRESLAU.

WRADUK, a town of Turkey in Europe, in Bosnia, to the E of Travnik, on the Bosna.

WRAGBY, a parish and market-town in the co. of Lincoln, 12 m. NE of Lincoln. Area 1,594 acres. Pop. in 1831, 601; in 1851, 610. The village stands in the middle of a fertile district on the high road from Lincoln to Horncastle. It is neatly built. The inhabitants are chiefly employed in agriculture. —Also a parish comprising the townships of West Hardwick, Hasle, Hill-top, and Hurstwick-with-Nostal, in the wapentake of Osgoldcross, and those of Ryhill and Winterset, in the W. R. of Yorkshire, 5 m. SW of Pontefract. Area 3,944 acres. Pop. in 1831, 756; in 1851, 762.

WRAHOWITZ, a village of Austria, in Moravia, in the circle of Olmutz, and seignory of Kralitz. Pop. 550.

WRAMPLINGHAM, a parish in Norfolk, 3 m. N by E of Wymondham. Area 845 acres. Pop. in 1831, 247; in 1851, 245.

WRANAN, a village of Austria, in Moravia, in the circle and 9 m. N of Brunn. Pop. 440. It has a fine collegiate church, in which are the tombs of the princes of Lichtenstein.

WRANAY, a town of Austria, in Bohemia, in the regency of Prague, 8 m. NNW of Schlan. It has an hospital.

WRANGA, a village of Austria, in Styria, in the circle of Marburg. Pop. 400.

WRANGEL, or VRANGEL (GREAT and LITTLE), two islands of the gulf of Finland, 9 m. NE of Revel. They are thickly covered with wood. Pop. 1,100.

WRANGLE, a parish in the co. of Lincoln, 8½ m. NNE of Boston. Area 9,780 acres. Pop. 1,196.

WRANIAN, a village of Austria, in Bohemia, in the regency of Prague, and bail. of Melnik. Pop. 240.

WRANOW, a village of Austria, in Bohemia, in the regency of Budweis, and bail. of Veneschan. Pop. 230.—Also a village of the regency of Pilsen. and bail. of Rokitzan. Pop. 260.

WRANOWEY, a village of Austria, in Bohemia, in the regency of Gitschin, and bail. of Semil, on the Iser. Pop. 260.

WRANOWITZ, a village of Austria, in Bohemia, in the regency of Pilsen, and bail. of Brzeznitz. Pop. 490.—Also a village of the same regency, in the bail. of Rokitzan. Pop. 360. It has vitriol works.—Also a village of Moravia, in the regency of Olmutz, and seignory of Plumenau. Pop. 385.

WRASCHKOW, a village of Austria, in Bohemia, in the reg. of Prague, and bail. of Melnik.

WRATH (CAPE), a celebrated head-land, in the parish of Durness in Sutherlandshire, forming the NW point of Scotland, in 58° 37′ N lat., and 5° W long. It is a fine promontory of granitic-gneiss, towering up in a pyramidal form to the height of 300 ft., and standing boldly out into the waves. "Nothing," says Macculloch, "can exceed the elegance and majesty of its form, declining towards the sea in a second and much lower pyramidal rock; the whole forming an outline as graceful as it is unexpected, and as grand as it is appropriate. No vessels approach this shore, as the rapidity and turbulence of the tide are extreme; and as this is esteemed both a difficult and a dangerous point to double." In 1828, a lighthouse was erected here at an expense of £14,000. It shows a white revolving light, which is elevated 400 ft. above high water, and is seen at the distance of 24 m. in clear weather. The Butt of Lewis on the SW, and the Hoyhead of Orkney towards the NE, can be seen in clear weather from the top of this lighthouse.

WRATISCHOW, a village of Austria, in Bohemia, in the regency of Budweis, and bail. of Pilgram. Pop. 280.

WRATKO, a village of Austria, in Bohemia, in the regency of Pardubitz, and bail. of Schwarz-Kosteletz. Pop. 250.

WRATNO, a village of Austria, in Bohemia, in the regency of Gitschin, and bail. of Jungbunzlau. Pop. 260.

WRATTENY. See FRATTING.

WRATTING (GREAT), a parish in Suffolk, 6½ m. WNW of Clare. Area 1,329 acres. Pop. 391.

WRATTING (LITTLE), a parish in Suffolk, 5¾ m. W by N of Clare. Area 936 acres. Pop. 212.

WRATTING (WEST), a parish in the co. of Cambridge, 5 m. N by E of Linton. Area 3,441 acres. Pop. in 1831, 763; in 1851, 868.

WRATZLAU, a village of Austria, in Bohemia, in the regency of Pardubitz, and bail. of Hohenmauth. Pop. 620.

WRATZOWITZ, a village of Austria, in Bohemia, in the regency of Budweis, and bail. of Veneschan. Pop. 380.

WRAWBY, a parish in the co. of Lincoln, 8 m. NE of Caistor, including the hamlets of Glandfordbrigg and Kettleby. Area 5,070 acres. Pop., with Kettleby, in 1831, 1,780; in 1851, 3,132.

WRAXALL, or WRAXHALL, a parish in Dorset, 8 m. E of Beaminster. It is divided into Higher and Lower W. Area 952 acres. Pop. in 1851, 87.

WRAXALL, a parish in Somersetshire, 6½ m. W by S of Bristol. Area 3,772 acres. Pop. 1,016.

WRAXALL (NORTH), a parish in Wilts, 7 m. W by N of Chippenham. Area 2,127 acres. Pop. in 1831, 415; in 1851, 450.

WRAXALL (South), a chapelry in the p. of Great Bradford, Wilts, 5 m. W by N of Melksham. Pop. in 1831, 389; in 1851, 356.

WRAY, a township in the p. of Melling, co.-palatine of Lancaster, 10 m. NE by E of Lancaster. Area, with Bolton, 6,506 acres. Pop. in 1831, 586; in 1851, 833. Nails are manufactured here to a considerable extent.

WRAZOW, a market-town of Austria, in Moravia, in the circle and 15 m. WSW of Hradisch. Pop. 2,050.

WRBCAN, a village of Austria, in Bohemia, in the regency of Pardubitz, and bail. of Kaurzim. Pop. 500.

WRBENTHUL. See Wurbenthal.

WRBICAN, a village of Austria, in Bohemia, in the reg. of Prague, and bail. of Schlan. Pop. 300.

WRBIETITZ, a village of Austria, in Moravia, in the circle of Hradisch. Pop. 330.

WRBITSCHAN, a village of Austria, in Bohemia, in the regency of Bohmisch-Leipa, and bail. of Leitmeritz. Pop. 330.

WRBITZ, a village of Austria, in Bohemia, in the regency of Gitschin, and bail. of Reichenau. Pop. 200.—Also a village of the same regency and bail. of Podiebrad. Pop. 430.—Also a village of the same regency in the bail. of Gitschin.—Pop. 400.—Also a village of the regency of Pardubitz, and bail. of Ghotiebor. Pop. 540. It has a castle.—Also a village of the regency of Pilsen, and bail. of Wolin. Pop. 240.—Also a village of the regency of Prague, and bail. of Melnik. Pop. 300.—Also a village of Moravia, in the circle of Brunn, and seignory of Göding. Pop. 800.

WRBKA, a village of Austria, in Bohemia, in the regency of Prague, and bail. of Schlan. Pop. 250.—Also a village of Moravia, in the circle of Hradisch, and seignory of Strassnitz. Pop. 750.

WRBNO, a village of Austria, in Bohemia, in the regency of Prague, and bail. of Melnik. Pop. 300.—Also a village of the regency of Pilsen, and bail. of Brzeznitz. Pop. 200.

WRBNO-AN-DEM-WALDERN, a village of Austria, in Bohemia, in the regency of Prague, and bail. of Laun. Pop. 370.

WRBOCE. See Werbau.

WRBOWE. See Mannersdorf. See also Verbo.

WRCHLABY. See Hohenelbe.

WRCHOSLAWITZ, a village of Austria, in Moravia, in the circle of Hradisch, on the Hanna. Pop. 500.

WRCHOWIN, a village of Austria, in Bohemia, in the regency of Gitschin, and bail. of Neustadt. Pop. 410.

WRCHOWINA (Nokitnay), a village of Austria, in Bohemia, in the regency of Gitschin, and bail. of Horzitz. Pop. 710.

WRCOW, a village of Austria, in Bohemia, in the regency and bail. of Budweis. Pop. 280.

WREAK, a river in Leicestershire, rising from various sources in the NE part of the co., and flowing SW by Melton Mowbray and Rotherby, to the Soar near Rothley. Near its efflux into the Soar it is united with the Leicester canal, and it is here also joined by its principal tributary streams from the SE.

WREAY, a chapelry in the p. of St. Mary, Cumberland, 5¾ m. SE by S of Carlisle. Area 1,088 acres. Pop. in 1831, 166; in 1851, 149.

WREBRONIG, a village of Austria, in Styria, in the circle of Marburg. Pop. 510.

WRECHEN, a village of the grand-duchy of Mecklenburg-Strelitz, circle of Stangard. Pop. 200.

WRECHOW, a village of Prussia, in the regency of Frankfort, and circle of Konigsberg. Pop. 310.

WREDEN, a town of Prussia, in the prov. of Westphalia, regency of Munster, and circle of Ahan, on the Berkel. Pop. in 1843, 2,596. It has a gymnasium, and possesses manufactories of linen and of chicory.

WREDENHAGEN, a bail. and village of the grand-duchy of Mecklenburg-Schwerin, in the circle of Windisch, on a small lake. Pop. 540.

WREIGH-HILL, a township in the p. of Rothbury, Northumberland, 5¼ m. W of Rothbury, on the river Coquet. This place was formerly populous, but, in 1665, its inhabitants were almost entirely swept away by the plague, and the pop. has ever since been small.

WREKIN (The), a noted mountain lying to the E of Shrewsbury in Salop, between the Watling-street and the Severn, 1 m. from Wroxeter. It was a famous station of the Romans. The steepest side of the hill is the eastern. Its height is 1,200 ft. The summit of this hill commands a scene on the W and N of great extent and grandeur, looking right into the central outer positions of the Principality. The views in the opposite directions command the busy mining district of Ketley, Donington, &c., and breaking away from Shiffnal, carry the eye into the very heart of the kingdom, over one of the richest and densest manufacturing and mining districts in the empire. The peculiar prominence and distinction of the W. arises from its sudden and abrupt elevation, in a wide and extensive flat county, and its pudding-like form, disconnected from any eminences, except the undulating elevations at Coalbroke-dale. The Shropshire coal-field lies in a triangle, with the Severn as the base, and the apex extending to Lilleshall, overlying the mountain limestone at the latter place and on one of the flanks of the Wrekin. The field is said to possess "ten beds of good workable coal," in some instances 16 ft. in thickness. The sandstone, which has been worked on the side of the W., discloses some curious fantastic forms, and is, in many places, converted into quartz.

WREKIN, a river in Lancashire, which falls into the Stour at Cussington.

WRELTON, a township in the p. of Middleton, N. R. of Yorkshire, 2¾ m. WNW of Pickering. Area 1,230 acres. Pop. in 1831, 172; in 1851, 232.

WREMEN, a village of Hanover, in the gov. of Stade, and state of Bremen, near Dorum. Pop. 545.

WRENBURY, a parish in the co.-palatine of Chester, 4⅝ m. SW by S of Nantwich, intersected by the Ellesmere canal. It contains the townships of Bromhall, Chorley, Woodcot, and Wrenbury-with-Frith. Area 11,665 acres. Pop. 2,060.

WRENINGHAM (Great), a parish in Norfolk, 4 m. SE by E of Wymondham. Area 1,528 acres. Pop. in 1831, 409; in 1851, 452.

WRENTHAM, a parish in Suffolk, 5¼ m. N by W of Southwold. Area 2,303 acres. Pop. 1,026.

WRENTHAM, a township of Norfolk co., in the state of Massachusetts, U. S., 23 m. SW by S of Boston, drained by branches of Taunton, Charles, and Neponset rivers, and intersected by the Norfolk co. railway. Pop. in 1840, 2,915; in 1850, 3,037. It has a village of the same name.

WRENTHORPE. See Stanley.

WRESCHEN, or Wreszma, a circle and town of Prussia, in the prov. and regency of Posen. The circle comprises an area of 12·6 German sq. m. Pop. 36,629. The town is 32 m. ESE of Posen. Pop. 3,144. It contains several churches, Lutheran and Catholic, and has manufactories of woollen and linen fabrics.

WRESCHIN, a village of Prussia, in the reg. of Bromberg, and circle of Czarnikau. Pop. 350.

WRESEN, a village of Austria, in Styria, in the circle of Cilli. Pop. 400.

WRESSEL, a parish in the E. R. of Yorkshire, 3¾ m. N of Howden, on the river Derwent, and in the line of the Hull and Selby railway, which is here carried across the river by a cast-iron bridge of 70 ft. span. It contains the townships of Newsham with Brind, and Wressel with Loftsome. Area 3,705 acres. Pop. in 1831, 386; in 1851, 378.

WRESTE, a village of Prussia, in the regency and circle of Oppeln. Pop. 270.

WRESTEDT, a village of Hanover, in the principality of Luneberg, and bail. of Bodenteich, near Welzen. Pop. 352.

WRESTLINGWORTH, a parish in Bedfordshire, 5½ m. ENE of Biggleswade. Area 1,620 acres. Pop. in 1831, 448; in 1851, 588.

WRETHAM (Great or East), a parish in Norfolk, 6 m. NE by N of Thetford. Area 6,442 acres. Pop. in 1831, 325; in 1851, 385.

WRETTON, a parish in Norfolk, 1 m. W of Stoke-Ferry. Area 1,154 acres. Pop. in 1851, 538.

WREXEN, a market-town of Waldek, in the district of the Diemel, on the Orpe. Pop. 1,030. It has a paper-mill and iron-works.

WREXHAM, a parish and market-town in the co. of Denbigh, 13 m. S by W of Chester. The parish extends into Flintshire, and, besides the town of Wrexham, contains the chapelries of Berse Drelincourt, Brymbo, and Minera, and the townships of Abenbury-fechan, Abenbury-vawr, Acton, Beiston or Boreston, Boras-bovah, Broughton, Esclusham-above, Esclusham-below, Gourton, Stansby, Wrexham-Abbot, and Wrexham-Regis. Area 15,879 acres. Pop. in 1831, 11,515; in 1851, 15,520. The church, an elegant structure, 178 ft. long and 72 ft. broad, with a tower 185 ft. high, consists of a chancel, pentagonal in shape, a nave, two collateral aisles, and a lofty quadrangular tower at the W end. The town consists of several streets, containing some handsome houses. It is a place of considerable importance, and is greatly benefited by its situation as a thoroughfare between Liverpool and Chester and North Wales. The inhabitants are partly employed in the flannel trade, also in paper-making, and carry on considerable traffic in coal, iron, and lead, which are mined in the vicinity.

WRHAWETZ, a village of Austria, in Bohemia, in the reg. of Pilsen, and bail. of Klattau. Pop. 270.

WRIETZEN, or Brietzen, a town of Prussia, in the regency of Potsdam, and circle of Oberbarnim, 36 m. ENE of Berlin, on an arm of the Oder. Pop. 6,450. It is enclosed by walls, with four gates, and has an hospital. It possesses manufactories of linen and woollen fabrics, and of tobacco, tanneries, several breweries, and important fisheries. The environs are adorned with gardens.—Also a village distinguished as Alt-Brietzen. Pop. 332.

WRIGHT, a county in the N part of the state of Iowa, U. S., comprising an area of 576 sq. m., drained by Borne river, and by the head branch of Iowa river. It has an undulating surface, and is generally fertile.—Also a county in the SW part of the state of Missouri, comprising an area of 913 sq. m., drained by Gasconade river and its branches. It has a finely diversified surface, and is very fertile. Pop. in 1850, 3,387. Its capital is Hartsville.

WRIGHTINGTON, a township in the p. of Eccleston, co.-palatine of Lancaster, 4 m. NW of Wigan. Area 3,876 acres. Pop. in 1851, 1,613.

WRIGHTSBORO, a village of Columbia co., in the state of Georgia, U. S., on a branch of Little river, 54 m. NE of Milledgeville. Pop. in 1850, 150.

WRIGHTS TOWN, a township of Bucks co., in the state of Pennsylvania, U. S., 7 m. SE of Doylestown, bounded on the S by Neshaming creek, and drained by its tributaries. Pop. in 1840, 708.—Also a village of Hanover township, Burlington, co., in the state of New Jersey, 23 m. S of Trenton. Pop. 250.

WRIGHTSVILLE, a town of York co., in the state of Pennsylvania, U. S., 24 m. SE of Harrisburg, on the W bank of the Susquehanna, opposite Columbia, with which it is connected by a bridge 5,690 ft. in length, and on the York and Wrightsville railroad. Pop. in 1840, 672; in 1850, 750.

WRINGTON, a parish, formerly a market-town, in the co. of Somerset, 7 m. NNE of Axbridge. It contains the chapelry of Burrington and the tything of Broadfield. Area 5,786 acres. Pop. 1,620.

WRISBERGHOLZEN, a village of Hanover, in the gov. and principality of Hildesheim and bail. of Winzenburg. Pop. 525. It has a castle, and possesses manufactories of earthenware.

WRISCHT, a village of Austria, in Moravia, in the circle of Iglau. Pop. 380. It has iron-works.

WRISSA, a town of Hindostan, in Sinde, in the district and 15 m. SW of Tasta, near the r. bank of the Sinde.

WRISWALL. See Wirswall.

WRITHLINGTON, a parish in Somersetshire, 7 m. NW by N of Frome, intersected by a canal and railroad communicating with Bath. Area 772 acres. Pop. in 1831, 245; in 1851, 296.

WRITTLE, a parish, formerly a market-town, in Essex, 2½ m. W by S of Chelmsford. Area 8,672 acres. Pop. in 1831, 2,341; in 1851, 2,423.

WRITZ, a commune of France, in the dep. of the Loire-Inferieure, and cant. of Saint-Mars-la-Gaille, 18 m. N of Ancenis. Pop. 1,706.

WROBLEWO, a village of Prussia, reg. of Posen, and circle of Samter. Pop. 360.

WROCKWARDINE, a parish in Salop, 2 m. W by N of Wellington, intersected by the Shrewsbury canal. Area 4,608 acres. Pop. in 1851, 3,107.

WRODAWEK, a town of Russia in Europe, in Poland, in the gov. of Warsaw, on the Wieschel, ENE of Brzesc. Pop. 1,480.

WROHE, or Wrau, a village of Denmark, in the duchy of Holstein, to the S of the Westen-see, and in the parish of that name. Pop. 408.

WROHM, a village of Denmark, in the duchy of Holstein, bail. of Norden-Ditmarschen, and parish of Zellingstedt. Pop. 352.

WRONCZIN, a village of Prussia, in the regency and circle of Posen. Pop. 300.

WRONIAWE, or Wroniowo, a village of Prussia, in the regency of Posen, and circle of Bomst, near Wollstein. Pop. 332.

WRONIN, a village of Prussia, in the regency of Oppeln, and circle of Kosel. Pop. 440.

WRONKE, or Wronki, a town of Prussia, in the prov., regency, and 33 m. NW of Posen, and circle of Samter, on the l. bank of the Warthe. Pop. 2,820. It has manufactories of cloth.

WRONKEN, a village of Prussia, in the regency of Gumbinnen, and circle of Goldau. Pop. 200.

WRONNEN (Gross and Klein), two villages of Prussia, in the regency of Gumbinnen, and circle of Lotzen, containing respectively 260 and 210 inhabitants.

WROOT, a parish in Lincolnshire, 8 m. NE by N of Bawtry. Area 3,246 acres. Pop. 349.

WROTHAM, a parish, formerly a market-town, in Kent, 24 m. SE by E of London. It contains the townships of Hale, Nepicar, Plaxtol, Winfield, and Roughway, and the hamlet of Borough-Green. Area 8,878 acres. Pop. in 1851, 3,184.

WROTKOW, a village of Prussia, in the regency of Posen, and circle of Krotossyn. Pop. 240.

WROTTESLEY, a hamlet in the p. of Tettenhall, Staffordshire, 3 m. NW of Wolverhampton.

WROUGHTON, a parish in Wilts, 3 m. SW by S of Swindon. Area 4,546 acres. Pop. 1,645.

WROXETER, a parish in Salop, 5¾ m. SE by S of Shrewsbury, on the river Severn. Area 4,774 acres. Pop. in 1831, 636; in 1851, 642. This is the site of the Roman station *Uriconium*.

WROXHALL, a parish in Warwickshire, 6 m. NW by N of Warwick. Area 1,735 acres. Pop. 179.

WROXHAM, a parish in Norfolk, 2½ m. SE of Coltishall, on the river Bure. Area 1,489 acres. Pop. in 1831, 368; in 1851, 429.

WROXTON, a parish in Oxfordshire, 3 m. WNW of Banbury. It contains the chapelry of Balscott. Area 2,350 acres. Pop. in 1851, 789.

WRSCHOWITZ, a village of Austria, in Bohemia, in the regency of Prague, and bail. of Carolinthal. Pop. 1,000.—Also a village of the same regency, and bail. of Laun. Pop. 320. It has a castle.

WRTSCHOW, a village of Austria, in Bohemia, in the regency of Pilsen, and bail. of Przestitz. Pop. 560.

WRUTEK. See RUDIG.

WRUTITZ, a village of Austria, in Bohemia, in the regency of Gitschin, and bail. of Nimburg. Pop. 450. It has a castle.

WRZCLOWICE, a town of Russia in Europe, in Poland, in the gov. of Lublin. Pop. 210.

WRZESIN, a village of Prussia, in the regency of Oppeln, and circle of Ratibor. Pop. 240.

WSCHECHLAP, a village of Austria, in Bohemia, in the regency of Budweis, and bail. of Veneschau. Pop. 280.—Also a village of the regency of Gitschin and bail of Nimburg. Pop. 240.

WSCHECHOWITZ, a village of Austria, in Moravia, in the circle of Prerau. Pop. 800. It has a castle, and possesses several mills and breweries.

WSCHEHERD, a village of Austria, in Bohemia, in the reg. of Pilsen, bail. of Kralowitz. Pop. 230.

WSCHEJAN, a village of Austria, in Bohemia, in the regency of Gitschin, and bail. of Nimburg. Pop. 200.

WSCHELIB, a village of Austria, in Bohemia, in the regency of Gitschin, bail. of Braunau. Pop. 300.

WSCHELIS, a village of Austria, in Bohemia, in the regency of Gitschin, and bail. of Nimburg. Pop. 360. It has a castle.

WSCHEMINA, a village of Austria, in Moravia, in the circle of Hradisch, and seignory of Lukow. Pop. 800.

WSCHEN, a village of Austria, in Bohemia, in the regency of Gitschin, bail. of Turnau. Pop. 300.

WSCHERAU, a village of Austria, in Bohemia, in the regency of Pilsen, and bail. of Tuschkau. Pop. 1,300. It has three mills.

WSCHESTAR, a village of Austria, in Bohemia, in the regency of Gitschin, and bail. of Königgrätz. Pop. 310.—Also a village of the regency of Prague, and bail. of Gule. Pop. 280.

WSCHETAT, a village of Austria, in Bohemia, in the regency of Prague, and bail. of Carolinenthal. Pop. 610.—Also a village of the same regency, and bail. of Rakonitz. Pop. 490.

WSCHETETZ, a village of Austria, in Bohemia, in the regency and bail. of Budweis. Pop. 215.

WSCHETUL, a village of Austria, in Moravia, in the circle of Hradisch, and seignory of Holleschau.

WSCHEWEIL, a village of Austria, in Bohemia, in the regency of Pilsen, and bail. of Brzetznitz. Pop. 290.

WSCHOWA. See FRANSTADT.

WSEMBORZ, a village of Prussia, in the regency of Posen, and circle of Wreschen. Pop. 300.

WSISKO, a village of Austria, in Moravia, in the circle of Olmutz. Pop. 290.

WSZOWA. See FRANSTADT.

WSZERUB. See SCHERAU.

WSZETIN, a town of Austria, in Moravia, in the circle and 35 m. NE of Hradisch. Pop. 3,200. It has manufactories of cloth.

WTELN, a village of Austria, in Bohemia, in the regency of Eger, and bail. of Brux. Pop. 300.

WTELNO, a village of Prussia, in the regency and circle of Bromberg. Pop. 458.—Also a village of Austria, in Bohemia, in the regency of Gitschin, and bail. of Jung-Bunzlau. Pop. 440.—Also a village of the regency of Prague, and bail. of Melnik. Pop. 450.

WTUREK, a village of Prussia, in the regency of Posen, and circle of Adelnau. Pop. 350.

WUBISER (GROSS and KLEIN), a village of Prussia, in the regency of Frankfurt, and circle of Konigsberg. Pop. 540.

WU-CHAE, a district and town of China, in the prov. of Shan-se, and div. of Ning-wu-fu.

WU-CHANG, a district and town of China, in the prov. of Hu-pih, and div. of Wu-chang-fu, in N lat. 30° 22′, and E long. 114° 48′.

WU-CHANG-FU, a division and town of China, in the prov. of Hu-pih. The div. comprises ten districts. The town is on the r. bank of the Yang-tsen-keang, opposite Han-yang, and nearly opposite the confluence of the Han-keang, 735 m. SSW of Peking, and 540 above the mouth of the river, in N lat. 30° 34′ 50″, and E long. 114° 13′ 30″. This town, the residence of the viceroy, yields to few cities in the empire in extent, populousness, and prosperity. It possesses, from its situation, an active entrepot trade; and, in the number of vessels that enter its port, it rivals the most flourishing emporium of the maritime provinces. The environs are noted for their tea, and the bamboo paper manufactured in the city is exported to all parts of the empire.

WUCHERN, a village of Austria, in Styria, in the circle of Cille. Pop. 250.

WU-CHIH-HEEN, a district and town of China, in the prov. of Ho-nan, and div. of Hwae-king-fu, in N lat. 35° 7′, and E long. 113° 31′.

WU-CHIN, a town of China, in the prov. of Keang-se. Although possessing no territorial importance, it is noted for its trade. It contains spacious and well-stocked magazines, excellent inns, richly decorated temples, and shops replenished with wares of every description. One of the temples dedicated to Kwang-sin-chu, the god of longevity is remarkable for the richness of its gilding and sculpture. In the exterior court are porcelain shops.—Also a district and town in the prov. of Kwantung, 225 m. SW of Canton, on the China sea.—Also a town of the Corea, in the prov. of King-ke, 60 m. SSE of Han-yang.

WU-CHING-HEEN, a district and town of China, in the prov. of Shan-tung, and div. of Lingtsing-chu, in N lat. 37° 15′, and E long. 116° 3′.—Also a district and town in the prov. of Che-keang, and div. of Hu-chu-fu.

WU-CHI-SHAN, a mountain of China, in the prov. of Kwang-tung, and near the centre of the island of Hai-nan. It is the loftiest in the island, and sends on all sides ramifications extending to the coast. It derives its name from its five summits.

WU-CHU-FU, a division and town of China, in the prov. of Kwan-se. The div. comprises five districts. The town is 135 m. SSE of Kwei-lin, on the l. bank of the Ta-keang, at the confluence of the Kwei-leang, in N lat. 23° 28′ 48″, and E long. 110° 51′ 15″. It possesses an active trade. The district which is mountainous, is noted for the Kwang-lang, the pith of which forms an important article of food.

WU-CHUEN, a district and town of China, in the prov. of Kwei-chu, and div. of Sze-nan-fu, in N lat. 28° 24′, and E long. 108° 11′ 19″.

WU-CHU-MUCHIN, or UDZEMERDEI, a Mongolian tribe, of Mongolia Proper, about 300 m. NE of Ku-pe-kheon, one of the gates of the great wall They inhabit a territory 105 m. in extent from E to W, and 120 m. from S to N. The r. wing extends to Mount Bakesurkhatai, and the l. to the Kouisun-tologaï. Its chief river is the Khulugur.

WU-CHU-SHAN, a mountain of China, in the prov. of Kan-suh, and dep. of Kwang-chang, in N lat. 35° 7′, and E long. 104° 5′. It is covered with perpetual snow.

WUDARGE, a village of Prussia, in the regency of Stettin, and circle of Saatzig. Pop. 250.

WUDD, a plain of Beluchistan, in the prov. of Jhalawaw, 6 m. in breadth from N to S, and upwards of that extent in length. It is in some parts fertile, and contains a village of the same name.

WUDICKE (GROSS and KLEIN), two villages of Prussia, in the regency of Magdeburg, and circle of Jerichow, containing respectively 380 and 168 inhabitants.

WUDISCHOFZEN, two villages of Austria, in Styria, in the circle of Marburg. Pop. 230.

WUDNI, a pergunnah and village of Sirhind, in Hindostan, 16 m. from Bussian. There is here a square brick fort, encircled by a deep ditch. The village consists of about 400 mud-houses, inhabited chiefly by Sikhs.

WUDZYN, a village of Prussia, in the regency and circle of Bromberg. Pop. 350.

WU-E, a district and town of China, in the prov. of Che-keang, and div. of Kin-wha-fu, in N lat. 28° 53′, and E long. 119° 50′.—Also a district and town in the prov. of Chih-le, and div. of Ching-ting-fu, in N lat. 37° 50′, and E long. 116°.

WUERDALE, or WEARDALE, a township in the p. of Rochdale, co-palatine of Lancaster, 2 m. NE of Rochdale, in the line of the Manchester and Leeds railway, and the Rochdale canal. Pop. with Wordle, in 1831, 6,754; in 1851, 7,855.

WU-GAN, a district and town of China, in the prov. of Ho-nan, and div. of Chang-tih-fu, in N lat. 36° 46′, and E long. 114° 24′.

WU-HEANG-HEEN, a district and town of China, in the prov. of Shan-se, and div. of Tsin-chu, in N lat. 36° 50′, and E long. 112° 50′.

WU-HEEN, a district and town of China, in the prov. of Keang-su, and div. of Su-chu-fu. This district, with those of Chang-chu and Yuen-ho, form the special district of the capital. Wu-heen was formerly capital of the kingdom of Wou, established by the emperor Chaou.

WU-HO-HEEN, a district and town of China, in the prov. of Jan-hwuy, and div. of Sze-chu, near the r. bank of the Yang-tsen-keang, and 15 m. SW of Tae-ping-fu, in N lat. 35° 12′, and E long. 117° 43′.

WUHR, a village of Austria, in Moravia, in the circle of Brunn, and seignory of Runstadt. Pop. 350.

WUHRDEN (LAND), a bail. of the duchy of Oldenburg, in the circle of Ovelgonne, on the r. bank of the Weser. Pop. 1,600. Its chief town is Dodesdorf.

WU-HU, a district and town of China, in the prov. of Jan-hwuy, and div. of Tae-ping-fu, to the SE of Lake Tsaou, and 85 m. NE of Nanking, in N lat. 31° 27′, and E long. 118° 21′. The city is enclosed by a wall, and is said to contain a pop. of 500,000; its trade, however, is at present entirely interrupted by the revolution. There are few objects particularly worthy of note, except the superb bay that stretches out for a distance of several miles just in front of the town; the great bend of the river at this point makes the view singularly grand. The mountains in the vicinity are said to contain coal in great abundance, and the time cannot be far distant when this mineral wealth will be fully developed, as an auxiliary to the steam navigation of the river. The American frigate Susquehannah was under steam 51 hours 20 minutes ascending from Wusung to Wuhu, and 31 hours descending the river from Wuhu to Shanghai.

WUILLY, or MISTELLACHERBERG, a mountain of Switzerland, in the cant. of Friburg, between the lakes of Morat and Neufchatel.

WU-KANG-CHU, a district and town of China, in the prov. of Che-keang, and div. of Hu-chu-fu. in N lat. 30° 33′, and E long. 119° 53′.

WU-KEANG, a district and town of China, in the prov. of Keang-su, div. and 12 m. S of Su-chu-fu, in N lat. 31° 12′, and E long. 120° 20′.—Also a river which has its source in the W part of the prov. of Kwei-chu; runs first E, then N; it enters the prov. of Sze-chuen; and, after a total course of about 600 m., joins the Yang-tsen-keang, on the r. bank, at the town of Pe.

WU-KEANG-HEEN, a district and town of China, in the prov. of Chih-le, and div. of Shin-chu, 144 m. SSW of Pe-king, in N lat. 38° 3′, and E long. 115° 56′.

WU-KEAOU, a district and town of China, in the prov. of Chih-le, and div. of Ho-keen-fu, in N lat. 37° 42′, and E long. 116° 34′.

WU-KUNG-HEEN, a district and town of China, in the prov. of Shen-se, and div. of Kan-chu, in N lat. 34° 20′, and E long. 108° 8′.

WU-KWAN, a district and town of China, in the prov. of Shan-se, and div. of Lu-gan-fu.

WU-KY-NU, or YUKIDO, an island of China, in the Lu-chu archipelago, 27 m. NNE of Great Lu-chu island, and in N lat. 26° 13′, and E long. 128° 42′.

WU-LAN-KHOTAN, a town of Mongolia Proper, 150 m. NNW of Pekin, in N lat. 41° 46′ 48″, and E long. 115° 54′ 30″.

WU-LAN-OBO, a lofty mountain of Mongolia, in the Khalkas country, 240 m. SE of Ourga. It rises in the midst of an extensive plain, and is composed of granite and quartz.

WULATAI, or OULATAI, a town of Mongolia, in the Khalkas country, on the l. bank of the Shishkit, the upper part of the Yenisei, 240 m. from the Russian frontier, and 600 m. WNW of Ourga. It is enclosed by a deep ditch lined on either side with palisades, and beyond is an entrenchment filled with stones and earth. It is the residence of the governor, and the place in which tribute is paid by the nomade tribes that frequent the region between Kiakhta and Dzongaria. It contains about 1,000 houses, and possesses considerable trade. Its inhabitants are chiefly Mongols, who form the military portion of the pop., and Chinese who are mostly artizans or merchants.

WU-LEI-TOU, a lake of Mongolia, in the Khalkas country, 390 m. W of Ourga. It discharges itself on the E by the Shilotie, which a little below takes the name of Selinga.

WU-LE-LEK, a town of Chinese Turkestan, in N lat. 37° 41′, and E long. 76° 39′ 20″, and 8 m. W of the river Didsan.

WULFEL, a village of Hanover, in the district of Kalenburg and bail. of Koldingen, near Hanover. Pop. 438.

WULFEN, a village of the duchy of Anhalt-Cothen, 6 m. NNW of Cothen. Pop. 800, of whom a large proportion are Jews.

WULFER, a village of Lippe-Detmold, and bail. of Schotmar. Pop. 530.

WULFERODE, a village of Hanover, in the district of Kalenberg and bail. of Hanover. Pop. 220.

WULFERSDORF, a market-town of Austria, in the circle of the Lower Manhartsberg, on the Zaga. Pop. 1,800. It has a castle.

WULFING, a village of Prussia, in the regency of Dusseldorf, and circle of Elberfeld. Pop. 430.

WULFINGEN, a village of Hanover, in the principality and bail. of Kalenberg, near Thiedenwiese. Pop. 498.

WULFINGERODE, a village of Prussia, reg. of Erfurt, and circle of Nordhausen. Pop. 442.

WULFINGEN, a village of Switzerland, in the cant. of Zurich, and bail. of Winterthur, on the Toss. Pop. 2,034.—Also a village of Bavaria, in the circle of Lower Franconia and presidial of Hassfurt.

WULFINGHAUSEN, a village of Hanover, in the principality and bail. of Kalenberg, near Elze. Pop. 200.

WULFRATH, a town of Prussia, in the reg. of Dusseldorf, and circle of Elberfeld. Pop. 1,108. It has manufactories of woollen fabrics and iron-works.

WULFTE, a village of Prussia, in the regency of Arnsberg, and circle of Brilon. Pop. 200.

WU-LIASSUTAI, or OULIASSOUTAI, a town of Mongolia, in the W part of the Khalkas country, in N lat. 47° 38′, and E long. 96° 32′, 9 m. N of a river of the same name, an affluent of the Dabkan, which flows into Lake Ike-Aral-Noor.

WU-LING-SHAN, a mountain of China, in the prov. of Chih-le, and div. of Shun-teen-fu, in N lat. 40° 43′, and E long. 117° 25′. It is covered with perpetual snow.

WULLEN, a village of Prussia, in the regency of Munster, and circle of Ahaus. Pop. 680.

WULLERSDORF, a town of Austria, in the prov. of Lower Austria, and lower circle of the Mannhartsberg, on the Schmeida, 24 m. NW of Korneuburg. Pop. 1,000. It has a castle.

WULLERSLEBEN, a village of Schwarzburg-Sondershausen, and bail. of Stadt-Ilm. Pop. 380.

WULLMERINGHAUSEN, a village of Prussia, reg. of Arnsberg, and circle of Brilon. Pop. 240.

WULLUKHAN, a village of Afghanistan, on the l. bank of the Turnak.

WULMSDORF, a village of Hanover, in the principality of Unter-Hoya, and bail. of Wexten, near Verden. Pop. 572.—Also a village of the principality of Luneburg, and bail. of Moisburg, near Harburg. Pop. 278.

WULPERODE, a village of Prussia, in the reg. of Magdeburg and circle of Halberstadt. Pop. 350.

WULSDORF, a village of Hanover, principality of Bremen and bail. of Stotel-Vreland. Pop. 484.

WULTENDORF, a market-town of Austria, in the prov. of the Lower Ens and presidial of Poisdorf. Pop. 600.

WULTSCH, a village of Prussia, in the regency, and circle of Liegnitz. Pop. 448.

WULTSCHKAN, a village of Prussia, in the reg. of Breslau, and circle of Neumarkt. Pop. 370.

WULTSCHUTZ, a village of Prussia, in the reg. of Breslau, and circle of Dels. Pop. 362.

WU-LUK-TAK, a chain of mountains on the S confines of Siberia, partly adjacent to the Chinese empire, and partly to Independent Tartary. It runs from E to W between the Little Altai and the Aitaon mountains, and makes part of the great NW branch of Asia, namely, the ridge which separates the bed of the Northern ocean from that of the W.

WU-LUNG, a district and town of China, in the prov. of Sze-chuen, and div. of Shun-king-fu, 240 m. ESE of Ching-lu-fu, in N lat. 29° 20′, and E long. 107° 44′.

WULUR, a lake of Kashmir, formed by an expansion of the Jailum. It is 21 m. in length from E to W, and 9 in width, and contains a small island, on which are the ruins of a Buddhist temple. It abounds with fish and aquatic birds.

WULVERINGEN, a village of Belgium, in the prov. of W. Flanders, and arrond. of Furnes.

WULZBURG, a town and fort of Bavaria, in the circle of Middle Franconia, and presidial of Weissenburg. Pop. 332.

WULZESHOFEN, a village of Austria, in the prov. of Lower Austria, and circle of Lower Mannhartsberg. Pop. 510.

WUMBACH, a village of Schwarzburg-Sondershausen, in the bail. of Gehren. Pop. 410.

WU-MI, or OUMI, a province of Japan, in the central part of the island of Nifon, to the S of the prov. of Getsisen, and to the W of those of Mino and Ise. Towards the middle of the prov. is a great lake, named Biwano-oumi or Oitz, which discharges itself on the S by the Gedogava, and the E bank of which is bristled with lofty mountains. This principality is well watered, and possesses considerable fertility, producing in great abundance rice, corn, legumes, tea, and excellent tobacco. The bamboo is abundant, but wood is generally scarce. This prov. is one of the most prosperous in Nifon, and is noted for the manufactures of silk and cotton fabrics, and of japanned iron-ware. It contains 15 districts. Its capital is Meaco.—Also a district in the same island, in the prov. of Gamatto.

WU-MINGAMI, a district of Japan, in the island of Nifon, and prov. of Kadsusa.

WUMKU, a mountain of Butan, near Buxadenar, entirely covered with trees, loaded with moss, and entwined with the wild vine. It is in some parts slate, but to a great extent consists of vitrifiable stone, with veins of talc and marble. The bamboo of large size abounds here as well as the banana.

WUMME, a river of Hanover, which has its source in the NW part of the principality of Luneburg, and separates from the gov. of Stade, traversing the marsh named the Grosse Moor. It afterwards enters the S part of the last named gov.; divides it for some distance from the N part of the gov. of Bremen; and, after a total course of 75 m., joins the Hamme; and thence takes the name of Lessum, an affluent of the Weser. The Wumme, in the middle part of its course, divides into several branches.

WU-MUNG-KINN-MIN-FU, a district and town of China, in the prov. of Sze-chuen, in N lat. 27° 20′ 24″, and E long. 103° 46′ 30″.

WU-NATO, a town of Japan, in the island of Nifon, and prov. of Monts, 240 m. NNE of Yedo.

WUNDERSLEBEN, a village of Prussia, in the regency of Erfurt, and circle of Weisensee, on the Unstrut. Pop. 475.

WUNDERTHAUSEN, a village of Prussia, reg. of Arnsberg, and circle of Wittgenstein. Pop. 472.

WUNDICHOW, a village of Prussia, reg. of Koslin, and circle of Stolpe. Pop. 200.

WUNDSCHUH, a village of Austria, in Styria, in the circle of Gratz. Pop. 240.

WUNDSCHUTZ, a village of Prussia, in the reg. of Oppeln, and circle of Kreutzberg, on the Stober. Pop. 430.

WUNG, a village of Afghanistan, 5 m. N of Mittunkote, on the road thence to Kohun.

WUNGA, a town of Sind, on the Purana, an arm of the Indus.

WUNGA, or WUNSEN, a volcano of Japan, in the E part of the island of Kiu-siu, and prov. of Fizen, near the W side of the bay of Simabara, and a little to the NW of the town of that name. It emits almost continually volumes of thick smoke, and the surrounding soil is warm, spongy, and exhales an offensive vapour. This volcano is noted in history as the scene of numerous Christian martyrdoms.

WU-NING, a district and town of China, in the prov of Keang-se, district and 72 m. NW of Nan-chang-fu, in N lat. 29° 15′ 56″, and E long. 115° 11′ 53″.

WU-NING-HEEN, a district and town of China, in the prov. of Hu-nan, and div. of Chang-tih-fu.

WUNKENDORF, a village of Bavaria, in the circle of Upper Franconia, and presidial of Weiss-main. Pop. 300.

WUNNENBERG, a town of Prussia, in the reg. and 57 m. SSW of Minden, and circle of Buren, on the After. Pop. 1,090. It is divided into an upper and lower town. It has paper and saw-mills.

WUNNESCHIN, a village of Prussia, reg. of Koslin, and circle of Lauenburg-Britow. Pop. 200.

WUNSCH, or WUNDSCH (NIEDER and OBER), two villages of Prussia, in the regency of Merseberg, and circle of Onerfurt, near Schafstadt, and containing respectively 318 and 332 inhabitants.

WUNSCHELBERG, or HRADECK, a town of Prussia, in the regency and 54 m. SSW of Breslau, and circle of Glatz. Pop. 1,400. It is enclosed by walls, with three gates, and has two suburbs, two Catholic churches, and an hospital. It possesses manufactories of cloth and linen, and several breweries.

WUNSCHENDORF, a village of Prussia, in the regency of Liegnitz, and circle of Lauban. Pop. 498.—Also a village of the circle of Lowenberg. Pop. 332.—Also a village of Saxony, in the circle of Dresden and bail. of Pirna. Pop. 282.—Also a village of the circle of Zwickau and bail. of Wolkenstein, near Zschopau. Pop. 631.—Also a village of Saxe-Weimar, in the bail. of Weida. Pop. 320.—Also a village of Austria, in Bohemia, in the reg. of Bohmisch-Leipa, and bail. of Friedland. Pop. 700.

WUNSCHENSUHL, a village of Saxe-Weimar, in the circle of Eisenach, and bail. of Gerstungen, on the Suhl. Pop. 360.

WUNSDORF (FERN and NACHST), two villages of Prussia, in the reg. of Potsdam, and circle of Teltow. Pop. 230.

WUNSIEDEL, a presidial and town of Bavaria, in the circle of Upper Franconia. The town is 24 m. E of Bayreuth, on the Roslau. Pop. 4,100. It is enclosed by walls with four gates. The streets are well laid out, but badly kept. It contains three churches, an hospital, and a school, and has extensive manufactories of linen and woollen fabrics, sugar-mills, and in the environs numerous iron-mills. It is noted as the birth-place of Jean-Paul-Richter.

WUNSTORF, a town of Hanover, in the principality of Kalenberg, 15 m. WNW of Hanover, on the Casp-Au, an affluent of the Leine. Pop. 1,980. It has an hospital and a convent, and contains manufactories of woollen fabrics and tanneries.

WU-PING, a district and town of China, in the prov. of Fo-keen, and div. of Ting-chu-fu, 225 m. WSW of Fuh-chu-fu, in N lat. 25° 4′ 48″, and E long. 116° 12′ 30″.

WU-PO, a district and town of China, in the prov. of Shen-se, and div. of Jen-gan-fu, 276 m. NNE of Se-gan-fu, on the r. bank of the Hwang-ho.—Also a district and town in the prov. of Ho-nan, 66 m. WNW of Kai-fung-fu.

WUPPENAU, a village of Switzerland, in the cant. of Thurgau and bail. of Topel, to the S of Gatristoch. Pop. 927.

WUPPER and WUPPERTHAL. See WIPPER.

WUPPERFELD, a town of Prussia, in the prov. of the Rhine, regency of Dusseldorf, circle and 3 m. NE of Elberfeld, on the Wipper.

WURA, a town of Nigritia, capital of the kingdom of Bergu, and 300 m. WNW of Kobbe. Its inhabitants are Mussulmans.

WURANG. See FORMOSA.

WURBENA, a village of Prussia, in the regency of Breslau, and circle of Ohlau. Pop. 562.—Also a village of the same regency, in the circle of Schweidnitz. Pop. 528.—Also a village of the regency of Oppeln and circle of Neisse. Pop. 240.—Also a village of the same regency, in the circle of Grottau. Pop. 210.—Also a village of Austria, in Bohemia, in the regency of Bohmisch-Leipa, and bail. of Friedland. Pop. 700.

WURBENTHAL, or WRBENTHAL, a town of Austria, in Silesia, in the circle and 27 m. NW of Troppau, on the Gold Fluss, an affluent of the Oppa. Pop. 2,100. It has extensive manufactories of linen and woollen fabrics, bleacheries, wire-mills, and in the environs important iron-mines, and blast furnaces.

WURBIS, a village of Saxony, in the circle and bail. of Bautzen. Pop. 215.

WURBITCH, a Japanese establishment in the island of Iturup, Kurile islands, in the SW part of the island, near a volcano. It has a port, defended by a fort. In 1807, it was attacked and pillaged by the Russians.

WURBITZ (DEUTSCH and POLNISCH), two villages of Prussia, in the regency of Oppeln, and circle of Kreutzburg, containing respectively 532 and 678 inhabitants.

WURCHA, a town of Abyssinia, in the state of Shoa and Efat, and prov. of Efat, 45 m. W of Ankober.

WURCHLAND, a village of Prussia, in the regency of Liegnitz, and circle of Glogau. Pop. 210. It has a castle.

WURCHOW, a village of Prussia, in the regency of Koslin, and circle of Neu-Stettin. Pop. 659.

WURCHWITZ, a village of Prussia, in the regency of Liegnitz, and circle of Glogau. Pop. 240. It has a castle.—Also a village of the regency of Merseburg, and circle of Zeitz. Pop. 200.

WURDA, a river of Hindostan, in the prov. of Gundwana, an affluent of the Wyne-Gunga, in E long. 79° 51′. Its principal affluent is the Payne-Gunga, a stream little inferior in magnitude.

WURDAN, a town of Lower Egypt, in the prov. and 39 m. NW of Gizeh, on the l. bank of the chief W branch of the Nile. It was long a resort of pirates.

WURDENHAIN, a village of Prussia, in the regency of Merseburg, circle of Liebenwerda. Pop. 200.

WURDINGHAUSEN, a village of Prussia, in the regency of Arnsberg, and circle of Olpe. Pop. 215. It has an iron-work.

WURDIZOWANG, a parish of Australia Felix, in the co. of Grant.

WURDWAN, a village of Cashmere, near the head of the Muru Wurdwan valley, on the E frontier of Cashmere.

WUREMATTA, an island of the Asiatic archipelago, in the group of the Banda islands, and to the W of Timor-Laut.

WURENLINGEN, a village of Switzerland, in the circle of Aargau, and bail. of Baden. Pop. 1,130.

WURENLOS, a village of Switzerland, in the cant. of Aargau, bail. and 5 m. SE of Baden, near the r. bank of the Limmat. Pop. 793. It contains some Roman antiquities.

WURFLACH, a market-town of Austria, in the circle of the Lower Ens, and bail. of Neu-Kirchen. Pop. 600.

WURGASSEN, a village of Prussia, in the regency of Minden and circle of Hoxter. Pop. 410.

WURGAU, a village of Bavaria, in the circle of Upper Franconia, presidial of Schesslitz. Pop. 247.

WURGENDORF, a village of Prussia, in the regency of Arnsberg, and circle of Siegen. Pop. 345.

WURGES, a village of Nassau, in the circle of Idstein. Pop. 940.

WURGLAH, a town of Algeria, in the prov. of Belud-ul-Jered, 135 m. SW of Taggart, near the Sahara.

WURGLEN, a village of Switzerland, in the cant. and 14 m. ENE of Zurich, and bail. of Kyburg. It has a cotton-factory.

WURGSDORF (NIEDER and OBER), two villages of Prussia, in the regency of Liegnitz, and circle of Volkenheim, containing respectively 1,024 and 588 inhabitants. They have extensive bleacheries.

WURGWITZ, a village of Saxony, in the circle and bail. of Dresden. Pop. 200.

WURIA, a river of Abyssinia, in the kingdom of Tigre, which, after a course of 75 m. in a generally W direction, joins the Tacazze on the W.

WURION, or WURKHOKH, a town of Senegambia, in the country of the Yolofs, 105 m. SE of St. Louis, in N lat. 15° 23′ 46″, and W long. 15° 16′. It is noted for its gum.

WURLITZ, a village of Bavaria, in the circle of Upper Franconia, presidial of Schesslitz. Pop. 200.

WURM, a village of Baden, in the circle of the Middle Rhine, and bail. of Pforzheim, on the Wurmbach. Pop. 574. It has a castle.—Also a village of Prussia, in the regency of Aachen, and circle of Geilenkirchen. Pop. 1,418.—Also a river of Würtemberg, an affluent of the Nagold.

WURMANSGUICK, a town of Bavaria, in the circle of Lower Bavaria, and presidial of Eggenfelden, 42 m. WSW of Passau. Pop. 406.

WURMBACH (OBER and UNTER), two villages of Bavaria, in the circle of Middle Franconia, and presidial of Gunzenhausen, containing respectively 143 and 368 inhabitants.

WURMBERG, a village of Würtemberg, in the circle of the Neckar, bail. of Maulbronn. Pop. 933.

WURMBRAND, a village of Austria, in the prov. of Lower Austria, and circle of the Upper Mannhardsberg, and presidial of Weitra. Pop. 280.

WURMERSHEIM, a village of the grand duchy of Baden, in the circle of the Middle Rhine, and bail. of Rastadt. Pop. 298.

WURMLINGEN, a village of Würtemberg, in the circle of the Schwarzwald, and bail. of Rottenburg. Pop. 1,061.—Also a village of the same circle, in the bail. of Tattlingen. Pop. 1,293.

WURM-SEE, a lake of Bavaria, in the circle of Upper Bavaria, and presidial of Sternberg, to the E of lake Ammer, and SW of Munich. It is 15 m. in length from N to S, and 3 m. in extreme breadth.—A small river of the same name issues from the N extremity, and falls into the Amper.

WURNITZ, a village of Austria, in the circle of the Wienerwald, on the Rusbach. Pop. 380. It has a castle.

WURNO, a town of Central Africa, about 15 m. NE of Sakatu, the present residence of the Emerel-Mumenin, the ruler of the Fellatah empire. Dr. Barth, who visited it in 1853, estimated its pop. at 12,000.

WURRICH, a village of Prussia, in the regency of Coblenz, and circle of Zeli. Pop. 200.

WURSCHEN, a village of Saxony, in the circle of and near Bautzen, and bail. of Lobau.

WURSELEN, a village of Prussia, in the regency and circle of Aachen, on the Wurm. Pop. 359.

WURTEMBERG, or WIRTEMBERG, a small kingdom in the SW of Germany, which takes its name from an old castle situated on a mountain not far from Unterturkheim, from the ancient proprietors of which the present dynasty dates its origin. This kingdom, which has the sixth place in the German confederacy, is bounded on the NE, E, and SE, by Bavaria; on the S, for a small distance, by the Lake of Constanz; and on the SW, W, NW, and N, by Baden. Its superficial extent is, according to Rosch, 348 German sq. m.; another calculation estimates it at 362·15 sq. m.; the *Almanach de Gotha* for 1855 states it at 354·14 sq. m. The census of 1817 returned the pop. at 1,397,451; that of December 1852, at 1,733,263.

Physical Features.] The surface of the country is mountainous; on the E runs the Suabian Alb, and on the western edge the Black Forest; and from both these chains several branches run off in all directions into the country. The Alb is not so high as the Black Forest, but of a rougher aspect; it presents limestone and sandstone, and several remarkable caves are found in it. None of the mountains in W. reach the snow-line. The highest point of that part of the Black Forest lying in W., is the Katzenkopf, or 'Cat's Head,' 3,603 ft. in height; and the highest point of the Alb is the Sternberg, 2,776 ft. above the level of the sea. There are no plains, but several large fertile valleys, of which those of the Neckar and Danube are the principal. The principal rivers are the Neckar, which receives the Kocher, the Jaxt, and the Enz; the Danube, which receives the Iller and the Blau; the Tauber; the Schussen; and the Arzen. The two latter fall into the lake of Constanz, which is the only large lake in the kingdom. The climate is mild and healthy, though in the higher parts the winters are very severe.

Soil and Produce.] A few small tracts excepted, W. is one of the most fertile and well-watered countries in Germany. It generally consists of champaign lands, and pleasant well-watered vales. Its fertility is such, that more grain is raised than suffices for internal consumption, and considerable quantities are exported; but this grain is chiefly that species denominated spelt, rye and wheat being much less cultivated here. Flax and hemp are also cultivated. The valleys are almost covered with forests of fruit-trees, which are also abundant in other parts of the country. In 1826 there were in this kingdom 597 vineyards, comprising 82,729 acres; and the total wine produced was estimated at 184,380 kilderkins, of the value of 3,990,831 florins. The wines are rich, palatable, and wholesome; and are generally denominated 'Neckar wines,' though each has a peculiar name of its own, received from the part where it is produced. Cherries are grown in great quantity in the districts of the Alb and the Black Forest, and used for making the celebrated strong liquor called *kirschwasser*. Game and poultry are abundant, and large herds of horned cattle are reared in various parts of the country. In the neighbourhood of Ulm, a particular branch of industry is the feeding of snails: millions of these animals are fattened here in autumn, and sent to Vienna and Italy.—The mountains abound in marbles of variegated hues, some of which are esteemed equal to those of Italy. Remarkably transparent alabaster, agate, crystalline pebbles, black amber, and fine millstones, are also procured; the other minerals are salt, cobalt, sulphur, coal, and porcelain earth. The salt-works at Sulz produce salt sufficient to supply the kingdom. There are mines of silver and copper near Freudenstadt and at Konigswart; of silver at Konigstein, and of copper, at Gullach, near Hornberg. Iron is also found, but this useful mineral was generally brought from Montbelliard, now belonging to France. There are many warm-baths and medicinal springs. Among the former, the most celebrated are those of Wildman. Heilbron is famous for its medicinal springs. There are also large salt-works in the territory of Suabian Hall, lately ceded to W.

Manufactures and Commerce.] W. is an agricultural and not a manufacturing country; only in a few towns are established manufactures of any extent; several branches of industry, however, such as spinning, weaving, and lace-making, are carried on in this country, besides agriculture. Distilleries and oil-mills of great extent are found in several towns. The articles of exportation are cattle, fat oxen, corn, wood, tar, potash, oil, and a few articles of industry; those of importation are colonial wares, silk, and different articles of foreign manufacture. The wages of the workmen in towns range from 1s. 8d. to 4s. 2d. a week, and in the country from 8d. to 20d. per week, beside board and lodging in each case. Workmen engaged by the year generally board and lodge with their employers, receiving in towns from £4 12s. to £5, and in villages from £1 15s. to £3 10s. per annum. Women and children can gain from £3 10s. to £4 10s. per year.

Inhabitants.] The inhabitants are—with the exception of about 9,000 Jews, and some families of Waldenses, and other colonists—all of German descent, speaking partly the Franconian, and partly the Suabian dialect. There were, in 1817, about 950,632 Lutherans, 432,616 Catholics, 2,308 of the Reformed creed, and 500 or 600 of other sects; the proportions of the different denominations remain, in all probability, much the same, although the gross pop. has considerably increased. There is no established religion. The kingdom of W. is one of the states whose inhabitants possess the right of receiving public assistance in distress. The pop. is divided into two classes, the freemen, and those who do not enjoy the *droit de cité*, called *bersitzers*, the latter form a tenth of the whole. The freemen who have acquired their title by purchase or birth, have the right to participate in the revenues of the district or parish of which they are members. The *droit de cité* may be obtained by payment of a sum inferior to that required for the right of freedom; but that payment does not entitle the purchaser to the privileges of the citizens. Nevertheless, whoever is unable to command the necessaries of life by trade or labour, or to obtain them by the assistance of friends, may claim the succour of the parish in which he has either of the rights just mentioned. If a man is too poor to purchase the lower right, he is assigned by the police to some one parish. The care of the poor is carried to a high degree of solicitude by the government, so that a death from want in time of dearth would entail the most rigorous punishment on the functionaries who might be found chargeable with neglect of the deceased. A number of parishes possess a fund called *pium corpus*, arising from voluntary contributions and other casual sources, but principally derived from lands which before the Reformation were held by the Catholic church, but which at that period, instead of undergoing confiscation by the government, were set apart for the relief of the poor. Most of the towns have their poor-houses for the reception of the aged and infirm; where those are wanting, the poor are received into all the houses of the inhabitants in turn, or farmed out in small establishments.

State of Education.] W. is one of the most enlightened countries in Germany; it has produced a number of scholars, and two of the most celebrated modern poets of Germany. Wieland and Schiller were natives of this country. Perhaps no country in Germany, of the same extent, has produced so many distinguished scholars, an honour for which it is indebted to the excellent establishments for education which have always existed here. There is one university, besides a number of gymnasiums, lyceums, and high schools, and several seminaries for Protestant and Catholic clergymen, and schoolmasters. Almost every village has its own school, which is carefully inspected, and provided with well-informed schoolmasters from the seminary at Eslingen. A special law directs that every child must attend school from 6 to 14 years of age, so that in old W., at least, there is scarcely one individual who cannot read and write.

Government.] W. is a constitutional monarchy. The constitution acknowledges equal civil and political rights in every citizen of the state: none are excluded by birth from any employment of the state. Personal liberty, security of property, and perfect liberty of conscience, is guaranteed to every citizen, besides liberty of emigration. The ministers and officers of state are responsible for their acts. The assembly of the state consists of two chambers, the upper one having for its members the princes of the royal family, and the chiefs of those noble families who formerly formed immediate states of the empire, besides certain hereditary members, and members for life, whom the king names; the lower chamber is elective. The transactions of both chambers are published, and the sittings of the lower are public. Without the consent of both chambers no laws can be enacted, changed, or abolished; complaints of faults or abuses in the administration are brought forward in them, and they have the right of granting the budget. The king swears to the constitution when he receives the oath of allegiance. His person is sacred; and he is of age at 18. The succession is hereditary in the male line; the female is only to succeed if the male shall be entirely extinguished. Justice is administered by different courts; the clergy stands under consistories; and the Catholics have a vicar-general. The 6 Catholic convents, which still exist here, are to be suffered to decay with the lives of their present inmates.

Revenue.] The revenue of W. amounted in 1826, according to Balbi, to 23,761,000 francs, or £990,041 13s. 4d. The debt was, in 1817, 30,000,000 florins, for the liquidation of which there existed a fund, which in 1824 had reduced it to 24,152,035 florins, or £2,717,103 18s. 9d. According to Balbi, it was further reduced in 1826 to 56,500,000 francs, or £2,354,166 13s. 4d. The budget for 3 years from 1852-3 to 1854-5 presented an estimated receipt of 36,524,815 florins; and of expenditure 36,530,249 florins. The debt on 30th June 1851 was stated at 48,423,718 florins.

Military force.] By the census of the Rhenish confederation, the military establishment of W. was fixed at 18,000 men, and its contingent at 12,000, which were all called forth at Napoleon's orders, to assist in the Russian campaign. The number of troops furnished to the allies, during the first and second invasions of France, amounted to upwards of 20,000 men; but under the auspices of more peaceful times, the sovereign reduced his military establishment to 8,200 men; and fixed it at 19,017 men, of whom 14,235 are infantry, 2,789 cavalry, and 1,963 artillery in time of war. Every citizen of the state, from the age of 21, is subject to conscription for military service, which lasts 6 years.

Topography.] This kingdom contains 130 towns, 128 boroughs, 1,115 parish villages, 2,410 smaller villages, 2,591 farms, and 296 castles. It is divided into 4 circles, which have their own administration, and are subdivided into bailiwicks, viz.

	Area in German sq. m.	Pop. in 1852.
Circle of the Neckar,	60·42	501,034
Black Forest,	86·72	443,872
Danube,	113·69	413,444
Jaxt,	93·31	374,913
	354·14	1,733,263

The circle of the Neckar is an extensive valley, through which the Neckar flows, on both sides bordered by mountains and hills which in some points rise to about 1,560 ft.: on the right side of the river are some smaller valleys, as the Remsthal, the Murrthal, and the Kocherthal. All these valleys are fertile and well-watered. Besides the Neckar, the principal rivers are the Ens, the Rems, and the Murr. The climate here is the mildest of the whole kingdom. The population, chiefly Lutheran, are mostly employed in agriculture. Stuttgard, or Stuttgart, the metropolis of the kingdom, lies near the Neckar, and in the middle of W. Proper on the Nesenbach, a few miles above its influx into the Neckar, nearly equidistant from the Danube and the Rhine, and 72 m. to the NE of Strasburg.—The circle of the Black Forest lies in and upon the Black Forest, which covers the whole W part, and spreads several of its branches into the interior; on the E the Alb stretches. It has not so mild a climate as the circle of the Neckar, although several of its valleys are fertile. The Neckar has its sources in this circle, and the Danube flows through a small part of it. Having less agriculture, this circle is more industrious than the former, though there are no extensive manufactures. Spinning and weaving is carried on in the country, and in several towns. Ehingen, on the Schmiecha, is one of the most industrious towns of the kingdom, and contains some manufactures of broad cloth, and hosiery. Reutlingen, on the Echaz, has some manufactures, and extensive tanneries. Tubingen, on the Neckar, with near 7,000 inhabitants, is the seat of a university, with 5 faculties, 38 professors, and a library of 30,000 volumes. The circle of the Danube contains 30 towns, 32 boroughs, and 2,015 villages and hamlets. The N part beyond the Danube is covered by the Rauhe Alb; the S part consists of ridges of small mountains and hills. The soil of this circle, except on the N, is more fitted than any other in W. for the purposes of the agriculturist. The richest districts are those between the Danube, the Iller, and the Lake of Constance. Fruit is here grown in great quantity. Ulm in this circle, is a considerable city, at the confluence of three rivers, the Blau, Iller, and Danube. Here the Danube becomes navigable by vessels of burden. The territory belonging to it, now transferred to Wurtemberg along with the city, is about 26 m. long, and 18 m. broad. The circle of the Jaxt is in most parts well cultivated, and produces corn, wine, and fruit; but its principal riches consist in cattle, of which a great number is exported. Sheep are also numerous, and bees are extensively reared in some parts.

History.] The first notice in history of the lords of W. is in 1139, where we find counts of Wurtemberg, or rather Wirtemberg, as the name of the old castle from which they dated their origin was Wirtinberck or Wirtinberg. From Ulrich, count of W., about the middle of the 13th cent., the history of this country runs down in an uninterrupted line to our times. Ulrich, the real founder of the present reigning house, was known as the most enterprising and the most gallant knight in Suabia. Eight times, says an old chronicle, he went to the field, and never was defeated. He died in 1265. His son and successor, Eberhard, reigned above 50 years. He was placed under the ban of the empire by Henry the VII., and driven from his country; but Henry having died in Italy, Eberhard returned home, and transferred his residence to Stuttgard, as the castle of W. had been dismantled during his absence. His son, and several of his successors, enlarged their territory, which, after having been for a short time divided under two lines of chiefs, was again united under another Count Eberhard, who was raised to the dignity of a duke by Maximilian I., in 1495 and who gave a kind of representative constitution to the country. Duke Ulrich introduced the Reformation in the middle of the 16th cent., and joined the Schmalkaldian league, which led him into a war with Austria. His son Christopher remodelled the legislation and administration of the country. Under his successors, in the 17th cent., the reigning house became divided into three lines. That of Stuttgard was continued by Eberhard, or Everhard III. W. suffered much in the Thirty Years' war. Duke Charles Alexander, who reigned in 1733, became a Catholic, and ruined the finances by the bad administration of a Jew, who was at the head of the finance department, and who was hanged by Charles's successor. He left three sons under age, who all reigned one after the other. The last was Frederic Eugene, who had served in the Seven Years' war under Frederic the Great, and educated his children in the Lutheran creed. He was succeeded by his son, Duke Frederic II., who, after an increase of territory occasioned by exchanges and acquisition, became an elector. He allied with Napoleon against Austria, and obtained, in the peace of Presburg, an augmentation of territory, and the title of king. He was one of the first German princes who entered the Rhenish confederacy on the 12th of July, 1806, when he again obtained an accession of territory. He was a firm adherent to Napoleon's system, but early after the battle of Leipzig entered into negotiations with the allies. He maintained his country on the *statu quo* in the congress of Vienna, but joined the German confederacy on the 1st of September, 1815. While the Congress was yet sitting, he summoned the States of his kingdom on the 11th of June, 1815, and laid before them the new constitution he intended to give to his territories. But the resistance of the States of the ancient part of W. prevented the adoption of this new constitution; and the king yielded in so far, that the ancient constitution was to remain in Old W., but not to be extended to the new acquisitions. During these transactions the king died suddenly, on the 28th of October, 1816. He was a man of no common talents and information, and of great firmness and strength of mind; but his strong passions often drove him to commit the most despotical actions. He was succeeded by his son, Wilhelm I., born on the 17th September, 1781. He was married to the princess of Bavaria, but this marriage was dissolved by mutual consent, upon which the princess married the emperor of Austria, and he married the grand-duchess of Russia, sister of the emperor. The present queen, his third wife, is his cousin, and a princess of W. He abolished bondage, and on the 31st of December, 1818, issued an excellent regulation for the organization of the municipalities, which was to serve as a ground-work for the new constitution. On the 10th of June, 1819, he summoned the States, and expressed his positive determination to establish a new constitution, which was unanimously accepted and proclaimed on the 25th September, 1819. The prince-royal, born in 1823, married in 1846, the grand duchess Olga, daughter of the Russian czar.

WURTHFLETH, a village of Hanover, in the gov. of Stade, and principality of Bremen, near Hagen. Pop. 318.

WURTING, a village of Bavaria, in the circle of Lower Bavaria, and presidial of Rothalmunster. Pop. 280.

WURTINGEN, a village of Würtemberg, in the circle of the Schwarzwald, and bail. of Urach, on the Alp. Pop. 807. It has a castle.

WURTSBOROUGH, a village of Mamakating township, Sullivan co., in the state of New York, U. S., 97 m. SW of Albany, on the Delaware and Hudson canal. Pop. in 1840, 200.

WURTSCH-HOLLE, a village of Prussia, in the regency of Liegnitz, and circle of Lüben. Pop. 340. It has several mills.

WURTZEN. See WURZEN.

WURZACH, a town of Würtemberg, in the circle of the Danube, and bail. of Leutkirch, on the Aitrach, 38 m. S of Ulm. Pop. 1,050. It has a seignorial castle, the residence of the princes of Waldburg-Zeil-Wurzach.

WURZBACH, a village of Reuss-Schleiz, in the district and to the W of Lobenstein, on the Gormitz, Pop. 1,550. It has several spinning-mills and iron-works.

WURZBACH (NIEDER and OBER), two villages of Bavaria, in the circle of the Pfalz, and cant. of Bliescastel, containing respectively 762 and 279 inhabitants.—Also a village of Würtemberg, in the circle of the Schwarzwald, and bail. of Calco. Pop. 280.

WURZBERG, a village of Hesse-Darmstadt, in the prov. of Starkenburg, and presidial of Michelstadt. Pop. 440.

WURZBURG, a town of Bavaria, capital of the circle of Lower Franconia, pleasantly situated on the Main, 63 m. WNW of Nuremburg, and 147 m. NW of Munich, at an alt. of 540 ft. above sea-level, in N lat. 49° 47′ 39″, and E long. 27° 55′ 15″. Pop. 28,000. It is surrounded by a lofty wall and deep ditch, and has six gates. It is divided by the Main, which is here crossed by a fine bridge, into two parts, of which that on the r. bank forms the town proper. The other, distinguished as the quarter of the Main, contains the castle of Marienburg, or citadel, situated on a rock 400 ft. in height. It is built without regard to regularity, but in some portions possesses considerable elegance. The principal edifice is the royal palace, forming the episcopal palace, a fine structure erected on the plan of that of Versailles, and possessing extensive gardens. It contains 33 churches, inclusive of the cathedral, an edifice of the 8th century, twelve hospitals, two infirmaries, an orphans' asylum, houses of correction, and of industry, a Catholic university, with about 700 students, and connected with it an hospital, a medical school, museum, and library. It has besides a gymnasium, a normal, an ecclesiastical, and other seminaries, three convents, a museum, and a theatre. It possesses manufactories of cloth, sealing wax, leather, hats, surgical and mathematical instruments, tobacco, saltpetre, vinegar, glauber's salts and colours. The environs afford fine promenades, and produce excellent wine. This town, which is one of the most ancient in Germany, was formerly the capital of Franconia, and one of the episcopal cities of the empire. It was ceded to Tuscany in 1803, and in 1815 was re-united to Bavaria.

WURZEH, a village of Afghanistan, in the district of Jelalabad, in a fertile valley of the Sufeid-Koh mountains.

WURZEN, a town and bail. of Saxony, in the circle and 17 m. E of Leipzig, on the r. bank of the Mulde. Pop. in 1837, 4,061; in 1843, 4,145; in 1849, 4,868; in 1852, 5,256. It is enclosed by walls, and has a suburb, a castle, three churches, and an hospital. It possesses manufactories of linen fabrics, hosiery, and tobacco, and several bleacheries.

WURZURKHAN, a fortress of Afghanistan, 50 m. W of Cabul.

WUSA, a town of Japan, in the island of Kiusiu and prov. of Bouzen, 90 m. NE of Nagasaki.

WU-SA-KINN-MIN-FU, a district and town of China, in the prov. of Sze-chuen, on the E of that of Wu-mung-kinn-min-fu.

WUSANSK, a town of Russia in Europe, in the gov. of Orenburg, and district of Sterlitamak. It has extensive iron-works.

WUSATA, a district of Japan, in the island of Nifon and prov. of Musasi.

WUSCHAN, a village of Austria, in Styria, in the circle of Gratz. Pop. 240.

WUSCHM - WOLOTSCHOK. See WISCHNEI-WOLOTSCHOK.

WUSCHUTEE, or MUCH MOUNTAINS, a range of mountains in Beluchistan, in the prov. of Mekran, in N lat. 28°. It is noted for its dates.

WUSCHWEIER, a colony of Prussia, in the reg. of Potsdam, and circle of Ober-Barunn. Pop. 516.

WUSCHWEILER, a village of Bavaria, in the circle of the Pfalz, and cant. of Rockenhausen. Pop. 345.

WU-SEIH, a district and town of China, in the prov. of Keang-su, and div. of Chang-chu-fu, a little to the N of Lake Tai-hu, 69 m. ESE of Nan-king, in N lat. 31° 38′, and E long. 120° 8′. It is 5 m. in circumference, exclusive of the suburbs, and is surrounded by a substantial wall, 25 ft. in height, and a deep ditch. It is well supplied with good water, and is noted for its porcelain.

WUSEN, a village of Prussia, in the reg. of Konigsberg, and circle of Braunsberg. Pop. 638.

WU-SEUEN, a district and town of China, in the prov. of Kwan-se, and div. of Tsin-chu-fu, in N lat. 23° 42′, and E long. 109° 21′.—Also a district and town in the prov. of Kwang-tung, and div. of Kaou-chu-fu, in N lat. 21° 19′, and E long. 110° 18′.

WU-SHAN, a district and town of China, in the prov. of Sze-chuen, div. and 21 m. E of Kwei-chu-fu, on the l. bank of the Yang-tsen-keang, in N lat. 31° 9′, and E long. 109° 52′.

WUSLACK, a village of Prussia, in the reg. of Konigsberg, and circle of Hutsburg. Pop. 478.

WUSSEEKEN, a village of Prussia, in the reg. of Koslin, and circle of Lauenburg-Butow. Pop. 220.—Also a village of the same regency, in the circle of Fürstentheim. Pop. 290.—Also a village of the same regency, in the circle of Schlawe, on the Wipper. Pop. 210.

WUSSEGEL, a village of Hanover, in the principality of Luneburg, and bail. of Hitzacker. Pop. 340.

WUSSENTIN, a village of Prussia, in the reg. of Stettin, and circle of Anclam. Pop. 300.

WUSSOW, a village of Prussia, in the reg. of Stettin, and circle of Randow. Pop. 330.—Also a village of the regency of Koslin, and circle of Lauenburg-Butow. Pop. 300.

WUST, a village of Prussia, in the regency of Magdeburg, and circle of Jerichow. Pop. 528.

WUSTEFELDE, a village of Hesse-Darmstadt, prov. of Fulda, and circle of Hersfeld. Pop. 240.

WUSTENBRAND, a village of Saxony, in the circle of Zwickau, and bail. of Chemnitz. Pop. 880.

WUSTENDORF, a village of Prussia, in the prov. of Silesia, regency and circle of Breslau. Pop. 384.

WUSTENHAMME, a village of Prussia, in the regency of Oppeln, and circle of Lublin. Pop. 220.

WUSTENMARK, a village of Mecklenburg-Schwerin, in the prov. of Mecklenburg, and bail. of Schwerin. Pop. 220.

WUSTENROTH, a village of Würtemberg, in the circle of the Neckar, and bail. of Weinsberg. Pop. 440.

WUSTENSACHEN, a market-town of Bavaria, in the circle of Lower Franconia, and bail. of Hilders, on the Alster. Pop. 1,160.

WUSTENSELBITZ, a village of Bavaria, in the circle of Upper Franconia, and presidial of Munchberg. Pop. 490.

WUSTENSTEIN, a village of Bavaria, in the circle of Upper Franconia, and presidial of Obermanstadt. Pop. 319. It has a paper-mill.

WUSTENZELE, a village of Bavaria, circle of Upper Franconia, and presidial of Homberg. Pop. 283.

WUSTERBARTH, a village of Prussia, in the regency of Koslin, and circle of Belgard. Pop. 390.

WUSTEREI, a village of Austria, in Bohemia, reg. of Gitschin, and bail. of Braunau. Pop. 610.

WUSTERHAUSE, a village of Prussia, in the reg. of Koslin, and circle of Neustettin. Pop. 360.

WUSTERHAUSEN, a town of Prussia, in the prov. of Brandenburg, regency of Potsdam and circle of Ruppin, on the Dosse, 15 m. WSW of Neu-Ruppin. Pop. 2,900. It is enclosed by walls with three gates, and has an hospital.

WUSTERHAUSEN (DEUTSCH), a village of Prussia, in the prov. of Brandenburg, regency of Potsdam, and circle of Teltow, on the Notte. Pop. 128.

WUSTERHAUSEN (KONIGS), a market-town of Prussia, in the prov. of Brandenburg, reg. of Potsdam, and circle of Teltow, on the Suhne. Pop. 300.

WUSTERHUSEN, a village of Prussia, reg. of Stralsund, and circle of Greifswald. Pop. 220.

WUSTERMARK, a village of Prussia, in the reg. of Potsdam, and circle of Osthavelland. Pop. 388.

WUSTERNITZ, a village of Prussia, in the reg. of Stettin, and circle of Kammin. Pop. 230.

WUSTERORESDORF, a village of Prussia, reg. of Liegnitz, and circle of Hirschberg. Pop. 615.

WUSTERWITZ, a village of Prussia, in the reg. of Koslin, and circle of Dramburg. Pop. 378.—Also a village of the same regency, in the circle of Schlawe. Pop. 334.—Also a village of the reg. of Frankfort, and circle of Soldin. Pop. 210.

WUSTERWITZ, GROSS and KLEIN (LUTJEN), two villages of Prussia, in the regency of Magdeburg, and circle of Jerichow, containing respectively 270 and 255 inhabitants.

WUSTE-WALTERSDORF, a village of Prussia, in the regency of Breslau, and circle of Waldenburg. Pop. 1,600. It has a castle, and possesses manufactories of cloth, and several bleacheries.

WUSTRA, a village of Austria, in Bohemia, circle of Bunzlau, and seignory of Lissa. Pop. 500.

WUSTRAU, a village of Prussia, in the reg. of Potsdam, and circle of Ruppin. Pop. 508.

WUSTROW, a town and bail. of Hanover, in the principality of Luneburg, on the Jeetze, at the confluence of the Damme. Pop. 700. It has a castle, and possesses manufactories of linen fabrics.—Also a village of the grand-duchy of Mecklenburg-Strelitz, in the circle of Stargard. Pop. 260.—Also a village of Prussia, in the regency of Frankfort, and circle of Konigsberg. Pop. 220.—Also a parish of Mecklenburg-Schwerin, in the circle of Justrow, 21 m. NE of Rustock, on the peninsula of Fischland, in N lat. 54° 20′ 48″, and E long. 30° 3′ 43″. Pop. 1,500.

WUSTRUT, a village of Bavaria, in the circle of Lower Franconia, and presidial of Geroldshofen.

WUSTUNG, a village of Austria, in Bohemia, in the reg. of Bohmisch-Leipa, and bail. of Friedland. Pop. 400.

WUSTWALLENROTH, a village of Hesse-Darmstadt, in the prov. of Hanau, and circle of Gelnhausen. Pop. 200.

WUSTWEILER, a village of Prussia, in the reg. of Treves, aad circle of Ottweiler. Pop. 495.

WUSTWERGK, a village of Prussia, in the reg. of Frankfort, and circle of Lubben. Pop. 300.

WUSTWESEL, or WUESTWSEEL, a department, commune, and village of Belgium, in the prov. and arrond. of Antwerp. Pop. of dep. 1,878, and of village, 441. It has a vitriol work.

WUTACH, a river of the grand-duchy of Baden, which issues from Lake Titi, in the W part of the circle of the See; runs first E to the vicinity of Bromberg; then SSW, and receives the Schlucht, near Thiengen; and, after a total course of about 45 m., falls into the Rhine, on the r. bank, between Waldshut and Kadelburg.

WU-TAE-HEEN, a district and town of China, in the prov. of Shan-se, div. and 90 m. NE of Tae-chu, in N lat. 38° 45′ 36″, and E long. 113° 24′.

WUTHENON, a village of Prussia, in the reg. of Frankfort, and circle of Soldin. Pop. 350.—Also a village of the regency of Potsdam, and circle of Ruppin. Pop. 260.

WUTICKE, a village of Prussia, in the reg. of Potsdam, and circle of Ostpriegnitz. Pop. 438.

WUTING, a bay of Great Lu-chu island, on the NW coast, to the E of a peninsula, in which is a town and port of the same name, in N lat. 26° 40′, and E long. 127° 55′. It is nearly circular, and is 5 m. in diameter. It contains several islands, one of them of considerable size, and at its entrance are several rocks. The port of Wu-ting is safe and commodious. The entrance is narrow, but easy of access.

WU-TING-CHU, a division and town of China, in the prov. of Yun-nan. The div. comprises two districts. The town is 39 m. NNW of Yun-nan, in a fertile locality, bordered by a range of arid mountains, in N lat. 25° 32′ 24″, and E long. 102° 32′ 30″.

WU-TING-FU, a division and town of China, in the prov. of Shan-tung. The division comprises ten districts. The town is in N lat. 37° 33′, and E long. 117° 41′.

WUTO, a town of Japan, in the island of Kiusiu, and prov. of Figo, near the E coast of Simabara bay, and 60 m. E of Nagasaki.

WUTOSCHINGEN, a village of Baden, circle of the Upper Rhine, and bail. of Waldshut. Pop. 480.

WUTRIENEN, a village of Prussia, in the reg. of Konigsberg, and circle of Allenstein. Pop. 330.

WUTSCHDORF, a village of Bavaria, in the circle of the Upper Pfalz, aud presidial of Amberg. Pop. 230.

WU-TSING, a district and town of China, in the prov. of Chih-le, and div. of Shun-teen-fu.

WU-TSIN-HEEN, a district and town of China, in the prov. of Keang-su, and div. of Chang-chu-fu.

WUTTENDORF, a village of Prussia, in the reg. of Oppeln, and circle of Kreutzburg. Pop. 220.

WU-TUNG-HUN, a district and town of China, in the prov. of Che-keang, and div. of Hu-chu-fu.

WUTZ, a district of Japan, in the island of Nifon, and prov. of Bichu.—Also a district of the same island, in the prov. of Gamatto.

WUTZ, a village of Prussia, in the reg. of Potsdam, and circle of West Priegnitz. Pop. 200.

WUTZETZ, a village of Prussia, in the reg. of Potsdam, and circle of Kuppin. Pop. 260.

WUTZIG, a village of Prussia, in the regency of Koslin, and circle of Dramburg. Pop. 320.—Also a village of the regency of Frankfort, and circle of Friedeberg. Pop. 260.

WUTZOW, a village of Prussia, in the regency of Frankfort, and circle of Konigsberg. Pop. 500.

WUWA, a district of Japan, in the island of Si kokf, and prov. of Iyo.

WU WANG CHU, a district and town of China, in the prov. of Hu-nan, and div. of Paou-king-fu.

WU-WEI-CHU, a district and town of China, in the prov. of Gan-hwuy, and div. of Leu-chu-chu, in N lat. 31° 23′, and E long. 117° 53′.

WU-WEI-HEEN, a district and town of China, in the prov. of Kan-suh, and div. of Leang-chu-fu.

WU-Y, a mountain of China, in the prov. of Fu-keen, and dep. of Keen-ning. It grows large quantities of the tea known in Europe as Bohea, and is noted for the number of temples, hermitages, and handsome private dwellings which are to be found in it.

WU-YANG, a district and town of China, in the prov. of Ho nan, and div. of Nan gan fu, in N lat. 33° 32′, and E long. 113° 38′.

WU-YIH-HEEN, a district and town of China, in the prov. of Chih-le, and div. of He-chu.

WU-YUEN, a district and town of China, in the prov. of Gan-hwuy, and div. of Hwuy-chu-fu.

WU-YUEN-HEEN, a district and town of China, in the prov. of Kwan-se, and div. of Sze-gan-fu.

WUYUK-LINGGA, or WYUK, a town of Tibet, in the prov. of Thsang, and 60 m. SW of Lassa, in a valley, watered by the Uyuk, an affluent of the Yaru-dzangbo-chu.

WYAKOMING, a lake in the S part of the state

of Wisconsin, U. S., in La Ponte co., which discharges itself, by a river of the same name, into the Croix river, of which it forms the principal head stream.

WYALUSING, a township of Bradford co., in the state of Pennsylvania, U. S., 103 m. NNE of Harrisburg, drained by a creek of the same name, an affluent of the Susquehanna. Pop. in 1840, 1,400.

WYANDOF, a central co. of the state of Michigan, U. S., in the N part of the peninsula, comprising an area of 576 sq. m., drained by Cheboygan and Skooawargurmish rivers, which are received into large lakes, in the N part of the co., and other considerable streams, the waters of which are discharged into Mackinan straits. It is generally level, and possesses considerable fertility.—Also a county in the NW part of the state of Ohio, comprising an area of 351 sq. m., drained by Sandusky and Tychmochte rivers, and intersected by Mad river and Lake Erie, the Finlay branch and the Ohio and Indiana railways.

WYANG, an island of the Asiatic archipelago, in the group of the Molucca islands, to the E of Gilolo.

WYBELSUM, a village of Hanover, in the prov. of Aurich, and district of West Friesland near Emden. Pop. 394.

WYBERTON, a parish in the co. of Lincoln, 2½ m. S of Boston. Area 3,231 acres. Pop. 647.

WYBRANOWKA, a market-town of Austria, in Galicia, in the circle and 33 m. WNW of Brzezany, on two small lakes.

WYBUNBURY, a parish in the co.-palatine of Chester, 3½ m. ESE of Nantwich, containing the chapelries of Doddington and Weston, and the townships of Bartherton, Basford, Blakenhall, Bridgemere, Checkley with Wrynchill, Chorlton, Hatherton, Hough, Hunsterson, Lea, Rope, Shavington with Gresty, Sound, Stapeley, Walgherton, Willaston, and W. The church is a spacious building, with a remarkable leaning tower. Area 18,414 acres. Pop. in 1831, 4,193; in 1851, 7,179.

WYCH. See Droitwich.

WYCHEN, a village of Holland, in the prov. of Gueldres, arrond. and 8 m. WNW of Nimeguen. It has a castle.

WYCHMAEL, a department and prov. of Belgium, in the prov. of Limburg, and arrond. of Hasselt, watered by the Dommel. Pop. of dep., 349; of com., 159.

WYCISTOWO, a village of Prussia, in the reg. of Posen, and circle of Schrimm. Pop. 200.

WYCK, a town of Holland, in the prov. of Limburg, on the r. bank of the Meuse, opposite Maëstricht, of which it may be considered a suburb.

WYCK-AND-BERG, a commune of Belgium, in the prov. of Antwerp, and dep. of Zoerleparwys. Pop. 252.

WYCK-HUYS, a commune of Belgium, in the prov. of East Flanders, and dep. of Cruyshantem. Pop. 1,974.

WYCLIFFE, a parish in the N. R. of Yorkshire, 5½ m. ESE of Bernard Castle. It contains the township of Thorpe. Wickliffe, the reformer, is said to have been born here in 1324. A fine portrait of this celebrated man, by Sir Antonio More, was given by Dr. Zouch, when rector of this parish, to his successors, the rectors of Wycliffe, who were requested to preserve it as an heir-loom to the rectory house. Area 2,162 acres. Pop. in 1831, 156; in 1851, 144.

WYCOMBE, a parish of Tasmania, in the co. of Devon, bordered on the S by the river Meander.

WYCOMBE (High, or Chipping), a parish, borough, and market-town in Buckinghamshire, 29 m. W by N of London. Area of p., 6,318 acres. Pop. in 1801, 4,248; in 1831, 6,299; in 1851, 7,179. The church is an ancient structure in the early style of English architecture, with a square embattled tower crowned with modern pinnacles. The borough first returned members to parliament 28° Edward I., since which time it has always returned 2 members. The new boundaries are much more extensive than the old, being coincident with those of the parish. The town is situated on the road from London to Oxford, and consists principally of one street, with some smaller ones branching off at right angles. A rivulet, called Wycombe stream, flows through it, and afterwards falls into the Thames a little below Marlow. The adjoining country is agreeably diversified with hills covered with wood. Wycombe-abbey, the seat of Lord Carrington, is situated in a bottom near the town, in a park of about 200 acres. Many of the inhabitants of W. are engaged in the manufacture of paper, which has been carried on here to a great extent, numerous mills on the Wick river and another stream in the manufacturing district, being kept in active operation. On Wycombe stream are also several mills for the grinding of corn. Lace is manufactured to a considerable extent, as also some articles of wood, such as chairs. The prosperity of Wycombe, however, has been chiefly sustained by the great traffic constantly passing through it, a source of emolument not benefited by the opening of the Great Western railway, which passes about 9 or 10 m. to the S of the town. This place is undoubtedly of great antiquity, and is supposed by some to have been founded by the Romans. The word *cheaping*, is the Saxon term for 'a market,' which proves it to have been occupied by that people, and the remains called Desborough-castle are generally supposed to have been of Saxon origin, and to have been intended to keep off the invasions of the Danes. W. gives the inferior titles of earl and baron to the marquess of Lansdowne.

WYCOMBE (West), a parish in Buckinghamshire, 2½ m. NW by W of High Wycombe. Area 6,340 acres. Pop. in 1831, 1,901; in 1851, 2,000.

WYCZECHOWO, a village of Prussia, in the reg of Danzig, and circle of Karthaus. Pop. 400.

WYDAWY, a village of Prussia, in the reg. of Posen, and circle of Kröben. Pop. 410.

WYDCIERCEWICE, a village of Prussia, in the reg. of Posen, and circle of Schrode. Pop. 210.

WYDDIALL, a parish in Hertfordshire, 1½ m. NE of Buntingford. Area 1,187 acres. Pop. 245.

WYDEN, a village of Switzerland, in the cant. of Aargau, and bail. of Bremgarten. Pop. 342.

WYDOW, a village of Prussia, in the prov. of Silesia, reg. of Oppeln, and circle of Zost. Pop. 320.

WYE, a river of Tasmania, an affluent of the Great Swan Port river, in the co. of Glamorgan.

WYE, a parish, formerly a market-town, in Kent, 4 m. NE of Ashford, on the river Stour, which is here crossed by a bridge of 5 arches. The town consists of two streets, running parallel, and two cross streets. The houses are, in general, well-built. Area 7,282 acres. Pop. in 1851, 1,724.

WYE, a river which rises in Cardiganshire, and, passing through Radnor, Hereford, Brecknock, and Monmouth, shires, falls into the Severn, below Chepstow: see Gloucestershire. This river is navigable almost to the Hay, and by a canal communicates with the Severn from Hereford to Gloucester. A railway, named the Lydney and Lidbrook new Severn and Wye railway, runs from the Severn at Lydney to the Wye at Ruarden, a length of 26 m. with branches to collieries. &c. It joins the Monmouth railway.

WYENAD, or Bynadu, also called Nellcala

and WYNATEL, a district of Hindostan, in the presidency of Rombay, forming a subdivision of the prov. of Malabar, extending between the 11° and 12° N lat., skirted on the W and S by the Western Ghauts, and comprising an area of 1,250 sq. m. It is generally hilly, and is in many parts densely overrun with jungle. It is noted for its cardamons. It contains a village of the same name, also called Panamburt Cotta, 40 m. NE of Calicut, in N lat. 11° 45′, E long. 76° 10′. Pop. in 1800, 8,070.

WYERKEN, a commune of Belgium, in the prov. of Limburg, and dep. of Lommel. Pop. 167.

WYERSDALE (NETHER), a township in the p. of Gartstang, co.-palatine of Lancaster, 4 m. NNE of Garstang, near the Preston and Lancaster railway. Area 3,640 acres. Pop. in 1831, 770; in 1851, 704.

WYERSDALE (OVER), a chapelry in the p. of Lancaster, co.-palatine of Lancaster, 6 m. NNE of Garstang. Area 16,938 acres. Pop. in 1851, 680.

WYFORDBY, or WYVERBY, a parish in Leicestershire, 3 m. E of Melton-Mowbray, intersected by the river Eye and the Oakham canal. It comprises the chapelry of Brentingby. Area 1,350 acres. Pop. in 1831, 98; in 1851, 161.

WYGANOW, a village of Prussia, in the reg. of Posen, and circle of Krotoszyn. Pop. 300.

WYGMAEL, a commune of Belgium, in the prov. of Brabant, and dep. of Herent. Pop. 555.

WYHAM, a parish in Lincolnshire, 7 m. NW by N of Louth. It comprises the hamlet of Cadeby. Area 1,880 acres. Pop. in 1831, 94; in 1851, 161.

WYHL, a village of the grand-duchy of Baden, in the circle of the Upper Rhine, and bail. of Kenzingen. Pop. 1,630.

WYHLEN, a village of the grand-duchy of Baden, in the circle of the Upper Rhine, and bail. of Lorrach. Pop. 1,080.

WYHRA, a village of Saxony, in the circle of Zwickau, and bail. of Glauchau. Pop. 569. It has mineral wells.

WYK. See WICK. See also BEVERWYCK.

WYK, or WIJK-BY-DUURSTEDE, a town of Belgium, in the prov. and arrond. and 14 m. SE of Utrecht, on the r. bank of the Leck, at its confluence with the Rhine, in N lat. 51° 58′ 23″, and E long. 23° 20′ 38″. Pop. 2,050. In the vicinity are the ruins of the castle of Duurstede.

WYKEHAM, a parish in the N. R. of Yorkshire, 6½ m. S by W of Scarborough. Area 7,535 acres. Pop. in 1831, 605; in 1851, 643.

WYKEHAM (EAST), a parish in the co. of Lincoln, 7 m. W by N of Louth. Area 560 acres. Pop. in 1851, 26.

WYKEN, a parish in Warwickshire, 3 m. NE by E of the city of Coventry. Area 670 acres. Pop. in 1831, 104; in 1851, 141.

WYKE-REGIS, a parish in Dorset, 1 m. WSW of Weymouth, on the coast at Chesil-bank. Area 2,062 acres. Pop. in 1831, 1,197; in 1851, 1,898.

WYKON, a village of Switzerland, in the cant. of Luzern, and bail. of Willisan. Pop. 959.

WYL, a town of Switzerland, in the cant. and 18 m. NW of St. Gall, and bail. of Gossan, near the Thur. Pop. 1,555.—Also a parish of the cant. and SE of Bern, and bail. of Konolfingen. Pop. 954.—Also a parish of the cant. of Zurich, and bail. of Bulach, on the r. bank of the Rhein. Pop. 932.—Also a parish of the cant. of Aargau, and bail. of Laufenburg. Pop. 636. See also WEIL.

WYLA, a parish of Switzerland, in the cant. of Zurich, and bail. of Pfaffikon, on the Toss. Pop. 1,131.

WYLAM, a township in the p. of Ovingham, Northumberland, 9 m. W of Newcastle-upon-Tyne, on the river Tyne, and in the line of the Newcastle and Carlisle railway, on which there is here a station. Area 930 acres. Pop. in 1851, 1,091.

WYLDECOURT, a tything in the p. of Hawkchurch, Dorset, 5 m. NE of Axminster. Pop. 267.

WYLEN, a village of Switzerland, in the cant. of Thurgau, and bail. of Tòbel. Pop. 292.

WYLER, a village of Prussia, in the regency of Dusseldorf, and circle of Geldern. Pop. 630.

WYLEROLTIGEN, a village of Switzerland, in the cant. of Bern, and bail. of Laupen. Pop. 451.

WYLHOF. See HOFWYL.

WYLIHOF, a village of Switzerland, in the cant. of Luzern, and bail. of Sursee. Pop. 266.

WYLLIAN. See WILLEN.

WYLP, a village of Holland, in the prov. of Gueldres, to the N of Zutphen.

WYLRE, a village of Holland, in the prov. of Limburg, WSW of Maëstricht. Pop. 1,588.

WYMEER, a village of Hanover, in the district of Ostfriesland, and bail. of Weener, near Bunde. Pop. 668.

WYMERING, a parish in the county of Hants, 4¾ m. W of Havant. Part of this parish is in Portsea island, and is connected with the mainland by a bridge. Great and Little Horsea islands are also in the parish, and it contains the hamlet of Hilsea. Area 4,307 acres. Pop. in 1831, 578; in 1851, 751.

WYMINGTON, or WINNINGTON, a parish in Bedfordshire, 6½ m. N of Harrold. Area 1,710 acres. Pop. in 1831, 257; in 1851, 296.

WYMONDHAM, a parish in Leicestershire, 6½ m. E of Melton-Mowbray. Area 2,852 acres. Pop. in 1831, 746; in 1851, 800.

WYMONDHAM, or WINDHAM, a parish and market-town in Norfolk, 9 m. WSW of Norwich. Besides the town, which forms the in-soken, the parish contains the divisions of Downham, Market-street, Silfield, Suton, Towngreen, and Wattlefield, which constitute the out-soken. Area 10,613 acres. Pop. in 1831, 5,485; in 1851, 5,177. The town is well-built. Bombazines and crapes are made here to a considerable extent.

WYMONDLEY, or WIMUNDESLEY (GREAT), a parish in Hertfordshire, 2 m. ESE of Hitchin. Area 1,120 acres. Pop. in 1831, 321; in 1851, 335.

WYMONDLEY, or WIMUNDESLEY (LITTLE), a parish in Hertfordshire, 2½ m. SE by E of Hitchin. Area 790 acres. Pop. in 1831, 226; in 1851, 500.

WYNDHAM, a town of Western Australia, in the co. of Plantagenet, on the shores of Port Two-people.

WYNDORP, or FOCO, a town of the island of Foco, in the gulf of Biafra, Western Africa.

WYNEGHEM, a department and commune of Belgium, in the prov. and arrond. of Antwerp. Pop. of dep., 958.

WYNE GUNGA, or WEIN GUNGA, a river of Hindostan, in the prov. of Gundwana, which has its source in the Mâhadee mountains, in the district of Seonee-Chawparrah, at an alt. of 1,850 ft. above sea-level; runs first E, then in a generally S direction to the Godavery, which it joins on the l. bank at Kalishwar, and about 10 m. ESE of Chinnur. It has a total course of about 439 m. Its principal affluents are the Khahan and Wurda, both of which it receives on the r. This river gives its name to a district of considerable extent, containing 13 pergunnahs, 2,111 villages, and a total pop. of 690,770 persons.

WYNGENA, a department and commune of Belgium, in the prov. of West Flanders, and arrond. of Bruges. Pop. 7,000. It has manufactories of linen.

WYNHUYZE, a commune of Belgium, in the prov. of East Flanders, and dep. of Essche-Saint-

Lieven. Pop. 497.—Also a commune of the same prov., dep. of Steenhuyze-Wynhuyze. Pop. 261.

WYNINGEN, a village of Switzerland, in the bail. and 4 m. NNE of Berthoud. Pop. 2,725. It has some Roman antiquities, and possesses manufactories of linen.

WYNKEL, a department and commune of Belgium, in the prov. of East Flanders, and arrond. of Ghent, watered by the Moervaerd and Sasschevaerd. Pop. of dep., 2,062.

WYNKELDRIES, a commune of Belgium, in the prov. of East Flanders, and dep. of Melden. Pop. 161.

WYNKEL-SAINT-ELOI, a department and com. of Belgium, in the prov. of West Flanders, and arrond. of Courtrai. Pop. of dep., 2,909.

WYNKOOP'S BAY, a bay on the S coast of the island of Java, in the Preanger regency, opening between Ragat and Payong points, which are 9½ m. distant from each other. It has a depth of 10 m., and is surrounded by high hills clothed with trees. On its NE side are government storehouses for coffee and salt, in S lat. 6° 59½', E long. 106° 35'.

WYNN, a parish of Durham co., New South Wales, bounded on the S by the river Hunter.

WYNORY. See PRACSA.

WYOCENA, a township of Columbia co., in the state of Wisconsin, U. S., 30 m. N by E of Madison, drained by Duck creek. Pop. in 1850, 506.

WYOMING, a county in the W part of the state of New York, U. S., comprising an area of 580 sq. m., drained by Allen, Cayuga, Tonawanda, Hueca, and other creeks, and intersected by the Buffalo and New York City, the Genesee Valley railways, and the Genesee Valley canal. Pop. in 1850, 31,981. Its capital is Warsaw.—Also a county in the NE part of the state of Pennsylvania, intersected by the Susquehanna river, and by the Lackawanna and Western railway, North Branch canal, &c. The surface is diversified, and generally fertile. Pop. in 1850, 10,655. Its capital is Tunkhannock.—Also a county in the W part of the state of Virginia, comprising an area of 1,342 sq. m., drained by the head branches of Guyandotte river, the Tug Fork of Sandy river and Camp creek. Its capital, Wyoming, is on the W bank of Allen's creek, 229 m. W of Albany. Pop. in 1850, 700.—Also a village of Middleburg township, Wyoming co., in the state of New York, on Allen's creek. Pop. in 1840, 600; in 1850, 700.—Also a village of Luzerne co., in the state of Pennsylvania, 131 m. NE of Harrisburg, on the W side of the Susquehanna. Pop. in 1840, 200.—Also a village of Starke co., in the state of Illinois, on the E side of Spoon river, 85 m. N by W of Springfield. Pop. in 1840, 75.—Also a township of Iowa co., in the state of Wisconsin, 32 m. W of Madison, drained by Pipe creek and other affluents of the Wisconsin. Pop. in 1850, 206.

WYRAGUR. See BYRAGUR.

WYRARDISBURY, or WRAYSBURY, a parish in Buckinghamshire, 3 m. SW by S of Colnbrook, on the NE bank of the Thames. Area 1,656 acres. Pop. in 1831, 682; in 1851, 701.

WYRE-PIDDLE, a township in the parish of Fladbury, Worcestershire, 2 m. NE by E of Pershore, on the river Avon. Pop. in 1831, 175; in 1851, 269.

WYRLEY (GREAT), a township in the p. of Cannock, Staffordshire, 6½ m. NNW of Walsall. Pop. in 1831, 541; in 1851, 824.

WYROW, a village of Prussia, in the regency of Bromberg, and circle of Wirsitz. Pop. 260.

WYRZEKA, a village of Prussia, in the reg. of Posen, and circle of Kostyn. Pop. 380.

WYSALL, or WISHON, a parish in the co. of Nottingham, 9 m. S by E of Nottingham. Area 1,360 acres. Pop. in 1831, 271; in 1851, 284.

WYSKITA, a village of Austria, in Bohemia, in the reg. of Budweis, and bail. of Pilgram. Pop. 550.

WYSMIERZYCE, a town of Russia in Europe, in Poland, in the gov. and 28 m. of Radom, on the r. bank of the Pilcia. Pop. 1,000.

WYSOCKO, two villages of Prussia, in the reg. of Posen, and circle of Adelnau. Pop. 550.

WYSOCZAN, a village of Austria, in Bohemia, reg. of Prague, and bail. of Karolinthal. Pop. 660.

WYSOGOTOWO, a village of Prussia, in the regency and circle of Posen. Pop. 200.

WYSOKA, a village of Prussia, in the regency of Pardubitz, and bail. of Hohenmauth. Pop. 280. —Also a village of the same regency, and bail. of Pardubitz. Pop. 650.

WYSOKIE, a town of Russia in Europe, in Poland, in the gov. of Lublin, circle and 24 m. WSW of Krasnistan. Pop. 300.—Also a town of the gov. of Augustowo. Pop. 890.

WYSOKI-MEYLO. See HOHENMAUTH.

WYSOKOW, a village of Austria, in Bohemia.

WYSOX, a township of Bradford co., in the state of Pennsylvania, U. S., 105 m. N by E of Harrisburg, drained by Wysox and Rumfield creeks. Pop. in 1840, 1,871; in 1850, 2,000.

WYSSECHRAD. See PLINTENBURG.

WYSSOGOTOWO, a village of Prussia, in the regency of Oppeln, and circle of Gross-Strelitz. Pop. 320. It has a castle, and possesses manufactories of pottery.

WYSTEMP, a village of Prussia, in the reg. of Oppeln, and circle of Ortelsburg. Pop. 210.

WYSTOCZ, a village of Prussia, in the regency of Posen, and circle of Kosten. Pop. 280.

WYSTYTEN, a town of Russia in Europe, in Poland, in the gov. of Augustowo, and 27 m. NNW of Suwalki, on the confines of Prussia, and to the N of a lake of the same name. Pop. 2,500.

WYSZANOW, a village of Prussia, in the reg. of Posen, and circle of Schildberg. Pop. 542.

WYSZECZIN, a village of Prussia, in the reg. of Danzie, and circle of Neustadt. Pop. with that of Wysz-Hutte, 200.

WYSZKI, a town of Prussia in the regency of Posen, and circle of Pleischen. Pop. 205.

WYSZKOW, a town of Russia in Europe, in Poland, in the gov. and 17 m. ESE of Plock, on the r. bank of the Bug. Pop. 1,290.

WYSZOGROD, a town of Russia in Europe, in Poland, in the gov. and 27 m. ESE of Plock, on the r. bank of the Vistula. Pop. 3,300. It has three churches, an ancient convent, and a poors'-house, and possesses manufactories of woollen and linen fabrics.

WYTEGRA. See VYTEGRA.

WYTHAM, or WIGHTHAN, a parish in Berks, 3 m. NW of Oxford. Area 1,670 acres. Pop. 195.

WYTHBURN WITH ST. JOHN CASTLERIGG, a chapelry in the p. of Crossthwaite, Cumberland, 8 m. SE by S of Keswick. Pop. in 1851, 558.

WYTHE, a county in the SW part of the state of Virginia, U. S., comprising an area of 578 sq. m., drained by New river and its branches, and by branches of Holston river, and intersected by the Virginia and Tennessee railway. It consists of an elevated table land, and possesses considerable fertility. Pop. in 1840, 9,375; in 1850, 12,024. Its capital is Wytheville.

WYTHEVILLE, a village of Wythe co., in the state of Virginia, U, S., 195 m. W by S of Richmond, and on the Virginia and Tennessee railroad. Pop. in 1840, 450; in 1850, 600.

WYTHOP, a chapelry in the p. of Lorton, Cumberland, 5 m. E by S of Cockermouth. Area 3,013 acres. Pop. in 1831, 121; in 1851. 119.

WYTIKON, a village of Switzerland, in the cant., bail., and 3 m. from Zurich. Pop. 328.

WYTON, a township in the p. of Swine, E. R. of Yorkshire, 5 m. NE of Hull. Area 788 acres. Pop. in 1331, 93; in 1851, 91.

WYTONSYL, a village of Prussia, in the regency of Posen, and circle of Buk. Pop. 290.

WYTSCHAETE, a commune of Belgium, in the prov. of W. Flanders, and arrond. of Ypres. Pop. 3,000. It has manufactories of table linen.

WYTSCHENSKE, a village of Prussia, in the reg. of Posen, and circle of Frankstadt. Pop. 200.

WYTUTAKE, an island of the Southern ocean, in Cook's archipelago.

WYVELL, a parish in Lincolnshire, 4 m. NW of Colsterworth. Area 1,670 acres. Pop. in 1831, 128; in 1851, 135.

WYVERSTONE, a parish in Suffolk, 6¾ m. N of Stow-market. Area 1,522 acres. Pop. 329.

WYZAYNY, a town of Russia in Europe, in Poland, in the gov. of Augustowo, 20 m. N of Suwalki, on the Prussian frontier. Pop. 1,000.

WYZONCHOW, a village of Prussia, in the reg. of Posen, and circle of Krotoszyn. Pop. 240.

X

⁎ For names beginning with this letter not found amongst the following, see under the letters G, H, J, or S.

XABEA, or Jabea, a town of Spain, in the prov. of Alicante, and partido of Denia, on the Mediterranean. Pop. 3,654. It has extensive fisheries.

XACURUNIA, a lake of Brazil, in the prov. of Mato-Grosso, which discharges itself by a river of the same name, into the Sumidouro, an affluent of the Arinos.

XAFALOTE, a mountain of Brazil, on the confines and forming one of the points of demarcation of the prov. of São-do-Pedro-Rio-Grande, and the Banda-Oriental.

XAGUA, a bay of the S coast of the island of Cuba, 45 m. WNW of Trinidad. It forms one of the best ports of the Antilles. Towards its centre, at the distance of 8 m. from the shore, springs of fresh water rise in ebullitions to the surface.

XAGUA, or Aguan, a river of the state of Honduras, which separates the districts of Comayagua and Taguzgalpa, and, after a course, in a generally NNE direction, falls into the gulf of Honduras, 33 m. WSW of Truxillo.

XAINTES. See Saintes.

XAINTRAILLE, or Saintraille, a town of France, in the dep. of the Lot-et-Garonne, and cant. of Lavardac, 7 m. NNW of Nerac. Pop. 700.

XALA, a town of Mexico, in the state of Michoacan, 210 m. WSW of Valladolid. Pop., chiefly Indians, 210.

XALANCE, or Jalance, a town of Spain, in the prov. of Valencia, and partido of Jarafuel, 42 m. WNW of San-Felipe-de-Jativa, near the r. bank of the Jucar, on a mountain crowned by a fortress. Pop. 1,300.

XALAPA. See Jalapa.

XALISCO, a market-town and port of Mexico, in the intendency and 159 m. W of Guadalajara, on the Pacific. Pop. 270. It contains a Franciscan convent, and has several tanneries. See Guadalajara.

XALON, or Jalon, a town of Spain, in the prov. of Alicante, and partido of Denia, on the r. bank of a river of the same name. Pop. 2,276. It has a parish-church, a custom-house, and a public granary, and possesses manufactories of linen and other fabrics.—The river X. has its source on the E side of the Iberian mountains, in the W part of the prov. of Calatayud; runs first NE past the town of that name; enters the prov. of Zaragoza; and, after a course of 120 m., joins the Ebro, on the r. bank, 12 m. below Zaragoza. Its principal affluent is the Xiloco.

XALPOTAN, a name by which the N part of the lake S. Cristoval, in Mexico, in the state of that name, is distinguished. It contains an island on which is a village of the same name.

XAMAPA, a river of Mexico, in the state of Vera-Cruz, which has its source in Mount Orizaba; runs E, and, after a course of about 90 m., throws itself into the gulf of Mexico, a little to the S of Vera-Cruz.

XAMILTEPEC, or Jamiltepec, a market-town of Mexico, in the state and 90 m. SW of Oaxaca, on the Chicometepec. Pop. 4,000. It has an active trade in honey, wax, cacao, cotton, and salt.

XANACI, or Tapiraquia, a river of Brazil, in the prov. of Mato-Grosso, an affluent of the Xingu.

XANDAY. See Xenday.

XANDRE (Saint), a commune of France, in the dep. of the Charente-Inferieure, cant. and 4 m. NE of La Rochelle. Pop. 1,126.

XANTEN, a town of Prussia, in the prov. of the Rhine, regency and 36 m. NNW of Dusseldorf, and circle of Geldern, near the l. bank of the Rhine. Pop. in 1843, 3,200. It has a fine Gothic church, and a gymnasium, and possesses manufactories of woollen, cotton, and silk fabrics, and tobacco. This town appears to occupy the site of a colony founded by Trajan, under the name of Colonia Trajana. Near it are still to be seen the traces of the Roman camp of Vetera Castra.

XANTHI, a range of mountains in Turkey in Europe, in Rumelia, in the sanj. of Gallipoli, to the NW of the gulf of Lagos.—Also a village 8 m. NW of Lagos.

XANTHUS, an ancient city of Asia Minor, situated on the Etchenchai, in N lat. 36° 21′, 20 m. SE of Macri, the site and remains of which were first explored, in modern times, by Sir C. Fellows in 1838. Its predecessor was the *Arina* of the Termilæ. Harpagus, a general of Cyrus the Great, destroyed this city B. C. 547, and X., erected on or near its site, by Lycian Greeks, was of little importance for some centuries. It revived, however, under Roman auspices a little before the Christian era.

but was destroyed by an earthquake towards the close of the 5th century.

XAPICO, a river of Brazil, in the S part of the prov. of São-Paulo, which has its rise in the W side of the Serra-do-Mar; runs W across the Campos-de-Vicaria, and along the confines of the prov. De-Rio-Grande-do-Sul; and, after a course of 180 m., joins the Rio-das-Pilotas, a head stream of the Uruguay.

XARA-DUKA, or KASAR-ADUKIN, a town of N. Africa, in Tripoli, 12 m. E of Marate, and near the SE coast of the gulf of Sidra. The surrounding territory is covered with ruins, one of the most remarkable of which is a castle, circular in form, cut out of the solid rock, and surrounded by a ditch.

XARA-HOTUN. See KARA-HOTUN.

XARAYES, an extensive plain of South America, stretching from the Brazilian prov. of Mato-Grosso, to the S of the city of that name, into Bolivia, a distance of 260 m. During the rainy season it is to a great extent inundated by the waters of the Paraguay, but at other periods these are confined to mere rills, and the soil is covered with vegetation.

XATIVA. See FELIPE-DE-JATIVA (SAN).

XAUJA, or JAUJA, a town of Peru, in the intendency and 78 m. N of Guancabellica, and 120 m. E of Lima, on a river of the same name, an affluent of the Mantaro. Pop. 3,000. It has productive silver mines, and carries on an extensive trade in grain, fruit, and cattle.

XAVEA, or JABEA, a town and fortress of Spain, in the prov. and 48 m. NE of Alicante, and partido of Denia, near Cape San Antonio. Pop. 3,654. It is well-built, and has a parish-church, two convents, and an hospital. The trade, which is considerable, consists chiefly in figs, raisins, grain, fruit, wine, silk, &c.

XAVIER, a small island of Brazil, in the prov. of Santa Catharina, in the Atlantic, near the S point of the island of Santa Catharina.—Also an island of Patagonia, in S lat. 47° 4′, W long. 74° 27′.

XAVIER (SAO), a town of the Argentine republic, in the state and 105 m. NE of Santa-Fe.—Also a town of the republic and 300 m. N of Monte Video, on a small affluent of the Ybicuy.—Also an island of the gulf of Penas, on the W coast of Patagonia, in S lat 47°, and W long. 79°.—Also a missionary settlement of Bolivia, in the dep. of Moxos, prov. and 165 m. NE of Santa-Cruz. There is another mission of the same name on the Mamora, 300 m. NNW of Santa-Cruz.

XAWEROW, a market-town of Russia in Europe, in the gov. of Volhynia, district and 38 m. SSE of Obrutch, on the l. bank of the Kamenka. Pop. 1,500.

XEA. See GEA.

XELSA, a market-town of Spain, in Aragon, in the prov. and 36 m. SE of Zaragoza, on the l. bank of the Ebro. Pop. 2,250. It has an hospital and several convents.

XENDAY, a town of Japan, on the E coast of the island of Nifon, in the prov. of Mouts, 255 m. NE of Yedo.

XENIA, a large and fertile township of Green co., in the state of Ohio, U. S., 47 m. WSW of Columbus, drained by Little Miami river, and intersected by the Little Miami, the Columbus, and Xenia, and the Dayton and Xenia railroads. Pop. in 1840, 4,913; in 1850, 7,056. It contains a village of the same name, with a pop., in 1840, of 1,200; and in 1850 of about 2,200 persons.

XENIL. See JENIL.

XEPEE. See JEPEE.

XERENTES, a tribe of Indians of Brazil, who inhabit the N part of the prov. of Goyaz.

XEREZ, or JERES, a town of Spain, in Andalusia, in the prov. and 39 m. E of Granada, and partido of Guadex, amid the chain of the Sierra-Nevada. Pop. 1,700. It has a parish-church, a custom-house, and a public granary.

XEREZ-DE-LA-FRONTERA, ZERES, or JERES, a judicial partido and town of Spain, in Andalusia, in the prov. of Cadiz. The town lies in a fertile district, 15 m. NE of Cadiz, and 2 m. from the r. bank of the Guadalete, on the road from Cadiz to Seville, in N lat. 36° 41′, and W long. 6° 7′ 20″. Pop. 33,104. It is intersected by a semi-circular line of old walls. The streets of the inner and ancient portion of the town are narrow and tortuous, those of the exterior, spacious and regular. It contains 8 parish churches, inclusive of a large collegiate church, containing a library and a museum of ancient coins, several chapels, numerous convents, a town-house—a fine edifice—4 hospitals, a foundling asylum, a college, two free-schools, a public granary, and barracks. The fortress is flanked with ancient towers. The wine stores are of great extent. The finest entrances to the town are towards Utrera and Porto-Santa-Maria. X. possesses little industry. Its only manufactures are coarse woollen fabrics, leather and soap for local use. The trade consists chiefly in wine grown in the surrounding district, amounting annually to 500,000 *arrobas*, of which more than a half is exported from the two ports of Cadiz and Santa-Maria to England. In the vicinity are quarries of white marble, and about 5 m. to the SE are the sulphureous baths of Jogonza, and 2 m. from it in the same direction, on the r. bank of the Guadalete, is an abbey, a fine edifice, adorned with pictures by the first masters. This town derives its surname from its situation near the Spanish frontiers. A signal victory over Don Rodrigo, in the vicinity, near the Guadalete, gained by the Moors in 1714, laid the foundation of the domination of that people in Spain. In 1264 this town was taken by Alphonso the Wise. X. is supposed by some authors to be the *Asta-Regia Cœsariana* of the Romans; by others, to have been only erected from its ruins, and identifying these with the adjacent Mesa-de-Asta.

XEREZ-DE-LA-FRONTERA, a market-town of Brazil, in the prov. of Mato-Grosso, 375 m. S of Cuyaba, and a little to the E of the Paraguay.

XEREZ-DE-LA-FRONTERA, a town of the state of Honduras, in the district of Tegucigalpa, 78 m. S of Comayagua, in the valley of Choluteca, on the Xerez, near its entrance into the gulf of Fonseca. It is reputed the most southerly and warmest town in the state. The gold-mine in the vicinity, formerly the richest in Guatemala, is now nearly exhausted. Cacao and indigo are extensively cultivated in the locality.

XEREZ, or JEREZ-DE-LOS-CABALLEROS, a judicial partido and town of Spain, in Estremadura, in the prov. of Badajoz. The partido comprises 10 pueblos. The town is 42 m. S of Badajoz. Pop. 5,628. It is enclosed by walls, and contains four churches, numerous convents, and two hospitals. It possesses manufactories of linen, leather, hats, pottery, and soap, and carries on an active trade in cattle. In the vicinity are mines of sulphur and silver.

XERICA, or JERICA, a town of Spain, in Valencia, in the prov. of Castellon-de-la-Plana, and partido of Viver, on the l. bank of the Palencia, and 39 m. NW of Valencia. Pop. 2,979. It has a parish-church, two convents, and contains some Roman antiquities. Agriculture, the manufacture of coarse linen fabrics, and distillation of brandy, form the chief branches of local industry.

XERTA, or CHERTA, a town of Spain, in Catalonia, in the prov. of Tarragona, partido and 20 m. N of Tortosa, on the r. bank of the Ebro. Pop. 2,433. It has a parish-church and a custom-house

XERTIGNY, a canton, commune, and town of France, in the dep. of the Vosges, and arrond. of Epinal. The cant. comprises 9 com. Pop. in 1831, 15,866; in 1846, 17,054. The town is 11 m. S of Epinal. Pop. in 1846, 3,871. It has manufactories of iron-ware and cutlery, forges and fineries.

XEUI, or Jejuhi, a river of Paraguay, which has its source in the Cordillera-de-Maracavu, runs W, and joins the Paraguay on the l. bank, 90 m. above Asuncion. Its principal affluents are the Aquarahi-guazu-mini on the r., and on the l. the Arrientes.

XHAVEE, a commune of Belgium, in the prov. of Liege, and dep. of Wandre. Pop. 215.

XHAWIRS, or Sur-les-Haies, a commune of Belgium, in the prov. of Luxemburg, and dep. of Xhendelesse. Pop. 180.

XHENDELESSE, a department and commune of Belgium, in the prov. of Liege, and arrond. of Verviers. Pop. of dep. 1,325; and of com., 320.

XHENDREMAEL, a department and commune of Belgium, in the prov. and arrond. of Liege. Pop. of dep. 748; and of com. 526.

XHIERFOMONT, a commune of Belgium, in the prov. of Liege, and dep. of Rahier. Pop. 119.

XHORIS, a department and commune of Belgium, in the prov. of Liege, and arrond. of Huy. Pop. 726.

XICALANQUES, a tribe of Indians who inhabit the Mexican state of La Puebla.

XICAQUES, a tribe of Indians who inhabit the state of Honduras.

XICOCO. See Sikokf.

XICOTLAN. See Zacatlan.

XIE. See Ixie.

XIHUQUILPAN, a village of Mexico, in the state of Michoacan, 60 m. W of Valladolid. It has a Franciscan convent, and possesses several sugar and salt-mills.

XILOCA, or Giloca, a river of Spain, in Aragon, which has its source, named Fuelda-de-Celda, in the prov. and to the NW of Teruel; runs NNW into the prov. of Zarragoza; and, after a course of 90 m. joins the Jalon at Calatayud.

XILOCASTRON, a market-town of Greece, in the Morea, in the dep. of Achaia, near the entrance of a small river of the same name, into the gulf of Lepanto, and 21 m. WNW of Corinth. Its trade consists chiefly in raisins.

XILOTEPEC, a market-town of Guatemala, in the state of that name, and valley of the same name. Pop. 4,000. It has several sugar-mills.

XIMA, or Sisio, a prov. of Japan, on the S coast of the island of Nifon, bathed on the E by the bay of Obari, and on the S by the ocean. It is generally mountainous. Fishing forms the chief employment of the inhabitants. Its chief town is Toriba.

XIMABARA. See Simabara.

XIMBINA, a tribe of Indians of Brazil, who inhabit the wilds of the prov. of Mato Grosso.

XIMENA, or Jimena, a market-town of Spain, in Andalusia, in the prov. and 11 m. E of Jaen, and partido of Mancha-Real, in the valley of Belval or Bedmar. Pop. 1,507. It has a parish-church, a custom-house, and a public granary.

XIMENA, or Gimena-de-la-Frontera, a town of Spain, in Andalusia, in the prov. and 66 m. ESE of Cadiz, and partido of San-Roque. Pop. 5,878. It has a parish-church, four convents, an hospital, a castle, and a public granary, and contains several tanneries. In the vicinity is an iron-mine.

XIMO. See Kiu-Siu.

XIMONOSCKI. See Simonoscki.

XINGO, a river of Brazil, which has its source in the Serra-de-Borracha; runs NE, and joins the São-Francisco, on the r. bank, on the W confines of the prov. of Sergipe.

XINGU, a river of Brazil, which has its source in the N side of the great mountain chain which forms, in the prov. of Mato-Grosso, the dividing line between the basins of the Amazon and La Plata, intersected by the prov. of Para, and, after a tortuous course, in a generally N direction of about 1,320 m., joins the Amazon, on the r. bank. 210 m. W of Para, and a little to the E of the island of Aquiqui. Its principal affluents are on the r. the Jangada, Bois, Fresco, Atoma, Ilabagua, Pacaxa, &c., and on the l. the Trubario, Barubo, Trahiras, Bacauris, Acariai, Turucuri, and Juraussu. The chief places on its banks are Souzel, Pombal, Veiros, Porto-de-Moz, and Boa-Vista. Its banks are salubrious and fertile, and generally well-populated.

XINGUTANIA, a district of Brazil, in the prov. of Para, bounded on the N by the Amazon; on the W by the Xingu, from which it derives its name, and by which it is separated from the district of Tapagonia; on the S by the prov. of Mato-Grosso; and on the E by the Tocantins. It is watered by the Anapu, Aratitu, Arias, Jacundaz, Pacajaz, and Tacanhunas; and contains in the N part the towns of Cameta, Gurupa, Melgoco, Oeiras, Pombal, Protel, Porto-de-Moz, and Veiros. The banks of the rivers are finely wooded, and abound with game; and the soil, which is generally fertile, produces rice, tobacco, sugar, &c. It is inhabited chiefly by Indians.

XIONDZ, or Xionz, a town of Prussia, in the prov. and regency and 27 m. SE of Posen, and circle of Schrimm, near the l. bank of the Warta. Pop. 1,150. It possesses a Lutheran and a Catholic church and a synagogue, and has manufactories of linen and of leather.

XIONZEK, a village of Prussia, in the reg. of Posen, and circle of Schrimm. Pop. 200.

XIONZENICE, a village of Prussia, in the reg. of Posen, and circle of Schildberg. Pop. 260.

XIPIXAPA, a town of Ecuador, in the dep. and 75 m. NW of Guayaquil, near the Pacific.

XIPOTO. See Chopoto.

XIQUEXIQUE, a small town of Brazil, in the prov. of Bahia, on the r. bank of the São-Francisco.

XIRIRICA, a tribe of Indians of Brazil, who inhabit a village of the same name in the prov. of Sao-Paula and comarca of Curitiba.

XIST (Saint), a village of France, in the dep. of the Lot-et-Garonne, and cant. of Beauville, 11 m. ENE of Agen. Pop. 750.

XIVRAY-MARVOISIN, a commune of France, in the dep. of the Meuse, and cant. of Mihiel, 15 m. E of Commercy. Pop. 515. It has manufactories of gloves and stockings.

XIXONA, or Jijona, a judicial partido and town of Spain, in Valencia, in the prov. of Alicante. The partido comprises ten pueblos. The town is 18 m. NNW of Alicante, on the side of a mountain of the same name. Pop. 4,795. It has a parish-church, an hospital, and two convents, and is noted for its almonds.

XOA. See Choa.

XOCHIMILCO, a village of Mexico, in the state and 9 m. SSE of Mexico, on the W bank of a lake of the same name. It has manufactories of iron-ware, and under the domination of the Aztecs was a town of considerable importance. The lake of X. is separated from that of Chalco, on the SE, by the dike of Tlahuac, and discharges itself on the N into Lake Tescuco.

XOLO. See Sulu.

XORULLO. See Jorullo.

XUAN-DAI, a town and port of Annam, on the E coast of Cochin-China, in the prov. of Phuyen, in N lat. 13° 22′, and E long. 109° 14′.

XUCAMEL, a lofty mountain of Central America, in the prov. of Vera-Paz, rising between Chichen and Chitzujay, with its summit towards the SE.

XUCAR, or JUCAR, a river of Spain, which has its source on the S side of the Sierra-di-Albarracin, on the N part of the province of Cuenca, and to the NNE of the town of that name; intersects in a N direction the S part of that prov., then turns E, flows into the prov. of Valencia, and after a total course of 210 m. falls into the Mediterranean, a little to the S of Lake Albufera, and 30 m. SSE of Valencia. Its principal affluents are the Gabriel and Magro or Requeno on the l., and on the r. the Albagna. Cuenca is the chief town on its banks.

XUCARAY, a river of Ecuador, in the dep. of Assuay, an affluent of the Amazon, which it joins on the l. bank, between the Pastaca and Tigre or Peguena.

XULLA. See ZULLA.

Y

*** For places not found under Y, look under E, I, or J

Y, HET-Y, or I, a gulf of Holland, in the prov. of N. Holland, extending in a WNW direction, between the Zuider-see and the Haarlem-meer. The chief towns on its banks are Amsterdam on the S coast, near the entrance to the Zuider-see, and Berdyk, at the NW extremity. This gulf was formed in the 13th cent., by an irruption of the sea.

YAAMBLE, a town of New South Wales, in the co. of Wellington, near Mudgee, on the Cudgegong river.

YA-ASAUA, an island of the S. Pacific, in the Fiji group, in S lat. 16° 50′, E long. 177° 23′. It is 10 m. in length, and attains an alt. of 780 ft.

YABAPIAS, an Indian tribe who inhabit the N part of Mexico, between the Gila and Colorado.

YABAQUE, a small island of the Bahama archipelago, in the group of the Lucayes, to the SE of the island of Mariguani, in N lat. 22° 30′.

YABARY, JAVARI, JABARI, or HIAVARI, a town of Brazil, in the prov. of Para, on the r. bank of the Amazon, near the confluence of a river of the same name. It is inhabited by Indians of the Cambelas and Tecunas tribes, who live chiefly by the chase, and by the culture of cacao and legumes. They carry on a small trade in sarsaparilla and cacao, which they barter for iron-ware.

YABBA, a town of New South Wales, in the district of Murray, on the Mitta-Mitta, near Albany.

YABO (EL), a town of Marocco, about a day's journey NW of the wells of Zenalyia.

YABLONEV, a town of Russia in Europe, in the gov. and 108 m. WNW of Poltava, district and 23 m. WSW of Lubni, on the r. bank of the Orjitza.

YABLONNE, a village of Poland, in the gov. of Masow, obwod and 11 m. N of Warsaw, on the r. bank of the Vistula. It has a castle with magnificent gardens.

YABLONNOI. See STANOVOI.

YABLONOV, a village of Russia in Europe, in the gov. of Kursk, district and 12 m. ENE of Karotcha, at the source of the Khasok. It was erected in 1617, and formerly bore the rank of a town.

YABLONOVKA, or CHENFELD, a village of Russia in Europe, in the gov. and 36 m. S of Saratov, district and 45 m. NNE of Kamishin, on the l. bank of the Volga. A German colony settled here in 1763.

YABLOTCHNAIA, a town of Russia in Europe, in the gov. and 45 m. NW of Kharkov, district and 15 m. NW of Bogodukhov, on the l. bank of the Rabina.

YABLUNKA, a town of Austria, in Moravia, in the gov. of Brünn, circle and 17 m. SSE of Telschen, on the l. bank of the Olsa. Pop. 1,780. It has extensive manufactories of linen. About 6 m. S of this town is a fortress of the same name, erected for the defence of the only pass between Silesia and Hungary.

YABRIN, a town of Arabia, in the prov. of Nedjed, 66 m. ENE of El-Haryk. The surrounding district abounds with salt.

YABTAR, a small town of Independent Tartary, in Badakshan, in N lat. 36° 8′, and E long. 73° 15′, on the l. bank of the Kudsa, 54 m. SE of Badakshan.

YABU, a state of Upper Guinea, on the Slave coast between Dahomey on the W, and Benin on the E. It is very fertile, and is inhabited by a vigorous people who employ themselves in agriculture and the rearing of cattle, sheep, and poultry. It is noted for its cotton fabrics.

YABUS, a river which has its rise in Abyssinia, on the W part of the country of the Galla; runs first NW, then N along the confines of Nubia to the Bahr-el-Azrek, which it joins in S lat. 11° 15′. Fadessi and Dolly are the chief places on its banks. Its waters abound with crocodiles and hippopotami. Russegger calls this stream the INBUSS.

YACANA, a Patagonian tribe, who inhabit the S part of Patagonia, near the banks of the Galligos.

YA-CHU-FU, a division and town of China, in the prov. of Sze-chuen. The div. comprises seven districts. The town is 90 m. SW of Ching-tu-fu, in N lat. 30° 3′ 30″, and E long. 103° 3′ 38″.

YACIPARANA, or JACIPARANA, a river of Brazil, in the prov. of Mato-Grosso, which has its source in the Sierra-Parecis, runs NW, and after a course of about 180 m. joins the Madeira, on the r. bank.

YACKABUTTA, a town of New South Wales, in the co. of Bligh, near Murruthera.

YACKADANDA, a creek of Australia Felix, in the district of Murray, which joins the river of that name near Albany.

YACK-JACQUES, a town of Upper Guinea, on the Ivory coast, on the Atlantic, 18 m. E of Lahu, and 135 m. WNW of Cape Three Points. It carries on an active trade in corn, yams and other productions.

YACOB (SAINT), a village of Austria, in Bohemia, in the circle and 8 m. S of Chrudim, on the l. bank of the Chrudimka. In the vicinity is a sulphur mine.

YACOB (BAY OF) OMENAK, or NORTH EAST BAY, an indentation of the W coast of Greenland, in Baffin's Bay, to the N of Waygal's strait, and to the NE of Disco island, and between the districts of Omenak and Rettenbeuks, in N lat. 71°, and W long. 52° 40'.

YACOBA, or JACOBA, a state of Nigritia, to the S of Houssa, and separated from Funda on the S by the Tshadda or Shary. It is covered with mountains of calcareous structure, and containing it is said antimony and silver. The inhabitants are named *yemyems* or cannibals by neighbouring states, and by the Mussulmans of Houssa the state is distinguished by the name of Bouchy, or country of the infidels. It contains a town of the same name, 270 m. S of Kano.

YACOVA, JACOVA, or DIACOVO, a district and small town of Turkey in Europe, in Albania, in the sanj. and 52 m. ENE of Scutari, on the r. bank of a small river of the same name. The district occupies the central part of the sanj. of Scutari, to the r. of the Drin, by which it is bounded on the S.

YACURZO, or JACURZO, a town of Naples, in the prov. of Calabria-Ultra, district and 8 m. S of Nicastro, on a lofty hill. Pop. 1,500.

YADE. See JAHDE.

YADJY, a town of Upper Guinea, on the Gold coast, in the state of Burum, 195 m. E of Kumassi.

YADKIN, a river of the state of North Carolina, U. S., which has its source in the Blue ridge; flows E and SE; enters the state of South Carolina, when it takes the name of Great Pedee; and falls into Winyard bay, near Georgetown. In Montgomery co. it passes through the 'Narrows,' where it is contracted from 600 to 100 ft. in width. It is navigable for small vessels throughout the greater part of its course.

YADRIN, or CHADRIN, a district and town of Turkey in Europe, in the gov. and 120 m. W of Kazan, on the l. bank of the Sura, and 33 m. above the embouchure of that river. The district, which lies in the E part of the gov., is populous and well-cultivated.

YAE-CHU, a district and town of China, in the prov. of Kwang-tung, and div. of Keun-che-fu, or island of Hainan, 123 m. SSW of Keun-che-fu, in N lat. 18° 22', and E long. 108° 48'.

YAENIS-JAERVI, a lake of Russia in Europe, in the grand-duchy of Finland, gov. of Viborg, and district of Norra-Kexholm, to the N of Lake Ladoga, and 15 m. NNE of Serdobal. It is 15 m. in length, and 9 in breadth, and discharges itself by the Leshkiloe, at the S extremity, into Lake Ladoga.

YAESKIS, a district of Russia in Europe, in the grand-duchy of Finland and central part of the gov. of Viborg. It contains in the W part a portion of Lake Saima. S. André is one of its chief places.

YAGAMI, a town of Japan, in the island of Nifon, on the sea of Japan, WNW of Meaco.

YA-GAN-HEEN, a district and town of China, in the prov. of Sze-chuen, and div. of Ya-chu-fu.

YAGNA, a town of New Granada, in the prov. and 55 m. S of Neiva.

YAGNACHE, a town of Ecuador, in the dep. and 26 m. NE of Guayaquil, on a river of the same name, which descends from the Andes, and joins the Guayaquil near the town of that name.

YAGNANIQUE, a port of the N coast of the island of Cuba, in N lat. 20° 42', and W long. 79° 38'.

YAGNARI, a town of the republic and 165 m. NNE of Monte-Video.

YAGNAROCHA, a lake of Ecuador, 3 m. from Ibarra. It is 5 m. in length, and very deep. It derives its name, which signifies 'lake of blood,' from the massacre of the inhabitants of Otabalo by the inca Huana-Cassac, which took place on its banks.

YAGO, or JAGO (SAN), or RIBEIRA-GRANDE, a town of the Cape Verde Islands, on the S coast of the island of San-Jago, and 8 m. WSW of Porto-Prayo, in a deep valley. It has a small port, and possesses some manufactures of cotton fabrics. It rapidly declined in importance since the removal of the seat of government to Porto-Prayo, and the castle and principal dwellings are falling into decay. Its pop. does not now number more than 60 families. On the NE coast of the island is another town also named S. Jago.

YAHNDI, a town of Upper Guinea, on the Gold coast, capital of the kingdom of Dagomba, and 225 m. NE of Komassi. It is said to be a large and flourishing place.

YAHORLYK. See JAGORLIK.

YAILA, or JAILA, a range of mountains in Russia in Europe, extending from SW to NE along the S coast of the Crimea, from Balaklava to Kaffa, and generally calcareous in structure, and presenting a rapid decline towards the S. They are intersected by valleys, in some parts spacious, and in others narrowed by projecting rocks, and watered by streams which descend to the sea. The vegetation is luxurious, and the hills, even to the summits, are studded with Tartar villages. The word *Yaila* signifies 'summer-quarters,' and is often given to any mountain-pasture at a considerable elevation.

YAILEGHERRY, a town of Hindostan, in the presidency of Madras, district and 73 m. NNE of Jalem.

YAIMIR, a town of Hindostan, in the presidency of Madras, and district of South Canara, 22 m. NE of Mangalore. It contains a colossal image of a Jain deity and numerous temples.

YAIR, a town of Upper Guinea, in Sherborough island, on the Sierra-Leone coast.

YAITCE, or JAITZE, a town of Turkey in Europe, in Bosnia, in the sanj. and 33 m. S of Bagna-Luka, and 7 m. E of Ghul-Hissar, on the l. bank of the Verbitza, which is here crossed by a stone bridge, and near the confluence of the Plieva. It is enclosed by a wall flanked with towers, and entered by two gates, and is defended by a fortress. It is noted as the burial-place of a Catholic bishop, who fell by the hands of the Turks 200 years ago, and whose memory is still held in sacred veneration by numerous devotees, who annually resort to the tomb. In the vicinity is a saltpetre work.

YAJGO, a town of Burmah, 150 m. NW of Ava, and to the W of the Khyendwen river.

YAKE-YAMA, a volcano of Japan, in the island of Nifon, and prov. of Monts, between Tanab and Obata.

YAKHSA, or ALBAZIN, a fortress of China, in Manchuria, in the prov. of He-lung-keang, on the l. bank of the Amour, 261 m. ENE of Nertchinsk, and near the Russian frontier. It was demolished after a long struggle with Russia, in 1715.

YAKHVA, a river of Russia in Asia, in the gov. of Tobolsk, which, after forming the dividing line between the districts of Bezerob and Tourinsk, enters the gov. of Tobolsk, traverses it in a NW direction, and joins the Konda on the l. bank, at Maïkoni, 90 m. NNW of Tobolsk, and after a total course in a generally S direction of about 150 m.

YAKIMA, a river of the Oregon territory, U. S., which has its source in the Cascade range; flows first SSE then E to the Columbia river, which it joins near the confluence of Lewis river.

YAKOB (Jisr-Beriat), a bridge of Turkey in Asia, in Syria, in the pash. and 57 m. SW of Damascus, between lakes Houle and Tabariyeh. This celebrated monument consists of three massive arches.

YAKOBENY, or Jakobeny, a village of Austria, in Galicia, in the circle and 69 m. SSW of Tschernowitz, on the l. bank of the Bisztritz. It is noted for its mineral springs. In the vicinity are valuable iron and copper-mines.

YAKOBSAU, or Jakubjau, a village of Hungary, in the comitat of Zips, 17 m. NNE of Leutschau. In the environs is an iron-mine.

YAKOBSHAGEN, or Jakobshagen, a town of Prussia, in the prov. of Pomerania, regency and 45 m. E of Stettin, and circle of Saazig, on a lake of that name at the junction of the Shua. Pop. 1,423.

YAKOBSTADT, or Jakobchtat, a town of Russia in Europe, in the gov. of Kourlande, and district of Alt-Jelburg, 84 m. ESE of Mitau, on the l. bank of the Dvina. Pop. 1,448. It has a Catholic and two Greek churches.

YAKOHALMA, a village of Hungary, in the district of the Yazyges, 5 m. ENE of Jaz-Bereny, and 48 m. E of Pesth, on the l. bank of the Tarna. Pop. 1,600.

YAKSA. See Yakhsa.

YAKUBOV, a town of Russia in Europe, in the gov. and 90 m. NE of Vilna, and 18 m. NNE of Vidzi. See Jakubova.

YAKULGUNJ, a village of Hindostan, in the district of Furruckabad.

YAKULPUR, a village of Hindostan, in the district of Alligurh.

YAKUNOSUNA, an island of Japan, 24 m. W of the island of Tanegasima, and 42 m. S of the island of Kiu-siu. It is 24 m. in length, and 9 in breadth, and according to Krusenstern, its N extremity is in N lat. 30° 42′ 30″, and E long. 131° 7′ 45″. It is perfectly flat, and so low that the trees with which it is covered scarcely seem to rise above the level of the water. On the E and W sides are two large bays. This island is frequented by the inhabitants of Nagasaki, who come hither in quest of wood.

YAKURGA, a village of Russia in Europe, in the gov. of Vologda, and district of Veliki-Oustioug, on the Dvina. It has extensive salt-works.

YAKUSI, a town of Japan, in the island of Nifon, and prov. of Dena, 135 m. N of Magani.

YAKUTSK, an extensive district of Asiatic Russia, originally forming one of the four provinces into which the gov. of Irkutsk was divided, but now composing a separate government. It extends N from the parallel of 43° 30′, or the boundary of Irkutsk-proper and of Nertshinsk, as far as the Frozen ocean. On the E it is bounded by the prov. of Okhotsk; on the S by China; on the SW by the gov. of Irkutsk; and on the W by that of Yeniseisk. From E to W it has a length of nearly 1,700 m.; from S to N it stretches above 1,000 m. The Lena traverses it through its whole extent from N to S. The other rivers are the Olonek, the Indeghirka, and the Kolyma. The sea-coast is low; but the surface rises towards the S and E, where numerous ridges from the Yablannoi mountains intersect the country. Timber attains a large size in the S; in the N the forests are chiefly of stunted pine and birch. The geological formation of the country seems to consist for the most part of carboniferous and other palæozoic deposits, which in the ridges of the Stanovoi hills, and near the coasts, are broken up and dislocated by the intrusion of rocks of igneous origin. Through nearly the whole of this vast country, the severity of the climate renders it impossible to raise grain to advantage; and if a crop of rye is sown, its ripening is extremely uncertain. In the neighbourhood of the town of Yakutsk, the ground, even in June, is frozen 3 ft. deep. The chief employment of the inhabitants, therefore, consists in hunting; and this region was at one time so abundant in valuable furs, as to be called the Peru of Siberia. The sables, however, being now nearly extirpated, the produce of the chase has greatly declined in value. The inhabitants of this dreary region bear the name of Yakuts or Jakuti. They are supposed to have been originally a Tartar race, and to have occupied the countries on the Angara and Upper Lena; but, forced to yield to the violence of the Mongols and Buriats, they have taken shelter in the frozen solitudes of the Lower Lena, from the Witim downwards. The territory occupied by them may be reckoned to extend 18° from N to S, and nearly 50° from E to W. The pop., however, of this region, does not bear any proportion to this extent. The number of males paying tribute in 1796 amounted to 34,979; and this, it was supposed, might be about the third of the entire number of both sexes. The Y. who inhabit the southern districts are tolerably rich; they possess the usual size, activity, and vigour; but those farther to the N form a stunted race, indolent and wretched. M. Ermann says that their countenance and language fully confirm the tradition of their descent from the Tartars. Their complexion is swarthy, and their hair black. "They are properly a pastoral people, whose chief riches consist in the number of their horses and horned cattle, on the produce of which they subsist almost entirely. But the abundance of fur-animals in their vast forests, and the profit which they can make by selling them to the Russians have turned a large part of their attention to the chase, of which they are often passionately fond, and which they follow with unwearied ardour and admirable skill. Accustomed from infancy to the privations incidental to their severe climate, they disregard hardships of every kind. They appear absolutely insensible to cold, and their endurance of hunger is such as to be almost incredible. Their food consists of sour cows' milk, and mares' milk, and of beef, and horse-flesh. They boil their meat, but never roast or bake it, and bread is unknown among them. Fat is their greatest delicacy. They eat it in every possible shape; raw, melted, fresh, or spoilt. In general they regard quantity, more than quality, in their food. They grate the inner bark of the larch, and sometimes of the fir, and mix it with fish, a little meal and milk, or by preference with fat, and make it into a sort of broth, which they consume in large quantities. They prepare from cows' milk what is called the Jakutian butter. It is more like a kind of cheese, or of curd, and has a sourish taste; it is not very rich, and is a very good article of food eaten alone. Both men and women are passionately fond of smoking tobacco. They prefer the most pungent kinds, especially the Circassian. They swallow the smoke, and it produces a kind of stupefaction which nearly resembles intoxication; and if provoked when in this state the consequences are dangerous. Brandy is also used, though the long inland carriage renders it extremely dear. The Russian traders know how to avail themselves of these tastes in their traffic for furs." They are one of the most superstitious races in existence. They reckon not less than thirteen tribes of aerial beings, to whom they pay homage. Their shamans, or magical impostors, enjoy unbounded influence.

Yakutsk, the capital of the above government.

is situated on the W bank of the Lena, on a plain surrounded by mountains, which enclose it at the distance of 10 or 12 m., and only elevated 270 ft. above sea-level. It displays a few wooden houses, and a number of huts plastered over with cow-dung, and windowed with ice; and contains about 7,000 inhabitants, chiefly Yakuts, whose language is spoken by all the Russian merchants here. Notwithstanding its desolate situation, Y. is a place of some trade. It forms the emporium of all the furs which are collected in the extensive regions to the north; and it imports the Russian and Asiatic commodities which are given in exchange. It forms also a species of *entrepot* for the Russian trade with Kamtchatka and the western coast of America. The cold is excessive; and during winter the inhabitants are almost entirely confined to their houses, the day-light not continuing above a few hours. This season is consumed by them almost entirely in sleeping and drinking. It is hardly conceivable how the love of gain could ever have induced men, acquainted with a milder climate, to take up their abode in this, the coldest spot on the earth's surface. The earth is always frozen, the summer's thaw never reaching below 3 ft. from the surface; and MM. Ermann and Middendorff are disposed to conclude that the subterranean ice or frozen subsoil has a depth of 600 ft.! Astounding as this statement may be, yet it seems to be confirmed by all that we know of the observed temp. of Y. For two months in every year, it has a medium temp. of —40°: in January, the therm. has been known to descend to —72°, or 18° below the degree of cold experienced by Captain Ross in his last expedition. Yet the inhabitants of Y., favoured by a warm, though short summer, reap wheat and barley, and cultivate successfully potatoes and a variety of hardy vegetables.

YALA (Cape), a headland of the island of Candia, on the E coast, to the S of the island of Carube, in N lat. 35° 3′, and E long. 26° 15′ 10″.

YALAO-SAKI (Cape), a headland of Japan, on the SE coast of the island of Sikokf, and 45 m. NNE of Cape Tosa.

YALBRET, a town of New South Wales, in the co. of Georgiana, near Lake Burrah-Burrah.

YALI, a small island of the Archipelago, near the SW coast of Anatolia, between the island of Piscopi on the NW, and that of Rhodes on the SE, in N lat. 36° 22′ 15″, and E long. 27° 28′ 35″.

YALLABUSHA, a county in the N part of the state of Missisippi, U. S., comprising an area of 897 sq. m., intersected from E to W by a river of the same name, and drained by its branches, and crossed by the Mississippi Central railroad. It has a finely undulating surface, and is generally fertile. Pop. in 1840, 12,248; in 1850, 17,258, of whom 8,597 were slaves. Its capital is Coffeeville.

YALLAH, a river of Jamaica, in the co. of Surrey, and parish of St. David. It has its source in the Blue mountains; runs SE; flows into a bay of the same name to the W of Cape Yallah.

YALLUNG, a river of Neypal, which has its source in the mountain of Kangchang, a summit of the Himalayas; flows SW; and, after a course of about 40 m., joins the Tambur, in N lat. 27° 21′. On its banks is a town of the same name.

YALO, a river of Mantchuria, which has its source in a range of mountains of the same name, in the prov. of Sagalin-Ula and district of the Solones; runs SE; forms for some distance the dividing line of Mongolia, and joins the Songari about 50 m. SW of Tsitsokar. The Yalo mountains form a portion of the great Khinghan or Siolki chain.

YALOFS. See Jolofs.

YAL-U, a river of Corea, in the prov. of Phingan, which has its source in Mount Golmin-shan-yan-alin, runs in a generally SW direction, and after a course of upwards of 300 m. enters the Yellow-sea, near the E confines of the prov. of Shin-king.

YA-LUNG-KEANG, or Yar-lung-keang, a river of the Chinese empire, which has its source on the confines of Tibet, and of the district of Koko-nor; forms the boundary line of Tibet, and of the Chinese prov. of Sze-chuen; enters that prov., and near its confines joins the Kiu sha-keang, the principal head-stream of the Jang-tse-keang. It has a total course in a generally SSE direction of about 750 m., and bears in Tibet the successive appellations of Tsachu, Tsitsirkana, and Miniakchu.

YAM, a mountain of Persia, in Azerdbijan, between the plain of Khoï and Lake Urumiyah. It has an alt. of 3,230 ft. above sea-level.

YAMADA, a town of Japan, in the island of Sikokf, and prov. of Sanoki, 54 m. N of Awa, on the strait of Mitsu-sima-nada.

YAMAGA, a town of Japan, in the island of Kiu-siu, and prov. of Fijo, 90 m. ESE of Nagasaki.

YAMAISOIL, a region of Japan, in the W part of the island of Nifon. It comprises 13 prov., inclusive of the islands of Ikia and Tsousima.

YAMA-ISOUKURI, a town of Japan, in the island of Nifon, and prov. of Monts, 72 m. NNW of Xendav, on the sea.

YAMALA, a town of Japan, in the island of Nifon, and prov. of Tutami, 102 m. WSW of Jedo.

YAMANASSIRO, a town of Japan, in the island of Nifon, and prov. of Kaï, 81 m. NW of Jedo.

YAMAN-TAU, a mountain-ridge of the Ural system, in the Russian prov. of Orenburg, near the upper valley of the Inzer, between the rivulets Kus and Kazaulu. Its principal peak presents the largest mass of rocks in the Southern Ural.

YAMASETA, a town of Japan, in the island of Kiu-siu, and prov. of Tsikongo, and 27 m. ESE of Sanga, on the gulf of Samabara.

YAMATFOURO, or Sansion, a prov. of Japan, in the SW part of the island of Nifon, and region of Yetsen, and to the W of the prov. of Omni, to which it formerly belonged. It is generally mountainous, but the valleys are fertile, producing rice, corn, legumes, cotton, and silk. Hara is the chief town.

YAMATTO, a prov. of Japan, in the island of Nifon, and region of Yetsen, to the N of the prov. of Kinokouni, and E of that of Kavatsi. It is intersected by mountains, with intervening valleys and plains of great fertility. Firao is the chief town.

YAMAYAS, an Indian tribe who inhabit the N part of Mexico, near the Colorado.

YAMAYO, a town of Japan, in the island of Kiu-siu, and prov. of Tsi-Koulen, 60 m. NE of Sanga.

YAMBARANGA, a town of New South Wales, on the river Lachlan, near Pomabil.

YAMBEESEE, a town of Sind, in the district and 68 m. NE of Hydrabad.

YAMBU. See Yembo.

YAMERINNA, a river of New South Wales, in the district of Wellington, an affluent of the Lachlan, near Mount Torrens.

YAM-HILL, a county of Oregon Territory, drained by a river of the same name, and by the Chehalem, affluents of the Willamette. Pop. in 1850, 1,512. Its capital is Lafayette.

YAMINA, a town of Nigritia, in the state of Bambarra, on the l. bank of the Joliba, 78 m. WSW of Sego.

YAMMIE, a town of Upper Guinea, on the Gold

coast, in the kingdom of Gaman, 90 m. NW of Kumassi.

YAMNAIA, or JAMNAIA, a town of Russia in Europe, in the gov. and 48 m. NW of Kharkov, and district of Bogodoukhov, on the l. bank of the Vorskla.

YAMNITZ, or GEMMEE, a town of Austria, in Moravia, in the circle and 26 m. NW of Znaym, on a mountain, near the l. bank of the Zeletawa. Pop. 1,100. It has a castle.

YAMPA, a town of Utah Territory, U. S., which has its source on the W side of the Rocky mountains, flows W, and falls into Green river.

YAMPARAES, a district or prov. of Bolivia, in the prov. of Charcas. It has a town of the same name, 15 m. NW of La Plata.

YAMPARICAS, an Indian tribe of Mexico, who inhabit the district to the NE of Lake Timpanagos, near the Sierra-Verde.

YAMPEE POINT, a headland of Australia, on the NW coast, between King's Sound and Collier Bay, and opposite the Buccaneer's archipelago.

YAMPOL, IAMBOL, or JAMPOL, a district and town of Russia in Europe, in the gov. of Podolia. The district occupies the central part of the gov., and is intersected by the Dniester. The soil is sandy, but is in some parts fertile, and towards the centre of the district is an extensive forest. The town is 87 m. ESE of Kamenetz, on the l. bank of the Dniester. It contains a lazaretto, and about 118 houses.—Also a town of Russia, in the gov. of Kharkov, and district of Izium. In the vicinity are several salt-springs.—Also a town of the gov. of Volhynia, district and 24 m. ESE of Kremenetz, on the l. bank of the Gorin.

YAMSAE, a town and parish of Russia in Europe, in the grand-duchy of Finland, gov. of Tavastehus, and district of Œfre-Sææxmæki, 57 m. NE of Tammerfors, and 69 m. NNE of Tavastehus, on the W bank of Lake Pæjjæne.

YAMSK, or JAMSK (BAY OF), or GOUBA-JAMSKAIA, a bay of Russia in Asia, formed by the sea of Okhotsk, on the coast of the district of that name, and to the SW of the bay of Ijighinsk, in N lat. 58° 50′, and E long. 174° 20′. It is about 54 m. in length from E to W, and 24 m. in width from N to S. To the N of its entrance is Cape Olemi, and at the W extremity, near the town of Jamsk, it receives the Yama. This bay forms a commodious port, but the entrance is rendered dangerous by sand-banks and sunken rocks.

YAMSK, or JAMSKOI-OSTROG, a town of Russia in Asia, in the district and 375 m. E of Okhotsk, and 225 m. SW of Ijighinsk, near the confluence of the Jama with the bay of Okhotsk. Pop. about 100. It is enclosed with palisades, and contains a church and about 30 houses. Fishing forms the chief employment of the inhabitants.

YAMSKAIA, a town of Russia in Europe, in the gov. and district and 2 m. ENE of Koursk.—Also a town in the gov. of Koursk, district and 1½ m. S of Staroï-Oskol.

YAN, a village of France, in the dep. of the Saone-et-Loire, cant. of Paray-le-Monial, near the r. bank of the Reconce, 13 m. W of Charolles. Pop. 718.

YAN, a range of mountains in Mongolia, to the N of China Proper, and forming part of the E ridge of the great central plateau of Asia.

YANA, a town of Japan, in the island of Nifon, and prov. of Mikawa, 165 m. W of Jedo.

YANA, a river of Russia in Asia, in the prov of Yakoutsk, and district of Verkhoïansk. It issues from a small lake to the SW of that district, 225 m. NE of Yakoutsk; runs N, and throws itself by several embouchures into the Arctic ocean, between Capes Darighin and Borkhaïa, in N lat. 71° 30′, and E long. 136° 20′, and 150 m. SE of the embouchures of the Lena. It has a total course of about 750 m. and receives the Adiga on the r., and on the l. the Doulgalakh, Tekan-Chemanova, and Boutaktaï. Verkhoïanst is the chief place on its banks. The Youkahirs are almost the exclusive inhabitants of its banks.

YANAGAVA, a river of Japan, in the island of Kiu-siu, and prov. of Tsikougo, which has its source in the NE part of the prov., and after a course of 60 m. throws itself into the bay of Simabara, near a town of the same name. This town is 21 m. E of Sanga.

YANA-KOURGAN, a small town of Independent Tartary, in the state of Kokan, and 72 m. from Namghan. It has a fort, with a garrison of 200 men.

YA-NANG-HEEN, a village of Burmah, near the l. bank of the Irawadi, 34 m. N by W of Patanago, celebrated for its wells of mineral oil or petroleum. These wells are scattered irregularly over a tract of rugged arid downs and hillocks about 3 m. from the river, to the number, it is said, of about 200; and 4 or 5 m. to the NE there are about 500 more. In making a well, the hill is cut down, so as to form a square table of 14 or 20 ft. for the crown of the well, and from this table a road is formed by scraping away an inclined plane for the drawers to descend, in raising the excavated earth from the well, and subsequently the oil. The shaft is sunk of a square form, and lined with cassia-wood staves. The soil or strata to be pierced is, first, a light sandy loam intermixed with fragments of quartz, silex, &c. Secondly, a friable sandstone easily wrought, with thin horizonal strata of a concrete of martial ore, talc, and indurated argil, at from 10 or 15 ft. from the surface, and also from each other, as there are several of these veins in the great body of freestone. Thirdly, at 20 cubits, more or less, from the surface, and immediately below the freestone, a pale blue argillaceous earth appears, impregnated with the petroleum, and smelling strongly of it. This is very difficult to work, and grows harder as they get deeper, ending in schist and slate, such as is found covering veins of coal in Europe. Below this schist, at the depth of 130 cubits, is coal. The oil is drawn pure from the wells in the liquid state; but in the cold season it congeals in the open air, and loses something of its fluidity. The oil is of a dingy green, and odorous: it is used for lamps, and, boiled with a little dammer (a resin of the country) for painting the timbers of houses, and the bottoms of boats, &c. which it preserves from decay and vermin. The average produce of each well per diem, in 1796 was 500 viss, or 1,825 lbs. avoird. Each well was worked by four men, and their wages regulated by the average produce of six days' labour. Mr. Cox estimated on the best data at which he could arrive, that there were 520 wells registered by government, and the gross produce of the whole per annum, was 56,940 viss, or 92,781 tons, 1,560 lbs., or 412,360 hogsheads; worth at the wells, at one and a quarter ticals per hundred viss, 711,750 ticals, or 889,687½ sicca rupees. From the wells the oil is carried in small jars, by coolies or on carts, to the river; where it is delivered to the merchant exporter. There were between 70 and 80 boats, average burthen sixty tons each, loading oil at the several wharfs, and others constantly coming and going while he was there. A number of boats and men also find constant employment in providing pots, &c. for the oil.

YANAON, or YANAN, a French settlement and village of Hindostan, in the presidency of Madras, district and 30 m. SE of Rajamundry, near the bifurcation of the Godavery and Coringa, and about

9 m. above the embouchure of the former river. Pop. 6,881. The territory connected with the French factory comprises an area of 8,147 acres.

YANA-URCU, a lofty summit of the Andes, in Ecuador, to the N of Quito. It is covered with perpetual snow.

YANAUZA, an island in the estuary of the Amazon, to the NW of the island of Caviana, in N lat. 0° 30′.

YANBO-EL-BAHR [*i. e.*, 'Yanbo of the Sea'], a town of Arabia, in the Hedjaz, 135 m. SW of Medina, and 134 m. NNW of Mecca, in a vast plain, on the shore of the Red sea, in N lat. 24° 7′ 6″, and E long. 37° 32′ 15″. Pop. 3,000. It has a good and commodious port, but difficult of access on account of the number of rocks which obstruct its entrance. The town is enclosed by walls, irregular in outline, but averaging a diameter of 746 yds. from E to W, and of about 426 yds. from N to S. The wall is 9 ft. in height, but does not exceed 15 in. in thickness. The fortifications consist chiefly of towers, one of which is strongly fortified with artillery. The central part of the town, which is on the harbour, is surrounded by a second enclosure of walls, of greater height, flanked with towers. The houses are low and flat-roofed. Yanbo-el-Bahr is the port of Medina, and carries on an active trade. The climate is excessively warm and insalubrious, and the environs present the aspect of an arid desert.

YANBO-EL-NAKHL [*i. e.*, 'Yanbo of the Palm-trees'], a town of Arabia, in the Hedjaz, about 30 m. ENE of Yanbo-el-Bahr, in the midst of mountains. It is well supplied with water, and possesses fine gardens and numerous palm-trees. The inhabitants are *sheriffs*, or descendants of the prophet, and are noted warriors.

YANCARDOOK, a parish of Australia Felix, in the co. of Bourke, bounded on the W by the Werribee river, and on the E by a creek, by which it is separated from Holden.

YANCEY, a county in the W part of the state of North Carolina, U. S., comprising an area of 687 sq. m., drained by Nolichucky river and its branches. It is generally mountainous, and has one summit, named Black mountain, which has an alt. of 6,476 ft. above sea-level. Pop. in 1840, 5,926; in 1850, 8,205. Its cap. is Burnsville.

YANCEYVILLE, a village of Caswell co., in the state of North Carolina, U. S., 60 m. NW of Raleigh. Pop. in 1840, 250; in 1850, about 300.

YANDABU, a town of Burmah, in the Mranma, on the l. bank of the Irawadi, 63 m. WSW of Ava Here, on the 26th February 1826, the treaty was concluded by which it was stipulated that all pretension by the Burmese to Cassay, Cachar, Jynteca, and Assam should be renounced, and that to the English should be ceded Aracan, the provinces of Ye, Tavoy, Mergui, and Tenasserim.

YANDOKA, a large village of Nigritia, in the Haoussa, and prov. of Kashena, between the town of that name and Zirmia. It is enclosed by walls.

YANDINSKOI-OSTROG, a town of Russia in Asia, in the gov. and 180 m. NNW of Irkutsk, and district of Nijni-Oudinsk, on the l. bank of the Angara. It is enclosed with a palisade.

YANEYA, a town of Western Africa, in the district of Benna, on the Kolantang, or N branch of the Great Scarcies river. It is in a dilapidated state; but the country around is fertile and well-cultivated.

YANG, a town of Burmah, in the state of Munipur, 99 m. SE of Nowgong.

YANG, an extensive mountainous district of Laos, lying to the N of the Keu-lung river.

YANGAIN-SHAIN-YAH, a town of Burmah, in Pegu, 42 m. NW of Rangoon, at the point of separation of the Panlang from one of the great branches of the Irawadi.

YANGAR, a lake of New South Wales, in the district of Murrumbidgee, between Lake Tala and Edward river.

YANG-CHING, a district and town of China, in the prov. of Shan-se, and div. of Tsih-chu-fu, in N lat. 35° 26′, and E long. 112° 36′.

YANG-CHU, a town of the Corea, in the prov. of King-ke, 9 m. E of Han-yang.—Also a town in the prov. of Chu-sin, 30 m. SW of Ku-fu.

YANG-CHU-FU, a division and town of China, in the prov. of Keang-su. The div. comprises eight districts. The town is situated on the Yu-ho canal, near the l. bank of the Yang-tsze-keang, and 51 m. NE of Nan-king, in N lat. 32° 26′ 32″, and E long. 119° 24′ 13″. It is about 6 m. in circumference, and contains several pagodas. Standing on the great thoroughfare on the Yang-tsze-keang, it possesses an active trade, especially in salt, and is remarkable for the miscellaneousness of its inhabitants, including even Jews and Christians, and notorious for their licentiousness. The surrounding country is romantic, and generally fertile, and in the vicinity is Kaou-min-she, an occasional residence of his imperial majesty.

YANGHECAKORRY, a town of Upper Guinea, on the Sierra Leone coast, and country of the Mandingoes.

YANGHIAKOURI, a town of Upper Guinea, on the Sierra Leone coast, and country of the Sousous.

YANGHIZ-TAGH, a group of mountains of Turkestan, in the country of the Kirghiz of the Middle horde, in N lat. 50° 30′, and E long. 62° 50′. The Tshaian, an affluent of the Tashlabulak, has its source in this chain.

YANG-HO, a river which has its source in Mongolia; runs SE; enters the Chinese prov. of Chihle, to the NW of Senen-hwa-fu; bathes the walls of that town; and, after a rapid course, joins the Sang-kan-ho, or San-cam-ho, to form the Hwan-ho.

YANG-HU, a district and town of China, in the prov. of Keang-su, and div. of Chang-chu-fu.

YANG-HUH, a district and town of China, in the prov. of Shan-tung, and div. of Yen-chu-fu, in N lat. 36° 9′, and E long. 115° 59′.

YANG-KANG, a town of China, in the prov. of Che-keang, 105 m. S of Hang-chu-fu, in N lat. 28° 58′, and E long. 120° 10′ 45″.

YANG-KEANG, a district and town of China, in the prov. of Kwang-tung, and dep. of Chaou-king-fu, on the l. bank of a river of the same name, near its entrance into the China sea, and 150 m. SW of Canton, in N lat. 22° 8′, and E long. 111° 33′.

YANG-KEAOU, a district and town of China, in the prov. of Shan-se, and div. of Tae-tung-fu.

YANG-LE-CHU, a district and town of China, in the prov. of Kwang-se, and div. of Tae-ping-fu, in N lat. 22° 54′, and E long. 106° 54′.

YANGMA, a river of Nepal, an affluent of the Tambur.

YANG-SHAN-HEEN, a district and town of China, in the prov. of Kwang-tung, and div. of Leen-chu, 105 m. NNW of Canton, in N lat. 24° 30′, and E long, 112° 24′ 30″.

YANG-SHUN, a district of China, in the prov. of Kwang-tung, on the Yang-keang, 144 m. SW of Canton.

YANG-SIN, a district and town of China, in the prov. of Shan-tung, and div. of Wu-ting-fu, 84 m. NNE of Tse-nan, in N lat. 37° 45′, and E long. 117° 38′.

YANG-SUH, a district and town of China, in the

prov. of Kwan-se, and div. of Kwei-lin-fu, in N lat. 24° 28′, and E long. 110° 20′.

YANG-TSZE-KEANG [*i. e.* 'son of the ocean'], called also Keang or Ta-keang [*i. e.* river, or great river], and often by Europeans Blue river, a river of China, one of the greatest in Asia, and scarcely inferior to any in America, formed by the confluence of the Kin-sha-keang and Ya-lung-keang, on the confines of the provinces of Sze-chuen and Yun-nan. Taking first a S direction, it runs along the confines of the above-named provinces, nearly in a line parallel with the meridian of 102° E, and then describes a curve to the E. On reaching 102° 42′ E long., it enters the prov. of Sze-chuen, and directs its course first NE, and afterwards N, skirting for some distance the prov. of Kwei-chu. In N lat. 28° 20′, it resumes a NE direction; passes Ma-hu-fu, the first large city on its banks, and soon after, at the confluence of the Yang-keang, a large river, Su-chu. In N lat. 29° 48′, and E long. 106° 40′, near Chung-king-fu, it receives the Kei-ho-tung-kean or Kea-ling-keang; and in its further course through Sze-chuen, passes many important cities, of which the principal, in addition to those already named, is Kwei-chu. On entering the prov. of Huh-pih, it runs SE till it reaches Tung-ling lake, with which it communicates, when it abruptly bends NE, a direction which it retains to the confluence of the Han-keang, near Han-yung-fu and Wu-chang-fu, in N lat. 30° 40′ and E long. 114° 10′. Then sweeping SE, it passes Wang-chu-fu, and through the N extremity of the prov. of Keang-si, in which it receives the outlet of the Po-yang lake. Again taking a NE direction, it traverses the prov. of Gan-hwuy, it waters Gan-king-fu, Che-chu-fu, and Tae-ping-fu, and enters that of Keang-su. Passing Keang-ning-fu or Nan-king, it directs its course ENE, crossing near Chin-keang-fu, the En-ho or Imperial canal, and, afterwards bending SE, enters the Tong-hai or Eastern sea, by an estuary about 60 m. wide, between the parallels of 31° and 32° N. It has a total course of 1,980 m. or to the source of the Ku-sha-keang 3,000 m. Its principal affluents are, on the r., the Ou-keang, the outlet of Lake Tung-ling, and the outlet of Lake Po-yang; and on the l., the Meu-keang, Kea-ling-keang, Han-keang, and the outlets of Lakes Tseaou and Kao-Yiou. The mountains of Pe-ling on the N, and those of Nan-king on the S, form, to a great extent, the basin of this river, which reaches on the W to the E coast of the great central plateau of Asia. It traverses a mountainous region in the upper part of its course, but towards the E it flows through vast plains of great fertility. The action of the tide reaches to Keon-keang, 450 m. from the sea, and vessels of large size can ascend to Lake Tung-ling, 300 m. higher. The Yang-tsze-keang is broad, deep, and abounding with fish. It is 21 m. wide above the island of Tsung-ming, at its mouth, and throughout a great part of its course is not less than 1½ m. in width. It forms numerous islands, on which reeds grow in great quantities, and being used for fuel, prove a useful production to the inhabitants of the adjacent banks.

YANGUAS, a town of Spain, in the prov. of Soria, and partido of Agreda, on the Cicados, and 30 m. SSE of Logrono. Pop. 838. It has a parish-church, an hospital, and a custom-house. It has several mills and bleachfields, and carries on an active trade in cattle.

YANGUITLAN, a village of Mexico, in the state and to the NW of Oaxaca. It has a fine Dominican convent, founded in the 16th century, by D. Fr. Las Casas, a companion of Cortez, and contains about 900 Indian families, who find their chief employment in the cultivation and manufacture of cotton. Y. was formerly a populous town.

YANG-WEI, a fortress of China, in the prov. of Shan-se, near the Great Wall, 198 m. NNE of Tae-quen-fu.

YANG-WU, a district and town of China, in the prov. of Ho-nan, and div. of Hwae-king-fu, in N lat. 35° 5′, and E long. 114° 8′.

YANG-YIN, a district and town of China, in the prov. of Ho-nan, and div. of Chang-tih-fu.

YAN-HEEN, a district and town of China, in the prov. of Shen-se, and div. of Han-chung-fu, on the Han-keang, 120 m. SW of Segan-fu, in N lat. 33° 5′, and E long. 107° 43′.

YAN-HO, a river of China, in the prov. of Shen-se, which runs ESE, and after a course of about 135 m., joins the Hwang-ho, on the r.

YANI, a central kingdom of Senegambia, to the r. of the Gambia, and near the central part of the course of that river. The territory is flat and fertile, and to a considerable extent covered with forests, containing in large numbers elephants and other wild beasts, and gazelles. The ass is the common beast of burden. Cattle and fowls are abundant. Corn is extensively cultivated in the vicinity of the towns, and there are also large plantations of tobacco, indigo, and cotton. Pisania and Yani are the chief towns.

YANIARI, a town of Turkey in Europe, in Albania, in the sanj. and 30 m. S of Delvino, and near the r. bank of the Kalamas.

YANIEVO. See YANOVA.

YANIKOW, a town of Poland, in the woiwodie, obwod, and 14 m. N of Sandomir. It contains about 51 houses.

YANINA, a town of Austria, in Dalmatia, in the circle, and 43 m. WNW of Ragusa, in the peninsula of Sabioncello.

YANINA. See JANINA.

YANIS, a group of islets of the Archipelago, a little to the W of Cape Sidero, the NE extremity of the island of Candia.

YANISHKE′LE′, a town of Russia in Europe, in the gov. of Vilna, district and 21 m. N of Ponieviej, and 38 m. ENE of Shavli.

YANISHKI, a town of Russia in Europe, in the gov. and 135 m. NNW of Vilna, and district of Shavli. Pop. 2,500, of whom a large proportion are Jews. It has manufactories of ironware.—Also a town in the gov., district, and 33 m. NNE of Vilna.

YANITZA, a town of Greece, in the Morea, in the gov. of Messenia, 3 m. E of Kalamata, on an affluent of the gulf of Koron. It contains about 300 houses.

YANKEE SPRINGS, a township of Barry co., in the state of Michigan, U. S., 44 m. W by S of Lansing, and drained by branches of Thorn-apple river. Pop. in 1840, 130; in 1850, 292.

YANKO-YANKO, a creek of New South Wales, in the district of Murrumbidgee, flowing between Billeboug creek and the Murrumbidgee.

YANKTONS, or YANCTHOANA, a branch of the Sioux Indians, in North America. It comprises eight tribes, all of whom are found near the sources of the St. Peter, in the vicinity of the Red river, and in the tract lying between these two rivers and the Missouri.

YANKYLILLY BAY, an indentation of the coast of South Australia, in the co. of Hindmarsh, and gulf of St. Vincent.

YANKUMASI, a town in the Assin territory of Western Africa, 16 m. S of Mansu.

YANNAMAROA, a town of Senegambia, on the r. bank of the Gambia, a little below Dami islands, in N lat. 13° 42′, and W long. 14° 58′.

YANONCHICHKI, a town of Russia in Europe,

in the gov. and 105 m. N of Vilna, and district of Vilkomir.

YANONCHOV, a town of Russia in Europe, in the gov. of Vilna, district and 33 m. ESE of Rossiena, on the r. bank of the Pevieja.

YANONCHPOL, a town of Russia in Europe, in the gov. of Volhynia, district and 36 m. SW of Jitomir. Pop. 1,785.

YANOV, or YANOVO, a town of Russia in Europe, in the prov. and 24 m. N of Bialistok, and district of Sokolka. Pop. 700.—Also a town in the gov. and 135 m. SSE of Grodno, and district of Kobrin. Pop. 700.—Also a town in the gov. of Podolia, district and 15 m. NE of Litin, on the l. bank of the Bug. Pop. 1,200.

YANOVA, or YANIEVO, a town of Turkey in Europe, in Albania, in the sanj. and 87 m. ENE of Scutari, and district of Pristina, on the l. bank of the Ibar.

YANOVA, or YANOW, a town of Russia in Europe, in the gov. and 51 m. NW of Vilna, and district of Kovno, on the r. bank of the Vilna.

YANOVITSHI, a town of Russia in Europe, in the gov. and 24 m. ENE of Vitebsk and district of Souraj.

YANOVKA, a town of Russia in Europe, in the the gov. of Volhynia, district and 26 m. SE of Kovel. Pop. 175.

YANOW, a town of Austria, in Galicia, in the circle and 14 m. WNW of Lemberg, on a lake abounding with fish.—Also a town in the circle and 27 m. SSE of Tarnopol, in an island formed by the Sered.—Also a town of Poland, in the woiwodie of Krakow, obwod and 35 m. N of Olkusz. Pop. 545.—Also a town in the woiwodie and 42 m. SSW of Lublin, obwod and 41 m. W of Zamosc. Pop. 290.—Also a small town in the woiwodie and 39 m. E of Siedlec and obwod of Bralo, near the l. bank of the Bug.—Also a village of the woiwodie of Sandomir, obwod and 20 m. SSE of Opoczno. Pop. 140. It has two blast furnaces.

YANOWICI, a town of Prussia, in the prov. and 38 m. NE of Posen, regency of Bromberg, and circle of Wongrowitz, on the Wilna. Pop. 298. It has a Catholic church.—Also a town of Poland, in the woiwodie of Sandomir, obwod and 36 m. ESE of Radom, near the l. bank of the Vistula. Pop. 465.

YANOWITZ, a village of Prussia, in the prov. of Silesia, regency and 26 m. SSW of Liegnitz, and circle of Hirschberg. Pop. 1,100. In the environs are mines of copper and lead.

YANOWO, a town of Poland, in the woiwodie and 72 m. NE of Plock, and obwod of Przasznic, on the Prussian frontier. Pop. 721.

YANSBORK, or JOHANNISBURG, a town of Poland, in the woiwodie of Augustowa, obwod and 48 m. NW of Marianpol, on the l. bank of the Niemen. Pop. 215.

YANTIC, a village of New London co., in the state of Connecticut, U. S., on a river of the same name, and on the New London, Willimantic, and Palmer railroad, and 50 m. from Palmer. The river Y. is formed by the junction of three branches in Lebanon co.; flows E and SE, and joins Thames river below Norwich.

YANWATH, a joint township with Eamont-Bridge, in the p. of Barton, Westmoreland, 2 m. S by W of Penrith, on the Eamont. Area 1,169 acres. Pop. in 1831, 327; in 1851, 356.

YANWORTH, or ENWORTH, a chapelry in the p. of Hazleton, Gloucestershire. Area 1,340 acres. Pop. in 1831, 123; in 1851, 89.

YANYANANT, a parish of Australia Felix, in the co., of Bourke, bounded on the W. by the river Barwon.

YANYEAN, a parish of Australia Felix, in the co. of Bourke.

YAOU. See YEOU.

YAOU, a town of Japan, in the island of Nifon and prov. of Isoumi, on the gulf of Osaka, 45 m. SW of Miaco. See also YO.

YAOU-CHU, a district and town of China, in the prov. of Shen-se, div. and 48 m. N of Yan-fu, in N lat. 34° 56′, and E long. 108° 53′.—Also a district and town of the prov. of Yun-nan, and div. of Tsu-heung-fu, 99 m. WNW of Yun-nan-fu, in N lat. 25° 32′ 20″, and E long. 99° 24′ 50″. It carries on an active trade in musk, the produce of the environs. In the vicinity is a salt lake, from which salt in quantity sufficient for the supply of the prov. is extracted.

YAOU-GAN-FU. See TSU-HEUNG-FU.

YAOURI, a kingdom of Sudan, bathed by the Joliba, and bounded on the N by Haussa; on the E by the state of Koton Kora; on the S by that of Nyffe or Tappa; and on the W by Borgu. The government is hereditary and absolute. The soil produces indigo, tobacco, corn, and rice, and affords pasturage to large numbers of horses, cattle, and sheep.—The capital, which bears the same name, is at some distance from the l. bank of the Joliba, and about 165 m. S of Sakkatu. It is large and populous, and is enclosed by a substantial clay-built wall of considerable height. It is nearly 24 m. in circumference, and is entered by eight large gates all well fortified. The inhabitants manufacture a coarse description of gunpowder, also saddlery and cloth. They grow indigo, tobacco, onions, wheat, and rice.

YAOUZA, a river of Russia in Europe, in the gov. and district of Moscow, which has its source to the N of that town; traverses one of its quarters, to which it gives its name; and, after a total course of 24 m., joins the Moskwa on the l. bank.

YAP, an island of the North Pacific, in the group of the Caroline islands, in 9° 35′ 30″ N lat., and 138° 8′ E long. Its length from N to S does not exceed 11 m. It is generally elevated, without having any mountains properly so called. It is watered by several small rivers, is well-cultivated, and has some fine wood. The inhabitants are energetic in character and industrious. This island, which formerly constituted a state, is now divided into several small principalities.—The name of Yap is sometimes extended to the adjacent group of Ngoli, Mogening, and Feis.

YAPEGU, a town of La Plata, in the prov. of Corrientes, on the r. bank of the Uruguay, a little above the confluence of the Ybicuy, and 300 m. NE of Santa Fe. The Jesuits have a fine college here.

YAPHAM, a chapelry in the p. of Pocklington, Yorkshire. Area 1,818 acres. Pop. in 1851, 151.

YAPO. See JAPA.

YAPOMSKOE. See JAPOMSKAIA.

YAPTON, a parish in Sussex, 5 m. SW of Arundel, intersected by the Arundel and Portsmouth canal. Area 1,699 acres. Pop. in 1851, 609.

YAPURA. See JAPURA.

YAQUE, a summit in the central part of the island of Haiti, in the range of the Cibao mountains. It gives rise to two rivers of the same name, distinguished as the Great and Little Yaque. The former runs first NNW, then WNW, traverses the plain of Santiago, and, after a total course of about 105 m., falls, by several embouchures, into the bay of Monte-Christi. The latter runs SSW, and, after a course of 45 m., joins the Neybe, on the l. bank.

YAQUI. See HIAQUI.

YAQUIN. See ACQUIN.

YARA, a town of Nubia, in the district of Fazoklo, in N lat. 11° 14′ 47″, on a hill, and about ¾ of a mile from the l. bank of the Bahr-el-Azrek.

YARACUY, a navigable river of Venezuela, in the prov. of Caracas, which descends from the mountains to the E of Barquisimeto, runs NE, and throws itself into the sea of the Antilles, to the NW of Porto-Cabello, and to the SE of the gulf of Trieste.

YARBA. See YARRIBA.

YARBOROUGH, or YARBURGH, a parish in Lincolnshire, 5 m. NE by E of Louth, in the line of the Louth navigation. Area 1,160 acres. Pop. 245.

YAR-BROK-YOU-MTHSO, YAMBRO-YOU-MTHSO, or PALTE, a lake of Tibet, on the confines of the provinces Thsang and Wei, to the SSW of Lassa, and near the r. bank of the Yaru-dzangbo-chu, by which it is separated from Mount Notdzing-Gangazang. It is 45 m. in length from N to S, and 30 m. in breadth from E to W. In form it is nearly circular, and in its centre is a large island, well cultivated, and containing several convents.

YARCOMBE, a parish in Devon, 5 m. W of Chard. Area 4,689 acres. Pop. in 1851, 780.

YARDLEY, a parish in Hertfordshire, 4½ m. WSW of Buntingford. Area 3,510 acres. Pop. in 1831, 599; in 1851, 1,210.—Also a parish in Worcestershire, near the line of the London and Birmingham railroad, 4½ m. E of Birmingham, containing the chapelries of Marston and Yardley-wood. Area 7,355 acres. Pop. in 1851, 2,753.

YARDLEY-GOBIONS, a hamlet in the p. of Potterspury, Northamptonshire, 3½ m. NNW of Stony-Stratford, near the Grand Junction canal, the river Tove, and the London and Birmingham railway. Pop. in 1831, 594; in 1851, 673.

YARDLEY-HASTINGS, a parish in Northamptonshire, 8 m. ESE of Northampton. Area 3,510 acres. Pop. in 1831, 1,051; in 1851, 1,210.

YARDLEYVILLE, a village of Bucks co., in the state of Pennsylvania, U. S., on the W bank of Delaware river, and 102 m. E of Harrisburg. Pop. in 1850, about 150.

YARE (THE). See NORFOLK.

YARGHIA-DZANGBO, a river of Tibet, in the N part of the prov. of Wei. It runs in a generally NE direction, but the termination of its course is still unknown.

YARKHILL, or YARCLE, a parish in Herefordshire, 7¼ m. E by N of Hereford, on the river Frome. Area, 1,666 acres. Pop. in 1831, 409; in 1851, 547.

YARGHUIT, or YARGUI, a town of Tibet, in the prov. of Wei, near the r. bank of the Yaru-dzangbo-chu, and 100 m. ESE of Lassa.

YARILL, a creek of New South Wales, in the co. of Stanley, an affluent of the Bremer.

YARIMOL, a town of New Granada, in the dep. of Cundinamarca, prov. and 54 m. NE of Antioquia, at an alt. of 123 toises above sea-level.

YARKAND, or GARKIANG. See ERGHEN.

YARKAND, YARKHAND, YERKIANG, or YARKEND, a city of Turkestan, in the Chinese government of Ele, in N lat. 40° 10′, 120 m. SE of Cashgar, on the route to Cashmere, and on a river of the same name. It is surrounded by stone walls about 3 m. in circuit; but the suburbs extend greatly beyond the walls. The houses are chiefly built of sun-dried bricks. There are numerous mosques, and 10 colleges within the walls; and several public buildings which also afford barracks for the Chinese garrison of 7,000. The population is variously estimated at from 40,000 to 200,000. An active trade is carried on here in horses, silk, and wool, and the customs are said to yield 45,000 dollars annually. This city was reunited to the Chinese empire in 1757. It is governed by the Mahommedan and the Chinese functionaries, who are replaced by new officers every third year. The vicinity is fertile, yielding wheat, barley, rice, fruits, and silk, and extensive pasturage.—The river of Y. rises on the N flank of the Karakoram mountains, flows NNW to the city of Y., whence it flows 300 m. to its junction with the Cashgar, the united streams forming the Tarym, which runs E into the Lob-nor. The *yesm* or jade-stone, a kind of jasper highly esteemed by the Chinese, is found chiefly in the bed of this river; and forms an important article of export.

YARKON, a town of Sudan, capital of a state of the same name in Borgu, and about 300 m. W of Sakkatu.

YARKLINGTON, a parish of Tasmania, in the co. of Monmouth.

YARLESIDE, a township in the p. of Dalton-in-Furness, co.-palatine of Lancaster. Pop. 561.

YARLINGTON, a parish in Somersetshire, 3½ m. W of Wincanton. Area 1,207 acres. Pop. 231.

YARM, or YARUM, a market-town and parish in the N. R. of Yorkshire, 44 m. NNW of York, on the river Tees, here crossed by a bridge, and 1 m. S of the Stockton and Darlington railway. Area 1,135 acres. Pop. in 1831, 1,636; in 1851, 1,647. The town has frequently suffered from inundation of the Tees.

YARMOUTH, a borough, parish, and sea-port, in the hund. of Flegg, co. of Norfolk, at the mouth of the Yare, 24 m. E of Norwich, with which it is now connected by railway. Area of the parish 1,510 acres. Pop. in 1801, 14,845; in 1831, 21,115; in 1851, 26,880. The borough, till 1832, consisted of the p. and hamlet of South-Town, or Little Y., in the parish of Gorlestone, Suffolk, opposite Great Y., being divided from it by the Yare, but connected by a drawbridge nearly opposite the centre of the town, and about 2½ m. from the mouth of the river. There is also an iron suspension-bridge over the North river. By the Reform act, the whole p. of Gorlestone, including the hamlet of Gorlestone, continuous with Little Y., was included within the borough boundaries. Great Y., the chief and ancient part of the town, stands on a narrow peninsular slip, running along the coast between the Yare on the W, and the sea on the E, upwards of 1 m. in length, and ½ m. in breadth. It is irregularly built, and contains four principal and numerous narrow streets. A wall, begun in the reign of Edward I., formerly extended on the E, N, and S sides, 6,720 ft. in aggregate length, with 10 gates and 16 towers. It is composed of flint, pebbles, and shingle, so strongly cemented together, that its demolition, in the course of improvement, is even still attended with no little labour. The suburb of Little Y. or South-Town, consists chiefly of neat and substantial private residences, extending along the W side of the high road leading from the bridge to Gorlestone. The space between that road and the river opposite to Great Y., is occupied by docks, timber-wharfs, building-yards, &c. The village or town of Gorlestone lies more to the S, near the entrance to the harbour. The guildhall is an ancient edifice. The borough-jail and house-of-correction is situated in the densest part of the town. The royal barracks on the beach are capable of accommodating 1,000 men. On the sands SE of the town, is a column 144 ft. high, erected to the memory of Nelson.—The harbour was formed with great difficulty and labour by the erection of piers and a jetty, and is kept in a state of repair at a very heavy cost. A N pier was first constructed with much labour, but this has been replaced by another, chiefly for the convenience of mooring ships in harbour. The S pier is at the mouth of the Yare. A continuation of it is carried up the river in a fine curve, forming an extensive and excellent wharf. This pier is 1,020 ft. long and 30 ft. broad. In addition to the piers, a jetty was constructed on the beach some way to the N in 1808: this extends into the sea upwards of 450 ft. and has a platform 21 ft. in width. This harbour affords a secure shelter to shipping; and vessels seek protection in it against easterly storms. The coast is shallow, full of sand-banks, and dangerous. Y. is the principal port, except London, for the English trade to the north of Europe; and besides fishing smacks, upwards of 500 vessels belong to the port. An extensive inland trade is conducted in the exchange of agricultural produce for coals, other heavy goods, and the produce of the continent and the colonies. A direct and extensive import timber trade is conducted with the Baltic; an important export trade in fish, is conducted with the West Indies and various foreign countries, particularly those bordering on the Mediterranean; and a regular communication is kept up interiorly with Norwich, and exteriorly with the chief ports of the east coast of England. The herring and mackerel fisheries are the chief source of trade, and are, in general, remarkably productive. In one year 15,000 barrels have been cured here for home consumption, and 70,000 for exportation. In 1853, 1,577 vessels = 133,249 tons, and 183 steam vessels = 25,358 tons, entered this port coastwise; and the tonnage belonging to the port was 328 vessels under 50 tons, and 289 vessels above 50 tons, making a total tonnage of 42,323 tons. Ship-building, and the various trades connected with it, are somewhat extensively conducted. A manufactory of crapes and other silk goods, employs about 500 persons.

YARMOUTH, a parish, and a disfranchised borough, and market-town, in the Isle of Wight, 10 m. W of Newport, at the mouth of the river Yar. Pop. in 1801, 343; in 1831, 586; in 1851, 572.—It regularly returned two members till the passing of the Reform act, when it was disfranchised. The town is situated on the slope of a hill, and formerly occupied more ground than it now does. It is well-built. The trade is now insignificant, and consists principally in the exportation of a fine white sand for the use of glass-makers, and in the importation of coals from Sunderland, and timber.

YARMOUTH, a township of Upper Canada, in the London district, bounded on the S by Lake Erie, and separated by Catfish creek from the township of Malahide, and from that of Southwold by Kettle creek. It is generally fertile and well cultivated, and has been long-settled. Pop. in 1842, 1,239. It contains two villages, St. Thomas and Port Stanley, and the settlements of Jamestown and Sparta.

YARMOUTH, a township of Cumberland co., in the state of Maine, U. S., 41 m. SSW of Augusta, drained by a small stream, and intersected by the Kennebec and Portland railroad. Pop. in 1850, 2,144. It contains a village of the same name.—Also a township of Barnstable co., in the state of Massachusetts, 63 m. SE of Boston, and extending to Cape Cod. The soil is sandy and unproductive. Pop. in 1840, 2,554; in 1850, 2,595.

YARMOUTH LITTLE. See South-Town.

YARMOUTH (North), a township of Cumberland co., in the state of Maine, U. S., 40 m. SW by S of Augusta, and intersected by the Atlantic and St. Lawrence railroad. Pop. in 1840, 1,221.

YARMOUTH PORT, a village of Barnstable co., in the state of Massachusetts, U. S., on a branch of Cape Cod bay.

YARMUK, or Jarmuk, a river of Turkey in Asia, in Syria, in the pash. of Damascus, which has its source near the fort of Azrak, runs first NW, then SW. and after a course of about 60 m., joins the Charia or Jordan, on the l. bank, 6 m. S of the lake of Tabarieh. It passes near the village of Adreat, and ruins of Gadara, and on its r. bank is a village of the same name. The Y. is the *Hieromax* of the ancients.

YARNESCOMBE, a parish in Devon, 6 m. NE by E of Great Torrington. Area 3,047 acres. Pop. in 1831, 498; in 1851, 479.

YARNTON, a parish in Oxfordshire, 4½ m. NW by N of Oxford. Area 1,613 acres. Pop. 317.

YAROOK, a parish of Australia Felix, in the co. of Bourke, bounded on the E by Merri-Merri creek.

YAROSLAV, or Jaroslav, an extensive government in the interior of European Russia; bounded on the N by Vologda; on the E by Kostroma; on the S by Vladimir; and on the W by Tver and Novogrod, and lying between the parallels of 56° 44′ and 58° 57′ N lat. Its territorial extent is 29,166 sq. versts. Its population was returned in 1846 at 1,008,100. The surface chiefly consists of large undulating plains, traversed by the Wolga, the Scheksna, the Mologa, the Kotoroste, the Sogoja, the Sita, the Pontna, the Vexa, the Kostroma, and several smaller rivers. The lakes are also numerous. The principal of them are Lake Rostof or Nero, the great Yagorskoie, the Kharlam, the Somskoie, and the Yarrobol. The soil is sand, clay, or black mould; but in many places it is marshy. The climate is not particularly severe; but agriculture is in so backward a state that it is often necessary to import corn from the neighbouring provinces. Of a total area of 3,040,349 sq. *desoiatines*, 1,065,881 were returned as arable in 1825. The inhabitants rear and export cattle: they trade also in wood to great extent, and have established some manufactures and fisheries. The live stock in 1825 was returned at 366,046 horses, 369,917 horned cattle, and 560,980 sheep. This government is divided into 10 circles, but does not comprehend the level and very slightly cultivated tract called the steppes of Yaroslav. It contains 11 towns, and 7,977 villages.

Yaroslav, the capital of the above government, is situated on the r. bank of the Wolga, 160 m. NNE of Moscow, and 360 ESE of St. Petersburg. Though built almost entirely of wooden houses, and with narrow irregular streets, it is a place of considerable importance. Its manufactures of linen, silk, and leather, are on a large scale; and paper, soap, ropes, brass, iron and tin wares, white lead, and oil are also manufactured here. There is a noted bell foundry; and, by means of the Wolga, a traffic of some extent is conducted. The pop. in 1846 was about 25,000. It is the see of an archbishop, and is said to possess 44 churches. An academy, founded by Prince Demidoff, for the education of youths of genteel family, a high school for the education of those of less elevated birth, and a theological seminary on a large scale exist here. Peter the Great founded the manufactures of this place; but they were brought to perfection chiefly by the fostering hand of John Ernest, duke of Courland.

YAROU-DZANGBO-TCHOU. See Dihong, and Irawaddy.

YAROUGA, a town of Russia in Europe, in the gov. of Podolia, district and 18 m. WNW of Yampol, and 72 m. ESE of Kamenetz, on the l. bank of the Dniester. Pop. 189.

YAROWA, a town of New South Wales, in the co. of Brisbane, on the Goulburn.

YARPOLE, a parish in Herefordshire, 5 m. NNW of Leominster. It contains the township of Bircher. Area 2,523 acres. Pop. in 1831, 651; in 1851, 645.

YARRALUMLA PLAINS, a level tract of New South Wales, in the co. of Murray.

YARRAMUNBAW, a creek of New South Wales, in the district of Liverpool plains.

YARRANE, a river of Australia Felix, in the district of Western Port, to the NE of the Granitic range.

YARRA-YARRA, a river of Australia Felix, in the co. of Bourke, which flows past Melbourne, and 5 m. below falls into Hobson's bay, on the NW side. This river, the largest that flows into Port Philip, is navigable only a little way above Melbourne.

YARRIBA, YARIBA, YOURIBA, or HIO, a state in the S part of Nigritia, to the E of Dahomey, W of Nyffe or Tappa, from which it is separated by the Quorra, and S of Borgu. It is intersected by the mountains of Kong, and by the rivers Niger and Cola. The surface is an elevated flat or table-land. It is finely wooded, and formed one of the most important kingdoms in Nigritia, until about the year 1817, when it was devastated by a civil war. From the ruins of above 145 towns of this once-flourishing country, the present city of Abbeokuta, on the Ogun, 60 m. from Lagos, has arisen; and it is reported that there are other four large towns within two or three days' journey of it, with an aggregate of 200,000 inhabitants. The people are chiefly agriculturists, but conduct a considerable commerce in gums, spices, dyewoods, ivory, gold dust, and palm-oil. They have long straight black hair, and rather European countenances. They speak a dialect of simple structure.

YARROW, an extensive district and parish in Selkirkshire, comprising several miles of the S side of the vale of the Tweed, about four-fifths of the vale of the Yarrow, and a portion of the central vale of the Ettrick. Its area of 71,142 acres, is chiefly occupied with sheep-walks, on which upwards of 50,000 sheep are pastured. Pop. in 1801, 1,216; in 1831, 1,221; in 1851, 1,294.—The river Y. rises at Yarrow cleuch, and runs in an ESE course of about 25 m. to the Ettrick, which it joins a little above Selkirk.

YARROW, a creek of New South Wales, on Beardy Plains, in the district of New England.

YARROW-PEAK, a summit of New South Wales, in the co. of Murray, between the Molonglo and Queanbegan rivers.

YARWELL, a parish in Northamptonshire, 1¼ m. S by W of Wansford, on the river Nen. Area 1,830 acres. Pop. in 1831, 369; in 1851, 450.

YARYCZOW, a town of Austria, in Galicia, in the circle and 12 m. ENE of Lemberg, on a marsh, near the l. bank of the Peltew. It has a United Greek church, and possesses manufactories of blankets.

YASASHNAIA, a river of Russia in Asia, in the prov. of Yakoutsk, and district of Sredne-Kolymsk, in the country of the Koriaks. It runs NE, and after a course of 180 m. joins the Kolyma, on the l. bank, in N lat. 65° 20′, and E long. 153° 20′.

YASASHNIA-PISMIANKA, a town of Russia in Europe, in the gov. and 225 m. NW of Orenburg, and district of Bougoulma.

YASELDA, or YATSOLDA, a river of Russia in Europe, which has its source in the gov. of Grodno and district of Volkovisk, near Novoï-Dvor; runs ESE, traverses Lake Sporovskoe; enters the gov. of Minsk, and after a course of 135 m., throws itself into the Pripet, on the l. bank, to the E of Pinsk. The Oglimski canal joins this river on the left bank, at the village of Velesnitse, and connects it with the Shtshara.

YASNITZA, a fortress of Turkey in Europe, in the sanj. and 54 m. W of Banja-Luka, and 21 m. SW of Novi.

YASNOGORODKA, a town of Russia in Europe, in the gov. and district and 24 m. WSW of Kiev. Pop. about 150.

YASON, a headland of Turkey in Asia, in the pash. of Sivas, and sanj. of Djanik, on the Black sea, to the W of Cape Vona, and 12 m. NE of Fatsa, in N lat. 41° 8′ 15″, and E long. 37° 39′ 20″. It is of little elevation, and is surrounded by rocks, which render it difficult of approach.

YASOR, or YAZOR, a parish in Hereford, 4 m. S of Weobly. Area 2,051 acres. Pop. in 1851, 222.

YASS, or YESS, a territory of Nigritia, in the Haoussa, and prov. of Zeg-Zeg.

YASS, a district of New South Wales, comprising the S part of the co. of King, and N part of that of Murray, comprising an area of 1,111,486 acres. Pop. 2,187. It contains a town of the same name, 179 m. from Sydney, and divided into two parts by the Yass or Yhar river. Pop. 274. The river has its source near Mount Ainslie, flows N and W between the cos. of King and Murray, and joins the Murrumbidgee, about 170 m. SW of Sydney; and to the S of the Yustron is a level tract to which it gives its name. This plain is from 9 to 12 m. in length, and from 5 to 7 in breadth, and is surrounded by forests. To the WNW is a remarkable hill called Ponni.

YASSA, a river of Turkestan, in the khanat of Kokan, which descends from the Kashgar-Divani; runs SW, and joins the Kafernikhan a little below Yaz, and 135 m. E of Samarkand.

YASSAKTCHI, a town of Turkey in Europe, in Bulgaria, in the sanj. and 107 m. NE of Silistria, and 22 m. W of Toultcha, on the r. bank of the Danube, which is here crossed by a ferry. The town is defended by a castle, and contains several mosques, baths, and a harem. It is very populous, and possesses an active trade. It was taken in 1790, and again in 1828 by the Russians.

YASSIH, or YASSAH, a town of Burmah, in the prov. of Mranma, on the l. bank of the Irawady, and 9 m. S of Pagahm-Mion.

YASSIN, a town of Nigritia, in Kordofan, to the E of Obeit.

YASSY. See JASSY.

YASVOIN, a town of Russia in Europe, in the gov. and 75 m. NW of Vilna, district and 36 m. ESE of Rossiena, near the r. bank of the Pevieje.

YATA, two rivers of Peru, distinguished as First and Second. They issue from Lake Roguaguado, run NE, and after a course of about 150 m. join the Mamore, on the l. bank. The First is the more easterly. Between them is the Tamaya Guibo.

YATASCO, a town of La Plata, in the prov. and 75 m. SSE of Salta.

YATCHENKA, a small river of Russia in Europe, in the gov. and district of Kalonga, which runs S, and after a course of 21 m. throws itself into the Oka, on the l. bank, 1½ m. W of Kalonga. On its banks the remains of an ancient town, said to have been the place of residence of a prince named Simeon, supposed to have been the son of Ivan-Danilovitch-Calita, and the same prince who reigned in Moscow and Novgorod.

YATE, a parish in Gloucestershire, 1 m. W of Chipping-Sodbury. Area 4,042 acres. Pop. 1,080.

YATE WITH PICK-UP-BANK, a joint township in the p. of Whalley, co.-palatine of Lancaster. Area 1,360 acres. Pop. in 1851, 1,208.

YATE, or JATTE, an island of the coast of Senegambia, to the S of the mouth of the Cacheo, and separated from the continent by a narrow channel, which bears the name of Jatte river, and appears to form a branch of Catherina river, by which the island is divided from the peninsula in the S. It possesses considerable fertility.

YATELEY, a parish in Southamptonshire, 4 m.

NE by E of Hartford-Bridge, on the river Blackwater, including the tythings of Cove and Hawley-with-Minley. Area 10,036 acres. Pop. 2,156.

YATES, a mountain of Western Australia, in the co. of Sussex.

YATES, a central co. of the state of New York, U. S., comprising an area of 320 sq. m., drained by Flint creek, West river, Crooked lake outlet and Big and Rock streams. It has a finely diversified surface, and is generally fertile, and intersected by the Canandaigua and Elmira railroad. Pop. in 1840, 20,444; in 1850, 20,590. Its capital is Penn-Yan.—Also a township of Orleans co., in the same state, 182 m. W by N of Albany, drained by Johnson's creek, an affluent of Lake Ontario, by which the township is bounded on the N. Pop. in 1840, 2,230; in 1850, 2,242.

YATESBURY, a parish in Wilts, 4½ m. E by N of Calne. Area 1,167 acres. Pop. in 1851, 251.

YATIR, a village of Syria, overlooking the plain of Tyre, and occupying an ancient site.

YATLAKA, a town of Hindostan, in the presidency of Madras, district and 69 m. E by S of Bellary.

YATOVA, a village of Spain, in the prov. of Valencia, and 1½ m. S of Bunol. Pop. 1,180.

YATREB. See MEDINA.

YATTABBAS, a tribe of the Bichuana family, in Caffraria.

YATTENDON, a parish in Berks, 6 m. SE of East Ilsley. Area 1,393 acres. Pop. in 1851, 263.

YATTON, a chapelry in the p. of Great Marcle, Herefordshire, 6 m. NE by N of Ross. Pop. 203.

YATTON (EAST and WEST), a parish in Somersetshire, 8 m. N of Axbridge, on the river Yeo, and in the line of the Bristol and Exeter railway. Area 5,374 acres. Pop. in 1831, 1,865; in 1851, 2,061.

YATTON-KEYNELL, a parish in Wilts, 4½ m. NW by W of Chippenham. Area 1,749 acres. Pop. in 1831, 419; in 1851, 516.

YAUCO, a town of the Antilles, in the S part of the island of Porto Rico, and 8 miles from the sea. Pop. 2,680. The surrounding district produces rice, maize, and tobacco.

YAUSH, a large town of Abyssinia, in the prov. of Gojam, near the river Yeda.

YAUTEPEC, a town of Mexico, in the state and 45 m. SSE of Mexico, in the fertile valley of the Istla. Pop. 2,000. It has a Dominican convent, and an hospital.

YAUYOS, or LLAULLOS, a district or prov. of Peru, in the dep. and to the SE of Lima. Pop. about 10,000. It is one of the most mountainous and coldest in Peru. The rearing of cattle forms the chief employment of the inhabitants. It contains a town of the same name.

YAVAHES, a tribe of Indians of Brazil, in the prov. of Goyaz, who inhabit the island of Santa Anna, formed by the Araguay.

YAVAL-COHOL, a mountain of Spain, in the prov. and 66 m. ENE of Granada, and 5 m. NNW of Baza. At its base are several abundant thermal springs.

YAVARY. See JAPURA.

YAVATA, a town of Japan, in the island of Nifon, and prov. of Foki, 96 m. WNW of Meaco.

YAVENESUREN, a town of Hindostan, in the presidency of Madras, district and 42 m. SE of Madura.

YAVERLAND, a parish in the Isle of Wight, 8 m. ESE of Newport. Area 1,834 acres. Pop. in 1851, 78.

YAVISA. See JAVISA.

YAVOM, a river of Russia in Europe, in the gov. of Novgorod, which issues from Lake Velce, in the district of Valdai; runs W, and throws itself after a course of 12 m. into the Pola on the r. bank.

YAXHAM, a parish in Norfolk, 2½ m. SE by S of East Dereham. Area 1,596 acres. Pop. 506.

YAXLEY, a parish, formerly a market-town, in the co. of Huntingdon, 14 m. N by W of Huntingdon, in the vicinity of Whittleseamere. Area 4,290 acres. Pop. in 1831, 1,140; in 1851, 1,445.—Also a p. in Suffolk, 1½ m. N of Eye. Area 1,238 acres. Pop. in 1831, 478; in 1851, 550.

YAYGEE, a town of Burmah, in the prov. of Pegu, on one of the branches of the Irawady.

YA-YAMA. See PATCHOW.

YAZOO, a central county of the state of Mississippi, U. S., comprising an area of 731 sq. m., drained by tributaries of Big Black and Yazoo rivers, and intersected by the Mississippi Central railroad. It has an undulating surface, and is generally fertile. Pop. in 1840, 10,480; in 1850, 14,418. Its capital is Benton.—Also a town of the above named county, situated on a low bluff, on the E side of Yazoo river, and 40 m. N by W of Jackson. Pop. in 1840, 700; in 1850, about 2,000. The river Y. is formed near the W confines of the co. of Carroll, by the confluence of the Tallahatchee and Yallabusha rivers, runs SSW, and after a winding course of about 160 m. joins the Mississippi near Vicksburg. It is navigable by steam-boats.

YBA, a village of the island of Luzon, Philippine archipelago, on the SW coast, 90 m. NW of Manilla.

YBARS (SAINT), or SAINT IBARS, a commune and town of France, in the dep. of Arriege, and cant. of La Fossat, 11 m. NW of Panners, on the Leze. Pop. in 1846, 2,328.

YBART (SAINT), a commune of France, in the dep. of Correze, cant. and 3 m. NW of Uzerche. Pop. 1,557.

YBERA. See IBERA.

YBICUY. See IBICUI.

YBOS, an island of the Philippine group, in the bay of Llana, to the S of the island of Mindanao. It is about 3 m. in circumference, and is a resort of pirates.

YCABAQUA. See ICABAQUA.

YCACOS. See ICACOS (CAPE).

YCAHUATES, an Indian people of Ecuador, who inhabit the banks of the Napo, about 100 m. above the confluence of that river with the Amazon.

YCO. See ICO.

YE, a province of British Indu-China, bounded on the N by the prov. of Amherst, on the E by the Siamese mountains, on the S by the prov. of Tavoy, and on the W by the Bay of Bengal, and is the smallest of the Tenasserim provinces. It is generally mountainous, and is to a great extent covered with wood or jungle. Its principal productions are rice and timber. It contains a river of the same name, the only one in the prov., and on its banks about 6 m. above its embouchure in the Bay of Bengal, is a town also called Ye. The prov. of Ye was ceded to the English in 1826.

YEADON, a township in the p. of Guisley, Yorkshire, 7 m. NNE of Bradford. Area 1,660 acres. Pop. in 1831, 2,761; in 1851, 4,109.

YEALAND-CONYERS, a township in the p. of Warton, co.-palatine of Lancaster, 2 m. WSW of Burton-in-Kendal, in the line of the Lancaster and Kendal canal. Area 1,577 acres. Pop. in 1851, 306.

YEALAND-REDMAYNE, a township in the p. of Warton, co.-palatine of Lancaster, 3 m. SW of Burton-in-Kendal. Area 2,099 acres. Pop. 226.

YEALM, a small river in Devonshire, rising in Ermington hundred, and flowing through an estuary into the English channel in Plympton hund.

YEALMPTON, or YALMTON, a parish, formerly a borough, in the co. of Devon, 4 m. SE by S of Earl's

Plympton, intersected by the river Yealm. Area 3,537 acres. Pop. in 1831, 1,262; in 1851, 1,155.

YEAROAH, a river of New South Wales, in the district of Darling Downs, near the source of the Boyne.

YEARSLEY, a township in the p. of Coxwold, Yorkshire, 6½ m. SSW of Helmsley. Area 2,764 acres. Pop. in 1831, 164; in 1851, 171.

YEAVAN-GHEOUN, YEAGHAIN, or SHONEGHAIN, a town of Burmah, in the prov. and 84 m. NW of Pegu, on the l. bank of the Irawady, in N lat. 20° 31′, and E long. 94° 43′.

YEAVELEY, a chapelry and township in the p. of Shirley, co. of Derby, 4½ m. S of Ashbourn. Pop. in 1831, 271; in 1851, 238.

YEBARA, a town of Japan, in the island of Nifon, and prov. of Mossasi, and 21 m. NNE of Yedo.

YEBENES, a town of Spain, in the prov. and 21 m. S of Toledo, and partido of Orgaz, at the foot of a mountain. Pop. 3,063. It is divided into two parts, and distinguished as Yebenes-de-Toledo, and Yebenes-de-S. Juan. It has a parish-church, a custom-house, an hospital, and a public granary. Its industry consists chiefly in the manufacture of coarse woollen fabrics and hosiery. Honey abounds in the surrounding country, and forms an important object of trade.

YEBES, a town of Spain, in the prov. and partido, and 9 m. SE of Guadalajara. Pop. 430.

YEBLERON, a commune of France, in the dep. of the Seine-Inferieure, and cant. of Fauville, 12 m. W of Yvetot. Pop. 1,817.

YEBRA, a town of Spain, in the prov. of Guadalajara, and partido of Pastrana, 33 m. SE of Madrid, amidst mountains, chiefly of limestone construction, and possessing considerable fertility. Pop. 1,290. It has a parish-church, a custom-house, and a public granary, and has manufactures of fine leather, and of haircloth.

YECHA, or JECHA, a village of the principality of Schwarzburg-Sondershausen, bail. and 1½ m. SE of Sondershausen, on the r. bank of the Wipper. Pop. 660. In the vicinity, on the mountain of Frauenberg, is the fortress of Jechaburg, a residence of the emperor Louis II., which was taken by the Huns in 933.

YECHLINGEN, a village of Baden, in the circle of the Rhine, and bail. of Breisach, on the Rhine. Pop. 850.

YECKENBACH, a village of Hesse-Homburg, in the bail. of Meissenheim. Pop. 390.—Also a village of Prussia, in the regency of Treves, and circle of the Wendel. Pop. 426.

YECLA, a judicial partido and town of Spain, in the prov. of Murcia. The partido comprises 2 pueblos. The town is 48 m. SE of Chinchilla, and at about an equal distance from Murcia, partly on the slope, and partly at the foot of a mountain, on the summit of which there formerly was a fortress. Pop. 9,933. It has two parish churches, a convent, six chapels, and an hospital, and possesses numerous distilleries of brandy, oil-mills, and tanneries. Of the ancient ramparts of Y. only the traces are now to be found. In the vicinity are the remains of a fortress, and near it some vestiges of a town, in which vases and funeral urns, medals of Roman emperors, especially those of Septimus Severus and Gallus, have been found. In the locality are also the foundations of a Catholic chapel, which has existed from before the invasion of the Moors.—Also a town of the prov. and 36 m. WSW of Salamanca, and partido of Lumbrales, in a fertile plain, near the confluence of the Huebra and Yeltes. Pop. 700. It has manufactories of coarse woollen and linen fabrics. Surrounding it are some traces of a Roman wall, in which numerous ancient coins and medals were found.

YECORA, a town of Spain, in the prov. of Alava, and partido of La Guardia, 8 m. N of Logrono, in an infertile district. Pop. 570.

YECZOW, a village of Prussia, in the regency of Koslin, and circle of Lauenburg-Bütow. Pop. 200.

YEDANCHOW, a village of Austria, in Bohemia, in the regency of Pardubitz, and bail. of Deutschbrot. Pop. 290.

YEDAUSOW, a village of Austria, in Bohemia, in the regency and bail. of Pardubitz. Pop. 230.

YEDDIE, a town of Nigritia, in Bornu, 24 m. S of Kuka. It is large, and walled, and is governed by a kaïd.

YEDDINGEN, a village of Hanover, in the gov. of Stade, and bail. of Rotenburg. Pop. 210.

YEDDINGHAM, a parish in the E. R. of Yorkshire, 8¼ m. NE of New Malton, on the river Derwent. Area 1,150 acres. Pop. in 1851, 104.

YEDE, a village of Austria, in Moravia, in the regency of Olmutz, and bail. of Hohenstadt. Pop. 1,600.

YEDENSPEIGEN, a market-town of Austria, in the district below the Ens, and bail. of Gross-Enzersdorf, on the March. Pop. 800. It has a castle.

YEDERITZ, a village of Prussia, in the regency of Potsdam, and circle of West Priegnitz. Pop. 270.

YEDERSDORF, a village of Austria, in the district below the Ens, and bail. of Waidhofen. Pop. 1,200.

YEDESHEIM, a village of Bavaria, in the circle of Swabia, presidial and near Illertissen. Pop. 590.

YEDI-BURON, or SEVEN CAPES, a mass of steep mountains of Turkey in Asia, in the sanj. of Meis, on the Mediterranean coast, between the gulf of Macri and the mouth of the Etchenchai.

YEDLAN, a village of Austria, in Bohemia, in the regency of Budweis, and bail. of Tabor. Pop. 320.—Also a village in the regency of Pardubitz, and bail. of Ledetsch. Pop. 200.

YEDLER, a village of Prussia, in the regency of Posen, and circle of Pleschen. Pop. 437.

YEDLERSDORF, a village of Austria, in the district below the Ens, and to the N of Vienna. Pop. 500. It has a castle.

YEDLESEN, a village of Austria, in the district below the Ens, on the Danube. Pop. 680. It has several breweries.

YEDLICE, a market-town of Austria, in Galicia, in the circle and 9 m. ESE of Jaslo, on the r. bank of the Jasiel, an affluent of the Wisloka.—Also a village of Prussia, in the regency and circle of Oppeln. Pop. 147.

YEDLIN, a village of Prussia, in the regency of Oppeln, and circle of Pless. Pop. 340. It has a castle and several mills.

YEDLINA, a village of Russia in Europe, in the gov. and to the SW of Vilna.

YEDLOWNITZ, a town of Prussia, in the regency of Oppeln, and circle of Rybnik. Pop. 218. It has several mills.

YEDNA, a town of Russia in Europe, in the gov. of Novgorod, and NE of Waldai.

YEDO, YEDDO, or JEDO, a city of Japan, the capital of the island of Nifon, on a fine insular gulf, on the SE side, in N lat. 35° 40′, E long. 139° 40′. It is represented as occupying an extensive plain, and bounded by a semicircular range of mountains. The houses are generally of one story, and there are few imposing public edifices. The pop. has been variously estimated at from 500,000 to 1,500,000. When the American squadron was in the bay of Yedo in 1854, an attempt was made to get as near the capital as possible, and the commodore ran up to within 8 m. of the city; but the imperial commissioners so warmly opposed the design, that it was relinquished. See article JAPAN.

YEDOGAVA, a river of Japan, in the island of Nifon, which issues from the S extremity of lake Biwano-Umi; runs SW, passing Meaco; and, after a course of about 73 m., falls into the bay of Osaka, near the town of that name. It is narrow, but deep, and is always filled with vessels.

YEDOW, a village of Austria, in Moravia, in the regency of Brünn, and bail. of Kroman. Pop. 208. It has several mills.

YEDOWNITZ, a market-town of Austria, in Moravia, in the circle and 12 m. NNE of Brünn. Pop. 825. It has an old castle. In the vicinity is the famous grotto of Maczocha, 950 ft. in depth.

YEDROWO, a town of Russia in Europe, in the gov. of Novgorod, SE of Waldai.

YEDTORRA, a town of Hindostan, in the state of Mysore, on the r. bank of the Cavary, 22 m. W by N of Seringapatam, in N lat. 12° 29′, and E long. 76° 26′.

YEDWABNO, a town of Russian Poland, in the gov. and 48 m. SW of Augustowo, and 12 m. NE of Lomza. Pop. 475.—Also a village of Prussia, in the regency of Konigsberg, and circle of Neidenburg. Pop. 256.

YEDWILLEITEN, a village of Prussia, reg. of Gumbinnen, and circle of Niederung. Pop. 280.

YEEBEN, a village of Prussia, in the reg. of Magdeburg, and circle of Salzwedel. Pop. 315.

YEETZE, or YETZE. See JETZE.

YEETZEL, a village of Hanover, in the principality of Luneburg, and bail. of Luchow. Pop. 210.

YEFLEREVSKAIA, a town of Russia in Europe, in the gov. of the Don Cossacks, and district of Onst-Medvieditza, on the r. bank of the Medvieditza, and 270 m. NE of Novo-Tcherkask. Pop. 1,000. The rearing of cattle forms the chief employment of the inhabitants.

YEFREN, a town of Northern Africa, on the Jebel range, in N lat. 32° 5′, E long. 12° 30′. There is a Turkish garrison of 500 men here, under the command of a Kaimakan, whose authority extends over Ghadamis, and the Jebel and Ghaurian ranges.

YEFREMOW, a district and town of Russia in Europe, in the gov. of Tula. The district is in the S part of the gov. It is but indifferently cultivated, and contains little wood. The town is 84 m. SSE of Tula, on the l. bank of the Metscha, which bathes the foot of the hill on which the town is situated. Pop. 7,366. It contains seven churches, all with one exception built of wood, and five suburbs. It was formerly enclosed by a wall, of which only a few traces now remain.

YEGENYE, a village of Austria, in Transylvania, to the WNW of Klausenburg. It has several mineral wells.

YEGGAN, a village of Prussia, in the regency of Magdeburg, and circle of Gardelegen. Pop. 271.

YEGGEN, a village of Hanover, in the district of Osnabruck. Pop. 306.

YEGHE, a town of China, in Manchuria, cant. and 135 m. SW of Kirin.

YEGINOE, or JEGINOE, an island of Denmark, in the Lym-fiord, diocese of Aalborg, and bail. of Thisted, in Jutland, and to the S of the island of Mors. It is 4 m. in length and 2 in extreme breadth, and contains a village of the same name.

YEGISTORF, a village of Switzerland, in the cant. and 8 m. NNE of Berne, and bail. of Fraubrunnen, on the road from Berne to Soleure. It has a fine castle belonging to the family of Erlach. Pop. of p. 2,438.

YEGODNEN (GROSS and KLEIN), two villages of Prussia, in the reg. of Gumbinnen, and circle of Lotzen. Pop. 240.

YEGORIEVSK, a district and town of Russia in Europe, in the gov. of Riazan. The district, which is in the N part of the gov., is woody and marshy, and contains several lakes. The culture of flax and hemp, and manufacture of linen, form the chief branches of local industry. The town is 63 m. NNW of Riazan, and 57 m. ESE of Moscow. Pop. 500. It contains two churches. The trade consists chiefly in hops, the produce of the environs.

YEGORLYK, a river of Russia in Europe, in the prov. of the Caucasus, and district of Slavropol, which has its source 12 m. SE of the town of that name, runs from S to N, and throws itself into the Manytsh, on the confines of the provinces of the Caucasus and of the Don Cossacks, about 30 m. W of Lake Bolcheï, and after a course of about 195 m.

YEGOUGIN, a town of Russia in Europe, in the gov. and 39 m. NW of Vilna, and district of Novo-Troki, on the Viliia.

YEH. See YE.

YEHYAPUR, a town of Hindostan, in the presidency of Bengal, and district of Sarun, 41 m. SW of Beltiah, in N lat. 26° 22′, and E long. 84° 7′.

YEIA, a river of Russia in Europe, in the prov. of the Caucasus, which has its source in the district and 84 m. NW of Stavropol; enters the district of the Cossacks of the Black sea; receives the Kongoi Yeïa; forms for the distance of about 21 m. the S confines of the gov. of the Don Cossacks; re-enters the former territory, and after a total course of 150 m., in a generally WNW direction, throws itself near Fort Yeïsk into the gulf of Yeïskoï-liman.

YEINVIK, a gulf of Russia in Europe, formed by the Baltic, on the W coast of the gov. of Esthonia, and district of Hapsal, at the mouth of the Kazarin. It is about 17 m. in length, and 6 in breadth.

YEISK, a fort of Russia in Europe, in the prov. of the Caucasus, in the territory of the Cossacks, near the mouth of the Yeïa, in the Yeïskoï-liman, and 135 m. N of Yekaterinodar.

YEISKOI-LIMAN, a gulf of the sea of Azof, at the mouth of the Yeïa, on the coast of the prov. of the Caucasus, in Russia in Europe. It is 14 m. in length, and 9 in breadth.

YEKALI, a town of Hindostan, in the prov. and 63 m. WNW of Hydrabad.

YEKATERINBURG, or EKATERINBURG, or JEKATERINBURG, a town of Asiatic Russia, in the gov. and 170 m. SE of Perm, in N lat. 56° 50′, E long. 60° 34′, at an alt. of 850 ft. above sea-level. It forms the capital of the mining districts of the Ural; and in external appearance reminds one of the manufacturing towns of Europe. Many of its houses are of stone; and the government magazines and offices are built in a very handsome style. The streets are straight, but unpaved, with logways at the sides for foot-passengers. The principal street runs parallel with the river Iset, which flows through the city. The principal branch of industry practised here, is the cutting and polishing of precious stones. The pop. has been stated at 15,000.

YEKATERINENSTADT, or YEKATERINSTADT, a village of Russia in Europe, in the gov. and 35 m. NE of Saratov, and district of Volsk, on the r. bank of the Volga. Pop. 646. It was founded by a German colony in 1763, who carried on an active trade in tobacco and cheese. The colonies of the Volga have here a general grain depot.

YEKATERINODAR, a town of Russia, in the country of the Tchernomaski Cossacks, on the l. bank of the Kuban, about 90 m. from its mouth. It is regularly built, and has a cathedral; but the streets are unpaved, and the houses mostly of wood.

YEKATERINOGRAD, a town of Russia, in the gov. of the Caucasus, on the l. bank of the Terek, in N lat. 43° 40′. It is an important military station.

YEKATERINOPOL, a town of Russia in Europe, in the gov. and 111 m. S of Kiev, district and 11 m. S of Zvenigorodka. It formerly bore the name of Kalniboloto.

YEKATERINOSLAV, or EKATERINOSLAV, a southern government of Russia, stretching between the parallels of 47° and 49° 30′ N; and bounded on the N by the govs. of Pultova, Charkof, and Voronetz; on the E by the territory of the Don Cossacks; on the S by Taurida; and on the W by Kherson. It has an area of 54,253 sq. versts, exclusive of Taganrog, and the territory of the Azof Cossacks; with a pop. in 1846 of 787,000. The river Dnieper intersects this gov. Donetz forms its NE boundary. That portion which lies to the E of the Dnieper is a sandy barren steppe; on the W side the soil is richer, and the country better cultivated. Wheat, barley, oats, flax, and hemp, are cultivated; but the chief attention is bestowed on grazing cattle, and rearing sheep. Some of the German colonists in this gov. are said to possess flocks of 17,000 head. Honey forms an important article of commerce, and many of the farmers keep several hundred hives. Fish and tallow are exported; but the manufacturing industry is nearly confined to articles of local consumpt.—The cap., which gives name to the gov., is situated on the r. bank of the Dnieper, in N lat. 48° 27′, about 250 m. NE of Odessa, and 170 NE of Oczatov. It has a pop. of 10,000, and is the seat of an archbishop. Its most important article of manufacture is fine woollen cloth.

YEKATERINSK, a mining village of Russia in Asia, in the gov. of Irkutsk, and district of Nertchinsk. It has a valuable silver-mine.

YEKATERINSKAIA, a town of Russia in Europe, in the gov. of the Don Cossacks, and district of the Don, 63 m. NNE of Novo-Tcherkask, on the l. bank of the Donetz.

YEKATERINSKAIA, or KHOLAN, a town of Russia in Europe, in the gov. and 75 m. SE of Koursk, and district of Novoi-Oskol.

YEKDER, or EKDER, a town of Turkey in Asia, in Anatolia, in the sanj. of Teke or Adalia, on a small river, near the W coast of the gulf of Adalia, and 23 m. SSW of the town of that name. To the E of Y. is Cape Arova, also called Ekdar, which forms the extremity of mount Taktalu, the *Solymus* of antiquity, and a little to the S at the village of Tekrova are the ruins of the ancient *Phaselis*.

YEKHOLM, a small island of Russia in Europe, in the gulf of Finland, and gov. of Esthonia, to the N of the port of Munkewick, in N lat. 59° 41′, and E long. 25° 55′. It has a lighthouse at the N extremity.

YEKIL-IRMAK, or KASALMAK [i. e. Green River], a river of Turkey in Asia, in the pash. of Sivas, which has its source in the Anti-Taurus range, in the sanj. of Devriki; runs first WNW under the name of Tozanlu, then NW, and afterwards NE, and after a total course of about 300 m., throws itself by several embouchures into the Black sea between Samsun and Thermeh. Its principal affluent is the Kelki, which it receives on the r. Tokat, Turkhal, and Amasia are the chief places on its banks.

YEKIMANIA, a town of Russia in Europe, in the gov. and 66 m. WNW of Vitebsk, district and 2 m. W of Polotsk, on the l. bank of the Dvina.

YEKISMOKORI, a town of Japan, in the island of Nifon, and prov. of Umi, 84 m. SSW of Meaco, on the W bank of Lake Umi.

YEKNAS. See EKNAS.

YEKTHOON, a town of Hindostan, in the prov of Hydrabad, and 18 m. WSW of Jaulna.

YELA, a town of Spain, in the prov. and 33 m. NE of Guadalajara, and partido of Brihuega, near the r. bank of the Tajuna. Pop. 350.

YELABUGA, a district and town of Russia in Europe, in the gov. of Viatka. The district contains extensive forests, and near the Kama is extremely fertile. Its inhabitants are chiefly Vostiaks and Tsheresnisses. The town is 222 m. SSE of Viatka, and 30 m. E of Mamadish, on the r. bank of the Kama. Pop. 3,500. It contains three churches, and has manufactories of cloth and of cotton fabrics. In the vicinity are the ruins of a temple.

YELAGONI, a river of Russia in Asia, in the W part of the gov. and district of Yeniseïsk. It runs from W to E, and after a course of about 225 m. joins the Yenisei, by several embouchures on the l. bank, between the confluence of the Podkameunaia-Toungouska and Nijniaia-Toungouska. Its principal affluents are the Tilgan and Pokatka, the source of the latter of which is near that of the Vakh, a tributary of the Oby.

YELAMOS-DE-ABAJO, a town of Spain, in the prov. and 15 m. from Guadalajara, and partido of Brihuega. Pop. 470.

YELAMOS-DE-ARRIBA, a town of Spain, in the prov. and 15 m. from Guadalajara, and partido of Brihuega, 1½ m. from Yelamis-de-Abajo. Pop. 670.

YELANDU-SINGUERI, a long chain of hills in Turkistan, in the country of the Kirghiz of the Middle horde, and to the E of the Yanghiz-tagh, between 50° and 51° N lat., and 63° 20′ and 64° 20′ E long.

YELANSKAIA, a town of Russia in Europe, in the gov. of the Don Cossacks, and district of Oust-Medvieditza, on the l. bank of the Don, and 192 m. NNE of Novo-Tcherkask.

YELATA, a town of Senegambia, in the state of Futa-Jalon, 39 m. N of Labbé, and near a river of the same name, which, after a course in a generally E direction of 45 m., joins the Gambia, on the l. bank.

YELATOM, or YELATMA, a district and town of Russia in Europe, in the gov. of Tambov. The district, which occupies the NW part of the gov., is hilly and ill-cultivated, and to a great extent covered with wood, amongst which the oak is distinguished for its magnitude. Agriculture and the rearing of cattle form the chief objects of local industry.—The town is 165 m. N of Tambov, and 17 m. ENE of Kasimov, on the l. bank of the Oka. Pop. 5,800. It contains ten churches, and has manufactories of linen, sulphur, vitriol, and copper, and carries on an active trade in tallow, hemp, corn, wax, and honey. In the vicinity are the great iron-works of Yeremschink.

YELBURGA, a town of Hindostan, in the presidency of Madras, and prov. of Hydrabad, 69 m. WNW of Bellary, in N lat. 15° 38′, and E long. 76° 4′.

YELDERSLEY, a hamlet in the p. of Ashbourn, co. of Derby, 3½ m. ESE of Ashbourn. Pop. 199.

YELDHAM (GREAT, or NETHER), a parish in Essex, 3¼ m. NW by N of Castle-Hedingham. Area 1,820 acres. Pop. in 1831, 673; in 1851, 716.

YELDHAM (LITTLE, or UPPER), a parish in Essex, 4 m. N of Castle-Hedingham. Area 938 acres. Pop. in 1831, 371; in 1851, 306.

YELDIATSK, a small fort of Russia in Europe, in the gov. of Orenburg, district and 51 m. NE of

Birsk, on the r. bank of the Oufa. It contains a church, and about 100 houses. The garrison consists of a half company of infantry, and about 100 Cossacks.

YELDURTY, a town of Hindostan, in the presidency of Madras, district and 21 m. SSW of Kurnool, in N lat. 15° 32′, and E long. 77° 59′.

YELENKA, a town of Russia in Europe, in the gov. of Tchernigov, district and 12 m. SSW of Starodoub.

YELETZ, a district and town of Russia in Europe, in the gov. of Orel. The district, which is in the E extremity of the gov., is hilly, partly woody, and partly cultivated, producing corn in considerable quantities, and pasturing large numbers of cattle.—The town is 120 m. ESE of Orel, on the l. bank of the Sosna, a little below the confluence of the Loutshka and Yeletz. Pop. 8,000. It is regularly built, and contains 14 churches, 12 of which are built of stone, and three charitable institutions. It has several soap-works, tanneries, and founderies, and carries on an active trade in cattle, corn, honey, and leather. In the vicinity are extensive ironworks. Y. is of great antiquity. Before the invasion of Batou-Khan, it formed an appanage of the princes of Riazan. In 1392 it was taken and destroyed by Tamerlane, and its sovereign made prisoner. The period of its falling under Russian domination is unknown. It has suffered frequently from the incursions of the Tartars of the Crimea, and in 1745 was reduced to ashes.

YELEZINSKAIA, a fort of Russia in Asia, in the prov. and 150 m. SE of Omsk, and 300 m. NW of Semipalatinsk. It makes part of the line of the Irtish, and contains a church and 164 houses, chiefly wood-built. The barracks, magazines, and house of the commandant, are of stone. The environs are fertile, and the plain towards the W affords excellent pasturage.

YELFORD, a parish in Oxford, 3 m. S of Witney. Area 305 acres. Pop. in 1851, 17.

YELINEWO, a town of Poland, in the gov. and obwod of Augustowo, 8 m. N of Suwalki. Pop. 516.

YELINKA-POSA, or ZLINKA, a town of Russia in Europe, in the gov. and 75 m. NNE of Tcheringov, and district of Novo-Zibcov.

YELISAVETGRAD, a district and town of Russia in Europe, in the gov. of Kherson. The district is in the NE part of the gov. Towards the N it is hilly, fertile, and well-cultivated, and covered with extensive forests, while in the S it presents the aspect of a steppe, in which numerous flocks and troops of wild horses roam at large. Saffron grows indigenously. Game is abundant. Millstone is extensively quarried, and forms an important article of commerce. The town is 138 m. N of Kherson, on the l. bank of the Ingoul, in N lat. 48° 30′ 17″, and E long. 30° 7′ 30″. Pop. 12,000. It has four suburbs, and regularly built and fortified. The streets are straight and lined with trees. It contains a large hospital, several magazines, and five churches, inclusive of one built by Potemkin for the rascolnikis or separatists, who form the majority of the population. An active trade is carried on with the adjoining governments.

YELISAVETINSKOI, a redoubt of Russia in Asia, in the gov. of Georgia, on the great road of the Terek, from Tiflis to Mozdok, and 15 m. NNW of Vladikavkas, on the l. bank of the Koumbalei.

YELISAVETPOL, GHENDIH, GANJA, or KANSAG, a district and town of Russia in Asia, in the gov. of Georgia. The district, which is situated in the S part of Georgia, is bounded on the N by the Kur; on the E by the Karabegh; and on the S by Persia; and comprises an area 60 m. in length and 30 m. in breadth. It is of all the districts of Georgia the warmest and most genial, and produces silk and cotton in great abundance. Its inhabitants are chiefly Turkomans, whose nomadic habits are unfavourable to agriculture. The remainder consists of Armenians, Jews, and Georgians.—The town is 75 m. SSE of Siguakhi, and 111 m. SE of Tiflis, on a small river of the same name, which divides into six branches, and about 18 m. below joins the Kur. It is fortified, and was formerly the residence of a khan, and is now of little importance. This town is of great antiquity. After the destruction of the kingdom of the Arsacides it fell into the power of the Aghovans, and under their dominion rose to considerable eminence. In 1088, it was taken by the Seljuks, and in 1235 fell into the hands of the Turks, who put its inhabitants to the sword. Some time after it was conquered by the Persians, and from them was taken by Russia.

YELJAL, a town of Hindostan, in the presidency of Madras, prov. and 40 m. S by W of Hydrabad, in N lat. 16° 49′, and E long. 78° 24.

YELL, a county in the NW part of the state of Arkansas, U. S., comprising an area of 1,027 sq. m., drained by Petite-Jean, and other affluents of the Arkansas, by which it is bordered on the NE. The surface is diversified, and possesses considerable fertility. Pop. in 1850, 3,341. Its capital is Danville.—Also a central county of the state of Iowa, comprising an area of 615 sq. m., drained by Des Moines river and its branches. It has a finely diversified surface, and is generally fertile.

YELL, one of the Shetland islands, the second of the group in point of size, and the most northerly in situation, excepting Unst. Yell sound, which is studded with isles and islets, separates Yell from the N part of Mainland. Its length from N to S is 19 m.; its greatest breadth is 9½ m.; and its superficial extent is about 94 sq. m. The tides on both sides of the island are very impetuous; and both on Yell and Blomel sounds, where they meet with obstructions, and often run at the rate of 9 or 10 m., they for continuous hours defy boat-navigation, and toss the sea, even during a calm, into foam and tumult. The surface of the island presents a heavy and cheerless aspect. Two nearly parallel ridges of gneiss rocks, of almost uniform outline, and only from 200 to 400 ft., traverse it nearly from end to end; sloping gradually toward the shores, and, in some places, connected by transverse ridges, running from E to W. Almost the whole soil is moss; occasionally, yet seldom, mixed with clay or sand. Agriculture is in a miserable condition. Only about 4,000 acres are enclosed; probably not above 1,000 are at any time under culture; and the spade is the only implement of tillage. The island, however, is an excellent fishing-station. Pop. in 1835, 2,693; in 1851, 3,397.—Y. is ecclesiastically distributed into three parishes, North, Mid, and South. North Y. is united to FETLAR: which see. Mid Y. and South Y. form one ministry, or united parish, comprehending also the island of Samphray, lying 1½ m. SSW of the S extremity of Y., and measuring about 360 acres in area, and the island of Bigga, lying 1¼ m. NW of the former, and measuring about 540 acres in area; exclusive of these, it has a superficial extent of 36,000 acres, and an extreme length and breadth of respectively 13 and 6½ m. Pop. in 1801, of North Y. and Fetlar, 1,389; in 1831, 1,812; in 1851, 1,656.

YELLAGODE, a town of Hindostan, in the presidency of Madras, district and 36 m. E by S of Kurnool, in N lat. 15° 44′, and E long. 78° 38′.

YELLANUR, a town of Hindostan, in the presidency of Madras, district and 86 m. ESE of Bellary, in N lat. 14° 40′, and E long. 78° 8′.

YELLAPUR, a town of Hindostan, in the presidency and 400 m. NW of Madras, and district of North Canara, in N lat. 14° 58′, and E long. 74° 46′.

YELLING, a parish in Huntingdonshire, 6 m. ENE of St. Neot's. Area 1,670 acres. Pop. in 1851, 386.

YELLNANDER, a town of Hindostan, in the prov. of Mysore, 36 m. SE of Seringapatam, in N lat. 12° 3′, and E long. 77° 4′.

YELLOW, a lake of Jefferson co., in the state of New York, U. S., 160 m. NW of Albany. It is 4 m. long, ½ m. in width, and abounds with fish.—Also a river of the state of Indiana, which has its source in the NW part of Kosciusco co.; flows WSW and W, and falls into English lake. It has a course of about 60 m., and is navigable at high water a distance of near 40 m. Its average width below Plymouth is 100 ft.

YELLOW-CREEK, a village of Northampton township, Summit co., in the state of Ohio, U. S., 135 m. NE of Columbus, on the W side of Cuyahoga river, and on the Ohio canal.

YELLOW-KNIFE, a river of British North America, in Great Slave Lake district, which, after a southerly course, in which it forms a series of small lakes, of about 210 m., falls into the N part of Great Slave Lake.

YELLOW SEA, or WHANG-HAI, a large gulf of the Pacific, on the NE coast of China, so called from the tinge of its waters by the large quantity of alluvium carried down into it by the great rivers the Hoang-Ho and the Yang-tsze-kiang. It lies between the parallels of 34° and 41° N, and the meridians of 118° and 127° E; but is of very irregular outline. Its mouth, measured from the SW extremity of the peninsula of Corea to the mouth of the Hoang-Ho, is 320 geog. m. wide; and from the parallel of the latter point, to the extreme N point of the gulf of Leaotong, it has a length of about 420 m. Between the promontory of Shan-tung on its W coast, and Hong-kin on the W coast of Corea, its breadth is contracted to 130 m.; while further N, between Tong-Tchu, on the coast of Shan-tung, and the extremity of the long peninsular projection, known as Regent's Sword or Chao-phing-tchu, it is contracted to 55 m. The W coasts of the gulf are formed by the Chinese provs. of Shan-tung and Petche-li; the N coast by Leaotong; and the E coast by Corea.

YELLOW-SPRING, a village of Blair co., in the state of Pennsylvania, U. S., 72 m. W by N of Harrisburg. In its vicinity are extensive iron-works.

YELLOW-SPRINGS, a village of Green co., in the state of Ohio, U. S., on the Little Miami railroad, and 72 m. from Cincinnati. It has a sulphur spring, and is a place of frequent resort.

YELLOW-WATER, a river which has its source in Covington co., in the state of Alabama, U. S., flows SSW into Florida, and discharges itself into a bay of the same name, in Pensacola bay.

YELLVILLE, a village of Marion co., in the state of Arkansas, U. S., on the N side of Crooked creek, a tributary of White river, and 102 m. N by W of Little Rock. Pop. in 1850, about 100.

YELNIA, a district and town of Russia in Europe, in the gov. of Smolensko. The district is situated in the SE part of the gov. and is traversed by the Desna. It is generally woody. The town is 51 m. ESE of Smolensko, and 30 m. SSE of Dorogobuj, on the Desna, at the confluenee of the Bedomoika and Gorodenka. Pop. 1,000. Agriculture forms the chief object of industry.

YELO, a modern town of Spain, in the prov. of Soria, and partido of Medinaceli. Pop. 448. It has a parish-church, and a custom-house.

YELOGONI. See YELAGONI.

YELOVKA, a river of Russia in Asia, in the district of Kamtchatka, which has its source on the E side of the chain of mountains by which the peninsula is intersected in its entire length; runs SE, and after a course of about 75 m., joins the Kamtchatka on the l. bank, 75 m. W of Nijne-Kamtchatsk. It is navigable almost up to its source. The natives who inhabit its banks are named Kootsh, and are considered more warlike and courageous than those of the Kamtchatka.

YELSI, a town of Naples, capital of a canton in the prov. of Sannio, district and 10 m. ESE of Campobasso, on the slope of a mountain. Pop. 2,200. It has a parish-church and a convent.

YELSK, or KAROLIN, a town of Russia in Europe, in the gov. of Minsk, district and 20 m. SSW of Mozir.

YELSOE, a parish of Norway, in the diocese of Christiansand, in the N part of the bail. of Stavanger, on the E part of the Bukkefiord. Pop. 1,908.

YELTES, a river of Spain, in the prov. of Salamanca, which has its source to the N of Pina-de-Francia; receives the Huebra, near Giela, and after a course, in a generally NW direction of about 60 m., joins the Duero, on the l. bank. The Y. is in summer every where fordable, but in winter swells to a large stream.

YELTON, a salt lake of Russia in Europe, in the gov. of Saratov, at the S extremity of the district of Kamishin, and 75 m. from the l. bank of the Volga. It is nearly circular in form, and is about 9 m. in diameter. It receives several small rivers, of which the principal is the Khirikhaza-Sclenai, which flows from the NW. It is of little depth, and its waters, when evaporated by the sun, become edged with incrustations of salt several inches in thickness. The quantity of salt thus obtained amounts to 100,000 tons annually, and employs about 10,000 men in its preparation and conveyance to Saratov.

YELVERTOFT, a parish in Northamptonshire, 10 m. N by E of Daventry, intersected by the Union canal. Area 2,080 acres. Pop. in 1851, 714.

YELVERTON, a parish in Norfolk, 6 m. SE by S of Norwich. Area 439 acres. Pop. in 1851, 66.

YELWALE, a town of Hindostan, in the presidency of Madras, and prov. of Mysore, 9 m. SW of Seringapatam, in N lat. 12° 20′, and E long. 76° 40′.

YEM. See EMBA.

YEMANGELKA, a fort of Russia in Asia, in the gov. of Orenburg, district and 60 m. NNW of Troitzk, on the NW bank of Lake Sarikoul.

YEMAMAH, a district and ancient town of Arabia, in the Nejid, 90 m. SSE of Derayah, in a plain, near the Aftan. Palm trees grow in the locality.

YEMBONNTUNG, a town of Burmah, about 100 m. NW of Ava, in N lat. 23°, and E long. 95° 9′.

YEMBO, or YAMBO, a town of Arabia, in Nedjaz, 130 m. SW of Medinah, on the N side of a capacious inlet of the Red sea. It is governed by a Turkish effendi, and forms the port of Medinah. The entrance of the harbour lies in N lat. 34° 4½′, E long. 38° 6′. It affords excellent water, and sheep, bullocks, fowls, and fruit. It has a pop. of about 2,000. Lieut. Burton, who recently travelled from Y. to Medinah, by the land-route, computes, that from Y. to Muzahhal the distance is 16 m.; from Muzahhal to El-Hamra, 48 m.; from El-Hamra to Suwaykeh, 46 m.; from Suwaykeh to Medinah, 20 m.

YEMEN, a district of Arabia, comprising the SW angle of the peninsula, and bounded on the N by the Hedjaz and Nedjid; and on the E by Hadramaut. It once formed the kingdom of Sabu; but the principal political chief now acknowledged in it is the imam of Sana. A strip of comparatively level ground, varying in width from 10 to 50 m. runs between the sea-shore and the interior table-land, which rises abruptly from the plain, and attains an elevation, it is supposed, of at least 3,000 ft. in the interior. The sheiks or sultans of Aden, Loheia, and Abu-arish, had independent sway in portions of the Tahema, and to some distance inland. In the Jebel, or highlands, nearly every chief maintains a virtual independence, although they will occasionally bear arms, for payment, in the service of the imam of Sana. Some of them appear to have formed a kind of confederacy for mutual defence, known as the Hashed-ul-Bekil. Ibrahim Pasha bestowed the province of Haiz on the sheikh who had governed Taas for the imam of Sana; and did much to foment the feuds between the petty chieftains of this quarter of Arabia, in the hope of profiting by their dissensions; but the people of Y. are not, generally speaking, attached to military life. Wheat, maize, coffee, and *khat* are the productions chiefly cultivated throughout Y. The latter is the *Celastrus* or *Catha edulis*. Its soft twigs and tender leaves are universally used as a restorative from fatigue, and produce an agreeable soothing excitement, which, when persevered in, becomes a slight intoxication. The Arabs of Y. sit chewing it, and talking whole nights together. The finer sorts are, however, expensive. The use of kat, which is said not to leave any disagreeable consequences, is older than that of coffee, and that cultivated on Mount Saber is the best, and a source of great wealth to the inhabitants, as the trade in this herb is very considerable. The trees are not stripped till four years old, and their first produce is of an inferior quality. It furnishes the best produce in the sixth year. The wild kat tree also furnishes a coarse and very intoxicating produce. Excellent grapes, apples, and medlars are raised. The pop. is in part Bedouin and Somali. The natives of Mount Saber, the most elevated portion of the inland plateau, are of superior physical development. The women of this district are said to be remarkably beautiful, with Italian features, fair complexion, and free graceful figures. The principal merchants are Hadramy, or natives of Hadramaut. See article ARABIA.

YEMGUM, or JEMGUM, a market-town and bail. of Hanover, in the gov. and 17 m. SSW of Aurich, on the l. bank of the Ems, which here forms a small port. Pop. 1,196. In 1568, Prince Louis of Nassau, brother of the Prince of Orange, was defeated with great loss near this town.

YEMHATTI, a town of Hindostan, in the presidency of Madras, 92 m. NE by N of Coimbatur, in N lat. 12° 10′, and E long. 77° 40′.

YEMOURTLINSKAIA, a town of Russia in Asia, in the gov. of Tobolsk, district and 33 m. S of Yalontorovsk, on the Yemourtla, a small affluent of the Tabol.

YEMTZA, a river of Russia in Europe, in the gov. of Archangel, which issues from Lake Konevskoi, in the district of Oneg; runs first NE, afterwards N; enters the district of Kholmogori; receives the Megrenga and Vaimontsha; and after a course of about 99 m., joins the Dwina on the l. bank, 51 m. S of Kholmogori. Wood in large quantities grows on its banks.

YEMWUNTOUNG, a town of Burmah, 100 m. NW of Ava, in N lat. 23°, and E long. 95° 9′.

YEN. See BORGU.

YEN, a town of China, in the prov. of Shen-se, on the Han-keang, 120 m. SW of Segan-fu.

YENANTSY, a town of Russia in Europe, in the gov. and district and 75 m. N of Vilna.

YENATZ, a village of Switzerland, in the cant. of the Grisons, and jurisdiction of Castels, 9 m. NE of Coire. It has several mineral springs and baths.

YEN-CHANG, a district and town of China, in the prov. of Shen-se, div. and 39 m. E of Yen-gan-fu, on the Yan-ho, in N lat. 36° 37′, and E long. 109° 58′.

YEN-CHA-TING, a district and town of China, in the prov. of Kan-suh, and div. of Ping-leang-fu.

YEN-CHIN-CHING, a town of China, in the prov. of Shan-tung, 54 m. ESE of Tse-nan-fu. It is noted for its manufacture of glass.

YEN-CHING, a district and town of China, in the prov. of Keang-su, and div. of Hwae-gan-fu, 120 m. NE of Nan-king-fu, in N lat. 33° 21′ 55″, and E long. 120° 0′ 22″.

YEN-CHING-HEEN, a district and town of China, in the prov. of Ho-nan, and div. of Hen-chu, 99 m. SSW of Kae-fung-fu, in N lat. 33° 38′ 20″, and E long. 114° 4′ 40″.

YEN-CHUEN, a district and town of China, in the prov. of Shen-se, and div. of Yen-gan-fu, in N lat. 36° 57′, and E long. 110° 8′.

YEN-CHU-FU, a division and town of China, in the prov. of Shan-tuy. The div. comprises 10 districts. The town is 81 m. S of Tse-nan-fu, in N lat. 35° 41′ 51″, and E long. 117° 3′ 30″. It is well situated between two rivers, and holds the rank of second town in the prov.—Also a div. and town in the prov. of Che-keang. The div. comprises 6 districts. The town is 66 m. SW of Hang-chu-fu, on the l. bank of the Tseen-tang-keang, in N lat. 29° 37′ 12″, and E long. 119° 32′ 47″. It is noted for the manufacture of paper. In the adjacent mountains are several copper-mines, and large numbers of trees yielding varnish are found in the environs.

YENDAN, a river of Senegambia, in the state of Sangara, which runs N and joins the Joliba on the r. bank, and near the confines of the state of Amana.

YENDELDI, a town of Nigritia, to the SW of Dar-fur. It is the residence of the sultan of Tonmourki.

YENDEN. See HING-KING, or GHIRIN.

YENDOVISHTSHA, a town of Russia in Europe, in the gov. and 9 m. WNW of Voronej, and district of Zemliansk, near the r. bank of the Don.

YEN-GAN-FU, a division and town of China, in the prov. of Shen-se. The div. comprises 10 districts. The town is on the Yan-ho, 180 m. NNE of So-gan-fu, in N lat. 36° 42′ 20″, and E long. 109° 28′. Petroleum, marble, and cinnabar are found in large quantities in the environs. Furs, especially those of the sable, form an important branch of local trade.

YENGBINE, a river of Indo China, in the Tenasserim provinces, an affluent of the Thaluyn-Myeet, which it joins in N lat. 17° 22′, and E long. 97° 48′.

YENGHI-HISSAR, sometimes written YINGKEIHSHAR, or INGASHAR, or TSEIH-GEN, a small town of Chinese Turkistan, on the r. bank of the Bildur, 60 m. SE of Kashgar, 135 m. NW of Yarkand, in a fertile territory, producing rice, wheat, flax, melons, &c.

YENGOKO, a town of Japan, in the island of Nifon, and prov. of Fitats, 60 m. ENE of Yedo.

*** YENI, names commencing with this syllable, not found amongst the following, are to be sought for under Jeni or Geni.

YENIBAZAR. See NOVIBAZAR.

YENIDJA, a lake of Turkey in Europe, in Rumelia, in the sanj. of Salonica, and to the S of Yemdja-Vardar. It is 3 m. in length from NE to SW, and 3 in breadth. It receives several rivers, the principal of which is the Vistritza, and discharges itself by the Caraismak into the gulf of Salonica. Its waters are saline, and in the incrustation of salt caused by evaporation on its banks, affords an important article of commerce. It is destitute of fish

YENIDJA-CARASU, a town of Turkey in Europe, in Rumelia, in the sanj. and 108 m. WNW of Gallipoli, and 30 m. ENE of La Cavale, in the midst of a fine plain, and near Lake Lafri, which discharges itself into the gulf of Lagos. Pop., chiefly Turks, 2,500. It has a mosque, baths, a bazaar, and several khans. The trade consists chiefly in tobacco, the produce of the locality, and noted as the finest in Turkey.—About 6 m. to the S of this town, on the banks of the sea, are the ruins of the ancient *Abdera*.

YENIDJA-KIZILAGHADJ, a town of Turkey in Europe, in Rumelia, in the sanj. and 21 m. NE of Tshirmen, and 33 m. N of Adrianople, in a plain, on the l. bank of the Tunja, which is here crossed by a bridge. Pop. 2,500. It has a mosque, a bazaar, a bath, and a khan.

YENIDJA-VARDAR, a town of Turkey in Europe, in the sanj. and 29 m. WNW of Salonica, 17 m. S of Mogrena, and a little to the N of a lake of the same name. Pop. about 6,000. It is open, and contains numerous mosques, Greek churches, khans, baths, and shops. It has manufactories of woollen fabrics, and carries on an active trade in tobacco, the produce of the locality, which is reckoned the best in Macedonia. It was conquered by Ornosbeg in 763, and has several foundations originated by him and his son. It contains the tomb of the former. A little to the S of this town are the ruins of the ancient *Pella*, the birthplace of Alexander the Great, and place of sepulture of Euripides.

YENI-GHEUL. See AINAGOUL.

YENIGUEDEL WELLS, a watering station of Marocco, in the prov. of Draha, to the N of Mimcina, visited by Caillé in 1828. They afford abundant supplies of excellent water.

YENIKALEH, JENIKALAH, or ENIKALE, a town of Russia, in the Crimea, on a cape 5½ m. to the E of Kertch, at the N entrance of the strait of Kertch, leading from the Black sea into the sea of Azof, in N lat. 45° 21′, E long. 36° 38′. It is a town of about 1,500 inhabitants, with a fortress on the highest point of the promontory. The town consists of two parts, one of which lies close to the water's edge, and is commanded by the high land above it. In the suburb stands the church, a handsome building in the Byzantine style. The other part of the town chiefly consists of a quadrangular fort, each side of which is about a ¼ m. in length. Within the walls of this quarter are the hospital, barracks, stores, and governor's house. The chief employment of the inhabitants is derived from the fisheries in the sea of Azof. A sandbank stretching from the NE extremity of the peninsula of Taman runs towards Y., and contracts the strait to the breadth of 1⅝ m. just before it opens into the sea of Azof, and there is not more than 13 ft. water in the channel opposite Y. On the bank, which is full of salt marshes, and from 2 to 3 m. broad in some places, there is a strong battery crossing fire with the principal battery on the side of Y. This place was taken by the allies in 1855, and is now held by them.

YENI-KENI, a town of Turkey in Europe, in Rumelia, in the sanj. of Viza, 9 m. NE of Constantinople, on a strait of the same name.—Also a village of the beglik and 21 m. SSE of Seres, and 6 m. NW of Orphano. It is built in the mountains encircling the gulf of Orphano, to the SE of Lake Takinos. The name, signifying new village, is common to many in Turkey in Asia. In the environs are the ruins of the ancient *Amphipolis*.

YENI-KHAN, or NOVIKHAN, a town of Turkey in Europe, in Rumelia, in the sanj. and 19 m. S of Sophia.

YENIKIM, or INIDJA, a village of Turkey in Asia, in Anatolia, in the sanj. and 60 m. NNE of Castamuna, and 24 m. E of Ineboli, on the Black sea. It has a good roadstead, and carries on an extensive trade with Constantinople, in the produce of the environs.

YENI-SAGHRA, or YENI-SARA, a town of Turkey in Europe, in Rumelia, in the sanj. and 42 m. N of Tshirmen, and 63 m. NNW of Adrianople, in a fertile plain on the r. bank of the Tunja.

YENISEI, a river of Asia, which has its source in the NW part of the Chinese empire in the country of the Khalkas. It issues under the name of Khouagem, from Lake Uru, a small lake separated by a ramification of the Tangnou mountains from Lake Koussongol; runs first in a westerly direction, taking, after the confluence of the Kamsara, the name of Outoukem. On reaching the N between the Saiansk and Kem-Kemtsik mountains, it bends N to the Little Altai chain, and thence assumes the appellation which it retains throughout the remainder of its course. It traverses longitudinally the gov. to which it gives its name, and falls into the Arctic ocean by a spacious estuary, frequently distinguished as the gulf of Yenisei. This immense river, which intersects Siberia in its entire breadth, has a total course of 1,950 m., of which 360 m. only are within the confines of the Chinese empire. It has numerous affluents, especially on the r., on which it receives the Tuba, Kan, Verkhniaïa-Tunguska, or Angara, the outlet of Lake Baikal, the Pit, the Podkamenna-Tunguska, Bakta, Nijnaia-Tunguska and Kursika. On the l. it receives the Kemt-chik, Abkan or Abakan, Sim, Elagni, and Turuka. Of all these rivers, the most important as the outlet of Lake Baikal, is the Verkhniaïa-Tunguska, or Upper Tunguska. The area of the basin of the Y., comprising that of Lake Baikal, and thus forming the greatest in Russia, is 2,100 m. in length, from the sources of the Orkhon in the Khangai mountains, to the Frozen ocean, and 1,200 in breadth from the source of the Angara to that of the Akban in the mountains of Tiletzk. These mountains, and the Kuznetzk range, a continuation of the former, form a portion of the dividing line between the basin of the Y. and that of the Oby on the W, while the mountains of Baikal, and a long chain still but little known, form the line of separation from the basin of the Lena. Towards the S the banks of the Y., generally well-peopled, are studded with villages and towns, amongst the latter of which are to be noted Minousinsk, Krasnoïarsk and Yeniseisk. Towards the N, Touroukhansk under the 66th parallel, is the last town on the Y., which here has a depth of 8 fathoms. The last village near its mouth is in 70° 30′ N lat. In some parts this river attains a considerable breadth, and a little above the confluence of the Podkammenaia-Tunguska expands to the dimensions of a lake. Its waters abound with fish.

YENISEI (GULF OF THE), or Gulf of the Seventy Islands, a gulf formed by the Arctic ocean, on the N coast of Russia in Asia, in the gov. of Yeniseisk, at the mouth of the river from which it takes its name, and of which it may properly be considered the estuary. It is 240 m. in length from N to S, 54 m. in extreme breadth, and 30 m. wide at the entrance, which is in N lat. 72° 30′, and E long. 84° 20′, and opposite to which is the island of Konskin. It derives the second of its names from the number of its islands. It does not open to navigation till the month of June.

YENISEISK, a government, district, and town

of Russia in Asia, situated in the western, but politically belonging to the eastern division of Siberia, lying between 50° 20′ and 78° 25′ N lat., and between 76° 50′ and 111° 40′ E long., bounded on the N by the Arctic ocean; on the E by the gov. of Irkutsk and the prov. of Yakutsk, from which it is partly divided by the Anabara; on the S by the Chinese empire, from which it is separated by the Little Altaï chain and the Saïansk mountains; and on the W by the gov. of Tomsk and Tobolsk, comprising an area 1,980 m. in length from N to S, and 840 m. in breadth from E to W, and a total superficies of 1,235,000 sq. m. The coast of the gov. of Y. forms on the NE Cape Sieverò-Vostochni, the most northerly point of the old continent, and presents several considerable indentations, of which the principal are the gulfs of Khatanga, Taimur, Yenisei, Varentzov, Tedanski and Tazovska. This vast district is almost entirely comprised within the basin of the Arctic ocean, and nearly all its waters are collected by the Yenisei. The principal affluents of that river in this gov. are the Verkhnaia-Tunguska, Podkamennaia-Tunguska, Bakta, Nijnaia-Tunguska and Kussika, on the r., and on the l. the Abakan, Sim, Elagui, and Turuka. In the NE are the Piasina and Katanska, which discharge themselves into the ocean; in the NW the Taz falls into Tazovsk bay; and on the W the Vakh, Tim, Ket, and Tchelun, all affluents of the Obi. In the N part are numerous lakes, the principal of which is Piasina. The E part is covered with hills of little elevation, while in the S are numerous ranges, ramifying from the great mountain chain, which run along the S and SW confines of the gov. This portion of the gov. is moderate in temperature, and is in some parts well-wooded. It is well-peopled, and contains several towns and villages built by the Russians and Cossacks. Culinary salt and iron are among the most important productions of this gov. About 30,000 poods of iron, and 130,000 poods of salt are annually made. Its aboriginal inhabitants consist of several tribes, amongst whom may be mentioned the Arintsis, Kamasintsis, Katchintsis, Sagaitsis, Koibalis, and Beltiris. The N portion, which is chiefly desert, is inhabited by Tunguses, Tchapogirs, Yakuts, Samoiedes, and Ostiaks. This gov. was formed in 1823 from the gov. of Tomsk, of which it comprises the greater portion. It is divided into 5 circles, viz., Atchinsk, Yeniseisk, Kansk, Krasnoiarsk, and Minousinsk. Krasnoiarsk is the capital. The number of its inhabitants in 1831 did not exceed 173,898. In 1846 they were returned at 191,500. The yearly increase of pop. has been estimated at 1547; while that by convicts amounts to 3,500.—The district of Yeniseisk occupies the N part of the gov., and possesses an area 750 m. in length from E to W, and 120 m. in breadth from N to S. It presents extensive sterile plains, intersected by the Yenisei, the Tunguskas, and the Khatanga, and inhabited by the Tunguses, Yakuts, Samoiedes, Ostiaks and Tchapogirs.—The town is 186 m. NNW of Krasnoiarsk, and 900 m. E of Tobolsk, in a fertile plain, on the l. bank of the Yenisei, in N lat. 58° 27′ 17″, and E long. 100° 58′ 30″. Pop. in 1831, 6,000. It is about 4 m. in circumference, and contains four churches, two convents, a custom-house, a powder-magazine, and provision store. It possesses an active trade, and a periodical fair resorted to by merchants from all the surrounding regions. Cattle and poultry are extensively reared in the vicinity.

YENI-SHEHR, a town of Turkey in Asia, in Anatolia, in the sanj. of Khodavenkiar, 15 m. S of Isnik, and 30 m. ENE of Brusah, and a little to the NE of a lake of the same name, which is about 6 m. long, and 2 m. in breadth.—Also a fort in the sanj. of Aidin, on a mountain, 12 m. E of Guzel-Hissar, and 90 m. SE of Smyrna, near the l. bank of the Gunjéré, a little above its confluence with the Bünek-Meindere. It occupies a portion of the ancient *Antiochia-ad-Mæandrum.*—Also a village of the same sanj., 18 m. WNW of Guzel-Hissar, and a little to the W of Inek-Bazar. It occupies a portion of the site of the ancient *Magnesia-ad-Mæandrum.*—Also a village in the sanj. and 75 m. WSW of Riga, on the Archipelago, a little to the SW of the channel of the Dardanelles, and a little to the NW of the site of the ancient Troy. It is inhabited chiefly by poor Greeks. On an adjacent hill was the acropolis or citadel of *Sigium.*

YENI-SU. See GATIPO.

YEN-KANG, a town of China, in the prov. of Che-keang, 105 m. S of Hang-chu-fu, in N lat. 28° 58′, and E long. 120° 10′ 45″.

YEN-KING, a town of China, in the prov. of Chih-le, a little beyond the Inner Great Wall, and 48 m. NNW of Peking, in N lat. 40° 29′ 5″, and E long. 116° 2′ 39″.

YEN-KOU, a name by which Chih-le was formerly known.

YEN-KWANG, or XU-YEN-KWANG, a maritime prov. of An-nam, in Tonquin.

YEN-LE, a town of China, in the prov. of Hupih, on the l. bank of the Yang-tse-keang, 114 m. SW of Wu-chang-fu.

YEN-LING, a district and town of China, in the prov. of Ho-nan, div. and 57 m. SSW of Kae-fung-fu, in N lat. 34° 10′, and E long. 114° 21′.

YENNA, a town of the Papal states, in the comarca and 38 m. E of Rome. Pop. 1,200.

YENNAN, an offset of the Irawady, in Burmah, in Mranma, which falls into the Sittang river, in N lat. 19° 54′, and E long. 96° 6′.

YENNE, a town of Japan, in the island of Nifon, prov. and 45 m. SSW of Noto.

YENNE, or HYENNE, a town of Sardinia, capital of a mandemento, in the prov. of Savoy, 15 m. NW of Chambery, at the confluence of the Rhone and Floss. Pop. 2,500. This town is supposed to have been the ancient *Epanna*, where a council was held in 517 by Sigismund, king of Burgundy.

YENNICOUL-GOOTA, a town of Hindostan, in the prov. of Mysore, 111 m. N of Seringapatam, in N lat. 14° 1′, and E long. 76° 52′.

YENOTAEVSK, a district and town of Russia in Europe, in the gov. of Astrakhan. The district extends along the Volga, in the central part of the gov., and contains vast plains or steppes studded with salt lakes. The town is 87 m. NW of Astrakhan, on the l. bank of the Volga. Pop. 500. It has a fort constructed of wood in 1741, for defence against the Kalmuks. The environs are generally sterile, and the surface generally is covered with fine shifting sand.

YEN-PING-FU, a division and town of China, in the prov. of Fo-keen. The div. comprises 5 districts. The town is well built, on the descent of a hill, near the l. bank of the Minho, and 90 m. NW of Fuh-chu-fu, in N lat. 26° 38′ 24″, and E long. 118° 17′ 50″. It is defended on all sides by inaccessible mountains. Although of no great extent, it is considered one of the finest cities in the empire, and it has a considerable trade.

YEN-SHAN [i. e. Salt Mountain], a district and town of China, in the prov. of Chih-le, and div. of Teën-tsin-fu, in N lat. 38° 7′, and E long. 117° 16′, and 144 m. SSE of Chih-le.

YEN-SZE, a district and town of China, in the prov. of Ho-nan, and div. of Ho-nan-fu, in N lat. 34° 45′, and E long. 112° 43′.

YENTELICHKI, a town of Russia in Europe, in the gov. of Vilna, district and 24 m. NW of Telsh, and 90 m. WSW of Mitau.

YEN-TING, a town of China, in the prov. of Tze-chuen, and dep. of Tung-chuen-fu, 90 m. NE of Ching-lu-fu, in N lat. 31° 14′, and E long. 105° 26′.

YEN-TSIN-HEEN, a district and town of China, in the prov. of Ho-nan, and div. of Wei-hwuy-fu, in N lat. 35° 10′, and E long. 114° 20′.

YEN-TSUNG, a town of China, in the prov. of Kwang-tung, on the Yang-keang, 144 m. SW of Canton.—Also a district of China, in the prov. of Kwei-chu, 165 m. ESE of Kwei-gang-fu, in N lat. 25° 57′ 36″, and E long. 109° 3′.

YENVILLE. See JANVILLE.

YEN-WEI, a fortress of China, in the prov. of Shan-se, near the Great Wall, 194 m. NNE of Tae-guen-fu.

YEN-YANG, a district and town of China, in the prov. of Shen-se, and div. of Tung-leang-fu.

YEN-YUEN [i. e. Salt-Spring], a district and town of China, in the prov. of Sze-chuen, and div. of Ning-yuen-fu, in N lat. 27° 20′, and E long. 101° 32′.

YEO. See IVEL.

YEO, a small river of Devonshire, tributary to the Cready, into which it falls to the E of Crediton.

YEOU, a river of Sudan or Nigritia, which has its sources in the mountains to the N of Bobyra; runs N through the state of Katagum, in which it is joined by the Shashum, and passes a town of the same name. It then sweeps easterly along the confines of the state of Aweak; passing Hadeiga, and through Bedu into Bornu, bending ENE discharges itself into Lake Chad, by an embouchure 900 ft. in width, a little below a town of the same name, and about 90 m. NNW of Kouka. It possesses considerable depth, and its average current is not less than 3 m. an hour. The chief towns on its banks in Bornu, are on the r. bank, Old Birnee at the junction of a stream of considerable size, and Bidekarsee, and on the l. Kabshary. Its bed consists of firm sand, the banks nearly perpendicular, and the waters, which are beautifully limpid, abound with fish. This river sometimes bears in its upper part the name of Gambaru, and was long considered a banch of the Joliba. The town of Yeou is on the r. bank of the river of the same name, about 10 m. above its entrance into Lake Chad. Pop. about 3,000. It is enclosed by a wall.

YEOU-LIU, or YOULO, a small island of the Chinese empire, in the group of the Lu-chu islands, to the N of Great Lu-chu island.

YEOU-LUN, or YOURUN, an island of Japan, to the SW of the island of Kiu-siu, and NE of the Lu-chu islands.

YEOU-NA-KOU-MI, or YOU-NEKORY, or KOUMI, an island of China, between the group of the Madjicosemah islands and Formosa. It is crescent-shaped, and about 8 m. in length from NE to SW.

YEOUNDAOUN, a town of Burmah, in the Mranma, on the r. bank of the Irawady, 18 m. NNW of Prome.

YEOUNGBENSAH, a town of Burmah, in Pegu, on the r. bank of the Irawady, 66 m. NW of Rangoon.

YEOUNG-LERAY, a town of Burmah, in Mranma, on the l. bank of the Irawady, on the confines of the prov. of Pegu, and 27 m. S of Prome.

YEOU-TSO-GO, or YOUSSA-GAFK, a mountain of the Chinese empire, in the island of Great Lu-chu, and prov. of Shan-nan, and near the SE extremity of the island.

YEOVIL, a parish and market-town in Somersetshire, 9½ m. SSE of Somerton. Area 4,056 acres. Pop. in 1831, 5,921; in 1851, 7,744. The town is well-built. Woollens were formerly made here to a great extent, but gloves are now the chief article of manufacture. Butter is sent in large quantities to London from the vicinity.

YEOVILTON, a parish in Somerset, 1½ m. E of Ilchester. Area 1,753 acres. Pop. in 1851, 329.

YEOWAH, a town of Burmah, on the r. bank of the Irawady, 108 m. SW of Ava.

YEPES, a town of Spain, in the prov. and 18 m. E of Toledo, and partido of Ocana, and at the W extremity of the fertile plateau of that name. Pop. 3,065. It is well-built, and has a parish church, two convents, two hospitals, a public granary, and a custom-house. A portion of the inhabitants live in caves skilfully excavated. Hosiery and basket-ware form the chief articles of local manufacture. The locality is noted for its white wine and asparagus.

YEPIFAN, a district and town of Russia in Europe, in the gov. of Tula. The district is in the E part of the gov. It contains some fine cultivated fields, but is to a great extent marsh or waste-land. The town is 51 m. SE of Tula, and 15 m. SE of Bogoroditzk, on the l. bank of the Don. Pop. 900. It contains eight churches, three of which are built of stone. It has manufactories of iron-ware.

YEPIFANOVKA, a town of Russia in Europe, in the gov. of Saratov, district and 24 m. SW of Staro-Belsk.

YERAGOLIA, or EYRAGOLY, a town of Russia in Europe, in the gov. and 90 m. NW of Vilna, district and 19 m. SE of Rossiena, on the r. bank of the Doubisa.

YERAPOLTSK, a village of Russia in Europe, in the gov. of Moscow, district and 9 m. NW of Voloklamsk, on the r. bank of the Lama. It has some good country houses.

YERBESTON, a parish in the co. of Pembroke, 4 m. SW of Narberth. Area 1,224 acres. Pop. 329.

YEREMCHINK, a mining village of Russia in Europe, in the gov. of Tambov, and district of Yelatom. Its works produce annually 40,535 lbs. of raw iron, and 25,162 in bars.

YEREMIEIEVKA, a town of Russia in Europe, in the gov. and 102 m. W of Poltava, district and 33 m. SE of Zolotonocha.

YEREMITCHI, a town of Russia in Europe, in the gov. of Grodno, district and 22 m. E of Novogrodek, near the l. bank of the Niemen.

YEREMSHA, a river of Russia in Europe, which has its source in the gov. of Nijny Novgorod, and district of Ardatov, near Semilova; runs SSE; enters the gov. of Tambov, and district of Temnikov; and after a course of 39 m., joins the Moksha, on the r. bank, 5 m. below Kadom. Near it are some extensive iron-works.

YERES, a river of France, which has its source in the dep. of the Seine-et-Marne, com. and 11 m. NNW of Provens, traverses the central and W part of the dep.; enters that of the Seine-et-Oise; and after a total course, in a generally W direction, of about 60 m., joins the Seine, on the r. bank, at Villeneuve-St.-Georges. It passes Rosay and Chaulmes, and a little to the S of Brie-Comte-Robert.—Also a small river in the dep. of the Seine-Inferieure, which has its source in the arrond. of Neufchatel, a little to the S of Foucarmont, flows thence into the arrond. of Dieppe, and, after a course to the NW of 30 m., falls into the English channel. Foucarmont, Grandcourt, and Criel, are the chief towns on its banks.

YERES, or YERRES, a commune and village of France, in the dep. of the Seine-et-Oise, cant. and 3 m. SSW of Boissy-St.-Leger, and 15 m. SE of Paris, in a valley on the r. bank of a river of the same name. Pop. 978. It contains many fine villas, and on an adjacent height is the Chateau-de-la-Grange, a handsome edifice, with an extensive park and gardens. There was formerly an abbey here, founded in 1122 by the Countess d'Etampes, sister of Louis-le-Gros.

YERESKI, a town of Russia in Europe, in the gov. and 38 m. NW of Poltava, district and 16 m. SE of Mirgorod, on the l. bank of the Psioul.

YERETZ. See ERETZ.

YERIKLINSK, a town of Russia in Europe, in the district and 45 m. ESE of Simbirsk, on the r. bank of the Tcheremshan.

YERING, a parish of Australia Felix, in the co. of Bourke, bounded on the N by the Yarra-Yarra river.

YERISCHMARKT. See GYERES.

YERJVILKI, a town of Russia in Europe, in the gov. and 120 m. WNW of Vilna, district and 17 m. SW of Rossiena.

YERKEL, a lake of Russia in Europe, in the gov. of Esthonia, district and 11½ m. SSE of Revel. It is 4 m. in length, and 2 m. in breadth, and discharges itself into the gulf of Revel.

YERKIM, or YERKIEN. See YARKAND.

YERMA, a town of Turkey in Asia, in Anatolia, in the sanj. of Sultan-Emir, on the r. bank of a small river of the same name, an affluent of the Sakaria, and 100 m. E of Kutaya. It has several mineral springs, and contains some remarkable ruins. Yerma, a Roman colonist, was noted for his sanctity; and it is said the emperor Justinian made a pilgrimage to the place in fulfilment of a vow.

YERNAGUDERN, a town of Hindostan, in the presidency of Madras, district and 63 m. NNE of Masulipatam, in N lat. 17°, and E long. 81° 33′.

YERO, JERO, or OLIEVIER, a port of Turkey in Asia, in Anatolia, on the SE side of the island of Mitylene. The entrance formed by Capes Agea-Maria and Patras is very narrow, but the interior is spacious, presenting an extent of about 6 m. in depth from N to S, and 4 m. in breadth from E to W.

YERRACURRON, a town of Hindostan, in the presidency of Madras, and district of Malabar, 42 m. NNE of Calicut.

YERRAMUMBA, a town of New South Wales, in the district of Liverpool Plains, on the Birebula river.

YERRAPILLY, a village of Hindostan, in the presidency of Madras, in the Carnatic. It has a copper-mine.

YERRARAS, a range of mountains of New South Wales, in the district of Lachlan, a ramification of the Peel range. See GOULBURN.

YERRIMONG, a creek of New South Wales, in the co. of St. Vincent, an affluent of Shoalhaven river.

YERRUMAUR, a town of Hindostan, in the presidency of Madras, and district of Malabar, 68 m. SE by E of Calicut, in N lat. 10° 40′, and E long. 76° 38′.

YERVILLE, a canton and commune of France, in the dep. of the Seine-Inferieure, and arrond. of Yvetot. The cant. comprises 19 com. Pop. in 1831, 12,530; in 1846, 13,244. The village is 8 m. ENE of Yvetot, and 20 m. NNW of Rouen. Pop. 1,495. It has manufactories of calico, cloth, linen, and hats, and several brick and lime kilns.

YESAN. See YESSAN.

YESD. See YEZD.

YESSAN, or YESAN, a mountain of Japan, in the island of Nifon, and prov. of Oumi, to the E of Lake Brivano-Oumi. It is held in sacred veneration by the inhabitants.

YES-SIN, or YESHIL, a lake of the S part of Chinese Turkestan, to the N of Kiria. It receives the Kiria and several other rivers.

YESTE, a judicial partido and town of Spain, in the prov. of Albacete. The partido comprises 16 pueblos. The town is 51 m. SW of Chinchilla, partly at the foot, and partly on the slope of a mountain, and in a fertile locality. Pop. 17,824. The district contains the remains of several fortresses, and towards its centre is a castle called Casa Palacio. It has manufactories of coarse linen and woollen fabrics, and carries on an active trade in honey. About 6 m. S of this town are mineral baths.

YESTER, anciently ST. BOTHAN'S, or ST. BATHAN'S, and popularly GIFFORD, a parish in the south of Haddingtonshire, having a superficial extent of nearly 14 sq. m. The village of Gifford stands in the middle of the N division of the strath. The other villages are Long-Yester and Long-Newton, at the foot of the Lammermoors. Pop. in 1831, 1,019; in 1851, 1,202. The family of Yester and Locherwart obtained the titles of Lord Yester in 1488, Earl of Tweeddale in 1646, and Marquis of Tweeddale and Earl of Gifford in 1694.

YETES, an Indian tribe, who inhabit the banks of the Amazon, in Ecuador.

YETHOLM, a parish on the NE border of Roxburghshire, having a superficial extent of nearly 6,000 acres. About 100 acres are covered with wood; about 200 are a wild moor, called Yetholm-common, on the boundary with England, and rather doubtfully belonging to Scotland; and the rest of the area is distributed into tillage and pasturage-grounds, in the mutual proportions of 13 to 15. Kirk-Yetholm and Town-Yetholm are the only villages. Pop. in 1831, 1,289; in 1851, 1,352.

YETHOLME, a parish of New South Wales, in the co. of Roxburgh, bounded on the S by Fish river.

YETKERN, a river of Marocco, in the prov. of Fez, which flows NW, and after a course of about 75 m. falls into the Atlantic, to the SW of Sali.

YETMINSTER, or YATEMINSTER, a parish in the co. of Dorset, 6 m. SW of Sherborne, containing the chapelries of Leigh and Chetnole. Area 4,321 acres. Pop. in 1831, 1,199; in 1851, 1,333.

YETSEGHEN, a central region of the island of Nifon, in Japan, comprising the provs. of Iga, Ise, Ovari, Mino, Yetsisen, Yetstsion, Kanga, Noto, and Sado (island).

YETSEN, a central region of the island of Nifon, in Japan, comprising the provinces of Oumi or Umi, Vakasa, Tango, Tamba, Yamatfouro, Sitzo, Farima, Kavatsi, Yamatto, Xima, Kinokoumi, Joumi, and Avadsi (island).

YETSINGO. See YETSONGO.

YETSISEN, a province of Japan, in the island of Nifon, and region of Yetseghen, to the N of the prov. of Oumi, and NE of that of Vakasa, and bounded on the NW by the sea of Japan. It is covered in the S by lofty mountains, while on the N it is level and fertile, producing corn, rice, potherbs, and silk, and affording pasturage to large herds of cattle. It contains the town of Yetz.

YETSONGO, or YETSINGO, a province of Japan, in the island of Nifon, and region of Guanto, bounded on the N by the sea of Japan, on the NE by the prov. of Dewa, on the E by that of Monts, on the S by the provinces of Kaodsouki and Sinano, and on the SW by that of Yetstsion. Its S part is mountainous, but the coasts are fertile, producing cotton, and silk, and, in small quantities, corn. Near the village of Kourougaou-Moura naphtha is found in great abundance, and is used by the inhabitants in their lamps.

YETSTSION, a province of Japan, in the island of Nifon, and region of Yetseghen, to the N of the prov. of Fida, and SE of that of Noto, and bathed on the N by the sea of Japan. It is generally mountainous, and is intersected by fertile and well-cultivated valleys, which produce in great abundance rice, grain, and other cereals. The forests afford excellent timber. The manufacture of pottery forms an important branch of local industry. Oyama is the chief town.

YETZ, a town of Japan, in the island of Nifon, and prov. of Yetsisen, 105 m. NNE of Meaco, and 21 m. W of Lake Biwano-Oumi.

YEU (Ile d'). See Dieu.

YEUROUKS, a people of Turkey in Europe, in the sanj. of Salonica. They are descendants of the Turkomans, and are Mussulmans. They inhabit small villages in the mountains, and employ themselves in agriculture and rearing sheep and goats.

YEVPATORIIA. See Eupatoria.

YEVSIEIVSKAI, a lake of Russia in Asia, in the N part of the gov. and of the district of Yeniseisk, in N lat. 68° 30′, and E long. 98° 20′. It is about 30 m. in length from N to S, and nearly equal in breadth.

YEVSIONGA, a town of Russia in Europe, in the gov. and 186 m. S of Voronej, district and 21 m. E of Staro-Bielsk.

YEVST, a river of Russia in Europe, which issues from the NE extremity of Lake Louban; runs first N to the confluence of the Pedetz; turns N; and after watering the S part of Livonia, enters the gov. of Vitebsk, and joins the Dwina on the r. bank, about 60 m. below Dunaburg. It has a total course of about 75 m.

YEVTOUCHKEVITCHI, a town of Russia in Europe, in the gov. and 135 m. SE of Minsk, district and 41 m. WNW of Retchitza, near the l. bank of the Spa.

YEW, a district and town of China, in the prov. of Hu-nan, div. and 90 m. SSE of Chan-sha-fu, in N lat. 27° 8′, and E long. 113° 3′.

YEW-KE, a district and town of China, in the prov. of Fo-keen, and div. of Yen-ping-fu, in N lat. 26° 15′, and E long. 118° 10′.

YEW-YEN-CHU, one of the eight cities of the Chinese prov. of Leaou-tung or Shing-hing.

YEW-YEN-TING, a district and town of China, in the prov. of Leaou-tung or Shing-king, and div. of Fung-teen-fu.

YEW-YUH-HEEN, a district and town of China, in the prov. of Shan-se, and div. of Suh-ping-fu, in N lat. 40° 12′, and E long. 111° 13′.

YEZD, a district and town of Persia, in Khorassan. The district which occupies the S part of Khorassan is in its greater extent sandy and sterile. The cultivated portions produce corn, wine, sugar, figs, melons, cotton, silk, and madder. Camels are here reared in large numbers. Y. is now the only part of Persia in which Guebres are still to be met with. The town is 250 m. ESE of Ispahan, and nearly the same distance NE of Shiraz, in a wide sandy plain. Pop. 30,000. It is about 1½ m. in extent from ESE to WNW, and about a third less in breadth, and is surrounded by gardens, the walls of which form its only enclosure. The gates, about 20 in number, open into as many streets. The citadel is enclosed by a deep ditch. The town contains 21 mosques, four colleges, and numerous caravanserais, and possesses manufactories of varieties of silk, cotton and woollen fabrics, brocades in silk, silver, &c., and shawls. It has also several sugar refineries, and manufactories of arms.

YEZD-ABAD, or Haz-abad, a town of Persia, in Khorassan, in the district and 24 m. NW of Yezd. Pop. 3,000. It contains two mosques.

YEZDIDI, a town of Persia, in Khorassan, in the district of Tubbes. It has a fortress. The surrounding country abounds in dates and oranges.

YEZDKHAST, a town of Persia, on the NE confines of the prov. of Fars, 90 m. SSE of Ispahan, and 120 m. NNW of Shiraz, near the l. bank of the Khonah, which is soon after lost in the sand. It stands on a rock, and is enclosed by a brick wall. From the exterior it has a striking aspect, but within it is rendered revolting by the accumulated impurities with which it abounds. In the environs is a caravanserai. Y. is noted for the excellence of its bread. Cotton and rice are extensively cultivated in the vicinity.

YEZIDIS, or Yeyids, a singular race of people in Russian Armenia, and scattered over the adjacent Turkish and Persian dominions. They speak a Kurdish dialect. Their principal stronghold is the Jebel-Sinjar, a solitary mountain, rising in the centre of the Mesopotamian desert, to the N of Mosul. Mr. Layard states that the Y. recognise one Supreme Being; but that, as far as he could learn, they do not offer up any direct prayer or sacrifice to him. "The name of the Evil spirit is never mentioned; and any allusion to it by others so vexes and irritates them, that it is said they have put to death persons who have wantonly outraged their feelings by its use. So far is their dread of offending the Evil principle carried, that they carefully avoid every expression which may resemble in sound the name of Satan, or the Arabic word for 'accursed.' Thus, in speaking of a river, they will not say *shat*, because it is too nearly connected with the first syllable in *sheitan*, the devil; but substitute *nahr*. Nor, for the same reason, will they utter the word *keitan*, thread or fringe. *Naal*, a horse-shoe, and *naal-band*, a farrier, are forbidden words, because they approach to *laan*, a curse, and *mâloun*, accursed. When they speak of the devil, they do so with reverence, as *Melek Taous*, King Peacock, or *Melek-el-Kout*, the mighty angel." Baron Haxthausen regards them as a religious sect, not a distinct race, and probably of Kurdish descent. He imagines that they were originally a Gnostic sect, early expelled from the Christian church. They do not exhibit any trace of polytheism, and their religion is evidently not of Parsi origin, for with them Satan is not a personification of the Evil principle. They believe Satan to be the chief of the angelic host, now suffering punishment for his rebellion against the Divine will; but still all-powerful, and to be restored hereafter to his high estate in the celestial hierarchy. He must be conciliated and reverenced, they say; for as he now has the means of doing evil to mankind, so will he hereafter have the power of rewarding them. Next to Satan, but inferior to him in might and wisdom, are seven archangels, who exercise a great influence over the world. They are Gabrail, Michail, Raphail, Azrail, Dedrail, Azrapheel, and Shemkeel. Christ, according to them, was also a great angel, who had taken the form of man. He did not die on the cross, but ascended to heaven. They hold the Old Testament in great reverence, and believe in the cosmogony of Genesis, the deluge, and other events recorded in the Bible. They do not reject the New Testament nor the Koran; but consider them less entitled to their veneration. Still they always select passages from the latter for their tombs and holy places. Mahommed they look upon as a prophet, as they do Abraham and the patriarchs. They expect the second coming of Christ, as well as the re-appearance of Imam Mehdi, giving credence to the Mussulman fables relating to him. An inextinguishable national, as well as a religious enmity, exists between the Y. and the Persians.

YEZIORI, a town of Russia in Europe, in the gov. and district and 15 m. E of Grodno, near the E bank of a lake formed by the Pira.

YEZOROS, a town of Russia in Europe, in the gov. and 84 m. NNE of Vilna, district and 24 m. NW of Vidzi.

YFENSBERG, or Yfilsberg, a hill of Switzerland, in the SW part of the cant. of Zurich, and bail. of Cappel.

YFFINIAC, a commune of France, in the dep. of the Côtes-du-Nord, cant. and 4 m. ESE of Brieux. Pop. in 1846, 2,213.

YGOS, a village of France, in the dep. of the Landes, and cant. of Arjuzan, 14 m. WNW of Mont-de-Marsan. Pop. 1,000.

YGUIRA, a town of New Granada, in the dep. of Cundinamarca, prov. and 30 m. SW of Neiva.

YGUISAN, a small island of the Philippine archipelago, near the N coast of the island of Panay, in N lat. 11° 35′, and E long. 122° 32′.

Y-HO, a river of China, which has its source in the central part of the prov. of Shan-tung, runs S into that of Keang-su, and after a course of about 210 m. joins the Hwang-ho, on the l. bank. It forms Lake Lo-ma-hu.

YIELDEN, a parish in Bedford, 3½ m. ESE of Higham-Ferrers. Area 1,912 acres. Pop. 328.

YIGDIG, or Ozernaia [*i. e.* River of the lake], a river of Russia in Asia, in the district of Kamchatka, near the S extremity of the peninsula of that name. It issues from the W side of Lake Kurilsko, runs W, and, after a course of about 30 m., throws itself into the sea of Okhotsk. On its banks is a village of the same name.

YIH, or Yi, a district and town of China, in the prov. of Ho-nan, and div. of Nan-gan-fu.—Also a town of the prov. of Che-keang, and div. of Yen-chu-fu, in N lat. 34° 53′, and E long. 117° 51′.

YIH-CHANG-FU, a division and town of China, in the prov. of Hu-pih. The div. comprises seven districts. The town is 195 m. W of Wu-chang-fu, on the l. bank of the Yang-tze-keang.

YIH-CHEUEN-HEEN, a district and town of China, in the prov. of Keang-se, and div. of Yuen-chu-fu.

YIH-CHING, or E-Ching, a district and town of China, in the prov. of Keang-su, and div. of Yang-chu-fu, in N lat. 32° 18′, and E long. 119° 9′, on the l. bank of the Yang-tze-keang, and 30 m. NE of Nan-king-fu.—Also a district and town in the prov. of Hu-pih, and div. of Fang-yang-fu, on the r. bank of the Han-keang, 150 m. NW of Wu-chang-fu, in N lat. 31° 40′, and E long. 112° 8′.

YIH-CHING, or E-Chang, a district and town of China, in the gov. of Hu-nan, and div. of Lin-chu, 180 m. S of Chan-sha-fu.

YIH-CHING-HEEN, a district and town of China, in the prov. of Shan-se, and div. of Ping-yang-fu, in N lat. 35° 37′, and E long. 111° 47′.

YIH-CHU, a division and town of China, in the prov. of Chih-le. The div. comprises two districts. The town is 66 m. SW of Pe-king.

YIH-CHU, or E-Chu, a district and town of China, in the prov. of Leaou-tung, or Shing-king, and div. of Kin-chu-fu, in N lat. 41° 30′, and E long. 112° 20′, on the Ta-lin-ho, 126 m. WSW of Fungteen-fu.—Also one of the eight cities of the same prov.

YIH-CHUEN-HEEN, a district and town of China, in the prov. of Shen-se, and div. of Yen-gan-fu, in N lat. 36° 8′, and E long. 110°.—Also a district and town in the prov. of Keang-se, and div. of Yuen-chu-fu.

YIH CHU-FU, a division and town of China, in the prov. of Shan-tung. The div. comprises six districts. The town is on the r. bank of the Y-ho, 150 m. SE of Tse-nan-fu, in N lat. 35° 8′, and E long. 118° 33′. It has a few shops and gardens; but the houses are generally poor, and the general aspect of the place that of neglect. The surrounding district is fertile and well-cultivated.—Also a district and town in the prov. of Chih-le, and div. of Paou-ting-fu, in N lat. 39° 24′, and E long. 115° 35′.

YIH-HEEN, a district and town of China, in the prov. of Shan-tung, and div. of Lae-chu-fu, 141 m. SSE of Tze-nan-fu.

YIH-HEEN, or E-Heen, a district and town of China, in the prov. of Gan-hwuy, and div. of Hwuy-chu-fu, in N lat. 30° 5′, and E long. 117° 58′.

YIH-HING, a district and town of China, in the prov. of Keang-su, and div. of Chang-chu-fu, in N lat. 31° 28′, and E long. 119° 49′.

YIH-HO-HEEN, or E-ho-heen, a district and town of China, in the prov. of Kan-su, and div. of Chin-se-fu.

YIH-HWANG-HEEN, a district and town of China, in the prov. of Keang-su, and div. of Fu-chu-fu.

YIH-KEUN-HEEN, a district and town of China, in the prov. of Shen-se and div. of Luh-chu, in N lat. 35° 28′, and E long. 109° 2′.

YIH-LEANG, a district and town of China, in the prov. of Yun-nan, div. and 30 m. SE of Yun-nan-fu, in N lat. 24° 58′, and E long. 103° 14′.

YIH-LING-CHU, a district and town of China, in the prov. of Hu-pih, and div. of King-chu-fu, in N lat. 30° 49′, and E long. 111° 10′ 20″.

YIH-LUNG-HEEN, a district and town of China, in the prov. of Sze-chuen, and div. of Shun-hing-fu, in N lat. 31′ 28′, and E long. 106° 25′.

YIH-MIM, a district and town of China, in the prov. of Yun-nan, and div. of Yun-nan-fu, in N lat. 24° 45′, and E long. 102° 19′.

YIH-NING, a district and town of China, in the prov. of Kwan-se, and div. of Kwei-lin-fu, in N lat. 25° 22′, and E long. 110′.

YIH-NING-CHU, a district and town of China, in the prov. of Keang-se, and div. of Nan-chang-fu.

YIH-PIN-HEEN, a district and town of China, in the prov. of Sze-chuen, and div. of Leu-chu-fu.

YIH-PING-TING, a district and town of China, in the prov. of Ho-nan, and div. of Kai-fung-fu, in N lat. 34° 55′, and E long. 115° 7′ 30″.

YIH-SHAN-HEEN, a district and town of China, in the prov. of Kwang-se, and div. of King-yuen-fu.

YIH-SHI, a district and town of China, in the prov. of Shan-se, and div. of Pu-chu-fu, in N lat. 35° 11′, and E long. 110° 43′.

YIH-SHWUY-HEEN, a district and town of China, in the prov. of Shan-tung, and div. of E-chu-fu, in N lat. 35° 46′, and E long. 119°.

YIH-SO, a district and town of China, in the prov. of Yun-nan, and div. of Keuh-tsing-fu, in N lat. 25° 20′, and E long. 104° 26′.

YIH-TU, a district and town of China, in the prov. of Hu-pih, and div. of King-chu-fu, in N lat. 30° 28′, and E long. 111° 19′.

YIH-WU-HEEN, a district and town of China, in the prov. of Che-keang, and div. of Kin-wha-fu.

YIH-YANG, a district and town of China, in the prov. of Keang-se, and div. of Kew-keang-fu.—Also a district and town of the prov. of Hu-nan, and div. of Chan-sha-fu, in N lat. 28° 35′, and E long. 112° 8′.—Also a district and town of the prov. of Keang-su, and div. of Kwang-sin-fu, in N lat. 28° 25′, and E long. 117° 33′.—Also a district and town in the prov. of Ho-nan, and div. of Ho-nan-fu, in N lat. 34° 31′, and E long. 112° 10′, 114 m. WSW of Kae-fung-fu. Copper is found in the adjacent mountains.—Also a district and town of the same prov. in the div. of Pi-chu, in N lat. 34° 12′, and E long. 112° 30′.

YIN, In-shan, Khinghan, or Siolki mountains, a general name for several mountain chains in Asia, viz.: 1st, The portion of the Yablonnoi or Stannovoi range, which runs along the frontier of the Russian prov. of Yakutsk and Chinese prov. of He-lung keang. 2d, A branch of the above-named chain which runs from N to S through the N part of the Chinese prov. of He-lung-keang. 3d, The mountains which form a continuation of the Gan mountains, and encircle the sources of the Amur in the Khalkas country; and lastly, the chain which traverses Shara-Mongolia and the SW part of the prov. of He-lung-keang. These mountains are but imperfectly known.

YING-CHING, a district and town of China, in the prov. of Ho-nan, and div. of Kwei-tih-fu.

YING-CHING-HEEN, a district and town of China, in the prov. of Hu-pih, and div. of Tih-gan-fu, in N lat. 31° 5′, and E long. 113° 27′.

YING-CHU, a division and town of China, in the prov. of Shan-se, and div. of Tae-ting-heen, in N lat. 39° 39′, and E long. 113° 13′.

YING-CHU-FU, a division and town of China, in the prov. of Gan-hwuy. The div. comprises 5 districts. The town is in N lat. 32° 58°, and E long. 115° 57′.

YINGGHEO, a town of Burmah, in the Mranma, on the r. bank of the Irawady, 42 m. SSW of Pagham-mew.

YING-KEANG-HEEN, a district and town of China, in the prov. of Kwei-chu, and div. of Sze-nan-fu.

YING-SHAN, a district and town of China, in the prov. of Sze-chuen, and div. of Shun-king-fu, in N lat. 31° 8′, and E long. 106° 35′.—Also a district and town in the prov. of Hu-pih, and div. of Tih-gan-fu, in N lat. 31° 40′, and E long. 113° 41′.—Also a district and town in the prov. of the Corea, 66 m. S of Kang.—Also a district and town in the prov. of Keang-se, and div. of Kwang-sin-fu, in N lat. 28° 15′, and E long. 117° 50′.

YING-SHANG-HEEN, a district and town of China, in the prov. of Gan-hwuy, and div. of Ying-chu-fu, in N lat. 32° 40′, and E long. 116° 13′.

YING-SHAN-HEEN, a district and town of China, in the prov. of Gan-hwuy, and div. of Luh-gan-chu, in N lat. 30° 45′, and E long. 115° 45′.

YING-TIH, a district and town of China, in the prov. of Kwang-tung, and div. of Shaou-chu-fu, in N lat. 24° 11′ 3″, and E long. 112° 55′.

YIOMATOUNG, POKHING-LOUNG, or ANU-PEC-TU-MU, a range of mountains which separates the Burman empire from Aralan.

YJUY, or HYJUHY, a river of the republic of Monte-Video, which has its sources in the Sierra Xavier, runs W, and joins the Uruguay, on the l. bank, in S lat. 27° 50′. It has a total course of 150 m. Its principal head-streams are distinguished as Yjuy Grande and Yjuy Poco.

Y-KI-MA, an island of the Madjicoseamah group, to the SW of the Lu-chu islands.

Y-KY, or IKE, an island of the Chinese empire, in the group and to the E of Great Lu-chu island. It is 15 m. in length, and 8 m. in breadth.

YLO, or ILO, a port of Bolivia, in S lat. 17° 37′, W long. 71° 23′. It is a small place of 300 inhabitants; but has a good roadsted.

YLO-COS, a tribe of aborigines of the island of Luzon, Philippine archipelago. It gives its name to one of the 15 prov. into which the island is divided, and the pop. of which is estimated at 862,000, of whom 362,000 are Indians.

YLOILO, a town and port of the Philippines, on the E coast of the island of Panay.

YLOPANGO, a lake in the state and 6 m. SE of San Salvador. It is 9 m. long, and 3 m. broad; and occupies apparently an extinct crater.

YLST, a town of Holland, in the prov. of Friesland, cant. and 2 m. WSW of Sneek. Pop. 1,200.

YNGAXEN, a lake of Sweden, in the prefecture of Nykoeping. It is 24 m. in length, and 6 m. in breadth, and discharges itself into the Baltic.

YNGEN, a lake of Sweden, in the prefecture of Karlstadt, which discharges itself into Lake Wener.

YNGOHAN, a town of the Corea, in the prov. of Khing-chang, 66 m, S of Kang.

YNGONA, a state of Nigritia, to the NW of Dogombah, on which it depends. It contains a town of the same name.

YNGUARAN, a mining village of Mexico, in the state of Mechoacan, 63 m. SW of Valladolid.

YNYS-CYNHAIARN, a parish in Carnarvonshire, 7 m. E of Tremadoc. Area 6,546 acres. Pop. 2,347.

YNYS-ENLLI. See BARDSEY-ISLE.

YO, or Jo, a town of Burmah, capital of a district of the same name, in the Mranma, 132 m. SW of Ava, and 120 m. NE of Aracan.—Also a mountain of China, in the prov. of Sze-chuen, noted for its precious stones.

YO-CHU-FU, a division and town of China, in the prov. of Hu-nan. The div. comprises 4 districts. The town is on the NE bank of Lake Tung-ting, at the point of egress of its outlet into the Yang-tsze-keang, and 90 m. N of Chan-sha-fu, in N lat. 36° 15′, and E long. 112° 6′.

YODO, a town of Japan, in the island of Nifon and prov. of Oumi, 6 m. SW of Meaco, on the l bank of the Yedogava, which is here crossed by a bridge of 40 arches, and 400 ft. in length. This town has two suburbs, and a castle. The houses are generally well-constructed, and the inhabitants are noted for their industry.

YOEC, a village of Beluchistan, 250 m. SW of Norshky, near the S confines of the desert of Afghanistan, in N lat. 28° 20′, and E long. 61° 56′. The surrounding district is hilly, and generally fertile and well cultivated.

YOF, a town of Senegambia, in the state of Kayor, on the N coast of the peninsula of Cape Verde, and 105 m. SSW of St. Louis.

YOHIBIDE, a mountain near the S confines of Bolivia, in S lat. 20° 30′, and E long. 62° 40′.

YOHN, or JOHN-NAUBISCHEN, a village of Prussia, in the regency of Gumbinnen, and circle of Heidefrug. Pop. 206.

YOHNSBACH, or JOHNSBACH, a village of Austria, in Bohemia, in the regency of Böhmisch-Leipa, and bail. of Böhmisch-Ramnitz. Pop. 380.—Also a village of Prussia, in the regency of Breslau, and circle of Frauenstein. Pop. 280.

YOHNSDORF, or JOHNSDORF, a village of Austria, in Bohemia, in the regency of Pardubitz, and bail. of Landskron. Pop. 690.

YOHNSDORF, or JOHNSDORF (OBER), a village of Austria, in Bohemia, in the regency of Pardubitz. Pop. 1,260.—Also a village of the regency of Böhmisch-Leipa, and bail. of Terschen. Pop. 1,100. It has several yarn-mills.—Also a village of the same regency, in the bail. of Karbitz.—Also a village of Moravia, in the circle of Olmutz, and seignory of Trübau. Pop. 309.—Also a village of the same circle, and bail. of Sternberg. Pop. 1,430. It has several iron-works.—Also a village of Prussia, in the regency of Leignitz, and circle of Landschut. Pop. 460.—Also a village of the same regency, in the circle of Lowenberg. Pop. 345.—Also a village of the same regency, in the circle of Sprottau. Pop. 352.—Also a village of the same regency, in the circle of Schonau. It has two mills.

YOHNSDORF (SCHON), a village of Prussia, in the regency of Breslau, and circle of Munsterberg. Pop. 350.—Also a village of the same regency, in the circle of Brieg. Pop. 220. It has a castle.

YOHNSDORF, or JOHNSDORF (ALT and NEU), two villages of Saxony, in the circle of Bautzen, and bail. of Löbau. United pop., 1,673.

YOINGT, or OINGT, a town of France, in the dep. of the Rhône, and cant. of Bois d'Yoingt, and 8 m. WSW of Villefranche. Pop. 500. It has the remains of an old church, the chapel of which is used as the parish church.

YOKA, a village of Hungary, in the comitat of Presburg. Pop. 1,970.

YOKAITZ, a town of Japan, in the island of Nifon, and prov. of Isé, 15 m. WSW of Nagoja, on Obari bay. Pop. 3,600.

YOKEFLEET. See YORKFLEET.

YOKKAS, a town and parish of Russia in Europe, in the grand-duchy of Finland, gov. of Kymmenegard, and district of Sawolax Nedredels, and 84 m. NE of Heinola.

YO-KEON, a town of the Corea, in the prov. of Chu-sin, 51 m. SSW of Hia-min.

YOKI, a town of Japan, in the island of Nifon, and prov. of Tasima, 66 m. NW of Meaco.

YOKOHANA, a town of Japan, on the coast of the bay of Yeddo, in the island of Nifon, about 10 m. below the city of Yeddo. The recent American expedition met with the commissioners of the emperor at this place in March 1854. They estimated the pop. at about 10,000.

YOKOMI, a small town of Japan, in the island of Nifon, and prov. of Monsasi, 27 m. N of Yedo.

YOLA, a town of Central Africa, the capital of Adamowa, 230 m. S by W of Kuka, in a swampy plain near the Benue.

YOLAY, or YOLET, a village of France, in the dep. of Cantal, cant. and 5 m. E of Aurilla.

YOLBACHI, a market-town of Turkey in Europe, in Rumelia, in the sanj. and to the N of Salonica.

YOLO, a county of California, occupying the tract between the Sacramento river and Putah creek and the coast range, and extending N to 39° 20′ N lat. It is drained by Cache and other creeks of Clear lake, and several other affluents of the Sacramento and Putah, and contains several lakes. Area 1,400 sq. m. Pop. in 1852, 1,307. Its capital is Freemont.

YOLOFS, WOLOFS, or JOLOFS, a people of Senegambia, who inhabit the vicinity of the French colony of Senegambia. Their state called Bourb-be-Ghiolof was formerly the most extensive in that region of Africa, and their chief still retains pre-eminence among the neighbouring coloured sovereigns. The Yolofs, in common with the Foulahs, inhabited, previous to the irruption of the Saracens, the fertile plains of Numidia and Mauritania, and when driven thence retired to the region they now occupy, which, although possessing little fertility, yield abundant returns to the exertions of industry. The forests abound with gum trees, but their produce is left chiefly to the Moors. Millet, cotton, and indigo abound in the fields. The Y. are large in stature, well-made, active, and hospitable, and their language is sonorous and easily understood. They manufacture cotton fabrics and articles in ivory. The gov., like that of the surrounding states, is feudal, but the king, owing chiefly to the number of his slaves, possesses despotic authority. Fetish paganism is more generally professed by the people than Mahommedanism.

YOLOMBO, a town of New Granada, in the dep. of Cundinamarca and prov. of Antioquia, 108 m. NNW of Santa-Fe-de-Bogota. Pop. 1,000.

YOMBERE, or JAMBER, an island in the S part of the Bissagos group, near the coast of Senegambia, 12 m. SE of the island of Kanybac, in N lat. 11° 6′, and W long. 15° 35′. It is 9 m. in length, and 6 m. in breadth. A little to the W are the small islands of Honey and Jument.

YONE-SAWA, or MI-TSE, a town of Japan, in the island of Nifon, and prov. of Dewa, 150 m. NNE of Yedo.

YONGE, a township of Upper Canada, in the Johnstown district, bounded on the S by the St. Lawrence, and containing several lakes, of which the principal is Gananque, in the W. Pop. in 1842, 4,036.

YONGUESVILLE, a village and station of Fairfield district, in the state of South Carolina, U. S., on Charlotte and South Carolina railroad, 32 m. N by W of Columbia. Population about 100.

YONKERS, a township and village of Westchester co., in the state of New York, U. S., 132 m. S of Albany, drained by Bronx river on the E, and bordered on the W by Saw-Mill or Nepperham river, an affluent of the Hudson, and intersected by the Hudson river railway. The village is noted for its salubrity, and is rapidly rising in importance. Pop. of township in 1840, 2,968; in 1850, 4,160.

YONNE, a river of France, which has its source in the dep. of the Nievre, in the Belleperche marshes, at the foot of Mount Beuvron, in the arrond. and to the SE of Chateau-Chinon, and bearing the name of Belleperche to the Etang d'Yonne, 12 m. below its source. After traversing the departments of the Nievre and Yonne, and a small part of that of the Seine and Marne, and running in a NNW direction past Corbigny, Clamecy, Auxerre, Joigny, Villeneuve-le-Roi, Sens, and Pont-sur-Yonne, it throws itself into the Seine, on the l. bank, at Montereau-Fault-Yonne, after a total course of about 192 m., of which 80 m. are navigable. Its principal articles of transit are timber, charcoal, wine, grain, &c., for the Paris market.—Its chief affluents are the Moine, Fachin, Proye, Tourou, Houssiere, Menage, Bruit, Anguison, Aussois, Beuvron, Ahdrie, Aire, Armançon, Vrin and Vannes. The canal de Nivernais forms a communication with the Loire, and that of Burgundy with the Saône.

YONNE, a department situated between 47° 22′ and 48° 25′ N lat., and between 2° 52′ and 4° 20′ E long., and bounded on the NW by the departments of the Seine-et-Marne; on the NE by that of the Aube; on the SE by the dep. of Côte d'Or; on the S by that of the Nievre; and on the W by the dep. of the Loiret. It is 105 m. in length from NW to SE, and 60 m. in extreme breadth from NE to SW, comprising a total area of 2,700 sq. m. Pop. in 1801, 320,596; in 1821, 332,905; in 1831, 352,487; in 1841, 362,961; in 1846, 374,856. The surface of this dep. is diversified with hills, valleys and plains, but contains no lofty summits. The principal heights are those which on the SW separate the basin of the Loire from that of the Seine, and form a continuation of the mountains of Morvan. With the exception of the Vrille, a small affluent of the Loire, all the waters of this dep. belong to the basin of the Seine. The principal river is the Yonne, which traverses the dep. from S to N, and receives on the r. the Cure, Armançon, and Vannes, and on the l. the Vrin, Loing, and its affluent the Ouanne. In the E part is Burgundy canal, which connects the Yonne and Saone. The soil is in some parts strong, but possesses great fertility, producing in great abundance, grain, wine, fruit, cider, hemp, and forage. The principal wines—which bear the general name of Lower Burgundy—are noted as the finest in the empire. Of the total number of arable hectares estimated at 709,268 in 1839, 88,223 were in wheat, 29,963 in maslin, 39,920 in rye, 38,476 in barley, 78,650 in oats, 531 in sarrasin, 6,822 in potatoes, 4,903 in legumes, 37,600 in the vine, 3,527 in gardens, 204 in beetroot, 2,468 in hemp, 29,154 in natural meadow, 42,699 in artificial meadow, 6,678 in orchards, and 162,299 in wood. Of these at the same period the estimated produce was 979,973 hectol. of wheat, 299,256 of maslin, 298,657 of rye, 355,319 of barley, 792,807 of oats, 4,882 of sarrasin, 584,576 of potatoes, 37,157 of legumes, 856,223 of wine, 32,234 of cider, 99,504 q. 1 m. of beetroot, 17,354 of hemp, 825,313 of grain, 606,370 q. m. of hay, 1,015,497 of forage, and 721,942 of timber. Of domestic animals the number was 28,163 horses, 3,787 mules, 11,876 asses, 139,218 head of cattle, 350,144 sheep, 35,489 pigs, and 3,172 goats. The principal mineral production is iron, the quality of which is esteemed equal to that of Berri. It yields also red and yellow ochre, granite, paving stone, marble, mill and lithographic stone, gun-flint, limestone, and pottery clay. The manufacturing industry of the dep. consists chiefly in iron, cloth, and some common woollen fabrics, blankets, cotton coverlets, casks, tiles, bottles, earthenware, and beetroot sugar. The principal exports are wine, grain, wood, charcoal, and staves. Fairs to the number of 406 are held in 89 com., lasting 421 days. In 1842 there were in the dep. 15 educational establishments, of which 6 were colleges at Auxerre, Avallon, Joigny, Noyers, Sens, and Tonnerre, 15 hospitals, and 71 other charitable institutions. The judicial divisions consist of 5 arrond., 35 cant., and 482 com. The capital is Auxerre. This dep. was formed in 1790, from the NW part of the prov. of Burgundy, the SW of that of Champagne, and a small portion of the Ile-de-France. This district of France was the territory of the ancient *Senones*, a Gallic people who spread themselves even to the heart of Italy, and who were powerful at the period of the Roman invasion of Gaul. Sens is one of the most ancient, and was for some time one of the most important towns in France.

YONY, a town of Sherborough island, on the Sierra Leone coast, in Upper Guinea. The convention by which the king of Sherborough and his tributaries ceded their territories to the English, was ratified here on the 5th Oct. 1825.

YOOANBA, a river of New South Wales, an affluent of the Manning.

YOODOMA. See JOUDOMO.

YOOT-SIMA, a small island of Japan, in the sea of that name, to the NW of Cape Noto. on a peninsula of that name, in the island of Nifon, in N lat. 37° 51', and E long. 137° 34'. It is flat and woody, and generally well peopled.

YOPARE', a town of the N coast of New Guinea, in S lat. 0° 53' 26", and E long. 134° 6' 18".

YORI, or YORA, a river of Russia in Asia, in Georgia, which has its source in the Caucasus, in the N part of the Kakhethi; runs first S, then SSE; and after a course of 210 m. joins the Abazani, on the r. bank, on the confines of Shirvan, and 18 m. NNE of Ganja. It flows through a mountainous country, in the upper part of its course, but afterwards traverses the steppe of Oupadar, and other sandy plains.

YORIBA. See YARRIBA.

YORIMAN, a numerous Indian tribe of Brazil, who inhabit the district to the N of the Hyapura.

YORK, a city, and ainsty or county in itself, and the capital of Yorkshire, is situated near the centre of the shire, at the point where the three ridings meet; at the confluence of the tidal rivers Ouse and Foss, in N lat. 53° 48', and W long. 4° 19'; at a mean alt. above the sea of 44 ft. Area of the city 2,720 acres. Pop. in 1841, 28,842; in 1851, 36,303. Pop. of the parliamentary borough, in 1841, 30,152; in 1851, 40,359. Area of the ainsty of the city, 49,720 acres. The ainsty was annexed to the West riding in 1837, as a wapentake, called Ainsty wapentake: but electors in the ainsty vote for members in the North riding. The whole of it lies on the W side of the city, and extends to a distance, in one direction, of 10 miles. The city and ainsty together form the co. of the city. York is 198 m. by coach-road, and 218½ by railroad, NNW of London; and 201 m. SSE of Edinburgh. From Leeds it is 22 m. by coach-road, and 30 m. by railroad; from Hull, 40 m. by coach-road, and 51 m. by railroad. The city is situated in the centre of one of the richest and most extensive valleys in England, having generally an alluvial clay subsoil. It is surrounded by walls, which still remain of the original height, and were renovated in 1831. The older and principal part of the city rises from the E side of the Ouse, with a gentle acclivity, reaching the summit level about 300 yds. from the river. Over the Ouse there is a handsome three-arched stone-bridge. The Foss—which is retained at a level 7 feet higher than that of the Ouse, by lock-gates—is crossed by four bridges. The streets are mostly broad and open, contain many elegant buildings, and are well-paved and lighted with gas. The castle stands in the angle between the Ouse and the Foss. The castle walls are 1,103 yds. in circumference. The county-hall and new county-prison buildings occupy a portion of the area which amounts to 4 acres. The prison is arranged on the radiated system. Extensive cavalry barracks were erected to the S of the city in 1796, at an expense of nearly £30,000. The guildhall comprises a very noble Gothic hall 96 ft. in length. A magnificent concert hall was built in 1825. The mayor's residence, in Blake-street, is a stately edifice. The city jail occupies a retired spot close to the city walls. There are public walks extending above 1 m. along the E bank of the Ouse; and the old city ramparts have been made into a dry flagged walk.—York minster, the most stately Gothic edifice in England, stands on the N side of the city, and, except on its northern side, is closely hemmed-in by other buildings. It was probably completed about the year 1410 or 1412, on the site of a still more ancient edifice. It is cruciform; and, in the order of its antiquity, consists of a south transept built in 1227, north transept built in 1260, nave finished in 1330, central tower begun in 1370, and choir extending from the west end to the east. Viewed in this order, it is perhaps the most perfect example extant of the history and progress of the Gothic style, during the lapse of nearly two centuries, which its construction occupied. The whole length of the edifice from E to W is 524½ ft.: breadth of E end 105; of W 109. Length of cross aisles from N to S 222 ft.: height of grand lantern tower 235: height of two western towers 196; height of nave or body of church 99: height of eastern window 75; breadth, 32 ft. The W front is adorned with two beautiful towers, between which, over the entrance, is a window of fine painted glass. The lantern steeple is ornamented with great taste, and has 8 windows, 45 ft. in height. The choir is divided from the rest of the church by a curious stone screen, over which is a noble and newly built organ; the front is adorned with statues of the kings of England, from William the Conqueror to Henry VI. There are 32 stalls for the prebendaries, all of marble; behind which, on each side of the choir, are ranges of carved wood work adorned with pinnacles. The ascent to the altar is by a flight of 16 steps, and the whole of the body of the church is paved in mosaic. Many of the ancient nobility, as well as archbishops, lie buried here, and some of the monuments are magnificent. The chapter house is a fine specimen of Gothic architecture. It is an octagon 63 ft. in diameter, and 67 ft. high; the roof, supported without pillars, is a cone of curious carved work covered with lead. Around the interior of the octagon are 44 beautiful marble stalls; and in each of the eight sides is a window except in one, through which is the entrance from the transept. On 2d February, 1829, a lunatic set fire to the choir of this venerable church, the stalls of which, with the organ, and part of the roof, were completely destroyed. A restoration was effected in 1833, under the direction of Mr. Smirke, at an expense of £65,000, which was defrayed by voluntary subscription. In May 1840, the SW tower was accidentally fired, and the roof of the nave became involved in the disaster. The expense of the restoration thus again rendered necessary, was estimated at about £20,000, and the repairs have now been completed and defrayed by voluntary subscription. In the southern tower a huge bell—the largest in England—has been put up. It weighs nearly 13 tons, and with the stock 16 tons; and is 9 feet 6 inches in height. The clapper, which is of wrought iron, weighs 4 cwt. It is rung with two wheels, each 14 ft. in diameter. It is heavier by 7 tons than 'Tom of Lincoln,' and by 5 tons than 'Old Tom of Oxford.'—The churches and dissenting chapels are much too numerous for minute description here. They are chiefly in the later style of

English architecture, with an occasional intermixture of the Norman and early English. The city returns two members to parliament. The franchise was conferred so early as 23° Edward I. Previous to the Reform act, the right of election was in the freemen. The number of electors registered in 1837, was 2,864, of whom 2,084 were freemen.

Diocese and see.] Until recently the diocese of York comprehended the archdeaconries of Cleveland, East riding, West riding, and Nottingham; but the ecclesiastical commissioners have transferred the archd. of Nottingham to the dio. of Lincoln; and part of the archd. of Cleveland, to the new dio. of Ripon, and part to the dio. of Durham: see these articles. The diocese, therefore, now consists of the archd. of Cleveland and the East and West ridings: see article YORKSHIRE. The number of benefices in the dio. returned in 1831, inclusive of sinecure rectories, but exclusive of benefices annexed to other preferments, was 891, besides 12 not returned. The average gross income of the returned benefices was £250. The average gross yearly income of the archiepiscopal see, in 1831, was £13,798: nett yearly income, £12,629. By order in council, of date 21st June, 1837, the income of the future archbishops of York is limited to £10,000 per annum. The chapter of York, besides the archbishop, consists of a dean, precentor, chancellor, subdean, 4 archdeacons, 28 prebends, a subchanter, 5 vicars, &c.

Commerce, &c.] York was anciently a port, and as such furnished, in 1298, one vessel to the fleet of Edward I. The tide, however, does not now rise so far as the city, being interrupted by Naburn lock about 5 m. distant. Its trade was anciently very considerable. In 1354, it obtained the staple for wool; and the woollen manufacture flourished here so late as the reign of Henry VIII., and indeed until the commencement of the 18th century, when this branch of trade removed to the West riding. The manufactures are now of small extent. Glass, carpets, white and red lead, linens, and ironmongery, are manufactured to a moderate extent. The trade is small, and consists chiefly in the importation of coal. Vessels of 90 tons can sail up to the Ouse bridge; and steamers ply to and from Hull in all states of the tide.

History.] In the time of Roman dominion, Y. was an important station; and in the year 70 was made the capital of the province of *Maxima Cæsariensis.* In 212, Severus died here. Under the heptarchy, Y. became the capital of Northumbria, and afterwards of Deira. Before the reign of the Conqueror it was called a second Rome, from the extent of ground which it occupied, and the magnificence of its buildings. The citizens joined the Scots and Danes against the Conqueror, who in consequence razed the city to the ground. In 1137 it had partially risen from its ashes, but was then destroyed by a fire, which burnt down the cathedral, the abbey of St. Mary, and forty parish churches, besides many streets and public buildings. The first meeting known by the name of a parliament was held here in 1160, by Henry II. In the beginning of the reign of Richard I. an extensive massacre of Jews took place here, in which it is supposed that not less than 2,000 Jews perished. In 1251, Alexander, king of Scotland, was married here to the daughter of Henry III. In 1349 the plague ravaged York, and again in 1390, in 1506, and in 1550. In 1604 the plague raged to such an alarming extent at York, that the citizens left the city and encamped on Hob moor and on the Horsefair. During the parliamentary war the city was held by the royalists till July 1644, when it fell into the hands of Fairfax. At the period preceding the Revolution, this city was noted for its opposition to the king; and in the very year of the Revolution, James II. took away its charter, and declared the office of mayor to be vacant. York has given title to many noblemen, especially to sons of sovereigns. The last who held the title of duke was Prince Frederick, second son of George III., by whose death, in 1827, the title became extinct.

YORK, a county and town of Western Australia. The co. is bounded on the N by the co. of Victoria; on the E by that of Howick; on the S by the co. of Grantham; and on the W by that of Perth. It is intersected from SSE to NNW by the Avon, which flows thence into the SW corner of the co. of Victoria, and there making an abrupt turn, re-enters and traverses the NW corner of the co. of York, under the name of Swan river. Its principal affluents are the Mackie and Mortlock on the r., and the Dale on the l., and towards the SW are the head-streams of the Helena, also affluents of the Swan. The chief summits are Leake, Ommaney, Moloine, Madie, Bakewell, and Matilda. On the W are extensive sandy downs, covered in some parts with cabbage trees, and in others with grass or scrub, while the W presents a more hilly aspect, and a predominance of red loam. Limestone is found in the SE towards Comely Peak. The chief towns are York, Beverley, and Northam. The town of Y. is on both sides of the Avon, which is here a mere chain of ponds, near the centre of the co., and about 60 m. E of Perth. It is surrounded by detached granitic hills, and beyond are immense tracts of slightly undulating surface, and densely covered with wood.—Also a mountain of New South Wales, in the co. of Cook, 72 m. WNW of Sydney. It forms the NW extremity of a range of mountains named King's table land, and has an alt. of 3,292 ft. above sea-level.—Also a parish of Tasmania, in the co. of Monmouth, bordered on the W by the Jordan.—Also a town in the co. of Devon, on the Tamar.—Also an island of the South Pacific, in the archipelago of New Britain. Port Hunter on the NE coast, is in S lat. 4° 7′, and E long. 6° 15′. This island is called by the natives Amacata.

YORK, a village of Upper Guinea, on the W coast of the peninsula of Sierra Leone, on a small river, and 9 m. N of Cape Shilling. It is well and regularly built, and contains some handsome houses. The environs are well-cultivated, and produce iguames, maize, and cacao. The inhabitants are chiefly sawyers, and carry on an active trade in timber.

YORK, a county of Upper Canada, in the Home district, bounded on the E by Lake Ontario, and watered by the Humber and Don rivers. It is tolerably fertile towards the rear, but near the lake is poor and sandy. Area 808 sq. m. Pop. in 1852, 79,719.—Also a village in the township of Huoon, on the Grand river, 5 m. from Caledonia. Pop. 150.—Also a county in the SW of New Brunswick, bordering on Maine, and intersected by the St. John's river.

YORK, a county in the SW part of the state of Maine, U. S., comprising an area of 822 sq. m., drained by Saco and Monsam rivers and their branches, affluents of Piscataqua river by which it is bordered on the SW, and intersected by the Portland, Saco and Portsmouth, and the York and Cumberland railroads. It has a finely diversified surface, with numerous lakes, and is in some parts very fertile. On the Atlantic are several good harbours. Pop. in 1840, 54,034; in 1850, 60,101. Its capital is Alfred.—Also a county in the SE part of the state of Pennsylvania, comprising an area of 864 sq. m., bordered on the NE by the Susquehanna, and drained by its tributaries, Cadoras, Conewago and Yellow Breeches creeks. It has a hilly surface, and is generally fertile. It is intersected by the York and Cumberland, the Baltimore and Susquehanna, the Hanover branch, and the York and Gettysburgh, and the Susquehanna and Tide Water Gettysburgh canal. Pop. in 1840, 47,010; in 1850, 57,450. Its capital is York.—Also a county in the SE part of the state of Virginia, in the centre of the peninsula between James and York rivers. It has a diversified surface, and possesses considerable fertility. Pop. in 1840, 4,720; in 1850, 4,460. Its capital is Yorktown.—Also a district of the N part of the state of South Carolina, comprising an area of 731 sq. m., drained by King's, Bullock's, Turkey, and Buffalo creeks, affluents of Broad and Catawba rivers, and intersected by the Charlotte and South Carolina, the King's mountains, &c., railroads. It presents a generally level surface, and possesses considerable fertility. Pop. in 1840, 18,383; in 1850, 19,433. Its capital is Yorkville.—Also a township and port of the co. of the same name, in the state of Maine, 94 m. SSW of Augusta. The surface is diversified, and the soil variable in quality. Pop. in 1840, 3,111; in 1850, 2,980. The village is in N lat. 43° 10′, and W long. 70° 40′. It has a harbour capable of receiving vessels of large size, and was formerly capital of the co.—Also a township of Livingstone co., in the state of New York, 220 m. W of Albany, bordered on the E by the Genesee river, and the Genesee Valley canal, and drained by branches of the river. Pop. in 1840, 3,049; in 1850, 2,785.—Also a township of York co., in the state of Pennsylvania, 21 m. SSE of Harrisburg, in N lat. 39° 58′, and W long. 76° 40′, intersected by the York and Cumberland, the Baltimore and Susquehanna, and the York and Wrightsville railroads. Pop. in 1840, 4,779; in 1850, about 6,000.—Also a township of York co., in the same state. Pop. in 1840, 1,294.—Also a township of Athens co., in the state of Ohio, drained by Hockhocking river and its tributaries. Pop. in 1840, 1,601.—Also a township of Belmont co., in the same state, bounded on the E by Ohio river, and drained by Captina creek. Pop. 1,294.—Also a township of Darke co., in the same state. Pop. 369.—Also a township of Medina co., in the same state. Pop. 781.—Also a township of Morgan co., in the same state. Pop. 1,032.—Also a township of Sandusky co., in the same state. Pop. 1,301.—Also a township of Tuscarawas co., in the same state. Pop. 865.—Also a township of Van Wert co., in the same state. Pop. 181.—Also a township of Union co., in the same state, drained by Rush and Bogues creeks, affluents of Sciota river, and 42 m. NW by N of Columbus. Pop. in 1840, 493; in 1850, 831.—Also a township of Washtenano co., in the state of Michigan, 61 m. SE of Lansing, drained by affluents of Raisin river. Pop. in 1840, 1,146; in 1850, 1,360.—Also a township of Switzerland co., in the state of Indiana. Pop. in 1840, 1,331.—Also a township of Steuben co., in the same state. Pop. 191.—Also a township of Noble co., in the same state. Pop. 289.—Also a village of Gibson co., in the same state, 120 m. SW by S of Indianapolis, near the Evansville and Illinois railroad.—Also a village of Crawford co., in the state of Illinois, on the W side of Wabash river, and 115 m. ESE of Springfield river. Pop. in 1840, 350; in 1850, 500. —Also a river in the state of Maine, in the S part of the township of York, which flows E and falls into the Atlantic by a wide embouchure, forming a capacious harbour.—Also a river of the state of Virginia, formed by the confluence of Pamunkey and Mattapony rivers, which unite in the SE extremity of King William co., and which, after a course of about 30 m., enters Chesapeake bay, by a mouth 3 m. in width. It forms the best harbour in Virginia.

YORK (Cape), a headland forming the NE point of Australia, advancing into Torres straits between the ocean and the gulf of Carpentaria, and opposite New Guinea, from which it is distant about 80 m., in S lat. 10° 41′, E long. 142° 34′. To the NE, about 7 m. from the continent, is a group of islands 3 in number, of the same name. The principal island is about 3 m. long, and has a conspicuous flat-topped hill. It is surrounded by coral reefs.

YORK (Gulf of), a sound or gulf of North-west Australia, to the S of Montague sound, and N of the embouchure of Prince Regent's river, in S lat. 14° 50′, and E long. 125° 5′. Its entrance, formed by Capes Pond and Hardy, is about 18 m. in width. It receives the waters of two rivers, viz., Roe river at the head, and Harper river on the N, and forms a spacious harbour, distinguished as Prince Frederick's harbour. It is bounded on all sides by precipitous rocks from 100 to 300 ft. in height, and contains numerous islands.

YORK (New), a village in the township of Murton, parish of Tynemouth, Northumberland. Pop. 253.

YORKE, a river of New South Wales, in the district of Liverpool Plains, an affluent of the Turrabeile.

YORKE PENINSULA, an extensive promontory of South Australia, extending between the Gulf of St. Vincent on the E, and Spencer's Gulf on the W, and separated on the S from Kangaroo island by Investigator's strait, and terminating in the SW in Cape Spencer or Berthier. Its shores are generally low and sandy, and, with the exception of the cape already named, and Corny Point and Cape Pierce, forming Hardwick bay on the W, present no important point of outline.

YORKFLEET, a township in the parish of Howden, E. R. of Yorkshire, 6¾ m. SE by E of Howden, at the entrance of the Ouse into the Humber. Area 1,597 acres. Pop. in 1851, 206.

YORK HAVEN, a village of York co., in the state of Pennsylvania, U. S., on the SW bank of the Susquehanna, at the junction of the Conewago, and intersected by the York and Cumberland railroad.

YORK MINSTER, a summit on the S side of Waterman island, Terra-del-Fuego, near the entrance to Christmas sound, in S lat. 55° 24′ 30″, and E long. 70° 1′ 50″.

YORK POINT, a headland of Patagonia, in Magalhaens strait, near a roadstead of the same name, in S lat. 53° 32′ 35″. The roadstead is between Port Gallant and Jerome channel, and off Bachelor's river.

YORKSHIRE, a northern maritime county, by far the largest in England, extending, in its extreme points, between 53° 18′ and 54° 40′ N lat., and between 2° 40′ W and 0° 10′ E long. It is bounded on the N by Westmoreland and Durham, divided from the latter by the river Tees; on the E by the German ocean; on the S by the Humber, dividing it from Lincolnshire, and by the shires of Nottingham, Derby, and Chester; and on the W by Lancashire. The form of the whole county is that of an irregular quadrangle, the longest diagonal extending from NW to SE, about 130 m., and the shortest from SW to NE about 90 m. Square area 3,635,820 acres. From its great extent, this co., exclusive of the city and ainsty of the city of York, has been distributed into three grand districts, called *ridings*, a name derived from a Saxon term signifying *thirds*. The north riding is divided into 11 wapentakes besides Whitby Strand liberty; the east into 7; and the west into 9. There are 613 parishes in the county, 1 archiepiscopal city, 17 parliamentary boroughs, and 59 market-towns. Inhabited houses in 1831, 261,281; in 1841, 307,372; in 1851, 360,188. The pop. of the co., with that of the city in 1801, was 858,892; in 1831, 1,371,359; in 1841, 1,591,584; in 1851, 1,797,995. Mr. Phillips, in an able chapter on the races of men in Yorkshire, says: "If, without regard to any real or supposed evidence of their national origin, we attempt to class the actual pop. of Yorkshire into natural groups, we shall find, independent of Irish emigrants, three main types, frequently distinct, but as often confused by interchange of elementary features. First, Tall, large-boned, muscular persons; visage long, angular; complexion fair, or florid; eyes blue, or grey; hair, light brown, or reddish. Such persons in all parts of the country form a considerable part of the population. In the North riding, from the eastern coast to the western mountains, they are plentiful. Blue-eyed families prevail very much about Lincoln. Second, Person robust; visage oval, full and rounded; nose often slightly aquiline; complexion somewhat embrowned, florid; eyes brown, or grey; hair brown, or reddish. In the West riding, especially in the elevated districts, very powerful men have these characters. Third, Persons of lower stature and smaller proportions; visage short, rounded; complexion unbrowned; eyes very dark, elongated; hair very dark. Such hair and eyes are commonly called black. Individuals having these characters occur in the lower grounds of Yorkshire, as in the valley of the Aire below Leeds, in the vale of the Derwent, and the level regions S of York. They are still more frequent in Nottinghamshire and Leicestershire, and may be said to abound amid the true Anglians of Norfolk and Suffolk. The physical characters here traced cannot be, as Dr. Pritchard conjectures in a parallel case in Germany, the effect of some centuries of residence in towns, for they are spread like an epidemic among the rural and secluded population as much as among the dwellers in towns. Unless we suppose such varieties of appearance to spring up among the blue-eyed races, we must regard them as a legacy from the Roman colonists and the older Britons, amongst whom, as already stated, the Iberian element was conjecturally admitted. Adopting this latter view, there is no difficulty in regard to the other groups. They are of North German or Scandinavian origin, and the men of Yorkshire inherit the physical organization, and retain many of the peculiarities of language of their adventurous sires. In the words employed, in the vowel sounds, the elisions, and the construction of sentences, the Yorkshire dialects offer interesting analogies to the old English of Shakspeare and Chaucer, the Anglo-Saxon of the Chronicles, and the Norse, as it is preserved to us by the Icelanders."

Climate.] The climate of the North riding is various. In the vale of York the air is mild and temperate, except near the moors. The Howardian hills are cold. Ryedale and the marshes on the skirt of the Derwent enjoy a mild but damp air. The great alt. of the eastern moorlands renders their climate extremely cold. The vale of Cleveland has a climate somewhat severe; but the dryness of the soil, and the frequent use of lime, concur to accelerate the harvest. The climate of the western moorlands is colder and more moist than that of the eastern. Some variations also exist in the East riding, it being colder on the eastern than on the western side of the wolds. The levels in the western part of this riding enjoy a mild climate. In the eastern parts of the West riding, towards the banks of the Ouse, the climate, as well as the face of the country, resembles that of the East riding on the opposite side of the river; and damps and fogs are somewhat prevalent. In the middle district the air is sharper, clearer, and more healthful. In the western the climate is cold, tempestuous, and rainy.

Surface.] The geographical features of this co. are strongly marked, and render the whole province one of the most interesting in the kingdom. The North riding is distinguished by a coast strikingly bold and rocky, and presenting cliffs which occasionally rise to an alt. of 500, and even 800 ft. There is an average loss of land, by encroachments of the sea on this coast, of about 2½ yds., or 33 acres annually. A little further inland, successive ranges of hills, rising one above the other, form the elevated tracts of the moorlands, presenting a dreary and desolate aspect of hopeless sterility. Further westward appear the extensive vales of Cleveland and York, and beyond them the western mountains and moorlands. The East riding is the least interesting in the county, and, with the exception of the wolds, displays no prominent variations in the appearance of the country. The West riding is irregular in its surface—so far westward as Sheffield, Bradford, and Oatley, it rises into hills and is beautifully variegated—farther westward still it becomes mountainous and rugged, occasionally intersected with beautiful valleys, and displaying, in Ingleborough, Whernside, Pennigent and Bowfell some of the loftiest passes in England.

Soils and produce.] In the North riding, the soil along the coast is brownish clay and loam: the hills abound with alum-shale. The district of Cleveland has a fertile clay and fine red sandy soil. The vale of York, both in soil and fertility, is very variable. Swaledale, on each side of the river Swale, is extremely fertile. The surface of some of the higher hills in the eastern moorlands is entirely covered with large masses of freestone and extensive morasses, and peat bogs highly dangerous to pass. In the West riding the soils vary from a deep strong clay or loam to the worst peat earth. In the East riding there is a light loam with a mixture of gravel in the sheep districts of the Wolds. The flat country extending between the wolds, the Ouse, and the Humber to Hull, towards the Spurn-head, along the side of the Humber, presents a soil of a strong nature. Throughout the greater part of the North riding, agriculture has, within the last few years, advanced as rapidly as in most parts of the kingdom, considering the circumstances of climate and soil. In the vale of York it is computed that one-third of the ground is in tillage, and two-thirds in grass. The enclosed lands, in many of the dales, are chiefly appropriated to meadow. The lower and better parts of the moors are mostly stinted pastures, on which the cattle are kept in summer. Barley is not much cultivated in the North riding, nor rye, except on poor and sandy soils. Meslin, or a mixture of wheat and rye, is very common, and of this the household bread is made. Ryedale is as remarkable for the culture of oats as Cleveland is for that of wheat. The extent of the woodlands is comparatively small. There is a considerable quantity of timber in the hedge-rows. The oak timber, though not large, is of an excellent quality, being produced slowly on sound, and often rocky ground. The breed of cattle is chiefly the short-horned. The northern part of the vale of York, and of Cleveland, are known by the name of Tees-water breed. This district is supposed to produce the largest cattle in England, and the breed has of late years been greatly improved. The Tees-water sheep are large, coarse boned, and slow feeders; and the wool is harsh and dry. But the stock has been greatly improved by a mixture of the Disley and Northumberland breeds. The native moorland sheep are small. Their wool is worked up into knitted hosiery, for which the dales are celebrated. On the western moors, however, the sheep are chiefly of the small Scotch breed. Yorkshire has long been famed for its horses, and the North riding is particularly distinguished for its breed. The Cleveland horses being cleanly made, strong, and active, are extremely well-adapted to the coach and to the plough; those of the northern part of the vale of York are, by the general introduction of the racing blood, rendered the most valuable breed for the saddle. The southern part of the vale, the Howardine hills, Ryedale, and the marshes, also produce a great number of horses both for the saddle and the coach. The dales of the eastern moorlands rear many horses, which, being of a smaller breed, are too low for the coach, but are a useful and hardy race. Horses also constitute a considerable part of the stock in the higher parts of the western moorlands. They are generally bred between the Scottish Galloways and the country breed, and are a hardy and very strong race, in proportion to their size: these are chiefly sold into the manufacturing parts of the W riding and Lancashire for the ordinary purposes. In the East riding greater improvements have been made in agriculture than in any other portion of the county. In the low grounds adjoining to the river Hull, such improvements have been made by drainage as less than a century ago would have been deemed impossible. On the wolds very great improvements have taken place. The valleys and declivities of the hills now wave with plentiful crops of wheat, and neither servants nor labourers will eat barley bread. The rabbit-warrens are rapidly disappearing, and the breed of sheep has been improved, especially by crosses from the Leicestershire. Enclosures have indeed become almost general. The whole of the level, extending from the foot of the wolds to the western limits of the riding, as well as the rest of the East riding, is now a corn country. Although there are no extensive woods between the wolds and the Ouse, there are perhaps a superabundance of plantations and trees in the hedge-rows of old enclosures. In the West riding almost all the arable land is enclosed with hedges or stone walls; the former in the eastern, the latter in the western parts. A greater part of this riding is exclusively kept in grass. The extensive commons of Hatfield, Stainforth, Thorne, &c., extending to about 12,000 acres, have been enclosed and rendered arable. In the arable lands of the West riding, a greater quantity of wheat is raised than of any other grain. The quantity of oak and ashwood is very considerable, and both meet with a ready market at the shipping and manufacturing towns. A large portion is also used at the mines and collieries. Draining is assiduously attended to in the West riding. In regard to cattle, less attention has been paid to the breeds in this district, than in either of the other two ridings. The horned cattle may be classed under three distinct heads: First, the short-horned or Holderness breed, which prevails in the eastern parts of the riding: second, The long-horned, or Craven breed, which are both bred and fed in the western parts; and, third, A cross-breed between the two above-mentioned. Immense numbers of Scotch cattle, also, are fed in this district, and sold to the butchers. The sheep have been so often crossed, as to confound all distinction. But a mixture of the Leicestershire breed has much improved them of late years.

Minerals, &c.] In the North riding the minerals consist chiefly of the alum mines, on the coast of Whitby, and the lead mines in Swaledale and the neighbouring valleys. Excellent freestone or grit is found in many parts of this riding. A long but narrow ridge, producing lime of a quality peculiarly excellent for agricultural purposes, extends along the edge of the eastern moorlands. Marble of various kinds is found in many parts of the western moorlands, but is converted to no other purposes than those of making lime or repairing roads. Coal is found, but the quantity is small and the quality bad. In the East riding, chalk is the principal fossil substance. The mineral productions of the West riding are of peculiar value, as they create and supply the manufactures of the district. They consist of coal, iron, and stone; and lead is extracted in great abundance from the mines of Grassington and Patelybridge. The West riding, indeed, yields in geological interest to no equal space in the kingdom. In this portion of the island four clearly marked divisions present themselves. The levels on the east rest on the stratum of red sand and clay, with gypsum or alabaster in varying quantity. The magnesian limestone range is one great plain rising from beneath the levels, and terminating toward the west in a regular well-defined edge, forming the partial summit of drainage. In the south is the great Yorkshire and Derbyshire coal field, which rivals, or even surpasses in importance, that of Northumberland. The mining district is, in some parts of the north, exceedingly variable in features, occupying either high or low ground, producing or not producing metallic ores.

Rivers.] See articles TEES and SWALE. The Swale waters the romantic tract called Swaledale; and passing by Richmond and Catterick, enters the Vale of York, and flows in that level country till it receives the Wiske, a small river rising near Osmotherley. It then continues its course till at Myton it joins the URE: which see. The united river then continues its course to about 6 m. below Boroughbridge, where it takes the name of the OUSE: which also see. The Nidd rises in Netherdale, and passes by Knaresborough, to Nun-Monkton, where it joins the Ouse. The Foss rises near Craike-Castle, and joins the Ouse at York. The Wharf rises at the foot of Craven hills, waters the beautiful district of Wharfdale, passing Tadcaster, and joins the Ouse at Nun-Appleton. The Derwent rises in the eastern moorlands, in the North riding, within about 4 m. of the sea; runs at first in a line almost parallel to the coast, by Ayton, to the foot of the Wolds, and afterwards in a westerly direction, till it receives the Rye from Helmsley; then south-westwardly by Malton, and thence by Gate-Helmsley, Stamford-bridge, &c. to the Ouse, near Barmby, from which it is navigable for vessels of 25 tons burthen to Malton. From its junction with the small river Hartford near its source, the Derwent forms the boundary between the North and East ridings, till it approaches near Stamford-bridge, where it enters the East Riding. The AIRE and CALDER are described in separate articles: which see. The Don rises in the western moors beyond Pennistone, flows south-east by Sheffield, north-east by Rotherham, Doncaster, &c., and north by Rawcliffe bridge to Goole, where it falls into the Ouse. In its course it is joined by the Hodbeck, the Wente, and other tributaries, and by several canals to be afterwards noticed. The Ouse is thus the great drain of all Yorkshire. See article HULL.

Canals.] The canals of Yorkshire are chiefly in the West riding.—1. The Barnsley canal; from the Calder, near Wakefield, to the Dearn and Dove canal near Barnsley, a distance of 15 m.—2. The Bradford canal: from Bradford to the Leeds and Liverpool canal at Windhill, 3 m.—3. The Dearn and Dove canal; from the river Don near Mexborough to the Barnsley canal, 9 m.—4 The Huddersfield canal; from Huddersfield to the Manchester and Oldham canal in Lancashire, 19½ m., passing through a tunnel 3½ m. in length.—5. The Leeds and Liverpool canal; beginning at Leeds, and leaving the co. near Colne, in Lancashire.—6. The Calder and Hebble canal, on the line of the river Calder from Wakefield to Halifax.—7. The Ramsden canal; from the Calder and Hebble navigation to the Huddersfield canal, 4 m.—8. The Rochdale canal; entering from Rochdale in Lancashire, and joining the Calder and Hebble navigation near Halifax.—9. The Stainforth and Keadley canal; partly in this co. and partly in the Isle of Axholme, county of Lincoln, branching from the navigation of the Don at Fishlake near Stainforth, and terminating in the Trent at Keadley, 15 m.

Railways.] The North Midland, the York and North Midland, and the Great North of England railways, forming part of the main line from the Midland and more southern parts of England, intersect this co. The North Midland railway enters from Derbyshire by crossing the small river Rother, and runs northwards by Musborough near Rotherham, Kilnhurst, Swinton, Darfield, Cudworth near Barnsley, Oakenham near Wakefield, Altofts near Normanton, Woodlesford, and Leeds. The York and North Midland joins the north Midland line at the Altoft station near Normanton, and runs NE 23¼ m., by Castleford, Burton-Salmon, Milford, Sherburne, Ulleskelfe, and Bolton-Percy, to York. The

Great North of England line joins the York and North Midland at York, and runs NNW by Shepton, Tollerton, Alne, Ruskelt, Sessay, Thirsk, Otterington, Northallerton, Danby, Wiske, East Cowton, and Croft. At Croft it enters the co. of DURHAM: which see. See also CROFT, DARLINGTON, and NORTHUMBERLAND. The Manchester and Leeds line of railway enters this co. at Todmorden, and terminates in the North Midland at Normanton: see article LANCASHIRE. This important line completes the railway communication across the island by its connections with the Leeds and Selby, and the Selby and Hull lines. The Leeds and Selby railway runs from Leeds by Garforth, Mucklefield, Milford, and Hamble, to Selby, crossing the York and North Midland line at Milford, where the general western terminus of the line may be said to be now placed. The Selby and Hull line runs from Selby by Cliff, Howden, Eastrington, Saddlethorpe, Brough, Ferriby, and Hessle, to Hull; a distance of 30¾ m. The Manchester and Sheffield railway enters this co. through a tunnel near Tintwhistle, and runs by Penistone, &c. to Sheffield, as described under article LANCASHIRE: which see. The Sheffield and Rotherham line runs from Sheffield by Holmes and Brightside to Rotherham, a distance of 5½ m., exclusive of a branch of 1½ m. to Greasborough. Leeds and Bradford are connected by railway. A continuation of the Stockton and Darlington to Middleburgh, traverses a portion of the northern boundary of the North riding of this county. See also articles DURHAM and MIDDLEBURGH. The Pickering and Whitby railway runs from Pickering NNE in rather a circuitous course by Newton-dale, &c., to Whitby, a distance of 24 m.

Commercial History.] Respecting the origin and the earlier history of the woollen manufactures of Yorkshire, we must here refer to article HALIFAX. The beginning of the 17th cent. was an important epoch in the history of the woollen manufacture, as till then the English were unable to dye and dress their own cloths, which were, therefore, usually sent to Holland to be dyed and dressed, and returned to England for sale. The introduction of the inventions of Arkwright and Watt towards the end of the 18th cent. and the beginning of the present, formed a still more important epoch, as these inventions were soon applied to the woollen and worsted manufacture, and produced an entire revolution in some of its seats in this county. The woollen trade is carried on chiefly in and around LEEDS and HUDDERSFIELD: which see; the worsted stuff at BRADFORD and HALIFAX; and blankets and carpets at DEWSBURY and HECKMONDWIKE: which also see. In 1838 the number of woollen-mills in this county was 606, employing 18,735 males, and 8,813 females; in 1850 they were 600 in number. The worsted-mills in 1838 were 348, employing 6,992 males, and 19,611 females; in 1850 they were 418. There are also an immense number of hand loom weavers in this trade, as well as in the linen, silk, and cotton manufactures. The number of flax-mills in the co. in 1838 was 91, employing 3,230 males, and 6,414 females; in 1850 the flax-mills were only 60 in number. In 1838 the number of cotton-mills was 169, employing 5,662 males, and 6,774 females; in 1850, the number of cotton-mills was 227, with 1,913,807 spindles, and 8,102 power looms. The next great manufacture in this county is that of cutlery: see SHEFFIELD.

Franchise and Government.] Previous to the passing of the Reform act, two members were returned to parliament for the county, and two each for the boroughs of Aldborough, Beverley, Boroughbridge, Heydon, Kingston-upon-Hull, Knaresborough, New Malton, Northallerton, Pontefract, Richmond, Ripon, Scarborough, Thirsk, and York. Under the Reform act two members are returned for each of the ridings. The boroughs of Aldborough, Boroughbridge, and Heydon, were disfranchised; those of Northallerton and Thirsk were deprived of one member each; Bradford, Halifax, Leeds, and Sheffield, were granted two members each, and Huddersfield, Wakefield, and Whitby, one each. The co. is included in the northern circuit. The assizes are held at York, where there is a county jail. This co. is in the province of York; and until recently was partly in the diocese of Chester, and partly in that of York. The former part consisted of the deaneries of Richmond, Catterick, and Boroughbridge, with part of Kirkby-Lonsdale, all in the archd. of Richmond and dio. of Chester; and the latter of the deaneries of Cleveland, Ridall, Bulmer, Ripon, and Ripon with Massam, all in the archdeaconry of Cleveland; the deaneries of Dickering, Buckrose, Herthill and Hull, and Holderness, all in the archd. of the East riding; and the deaneries of Craven, York Ainsty, York city, Pontefract, and Doncaster, all in the archd. of York, or West riding. By orders in council, however, of recent date, those parts of the co. previously in the dio. of Chester, together with the deaneries of Ripon, Ripon with Massam, Craven, and parts of York Ainsty, and of Pontefract, have been transferred from their respective dioceses in order to form the new diocese of RIPON: see that article.

YORKSHIRE, a township of Cattaraugus co., in the state of New York, U. S., 250 m. W of Albany, drained by Cattaraugus creek and its branches. Pop. in 1840, 1,292; in 1850, 2,010.

YORK SULPHUR-SPRINGS, a village of Adams co., in the state of Pennsylvania, U. S., 20 m. SSW of Harrisburg.

YORKTOWN, a village of Delaware co., in the state of Indiana, U. S., 48 m. NE of Indianapolis, on the Indianapolis and Bellefontaine railroad, and near the W fork of White river. Pop. in 1850, about 300.—Also a township of West Chester co., in the state of New York, 94 m. S of Albany, drained by Croton river and its branches, from which, by means of a dam 250 ft. long and 40 ft. in breadth, is formed Croton Pond, which supplies New York with water. Pop. in 1840, 2,819; in 1850, 2,273.—Also a village and port of York co., in the state of Virginia, on the S side of York river, 11 m. from its mouth and 70 m. ESE of Richmond, in S lat. 37° 13′, and W long. 76° 34′. Pop. in 1840, about 300. It is noted as the place of the surrender of Lord Cornwallis' troops to General Washington, on October 19th, 1781.

YORKVILLE, a village of Upper Canada, 2 m. N of Toronto. Pop. 1,750.—Also the cap. of York district in S Carolina, U. S., 86 m. N of Columbia.

YORKVILLE, a village of New York co., in the state of New York, U. S., on the Harlem railway, 5 m. NNE of the City Hall, New York.—Also a village of York district, in the state of South Carolina, on a branch of Broad river, and 63 m. N by W of Columbus. Pop. in 1840, 400; in 1850, 850.—Also a township of Racine co., in the state of Wisconsin, 78 m. ESE of Madison, drained by Des Plaines river. It has an undulating surface, and is generally fertile. Pop. in 1850, 998.—Also a village of Gibson co., in the state of Tennessee, 143 m. W of Nashville, on a branch of Obion river. Pop. in 1840, about 100.

YORUBA. See YARRIBA.

YOSEPHEN, or JOSEPHEN (SAINT), a village of Switzerland, in the cant. of St. Gall, and bail. of Gossau, on the Sitter. Pop. 1,269.

YOSEPHIDORF, a village of Austria, in Bohemia, in the regency of Bohmisch-Leipa, and bail. of Rumburg. Pop. 269.

YOSEPHIN, a village of Prussia, in the regency and circle of Bromberg. Pop. 212.

YOSEPHOVA, a village of Hungary, in the comitat of Torontal, on the Theiss. Pop. 1,430.

YOSEPHSDORF, a village of Austria, in Bohemia, in the regency of Pardubitz, and bail. of Czaslau. Pop. 340.—Also a village in the regency of Bohmisch-Leipa and bail. of Haida. Pop. 270.—Also a village of Hungary to the NW of Peterwardein, on the Theiss. Pop. 3,050

YOSEPHSLAG, a village of Austria, on the Enns, and bail. of Gmund. Pop. 228.

YOSEPHSRUHE, a village of Prussia, in the regency of Bromberg, and circle of Chodziesen. Pop. 245.

YOSEPHSTADT, a village of Moravia, in the circle of Brunn. Pop. 980. See also JOSEPHSTADT.

YOSEPHSTHAL, a village of Austria, in Bohemia, in the regency of Bohmisch-Leipa, and bail. of Yablonz. Pop. 460. It has a glass-work and several mills.—Also a village of the regency of Budweis, and bail. of Ober-Plau. Pop. 150. It has a glass-work.—Also a village of the Military confines of Croatia, to the S of Ogulin. Pop. 1,150.

YOSIDA, a town of Japan, in the island of Nifon, and prov. of Mikava, 162 m. WSW of Yedo. Pop. 1,350. It has two suburbs, a castle, and extensive iron-works.

YOSSA, or JOSSA, a village of Hesse, in the prov. and circle of Fulda. Pop. 500.

YOSS-EN-YOSSA, a village of Hesse, in the prov. of Hanau, and circle of Schluchtern. Pop. 420.

YOSSERNDORF, or JOSSERNDORF, a village of Bavaria, in the circle of Lower Franconia, and presidial of Ebern. Pop. 260.

YOSSHOFEN, a village of Bavaria, in the circle of Swabia, and presidial of Neuburg. Pop. 212.

YOSTHAL, or JOSTHAL, a village of the grand-duchy of Baden, in the circle of the See, and bail. of Neustadt. Pop. 560.

YOSVAFEO, or JOSVAFEO, a village of Hungary, in the comitat of Torna. Pop. 900.

YOSYVARA, a town of Japan, in the island of Nifon, and prov. of Souronga, 72 m. SW of Yedo, at the mouth of a river. In has an active transit trade in iron-wood. In the vicinity is a harbour formerly a station for the imperial vessels.

YOSZ, or JOSZ, a market-town of Hungary, in the gsp. of Abauj, near Raschau.

YOUDGIA-KIRTA. See JOCJACARTA.

YOUDIA and YONDRA. See SIAM.

YOUDRA-SKAN. See CAMBOGE.

YOUGHAL, a rivulet in the p. of Youghalarra, co. Tipperary, which rises among the Arra mountains, and flows 5 m. NE to a point near the head of the SW side of Youghal bay.

YOUGHAL, or ST. MARY'S OF YOUGHAL, a parish and borough of co. Cork. Area of p. 4,830 acres. Pop. in 1831, 11,327; in 1851, 11,311. The surface extends along the Blackwater and its estuarial expansion, to the middle of the W side of Youghal bay. The inner portion of Youghal bay constitutes the harbour, and comprises the small estuary of the Blackwater. At the entrance of the harbour may be seen the remains of the foundation of a mill standing on a rock, which shows that the ocean has greatly exceeded its limits on this shore.—The burgh of Youghal stands on the estuary of the Blackwater, 8½ m. E by N of Castle-Martyr, 12½ S by E of Lismore, and 23½ NE of Cork. It extends between the W margin of the estuary of the Blackwater, and the E base of a long and steep hill. The principal street is about ¾ m. in length. Vestiges of the old town walls enclose the central district of the town, and an adjacent portion of the hill slopes. The suburbs are large, and extend in every direction up the hill, behind the old town wall, and contain many very miserable cabins. A considerable improvement upon the town has recently been effected, in the re-edification of several old and tottering buildings, the erection of a good inn, the building of one or two good public edifices, the introduction of gas, and the regular cleansing and paving of the streets. The church, situated in the NW part of the town, and usually called the collegiate church, displays an incongruous mixture of parts and tastes, but was in its original condition one of the finest specimens in all Ireland of the pointed architecture now technically known as the decorated English style. The pop. of the burgh in 1841 was 7,372. It sends one member to parliament. Its only manufactures are bricks, cordage, and malt liquors.

YOUGHAL BAY, a bay on the E side of Lough Derg, co. Tipperary, projecting from the main body of the lough at a point E by N of the opposite projection of Scarriff bay. It commences with a breadth of rather more than 1 m., and gradually contracts to a breadth of ⅓ m., and penetrates the land to the extent of 2 m.

YOUGHIOGHENY, a river which has its source in Preston co., in the state of Virginia, U. S., passes through Alleghany co., in the state of Maryland, and enters that of Pennsylvania, between Somerset and Fayette cos. flows N and NW, and enters the Alleghany river 18 m. SE of Pittsburg. It is navigable to Ohio-pyle falls, 30 m. above its embouchure, where it has a perpendicular fall of 20 ft.

YOUGOURTAI. See ERGETU.

YOUITOS, an Indian tribe of Ecuador, who inhabit the territory lying between the Tigre, Napo, and Amazon.

YOUKAHIRS. See JOUKAHIRS.

YOUKOUSKI, a town of Russia in Europe, in the gov. of Vilna, district and 27 m. N of Chavli, and 34 m. SW of Mitau.

YOUL. See AION.

YOULGRAVE, an extensive parish in the co. of Derby, 3 m. SW by W of Bakewell, containing the chapelries of Birchover, Stanton, and Winster, the township of Youlgrave, and the hamlet of Gratton, all in the hund. of High Peak; and the chapelry of Elton, and township of Middleton with Smerril, in the hund. of Wirksworth. Pop. in 1851, 3,764.

YOULTHORPE, a township in the p. of Bishop's Wilton, Yorkshire, 5 m. NW by N of Pocklington. Area 1,350 acres. Pop. in 1831, 106; in 1851, 105.

YOUNG, an island off the NE coast of Australia, in S lat. 12°.

YOUNGAI, a town of the Tennasserim provinces, in the district of Tavoy, 114 m. SSE of Amherst, in N lat. 14° 30′, and E long. 98° 18′.

YOUNG-SESTERS, NEETU, or NEW SESS, a town of Upper Guinea, in Liberia, about 5 m. NW of Trade Town and 80 m. SE of Monrovia, near the mouth of a river of the same name.

YOUNGSTOWN, a village and port of Niagara co., in the state of New York, U. S., on the E side of Niagara river, near its entrance into Lake Ontario, and 276 m. W by N of Albany. It is connected by a ferry with Newark in Canada. Pop. in 1840, 400; in 1850, 550.—Also a township of Mahoning co., in the state of Ohio, 138 m. NE by E of Columbus, drained by Mahoning river and its branches, and near the Pennsylvania and Ohio canal. Pop. in 1840, 654; in 1850, 3,002.—Also a town of Westmoreland co., in the state of Pennsylvania, between the head branches of Loyalhanna, and 127 m. W of Harrisburg. Pop. in 1840, 415; in 1850, about 600.

YOUNGSVILLE, a village of Warren co., in the state of Pennsylvania, U. S., on Broken Straw creek, and 167 m. NW by W of Harrisburg. Pop. in 1840, 200; in 1850, about 400.—Also a village of Tullapoosa co., in the state of Alabama, 139 m. ESE of Tuscaloosa. Pop. in 1840, about 75.

YOUNG WILLIAM'S GROUP, a group of islands in the Pacific, about 60 m. in circumf., and consisting of a number of islands connected by coral reefs. The two southernmost are each about 8 m. in length, and a ½ m. in breadth. The NW point of the group is in N lat. 5° 27′, E long. 153° 24′; the SW, in N lat. 5° 8′, E long. 153° 38′. These islands are covered with cocoa-nut and bread-fruit trees, and are thickly peopled by a light-complexioned able-bodied race.

YOUNG-WOMANSTOWN, a village of Clinton co., in the state of Pennsylvania, U. S., on the N side of the W branch of the Susquehanna, and 85 m. NW by N of Harrisburg. Pop. in 1840, about 40.

YOU-NO-DAKE, or KOKENSAM, a mountain of Japan, in the island of Nifon, and prov. of Bingo. It has several hot springs.

YOUN-ZERAY, a town of Burmah, in the prov. of Pegu, on the Irawady, 18 m. S by W of Prom, in N lat. 18° 31', and E long. 95°.

YOURI. See YAOURI.

YOURIBA. See YARRIBA.

YOURINGI, a town of Japan, in the island of Nifon, and prov. of Sangami, 54 m. WNW of Yedo.

YOUROUNG-KASHI OF KHOTEN, a river of Chinese Turkestan, in the district of Khoten, which has its source in the Karangoui mountains; runs N; divides into 2 branches, one of which passes Khoten on the l., and the other near Youroung-Kash on the r.; reuniting, they pursue their course to the Yarkand, which they join on the r., and after a total course of 300 m. These rivers abound with jasper.—The town of Y. is about 5 m. ESE of Khoten, in N lat. 36° 52', and E long. 80° 50'.

YOUX, a village of France, in the dep. of the Puy-de-Dôme, cant. and 3 m. S of Montaigu. Pop. 500. It has a copper-mine.

YOWEN, a mountain of Australia Felix, in the district of Western Port, near the Avoca river.

YOXFORD, a parish in Suffolk, 23½ m. NE of Ipswich. Area 2,724 acres. Pop. in 1851, 1,272.

YOXHALL, a parish in Stafford, 7½ m. NNE of Lichfield. Area 4,813 acres. Pop. in 1851, 1,496.

YPANE-GNASSU, or YPANE-GNASSU, a river of Paraguay, which has its source in the W side of the cordillera of Anambahy; runs first SSW, then W; and after a course of about 120 m., joins the Paraguay on the l. bank, a little below Villa-Real-de-la-Concepcion.—At some distance from the N bank is a village named Ypane.

YPANE-MINI, a river of Paraguay, which runs SW, and joins the Paraguay on the l. bank, 50 m. below the confluence of the Ypane-Gnassu.

YPEREN. See YPRES.

YPERLEE, a river of Belgium, in the prov. of West Flanders, which has its source in the S part of the dep. of Ypres; flows N past the town of that name, when it becomes navigable near Dixmunde and Nieuport; and 3 m. below the latter throws itself into the North sea. It has a total course of 39 m., and is defended at the mouth by a double flood-gate. Its chief affluent is the Ysser.

YPITA, a town of Paraguay, near the l. bank of the Parana, and 60 m. NNE of Assomcion.

YPORT, a village of France, in the dep. of the Seine-Inferieure, cant. of Focamp, and com. of Criquebeuf-en-Caux. Pop. 1,000.

YPRES, a department, arrondissement, commune, and town of Belgium, in the prov. of West Flanders. Pop. of dep. 15,455.—The town is situated in a fertile plain, on the Yperlee, 30 m. S of Bruges, and 18 m. NNW of Lille, in N lat. 50° 51' 10", and E long. 2° 52' 49". Pop. 13,281. It is well-built, and contains a splendid town-hall, five churches, a bank, a royal college, an hospital, a public library, and an academy of painting. The principal buildings are the cloth and town-hall, both within one large Gothic edifice, founded in 1342, and surmounted by a fine tower, the ancient church of St. Martin, formerly the cathedral, and the castellany, which has a richly ornamented façade. In the 14th century, Y. was noted for its cloth. Lace now forms its staple manufacture; but it has also several spinning-mills, bleacheries, oil-works, dye-works, tanneries, a salt-refinery, soap-works, and manufactories of a variety of linen and woollen fabrics, tape, &c. This town, formerly one of considerable importance, was, previous to the year 800, a mere fortress, which was destroyed by the Normans, but rebuilt in 960 by Baudonin III., count of Flanders. A town gradually rose around it, which was aggrandised by Count Thierry d'Alsace. In 1128 it was taken by Louis VI., king of France; in 1213 by Philip Augustus; in 1297 by Philip le-Bel; and, in 1383, sustained an unsuccessful siege by the Gauls. In 1388, it was enlarged and walled round by Phillipe-le-Hardi, duke of Burgundy. Y. was taken in 1577, and again in 1648, by the Duke-de-Conde; by the archduke Leopold the following year; by Turenne in 1658; and, in 1678, by Louis XIV. By the treaty of Nimeguen, it was ceded to the French, and became one of the strongest places in Flanders. Under the French empire, it was the capital of an arrond. of the dep. of the Lys.

YPSILANTI, a township and village of Washtenan co., in the state of Michigan, U. S., 60 m. SE of Lansing, and intersected by the Michigan Central railroad. Pop. of township, in 1840, 2,419; in 1850, 3,052; and of village about 2,000. The village is pleasantly situated on the Huron river.

YQUILAO, a small island of Chili, to the S of that of Chiloe. It affords good pasturage.

YRABO. See ERABU.

YRANCY. See IRANCY.

YRIEIX (SAINT), or SAINT YRIEIX-DE-LA-PERCHE, an arrondissement, canton, commune, and town of France, in the dep. of the Haute-Vienne. The arrond. comprises an area of 90,809 hect., and contains 4 cant. Pop. in 1831, 41,788; in 1836, 42,260; in 1841, 41,298; in 1846, 44,732. The cant. comprises 4 com. Pop. in 1846, 13,364.—The town is on the Lone, near the source of that river, 24 m. S of Limoges, and near the confines of the departments of the Dordogne and Correze. Pop. in 1789, 6,110; in 1821, 6,413; in 1831, 6,542; in 1841, 7,110; in 1846, 7,715. It contains five churches, one of which is a fine Gothic structure, and an alms-house. It possesses manufactories of linen and woollen fabrics, porcelain and earthenware, colour-mills, tanneries, and several forges and iron-works, and carries on an active trade in kaolin, petunze, and porcelain, the produce of adjacent quarries, leather, hides, hemp, tobacco, cattle, and pigs. This town, which is ill-built, owes its origin to a monastery founded in the locality in the 15th century, by a seigneur of the name of Yrieix, who became its first abbot. During the wars of the league, it sustained a long siege. It is sometimes called Saint Yrieix-la-Montague.

YRIEIX-LA-MONTAGUE (SAINT), a commune of France, in the dep. of the Creuse, and cant. of Felletin, 11 m. SW of Aubusson. Pop. 1,322.

YRIEIX-LE-DEGALET (SAINT), a commune of France, in the dep. of the Correze, and cant. of Egletons, 18 m. NE of Tulle. Pop. 1,001.

YRIEIX-LES-BOIS (SAINT), a commune of France, in the dep. of the Creuse, and cant. of Ahun, 4 m. SE of Gueret. Pop. 1,109.

YRONDE, a commune of France, in the dep. of the Puy-de-Dome, and cant. of Vic-le-Comte, 17 m. SSE of Clermont-Ferrant. Pop. 1,301.

YROUN, a town of Nigritia, in the SE part of Bornu, on the Yambalaroum. It is governed by a kaïd.

YRUELA (LA). See IRUELA.

YSABAL, a town of Nicaragua, on a gentle elevation on the banks of the Golfo-Dolce. It is 90 m. NE of Guatimala. Pop. 1,500.

YSABEL (Santa). See Isabella (Saint).

YSCEIFIOG, a parish in Flintshire, 3 m. SW of Holywell. Area 5,905 acres. Pop. in 1851, 2,397.

YSCLYDACH, a hamlet in the p. of Llywel, co. of Brecon. Pop. in 1831, 362; in 1851, 327.

YSER, a river which has its source in France, in the dep. of the Nord, to the E of St. Omer, flows NE into Belgium, traverses the W part of the prov. of West Flanders, and after a course of about 39 m. joins the Yperlee, on the l. bank, at the foot of Knocke. Its principal affluents are the Peene in France, and in Belgium the Vleter. It is also joined by the Loo and Boesinghe, the Nieuport and the Furnes canals.

YSGWYDDWYN, a hamlet in the p. of Gelligaer, co. of Glamorgan. Pop. in 1851, 112.

YSPYTTY-EVAN, a parish partly in Denbighshire, and partly in Carnarvonshire, 10 m. S by E of Llanrwst. Area 4,768 acres. Pop. in 1851, 892.

YSPYTTY-RHIW-YSTWYTH, a parish in Cardiganshire, 13 m. SE of Aberystwith. Area 5,544 acres. Pop. in 1831, 472; in 1851, 718.

YSSANDON, a commune of France, in the dep. of the Correze, and cant. of Ayen, 12 m. WNW of Brèves. Pop. 1,254.

YSSEL (Alte), or Oude Yssel, a river which has its source in the Prussian prov. of Westphalia, near Borken; runs WNW; enters, near Anholt, the Dutch prov. of Gelderland; passes near Doetinchem; and after a course in a generally WNW direction of about 45 m., joins the Neue-Yssel at Doesburg, whence the united streams take the name of Ober-Yssel.

YSSEL (Neder), or Hollandisch-Yssel, a river of Holland, a branch of the Lech, from which it separates in the S part of the prov. of Utrecht, to the W of Vianen. It passes Ysselstein and Montfoort, enters the S part of the prov. of Holland, bathes Oude water and Gonda, and joins the Maas on the r. at Stormpold, and opposite the island of Ysselmonde, 3 m. below Rotterdam. It has a total course of about 30 m., and is connected by canals with the old Rhine.

YSSEL (Neue), a branch of the Rhine, which occupies the bed of a canal constructed by Drusus, in the Dutch prov. of Gueldres. It separates from the river a little above Arnsheim, flows NE, and joins the Alte-Yssel at Doesburg, whence the united streams take the name of Yssel or Ober-Yssel. It has a course of about 18 m.

YSSEL (Ober), or Yssel, a river of Holland, formed by the junction of the Alte and Neue Yssel, which unite in the prov. of Gelderland at Doesburg; flowing NNE, it passes Zutphen, forms for some distance the line of separation between the provinces of Gelderland and Ober-Yssel; passes Deventer; enters the last-named prov., and to the W of Zwolle, and a little below Kampen, throws itself into the Zuyder-zee, opposite the island of Schokland, and after a course in a generally N direction of about 60 m.

YSSELMONDE, an island of Holland, in the prov. of South Holland, arrond. of Dordrecht, and cant. of Ridderkerk, formed by the two arms of the Maas, which receives opposite this island on the N the Leck and Neder-Yssel, and is 3 m. ESE of Rotterdam. Pop. 1,500. It is 16 m. in length from E to W, and 5 m. in extreme breadth. It contains a village of the same name.

YSSELSTEIN, a town of Holland, in the prov. and arrond. and 7 m. SW of Utrecht, on the l. bank of the Neder-Yssel. Pop. 3,000.

YSSENGEAUX, or Issengeaux, an arrondissement, canton, commune, and town of France, in the dep. of the Haute-Vienne. The arrond. comprises an area of 119,337 hect., and contains 6 cant. Pop. in 1831, 81,664; in 1836, 81,785; in 1841, 82,571; in 1846, 87,079. The cant. comprises 7 com. Pop. in 1831, 18,015; in 1846, 20,374.—The town is 15 m. NE of Puy, on an affluent of the Lignon. Pop. in 1789, 6,240; in 1821, 7,160; in 1831, 7,166; in 1841, 7,408; in 1846, 7,707. It has manufactories of silk, ribbon, lace, cheese, and several tanneries, and carries on an active trade in timber, cattle, and grain.

YSSER. See Isser.

YSSOIRE. See Issoire.

YSTAD, a town of Sweden, in the län and 39 m. ESE of Malmoe on the Baltic, in N lat. 55° 25′ 31″, and E long. 13° 48′ 15″. Pop. 4,108. It is fortified, and has a new port, the old harbour being small and possessing little shelter. It contains two churches. It has manufactories of cloth, and other woollen fabrics, and carries on an active trade in corn, cattle, alum, potash, iron-ware, &c. Y. has regular steam communication with Stralsund.

YSTLAN, a village of Mexico, in the state of Mechoacan, in a widely extending valley, enclosed by low woody hills, and covered in many parts with the efflorescence of the muriate of soda, which forms an important object of local trade. In the valley are several hot springs.

YSTRAD, a hamlet in the p. of Llandingad, Carmarthenshire. Pop. in 1831, 223; in 1851, 175.

YSTRAD-DYFODWG, a parish in Glamorganshire, 9 m. NW of Llantrisaint. Area 24,515 acres. Pop. in 1831, 1,047; in 1851, 1,998.

YSTRAD-FELLTE, a parish in Breconshire, 15 m. SE of Brecon. Area 19,025 acres. Pop. 711.

YSTRAD-GYNLAIS, a parish in Breconshire, 10 m. N of Neath. Area 21,954 acres. Pop. in 1851, 3,758.

YSTRAD-OWAIN, a parish in Glamorganshire, 3 m. NE of Cowbridge. Area 1,494 acres. Pop. in 1831, 233; in 1851, 221.

YTAIRE (Saint), or Ithaire, a commune of France, in the dep. of the Saône-et-Loire, cant. and 5 m. SW of Gengoux-le-Royal. Pop. 1,032.

YTAPNA, a town of Paraguay, on the r. bank of the Parana, 195 m. SSE of Villa-Rica-de-la-Assomcion. This was formerly the only point at which the Brazilians could maintain commercial relations with Paraguay. The environs are noted for their yerva-mete, or tea of Paraguay.

YTATI, a town of La Plata, in the prov. and 36 m. E of Corrientes, to the NW of the lagune of Ybera, and near the l. bank of the Parana.

YTERSEN. See Utersen.

YTHAN, or Ithan, a river of Aberdeenshire, which rises from two springs, called the wells of Ythan, in the uplands of Forgue, about 1 m. W of the boundary of that parish with Auchterless; and flows to the sea a little below the village of Newburgh, in a NE, and then SE direction, a distance of about 31 m. It is navigable for river-craft to Ellon, and for vessels of 150 tons to the distance of about 1 m. from the sea.

YTRAC, a commune of France, in the dep. of Cantal, cant. and 4 m. W of Aurillac, on the l. bank of the Dantre. Pop. 1,742.

YTTERNAES, a good port of Russia in Europe, in the grand-duchy of Finland, on the coast of the island of Aland, in the archipelago of that name off Iomala.

YTTEROEN, a parish of Norway, in the diocese of Drontheim, and bail. of North Drontheim. Pop. 2,090.

YTTRE-DOLMEDAL, a parish of Norway, in

the diocese of Bergen, and bail. of North Bergenshuus. Pop. 2,300.

YTTRE-SULEN, an island of the North sea, off the W coast of Norway, in the diocese of Bergen, and bail. of North Bergenshuus, in N lat. 61°, and E long. 4° 43′.

YTUCALES, a tribe of Indians of Ecuador, who inhabit the district extending between the Amazon on the S, and the Chambira on the N, and Pistaca on the W.

YUAB, or Juab, a county of the state of Utah, U. S., which lies between Toule and Millard counties, and between San Peto co. and the California line. It contains several lakes, of which Nicollet is the largest, and is drained by the Carson and the Nicollet rivers. Its chief town is Nephi.

YUANES, a tribe of Indians of La Plata, who inhabit the E part of the prov. of Tucuman, and S part of that of Chaco.

YUBA, a river of the state of California, formed by the confluence of three streams, in the co. of the same name, and flowing SSW, falls into Feather river. The gold diggings on Y. river are among the most profitable in the state.—Also a county of the same state, drained by Bear and Yuba, and Middle Yuba rivers, Dry, Indiana, Deer, Canon, and Slate creeks, Clark's run, &c., and intersected by the Benecia and Marysville railroad. Its highest summit, Oregon-hill, has an alt. of 2,800 ft. above sea-level. Pop. in 1852, 22,005, of whom 17,299 were whites. The chief towns and settlements are Marysville, the capital, Ousley's Bar, Kennebec, Long's Bar, and Park's Bar.

YUCA, a town of New Granada, in the dep. of Cundinamarca, prov. and 21 m. N of Bogota.

YUCARES, a tribe of Indians of Bolivia, who inhabit the dep. of Cochabamba.

YUCATAN, or Merida, a peninsular region of America, situated between the gulf of Mexico and the bay of Honduras, and between 17° and 22° N lat., and 87° and 94° W long.; having the territory of British Honduras on its SE coast; and the Mexican prov. or state of Tabasco on the W. The gulf of Mexico stretches along its W side, the gulf of Honduras on the E. Its length from N to S is about 250 m.; and its average breadth 200 m. Area 79,500 square miles. The pop. is stated at 668,623 in 1854. The central part of the peninsula is occupied by a ridge of high ground, which becomes gradually lower as it advances to the N; at the northern extremity its elevation is about 3,000 ft., but near Cape Catoche it sinks to a few hundreds. The ridge in the W is skirted by an extensive plain, which towards the N is about 100 m. wide, and becomes narrower towards the S. Its surface, however, is so sandy and arid, that from the bay of Campeaché to Cape Catoche there is not a single spring of fresh water along the coast. To the S of Cape Catoche, on the eastern side of the peninsula, and also on the west coast, to the S of Rio Francisco, near Campeaché, as far as the mouth of the river Usamasinta, the country is undulating and even hilly. The soil, except on the very shores, is less sandy, and the country is chiefly covered with lofty forest trees. The climate is hot and unhealthy along the coasts, but in the interior the atmosphere is said to be salubrious and the heat more tempered. Mr. Ward says, "Yucatan is one of the poorest states in the Mexican federation. On parts of it maize, cotton, rice, tobacco, pepper and the sugar cane are produced; with dyewoods, hides, soap, &c.; but the scarcity of water in the central parts of the peninsula, where not a stream of any kind is known to exist, and the uncertainty of the rainy season, render the crops very variable; and years frequently occur in which the poorer classes are driven to seek a subsistence by collecting roots in the woods, when a great mortality ensues in consequence of their exposure to a very deleterious climate. Y. has no mines. An active intercourse was formerly carried on with Havana, which Y. supplied with Campeaché wood, salt, hides, deer skins, salted meat and the *jenequen*, a plant from which a sort of coarse thread was made and wrought up into sacking, cordage and hammocks. This trade was cut short by the war; and as few foreigners have been induced to settle in Y., the inhabitants have derived but little advantage from the late change of institutions." Mr. W. P. Robertson agrees with Mr. Wood in describing Y. as a poor country. "To Europe," he says, "it only sends dye woods, (Palo de Campéché;) but it supplies Mexico, New Orleans, Belize, and the Havana with a great variety of minor articles, as salt, hemp, hides, bags, cordage, cocoanuts, fruits, and other things; while, for the consumption of the country, the Yucatecos seem to have all the necessaries and some of the luxuries of life. The fisheries are very extensive, and the poorer classes, Creole and Indian, on the coast, chiefly subsist by barter or consumption of the produce of the sea, the fish being at once good and abundant. They have all the tropical fruits, and many peculiar to the Spanish American colonies, as the *chirimoya*, *palta*, and others; while they also produce Indian corn, rice, coffee, sugar, rum, tobacco, &c. Horses, horned cattle, and sheep they possess in moderate abundance; they reckon up 1,388 *haciendas*, something between a farm and an estate, now the one, now the other; and 2,040 *ranchos* and *sitios*, which may be likened to the huts and hovels of the Irish peasantry, without the starvation, or to the holdings of the Highland cotters in Scotland."—It is administratively divided into 5 districts: viz. 1st, Merida, with a pop. of 118,839; 2d, Yzmal with 72,096; 3d, Valladolid, 97,468; 4th, Tekaz, 134,000; 5th, Campeachy, 82,232. The cities called after the above districts have respectively, 1st, the capital Merida, 24,090 inhabitants; 2d, Yzmal, 5,335; 3d, Valladolid, 11,457; 4th, Tekaz, 4,000; 5th, Campeachy, 15,357. There are of secondary towns and villages 253, with populations ranging from 7,551 down to 924. The townspeople of Campeachy claim a much higher pop. than here stated, some calling it 30,000. Bacalar is a town on a small lake near the Rio-Hondo, N of Balize. Considerable quantities of logwood, mahogany, dye-woods, with some valuable drugs and medicinals, form its exports, which are carried down the river to the bay of Honduras, and principally shipped to England and the United States. Vittoria lies on the coast near the SW extremity of the peninsula. Valladolid is an inland town on the N, and has a communication by river to the Caribbean sea.—Yucatan presents many evidences of having been the abode in former times of a race of highly civilized people, but whence they came and whither they have gone will ever remain unknown. Their having existed is proved by numerous ruins of towns and villages, which exhibit much architectural beauty. The Indians of the present day are incapable of such works, nor indeed was their existence known to them previous to the researches of the European and American archæologists, whose works have created such a lively interest. In the eastern part of the state, and particularly to the S of Merida, there are the remains of several ancient stone structures; one of which, called by the natives Uxmal, is still in good preservation; it is about 600 ft. square. The rooms, corridors and pillars are adorned with figures, in half relief, of serpents, lizards, &c., in stucco. There are also figures of men in the attitudes of dancers, and resembling in every respect those which are found in the ruins of Palenque, which proves that the same race has swayed the destinies of all the lower province of Mexico as well as Yucatan. See articles UXMAL and PALENQUE.

History.] The east coast of Yucatan was first seen in 1506, by Juan Diaz-de-Solis and Vicente Munoz-Penzou, who had accompanied Columbus in his last voyage; but the country, properly speaking, was discovered by Francisco Hernandez de Cordova, who landed in 1517, from the Havana, at Cape Catoche, the extreme NE point of the country. He went on to Campeachy, and thence to Champoton, opposed throughout by the Indians, who at last succeeded in beating him and his small force off, as they also did, next year, the expedition of Don Juan-de-Grajalon. This captain had sailed along the coast of Y. till he reached the Mexican prov. of Tabasco. Here the report of the riches of Mexico first met the greedy ears of the Spaniards; and the eventual result was the conquest of Mexico by Hernan Cortes. One of his chief captains, Don Francisco-de-Montejo, in 1526, undertook the conquest of Y.: and he obtained, moreover, from Charles V., a commission, or royal grant, appointing him *adelantado*, or governor, in perpetuity, for himself and his descendants, with various other privileges and immunities accorded to him and his followers. After going through incredible hardships, which were met with unflinching fortitude, and overcome by the prowess which was in these conquests everywhere displayed, the Montejos, first the father, and then the gallant son, succeeded in conquering the indigenous possessors of Y., so that, in 1542, it was raised to a captain-generalship of Spain, independent of Mexico, as well as of Guatimala. On the consummation of the independence of Mexico, Y. became an integral portion of that nation. From the dissolution of the empire, founded by Don Augustin Iturbide, until the formation of the pact which bound together the United Mexican states, in October 1824, Y maintained an independent position, administering her own government, in the meantime, on republican principles. The compact of the confederation having been accepted by her, she became incorporated in the Mexican republic, and so remained until the consolidation of the states by Santa Anna in 1835, when a *quasi* separation took place, Y. standing aloof from all connection with the central power. In 1840 an actual secession was consummated, and Y. was proclaimed an independent republic, and the people soon after formed for themselves a constitution or fundamental law. Mexico in vain endeavoured to persuade or coerce Y. into submission; but finally the new republic wrested from that government a peace on its own terms. This peace was based on the convention of 14th December, 1843, which secured to Y. many advantages she had not before enjoyed, and relieved her from many oppressions from which she had, since her first incorporation, severely suffered. By the conditions of this peace, she became again incorporated with Mexico, and so remained until 1st January 1846, when she again solemnly renounced the connection and declared anew her independence. The chief cause which led to these results was the violation by Santa Anna of the convention of December 1843, causing, by a decree of 21st February 1844, the ports of the republic to be closed against the productions of Y., and subjecting that state to many harassing vexations. Another cause was the refusal of the chamber of deputies to recognise the convention as of any effect. These proceedings, and a renewal of the hardships which she had before endured, roused Y. to throw off the yoke of the supreme government, and to convoke a congress for the purpose of organizing an independent state. Towards the close of 1846, a frightful social war broke out with the Indian part of the pop who had supported the Yucatecos in their struggle with Mexico, but were left unrewarded by the whites. Emboldened by political dissensions between the people of Merida and those of Campeachy, the Indians fell upon the provinces of Tekaz and Valladolid, took the towns, and, with fire and sword, devastated the country. A war was now begun, not only of extermination, but of pitiless cruelty, of savage and ferocious revenge. Men, women, and children were massacred with every variety of ingenious torture. Villages, towns, and cities were burnt or demolished; and the whole north-eastern division of the republic presented one vast scene of devastation or ruin. The troops of the government became panic-stricken, and everywhere dispersed. Terror extended itself on all sides: the people, abandoning their homes and their strongholds, sought for safety in a flight to the sea-coast. Valladolid and Tekaz were abandoned, and gradually the whole interior seemed to be returning to its original owners. While the Indians were thus gradually overpowering the white or Creole population of Y., the people sought for aid from their surrounding neighbours. But both the United States and Mexico were too busy with their own war to embarrass themselves with the troubles of Y. Mexico having at length made peace with the United States, admitted the offer of the commissioners sent to propose the incorporation of Y. with the Mexican nation. Arms and war-munitions, as well as money, were sent, and the whole white population rose up and armed. They attacked the Indians, routed them in successive engagements, dislodged them from their principal points of advance, and drove them towards their fastnesses in the east, bordering on the British settlement of Belize. But the war is by no means ended as yet. An obstinate struggle, on the part of the aborigines, still goes forward; and, with public opinion still divided, Y. is once more an integral part of the Mexican federation.

YU-CHING, a district and town of China, in the prov. of Shan-tung, div. and 30 m. NW of Tse-nan-fu, in N lat. 37° 2′ 30″, and E long. 116° 51′.—Also a district and town of the prov. of Ho-nan, and div. of Kwei-tih-fu, near the l. bank of the Hwang-ho, in N lat. 34° 38′ 35″, and E long. 116° 9′.

YU-CHU, a district and town of China, in the prov. of Ho-nan, and div. of Kae-fung-fu, in N lat. 34° 15′, and E long. 113° 35′.—Also a district and town of the same prov., in the district of Nan-gan-fu, in N lat. 33° 23′, and E long. 113° 1′.

YU-CHU, or WEI-CHU, a district and town of China, in the prov. of Chih-le, and div. of Senen-hwa-fu, in N lat. 39° 50′ 54″, and E long. 114° 36′.

YUDI, or ITIDY, a river of Paraguay, which descends from the mountain-chain by which the prov. is intersected, runs E, and joins the Parana, about 20 m. below Salto-de-Iguazu, on the confines of Brazil.

YUEN-CHING-HEEN, a district and town of China, in the prov. of Chih-le, and div. of Ta-ming-fu.

YUEN-CHU-FU, a division and town of China, in the prov. of Keang-se. The div. comprises 4 districts. The town is 114 m. WSW of Nan-chang-fu, on the Yu-ho, an affluent of the Kan-keang, in N lat. 27° 51′ 32″, and E long. 114° 22′ 26″. Vitriol and alum are found in large quantities in the environs.—Also a division and town in the prov. of Hu-nan. The div. comprises 3 districts. The town is on the Yung-keang, 240 m. WSW of Chan-sha-fu, in N lat. 27° 24′ 30″, and E long. 109° 25′ 10″.

YUEN-GAN-HEEN, a district and town of China, in the prov. of Hu-pih, and div. of King-mun-chu, in N lat. 31° 10′, and E long. 111° 30′.

YUEN-HO, a district and town of China, in the prov. of Keang-su, and div. of Su-chu-fu.

YUEN-KEANG, a division and town of China, in the prov. of Hu-nan, and div. of Chang-tih-fu, near the S bank of lake Tung-ting, 54 m. NW of Chan-sha-fu, in N lat. 28° 45′ 30″, and E long. 112° 13′ 30″.

YUEN-LING-HEEN, a district and town of China, in the prov. of Hu-nan, and div. of Shin-chu-fu.

YUEN-MING-HEEN, a district and town of China, in the prov. of Sze-chuen, and div. of Ning-yuen-fu.—Also an imperial palace in the prov. of Chih-le, 36 m. N of Pe-king, in a magnificent park.

YUEN-MOW-HEEN, a district and town of China, in the prov. of Yun-nan, and div. of Wu-ting-chu, 66 m. NW of Yun-nan-fu, in N lat. 25° 38′, and E long. 102° 3′.

YUEN-SHAN-HEEN, a district and town of China, in the prov. of Keang-se, and div. of Kwang-sin-fu.

YUEN-WU, a district and town of China, in the prov. of Hu-nan, and div. of Hwae-king-fu, in N lat. 35° 6′, and E long. 113° 56′.

YUE-TSUY-TING, a district and town of China, in the prov. of Sze-chuen, and div. of Ning-yuen-fu.

YU-FUNG-SHAN, a mountain of China, in the prov. of Hu-nan, in N lat. 26° 56°, and E long. 109° 32′. It is covered with perpetual snow.

YU-GHIA-LAIDZA-DZOUNG, a town of Tibet, in the prov. of Wei, to the E of Lake Yar-brok-gon-mthso, and 90 m. SSW of Lassa.

YU-HANG, a district and town of China, in the prov. of Che-keang, and div. of Hang-chu-fu, in N lat. 30° 17′, and E long. 119° 52′.

YU-HEEN, a district and town of China, in the prov. of Shen-se, and div. of Se-gan-fu.—Also a district and town of the prov. of Shan-se, and div. of Ping-ting-chu, in N lat. 38° 7′, and E long. 113° 22′.—Also a district and town of the prov. of Chih-le,

78 m. E of Pe-king, in N lat. 39° 56′, and E long. 117° 45′ 40″.

YUH-HWAN-TING, a district and town of China, in the prov. of Che-keang, and div. of Wan-chu-fu.

YUH-LIN-CHU, a division and town of China, in the prov. of Kwan-se. The division comprises four districts. The town is 198 m. SSW of Kwei-lin-fu, in N lat. 22° 41′, and E long. 109° 45′.

YUH-NUN-HEEN, a district and town of China, in the prov. of Kan-suh, and div. of Yan-se-chu.

YUH-PING-HEEN, a district and town of China, in the prov. of Kwei-chu, and div. of Sze-chu-fu.

YUH-SHAN, a district and town of China, in the prov. of Keang-se, and div. of Kwang-sin-fu, 168 m. E of Nan-chang-fu, on the confines of the prov. of Che-keang, in N lat. 28° 45′, and E long. 118° 28′. It is situated on a small stream which is here crossed by a substantial stone bridge, and is enclosed by walls of red sandstone. The houses are tolerably well built. It contains several triumphal arches and a pagoda.

YUH-SHE, a district and town of China, in the prov. of Sze-chuen, and div. of Shun-king-fu, in N lat. 30° 35′, and E long. 106° 26′.

YUH-TSEEN, a district and town of China, in the prov. of Che-keang, and div. of Hang-chu-fu, in N lat. 30° 14′ 27″, and E long. 119° 22′ 57″.

YUIQUIS, a tribe of Indians of Bolivia, who inhabit the dep. of Moxos, between the San-Miguel and the Mamore.

YU-KAN, a district and town of China, in the prov. of Keang-se, and div. of Faou-chu-fu, 51 m. E of Nan-chang-fu, in N lat. 28° 41′, and E long. 116° 40′.

YU-KEANG, a district and town of China, in the prov. of Shan-se, and div. of Pu-chu-fu.

YU-KING-HEEN, a district and town of China, in the prov. of Kwei-chu, and div. of Ping-yin-chu, 93 m. NE of Kwei-yang-fu, in N lat. 27° 6′, and E long. 107° 34′.

YUKON, or KWICHPACK, a river of North America, which rises to the W of the Rocky mountains, not far from the head-branches of the Pelly; flows first N; and then, after receiving a large tributary named the Porcupine, turns W, and falls into Behring's sea, in N lat. 66°, W long. 147° 30′, after a course of above 1,000 m.

YULA, a river of Hindostan, in Kunawur, in the district of Bussahir, which has its source in the Damuk-chu mountains, on the S confines of Ladak, flows S, and joins the Sutlej on the S side, in N lat. 31° 31′, and after a course of about 15 m.

YULANG, a mountain-stream of Hindostan, in Kunawur, in the district of Bussahir, an affluent of the Li, or Spiti, which it joins on the r. bank, in N lat. 31° 55′.

YU-LAY, or EIL-RAY, a town of China, in Great Lu-chu island, in the prov. of Tung shan.

YULDAZ (GREAT), a lofty summit of the Thian-chan, or Teen-shan chain, on the SE confines of the prov. of Ili, between Sungaria, and Eastern Turkestan. It is upwards of 30 m. in circumference. It affords excellent pasturage, and has many fine springs of water.

YULDAZ (LITTLE), a summit of the Chinese empire, in the Teen-shan chain, on the SE confines of Sungaria and NNE of Great Yuldaz. Between these two summits, near the source of the Kaidon, is a village of the same name.

YU-LIN-FU, a division and fortified town of China, in the prov. of Shen-se. The div. comprises 5 districts. The town is situated near the great wall, 324 m. N of Se-gan-fu, in N lat. 38° 18′ 8″, and E long. 109° 22′ 30″.

YU-LING, a chain of mountains in China, which runs from NW to SE, through the S part of the prov. of Yun-nan, and the W part of those of Kwan-se and Kwang-tung, and terminates on the shore of the China sea.

YU-LIN-HEEN, a district and town of China, in the prov. of Shen-se, and div. of Yu-lin-fu.

YUMA. See LONG ISLAND.

YUMADOUNG MOUNTAINS, a range of mountains in India beyond the Ganges, forming a continuation of the great mountain-chain extending from the S part of Assam, parallel to the Irawady, running NNE, from Cape Negrais in the SW of the prov. of Pegu, in N lat. 16° 22′ 20″, and forming a natural barrier between Arracan and Ava. Its summits, which bear different names, vary from 200 to 8,000 ft. in alt. above sea-level. These mountains give rise to several rivers, the principal of which is the Arracan.

YUMBA, a bay of the SE coast of the island of Hayti, between the island of Saona and cape Espada. See also LONG ISLAND.

YUMEN-HEEN, a town of Chinese Turkestan, 60 m. W of Kia-gu, situated in the NW angle of China.

YUMETOS. See JUMENTOS (LOS).

YUNA, a river of Hayti, in the E part of the island, which has its source in the Sierra-Cibao, runs first NE, then E, and after a course of about 75 m. throws itself by several embouchures into the bay of Samana. It traverses, in the lower part, the plains of La Vega. Its principal affluent is the Comon, which it receives on the l.

YU-NA-CHING, or YO-NA-TSIO, a town of the Chinese empire, in Great Lu-chu island, in the prov. of Tung-shan, 11 m. NNE of Zieuly.

YUN-CHING-HEEN, a district and town of China, in the prov. of Shan-ting, and div. of Tsaou-chu-fu, in N lat. 35° 45′, and E long. 116° 14′.

YUNCILLOS, a town of Spain, in the prov. and 9 m. N of Toledo, and partido of Illescas. Pop. 600.

YUNCLER, a town of Spain, in the prov. and 12 m. from Toledo, and partido of Illescas. Pop. 800.

YUNCOS, a town of Spain, in the prov. of Toledo, and partido of Illescas. Pop. 400.

YUNG, a district and town of China, in the prov. of Sze-chuen, and div. of Kea-tsing-fu, in N lat. 29° 30′, and E long. 104° 30′.—Also a district and town in the prov. of Chih-le, and div. of Paou-ting-fu.—Also a district and town of the prov. of Kwan-se, and div. of Wu-chu-fu, in N lat. 22° 53′, and E long. 110° 3′.—Also a district and town of the same prov., in the div. of Lew-chu-fu, in lat. 24° 59′, and E long. 108° 58′.

YUNGAS, or YUNCAS, a tribe of Indians of Bolivia, who inhabit the district of La Paz.

YUNG-CHANG-FU, a division and town of China, in the prov. of Yun-nan. The div. comprises four districts. The town is in a mountainous locality, 225 m. WSW of Yun-nan-fu, in N lat. 25° 4′ 40″, and E long. 99° 25′ 55″. It is very populous, and carries on an active trade. The surrounding territory affords in considerable quantities gold, marble, amber, honey, wax, flax, and silk. This town is supposed to be the Unshiam of Marco Paulo.

YUNG-CHANG-HEEN, a district and town of China, in the prov. of Kan-suh, and div. of Leang-chu-fu, in N lat. 38° 20′, and E long. 102° 10′.—Also a district and town in the prov. of Sze-chuen, and div. of Chun-king-fu, in N lat. 29° 34′, and E long. 105° 38′.

YUNG-CHING, a district and town of China, in the prov. of Chih-le, and div. of Paou-ting-fu, in N lat. 39° 4′, and E long. 115° 58′.—Also a district and town in the prov. of Ho-nan, and div. of Kwei-tih-fu, in N lat. 34° 5′, and E long. 116° 27′.

YUNG-CHING-HEEN, a district and town of China, in the prov. of Shan-tung, and div. of Tsaou-chu-fu, in N lat. 35° 45′, and E long. 116° 15′.

YUNG-CHU, a district and town of China, in the prov. of Yun-nan, and div. of Shun-ning-fu, in N lat. 24° 25′, and E long. 100° 20′.—Also a district and town in the prov. of Shen-se, div. and 66 m. NW of Sin-gan-fu, in N lat. 34° 48′, and E long. 108° 3′.—Also two islands of the roadstead of Macao, in N lat. 22° 11′, and E long. 113° 40′.

YUNG-CHUEN, a district and town of China, in the prov. of Sze-chuen, and div. of Chun-king-fu, in N lat. 29° 30′, and E long. 106°.

YUNG-CHU-FU, a division and town of China, in the prov. of Hu-nan. The div. comprises eight districts. The town is 180 m. SSW of Chan-sha-fu, on the r. bank of the Tu-keang, near the confluence of the Seang-keang, in N lat. 26° 8′ 24″, and E long. 111° 34′ 50″.

YUNG-CHUN, a district and town of China, in the prov. of Kwang-ping, and div. of Chaou-king-fu.—Also a district and town in the prov. of Kwan-se, and div. of Nan-ning-fu, in N lat. 22° 41′, and E long. 108° 33′.

YUNG-CHUN-CHU, a division and town of China, in the prov. of Fo-keen. The div. comprises two districts. The town is 57 m. SW of Fuh-chu-fu, in N lat. 25°, and E long. 118° 28′.

YUNG-FUH, a district and town of China, in the prov. Fo-keen, div. and 36 m. SW of Fuh-chu-fu, in N lat. 25° 46′ 48″, and E long. 118° 1′ 50″.—Also a district and town in the prov. of Kwan-se, and div. of Kwei-lin-fu, in N lat. 24° 58′, and E long. 109° 51′.

YUNG-FUNG, a district and town of China, in the prov. of Keang-se, and div. of Kein-gan-fu, in N lat. 27° 25′, and E long. 115° 20′.—Also a district and town of the same prov., in the div. of Kwang-sin-fu, in N lat. 28° 31′, and E long. 118° 20′.

YUNG-GAN, a district and town of China, in the prov. of Kwang-tung, and div. of Hwuy-chu-fu, in N lat. 23° 42′, and E long. 115° 8′.—Also a district and town of the prov. of Fo-keen, and div. of Yen-ping-fu, in N lat. 25° 54′, and E long. 117° 33′.—Also a district and town of the prov. of Kwei-chu, and div. of Ping-yue-chu.

YUNG-GAN-CHU, a district and town of China, in the prov. of Kwan-se, and div. of Ping-lo-fu, in N lat. 24° 1′ 12″, and E long. 110° 18′ 10″.

YUNG-HING-HEEN, a district and town of China, in the prov. of Hu-nan, and div. of Lin-chu-fu, 165 m. S of Chan-sha-fu, in N lat. 26° 4′ 48″, and E long. 112° 44′ 51″.

YUNG-HO-HEEN, a district and town of China, in the prov. of Shan-se, and div. of Pu-chu-fu, on the l. bank of the Hwan-ho, 225 m. SW of Tae-yuen-fu, in N lat. 35° 23′, and E long. 110° 25′.

YUNG-KANG, a district and town of China, in the prov. of Che-keang, and div. of Kin-wha-fu, in N lat. 28° 58′, and E long. 120° 11′ 45″.

YUNG-KANG-CHU, a district and town of China, in the prov. of Kwan-se, and div. of Tae-ping-fu, in N lat. 22° 56′, and E long. 107° 36′.

YUNG-KEEN, a district and town of China, in the prov. of Che-keang, and div. of Wan-chu-fu.

YUNG-KHE-SHAN, a mountain of China, in the prov. of Yun-nan, and dep. of Keo-tsing, in N lat. 25° 24′, and E long. 103° 9′. It is covered with perpetual snow.

YUNG-KING-HEEN, a district and town of China, in the prov. of Sze-chuen, and div. of Ya-chu-fu, in N lat. 29° 52′, and E long. 102° 55′.

YUNG-KO-TSIN [*i. e.* Parroquet's Beak], the S point of the island of Hainan, in the Chinese prov. of Yuang-tung, at the extremity of the E cape of the bay of Yu-lin-keang, in N lat. 18° 9′ 35″, and E long. 109° 34′ 15″.

YUNG-LEEN, a district and town of China, in the prov. of Sze-chuen, and div. of Sin-chu-fu, in N lat. 28° 6′, and E long. 104° 20′.

YUNG-MING, a district and town of China, in the prov. of Hu-nan, and div. of Yung-chu-fu, in N lat. 25° 20′, and E long. 111° 15′.—Also a district and town of China, in the prov. of Keang-se, and div. of Keih-yan-fu, in N lat. 26° 45′, and E long. 114° 2′.

YUNG-NEEN-HEEN, a district and town of China, in the prov. of Chih-le, and div. of Kwang-ping-fu.

YUNG-NING, a district and town of the prov. of Ho-nan, and div. of Ho-nan-fu, in N lat. 34° 22′, and E long. 111° 43′.—Also a district and town of the prov. of Chih-le, and div. of Yen-ping-chu.—Also a town of Tibet, in the prov. of Kam, near the confines of the Chinese prov. of Yun-nan, 231 m. NW of Yun-nan-fu, in N lat. 27° 48′ 28″, and E long. 100° 46′ 10″. To the E of this town is a fine lake containing four islands.

YUNG-NING-CHU, a district and town of China, in the prov. of Shan-se, and div. of Fun-chu-fu, in N lat. 33° 33′ 36″, and E long. 111° 4′.—Also a district and town in the prov. of Kwan-se, and div. of Kwei-lin-fu, in N lat. 25° 7′ 12″, and E long. 109° 36′ 10″.—Also a district and town of the prov. of Kwei-chu, and div. of Gan-shun-fu, in N lat. 27° 52′ 48″, and E long. 105° 23′ 20″.

YUNG-NING-HEEN, a district and town of China, in the prov. of Sze-chuen, and div. of Sen-yung-ting.

YUNG-NING-KEEN, a town of China, in the prov. of Shing-king, on the E side of the gulf of Leao-tung, 165 m. SSW of Fung-thian.

YUNG-PI, a district and town of China, in the prov. of Yun-nan, in a mountainous locality, on the Koen-tin-ho, 165 m. NW of Yun-nan-fu, in N lat. 26° 42′, and E long. 100° 58′ 10″.

YUNG-PING, a district and town of China, in the prov. of Yun-nan, and div. of Yun-chang-fu.—Also a district and town of the same prov., in the div. of Yung-chang-fu, in N lat. 25° 30′, and E long. 99° 48′.

YUNG-PING-FU, a division and town of China, in the prov. of Chih-le. The div. comprises seven districts. The town is on the Lan-ho, 141 m. E of Pe-king, in N lat. 39° 56′ 10″, and E long. 118° 53′ 58″. It has manufactories of paper, and carries on an active trade. Tin is found in the adjacent mountains.

YUNG-SHEN, a district and town of China, in the prov. of Yun-nan, and div. of Tung-chuen-fu.

YUNG-SHOW-HEEN, a district and town of China, in the prov. of Shen-se, and div. of Kan-chu.

YUNG-SHUN-FU, a division and town of China, in the prov. of Hu-nan. The div. comprises four districts. The town is in N lat. 28° 58′, and E long. 109° 51′.

YUNG-SHUN-HEEN, a district and town of China, in the prov. of Hu-nan, and div. of Yung-shun-fu.

YUNG-SIN, a district and town of China, in the prov. of Keang-se, and div. of Keih-gan-fu, 174 m. SW of Nan-chang-fu, in N lat. 26° 55′, and E long. 113° 7′.

YUNG-SUH, a district and town of China, in the prov. of Kwan-se, and div. of Kwei-lin-fu.

YUNG-TING, a district and town of China, in the prov. of Hu-nan, and div. of Fung-chu, 165 m. WSW of Chan-sha-fu.—Also a district and town of the prov. of Fo-keen, and div. of Ting-chu-fu, 192 m. SW of Fuh-chu-fu, in N lat. 24° 44′ 54″, and E long. 116° 52′ 30″.—Also a town of the Corea, in the prov. of Ping-gan, 24 m. SW of Lisan.—Also an island near the S coast of the Corea, in the prov. of Tsuen-lo, opposite Yo-ping.

YUNG-TSE, a district and town of China, in the prov. of Ho-nan, div. and 51 m. W of Kae-fung-fu, near the r. bank of the Hwang-ho, in N lat. 34° 56′ 40″, and E long. 113° 44′.

YUNG-TSING, a district and town of China, in the prov. of Chih-le, and div. of Shun-teen-fu, on the r. bank of the Hwan-ho, 42 m. S of Pe-king, in N lat. 39° 20′, and E long. 116° 35′.

YUNG-TSUNG, a district and town of China, in the prov. of Kwei-chu, and div. of Le-ping-chu.

YUNG-YANG, a district and town of China, in the prov. of Ho-nan, div. and 60 m. W of Kae-fung-fu, in N lat. 34° 52′ 40″, and E long. 113° 32′ 30″.—Also a district and town of the prov. of Fo-keen, and div. of Yen-ping-fu.

YUN-HEEN, a district and town of China, in the prov. of Hu-pih, and div. of Yun-yang-fu.

YUN-LEANG-HO, or YUN-HO. See YU-HO.

YUN-LING, a chain of mountains in the central part of China. They extend into the E part of Tibet, and the country of Koko-nor, and are supposed to form a part of the great central chain of Asia.

YUN-LUNG-CHU, a district and town of China, in the prov. of Yun-nan, div. and 48 m. WNW of Ta-le-fu, on the l. bank of the Lan-thsang-ho, in N lat. 25° 54′, and E long. 99° 36′.

YUN-MANG, a district and town of China, in the prov. of Hu-pih, and div. of Tih-yan-fu, 48 m. NW of Wu-chang-fu, in N lat. 31° 5′, and E long. 113° 41′.

YUN-NAN, a province of China, bounded on the N by Tibet and the Chinese prov. of Sech-wen; on the NE by Sech-wen and Kwei-chu; on the E by Kwei-chu and Quang-si; on the S by Tonquin, Laos, and the Birman territories; and on the W by the Birman territories and the Bong country. It lies between the parallels of 22° and 28° N; and the meridians of 97° and 106° E; and has an area of 110,000 sq. m. Its surface is that of a highly rugged table-land, sloping from N to S; in the N rising in the Yung-ling, into very elevated snowy peaks; but on the S sinking into gently undulated level tracts, approaching in character to those which border the gulfs of Siam and Bengal. The Yang-tse-keang skirts a portion of the northern frontier in a SE course; crosses a projecting portion of this frontier; forms for some distance the common boundary with Sech-wen; and, a little to the E of the meridian of 102°, bends northwards, and leaves the prov. of Y. The Hong-keang rises to the S of the cap. of the prov., and running E, enters the prov. of Quang-si; the S and W portions of the prov. are drained by the head-waters of the great river of Tonquin, by the Meinam-kong, the Menam, the Thaleyn, and some affluents of the Irawady. High mountain-ridges separate the head-waters of these streams from each other. In the E part of the prov. and to the S of the capital, is the large lake called Sien, a feeder of the Yang-tse-keang; and to the S of it, and separated by a high ridge of mountains, lies the Serin lake, which sends its waters towards the Hong-keang. In the NW is Lake Tchao, a feeder of the Meinam-kong. The climate is rude, and there is comparatively but a small portion of the surface under cultivation; but this prov. is reputed to be rich in minerals, and gold, silver, tin and copper are wrought. The pop. has been estimated at 5,500,000, of which a large proportion appear to be affiliated to the Burmese.—The cap., which gives name to the prov., lies at the N extremity of Lake Chin, in N lat. 25° 6′, W long. 13° 36′. It is said to be a large place, and to have considerable manufactures of carpets, woollens, and silk goods; but the chief trade with Burmah and India is conducted between the towns of Tsan-tah in the SW of the prov., and Bhaumt on the Irawady.—Also a district and town in the same prov., in the div. of Ta-le-fu, 114 m. WNW of Yun-nan-fu.

YUN-NAN-FU, a division and town of China, in the prov. of Yun-nan. The div. comprises nine districts. The town is on the N bank of a lake, in N lat. 25° 6′, and E long. 102° 51′ 40″, 1,425 m. SW of Pe-king, and 720 m. WNW of Canton. It is intersected by canals, and carries on an active trade, especially in metals. It is noted for its carpets, and for a species of silk fabrics named touhaitoanese. It formerly contained many fine edifices, most of which have been destroyed during the Tartar invasions. In the environs are quarries of azure stones and of marble. The lake of Y. is 30 m. in length from N to S. and 9 m. in extreme breadth.

YUN-PIN-OUSHI, or OUSHI, a canton and town of Chinese Turkestan, in the prov. of Sin-keang. The cant. is mountainous, extending N to the glaciers of the Teën-shan chain, and is intersected with fertile valleys, and extensive plains covered with reeds. Its chief rivers are the Aortaksi and Kumarih, and is frequently visited by the nomade Khirgese. The town is 69 m. W of Aksou, and 300 m. NW of Koutche, in N lat. 49° 9′, and E long. 78° 27′, on an inosculation of the Aortaksi. A rate of 10 per cent. is levied upon all merchandise brought by strangers to this town. It was formerly a flourishing place.

YUNQUERA, a town of Spain, in Andalusia, in the prov. and 21 m. WNW of Maley, and partido of Ronda, in a fine plain. Pop. 2,726. It has a parish church, an hospital, several convents, a customhouse, and a public granary, and possesses manufactories of cloth, numerous fulling-mills, and distilleries of brandy. Metals of different kinds, and jasper are found in the surrounding mountains.—Also a town of the prov., partido, and 6 m. N of Guadalajara, in a fine plain. Pop. 983. It has a parish church, an hospital, a custom-house, and a public granary.

YUN-SE, a division and town of China, in the prov. of Hu-pih, and div. of Yun-yang-fu, in N lat. 32° 56′, and E long. 110° 23′.

YUN-SHAN, a town of the Corea, in the prov. of Hwang-hai, 51 m. NE of Han-chu.

YUNTA (LA), a town of Spain, in the prov. of Zaragoza, 27 m. S of Calatayud, on the Mesa. Pop. 550.

YUN-TAE-SHAN, an island of China, off the coast of the prov. of Keang-su, opposite Hae. It is well-cultivated, and contains the small town of Hin keou ching.

YUN-YANG, a district and town of China, in the prov. of Sze-chuen, and div. of Keu-chu-fu, in N lat. 31° 5′, and E long. 109° 6′.

YUN-YANG-FU, a division and town of China, in the prov. of Hu-pih. The div. comprises six districts. The town is on the l. bank of the Han-keang, in an extensive plain, 255 m. NW of Wu-chang-fu, in N lat. 32° 49′ 20″, and E long. 110° 51′ 41″.

YUPATA. See UPATA.

YUPI, a people of Chinese Tartary, who inhabit the banks of the Oussouri, in the SE part of Mandshuria. They are pacific in character, but lazy, stupid, and without any species of religious worship. The country is covered with impenetrable forests. They subsist chiefly on fish, and use the skins, skilfully prepared and sewed together for the purpose, as garments. They form boats of the bark of trees, and employ dogs in their sledges.

YURIRAPUNDARO, a village of Mexico, in the

state of Guanaxuato, containing about 500 Indian families. It has a fine convent.

YURRE, a town of Spain, in the prov. of Vizcaya, and partido of Durango. Pop. 884. It has a parish church.

YURUARI, a river of Venezuela, a branch of the Cuyuni, flowing between the Sierra-Usupama and the Sierra-Imataca, and communicating with the Caroni in the rainy season.

YURUBAXA, a river of Brazil, in the prov. of Para, forming a branch of the Rio-Negro, from which it separates 75 m. above Thomar, runs SSW, forms Lake Maraho, and after a course of about 150 m., joins the Japura on the l. bank.

YU-SHAY-HEEN, a district and town of China, in the prov. of Shan-se, and div. of Leaou-chu, in N lat. 37° 4', and E long. 113° 1'.

YU-SHE, a district and town of China, in the prov. of Ho-nan, and div. of Kae-fung-fu, in N lat. 34° 30', and E long. 114° 23'.

YUSTE, a monastery of Spain, delightfully situated in a finely wooded valley, 2 leagues W of Xarandilla in Estremadura, celebrated as the spot to which the emperor Charles retired in February 1557, and where, in the month of September of the following year, he breathed his last.

YUSU, a village of Hindostan, in Kunawur, in the district of Bussapir, near a pass of the Himalaya chain of the same name. This pass, in N lat. 31° 24', and E long. 78° 9', is 15,877 ft. above sea-level.

YU-SZE, a district and town of China, in the prov. of Shan-se, and div. of Tae-yuen-fu, in N lat. 37° 42', and E long. 112° 44'.

YU-TAE-HEEN, a district and town of China, in the prov. of Shan-tung, and div. of Tse-ning-chu, 126 m. S of Tse-nan-fu, and to the W of Lake Tu-shan-hu, in N lat. 35° 7' 21", and E long. 116° 46' 30".

YUTAS-TABEQUACHIS, a tribe of Indians of Mexico, who inhabit the district in the vicinity of the source of the Colorado.

YU-TEEN-HEEN, a district and town of China, in the prov. of Chih-le, and div. of Tsun-hwa-chu, in N lat. 39° 56' 10", and E long. 117° 46' 40".

YU-TSEEN, a district and town of China, in the prov. of Che-keang, div. and 51 m. W of Hang-chu-fu, in N lat. 30° 14' 27", and E long. 119° 21' 57".

YU-TU-HEEN, or Heu-tu-heen, a district and town of China, in the prov. of Keang-se, and div. of Kan-chu-fu, in N lat. 26°, and E long. 115° 21'.

YUTZ-BASSE, a commune of France, in the dep. of the Moselle, cant. and ¾ of a mile E of Thionville, on the r. bank of the Moselle. Pop. 1,337. It has an oil-mill, a distillery, manufactories of beet-root sugar, and tile-kilns.

YU-YAOU-HEEN, a district and town of China, in the prov. of Che-keang, and div. of Shaou-hing-fu, in N lat. 30° 7', and E long. 121° 3'.

YVECRIQUE, a village of France, in the dep. of the Seine-Inferieure, cant. and 2 m. S of Dondeville, 7 m. NNE of Yvetot. Pop. 1,000.

YVERDON, or Yferten, a district and town of Switzerland, in the cant. of Vaud. Pop. of district, 9,000. The town is 20 m. N of Lausanne, and 53 m. NE of Geneva, on an island of the Thiele, a little above the entrance of that river into the lake of Neuchâtel. Pop. 3,461. It has several suburbs with which it is connected by bridges, and on the lake is a fine port. The town, which is generally well-built, consists of three principal parallel streets, terminating in a handsome square, lined with fine edifices, amongst others the castle built in the 12th century by Conrad of Zähringen, and in which in 1805 the celebrated Institute of Pestalozzi was established, the church and town house, all of modern erection. It has besides a college, an hospital, a public library, and several benevolent institutions. Y. possesses an active trade, and is an entrepot for the wines brought by the canal of Entreroches, and destined for the German Swiss market. The environs are adorned with numerous villas, and extending along the banks of the lake is a fine promenade. In the vicinity are several mineral baths. This town, which is the *Ebrodunum* of the Romans, belonged successively to the kings of Burgundy and dukes of Zähringen. In 1259 it was taken by Peter of Savoy, and in 1473 by the Swiss, but three years after was recovered by the former, and retained till 1536, when it again fell into the hands of the Bernese. It presents, in the number of the ruins which are still scattered around it, evidence of its former extent and importance.

YVES-GOMEZEE, a department, commune, and town of Belgium, in the prov. of Namur, and arrond. of Dinant. Pop. 1,552.

YVETOT, an arrondissement, canton, commune, and town of France, in the dep. of the Seine-Inferieure. The arrond. comprises an area of 115,567 hect., and contains 10 cant. Pop. in 1831, 138,429; in 1841, 142,349; in 1846, 141,412. The cant. comprises 11 com. Pop. in 1831, 18,508; in 1846, 18,574. The town is 21 m. NW of Rouen. Pop. in 1789, 8,294; in 1821, 9,758; in 1831, 9,021; in 1841, 9,083; in 1846, 9,183. The houses are generally old, and surrounded with gardens. There is a fine promenade called L'Etoile. From the situation of the town on a plateau, its only supply of water is by means of wells and cisterns. It has an alms-house and a prison, and has manufactories of woollen, cotton and linen fabrics, velvet, hosiery, tape, &c. Grain and wool are its chief articles of trade. The lords of Y. formerly bore the title of kings, and had, it is said, the privilege of using their own coin.—Also a commune in the dep. of the Manche, cant. and 2 m. SW of Valognes. Pop. 1,235.

YVETTI, a river of France, in the dep. of the Seine-et-Oise, which has its source to the NE of Rambouillet, in the cant. of Essarts, passes Chevreuse, waters afterwards the S part of the arrond. of Versailles, bathing Orsay and Palaiseau, enters that of Corbeil, and 9 m. NW of the town of that name, joins the Orge on the l. bank, and after a course in a generally E direction of about 39 m. On its banks are quarries of excellent paving stone.

YVI (Saint), a commune of France, in the dep. of the Finistere, cant. of Rosporden, 12 m. SE of Quimper. Pop. 1,169.

YVIAS, a commune of France, in the dep. of the Côtes-du-Nord, cant. and 5 m. of Paimpol. Pop. in 1846, 2,400.

YVIERS, a commune of France, in the dep. of the Charente, and cant. of Chalais, 20 m. SE of Barbezieux. Pop. 1,051. It has a tannery.

YVIGNAC, a commune of France, in the dep. of the Côtes-du-Nord, cant. and 5 m. NE of Broons. Pop. 1,784.

YVOIR, a department and commune of France, in the prov. of Namur, and arrond. of Denant, watered by the Boue. Pop. of dep., 673; and of com., 254. In the locality are extensive iron-works.

YVOIRE, a town of Sardinia, in the prov. of Chablais, 14 m. W of Thonon, on the S side of the lake of Geneva.

YVONAND, a fine village of Switzerland, in the cant. of Vaud, district and 5 m. ENE of Yverdun, and circle of Molondin, near the r. bank of the Mantua, a little above its entrance into the lake of Neuchâtel.

YVORNE, a small village of Switzerland, in the cant. of Vaud, district and circle and 1½ m. N of Aigle. It was in 1584 overwhelmed by the fall of an adjacent mountain.

YVOY. See CARIGNAN.

YVOY-LE-PRE. See IVOY-LE-PRE.

YVOY-LE-POLIN, a commune of France, in the dep. of the Sarthe, and cant. of Pontvallain, 15 m. ENE of La Fleche.

YVRE-L'EVEQUE, a commune of France, in the dep. of the Sarthe, cant. and 4 miles E of Le Mans, on the r. bank of the Huisne. Pop. in 1846, 2,192.

YZENDYKE, a fortified town of Holland, in the prov. of Zeland, 14 m. S of Middelburg, in the island of Cassandria. Pop. 1,150.

YZER, a commune of Belgium, in the prov. of Brabant, and dep. of Duysburg. Pop. 327.

YZER-BERGEN [i. e. mountains of iron], a range of mountains in the island of Celebes, which extend along the E coast, and terminate in Cape Tulabo.

YZERNAY, a commune of France, in the dep. of the Maine-et-Loire, cant. and 10 m. E of Chollet. Pop. 1,601.

YZEURE, a commune of France, in the dep. of the Allier, cant. and 2 m. NE of Moulins-sur-Allier. Pop. in 1846, 2,610.

YZEURES, a commune of France, in the dep. of the Indre-et-Loire, cant. and 6 m. SSW of Preuilly, on the r. bank of the Creuse. Pop. 1,690.

Z

*** For names beginning with Z not found amongst the following, see under C, J, or S.

ZAAB. See ZAB.

ZAADDIK, a town of Holland, in the prov. of North Holland, and arrond. of Amsterdam, 2 m. N of Saardam, on the Zaan. Pop. 1,800. It has numerous paper-mills.

ZAAHN, a village of Prussia, in the regency of Liegnitz, and circle of Grunberg. Pop. 235.

ZAAN, a river of Holland, in the prov. of North Holland, which issues from the Schermer-meer, runs S, and after a course of about 12 m., throws itself into the Y, on the N bank, near Zaandam.

ZAANDAM, ZAANREDEM, SAARDAM, or SARDAM, a town of Holland, in the prov. of North Holland, arrond. and 6 m. NW of Amsterdam, on the Zaan, near its confluence with the Y. Pop. 11,000. It consists of two parts separated by the river, and distinguished as Oost Z. and West Z. It is an "eminently Dutch town," and has a generally picturesque appearance, the houses, which are nearly all of wood, being fancifully painted, each having a garden attached, enclosed by a ditch. It contains about 1,000 mills, comprising manufactories of paper, sail-cloth, cinnabar, mustard, powder, pitch, starch, glue, saw-mills, &c. It has a port, and extensive building-docks, noted in the history of Peter the Great, who, in 1696, here wrought, under the name of Peter Mikhailov, as a common ship-wright.

ZAAP, a market-town of Austria, in Bohemia, in the regency of Prague, and bail. of Carolinenthal. Pop. 480.

ZAASCH, a village of Prussia, in the regency of Merseburg, and circle of Delitzsch. Pop. 350.

ZAATSHA, an oasis and village of Algeria, in the Zab-Daari, or Northern Zab, about 20 m. from Biskara. It is a fortified village of about 150 inhabitants. It is of solid construction, surrounded by thick and high walls, and a deep moat, which offered formidable resistance to the French troops under General Herbillon. It is built in the centre of large gardens which contained 17,500 palm-trees and 1,500 fruit-trees, each standing 20 ft. from the other; the different enclosures were separated by tolerably high walls, and by wide drains for irrigation; moreover, its foundations are upon very solid ground, and, however slightly the ground may be excavated, water immediately shows itself in great abundance. Two oases—Lichana, in the SW, an oasis with 50,000 palm-trees, and a village containing 1,800 inhabitants; and Zoonia, with about 5,000 trees, are united to Z. by their gardens. When attacked by General Herbillon's troops in 1852, every night new defenders from Lichana and Zoonia penetrated into this place, while profiting by the obstacles and shelter which the peculiar nature of the terrain affords, other parties kept up a murderous fire on the troops. Behind their ambuscades, these marksmen decimated the French soldiers, without their ever being able to perceive from whence the ball proceeded which proved fatal to them. Siege operations were undertaken; but the rising waters retarded their execution; then Zoonia and its palm-trees were destroyed; then an endeavour was made to isolate Z. from the gardens of Lichana by means of wide ditches: but it soon became clear that to isolate this vast triangle, would require a force of 50,000 men. The works advanced but slowly and with difficulty. Zoonia was destroyed, and Lichana then became the sole rallying-point for the allies, who hastened from all parts to the support of the defenders of Z. Their numbers were supposed to be not less than 4,000. Ultimately, after immense exertions, the fortress was taken.

ZAATZKE, a village of Prussia, in the regency of Potsdam, and circle of Ost-Priegnitz. Pop. 300.

ZAB, a district of Algiers, in the S part of the gov. of Algiers, to the S of Bescara. It is watered by the Jiddi, and comprises ten oases or fertile and cultivated spots.

ZAB, ZARB-SOUI, or GREAT ZAB, a river of Turkey in Asia, in Kurdistan, which has its source near the Prussian frontier; runs SW; and joins the Tigris on the l. bank, 57 m. SSE of Mosoul, and after a total course of about 180 m. It is fordable during the dry season, but at other periods forms an impetuous torrent. This river is the *Zabatrus*, *Zerbis*, or *Lycus* of antiquity.

ZAB, ALTUN, or LITTLE ZAB, a river of Turkey in Asia, in Kurdistan, which has its source on the Persian frontier; runs SW; and, after a course of about 135 m., falls into the Tigris, on the l. bank, 45 m. below the confluence of the Great Zab.

ZABAKUCK, a village of Prussia, in the regency

of Magdeburg, and circle of Jerichow, on the Stremme. Pop. 330.

ZABANAGO, or LABBHOEHNAGO, a town of Burmah, in the Mranma, on the l. bank of the Irawady, 60 m. N of Ummerapura. Rubies are found in the vicinity.

ZABANOS, a town of Austria, in Bohemia, in the circle of Kaurzim, and seignory of Radim. Pop. 350.

ZABARAH (JEBEL), or EMERALD MOUNTAINS, a mountain-range of Upper Egypt, near the shore of the Arabian gulf, in N lat. 24° 40′, and E long. 25° 5′. It was formerly noted for its emeralds.

ZABARTOWA, a village of Prussia, in the regency of Bromberg, and circle of Wirsitz. Pop. 230.

ZABBANAGO. See ZABANAGO.

ZABEDOW, a village of Austria, in Bohemia, in the regency of Gitschen, and bail. of Neu Bidschow. Pop. 230.

ZABEHLA, a village of Austria, in Bohemia, in the reg. of Pilsen, and bail. of Brzeznitz. Pop. 300.

ZABELKAU, a village of Prussia, in the regency of Oppeln, and circle of Ratibor. Pop. 562.

ZABELSDORF, a village of Prussia, in the regency of Potsdam, and circle of Templin. Pop. 250.

ZABELTITZ, a village of Saxony, in the circle of Dresden, and bail. of Grossenham. Pop. 510.

ZABENSTEDT, a village of Prussia, in the reg. of Merseburg, and circle of Mansfeld. Pop. 240.

ZABER, a small river of Würtemberg, in the circle of the Neckar, which has its source near Zaberfeld, and falls into the Neckar.

ZABERFELD, a village of Würtemberg, in the circle of the Neckar, and bail. of Brackenheim, on the Zaber. Pop. 784. It has a castle.

ZABERN, or GLSASS-ZABERN. See SAVERNE.

ZABICZYN, a village of Prussia, in the regency of Bromberg, and circle of Wongrowice. Pop. 200.

ZABIEHLITZ, a village of Austria, in Bohemia, in the regency of Prague, and bail. of Carolinenthal. Pop. 730. It has a castle.

ZABIERZAU, a village of Prussia, in the regency of Oppeln, and circle of Neustadt. Pop. 200.

ZABINKEN, a village of Prussia, in the regency of Gumbinnen, and circle of Augerburg. Pop. 260.

ZABLAT, a village of Austria, in Bohemia, in the regency and bail. of Budweis, and seignory of Frauenberg. Pop. 270.—Also a village of the regency of Neuhaus, and seignory of Wittingau. Pop. 280.—Also a village of Silesia, in the regency of Troppau, and seignory of Schwarzwasser. It has 3 mills.

ZABLATOW, a market-town of Austria, in Galicia, circle and 15 m. E of Kolomea, near the l. bank of the Pruth.

ZABLATY, a village of Austria, in Moravia, in the circle of Znaim, and seignory of Namiescht. Pop. 230.

ZABLJAK, a village and frontier fortress of Albania, at the head of the lake of Scutari, of considerable strategetic importance, in the district of Moraka. The village, which represents the former capital of the old princes of Servia, only contains about 50 houses. It was built in 1423 by Stephen the Black, lord of Montenegro and Zeuta. His son, Ivan, resided in the stronghold as governor of Montenegro, and fought on the side of the Venetians, in 1478, against the Turks. In 1480 the Turks, already in possession of Scutari, revenged themselves by taking Z. by storm. In 1842 Ivan recovered it; but in the same year it was besieged anew by the Turks. Ivan now sought aid of the republic of Venice, pointing out how much he was suffering through their alliance: the senate, however, excused itself from rendering assistance, by alleging the separate peace it had made with the Sultan Bajazet. Hereupon Ivan abandoned the fortress, and fled to Cettinge, in Montenegro, and Z. remained in the hands of the Turks until 1835. In that year, the Montenegrins agreed with two Albanian sentinels of the Greek religion, who opened the gates of the fortress on the night of March 25, and allowed them to take possession of the place. Hafiz Pasha, commandant of Scutari, hastened to the fort with numerous regular and irregular troops, whereupon the Montenegrins again abandoned it. In 1853, the Montenegrins again got possession of it by a *coup-de-main;* but were driven from it by Omer Pasha.

ZABLOTZ, a town of Russia in Europe, in the gov. and 48 m. NE of Grodno, and district of Lida.

ZABLUDOV, a town of Russia in Europe, in the gov. of Grodno, 9 m. SSE of Bialistok. Pop. 2,000.

ZABNO, a market-town of Austria, in Galicia, in the circle and 12 m. NNW of Zarnow.—Also a village of Prussia, in the regency of Posen, and circle of Schrimm. Pop. 240.

ZABOKREKY. See ZSAMBOKRETH (NYITRA).

ZABOLA, a market-town of Austria, in Transylvania, in the country of the Szeklers, and stuhl of Haromszek.

ZABOR, a village of Austria, in Bohemia, in the regency of Pardubitz, and bail. of Hohenmauth. Pop. 700.—Also a village of the same regency, and bail. of Deutschbrod. Pop. 350.—Also a village of the regency of Gitschen, and bail. of Königinof. Pop. 290.—Also a village of the regency of Pilsen, and bail. of Pisek. Pop. 320.

ZABORESK, a village of Austria, in Bohemia, in the reg. of Prague, and bail. of Schlan. Pop. 330.

ZABORJ, a village of Austria, in Bohemia, in the regency of Pilsen, circle of Prachin, and seignory of Proitiwin. Pop. 360.

ZABORNA, a village of Austria, in Bohemia, in the regency of Pardubitz, and bail. of Deutschbrod. Pop. 330.

ZABOROWO, a town of Prussia, in the regency of Posen, and circle of Fraustadt. Pop. 860.—Also a village of the same regency, in the circle of Schrimm. Pop. 280.

ZABORZE, a village of Prussia, in the regency of Oppeln, and circle of Benthen. Pop. 410.

ZABRD, a village of Austria, in Bohemia, in the reg. of Pilsen, and bail. of Winterberg. Pop. 300.

ZABRECH, a village of Austria, in Moravia, in the circle of Prerau. Pop. 770.

ZABREZ, a town of Russia in Europe, in the gov. of Vilna, district and 30 m. SE of Ochmiana.

ZABROD, a village of Austria, in Bohemia, reg. of Gitschen, and bail. of Braunau. Pop. 200.

ZABROW, a village of Austria, in Bohemia, in the reg. and bail. of Budweis. Pop. 300.

ZABROZAU, a village of Prussia, in the prov. of Silesia, regency of Oppeln, and circle of Neustadt. Pop. 150.

ZABRZE, a village of Prussia, in the prov. of Silesia, regency of Oppeln, circle and 12 m. E of Fraustadt. Pop. 912. It has manufactories of cloth.

ZABRZEG, a village of Prussia, in the regency of Oppeln, and circle of Pless. Pop. 200.—Also a village of the same regency, in the circle of Ratibor. Pop. 340.

ZABU, a village of Egypt, in the W part of the oasis of El Bahryeh, in N. lat. 28° 21′ 47″, and E long. 28° 54′. Pop. 400. It is enclosed by walls with three gates. In the vicinity are several mineral springs.

ZACAPA, a village of Guatimala, capital of a district in the state of that name, on the SW side of the gulf of Dulce, on which it has a small port, and 135 m. ENE of Guatimala. Pop. 3,000.

ZACAPULAS. See SACAPULAS.

ZACATA, a town of Peru, in the dep. of Cuzco, on the slope of the Andes, 51 m. SW of Chucuyto.

ZACATAN, a river of Guatimala, in the state of that name, and dep. of Verapaz, which runs SE, and after a course of about 90 m., throws itself into the gulf of Honduras.

ZACATECAS, a central state of the republic of Mexico; bounded on the N by Cohahuila; on the NE by Nuevo-Leon; on the E by San-Luis-Potosi; on the S by Guanaxuato; on the W by Guadalaxara; and on the NW by Durango. It has a triangular outline, with an area of 20,000 geog. sq. m., and a pop. of 360,000. It belongs to the great plateau of Mexico, and is, in general, an arid uncultivated waste. As a mining district, however, it is of considerable importance; and on the N and E are vast cattle and horse rearing estates.—Its capital, of the same name, is situated in a narrow valley, 290 m. NW of Mexico. It has a pop. of 25,000. Its streets are well-paved, but narrow and crooked, owing to the city being closely hemmed in by mountains. Its post-office and custom-house are substantial buildings, and the government palace, on the S side of the Plaza, is a very handsome edifice. The church of La Parogia is nearly the size of the great cathedral of Mexico. There is a large mint here which is said to coin from £5,000,000 to £8,000,000 per annum.

ZACATEPEQUEZ. See SACATEPEC.

ZACATLAN, or XICOTLAN, a village of Mexico, in the state and 45 m. NNE of La Puebla. Pop. 750. It has extensive salt-works.

ZACATOTLAN, formerly a district of Mexico, on the coast of the Pacific, now comprised in the W part of the states of Mexico and Michoacan. It contains the town of Zacatula.

ZACATULA, MUSCALA, or BALSAS, a river of Mexico, in the state of that name, which has its source in the Great Cordillera of Anahuac, to the SE of Cuernavaca; runs W; and after a total course of about 300 m., throws itself into the Pacific, near a town of the same name. The embouchure forms a port, which bears the same name. A recent notice of this river, by one of an exploring party of Americans, states, that starting from Mesquitelan, near the frontier of the state of Puebla and the first village in the state of Guerrero, on the river-bank, the downward course, for 4 leagues, lay to the S, and then W between high mountains, some of them nearly 1,000 ft. high, whose sides were covered with a dense forest, in which were noticed Brazil and Nicarauag wood, mahogany, and many other dye trees. The river was about 100 yds. wide, and 7 ft. deep, with a three-knot current. The rapids and rocks were but trifling impediments to steam-boat navigation. About one league below the village of Slal-co-so-titlan, the river Atanango enters from the N, and is as large as the Muscala. "All map-makers have made the mistake of marking the Atanango for the Muscala, and above the mouth of the Atanango they have made the Muscala flow to the S and E, instead of to the W and S. Disturnell's map of Mexico, of 1850, is the only correctly printed map of the course of the Muscala or Zacatula. Below the mouth of the Atanango the depth of water varies from 12 to 20 ft., without increasing in width. We are now fully satisfied," continues this writer, "that this river is navigable for steam-boats at all stages of water for 450 m. This will bring us within 100 m. of Puebla, and 170 m. more to Vera Cruz; all excellent stage road, which can be driven over in 36 hours; so that you may calculate as follows:—six days from New York to Vera-Cruz in steamer, 36 hours across land to the Muscala river, 48 hours down to the mouth of the river; and 5 days thence to San Francisco—being less than 15 days from New York to San Francisco. The river Muscala is the largest and longest river exclusively in the republic of Mexico. The port of Zacatula, at its mouth, is the oldest port on the Pacific ocean, being the place where Cortez built his first ships on that sea. Among the old archives a record is made of the fact that Cortez transported iron, cordage, and other articles from Vera-Cruz to the head waters of the Muscala, and floated them down the river, so as to equip the first fleet built on the Pacific at the port of Zacatula. The hero Morelos, in the wars of independence, improved on this information, and brought up on rafts his artillery to Mesquitelan, near which village he was captured. The survey made by order of the state of Puebla in the dry season in 1850 reports 3½ ft. water in the channel from Mesquitelan to the Atanango river; and 4½ ft. thence for 15 leagues; then 6 ft., then 20 ft.; then 40 ft., and, for the last 100 m., 60 ft. of water in the channel. A beautiful bay exists near the mouth, and Humboldt, in his work on New Spain, frequently notices the port of Zacatula. Yet, with all this, the impression has prevailed that the river was not navigable, first originating in the fact, that it is not adapted to sailing vessels, without reflecting that it might be so for steam-boats. In the second place, the river has never been properly laid down on the charts, and all information of the courses, distances, depths, and currents was confined to the simple natives on the banks, who only knew it from village to village, and wherever it ran between high mountains it was said to contain falls. Again, the elevation of Puebla above the sea is known to be 7,000 ft., and many intelligent persons imagined that in 700 m., by the course of the river, that descent could not be overcome, without knowing the fact, that within the first 100 m. immediately below Puebla the descent is more than 3,000 ft. The elevation of Mesquitelan above the sea is only 2,000 ft., and in the village of Muscala Humboldt found the elevation to be less than 1,600 ft. Here is a river which has its mouth on the Pacific ocean, directly west from Vera-Cruz, and navigable for more than half the distance from port to port, and lying directly in the course, diminishing the distance by one-half from New Orleans to California, and the expense in the same proportion."—The town of Z. is 225 m. SW of Mexico, and 340 m. NW of Acapulco, on the l. bank of the river of the same name. Pop. 800.

ZACH, a mountain of Japan, in the SW part of the island of Nifon, and prov. of Isoumo, near the coast of the Sea of Japan.

ZACHAN, a town of Prussia, in the prov. of Pomerania, regency of Stettin, and circle of Saazig, 12 m. ESE of Stargard, on the Krasbach. Pop. 1,159.

ZACHARCZOWITZ, a village of Prussia, in the reg. of Oppeln, and circle of Tost. Pop. 260.

ZACHARIE (SAINT), a commune of France, in the dep. of the Var, cant. and 10 m. SW of Maximin, on the Verne. Pop. in 1841, 1,622. It has a glass-work and manufactories of pottery.

ZACHARIN, a village of Prussia, reg. of Marienwerder, and circle of Deutsch-krone. Pop. 230.

ZACHARZOW, a village of Prussia, in the reg. of Posen, and circle of Adelnau. Pop. 370.

ZACHASBERG, a village of Prussia, in the reg. of Bromberg, and circle of Chodziesen. Pop. 570.

ZACHENBERG, a village of Bavaria, in the circle of Lower Bavaria, and presidial of Viechtach. Pop. 200.

ZACHGLUM, a village of Austria, in Bohemia, reg. of Gitschen, and bail. of Reichenau. Pop. 410.

ZACHIGOV-JER, a town of Russia in Europe, in the gov. of Irkutsk, district and 6 m. N of Nertchinsk, on a height, near the Nertcha.

ZACHITA, a village of Mexico, in the state and valley of Oaxaca. It is noted as the place in which the ancient Tzapotec kings held their court.

ZACHIVERSK, a district and town of Russia in Asia, in the prov. of Yakutsk. The town is 510 m. NE of Yakoutsk, on the r. bank of the Indighirka. The district presents a dreary aspect, and is surrounded by sterile mountains. Fish is plentiful in the streams, and hares, elans, bears, reindeer, and game are abundant.

ZACHKEVITCHI, a small town of Russia in Europe, in the gov. of Vilna, district and 30 m. E of Ochmiana.

ZACHOROWITZ, a village of Austria, in Moravia, in the circle of Hradisch. Pop. 800.

ZACHOTIN, a village of Austria, in Bohemia, in the regency of Budweis, circle of Tabor, and seignory of Roth-Rzeczicz. Pop. 300.

ZACHOW, a village of Prussia, in the regency of Potsdam, and circle of Westhavelland. Pop. 380.—Also a village in the reg. of Frankfort, and circle of Konigsberg. Pop. 280.—Also a village of the regency of Stettin, and circle of Regenwalde. Pop. 240.

ZACKEN, a river of Prussia, in the prov. of Silesia, and regency of Liegnitz, which descends from the Riesen-Gebirge; runs NE, and joins the Bober, near Hirschberg, and after a course of about 10 m. Its banks are extremely picturesque.

ZACKERICK, a village of Prussia, in the regency of Frankfort, and circle of Konigsberg. Pop. 1,155.

ZACKERSDORF, or ZACKINGEN. See CSAKANG.

ZACKRZEWO, a town of Poland, in the woiwodie of Masow and obwod of Kujavie, 20 m. NW of Brzesc. Pop. 500. It has an ancient Carmelite convent.

ZACUALCO, a village of Mexico, in the state of Xalisco, on a lake of the same name.

ZACUALPAN, a market-town of Mexico, in the state of that name, SSE of Saltepec.

ZACUALTIPAN, a village of Mexico, in the state of that name, near the confines of Vera-Cruz.

ZACYNTHE. See ZANTE.

ZADACATA, or SZADEKE, a halting-place of the Arabian Desert to the E of the mountains of El Araba, at the junction of the caravan route from Suez with that running N from Akaba. It has a small stream.

ZADAICA, one of the principal villages of the island of Zerbi, near the coast of the kingdom of Tunis.

ZADEL, a village of Saxony, in the circle of Dresden, bail. and near Meissen. Pop. 445.—Also a village of Prussia, in the regency of Breslau, and circle of Frauenstein. Pop. 815.

ZADELOW, a village of Prussia, in the regency of Stettin, and circle of Saatzig. Pop. 350.

ZADELSDORF, a village of Saxe-Weimar, circle of Neustadt, and bail. of Weida. Pop. 230.

ZADEN (HOHEN and NIEDER), two villages of Prussia, in the regency of Stettin, circle of Randow. Pop. 600.

ZADONSK, a district and town of Russia in Europe, in the gov. and 63 m. NNW of Voronej, on the l. bank of the Don. Pop. 2,000.

ZADORRA, a river of Spain, in the prov. of Alava, formed by the confluence of two streams, which descend from the mountains of Salvatierra, and unite near the town of that name; flows SW, and after a total course of about 36 m. joins the Ebro, on the l. bank, opposite Ircio, and near Zembrana. It is the largest river in the prov., although fordable at all parts of its course. Its waters abound with fish.

ZADORY, a village of Prussia, in the regency of Posen, and circle of Kosten. Pop. 200.

ZADRASCHAN, a village of Austria, in Bohemia, in the regency of Gitschin, and bail. of Neu Bidschow. Pop. 300.

ZADRIM, ZADRA, or ZADRINA, a town of Turkey in Europe, in Albania, in the sanj. and 15 m. SSE of Scutari, on the l. bank of the Drino, near the confluence of the Dibra.

ZADROORITZ, a village of Austria, in Moravia, in the circle of Hradisch, and seignory of Wisowitz. Pop. 850.

ZAFARINES, or ZAFFARINA, a group of islets in the Mediterranean, near the coast of Marocco, and prov. of Fez, to the E of Melilla, and NW of the embouchure of the Moulouïa, in N lat. 36° 17′.

ZAFFARANA, a headland of Sicily, in the district and 11 m. E of Palermo. It consists of a steep, conical-shaped mountain.—Also a commune and market-town in the prov. and 16 m. N of Catania, on the E declivity of Ætna. Pop. of com., 3,000. It is a place of considerable industry, and the vicinity produces corn, wine, charcoal, fire-wood, and forage. It was considerably damaged by a lava stream from Mount Ætna in Sept. 1852.

ZAFFARGHUR, a town of Hindostan, in the territory of the Nizam, 72 m. ENE of Hydrabad, in N lat. 17° 46′, and E long. 79° 31′.

ZAFFERAN, a town of Tripoli, on the gulf of Sidra, 144 m. SE of Mesurata, in N lat. 31° 12′ 10″, and E long. 16° 41′ 14″.

ZAFFERANLU, or ZAFFARNLU, a Kurd tribe who inhabit the the Persian prov. of Khorassan. They number about 10,006 individuals.

ZAFRA, a judicial partido and town of Spain, in the prov. of Badajoz. The partido comprises 10 pueblos. The town is 42 m. SE of Badajoz. Pop. 4,894. It is clean and regularly built, and contains two public squares, surrounded with porticoes. In a commanding situation is the magnificent palace of the dukes of Medina-Celi. It possesses manufactories of gold-ware, hosiery, and pottery, and several tanneries. In the vicinity is a mineral spring. This town was founded by the Celts 580 years before the Christian era. It was greatly extended by Julius Cæsar, and received from him the name of Restituta Julia. Its present name is said to be of Arabic origin. It was taken by the Moors in 1240.—Also a town of the prov. and 24 m. SW of Cuenca, on a hill. Pop. 760. It has a fort in ruins, and an hospital.

ZAFRANEH, or ZAFFARANNA, a headland of Egypt, in the gulf of Suez, in N lat. 29°.

ZAFURABAD, a town of Hindostan, in the prov. and 6 m. SE of Jounpur, on the r. bank of the Joomtee, in N lat. 25° 42′, and E long. 82° 47′.

ZAFYR (EL), an Arab tribe in the district of Nedjid. It is governed by a sheik, and is said to number 3,200 who carry arms.

ZAFZAF, a caravan station of the Sahara, to the W of Fezzan.

ZAGALE, a town of Russia in Europe, in the gov. of Minsk, district and 51 m. SW of Retchitza.

ZAGARA. See PALÆOVUNI.

ZAGARISE, a market-town of Naples, in the prov. of Calabria-Ultra, district and 9 m. NNE of Catanzaro, and cant. of Soveria, on the slope of a hill, the base of which is washed by the Moloiano. Pop. 1,200. It has a collegiate church.

ZAGAROLO, a market-town of the Papal states, in the comarca and 18 m. E of Rome. Pop. 1,080.

ZAGETSCHITZ, a village of Austria, in Bohemia, in the regency of Pardubitz, and bail. of Chrudim. Pop. 910.

ZAGEZD, a village of Austria, in Bohemia, in the regency of Gitschin, and bail. of Senstenberg. Pop. 270.

ZAGEZDEZ, a village of Austria, in Bohemia, in the regency of Pardubitz, and circle of Chrudim. Pop. 450. It has a synagogue and several breweries.

ZAGHOUTH, a town of Tripoli, in the E part of Tripoli Proper, 104 m. W of Augela.

ZAGLAWA, an island of Austria, in Dalmatia, to the W of the island of Cherso.

ZAGLINKA, a village of Russia in Asia, in Georgia, near Tiflis. In the vicinity is an extensive alum-work.

ZAGORA, a market-town of Turkey in Europe, in Thessaly, on a small river which descends from the mountain of Plesnid-oro or Zagora, and about 2 m. from the shore of the Archipelago. The surrounding district is fertile, and produces silk of good quality. To the NE of Z. is a cape of the same name.—Also a town of Rumelia, on a small lake which communicates with the Black sea, and about 10 m. SW of Burgas. This town is the *Debultum* of the ancients.

ZAGORA. See Liakuri.

ZAGORI, a village of Turkey in Europe, in Albania, in the S part of the sanj. of Valona, and near the source of the Zagoria.

ZAGORIA, a river of Turkey in Europe, in Albania, which has its source in the NW part of the sanj. of Janina, flows NW, and joins the Voiussa or Vojutza, on the l. bank, about 7 m. NNE of the confluence of the Deropuli.

ZAGOROWO, a town of Russia in Europe, in Poland, in the gov. and 33 m. NNW of Kalisch, and obwod of Konin. Pop. 1,650.

ZAGRAB. See Agram.

ZAGROS. See Tak.

ZAGUALTIPAN, a town of Mexico, in the state of Queretara, 87 m. NE of Mexico. Pop. 6,000. It is substantially built of brick and stone, and contains several churches and other public edifices.

ZAGUANANAS, a name by which the Rio Colorado is distinguished in the upper part of its course.

ZAGURA, a river of Turkey in Europe, in Albania, which has its source towards the centre of the sanj. of Janina, and to the SW of the town of that name, runs first SW, then SE to the confluence of the Romandes. About 3 m. below Kako-Souli it is joined by the Mavro, whence it takes the name of Mavro-Polaino, which see.

ZAGVAN, a town of Tunis, on a river, and near the foot of a hill of the same name. It has several dye-works and bleacheries, and is surrounded by fine gardens. In the vicinity are the ruins of the ancient Carthage.

ZAHARA, a town of Spain, in Andalusia, in the prov. of Cadiz, and partido of Olvera, on the slope of a lofty mountain, near the source of the Guadalete. Pop. 1,151. It has a parish church, a custom-house, and a public granary.

ZAHARA. See Sahara.

ZAHAY, a village of Austria, in Bohemia, in the regency of Pardubitz, and bail. of Leitomischl. Pop. 660.

ZAHINOS, a town of Spain, in the prov. of Badajoz, partido and 8 m. NW of Xerez-de-los-Cabelleros. Pop. 1,000. It has manufactories of linen fabrics.

ZAHLBACH, a village of Bavaria, in the circle of Lower Franconia, and presidial of Kissengen. Pop. 591.

ZAHLENITZ, a village of Austria, in Moravia, in the regency of Olmutz, circle of Hradisch, and seignory of Napagedl. Pop. 490.

ZAHLING, a village of Bavaria, in the circle of Upper Bavaria, and presidial of Aichach. Pop. 200.

ZAHMEN, a village of Hesse, in the circle of Ober-Hesse, and presidial of Altenschlirf. Pop. 200.

ZAHNA, a town of Prussia, in the prov. of Saxony, regency of Merseburg, circle and 7 m. ENE of Wittenberg. Pop. 1,345. It is enclosed by lofty walls, and contains two churches and an hospital. Pop. in 1843, 2,134.

ZAHNASCHOWITZ, a village of Austria, in Moravia, in the circle and 17 m. E of Hradisch. Pop. 340.

ZAHOHILLO, an island of Guatemala, in the SE part of Lake Nicaragua, and in the state of that name, to the SE of the island of Zapone.

ZAHOR, a village of Austria, in Bohemia, in the regency of Budweis, circle of Tabor, and seignory of Chotowin. Pop. 270.

ZAHORAN, a village of Austria, in Bohemia, in the regency of Pilsen, circle of Plattau, and seignory of Kanth. Pop. 560.

ZAHOREITZ, a village of Austria, in Bohemia, in the regency of Pilsen, circle of Prachin, and seignory of Worlik. Pop. 200.

ZAHORJ, a village of Austria, in Bohemia, in the regency of Pilsen, and bail. of Prachatitz. Pop. 200.—Also a village of the same regency, and bail. of Jung-Woschitz. Pop. 260.

ZAHORNITZ, a village of Austria, in Bohemia, in the regency of Gitschin, and bail. of Neustadt. Pop. 340.—Also a village of the same regency, and bail. of Podiebrad. Pop. 660.

ZAHRADKA, a market-town of Austria, in Bohemia, in the regency of Pardubitz, and bail. of Lodetsch. Pop. 1,100. It has a paper-mill.—Also a village of the regency of Budweis, and bail. of Nachaus. Pop. 400.—Also a village of the regency of Pilsen, and bail. of Przestik. Pop. 260.

ZAHR-EL-AKABA, a village of Arabia, in the desert, extending between Nedjed, Hedjaz, and Asiatic Turkey, and forming one of the stations of the great caravan to Mecca. It is on the slope of the mountains of Akaba, which are here crossed by a road always difficult, and in winter impassable.

ZAHRENSDORF, a village of Mecklenburg-Schwerin, circle of Wendisch, and bail. of Boitzenburg. Pop. 200.

ZAHRENSTORF, a village of Mecklenburg-Schwerin, in the district of Schwerin, and bail. of Tempzin. Pop. 330.

ZAHZINGEN, a village of the grand-duchy of Baden, in the circle of the Upper Rhine, bail. and 2 m. N of Freiburg, at the foot of a mountain on which are the ruins of the celebrated castle of the same name. Pop. 690.

ZAI, a river of Russia in Europe, in the W part of the gov. of Orenburg, which has its source in the environs of Bugulma, runs NNW, and joins the Kama on the l. bank, 9 m. SE of Mamadich, and after a total course of about 120 m.

ZAIDIN, a town of Spain, in Aragon, in the prov. of Huesca, and partido of Fraga, 33 m. from Barbastro, on the l. bank of the Cinca. Pop. 1,610. It has a parish church, a custom-house, and a public granary. Linen is here the chief article of local manufacture.

ZAIERZHOFEN, a village of Bavaria, in Swabia, in the presidial of Boggenburg. Pop. 250.

ZAIKANY, or Zajkany, a village of Transylvania, in the comitat of Hunyad, 20 m. SE of Eisenmarkt, and 14 m. SW of Hatszeg, near the famous pass called Vaskapu or Irongate. It is noted for the victory which was here for the third time gained by Trajan over Decébalus, king of the Dacians, treasure belonging to whom was found in the locality in 1543.

ZAINAH. See TAGGAH.

ZAINIEN, a village of Würtemberg, in the circle of the Schwarzwald, and bail. of Neuenburg. Pop. 250.

ZAININGEN, a village of Würtemberg, in the circle of the Schwarzwald, and bail. of Urach. Pop. 956.

ZAINSK, a market-town of Russia in Europe, in the gov. of Orenburg. Pop. 1,700.

ZAIONCZKI, a village of Prussia, in the regency of Posen, and circle of Schildberg. Pop. 260.

ZAIONCZKOWO, a village of Prussia, in the regency of Posen, and circle of Samter. Pop. 250.

ZAIRE, ZAHIR, or CONGO, a large river of Southern Africa, which appears to have one of its sources somewhere under the parallel of 2° N, and between the meridians of 17° and 18° E. This head-stream pursues at first a course to the W of S, passing through the Anzeco and Wamba territories. Below Konkobale, it unites with a very large river coming from the SSE, which is probably the true main stream of the Zaire, which then flows in a prevailingly W course to the ocean, into which it falls with a strong current in S lat. 6°, between the Kongo and Kakongo territories, by an estuary 10 m. in width. The N head-stream of the Z. was identified by some geographers with the Niger, and others even connected it with the imaginary fountains of the Nile. Mr. Cooley, in his recent work entitled 'Inner Africa laid open' [Lond. 8vo. 1852], has thrown great light on the higher and true sources of this great river, he states that "in 1802, F. H. Da Costa, superintendent of the fair or factory of Casange, sent two *pombeiros*, or native mercantile travellers, into the interior, with instructions to cross the continent, if possible, to the Zambeze. But a principal object of their mission was to endeavour to establish relations of amity and intercourse with the *muropúe*, or king of the Moluas (as they are called by the Portuguese), who was known to dwell beyond Casange, towards the NE or NNE. The starting point of the pombeiros was the Mucari or domain containing the factory for Casange, within a day's journey, probably, of Pungo-Andongo, in about lat. 9° 30′ S, long. 15° 34′ E. Nearly 14° further E, and in the same parallel, stands Lucenda, the Cazembe's capital. Between these two points we have to arrange a route of 150 days' march, made by experienced travellers, who halted often and long for rest, and whose daily route may be taken at 10 miles. In order to avoid Casange on the l., they went along the r. bank of the Quanza, which rises in Bíhê, at a distance of 15 days SE, perhaps, from Pungo-Andongo. They thus came to the estates of Bomba, between the rivers Quanza and Quango, which are said to be but 7 days asunder. The titles of this chief are thus enumerated by Francesco Honorato: 'Seculo Bomba, Cambambi, Camasaca, and Mugumbo Acalunga, Ruler and Lord of all the Songo and passage to the interior.' The pombeiros appear to have entered the territory of Bomba, when they crossed the river Jumbo on the 12th day of their march. On the 22d, they arrived at the town of the Secúlo, or, as the Portuguese would say, of Duke Bomba. At a distance of 4 days from this they came to Pepumdi (?) Songo, also on a river Jumbo, and in 3 days more (29 in all) crossed the Quango. Five days beyond the Quango, the travellers crossed a desert nine or ten days in extent, to the town of Cabungi, on the frontier of the Muáta-ya-Nvo. The desert here mentioned, in which four rivers were met with, the road going along one of them (the Quihubue) for 3 days, extends probably over the dividing ridge between the valley of the Quango and that of the great rivers whereon lie the dominions of the Muáta ya Nvo. Nine days from Cabungi, they passed through a village forming part of the estates of Luconquéso, the queen-mother, and two days afterwards crossed the great river Casais (more probably Casézi) in a canoe. Again, in ten days, traversing another dividing ridge, marked by a two days' desert, they crossed the Lulúa also, the chief river of this region, in a canoe. In 12 days more, they arrived at the residence of the Muáta-ya-Nvo, or muropúe." It is to be borne in mind that from Pungo-Andongo to the abode of 'Duke Bomba,' the travellers had proceeded up the valley of the Quango, in a SE direction, that, soon after leaving this spot they turned to the NW, until they had reached the summit of the mountain-ridge which divides the waters that feed the Quango from those that fall into the Lulúa. Hence they proceeded due N, till they arrived at the capital of the muropue. "Leaving the town of this chief," Mr. Cooley proceeds, "at the end of May 1806, on their way to the Cazembe, they tell us that they had the sun (rising) full on the left side, which implies a course about two point E of S; and in this course they persisted for about 37 days. The frequent mention of rain in this part of the journal (from May to September), shows that the wet season had set in. The rivers were numerous, and many of them waist deep. The chief were the Izabuígi, the Calalímo, Roando, Rova, Cazelle, the Caginrígi, crossed in a canoe; the Reu, Ropoege, and Lubúri, 80 yds. wide, forded on the 34th day, and where the muropúe's dominion terminates. These rivers, together with other and large streams further on, to a distance of 12 days, all flow into the Lulúa." It is clear, from these facts, and the features of the country, that the Lulúa is the main stream, and its valley the cradle, of the Zaire or Congo. The real origin of this river is, therefore, to be found far to the S, instead of the N of the equator: the greater part of its course runs in a NW instead of a SW direction; nor is it until it has performed more than five-sixths of its allotted journey, and has united itself with the Quango also flowing from the SE, as well as with the northern stream hitherto regarded as the true Zaire, that it adopts that name which has been supposed to characterise it throughout. The muropúe appears to be a sovereign of considerable importance. On both sides of the valley of the Lulúa his empire extends to a distance of 34 days' journey, though on the W bank this measure must be taken as representing less than on the E.—At about 140 m. above its mouth, the width of the Z. contracts to from 300 to 500 yds., and it flows for about 40 m. between lofty rugged banks, over a rocky channel. Above this, it resumes a majestic breadth, varying from 1 to 3 m., and flows with a current of from 2 to 3 m. an hour, except when highly flooded. The waters of the Z. retain their freshness 13 m. to the WNW of the mouth of the stream, and their effect is perceived at 36 m. distance. Like most other rivers upon this coast, it appears to have a bar across its entrance, but with perhaps 20 fath. water upon it.

ZAISAN, or KONG-KO-LU, a lake of China, in the NE part of Sungaria, near the frontier of the Russian gov. of Omsk. It covers an area 75 m. in extreme length from E to W, and 27 m. in breadth from N to S, extending between 47° and 48° N lat., and between 83° and 85° E long. It is formed by an expansion of the Irtish, and receives also the waters of the Oulasly, Kara-bonga and Bazar on the S, and the Kokboukhty or Kupkak-Bugutchik on the W.

ZAISENHAUSEN, a market-town of the grand-duchy of Baden, in the circle of the Middle Rhine, and bail. of Bretten. Pop. 1,050. It has salt-works.—Also a village of Würtemberg, in the circle of the Jaxt, and bail. of Kunzelsan. Pop. 306.

ZAISERSWEIHER, a village of Bavaria, in Swabia, in the presidial of Türkheim. Pop. 584.

ZAITAN (CAPE), a headland of Turkey in Europe, in Rumelia, and sanj. of Kirk-Kilissia, on the Black sea, a little to the N of the embouchure of the Kanarahdéré, at the E extremity of Mount Shaïtan, in N lat. 42° 17′ 55″, and E long. 27° 47′ 20″.

ZAITUN, or ZETOWNA, a town of Fezzan, 30 m. E of Mourzouk.

ZAITZKOFEN, a village of Bavaria, in the circle of Bavaria. Pop. 138. It has a castle.

ZAIZEHS, a commune of Belgium, in the prov. of Hainault, and dep. of Roux. Pop. 689.

ZAKLICZAN, a market-town of Austria, in Galicia, in the circle of Bochnia, on the r. bank of the Dunajec.

ZAKLIKOW, a town of Russia in Europe, in Poland, in the gov. and 42 m. SW of Lublin, and obwod of Samosc, on the l. bank of the Sanna. Pop. 1,000.

ZAKOPANI, a village of Galicia, in the circle of Sandec, on the White Dunajec.

ZAKOTNAIA, a town of Russia in Europe, in the gov. of Voronej, district and 24 m. N of Starobielsk.

ZAKRAN, a village of Austria, in Moravia, in the circle of Brunn, and seignory of Rossitz. Pop. 290.

ZAKREZEWO, a village of Prussia, in the regency of Marienwerder, and circle of Flatow. Pop. 499.—Also a colony distinguished as Neu-Zakrezewo, in the same regency and circle. Pop. 200.

ZAKROCZYN, a town of Russia in Europe, in Poland, in the gov. and obwod of Plock, on a height near the r. bank of the Vistula, and 27 m. NW of Warsaw. Pop. 1,200.

ZAKRZEWKE, a village and colony of Prussia, in the regency of Marienwerder, and circle of Flatow, possessing respectively 280 and 400 inhabitants.

ZAKRZEWKE, a village of Prussia, in the regency of Posen, and circle of Pleschen. Pop. 240.—Also a village of the same regency, in the circle of Schrimm. Pop. 230.—Also a village of the same regency, in the circle of Posen. Pop. 230.—Also a village in the same regency, in the circle of Kröben. Pop. 390.

ZAKUK, the most northerly inhabited spot of the Wady-Natrun in Egypt, 12 hours' march from Temneh. The road passes over the ruins of an ancient town, which have been turned up in every direction, for the sake of the nitre that abounds in the mounds of all the old towns of Egypt. On reaching the summit of the low hills that skirt the desert, the road continues upon a high plain, with a slight ascent towards the W, varied by occasional undulations, for about 20 m.: after which it descends towards the Natrun valley, the bottom of which is lower than the valley of the Nile. The village of Z., founded about 20 years since by Europeans, who established works here for the drying of the natron, now consists of about 60 huts, with a pop. of about 200. The natron is found both in the plain and in two or three of the lakes; some of the lakes contain only muriate of soda; most of the lakes contain water all the year round, and some are dried up in summer. In those lakes which contain both common salt and natron, these crystallize separately, the former above, in a layer of about 18 inches thick, and the latter beneath, in a layer of about 27 inches. All the lakes contain muriate of soda, though few produce natron. When the water of the salt and natron lakes has evaporated, a firm incrustation is left, when the natron called *sottanee* is collected. The natron is of two kinds, the white and the sottanee: the latter taken from the lakes, the former from the ground surrounding the lakes, and which is not inundated. The latter is of the best quality. There are several springs of fresh water in the Natrun valley, the purest of which is near the convents to the S; that of Dayr Baramoos being slightly salt. Sir Gardner Wilkinson is of opinion, that the fresh water found here and at the oases, filters beneath the mountains that separate the Wady Natrun from the Nile, and remains fresh from its not meeting with any of the salts on its passage. The process of infiltration is very slow, as the water of the lakes does not rise till three months after the rise of the Nile. The dip of the strata that borders the Natrun valley is towards the NE. Besides the population before stated of the village of Z., the four convents reckon together 72 inhabitants. The valley produces, in addition to the natron, rushes used for making the well-known mats of Egypt. The best rushes, however, those used in the mats called Menoofee, come from the valley called Wady-è-Sumar, to the SW of the Natrun valley three days' journey. Tamarisk and stunted palms, with the usual herbs of the desert, are the only vegetable productions of these valleys, besides the rushes. Gazelles, jerboas, foxes, and a few others common to the Libyan hills, are the only animals. The length of the Wady-Natrun is 22 m.; its greatest breadth 5½ from the brow of the hills, the bottom being only 2 m. The hills and the undulating banks of the Wady are covered with rounded siliceous pebbles, and pieces of petrified wood. Of the Bahr-el-Fargh, or Bahr-bila-ma ['river without water'], Sir Gardner Wilkinson is disposed to deny its ever having been a water-course; he says it has none of the characters of one, and asks what has become of the alluvial deposit which should be found, if in reality a portion of the waters of the Nile has ever flowed here.

ZALA, a village of Tripoli, to the S of Makdar, on the shore of the Great Syrtis.

ZALAD, or ZALA. See SZALAD.

ZALAHN. See WALTENBERG.

ZALAKNA. See ZALATHNA.

ZALAMEA, a market-town of Spain, in Estremadura, in the prov. and 57 m. ESE of Badajoz, and partido of Castuera, near the Sierra Morena, on a granitic rock. Pop. 3,980. It has a parish church, a convent, a custom-house, a public granary, and an hospital. It contains the ruins of a castle, and some Roman antiquities, and in the vicinity are mines of silver.

ZALAMEA-LA-REAL, a town of Spain, in Andalusia, in the prov. of Huelva, and partido of Cerro, 45 m. WNW of Seville, on a height. Pop. 4,180. It has a parish-church, an hospital, several schools, a public granary, and contains some well-built houses. Candles, and other articles in wax, linen and woollen fabrics, leather, and brandy, are its chief articles of manufacture. In the vicinity are the famous mines of Rio-Tinto.

ZALATHNA, KLEIN-SCHLATTEN, GOLDENMARKT, ZALAKNA, or ZLAKNA, a town of Transylvania, in the gsp. of Unter-Alba, 18 m. WNW of Karlsburg, on the Ompoly or Ampoly, a small affluent of the Maros. Pop. 5,000, consisting chiefly of Hungarians, Germans, and Wallachians. The environs are noted for their mines of gold, silver, lead, copper, and quicksilver.

ZALATHNA (SZASZ), SCHLATT or SCHLATTENDORF, ZALAKNA or ZLAKNA, a town of Transylvania, in the comitat of Upper Weissenburg, and 12 m. SE of Medwisch.

ZALDIVIA, or ZALDIVAR, a town of Spain, in the prov. of Guipuzcoa, partido of Tolosa, and 6 m. from San-Sebastian, in a fertile valley. Pop. 1,015. It has a parish church. It possesses manufactures of common cloth, and carries on an active trade in cattle and pigs.

ZALDUENDO, a town of Spain, in the prov. of Alava, partido and 3 m. E of Salvatierra, in a mountainous locality. Pop. 400. It has an hospital.

ZALEC, a village of Afghanistan, at some distance from the r. bank of the Helmund.

ZALEMAH, a group of islands in the strait of Macassar, to the W of Celebes, in N lat. 5° 30′.

ZALENSCHE, a village of Prussia, in the regency of Oppeln, and circle of Benthen. Pop. 600.

ZALES, a village of Austria, in Bohemia, in the regency of Pilsen, and bail. of Wolin. Pop. 300.

ZALESCH, a village of Austria, in Bohemia, in the regency of Pardubitz, and bail. of Leitomischl. Pop. 260.

ZALESIN, a village of Prussia, in the regency of Marienwerder, and circle of Konitz. Pop. 250.—Also a village of the regency of Posen, and circle of Kröben. Pop. 400.

ZALESIN (GROSS and KLEIN), two villages of Prussia, in the regency of Posen, and circle of Krotoszyn. Pop. 476.

ZALESZCZYKY, a town of Austria, in Galicia, in the circle and 24 m. S of Czortkow, and 120 m. SE of Lemberg, on the l. bank of the Dniester, in N lat. 48° 43′, and E long. 25° 45′ 50″. Pop. 5,500, of whom 3,200 are Jews. It has a convent and a school, and possesses considerable trade. The circle of Czortkow formerly bore the name of Zaleszczyky.

ZALLEN, a village of Austria, in Bohemia, in the reg. and bail. of Budweis. Pop. 220.

ZALLENFELDE, a village of Prussia, in the regency of Konigsberg, and circle of Preussisch-Holland. Pop. 380.

ZALOSZE, or ZALOSCE, a town of Galicia, in the circle of Zloczow, 21 m. SSE of Brody, on the Sered. It has a castle, Catholic and United Greek churches, and an ancient convent. It has manufactories of fine cloth.

ZALOW, a village of Austria, in Bohemia, in the reg. of Prague, and bail. of Schmichow. Pop. 220.

ZALSCHY, a village of Austria, in Bohemia, in the regency of Budweis, and bail. of Tabor. Pop. 330. It has a castle. Potash is its chief article of manufacture.

ZALT-BOMMEL. See BOMMEL.

ZALWAALD. See SAALFELDT.

ZAMAYON, a town of Spain, in the prov. and 18 m. NW of Salamanca. Pop. 300.

ZAMBA, or ZEMBA, a town and port of New Granada, in the dep. of the Magdalena, in the prov. and 30 m. NE of Carthagena, on a bay of the sea of the Antilles, which forms a spacious harbour. A little to the N is the island of Zemba, or Galera-de-Zamba.

ZAMBALES, a prov. of the Philippine islands, in the island of Luçon. Pop. 24,500, of whom about 500 are aborigines.

ZAMBERK. See SENFTENBERG.

ZAMBEZE, or ZAMBEZI, a river of Africa, which now appears to have its sources about the parallel of 10° S, near those of the Zaire, and also of the chief streams flowing into the great lake Nyassi, on the high land of Lobale or Lovar. The main stream flows in a prevailingly SSW direction, until it has crossed the 15th parallel, shortly after which it bends round towards the SSE, between the meridians of 20° and 25° E; then flows to the N of E, to its junction with the Mamutshi, in about 16° 30′ S lat., and 31° E long. Below this point, at Zumbo, it receives the Aranagoa, coming from the N. It then flows E by S to Tete, near which it receives a considerable stream from the N, and another from the S, after which it breaks a passage through the Lupata chain, and flowing SSE, passes Sena, and falls into the ocean by three principal deltoid streams, the Quilimane, the Cuama, and the Luaba.—Another large stream, also called the Zambezi, or Lecambye, appears to rise under the parallel of 12° S, near the source of the Aranagoa branch of the river just described, and to flow in a W direction, towards the meridian of 27° E, when it bends round to NNE, and flowing past Pemba and Lucenda, discharges itself into the great lake of Nyassi, on the W bank, and under the parallel of 8° 30′ S.

ZAMBI, or SUMBI, a mountain of Lower Guinea, in the N part of Libolo. It is an active volcano.

ZAMBOANGA. See SAMBOANGA.

ZAMBORST, a village of Prussia, in the regency of Koslin, and circle of Neustettin. Pop. 260.

ZAMBROW, a town of Russia in Europe, in Poland, in the gov. of Augustowo, obwod and 17 m. SE of Lomza. Pop. 600.

ZAMES, a village of Austria, in Bohemia, in the regency and bail. of Gitschin. Pop. 210.

ZAMFARA, or ZAMFRA, a district of Nigritia, in the W part of Houssa, bordered on the E by the Guarrama, and intersected by the Fulshi. Its capital is Zirmie.

ZAMIANSK, or ZAMIANOVSKAIA, a town of the Cossacks of Astrakhan, in Russia in Europe, district and 45 m. NW of Astrakhan, on the r. bank of the Volga.

ZAMIEL, a village of Austria, in Bohemia, in the reg. of Gitschin, and bail. of Reichenau. Pop. 760.

ZAMIKHOV, a town of Russia in Europe, in the gov. of Podolia, and district of Ouchitza, 33 m. NE of Kamenetz.

ZAMLEKAN, a village of Austria, in Bohemia, in the reg. of Pilsen, and bail. of Planitz. Pop. 380.

ZAMMARO, a market-town of Naples, in the prov. of Calabria-Ultra, district, cant. and 1½ m. S of Monteleone, on the slope of a hill. Pop. 500. It was to a great extent destroyed by an earthquake in 1783.

ZAMOCHA, a town of Russia in Europe, in the gov. of Minsk, district and 45 m. W of Disna.

ZAMORA, a province of Spain, in the kingdom of Leon, bounded on the N by the prov. of Leon; on the E by Valladolid; on the S by Salamanca; and on the W by Orense and by Portugal. It has an area of 167·73 German or 3,560 English sq. m. Pop. in 1849, 180,000. The surface, except in the NW, is generally level. The principal rivers are the Douro, and the Esla, the Cea, the Orbiga, and the Tera. The soil is in general fertile, but poorly cultivated. Nitre and potash are obtained in some districts. The prov. is administratively subdivided into 6 partidos.—Its cap., of the same name, stands on the r. bank of the Duero, 130 m. NW of Madrid. It contained, in 1840, 2,179 houses, with 8,926 inhabitants. It is a walled town, and occupies a site of strategetic importance; but its fortifications have been neglected. It has an ancient cathedral, and several other churches, and is the seat of a college or high school with several professors. Its manufactures consist of hats, leather, brandy, and coarse woollens.

ZAMÓRA, a market-town of Mexico, in the state of Michoacan, 66 m. WNW of Valladolid, in a valley, surrounded with dense forests. Pop. 2,000. Its has two convents.

ZAMORA. See Santiago.

ZAMORZE, a village of Prussia, in the regency of Posen, and circle of Samter. Pop. 200.

ZAMOSC (Slari), a town of Russia in Europe, in Poland, in the gov. of Lublin, obwod and 9 m. NNW of Zamosc. Pop. 500. It has a lyceum.

ZAMOSC, or Zamosz, a town and fortress of Russia in Europe, in Poland, capital of an obwod of the same name, in the gov. and 51 m. SE of Lublin, on the r. bank of the Wieprz, in N lat. 50° 42′ 50″, and E long. 23° 15′ 10″. It is built in the Italian style, and has a large and handsome castle, a fine town-house, four churches, two convents, an hospital, a theatre, and a gymnasium. It possesses manufactories of linen fabrics, wax-ware and several tanneries.

ZAMOST, a market-town of Austria, in Bohemia, in the regency and bail. and 6 m. N of Budweis.

ZAMOW, a village of Prussia, in the regency of Stettin, and circle of Greifenberg. Pop. 200.

ZAMPELHAGEN, a village of Prussia, in the regency of Stettin, and circle of Nangardt. Pop. 200.

ZAMRST, a village of Austria, in Bohemia, in the regency of Pardubitz, and bail. of Hohenmauth. Pop. 430. It has a castle, several paper-mills, breweries, &c.

ZAMS, a village of Austria, in the Tyrol, in the reg. of Innspruck, and circle of Imst. Pop. 870. It has manufactories of woollen and linen fabrics.

ZAMUCOS, a tribe of Indians of La Plata, who inhabit the prov. of Chaco.

ZAMZOW, a village of Prussia, in the regency of Stettin, and circle of Saatzig. Pop. 260.

ZANA. See Dembea.

ZAND, a commune of Belgium, in the prov. of Brabant, and dep. of Cortenaeken. Pop. 208.—Also a com. in the prov. of Limbourg, and dep. of Lille-Saint-Hubert.

ZANDACAO, a village of Venezuela, in the dep. of the Orinoco, to the SE of Barceloneta.

ZANDBERGEN, a com. of Belgium, in the prov. of East Flanders, and dep. of Zele. Pop. 760.

ZANDE, a department and commune of Belgium, in the prov. of West Flanders, and arrond. of Bruges. Pop. 584.—Also a com. in the prov. of East Flanders, and dep. of Hansbeke. Pop. 314.

ZANDEN, a commune of Belgium, in the prov. of East Flanders, and dep. of Grembergen. Pop. 700.

ZANDLOOPER, a commune of Belgium, in the prov. of West Flanders, and dep. of Vracene.

ZANDOLENGE, a commune of Belgium, in the prov. of West Flanders, and dep. of Ruysselede. Pop. 1,016.

ZANDOLEUGE, a commune of Belgium, in the prov. of East Flanders, and dep. of Eecloo. Pop. 110.

ZANDOWITZ, a village of Prussia, in the regency of Oppeln, circle of Gross-Strelitz. Pop. 1,000.

ZANDSTRAET, a commune of Belgium, in the prov. of East Flanders, and dep. of Sulsique.

ZANDT, a commune of Belgium, in the prov. of East Flanders, and dep. of Saint-Paul.

ZANDT, a village of Bavaria, in the circle of the Middle Rhine, and presidial of Ripfenberg, 27 m. WSW of Ratisbon. Pop. 531. It has a quarry of stone, and manufactories of potash.

ZANDVLIET, a town of New South Wales, in the parish of Alexandria, co. of Cumberland, 8 m. from Sydney.

ZANDVOORDE, a department and commune of Belgium, in the prov. of West Flanders, and arrond. of Bruges. Pop. 543.

ZANDVOORT, a village of Holland, in the prov. of North Holland, and near Haarlem.

ZANEN (Oost) a village of Holland, and prov. of North Holland, and arrond. of Hoorn, 6 m. N of Amsterdam. Pop. 1,160.

ZANEN (West), a village of Holland, in the prov. of North Holland, arrond. and 8 m. NE of Haarlem, near the N coast, on the Y. Pop. 2,350.

ZANESVILLE, a village of Logan co., in the state of Ohio, U. S., 43 m. NW by W of Columbus, on a head branch of Mad river. Pop. in 1840, 150; in 1850, 250.—Also a village of Muskingum co., in the same state, on the E bank of Muskingum river, at the confluence of Licking river, 51 m. E of Columbus, and intersected by the central Ohio, the Cincinnati, Wilmington, and Zanesville, and the Cleveland, Zanesville, and Cincinnati railroads. Pop. in 1840, 4,766; in 1850, 7,791.

ZANG, a village of Würtemberg, in the circle of the Jaxt, and bail. of Heidenheim. Pop. 650.

ZANGARA, a river of Spain, in New Castile, an affluent of the Guadiana.

ZANGARONE, a market-town of Naples, in the prov. of Calabria-Ultra, district and cant. and 3 m. E of Nicastro, in a fertile plain. Pop., chiefly Greek, 700. Marble of different colours is wrought in the environs.

ZANGEIA, a town of Nigritia, in Houssa, on the road from Kano to Katagum, and near the SW extremity of Duchee mountains. This town appears to have formerly been a place of considerable importance, and is still surrounded by extensive plantations of cotton, tobacco, and indigo. It now consists of an assemblage of villages scattered over a considerable territory. Within the walls is a range of stone in large masses, rising to the height of about 200 ft., and extending to the adjacent hills. Z. is the residence of a governor depending upon the sultan of Kano.

ZANGENBURG, a village of Prussia, in the regency of Merseburg, and circle of Seitz. Pop. 380.

ZANGENHAUSEN, a village of Bavaria, in the circle of Upper Bavaria, and presidial of Bruck. Pop. 200.

ZANGFARA. See Zamfra.

ZANGNANE. See Natal.

ZANGUEBAR, a term which, equivalent to Bar-es-Zenug, signifies literally 'the country of the Zenug,' or 'Blacks,' vaguely applied to a large extent of the eastern coast of Africa between Mozambique and the Red sea. The inhabitants consist of Negroes, with a number of Arabs settled in the towns. The cities of Quiloa, Mombaza, Melinda, and some others, have been visited by Europeans; but the interior is entirely unknown. During the prosperity of the Portuguese, their empire extended over the greater part of this coast; but they have now been generally obliged to give place to the imam of Muscat.

ZANIESMYSHL. See Santomysl.

ZANNONE, or Zanone, an island of the Tyrrhenian sea, in the group of the Pouza islands, of which it is the most northerly, and depending upon the kingdom of Naples, in the prov. of the Terra-di-Lavoro, and district of Gaita. It is equilateral in form, elongated from N to S, and is partly volcanic and partly calcareous in structure. It is destitute of water, and desert. It formerly had a Benedict convent.

ZANOW, a town of Prussia, in the regency of Koslin, and circle of Schlawe, on the E slope of the Gallenberg. Pop. 1,278. It has several distilleries of brandy and breweries.

ZANSKAR, a division of Ladak, comprising the country lying along the two great branches of the river of the same name; and bounded by Ladak Proper on the N; by Rukchu on the E; by Lahul on the S; and by Purik and Waula on the W. The

Himalaya forms its S boundary; the Trans-Himalayan chain its N. It has an area of 3,000 sq. m., and a mean elevation of 13,154 ft.—The head-waters of the Z. river, the Yunam, the Sachu, and the Cherpa, all rise to the N of the Himalaya range. From the source of the Cherpa, in N lat. 32° 40′, to the town of Phadam, the distance is 130 m.; and the fall of the Z. 4,000 ft. From Phadam the stream flows due N 80 m., to its junction with the Indus opposite Nijimo. Its entire length is 210 m.; total fall, 6,000 ft.

ZANT, a village of Bavaria, in the circle of Lower Bavaria, and presidial of Kotzting. Pop. 260. It has a castle.

ZANTE, the ancient *Zacynthus*, an island of the Mediterranean, one of the Ionian group, situated 9 m. S of Cephalonia, and 15 m. W of the Morea. Its form is irregular. Its length is 25 m.; its breadth above 10 m.; and its circumference 56 m. In its aspect it is the finest of the Ionian islands, presenting, when viewed from above the town of Zante, a series of vales and eminences richly cultivated, and covered with hamlets or villages, embosomed in olive plantations. The W coast is high and cliffy. A chain of mountains extends throughout its whole length on the W side. The highest hill in the island [alt. 2,200 ft.] is called Skopo. The plain of Chierri occupies nearly half the surface of the island in the N and E parts. Z. has no large rivulets; and in summer, considerable inconvenience is experienced from the drying up of the springs and wells. Near the SE extremity of Chierri bay is an extensive morass. The whole surface of the island presents the traces of subterraneous fire, discovered in some parts by warm sulphureous springs, and occasionally by the occurrence of earthquakes; springs of petroleum and mineral tar, known in the days of Herodotus, are still worked to advantage. The coast in many places forms recesses, answering, in some degree, the purpose of harbours. The climate, though extremely hot in summer, is not unwholesome, nor so variable as that of the other Ionian islands. Snow never lies except on the mountain-tops. The winds generally blow from NW and N in spring and summer. The chief products of Zante are currants, olives, and other fruits of a warm latitude. Currants have long formed a great article of export to England and Holland; and the extent of ground occupied by their cultivation exceeds 6,500 acres. Of olive oil, the annual produce is fully 32,000 barrels, of about 130 lbs each. On the other hand, the corn raised is hardly equal to four months' consumption; and the deficiency is supplied by imports from the Morea, whither the peasantry of Z. resort in the harvest to the number of 4,000 or 5,000, receiving payment for their labour in corn, of which they bring back with them on an average, not less than 50,000 bushels. Cotton and silk are cultivated to a small extent; pasturage is very scanty, in consequence of the droughts of summer. Goats are the only live-stock. The soil of this island being so fertile, the pop. is much more dense than in other parts of Greece. The total in 1851 was 39,103. Of these above 18,000 reside in the town of Zante; the rest in villages, hamlets, and scattered habitations, many of which are pleasantly situated along the coast. The language of the town, for the purpose of intercourse with foreigners, is Italian; but the current language is Greek. The island fell into the hands of the British in October 1809.

Zante, the capital of the foregoing island, is situated on the SE coast, about 12 m. W of Cape Tornese in the Morea, at the bottom of a small bay, and on a hill of gentle declivity, descending in an almost imperceptible slope to the surface of the sea. In its appearance, Z. resembles an Italian town. The principal street, which traverses it in its whole length, is broad and handsome: it is bordered with well built houses and churches, some supported by colonades, others provided with piazzas. It has likewise the advantage, so rare on the continent, of a foot-pavement. The market-place, situated near the sea, is spacious and well laid out. The houses are built partly of brick, partly of wood; and, on account of the frequency of earthquakes, seldom exceed one or two stories in height: the last visitation of this kind, in 1820, overthrew several hundred houses, without, however, causing a loss of lives. Z. partakes of the character of a commercial town, comfort being more studied than elegance, and there being no public amusements except billiards, and occasionally a theatrical representation by Italian players. Of the churches, several are well built, but none particularly striking. The harbour is capacious.

ZANTKAU, a village of Prussia, in the regency of Breslau, and circle of Trebnitz. Pop. 260. It has a castle.

ZANTOCH, a village of Prussia, in the regency of Frankfort, and circle of Landsberg. Pop. 868.—Also a village of the regency of Breslau, and circle of Dels. Pop. 278.

ZANZHAUSEN, a village of Prussia, in the prov. of Brandenburg, regency of Frankfurt, circle and 11 m. NE of Landsberg, on the Zanze. Pop. 555. It has extensive iron-works.

ZANZIBAR, an island in the Indian ocean, situated off the E coast of Africa, between the 6th and 7th parallels of S lat. It is nearly 50 m. in length, and 20 m. in breadth; and is distant from the continent, along which it stretches in a NE direction, about 45 m. Between the continent and it there is, however, no navigable passage for large vessels, except through the harbour, as a reef runs obliquely across from the African shore to the small islands which lie close to the W side of Zanzibar. The appearance of the island is extremely delightful; and the climate is similar to that of India, only the monsoon, or rainy season, sets in sooner. The town is situated on the W side of the island, and is both large and populous. The sovereignty of the island belongs to the imam of Muscat, who appoints the *haikim*, or governor, and to whom the revenue derived from its commerce and land tenure devolves. The imam maintains no kind of military force; but the *haikim's* slaves, amounting to 400 or 500 men, are armed, and serve as soldiers under three Arab officers. The population of the island may be estimated at 200,000, three-fourths of whom are slaves. The principal articles of export are slaves and ivory, with a small quantity of drugs; the number of slaves annually sent to Muscat, India, and elsewhere, is estimated at not less than from 6 to 10,000. The imports are chiefly from India and the neighbouring coast of Africa, and are of considerable importance; judging from the amount of customs, their value cannot be under 300,000 dollars annually. The foreign commerce of Zanzibar is considerable, and almost entirely in the hands of the Americans and English. The American trade bears to the English a proportion of five to one, and to that of any other nation, about twenty to one. The American business is generally held by the merchants of Salem. The imports from the United States, are domestic cottons, and various other manufactured goods, as ammunition, household furniture, &c. By a late treaty between the sultan of Muscat and the United States, these imports are introduced subject to five per cent. duty, while exports in American ships are freed of the five per cent. duty imposed on those conveyed in the ships of other nations. As in most eastern countries, however, the sultan farms out the customs for a specific sum—here 150,000 dollars per annum.—The cap. of the island is situated on a sandy peninsula projecting from its W side, in S lat. 6° 10′. It is a miserably built place, with a pop. of perhaps 25,000. On the SW side of the island is another considerable town.

ZANZIN, a village of Prussia, in the regency of Frankfurt, and circle of Landsberg. Pop. 584.

ZANZTHAL, a village of Prussia, in the regency of Frankfurt, and circle of Landsberg. Pop. 425.

ZAOREJAS, a market-town of Spain, in the prov. and 48 m. from Cuenca. Pop. 700. It has an aqueduct and an hospital.

ZAORLA, a village of Prussia, in the regency of Posen, and circle of Kröben. Pop. 220.

ZAPADNAIA, a town of Russia in Europe, in the gov. of the Don Cossacks, 33 m. E of Tcherkask.

ZAPARA, an island off the N coast of Venezuela, in the dep. of Sulia, and prov. of Coro, on the S part of the gulf of Maracaibo, 30 m. N of the lake of that name. It is about 15 m. in length from E to W.

ZAPARDIEL, a river of Spain, which has its source in the prov. and to the NW of Avila, traverses the N part of that prov. and the S of that of Valladolid, and after a course in a generally S direction of about 75 m., joins the Duero on the l. bank, 3 m. WSW of Tordesillas. Its waters are sluggish and impure, but abound with fish, and its banks unhealthy.

ZAPATERA, an island in the lake of Nicaragua, in Central America. It is mountainous, rising to a height of nearly 2,000 ft.; and is the largest in the lake, but is uninhabited. Squiers thus describes the islands of this lake generally, and that of Z. in particular. Perhaps a more singular group of islets cannot be found. They are all of volcanic origin, generally conical in shape, and seldom exceeding three or four acres in area. "All are covered with a cloak of verdure; but nature is not always successful in hiding the black rocks which start out in places, as if in disdain of all concealment, and look frowningly down in the clear water, giving an air of wildness to the otherwise soft and quiet scenery of the islands. Trailing over these rocks, and dropping in festoons from the over-hanging trees, their long pliant tendrils floating in the water, are innumerable vines, with bright and fragrant flowers of red and yellow, mingled with the inverted cone of the 'gloria de Nicaragua,' with its overpowering odour, with strange and nameless fruits, forming an evergreen roof so dense that even a tropical sun cannot penetrate. Many of these islands have patches of cultivated ground, and on such, generally crowning their summits, relieved by a dense green background of plantations, and surrounded by kingly palms and the papaya with its great, golden fruit, are the picturesque cane huts of the inhabitants. Groups of naked, swarthy children in front,—a winding path leading beneath the great trees down to the water's edge,—an arbour-like miniature harbour, with a canoe lashed to the shore,—a woman naked to the waist with a purple skirt of true Tyrian dye (for the famous *murex* is found on the Pacific shores of Nicaragua), her long, black, glossy hair falling over neck and breast, reaching almost to her knees,—a flock of noisy parrots in a congressional squabble among the trees,—a swarm of parroquets scarcely less noisy,—a pair of vociferating macaws like floating fragments of a rainbow, inquisitive monkeys hanging among the vines,—active iguanas scrambling up the banks,—long-necked and long-legged cranes in deep soliloquy at the edge of the water, their white bodies standing out in strong relief against a background of rock and verdure,—a canoe glancing rapidly and noiselessly across a vista of water,—all this, with a golden sky above, the purple sides of the volcano of Momobacho overshadowing us, and the distant shores of Chontales molten in the slanting sunlight,—these were some of the elements of the scenery of the islands—elements constantly shifting, and forming new and pleasing combinations." After toiling for a long time, Mr. Squiers continues, "we came suddenly upon the edge of an ancient crater of great depth, at the bottom of which was a lake of a yellowish green or sulphurous colour, the water of which Manuel assured me was salt. This is probably the fact, but I question much if any human being ever ventured down its rocky and precipitous sides. Manuel now seemed to recognize his position; and turning sharp to the left, we soon came to a broad level area, covered with immense trees, and with a thick undergrowth of grass and bushes. There were here some large, regular mounds composed of stones, which I soon discovered were artificial. Around these Manuel said the *freyles* were scattered, and he commenced cutting right and left with his machete. I followed his example, and had not proceeded more than five steps, when I came upon an elaborately sculptured statue, still standing erect. It was about the size of the smaller one discovered at Pensacola, but was less injured, and the face had a mild and benignant aspect. It seemed to smile on me as I tore aside the bushes which covered it, and appeared almost ready to speak. In clearing further I found another fallen figure, but a few feet distant. The first monument which claimed our attention was a well-cut figure, seated crouching on the top of a high ornamented pedestal. The hands were crossed below the knees, the head bent forward, and the eyes widely opened as if gazing upon some object on the ground before it. A conical mass of stone rose from between the shoulders, having the appearance of a conical cap when viewed from the front. It was cut with great boldness and freedom, from a block of basalt, and had suffered very little from the lapse of time. It is impossible, without engravings and plans, to give any clear comprehension of these monuments, and I shall not attempt a detailed account of them. They are very different from those discovered by Mr. Stephens at Copan. Instead of the heavy and incongruous mass of ornament with which those were loaded, most of these are simple and severe,—and though not always elaborately finished, are cut with great freedom and skill. There is no attempt at drapery in any of the figures. Some are erect, others seated, and still others are in crouching or reclining postures. The material, in every case, is a black basalt. A few of the figures, from defects of the stone, have suffered somewhat from the weather, but less from this cause than from the fanaticism of the conquerors. They all bear marks of the heavy sledges, or other instruments, with which the Catholic zealots endeavoured to destroy them; but the task was not an easy one, and fortunately for the archæologist, the massive stones resisted their assaults."

ZAPATOSA, a lake of New Granada, in the dep. of the Magdalena, and prov. of Mompox. It is formed by the Cesare, an affluent of the Magdalena, to the E of the river of that name, and of Mompox. It is about 30 m. in length from E to W, and 27 m. in breadth.

ZAPEL, a village of Mecklenburg-Schwerin, in the prov. of Mecklenburg, and bail. of Erivitz. Pop. 310.

ZAPENDOWO, a village of Prussia, in the regency of Marienwerder, and circle of Konitz. Pop. 200.

ZAPES, a people of Western Africa, who inhabit the W part of Senegambia near Bagos.

ZAPFENDORF, a village of Bavaria, in the circle of Upper Franconia, and presidial of Schesslitz. Pop. 597.

ZAPOLIANSKAIA, a town of Russia in Europe, in the gov. of the Don Cossacks, on the r. bank of the Medoiaditza, 279 m. NE of Novo-Tcherkask. It is surrounded with forests, which are much infested with bears.

ZAPONE, an island of Central America, in the state of Nicaragua, and in the SE part of the lake of that name.

ZAPOROGUES. See COSSACKS.

ZAPOTITLAN, a headland of Central America, in the state and 105 m. SE of Vera Cruz, on the E side of a small islet formed by the gulf of Mexico.—Also a village of Mexico, in the state of Xalisco, about 20 m. SSE of Guadalajara, and near the l. bank of the Rio Grande Santiago.

ZAPOTLAN-EL-GRANDE, or ZAPOTITLI, two large villages of Mexico, in the state of Xalisco, to the N of the volcano Colima.

ZAPPENDORF, a village of Prussia, in the regency of Merseburg, and circle of the See. Pop. 510.

ZAPPLAU, a village of Prussia, in the regency of Breslau, and circle of Guhran. Pop. 375.

ZAPPLIN, a village of Prussia, in the regency of Stettin, and circle of Greifenberg. Pop. 380.

ZAPUTERA. See ZAPATERA.

ZARA, a circle of Austria, in Dalmatia, bounded on the N by Sclavonia, on the E by Turkey in Europe, on the S by the circle of Spalatro, and on the W by the Adriatic. It is formed from the ancient county of Zara or Kotar, Bukovizza, the counties of Sebenico, Knin, and Nona, the territory of Novigrad, the islands of Pago and Arba, and some others of smaller extent, comprising a total area of 840 sq. m. Pop. in 1839, 137,510. See also DALMATIA.

ZARA, or ZARDRU, a town of Austria, capital of the kingdom or ldgb., of Dalmatia, and of the circle of the same name, 180 m. SE of Venice, and 321 m. S of Vienna, on the Adriatic, opposite the island of Uglian, from which it is separated by a strait of the same name, in N lat. 44° 6′ 51″, and E long. 15° 13′ 33″. Pop. in 1843, 6,850. It is situated on a tongue of land, separated from the continent by a deep ditch, crossed by a bridge, and is fortified. It is well built, and contains a cathedral, Catholic gymnasium, a lyceum, an archiepiscopal seminary, a martime hospital, several schools, and a theatre, It has manufactories of silk and woollen fabrics, and tanneries; and is noted for its manufactory of *maraschino*, a liqueur made from the *marasca*, or black cherry, which is grown mostly in the neighbourhood of Almissa, between Spalato and Macarsca. Bordeaux is not more famous for its wines than Z. for its liqueurs. There are, altogether, about a dozen distilleries in the town. Its trade consists chiefly in wine and figs. Z. is destitute of springs of water, and receives its only supply by means of cisterns. It has the remains of a Roman aqueduct. The port is spacious, but wanting in depth, and open to the N wind, which here frequently blows with great violence.

ZARACCA, VULSI, or STYMPHALUS, a lake of Greece, in the Morea, in the province and 22 m. SW of Corinth. It is nearly 5 m. in length from NW to SE, and has no apparent outlet. At its NE extremity is a village of the same name.

ZARAGOZA. See SARAGOSSA.

ZARAGOZA, a town of New Granada, in the dep. of Cundinamarca, prov. and 120 m. NE of Antioquia, near the confluence of the Porce and Nechi. Pop. 2,000. The district is extremely unhealthy.

ZARAISK, a town of Russia in Europe, in the gov. and 36 m. NW of Riazan.

ZARAKOULA, a village of Greece, in the Morea, diocese and 28 m. WSW of Corinth, at the NE extremity of Lake Phonia.

ZARA-LAKALIA, or ZERA-LACALIA, an island of Tunis, in the gulf of Cabes, the most westerly of the Kerkima islands, in N lat. 34° 39′ 50″, and E long. 10° 57′.

ZARAND, a town of Hungary, in the comitat and 24 m. NE of Arad, at the confluence of the Schwartze-koros and Csiger. Pop. 2,827. Tobacco is cultivated in the environs.—Also formerly a province of Transylvania, now comprised in the comitat of Arad.

ZARANG. See JELLALABAD.

ZARATA, a town of Bolivia, in the dep. and 111 m. N of La Paz, near the l. bank of the Beni.

ZARATAN, a town of Spain, in the prov., partido, and 3 m. from Valladolid, near the Pisuerga. Pop. 896. It is noted for its bread.

ZARA-VECCHIA, or ALT-ZARA, BIOGRAD or BIELOGRAD, a town of Austria, in Dalmatia, in the circle and 21 m. SE of Zara, on the Adriatic, in N lat. 43° 56′ 19″, and E long. 15° 26′ 18″. Pop. 1,348. It has a port, and was formerly a flourishing place, and the residence of several of the kings of Croatia. It was the Colonia-Claudia-Augusta-Felix, Blandona, and Alba Maris of the Romans. During the war of the Venetians with the kings of Hungary, it was suffered by the doge Faliero to fall into ruins. In the 18th century it was destroyed by the government of Venice.

ZARAUZ, or SARAUS, a town of Spain, in the prov. of Guipuzcoa, and partido of Vergara, 12 m. W of San Sebastian, at the foot of the mountain of Santa Barbara, and on the coast of the gulf of Gascogny. Pop. 969. It is well-built, and has a parish church and several convents. It has a small fishing port.

ZARAZITZ, a village of Austria, in Moravia, in the circle of Hradisch, and seignory of Wesselt. Pop. 660.

ZARBEN, a village of Prussia, in the regency of Stettin, and circle of Greiffenberg. Pop. 400.

ZARBRAK, ZUURBRAK, or MOSES KRAAL, a Dutch village and English missionary settlement, in the gov. of the Cape of Good Hope, prov. and 15 m. WNW of Zwellendam, on the r. bank of Buffeljagts river, and near its sources in the Zwarte-Berg.

ZARB-SUI. See ZAB.

ZARCYN, a village of Prussia, in the regency of Bromberg, and circle of Wongrowiec. Pop. 260.

ZAREC, a village of Austria, in Bohemia, in the regency of Pilsen, and bail. of Horazdiowitz. Pop. 400.

ZARED, SZAFYE, HUSSAN, or AHSA, a river of Turkey in Asia, in Syria, which has its sources on the confines of Arabia, runs NW, and flows into the SE extremity of the Dead sea, a little below the hamlet of Szafye.

ZAREVOKOKSCHAISK, a town of Russia in Europe, in the gov. of Kasan, on the Maloi Kolschaja. Pop. 5,000.

ZAREVOSANTSCHURSK, a town of Russia in Europe, in the gov. of Viatka. Pop. 2,000.

ZAREWO, a town of Russia in Europe, in the gov. of Arkhangel. Pop. 4,000.

ZARGUN, or ZERGUN, a town of Persia, in the prov. of Fars, 15 m. NE of Huran, at the foot of a chain of low hills. It is enclosed by an earthen wall, and contains about 300 houses, most of which have wells of good water. It has a caravanserai, but of the meanest description. The industry of the inhabitants consists chiefly in the manufacture of a species of blue cloth. The environs are fertile, producing corn, sesame, and cotton.

ZARIA, a town of Nigritia, in Houssa, capital of the state of Zegzeg, and 75 m. SW of Kano. Pop. chiefly Felatahs, 50,000. The environs are beautiful and well-cultivated. It was visited by Clapperton in 1826.

ZARICZ, a village of Austria, in Bohemia, in the regency of Budweis, and bail. of Tabor. Pop. 300.

ZARICZAN, a village of Austria, in Bohemia, in the regency of Pardubitz, and bail. of Czaslau. Pop. 420.

ZARIGRAD, a mountain of Turkey in Europe, between Bulgaria and Servia, on the confines of the sanjaks of Kruchovatz and Sophia. It belongs to the N branch of the Balkan, and lies between the basins of the Morava and Isker.

ZARITZ, a village of Austria, in Moravia, in the circle of Prerau, and seignory of Kremster. Pop. 500.

ZARITZIR. See Tzaritzin.

ZARJECS, a village of Hungary, in the comitat and 3 m. WSW of Trentsen, on the r. bank of the Waag. It is noted for its cutlery.

ZARKI, a town of Russia in Europe, in Poland, in the gov. of Lublin, obwod and 29 m. NNW of Olkusz. Pop. 1,980. It has three churches, a convent, and a synagogue. In the vicinity are mines of iron.

ZARKO, a village of Turkey in Europe, in Thessaly, 30 m. E of Trikali, and near the N bank of the Cachia.

ZARMIZAGELHUSA. See Varhely.

ZARNEFAUX, a village of Prussia, in the regency of Koslin, and circle of Belgard. Pop. 200.

ZARNEKOW, a village of the grand-duchy of Mecklenburg-Schwerin, circle of Wendisch, and bail. of Dargun. Pop. 200.

ZARNEWANZ, a village of the grand-duchy of Mecklenburg-Schwerin, circle of Wendisch, and bail. of Lessin. Pop. 200.—Also a village of Prussia, in the regency of Stralsund, and circle of Grimmen. Pop. 200.

ZARNGLAS, a village of Prussia, in the regency of Stettin, and circle of Kammin. Pop. 220.

ZARNIKOW. See Czarnikau.

ZARNINKOW, a village of Prussia, in the reg. of Stettin, and circle of Saatzig. Pop. 200.

ZARNO, a town of Austria, in Galicia, in the circle of Tarnow. Pop. 1,500.

ZARNOW, a village of Prussia, in the regency of Stettin, and circle of Pyritz. Pop. 240.—Also a village and colony of the same regency, in the circle of Greiffenhagen. Pop. 430.—Also a town of Russia in Europe, in Poland, in the gov. of Radom, obwod and 11 m. SW of Opoczno. Pop. 730.

ZARNOW, or Czarnow, a village of Poland, in the gov. of Radom, obwod and 1½ m. W of Kielce. Pop. 320. On an adjacent mountain are the remains of a Benedictine convent.

ZARNOWICE, a town of Russia in Europe, in Poland, in the gov. of Kielce, obwod and 15 m. NNW of Miechow, near the Pilica. Pop. 1,638. It has the ruins of a castle, noted as the residence of the spouse of Casimir the Great.

ZARNOWITZ, a village of Prussia, in the regency of Danzig, and circle of Neustadt. Pop. 360.

ZAROSCHITZ, a village of Austria, in Moravia, in the circle of Brunn. Pop. 980. It has manufactories of potash and several breweries.

ZARPANA. See Rota.

ZARPEN, or Sarpen, a parish of Holstein, to the N of Reinfeld. Pop. 570.

ZARRA, a market-town of Spain, in the prov. and 51 m. SW of Valencia, and partido of Jarafuel, on a calcareous mountain. Pop. 900. Wine and oil are cultivated in the vicinity.

ZARRA, a town of Tunis, 9 m. SE of Cabes, on the shore of the gulf of that name.

ZARRAH, or Zirrah. See Hamun.

ZARREN, a village of Belgium, in the prov. of West Flanders, and arrond. of Furnes. Pop. 1,668.

ZARRENDORF, two villages of Prussia, in the reg. of Stralsund, and circle of Grimmen. Pop. 480.

ZARRENTIN, a market-town of the grand-duchy of Mecklenburg-Schwerin, in the circle and 23 m. WSW of Schwerin, on the SW bank of Lake Schaal. Pop. 1,000.

ZARSHOE, a village of Afghanistan, in the district of Kohistan, on an affluent of the Pinyshreen, in N lat. 35° 7′.

ZARSKOJE-SELO. See Sofia.

ZARTEN, a village of the grand-duchy of Baden, circle of the Middle Rhine, and bail. of Freiburg. Pop. 428.

ZARTLESDORF, a village of Austria, in Bohemia, in the regency of Budweis, and bail. of Kaplitz. Pop. 250. It has two castles.

ZARUDZIE, a town of Austria, in Galicia, in the circle of Zloczow, on the Strypa.

ZARUMA, or Zuruma, a town of Ecuador, in the dep. of Guayaquil, 30 m. NW of Loxa, in the Andes, at an alt. of 4,921 ft. above sea-level, and on the Tumbez. Pop. 5,000. It had formerly important gold-mines in its vicinity.

ZARUS, a village of Afghanistan, on the l. bank of the Helmund, in N lat. 30° 21′.

ZARYB, a village of Austria, in Bohemia, in the regency of Prague, and bail. of Melnik. Pop. 240.

ZARYN, a mountain of Mongolia, in the district of Sounit, and desert of Gobi.

ZARZA (La), a town of Spain, in New Castile, in the prov. of Cuenca, and partido of Ucles, 3 m. W of Tarancon. Pop. 355. Cordage is its chief article of manufacture. It is of great antiquity, and was formerly very populous. The *Vilismea* of the Romans was situated in its vicinity.

ZARZA CAPILLA, a town of Spain, in Estremadura, in the prov. of Badajoz, and partido of Puebla-de-Alcocer, 27 m. from Villanueva-de-la-Serena, in a mountainous locality. Pop. 1,800.

ZARZA-DE-ALANGE, or Zarza-pinte-alanje, a town of Spain, in Estremadura, in the prov. of Badajoz, partido and 11 m. SE of Merida, on the l. bank of the Guadiana, between two rocks. Pop. 3,127. It contains a parish church and a custom-house, and has manufactories of linen and coarse woollen fabrics. The environs are noted for their excellent wine.

ZARZA-DE-MONTANCHEZ, a town of Spain, in Estremadura, in the prov. of Caceres, partido and 6 m. WSW of Montanchez, in a fertile valley, about ¾ of a mile from the Sierra-de-San-Cristoval. Pop. 1,200. It has manufactories of soap, blankets, pottery, tiles, and bricks, and is noted for its hams. Wine is extensively grown in the environs.

ZARZA-LA-MAYOR, a town of Spain, in Estremadura, in the prov. of Caceres, partido and 12 m. N of Alcantara, in a spacious valley. Pop. 2,313. It has a parish church, several convents, and a custom-house, and is tolerably well built.

ZARZALEJO, a town of Spain, in New Castile, in the prov. and 24 m. NW of Madrid, and 3 m. SW of the Escorial, between two mountains. Pop. 750.

ZARZIG, a village of Prussia, in the regency of Stettin, and circle of Saatzig. Pop. 310.

ZARZIS, a village of Tunis, on the SW coast, on the confines of Tripoli.

ZARZOSA, a market-town of Spain, in Old Castile, in the prov. and 30 m. from Soria. Pop. 560. It has manufactories of common cloth, and several dye-works.

ZARZUELA, a market-town of Spain, in New Castile, in the prov. and partido and 6 m. from Cuenca. Pop. 500.—Also a royal residence in the prov. and 6 m. N of Madrid.

ZARZYDSCHE, a village of Prussia, in the regency of Oppeln, and circle of Pless. Pop. 300.

ZASADA, a village of Austria, in Bohemia, in the reg. of Gitschin, and bail. of Eisenbrod. Pop. 800.

ZASCHAN, a village of Austria, in Moravia, in the circle of Olmutz, and bail. of Walachisch-Meseritsch. Pop. 1,200.

ZASCHOWITZ, a village of Austria, in Moravia, in the circle of Iglau. Pop. 240.

ZASCINOSCZE, a village of Galicia, in the circle of Tarnapol. It has extensive quarries of sharpingstone.

ZASENBECK, a village of Hanover, in the principality of Luneberg, bail. of Knesebeck. Pop. 200.

ZASIGNANE, a name by which the aborigines of the Aleutian islands distinguish the group consisting of the Aleutian islands, properly so called, and Copper and Behring's islands.

ZASLAV, Zaslaw, or Iziaslav, a district and town of Russia in Europe, in the gov. of Volhynia, 93 m. W of Jitomir, on the Gorin. Pop. 8,200, of whom a large proportion are Jews. It has a synagogue and a convent. This town was founded by Vladimir the Great, and named by him, in honour of his son, Iziaslav.

ZASMUK, a market-town of Austria, in Bohemia, in the regency of Pardubitz, bail. and 5 m. SE of Kaurzim. Pop. 1,760. It has a castle.

ZASTRIZL, a village of Austria, in Moravia, in the circle of Hradisch. Pop. 350.

ZATAS. See Sorraya.

ZATAU, a village of Austria, in Bohemia, in the regency of Pilsen, and bail. of Pisek. Pop. 250.

ZATECZ. See Saatz.

ZATKOW, a village of Prussia, in the regency of Koslin, and circle of Belgard. Pop. 300.

ZATMEOS, an Indian tribe of La Plata, in the prov. of Chaco.

ZATOR, a town of Austria, in Galicia, in the circle and 9 m. NW of Wadowice, on the Skawa, near the r. bank of the Vistula. Pop. 1,466, of whom 243 are Jews. It has a castle.

ZATTEN, a village of Prussia, in the regency of Frankfort, and circle of Arnswalde. Pop. 270.

ZATTIG, a village of Austria, in Silesia, in the circle of Tropau. Pop. 470.

ZATUNA, a town of Greece, in the Morea, in the gov. of Gortynia, and 17 m. WNW of Tripolitza.

ZATZENHAUSEN, a village of Würtemberg, in the circle of the Neckar, and bail. of Cannstadt. Pop. 385.

ZAUBACH, two villages of Bavaria, in the circle of Upper Franconia, and presidial of Stadtsteinach. Pop. 290.

ZAUCEJO, a town of Spain, in Estremadura, in the prov. of Badajoz, partido and 36 m. from Villanueva-de-la-Serena. Pop. 300.

ZAUCH-BELZIG, a circle of Prussia, in the reg. of Potsdam. Pop. 43,155.

ZAUCHE, a village of Prussia, in the reg. of Frankfort, and circle of Lubben. Pop. 400.

ZAUCHE (Gross), a village of Prussia, in the reg. of Breslau, and circle of Trebnitz. Pop. 300. It has a castle.—Also a village of the regency of Liegnitz, and circle of Grunberg. Pop. 330.

ZAUCHE (Neu), a village of Prussia, in the reg. of Frankfort, and circle of Lubben. Pop. 400.

ZAUCHE (Ober), a village of Prussia, in the reg. of Liegnitz, and circle of Glogau. Pop. 360.—Also a village of the same regency, and circle of Sprottau. Pop. 350. It has a castle.

ZAUCHTEL, Zauchtenthal, or Segdal, a village of Austria, in Moravia, in the circle of Prerau, and seignory of Runewald. Pop. 1,140.

ZAUCHWITZ, a village of Prussia, in the reg. of Oppeln, and circle of Leobschutz. Pop. 705.—Also a village of the regency of Potsdam-Beutz. Pop. 285.

ZAUDITZ, a market-town of Prussia, in the reg. of Oppeln, circle and 8 m. SW of Ratibor, on the Muhlbach. Pop. 1,000. It has a castle. It has manufactories of cloth, shoes, and leather, and several dye-works.

ZAUGGENRIED, a village of Switzerland, in the cant. of Bern, and bail. of Fraubrunnen. Pop. 377.

ZAUGHALS, a village of Prussia, in the reg. of Breslau, and circle of Glatz. Pop. 260. It has a castle.

ZAUKERODE, a village of Saxony, in the circle and 4 m. SSW of Dresden. Pop. 610. Coal is extensively wrought in the environs.

ZAUMGARTEN, a village of Prussia, in the reg. and circle of Breslau. Pop. 260.

ZAUNRODEN, a village of Prussia, in the reg. of Erfurt, and circle of Worbis. Pop. 260.

ZAUSKAR, a canton of Ladak or Little Tibet, extending between the Sechu or Lingtu, on the E, and the Himalaya chain on the W.

ZAUSSWITZ, a village of Saxony, circle of Leipsig and bail. of Aschak, near Strehla. Pop. 325.

ZAUTKE, a village of Austria, in Moravia, in the circle of Olmutz, and seignory of Hohenstadt, on the March. Pop. 560.

ZAVARA. See Inhaga.

ZAVATERILLO, a village of Sardinia, in the div. of Genoa, prov. and 11 m. NW of Bobbio. Pop. 1,700.

ZAVEDIN, a mountain of Turkey in Europe, in the W part of Herzegovinia, near the Dalmatian frontier, and joining the Prolog chain, on the NW.

ZAVELSTEIN, a town of Würtemberg, in the circle of the Schwarzwald, bail. and 3 m. S of Calw, and 23 m. NW of Tubingen. Pop. 357. It contains the ruins of a castle, and has a suburb.

ZAVERDA (Gulf of), an indentation of the coast of Turkey in Europe, on the W coast of Livadia, opposite the island of Santa-Maura. It receives a river on the N, on which is a village of the same name.

ZAWADA, a village of Prussia, in the regency of Oppeln, and circle of Tost. Pop. 270.—Also a village of the same regency, in the circle of Pless. Pop. 460.—Also a village of the same regency, in the circle of Ribnik. Pop. 250.

ZAWADILKA, a colony of Austria, in Moravia, in the circle of Olmutz and seignory of Laschkau. Pop. 200.

ZAWADKA, a village of Prussia, in the reg. of Oppeln, and circle of Pless. Pop. 200.

ZAWADOW, a town of Austria, in Galicia, in the circle of Lauberg. It has a paper-mill.

ZAWICH. See Zouvaya.

ZAWICHOST, a town of Russia in Europe, in Poland, in the gov. of Radom, obwod and 11 m. NNE of Sandomir, on the l. bank of the Vistula. Pop. 2,500.

ZAWIDOWICE, a village of Prussia, in the reg. of Posen, and circle of Pleschen. Pop. 300.

ZAWIESCHIN, a village of Austria, in Bohemia, regency of Pilsen and bail. of Blatna. Pop. 300.

ZAWISCZ, a village of Prussia, in the regency and circle of Oppeln. Pop. 240.

ZAWITZ, a village of the duchy of Anhalt-Dessau, in the prov. of Kothen, and bail. of Wulffen, on the Fuhne. Pop. 220.

ZAWOBRESK, a village of Austria, in Bohemia, reg. of Budweis, and bail. of Kruman. Pop. 330.

ZAWODZIE, a village of Prussia, in the reg. of Posen, and circle of Wreschen. Pop. 200.

ZAYANE, a district and town of Marocco, 125 m. ESE of the town of that name, in the prov. of Tafilet, on the E side of the Atlas mountains, and on one of the head streams of the Draha. Pop. of district, 100,000.

ZAYAS-DE-TORRE, a town of Spain, in Old Castile, in the prov. and 51 m. SE of Burgos. Pop. 300. On an adjacent hill are the ruins of a fortress.

ZAYENDIH-RUD, or ZENDERUD, a river of Turkey in Asia, in Irac-Ajemi, which has its source in the Koh-i-Zerd, a summit of the mountains of Awa; runs NE to Ispahan, where it is joined by a considerable stream from the NW; and bending ESE, is gradually lost in the sand and marshes in the SW corner of Khorasan, and after a total course of about 180 m.

ZAYLA, a town of Peru, in the dep. of Guamanga, on the r. bank of the Pausa, and 24 m. SW of the town of that name.

ZAZRIVA, a village of Hungary, in the comitat of Arva, 11 m. NW of Kubin. Pop. 2,300.

ZAZUAR, a town of Spain, in Old Castile, in the prov. and 40 m. S of Burgos, and partido of Aranda-de-Duero, near the confluence of the Arandilla with the Duero. Pop. 700.

ZBANSZYN. See BENTSCHEN.

ZBANITZ, a village of Austria, in Moravia, in the circle of Brunn, and seignory of Obrowitz. Pop. 360.

ZBARASZ, a town of Austria, in Galicia, in the circle and 12 m. NE of Tarnopol, on the Ikwa. Pop. 5,700. It has an ancient castle, a Catholic and two Greek churches, and a Bernardine convent. The defence of this place, in 1652, against Khmielnitski at the head of the revolted Zaporogues, by Andrew Firloy, a man of English descent, but at that time a Polish noble, and castellan of Betz, preserved Poland for the time, and threw the Cossacks into the hands of Russia.

ZBEGSCHOW, a village of Austria, in Moravia, in the circle of Brunn, and seignory of Obrowitz. Pop. 360.

ZBELITOW, a village of Austria, in Bohemia, in the regency of Budweis, and bail. of Mühlhausen. Pop. 230.

ZBENCHY, a village of Prussia, in the regency of Posen, and circle of Kosten. Pop. 260.

ZBENITZ, a village of Austria, in Bohemia, in the regency of Pilsen, and bail. of Wirowitz. Pop. 310. It has a castle, and possesses manufactories of potash and breweries.

ZBESCHAU, a village of Austria, in Moravia, in the circle of Brunn, seignory of Oslawan. Pop. 360.

ZBETSCHNICK, a village of Austria, in Bohemia, in the regency of Gitschin, and bail. of Braunau. Pop. 800.

ZBETSCHNO, a village of Austria, in Bohemia, in the reg. of Prague, bail. of Rakonitz. Pop. 680.

ZBICZNO, a village of Prussia, in the regency of Marienwerder, and circle of Strasburg. Pop. 240.

ZBIRON, a market-town of Austria, in Bohemia, in the regency of Prague, and bail. of Horzowitz. Pop. 800.

ZBISLAU, a village of Austria, in Bohemia, in the reg. of Pardubitz, bail. of Czaslau. Pop. 500.

ZBISLAW, a village of Austria, in Bohemia, in the regency of Pilsen, and bail. of Planitz. Pop. 260.

ZBITOW, a market-town of Austria, in Bohemia, in the regency of Prague, and circle of Beraun. Pop. 1,000. It has a castle and iron-works.

ZBLEW, a village of Prussia, in the regency of Danzig, and circle of Stargard. Pop. 500.

ZBONIN, a village of Austria, in Bohemia, in the regency of Pilsen, and bail. of Wirowitz. Pop. 300.

ZBORO, or ZBOROW, a market-town of Austria, in the comitat of Saros, 7 m. NNE of Bartfeld, on the r. bank of the Breznik. It has a tile-kiln and paper-mill.

ZBOROW, a village of Austria, in Bohemia, in the regency of Budweis, and bail. of Wittengau. Pop. 250.—Also a village of the regency of Pilsen, and bail. of Planitz. Pop. 500.—Also a village of Moravia, in the circle of Olmutz, and seignory of Hohenstadt. Pop. 460.—Also a market-town of Galicia, in the circle of Zloczow. It has a tobacco-factory.

ZBOROWITZ, a village of Austria, in Bohemia, in the regency of Pilsen, and bail. of Strakonitz. Pop. 500.—Also a village of Moravia, in the circle of Hradisch. Pop. 600. It has a castle.

ZBOROWSKY, a village of Prussia, in the prov. of Silesia, regency of Oppeln, circle and 9 m. N of Lublinitz. Pop. 500.

ZBOSCH, a village of Austria, in Bohemia, in the regency of Gitschin, and bail. of Neu-Paka. Pop. 480.

ZBOZE, a village of Prussia, in the regency of Marienwerder, and circle of Flatow. Pop. 220.

ZBOZICKO, a village of Austria, in Bohemia, in the reg. of Gitschin, and bail. of Nimburg. Pop. 200.

ZBOZJ, a village of Austria, in Bohemia, in the reg. of Gitschin, and bail. of Podiebrad. Pop. 300.

ZBRACHLIN, a village of Prussia, in the regency of Marienwerder, and circle of Schwetz. Pop. 360.

ZBRASCHAU, a village of Austria, in Moravia, in the circle of Olmutz, and seignory of Weisskirch. Pop. 220.

ZBRASLAU, a village of Austria, in Moravia, in the circle of Brunn. Pop. 580.

ZBRASLAWITZ, a market-town of Austria, in Bohemia, in the regency of Pardubitz, and bail. of Ruttenberg. Pop. 1,160. It has a castle, a synagogue, and a manufactory of potash.

ZBRUDZEWO, a village of Prussia, in the regency of Posen, and circle of Schrimm. Pop. 280.

ZBUCZYN, a market town of Russia in Europe, in Poland, in the gov. of Podlachie, obwod and 9 m. SE of Siedlec. Pop. 450.

ZBUDAU, a village of Austria, in Bohemia, in the regency and bail. of Budweis. Pop. 300.

ZBUZAN, a village of Austria, in Bohemia, in the reg. of Prague, and bail. of Rakonitz. Pop. 290.

ZBYKI, a village of Prussia, in the regency of Posen, and circle of Pleschen. Pop. 250.

ZBYSCEWICE, a village of Prussia, in the regency of Bromberg, circle of Chodziesen. Pop. 210.

ZCHINWALI, a town of Russia, in Georgia, in Middle Kartalinia, at the mouth of the Liachwa.

ZCHORI, or SCHORI, a town of Russia, in Trans-Caucasia, and district of Abasia, on the Black sea, 40 m. SE of Soukoum Kall.

ZDANITZ, a village of Austria, in Bohemia, in the regency and bail. of Pardubitz. Pop. 580.—Also a village of Moravia, in the circle of Iglau, and seignory of Bistritz. Pop. 290.

ZDARETZ, a village of Austria, in Bohemia, in the regency of Gitschin, and bail. of Richenburg. Pop. 320.

ZDAUNEK, a village of Austria, in Bohemia, in the regency of Pardubitz, and bail. of Chrudim. Pop. 600. It has a castle and a synagogue.

ZDAUNEK, or Zdannky, a town of Austria, in Moravia, in the regency of Brunn, circle and 24 m. WNW of Ungarisch-Hradisch. Pop. in 1834, 937.

ZDELOW, a village of Austria, in Bohemia, in the regency of Gitschin, and bail. of Adlerkosteletz. Pop. 520.

ZDENKAU, a village of Austria, in Moravia, in the circle of Iglau. Pop. 210.

ZDESLAWITZ, a village of Austria, in Bohemia, in the regency of Pardubitz, and bail. of Ruttenberg. Pop. 210.

ZDIAR, a village of Austria, in Bohemia, in the regency of Budweis, and bail. of Pilgram. Pop. 270.—Also a village of the same regency, in the bail. of Tabor. Pop. 370.—Also a village in the regency of Gitschin, and bail. of Adlerkosteletz. Pop. 470.—Also a village of the same regency, in the bail. of Gitschin. Pop. 230.—Also a village of the same regency, in the bail. of Braunau. Pop. 500.—Also a village of the regency of Bohmisch-Leipa, and bail. of Danpa. Pop. 500.—Also a village of the regency of Pilsen, and bail. of Blowitz. Pop. 210.—Also a village of the same regency, in the bail. of Wodniau. Pop. 330.—Also a village of the same regency, in the bail. of Pilsen. Pop. 260. —Also a village of Moravia, in the circle of Brunn, and seignory of Raitz. Pop. 470.—Also a village of Hungary, in the comitat of Zips, to the NE of Kasmark. See also Saar.

ZDIAREK, a village of Austria, in Bohemia, in the regency of Gitschin, and bail. of Braunau. Pop 600.

ZDIARETZ, a village of Austria, in Bohemia, in the regency of Pardubitz, and bail. of Hohenmauth. Pop. 220.—Also a village of Moravia, in the circle of Brünn, and seignory of Tischnowitz. Pop. 370.

ZDIARNA, a village of Austria, in Moravia, in the circle of Brünn, and seignory of Boskowitz. Pop. 300. It has a castle.

ZDIBY, a village of Austria, in Bohemia, in the regency of Prague, and bail. of Karolinenthal. Pop. 220. It has a castle.

ZDIECHOW, a village of Austria, in Moravia, in the circle of Hradisch, and seignory of Wsetin. Pop. 950.

ZDIETIN, a village of Austria, in Moravia, in the circle of Olmutz, and seignory of Plin. Pop. 250.

ZDIKAU (Gross and Klein), two villages of Austria, in Bohemia, in the regency of Pilsen, and bail. of Winterberg. Pop. 1,200. It has a castle.

ZDIRETZ, a village of Austria, in Bohemia, in the regency of Pardubitz, and bail. of Chotiebor. Pop. 250.

ZDIRNITZ (Norder and Hinter), two villages of Austria, in Bohemia, in the regency of Gitschin, and bail. of Horzitz. Pop. 580.

ZDISLAWITZ, a market-town of Austria, in Bohemia, in the regency of Budweis, and bail. of Wlaschin. Pop. 800.—Also a village of the regency of Pardubitz, and bail. of Humpoletz. Pop. 220.

ZDISSLAWITZ, a village of Austria, in Moravia, in the circle of Hradisch. Pop. 420.

ZDITZ, a village of Austria, in Bohemia, in the reg. of Prague, and bail. of Horzowitz. Pop. 1,000.

ZDUNEK, or Zdunik, a market-town of Austria, in Moravia, in the circle and 15 m. NNW of Hradisch. Pop. 600. It has a castle.

ZDUNSKA-WOLA, a town of Russia in Europe, in Poland, in the gov. of Kalisch, and obwod of Sieradz. Pop. 2,920.

ZDUNY, a town of Prussia, in the prov. and regency and 60 m. SSE of Posen, circle and 5 m. SSW of Krotosczyn. Pop. in 1843, 3,406. It has a Catholic and a Lutheran church, a Jewish school, and a Lutheran gymnasium. Cloth and linen are its staple manufactures.

ZEA, Cea, Ceos, or Kea, an island of Greece, in the diocese of Syra, in the Archipelago, between Kythnos and Makronisi, and 12 m. SE of Cape Colonna, in Livadia. Its highest summit, Mount St. Elias, is in N lat. 37° 37′ 18″, and E long. 24° 21′ 25″. It is about 15 m. in length from NNE to SSW, and 9 m. in extreme breadth, comprising a total area of 100 sq. m., of which about ⅜ are under cultivation. Pop. 3,500. The coasts are low, but towards the centre the elevation is considerable. The soil is chiefly calcareous, and extremely fertile, producing corn, wine, silk, cotton, figs, citron, pomegranates, and the climate salubrious. The island contains 34 churches, and 5 convents.—The chief town, which bears the same name, rises above a series of terraces, near the centre of the island, overlooking a deep ravine, and 3 m. from its port of Raphtilemani. It is partly built on the ancient massive walls.

ZEALAND (New). See New Zealand.

ZEAL-MONACHORUM, a parish in Devonshire, 1¾ m. N of Bow. Area 3,264 acres. Pop. in 1831, 747; in 1851, 578.

ZEBAN, a village of Austria, in Bohemia, in the regency of Eger, and bail. of Weseritz. Pop. 360. It has a castle and a synagogue.

ZEBAYR, a group of islands in the Red sea, in N lat. 15° 7′, E long. 42° 12′. They are all of volcanic origin.

ZEBBAR, or Zabbar, a market-town of Malta, about 3 m. S of Valetta. Pop. 300.

ZEBBIN, a village of Prussia, in the regency of Stettin, and circle of Kammin. Pop. 250.

ZEBEDANY, a large village in the pash. of Damascus, on the N side of a fine plain bordered on the E by the Anti-Lebanon, on the route from Damascus to Baalbek.

ZEBEE, or Zebeh, a river of Africa, which has its source near the S confines of Abyssinia, to the SW of Bosham, waters the district of Gingero, and flows into the Indian ocean, on the Zanguebar coast, near the town and island of Lamu, and about 2° 30′ S lat. Its intermediate course is unknown.

ZEBEIR. See Zobeir.

ZEBEN. See Seben.

ZEBIBI (Cape), a headland of Tunis, on the N coast, to the NW of the embouchure of the Majerda, in N lat. 37° 10′, and E long. 28° 17′.

ZEBICO-NABERO, a town of Spain, in the prov. and to the E of Palencia.

ZEBICO-DE-LA-TORRE, a town of Spain, in the prov. and 9 m. SE of Palencia, in the valley of Cerrato.

ZEBID, a town of Arabia, in Yemen, in the district of Tehama, 21 m. S of Beit-el-Fakih, and 105 m. SW of Sana, near the greatest and most fertile valley of the Tehama, and about 15 m. from the shore of the Arabian gulf. The valley during the rainy season receives so large a body of water from the adjacent mountains, as to become the bed of a great river, named also Zebid. The town is enclosed by low walls, and had formerly three gates, two of which were destroyed by inundations. The wall describes a considerable circuit, but scarcely a half of the enclosure is now occupied by the town. It contains numerous mosques, an academy for the instruction of youth of the Sunnite sect in the sciences cultivated by Mahommedans, and possesses important manufactories of silk and cotton fabrics. The houses are built of a dark coloured brick, but with great irregularity, and with few interior means of comfort. The use of glass in casements is almost unknown, and the only local substitute fish-skin renders the habitations gloomy in the extreme. This town was formerly the most commercial in the Tehama, but the port of Ghalef'ka having become unfit for use, the commerce to which Z. formerly owed its prosperity has passed to Mocha, Hodeida, Loheïa, and Beit-el-Fakih. The pop. is at present about 7,000.

ZEBLIN, a village of Prussia, in the regency of Koslin, and circle of Furstenthum, near Bublitz. Pop. 215.

ZEBNITZ, a village of Austria, in Bohemia, in the reg. of Pilsen, and bail. of Kralowitz. Pop. 400.

ZEBRAK, a town of Austria, in Bohemia, in the reg. and 28 m. WSW of Prague, and circle of Beraun. Pop. 1,700. Coal is wrought in the environs.

ZEBRERO, a town of Spain, in Galicia, in the prov. and 36 m. SE of Lugo. The environs are noted for their chase.

ZEBU, an island of the Philippine archipelago, between the island of Negros on the W, and those of Leyte and Bohol on the E, and of Mindanao on the S, extending between 9° 25', and 11° N. It is about 60 m. in length from NNE to SSW; but is comparatively narrow, not exceeding 240 m. in circumference. Pop. 170,000. It is generally mountainous, but is well watered, and is in many parts extremely fertile, producing tobacco, cotton, flax, and a species of grain named barona. The forests afford ebony and several kinds of dyewood, and gold is found in some of the rivers. Birds' nests form one of its chief articles of trade. The inhabitants, who are chiefly Bissayas, are enterprising, active, and industrious. This island, which was discovered by Magellan in 1521, was the first of the Philippines of which the Spaniards rendered themselves masters. The Chinese, who had long been established in the island, united with the aborigines in their efforts to resist the Spaniards, and it was not until the 27th attempt of the latter to make good their footing on its shores, that they achieved success.—The town, also called Zebu, is situated on the E coast of the island. Pop. 25,000. It is the residence of a Spanish governor, and has an earthen fort and good harbour. Magellan here lost his life by the hand of the king of Zebu.

ZEBUEBEIN, a town of Burmah, on the Irawady, 121 m. SW of Ava, in N lat. 20° 41', and E long. 94° 34'.

ZEBULON, a village of Pike co., in the state of Georgia, U. S., 68 m. W of Milledgeville. Pop. in 1840, 200; in 1850, 300.—Also a village of Pike co., in the state of Arkansas, 90 m. WSW of Little Rock. Pop. in 1840, 75; in 1850, 150.

ZECHAAN, or ZEEHAN, a summit of Van Diemen's Land, near the W coast, to the SE of Mount Heemskirk, in S lat. 44° 55', and E long. 145° 18'.

ZECHAN (ZECHANOW), a village of Austria, in Moravia, in the circle of Olmutz, and seignory of Gulenberg. Pop. 280.

ZECHAU, a village of Saxe-Altenburg, and bail. of Altenburg. Pop. 280.

ZECHEN, a village of Prussia, in the reg. of Breslau, and circle of Guhrau. Pop. 300.

ZECHENDORF, a village of Prussia, in the reg. of Koslin, and circle of New Stettin. Pop. 260.

ZECHIN, a village of Prussia, in the regency of Frankfort, and circle of Kustrin. Pop. 1,258.

ZECHITZ, a village of Austria, in Moravia, circle of Olmutz and seignory of Culenberg. Pop. 440.

ZECHLAVIN, a town of Spain, in the prov. and 27 m. NW of Caceres.

ZECHLIN, a market-town of Prussia, in the reg. of Potsdam, and circle of Ost Priegnitz, on the W bank of a lake of the same name, and 20 m. E of Wittstock. Pop. 560.—Also a village of the same circle. Pop. 300.

ZECHOW, a village of Prussia, in the regency of Frankfort, and circle of Landsberg. Pop. 310.

ZECHOWITZ, a village of Austria, in Bohemia, in the reg. of Gitschin, and bail. of Neu-Bidschow. Pop. 240.—Also a village of the reg. of Pilsen, and bail. of Wolin. Pop. 280.

ZECHSDORF, two villages of Austria, in Silesia, in the circle of Troppau. Pop. 800.

ZECKENDORF, a village of Bavaria, in the circle of Upper Franconia, and presidial of Schesslitz. Pop. 325, of whom 166 are Jews. It has a synagogue.

ZECKERIN, a village of Prussia, in the reg. of Frankfort, and circle of Luckau. Pop. 350.

ZECKERN, a village of Bavaria, in the circle of Upper Franconia, and presidial of Herzogenaurach. Pop. 200.

ZECKRITZ, a village of Prussia, in the reg. of Merseburg, and circle of Torgau. Pop. 350.

ZEDA (MONTE-DELLA), a mountain of Sardinia, in the division of Novara, and prov. of Domo d'Issola, to the W of Lake Maggiore, in N lat. 46° 2' 25", and E long. 8° 32' 2".

ZEDDEMICK, a village of Prussia, in the reg. of Magdeburg, and circle of Jerichow. Pop. 220.

ZEDDIN, a village of Prussia, in the regency of Potsdam, and circle of West Priegnitz. Pop. 330.

ZEDEL, a village of Prussia, in the regency of Frankfort, and circle of Gorau. Pop. 320.

ZEDELGHEM, a department and commune of Belgium, in the prov. of W. Flanders, and arrond. of Bruges. Pop. of dep. 2,556; and of com. 244.

ZEDELSDORF, a village of Prussia, in the reg. of Liegnitz, and circle of Sagan. Pop. 330.

ZEDIK, a town of Persia, in the prov. of Ghilan, 5 m. W of Resht. Pop. 3,000.

ZEDLACH, a village of Austria, in the Tyrol, in the circle of Brixen, and bail. of Lienz. Pop. 250.

ZEDLIN, a village of Prussia, in the regency of Stettin, and circle of Grafenberg. Pop. 398.

ZEDLITZ, a village of Saxony, in the circle of Leipsig, and bail. of Vorna. Pop. 450.—Also a village of Prussia, in the regency of Breslau, and circle of Ohlau. Pop. 545.—Also a village of the same regency, and circle of Schweidnitz. Pop. 389. —Also a village of the same regency, and circle of Steinau. Pop. 480.—Also a village of the same regency, in the circle of Trebnitz. Pop. 278.—Also a village of the regency of Oppeln, and circle of Grotkau. Pop. 195.—Also a village of the reg. and circle of Oppeln. Pop. 250.

ZEDLITZ (NIEDER and OBER), two villages of Prussia, in the reg. of Posen, and circle of Fraustadt, containing respectively 472 and 389 inhabitants.

ZEDLITZ (ALT or ALT CZETLICZE), a town and castle of Bohemia, in the circle of Pilsen, 3 m. NW of Hayd. It has manufactories of hosiery, and a considerable trade in cotton and in cattle.

ZEDLITZHEIDE, a village of Prussia, in the reg. of Breslau, and circle of Waldenburg. Pop. 550.

ZEDTWITZ, a village of Bavaria, in the circle of Upper Franconia, and presidial of Hof. Pop. 450.

ZEEGENDORF, a village of Bavaria, in the circle of Upper Franconia, and presidial of Bamberg. Pop. 230.

ZEEKOE, a river of South Africa, in the district of Graaff-Reynet, which has its source in Compass mountain, a summit of the Sneeuw-Berg or Snow mountains; runs N and NE, after a course of about 150 m.; joins the Nu-gariep, Black or Cradock river, on the l. bank, and on the confines of Hotentolia.

ZEELAND, a province of Holland, lying between

51° 14′ and 51° 45′ N lat., and between 3° 13′ and 4° 7′ E long., and bounded on the N by the arm of the Maas, distinguished by the names Krammer and Grevelingen, by which it is separated from Holland; on the E by the prov. of North Brabant, and by Belgium; on the S and SW by Belgium; and on the W by the North sea. It is 42 m. in length from N to S, and nearly equal in breadth, and comprises a total area of 225 sq. m. The S part, improperly distinguished as the island of Cassandria, taken from Dutch Flanders, belongs to the continent, its isolation being merely by canals. Z., properly so called, consists of islands formed by two main and several minor arms of the Schelde, arms of the Maas and the North sea, viz. North and South Beveland and Walcheren on the S, and Tholen, Duiveland, and Schouwen on the N. The surface, with the exception of some small parts of the islands of Walcheren and Schouwen, where the sea has thrown up sand banks or downs, locally named tarpen, is level throughout, and lies so low as to require for its defence against inundation large dykes, which run along the coast and the sides of rivers, and are kept up at great labour and expense. These dykes are from 20 to 30 yds. in breadth at the bottom, and of sufficient width at the top for two carriages to pass abreast. Yet different parts of this prov. have at times been exposed to heavy calamities by the sea breaking over the dykes in storms at high tides, particularly in 1302, 1309, 1522, 1532, and 1548. In these catastrophes whole towns and districts have been overflowed and abandoned; and though part of them have been subsequently recovered, several islands have undergone a permanent reduction; in particular, that of Schouwen, which is said to have once been 60 m. in circumference, has now only 25. In the river Scheldt, between the towns of Bergen-op-Zoom and Antwerp, there exist on both sides large tracts of land, to the extent of several miles, which at high water are covered by the sea, and at low water present a varying surface of several feet in depth of the richest alluvial soil, ever on the increase by the daily action of the tides. To redeem and dispose of this valuable land, a company has been formed, and the necessary powers granted by a concession for 99 years, under which the company may recover and enclose land to the extent of 35,000 acres, the line of operation marked out being almost identical with that planned by the Emperor Napoleon for the same design. The reclamation of the land will cost £20 per acre; the land reclaimed, judging from the price realized for adjacent plots similarly recovered, will be worth from £60 to £70 per acre. Indeed, the alluvial deposits in this province of Zeeland are of so peculiarly rich and valuable a quality that it is affirmed by competent judges such lands will bear crops for 20 years in succession without the application of manure. The 3,000 acres first enclosed are formed of alluvial deposits, of which the lowest parts are about 8 ft. above the level of low tide. The soil of Z. is a rich black mould, excellent for pasturage and the culture of such crops as madder, flax, and cole seed, which require a very heavy soil. Wheat is raised chiefly in South Beveland, and madder in large quantities in the island of Schouwen. The climate is damp and variable. To foreigners it is productive of bilious complaints and agues, as was so severely experienced by our troops in 1809, and affects even the health and longevity of the natives. Frost and snow are of less frequent occurrence than on the continent. The principal productions are cattle, fowls, game, fish, grain, fruit, legumes, melons, potatoes, flax, rape-seed, madder, and mulberries. The sheep are small, and the wool inferior in quality. The horses are strong, but ill-made. The manufacturing industry of the prov. consists chiefly in the production of fine yarn, and some species of linen and woollen fabrics, oil, and gin. The trade is considerable, the exports consisting chiefly in articles of local production.—The Zeelanders on the islands are strictly Dutch; on the mainland they are a mixture of Dutch and Flemings. They resemble the inhabitants of the rest of the United Provinces, in their industry, perseverance, and phlegmatic tendencies, and are remarkable for their adherence to the ancient manners and customs. The men, when in full costume, wear short velvet jackets with waistcoats, and amplitudinous trousers of the same material, bound round the waist with a girdle whose buckle, or buckles, is generally of pure silver. The head-dress is a black cap, or loose hat, producing very much the effect of the Andalusian cap. The faces of the women are intelligent, and many of them are well-looking; their forms being the very reverse of that which we understand by the term Dutch-built. In full dress, they have slips of striped velvet, or silk, or stuff, with boddices of silk or velvet, and a sort of neckerchief of many-coloured ribands run together. But their great glory is the singular but effective gold or gilt plates, prominent just where the ear meets the forehead; some to these hang other plates, and from these, again, all sorts of trinkets; a few may be observed with, in addition, a broad plate or band of gold, altogether covering the forehead between the two plates first described, very like the antique Egyptian forehead decoration. They wear silver and gold brooches at the neck, and at the waist, and huge plates of the same on the shoes.—This prov. sends three deputies to the states general. Its provincial states consist of 44 members, of whom six are for the knights, 20 for the towns, and 18 for the country. It is divided into three arrond., viz., Goes, Middelburg, and Zierikzee. Its capital is Middelburg. The general aspect of the country seems to have been singularly changed by the action of its waters since the period of Roman domination, when the mouths of the Schelde and Maas were occupied by the Caninefates. After numerous vicissitudes, Z. passed under the domination of Spain, but was one of the provinces which shook off the yoke of Philip II. in the 16th century. Till 1798, it formed one of the United Provinces. In 1810, it was united to the French empire, and became the dep. of the Bouches-de-l'Escaut, and was finally attached to Holland at the close of the revolution of 1830.

ZEELANDIA, a fort of Dutch Guayana, in the Surinam.

ZEELBOURG (ALT). See ZELBOURG (ALT).

ZEELE. See ZELE.

ZEELHEM, a village of Belgium, in the prov. of Limburg, near Hasselt. Pop. 2,650.

ZEELST, a village of Holland, in the prov. of North Brabant, and arrond. of Emdhoven, on the Gender. Pop. 1,000. It has important manufactories of linen.

ZEEMANIA, or NUSRUTHPUR, a town of Hindostan, in the district and 11 m. S of Ghazeepur, in N lat. 25° 23′, and E long. 83° 38′.

ZEERAPUR, a town of Hindostan, in the territory of Holkar, 103 m. NNE of Indur, in N lat. 24° 3′, and E long. 26° 26′.

ZEERUK, a station of Afghanistan, 96 m. NW of Candahar, on the road thence to Herat. It is well supplied with water, but the surrounding district possesses little fertility.

ZEESTOW, a village of Prussia, in the regency of Potsdam, and circle of Ost Havelland. Pop 250.

ZEFTI, a town of Lower Egypt, in the prov. of

Garbieh, on the l. bank of the principal E branch of the Nile, and 54 m. N of Cairo. It has three mosques.

ZEGAMA, a town of Spain, in the prov. of Guipuzcoa, 24 m. SSW of San Sebastian.

ZEGENDORF (Gross), Nagy-Czeg, or Cragu, a village of Transylvania, in the comitat and 33 m. E of Klausenburg. It has several mineral springs.

ZEGGERS-CAPPEL, a commune of France, in the dep. of the Nord, cant. and 5 m. W of Wormhout. Pop. 1,775.

ZEGHAWA, a town of Sudan, in Darfur, 60 m. N of Kobbe, on the Sudan and Cairo caravan route.

ZEGHEN, or Zedjhan, a town of Tripoli, in Fezzan, 135 m. NNE of Murzuk, in N lat. 27° 26′, and E long. 16° 26′. It is surrounded with plantations of palm-trees.—This town gives its name to a range of mountains running E and W, and joining the Harutsch mountains on the W.

ZEGLINGEN, a village of Switzerland, in the cant. of Basle, and bail. of Sissach. Pop. 586.

ZEGROWO, a village of Prussia, in the regency of Posen, and circle of Kosten. Pop. 200.

ZEGRZ, a town of Russia in Europe, in the gov. of Plock, on the Bug.

ZEG-ZEG, an extensive province of Nigritia, in Houssa. It is intersected by the head-streams of the Yeou and Makamie, and contains several towns, of which Zaria is the principal.

ZEHDEN, a town of Prussia, in the regency of Frankfort, and circle of Königsberg, on the Muglitz. Pop. 1,530.

ZEHDENICK, a town of Prussia, in the regency of Potsdam, and circle of Templin, on the r. bank of the Havel. Pop. in 1843, 2,858. It has three gates, and the same number of suburbs, two churches, an hospital, and a convent, and possesses manufactories of cloth, tobacco, and potash.

ZEHDINE. See Sehdine.

ZEHENSDORF, a village of Austria, in Styria, in the circle of Gratz. Pop. 260.

ZEHISTA, a village of Saxony, in the circle of Dresden, and bail. of Pirna. Pop. 290.

ZEHLENDORF, a village of Prussia, in the regency of Potsdam, and circle of Niederbarnim, near Dranienburg. Pop. 590.

ZEHLENDORF (Alt), a village of Prussia, in the regency of Potsdam, and circle of Teltow. Pop. 235.

ZEHMA, a village of the duchy of Saxe-Altenburg, bail. and 4 m. S of Altenburg. It has a limestone quarry.

ZEHMEN, a village of Saxony, in the circle and bail. of Leipsig. Pop. 300.

ZEHNA, a village of Mecklenburg-Schwerin, circle of Windisch, and bail. of Gustrow. Pop. 210.

ZEHNGERICHTENBUND, a state or political subdivision of the Swiss cant. of the Grisons, comprising the 7 nochgerichts of Davos, Klosters, Castels, Schiersh and Seewis, Marienfeld and Malans, Belfort, Schansigg, and Langwies.

ZEHNHAUSEN, a village of Nassau, in the bail. of Keemerod. Pop. 200.

ZEHNITZ, a village of Austria, in Bohemia, in the circle of Prachin, and seignory of Stickna. Pop. 660.

ZEHOLFING, a village of Bavaria, in the circle of Lower Bavaria, and presidial of Landau. Pop. 290.

ZEHREE, a district and town of Beluchistan, in the prov. of Jhalawan. The town, which is the residence of the sirdar or chief of the prov., is of considerable extent, and is surrounded by a mud wall.

ZEHREN (Nieder), a village of Prussia, in the regency and circle of Marienwerder. Pop. 615.

ZEHREN (Nieder and Ober), a village of Saxony, in the circle of Dresden, and bail. of Meissen. Pop. 382.

ZEHROWITZ, a village of Austria, in Bohemia, in the regency of Prague, and bail. of Rakonitz. Pop. 490.

ZEHRTEN, a village of Prussia, in the regency of Stettin, and circle of Saatzig. Pop. 200.

ZEIDELBORN, a village of Bavaria, in the circle of the Upper Pfalz, and seignory of Worth. Pop. 300.

ZEIDEN, Schwarzburg, or Fekete-Kalom, a market-town of Austria, in Transylvania, in the Saxon territory, district and 8 m. NW of Cronstadt, at the foot of a mountain of the same name. Pop. 3,220. It has Greek and Lutheran churches.

ZEIDLER, a village of Austria, in Bohemia, in the regency of Bohmisch-Leipa, and bail. of Schluckenau. Pop. 1,600.

ZEIER, a village of Prussia, in the regency of Danzig, and circle of Elbing. Pop. 910.

ZEIERS (Nieder and Vorderhampen), two villages of Prussia, in the regency of Danzig, and circle of Elbing, containing respectively 300 and 380 inhabitants.

ZEIERS-ROSENGARTH, a village of Prussia, in the regency of Danzig, and circle of Elbing. Pop. 850.

ZEIHEN (Nieder), a village of Switzerland, in the cant. of Aargau, and bail. of Laufenburg. Pop. 457.

ZEIL, or Zeyl, a town of Bavaria, in the circle of Lower Franconia, and presidial of Ellmann, on the r. bank of the Main, 38 m. NE of Wurtzburg. Pop. 1,310. It has a castle, and possesses manufactories of cloth, paper, potash, &c.

ZEIL (Ober), a seignory and castle of Würtemberg, in the circle of the Danube, and bail. of Leutkirch, belonging to the princes of Waldburg-Zeil-Tranchburg. It contains the villages of Unter and Ober Zeil.

ZEIL (Unter), a village of Austria in Styria, in the circle of Gratz. Pop. 380.

ZEILAH, or Zeyla, a town and port of the Somauli territory, on a low sandy cape, called Ras Mahmaler, 150 m. NNE of Hurrur, and about 200 m. WNW of Burburra. It is enclosed by a wall in a ruinous and dilapidated condition, and contains about 200 huts, and a few stone-buildings. It is supplied with water from the bed of a water course, about 4 m. to the SW named Tacushah, where there is a small round tower and an Arab guard to protect the watering place. There are a few Arab merchants residing in the town, a few also of the Eessah-Gudu-burie Somaulis. Coffee, dye, ghee, ivory and ostrich feathers form the chief articles of export, and the slave trade is also carried on here to a considerable extent. The harbour, which is the only port on the Eessah coast, is bounded on the W side by a range of sand-banks, extending from Ras Tacushah in a NNE direction towards Taddickdeen island, having three channels for small ships, each about 300 yds. wide, and 18 ft. deep at low water. This town is believed to be the *Mossyllon* of Ptolemy.

ZEILBACH, a village of Hesse, in the prov. of Upper Hesse, and presidial of Grünberg. Pop. 250.

ZEILERN, a market-town of Austria, in the circle of the Upper Weinerwald, and 12 m. N of Waidhofen. Pop. 500.

ZEILERVIERTL, a village of Austria, in Styria, in the circle of Gratz, and bail. of Pollau. Pop. 500.

ZEILFELD, a village of Saxe-Meiningen, in the bail. of Romhild. Pop. 260.

ZEILHARD, a village of Hesse, in the prov. of Starkenburg, and presidial of Lichtenberg. Pop. 300.

ZEILSHEIM, a village of Nassau, in the bail. of Hochst. Pop. 400.

ZEIMIKE, a village of Prussia, in the regency of Stettin, and circle of Saatzig. Pop. 370.

ZEINHADI, a village of Persia, in Farsistan, to the SE of Yezd.

ZEININGEN, a village of Switzerland, in the cant. of Aargau, and bail. of Reinfelden. Pop. 983.

ZEIPPERN, a village of Prussia, in the regency of Breslau, and circle of Guhrau. Pop. 250.

ZEIRING (Ober and Unter), a village of Syria, in the circle and 14 m. NW of Judenburg, on the Pols. It has a castle, and possesses mines of argentiferous lead and of iron.

ZEISDORF, a village of Prussia, in the regency of Merseburg, and circle of Eckartsberga. Pop. 200.

ZEISELMAUER, a town of the archduchy of Austria, in the country below the Ens, and upper circle of the Wienerwald, 29 m. ENE of St. Pölten.

ZEISELWITZ, a village of Prussia, in the regency of Oppeln, and circle of Neustadt. Pop. 380.

ZEISEN, a village of Prussia, in the regency of Gumbinnen, and circle of Lyk. Pop. 230.

ZEISGENDORF, a village of Prussia, in the regency of Danzig, and circle of Stargard. Pop. 450.

ZEISKAM, a village of Bavaria, in the circle of the Pfalz, and cant. of Germersheim. Pop. 1,703.

ZEISKE, a village of Austria, in Silesia, in the circle of Troppau. Pop. 470.

ZEIST, a village of Holland, in the prov. and 6 m. E of Utrecht, and arrond. of Amersfoort. Pop. 2,300, of whom about 300 are Herrnhuters or Moravian brethren, who form a sort of colony, and manufacture varieties of articles in gold, silver, ivory, and japan, gloves, wax-candles, &c., besides employing themselves as agriculturists and gardeners.

ZEISTSDORF, a village of Prussia, in the regency of Liegnitz, and circle of Sprottau. Pop. 240. It has a castle.

ZEISZAU, a village of Prussia, in the regency of Liegnitz, and circle of Sagan. Pop. 230.

ZEISZIG, a village of Prussia, in the regency of Liegnitz, and circle of Hoyerswerda. Pop. 200.

ZEITH, a village of Prussia, in the regency of Arnsberg, and circle of Siegen. Pop. 200.

ZEITHAM, a village of Saxony, in the circle of Dresden, bail. and near Grossenheim. Pop. 560.

ZEITHUN, a district of Asia Minor, chiefly consisting of an elevated plateau, stretching between the parallels of 34° and 35°; and bounded on the N by the Kataonian Taurus; on the E and S by the river Pyramus; on the W by Karmesdagh. It is watered by affluents of the Teker-su. It contains about 3,000 Armenian dwellings, and not above 20 Turkish houses. The nominal governor of the territory is the pasha of Adana.

ZEITLARN, a village of Bavaria, in the circle of the Ober Pfalz, and presidial of Regenstauf, on the Regen. Pop. 308.

ZEITLITZ, a village of Prussia, in the regency of Stettin, and circle of Regenwalde. Pop. 240.

ZEITLOSS, a town of Bavaria, in the circle of Lower Franconia, and presidial of Bruckenau. Pop. 740. It has a castle, and possesses manufactories of woollen fabrics.

ZEITSCHACH, a village of Austria, in Syria, in the circle of Gratz. Pop. 260.

ZEITUN, Isdin, or Lamia, a town of Greece, in Livadia, capital of the diocese of Phiotide, 5 m. NW of the gulf of Zeitun, and near the l. bank of the Hellada, 54 m. NNW of Livadia, and 99 m. NW of Athens. Pop. 4,000. It is the see of an archbishop, and has a custom-house and printing press. About 9 m. SE of Z. is the celebrated pass of *Thermopylæ*.—The gulf of Z., also called Molo, forms the apex of the angle described by the channels of Talanda and Trikeri, by which the island of Egripo is separated from the continent. It is 18 m. in depth from E to W, and 9 m. in breadth from N to S, and receives the Ellada on the SW, near the defile of Thermopylæ.

ZEITUN-BURUN, or Santa Agia Maria, a headland of Turkey in Asia, at the E extremity of the island of Mitylene, and to the SE of the town of that name.

ZEITZ, a circle and town of Prussia, in the prov. of Saxony, and regency of Merseburg. Pop. of circle 35,360. The town is on the r. bank of the Elster, 21 m. SSE of Merseburg. It is enclosed by walls, and is divided into two parts, distinguished as the Upper and Lower town, and has an old and a new castle, a public library, four churches, two schools, an orphans' asylum, a poor-house, and an infirmary. It possesses manufactories of woollen fabrics, shoes, pottery, and several breweries, distilleries, and wax bleacheries.

ZEKANOVITZ, a market-town of Russia in Europe, in the gov. of Bialystock. Pop. 1,500.

ZEKINOWKA, a town of Russia in Europe, in the gov. of Podolia, on the Dniester. Pop. 1,600.

ZELANDIA, a fort of Dutch Guyana, on the l. bank of the Surinam, a little below Paramaribo.

ZELANZ. See Klagenfurt.

ZELASEN, a village of Prussia, in the reg. of Koslin, and circle of Lauenburg-Britow. Pop. 200.

ZELASNA, a village of Prussia, in the reg. and circle of Oppeln, on the Oder. Pop. 370.

ZELASZKOWO, a village of Prussia, in the reg. of Bromberg, and circle of Gnisen. Pop. 220.

ZELATAWA. See Scheletau.

ZELATOWITZ, a village of Austria, in Moravia, in the circle of Prerau. Pop. 410.

ZELAYA, a town of Mexico, in the dep. and 42 m. SE of Guanaxuato, on the r. bank of the Rio-Grande, in a fertile territory, and at an alt. of 6,020 ft. above sea-level. Pop. 2,500. It is well-built, and has two churches, one of which is reputed the finest in Mexico, five convents, a college, and an hospital. It has an active trade, and the environs produce in great abundance sugar, wine, and olives.

ZELBURG (Alt), a town of Russia in Europe, in the gov. of Courland, on the l. bank of the Dwina, 58 m. E of Mittau.

ZELCK-HAUTZUM, a commune of Belgium, in the prov. of Limburg, and dep. of Haelen. Pop. 330.

ZELDONCK, a commune of Belgium, prov. of W. Flanders, and dep. of Oedelem. Pop. 480.

ZELE, a department and commune of Belgium, in the prov. of E. Flanders, arrond. and 5 m. NW of Termonde, and watered by the Schelde and Durme. Pop. of dep. 10,838; and of com. 1,250. It has manufactories of linen, blankets, oil, starch, soap, distilleries and breweries.

ZELECHOW, or Zelechowo, a town of Poland, in the gov. of Podlachie, obwod of Lukow, near the Wolga. Pop. 2,300.

ZELENDEN, a town of Peru, in the dep. and 114 m. NNE of Truxillo, and near the l. bank of the Maranon.

ZELENETZ, a village of Austria, in Bohemia, in the regency of Prague, and bail. of Karolinenthal. Pop. 350.

ZELENITZ, a village of Austria, in Bohemia, in the reg. of Prague, and bail. of Schlan. Pop. 220.

ZELEZNA, a village of Austria, in Bohemia, reg. of Prague, and bail. of Schmichow. Pop. 370.

ZELEZNIK, a mountain of Hungary, in the central part of the comitat of Jomor, near the village of Szink, and to the W of Jolsva.

ZELIENOPLE, a village of Butler co., in the

state of Pennsylvania, U. S., on the S side of Conequennessing creek, and 164 m. W by N of Harrisburg. Pop. in 1850, about 300.

ZELL, a village of the grand-duchy of Baden, in the circle of the Middle Rhine, and bail. of Buhl. Pop. 370.—Also a village of the same circle, in the bail. of Offenburg. Pop. 540.—Also a village of Bavaria, in the circle of Lower Bavaria, and presidial of Pfaffenhofen. Pop. 260.—Also a market-town in the circle of Upper Franconia, and presidial of Münchberg, on the Saxon Saale, near its source, and 20 m. NNE of Bayreuth. Pop. 850.—Also a village of the same circle, in the presidial of Helpolstein. Pop. 240.—Also a village of the same circle, in the presidial of Riedenburg. Pop. 240.—Also a village of the circle of the Pfalz, and cant. of Gollheim. Pop. 410.—Also a village of the circle of Swabia, and presidial of Gronenbach. Pop. 360.—Also a village of the same circle, in the presidial of Neuburg. Pop. 640.—Also a village of the circle of Lower Franconia, and presidial of Schweinfurt. Pop. 200.—It has manufactories of cloth and of potash.—Also a village of Kurhessen, in the prov. and circle of Fulda. Pop. 200.—Also a village of Hesse, in the prov. of Upper Hesse, and presidial of Alsfeld. Pop. 770.—Also a village of the prov. of Starkenburg, and presidial of Zwingenberg. Pop. 610.—Also a village of the same prov., in the presisidial of Michelstadt. Pop. 380.—Also a village of Austria, in the country above the Ens, and circle of Hausruck, on the Zellerbach. Pop. 480.—Also a circle of Prussia, in the prov. of the Rhine, and reg. of Coblenz. Pop. 29,090. It has a town of the same name, 45 m. SW of Coblenz, on the Moselle, at the confluence of the Zellerbach. Pop. 2,310. It has an active trade in wine, and other articles of local produce.—Also a village of Würtemberg, in the circle of the Neckar, and bail. of Backnang. Pop. 280.—Also a village of the same circle, in the bail. of Esslingen. Pop. 730.—Also a village of Switzerland, in the cant. of Zurich, bail. and 6 m. SE of Winterthur, on the Toss. Pop. 1,855.—Also a village of the cant. of Luzern, bail. and 4 m. WNW of Willisau. Pop. 1,355.

ZELL (OBER) a village of Bavaria, in the circle of Lower Franconia, near Wurtzburg.

ZELL-AM-EBERSBERGE, a village of Bavaria, in the circle of Lower Franconia, and presidial of Eltmann. Pop. 460.

ZELL-AM-HALLBACH, a market-town of Austria, in the country below the Ens and Lower Wienerwald. Pop. 1,000.

ZELL-AM-HARMERSBACH, a market-town of Baden, in the circle of the Middle Rhine, and bail. of Gengenbach, 11 m. SE of Offenburg, on the Harmersbach, and near the Kinzig. Pop. 1,400. It was formerly a free imperial town, and is enclosed by walls with three gates. It possesses manufactories of potash, cloth, paper, and several ironworks.

ZELL-AM-MOOS, a village of Austria, in the country above the Enns, and circle of Hausruck. Pop. 740.

ZELL-AM-SEE, a market-town of Austria, in the circle of Salzburg. Pop. 740. It has a castle.

ZELL-AN-DER-BRAMBACH, a market-town of Austria, in the circle of the Inn, and bail. of Schwerding. Pop. 860.

ZELL-AN-DER-YBBS, a market-town of Austria, in the country below the Enns, and circle of the Oberwienerwald. Pop. 680.

ZELLA, a village of Saxony, in the circle of Zwickau, and bail. of Schwarzenberg, near Aue. Pop. 557.—Also a village of the circle of Leipzig, and near Nossen. Pop. 96.—Also a village of Prussia, in the regency of Erfurt, and circle of Muhlhausen. Pop. 350.—Also a village of Kurhessen, prov. of Oberhessen, circle and bail. of Ziegenhain. Pop. 380.

ZELLA ST. BLASII, or BLASIEN-ZELLA, a town of the duchy of Saxe-Coburg-Gotha, principality and 24 m. SSW of Gotha, on the Gemeinbach, in the midst of the mountains of Thuringia. Pop. 1,500. It has manufactories of arms and of ironware.

ZELLBERG, a village of Austria, in the Tyrol, circle of Schwatz, and presidial of Zell. Pop. 680.

ZELLE. See CELLE.

ZELLE (NEU), a village of Prussia, in the reg. of Frankfort, and circle of Guben. Pop. 240.

ZELLENBERG a village of France, in the dep. of the Haut-Rhin, arrond. and 8 m. NNW of Colmar, and cant. of Kaysersberg, on a mountain. Pop. 580. It has two forts.

ZELLERFELD, or CELLERFELD, a town of Hanover, in the Harz, in the territory of Klausthal, and a little to the N of the town of that name, from which it is separated by a small river named Zellerbach, and 36 m. SSW of Brunswick. Pop. 4,100. It is tolerably well built, with wide streets lined with trees, and a large market-place. It has a gymnasium and a bank.

ZELLERHORN, a summit of the mountains of Hohenzollern-Hechingen.

ZELLER-SEE, or UNTER-SEE, a lake or river-expansion, partly in the grand-duchy of Baden, and partly in the Swiss cant. of Thurgau, formed by the Rhine, above Stein, and 3 m. below Lake Constanz, under the name of which it is often comprehended. It is 11 m. in length, and contains the island of Reichenau, on which are two villages distinguished as the Ober and Unter Zell.

ZELLGNIEWO, a village of Prussia, in the reg. of Bromberg, and circle of Chodziesen. Pop. 470.

ZELLHAUSEN, a village of Hesse, prov. of Starkenburg, and presidial of Geligenstadt. Pop. 630.

ZELLHOF, a village of Austria, in the country above the Enns, and circle of the Muhl. Pop. 240. It has a castle.

ZELL-IM-WIESENTHAL, a market-town of Baden, in the circle of the Upper Rhine, and bail. of Schonau. Pop. 5,000.

ZELL-IM-ZILLERTHAL, a market-town of Austria, in the Tyrol, in the circle of Innspruck, on the Zeller. Pop. 1,000.

ZELLIN, a market-town of Prussia, in the reg. of Frankfort, and circle of Konigsberg, on the r. bank of the Oder. Pop. 1,800.

ZELLINE, or ZELLINA, a mountain torrent of Austria, in Lombardy, in the prov. of Udine, which has its source in the NW part of that prov. between the mountains of Costomo, Costa Bluda, and Aosaip; runs first SW, then SE; and, after a course of 45 m., joins the Meduna, near Pordenone.

ZELLINGEN, a village of Bavaria, in the circle of Lower Franconia, and presidial of Carlstadt, on the Main. Pop. 1,985.

ZELLNITZ, a village of Austria, in Syria, in the circle of Marburg. Pop. 370.

ZELL-UNTER-AICHELBERG, a market-town of Würtemberg, in the circle of the Danube, and bail. of Kirchheim. Pop. 748.

ZELLWILLER, a commune of France, in the dep. of the Bas-Rhin, and cant. of Obernai, 12 m. N of Schelestadt. Pop. 1,137.

ZELO-BUON-PERSICO, a village of Austria, in Lombardy, in the prov. of Lodi and Cremona, 8 m. NNW of Lodi.

ZELODI, an island of the Indian ocean, near the W coast of Sumatra.

ZELS, a village of Prussia, in the reg. of Frankfort, in the circle of Sorau. Pop. 210.

ZELTINGEN, a market-town of Prussia, in the reg. of Treves, and circle of Berncastel, on the r. bank of the Moselle. Pop. 1,508.

ZELTSCH, a village of Austria, in Moravia, in the circle of Brunn. Pop. 760.

ZELVA, a river of Russia in Europe, in the gov. of Grodno, which has its source to the SE of Volkovisk; runs N; and, after a course of about 75 m., joins the Niemen, on the l. bank, to the NNE of the town of that name. On its banks is a town also named Zelva. Pop. 1,000.

ZEMANI. See CASAMANSA.

ZEMBERE. See ZAIRE.

ZEMBI, a mountain of Greece, in the N part of the Morea, to the W of Mount Velia, and WSW of Calavrita.

ZEMBIN, a town of Russia in Europe, in the gov. of Minsk, district and 15 m. NW of Borisov.

ZEMBKOW, a town of Prussia, in the reg. of Posen, and circle of Adelnau. Pop. 290.

ZEMBLA (NOVA). See NOVA ZEMBLA.

ZEMBLIANK, or ZEMLIANSK, a town of Russia in Europe, in the gov. and 27 m. NW of Voronej.

ZEMBOWITZ, a village of Prussia, in the reg. of Oppeln, and circle of Rosenburg. Pop. 420. It has a castle. Potash is its chief article of manufacture.

ZEMBOWO, a village of Prussia, in the reg. of Marienwerder, and circle of Schwetz. Pop. 260.—Also a village of the reg. of Posen, and circle of Buk. Pop. 470.

ZEMBRA, an island of Tunis, near the entrance of the gulf of that name.

ZEMBSCHEN, a village of Prussia, in the reg. of Merseburg, and circle of Weissenfels. Pop. 210.

ZEMLIN. See SEMLIN.

ZEMLING, a village of Austria, in the country below the Enns, in the Lower Mannhartsberg, and presidial of Limburg. Pop. 270.

ZEMMEN, a village of Prussia, in the reg. of Koslin, and circle of Lauenburg-Butow. Pop. 300.

ZEMMER, a village of Prussia, in the reg. and circle of Treves. Pop. 500.

ZEMMIN, a village of Prussia, in the reg. of Koslin, and circle of Neustettin. Pop. 200.—Also a village of the same regency, in the circle of Stolpe. Pop. 290.

ZEMNO. See ZEMPLIN.

ZEMOH, a town of Burmah, in the prov. of Pegu, on the l. bank of the Irawady, 99 m. NW of Rangoon.

ZEMONICO, a market-town of Austria, in Dalmatia, in the circle of Zara. It has an active trade with Bosnia.

ZEMPCZ. See SZEMPCZ.

ZEMPELBURG, or ZEMPLEN, a town of Prussia, in the reg. of Marienwerder, and circle of Flatow, on a small lake of the same name. Pop. 3,300. It has Catholic and Lutheran churches, and a synagogue, and possesses manufactories of cloth, linen, and lace, and a fulling-mill.

ZEMPIN, a village of Prussia, in the prov. of Pomerania, reg. and 51 m. NW of Stettin, and circle of Usedom-Wollin, on the island of Usedom. Pop. 80.

ZEMPLIN, SEMPLIN, or ZEMPLEN, a gespenschaft or comitat and a town of Hungary, in the circle beyond the Theiss. The comitat is bounded on the N by Galicia; on the E by the comitats of Unghvar and Szabolcs; on the S by the latter comitat; and on the W by those of Borsod, Abauj, and Saros, containing an area 108 m. in extent from NNE to SSW, and 30 m. in extreme breadth, with a total superficies of 900 sq. m. Pop. in 1839, 347,100; in 1850, 350,000. It is covered in the N part witn the Carpathian chain, a ramification of which, named the Heggalla, is noted for a species of wine, known as Tokay. The principal rivers are the Theiss, which forms to a great extent the S confines of the comitat; the Hernad on the SW; and in the interior the Bodrog and its affluents the Labonez, the Ondava, and Topla. The climate in the N is cold, but towards the S is mild and genial. The valleys are extremely fertile, and produce in great abundance corn, tobacco, hemp, and fruit. Wine is grown on the hills, and the mountains afford timber, iron, petroleum, alum, jasper, and saltpetre. Large number of horses, cattle, and sheep are reared on the mountain pastures. The comitat of Z. is divided into six marches, and contains 28 towns and 119 villages. Ujhely-Satorallia is the chief town.

ZEMPLIN, SEMPLIN, ZEMPLEN, or ZEMNO, a town of Hungary, in the comitat of the same name, on the r. bank of the Bodrog, 9 m. ENE of Ujhely-Satorallia. It has an ancient castle.

ZEMSCHEN, a village of Austria, in Bohemia, in the reg. of Pilsen, circle of Klattau, and seignory of Hostau. Pop. 260.

ZEMY, TCH'HAING-MAI, SAYM-MAY, or JAUGOINA, a town of Burmah, capital of the prov. of Yun-shan, on the r. bank of the Meinam, and 330 m. ESE of Ava.

ZEMZEN (WADY), a valley of Tripoli, 78 m. S of Mesurata, on the road to Murzuk. It has several springs of water.

ZENAGA. See SENEGAL.

ZENATI, a name by which the Seibus, a river of Barbary, is distinguished in the upper part of its course.

ZENATYIA, a well-watered tract of Marocco, in the prov. of Tafilet.

ZENDEJAS-DE-LA-TORRE. See CENDIJAS-DE-LA-TORRE.

ZENDERO. See GINGIRO.

ZENDEROND. See ZAYENDEH-RUD.

ZENDS, a nomade tribe of Persia, in the prov. of Irak-Ajemi, in the environs of Ispahan, and N part of Farsistan, and comprising about 12,000 individuals.

ZENEGGEN, a village of Switzerland, in the cant. of Wallis, and bail. of Visp. Pop. 210.

ZENFARAH. See ZAMFRA.

ZENG, ZENGG, SEGNA, SEGNIA, or SZENY, a free royal town of Hungary, on the military confines of Croatia, generalat and 51 m. SW of Carlstadt, and regimentary district of Ottochaow, on the Adriatic, opposite the island of Veglia, from which it is separated by the canal of Morlacca, and 39 m. SE of Fiume. Pop. 5,000. It has two Catholic churches, three chapels, a Franciscan convent, an Episcopal lyceum, and a theological seminary. Its trade consists chiefly in the corn of Hungary, the wine of Dalmatia, and in fish. The port, which was declared free in 1785, possesses few advantages, being open to the NW. In the vicinity of this town is the road over the mountains of Capella, formed by Joseph II., and named by him the Josephine road.

ZENGHIAN, a town of Persia, in the prov. of Irak-Ajemi, capital of the district of Khamsah, and 180 m. NW of Teheran. Pop. 15,000. It is enclosed with a wall flanked with towers, is well-built, and has a fine bazaar, to which the Avchars, a nomade tribe, come to barter carpets, skins, and woollen fabrics for cloth, arms, powder, and lead. The palace of the khan is a spacious edifice, and possessing considerable elegance of architecture. This town was destroyed by Tamerlane.

ZENITZA. See SIENITZA

ZENKOV, a town of Russia in Europe, in the gov. and 45 m. N of Poltava, on the Gronnia. It has eight churches.

ZENNERN, a village of Kurhessen, in the prov. of Niederhessen, circle and bail. of Fritzlar. Pop. 560.

ZENNOR, a parish in Cornwall, 6 m. WSW of St. Ives. Area 4,229 acres. Pop. in 1851, 918.

ZENONE (San), a town of Austria, in Lombardy, in the prov. and 14 m. ESE of Pavia, and district of Corte-Olona, on the Olona, which is here crossed by a substantial bridge, and a little above its confluence with the Po. Pop. 1,150.

ZENTA. See Szenta.

ZENTAWA, a village of Prussia, in the regency of Oppeln, and circle of Gross-Strehlitz. Pop. 385.

ZENTBECHOFEN, a village of Bavaria, in the circle of Upper Franconia, and presidial of Höchstadt. Pop. 260.

ZENTCHA. See Znatchka.

ZENTENDORF, a village of Prussia, in the reg. of Liegnitz, and circle of Gorlitz. Pop. 230.

ZENZ, a village of Prussia, in the regency of Magdeburg, and circle of Kalbe. Pop. 290.

ZENZEGUAJES, an Indian tribe, who inhabit the S part of New Granada, near the Amazon.

ZEPADA. See Cepada.

ZEPEE. See Jepee.

ZEPITA, a town of Bolivia, in the dep. and 51 m. WNW of La Paz, on the S bank of Lake Vinamarca, a branch of Titicaca.

ZEPKOW, a village of Mecklenburg-Schwerin, circle of Wendisch, and bail. of Wredenhagen. Pop. 260.

ZEPPENFELD, a village of Prussia, in the regency of Arnsberg, and circle of Siegen. Pop. 580.

ZEPPEREN, a village of Belgium, in the prov. of Limburg, cant. and 3 m. ENE of St. Trond. Pop. 1,288.

ZEPPERNICK, a village of Prussia, in the reg. of Magdeburg, and circle of Jerichow. Pop. 240.

ZERA, a village of Marocco, on the N coast, to the W of Milella.

ZER-AFCHAN, Kohek, Sogd, or Konan-Daria, a river of Independent Tartary, in Bokara, which descends from the range of the Kara-tagh, runs W, passes Samarcand and Bokara, and after a course of 300 m., loses itself in Lake Dengis. Its waters are extensively employed in irrigation.

ZERARA, a district of the empire and prov. of Marocca, to the N of the town of that name.

ZERAWIN, a village of Austria, in Moravia, in the circle of Hradisch, and seignory of Ungarisch-Ostra. Pop. 260.

ZERAWITZ, a market-town of Austria, in Moravia, in the circle of Hradisch. Pop. 400.—Also a village of the circle of Prerau. Pop. 420.

ZERBA. See Jerbeh.

ZERBAU, a village of Prussia, in the regency of Liegnitz, and circle of Glogau. Pop. 395.

ZERBEN, a village of Prussia, in the regency of Magdeburg, and circle of Jerichow. Pop. 300.

ZERBST, a town of the duchy of Anhalt-Dessau, capital of an amt or bail. of the same name, on the Nuthe, 14 m. NW of Dessau. Pop. 11,600. It is enclosed by walls, with three gates, and has a suburb, an old castle, a Lutheran and two Reformed churches, two hospitals, an orphans' asylum, a house of correction, and a school called the Franciseum. It has manufactories of gold and silver ware, earthenware, cloth, parchment, and tobacco, several breweries, bleacheries of wax, &c. Legumes and fruit are extensively cultivated in the environs. This town is noted as the birthplace of the Russian empress Catherine II.

ZERD-KOH, or Koh-i-Zerd, a summit of the Awa chain, in Persia, in Irak-Ajemi, 70 m. SW of Ispahan. It gives rise to the Zindarud.

ZERELITZ, a village of Austria, in Bohemia, in the regency and bail. of Gitschin. Pop. 540.

ZERERE, a small river of Brazil, in the prov. of Mato-Grosso, which has its source in the Serra Santa Barbara, runs NE, and joins the Rio Embotetiu or Mondego on the l. bank.

ZEREZUELA, a town of New Granada, in the dep. of Cundinamarca, and prov. of Bogota, 9 m. SE of Santa Fe-de-Bogota.

ZERF (Nieder and Ober), two villages of Prussia, in the regency of Treves, and circle of Saarburg, containing respectively 332 and 380 inhabitants.

ZERGUN, a village of Persia, in Farsistan, to the N of Shiras.

ZERIND, a village of Hungary, in the comitat of Arad, and NE of Siklo, on the Koros.

ZERINZA. See Serinza.

ZERKITZA, a mountain on the confines of Turkey in Europe and Dalmatia, between the Herzegovine and the district of Cattaro.

ZERKOWITZ, a village of Austria, in Moravia, in the circle of Znaim, and seignory of Hösling. Pop. 300.

ZERKOWO, a town of Prussia, in the regency and 39 m. SE of Posen, and circle of Wreschen. Pop. 1,200. It has several breweries and distilleries of brandy.

ZERKWITZ, a village of Prussia, in the regency of Frankfurt, and circle of Kalau. Pop. 300.

ZERLACH, a village of Austria, in Styria, in the circle of Gratz. Pop. 420.

ZERLANG, a village of Prussia, in the regency of Potsdam, and circle of Ruppin. Pop. 240.

ZERMAGNA, a market-town of Austria, in Dalmatia, in the circle of Zara, on a river of the same name.

ZERMATT-THAL, a valley of Switzerland, in the cant. of the Valais, descending from the Alps at Ft. S. Theodule, and joining the Visp Thal, a little below the village of Zermatt or Praborn. Pop. 411.

ZERNAMEL. See Tschernembel.

ZERNETZ. See Cernetz.

ZERNEWO. See Czerniejewo.

ZERNIKI. See Schernik.

ZERNOWITZ. See Zsarnocza.

ZERNOWNA, a village of Austria, in Moravia, in the circle of Brunn, and seignory of Czernahora. Pop. 210.

ZERNOWNIK, a village of Austria, in Moravia, in the circle of Brunn, and seignory of Czernahora. Pop. 240.

ZERO, a river of Austria, in Lombardy, which has its source in the prov. of Treviso, and district of Castelfranco, separates that prov. from that of Padua, and enters the prov. of Venice, joins the Dese, and after a course of about 60 m., loses itself in the Lagunes to the N of Venice. Its waters are limpid, and it is navigable for a short distance.

ZERO, a mountain of Australia Felix, in the district of Wimmera, at the N extremity of the Grampian chain. It consists of micaceous sandstone.

ZEROTIN, a village of Austria, in Bohemia, in the reg. of Prague, and bail. of Rakonitz. Pop. 360.

ZEROTITZ, a village of Austria, in Moravia, in the circle of Znaim. Pop. 400.

ZEROWITZ. See Serowitz.

ZERUTEK, a village of Austria, in Moravia, in the circle of Brünn, and seignory of Lissitz. Pop. 200.

ZERYBE, a village of Algeria, in the prov. of Constantine, on the Serka.

ZESCHA, a village of Saxony, in the circle and bail. of Bautzen. Pop. 280.

ZESCHNIG, a village of Saxony, in the circle of Dresden, and bail. of Hohnstein. Pop. 200.

ZESCHWITZ, a village of Saxony, in the circle of Leipzig, and bail. of Pegau. Pop. 266.

ZESELBERG, a village of Bavaria, in the circle of the Pfalz, and cant. of Waldsischbach. Pop. 332.

ZESSEL, a village of Prussia, in the regency of Breslau, and circle of Münsterberg. Pop. 270.

ZESSENDORF, a village of Prussia, in the reg. of Liegnitz, and circle of Sagan. Pop. 370.

ZETEL (Ostseite and Westseite), two villages of the duchy and to the NW of Oldenburg, and bail. of Bockhorn. Pop. 1,590.

ZETHLAU, a village of Saxony, in the circle of Dresden, and bail. of Freiburg. Pop. 1,590.

ZETHLINGEN, a village of Prussia, in the regency of Magdeburg, and circle of Salzwedel. Pop. 300.

ZETHUN, a village of Prussia, in the regency of Koslin, and circle of Furstenthum. Pop. 210.

ZETLAND. See Shetland.

ZETSCHOWITZ, a village of Austria, in Bohemia, in the regency of Pilsen, and bail. of Bischofteinitz. Pop. 270.

ZETTELSDORF, a village of Bavaria, in the circle of Upper Franconia, and presidial of Burgebrach. Pop. 220. It has a castle.

ZETTWEIL, a village of Prussia, in the regency of Merseburg, and circle of Zeitz. Pop. 290.

ZETTWING, a market-town of Austria, in Bohemia, in the regency of Budweis, and bail. of Kaplitz. Pop. 700.

ZEUBELRIED, a village of Bavaria, in the circle of Lower Franconia, and presidial of Ochsenfurt. Pop. 230.

ZEUCHING, a village of Bavaria, in the circle of Lower Bavaria, and presidial of Kotzing. Pop. 240.

ZEUGFELD, a village of Prussia, in the regency of Merseburg, and circle of Ouerfurt. Pop. 200.

ZEULENRODA, a town of the principality of Reuss-Greitz, 10 m. WSW of Greitz, in a woody locality. Pop. 5,400. It is enclosed by walls, and has a burgess school and an hospital. It possesses manufactories of woollen fabrics, hosiery, hats, and watches, and carries on a considerable trade also in cattle.

ZEULN, or Marktzueln, a market-town of Bavaria, in Lower Franconia, near Schweinfurt. Pop. 370. It has several breweries, and carries on an active trade in wood.

ZEUNDORF, a village of Anhalt-Dessau, in the circle of Rothen, on the Fuhne. Pop. 240.

ZEUTERN, a village of the grand-duchy of Baden, in the circle of the Middle Rhine, and bail. of Bruchsal. Pop. 1,490.

ZEUTSCH, a village of the duchy of Saxe-Altenburg, and bail. of Rahla. Pop. 300.

ZEUTZLEBEN, a village of Bavaria, in Lower Franconia, in the presidial of Werneck. Pop. 584.

ZEVAL, a village of Turkey in Europe, in Bulgaria, in the sanj. of Widden, on the Danube.

ZEVECOTE, a village of Belgium, in the prov. of West Flanders, and arrond. of Bruges. Pop. 620.

ZEVEN, or Kloster-Zeven, a market-town and bail. of Hanover, in the gov. and 27 m. SSW of Stade, duchy and 26 m. NE of Bremen, on the r. bank of the Aue. Pop. 900. A convention between the French and Hanoverians was concluded here in 1757.

ZEVENAAR, a town of Holland, in the prov. of Gelderland, arrond. and 9 m. ESE of Arnheim, near the r. bank of the Rhine. Pop. 900.

ZEVENBERGEN, a market-town of Holland, in the prov. of North Brabant, arrond. and 9 m. WNW of Breda, on a canal which branches from the Merk at Hollands-Diep. Pop. 3,800.

ZEVENECKEN, a commune of Belgium, in the prov. of West Flanders, 8 m. ENE of Ghent. Pop. 2,500. It has manufactories of fine linen and cotton fabrics, and carries on an active trade in grain, flax, linen, and cattle.

ZEVEREN, a commune of Belgium, in the prov. of East Flanders, arrond. and 12 m. W of Ghent. Pop. 1,000. It has manufactories of linen and cotton fabrics.

ZEVENHUIZEN a parish of Holland, in the prov. of South Holland. Pop. 1,200.

ZEVIO, a market-town of Austria, in Lombardy, in the prov. and 8 m. SE of Verona, near the r. bank of the Adige. Pop. 1,748. In the vicinity is a fine castle, surrounded with a deep ditch.

ZEWELIN, a village of Prussia, in the regency of Koslin, and circle of Furstenthum. Pop. 250.

ZEWEN, a village of Prussia, in the regency and circle of Treves, on the Mosel. Pop. 660.

ZEWITZ, a village of Prussia, in the regency of Koslin, and circle of Lauenburg-Buton. Pop. 200.

ZEYERN, a market-town of Bavaria, in Upper Franconia, in the presidial of Kronach. Pop. 459.

ZEYRING, a market-town of Austria, in Styria, in the circle of Judenburg. Pop. 560.

ZEYRINGGRABEN, a village of Austria, in Styria, in the circle of Bruck. Pop. 260.

ZEYST. See Seist.

ZEYTOUN, a village of Egypt, in the oasis and 9 m. E of Siwah, on the caravan route from Murzuk to Cairo. Olives abound in its vicinity.

ZEZARINE, Zazarine, or Farsey, a small island of the Persian gulf, to the NW of Cape Kenn, in Farsistan, and about midway between the coasts of Persia and Arabia.

ZEZENOW, a village of Prussia, in the regency of Koslin, and circle of Stolpe. Pop. 350.

ZEZERE, a river of Portugal, in the prov. of Beira, which has its source in the Serra-de-Estrella, at Paulo-Martius, in the territory of Manteigas; runs from NE to SW; and joins the Tagus on the r. bank, between Punhete and Paio-de-Pelle, and after a rapid course of 120 m. Its waters are turbid. Its principal affluents are the Meimoa and Naboa.

ZEZITOU, a village of Switzerland, in the cant. of Thurgau, and bail. of Kulm. Pop. 1,149.

ZEZOWA, a town of Tripoli, in the Fezzan, 15 m. E of Murzuk, and on a caravan route thence to the S.

ZEZWEIL, a village of Switzerland, in the cant. of Aargau, and bail. of Kulm. Pop. 1,149.

ZGIERZ, a town of Russia in Europe, in Poland, in the gov. of Warsaw, obwod and 18 m. SE of Lenczy. Pop. 3,200.

ZGNILOBLOTT, a village of Prussia, in the regency of Marienwerder, and circle of Strasburg. Pop. 320.

ZGOIN, a village of Prussia, in the regency of Oppeln, and circle of Pless. Pop. 498.

ZHOBE, a river of Afghanistan, in Sewestan, which has its source near the NE base of the Toba mountains, and after a course in a generally NE direction of about 170 m., joins the Gomul, in N lat. 32° 2′, and E long. 69° 40′.

ZHOR, a village of Austria, in Bohemia, in the regency of Budweis, and bail. of Tabor. Pop. 270.—Also a village of the same regency, in the bail. of Muhlhausen. Pop. 230.—Also a village of the reg. of Pardubitz, and bail. of Ruttenberg. Pop. 460.—Also a village of the same regency, in the bail. of Hohenmauth. Pop. 410.

ZHOREZ, a village of Austria, in Bohemia, in

the regency of Budweis, and bail. of Pilgram. Pop. 300.

ZHORZ-STRANEZKZKA, a market-town of Austria, in Moravia, in the circle of Iglau. Pop. 600.

ZIA, THIKIRI, or ZINGHIRI, a river of Mandshuria, which has its sources on the S side of the Stannovo mountains; runs SSE; receives the Silimpdi on the l., and after a rapid course of about 450 m., joins the Amoor on the l. bank, a little above Sagalu-Oola, and by an embouchure about 1½ m. in width.

ZIABNA, a village of Austria, in Moravia, in the circle of Prerau. Pop. 320.

ZIADOWITZ, a village of Austria, in Moravia, in the circle of Hradisch. Pop. 550. It has a castle.

ZIAKAU, a village of Austria, in Bohemia, in the reg. of Pilsen, and bail. of Blowitz. Pop. 400.

ZIAKOWITZ, a village of Austria, in Moravia, in the circle of Prerau, and seignory of Bistritz. Pop. 300.

ZIALKOWITZ, a village of Austria, in Moravia, in the circle of Olmutz, and bail. of Kremster. Pop. 570.

ZIAMNA-GUNTZO, a lake of Tibet, in the prov. of Wei, 120 m. NE of Lassa, from which the Niangchu issues in the S, runs SSE, and it is supposed to join the Yarou-dzang.

ZIAN, a village to the N of Kashmir, near the r. bank of the Kishengunga river, in N lat. 34° 35′, and E long. 74° 42′.

ZIANDOWITZ, a village of Prussia, in the reg. of Oppeln, and circle of Gross Strehlitz. Pop. 1,200.

ZIANGAMRING, a village of Tibet, in the prov. of Dzang, 150 W of Lassa, on the Chang, an affluent of the Yarou-dzang.

ZIAROWITZ, a village of Austria, in Moravia, in the circle of Olmutz, and seignory of Plumenau. Pop. 300.

ZIAZ, a village of Prussia, in the regency of Magdeburg, and circle of Jerichow Pop. 350.

ZIBELLE (MITTEL, NIEDER and OBER), a market-town of Prussia, in Silesia, in the regency of Liegnitz, and circle of Rothenburg. Pop. 650.

ZIBELLO. See GIBELLO.

ZIBEL-TEIR, or GEBEL-TUR, a small volcanic island of the Arabian gulf, 48 m. from the shore of Yemen, in Arabia, and 6 m. WSW of Loheia, in N lat. 15° 30′, and E long. 41° 45′.

ZIBIB. See ZEBIBI.

ZIBIN, or CIBIN, a river of Transylvania, in the Saxon territory, which has its sources in the Sztrimba and Froma mountains, on the W confines of the sea of Hermanstadt, runs NE to the town of that name, then SE, and after a course of about 36 m. joins the Hartbach.

ZIBKOV (NOVO), a town of Russia in Europe, in the gov. and 84 m. NE of Tchernigov.

ZIBOBLAW, a village of Austria, in Bohemia, in the regency of Pardubitz, and bail. of Neukolin. Pop. 200.

ZIBREIRA, or CIBREIRA, a market-town of Portugal, in the prov. of Beira, comarca and 30 m. SE of Castello-Branco, on the Eljas, near the Spanish frontier. Pop. 800. It has a fort and an hospital.

ZICAVO, a canton, commune, and town of France, in the dep. of Corsica, and arrond. of Ajaccio. The cant. comprises 9 com. Pop. in 1830, 4,925; in 1846, 5,474.

ZICHER, a village of Prussia, in the regency of Frankfort, and circle of Kustrin. Pop. 860.

ZICHTAU, a village of Prussia, in the regency of Magdeburg, and circle of Gardelegen. Pop. 300.

ZICKELHID. See SZEKELYHID.

ZICKER, a village of Prussia, in the regency of Koslin, and circle of Neu-Stettin. Pop. 530.

ZICKERKE, a village of Prussia, in the regency of Stettin, and circle of Naugardt. Pop. 300.

ZICKHUSEN, a village of Mecklenburg-Schwerin, in the prov. of Mecklenburg, and bail. of Schwerin. Pop. 215.

ZIDDORF, a village of Mecklenburg-Schwerin, in the circle of the Wendisch, and bail. of Teterow. Pop. 200.

ZIDEN, a village of Prussia, in the regency of Gumbinnen, and circle of Lyk. Pop. 200.

ZIDLOCKOWICE. See SELOWITZ.

ZIEBENDORF, a village of Prussia, in the reg. of Liegnitz, and circle of Luben. Pop. 374.

ZIEBENGEN, a village of Prussia, in the reg. of Frankfort, and circle of Sternberg. Pop. 1,190.

ZIEBERN, a village of Prussia, in the regency of Liegnitz, and circle of Glogau. Pop. 300.

ZIECHANOWITZ, a village of Russia in Europe, in the gov. of Bialystock, to the S of Bransk.

ZIECHAU, a village of Prussia, in the regency of Potsdam, and circle of Angermunde. Pop. 300.

ZIECKAU, a village of Prussia, in the regency of Frankfort, and circle of Luckau. Pop. 230.

ZIEDER (NIEDER and OBER), two villages of Prussia, in the regency of Liegnitz, and circle of Landshut, containing respectively 525 and 478 inhabitants.

ZIEGELANGER, a village of Bavaria, in the circle of Lower Franconia, and presidial of Altmaun. Pop. 260.

ZIEGELBACH, a village of Würtemberg, in the circle of the Danube, and bail. of Waldsee. Pop. 74.

ZIEGELERDEN, a village of Bavaria, in Upper Franconia, in the presidial of Kronach. Pop. 240.

ZIEGELHAUSEN, a village of Baden, in the circle of the Lower Rhine, and bail. of Heidelberg, on the Neckar. Pop. 1,545. It has three churches, and possesses several paper-mills, iron-works, linen factories, breweries, and tile-kilns. In the vicinity is the ancient convent of Neuburg.

ZIEGELHEIM, a village of Saxony, in the circle and bail. of Zwickau, near Waldenburg. Pop. 589.

ZIEGELLACK, a village of Prussia, in the reg. and circle of Marienwerder. Pop. 230.

ZIEGELRODA, a village of Prussia, in the reg. of Merseburg, and circle of Oberfurt. Pop. 515.—Also a village of the same regency, in the circle of Mannsfeld. Pop. 280.

ZIEGELSCHENNE, a village of Prussia, in the regency of Posen, and circle of Meseritz. Pop. 200.—Also a village of the regency of Breslau, and circle of Militsch. Pop. 280.

ZIEGELSTEIN, a village of Bavaria, in Middle Franconia, in the presidial of Nurnberg. Pop. 200.

ZIEGENBURG, a village of Austria, in Styria, in the circle of Gratz. Pop 250.

ZIEGENDORF, a village of Mecklenburg-Schwerin, in the prov. of Mecklenburg, and bail. of Grabon. Pop. 430.

ZIEGENFUSS, a village of Austria, in Bohemia, in the regency of Pardubitz, and bail. of Landskron. Pop. 250.

ZIEGENHAGEN, a village of Hesse-Cassel, in the prov. of Nieder Hessen, circle and bail. of Witzenhausen. Pop. 422.—Also a village of Prussia, in the regency of Stettin, and circle of Saatzig. Pop. 250.

ZIEGENHAIN, a village of Saxony, in the circle of Dresden, and bail. of Meissen. Pop. 237.—Also a village of Saxe-Weimar, in the prov. of Weimar, and bail. of Jena. Pop. 320.—Also a town and fortress of Hesse-Cassel, capital of a circle and bail. of the same name, in the prov. of Oberhessen, on the

Schwalin. Pop. 1,777. The fortress, which belongs to the counts of Z., was founded in the 12th century, and was formerly capital of the co. of the same name. Its fortifications were destroyed by the French in 1806.—The circle of Z. comprises an area of 360 sq. m., of which 100 sq. m. are arable. Pop. in 1840, 34,298. It is mountainous, but comprises some fertile tracts. The culture of flax and the rearing of cattle form the chief branches of local industry. The circle comprises 4 Justiz-amt, and contains 6 towns and 74 villages.

ZIEGENHALS, a town of Prussia, in the reg. of Oppeln, circle and 12 m. SSE of Neisse, on the Biela. Pop. 2,750. It is partly enclosed with walls, with two gates, and has two Catholic churches, and an hospital.

ZIEGENORT (Gross and Klein), a village of Prussia, in the regency of Stettin, and circle of Uckermunde, at the confluence of the Oder with the Pommersche-Haff. Pop. 1,100.

ZIEGENRUCK, a village of Austria, in Bohemia, in the regency of Pilsen, and bail. of Schuttenhofen. Pop. 300.—Also a town of Prussia, in the prov. of Saxony, capital of a circle of the same name, in the regency and 39 m. SE of Erfurt, on the Saale and Drebebach. Pop. 800. In its environs is a slate-quarry. The circle consists of districts, in the states of Reuss, Schwarzburg-Rudolstadt, and Saxe-Meiningen.

ZIEGLERSHUBEN, a village of Prussia, in the regency of Marienwerder, and circle of Stuhm. Pop. 250.

ZIEGRA, a village of Saxony, in the circle of Leipsig, and bail. of Leisnig. Pop. 275.

ZIEKO, a village of the duchy of Anhalt-Bernburg, and bail. of Roswig. Pop. 200.

ZIELAZNA, a village of Prussia, in the reg. of Bromberg, and circle of Wirsitz. Pop. 260.

ZIELAZNO, a village of Prussia, in the reg. of Posen, and circle of Koslin. Pop. 230.

ZIELCHOWITZ, a village of Austria, in Moravia, in the circle of Olmutz and bail. of Sternburg. Pop. 350.

ZIELENICE, a village of Prussia, in the reg. of Posen, and circle of Wreschen. Pop. 300.

ZIELENZIG, a town of Prussia, in the prov. of Brandenburg, capital of the circle of Sternburg, in the reg. and 27 m. ENE of Frankfort, on the Poste. Pop. in 1851, 4,800. It is enclosed by walls, has two suburbs, and possesses manufactories of cloth, linen, hosiery, and several tanneries. Coal and alum are found in the environs.

ZIELITZ, a village of Prussia, in the reg. of Magdeburg, and circle of Wollmirstadt. Pop. 300.

ZIELKAU, a village of Prussia, in the reg. of Marienwerder, and circle of Lobau. Pop. 310.

ZIELLECHOWITZ, a village of Austria, in Moravia, in the circle of Hradisch and seignory of Wisowitz. Pop. 740.

ZIELONA, a village of Prussia, in the reg. of Oppeln, and circle of Lublinitz. Pop. 500.

ZIEMETSHAUSEN, a market-town of Bavaria, in Swabia, in the presidial of Krumbach, on the Zasam. Pop. 849.

ZIEMIENZITZ, a village of Prussia, in the reg. of Oppeln, and circle of Tost. Pop. 400.

ZIEMIN, a village of Prussia, in the reg. of Posen, and circle of Kosten. Pop. 250.

ZIEMLIN, a village of Prussia, in the reg. of Posen, and circle of Kroben. Pop. 210.

ZIENAGA, a town of New Granada, in the dep. of the Magdalena, prov. and 111 m. S of Carthagena. —Also a bay of Venezuela, in the dep. of Sulica, in the NW part of Lake Maracaibo.

ZIER (Nieder and Ober), two villages of Prussia, in the regency of Aachen, and circle of Duren. Pop. respectively 848 and 520.

ZIERANOWITZ, a village of Austria, in Moravia, in the circle of Hradisch. Pop. 440. It has a castle.

ZIERDE, a village of Austria, in Bohemia, reg. of Bohm-Leipa, and bail. of Auscha. Pop. 230.

ZIERENBERG, a town and bail. of Electoral Hesse, in the prov. of Lower Hesse, circle and bail. of Wolfhagen, on the Wurme, 10 m. WNW of Cassel. Pop. 1,601. It is enclosed by walls, with three gates. In the vicinity are the ruins of the ancient castles of Schartenberg and Gudenburg. Z. was taken by the Duke of Brunswick in 1760.

ZIERIKZEE, an arrondissement, canton, and town of Holland, in the prov. of Middelburg. The arrond. comprises three cants. Pop. 23,000. The town is 17 m. NE of Middelburg, in the SE part of the island of Schouwen, and near the r. bank of the E. Schelde, with which it is connected by a port. Pop. 6,900. It is well fortified, and contains several churches. The ancient cathedral was destroyed by a storm in 1832. The manufacture of salt and hydromel and fishing form the chief objects of local industry. Near the cathedral is a cistern capable of containing 2,000 tons of water. This town, the most ancient in Zeeland, was built and surrounded with walls in 859. It sustained an unsuccessful attack by the Flemings under Guy-de-Dampier, Count of Flanders, in 1303, but was taken and for some time held by the Spaniards in 1576.

ZIERKE, a village of Mecklenburg-Strelitz, and circle of Stargard. Pop. 270.

ZIEROLSHOFEN, a village of the grand-duchy of Baden, in the circle of the Middle Rhine, and bail. of Rhein-Bischofsheim. Pop. 345.

ZIEROTEIN, a village of Austria, in Moravia, in the circle of Olmutz. Pop. 470.

ZIERSDORF, a village of Austria, in the district below the Enns, and circle of the Lower Mannhardsberg. Pop. 500.

ZIERSPIENTEN, a village of Prussia, reg. of Gumbinnen, and circle of Lensburg. Pop. 290.

ZIERTHEIM, a village of Bavaria, in Swabia, in the presidial of Laningen. Pop. 441.

ZIERUTEK, a village of Austria, in Moravia, in the circle of Znaim. Pop. 240.

ZIERWIENZ, a village of Prussia, in the reg. of Koslin, and circle of Stolpe. Pop. 230.

ZIERZOW, a village of Mecklenburg-Schwerin, in the prov. of Mecklenburg, and bail. of Grabow. Pop. 290.

ZIESAR, a town of Prussia, in the prov. of Saxony, in the reg. and 30 m. ENE of Magdeburg, and circle of Jerichow. Pop. 2,830. It has two churches, and possesses manufactories of cloth, linen, and hosiery.

ZIESCHOW, a village of Austria, in Moravia, in the circle of Olmutz. Pop. 450.

ZIESCHT, a village of Prussia, in the reg. of Potsdam, and circle of Juterbogk. Pop. 300.

ZIESENDORF, a village of Mecklenburg-Schwerin, in the circle of Wendisch, and bail. of Schwaan. Pop. 200.

ZIESERWITZ, a village of Prussia, in the reg. of Breslau, and circle of Neumarkt. Pop. 580. It has a castle.

ZIETHEN (Hohen), a village of Prussia, in the reg. of Frankfort, and circle of Soldin. Pop. 320.

ZIETHEN (Gross and Klein), two villages of Prussia, in the reg. of Potsdam, and circle of Angermunde, containing respectively 440 and 280 inhabitants.

ZIETHEN (Gross), a village of Prussia, in the reg. of Potsdam, and circle of Osthavelland. Pop.

310.—Also a village of the duchy of Mecklenburg-Strelitz, and circle of Stargard. Pop. 330.

ZIETLOW, a village of Prussia, in the reg. of Koslin, and circle of Belgard. Pop. 200.

ZIEVERICH, a village of Prussia, in the reg. of Cologne, and circle of Rergheim. Pop. 210.

ZIEZA, a town of Spain, in the prov. and 30 m. NW of Murcia, on the l. bank of the Segura.

ZIEZENOW, a village of Prussia, in the reg. of Koslin, and circle of Belgard. Pop. 250.

ZIEZENOW, a village of Prussia, in the regency of Koslin, and circle of Belgard. Pop. 250.

ZIEZOW, a village of the grand-duchy and duchy of Mecklenburg-Strelitz, and bail. of Mirow.

ZIFFER. See CZIFFER.

ZIFFLICH, a village of Prussia, in the regency of Dusseldorf, and circle of Kleve. Pop. 470.

ZIGABNEN, a village of Prussia, in the regency and circle of Marienwerder. Pop. 370.

ZIGANKENBERG, a village of Prussia, in the regency and circle of Danzig. Pop. 500.

ZIGGELMARK, a village of Mecklenburg-Schwerin, in the prov. of Mecklenburg, and bail. of Wittenburg. Pop. 200.

ZIGNA, or ZIKNA, a town of Turkey in Europe, in the beglik and 15 m. ESE of Seres, on a small river of the same name, an affluent of Lake Takinos.

ZIGNANO, a village of Sardinia, in the gov. of Genoa, and prov. of Levante. Pop. 1,300.

ZIGNITZ, a village of Prussia, in the regency of Koslin, and circle of Schlawe. Pop. 210.

ZIGOS, a mountain of Greece, in Livadia, in Ætolia, to the N of Mesolonghi, extending between the Aspropotamo and Fidari, at the distance of about 5 m. from the coast.—Also a mountain of Turkey in Europe, in Albania, in the sanj. of Janina, near the point of junction of Macedonia and Thessaly. It makes part of the Mezzovo range in the great Hellenic chain.

ZIGUINES, ZIGUENEZ, or BOHEMIANS, a nomade people who are found in almost every part of Europe, in the W of Asia, and N of Africa. Their history is still a matter of doubt; but the most probable opinion is that they are of Hindostani origin. On some authorities, they are descended from the *Sigynnæ*, a people, according to Herodotus, settled near the Getæ and Thracians. By some they are traced to Egypt, and hence derive the name *Gipsy*, by which they are commonly known in England, and that of *Pharaons*, their common appellation in Hungary, where they are also named *Tzinganu*. By the Greeks they are named *Athinganoi*; by the Wallachians and Moldavians, *Cyganis*; by the Italians, *Zingani* or *Zingari*; and by the Dutch, *Heiden*. In Syria, and by the Turks, they are called *Djinganes* or *Tchinghenes*; in Syrmia, *Madjub*; in Bokhara, *Diaji*; in Barbary, *Slaubara*; amongst the Arabs, *Courbad*; and in Egypt, *Ghasie*. By themselves they are named *Roma*, *Sinte*, and *Mallele-Tchehl*. Their number in Europe is estimated at 340,000, of whom 50,000 are in Austria; 150,000 in Turkey; 10,000 in France; 20,000 in England; and 20,000 in Russia. The independent and vagabond life led by the Z., the misery and degradation of their condition, seem to ally this people with the Gitanos of Spain, a tribe, however, with which they cannot properly be identified.

ZIHLSCHLACHT, a village of Switzerland, in the cant. of Thurgau, and bail. of Bischofszell. Pop. 485.

ZIHOBETZ, a village of Austria, in Bohemia, in the regency of Pilsen, and bail. of Schutenhofen. Pop. 600.

ZIKANKA, a village of Bohemia, in the circle of Chrudim. It has a paper-mill.

ZIKKAIN, or SAIGAING, a town of Burmah, in the Mranma, on the r. bank of the Irawady, 3 m. NNW of Ava, in N lat. 21° 25′, and E long. 96°. It is situated partly at the base, and partly on the side of a range of hills, most of the summits of which are crowned with temples. The number of these temples, ancient and modern, forms an extraordinary assemblage of objects, the effect of which is increased by their being carefully whitewashed and kept in repair. The houses are scattered amid gardens and plantations, in which are many fine old tamarind trees. The Irawady contracts considerably between Zikkaïn and the citadel of Ava, but expands again below the latter, and is always covered with numerous boats plying between the two banks.

ZILEH, a town of Turkey in Asia, in the pash. and sanj. and 30 m. NW of Sivas. It is the *Zela* of the ancients, and is noted as the scene of the defeat of Pharnaces by Julius Cæsar.

ZILINA. See SZOLNA.

ZILITEN, or ZLITUN, a market-town of Tripoli, in the kingdom and 30 m. ESE of Tripoli, in N lat. 32° 32′ 54″, and E long. 14° 34′ 20″, on the Mediterranean. It is inhabited by Jews and Marabouts. In the vicinity is the tomb of a celebrated Mahommedan saint, named Sidi-Abd-el-Salam.

ZILLAH, or ZILLENMARKT. See WALTENBERG.

ZILLAHPUR, a village of Hindostan, in the district of Cawnpore, 57 m. SE of Etawa, on the road thence to Allahabad, in N lat. 26° 16′, and E long. 79° 46′.

ZILLBACH, a village of the grand-duchy of Saxe-Weimar, in the circle of Eisenach, and bail. of Kalten-Nordheim. Pop. 368. It has a grand ducal hunting-lodge.

ZILLER, a river of Austria, in the Tyrol, an affluent of the Inn.

ZILLERTHAL (HOHEN, MITTEL, and NIEDER), a village of Austria, in the regency of Liegnitz, and circle of Hischberg, and near Erdmannsdorf. Pop. 350.

ZILLHAUSEN, a village of Würtemberg, in the circle of the Schwarzwald, bail. and 3 m. E of Balingen, on the Bach, which in the vicinity forms a fine cascade. Pop. 713.

ZILLINGDORF, a market-town of Austria, in the circle of the Lower Wienerwald, on the Leitha. Pop. 950.

ZILLIS, a village of Switzerland, in the cant. of the Grisons, and bail. of the Hinter-Rhein, to the S of Thusts. Pop. 306.

ZILLISHEIM, a commune of France, in the dep. of the Haut-Rhin, and cant. of Mulhausen, 9 m. N of Altkirch. Pop. 1,176.

ZILLY, a village of Prussia, in the regency of Magdeburg, and circle of Aschersleben. Pop. 768.

ZILMITZ, a village of Prussia, in the regency of Koslin, and circle of Schlawe. Pop. 210.

ZILSHAUSEN, a village of Prussia, in the reg. of Coblenz, and circle of Kochem. Pop. 345.

ZILTENDORF, a village of Prussia, in the reg. of Frankfort, and circle of Guben. Pop. 660.

ZIMANRO, a strait by which the island of Sabrao is separated from the islands of Lomblem and Solor, in the Sunda archipelago.

ZIMAPAN, a town of Mexico, in the dep. and 105 m. of Queretaro, on the elevated plateau of Mextillan. Pop. 5,000. It has important silver-mines.

ZIMATLAN, a village of Mexico, in the dep. and 12 m. S of Oaxaca, on the r. bank of the Rio Verde. Pop. 3,075. It has a Dominican convent.

ZIMBAO, or ZIMBAOE [royal residence], a town of Caffraria, capital of Monomotapa, and residence of the Quiveto, on the r. bank of the Zambèze, near the confluence of the Manzora, between Tête and Sena. It was formerly capital of an empire of the same name.—Also a town in the state of Manica, on the Sofala.—Also a town to the SW of Monomotapa, in the Botongas territory.

ZIMBEBAS. See CIMBEBAS.

ZIMBES. See MICZIMBES.

ZIMBO, a headland of Brazil, in the prov. of Santa Catharina.—Also a river of the prov. of Bahia, in the comarca of Caravellas, which descends from the Cordilheira-dos-Aimores, runs about 30 m., and falls into a gulf named the Rio-de-Caravellas.

ZIMBRE, a small island of the Mediterranean, near the N coast of Tunis, and in the E part of the gulf of that name, in N lat. 37° 8′ 30″, and E long. 10° 48′ 10″. It is frequented by corsairs.

ZIMDARS, a village of Prussia, in the regency of Stettin, and circle of Greifenburg. Pop. 270.

ZIMITI, a town of New Granada, in the dep. of the Magdalena, prov. and 90 m. SSE of Mompox, on the S bank of a small lake.

ZIMITZ, a village of Austria, in Bohemia, in the regency of Pilsen, and circle of Prachin. Pop. 480.

ZIMMAY, a petty state of India, lying to the N of the Tenasserim provinces, on the E bank of the Salueen, inhabited by a wild, barbarous, but independent tribe of mountaineers called Red Kayens. The town of Zimmay, or Changmai, is situated in N lat. 20°, and E long. about 99°. A town called Labong is distant from it only 10 m. SE; and another called Lagon about 50 m. SE. These three towns are respectively the capitals of three small states of the same name, our information respecting which has been chiefly gleaned by Dr. D. Richardson, who visited them in 1830. Their respective boundaries are not well defined, but, together with those of Mung-pay and Mung-nam, they appear to be the patrimony of one family, who, when the whole of this country was under the dominion of Ava, succeeded, with the assistance of Siam, in throwing off the Burmese yoke, ejected the latter from the above named towns, and having been confirmed in the government of them by Siam, have continued tributary to that kingdom, and successfully resisted all the attempts of Ava to regain possession. The elder brother was invested with the title of *Chowtchee-Weet*, or 'Lord of Life,' with the supreme authority over the others, and the title descended to each brother successively alive. Recent revolutions have probably altered the relations of these petty states. Of the original inhabitants of this country but a very small portion—perhaps not above one-third of the whole—now remains in the territory, owing to the great number carried off by the kings of Ava when they overran the country. The remainder consists chiefly of Burmese, Peguers and Shans, either refugees or slaves. Dr. Richardson was unable to form an opinion as to the amount of the pop. He was told that the towns of Zimmay, Lagon, Mung-pay and Mung-nam contain each about 20,000, and Labong 14,000, but he thinks these numbers exaggerated. In person the Shans bear a great resemblance to their Burmese and Siamese neighbours, though somewhat fairer. They are muscular, well-formed, and healthy in their appearance: eyes moderately linear; nose small rather than flat; the mouth large, and disfigured by black teeth and gums, which they cherish as a beauty; the hair is long, straight, lank, and almost always black. They tatoo the lower limbs, but to less extent than the Burmese. Their dress consists of a cotton *putso* or cloth round the loins, generally blue, a blue cotton jacket reaching well over the hips, and a coarse red cotton turban; though many go uncovered. The dress of the chiefs is of the same description, but the materials are more costly,—Chinese crape or satin jackets, with gold or silver lace, the putso of silk. The women are fair and good-looking. Many are disfigured with goitre, but it seldom attains a large size. They are a quiet, mild, good-humoured race. Their religion is that of Budh. Speaking of the soil and cultivation, Dr. Richardson says—"The soil in the low paddy and garden lands is a rich black loam apparently inexhaustible, in which the crops follow each other in uninterrupted succession. Cultivation is conducted with a good deal of care by irrigation and transplanting, and the return on good lands is about 120 and on inferior about 75 fold. The general seed-time is in July and August, and the harvest in December and January, though some descriptions of paddy are perfected in three months. Their plough closely resembles those used in the south of Scotland without the coulter or ploughshare, and is often drawn by one buffalo. The principal articles of cultivation are the cauth-brier, or glutinous rice, cotton, maize, sugar-cane, tobacco, ground nut, chillies, several kinds of pulse, radishes, and turnips. Of grain, wheat, and other corn they have none. Pepper, cloves, and all the finer species of spices are unknown." Tin, iron, and lead ores appear to be abundant, though it does not appear that the extraction of the metal is engaged in to any extent. Dr. R. was informed that the tin ore yields 80 per cent., and some iron ore that was shown him seemed equally rich. There are extensive teak forests in that part of the country immediately adjoining our provinces, but there is difficulty in bringing the timber down the Salueen river in consequence of rocks and rapids. The trade of the country is unimportant. An annual caravan visits it from the frontier provinces of China, bringing silks, satins, velvet and woollens, (the latter chiefly English,) cooking vessels, musk, and trifling articles of Chinese manufacture; and carrying away chiefly raw cotton, the produce of the country. From Maulmein are received British piece goods, chintzes, muslins, hardware, &c., in return for cattle, ivory, and a small quantity of stick-lac. With the hill tribes on the W bank of the Salueen they exchange cattle, grain and betelnut, for slaves, tin, lead, and stick-lac.—The walls of the inner town of Zimmay are 800 fathoms from E to W, and 1,000 from N to S; all of brick, with a ditch and rampart all round. The outer wall, which reaches from the NE to the SW corner, is circular and upwards of 1,800 fathoms, one-half of brick, the other of wood with a rampart round the brick part, and a ditch surrounding the whole. The ditches when in repair can be filled from the river. The town is situated 4 or 5 m. from the eastern foot of the Bya-tha-Dyk hill, the highest in the range, and between it and the hill is another small single-walled fort about the size of Labong, called Mung-Sun-dank. All the houses in Zimmay above the poorest of the people are surrounded by compounds fenced in, in which are cocoa-nut, areca, betel, bamboo, and other useful trees, with a great variety of flowers and flowering shrubs, which are watered by a stream of clear water brought from the hill. The valley in which this town and Labong both stand, is little less than one day from E to W, and little more than three from N to S. Much of the valley near the town is under cultivation, which is all prepared by irrigation, and the grain is transplanted, yielding upwards of one hundred-fold, though the fields are never left fallow. At Zimmay Dr. R. found a caravan of Chinese traders, consisting of 200 mules and horses. Three hundred more were said to be at Mung-nam, where cotton is abundant. They had arrived in the country a considerable time before, and were preparing shortly to return home. The imports by these caravans consist of copper and iron vessels, silk, (raw and manufactured,) satins, gold and silver thread and lace, musk walnuts, carpets, and vermilion. They export from the Shan country cotton, ivory, skins, horns, &c From the information which Dr. R. could collect the caravan assembled at Mungku, distant from Zimmay about two months' journey. Their goods are conveyed by mules, and they would appear t travel rapidly; as they asserted they would not be more than 12 days from Zimmay to Maulmein.

ZIMMERBACH, a village of France, in the dep. of the Haut-Rhin, and cant. of Wintzenheim, 7 m. W of Colmar, near the Fecht. Pop. 424.

ZIMMERBUDE, a village of Prussia, in the regency of Konigsberg, and circle of Fischhausen. Pop. 400.

ZIMMERHAUSEN, a village of Prussia, in the reg. of Stettin, and circle of Regenwalde. Pop. 400.

ZIMMERHOLZ, a village of Baden, in the circle of the See, and bail. of Engen. Pop. 320.

ZIMMERN, a village of Bavaria, in the circle of Lower Franconia, and seignory of Rothenfels. Pop. 400.—Also a village of Prussia, in the principality of Hohenzollern. Pop. 353.—Also a village of the regency of Erfurt, and circle of Langensalza. Pop. 350.—Also a village of Saxe-Weimar, in the prov. of Weimar, and bail. of Vieselbach. Pop. 820.

ZIMMERN (GROSS), a town of the grand-duchy of Hesse-Darmstadt, in the prov. of Starkenburg, bail. and 2 m. SSW of Dieburg, on the l. bank of the Gersprienz. Pop. 2,754, of whom 1,738 are Lutherans, 883 Catholics, and 133 Jews.

ZIMMERN (NIEDER), a village of Saxe-Weimar, prov. of Weimar, and bail. of Vieselbach. Pop. 820.

ZIMMERN (SUPRA), a village of Prussia, in the regency and circle of Erfurt. Pop. 370.

ZIMMERN-BEI-GMUND, a village of Würtemberg, in the circle of the Jaxt, and bail. of Gmund, on the Rems. Pop. 200.

ZIMMERN-BEI-GRUNSFELD, a village of Baden, in the circle of the Lower Rhine, and bail. of Gerlachsheim. Pop. 400.

ZIMMERN-BEI-OSTERBURKEN, a village of Baden, in the circle of the Lower Rhine, and bail. of Abelsheim. Pop. 350.

ZIMMERN-BEI-URLOFFEN, a village of Baden, in the circle of the Middle Rhine, and bail. of Offenburg. Pop. 600.

ZIMMERN HEILIGEN, a village of Prussia, in Hohenzollern, prov. of Sigmaringen, and bail. of Haigerloch. Pop. 600.

ZIMMERN-OB-ROTTWEIL, a village of Würtemberg, in the circle of the Schwarzwald, and bail. of Rottweil. Pop. 550.

ZIMMERN-UNTER-DER-BURG, a village of Wurtemberg, in the circle of the Schwarzwald, and bail. of Rottweil, on the Schwarzbach. Pop 530.

ZIMMERS, a village of Electoral Hesse, in the prov. of Lower Hesse, circle of Fritzlar, and bail. of Jesberg. Pop. 535.

ZIMMERWALD, a village of Switzerland, in the cant. of Bern, and bail. of Seftigen. Pop. 810.

ZIMNAWODA, a village of Prussia, in the reg. of Posen, and circle of Krotoszyn. Pop. 280.

ZIMNICEA, a town of Turkey in Europe, in Wallachia, capital of the district of Teleorman, and 90 m. SW of Bukharest, on the l. bank of the Danube, opposite Sistov.

ZIMNIE. See ATARAM.

ZIMONY. See SEMLIN.

ZIMROWITZ, a village of Austria, in Silesia, in the circle of Troppau. Pop. 360.

ZIMUR-SHELLUKS, a tribe of Berebbers, who inhabit Marocco.

ZINARI, or KINARO, a small island of the Archipelago, to the NE of Amorgo, and W of Levitha. Its highest summit is in N lat. 36° 58′ 42″, and E long. 26° 17′ 18″.

ZINCKEN, a mountain of Austria, in Styria, on the confines of the circles of Judenburg and Bruck, and to the NNE of Judenburg.

ZINDEL, a village of Prussia, in the regency and circle of Breslau. Pop. 435.—Also a village of the same regency, in the circle of Brieg. Pop. 440.—Also a village of the regency of Oppeln, and circle of Grottkau. Pop. 250.

ZINDER, (GROSS and KLEIN), two villages of Prussia, in the regency and circle of Danzig, containing respectively 675 and 390 inhabitants.

ZINGA, a town of Peru, in the dep. and 165 m. NNW of Tarma, and near the l. bank of the Marañon.

ZINGARI. See ZIGUINES.

ZINGAS, or GINGAS, a name which the Portuguese of Angola, in Lower Guinea, usually gave to the inhabitants of Mattemba.

ZINGI, a village of Turkey in Asia, in Kurdistan, in the pash. of Mosul.

ZINGICHOR, or ZINGHINCHOR, a Portuguese establishment in Senegambia, in the Baznyans' Territory, on the S side of the Casamanza, and 30 m. NNW of Cachea-Fort.

ZINGILLA, a village of Abyssinia, in the prov. of Amhara, and district of Lasta, 15 m. NW of Azof.

ZINGSHEIM, a village of Prussia, in the reg. of Aachen, and circle of Schlieden. Pop. 498. It has a cloth-factory.

ZINGST, an island of the Baltic, off the coast of the Prussian prov. of Pomerania, regency of Stralsund, and circle of Franzburg, in N lat. 54° 25′, and E long. 12° 50′. On the S it is separated from the coast of Barth by the Binnen-see. Pop. 1,160. It is 15 m. in length from E to W, and 3 m. in medium breadth, and contains several villages, one of which bears the same name.

ZINGUENEH, a Luri tribe of Persia, who inhabit the environs of Kirmanshah, in Kurdistan. They number about 6,000 individuals.

ZINGUES (COUNTRY OF THE). See ZANGUEBAR.

ZINGUN (BENDER), a town of Persia, in Farsistan, on the Persian gulf, 150 m. SSW of Shiraz.

ZINKENDORF (GROSS), or NAGY-CZENK, a market-town of Hungary, in the comitat and 8 m. SE of Oedenburg, on the Ikva, near the S bank of Lake Neusiedel. Pop. 800.

ZINKENDORF (KLEIN), or KIS-CZENK, a village of Hungary, in the comitat of Oedenburg, a little to the ENE of Gross-Zinkendorf, on the Ikva. It has a castle and an hospital.

ZINKOV, a town of Russia in Europe, in the gov. of Podolia, and district of Lelitchev, 33 m. NNE of Kamenetz, on the Ouchitza.

ZINNA, a town of Prussia, in the regency of Potsdam, and circle of Juterbogk, on the Nuthe. Pop. in 1843, 1,745. It has manufactories of linen and woollen fabrics, and tanneries. In the vicinity is an abbey of the same name founded in 1171. The town owes its foundation to Frederick the Great, having been built towards the end of the 18th century.—Also a village of the regency of Merseburg, and circle of Torgau. Pop. 360.—Also a village of the regency of Potsdam. Pop. 285.—Also a village of the same regency, in the circle of Niederbarnim. Pop. 265.

ZINNAH, a village of the Punjab, on an offset and 6 m. E of the Indus.

ZINNDORF, a village of Prussia, in the regency of Potsdam, and circle of Niederbarnim. Pop. 290.

ZINNITZ, a village of Prussia, in the regency of Frankfurt, and circle of Kalau. Pop. 230.

ZINNOWITZ, a village of Prussia, in the reg. of Stettin, and circle of Usedom. Pop. 240.

ZINNSDORF, a village of Prussia, in the regency of Merseburg, and circle of Liebenwerda. Pop. 200.

ZINNWALD (NEU or SACHSISCH), a market-town of Saxony, in the circle and 24 m. S. of Dresden, on the Rothen Weiseritz, by which it is separated from a town of the same name in Bohemia, and at an alt. of 2,755 Parisian ft. above sea-level. Pop. 697. It has mines of tin, several mills and manufactories of lace.

ZINNWALD, or BOHMISCH-ZINNWALD, a town of Bohemia, in the circle and 24 m. NW of Leitmeritz, on the Rothen Weiseritz. It has mines of tin, copper, and silver.

ZINOTTEN, a village of Austria, in Bohemia, in the regency of Budweis, and bail. of Neuhaus. Pop. 550.

ZINSAT, a village of Austria, in Styria, in the circle of Marburg. Pop. 500.

ZINSERBUTTEN, a village of Prussia, in the regency and circle of Treves. Pop. 240.

ZINTEN, a town of Prussia, prov. of East Prussia, in the regency and 20 m. SSW of Konigsberg, and circle of Heiligenbeil, on the Straddig. Pop. in 1843, 2,458. It has manufactories of cloth and hats, and several tanneries.

ZINTI, a town of Bolivia, in the dep. and 135 m. SE of Potosi.

ZINTZEL, a river of France, in the dep. of the Lower Rhine, formed by the confluence of several streams, which have their sources in the dep. of the Meurthe, to the NW of Phalsburg, waters the W part of the dep. of the Lower Rhine, and after a course of about 18 m. throws itself into the Zorn, on the l. bank, a little below Steinbourg.

ZINN. See SINN.

ZINZENZELL, a village of Bavaria, in Lower Bavaria, in the presidial of Mitterfels. Pop. 200.

ZINZEREN, a town of Turkey in Europe, in Little Wallachia, and district of Mehenditzi, near the l. bank of the Schiul.

ZINZILI. See INZELI.

ZINZIRENI, a village of Turkey in Europe, in Upper Wallachia, on the Danube.

ZINZON, a district of Japan, in the central part of the island of Nifon, and prov. of Sinano. It contains the volcano of Asamaga-daki, the last eruption of which occurred in 1823.

ZINZWEILER, a village of France, in the dep. of the Lower Rhine, cant. and 5 m. W of Niederbronn, and 20 m. SW of Wissemburg. Pop. 930. It has iron-mines and furnaces.

ZION, a village of Centre co., in the state of Pennsylvania, U. S., 65 m. NW of Harrisburg. It has an extensive charcoal furnace.

ZIOPP, a village of Austria, in Moravia, in the circle of Olmutz, and seignory of Holleschau.

ZIPANGRI, or CIPANGU, the name by which Japan was originally known to the Europeans.

ZIPAQUIRA, a town of New Granada, in the dep. of Cundinamarca, and prov. of Bogota, 11 m. NNE of Santa-Fe-de-Bogota. Pop. 4,000. It has extensive salt-mines.

ZIPKOW, a village of Prussia, in the regency of Koslin, and circle of Stolpe. Pop. 300.

ZIPPELZERBST. See ZORBIG.

ZIPPLINGEN, a village of Würtemberg, in the circle of the Jaxt, and bail. of Ellwangen. Pop. 518.

ZIPPNOW, a village of Prussia, in the regency of Marienwerder, and circle of Deutsch-Krone. Pop. 1,200.

ZIPREIN, a village of Austria, in Styria, in the circle of Gratz. Pop. 500.

ZIPS, or SZEPES-VARMEGYE, an administrative province and comitat of Hungary, in the circle beyond the Theiss, bounded on the N by Galicia; on E by the comitat of Saros; on the S by those of Abauj, Torna, and Gomor; and on the W by the comitat of Lyptau. It is 69 m. in length from NW to SE, and 24 m. in medium breadth, comprising an area of 562 sq. m. Pop. in 1837, 224,510, of whom the majority are Sclaves. This comitat may be regarded as the Switzerland of Hungary, being to a great extent covered by the Carpathian mountains, which here form the northern boundary with Poland, and reach their highest point in the vast Tatra group, the lowest of which rises nearly 7,000 ft. above the level of the sea. "These granite giants, though unadorned by nature's varied tapestry, still, with their snow-clad peaks, rugged perpendicular cliffs, unfathomable precipices and roaring cataracts, vie with the most majestic Alpine scenery, unfolding to the beholder a picture of wild beauty and stern grandeur that at the same time awes and delights." The central part is watered by the Poprad, the S by the Hernad and Gobnitz. The N confines are for some distance traced by the Dunajec. The climate is cold, unsuited to the culture of the vine, but the soil yields barley, flax, pulse, fruit, timber, and in small quantities grain. Cattle, sheep, pigs, and fowls are reared extensively, and game and bees are found in large quantities in woods, and fish abound in the streams. The comitat is noted for the variety and abundance of its mineral wealth. It has mines of gold and silver, copper, and iron, and vitriolic springs. Brandy, linen, pottery, and leather, are its chief articles of manufacture. The com. of Z. comprises 38 towns, two of which are royal, and 175 villages, one of which bears the same name. These towns form a district independent of the jurisdiction of the comitat, and enjoy several privileges, amongst others that of being governed by a count of their own election. The chief town is Leutschau. The village of Z. is near Kirchdorf. It contains the fine ruins of an ancient royal castle of the same name. "The Zipsers are an enduring and vigorous race of people, honest and simple in their habits, and remarkable alike for their general mental culture and ardent attachment to liberty and the Protestant faith. Of this they gave countless proofs amidst the vicissitudes their country was subjected to under the Austrian rule. Their devotion, however, reached its height during the memorable period of 1848, when they placed their savings at the disposal of the national government, and sent thousands of their bravest sons to the plains on the banks of the Theiss against the Austrian and Russian invaders. Hungarians in heart and from conviction, the Zipsers were so deeply affected by the tragic conclusion of the war, that, when the remnant of their gallant bands returned to their mountain fastnesses and related the terrible tidings of their country's downfall, their despair knew no bounds."

ZIPSENDORF, a village of Prussia, in the regency of Merseburg, and circle of Zeitz. Pop. 330.

ZIPSER-HANS, SZEPES-VAR, or SPISSKY-ZAMEK, a castle of Hungary, in the comitat of Zips, 1½ m. E of Kirchdrauf, on a lofty rock. It belongs to the counts of Csaky, and gives its name to the comitat. Near it is the fine church of St. Martin.

ZIPUEZA, an ancient Indian town of New Granada, in the dep. of the Magdalena, 30 m. SE of Mompox, on the SW bank of Lake Zapatosa. It is now a mere village.

ZIRAFFE, a town of Tripoli, on the gulf of Sidra, 126 m. SE of Mesurata.

ZIRCHOW, a village of Prussia, in the regency of Koslin, and circle of Stolpe. Pop. 220.

ZIRCZ, or ZIRTZ, a town of Hungary, in the comitat and 12 m. N of Veszprim.

ZIRGESHEIM, a village of Bavaria, in the circle of Swabia, and presidial of Donauwörth. Pop. 393.

ZIRK, a village of Austria, in Bohemia, in the reg. of Eger, and bail. of Pfraumberg. Pop. 260.

ZIRKAU, a village of Prussia, in the regency of Liegnitz, and circle of Sprottau. Pop. 260.

ZIRKE, or SIRAKOW. See SIRAKOW.

ZIRKENDORF, two villages of Bavaria, in the circle of the Oberpfalz, and presidial of Eschenbach. Pop. 350.

ZIRKLACH, a village of Illyria, in the gov. and circle and 17 m. NNW of Laybach.

ZIRKNITZ, a market-town of Austria, in Illyria, in the gov. of Laybach, and circle of Adelsberg, to the N of a lake of the same name. Pop. 1,300. It has a considerable trade in salt. Lake Z. is 3 m. long from N to S, and 1½ in breadth. It lies amid lofty mountains, and the rocks by which it is surrounded are naked, sterile, and uniform. It is remarkable for its flux and reflux. Towards the middle of summer, when the snow disappears from the mountains, the waters decrease, and the bed becoming dry, is rapidly covered with rich vegetation, and abundant crops of clover, sainfoin, luzern, and rice reward the diligence of its cultivators.

ZIRKWITZ, a village of Prussia, in the regency of Breslau, and circle of Treb. Pop. 540.—Also a village of the regency of Marienwerder, and circle of Flatow. Pop. 390.—Also a village of the reg. of Stettin, and circle of Greiffenberg. Pop. 250.

ZIRL, a village of Austria, in the Tyrol, and cir- of Oberhinthal, on the Inn, 5 m. W of Innspruck. Pop. 1,700.

ZIRLAU, a village of Prussia, in the regency of Breslau, and circle of Schweidnitz. Pop. 925.

ZIRMIE, or ZIRMI, a town of Nigritia, in Haoussa, to the E of Sakatu, capital of the prov. of Zamfra. It occupies a peninsula formed by the Guarrama, the banks of which rise precipitously, covered with brushwood, through which a path leads to the town, which is surrounded with a mud wall and ditch. The inhabitants are noted for their rapacity.

ZIRMONY. See JERMOUN.

ZIRNAU, a village of Austria, in Bohemia, in the regency and bail. of Budweis. Pop. 1,000.

ZIRNDORF, a village of Bavaria, in the circle of Middle Franconia, and presidial of Nürnberg. Pop. 1,689, of whom about 100 are Jews. It has a synagogue, and possesses manufactories of tobacco and breweries.

ZIRNELSCHLAG, a village of Austria, in Bohemia, in the regency of Budweis, and bail. of Kaplitz. Pop. 400.

ZIRONA, an island of the Adriatic, near the Dalmatian coast, in the circle of Spalatro, and to the NW of the island of Bua. It is about 3 m. in length from N to S, and contains a village.

ZIROW, a village of Austria, in Bohemia, in the regency of Budweis and bail. of Pilgram. Pop. 290.

ZIRRA, a river of Persia, in Farsistan, which has its source to the E of Guiere, and to the SW of Shiraz; runs in a generally S direction; and after a course of about 120 m., throws itself into the Persian gulf, between Bender-Rick and Abushehr.

ZIRZOW, a village of Mecklenburg-Strelitz, in the circle of Stargard. Pop. 280.

ZISCHKOW, a village of Austria, in Moravia, in the circle of Brunn. Pop. 480.

ZISCHOW, a village of Austria, in Bohemia, in the reg. of Budweis, bail. of Wittengau. Pop. 200.

ZISSEN, a village of Prussia, in the regency of Coblenz, and circle of Ahrweiler. Pop. 750.

ZISSERSDORF, a village of Austria, in the country below the Enns, in the Upper Mannhartsberg, and presidial of Drosendorf. Pop. 500.

ZISTENDORF (NIEDER and OBER), two villages of Prussia, in the regency of Coblenz, and circle of Ahrweiler, containing respectively 580 and 430 inhabitants.

ZISTERSDORF, a town of Austria, in the district below the Enns, and lower circle of Mannhartsberg, 33 m. NW of Vienna. Pop. 2,000. It has a castle.

ZISWINGEN, a village of Bavaria, in Swabia, in the presidial of Nordlingen. Pop. 250.

ZITANG. See ZITTANG.

ZITAQUARO, a town of Mexico, in the state of Michoacan, 54 m. ESE of Valladolid. Pop. 2,000. It has important silver-mines.

ZITARA, a town of New Granada, in the dep. of the Cauca, and prov. of Choco, on the Atuato, 285 m. N of Popayan. It has mines of gold and platina.

ZITATE, or CITATE, a village of Wallachia, near the l. bank of the Danube, and 10 m. NNE of Widin. It stands on the brow of a hill which overlooks the surrounding plain for many miles, and is flanked on two sides by ravines. That towards the E is of considerable depth, broken and *escarpé*, and shelving at the bottom into a lake, beyond which a plain extends to the Danube at a little distance; the other is less formidable, and winds towards the top of the hill in the rear of the village, forming a sort of hollow or covered way, which can be traversed from one end to the other without difficulty. The village, like all those of Wallachia and Bulgaria, is composed of scattered houses, all open and unprotected, except by slight fences, designed to secure the corn and hay. The road from Kalafat towards Saltsha, Gruja, and Tschernetz, passes through the centre in a NW direction, after having ascended between the ravines. A hard-fought battle took place in the neighbourhood of this village in the late war with Russia, on the fruitful and picturesque plain which extends W, in a SW direction from Radovan to Cioroin, and thence to Z.; and from the latter place N to Plenizza.

ZITTANG, SETANG, or PAULAN, a river of Burmah, a branch of the Irawady, from which it separates in the Mranma, between Ava and Ummerapura; runs first SE; then SW and S; and falls at a little below a town of the same name, into the gulf of Martaban, to the E of Rangoon, and NW of Thaleayn. It is navigable to Taungu, a distance of 180 m. from its mouth. The latter is much obstructed with sand-banks, and is liable to sudden swelling of the waters. The Z. has a total course of about 480 m. At Graungrue, where it forms a lake, it communicates by the Mobia with the Thaluyn.

ZITTAU, a town of Saxony, in the circle and 29 m. SE of Bautzen, on the Altwasser and Mandau, an affluent of the Niesse. Pop. in 1837, 8,674, in 1843, 9,268; in 1852, 11,000. It is enclosed by a substantial double wall with four gates, and has two suburbs. It contains eight churches, a gymnasium, an hospital, an orphans' asylum, a theatre, a library, a museum, and possesses manufactories of woollen and linen fabrics, carding machines, pottery, bleacheries, a paper-mill, dye-works, and breweries, and carries on an active trade in flax and grain. Z. was erected into a town in 1255, and enclosed with walls in 1287. In 1757 it was taken by assault by the allies of the elector of Saxony. In the vicinity are the mineral springs and baths of Augustusbad.

ZITTNA, a village of Prussia, in the reg. of Oppeln, and circle of Ribnitz. Pop. 240.

ZITTOW, a village of Mecklenburg-Schwerin, in the circle and bail. of Schwerin. Pop. 200.—Also a village of Austria, in Moravia, in the circle of Prerau, on the March. Pop. 600. It has a castle.

ZITZ, a village of Prussia, in the reg. of Magdeburg, and circle of Jerichow. Pop. 290.

ZITZENHAUSEN, a village of Baden, in the

circle of the See, and bail. of Stockach, on the Aach. Pop. 1,030.

ZITZEWITZ, a village of Prussia, in the reg. of Koslin, and circle of Stolpe. Pop. 270.

ZITZISHAUSEN, a village of Würtemberg, in the circle of the Schwarzwald, and bail. of Nurtingen, on the Neckar. Pop. 397.

ZITZLAU, a village of Austria, in the district above the Enns, and bail. of Linz. Pop. 370.

ZITZMIN, a village of Prussia, in the reg. of Koslin, and circle of Schlawe. Pop. 420.

ZITZOW, a village of Prussia, in the reg. of Koslin, and circle of Schlawe. Pop. 290.

ZITZCHEN, a village of Prussia, in the reg. and circle of Merseburg. Pop. 320.

ZITZSCHEWIG, a village of Saxony, in the circle and bail. of Dresden. Pop. 1,020.

ZIVOTOV. See JIVOTOV.

ZIWENITZ, a village of Austria, in Bohemia, in the reg. and bail. of Pardubitz. Pop. 480.

ZIZ, or SEGELMESA, a river of Marocco, in the prov. of Tafilet, which descends from the S side of the Great Atlas; runs SE, and, after a course of 300 m., loses itself in the desert.

ZIZELITZ, or ZIZELICZE, a market-town of Austria, in Bohemia, in the reg. of Gitschin, and bail. of Ehlumetz, on the Ezidlina. Pop. 2,000.

ZIZELOWES, a village of Austria, in Bohemia, in the reg. of Gitschin, and bail. of Koniginhof. Pop. 380.

ZIZERS, a market-town of Switzerland, in the cant. of the Grisons and Ligne-de-la-Maison-de-Dieu, near the r. bank of the Rhine, and 8 m. N of Coire. Pop. 1,115. It has an alms-house and mineral baths.

ZLABERN, a village of Austria, in the district below the Enns, and lower circle of the Mannhartsberg. Pop. 500.

ZLABINGS, ZLABINCS, or SLAWONICE, a town of Austria, in Moravia, in the circle and 30 m. SSW of Iglau. Pop. 2,300. It has a manufactory of cloth, and possesses mineral wells.

ZLACZEW, or ZLUCZEWO, a town of Russia in Europe, in Poland, in the gov. and 35 m. SE of Kalisch, and obwod of Jieradz. Pop. 800. It has manufactories of cloth, hats, and hosiery.

ZLADKOW, a village of Austria, in Moravia, circle of Brunn, and seignory of Pernstein. Pop. 230.

ZLAKOWITZ, a village of Austria, in Bohemia, in the reg. of Pilsen, and bail. of Brzeznitz, on the l. bank of the Moldau. Pop. 540.

ZLAMANKA, a village of Austria, in Moravia, in the circle of Hradisch. Pop. 260.

ZLATNIK, a village of Austria, in Bohemia, in the reg. of Prague, and bail. of Eul. Pop. 330.

ZLATTEN, a village of Austria, in Styria, in the circle of Bruck. Pop. 250.

ZLAUKOWITZ, a village of Austria, in Bohemia, in the reg. of Prague, and bail. of Rakonitz. Pop. 310.

ZLEB, a market-town of Austria, in Bohemia, in the reg. of Pardubitz, and circle of Czaslau. Pop. 1,300. It has a castle, and possesses manufactories of saltpetre, potash, and woollen fabrics and bleacheries.

ZLECHAU, a village of Austria, in Moravia, in the circle of Hradisch. Pop. 600.

ZLEM, a village of Austria, in Styria, in t e circle of Judenbruck. Pop. 280.

ZLIN, a market-town of Austria, in Moravia, in the circle and 15 m. NE of Hradisch, on the Drewnitza. Pop. in 1834, 2,630. The manufactures of boots and of pottery form important branches of local industry.

ZLINKA, or JELINKA-POSA, a town of Russia in Europe, in the gov. and 75 m. NNE of Tchernigov, and district of Novo-Zibcov.

ZLITOUN. See ZILITEN.

ZLITSCH, a village of Austria, in Bohemia, in the reg. of Gitschin, and bail. of Braunau. Pop. 300

ZLIW, a village of Austria, in Bohemia, in the reg. and bail. of Budweis. Pop. 260.

ZLOBITZ, a village of Austria, in Moravia, circle of Olmutz, and seignory of Kreinster. Pop. 440.

ZLOCZOW, a circle and town of Austria, in Galicia. The circle is bounded on the N and NE by Russia; on the SE by the circle of Tarnapol; on the SE by that of Brzezany; and on the W by the circles of Lemberg and Zolkiew. It is 63 m. in length from NW to SE, and 39 m. in extreme breadth. Except in the SE it presents a generally flat surface. It gives rise to the Bug and Sered, and has in some parts a tolerably fertile soil. Agriculture and the rearing of cattle form the chief pursuits of the inhabitants. Pop. in 1837, 232,964; in 1843, 245,046, of whom 26,000 are Jews.—The town is 36 m. E of Lemberg. Pop. 3,958, of whom 1,850 are Jews. It has a Catholic, Greek, and a United Greek church, and an ancient castle. Sail-cloth forms its chief article of manufacture.

ZLONIN, a village of Austria, in Bohemia, in the reg. of Prague, and bail. of Rakonitz. Pop. 250.

ZLONITZ, a village of Prussia, in the reg. and circle of Oppeln. Pop. 480. It has a fine castle.

ZLONITZ, a market-town of Austria, in Bohemia, in the circle of Rakonitz. Pop. 770. It has a castle belonging to the princes of Klinsky, and an hospital.

ZLOTNIKI, a village of Prussia, in the reg. and circle of Posen. Pop. 240.

ZLOTOWO, a village of Prussia, in the reg. of Marienwerder, and circle of Lobau. Pop. 560.

ZLOTTERIE, a village of Prussia, in the reg. of Marienwerder, and circle of Thorn. Pop. 460.

ZLOTTNIK, a village of Prussia, in the regency and circle of Oppeln. Pop. 320.

ZLUCZEWO. See ZLACZEW.

ZMATYN. See SMATYN.

ZMEEV, or ZMIEV, a town of Russia in Europe, in the gov. and 24 m. SSE of Kharkov, on the Sievernoï-Donetz. Pop. 5,000. It has two churches.

ZMEINOGOROSKOI, a fortified town of Russia in Asia, in the gov. of Tomsk, and district of Büsk, on the Smeïevka, at the foot of Mount Schlaugenberg. Pop. 3,000. Gold is found in the adjacent streams.

ZMIETSCH, a village of Austria, in Bohemia, in the reg. of Budweis, and bail. of Krumau. Pop. 500.

ZMIEWO, a village of Prussia, in the regency of Marienwerder, and circle of Strasburg. Pop. 280.

ZMISCHBEKSCH, a village of Russia, in the gov. of the Caucasus, on the Kuban.

ZMOLLENITZ, a village of Austria, in Styria, in the circle of Gratz. Pop. 260.

ZMYGROD, a market-town of Austria, in Galicia, in the circle and 11 m. S of Yaslo, on the r. bank of the Wisloka. It has manufactories of linen.

ZMYSLONA, a village of Prussia, in the regency of Posen, and circle of Schildberg. Pop. 250.

ZNATCHKA, a town of Russia in Europe, in the gov. of Poltava, district and 12 m. SE of Lonbuy.

ZNAYM, a circle of Austria, in the SW part of Moravia, bounded on the E and NE by the circle of Brünn; on the S by the archduchy of Austria; and on the W and NW by the circle of Iglau. It is 51 m. in length from E to W, 45 m. in extreme breadth from N to S, and comprising an area of 501 sq. m. Pop. in 1837, 157,752; in 1843, 167,406. The principal river is the Taya, by which it is watered in the S. It is intersected also by the Iglawa, Rokitna,

Jaromirszka, and Oslawa, and contains several ponds. The N part is mountainous, but in other directions it consists of fertile plains, producing in great abundance corn, fruit, wine, flax, anise, mustard, and fennel. It has extensive forests, and possesses large numbers of cattle.—The chief town, called also Znaym, Znaymo, or Znogmo, stands on a hill, the base of which is washed by the Taya, 38 m. SW of Brünn, in N lat. 48° 51′ 15″, and E long. 16° 2′ 36″. Pop. in 1834, 5,010. It is enclosed by walls, is well built, and has several suburbs. It contains an ancient fortress, now employed as a military fortress, formerly the residence of the Moravian princes of the Znaym dynasty, several convents, several churches, a gymnasium, and an educational establishment for the education of children of the military. An engagement took place in the vicinity of this town, between the French and Austrians, in June 1809.

ZNETINEK, a village of Austria, in Moravia, in the circle of Iglau, and seignory of Gross-Meseritsch. Pop. 270.

ZNIATKA, a village of Austria, in Moravia, in the circle of Znaim, and seignory of Namiescht. Pop. 220.

ZNIN, or SCHNIN, a town of Prussia, in the reg. of Bromberg, and circle of Schubin. Pop. 1,618.

ZNIO-VARALLYA. See VARALLYA (ZNIO).

ZNOGMO. See ZNAYM.

ZNOROW, a village of Austria, in Moravia, in the circle of Hradisch, and seignory of Weselh.

ZOAGLI, a village of Sardinia, in the div. and on the gulf of Genoa. Pop. 3,900.

ZOAR, a village of Tuscarawas co., in the state of Ohio, U. S., on the E bank of the Tuscarawas river, and 90 m. ENE of Columbus. Pop. in 1850, 249. It was formed by a German communal colony, called Separatists, under the government of a patriarch.

ZOAR (EL), a village of Turkey in Asia, in Syria, near the SW extremity of the Dead sea, and near the frontier of Arabia Petræa.

ZOARA, a town of Tripoli, on the Mediterranean, 75 m. W of Tripoli.

ZOBBENITZ, a village of the duchy of Brunswick, in the circle of Helmstadt. Pop. 360.

ZOBEIR, a town of Turkey in Asia, in the pash. and 9 m. SW of Bassora. It is well-built, with straight and well-kept streets. This town was founded on the ruins of the ancient *Bassora*. It derived its name from one of the first disciples of Mahomet, whose tomb is still to be seen in the vicinity, amid the ruins of the ancient city.

ZOBEL, a village of Prussia, in the regency and circle of Liegnitz. Pop. 200. It has a castle.

ZOBELITZ. See ZOBLITZ.

ZOBERITZ, a village of Prussia, in the regency of Merseburg, and circle of Bitterfeld. Pop. 200.

ZOBERSDORF, a village of Prussia, in the reg. of Merseburg, and circle of Liebenwerda. Pop. 240.

ZOBES, a village of Saxony, in the circle of Zwickau, and bail. of Plauen. Pop. 385.

ZOBIGKER, a village of Saxony, in the circle and bail. of Leipsig. Pop. 312.

ZOBING, a town of Austria, in the upper circle of the Mannhartsberg, on the r. bank of the Great-Kemp, 8 m. NE of Krems.

ZOBINGEN, a village of Wurtemberg, in the circle of the Jaxt, and bail. of Ellwangen. Pop. 775.

ZOBLITZ, or ZOBELITZ, a town of Saxony, in the circle of Zwickau, bail. of Lauterstein, and 18 m. SE of Chemnitz. Pop. in 1837, 1,415; in 1843, 1,545. It is well-built, and possesses manufactories of linen and cotton fabrics, and lace, and several breweries.—Also a village of Prussia, in the reg. of Liegnitz, and circle of Rothenburg. Pop. 280.

ZOBRA, a town of Arabia, in Yemen, 63 m. NW of Sana.

ZOBTEN, a town of Prussia, in the prov. of Silesia, regency and 27 m. SW of Breslau, and circle of Schweidnitz, at the foot of a mountain of the same name, on which is a quarry of fine marble. It has two churches and an hospital.—Also a village of the regency of Liegnitz, and circle of Lowenberg. Pop. 650. It has a castle.

ZOBTENBERG, a mountain of Prussia, in Silesia, in the regency of Breslau, and circle of Schweidnitz. It has an alt. of 2,320 ft. above sea-level.

ZODEL, a village of Prussia, in the regency of Liegnitz, and circle of Gorlitz. Pop. 700. It has two castles.

ZODIN. See JODIN.

ZOEN, a town of Norway, in the diocese and bail. of Aggershuus, on the gulf of Christiania.

ZOERLEPARWYS, a department and commune of Belgium, in the prov. of Antwerp, and arrond. of Turnhout. Pop. 431.

ZOERSEL, a department and commune of Belgium, in the prov. and arrond. of Antwerp. Pop. 944.

ZOEST, a village of Holland, in the prov. and 17 m. ENE of Utrecht, and arrond. of Amersfoort. Pop. 1,290. To the NW of this village is the castle of Zoestdyk.

ZOETENAEY, a department and commune of Belgium, in the prov. of West Flanders, and arrond. of Furnes.

ZOETERMEER, a parish of Holland, in the prov. of South Holland, and S of Haag. Pop. 960.

ZOFINGEN, a bail. and village of Switzerland, in the cant. of Aargau, on the r. bank of the Wigger, 11 m. SSW of Aarau. Pop. 3,175. It has a public library, and possesses manufactories of cotton fabrics, ribbon, velvet, &c. This town is of great antiquity. The district of Z. comprises 5 circles.

ZOGANY, a town of Nigritia, in Kanem, on the N bank of Lake Tchad. It stands in a flat locality, covered with a species of low brushwood, locally named Kaneskia.

ZOGERSDORF, a village of Austria, in the district below the Enns, and lower circle of the Mannhartsberg. Pop. 200.

ZOGGE, a commune of Belgium, in the prov. of East Flanders, and dep. of Hamme. Pop. 408.

ZOGNO, a village of Austria, in Lombardy, in the deleg. and 8 m. N of Bergamo, on the r. bank of the Brembo. Pop. 2,000.

ZOHAB, a town and district of Asia, the political relations of which are matter of dispute between the governments of Turkey and Persia. The district is bounded on the NW by the river Diyalah; on the E by the mountain-range of Zagros; and on the S by the river Holwan. The soil is rich, and where irrigated extremely fertile in rice and corn, which is principally disposed of to traders from Bagdad. The town, which lies about 100 m. NE of Bagdad, from its frontier position between Turkey and Persia, is exposed to constant spoliation, and in 1836 was a mass of ruins, with scarcely 200 inhabited houses. Immediately overhanging the town to the E, is the fortress of Bán-Zardah, or Kalahi-Yezdijird, the stronghold of Holwán, to which Yezdijird, the last of the Sasanian kings, retreated after the capture of Ctesiphon by the Arabs. It is a noble specimen of the labour which the monarchs of those ages bestowed upon their royal buildings. It is formed by a shoulder projecting westward from the mountain of Dáláhú, girt upon three sides by an inaccessible scarp, and defended upon the other, where alone it admits of attack, by a wall and dry ditch of colossal dimensions, drawn right across from one scarp to the other, a distance of above 2 m. The wall is now in ruins, and the debris has fallen down into the ditch at its foot, but it still presents a line of defence of no ordinary description. It is flanked by bastions at regular intervals, and it would seem to have been about 50 ft. in height and 20 ft. in thickness; the edge of the scarp has also been faced all round with a wall of less dimensions. The hill itself is elevated very considerably above the plain of Z., perhaps 2,000 ft.: the slope from the plain is abrupt, and it is everywhere crowned by a scarp varying from 300 to 500 ft. The N side of the hill is higher than the S, and the table-land of the fort, containing about 10 sq. m., presents an inclined surface throughout. At the NE angle, where the scarp rises in a rocky ridge to its highest point, and joins the mountain of Dáláhú, there is a pass which conducts into the fort, the ascent rising gradually along the shoulder; the whole way from the town of Z. is easy enough, but the descent on the other side into the table-land of the fort is by a precipitous and difficult gorge. A straight line drawn from Mount Ararat to the mouth of the Tigris will fall in pretty well with the frontier boundary of Turkey and Persia, as it is at present laid down in our maps, and as it has remained from the time of Nadir Shah to the present day. About the northern half of this line there is no dispute. Zohab, one of the disputed districts, falls about its middle; and about half of the southern part of the same line will pass through the disputed district of Mohamran. According to the decision of the conferences at Erzerum, Zohab was assigned to the Turks, and Mohamran to the Persians. The district of Mohamran has an extensive littoral upon the Persian gulf, stretching from the mouth of the Tigris towards Bushire, and a convenient port near the river's mouth. The Turks lay much more store by Mohamran than Zohab, a territory comparatively fruitless to them by the intractable nature of the Kurdish tribes which inhabit it, and from which tribute can only be extracted at the sword's point.

ZOHLOW, a village of Prussia, in the regency of Frankfurt, and circle of Lebus. Pop. 290.

ZOHSE, a village of Austria, in Bohemia, in the regency of Pardubitz, bail. of Landskron. Pop. 650.

ZOLCHOW, a village of Prussia, in the regency of Potsdam, and circle of Prenzlau. Pop. 200.

ZOLDEKOW, a village of Prussia, in the regency of Stettin, and circle of Kammin. Pop. 215.

ZOLDER, a department and commune of Belgium, in the prov. of Limburg, and arrond. of Halselt. Pop. of dep., 2,378; of com., 1,234.

ZOLENDNICE, a village of Prussia, in the regency of Posen, and circle of Kroben. Pop. 200.

ZOLES (SAN), a town of Spain, in the prov. and 14 m. from Zamora, in a fertile locality. Pop. 800.

ZOLKENDORF, a village of Mecklenburg-Schwerin, in the prov. of Mecklenburg, and bail. of Ivenack. Pop. 200.

ZOLKIEW, a circle and town of Austria, in Galicia. The circle is bounded on the NW by Poland; on the NE by Russia; on the E by the circle of Zloczow; on the S by that of Lemberg; and on the SW by the circle of Przemysl. It comprises a surface, generally flat, of 75 m. in length from E to W, and 48 m. in medium breadth. Its chief river is the Bug. It produces grain, flax, hemp, and legumes, and contains extensive forests. Pop. in 1837, 210,043; in 1843, 224,027. It contains 17 towns and 267 villages.—The town is 15 m. N of Lemberg, on a small river, in N lat. 50° 4', and E long. 24°. It has a castle, a convent, a school, and a military hospital. Pop. 3,800. It has manufactories of cloth, and of china, and tanneries. In the vicinity is an ancient castle belonging to the Sobieski family.

ZOLKIEWKA, a town of Russia in Europe, in Poland, in the gov. and 27 m. SSE of Lublin, and obwod of Krasnistan. Pop. 200.

ZOLKOW, a village of Mecklenburg-Schwerin, in the prov. of Mecklenburg, and bail. of Lubz-on-the-Warnow. Pop. 300.

ZOLLCHAU, a village of Prussia, in the reg. of Magdeburg, and circle of Jerichow. Pop. 320.

ZOLLDORF, a village of Austria, in Bohemia, reg. of Gitschin, and bail. of Jungbunzlau. Pop. 340.

ZOLLEN, a village of Prussia, in the reg. of Frankfort, and circle of Soldin. Pop. 260.

ZOLLGRUN, a village of Reuss-Schleiz, in the prov. of Schleiz. Pop. 410.

ZOLLIKON, a village of Switzerland, in the cant., bail., and SW of Zurich, on the Zurcher-See. Pop. 1,316.

ZOLLIKOFEN, a village of Switzerland, in the cant. and bail. of Bern. Pop. 1,045.

ZOLLING, a village of Prussia, in the reg. of Liegnitz, and circle of Friestadt. Pop. 490. It has a castle.

ZOLLING (GROSS and KLEIN), two villages of Prussia, in the reg. of Breslau, and circle of Dels. Pop. 601 and 303.

ZOLLNITZ, a village of Saxe-Altenburg, in the bail. of Roda. Pop. 300.

ZOLLINO, a market-town of Naples, in the prov. of the Terra-d'Otranto, district and 12 m. SSE of Lecce. Pop. 560.

ZOLLWITZ, a village of Saxony, in the circle of Leipsig, and bail. of Golditz. Pop. 252.

ZOLONDOWO, a village of Prussia, in the reg. and circle of Bromberg. Pop. 240.

ZOLOTCHEV, a town of Russia in Europe, in the gov. and district and 24 m. NNW of Kharkov, on the Ouda. Pop. 5,000. It has some dilapidated fortifications.

ZOLOTOI-OSTROV, an island of Russia in Europe, in the gov. of Astrakhan, formed by two of the embouchures of the Ural, in the Caspian sea. It is surrounded with sand-banks, which now prevent all approach except in small boats.

ZOLOTONOCHA, a town of Russia in Europe, in the gov. and 126 m. W of Poltava, on a small river of the same name, an affluent of the Dnieper. Pop. 1,000. It has two convents, and carries on an active trade in cattle.

ZOLOTOPOL, a town of Russia in Europe, in the gov. of Kiev, district and 45 m. SE of Zvenigorodka.

ZOLTKAMP, a fortified village of Holland, in the prov. and 14 m. NW of Groningen, and 27 m. ENE of Leeuwardin, on the r. bank of the embouchure of the Hunse.

ZOLUDEK, or JOLUDEK, a town of Russia in Europe, in the gov. and 51 m. ESE of Grodno, and district of Lida.

ZOLVERN, a village of Holland, in the prov. of Luxemburg, and cant. of Esch, on the Alzette. Pop. 360. It has a castle.

ZOLY. See SOHL.

ZOLYOM. See SOHL. See also SOHL (ALT).

ZOMA (LA), a village of Spain, in the prov. and 63 m. from Saragosa, and 27 m. from Alcaniz. Pop. 140. In the vicinity are mines of iron and lead.

ZOMBER, or SOMBOR, a royal town of Hungary, capital of the gov. or comitat of Bags-Bodrogh, 29 m. NE of Eszek, on the Mosztunka, near the French canal, which connects the Danube and Theiss. Pop. 22,000, of whom upwards of two-thirds are Catholics. It has a Greek and a Catholic school, and a Greek gymnasium and school. The trade consists chiefly in corn and in cattle.

ZOMOEIVECK, a colony of Prussia, in the reg. and circle of Bromberg. Pop. 220.

ZONAS, a mountain of Greece, in Livadia, on the SW confines of the hypodiocese of Doriale, and 30 m. NE of Lepanto. It forms a part of the Pindus chain, rises between the mountains of Vardisio and Elato, and to the NE of Mount Oxas.

ZONDAG, or SUNDAY, a river of South Africa, which has its source on the SW side of the Snieuw-Berg or Snow mountains; runs S; forms for some distance the line of separation with the district of Uitenhagen; traverses the latter, and, after a total course of 150 m., enters Algoa bay. Its fall is considerable, and its bed rocky. Its tributaries are chiefly periodical streams.

ZONEBAD, a river of Persia, in the prov. of Azerdbijan, an affluent of Lake Urmina or Shahee.

ZONHOVEN, a department and commune of Belgium, in the prov. of Limburg and arrond. of Hasset. Pop. of dep. 2,556; of com. 244.

ZONNE, a commune of Belgium, in the prov. of East Flanders, and dep. of Saint-Jean-d'Eremo. Pop. 161.

ZONNEBEKE, a department and commune of Belgium, in the prov. of West Flanders, and arrond. of Ypres. Pop. 2,421.

ZONNEGHEM, a department and commune of Belgium, in the prov. of East Flanders, and arrond. of Audenarde. Pop. of dep., 385; of com., 166.

ZONS, a town of Prussia, in the prov. of the Rhine, regency and 9 m. SSE of Dusseldorf, and circle of Reuss, near the l. bank of the Rhine. Pop. 900. It has manufactories of cloth, tanneries, and distilleries of brandy. This town, under the name of *Gesonia*, was one of the 50 castles of Drusus. In the ancient maps it is called Sontenum. It was formerly bathed by the Rhine. It was fortified by Sigfried, and in 1645 withstood an attack by the French.

ZONZONATE. See SONSONATE. See also TRINIDAD.

ZOOGO, a town of Africa, in Sudan, to the W of Borgu.

ZOOTZEN, a village of Prussia, in the reg. of Potsdam, and circle of Ost Priegnitz. Pop. 220.

ZOPEN, a village of Prussia, in the regency of Konigsberg, and circle of Wehlau. Pop. 200.—Also a village of Saxony, in the circle of Leipsig, bail. and near Vorna. Pop. 389.

ZOPPOK, a village of Prussia, in the regency of Danzig, and circle of Neustadt, on the Baltic. Pop. 598. It has manufactories of cloth and of paper.

ZOPPOLHEN, a village of Reuss-Greiz, in the bail. of Burgk. Pop. 580.

ZOPTAU, a village of Austria, in Moravia, in the circle of Olmutz, and seignory of Wiesenberg. Pop. 1,140.

ZOPTEN, a village of Saxe-Meiningen, in the bail. of Grafenthal. Pop. 220.

ZORA, a village of Turkey in Asia, in the pash. and 39 m. NNE of Bagdad, near a canal drawn from the Diala. Pop. 1,000. The environs produce dates and silk, and pasture large herds of buffaloes.

ZORBIG (KLEINZERBST, or ZIPPERZERBST), a town of Prussia, in the regency of Merseburg, and circle of Bitterfeld, on the Fuhne, 12 m. NNE of Halle. Pop. in 1843, 2,856. It has a suburb, an hospital, and an ancient castle. Tobacco is cultivated in the environs.

ZORE, or ZUR, a town and port of Arabia, in Oman, to the W of Ras-el-Had, and 105 m. SE of Muscat. It carries on a considerable trade with Persia and Sinde.

ZORGE, a town of Brunswick, in the circle and 27 m. SW of Blankenburg, on a small river of the same name. Pop. 1,268. It has important iron-works.

ZORITA, a town in the prov. of Caceres, and partido of Logrosan. Pop. 2,273. It has a parish church and a custom-house.—Also a market-town of Spain, in the prov. of Castellon-de-la-Plana, and 69 m. N of Valencia, on a height near the l. bank of the Bergante. Pop. 600. On the r. bank of the river, in a grotto, is a curious chapel.

ZORITA-DE-LAS-CANES, a town of Spain, in the prov. and 42 m. E of Madrid, on the l. bank of the Tagus. Pop. 150.

ZORN, a river of France, which has its source on the E side of the Vosges, in the dep. of the Meurthe, flows thence N into the dep. of the Lower Rhine, which it traverses from W to E, and after a course of about 60 m., throws itself into the Moder, on the r. bank, a little below Rohrweiler. Its principal affluents are the Zintzel, Moselle, and Rohrbach.

ZORNDORF, a village of Prussia, in the prov. of Brandenburg, regency of Frankfurt, and circle of Kustrin. Pop. 500. An engagement between Frederick II. and the Russians took place here in 1758.—Also a village of Hungary, in the comitat of Wieselburg, on an arm of the Leitha, to which it gives its name. Pop. 1,267. It has a considerable trade in grain, hay, cattle, and wool, and is noted for its cray fish.

ZORNÈTING, or ZORNEDING, a village of Bavaria, in the circle of Upper Bavaria, and presidial of Ebersberg. Pop. 454.

ZORNHEIM, a village of Hesse, in the prov. of Rhein-Hessen, and circle of Mainz. Pop. 680.

ZORNISZCZE. See JORNICHTCHI.

ZORNOSA, a village of Spain, in the prov. of Vizcaya, between Bilboa and Vittoria. An engagement took place here between the Spaniards and French on the 31st Oct. 1808.

ZORRAQUIN, a market-town of Spain, in the prov. and 15 m. E of Burgos. Pop. 200.

ZORRICA, or ZURRICO, a town of Malta, 7 m. SW of La Valetta, and about ½ a mile from the S coast of the island. Pop. 3,500.

ZORRO, or SORRO, a river of Portugal, in the prov. of Alemtejo, which has its source to the NW of Alpalhao, near the S bank of the Tagus, runs W and SW, and after a course of about 60 m. joins the Sorraya, opposite Conco.

ZOSCHEN, a village of Prussia, in the regency and circle of Merseburg, on the Luppe. Pop. 580.

ZOSCHINGEN, a village of Bavaria, in Swabia, in the presidial of Lauingen. Pop. 596.

ZOSNITZ, a village of Austria, in Bohemia, in the regency of Bohmisch-Leipa, and bail. of Auscha. Pop. 260.

ZOSSEN, a town of Prussia, in the regency of and circle of Teltow, on the Notte, 21 m. S of Potsdam. Pop. in 1843, 1,919. It is enclosed by palisades and ditches, and has three suburbs, a castle, and an hospital. The manufacture of linen, and adjacent fisheries form the chief branches of local industry.—Also a village of Austria, in Silesia, in the circle of Troppau. Pop. 980.

ZOTORSKAIA, a town of Russia in Europe, in the gov. of the Don Cossacks, 240 m. NE of Novo-Tcherkask, on the left bank of the Khoper.

ZOTTANCE, a village of Austria, in Galicia, in the circle of Zolkiew. Pop. 2,000. It has a Catholic and a United Greek church.

ZOTTELSTADT, a village of Saxe-Weimar, in the prov. of Weimar, and bail. of Rossla. Pop. 400.

ZOTTEWITZ, a village of Saxony, in the circle of Dresden, and bail. of Grossenhain. Pop. 200.

ZOTTIG, a village of Austria, in Silesia, in the circle of Troppau. Pop. 550.

ZOTTISHOFEN, a town of Würtemberg, in the circle of the Jaxt, and bail. of Künzelsau. Pop. 2?0.

ZOTTKUTTEL, a village of Austria, in Moravia, in the circle of Olmutz, and seignory of Hohenstadt. Pop. 1,250.

ZOTTWITZ, a village of Prussia, in the regency of Breslau, and circle of Ohlau. Pop. 730.

ZOTZENBACH, a village of Hesse, in the prov. of Starkenburg, circle of Hippenheim, and presidial of Furth. Pop. 580.

ZOTZENHEIM, a village of Hesse, in the prov. of Rhein-Hesse, and circle of Bingen. Pop. 330.

ZOUBLOV, a town of Russia in Europe, in the gov. and 75 m. SW of Tver, on the Volga, at the confluence of the Vazowza. Pop. 1,000. It contains four churches and an alms-house.

ZOUCHA, a river of Russia in Europe, which has its source in the gov. of Toula, and district of Novosil, enters the gov. of Ozel, traverses Mtzensk, and after a course in a WNW direction of about 75 m., joins the Oka, on the r. bank, a little to the N of Dechkin, and on the confines of the governments of Ozel and Tula.

ZOUT, or SALT-RIVER, a river of South Africa, which has its sources in the district of Beaufort, in the Nieuwveld-Bergen, runs along the E confines of the district, forming the dividing line with that of Graaf-Reynet, and at the point of junction with the districts of Georgia and Uitenhage, joins the Karreeka, and thence takes the name of Camtoos or Great river. It has a total course of about 90 m.

ZOVAN. See ZAGRAN.

ZOVANY, a village of Transylvania, in the comitat of Kraszna, on the r. bank of the Berettyo, and 6 m. W of Somlyo.

ZOVAR, a halting-place in the Sahara, in the district of Bergu, 24 m. NW of the town of that name. It has good wells.

ZOVAYA, or ZAWIEH, a town of Tripoli, 9 m. W of Mesurata. The houses are well built, and are surrounded with gardens.

ZOW, or ZO, an oasis of the Sahara, in the country of the Tibbus, to the S of Bilma, on the road from Murzuk to Kouka. It is well-supplied with water. The name, signifying 'difficult,' is derived from the dangerous character of the route in which it is situated. A little to the N of Z. is another oasis named Zow-Seghrir or Little Zow, near a group of hills named Kakoroum.

ZOWAT, ZAWAT, or BRABEESH, a Moorish tribe of Nigritia, who inhabit the territory to the N of Timbuctu. Their chief towns are El Arawan and Bousbeya. Major Laing was assassinated by order of the chief of this territory in 1827.

ZRUCZ, a town of Bohemia, in the circle and 20 m. SW of Czaslau, on the Sarawa.

ZRUTSCH, a market-town of Austria, in Bohemia, in the regency of Pardubitz, and bail. of Lebetsch. Pop. 800. It has a castle.

ZSAMBEK, a market-town of Austria, in Hungary, in the comitat of Pesth, 15 m. WNW of Ofen.

ZSAMBOKRETH (NYITRA), or ZABOKREKY, a market-town of Austria, in Hungary, in the comitat and 24 m. NNE of Neutra, and near the r. bank of the river of that name.

ZSAMBOLY. See HATZFELD.

ZSARNOCZA, ZERNOWITZ, or ZARNOWICE, a town of Hungary, in the comitat of Bars, on the r. bank of the Gran, and 6 m. NE of Uj-Banya.

ZSCHAGAN, a village of Prussia, in the regency of Merseburg, and circle of Torgau. Pop. 290.

ZSCHAITEN, a village of Saxony, in the circle of Dresden, and bail. of Grossenhain. Pop. 280.

ZSCHAITZ, a village of Saxony, in the circle of Dresden, and bail. of Meissen. Pop. 285.

ZSCHEPA (GROSS and KLEIN), a village of Saxony, in the circle of Dresden, and bail. of Grossenhain. Pop. 320.

ZSCHEPPLIN, a village of Prussia, in the reg. of Merseburg, and circle of Delitzsch, on the Mulde, near Eilenburg. Pop. 628.

ZSCHERBEN, a village of Prussia, in the reg. of Merseburg, and circle of the Saal. Pop. 310.

ZSCHERNGOSTA, a village of Prussia, in the prov. of Brandenburg, regency of Frankfort, on the Oder, circle and 18 m. SSW of Kalau. In its vicinity is an extensive glass manufactory.

ZSCHERNITZ, a village of Prussia, in the reg. of Merseburg, and circle of Delitzsch. Pop. 350.

ZSCHERNITZSCH, a village of Saxe-Altenburg. Pop. 360.

ZSCHERNSKE, a village of Prussia, in the reg. of Liegnitz, and circle of Rothenburg. Pop. 200.

ZSCHIEDGE, a village of Saxony, in the circle and bail. of Dresden. Pop. 200.

ZSCHIEREN, a village of Saxony, in the circle of Dresden, and bail. of Pirna. Pop. 310.

ZSCHIESEWITZ, a village of Prussia, in the reg. of Merseburg, and circle of Bitterfeld. Pop. 200.

ZSCHIRLA, a village of Saxony, in the circle of Leipsig, and bail. of Golditz. Pop. 285.

ZSCHOCHAU, a village of Saxony, in the circle of Dresden, and bail. of Meissen, near Döbeln. Pop. 436.

ZSCHOCKEN (OBER and UNTER), a village of Saxony, in the circle of Zwickau, near Hartenstein. Pop. 1,093.

ZSCHOLLAU, a village of Saxony, in the circle of Leipsig, and bail. of Oschatz. Pop. 200.

ZSCHOPAU, a river of Saxony, which has its source on the N side of the Erzgebirge, in the S part of the circle of that name, runs N, waters the E part of the circle of Leipsig, and after a course of about 75 m. joins the Mulde, at Treeburgus, 3 m. WNW of Dobeln. Wolkenstein, Zschopau, Mettweyda, and Waldheim, are the chief places on its banks.—Also a town in the circle of Zwickau, and bail. of Augustusburg, on the l. bank of the Zschopau, at the foot of the Zschopenberg. Pop. in 1837, 5,813; in 1843, 6,169; in 1849, 6,938. It has a castle and two churches, and possesses manufactories of linen and cotton fabrics, hosiery, parchment, and pottery, bleacheries, and several breweries.

ZSCHOPPENITZ, a village of Saxe-Altenberg, and bail. of Altenberg. Pop. 200.

ZSCHORLAU, a village of Saxony, in the circle of Zwickau, and bail. of Kirchberg, near Schneeberg. Pop. 2,065. It has manufactories of lace, and iron-ware.

ZSCHORNA, a village of Saxony, in the circle of Leipsig, and bail. of Wurzen. Pop. 336.

ZSCHORNEGOSDA, a village of Prussia, in the regency of Frankfort, near Kalau. Pop. 310.

ZSCHORNEWITZ, a village of Prussia, in the regency of Merseburg, and circle of Bitterfeld. Pop. 220.

ZSCHORTAU, a village of Prussia, in the reg. of Merseburg, and circle of Delitzsch, on the Loberbach. Pop. 500.

ZSELLELZ, a village of Hungary, in the comitat of Bars, on the Gran.

ZSOGOD, or SOGOD, a village of Transylvania, in the Szeklers territory, on the l. bank of the Aluta, 3 m. ESE of Csik-Szereda.

ZSOLNA. See SZOLNA.

ZSOLT. See SOLT.

ZSOMBOLY. See HATZFELD.

ZUBA, a village of Syria, 15 m. ESE of Jaffa.

A signal victory was here gained over Antiochus by Judas Machabeus.

ZUBAKI, a river of Japan, in the island of Nifon, and prov. of Dewa, which descends from the Oraxi mountains, runs NW, and after a course of about 60 m., throws itself into the sea of Japan.

ZUBEN, a village of Switzerland, in the cant. of Thurgau, and bail. of Gottlieben. Pop. 225.

ZUBERWANGEN, a parish of Switzerland, in the cant. of St. Gall, and bail. of Wyt. Pop. 1,127.

ZUBIENA, a market-town of Sardinia, in the gov. of Turin, SSW of Biella. Pop. 5,000.

ZUBIETA, a village of Spain, in the prov. and 5 m. from San Sebastian. Pop. 158.—Also a market-town in the prov. of Navarre, and 30 m. N of Pamplona, in the valley of S. Esteban-de-Lerin.

ZUBOV, a mountain of Turkey in Europe, in the W part of Servia, near the sources of the Lieg and Kamenitza.

ZUBRI, a village of Austria, in Moravia, in the circle of Olmutz, and bail. of Walachisch-Meseritsch. Pop. 2,060.

ZUBRY, a village of Austria, in Moravia, in the circle of Iglau. Pop. 440.

ZUBSUCH, a village of Austria, in Galicia, in the circle of Sandie. Pop. 2,000.

ZUCAPA, a town of Guatimala, in the state of that name, midway between the coast and Guatemala-la-Vieja. Pop. 8,000.

ZUCARELLO, a town of Sardinia, in the division of Genoa, prov. and 6 m. NW of Albengo, on the Arossia Pop. 1,500.

ZUCHAU, a village of Prussia, in the reg. of Magdeburg, and circle of Kalbe. Pop. 465.

ZUCHEN, a village of Prussia, in the reg. of Koslin, and circle of Neustettin. Pop. 240.

ZUCHERING, a village of Bavaria, in Swabia, in the presidial of Neuburg. Pop. 622.

ZUCHOW, a village of Prussia, in the regency of Koslin, and circle of Dramburg. Pop. 328.

ZUCHWEIL, a village of Switzerland, in the cant. of Soleure, and bail. of Bucheggberg, near the confluence of the Emmen with the Aar. Pop. 435.

ZUCKAU, or ZUCKOW, a village of Prussia, in the regency of Danzig, and circle of Karthaus, on the Radaune, a little above the confluence of the Stolpe. Pop. 580. It has a paper-mill.

ZUCKERHANDL, a village of Austria, in Moravia, in the circle of Znaim. Pop. 260.

ZUCKERHU, an island of New Zealand, to the N of Cape Borell, in the island of Eahenomauwe.

ZUCKERS, a village of Prussia, in the regency of Koslin, and circle of Rummelsburg. Pop. 240.

ZUCKLAU, a village of Prussia, in the regency of Breslau, and circle of Dels. Pop. 470.

ZUCKMANTEL, a town of Austria, in Silesia, in the circle and 12 m. NW of Troppau, and duchy of Neisse. Pop. 3,200. It has an hospital, and possesses manufactories of linen and cotton fabrics.

ZUDA, a village of Arabia, in Yemen, 114 m. NW of Sana.

ZUELA, a town of Tripoli, in Fezzan, 114 m. ENE of Murzuk. It was formerly capital of Fezzam.

ZUERA, a town of Spain, in the prov. and 18 m. NE of Zaragoza, on the r. bank of the Gallejo. Pop. 2,000. It is well-built, and has a collegiate church, convent, and almshouse.

ZUEROS, a town of Spain, in the prov. and 36 m. SE of Cordova, and partido of Cabra. Pop. 2,024. It has a parish-church, a school and a public granary.

ZUF, GUR, JANGARIE, or CHING HEHIRAN, a town of Afghanistan, capital of the Ghorat district, 170 m. SSW of Balkh, and 130 m. NNW of Candahar. This town, which was formerly capital of a state, was taken by the Khan of Kharism, and sacked by the armies of Denghiz Khan and Tamerlane. Little now remains but the name.

ZUFFERABAD, a town of Hindostan, in the presidency of Bengal, and prov. of Allahabad, 5 m. N of Juanpur. Pop. 20,000.

ZUFFONNE, or MERS-EL-FAUM, a town of Algeria, 75 m. E of Algiers, and 48 m. W of Bougie, in the prov. of Constantine, on the Mediterranean. Coal is found in the environs.

ZUFRA, a town of Spain, in the prov. of Huelva, and partido of Aracena, 45 m. NNW of Seville, at the foot of the Sierra-Morena. Pop. 730.

ZUG, a canton of the Swiss confederation, situated between 47° 5′ and 47° 13′ N lat., and between 8° 24′ and 8° 39′ E long., bounded by the canton of Zurich—from which it is separated by the Sihl—on the N; by Schwytz—from which for several leagues it is divided by Lake Zug, the line of demarcation from which in its entire length is described by the Reuss—on the E and W; and by Lucerne and Aargau on the S. It comprises an area 14 m. in length from E to W, and 9 m. in extreme breadth, and is, next to Geneva, the smallest canton in Switzerland. Pop. in 1837, 15,322, of whom 14,193 were citizens of the cant., 1,019 those of other cants., and 110 strangers. Pop. in 1850, 17,461, of whom 17,346 were Catholics. The German language is that commonly spoken. Towards the NW this cant. consists chiefly of plains. The Zugerberg rises to the S of Lake Zug, and in the SE are several mountain-ranges. Except on the E frontier, traversed by the Limmat, this cant. is comprised entirely within the basin of the Reuss. It contains the greater part of the lake of the same name, and the whole of that of Egeri. Its principal rivers are the Reuss and Sihl, on the E and W-frontiers; and, in the interior, the Lorze, an efflux from Lake Egeri, and affluent of Lake Zug, and ultimately the Reuss. The climate is mild, especially in the vicinity of the lake. The soil is fertile, producing grain, fruit, cider, and in considerable quantity, but inferior in quality, wine. Butter and cheese are among the most important agricultural productions. The manufactures consist chiefly in paper leather, and flocs-silk. Cattle, timber, kirchwasser, cider, chestnuts, fruit, and butter, are the chief articles of trade. A small school at Zug, and a Latin school at Zaar, are its only educational establishments. The cant. of Z. ranks eighth in the order of the Helvetian confederation. Its contingent for the federal army is 456 men, and 16 horses, and 2,295 francs in silver. Its receipts in 1843 amounted to 90,000 francs. It depends upon the diocese of Coire, and has an Episcopal commission. It is a democratic republic, the citizens of which enjoy political privileges, and are all soldiers on attaining the age of 19. In 1837 the number of active citizens was estimated at 3,500. The sovereignty belongs to the people, and is exercised either directly in the national, or in the communal assemblies, or by representatives forming the councils or constituted bodies. The *landsgemeinde*, or National Assembly, meets once a-year to receive the reports of the administrators of public affairs, and to appoint the public functionaries. The chief of these, the *vorgesetzten*, is named for life. He presides at the landsgemeinde, and is assisted in the exercise of the executive power by the *kantonsrath*, or council of the canton, of which also he is president. This council consists of the vorgesetzten and 54 members elected for two years by the communal assemblies, with liberty to represent their respective localities rather than the country generally. The legislative power, and the right of electing deputies to the diet, were delegated by the constitution in 1814 to the *landrath*, or council of the country, consisting of a number triple to that of the kantonsrath, and hence called the great or triple council, and which meets three times a-year. The cant. is divided into two amts,—viz., the innere-amt, containing 6 communes, and the aussere-amt, which comprises 5.—The chief town, which bears the same name, is on the E bank of the lake of the same name, at the foot of the Zugerberg, and 17 m. S of Zurich, in N lat. 47° 9', and E long. 8° 35', at an alt. of 1,443 ft. above sea-level. Pop. 3,302. This town, which was founded in the 15th century, still preserves some of its ancient fortifications, and its massive walls and towers. It is well and regularly built, and has an ancient cathedral, in which are the tombs of the Zursanben family, a capuchin convent, a gymnasium, a public library, an arsenal, and a poors'-house. It has manufactories of paper and of leather, and carries on an active trade in cattle, wood, kirchwasser, cider, chestnuts, and dried fruit. In 1435 two of the streets sunk into the lake, and in 1594 several houses shared the same fate. In 1795 it suffered extensively by fire.

ZUG (Lake of), a lake of Switzerland, divided into two parts by an angle formed near its centre by the cant. of Lucerne, the northern of which is comprised in the cant. of the same name, and the southern in that of Schwytz. It is connected with the basin of the Rhine, or of its affluents, by the Lorze, Reuss, and Aa, and is bordered on the E and S by the Zugerberg and Righi mountains. It is 12 m. in length from N to S, and 3 m. in extreme breadth, and is 1,340 ft. in alt. above sea-level. Its principal affluent is the Lorze, an efflux of Lake Egeri, which it receives on the N at Chain, and which afterwards conveys its waters to the Reuss. The E bank of this lake is the most fertile. It abounds with fish, and is said to contain carp of 90 lbs weight.

ZUGA, or Zouga, a river of South Africa, which flows from the SE extremity of Lake Ngami, in a stream which, for two or three days from the lake, is from 300 to 500 yds. wide, and pursues an E, S, SE, and then NE course, becoming narrower and shallower as it proceeds, until it divides into a number of small rivulets, which are finally absorbed in extensive salt marshes in the great Kalahari desert. Its water is clear, soft, and cold. Its volume of water is largest in October, and it then gradually decreases till June. Its banks, for some distance from the lake, are flat and swampy; they then rise, and are generally of calcareous tufa. They are lined towards the lake by gigantic trees, among which are the baobab, the Palmyra palm, and the banyan. Alligators and hippopotami are numerous in its waters. The Bakoba are the principal dwellers on the islands and banks of the river.

ZUGAR (Jebel), a volcanic island in the Red sea, having its NE point in N lat. 14° 5', E long. 42° 48'. It is 28 m. in circumf., and nearly 9 m. long from N to S; and is composed of a series of lofty barren hills. A sprinkling of timber occurs on the N and S parts of the island, and a little grass in some of the valleys affords subsistence to a few antelopes; but the island is only occasionally resorted to by fishermen.

ZUGDAMM, a village of Prussia, in the reg. and circle of Danzig, on the Werder. Pop. 430.

ZUGERBERG, a group of mountains in Switzerland, in the cant. of Zug, to the E of the town of that name. It presents a generally fertile surface, and has fine fruit trees. Its principal summit, the Kamstal, has an alt. of 912 Paris ft. above the level of the lake.

ZUGLIANO, a village of Austria, in Lombardy, in the deleg. and to the S of Udine.

ZUG-MEKAYIL, a town of Syria, capital of a district of the same name, in the N part of Kesrawan, and about 5 m. from Nahr-el-Kelb. It is of recent foundation, is well-built, and has a handsome church dedicated to St. Michael, two palaces, one belonging to the sheik, with fine gardens, and two convents. Pop. 12,000. The environs are noted for their silk and wine.

ZUHLEN, a village of Prussia, in the reg. of Potsdam, and circle of Ruppin. Pop. 368.

ZUHLSDORF, a village of Prussia, in the reg. of Potsdam, and circle of Nieder-Barnim. Pop. 200.—Also a village of the reg. of Frankfort, and circle of Arnswalde. Pop. 338.

ZUHLSHAGEN, a village of Prussia, in the reg. of Koslin, and circle of Dramburg. Pop. 320.

ZUHR, a village of Mecklenburg-Schwerin, prov. of Mecklenburg, and bail. of Wittenburg. Pop. 260.

ZUIDER-ZEE, Zuyder-Zee, or Sudersee [i. e. 'South sea'], an inlet of the German ocean, on the coast of Holland, enclosed by the prov. of North Holland on the W, those of Utrecht and Gelderland on the S, the prov. of Over Yssel on the E, and that of Friesland on the NE. Stretching across its entrance on the N, are the islands of Ameland, Schelling, Vlieland, and Texel, and to the SE of these near the opening of the gulf is the island of Wieringen. Between Stavoren, the W extremity of Friesland, and Enkhuizen in Holland, the Zuider-Zee does not exceed 11 m. in breadth; further S it expands into a circular basin, forming the Zuider-Zee, properly so called, and presents a diameter of 45 m. The total area is estimated at 1,207 sq. m. To the SW it is connected by the Pampus channel and Het-Y, on the S bank of which is Amsterdam, with the Haarlem Meer. The entrance to the gulf is encumbered with sand-banks. The Zuider-Zee, properly so called, has, except near the shores, comparatively few. It contains the islands of Schokland, on the E of Urk towards the centre, and of Marken in the SW. The Yssel, a branch of the Rhine, and the Vecht with its affluent, the Zwarte-water, in the E, and another Vecht, a branch of the Rhine, on the S, are its chief tributaries. The southern part of the gulf was formerly a lake named Flevo. It received the Yssel, and discharged a stream into the sea, then 50 m. distant. In 1170, Lake Flevo rose to such a height, that the water spread over the land as far as Utrecht, and the people of that town caught fish from the walls. In 1230, 100,000 persons were drowned in Friesland; and from this date encroachments were rapid: the barrier-islands on the sea-board were diminished in size, and the channels between them deepened. By the wasting action of successive inundations, the Zee grew to nearly its present extent by the beginning of the fifteenth century. On some occasions the water has stood 8 ft. above the surface of the dikes in Friesland and Gelderland. The wind most dreaded on the shores of the Z. is one which beginning at the S, sweeps gradually round, and blows for a time from the NW. This pours in such a heaped-up mass of water from the German ocean, that the dams, except under extraordinary circumstances, are too low to repel it, and meadows, farms, towns, and villages, are overwhelmed. The tide runs with great velocity through the channels between the islands, particularly at the Helder, or 'Hell's-door,' as the Dutch name signifies. The growth of the Z. has resembled that of the Haarlemmer-meer, but on a grander scale. The average depth of the two is about the same, excepting certain cavities in the Z. which are very profound; and the formation of a barrier-dike in water not more than 12 ft. deep appears to be a practicable undertaking. The project is to construct a dike, with buttresses and flood-gates across the narrowest part of the Z., from Medemblik or Enkhuysen, on the peninsulated portion of N. Holland, to Stavoren in Friesland. The cost, as estimated, would be £5,000,000. In the spring of 1856, the Z. in consequence of the long continuance of northerly and easterly winds, was almost dry. Between Geneminden, Blockzil, Lemmer, Kampen, and Harderwyk the bed of the sea might be crossed dry footed, and steamboats and other vessels were everywhere high on the sands.

ZUIDHORN, a village of Holland, in the prov., arrond., and 7 m. W of Groningen. Pop. 1,400.

ZUIDLAND, a village of Holland, in the prov. of S. Holland, to the E of Helvoetsluis. Pop. 1,200.

ZUIDSCHOTE, a village of Belgium, prov. of W. Flanders, and arrond. of Ypres. Pop. 590.

ZUILWOLDE, a village of Holland, in the prov. and to the NE of Groningen.—Also a village of the prov. of Drentheim, to the SE of Meppen. Pop. 1,120.

ZUJAR, a town of Spain, in Andalusia, in the prov. and 57 m. NE of Granada, partido and 6 m. N of Baza, near the Barbata. Pop. 1,655. It has a parish-church, a custom-house, and a public granary. In the vicinity are mines of copper, and sulphureous springs and baths.

ZUKIN, a village of Afghanistan, on the l. bank of the Turnah, 8 m. SE of Kandahar.

ZUKLEBE, a village of Prussia, in the reg. of Frankfort, and circle of Sorau. Pop. 280.

ZUKON. See Zuckau.

ZULAUF. See Sulau.

ZULB, a village of Austria, in Moravia, in the circle of Znaim. Pop. 850.

ZULCHOW, a village of Prussia, in the reg. of Stettin, and circle of Randow. Pop. 450.

ZULIA. See Sulia.

ZULICHENDORF, a village of Prussia, in the reg. of Potsdam, and circle of Juterbogk. Pop. 200.

ZULIN, a village of Galicia, in the circle of Stry, on a river of that name. It has a Catholic and a United Greek church.

ZULKENHAGEN, a village of Prussia, in the reg. of Koslin, and circle of Neustettin. Pop. 260.

ZULKOWO, a village of Prussia, in the reg. of Posen, and circle of Kroben. Pop. 240.—Also a village of the circle of Wreschen. Pop. 370.

ZULLA, or Xulla, a town of Abyssinia, in Dankali, in the Hazortas territory, on the S side of Annesley bay, and about 30 m. SE of Arkeeko. In the vicinity are the remains of the ancient *Aduli*.

ZULLA, or Xulla, a group of islands, four in number, in the Molucca strait, to the E of the island of Celebes, extending between 1° 55′ and 2° 30′, and between 124° 40′ and 126° 30′ E long. They are high, well-wooded, and populous. One of them, named Xulla-Bessey, is a Dutch factory.

ZULLICHAU, a circle and town of Prussia, in the prov. of Brandenburg and reg. of Frankfort. The circle comprises an area of 350 sq. m. Pop. 39,810.—The town is 24 m. ESE of Frankfort-on-the-Oder, in a fertile plain. Pop. in 1843, 4,898. It is enclosed by walls and ditches, and has four suburbs, a castle, a Lutheran and a Reformed church, an orphans' asylum, and a royal college. It possesses manufactories of fine cloth, hats, hosiery, linen, brandy, and vinegar, and several tanneries.

ZULLSDORF, a village of Prussia, in the reg. of Merseburg, and circle of Torgau. Pop. 348.

ZULLWEIL, a village of Switzerland, in the cant of Soleure and bail. of Thierstein. Pop. 340.

ZULPICH, a town of Prussia, in the reg. of Cologne, and circle of Enskirchen. Pop. 1,340. It has manufactories of woollen fabrics and tanneries. This town was fortified by the Romans, and called by them Tolbiacum. In the vicinity Clovis successfully resisted the Germans in 496, and in 612 the same locality was the theatre of an engagement between the sons of King Childebert and Thierry II., king of Burgundy.

ZULTE, a department and commune of Belgium, in the prov. of E. Flanders, arrond. and 17 m. WSW of Ghent. Pop. of dep. 2,002; of com. 296.

ZULTEPEC, or Sultepec, a village of Mexico, in the state of Mexico, 30 m. SW of Toluca. Pop. 2,000. The surrounding district is named La-Plata, from its silver mines.

ZULUS, a remarkable tribe of Kaffirs, occupying the country to the N of Natal, and separated from it by the Tuguela river Their territory is fertile, yielding dhurra and millet, and good pasturage; the greater part of the surface is covered with jungle, and infested by wild animals and venomous reptiles. The Z. are a tall, athletic, well-proportioned race; warlike in their habits, and surpassing most African tribes in intelligence, but improvident, superstitious, and cruel. About seventy years ago, the Z. were an inconsiderable race, finding a scanty livelihood on the banks of the Mozambique channel, and to the N and E of what is now Natal. In this tribe, however, rose a family the members of which for several generations governed the Z. with a rod of iron. The first of these chiefs was named Essenzingercona. Warlike and ambitious, he led the tribe from the swamps of the Mozambique into the more fertile regions of the interior, and, falling upon the inhabitants, either massacred or drove them out of their possessions. Chaka—a man whose name is as terrible in Zulu annals as Domitian or Nero in Roman, was sent by his father Essenzingercona, to live with the king of a neighbouring tribe. Shortly afterwards the monarch died, and a younger brother of Chaka reigned in his stead. Chaka spread a report of his own death, and returning in disguise, with the aid of another of his brothers, murdered the new monarch and proclaimed himself king, his first care being to exterminate all Essenzingercona's councillors and friends, and then to assassinate his accomplice brother. He then attacked the king who had shown him hospitality, killed him, exterminated half his people, and made the rest swear fidelity. A rapid round of similar expeditions took place. Chaka became the sole, absolute, and despotic monarch and master of what might be called a kingdom. Isaacs—now a Natal settler, and who was forced to accompany Chaka on several of these dreadful raids—witnessed a massacre which Chaka instituted as a sort of sacrifice to the manes of his mother, which lasted a fortnight, and in which it was estimated that upwards of 10,000 people were slain! Chaka, beginning to grow old, had that done to himself by his brother Dingarn which he had done to the brother of both. Dingarn was a modified sort of Chaka. He did not kill so many people as Chaka—for European influence was beginning to assert itself, and the missionaries were saving many lives; but Dingarn, in his heart, was as cruel and as treacherous as his predecessor, and at length, incensed at the trespasses of the Dutch boers upon what he considered his dominions, he invited upwards of 60 of the principal of these farmers to a dinner of friendship. They went, unhappily, unarmed. In the middle of the feast each boer was struck down senseless from behind, and then, before death, impaled upon spikes prepared for the purpose. As might have been expected, the countrymen of the victims took a fearful and a bloody revenge. Every man of white blood, for hundreds of miles around, obeyed the call with all his trusty servants; and about half the tribe of Z. revolted against Dingarn, and joined the Dutch. Dingarn led the remainder of his warriors, and a desperate encounter ended in the rout of Dingarn's army, and his own flight to a neighbouring chief, who received him courteously, and murdered him in the night. This expedition and battle is called by the Dutch people 'the Volks Rald,' and is the Waterloo of Natal and the Cape colony. It broke the power of the Z., and their present king, Panda, has the good sense to prefer trading to fighting, and, by the advice of the colonial authorities, now his neighbours, has abolished the murderous and despotic system of his ancestors. Zulu is, however, far from being yet in a satisfactory state of civilization, but it is immeasurably better than when subjected to the tender mercies of Chaka and Dingarn. Nearly 100,000 Zulus fled during the last reign to Natal.

ZULZ, or Biala, a town of Prussia, in the prov. of Silesia, reg. and 24 m. SSW of Oppeln, and circle of Neustadt, on the Biala. Pop. in 1834, 2,739. It has a castle, three churches, and a synagogue. Lace is its chief article of manufacture.

ZULZENDORF, a village of Prussia, in the reg. of Breslau, and circle of Nimptson. Pop. 468.

ZUMAMPA, a town of La-Plata, in the prov. and 150 m. NE of Cordova.

ZUMARRAGA, a town of Spain, in the prov. of

Guipuzcoa, and partido of Vergara, 21 m. SW of San Sebastian, near the r. bank of the Urola, over which is a bridge, which forms a communication between this town and Villa-Real. Pop. 969.

ZUMAYA, a town of Spain, in the prov. of Guipuzcoa, and arrond. of Azpeitia, 18 m. W of San-Sebastian, at the foot of a mountain, and at the confluence of the Urola with the gulf of Gascogny. It has a small fishing port, and is well and regularly built. Pop. 1,500. It has a town-house, a handsome edifice, a square, several fountains, a convent, and two hospitals. The adjacent fisheries are important, and in the vicinity are quarries of jasper and limestone.

ZUMAYAQUE, a village of Guatimala, in the state of that name. It has manufactories of cotton fabrics. Cacao is extensively cultivated in the vicinity.

ZUMBO, a settlement of Mozambique, on an island of the Zambese, about a month's journey WSW of Tite. The Portuguese have a factory here. The trade consists chiefly in gold from the mines of Abutua, Pamba, and Murusura, ivory from the banks of the Orange river, and iron, copper, and coal of the vicinity. The surrounding district is richly wooded.

ZUM-BUM, a lake of Nigritia, in the district of Bedee, to the E of Houssa, and 12 m. ENE of Bedeguna.

ZUMEL, a town of Spain, in the prov. and 11 m. NW of Burgos, on the Urbel. Pop. 200.

ZUMHOF, a village of Würtemberg, in the circle of the Jaxt, and bail. of Welzheim. Pop. 300.

ZUMHOLZ, a village of Switzerland, in the cant. of Freiburg. Pop. 247.

ZUMIKON, a village of Switzerland, in the cant. of Zurich, and bail. of Meilen. Pop. 711.

ZUMILIANSK, a village of Russia in Europe, in the gov. of the Don Cossacks, on the Don, to the NE of Novo-Tcherkask.

ZUMMERN, a village of Austria, in Bohemia, in the regency of Eger, and bail. of Tachan. Pop. 270.

ZUMPANGO, a town of Guatimala, in the state of that name, and dep. of Sacatepec and Chimaltenango, and district of Texar. Pop. 4,800.—Also a small lake of Mexico, in the state of Mexico, and to the NW of the lake of San Christoval. It is divided by the dike of the Calzada-de-la-Cruz-del-Rey into two basins, distinguished by the names of Zitlaltepec and Cayotepec. It receives the Pachuca on the NE. The Guantitlan, which formerly flowed into it on the W, now joins the Zula. On the NE bank of the lake is a town of the same name. Pop. 4,000.

ZUNDER (GROSS and KLEIN), two villages of Prussia, in the reg. and circle of Danzig. Pop. 930.

ZUNDERSBACH, a village of Bavaria, in the circle of Lower Franconia, and presidial of Bruckenau. Pop. 404.

ZUNDERT, a parish of Holland, in the prov. of North Brabant, and SW of Breda. Pop. 3,500.

ZUNEDA, a town of Spain, in Old Castile, in the prov. and 36 m. NE of Burgos. Pop. 225.

ZUNGOLI, a town of Naples, in the prov. of the Principato-Ultra, district and 7 m. SE of Ariano, and cant. of Villanova, on a hill. Pop. 1,800. It has a collegiate church. This town was founded by the Normans, by whom a fortress was erected in the locality to repress the incursions of the Greeks.

ZUNIGA, a town of Spain, in the prov. of Navarre, 36 m. SW of Pamplona, in the valley of the Solana. Pop. 260.

ZUNIGA, a town of New Granada, in the dep. of the Magdalena, prov. and 42 m. S of Carthagena.

ZUNIL, a market-town of Guatimala, in the district of Quezaltenango. Pop. 2,900.

ZUNSWEYER, a village of Baden, in the circle of the Middle Rhine, and bail. of Offenburg. Pop. 1,430.

ZUNTERSBACH, a village of Electoral Hesse, in the prov. of Henau, and circle of Schluchtern, and bail. of Schwarzenfels. Pop. 410.

ZUNZGEN, a village of Switzerland, in the cant. of Basle, and bail. of Sissach. Pop. 633.

ZUNZINGEN, a village of Baden, in the circle of the Upper Rhine, and bail. of Mullheim. Pop. 230.

ZUPPA-XUPPA ['Four Counties'], a district of Austria, in Dalmatia, in the circle of Cattaro, consisting of four counties, viz.: Lazzarovich, Boicovich, Clubanovich, and Tuicovich. It is very populous, and its inhabitants, although employed chiefly in agriculture, are distinguished in many of their customs and manners from the surrounding people.

ZUR (RAS-EL), a headland of Arabia, in the Persian gulf, to the S of Erzug bay.

ZURAWIA, a village of Prussia, in the regency of Bromberg, and circle of Schubin. Pop. 280.

ZUR-BURG. See BURG.

ZURGENA, a town of Spain, in the prov. and 48 m. NE of Almeria, partido of Huercalovera, on the r. bank of the Almanzor. Pop. 2,930. It has a parish church and a custom-house, and possesses manufactories of linen and woollen fabrics, and saltpetre.

ZURGHEIM, a Moorish tribe of the W part of the Sahara, in the vicinity of Mongearts.

ZURI, a small island of the Adriatic, dependent on Dalmatia, in N lat. 43° 38′ 30″, and E long. 15° 40′ 20″. It has coral fisheries, and contains a village.

ZURICH, a northern canton of Switzerland; bounded on the N by Baden and by the territory of Schaffhausen; on the E by the cantons of Thurgau and St. Gall; on the S by Schwyz and Zug; and on the W by Aargau. It has an area of 690 sq. m.; and is in length from N to S about 32 m.; in breadth 25 m. It is somewhat smaller than the average size of an English county; but its pop., 250,698 in 1850, is such as to rank it next to Geneva in density, and among the best-peopled tracts of the continent of Europe: the number of inhabitants being on an average 3,499 to a sq. league; while the average for all Switzerland, in 1850, was 1,370 to a sq. league. The density of the population of this canton is owing less to the fertility of the soil, than to the long continued enjoyment of good government, and consequent habits of industry. The general aspect is pleasant, though without that bold and magnificent scenery which marks the interior and south of Switzerland. The hills do not rise above 3,200 ft., and are separated by beautiful valleys and lakes. There are no glaciers in this cant., the highest mountains being covered with forests and constant verdure. The principal summits are the Allmannskette, the Lägernkette, and the Albikette.—The principal rivers are the Rhine, the Reuss, and the Limmat; all pursuing a W or NW course. The smaller streams are the Thur, the Töss, the Sihl or Sil, and the Glatt. The Limmat is navigable, and flowing through the lake of Zurich, unites with the Aar below Baden. The other lakes are, the Greiffen-see, and the Pfeffikon-see, on the E of Zurich; the Katzen-see on the NW, and the Türler-see on the SW. Among the mineral springs are those of Nidel or Nydelbad, the two Geirenbads, Röslibad, and the baths of Zollikhof, Arnheim, and Drathschmittlein.—The climate is mild, the greatest cold never exceeding 7° 25′, and the greatest heat 70° of Fahr. It rains upon an average 100 days in the year. The soil is not naturally fertile, but rendered productive by careful cultivation. In no part of Switzerland is irrigation, manuring, and the mixture of different soils, better understood: rich pastures and extensive orchards meet the eye of the traveller in every direction. The cattle are in general of a good breed. Wine is cultivated; and the quality, in certain tracts of favourable exposure, is good; but in general it is thin and sharp. The cyder and cherry-brandy are excellent. The manufactures of this canton are mostly carried on by weavers in their own houses, to which are attached patches of land. The fabrics are chiefly cotton and light silk stuffs; also linen, woollen, and leather. The iron-steam-boats which ply on the lake are all built here. Corn is imported, the produce being inadequate to the consumption.—In its mode of government, this canton was long aristocratic; but since 1831 the government, as far as election goes, has been democratic. The great council or legislative body of 213 members is elected by universal suffrage. The session of this council is short, the executive functions being committed to 19 of its members. The revenue of the cant. is about £50,000; its contingent to the military force of Switzerland is 3,858 men. The inhabitants are, with the exception of two Catholic parishes, almost all Calvinists. Having joined the Swiss confederacy soon after 1315, this cant. was long the principal state, and had the honour of giving a president to the diet, as well as of receiving at Zurich the ambassadors of foreign powers. At present it ranks in the diet next to the cant. of Berne. Its territory was the scene of important military operations in 1798 and in 1799.

Zurich, the capital of the above cant., stands on the river Limmat, at its outlet from the N extremity of the lake of Zurich, in a narrow valley between hills, 46 m. ESE of Basle, and 60 m. ENE of Berne. It was fortified with a wall and ditch, and is neat and clean, though most of the houses are old fashioned, and the ramparts have been levelled since 1833. The population, 10,475 in 1812, and 17,040 in 1850, are almost all Calvinists. Z. contains few public buildings of note. The *rath-haus* or town-hall is large and well situated, but inelegant; the cathedral is an old massive structure; of the four other churches, none are entitled to notice; on the other hand, the beauty of scenery is striking. The archery-ground, which extends along the banks of the Limmat to the spot where that river is joined by the Sil, a copious stream flowing from the southward, is one of the finest walks in Switzerland; and there are a number of fine promenades around the town. Z. having in its lake and river the advantage of water-communication, has long been a place of manufacture and trade, and is now celebrated for its silk and cotton manufactures, and calico-printing. So early as the 13th cent., we read of fabrics of woollens, linen, and leather, in this place; here were also silk-works, small in extent but of note in an age when manufacturing industry was backward. They were soon surpassed by those of Tours and Lyons, and have been latterly replaced by cotton-works, which are carried on to a considerable extent. Few places of the size of Z. have surpassed it in the cultivation of literature. It contains extensive collections of books, manuscripts, paintings, and antiques. The town-library founded in 1628, contains 40,000 vols.; that of the cathedral is also considerable. Of schools and literary establishments, great and small, there are here no less than 24. Philosophy, divinity, and classics, are taught in the *Collegium Carolinum*, but most young divines are educated in Germany. There are here also a military and a medical seminary, a botanical garden, and a society of physics, economics, and natural history. Z. is the seat of the *vorort* or Swiss diet, alternately with Berne and Lucerne. A railway connects Z. with Baden or Aargau.

ZURICHER-SEE, a lake of Switzerland, extending in the form of a crescent, chiefly through the canton of Z., but partly also between those of Schwyz and St. Gall. It is divided into an upper and lower lake by the strait at Rapperschwyl, which, the breadth being little more than a quarter of a mile, is crossed by a long wooden bridge; in other places the breadth varies to the extent of nearly 3 miles. Its entire length is about 30 m. in a long irregular course. Its greatest depth is 600 ft. This lake, without rivalling those of Geneva or Lucerne in awful sublimity of scenery, is still one of the finest in Europe, being surrounded by a populous and well cultivated country, and the prospects on its banks being richly varied. It abounds in fish, and affords the means of extensive water-carriage. By the Linth it is connected with the lake of Wallenstadt.

ZURITE, a town of Peru, in the dep. and 30 m. NW of Cuzco.

ZURMATT, a village of Switzerland, cant. of Valais, and bail. of Visp, in the Matterthal. Pop. 369.

ZURMIE, or Zirmie, a town of Nigritia, in Houssa, on the r. bank of the Quarama or Kambarra, 100 m. ESE of Saccatu, in a fine plain. This town was visited by Clapperton in 1826.

ZURNAE, a fortress of Afghanistan, 44 m. SW of Ghuznee, on the road thence to Cabul, and at some distance from the r. bank of the Logurk.

ZUROMIN, a town of Russia in Europe, in Poland, in the gov. and 23 m. NNE of Plock, and obwod of Mlawa. Pop. 1,360.

ZUROW, a village of the grand-duchy of Mecklenburg-Schwerin, prov. of Mecklenburg, and bail. of Schwerin. Pop. 220. It has glass and tile-works.

ZUROW, or Zuroiona, a town of Galicia, in the circle and 18 m. E of Stry, at the confluence of the Swica and Dniester.

ZURRAH, Zereh, Zarang, Durrah, Hamun, Khashek, or Loukh, a lake of Afghanistan, in Seistan, near the Persian frontier. It is still little known, but is stated to be 105 m. in length from ENE to WSW, and 30 m. in breadth. It receives the Helmund at the E extremity, and the Furrah-Rud on the NE, but has no apparent outlet. Its waters are said by Ebn-Haukal to be sweet and abounding in fish, rising and falling periodically with the season, and bordered with reeds.: Elphinstone, on the other hand, pronounces the waters to be scarcely drinkable, and that they annually occasion extensive inundations. Near its centre is an island named Kohizor or Killahi, on which is Fort Rustam. This lake is the *Aria Palus* of ancient geographers.

ZURSACH, a village of Austria, in the country below the Ens, and upper circle of the Wienerwalde, near Gottweig. Garnets of fine quality are found in the environs.

ZURSZYM, a market-town of Austria, in Galicia, in the circle of Sanok.

ZURUKHAITNEWSK, a village of Russia in Asia, in the gov. of Irkutsk, SE of Nertchinsk, on the confines of China.

ZURUMA, or Zaruma, a town of Ecuador, in the dep. of Assuay, prov. and 30 m. NW of Loxa, in the midst of the Andes, at an alt. of 10,000 ft. above sea-level. Pop. 6,000. It was previous to the exhaustion of the rich gold mines in the vicinity extremely populous.

ZURUMILLO, a town of Peru, in the dep. and 105 m. E of Truxillo, near the r. bank of the Maranon.

ZURZACH, a market-town of Switzerland, capital of a circle in the cant. of Aargau, on the l. bank of the Rhine, 18 m. NE of Aarau. Pop. 700. Near the town are the ruins of the castle of Kussenberg, and the site of the *Forum Tiberii* of the Romans. Pop. of district 11,500.

ZUSAM, a river of Bavaria, in the circle of Swabia, which has its source to the SW of Augsburg; runs NNE; and, after a course of about 45 m., joins the Danube on the r. bank, opposite Donauworth.

ZUSAMALTHEIM, a village of Bavaria, in the circle of Swabia, and presidial of Wertingen, on the Zusam, to the NW of Augsburg. Pop. 490.

ZUSAMZELL, a village of Bavaria, in Swabia, in the presidial of Wertingen. Pop. 200.

ZUSCH, a village of Prussia, in the reg. and circle of Treves. Pop. 634.

ZUSCHEN, a village of Prussia, in the reg. of Arnsberg, and circle of Brilon. Pop. 440.—Also a town of Waldeck, in the district of the Eder. Pop. 799.—Also a town of the principality of Waldeck, and bail. of Eder, on the Elbe, 17 m. SW of Cassel. Pop. 1,000.

ZUSCHERING, a village of Bavaria, in the circle of Swabia and presidial of Neuburg. Pop. 630.

ZUSENHOFEN, a village of Baden, circle of the Middle Rhine, and bail. of Oberkirch. Pop. 615.

ZUSMARSHAUSEN, a presidial and town of Bavaria, in the circle of the Middle Rhine. Pop. 15,800. The town is on the Zusam, 15 m. WNW of Augsburg. Pop. 931. It has a castle and an hospital.

ZUSOW, a village of the grand-duchy of Mecklenburg-Schwerin, seignory of Wismar and bail. of Warin. Pop. 220.

ZUSSDORF, a village of Würtemberg, in the circle of the Danube, and bail. of Ravensburg, on the confines of Baden. Pop. 206.

ZUTKUERQUE, a commune of France, in the dep. of the Pas-de-Calais, and cant. of Audriucq, 12 m. NW of the Omer. Pop. 1,862.

ZUTPHEN, a town of Holland, capital of an arrondissement and canton of the same name, in the prov. of Gelderland, 9 m. S of Deventer, and 17 m. ENE of Arnheim, on the Yssel, by which it is divided into two parts, and at the confluence of the Berkel, in N lat. 52° 8′ 24″, and E long. 10° 11′ 39″. Pop. 10,000. It is well-built and fortified, and contains an ancient cathedral, or *groote-kerk*, Saint-Walburga, with a lofty tower, a town-house, and council buildings, with a peel of 36 bells, and a Latin school. It has paper, oil, tan, and fulling-mills and tanneries, and carries on an active trade in grain. The view of Z. across the river Yssel is striking: its bastions, bridges, and the great tower of Saint Walburga contrasting remarkably with the vast plains through which the Berkel and the Yssel flow to their point of confluence here. The environs are pleasant, and contain considerable quantities of salt. Z. is of great antiquity, and was formerly a Hanse town. In 1572, it was pillaged by the Spaniards under Frederick of Toledo, son of the Duke of Alba, and again in 1583 by the Duke of Parma. In 1591 it was taken by Prince Maurice of Nassau, and in 1672 was captured and dismantled by the French. —The arrond. comprises 14 cant. Pop. 80,000.

ZUTTAU, a village of Austria, in Moravia, circle of Hradisch and seignory of Napagedl. Pop. 580.

ZUTTLINGEN, a village of Würtemberg, circle of the Neckar, and bail. of Neckarsulm. Pop. 490.

ZUTZ. See Zuz.

ZUTZEN, a village of Prussia, in the reg. of Frankfort, and circle of Luckau. Pop. 380.—Also a village of the reg. of Potsdam, and circle of Angermunde. Pop. 220.

ZUTZER, a village of Prussia, in the reg. of Marienwerder, and circle of Deutschkrone. Pop. 450.

ZUTZKEN, a village of Prussia, in the reg. of Konigsberg, and circle of Gerdauen. Pop. 200.

ZUTZLAN, a village of Austria, in Bohemia, reg. of Gitschin, and bail. of Koniggratz. Pop. 340.

ZUURBERG, a district of South Africa, to the N of the co. of Somerset, bordering on Hottentotia. It is intersected by a range of mountains of the same name, forming a continuation of the Snowy mountains, and is divided from the district of Graaf Reynet by a river also called Zuurberg, an affluent of the Nu-Gariep.

ZUURBRAK, or Moses Kraal, an English Missionary establishment and village of South Africa, in the district and 15 m. ENE of Zwellendam, on the Buffeljagts river.

ZUUREVELD, a village of South Africa, in the district of Albany or Zuureveld, to the NE of Bathurst. See also Albany.

ZUVIA, a town of Spain, in the prov. and partido and 3 m. S of Granada, on the l. bank of the Genil. Pop. 3,095. It has a parish-church, a convent, a custom-house, and a public granary.

ZUYDER-ZEE. See Zuider-Zee.

ZUZ, or Zutz, a market-town of Switzerland, in the cant. of the Grisons, in the Haute-Engadine, on the l. bank of the Inn, and 27 m. SE of Coire. Pop. 423. Romani is chiefly spoken by the inhabitants.

ZUZELLA, a town of Prussia, in the reg. and circle of Oppeln. Pop. 460.

ZUZENHAUSEN, a village of the grand-duchy of Baden, in the circle of the Upper Rhine, bail. of Enisheim, on the Elseuz. Pop. 1,040.

ZUZGEN, a village of Switzerland, in the cant. of Aarau, and bail. of Rheinfelden. Pop. 775.

ZUZWEIL, a village of Switzerland, in the cant. of St. Gall, and bail. of Wyl. Pop. 1,127.

ZVENIGOROD, a town of Russia in Europe, in the gov. and 36 m. W of Moscow, on the l. bank of the Moskva, in a pleasant and picturesque situation. Pop. 1,000. It has a citadel, situated on a height, and three churches, and contains a salt magazine, a brandy store, and manufactories of silk-fabrics. In the vicinity is the convent of Savinskoi-Storojevskoi.

ZVENIGORODKA, a town of Russia in Europe, in the gov. and 102 m. SSE of Kiev, on the Guiloi-Tikitch. Pop. 800.

ZVERINOGOLOVSKAIA-KREPOST, a town and fort of Russia in Asia, in the gov. of Tobolsk and district of Kourgan, on the r. bank of the Tobol. Pop. 400. It has two churches.

ZVERLIK, or Ispervik, a fortress of Turkey in Europe, in Bulgaria, in the sanj. and 51 m. SW of Widdin.

ZVINGI. See Sving.

ZVORNIK, or Isvornik, a town of Turkey in Europe, in Bosnia, 39 m. SW of Czabatz, and 57 m. NE of Bosna-Serai, on the l. bank of the Drina. Pop. 15,000. It is fortified, and has two castles, several mosques, and Greek and Catholic churches. It possesses several manufactories, and carries on an active trade, especially in timber, with Belgrade and Semlin. In the vicinity are mines of argentiferous lead.

ZWAINDORF, a village of Austria, in Bohemia, in the reg. of Budweis, and bail. of Kaplitz. Pop. 280.

ZWALBACH, a village of Prussia, in the reg. of Ireoes, and circle of Merzig. Pop. 200.

ZWALLINNEN, a village of Prussia, reg. of Gumbinnen, and circle of Johannisberg. Pop. 200.

ZWALUWE, a village of Holland, in the prov. of North Brabant. Pop. 2,000.

ZWAMMERDAM, a parish of Holland, in the prov. of South Holland, on the Rhine. Pop. 1,120.

ZWANOWITZ, a village of Austria, in Bohemia, in the reg. of Pardubitz, and bail. of Schwarzkostelitz. Pop. 410.

ZWARTDOON, a river of South Africa, in the district of Clan-William, which has its source in the Kamiesbergen; directs its course first S; then bends W; and joining Green river, falls into the ocean in S lat. 30° 55′.

ZWARTE-BERG, [*i. e.* 'Black Mountain',] a mountain of South Africa, in the district of Swellendam, which it intersects from W to E.—Also a mountain in the N part of the district of Stellenbosch.

ZWARTE-BERG (GROOTE), [*i. e.* 'Great Black Mountain',] a range of mountains in South Africa, extending from W to E along the confines of the districts of Worcester and Zwellendam, Beauford, and George, and presenting a total length of about 270 m. They average from 2,500 ft. to 3,500 ft. in height, and contain mineral springs.

ZWARTEGAT, a commune of Belgium, prov. of W. Flanders, and dep. of Clercken. Pop. 100.

ZWARTELINJES, a river of South Africa, in the district of Clan-William, which has its source in the Kamiesbergen; runs W, and falls into the ocean in S lat. 29° 56′.

ZWARTENBROCK, a commune of Belgium, in the prov. of E. Flanders, and dep. of Hundelgen. Pop. 335.

ZWARTESLUIS, a fortified town of Holland, in the prov. of Ober-Yssel, and cant. of Hasselt, and 11 m. N of Zwolle, on the r. bank of the Zwarte-water, at the confluence of the Havelter-Aa, and at some distance from the Zuider-Zee. Pop. 2,500.

ZWARTE-WATER [*i. e.* 'Black Water',] a river of Holland, in the prov. of Ober-Yssel, and arrond. of Zwolle, which is formed, a little above the town of that name, by the confluence of four streams, which unite at Vecht, and which, after a course of about 15 m., in a generally NNW direction, falls into the Zuider-Zee, near Geneminden and Zwartesluis.

ZWARTINHOEK, a commune of Belgium, in the prov. of Limburg, and dep. of Guaed-Mechelen. Pop. 158.

ZWARTKLIP, or OORLOGSKLOOF, a river of South Africa, in the district of Clan-William, which has its source in the W side of the Roggeveld-Bergen; runs W, and after a course of about 120 m., joins the Great Doom or Thorn river, about 20 m. above the confluence of that river with Elephants or Olifants river.

ZWARTVELD, a commune of Belgium, prov. of W. Flanders, and dep. of Markegem. Pop. 100.

ZWATZEN, a village of Saxe-Weimar, in the prov. of Weimar, and bail. of Jena. Pop. 370.

ZWECKEL, a village of Prussia, in the reg. of Munster, and circle of Recklinghausen. Pop. 170.

ZWEEL, a parish of Holland, in the prov. of Drenthe, SW of Emmen. Pop. 800.

ZWEHRENBERG, a village of Würtemberg, in the circle of the Schwarzwald, and bail. of Calw. Pop. 279.

ZWEIBRODT, a village of Prussia, in the reg. and circle of Breslau. Pop. 300.

ZWEIBRUCKEN, or ZWEI-BRUGGEN, or DEUX-PONTS, a canton and town of Bavaria, in the circle of the Pfalz, pleasantly situated on the Erlbach, 52 m. W of Spire, and 66 m. SW of Mayence. Pop. 7,920. It is of small extent, but is well and regularly built. It has a fine castle, formerly the residence of the ducal family of Z., a Calvinist and two Lutheran churches, a gymnasium, and an orphans' asylum. It has manufactories of woollen fabrics, steel-ware, tobacco, gypsum, oil-mills, and tanneries. The printing press of this town, noted for its fine editions of the classics, was, during the French occupation, removed to Strasburg. In the environs are the remains of the chateau built by Stanislaus, king of Poland, and of the castle of Carlsberg belonging to the Duke de Deux-Ponts, and destroyed during the French Revolution.—The duchy of Zweibrucken was formerly a county-fief of the bishopric of Metz. Half of it was sold to Stephen, count-palatine of the Rhine, third son of the emperor Robert, and the other half passed to Philip V., count of Hanau, by his marriage with the daughter of James, last duke of Z. These two portions were reunited by Louis-le-Novi, the first duke, and the duchy was ever held by a descendant of the counts-palatine. It fell by succession to Charles XI., king of Sweden, but the king of France, desirous of reviving the rights of sovereignty on the duchy, took it in 1676, and held it till after the peace of Ryswick. Charles XII. leaving no heir, the duchy fell to Gustavus Emmanuel, and thence to John Casimir. The duke of Z. had a sitting and voice in the diets of the empire, when he took rank in the college of princes. His revenue was estimated at 1,375,000 francs. This duchy was repeatedly possessed by the French between 1792 and 1794, and in 1802 was ceded by the treaty of Luneville to France. In 1814 it was granted in exchange for other territory to Bavaria.

ZWEIBRUGGEN, a village of Prussia, in the regency of Aachen, and circle of Geilenkirchen. Pop. 200.

ZWEIDORF, a village of Mecklenburg-Schwerin, in the circle of Wendisch, and bail. of Boitzinburg. Pop. 240.—Also a village of the duchy and circle of Brunswick. Pop. 350.

ZWEIFALTEN, a town of Würtemberg, in the circle of the Danube, and bail. of Munsingen, on the Aach, 30 m. SW of Ulm. Pop. in 1840, 306, of whom 277 were Catholics. It has a fine convent, now used as a lunatic asylum, and in the vicinity the chapel of Notre-Dame-de-Loretto, a much frequented place of pilgrimage. This town was united to Wurtemberg in 1802.

ZWEIFEILE, a village of Prussia, in the regency of Aachen, and circle of Montjou, on the Vicht. Pop. 580. It has manufactories of cloth and of iron-ware.

ZWEIFLINGEN, a village of Würtemberg, in the circle of the Jaxt, bail. of Oehringen. Pop. 260.

ZWEIMEN, a village of Prussia, in the regency and circle of Merseburg. Pop. 230.

ZWEINAUNDORF, a village of Saxony, in the circle and bail. of Leipsig. Pop. 327.

ZWEINERT, a village of Prussia, in the reg. of Frankfort, and circle of Lebus. Pop. 200.

ZWEISIMMEN, a village and parish of Switzerland, in the cant. and 30 m. S of Berne, and bail. of Ober-Simmenthal, at the confluence of the Great and Little Simmen. Pop. 2,128.

ZWEIT-BAUENSCHAFT, a village of Prussia, in the regency of Dusseldorf, and circle of Geldern. Pop. 250.

ZWELLENDAM, a district and town of South Africa. The district is bounded on the N by the district of Worcester; on the E by that of George; on the S by the Atlantic; and on the W by the district of Stellenbosch. It is intersected from W to E by the Zwarte-Berg, and running along the N confines is the Groote Zwarte-Berg. Pop. 13,800. The principal rivers are the Breede, with its affluents the Zondereinde, Beffelgots, and Slange, which falls into St. Sebastian bay, and the Gauritz, which receives the waters of the Buffels and Taun in this district, and forms its line of separation from the district of George. St. Sebastian bay contains Beaufort port, formed by the embouchure of the Breede, and is enclosed on the W by Cape Infanta. Further WSW is Struys bay, and thence SW Cape Lagullas, the S extremity of Africa. Towards the W extremity of the district, and to the N of Danger Point, is Walker's bay. Its chief towns are Zwellendam and Caledon.—The town of Z. is situated to the S of the Zwarte-Berg, about 6 m. N of Breede river, and 120 m. E of Cape Town. Pop. 1,800.

ZWENKAU, a town of Saxony, in the circle and to the S of Leipsig, and bail. of Pegau, on the r. bank of the Flossgrabens. Pop. in 1843, 2,612. It has a suburb, and possesses manufactories of cloth and saltpetre.

ZWENTENDORF, a village of Austria, in the country below the Enns, and circle of the Oberweirwald, on the Danube. Pop. 550.—Also a village of the Lower circle of the Mannhartsberg, and bail. of Ernstbrunn, on the Zapâ. Pop. 360.

ZWERGEN, a village of Electoral Hesse, in the prov. of Lower Hesse, circle and bail. of Hofgeismar. Pop. 628.

ZWERINCK, a village of Austria, in Bohemia, in the regency of Gitschin, and bail. of Podiebrad. Pop. 200.

ZWERNDORF, a village of Austria, in the country below the Enns, Lower district of the Mannhartsberg, and bail. of Marchegg. Pop. 360.

ZWESTEN, a village of Electoral Hesse, in the prov. of Lower Hesse, circle of Fritzlar, and bail. of Jesberg, on the Wezelbach. Pop. 1,156.

ZWETHAU, a village of Prussia, in the regency of Merseburg, and circle of Torgau. Pop. 330.

ZWETKOFZEN, a village of Austria, in Styria, in the circle of Marburg. Pop. 300.

ZWETTEL, or ZWELL, a town of Austria, in the country below the Enns, and Upper circle of the Mannhartsberg, on the l. bank of the Great Kamp, at the confluence of the Zwettel, and 26 m. NW of Krems. Pop. 1,092. It has a Cistercian abbey, founded in 1138, with a valuable library. It possesses manufactories of woollen, linen, and cotton fabrics, ribbon, and several tanneries, and carries on an active trade in flax, hemp, wool, grain, and cattle.

ZWETTEL (KURZEN), a town of Austria, in the country above the Enns, regency of Linz, and circle of Muhl, on the Great Rottel, 11 m. W of Rottel. It has extensive manufactories of linen.

ZWEYBRUCKEN. See ZWEIBRUCKEN.

ZWICKAU, a kreis or circle, amt and town of Saxony. The circle comprises an area of 84.25 German sq. m. Pop. in 1852, 735,557. The town is on the l. bank of the Mulde, 20 m. WSW of Chemnitz, at an alt. of 850 ft. above sea-level. Pop. in 1837, 7,239; in 1840, 8,806; in 1852, 12,000. It was formerly fortified, and has several suburbs, four churches, a gymnasium, with a library of 16,000 volumes, an hospital, and two military stores, and a house of correction, formerly the castle of Osterstein. It possesses manufactories of woollen and cotton fabrics, hosiery, nails, sealing-wax, and cards, bleacheries, and dye-works, and has an important entrepot trade in salt. Coal is wrought in the environs.—Also a small town of Austria, in Bohemia, in the circle of Bunzlau, 28 m. NNW of Jung-Bunzlau, on the Bober. Pop. 3,800.

ZWIEFALLENDORF, a village of Würtemberg, in the circle of the Danube, and bail. of Riedlingen, on the Danube. Pop. 380. It has a castle.

ZWIELIPP, a village of Prussia, in the regency of Koslin, and circle of Furstenthum. Pop. 270.

ZWIERCYNICE, a village of Russia in Europe, in Poland, in the gov. of Lublin, and obwod of Zamos. It has a large brewery.

ZWIESEL, a market-town of Bavaria, in the circle of Lower Bavaria, and presidial of Regen, on the Schwarzen-Regen. Pop. 2,245. It has an hospital, and possesses manufactories of wooden-ware, and extensive breweries.

ZWIESELAN (OBER and UNTER), a village of Bavaria, in the circle of Lower Bavaria, and presidial of Regen. Pop. 324.

ZWIESELDORF, a village of Switzerland, in the cant. of Bern, and bail. of Thun. Pop. 237.

ZWIESIGKO, a village of Prussia, in the reg. of Merseburg, and circle of Schweinitz. Pop. 230.

ZWIESTON, a village of Austria, in Bohemia, in the regency of Budweis, and bail. of Wotitz. Pop. 440.

ZWIKAU, a market-town of Austria, in Bohemia, in the reg. and bail. of Bohmisch-Leipa. Pop. 4,380. It has manufactories of linen.

ZWIKOWETZ, a village of Austria, in Bohemia, in the regency of Prague, and bail. of Horzowitz. Pop. 480.

ZWIN, a small arm of the North sea, on the confines of the Belgian prov. of West Flanders, and the Dutch prov. of Zeeland, to the W of the island of Cassandria, and to the N of L'Ecluse. The canal of Bruges terminates here.

ZWINGEN, a village of Switzerland, in the cant. of Bern, and bail. of Laufen. Pop. 370.

ZWINGENBURG, a town of the grand-duchy of Hesse-Darmstadt, and prov. of Starkenburg, and bail. of Bensheim, 13 m. S of Darmstadt. Pop. 1,300.—Also a village of the grand-duchy of Baden, in the circle of the Lower Rhine, and bail. of Eberbach, on the r. bank of the Neckar. Pop. 380.

ZWINGENDORF, a village of Austria, in the country below the Enns, and Lower circle of the Mannhartsberg, on the Palka. Pop. 500.

ZWINOMASS, a village of Austria, in Bohemia, in the regency and circle of Pilsen. Pop. 240.

ZWINZEN, a village of Austria, in Bohemia, in the regency of Budweis, and bail. of Kruman. Pop. 400.

ZWIPPENDORF, a village of Prussia, in the regency of Frankfort, and circle of Sorau. Pop. 220.

ZWIRKOWITZ, a village of Austria, in Bohemia, regency of Budweis, and bail. of Moldauthein. Pop. 240.

ZWIRSCHEN, a village of Austria, in Bohemia, in the reg. of Pilsen, and bail. of Hostau. Pop. 220.

ZWISCHENAHN, a parish and village of the duchy and circle of Oldenburg, on the S bank of a small lake of the same name, and 11 m. WNW of Oldenburg. Pop. of bail., 2,953; of village, 280.

ZWISCHENWASSER, a village of Austria, in Tyrol, in the Vorarlberg, and presidial of Feldkirch. Pop. 780.

ZWISCHENWASSERN, formerly HOCKSTEIN, a village of Illyria, in the circle of Klagenfurt, at the confluence of the Olkza and Gulk. It has a fine castle belonging to the bishop of Gurk.

ZWITTAU, or ZWITTAWA, a town of Austria, in Moravia, in the circle and 39 m. WNW of Olmutz, near the source of a river of the same name, an affluent of the Schwarza. Pop. 3,699. It is enclosed by walls and ditches, is well-built, and contains three churches. It has manufactories of woollen and linen fabrics, and carries on an active trade also in flax and wool.

ZWITTAWKA, a market-town of Austria, in Moravia, in the circle of Brunn, on the Zwittau.

ZWITTE, a village of Austria, in Bohemia, in the reg. of Bohmisch-Leipa, and bail. of Haida. Pop. 280.

ZWOCHAU, a village of Prussia, in the reg. of Merseburg, and circle of Delitsch. Pop. 330.

ZWOISCHEN, a village of Austria, in Bohemia, in the reg. of Pilsen, and bail. of Vergreichenstein. Pop. 260.

ZWOLEN. See ALT-SOHL.

ZWOLENSKA-STOLICE. See SOHL.

ZWOLIN, a town of Russia in Europe, in Poland, in the gov. of Radom. Pop. 2,026.

ZWOLL, a village of Austria, in Bohemia, reg. of Gitschin, and bail. of Koniginhof. Pop. 450.

ZWOLLE, or ZWOL, an arrondissement, canton, and town of Holland, in the prov. of Over-Yssel. The arrond. comprises five cant. Pop. 50,000. The town is on the Vecht, at the confluence of the Soest, 54 m. ENE of Amsterdam, 9 m. from the Zuider-Zee, and at some distance from the r. bank of the Yssel, with which it is connected by a canal. Pop. 16,000. It is fortified, regularly built, and has several spacious streets, intersected with canals. Three gates connect the suburbs with the town properly so called. Z. possesses eight churches, inclusive of the ancient cathedral, and a house-of-correction, and contains manufactories of pins, linen, salt, soap, and sugar refineries, and bleacheries, a vinegar-work, and tanneries. Horses, cattle, butter, and leather are exported from it. In the vicinity is Mount St. Agnes, on which was formerly an Augustine convent, noted as numbering amongst its inmates the celebrated Thomas-a-Kempis.—Also a village of Austria, in Bohemia, in the reg. and bail. of Pilsen. Pop. 280.

ZWOLNO, a village of Prussia, in the reg. of Posen, and circle of Schrimm. Pop. 250.

ZWONITZ, a town of Saxony, in the circle of Zwickau, and bail. of Grunhain, on a river of the same name. Pop. in 1843, 2,012; in 1849, 2,400. It has manufactories of linen and cotton fabrics, blankets, and lace.

ZWONITZ (NIEDER), a village of Saxony, in the circle of Zwickau, and bail. of Grunhain, near the Zwonitz. Pop. 1,200. It has two churches, and possesses manufactories of cotton fabrics, paper, saw and oil-mills. In the vicinity are beds of porcelain clay, and two mineral springs.

ZWORNIK. See ZVORNIK.

ZWOTA, a village of Saxony, in the circle of Zwickau, and bail. of Voigtsberg, near Klingenthal. Pop. 1,089.

ZWOTZEN, a village of Reuss-Schleiz, in the prov. of Gera, on the Elster. Pop. 260.

ZWYCHLIN, a town of Poland, in the gov. and 63 m. W of Warsaw. It has several tanneries, and manufactories of gold-ware.

ZWYNAERDE, a commune of Belgium, in the prov. of E. Flanders, arrond. and 4 m. S of Ghent, on the Schelde. Pop. 1,802.

ZWYNDRECHT, a department and commune of Belgium, in the prov. of E. Flanders, and arrond. of Termonde, watered by the Schelde. Pop. of dep. 1,935; of com. 1,649.

ZWYNENBERG, a commune of Belgium, in the prov. of Brabant, and dep. of Lennick-Saint-Martin. Pop. 102.

ZYDACZOW, a town of Austria, in Galicia, in the circle and 15 m. NE of Stry, on a river of the same name, an affluent of the Dniester. Pop. 2,141, of whom 524 are Jews. It has three churches.

ZYDARKO, a village of Prussia, in the reg. of Bromberg, and circle of Gnesen. Pop. 200.

ZYDLOCHWITZE. See SCLOWITZ.

ZYDOWO, a town of Prussia, in the regency of Bromberg, and circle of Gnesen. Pop. 300.

ZYFEN, a village of Switzerland, in the canton of Basle, and presidial of Liestal. Pop. 910.

ZYFFLICH, a village of Prussia, in the regency of Dusseldorf, and circle of Cleves. Pop. 470.

ZYGHOR, or JAYGHUR, a port of Hindostan, in the presidency of Bombay, prov. and 174 m. W of Bejapur, on the Concan coast, at the mouth of a river of the same name, which forms a small bay. This port is well sheltered, but is difficult of access. It has a considerable trade in pepper, salt, and hemp.

ZYGLIN, a village of Prussia, in the regency of Oppeln, and circle of Beuthen. Pop. 290.

ZYGOS, a mountain-range on the frontier of Thessaly, connecting the northern and southern range of Pindus, and over which the road leads from Metzovo to Trikala. The ascent, of three hours, is steep and difficult. The view from the summit of the pass extends from the snowy top of Mount Olympus to the Olytzka and the Suli range. On the side of Albania the eye beholds rugged peaks covered with dark firs; on that of Thessaly it rests upon a succession of beautiful hills and plains, covered with forests of noble birch, oak, and plane-trees.

ZYLDEGHEM, a com. of Belgium, in the prov. of East Flanders, dep. of Cruyshautein. Pop. 1,090.

ZYMINO, a village of Prussia, in the regency of Posen, and circle of Schroda. Pop. 240.

ZYORY. See SORAU.

ZYPE, a canton of Holland, in the prov. of North Holland, and arrond. of Alkmaer, nearly equidistant from Helder and Alkmaer. Pop. 2,400. It is marshy, but affords good pasturage. It is noted for its sheep. Its chief town is Schagerburg.

ZYRANKA, a river of Russia in Asia, in the E part of the prov. of Jakutsk, an affluent of the Kolyma, which it joins to the N of Verkhne-Kolymski, and after a course in a generally ENE direction of 180 m.

ZYRIA, a mountain of Greece, in the N part of the Morea, near Trikala, a part of the *Cyllene* of the ancients. It has an alt. of 7,605 ft. above sea-level.

ZYRIANES. See SIRIANES.

ZYROWA, a village of Prussia, in the regency of Oppeln, and circle of Gross Strelitz. Pop. 380.

ZYSKALENFIELD, a mountain of Norway, on the confines of the dioceses of Nordland and Drontheim. It is remarkable for the pyramidal rocks of Torgehatten, which rise to the height of 3,000 ft.

ZYTNO, a town of Poland, in the woiwodie of Kalisch, obwod and 12 m. S of Petrikau. Pop. 260.

ZYTOMIERZ. See JITOMIR.

ZYWIEC, or SAYBUSZ. See SEYPUSCH.

ZYWIELSK, a town of Russia in Europe, in the gov. of Kasan, on a river of the same name. Pop. 1,300

ZYWODCZYTZ, a village of Prussia, in the regency and circle of Oppeln. Pop. 330.

ZZANG-TSEOU. See KALDYAO-MURAN.

ZZUBIN. See SCHUBIN.

APPENDIX.

A LIST OF GEOGRAPHICAL NAMES

SHOWING THE

ANCIENT, MEDIÆVAL, AND MODERN DESIGNATIONS BORNE BY THE SAME PLACE.

*** It will be observed from the following List that the Latinized forms of modern geographical nomenclature, as well as the modern forms of ancient names, have been chiefly effected by the substitution of the following terminations for each other.

MODERN.	ANCIENT.	MODERN.	ANCIENT.	MODERN.	ANCIENT.
ach or *ack*	*acum* or *achium*	*ent*	*entium*	*land*	*landia*
ad	*adum*	*entre*	*inter*	*leben*	*laba* or *leba*
agne	*ania*	*er*	*era*	*mate*	*mates*
ailles	*alia*	*es* or *eis*	*ai*	*mold*	*moldia*
ain or *eine*	*ania* or *anium*	*euil*	*olium*	*mond, mons,* or *mont*	*montium*
al	*alium*	*ey*	*ia* or *eja*	*mouth*	*muthum*
alta	*altus*	*feld*	*felda*	*munde*	*munda*
am	*amum*	*fels*	*felsa*	*o*	*um*
an	*ania* or *anum*	*ford*	*fordia*	*oglio*	*oleum*
ant, anz	*antia*	*frith*	*fretum*	*ogne*	*onia*
ar	*aria* or *ara*	*furt*	*furtum*	*oise*	*osia*
atz	*atium*	*gart, garten, gorod, grod*	*gardia*	*oping*	*opia*
at	*atum*	*gau*	*govia*	*oux*	*osum*
au	*avia* or *ovia*	*gen*	*gia* or *ga*	*ow*	*ovia*
aux	*atium*	*hausen* or *husen*	*husium*	*pel* or *poli*	*polis*
berg	*berga*	*hava* or *haff*	*hama*	*por*	*pons*
born	*borna*	*haven*	*havia*	*que*	*ça*
burg	*burgum*	*heim*	*hemium*	*sand*	*sanda*
c, n, r, t	*iacum*	*hofen*	*hofa, horra,* or *hovium*	*stadt*	*stadium*
caster	*castrum*	*holm*	*holmia*	*stein*	*steinium*
cester	*cestria*	*holz*	*holtia*	*stock*	*stockium*
citta	*civitas*	*horst*	*horstium*	*thal*	*dalia* or *thalia*
dorf	*dorfium*	*hut*	*hutum*	*vic, wy,* or *wyk*	*vicus*
dorp	*dorpium*	*ich, ig, ik*	*icum*	*ville*	*villa*
dun	*dunum*	*ie*	*ia*	*wegen*	*vegia*
e	*a* or *as*	*igno*	*inium*	*weiss* or *weissen*	*albis*
eck	*ecca*	*ili*	*olia*	*worth*	*verda*
eglia	*ella*	*im*	*imum*	*s*	*ium*
eil or *eille*	*elia*	*in*	*inum*	*zell*	*cella*
ein	*inum*	*itz* or *itsch*	*icium* or *itium*	*zza*	*ssa*
em	*emum*	*kerchen*	*querca*		
en	*ena*				
ence, en, enz	*entia*				

AA, or Œ, a river of Holland, the AA of Overyssel; also the AADE of North Brabant.

AAMANTIA, or AMANTIA, a fort on the coast of Illyria, PORTO-RAGUSIO.

AARA, ABRINCA, ABRICCA, or OBRINGA, the river AAR, AER, or AHR of Rhenish-Prussia.

AARDALINUS SINUS, the Norwegian bay of BUCKER-FORD.

AARHUSIA, or ARHUSIA, the Danish town of AARHUS or ARHUS.

AARHUSIUS, or ARHUSIUS-COMITATUS, the Danish province of AARHUS or ARHUSEN.

AASA, the Norwegian port of AAS.

ABA, or ABUS, a mountain in Armenia, by some identified with the modern ARARAT.

ABÆ, a city of Greece, identified with ruins near the village of EXARCHO.

ABACENA, or ABACÆNUM, a town of Sicily, the SANCTA-BIGENIS in the Val-di-Demona.

ABACUM, the town of ALBACH or ABACH in Lower Bavaria.

ABALLABA, or ABALABA, the town of APPLEBY in Westmoreland.

ABALLO, or ABALLIUM, the town of SAINT-AVALON in the French dep. of Yonne.

ABALUS, or RANTOMANNIA, an island in the Frisch-haff of Prussia.

ABANA, AMANA, BARDINES, or CHRYSORRHOSA, river of Syria, the BARADA.

ABANTIAS, an island off the coast of Greece, EUBŒA, or NEGROPONTE.

ABANTONIUM, the modern ST. AUBENTON in the French dep. of Aisne.

ABASCIA, the territory of ABASSIA on the coast of the Black sea.

ABASINIA, the territory of ABYSSINIA or ABESSINIA.

ABATOS, an island in the Nile, generally identified with BIGGEH near Phila.

ABBATIA, the modern SANCTA-BADIA in the Polesina of Lombardy.

ABBATIA ABUNDANTIÆ, the abbey of NOTRE-DAME-D'ABONDANCE in Savoy.

ABBATIA MARIÆ LUCEDIÆ, the abbey of LUCEDIO in Montferrat.

ABBATIA MARIÆ DE ROSIS, the abbey of ROOSEN, near St. Alost in West Flanders.

ABBATIA BINTENSIS, or HORTUS FLORIDUS, the abbey of BAINDT in Würtemberg.

ABBATIA CÆSARIENSIS, the abbey of KAISERSHEIM in Bavaria.

ABBATIA LORCENSIS, the town and abbey of LOCKUM in Hanover.

ABBATIA ROSACIS, or ABBATIA SANCTI JACOBI, the abbey of ST. JAKOB in Illyria.

ABBATIA SORETHANA, or SORETHIUM, or SORETUM, the cloister of SCHUSSENRIED in Würtemberg.

ABBATIS CELLA, or ABBATUM CELLA, the town of APPENZELL in Switzerland.

ABBATIS VILLA, or ABBAVILLA, or ABBATICO VILLA, the town of ABBEVILLE in France.

ABBEFORTIA, the town of ABBEFORT in Norway.

ABBENDONIA, the town of ABINGDON in Berkshire.

ABDERA, or ABDARA, or AUDERA, the town of ADRA in Andalusia, on the Mediterranean.

ABDERA, a once opulent city of Thrace, the Polystylus of the middle ages, the ruins of which are said to exist near Cape Balustra in Romelia.

ABDUA, the town of LODI, on the Adda in Austrian Italy.

ABDUA, or ADDUA, the ADDA, a river of Italy, falling into the Po near Cremona.

ABELLA, or AVELLA, the AVELLA-VECCHIA of the Neapolitan prov. of Terra-di-Lavoro.

ABELLABA, or ABALLABA, a station on Hadrian's wall in Northumberland, identified with WATCHCROSS.

ABELLINUM, a city of the Hirpini, represented by the modern ABELLINO or AVELLINO in Naples.

ABELLINUM-MARSICUM, a town of Lucania, corresponding to MARSICO-VETERE in the Neapolitan prov. of Basilicato.

ABENSPERGA, ABUSINA, AVENTINUM, or ARUSENA, the town of ABENSBERG in Bavaria.

ABERAVONIUM, ABERAVON in Glamorganshire.

ABERCONVONIUM, ABERCONWAY in Carnarvonshire.

ABERDONA, ABERDONIA, or ABEDONIUM, the city of ABERDEEN in Scotland.

ABERDONIA-VETUS, or DEVONA, the town of OLD ABERDEEN.

ABERDONIA-NOVA, or ABERDEA, the town of NEW ABERDEEN.

ABERFRAVIA, or GADIVA, ABERFRAW in Anglesea.

ABERGONIUM, or GOBANIUM, ABERGAVENNY in Monmouthshire.

ABERISTIVUM, ABERYSTWITH in Cardiganshire.

ABERNETHACA, or ABRONIA, the village of ABERNETHY in Forfarshire.

ABESTA, the port of BAST in the Persian prov. of Segistan.

ABEXIA, the modern ABEX, or HABESH, stretching along the coast of the Red sea.

ABIA, the river ABRUST in Upper Bavaria.

ABILA, or ABILLA, a town of Cœle-Syria, the principal city of Abilēne, now identified with the village of SUK on the Barada.

ABINDONIA, or ABINTONIA, the town of ABINGDON in Berkshire.

ABLONIUM, the village of ABLON on the Seine in France.

ABNOBA MONS, a name given by Pliny and by Tacitus to a mountain on the SE side of the Schwarzwald, or Black Forest, in which the Danube takes its rise.

ABOA, the town of ABO in Finland.

ABOBRICA, or AOBRIGA, a town of Spain, the modern BAYONA on the coast of Galicia.

ABADIACUM, ABADIACON, or ABUDIACUM DANUBIANUM, the modern town of ABACH or ABBACH in Bavaria.

ABOLUS, the modern CANTERA in Sicily between Catania and Syracuse.

ABONITICHOS, ABONI-CASTRUM, or IONOPOLIS, the modern INOBOLI or INEBOLI on the coast of the Black sea.

ABONA ÆSTUARIUM, the DORNOCH FRITH on the N coast of Scotland.

ABONE, or AD ABONAM, the present BITTON in Gloucestershire.

ABORAS, ABORRAS, or CHABORAS, a river of Mesopotamia, the modern KHABUR, an affluent of the Euphrates, called by Xenophon the ARAXES.

ABOTIS, a town of the Thebaid in Egypt, represented by the modern ABUTIGE.

ABOXIS, or ABOCIMPOLIS, the hamlet of IPSAMBUL or ABUSAMBUL, on the W bank of the Nile.

ABRAVANNUS SINUS, LUCE BAY on the coast of Galloway.

ABRINCA. See AARA.

ABRINCÆ, the French town of AVRANCHES in the dep. of Manche.

ABRONTIUM, the modern ABRANTES on the coast of the Portuguese prov. of Estremadura.

ABROTONUM, an ancient port on the Syrtic coast of N. Africa, identified by Mannert with Sabrata, perhaps the modern TRIPOLI-VECCHIO.

ABSORUS, or AUSORIENSIS-INSULA, the principal island in the group of the Absyrtides, the modern OSERO in the Adriatic.

ABSYRTIDES-INSULÆ, or APSYRTIDES, a group of islands in the Adriatic, off the coast of Illyria, in the Sinus Flanaticus, off the Promontorium Polaticum. They are four in number, and known in modern geography as CHERSO, OSERO, FEROSINA, and CHAO.

ABULA, AVILA in the Spanish prov. of Old Castile.

ABUS, the river HUMBER in England. The name ABUS ÆSTUARIUM is more peculiarly appropriate to the Humber, which is, strictly speaking, the estuary of the united streams of the Trent and the Ouse; while the Ouse itself is designated ABUS FLUVIUS.

ABUSINA. See ABENSPERGA.

ABYDOS, a city of Upper Egypt, now represented by ARABAT-EL-MATFUN. See article ABYDOS.

ABYDUS, ABYDUM, or ABYDONA URBS, a once rich and flourishing town of Asia Minor, on the shore of the Hellespont, on the point of land now known as Cape Nagara. See article DARDANELLES.

ABYLA, or ABYLA-COLUMNA, a mountain on the N coast of Africa, one of the pillars of Hercules, the extreme point of the SIERRA-DE-LAS-MONAS. See article ABYLA.

ACA, ACE, ACCO, PTOLEMAIS, COLONIA PTOLEMAIS, a coast town of Syria, the modern ACRE or ST.-JEAN-D'ACRE.

ACALANDRA, a town of Lucania, now represented by the town of SALANDRA in the Neapolitan prov. of Basilicata.

ACALANDRUS, a river of Italy, the SALANDRELLA, falling into the gulf of Tarentium.

ACAMUS PROMONTORIUM, Cape Amant or Salizano on the NW extremity of the island of Cyprus.

ACANTHUS-ÆGYPTIA, a town of the Egyptian Heptanomis, to the S of Memphis, now Dashur.

ACANTHUS MACEDONICA, or Acanthos, a coast town on the Sinus Singiticus in Macedonia, near the modern village of Cheriasa, or Hierisos.

ACAPULCUM, or Portus Aquæ Pulchræ, the port of Acapulco in Mexico.

ACARNANIA, a region of Central Greece, mainly represented by the modern Karlelia or Karnia. The ancient region was bounded by the Sinus Ambrasius, or gulf of Arta, on the N; on the E by the river Achelous; on the SW and W by the Mare Ionium, or Ionian sea. See article Acarnania.

ACAUNUM, Agaunum, or Mauritianum, a town of Gallia Narbonensis, the modern St. Maurice or Moritz, in the Swiss cant. of Valais.

ACCADIA, the peninsula of Acadie, on the frontier of Canada.

ACCI, Colonia, Accitana Gemellensis, in the vicinity of the town of Guadix-el-viejo in the Spanish prov. of Granada.

ACCIPITRUM INSULÆ, Flandricæ, or Terceres Insulæ, the modern group of the Azores.

ACCO. See Aca.

ACCELLIUM, or Aveglio, represented by Arona on the Lago Maggiore, in Sardinia.

ACELUM, or Acilium, a town of Italy, represented by Asola in the Lombardian prov. of Treviso.

ACEMUM, or Achemum, the modern Acheen in Sumatra.

ACCERRÆ, a town of the Insubres in Gallia Cisalpina, now represented by Gherra.—Also a town of Campania, near the sources of the Clanius, the modern Acerra in Naples.

ACESINES, a tributary of the Indus, the modern Chenab.—Also a river of Sicily, the modern Cantera.

ACESTA, Ægesta, Egesta, Segesta, a town of Sicily, in the vicinity of the modern Castell-a-Mare-di-Golfo.

ACHADA, Achata, or Achonrita, the modern Achonree in co. Sligo.

ACHÆORUM PORTUS or Achæon, a haven on the coast of the Troad, at the mouth of the Xanthus or Ergentchai.

ACHAIA, a district of Greece, inhabited by a Pelasgic race, who became afterwards blended with an Ionian colony, when the name of the country was changed from Ægialus to Ionia. About B. C. 1104, a body of Achæi, driven before the Dorians, expelled the Ionians, and gave their name to the country. The Roman prov. of Achaia, after B. C. 146, comprehended all the Peloponnesus and all Northern Greece S of Thessaly. Achaia, including Sicyonia, was bounded on the N by the Sinus Corinthiacus, or gulf of Corinth; on the E by Corinthia; on the S by Arcadia and Elis; and on the W by the Mare Ionium or Ionian sea. See article Achaia.

ACHARNÆ, a town or village of Attica, the ruins of which are near the modern Menidi, on the route from Athens to Phyle, by Khassia.

ACHATES, a river of Sicily, the modern Drillo of the Val-di-Noto.

ACHATIUS, the Echitz of Würtemberg, an affluent of the Neckar.

ACHELOUS, a river of Central Greece, the modern Aspropotamos, or Aspropotamo.

ACHERON, a river of Epirus, the modern Suli, descending from the mountains to the W of Pindus, and falling into the Ionian sea near Parga. See article Acheron.—Also a river of Magna Græcia, in the Roman prov. of Bruttium, now the Maresanto or Arconti, which unites with the Crathis near Cosenza in Naples.—Also a river of the Peloponnesus, an affluent of the Alpheus, now known as the Savuto.

ACHERONTIA, a town of Apulia Peucetia or Daunia, now Acerenza in Naples.

ACHERONTINA PROVINCIA, or Lucania, a district of Magna Græcia, now represented by the Neapolitan provs. of Basilicata and Principato-Citra.

ACHERSUNDIA, the Norwegian island of Ackersund.

ACHERUSIA PALUS, a marsh through which the river Acheron Epiroticus flows in the early part of its source.—Also a lake of Campania, by some identified with the Lacus Lucrinus; by others with the Lacus Avernus. Probably the modern Lago Fusaro.

ACHILA, or Achilla, a town of Byzacium or Africa Propria, near Elalia.

ACHILLEA, or Achillis Insula, or Leuce, an island in the Black sea, the modern Fidonisi or Ylan Adassi, 24 m. E by N of the Sulina mouth of the Danube.

ACHMETA. See Ecbatana.

ACHRIS, or Lychnidus, a town of Macedonia, the modern Ochrida or Ghiustendil.

ACHSIB, or Ecdippa, a town of Upper Galilee, represented perhaps by the modern Dsib or Zib between Acre and Tyre.

ACHYRUM, the town of Achtyrka, in the Roman gov. of Charkof.

ACILIA AUGUSTA, the modern Straubingen in Bavaria.

ACILIO, a town in Gallia Aquitanica, represented by Aiguillon, in the French dep. of Lot-et-Garonne.

ACIMINCUM, or Acumincum, a town of Pannonia Inferior, near the modern Peterwardein in Hungary, on the Danube.

ACINCUM, Acincus, or Aquincum, a town of Pannonia Inferior, now Alt-Ofen, on the Danube.

ACINIPO, or Acinippo, a town of Hispania Bœtica, represented by Ronda-la-Vieja, on the Sierra-de-Ronda.

ACIRIS, a river of Græcia Magna, the Agri in the Neapolitan territory, falling into the gulf of Tarentum.

ACIS, a river of Sicily, the modern Jaci, falling into the gulf of St. Thecla.—Also one of the Cyclades, now Serpho.

ACONA, or Aquæ Saxonicæ, the modern St. Achen, on the Elbe, in the Prussian gov. of Magdeburg.

ACRAGAS, or Agrigentum, the modern Girgenti, in Sicily.

ACRICONIUM, the town of Hereford in Herefordshire.

ACRITAS PROMONTORIUM, a cape on the S coast of Messenia, now Cape Gallo.

ACROATHON, Cape Monte-Santo, on the W coast of Asiatic Turkey.

ACROCERAUNIA, or Acroceraunii Montes, a range of mountains in Epirus, the Monti-della-Chimera, or Kimaria of Upper Albania.

ACROCERAUNIUM PROMONTORIUM, now Cape Linguetta.

ACROLISSUS, a fortress in Dalmatia, now Alessio, near the mouth of the Drin.

ACRONIUS LACUS, the Uberlinger arm of the lake of Constanz.

ACROPOLIS, the modern Agropoli, in the Neapolitan prov. of Principato-Citra.

ACSTEDA, the village of ACKSTEDT, a little to the N of Bremen.
ACTANIA, the island of SCHELLING, in the German ocean, off the coast of Friesland.
ACTE, the peninsula of MOUNT ATHOS in Asiatic Turkey.
ACTIUM, a point of land at the entrance of the Sinus Ambrasius now the bay of Previsa, on the coast of Greece, by some identified with LA PUNTA, by others with AZIS.
ACTIUM PROMONTORIUM, CAPO-D'ACCIAGUALA, on the W side of Sardinia.
ACULA, or AQUÆ TARINÆ, the town of ACQUAPENDENTE in the Papal states.
ACUSIO, the village of ST. ANCONA, on the Rhone, in the French prov. of Dauphiny.
ACUTUS, the river AGOUT, in the French dep. of Tarn-et-Garonne.
ACUUM CAPUT, CAPE AGULHAS, the S point of Africa.
ADAMANTIA, or AMANTEA, the town of SANTA AMANTEA in Calabria-Citra.
ADAMAS, a river of Macsolia, probably the BRAMNI, an affluent of the Ganges.
ADANA, the modern ADEN.
AD AQUAS, or BELGA UVELLA, the modern WELLS in Somersetshire.
AD AQUAS, the modern ST. SCIACCA, near Girgenti in Sicily.
AD AQUAS HELVETICAS, the modern BADEN in the Swiss cant. of Aargau.
AD AQUILAS, the village of EIGELL or ST. AIGLE in the Swiss cant. of the Valais.
ADATA, a town of Commagene, in Syria, now ARAH.
ADAX, or ATTAGUS, a river of Gallia Narbonensis, the AUDE of Carcassone in France.
AD CALLENO, or AD CALE, a town of Umbria, the modern ST. CAGLI in the Papal deleg. of Fermo.
ADDUA, or ADDUAS, the river AIN, an affluent of the Rhone.
AD DURIUM, a town of the British tribe of the Damnonii, probably the modern TOTNESS.
ADEBA, the town of AMPOSTA in Spain.
ADELLUM, a town of Hispania Tarraconensis, the modern ST. ELDA in Valentia.
ADELUM REGNUM, the territory of ADEL, on the E coast of Africa.
ADENUM PROMONTORIUM, CAPE ADEN.
ADERBORNA, ADERBORN in Pomerania.
AD FAVARIAS, or FAVARIUM, the town of PFEFFER in the Swiss cant. of St. Gall.
AD FINES, the town of PFIN in the Swiss cant. of Zurich.
AD FLEXUM, ANTIQUUMBURGUM, or OVARIA, a town in Pannonia Superior, the modern ALTENBURG or OBER of Hungary.—Also a town of Gallia Cisalpina, the modern RIVOLTELLA.
AD HORREA, or HORREÆ CELLÆ, a town of Byzacium, the modern ERCLEA.—Also the modern CANNES in the French dep. of Var.
ADILBEGIANA, or ADERBEIZANA, the Persian prov. of AZERDBIJAN.
ADISATHRUS, the GHAUTS of peninsular Hindostan.
ADJACII SINUS, the gulf of AJACCIO on the W coast of Corsica.—Also the gulf of ALEXANDRETTA or SCANDERUN, on the coast of Syria.
ADJACIUM, Issuz, or NICOPOLIS, the town of AJAZ or AYAS, on the coast of Asiatic Turkey.
AD LACCUM the town of LACHEN on Lake Zurich, in the Swiss cant. of Schwytz.
ADMIRALES INSULÆ, the SEYCHELLES, off the coast of Zanguebar.—Also the ADMIRALTY GROUP in the S. Pacific.
AD MONILIA, the town of MONEGLIA in Genoa.
AD MONTES, or ADMONTIUM, the town of ADMONT on the Enns in Styria.
AD NOVAS, the town of RUTSCHUK in Bulgaria.
ADONCUM, ADONCO on the lake of Como.
ADORFIUM, ST. ADORF on the Elster in Saxony.
AD PONTEM, a town of the Coritani, SOUTHWELL on the Trent.
AD PONTEM ISIS, or IBISSA, the modern ST. IPS or YBBS, a village at the influx of the Ips into the Danube in Austria.
AD PONTEM MURI, the village of ST. MURAU on the Muhr in Styria.
ADRAMYTTENUS SINUS, the GULF OF ADRAMYTI or ADRAMITI, on the W coast of Asiatic Turkey.
ADRAMYTTIUM, a city at the head of the gulf of Adramiti, represented by the modern ADRAMYTI.
ADRIANA, or ADRANUS, the EDDER or Edda, a branch of the Fulda in Cassel.
ADRANUM, ADERNO in Sicily.
ADRIA, or HADRIA, the modern ADRIA, in the Venetian prov. of Polesina.—Also ALTRI, in the Neapolitan prov. of Abruzzo-Ulteriore 1ma.
ADRIA, ADRIA SINUS, ADRIATICUS SINUS, ADRIACUS PONTUS, MARE HADRIANUM, MARE HADRIATICUM, MARE SATURNIUM, or MARE SUPERUM, that eastern arm of the Mediterranean which extends between Italy and the shores of Illyricum, Epirus, and Greece, comprising the modern GULF OF VENICE, and the IONIAN SEA or ADRIATIC.
ADRIÆ SCOPULUS, the island of PELAGOSA in the gulf of Venice.
ADRIANOPOLIS, or HADRIANOPOLIS, the city of ADRIANOPLE in Turkey.
ADROBICA, the town of CORUNNA on the N coast of Spain.
ADRUMETUM, ADRUMETTUS, ADRYMETUS, HERACLEA or HADRUMETUM, the modern HAMAMET in Tunis, or perhaps HERKLA.
AD SABRINAM, PORTISHEAD in Somersetshire.
ADULA, ALPES SUMMÆ, ARULA, the highest summit of the Alps, in the NW of Italy, on the borders of Helvetia, applied to the FURKA ST. GOTHARD and LUCMANIERBERG.
ADULE, or ADULES, the port of Axume, on the Sinus Arabicus, supposed by D'Anville to be the modern ARKIKO, but more probably ZULLA.
ADURA, or ADURUM, or CIVITAS ATMENSIUM, represented by AIRE on the Lys, in the French dep. of Pas-de-Calais.
ADVERSA, AVERSE in Naples.
ADVICENAS, VINCENNES in France.
AD VIGESIMUM, CASTLE FLEMISH in South Wales.
ADZELIA, ADZEL or ABZEL in Livonia.
ÆA, a town of Colchis, perhaps MALE on the Phasis.
ÆANTIUM PROMONTORIUM, CAPE TRIKERI or VOLO on the coast of Greece.
ÆBUDÆ, EBUDÆ, HEBRIDÆ, or HEBRIDES OCCIDENTALES, the HEBRIDES of Scotland.
ÆCLANUM, or ÆCULANUM, a city of the Hirpini, perhaps FRICENTI on the Tripalta, in the Neapolitan prov. of Principato-Ultra.
ÆDEPSUS, DIPSO on the coast of Negroponte.
ÆGÆ, or EDESSA, the capital of Emathia, perhaps the modern VODINA in Northern Greece.—Also a place in Eubœa celebrated for the worship of Neptune, now represented by AKIO on the coast of Negroponte.
ÆGÆUM MARE. See MARE ÆGÆUM.
ÆGATES INSULÆ, the islets of TAVAGNANA, MARETIMO, and LEVENZA, off the W coast of Sicily.

ÆGESTA, a town of Sicily, whose ruins exist near the modern ALCAMO.

ÆGIDA, or JUSTINOPOLIS, a city of Histria, now CAPO-D'ISTRIA, at the mouth of the Fornico in Illyria.

ÆGILIUM, or IGELIUM, the island of GIGHO off the coast of Tuscany.

ÆGILON, or CAPRARIA, the island of CAPRAIA off the coast of Tuscany.

ÆGINA, ÆNONE, ÆNOPIA, MYRMIDONIA, the island of EGINA or ENGHIA off the coast of Greece.

ÆGINIUM, a town of Thessaly, the KALABACHI of the Turks according to Leake; but supposed by Cramer to be represented by the modern MOCOSSI.

ÆGIRA, a town of Achaia, the ruins of which still exist near BLUBUKI to the NW of Trikala.

ÆGIRCIUS, the town of GERS in the French dep. of Hautes-Pyrenees.

ÆGISSUS, or ÆGYPSUS, a town of Mœsia Inferior, the modern TATZA near the mouth of the Danube.

ÆGITNA, a town of CANNES in the French dep. of Var.

ÆGIUM, a town of Achaia, the modern VOSTIZZA.

ÆGLESBURGUS, AILESBURY in Buckinghamshire.

ÆGOS FLUMEN, or ÆGOSPOTAMOS, or CAPRÆ-FLUMINA, a river and town of the Chersonese, the port of which is perhaps represented by the modern GALATA.

ÆGUSÆ. See ÆGATES.

ÆGYPTI FLUVIUS, the modern WADY-EL-ARISH on the S frontier of Palestine, meeting the sea at Rhinocolura.

ÆGYPTUS, MIZRAIM, RAHAB, or CHAMI, the ancient EGYPT, bounded on the N by the Mediterranean; on the E by Arabia Petræa and Palestine; on the S by Æthiopia; and on the W by the Libyan desert.

ÆGYPTUS INFERIOR, the DELTA of Egypt.

ÆGYPTUS MEDIA, or HEPTANOMIS, a name originally applied to that part of Egypt extending from Abydos to the fork of the Delta at Cercasorum. Under the Eastern empire it took the name of ARCADIA.

ÆGYPTUS SUPERIOR, THEBAIS, or PATHROS, that portion of Egypt which extended from Abydos to the N frontier of Æthiopia.

ÆJOPOLIS, the modern HILLAH on the Euphrates.

ÆLANA, AILA, AILAS, ELANA, or ELATH, at the head of the Sinus Ælanītes, the modern AKABA.

ÆLBURGUM, or ALBURGUM, ALBORG in Jutland.

ÆLI, HALI in the Arabian prov. of Yemen.

ÆMILIANUM, or MILLIADUM, ST. MILHAUD in the French dep. of Tarn.

ÆMINIUM, AGUEDA in the Portuguese prov. of Beira.

ÆMODÆ, EMODÆ, EMUDÆ, EMUDES, ATMODES, or ACMODÆ, a group of islands in the Oceanus Deucaledonius, probably the SHETLAND group to the N of Scotland.

ÆMONA, EMANA, EMA, or HÆMA, the town of LAYBACH in Illyria.

ÆMONIA NOVA, CITTA NUOVA in the circle of Trieste on the Adriatic.

ÆNARIA, INARIMA, or PITHECUSA, an island off the coast of Campania, the modern ISCHIA.

ÆNEA, a town on the E shore of the Sinus Thermaicus, near CAPE PANOMI, on the coast of Macedonia.

ÆNI PONS, AD ENUM, or ÆNIPONS, the modern INSBRUCK in Tyrol.

AENNUM, PHILOTERA, PHILOTERÆ PORTUS, or PHILOTERIS, the modern COSSEIR on the Red sea.

ÆNONA, NONA on the gulf of Venice.

ÆNOS, ÆNUS, or ABSYNTHUM, a town of Thracia, at the mouth of the Stentoris Palus, the modern ENOS.

ÆNUS, or ŒNUS, INN in Tyrol.

ÆOLIÆ INSULÆ, ÆOLIDES, VULCANIÆ HEPHÆSTIADES, LIPAREÆ, LIPAREORUM INSULÆ, or PLATÆ, a group of islands off the N coast of Sicily, embracing Lipăra, Osteodes, Heraclea, Didyme, Phœnicusa, Hiera, and Strangile, now known as the LIPARI group.

ÆONA, ÆONS, or MYSIA, a district of Asia Minor, settled by colonies from Greece, about the year B.C. 1124; in the largest sense of the term embracing the W and a portion of the NW coast of Asiatic Turkey.

ÆPEA, or ÆPEIA, a town of Messenia, the modern PETALIDI, about 15 m. N of Coron.—Also a town on the coast of Cyprus, near the modern SOLIA.

ÆQUINOCTIUM, the town of FISCHAMUND at the influx of the Fischa into the Danube.

ÆQUOLESIMÆ, or ENGULISMA, a town in Gallia Aquitania, near the modern ANGOULEME in the French dep. of Charente.

ÆQUOR TUSCUM, the TUSCAN sea.

ÆRA, or ÆREA, the town of AYR in Ayrshire.

ÆREUS, the river AYR in Ayrshire.

AERIA, VAISON in the French dep. of Vaucluse.

AERIA, AETHRIA, CHRYSA, THASUS, or THASSUS, an island of the Chersonesus, the modern TASSO.

ÆSEPUS, a river of Mysia Minor, the modern SATAS-DERE in the Troad.

ÆSERNIA, a city of Samnium, now ISERNIA in the Neapolitan prov. of Sannio.

ÆSIA, or ŒSIA, the river OISE of France.

ÆSICA, the SOUTHERN ESK of Forfarshire.—Also a station on Hadrian's Wall, represented by GREAT CHESTERS in Northumberland.

ÆSTHONIA, or ÆSTIA, the town of REVEL on the gulf of Finland.

ÆSTIVÆ INSULÆ, or SOMMERIÆ INSULÆ, the BERMUDA group.

ÆSTUARIUM AVUM, the HUMBER in England.

ÆSTUARIUM ST. GEORGII, SINUS SABRINÆ or SABRIANA, or MUNICA BRISTOLIENSIS, the BRISTOL CHANNEL.

ÆSTUARIUM VARÆ, the FRITH OF CLYDE in Scotland.

ÆTHALIA, or ILVA, the island of ELBA.

ÆTHIOPIA, a name of somewhat undefined application in ancient geography, but applied to the region lying between Egypt on the N; the Sinus Arabicus and Mare Erythræum on the E; and Libya Interior on the W. Its S boundaries were unknown. It included the modern NUBIA, SENNAAR, and a part of ABYSSINIA. The name is given, by later writers, more particularly to Abyssinia.

ÆTHIOPICUS SINUS, or OCEANUS MERIDIONALIS, the GULF OF GUINEA.

ÆTHONIA, or ETONA, ETON in Buckinghamshire.

ÆTNA, or INNESA, a town on the S declivity of Ætna, now represented by CASTRO.

ÆTNA MONS, a mountain of Sicily, the modern ETNA or MONGIBELLO.

ÆTOLIA, a district of ancient Greece, bounded on the N by Aperantia and Æniana; on the E by Doris, and the territory of the Locri Ozŏlæ; on the S by the Sinus Corinthiacus; and on the W by the river Achelous, separating it from Acarnania. Under the Turkish empire this region formed part of LIVADIA, it is now included within the frontiers of Greece. The northern and more rugged portion of the ancient territory was distinguished as ÆTOLIA EPICTETUS; that section which extended along the coast, from the mouth of the

Achelous eastwards to the narrow entrance of the gulf of Corinth, was ÆTOLIA ANTIQUA. Æ. formed part of the Roman province of Achaia.

ÆXONE, a town of Attica, near the modern hamlet of AXAONA.

AFRICA, a name originally given by the Romans to the territory around the city of Carthage, and said to have been derived by them from a small Carthaginian district. When the name had become applied to the whole continent of Africa, the Roman geographer still distinguished AFRICA PROPRIA, or AFRICA VERA, a region corresponding to the modern TUNIS and part of TRIPOLI. Herodotus considered Libya or Africa, as terminating N of the equinoctial line, and as surrounded by water, except at the isthmus of Suez. It is supposed that the Carthaginian Hanno had passed to the S of the mouth of the Senegal river, in his exploratory voyage made probably before B. C. 500. 'The Periplus of the Erythrean sea,' a work compiled probably about the time of Pliny the elder, mentions Rhapta as the extreme S point reached on the African coast, and this has been identified with QUILOA. In time of Ptolemy, the W coast of Africa was known as far as to the parallel of 11° N; and the probability is, that the Romans were not altogether ignorant of the upper course of the NIGER. See that article.

AGANA, PAGANA, INSULA ST. JOANNIS, the island of GUAHAM or GUAM in the Ladrones.

AGARA, a town of India, to the NW of Palibothra, on the S bank of the Jomanes, now represented by AGRA in Hindostan.

AGARUS, a river of Dacia, now the BERDA.

AGASUS PORTUS, a town of Daunia, probably PORTO-GRECO in Naples.

AGATHA, AGATHE, or CYRTA, a town at the mouth of the Arauris, now AGDE in the French dep. of Herault.

AGATHA DÆMONOS, an island in the Indian ocean, probably one of the Nicobar group.

AGATHOPOLIS, or FORUM SANCTÆ AGATHÆ, now the town of AGATA-DI-GOTI, in the Neapolitan prov. of Principato-Ultra.

AGATHYRNA, or AGATHYRNUM, a town on the N coast of Sicily, near the modern SANTA AGATHA.

AGBATANA. See ECBATANA.

AGEDINCUM, AGENDICUM, AGREDICUM, or LUGDUNENSIS QUARTÆ, a town of the Senones, on the r. bank of the Acauna, now SENS in the French dep. of Yonne.

AGEDUNUM, AHUN in the French dep. of Creuse.

AGEIUM, or ANESIACUM, the town of AI or AY in the French dep. of Marne.

AGELOCUM, or SEGELOCUM, a town of the Coritani, perhaps LITTLEBOROUGH in Essex.

AGENNO, AGENNUM, or CIVITAS AGINNENSIUM, a town of Gallia Antiquitana Secunda, on the Garumna, now AGEN in the French dep. of Lot-et-Garonne.

AGERANA VALLIS, the valley of the ELAN in the Swiss cant. of Valais.

AGER MARIANUS, the town of MARIAGER in Jutland.

AGESINATES, a town of Gallia, now LUSIGNAN.

AGGER GANDAVENSIS, the town of SAS-VAN-GAND in the Dutch prov. of Zieland.

AGGERHUSIUM, AGGERHUUS in Norway.

AGGERIPONTUM, the town of THOMASBRUCK or THAMESBRUCK in the Prussian circle of Erfurt.

AGINO, AGNIUS, or EUNENO, the river AA in the French dep. of Manche.

AGISYMBA, or AGIZIMBA, a region in Ethiopia, S of the equator, comprising the whole of Central Africa as known to Ptolemy.

AGMUNDESHAMUM, or AMERSHAMUM, AMERSHAM in Buckinghamshire.

AGNIADELLUM, the town of AGNADEL in Lombardy.

AGORUM, AGORA in the Venetian prov. of Belluno.

AGRAMONTIUM, ST. AGRA in Catalonia.

AGRIA, the town of ERLAU or EGER in the Hungarian prov. of Hevesch.—Also the river ERLAU, an affluent of the Theiss in Hungary.

AGRIANES, a river of Thrace, a tributary of the Hebrus, now the ERGENE.

AGRIGENTUM, the modern GIRGENTI in Sicily.

AGRIMUM, the town of AGHRIM in co. Roscommon.

AGRIPPINA, or COLONIA AGRIPPINA UBIORUM, a town of Gallia Belgica, now COLOGNE in the Prussian prov. of the Rhine.

AGROPOLIS, or NOVOMARCHIA, the town of MAROS-VASARHELY in Transylvania.

AGUILARIA, ST. AGUILAR in the Spanish prov. of Navarre.

AGUILARIA-CAMPESTRIS, ST. AGUILAR-DEL-CAMPO in Old Castile.

AGUNTUM, INNICHEN on the Drave in Tyrol.

AGYRIUM, or AGYRENA, a town of Sicily to the W of Etna, the ruins of which are near ST. FILIPPO-D'AGYRO.

AHUSIA, AHUUS in the Swedish territory of Blekingen.

AIANIA, the AYAN coast of NE. Africa.

AICHA, AICHACH or AICH in Bavaria.

AICHSTADIUM, ARENTUM, DRYOPOLIS, the modern AICHSTADT or EICHSTADT on the Bavarian Altmuhl.

AIRIACUM, the town of AIRY in the French dep. of Yonne.

AKHASSA REGIO, ACHASA, or CESI, a region of India, which Ptolemy places between the Chatæ Scythæ on the N, and the Chauranæi Scythæ on the SE, and to the eastward of the Byltæ, and which Major Cunningham identifies with LADAK.

ALA, ALENA, or OLA, now AALEN or ALEN in Würtemberg.

ALA, or AQUILEGIA, the town of AELEN or AIGLE in the Swiss cant. of Vaud.

ALABANDA, or ANTIOCHIA AD MÆANDRUM, a town of Caria, identified by Leake with ARABI-HISSAR.

ALABASTRITES MONS, a portion of the JEBEL-MOKATTAM on the E side of the valley of the Nile.

ALABON, or ALLABONA, the modern ALAGOA, near the influx of the Xalon into the Ebro in the Spanish prov. of Aragon.

ALABUS, or ALABON, the river CANTERA in Sicily.

ALÆNUS, or ALAUNUS, perhaps the river BRIT or BRIDE of Dorsetshire.

ALÆSUS, a river on the N coast of Sicily, the PATTINEO.

ALALIA, or ALERIA, a town of Corsica, the modern ALERIA.

ALANDIA, the island of ALAND in the mouth of the gulf of Bothnia.

ALANGUERA, or ALANGUERUM, the town of ALANGUER in Estremadura.

ALANI, or ALAUNI MONTES, or ALAUNUS MONS, a mountain of Sarmatia Europæa, now known as the WALDAI range between the Dnieper and the Don, in Russia.

ALANTIA, ALANCHES in the French dep. of Cantal.

ALANUS, the river ALAN in Cornwall.

ALARA, the ALLER in Lower Saxony.

ALATA CASTRA, CASTRA PUELLARUM, EDINUM, EDINBURGUM, BURGUS ALATUS, BURGUS ALARUM, the city of EDINBURGH, the capital of Scotland.

ALATRIUM, ALATRI in the Campagna-di-Roma.

ALAUNA, or ALAUNIUM, a town of Gallia Lugdunensis, now ST. QUIMPER in the French dep. of Finistere.

ALAUNUS, or ALÆNUS, the river ALNE in Northumberland.

ALBA, the river AUBE, a tributary of the Seine in France.—Also the Spanish town of ESTELLA on the Ega in Navarre.

ALBA, or ALBA BULGARICA, or ALBA GRÆCA, the town of BELGRADE in Servia.

ALBA, or HELVÆ, the town of ELVAS in the Portuguese prov. of Alentejo.

ALBA, ALBIA, or ALBIS, the river ELBE in Germany.

ALBA, or TEZERUS, the river TER in the Spanish prov. of Catalonia.

ALBA AUGUSTA, ALBA HELVIORUM, or VIVARIUM, the capital of the Helvii in Gallia, now ALPS or APS in the French dep. of Ardeche, or perhaps VIVIERS.

ALBA CAROLINA, ALBA JULIA, APULUM, COLONIA APULENSIS, MUNICIPIUM APULUM AUGUSTUM, a town of Dacia, now CARLSBERG or WEISSENBURG in Transylvania.

ALBA DOMINARUM, the cloister of FRAUENALB on the Alb, in the grand-duchy of Baden.

ALBA DOMINORUM, the town of HERRENALB in Würtemberg.

ALBA FUCENTIA, or ALBA MARSORUM, near ALBA in the Neapolitan prov. of Abruzzo-Ultra 2da.

ALBA INGANNORUM, ALBIUM INGAUNUM, or ALBINGA, now ALBIENGA in the prov. of Genoa.

ALBA LONGA, or ALBONA, a town of Latium, now represented by PALAZZOLO.

ALBA MALA, ALBAMARLA, the town of ST. AUMAL in the French dep. of Seine-Inferieure.

ALBA MARIS, or BLANDONA, a port near ST. BROGRAD in Dalmatia.

ALBA MONS, ALPAGORIA, ALPES SUEVIÆ, ALPENSIS PAGUS, the RAUHE-ALP in Würtemberg, extending from the source of the Danube, along its N bank, to the junction of the Licus or Lech.

ALBANIA, a region of Asia, bounded on the N by the Caucasus; on the E by the Caspian; on the S by Armenia; and on the W by Iberia; corresponding to the Persian provs. of DAGHESTAN and LAZISTAN. Its inhabitants were of Scythian origin.

ALBANIA, or EPIRUS, the modern pashalik of ALBANIA in Turkey.

ALBANIA NOVA, the town of ALBANY in the state of New York, U. S.

ALBANIÆ PORTÆ, or PYLÆ, or PORTA FERREA, the pass of DERBEND in the Persian prov. of Daghestan.

ALBANIUM, or ALBA REGALIS, the town of STUHLWEISSENBURG in Lower Hungary.

ALBANOPOLIS, a town of Illyris Græca, the modern ALBANOPOLI on the Drino.

ALBANUM, a town in Latium, the modern ALBANO in the Campagna-di-Roma.

ALBANUS, a river of Asia, the SAMURA, or perhaps the BILBANA, falling into the Caspian.

ALBANUS, or ALBIUS MONS, the modern MONTE-DEL-CORSO in Carinthia.

ALBANUS LACUS, the modern LAGO-CASTELLO or LAGO-DI-CASTEL-GONDOLFO in the Papal states.

ALBA PAUPEIA, the modern ALBA in the duchy of Montferat.

ALBARICINUM, or LUBETUM, the town of ALBARCIN on the Guadalaviar, in the Spanish prov. of Aragon.

ALBA SELUSIANA, or SELUSIUM, the town of WEISSENBURG, on the Lauter, in the French dep. of Bas-Rhin.

ALBA TERRA, AUBETERRE in the French dep. of Charente.

ALBECE, REGIUM, REGII APOLLINARIS, a town of Gallia Narbonensis, represented by the modern RIEZ in the French dep. of Basses-Alpes.

ALBERTONIA, ALVERTON in Yorkshire.

ALBIA, or ALBIGA, a town of Gallia Aquitania, now ALBI in the French dep. of Tarn.

ALBIANUM, KUFSTEIN in Tyrol.

ALBIATE, ALBIATUM GRASSUM, or ABIAGRASSA, the modern BIAGRASSA in Austrian Lombardy, on the Canal Naviglio.

ALBICASTRUM, or CASTROBRACUM, on the river Liria, the modern CASTELLO BRANCO in the Portuguese prov. of Beira.

ALBIGA. See ALBIA.

ALBIMONTIUM, the town of BLEMONT on the Vezouze, in the French dep. of Meurthe.

ALBION, or BRITANNIA, the island of Britain. The name Albion is derived from the Celtic *alp* or *alb*, and may have reference either to the lofty coasts of the island as beheld from the opposite shores of France, or to the white chalky cliffs of Britain.

ALBION NOVA, the states of NEW ENGLAND on the American continent.

ALBIPOLIS, the HOHENELBE in the Bohemian circle of Bitschow.

ALBIS. See ALBA.

ALBIUM INGAUNUM, the chief city of the Ingauni in Liguria, the modern ALBENGA.

ALBIUM INTEMELIUM, or ALBINIMINIUM, a city of the Intemilii in Liguria, now VINTIMIGLIA, at the mouth of the Roya, in the prov. of Genoa.

ALBIUS MONS. See ALBANUS.

ALBONA, or ALBUNEA, the river AGOGNA in Piedmont.

ALBONA, ALVONA, or ALVUM, the modern ALBONA in Istria.

ALBORIUM, the town of ALVAR in the Spanish prov. of Algarve.

ALBRETUM, ALBRET, or LA BRITT in the French dep. of Landes.

ALBUCELLA, or ARBUCULA, the largest city of the Vaccæi, now VILLA FAGILA in the Spanish prov. of Asturias.

ALBUCIUM, or ALBUCUM, the town of AUBUSSON in the French dep. of Creuse.

ALBULA, TIBERIS, TIBRIS, TYRRHENUS AMNIS, or THYMBRIS, the TIBER of the Papal states. Albula was probably the old Latin name of this river; Tiberis, the Tuscan one.

ALBULA, the river WEISSERITZ of Bohemia, which joins the Elbe near Dresden.

ALBULÆ AQUÆ, or ALBUNEA FONS, probably represented by the LAGO-DI-ZOLFO, and other two sulphureous lakes, on the road from Rome to Tivoli.

ALBUM MARE, or SINUS GRANVICUS, the WHITE SEA, a gulf of the Arctic ocean, on the coast of the Russian gov. of Archangel.

ALBUQUERCUM, ALBUQUERQUE in the Spanish prov. of Estremadura.

ALBUQUERCUM AMERICANUM, ALBUQUERQUE in New Mexico.

ALBURACIS, or AURIGERA, the river AURIGE in the French dep. of Arriege.

ALBURNUS MONS, a ridge of mountains on the N frontier of Lucania, near the junction of the Silarus and Tanager, now MONTE-DI-POSTIGLIONE, or MONTE ALBRUZZO, in Naples.

ALBUS PORTUS, ALGECIRAS on the S coast of Spain.

ALCALA REGALIS, or ALCALA REGIA, the town of ALVALE-REALE in the Spanish prov. of Jaen.

ALCAMUS, ALCAMO in Sicily, SW of Palermo.

ALCANITIUM, ALCANIZ on the river Guadalupe in Aragon.

ALCANTARA, ALCANTARILLA in Andalusia.

ALCARATIUM, a town of the Celtiberi, now represented by ALCARAZ in New Castile.

ALCASARIUM MAGNUM, or CÆSAREA MAGNA, the town of ALCAZAR-QUEVIR, on the river Luco, in Fez, in Northern Africa.

ALCASARIUM PARVUM, the coast-town of ALCAZAR-COGUER in Fez.

ALCASARIUM SALINARUM, or SALACIA IMPERATORIA, the town of ALCAÇAR-DO-SAL in Portuguese Estremadura.

ALCASARIUM SANCTI JOANNIS, the town of ALCAÇAR-DE-SAN-JUAN in the Spanish prov. of Toledo.

ALCEJO, or ALISO, the town of ALZEY or ALTZOY in Rhein-Hessen.

ALCIACUM, AUXI-LE-CHATEAU in the French dep. of Pas-de-Calais.

ALCIATUM, ALZATO in the Lombardian prov. of Como.

ALCIMOENNIS, SAMULACENÆ, or ULMA, the town of ULM in Würtemberg.

ALCINOI INSULA. See CORCYRA.

ALCIRA, or SETABICULA, the town of ALVIRE in the Spanish prov. of Valencia.

ALCMARIA, or ALCMARIUM, the town of ALKMAER in North Holland.

ALCOBATIA, or EBUROBRITIUM, the town of ALCOBAZA in Spanish Estremadura.

ALCOLA, AREA, or FLAVIUM ARVENSE, the town of ALCOLE in Seville.

ALDEA MURI, ALDEA-EL-MURO in Old Castile.

ALDEA RIVI, ALDEA-EL-RIVO in Seville.

ALDENARDA, OUDENAARDE in East Flanders.

ALDENARIA, ALDENHAR in Rhenish Prussia, NW of Bonn.

ALDERGENIUM, AUVEGHEM in East Flanders.

ALDNUBIS, or ALDUADUBIS, a town of Gallia Lugdunensis, now represented by DOUBS in the French dep. of Doubs.

ALECTA, or ELECTA, ALETH in the French dep. of Aube.

ALEMANNIA PROPRIA, or ALGEA, or ALMANGOVIA, the lordship of ALGAU in Suabia.

ALEMANNI MONASTERIUM, the monastery of ALTMUHL-MUNSTER in Bavaria.

ALEMANNUS, ALEMO, or ALMONA, the river ALTMUHL in Bavaria.

ALENCONIUM, ALENCON in the French dep. of Orne.

ALEPUM, ALEPPUM, BEREA, BERHÆA, or CHALEP, the city of ALEPPO in Asiatic Turkey.

ALERE, ANDRIA, ANGER, or INGERIS, the river INDRE in France.

ALEREA, a town of the Bituriges Cubi, now ST. VINCENT-D'ARDENTES on the Indre.

ALESIA, ALISIA, or ALESSIA, now represented by the village of ALISE, at the base of Mount Auxois, in the French dep. of Cote-d'Or.

ALESIA, or ALESIUM, the town of ALAIS in the French dep. of Garde.

ALESIA, or USELLIS, a town of Sardinia near the modern ALES.

ALESUS, the modern river SANGUINARIA in the Papal states.

ALETHA MACLOVIA, or MACLOPOLIS, the island of ST. MALO in the French dep of Ille-et-Vilaine.

ALETUM, a town of Gallia Lugdunensis near ST. MALO.

ALEXANDREA, ALEXANDRIA, ALEXANDRIA TROAS, ANTIGONEA, COLONIA TROAS a town founded by Antigonus, on the coast of the Troad, in Asia, the ruins of which appear at ESKI-STAMBUL.

ALEXANDREA, or ALEXANDRIA, a city founded by Alexander the Great, on the W of the Delta in Egypt, now ALEXANDRIA.

ALEXANDREA ARIANA, or ALEXANDRIA APUD ARACHOTOS, the ancient capital of Aria in Asia, founded by Alexander, probably near the modern city of HERAT.

ALEXANDREA-AD-ISSUM, or ALEXANDRIA CATISSON, or ALEXANDRIOLA, now represented by ALEXANDRETTA or SCANDERUN in Syria.

ALEXANDRIA A PALEA, or ALEXANDRIA STATIELLORUM, the town of ALESSANDRIA-DELLA-PAGLIA in Piedmont.

ALEXANDRIA AD CONFLUENTES ACESINIS, the city of LAHORE in the Punjab.

ALEXANDROVIUM, ALEXANDROFSK, in the Russian gov. of Yekaterinoslav.

ALEXANI CIVITAS, or ALEXANUM, ALESSANO in the Neapolitan prov. of Terra-d'Otranto.

ALEXODUNUM, HEXHAM in Northumberland.

ALFACHUSA, ALFAQUES in Tunis.

ALFELDA, ALFELDEN in Hanover.

ALFORDIA, ALFORD in Lincolnshire.

ALFRADONIUM, ALFRETON in Derbyshire.

ALGARBIA, the Portuguese prov. of ALGARVE.

ALGERIA, CORAX, or PORTUS CARACODES, the town of ALGHIWAS on the W coast of Sardinia.

ALGERIA, or ALGERIUM, the city of ALGIERS in Africa.

ALGERIANUM REGNUM, the kingdom of ALGIERS.

ALGIA, the French port of HONFLEUR.

ALGIDUS MONS, a chain of hills in the Papal states, stretching from the Albon mount to the Tusculan hills.

ALIACMON, or HALIACMON, a river of Macedonia, flowing into the Sinus Thermaicus or gulf of Saloniki, identified by Leake with the IRJE-KARASU.

ALICANTA, ALICANTIUM, or LUCENTUM, ALICANT in Valencia.

ALICANUM, HALICANUM, or RACIATANUM, the village of RUCKSBURG, on an island of the Muhr in Styria.

ALIETUM, ISOLO in the Illyrian circle of Trieste.

ALIFA, ALLIFE, ALLIPHA, or ALLIFÆ, ALIFA in the Neapolitan prov. of Terra-di-Lavoro.

ALILÆI, SOKIA in the prov. of Yemen in Arabia.

ALINDA, the capital of Caria, represented by the modern MUGHLAH.

ALINGO, or PORTUS ALINGONIS, a town on the river Gammua, probably near LANGON in the French dep. of Gironde.

ALIONE, ALONA, or LANCASTRIA, a town of the Brigantes, the modern LANCASTER in Lancashire.

ALIPHERA, a town of the Eleans, in Arcadia, corresponding to the modern NEROVITZA.

ALISIUM, or HEILBRONNA, HEILBRON on the Neckar in Würtemberg.

ALISO, or ELISO, the ALME, an affluent of the Lippe in Westphalia; or perhaps the LISE.

ALISO, ALISUM, or VESALIA INFERIOR, NIEDERWESAL at the influx of the Lippe into the Rhine.

ALISO ROMANORUM, perhaps represented by the village of LISBORN in Westphalia; but identified by some with ELSEN at the junction of the Alme and Lippe.

ALISONTIA, ALIZUNTA, or ELZA, the ELZE, an affluent of the Sure passing Luxemburg.

ALLÆ CASTRUM, or ALLENBURGUM, ALLENBURG in Brandenburg.

ALLATURA, ALATYR on the Sura, in the Russian gov. of Nijny-Novgorod.

ALLECTUM, DONUM DEI, or TUADANUM, the town of DUNDEE in Forfarshire.

ALLENDORFIUM, ALLENDORF on the Werra, in Kur-Hessen.—Also ALLENDORF in Hessen-Darmstadt.

ALLENSTENIUM, ALLENSTEIN in the Prussian prov. of Konigsburg.

ALLIA, a river of the Sabini, an affluent of the Tiber, on which Brennus defeated the Romans B. C. 390, by some identified with the RIO-DEL-MOSSO or FOSSO-DI-PRADARONI; by others with the SCOLO-DEL-CASALE; by others with the FOSSO-DI-CONCA, which flows into the Tiber at Malpasso.

ALLIENI FORUM, FENARA, or TRIGABOLI, a town of Gallia Cispadana, the modern FENARA.

ALMA, or ALMUS MONS, a mountain in Pannonia Inferior, perhaps the WEINBERG.

ALMAGRA, ALMAGUER in New Grenada.

ALMAGRUM, the ALMAGRA in New Castile.

ALMANDRALEGIUM, ALMANDRALEJO in Spanish Estremadura.

ALMANTICA, ALMANZA in New Castile.

ALMAZANUM, ALMAZAN in Old Castile.

ALMELOA, ALMELO in the Dutch prov. of Over-Yssel.

ALMERINUM, ALMERIA in Portuguese Estremadura.

ALMINIUM, or PEGUNTIUM, the modern ALMISSA in Dalmatia.

ALMO, a river of Latium, the FOSSO-SCARICATORE, rising in the Albion mount, and flowing towards the Tiber.

ALMODAVURIA CAMPESTRIS, ALMODAVUR in New Castile.

ALMONDBURIUM, ALMONDBURY in Yorkshire.

ALOCETUM, or LANNAJAM, the town of LANNOY in the French dep. of Nord.

ALOCIÆ INSULÆ, a group off the coast of Jutland, MORS, THYE, and VENDYSSEL.

ALONE, or ALONÆ, perhaps TORRE-DE-SALINAS at the mouth of the Tader.

ALONE BRITANNICA, a town of the Brigantes to the SE of Galava, now AMBLESIDE.

ALONTA, a river of Sarmatia Asiatica, the modern TEREK flowing into the Caspian.

ALOPECE, or ALOPECIA, the village of KRABATO in the vicinity of Athens.—Also an island in the mouth of the Don.

ALORUS, a town of Macedonia, perhaps KAITCO CHORI.

ALOSTUM, ALOST in East Flanders.

ALPES, ALPINA JUGA, ALPINI MONTES, AGGERES ALPINI, the extensive mountain-system which, under various names, separates Gallia, Helvetia, and Germania from Italia, extending in a crescent form from the Sinus Flanaticus or gulf of Quarnero in the Adriatic, and the sources of the Kolapis or modern KULPE, to Vada Sabatia, now SAVONA on the Sinus Lingusticus, or gulf of Genoa.

ALPES, or ALPIUM URBS, the modern AULPS or AUPS in the French dep. of Var.

ALPES BASTARNICÆ, that part of the Carpathian chain which passes through Transylvania.

ALPES CARNICÆ, or JULIÆ, that portion of the Alps which intersects the S of Styria, the modern CARNIC ALPS.

ALPES COTTIÆ, the modern COTTIAN ALPS extending from Monte-Viso to Mont-Cenis.

ALPES ENGANÆ, a branch of the Carnic Alps.

ALPES GRACÆ, or GRAJUS MONS, the GRAIAN ALPS, stretching from Mont-Cenis to the Little St. Bernard. The origin of the name has been traced by some to the Celtic *craig* or *carraig*, signifying 'a rock.'

ALPES JULIÆ, ALPES VENETÆ, or ALPES PANNONIÆ, a portion of the Carnic Alps, comprising the BIRNBAUMER WALD.

ALPES LEPONTIÆ, or ALPES LEPONTINÆ, the LEPONTIAN ALPS sometimes called the HELVETIAN ALPS, in Western Switzerland, extending from Monte Rosa to the St. Gothard.

ALPES MARITIMÆ, or ALPES LITTOREÆ, the MARITIME ALPS, extending from Nice to Monte-Viso.

ALPES NORICÆ, the NORIC ALPS, extending from the frontiers of Carinthia through Salzburg and Styria.

ALPES PANNONIÆ. See ALPES JULIÆ.

ALPES PENNINÆ, the PENNINE ALPS, skirting the S side of the Swiss cant. of Valais, and extending from Mont Blanc to the Simplon.

ALPES RHÆTICÆ, ALPES RÆTICÆ, or JUGA RHETICA, the RHÆTIAN ALPS, separating Italy from Rhætia, and extending from the sources of the Rhone, through the Grisons and Tyrol, to the Drieherrnspitz on the borders of Carinthia.

ALPES SUEVIÆ. See ALBA.

ALPES SUMMÆ, the modern ST. GOTHARD.

ALPES TRIDENTINÆ, or JUGA TRIDENTINA, a portion of the Rhætian Alps.

ALPES VENETÆ. See ALPES JULIÆ.

ALPHA, the AA of the Swiss cant. of Unterwalden. —Also the AA of Munster.

ALPHANUM TRAXINUM, ALPHANO in Portuguese Alentejo.

ALPHEUS, a river of Elis, passing to the S of Olympia, the modern KARITENA or RUFIA.

ALPIS, the river ALBIN falling between Trieste and Capo-d'Istria, into the gulf of Venice.

ALPIS COTTIA, MONTE VISO and CENIS.

ALPIS GRAIA, MOUNT ST. BERNARD.

ALPUXARI MONTES, the ALPUXARAS in Granada.

ALSA, or ALSENA, the island of ALSEN in the Little Belt.

ALSA, a river of Gallia Transpadana, perhaps the AUSA which falls, near the Rimini, into the gulf of Venice.

ALSADAMUS, a mountain of Trachonitis, perhaps the KELB-HAURAN a summit of the Jebel-Hauran in Syria.

ALSÆ FRETUM, the ALSEN-SUND on the coast of Schleswig.

ALSATIA, or ELISATIA, ALSACE.

ALSFELDA, ALSFELD to the W of Marburg in Prussia.

ALSIETANUS LACUS, or ALSIA AQUA, now LAGO-DI-MARTIGNANO near Trivignano in the Papal states.

ALSTADIUM, ALSTADT in Thuringia.

ALSTERA, the ALSTER, an affluent of the Elbe.

ALTAIUS MONS, ALTAII MONTES, or ANNIBI MONTES, a mountain-range in Scythia extra Imaum, the modern ALTAI.

ALTA MAUTA, HOHENMAUTH in Bohemia.

ALTA REPA, ULTENREIF in the Swiss cant. of Freiburg.

ALTA SPECULA, or SUMMONTORIUM, HOHENWARTH in Bavaria.

ALTENACHIUM, or ALTENACUM, ALTENA in the Prussian gov. of Arensburg.

ALTENBURGUM, or PALÆOPYRGUM, ST. ALTENBURG on the Pleisse.

ALTENHOVIA, or CURIA VETUS, ALTENHOFEN on the Danube in Upper Austria.

ALTERPRETUM, or ALTSTETTUM, ALTSTAETTEN in the Swiss cant. of St. Gall.

ALTHÆA, or CARTERA, perhaps ORGAZ in New Castile.

ALTILIA, the river AUTHIE in the French dep. of Pas-de-Calais.
ALTINÆ, or ELTENUM, ELTEN in the Prussian prov. of the Rhine.
ALTISIADARUM, or AUTESIODORUM, the town of AUXERRE in the French dep. of Yonne.
ALTISOLIUM, or SOLIUM VETUS, ALTSOHL in LOWER Hungary.
ALTORFIUM, or VICUS VETUS, ALTORFF in the Swiss cant. of Uri.
ALTOVADUM, or VADUM ALTUM, HOHENFURT on the Muldau.
ALTUM CASTRUM, ALTA ARÆ, or VICEGRADUM, WISCHEGRAD in Lower Hungary near Ofen.
ALUTA, ALVATA, or ALVATUS, a river of Dacia, the modern ALT or OLT.
ALVERNIA, AUVERGNE in France.
ALVERNIA SUPERIOR, UPPER AUVERGNE in France.
ALVINAUM, WINZENDORF on the Marosch in Transylvania.
ALVONA. See ALBONA.
AMACAOUM, the modern Portuguese settlement of MACAO on the Chinese peninsula of Hiang-shan.
AMADES, or AMISIUM, EMBS or HOHENEMBS in Tyrol.
AMAGETOBRIGA, a town of the Sequani in Gallia Lugdunensis, perhaps BINGEN.
AMAGRIA, the island of AMAK off the coast of Danish Sieland.
AMALCHIUM MARE, the BALTIC.
AMALIA, AMAL on Lake Wener in Sweden.
AMALLOBRIGA, a town of Vaccæi, now MEDINA-DEL-RIO-SECO in Spain.
AMANA, a portion of the Anti-Libanus in Syria, N of Damascus.
AMANA. See ABUNA.
AMANA, the OHEN in Hessen, an affluent of the Lahn.
AMANDOPOLIS, ST. AMAND in the French dep. of Nord.
AMANIA, the modern YEMEN in Arabia.
AMANISÆ PYLÆ, AMANI MONTIS PORTÆ, or AMANIDOS PYLÆ, a mountain pass in Cilicia, to the E of Issus, leading through the range of Amanus, the modern DEMIR-KAPU or KARA-KAPU of the Jawur-Dagh in Syria.
AMANTIA, AMENTIA, or ADAMANTIA, AMANTEA in Neapolitan prov. of Calabria Citra.
AMANUS MONS, a mountain on the E of Cilicia, now the JAWUR-DAGH skirting the gulf of Scanderun in Syria.
AMARDUS, or MARDUS, a river of Northern Media, now the SAFID-RUD of Ghilan in Persia.
AMARI LACUS, a bitter lake in the Delta of Egypt, near Arsinoe, now known as the SCHEIB.
AMARINUM, ST. AMARIN or EMMERIN on the Thur, in the French dep. of Haut-Rhin.
AMASEA, AMASIA, or AMASIA HADRIANA, a city on the river Iris in Pontus, the modern AMASIAH in Anatolia.
AMASENUS, or BADINUS, a river of the Volsci, the AMASENO of the Papal States.
AMASIA, AMASIUS, AMISIA, or AMISIS, the EMS flowing into the Dollart bay.
AMASIA, EMBDA, or EMDA, the modern EMBDEN in East Friesland.
AMASIA CATTORUM, the modern MARBURG in Ober-Hessen.
AMASIÆ PRÆFECTURA, the pashalik of SIVAS in Asiatic Turkey.
AMASTRA, MISTRETTA in the Val-di-Demona in Sicily.
AMASTRIS, AMASTRIANORUM CIVITAS, AMASTRIANA URBS, or SESAMUS, the modern AMASERAH, AMASTRO, or BADISTAN, near Cape Diwan-Burni on the coast of the Black sea.
AMATHA, HAMATH, CHAMATH, AMATHIS, EPIPHANEA, or EPIPHANIA, now represented by HAMATH on the Orontes in Syria, and the fertile district surrounding it.
AMATHUS CYPRIA, a town of Cyprus, the ruins of which exist in the vicinity of LIMASOL.
AMATIQUA, or JANUM ST. THOMÆ DE CASTELLA, AMATIQUE in Guatimala.
AMATRICA, AMATRICE in Abruzzo Ultra.
AMAVILLA, AMEVILLE on the Doria in Piedmont.
AMAZONIDUM, or AMAZONUM FLUVIUS, the river AMAZON or MARANON.
AMBACIA, ST. AMBOISE in the French dep. of Indre-et-Loire.
AMBERGA, AMBERG in Bavaria.
AMBITARINUS VICUS, or CAPELLA, CAPELLE in Rhenish Prussia, according to Cluver; but REUSE according to Ukert.
AMBITIVUS PAGUS, MAINFELD in the Prussian prov. of Coblenz.
AMBLAVIA, the AMBLEVE, an affluent of the Ourthe, in Liege.
AMBLETOSA, AMBLETEUSE in the French dep. of Pas-de-Calais.
AMBRA, the river EMMER in Westphalia.
AMBRACIA, ARTA in Northern Greece.
AMBRACIUS SINUS, the gulf of ARTA between Epirus and Acarnania in Northern Greece.
AMBRÆ LACUS, or AMBRONIS LACUS, the AMMER-SEE in Bavaria.
AMBRESBURIA, or AMBROSII VITUS, AMERSBURY in Wiltshire.
AMBRONIACUM, AMBRONAY in the French dep. of Ain.
AMBRONICUS PAGUS, the Swiss cant. of SOLOTHURN.
AMBROSIOPOLIS, or SAXOPOLIS, BROOS in Transylvania.
AMBRYSUS, or AMBRYSSUS, near DISTOMO in Northern Greece.
AMELANDIA, the island of AMELAND off the coast of Friesland.
AMENA, AYMERIES in the French dep. of Nord.
AMERIA, AMELIA in the Papal deleg. of Spoleto.
AMERICA, or INDIA OCCIDENTALIS, or TERRA NOVA, the continent of AMERICA.
AMERIOLA, a town of the Sabines, the ruins of which are situated on the N side of MONTE ST. ANGELO, in the Corniculan range, in the Papal states.
AMERSFORDIA, AMERSFORT on the Ems, in the Dutch prov. of Utrecht.
AMIDA, DIYARBEKER in Asiatic Turkey.
AMILHANUM, or MILHADUM, MILHAU in the French dep. of Aveyron.
AMILIANUM, MILHAUD in the French dep. of Gard.
AMISUS, a city of Pontus, the modern SAMSUN in the Turkish prov. of Trebisond.
AMITERNUM, a city of the Sabini, near the modern VITTORINO in Naples.
AMMA, or EMMANA, the river EMMA in the Swiss cant. of Bern.
AMMONIUM, AMMONIACA REGIO, or HAMMONIACUS NOMOS, the oasis of SIWAH in the Sahara.
AMNIAS, a river of Paphlagonia, the modern KARAZU, an affluent of the Halyz or Kizel-Irmak.
AMŒNUM STAGNUM, the lake of ALBUFERA in Valencia.
AMOLIA, AMOL or AMUL in the Usbek territory.
AMORGUS, or AMARGUS, the island of AMORGO or MARGO in the Grecian Archipelago.
AMPELUSIA, or COTTES, CAPE SPARTEL on the Straits of Gibraltar.

AMPELUS MONS, AMBELONA in the island of Samos.
AMPELUS PROMONTORIUM, CAPE FALSO on the coast of Asiatic Turkey.
AMPHIALE PROMONTORIUM, CAPE DAPHNE on the W coast of Attica.
AMPHIMALIA, a town of Crete, near the modern ARMIRA.
AMPHIPAGUS, or AMPHIPAGUM, probably the modern CAPE SAN ANGELO in the island of Corfu.
AMPHIPOLIS, or CHRISTOPOLIS, an important city of Thrace, near the hamlet of JENIKEUI in Northern Greece.
AMPHISSA, a city of the Ozolæ, now represented by SALONA on the gulf of that name.
AMPHYSUS, or AMPHRYSSUS, a river of Phthiotis, the modern ARMYRO of Greece.
AMPLIPUTEUM, AMPLEPUIS in the French dep. of Rhone.
AMPSAGA, or FLUVIUS CIRTENSIS, a river of Numidia Massylorum, the modern EL-KIBIR or RUMMEL of Algeria.
AMSANCTI LACUS, a lake in the territory of the Hirpini, the modern MEFITI in the Campagna-di-Roma.
AMSTELA, the river AMSTEL on which Amsterdam is situated.
AMSTELŒDAMUM, AMSTELODAMUM, AMSTERODAMUM, or AMSTERADAMA, the city of AMSTERDAM in Holland.
AMSTRUTTERA, or ANSTRUTTERA, the town of ANSTRUTHER in Fifeshire.
AMYCLA, a city of Laconia, near AIA KYRIAKI in Northern Greece.
AMYCLÆUS MONS, the modern TAYGETUS or ST. ELIAS in Northern Greece.
ANACTORIA VONITZA, on the gulf of ARTA in Greece.
ANAGNIA, a city of the Hernici, now ANAGNI in the Papal states.
ANANIA, NONSBERG in Tyrol.
ANAPHE, the island of ANAPHI or NAMFIO in the Ægean sea.
ANAPHLYSTUS, or PANIS FANUM, ANAPHISO in Attica.
ANAPUS, the river ÆTES, a tributary of the Aspropotamo of Greece.—Also the river ALFEO in Sicily.
ANAREI MONTES, a branch of the ALTAI.
ANARISMUNDI, or ANDRASIMUNDI PROMONTORIUM, perhaps the modern CALPENTYN on the W coast of Ceylon.
ANAS, a river of Spain, the modern GUADIANA.
ANASSIANUM, ANASUM, or ANISIA, ENS on the Danube, in Austria.
ANASUS, ANESUS, or ANISUS, the river ENS, an Austrian affluent of the Danube.
ANATHO, the town of ANAH on the Euphrates.
ANATILIORUM URBS, MARITIMA COLONIA, or MARITIMA AVATICORUM, the town of MARTIGUES in the French dep. of Bouches-du-Rhone.
ANATIS, or CUBA, a river of Mauritania Tingitana, the OMMIRABI of Marocco.
ANATOLIA, or NATOLIA, the modern ASIATIC TURKEY. See ASIA MINOR.
ANCANICUM, the river ALANIS in Andalusia.
ANCIACUM, ANCY-LE-FRANC in the French dep. of Yonne.
ANCON, or ANCONA, a city of Picenum, the modern ANCONA.
ANCYRA, a city of Galatia, the modern ANGORA or ANGURI.
ANDECAMULUN, a town of the Lemovices, represented by the modern RANÇON in the French dep. of Haut-Vienne.
ANDECAVIUM, or ANDETRIUM, now represented by OLISSA in Dalmatia.
ANDEGAVA, CIVITAS ANDICAVORUM, ANDEGAVUM or JULIOMAGUS, the town of ANGERS in the French dep. of Marne-et-Loire.
ANDEGAVIA, DUCATUS ANDEGAVENSIS, or AGER ANDEGAVENSIS, the territory of ANJOU in France.
ANDELAGUS, ANDELIUM, or ANDELAUS, the town of ANDELYS in the French dep. of Eure.
ANDELAUS, the town of ANDELOT in the French dep. of Haut-Marne.
ANDELUS, or POMPEIOPOLIS, PAMPELUNA in Spanish Navarre.
ANDERITUM, a town of the Gabali, now JAVOUX in the Cevennes.
ANDERLACUM, ANDERLECH in Brabant.
ANDETHANA, ECHTERNACH or ECHTERN in Luxemburg.
ANDIATROCUS, LAKE ONTARIO in North America.
ANDOCERA, ANDOVER in Hampshire.
ANDOMATUNUM, ANTEMATUNNUM, or CIVITAS LINGONUM, represented by LANGRES in the French dep. of Haut-Marne.
ANDREOPOLIS, ST. ANDREWS in Fifeshire.
ANDRIACE, a town of Lycia, now ANDRAKE.
ANDROS, HYDRASSA, LASIA, or ANTANDRUS, the modern ANDROS in the Ægean sea.
ANDUSIA, ANDUZE in the French dep. of Gard.
ANECIUM, ANNECIUM, or ANNESIUM, ANNECY in Savoy.
ANEMURIUM PROMONTORIUM, CAPE ANAMUR, the S extremity of Asiatic Turkey.
ANETUM, ANET in the French dep. of Eure-et-Loire.
ANGADIVA, the island of ANKEDIVA off the coast of the Deccan.
ANGARIA, or ANGIRA, ENGER in Bavaria.
ANGELOPOLIS, or CIVITAS ST. ANGELI, SANTO-ANGELO in the Capitanata of Naples.—Also SANTO-ANGELO in Benevento.
ANGELOPOLIS-AD-LOMBORDAM, ST. ANGELO-DE-LOMBARDI in the Neapolitan prov. of Principato-Ultra.
ANGELOPOLIS AD METARUM, ST. ANGELO-DE-VADO in the Papal states.
ANGELOPOLIS AMERICANA, LA-PUEBLA-DE-LOS-ANGELOS in Mexico.
ANGELOSTADIUM, or INGOLSTADIUM, the town of INGOLSTADT in Bavaria.
ANGENACUM, ST. JEAN-d'ANGELI in the French dep. of Charente-Inferieure.
ANGERMANNIA, the old Swedish prov. of ANGERMANLAND.
ANGERMANNUS FLUVIUS, the river ANGERMANN in Swedish Lapland.
ANGIA, or ANGIANUM, ENGHIEN in Belgium.
ANGLA, ANGLE in the French dep. of Vendee.
ANGLA MONTANA, ANGLE in the French dep. of Herault.
ANGLERIA, ANGHIERE in Milan.
ANGLIA. The Angli, a tribe of the Suevi, are first noticed by Tacitus, when they appear to have been located to the E of the Elbe. Ptolemy places them on the W of that river, in what is now Magdeburg. D'Anville represents them as occupying in the 5th cent. the greater portion of Schleswig and part of Holstein, having the German ocean on the W, the Saxons on the S, the Vacini on the SE, and the Jates on the N. In 449, a large body of Saxons and Angles, led by Hengist and Norsa, settled themselves in England; and it would appear that the Angles were of predominating influence in this expedition, for they gave their own name to the newly acquired country,—ANGELLAND or ANGLIA.

ANGLIA MINOR, the lordship of ANGLEN or ANGELN in the duchy of Schleswig, now chiefly occupied by small peasant proprietors, speaking a dialect of the Platt-Deutsch. See article ANGELN.
ANGLOMONASTERIUM, INGELMUNSTER in W Flanders.
ANGLORUM INSULA, MONA, or MONNA, the modern ISLE OF ANGLESEY.
ANGOTINUM REGNUM, ANGOT in Abyssinia.
ANGULUM, ANGULUS, or ANGELOPOLIS AD MARE, SANTO-ANGELO at the mouth of the Aternus, in Abruzzo Citra.
ANGULUS ALPIUM, ALBEK in Suabia.
ANHALTINUM VETUS, ANHALT in Anhalt-Bernburg.
ANICIUM, VELLAVA URBS, or URBS VELLAVORUM, the modern PUY in the French dep. of Haut Loire.
ANIENE, or ANIO, the river TEVERONE which falls into the Tiber below the site of Antoninæ.
ANISIARUM AD LIGERIM, ST. AY in the French dep. of Loiret.
ANNABERGA, or ANNÆMONTIUM, ANABERG in the kingdom of Saxony.
ANNANDIA, or ANNANDI VALLIS, ANNANDALE in Dumfries-shire.
ANNANDUM, ANNAN in Dumfries-shire.
ANNANDUS, the river ANNAN in Dumfries-shire.
ANNAPOLIS, PORT ROYAL in Jamaica.
ANNECIUM VETUS, ANNECY-LE-VIEUX in Savoy.
ANNONACUM, ANNONAY in the French dep. of Ardeche.
ANONIUS, the river MONS in Tyrol.
ANOTHIA, ANNOT in the French dep. of Hautes-Alpes.
ANSA, ANTIUM, or ANSA PAULINI, ANSE in the French dep. of Rhone.
ANSER, or SERCULUS, the river SERCCHIO in Lucca.
ANSLAGA, or CHRISTIANIA, the town of ANSLO or ORSLO in Norway.
ANSTRUTIERA. See ARNSTRUTIERA.
ANTÆOPOLIS, a town of Upper Egypt, near the present village of KAU.
ANTANDRUS, or ANTANDROS, a town of Mysia Major, now ANTANDRO at the foot of Mount Ida in the Troad.
ANTARADUS, a town of Lycia, the modern TORTOSA in the pash. of Tarablus.
ANTEA, or ANTEIS, AMPUIS in the French dep. of Rhone.
ANTECARIA, ANTICARIA, or ANTEQUARIA, a town of Hispania Bœtica, the modern ANTEQUERA in the Spanish prov. of Granada.
ANTEMNÆ, near the present TORRE-DI-QUINTO at the junction of the Anco and the Tiber.
ANTIA, the river ANZA, an affluent of the Tosa in Lombardy.
ANTIBARUM, ANTIVARI in Dalmatia.
ANTICARIA NOVA, ANTEQUERA in the Mexican prov. of Guanaxuato.
ANTICOSIA, ANTISCOTIA, or ASSUMTIONIS INSULA, the island of ANTICOSTI in the St. Lawrence.
ANTILIA, AMPTHILL in Bedfordshire.
ANTILIBANUS, AMANA, HERMON, or SENIR, the southernmost of the two parallel chains of Lebanon in Syria.
ANTILLÆ, ANTILIÆ, CAMERCANÆ INSULÆ, CARIBES or CARIBANES, the CARIBBEAN ISLANDS or ANTILLES in the gulf of Mexico.
ANTIMELOS, ANTIMILO in the Grecian archipelago.
ANTIMONASTERIUM, EYMOUTIERES in the French dep. of Haut-Vienne.
ANTINOE, ANTINOI URBS, or BESA, a town in the Egyptian delta, the ruins of which are near SHEIKH-ABADEH.
ANTIOCHEA SUPER CRAGUM, a town in Cilicia Trachea, now ANTIOCHETA in Cyprus.
ANTIOCHIA, ANTIOCHEA AD ORONTEM, ANTIOCHEA TETRAPOLIS, ANTIOCHIA APUD DAPHNEN, ANTIOCHIA THEOPOLIS, the modern ANTIOCH or ANTAKIA in Upper Syria.
ANTIOCHIA AD MÆANDRUM, or ANTIOCHIA CARLÆ, NISA, or NYSSA, the modern JENISHEHR in Asiatic Turkey, near the Mendere-su.
ANTIOCHIA AD PISIDIUM, or COLONIA CÆSAREA ANTIOCHIA, a town in Phrygia Major, the modern AKSHEHR in Asiatic Turkey.
ANTIOCHIA AD TAURUM, perhaps the modern frontier fortress of BAHASNA in Northern Syria.
ANTIOCHIA MARGIANA, or ANTIOCHIA ARIDA, a town in Margiana, according to Mannert, MEMRUD or MANUSHAK, on the Murghab river in Northern Asia; according to others, MAWRI or SHAHJEHAN.
ANTIOCHIA PISIDIA, YALOVATCH in Asiatic Turkey.
ANTIPATRIS, in Judæa, ARSUF, or perhaps KAFFR-SABA in Syria.
ANTIPOLIS, ANTIBES in the French dep. of Var.
ANTIQUA, ANTIGUA in the Antilles.
ANTIRRHIUM PROMONTORIUM, a headland on the coast of Ætolia, at the entrance of the Sinus Corinthiacus, now CAPE RUMELI opposite Cape Drepano.
ANTITAURUS, a branch of the Taurus, passing through Cappadocia, the modern ALIDAGH.
ANTIUM, a city of the Volsci, now CAPO-D'ANZO, near Nettuno in the Papal states.
ANTIUM, or HANTIUM, HANZ in the Grisons.
ANTIVESTÆUM PROMONTORIUM, CAPE ST. IVES on the W coast of Cornwall.
ANTONA, AUFONA, or AVONA, the river AVON falling into the Bristol channel.
ANTONIA, ANTOING on the Scheld.
ANTONIA, or TRAJECTUM AD RHENUM, TRAJECTUS RHENI or ULTRAJECTUM, UTRECHT in Holland.
ANTORICUM, or ANTRICUM CARNUTUM, the town of CHARTRES in the French dep. of Eure-et-Loir.
ANTRINUM, ANTRIM in the Irish prov. of Ulster.
ANTROS, probably CORDUAN on the river Garonne.
ANTUNACUM, or ANDERNACUM, ANDERNACH on the Rhine.
ANTVERPIA, ANTWERP in Belgium.
ANUBINGARA, probably NEGOMBO in Ceylon.
ANXANUM, LANCIANO-VECCHIO in Abruzzo-Citra.
ANXELLODUNUM, EXCLODUNUM, or OSOLDUNUM, YSSODUN in the French dep. of Indre.
ANXIA, ANZI in the Neapolitan prov. of Basilicata.
ANXUR, or TARRACINA, the modern TERRACINA in the Papal deleg. of Frosinone.
APAMEA, the capital of Apamene, now KALAT-EL-MEDIK or FAMIEH.
APAMEA, in the S extremity of the island of Mesene, the modern KORNAH on the Tigris.
APAMEA CIBOTOS, APAMEA PHRYGIÆ, or APAMEA-AD-MÆANDRUM, a town of Southern Phrygia, near the modern DINARE.
APAMIA, PAMIERS in the French dep. of Arriege.
APENESTÆ, or VIESTÆ, a town in Apulia Daunia, now VENESTI on the coast of the Neapolitan prov. of Capitanata.
APENNINUS, APENNINI JUGA, APENNINUS MONS, the modern APENNINES. The true form is Apenninus, not Appeninus, nor Apeninus.
APENROA, APENRADE in Schleswig.
APERIASCIO, or EPERIÆ, EPERIES in Transylvania.
APEROPIA, a town of Argolis, now DHOKO, or perhaps HYDRON.
APHACA, AFKA in Northern Syria.
APHRODISIAS, in Caria, now GHEIRA.

APHRODISIUM PROMONTORIUM, or PYRENÆ PROMONTORIUM, the modern CAPE CREUX on the coast of the Mediterranean.
APHRODILOPOLIS, ASFUN in Upper Egypt.—Another place of this name was in the vicinity of the modern ATFIEH; and another stood near the modern TACHTA in Upper Egypt.
APIARIUM, BEJAR in the Spanish prov. of Murcia.
APOCREMNUS PROMONTORIUM, in Ionia, now CAPE ESOMENO.
APOLLINARIUM, a town of the Volcæ Arecomici, now AUBENAS in the French dep. of Ardeche.
APOLLINIS PROMONTORIUM, CAPE MOSTAGAN on the N coast of Africa.
APOLLINOPOLIS MAGNA, EDFU in Upper Egypt.
APOLLONIA CYRENAICA, near MARSA-SUZA on the N coast of Africa.
APOLLONIA ILLYRIA, POLLINA near the mouth of the Vojussa.
APOLLONIA MACEDONICA, a town on the Egnatian way, near POLLINA.
APOLLONIA THRACIA, or SOZOPOLIS, now SIZEBOLI.
APOLLONIATIS LACUS, a lake in Mysia Minor, the modern ABULLIONTE.
APONIANA INSULA, EGUSA off the W coast of Sicily.
APONUM, ABANO or ALBONO in Lombardy.
APOSTOLORUM PORTA, or APOSTOLORUM MONS, POSTELBERG in Bohemia.
APPIA, or VIA APPIA, a celebrated Roman road, leading from Rome to Capua, a distance of more than 1,000 stadia; and afterwards prolonged to Brundusium. It began at the Porta Capena of Rome, near the modern hill of S. Balbona, where the Via Latina was united with it, and joins the modern road from Velletri to Cisterna, at the 32d mile-stone, whence they run nearly together to Terracina.
APRIMONASTERIUM, or NOVIENTIUM, the abbey of EBERSHEIMMUNSTER on an island in the Ill in the French dep. of Bas-Rhin.
APROSITOS, the island of FUERTEVENTURA in the Canary group.
APRUSA, the river AUSA in the Papal states.
APRUTIUM, ABRUZZO in Naples.
APRUTIUM PROVINCIA AQUILENSIS, the Neapolitan prov. of ABRUZZO CITRA.
APSARUS, or APSORUS, the river OSERO flowing into the gulf of Trieste.
APSUS, a river of Illyria Græca, the modern BARETINO or ERCENT flowing into the Adriatic.
APTA JULIA, APTA, or OURTAI APTENSIUM, the modern APT in the French dep. of Vaucluse.
APUA, or PONS TREMULUS, the town of PONTREMOLI in Tuscany.
APULIA, APULIA DAUNIORUM, or DAUNIA, a district of Græcia Magna, now PUGLIA or APUGLIA in Naples. The name Apulia was originally applied to a small tract of country immediately to the SE of the Frentani; but in the reign of Augustus it included not only Daunia but also Messopia.
AQUA ANTIQUA, ALTWASSER in Silesia.
AQUABELLETTA, or AQUÆ BELLÆ, AIGUEBELLE in Savoy.
AQUABURGUM, WEISSERBURG in Bavaria.
AQUA CAMPESTRIS, AGUILAR-DEL-CAMPO in Old Castile.
AQUÆ, BADENA, CASTELLUM AQUARUM, or THERMÆ INFERIORES, the town of BADEN in Baden.
AQUÆ, AQUISGRANUM, GRAND PALATIUM, or URBS AQUENSA, AACHEN or AIX-LA-CHAPELLE in Rhenish Prussia.
AQUÆ ALLOBROGUM, or AQUÆ GRATIANÆ, AIX in Savoy.
AQUÆ AUGUSTÆ, AQUÆ TURBELLICÆ, or TURBELLA CIVITAS, near ACQO in the French dep. of Landes.
AQUÆBELLICUS in Augusta Trevirorum, the modern WASSERBILLICH at the conflux of the Moselle and the Sauer.
AQUÆ BIGERRONUM, or VICUS AQUENSIS, the town of BAGNERES-EN-BIGORRE in the French dep. of Hautes-Pyrenees.
AQUÆ BILBITANORUM, or AQUÆ BILBICITANÆ, near ALBANIA in Spanish Aragon.
AQUÆ BORMONIS, or BORBONIUM ACCIMBALDI, the town of BOURBON-L'ARCHAMBAULD in the French dep. of Allier.
AQUÆ BORVONIS, the town of BOURBONNE-LES-BAINS in the French dep. of Marne.
AQUÆ CALENTES, the town of CHAUDES-AIGUES in the French dep. of Cantal.
AQUÆ CALIDÆ, AIGUES-CHAUDES in the French dep. of Basses-Pyrenees.
AQUÆ CILINORUM, AMPHIOCHIA, or AQUÆ CELENIÆ, ORENSE in Spanish Galicia.
AQUÆ DURÆ, ALCALA-DEL-RIO in Seville.
AQUÆ FLAVIÆ, near CHAVES in the Portuguese prov. of Tras-os-Montes.
AQUÆLUPÆ, GUADELUPE in New Castile.
AQUÆ MATTIACÆ, MATTINEI FONTES, or THERMÆ WISBADENSES, the town of WIESBADEN in Nassau.
AQUÆ MORTUÆ, AIGUES-MORTES in the French dep. of Gard.
AQUÆ NERÆ, NERIS in the French dep. of Allier.
AQUÆ ONESIORUM, BARREGES-LES-BAINS in the French dep. of Hautes-Pyrenees.
AQUÆ PANNONICÆ, THERMÆ AUSTRIACÆ, or THERMÆ PANNONIÆ, the town of BADEN in the archd. of Austria.
AQUÆ REGIÆ, AEGERI or EGERI in the Swiss cant. of Zug.
AQUÆ RUBEÆ, or RUFFACUM, RUFFACH in the French dep. of Haut-Rhin.
AQUÆ SEXTIÆ, or COLONIA AQUENSIS, AIX in the French dep. of Bouches-du-Rhone.
AQUÆ SPADANÆ, SPA in Liege.
AQUÆ STATIELLÆ, ACQUI in the duchy of Montferrat.
AQUÆ TARBELLICÆ, the modern DAX in the French dep. of Landes.
AQUÆ TIBILITANÆ, the springs now known as HAMMAM-EL-BERDÆAH in Algeria.
AQUÆ VETERES, OUDEWATER in the Dutch prov. of Utrecht.
AQUÆ VIVÆ, AIGUES-VIVES in the French dep. of Var.
AQUALATUM, YGNALADA in Catalonia.
AQUARIA, YVOIRE on the lake of Geneva.
AQUA SPARSA, or AQUÆ CALIDÆ, AIGUEPERSE in the French dep. of Puy-de-Dome.
AQUIANUM, EVIAN on the lake of Geneva.
AQUILA IN VESTINIS, or AQUILIA, AQUILO in the Neapolitan prov. of Abruzzo-Ultra 2da.
AQUILANA PROVINCIA, or ABRUTIUM ULTERIUS, the Neapolitan prov. of ABRUZZO-ULTRA.
AQUILEJA, AQUILEJA COLONIA LATINA, the modern AGLAR or AQUILEJA on the gulf of Trieste.
AQUILEJA. See ACULA.
AQUILONIA, CEDOGNA in the Neapolitan prov. of Principato-Ultra.
AQUINUM, AQUINO in Naples.
AQUITANIA, one of the four provinces of Gaul under the Romans, comprising the territory of the Aquitani reaching up to the Garumna or Garonne, and that portion of Celtic Gaul which

lay between the Garumna and the Legeris or Loire. It was subdivided into Novem Populana, Aquitania Prima, and Aquitania Secunda. The division of Novem Populana comprised the original Aquitania, or the country between the Pyrenees and the Garonne.

AQUITANIA PRIMA, that portion of Gallia Celtica which Augustus added to Aquitania Proper, comprising the modern French departments of CHER, INDRE, ALLIER, CANTAL, PUY-DE-DOME, TARN-ET-GARONNE, CAREZE, and HAUTE-VIENNE. Its cap. was Civitas Bituriges now Bourges.

AQUITANIA SECUNDA comprised the modern French deps. of VENDEE, DEUX-SEVRES, VIENNE, CHARENTE-INFERIEURE, and a portion of LOT, TARN-ET-GARONNE, and GIRONDE. Its cap. was CIVITAS BURDIGALENSIUM now Bourdeaux.

AQUITANICUS OCEANUS, the GULF OF GASCOGNY.

ARA BACCHI, or BACCHARACUM, BACCHARACH in Rhenish Prussia.

ARABIA, a region bounded in ancient, as in modern geography, by Palestine and the desert of Syria on the N; by the Sinus Persicus or Persian Gulf on the NE; by the Mare Erythræum or Indian Ocean on the E and SE; by the Sinus Arabicus on the SW; and by the isthmus of Suez on the W. The name *Arab* signifies 'an inhabitant of the West,' that is, one dwelling to the west of the river Euphrates. A. was divided by the Greeks and the Romans into Arabia Deserta, Arabia Petræa, and Arabia Felix.

ARABIA DESERTA, the interior of the Arabian peninsula, more particularly the N part of it, bordering on the Syrian desert, and roamed over by tribes whom the Greeks distinguished as *Arabes Scenitæ*, that is, 'tent-dwelling Arabs.'

ARABIA FELIX, that portion of Arabia which borders on the Sinus Arabicus or Arabian gulf, below the Sinus Ælanites or gulf of Akaba, and on the Mare Erythræum or Red sea, and the Sinus Persicus or Persian gulf. It answers to the modern YEMEN, HEDJAZ, HADRAMAUT, and OMAN.

ARABIA PETRÆA, the NW portion of the Arabian Peninsula, from Palestine to the Sinus Arabicus, including the country around the Sinus Ælanites. This portion of A. was so called from its capital, Petra. It was the IDUMÆA or EDOM of Scripture, and is now nearly identical with the Great desert of EL-TYH.

ARABICI MONTES, the JEBEL MOKATTIM forming the E side of the valley of the Nile.

ARABICUS SINUS, or MARE RUBRUM, the RED SEA.

ARABISSUS, perhaps GURUM on the Tokmah-su, in Armenia.

ARABO, RABUS, or AQUA NIGRA, the river RAAB in Hungary.

ARACCA, the ancient ARACH or EDESSA, perhaps ORFA or URFA in the Asiatic pash. of Diyarbekir.

ARACEME. See PETRA.

ARACHOSIA, or COPHEN, identified by Major Rawlinson with ULAN-ROBAT or SHAHRI-ZOHAK in the Ghilzizeh country, SE of Kandahar.

ARACILLUM, ARAQUIL in the Spanish prov. of Pampelona.

ARACOSIA, ARCOZ in Andalusia.

ARADIENSIS COMITATUS, the com. of ARAD in Hungary.

ARADUS, the island of RUAD off the coast of Syria.

ARÆ FLAVIÆ, ERYTHROPOLIS, or AURICUM, NORDLINGEN in Bavaria.

ARÆGENUS, or CIVITAS-BAJOCASSIUM, BAYEUX in the French dep. of Calvados.

ARAGNUM, the river ALMEN in the Swiss cant. of Valais.

ARAGONIA, CELTIBERIA, or TERRA IBERIA, the kingdom of ARAGON in Spain.

ARAM, the ancient Hebrew name of SYRIA.

ARAM-BETH-RECHOB, probably the district of ARD or EL-HULEH in Syria, near the source of the Jordan.

ARAM-DAMESEK, or SYRIA DAMASCENA, DAMASCENE-SYRIA.

ARAM-MAACHAH, a district of Syria, extending southwards from Damascus.

ARAM-NAHARAIM, PADAN-ARAM, or CAMPI-MESOPOTAMIA, the modern EL-JEZIREH in the pash. of Bagdad.

ARAM-ZOBU, or ARAM-SABU, probably a region of Cœlo-Syria.

ARANDA DURII, ARANDO-DE-DUERO in Old Castile in Spain.

ARANDI, a town of the Tardetani, now OURIQUE in Spain.

ARANIA, the island of ARRAN in the Frith of Clyde.

ARANIA VALLIS, the valley of ARAN in the Pyrenees.

ARANIÆ INSULÆ, the ARRAN islands off the W coast of Ireland.

ARANUM, ARANNUM, or ARANCI, near AVRANCHES in the Portuguese prov. of Alentejo.

ARANUS, ARANYOS in Transylvania.

ARAR, ARARIS SAGANA, or SAUGONNA, the river SAONE in France.

ARARAT, MONTES ARMENII, TERRA ARMENIORUM, KARDU, a mountain of Armenia, still generally known as MOUNT ARARAT, the MASIS or MASEES of the Armenians, the KOH-I-NU of the Persians, the DAGHERDAGH or AGHRIDAGH of the Turks.

ARARUS, perhaps the ALUTA, a branch of the Danube.

ARASSIUM, ARASSI in the duchy of Genoa.

ARA UBIORUM, BONN in Rhenish Prussia.

ARAUGIA, ARAVIA, or AROVIUM, ARAU in the Swiss cant. of Aargau.

ARAURIS, or ERAVUS, the HERAULT or GRACE, falling into the Mediterranean near Agde.

ARAUSIO, or COLONIA ARAUSIO, or CIVITAS ARAUSICORUM, a city of the Cavares, ORANGE in the French dep. of Vaucluse.

ARAUSIONENSE CASTRUM, ORANIENBURG in Prussia.

ARAUSIONENSIS PRINCIPATUS, the principality of ORANGE in the French dep. of Vaucluse.

ARAXES, a river of Armenia Major, now the ARAS.

ARAXES, a river of Persia, now the BEND-EMIR, flowing into the lake of Bakhtegan.

ARAXES, the Oxus falling into the Caspian.

ARAXUS PROMONTORIUM, CAPE PAPAS on the NW coast of Greece.

ARBA, the island of ARBE in the Adriatic.

ARBACALA, VILLENA in the Spanish prov. of Murcia.

ARBELA, ARBEL or ERBIL in Kurdistan.

ARBERGA, or ARLABURGUM, AARBERG in the Swiss cant. of Berne.

ARBONA, or ARBOR FELIX, ARBON in the Swiss cant. of Thurgau.

ARBOREA, or ORISTANA, ORISTANO on the W coast of Sardinia.

ARBOSCA, ARBOIS in the French dep. of Jura.

ARBUDA, KNIN or TINEN in Dalmatia.

ARBURGUM, ARALÆBURGUM, AARBURG in the Swiss cant. of Aargau.

ARCA, ARQUES in the French dep. of Seine-Inferieure.

ARCADES, or ARKADE, ARCADION in the island of Candia.

ARCADIA, a mountainous region, the Switzerland of Greece, now embracing the eparchies of MEGALOPOLIS, MANTINEA, GORTYNE, and KYNURIA.

ARCATIF, or ARCATUS, near the modern ARCOL on the Coromandel coast of India.

ARCANNUM, BRACCIANUM, or BRYGIANUM, BRACCIANO in the Papal states.

ARCES AD AUGUSTIAS HELLESPONTI, or ARCES DUÆ AD HELLESPONTIUM, the DARDANELLES.

ARCHA, HERCK in Liege.

ARCHANGELOPOLIS, or FANUM, ST. MICHAELIS ARCHANGELI, ARCHANGEL in Russia.

ARCHIPELAGUS, or MARE ÆGEUM, the ÆGEAN SEA OR GRECIAN ARCHIPELAGO.

ARCI, ARCOS-DE-LA-FRONTERA in Seville.

ARCIDOVA, SLATINA in Wallachia.

ARCISSA, VAN or WAN in Armenia.

ARCIUS, the river ARC or ARCHE in Savoy.

ARCLOA, ARKLOW in co. Wicklow.

ARCOBRIGA, ARCOS on the Xalon in Old Castile.

ARCTOPOLIS, or BERNA, BERN in the Swiss cant. of that name.

ARCTOPOLIS, BIERNEBURGUM, URSORUM CASTRUM, BIORNEBURG in Finland.

ARCTOPOLIS AD SALAM, or URSOPOLIS, BERNBURG in Anhalt-Bernburg.

ARCTORIS OCEANUS, OCEANUS BOREALIS, OCEANUS SEPTENTRIONALIS, MARE GLACIALE, the ARCTIC OCEAN.

ARCUS JULIANUS, ARCUEIL in the French dep. of Seine.

ARDARTUM, ARDFERT in co. Kerry.

ARDEA, the capital of the Rutuli, the village of ARDEA, 22 m. from Rome.

ARDEA, ARDRA, or ARDRESIUM, ARDRES in the French dep. of Pas-de-Calais.

ARDABILA, ARDEBIL in the Persian prov. of Azerdbijan.

ARDELICA, PESCHIERA in the Venetian prov. of Mantua.

ARDEVICUM, or HARDEROVICUM, HARDERWYK on the Zuyder-see in Holland.

ARDIMACHA, or ARMACHA, or ARMACANUM, ARMAGH in Ulster.

ARDONCÆ, or HERDONIA, CEDONA or CEDOGNA in the Neapolitan prov. of Principato-Ultra.

ARDROSA, ARDROSSAN in Ayrshire.

ARDUENNÆ SILVÆ, a forest in Gallia Belgica, the modern ARDENNES.

AREÆ, or OBIA, or OLBIA, HAJERES in the French dep. of Var.

AREBURGIUM, or AREMONTIUM, AREMBERG on the Aar in Rhenish Prussia.

ARECANUM, ARNEMIUM, or ARNHEMIA, ARNHEIM in Holland.

ARECONIUM, or HARFORDIA, HERTFORD in Hertfordshire.

AREDATÆ, AREDATUM, AURELIANUM, GESADUNUM, LENTIA, LINCIA, or LINCIUM, LINZ on the Ens in Austria.

AREFLUCTUS, HARFLEVIUM, or HARFLONIUM, HARFLEUR in the French dep. of Seine-Inferieure.

ARELAS, GALLULA ROMA, ARELATE, ARELATUM CONSTANTINA, COLONIA JULIA PATERNA ARELATE, the town of ARLES in the French dep. of Bouches-du-Rhone.

AREMORICA, ARMORICÆ ORÆ, ARMORICA, ARMORICANUS TRACTUS, BRITANNIA MINOR, NEUSTRIA, now represented by the French provs. of BRETAGNE and NORMANDY.

ARENÆ MONTES, the ARENAS GORDAS in Spain.

ARENÆ OLONENSIS, SABLES D'OLONNE in the French dep. of Vendee.

ARENSBERGE, ARENSBERG in Westphalia.

ARENSEUM, ARLEUX in the French dep. of Nord.

ARESCHOTTUM, or ARSCHOTTUM, ARSCHOTT in Belgium.

ARETHON. See ARACTHUS.

ARETIUM, or MUNICIPIUM ARRETIUM, AREZZO in Tuscany.

ARECA, ARLANZON in Old Castile.

AREVACORUM URBS, SEGOVIA, or SEGUBIA, SEGOVIA in Old Castile.

ARGANTOMAGUS, near ARGENTON in the French dep. of Creuse.

ARGATHELIA, ARGYLESHIRE in Scotland.

ARGELIA, or FORGAVIA, FORGAN on the Elbe.

ARGENTANUM, SANTO MARCO IN LAMIS, a prov. of Calabria-Citra.

ARGENTARIA, or ARGENTOVARIA, near HARBURG in the French dep. of Haut-Rhin.

ARGENNUM PROMONTORIUM, CAPE BIANCO on the W coast of Asiatic Turkey.

ARGENTARIA, ARGENTIERE in the French dep. of Hautes-Alpes.

ARGENTEA REGIA, the modern Indian province of ARACAN.

ARGENTEUS MONS, the SIERRA-SEGURA in Spain.

ARGENTINA, JORNICK in Bosnia.

ARGENTINA CIVITUS, ARGENTORATUM, STRATEBURGUM, STRATÆBURGUS, or TRIBOCORUM URBS, the city of STRASBURG in the French dep. of Bas Rhin.

ARGENTOLIUM, ARGENTEUIL in the French dep of Yonne.

ARGENTOLIUM AD SEQUANUM, ARGENTEUIL in the French dep. of Seine.

ARGENTOMUM, ARGENTAN in the French dep. o Orne.

ARGI, ARGOLICA URBS, ARGOS, DEPSIUM, ARGO near the gulf of Nauplia in Greece.

ARGOLICUS SINUS, the gulf of NAUPLIA or NAPOLI-DI-ROMANIA.

ARGOLIS, an ancient territory of Greece, bounded on the N by Corinthia and Sicyonia; on the E by the Sinus Saronicus or gulf of Engia; on the S by Laconia and the Argolicus Sinus; and on the W by Arcadia; now composing with Corinth the eparchies of NAUPLIA, ARGOS, CORINTH, HYDRA HERMIONE and TRŒZENIA.

ARGOS AMPHILOCHICUM in Acarnania, now NEOKHORI near the river Ariadha; or perhaps KARAVASARA on the gulf of Arta.

ARGOUS PORTUS, PORTO-FERRAJO in Elba.

ARIA, a region of Asia, bounded on the N by Margiana and Bactriana; on the E by India; on the S by Gedrosia; and on the W by Parthia. It included within these boundaries the Paropamisadæ and Arachosia; and probably corresponds to the modern KHORASAN, SEISTAN, and AFGHANISTAN.

ARIA, HERAT in Khorasan.

ARIA, Lake DURRAH or ZEREB in Khorasan.

ARIANI MONTES, ARIORUM MONS or MARIANUS MONS, the SIERRA-MORENA in Spain.

ARIAS, HERI in Khorasan.

ARICA, EBODIA, or ORINIACUM, the island of ALDERNEY in the English channel.

ARICIA NEMORALIS, NEMUS, now ARICCIA or LA RICCIA in the Campagna-di-Roma.

ARICINUS LACUS, VALLERICCIA an empty crater below La Riccia in the Campagna-di-Roma, but by some identified with the LAGO-DI-NEMO

ARIGÆUM, CABUL in Afghanistan.

ARIMATHÆA, RAMA, RAMULA, the modern RAMLEH or RAMA of Syria.

ARIODUNUM, ERDING in Bavaria.

ARIOLA, or AYROLUM, AIROLO or ORIENZ in the Swiss cant. of Ticino.

ARIOVICA, PONS DUBIS, or PONTARIUM, PONTARLIER in the French dep. of Doubs.

ARISTADIUM, ARNESTADUM, or ARNSTADIUM, ARMSTADT near Erfurt.

ARIUS, a river of Aria, probably the HERIRUD.

ARLUNUM, ARALAUNUM, or ORALUNUM, ARLON in Luxemburg.

ARMADATUM, or HARIMEDABATUM, AMADABAD in Gujerat.

ARMASANICÆ, AYMARQUE in the French dep. of Gard.

ARMENIA, a region of Asia, lying between the Caucasus on the N, and the chain of Taurus on the S; but of less easily defined limits towards the E and the W. By the Greeks, Romans, and Armenians, it was divided into A. Major and A. Minor.

ARMENIA MAJOR, or ARMENIA PROPER, was bounded on the N by Colchis, Iberia, and Albania; on the NE and E by Albania; on the S by Mesopotamia, Assyria, and Media; and on the W by Armenia Minor. It constitutes the existing, and by far the larger division of Armenia; and is separated from Armenia Minor by the Euphrates; and from Mesopotamia and Kurdistan by Mount Taurus. It now belongs politically to Russia, Turkey, and Persia.

ARMENIA MINOR, ARMENIA PARVA, or ARMENIA PARTHICA, was bounded on the N by Pontus; on the E by the Euphrates; on the S and W by Cappadocia. It includes portions of the modern pashaliks of KARAMANIA, MERASH, and S VAS.

ARMENIACENSIS CIVITATUS, ARMAGNAC in the French dep. of Gers.

ARMENISTA, ARNINA, or FLORA, the river FIORE in Tuscany.

ARMENTARIÆ, ARMENTIERES in the French dep. of Nord.

ARMENTIO, the river ARMANCON in France.

ARMORICA, according to Pliny and Mannert, the early name of AQUITANIA PROPER, on the SW of France; according to Ukert, the tract of country stretching along the Atlantic between the mouths of the Loire and the Seine.

ARMORIUM CAPUT, or RHEGIUM PROMONTORIUM, the promontory of CAPO-DELL-ARMI on the coast of Calabria-Ultra.

ARMUZIA, or HARMOZIA, the island of HORMUZ in the Persian gulf.

ARNAPHA, ERFTA, or ERVATIS, ERFT in Rhenish Prussia.

ARNEMUDA, ARNUYDEN in the island of Walcheren.

ARNON, the MODJEB, MUDJEB, or WADY-EL-MUJEB flowing into the E side of the Dead sea.

ARNUS, the modern river ARNO of Tuscany.

AROCHA, a river of Bruttium, the modern CROCHA or CROCCHIO in Naples.

AROLA, or ARULA, the Swiss river AAR.

AROMATA PROMONTORIUM, CAPE GUARDAFUI on the coast of Arabia.

ARANATA, perhaps the modern ZEILAH on the Adel coast of Africa.

AROSIS, the river ARAS of Persia.

AROTTIA, the town of AROLSEN in Waldeck.

ARPI, a city of Daunia near MANFREDONIA.

ARPINUM, a town of the Volsci, now ARPINO in Naples.

ARQUATUM, ARQUA in Lombardy.

ARRETIUM, or ARRETIUM VETUS, AREZZO in the Papal states.

ARRETIUM FIDEUS, CASTIGLIONE in Tuscany.

ARRETIUM JULICUM, SABLIANO in Tuscany.

ARRIA, the island of ARROE in the Little Belt.

ARRIACA, GUADALAXARA in New Castile.

ARROSIUS, the river ARON in France, a tributary of the Loire.

ARSACIA, or RAGA, KASBIN in Persia.

ARSAMOSTA, or ARMOSATA, a town of Armenia Major, now SEMSAT.

ARSARATA, the modern ASTRABAD in Persia.

ARSIA, the river ARSIA in Dalmatia.

ARSIGNANUM, or ARSÆ LANÆ, ARSIGNANO in Lombardy.

ARSINARIUM PROMONTORIUM, or CAPUT VIRIDE, CAPE VERDE on the W coast of Africa.

ARSINOE, CLEOPATRIS, or PISIDIUM, now ASSAB on the NE coast of Africa.

ARSINOE, or CROCODILOPOLIS, a town of the Heptanomis of Egypt, near MEDINET-EL-FAYUM.

ARSINOE, or FAMA AUGUSTA, the modern FAMAGOSTA in Cyprus.

ARSINONE, or TAUCHIRA, the modern TEUCHIRA on the coast of Barbary.

ARSISSA PALUS, Lake VAN in Armenia.

ARTABRUM PROMONTORIUM, or PROMONTORIUM CELTICUM, CAPE FINISTERE, the NW point of Spain.

ARTACOANA, the ancient capital of Aria, represented by the modern HERAT in Khorassan.

ARTAMIS, a river of Bactria, now the DAKASH, an affluent of the Zohrab.

ARTAUNUM, HERBIPOLIS, MACROPOLIS, or WURCEBURGUM, the town of WURZBURG in Bavaria.

ARTAXATA, or NERONIA, the capital of Armenia-Major, now perhaps ARDASHAT.

ARTEMISIA, or DIANIUM, an island near the mouth of the Tiber, now GIANUTI or JANUTI off the coast of Tuscany.

ARTEMISIUM, DIANIUM, or HEMOROSCOPION, the town of DENIA in Valencia.

ARTEMISIUM PROMONTORIUM, DIANIUM PROMONTORIUM, or FERRARIA, at the NE extremity of the Spartarius Campus in Spain, now CAPE ST.-MARTIN.

ARTEMISIUM PROMONTORIUM, or FANUM DIANÆ, CAPE SYROCHORI on the N coast of the island of Eubœa.

ARTEMISIUS, a mountain range stretching between Monte Cave or Monte Albano and Velletri in the Papal states, near MONTE ARICANO.

ARTEMITA, a town of Assyria, now represented by SHEHRBAN.

ARTENA VEIENTIUM, probably BOCCEA near the river Arrone in the Papal states.

ARTENA VOLSCIRUM, now MONTE FORTINO in the Papal states.

ARTERA, ARTERN on the Unstrut in Prussia.

ARTESIA, or ATREBATENSIS COMITATUS, the lordship of ARTOIS in French Flanders.

ARTIGI, a town in Hispania Bœtica, now the modern ALHAMA in Granada.

ARTISCUS, a river of Thracia, the modern BUJUKDERE, a confluent of the Ergene.

ARTOBRIGA, a town in Noricum, identified by some with LAUFEN in the Salz-Kammergutt of Austria; by Mannert, with TEISSENDORF in the circle of Salzburg.

ARUCCI, or ARUCÆ, in Hispania Bœtica, perhaps MOURA in the Portuguese prov. of Alentejo.

ARULA, the ARLBERG in Suabia.

ARUNDA, the town of RONDA in Granada.

ARUNDÆ MONTES, the SIERRA-DE-RONDA in Granada.

ARUNTINA, ARUNDEL in Sussex.

ARUPENUM, a town of Illyricum, now MODRUSH in Dalmatia.

ARUSINI CAMPI, in Samnium, now ARUSIA.

ARVA, ARAVA or ORAWA in Hungary.

ARVERNA, ARVERNARUM CIVITAS, CLARUS MONS, or CLAROMONTIUM, the modern CLERMONT or CLER-

MONT-FERRANT in the French dep. of Puy-de-Dome.

ARVONIA, ARVONIENSIS COMITATUS, or SNAUDONIA, CARNARVONSHIRE in Wales.

ARX ARMINII, HERMANNSBURG in Waldeck.

ARX BATAVORUM, BATEMBURG on the Maas, in the Dutch prov. of Gelderland.

ARX BRITANNICA, BRITTENBURG in the prov. of South Holland.

ARX FONTANA, FUENTES at the influx of the Adda into Lake Como.

ARX GANDULFI, CASTEL GANDOLFO on the Lago-di-Castello in the Papal states.

ARX HERCULES MONŒCI, now MONACO in Sardinia.

ARX IPHIA, ARX FAXIANA, or CASTELLUM IPHIUM, the CHATEAU-D'IF at the mouth of the Rhone.

ARX KELLINA, ENNISKILLEN in Ulster.

ARX LUDOVICI AD SARUM, FANUM ST. LUDOVICI, or SARAVUS LUDOVICI, SAARLOUIS in Prussia.

ARX MINERVÆ, CASTRUM MINERVÆ, or PORTUS VENECIS, CASTRA in Calabria.

ARX NOVA, or CASTRUM NOVUM, NYSSLOT in Russian Finland.

ARX NOVA, NEOSELIUM, or UJAVERINUM, NEUHAUSEL on the Neutra, in Hungary.

ARX REGIA, FORT ROYAL in Martinique.

ARX RUBRA, the ROTHERTHURM in Transylvania.

ARX SANCTI HOSPITII, ST. OSPITIO in Sardinia.

ARX SANCTI JULIANI, SAO JOAN at the mouth of the Tagus.

ARX SANCTI MARTINI, or FANUM ST. MARTIN, MARTINSBERG on the Raab in Hungary.

ARX SCEPUSIENSIS, or ARX CEPUSIUM, ZIPZERHAUS in Hungary.

ARX SCHENKIANA, or PROPUGNACULUM SCHENCKIARUM, SHENKENHAUS at the influx of the Rhine into the Waal.

ARX SEPTENILIA, SETENILIA, or SEPTENILIUM, SETTENIL in Granada.

ARZEN, ARZES, or ATRANUTZIN, now ERZERUM in Turkish Armenia.

ARZILLA, or COLONIA AUGUSTA JULIA CONSTANTIA ZILI, or ZILIA, a town in Mauritania Tongitana, probably ARVILLE in Fez.

ASCALINGIUM, BENNOPOLIS, HILDESIA, or HILDESHUMIUM, HILDESHEIM in Hanover.

ASCALO, ASCALON, the modern ASKALON in Syria.

ASCANIA, or ASCHARIA, ASCHERSLEBEN in Prussia.

ASCANIA, or TUSCANIA, TOSCONELLA in the Papal deleg. of Viterbo.

ASCANIA REGIO, or PRINCIPATUS ANHALTINUS, the principality of ANHALT in Upper Saxony.

ASCANIUS, the river AKSU in Asiatic Turkey.

ASCANIUS LACUS, ASCANIUS SINUS, ASTACENUS SINUS, or OLBIANUS SINUS, now the ISNIK-GOL in the sanjak of Khodavendikar in Asiatic Turkey.

ASCHAVIA, ASCHACH in Bavaria.

ASCIBURGIUM, a town of Gallia Belgica, now ASBURG, or perhaps ESSENBURG.

ASCIBURGIUS, or ASCIBURGII MONTES, or GIGANTES MONTES, the RIESENGEBIRGE in Upper Lusatia.

ASCIBURGUM, EMBRICA, or EMERICA, EMMERICH in Rhenish Prussia.

ASCIBURGUM SCHAFNABURGUM, ASCHAFFENBURG in Bavaria.

ASCRIVIUM, CATTARA, or CATARÆ, CATTARO in Dalmatia.

ASCULUM APULUM, a town in Daunia, the modern ASCOLI-DE-SATRIANO in the Capitanata of Naples.

ASCULUM PICENUM, or COLONIA ASCULANORUM, the town of ASCOLI in the Papal states.

ASCURIS LACUS, now the MAVRO - LIMNE or EZERO in Northern Greece.

ASHTAROTH, in the vicinity of MEZARIB, 3 days' march SSW of Damascus.

ASIA, a quarter of the globe which, as known to the ancients, was bounded on the N by the Great Northern ocean; on the E by the Great Eastern ocean; on the S by the Oceanus Indicus or Indian ocean; on the SW by the Sinus Arabicus or Red sea, which separated it from Egypt; and on the W by the Mediterranean and Ægean seas, and a line drawn through the Hellespontus or Dardanelles, the Propontis or Sea of Marmora, the Pontus Euxinus or Black sea, the Palus Mæotis or Sea of Azof, and the river Tanais or Don. The older geographers considered Egypt as belonging either partially or entirely to Asia; but they were unacquainted with the extreme N and E portion of Asia. The conquests of Alexander added somewhat to their knowledge of India. The Romans, in their wars with Mithridates, king of Pontus, became acquainted with the Caucasus; and pushing forward to the shores of the Caspian, they got information of a commercial route through Bactria to India; and soon afterwards of another route over the high table-land of Upper Asia, to the Seres or Chinese. The Periplus of Arrian contains a brief description of the Pontus Euxinus or Black sea; and Ptolemy knew the Caspian to be an inland sea.

ASIA CIS TAURUM, ASIA CITERIOR, ASIA INTRA HALYN, or ASIA PROPRIA, the western half of Asia Minor.

ASIA EXTRA HALYN, ASIA EXTRA TAURUM, or ASIA MAJOR, the SE part of Asia as known to the Romans.

ASIA PROCONSULARIS, comprising Mysia, Lydia, Caria, and Phrygia, with the exception of Lycaonia.

ASINÆUS SINUS, the GULF OF CORON in the Morea.

ASINARUS, the river FREDDO in Sicily.

ASINDO, ASSIDONIA, or METHYMNA SIDONIA, now MEDINA-SIDONIA in Spain.

ASISIUM, ASSISSI in the Papal deleg. of Perugia.

ASIUS, the river ASI which flows into the Tiber at Orvieto.

ASIUS, or ŒASIUS, the river CHIASCIO, an affluent of the Tiber.

ASMIRÆA, perhaps HAMEL or HAMI in Bokhara.

ASMONÆI MONTES, the western part of the modern DAOURI chain in Bokhara.

ASNESUM, ASSENSEE in the Danish island of FUNEN.

ASNIDIA, or ASSINDIA, or ESSENDIA, ESSEN ABBEY in Prussia.

ASOPUS, a river of Bœotia, the modern Asoro.

ASOVIA, ASSOFFIUM, or TANAIS, the town of AZOF on the sea of that name.

ASPA, ASPADANA, or ASPAHAMUM, the city of ISPAHAN in Persia.

ASPALATOS, SALONA NOVA, or SPALATIUM, the town of SPALATRO in Dalmatia.

ASPALUCA, ACOUS in the French dep. of the Basses-Pyrenees.

ASPERA, ASPERN in the prov. of S. Holland.

ASPEROSA, or ASTRIZZA, ASPEROSA in Romania.

ASPHALTITES LACUS, MARE MORTUUM, MARE ORIENTALE, MARE SALINARUM, SODOMITUS LACUS, the DEAD SEA.

ASPIS, the town of ASPE in Catalonia.—Also a small harbour in Zeugitana, now CLYBEA in Barbary.

ASPRENCIA, ESPERNÆUM, or SPARNACUM, EPERNAY in the French dep. of Marne.

ASPRICOLLIS, or MONS ACUTUS, SCHERPENHEUVEL in S. Brabant.
ASPRIMONTIUM, APREMONT or ASPREMONT in the French dep. of Vendee.
ASSUMPTIONIS CIVITAS, ASSUMCION in Paraguay.
ASSUS, a Lesbian city in the Troad, now BERIAM KALESI.
ASSYRIA, in the most limited sense of the term, a province of Babylonia, now answering to a part of KURDISTAN. It was bounded on the N by Armenia; on the E by Media and Susiana; on the S by Babylonia; and on the W by Mesopatamia and part of Babylonia. Its subdivisions were Aturia to the NW of the Greater Zab; Adiabene between the Greater and the Lesser Zab; Apolloniatis to the S of the Lesser Zab; Chalonitis to the E of Apolloniatis; Sittacene around the city of that name; and Satrapene in the extreme S.
ASSYRIUM STAGNUM, the LAKE of GALILEE.
ASTABORAS, or ASTOBORRAS, the river ATBARA or TACAZZE in Abyssinia.
ASTACUS, a town of Acarnania, probably TRAGAMESTI, near the mouth of the Aspropotamo.
ASTÆ, ASTA POMPEIA, ASTENSIA CIVITAS or HASIA, ASTI in Piedmont.
ASTAPA, a town of Hispania Bœtica, near ESTEPA in Seville.
ASTAPES, or ASTAPUS, the modern ABAI or BAHR-EL-AZREK, Abyssinia.
ASTA REGIA, XERES-DE-LA-FRONTERA in Spain.
ASTASABAS, or ASTOSABUS, the BAHR-EL-ABIAD, or the true head-branch of the Nile.
ASTERABATIA, a town of Hircania, ASTERABAD on the Caspian.
ASTIJI, ASTYGIS, or COLONIA JULLÆ AUGUSTA FIRMA, in Hispania Bœtica, now ECIJA on the Xenil in Seville.
ASTRACANUM, ASTRAKHAN on the Volga.
ASTRACUS, a river of Macedonia, the modern VOSTRIZZA.
ASTUIA, or BOBENHUSA, BOBENHAUSEN near Frankfort-on-the-Main.
ASTURA, now TUERTO in the Spanish prov. of Asturias.
ASTURICA AUGUSTA, or ASTURGIA, a town in Hispania Tarraconensis, now ASTORGA in Leon.
ASTURUM LUCUS, now OVIEDO in Spain.
ASTYPALACA, or PYRRHA, or PYLÆA, the island of STAMPALIA in the Grecian archipelago.
ASTYPALÆA PROMONTORIUM, CAPE ANAPHISO in Greece.
ASYPHUS, a mountain in Libycus Nomus, now LA RABBIA.
ATAGIS, ERSACUS, or ITARGUS, the river EISACH, an affluent of the Adige.
ATALANTA, an island in the strait of Euripus, now TALANTA, or perhaps TATENDI.
ATALANTA, an island off the coast of Attica, now TALANTUS.
ATELLA, or ATELA, a town of the Osci in Campania, near AVERSA in Naples.
ATARNEA, a town of Æolia, now DIKALIK.
ATERNUM, now PESCARA in Naples.
ATESTE, a town of the Veneti in Gallia Transpadana, now ESTE in the Lombardian prov. of Padua.
ATHA, ATH in Belgium.
ATHANAGIA, a city of the Ilergetes, perhaps AGRAMANT in the Spanish prov. of Lerida.
ATHENÆ, or CECROPIA, the modern ATHENS.
ATHENÆ SALANÆ, JENA in Weimar.
ATHENÆON, or SUGDAJAH, SUDAK in the Crimea.
ATHENOPOLIS, or SAMBRACIA, GRIMAUD in the French dep. of Var.
ATHENRIA, ATHENRY in co. Galway.
ATHERDA, ATHERDEE in co. Louth.
ATHESIA, or ATHESINUS AGER, ETSHLAND in Tyrol.
ATHESIS, ATESIA, or ATISON, the river ADIGE of Italy.
ATHISO, the river TASA in the Swiss cant. of Valais.
ATHLONA, ATHLONE in co. Roscommon.
ATHO, the modern MONTE-SANTO or HAGIOS-OROS on the coast of Asiatic Turkey.
ATHOLIA, the district of ATHOL in Perthshire.
ATHRIBIS, or ATHARRABIS, represented by the village of ATRIB in the delta of Egypt.
ATHRULLA, MEDINAH-EL-NABI on the Arabian Hedjaz.
ATHURNUS, or VOLTURNUS, a river of Campania, the modern VOLTURNO of Naples.
ATHYRAS, or PYDARAS, a river of Thracia, the modern KARASU.
ATINUM, a town of Lucania, ATINO in Naples.
ARTIA, ALTENBURG in Bavaria.
ATLANTICÆ INSULÆ, ATLANTIDES, FORTUNATÆ INSULÆ, HESPERIDUM INSULÆ, or CANARIÆ INSULÆ, the CANARY group.
ATLANTICUM MARE, the ATLANTIC.
ATLANTIS, an island said to have existed at a very early period in the Atlantic ocean, over against the Columns of Hercules. Plato, who tells us that he obtained his information from the priests of Egypt, represents it as larger than Africa and Asia taken together. Some writers suppose AMERICA to have been known under this name. M. l'Abbé Jolibois, in a recent dissertation, attempts to show that this region must have been composed of the chain of Mount Atlas, the Spanish peninsula in whole or in part, and a region now submerged, situated between the Cape Verd, Canary, and Azore islands.
ATLAS MONS, DYRIN, or DARAH, the ATLAS chain in Southern Africa.
ATRIA. See ADRIA.
ATRIANUS, the TARTARO, an affluent of the Po.
ATROPATIA MEDIA, or ATROPATENE, the Persian prov. of ADERBEIDJAN.
ATTALIA, ANTALI, or SATALIA, the modern SATALIYAH in Asiatic Turkey.
ATTIDIUM, in UMBRIA, now ATTIGIO.
ATTINIACUM, now ATTIGNY in the French dep. of Ardennes.
ATTOVRIGA, VALENCIA, or MONASTERIUM VELTENBURGIUM, WITTENBURG in Bavaria.
ATURIS, or ATURIUS, the ADOUR falling into the gulf of Gascogny.
ATURUS SILVENSIS, the ADOUR-DE-SUEBE, an affluent of the Adour.
AUCENSIS FLUVIUS, the river OKA in the Russian gov. of Orel.
AUDOMAROPOLIS, AUDOMARUM, or FANUM DIVI AUDOMARI, SAINT OMER in the French dep. of Pas-de-Calais.
AUDURA, AUTURA, or EBURA, the river EURE, an affluent of the Seine.
AUFIDENA, ALFIDENA in Naples.
AUFIDUS, the river OFANTO in Naples.
AUFINA, or AUFINIUM, OFENA in Abruzzo-Ultra.
AUFONA, ALTONA, or AUVONA, the river AVON of Warwickshire.
AUGA, AUGUR, or AUGIUM, EU on the Breste in Normandy.
AUGIA ALBA, or WISSENAVIA, WEISSENAU in Suabia.
AUGIA BRIGANTINA, or AUGIA MAJOR, the abbey of MEHRERAN, on the lake of Constanz.
AUGIA DOMINA, HERISAN in the Swiss cant. of Appenzell.

AUGIA RHENI, RHEINAU on the Rhine near Schaffhausen.
AUGIA SACRA, ELGG in the Swiss cant. of Zurich.
AUGIA VIRGINUM, MAGDENAU in the Swiss cant. of St. Gall.
AUGUSTA, AGOSTA in Sicily.
AUGUSTA. See LONDINIUM.
AUGUSTA, NEOMAGUS, NEODUNUM, NOVIDUNUM, or NIVIODUNUM, near the modern NYONS or NEUS in the Swiss cant. of Vaud.
AUGUSTA ALLOBROGUM, COLONIA ALLOBROGUM, OPPIDUM ALLOBROGUM, CEBANUM, GEBENNA, GENAVA, JANOBA, or JANUA, the city of GENEVA in Switzerland.
AUGUSTA ANTONINI, GASTANIUM, or GASTINUM, GASTEIN in the circle of Salzburg.
AUGUSTA AUSCIORUM, or CLIMBERRUM, the town of AUCH in the French dep. of Thoulouse.
AUGUSTA BATIENNORUM, a town in Gallia Cisalpina, near BASSIGNANA in Lombardy.
AUGUSTA BILBILIS, a town of Hispania Tarraconensis, BAMBOLA near Calatayud in Spain.
AUGUSTA BRACARA, AUGUSTA BRACARUM, BRACARA, or BRACARAUGUSTA, the modern BRAGA in Portugal.
AUGUSTA CASTRA, AZELBURG in Bavaria.
AUGUSTA DACICA, COLONIA SARMIZEEGETHUSA, or ZARMIGETHUSA, the town of VARHELY in Hungary.
AUGUSTA DEA, CIVITAS DEENSIUM, or DEA, near ST. DIE in the dep. of Drome.
AUGUSTA EMERITA, or EMERITA, MERIDA in Estremadura.
AUGUSTA JULIA GADITANA, GADITANA CIVITAS, or GADES, the modern CADIZ in Spain.
AUGUSTA NEMETUM, or NOVIOMAGUS, SPIRES in Bavaria.
AUGUSTA NOVA, TORQUEMADA in the Spanish prov. of Palencia.
AUGUSTA PRÆTINA, AOSTA in Piedmont.
AUGUSTA RAURACORUM, or RAURICUM, now AUGST on the Rhine.
AUGUSTA ROMANDUORUM, LUCIBURGIUM, or LUXEMBURGUM, the modern LUXEMBURG.
AUGUSTA SUESSONUM, or NOVIODUNUM, in Gallia Belgica, now SOISSONS in the French dep. of Aisne.
AUGUSTA TAURINORUM, or COLONIA TAURINA, TURIN in Piedmont.
AUGUSTA TIBERII, REGINUM, CASTRA REGINA, RATISBONA, HIEROPOLIS, or TIBURNIA, REGENSBURG in Bavaria.
AUGUSTA TREVIRORUM, or AUGUSTA IN TREVIRIS, TREVERIS, or TREVERORUM URBS, TREVES in Prussia.
AUGUSTA TRICASTINORUM, or FANUM ST. PAULI TRICASTI, ST. PAUL DE TROIS CHATEAUX in the French dep. of Drome.
AUGUSTA TRINOBANTUM, LONDINIUM, LUNDINIUM, or LUNDONIA, the city of LONDON.
AUGUSTA VAGIENNORUM, SALUTIÆ, or SALUTIUM, SALUZZO in Piedmont.
AUGUSTA VANGIONUM, BORBITOMAGUS, CIVITAS VANGIONUM, WARMATIA, GARMATIA, WARMATIA, or WORMACIA, the city of WORMS in Prussia.
AUGUSTA VEROMANDUORUM, QUINTINOPOLIS, SANQUINTINUM, or FANUM ST. QUINTINI, ST. QUENTIN in the French dep. of Aisne.
AUGUSTA VESUNNA, VESONNA, PETRICORDIUM, PETROCORA, or PETROGORICUM, PERIGUEUX in the French dep. of Dordogne.
AUGUSTA VINDELICORUM, COLONIA AUGUSTA RACTORUM, or VINDELICA, the city of AUGSBURG in Bavaria.
AUGUSTAMNICA, a district in Egyptus Inferior, the modern ER-RIF on the borders of Arabia.
AUGUSTANA PRÆTORIA, a town of Dacia, ROMAN on the Moldau.
AUGUSTI LUCUS, LUGO in the Spanish prov. of Gallicia.
AUGUSTI MURI, PONS SARAVI, or SARÆ PONS, in Gallia Belgica, now SAARBRUCK.
AUGUSTI PORTUS, the exterior port of Portus, now PORTO on the Tiber.
AUGUSTI TROPÆA, TORBIA in Nice.
AUGUSTOBONA, AUGUSTOMANA, or CIVITAS TRICASSIUM, TROYES in the French dep. of Aube.
AUGUSTOBRICA, PUENTE-DEL-ARZOBISPO in New Castile.
AUGUSTOBRIGA, AGREDA in Old Castile.
AUGUSTODUNUM, CIVITAS ÆDUORUM, HEDUA, or BIBRACTE, a town of the Ædui, now AUTUN in the French dep. of Saone-et-Loire.
AUGUSTOMAGUS, CIVITAS SILVANECTUM, or RATOMAGUS, SENLIS in the French dep. of Oise.
AUGUSTORITUM, LEMOVICUM, or LEMOVICÆ, LIMOGES in the French dep. of Vienne.
AUGUSTUS, AUGST in the French dep. of Somme.
AULA BONA, AUBONNE in the Swiss cant. of Vaud.
AULA MAGNI MAGISTRI ORDINIS TEUTONICI, MARIÆ DOMUS, or MERGENTHEMIUM, MERGENTHAL in Würtemberg.
AULA NOVI, EBERSDORF on the Danube.
AULA QUIRIACA, GUERANDE in the French dep. of Loire-Inferieure.
AULA REGIA, KONIGSSAAL in Bohemia.
AULA REGIA AD LYNUM, or AULICA, ELZE in Hildesheim.
AULÆI TICHOS, in the Chersonesus, now KURUDERE.
AULIS IN BŒOTIA, the modern port of OATI.
AULON, the modern TERRA-DI-MELONE in Naples. —Also VALONA in Albania.
AUNUS, ANNEAU in the French dep. of Eure-et-Loire.
AURACIUM, AURAY in the French dep. of Nord.
AURACUM, or URACUM, AURACH or URACH in the Black Forest.
AURACUM DUCIS, or AURIACUM, HERZOGEN-AURACH in Bavarian Bamberg.
AURANITIS, or HAVRAN, the HAOURAN in Northern Syria.
AURARIA, AURARIACUM, or ABROBANIA, GROSS SCHLATTEN in Transylvania.
AURARIA PARVA, KLEIN SCHLATTEN or ZALATHNA on an affluent of the Maros.
AURASIUM, AURAS in Lower Silesia.
AUREA CHERSONESUS, or AUREA TERRA, probably the peninsula of Malacca.
AUREA TEMPE, AUREA PLANITIES, or AUREUM ARVUM, the AUE or ROSENAUE, an affluent of the Unstrutt.
AUREA VALLIS, AIRVAUX in the French dep. of Deux-Sevres.
AURELIA, AURELIANORUM CIVITAS, or CENABUM, the modern ORLEANS.
AURELIA AQUENSIS. See AQUÆ.
AURELIACUM, or AURILIACUM, AURILLAC in the French dep. of Cantal.
AURELIANENSIS AGER, the French prov. of ORLANOIS.
AURELIANENSIS INSULA, the city of ORLEANS on the Mississippi.
AURELIANUM, or ORIGANUM, ORIGANO in Lombardy.
AURICHUM, AURICH in E. Friesland.
AURIMONTIUM, GOLDBERG at the foot of the Riesengebergen.

AURISIUM, ROTH in Bavaria.
AURUNCA, SUESSA, or COLONIA JULIA CLASSICA SUESSA, a town of the Arunci, now SESSA or SOZZA in the Neapolitan prov. of Terra-di-Lavoro.
AUSA, AUSA NOVA, or AUSONA, VICH or VIQUE in Catalonia.
AUSER, or AUSUR, the river SERCHIO, an affluent of the Arno.
AUSIMUM, or AUXUMUM, OSMO or OSIMO near Ancona.
AUSSONA, or AUXONIA, AUXONNE on the Saone, in the French dep. of Cote-D'Or.
AUSTA, AUSTIA, or USTA, AUSSIG on the ELBE.
AUSTERBATIUM, OSTREVEND on the Scheldt.
AUSTRALIS TERRA, or INDIA AUSTRALIS, the AUSTRALIAN continent with POLYNESIA.
AUSTRASIA, or WESTERREICH, a name given to a territory composed, in the Middle Ages, of Brabant, Liege, Luxemburg, Lothringia, the Eifel, and Treves.
AUSTRAVIA, or CASTRA PRÆTENSIA, OSTERHOFEN in Bavaria.
AUSTRIA, or FANUM ST. LUCERII, ST. LIZIER DE CONSERANS in the French dep. of Ariege.
AUSTRIÆ ARCHIDUCATUS, the archduchy of Austria, comprising Austria Inferior and Austria Superior.
AUTIANÆ, or BATHIA, BACHIA or BATHA in Hungary.
AVALITES EMPORIUM, the modern ZEILAH on the coast of Aden.
AVALLONIA, or ABALLONIA, the district of AVALON in Nova Scotia.
AVALONIA, GLASTONBURY in Kent.
AVANTICUM, in Gallia Lugdunensis, the modern AVRANCHES in Bern.
AVARA, the river EVRE in the French dep. of Cher.
AVARICUM, BOURGES in the French dep. of Cher.
AVARIO, the river AVEYRON, an affluent of the Garonne.
AVARUM PROMONTORIUM, or VIANÆ CAPUT, near VIANA in Portugal.
AVATICORUM STAGNUM, or MASTRAMELUM STAGNUM, the MER-DE-MARTIGNES in the French dep. of Bouches-du-Rhone.
AVENACUM, AVENAY in the French dep. of Marne.
AVENDI CASTRUM, REMIREMONTIUM, or ROMERICUM CASTRUM, REMIREMONT in the French dep. of Vosges.
AVENIO, a town of the Cavares, near AVIGNON in the French dep. of Vaucluse.
AVENIONETUM, AVIGNONET in the French dep. of Aude.
AVENIONIS CASTRUM, NAPOULE in the French dep. of Var.
AVENLIFNIUS, or LIBNIUS, the river LIFFEY in Ireland.
AVENNÆ, AVESNES in the French dep. of Nord.
AVENTICENSIS LACUS, the MURTNERSEE in the Swiss cant. of Freyburg.
AVERIUM, ERCOVRIGA, or TALABRICA, AVEIRO in Portugal.
AVERNUS, AVERNO in Naples.
AVERSBERGA, AVERSBERG in Carinthia.
AVIARUM, or PLUVERIUM, PITHIVIERS in the French dep. of Loiret.
AVICULA, the VOGELBERG in the Rhætian Alps.
AVILIANA, or VILLIANA, AVEGLIANA in Piedmont.
AVILLA, AVILES in Spanish Asturias.
AVIMONS, OISEMONT in the French dep. of Somme.
AVISIUM, AVIS in the Portuguese prov. of Alentejo.
AVIUM INSULA, the Bird Islands off the Buenos-Ayres coast.
AXA, AXBRIDGE in Somersetshire.
AXALITA, LORA in Seville.
AXELLA, AXAL in Sieland.
AXELVELUNUM, HEXHAM in Northumberland.
AXIACA, OCZAKOV in the Russian gov. of Cherson.
AXIACES, the river TELIGOL in Podolia.
AXIOPOLIS, GALATZ in Moldavia.
AXIUS, or BARDARIUS, a river of Macedonia, the modern VARDAR.
AXONA, or AUXUNUUS, the AISNE an affluent of the Oise in France.
AYMONTIUM, AYAMONTE in Andalusia.
AZANIA, the AJAN territory on the NE coast of Africa.
AZELUM, ASOLO in Lombardy.
AZETIUM, a town of Peucetia now ROTEGLIANO.
AZORUS, a town of Pelasgiotis, now VUVALA in Asiatic Turkey.
AZOTUS, now ES-DUD in Syria.

B

BAALATH, or BAAL-GAD. See article BAALBEK in the body of this work.
BABA, the village of BENI-TEUDI in Fez.
BABARDIA, BAUDOBRICA, BAPPARDIA, or BOTOBRIGA, the town of BOPPARD on the Rhine.
BABBA, COLONIA JULIA BABBA, or BOBBA, near NARANJAH in Marocco.
BABEBERGA, BAMBERGA, or BERGIUM, the town of BAMBERG in Bavaria.
BABECILLUM, or BARBIZILLUM, the town of BARBEZIEUX in the French dep. of Charente.
BABELMANDALIA INSULA, a small island in the straits of Bab-el-Mandib.
BABELMANDELUM FRETUM, the STRAITS OF BAB-EL-MANDIB leading into the Red sea.
BABIA, BALBIA, ALTOMONTIUM, or MONS ALTUS, the town of MONTALTO in the Neapolitan prov. of Calabria-Citra.
BABINA, BABASEK in Hungary.
BABYLON. See article BABYLON *ante*.
BABYLONIA, a region of Asia, which, as known to the ancients, was divided into Babylonia Proper and Chaldæa; the former comprising the country extending southward from Mesopotamia, and enclosed between the Euphrates and Tigris; while Chaldæa was the district lying along the r. bank of the Euphrates, and extending as far into the Desert on the W, as the soil could be rendered fertile by irrigation. The whole territory—now corresponding chiefly to IRAK-ARABI—was bounded on the N by Mesopotamia; on the E by Susiana and Assyria; on the S by the Sinus Persicus or Persian gulf; and on the W by the Arabian Desert.
BACACUM NERVIORUM, or BAVACUM, a town of the Nervii, now BAVAY in the French dep. of Nord.
BACARI, CALICUT in Hindostan.
BACASARA, or BACASARIUM, BAKSHI-SERAI in the Crimea.
BACCÆ, or BACCIUM, BEX in Switzerland.
BACCANÆ, near BACANO in the Papal states.
BACCHARACUM. See ARA BACCHI.
BACCHILIO, or MEDOACUS MINOR, now the river BACCHIGLIONE falling into the gulf of Venice.
BACEMUM, BASREIN on the Concan coast of Hindostan.
BACENÆ, or BINASCUM, BINASKO in Lombardy.
BACENIS SYLVA, or BACHONIA, now the THURINGERWALD chain in Germany.
BACHIA, BACH in Lower Hungary.
BACODURUM, BATAVA CASTRA, PASSAVIA, or PA

TAVIA, a town of Vindelicia, now PASSAU in Bavaria.

BACQUIANUM, the island of BACHIAN in the Moluccas.

BACSIENSIS COMITATUS, the county of BATSCH in Lower Hungary.

BACTRA, the capital of Bactria, now BALKH.

BACTRIA, or BACTRIANA, a region of India, bounded on the N by Sogdiana, from which it was separated by the river Oxus; on the E and S by Aria; and on the W by Margiana. It answers now to the khanate of BALKH.

BACTRUS, or ZARIASPIS, the modern ANDERAB, flowing into the river Oxus.

BACUNTIUS, a river of Pannonia Inferior, the modern BOSSET an affluent of the Save.

BADACUM, BIDACUM, or BEDAJUM, a town in Noricum, now BUDACH in Bavaria.

BADENACHA, the district of BADENOCH in Inverness-shire.

BADENVILLA, or BADENWEILLERIUM, BADENWEILER in Baden.

BADERA, in Gallia Narbonensis Prima, now BAZIEGES in the French dep. of Haut Garonne.

BADESA, or BEDESA, in Hispania Tarraconensis, now ST. JOANNES-DE-LAS-BADESAS.

BADIA, or BATHEA, or PAX AUGUSTA, a town of Hispania Bœtica, now BADAJOZ in Estremadura.

BADRINUS, or VATRENUS, a river in Gallia Cispadana, the modern SANTERNO, an affluent of the Po.

BAECULA, in Gallia Tarraconensis, perhaps RODA on the Ter, in Catalonia.

BAELA CLAUDIA, BAELON, or BELA, a town of Hispania Bœtica, BARBATO in Andalusia.

BÆONES, or BARACE, the island of DIU in Gujerat.

BAETANA, a town in Ariaca or India extra Gangem, perhaps BEDER in Golconda.

BAETERRA SEPTIMANORUM, BETERRÆ, BACTIRÆ, or BLITERÆ, the town of BEZIERS in the French dep. of Herault.

BAETES, BETIS, or TARTESSUS, a river in Hispania Bœtica, the GUADALQUIVIR.

BÆTULA, or BETULO, BADALONIA in Catalonia.

BAGA, VAGA, or VACCA, a town in Numidia Marsylorum, near the modern VEGJA.

BAGAMEDICUM REGNUM, the state of BEGEMDER or BIGEMIDIR in Abyssinia.

BAGAUDARUM CASTRUM, the town of SAINT MAUR-DE-FOSSEZ in the French dep. of Seine.

BAGDADUM, the city of BAGDAD on the Tigris.

BAGENNÆ, or BENNA, the village of BENA in Piedmont.

BAGNERIÆ, or ONESIÆ THERMÆ, the town of BAGNERES-DE-LUCHON in the French dep. of Haute-Garonne.

BAGOUS MONS, the GAUR mountain in the Paropamisan chain of Northern India.

BAGRADA, the river MEJERDAH in Tunis.

BAHARENIA, the port of BAHREIN on the Persian gulf.

BAHUSIA, or BAHUSIUM, the town of BAHUUS in Sweden.

BAIÆ, BAIA in Naples.

BAIMOCIUM, or BAIMOZÆ, the town of BAINITZ, on the Neutra, in Hungary.

BAJANUM, or TORNACUM, the city of TOURNAY in Belgium.

BAJOARIA, or BOJORIÆ, the kingdom of BAVARIA.

BAJONA, or LAPURDUM, the town of BAYONNE in the French dep. of Basses-Pyrenees.

BAJONICÆ, BAYONNE in New Castile.

BAJONIUM, the town of BAYON in the French dep. of Moselle.

BAJONNA AD MINIUM, BAYONNE in Galicia.

BALAEPATNA, or BALIPATNA, the city of PATAN, in the state of Ajmir, in Hindostan.

BALAGNIA, ALGAIOLA in Corsica.

BALAMBUANUM, PALIMBUAM on the E coast of Java.

BALBASTRUM, BALBASTRO in Aragon.

BALEARES, the BALEARIC ISLANDS.

BALGENTIACUM, BAUGENCY in the French dep. of Loiret.

BALGIACUM, BAUGE in the French dep. of Saone-et-Loire.

BALISTELLA, BALSTAL in the Swiss cant. of Soluthurn.

BALIUM REGNUM, the DANAKIL territory on the coast of the Red sea.

BALLEGARIUM, BELLEGARIUM, or VALAGNERIA, the town of BALAGUER in Catalonia.

BALLIOLA, or BALLIOLUM, BELLE in the French dep. of Nord.

BALMA, or BAPALMA, the town of BAPAUME in the French dep. of Pas-de-Calais.

BALMA, or PALMA, BAULME in the French dep. of Doubs.

BALNEA, or BALNEALIS, BAGNOLS in the French dep. of Gard.

BALNEA REGIA, or BALNEOREGIUM, BAGNAREA in the Papal deleg. of Viterbo.

BALNEOLIUM, BAGNOLO in the Neapolitan prov. of Principato-Ultra.

BALNEUM MARIÆ, MARIENBAD in Bohemia.

BALSA, TAVIRA in the Portuguese prov. of Algarve.

BALSIO, in Hispania Tarraconensis, now BOVIA in Aragon.

BALTIA, BASILIA, SCANDIA, SCANIA, SCANDINAVIA, or SCHONIA, the Scandinavian peninsula, embracing SWEDEN, NORWAY, LAPLAND, and FINMARK.

BALTICUM FRETUM MAJUS, the arm of the Baltic known as the GREAT BELT.

BALTICUM FRETUM MINUS, the LITTLE BELT.

BALTICUM MARE, the BALTIC.

BALTINGLASIA, BALTINGLASS in co. Wicklow.

BALTIONA, BELENIZONA, or BILITIO, BELLINZONA in the Swiss cant. of Ticino.

BALUCLAVIA, JAMBOL or BALAKLAVA in the Crimea.

BAMBYCE, HIERAPOLIS, or NINUS VETUS, in Cyrrhestica, near the modern MAMPEJ in Asia Minor.

BANASA, or VALENTIA, near MAMORA on the Seibus in Marocco.

BANAVASI, the modern PUNAH in Hindostan.

BANBURIA, BANBURY in Oxfordshire.

BANDER-ABASSIUM, MERCATUS-ABASII, or GOMROMA, the port of BENDER-ABASSI on the Persian gulf.

BANDERA CONGI, or CONGENIS, BENNDER-CONGO in Farsistan.

BANDUM, the Hindu territory of AJMIR.

BANEA VALLIS, BAMENTHAL in the Swiss cant. of Valais.

BANFIA, BANFF in Elginshire.

BANGERTIUM, BANGOR in Carnarvonshire.

BANIASCUM, BAGNASCO in Piedmont.

BANTUM, the Benedictine abbey of BANZ in Bavaria.

BARANIVARIUM, BARANYA in Hungary.

BARAVELLUM, BARAVELLO on Lake Canso.

BARBANSONIUM, BARBANÇON in the French dep. of Nord.

BARBARIA, BERBERORUM TERRA, a name applied by Ptolemy to the AJAN and ZANGUEBAR coasts of Africa; and by the author of the 'Periplus' to that part of the NE coast of Africa which lies between the straits of Dire or Bab-el-Mandeb, and the Aromata-Promontorium.

BARBARIÆ MARE, or AFRICUM MARE, that part

of the Mediterranean which bathes the coasts of Tunis and Algeria.

BARBARICUS CAMPUS, the Syrian desert to the W of the Euphrates.

BARBARICUS SINUS, a gulf on the E coast of Africa, probably identical with the GULF OF ZANGUEBAR.

BARBARIUM PROMONTORIUM, CAPE ESPICHEL in Spain.

BARBATA, the island of BARBADOES.

BARBECILLUM, or BABEZILLUM, the town of BARBEZIEUX in the French dep. of Charente.

BARBELLUM, or SACER PORTUS, the abbey of BARBEAUX in the French dep. of Seine.

BARBESOLA, a town in Hispania Bœtica, now represented by GUADIARO.

BARBIUM, BARBY in Saxony.

BARCÆORUM PORTUS, or BARCENSICUM URBS, a town in Cyrenaica, now BARCA to the E of Tripoli.

BARCELORIUM, BARCELOR on the Malabar coast.

BARCELUM, BARCELOS in Portugal.

BARCHERIA, or BERCHERIA, BERKSHIRE in England.

BARCHONIUM, BARCHON in New Castile.

BARCINA, BARCINON, or COLONIA FAVENTIA, now represented by BARCELONA in Catalonia.

BARCINO VESCONIÆ, BARCELONE in the French dep. of Gers.

BARCINONA, or BARCINO NOVA, BARCELONETTE in the French dep. of Basses-Alpes.

BARCOVICUM, BARAVICUM, or BARVICUM, the town of BERWICK-UPON-TWEED.

BARDEATIS, the Indian state of BIKANIR.

BARDENSIS PRINCIPATUS, BARTH in Pomerania.

BARDENUVICUM, or BARDEVICUM, BARDOWICK in Luneburg.

BARDIA, BARTHE in Pomerania.

BARDINETIUM, the BARDINEZZA, an affluent of the Po.

BARDIUM, BARDI on the river Cano in Parma.

BARDUM, BARDO in Piedmont.

BAREA, or BARIA, the town of VERGA in Granada.

BAREDGINA VALLIS, the valley of BARREGE in the French dep. of Hautes-Pyrenees.

BARETIUM, or VARESIUM, VARESE in Lombardy.

BARGEMONTIUM, or BARGEMONUM, BARGEMONT in the French dep. of Var.

BARIANUS, BARITANUS, or BARIENSIS AGER, the Neapolitan prov. of TERRA-DI-BARI.

BARISSUS, BARASSUS, or BARTHA, the river BARTSCH an affluent of the Oder.

BARIUM, or BARRIUM, a town in Apulia Pucentia, now represented by BARI in Naples.

BARIUM, BAR in Podolia.

BARIUM AD ALLULUM, BAR-SUR-AUBE in the French dep. of Aube.

BARIUM AD SEQUARIUM, BAR-SUR-SEINE in the French dep. of Aube.

BARIUM DUCIS, or BARRA DUCUM, the town of BAR-LE-DUC in the French dep. of Meuse.

BARNE, DIONYSOPOLIS, or MATIOPOLIS, the modern VARNA in Bulgaria.

BAROCA, or BROCHIA, BAROACH in Gujerat.

BAROCCA, BAROCHE in the French dep. of Orne.

BAROFLUCTUM, BARFLEUR in the French dep. of La Manche.

BAROLUM, BARULUM, or BARUM, BARLETTA in Naples.

BARPANA, or HARPONA, the island of CERBOLI in the Tuscan sea.

BARRANA ARX, BARRAUX in the French dep. of Isere.

BARROJUS, or BIRGOS, the river BARROW in Leinster.

BARSCHIENSIS COMITATUS, the Hungarian comitat of BARS or BARSH.

BARTONIA, or BARTONIUM, BARTUS in Brandenburg.

BARUCA, BAKU in Shirwan.

BARUSSÆ INSULÆ, the ANDAMAN ISLANDS.

BARUTHUM, or BYRUTHUM, BAIREUTH in Bavaria.

BARVIA, the town of BERVIE in the Mearns.

BARYGAZENUS SINUS, the GULF OF CAMBAY.

BASENTINUS, the river BASIENTO in Naples.

BASIANA, a town in Pannonia Inferior, perhaps SABATZ on the river Sabu, but by some identified with POSEGA.

BASILEA, BASILACA, or COLONIA MUNATIANA, the town of BASLE in Switzerland.

BASILCENSIS PAGUS, the Swiss coast of BASLE.

BASINIUM, BAZINGA, or BAZINIUM, PŒSING in Lower Hungary.

BASILUSSA, or HERCULIS INSULA, the island of BASILUZZA in the Lipari group.

BASSACUM, BASSAI in the French dep. of Charente.

BASTARNORUM REGIO, RUSSIA NEGRA, or RUSSIA RUBRA, the district of PODOLIA in Russia.

BASTI, BAZA in Granada.

BASTONACUM, or BASTONIA, BASTOGNOCK in Luxemburg.

BATANA, or BATINA, a town in Media Magna, the modern SULTANIYEH in Persia.

BATAVIA, BATAVIUM, BATAVORUM AGER, or HOLLANDIA, the kingdom of HOLLAND.

BATAVODURUM, or DUERSTADIUM, a town of Gallia Belgica, now represented by WYK-DE-DUARSTEDE in the Dutch prov. of Utrecht.

BATHENIS COMITATUS, the Hungarian comitat of BACH.

BATIA, or BIETIA, BAEZA in Andalusia.

BATIANA, in Gallia Narbonensis, now BAIS on the Rhone.

BATUA, or BULUA, BUDOA in Dalmatia.

BAUCONICA, or BANCONICA, by some identified with BOCKENHEIM in Hainau, by others with OPPENHEIM in Rhein-Hessen.

BAUDOBRICA, or BAULOBRIGA, a town of the Treviri, probably near BOPPART.

BAUDRIA, BAUDRY in Neufchatel.

BAUJOVIUM, BELLIJOCUM, or BELLOJOVIUM, BEAUJEU in the French dep. of Rhone.

BAUTÆ, or BONAVILLA, BONNEVILLE in Savoy.

BAUTISUS, a river in Senia, probably the HOANG-HO of China.

BAUZANUM, BOTZANUM, or BAUXARE, BOTZEN in Tyrol.

BAVA, or BOVO, the town of BUA in Dalmatia.

BEALTA, or BULLACUM, BWELT in Brecknockshire.

BEARINA, BENECHARNUM, or BENCHARNIA, the district of BERNE in the Pyrenees.

BEATIA, or BIATIA, a town of Hispania Bœtica, now BAEZO in Andalusia.

BEBIANA, BABENHUSEN in Bavaria.

BEBRIACUM, BEDRIACUM, or VETRIACUM, a town of Gallia Cispadana, probably near CANETO in Lombardy.

BECHINA, or BECHINUM, BECHIN in Bohemia.

BECKENA, BECKUM in Westphalia.

BEDA, BITBURG in Luxemburg.

BEDAJUM, LAUFEN in Bavaria.

BEDFORDIA, BEDFORD in Bedfordshire.

BEDFORDIENSIS COMITATUS, BEDFORDSHIRE.

BEDORUM, MENDRA in Fezzan.

BEDOVINUM, BEDWIN in Wiltshire.

BEFORTIA, or BEFORTIUM, BELFORT in the French dep. of Haut-Rhin.

BEGA, BOEGE in Savoy.
BEGRA, BEJARA, or PAX JULIA, BEJA in the Portuguese prov. of Alentejo.
BEICHLINGA ARX, or BEICHLINGIUM, BACHLINGEN in Saxe-Weimar.
BEIJA, or BULLA REGIA, BEJA in Tunis.
BEJARA MELENÆ, BEXAR-DE-MELENA in Andalusia.
BELACUM, BELLAC in the French dep. of Haut-Vienne.
BELALCAZARIUM, BEL-ALCAZAR in the Spanish prov. of Cordova.
BELBUS, the river BELBO, an affluent of the Po.
BELCASTRUM, or BELLICASTRUM, BELCASTRO in Calabria Ultra 2da.
BELEGRA, or BEREGRA, a town of the Piceni, now CIVITELLA in Abruzzo-Ultra.
BELEMUM, BELEM on the Tagus.
BELENDI, BELIN in the French dep. of Gironde.
BELERIUM, or BALERIUM, probably the LAND'S END in Cornwall.
BELESMUM, or BELISMUM, BELESME in the French dep. of Orne.
BELESUS, BELEZ in Catalonia.
BELFORDIUM, BELFORTE in Calabria Ultra.
BELFORTIUM, BELFORTE in Parma.
BELGICA, or GALLIA BELGICA, a name given by the Romans to the old territory of the Belgæ, and that portion of Celtic Gaul which lay to the E of the Arar or Saone, and of the Vosges range, and extended to the Rhine. It was afterwards, by successive dismemberments, divided into five provinces, viz: (1) BELGICA PRIMA, comprising the modern duchy of TREVES, part of LUXEMBURG, and LORRAINE; (2) BELGICA SECUNDA, comprising the northern part of CHAMPAGNE, the territory of TOURNAY, the N half of the ISLE-OF-FRANCE, PICARDY, ARTOIS, and FRENCH HAINAULT; (3) GERMANIA PRIMA, comprising all the country on the l. bank of the Rhine from Mount Vocetius to the confluence of the Ovringa; (4) GERMANIA SECUNDA, comprising the l. bank of the Rhine, from the confluence of the Ovringa to the Vahalis; and (5) MAXIMA SEQUANORUM, corresponding to FRANCHE COMTE, the W portion of SWITZERLAND, and SOUTHERN ALSACE.
BELGICA, a town of the Ubii, represented by BALCKHAUSEN in the Prussian prov. of the Rhine.
BELGIUM, BELGICA, or GERMANIA INFERIOR, the present kingdom of BELGIUM.
BELGIUM AUSTRIACUM, or B. HISPANICUM, the AUSTRIAN NETHERLANDS.
BELGIUM CONFŒDERATUM, the DUTCH NETHERLANDS.
BELGIUM GALLICUM, part of FLANDERS, HENNEGAU, NAMUR, and LUXEMBURG.
BELGIUM NOVUM, the state of NEW YORK in North America.
BELGRADUM, or BELLOGRADUM, the fortress of BELGRADE.
BELICA, or BELLICUM, BELLAY in the French dep. of Ain.
BELIGRADA, or ZARA VETUS, ZARACHA in the Morea.
BELIO, BELION, LIMAEA, LIMIUS, or OBLIVIONIS FLUMEN, a river in Hispania Tarraconensis, now the LIMA of Portugal.
BELISAMA, the river RIBBLE in Lancashire.
BELJOCUM, BELJOCO in Aragon.
BELLA AQUA, the abbey of BELLE-AIGUE in Auvergne.
BELLA BRANCA, the abbey of BELLE-BRANCHE in Maine.
BELLA PERTICA, the abbey of BELLE-PERCHE on the Garonne.
BELLA REPARIA, BEAUREPAIRE in the French dep. of Isere.
BELLA STELLA, the abbey of BELLE-ESTOILE in Normandy.
BELLA VILLA, BELLEVILLE on the Saone, in the French dep. of Rhone.
BELLAMANTIUM, or BELLOMANTIUM, the town of BEAUMONT-LE-ROGER in the French dep. of Erne.
BELLEFORDIA, BEAUFORT in the French dep. of Mayenne-et-Loire.
BELLEGARDIA, or BELLEGARDUM, BELLEGARDE in the French dep. of Cote-d'Or.
BELLELAGICUM, the abbey of BELLELAY in the Swiss cant. of Basle.
BELLICADRUM, or BELLOQUADRUM, BEAUCAIRE in the French dep. of Gard.
BELLIJOCENSIS, or BELLAJOVIENSIS AGER, the territory of BEAUJOLAIS in Burgundy.
BELLILOCUS, BEWDLEY in Worcestershire.
BELLIOLA. See BALLIOLA.
BELLIPRATUM, BEAUPRE in Flanders.
BELLISMUM, BELLESME in the French prov. of Perche.
BELLOFONTANUM, FONS BLIANDI, or FONS BELLAQUENS, FONTAINEBLEAU in the French dep. of Seine-et-Marne.
BELLOMARISCUS, BEAUMARIS in Anglesea.
BELLOMONTIUM, BEAUMONT in the French dep. of Haute-Garonne.
BELLOMONTIUM, or BELLUS MONS, BEAUMONT in Hainault.
BELLOMONTIUM, BELMONTE in Calabria-Citra.
BELLOMONTIUM in Argona, BEAUMONT-EN-ARGONE in the French dep. of Ardennes.
BELLOMONTIUM VICECOMITES, BEAUMONT-LE-VICOMTE in the French dep. of Sarone.
BELLOPRATUM, BEAUPREAU in the French dep. of Eure.
BELLOVACI, BELVACUS, or CÆSAROMAGUS, BEAUVAIS in the French dep. of Oise.
BELLUM VADUM, BILBICUM, or FLAVIOBRIGA, BILBOA in Biscay.
BELLUNENSIS AGER, BELLUNESE in Lombardy.
BELLUS DECUS, the abbey of BEAUBEC in Normandy.
BELLUS LACUS, BEAULIEU in Vendee.
BELLUS LACUS AD DARANIUM, BEAULIEU in the French dep. of Correze.
BELLUS LACUS AD INGERIM, BEAULIEU in the French dep. of Indre.
BELLUS LACUS AD LIGERIM, BEAULIEU on the Loire.
BELLUS MONS, BEAUMONT-SUR-OISE in the French dep. of Seine et Oise.
BELNA, or BELNUM, BEAUNE in the French dep. of Cote-d'Or.
BELNA, BEAUNE in the French dep. of Loiret.
BELOGRADUM, BIELGOROD in the Russian gov. of Kursk.
BELSIA, BEAUCE in the French dep. of Eure-et-Loir.
BELZA, or BELZIUM, BELZ in Austrian Galicia.
BENACIUM, GARDA on the Lago-di-Garda.
BENACUS LACUS, the LAGO-DI-GARDA.
BENALGIÆ, BENANGES in the French dep. of Gironde.
BENDIMIRUS FLUVIUS, the ARAXES or BEND-EMIR.
BENEHARNUM, LASCAR in the French dep. of Basses-Pyrenees.
BENEVENTUM, or COLONIA BENEVENTANA, BENEVENTO in the Papal States.
BENGALÆ REGNUM, the province of BENGAL in India.

BENGALÆ SINUS, or GANGETICUS SINUS, the GULF OF BENGAL.
BENIBOCHULA, BENIBOCHUL in Fez.
BENIMARASIUS MONS, BENIMARAZ in Fez.
BENINUM, BENIN on the Slave coast.
BENINUS SINUS, the GULF OF BENIN.
BENNAVENNA, or BENNAVENTA, DAVENTRY in Northamptonshire.
BENTHEMIUM, BENTHEIM in Hanover.
BENTOLIERA, the island of VENTOLIERE off the Neapolitan coast.
BERARUM REGNUM, the state of BERAR in Hindostan.
BERAUNA, or VERONA, BERAUN in Bohemia.
BERCIZOMA, or BERGA-AD-ZOMAM, BERGEN-OP-ZOOM in Holland.
BERCORIUM, or BRESSUIRA, BRESSUIRE in the French dep. of Deux-Sevres.
BERENICE, HESPERIS, or HESPERIDES, a city of the Pentapolis in Cyrenaica, near the modern BENEGASI.
BERENICE PANCHRYSOS, or B. JUXTA SABAS, a town in Troglodytica, perhaps ATALAKI on the Arbian gulf.
BERGA, BERGEN in Norway.
BERGÆ DIVÆ GERTRUDIS, or GERTRUDEBERGA, GERTRUIDENBERG in North Brabant.
BERGÆ SANCTI VINOXI, or VINOCIBERGA, BERG-SAINT-VINOX in the French dep. of Nord.
BERGAMENSIS AGER, or BERGAMENSE TERRITORIUM, the province of BERGAMO in Austrian Italy.
BERGENSIS PREFECTURA, the prov. of BERGENHUUS in Norway.
BERGERACUM, BERGERAC in the French dep. of Dordogne.
BERGIDUM, BERGIUM, or VIRGIUM, BENAVARRI in Aragon.
BERGINTRUM, or FANUM SANCTI MAURITIS, SAINT-MAURICE in Savoy.
BERGOMUM, BERGAMO in Lombardy.
BERGULÆ, BERGULA, or BERGULIUS, a town in Thracia, now BURGHAZ in Bulgaria.
BERGUSIUM, or BARGUSIA, BOURGOIN in the dep. of Isere.
BERISA, a town in Pontus, near TOCAT in the pash. of Siwas.
BERLENGA, or VELERANICA, BERLENGA in Old Castile.
BERNAYUM, BERNAY in the French dep. of Eure.
BERNENSIS PAGUS, or BERNENSIUM, the Swiss cant. of BERNE.
BEROBE, the state of MERGH in India.
BEROLINUM, BERLINIUM, or BERLINUM, the city of BERLIN in Prussia.
BEROLSTADIA, or BERNSTADIUM, BERNESTADT on the Weida in Silesia.
BEROLSTADIA LUSATIÆ, BERNSTADT on the Pliesnitz.
BERONICUM, or VERNULACUM, BERNRIED in Bavaria.
BEROSTONUM, BERNSTEIN in Prussia.
BEROSUS MONS, the CHATYR-DAGH in the Crimea.
BEROTHA, BERETHUS, BERYTUS, or COLONIA FELIX JULIA, BEIRUT in Syria.
BERSINUM, or LOMBARIA, LOMBEZ in the French dep. of Gers.
BERTISCUS MONS, the JAMUS-DAGH in Albania.
BERTRANOPOLIS, or FANUM SANCTI BERTRANDI, BERTRAND in the French dep. of Haut-Garonne.
BERUBIUM, or BERUVIUM, DUNCANSBY HEAD on the coast of Carmarthenshire.
BESCIA, or VESCIA, VESCINO in Naples.
BESELDUNUM, BESALU in Catalonia.
BESIDLÆ, BISIGNANO in Calabria-Citra.
BESIGHAMIUM, or BESSIGHEMIUM, BESSIGHEIM in Würtemberg.
BESIPPO, VEGEL in Andalusia.
BESONTIUM, CHRYSOPOLIS, or VISONTIO, a town of Gallia Belgica, BESANÇON in the French dep. of Doubs.
BESYNGA, or BABYSENGA, the city of PEGU in India.
BETH-SEMES, MATARIEH in Egypt.
BETHULIA, or BETYLUA, SAFAT in Syria.
BETHUNIA, BETHUNE in the French dep. of Pas-de-Calais.
BETTIGUS, or GATES, the GHAUTS in Hindostan.
BETULA, UBEDA in the Spanish prov. of Jaen.
BETULUS, BELES in Catalonia.
BEUCINUM, BUCEPHALEA, or BUXONIUM, BUTZOW in Mecklenburg-Schwerin.
BEVELANDIA AUSTRALIS, ZUID BEVELAND in the Dutch prov. of Zeeland.
BEVELANDIA SEPTENTRIONALIS, NORD BEVELAND in Zeeland.
BEVERA, the river BEVER, an affluent of the Weser.
BIALOQUERCA, BIALOCERKIEF in the Russian gov. of Kief.
BIARMIA, or PERMIA, the Russian gov. of PERM.
BIBACUM, VICHTACH in Bavaria.
BIBERACUM, or BIBRACUM, BIBERACH in Würtemberg.
BIBERAHA, BIBRA in Saxony.
BIBERVACUM, BIBERBACH in Bavaria.
BIBISCUM, or VIVIACUM, VEVAY in the Swiss cant. of Vaud.
BIBIUM, CARLSTADT in Croatia.
BIBONIUM, BÖCHLINGEN in Würtemberg.
BICONA, or PICINA, PITSCHEN in Prussia.
BICONA, or BICHSEUM, BITSCH in the French dep. of Moselle.
BICORNIS, FURCA, FURCELLA, or FURCULA, the FURKA or GABELBERG in the Swiss cant. of Valais.
BIDBURGUM, BIEDBOURG on the river Nyms in Luxemburg.
BIDGROTIA, or BROMBERGA, BROMBERG in Prussia.
BIDINUM, VIDINUM, or BYDENA, WIDDIN in Bulgaria.
BIDOSSA, VEDASUS, or VIDASSUS, the river BIDASSOA on the Spanish frontier.
BIDRUNTUM, BITUNTUM, or BUDRANTUM, BITONTO in the Neapolitan prov. of Terra-di-Bari.
BIELA, or EBILLINUM, BIEL in Aragon.
BIELCA, BIELSK in the Russian gov. of Bialystock.
BIELCENSIS PALATINATUS, the Polish territory of PODLACHIA.
BIELLA, BIELLUM, BIENNA, or BIPENNIUM, BIEL or BIENNE in the Swiss cant. of Bern.
BIENNENSIS LACUS, the BIELER-SEE in the Swiss cant. of Bern.
BIENTINUM, BIENTINA in Lucca.
BIGASTRUM, or ORIOLA, ORIHUELA in Valencia.
BIGENA, VILLENA in Murcia.
BIGONENSIS COMITATUS, the territory of BIGORRE in the French dep. of Hautes-Pyrenees.
BIHACIUM, BIHACZ in Croatia.
BIHARIENSIS COMITATUS, the Hungarian comitat of BIHAR.
BIHARIUM, BIHAR in Hungary.
BILBILIS, or SALO, a river of Hispania Tarraconensis, an affluent of the XALON in Aragon.
BILBILIS NOVA, or CALATAJUBA, CALATAYUD in Aragon.
BILCHÆ, BILCHES in the Spanish prov. of Jaen.
BILEDULGENDIA, BILEDULGEND in Barbary.
BILEFELDIA, BIELEFELD in Prussia.

BILHOMUM, or Billemum, Billom in the French dep. of Puy-de-Dome.
BILINA, Bilin in Bohemia.
BILITIUM, Bilitz in Silesia.
BILLERBECCUM, Billerbeck in Westphalia.
BILOCERENSIS DUCATUS, Bielo-Osero in the Russian gov. of Novgorod.
BIMINACIUM, a town of Mœsia Superior, probably near Widdin in Bulgaria.
BIMONIUM, or Binonium, Binchester in Durham.
BINÆ, or Binascum, Binasco in Lombardy.
BINARUSIA, Binaros in Valentia.
BINCHIUM, Binche in Hainault.
BINGIUM, Bingen in Rhein-Hessen.
BIORA, Biota in Sardinia.
BIPONTIUM, Bipontum, or Geminus Pons, Zweybrucken in Bavaria.
BIRCA, the island of Biorkoe in Sweden.
BIRCOFELDA, Birkenfeld in Rhenish Prussia.
BIRFLITUM, Biervliet in Dutch Flanders.
BIRICIANA, the river Wernitz in Bavaria.
BIRTHA, Bir in the pash. of Diyarbekir.
BIRTHALBINUM, Birthelm in Saxony.
BISANICUS, the river Bisagno in Genoa.
BISANTHE, or Rhædestus, Rodosto in Rumelia.
BISCAJA, or Cantabria, the Spanish prov. of Biscay.
BISCARGIS, Flix in Catalonia.
BISSONUM, Bessan in the French dep. of Herault.
BISTRICIA, or Bistricium, Bistritz in Saxony.
BISULA. See Visula.
BITECTUM, Bitetto in the Neapolitan prov. of Terra-di-Bari.
BITURGIA, Borgo-di-San-Sapolero in Tuscany.
BITURICENSIS DUCATUS, or Biturigum Provincia, the duchy of Berry in France.
BITURIGES VIBISCI, Medoc in the French dep. of Gironde.
BIZYA, in Astica, probably Vizia in Romania.
BLABIA, or Portus Ludovici, Port Laus in the French dep. of Morbihan.
BLAESÆ, Blesum, or Blæsia, Blois in the French dep. of Loire-et-Cher.
BLANCOBERGA, Blankenburg in Brunswick.
BLANCOFORTIUM, Blanquefort in French Guienne.
BLANCOHAYNIUM, Blankenhayn in Weimar.
BLANDA, Blanes in Catalonia.
BLANDENA, or Blandenona, in Gallia Cisalpina, near Broni in Piedmont.
BLANDONA, Zara-Vecchia in Dalmatia.
BLARA, Blair Athole in Scotland.
BLASCON, or Bresconia, the island of Brescon in the gulf of Lyons.
BLATOBULGIUM, or Blatum Bulgium, a station of the Brigantes, probably Bowness in Cumberland.
BLAVIA, or Blavatum, Blaye in the French dep. of Gironde.
BLECHINGIA, or Blekinga, the Swedish Bleckingen.
BLENAVIUM, Blesnau in the French dep. of Yonne.
BLEONIS VILLA, Blainville, in the French dep. of La Manche.
BLEONIS VILLA AD MASAM, Blainville on the Maase.
BLESSA, Blies in the Prussian circle of Saarbruck.
BLETERUM, Bletteraus in the French dep. of Jura.
BLETISA, Ledesma in Salamanca.
BOA, the island of Bua in the Adriatic.
BOANDUS, Boina, or Buvinda, the river Boyne in Ireland.
BOARIS, the town of Toro on the coast of Sardinia.
BOBIANUM, or Boianum, Bojano in Naples.
BOBIUM, or Bobbium, Bobio on the Trebia in Tuscany.
BOCARDI INSULA, Isle Boachard in the French dep. of Indre.
BOCCHOTTIA, Bachott on the Aa in Prussia.
BOCHANIUM, or Bachanium, Bouchain in the French dep. of Nord.
BOCKANICA, Bockenheim in the French dep. of Bas-Rhin.
BODA, or Hercynidus, the river Bode, an affluent of the Saale.
BODAMI CASTRUM, Bodungo, or Padium, the castle of Bodmen on the Boden-see.
BODAMICUS LACUS, Brigantiæ Lacus, Brigantinus Lacus, or Suevicus Lacus, the Boden-see or Lake of Constanz.
BODINCONIGUM, Bodincomagus, Industria, or Casalium, Casale in the duchy of Montferrat.
BODINCUS, Bodingus, or Eridanus, the river Po.
BODIUS MONS, Monte Alboddo in the marquisate of Ancona.
BODOTRIA, or Bodena Æstuarium, the Frith of Forth in Scotland.
BODROGENSIS COMITATUS, the Hungarian comitat of Bodrog.
BODROGUM, Bodrog in Hungary.
BODROGUS, the river Bodrog, an affluent of the Theiss in Hungary.
BOEA, or Boiæ, Vatika in the Morea.
BOEBICUS SINUS, the Gulf of Vatika.
BOEMIA, Bojemum, Bojohemia, or Boamia, the kingdom of Bohemia.
BŒOTIA, a district of Greece, bounded on the N and NE by the Opantius Sinus and Euripus; on the SE by Athens; and on the SW by the Sinus Corinthiacus, now part of Livadia.
BOEROSIA, Boraas in Swedish Gothland.
BOETONOMACUM, Ribchester in Lancashire.
BOEUM, or Boium, in the Dorian Tetrapolis, now Mariolates, or perhaps Astaka.
BOJATUM, Buchsium, or Buxium, Buch in the French dep. of Gironde.
BOJOBINUM, or Boviasmum, a town of the Marcomanni, now Prague in Bohemia.
BOJODURUM, or Bolodurum, a town in Noricum Ripense, now Innstadt.
BOJORUM AGER, the Bourbonnois, or dep. of Allier in France.
BOJORUM DESERTA, the Wienerwalde in Germany.
BOJUS AGER, Boicus Ager, Lombardia, or Longobardia, Lombardy in Northern Italy.
BOLAGASUS, Volgesia, or Vologessia, Meshid-Hosein on the Euphrates.
BOLBE, or Prasias, Siderokapsa on the gulf of Contessa.
BOLBITINA, Rosetta in the Delta of Egypt.
BOLBITINUM OSTIUM, or Bolbiticum Ostium, one of the seven ancient deltoid branches of the Nile, now known as the Rosetta arm.
BOLCONES FANUM, Bolkenhayn in Silesia.
BOLESLAI FANUM NOVUM, or Boleslavia Bohemiæ, Jung-Bunzlau in Bohemia.
BOLESLAI FANUM VETUS, Alt-Bunzlau in Bohemia.
BOLESLAVIA, or Baleslavia Silesiæ, Bunzlau on the river Bober in Silesia.
BOLEUM, or Bovillium, Boglio in Piedmont.
BOLIASCUM, Bogliasco on the coast of Genoa.
BOLONIA, Bononia, Gessoriacum, or Moriorum Portus Britannicus, Boulogne-sur-Mer in the French dep. of Pas-de-Calais.
BOLONIA, or Bononia Vasconiæ, Boulogne in the French dep. of Haut-Garonne.

BONA CELLA, or DEI CELLA, GOTTESZELL in Würtemberg.
BONA DIES, GORTENTAG in Silesia.
BONA FORTUNA, the island of GORT-GLUCK or FORTUNA off the coast of Sumatra.
BONA VALLIS, or BONNOVALLIS, BONNEVILLE in the French dep. of Loire.
BONACUMBA, the abbey of BONNECOMBE in the French dep. of Avignon.
BONÆ FORTUNÆ INSULA, the larger ANDAMAN island in the gulf of Bengal.
BONÆ SPEI INSULA, the island of GOOD HOPE in the Navigator's group.
BONÆ SPEI PROMONTORIUM, the CAPE OF GOOD HOPE.
BONDELIA, BAGNONE in Tuscany.
BONI AERIS INSULA, BONAIR in the Little Antilles group.
BONI DESIDERII PROMONTORIUM, Cape BUENO-DESEO in New Guinea.
BONIFACII CARTAS, or BONIFACIUM, BONIFACIO on the S coast of the island of Corsica.
BONIFACII SINUS, the GULF OF BONIFACIO.
BONNA, BONNA AD RHENUM, or CASTRA BONENSIA, BONN on the Rhine.
BONNIUM, BONNY in the French dep. of Loire.
BONONIA, BONONIENSIS COLONIA, or BONONIA PINGUIS, the city of BOLOGNA.
BONONIA, a town of Pannonia Inferior, probably ILLOK on the Danube.
BONONIENSIS AMNIS, or RHENUS, the river RENO, an affluent of the Po.
BONOPPIDUM, the Prussian town of GUTSTADT on the Alle.
BONUS AER, or FANUM SANCTI TRINITATIS, the city of BUENOS-AYRES on the Rio-de-la-Plata.
BORA, PRILIPPO in Macedonia.
BORBONIUM ANSELMIUM, BARBA ANCELLI, or BAIA, BOURBON-LANCY in the French dep. of Saone-et-Loire.
BORCKELAA, BARCHELO in Gelderland.
BOREALE ARCTOUM, SEPTENTRIONALE MARE, or SEPTENTRIONALIS OCEANUS, the NORTH SEA.
BOREALE CAPUT, the NORTH CAPE in Norway.
BOREALIS FLUVIUS, the RIO BRAVO-DEL-NORTE of America.
BORECA, JERKAN in Bohemia.
BOREUM PROMONTORIUM, on the coast of Cyrenaica, CAPE TEJORES on the Barbary coast.
BOREUM PROMONTORIUM, CAPE KALASNAIN in Ceylon.
BORINGIA, or BORNHOLMIA, the Swedish island of BORNHOLM.
BORMANICIUM, MOMESQUE in Provence.
BORMIANA JUGA, or BRAULIUS MONS, MOUNT BRAULIO in Tyrol.
BORMIANÆ THERMÆ, WURMSER-BAD in the VALTELINE.
BORMIUM, BORMIO on the Adda in Lombardy.
BORONTIUM, BURANZO in Piedmont.
BORSODIENSIS COMITATUS, the comitat of BORSOD in Hungary.
BORTINA, or BURTINA, ALMODEVAR in Aragon.
BORUSSIA, or PRUSSIA, the kingdom of PRUSSIA.
BORYSTENES, or DANAPUS, the river DNIEPER.
BORYSTHENIS, ORBIA, OLBIOPOLIS, MILETOPOLIS, or SAVIA, KUDAK on the Dnieper, in the Russian gov. of Kief.
BOSCOBELLUM, or HENRICOMONTIUM, BOISBELLE in the French dep. of Cher.
BOSCODUCUM, BOSCUM DUCIS, or SYLVA DUCIS, HERZOGENBERSCH in North Brabant.
BOSNIA, or BOSINA, the kingdom of BOSNIA.
BOSPHORUS, OCHSENFURTUM, or OXONIUM, OCHSENFURT in Bavaria.
BOSPORUS, BOSPORUS CIMMERIUS, or FRETUM BOSPORANUM, the STRAITS OF KAFFA.
BOSPORUS THRACIÆ, or BOSPORUS THRACIUS, the STRAITS OF CONSTANTINOPLE.
BOSTAMPIUM, POSTAMPIUM, POSTERNUM, or POTTSTAMPIUM, the city of POTSDAM.
BOSTONIUM, BOSTON in Lincolnshire.
BOTHNIA, or BOTNIA, BOTHNIA in Sweden.
BOTHNICUS SINUS, or BOTNICUS SINUS, the GULF OF BOTHNIA.
BOVINÆ, or BOVIAICECUM, BAVOIGNES in Namur.
BOVINUM, the island of BOUM at the mouth of the Loire.
BOXTEHUDA, BUXTEHUD in Bremen.
BOYNUM, or PAYNUM CASTRUM, PEINA in Hanover.
BOYOLUM, BOYOLO in Lombardy.
BOZOKIENSIS COMITATUS, the comitat of BOSOGH in Hungary.
BRABANTIA, BRABANT in the Netherlands.
BRABONIACUM, BROUGHAM in Westmoreland.
BRACANETUM, BRIQUENAY in Champagne.
BRACCHIA, BRACHIA, or BRATTIA, the island of BRAZZA in the Adriatic.
BRACCHION, an island in the Syrtis Minor, now GERBEH.
BRACHME, SERINGAPATAM in Hindostan.
BRACLAVIA, or BRASLAVIA AD HYPANUM, BRASLAU in Podolia.
BRACLEJUM, BRACKLEY in Northamptonshire.
BRADANUS, the river BRADANO falling into the gulf of Tarento.
BRÆA, or BREA, BREY in Liege.
BRAGANTIA, or BRIGANTIA, BRAGANZA in Portugal.
BRAILOVIA, BRAILOW in Podolia.
BRAJA, or BRIA COMITIS ROBERTI, BRIE-SUR-YERES in the French dep. of Seine-et-Marne.
BRAMENIUM, or BREMENIUM, BRAMPTON in Cumberland.
BRAMOSANTUM, BRAMASANT in the Swiss cant. of Berne.
BRANA, BRENNIA AD VIDULAM, BRENNACUM, or BRONA, BRENNA in the French dep. of Aisne.
BRANA, or BRENNIA ALLODIENSIS, BRAINE-D'ALEN in South Brabant.
BRANCASTRUM, or BRANCIODUNUM, BRANCASTER in Norfolk.
BRANDEBURGIUM, or BRENNOBURGUM, BRANDENBURG on the river Havel in Prussia.
BRANDENBURGUM, BRANDENBURG on the Frische-haff in Prussia.
BRANDENBURGUM NOVUM, NEW BRANDENBURG in Mecklenburgh-Strelitz.
BRANDESIUM, or BRANDUSIUM, BRANDEIS in Bohemia.
BRANECIUM, BRANECUM, or BRUNOPOLIS, BRUNEGG in the Pusterthal.
BRANESIA, OLDENBURG in Holstein.
BRANGONIA, BRANONIUM, BRANOVIUM, or VIGORNIA, the city of WORCESTER in England.
BRANSKUM, BRANSK in Bialystock.
BRANTOSONIUM, BRANTHANE in the French dep. of Dordogne.
BRASILIA, the empire of BRAZIL.
BRASLAVIA, or BATISLAVIA, BRASLOW in the Russian gov. of Wilna.
BRASLOVIA, BRASLOW in Wallachia.
BRASSOVIA, CORONA, or STEPHANOPOLIS, CRONSTADT in Transylvania.
BRAUNODUNUM, BRAUNAU in Bavaria.
BRAUNSBERGA, or BRUNSBERGA, BRAUNSBERG in Prussia.
BRAURON, or BRAURONA, VRANA in Rumelia.
BRAVIARUM AD SAMARAM, BRAY in Picardy.
BRAVINIUM, BRAMFIELD in Shropshire.

BREANIA, BRENIA, or CAVANUM, CAVAN in co. Cavan.
BRECECHA, BRESACUM, or BRISCACUS MONS, BREISACH in Baden.
BRECHINIA, BRECKNOCK in Brecknockshire.
BRECHINIENSIS AGER, BRECKNOCKSHIRE.
BRECHINIUM, BRECHIN in Forfarshire.
BRECISLABURGUM, POSONIUM, or PRESBURGAM, PRESSBURG in Lower Hungary.
BREDANA, or BAROCHIA, BREDA in North Brabant.
BREDEFORTIA, or BREFORTIUM, BREEDEVOORT in Gelderland.
BREGA, or BRIGA, BRIEG in Silesia.
BREGENSA THERMÆ, BRIEG in the cant. of Valais.
BREGENTIA, or BRIGANTIA, BREGENZ on the Lake of Constanz.
BREGENTIUM, or BRIGANTIA, BREGENZ in Tyrol.
BRELIUM, BREGLIO in Piedmont.
BREMA, the city of BREMEN on the Weser.
BREMBUS, the BREMBO an affluent of the Adda.
BREMERFURDA, or BREMERVORDA, BREMERWORDE in Hanover.
BREMETUM, or BREMMA, BREMA in Piedmont.
BREMOGARTUM, BREMGARTIN in the Swiss cant. of Aargau.
BRENA, BREANA, or CASTRUM BRIENNENSE, BRIENNE in the French dep. of Aube.
BRENDULUM, or BRENTULA, BRENDOLA in Lombardy.
BRENIANUM, BREGNANO on the Sesia.
BRENNÆ VALLES, BREGNA in the Swiss cant. of the Grisons.
BRENNIA CASTRENSIS, BRAIN-LE-CHATEAU in Hainault.
BRENNIA COMITIS, or BRONIUM, BRAIN-LE-COMTE in Hainault.
BRENNOVICUM, or MAURIANA, SAINT-JEAN-DE-MAURIENNE in Savoy.
BRENNUM, BRENO in Lombardy.
BRENNUS, the BREGNA, an affluent of the Bessin, in the Swiss cant. of the Grisons.
BRENNUS MONS, the BRENNER-BERG in Tyrol.
BRENTA, the BRENZ, an affluent of the Danube.
BRENTESIA, or MEDOACUS MAJOR, the river BRENTA in Northern Italy.
BRESEGELLA, BRISIGELLA in the Papal states.
BRESSIA, or SEBRISIANUS, BRESSE in the French dep. of Ain.
BRESTA, or BRESTIA CUJAVIÆ, BRZESE in Poland.
BRESTIA AD HYPANIM, BRZESC in the Russian gov. of Grodno.
BRETELIUM, or BRETOLIUM, BRETEUIL in the French dep. of Eure.
BRETTA, or VENERBURGUM, WENERSBORG in Swedish Gothland.
BREUCOMAGUM, or BROCOMAGUS, BRUMFT 12 m. NW of Strasburg.
BREUTHE, KARITENA in the Morea.
BREVIODURUM, PORT-AUDEMER in the French dep. of Eure.
BREXALIA VALLIS, the BERGELTHAL in the Grisons.
BREXIA, or BRIXIA, a town of the Cenomanni, now BRESCIA in Lombardy.
BREZINA, BRESINI in Poland.
BRIANZONIA ARX, BRIANÇONET in the French dep. of Isere.
BRIARIA, BRIBODURUM, or BRIDOBORUM, a town in Gallia Lugdunensis, near BRIARE in the French dep. of Loiret.
BRICEJUM, or BRIEUM, BRIEY in the French dep. of Moselle.
BRIDLINGTONIUM, or BRILLENDUNUM, BRIDLINGTON in Yorkshire.
BRIELA, BRIEL in Holland.
BRIENSIS AGER, or BRIGENSIS PAGUS, the lordship of BRIE in the French dep. of Seine-et-Marne.
BRIEZA FIDA, BRIEZEN in Prussia.
BRIGA, the BREG an affluent of the Danube.
BRIGABANNIS, or BRIGOBANNA, BRULINGEN in Baden.
BRIGALCIUM, near BENEVENTE in the Spanish prov. of Leon.
BRIGANCONIA, BREGANÇON in the French dep. of Var.
BRIGANTIA, or VIRGANTIA, BRIANÇON in the French dep. of Hautes-Alpes.
BRIGANTINUS AGER, the lordship of BRIANÇON in France.
BRIGANTINUS COMITATUS, BREGENZ on the Lake of Constanz.
BRIGIA, BRIG or BRUG in the Swiss cant. of Valais.
BRIGIANA, BRIGASH in Baden.
BRIGNÆSIUM, or PRIGNESIUM, BRIGNAIS in the French dep. of Rhone.
BRIGOLIUM, BREGUEIL in the French dep. of Charente.
BRIMEUM, BRIMEN in Picardy.
BRINNEUM, BRINNUM, or BRUNNA, BRUNN in Moravia.
BRINOLIUM, or BRINONIA, BRIGNOLE in the French dep. of Var.
BRIOCENSE OPPIDUM, BRIOCUM, or OPPIDUM SANCTI BRIOCI, the town of ST. BRIEUX or ST. BRIEUCE in the French dep. of Nord.
BRIONIA, BRIONNE in the French dep. of Eure.
BRIOVERA, or OPPIDUM SANCTI LAUDI, SAINT LO in the French dep. of La Manche.
BRIPIUM, BRIVIO in Lombardy.
BRIQUERASCUM, BRIQUENSAS in Piedmont.
BRISACENSIS VICUS, BRISACH in Prussia.
BRISACUM AUDEGAVENSE, BRISSAC in the French dep. of Maine-et-Loire.
BRISGAVIA, BRISGOIA, or BRISGOVIA, the BREISGAU in the Black Forest.
BRISTOLIA, or BRISTOLIUM, BRISTOL in Somersetshire.
BRISTOLIENSIS MANICA, SAINT GEORGE'S CHANNEL.
BRITANNIA, or BRITANNIA MAGNA, the kingdom of GREAT BRITAIN.
BRITANNIA BARBARA, CALEDONIA, or CALIDONIA, the kingdom of SCOTLAND.
BRITANNIA INFERIOR, VALENTIA, or BRITANNIA SEPTENTRIONALIS, that portion of Britannia Romana which comprised the southern counties of Scotland and the LOTHIANS, with NORTHUMBERLAND, CUMBERLAND, WESTMORELAND, LANCASHIRE, YORK, and DURHAM.
BRITANNIA PRIMA, that part of Britain, under the Romans, which lay to the S of the Thames and of the Bristol channel.
BRITANNIA ROMANA, BRITANNIA PROPRIA, or BRITANNIA MAJOR, that part of the island of Britain which was under Roman sway. It was divided into five provinces, viz.: BRITANNIA PRIMA, BRITANNIA SECUNDA, FLUVIA CÆSARIENSIS, MAXIMA CÆSARIENSIS, and VALENTIA or VALENTIANA.
BRITANNIA SECUNDA, that part of Britain, under the Romans, which was separated by the Sabina or Severn, and the Deva or Dee, from the rest of Britain, in other words, WALES, HEREFORDSHIRE, MONMOUTHSHIRE, part of SALOP, and GLOUCESTERSHIRE and WORCESTERSHIRE.
BRITANNICÆ INSULÆ MINORES, VECTIS or the ISLE OF WIGHT; TANETOS or the ISLE OF THANET; MONA (of Cæsar) or the ISLE OF MAN;

MONA (of Tacitus) or the ISLE OF ANGLESEY; the ÆBUDÆ or HEBRIDES; ÆMODÆ, or SHETLAND ISLES; ORCADES or ORKNEY ISLANDS; and CASSITERIDES or SCILLY ISLES.
BRITANNICUS OCEANUS, or BRITANNICUM MARE, the ENGLISH CHANNEL.
BRITANNODUNUM, CASTRUM BRITONUM, or DUMBRITONIUM, DUMBARTON in Scotland.
BRITEXTA, BRIATESTE in the French dep. of Tarn.
BRITONUM CAPUT, or INSULA REGALIS, the island of CAPE BRETON.
BRITZNA, BRIES in Hungary.
BRIVA CURETIA, or BRIVITENSIS VICUS, BRIVE-SUR-LA-CORREZE in the French dep. of Correze.
BRIVA PARÆ, or ŒSIENSIS PONS, PONTOISE in the French dep. of Seine-et-Oise.
BRIVENES, BRENZ in Würtemberg.
BRIVUS, or BRIVOTENSIS VICUS, BRIOUDE in the French dep. of Allier.
BRIXELLUM, or BREXILLUM, BRESELLO in Lombardy.
BRIXIA, or BRIXINA, BRIXEN in Tyrol.
BRIXIANIS AGER, BRESCIANO in Lombardy.
BROAGIUM, or BURAGIUM, BROGE in the French dep. of Charente-Inferieure.
BROBURGUM MORINORUM, BOURBOURG in the French dep. of Nord.
BROCKMERIA, BROCKMERLAND in East Friesland.
BRODA, BROD in Transylvania.
BRODA BOHEMICA, BROD in Bohemia.
BRODA TEUTONICA, BROD-NAMETZKI in Bohemia.
BRODNICA, or STRATIOBURGUM, STRASBURG in Prussia.
BRUBACUM, BRAUBACH in Nassau.
BRUCHIUM AD RUVAM, BRUCK-AN-DER-ROER in Westphalia.
BRUCTERUS MONS, the BROCKEN in Germany.
BRUGA, or ARULÆ PONS, BRUCK in Switzerland.
BRUGÆ, BRUGES in Belgium.
BRUGÆ BEANUÆ, BRUGES in the French dep. of Basses-Pyrenees.
BRUGNATUM, or BUNNICIUM, BRUGNATO in Sardinia.
BRUMIA, or PRUMIA, PRUM in Prussia.
BRUNDISCA, or BRUNDUSCA, PORENTRUY in the Swiss cant. of Berne.
BRUNDULUS PORTUS, BRONDOLO in Lombardy.
BRUNONIS VICUS, BRUNOPOLIS, or BRUNSIRGA, the city of BRUNSWICK.
BRUNSBRITTA, BRUNSBRITTEL in Holstein.
BRUNSVICIENSIS DUCATUS, the duchy of BRUNSWICK.
BRUSCA, the BRUSK, an affluent of the Ill.
BRUTTIUM PROMONTORIUM, CAPO-DELL-ARMI on the Neapolitan coast.
BRUVENHAVIA, or BRUVARI PORTUS, BROWERSHAVEN in the Dutch prov. of Zeeland.
BRYSTACIA, UMBRIATICO in Calabria.
BUBASTICUS FLUVIUS, the extreme eastern deltoid branch of the Nile.
BUBINDA, CARLINGFORD in Ireland.
BUCA, a town in SAMNIUM, represented by TERMOLI in the Neapolitan prov. of Capitanata.
BUCARESTA, BUCHAREST in Walachia.
BUCCINA, LEVENZA on the coast of Sicily.
BUCELLUM, CIVITA BORELLA in Abruzzo-Citra.
BUCEPHALA, KAVALA in Macedonia.
BUCEPHALA, or BUCEPHALUS, a town of India, perhaps MULTAN, but by some regarded as being now represented by LAHORE.
BUCHANIA, the prov. of BUCHAN in Aberdeenshire.
BUCHAVIA, BUCHAU in Bavaria.—Also BUCHAU in Bohemia.
BUCKINGHAMENSIS COMITATUS, BUCKINGHAMSHIRE.
BUCKINGHAMIA, the town of BUCKINGHAM.
BUDA, or SALINUM, OFEN in Hungary.
BUDINA, BUDIN in Bohemia.
BUDINGA, BUDINGEN in Upper Hesse.
BUDISSA, or BUDISSINA, BAUTZEN in Saxony.
BUDOVICUM, or BUDOVISSA, BUDWEIS in Bohemia.
BUGELLA, or GAUMELLUM, BIELLA in Piedmont.
BUGUSTIA, or PATISCUM, BAUZKE in Prussia.
BULGARIA, VULGARIA, or MŒSIA INFERIOR, the Turkish prov. of BULGARIA.
BULLA, BEJE in Tunis.
BULLIO, or BULLONIUM, BOUILLON in Luxemburg.
BUNDIUM, BONDO in the Swiss cant. of the Grisons.
BUNIA, the BOGNA, an affluent of the Tosa in Lombardy.
BUNITIUM, ROSTOCHIUM, or ROSARUM URBS, ROSTOCK in Mecklenburg-Schwerin.
BURA, BUREN in Prussia.
BURCHANIA, BORKUM in East Friesland.
BURCHUSA, BURGHAUSEN in Bavaria.
BURCINALIUM, CRAUENBERG in Prussia.
BURDEGALA, BOURDEAUX in France.
BURGAVIA, BURGAU in Bavaria.
BURGETUM, BORGHETTO in Lombardy.
BURGETUM AD NUNCIUM, BORGHETTO in the Lombardian prov. of Verona.
BURGI, BURGOS in Old Castile.
BURGLA, or VENSILIA, WENSUSSEL in North Jutland.
BURGOMANERUM, BORGOMANERO in the Sardinian prov. of Novara.
BURGUM, BURG in Prussia.
BURGUM BONÆ GELENÆ, BARLENE in the French dep. of Vaucluse.
BURGUM FRANCUM, BORGOFRANCO in Lombardy.
BURGUM NOVUM, BOURGNEUF in the French dep. of Charente-Inferieure.
BURGUM NOVUM AD LIGERIM, BOURGNEUF in the French dep. of Loire-Inferieure.
BURGUNDIA, BURGUNDY in France.
BURGUNDIA INFERIOR, or BURGUNDIÆ DUCATUS, LOWER BURGUNDY.
BURGUNDIA SUPERIOR, or BURGUNDIA COMITATUS, UPPER BURGUNDY.
BURGUS, BOURG-SUR-MER in the French dep. of Gironde.
BURGUS ANDEOLII, BOURG-ST.-ANDEOL in the French dep. of Ardeche.
BURGUS ARGENTALIS, BOURG-ARGENTHAL in the French dep. of Loire.
BURGUS BRESSIÆ, BOURG-EN-BRESSE in the French dep. of Ain.
BURGUS FORTIS, BORGOFORTE in Lombardy.
BURGUS SANCTI ANGELI, BORGO-DI-SAN-ANGELO in Malta.
BURGUS SANCTI DONNINI, BORGO-DI-SANTA-DONNINO in Parma.
BURGUS SANCTI SEPOLCRI, BORGO-DI-SAN-SEPOLCRO in Tuscany.
BURGUS SASSITES, BORGO-DI-SESIA in Lombardy.
BURICHUM, or BUDERCHIUM, BURICH in Prussia.
BURNONIA SILVA, or NISTRIA, the WESTERWALD in Germany.
BURNONIS MONS, BOURMONT in the French dep. of Haute-Marne.
BUSIRIS, ABUSIR in Lower Egypt.
BUSIRIS, in the Heptanomis, BUSIR near Lake Mœris in Egypt.
BUSSUS, the river Bozzo falling into the Lago-di Gavira in Northern Italy.

BUSUM PROMONTORIUM, Cape Buso in Crete.
BUTHRORUM, Butrinto in Albania.
BUTROTUS, the river Bucorta in Naples.
BUVINDUM, or Carlingfordia, Carlingford in Ireland.
BUXENTIUS, or Pyscus, the river Bussemto in Naples.
BUXENTUM, Bolicastro in Naples.
BUXOVILLA, Buchswieler in the French dep. of Bas-Rhin.
BYBLOS, Jebele in Syria.
BYRRA, or Bynus, the river Rienz in Tyrol.
BYSTRICIUM, Bistricza in the Russian gov. of Grodno.
BYZANTIUM, Lygos, or Constantinopolis, the city of Constantinople.

C

CABALACA, or Chabala, Carlasvur in Georgia.
CABALLIACENSIS DUCATUS, or Cabellicus Ager, the territory of Chablais.
CABALLIO, or Cabellio, the capital of the Cavares, now Cavaillon on the Durance, in the French dep. of Vaucluse.
CABAR-SUSIS, Susa in Turin.
CABELLONENSIS AGER, the Chalonnais in France.
CABIRA, Diopolis, Sebasta Armeniæ, Augusta, or Sebastopolis, a town in the interior of Pontus, perhaps represented by the modern Siwas.
CACHALES, the modern Kako-Rheuma, an affluent of the Cephisus in Phocis.
CACOSANA, Aracan in India.
CACUS, or Caunus Mons, the Sierra-Moncajo in Aragon.
CACYPARIS, now the Cassibele, a river of Sicily.
CADACHENUM, Cadagnes in Catalonia.
CADANUM, Kaden in Bohemia.
CADAVUS, the river Cadavo in Galicia.
CADI, a town in Phrygia Epictetus, now Kedus.
CADMEA, or Thebæ, in Bœotia, now Theba or Stiva.
CADOMUM, or Cadomus, Caen in the French dep. of Calvados.
CADROTIUM, Caudrot in the French dep. of Gironde.
CADURCA TERRA, or Cadurcorum Regis, the territory of Quercy in the French dep. of Lot.
CADURCI, Cadurcum, Divana, Bibona, or Divonna, Cahors in the French dep. of Lot.
CADYNA, or Cadena, a town of Isauritis, Nighdeh in the pash. of Koniyeh.
CÆA, the island of Zea in the Grecian Archipelago.
CÆCILIA CASTRA, according to D'Anville, Caceres in Estremadura; but by some other geographers identified with San-Maria di Guadalupe.
CÆLANUM, Celanum, or Cœlanum, Celano in Naples.
CÆLIANUM, Cigliano in Naples.
CÆLINA, the river Celina in Lombardy.
CÆLIUS, or Cælius Mons, probably Kellmunz in Bavaria.
CÆNE, Nova Urbs, or Neapolis, Kenneh or Ghenneh in Egypt.
CÆNUS, the river Arc in France.
CÆNYS PROMONTORIUM, the Punta-del-Pezzolo, or Coda-del-Volpe on the coast of Calabria-Ultra, opposite the Sicilian promontory of Pelorus.
CÆRE, Cacistum Municipium, Agylla, or Agyllina Urbs, Cerveteri in the Papal States.
CÆRETANORUM EMPORIUM, a town of Emeria, Santa-Severa in the Papal States.
CÆSARAUGUSTA, Colonia Cæsarea Augusta, or Salduba, a town of the Edatani, in Hispania Tarraconensis, now Saragossa in Aragon.
CÆSAREA, the island of Jersey.
CÆSAREA AD ARGÆUM MONTEM, Cæsarea Metropolis, Mazaca, or Eusebia Cæsarea, now Kaisarieh in the pash. of Karieh.
CÆSAREA, or Mauritaniæ Tingitanæ, Tingis, Tiga, or Julia Traducta, now the city of Tangiers in Marocco.
CÆSAREA BOJORAM, Kotzing in Bavaria.
CÆSAREA LUTRA, or Cæsaro Luthera, Kaiserlautern in Rhenish Prussia.
CÆSAREA PALÆSTINÆ, Cæsarea Stratonis, or Colonia Flavia Cæsarea, the modern Kaisarieh in Samaria.
CÆSAREA PANEAS, Cæsarea Philippi, or Neronias, near Banias or Panias at the sources of the Jordan in Syria.
CÆSAREANUM, Cere in the French dep. of Lot.
CÆSAREOPOLIS, Kaisersmark in the Hungarian comitat of Zips.
CÆSARIS BURGUS, Cæsaroburgus, or Caroburgum, Cherbourg in Normandy.
CÆSARIS INSULA, or Insula Rheni, Kaiserswerth in Rhenish Prussia.
CÆSARIS MONS, Kaisersberg in the French dep. of Haut-Rhin.
CÆSARIS TRIBUNÆ, or Forum Tiberii, Kaiserstuhl in the Swiss cant. of Aargau.
CÆSARO-FELDA, Kaiserfeld in Illyria.
CÆSARODUNUM, Metropolis Civitas Turonorum, or Turonica Urbs, Tours in the French dep. of Indre.
CÆSAROMAGUS, or Baromacus, Chelmsford in Essex.
CÆSAROTIUM, or Gisortium, Gisors in the French dep. of Eure.
CÆSENA, Curoa Cæsana, Cesena in the Papal States.
CÆSIA SYLVA, now the Heserwald in the Prussian reg. of Munster.
CÆTOBRIX, Catobriga, or Setubalia, Setuval in Spanish Estremadura.
CAFERONIANUM, or Castellum Novum Garfunanæ, now Castel-Nuovo-di-Carfaguana in the duchy of Modena.
CAFFARIA, Caffraria.
CAINO, or Chininium, Chinon in the French dep. of Indre-et-Loire.
CAIRUS MAGNA, or Cairum, Cairo in Egypt.
CAJARCUM, Cajaco in the French dep. of Lot.
CAJETA, Gaeta in Naples.
CALAA, in Mauritania Cæsariensis, now Kalaah in Algeria.
CALABRIA, Japygia, Messapia, or Salentina, the Neapolitan prov. of Terra-d'Otranto.
CALACTA, Calacte, or Cale Acte, probably Santo-Marco on the N coast of Sicily.
CALADUNUM, Mirandella in the Portuguese prov. of Tras-os-Montes.
CALAGORINA, Calagunis, or Municipium Calaguni Julia, Calahorra in Old Castile.
CALAGORRIS, Caçeres in the French dep. of Garonne.
CALAMÆ, in Messenia, near Kalamata in Northern Greece.
CALANCORUM, Gorlitz in Prussia.
CALARIS, or Carala, Cagliari in Sardinia.
CALAROGA, Calaruega in Old Castile.
CALATA, or Calacia, Galati in Sicily.
CALATIA in Campania, now Cajazzo in the Neapolitan prov. of Terra-di-Lavoro.
CALATIS, Callatis, or Callatia, Caliaca in Bulgaria.

CALAURIA, HYPEREIA, or POSSIDONIA, now the island of PORO or POROS, off the coast of the Morea.
CALBA, or CALVA, CALBE in Würtemberg.
CALBA AD SALAM, CALBE in Prussia.
CALCARIA, TADCASTER in Yorkshire.
CALCARIUM, KALKAR in Rhenish Prussia.
CALCIATA, or CALZIATA, CALZADA in Old Castile.
CALENA, or CALLEVA ATTREBATUM, WALLINGFORD in Berkshire.
CALDARIUM, CALDEM in Tyrol.
CALDEBECCUM, or CALIDUM BECCUM, CAUDIBEC in the French dep. of Seine-Inferieure.
CALE, or PORTUS CALE, OPORTO in Portugal.
CALE, or MUNICIPIUM CALENUM, CALVI in the Neapolitan prov. of Terra-di-Lavoro.
CALECUTUM, CALICUT on the Malabar coast.
CALEDONIA, CASTRUM CALEDONIUM, or DUNCALEDONIA, DUNKELD in Perthshire.
CALEDONIUS OCEANUS, DEUCALEDONIUS SINUS, LAELAMNONIUS SINUS, or CALEDONIUM MARE, that portion of the ocean extending between the NW and N coast of Scotland and the Orkney islands.
CALEGIA, LEUCOREA, or WITTEBERGA, WITTENBERG in Prussia.
CALENUM, CARINOLA in Naples.
CALETORUM PROMONTORIUM, CAPE-DE-CAUX, the N coast of France.
CALETUM, CALESIUM, or ICTIUS PORTUS, CALAIS on the N coast of France.
CALICULA, HUESCAR in the Spanish prov. of Granada.
CALIDUM, CHAUD in Savoy.
CALINGON PROMONTORIUM, probably PUNTA-DE-PALMEIRAS near the mouth of the Ganges.
CALINIPAXA, or CANOGIZA, CANOGA in the Indian prov. of Agra.
CALIPUS, or SADANUS, the river ZADAON in Spanish Algarvia.
CALISIA, or CALISSIUM, KALISCH in Poland.
CALLICULA MONS, near CAJANELLO in Naples.
CALLIDROMUS, the highest summit of MOUNT KATAVOTHRA in Greece.
CALLIENA, CORINGA in Guzerat.
CALLIPOLIS, or ANXA, GALLIPOLI in Naples.
CALLIPOLIS, GALLIPOLI in Rumelia.
CALLISTE, or THERA, the island of SANTORIN in the Grecian archipelago.
CALLUNDA, KALUNDBORG in Denmark.
CALMARIA, CALMAR in Sweden.
CALMONTIUM, CHAUMONT in Savoy.
CALNERIA, CHAUNES in the French dep. of Somme.
CALNIACUM, CHAUNY in the French dep. of Aisne.
CALOW in Germania Inferior, perhaps KALENHUYSEN.
CALONESUS, or PULCHRA INSULA, the island of BELLE-ISLE off the coast of Brittany.
CALOR, the river CALAE in Naples.
CALPE, CALPIA, COLONIA JULIA CALPE, or GIBRALTARIA, GIBRALTAR in Spain.
CALPE PROMONTORIUM, in Hispania Bœtica. the JEBEL-EL-TUARIK.
CALPE, KIRPEH on the coast of the Black sea.
CALPURNIANA, CARPIO in the Spanish prov. of Cordova.
CALUNIA, VARVICUM, or VEROVICUM, WARWICK in Warwickshire.
CALVENCIACUM, CHAVANCY in Luxemburg.
CALVENZANUM, CALVENZANA in Lombardy.
CALVIMONTIUM, or CALVUS MONS, CHAUMONT in the French dep. of Oise.
CALVINIACUM, CHAUVIGNY in the French dep. of Vienne.
CALVIUM, CALVI in Corsica.
CALVO MONS, CALMONT in the French dep. of Lot-et-Garonne.
CALVO MONS NORMANIÆ, CAUMONT in the French dep. of Calvados.
CALVO MONS PROVINCIÆ, CAUMONT in the French dep. of Vaucluse.
CALVO MONS VASCONIÆ, CHAUMONT in the French dep. of Haut-Garonne.
CALYCADINIS, a river in Cilicia, now the GHUIK-SU.
CAMALDULUM, CAMALDOLI in Tuscany.
CAMALODUNUM, or CAMUDOLANUM, MALDEN in Essex, or perhaps COLCHESTER.
CAMAR, CAMARA, or CANAR, near CANORIN in Ceylon.
CAMARACUM, or CAMARACENSIUM CIVITAS, in Gallia Belgica, probably CAMBRAY on the Schelde.
CAMARICA, VITTORIA in the Spanish prov. of Alava.
CAMARINA, or HYPERIA, CAMARANA on the coast of Sicily.
CAMARINA PALUS, the LAGO-DI-CAMARANA in Sicily.
CAMARS, or TUSCIA, now CHIUSI.
CAMBAJA, or CAIRUS INDIÆ, CAMBAYA in Hindostan.
CAMBAJÆ REGNUM, CAMBAYA in Guzerat.
CAMBERIACUM, CAMLERIACUM, CHAMARIUM, or CHAMBURIACUM, CHAMBERY in Savoy.
CAMBODUNUM, CAMPODUNUM, or DRUSOMAGUS, KEMPTEN in Bavaria.
CAMBODUNUM, WEILHEIM in Bavaria.
CAMBORICUM, CAMBORITUM, COMIVADUM, or CANTABRIGIA, CAMBRIDGE in Cambridgeshire.
CAMBORITUM, or CAMBORIUM, CHAMBORD in the French dep. of Loir-et-Cher.
CAMBRATONIUM, BRETENHAM in Suffolk.
CAMBRIA, CAMBRO BRITANNIA, or VALLIA, the principality of WALES.
CAMBUS, the KAMP or CAMB, an affluent of the Danube.
CAMBUS BAVARICUS, the CHAM, an affluent of the Regen.
CAMBYSÆ ÆRARIUM, or CAMBUSIS, MOCHO in the district of Meroe in Upper Egypt.
CAMBYSIS, the river ASTARA or ZARA in Armenia.
CAMECHIA, SHAMAKI in Albania.
CAMENECIA, or CAMIENICUM, KAMIENETZ in Russian Podolia.
CAMENGRADUM, CAMENGRAD in Bosnia.
CAMENTIA, CAMENZ in Saxony.
CAMENTIA AD NISSAM, CAMENZ in Prussia.
CAMERACENSIS AGER, the lordship of CAMBRAY in the French dep. of Nord.
CAMERACUM AD ALBIM, KEMBERG in Prussia.
CAMERINUM, CAMERINO in Lombardy.
CAMICUS, the river CAMIA in Sicily.
CAMILIACUM, CHAMBLY in the French dep. of Oise.
CAMINA, KAINS in Tyrol.
CAMINIUM, CAMINA in Prussia.
CAMIROS, FERUCCHIO in the island of Rhodes.
CAMIRUS, CYRBA, HIERA, HIERAPYTRA, or HIERA PETRA, GIRAPETRA in the island of Crete.
CAMPANIA, REGIO FELIX, or LABORINUS AGER, the Neapolitan prov. of TERRA-DI-LAVORO.
CAMPANIA, CAMPAGNA in the Neapolitan prov. of Principato-Citra.
CAMPANIA AD ISALAM, CAMPEN in Holland.
CAMPANIA FRANCICA, or CAMPANIA FRANCO-GALLICA, the prov. of CHAMPAGNE in France.
CAMPI CANINI, the canton of the GRISONS in Switzerland.
CAMPI MACRI, the VAL-DI-MARTIRONE in Parma

CAMPI PHLEGRÆI, the SOLFATERRA in the Neapolitan prov. of Terra-di-Lavora.
CAMPIANIUM, CAMPIANO in Sicily.
CAMPILILIUM, LILIENFELD in Austria.
CAMPIMONTIUM, CHAUMOUNY in Savoy.
CAMPINIA, COMPIEGNE in Liege.
CAMPIPORCHERIUM, CHIAMPORRIERO in Piedmont.
CAMPIVERIA, CAMPOVERIA, or VERA, TER-VERE in the Dutch prov. of Zieland.
CAMPLUM, CAMPOLI in Abruzzo-Ultra.
CAMPSUM, or CAMSO, ZAMS in the Swiss cant. of St. Gall.
CAMPUS, CAMPO in the Sardinian prov. of Genoa.
CAMPUS CAROLINUS, CARLOPAGO in Dalmatia.
CAMPUS CASSOBUS, CASSOVIUS, or CAMPUS MERULÆ, AMSELFELD in Servia.
CAMPUS DULCINUS, CAMPOLSHIN in the Swiss cant. of the Grisons.
CAMPUS LONGUS, CAMPO LONGO in Wallachia.
CAMPUS REGIUS, KONIGSFELDEN in the Swiss cant. of Bern.
CAMPUS ROTUNDUS, or EGOSA, CAMPREDON in Catalonia.
CAMUDOLANUM. See CAMALODUNUM.
CAMUNORUM VALLES, the VAL-DI-CAMONICA in Lombardy.
CANA, or CAPELLA, now CAPE COLONI on the coast of Greece.
CANA, or CHANNEH, KASCHIM in the Arabian prov. of Yemen.
CANA, or CANSTADIUM, CANSTADT in Würtemberg.
CANADA, the British province of CANADA in North America.
CANADIENSIS COMITATUS, the palatinate of CHANAD or CSANAD in Hungary.
CANAGORA, CASHMIR in Afghanistan.
CANALOVII MONTES, a range on the confines of Illyricum, now CRASTÆ.
CANAMA, in Hispania Bœtica, VILLA-NOVA-DEL-RIO in Andalusia.
CANARIA, or CANARIA MAGNA, the GREATER CANARY ISLAND.
CANASTRÆUM, KASSANDRA in Macedonia.
CANCALLIUM, CANCALE in the French dep of Isle-et-Vilaine.
CANCIUS, or QUENTIA, CANCHE in the dep. of Pas-de-Calais.
CANDACE, perhaps THAK in Cabulistan.
CANDIDA CASA, WHITHORN in Galloway.
CANDIDUM PROMONTORIUM, RAS-EL-ABIAD on the coast of Barbary
CANDUUM, CAMBURG in Altenburg.
CANE, or CANIUM, CAPE LITHODA on the coast of Greece.
CANELATA, FURENZA in Corsica.
CANENTULUS, a river of Gallia Aquitanica, the SEVRE according to Mamert.
CANISIA, or CALISSUM, KALISCH in Prussia.
CANISIA AD DRAVUM, CANISCHA in Lower Hungary.
CANOPUS, or CANOBUS, an ancient city in the Egyptian delta, near the modern village of ABUKIR.
CANORGIA, CANOURGUE in Languedoc.
CANORICUM, or CHANORICUM, CHANONRY in Ross-shire.
CANTABRIA, BISCAY in Spain.
CANTABRICUS OCEANUS, the BAY OF BISCAY.
CANTABRIGUENSIS COMITATUS, CAMBRIDGESHIRE.
CANTASIUM, or CANTAZARUM, CANTAZARO in the Neapolitan prov. of Calabria-Ultra.
CANTECRUCIUM, CANTECROIX in S. Brabant.
CANTERIUS MONS, the MONTE-DI-GIOVANNE in the Papal states.
CANTHI COLPUS, the GULF OF KERTCH.
CANTIERA, KINTYRE in Scotland.
CANTIERÆ FRETUM, the MULL OF KINTYRE.
CANTIERÆ ROSTRUM, or EPIDIUM PROMONTORIUM, KINTYRE-HEAD.
CANTII LITORA, the KENTISH DOWNS.
CANTIOCHIS, HANBUCH in Bavaria.
CANTIUM, KENT in England.
CANTUARIA, DARVERNUM, DUROVERNUM, or DOROVERNIA, CANTERBURY in Kent.
CANUSIUM, CANOSA in Naples.
CAPEDUNUM, KAPFENSTEIN in Hungary.
CAPENA, or COLONIA CAPES, CIVITELLA in the Papal states.
CAPHA CAVUM, or THEODOSIA, CAFFA or FEODOSIA in the Crimea.
CAPHAREUS, or ZYLOPHAGUS, the modern CABO-DELL'-ORO in Eubœa.
CAPISA, perhaps CANDAHAR in India.
CAPITOLIAS, a town in Cœlosyria, probably near BEIT-EL-RAS.
CAPPADOCIA, that portion of Asia Minor lying to the E of Lake Tatta, and W of the Euphrates, embracing a large portion of the pashaliks of BOZUK and KARAMANIA.
CAPRA DORSUM, ZIEGENRUCK in Prussia.
CAPRARIA, the island of PALMA.—Also the island of CABRERA in the Balearic group.
CAPRARIUM INSULA, the island of TRAJONISI in the Greek archipelago.
CAPREÆ, or CAPRIA, the sland of CAPRI in the Gulf of Naples.
CAPRUS, a river of Assyria, probably the LITTLE ZAB.
CAPRUS, a haven on the Sinus Strymonicus, now SGORA.
CAPRUSIUM, CHEVREUSE in the French dep. of Seine.
CAPSA, CAPA in Turin.
CAPUA, now the modern CAPUA in the Neapolitan prov. of Terra-di-Lavoro.
CAPUNGUM, or CAUFUNGA, KAUFUNGAS in Lower Hesse.
CAPUT ACCIUM, CAPE AGULHAS in Lower Africa.
CAPUT ARENARIUM, CAPE SABLE in Newfoundland.
CAPUT AROMATUM, CAPE GUARDAFUI.
CAPUT CORSUM, CAPE CORSO in Corsica.
CAPUT CORSUM AFRICANUM, CAPE CORSO on the Gold Coast.
CAPUT HORNANUM, CAPE HORN.
CAPUT OENI, the INNTHAL or ENGADINE in the Grisons.
CAPUT RISUM, CAPE RIZO in Cyprus.
CAPUT THYRSI, THYRSO in Sardinia.
CARABUSSA, the island of GARABUSA off Candia.
CARACOTINUM, or PORTUS GRATIÆ, HAVRE on the N coast of France.
CARADRINA, the river DRIN or DRINO in Rumelia.
CARALITANUM PROMONTORIUM. CAGLIARI in Sardinia.
CARALITANUS SINUS, the GULF OF CAGLIARI.
CARAMBACIS, the river DWINA in Russia.
CARAMBICE, NOVA ZEMBLAIA.
CARAMBIS, or CARAMBICUM PROMONTORIUM, CAPE KEREMPE in the Crimea.
CARAMBUCIS, or OBIUS, the river OB or OBI in Asiatic Russia.
CARANCEBÆ, CARANSEBES in Hungary.
CARANTONUS, a river of Gallia Aquitanica, the modern CHARENTE.
CARANUSCA, SARANUSCA, or SARBURGUM, SAARBURG in Prussia.
CARAVACIUM, or CARAVAGIUM, CARAVAGIO in Lombardy.

CARBANTORIGUM, CAERLAVEROCK in Galloway.
CARBONARIA, PORTO-DI-GORO, one of the mouths of the Po.
CARBONARIA SYLVA, the KOHLENVALD in the Ardennes.
CARCASO, CARCASSO, or CARCASSUM, CARCASSONE in France.
CARCHA, ESKI-BAGDAD on the Tigris.
CARCHESIUS, CIRCESIUM, or KARKESIA, KIRKESSIA on the Euphrates.
CARCINE, at the mouth of the Hypacacis, perhaps AKMETCHET on the gulf of Perekop.
CARCINITES, or PACYRIS, the KANIKSHAK flowing into the gulf of Karkinite.
CARCINITES SINUS, the gulf of KARKINITE or PEREKOP.
CARCOVIACA, KIRKWALL in Orkney.
CARDABIANCA, FRIBURG in Styria.
CARDAMYLE, SKARDAMULA in Northern Greece.
CARDANIA, CARDAGNE in the Pyrenees.
CARDANUM, CARDANE in Lombardy.
CAREA, CARIUM, or CHIERIUM, CHIERI in Piedmont.
CARENTO, CARENTAN in Normandy.
CARENTONIUM, CHARENTON in the French dep. of Isle-de-France.
CARENTONIUM BITURICENSE, CHARENTON in the French dep. of Cher.
CARES, now PUENTA-DE-LA-REYNA in Pampeluna.
CARESSUS, now CABIA in the island of Ceos.
CARICTA, or CARRICKTA, CARRICK in Ayrshire.
CARILOCUS, CHARLIEU in Burgundy.
CARINIANA VALLIS, or BABOLIZA, BABOLITZA in Slavonia.
CARINIANUM, CARIGNANO in Lombardy.
CARIS, CARUS, or CHARES, the river CHER in France.
CARISIACUM, CRE'CY or CRESSY in the French dep. of Somme.
CARITACUM, or CHARITAS, LA CHARITE' in the French dep. of Nievre.
CARLEOLUM, LUGOVALLUM, LUGUVALLIA, or LUGUVALLUM, CARLISLE in Cumberland.
CARMANICA, KERMAN with the eastern portion of LARISTON and MOGHISTAN in Persia.
CARMANIOLA, CARMAGNOLA in Piedmont.
CARMELUS, MOUNT CARMEL in Syria.
CARMENE, CARMONE in Seville.
CARNAIRUS SINUS, the GULF OF QUARNERO in the Adriatic.
CARNIA, CARNIOLA.
CARNIOBURGUM, KRAINBURG in Illyria.
CARNOVIA, JÆGERNDORF in Silesia.
CARNUNTUM, or CARNUS, HAINBURG on the Danube, in the archd. of Austria.
CAROCELIA VALLIS, the MAURIENNETHAL in Savoy.
CARODUNUM, or CRACOVIA, CRACOW in Poland.
CAROLESIUM, or QUADRELLENSIS, the CHAROLAIS in the Bourbonnois of France.
CAROLI, or CORONA, CARLSCRONA in Sweden.
CAROLI HESYCHIUM, CAROLINA HESYCHIA, CAROLSRUHA, CARLSRUHE in Baden.
CAROLI PORTUS, CARLSHAMN in Sweden.
CAROLI PORTUS AD VISURJUM, CARLSHAFEN in Hesse-Darmstadt.
CAROLIÆ, CHAROLLES in Burgundy.
CAROLINA ANTIQUA, ALT-CARLEBY in Finland.
CAROLINA NOVA, or CAROLOPOLIS, NEW CARLEBY in Finland.
CAROLIUM, KAROLY in Hungary.
CAROLOLESIUM, or CAROLOREGIUM, CHARLEROY in Belgium.
CAROLOMONTIUM, CHARLEMONT in Belgium.
CAROLOMONTIUM HIBERNICUM, CHARLEMOUNT in co. Armagh.
CAROLOPOLIS, CHARLEVILLE in Belgium.
CAROLOPOLIS AMERICANA, CHARLESTOWN in Maryland, U. S.
CAROLOSTADIUM, CARLSTADT in Bavaria.
CAROLOSTADIUM SCUECICUM, CARLSTADT in Sweden.
CAROLOVICIA, or CAROLOVITIUM, CARLOWITZ in Hungary.
CARONIUM, CORUNNA in Spain.
CARPASIA, or CARPASIUM, JALUSSA in Cyprus.
CARPATES, ALPES BASTARNICÆ, SARMATICI MONTES, or CARPATICI MONTES, the CARPATHIAN MOUNTAINS, or great Hercynian system which is separated from the system of the Alps and of Hæmus by the basin of the Danube.
CARPATHIUM MARE, CARPATHIUM PELAGUS, or ASIATICUM MARE, the GULF OF SCARPANTO.
CARPELLA, CAPE JASK on the Persian Gulf.
CARPENTORACTE, now CARPENTRAS on the river Auzon in France.
CARPIO, KERPEN in Prussia.
CARPIS, GARBOS on the coast of Tunis.
CARPIS, the river DRAVE.
CARPIUM AD ATHESIN, CARPI on the Etsch in Lombardy.
CARREA, or POLLENTIA, POLENZA in Piedmont.
CARRECTANUM, CARETTO in Sardinia.
CARRIO COMITUM, CARRION-DE-LOS-CONDES in the Spanish prov. of Leon.
CARRODUNUM, a town of the Lygii, by Cellarius thought to be represented by RADOM in Poland; by Kruse identified with ZAMOWICE.
CARRODUNUM, KRAINBURG in Bavaria.
CARSE, the city of KARS in Asiatic Turkey.
CARSIDAVA, KARYMA in Wallachia.
CARSIUM, or CARSUM, KERSCHOWA in Bulgaria.
CARTEJA, CARTEGIA, or CARPESSUS, perhaps ROCADILLO near Gibraltar in Spain; but by some identified with TARIFA.
CARTENNA, perhaps MOSTAGHAN in Algeria.
CARTHAGO, CARTHAGO MAGNA, COLONIA CARTHAGO, or TYRIA URBS, near MERSAH on the coast of Tunis.
CARTHAGO NOVA, CARTHAGO POENORUM, or CARTHAGO SPARTARIA, CARTHAGENA in Murcia.
CARTHAGO VETUS, a town of Hispania Tarraconensis, now CANTAVIEJA in Aragon.
CARTHUSIA MAGNA, the abbey of CHARTREUX, or LE GRANDE CHARTREUX in Dauphiny.
CARTRIS, CIMBRORUM PROMONTORIUM, JUTIA, or JUTLANDIA, the Dutch peninsula of JUTLAND or NORTH JUTLAND.
CARUO, a town in Gallia Belgica, near KUILENBERG in the prov. of South Holland.
CARURA, KARUR in Southern Coimbatore in India.
CARUSSA, CHAROUSSE in Savoy.
CARYÆ, or CARYA, a town in Laconica, now KRAVADA.
CARYSTOS, CARISTO or CASTEL-ROSSO in the island of Negroponte.
CASA CANDIDA. See CANDIDA.
CASA DEI FŒDUS, the ENGADINE in the Swiss cant. of the Grisons.
CASALAQUEUM, CAZALEGAS in New Castile.
CASALE MAJUS, CASALE-MAJORE on the Po.
CASANA, or CASANUM, KAZAN in Russia.
CASANDRIA, CADSAND in the Dutch prov. of Zeeland.
CASANUM AD ADDUAM, CASANO on the Adda in Lombardy.
CASCALE, CASCAOS in Estremadura.
CASCHOVIA, CASSOVIA, or BORMANUM, CASCHAU in Hungary.
CASHILIA, or CASSILIA, CASHEL in Ireland.
CASIA, KASHGAR in India.

CASII MONTES, a part of the KHARA range in Mongolia.
CASIMIRIA, CASIMIR in Prussia.
CASINOMAGUS, perhaps COULOGNE in Gascony.
CASINUS, or CASSINUS MONS, CASINO in Naples.
CASIUS, the river AMUR.
CASLETUM, or CASLETTUM MORINORUM, CASSEL in the French dep. of Nord.
CASPERIA, or CASPERULA, a town of the Sabine, now ASPRA near Tivoli in the Papal States.
CASPIÆ PORTÆ, or CASPIÆ PYLÆ, a pass between Media and Hyrcania, in the ELBURZ range between the Hurka-Koh and the Siah-Koh.
CASPINGIUM, ASPEREN in Holland.
CASPIRIA, CASHMIR in India.
CASPIUM MARE, CASPIUM PELAGUS, CASPIUM SINUS, or HYRCANIUM MARE, the CASPIAN SEA.
CASPIUS MONS, the ELBURZ.
CASSANDA, CADSAND in the Dutch prov. of Zeeland.
CASSANUM, CASSANO in Naples.
CASSELETUM, CHATELET in the Belgian prov. of Hainault.
CASSELLA, CASSELIA, CASSILIUM, or CASTELLUM CATTORUM, CASSEL in Germany.
CASSINUM, MONTE-CASINO in Naples.
CASSIOPE, JANINA in Albania.—Also CASSOPO in Corfu.
CASSITERIDES, CASSITERIDUM INSULA, SYLINA INSULA, or HESPERIDES, the SCILLY ISLANDS.
CASTAGNEDOLUM, CASTAGNEDOLI in the duchy of Genoa.
CASTANA, or CASTANEA, now KASTANIA in Thessaly.
CASTANIA, CASTELLANATA in Naples.
CASTANOVITIUM, KOSTANITZA in Croatia.
CASTELHOLMIUM, CASTLEHOLM in Sweden.
CASTELLA, or CASTILIA, CASTILE in Spain.
CASTELLA AUREA, or TERRA FIRMA, NEW GRANADA in South America.
CASTELLA NOVA, NEW CASTILE in Spain.
CASTELLA VETUS, OLD CASTILE in Spain.
CASTELLANUS, CASTELL in Bavaria.
CASTELLARIUM, CHASTELLAR in Savoy.
CASTELLIO, CHATILLON in the French dep. of Bouches-du-Rhone.
CASTELLIO AD AVARIM, CHATILLON-SUR-SAONE in France.
CASTELLIO AD AUGERIM, CHATILLON-SUR-L'INDRE.
CASTELLIO AD CARIUS, CHATILLON-SUR-CHER.
CASTELLIO AD LIGERIM, CHATILLON-SUR-LOIRE.
CASTELLIO AD LUPIAM, CHATILLON-SUR-LOING.
CASTELLIO AD MATRONAM, CHATILLON-SUR-MARNE.
CASTELLIO AD SEPARAM NANNETENSEM, CHATILLON-SUR-SEVRE.
CASTELLIO AD SEQUANAM, CHATILLON-SUR-SEINE.
CASTELLIO BURGUNDIÆ, CHATILLON-LES-DOMBES.
CASTELLIO MEDULCI, CHATILLON-LE-MEDOC.
CASTELLIO MORMENSIS TRACTUS, CHATILLON-EN-BAZORI.
CASTELLIONUM, CASTIGLIONE in Piedmont.
CASTELLIO PEDIMONTII, CHATILLON in Piedmont.
CASTELLIO PICTAVIÆ, CHATILLON-SUR-SEVRE.
CASTELLIO PISCARA, CASTIGLIONE in Tuscany.
CASTELLODUNUM, or CASTRODUNUM, CHATEAUDUN in the French dep. of Eure-et-Loir.
CASTELLUM, CASTELBERG in Lichtenau in Prussia.
CASTELLUM ANGELI, or MOLES HADRIANI, the castle of SAN ANGELO in Rome.
CASTELLUM ARIANORUM, CASTEL-NAUDRY in the French dep. of Aude.
CASTELLUM ASENSE, CASTEL D'ASENS in Catalonia.
CASTELLUM BALDUM, CASTEL-BALDO in Lombardy.
CASTELLUM DURANTIUM, CASTEL-DURANTE in the Papal states.
CASTELLUM EPISCOPI, BISHOPCASTLE in Shropshire.
CASTELLUM HANNO UM, CASTELLAUN in Rhenish Prussia.
CASTELLUM HERALDI, CHATELHERAULT in the French dep. of Vienne.
CASTELLUM ICTIUM, ISLE-JOURDAIN in the French dep. of Gers.
CASTELLUM MAJUS, CASTLELIMAGGORE in Lucca.—Also CASTELMAGNO in Piedmont.
CASTELLUM MOSELANUM, or TABERNÆ MOSELLANICÆ, BERNCASTEL in Prussia.
CASTELLUM NOVUM, CASTEL-NUOVO in Dalmatia.
CASTELLUM ORIENTIS, LEVANTA in the principality of Genoa.
CASTELLUM PETRÆ, CASTELL-DELLA-PIETRA in Tyrol.
CASTELLUM SALINARUM, CHATEAU-SALINS, in the French dep. of Meurthe.
CASTILIO CALABRIÆ, CASTIGLIONE MARITIMO in Naples.
CASTILIO CONSENTINA, CASTIGLIONE-DI-COSENZA in Calabria-Citra.
CASTILIO MANTUANA, CASTIGLIONE-MANTUANA in Lombardy.
CASTINETUM, KESTENHOLZ in Rhenish Prussia.
CASTIO, or CASTRIUM STILICONIS, CASTIONE in Milan.
CASTIODUM, OESCH in the cant. of Bern.
CASTRA, ARPAGIN in the French dep. of Aveyron.
CASTRA, or CASTRA RHETICA, GASTER in the cant. of Glarus.
CASTRA AD GARUMNAM, CASTRES in the French dep. of Gironde.
CASTRA CONSTANTINA, or CONSTANTIA, COUTANCES in Normandy.
CASTRA CORNELIA, or SUPERBUS VALLUM, a town in Zengitana, near GELLAH in Tunis.
CASTRA EXPLORATORUM, probably NETHERBY in Cumberland.
CASTRA FABIANA, near BABENHAUSEN in Baden.
CASTRA HANNIBALIS, CASTELLETE in Calabria-Ultra.
CASTRA HERCULIS, near KESTEREN in Holland.
CASTRA HORDEANI, or ORDINGA, UERDINGEN in Prussia.
CASTRA JULIA, or TURRIS JULIA, TRUXILLO in Estremadura.
CASTRA METELLINA, or METELLINUM, MEDELIN in the prov. of Estremadura.
CASTRA NOVA, CARACAL in Wallachia.
CASTRA POSTUMIANA, CASTRO-DEL-RIO in Cordova.
CASTRA TRAJANA, RIBNIK in Wallachia.
CASTRA TRICORNIA, or TRICORNIUM, KOLUMBUZ in Servia.
CASTRA ULPIA, COLONIA TRAJANA, OTTRICENSIMÆ OPPIDUM, KELLER in Prussia.
CASTRA VARIA, or VARIANA, VELIHA in Slavonia.
CASTRA VITERA, or SANTENA, near SANTEN in Rhenish Prussia.
CASTRENSIS DUCATUS, CASTRO in the Papal States.
CASTRIFENENSE OPPIDUM, SARVUR or KOTHBURG in Hungary.

CASTRUM AD LÆDUM, Chateau-du-Lar in the French dep. of Sarthe.
CASTRUM ALARICI, Alairac in the French dep. of Aude.
CASTRUM ALBRAISIUM, Castres in the French dep. of Tarn.
CASTRUM ARAGONENSE, Castello Aragonese in Sardinia.
CASTRUM BELLUM, Castel-Bell in Tyrol.
CASTRUM BIGONENSE, or Tarba, Tarbes in the French dep. of Haut-Pyrenees.
CASTRUM BONONIENSE, Castel-Bolognese in the Papal States.
CASTRUM BRIENTII, Chateau-Briant in the French dep. of Cher.
CASTRUM CÆSARIS, or Castrum Sigenci, Castro-Ceras in Burgundy.
CASTRUM CAMARACENSE, Chateau-Cambresis in the French dep. of Nord.
CASTRUM CANINUM, Chateau-Chinon in the French dep. of Nievre.
CASTRUM CARBONARIÆ, Charboniere in Savoy.
CASTRUM CELSUM, Champtoceaux in the French dep. of Mayenne-et-Loire.
CASTRUM CORNU, Castelcorn in Tyrol.
CASTRUM DE ARCURUS, Les Arcs in the French dep. of Var.
CASTRUM DOLENSE, Deves in the French dep. of Cher.
CASTRUM DUNI, or Regiodunum, Dun-le-Roi in the French dep. of Cher.
CASTRUM EBREDUNENSE, Eborodunum, or Ebredunum, Embrun in the French dep. of Hautes-Alpes.
CASTRUM FERREUM, Eisenburg in Hungary.
CASTRUM FONTARABIÆ, Andaya in Navarre.
CASTRUM FORO-JULIANUM, Forum Julii, or Castrum Forojuliense, Ciudad-de-Friuli in Lombardy.
CASTRUM FRANCORUM, Castelfranco in Lombardy.
CASTRUM GAILLONIS, or Gallio, Gaillon in the French dep. of Eure.
CASTRUM GELOSUM, Castel-Geloux in the French dep. of Lot-et-Garonne.
CASTRUM GONTERII, Chateau-Gontier in the French dep. of Mayenne.
CASTRUM LANDONIS, Chateau-Landon in the French dep. of Seine-et-Marne.
CASTRUM LEONIS, Castel-Leone in Lombardy.
CASTRUM LEONIS AD MARE, Castel-Leone in the Neapolitan prov. of Terra-di-Lavoro.
CASTRUM LUCII, Chalas in the French dep. of Haut-Vienne.
CASTRUM MONTIS CALERII, Montcalvo in the duchy of Montferrat.
CASTRUM NOVUM, Chateauneuf in the French dep. of Eure-et-Loire.
CASTRUM NOVUM AD ADDUAM, Castel-nuovo in Lombardy.
CASTRUM NOVUM AD CARIM, Chateauneuf-sur-Cher in the French dep. of Cher.
CASTRUM NOVUM AD LIGERIM, Chateau-neuf-sur-Loire.
CASTRUM NOVUM AD MARTAM, or Corne-tum, Corneto in the Papal deleg. of Viterbo.
CASTRUM NOVUM AD SARTAM, Chateauneuf on the Sarthe.
CASTRUM NOVUM AREMORICÆ, Chateau-neuf in the French dep. of Ille-et-Vilaine.
CASTRUM NOVUM DERTONENSE, or Nuova Tortonese, the Sardinian prov. of Tortona.
CASTRUM NOVUM INSULISMENSE, Chateau-neuf in the French dep. of Charente.
CASTRUM OSCÆ, Usk in Monmouthshire.
CASTRUM PIPONI, Bipp in the Swiss cant. of Bern.
CASTRUM PONZONIS, Castel-Ponzone in Lombardy.
CASTRUM RADULPHI, or Castrum Rufum, Chateau-Roux in the French dep. of Indre.
CASTRUM REGALE, Castroreale in Sicily.
CASTRUM REGINALDI, Chateau-Regnault in the French dep. of Indre-et-Loire.
CASTRUM RUPTUM, Castelmo in Tyrol.
CASTRUM SANCTI ANEMUNDI, Chaumont in the French dep. of Loire.
CASTRUM SANCTI GUELPHI, Guelfo in the duchy of Parma.
CASTRUM SANCTI JOANNIS, Giovanni in the duchy of Piacenza.
CASTRUM SANCTI MACAIRI, Saint-Macaire in the French dep. of Gironde.
CASTRUM SANCTI MICHAELIS, Sankt Michael on the Muhr in Austria.
CASTRUM SANCTI SECUNDI, Santo-Secondo in Parma.
CASTRUM SANCTI SEVERI, Sancto-Severo in Naples.
CASTRUM SEDUNUM, Suss in the Swiss cant. of Valais.
CASTRUM SINEMURUM BRIENNENSE, Semur in the French dep. of Saone-et-Loire.
CASTRUM THEODORICI, Chateau-Thierry in the French dep. of Aisne.
CASTRUM UCOCENSE, or Ucetia, Uzes in the French dep. of Gard.
CASTRUM VALERIANUM, Bingenheim in Hessen-Homberg.
CASTRUM VETRIUM, Caulonia, or Aulonia, Castel-Vetere in Calabria-Ultra.
CASTRUM VICECOMITUM, Castel-Visconte in Lombardy.
CASTRUM VILLANUM, Chateau-Vilain in the French dep. of Haute-Marne.
CASTRUM VINDONICUM, Vendome in the French dep. of Loir-et-Cher.
CASTRUM VULPINUM, Chateau-Renard in the French dep. of Bouches-du-Rhone.
CASTUA, Kastau in Illyria.
CASTULA, Castiglione-Aretino in Tuscany.
CASTULO, or Castalon, perhaps Caslona in Andalusia.
CASTULONENSIS SALTUS, Castona-la-Vieja in Spain.
CASUENTUS, the river Basento flowing into the gulf of Tarentum.
CASULÆ, Casoli in Abruzzo-Citra.
CASURGIS, a town of the Lygii, near Kaurzim in Silesia.
CATABAMA, Hadramaut in Arabia.
CATABANUM, Shibam in Arabia.
CATACCUM, Catanzaro in Naples.
CATADA, the river Milianah in Tunis.
CATADUPA, or Cataractes Minor, the Cataracts of the Nile near Philo.
CATAJA, the northern portion of China.
CATALAUNI, or Catalaunum, Chalons-sur-Marne in the French dep. of Haut-Marne.
CATALAUNIA, or Gothalonia, Catalonia in Spain.
CATANA, Catania in Sicily.
CATARACTA RHENI, the falls of the Rhine at Schaffhausen.
CATARRACTES, a river of Pamphylia now the Ampadere.
CATARRACTONIUM, Catterick Bridge in Yorkshire.
CATERÆ ARDIUM, or Montes Nivosi, the Cordillera-de-los-Andes.

CATERLOGUM, CARLOW in co. Carlow.

CATHENASIA, or CATHANASIA, CAITHNESS in Scotland.

CATIGARA, or CAITAGHORA, by some supposed to be SUCCADANA on the coast of Borneo; by others, MERGUI on the Birman coast.

CATOLACUM, CATULLIACUM, or DIONYSIOPOLIS, SAINT DENIS in the French dep. of Seine.

CATONEUM, CADOGNO in Lombardy.

CATTIMELIBOCENSIS COMITATUS, the lordship of CATZENELNBOGEN in Nassau.

CATTORUM VICUS, CATTWYK in Holland.

CATTUS, KATZBACH in Prussia.

CATUACUM, DOUAY in the French dep. of Nord.

CATUALIUM, HALEN in Liege.

CATUIACA, CARLUEC in Provence.

CATULI ARA, GATTINARA in Piedmont.

CATURIGÆ, or CATURIGOMAGUS, CHORGES in the French dep. of Hautes-Alpes.

CAUCA, or CORU, MEDINA-DEL-RIOSECO in the Spanish prov. of Valladolid.

CAUCASIÆ PORTÆ, or IBERIÆ PORTÆ, the pass of VLADI-KAUKAS in the Caucasus.

CAUCASUS, or CAUCASII MONTES, the CAUCASUS or KAUKAS range in Asia.

CAUDA BOVIS, CLIDES, or DINARCTUM, CAPE SAINT ANDRE in Cyprus.

CAUDA VULPIS, the promontory of CODO-DI-VOLPE on the Neapolitan coast.

CAUDERIÆ, CAUDIEZ in Languedoc.

CAULUM, CHAOUL in the Deccan.

CAURIA, CORIA in Spanish Estremadura.

CAURZIMIUM, KAURZIM in Bohemia.

CAUSENNÆ, KESWICK in Cumberland.

CAVANENSIS COMITATUS, the county of CAVAN in Ireland.

CAVORTIUM, CAVOURS in Piedmont.

CAYSTER, or CAYSTRON, the river KARASU in Asiatic Turkey.

CAZORLA, CACORLO in Andalusia.

CEA, CEOS, or HYDRUSSA, the island of ZEA in the Grecian archipelago.

CEBA, CEVA in Piedmont.

CEBENNA, GEBENNICI MONTES, or CEMMENUS MONS, the chain of the CEVENNES.

CECINA, or CECINNA, the river CESINA in Naples.

CEDROSIA, or GEDROSIA, the prov. of MEKRAN in Persia.

CELA, the river FIUME-DI-TERRA-NUOVO in Sicily.

CELEJA, a town in Noricum, CILLI or ZILLI in Styria.

CELEMANTIA, KALMINZ near Komorn in Hungary.

CELENDERIS, in Cilicia Aspera, by some identified with CANDEROLA, by others with CELINDRO, or with CENCHREO.

CELETRUM, CASTORIA in Greece.

CELEUSA, the river KELS in Swabia.

CELEUSUM, CELEUSUS, or CELTÆ DOMUS, by some supposed to be represented by KELHEIM in Bavaria, by others identified with NEUSTADT, and by some with OETLING.

CELIA, or CŒLIUM, CEGLI in the Neapolitan prov. of Terra-di-Bari.

CELIDNUS, or CELYDNUS, a river of Epirus, the modern SALMIEH.

CELLA, CELLÆ, CESLA, or SKAESLA, CELLE in Hanover.

CELLA AD ISIUS, ZELL-OB-DER-IPS in Austria.

CELLA AD LACUM INFERIOREM, ZELL-AM-ZELLARSEE in Austria.

CELLA AD MOSELLAM, ZELL-IM-HAMM in Rhenish Prussia.

CELLA BADENSIS, ZELL in Baden.

CELLA DEI INFERIOR, the abbey of UNTER ZELL in Bavaria.

CELLA DEI SUPERIOR, HOFVERZELL or OBER ZELL in Bavaria.

CELLA FRANCONICA, ZELL on the Saale in Bavaria.

CELLA RUDOLPHI, ZELL-AM-HARMERSBACH in Baden.

CELLA SANCTI BLAZII, BLASIENZELLA in Thuringia.

CELLA SANCTI CANICI, KILKENIA, KILKENNA, or OSERIA, KILKENNY in Ireland.

CELLA SALÆ, SAHLENHOFEN in Bavaria.

CELLA TIROLENSIS, ZELL in Tyrol.

CELNIUS, by some antiquaries identified with the river DEVERON in Banffshire, by others with the SPEY.

CELSONA, or CALEA, SOLSONA in Catalonia.

CELTIBERIA, a province of Hispania Tarraconensis, comprising the modern provinces of ARAGON, NAVARRE, and the CASTILES in Spain.

CELTORUM MONTES, the CANTAL range in Auvergne.

CELURCA, or MONS ROSARUM, MONTROSE in Scotland.

CEMA, MONTE-CAMELIONE in the Maritime Alps.

CEMELIUM, CIMIEZ in the Sardinian prov. of Nice.

CENEBUM, GENNAPPE in Limburg.

CENESTUM, SANTA LUCIA in Corsica.

CENETUM, CERRETO in the Neapolitan prov. of Terra-di-Lavoro.

CENIO, or CENION, the river FAL of Cornwall.

CENISIUS MONS, MONT CENIS in Savoy.

CENNA, or CINNA, LANGENZENN in Bavaria.

CENNACUM, or CHINEIUM, CHINEY in Namur.

CENOMANIA, LE MANS in the French dep. of Sarthe.

CENTA, the river CONTA in Piedmont.

CENTRITES, or NICEPHORIUS, a river in Armenia Magna, probably the KHABUR.

CENTRONUM REGIO, or TARRENTASIÆ COMITATUS, the TARENTAISE in Savoy.

CENTRONUM CIVITAS, DARANTUSIA, or FORUM CLAUDII, MOUTIERS in Savoy.

CENTULUM, SAINT RICQUIER in the French dep. of Somme.

CENTUM CELLÆ, or TRAJANI PORTUS, CIVITA VECCHIA in the Papal states.

CENTUM COLLES, HUNDERTSBUHEL in Transylvania.

CENTUMVALLES, CENTOVALLI in Lombardy.

CEPARUM PROMONTORIUM, CAPE CORMACHETTI in Cyprus.

CEPELIA, the island of CSEPEL in the Danube.

CEPEROSUM, FORT LOUIS in Guyana.

CEPHALÆ PROMONTORIUM, or CAPITA, CAPE MESURATA in Tunis.

CEPHALENIA, CEPHALLENE, CEPHALONIA, SAMUS or MELÆNA, the island of CEFALONIA in the Ionian group.

CEPHALOEDES, CEPHALOEDIUM, or CEPHALEDUM, CEFALU in Sicily.

CEPHISSUS, or CEPHISOS, the river PODHONISTA, falling into the gulf of Enghia.—Also the MAROSPOTAMO flowing into Lake Copais.

CEPIANA, or PINELUM, PINHEL in Portugal.

CERAMICUS SINUS, the GULF OF COS in the German archipelago.

CERAMIN, SEPOLIA in Attica.

CERASUS, or PHARNACEA, KERESUN in Anatolia.

CERAUNIA, CERINES in Cyprus.

CERAUNI MONTES, the MONTI-DELLA-CHIMERA or ACROCERAUNIAN MOUNTAINS of Epirus.

CERBALUS, or CERVARUS, the river CERVARO in Naples, an affluent of the Candelaro.

CERBICA, SBEKHAH in Tunis.
CERCEDIUS, the river BAVANO in Sicily.—Also the river LIAMONE in Corsica.
CERCINA, or CERCINNA, the island of KERKINE in the Syrtis Minor off the coast of Barbary.
CEREBELIACA, CHABEUIL in the dep. of Drome.
CERENTHIA, GERUNTIA in Calabria-Citra.
CERESIUS, the river TRESA in Lombardy.
CERESIUS LACUS, the LAGO-DI-LUGANO.
CERETICA, CARDIGAN in Wales.
CERETICENSIS COMITATUS, CARDIGANSHIRE.
CERFENNIA, POPOLI in Abruzzo-Citra.
CERNE, by some identified with MADAGASCAR, by others supposed to be MADEIRA or PORTO-SANTO.
CERNIA, CERINES in Cyprus.
CERTERATÆ, or CORTACUM, COUTRAS in the French dep. of Gironde.
CERTIACUM, or ZURZACUM, ZURZACH in the Swiss cant. of Aargau.
CERVARIA, CERVERA in Catalonia.
CERVIA, CHIEVRES in Hainault.
CERVIMONTIUM, or HIRSCHBERGA, HIRSCHBERG in Silesia.
CERVIMONTIUM AD SALAM, HIRSCHBERG on the Saale.
CERVIMONTIUM WESTFALIÆ, HIRSCHBERG in Prussia.
CERVIUM, CIERF in the Swiss cant. of Grisons.
CESTRIA, DEVA, or DEUNA, CHESTER in Cheshire.
CESTRIENSIS COMITATUS, CHESHIRE.
CETANA, SCOPELLO in Sicily.
CETIUM, or AELIUM CETIUM, a town in Noricum, according to Kruse, MAUTERN in the Wienerwalde, according to others ST. POLTEN.
CETIUS MONS, in Pomerania, the KALLENBERG.
CHABERIS, CAVERYPUTNAM in Hindostan.
CHABERIS, the river CAVERY in Hindostan.
CHABRIUS, a river in Chalcedice, the modern CHABRIA or CILABRO.
CHADRACH, HADRACH in Cœle-Syria.
CHALCIS EUBŒÆ, the island of NEGROPONT.
CHALCEDON, or CALCHEDON, now KADI-KEUI on the shore of Asia-Minor opposite Constantinople.
CHALDÆA, the SW part of Babylonia, now IRAK-ARABI, which, stretching along the r. bank of the Euphrates, extended as far into the desert on the W as could be rendered habitable by irrigation. The oldest name of this region was SHINHAR or SHINGAR. The names CHALDÆA and BABYLONIA are, however, used interchangeably by early writers.
CHALUS, or CHALYS, a river in Cyrrhestica, the modern KAWIK or KOVAIK.
CHALUSUS, or TRAVA, the river TRAVE in Holstein.
CHANTILIACUM, CHANTILLY in the French dep. of Oise.
CHAONIA, in Epirus, now CANINA in Albania.
CHARADRUS, a river in Epirus, now the ZALONGOS.
CHARAX, now KARA-KAJA in the Crimea.
CHARIATUM, CARIATI in Naples.
CHARICIS, or CHARISTUS, a river in Colchis, the modern TAMUSA.
CHARIDEMI PROMONTORIUM, CAPE GATA on the coast of Granada in Spain.
CHARMUTHAS, probably YEMBO on the E coast of the Arabian gulf.
CHELEDONIUM PROMONTORIUM, SACRUM PROMONTORIUM, or TAURI PROMONTORIUM, CAPE CHELEDONI on the S coast of Anatolia.
CHELONITES PROMONTORIUM, the modern CAPE TORNESE.
CHELONOTICUS SINUS, the GULF OF ARCADIA.
CHEMONTIUM, CHEMNITZ in Transylvania.
CHERRONE, CHERONESUS HEROCLEA, or HERACLEOTARUM OPPIDUM, probably EUPATORIA on the coast of the Black sea, in the Crimea.
CHERSO, CHERSON on the Dnieper in Russia.
CHERSONESUS, CAPE CHERSONESI on the coast of Greece.
CHERSONESUS, the peninsula of KORIPATAM in the Concan in India.
CHERSONESUS CIMBRICA, the Danish territory of JUTLAND, SCHLESWIG, and HOLSTEIN.
CHERSONESUS TAURICA, TAURICA, TAURICA TERRA, CHERSONESUS SCYTHICA, CHERSONESUS MAGNA, CHERSONESUS CRIMÆA, or TARTARIA CRIMEA, the CRIMEA in the Russian gov. of Taurida.
CHERSONESUS THRACIÆ, the peninsula of the DARDANELLES or GALLIPOLI.
CHIAGRA, the river CHAGRES in Central America.
CHIEMUS LACUS, the CHIEM-SEE in Bavaria.
CHILCA, or INSULA SANCTI COLUMBIUM, ICOLMKILL in the Scottish Hebrides.
CHILI REGIO, CHILE or CHILI in S. America.
CHILMORIA, or KILMORA, KILMORE in co. Cavan.
CHILONIUM, KIELA, KIELANUM, KILANA, or KILONIUM, KIEL in Denmark.
CHIMACUM, CHIMAY in Hainault.
CHIMÆRA, now KIMARA in Albania.
CHIMÆRA MONS, in Lycia, probably the modern CRAGUS.
CHINA, SINA, or SINARUM REGIO, the modern CHINA.
CHINAPHAL, or SARDABAL, a river of Mauritania Cæsariensis, probably the modern SHELLIFF.
CHINIACUM, CHINY in Luxemburg.
CHINNERETH MARE GENNESAR LACUS, GENNESARITIS LACUS, GALILÆÆ MARE, or TIBERIDIS LACUS, the BAHR-EL-TABERIA, or LAKE OF TIBERIAS of Syria.
CHION, MACRIS, or PITYUSA, the island of SCIO in the Grecian archipelago.
CHIOVIA, or KIOVIA, the gov. of KIEV in Russia.
CHIOVIENSIS PALATINATUS, the gov. of KIEV in Podolia.
CHIPPENHAMUM, CHIPPENHAM in Berks.
CHOASPES, the river ATTOCK, an affluent of the Ganges.—Also a river of Susiana, the modern KERAH or KARASU.
CHOBUS, a river of Colchis, perhaps the KEMKHAL, which flows into the Black sea to the N of the Phasis.
CHOINITIA, or KOMZA, KAUNITZ in Prussia.
CHORSA, a town of Armenia Major, now KARS.
CHOTIMIA, or CHOTINUM, CHOTYM in Bessarabia.
CHREMISSÆ MONASTERIUM, or CREMISSANUM, KREMSMUNSTER in Austria.
CHREPSA, or CRESSA, the island of CHERSO in the Adriatic.
CHRISTA, CRESTIDIUM, or CRESTUM, CREST in the French dep. of Drome.
CHRISTIANI MUNITIO, FRIEDRICHSORT, or CHRISTIANSPRIESS in Schleswig.
CHRISTIANOPOLIS, or CHRISTIANSTADIUM, CHRISTIANSTADT in Sweden.
CHRISTINÆ PORTUS, CHRISTIANHAUSEN on the Wener lake in Sweden.
CHRISTOPOLIS, in Thrace, JENIKOI near the river Strymas.
CHRONOPOLIS, or TILSA, TILSIT in Prussia.
CHRONUS, PREGELLA, or MEMELA, the river MEMEL in Prussia.
CHRUDIMA, CHRUDIM in Bohemia.
CHRYSA, GARDEROGNISSA in Crete.
CHRYSII AURANIA, ALTENBURG in Hungary.
CHRYSIUS, the river KOROSCH in Hungary.
CHRYSOPOLIS, SCUTARI in Albania.
CHYDAS, the river ROSAMARINA in Sicily.
CIABRUS, CIAMBRUS, or CEBRUS, the river JOCA or ZEBRIS in Bulgaria.

CIBONIENSIS COMITATUS, the territory of HERMANNSTADT in Transylvania.
CIBONIUM, or HERMANNOPOLIS, HERMANNSTADT in Transylvania.
CIBURIUM MUNIS, ZEBEN or SABINOW in the Hungarian comitat of Zarosch.
CICESTRIA, CHICHESTER in Sussex.
CILIANUM, CIGLIANO in Piedmont.
CILICIA, BIALA in Silesia.
CILICIA, a district of Asia Minor, comprising the portions of the Turkish pashaliks of KARAMANIA and ADONIA. The W division of this region was known as C. TRACHEA; the E as C. CAMPESTRIS.
CILNIANA, ESTEPONA in the Spanish prov. of Granada.
CILURNUM, perhaps COLLERTON in Northumberland.
CIMACULUM, or CAMACLUM, COMACCHIO in Lombardy.
CIMARUM PROMONTORIUM, CAPE SPADA in Crete.
CIMBRIA PARVA, POMERANIA in Prussia.
CIMBRORUM PORTUS, CIMBRISHAM in Sweden.
CIMINIUS LACUS, LAGO-DI-VICO in the Papal deleg. of Viterbo.
CIMMERIUM, ESKI-KRIM or LEUKOPOL in the Russian gov. of Taurida.
CIMOLIS, KINOLI on the Black sea.
CIMOLUS, or ECHINUSSA, the island of CIMOLI in the Cyclades.
CINGA, CINCA in Aragon.
CINGULUM, CINGOLI in the Papal States.
CINAMONIFERA REGIO, a name given by Ptolemy to the NE angle of Africa, or of the Somali territory.
CINNIBANTUM, KIMBOLTON in Huntingdonshire.
CIOS, or OESCUS, a river of Mœsia Inferior, perhaps the ESKER or ISKRA.
CIOS, or PRUSCIAS, GHIO in Anatolia.
CIRCÆUM PROMONTORIUM, MONTE-CIRCELLO in the Campagna-di-Roma.
CIRCIDIUS, a river of Corsica, the modern RIPE.
CIREOLA, ZIRL or CIRLE in Tyrol.
CIRRADIA, SILHET in Bengal.
CIRRHA, SALONA on the gulf of Salona.
CIRTA, or CIRTA JULIA, CONSTANTINA in Algeria.
CISAMUS, CHISAMO in Crete.
CISIMONS, CSIMONE in Lombardy.
CISTERCIUM, CITEAUX or CISTERZ in the French dep. of Cote-d'Or.
CISTHENE, the island of CASTEL-ROSSO off the coast of Caramania.
CITHÆRON in Bœotia, the modern MOUNT ELATEIA.
CITHRUM, ZOTTIWAR in Thessaly.
CITIUM, CHIETI in Cyprus.
CITIUM, or ZIZE, ZEITZ in Prussia.
CITUATUM, or CITUORUM INSULA, the island of SCHUTT in the Danube.
CIVITAS, CIVEDA in Lombardy.
CIVITAS CASTELLANA, CITTA-DI-CASTELL in the Papal deleg. of Perugia.
CIVITAS CURIOSOPITUM, or CURIOSOLIMAGUS, by some identified with QUIMPER, by others with CORSEULT, in the French dep. of Finesterre.
CIVITAS DINIENSIUM, or DINIA, DIGNE in the French dep. of Basses-Alpes.
CIVITAS DUCALES, CIVITA-DUCALE in Naples.
CIVITAS EQUESTRIUM, COLONIA EQUESTRIS, or NOVIODUNUM, NYONS in the Swiss cant. of Vaud.
CIVITAS GOSLARIA, GOSLAR in Hanover.
CIVITAS LACTORATIUM, or LACTURA, LECTOURE in the French dep. of Gers.
CIVITAS LEXOVIORUM, or NOVIOMAGUS, LISIEUX in the French dep. of Calvados.
CIVITAS LUTEVENSIUM, LUTEVA, or FORUM MERONES, LODEVE in the French dep. of Herault.
CIVITAS MELDORUM, or MELDUNUM, MEAUX in the French dep. of Seine-et-Marne.
CIVITAS MONTIS GRACCENSIS, or ZAGRAVIA, AGRAM in Croatia.
CIVITAS NAMNETICA, CONDIVICNUM, or NAMNETES, NANTES in the French dep. of Loire-Inferieure.
CIVITAS OCENSIS, OCA, COLONIA OSA, or EVA, TRIPOLI in Barbary.
CIVITAS POMPEJONENSIS, PAMPELUNA in Andalusia.
CIVITAS REGALIS, or CIVITAS REGIA, CIUDAD-REAL in La Mancha.
CIVITAS REMORUM, DURICORTORA, or DUROCORTURUM, a town of the Remi in Gallia Belgica, now represented by Rheims in the French dep. of Marne.
CIVITAS RODERICI, or RODERICOPOLIS, CIUDAD-RODRIGO in the Spanish prov. of Leon.
CIVITAS SALMARUM, or CIVITAS SALMIENSIUM, CASTELLANE in the French dep. of Basses-Alpes.
CIVITAS SANCTI ANGELI, CIVITA-DI-SANTO-ANGELO in Abruzzo-Ultra.
CIVITAS SANCTI MINIATIS AD TEDESCUM, NUMATO-AL-TODESCHO in Tuscany.
CIVITAS SANICIENSIUM, SANICIO, or SANECIUM, SENEZ in the French dep. of Basses-Alpes.
CIVITAS SEGESTERORUM, SEGUSTERO, or SISTARICUM, a town of the Memini in Gallia Narbonensis, near SISTERON in the French dep. of Basses-Alpes.
CIVITAS TRIDENTINA, TRIDENTUM, or TRIDENTONUM CASTELLUM, TRIENTE or TRENTO in Tyrol.
CIVITAS VALENTINORUM, VALENCE in the French dep. of Drome.
CIVITAS VALLENSIUM, OCTODURUS, or VALINSA, MONTIGNI in the Swiss cant. of Valais.
CIVITAS VAPPINCENSIUM, or VAPINIUM, probably GAP in the French dep. of Hautes-Alpes.
CIVITAS VASATICA, COSSIO, or COSSIUM, BAZAS in the French dep. of Gironde.
CIVITAS VENETORUM, CIVITAS VENETICA, or VENETRA, VANNES in the French dep. of Morbihan.
CIVITAS VIDUCASSIUM, VIEUX in the French dep. of Nord.
CLADRUBIUM, KLADRAU in Bohemia.
CLAGENFURTUM, CLAGENFURT in Illyria.
CLAMOCIACUM, CLAMECY in the French dep. of Nievre.
CLAMERGANIA, GLAMORGAN in Wales.
CLAMPETIA, or LAMPETIA, SANTO-LUCILO in Calabria.
CLANIS, GLANIS, or GLANUS, a river of Etruria, the modern CHIANA.
CLANIS, or LIRIS, a river of Latium, the modern GARIGLIANO.
CLANIUS, CLANUS, GLANIS, or LITERNUS, a river of Campania, the modern AGNO or CLANIO.
CLARA, or MAGNANA, WELIKA in Croatia.
CLARA, co. CLARE in Ireland.
CLARÆ VALLES, CLARVEAUX in the French dep. of Aube.
CLARA VALLIS, CLAIRVEAUX in Luxemburg.
CLARA VALLIS, ZWETL in Austria.
CLARA WERDA, SCHONEWERD in Switzerland.
CLARAMONTIUM, CLERMONT in Savoy.
CLARASCUM, CHERUSCA in Piedmont.
CLARENNA, a town in Rhetium, by some supposed to be represented by EHINGEN on the Danube, by others identified with KIRCHHEIM.
CLARENTIA, CLARE in Suffolk.
CLARENTIA, CHIARENZA on the gulf of Lepanto.

CLARIACUM AD LIGERIM, Clery in the French dep. of Loiret.
CLARIACUM AD OLDAM, Clairac in the French dep. of Lot-et-Garonne.
CLARITAS JULIA, Espejo in Cordova.
CLARIUM, Chiari in Milan.
CLAROFONTARUM PALATIUM, Hellebronn in Salzburg.
CLAROMONS, Chiaromonte in Sicily.
CLAROMONTIUM LUTEVENSE, Clermont-di-Lodeve in the French dep. of Herault.
CLARUS FONS, Sherborne in Dorset.
CLARUS MONS BELLEVOISENSIS, Clermont-en-Beauvoisis in the French dep. of Oise.
CLARUS MONS LOTHARINGIÆ, Clermont-en-Argonne in the French dep. of Marne.
CLATEMA, Varignano in the Papal states.
CLAUDIA FOSSA, Chioggia in the Adriatic.
CLAUDIA, Municipium Claudium, or Claudonium, Clagenfurt.
CLAUDIA CASTRA, Claudiocestra, Clevum, or Glocestria, the city of Gloucester in England.
CLAUDIANOPOLIS, Claudiopolis or Colusvaria, Klarisenburg or Kolosvar in Hungary.
CLAUDOMERUM, Muros in Spanish Galicia.
CLAUSA, or Veruca, Chiusa in Lombardy.
CLAUSENTUM, or Southantonia, Southampton in England.
CLAUSINA, or Clausium, Clausen in Tyrol.
CLAUSTRUM CAVALI, Kafel in Tyrol.
CLAUSULÆ, Clusæ, or Slusa, Sluys in the Dutch prov. of Zieland.
CLAUSULA NIGRÆ, Swarte-Sluys in the Dutch prov. of Overyssel.
CLAVASIUM, Chiavasco or Chivaria in Piedmont.
CLAVUNIA, Chiavenna in Lombardy.
CLAZOMENÆ, Vourla in Macedonia.
CLEPIDAVA, Kamieniec in Podolia.
CLESIUS, Clausius, or Clusius, the river Chiese an affluent of the Oglio in Lombardy.
CLIMAX, a mountain in Cœle-Syria, the modern Kastrovan.
CLINIACUM, or Clunineum, Cluny in the French dep. of Saone-et-Loire.
CLIPPIACUM, or Cligiagum, Clichy in the French dep. of Seine.
CLISOBORA, Delhi in Hindostan.
CLITERNIA, Civita-Mare in Naples.
CLITOR, or Clitorium, Katzenes in Northern Greece.
CLITUMNUS, the river Clitumno in the Papal states.
CLIVIA, Cleve in Rhenish Prussia.
CLIVUS SANCTI LUCII, Luciensteig in the Swiss cant. of Grisons.
CLODRANUS, the river Fluvia in Catalonia.
CLONA, Clones in co. Fermanagh.
CLONFERTIA, Clonfert in co. Galway.
CLUANUM, Cloyne in co. Cork.
CLUDANUS AMNIS, Cluda, Clota, or Glota, the river Clyde in Scotland.
CLUDEM, Klundert in the Dutch prov. of N. Brabant.
CLUIDA, the river Clwyd in Wales.
CLUNIA, Feldkirch in Rhenish Prussia.
CLUNIA, or Colonia Clunia, Corunna-del-Conde in Old Castile.
CLUNIUM, or Fanum Sanctæ Catharinæ, Catharina in Corsica.
CLUPEDA, Memelia, or Memelium, Memel in Prussia.
CLUSÆ VALLES, Vaucluse in the French dep. of Vaucluse.
CLUSONIUS, the river Clusone an affluent of the Po.
CLYSMA, Kocsum near Suez in Egypt.
CNAPDALIA, Knapdale in Argyleshire.
CNEDUS, Gnidus, Triopia, or Pegusa, a city of Caria, near Porto-Genovese on the SW coast of Asia Minor.
CNESSUS, Guiosa in Crete.
COAGIA, Kioege in Danish Sieland.
COAGIUM, Cassaro in Sicily.
COBURGUM, Coburgium, or Devona, Coburg on the Itz, in the duchy of Coburg.
COCCYMUTIUM, or Cocermutium, Cockermouth in Cumberland.
COCINTIA, or Consilinum, Consentia in Calabria.
COCINTUM PROMONTORIUM, Cape Stilo in Calabria.
CODANIA, Codonia, Hofnia, or Haunia, the city of Copenhagen in Denmark.
CODANONIA, Sieland in Denmark.
CODANUS SINUS, Codanum Mare, or Suevicum Mare, the Cattegat.
CŒLA in the Thracian Chersonese, Kilia.
CŒLE PERSIS, Merdasht in Persia.
CŒLESYRIA, or Phœnicia Libanesia, the valley between the chains of Libanus and Antilibanus in Syria, now El-Bekach.
CŒLICORONA, Himmelskron in Bavaria.
CŒLIUS MONS, Kellmunz in Bavaria.
COGNACUM, or Conacum, Cognac in the French dep. of Charente.
COILA, or Covalia, Kyle in Ayrshire.
COLA, the island of Coll off the W coast of Scotland.
COLA LAPPORUM, Kola at the mouth of the Lutto in Russian Lapland.
COLANIA, Lanark, or perhaps Coldingham, in Scotland.
COLAPIS, the river Kulpa in Illyria.
COLASÆ, or Colossæ, a town in Phrygia Magna, near the village of Khonas in Asia Minor.
COLATIO, or Lithopolis, a town of Noricum, Stein or Kamneck in Illyria.
COLBERGA, Colberg in Prussia.
COLCESTRIA, or Colonia, Colchester in Essex.
COLCHIS, a district of Asia, bounded on the N by the Caucasus, on the E by Iberia, on the S by Armenia, and on the W by the Euxine, comprising the modern Mingrelia and a part of Imiretia.
COLDANIA, Coldana, or Coldingamum, Coldingham in Berwickshire.
COLDINGA, Koldingen in Jutland.
COLDITIA, Colditz in Saxony.
COLEDA, or Colonia ad Windam, Coln-an-der-Losse in Prussia.
COLINUM, Collin in Bohemia.
COLLATIA, a town of the Sabines, near Corcollo in the Papal states.
COLLIPPO, Leiria in Estremadura.
COLLIS, Colle in Tuscany.
COLLIS MONTES, Colmars in the French dep. of Basses-Alpes.
COLLIS PEREGRINORUM, or Marbacum, Marbach in Würtemberg.
COLLOPS MAGNUS, or Chulla, Collo or Colle in Tunis.
COLMAR, Coloburgum, or Columbanum, Colmar in the French dep. of Haut-Rhin.
COLMOGARO, Kolmogrod in the Russian gov. of Wologda.
COLOCIA, or Colozza, Kolotsha in Hungary.
COLOMENA, Coulommiers in the French dep. of Seine-et-Marne.
COLONE, or Colonides, Coras in Greece.

COLONIA AD SPREAM, or COLONIA MARDUCA, COLN-ON-THE-SPREE in Prussia.

COLONIA AUGUSTA AROE PATRENSIS, PATROS in Greece.

COLONIA AUGUSTA PANHORMITANORUM, PANHORMUS, or PANORMUS, PALERMO in Sicily.

COLONIA AUGUSTA TAUROMENITANA, TAORMINA in Sicily.

COLONIA AUGUSTA VERONA NOVA GALLIENIANA, VERONA in Lombardy.

COLONIA AURELIA ANTONIANA, or OVILIA, WELS in Austria.

COLONIA AURELIANA LAUREACENSIS, LAURACH in Austria.

COLONIA CLAUDA AUGUSTA FELIX JUDERA, ZARA-VECCHIA in Dalmatia.

COLONIA CLAUDIA SAVARIA, STEIN-AM-ANGER in Hungary.

COLONIA EBORACENSIS, or COLONIA EBORACUM, YORK in Yorkshire.

COLONIA FERENTENENSIS, or FERENTINUM, MONTE-FIASCONE in the Papal states.

COLONIA GEMELLA JULIA HIPPONESNIS PIA AUGUSTA, or HIPPO REGIUS, BONA in Algeria.

COLONIA ITALICENSIS, or TRAJANI CIVITAS, SEVILLA-LA-VIEJA in Spain.

COLONIA JULIA AUGUSTA DERTONA, TORTONA in Piedmont.

COLONIA JULIA AUGUSTA PARMA, PARMA in Italy.

COLONIA JULIA AUGUSTA SINOPE, SINAB or SINOPE on the Black sea.

COLONIA JULIA FANESTRIS, or FANUM, FANO in Lombardy.

COLONIA JULIA OCTAVIONORUM, or COLONIA PACENSIS CLASSICA, FREJUS in the French dep. of Var.

COLONIA JULIA PARIANA, or PARIUM, a town in Mysia Minor, according to Kruse KAMARISS.

COLONIA JULIA SALONA, SALONA, or SALO, SALONA in Dalmatia.

COLONIA LAODICENA, LAODICEA CABIOSA, LAODICIA, or LAODACEA AD LIBANUM, near JUSHIA in Cœle-Syria.

COLONIA LIBISOSANORUM, or LIBISOSA, LIZUZA in New Castile.

COLONIA NAPOCENSIS, or NAPOCA, a town in Dacica, by Spon identified with KLAUSENBERG, by D'Anville with DOBOCA.

COLONIA NEPENSIS, or NEPETE, NEPI in the Papal deleg. of Viterbo.

COLONIA OSTIA, OSTIA near the mouth of the Tiber.

COLONIA PATRICIA CORDUBENSIS, or CORDUBA, CORDOVA in Spain.

COLONIA ROMULEA, COLONIA ROMULENSIS, or HISPALIS, SEVILLA in Andalusia.

COLONIA SABRATA, SABRATHA, or TRIPOLI AFRICANA, TRIPOLI-VECCHIA in Tripoli.

COLONIA SEPTIMIA RESAINESIORUM, RESAINA, or RESINA, RAS-AIN in Irak-Arabi.

COLONIA SOLMONTINA, SOLMONA, or SULMO, SALMONA in Abruzzo-Ultra.

COLONIA SUTRINA, or SUTRIUM, SATRI in Viterbo.

COLONIA TOLOSA, TOLOSATIUM CIVITAS, TOLOSA PALLADIA, or THOLOSA, TOULOUSE in the French dep. of Haut-Garonne.

COLONIA VENEREA NABRISSA AUGUSTA, LEBRIJA in Seville.

COLONIA ZEMENSIUM, CZEMETZ on the Danube in Wallachia.

COLOPHON, a city of Ionia, near CHILLE on the Halys.

COLOSIA, or COLOSIUM, KOLOSCH in Transylvania.

COLOSVARIENSIS COMITATUS, the co. of CLAUSENBURG in Transylvania.

COLRANA, COLERAINE in Ireland.

COLRANENSIS COMITATUS, co. LONDONDERRY in Ireland.

COLUBRARIA, the island of OPHIUSA in the Mediterranean.

COLUMBARIA, or PELIUS, the island of COLUMBARI-DE-TREPANI off the W coast of Sicily.

COLUMBIANA, COLUMBATZ in Servia.

COLUMELLA, CATHONE in Calabria.

COLUMNÆ HERCULES, CALPE and ABYLA on the opposite sides of the straits of Gibraltar.

COLUMNARUM CAPUT, CAPE COLONNA in Calabria-Ultra.

COLURNUM, COLORNO in Parma.

COMACENUS LACUS, LARIUS LACUS, or COMENSIS LACUS, the LAGO-DI-COMO in Italy.

COMAGENUS, or COMIANUS, a town in Pannonia Superior, near KAUMBERG in Austria.

COMANA PONTICA, a town of Pontus, near TABACHZAN.

COMARA, COMARNUM, or CRUMENUM, COMORN in Hungary.

COMARIA, CAPE COMORIN in Hindostan.

COMARUS, CAMERINO in Paxos.

COMARUS, or MONACENIS, a Roman station in Lower Egypt, near SHAMAM-EL-GHARBIEH.

COMESATIUM, COMESAZZO in Lombardy.

COMIDAVA, in Dacia, now SALZDORF in Transylvania.

COMIDAVA AD SAMOSIUM, DIS-DIESCH in Transylvania.

COMINEUM, COMMINIUM, or COMMINSICUM, COMINES in West Flanders.

COMINUM, or HEPHÆSTIA, the island of COMINO in the Mediterranean.

COMITATIS MALA, GREVESMUHLEN in Mecklenburg-Schwerin.

COMMAGENE, or EUPHRATESIA, SHEMISATH in Northern Syria.

COMMENASES, the river GOGRA in India.

COMMERCIACUM, COMMERCY in the French dep. of Meuse.

COMMOTOVIA, or COMMOTOVIUM, KOMMOTAU in Bavaria.

COMPITUM ANAGNINUM, SARVIGNANO in the Papal States.

COMPLUTUM, ALCALA-DE-HENARES in New Castile.

COMPSA, CONZA in Principato-Ultra.

COMPULTERIA, MARIA-DE-COVULTERE in Holstein.

COMUM NOVOCOMUM, or NOVUM COMUM, COMO in Lombardy.

CONA, CONDIDA, or CONIUM, COSNE in the French dep. of Nievre.

CONATIA, TYRCONNEL in Ireland.

CONCA, CONCHA, or CONCIA, CUENCA in Spain.

CONCANA, CONGAS-DE-ONIS in Asturias, or perhaps SANTILLANA in Asturias.

CONCANGIUM, KENDAL in Westmoreland.

CONCARNEUM, CONCARNEAU in the French dep. of Finisterre.

CONCEPTIONIS SINUS, CONCEPTION BAY in Newfoundland.

CONCHÆ, CONCHES in the French dep. of Eure.

CONCORDIA, TOMAR in Estremadura.

CONCORDIA, a town of the Nemetes, supposed by Simler to be KICHERSBERG in Lower Alsace; by others identified with ALTSTADT on the Lutter; by others with DRUSENHEIM.

CONCORDIA SALTUS, or CONCRESSANUM, CONCORSAULT in the French dep. of Cher.

CONDÆUM, CONDATUM, or CONDITUM, CONDE in the French dep. of Nord.
CONDÆUM AD NOVALLUM, CONDE-SUR-NOIREAU in the French dep. of Calvados.
CONDATE, KONGEN in Würtemberg.
CONDATE, CONDE in the French dep. of Eure.
CONDATE, a town of the Carnavii, by some identified with NORTHWICH, by others with CONGLETON in Cheshire.
CONDATE, or REDONUM CIVITAS, RENNES in the dep. of Isle-et-Vilaine.
CONDATE, or MONASTERIUM SENONUM, MONTEREAU-FAULT-YONNE in the dep. of Seine-et-Marne.
CONDERCUM, a town of the Brigantes, BENWELL in Northumberland.
CONDOCHATES, probably the river GUNDUK in Hindostan.
CONDOJANUM, CONDOJANI in Calabria-Ultra.
CONDOMIUM VASCONUM, or LONDOMUM, CONDOM in the dep. of Gers.
CONDRIACUM, CONDRIUM, or CONDRUSIUM, CONDRIEUX in the dep. of Rhone.
CONDRUSORUM REGIO, or CONDRUSTO, CONDROS in Liege.
CONDURA, CONDUR in Bissenagur.
CONEDRACIUM, or CONNARÆUM, CONNERE in the French dep. of Sarthe.
CONELIANUM, or CONEGLIANUM, CONEGLIANO in Lombardy.
CONEUM, or CUNEJUM, CONI in Piedmont.
CONFLUENTES, COBLENZ in Rhenish Prussia.
CONFLUENTES, or CONFLUENTIA, CONFLANS in the French dep. of Haut-Saone.
CONFLUENTES, or CONFLUENTIA CENTRONUM, CONFLANS in Savoy.
CONFLUENTES, or CONFLUENTIA HELVETIORUM, COBLENZ in Switzerland.
CONFLUENTES, or CONFLUENTIA INSULÆ FRANCIÆ, CONFLANS-ST.-HONORINE in the French dep. of Seine-et-Oise.
CONFLUENTES, or CONFLUENTIA LOTHARINGIÆ, CONFLANS-EN-JAMY in the French dep. of Meuse.
CONGAVATA, STANWICKS on the river Eden in Northumberland.
CONGELATUM MARE, GLACIALE MARE, CONCRETUM MARE, PIGRUM MARE, SCYTHICUM MARE, or MOSCOVITICUM MARE, the ARCTIC OCEAN.
CONGIUM, CABEÇON in Spain.
CONGLETONIUM, CONGLETON in Cheshire.
CONGUM, CONGO in Lower Guinea.
CONIA, COGNA in Piedmont.
CONILIUM, CONIL in Seville.
CONIMBRA, CONIMBRICA, or COLIMBRIA, COIMBRA in the Portuguese prov. of Beira.
CONNACHTIA, CONNAUGHT in Ireland.
CONNOBURGUM, CONNEBURG in Prussia.
CONNOVIUS, or TISOBIS, CONWAY in Wales.
CONOVIUM, ABERCONWAY, or perhaps CAER-RHYN near Conway in Wales.
CONSABRUM, CONSUEGRA in the Spanish prov. of Toledo.
CONSENTIA, or CONSENTINA CIVITAS, COSENZA in Calabria-Citra.
CONSENTINA PROVINCIA, CALABRIA-CITERIORE.
CONSTANTIA, or GANODURUM, CONSTANZ in Baden.
CONSTANTIA, CONSTANTINA, or NICEPHORIUM, RAKKA in Mesopotamia.
CONSTANTIANA, KONSTENZA in Bulgaria.
CONSTANTIENSE TERRITORIUM, COTENTIN in the French dep. of La Manche.
CONSTANTINÆ PROVINCIA, or CONSTANCIA, the prov. of CONSTANTINE in Algeria.
CONSTANTINOPOLIS, CONSTANTINOPLE.
CONTEGIUM, CUNDIS in the Swiss cant. of Valais.
CONTIACUM, or CONTIUM, CONTY in the French dep. of Somme.
CONTILIANUM, CONTIGLIANO in the Papal States.
CONTINUM, CONCONUM, or MIRANDA DURII, MIRANDA-DE-DUERO in Portugal.
CONTRA ACINIUM, or TRANSACINCUM, PESTH in Hungary.
CONTRIBUTA, MEDINA-DE-LAS-TORRES in Estremadura.
CONVALLIS, CONVALLIA, or NIVARIA, the island of TENERIFFE.
CONVENNAS, CONVEY in Essex.
CONVENTRIA, or COVENTRIA, COVENTRY in Warwickshire.
CONVERSANUM, or CAPERSANUM, CONVERSANO in Naples.
COPAIS, CEPHINIS, HALIARTIUS, the LAGO-DI-TAPOGLIA or LAGO STIVA in Livadia.
COPHEN, COAS, or CHOES, the river HELMUND in Afghanistan.
COPID, SYBARIS, or THURIUM, TERRANUOVO near the gulf of Tarentum in Naples.
COPINGA, KOPING on the Mälarsee in Sweden.
COPRANITIA, COPRANITZ on the river Drave.
COPRANITIA, KAPRONEZA in Croatia.
COPTOS, COFT or KEPHT in Egypt.
COQUEDA, the river COQUET in Northumberland.
CORACODES PORTUS, PORTO-CONDE in Algeria.
CORASESIUM, in Cilicia, now ALAYA.
CORASIÆ, or CORASSIÆ, the island of DRAGONISE in the Ægean sea.
CORAXII MONTES, the ELBURZ.
CORAX PROMONTORIUM, CAPE KERKINOS-BORNU in the Crimea.
CORAX, or CROTALUS, the river CORACE in Naples.
CORBARIA VALLIS, CORBIERES in Languedoc.
CORBEJA, CORBIE in the French dep. of Somme.
CORBEJA, CORVEY in Westphalia.
CORBELIUM, CORBELLUM, or CORBONIUM AD SEQUANUM, CORBEIL in the French dep. of Seine-et-Oise.
CORBENIACUM, CORBENY in the French dep. of Aisne.
CORBIENA, or CARBIENE, KHORUMABAD in Irak-Arabi.
CORBINIACUM, CORBIGNY in the French dep. of Marne.
CORBINIACUM NIVERNENSE, CORBIGNY in the French dep. of Nievre.
CORBULONIS MONUMENTUM, GRONINGEN in Holland.
CORCAGIA, or CORCANIA, CORK in Ireland.
CORCAGINENSIS COMITATUS, co. CORK in Ireland.
CORCORAS, or NAUPACTUS, LAIBACH in Illyria.
CORCYRA, DREPANE, PHÆACIA, or PHÆACUM INSULA, the island of CORFU in the Mediterranean.
CORCYRA, CORZOLA in the island of Corzola.
CORCYRA NIGRA, or CORCYRA ILLYRIÆ, the island of CORZOLA off the Dalmatian coast.
COREA, the peninsula of KOREA.
CORETUS SINUS, the GIULOE-MARE, or W part of the Palus Mæotis.
CORFINIUM, SANTO-PELINO in Abruzzo-Citra.
CORICEUM PROMONTORIUM, CAPE KORAKA on the coast of Asia Minor.
CORINIUM, DUROCOMOVIUM, or CIRCESTRIA, CIRENCESTER in Gloucestershire.
CORINTHIACUS SINUS, the GULF OF LEPANTO.
CORINTHUS, or EPHYRA, the city of CORINTH in Greece.
CORIOLANUM, or CORIOLUM, CARIGLIANO in Calabria-Citra.
CORIOVALLUM, FALCOBURGUM, or VALCONIS MONS, FALKENBERG in Limburg.

CORITICUM, KRINGA in Carinthia.
CORLINUM, CORLIN in Prussia.
CORNACUM, ZATA in Lower Hungary.
CORNELIA, or WIMPINA, WIMPFEN on the Neckar.
CORNELIANUM, CORNEILLAN in the French dep. of Gers.
CORNELII MONASTERIUM, CORNELIUSMUNSTER in Prussia.
CORNETUM, GORNETO in Naples.
CORNI, CORNETO in Sardinia.
CORNUBIA, CORNUWALLIA, or DUMNONIORUM REGIO, CORNWALL in England.
CORNUVIENSIS, JAGERNDORF in Silesia.
COROCONDAME, FANAGORIA or TAMAN in the Crimea.
CORODAMUM, CAPE KURIAT on the Arabian gulf.
COROMANDELA, or SORARUM REGIO, COROMANDEL in Hindostan.
CORONA, the river CORONE in Lombardy.
CORONÆBURGUM, CRONENBURG or KRONBERG in Denmark.
CORONIA, LANDSKRONA in Sweden.
CORONTA, near ASPRO in Northern Greece.
COROPASSUS, KON-HISSAR in Asia Minor.
COROS, the river BENDIMER in Persia.
CORREGIUM, CORREGIO in Modena.
CORSILIANUM, or CORSIANUM, PIENZA in Tuscany.
CORSTORPITUM, or MORSTORPITUM, according to Camden, MORPETH; but by some supposed to be CORBRIDGE in Northumberland.
CORTENACUM, COURTENAY in the French dep. of Loiret.
CORTENACUM, CORTRACUM, or CORTORIACUM, COURTRAY in W. Flanders.
CORTYNA, CORTYNIA, or GORTINA, near NOVI-CASTELL in Crete.
CORU, CAPE ORTEGAL in Spain.
CORVI INSULA, CORVO in the Canary group.
CORYCEON PROMONTORIUM, or CORYCUS MONS, CAPE COURCO on the coast of Asia Minor.
CORYCUM PROMONTORIUM, CAPE KARACUS in Crete.
CORYPHASIUM, CAPE ZONCHIO on the coast of Greece.
COS, COOS, CEOS, CEA, CARIS, or MEROPIS, the island of STANCHIO in the Mediterranean.
COSA, or COSSA, ANSIDONIA in Tuscany.
COSAMBA, BALASORE in Hindostan.
COSANUM, SANTO-STEFANO in Tuscany.
COSELIA, KOSEL in Prussia.
COSFELDIA, COESFELD in Hanover.
COSLINUM, COSLIN in Prussia.
COSSIACUM, COSSE-LE-VIVIERS in the French dep. of Mayenne.
COSSOVOPOLITANA, AMSELFELD in Servia.
COSSYRA, COSYRA, or COSYRUS, the island of PANTALARIA in the Mediterranean.
COSSYRA, OPPEDOLO in the island of Pantalaria.
COSTRINUM, or CASTRINUM, CASTRIN in Prussia.
COTBRISIUM, COTTBRIS in Prussia.
COTHA, COTHENA, or COTHENIUM, KOTHEN in Anhalt-Kothen.
COTIARIS, probably the river MEIKONG in Burmah.
COTTIÆ, Cozzo in Piedmont.
COTTIARA, COCHIN in Travancore.
COTTOBARA, MASULIPATAM in Hindostan.
COTTONARA, CANARA in Hindostan.
COTYÆUM, KUTAHIYAH in Asiatic Turkey.
COTYORA, BUYUK-KALEH in Asiatic Turkey.
COVARIA, KOEVAR in Transylvania.
COVELIACÆ, KOCKEL in Bavaria.
COVELIACUS, the river KOCKEL in Transylvania.
COVORDIA, KOEVORDEN in Holland.
CRACOVIA, CRACOW.
CRAGUS, a mountain in Lycia, MONTE-DI-GORONTE.
CRAMBUSA, the island of CRAMBUSSA in the Mediterranean.
CRANÆ, the island of MARATHONISI in the Mediterranean.
CRANON, according to Kruse, SARLIKI in Thessaly.
CRATHIS, or CRATHUS, the river CRATI in Calabria.
CRATUMNUM, CRAON in the French dep. of Mayenne.
CREÆ ÆSTUARIUM, the FRITH OF CREE in Galloway.
CREDILIUM, CROIL in the French dep. of Oise.
CREDULIO, CRILLON in the French dep. of Vaucluse.
CREMA, CREMA in Lombardy.
CREMENECUM, KREMINIEK in Volhynia.
CREMENSIS AGER, CREMASCO in Lombardy.
CREMESIA, KREMS in Austria.
CREMNICIUM, KREMNITZ in Hungary.
CREPACORIUM, CREVACOURE in Piedmont.
CREPIACUM, or CRESPIACUM, CRESPY in the French dep. of Oise.
CREPIACUM, CREPY in the French dep. of Aisne.
CREPICORDIUM, CREVECŒUR in N. Brabant.
CRETA, the island of CRETE.
CREUTZBERGA, KREUTZBERG in Silesia.
CREVANTIUM, CREVANT in the French dep. of Yonne.
CRIMISA, or CRIMISSA, CAPE D'ALICE in Calabria-Citra.
CRIMISUS, CRIMISSUS, or CRINISUS, the river BELICA-DESTRO in Sicily.
CRISIUM, KREUTZ in Croatia.
CRISSACUS SINUS, the GULF of SALONA.
CRISTOLIUM, CREIL in the French dep. of Oise.
CRIUMETOPUM, CAPE ORIO in Crete.
CROATIA, or CHROVATIA, CROATIA.
CROCIATONUM, CARENTAN in the French dep. of La Manche.
CROCILLIACA, CROISILLE in Savoy
CROCODILORUM INSULA, the greater CAYMAN island off the Gulf of Mexico.
CROCODILORUM OPPIDUM, MEDINET-FAYUM in Egypt.
CROLLEJUM, or CURLEJUM, CREVILLY in the French dep. of Calvados.
CROMARTINUS COMITATUS, CROMARTYSHIRE in Scotland.
CROMARTIUM, CROMARTY in Scotland.
CROMENA, KURMAN in Bohemia.
CROSA, the river CREUSE in France.
CROSNA, CROSSEN in Prussia.
CROSNA AD ELYSTRUM, CROSSEN in Prussia.
CROTALUS, the river CORACE in Calabria.
CROTO, CROTONE in Naples.
CROVIACUM, CROY in the French dep. of Seine-et-Marne.
CROVUS, CROW in the French dep. of Seine-et-Oise.
CRUA, or CROJA, AKHISSAR in Romania.
CRUCENACUM, or CRUCINACUM, KREUTZNACH in Prussia.
CRUCIBURGUM AD VIERNAM, KREUTZBURG in Weimar.
CRUCIBURGUM VENEDICUM, KREUTZBERG in Prussia.
CRUCISORA, CORSOER in the Great Belt.
CRUMLAVIA, or CRUMLOVIUM, KRUMLAU in Moravia.
CRUSTUMERIA, MARIGLIANO-VECCHIO in the Papal states.
CRYBENSTENIUM, the castle of KRIEBENSTEIN in Prussia.
CRYPTA, GROTTA in Lombardy.

CRYPTA AUREA, GROTTAGLIA in Naples.
CRYPTA NEAPOLITANA, MONTE-DI-PAUSILIPPO in the Neapolitan prov. of Terra-di-Lavoro.
CSANADIENSIS COMITATUS, the comitat of TSANAD in Hungary.
CSONGRADIENSIS COMITATUS, the comitat of TSONGRAD in Hungary.
CTESIPHON, or CHALNO, in Assyria, near AL-MODAIN.
CUBITUS, or ELNBOGA, ELNBOGEN in Bohemia.
CUCULUM, CUCULLO in Abruzzo.
CUDA, COA in Portugal.
CUFFINSTANIUM, KOSTHEIM on the Main.
CULARO, GRENOBLE in France.
CULEMBACENSIS PRINCIPATUS, the lordship of CULMBACH in Bavaria.
CULMBACHIUM, or CULMBACUM, CULMBACH in Bavaria.
CULMEN URSI, or URSULUS, the SPLUGENBERG in the Grisons.
CUMA, or CYMA, a city in Æolis, near SANDERLI.
CUMBERLANDIA, CUMBERLAND.
CUMEOBURGUM, CUMMEBURG in Fulda.
CUMETENSE TERRITORIUM, COMO in Lombardy.
CUMILLUM MAGNUM, CIGOGNOLA in Lombardy.
CUNCIANUM, GOCIANO in Sardinia.
CUNETIO, or MARGABERGA, MARLBOROUGH in Wilts.
CUNEUM PROMONTORIUM, CAPE SANTA-MARIA in Portugal.
CUNEUS AUREUS, or SPELUGA, the SPLUGEN in the Grisons.
CUNIGAMIA, or CUNNINGHAMIA, CUNNINGHAM in Ayrshire.
CUPA, the river COPA in Lombardy.
CUPRA, or CYPRÆ FANUM, BENEDETTO in the Papal states.
CUPRA, or CUPRUM, CUPAR in Fife.
CUPRIMONTIUM, KUPFERBERG in Silesia.
CUPRIMONTIUM AD ALCMONAM, KUPFERBERG in Bavaria.
CUPRIMONTIUM BOHEMIÆ, KUPFERBERG in Bohemia.
CUPRIMONTIUM GESTRICIÆ, KUPFERBERG in Norway.
CURDIA, or CURDISTANIA, KURDISTAN.
CURENSIS AMNIS, the river CORRESE, an affluent of the Tiber.
CURES, CORREZE in the Papal states, or perhaps VESCOVIA.
CURETIA, the river CORREZE in France.
CURETIUM, the town of CORREZE in France.
CURIA, or CURIA RHÆTORUM, CHUR or COIRA in the Grisons.
CURIA, probably LANARK in Scotland.
CURIA, CORTE in Corsica.
CURIA BAVARICA, or CURIA REGENTIANA, HOF in Bavaria.
CURIA MAJOR, CORTE-MAGGIORE in PARMA.
CURIA MORAVICA, HOF in Moravia.
CURIA MORICI, HOF in Austria near Salzburg.
CURIA PIETRA, CORBETTE in Lombardy.
CURIA REGIA, KONIGSHOFEN in Bavaria.
CURIA REGIS AD ALBIUS, KONIGSHOF in Bohemia.
CURIA REGIS BADENSIS, KONIGSHOF in Baden.
CURIANUM PROMONTORIUM, CAPE FERET in Guyenne.
CURIAS, CAPE GAVATA in Cyprus.
CURIETA, the island of VEGLIA in the Adriatic.
CURIETUM, VEGLIA in the island of that name.
CURIOSOLIMAGUS, CORSENTI in the French dep. of Nord.
CURLANDIA, or CURONIA, COURLAND in Russia.
CURLANDIÆ SINUS, the CURISCHE-HAFF in Prussia.
CURSUS ACHILLIS, the island of FIDONISI in the Black sea.
CURTISMILIUM, CORTEMIGLIA in Piedmont.
CUSSENACUM, KUSSNACHT in the cant. of Schwyz.
CUSUS, the river WAAG in Hungary.
CUTACIUM, KUTAIS in Georgia.
CUTILIA, CONTIGLIANO in the Papal States.
CUTNA, CUTTNA, or KUTTENBERGA, KUTTENBERG in Bohemia.
CUXHAVIA, KUXHAVEN in the Hamburg territory.
CYANEÆ INSULÆ, the PAVONAKE islands near the mouth of the Bosphorus.
CYBISTRA, a town of Cappadocia, according to D'Anville, COSTEREH; according to Leake, KARAHISSAR in Asiatic Turkey.
CYCLADES, the group of islands in the Grecian archipelago which encircled DELOS.
CYCLOPUM SCOPULI, the island of FARIGLIANI off the E coast of Sicily.
CYDAMUS, in Regio Syrtica, perhaps GHADAMES.
CYDNUS, a river of Cilicia Campestris, the KARASU.
CYGNEA, CYNAVIA, or ZWICKAVIA, ZWICKAU on the river Mulde in Lunenburg.
CYLINDRINA, the prov. of SIRINAGUR in Hindostan.
CYLISTARNUS, the river ROCANELLO in Calabria.
CYLLENE, CHIARENZA or CLARENZA in Greece.
CYMÆUS SINUS, the GULF OF SANDALI on the coast of Asiatic Turkey.
CYMINUS LACUS, the LAGO-DI-PICO in the Papal States.
CYNOPOLIS, or CANUM URBS, near SAMULLAT on the Nile.
CYPARISSIUM PROMONTORIUM, CAPE CANELLO on the W coast of Greece, or perhaps CAPE APIDAGLIA.
CYPARISSIUS SINUS, the GULF OF RONCHIO at the mouth of the Alpheius in Greece.
CYPRESSETA, SORGUE in the dep. of Vaucluse.
CYPRUS, ACAMANTHIS, AMATHUSIA, or COLINIA, the island of CIPRO or CYPRUS.
CYPSELA, or CYPSELLA, a town in Thrace, according to Kruse, IPSALA.
CYRENAICA, CYRENÆA, or PENTAPOLIS LIBYA, the territory of BARCAH in Barbary.
CYRENE, KAIRWAN in Tunis.
CYRRHESTICA, HALEB or ALEPPO in Asiatic Turkey.
CYRUS, KORO, or CYSSUS, the river KUR in Armenia.
CYTA, KUTAIS in Imiretia.
CYTÆUM, SETTIA in Crete.
CYTHERA, the island of CERIGO.
CYTHNUS, or OPHIUSI, THERMIA one of the Cyclades.
CYZICUS, or DOLIONIS INSULA, CHIZICO in the sea of Marmora.
CZASLAVIA, CZASLAU in Bohemia.
CZERDINUM, CZERDIN in the Russian gov. of Perm.
CZERNICHOVIA, the Russian gov. of CHERNIGOF.
CZYRCASSIUM, CZIRKASSI in the Volhynia.

D

DABRONA, the river BLACKWATER in co. Waterford.
DACHANUM, DACHAU in Bavaria.
DACHINABADES, GOLCONDA in Hindostan.
DACIA, the ancient name of the country bounded on the N by European Sarmatia; on the E by the Euxine; on the S by the Danube; and on the W

by the river Theiss. As a Roman prov. it was subdivided into DACIA-RIPARIA or WALLACHIA, stretching along the Danube; DACIA MEDITERRANEA, the central part, or TRANSYLVANIA; and DACIA ALPENSIS in the vicinity of the Carpathians.

DACTONIUM, MONTFORTE-DE-LEMOS in Spanish Gallicia, or perhaps RIBADEO.

DÆA, DEA, DEVA, or DIVA, the river DEE in Galloway.

DAGHOA, DAGOE, an island in the gulf of Finland.

DAGOBERTI SAXUM, DACHESTEIN, in the French dep. of Bas-Rhin.

DAHARUM REGIO, DAGHESTAN in Persia.

DAIRAGUM, DAIRAGO in Lombardy.

DAIX, or OARUS, the river JAIK or URAL flowing into the sea of Aral.

DALECARLIA, or DALIA, the Swedish prov. of DALARNA.

DALECARLIUS, the DAL-ELF in Sweden.

DALKETHUM, DALKEITH in Mid-Lothian.

DALMATIA, a district of Illyricum, nearly corresponding with the Austrian territory of the same name.

DALMIUM, DELMINO in Bosnia.

DAMASCENORUM OPPIDUM, or COLONIA DAMASCUS METROPOLIS, DIMASCHK, or DARMSUK, the city of DAMASCUS in Syria.

DAMASIA, DIESSEN in Bavaria.

DAMGARDIA, DAMGARTEN in Prussia.

DAMIATA, or DIMIAHA, DAMIETTA in Egypt.

DAMMONA, or DAMUM, APENGADAM in Holland.

DAMMONUS, DAMSSERDIEP in Holland.

DAMNONIUM PROMONTORIUM, the LIZARD-POINT in Cornwall.

DAMNUM MARTINUM, or DAMMARTINUM, DAMMARTIN in the French dep. of Seine-et-Marne.

DAMOVILLA, DOMVILLE in the French dep. of Eure.

DAMPUILLERIUM, DAMVILLIERS in the French dep. of Meuse.

DANASTER, DANASTRUS, or TYRA, the river DNIESTER.

DANGCLONIUM, DANJEAU in the French prov. of Perche.

DANIA, DENMARK.

DANICA SYLVA, the DANISCHWALD in Schleswig.

DANICUM FRETUM, or ORESUNDICUM FRETUM, the OERSUND between the Danish islands of Sieland and Schonen.

DANNEBERGA, DANNEBERG in Hanover.

DANTISCUM, or GEDANUM, DANTZIG in Prussia.

DANUBII INSULA, DONABERGA, or VERTIA, DONAUWERTH in Bavaria.

DANUBIUS, DANOVIUS, or ISTER, the river DANUBE.

DANUM, DONCASTER in Yorkshire.

DANUS, or INDUS, the river AIN or ENS an affluent of the Rhone.

DAONES, the river SANG-PO, or perhaps the IRAWADDI in India.

DAPHNITIS INSULA, the island of DOLLACA or DAHALEG on the NE coast of Africa.

DARADUS, or DARAS, the river SENEGAL.

DARDANELLÆ, the DARDANELLES.

DARDANIA, the sanjak of USCOP in Servia.

DARDANIUM PROMONTORIUM, CAPE BURUN on the Black sea.

DARDANUS, PENNA-DI-ORDUNA in Biscay.

DARGOMANES, or ORGOMANES, a river of Bactriana, probably the BAMIAN, an affluent of the Oxus.

DARMSTADIUM, DARMSTADT in Hessen-Darmstadt.

DARNASIA, or DIESENHOFIA, DIESSENHOFEN in the cant. of Thurgau.

DARNIS, a town of Cyrenaica, represented by DERNEH in Barca.

DARVANIUM, DRAAS in Transylvania.

DARVENTUS, or DERVENTUS, the river DERWENT in Derbyshire.

DASCYLEUM, or DASCYLOS, a coast town of Bithynia, now DIASKILLO.

DATUM, DATOS, in Thracia, the modern CAVULLA.

DAUDYANA, DIADIN in Armenia.

DAULIS, DAULIA, or DAULIORUM URBS, a town in Phocis, near the modern DAULIA.

DAUNIA, or DAUNIAS, TERRA-DI-BARI in Naples.

DAVENTRIA, or DEVONTURUM, DEVENTER in Holland.

DAVIUM SACELLUM, FALKIRK, or CARRON in the vicinity of Falkirk, in Scotland.

DEANENSIS SYLVA, DEAN FOREST in Gloucestershire.

DEBRECINUM, or DEBRETTINUM, DEBRECZIN in Hungary.

DECASTADIUM, CASTIDIO in Calabria.

DECEM PAGI, DIENZE in the French dep. of Meurthe.

DECENTIANUM, DESENZANO in Lombardy.

DECETIA, DECIZE in the French dep. of Nievre.

DECIA, or DOZZA, DIETZ in Nassau.

DECIDAVA, DIEMBRICH in Transylvania.

DEDESSUS, DEDES in Tripoli.

DEI FACIES, THEU-PROSOPON, or EUPROSOPON, near CAPE CAROGE on the coast of Syria.

DEI MONS, DIESTALDEN in the Swiss cant. of Valais.

DEINSA, DEINSE in East Flanders.

DELAS, or SILLA, the river DIALA in Persia.

DELAVARUS, the river DELAWARE in N. America.

DELBRUGGIA, DELBRUCK in Prussia.

DELEMONTIUM, or TELAMONTIUM, DELMONT in the Swiss cant. of Bern.

DELFI, DELFUM, DELFTA, or DELPHIUM, DELFT in Holland.

DELGOVITIA, or GODNUNDHAM, WIGTON in Cumberland.

DELITIUM, DELITZSCH in Saxony.

DELLINA, DELINA, or DELBRIA, the river DALKE, an affluent of the Ems in Germany.

DELLIUM, SURASENORUM REGIO, the prov. of DELHI in Hindostan.

DELMENHORSTIUM, DELMENHORST in Oldenburg.

DELMINIUM, DALMINO in Bosnia.

DELOS, ASTERIA, CHLAMIDIA, CYNTHIA, LAGIA, ORTYGIA, or PELASGIA, the island of DELOS, DILI, or SEDILLI in the Mediterranean.

DELPHI, PYTHIA, PARNASSIA VALLIS, now CASTRI on Mount Liakura in Greece.

DELPHINATUS, the French prov. of DAUPHIGNY.

DELPHINI PORTUS, PORTO-FINO in the Sardinian prov. of Geneva.

DELPHINIUM, DELFINO in Chios.

DELPHOLANDIA, or DELPHORUM AGER, DELFTLAND in S. Holland.

DELPHORUM PORTUS, DELFTSHOFEN in Holland.

DELTA, that portion of Egypt which, in ancient times, was enclosed between the Pelusian and the Canopic branches of the Nile, which branched off at Cercassorum. The apex of this triangular space—whence its name—is now much less obtuse than in ancient times; and its triangular figure is determined by the Rosetta branch, the old Canopic arm, and the Phatnitic or Damietta branch.

DEMETRIOVITIA, MITROWITZ in Slavonia.

DEMETRIOVITIA SARMATICA, DEMETRIOWITZ in the Russian gov. of Smolensk.

DEMITRIOVIA, DEMITRIOFF in the Russian gov. of Wologda.

DEMMINUM, Demmin in Prussia.
DEMONNESI, or Demonesi, Princes Islands in the sea of Marmora.
DENBIGENSIS, Denbighshire in North Wales.
DENBIGHUM, or Denbiga, Denbigh in Denbighshire.
DENIGOTIA, Denigen in Bulgaria.
DEOBRIGA, Mirando-di-Ebro in the Spanish prov. of Burgos.
DEOMANT, Donandium, or Dinoantum, Dionant in Belgium.
DEORUM CURRUS, or Theorum Ochema, probably Sierre-Leone on the coast of Africa.
DEORUM INSULÆ, a group of islands off the coast of Galicia in Spain.
DEORUM PORTUS, or Divini Portus, perhaps Marz-al-Quibir in Barbary.
DEPPA, Dieppa, or Juliabona, Dieppe in France.
DERBATUM, Derpatum, or Dorpatum, Dorpt in the Russian gov. of Riga.
DERBICENSIS COMITATUS, Derbyshire.
DERE, or Dire, perhaps Ras-Bel on the coast of the Red sea.
DERGUS LACUS, Lough Dergh in Galway.
DERIA, Derra, or Roboretum, Londonderry in Ireland.
DERNUS, or Ernus Lacus, Lough Earne in Ireland.
DERSIA, Hessia, or Hessiga, Hessen in Kur-Hessen.
DERTONENSE TERRITORIUM, the prov. of Tortona in Piedmont.
DERTOSA, Tertosa in the Spanish prov. of Catalonia.
DERVENTIA, Derby in Derbyshire.
DERVENTIO, perhaps Auldby in Yorkshire.
DESERTINA, Discentium, or Dissertinum, Disentis in the Swiss cant. of Grisons.
DESERTUM, or Dysertum, Dysart in Fifeshire.
DESIDERII MONS, Mondidier in Spanish Galicia.
DESSAVIA, Dessau in Anhalt-Dessau.
DETHMOLDA, or Diethmolium, Detmold in Lippe-Detmold.
DEUCALEDONIUS. See Caledonius.
DEVA, or Setejà, the river Dee in North Wales.
DEVA, or Diva, the river Dee in Scotland.
DEVA, or Diva, the river Dive an affluent of the Loire in France.
DEVALES, the Deva in Asturias in Spain.
DEVELTIN COLONIA, Develtum, or Debeltus, Zagora or Develto in Romelia.
DEVENA, Dowina, or Thebena, Deven or Theben in Hungary.
DEVIOTIA, Teviotia, or Ripa Alta, the district of Teviotdale in Roxburghshire.
DEVONA, Schweinfurt in Coburg.
DEVONA VETUS, Old Aberdeen in Scotland.
DEVONIA, or Devoniensis Comitatus, Devonshire.
DIA, the island of Standia in the Mediterranean.
DIABLINTES, Novodunum, or Novigentum Rotrudum, Nogent-le-Routron in the French dep. of Eure-et-Loire.
DIAMPOLIS, a city of Thracia, near the modern Jamboli.
DIANA, Zamah in Algiers.
DIANÆ FONS, Favara in Sicily.
DIBIO, Divio, or Diviodunum, Dijon in the French dep. of Cote-d'Or.
DICÆARCHIA, or Colonia Augusta, Puzzuolo in Naples.
DICIACUM, Duciacum, or Duodeciavum, Douzy in the French dep. of Somme.
DICTAEUS MONS, Lasthi in Crete.
DICTAMNUM, Dictamo in Crete.
DIDATTIUM, Dole in the French dep. of Jura.
DIDIGNA, Digna, or Digba, Korna in the pash. of Bagdad.
DIDYME, the island of Salini in the Lipari group.
DIDYMI, Didymo in Greece.
DIDYMI MONTES, Rás-el-Hhad in Persia.
DIDYMOTICHOS, Dimotika in the sanj. of Gallipoli.
DIEGI VILLA, Diego in Old Castile.
DIEMENI INSULA, Van Diemen's Land or Tasmania.
DIEPHOLTA, Diepholz in Brunswick.
DIESTA, Diesthemium, or Distemium, Diest in South Brabant.
DILA, Dilis, or Verdonia, Verdun in the dep. of Bouches-du-Rhone.
DILLINGA, Dillingen in Bavaria.
DILURON, or Iluro, Mataro in Catalonia.
DIMOLA, Diemel in Kurhessen.
DINANTIUM, Dinan in the French dep. of Nord.
DINCKELSPUHLA, Tricollis, or Zeapolis, Dinkelsbuhl in Bavaria.
DINGLIA, Dingle in co. Kerry.
DINGOLVINGA, or Dingelfinda, Dingelfingen in Bavaria.
DIODEA, Medon in Albania.
DIODORI INSULA, the island of Perim in the straits of Bab-el-Mandeb.
DIOMEDEA, Diomedia, or Trimetus, Tremiti in Naples.
DIOMEDIS PROMONTORIUM, Piança in Illyria.
DIONYSIAS, near Schohba in Middle Egypt.
DIONYSIOPOLIS, Nagara in India.
DIOSCORIAS, Sebastopolis, or Soteriopolis, Iskurieh in Mingrelia according to some geographers, but generally identified with Sebastopol on the W coast of the Crimea.
DIOSCORIDIS INSULA, Socotora in the Indian ocean.
DIRIGOTA, or Dinogetia, or Dinogatria, Drimaço in Bulgaria.
DISIBODENGENSE CŒNOBIUM, Diesenberg in Rhenish Prussia.
DISMUDA, or Dixmuda, Dixmuyden in West Flanders.
DITMARSIA, or Dithmarsia, Ditmarschen in Holstein.
DITTAINUS, Dittaino in Sicily.
DIUM PROMONTORIUM, Cape Sossoso in Crete.
DIUSBURCH, or Diusburgum, Diusburg in Rhenish Prussia.
DIUZA, or Diutia, Deutz in Rhenish Prussia.
DIVA, the river Deva in Biscay.
DIVI LAURENTII INSULA, Lunæ Insula, or Menuthias, the island of Madagascar.
DIVODURUM, Mettis, Mediomatricum, or Metensis Urbs, Metz in the French dep. of Moselle.
DOADUM, Dovacum, or Theodoadum, Douay in the French dep. of Mayenne-et-Loire.
DOANAS, or Daona, according to D'Anville, the Indian territory of Tenasserim.
DOBERANUM, or Dobranum, Doberan in Mecklenburg-Schwerin.
DOBOCA, Doboka in Transylvania.
DOBRINIA, Dobrinum, or Debieconium, Doerzyn in Poland.
DOCCOMIUM, Dochzatum, or Doccumum, Dokkum in Friesland.
DOESBURGUM, or Drusoburgum, Doesberg in Holland.
DOLA, Deal in Kent.
DOLA, or Dolum, Dol in the French dep. of Ille-et-Vilaine.

DOLA, DOLE in the French dep. of Doubs.
DOLIANUM, DOGLIANI in Piedmont.
DOLLARIUS SINUS, or EMDANUS SINUS, the DOLLAERT at the mouth of the Ems.
DOMINICOPOLIS, DOZIER in the French dep. of Haute-Marne.
DOMITIUM, DOEMITZ in Saxony.
DOMODUSCELLA, DOMO-D'OSSOLA in Lombardy.
DONASTIENUM, SAN-SEBASTIAN in Spain.
DONCANONIUM, DUNCANNON in co. Wexford.
DONESCHINGA, DONESCHINGEN in Baden.
DONNIFRONS, or DAMFRONIUM, DOMFRONT in the French dep. of Orne.
DORA BALTEA, or DURIA MAJOR, the river DORA in Piedmont.
DORCESTRIA, DURNOVARIUM, or DORCUNÆ CIVITAS, DORCHESTER in Dorset.
DORDANUM, or DORDINGA, DOURDAN in the French dep. of Seine.
DORDRACUM, DORDRECHT in the prov. of S. Holland.
DORICUS SINUS, the GULF OF SYMO.
DORIUS, the river PEGU in India.
DORMUNDA, or TREMONIA, DORTMUND in Prussia.
DORNACUM, DORNECK in the Swiss cant. of Soluthurn.
DORNBURGUM, DORNBURG in Weimar.
DORNOCUM, or DORNOTUNUM, DORNOCK in Sutherlandshire.
DOROSTENA, or DOROSTORUM, SILISTRIA in Bulgaria.
DORSETIA, DORCESTRIA, or DORCESTRIENSIS COMITATUS, DORSETSHIRE.
DORYLÆUM, or DORYLLEIUM, ESKI-SHEHR in Asia-Minor.
DOTECUM, DENTICHEM in Guelderland.
DRACENÆ, or DRACENUM, DRAGUIGNAN in the French dep. of Var.
DRACHONUS, or DRACONUS, the river TRAEN, an affluent of the Moselle.
DRACO, the river DRAGONE in Naples.
DRACOMONTIUM, TRACHENBERG in Silesia.
DRACONIS MONS, DRAGON in Provence.
DRACUINA, EHINGEN in Würtemberg.
DRAGAMUNTINA, or TRAVEMUNDA, TRAVEMUNDE in Lubeck.
DRANGIANA, SISTAN in Cabulistan.
DRANSA, the river DRANSE in Switzerland.
DRAVOBURGUM, DRABURG in Illyria.
DRAVUS, DRABUS, or DRAUS, the river DRAVE.
DRENTHIA, the Dutch prov. of DRENTHE.
DREPANUM, TRAPANI in Naples.
DREPANUM PROMONTORIUM, CAPE DRAPANO in Naples.—Also CAPE BIANCO in Cyprus.—Also CAPO-DI-SANTO-ALESIO in Sicily.—Also CAPE TRAPANI in Sicily.
DRESDA, or DRESENA, the city of DRESDEN.
DRESSENIUM, or DRIESENA, DRIESEN in Prussia.
DRIBURGUM, DRIBURG in Prussia.
DRILO, or DRINUS ALBUS, the river DRINO-BIANCO in Illyria.
DRILONIUS SINUS, the GULF OF DRINO on the coast of Albania.
DRINUS NIGER, the DRINO-NEGRO in Illyria.
DRINUS, or DRILO MINOR, the DRINA in Bosnia.
DROCÆ, DROCUM, or DUROCASSIUM, DREUX in the French dep. of Eure.
DROGHDÆA, or DROGEDA, DROGHEDA in co. Louth.
DROMARIA, or DRUMORIA, DROMORE in Ulster.
DRUBETIS, CRAJOWA in Wallachia.
DRUENTIA, the river DURANCE an affluent of the Rhone.
DRUMA, or DRUNA, the DROME an affluent of the Rhone.
DRUSENHEMIUM, DRUZENHEIM in the French dep. of Bas-Rhin.
DRUSIANA URBS, or FRAUENBURGUM, FRAUENBURG in Prussia.
DRUSANAGUS, DRUISHEIM in Prussia.
DUALA GALLICA, or GALLIVENSIS COMITATUS, co. GALWAY in Ireland.
DUBA, or DUBENA, DUBEN in Prussia.
DUBLINUM, or DUBLINIUM, the city of DUBLIN.
DUBRÆ, DUBRIS, or DAVONIA, DOVER in Kent.
DUDERSTADIUM, DUDERSTADT in Hanover.
DUINA, or DUNA, the river DWINA in Russia.
DULCIS AQUA, DOLCE-AQUA in Piedmont.
DULECUM, or DAMLIAGUM, DULECK in Leinster.
DULICHIA, DULICHIUM, or DOLICHNA, by some supposed to be the island of NEOCHORI in the Mediterranean; by others that of PETALA, the largest of the Echinades.
DUMBÆ, or DUMBORUM, DOMBS in the French dep. of Ain.
DUMBARIUM, DUNBAR in East Lothian.
DUMBLANUM, DUNBLANE in Perthshire.
DUNELMENSIS COMITATUS, the palatinate of DURHAM.
DUNELMUM, or DUREMUM, the city of DURHAM.
DUNENSIS COMITATUS, co. DOWN in Ireland.
DUNESTORIUM CASTRUM, DUNSTER in Somersetshire.
DUNFREJA, or DUMFREGA, DUMFRIES in Dumfriesshire.
DUNGA, or DUNPA, GOA in Hindostan.
DUNGALENSIS COMITATUS, co. DONEGAL.
DUNKELDINUM, DUNKELD in Perthshire.
DUNKERANUM, DUNDALK in co. Louth.
DUNQUERCA, or DUNKERKA, DUNKIRK in the French dep. of Nord.
DUNSIUM, or DUSIUM, DUNSE in Berwickshire.
DUNUM, DOWNPATRICK in Ireland.
DUNUM ÆSTUARIUM, WHITBY BAY on the coast of Yorkshire.
DUPLICES AQUA, ZWIEFALTEN in Würtemberg.
DURA, DUREN in Prussia.
DURACIUM, THOUARS in the French dep. of Dieux-Sevres.
DURANIUS, or DRONA, the river DORDOGNE in France.
DURASTELLUM, DURETAL in the French dep. of Loire.
DURBIS, DURBY in the Belgian prov. of Liege.
DUREGUM, THUREGUM, or TURIGUM, ZURICH in Switzerland.
DURFOS, DEVEREM in North Brabant.
DURIA, the river DUR or THUR, an affluent of the Rhine.
DURIAS, TURIA in Valencia.
DURIUS, or DORIUS, the river DUERO or DOURO in Portugal.
DUROBRIVÆ, a town of the Coritani, BRIGH-CASTERTON in Northamptonshire.
DUROBRIVIS, a town of the Cantii, ROCHESTER in Kent.
DUROCOBRIVÆ, a town of the Catyeuchlani, by some supposed to be HERTFORD; by others GREAT BERKHAMPSTEAD; and by others DUNSTABLE.
DUROLI PONS, perhaps GODMANCHESTER in Huntingdonshire; but by Mannert supposed to be CAMBRIDGE.
DUSSELDORPIUM, DUSSELDORF in Luneburg.
DUTLINGA, TUTLINGEN in Würtemberg.
DUVELANDIA, DUIVELAND in Holland.
DYARDANES, or OLDANES, the river BRAHMAPUTRA.
DYLA, the river DYLE in Brabant.
DYME, DYMÆ, or PALEA, a town of Achaia, perhaps the modern KANIMITZA.

DYRRACHIUM, or EPIDAMNUS, DURAZZO in Albania.
DYSTUS, in Eubœa, now DISTO.

E

EASO, ŒASA, or OLARSO, OIARSO in the Spanish prov. of Guipuscoa.
EAUNA, EONA, or JAUNA, YENNE in Savoy.
EBA, or MARANUS MONS, MONTE-MARANO in the Neapolitan prov. of Principato-Ultra.
EBELTOFTIA, or POMAGNUM, EBELTOFT in North Jutland.
EBERACIUM, or EBRIACUM, ERBACH in Germany.
EBERSBERGA, EBERSBERG in Germany.
EBERSDORFIUM, EBERSDORF in Germany.
EBERSTENIUM, EBERSTEIN in Germany.
EBESHANIUM, EPSOM in Surrey.
EBILINGUA, WEINBLINGEN in Würtemberg.
EBLANA, by some supposed to be DUBLIN, by others DUNDALK, in Ireland.
EBOLUM, or EBOLUS, EBOLI in the Neapolitan prov. of Principato-Citra.
EBORA, or LIBERALITAS JULIA, EVORA in Portugal.
EBORA, a town in Hispania Tarraconensis, perhaps MUROS in Galicia.
EBORACENSIS COMITATUS, YORKSHIRE.
EBORACENSIS NOVA CIVITAS, or BELGIUM NOVUM, NEW YORK in North America.
EBORACUM, YORK in Yorkshire.
EBORESHEMIUM, EBERSHEIM in the French dep. of Bas-Rhin.
EBORICA, EBROICUM, or MEDIOLANUM, EVREUX in the French dep. of Eure.
EBRODUNUM, YVERDON in the Swiss cant. of Vaud.
EBROLIUM, EBREUIL in the French dep. of Allier.
EBRONIUM, EVRON in the French dep. of Mayenne.
EBUDA ORIENTALIS, the island of SKY.
EBUDA INSULÆ, the Scottish HEBRIDES.
EBUROBERGOMUM, EBERSPERGA in Bavaria.
EBURUM, or OLMUTIUM, OLMUTZ in Moravia.
EBUSUS, EBUSIA, or EBYSSOS, the island of IVICA in the Mediterranean.
ECBATANA, AGBATANA, or APOBATANA, the capital of Media, now HAMADAN in Persia.
ECBATANA, CAIFA or KAFFA on the coast of Syria.
ECCLESIA, or METHYMNA CŒLESTIS, MEDINA CŒLI in the Spanish prov. of Soria.
ECCLESIÆ, IGLESIAS in Sardinia.
ECDIPPA, SIB or ACHZIB in Syria.
ECHE, EYCH in Limburg.
ECHETLA, AQUILA in Sicily.
ECHINÆ, the island of CURZOLARI in the Mediterranean.
ECHINUS, ECHINO on the coast of Thessaly.
ECKESIŒA, or EKESIUM, EKSJO in Sweden.
ECOLESIMUS, or ECOLIMENSIS PAGUS, the prov. of ANGOUMOIS in France.
EDELBERGA, HEIDELBERGA, or MYSTILLORUM MONS, HEIDELBERG in Baden.
EDESSA, VODINA in Macedonia.
EDESSA CALIRRHOE, or ANTIOCHIA, ORFAH or URFAH in the pash. of Mosul.
EDETA, LEIRIA or LIRIA in the Spanish prov. of Valencia.
EDINUM, or EDENBURGUM, the city of EDINBURGH, the capital of Scotland.
EDRUM, or IDRUS, IDRO in Lombardy.
EDRUS, or AUDROS, perhaps the island of BARDSEA in the Irish Channel.
EDULUM, EDULO in Lombardy.
EGABRA, or AGABRIA, CABRA in Andalusia.
EGEDORA, the river EYDER in Denmark.
EGILIUM, GIGLIO off the coast of Tuscany.
EGLIS, or THELIS, the river OGLY in Languedoc.
EGLISAVIA, EGLISAU in the Swiss cant. of Zurich.
EGONUM VICUS, VICOVENZA in the Papal States.
EGRA, or AGARA, the river EGER in Bohemia.
EGRA, or OEGRA, EGER in Bohemia.
EHRENBERTI SAXUM, EHRENBREITSTEIN on the Rhine.
EICHSFELDIA, EICHSFELD in Prussia.
EICHSTETUM, EICHSTADT in Prussia.
EIFFALIA, the EIFEL in Rhenish Prussia.
EIGARUS, AIGARUS, or ISARUS, the river AIGUES in Dauphiny.
EINDOVIA, EINDHOFEN in North Brabant.
EION, a town of Edonia, by some supposed to be represented by the modern CONTESSA; by others, by RENDINA, or by PONDINO.
EISENBERGA, EISENBERG in Bohemia.
EISEOBERGA, or EISENBERGA, EISENBERG in Altenburg.
EKENESIUM, or QUERCUUM PENINSULA, EKENAAS on the gulf of Finland.
ELÆA, in Æolis, perhaps AYASMON on the gulf of Tchandeli.
ELÆUS, in the Thracian Chersonese near the modern KRITIA.
ELAITICUS SINUS, the GULF OF TCHANDELI on the coast of Asiatic Turkey.
ELANITICUS SINUS, or ALAINITES SINUS, the GULF OF AKABA.
ELAPHONESON, ELAPHONNESOS, or PROECONNESOS, the island of MARMORA off the coast of Anatolia.
ELARONA, ELORO, or ILURO, OLERON in the French dep. of Basses-Pyrenees.
ELBERFELDIA, ELBERFELD in Prussia.
ELBII LACUS, the LAGO-DI-VICO in the Tuscan prov. of Pisa.
ELBINGA, ELBING in Prussia.
ELBORA, or LIBORA, TALAVERA-DI-LA-REYNA.
ELBOVIUM, or ELLEBOVIUM, ELBEUF in the French dep. of Seine.
ELBURGUM, ELBURG in Guelderland.
ELCEBUS, HELCEBUS, or HELVETUS, a town of the Tribocci, ZELZENHEIM in Prussia.
ELDANA, SALDANNA in Old Castile.
ELEA, ELLA, HELIA, or VELEA, a town on the coast of Lucania, probably represented by the modern CASTELLAMARE-DELLA-BRUCA on the gulf of Salino.
ELEA, or HALES, PISCIOTTA in the Neapolitan prov. of Principato-Citra.
ELEGIA, a town in Armenia Magna, now ILIJA on the Euphrates.
ELEGIUM, in Noricum Ripense, perhaps ERLACH in Upper Austria; or according to some STRENOBERG.
ELENÆ, HELENÆ VICUS, or LENDUM, LENS in the French dep. of Pas-de-Calais.
ELEPHANTARIA, TEMPI in Sardinia.
ELEPHANTIACUM, or ELWANGA, ELLWANGEN in Germany.
ELEPHANTIUS, or ELEPHANTUS, JEZEREH-EL-SAG on the river Nile.
ELEPTA, NIEBLA in the prov. of Seville.
ELETHYIA, or LEUCOTHEÆ OPPIDUM, ELEITS in the prov. of Said, in Egypt.
ELEUSA, the island of LEOVUSA in the Saronic gulf.
ELEUSIN, near LESSINA in Attica.
ELEUTHEROPOLIS, FREYSTADT in Silesia.
ELEUTHEROPOLIS AD VAGUM, FREYSTADT on the Waag in Hungary.
ELEUTHERUS, a river of Phœnicia, probably the NAHR-EL-KEBIR of Syria; but by some supposed

to be the BARADA.—Also a river of Sicily, the modern BAJAIRA.
ELGINA, or ELGIS, ELGIN in Scotland.
ELGOVIA, or ELIGOVIA, ELCK or ELCKOW in the Swiss cant. of Zurich.
ELIBANUS MONS, MONTE-FISARDO in Calabria-Ultra.
ELIBERI, ILIBERI, or ELIBERIS, by Mentelle identified with GRANADA in Spain; by others, with ELVIRA.
ELIM, the WADY-GHARENDAL in the peninsula of Sinai.
ELIOCRATA, LOCRA in the Spanish prov. of Murcia.
ELIS, a town of Elis in Greece, near BELVEDERE.
ELISTER, or ELSTERA, the river ELSTER.
ELIZATIUM, or SALESIA, SELTZ in Rhenish Prussia.
ELLA, ELLUS, or ALSA, the river ELL or ILL in Rhenish Prussia.
ELLA, or ILLA, the river LILLE, an affluent of the Dordogne.
ELLEBOGIUM, or MALMOGIA, MALMOE in the Swedish island of Schonen.
ELMATICA, or SALMANTICA, SALAMANCA in the Spanish prov. of Leon.
ELORUM, the river ATELLARO in Sicily.
ELORUS, or HELORUS, NOTO in Sicily.
ELRICA, or ALRICA, the river ELRICK, an affluent of the Tweed.
ELRICUM, ELLRICH in Prussia.
ELSENORA, or HELSINGORA, ELSINORE in Sieland.
ELTZIA, ELZ in Rhenish Prussia.
ELURO, PINEDA in Catalonia.
ELYMAIS, LURISTAN in Persia.
ELYMEA, in Macedonia, GRENOW, or perhaps CANINA.
ELYSTER, the river ELSTER or WHITE ELSTER in Saxony.
EMBA, the river JEMBA which falls into the Caspian.
EMBDAMUS COMITATUS, or FRISIA ORIENTALIS, the Hanoverian prov. of EAST FRIESLAND.
EMERITA, CASTELLO-MENDO in the Portuguese prov. of Beira.
EMERITA JUCATANIENSIS, MERIDA in Yucatan.
EMESA, EMESUS, or EMISA, HOMS in the pash. of Damascus.
EMISSARIUS, the river DESAGUADERO in Chili.
EMMENÆ VALLIS, the EMMENTHAL in the Swiss cant. of Berne.
EMMERA PARVA, the EMMER in Westphalia.
EMODI MONTES, or EMODUS, the HIMALAYA mountains.
EMPORIÆ, AMPURIAS in Catalonia.
EMPORIUM, or EMPOLIA, EMPOLI in Tuscany.
EMPORIUM ARABIÆ, probably MAKULLAH in Arabia.
EMPULUM, AMPIGLIONE in the Papal States.
ENARGIRUM, ORGON in the French dep. of Bouches-du-Rhone.
ENCHUSA, or ENCHUSIA, ENKHUYZEN in North Holland.
ENCOPIA, ENCOPING in Sweden.
ENDELAVIA, ENDELAVE in Jutland.
ENINGIA, FENNINGIA, FINONIA, or FINNLANDIA, the region of FINLAND.
ENJEDINUM, ENYEDIMUM, or AGNETTINUM, STRASBURG in Transylvania.
ENNA, near CASTROGIOVANNE in Sicily.
ENOSIS, ANTIOCO on the SW coast of Sardinia.
ENSDORPIUM, ENSDORF in Bavaria.
ENSISHEMIUM, or ENSHENIUM, ENSHEIM in French Alsace.
ENTELLA, the river LAVAGNA in the prov. of Genoa.
EPAUNENSIS CIVITAS, or EPAONA, according to some PAMIERS; according to others, PAU in Burgundy.
EPERODIA, or EBOREGIA, IVREA in Piedmont.
EPHESUS, AIA-SALUK in Asiatic Turkey.
EPICARIA, PAPADORAS in Albania.
EPIDAURUM, RAGUSA-VECCHIA in Dalmatia.
EPIDAURUS LIMERA, or MALVASIA, a town of Laconica, NAPOLI-DI-MALVASIA in Asiatic Turkey.
EPIDIUM, or EPIDIA, the island of ISLAY in the Scottish Hebrides.
EPINABURGUM, the abbey of BIBURG in Bavaria.
EPIRUS, the name anciently borne by the NW portion of Greece, between the Pindus range and the Ionian gulf, and stretching from the Acroceraunian mountains to the river Achelous. It nearly corresponds to the modern pashalik of ALBANIA.
EPISCOPATUS, VISCOVATO in Lombardy.
EPISCOPI CASTRUM, BISHOP'S CASTLE in Shropshire.
EPISCOPI CELLA, BISCHOFZELL in the Swiss cant. of Thurgau.
EPISCOPI INSULA, BISCHOFSWERDA in Saxony.
EPISCOPI VILLA, BISCHWEILER in Rhenish Prussia.
EPOISSUS, or IVODIUM, YVOIX in Luxemburg.
EPONA, or PHOERINGA, PFORING in Bavaria.
EPORA, or RIPEPORA, perhaps MONTONE in Andalusia.
EPTEMACUM, EPTEMACH in Luxemburg.
EQUITANIA, IDANHA-VELHA in the Portuguese prov. of Beira.
EQUUS TUTICUS, EQUUS MAGNUS, or TUTICUM, ACANO in Naples.
ERANA, NAVARINO in Greece.
ERANNOBOA, the river SONE in India.
ERASINUS, ERASINO on the gulf of Corinth.
ERBITA, NICOSIA in Sicily.
ERCTA, or ERCTE, MONTE-PELEGRINO in Sicily.
ERCURIACUM, ECRI in the French dep. of Ain.
ERDODIUM, ERDOD in Hungary.
EREBANTIUM PROMONTORIUM, CAPO-DELLA-TESTA on the coast of Sardinia.
EREMITARUM CŒNOBIUM, or MEGINRADI CELLA, EINSIEDELN in the Swiss cant. of Schwytz.
ERENBERTI SAXUM, EHRENBREITSTEIN in Prussia.
ERENEUM, ERNEE in the French dep. of Mayenne.
ERESBURCH, ERISBURGUM, ÆRESBERGUM, or MERESBURG, MARSBERG in Prussia.
ERETUM, near MONTE-ROTONDO in the Papal States.
ERFA, ERFF in Saxony.
ERFORDIA, ERFESFURDUM, or HIERANOFORDIA, ERFURT on the Gera, in Saxony.
ERGAVIA, IGUALADA in Catalonia.
ERGAVICA CELTIBERUM, ALCANIZ in Aragon.
ERGITIA, the river ERGERS an affluent of the Ill.
ERIBOLUM, or HYRIBOLUM, EREKLI on the sea of Marmora.
ERIDANUS, the river RODAUN an affluent of the Vistula.
ERLA, ERLAU in Hungary.
ERLACUM, ERLACH in the Swiss cant. of Bern.
ERLANGA, ERLANGEN in Bavaria.
ERLAPHIN, or ARLAPA, ERLAF in Hungary.
ERMÆA, or HERMÆA, TAVOLARA in Sardinia.
ERMANDICA, HELMANTICA, or VETTONUM URBS, SALAMANCA in Spain.
ERMSLEBIA, ERMSLEBEN in Prussia.
ERNODUNUM, AMBROISE-SUR-L'ARNON in the French prov. of Berry.
EROLATIA, HAAL in the Authrian circle of France.

ERPACHIUM, ERBACH in Darmstadt.
ERVANUM, or TERVA, ERIVAN in Armenia.
ERYCIS PORTUS, LERICE on the gulf of Genoa.
ERYMANTUUS, the river DIMITZANA an affluent of the Alpheus in Greece.
ERYTHIA, APHRODISIAS, or JUNONIS INSULES, probably an island which adjoined the present peninsula of Cadiz in Spain.
ERYTHRÆUM MARE, AUSTRALE MARE, or INDICUS OCEANUS, the PERSIAN GULF.
ERYTHRÆUM PROMONTORIUM, CAPE ROSSO in Crete.
ERYTHRUM, BARCA on the coast of Barbary.
ERYX, TRAPANI-DEL-MONTE in Sicily.
ESCAMUS, ESCUS, or OESCUS, the river ISCA, an affluent of the Danube.
ESCIA, ESKDALE in Scotland.
ESCUNACUM, ESCURIALE, or SCORIALE, the ESCURIAL in Spain.
ESCUS, or ISCUS, ISCHA in Bulgaria.
ESENA, ESENS in the Hanoverian prov. of East Friesland.
ESEPUS, or ÆSEPUS, a river of Asia Minor, according to Kruse the modern SATAL.
ESESFELTH, ESSESFELD, or ESSEVELDOBURG, ITZEHOE in the duchy of Holstein.
ESIA, OESIA, or OISIA, the river OISE in the French dep. of Ardennes.
ESPERNÆUM. See ASPRENTIA.
ESPINOIUM, or SPINETUM, ESPINOY in the French dep. of Pas-de-Calais.
ESSEXIA, or TRINOBANTUM REGIO, ESSEX.
ESTEVA, ESTEVACA, or STAVIACUM, ESTEVAY in the Swiss cant. of Friburg.
ESTHONIA, ESTONIA, or ESTENIA, the Russian gov. of REVEL.
ESTIA LACUS, or AQUÆ DULCIS SINUS, the DUMMER-SEE in Hanover.
ESTIONUM MONS, the ESTHENBERG in Tyrol.
ESTIVALE, or STIVALE, ESTIVAL in the French dep. of Maine.
ESTOLA, the Spanish river ESLA.
ESURIS, or XEREZ-DE-LA-FRONTECA, ST. FARO in Algarve.
ETOBEMA, ETOBESA, ETOVISSA, or SEGORVIA, SEGORBE in Valencia, or perhaps OROPESA.
ETOCETUM, or LICHFELDIA, LICHFIELD in Staffordshire; according to others UTCESTER.
ETRICULUM, LATTARACO in Calabria.
ETRURIA, ÆTRURIA, TUSCIA, or TYRRHENIA, a district of Italy, or a Roman province, bounded by the river *Macra*, the modern Magra, on the N; by the Apennines and the river Tiber on the E; by *Latium* on the SE; and by the *Mare Inferum*, or Tuscan Sea, on the W and SW. It nearly corresponds to the modern TUSCANY.
ETRUSCUM MARE, or TYRRHENUM MARE, the TUSCAN OR ITALIAN SEA.
ETTERSBURGUM, ETTERSBURG in Weimar.
ETYMANDER, the river HELMUND in Cabul.
EUANTHIA, or ŒANTHE, according to Kruse, GALAXIDI on the gulf of Lepanto.
EUGANEA VALLIS, the VAL SAGAN in Tyrol.
EUGENII INSULA, the peninsula of INNISHOWEN in Ulster.
EULÆUS, or ULAI, a river in Susiana, identified by some writers with the CHOASPES, by others with the modern KURAN.
EULISIA, the modern KABARDAH in the Caucasian isthmus.
EUMENIA, or EURYMNA, a town of Lydia, the modern TRALLES.
EUMENIA, a town of Phrygia Major, on the river Cludros near ASHKLI.
EUMENIA, OMENIA, or OMEGNA, NOVARA in the duchy of Milan.
EUPATORIA, a town on the west coast of the Tauric Chersonese, now KOSLOF, EUPATORIA, or JUPATORIA in the Russian gov. of Taurida.
EUPATORIA, in Pontus, the modern KULKHISSAR.
EUPILIS, a lake in Cisalpine Gaul, the LAGO-DI-PUSCIANA.
EURA, the river AURE in Normandy.
EURIPUS, the channel of NEGROPONTE, between the E coast of Greece and the island of *Eubœa*.
EUROPA, EUROPE, a name of Phœnician origin probably. In the time of Herodotus, this name was given to that portion of the old continent which extended to the river Phasis on the E; or perhaps to the Tanais. In the time of Ptolemy its boundaries were the Icy sea on the N; the *Tanais* or Don, the *Palus Meotis* or sea of Azof, the *Pontus Euxinus* or Black sea, the *Propontes* or sea of Marmora, the *Hellespont* or Dardanelles, and the Ægean sea on the E and SE; the Mediterranean on the S; and the Atlantic on the W.
EUROPUS, or RHAGES, near the modern VERAMIN in Persia.
EUROTAS, a river of Laconia, the modern BASILIPOTAMO OR IRI.
EUROTAS, a river of Calabria, the modern BAGRADA OR TARA.
EURYALUS, BELVEDERE on the W coast of Sicily.
EURYMEDON, a river of Pamphylia, the modern KAPRI-SU.
EUXINUS PONTUS, the BLACK SEA.
EVENUS, a river of Ætolia, the modern FIDARI.
EVONIUM, DUNSTAFFNAGE on the coast of Argyle.
EX, SEX, or SIXUS, a town on the S coast of Spain, ALMUNESSAR, according to others VELEZ-MALAGA.
EXA, or ISEA, the river Ex, which falls into the English channel near Exmouth.
EXCLODUNUM, ISSUNDUN in the French prov. of Berry.
EXIDOLIUM, EXIDEUIL in the French dep. of Dordogne.
EXILISSA, or LISSA, in Mauritania Tingitana, probably CEUTA.
EXONABA, STOMBAR in Algarve.
EXONIA, or ISCA, the modern EXETER.
EXTREMA, or STREMONTIUM, ESTREMOZ in the Portuguese prov. of Alentejo.
EXTREMA DURII, the prov. of ESTREMADURA in Portugal.
EXTREMADURA, or EXTREMADURA LEGIONENSIS, the Spanish prov. of ESTREMADURA.
EXTREMA MINII, the prov. of ENTRE-DUERO-E-MINHO in Portugal.
EZELINGA, or ESSLINGA, ESSLINGEN in Würtemberg.
EZRA, a town in Syria, ZARAVA.

F

FABARIA, the village of PFÆFERS near the Rhine.
FABARIS, FABER, or FARFARUS, the FARFA, a river flowing into the Tiber.
FABIA PRISCA SERPENSIS, the modern SERPA in the Portuguese prov. of Alentejo.
FABIRANUM, according to Kruse, now BEVERSTEDT in Bremen.
FABRATERIA, a town of Latium, now FALVATERA, in the Neapolitan prov. of Terra-di-Lavoro.
FABRICIUS PONS, a bridge connecting an island in the Tiber with Rome, now PONTE-DI-QUATRO-CAPI.
FÆNIANA PINIANA, FINNINGEN in Bavaria.
FÆROÆ, or FAROENSES INSULÆ, the FAROE ISLANDS.

FÆSULÆ, a town of Tuscany, now FIESOLE.
FAGARIA, the district of BUCHENAU in the German duchy of Fulda.
FAGONIUM, the FELDSPERG in the Swiss cant. of Grisons.
FAGUS, FOUG or FAU in the French dep. of Bar.
FALAZA, FALESIA in the French dep. of Calvados.
FALCOMONTIUM, FALCKENBERG in Sleswick.
FALCOPIA, the town of FALKÖPING in Gothland.
FALISCA, a town of Tuscany, supposed to be represented by the modern GALLESE.
FALISCORUM MONS, a mountain of Latium, now MONTE-SANCTI-ORESTE.
FALMUTHUM, VOLEMUTHIUM, or CENIONIS OPPIDUM, FALMOUTH in Cornwall.
FALSTRIA, the Danish island of FALSTER.
FANANUM, FANANO in the duchy of Modena.
FANESIORUM INSULA, or WOLLINIA, the island of WOLLIN.
FANIA, or FANIA SYLVA, the district of LA FAIGNE in the prov. of Liege in Belgium.
FANIOLUM, FAGNAUX in Languedoc.
FANUM AD TAFFUM, or LANDAVA, LANDAFF in Wales.
FANUM DAVIDIS, or MENEVIÆ, ST. DAVID'S in Wales.
FANUM FERONIÆ. See LUCOFERONIA.
FANUM FERONIÆ, PIETRA SANCTA in Tuscany.
FANUM FORTUNÆ, or TYCHOPOLIS, GLUCKSTADT in the duchy of Holstein.
FANUM JOVIS, DIOSHIERITÆ in Lydia.
FANUM JOVIS, the town FANJEAUX in Languedoc.
FANUM MARIÆ LAURETANÆ, or LAURETUM, ST. LORETTO in Ancona.
FANUM MARTIS, FAMARS in Hennegau.
FANUM MARTIS, a district of Normandy, MONT MARTIN.
FANUM ST. CRUCIS, the town of ST. CRUX in Cuba.
FANUM ST. CRUCIS, the village of HEILAGEN-KREUZ in Waldthale.
FANUM ST. EULALIÆ, ST. OLOLA in New Castile.
FANUM ST FIDEI, ST. FE in Granada.—Also ST. FOY LA GRANDE in the French dep. of Guienne. Also ST. FE in New Mexico.
FANUM ST. JULIANÆ, the Spanish SANTILLANA in Asturias.
FANUM ST. MENEHILDIS, ST. MENEHOULD in the French dep. of Champagne.
FANUM ST. ÆGIDII, ST. GILLES in the French dep. of Gard.
FANUM ST. ÆGIDII PICTAVIÆ, ST. GILLES in Poitou.
FANUM ST. ALBANI, ST. ALBANS in Hertfordshire.
FANUM ST. ALBINI, ST. AUBIN DU CORMIER in Bretagne.
FANUM ST. AMATORIS, ST. AMOUR in the French dep. of Jura.
FANUM ST. ANDEOLI, or ANDOCHI, ST. ANDEOL in the French dep. of Rhone.
FANUM ST. ANDREÆ, SANTANDER in Burgos in Spain.
FANUM ST. ANDREÆ AD DANUBIUM, ST. ANDRE in the Hungarian com. of Pesth.
FANUM ST. ASAPHI, or ELWA, ST. ASAPH in Flintshire.
FANUM ST. AUGENDI, or EUGENII, ST. CLOUD in the French dep. of Jura.
FANUM ST. BASSANI, ST. BASSANO in the duchy of Milan.
FANUM ST. BEATI, ST. BEAT in the French dep. of Haute-Garonne.
FANUM ST. BELLINI, ST. BELLINO in Lombardy.
FANUM ST. BONIFACIO, ST. BONIFACIO in Corsica.
FANUM ST. CHRISTOPHERI, HAVANNA in Cuba.
FANUM ST. DESIDERII, or DOMINICOPOLIS, ST. DIDIER in the French dep. of Bouches-du-Rhone.
FANUM ST. DIODATI, ST. DIE, or DIEZ, on the river Meurthe.
FANUM ST. EUTROPII, ST. TROPEZ in Provence.
FANUM ST. FACUNDI, ST. FANGON in Leon in Spain.
FANUM ST. FLORI, ST. FLOUR in Auvergne.
FANUM ST. FRANCISCI, QUITO in New Granada.
FANUM ST. GALLI, the Swiss ST. GALL.
FANUM ST. GEORGII, ST. GEORGE in the com. of Presburg.
FANUM ST. GERMANI IN LAJA, ST. GERMAIN-EN-LAYE in the Isle-de-France.
FANUM ST. GISLENI, ST. GUILIN in Hennegau.
FANUM ST. GOARI, ST. GOAR in Rhenish Prussia.
FANUM ST. GOTTHARDI, ST. GOTTHARD in the com. of Eisenburg.—Also ST. GROT in the com. of Czalad.
FANUM ST. HIPPOLYTI, ST. POLTEN in the archd. of Austria.—Also the French town of ST. HIPPOLYTE in Alsace.
FANUM ST. JACOBI, ST. JAMES in Barbadoes.
FANUM ST. JACOBI LEGIONENSIS, ST. JAGO DE LEON in Venezuela.
FANUM ST. JACOBI MAJORIS, ST. JAGO DE CHILI.
FANUM ST. JOANNIS, ST. JEAN.
FANUM ST. JOANNIS AD FAVUM, the Scottish town of PERTH.
FANUM ST. JOANNIS LAUDONENSIS, or ST. JEAN DE LAUNE, LOSNE in the French dep. of Cote d'Or.
FANUM ST. JOANNIS LUISII, ST. JEAN DE LUZ in the French dep. of Basses-Pyrenees.
FANUM ST. JOANNIS PETEPORTUENSIS, ST. JEAN PIE DE PORT in the French dep. of Basses-Pyrenees.
FANUM ST. JOANNIS PORTUS DIVITIS, ST. JUAN DE PUERTO RICO in the West Indies.
FANUM ST. JOBI, ST. JOB in the com. of Gross-Wardein.
FANUM ST. LEONIS, ST. LEO in the Papal deleg. of Urbino.
FANUM ST. LUCIFERI, ST. LUCAR DE BARRAMEDA in Seville.
FANUM ST. MARCELLINI, ST. MARCELLIN in Dauphiny.
FANUM ST. MARCI. See ARGENTANUM.—Also ST. MARCO in Sicily.
FANUM ST. MARIGNI, ST. MARINI in the Papal deleg. of Romagna.
FANUM ST. MARTINI, SZENKT MARTIN in the Hungarian com. of Thurocz.—Also SZENKT MARTEIN in the com. of Œdenburg.—See also ARX ST. MARTINI.
FANUM ST. MAURITII, perhaps the same as TARNATES.
FANUM ST. MAXENTII, ST. MAIXENT in Poitou.
FANUM ST. MICHÆLIS, MIHALY in the Hungarian com. of Semplin.—Also ST. MIGUEL in Guatimala.—Also ST. MICHAEL near Verona.—Also ST. MICHAELSTON in Barbadoes.—Also ST. MIHIEL in Belgium.
FANUM ST. MICHAELIS ARCHANGELI, ARKHANGEL in Russia.
FANUM ST. NABORIS, the abbey of ST. AVO in Lothringen.

FANUM ST. NICOLAI, St. Miklos in the Hungarian com. of Liptau.—Also St. Niklas, or St. Nicolo du Port.
FANUM ST. PALATII, St. Palais in the French dep. of Basses-Pyrenees.
FANUM ST. PAPOLI, St. Papoul in Languedoc.
FANUM ST. PAULI, Alexandrien or St. Paul in Virginia.
FANUM ST. PAULI LEONENSIS, or Leonum, St. Paul-de-Leon in the French dep. of Finisterre.
FANUM ST. PONTII TOMERIARUM, St. Pons-de-Thomieres in Languedoc.
FANUM ST. PORTIANI, St. Pourcain in the Bourbonnois.
FANUM ST. REMIGII, St. Remi in Provence.
FANUM ST. REMULI, St. Remo in Genoa.
FANUM ST. SALVATORIS, St. Salvador in the prov. of Nice.
FANUM ST. SEBASTIANI, St. Sebastian in Gomera of the Canary group.
FANUM ST. SEVERI, St. Sever in the French dep. of Landes.
FANUM ST. SPIRITUS, or Kedziensa, St. Kesdy-Lelek in Transylvania.
FANUM ST. STEPHANI, Launceston in Cornwall.—Also St. Etienne in the French dep. of Loire.—Also St. Acoma in New Mexico.
FANUM ST. THOMÆ, or Amatica, St. Amatique in Guatimala.
FANUM ST. TRUDONIS, St. Trond, or St. Trou, St. Truyen in Limburg.
FANUM ST. URSICINI, St. Ursane or Ursitz in the Swiss cant. of Bern.
FANUM ST. VALERII, St. Valery in Picardy.
FANUM ST. VENANTII, St. Venant in Artois.
FANUM ST. VITI, or Vitopolis, St. Veit in the circle of Klagenfurt.
FANUM ST. VITI FLAMONIENSIS, Fiume in the gulf of Quamero.
FANUM SPIRITUS SANCTI, Pont St. Esprit in Languedoc.
FANUM SPIRITUS SANCTI VASCONIÆ, St. Esprit in Gascogne.
FANUM VACUNÆ, Vocone in the del. of Rome.
FANUM VOLTUMNÆ, or Vitersium, Viterbo in the deleg. of Viterbo.
FARA, or Ferra, Fere in Picardy.
FARDIUM, or Verdia, Verden in Hanover.
FARINGA, Pforingen in Bavaria.
FARNESIUM CASTRAM, Farnese in the Papal deleg. of Viterbo.
FAUCES NORICARUM, or Scæva Viennæ, Schœdtwein in the archd. of Austria.
FAUCES PERTUSÆ, Pertus, or the Col-de-Pertuis, a pass over the Pyrenees.
FAUSTINI VILLA, Edmundsbury in Suffolk.
FAVENTIA, Faenza in Ravenna.—Also Fayence in Provence.
FAVERGA, Faverge in Savoy.
FAVONII PORTUS, Porto-Favona in Corsica.
FEBIANA CASTRA, Bebenhausen in Würtemberg.
FEGE, or Fara, Pharphar, a river in Cœle-Syria.
FELICITAS JULIA OLISIPO, Ulyssia, or Lissabona, Lisbon in Portugal.
FELINUM, Fellin in the Russian gov. of Riga.
FELTRIA, or Feltuum, Feltre in the Lombardian prov. of Belluno.
FENESTRELLÆ, Fenestrelles in Dauphiny.
FERA, or Ferrea Vallis, Freel in the Swiss cant. of Grisons.
FERDA, or Fardium, Verden in Hanover.
FERDINANDI INSULA, Fernando Po on the coast of Guinea.
FERENTUM, Forenza in the Neapolitan prov. of Terra d'Otranto.
FERGUSII RUPES, Carrickfergus in Antrim.
FERMAGANENSIS COMITATUS, Fermanagh in Ulster.
FERNAMBOCUM, Pernambuco in Brazil.
FERONIA, in Sardinia, or perhaps Orosai.
FERRARIÆ, perhaps Ferrieres in the French dep. of Isle-de-France.
FERRARIUS PORTUS, Porto Ferrago in Elba.
FERRETA, or Ferrete, Pfrit in Upper Rhine.
FERRATUS MONS, supposed to be the Jurjura of Algiers.—Also the Frankenberg in Alsace.
FERRERA, Ferrieres in the French dep. of Lot.
FERRI INSULA, or Ferro, Hiero in the Canary group.
FERTINA, the Illyrian Island Berwitsch.
FESCENNIA, Fescennini in Tuscany.
FESSA, or Fezzanum Regnum, Fez in Africa.
FEVUS, Vraita in Lombardy.
FIBRENUS, a river in Latium, the Fiume della Posta.
FICARIA, an island near the north coast of Sardinia, perhaps Cortelazzo.
FICARIA, Figueras in Catalonia.
FICARIUS, a river in Corsica, the Ficari, according to others the Valença.
FICELLA, Oberwesel in Rhenish Prussia.
FICOCLE, Cervia in the Papal deleg. of Ravenna.
FIDENTIUM, Fezenza in Gascony.
FIGULNENSIS PORTA, Porta-di-Santa-Agnese.
FILICERIÆ, Fougeres in Brittany.
FINARIUM, Finale or Finaro in Genoa.
FINIS TERRÆ CAPUT, Finisterre in Gallicia.
FINIS VALLI, or Vindomoro, thought by some to be Wallsend; by others, Newcastle on the Tyne.
FIRMANUM, a town in Umbria, Formignano.
FIRMITAS, Ferte', a town in Luxemburg.
FIRMITAS AD ALBULAM, Ferte'-sur-Aube in Champagne.
FIRMITAS AD JOTRUM, Ferte'-sous-Jouarre in the French dep. of Seine-et-Marne.
FIRMITAS ADELHEIDIS, or Alesia, Ferte Alais, Aleps in the French dep. of Isle-de-France.
FIRMITAS AUCULPHI, or Qualquarii, Ferte'-Guacher in the French dep. of Seine-et-Marne.
FIRMITAS AURANIENSIS, Ferte' Aurin in the French dep. of Loire-et-Cher.
FIRMITAS BERNHARDI, Ferte Bernard in Maine.
FIRMITAS MILONIS, Ferte' Milon in the French dep. of Isle-de-France.
FIRMIUM, or Firmus, Fermo in the Papal States.
FIRMIUM JULIUM, perhaps the same as Lexi Motril in Granada.
FISCA, or Fiskaha, the Grosse-Fischa, an Austrian affluent of the Danube.
FISCAMNUM, or Fecamp, Fescamp in Normandy.
FISCELLUS, by some identified with Monte-Fiscello in Sabina, by others with Monte-della-Sabilla in Umbria.
FISCUS ISIACENSIS, or Isiacus, Issy near Paris.
FISERA, Ficari in Corsica.
FIVELGONIA, or Fivelingo, the Dutch prov. of Groningen.
FIXA, or Flexia, La Fleche in Anjou.
FLANATICUS SINUS, the Gulf of Quarnero.
FLANDRIA, Flanders.
FLANONA, Fianona on the W side of Quarnero.
FLAVIA CONSTANS, Spello in Umbria.
FLAVIA GALLICA, Fraga in Aragon.
FLAVIA IRIA, Padrono or El-Padron in Gallicia.

FLAVIA LAMBRIS, RIVADIA in Gallicia.
FLAVIA SOLVA, a town of Austria, perhaps SOLFIELD.
FLAVIACUM, FLAVIGNY in Burgundy.
FLAVIÆ ARÆ, IMENDINGEN in Baden.
FLAVIANA CASTRA, or VINDOBONA, VIENNA.
FLAVIANUM, FIANO in Viterbo.
FLAVIONANIA, or FANUM ST. ANDREÆ, FUANES in Biscay.
FLAVIONIA, COMPOSTELLO in Gallicia.
FLAVIOPOLIS, or FLAVIAS, BOLI in Asia Minor.
FLAVIUM ARGITANIUM, JAEN in Andalusia.
FLAVIUM BRIGANTIUM, BETANZOS in Gallicia.
FLAVIUM INTERAMNIUM, PONFERRADA in the Spanish prov. of Leon.
FLAVIUM LAMINITANUM, the ALHAMBRA in Spain.
FLEMARUM VALLIS, the valley of FLEIMS in Tyrol.
FLEMIUM, the village of FLIMS in Switzerland.
FLEMMA, the town of FLUMS in the Swiss cant. of St. Gall.
FLENIUM, VLAARDING in Holland.
FLENOPOLIS, FLENSBURG in Schleswig.
FLESINGA, VLIESSINGEN in the Netherlands.
FLETIO, FLEUTEN in the Dutch prov. of Utrecht.
FLEVO INSULA, URK, an island in the Zuyder-See.
FLEVO LACUS, the ZUYDER-SEE.
FLEVOLANDIA, VLIELAND, an island of Holland.
FLEVUM (OSTIUM), the N mouth of the Rhine, joining the Zuyder-See with the German ocean (now the HET-VLIE).
FLEVUM CASTELLUM, or PHLEUM, now WIESFLIET in Holland, SW of the mouth of the Unse
FLEXUM AD FLEXUM, WIESELBURG in Hungary.
FLORA, the river FIORE in Tuscany.
FLORENTIA, FLORENCE in Tuscany.—Also FIORENZUOLA, a town in Parma.
FLORIACUM AD OSCARUM, FLEURY in Burgundy.
FLORIACUM MONASTERIUM, FLEURY in the Hennegau.
FLORIACUS, the French town and abbey of FLEURY, ST. BENEDICT in the Orleannois.
FLORICHINGÆ. See FLENIUM.
FLORIDA VALLIS, the abbey of FLORIVAL in Brabant.
FLORIMONTIUM, BLUMBERG in Upper Alsace.
FLORINKINGÆ, FLORENNES in Namur.
FLORIUS, a river of Spain, the RIO-DE-CASTRO.
FLORIVALLIS, the Swiss BLUMENTHAL in the cant. of Grisons.
FLORUM INSULÆ, FLORES, one of the Azores.
FLUETUM, VLUETEN in the Dutch prov. of Utrecht.
FLUMEN SALVUM, an arm of the river Euphrates, near its mouth, now the island of JEZIR-ET-KHADER.
FLUMENTANA PORTA, the PORTA-DEL-POPULO in Rome.
FLUMETUM, FLUMET in Faucigny.
FLUSOR, supposed to be the CHIENTI, a river of Italy.
FLUVIUS FŒDERATORUM, a river of Granada, the RIO-GUADALMEDINA.
FOCHIUM, FOQUI on the Japanese island of Niphon.
FOLIA, ISAURUS, or PISAURUS, now the FOGLIA, a river of Italy.
FONS ADMIRATIONIS, the source FONTAINE-DE-MERVEILLE in Savoy.
FONS BEATÆ VIRGINIS, a town and ancient nunnery, FRAUENBRUNNEN in the Swiss cant. of Berne.
FONS BELLUS, the castle of SCHONBRUNN in the archd. of Austria.
FONS EBRALDI, FONTEVRAULD in Anjou.
FONS EPISCOPI, FONTAINE L'EVEQUE, in the Hennegau.
FONS GOMBALDI, FONCOMBAUT in Berry.
FONS IBERE, FONTIBRE in Old Castile.
FONS LATIUS, LATZFASS in the territory of Brixen.
FONS PIGI, SOMMEPI in Champagne.
FONS RAPIDUS, FUENTE RABIA or FONTARABIA in Guipuscoa.
FONSKESONUS, the BULLERBORN in Westphalia.
FONS ST. ANDREÆ, FONTAINE ANDRE in Neufchatel.
FONS SANUS, FOSSANO in Piedmont.
FONTANELLA, the abbey of FONTENELLE in Flanders.
FONTANETUM, FONTANETO, a town of Milan.
FONTANEUM, FONTENAEI in Burgundy.
FONTANEUM COMITUM, FONTENAI LE COMTE in Poitou.
FONTES, FONTAINE in Burgundy.
FONTES BELGÆ, WELLS in Somerset.
FONTES PADERÆ, PADERBORN in Prussia.
FONTIA, or PONZA, the TUSCAN SEA.
FORA, the island of FOHR on the W coast of Sweden.
FORAHEIM, FORCHEIM in Bavaria.
FORCALQUERIUM, FORCALQUIER, a town of Provence.
FORDUNIUM, FORDUN in Kincardineshire.
FORGIÆ, FORGES-LES-CAUX in Normandy.
FORMIÆ, supposed to be MOLA-DI-GAETA in Naples.
FORMIO, a river in the N of Istria, the RESINO.
FOROJULIENSIS CIVITAS, CIVIDALE-DEL-FRIULI in Austrian Lombardy.
FORTALITIUM, or FORTIS MONS, LA-FORZA in the Val-di-Demona, Sicily.
FORTALITIUM LUDOVICI, FORT LOUIS in Alsace.
FORTALITIUM ST. CHRISTOPHORI, ST. CHRISTOVAL in Brazil.
FORULI, CIVITA THOMUSSA in Naples.
FORUM ADRIANI, VORBURG in Holland.
FORUM APPII, ST. DONATA in the States of the Church.
FORUM AURELII, MONTE ALTO in Tuscany
FORUM BIBALORUM, FOMILLO in Spain.
FORUM CALVISII, CALVISANO in Austrian Lombardy.
FORUM CASSII, ST. MARIA FORCASSI in Tuscany.
FORUM CLAUDII, ORIOLO on the coast of Tuscany.
FORUM CORNELII, IMOLA in the States of the Church.
FORUM DIUGUNTORUM, PIZZIGHETTONE in Austrian Lombardy.
FORUM DOMITII, FRONTIGNAN, but, according to some, FABREGNES in the French dep. of Herault.
FORUM EGURRORUM, MEDINA-DEL-RIOSECA in Valladolid; but, according to others, SALAS in Asturias.
FORUM FLAMINII, PONTE CENTESIMO in the States of the Church.
FORUM FULVII, VALENZA in Piedmont.
FORUM LIBRICORUM, BORGO LAVIZARA in Milan.
FORUM LICINII, or PIEVE D'INCINO, the BERLASINA in Milan.
FORUM LIMICORUM, or LIMIA, PONTE-DE-LIMA in Portugal.
FORUM LIVII, FORLI in the States of the Church.
FORUM NERONIS, BOURG D'OISANS in the French dep. of Isere.

FORUM NOVUM, FORNOVO in Austrian Lombardy.
FORUM POPULI, FORLIMPOPULI in the deleg. of Romagna.—Also LA FOSSA in Naples.
FORUM SEGUSIANORUM, FEUR in the French dep. of Loire.
FORUM SEMPRONII, FOSSOMBRONE in the deleg. of Urbino.
FORUM STATIELLORUM, perhaps VALLADASO in Lombardy.
FORUM TRAJANI, PORDONGIANO in Sardinia.
FORUM VIBII, BUBIENA in Piedmont; but, according to others, CASTEL FIORI.
FORUM VOCONII, perhaps CANET in Haute-Garonne; according to others, LUZ in the dep. of Hautes-Pyrenees.
FORUM VULCANI, SALFATARA in Naples.
FOSSA CLODIA, CHIOZZA in Venice.
FOSSA CORBULONIS, or VLIE, VLIESTROM in Holland.
FOSSA MESSANICA, CANAL DI ST. ALBERTA in the deleg. of Ravenna.
FOSSA PAPIRIANA, or VIAREGGIO, VAREGGIA in Genoa; but, according to others, FOS-DI-NOVO.
FOSSÆ, FOSSE in Namur, Belgium.
FOVEA, FOGGIA in the Capitanata, Naples.
FOXUM, FOIS in Languedoc.
FRAGIZA, perhaps RAJIK in Syria.
FRANCIACUM, FRONZAC in Guienne.
FRANCISCOPOLIS, HAVRE-DE-GRACE.
FRANCODALIA, FRANKENTHAL in Bavaria.
FRANCOFURTUM AD MŒNUM, FRANKFORT-ON-THE-MAIN.
FRANCOFURTUM AD ODERAM, FRANKFORT-ON-THE-ODER.
FRANCONATUS, HET VRYE LAND, a district in Old French Flanders.
FRANCONATUS BRAGARUM, LA-FRANCE-DE-BRUGES in the Netherlands.
FRANCONATUS ORIENTALIS, the district of OOST VRYE in Flanders.
FRANCOPOLIS, VILLE FRANCHE (DE ROVERGNE) in the French dep. of Aveyron.
FRANCOSTEINIUM, FRANKENSTEIN in Prussian Silesia.
FRANECHERA, FRANECKER in Friesland.
FRASEAROLUM, FRESEARNOTO in the prov. of Milan.
FRATRUERTIUM, LA FEZZA in Terra d'Otranto, Naples.
FRAUSTADIUM, FRAUSTADT in Posen, Prussia.
FRAXINETUM, GARDE FREMET, in the French dep. of Var.
FRAXINIA, FRESSING in Bavaria.
FREGELLÆ, PENTECORVO in the Terra-di-Lavoro; according to others, CEPRANO in Naples.
FREGENÆ, perhaps MACARESE in Tuscany.
FREMICURIA, FREMINCOURT in the French dep. of Isle-de-France.
FRENTO, the river FORTORE in the Capitanata, Naples.
FREQUENTUM, FRICENTI in the Principato-Ultra, Naples.
FRESNACUM, FRESNAY in Maine.
FRETUM ANJANUM, the strait of ANIAN between California and Jedso.
FRETUM BRITANNICUM, FRETUM CALETANUM, or EURIPUS, the STRAITS OF DOVER.
FRETUM DAVISII, DAVIS' STRAITS, Greenland.
FRETUM FORBRISSERI, FORBISHER STRAITS in Greenland.
FREYBERGA, FREYBERG in Saxony.
FRIBURGENSIS PAGUS, the Swiss canton of FRIBURG.
FRIBURGUM, FREYBERG in Baden.
FRIBURGUM AD WINDAM, FREYBERG in Prussia.
FRICDISLARIA, FRIZLAR in Hesse-Cassel.
FRIDERICI COLLIS, FRIEDERICHSBUHEL in Bavaria.
FRIDERICI ODA, FREDERICIA in Jutland.
FRIDERICI PORTUS, FREDERICKSHAVN in Finland.
FRIDERICIANA VALLIS, FRIEDERICHSTHAL in Zieland.
FRIDERICOPOLIS, FREDERICSTADT in Schleswick.
FRIEDBERGA, FRIEDBERG in Bavaria.
FRIEDLANDIA, FRIEDLAND in Prussian Silesia.
FRIGENTIUM, FRIGENTO in the Principato-Ultra, Naples.
FRIGIDA, FRIAS in Old Castile.
FRIGIDA VALLIS, FROIDEVAL in Alsace.
FRIGIDA VALLIS, VALLE-FREDDA in Naples.
FRIGIDUS, the river VYPAO in Trieste.—Also the FREDDO, a river in Tuscany.
FRIGIDUS MONS, the abbey FREMONT in the French dep. of Isle-de-France.
FRISIA, FRISA, or FRESIA, FRIESLAND in Holland.
FRISIA CIMBRICA, FRIESLAND in Schleswick.
FRISIA ORIENTALIS, the district of EMBDEN.
FRITTOLANÆ AQUÆ, FRITTOLA in Naples.
FRIZACUM, FREISAC in Illyria.
FRONTENSIS VILLA, FRONT in Savoy.
FRUSINO, FROSINONE in the Papal States.
FUCINUS LACUS, the LAGO-DI-CELANO in Abruzzo-Ultra 2da.
FULCHERIA, the GHIERRA D'ADDA in the prov. of Milan.
FULDENSE CÆNOBRUM, FULDA in Hesse.
FULFINIUM, perhaps WELKA in the island of Vegia.
FULGINIA, FOLIGNO in the Papal States.
FULINUM, FEVILLANS in Languedoc.
FUMACUM, FUMAY in the French dep. of Ardennes.
FUNDI, FONDI in Naples.
FUNDUS MARUCANUS, MAZOUNAH in Algeria.
FURA, FERVUEREN in the Netherlands.
FURCÆ CAUDINÆ, FURCHLÆ in the valley of Arpaia, in the Neapolitan prov. of Principato-Ultra.
FURCONA, FORCONE in Abruzzo.
FURNA, or FOURNES, VEURNE in the Netherlands.
FUSNIANUM, FOIGNY, an abbey in Picardy.
FUSUS, the FUSE, an affluent of the Aller, in Hanover.

G

GABA, perhaps the same as GAMALA.—Also GIB in Samaria.—Also GIBEA in the same.
GABÆ, DARABGHERD in the south of Persia.—Also KABAS in Sogdiana.
GABALA, DJBILE or GEBILEH in Syria.
GABARUS OLERONENSIS, the river GAVE D'OLERON in the French dep. of Basses-Pyrenees.
GABARUS PALENSIS, the river GAVE DE PAU in Basses-Pyrenees.
GABATHON, or GIBETHON, GEBATH in Judea.
GABELLUS, the river SECCHIA in the duchy of Modena.
GABII, PONTANO in the Papal States.
GABINUS LACUS, the LAGO-DI-CASTIGLIONE in the Papal States.
GABLONA, or JABLONA, GABEL in Prussian Silesia.
GABRANTOVICORUM SINUS, BRIDLINGTON BAY on the coast of Yorkshire.
GABRIS, TAURIS or TABRIS in the prov. of Ghilan

in Persia.—Also CHABRIS in the French prov. of Berry.
GABROMAGUS, LEONHARDY-AM-FORST in Austria.
GABROSENTUM, or CAPRA CAPUT, GATESHEAD in Durham, or perhaps DRUMBURGH.
GADAMUS, perhaps GIODDA in Arabia Felix.
GADARA, near OM-KEIS in Palestine.
GADIRTHA, EL-DER in Syria; according to others RABABEH.
GADITANUM FRETUM, COLUMNARUM FRETUM, or HERACLEUM FRETUM, the STRAITS OF GIBRALTAR.
GADITANUS PORTUS, the BAY OF CADIZ.
GADIVA, ABERFRAW in Anglesea.
GAGUETA, CAKET in Gurgistan.
GAITIA, the river JAIEZE in Bosnia.
GALAAD, MOUNT GILEAD in Palestine.
GALABA, perhaps GIALAB in Mesopotamia.
GALABER, the river GALAURE in Dauphiny.
GALACUM, KENDAL in Westmoreland.
GALARIA, GAGLIARO in Sicily.
GALATA, GALITA or JALTA in the Mediterranean.
GALATIA, GALLIA, GALLIA TRANSALPINA, GALLIA PROPRIA, GALLIA ULTERIOR, or GALLIA ULTIMA, names given by the Romans to that region of Europe bounded on the N by the German ocean and the Rhine; on the E by the Rhine, the Alps, and the Varus; on the S by the Pyrenees and the Mediterranean; and on the W by the Atlantic, corresponding nearly to the modern countries of the Netherlands, Switzerland, and France.
GALATIA, the Spanish GALICIA.
GALATIA, GALLIA PARVA, or GALLOGRÆCIA, a district of Asia Minor, lying between Pontus on the E, and Bithynia on the W, now mainly comprised in the pash. of Angora.
GALEPSUS, ST. KIRILI on the Toronean gulf.
GALICIA, HALICZ in Galicia.
GALILÆUM MARE, the LAKE of TIBERIAS in Palestine.
GALINDIA, AMUL in Persia.
GALLA, AMUL in the Persian prov. of Azerdbijan.
GALLIA AQUITANIA, that part of GALLIA which the river GARUMNA or GARONNE divided from Celtic Gaul.
GALLIA CELTICA, or GALLIA LUGDUNENSIS, that division of GALLIA which lay between the Marne and the Seine on the N, and the Loire on the S.
GALLIA CISALPINA, GALLIA CITRA ALPA, ITALIA SUBALPINA, or PROVINCIA CITERIOR, that part of GALLIA which comprised AUSTRIAN LOMBARDY, PIEDMONT, and SAVOY.
GALLIA NARBONENSIS, or PROVINCIA ROMANA, that region in the SE of Gallia which the river Varus and the Alps divided from Gallia Cisalpina.
GALLIANUM, GALLIANO in Milan.
GALLICUS, the GALLEGO in Aragon.
GALLICUS SINUS, the GULF OF LYONS.
GALLINARIA, URGO, or GORGONA, GALLINARA, according to others, ISOLA D'ALBIENGA in the Gulf of Lyons.
GALLITARUM OPPIDUM, GUILLIESTRE in Dauphiny.
GALLIVA, GALWAY in Ireland.
GALLUS, GALIPO, a river in Phrygia.
GALTHERA, the DENDER in Flanders.
GALTIS, GALULA in Assyria.
GALVEJA, GALLOWAY in Scotland.
GALYMNA, KALYMNO, according to others, KOLMONE in the Cyclades.
GAMALA, HOSN in Palestine.
GAMANODUNUM, GROBMING in Austria.
GAMBLATUM, GAMBOLO in Milan.
GAMBRACUS SINUS, GRIMAND, or the GULF OF ST. TROPEZ in Provence.
GAMMUNDIA, GEMUND in Würtemberg
GAMUNDIUM, CASTELAZZO in Milan.
GANGES, the river GANGA, or GONGA in India.
GANGETICUS SINUS, the BAY OF BENGAL.
GANGRA, KANGREH in PAPHLAGONIA.
GANNATUM, GANAT in Bourbon.
GANNODURUM, LAUFENBERG in the Swiss cant. of Aargau.
GANUS, KANOS in Romelia.
GARACTUM, GUERET in the French dep. of Creuse.
GARAMA, GERMA in Africa.
GARAPHI MONTES, GHIBB-EL-ZICKAR in Fez.
GARDELEGIA, GARDELEBEN in Prussia.
GARDICE LACUS, GUARDE in Sudan, Africa.
GARETIUM, GARESSIO in Piedmont.
GARGANUM PROMONTORIUM, PROMONTORIO-DEL-MONTE-GARGANO in Naples.
GARGANUS, or MONTE-DI-ST.-ANGELO, GARGANO in Naples.
GARGARA, TSCHEPINI in Mysia.
GARGARCI, perhaps GARZARO in the Troad.
GARGARIUS, ST. JEAN-DE-GARGUIER in Languedoc.
GARGOVIUM, JARGEAU in Orleanois.
GARI, GRISCH in Cabul.
GARIANONUM, YARMOUTH in Norfolk.
GARIENIS, the YARE in Norfolk.
GARNÆ PORTUS, RODIA in Naples.
GARONA, the GARON in the French dep. of Isere.
GARONNA, or GARUMNA, the GARONNE in Gironde.
GARRA, LONA in Fez.
GARREJENUS, CLEY in Norfolk.
GARSA, GARZ in Stettin.
GARSAVORA, AK-SERAI in Asia Minor.
GARTIA, the GARZA in Milan.
GASGAMIRA, CASHMIR in Northern India; according to others, JESSULMIR.
GASCOGNA, or VASCONIA, the French prov. of GASCOGNY.
GASORUS, ALESTEROFORI in Thrace.
GATH, MESHED in Palestine.
GAUDIA MUNDI, GEMUNDEN in Bavaria.
GAUDIOSA, JOYEUSE in Languedoc.
GAUDOS, GOZO, an island south of Crete, according to some GIADUROGISSA; others, PORTO GABOSO.
GAUGAD, GOZA, an Æthiopian island.
GAUGAMELA, or KARMELIS, JERMELY in Assyria.
GAULOS, GOZO in the Mediterranean.
GAUNODURUM, STEIN in the Swiss cant. of Schaffhausen.
GAURUS, MONTE-DI-GAURO in Naples.
GAUZACA, GHIZNI in Cabul.
GAUZANIA, SOFIAN in Persia, in Ghilan.
GAVIODORUM, EBENHEIM in Alsace.
GAVIRATIUM, GHIVIRA in Milan.
GAZA, TAURIS in Ghilan.
GAZENA, GAZARAT in Phrygia.
GEDDA, JEDDA in Arabia.
GEDROSIA, the Persian prov. of MEKRAN.
GELA, TERRA-NUOVA in Sicily.
GELDUBA, GELB in Prussia.
GELURNUM, GLURENZ in Tyrol.
GEMBLACUM, GEMBLOURS in the Netherlands.
GEMELLÆ, IMMELACH in Numidia.
GEMENICIUM, JAMNITZ in Moravia.
GEMETICUM, JAMIEGES in Normandy.
GEMUNDA AD NICRUM, NECKAR-GEMUND in Nassau.
GEMUNDA AD TRAUNUM, GEMUND in Illyria.
GENA, GROSS-JENA in Weimar.
GENABUM, the city of ORLEANS in France.
GENCHOA, CHENCOA in Sudan.

GENEBRA, a mountain, GENEVE in Dauphiny.
GENGILIACUM, GENGENBACH in Baden.
GENESARITIS LACUS, the SEA OF GALILEE.
GENONIA, SINUNIA in Parthia.
GENOSIA, GENSEN in Bavarian Germany.
GENTIANUM, GENTANO in the Papal States.
GENTIFORUM, VOKELMARKT in Illyria.
GENTILIACUM, GENTILLY in the French dep. of Seine.
GENUA, the city of GENEVA.
GENUA URSORUM, or URSO, OSSUNA in Seville.
GENUSIUM, GENOSA in the Neapolitan prov. of Capitanata.
GENUSUS, the river SCOMBI in Greece.
GEOFANUM, GIFANI in the Neapolitan prov. of Principato-Citra.
GEPHYRA, perhaps SIRMIA in Syria.
GER, the DRAB or DARABA in Marocco.
GERA, or GERAHA, SCHWARTZ in E. Weimar.
GERÆSTUM, CAPO-MONTELLO in Negropont.
GERÆSTUS, CARYSTO or GERESTO in Negropont.
GERANIA, or MACRIPLAI, PALEOVOUNI, a mount in Greece.
GERANTHRÆ, JERAKI in Greece.
GERARIDI MONS, or GRAMMONT, GEENTSBERGE in the Netherlands.
GERASA, DSCHERRASCH in Decapolis.
GERBERACUM, GERBEROY in the French prov. of Isle-de-France.
GERENA, GEHREN in Switzerland.
GERENIA, ZARNATA in Greece.
GERGOVIA, a town of the Arverni, near CLERMONT in France.
GERION, DRAGONARA in Capitanata, Naples.
GERISA, GHERZE in Tripoli.
GERLOCURIA, GEROLSHOFEN in Bavaria.
GERMA, GERMHASTI in Asia Minor.
GERMANIA CISRHENANA, that part of *Gallia* lying immediately south of the Lower Rhine.
GERMANIA TRANSRHENANA, or GERMANIA MAGNA, that part of Europe bounded by the *Mare Suevicum* or Baltic on the N; the *Montes Sarmatici* or Carpathians on the E; the Danube and the Rhine on the S; and the Rhine and the German ocean on the W. Its boundaries eastward were nominal; to the N it included Denmark.
GERMANICOPOLIS, ZETERAMBOL in Paphlagonia.
GERMANICUM MARE, the NORTH SEA.
GERMANICUM OPPIDUM, VOBURG in Bavaria.
GERMINIACUM, GERMIGNY in Orleanois.
GERNE, the LAGO-DE-VARANA in Capitan ta, Naples.
GERNINGERODA, GERNRODE in Anhalt, Germany.
GERNIUM, CIREGNOLA in Capitanata, Naples.
GERONTIA, JURA in the Ægean sea.
GERRA, GERRUM MASELI in Lower Egypt.—Also EL-KATIF in Arabia.—Also SERUK in Syria.—Also the island of GIRBA.
GERRAICUS SINUS, EL-KALIF bay in the Persian gulf.
GERRHUS, KOISU, according to others SULAK in Albania.—Also TRUZKOY in Russia.
GERSOVIA, GERSAU in Switzerland.
GERULATA, perhaps OROSVAR in Hungary.
GERUNDA, GERONA in Catalonia.
GESCHUR, SHIR-BENIR-JAKUB in the Syrian pash. of Damascus.—Also SHIR-SHOGR on the river Orontes.
GESECENA, GESEKE in Prussia.
GESIA, GEX in the French dep. of Aisne.
GESORIACUM. See BONONIA.
GESSORIACUM, BOULOGNE in France.
GESTESIA, GOSTESCH in Hungary.
GESTRICIA, GESTRIKLAND in Norway.
GETA, GETS in Brabant.
GETARA, ABCHERON in Albania.
GETARUM DESERTUM, BESSARABIA.
GEVALIA, GEFLE in Sweden.
GEVALIENSIS PROVINCIA, the prov. of GEFLEBORG in Norway.
GEYLA, the river GEIL in Tyrol.
GIANUM, GIEN in the French dep. of Loiret.
GIBEATH PHINEHAS, GIBEA in Palestine.
GIBRALTERIA, GIBRALTAR.
GIESSA, GIESSEN in Hesse-Darmstadt.
GIGONUS, PANOMI in Rumelia.
GIJIA, GIJON in Asturias.
GILAVIA BORUSSICA, EYLAU in Prussia.
GILAVIA GERMANICA, ILAWA in Prussia.
GILDONACUM, or GELDENACKEN, JUDOIGNE in Brabant.
GILFORDIA, GUILFORD in Surrey.
GILOVIA, EGLAU in Bohemia.
GIMO, GIMONT in the French dep. of Gers.
GINÆA, DSCHENNIN, or GENIN, GILIN in Palestine.
GINSIUM, GUNZ or KOZEG in Bavaria.
GIPPEVICUM, IPSWICH in Suffolk.
GIR, the WADI-ADJEDEC or DHGIR in Central Africa.
GIRBA, JERBA, or ZERBI, an island N of Tripoli.
GIRGIRIS, or GYRIS, EYREH, a mount in Central Africa.
GIRVIUM, or YARUM, YARM in York.
GISNÆ, GUISNES in the French dep. of Pas-de-Calais.
GISONIS CASTRA, GEISENFIELD in Bavaria.
GITLUI, or APFAX, EL-CALLAH in Algeria.
GITMIACINUM, GITSCHIN in Bohemia.
GIUF, supposed to be MESSHERGA in Zeugitana.
GLACENSIS COMITATUS, the co. of GLATZ in Silesia.
GLACIALE PROMONTORIUM, ICY CAPE in Tartary.
GLACIALIS OCEANUS, the ARCTIC OCEAN.
GLACIUM, the town of GLATZ in Silesia.
GLAMNATEVA CIVITAS, or GLANDATUM, GLANDEVES in the S of France.
GLAMORGANIENSIS COMITATUS, GLAMORGANSHIRE.
GLANAVENTA, ELLENBOROUGH in Cumberland.
GLANDOMIRUM, or TAMARA, PIERTE BRANDOMIL in Spanish Galicia; or perhaps MUROS at the mouth of the Tambre.
GLANEATUM, or GALERATUM, GALLARA in Milan.
GLARIZIUM, GLARIS in Switzerland.
GLARONENSIS PAGUS, the Swiss cant. of GLARIS or GLARUS.
GLASCONIA, GLASTONBURY in Somerset.
GLASCOVIUM, GLASCUM, or GLASCUA, GLASGOW in Scotland.
GLAUCHUS, the LEUCOS or LEUKA, a river in the Morea.
GLAUCUS, the river CHORAK in Mingrelia.
GLAUCUS SINUS, the GULF OF MACRI in Asia Minor.
GLEMONA, GEMONA in Friuli.
GLESSARIA, NORDSTRAND, or perhaps AMELAND, an island of Denmark.
GLOCESTRIA, GLOUCESTER.
GLOGOVIA MAJOR, GLOGAU in Prussian Silesia.
GLOGOVIA MINOR, OBER-GLOGAU in Prussian Silesia.
GLORIUM, or GLURNIUM, GLURNS in Tyrol.
GLOTA INSULA, the island of ARRAN in the Clyde.
GLOTÆ ÆSTUARUM, the FRITH OF CLYDE.
GLOTTANA VALLIS, or CLIDESDALIA, CLYDESDALE in Lanarkshire.

GLUNICENSE CŒNOBIUM, GLEINK in the Fraunviertel.
GNATIA, GIOVENAZZO in the Neapolitan prov. of Terra-di-Bari.
GNESNA, GNEZEN in the Prussian reg. of Posen.
GNEVUM, MEWE in Prussia.
GNOSUS, CANDIA in Crete.
GOARIS, the river TAPTI in Hindostan.
GOBANNIUM, ABERGAVENNY.
GOBANNINO, the river GAVENNY.
GOBARUM, or GOBÆUM PROMONTORIUM, CAPE ST. MAHE in the French dep. of Finistere.
GODERA, GEDERN in Hesse-Darmstadt.
GODEREA, GOEREE, an island at the mouth of the Maese in Holland.
GOEMOERIA, GOMOR in Hungary.
GOETTINGA, GOTTINGEN in Germany.
GOFNA, now JIFNA in Syria.
GOGANA, perhaps KONKUN in Persia.
GOLNOVIA, GOLLNOW in the Prussian reg. of Stettin.
GOMADÆORUM INSULÆ, the islands of DARADATA and DOLCOFALLAR off the coast of Barbary.
GOMERIA, GOMER in Hungary.
GOMESIANORUM CONVENTUS, the district of CONCHES in the Swiss cant. of Valais.
GOMPHI, a town in Thessaly, now STAGI; or according to Kruse, KALABAKI; according to Cramer, STAGOUS; and by Leake identified with EPISCOPI.
GONNI, or GONUSA, GONIGA in Thessaly.
GOPPINGA, GOPPINGEN in Austria.
GORCOMIUM, GORKUM in S. Holland.
GORDENIA, or GORTYNIA, according to Kruse GRADISCA in Macedonia.
GORDIANI MONUMENTUM, perhaps ZOXO-SULTAN in Mesopotamia.
GORDITANUM PROMONTORIUM, according to Cluvier, CAPO-DI-MONTE-FALCONE; according to others, CAPO-DI-ASMARA in Sardinia.
GORDYÆI MONTES, a portion of MONS MASIUS in Kurdistan.
GORGADES INSULÆ, the BISSAGOS on the W coast of Africa.
GORGO, URGENZ in Turkistan.
GORGUS, DIALA in Mesopotamia.
GORITIA, GORZ in Illyria.
GORNACUM, GOURNEY in Normandy.
GORSENIUM, GORZAGNA in Piedmont.
GORTYNA, or GORTYS, according to Kruse, ATELUCELO in Arcadia.
GOSAN, the KISIL-OSEN in Persia.
GOTHLANDIA, the island of GOTHLAND in the Baltic.
GOTHIA, or GATHIA, GOTHLAND in Sweden.
GOTHOBURGUM, GOTHENBURG in Sweden.
GOTTORPIA, or GOTTORPIUM, GOTTORP in Schleswig.
GRABOVIA, GRABOW in Mecklenburg.
GRACA, GRASSÉ in Provence.
GRACCURIS, CORELLA in Navarre.
GRADACIUS, MONTE-GRADACCIO in Corsica.
GRADICUM, GRAY in Franche-Comte.
GRADIUM REGINÆ, or KONIGRATTZ, GRADECZ in Bohemia.
GRADUS, GRADO, an island in the Adriatic.
GRÆA, HOHENKRAHN in Baden.
GRÆCIA, or HELLAS, excluding Macedonia and Epirus, a region of Europe, bounded in ancient times on the N by *Macedonia;* on the E by the *Ægeum Mare* or Ægean sea; on the S by the *Mare Mediterraneum* or Mediterranean; on the W by the *Ionium Mare* or Ionian sea; and on the NW by *Epirus.*—Its main divisions were GRÆCIA PROPRIA and PELOPONNESUS; nearly corresponding to the NORTHERN GREECE and the MOREA of the present day.
GRÆCIA MAGNA, a name given by the Romans to the whole of SOUTHERN ITALY, from the river *Silarus* on the W, to the *Frento* on the E.
GRAECIUM, GREIS in Jutland.—Also GRATZ in Styria.
GRAMMATUM, GRANDVILLARS in the French dep. of Haut-Rhin.
GRAMMONTIUM, GRAMMONT in the French dep. of Basses-Pyrenees.
GRAMPIUS MONS, the GRAMPIAN MOUNTAINS in Scotland.
GRANADA, GRANATA, or GRANATUM, the city of GRANADA in Spain.
GRANCEGUM CASTRUM, GROMCEY-EN-MONTAGNE in the French dep. of Cote-d'Or.
GRANDIMONTUM, GRAMONT in the French dep. of Gers.
GRANDIPARTUM, GRANDPRE in the French dep. of Ardennes.
GRANDISONUM, GROMSON in the Swiss cant. of Vaud.
GRANDISVILLA, GRANVILLE in the French dep. of Manche.
GRANDIVALLIS, GRANFELT in Switzerland.
GRANIANUM, CAPO-D'ERBICARIA in Corsica.
GRANICUS, the river GUSTEVOLA or OUSVOLA in Mysia; according to others, the SUSUGHIRLI; and according to Diez, the GRANIFARA or GRANIQUE.
GRANIS, the river DARABYE or BOSHAVIR in Persia.
GRANNONA, PORT-EN-BESSIN in Normandy.
GRANUA, the river GRAN an affluent of the Danube.
GRATIÆ MONS, the GNADENBERG in Bavaria.
GRATIANOPOLIS, GRENOBLE in France
GRATIARUM VALLIS, the GNADENTHAL in Switzerland.
GRATICULA, ST. LORENZO in the Papal deleg. of Viterbo.
GRAUDENTIUM, GRAUDENZ in Prussia.
GRAVELINGIA, GRAVELINES in the French dep. of Nord.
GRAVESCENDA, GRAVESEND in Kent.
GRAVIA, GRAVE in the Netherlands.
GRAVIACI, MURAN in Hungary.
GRAVISCA, EREMO-DI-ST.-AUGUSTINO in Etruria.
GREGORII ST. MONASTERIUM, GREGORSMINSTER in Alsace.
GRIGNIACUM, GRIGNAN in Dauphiny.
GRIMUS, GRIM in Prussian Stralsund.
GRINARIA, GRUNINGEN in Würtemberg.
GRINNES, SAINT ANDRIES in the Netherlands.
GRIPESWOLDA, GRIEFSWALDE in Prussian Stralsund.
GRISELUM, GREOULX.
GRISONIA, GRANBUNDEN in Switzerland.
GRISSOVIUM, GRISSAU in Prussian Silesia.
GROENINGA, GRONINGEN in the Netherlands.
GRONAICUM, GREENWICH in Kent.
GROTGOVIA, GROTTKAN in Prussian Silesia.
GRUARII PORTUS, PORTO-GRUARO in Venice.
GRUBENHAGENSIS PRINCIPATUS, the principality of GRUBENHAGEN or GOTHINGEN in Hanover.
GRUERIA, GRIERS or GRUYERE in Switzerland.
GRUMENTUM, AGRIMONTE in Naples, or near SAPONERA.
GRUMUM, GRUMO near Palo in the Papal States.
GRUNUM, the Swiss town GRON.
GRYNEUM, GLISSELIK in Greece.
GRYPHÆUM, GREIFEN-SEE in the Swiss cant. of Zurich.
GRYPHÆUS LACUS, the GREIFEN-SEE in the cant. of Zurich.
GRYPHIBERGA, GREIFENBERG in Prussian Stettin.

GRYPHIMONTIUM, GREIFENBERG in Prussian Silesia.
GRYZELIUM, GREOULS in Narbonne.
GUADELA, GOADEL, KHEVADIR in Gedrosia in the Indian sea.
GUALACRA, WALCHEREN in Holland.
GUARDISTALLUM, GUASTALLO in Modena.
GUELDRIA, GELDERN in the grand-duchy of the Rhine.
GUELFERBYTUM, WOLFENBUTTEL in Brunswick.
GUERCHIA, GUIERCHE in the French dep. of Ille-et-Vilaine.
GUERICA, GUIERCHE in Touraine.
GUESTA, WESEN in the Swiss cant of St. Gall.
GUESTFALIA, WESTPHALIA.
GUILIELMOSTADIUM, WILHELMSTADT in N. Brabant.
GUIMARANUM, GUIMARANES in Portugal.
GUINÆ, GUINES in the French dep. of Calais.
GUINEA NOVA, NEW GUINEA in the Pacific.
GUISIA, or GUSGIA, GUISE in Picardy.
GULIA, GEULE in W. Flanders.
GULPIA, GULPE in Limburg.
GUNDOLFI-VILLA, GONDREVILLE in the French dep. of Meurthe.
GUNDULPHI CURIA, GONDRECOURT in the French dep. of Meuse.
GUNTIA, the river GUNS in Bavaria.—Also GUNZBURG in Bavaria.
GURGES ATLANTICUS, the ATLANTIC.
GURGES CARPATHIUS, the ÆGEAN SEA.
GUSA, or GUSIA, GOES, an island in S. Holland.
GUTTALUS, the river PREGEL in E. Prussia.
GYARA, or GYARUS JOURA, GHIURA in the group of the Cyclades.
GYGÆA PALUS, LAKE MERMERE in Asia Minor.
GYGANEUM, GUGNIE in Colchis.
GYMNIAS, COMASAR, KUMAKIE in Armenia.
GYNÆCOPOLIS, SELAMON in Lower Egypt.
GYNÆPEDIUM, FRAUENFELD in the Swiss cant. of Thurgau.
GYNDES, the river KARA-SU or SALEH in Assyria.
GYRIENSE CŒNOBIUM, GYRACH in Syria.
GYRTON, TARCHI-VOLICATI in Thessaly.
GYTHANÆ, DELFINO in Upper Albania.
GYTHEUM, PALEOPOLIS in Greece.
GYTHIUS, according to some, KOLO-KYTIA in Northern Greece.

H

HA-ARABAH, EBENE, AULON, EL-GHOR in Palestine.
HABITACULUM MARIÆ, MARIBOE in Denmark.
HABOLA, the HAVEL, a river in Prussian Brandenburg.
HABUS, the FRISCHE-HAFF in Prussia.
HADELIA, HADELN in Bremen.
HADEMARA, HADAMAR in Nassau.
HADINA, HADDINGTON in East Lothian.
HADRIA, or HATRIA, ADRIA in Lombardy.
HADRIANE, BENGAZI in Tripoli.
HADRIANI MONS, ENGELSBURG, a citadel in Rome.
HADRIANOPOLIS, in Macedonia, now TONJA.
HADRIANOPOLIS BITHYNIÆ, BOLI in Asiatic Turkey.
HADRIATICUM MARE, the ADRIATIC, comprising both the Gulf of Venice and the Ionian sea.
HADRUMETUM, NABAL in Barbary.
HÆDICOLLIS, KITZBIEHL in Tyrol.
HÆMI EXTREMA, the EMINEH-DAGH in Turkey.
HÆMUS MONS, the BALKAN in Turkey.
HAGA, the HAGUE in Holland.
HAGA AURELIANENSIS, LA HAYE in the Orleanois.
HAGANOA, GROSSENHAYN, HAYN in Saxony.
HAGENOA, HAGENAU in the French dep. of Bas-Rhin.
HAINANA, the Chinese island of HAINAN or FORMOSA.
HAINORIA, HAINAU, HEYNAU in Prussian Silesia.
HALA, HALA MAGDEBURGICA, or HALA SALICA, HALLE in Prussian Saxony.
HALA-AD-ŒNUM, HALL in Tyrol.
HALA SUEVICA, HALLA SUEVORUM, HALL in Würtemberg.
HALÆ, according to Kruse, ALACHI in Greece.
HALANDIA, the district of HALMSTADT in Sweden.
HALBERSTADIA, HALBERSTADT in Prussian Saxony.
HALESA, TUSA in Sicily.
HALESUS, the river HALYS in Asiatic Turkey.
HALEX, the river ALECE in Calabria Ultra.
HALIACMON, the JUJE-KARA-SU, a river of Macedonia.
HALIARTUS, MAZI, or perhaps TRITONNI, in Livadia.
HALICA, HALIEIS, according to Kruse, CHILADIA in Argolis.
HALICANUM, TSCHAKATURA.
HALICIA, HALICZ in Galicia, Austria.
HALICAMASSUS, BODRUN in Asiatic Turkey.
HALIFACIUM, or HORTONIUM, HALIFAX in York.
HALIOLA, or HALLULA, HALLE in Austria.
HALIUSA, CARAVI in the Ægean sea.
HALMOSTADIUM, HALMSTADT in Sweden.
HALMYDESSUS, MIDJEH on the coast of the Black sea.
HALMYRIS, the CARASU, a river of Bulgaria.
HALONÆ, SALBAZAR in Ionia.
HALONE, ALONI, an island in the sea of Marmora.
HALONESUS, or DROMI, the island of KHELIDROMI in the Ægean sea.
HALUNTIUM, PHILADELFO in the Val-di-Demona, Sicily.
HALYCUS, the river PLATANI in Sicily.
HALYS, the river KIZIL-IRMAK in Asiatic Turkey.
HALYZEA, CANDILI, ALCIPO in Greece.
HAMAXITUS, KURAN-KEVI in Asia Minor.
HAMBURGUM, or HAMMONIA, HAMBURG in Hanover.
HAMELA, HAMELN in Hanover.
HAMETA, HAM in Picardy.
HAMILTONIUM, HAMILTON in Lanark.
HAMMARIA, HAMMER in Norway.
HAMMONA, HAMM in Westphalia.
HAMONS, HAMONT in Limburg.
HAMPTONI CURIA, HAMPTON-COURT in Middlesex.
HANES, THENNIS.
HANNEBOTUM, HANEBONT, HENNEBON in Brittan .
HANNIBALIS PORTUS, near ALBOR in Algarve.
HANNIBALIS SCALÆ, LES-COSTES-DE-CARRAF, a mountain in N. Spain.
HANNIBALIS TURRIS, MAHADIAH in Tunis.
HANNONIS INSULA, MADAGASCAR.
HANNUVIUM, HANNUT, HANNUYE in the Netherlands (in Liege).
HANOVIA, HANAU in Germany.
HAPSELIA, HABSAL in the Russian gov. of Esthonia.
HARÆ, according to D'Anville, JARECCA in Syria.
HARBURGUM, HAARBURG in Hanover.
HARCURTIUM, HARCOURT in Normandy.
HARISTALLE, HERSTALL in Liege.
HARISTELLUM, HERSTELLE in Germany.
HARLEMUM, HAARLEM in Holland.

HARLINGA, Harlingen in Friesland.
HARMAZIA, Ormus in the Persian gulf.
HARMENE. See Armene.
HARMOZON PROMONTORIUM, Cape Kuhestek on the Persian gulf.
HARPASA, Arabhissar in Asiatic Turkey.
HARPASUS, the Jina, a river in Asiatic Turkey.—Also the Harpasu a branch of the Aras.
HARPESSUS, the Arda, a river in Thrace.
HARPIS, Adshud in Bulgaria; according to others, Licostomo.
HARPONIUM, Cerchiera in Calabria.
HARSEFELDUM, Harsefeld in Bremen.
HARTFORDIA. See Areconium.
HARTFORDIENSIS COMITATUS, Hertfordshire.
HARTIANA, or Nigra Sylva, the Schwarz-wald in Germany.
HARUDORUM PAGUS, Halberstadt.
HARVIACUM, Harwich in Essex.
HASA, the river Hase in Osnaburg, Hanover.
HASELIA, or Haselina Vallis, the valley of Hasli in the canton of Bern.
HASICON, Hasik in Yemen in Arabia Felix.
HASSELETUM, Hasselt in the Dutch prov. of Overyssel.
HASSELETUM-AD-DEMERAM, Hasselt in Limburg.
HASSIA, Hessen in Germany.
HASTIA, Voltri in Tuscany.
HATRÆ, or Atræ, Atreni in Mesopotamia.
HATTEMIUM, Hattem in the Dutch prov. of Gelderland.
HATTONIS CASTRUM, Hattonchatel in Belgium.
HATUANUM, Hatvan in Hungary.
HAUGASTALDIUM, or Sancta Insula, Holy Island off the coast of Northumberland.
HAVREA, Havre in Belgium.
HAZOR, Kassr Antar in Syria.
HEBRIDES. See Ebudæ.
HEBROMAGUS, Brane in Languedoc.
HEBRUS, the Maritza, a river of Romelia.
HECATONNESI, Musconisi, Miosconisi, a group of islands NE of Mitylene.
HEDDEMHENUUM, Heddemhein in Nassau.
HEDENA, Hedin or Hesdin in the French dep. of Pas de Calais.
HEDYPHON, or Hedypnus, a river of Susiana, by some identified with the modern Jeraki.
HEGENENSE MONASTERIUM, Hunningen in Leiningen.
HEGETMATIA, Massel in Prussian Silesia.
HEGONIUM PROMONTORIUM, Cape Karaburon on the Black sea.
HEGOVIA, Hegau on the Boden-see.
HEIDEBA, or Slesvicum, the Duchy of Schleswick in Denmark.
HEIDELBERGA, Heidelberg.
HEIGERA, Hager in Nassau.
HEILBRONNA, Heilbronn in Nassau.
HEILIGENSTADIUM, Sanctorum Urbs, Heiligenstadt in Prussian Saxony.
HEIMBODESHEIM, Heimsen in Würtemberg.
HEINSILIANUS MONS, the Heinzenberg in the Swiss cant. of the Grisons.
HELAUSE, or Eleusa, an island off the coast of Asiatic Turkey; White according to Kruse.
HELCIPOLIS, Commotan in Bohemia.
HELELLUM, or Helvetum, Shlettstadt in the French dep. of Bas-Rhin.
HELENA, an island of Greece, now Macronisi.—Also Elne in the French dep. of Pyrenees-Orientales.
HELENOPOLIS. See Drepane.
HELES, or Elees, the river Alento in Italy.
HELIA, or Elia, Ely in Cambridge.
HELICE, Trypia in Achaia; according to others, Niora.
HELICON, Mount Palæovuni or Zagora in Greece.—Also the Baphyra, a river of Bœotia.—Also the Olivero in Sicily.
HELIOPOLIS, Matarieh in Lower Egypt.—Also Baalbec in Syria.
HELISSON, the river Leondari in the Morea.
HELIUM, the mouths of the Meuse and Waal, Brihel and Helfoek.
HELLA, according to Kruse, Castritza in Albania.
HELLAS, a name given by the ancients not only to Greece, but also to all the regions colonized by the Hellenes in Asia Minor, Africa, Italy, and Sicily.
HELLESPONTUS, the Straits of the Dardanelles, Stambul-Denghiz.
HELLOMENUM, now Climeno in Acarnania.
HELLOS, according to Kruse, Helles in Laconia; according to others, Tsili.
HELMONTIUM, Helmond in N. Brabant.
HELMSTADIUM, Helmstedt in Brunswick.
HELORUM, Muri-Ucci in Sicily.
HELORUS, the Atellaro, Acellaro, or Abisso, a river in Sicily.
HELSINGA, Helsingborg in Sweden.
HELSINGFORDIA, Helsingfors in Russian Finland.
HELSINGIA, Helsingland in Norway.
HELVETIA, Switzerland.
HELVETIUM, Helvaux or Hevaux in the French dep. of Haute-Vienne.
HELVILLUM, Sigello in the Papal deleg. of Urbino.
HELVINUS, the Salinello, a river in Abruzzo.
HEMEROSCOPIUM, Denia in Valencia.—Also Cape St. Martin.
HEMIPYRGUM, Halb-Thurn in Hungary.
HENA, Anah in El-Jezireh in the pash. of Bagdad.
HENIUS, the river Haisne in the French dep. of Nord.
HENNIACUM LITARDI, Henin-Lietard in the French dep. of Pas-de-Calais.
HENRICI PAGUS, Einrich in Wiesbaden.
HEPHA, or Kaifa, Caffa on the Syrian coast.
HEPHÆSTIA, Sotiri in Lemnos; according to others, Cochino.
HEPTANOMIA, or Heptanomis, Central Egypt, or Vostani.
HERACLEA, Felicuda in Sicily.—Also Raclia, an island in the Archipelago.
HERACLEA ÆOLIDIS, according to Kruse, Kidonia in Asia Minor.
HERACLEA CACCABARIA, Fanum St. Eutropii.
HERACLEA CASSIOTIDIS, Meinta Posseda in Syria.
HERACLEA LUCANIÆ, Policoro in the prov. of Basilicata in Naples.
HERACLEA MEDIÆ, Burugkerd in Armenia.
HERACLEA MINOA, Torra-di-Capo-Bianco, or Castel-Bianco in Sicily.
HERACLEA PELAGONIÆ, according to Kruse, Perlipe in Macedonia.
HERACLEA PONTICA, Harakie or Eregri in Anadolia; according to others, Penderachi.
HERACLEA SENTICA, Iscar or Zenorva in Macedonia.
HERACLEA THRACIÆ, Heraklitza on the sea of Marmora.
HERACLEOPOLIS MAGNA, Ahnas in Lower Egypt.
HERACLEOPOLIS PARVA, Delbom in Lower Egypt.

HERACLEOTES, BAZILUZZO, an island near Sicily.
HERACLEUM, CAPE TSHERSHEMBI on the coast of Asiatic Turkey.—Also CANDIA in Crete; but according to some MIRABELLO, or SPINALONGA.—Also ARABAT in Russian Taurida.
HERACLIUS, the river BULLA in Livadia.—According to others, the HERACE.
HERÆA, according to Kruse, AGIANI or AIANI in the Morea; according to others, IRI.
HERÆI MONTES, MONTI-SORI in Sicily.
HERBANUM, URBIBENTUM, or URBEVETUM, ORVIETO in the Papal deleg. of Viterbo.
HERBERIA, or RUBERIA, RUBIERA in Modena.
HERBITA, NICOSIA in the Val-di-Demona, Sicily.
HERCULANEUM, or HERCULANUM, a buried city in the vicinity of PORTICI in Naples.
HERCULEUM, or ERKELENS, HERKELENS in Prussia.
HERCULIS CASTRA, HERVELT.
HERCULIS FANUM, MASSA-CARRARA in Tuscany.—Also CASTILLO in Andalusia.—Also MARZASTROCCO in Malta.
HERCULIS INSULÆ, ZAVARA and ASINARA, islands near Sardinia.
HERCULIS LUCUS, the SUNTELGEBIRG, near Minden, in Germany.
HERCULIS MONOECI, MONACO or MOURGUES in Piedmont.
HERCULIS PORTUS, TROPÆA in Calabria-Ultra.
HERCULIS PROMONTORIUM, HARTLAND-POINT in Devonshire.—Also CAPE SPARTIVENTO in Calabria-Ultra.—Also CAPE GIR in Marocco.
HERCULIS TURRIS, LA-TORRE-DE-CAMIRA in Barbary.
HERCYNIA SYLVA, H. SALTUS, H. LUCUS, MARCIANA SYLVA, ABNOBA MONS, SEMANA SYLVA, MELIBOCUS MONS, SUDETI MONTES, SARMATICI MONTES, a name embracing the modern ODEN-WALD, the SCHWARZ-WALD, the RHONG-EBIRG, the THURINGER-WALD, the HARZ, and the BOHMER-WALD in Baden.
HERDALIA, HERGEDALEN in Swedish Gefleborg.
HERDONIA, ORDONA in the Capitanata of Naples.
HEREFORDIA, HEREFORD in England.
HERESFELDA, HERSFELD in Germany.
HERFORDIA, HERFORD or HERVORDEN in Westphalia.
HERIBODESHEIM, HERBITZHEIM in Lothringen.
HERICURIA, HERICOURT in the French dep. of Haute-Saone.
HERICUS, the island of HEIS in the French dep. of Vendee.
HERINUM, REINA in Calabria.
HERISLÆ NOVÆ, NEIN-HERSE in Westphalia.
HERISPICH, HEERWEN in Gelderland.
HERIUS, the RIVIERE-D'AURAY in Brittany; or, according to d'Anville, the VILAINE.
HERIVALLIS, the abbey of HERIVAL in Lothringia.
HERLEFSHOLMIA, HERLEFSHOLM in Denmark.
HERMÆA EXTERNA, RIPA ALBA, a cape on the coast of Barbary.
HERMÆUM PROMONTORIUM, PONTA-DI-TRIPITI, a cape in Crete.—Also CAPE BON in Tunis.
HERMINIUS, or ARMINNO MONS, SIERRA-DE-LA-ESTRELLA in Beira, Portugal.
HERMIONE, CASTRI; according to others, MARIA in the Morea.
HERMONASSA, PLATANA in Pontus.—Also MATRIGA in Halbinsel.
HERMONTHIS, ERMENT in Upper Egypt.
HERMOPOLIS MAGNA, ACHMUNEIN in Central Egypt.
HERMOPOLIS PARVA, DEMENHUR in Lower Egypt.
HERMOPOLITANA CUSTODIA, BAHR-JUSEF in Central Egypt.
HERMUS, a river in Asia Minor, now the SARABAT; according to d'Anville, the REDUS.
HERNADUS, the river HERNATH in Upper Hungary.
HEROCAMPIA, or HEROFELDA, HERSFELD in Hesse-Cassel.
HEROOPILIS, a town in Lower Egypt, now ABUKEHEID.
HEROOPOLITINUS SINUS, or SINUS ALANITES, the GULF OF SUEZ, or BAHR-EL-KOLSUM.
HESIDRUS, the river KEHKER in India.
HESPERIA, SPAIN.
HESPERIUM PROMONTORIUM, CAPE VERD.
HESTRACA, OREO in Negropont.
HESYDRIUS, the river SETLEDJE or SUTLUIJ in India.
HETHLANDIA, the MAINLAND in Shetland.
HETHLANDICÆ INSULÆ, the SHETLAND ISLANDS.
HETRICULUM, LATARACO in Calabria.
HETRURIA MAGNUS DUCATUS, TUSCANY.
HEUDENA, HEUSDEN in N. Brabant.
HEVESIA, HEVES-SZOLNOK in Hungary.
HIBERNA REGIA, KONIGSWINTER in Prussia.
HIBERNIA, IBERNIA or SCOTIA MAJOR, IRELAND.
HIBERNICUM MARE, the IRISH SEA.
HICESIA, PANARI, one of the Lipari islands.
HIENIPA, ALCALA-DE-GUADAIRA in Andalusia.
HIERA, MARETIMO, an island near Sicily.—Also VOLCANO, one of the Lipari islands, or CAMENI, one of the Cyclades.—Also PORTO-GERO in Lesbos.
HIERA, or GERME, GHERMASTI in Mysia.
HIERACIUM, GIERACI in Calabria-Ultra.
HIERACUM INSULA, ISOLA-DI-ST.-PIETRO near Sardinia.
HIERAPOLIS, BAMBUK-KALESSI, or BAMBURKESSER in Anadolia.
HIERASUS, the river PRUTH between Turkey and Russia.
HIERATIS, DEKIAN or KIERASIN, a river in Persia.
HIERICUS, JERICHO, or PALMARUM CIVITAS, now RIBAH in Palestine.
HIEROCÆSAREA, SEMEH in Anadolia.
HIEROMIACE, or HIEROMAX, a river in Palestine, the YARMUK, or SHERIAT-EL-MANDUR.
HIERON PROMONTORIUM, CAPE CARNSORE in Leinster.
HIERON OROS, MOUNT TEKMAN in Pontus.
HIEROSOLYMA, JEBUS, AELIA CAPITOLIA, DAVIDIS URBS, SALEM, JERUSALEM in Palestine.
HIERSPERGA, HIRSPERG in Bavaria.
HIERUS, the river ORBO in Corsica.
HILARA, the river KLER in Bavaria.
HILARIA, the abbey of WILLERING.
HILICANUM, SZERDAHELY in Hungary.
HILPERSHUSIA, HILDBURGHAUSEN in Meiningen, Germany.
HIMELLA, the river AJA in Sabina.
HIMERA, near TERMINI in Sicily.
HIMERA, the FIUME-GRANDE in Sicily.
HIMERENSES THERMÆ, TERMINI in Sicily.
HIOVIA, HJO in Swedish Westgothland.
HIPPAGRETA, HIPPO ZARYTUS in Numidia.
HIPPARENUM, SEPHARVAIM, BARSITA or BORSIPPA in Mesopotamia.
HIPPARIS, CARINA in Sicily.
HIPPO, BIVONA in Calabria-Ultra.
HIPPOCURA, or BAGNAGUR, GOLCONDA in India.
HIPPO DIARRHYTUS, BISERTA in Tunis.
HIPPOLA, MAINA in Laconia.
HIPPONIATES SINUS, GOLFO-DI-SAN-EUFEMIO on the coast of Genoa.

HIPPONIUM, MONTE-LEONE.
HIPPO REGIS, near BONA in Algeria.
HIPPORUM, FELO in Calabria.
HIPPOS, SZAMMERA in Palestine.
HIPPURIS, FIDULSA, an island of the Cyclades.
HIRA, MEZJED-ALI or IMAN-ALI in the pash. of Bagdad.
HIRMINIUS, the river MAULO or FIUME-DI-RAGUSA in Sicily.
HIRSANGIA, HIRSAN or HIRSCHAN in Würtemberg.
HIRSCHAVIA, HIRSCHAU in Bavaria.
HISORIS, IBSON in Upper Egypt.
HISPALIS AUREA, SEVILLE in Spain.
HISPANIA, SPAIN and PORTUGAL.
HISPANIA BŒTICA, a division of Hispania Ulterior, comprising the modern ANDALUSIA, part of the Portuguese prov. of ALENTEJO, that part of Spanish ESTREMADURA which lies to the S of Anas, and a large part of LA MANCHA.
HISPANIA LUSITANIA, a region in Roman geography embracing PORTUGAL, the NW of ESTREMADURA, the S of LEON, and the SW of OLD CASTILE.
HISPANIA NOVA, MEXICO.
HISPANIENSIS OCEANUS, the IBERIAN SEA.
HISPANIOLA, the island of ST. DOMINGO or HAYTI.
HISTIÆA, near OREOS in Negroponte.
HISTONIUM, VASTO-D'AMMONE in Abruzzo-Citra.
HISTRIA, HISTRI in Trieste.
HISTRIOPOLIS, CHIUSTANGE or PROSCHLOVIZA in Bulgaria; or, according to others, KARAHIRMEN.
HITONA, AITONA in Spanish Catalonia.
HIZGERA, HITZACKER in Hanover.
HLIDBEKI, LIDBACH in Westphalia.
HLIUNI, LUENE in Hanover.
HOCSEBURGUM, HOCHSEEBURG or SEEBURG in Prussian Saxony.
HODINGÆ, ALT-OETTINGEN in Bavaria.
HOECHSTA, HOCHST in Nassau.
HOHBUOCHI, BUCHEN, a Danish village.
HOINBORCHUM, the convent of ODILIENBERG in the French dep. of Bas-Rhin.
HOLA, HOLAR in Iceland.
HOLBECA, HOLBECK in the island of Zieland.
HOLDISTEDE, HOLLINSTED in Hanover.
HOLDSTEBROA, HOLSTEBROE in Jutland.
HOLIS, HOLEN in Iceland.
HOLMIA, STOCKHOLM in Sweden.
HOMELEA, the river HUMBLE in Hantshire.
HOMOLIUM, near FTERI in Thessaly.
HOMONA, or HOMMADA, ERMENAK in Asia Minor.
HONFLEVIUS, HONFLEUR in Normandy.
HONOSCA, JOYOSA in Valencia.
HORDEANI, URDINGEN in the Prussian prov. of the Rhine.
HOREB, now SINAI, a mountain in Arabia Petræa.
HORNA, HOORN, a town in N. Holland.
HORNANUM CAPUT, CAPE HORN in S. America.
HORREA MARGI, MORAVA-HISSAR in Servia.
HORRISONUS MONS, the HORSELBERG in Germany.
HORSNESIA, HORSENS in Jutland.
HORTA, ORTA in the Papal deleg. of Viterbo.
HORTUS FLORIDUS, the ancient abbey of BAINDT in Schwaben.
HOSEMUM, HUSUM in Schleswig.
HOSPITELLUM, SOSPELLO in Sardinia.
HOSTA, the river OSTE in Hanover.
HOSTILIA, OSTIGLIO in Lombardy.
HOSTUNUM, OSTUNI in Neap. Terra-d'Otranto.
HUCULBI, PETERSHAGEN in Westphalia.
HUDWICKSOVALDUM, HUDICKSWALL in Helsingland.
HUENA, WEEN, a Swedish island.
HUGONIS CURIA, HUGSHOFEN in Alsace.
HULTONIA, ULSTER in Ireland.
HUNINGA, HUNINGEN in the French dep. of Haut-Rhin.
HUNNICURIA, the town and abbey of HONNECOURT in Picardy.
HUNNICUS PAGUS, HUNDSRUCK in Bavaria.
HUNNOBRODA, BROD in Moravia in Austria.
HUNNUM, SCAVEN-SHALE in Northumberland.
HUNYADENSIS COMITATUS, HUNYAD in Hungary.
HUUM, HUYEMUM, HUY, HOYE in Liege in the Netherlands.
HUXARIA, HOXTER, HUXTOR in Prussian Westphalia.
HYALA, LAHEVI in India.
HYAMPEA, NAUPLIA or NAPOLI-DI-ROMANIA in the Morea.
HYAMPOLIS, according to Kruse, BOGDANA in Greece.
HYAROTIS, the river IRRAWADY in India beyond the Ganges.
HYBLA MAJOR, PATERNO in the Val-di-Demona, Sicily.
HYBLA MINOR, RAGUSA in the Val-di-Noto, Sicily.
HYCCARA, BICCARI, in the Val-di-Mazzara, Sicily.
HYDASPES, the river BEHAT or JELUM in India.
HYDRÆA, or HYDREA, the island and town of HYDRA in the Ægean sea.
HYDRALIS, BELGRADUS, BELGRADO in Rumelia.
HYDRAOTES, the RAVI in the Punjab.
HYDRIACES, the river GATTER in Persia.
HYDROPOLIS, FEUCHTWANG in Bavaria.
HYDRUNTUM, or HYDRUS, OTRANTO in Naples.
HYELE, or VELIA, near CASTELLAMARE-DELLA-BRUCA in Naples.
HYGRIS, the Russian river DONETZ in the gov. of Charkof.
HYLA, YLA, YIL, ISLA in the Hebrides of Scotland.
HYLÆA, perhaps the NOGAI STEPPES in the S of Russia.
HYLIAS, a river in Calabria, now the AQUANILE.
HYLICE (LACUS), now LIMNE-STIVA or SENZINA, a lake in Livadia.
HYMETTUS, MONTE IMETTO or MATIO in Livadia.
HYPACARIS, the river DESNA.
HYPÆA, the ISLE-DU-TITAN off the S coast of France.
HYPÆPA, BIRGHE, BERKI, TOPOÏ, or IPEPA ruins in Anadolia.
HYPANIA, ACCOMBA in the Morea.
HYPANIS, the river BOG or BOUG in the S of Russia.—Also the KUBAN in Taurida.—Also the BEYPASHA in India.
HYPATA, NEOPATRA, CASTRITZA in Thessaly.
HYPER LABACUM, UPPER LAIBACH in Crain.
HYPERBOREI MONTES, the URAL MOUNTAINS.
HYPERGRÆCIA, OBERKIRCH in Germany.
HYPEROSIA, BLUBUKI in Greece.
HYPIA, the LIPPO, a river in Asiatic Turkey.
HYPIUS, the river SKENI in Asia Minor.
HYPSA, the DRAGO, or more likely the BELICE in Valdi Mazzara, Sicily.
HYPSELE, now SIUT in Upper Egypt.
HYRCANIA, SEBSWAR in Khorasan; MARMARA in Asiatic Turkey.
HYRGIS, according to some the, SEVIERSKY.
HYSSUS, a river in Colchis, now the HORCHID.

I

IADERA, ZARA in Illyria.
IALYSUS, IALISO in Rhodes.
IAPYGIA, the TERRA-D'OTRANTO in Naples.

IAPYGIUM PROMONTORIUM, CAPO-DI-LEUCA in Naples.
IASONIUM PROMONTORIUM, CAPE JASAN in Asiatic Turkey.
IASSAS, HASSAN-KALESI in Asiatic Turkey.
IASSIUS SINUS, the GULF OF ALESSI.
IATRAS, the river IANTRA in Asiatic Turkey.
IBERA, perhaps TORTOSA in Spain.
IBERIA, SPAIN.
IBERIA, GEORGIA or GRUSIA and IMIRETIA in Asiatic Turkey.
IBERIUM, IVRY in Normandy.
IBERNIUM, BERE-REGIS in Dorsetshire.
IBERUS, the river EBRO in Spain.—Also the JINTO in Seville.
IBLIODURUM, BEAUVILLE; according to others. CONFLANS-EN-JARNISI in the French dep. of Moselle.
IBOLA, the river YBOL or YPOL in Hungary.
IBURINGA, UEBERLINGEN in Baden.
ICARIA, NIKARIA or ACHIKRIA, an island of the Cyclades.
ICARIUM, an island in the Persian gulf, the modern CARAI or GARGE.
ICAUNA, or YTUMNA, the river YONNE, an affluent of the French Seine.
ICARUS, PELUDJI, an island in the Persian gulf.
ICCIUS PORTUS, or ITIUS PORTUS, or WISANTI PORTUS, WISSAND or WISSAN in the French dep. of Pas-de-Calais, near Cape Griz-Nez.
ICENORUM OPPIDUM, IKSWORTH in Suffolk.
ICHNUSA, the island of SARDINIA.
ICHTHYS PROMONTORIUM, CAPE ZANCHI or CATACOLO in the Morea; according to others, CAPE TORNESE.
ICIANI, perhaps ICHOROW in Norfolk, Chesterford.
ICIDMAGUS, YSSENGEAUX in the French dep. of Haute-Loire.
ICINIACUM, according to Kruse, LECH-GEMUND in Bavaria.
ICIODURUM, ISSOIRE in Auvergne.
ICIUM PROMONTORIUM, CAPE GRIS-NEZ in the French dep. of Pas-de-Calais.
ICONIUM, COGNI, KONIEH in Asiatic Turkey.
ICOSIUM, ORAN in Algeria.
ICTIMULI, VICTIMOLO in Piedmont.
ICTIUM CASTRUM, L'ILE JOURDAIN in Gascogny.
ICTODURUM, AVANCON or GUTENBURG.
ICUS, the island of SKIPHE or SCYRO PULO in the Archipelago.
IDA, PSILORITI or MONTE GIOVE in Crete.—Also the KAZ-DAGHI in Asia Minor.
IDALIUM, DALIN in Cyprus.
IDEX, the river L'IDICE in Bologna.
IDICARA, HIT or HAITA in the pash. of Bagdad.
IDOMENE, IDOMENI in Macedonia.
IDONIA, VINCA, the river HUISNE in the French dep. of Sarthe.
IDRINUM, IDRO in the cant. of Bresciano in Lombardy.
IDUBEDA MONS, the SIERRA-D'OCA or SIERRA-SEGMA in Old Castile.
IDUMANIA, or SIDUMANIS, the river CHELMER or BLACKWATER in Essex.
IDUMÆA, EDOM and a portion of SOUTHERN PALESTINE.
IDUNUM, JUDENBURGUM in Hungary.
IERNE. See HIBERNIA.
IERNIS, DUNCANNON in Ireland.
IGÆDITA, IDANHA-VELHA.
IGENIA, TEGENGILL in Wales.
IGILGILIS, JIGEL, JIGELI or GIGERI on the coast of Barbary.
IGILIUM. See ÆGILIUM.
IGLAVIA, or GIGLOVA, IGLAU in Moravia.
IGLOVIA, or NEOCOMIUM. IGLO or NEUDORF in the Hungarian com. of Hernad.
IGNIS TERRA, TJERRA-DEL-FUEGO in S. America.
IGNIUM INSULA, ILHA-DE-FUEGO.
IGUVIUM, ENGABBIO or GABBIO in the Papal States.
ILA, GILLA a town in Persia.
ILARCURIS, CAROS-DE-LOS-INFANTOS in New Castile.
ILARGUS, the river ILER or ILLER in Bavaria.
ILDUM, SAN-MATTHEO in Valencia; according to others, SALSADELLA.
ILEA, or ILA, WICK in Caithness.
ILEBURGUM, EILENBURG in Prussian Sachsen.
ILEI, ELEO in the Morea.
ILERCAO, or ILLERCO, ALARCON in New Castile.
ILERDA, LERIDA in Catalonia.
ILICIS, ILLICI or ILLICE, ELCHE in Spain.
ILIPA, probably the same as ALCOLA; or perhaps NIEBLA in Spain.
ILIPULA, or ILIPULA MAGNA, NIEBLA in Seville; according to others, VILLANUEVA-DEL-RIO.
ILIPULA MONS, perhaps the SIERRA-NEVADA, or the SIERRA-D'ALHAMILLA in Spain.
ILISSUS, the river ILS or ILZ between Bavaria and Bohemia.—Also the ILISSE in Attica.
ILITURGIS, ANDUJAR-DEL-VEJO in Andalusia; according to others BÆZA.
ILIUM NOVUM, KALAFATLI near the lower castle of the Dardanelles.
ILIUM VETUS, BUNAR-BASHI in Asia Minor.
ILLICE. See ILICIS.
ILLICITANUS SINUS, the BAY OF ALICANTE in Illyria.
ILLYRICUM, comprising ILLYRIA BARBARA and ILLYRIA ROMANA, the country lying between the river *Arsia*, the modern Arsa; the *Savus* or modern Save; the *Drinus* or modern Drino; the Adriatic and the *Aoas*, or Vojussa; a region stretching from the confines of Italy to Epirus.
ILMA, or ILMUS, the river ILM in Germany.
ILMA, ILM or ILMSTADT in Germany.
ILMENAVIA, ILMENAU in Saxe-Weimar.
ILMENUS LACUS, LAKE ILMEN in the Russian gov. of Novgorod.
ILMETUS, the river ILMENT in Sistan.
ILMI MONASTERIUM, ILLMUNSTER in Bavaria.
ILORCI, LORCA in Spanish Murcia.
ILOSTUM, or ILSTA, YLST, or DRIELST in Friesland.
ILURO, PALAMOS, or perhaps ALORA in Catalonia.
IMACHARA, perhaps TRAINA in Sicily.
IMAUS MONS, the MUS-TAG or MUZ-DAGH in Tibet, and the HIMALAYA chain of India.
IMBRUS, the island of EMBRO in the Archipelago.
IMMADRAS, probably the ISLE-DE-MAIR off the French coast.
IMMUNDUS SINUS, FOUL GULF in the Red sea.
IMPATIS MINISTERIUM, OBERHALBSTEIN in Switzerland.
IMPERIUS, the river IMPERIALE in Genoa.
IMUM CASTRUM, FIEFENCASTEL in the Swiss cant. of the Grisons.
INA, the Prussian river IHNA, an affluent of the Oder.
INACHUS, the river NAJO, SPLANISSA, ZERIA, in the Morea.—Also the KRIKELI in Carnia.
INATUS, EIPATOS in Crete.
INCARUS, CARRI or CURI in Provence.
INDIA, a region of Asia which, as known to the ancients, was bounded on the N by *Scythia extra Imaum*, and part of *Serica* or China; on the E by the *Sinæ* or Cambodia, and the Eastern ocean; on the S by the *Mare Erythræum;* and on the W by the *Mare Erythræum* and *Aria*. Ptolemy divided it into *India intra Gangem* or India W of

the Ganges; and *India extra Gangem* or India E of the Ganges. Of the latter region very little was known to the ancients.
INDIA OCCIDENTALIS, the WEST INDIES.
INDIA ORIENTALIS, the EAST INDIES.
INDIBILE, XERTA in Catalonia.
INDI OSTIA, the mouths of the river INDUS in India.
INDRISTA, the river INNERSTE in Hanover.
INDUS, the river INDUS, MEHRA, NILAB, OTTOK, SHINDU, SIND, or SUR in India.
INFERUM MARE, FUSCUM ÆQUOR, or TYRRHENUM MARE, the LIGURIAN SEA.
INFRA PORTUM, UNTER PORTA in Switzerland.
INGELBERGA, or ANGELORUM MONS, the abbey of ENGELBERG in Switzerland.
INGERMANNIA, ST. PETERSBURG.
INGHILINHAIM, INGELHEIM in Germany.
INICERUM, CZERNEK in Hungary.
IN MONTIBUS, ORIS MONS, or MONTI, MUND in Switzerland.
INNERNIUM, or INVERNIUM, INVERNESS in the N of Scotland.
INOLOCZA, or INOULADA, INOWOLODZ in Poland.
INSULA, the river AYE, EAY, or EYE in Suffolk.—Also the YSEL in Tyrol.—Also the L'ISLE in the French dep. of Gironde.—Also the LILLE or RYSSEL in Flanders.
INSULA ALBIGENSIS, the French town ISLE-EN-ALBIGEOIS in the dep. of Tarn.
INSULA ASCENSIONIS, the island of ASCENSION in the Atlantic.
INSULA BRUTTIORUM, ISOLA in Calabria-Ultra.
INSULA BURBONIA, the island of BOURBON.
INSULA CANUM, DOG-ISLAND off the Australian coast.
INSULA CANUM MARINARUM, DOG-ISLAND in the West Indies.
INSULA CATHARINÆ, ST. CATHARINA in Lower Guinea.
INSULA DEI, the island of DIEU, ISLE DIEU, off the coast of France.—Also the ancient convent of HOLME, afterwards BRAHE-FROLLENBORG in Funen.—Also NOIRMONTIERS at the mouth of the Loire in France.
INSULA DIABOLORUM, DROGEO, an island near Jamaica.
INSULA ERRORIS, ALBORAN, an island near Fez in Africa.
INSULA SANCTÆ CATHARINÆ, ST. CATALINA, an island of South America.
INSULA ST. CRUCIS, HEILIGENKREUZ among the Antilles.
INSULA ST. HELENÆ, ST. HELENA in the Atlantic.
INSULA ST. MARGARETHÆ, CSEPEL HASEN in Silesia.
INSULA ST. MARIÆ, the convent of DINDORF in Prussian Saxony.
INSULA ST. ANDREÆ, ST. ANDREAS in Hungary.
INSULA ST. DOMINICI, HISPANIOLA, ST. DOMINGO, or HAYTI in the East Indies.
INSULA ST. GENESII, the island of ST. GENNAIS in Provence.
INSULA ST. GEORGII MAJORIS, ST. GEORGE near Venice.
INSULA ST. JACOBI, the island of ST. JAGO, W of Cape Verd.
INSULA ST. JOANNIS, ST. JEAN in the West Indies.
INSULA ST. JOANNIS PORTUS DIVITIS, PORTO RICO in the West Indies.
INSULA ST. LAURENTIIS, the island of MADAGASCAR.
INSULA ST. STEPHANI, ST. ETIENNE on the coast of Provence.
INSULA ST. THOMÆ, ST. THOMAS on the coast of Guinea.
INSULÆ CUNICULARES, the BORROMEAN ISLANDS in Lake Maggiore.
INSULÆ INFRA VENTUM, SOTTOVENTO in the West Indies.
INTERAMNA, TERNI in the Papal States.
INTERAMNA FRENTANA, TERMOLI.
INTERAMNA PALESTINÆ, TERANO in Abruzzo.
INTERAMNIS, ENTRAMES in the French dep. of Mayenne.—Also ENTRAINS in the dep. of Nievre.
INTERAMNIUM FLAVIUM, perhaps PONS FERRATUS or PONFERRADO in Spanish Leon.
INTERAQUÆ, ENTRAIGUES in the French dep. of Aveyron.
INTERCISA, RACKSKENEY, a village in Hungary.
INTERLACUS, the UNTER-SEE in the Swiss cant. of Bern.
INTERVALLIS, ENTREVAUX in Provence.
INTRA FLUVIUS, JOANVILLA or JOINVILLE in Champagne.
INUTRIUM, MITTENWALD in Bavaria.
INYCUM, CALTA BELOTA in Sicily.
IONIA, that part of the ancient *Lydia* which stretched from the *Sinus Camœus* or GULF OF CHANDERLI, to the *Sinus Bargylieticus*.
IONIUM MARE, the ADRIATIC.
IONOPOLIS, INEBOLI.
IOS, the island of NIO in the Archipelago.
IPRA, YPRES or YPERN in W. Flanders.
IRIA, VOGHERA in Piedmont.
IRIA FLAVIA, EL PADRON in Spanish Galicia.
IRINUS SINUS, the GULF OF CUTCH.
IRIS, the KISEL-IRMAK in Asiatic Turkey.
IRSINGUM, the abbey of IRSINGEN or YRSEL in Swabia.
IS, HIT in Algezira, in Asiatic Turkey.
ISABELLÆ ARX, FORT ISABELLE in N. Brabant.
ISACA, the EX in Somersetshire.
ISALA, the ISSEL or YSSEL in Gelderland.
ISAMNIUM PROMONTORIUM, ST. JOHN'S POINT in county Down.
ISANA, the river ISEN in Bavaria.
ISAPIS, the LAVIO or ALPS in the Papal deleg. of Urbino.
ISAR, the ISERE in Dauphiny.
ISARA, the ISAR in Bavaria.
ISARA, the French river OISE.
ISARGUS, the river ISER.
ISARLONIA, ISERLOHN in Prussia.
ISARUS, the FOGLIA in the Papal deleg. of Urbino.
ISBURUS, CALATABELLOTA in the Val-di-Mazzara, Sicily.
ISCA DUMNONIORUM, EXETER in Devonshire.
ISCA SILURUM, CAER LEON in Monmouthshire.
ISCALIS, ILCHESTER in Somersetshire.
ISCOPOLIS, TRIPOLI or TARABLUS in Asiatic Turkey.
ISEGENIUM, ISEGHEM in W. Flanders.
ISENACUM, EISENACH in Weimar.
ISENDICUM, YSENDIEK in Zieland.
ISION, the RAS-AL-DAWER in Nubia.
ISINISCA, the ISEN in Bavaria.
ISIS, the IPS in Hungary.—Also the CHOROK in Colchis.
ISIUS MONS, RAS-EL-DHAR on the E coast of Africa.
ISLEBIA, EISLEBEN in Prussian Thuringia.
ISMARUM PROMONTORIUM, CAPE MAROGNA on the coast of Thrace.
ISMENUS, the river ISMENO flowing into the lake of Likari in Northern Greece.
ISONTIUS, ISONZO, LISONZO, LIESNITZ in Illyria.

ISPINUM, Yepes in Spanish Toledo.
ISSA, Lissa in the Adriatic sea.
ISSEDON SCYTHICA, Haracar or Karacorum in Chinese Tartary.
ISSEDON SERICA, Kantschu, Cantcheou in Chinese Tartary.
ISSICUS SINUS, Scanderun Gulf.
ISSUS, Alexandretta or Scanderun.
ISTADIUM, Ystadt, Œgestadt in Sweden.
ISTER, the river Danube.
ISTROPOLIS, Chiustange at the mouth of the Danube
ISURIUM, Aldborough in Suffolk.
ITABYRIOS MONS, Tabor in Galilee.
ITALIA, Hesperia, Ausonia, Saturnia, or Œnotrea, a name originally given to that portion of the S extremity of Italy which lies below the *Sinus Scylleticus* or Gulf of Squillace, and the *Sinus Lameticus* or Gulf of St. Euphemia. It was afterwards extended to all the country S of the river *Laus* on the W, and *Metapontum* on the E. After 278, it comprised the southern and middle part of the peninsula as far N as the river *Tiber;* and was finally extended to the rivers *Rubicon* and *Macra*. In the reign of Augustus, Italy was bounded by the Alps on the N; the *Mare Superum* or Adriatic on the NE; the *Mare Ionium* or Ionian sea on the S; the *Mare Inferum* or Sea of Italy on the SW; and the *Alpes Maritimæ*, and river *Varus* or Var, on the NW.
ITALICA, Sevilla-la-Vieja in Spain.
ITALICA PELIGNORUM, near Popoli in the Papal States.
ITANUM PROMONTORIUM, Cape Xacro in Crete.
ITANUS, Sitano in Crete; according to others, Palaco Castro.
ITENEDIUM MENUTHESIAS, Monbusa in the SE of Africa.
ITHACA, Theaki, Cefalonia Piccola in the Ionian sea.—Also Aita, Vaitha in the island.
ITHOME, Monte Volcano in the Morea.
ITHORIA, Angelo-Castro in W. Livadia.
ITRIUM, Lamuranum, Itri in Terra-di-Lavoro, Naples.
ITUNA, the Eden flowing into the Solway Firth.
ITUNÆ ÆSTUARIUM, the Solway Firth.
ITURÆA, the district of Jedur in Palestine.
ITURISCA, St. Stephan; according to others, Sanguessa in Spanish Navarre.
ITYS, the Assin (Carron?) in the N of Scotland.
IVERNA. See Hibernia.
IVIA, Juvia, a river in Spanish Galicia.
IXARIUM, Hijar, Ixar in Spanish Aragonia.

J

JABADII INSULA, Sumatra or Banca in the Indian ocean.
JABNE, Jabneel, Ibne, Gebne in Judæa.
JABOK, the Wadi-Zerka in Palestine.
JABRUDA, perhaps Hebud in the p. of Damascus.
JACOBIPOLIS, Jamesborough in Leinster.
JACOBIPOLIS AMERICANA, Jamestown in Virginia, U. S.
JADA, the Jahde, a river of Oldenburg.
JADER, the Salone in Dalmatia.
JAGATH, Tetuan or Tetwen in Marocco.
JALYSUS, Jaliso or Uxilica in Rhodes.
JAMNO, Ciudadella in Minorca.
JAMPHORINA, Nerocop in Rumelia.
JAPONIA, the empire of Japan.
JAPYGIUM PROMONTORIUM, Cape-di-Leuca or Cape Finisterre in Terra d'Otranto.
JASENOCIUM, Jesenowitz in Hungary.
JASONIUM PROMONTORIUM, Cape Vono in Asiatic Turkey.
JASSIORUM DACORUM MUNICIPATUM, Jaseh or Jassy in Moldavia.
JASSUS, Assim-Kalesi in the SW of Anatolia.
JASTUS, the Kressel in Tartary.
JATRUS, the Otzuma in Bulgaria.
JAURAVIA, or Javoria, Jauer in Prussian Silesia.
JAURINUM, Raab in Hungary.
JAVENNUM, Giavenno in Piedmont.
JAXARTES, the Sihon, Sir or Sir-Daria in Great Bucharia.
JAZER, the Wady-Szyr in S. Damascus.
JECORA, the Jecker in Liege.
JEDUM, Jeddo in Japan.
JENA ÆSTUARIUM, Wigton Bay.
JENECOPIA, Joencoping in Sweden.
JENISIA, the Yenisei in Asiatic Russia.
JERACUM INSULA, Sabæa in the Arabian Gulf.
JERESLAVIA, the Russian gov. of Yaroslav.
JERNUS, the river Kilmare or the Suir in Ireland.
JESEDA, Jezd, Yesd in Persian Farsistan.
JESNA, Deszna, Jeno in Hungary.
JESSENA, Jessen in Prussian Thuringia.
JETÆ. Juto in the Val-di-Mazzara in Sicily.
JOACHIMICA VALLIS, Joachimsthal in Prussian Brandenburg.
JOANNIS GEORGII OPPIDUM, Johann-Georgenstadt in Saxony.
JOANNIS PONS, Hidvegh in Hungary.
JOCOSA, Freudenthal, Brunthal in Sleswick.
JODRUM, Adiodrum, Jouare in the French dep. of Seine-et-Marne.
JOGA, St. Gioja in the Neapolitan prov. of Terra-di-Bari.
JOGALIA, Youghall in Cork, Ireland.
JOLCUS. or Boritza, Vola in Thessaly.
JOMANES, the river Jumna in Hindostan.
JOMNIUM, Temen in Algeria; or perhaps Skurrfah.
JONNOSIA, Villa Loysa in Valencia.
JOPE, Jaffa in the pashalic of Damascus.
JOPILA, Jupil in Liege.
JORDANES, the Nahr-el-Arden, or Jordan in Palestine.
JORDANES MINOR, Hasbani, Mojet-Hasbeya.
JORDANIS PLANITIES, the Ghor in Syria.
JORNACUM, the river Irnis in Switzerland.
JOTAPE, Castel-Lombardo in Asiatic Turkey.
JOVALIA, Valpo in Hungary.
JOVARUS, the Salza in Austria.
JOVIA, perhaps Legrad in Lower Hungary.
JOVIACUM, according to Kruse, Pied in Austria.
JOVINIACUM, Juiniacum, Joigny in the French dep. of Yonne.
JOVIS URII FANUM. See Dios Sacra.
JOVIUM, Juca, Joux in the French prov. of Franche Comte.
JUBERG, Iburg in Hanover.
JUDECA, a Venetian island, now Giudeca, Zuecca.
JUENNA, Jaunstein in Carinthia, more likely Neu-Marktl-Terzesch in Laibach.
JUERNIS, Dunkerron in Munster; or according to d'Anville, Cashel.
JUERNUS, the Shannon in Munster.
JUFICUM, Sassoferrah in the Papal deleg. of Urbino.
JUGA SUIVONIS MONTIS, the Kolen mountains in Norway.
JUIA, Juvia in Spanish Asturias.
JULIA, Gyula in Hungary.
JULIA, or Zea, the Gail in Tyrol.
JULIA JOZA, or Transducta, Tarifa in Andalusia

JULIA LIBYCA, LLIVIA or ST. JULIA in Catalonia.
JULIA MYRTILIS, MERTOLA in Portuguese Alentejo.
JULIA OPTA, HUETE or GUETE in New Castile.
JULIA SCARABANTIA, OEDENBURG in Hungary.
JULIACUM, JULICH in Prussia.
JULINUM, WOLLIN in Prussian Pomerania.
JULIOBONA, LILLEBONNE in Normandy.
JULIOBRIGA, FUENTO-DI-IVERO or RETORTILLO in Spanish Burgos.
JULIOBURGUM, JULIUSBERG in Prussian Silesia.
JULIODUNUM, LOUDUN in Poitou.
JULIOLA, CASTRO DORIA in Sardinia.
JULIOMAGUS, STUHLINGEN in Baden.—Also the river WUTACH in Baden.
JULIUM CARNICUM, GUILDORF in Laibach.
JULIUS MONS, the JULIERBERG in the Swiss cant. of the Grisons.
JULIUS VICUS, GEMERSHEIM in Bavaria.
JUNCARIA, JUNQUERA in Catalonia.
JUNCETANA DOMUS, ALTENBIESEN in Limburg.
JUNIANELLUM, VIGNIANELLO in Tuscany.
JUNIANUM LIGANUM, LAUIS, LAUWERZ, or LUGANO in the Swiss cant. of Tessin.
JUNIVLADISLAVLENSIS PALATINATUS, INOWOLOCZ or INOWLADISLAW in Poland.
JUNNANIA, the Chinese prov. of YUNNAN.
JUNONIA, PALMA, one of the Canary islands.
JUNONIS INSULA, CADIZ in Spain.
JUNONIS PROMONTORIUM, CAPE TRAFALGAR on the coast of Spain.
JURENSIS URBS, ST. RAMBERT-LA-JOUX in the French dep. of Ain.
JUTA, JITTA in Judea.
JUTREBOCUM, JUTERBOCK in Prussian Brandenburg.
JUVANTUS, the TRONTE in the Papal deleg. of Ravenna.
JUVAVIA, or JOVAVUM, SALZBURG in Austria.
JUVENACIA, GIOVENAZZO in Neapolitan Bari.

K

KALMUNDA, KALMUNE in Bavaria.
KEDES, KISJON in Samaria.
KEKKOJA, KEKKOE in Hungary.
KER, the river AKKER or AKER in Syria.
KESDIENSIS SEDES, KESD in Transylvania.
KEXHOLMIA, KEXHOLM or KORDSKIGOROD in Finland.
KILLOCIA, KILMALON in Leinster.
KIMPERLÆUM, QUIMPERLAY in Brittany.
KIPHUSA, KIFHAUSEN in Prussia.
KIPHUSANUS MONS, MOUNT KUFFHAUSER in Prussian Thuringia.
KIR-MOAB, KARRAK in Palestine.
KIRITIUM, KYRITZ in Prussian Brandenburg.
KIRJATHAIM, EL-FEYM in the Arabian prov. of Nedjed.
KISDEMUM, KAYST in Transylvania.
KISON, the river MUKATTA in Syria.
KISTNIA, GUSTROW in Mecklenberg.
KIZINGA, CHIZICHEMIUM or KISSINGEN in Bavaria.
KLITSOVIA, KLITSCHOW in Poland.
KYRIOPOLIS, HERRNSTADT in Prussian Silesia.

L

LABACHUS, the river LAIBACH in Illyria.
LABACUS, the river LAWERS in Dutch Groningen.
LABARUS, the LAMBRO, a river in Lombardy.
LABBANA, MOZUL in the pash. of Bagdad.
LABEATIS, LAGO-DI-SCUTARI in Albania.
LABELLUM, LAVELLO in the Neapolitan prov. of Basilicata.
LABERUS, KILDARE in Leinster; according to others, KELLS in co. Meath.
LABIAVIA, LABIAU in the Prussian reg. of Konigsberg.
LABICUM, COLONNA in the Campagna-di-Roma; according to others, MONTE-COMPATRO.
LABIENI CASTRA, LOBBE, an abbey in the Hennegau.
LABINIUS, the LAVINO, a river in the Papal States.
LACCOBRIGA MUNICIPIUM, or LACOBRIGA, BURGOS in Old Castile.
LACEDÆMON, PALÆO-CHORI; according to Kruse, PALAIO-CASTRO in the Morea.
LACEREA, LAKE CARLO in Northern Greece.
LACIACA, FRANKENMARKT; according to others, MONDSEE in Austria.
LACIBURGUM, RATZEBURG or WISMAR in Meklenberg.
LACINIUM PROMONTORIUM, CAPE-DELLE-COLONNE in Calabria Citra.
LACIPPO, PORCUNA in Andalusia.
LACMON, MONTE-LIACA in the N of Thessaly.
LACOBRIGA, LAGOS on the coast of Portugal.
LACONICUS SINUS, the GULF OF KOLOKYTHIA in the Morea.
LACONIMURGI, COLMENAR-DE-OREJA in Toledo.
LACOTENA, MARHES or MERASCH in Armenia.
LACTIS MONS, MONTE-LATTARIO in Campania.
LACTODUNUM, BEDFORD; according to others, TOWCESTER in Northampton.
LACUFELICIS, OBER-WELS in Austria.
LACUS ALBUNEUS, the LAGO-DI-BAGNI in Campagna-di-Roma.
LACUS ALTINUS, the ALTIN-NOR in Asiatic Russia.
LACUS AMPSANCTI, LAKE MEPHITI in Naples.
LACUS ANDURIANUS, or SALSUS, LAGO-SALSO, or D'ANDORIA in Naples.
LACUS ARICIUS, the LAGO-DI-NEMI in the Campagna-di-Roma.
LACUS BIELLENSIS, or BIENNENSIS, the LAKE OF BIENNE in Switzerland.
LACUS CERUSIUS, LAKE LAVIC or LUGANO in Switzerland.
LACUS EDRINUS, LAKE IDRO in Lombardy.
LACUS INFERIOR, or VENETUS, the LAKE OF ZELL in the NE of Switzerland.
LACUS ISEUS, LAKE ISEO in Lombardy.
LACUS ITALICUS, or VALLENSIS, the VALLEN-SEE in Bavaria.
LACUS RUBRACENSIS, ROBRINO LAKE in Languedoc.
LADESIA, or LAGUSTA, LASTRE on the coast of Dalmatia.
LADICUS MONS, CODOS-DE-LADOS in Spanish Leon.
LADIGARA, GUADIGA in Algeria.
LÆDUS, the LEZ in Languedoc.
LÆLIA, ARACENA in Spanish Huebla.
LÆMOCOPIA, BAGAZASAR in Rumelia.
LÆPA, LEPE in Andalusia.
LÆROS, the LERIZ, a river in Spanish Galicia.
LÆTIÆ, or LÆTITLÆ, the abbey of LIESSE in the French prov. of Isle-de-France.
LÆVEFANUM, VIANEN in Holland.
LAGAHOLMIA, LAHOLM in Sweden.
LAGANA, LAUGONA, or LAHANA, the LAHN in Nassau.
LAGARIA, ROCCA IMPERIALE in the Neapolitan prov. of Basilicata.
LAGECIUM, CASTLEFORD in Yorkshire.
LAGINA, LAKENO in Asia Minor.
LAGIUM, LAGOS on the coast of Algarve.

LAGUEDONIA, Lacedogna in the Neapolitan prov. of Principato-Ultra.
LAGURINA VALLIS, the Lager-Thal in Tyrol.
LAGUSA, Panagia-di-Cordialissa, an island of Asia Minor.
LAGUSSÆ, Taochan-Adasi, an island in the Archipelago.
LAGYRA, Belbeck in the Crimea.
LAHA, Lava, La, or Laab, Laba in Holland.
LAIBNITIA, Leibnitz in Austrian Styria.
LALASIS, Calendo in Asia Minor.
LALINUM, Lalain in Dutch Hennegau.
LAMA, or Lameca, Lamego in the prov. of Beira in Portugal.
LAMBESA, Tezzute in Algeria.
LAMECUS, or Amatius, Amato or Lamato in the prov. of Calabria-Ultra.
LAMELLUS, Mallora, an island of Tuscany.
LAMETIA, St. Eufemia in Calabria-Ultra.
LAMETINUS SINUS, the Gulf of St. Eufemia in Calabria-Ultra.
LAMIA, Zeutuni in Thessaly.
LAMIDA, Medea in Marocco.
LAMINIUM, Montiel in New Castile; according to others, Alhambra.
LAMPEA, Elanda, a mount in the Morea.
LAMPSACUS, Lepsek, Lampsaki; according to Kruse, Chardak in Asia Minor.
LAMUM, Marino in the States of the Church.
LANCASTRIA, or Longovicum, Lancaster in Lancashire.
LANCIA, Penna-di-Francia in Spanish Leon.
LANCIA OPPIDANA, Garda in Portugal.
LANCICIA, Lencig or Lentschitz in Poland.
LANCIOBURGUM, Laneburg in Savoy.
LANDÆ, Landen in Luttich.
LANDAVIA, Landau in Bavaria.
LANDULFESDORF, Leudesdorf in Germany.
LANGARUS, the river Lanquart in Switzerland.
LANGIACUM, Langeac in Auvergne.
LANGOBRIGA, Feira in Portuguese Beira.
LANNOGUM, Lannoy in Flanders.
LANUVIUM, Civita-Lavinia in the Campagna-di-Roma.
LANZONIS MONS, Roccalanzone in Parma.
LAODICEA, Laodicea ad Lycum, or Diocesarea, Eski-Hissar in Asiatic Turkey.
LAODICEA COMBUSTA, Ladik in Karamania.
LAODICEA SYRIÆ, Ladikieh in Syria.
LAONA, Killaloa, Killaloe, or Labu in co. Clare.
LAPATHOS, Lapito or Lapta in the island of Cyprus.
LAPIDARIA, or Sexamnis Vallis, the Shamser-Thal in the Swiss cant. of Grisons.
LAPIDEUM LITUS, La Gran Ferriere in Provence.
LAPIS REGIUS, Konigstein in Nassau.
LARENUSIÆ, the island of Mollicorno near Tunis.
LARES, Larbuss in Tunis.
LARIA, Lar in Persia.
LARISSA, Sheizar in Syria.—Also Vesrach-Kevi in Asia Minor.
LARISSA PENSILIS, Larizo in Crete.
LARISSA THESSALA, Jenisher or Larissa in Thessaly.
LARISSUS, Risso, a river in the Morea.
LARNUM, Fordera in Catalonia.
LARUS, the Arone, a river in the Papal States.
LARYMNA, Larnes in Livadia.
LAS, Scatari in the S of the Morea.
LASSIRA, Sarione in Spanish Aragon.
LASTIGI, Zahara in Seville.
LATAMEDA, the Moree, a river in India beyond the Ganges.
LATERA, Lette or Le-Chateau-de-Latties in Languedoc.
LATINIACUM, Lagny in the Isle-de-France.
LATMUS, Il-Monte-di-Palatshia in Asia Minor.
LATO, or Latopolis, Asna, Esneh in Upper Egypt.
LATONÆ CIVITAS, Errahue in Lower Egypt.
LATOPOLIS, Esneh in Upper Egypt.
LATRIPPA, Jathrib in Arabia.
LATRIS, the same as Osilio, Oesel island in the gulf of Riga.
LATRONUM INSULÆ, the Ladrones in the Pacific ocean.
LATURUS SINUS, the Gulf of Sereni in Numidia.
LATVERUM, Laviers in Picardy.
LATYMNUS, the Monte-di-Cotrone in Calabria-Citra.
LAUCOSTABULUM, Lichstall, Liestall, in the Swiss cant. of Basle.
LAUDERA, Lutra, the town and abbey of Luders in Alsace.
LAUDUM, Laus Pompeja Nova, Lodi in Lombardy.
LAUGINGA, Lauingen in Bavaria.
LAUMELLUM, Lomello in the Sardinian States.
LAUREACENSE MONASTERIUM, Lorch in Würtemberg.
LAURENS CASTRUM, Torre-di-Paterno in the Campagna-di-Roma.
LAURIUM, or Lorium, Castel-Guido in Tuscany.
LAUS POMPEJA, Lodi-Vecchio in Lombardy.
LAUS, Laino in the prov. of Basilicata in Naples.
LAUSANIUS LACUS, the Lake of Lausanne in Geneva.
LAVENTINA, Lavemunde in Illyria.
LAVENTUS, the river Lavant in Illyria.
LAVIACUM, or Lupha, Lauffen in Würtemberg.
LAVINIUM, Patrica in the Campagna-di-Roma.
LAVUS, the river Lohe in Silesia.
LEA, Savenus, the Lee in Ireland.
LEANITES SINUS, the Bay of Kadhema in the Persian gulf.
LEBADEA, Livadia in Greece.
LEBEDUS, Lebedigli in Asiatic Turkey.
LEBEN, Laonda in the island of Crete.
LEBINTHUS, Levintho, Leuda in the Archipelago.
LEBONA, Leban in Samaria.
LEBONIA, Levagna in Genoa.
LECCA, the river Leck in Holland.
LECHLINIA, Leighlinbridge in co. Carlow.
LECTUM PROMONTORIUM, Cape Baba or St. Maria in Asia Minor.
LECYTHUS, Kiriaki in Macedonia.
LEDESIA, Leeds in Yorkshire.
LEDI, Liers in Belgium.
LEDIA, Laye in the Isle-de-France.
LEDUM SALARIUM, or Lonsalinum, Lons-le-Saunier in Franche Comté.
LEGECESTRIA, Leogara, Leicester in England.
LEGEDIA, Havre-de-Longeville in Normandy.
LEGIA, the Lye, Lys in West Flanders.
LEGIO SEPTIMA GERMANICA, Leon in the Spanish prov. of Leon.
LEGUNTINA VALLIS, the Lugnetzer-thal in Switzerland.
LEHERICI MONS, Montlehery in the French dep. of Seine-et-Oise.
LEINIUS, or Lynius, the Leine in Hanover.
LEISNICIUM, Leissnig in Saxony.
LEITÆ PONS, Bruck on the Leythe in Germany.
LELAANONIUS SINUS, Loch Fyne in Argyleshire.
LEMANIS, perhaps the same as Novus Portus, Lime, Lyme in Dorsetshire.

LEMINCUM, Chambery in Savoy.
LEMNOS, Stalimene, Lemno in the Archipelago.
LEMOIGA, Lemvicum, Lemvig in Jutland.
LEMURIS, the Lemo in Genoa.
LENCIA, Linza, a river in Italy.
LENGELFELDENSIS ARX, Burglenkenfeld in Bavaria.
LENTUDUM, Luttenberg in Austrian Styria.
LEOBUSIUM, Leubus, Lubens in Silesia, Prussia.
LEOGUS, Lewis in the Hebrides.
LEON, the river Auleh in Syria.
LEONIACUM, Legnano in the Venetian prov. of Verona.
LEONICA, Aliaga in Spanish Aragon.
LEONICÆ, Lorgues in Provence.
LEONIS CASTRUM, Hort in Denbighshire.
LEONIS MONASTERIUM, Lemster in Hereford.
LEONIS MONS, Monte Leone in Calabria-Ultra.
LEONTOPOLIS, Tel-Essabe in Lower Egypt.
LEOPOLDINUM, Leopoldstadt in Hungary.
LEOPOLIS, Lemberg, Lowenberg in Silesia, Prussia.—Also Lemberg in Austrian Galicia.
LEOVALLIS, Liebenthal, Loewenthal in Prussian Silesia.
LEOVARDIA, Leewarden in Dutch Friesland.
LEPONTINA VALLIS, Livinerthal in the Swiss cant. of Tessin.
LEPORACENSIS VALLIS, Leberthal in Alsace.
LEPREUM, Strobitzi in the Morea.
LEPROSIUM, Levroux in French Berri.
LEPSIA, Lipso in the Archipelago.
LEPTE, a cape, Ras-el-Enf in Upper Egypt.
LEPTIS MAGNA, Lebida in African Tripoli.
LEPTIS MINOR, Lemta in African Tunis.
LERDAMUM, Leerdam in S. Holland.
LERIA, Lero, one of the Cyclades.
LERINA, St. Honorat, an island south of Provence.
LERINÆ, or Stoechades Insulæ, the Hieres Islands south of Provence.
LERNA, Petrina in the Morea.
LERNA PALUS, Molini or Mullenteich in the Morea.
LERO, St. Marguerite, one of the Hieres Islands.
LESA, St. Gadina in Sardinia.
LESBOS, Metelina in the Archipelago.
LESINIACUM, Lesignac in Languedoc.
LESORA, Losere, a mountain in Languedoc.
LESSINA, Lessines in Belgian Hainault.
LESUA, Lewes in Sussex.
LESURA, the Leser in German Treves.
LESTARUM REGIO, Siam in India beyond the Ganges.
LESTINÆ, Letines in Belgian Hainault.
LETANE, the Casmy, Casimir in Damascus.
LETHA, Leith in Scotland.
LETIA, the Lesche, a river in Belgian Namur.
LETOA, Gaiduronisia, an island S of Candia; according to others, Christina.
LETRINI, Pyrgo in the Morea.
LETSCHIA VALLIS, the Lettscherthal in Switzerland.
LETTERANUM, or Lycteræ, Lettere in Naples.
LETUSA, Leuse in Belgian Hainault.
LEUCA, St. Maria-di-Leuca in the Neapolitan prov. of Terra-d'Otranto.—Also Capo-di-Leuca.
LEUCADIA, St. Maura or Lefkhæda, an island in the Ionian sea.
LEUCARISTUS, Konstad in Prussian Silesia.
LEUCAS, Amaxechi in Santa-Maura.
LEUCASIA, Leucothia, Licoso, an island in the gulf of Salerno.
LEUCATE PROMONTORIUM, Capo Ducato in Santa-Maura.
LEUCE, Achillis Insula or Fidonisi.
LEUCE ACTE, Chiflik in Rumelia.
LEUCE COME, Hauran or Sherm Janbo in Arabia Felix.
LEUCE PORTUS, Chaona in Arabia Petræa.
LEUCENSES, or Leuceræ Thermæ, Leukerbad in Switzerland.
LEUCI MONTES, Alprovoana in Crete.
LEUCIA, Leuk in the Swiss cant. of Wallis.
LEUCIMNA, Cape Bianco in the island of Corfu.
LEUCOLLA, a town of Cyprus, now Armida.
LEUCONIUM, Leutschau in Hungary.
LEUCOPETRA, Cape-del-Armi in Calabria-Ultra.
LEUCOPIBIA, Wigton in Scotland.
LEUCOSIA, Nicosia, Lefcosia in Cyprus.
LEUCOSTABULUM, Lichstal in the Swiss cant. of Basle.
LEUCOSYRUS, the Jeihun in Karamania.
LEUCTRA, according to Kruse, Lefka; according to others, the ruins Eremo-Castro in Livadia.—Also Istechia or Londari in the Morea.
LEUCUM, Lecco in Milan.
LEUPHANA, Lubtheen, or perhaps Luneberg in Mecklenburg.
LEUTEVA, Ludeve.
LEVIA, Lewensz in Hungary.
LEVIDONA, Alvidona in Calabria-Ultra.
LEVINIA, Dumbarton in Scotland.
LEVITANIA, Lavedan.
LEZINIACUM, Lusignan in Poitou.
LIBA, Libau, Liveja in Russian Kourland.
LIBANUS MONS, the Jebel-el-Garbi and Jebel-el-Libnan in the pash. of Damascus.
LIBARNA, Lerna in Genoa.
LIBERNIDES, the Liburnian Islands off the coast of Dalmatia.
LIBETHRA, Nesivo in Macedonia.
LIBISSONIS TURRIS, Porte-de-Torre in Sardinia.
LIBORA, Talavera-de-la-Reina in Spanish Toledo.
LIBYA, a region of Barbary extending from the gulf of Syrtis to the Egyptian frontier.
LIBYSSA, Gebse in Anatolia.
LICHADES, the Ponticonesi islands.
LICHUS, or Licus, the river Lech in Switzerland.
LICOPIA, Lidkoping in Swedish Mariestad.
LIDALIA, Liddesdale in Scotland.
LIGER, the river Loire in France.
LIGERULA, the Loiret in Orleannois.
LIGNIACUM, Ligne in Belgian Hainault.
LIGNITIUM, Leignitz in Prussian Silesia.
LIGUSTICUM MARE, or Ligusticus Sinus, the gulf of Genoa.
LILLERTIUM, Lillers in Artois.
LILYBÆUM, Marsala in the Val-di-Mazara, Sicily.
LILYBÆUM PROMONTORIUM, Cape Bæo in Sicily.
LIMAGA, or Lindemagus, the Swiss river Limmat.
LIMANIA, or Alimania, Limagne.
LIMIOSALEUM, Lissa in Prussian Posen.
LIMNÆA, Loutra in Livadia.
LIMNUS, Dalkey in the Irish sea.
LIMODIA, Limegos in Gascony.
LIMONUM, Poitiers in French Poitou.
LIMOSUM, Limoux in Languedoc.
LIMUSA, Istvandi in Lower Hungary.
LIMYRICA, Goa in Hindostan.
LINCOLONIA, Lindum, Lincoln in England.
LINCOPIA, Lindkoping in Sweden.
LINDANGIA, or Lintona, Lindau in Bavaria.
LINDUM, Lincoln.
LINDUM DAMNIORUM, Kirkintilloch in Scotland.

LINGO, LINGEN in Hanover.
LINTERNA PALUS, LAGO-DI-PATRIA in Naples.
LINTERNUM, TORRE-DI-PATRIA in Lavoro, Naples.
LIONUM, LIONS in French Normandy.
LIPAREÆ, the LIPARI ISLANDS.
LIPPEBRUNE, or LIPPLÆ FONTES, LIPPSPRING in Prussian Minden.
LIPPIA, the river LIPPE in Westphalia.
LIPSIA, LEIPZIG in Saxony.
LIPSTADIUM, LUPIAS in Prussian Minden.
LIQUENTIA, LIVENZA, a river in Venice.
LIRIMIRIS, OLDENFELD in Mecklenburg; according to Kruse, WISMAR.
LISSUS, ALESSIO in Albania.
LITAHA, LEITHA, a river in Hungary.
LITANA SILVA, SILVA-DI-LUGE in Modena.
LITERNA PALUS, LAGO-DI-PATRIA in the Papal States.
LITERNUM, the TORRE-DI-PATRIA in the Campagna of Rome.
LITOMERICIUM, LEUTMERITZ in Bohemia.
LITOMISLIUM, LEUTOMISCHL in Bohemia.
LIXA, LARACHE in Fez.
LOBAVIA, or LOEBAU, LIEBE in Saxony.
LOBDUNUM, or LATINOBURGUM, LADENBURG in Baden.
LOBETUM, REQUENA in New Castile.
LOCHENITIUM, LOCKNITZ in Prussian Stettin.
LOCHIA, LOCHE in French Touraine.
LOCOPOLIS, BISCHOFSLACK in Carinthia.
LOCORITIUM, LOHR, a town in Bavaria; according to Kruse, LAURINGEN.
LOCRA, a river in Corsica, now LIAMONE; according to others TULAVO.
LOCRI, MOTTA DI BURZANO in Calabria-Ultra.
LODA, OLDA, or OLITIS, the French river LOT in Guienne.
LOGANA, the river LAHN, an affluent of the Rhine.
LOGIA, the river LAGAN in Ireland.
LOJA, LOYES in Burgundy.
LONCIUM, LIENZ in the Tyrol.
LONDINIUM, LONDON.
LONDINUM, the Swedish town LUND in Malmo.
LONGA SALINA, SALO in Milan.
LONGATICUM, LOHITSCH in Hungary.
LONGOSALISSA, LANGENSALZA in Prussian Erfurt.
LONGOVICUS, LONGWY in Lorraine.
LONGUM GEMELLUM, LONJUMEAU in the Isle-of-France.
LONGUS ÆSTUARIUM, the LINNHE LOCH in Argyleshire.
LOPADUSA, LAMPADUSA, an island in the Mediterranean.
LOPOSAGIUM, LUXIOL, near Beaume in France.
LOPSICA, LOPSI in Croatia.
LORIACUM, LORRIS in French Orleannois.
LORIUM, or VALCOLERIA, VAUCOLEURS in Lorraine.
LOSONTIUM, LOSCHONZ in Hungary.
LOSSA, LOON in Belgian Limburg.
LOTITIA PARISIORUM, LEUCOTERIA, LUTETIA PARISIORUM, or LUCOTOCIA, PARIS in France.
LOVANIA, LUVANIUM, or LOWEN, LEUVAN in S. Brabant.
LOVITIUM, LOWIEZ or LOWITSCH in Poland.
LOXA, the river LOSSIE in Elgin; according to Mannert, the STRUTH.
LOYSA, the LOISACH in Bavaria.
LUANUM, or LUANO, LOVANO in Piedmont.
LUBECA, LUBECK in Mecklenburg.
LUBENA, LUBBEN in Prussian Frankfort.
LUBLAVIA, LUBLAU in Hungary.
LUCCAVIA, LUCCA or LUCKAU, in Sachsen-Altenburg.
LUCENA, LUTZEN in Prussian Merseburg.
LUCENTUM, the same as ALICANTA.
LUCERIA, LUCERN in Switzerland.—Also LOUCERIA, NUCERIA APULORUM, LUCERA in Basilicata.
LUCINA, EL-KAB in Upper Egypt.
LUCIO, LUCON in Poitou.
LUCOMONIS MONS, the LUKMANIER, a mountain in Switzerland.
LUCOPIBIA, NEWTON-STEWART in Wigtonshire.
LUCRINUS LACUS, LAGO-LUCRINA in the Terra-di-Lavoro of Naples; according to others, MONTE-NUOVO-DI-CENERE.
LUCUS ASTURUM, OVIEDO in Spanish Asturias.
LUCUS AUGUSTI, or SEPTIANÆ ARÆ, LUGO in Spanish Galicia.—Also LUC in Provence.
LUDA AD AMBRAM, LUSDUM, LUYDA or LUGDE in the Prussian reg. of Minden.
LUDOVICI ARX, LUDWIGSBURG in Würtemberg.
LUDOVICI MONS, MONT LOUIS in Rousillon.
LUDUM, or LUTHA, LOUTH in Lincolnshire.
LUGDUNUM, LYON in French Lyonnais.
LUGDUNUM BATAVORUM, LEYDEN in Holland.
LUGOSIUM, LUGOS in Hungary.
LUGUIDONIS PORTUS, LOCONI in Sardinia.
LUGUVALLUM, CARLISLE in Cumberland.
LULEA, LULA in Sweden.
LUNA, LUNEGIANO in Tuscany.
LUNÆBURGUM, or SELENOPOLIS, LUNEBURG.
LUNÆ MONTES, the JEBEL-EL-KUMRI or MOUNTAINS OF THE MOON in Central Africa.
LUNÆ PROMONTORIUM, CAPE ROCCO in Portugal.
LUNÆ SYLVA, the MANHARTSBERG.
LUNARIS VALLIS, LUNEVILLE in Lorraine.
LUNARIUM PROMONTORIUM, CAPE-FELIX-DE-GUISOLA.
LUNATE, LUNILIUM in Languedoc.
LUPARIA, LOUVIERS in Normandy.
LUPATIÆ, ALTAMURA in the Neapolitan prov. of Terra-di-Bari.
LUPELLI MONS, MONLUEL in Burgundy.
LUPIA, LECCE in the Neapolitan prov. of Terra-d'Otranto.—Also LOING in Nivernois.
LUPINUM, or MAJÆ VILLA, MAYENFIELD in Switzerland.
LUPPIA, the LIPPE, an affluent of the Rhine.
LUQUIDO, LUGODORI in the island of Sardinia.—Also LAGIOLASTO in Corsica.
LUSIO, BATA; according to others FOLVAR in Hungary.
LUSITANIA, a division of the ancient Hispania, comprising the present kingdom of PORTUGAL, with the exception of that portion of it which lies N of the Douro, and likewise the greater part of SPANISH ESTREMADURA, and SALAMANCA, and part of NEW CASTILE and TOLEDO.
LUSIUS, a river of Greece, probably the modern ATZIKOLO.
LUSSONIA INSULA, LUÇON, or LUZON, MANILLA in the Philippines.
LUSSONIA INSULÆ, the PHILIPPINE ISLANDS or MANILLAS in the Pacific ocean.
LUTERA, LURE or LUDERS in Franche-Comte.
LUTERA REGIA, KONIGSLUTTER in Brunswick.
LUTETIA PARISIORUM, the city of PARIS.
LUTITIA, LOITZ in Prussian Stralsund.
LUTOSA, the river LOSE in France.
LUTRÆ CASTRUM, LAUTERBURG in Germany.
LUTTERVORTIUM, LUTTERWORTH in Leicester.
LUTUDARUM, perhaps LEEDS in Yorkshire.
LUTUM, LOUTH in Ireland.
LUVIA, the river LOUAIN in the French dep. of Seine-et-Marne.
LUXIA, ODIEL in Andalusia.
LUXOVIUM, LUXEVIL or LUXEN in Franche Comte.
LYCÆUS, TETRAGI, a mount in the Morea.

LYCHNIDUS, OCHRIDA or ACHRIDA, a town in Albania.
LYCHNITIS PALUS, the LAKE OF OCHRIDA.
LYCIA, a region of Asia Minor, comprising the basin of the river *Xanthus*, the modern ETCHEN-SAI.
LYCIORUM CAMPUS, LECHFELD in Augsburg.
LYCOPOLIS, SIUT or OSSIUT in Upper Egypt.
LYCOREA, LIACURA in Livadia.
LYCTUS, LYTTUS in Crete; according to others, LASSITI.
LYCURIA, PHONIA; according to others, LYKURIO in the Morea.
LYCUS, the GREAT ZAB, a river in Kurdistan; or, according to Kruse, the MIKALITZA in Anatolia.
LYCUS, the river NAHAR-KELB in Damascus.—Also the JESHIL-IRMAK in Armenia.—Also the ZAB or ZAB-ALA in Asiatic Turkey.
LYDIA, a region of Asia Minor, corresponding to the modern sanjak of SARAKHAN, and the N part of SIGHLA.
LYMNÆA, LOUTRA in Livadia.
LYNCESTINUS AMNIS, the VENETICO in Rumelia.
LYNCUS, according to Kruse, VODONICA; perhaps HERACLEA in Rumelia.
LYNUM REGIS, LYNN, or KING'S LYNN in Norfolkshire.
LYRA, LIER or LIERRE in Antwerp.
LYSINIA, AG-JALON in Asiatic Turkey.
LYSTRA, near BINKIR-KALISSEH in Asiatic Turkey.
LYSTRA, now LATIK in Karamania.
LYTARMIS PROMONTORIUM, CAPE OBY in Siberia.

M

MAAGRAMMUM, CANDY in Ceylon.
MAALINÆ, MECHLINIA, MECHELN in Antwerp.
MABARTHA, SYCHAR, FLAVIA NEAPOLIS SAMARIÆ, or NEAPOLIS, NABLUS in Samaria.
MACEDONIA, a region of Europe, bounded on the N by *Mœsia;* on the E by *Thracia*, from which it was separated by the river *Strymon*, and subsequently by the *Nestus;* on the S by *Thessaly*, and on the W by *Illyricum* and *Epirus*, from which it was separated by the chains of *Scardus* and *Pindus*. It nearly corresponds to the MAKEDONIA or FILIBA-VILAJETI of Turkey.
MACEDONICUM MARE, or THERMAICUS SINUS, the GULF OF SALONICA.
MACEPRACTA, MASJEB-KAN; according to others, MAI-FANEKIN in Mesopotamia.
MACERIÆ, or MADERIACUM, MEZIERES, a town of Champagne, France.
MACESTUS, the SUSUGHERLI in Asia Minor.
MACETA, or ARSABORUM PROMONTORIUM, CAPE MOCADON in Arabia Felix.
MACHERA COMITIS, GREVENMACHERN in Luxemburg.
MACHERA REGIS, KONIGSMACHERN in Luxemburg.
MACHICOLIUM, MACHECOUL in Brittany.
MACHMAS, or MICHMASH, EL-BIR in Judea, Palestine.
MACHORBE, MASKAT or MESKIET in Arabia Felix.
MACISTUS, a town of Triphylia, now MOFKITZA.
MACNA, MAGNE on the Red Sea.
MACOLICUM, KILMALOK in Limerick; according to others, DONERAILE in Cork.
MACORABA, perhaps MECCA.
MACRA, the MAGRA, a river in the Sardinian prov. of Genoa.
MACRINA, now VETERI, a town in the States of the Church.
MACRINUS, MATRINUS, or VIBATINUS, the TORDINO, PIOMBA in Naples.
MACRUM, MARO in Piedmont.
MADASCONA, or MATASCO, MAÇON in Burgundy.
MADIA, MAÏS in Armenia.—Also the Swiss river MAIN or MAGGIA.
MADIÆ VALLIS, or MAINTHAL, MAGGIA in the Swiss cant. of Tessin.
MADUS VAGNIACÆ, MAIDSTONE in Kent.
MADYTOS, MAITO in Rumelia, Turkey.
MÆANDER, the MINDER-SU or MENDERE-SU, a river of Asiatic Turkey.
MÆNALUS MONS, a range in Northern Greece, the modern ROINO, or perhaps AIDIN.
MÆNARIA, MELORA, an island on the coast of Tuscany.
MÆNOBA, the river VELEZ in Spanish Granada.
MÆNUS, MŒNIS, or MOHIN, the MAIN in Germany.
MÆOTIS PALUS, the SEA OF AZOV, or KABACHE in Russia.
MÆSANITUS SINUS, an inlet in the Persian gulf, now ADILLA.
MÆSOLUS, or MESOLUS, the river KISTNAH or KRISHNA in Hindostan.
MAGETOBRIGA, perhaps the same as BELLICARDUS or PILIGARDÆ MONS, MOMPELGARD in Franche Comté, or MOIGTE-DE-BROIE in the vicinity of Pontarlier.
MAGIOVINIUM, FENNY-STRATFORD in Bucks.
MAGNA, CARVONAN in Cumberland.
MAGNA, KENTCHESTER near Weston.
MAGNESIA AD MÆNADRUM, INEKBAZAR in Asiatic Turkey.
MAGNESIA AD SIPYLUM, now MANISSA in Asiatic Turkey.
MAGNESIÆ PROMONTORIUM, CAPE ST. GEORGE in Greece.
MAGNIACUM, MAYEN in Coblenz.
MAGNUM PROMONTORIUM, CAPO-DE-BRAGU.—Also CABO-DA-ROCA in Portugal.—Also ROMANIA POINT.
MAGNUS PORTUS, or PORTSMUTHUM, PORTSMOUTH in England.—Also ARZEN in Marocco.
MAGNUS SINUS, the GULF OF MARTABAN in the Bay of Bengal.
MAGO, MAHON in the island of Minorca.
MAGON, the RAMGUNGA river in Hindostan.
MAGONTIA, or MOGUNTIACUM, MAINZ in Hesse-Darmstadt.
MAGORAS, the NAHR-BEIROUT in Damascus.
MAGORUM INSULA, the island of DAHALAG in the Red Sea.
MAGRADA, the river BIDASSAO; or, according to others, the URUMEA in Biscay.
MAGUDA, MAKESSIN in the pash. of Bagdad.
MAGUSA, MEGARISCH-UZZIS in Arabia-Petræa.
MAININGA, or MINIMINGA, MEININGEN in Germany.
MAIS, the MAHI or MYHIE in Hindostan.
MAJORIS MONASTERIUM, MARMOUTIER in Touraine.
MALACA, MALAGA in Granada.
MALÆI COLON, CAPE ROMANIA in India beyond the Ganges.
MALAMANTUS, the river KAMEH in Hindostan.
MALANGA, MADRAS in Hindostan.
MALDUNENSE CŒNOBIUM, MALMSBURY in Wiltshire.
MALEA, ADAM'S PEAK in Ceylon.—Also CAPE ST. ANGELO in the Morea.—Also CAPE ST. MARY in the island of Lesbos.
MALEAS INSULA, MULL in the Scottish Hebrides.
MALEUS, MOUNT YATES in Hindostan.

MALEVENTUM, or BENEVENTUM, SABBATO in the Papal States.
MALIACUS SINUS, the GULF OF ZEITUN in Thessaly.
MALIARPHA, ST. THOMAS or MALIAPUR in Hindostan.
MALICHI INSULÆ, the SOKAR islands in the Red Sea.
MALLEO, MAULEON in the French dep. of Basses-Pyrenees.
MALLORUM METROPOLIS, probably MULTAN in Afghanistan.
MALMUNDARIÆ, MALMEDY in Prussia.
MALOBODIUM, MAMBEUGE in the French dep. of Nord.
MALUS, the MALE in Karamania.
MAMALA, HALI in Arabia Felix.
MAMERCIÆ, MAMERS in the French dep. of Sarthe.
MAMERTIUM, OPPIDO in Calabria-Ultra; according to others, MARTORANO in Calabria-Citra.
MAMMA, BEGHUI on the coast of Barbary.
MANAPIA, WEXFORD in Ireland.
MANCUNIUM, or MANDUESSEDUM, MANCHESTER in Lancashire.
MANDAGORA, DABUL in the prov. of Bejapur in Hindostan.
MANDAS, the river MAHANADDY in Hindostan.
MANDELA, now BARDELA in the Papal States.
MANDONIUM, MANDURIA, CASAL-NUOVO in the Neapolitan prov. of Terra-d'Otranto.
MANDRAS, MOUNT KONG in Soudan, Africa.
MANDUESSEDUM. See MANCUNIUM.
MANESCA, MANOSQUE in the French dep. of Lower Alps.
MANGARUTH, MANGALORE in Canara, Hindostan.
MANHEMIUM, INTERAMNIUM, MANHEIM in Baden.
MANIANA, MALLIANO, MALIANA in Barbary.
MANIOLÆ INSULÆ, the ANDAMAN ISLANDS in the Bay of Bengal.
MANLIANA, MAGLIANA in Tuscany.
MANNARICIUM, MAURIK in the Dutch prov. of Utrecht.
MANTALA, MONTAILLEAU in Savoy.
MANTIANA PALUS, LAKE VAN in Armenia.
MANTINEA, according to Kruse, PALEOPOLI in the Morea.
MANTINORUM OPPIDUM, BASTIA, a town in Corsica.
MANTUA HISPANICA, MADRID, the capital of Spain.
MARA, AIJUN-MUSSA in the Arabian Desert of Shur.
MARABODUI CASTELLUM, KONIGSWART or KUNEZWART in Bohemia.
MARACANDA, SAMARCAND in Independent Tartary.
MARACLEA, MERAKIAH in Syria.
MARANTIUM, MARANS or MARENNES, in the French dep. of Charente-Inferieure.
MARANTIUM, MARANT in Turkey in Asia.
MARANUM, MERAN in the Tyrol.
MARATHOS, MERAKIN in Syria.
MARCERUM, MERZIG in Prussia, Rhenish Province.
MARCHENIUM, or ROSBURGUM, ROXBURGH in Scotland.
MARCHIA FAMINA, MARCHE-EN-FAMINE in the Belgian prov. of Liege.
MARCHIANÆ, MARCIANÆ, MARCHIENNES in French Flanders.
MARCHUS, the river MARCH in Austrian Moravia.
MARCI, MARQUISE in Artois, France.
MARCIANA, MARCENA, or MARIANA CASTRA, MARBURG in Austrian Syria.
MARCIANA SYLVA, the SCHWARZWALD in Baden.
MARCIANOPOLIS, PEREJASLAW or PRESLAW in Bulgaria.
MARCILLIACUM, MARCILLAC in the French dep. of Puys-de-Dome.
MARCINA, VETERI in Naples.
MARCINIACUM, MARCIGNY in Burgundy.
MARCODURUM, or MARCOMAGUS, DUEREN in the Prussian prov. of the Rhine.
MARCOMAGUS, AACHEN or MARKMAGEN in Rhenish Prussia.
MARE ÆGEUM, the ARCHIPELAGO.
MARE AMALCHIUM, the DUTCH SEA, or NORTH SEA.
MARE EUXINUM, or PONTUS EUXINUS, the BLACK SEA.
MARE EXTERNUM, or ATLANTICUM MARE, the ATLANTIC.
MAREIA LACUS, MARIUT, MARUT, in Lower Egypt, the same as MAREOTIS LACUS.
MARE INTERNUM, or MEDITERRANEUM MARE, the MEDITERRANEAN.
MAREOTIS LACUS, the BUHET-MARIUT in Lower Egypt.
MARGINIA, MEIMARK in Great Bucharia.
MARGUM, PASSAROWITZ in Servia.
MARGUS, the river MORAWA in Servia.—Also the MARIAB in Turcomania.
MARIA-THERESIANOPOLIS, THERESIOPEL in Hungary.
MARIABA BARAMALACUM, MAREB in Arabia Felix.
MARIACELLA, MARIAZEL or GRASSMARIAZELL in Austria.
MARIACHELMUM, CULM, or MARIENCULM, in Bohemia.
MARIÆ DOMUS, MERGENTHEIM in Würtemberg.
MARIÆ VERDA, MARIANA INSULA, MARIENWERDER in Prussia.
MARIANA, GOLO in Corsica.
MARIANUM, MARANO in Venice.
MARIANUS MONS, the SIERRA-MORENA in Spain.
MARIDUNUM, CAERMARTHEN or CAERVYRDEN in Wales.
MARINUM PROMONTORIUM, CAPO-DI-CASA-BARBARICA in Corsica.
MARIONIS, the same as HAMBURGUM.
MARIONIS ALTERA, the same as LUBECCA.
MARISCUM, MARASH, MARAZ in Turkey in Asia.
MARISUS, the river MAROSCH or MAROS in Hungary.
MARMARICA, the modern district of BARCA, and the OASIS OF AMMON in Barbary.
MARNAMANIS PORTUS, MARNER in Bavaria.
MARO, MOUNT MANDONIA in Sicily.
MAROBUDUM, KONIGSBERG in Bohemia.
MARONEA, MAROGNO in Romelia.—Also CAMPO-MARANO in Italy.
MAROSIENSIS SEDES, MAROS in Transylvania.
MARPURGUM, MADEN in Hessen-Cassel.
MARRUVIUM, near SAN-BENEDETTO in the Papal States.
MARSIBURGUM, or MARSIPOLIS, MERSEBURG in Prussia.
MARSILINUM, MARSCHLINS in Switzerland.
MARTELLI CASTRUM, MARTEL in the French dep. of Lot.
MARTIA FAMINA, MARCHE-EN-FAMINE in Belgium.
MARTIA VILLA, MARVILLE in the French dep. of Meuse.
MARTIS STATIO, OULX, HOULX in Piedmont.
MARUCA, MERON-ERRUD in Khorasan.
MARUS, the MARCH, a tributary of the Danube.
MASA, the river MAAS in Holland.
MASADA, SEBBEH near the Dead Sea in Syria.

MASECA, Maaseyk in Limburg.
MASES, Psate in the Morea.
MASIACUM, Schwas in the Tyrol.
MASIUS MONS, the Karajeh-Dagh in Mesopotamia.
MASONIS MONASTERIUM, Masevaux or Masmunster in Alsace.
MASSA, Kaina in the Morea.—Also Achnagar in Hindostan.
MASSA LUBIENSIS, Massa-di-Sorrento or Lubrense in Naples.
MASSA VETERNENSIS, Massa in Tuscany.
MASSILÆ, Masjebkan in Irak-Arabi.
MASSICUS MONS, Mount Mondragone in the Neapolitan prov. of Terra-di-Lavoro.
MASSILIA, Marseilles in France.
MASTE, Gongar in Abyssinia.
MASTRAMELUS LACUS, the Lake of Martrigues in Provence or Maintenon in Orleannois.
MASTUSIA PROMONTORIUM, Cape Griego in Rumelia.
MATAVONIUM, Cabasse in Provence.
MATERENSE OPPIDUM, Matter in Tunis.
MATILO, Rheinsburg in South Holland.
MATINUM, Matinaca in Naples.
MATREJUM, Matray in the Tyrol.
MATRICA, Theten in Hungary.
MATRINUM, Porto-d'Atri in Abruzzo.
MATRONA, the river Marne in Champagne.
MATTHÆI VILLA, Matsdorf in Hungary.
MATTIUM, Marburg in Hessen-Cassel.
MAURECIACUS, or Morentiacus Mons, Montmorency in Isle-de-France.
MAURIANUM, Majus, Aquilejense Monasterium, Marmoutier in the French dep. of Bas-Rhin.
MAURILIACUM, Milly in the French dep. of Seine-et-Oise.
MAURONTI VILLA, Merville, Merghem in the Netherlands.
MAUSILIUM, Nova Ninus, Mozul in the pash. of Bagdad.
MAXENTII FANUM, Maixent in Poitou.
MAXILLA ASINI, Onugnachos, Cape Cervi in the Morea.
MAXIMIANOPOLIS, now Nekadeh, Neghade in Upper Egypt.
MEARNS, the river Mero in Spanish Galicia.
MECHLINIA, Malines, Mechelin in Belgian Antwerp.
MEDAMA, Medme, now Rosarno in Calabria-Ultra.
MEDANA, Meduana, now Mayenne in Maine, France.
MEDDEDACUM, Middlaer in Dutch Gelderland.
MEDELICIUM, Melicium, Melck, Molk in Hungary.
MEDEMELACUM, Medenblick, Memelick in Holland.
MEDERIACUM, Brugen, Bruck in Rhenish Prussia.
MEDIA, a region bounded by *Armenia* on the N; on the E by *Parthia;* on the S by *Susiana* and *Perses;* and on the W by *Assyria;* corresponding to the provinces of Azerbijan, Shirvan, Ghilan, the W part of Mazanderan, and the N part of Irak in modern Persia.
MEDIANIS, Minden in Westphalia.
MEDIESIS, Mediasch, or Medwisch in Transylvania.
MEDINA CELIA, Medymna Cœlestus, Medina Celi in the Spanish prov. of Old Castile.
MEDIOBURGUM, Middelburg in Holland.
MEDIOLANIUM, Meteln in Germany.
MEDIOLANUM, Milan in Lombardy.—Also Chateau-Mellant in the French dep. of Cher.—Also Moyland in the duchy of Cleves.
MEDIOLANUM SANTONUM, Santones, Saintes, Xaintes in Saintonge.
MEDIOLARIUM. See Ad Tres Lares.
MEDIUM, Walschmetz in the Tyrol.
MEDOBRIGA, Marvao in Portugal.
MEDOBRIGA, Medubrica, Montemora Villa, or Marvao, the Portuguese prov. of Beira.
MEDOSLANIUM, Meissau in Austria.
MEDUACUS MAJOR, the river Brenta in Lombardy.
MEDUACUS MINOR, the river Bachiglione.
MEDUANA. See Medana.
MEDULLAS MONS, Las Medulas in Portugal.
MEDUS, the river Abi-Kuren in Persia.
MEGALOPOLIS, Mecklenburg in Prussia.—Also Sinano in the Morea.
MEGIDDO, or Legio, Leijun in Syria.
MELÆ, Molise in Naples.
MELÆNA PROMONTORIUM, Cape-Kara-Boron in the Black Sea.
MELAMPHYLOS, Samos, Dryusa, or Anthemusa, Sussam in the Archipelago.
MELANGE, Madras in Hindostan.
MELAS, the Gensin Koremoz in Asia Minor.—Also the Mavro-Nero in Livadia.
MELAS SINUS, the Gulf of Saros in Rumelia.
MELDUNUM, or Minnodunum, Milden in the Swiss cant. of Bern.
MELFITUM, Molfetta in the prov. of Terra-di-Bari in Naples.
MELIBOCUS MONS, the Harz Mountains in Germany.
MELITA, Gaulos, or Lampas, the island of Malta in the Mediterranean.
MELITA, Civita-Vecchia in the States of the Church.
MELITENE, Malathija or Malatia in Asiatic Turkey.
MELLINDUM, or Mulancum, Meulan in the French dep. of Seine-et-Oise.
MELODUNUM, Metiosedum, or Milidunum, Melun in Isle-de-France.
MELOS, Milo in the Archipelago.
MELPES, the Mingardo in Naples.
MELPHIA, Melfi or Amalfi in Naples.
MELPIS, the Melfa in the Terra-di-Lavoro of Naples.
MELSUS, the river Abano in Spain.
MEMMALE, Mende in Languedoc.
MEMPHITIS NOMOS, Gizeh in Egypt.
MENÆ, Mineo in the Val-di-Noto in Naples.
MENAPIA, St. David's in Wales.—Also Wexford in Ireland.
MENAVIA, Monabia, Monœda, or Mona, the Isle of Man in the Irish sea.
MENBA, Menbigium, or Hierapolis, Menbigz in Syria.
MENDES, Achmun-Tanah in Egypt.
MENDICULEIA, Monzon in Spanish Aragon.
MENDRISIO, or Mendrium, Mendrees in Lombardy.
MENESTHEI PORTUS, Puerto de Santa Maria in Andalusia.
MENEVIA. See Menapia.
MENISI, the island of Gerbeh or Jerbah off the coast of Tunis.
MENITHINNA, Menden in Westphalia.
MENLASCUS, the river Bidassao in Biscay.
MENNIS, Dus Churmahk in Kurdistan.
MENOBA, or Mænoba, Velez Malaga in Andalusia.—Also Menuba, now Guadiamar in Andalusia.—Also the Velez river.
MENOSCA, Sumaya or St. Sebastian in Biscay.

MENUTHIAS, or MEMUNIAS, the island of MADAGASCAR in the Indian ocean.
MEPHITIS, MUFITI or MUFTI in Naples.
MERANIA, MORAN in the Tyrol.
MERCATORUM PORTUS, COPENHAGEN in Denmark.
MERCURIALE, MERCOGLIANO in the Terra-di-Lavoro of Naples.
MERCURII CURTIS, MIRECOURT in the French dep. of Vosges.
MERCURII INSULÆ, the TAVOLARO islands, near Sardinia.
MERGABLIUM, BEYER-DE-LA-MIEL in Andalusia.
MERGUM, AMERGO in Fez.
MERINIANUM, MARIGNANO or MELIGNANO in Milan.
MERMODAS, the river EGORLIK in Arabia.
MEROE, MERAWE in Nubia.
MEROPIA, the SIPHNUS or SIFANTO in the Archipelago.
MEROS, MEROTH in Damascus.—Also KUTAHIEH in Anatolia.
MERSABURGUM, or MARTIOPOLIS, MERSEBURG in Prussia.
MESAMBRIA, or MENEBRIA, the modern MISSAVIRIA in Macedonia.
MESE, PORTECROZ, one of the Hieres Islands in the Mediterranean.
MESENE, DIGEL in the pash. of Bagdad.
MESERIA, MEZIERES in Champagne.
MESONIA, the river MEDANIA in Sicily.
MESOPOTAMIA, the AL-JEZIRAH in the pash. of Bagdad.
MESSANA, MESSENE, MESSINA in Sicily.
MESSAPIA, MESSAGNA in Naples.
MESSENE, MAVRE-MATIA in the Morea.
MESSENIACUS SINUS, the GULF OF KORONE or KALAMATA in the Morea.
MESSUA COLLIS, now CETTE in Languedoc.
MESTUS, the river MESTRO in Rumelia.
MESUIUN, according to Kruse, MARSIBURGUM; according to others, MAGDEBURG in Prussian Saxony.
METACOMPSO, CONZO in Nubia.
MATACUM, CALAIOTE in Arabia Felix.
METAGONIUM PROMONTORIUM, CAPE-DE-TRES-FORCAS or SIGARI in Algiers.
META LEONIS, MATALONE in Naples.
METALLA, CIVITA-DE-GLISSA in Sardinia.
METALLINUM, MEDELLIN in Portugal.
METAPA, MEDENICO in Livadia.
METAPINUM, a mouth of the Rhone, now GRAS-D'ORGON.
METARIS ÆSTUARIUM, the WASH in England.
METARUS, or MEAROS, the MERSE.
METAURUM, GIOJA in Calabria-Ultra.
METAURUS, the METRO in the States of the Church.—Also the MARRO in Calabria-Ultra.
METELLI CASTRUM, MIDDLEBURG in Dutch Zieland.
METHAMAUCUM, MALAMOCO in Venice.
METHONE, MODON in the Morea.
METHYMNA, MOLIWA in Mitylene.
METHYMNA ASIDONIA, MEDINA SIDONIA in Spanish Seville.
METHYMNA CAMPESTRIS, MEDINA-DEL-CAMPO in Spanish Leon.
METHYMNA CETIA, MEDINA-CELI in New Castile.
METHYMNA SICCA, MEDINA-DE-RIO-SICCO in Spanish Leon.
METHYMNA TURRIUM, MEDINA-DE-LOS-TORRES in Spanish Estremadura.
METINA, or METAPINA, now COMMARGUE in Provence.
METOPON, ACRA SPANDONO.
METROPOLIS, now TIRIA in Asia Minor.
METUBARRIS, the OTOZECZ islands in the Gave.
METULUM, now METLING in Austrian Carniola.—Also MELLE in Poitou.
MEURSIA, MEURS or MORS in Prussian Dusseldorf.
MEVANIA, BEVANIA in the States of the Church.
MICHAELIA, ISOLA-DI-SAN-MICHELE in the Adriatic sea.
MICHAELOPOLIS, or ARCANGELOPOLIS, ARCHANGEL in European Russia.
MICHILINSTADIUM, MICHELSTADT in Hesse-Darmstadt.
MICULI MONS or PROMONTORIUM, PONTA MICHA, a cape in Dalmatia.
MIDÆUM, or MIGDUM, now SEID GAZA in Anatolia.
MILETOPOLIS, BELI-KESSR in Asiatic Turkey.
MILETUS, now PALATSCHA in Anatolia. — Also MILETONI in Crete.
MILEUM, or MILLEVIS, now MELA in Algiers.
MILIZIGARIS, or MELIZGERIS, now ARNAUL in India.
MILLÆ, MILLAS in the French dep. of Pyrenees-Orientales.
MIMALLIS, the island of MILO.
MIMATUM, MENDE in the French dep. of Lot.
MINAGARA, MUNICAPATAM in Hindostan.
MINARII MONTES, the MENDIP HILLS in Somersetshire.
MINCIUS, the river MINCIO in Lombardy.
MINERVÆ ARA, CHIFALE, an island in the Red sea.
MINERVÆ PROMONTORIUM, or ATHENÆUM, CAPE CAMPANILLA in Naples.
MINERVIUM. See ARX MINERVÆ.
MINIATUM, MIMATUM, or TEUTONIS, MINIATO-AL-FEDESCO in Tuscany.
MINIO, the river MIGNIONE in Tuscany.
MINIUS, the river MINHO in Portugal.
MINNIDUNUM, MOUNDON or MILDEN in the Swiss cant. of Vaud.
MINOA, MOLETIO on the coast of Crete.
MINORISSA, MANRESSA in Spanish Catalonia.
MINTHE, MOUNT MITTENA in the Morea.
MIRA, NURA, AYRE in the Balearic group.
MIROBRIGA, CAPILLA in Andalusia.—Also SINES or ODEMIR in Portugal.
MIRTILIS, MERTOLA in Portugal.
MISA, NIGOLA in the Papal deleg. of Urbino.
MISENUM PROMONTORIUM, MISENO in Naples.
MISNIA, MEISSEN in Saxony.
MISYNUS, MESURATA, an island off the coast of Tripoli.
MITYLENE, METELINO in the island of Mitylene.
MNEMIUM, CAPE RASEL DOAR in the Arabian gulf.
MOAB, EL-KURAR in Palestine.
MOBA, CHARAK M., KHARRAK in the pash. of Damascus.
MOCA, MOCHA in Arabia.
MOCENIA, MOTZING in Bavaria.
MODICIA, or MOGONTIA, MOZZA MONZA in Lombardy.
MODIM, ZUBA in Damascus.
MODOGULLA, MUDGULL in India.
MODONUS, the LIFFEY in co. Wicklow, according to Mannert; but more probably the SLANEY in co. Wexford, Ireland.
MODURA, BISNAGAR, or perhaps MADURA in Hindostan.
MODUTI, MOLADIVA in Ceylon.
MŒNUS, the river MAIN, at affluent of the Rhine.
MŒRIS (LACUS), or MŒRIDIS (LACUS), the BIRKET-EL-GARUM in Egypt.

MŒSIA, a Roman province corresponding to the modern SERVIA and BULGARIA.
MŒTONIUM, ROHATYN in Russia.
MOGELINA, MOGELN or MUGELN in Saxony.
MOGETIANA, ZICKA in Hungary.
MOGONTIACUM, or MAGONTIACUM, MAYENCE or MAINZ on the Rhine.
MOLBURIUM MONASTERIUM, MAUBEUGE in the French dep. of Nord.
MOLIBERGA, MUHLBERG in Prussian Merseburg.
MOLINUM, MOULINS in the French dep. of Allier.
MOLIUM. See AMANUS MONS.
MOLOCHAT, the MULUCHA in Barbary, Africa.—Also ALEABASSA, the town.
MOLOSSIS, JANINA in Albania.
MOLYBODES, ISOLO-DEL-FORO, off the E coast of Sardinia.
MOLYCRIUM, near the castle of RUMELIA in Northern Greece.
MOMEMPHIS, MENUF in Lower Egypt.
MONA. See MENAVIA.
MONACUM, MUNICH in Bavaria.
MONALUS, the POLLINA, a river in Sicily.
MONATIUM, MONETIUM, MANSPURG or MONSPURG in Carinthia.
MONDA, the river MONDEGA in Portugal
MONEMBASIA, MALVASIA in the Morea.
MONILIA (AD), MONEGLIA in Genoa.
MONŒCIA PORTUS, MONACCO in Tuscany.
MONOGLOSSUM, CAMBAY in Hindostan.
MONOSGADA, MITTENWALDE in Prussian Silesia.
MONS ACUTUS, MONTAIGUT in the French dep. of Tarn-et-Garonne.
MONS ALBANUS, or MONS AREOLUS, MONTAUBAN in Guienne.
MONS ALBUMUS, MONTE-DI-POSTIGLIONE in Naples.
MONS ALCINOUS, MONTALCINO in Tuscany.
MONS ALTUS, MONTALTO in the Papal States.
MONS ATER, the BLACK HARATSH in Fezzan.
MONS BRISIACUS, BREISACH in Baden.
MONS CHRISTI, the island of MONTE-CHRISTO off the coast of Tuscany.
MONS COMIANUS, KONIGSSTADTEN. See COMAGENÆ.
MONS DOLOROSUS, STIRLING in Scotland.
MONS EDULIUS, MONTSERRAT in Catalonia.
MONS GLABER, MADARE in Palestine.
MONS GOMERICUS, MONTGOMERY in Wales.
MONS JOVIS, MONGRI and MONTJOU in Catalonia.
MONS MARIORUM, MARINES in Andalusia.
MONS MARTYRUM, MERCURII, MONTMARTRE near Paris.
MONS MASSICUS, MONTE-DRAGONE in the Papal states.
MONS MEDIUS, M. MALEDICTUS, now MONMEDY in Lorraine.
MONS PESSULANUS, MONTPELIER in Languedoc.
MONS PHYSCON, MONTE-FIASCONE in the Papal states.
MONS REGALUS, MONTREALE in Sicily.
MONS REGIUS, KONIGSBERG in Prussia.
MONS ROSARUM, MONTROSE in Forfarshire.
MONS SACER, PUERTO-DE-RABANON in Spanish Galicia.
MONS SELEUCI, perhaps MONTMAUR in Languedoc.
MONS SILICIS, MONTELESE in Venice.
MONS SOLIS, M. BADONIA, BATH in Somerset.
MONS TABURNUS, MONTE-TABURNO in the Papal states.
MONS TIFATA, the MADDALONI spur of the Apennines.
MONS VICI, MONDOVI in Piedmont.
MONS VULTUR, MONTE-VULTURE in the Papal states.
MONTABORIUM, MONTABAUR in Nassau.
MONTES, MONS, BERGEN in Hainault.
MONTES ACUTI, SPITSBERGEN in the Arctic ocean.
MONTES ÆRII, MONTE ÆTÈSINO in Sicily.
MONTILIUM ADEMARII, MONTELIMART in Dauphiny.
MONUMETHIA, MONMOUTH in England.
MOPHIS, the river MAHE or MYHIE in India.
MOPSU ESTIA, MAMPSYZTA, now MESSIS in Karamania.
MORA, the MORAWA in Hungary.
MORADUNUM, WERDEN in Prussian Dusseldorf.
MORATUM, MURTEN in Switzerland.
MORBIUM, MORESBY in Cumberland.
MORDULA, MORDI PORTUS, BATICALO in Ceylon.
MORGENTIA, ST. GEORGIO in Calabria.
MORGUS, the L'ORCO in Italy.
MORICAMBE ÆSTUARIUM, MORECAMBE BAY on the coast of Lancashire and Cumberland.
MORIDUNUM, SEATON on the coast of Devonshire.
MORIMARUSA, the ARCTIC OCEAN.
MORITANIA, MORTAGNE in French Flanders.—Also a town in Marne.
MORON, MONTALVAO on the Tagus.
MOROSGI, ST. SEBASTIAN in Biscay.
MORTUUM MARE, the DEAD SEA.
MORUNDA, MARAND, MEHRAND in Persia.
MOSA, the river MAAS or MEUSE in Holland.
MOSCHA, MASCAT in Arabia Felix.
MOSCHICI MONTES, the MESIDJIRDAGH in Armenia.
MOSCHIUS, the river IBAR in Servia.
MOSELLA, the river MOSELLE.
MOSOMOGUM, MOUSON in Champagne.
MOSTORPITUM, MORPETH in Northumberland.
MOTENUM, MUTENUM, BRUCK on the Leitha in Lower Austria.
MOTYCA, MOTYCHANUS, MODICA in Sicily.
MUCIALLA, MUGELLO in Tuscany.
MULRICHA, the river MULVIAH in Algeria.
MUNDA, the river MONDEGO in Portugal.
MUNDA RURÆ, ROERMONDA in Limburg.
MUNDA TENERÆ, DENDERMONDE in E. Flanders.
MUNDA VISTULÆ, WEICHSELMUNDE in Prussian Dantzig.
MUNDU, the island and town of METE in the Arabian gulf.
MUNYCHI, STRATIOTICA, a port of Athens.
MURANUM, SUMMURANUM, HIERAMUM, MURANO in Calabria.
MURATUM ALVERNIÆ, MURAT in Auvergne.
MURGIS, ALMERIA in Spain.
MUROELA, now MUREK in Styria, Austria.
MURSA, ESSEK in Slavonia, Austria.
MURTA, LA MEURTE, a river in French Lorraine.
MURUS, MUR or CASTEL-MUR in Austria.—Also MORTALES in Spain.—Also MURO in Basilicata, Naples.
MUSA, MAUSCHID in Arabia.
MUSIS, the river ARPASU in Armenia.
MUSONIUM, MUSSON or MUZON in Hungary.
MUSOPALE, VISAPUR in India.
MUSSIPONTUM, PONSCAMASONIS or PONTA MOUSSON in Lorraine.
MUSTI, FESHURE in Algiers.
MUTARENSIS CIVIT., MUTAREM, now MAUTERN in Austria.
MUTILA, MODELO in the Papal states.
MUTINA, MOTINA, MODENA in Italy.
MUZIRIS, MODIRIS, MIRZNO or MIRDSCHNO in India.
MYCENÆ, near KRABATA in Greece.
MYCLIUS, HAGIOZ LUKAS on the coast of Greece.
MYENUS MONS, MOUNT ALPHION in Livadia.

MYGDONIUS, the river HERMAS or SINDSCHAR in the pash. of Bagdad.
MYLÆ, MARCELLO in Sicily.
MYLÆ, GNIOSA and LISTA, two islands near Candia.—Also MELAZZO in Sicily.
MYLASA, MELASSA, MELAZZA in Asia Minor.
MYNDUS, according to Leake, GAMISHLU; according to Cramer, MONTESHA in Asiatic Turkey.
MYOS HORMOS, KOSSEIR, COSSEIR in Egypt.
MYRA, MACRE, STRAMITA in Asia Minor.
MYRCINUS, perhaps ORPHANO in Thessaly.
MYRINA, SAMDARLIK in Asia Minor.—Also CASTRO or PALEO CASTRA in the island of Lemnos.
MYRIOPHYTUS, MEIROFITO in Otranto, Naples.
MYRIS PORTUS, or MYOS HORMOS, COSSEIR on the Red sea.
MYRONOS, the island MARATA in the Red sea.
MYRRHINUS, MERENDA in Livadia.
MYRTILIS, MERTOLA in Spain.
MYSIA, a district of Asia Minor, comprising the SW part of Asiatic Turkey.
MYSOCARAS, PORT AMAN in Marocco, Africa.
MYUS, PALATSHA in Asia Minor.

N

NABÆUS, NABARUS, the river DURNESS in Sutherlandshire.
NABALIA, the WAAL or NORDWAAL in Holland.
NABIUS, IVIA or NARIUS, now the JUVIA in Spain.
NABLIS, NAVUS, the river NAABE or NAB in Bavaria.
NABRISSA, VENERIA, now LEBRIGA in Andalusia.
NACCARORUM PALUS, ESTANQUE-DE-ALBELATE, a lake in Catalonia.
NACOLEIA, KODSHA SHEHR in Anatolia.
NACRASA, BOK-HAIR in Anatolia.
NADRA, the NARA in Parma.
NÆBIS, the CAVADO-NEYA in Portugal.
NAGARA, NATARURA, NAGARAM in Hindostan.
NAGNIA, NARI in Urbino.
NAISSUS, NEZZA or NISSA in Servia.
NAMADUS, LAMNÆUS, the river NARBUDA, NERBUDDA in India.
NAMARE, MAUER in Lower Austria.
NAMON, NAMURCUM, NAMUR in Belgium.
NANAGUNA, the river KRISHNA or the TAPTI in India.
NANASA, the ASON NANSA in Spanish Asturias.
NANCEGUM, NANCY in Lorraine.
NAPARIS, PROAVA in Wallachia.
NAPITINUS SINUS, the GULF of ST. EUPHEMIA on the coast of Naples.
NAPOCA, or COLONIA NAPOCENSIS, SAMOS-UJVAR in Transylvania.
NAR, the NERA in the Papal states.
NARABON, ARABON, ARRABONA, the RAAB in Hungary.
NARBO, NARBONNE in the French dep. of Aude.
NARES LUCANIÆ, MONTE-NERO in Naples.
NARNIA, NARNI in the Papal states.
NARO, the NARENTA in Dalmatia.
NARTHACIUM, AARIKA in Thessaly.
NARTHECUSA, CHESINUS near Rhodes.
NARYCIUM, GERACE in Greece.
NASICA, NASSUK in the prov. of Aurungabad in Hindostan.
NASIUM, perhaps GRAND-NANCY in Orleans.
NASTUS. See NESTUS.
NATIOLUM, GIOVENASSO in Bari, Naples.
NATISO, the river NATISONE in Northern Italy.
NAUBARUM, NAVARI or NAVARRI in Russia.
NAUCRATIS, near SALHAJAR in Lower Egypt.
NAUPACTUS, LEPANTO in Livadia.
NAUPLIA, the GULF OF NAPOLI-DI-ROMANIA in Greece.
NAUPORTUM, UPPER LAYBACH in Hungary.
NAUS PROMONTORIUM, CAPE COLONNA or CAPE NAU in Calabria-Ultra.
NAUSTATHMUS, ASPARANETTA in Sicily.—Also BALIREH in Anatolia.
NAUTACA, KECH in Turkistan.
NAVA, the NAHE, a tributary of the Rhine in Lower Rhine.
NAVILUVIO, the EO, NAVIA, or NALON in Spanish Asturias.
NAXOS, STRONGYLE, or DIA, the island of NAXIA in the Archipelago.—Also SCHISSO in Sicily.
NAXUANA, NAKSHIVAN in Armenia.
NAZARA, EL NACHAR in Palestine.
NEÆTHUS, the river NETO in Calabria-Ultra.
NEAPOLIS, NAPLES in Italy.—Also NAPOLI in Sardinia.—Also NABAKI in the Crimea.—Also SKALA-NOVA in Asia Minor.—Also GEDIDA in the pash. of Bagdad.—Also ANAKLIA in Mingrelia.—Also NABAL in Tunis.
NEAPOLIS DANICA, or NICOPIA, NIKIOPING in Denmark.
NEAPOLITANUS SINUS, the GULF OF HAMMAMET in Tunis.
NEBRISSA, LEBRIJA in Spain.
NECHESIA, GUALIBO in Egypt.
NEDA, the river BOUTZI in the Morea.
NEDUM, NOTO in Sicily.
NEGRA, NOKRA in Persia.
NELO, or NELUS, the RIO-DE-LA-PUENTE in Spanish Galicia.
NEMAUSUS, NIMES in the French dep. of Gard.
NEMEA, TRISTENA in the Morea.—Also the river LANGIA.
NEMESA, the NYMS in Luxemburg.
NEMETOBRIGA, MENDAYA in Spanish Asturias.
NENTIDAVA, NOSENSTADT in Austria.
NEOCÆSAREA, NIKSARA in Asiatic Turkey.—Also KALAT-EL-NEDSJUR in Aleppo.
NEOCOMUM, or COMUM, COMO on the Lago-di-Como.
NEON, or TITHAREA, near VELITZA in Greece.
NEONTICHOS, AINADSJIK in Anatolia.
NEPHERIS, MORAISAH in Africa.
NEPTUNIUM, or POSIDIUM PROMONTORIUM, BOS BURUN in Asia-Minor.
NERIGON, or NERIGOS, NORWAY.
NERITUM, NARDO in the Terra-d'Otranto of Naples.
NERIUM PROMONTORIUM, CAPE FINISTERE.
NERTOBRIGA, VALERA-LA-VIEJA in Spanish Aragon.
NERULUM, CASTELLUCIO, ORINOLO in Calabria.
NERVA, the river ORDUNA in Biscay.
NESCANIA, EL-VALLE-DE-ABDELACIZ in Andalusia.
NESIS, NISITA in the gulf of Naples.
NESTUS, or NASTUS, the river KARA-SU, or MESTO in Rumelia.
NETEGA, or NITHEGA, the river NETTE in the Prussian reg. of Minden.
NETUIM, NOJA in Naples.
NEVE, NEVA in Palestine.
NIA, BAMBOTUS, the SANAGA, SENEGAL in Africa.
NICÆA, NAGGUR in India.—Also ISNIK in Asia Minor.—Also NICE in the kingdom of Sardinia.—Also NIOLO in Corsica.—Also NEGA-PATNAM in India.
NICAR, NICER, or NIGER, the river NECKAR in Würtemberg.
NICATORIUM, the KARADJAG in Kurdistan.
NICE, KULELI, KUSELI, or KULULEU in Rumelia.
NICEPHORIUM, RACCA in the pash. of Bagdad.

NICIA, or NIGILLA, the CROSTOLO or ONGINO a tributary of the Po.
NICOMEDIA, ISMID in Anatolia.
NICOPOLIS, DIORIGUI in Armenia.—Also PREVESA in Livadia.—Also NICOPOLI in Rumelia.—Also NICOPOLI in Asia Minor.—Also KARS or KIASSERA in Egypt.
NICOPSIS, KAPARTIS on the coast of the Black sea.
NIDUM, NEATH on the Nedd in Glamorganshire.
NIDUS, the river NEDD in Glamorganshire.
NIELLA, NIVELLE or NIVELLES in S. Brabant.
NIEMA, NIHEIM in Prussian Minden.
NIGER. See NECKAR.
NIGIRA, GANA in Africa.
NILUS, the river NILE in Africa. Its mouths are (1.) BOLBITINUM OSTIUM, now the ROSETTA MOUTH; (2.) SEBENNYTICUM OST, now BURLOS; (3.) PHATNITICUM OST, the DAMIETTA MOUTH; (4.) MENDESIUM OST, now FUM-DIBEH; (5.) TANITICUM OST, now FUM-FAREDJEH; (6.) PELUSIACUM OST, or the TINEH MOUTH.
NINGUM, HUMAGO, MONTANA, GRISIGNANA, a town in Istria.
NINIA, KNIN in Illyria.
NINUS, or NINEVEH, now NIMRUD, KOYUNJIK, and KHORSABAD in the pash. of Bagdad.
NISÆA, HERAT in Georgia.
NISIBIS, NISEBIN in the pash. of Bagdad.
NISYRUS, NISIRO in the Archipelago.
NIUSA CASTRA, NEUS in the Prussian reg. of Dusseldorf.
NIVERNIUM, the NEVERS in the French dep. of Nievre.
NIVERIS, or NEVERIS, the river NIEVRE in France.
NIVARIA, perhaps the island of TENERIFFE.
NOÆ, NOARA in Sicily.
NOELA, NOJA in Spanish Galicia.
NOEOMAGUS VADICASSIORUM, either CHATEAU-THIERRY in the French dep. of Aisne, or CHALONS in the dep. of Marne.
NOLIBA, BOLANOS in Spain.
NOMENTUM, MENTANA in the States of the Church.
NOMISTERIUM, NIEMES in Bohemia. — Also NIMPTSCH in Sweden.
NONACRIS, NAUKRIA in the Morea.
NONYMA, NAUNI in Sicily.
NOORTA, ORTA CHAN in Al Gezira.
NORA, TORRA-FORCADIZZO in Corsica.
NORDEDI PAGUS, NORDEN in Holland.
NOREIA, FRIESACH in Austrian Styria.
NORIMBERGA, or NORICA, NURNBERG in Bavaria.
NOROSSUS, the GORI BISTAN mountains in the Ural chain.
NOTIUM PROMONTORIUM, MIZEN HEAD in the south of Ireland.—Also the CAPE OF CAMBOIA in Siam.
NOVA CASTELLA, NEUFCHATEAU in Luxemburg. —Also NEUFCHATEL in the French dep. of Seine-Inferieure.
NOVANA, MONTE-DI-NOVE in the States of the Church.
NOVANUS, the river GENGA in Italy.
NOVEMPAGI, BRACCIANO in Tuscany.
NOVESIUM, NEUS in Prussian Dusseldorf.
NOVIDUNUM, NOGENT-SUR-SEINE in the French dep. of Aube.
NOVIOMAGUM, NIMWEGEN in Holland.
NOVIOMAGUS, CAEN in the French dep. of Calvados; according to others, LISIEUX.—Also CASTILLON, or CASTELNAU-DE-MEDOC in the dep. of Gironde.—Also WOODCOTE in Surrey.—Also NEUMAGEN in Germany.—Also NOYON in the French dep. of Oise.
NOVIOREGUM, ROYAN in the French dep. of Charente-Inferieure.
NOVIUM, PORTO-MOURO in Spanish Galicia.
NOVIUS, the river NITH in Scotland.
NOVODUNUM, JUBLENS or JUBLEINS in Maine.
NOVUM OPPIDUM, NAY in the French dep. of Basses-Pyrenees.
NOVUS PORTUS, LYME in Dorsetshire.
NUBA, LAKE HADABA in Central Africa.
NUCARIA PALLARIENSIS, the NOGUERRA-PALERESA in Catalonia.
NUCARIA RIPACURTIA, the NOGUERRA-RIBAGORZANA in Catalonia.
NUCERIA ALFATERNA, NOCERA in Principato-Citra.—Also NOCERA in Urbino.
NUERIOLA, MASCANI in Naples.
NUMISTRO, NUSTRO, in Principato-Ultra.
NURSIA, NORCIA in the Papal States.
NYGDOSORA, NAGPUR in Hindostan.
NYMPHÆUM, JUAN-DE-MEDUA in Albania.
NYMPHÆUS, SANTIMBENIA in Sardinia.—Also the river NINFA in Campagna-di-Roma.
NYRAX, NIORT in the French dep. of Dieux-Sevres.
NYSA, NAZLI or NOSLI in Anatolia.—Also KOUS SHEHR in the Turkish pash. of Koniyeh.

O

OANOS, the river FRASCOLARI in Sicily.
OASIS MAJOR, EL-VAH, or AUGILEH, the Greater Oasis.
OASIS MINOR, EL-GHARBI, or EL-KASSAR, the Lesser Oasis.
OAXES, the PETRÆA, a river in Crete.
OBILA, AVILA, OVILA in Old Castile.
OBOCA, the river BOYNE in co. Meath, Ireland.
OBRIMA, the BUR-BASHA in Anatolia.
OBRIS, the river ORBE in Languedoc.
OBRINGA, the river AHR or AAR in Rhenish Provence.
OBUCULA, MONCLOVA in Andalusia.
OBULCO, PORCUNA; according to others, BUJALANCE in Andalusia.
OCEANUS ARCTORIS, the ARCTIC OCEAN.
OCEANUS EOUS, or INDICUS OCEANUS, the EASTERN OCEAN.
OCEANUS HYPERBOREUS, or CONGELATUM MARE, the ICY SEA.
OCEANUS SEPTENTRIONALIS, the NORTH SEA.
OCELLUM PROMONTORIUM, SPURNHEAD in Yorkshire.
OCELUM, FERMOSELLE in Spanish Leon. — Also AVIGLIANA in Piedmont.
OCETIS, SOUTH RONALDSHAY in the Orkney islands.
OCHUS, the river DEHASCH in Turkistan.—Also DAHR-ASBAN in Persia.
OCINARUS, the BAGNI, a river in Calabria-Ultra.
OCRA, BIRNBAUMER-WALDES in Austria, the same as the Julian Alps.
OCRICULUM, OTRICOLI in Urbino.
OCTAPITARUM, ST. DAVID'S HEAD in Wales.
OCTODURUS. See CIVITAS VALLENSIUM.
OCTOGESA, MEQUINENZA in Spanish Aragon.
ODESSUS, VARNA in Bulgaria.—Also ODESSA in the south of Russia.
ODYSSEA, CAPE MARZO in Sicily.
ŒASO, OYARZUN, or OYARCO, FONTARABIA in Biscay.
ŒASO PROMONTORIUM, CAPO-DEL-HIGUER in Biscay.
ŒBALIÆ TURRES. See TARENTUM.
ŒCHARDUS, the river ORKHON in Mongolia.
ŒN ADÆ, TRIGARDON in Livadia.
ŒNOE, UNISH, a river and port in Trebizonde.

ŒNI PONS, MUHLDORF on the river Inn.
ŒNONE, ŒNOPIA. See ÆGINA.
ŒNUSSÆ, the islands of SAPIENZA, KABRERA, &c., off the S coast of the Morea.—Also the SPERMADORI in the Archipelago.
ŒSCUS, ORESOVITZ in Bulgaria.—Also the river ESKER in Bulgaria.
ŒSTRYMNIDES INSULÆ, the CASSITERIDES or SCILLY ISLANDS.
ŒSTRYMNIS PROMONTORIUM, the LAND'S END in Cornwall.
ŒSYME, ESKI-KAVULLA in Thessaly.
ŒTA, MOUNT KUMAYTA; according to Kruse, KATAVOTHRA, VANNO in Greece.
ŒTYLOS, VITYLO or VITULO in the Morea.
OGIA, the island of D'YEU or DIEU off the French coast.
OGLASA, the island of MONTE-CHRISTO off the coast of Tuscany.
OGYRIS, perhaps the island of KISHM in the Persian gulf.
OLABUS, DSJUBBE, islands on the Euphrates.
OLANIGE, the island of ALNEY in the Severn.
OLARION, OLERON, an island in the Bay of Biscay.
OLBIA, TERRA-NOVA in Sardinia.—Also the island of ŒBES near Hieres.—Also KUDAK in the Russian gov. of Kief.
OLCACHITIS SINUS, the GULF OF STORA on the coast of Barbary.
OLCINIUM, DULCIGNO in Albania.
OLEAROS, ANTIPAROS in the Archipelago.
OLEASTRUM, BALAGUER in Catalonia.
OLENACUM, ELENBOROUGH, HELENSBURGH in Scotland.
OLGASSYS, ELKAS, ULGHUZ DAGH in Anatolia.
OLIBA, OLITTE in Navarre.
OLICANA, IKLEY in Yorkshire.
OLIMACUM, LIMBACH in Hungary.
OLINA, the river ORNE in France.
OLINS, HOLE in Baden.
OLINTHA. See OLYNTA.
OLINTIGI, PALOS; according to others, MOGUER in Andalusia.
OLIOS HIPPO, or OLISIPPO, the city of LISBON.
OLIVULA PORTUS, ST. HOSPICIO in Piedmont.
OLLIUS, the river OGLIO in Lombardy.
OLMIÆ PROMONTORIUM, MALANGARA, a cape in Greece.
OLONNA CURTIS, CORTE OLONO in Milan.
OLOOSSON, ALASSONA in Thessaly.
OLTIS, the river LOT in Guyenne.
OLUS, MIRABELLO in Crete.
OLYMPIA, MIRAKA in Thessaly.
OLYMPUS, the ANATOLÆ-DAGH or KESHISCH-DAGH in Asiatic Turkey.—Also MOUNT ELIMBE in Greece.—Also LACHA in Rumelia.—Also CAPE SANTA-CROCE in Cyprus.
OLYNTA, SOLTA, an island in the Adriatic sea.
OLYNTHUS, perhaps AGIO-MAMA in Rumelia.
OMANA, the GULF OF KATFAT in the Indian ocean.
OMBRI, KUM-OMBOS in Upper Egypt.
OMBRIO, or OMBRION, PALMA in the Canary islands.
OMENOGARA, AMEDNAGAR in Hindostan.
ONCHESTUS, ALAZARAKI in Livadia.
ONOBA, HUELVA, or perhaps ODIEL in Andalusia.
ONOBALAS, the ALCANTARA, a river of Sicily.
ONUGNATHOS PROMONTORIUM, CAPE XYLI in the island of Cervo.
ONUPHIS, BANUB in Egypt.
OORACTA, the island of KISHMEH in the Persian gulf.
OPHIS, the river OUF near Trebizond.
OPHIUSA, LAS COLUMBRETES or MONTCOLIBRE, FORMENTERA, an island of the Mediterranean.
OPHLA. See HIEROSOLYMA.
OPHRYNIUM, RONKIOV, or RENN-KEVI in Asiatic Turkey.
OPIÆ, OPIE, BOPFINGEN in Bavaria.
OPINUM, OPPIDO in Basilicata.
OPITERGIUM, ODEZZO in Venice.
OPIUS, OPH, a town in Asia Minor.
OPONE PROMONTORIUM, CABO-DEL-GADA on the coast of Africa.
OPPIDUM NOVUM, OSSUN in Gascony.—Also NARANDSHA in Marocco.
ORACANA, BALFRUSCH in Persia.
ORBELUS, MONTE-ARGENTARO in Turkey.
ORBITANIUM, VITOLANO in the Papal States.
ORCADES, the ORKNEY ISLANDS.
ORCAS, DUNNET-HEAD or DUNCANSBY HEAD in Scotland.
ORCELIS, ORIHUELA in Spanish Valencia.
ORCHOE, BASSORA in the Persian gulf.
ORCHOMENUS, SCRIPU in Livadia.—Also KALPAKI in the Morea.
ORDESSUS, the river in SERETH in Moldavia.
ORDIA, ORDIAGO in the north of Italy.
ORESTIA, or ORESTIS. See MEGALOPOLIS.
ORETHUS, the river OFETO in Sicily.
ORETUM GERMANORUM, ORETO in New Castile.
ORGANO, KHISHME in the Persian gulf.
ORGAS, the BURBASHA in Asia Minor.
ORGELLIS CIVITAS, SEO-D'URGEL in Aragon
ORGESSUM, ARGOVA in Turkey.
ORGIA, ARGAGNA in Spanish Aragon.
ORINE, DAHLAK, an island.
ORIPPO, VILLA-DE-DOS-HERMANOS in Andalusia.
ORIZA, SUKUEH in Syria.
ORNEON PROMONTORIUM, CAPE KARADIVA in Ceylon.
OROATES, the TAB, RASAIN in Persia.
OROLAUNO, ORLON in Luxemburg.
ORONTES, or TYPHON, the AASSI in Syria.
OROPUS, ROPO in Greece.
OROSPEDA MONS, the SIERRA-DE-MUNDO, DE-ALKARAS, and DE-RONDA in Murcia, Spain.
ORTHOSIA, ORTOSA in Anatolia.
ORTONA, ORITOLO in the Campagna-di-Roma
ORTOPLA, STARIGRAD in Croatia.
ORTOSPANO, perhaps CABUL in Afghanistan.
ORUBIUM PROMONTORIUM, CAPE CORROVEDO, CAPE SILLEIROS in Galicia in Spain.
ORURCOS, GORAX in Al-Gezira.
OSÆA, TORRE-DI-ORESTANO in Sardinia.
OSCA, HUESCA in Aragon.—Also HUESCA in Andalusia.
OSCARUS, ONCHE in the French dep. of Cote-d'Or.
OSCELLA, DOMO-D'OSSOLA in Piedmont.
OSERICTA, perhaps OESEL and the OELAND ISLANDS in the Baltic.
OSIANO, JEUZGAT in Roum, Turkey in Asia.
OSICERDA, OSSERA, IXAR in Aragon.
OSSA, MONTE-KISSAVO in Thessaly.
OSSET, CASTELLO-DE-DA-CUESTO or JUAN-DE-ALFARACHE in Andalusia.
OSSIGI LACONICUM, MAQUIZ in Andalusia.
OSSONOBA, ESTOY in Portugal.
OSTRA, ORZIANA in the Papal deleg. of Urbino.
OSTRÆ LACUS, the STAGNO-DI-LEVANTE in the Papal States.
OTHOCA, ORISTANO in Sardinia.
OTHONA, OTTERTON in England.
OVETUM, OVIEDO in Spanish Asturias.
OVILABIS, LAMBACH, according to others, WELS in Austria.
OXIÆ, the CURZOLARI ISLANDS in the Ionian sea.
OXIANA, TERMED in Turcomania.
OXIMAGIS, BOGMUTTY, a river in India.

OXINAS, or ACHERON, the river KUSSEH in Asiatic Turkey.
OXUS, the river AMU-DAIJA or ABU-AMU.
OXYNIA, MOKOSSI in Thessaly.
OXYRYNCHUS, BENESCH in Egypt.
OZENE, UZEN in Hindostan.
OZOGARDANA, PUCORIA in Al-Gezira.

P

PAALA, the SAVENA in the north of Italy.
PABLIA, the PAGLIA in Tuscany.
PACHNAMUNIS, HANDAHUR in Egypt.
PACHYNIPORTUS, PORTO-DI-PALO.
PACHYNUM, CAPE PASSARO in Sicily.
PACONIA, PONTALERIA, PORCELLI, an island of Sicily.
PACTIUS, CANALE-DI-TERZO in the prov. of Terra-d'Otranto in Naples.
PACTOLUS, the river SARABAT in Asia Minor.
PACTYA, BULAIR in Romelia.
PADINUM, BONDINO in Modena.
PADUS, the river Po, with its mouths, OSTIUM or FLUVIUM VOLANA, now the PORTO-DI-VOLANO; SPINETICUM or ERIDANUM OSTIUM, now the PORTO-DI-PRIMARO; OSTIUM CAPRASIÆ, now the PORTO-INTERITO-DI-BELL' OCHIO; and OSTIUM SAGIS, now the PORTO-DI-MAGNAVACCA.
PÆSTANUS SINUS, the GULF of SALERNO in Naples.
PÆSTUM, or POSIDONIA, PESTI or PESTO in Principato-Citra.
PÆSULA, EL-POZUELO in Andalusia.
PAGÆ, PACHE, now PSATO in Livadia.
PAGASITICUS SINUS, the GULF OF VOLO in Thessaly.
PAGRÆ, BAGRAS in Northern Syria.
PALÆBYBLOS, BALBYBLUS, ALCOBILE, now JAFR-JUNI in Damascus.
PALÆPAPHOS, CONCLIA in Cyprus.
PALÆPHARUS, PALÆPHARSALUS, KALABAKI in Thessaly.
PALÆPOLIS, or NEAPOLIS, the city of Naples.
PALÆRUS, near ZAVEDRA in Greece.
PALÆSTA, near PALEASSA in Greece.
PALÆSTINA, the modern SYRIA.
PALÆTYRUS, or TYRUS, near the modern SUR on the coast of Syria.
PALANTA, BALAGNA in Corsica.
PALANTIA, PALENCIA in Aragon.—Also VALENTIA-DE-DON-JUAN in Asturias.
PALATIUM, POLEGIA in the Papal states.—Also P. ADRIANI, PALAZZO in Sicily.—Also P. DIOCLETIANI, SPALATRO in Dalmatia.
PALFURIANA, VENDRELL in Catalonia.
PALIBOTHRA, PATELPUTER or PATALIPUTRA, Hindostan, ALLAHABAD, or perhaps PUTNA.
PALICA, PALAGONIA in Sicily.
PALINDROMOS, CAPE BAB-EL-MANDEB in Arabia.
PALINURUM PROMONTORIUM, CAPE SPARTIVENTO in Naples.
PALLA, PORTO-SAN-JULIANO in Corsica.
PALLANTIA, PALENCIA in Spain.
PALLANTIAS, the river PALANTIA in Spain.
PALLANUM, POLLUTRO, PAGLIETTA in Abruzzo-Citra.
PALLE, PALIO in Cephalonia.
PALLURA, PALI-KOIL in India.
PALMARIA, PALMARNOLA off the coast of Italy.
PALMATA, KUTSCHAK or KAINARISJIK in Bulgaria.
PALTUS, BALDO in Syria.
PALUDES PONTINÆ, the PONTINE MARSHES in Italy.
PALUMBINUM, PALOMBARO in the Papal states.
PALUS MÆOTIS, the SEA OF AZOF.
PALUS TRITONIS, the SHEBKAH-EL-LOWDIAH on the coast of Tunis.
PAMBOTIS PALUS, the LAKE of JANINA in Albania.
PAMISUS, PANISUS, the PIRNATZA in the Morea.
PAMPHYLIA, the district of TEKE-ILI in Karamania.
PANACHAICUS, MOUNT BOIDIA or VODA in the Morea.
PANAGRA, SEMAGDA in the north of Africa.
PANDATARIA, VANDATINA, an island on the coast of Naples.
PANDOSIA, near ANGLONA in Naples.
PANDOSIA BRUTTIORUM, near MENDOCINO in Naples.
PANGÆUS, CASTAGNATZ or the PUNDAR-DAGH in Rumelia.
PANIONIUM, PANIUM, DSCHANGLI in Anatolia.
PANNONIA, a region in ancient geography bounded on the N and E by the Danube; on the S by *Illyricum* and *Mœsia;* and on the W by *Noricum;* and corresponding to SLAVONIA, parts of HUNGARY, LOWER AUSTRIA, STYRIA, CROATIA, and those parts of BOSNIA and SERVIA which touch upon the Save.
PANOPE, ST. BLASIOS in Livadia.
PANOPOLIS, AKHENYN in Egypt.
PANORMUS, PANORMO on the W coast of Asiatic Turkey.—Also PORTO-RAPHTI in Livadia.—Also TEKET in the Morea.
PANTAGIUS, the river PORCARI in Sicily.
PANTANUS LACUS, the LAGO-DI-LESINA in the Papal states.
PANTICAPEA, BOSPHORUS, KERTCH or WOSPOR in the Crimea.
PANTICAPES, the river SAMARA in Russia.
PANTICHIUM, PANTIK in Anatolia.
PANYASUS, the SPIRNAZZA, a river in Albania.
PAPIÆ INSULÆ, the SOHAR ISLANDS in the Persian gulf.
PAPIRA, AJASSE in Anatolia.
PARACHOATHRAS, the ALVEND in Persia.
PARÆTONIUM, AL-BARETUN in Barbary.
PARALISSUM, NAGY-BANJA in Hungary.
PARAMBOLE, DEBOT in Egypt.
PARENTIA, PARENZO in Istria.
PARIENNA, BARIN in Hungary.
PARIETINA, VELEZ-DE-GOMERA in Marocco.
PARISII, the city of PARIS.
PARMÆ CAMPI, CHAM in Bavaria.
PARNASSUS, MOUNT LIAKURA in Greece.—Also the BASHA-DAGH in Asiatic Turkey.—Also BAZARDSJANLU in the pash. of Koniyeh.
PAROPUS, PARCO in Sicily.
PARRADUNUM, PARTENKIRCH in the Tyrol.
PAROPAMISUS MONS, the HINDU-CUSH in Afghanistan.
PARTHENIA, now SAMOS.
PARTHENICUM, PALAMITA in Sicily.
PARTHENIUM PROMONTORIUM, CAPE FELENKBURNON on the SW of the Crimea.
PARTHENIUS, the BARTIN or PARTHINE, a river in Anatolia.
PARTHENIUS MONS, BARBENIA in the Morea.
PARTHENIUS PORTUS, CETRARO in Calabria.
PARTHENOPOLIS, HADSJI-OGLU-BAZARDSJIK in Turkey.
PARTISCUM, RAOZ BECSE or FELEGYHAZA in Hungary.
PASAGARDA, PASA or FASA in Persia.
PASSA, PAPASLI, a town in Rumelia.
PASSALON, SHEICH-EL-HARDY in Upper Egypt.
PATAVISSA, MAR-UJVAR in Transylvania.

PATAVIUM, Padova in Venice.
PATERNIANA, Pedernoso in Spanish Estremadura.
PATERNUM, Torre-di-Fiumenica in Calabria.
PATHYSSUS, Tibissus, Tisianus, the river Theiss in Hungary.
PATMOS, Pathmos, Patino, Palmosa in the Archipelago.
PATRICIA. See Colonia Pat. Cerdubensis.
PATUMUS, Belbeis in Egypt.
PAUCA, Casa-di-Valineo in Corsica; according to others, Pocognano.
PAULON, the Poglion in the duchy of Nice.
PAUSILIPUM, the Grotte-del-Monte-di'-Posilipo in the Terra-di-Lavoro, Naples.
PAUSINUS, the Possidaria in Croatia.
PAUSULÆ, Monte Elpare, Grotta Azolino in the Papal states.
PAXÆ, the islands Paxo and Antiparo in the Ionian sea.
PAX JULIA, Beja in Portugal.
PEDALIUM, Cape Grega in Cyprus.
PEGUNTIUM. See Alminium.
PEGUSA. See Cnidus.
PEISO, Pelso, the Balaton or Platon See in Hungary.
PELACAS, Mount Daumakli in Anatolia.
PELINÆUS MONS, Mount Elias in Scio.
PELION, according to Kruse, Lagora in Thessaly. —Also Biklista in Rumelia.
PELLA, Palatisia, Alla-Kilissa in Rumelia, Bellue in Palestine.
PELLENE, near Trikala in Greece.
PELODES PORTUS, Butrinto in Epirus.
PELONTIUM APLANS, Pola-de-Lena or Conceja-de-Pilonna in Castile.
PELORUM PROMONTORIUM, Cape Peloro or Faro in Sicily.
PELTUINUM, La Civitella or Monte-Bello in Abruzzo-Ultra.
PELUSIUM, Tineh in Lower Egypt.
PELVA, Livono in Turkish Herzegovina.
PEME, Bembe in Egypt.
PENEUS, the river Salembria in Thessaly.—Also the Igliaco in the Morea.
PENNINUS MONS, the Great St. Bernard in Switzerland.
PENNOCRUCIUM, near Strettin; or, according to others, Penkridge in Shropshire.
PENOXULLUM PROMONTORIUM, Ord-Head in Caithness-shire.
PENTEDACTYLUS, Ras-al-Anf in Nubia.
PENTAPOTAMIA, the Punjab in India.
PEOS ARTEMIDOS, Beni-Hasan in Upper Egypt.
PEPARETHUS, the island of Scopelo in the Archipelago.
PEPHNOS, Pekno in the Morea.
PERGAMUS, or Pergamum, Bergma in Asiatic Turkey.
PERGANTIUM, Bregen in the kingdom of Sardinia.
PERGE, Karahissar in Anatolia.
PERIMUDA, Bombay in Hindostan.
PERINCARI, perhaps Peringary in Hindostan.
PERINTHUS, or Heraclea Thraciæ, Heraclitza.
PERMESSUS, Panitza in Livadia.
PERNICIACUM, Bertrais in Luxemburg.
PERSEIS, Yersele in Rumelia.
PERSEPOLIS, Tchil-Minar in Persia.
PERSIA, or Persis. See article Persia in the body of this work.
PERUSIA, or Pirusio, Perugia in Tuscany.
PESSIDA, Timbuctoo or Tombuctu in Central Africa.
PESSINUS, Bosan in Anatolia.
PESSIUM, Pesth in Hungary.
PETALIA PROMONTORIUM, Cape Carysto in Negroponte.
PETELIA, or Petilia, the modern Strongoli in Calabria-Ultra.
PETELIANA, Cataldo in Sicily.
PETERGALA, Bidur in India.
PETOBIO, Petau in Hungary.
PETRIANA, Old Penrith or Castle-Steeds in Cumberland.
PETRODAVA, perhaps Jassy in Moldavia.
PETROMANTALUM, Magny or Banteln in the Isle-de-France.
PETUARIA, Peterborough in Northamptonshire.
PEUCE, Piczina, an island at the mouth of the Danube.
PHAGRE, Orfan in Rumelia.
PHALACRUM PROMONTORIUM, Cape Sidari in Corfu.
PHALANNA, Karajoli in Thessaly.
PHALASIA, Falasia in Negropont.
PHALERIA, Phanari in Thessaly.
PHALERUM, Porto-Fanari near Athens.
PHANÆ PROMONTORIUM, Cape Mastico in Scio.
PHANOTE, Vuno or Gardiki in Albania.
PHARAN, Cape Ras-Mohammed in the Red sea.
PHARBÆTHITES NOMOS, Farbeit in Egypt.
PHARSALUS, Farsa in Thessaly.
PHARUS, Farillon, an island in Egypt.—Also Lesina or Hoar on the coast of Dalmatia.
PHARYGIUM PROMONTORIUM, Cape Aghia in Greece.
PHASÆLIS, Phozzeyil in Damascus.
PHASELIS, Takhova in Anatolia.
PHASIS, the river Rion or Rioni in Mingrelia, or the Poli near it.
PHATISANE, Vatisa or Fatsa in Trebizond.
PHEA, Castell-Tornese in the Morea.
PHENEOS, Phonea in the Morea.
PHERÆ, Palæa-Choro in the Morea.—Also Firino in Thessaly.
PHIGALIA, Paulitza or Phanari in the Morea.
PHILADELPHIA, Malatzkert in Karamania.—Also Allah-Shehr in Anatolia.—Also Amman in the pash. of Damascus.
PHILÆ, Jeziret-el-Birbe in Upper Egypt.
PHILECIA, Policza, Fulnek or Olmutz in Moravia.
PHILIA PROMONTORIUM, Cape Philine or Emineh in Rumelia.
PHILIPPOLIS, Eumolpius, Poneropolis, Filibe in Rumelia.
PHILISCUM, Blis in Syria.
PHILOMELIUM, Ilgua in Asiatic Turkey.
PHINOPOLIS, or Phinea, Dercus on the Bosporus.
PHINTONIS, Isola-de-Figo between Sardinia and Corsica.
PHOCÆA, Fokia in Anatolia.
PHOCARUM, Firan, an island in the Arabian gulf.
PHŒNICE, the river Zerka in Syria.
PHŒNICUS PORTUS, Dschesme in Anatolia.
PHŒNICUSA, Dattolo, one of the Lipari islands.
PHŒNIX, the river Salmenico in Greece.
PHOLIGANDROS, Policandro in the Archipelago.
PHOTICE, Soroto in Albania.
PHRICIUM, Urtziani in Thessaly.
PHRUDIS, the river Somme in Picardy; according to Mannert, the Bresle in Normandy.
PHTHURIS, Sasef in Egypt.
PHURGISATIS, Klingenberg in Moravia.
PHYCUS, Ras Sem in Tripoli, Africa.
PHYLE, Argivo-Castro in Livadia.

PHYSCUS, Odornek in the pash. of Bagdad.—Also Pozzi in Greece.

PIACUS, Piazza in Sicily.

PICENTIA, Piconza in Tuscany.

PICINIANA, Aragona in Sicily.

PICTONIUM PROMONTORIUM, Cape Les-Sables-d'Olone; according to others, Point-de-Boisvinet in Poitou.

PINDUS, Mount Agrafa in the west of Thessaly.

PINETUS, Pinhel in Portugal.

PINGUS, the river Ipek in Anatolia.

PINNA, Civita-di-Penna in Abruzzo-Ultra.

PINTIA, Valladolid in Spanish Leon.

PIRACEUS, or Piræus, Porto-Dracone in Greece.

PIRINA, Pire in Sicily.

PIRUM (Ad), the Birnbaumer-Wald among the Alps.

PISAURUM, Pesaro in Urbino.

PISAURUS, Isaurus, the Foglia in Urbino.

PISCENÆ, Pezenas.

PISORACA, Pisuerga in Spanish Leon.

PISTORIA, Pistoja in Tuscany.

PITANE, Sandarlik in the Morea.

PITINUM, Vasteo in Abruzzo-Ultra.

PITYNDA, Bagnagar in India.

PITYUS, Drandar in Mingrelia.

PITYUSA. See Chios, Demonesus, Ebusus, Lampsacus, Miletus, Opius.

PLACENTIA, Piacenza in Parma.

PLACIA, Panermo in Anatolia.

PLANASIA, Pianosa, an island near Corsica.

PLANESIA, Isola Plana in the east of Spain.

PLATÆA, Palæo Castra in Kokla, Greece.

PLAVIS, the river Piave in Venice.

PLEMMYRIUM PROMONTORIUM, Punta-di-Gigante, a cape in Sicily.

PLESTINA, Peschiolo in Abruzzo-Ultra.

PLITHANA, Pultana in Hindostan.

PLOTÆ INSULÆ, the Strivali group of islands in the Ionian sea.

PLUTIUM, Monte-Pulciano in Tuscany.

PNUPS, Aamara in Nubia.

PODIUM CERETANUM, Puicerda in Catalonia.

PODOPERURA, Barcelor in India.

POLATICUM PROMONTORIUM, Ponta-di-Promontoria in Istria.

POLEMONIUM, Fatisa, Fatsa in Trebizonde.

POLITORIUM, Pocigliano in the Papal States.

POLLENTIA, Pollenza in the island of Majorca.—Also Urbisaglia in the Papal States.

POLLUPICE, Finale in the duchy of Genoa.

POLYÆGOS, Polino in the Archipelago.

POLYTIMETUS, the river Sogd in Turcomania.

POMPELON, Pampeluna in Spanish Navarre.

POMPTINÆ PALUDES, the Pontine Marshes in the Campagna-di-Roma.

PONEROPOLIS, or Philippopolis, Filibe.

PONTES, Ponches in Picardy.

PONTIA, Isola-di-Ponza in the bay of Naples.

PONTUS EUXINUS, Sythicus Sinus, Pontus Tauricus, Mare Cimericum, or Sarmaticum, the Black Sea.

PORAS, or Pyretus, the river Pruth.

PORCA, Pforz or Alt-Pforz; according to others, Borgen.

PORCIFERA, Polcevera in the duchy of Genoa.

PORPAX, the river Birgi in Sicily.

PORTA AUGUSTA, Torquemada in Spanish Leon.

PORTÆ AMANI MONTIS, the same as Amanæ Pylæ.

PORTHMUS, Porto Bufalo in Negropont.

PORTOSPANA, Terpez in Persia.

PORTUNATA, Pontedura, an island in the Adriatic.

PORTUS CITARISTÆ, Toulon.

POSIDIUM, Posseta in Syria.—Also Castro in the island of Samos.

POSIDIUM, Cape Tschautchse-Aghisi in Anatolia.—Also Puntadella Licosa.

PRACTIUS, Borgas, a river in Anatolia.

PRÆNESTE, Polystephanon or Stephane Palæstrina in Campagna-di-Roma.

PRÆSIDIUM, Basiellica in Corsica.

PRÆSIDIUM CORNAVIORUM, Warwick in Warwickshire.

PRÆSIDIUM JULIUM, the same as Scalabis, Santarem.

PRÆTORIUM, Hebberstow in Lincolnshire.

PRÆTUTIANA REGIO, Teramo in Italy.

PRASLÆ. See Braslæ.

PRASUM, Cape-del-Gada in the SE of Africa.

PREPESINTHUS, Strongyla in the Archipelago.

PRIAPUS, Karaboa in Anatolia.

PRIENE, Samson-Kalesi in Anatolia.

PRIMIS MAGNA, I-Brim in Nubia.

PRIMIS PARVA, Old-Dongola in Nubia.

PRIVERNUM, Piperno Vecchio in the Campagna-di-Roma.

PROCHYTA, Procida, an island on the coast of Lavoro.

PROLAQUE, Pioraca in the Papal States.

PROMONTORIUM NERIUM. See Artabrum Promontorium.

PRONÆA, the Prum in Rhenish Prussia.

PROPONTIS, the Sea of Marmora in the south of Europe.

PROTE, Prodano, an island near the Morea.

PRUSA AD OLYMPUM, Bursa or Brusa in Anatolia.

PRYMNESIA, or Seid-Gazi, Amphion-Karahissar in Anatolia.

PSACUM, Cape Spada in Crete.

PSELCHE, or Pesla, Dakke in Nubia.

PSEUDOCELIS, Mocha in Arabia.

PSEUDOSTOMUS, the Cauvery in S. Hindostan.

PSYRA, Ipsara in the Archipelago.

PSYTTALA, Lipsocatalia in the Archipelago.

PTOLEMAIS, Tolometa in Tripoli, Africa.

PTOLEMAIS HERMII, Mensieh in Upper Egypt.

PTOLEMAIS THERON, now Mirza-Mombarik in Nubia.

PTUCCI, Tejada in Andalusia.

PTYCHIA, Scoglio di Vido, an island in the Ionian sea.

PUCINUM, Duino in Istria.

PUDNI, Dschesan in Arabia.

PUPULUM, or Porto Paglia, S. Giovanni-di-Pula in Sardinia.

PURA, Karkend or Khia in Beluchistan.

PURPURIÆ INSULÆ, the islands of Madeira near Marocco

PUTEOLANUS SINUS, or Cumanus Sinus, the Gulf of Pozzuoli in Naples.

PUTPUT, Hamamet in Tunis.

PYDNA, Kitros in Macedonia.

PYLOS, Old Navarino in the Morea.

PYRGI, St. Severa in Tuscany.

PYRRHA, Caloni in the island of Mitylene.

Q

QUINDA, Nembrot in Turkish Karamania.

QUINTANA, or Kinzen, Wischelburg in Austria

R

RABBATH MOAB, or Ar, Mehalet-el-Hadj in Syria.

RADANTIA, the REDNITZ or RETZAT in Bavaria.
RADIS, the island of RE near the mouth of the Garonne.
RÆTIARIA, ARZER-PALANCA in Bulgaria.
RAGANDONE, or RAGONDO, WINDISH FEISTRIZ in Styria.
RAPHIA, RETHA in Judea.
RAPTUM PROMONTORIUM, CAPE FORMOSA in Africa.
RARAPIA, FERREIRA in Portugal.
RATÆ, or RATIS, LEICESTER in Leicestershire.
RATIASTUM, MACHECOU in the French dep. of Loire-Inferieure.
RATOSTATHYBIUS, the river USK, or the NEATH in Wales.
RAUDA, ROA, a town in Old Castile.
RAURANUM, RAUM or ROM in Poitou.
RAVIUS, LOUGH ERNE in Ireland.
REATE, RIETI in the Papal States.
REATINUS LACUS, the LAGO-DELLE-MARMORE in the Papal States.
RECHOBOTIR, RAHABA in Asiatic Turkey.
REFUGIUM APOLLINIS, FANO in Sicily.
REFUGIUM GELA, near TERRA-NUOVA in Sicily.
REGANASBURCH, RAINESBURGUM, RATISBONA, or AUGUSTA TIBERII, the city of RATISBON.
REGIANA, VILLA-DE-REYNA in Andalusia.
REGILLUS LACUS, IL LAGHETTO-DELLA-COLONNA in the Campagna-di-Roma.
REGIO, KONTSHUK TSHEKINESH in Rumelia.
REGNUM, RINGWOOD in Southampton.
REGULBIUM, RECULVER in Kent.
REMISIANA, MUSTAPHA - PALANKA in Turkish Servia.
RERIGONIUM, STRATHNAVER in Scotland.
RESAPHA, ARSOFFA in Syria.
RETINA, RESINA in Neapolitan Lavoro.
REUNIA, RAGOGNA in Venice.
RHABANA, TAVOY in India.
RHABON, the river SYL in Walachia.
RHÆDIAS, the WHITE DRINO in Albania.
RHAMNUS, TAURO-CASTRO in Livadia.
RHAUSIUM, RAGUSA in Dalmatia.
RHEBAS, the river RIWA in Anatolia.
RHEGIUM, REGGIO in Calabria-Ultra.
RHEGIUM LEPIDI, or CIVITAS REGIUM, REGIO in Modena.
RHENUS, the river RHINE in Germany.
RHETICO MONS, the RHÆTIAN ALPS in the Tyrol.
RHIGODUNUM, RIBBLE-CHESTER; according to Mannert, RICHMOND in Lancashire.
RHINOCLURA, or RHINOCORURA, EL - ARISH in Lower Egypt.
RHITHYMNA, RETIMO in Crete.
RHIUM, CAPE-DI-FANO in Corsica.
RHIZON, RESINUS or RESANO in Turkish Montenegro.
RHIZONICUS SINUS, the GULF OF CATTARO in the Adriatic sea.
RHIZUS, RIZEH or IRRISCH in Trebizonde.
RHOBODUNUM, or EBORODUNUM, BRUNN in Moravia.
RHODA, ROSAS in Spanish Catalonia.
RHODANUS, the RHONE in France.
RHODE, the river SARIGOL in Russia.
RHODOPE, MOUNT DESPOTO or the DESPOTI-DAGH in Rumelia.
RHOE, KIRPE, a river in Anatolia.
RHOSSICUS SCOPULUS, CAPE KHYNZYR or TORTOSA in Syria.
RHOTANUM, the river TAVIGNANO in Corsica.
RHUBON, the WINDAU, a river in Russia; according to Gosselin, the NIEMEN.
RHUGIUM, REGENWALDE in Prussian Pomerania.
RHYBDUS, RIESI in Sicily.
RHYMNICI MONTES, the URAL MOUNTAINS.
RHYMNUS, the river GASURI in Russia.
RICINA, RECCO in the grand-duchy of Genoa.
RIGOMAGUM, RHEINMAGEN in Germany.
RIOBE, or ORBE, PROVINS in French Champagne.
RIRA, the river KAMOZIK in Rumelia.
RISARDIR, SAFI or AZAFFI in Marocco.
RITUMAGUS, or RODEPONT, RY in Normandy.
RITUPÆ, or AD RITUPIS PORTUM, RICHBOROUGH in English Kent.
RIVUS, PONTE-DELLA-RIVA or PONTE-DI-RIVALI in Venice.
ROBOGDIUM PROMONTORIUM, FAIRHEAD or BENNET HEAD in Antrim, Ireland.
ROBORETUM, TORRE-DE-MONCORVO in Portugal.
ROBRICA, PORTS-DE-LONGUE in the French dep. of Maine-et-Loire.
ROBUR, HOERBURG in Baden.
RODIUM, ROIGISE, ROYE in Normandy.
RODUMNA, ROANNE in the French dep. of Loire.
ROMULA, KARLSTADT in Croatia.
ROMYLIA, SUB-ROMULA in Naples.
ROSCIANUM, ROSSANO in Calabria-Citra.
ROSOLOGIACUM, or ROSOLODIACUM, TSCHERTSCHEM-KUPRI in Anatolia.
ROSTRUM NEMOVIÆ, MINDELHEIM in Bavaria.
ROTOMAGUS, ROUEN in Normandy.
RUBI, RUVO in Neapolitan Bari.
RUBICON, the river PISATELLO (FIUMICINO) in the Papal States.
RUBRAS, CABEZAS RUBRAS in Andalusia.
RUBRENSIS LACUS, L'ETANG-DE-SIGEAN, a lake in Languedoc.
RUBRICATA, OLESA in Catalonia.
RUBRICATUS, the river LLOBREGAT in Catalonia.—Also the SEIBUS in Algeria.
RUDIÆ, or RHUDIÆ, ROTIGLIANO in the south of Naples.
RUESIUM, or RIEUX, ST. PAULIEN in the French dep. of Morbihan.
RUFRÆ, LACOSTA RUFARIA in the Neapolitan prov. of Terra-di-Lavoro.
RURA, the river ROER in Prussian Aachen.
RUSADIR, or RYSSADIRUM, MELILLA in Marocco.—Also CAPO-DI-TRES-FORCAS.
RUSCINO, LA-TOUR-DI-ROUSSILLON in the south of France.—Also ROSCHINUS, TELIS, or VERNODUBRUM, the river TET in French Roussillon.
RUSELLÆ, ROSELLO in Tuscany.
RUSIBIS, MAZAGAN in Marocco.
RUSICADA, STORA on the coast of Barbary.
RUSICIBAR, RUSSUBICARI, or RUSIBRICARI, TEFESSAD in Algiers.
RUSSIPPIR, PORTOGAVEDO in Algiers.
RUSUCCOREÆ, or RUSUCURUM, COLEAH in Algiers.
RUTUBA, the river ROVA or ROYA in Nice.
RUTUNIUM, ROWTON in Shropshire.
RYSSADIUM, CAPE BLANCO on the west of Africa.

S

SABA, SAADA or SAADE in Arabian Yemen.—Also SABBEA in the same.—Also ASSAB on the coast of Abyssinia.
SABADIBÆ, LINGAN; according to others, PULO-WAY in the Malay archipelago.
SABARA, ARACAN in Further India.
SABARIA, the river MUR in Austria.
SABARUS, the river GAGRA in India.
SABATA, or SABATORUM VADA, SAVONA in the kingdom of Sardinia.
SABATE, SABATIA STAGNA, or SABATINUS LACUS, the LAGO-DI-BRACCIANO in Tuscany.

SABATINCA, St.-Johann-in-Traum in Upper Austria.
SABBATUS, the river Savato in Calabria.
SABIS, the river Sambre in Flanders.—Also the Torre-del-Savio in N. Italy.
SABRIANA, the river Severn in England.
SACATIA, Hodeida in Arabian Yemen.
SACHALITES SINUS, the Gulf of Sadshar, Seger, or Shahr in Arabia.
SACILI, Alcorucen in Andalusia.
SACIS AD PADUM, Comacchio in the Papal state of Ferrara.
SACRUM PROMONTORIUM, Cape St. Vincent in Portugal.—Also Cape Corso in Corsica.
SÆPINUM, Sepino in the Neapolitan prov. of Molise.
SÆTABICULA, Alzira in Valencia.
SÆTABIS, the Montesa river in Spain.—Also the town of Alcoy.—Also Xativa in Valencia.
SAGALASSUS, Aglason Bey in Anatolia.
SAGARICUS SINUS, Teligol, or the Gulf of Berezen in S. Russia.
SAGDIANA, or Hinderabi, Busheab, an island in the Persian Gulf.
SAGEDA, Schagepur, Singpur, or Sagur in India.
SAGRAS, the river Sagriano in Calabria.
SAGUNTIA, Xigonza or Ergonza in Andalusia.—Also Siguenza in New Castile.
SAGUNTUM, Murviedro in Valencia.
SALA, the river Saale in Prussian Merseburg.—Also the Szala or Beni-Tamar in Marocco.—Also the town of Selle or Sallee.—Also Setida in Andalusia.
SALACIA, Alsaçedo Sal in Portugal.
SALAMIAS, Salemjat in Syria.
SALAMIS, Porto-Constanza in Cyprus.—Also Colouri, an island near Greece.
SALANIANA, St. Jago-de-Villela in Portugal.
SALAPIA, Salpi in Capitanata.
SALAPINA LACUS, the Lago-di-Salpi in Capitanata.
SALARIA, Sabiote in Spanish Jaen.
SALAURIS, perhaps Puerto-de-Salen in Spain.
SALCHA, Salkhat in Damascus.
SALDÆ, Tedeles in Marocco.
SALDUBA, the Rio-Verde in Andalusia.—Also the town of Marbella.
SALEBRO, Scarlina in Tuscany.
SALETIO, Selz in the French dep. of Bas-Rhin.
SALGANEA, St. George in Greece.
SALIA, the river Sella in Spanish Asturias.
SALICE, Salæ or Taprobane, the island of Ceylon.
SALINÆ, Torre-delle-Saline in the Neapolitan prov. of Capitanata.—Also Torda in Transylvania.
SALIOCLITA, Saclas in the French dep. of Seine-et-Oise.
SALISSO, Sulzbach in Bavaria.
SALLUNTUM, Eski-Slana in Dalmatia.
SALMONA, the Salm, an affluent of the Rhine.
SALMONE, Cape Sidero in Crete.
SALMORUS, or Halmyris, Jeni-Sala in Bulgaria.
SALODURUM, the cant. of Solothurn or Soleure in Switzerland.
SALOPIA, Shrewsbury in England.
SALPESA, Facialcazar in Andalusia.
SALSOVIA, Tulcze in Bulgaria.
SALSULÆ, Salces in the French dep. of Pyrenees-Orientales.
SAMACHONITIS, or Aquæ Merom, the Bahr-el-Haleh in Syria.
SAMARA, the same as Ambiani.
SAMBRA, Santatoli in India.
SAMBUS, the river Sambul or Tshambul in India.
SAMICUM, Neocastro in the Morea.
SAMOSATA, Samisat in Asiatic Turkey.
SANCTIO, Seckingen in Baden.
SANDA, Miera in Spanish Oviedo.
SANGARIUS, the river Sakaria in Anatolia.
SANISERA, Alajor in the island Minorca.
SANTICUM, Wasserleonburg in Austria.
SANTONUM PORTUS, Rochelle, or Tonnay-Charente in French Poitou.
SAPIRENE, Sheduan, an island in the Arabian gulf.
SARABRIS, Toro in Spanish Leon.
SARALAPIS, Burgus in Sardinia.
SARAPANA, or Zadris, Sharapani in Georgia.
SARAVUS, the river Saar in France and Prussia.
SARDIS, Sart in Asia Minor.
SARDOPATRIS FANUM, St. Honorato in Sardinia.
SAREPTA, Serphant or Ssarfand in Syria.
SARIPHI MONTES, Mount Sahar or Shar in Persian Khorassan.
SARMATICI MONTES, Mount Tatra.
SARMIA, Guernsey in the English channel.
SARNADE, Ober-Unnacz in Turkish Croatia.
SARONICUS SINUS, the Gulf of Egina in Greece.
SARS, the river Sar or Ezaro in Spain.
SARUS, the river Seihan in Asia Minor.
SASINA, Porto-Cesarea in Calabria.
SASO, Saseno in the Adriatic.
SATACHTHA, Korti or Ambucote in Nubia.
SATALE, Sakme in the pash. of Erzerum.
SATANACUM, Stenacum in Champagne.
SATICOLA, Savignano in Capitanata, Naples.
SATRICUM, Pratica in Campagna-di-Roma.
SATURÆ PALUS, Lago-di-Paola in Calabria-Citra.
SATYRORUM INSULÆ, the islands of Anamba in the Chinese sea.
SAUNIUM, the river Saja in N. Old Castile.
SAVO, the river Saona in the Neapolitan prov. of Terra-di-Lavoro.
SAVUS, the river Sau or Save in Austria.
SAXETANUM, Motril in Spanish Granada.
SAXONES INSULÆ, the islands Fohr, Syll, and Romsoe near Denmark.
SCAIDAVA, Scedaba, Orotshuk or Rutshuk in Bulgaria.
SCALDIA, the river Schelde in France and the Netherlands.
SCANDILE, Skangero in the Ægean sea.
SCAPTE HYLE, Skipsilar in Rumelia.
SCARBIA, Scharnitz in the Tyrol.
SCARDONA, Arba in the Adriatic sea.
SCARDUS, Mount Argentaro between Servia and Macedonia.
SCARNIUNGA, the river Leitha in Hungary.
SCARPONA, Charpagne in the French dep. of Meurthe.
SCENA, the river Shannon in Ireland.
SCEPSIS, Eskiupsi in Asia Minor.
SCHERA, Calagero in Sicily.
SCHINUSSA, Skinosa in the Archipelago.
SCIATHUS, Skiatho in the Archipelago.
SCIDRUS, Sideroni, Siderno in Calabria-Ultra.
SCINGOMAGUS, Sezame in the Sardinian states.
SCODRA, Scutari in Albania.
SCOMBRARIA, the island of Islote in the bay of Carthage.—Also Capo-de-Palos in Spanish Murcia.
SCOMIUS, Scombros, Witoscha and Rulla, mountains in Turkey.
SCOPI, Uskup in Macedonia.

SCULTENNA, the river PANARO in the duchy of Modena.
SCURGUM, CÖSLIN in Prussian Pomerania.
SCYDRA, SIDERO-KAPSA in Macedonia.
SCYLACE, SIKI in Anatolia.
SCYLACEUM, SQUILLACE in Calabria-Ultra.
SCYLACEUS, or SCYLLETICUS SINUS, the GULF OF SQUILLACE.
SCYLLÆUM, SCYLLA in Calabria-Ultra.
SCYLLÆUM PROMONTORIUM, CAPE SCYLLA in the Morea.
SCYRUS, SKYRO in the Ægean sea.
SEBASTIA, SIWAS in Roum, Asiatic Turkey.
SEBATUM, SEBEN in Hungary.
SEBENNYTUS, SEMMENED in Lower Egypt.
SEBETHUS, FIUME-DELLA-MADALENA in Lavoro, Naples.
SEBINUS LACUS, LAKE SEO or ISEO in Lombardy.
SECERRÆ, now ST. PERE-DE-SERCADA, ST. CHELONI, or S. COLONIA SAJERRA in N. Spain.
SECOR, SABLES-D'OLONNES in the French dep. of La-Vendee.
SEDINUM, STETTIN in Prussian Pomerania.
SEGEDO ANGURINA, ST. JAGO-DELLA-HIGUERA in Andalusia.
SEGEDUNUM, COUSIN'S HOUSE in Northumberland.
SEGESAMUNCLUM, ST. MARIA-DE-RIBAREDONDA in N. Spain.
SEGESSERA, BAR-SUR-AUBE in Champagne.
SEGISA, a town of Spanish Murcia, now CEHOGIN.
SEGOBODIUM, SEVEUX in Franche Comte.
SEGOBRIGA, SEGORBE in Spanish Valencia.—Also PRIEGO in Cordova.
SEGODUNUM, RHODEZ, RODEZ in the French dep. of Aveyron.—Also SIEGEN; according to others, WURTZBURG in Westphalia.
SEGONA, the SAONE, a river of France.
SEGONTIA, SIGUENZA in Old Castile.
SEGONTIUM, CARNARVON in Wales.
SEGORA, SAUMUR in the French dep. of Maine-et-Loire.
SEGOSA, ESCOUSSE in France.
SEGUSIA, SUSA in Piedmont.
SELAMBINA, SALABRENNA in Andalusia.
SELE, SALAKIEH in Lower Egypt.
SELESTADIUM, SCHLETTSTADT in the French dep. of Bas-Rhin.
SELEUCIA, now KEPSE a village of Syria.—Also SELEFKIEH in Asia Minor.—Also UL MODAIN in Irak-Arabi.
SELEUCUS MONS, MONT SALEON in the French dep. of Hautes-Alpes.
SELINUS, SELENTI in Asia Minor.
SELLEIS, the river PACHISTA in the Morea.
SELYBRIA, SILIVRIA in Rumelia.
SEMANA SYLVA, the THURINGIAN FOREST.
SEMIRUS, SIMARI, a river in Calabria.
SEMNE, MANGALORE in Hindostan.
SEMPRONII FORUM, FOSSOMBRONE in the Papal deleg. of Urbino.
SEMPRONIUM, OEDENBURG in Austria.
SEMULOCENIS, ULM in Würtemberg.
SENA, SAIN, an island off the W coast of France.—Also SINIGAGLIA in the deleg. of Urbino.
SENIA, SEGNIA or ZENG in Croatia.
SENTIACA VILLA, SINZIG in the Prussian prov. of the Rhine.
SENUS, the MAY, a river of Chiampa.
SEPELACI, BURRIANA, a town of Spanish Valencia.
SEPHAR, DAFAR in Syria.
SEPIAS, CAPE ST. GIORGIO in Thessaly.
SEPINUM, SIPICCIANO in Central Italy.
SEPTEM FRATUS, the AFFEN-BERGE mountains in Algiers.
SEPTEMIACIS, MEMMINGEN in Bavaria.
SEPTEMPEDA, ST. SEVERINO in Central Italy.
SEPTIMANCA, SIMANCA in Spanish Valladolid.
SEQUANA, the SEINE, a river of France.
SERA, SINGAN-FU in Siam.
SERETIUM, SORICH in Dalmatia.
SERGENTIUM, ARTESINA MONTE in Sicily.
SERIANE, SAHARIDSJE or SERIA in Syria.
SERION, RIONS in the French dep. of Gironde.
SERMYLE, REVEL in Russian Esthonia.
SEROTA, VEROCZE or VEROVITO in Hungary.
SERRÆ, SERES in Turkish Macedonia.
SERUS. See SOBANNUS.
SERVITIUM BANIALUKA, SIEVEROUCZI in Turkish Bosnia.
SESSITES, SESIA, or SESSIA, a river in Piedmont.
SESTIANÆ ARÆ, CAPE VILANO on the NW coast of Spain.
SESTIARIA, CABO QUILATES in Marocco.
SESTUS, JALOWA in Rumelia.
SETELSIS, SOLSONA in Spanish Catalonia.
SETIA, SETENIL in Andalusia.
SETIA, SESSE or SEZZA in the Neapolitan prov. of Terra-di-Lavoro.
SETIDAVA, ZYDOWO in Prussian Posen.
SETIUM, CETTE in Languedoc.
SETUCI, CAYEUX in the French dep. of Somme.
SETUIA, SETIVA, SEVIA, SEGOVIA, or KESMARK, SYDZINA in Hungary; according to Kruse, CZYCHE.
SEUMARA, SAMTHAURO in Georgia.
SEVERUS MONS, VISSA in the Papal States.
SEVO MONS, the KIOLEN MOUNTAINS in Norway.
SEX, SEXI, or SEXITANUM, MOTRIL in Spanish Granada.
SIARUM, SARACATIN in Andalusia.
SIAZUR, SHAHRASOUR, SCHEREZUR, in Irak-Arabi.
SIBARIA, or SANTIZ, FUENTE-DE-SIBARRA in Spain.
SIBARIS, the river KOISMIR or ALADAN in Asia Minor.
SIBERENE, SEVERINA in Naples.
SICAMBRIA, ALT-OFEN in Hungary.
SICCA VENERIA, KEFF in Tunis.
SICILIA, the island of SICILY.
SICILIBRA, BAZILBAH in Tunis.
SICINUS, SIKINO in the Ægean sea.
SICORIS, SEGRE in Spanish Catalonia.
SICULIO. See TIBUR.
SICULUM FRETUM, the STRAITS OF MESSINA.
SICUM, SEBENICO in Dalmatia.
SICYON, ÆGIALIA, or MYCONE, now BASILICO in the Morea.
SIDA, ESKI in Anatolia.
SIDOLONCUM, or SEDELAUCUM, SAULIEU in the French dep. of Cote d'Or.
SIDON, SSEIDA or SAIDA on the Syrian coast.
SIEDUS, SOUSA in Greece.
SIGA, TAFNA, a river in Algiers.—Also NEDROMA, a village of Algiers.
SIGÆUM, JENISHEHR in Anatolia.
SIGMANUS, the river BOUCAUT VON REMISAN, PUECHE-D'ARCACHON in France.
SIGNIA, SEGNI in the Campagna-di-Roma.
SIGRIUM, CAPE SIGRI in Lesbos.
SIGRUS, TEMLUK in Algeria.
SIHOR, the river BESOR in S. Palestine.
SILARUM, the river SILARO.—Also CASTEL-ST.-PIETRO in the Papal States.
SILARUS, the river SELE in Naples.
SILBIUM, ad SILVIANUM, GARAGNONE in Bulgaria.
SILE, SALEHIEH in Egypt.
SILVINIACUM, SAUVIGNY in the French dep. of Loire et Cher.
SIMBRUINA STAGNA, or SIMBRIVIUM, SUBIACO in Central Italy.

SIMYLLA, CAPE ST. JOHANN in India.
SIMYRA, SUMRE in Egypt.
SINAI, MOUNT CATHARINE and SHAB-EL-MUSA in the Arabian desert.
SINCIACUS PAGUS, SINZIEH or SINSIEH in the Prussian prov. of the Rhine.
SINDÆ, the CELEBES ISLANDS in the Indian sea.
SINDICUS PORTUS, SUNDJICK in Asiatic Russia.
SINGARA, SIND-SHAR in the prov. of Al-Gezirah in Persia.
SINGAS, the river SIMEREN in Syria.
SINGIDAVA, or DEVA, SZEGEDIN in Transylvania.
SINGONE, TRENTSIN in the NW of Hungary.
SINNIUS, the river SENIO in the Papal States.
SINONIA, ZANNONE, an island in the sea of Tuscany.
SINUESSA, or SOESSA, ROCCA-DI-MONDRAGONE in the Neapolitan prov. of Terra-di-Lavoro.
SINUS FLANATICUS, the BAY OF QUARNARO in the Adriatic.
SIPHNUS, SIFANTO in the Cyclades.
SIRBITUM, SENNAAR, the capital of Sennaar in Africa.
SIRBON, or SIRBONIS LACUS, SEBAKET-BARDVIL in Egypt
SIRENUM PETRÆ, the Neapolitan island of LUNGA.
SIRIO, SERIO, SIRON, or PONT DE SIRON; according to others, RIONS in the French dep. of Gironde.
SIRMIO, SERMIONE in Venice.
SIRIS, TORRE-DI-SENNA in Naples.
SISAPON, or SISIPO, now ALMADAN or GUADALCANAL in Andalusia.
SISAR, AJEBBI in Algeria.
SISCIA, SISSEK in Hungary.
SITACE, ESKI-BAGDAD in Persia.
SITIFIS, SETIF in Algeria.
SITILLA, THIEL in France.
SITIOENTA, SALSOVIA in Bulgaria.
SITOMAGUS, THETFORD in Cambridgeshire.
SITTOCATUS, the river SIND in India.
SLIESTHORP, SCHLESWIG in Denmark.
SOATRA, PRAVADI in Asia Minor.
SOBANNUS, the river MEINAM in Further India.
SOCANNAA, the ABI-ATRAK in N. Persia.
SOETA, SAGA in Asiatic Russia.
SOLCI, PALMA-DI-SOLO in Sardinia.
SOLETUM, SOLITO in Naples.
SOLIA and SOLLURCO, SAN-LUCAR-LA-MAYOR in Andalusia.
SOLICIUM, SCHWETZINGEN in Holland.
SOLIMARIACA, SOULOSSE in Belgium.
SOLIS MONS, or SOLOCIS, CAPE CANTIN in Marocco.
SOLLINIENSIUM CIVITAS, SOLLIES in the French dep. of Var.
SOLOENTIA, CAPE BOGADOR in Africa.
SOLORIUS MONS, the SIERRA-NEVADA or SIERRA-DE-LOS-VERTIENTES in Spain.
SOLUS, CASTELLO-DI-SOLANTO in Sicily.
SONUS, the river SONE or SON, a tributary of the Ganges.
SOPIANÆ, SOPPAN or ZOPPIA in Hungary.
SORACTES, MONTE-DI-ST.-ORESTE in Tuscany.
SORBIODUNUM, OLD SARUM in Wiltshire.
SORDICE, LAKE LEUCATE in the French dep. of Aube.
SORIPHÆA, SERPHEAT in Judæa.
SOSSIUS, the river ARENA in Sicily.
SPALATUM, SPALATRO in Dalmatia.
SPATANA, TRINKOMALE in Ceylon.
SPAUTA, URMI or URMIAH, an ancient city of Persia.
SPELUNCA, SPERLONGA in the Campagna-di-Roma.
SPELUNCÆ, OSTUNI in Calabria.
SPERCHIUS, AGRAMELA or ELLADA in Greece
SPHETZANIUM, PECCIANA in Turkey.
SPINA, near ARGENTA on the Po.
SPINÆ, SPEEN in Berkshire.
STABIÆ, CASTEL-A-MARE in Naples.
STABULUM NOVUM, SOLIVELA or SAGARRE in Spain.
STACHIR, ST. JEAN, a river in Africa.
STAGIRA, STAVRO in Macedonia.
STALIOCAÑUS PORTUS, SLIOKAN in Brittany.
STANACUM, SCHARDING in Austria.
STANEDUM, CZACNITZA in Dalmatia.
STATONIA, CASTRO in Tuscany.
STENIMACHUM NICETAS, ICHTIMA in Turkey.
STENTORIS PALUS, the GULF OF ENOS.
STEPHANE, or STEPHANIO, ISTIPHANIE in Asia Minor.
STEREONTIUM, CASSEL; according to Reichardt, STEINFURT in Germany.
STIBOETES, or ZIOBERIS, the river ADJI-SU in Persia.
STILIDA, CAPE-DI-STILO in Calabria.
STLUPI, SLUIN in Croatia.
STOBI, ISTIB in Macedonia.
STŒCHADES, or MASSILIENSIUM INSULÆ, the HIERES ISLANDS off the S coast of France.
STRAGONA, STREHLEN, STRICGAU, or SCHINTAU in Prussian Silesia.
STRATONICA, FRANCO CASTRA in Macedonia.
STRATONICEA, ESKIHISSAR in Anatolia.
STRATUS, PORTA in Grecian Livadia.
STREVINTA, TREBITSCH in Moravia; according to Kruse, QUINTEN.
STRONGYLE, STROMBOLI, one of the Lipari islands.
STRYMONICUS SINUS, the GULF OF CONTESSA in the Ægean sea.
STUCIA, TUCCIA, now the DUFFI or YSTWITH, a river in Wales.
STURA, the STARA, a river in Italy.
STURIA, the river STOR, in Danish Holstein.
STYMBARA, STUBERA, STOBI in Macedonia.
STYMPHALUS, KATHOLIKON or KIONEA in the Morea.
SUASA, ST. LORENZO in the Papal deleg. of Urbino.
SUBI, the river FRANCOLI in Catalonia.
SUBLAQUEUM, or SUBLACUS, SUBIACO in the Campagna-di-Roma.
SUBRITA, perhaps GRUTES in Crete.
SUBUR, SITGES VILLANUEVA or SEGUR in the French dep. of Aveyron.
SUCHE, SUAKEM in Nubia.
SUCIDAVA, OSENIK or ASENIK in Austrian Galicia.
SUCRO, the river XUCAR in Spain.
SUDERNUM, SORTEANO in Tuscany.
SUDETÆ, or SUDITI MONTES, the ERZGEBIRGE and THURINGER WALD in Germany.
SUESTASIUM, SUISSATIUM in Spanish Biscay.
SUEVICUS LACUS, the same as BODAMICUS LACUS.
SULLIACUM, or SALLIACUM, SULLY-SUR-LOIRE in the French dep. of Loiret.
SULLIONACÆ MONTES, the BROCKLEY HILLS in the south of England.
SULMO, SERMONETA in the Campagna-di-Roma.
SUMMUM PYRENÆUM, SUMPORT in Spain.
SUMMUS LACUS, SAMMOLICO in Italy.
SUNIUM PROMONTORIUM, CAPE COLONNI in Attica.
SUPERÆQUANA COLONIA, or SUPEREQUUM, CASTEL VECCHIO SUBREQUO in Central Italy.
SURA, GJABAR or JABAR in Asiatic Turkey.
SURENTINUM PROMONTORIUM, CAPE VERDE in Africa.
SURRATHA, UM-ES SURRAB in Arabia Petræa

SURRENTUM, SORRENTO in the Neapolitan prov. of Lavoro.
SUSA, TOSTER, or SHUSTER in Persian Khusistan
SUSALEUS VICUS, STAGNO-DI-CAGLIARI, UTTA in Sardinia.
SUSUDATA, the city of BERLIN in Prussia.
SYBARIS, SIBARI or COSCILE, a river of Naples, in Calabria-Citra.
SYCURIUM, SIGURO in Thessaly.
SYMBOLON, BALAKLAVA in the Crimea.
SYME, SIMMI, an island in the Ægean sea.
SYNAOS, perhaps SEGULA in Asia Minor.
SYNNADA, SAID GAZELLE in Asia Minor.
SYPICIUS, CALA SISINA in Sardinia.
SYRACUSANUS PORTUS, PORTO-VECCHIO in Corsica.
SYRIAS, INDSJE, a cape on the Black sea.
SYROS, SIRA in the Ægean sea.
SYRTIS MAJOR, the GULF OF SIDRA on the N coast of Africa.
SYRTIS MINOR, the GULF OF CABES on the N coast of Africa.

T

TABERNÆ, RHEIN-ZABERN in French Alsace.—Also BERG-ZABERN in Bavaria.—Also BERN-CASTEL in the prov. of the Lower Rhine.
TABLA, or TABLÆ BATAVORUM, ALBLAS, DELFT in Holland.
TABRACA, THABRACA, TABARCA in Tunis.
TABURNUS MONS, ROCCA RAINOLA, or MONTE VERGINE, TABURNO in Central Italy.
TACATUA, TAKUSEH in Africa.
TADER, or TEREBRIS, the SEGURA, a river in Spain.
TÆNARUM, MAINA in the Morea.
TÆZALUM, KINNAIRD'S HEAD on the E coast of Scotland.
TAGÆ, DAMEGHAN, a town in Persia.
TAGARA, DEOGHIR, now DOWLATABAD in India.
TAGONIUS, HENARES, TAJUNNA, a river of Spain.
TAGRUS, TUNTO or MONTE-DI-SINTRA in Portugal.
TAGUS, the river TAGO or TEJO in Spain.
TALABRIGA, TALAVERA-DE-LA-REYNA or AVEIRO in Portugal.
TALARIA, TATRIA in Sicily.
TALIA, TATALIA in Turkey.
TALMENA, SCHAIMA in S. Persia.
TALMIS, KALABSCHE in Egypt.
TALUBATH, TAFILET in Africa.
TAMALLUMA, CALLA in Algeria.
TAMARA, the TAMBRE, a river in Spain.—Also TAMERTON in the south of England.
TAMARICIO PALMÆ, TORRE-DI-PALMA, PAGLIARA in Sicily.
TAMARUS, the river TAMAR in Cornwall.
TAMESUS, the river THAMES in England.
TAMNUM, MORTAGNE in the French dep. of Charente-Inferieure.
TAMYRAS, the river DAMAR in Syria.
TANAGER, the river NEGRO in Italy.
TANAGRA, SCAMINO, a town in Greece.
TANAIS, the river DON in Russia.
TANATIS, the island of THANET in England.
TANETIS VICUS, or TANETO, ST. ILLARIO in N. Italy.
TANGALA, DINDIGULL in India.
TANIS, SAMMAH or SAN in Upper Egypt.
TAOCE, BENDERIGK in Persia.
TAPHIÆ, MEGANISI, an island in the Ionian sea.
TAPHIS, TAFA in Nubia.
TAPHROS, FOSSA, or FRETUM GALLICUM, the STRAITS OF BONIFACIO.
TAPHRURA, the town of SFAX in Marocco.
TAPODIZUS, in Thrace, the town KIRKLISSA, KODSJE-TARLA.
TARANTASIA, MONSTIERS in Savoy.
TARENTUM, TARENTO in Naples.
TARGINES, TACINA in Greece.
TARNANTO, ALTEN-THUAN in Austria.
TARNIS, TARN, a river in the French dep. of Tarn.
TARODUNUM, MARK ZATERN or DERENDFORF in Germany.
TARRACON, TARRAGONA in Spain.
TARRAGA, LARRAGA in Spain.
TARSIUM, TERSAIZ in Hungary.
TARSUS, TARSO in Asia Minor.
TARVANNA, THEROUENNE in the French dep. of Pas-de-Calais.
TARVESIUM, TREVISO in Lombardy.
TASINEMETUM, KAPPEL, a town in Austria.
TATHIS, TAUD in Nubia.
TATTA, TUSLAG in Asia Minor.
TAUM ÆSTUARIUM, the FRITH OF TAY in Scotland.
TAURANIA, TORETTO in the Neapolitan prov. of Lavoro.
TAURESIUM, GIUSTENDIL in Wallachia.
TAURIS INSULA, TORKOLA in the Gulf of Venice.
TAURUNUM, BELGRAD or SEMLIN in Austrian Slavonia.
TAURUS, CAPO-DI-ST.-CROCE in Sicily.—Also the ALA-DAGH or AL-KURUN in Asia Minor.
TAVA, TAUS. See TAUM ÆSTUARIUM.
TAVIA, GOUKOURTHOY in Asia Minor.
TAXGÆTIUM, DAXWANG in Austria.
TAYGETUM, PORTAIS, MONTE-DI-MAINA, BRACCIO-DI-MAINA, a mountain in the Morea.
TEANUM APULORUM, PONTE ROTTO, CIVITARE in Italy.
TEANUM SIDICINUM, TEANO in the Neapolitan prov. of Terra-di-Lavoro.
TEARI JULIENSES, TRAYGUERA in Spain.
TEATE MARRUCINORUM, TEATEA or CHIETI in Central Italy.
TECELIA, OLDENBROOK or ELSFLETH in Germany.
TECUM, the river TECH in Languedoc.
TEDANIUS, the TERMANIA, a river of Dalmatia.
TEGANUSA, ISOLA-DI-CERVI, an island off the S coast of the Morea.
TEGEA, PALEO-EPISCOPI in the Morea.
TEGULA, TEULADA, a town of Sardinia.
TEJUM, TILIOS or TIOS in Asia Minor.
TELAMON, the village of TELAMONE in Tuscany.
TELESIA, TELESE in Central Italy.
TELLONUM, LUC in the French dep. of Var.
TELMESSUS, MACRI in Asia Minor.
TELOBIS, MARTORELL, VILLA DELLOPO in Spain.
TELO MARTIUS, TOULON in Languedoc.
TELONIUS, TURANO in Central Italy.
TELONNUM, or TELUMNUM, TOULON-SUR-ARROUX in the French dep. of Saone-et-Loire.
TELOS, PISCOPIA DILOS, one of the Cyclades.
TEMNUS, TSCHUMUS-DAG in Thessaly.—Also MENIMEN, a town in Asia Minor.
TENEBRIUM, CAPE TORTOSA in Catalonia.
TENEDO, THIENGEN in Austria.
TENOS, TENO or TINO, one of the Cyclades.
TEOS, SEDSHIDSHIEK in Asia Minor.
TEPHLIS, TEFLIS or TIFLIS in Persia.
TERGESTE, or TEGESTRA, the city of TRIESTE in Illyria.
TERGOLAPE, VOGGLBRUCK or LAMBACH in Austria.
TERIAS, LENTINI, a river in Sicily.
TERINA, the GULF OF EUPHEMIA in Italy.
TERMESUS, SCHENET in Asia Minor.
TERMOS, or TERMO, the SERRA, a river of Sardinia.
TERNODORUM, TONNERE in Champagne.

TERPONUS, Tschernemble in Carniola.
TESANA, Tesino in Austria.
TETARIUM, Aksher in Asia Minor.
TETRAPYRGIA, Ingesu in Asia Minor.
TETUS, Trieux in France.
TEUCERA, Thievres in Belgium.
TEUDERIUM, Detern in Germany.
TEURNIA, Villach, Larnfield in Austrian Illyria.
TEUTOBURGIUM, Bieloberdo in Hungary.
THABBA, Dobber in Arabia.
THALAME, Prastius in the Morea.
THAPSACUM, Deer or Deir; according to others, Zenobia in Syria.
THARO, Scharedsje or Zare, an island in the Persian gulf.
THARSANDALA, Czataleza in Thrace.
THAUMACI, Taimaki in Thessaly.
THEBÆ, Thiba or Thiva in Greece.—Also Armiro in Thessaly.—Also Duloma on the coast of the Arabian gulf.—Also Diospolis Magna, Karnak, Luxor, Gurnu, and Medinet-Abu in Egypt.
THEIPHALIA, Tiffange in French Poitou.
THEMISONIUM, Denisley in Asia Minor.
THEODONIS VILLA, Thionville in the French dep. of Moselle.
THEOTMALA, Detmold in Lippe-Detmold.
THERMÆ SELINUNTIÆ. See Ad Aquas.
THERMAICUS SINUS. See Macedonicum Mare.
THESPIA, Eremo Castro in Greece.
THESSALONICA, Saloniki in Turkey.
THEUDARIA, Todoriano in Albania.
THEUDURUM, Tudder in Belgium.
THILATICOMUM, Scharmely in Asiatic Turkey.
THILSAPHATA, Tellaafar, Tal-el-Hava in Asiatic Turkey.
THIMONEPSI, Tinna in Egypt.
THINÆ, Tana-Serim, Schunnin in Chinese Yunnan.
THISBE, Kakosia in Greece.
THORONOS, Isola Melere in the Ionian sea.
THOSPITES, the lake of Erzen or Arzen in Armenia.
THROASCA, Djirost or Girest in Persia.
THULE, the Mainland of Shetland.
THUMELITHA, Kanem or Kano in Bornou.
THYAMIS, Kallama in Albania.
THYATIRA, Akhissar in Asia Minor.
THYLIA, the river Dyle in the Netherlands.
THYNIA, Kirpeh, an island in the Black sea.
THYNIAS, Kiada Burun, a cape and town in Turkey.
THYRÆA, Tyros in the Morea.
THYSDRUS, El-Jemme in Africa.
TIARANTUS, the river Syl in Russia.
TIASA, Misitra, a river in the Morea.
TIBERIACUM, Bergen in Germany.
TIBERIOPOLIS, Ilgiun in Asia Minor.
TIBISCUM, Caravan in Turkey.
TIBULA, Longo-Sardo in Sardinia.
TIBUR, Tivoli in Central Italy.
TICARIUS, Tigari, a river in Corsica.
TICENUM, Pavia in Lombardy.
TICHIS, the Muga or Llobregat Minor in Catalonia.
TICINUS, the Tecino, a tributary of the Po.
TIFERNUM METAURENSE, St.-Angelo-in-Vado in Central Italy.
TIFERNUS, the Biferno in the Neapolitan prov. of Molise.
TIGAVA, El-Herba in Barbary.
TIGES, Tegusee in Tunis.
TIGRANAANA, Tauris in Armenia.
TIGULIA. See Segesta Tiguliorum.
TILAVENTUM, the river Tagliamento in Venice.
TILLIUM, Tulla in Sardinia.
TILOX, Cavallada, Scandola, a cape in Corsica.
TIMACHUS, the Timok in Turkey.
TIMACUM, Isperik in Bulgaria.
TIMAGENIS INSULA, Hassane in the Red sea.
TIMETHUS, Patti, a river in Sicily.
TINA, the Eden in Fifeshire.
TINIA, Timia, a river in Central Italy.
TINURTIUM, Tournous in the French dep. of Saone-et-Loire.
TIPARENUS, Spezzia, an island of Greece.
TIPASA, Damus in Algiers.
TISSA, Randazzo in Sicily.
TITHOREA, Velizza in Greece.
TITIANUS PORTUS, Torre-di-Campo-Moro in Corsica.
TITIUM, Kerka in Dalmatia.
TIUM, Tilios or Tios in Asia Minor.
TMOLUS, the Boz-Dagh, a mountain in Asia Minor.
TOBIUS, the river Towy in Wales.
TOCASANNA, the river Aracan in Bengal.
TOGISONUS, the river Concone in Venice.
TOLBIACUM, Zulch or Tulpich in Belgium.
TOLENUS, Turano in Central Italy.
TOLETUM, Toledo in Spain.
TOLIAPIS, the island of Sheppey in Kent.
TOLLEGATÆ, Telgate in Austrian Lombardy.
TOMARA, Mararco in India.
TOMERUS, the Hor, a river in Persian Mekran.
TOMI, Tomisvar in Bulgaria.
TONZUS, the river Tunsa or Tuncza in Rumelia.
TORNODURUM, Tonnere in French Champagne.
TORNOMAGENSIS VICUS, Tournon in France.
TOXIANDRIA, Tessenderloo in Belgium.
TRAEIS, the Trionto, a river in Naples.
TRAGURIUM, Trau in Dalmatia.
TRAJANA LEGIO, Coblenz in the Prussian prov. of the Rhine.
TRAJECTUM, or Trajectus Rheni, Utrecht, a city of Holland.
TRAJECTUM, Trajectus Mosæ, or Trajectum Tungrorum, Maestricht in Limburg.
TRALLIS, Sultan-Hissar in Anatolia.
TRANSMARISCA, Tromarisca, Turtukai in Turkish Bulgaria.
TRAPEZOPOLIS, Karadoghbru in Anatolia.
TRAPEZUS, Taraposan, Trebizond in Asia Minor.
TRASIMENUS LACUS, Lago-di-Perugia in Tuscany.
TRAUNUS, the river Traun in Upper Austria.
TREBA, Trevi in the Campagna-di-Roma.
TREBIÆ, Trevis, now Trevi in the Papal deleg. of Urbino.
TRES TABERNÆ, Borghetto in the Lombardian prov. of Lodi.
TRETUM PROMONTORIUM, Sebbakus, a cape in Algeria.
TREVA, Travendahl or Travemunde in Danish Holstein.
TRIBUNCI, Drusenheim in the French dep. of Bas-Rhin.
TRICA, Tricarico in the Neapolitan prov. of Basilicata.
TRICCA, Tricala in Thessaly.
TRIERUM PROMONTORIUM, Cape Mesurata in Tunis.
TRIGLYPTUM, Triglyphon, or Trilingum, Pegu in India.
TRINIUM, Trigno, a river in Neapolitan Molise.
TRIOBRIS, the river Obris, Truyere in the French dep. of Cantal.
TRIPOLI, Tripolitza in the Morea.—Also Triboli a town of Asiatic Turkey.—Also Ostraven or Tribul in Anatolia.

TRISANTON, the river ARUN in Sussex.
TRITÆA, TRITIA in the Morea.
TRITIUM MOTRICORUM, MOTRICO in Spanish Guipiscoa.
TRITON, TRITONIS LACUS, SHIBKAH-EL-LOWDIAH in Tunis.
TROEZEN, DAMALA in the Morea.
TRONUM, BUDIMIR in Dalmatia.
TUBUNÆ, TUBNAH in Algeria.
TUBUSUPTUS, BURG in Algeria.
TUDÆ, TUY in Spanish Galicia.
TUDER, TODI in the States of the Church.
TUEROBIS, the river TEIFY in Wales.
TUGIENSIS SALTUS, the SIERRA-DI-CAZORLA in Andalusia.
TULCIS, FRANCOLI in Spanish Catalonia.
TULLONIUM, TUDELA in Spanish Navarre.
TULLUM, TOUL in the French dep. of Meurthe.
TUOLA, GOLO in Corsica.
TURBULA, TOVARRA in Spanish Murcia.
TURENUM, TRANI in the Neapolitan prov. of Bari.
TURIA, the river GUADALAVIAR in Spain.
TURIASO, TARRAZONA in Spanish Aragon.
TURNACUM, TOURNAY in Belgian Hainault.
TURRES, TORRECILLAS in Portugal.
TURRES AURELIANÆ, or TURRES CÆSARIS, POLIGNANO in the Neapolitan prov. of Bari.
TURRUS, TORRE, TORRENTE, a river in Venice.
TURULIS, the river UDUBA or SÆTABIS in Spanish Aragon.
TURUNTUS, DUNA, WINDAU, a river in Russian Kurland.
TUSCA, WADI-EL-BERBER, a river in Tunis.
TUSCANA, TOSCANELLA in Tuscany.
TYANA, NIKDEH in Asia Minor.
TYBA, TAIBE or THAIBE in Syria.
TYLÆ, KILIOS in Rumelia.
TYRUS, SUR on the coast of Syria.
TZURULUM, SYRALLUM in Rumelia.

U

UCENSE CASTRUM, UZES in the French dep. of Gard.
UCIENSE, MARMOLEJO, ANDUJAR in Andalusia.
UDUBA, the river MIJARES in Spanish Aragon.
UDURA, CARDONA in Aragon.
UFFUGUM, FOGNANO in Naples.
UGERNUM, BEAUCAIRE in Lower Languedoc.
UGGADIS, PONT DE L'ARCHE in the French dep. of Eure.
UGIA, LAS CABEZAS in Andalusia.
ULCISIA CASTRA, SZENT-ENDRE in Hungary.
ULDA, AOUST, OUST in the French dep. of Arriege.
ULIA, MONTEMAYOR in Andalusia.
ULIARUS, OLERON, an island on the coast of France.
ULPIANUM, KOSTENDIL in Transylvania.
ULYSSIS PORTUS, S. MESSIO in Sicily.
UNA, the river IGUZUL in Marocco.
URBA, ORBE, ORBEN, or URBACH in the Swiss Pays-de-Vaud.
URBIACA, ALBAROCHES in New Castile.
URBIS, the river ORBE in Piedmont.
URBS VETUS, ORVIETO, a town in Tuscany.
URCAO, ARJONA in Andalusia.
URCESA, REQUENA, VELAS, ORGAS in New Castile.
URCINIUM, AJACCIO in Corsica.
URGO, GORGONA, an island of Tuscany.
URIA, ORIA in the TERRA D'OTRANTO of Naples.
URIAS SINUS, the GULF OF MANFREDONIA in Naples.
URICONIUM, or VIRONOCONIUM, WROXETER in Salop.
URIUM, the river TINTO in Spain.
URSOLÆ, ROUSSILLON in the French dep. of Isere.
USADIUM, CAPE OSEM in Marocco.
USCANA, near ISTURGA in the valley of the Drino.
UTHINA, UDINE in African Tunis.
UTICA, PORTO-FARINA in Tunis.
UTIS, the river MONTONE in Italy.
UXAMA, OSMA in Old Castile.
UXAMBRACA, OSMA in Biscay.
UXANTIS, OUESSANT or USHANT, an island off the coast of Brittany.
UXELA, EXETER in Devonshire.
UXELLODUNUM, PUECHE D'ISSOLU.
UXENTUM, UGENTO in the Neapolitan prov. of Terra-d'Otranto.

V

VACONTIUM, VAGY-VASONY in Hungary
VACORIUM, BISCHOFFSHOFEN or WAGREIN in Austria.
VACUA, VOUGA, a river of Portugal.
VADIMONIS LACUS, the LAGODI-BASSANO in Tuscany.
VAGEDRUSA, MANOMUZZA in Sicily.
VAGNIACÆ, NORTHFLEET in Kent.
VAHALIS, the river WAAL in Holland.
VALDANUS, LARWITZ in Bosnia.
VALENA, VALBACH in Hungary.
VALVATA, FORNACETTE in Tuscany.
VANDOGARA, PAISLEY in Scotland.
VANNIA, VENZONE in Venice.
VAPANES, VALPAJELA in Corsica.
VARA, the FIRTH OF CROMARTY in the north of Scotland.
VARÆ, near BODFARI in Wales.
VARDANUS, the KUBAN, a river of Asiatic Russia.
VASIS, FORCALQUIERE in the French dep. of Basses-Alpes.
VECTA, the ISLE OF WIGHT.
VEDRA, the river TYNE, or the WEAR in Durham.
VEGGIUM, VEZZO in Austria.
VELINUS LACUS, PIE-DI-LUCO in the Papal States.
VELITRÆ, VELLETRI in the Papal States.
VELLAUNODUNUM, BEAUNE in Burgundy.
VEMANIA, IRMENSTADT, or perhaps WANGEN in Bavaria.
VENAFRUM, VENAFRI in the Papal states.
VENDUM, VENDO in Dalmatia.
VENEDICUS SINUS, the GULF OF RIGA in the Baltic.
VENETIA, the city of VENICE.
VENIATIA, VINHAES in Portugal.
VENICNIUM PROMONTORIUM, the BLOODY FORELAND in Kent.
VENTA BELGARUM, WINCHESTER in Hampshire.
VENTA ICENORUM, CAISTER-ST.-EDMUNDS in Norfolk.
VENUSIA, VENOSA in the Papal states.
VERBANUS LACUS, the LAGO-MAGGIORE in Italy.
VERCELLÆ, VERCELLI in Piedmont.
VEREASUECA, PUERTO-DE-SAN-MARTIN in Spain.
VERLUCIO, perhaps LECKHAM in Warwickshire.
VEROMETUM, near WILLOUGHBY in Leicestershire.
VERULAMIUM, ST. ALBANS in Herefordshire.
VESONTIO, BESANÇON in France.
VESUNNA, or PETROCORII, PERIGUEUX in France.
VEXALLA ÆSTUARIUM, BRIDGEWATER BAY at the mouth of the Severn.
VIADRUS, the river ODER in Germany.
VICENTIA, VICENZA in Lombardy.

VICTORIA, KINROSS in Scotland.
VIDRUS, the VECHT, an arm of the Rhine, falling into the Zuyder-Zee.
VIGENNA, the river VIENNE, an affluent of the Loire.
VINDELIA PROMONTORIUM, PORTLAND BILL on the coast of Dorset.
VINDO, the river WERTACH in Bavaria.
VINDOBONA, the city of VIENNA.
VINDOCLADIA, WIMBOURNE, or perhaps GASSAGE in Dorset.
VINDOLANA, LITTLE CHESTERS in Northumberland.
VINDOMARA, EBCHESTER in Yorkshire.
VINDONISSA, WINDISCH in the Swiss cant. of Berne.
VINDONUS, FINCKLEY hamlet in Hampshire.
VINOVIA, BINCESTER in Yorkshire.
VIRIBALLUM PROMONTORIUM, CAPO-DE-GARGALO in Corsica.
VIRODUNUM, VERDUN in France.
VIRVEDIUM PROMONTORIUM, DUNCANSBY-HEAD on the coast of Scotland.
VOGESUS MONS, the chain of the VOSGES.
VOLATERRÆ, VADA on the coast of Tuscany.
VOLOGESIA, perhaps CUFAH in the pash. of Bagdad.
VOLSAS SINUS, CALVA BAY on the coast of Caithness.
VOLSINIENSIS LACUS, the LAGO-DI-BOLSENA in Italy
VOLUSTANA, VOLUTZA in Thessaly.
VOMANUS, the river VOMANO in Italy.
VULCEIUM, BUCCINO in the Neapolitan prov. of Apuglia.
VULTUMUM, CASTELLO-DI-VOLTUMA in the Papal states.

X

XANTHUS, the river ETCHEN-SHAI in Asiatic Turkey.
XEREDRUS, the river SUTLEDGE in the Punjab.

Z

ZABATUS, the ZAB-ALA river or GREATER-ZAB in the pash. of Bagdad.
ZACYNTHUS, the island of ZANTE.
ZAMA, ZAINAH in Marocco.
ZAO PROMONTORIUM, perhaps the BOC-DE-SORMIUS on the S coast of France.
ZELA, ZELEH in Asiatic Turkey.
ZEPHYRIUM PROMONTORIUM, CAPO-DI-BRUZZANO on the coast of Naples.
ZEUGMA, near BIR, on the Euphrates.
ZINGES PROMONTORIUM, CAPE DELGADO on the coast of Africa.
ZOSTER PROMONTORIUM, CAPE HALIKES on the coast of Greece.

ARTICLE II.

REVERSED MODERN, ANCIENT, AND MEDIÆVAL INDEX.

A

AA, a river in France, Agino.
AA, a river in Westphalia, Alpha.
AA, a river in Switzerland, Alpha.
AACHEN, Aquisgranum.
AADAYA, Castrum Fontarabiæ.
AALEN, Alena, Ala.
AAMARA, Pnups.
AAR, Arola, Arula.
AAR, Abrinca.
AARAKI, Narthacium.
AARAU, Araugia, Aravia.
AARBURG, Arlaburgum, Aroloburgum.
AARDAL-FIORD, Ardalius sinus.
AARGAU, Argovia.
ABACH, Abacum.
ABANCAY, Abancatus.
ABANO, Aponum, Aponus.
AB-ARACH, Saraca.
ABARANER, Abaranum.
ABASCIE, Glaucus.
ABASSI, Baetius.
ABAZES, Abkhasia.
ABBATIA, Aquæ Aponi.
ABBEVILLE, Abbatis villa, Abbavilla.
ABCHERON, Getara.
ABECOURT, HAPECOURT, Alba citria.
ABENSPERG, Abusina, Aventinum.
ABENST, ABENSEE, Ampla.
ABERDEEN (NEW), Aberdona.
ABERFORTH, Calcaria.
ABERFRAW, Gadiva.
ABERGAVENNY, Abergonium.
ABERNETY, ABERDORN, Abernethaea.
ABERYSTWITH, Aberistivum.
ABESH, ABEX, Abaxia ora.
ABIAGRASSO, BIAGRASSA, Albiate.
ABI-ATRAK, Socanaa.
ABI-KUREN, Medus.
ABIVERD, Apavarcticena.
ABO, Aboa.
ABONDANCE, Abbatia, Abundantia.
ABRANTES, Abrantium.
ABROBANIA, GROSS-SCHLATTEN, Auraria.
ABRUCENA, Urci.
ABRUST, Abia.
ABRUZZO, Aprutium.
ABSEL, ADZEL, GAUJENE, Adzelia.
ABU-AIJAN, Copar.
ABUKIR, Canopus.
ABUKSHEID, Heropolis.
ABUSIR, Busiris.
ABUTISCH, ABUTIG, Abotis.
AB-ZAL, Copratas.
ACAPULCO, Portus Aquæ pulchræ, Acapulcum.
ACCAR, Demetrias.
ACCOMBA, Hypania.
ACEGLIO, Acellium.
ACERE, Acerræ.
ACERENZA, Acherontia.
ACERNO, ACIERNO, Acernum.
ACERRA, Acerræ.
ACHEM, Acemum.
ACHIN, ANCHIN, Aquiscinctum.
ACHINA, ECHINA, ECHINO, Echinus.
ACHMUNEIN, ACHMUNIM, Hermopolis Magna.
ACHONRI, Achada.
ACHRIDA, OCHRIDA, GHIUSTENDIL, Achris.
ACHRIDA, Achridis Lacus.
ACHSTEDT, ACKSTEDT, Acsteda.
ACHTYRKA, Achyrum.
ACKEN, Aquæ Saxonicæ.
ACLE, Aclea.
ACLYBIA, Clypea.
ACOBA, Elcobatia.
ACQS, DAX, Aquæ Augustæ.
ACQUAPENDENTE, Acula.
ACQUI, Aquæ Statiellæ, Acquæ.
ACRA SPANDANO, Metopon.
ACRE, ST. JEAN D'ACRE, Aca, Acco.
ACSARAI, Anazarba.
ACSU, the river Ascanius.
ACSU, Ascanius Lacus.
ACTAMAR, WAN, Arsissa.
ACTAMAR, WAN, Artemita.
ADAMANB, Maniolæ.
ADAMSPIK, Malea.
ADDA, Abdua, Addua.
ADEL, ZEILA, Adolum Regnum.
ADELSBERG, Pastoina.
ADEN, ADEM, Adana, Adanum.
ADENBURG, ALDENBERG, Brannesia.
ADERNO, Adranum.
ADIAZZO, AJAZZO, AYAS, Adjacium.
ADIGE, ETSCH, Athesis, Alesius.
ADILLA, Mæsanites Sinus.
ADLERSBERG, Aquila montium.
ADMONT, Ad Montes.
AD LEMANUM, Allaman.
ADON, Adonum.
ADONCO, Adoncum.
ADOUR, ATUR, Aturis, Aturus.
ADOUR-DE-BAUDEAU, Aturus Baudeanis.
ADOUR-DE-SUEBE, Aturus Silvensis.
ADRA, Abdara.
ADRAMIT, Adramytteos.
ADRANA, Andrinople.
ADRIATIC, Adria.
ADRIBE, Crocodilopolis.
ADSCHI-SU, Stiboetes.
ADSJERUD, Arsinoe.
ÆGERI, or EGERE, Aquæ Regiæ.
ÆGER, LAKE EGERI, Egerius Lacus.
ÆLEN, AIGLE, Ala.
ÆRDING, ERDING, Ariodunum.

AERNON, AERNEN, Aragnum.
ÆROE, Arria, Arroa.
AERSCHOTT, Arescottum.
ÆTHIOPIC SEA, Æthiopicus Sinus.
AFFENBERGE, Septem Fratres.
AFKA, Aphaca.
AGAY, Agathenæ portus.
AGDE, Agatha.
AGEDABIA, Nasamones.
AGEN, Agenno.
AG-JALON, Lysinia.
AGHER, Aghera.
AGHIRMISCH-DAGH, Cimmerius mons.
AGIA, Dium.
AGIO-LINDI, Epidelium.
AGIO-MAMA, Olynthus.
AGITAKI, ALDEA-DES-GORRES, Commendum.
AGLA, Aeglæ.
AGLAR, AQUILEJA, Aquileja.
AGLASON-BEY, Sagalassus.
AGNADEL, Agniadellum.
AGNANO, Anianæ Thermæ.
AGNANO, Anianus Lacus.
AGNO, CLANIO, Clanius.
AGOBEL, Victoria.
AGOGNA, GOGNA, Albona.
AGORA, Agorum.
AGOSTA, Augusta.
AGOUT, Acutus, Agotius.
AGRA, Agara.
AGRAM, ZAGRAB, SAGABRIA, Civitas Montis Graccensis.
AGRAMELA, Sperchius.
AGRAPHA-GEBIRGE, Pindus.
AGREDA, Augustobriga.
AGREH, Artane.
AGRI, Aciris.
AGRIA, Abieta.
AGRI-DAGH, Ararat.
AGRIMONTE, Grumentum.
AGROPOLI, Acropolis.
AGUARICO, Ahuarico.
AGUAS, Aguarum Provincia.
AGUEDA, Æminium.
AGUERIA, Æminium.
AGULHAS, Acuum Caput.
AHUN, Agedunum.
AI, AY, Ageium.
AICHACH, AICH, Aicha.
AICHSTADT, Aureatum.
AIELO, Thyella.
AIGLE, EIGELL, in Switzerland, Aquilegia, Ala.
AIGLE (POINT), Catharistes.
AIGLE, in France, Ad Aquilas.
AIGUEBELLE, Aqua Pulchra.
AIGUEBELLETE, Aquæ Bellæ.
AIGUE-PERSE, Aqua Sparsa.
AIGUES, EIGUES, Eigarus.
AIGUES-CHAUDES, Aquæ Calidæ.
AIGUES-MORTES, Aquæ Mortuæ.
AIGUES-VIVES, Aquæ Vivæ.
AIGUILLES, Acuum Caput.
AIGUILLON, Acilio.
AILAH, AILE, Aelana.
AILESBURY, Aeglesburgus.
AIME, Axima.
AIN, DAIN, ENS, Danus.
AINADJIK, Neontichos.
AIN-CAREM, Aenon.
AINDAB, Deba.
AINGHUL, Modra.
AINOD, Aineda.
AINTAB, Antiochia ad Taurum.
AIR, AYR, Aera.
AIRE, in the French dep. of Pas-de-Calais, Aeria.
AIRE, in the French dep. of Landes, Adura.
AIROLO, ERIELS, ERIOLS, ORIENT, ORIENZ, Ariola.
AIRVAULT, OIRVAUX, Aurea vallis, Aravallis.
AIRY, Airiacum.
AISNE, Axona, Esna.
AITONA, Hitona.
AIX, in the French dep. of Bouches-du-Rhone, Aquæ Sextiæ.
AIX, Aquæ Gratianæ.
AJA, Himella.
AJA, KANDJES-BORUN, Criumetopon.
AJACCIO, Urcinium.
AJACCIO, Adjacium.
AJACCIO (GULF OF), Adjacii Sinus.
AJA-SALUK, AJA-JUNY, FESO, FIGERA, Ephesus.
AJASSE, Papyra.
AJEBBI, Sisar.
AKABAH, Elamiticus Sinus.
AKABET-ASSALOM, Catabathmus.
AKELO, Anchialus.
AKERMAN, Alba Julia
AKHENYN, Panopolis.
AKHISSAR, CROJA, Crua.
AKHISSAR, Thyatira.
AKKER, Ker.
AKLIBIA, Clupea.
AKMESCHID, Carionites Sinus.
AKMIN, Chemnis.
AK-SERAI, Garsavora.
AK-SHEHR, Antiochia ad Pisidiam.
AKSHER, Tetarium.
AKSU, Equius.
ALA, Sarnæ.
ALACHI, Halæ.
ALA-DAGH, Taurus.
ALAGOA, Alabon.
ALAIRAC, Castrum Alarici.
ALAIS, ALEZ, Alesia.
ALAJOR, Sanisera.
ALANCHES, Alantia.
ALANIS, Ancanicum.
ALARCON, Ilercao.
ALASSONA, Olusson.
ALATRI, Alatrium.
ALATYR, ALATUR, Allatura.
ALAYAH, Coracesium.
ALBA, Alba Fucentia.
AL-BAHRI, DELTA, Ægyptus Inferior.
ALBANI, Albania Nova.
ALBANIA, Illyria.
ALBANO, Albanum, Alba Longa.
ALBAROCHES, Urbiaca.
ALBASONA, Scampa.
ALBAXIN, Albasinum.
ALBE, ALBA, Alba Pompeja.
ALBA JULIA, CARLSBERG, WEISSENBURG, Alba Carolina.
ALBEGNA, Albinia.
ALBEK, Angulus Alpium.
ALBEN, MONTE-DEL-CARSO, Albanus.
ALBEN, Alpis.
ALBENGUE, ALBIENGA, Alba Ingaunorum
ALBERTON, Paraetonium.
ALBI, Albigæ.
ALBI, Alba Mursorum.
ALBIS, Albius, Albisus.
ALBON, S.-ROMAIND', Castrum Albonis.
ALBONOL, Sexitanum.
ALBORAN, Insula Erroris.
ALBORG, Alburgum.
ALBRET, LABRIT, Leporetum, Lebretum.
ALBUFERA, Amœnum Stagnum.
ALBUQUERKE, ALBUQUERQUE, Albuquercum Americanum.
ALBUQUERQUE, Albuquercum.

ALCACAR CEGUER, Alcasarium Parvum.
ALCACAR-DO-SAL, Alcasarium Salinarum.
ALCACAR-DE-SAN-JUAN, Alcasarium St. Joannis.
ALCACAR-QUIVIR, Alcasarium Magnum.
ALCACEDO, SAL, Salacia.
ALCAHIR, CAIRO, Cairus Magna.
ALCAI, Alcaja.
ALCALA-DE-GUADAIRA, Hienipa.
ALCALA-DE-HENARES, Complutum.
ALCALA-DEL-RIO, Aquæ Duræ.
ALCALA-LA-REALE, Alcala Regalis.
ALCANIZES, Alcanitium.
ALCANTARA, Norba Cæsarea.
ALCANTARA, Onobalas.
ALCANTARILLA, Alcantara.
ALCARAZ, Alcaratium.
ALCIRA, Algerisa.
ALCOBAZA, Eburobritium.
ALCORRUCEN, Sacili.
ALCOY, Sætabis.
ALDBOROUGH, Isurium.
ALECE, Halex.
ALEN, AALEN, Ala, Alena, Julia Alensis.
ALENTEJO, Provincia Transtagana.
ALEPPO, Alepum, Beræa.
ALERIA, Alalia.
ALES, Alesia.
ALESSANDRIA, Alexandria.
ALESSANO, Alexanum.
ALESSIO, Acrolissus, Lissus.
ALESTEROFORI, Gasorus.
ALETH, Alecta.
ALEXANDRETTA, SCANDERUN, Alexandria ad Issum.
ALEXANDRIA-DE-LA-PAILLAE, ALESSANDRIA DELLA PAGLIA, Alexandria Statiellorum, a Palea.
ALEXANDROWSK, Alexandrovium.
ALFACHS, Tenebrium.
ALFAQUES, ESFAGUES, Alfachusa, Ruspe.
ALFIDENA, Aufidena.
ALFORD, Alfordia.
ALFREDON, Alfradonium.
ALGAIOLA, Balagnia.
ALGARIA, Algarica.
ALGAU, Alemannia propria.
ALGEOIRAS, Algezira.
ALGHIER, ALGHER, Algeria.
ALGIERS, Algeria, Julia Cæsarea.
ALGIRO, Estiæ.
ALHAMA, in Aragon, Aquæ Bilbitanorum.
ALHAMA, in Granada, Artigi.
ALIAGA, Leonica.
ALICAD, ALCUDI, Pallentia.
ALICANTE, Lucentum.
ALICUR, Osteodes.
ALIDINELLA, Caria.
ALIOA, ALIOLA, Aliadora.
ALISE, ST. REINE D'ALISE, ALISO, Alesia.
ALIXAN, Alexianum.
ALIZETTE, Alisontia.
AL-KHABUR, CHABUR, Aboras.
ALKMAAR, Alcmaria.
ALLA-SHEHR, Philadelphia in Asia Minor.
ALLENBURG, Allæ Castrum.
ALLENJOIE, Allanum Jovis.
ALLER, Alara.
ALLIER, Elaver.
ALLONES, Vir.
ALMACARANO, Almacharana.
ALMADAN and GUADALCANAL, Sisapon.
ALMAGRO, Almagrum.
ALMAGUER, Almagra.
ALMANDRALEJO, Almandralegium.
ALMANZA, Almantica.
ALMARAZO, Lama.
ALMAS, Alisca.
ALMEDA, Alsena.
ALMELO, Almeloa.
ALMISSA, Alminium.
ALMONS, Camana Pontica.
ALMUDEVAR, Bortina.
ALMUNESSAR, Almunecara.
ALMUNIA, Nertobriga.
ALNE, Alaunus.
ALP, Alba Mons.
ALPEN, Alpes.
ALPHEN, Albiniana Castra.
ALPNACH, Alpinæum.
ALPROVOANA, Leuci Montes.
ALPUJARRAS, Alpuxari Montes.
ALRICK, ELRICK, Elrica.
AL-SAID, Ægyptus Superior.
ALSEN, Alsa.
ALSEN-SUND, Alsæ Fretum.
AL-SHEDUR, Gadaris.
ALT, OLT, Aluta.
ALTAI, Altajus Mons, Anarci Montes.
ALTAMURA, Lupatiæ.
ALT-BAGDAD, Sitace.
ALT-BUNZLAU, Boleslai Fanum Vetus, Boleslavia vetus.
ALT-CARLEBY, Carolina Antiqua.
ALT-CASTILIEN, Castella Vetus.
ALT-KATZENELNBOGEN, Cattimelibocum.
ALT-DONGALA, Primis Parva.
ALTEN, Alala-Joca.
ALTENA, Altenachium.
ALTENAU, Altenavium.
ALTENBIESEN, Juncetana domus.
ALTENBURG, in Bavaria, Atilia.
ALTENBURG, Altenburgum.
ALTENBURG, Korös Bangä.
ALTENBURG (OBAR), Ad Flexum.
ALTENHOVEN, Arrianum, Vetus Curia.
ALTENRIF, Alta Ripa.
ALTEN-THAN, Tarnanto.
ALTHEIM, Bragodunum.
ALT-MAMORA, Banasa.
ALT-MEDINGEN, Mesuium.
ALTMUHL, Alemannus, Alimona.
ALTMUHL-MUNSTER, Alemanni, Monasterium.
ALTO-BOSCO, Colophon.
ALTO-BOSCO (LAKE), Selenusia Palus.
ALTRANSTADT, Antiqua Ranstadt.
ALT-OETTINGEN, Hodingae.
ALT-OFEN, Acincum.
ALTREU, Alta-Riva.
ALTRIP, Alta Ripa.
ALTSTATTEN, Alterpretum.
ALVA-DE-TORMES, Alba.
ALVERTON, Albertonia.
ALVIDONA, Levidona.
ALWEND, Parachoathras.
ALZATO, Alciatum.
ALZEY, ALZHEIM, Alceja.
ALZIRA, Sætabicula.
AMACK, Amagria.
AMADABAD, Armadabatum.
AMADAM, HAMADAN, Ecbatana, Amadia.
AMAL, Amalia.
AMALFI, MELFI, Melphia.
AMANTEA, Adamantia, Amantia.
AMANZE, a river in the French dep. of Haut-Saone, Emaus, Amans.
AMANZE, Esmantia.
AMARANTE, Araducta.
A MARE, Cupra Montana.
AMASIA, Amasea.
AMASSERAH, AMASTRO, Amastris.

AMATO, Lamecus.
AMAXICHI, Leucas.
AMAYA, Amagia, Aregia.
AMBIERTE, Amberta
AMBLETEUSE, Ambletosa.
AMBLEVE, Amblavia, Amblava.
AMBOISE, Ambacia.
AMBON, Amboina.
AMBOURNAY, AMBRONAY, Ambroniacum.
AMEDNAGUR, Omenogara.
AMELIA, Ameria.
AMERGO, Mergum.
AMERSFORT, Amifortia.
AMERSHAM, Agmundeshamum.
AMESBURY, Ambrosii Vicus.
AM-HOF, Curia Norici.
AMIENOIS, Ambianensis ager.
AMIENS, Ambianum, Samarobriva.
AMMERLAND, Ambria.
AMMER-SEE, Ambræ Lacus.
AMOL, Galla, Amolia.
AMONA, Anemo.
AMONDE, Almon.
AMONEBURG, Castrum ad Amanam, Ohmam.
AMPADERE, Catarractes.
AMPIGLIONE, Empulum.
AMPLEPUIS, Ampliputeum.
AMPUGNANI, Ampuniana.
AMPUIS, Antea.
AMPURDAM, Emporiensis tractus.
AMPURIAS, Emporiæ.
AMSEL-, AMSERFELD, Campus Cassobus.
AMSTEL, Amstela.
AMSTERDAM, Amstelædamum, Amsterodamum.
AMU-DEREJA, ABU AMU, Oxus.
AMUR, Casius.
AMY, Dædalia.
ANAGHELONE, Anagelum.
ANAH, Hena, Anathan, Anathon.
ANAMBA, Satyrorum insula.
ANAPHI, Anaphe.
ANCENIS, Angenisium.
ANCHEDIVA, Angadiva.
ANCONA, Marchia Anconitana.
ANCONE, Acusio.
ANCRE, Ancora.
ANCY-LE-FRANC, ANCYLE, Anciacum.
ANDAGUILAS, Andaguili.
ANDALUSIA, Vandalitia.
ANDAMANS, Barussæ Insulæ.
ANDANCE, Andancia.
ANDECHS, Andex.
ANDELLE, Andelejus Fluvius.
ANDELOT, Andelous.
ANDELYS, Andelajus, Andelejum.
ANDENNE, Andana.
ANDEOL, Burgias.
ANDERAB, Bactrus.
ANDERNACH, Antunnacum, Antenacum.
ANDES, Catenæ Andium.
ANDORIA, LAC D', Lacus Andurianus, L. Salsus.
ANDRINOPLE, Adrana.
ANDROSEN, Ardrosa.
ANDUJAR, Iliturgis.
ANDURVIA, Caicandros.
ANDUZE, Andusia.
ANF-AL-HAGIAR (CAPE), BEL-MONTE, Dei Facies.
ANGELO-CASTRO, Ithoria.
ANGERMANNFLODT, Angermannus fluvius.
ANGERMANNLAND, Angermannia.
ANGERS, Andegava, Andegavum.
ANGHIARI, Anglarium, Angleria.
ANGLE, Angla Montana.
ANGLESEA, Anglorum Insula.
ANGLEN, Anglia Minor.
ANGLURE, Angledura.
ANGOULEME, Æquolesima, Iculisma, Angolemma.
ANGOUMOIS, Ecolesimus pagus.
ANGOURY, Ancyra.
ANGUILLA, Anguis.
ANHALT (DUCHY OF), Ascania Regio.
ANHALT, Anhaltinum Vetus.
ANJOU, Castrum de Anjove.
ANJOU (PROVINCE OF), Andegavia.
ANNAGIOUS, Annacingi.
ANNANDALE, Annandia.
ANNEBON, Annaboni Insula.
ANNECY (LAKE), Anecius Lacus.
ANNECY-LE-VIEUX, Annecium vetus.
ANNECY, in Savoy, Anecium, Annesiacum.
ANSBACH, Onolsbacum.
ANSE, Assa-Paulini.
ANSEDONIA, Cosa.
ANSENNA, Angira.
ANSGA, Ansita.
ANSLO, Ansloga.
ANTAKHIA, Antiochia.
ANTALI, Attalia.
ANTEQUERA, Antecaria.
ANTHIL, Antilia.
ANTIBES, Antipolis.
ANTIGOA, Antiqua.
ANTILODI, Coryne.
ANTIMILO, Antimelos.
ANTIOCHETA, Antiochea super Cragum.
ANTIPAROS, Olearus.
ANTIVARI, Antibarum.
ANTOING, Antonia.
ANTRAINS, ENTRAINS, Interamnes.
ANTRIM, Antrinum.
ANTRODOCO, Interocrea
ANVERS, Antverpia.
ANZA, Antia.
ANZI, Anxia.
AOSTA, Augusta Prætoria.
AOUST, Ulda.
APENRADE, Apenroa.
APLANS, Pelontium.
APPENZELL, Abbatis Cella.
APPLEBY, Aballaba.
APREMONT, Asprimontium.
APS, Alba Helveturum.
APT, Apta Julia.
AQUANILE, Hylias.
AQUAPENDENTE, Aquula, Acula.
AQUARA, Acquara.
AQUATACCIO, Almo.
AQUILA, Avella, Aquila in Vestinis, Amiternæ.
ARABA-MADFUN, Abydus.
ARABAT, Heracleum.
ARABHISSAR, Harpasa.
ARACAN (PROVINCE OF), Argentia Regio, Sahara
ARACAN, Cacosana, Tacosanna.
ARACENA, Lælia.
ARA-CLUDIEH, Claudias.
ARAK, KARAK, HADSCHE, Araceme.
ARANYAS, Aranyensis Sedes.
ARANYOS, Aranus.
ARAQUIL, Aracillum.
ARAS, Araxes.
ARAU, Aravia.
ARAVA, Arva.
ARBE, Scardona.
ARBERG, Mons Arolæ.
ARBOIS, Arbosia, Arborosa.
ARBON, Arbor felix.
ARBURG, Arolæ Burgus.
ARC, Cænus.
ARC, ARCHE, Arcius.

ARCADIA, Chelonaticus, Cyparissus Sinus.
ARCADION, Arcades.
ARCHANGEL, Archangelopolis, Michaelopolis, Arcangelo.
ARCHIPELAGO, Mare Ægæum.
ARCOS, in Seville, Aracosia.
ARCOS, in Aragon, Arcobriga.
ARCOS-DE-LA-FRONTERA, Arci.
ARCS (LES), Castrum de Arcubus, Arcolium.
ARCUEIL, Arcus Julianus.
ARDA, Harpessus.
ARDEK, Artaxarta.
ARDENNES, Arduenna Silva.
ARDFEART, Ardartum.
ARDRES, Ardea.
AREMBERG, Areburgium.
ARENA, Sossius.
ARENE, Erana.
AREZZO, Aretium.
ARGAGNA, Orgia.
ARGAU, Argoja.
ARGENS, Argenteus.
ARGENTAN, Argentomum.
ARGENTARO, Orbelus, Soardus.
ARGENTEUIL, in the French dep. of Seine, Argentolium ad Sequanam.
ARGENTEUIL, in the French dep. of Yonne, Argentolium.
ARGENTIERE, Argentaria.
ARGENTON, Argantomagus.
ARGIVO CASTRO, Phyle.
ARGO, Argi.
ARGOVA, Orgessum.
ARGUENON, Argenus.
ARGYLE, Argathelia.
ARHUS, Aarhusius Comitatus.
ARIANO, Equus Tuticus.
ARICCIA, LA RICCIA, Aricia Nemoralis
ARICHOVO, Trajanopolis Thraciæ.
ARISCH, Rhinocolura.
ARITZA, Ratiara.
ARJONA, Urcao.
ARKEL, Herculis Tractus.
ARKIKO, ARGUER, Adulis.
ARKLOW, Arcloa.
ARLBERG, Arula.
ARLES, in the French dep. of Bouches-du-Rhone Arelas, Arelate.
ARLES, in the French dep of Pyrenees Orientales, Arulæ.
ARLEUX, Arensium, Arlodium.
ARLON, Arolanum.
ARMAGH, Ardimacha.
ARMAGNAC, Armeniacensis Comitatus.
ARMANCON, Armentio.
ARMENIENSTADT, Armenopolis.
ARMENTIERES, Armentariæ.
ARMINACHA, EL BOSTAN, Comana Cappadociæ.
ARMIRO, Eretria Phthiotidis, Thebæ.
ARMUIDEN, Arnemuda.
ARNA, Andros.
ARNAUD BELGRAD, Elyma.
ARNAUL, Milizigeris, Melizigera
ARNAUT VILAJETI, Albania.
ARNAY-LE-DUC, Arnetium, Arnejum Ducis.
ARNHEIM, Arecanum.
ARNO, Arnus.
ARNSTADT, Aristadium, Arnestadium.
AROKHAGE, Arachosia.
AROLSEN, Arothia.
ARONCHES, Aranum, Arunci.
ARONDEL, ARUNDEL, Aruntina.
ARONE, Larus.
AROSEN, WESTERAES, Arosia.
AROUN, Trisanton.
ARPAJON, Castra.
ARPASU, Musis.
ARPI, Argyripa, Arpi.
ARQUA, Arquatum.
ARQUES, Arca.
ARRAN, Glota.
ARRAS, Atrebates, Atrebatum.
ARROE, Arria.
ARROUX, Arrosius.
ARSA, Arsia.
ARSOFFA, Resapha.
ARSUF, Antipatris.
ARTA, Ambrakia, Actium.
ARTA, Ambracius sinus.
ARTESINA MONTE, Sergentium.
ARTINGARI, Agari Urbs.
ARTINGARI, Argaricus sinus.
ARTOIS, Atrebatensis comitatus.
ARZEN, Magnus Portus.
ARZER-PALANCA, Ractiaria.
ARZERUM, Arzen.
ARZIGNANO, Arsignanum.
ASBEN, Agisimba.
ASCENSION, Insula Ascensionis.
ASCHACH, Aschavia.
ASCHAFFENBURG, Asciburgum, Schafnaburgum.
ASCHERSLEBEN, Ascaria.
ASCHNAGAR, Massa.
ASCHWELL, Magiovinium.
ASCIANO, Aqua Santa.
ASCOLI, Asculum Picenum.
ASCOLI-DE-SATRIANO, Asculum Apulum.
ASFUN, Aphroditopolis.
ASIM KALESI, Jassus.
ASIM KALESI, Jasius sinus.
ASINARA, Herculis insula.
ASNAH, Euphranta Turris.
ASOLA, Acelum.
ASOV, Mæotis.
ASOW, Asovia, Tanais.
ASPARANETTO, Naustathmus.
ASPE, in France, Aspaluca.
ASPE, in Spain, Aspis.
ASPEREN, Caspingium.
ASPIDO, Aspis.
ASPRA, Casperia.
ASPRASPITI, Cirrha.
ASPRO, Coronta.
ASPRONISI, Therasia.
ASPROPITI, Cyparissa.
ASPROPOTAMO, Achelous.
ASSAB, Adulis, Saba.
ASSENS, Asnesum.
ASSIMSHIRE, Assinum.
ASSIN, Itys.
ASSISI, Asisium.
ASSOMPTION, Assumptionis Civitas.
ASSUMPTION (ISLAND OF), Assumtionis Insula.
ASTACO, Boium.
ASTARABAD, Asterabatia.
ASTI, Asta.
ASTORGA, Asturica Augusta.
ATBARA, TACAZZE, Astaboras.
ATELLARO, Elorum, Helorus.
ATELUCOLO, Gortyna.
ATH, Athum.
ATHANIA, Aquæ Bilbilitanæ.
ATHENRY, Athenria.
ATHERDEE, Atherda.
ATHIES, Atheiæ.
ATRI, Adria.
ATRIB, Athribis.
ATTANEAL, Attinga.
ATTARUS, Nebo.

ATTIGIO, Attidium.
ATTIGNY, Attiniacum.
ATTOK, Choaspes.
AUBAGNE, Albinia.
AUBE, Alba.
AUBENAS, ALBENAS, Albenacium.
AUBENTON, Aubanton, Albantonium.
AUBETERRE, Alba Terra.
AUBIGNY-VILLE, Albiniacum.
AUBONNE, Aula Bona, Alpana.
AUBUSSON, Albucium.
AUCH, Augusta Ausciorum.
AUDE, Adax.
AUDENARDE, Aldenarda.
AUFNAU, Augia.
AUGE, Algia.
AUGSBURG, Augusta Vindelicorum.
AUGST, Augusta Rauracorum.
AULPS, Alpes, Alpium urbs.
AUMALE, Alba Mala.
AUNEAU, Aunus.
AUNIS, Alaitensis Tractus.
AURAN, Laurana.
AURAS, Aurasium.
AURAY, Auracium.
AURE, Eura.
AUREILLE, Teritias.
AURICH, Auricum.
AURIEGE, Alburacis, Aurigera.
AURIGNY, ORIGNY, Riduna, Arica.
AURILLAC, Aureliacum.
AUSE, Alsa.
AUSSIG, Austa.
AUTHIE, Altilia.
AUTUN, Augustodunum.
AUVERGNE, Alvernia.
AUWEGHEM, Aldergemum.
AUXERRE, Autesiodorum.
AUXI-LE-CHATEAU, Alciacum.
AUXONNE, Aussona.
AUZOM, Alsona.
AVA, Chavilah.
AVA, a river, Daonas.
AVADOS-RU, Ethiopia.
AVALON, Aballo.
AVALON (PROVINCE OF), Avallonia.
AVANCHE, Avanticum.
AVAUX, Avallis villa.
AVEIRO, Averium.
AVEIRON, Avario.
AVENAY, Avenacum.
AVENZA, Aventia.
AVERSE, Atella.
AVESNES, Avennæ.
AVIDO, Abydus.
AVIGLIANA, Aviliana, Ocelum.
AVIGNON, Avenio.
AVIGNONET, Avenionetum.
AVILA, Abula, Albicella.
AVILES, Avilla.
AVIS, Avisium.
AVON, Antona.
AVRANCHES, Abrincæ.
AWIRS, Aigremontium.
AWLE, Leon.
AWLON, Aulon.
AXBRIDGE, Axa.
AY, Evus.
AYAMONTE, Aymontium.
AYAN, Aiania.
AYAS, Ægæ.
AYE, EAY, EYE, Insula.
AYERBE, Ebillinum.
AYMARQUE, AYMARGUES, Armasanicæ.
AYMERINS, Ameria.
AYR, Aereus.
AYTON, Calydon.
AZELBURG, Augusta Castra.
AZORISCHE INSELN, Accipitrum Insulæ.
AZREK, Astappus.

B

BAALBECK. See Balbec.
BAASSON, Bazna.
BABA, Elatea.
BABA-DAGH, Vallis Domitiana.
BABASSEK, Babina.
BABEL, Byblus.
BAB-EL-MANDEB (STRAITS OF), Babelmandelum Fretum.
BAB-EL-MANDEB (CAPE), Palindromos.
BAB-EL-MANDEB (ISLAND OF), Babelmandelia Insula.
BABENHAUSEN, Castra Febiana.
BABOLITZA, Cariniana Vallis.
BABUROW, Bauerwitz.
BACAIM, Bacemum.
BACANO, Baccanæ.
BACASARA, Bakhtchisserai.
BACCHIGLIONE, Medoacus minor.
BACCIUM, Bex.
BACHARACH, Ara Bacchi.
BACHDA, Basistis.
BACHIA, Autianæ.
BACHIAN, Bacquianum.
BACODORUM, Passau.
BACTRA, Balkh.
BACU, Baruca.
BADA, Baden.
BADAJOZ, Badia, Pax Augusta.
BADALONA, Bætulo.
BADCO REGIUM, Jiddah.
BADEN, in Austria, Aquæ Pannonicæ, Therm Austriacæ.
BADEN, in Baden, Aquæ Badena.
BADEN, in Switzerland, Ad Aquas Helvetia Thermæ (Aquæ) Helvetiæ.
BADENOCH, Badenacha.
BADENWEILER, Badenvilla.
BADIA, Abbatia.
BADIAH, Arabia Deserta.
BADKIS, Bitaxa.
BÆZA, Batia, Biatia.
BAFFA, Paphos.
BAGA, Bacasis.
BAGAZASAR, Læmocopia.
BAGNA CAVALLO, Ad Caballos.
BAGNAGAR, Pitynda.
BAGNAGUR, HYDERABAD, or GOLCONDA, Hipp cura.
BAGNAREA, Balnea Regia, Balneum regis.
BAGNASCO, Baniascum.
BAGNERES, Aquæ Convenarum.
BAGNERES-DE-BIGORRE, Aquæ Bigerronum
BAGNERES-DE-LUCHON, Bagneriæ Thermæ.
BAGNI, Ocinarus.
BAGNI-D'ALBANO, Fons Aponi.
BAGNI-DI-BALLICANO, Aquæ Calidæ.
BAGNI-DI-SAN-CANTINO, Clusini Fontes.
BAGNI-DI-TRIDOLINO, Ciceronis Aquæ.
BAGNI (LAGO DI), Lacus Albuneus.
BAGNOLO, Balneolum.
BAGNOLS, Balnea.
BAGNONE, Bondelia.
BAGRADOS, Mejerdah.
BARGRAS, Pagræ.
BAHHARET TABERIA, Genesaritis lacus.
BAHRAIM, Baharima.

BAHR-BELA-MAH, Anydros, Sahara.
BAHREIN, Baharema.
BAHR-EL-ABIAD, Astasobas.
BAHR-EL-AKABA, Elaniticus Sinus.
BAHR-EL-AZREK, Astapes.
BAHR-EL-KOLSUM, Heroopoliticus Sinus.
BAHR-JUSEF, Hermopolitana Custodia.
BAHR-LOUT, Mare Mortuum.
BAHR-NOA, Borne.
BAIA, Bajæ.
BAIA-DE-CADIX, Gaditanus Portus.
BAIARIA, ADMIRATI, Eleutherus.
BAIGNEUX, Balneolum.
BAILLEUL, Balliola.
BAIMOT, Baimocium.
BAINBRIG, Glannobanta.
BAINDT, Hortus Floridus, Abbatia Bintersis.
BAINS D'HERCULE, Mehadia.
BAIREUTH, Baruthum.
BAJA, São Salvador.
BAJEUX, Bajocum, Bajocassium civitas, Juliobona.
BAJON, Bagyona.
BAJONNE, in Spain, Bajonna ad Minium.
BAJONNE, in France, Bajona.
BAIS, Batiana.
BAKTSCHISSERAI, Baccasara.
BALAGAT, Balaguatæ Regnum.
BALAGNA, Palanta.
BALAGUER, Oleastrum, Bellegarium, Valagnaria.
BALAKEA, Wallachia.
BALAKLAVA, Symbolon.
BALASFALVA, Blasii (Villa Sancti).
BALASOR, Cosamba.
BALATON, Balatinus Sinus, Volcea.
BALBEC, Baal-Hamon, Heliopolis.
BALCKHUSEN, Belgica.
BALDENAU, Belginum.
BALFRUSCH, Oracana.
BALK, Dargidus.
BALKAN- or TSCHENGJE-GEB., Hæmus.
BALKH, Alexandria Bactriana, Bactra.
BALLAIGNES, Bellæ Aquæ.
BALLIADA, Corone.
BALNEA, Bagnolles.
BALNEAS, Balanææ.
BALSTAL, Balistella.
BALTIC SEA, Mare Balticum.
BAMBERG, Babeberga, Bergium.
BAMBIG, or MAMBEDSCH, Bambyce.
BAMBOLA, Augusta Bilbilis.
BAMFF, Banfia.
BANAL, Ban.
BANBECULA, Benbecula.
BANDA-ORIENTAL, Uruguay.
BANGOR, Bangertium.
BANJA-CONGO, San Salvador.
BANNOLAS, Aquæ Calidæ.
BANUB, Onuphis.
BANZ, Bantum.
BAPAUME, Bapalma.
BAR-LE-DUC, Barium Ducis.
BAR-SUR-AUBE, Barium ad Albulam.
BAR-SUR-SEINE, Barium ad Sequanam.
BARABON, Berabonna.
BARACUN, Baracura.
BARADAH, Abana, Amana.
BARANYA, Baranivarium.
BARBADOS, Barbata.
BARBEAUX, Barbellum.
BARBEZIEUX, Barbecillum.
BARBY, Barbium and Barbiensis Comitatus.
BARCA, Barcana Provincia.
BARCA, Barcæorum Portus.
BARCELONA, Barcino.
BARCELONE, Barcino Vasconiæ.
BARCELONETTE, Barcinona.
BARCELONETTE, (VALLEY,) Barcinonensis Vallis.
BARCELOR, Podoperura.
BARCELOS, Barcelum.
BARDEWICK, Bardenuvicum.
BARDIANO, Bardacima.
BARDINEZZA, Bardinetium
BAREGE, Baredgina Vallis.
BARFLEUR, Barofluctum.
BARGAS, Pagræ.
BARGENY, Berigonium.
BARIN, Parienna.
BARJOLS, Barjolium.
BARKUL, Bargun.
BARLENGA, Berlenga.
BARLETTE, Barolum.
BAROCHE, in Hindostan, Baroca.
BAROCHE, in France, Barocca.
BAROUTH, Barygaza.
BARRA, Barray.
BARRAUX, Barrana Arx.
BARROIS, Barrensis Ager.
BARROW, Barrojus.
BARS, Barschiensis Comitatus.
BARTEN, Bartonia.
BARTH, Bardensis Principatus.
BARTKE, Bardia.
BARTSCH, Barissus.
BARUTH, Berytus.
BARWIC, Barcovicum.
BAS, Barsa.
BASEL, Basilea.
BASENTO, Casuentus.
BASIELLICA, Præsidium.
BASIENTO, Basentinus.
BASILICA, Sicyon.
BASILICATA, Lucania.
BASILIPOTAMO, Eurotas.
BASILUZZA, Basilussa.
BASRA-FERATH-MELLENE, Forath.
BASSAI, Bassacum.
BASSAIN, Bassenum.
BASSANO (LAGO DI), Vadimonis lacus.
BASSENTO, Casuentum.
BASSIGNANA, Augusta Batiennorum.
BASSIGNI, Bassinia.
BASSIGNY, Bassiniacum.
BASSORA, Orchoe.
BASTA, Bubastis.
BASTAN, Bithynium.
BASTI, Bassea.
BASTIA, Mantinorum Oppidum, Martinum.
BASTOGNE, Bastonia, Bastonacum.
BATECALO, Mordula.
BATEMBURG, Arx Batavorum.
BATH, Aquæ Solis, Bathonia.
BATMONSTER, Bathmonasterium.
BATON, Batonia Insula.
BATRON, Bostra.
BATSCH, Bacsiensis Comitatus.
BAUCA, Bovillæ.
BAUGE, in the French dep. of Maine-et-Loire, Bagiacum.
BAUGE, BEAUGE, Balgium.
BAUGENCY, Balgentiacum.
BAUGERAIS, Baugerium.
BAUME, BAUMES-LES-DAMES, Balma.
BAURON, Dicæa.
BAUTZEN, Budissa.
BAUVIGNES, Bovinæ.
BAUX, Baltium.
BAVANO, Cercidius.
BAVAY, Bacacum Nerviorum, Bagacum.
BAYEUX, Aræegenus, Bajocæ.
BAYONA, Abobrica, Aobriga.

BAYONNE, Lapurdum, Bajona.
BAYONNE ISLANDS, Deorum Insulæ, Bajonæ Ins.
BAZA, Basti.
BAZADOIS, Vasatensis ager.
BAZAS, Civitas Vasatica, Vasatæ.
BAZIEGES, Badera.
BAZILBAH, Sicilibra.
BAZILUZZO, Heracleotes.
BEARN, Benecharnia.
BEAUBEC, Bellus Becus.
BEAUCAIRE, Belloquadra.
BEAUCE, Belsia.
BEAUFORT, Bellefordia.
BEAUJEU, Baujovium, Bellus jocus.
BEAUJOLOIS, Bellijocensis Ager.
BEAULIEU, Bellus Locus ad Ligerim.
BEAULIEU, in Vendee, Bellus Locus.
BEAULIEU, in Correze, Bellus Locus ad Duraniam.
BEAULIEU, in Indre, Bellus Locus ad Ingerim.
BEAUMARIS, Bellomariscus.
BEAUMONT, Bellomontium.
BEAUMONT-EN-ARGONE, Bellomontium in Argona.
BEAUMONT-LE-ROGER, Bellomontium Rogerii.
BEAUMONT-LE-VICOMTE, Bellomontium Vicecomitis.
BEAUMONT-SUR-OISE, Bellus Mons.
BEAUNE, Belna, Vellaudunum.
BEAUNE, Belna.
BEAUPORT, Formosa Statio.
BEAUPRE, Bellipratum.
BEAUPREAU, Bellopratum.
BEAUREPAIRE, Bella Reparia.
BEAUVAIS, Bellovacensis Ager.
BEAUVAIS, Bellovacum.
BEBENHAUSEN, Febiana Castra.
BECKEM, Beckena.
BEDARIDES, Bituritæ.
BEDEAH, Hahiroth.
BEDI, Bætana.
BEDFORD, Lactodurum.
BEDUR, Petirgala.
BEGHUI, Byzacia.
BEHESNI, Pindenissus.
BEHNESE. See Benesch.
BEIBAZAR, Lagania.
BEILSTEIN, Bilestinum.
BEIT-EL-RAS, Capitolias.
BEJA, Begia, Pax Julia.
BEJAR, Apiarium.
BEJER-DE-LA-MIEL, Mergablum.
BEJIE, Bulla.
BELA-BANGA, Dilna.
BELBECK, Lagyra.
BELBEYS, Patumus.
BELCHITE, Belia.
BELFORTE, Diomedis Insula.
BELGIUM, Gallia Belgica.
BELGRAD, Alba Bulgarica, Taurunum.
BELGRADO, Hydralis.
BELI-KESSR, Miletopolis.
BELICE DESTRO, Crimisus.
BELIN, Belendi.
BELLAC, Belacum.
BELLAY, Belica.
BELLE-AIGUE, Bella Aqua.
BELLE-BRANCHE, Bella-Branca.
BELLE-ESTOILE, Bella Stella.
BELLEGARDE, Bellogardia.
BELLE-ISLE, Calonesus.
BELLELAY, Bellelagium.
BELLENZ, Baltiona, Bilitionum.
BELLE-PERCHE, Bella Pertica.
BELLEVUE-LES-BAINS, Borbonium Anselmium.
BELLEY, Belencum.
BELLINGER-WOLDER-SCHANZ, Bellingvoldium.
BELLOWERD, Clara Werda.
BELLUE, Pella.
BELLUNEZE, Bellunensis Ager.
BELLUS, Belussa.
BELMONTO, Bellomontium.
BELOVAR, Bellovarium.
BELOZERA, or Bielo-Osero, Bilocerensis Lacus.
BELT, Balticum Fretum.
BELTURBET, Rhæba.
BELUTEN, Comedæ.
BELVEDERE, Euryalus.
BELVEDERE, Caloscopi, Elis.
BELVEDERE, Elis.
BEMBE, Peme.
BEMBINADIA, Nemea.
BENA, Bagennæ.
BENAUGE, Benalgiæ.
BENAVARRI, Bergidum.
BENAVENTE, Brigæcium.
BEND-EMIR, Araxes.
BENDERRIGK, Taoce.
BENDUREN, Ebodurum.
BENESEH, Oxyrynchus.
BENGAL (Gulf of), Gangeticus sinus.
BENGARI, Berenice.
BENGAZI, Hadriane.
BENI-ASSER, Contra Laton.
BENIGEBARA, Beniguelidus Mons.
BENIGUMI, Benigomia.
BENI-TEUDI, Baba.
BENWEL, Condercum.
BERAUN, Berium.
BERBACH, Ripa prima.
BERBERES, Gætuli.
BERBEREY, Barbaria Terra.
BERG, Montanus Ducatus.
BERG-SAINT-VINOX, Bergæ St. Vinoxi.
BERGA, Berginium.
BERGAMASCA, Bergamensis Ager.
BERGAMO, Bergomum.
BERGEL-THAL, Brexalia Vallis.
BERGEN, Tiberiacum.
BERGENHUUS, Bergensis Præfectura.
BERGENLU, Parnassus.
BERGEN-OP-ZOOM, Bercizoma, Berga ad Zomam.
BERGZABERN, Tabernæ.
BERLENGA, Erythia.
BERLIN, Berolinum.
BERMEO, Vesperies.
BERMUDEN, Æstivæ Insulæ.
BERN, Arctopolis.
BERNAY, Bernayum, Bernacum.
BERNBURG, Arctopolis ad Salam.
BERNCASTEL, Castellum Mosellanum, Tabernarum, Tabernæ Mosellanicæ.
BERNHARDIN, Culmen St. Bernhardini.
BERNRIED, Beronicum.
BERNSTADT, Berolstadia.
BERNSTADT, Berolstadia Lusatiæ.
BERNSTEIN, Berostonum.
BERRY, Bituricensis Ducatus.
BERSEL, Barsalium.
BERSELLO, Brixellum.
BERTINORO, Britinorium.
BERTRAIS, Perniciacum.
BERWICK, Barcovium.
BERWITSCH, Fertina.
BERZETO, Barcetum.
BESAGE, Vosagus.

BESAGNO, Bisamnis.
BESALU, Beseldunum.
BESANÇON, Besontium, Vescontio.
BESES, Betulus, Batulo.
BESOR, Sihor.
BESSAN, Bissonum.
BESSARABIEN, Getarum Desertum.
BESSIGHEIM, Besighamium.
BESSIN, Bagisinus Ager.
BETANSOS, Flavium Brigantium.
BETHELSDORF, Bellenfalva.
BETHUNE, Bethunia.
BETHYSSI (Bethisy), Betisiacum, Bestum.
BETTONA, Vettona.
BETUVE, Batavorum Insula.
BEUIL, Boleum.
BEUTHEN, Bethania, Bythomia.
BEVAGNA, Mevania.
BEVERLEY, Betuaria.
BEVERSTEDT, Fabiranum.
BEVERWICK, Beverovicum.
BEWDLEY, Bellilocus.
BEX. Baccæ.
BEYPASHA, Hypanis.
BEZIERS, Bæterra Septimanorum, Bliterræ.
BIAGRASSA. See Abiagrasso.
BIALA-CESTRO, Decelia.
BIALOCERKIEW, Bialoquerca.
BIBRA, Biberaha.
BIBURG, Epinaburgum.
BIBURG, Epinaburgum ad Vilsam.
BICANIR, Bardeatis.
BICCARI, Hyccara.
BIDASSAO, Menlascus, Vedasus, Bidosso
BIDSCHOW, Bytziovia.
BIECZ, Becia.
BIELE (Lake), Biellensis Lacus.
BIELGOROD, Belogradum.
BIELLA, Bugella, Gaumellum.
BIELOBERDO, Teutoburgium.
BIELO-OSERO, Bilocerensis Ducatus.
BIELO-OSERO, Biela-Ozera
BIELSK, Bielca.
BIERVLIET, Birfletum.
BIESBOSCH, Bienboscum.
BIES-SCADI, Bieccadius Mons.
BIFERNO, Tifernus.
BIGENIS, Abacæna.
BIGORRE (Le), Bigerrensis Comitatus.
BILBAO, Bellum Vadum.
BILIN, Bylina.
BILLE, Billena.
BILLOM, Billomum, Billemum.
BILSEN, Belisia.
BINAROS, Binarusia.
BINASCO, Binæ, Bacenæ.
BINCHE, Binchium, Bintium.
BINCHESTER, Bimonium, Vinnovium.
BINGEN, Bingium.
BINGENHEIM, Castrum Valerianum.
BINOESCA, Visontium.
BIO (Lake), Bœbeis Lacus.
BIOGRAD, Zaravecchia, Alba Maris.
BIORCO, Birca.
BIORNEBURG, Arctopolis, Bierneburgum
BIPP, Castrum Pipini.
BIR, Birtha.
BIR SZABEA, Beer-Seba.
BIRGHE, Hypæpa.
BIRGI, Acithenis.
BIRKET-AL-GARUM, Mœris.
BIRNBAUMER WALD, Ad Pirum.
BIRTHELM, Birthalbinum.
BISCAY, Cantabria, Biscaja.
BISCAY (Bay of), Cantabricus Oceanus.
BISCHAH. Castra Germanorum.
BISCHOFSLACK, Locopolis.
BISCHOFSWEILER, Episcopi Villa.
BISCHOFSWERDA, Episcopi Insula.
BISCHOFSZELL, Episcopi Cella.
BISEGLIA, Vigiliæ.
BISENTIO, Vesentium, Visentium.
BISERTA, Hippo Diarrhytus.
BISHOPSCASTLE, Episcopi Castrum.
BISIGNANO, Besidiæ, Bisinianum.
BISNAGAR, Modura.
BISTRITZ, Bistricia.
BITCHE, Bicina, Bidiscum.
BITEJA, Buteium.
BITOLIA, Octolophum.
BITONTO, Bidruntum, Biluntum.
BITSCH, Bidiscum.
BITTBURG, Beda.
BIUTHERE, Bathycolpus.
BIZERTA, Benzerta.
BLACKWATER, Dabrona.
BLÆSOIS, Blesensis Ager.
BLAINVILLE, Bleonis Villa ad Mosam.
BLAINVILLE, Bleonis Villa.
BLAIR, Blarensis Ducatus.
BLAIR. Blara.
BLAISE. Blazi (Sanct).
BLAMONT, Albimontium.
BLANC (Le), Oblincum.
BLANDIN, Abbatia St. Petri in monte Blandinio.
BLANES, Blanda.
BLANGY. Blangiacum.
BLAUBEUERN, Blabira.
BLAVET, Blabius.
BLAVET, Blabia.
BLAYE, Blavia.
BLECKINGEN, Blechingia.
BLEEDA, Bida.
BLENIO, Bellenzer-Thal.
BLERICK, Blariacum.
BLESNEAU, Blenavium.
BLETTERANS, Bleterum.
BLIES, or Blise, Blessa.
BLIS. Philiscum.
BLOIS, Blæsæ, Blesæ.
BLOTZHEIM, Bladolsheim.
BLUMBERG, Florimontium.
BLUMEN-THAL, Florivallis.
BOBENHAUSEN, Astvia.
BOCCA-DI-BONIFACIO, Bonifacii Sinus.
BOCHOLT, Bocholtia.
BOCINO, Bucinum.
BOCKENHEIM, Boconica.
BODEN-SEE, Bodamicus. Brigantinus Lacus.
BODMEN, Bodami Castrum.
BODMER-SEE, Acronius Lacus.
BODMIN, Bosuenna.
BODOK, Bodoxia.
BODRU. Halicarnassus.
BODVAR, Varæ.
BOEBLINGEN, Bibonium.
BOEGE, Bega.
BOG, Hypanis.
BOGDANA, Hyampolis.
BOGLIASCO, Boliascum.
BOGMUTTY, Oximagis.
BOGNA. Bunia.
BOHMEN, Bojæmum. Bohemia.
BOHMISCH-BROD, Broda Bohemica
BOIDIA, or Voda, Panachaicus.
BOISBELLE, Boscobellum.
BOIS-COMMUN, Commeranum.
BOJANA, Barbana.
BOJANO, Bobianum.
BOK HAIR, Nacrasa.

BOLANOS, Noliba.
BOLGEN, Beleus.
BOLI, Hadrianopolis Bithyniæ.
BOLKENHAIN, Bolconis Fanum.
BOLLENZA, Breunia Vallis.
BOLOGNA, Bononia.
BOLSENA, Volsinium, Volsinii.
BOMARZO, Polimartium.
BOMBAY. Perimuda, Boa, Bahia.
BONA, Colonia Gemella Julia Hipponensis Pia Augusta.
BONADUZ, Beneducium.
BONAIR, Boni Æris Insula.
BONDENO, Padinum.
BONIFACIO, Bonifacii Civitas.
BONLIEUX, Bonilii.
BONN, Bonna.
BONNEVAL, Bona vallis.
BONNEVILLE, Bautæ.
BONNEVILLE, Bonavilla.
BONNY, Bonnium.
BONZIEDEN, Bonazida.
BOPFINGEN, Opiæ.
BOPPART, Baudobrica.
BORAES, Bœrosia.
BORDEAUX, Burdigala.
BORGAS, Practius.
BORGHETTO, Burgetum ad Mincium.
BORGO-LAVIZARA, Forum Libricorum.
BORGO-DI-ST.-ANGELO, Burgus St. Angeli.
BORGO-DI-ST.-DONINO, Fidentia.
BORGO-DI-SAN-SEPOLCRO, Biturgia.
BORGO-DI-SESSIA, Burgus Sessites.
BORGO-DI-VAL-DE-TARO, Burgus Tari Vallis.
BORGOFORTI, Burgus Fortis.
BORGOFRANCO, Burgum Francum.
BORGOMANERO, Burgomanerum.
BORIA, Balsio.
BORITZA, Jolcus.
BORKUM, Burchania.
BORNHOLM, Boringia.
BORROMEAN ISLANDS, Insulæ Cuniculares.
BOSAN, Pessinus.
BOS-BURUN, Neptunium.
BOSCO, Boschi, Mesia sylva.
BOSSET, Bacuntius.
BOSSY, Buxudis.
BOST, Abesta.
BOSTAN. See Arminacha.
BOTZEN, Bolsanum, Bolzanum.
BOUCHAIN, Bochanium.
BOUCHALOT, Vulchalon.
BOUCHARD, Bocardi Insula.
BOUDRY, Baudria.
BOUILLON, Bullio.
BOUIN, Bovinum.
BOULENC, Burgum Bonæ Gelenæ.
BOULNESS, Tunnocelum.
BOULOGNE, in the French dep. of Haute-Garonne, Bononia Vasconiæ.
BOULOGNE, in the French dep. of Pas-de Calais, Bononia Gesoriacum.
BOULON, Ad Stabulum.
BOULONNOIS (Le), Boloniensis Ager.
BOURBON, Insula Burbonia.
BOURBON-L'ARCHAMBAUD, Aquæ Bormonis, Burbo Arcimbaldi.
BOURBONNE LES BAINS, Aquæ Borvonis.
BOURBONNOIS (Le), Bojorum Ager.
BOURBOURG, Broburgum Morinorum.
BOURG-ARGENTAL, Burgus Argentalis.
BOURG DE HIEROCHIPE, Hierocepia.
BOURG D'OISANS, Forum Neronis.
BOURG-EN-BRESSE, Burgus Bressiæ.
BOURG-ST.-ANDEOL, Burgus Andeoli.
BOURG SUR MER, Burgus.
BOURGES, Bituriges, Avaricum.
BOURGEUIL, Burgolium.
BOURGNEUF, Burgum Novum.
BOURGNEUF, Burgum Novum ad Ligerim.
BOURGOGNE, Burgundia.
BOURGOIN, Bergusium.
BOURMONT, Burnonis Mons.
BOUSSAC, Bussatium.
BOUTON, Baton, Batonia insula.
BOUTZI, Neda.
BOVINES, Boviniacum.
BOYNE, Boandus, Buvinda.
BOZBERG, Vocetius.
BOZ-DAGH, Tmolus.
BOZGIA, Tenedos.
BOZZO, Bussus.
BRA, Barderate, Braida.
BRACCIANO, Arcennum.
BRACCIANO (Lago di), Sabate.
BRADANO, Brunduno.
BRAGA, Augusta Bracara.
BRAGANZA, Brigantia.
BRAIN-LE-CHATEAU, Brennia Castrensis.
BRAIN-LE-COMTE, Brennia Comitis.
BRAINE, Brennia ad Vidulam.
BRAINE-D'ALEU, Brennia Allodiensis.
BRAMANT, Brammovicum.
BRAMFIELD, Bravinium.
BRAMNI, Adamas.
BRAMPOR, Barampura.
BRAMPTON, Bramenium.
BRANDEIS, Brandusium.
BRANDEMIR, Coros.
BRANDENBURG, Marchia Brandenburgica.
BRANDENBURG, in Prussia, Brandeburgium.
BRANDENBURG, Bradeburgum.
BRANE, Hebromagus.
BRANTOME, Brantosomum.
BRASLAW, Bratzlaviensis Palatinatus.
BRASLAW, Braclavia.
BRASLOW, Braslovia.
BRATA, Brathia.
BRAUBACH, Brubacum.
BRAULIO, Bormiana Juga.
BRAUNAU, Brunodunum.
BRAUNEGG, Branecium.
BRAUNSBERG, Brunonis mons.
BRAUNSCHWEIG, Brunsvicensis Ducatus.
BRAUNSCHWEIG, Brunonis Vicus, Brunsviga.
BRAY, Braviarum ad Samaram, Braccius.
BRAYCHIPULT, Braychiputium Caput.
BRAZZA, Bracchia.
BRECHIN, Brechinum.
BRECKNOCK, Brechinia.
BRECKNOCKSHIRE, Brechiniensis Ager.
BREE, Bræa.
BREEDEVOORT, Bredefortia.
BREG, Briga.
BREGANÇON, Briganconia.
BREGEN, Pergantium.
BREGENZ, Brigantinus Comitatus.
BREGENZ, Bregentium, Bregantum.
BREGENZ, a river, Bregentia.
BREGLIO, Brelium.
BREGNA, Brennæ Vallis.
BREGNA, Brennus Fluvius.
RREGNANO, Brenianum.
BREISACH, Mons Brisiacus.
BREISGAU, Brisgovia.
BREITENECK, Britenecca.
BREMBATO-DI-SOTTO, Brembatum Inferius.
BREME, Bremetum.
BREMERWORDE, Bremerfurda.
BREMGARTEN, Bremogartum.

BRENDOLA, Brendulum.
BRENNER, Brennus Mons.
BRENTA, Medoacus Major, Brentesia.
BRENTONEGO, Brentonicum.
BRENZ, a river, Brenta.
BRENZ, Brivenes, Brenta.
BRESCIA, Brixia.
BRESCIANO, Brixianus Ager.
BRESCON, Agatha, Blascon.
BRESINI, Brezina.
BRESLAU, Vratislavia.
BRESSE, Bressia.
BRESSUIRE, Bercorium.
BREST, Brivates portus.
BRESTON, Calatum.
BRETAGNE, Britannia minor.
BRETENHAM, Cambretonium.
BRETEUIL, in the French dep. of Eure, Bretallium.
BRETEUIL, in the French dep. of Oise, Bretelium.
BRETIGNI, Bretiniacum.
BRETON, Britannicum Fretum.
BRETON (Cape), Britonum Caput.
BREUSCH, Brusca.
BREYN, Bruna.
BRIANÇON, Brigantia.
BRIANÇONET, Brianzonia Arx.
BRIANÇONNOIS (Le), Brigantinus Ager.
BRIANZA, Brigantini Montes.
BRIARE, Brivodurum.
BRIATESTE, Britexta.
BRIDGEWATER-BAY, Vexalla.
BRIDLINGTON, Brillendunum.
BRIE, Braja, Bria.
BRIEG, Bregenses Thermæ, Vibericus pagus.
BRIEG in Silesia, Brega, Briga.
BRIEG, Brigensis principatus.
BRIEL, Helium.
BRIENNE, Brena.
BRIES, Britzna.
BRIESCIA, Brisciensis Palatinatus.
BRIESCIA, Briscium.
BRIEY, Briocjum.
BRIEZEN, Brieza Fida.
BRIGACH, Brigiana.
BRIGNOLES, Brinolium, Brinonia.
BRIGUEIL, Brigolium.
BRIHUEGA, Driaca.
BRINDISI, Brundisium.
BRIOU, Brigeosum.
BRIOUDE, Brivas.
BRIQUENAY, Bracanctum.
BRIQUENRAS, Briquoraceum.
BRISACH, Brisacensis Vicus.
BRISIGHELLA, Bresegella.
BRISTOL CHANNEL, Æstuarium St. Georgii.
BRIVE-LA-GAILLARDE, Briva Curretia.
BRIVIESCA, Virovesca.
BRIVIO, Bripium.
BRIXEN, Brixina, Brixinum.
BROAD, Broadius Lacus.
BROCKEN, Bructerus Mons.
BROCKLEY-HILLS, Sullionacæ.
BROCKMERLAND, Brockmeria.
BROD, Hunnobroda.
BROGE, Brouage.
BROGLIO, Broilum.
BROMBERG, Bidgostia.
BRONDOLO, Brandulus Portus.
BRONDRUT, Pruntrut, Porrentrui, Brundisia.
BRONI, Blandeno.
BROSZ, Ambrosiopolis.
BROUAGE, Broagium.
BROUGHAM, Braboniacum.

BROUWERSHAVEN, Bruvenhavia.
BROWERS, Broveri Fretum.
BRUCK, Brugg an der Aar, Pons Arolæ.
BRUCK-AN-AMBER, Ambra, Ambarum.
BRUCK-AN-LEYTHA, Leitæ Pons, Motenum.
BRUCK-AN-MUR, Prugg, Pons Muræ.
BRUCK-AN-ROER, Bruchium ad Ruram.
BRUGES, Brugæ Bearniæ.
BRUGG, Bruga.
BRUGGE, Brugæ.
BRUGGEN, Mederiacum.
BRUGGERFAHRT, Brugensis Fossa.
BRUGH, Crococalanum.
BRUL, Brielium.
BRUMPT, Breucomagum.
BRUNN, Brinnium, Brunna.
BRUNNER KREIS, Brunnensis Circulus
BRUNNSTADT, Uruncæ.
BRUNSBUTTEL, Brunsbutta.
BRUS, Bursa.
BRUSSEL, Bruxella.
BRZESCIE, Brestiensis Palatinatus.
BRZESK, Brestia ad Hypanim.
BRZESK, in Poland, Bresta.
BUA, Boa, Bavo.
BUCH, Bojatum.
BUCHAN, Bodom.
BUCHAU, Buchavia Bohemica.
BUCHAU, in Bavaria, Buchavia Bavarica.
BUCHAU, Fagonia.
BUCHEN, Hohbuochi.
BUCHSWEILER, Buxovilla.
BUCKER-FIORD, Aardalinus Sinus.
BUCKOR, Buchoria.
BUCORTA, Butrotus.
BUDACH, Badacum.
BUDIMIR, Tronum.
BUDINGEN, Budinga.
BUDNOCK, Butnocum.
BUDOA, Batuo.
BUDWEIS, Budovicium.
BUENO-DESEO, Desiderii Promontorium.
BUENOS-AYRES, Bonus Aer.
BUFLES, Bubalorum Lacus.
BUGEY, Beugesia, Bugia.
BUGGIANO, Bellavista.
BUGIE, Chobæ.
BUIS, Busium, Buxium.
BUITRAGO, Blitabrum.
BUJUK-KALEH, Cotyora.
BULAWADNI, Philomelium.
BULLE, Bulium.
BULLERBORN, Fons Resonus.
BULLES, Bubulæ.
BULNESS, Blatobulgium.
BUNARBASCHI, Ilium Vetus.
BUNGO, Bundium.
BUNZLAU, Boleslavia.
BUQUOI, Buquojum.
BURBASCHA, Obrima, Orgas.
BUREN, Bura.
BURG, Tubusuptus.
BURGAS, Bergulæ.
BURGAU, Burgavia.
BURGHAUSEN, Burchusa.
BURGOS, Burgi.
BURGSCHEIDUNGEN, Dispurgum
BURGUND, Bourgogne.
BURGUNDIER, Mugilones.
BURICH, Burichum.
BURLOS, Buticus Lacus.
BURONZO, Borontium.
BURRAMPUTER, Dyardanes.
BURRIANA, Sepelaci.
BURSA, Prusa.

BURUGKERD, Heraclea Mediæ.
BURUZ, Cibyra.
BURY, Faustini Villa.
BURZENLAND, Barcia.
BUSETO, Buxetum, Bussetum.
BUSO, Busum Promontorium.
BUSSENTO, Buxentius.
BUSSIERE-POITEVINE, Buxerium Pictonum.
BUSTEREH, Cybistra.
BUSTO-GRANDE, Bustum Magnum.
BUTERA, Buterium.
BUTOW, Butavia.
BUTRINTO, Buthrotum.
BUTRIO, Butrium Umbrorum.
BUTZOW, Beucinum.
BUZANÇOIS, Buzancæum.
BYSTRZYCA, Bystricium.

C

CABEÇON, Congium.
CABES, Syrtis minor.
CABEZAS-RUBIAS, Ad Rubras.
CABIA, Caressus.
CABLASVAR, Cabalaca.
CABO. See Capo.
CABRA, Egabra.
CABRERA, Capraria.
CABRIERES, Capraria.
CABUL, Arigæum.
CABYLAN, Cabyle.
CACORLO, Cazorla.
CADAGNES, Cadacherium.
CADDE, Casæ Dei Fœdus.
CADIZ, Augusta Julia Gaditana, Gades, ium.
CADIZ (Gulf of), Fretum Gaditanum.
CADORE, Cadubrium.
CADORINO (Il), Cadorinus Ager.
CADSAND, Cassanda.
CAEN, Cadomum.
CAER-GWEND, Venta Silurum.
CAER-LEON, Caerlion, Isca Silurum.
CAERMARTHEN, Maridunensis Comitatus.
CAERMARTHEN, Maridunum.
CAERNARVON, Arvonia.
CAERWENT, Ventidunum.
CAFFA, Capha.
CAFZA, Capsa.
CAGLI, Ad Callem, Calium.
CAGLIARI, Caralitanum Promontorium.
CAGLIARI, Calaris.
CAGLIARI (Gulf of), Caralitanus Sinus.
CAHORLE, Caprulæ.
CAHORS, Cadurci, Divona Cadurcorum.
CAIFFA, Ecbatana.
CAILLY, Bibe.
CAIMAN ISLAND, Crocodilorum Insula.
CAIRO, Canalicum.
CAITHNESS, Cathenesia.
CAJANELLO, Callicula.
CAJAZZO, Gajazzo, Calatia.
CAKET, Gaguetia.
CALA-LONGA, Cinium.
CALAAH, Calaa.
CALABRIA, Bruttium.
CALABRIA CITERIORE, Consentina Provincia
CALAC, Calabrum.
CALAGERO, Schera.
CALAHORRA, Calagorina, Calaguris.
CALAIOTE, Metacum.
CALAIS, Caletum.
CALAMAT, Celama.
CALANDRA, Mendæ.
CALARUEGA, Calaroga.
CALA-SISINA, Sypicius.
CALATABELLOTA, Isburus.
CALATAGIRONE, Calata Hieronis.
CALATAYUD, Bilbilis Nova.
CALBE, Calba ad Salam.
CALBE, Calba Palæo — Marchiæ.
CALDERN, Caldarium.
CALENDO, Lalasis.
CALEPIO, Calepium.
CALIACA, Calatis.
CALICUT, Bacari.
CALIFORNIAN GULF, Mare Rubrum, or Purpureum.
CALISSIN, Rosbelum Caput.
CALLAH, Tamalluma.
CALLEANI, Calligeris.
CALLEN, Calanum.
CALLIAMERE, Calligicum.
CALLISANE, Calcaria.
CALMAR-SUND, Fretum Calmariense.
CALMINE, Calymna.
CALMONT, Calvo-Mons.
CALONI, Pyrrha.
CALPENTINE, Anarismundi.
CALTA-BELLOTA, Inycum.
CALUI-ACRA, Calinacron.
CALVA-BAY, Volsas.
CALVI, Cale.
CALVISANO, Forum Calvisii
CALW, Calba.
CALZADA, Calciata.
CAMARANA, Camarina.
CAMARASA, Cyssum.
CAMARGUE, Camaria.
CAMASCH, Comagene.
CAMBAY, Monoglossum.
CAMBAY (Gulf of), Barygazenus Sinus.
CAMBRAY, Camaracum, Cameracum.
CAMBRESIS, Cameracensis Ager.
CAMBRIDGE, Cantabrigiensis Comitatus.
CAMBRIDGE, Camboricum, Cantabrigia.
CAMELIONO, Cema.
CAMENI, Automate, Hiera.
CAMENZ, Camentia ad Nissam.
CAMERA-LA-TORRE, Herculis Turris.
CAMERINO, Comarus.
CAMERINO, Camerinum.
CAMIN, Caminium.
CAMIN, Caminum.
CAMISANO, Camissanum.
CAMPAGNA, Campania.
CAMPAGNA-DI-ROMA, Latium.
CAMPEN, Campania ad Isalam.
CAMPI-DOGLIO, Capitolium.
CAMPIGNE, Campinia.
CAMPOLI, Camplum.
CAMPOLSCHIN, Campus Dulcinus.
CAMPREDON, Campus rotundus.
CANAL-DE-BRIARE, Brivodurense Incile.
CANAL-DI-FERISINA, Farisinensis Sinus.
CANALE-DI-ST.-ALBERTO, Fossa Messanica.
CANARA, Cottonara.
CANARY ISLANDS, Atlanticæ Insulæ, Insulæ Canariæ, Fortunatæ.
CANAVESE, Canapitium.
CANCHE, Cantius.
CANDAHAR, Capissene.
CANDE, Candæum.
CANDI, Maagrammum.
CANDIA, Aëria, Creta.
CANDRIA, Didaucana.
CANEA, Cydonia.
CANES, Ad Horrea, Ægitna.
CANETO, Bebriacum.
CANGAS-DE-ONIS, Concana.

CANI, Dracontia.
CANIKSHACK, Carcinites.
CANINA, Chaonia.
CANISCHA, Canisia ad Dravum.
CANISTRO, Ampelusia.
CANNA, Camicus.
CANNONDEN, Cononium.
CANO, Cercura.
CANOSA, Canusium.
CANOSSA, Canosium.
CANOURGUE, Canorgia.
CANOVIA (La), Candavia.
CANSTADT, Cana, Cantaropolis.
CANTAL, Celtorum Montes.
CANTAVIEJA, Carthago Vetus.
CANTAZARO, Cantacium.
CANTECROIX, Cantecrucium.
CANTELLE-LA-VIEILLE, Cantilia.
CANTERA, Abolus, Acesines.
CANTERBURY, Cantuaria, Durovernum.
CANTIRE, or Kintyre, Cantiera.
CANVEY, Convennos.
CANZO, Cantium.
CAOURS, Cavortium.
CAPACCIO, Caput Aqueum.
CAPE, Capes, Tacapa.
CAPE AGUILHAS, Caput Acuum.
CAPE BABA, Lectum.
CAPE BERBIERI, Dardanis.
CAPE BIANCO, Drepanum Prom.
CAPE BLANCO, Ryssadium Prom.
CAPE BOEO, Lilybæum Prom.
CAPE BOJADOR, Solventia.
CAPE BONO, Hermæa.
CAPE BRETON, Britonum Caput.
CAPE CALASNANE, Boreum Pr.
CAPE CAMORIN, Comaria.
CAPE CANAN, Cephalæ Pr.
CAPE CANTIERE, Epidium Pr.
CAPE CANTIN, Solis Mons.
CAPE CAVATI, Curias.
CAPE CHELIDONI, Chelidonium Pr.
CAPE CHERSONESI, Chersonesus.
CAPE COLONNA, Columnarum Caput.
CAPE CORUSORE, Hieron Pr.
CAPE CRIO, Criumotopon.
CAPE DREPANO, Derris.
CAPE DUCATO, Leucata.
CAPE ELENI, Callinusa.
CAPE ESPARTEL, Ampelusia.
CAPE FERET, Curianum.
CAPE FINISTERRE, Artabrum Pr.
CAPE FORMOSA, Raptum Pr.
CAPE GARDAFUI, Caput Aromatum.
CAPE GATA, Charidemi Pr.
CAPE GEER, Herculis Pr.
CAPE GREGA, Pedalium.
CAPE HARTLAND, Herculis Pr.
CAPE HORN, Caput Hornanum.
CAPE JASQUES, Carpella.
CAPE KINKINOS-BURNU, Corax.
CAPE KORAKA, Coriceum Pr.
CAPE KUHESTEK, Armozon Pr.
CAPE KURIAT, Corodamum.
CAPE LEZARD, Dammonium.
CAPE LOPEZ-DE-GONSALVA, Caput Lupi Gundisalvi.
CAPE MARTIN, Artemisium Pr.
CAPE MASTICO, Phanæ.
CAPE MATAPAN, Tænaria.
CAPE MOCADON, Maceta.
CAPE NON, Chaunaria.
CAPE OBY, Lytarmis Pr.
CAPE ORFUI, Apocopa.
CAPE ORTEGAL, Coru.
CAPE OSEM, Usadium.
CAPE PAPA, Araxos.
CAPE PAPERNESS, Cantium.
CAPE ROCCO, Lunæ Pr.
CAPE ROCCO DI SINTRA, Magnum Pr.
CAPE ROMANIA, Malæi Colon.
CAPE SALON, Sellus.
CAPE SIDARI, Phalacrum.
CAPE SIDERO, Salmone.
CAPE SIGRI, Sigrium.
CAPE SKYLLO, Scyllæum Pr.
CAPE SOSSOSO, Dium Pr.
CAPE SPADA, Cimarum.
CAPE STILO, Cocintum.
CAPE TRAFALGAR, Junonis Pr.
CAPE TRAPANI, Drepanum Pr.
CAPE VERDE, Arsinarium Pr., Surrentinum Pr.
CAPE VONO, Jasonium Pr.
CAPE XAGRO, Stanum Pr.
CAPE ZONCHIO, Coryphasium.
CAPE CANTIRE, or Kintyre, Cantieræ Rostrum
CAPE DE-CAUX, Caletorum Pr.
CAPE DE-CREUS, Aphrodisium Pr.
CAPE DE-ESPICHEL, Barbarium Pr.
CAPE DE-LA-CROISETTE, Zoa.
CAPE DE-NORD, Boreale Caput.
CAPE D'OROS, Caphareus.
CAPE DE-SABLE, Caput Arenarum.
CAPE DELL'-ARMI, Leucopetra.
CAPE DEL-CAS, Casius.
CAPE DEL-GADA, Prasum.
CAPE DELLE-COLONNE, Lacinium
CAPE DI-LEUCA, Japygium Pr.
CAPE DI-TEJONES, Boreum Pr.
CAPE ST. ANDRE, Cauda Bovis.
CAPE ST. ANGELO, Malea.
CAPE ST. GIORGIO, Sepias.
CAPE ST. JOHANN, Simylla.
CAPE ST. MARIA, Cuneum.
CAPE ST. MARIE, Malea.
CAPE ST. MARTIN, Hemeroscopium
CAPE ST. VINCENT, Sacrum Pr.
CAPE ST. YVES, Antivestæum.
CAPELLE, Duronum.
CAPELLE, Ambitarinus vicus.
CAPESTAN, Caput stagni
CAPILLA, Mirobriga.
CAPIZZI, Capitina civitas.
CAPO CAMPANILLA, Minervæ Pr.
CAPO COLONNI, Sunium Pr.
CAPO CORMACHETI, Ceparum Pr.
CAPO CORSO, Caput Corsum Africanum.
CAPO CORSO, Sacrum Pr., Caput Corsum.
CAPO MARZO, Odyssea.
CAPO MESURATA, Trierum Pr.
CAPO MISENO, Misenum.
CAPO PASSARO, Pachynum.
CAPO QUILATES, Sestiaria.
CAPO SPADA, Psacum.
CAPO SUPRANO, Plaga Calvis.
CAPO TURGLIO, Viriballum.
CAPO VILLANO, Sestianæ aræ.
CAPO XACRO, Erythræum Pr.
CAPO D'ALBER, Catabathmus.
CAPO D'ERBICARIA, Granianum.
CAPO D'ISTRIA, Aegida, Justinopolis.
CAPO DE PALOS, Scambraria.
CAPO DE TRES FORCAS, Metagonium.
CAPO DEL HIGUER, Oeasso.
CAPO DELL'-ARMI, Bruttium Pr., Armorum Caput.
CAPO DELLA TESTA, Erebantium.
CAPO DI-BRUSSANO, Zephyrium.
CAPO-DI-CASA-BARBARICA, Marinum.
CAPO DI-FARO, Pelorus

CAPO DI-FENO, Rhium.
CAPO DI-SPARTIVENTO, Herculis Pr.
CAPO DI-ST.-CROCE, Taurus.
CAPO DI-STILO, Stilida.
CAPO DI-TRES-FORCAS, Ras Adir.
CAPRAJA, Capraria.
CAPRI, Capreæ.
CAPRISTANO (LAGO DI), or LAGO DI CELANO, Fucinus lacus.
CAPSE, Carpudæmon.
CARA-CAJA, Charax.
CARACAL, Castra Nova.
CARACARAN, Lariagara.
CARADIVA, Orneon.
CARANSEBES, Carancebæ.
CARASU, Cydnus.
CARASU, Athyras.
CARAVAGGIO, Carayacium.
CARAVELLE, Daunus.
CARAVI, Haliusa.
CARBON, Alpheus.
CARCASSONE, Carcaso, Carcasum.
CARDEN, Cardonia.
CARDIGAN, Ceretica.
CARDIGANSHIRE, Ceretícensis Comitatu
CARDONA, Udura.
CAREATAIN, Cariatha.
CARENTAN, Carento, Carentonum.
CARFAGNANA, Caferoniana.
CARI, Criassus.
CARIATI, Chariatum.
CARIDIA, Cardia.
CARIFE, Callifæ.
CARIGNAN, Cariniacum.
CARIGNANO, Carinianum.
CARINA, Hipparis.
CARINOLA, Calinula, Calenum.
CARISSA, Enipeus.
CARIZA, Carissa Regia.
CARLAVEROCK, Carbantorigum.
CARLEBY, Carlopolis, Carolina.
CARLINGFORD, Buvindum.
CARLISLE, Carleolum.
CARLOPAGO, Campus Carolinus.
CARLOW, Caterlogum.
CARLSHAFEN, Caroli Portus.
CARLSHAFEN, Caroli Portus ad Visurgim.
CARLSKRONA, Caroli Corona.
CARLSRUHE, Caroli Hesychium.
CARLSSTADT, Bibium.
CARLSSTADT, Carolostadium.
CARLUEC, Catuiaca.
CARMAGNOLA, Carmaniola.
CARMAING, Carmanum.
CARMONE, Carmene.
CARNAK, Thebæ.
CARNAL MANAZIL, Carana.
CARNARVON, Segontium.
CARNERO, Carnarius Sinus.
CAROMB, Carumbus.
CARONIA, Calacta.
CAROS-DE-LOS-INFANTES, Ilarcuris.
CAROSA, Cyptasia.
CARPATHEN, Carpates.
CARPENTRAS, Carpentoracte.
CARPI, Carpium ad Athesin.
CARPI, Carpium.
CARPI, Carpiensis principatus.
CARPIO, Calpurniana.
CARETTO, Carectanum.
CARRI, Incarus.
CARRICK, Caricta.
CARRICKFERGUS, Fergusii Rupes.
CARRION DE LOS CONDES, Carrio Comitum.
CARROUX, Carrofum.
CARSO, Caristum.
CARTHAGENA, Carthago Nova.
CARUSU, Halmyris.
CARVAJALES, Vicus Aquarius.
CARVAR, Aramagara.
CARVASIDAY, Corax.
CASAL, Bodincomagum.
CASAL-NUOVO, Mandonium.
CASALE-MAGGIORE, Casale Majus.
CASARICTON, Casarætonia.
CASBIN, Arsacia.
CASCAES, Cascale.
CASCANTE, Cascantum.
CASCHAU, Cassovia.
CASHEL, Cassilia.
CASIGLIANO, Carsulæ.
CASINO, Casinus Mons.
CASLONO, Castulo.
CASHMIR, Canagora.
CASHMIR, Caspiria.
CASMY, Letane.
CASOLI, Casulæ.
CASPE, Caspium.
CASPIAN SEA, Caspian Mare, Mare Hyrcanum.
CASSAND, Casandria.
CASSANO, Cassanum ad Adduam.
CASSARO, Caeyrum.
CASSEL, Cassela, Castellum Cattorum.
CASSEL, Castellum.
CASSEL, Casletum.
CASSIO, Cursula.
CASSIS, Carcicis Portus.
CASSO, Casus.
CASTAGNATZ, Pangæus.
CASTEL-A-MARE, Stabiæ.
CASTEL-ARAGONESE, Emporiæ, Castrum Aragonense.
CASTEL-BALDO, Castellum Baldum.
CASTEL-BELL, Castrum Bellum.
CASTEL-BOLOGNESE, Castrum Bononiense.
CASTEL-DURANTE, Castrum Durantium.
CASTEL-FRANCO, Castrum Francorum.
CASTEL-GANDOLFO, Arx Gandulfi.
CASTEL-GELOUX, Castrum Gelosum.
CASTEL-GUIDO, Laurium.
CASTEL-LEONE, Castrum Leonis.
CASTEL-LEONE, Castrum Leonis ad Mare.
CASTEL-LOMBARDO, Jotape.
CASTEL-MAGGIORE, Castellum Majus.
CASTEL-NAUDARY, Castellum Arianorum.
CASTEL-NUOVO, Castellum Novum.
CASTEL-NUOVO, Castrum Novum ad Adduam.
CASTEL-NUOVO-DI-CARFAGNANA, Caferonianum.
CASTEL-NUOVO-TORTONESE, Castrum Novum Dertonense.
CASTEL-PONZONE, Castrum Ponzonis.
CASTEL-ROSSO, Cisthene.
CASTEL-ROSSO, Carystos.
CASTEL-ST. PIETRO, Silarum.
CASTEL-TORNESE, Phea.
CASTEL-VECCHIO-SUBREQUO, Superæquana Colonia.
CASTEL-VETERE, Castrum Vetrium.
CASTEL-VISCONTE, Castrum Vicecomitum.
CASTEL-VOLTURNO, Volturnus.
CASTELBERG, Castellum.
CASTELBRANCO, Albicastrum.
CASTELCORN, Castrum Cornu.
CASTELHOLM, Castellum Holmium.
CASTELL, Castellanus Comitatus.
CASTELL-A-MARE-DELLA-BRUCA, Elea.
CASTELLANA, Æquum Faliscum.
CASTELLANE, Civitas Salinarum, Salinæ.
CASTELLANETA, Castania.

CASTELLAUN, Castellum Hunnorum.
CASTELLAZIO, Eubœa.
CASTELLAZZO, Gamundium.
CASTELLETE, Castra Hannibalis.
CASTELLO-ARAGONESE, Castrum Aragonense.
CASTELLO-DELLA-PIETRA, Castellum Petræ.
CASTELLO-DI-LIVIZANO, Sicinima.
CASTELLO-DI-SOLANTO, Solus.
CASTELLO-MENDO, Emerita.
CASTELLO-VILLONIO, Bilbium.
CASTELMAGNO, Castellum Majus sc. Pedemontii.
CASTELRUT, Castrum Ruptum, Castelrotta.
CASTER, Venta Icenorum.
CASTIGLIONE, Castellio Piscaria.
CASTIGLIONE, Castellionum.
CASTIGLIONE, Castilio Luccensis.
CASTIGLIONE-ARETINO, Castula.
CASTIGLIONE-MANTUANO, Castilio Mantuana.
CASTIGLIONE-MARITIMO, Castilio Calabriæ.
CASTIGLIONE-DELLA-PESCAJA, Aprilis Lacus.
CASTIGLIONE-DELLA-STIVIERE, Castilio Stiverorum.
CASTIGLIONE-DI-COSENZA, Castilio Consentina.
CASTILIEN, Castella.
CASTILLO, Herculis Fanum.
CASTIONE, Castio.
CASTLEFORD, Lagecium.
CASTONA-LA-VIEJA, Castulonensis Saltus.
CASTORIA, Celetrum.
CASTRALTA, Castrum Altum.
CASTRES. Castra ad Garumnam.
CASTRES, Castrum Albiensium.
CASTRI, Delphi.
CASTRITZA, Hella.
CASTRO, Myrina, (of Lemnos).
CASTRO, Castrensis Ducatus.
CASTRO, Castricomium.
CASTRO, Arx Minervæ.
CASTRO-DEL-RIO, Castra Postumiana.
CASTRO-DORIA, Juliola.
CASTRO-GERIZ, Castrum Cæsaris.
CASTRO-GIOVANNE, Enna.
CASTRO-GIUBILEO, Fidena.
CASTRO-REALE, Castrum Regale.
CATALDO, Petilliana.
CATALONIA, Catalaunia.
CATANIA, Catana.
CATANZARO, Catacium, Catancium.
CATAYA, Cataja.
CATELET, Casteletum.
CATHONE, Columella.
CATTARO, Ascrivium.
CATTEGAT, Codanus, Scagensis Sinus.
CATTWYK, Cattorum Vicus.
CATZENELNBOGEN, Cattimelibocensis Comitatus.
CAUB, Cuba.
CAUDEBEC, Calidobeccum.
CAUDIEZ, Cauderiæ.
CAUDROT, Cadrotium.
CAUMONT, Calvo-Mons Provinciæ.
CAUMONT, Calvo-Mons Normanniæ.
CAUSSADE, Calciata.
CAUX, Caletensis Ager.
CAVAILLON, Caballio.
CAVALA, Bucephala Peloponnesiaca.
CAVALLA, Datum.
CAVAN, Breania.
CAVERY, Chaberis.
CAVERY-PATNAM, Chaberis.
CAVERY-PORUM, Purata.
CAW, Choës.
CAYEUX, Setuci.
CAYRON, Cyrene.
CAZALEJAS, Casalaqueum.
CAZERES, Calagorris.
CEA, Cæa.
CEDOGNA, Ardoneæ.
CEFALINO, Archidemia.
CEFALONIA, Cephalenia.
CEFALU, Cephalœdis.
CEGLI, Celia.
CEHEGIN, Segisa.
CELONÆ, Colonæ.
CENEDA, Acedes.
CENIA, Sætabis.
CENIS, Cenisius Mons.
CENTAPOR, Centapora.
CENTORBE, Centuripa.
CENTOVALLI, Centumvallis.
CEPEROUX, Ceperosum.
CEPPOI, Cappæ.
CERA, Sera, Xera.
CERANO, Ceredanum.
CERASONDE, Cerasus.
CERBOLI, Barpana.
CERCHIALA, Harponium.
CERDAGNE, Ceretania.
CEREA, Cerera.
CERELO, Sirpium.
CERENZIA, Cerenthia.
CERIGO (Island of), Cythera.
CERIGO, Cythera.
CERIGOTO, Aegilia.
CERINES, Ceraunia.
CERISOLLES, Cerisolæ.
CEROSICKA, Cucci.
CERRERA, Ceressus.
CERRETO, Cenetum.
CERT, Carthus.
CERUGNONE, Gincunum.
CERVARO, Cerbalus.
CERVERA, Cervaria.
CERVETERRE, Cervetro, Cære, Agylla.
CERVI, Maxilla Asini.
CERVIA, Ficocle.
CERZUN, Cercunum.
CESANA, Cæcinum.
CESERIEUX, Casuaria.
CESINA, Cecina.
CETRARO, Parthenius Portus.
CETTE, Messua Collis, Setium.
CEVA, Seba, Ceba.
CEVENNES, Cebenna.
CEUTA, Septa.
CEYLON, Taprobane.
CHABEVIL, Cerebeliaca.
CHABLAIS, Caballiacensis Ducatus, Caballicus ager.
CHABLIS, Cabelia, Cabelium.
CHABRIA. See Cilabro.
CHABRIS, Gabris, Carobriæ.
CHADSAIE, Casiotis.
CHAGRA, Chiagra.
CHAILLI, Calagum.
CHAISE-DIEU (La), Chasse Dieu, Casa Dei.
CHALANT, Challandium.
CHALDAS, Chaldæa.
CHALONNE, Calonna
CHALONOIS, Cabillonensis Ager.
CHALONS-SUR-MARNE, Catalauni, Catalaunum.
CHALONS-SUR-SAONE, Caballinum, Cabillonum.
CHALUS, Castrum Lucii.
CHAM, Cambus Bavaricus.
CHAM, Cambum.
CHAMBERY, Camberiacum, Camberium.
CHAMBLY, Camiliacum, Cambliacum.

CHAMBORD, Camboritum.
CHAMBRE, Camera.
CHAMBRE (LA), Camera sc. Britanniæ Minoris.
CHAMBRE-FONTAINE, Camera Fontis.
CHAMOND (SAINT), Oppidum Sancti Annemundi, Castrum St. A.
CHAMOUNIS, Campimontium.
CHAMPAGNE, Campania Francica.
CHAMPIGNY, Campiniacum.
CHAMPTOCEAUX, Castrum celsum.
CHANGREH, Germanicopolis.
CHANONRY, Canoricum, Chanrea.
CHANTEL-LE-CHATEL, Cantillia.
CHANTILLY, Chantilliacum.
CHAONA, Leuce Portus.
CHAOURCE, Chaorcium.
CHAOURS, Catusiacum.
CHAPELLE, Capella.
CHAPELLE-D'ANGILLON (LA). Capella Domini Gilonis.
CHARBONNIERE, Castrum Carbonariæ.
CHARENTE, Carantonus.
CHARENTON, in the dep. of the Seine, Carentonium, Carentonus.
CHARENTON, in Cher, Carentonium Bituricense.
CHARITE (LA), Charitas.
CHARKI, Chalce.
CHARLEMONT, Carolomontium.
CHARLEMOUNT, Carolomontium Hibernicum.
CHARLEROY, Carolo regium.
CHARLESTOWN, Carolopolis Americana.
CHARLEVILLE, Carolopolis.
CHARLIEU, Carilocus.
CHARMANS, Sarrum.
CHAROLLES, Caroliæ, Quadrigellæ.
CHAROLOIS, Carolesium.
CHAROST, Carophium, Carovium.
CHAROUSSE, Carussa.
CHARPAGNE, Scarpona.
CHARTRAIN, Carnutensis Ager.
CHARTRES, Antoricum, Carnutes, ium.
CHASSENON, Cassinomagus.
CHASTELAR, Castellarum.
CHATEAU-BRIANT, Castrum Brientii.
CHATEAU-CAMBRESIS, Castrum Cameracense.
CHATEAU-CHINON, Castrum Caninum.
CHATEAU-DAUPHIN, Castrum Delphini.
CHATEAU-D'IF, Arx Iphiæ.
CHATEAU DU LOIR, Castrum ad Lædum, Castrum Lidi.
CHATEAU-DUN, Castellodunum.
CHATEAU-GONTHIER, Castrum Gonterii.
CHATEAU-LANDON, Castrum Landonis.
CHATEAU-MEILLANT, Mediolanum, Castrum Mediolani.
CHATEAU-PORCIEN, Castrum Portianum.
CHATEAU-RENARD, Castrum Vulpinum.
CHATEAU-RENAULD, Castrum Reginaldi.
CHATEAU-ROUX, Castrum Radulphi, Rudolphinum.
CHATEAU-SALINS, Castrum Salinarum.
CHATEAU-THIERRY, Castrum Theodorici.
CHATEAU-VILAIN, Castrum Villanum.
CHATEAUNEUF, in the French dep. of Mayenne-et-Loire, Castrum Novum ad Sartam.
CHATEAUNEUF, in the French dep. of Ille-et-Vilaine, Castrum Novum Aremoricæ.
CHATEAUNEUF, in the French dep. of Eure-et-Loire, Castrum Novum.
CHATEAUNEUF, in the French dep. of Charente, Castrum Novum Inculismense.
CHATEAUNEUF-SUR-CHER, Castrum Novum ad Carim.
CHATEAUNEUF-SUR-LOIRE, Castrum Novum ad Ligerim.
CHATELET, Casseletum.
CHATELLERAUD, Castellum Heraldi.
CHATILLON, in Piedmont, Castellio Pedemontii.
CHATILLON, in the French dep. of Bouches-du-Rhone, Castellio.
CHATILLON, in the French dep. of Deux-Sevres, Castellio ad Separam Nannetensem.
CHATILLON-DE-MEDOC, Castellio Medulci.
CHATILLON-EN-BAZOIS, Castellio Nivernensis Tractus.
CHATILLON-LES-DOMBES, Castellio Burgundiæ.
CHATILLON-SUR-CHER, Castellio ad Carim.
CHATILLON-SUR-L'INDRE, Castellio ad Augerim.
CHATILLON-SUR-LOING, Castellio ad Lupiam.
CHATILLON-SUR-LOIRE, Castellio ad Ligerim.
CHATILLON-SUR-MARNE, Castellio ad Matronam.
CHATILLON-SUR-SAONE, Castellio ad Ararim.
CHATILLON-SUR-SEINE, Castellio ad Sequanam
CHATILLON-SUR-SEVRE, Castellio Pictaviæ.
CHATONIE, Beberacus Lacus.
CHAUD, Calidum.
CHAUDES-AIGUES, Aquæ Calentes.
CHAUL, Caulum.
CHAULEBERNAU, Eleæ Promontorium.
CHAULNES, Calneria, Calviacum.
CHAUMERAC, Calmeracum.
CHAUMONT, in Savoy, Calmontium.
CHAUMONT, in the French dep. of Haute-Garonne, Calvo-Mons Vasconiæ.
CHAUMONT, in the French dep. of Oise, Calvimontium.
CHAUMONT-EN-BASSIGNI, Calmontium Bassiniæ.
CHAUNY, Calniacum.
CHAUVIGNY, Calviniacum.
CHAVANCY, Chavanciacum, Calvinciacum.
CHAVES, Chiaves.
CHEDUBA, Bazacata.
CHEGE, Chegensis Comitatus.
CHELADIA, Halica.
CHELLES, Cala.
CHELM, Chelmensis Palatinatus.
CHELMERS, Chelmerus.
CHEMACH, Camachus.
CHEMALI, Colonæ.
CHEMNITZ, Chemnitium, Chemn. Hermundurorum.
CHENCOA, Genchoa.
CHER (LE), Caris, Carus.
CHERASCO, Clarascum.
CHERBOURG, Cæsaris Burgus, Caroburgum.
CHERIASA, Acanthus.
CHERMES, Sermanicomagus.
CHERSO, Chrepsa, Cressa.
CHERSON, Cherso.
CHESINUS, Narthecusa.
CHESTER, Cestria, Deva.
CHESTER, Cestriensis Comitatus.
CHESTERFIELD, Cesterfida.
CHESTER-LE-STREET, Ermine-Street.
CHESU-MANSSUR, Charmodara.
CHEVREUSE, Caprusium.
CHEXBRES, Cubirasca
CHIAMPORRIERO, Campiporcherium.
CHIANA, Clanis.
CHIANA PALUDE, Clusina Palus.
CHIARAMONTE, Claromons.
CHIARENZA (MOUNT), Cyllenius Mons.
CHIARENZA, Clarentia.
CHIARI, Clarium.
CHIASCIO, Asius, Clasius.
CHIASTEZZO, Clastidium.

CHIAVARI, Clavarum, Claverium.
CHIAVASCO, Clavasium.
CHIAVENNA, Clavenna.
CHIAVES, Aquæ Flaviæ.
CHICHESTER, Cicestria.
CHIEMSEE. Chiemus Lacus.
CHIENTI, Flusor.
CHIENTO. Cluana.
CHIERI, Carea, Cherium.
CHIESE, Clusius.
CHIETI, Citium, Teate, Teatea.
CHIEVRES, Cervia.
CHIFALE, Minervæ ara.
CHIFLIK, Leuce Acte.
CHILE, Chili Regio.
CHIMAY, Chimacum, Cimacum.
CHIMERA, Chimæra.
CHIMERA MONS, Acroceraunia, or Ceraunii Montes.
CHINA, Sina.
CHINEY, Cennacum.
CHINON, Caino.
CHINY, Chiniacum, or Chisneium.
CHIOGGIA, Claudia.
CHIOZZA, Fossa Clodia.
CHISAMO, Cisamus.
CHIURO, Clurium.
CHIUSA, Clausa.
CHIUSI, Camars, Clusium.
CHIZICO, an island, Cyzicus.
CHIZICO. Cyzicum.
CHOCZIM, Chotimia, Chotinum.
CHOISEUL, Casœlum.
CHOISSY, Cauciacum.
CHOLM. Columna.
CHOLMKIL, Chilca, s. Columbani ins.
CHONAD (the com. of), Canadiensis Comitatus.
CHONAD, Canadium.
CHORASAN, Corasania.
CHORGES, Caturigæ, Caturigomagus.
CHREISSAN, Jabes.
CHRISSEE, Crusina.
CHRISTIANIA, Ansloga.
CHRISTIANIA, Aggerhusia Præfectura.
CHRISTIANSPREIFS, Christiani Munitio.
CHRISTIANSTADT, in Prussia, Christianostadium ad Boberam.
CHRISTIANSTADT, in Sweden, Christianopolis.
CHRISTINEHAMM, Christinæ Portus.
CHRISTINESTAD, Christina.
CHUR, Curia Rhætorum.
CHURWALDEN, Corvantiana Vallis.
CHURWALDEN, Corvantiense Monasterium.
CHUSISTAN, Susiana.
CIANIDZKHALI, Cianesus.
CIANIS, Cyaneus.
CIEN-POCUELOS, Centum Puteoli.
CIERF, Cervium.
CIGLIANO, in Naples, Cælianum.
CIGLIANO, Cilianum.
CIGUATEO, Ciguata.
CILABRO, Chabrius.
CILLY (circle of), Celejensis Comitatus.
CILLY, Claudia, Celeja.
CIMBRISHAMM, Cimbrorum Portus.
CIMIEZ, Cemelium.
CIMOGOL, Cumillum Magnum.
CIMOLI, Cimolus.
CIN, Cinum.
CINCA, Cinga.
CINEY, Cennæum.
CINGOLI, Cingulum.
CIOTAT, Carsici, Civitas.
CIPIONA, Cæpionis Turris.
CIRCARS, Masalia.
CIRCELLO, Circeji.
CIRCELLO (CAPE), Circæum Promontorium.
CIRCESTER, Corinium.
CIRELLA, Cerilli.
CIRENZA, s. Acerenza.
CIRIE, Ciriacum.
CIRISANO, Cytherium.
CIRKASSIER, Cercetæ.
CISMONE, Cisimons.
CISTEAUX. Cistercium.
CITADELLE, Civitas Tuta, Civitatula.
CITTA-DELLA-PIEVE, Civitas Plebis.
CITTA-DI-CASTELLO, Civitas Castellana, Tifernum.
CITTA-NUOVA, Æmonia nova.
CIUDAD-DE-LAS-CASAS, Villa Viciosa.
CIUDADELLA, Jamno.
CIUDAD-REAL, Civitas Regia, Philippopolis.
CIUDAD-RODRIGO, Civitas Roderici, Roderico polis.
CIVEDA. Civitas.
CIVIDAD-DI-FRIULI, Castrum Foro-Julianum, Forum Julii.
CIVIDALE-DEL-FRIULI. Forojuliensis Civitas.
CIVITA-BORELLA. Bucellum.
CIVITA-CASTELANA, Faliscæ.
CIVITA-D'ANTIA, Anxantium.
CIVITA-DE-GLISSA, Metalla.
CIVITA-DELLA-PIEVE, Civitas Plebis.
CIVITA-DI-PENNA, Pinna.
CIVITA-DI-ST.-ANGELO, Civitas St. Angeli.
CIVITA-DUCALE. Civitas Ducalis.
CIVITA-LAVINIA. Lanuvium.
CIVITA-A-MARE, Cliternia.
CIVITA-THOMUSSA, Foruli.
CIVITA-VECCHIA, Centum Cellæ, Civitas vetus.
CIVITELLA, Capena.
CIVITELLA, in Naples, Belegra.
CIZE, Cisiensis Ager.
CLAEVNER, CLEVNER-SEE, Clavennensis Lacus.
CLAGENFURT, Claudia, Clagenfurtum.
CLAIN, Clanis.
CLAIR, Clarius.
CLAIRETS, Claretum.
CLAIRMONT, Claramontium.
CLAIRVAUX, in the French dep. of Aube, Claræ Vallis.
CLAIRVAUX, Clara Vallis
CLAMECY, Clameciacum, or Clamiacum.
CLARE, CLARENCE, Clarentia.
CLARE, CHOMOND, Clara, Chomondia.
CLATEA, Cithæron.
CLAUSEN, Clausina.
CLAUSENBURG, Claudianopolis, Colosvaria.
CLAUSENBURG (the com. of), Colosvariensis Comitatus.
CLEANO, Cliternum.
CLERAC, Clariacum ad Oldam.
CLERMONT, Arverna.
CLERMONT-DE-LODEVE, Claromontium Lutevense.
CLERMONT-EN-ARGONNE, Clarus Mons Lotharingiæ.
CLERMONT-EN-BEAUVOISIS, Clarus Mons Bellovacensis.
CLERY, Cleriacum ad Ligerim.
CLEVE, Clivia.
CLEVE (dukedom of), Cliviensis Ducatus.
CLEVEN, s. Chiavenna.
CLEY, Garrejenus.
CLICHY, Clippiacum.
CLISSA, Andecrium.
CLISSELIK, Grynia.
CLITUNNO, Clitumnus.
CLONEY, Clona.

CLOUD (Saint), Novigentum.
CLOYNE, Cluanum.
CLUGNY, Cluny, Cliniacum, Cluniacum.
CLUNDERT, Cludera.
CLUSON, Clusonia Vallis.
CLUSON, Clusonius.
CLUYD, Cluida.
CLY, Clitum.
CLYDE, Cludanus Amnis.
CLYDE (Frith of), Æstuarium, Varæ Glotæ Æstuarium.
CLYDESDALE, Glotana Vallis.
CNIDO, Cnidus.
COA, Cuda.
COBLENZ, Confluentes, sc. Helvetiorum.
COBLENZ, Confluentia, Confluentes.
COCA, Cauca.
COCAGLIA, Colicaria.
COCHEM, Kochheim, Cochima.
COCHIN, Cottiara.
COCKERMOUTH, Coccymutium, Novantum.
CODA-DI-VOLPE, Cauda Vulpis.
CODES, Cogedus.
CODI-GARO, Neronia.
CODINIAC, s. Cadenac.
CODOGNO, Catoneum.
CODOS-DE-LADOC, Ladicus Mons.
CODROPIO, Codropolis.
COELLEDA, Coleda.
COELN-AN-DER-SPREE, Colonia ad Spream.
COERLIN, Corlinum.
COESFELD, Cosfeldia.
COESLIN (regency of), Cosliniensis Comitatus.
COESLIN, Coslinum.
COESNON, Coetenus.
COEVORDEN, Covordia.
COFT, Coptos.
COGNA, Conia.
COGNAC, Cognacum.
COGNI, Iconium.
COGORETO, Coguretum.
COIMBRA, Conimbra, Conimbrica.
COKET, Coqueda.
COL, Cola.
COLCHESTER, Colcestria, Coldonia.
COLDINGHAM, Coldania.
COLEAH, Rusuccoræ.
COLECHI, Colchi.
COLERAINE, Colrana.
COLERAINE, Colranensis Comitatus.
COLIGNY, Coloniacum.
COLLE, in Tunis, Collops Magnus.
COLLE, in Tuscany, Collis.
COLLERTON, Cilurnum.
COLLIN, Colinum.
COLLIOURE, Caucoliberum, Cocoliberis.
COLMAR, Cholembra.
COLMAR, Collis Martis, Coloburgum.
COLMENAR-DE-OREJA, Laconimurgi.
COLMOGROD, Colmogora.
COLOBIANO, Colobianum.
COLOGNA, Coloniola.
COLOGNE, Colonia Agrippina.
COLO-KYTIA, Gythius.
COLOMBARI-DI-TREPANI, Columbaria.
COLOMBO, Arabingara.
COLOMEA, Colonia.
COLOMIERS, Colomeria.
COLONI, Cana.
COLONIA (La), Eclanum.
COLONNE, Lacinium.
COLORNO, Culurnum.
COLRAINE, Colranensis Comitatus.
COLUMBATS, Galumbaz, Columbraria.
COLUMBI, Columbarum Insula.
COLUMBRETES (Las), Ophiusa.
COLUMBTON, Columbus.
COLURI, Salamis.
COM, Koum, Chaona.
COMANA, Cromna.
COMARIN, Camar.
COMBRAILLES, Combralia.
COMBRONDE, Oppidum Candidobrinse.
COMBURGOS, Callum.
COMER-SEE, Comacenus, Larius lacus.
COMESAZZO, Comesatium.
COMINES, Comineum, Comminium.
COMINGES, Comminges, Convenæ, Convenensis Tractus.
COMINO, Cuminum, Hephæstia.
COMMACHIO, Cimaculum, Comacium.
COMMARGUE, Metina.
COMMERCY, Commerciacum.
COMMOTAU, Helcipolis.
COMO (Lago-di), Cumanus Lacus.
COMO, Cumetense Territorium.
COMO, Comum.
COMORN, Comara, Crumenum.
COMORN (comitat of), Comaroniensis Comitatus.
COMOTAY, Comotæa.
COMPIEGNE, Compendium.
COMPOSTELL, Flavionia, Fanum S. Jacobi.
CONCA, Crustumium.
CONCABAR, Elymais.
CONCARCEAUT, Concordiæ Saltus.
CONCARNEAU, Concarneum.
CONCEPTION, Civitas Conceptionis.
CONCEPTION, Conceptionis Sinus.
CONCHES, Conchæ.
CONCRESSANT, Concurcalium.
CONDE, Condæum, Condatum.
CONDE, Condate.
CONDE-SUR-NOIREAU, Condæum ad Norallum.
CONDOJANI, Condojanum.
CONDOM, Condomium Vasconum.
CONDRIEUX, Condriacum.
CONDROS, Condrusorum Regio.
CONDUR, Coddura.
CONEGLIANO, Conelianum.
CONFLANS, Confluentia.
CONFLANS, Confluentes Centronum.
CONFLANS, Confluentes Burgundiæ Superioris.
CONFLANS-EN-JARNY, Confluentes Lotharingiæ.
CONFLANS-ST.-HONORINE, Confluentes Insulæ Franciæ.
CONGLETON, Congletonium.
CONI, Coneum, Cunejum.
CONIGLIERI, Insulæ Pelagiæ.
CONITZ, Choinitia.
CONNAUGHT, Connachtia, Connacia.
CONNEBURG, Connoburgum.
CONNERAY, Conedracium.
CONQUET, Conquestus.
CONSERANS, Conseranensis Ager.
CONSTANTIA, Constantine, Cirta.
CONSTANTINOPLE, Constantinopolis, Byzantium.
CONSTANTINOPLE (Straits of), Bosporus Thraciæ, Os Ponti, Fretum Constantinopolitanum.
CONSTANZ, Costnitz, Constantia.
CONSUEGRA, Consabrum.
CONTESSA, Sinus Strymonicus.
CONTIGLIANO, Contilianum, Cutilia.
CONTY, Contiacum.
CONUCHA, Cythera.
CONVERSANO, Conversanum, Cupersanum.
CONWAY, Conovium.
CONZA, Compsa.
COPA, Cupa.

COPENHAGEN, COPPENHAGEN, Codania, Hafnia.
COPRANITZ, Copranitia.
COPREINITZ, Coprinitia.
CORACE, Crotalus, Corax.
CORACOJAN, Caryanda.
CORAKE, Corax.
CORANGA, Calliena.
CORBEIL, Corbelium, Corbolium.
CORBENY, Corbeniacum.
CORBERG, Corberia.
CORBETTE, Curia Pietra.
CORBIE, Corbeja.
CORBIERES, Corbaria Vallis.
CORBIGNY, Corbiniacum.
CORBIGNY, Corbiniacum Nivernense.
CORBON, Corbo.
CORBRIDGE, Corstorpitum.
CORCHUELA (LA), Rusticiana.
CORDES, Cordua.
CORDON, Cordona.
CORDOVA, Colonia Patricia Cordubensis, Corduba.
CORDUAN, Antros.
CORELLA, Graccuris.
CORFU (ISLAND OF), Corcyra.
CORFU, Corcyra.
CORI, in the Papal states, Cora.
CORI, in Dalmatia, Corinium.
CORIA, Cauria.
CORIGLIANO, Coriolanum Calabriæ.
CORIGLIANO, Coriolanum in Calabria Citra.
CORINTH, Corinthus.
CORK, Corcaginensis Comitatus.
CORK, Corcagia.
CORLAY, Vorgium.
CORMEILLE, Curmiliaca.
CORMERY, Cormaricum, Cormeriacum.
CORMICY, Cormiciacum, Calmisciacum.
CORMONS, Cormones.
CORNEILLAN, Cornelianum.
CORNETO, Castrum Novum ad Martam, Cornetum.
CORNETO, in Sardinia, Corni.
CORNEVILLE, Cornavilla.
CORNICO, Corycus.
CORNIGLIONE, Cornilianum.
CORNIO, Ciconium Promontorium.
CORNWALLIS, Cornubia.
COROMANDEL, Coromandela.
CORON, Colonides.
CORONE, Corona.
CORREGGIO, Corregium.
CORRESE, CUREZE, Cures.
CORRESE (the Italian river), Curensis Amnis.
CORREZE (the French river), Curetia.
CORREZE, Curetium.
CORSOER, Crucisora.
CORTE, Curia.
CORTEGANA, Corticata.
CORTELAZZO, Ficaria.
CORTE-MAGGIORE, Curia Major.
CORTEMIGLIA, Curtismilium.
CORTE OLONO, Olonna Curtis.
CORTONA, Coriti Arx, Corythus.
CORTRYCK, Courtray.
CORUGNA, Clunia.
CORUNNA, Brigantium, Coronium.
CORVEY, Corbeja, Corbeia.
CORVO, CUERVO, Corvi Insula.
CORZOLA, Corcyra.
COS, s. Stanchio.
COSCAN, Cocusos.
COSENZA, Consentia.
COSMIN, Coccoranagora.
COSNE, Cona, Condate.
COSSE-LE-VIVIERS (VIVOIN), Cossiacum.
COSTANZA, Constantia.
COSTHEIM, Cuffinstanium.
COTBUS, s. Cottwitz.
COTENTIN, Constantiense Territorium.
COTES (LES), DE GARRAF, Hannibalis Scalæ.
COTRONE, Croto.
COTTE (LA), ST. ANDRE, Clivus St. Andreæ.
COTTWITZ, Cotbusium.
COUBEBI, Emaus.
COUCY, Cociacum, Codiciacum.
COULANGES-LES-VINE SES, Coloniæ Vinosæ.
COULANS, Villa Colonia.
COULOGNE, Casinomagus, Colonia.
COULONS, Colombæ.
COUPER, Cupra.
COURPIERRE, Curtipetra.
COURTENAY, Cortenæum, Cortiniacum.
COURTRAY, CORTRYCK, Cortracum, Corteriacum.
COUSINS-HOUSE, Segedunum.
COUTANCES, Castra Constantina.
COUTRAS, Certeratæ, Cotracum.
COVENTRY, Conventria.
COVOLI, Claustrum Cubali.
COWBRIDGE, Bovium.
COZZO, Cottiæ.
CRACAU, Carodunum, Cracovia.
CRACAU, Cracoviensis Palatinatus.
CRAIN, s. Krain.
CRAINBURG, Cariniobnrgum.
CRAJOWA, Drubetis.
CRANENBURG, Burcinalium.
CRANGANOR, Baris.
CRAON, Cratumnum.
CRAONOIS, Credonensis Ager.
CRASTA, Candavii Montes.
CRATI, Crathis.
CRAU, Campi Lapidei, Campus Herculeus.
CRAVANT, Crevantium.
CRECI, Gatiacum, Cressiacum.
CRECY, Carisiacum.
CREE-FIRTH, Creæ Æstuarium.
CREIL, Credilium, Creolium.
CREMASCO, Cremensis Ager.
CREMIEU, Cremiacum.
CREMINIECK, Cremenecum.
CREMNITZ, Cremnicium.
CREMONA, Cremonensis Ager.
CREMS, Cremesia.
CREMSMUNSTER, Chremissæ Monasterium.
CREPY, CRESPY, Crepiacum.
CREPY, CRESPY, Crepiacum Laudunense.
CREQUY, Crequium.
CREST, Christa.
CREUSE, Crosa.
CREUSSE, Crusena.
CREUTZ, Crisium.
CREUTZBURG, in Konigsberg, Cruciburgum Venedicum.
CREUTZBURG, in Saxe-Weimar, Cruciburgum ad Vierram.
CREUTZBURG, in Silesia, Creutzberga.
CREUTZENACH, Cruciniacum.
CREVACOURE, Crepacorium.
CREVASTA, Apsus.
CREVECŒUR, Crepicordium Cameracense.
CREVECŒUR, Crepicordium.
CREVILLY, Crollejum.
CRICHINGEN, Creance.
CRILLON, Credulio.
CRISTONIA, Creston.
CRITIA, Elæus.
CROCHA, Arocha.
CROISIL, Crociliacum.
CROISILLE, Crocilliaca.
CROMARTY, Cromartium.

CROMARTYSHIRE, Cromartinus Comitatus.
CRONENBURG, Coronæburgum.
CRONSTADT, Brassovia, Stephanopolis.
CROSSEN, Crosna ad Elystrum.
CROSSEN, Crosna.
CROSSEN, Crosna Sarmatica.
CROTOIS, Carocotinum.
CROU, CROULD, Crovus.
CROUY, Croviacum.
CRUZ DE LA ZORZA, Vicus Caminarius.
CSEPEL, Insula St. Margarethæ, Cepelia.
CUBU, Subur.
CUENCA, Conca.
CUESTRIN, Costrinum.
CUGIONE, Cusionum.
CUIVIN, Covinum.
CULLERA, Sucro.
CULLO, Calliga.
CULM, Mariæchelmum.
CULM, Culmia.
CULMORE, Vidua.
CUMBERLAND, Cumbria.
CUNDIS, Contegium.
CUNNINGHAM, Cunigamia.
CUPAR, Cupra.
CURDISTAN, Curdia.
CURDO, Curdius Mons.
CURE, Chora.
CURISCHE-NEHRUNG, Curonensis Peninsula.
CURISCHE-HAFF, Curonensis Lacus.
CURLAND, Curonia, Curlandia.
CURTAKOES, Curta.
CURZOLA, Corcyra Nigra.
CURZOLARI, Echinæ, Echinades.
CUTAIS, KUTAIS, Cyta.
CYPERN, Cyprus.
CZACKENTHURN, Chactornia.
CZACNITZA, Staneclum.
CZAKOTORN, Carrodunum (Carnodunum).
CZASLAU, Czaslavia.
CZASLAU, Czaslaviensis Circulus.
CZATALCZA, Tharsandala.
CZENSTOCHAU. Czenstochovia.
CZERHAT-FATRA, Sarmatici Montes.
CZERNEK, Inicerum.
CZERNETZ, Zerna, or Colonia Zernensium.
CZERSK, Ciricium, or Czercum.
CZIRKNITZ (LAKE), Circoniensis Lacus.
CZIRKNITZ, Czernicum, Circonium.
CZONGRAD, Celadium.

D

DABARNA, Coloë.
DABORA, Dabira.
DABUL, Mandagora.
DACHAN, Dachanum.
DACHSTEIN, Dagoberti Saxum.
DADI, Drymæa.
DAENEMARK, Dania.
DAENISCHWALD, Danica Sylva.
DAFAR, Sephar.
DAHALAG, Daphnitis Insula.
DAHLAK, Orine.
DAKKE, Pselche.
DALARNE, Dalecarlia.
DALCKE, Dellina.
DALELBE, Dalecarlius.
DALIN, Idalium.
DALKEITH, Dalkethum.
DALKEY, Limnus.
DALMINO, Delminium.
DAM, Dammona.
DAMAN, Camani.
DAMAR, Leontopolis.
DAMEGHAN, Tagæ.
DAMER, Tamyras.
DAMERY, or DAMERA, Dameriacum.
DAMMARTIN, Damnum Martinum, Dominium Martini.
DAMME, Damma.
DAMMERSEE, Estia.
DAMVILLE, Damovilla.
DAMVILLERS, Dampuillerium, Damvillerium.
DANA-PLU, Daona.
DANAR, Eleutherus.
DANGEAU, Dangellum.
DANZIG, Dantiscum, Gedanum.
DARABGHERD, Gabæ.
DARABIN, Dara.
DARABYE, Granis.
DARDANELLES, Arces ad Hellespontum.
DARDANELLES (STRAIT OF), Hellespontus, Fretum Hellesponticum, Gallipolitanum.
DASCARA-EL-MELIK, Dastagerda.
DATSCHITZ, Daczitium.
DATTENRIED, DELLE, Datira.
DATTOLO, Phœnicusa.
DAULIA, Daulis.
DAUMAKLI, Pelecas.
DAUN, Dumnus.
DAUPHINE, Delphinatus.
DAVENTRY, Bennavenna.
DAVIS' STRAIT, Fretum Davisii.
DAX. See ACQS.
DAXWANG, Taxgætium.
DEAL, Dola.
DEBOT, Parambole.
DEBRECZIN, Debrecinum, Debrettinum.
DECIZE, Decetia.
DEDES, Dedessa.
DEE, Devæ Æstuarium.
DEE, Deva, Dæa.
DEES AKKA, Comidava.
DEHASCH, Ochus.
DEKIAN, Hieratemis.
DELAWARE, Delavarus.
DELBOM, Heracleopolis Parva.
DELBRUCK, Delbruggia.
DELEBIO, Alebium, Dalebium.
DELEMONT, Telemontium.
DELFINO, Delphinium.
DELFINO, Gythanæ.
DELFT, Delfi, Delphi, Delphium.
DELFTLAND, Delpholandia.
DELFTSHAFEN, Delphorum Portus.
DELFZYL, Delfzilia Arx.
DELHI, Clisobora.
DELHI, Dellium.
DELICHI, Acheron.
DELITSCH, Delitium.
DELLE COLONNE, Naus.
DELMINO, Dalmium.
DELSBERG, Delemontium, Telamontium.
DEMENHUR, Hermopolis Parva.
DEMENYFALVA, Demanowa.
DEMER, Tabuda.
DEMIRCARPI, Cataractæ Danubii.
DEMITRIOFF, Demitriovia.
DEMONA (VALDI), Vallis Demonæ, Nemorensis.
DEMOTICA, Didymotichos.
DENBIGH, Denbiga.
DENDER, Galthera, Dendera.
DENDERAH, Tentyra.
DENDERMONDE, Munda Teneræ, Teneramunda
DENEUVRE, Danubrium.
DENIA, Artemisium, Dianium.
DENIGU, Denigotia.
DENIS (SAINT), Dianiscopolis.

DENISLEY, Thomisonium.
DENMARK, Danimarca.
DENSEN, Desena.
DENSEN, Dumnissus.
DEOGHIR, Tagara.
DEOLS, Castrum Dolense.
DERAS, Deris.
DERBEND, Albaniæ Portæ.
DERBEND-PUST, Nazada.
DERBY, Derventia.
DERBYSHIRE, Derbicensis Comitatus.
DERCUS, Dercon.
DERDASSO, Deboma.
DERGH, Dergus.
DERNE, Darnis.
DERRY, Deria.
DERWENT, Darventus.
DESAGUADERO, Emissarius.
DESENZANO, Decentianum.
DESIDERADE, Desiderata.
DESITZE, Deciatum.
DESMOUND, Desmonia comitatus.
DESNA, Hypacaris.
DESPOTO-DAGH, Rhodope.
DESSAU, Dessavia.
DESZNA, Jesna.
DETERN, Teuderium.
DETMOLD, Theotmala, Detmoldia
DEUPRAY, Catadupa.
DEUSCHBERG, Dei Mons.
DEUTICHEM, Dotecum.
DEUTSCH-BROD, Broda Teutonica.
DEUTSCHENDORF, Popradum.
DEUTSCHLAND, Germania.
DEUTZ, Diuza, Diutia.
DEVA, Decidava.
DEVA, Diva, Devales.
DEVENTER, Daventria, Devontarum.
DEVEREM, Durfos.
DEVONSHIRE, Devonia.
DEVRET-HISSAR, Deborus.
DEZISE, Decetia.
DIADIN, Daudyana.
DIALA, Delas.
DIARBEKIR, Amida.
DIAR MODHAR, Osrœne
DIASKILLO, Dascyleum.
DIBAN, Dibon.
DIBRES, Dibræ.
DICOMANO, Decumanum.
DICTAINO, Dittainus.
DICTAMO, Dictamnum.
DIDSCHLE, Tigris.
DIDYMO, Didymi.
DIE, Augusta Dea.
DIEDENHOFEN (Thionville), Theodonis villa.
DIEMEL, Dimala, Dimona.
DIEPHOLZ, Diepholtanus Comitatus.
DIEPHOLZ, Diepholta.
DIEPPE, Deppa.
DIESDORF, Insula St. Mariæ.
DIESENBERG, Disibodengense Cœnobium, Desenberga.
DIESSEN, Damasia.
DIESSENHOFEN, Darnasia.
DIETZ, Decia.
DIEU, Insula Dei.
DIEULOUARD, Deslonardum.
DIEUZE, Decem Pagi.
DIGEL, Mesene.
DIGNE, Civitas Diniensium, Dinia.
DIJON, Dibio, Divio, Diviodunum.
DIJONNOIS, Diviodunensis Pagus.
DIKELIK, Atarnea.
DILEM, Catadorbis.
DILI, Delos.
DILLSBOO, Delisboa.
DIMITZANA, Erymanthus.
DI MOLISE, Samnium.
DINANT, Dinantum.
DINDIGULL, Tangala.
DINGELFINGEN, Dingelvinga.
DINGLE, Dinglia.
DINKELSBUHL, Zeacollis.
DIOIS, Diensis Tractus.
DIONANT, Deomant, Sanctæ Mariæ Monastorium.
DIORIGUI, Nicopolis.
DIPRAG, Devaprayaga.
DIPSO, Ædepsum.
DISENTIS, Desertina, Disentium.
DIVA, Deva.
DJIROST, or Girest, Throasca.
DNIEPER, Borysthenes.
DNIESTER, Danaster, Tyras.
DOBBER, Thabba.
DOBERAN, Dobranum.
DOBRZYN, Dobrinia, Dobrinum.
DOCKUM, Doccomium, Dochzetum.
DOEMITZ, Domitium.
DOG-ISLAND, Insula Canum.
DOGHOE, Daghoe.
DOGLIANI, Dolianum.
DOLCEAQUA, Dulcis Aqua.
DOLE, Didattium, Dola.
DOLLART, Sinus Emdanus, Dollarius.
DOMBES, Dumbæ.
DOMBESCHGERTHAL, Domestica Vallis.
DOMFRONT, Donnifrons, Dumfronium.
DOMNISSAN, Dumnotinus.
DOMO-D'OSSOLA, Oscella, Domoduscella.
DOMOCHI, Demonicus.
DON, Tanais.
DONATO, Evoria.
DONAU, Danubius.
DONAUESCHINGEN, Doneschinga.
DONAUWORTH (Danubii Insula), Vertia, Donaverda.
DONAWITZ, Donau-Arm, Danubii Brachium.
DONCHERY, Doncherraeum.
DONEGAL, Dungalensis Comitatus.
DONEGAL, Dungalia.
DONEZ, Hygris.
DOOR, Ravius.
DOORNE, Turnium.
DORCHESTER, Dorcestria, Dunium.
DORCHESTER, Dorciniæ Civitas, Dorscia, Dorcestria.
DORDOGNE, Duranius, Turanius.
DORMAGEN, Durnomagus.
DORMOIS (LE), Dulmense.
DORNIK, or Tournay, Turnacum.
DORNOCK, Dornodunum.
DORO, Oboca.
DORPAT, Derbatum, Dorpatum.
DORSET, Dorsetia.
DORTMUND, Dormunda, Tremonia.
DORTRECHT, Dordracum.
DOS-DI-TRENT, Dorsum Tridentinum.
DOTIS, Deotatum, Theodata.
DOUARNENES, Dovarnena.
DOUAY, Catuacum, Douacum.
DOUBS, Alduabis.
DOUE, Doadum, Theodoadum.
DOULENS, Donincum.
DOURDAN, Dordanum.
DOURIERS, Duroicorejum.
DOUX, Dubis.
DOUZE, Dusa.
DOUZERE, Ducera.

DOUZY, Diciacum.
DOVER, Dubræ, Dubris.
DOWN, Dunensis Comitatus.
DOWNE, Dunum.
DRAA, Edrei, Adra.
DRAAS, Darocinium.
DRABURG, Dravoburgum.
DRAGON, Draconis Mons.
DRAGONARA, Gerion, Geronium.
DRAGONE, Draco.
DRAGONISI, Corasiæ.
DRAGUIGNAN, Dracenæ.
DRAMA, Drabescus.
DRAMISI, Delium.
DRANDAR, Pityus.
DRAPANO, Drepanum Promontorium.
DRAU, or DRAVE, Dravus.
DRAUSEN, Drusis.
DREUX, Drocæ, Drocum.
DRIESEN, Dressenium.
DRILLO, Achates.
DRIMAGO, Dirigota.
DRIN, Caradrina.
DRINA, Drinus.
DRINO-BIANCO, Drilo.
DRINO-NEGRO, Drinus.
DRINO, Drilonius Sinus.
DRINOVAR, Drinopolis, Trinopolis.
DROGEO, Insula Diabolorum.
DROGHEDA, Droghdæa, Pontana.
DROME, Druna.
DRONERO, Draconerium.
DRONTHEIM, Nidrosia.
DRUIDA, Diruta.
DRUMLANRIG, Uxellum
DRUSEN, Drusomagus.
DSCHANGLI, Panionium.
DSCHEBAIL, Byblos.
DSCHEDUR, Ituræa.
DSCHENNIN, Ginæa.
DSCHESAN, Pudni.
DSCHESIRA-EL-BIRBE, Philæ.
DSCHESIRAT-EL-ARAB, Arabia Felix.
DSCHESME, Phœnicus Portus.
DSCHIBBEL-EL-GHARBI, Libanus.
DSCHISSER-BENI-JAKUB, Jacobi Pons, Geschur.
DSCHOLAN, Gaulanitis.
DSJEDIE, Schedia.
DSJINA, Harpasus.
DSJISME, Organa.
DSJUBBE, Olabus.
DUBEN, Duba.
DUCHS, Duxonum.
DUERO, Durius.
DUESMOIS, Duesmensis tractus.
DUINO, Pucinum.
DUISBURG, Duisburgum, Tuiscoburgum.
DUIVELAND, Duvelandia.
DULCIGNO, Olchinium.
DULECKE, Dulecum.
DULOMA, Thebæ.
DUMATH, Duma.
DUMBARTON, Levinia.
DUMBARTON, Britannodunum, Dumbritonium.
DUMFRIES, Dunfreja.
DUMFRONT, Dumfronium.
DUMMERSEE, Dummera.
DUMNOTYR, Dumnoricum, Dunnovortium.
DUNBAR, Vara, Dumbarum.
DUNBERG, Dunus, Dunum.
DUNBLANE, Dumblanum.
DUNCANSBY-HEAD, Orcas Prom.
DUNCASTER, Danum.
DUNDALK, Dunkeranum.
DUNDEE, Allectum, Donum Dei.
DUNINGEN, Dunum.
DUNKELD, Caledonia.
DUNKERAN, Ivernis.
DUNKIRK, Dunquerca.
DUN-LE-ROI, Castrum Duni, Regiodunum.
DUNOIS, Dunensis ager.
DUNSTAFFNAGE, Evonium, Stephanodunum.
DUNSTER, Dunestorium Castrum.
DUNWICH, Dunmocha.
DURANCE, Druentia.
DURAS, Duracii, Duracium.
DURAZZO, Dyrrachium.
DURBUY, Durbis.
DUREN, Marcodurum, Dura.
DURETAL, Durastellum, Durstallum.
DURHAM (Palatinate of), Dunelmensis Comitatus
DURHAM, Dunelmum, Dunhelmum, Duremum.
DURRAH, Aria.
DUS-CHUR-MALU, Mennis.
DUSSELDORF, Dusseldorpium.
DUX, Duxovium.
DWINA, Carambacis, Duina.
DYLE, Thilia.
DYSBORCH, or DYSBORG, Dispargum.
DYSS, Desertum.

E

EARNE, Dernus.
EASTMEATH, Media.
EATON, Æthonia, Etona.
EAUZE, Elusa.
EBENHEIM, Gaviodurum.
EBERACH, Eberacum.
EBERSBERG, Eburobergomum.
EBERSDORF, Aula Nova.
EBERSHEIMMUNSTER, Aprimonasterium.
EBREUIL, or EBREVILLE, Ebrolium.
EBRO, Iberus.
ECHITZ, Achatius.
ECHTERNACH, Andethanna?
ECIJA, Astigi.
ECLUSE, Sluzæ.
ECOUIS, Escovium.
EDEN, Ituna.
EDER (the river), Adrana.
EDER, Eudrapa.
EDERINGTON, Adurni Portus.
EDFU, Apollinis Urbs Magna.
EDINBURGH, Alata Castra, Edinum
ED-SOPH, Aphroditopolis.
EDULO, Edulum.
EGER, Egra.
EGER (the river), Egra, Agara.
EGINA (GULF OF), Saronicus Sinus.
EGLISAU, Eglisavia, Eglisonia.
EGLON, Eglon.
EGLY, Eglis.
EGORLIK, Mermodus.
EGRI-PATANKA, Bylazora.
EGRIBOS, Euripus.
EGYPT, Ægyptus.
EHINGEN, Dracuina.
EHRENBREITSTEIN, Ehrenberti Saxum.
EICH, Echa.
EICHSTADT, Sedatum, Eistetta.
EIDER, Egidora, Eidora.
EIDERSTÆDT, Epidorensis Præfectura.
EIFFEL, Eiffalia.
EILENBURG, Ileburgum.
EINDHOFEN, Eindovia.
EINRICH, Henrici pagus.
EINSIEDELN, Eremitarum Cœnobium in Helvetiis.

EIPATOS, Inatus.
EISACH, Atagis.
EISENACH, Isenacum.
EISENBERG, Eiseoberga.
EISENBURG, Castrum Ferreum.
EISENBURG (Comitat of), Castriferrensis Comitatus.
EISENBUTTEL, Isenbuttella.
EISENSCHNIPPE, Fauces Ferreæ.
EISLEBEN, Islebia.
EKESJOE, Eckesiœa.
ELALIA, Achila.
ELANDA, Lampea.
EL-ARCAS, Cercasorum.
EL-ARISCH, Rhinocolura.
ELBA, Æthalia, Ilva.
EL-BAISAN, Bethsan.
ELBE, Albis.
ELBEUF, Elbovium.
EL-BIR, Machinas.
EL-BOKAH, Cœle-Syria?
EL-BOTTHIN, Basan.
EL-CALLAH, Gitlui.
ELCHE, Illice, Illici.
EL-COME, Cholle.
ELDA, Adellum.
ELDER, Gadirtha.
ELEFTA, Elatea.
ELEITZ, Elethvia.
ELENBOROGH, Olenacum.
ELEO, Eilei, Ilei.
ELEVANGON, Elephantiacum.
EL-HERBA, Tigava.
ELIASBERG, Pelinæus.
ELIMBO, Olympus.
EL-JEHUDIEH, Castra Judæorum.
EL-KAB, Lucina.
ELKAS, Olgassus.
EL-KATIF, Gerra.
EL-KHABUR, Chaboras.
EL-KIBIR, Ampsaga.
EL-KURAH, Moab.
ELL, Elcebus.
EL-LAHUM, Ptolemais.
ELLERENA, Regiana.
ELLGOW, Elgovia, Augia sacra.
ELL-JEMME, Thysdrus.
ELLRICH, Elricum.
EL-MADAIN, Seleucia ad Tigrim.
EL-NACHAR, Nazara.
ELNBOGEN, Cubitus.
ELNE, Helena, Eliberri.
ELPISARA, Ephaliga.
EL-POZUELO, Pæsula.
ELSASS, Alsatia.
ELSASS-ZABERN, Tabernæ Tribocorum, Alsaticæ.
ELSEN, Aliso.
ELSFLETH, Alisni.
ELSLOO, Haslacum.
ELSTER, Elister.
ELTEN, Altinæ.
EL-VALLE-DE-ABDELACIZ, Nescania.
ELVAS, Alba, Heluæ.
ELVIRE, Eliberi, Illiberis.
ELWANGEN, Elephantiacum.
ELY, Helia, Elia.
ELZ, Eltzia.
ELZE (the river), Alisontia.
ELZE, Aula Regia ad Lynum.
EM, Emma.
EMB, Emba.
EMBDEN, Amasia.
EMBOLI, Amphipolis.
EMBRO, Imbrus.
EMBRUN, Castrum Ebredunense, Eborodunum.
EMBS, in the Tyrol, Amades.
EMBS, in Nassau, Embasis.
EMELY, or Awn, Emelia.
EMLEY, Jemelaca.
EMMAT, Amma.
EMMEN, Emana.
EMMENTHAL, Emmanæ Vallis.
EMMER, Ambra.
EMMERICH, Asciburgum, Emmerica, Embrica.
EMPOLI, Emporium.
EMS, Amasia, Amasius.
ENDARO, Enderum.
ENGADIN, or Innthal, Caput Œni, Engiadina.
ENGELBERG, Ingelberga.
ENGELSBURG, Hadriani Moles, Castellum Angeli.
ENGER, Angaria, Angria.
ENGHIEN, Angia.
ENGIA, Ægina.
ENGLAND, Britannia, Anglia.
ENIED, Enjedinum.
ENKHUIZEN, Enchusa, Enchusia.
ENKOEPING, Encopia, Enecopia.
ENN, Endidæ.
ENNISKILLEN, Arx Kellina.
ENOS, Ænos.
ENS, Anassianum, Anisia.
ENS, Anasus, Anisus.
ENSDORF, Ensdorfense Monasterium.
ENSINEH, Antinoe.
ENTRAIGUES, Interaquæ.
ENTRAINS, Interamnis.
ENTRAMES, Interamnis.
ENTRE-DUERO-E-MINHO, Extrema Minii.
ENTREMONT, Intermontium.
ENTREVAUX, Intervalles, Intervallum.
ENVERMEU, Envermodium.
EPERIES, Aperiascio, Eperiesinum.
EPERNAY, Asprencia, Sparnacum.
EPERNON, Sparno.
EPINAL, or Espinal, Spinalium.
EPINOY, Espinojum, Spinetum.
EPSHAM, Ebeshamum, Thermæ Ebeshamenses.
ERBACH, Erpachium.
ERCOCA, Erquicum.
ERCSIN, Salinum.
EREKLI, Heraclea Thraciæ.
EREMO-DI-ST.-AUGUSTINO, Gravisca.
ERETIA, Cnemides.
ERETRI, Erythræ.
ERFT, Arnapha.
ERFURT, Erfordia.
ERGERS, Ergitia.
ERI, Cæretanus amnis.
ERIAH-NAHR-EBRAHAM, Adonis.
ERINEO, Erinæum.
ERISSI, Eresos.
ERIWAN, Eroanum.
ERIWAN (Prov. of), Turcomania orientalis.
ERKELENS, Herculeum, Herculanum.
ERKLIA, Ad Horrea.
ERLACH, Elegium, Elacum.
ERLACH, or Erlaf, Erlaphus, Arlapa.
ERLAU, Agria.
ERMENAK, Homona.
ERMENT, Hermonthis.
ERMITA-DE-LOS-PALACLOS, Flavium Vivitanum.
ERMITA-DE-NUESTRA-SENORA-DE-TIERMES, Termantia.
ERNE'E, Ereneum.
ERRAHUE, Latonæ Civitas.
ERRIF, Augustamnica.
ER-ROHA, Edessa.

ERZGEBIRGE, Metalliferi Montes.
ESARO, Æsaris.
ESBET, Calus.
ESCHBORN, Asgabrunnum villa.
ESCHWEGE, Schuvegia.
ESCORIAL, Escuriacum, Scoriale.
ESCOUIS, Escovium.
ESCOUSSE, Segosa.
ESDUD, Asdod.
ESENS, Esena.
ESINO, Æsis.
ESKDALE, Escia.
ESKER, Œscus.
ESKER, Cios.
ESKI, Sida.
ESKI-FOROS, Dandaca.
ESKI-HISSAR, Laodicea.
ESKIHISSAR, Stratonicea.
ESKI-SHEHR, Dorylæum.
ESKI-SLANA, Salluntum.
ESKI-STAMBUL, Alexandrea
ESLA, Estola.
ESPEJO, Claritas Julia.
ESSECK, Mursa, Essecium.
ESSEN, Essendia, Asnidia.
ESSENBERG, Asciburgium.
ESSENGEAUX, Ensigausium.
ESSENIDE, Xanthus.
ESSLINGEN, Ezelinga.
ESTAGEL, Stagellum.
ESTAIN, or ETAIN, Stagnum.
ESTAMPES, or ETAMPES, Stampæ.
ESTANFORT, Darinum.
ESTANQUE-DE-ALBELATE, Naccarorum palus.
ESTAPLES, Stapulæ.
ESTE, Ateste.
ESTE, Escheda, Ateste.
ESTELLA, Alba, Stella.
ESTEPA, Astapa.
ESTEPONA, Cilniana.
ESTHENBERG, Estionum Mons.
ESTHLAND, Esthonia.
ESTIVAL, Stivagium.
ESTOY, Ossonoba.
ESTRECHY, Scripanicum.
ESTREMADURA, Extrema Durii, Extremadura.
ESTREMOZ, Extrema.
ETANG-DE-SIGEAN, Rubrensis Lacus.
ETHAL, Coveliacæ.
ETSCH, s. Adige.
ETSCHLAND, Athesia, Athesius ager.
ETSED, Echedum.
ETTERSBERG, Mons Æthereus.
EU, Auga, Augium.
EUFEMIA, Lametia.
EUFEMIA (GULF OF), Vibonensis Sinus.
EUGENIUSBERG, Eugenius Hyge.
EUGUBIO, Aguvium, Inginium.
EUPHRAT, or PHRAT, Euphrates.
EURE, Audura, Ebura.
EUTIN, Oittinum.
EVIAN, Aquianum.
EVORA, Ebora, Liberalitas Julia.
EVORAMENTE, or EVORA-DE-MONTE, Ebora Alta.
EVRE, Avara.
EVREUX, Ebroicæ, Ebroicum.
EVRON, Ebronium, Aurio.
EXE, Exa.
EXETER, Isca Dumnoniorum, Uxela.
EXIDEUIL, Exidolium.
EYDER, Ægidora, Egidora.
EYLAU, Gilavia Borussica.
EYMOUTIERES, Antimonasterium.
EYREH, Girgiris.
EZARO, Sars.

F

FABREGUES, Forum Domitii.
FACIALCAZAR, Salpesa.
FAENZA, Faventia.
FAGNEAUX, Faniolum.
FAIGNE (LA), Fania.
FAIMINGEN, Pomo.
FAINS, Fanis.
FAIR HEAD, Robogdium promont.
FALAISE, Falaza, Falesia.
FALASIA, Phalasia.
FALCKENBERG, in Sweden, Falcoberga.
FALCKENBERG, in Silesia, Falcomontium.
FALCKENSTEIN, Falconis Petra, Falcostenium.
FALCO, Palandas.
FALKENBERG, Coriovallum.
FALKENSTEIN, Falchenstein.
FALKIRK, Davium Sacellum.
FALKOPING, Falcopia, Focopia.
FALLERONI, Falaria.
FALMOUTH, Falmuthum, Volemutnum, Volubæ Portus.
FALSTER, Falstria.
FALVATERA, Fabrateria.
FAMAGOSTA, Arsinoe, Fama Augusta.
FAMARS, Fanum Martis.
FAMENE, Pæmani.
FAMIEH, Apamea Syriæ.
FAMINE, Famiensis Tractus.
FANAGORIA, Corocondame.
FANAS, Fæntejum.
FANJEAUX, Fanum Jovis.
FANO, Refugium Apoliinis.
FANO, or FANEK, Vennum.
FANO, Colonia Julia Fanestris.
FARBEIT, Pharbæthus.
FARFA, Fabaris.
FARIGLIANA, Cyclopum Scopuli.
FARMOUTIERS, Farense monasterium.
FARNASU, Erithini Scopuli.
FARNESE, Farnesium Castrum.
FARNHAM, Vindomum.
FARO-DI-MESSINA, Siculum Fretum.
FAROER, Færoæ Insulæ.
FARRA, Parsia.
FARSA, Pharsalus.
FASSA, Fascia.
FASTON, Villa Faustini.
FAUCIGNY, Faciniacum.
FAUCIGNY, Fossiniacum.
FAULC BAY, Immundus Sinus.
FAUQUEMONT, Falconis Mons.
FAVAGNANA, Ægusa.
FAVARA, Dianæ Fons.
FAVARA, Dianæ Fanum.
FAVERGES, Faverga.
FAVERNAY, Favernajum.
FAYENCE, Faventia.
FECAMP, or FESCAMP, Fiscamnum, Fisci campus.
FEDER-SEE, Lacus plumarius.
FEIRA, Langobriga.
FELDSPERG, Fagonium.
FELENK-BOURNON, Parthenium.
FELICUDA, Heraclea.
FELK, Fulka.
FELLES, Elephas.
FELLIN, Felinum.
FELO, Hipporum.
FELTRE, Feltria.
FELTRO, Feretranus Ager.
FEMERN, Cimbria Parva, Fimbria.
FEMERSUND, or FEMERSEE, Fimbriæ Fretum.
FENESTRANGE, Vestringium.

FENIX, Fenisium.
FENOUILLEDES, Feniculetum.
FERACHIO, Camiroa.
FERE (La), Farra.
FERINA, Erigon.
FERMANAGH, Fermaganensis Comitatus.
FERMO, Firmium, Firmum Picenum.
FERNANDO-DEL-PO, Ferdinandi Insula.
FERRAJA-PORTO, Portus Ferrarius.
FERRARA, Allieni Forum, Ferraria.
FERREIRA, Rarapia.
FERRETE, Ferrata.
FERRIERES, Ferrariæ, Ferrera.
FERRO, Ferri Insula.
FERTE, Firmitas.
FERTE-ALAIS, Firmitas Adelheidis, Adelahidis.
FERTE-AURAIN, Firmitas Auraniensis, Aureni.
FERTE-BERNARD, Firmitas Bernhardi.
FERTE-GAUCHER, Firmitas (Auculphi) Gualtieri.
FERTE-LOUPIERE, Firmitas Lupera.
FERTE-MILON, Firmitas Milonis.
FERTE-SOUS-JOUARRE, Firmitas (ad Iotrum) Auculphi.
FERTE-SUR-AUBE, Firmitas ad Albulam.
FERVUEREN, Fura.
FESENSAC, Fidentiacum.
FEUCHTWANG, Hydropolis.
FEUERLAND, Ignis Terra.
FEUILLANS, Fulinum, Felium.
FEURS, Forum Segusianorum.
FEYSTRIZ, Bistricia.
FEZ, Fezzanum Regnum.
FEZ, Fessa, Fezza Volubilis.
FEZZAN, Phazania.
FIACCONE, Alianus.
FIANO, Flavianum.
FIANONA, Flanona.
FIASCONE, Colonia Ferentinensis.
FICARI, Fisera.
FICHTELBERG, Mons pinifer.
FIDARI, Evenus.
FIDNEK, Filekia.
FIDONISI (Island of), Leuce.
FIDONISI, Cursus Achillis.
FIDULSA, Hippuris.
FIESOLE, Fæsulæ.
FIFE, Otholonia.
FIGEAC, Figiacum.
FIGUEIRAS, Ficaria.
FILEK, Filekiensis Processus.
FILIBE, Philippopolis.
FILIPPO-D'ARGIRO, Agyrium.
FINALE, Finarium.
FINISTERRE (Cape), Finis Terræ Caput, prom. Artabrum.
FINLAND, Eningia, Finnia.
FINMARK, Finnmarchia.
FINNEN, Fenni, Finni.
FINNINGEN, Fæniana, Piniana.
FIORE, Flora, Armenita.
FIORENZUOLA, Fidentiola, Florentia.
FIRINO, Pheræ.
FIRMIANO, Firmanorum Castellum.
FIRTH OF CLYDE, Glotæ Æstuarium.
FIRTH OF FORTH, Bodotria.
FIRTH OF SOLWAY, Itunæ Æstuarium.
FIRTH OF TAY, Taum Æstuarium, Tava.
FISARDO, Elibanus Mons.
FISCHAMEND, Æquinoctium.
FISMES, Fimæ, Fima.
FIUME, Flavoniensis Circulus, Fanum S. Viti Flumoniensis.
FIUME-DELLA-CANE, Cena.
FIUME-DELLA-MADALENA, Sebethus.
FIUME-DELLA-POSTA, Fibrenus.
FIUME-DI-RAGUSA, Motycanus.
FIUME-DI-TERRA-NUOVA, Cela, Gela.
FIUME-GRANDO, Himera.
FIVEL, Dammonus.
FIVELINGO, Fivelgonia.
FLÆSCH, Falisca.
FLANDERS, Flandria, Flandriæ comitatus.
FLARCHHEIM, Fladichhemium.
FLAVIGNY, Flaviacum, Flaviniacum.
FLECHE (La), Fixa, Flexia.
FLEIMS, Flemarum Vallis.
FLENSBURG, Flenopolis.
FLENSBURGER-WYK, Fleni Sinus.
FLERS, Bratuspantium.
FLEURUS, Floriacum Monasterium.
FLEURY, in Burgundy, Floriacum ad Oscarum.
FLEURY, in the dep. of Loiret, Floriacus.
FLEUTEN, Fletio.
FLIESSINGEN. See Vliessingen.
FLIMS, Flemium.
FLIX, Biscargis.
FLORENMES, Florinæ.
FLORES, Florum Insula.
FLORIVAL, Florida Vallis.
FLUETEN, Fluetum.
FLUMS, Flemma.
FLUVIA, Clodianus.
FOGARASCH, Fogarasinensis Ager.
FOGARASCH, Fogarasinum.
FOGGIA, Fovea.
FOGLIA, Isaurus, Pisaurus.
FOGNANO, Ustugum.
FOHR, Fora.
FOIGNY, Fusnianum.
FOIX, Foxum, Fuxum.
FOJANO, Fogianum.
FOKIA, Phocæa.
FOLGARIA, Fulgarida.
FOLIGNO, Fulginia.
FOMILLO, Forum Bibalorum.
FONCOMBAUT, Fons Gombaldi.
FONDI, Fundi.
FONTAINE-ANDRE, Fons St. Andreæ.
FONTAINEBLEAU, Bellofontanum, Fons Bellaqueus.
FONTAINE-DE-MERVEILLE, Fons Admirationis.
FONTAINE-L'EVEQUE, Fons Episcopi.
FONTAINES, Fontes.
FONTENAY, Fontaneum.
FONTENAY-LE-COMTE, Fontenacum Comitum.
FONTENELLE, Fontanella.
FONTEVRAULD, Fons Ebraldi.
FONTIBRE, Fons Iberi.
FOOSHT, Cardamine.
FOQUI, Fochium.
FORAT, Euphrates.
FORBISHERS' STRAIT, Fretum Forbisseri.
FORCALQUIER, Forcalquerium, Forum Calcarium.
FORCELLI, Erythrum.
FORDON, Fordunium.
FORELLENBACH, Mediana.
FORENZA, Ferentum.
FOREZ, Forensis Provincia.
FORGES-LES-EAUX, Forgiæ.
FORLI, Forum Livii.
FORLIMPOPOLI, Forum Populi, Pompilii.
FORMIGNANO, Firmanum.
FORMOSELLE, Ocelum.
FORNOVO, Forum Novum.
FORON, Furonis.
FORSTENBERG, Vetera castra.
FORTAVENTURA, Aprositos.

FORT-DE-FRANCE, Arx Regia.
FORT-LOUIS, Fortalitium Ludovici, Castrum Ludovici.
FORTORE, Frento.
FORT-ST.-MIGUEL, Fortalitium St. Michaelis.
FORZA (LA), Fortalitium.
FOSSA (LA), Forum Popilii.
FOSSANO, Fons Sanus, Fossanum.
FOSSE, Fossæ.
FOSSOMBRONE, Forum Sempronii.
FOUG, Fagus.
FOUGERES, Filiceriæ, Fulgeriæ.
FOURNES, Furna.
FRAGA, Flavia Gallica.
FRANC (LE), Franconatus.
FRANCE-DE-BRUGES (LA), Franconatus Brugarum.
FRANCHE-COMTE, Burgundia Superior.
FRANCO-CASTRA, Stratonica.
FRANCOLI, Subi.
FRANCOLISE, Virena.
FRANECKER, Franequera.
FRANKEN, Francia Orientalis, Franconia.
FRANKENBERG, Ferratus Mons.
FRANKENTHAL, Francodalia.
FRANKFORT-ON-THE MAINE, Francofurtum ad Mœnum.
FRANKFORT-ON-THE-ODER, Francofurtum ad Oderam.
FRANCE, Gallia, Franco-Gallia.
FRANZA-CORTA, Francia Parva.
FRASCATI, Tusculum.
FRASCOLARI, Oanos.
FRAUBRUNNEN, Fons Beatæ Virginis.
FRAUENALB, Alba Dominarum.
FRAUENBURG, Drusiana Urbs.
FRAUENFELD, Gynæpedium.
FREDDO, in Tuscany, Frigidus.
FREDDO, in Sicily, Asinarus.
FREEL, Fera Vallis.
FREISACH, Frisacum.
FREJUS, Colonia Julia Octavianorum, Forum Julii.
FREMINCOURT, Fremicuria.
FREMONT, Frigidus Mons.
FRESCARUOLO, Frascarolum.
FRESNAY, Fresnacum, Frederniacum.
FREUDENTHAL, Jocosa Vallis.
FREYBERG, Friberga Hermundurorum.
FREYBURG, in Baden, Friburgum Brisgoviæ.
FREYBURG, in Switzerland, Friburgum Nuithonum.
FREYBURG, in Prussia, Friburgum ad Windam.
FREYENWALDE, Frienwalda.
FREYSINGEN, Fraxinia, Frisinga.
FREYSTADT, Eleutheropolis.
FREYSTADT, in Austria, Eleutheropolis Tessinensis.
FREYSTADT, in Hungary, Eleutheropolis ad Vagum.
FRIAS, Frigida.
FRIAUL, Fori Julii Ducatus, Forum Julium.
FRICENTI, Frequentum.
FRIEDERICHSBUHEL, Friderici Collis.
FRIEDERICIA, Friderici Oda.
FRIEDLAND, Irenopyrgus.
FRIEDRICHSBURG, Fridericoburgum.
FRIEDRICHSHAMM, or FRIEDRICHSHAVN, Friderici Portus.
FRIEDRICHSSTADT, Fridericopolis.
FRIEDRICHSTHAL, Fridericiana Vallis.
FRIEDRICHSWERTH, Erfa.
FRIESACH, Noreia, Frisacum.
FRIESLAND, Frisia orientalis and occidentalis.
FRIGENTO, Frigentium.
FRISCHE-HAFF, Habus, Sinus Venedicus.
FRITTOLA, Frittolanæ Aquæ.
FRITZLAR, Friedislaria, Fritzlaria.
FROIDEVAL, Frigida Vallis.
FRONSAC, Fronciacum.
FRONT, Frontensis Villa.
FRONTIGNAN. Forum Domitii, Frontinianum.
FRUSILONE, Frusino.
FUEGO, or ISLE-DE-FUEGA, Insula Ignium.
FUENGIROLA, Suel.
FUENTE-OVEJUNA, Mellaria.
FUENTE-RABIA, Fons Rapidus.
FUENTES, Arx Fontuna.
FUESSEN, Abodiacum.
FULIGNO, Fulginio.
FUMAY, Fumacum.
FUNEN, Fionia.
FURCA, Bicornis, Adula.
FURCHE, Caudium.
FUSANO, Archiphlegethon.
FUSCHUA-BAY, Dioscorum portus.
FUSE, Fusus.

G

GABAON, Gabao.
GABEL. or GABLON, Jablona.
GABIANO, Gabienum.
GADEBUSCH, Dei Lacus.
GADELONIS, Claude.
GAESBECK, Gasbeca.
GAESTRIKLAND, Gestricia.
GAETA, Cajeta.
GAGLIANO, Galaria.
GAGRA, Sabarus.
GAIDEROGNISSA, Chrysa.
GAIL, Julia.
GAILLAC, Galliacum.
GAILLON, Castrum Gaillonis.
GAISS, Casa.
GAJAZZO. See CAJAZZO.
GAJOLA, Euplœa.
GALASO, Galesus.
GALATA, Chalcis.
GALATI, Galata.
GALATZ, Axiopolis.
GALAURE, Galaber.
GALAXIDI, Euanthia.
GALERATO, Glaneatum.
GALICIA, Callæcia, Galatia.
GALINDER-LAND, Galindia.
GALL (SAINT), Gallum Santo.
GALLEGO, Gallicus Fluvius.
GALLESE, Falisca.
GALLEVESE, Gallovissia.
GALLIPOLI, in Naples, Callipolis.
GALLIPOLI, in Turkey, Callipolis, Gallipolis.
GALLO, Acritas.
GALLOWAY, Gallovidia.
GALULA, Galus, Halus.
GALWAY (co. of), Duaca Gallica.
GALWAY, Galliva.
GAMBARARO, Gambraria.
GAMBOLO, Gamblatum.
GAMS, Campsum Stiriæ.
GAMS, in Switzerland, Campsum.
GAND, Gantum.
GANDERSHEIM, Gandersium, Gandesium
GANDEWARY, Ganges.
GAP, Civitas Vappincensium, Vapincum.
GAPS, Tacape.
GARABUSA, Carabussa.
GARAGNONE, Silbium.
GARBIN-ESSUEN, Contra Suenen.

GARDA, Benacum.
GARDA (Lago-di), Benacus Lacus.
GARDE-FREMET, Fraxinecum.
GARDON, Vardus.
GARESSIO, Gâretium.
GARIB, Drepanum Promontorium.
GARIGLIANO, Clanis, Liris.
GARIS, Carasa.
GARON, Calarona.
GARONNE, or Gironde, Garumna, Garonna.
GARZ, Garsa.
GARZA, Gartia.
GARZARA, Gargarci.
GASCOGNE, Vasconia.
GASCOGNY (Gulf of), Aquitanicus Oceanus.
GASTEIN, Augusta Antonini, Gastanium.
GASTER, Castra Rhætica.
GASTINOIS, Gastinensis Pagus, Vastinium.
GASTUNIA, Peneus.
GASURI, Rhymnus.
GATES, Maleus.
GATES, Adisathrus, Bettigus.
GATESHEAD, Gabrosentum.
GATIPO, Gallus.
GATTER, Hydriaces.
GATTINARA, Catuli Ara.
GAVE-D'OLERON, Gabarus Oleronensis.
GAVE-DE-PAU, Gabarus Palensis.
GAVRIOS, Cephissus.
GAZARAT, Gazena.
GEBSE, Libyssa.
GEDIDA, Neapolis.
GEENTSBERGE, Gerardi Mons.
GEFLE, Gevalia.
GEFLEBORG, Gevaliensis Provincia.
GEHREN, Gerena.
GEIHUN, Leucosyrus.
GEISENFELD, Gisonis Castra.
GELB, Gelduba.
GELDENACKEN, Gildonacum.
GELDERS, Gueldria, Geldria.
GELLAH, Castra Cornelia.
GELOSIA, Mistia.
GELVES, or Gines, Vergentum.
GEMBLOURS, Gemblacum.
GEMONA, Glemona.
GEMUNDEN, Gaudia Mundi.
GEMUNDER SEE. See Traunsee.
GENEVA, or Genf, Augusta Allobrogum, Geneva.
GENEVE, Genebra.
GENEVOIS, Gebennensis Ducatus, Gebennesium.
GENF. See Geneve.
GENF, Genavensis Comitatus.
GENFER-SEE, Lausanius Lacus, Lacus Lemanus.
GENGA, Novanus.
GENGENBACH, Gengibacum.
GENNEP, Cenebum.
GENOA (Golf di), Ligusticum mare.
GENSIN, Melas.
GENT, Ganda, Gandavum.
GENTILLY, Gentiliacum.
GENZANO, Gentianum, Cyntianum.
GEORGIA, Cambysene.
GERACE, Hieracium.
GERBE, Bracchion.
GERBEROY, Gerberacum, Gerboredum.
GERGEAU, Gargovium.
GERGENTI, Acragas, Agrigentum.
GERMERSHEIM, Jullus Vicus.
GERMHASTI, Germa.
GERMIGNY, Germiniacum.
GERNRODE, Gerningeroda, Geronisroda.
GERNSHEIM, Gerinesheim.
GEROLA, Girola.
GEROLSHOFEN, Gerlocuria.
GERONA, Gerunda.
GERS, Ægircius.
GERSAU, Gersovia.
GERTRUIDENBERG, Bergæ Divæ Gertrudis.
GESEGNETER, Garizim.
GEUDERN, Godera.
GEULE, Gulia.
GEURGOWATZ, Timacum minus.
GEUSEN, Genosia.
GEVAUDAN, Gabalensis provincia, Gabalicus pagus.
GEX (co. of), Gesiensis Ager.
GEX. Gesia, Gesium.
GEYRACH, Gyriense Cœnobium.
GEZIRET-ASSUAN, or Elephantine, Elephantine.
GHDIBB-EL-ZICKAR, Garaphi Montes.
GHEDEMEZ, Cydamus.
GHERMA, Garama.
GHERMASTI, Hiera Germe.
GHERZE, Gerisa.
GHIERRA-D'ADDA, Fulcheria.
GHILAN, Gosan.
GHIVIRA, Saviratium.
GHURIA, Colchis.
GIALAB, Galaba, Calaba.
GIANUTI, Artemisia, Dianium.
GIAR. Duus.
GIAVENO, Javennum.
GIB, Gaba.
GIBRALTAR, Calpe, Gibraltaria, Gibilterra.
GIBRALTAR (Straits of), Fretum Gaditanum.
GIEN, Gianum.
GIERAPIETRA, Camirus.
GIFANI, Geofanum.
GIGLIO, Egilium, Ægilium.
GIGNAC, Gigniacum.
GIHAUD, Ubimum.
GILLA, Ila.
GILOWEY, Gilovia.
GIMONT, Gimo.
GINOSA, Cnossus.
GIODDA, Gadamus.
GIOJA, Taurianum.
GIOVENAZZO, Juvenacia, Gnatia.
GIRACE, Hieracium.
GIRGENTI, Acragas.
GIRGIO, Giorgiensis Ager.
GIRMASTI, Caicus.
GIRONDE, Girondia.
GISORS, Cæsarotium, Gisortium.
GITSCHIN, Redintuinum, Gilmiacinum.
GIUDECA, Judeca.
GIUSTENDIL, Tauresium, Justiniana prima.
GJABAR, Sura.
GJEDIS (Dschedis), Hermus.
GLAMORGAN, Clamorgania, Glamorganiensis civitas.
GLAN, Clanis.
GLANDELAGH, Glendelacum.
GLANDEVE, Glanateva.
GLANIO-VECCHIO, Glanis.
GLARUS, Glaronensis Pagus.
GLARUS, Glarizium, Glarona.
GLASGOW, Glascovium.
GLASTONBURY, Glastonia.
GLATZ, Glacensis comitatus.
GLATZ, Glacium.
GLEINK, Glunicense Cœnobium.
GLENDALOUGH, Glendelacum.
GLOGAU, Glogovia Major.
GLOUCESTER, Claudia Castra, Glocestria.
GLUCKSTADT, Fanum Fortunæ, Tychopolis.
GLURENZ, Glorium, Gloriæ Vallis.
GLYKEON, Glykys Limen, Dulcis Portus.

GNADENBERG, Gratiæ Mons.
GNADENTHAL, Gratiarum Vallis.
GNADENZELL, Gratiæ Cella.
GNESEN, Gnesna.
GNIOSA AND LISTO, Mylæ.
GNOIEN, Cœmœnum.
GOADEL, Guadela.
GOCIANO, Cuncianum.
GODMONHAM, Delgovitia.
GOEDEREEDE, Goderea.
GOERDUK, Lycus.
GOERLITZ, Calancorum, Gorlicium.
GOERZ, Goritia.
GOES, Gusa.
GOETHALAND, or GOTHLAND, Gothia.
GOGNA, Novaria.
GOLCONDA, Dachinabades.
GOLDBERG, Aurimontium.
GOLDENMARK, Auraria Parva.
GOLDNE-AUE, Aurea Tempe.
GOLDSTHAL, Kolitschia Vallis.
GOLLNOW, Golnovia.
GOLO, Tuolo.
GOMBS, Gomesianorum Conventus.
GONDRECOURT, Gundulphi Curia.
GONDREVILLE, Gundulphi Villa.
GONGAR, Maste.
GONIGA, Gonni.
GONNINGEN, Agennum.
GOOD-FORTUNE, Bona Fortuna.
GOOD-HOPE, Bonæ Spei Insula.
GORAR, Oruros.
GORBAGA, Gorbeum.
GORGO, Verrugo.
GORGONA, Urgo, Gorgon.
GORKUM, Gorcomium, Gorichemium.
GORNETO, Cornetum sc. Lucaniæ.
GORZAGNA, Gorsenium.
GOSLAR, Civitas Imperialis ad Gosam.
GOTTESZELL, Bona Cella.
GOUKOURTHOY, Tavia.
GOURNAY, Gornacum.
GOVERNOLO, Castellum Gubernium.
GOZO, Gaulos.
GOZZANO, Gaudianum.
GRADISCA, Gordenia.
GRAESIVAUDAN, Gratianopolitanus Pagus.
GRAEZ, Græcium.
GRAMMONT, Grandimontium.
GRAMPIAN, Grampius Mons.
GRAN, Ad Herculem, Strigonium.
GRAN, Granua.
GRANCEY-EN-MONTAGNE, Grancejum Castrum.
GRANDPRE, Grandipratum.
GRANDSELVE, Grandis sylva.
GRANDSON, Grandisonum.
GRANDVILLARS, Grammatum.
GRANDVILLE, Grandisvilla.
GRANFELT, Grandivallis.
GRAS-D'ORGON, Metapinum.
GRASSE, Graca.
GRAUBUNDEN, or cant. of GRISONS, Grisonia, Rhætia superior.
GRAUPEN, Crupna.
GRAVELINES, Grævelingia.
GRAVENMACHERN, Machera Comitis.
GRAVESAND, Gravescenda.
GRAVINA, Blera.
GRAY, Gradicum, Grajum.
GREAT ANDAMAN, Bonæ Fortunæ Insula.
GREAT BRITAIN, Britannia magna.
GREATER-ST.-BERNHARD, Penninus Mons.
GREENLAND, Gronia, Gronlandia.
GREENWICH, Gronaicum.
GREGORSMUNSTER, Gregorii St. Monasterium.
GREIFENBERG, Gryphiberga.
GREIFENSEE, Gryphæum.
GREIFEN-SEE, Gryphæus Lacus.
GREIFSWALDE Gripeswolda, Gryphiswalda.
GRENADILLEN, Grenadinæ Insulæ.
GRENESAY, Sarmia.
GRENOBLE, Cularo, Gratianopolis.
GREOULS, Gryzelium.
GRETLAN, Cambodunum.
GREVESMUHLEN, Comitatis Mola,
GRIECHENLAND, Græcia.
GRIERS, Grueria.
GRIGNAU, Grigniacum.
GRIMAUD, Gambracius Sinus.
GRIMAUD, Athenopolis.
GRIMM, Grimus.
GRIS-NESS, Icium.
GRISANO, Ctemenæ.
GRISCH, Gari.
GRISONS. See GRAUBENDEN.
GRISSAU, Grissovium.
GROBMING, Gamanodunum.
GRON, Grunum.
GRONDEL, Daneon Portus.
GRONINGEN. Corbulonis Monumentum.
GROSS-CUMANIEN, Cumania Major.
GROSS-DELOS, Rhene.
GROSSENHAYN, Haganoa.
GROTTA, Crypta.
GROTTAGLIA, Crypta Aurea.
GROTTA-ROSSA, Crypta Rosaria.
GROTTE (LA), Erbessus.
GROTTE, Crypta Ferrata.
GROTTE-DEL-MONTE-DI-PAUSILIPPO-DI NAPOLI, Pausilyppum. Crypta Neapolitana.
GROTTKAU, Grotgavia.
GRUB, Fovea.
GRUSCH, Crucium.
GRUTES, Subrita.
GRUYERES, Grueria.
GUADALAVIAR, Durias, Turia, Turium.
GUADALAXARA, Arriaca, Carraca.
GUADALQUIVIR, Bætis.
GUADELOUPE, Aquælupæ.
GUADIAMAR, Menoba.
GUADIANA, Anas.
GUADIARO, Barbesola.
GUADIX, Acci, Guadicum.
GUAGIDA, Ladigara.
GUAHAM, Agana.
GUALDO, Ptanias.
GUALIBO, Nechesia.
GUARDA, Lancia Oppidana.
GUARDAFUI, Aromata.
GUARDE, Gardiæ Lacus.
GUASTALLA, Guardistallum.
GUASTECAM, Guasteca.
GUDELAF, Cusum, Cusis.
GUERANDE, Aula Quiriaci.
GUERDEN, Ecbatana.
GUERET, Garactum, Gueretum.
GUERSELBIN, Cillaba.
GUESCAR, Calicula.
GUEVIN, Jovem.
GUGNIE, Gyganeum.
GUICHE, Guissunum.
GUIENNE, Aquitania.
GUIERCHE, Guerica.
GUILFORD, Gilfordia.
GUILLESTRE, Gallitarum Oppidum.
GUILOE-MARE, Coretus Sinus.
GUINES, Gisnæ.
GUIPUSCOA, Jupuscoa.
GUISE, Guisia.

GUISSONA, Cissa.
GUNDUK, Condochates.
GUNS, Ginsium.
GUNTRING, Ferrariæ Carnorum.
GUNZ, Guntia.
GUNZBURG, Guntia.
GURBOS, Carpis.
GURCK, Noaras.
GURIEL, Guria.
GURKFELD, Novidunum.
GUTENBURG, Ictodurum.
GUTENECK, Boneccia.
GUTENTAG, Bona Dies.
GUTSTADT, Bonoppidum.
GUZURAT, Orrotha.
GUYER, Guivia.
GYULA, Julia.

H

HAAG, Haga Comitum.
HAAL, Ernolatia.
HABRUN, Hebron.
HABSAL, Hapselia.
HADABA, Nuba.
HADDINGTON, Hadina.
HADRAMAUT, Catabania.
HADSJI-OGLU-BAZARDSJIK, Parthenopolis.
HAEGER, Heigera.
HAFEN, Athenacon.
HAFNERZELL, Cella Dei Superior.
HAGENAU, Hagenoa.
HAIMBURG, Hamburgum Austriæ.
HAINAU, Hainovia, Hannonia.
HAISNE, Henius.
HALB-THURN, Hemipyrgum.
HALEB. See ALEPPO.
HALEN, Catualium.
HALENTE, Hales.
HALICZ, Galicia, Halicia.
HALL, in Würtemberg, Hala Suevica, Suevorum.
HALL, in Tyrol, Hala ad Œnum.
HALLE, or HALLER-SAALE, Hala, Hala-Saxonum, Magdeburgica.
HALLEIN, Haliola.
HAM, Hametum, Hamona.
HAMAMET, Putput, Adrumetum.
HAMATH, Amatha.
HAMBURG, Hammonia, Hamburgum.
HAMEL, Asmiræa.
HAMELN, Hamela.
HAMISE, Tibilis.
HAMM, Hammona.
HAMMA, Aquæ Calidæ.
HAMONT, Hamons.
HAMPSHIRE, Hanonia.
HAMPTON-COURT, Hamptoni-Curia.
HANAU, Hanovia.
HANDAHUR, Pachnamunis.
HANEBONT, or HENNEBON, Hannebotum.
HANNUT, or HANNUGE, Hannuvium.
HANZ, Antium.
HARAD, Ader.
HARBERT, Salamboria.
HARBURG, Biriciana.
HARBURG, Argentaria, Argentuaria.
HARCOURT, Harecortis, Harcurtium.
HARDEGO, Harudorum Pagus.
HARDERWYCK, Ardevicum.
HARFLEUR, Arefluctus, Harflorium.
HARFORD, Durocobrivæ.
HARLEM, Harlemum.
HARRAN, Carræ.
HARTBERG, Heortis Mons.
HARTENBERG, Duroburgum.
HARTENFELD, Durus Campus.
HARTLAND-POINT, Herculis prom.
HARWICH, Harviacum, Harvicum.
HARZ, Hercynia sylva.
HASBEIA, Dan.
HASE, Assa, Hasa.
HASEK, Hasicon.
HASLE, Haselia Vallis, Hasselia.
HASSANI, or ED-SJEDERU-ADUN, Timagenis insula
HASSELT, Hasseletum.
HASTINGS, Othna, Astingua.
HATDER, Hatra.
HATTONCHATEL, Hattonis Castrum.
HATVAN, Hatuanum.
HAURAN, Leuce Come.
HAUTECOMB, Altacumba.
HAVANNA, Fanum St. Christophori.
HAVEL, Habela.
HAVRE, Havrea.
HAVRE-DE-GRACE, Caracotinum, Gratiæ portus, Franciscopolis.
HAVRE-DE-LONGUEVILLE, Legedia.
HAY, Dumna.
HEBBERSTOW, Prætorium.
HEBRIDES, Ebudæ Insulæ.
HEBUD, Jabruda.
HEERWEN, Herispich.
HEGAU, Hegovia.
HEIDELBERG, Edelberga, Heidelberga.
HEIDENHEIM, Aquileja.
HEILBRONN, Alisium.
HEILIGEN-KREUZ, Fanum St. Crucis.
HEILIGENKREUZ, Insula St. Crucis.
HEILIGER-BERG, Pirus.
HEIMSHEIM, Heimbodesheim.
HEINZENBERG, Heinzilianus Mons.
HEIS, Hericus.
HELIGOLAND, Insula sancta.
HELLA, Æjopolis.
HELLEBRUNN, Clarofontanum.
HELONI-MON, Dodona.
HELSINGBORG, Helsinga.
HELSINGOR, Elsenora.
HELVAUX, Helvatium.
HEMS, Emesa.
HENARES, Tagonius.
HENDU-KHOS, Paropamisus.
HENNEGAU, Haginoia.
HERAKLITZA, Heraclea Chersones, Thraciæ.
HERAT, Aria.
HERAT, Nisæa.
HERAULT, Arauris, Araura.
HERBITZHEIM, Heribodesheim.
HERCK, Archa.
HERCOLANO, Herculaneum.
HERDEN, Aredunum.
HERDNERBERG, Aredunum.
HERI, Arias.
HERISAU, Augia Domini.
HERISSON, Irritio.
HERJEDALEN, Herdalia.
HERKLA, Adrumetum.
HERMANNSBURG, Arx Arminii.
HERMANNSTADT, Cibinium, Hermannopolis.
HERRENALB, Alba Dominorum.
HERRENGRUND, Dominorum Vallis.
HERRNSTADT, Kyriopolis.
HERSFELD, Herocampia.
HERSTALL, Haristalle, Heristallum.
HERSTELLE, Haristellum.
HERTFORD, Areconium.
HERVELT, Herculis Castra.
HERZOGENBUSCH, Boscoducum, Sylva Ducis
HESDIN, Hedena.

HESER-WALD, Cæsia Sylva.
HESSEN, Hassia.
HESSN-KEIFA, Cephæ.
HET-VLIE, Flevum.
HEUSDEN, Heudena.
HEVES-SZOLNOK, Hevesia.
HEXHAM, Alexodunum.
HIALI, Ægiale.
HIDVEGH, Joannis Pons.
HIERES (De), Stœchades.
HIESMES, Oximium.
HIJAR, Ixarium.
HILDBURGHAUSEN, Hilpershusia.
HILDESHEIM, Ascalingium, Hildesia.
HIMMALAYA, Emodi Montes.
HIMMELSKRON, Cœli-Corona.
HINDMEND, Etymander.
HIR, Cophen.
HIRSAU, Hirsaugia.
HIRSCHBERG, in Silesia, Cervimontium.
HIRSCHBERG, in Reuss, Cervimontium ad Salum.
HIRSCHBERG, Cervimontium Westfaliæ.
HIRSCHBERG, in Bohemia, Dona.
HIT, Idicara.
HITA, Cesada.
HITZACKER, Hizgera, Hidonisager.
HJO, Hiovia.
HOANG-HO, Bautisus.
HOBA, Chobah.
HOCKERLAND, Hocheria.
HODEIDA, Sacatia.
HOF, Curia Variscorum.
HOF, Curia Inferioris Pannoniæ.
HOF, Curia Moravica.
HOF (Stadt-am), Curia Bavarica.
HOHE, Taunus.
HOHEMAUTH, Alta Mauta.
HOHENELBE, Albipolis.
HOHENFURT, Altovadum.
HOHENKRAHN, Græa.
HOHENTWIEL, Duellium.
HOHENWARTH, Alta Specula.
HOLEN, Holis.
HOLLAND, Batavia, Hollandia.
HOLLENSTEDT, Holdistede.
HOLME, Insula Dei.
HOLSTEIN, Holsatia.
HOLY-ISLAND, Insula sancta.
HON, Diospolis.
HONFLEUR, Honflevius, Honflorium.
HONNECOURT, Hunnicuria.
HOORN, Horna.
HOR, Arabis, Tomerus.
HORA, Gosen.
HORBURG, Robur.
HORCHID, Hyssus.
HORN (Cape), Hornanum caput.
HORNSEY, Gabrantonicorum Sinus.
HORRE-HEAD, Venicnium.
HORSELBERG, Horrisonus Mons.
HORSENS, Horsnesia.
HORT, Leonis Castrum.
HOSN, Gamala.
HOSZKZU-MEZOE, Campus Longus.
HOUDAN, Hosdencum.
HOWE-LEMUR, Sabææ Aræ.
HOXTER, Huxaria.
HRADISCH, Hradisca.
HUELVA, Onoba.
HUESCA, Osca.
HUESCAR, Lacuris.
HUETE, Julia Opta.
HUFINGEN, Brigobanne.
HUISNE, Idonia.
HUMBER, Æstuarium Abum.
HUMBER, Abus.
HUNDERTSBUEL, Centum Colles.
HUNDSCAP, Rhosus.
HUNDSRUCK, Hunnicus Pagus, Hunnorum tractus.
HUNNINGEN, Hegenense Monasterium.
HUNSINGO, Hunesgonia, Hunsingia.
HUREPOIX, Hurepœsium, Hurepoisius tractus.
HUSBAN, Esebon.
HUSUM, Hosemum.
HUTALIDSCH, Utus.
HUY, Huum.
HUYTS, Arx Britannica.
HWEEN, Huena.
HYDRA, Hydrea.
HYERES, Arcæ.

I

IASZAG, Iazygia.
IAUER, Iavoria.
IBERIAN SEA, Hispaniensis Oceanus.
IBNE, Jabne.
IBRIM, Primis Magna.
IBSON, Hisoris.
IBURAR, Cibyra Minor
IBURG, Juberg.
ICBOROW, Iciani.
ICELAND SEA, Deucaledonius Oceanus.
ICHTIMA, Stenimachum.
IDANHA-A-VELHA, Egiditania, Equitania
IDICE (L'), Idex.
IDRIA, Idrizza.
IDRO, Edrum, Idrinum.
IDRO, Lacus Edrinus.
IENISA (Lago di), Ezerus.
IF, Sphia, Hypæa.
IFERTEN, Eborodunum.
IFFIZHEIM, Bibium.
IGLAU, Iglavia.
IGLESIAS, Ecclesiæ.
IGLO, Iglovia, Neocomium.
IGUZUL, Una.
IKSWORTH, Icenorum Oppidum.
ILA, Epidium.
ILCHESTER, Iscalis, Ischalis.
ILGIUN, Tiberiopolis.
ILHA-DEL-FUEGO, Ignium Insulæ
ILIGA, Elegia.
ILKLEY, Olicana.
ILL, Licia.
ILL, or Ell, Elsus, Alfa, Ella.
ILLE, Insulæ.
ILLER, Ilargus, Lictus.
ILLESCAS, Tiruacia.
ILLMUNSTER, Ilmi Monasterium.
ILLOK, Bononia.
ILLYRIA, Illyricum, Illyria.
ILS, Ilissus.
ILSSTADT, Civitas Ilzensis.
IMMELACH, Gemellæ.
IMMENSTADT, Vemania.
IMOLA, Forum Cornelii.
IMPERIALE, Imperius.
INCISA, Ad Incisa Saxa.
INDIAN OCEAN, Erythræum Mare.
INDRE, Alere, Ingeris.
INDSJE, Syrias.
INEBAZAR, Magnesia ad Mæandrum.
INGELMUNSTER, Anglomonasterium.
INGERMANNLAND, Ingria.
INGESU, Tetrapyrgia.
INN, Ænus, Œnus.
INNERSTE, Indrista.

INNICHEN, Aguntum.
INNISOWEN, Eugenii Insula.
INNSBRUCK, Æni Pons, Œnipons.
INNSTADT, Bojodurum.
INOBOLI, Aboni-Tichos.
INOWLOCZ, Junicladislaviensis Palatinatus.
INOWLOCZ, Inolocza.
INVERNESS, Innernium, Invernium.
IPEK, Pingus.
IPF, Opie.
IPS, Isis, Ibissa.
IPS, Ad Pontem Isis.
IPSALA, Cypsela.
IPSWICH, Gippevicum.
IRELAND, Hibernia.
IRISH SEA, Hibernicum Mare.
IRNIS, Jornacum.
IS-BARTEH, Baris.
ISCA, Escamus, Escus.
ISCHA, Escus.
ISCHIA, Ænaria.
ISEGHEM, Isegenium.
ISEN, Isana.
ISEN (the river), Isinisca.
ISEO, Lacus Iseus, Sebinus Lacus.
ISER, Isra.
ISER, or ISAR (the Bavarian river), Isara, Urusa.
ISERE, Isara.
ISERLOHM, Iserlonia.
ISERNIA, Æsernia, Esernia.
ISLAND, Gardari.
ISLAY, Isla.
ISLE (L'), Insula.
ISLE-DE-FRANCE, Insula Franciæ.
ISLE-DE-MAIR, Immadras.
ISLE-DEL-AYRE, Mira.
ISLE-DU-LYS, Igilium.
ISLE-DU-TITAN, Hypæa.
ISLE-EN-ALBIGEOIS, Insula Albigensis.
ISLE-JOURDAIN (L'), Ictium Castrum, Castellum Ictium.
ISLE-ROUSSE, Isola-Rossa.
ISLOTE, Scombraria.
ISMIR, Smyrna.
ISNIK, Nicæa.
ISOLA, Alietum.
ISOLA, Insula Bruttiorum.
ISOLA DE FIGO, Phintonis Insula.
ISOLA-DEL-TORO, Molybodes.
ISOLA-DI-CERVI, Teganusa.
ISOLA-DI-PONZA, Pontia.
ISOLA-DI-SAN-MICHELO, Michaelia.
ISOLA-DI-ST.-PIETRO, Hieracum.
ISOLA-MELERE, Thoronos.
ISOLA-VANDOTINA, Pandataria.
ISOLE-DELL'-ARCHIPELAGO, Cyclades.
ISONZO, Sontius, Isontius.
ISPAHAN, Aspa.
ISPERIK, Timacum.
ISSEL, or YSSEL, Isala, Sala.
ISSEL, or YSSEL, NEUE, Fossa Drusiana.
ISSENGEAU, Icidmagus.
ISSNY, Isna.
ISSODUN, or ISSOUDUN, Anxellodunum, Exoldunum.
ISSOIRE, Iciodurum, Issiodorum.
ISSY, Fiscus Isiacensis.
ISTIB, Stobi.
ISTRIA (CAPO D'), Justinopolis.
ISTRIGA, Scirtiana.
ISTVANDI, Limusa.
ISVORNIK, Argentina.
ITALA, Attalea.
ITRI, Itrium.
ITZEHOE, Esesfelth.
IVIÇA, Ebusus.
IVREA, Eporedia, Eperodia.
IVRY, Iberium, Iberiacum.
IXWORTH, Icenorum Oppidum.

J

JACA, Edulius Mons.
JAEN, Flavium, Gienna, Gienum.
JAGERNDORF, Carnovia.
JAGODNA, Jagodina, Januaria.
JAHEL, Alyi.
JAICZE, Gaitia, Jaitza.
JAIX, Daix.
JAKOVO, Psophis.
JALOWA, Sestus.
JALTA, Galata.
JAMAGOROD, Jama.
JAMBO, Charmuthas.
JAMBOLI, Jambolensis Ager.
JAMBOLI, Diampolis.
JAMESBOROUGH, Jacobipolis.
JAMESTOWN, Jacobipolis.
JAMEZ, or JAMETS, Gemmacum, Gemmatium.
JAMNITZ, Gemenicium.
JAMTLAND, Jemtia.
JANINA, Epirus.
JANINA, Cassiope.
JAPAN, Japonia.
JAPARA, Parnassus.
JARECCA, Haræ.
JARGEAU, or GERGEAU, Gargogilum.
JAROMIERZ, Jaromirium.
JASSY, Jassium, Jassiorum Dacorum Municipium, Petrodava.
JATHRIB, Latrippa.
JATI, Bathys.
JATIVA, Sætabis.
JAUER, Jauravia, Juravia.
JAVOUX, Anderidum.
JEDDO, Jedum.
JEDEREN, Jadrensis Regio.
JEGNI-SALA, Salmorus.
JEHUDIEH (EL), Castra Judæorum.
JENI-KOI, Christopolis.
JENISCHEHER, Larissa Thessala.
JENISCHEHR, Sigeum.
JENISEI, Jenisia.
JERAKI, Geranthræ.
JERAN, Jovis Urii Fanum.
JERSEY, Cæsarea.
JERUSALEM, Hierosolyma.
JESD, Jessedа.
JESENOWITZ, Jasenocium.
JESI, Æsis.
JESOLA, Equilium.
JEUZGAT, Osiana, Soanda.
JITTA, Juta.
JOACHIMSTHAL, Joachimica Vallis.
JOBAB, Jotab.
JOIGNY, Joviniacum.
JOINVILLE, Joanvilla, Intra Fluvios.
JONCOPING, Jenecopia.
JONQUERAD, Egorigium.
JORKAU-BORECK, Boreca.
JOUARE, Jodrum, Jovis, Ara.
JOUSCHIA, Colonia Laodicena.
JOUX, Jovium.
JOUY-SUR-MORIN, Gaudiacus.
JOYEUSE, Gaudiosa, Joyosa.
JOYOSA, Honosca.
JUBLEINS, Diablintes, Novodunum.
JUHORSKY, Jughoria.
JUINE, Junna.

JULICH, Juliacum.
JULIERBERG, Julius Mons.
JULIS, Julis.
JULIUSBERG, Julioburgum.
JUMIEGE, Gemmeticum.
JUNG-BUNZLAU, Boleslai Fanum Novum.
JUNNAN, Junnania.
JUNQUERA, Juncaria.
JURA, Gerontia.
JUREKIAM-LADIK, Laodicea Combusta.
JURJURA, Ferratus Mons.
JUTERBOCK, Jutrebocum.
JUTLAND, Cartris.
JUTLAND AND SCHLESWICK, Chersonesus Cimbrica, Jutia.
JUVIA, Ivia, Nabius, Juvia.

K

KAADEN, or KADEN, Cadanum.
KABARDA, Eulisia.
KABAS, Gabæ.
KAFAR-TUTHA, Maurorum Castra.
KAFFERLAND, Caffraria.
KAHEM, Agamna.
KAHLENBERG, Cetius Mons, Mons Calvus.
KAIFA, Hepha.
KAIHWARES, Cæne.
KAINA, Massa.
KAINS, Camina.
KAIRWAN, Vicus Augusti.
KAISAR, Carissa.
KAISARIEH, Cæsarea.
KAISERSBERG, Cæsaris Mons.
KAISERSFELD, Cæsaro-Felda.
KAISERSHEIM, Abbatia Cæsariensis.
KAISERSLAUTERN, Cæsarea Lutra.
KAISERSTUHL, Cæsaris Tribunal, Forum Tiberii.
KAISERSWERTH, Cæsaris Insula, Cæsaris Verda.
KAKOSIA, Thisbe.
KALABAKI, Palæpharus.
KALABSCHE, Talmis.
KALAT-EL-HORSA, Chorazin.
KALAT-EL-NEDSJUR, Neocæsarea.
KALBA, Cabana.
KALEMON, Calamos.
KALENHUYSEN, Calon.
KALINGA-PATAM, Dandagula.
KALISCH, Calisia, Canisia.
KALLAMA, Thyamis.
KALLIVIA, Elis.
KALLUNDBORG, Callunda.
KALMINZ, Celemantia.
KALMIUS, Lycus.
KALPAKI, Orchomenus.
KALYMNO, Galymna.
KAMARI, Demetrias.
KAMEH, Malamantus.
KAMINIEK, Camenecia, Caminicium.
KAMINITZA, Dymæ.
KAMOZIK, Rira.
KAMP, Cambus.
KAN, Caina.
KANEM, Thumelitha.
KANISCH, Canisia.
KANOGE, Calinipaxa.
KANOS, Ganus.
KANUATH, Canatha, Kenath.
KAPARTIS, Nicopsis.
KAPFENSTEIN, Capedunum.
KAPPEL, Tasinemetum.
KAPSTADT, Castrum bonæ spei.
KARA, Carra.
KARABOA, Priapus.
KARA-DERRE, Dara.
KARADSCHEFU, Trapezopolis.
KARADSJAG, Nicatorium.
KARAHISSAR, Cestros.
KARAISSAR, Perge.
KARAK-SARAI, Damalis.
KARASU, Nestus.
KARAT, Ocorura.
KAR-DANAH, Beleus.
KARE, Arachotus.
KAREM, Charax Pasini.
KARITHENE (the river), Breutheates.
KARITHENE, Breuthe.
KARK, Carcha.
KARKEND, Pura.
KARKUF, Sittace.
KARLE-SATZAK, Ætolia Adjecta.
KARLINGFORD, Buvindum.
KARLSBAD, Thermæ Carolinæ.
KARMELIS, Gaugamela.
KARNICOBAR, Agathu Dæmonos sc. Insula.
KARNTHEN, Carinthia.
KAROLY, Carolium.
KARPFEN, Carpona.
KARRAK, Moba, Kir Moab.
KARS, in Asiatic Turkey, Chorsa, Carse.
KARS, in Egypt, Nicopolis.
KARST, Carujadius.
KARTPURT, Ziata.
KARTZEN, Casurgis.
KASALMAK, Iris.
KASANKA, Casancius.
KASBIN, Vesaspe.
KASCHAU, Cassovia.
KASCHEM, Cana.
KASCON, Cogæonos.
KASEMIESCH, Eleutherus.
KASMARK, Cæsareopolis, Cæsareo-Forum.
KASPIAN SEA. See CASPIAN SEA.
KASSR-ANTAR, Hazor.
KASTANIA, Castana.
KASTRAVAN, Climax.
KATHOLIKON, Stymphalus.
KATIEH (EL), Cassius Mons.
KATO-ACHAIA AND APANO-ACHAIA, Pharæ
KATRINA, Colonia Diensis, Dion.
KATTEGAT. See CATTEGAT.
KATTENBERGERDORF, Citium.
KATZBACH, Cattus.
KATZENELNBOGEN. See CATZENELNBOGEN.
KAUFUNGEN, Capungum.
KAUMBERG, Cumeoburgum, Comagenus.
KAUSCHAU, Gauzanitis.
KAWIK, Chalus.
KAYST, Kisdemum.
KEBRINAZ, Cremna.
KECH, Nautaca.
KEDUS, Cadi.
KEFF, Sicca Veneria.
KEHKER, Hesidrus.
KEICH, Catæa.
KEKKOE, Kekkoja.
KELB-HAURAN, Alsadamus.
KELLEN, Castra Ulpia.
KELLMUNZ, Cœlius Mons.
KELS, Celeusa.
KELSKEMET, Egopolis.
KEMBERG, Cameracum ad Albim.
KEMKHAL, Chobus.
KEMNADE, Caminata.
KEMPTEN, Campodunum, Campitona.
KENCHESTER, Ariconium.
KENDAL, Concangium.
KENGHEVER, Concobar.
KENKRIE, Cenchreæ.

KENNE, Cæne.
KENYER-MEZOE, Campus Panis.
KEPIL, Cepi.
KEPSE. Seleucia.
KERATHA. Coreathe.
KERBEH, Calpas.
KERCI, Cercum.
KEREMP, Carambis.
KEREND. Carine.
KERKA, Tilium.
KERKH, Carmania Deserta.
KERLBURG, Gerulata.
KERMAN, Carmania.
KERMAN, Carmana.
KERPEN, Carpio.
KERRAH, Choaspes.
KERSCHOWA, Carsium.
KERTSCH, Panticapæa.
KERZERS, Ad Carceres.
KESCHIN, Eristhe.
KESCHING. Ælia Flavia Cæsarea, Germanicum.
KESSEL, Castellum Menapiorum.
KESTENHOLZ, Castinetum.
KESTEREN. Castra Herculis.
KESWICK, Causennæ.
KETSKEMET, Ægopolis.
KEUPRELI, Europus ad Axium.
KHABUR, Centrites.
KHORREM-ABAD, Corbiena.
KIADA-BURUN, Thynias.
KIDJEH, Chodda.
KIDONIA, Heraclea Æolidis.
KIEF, Chiovia.
KIEIDANI, Cajodunum.
KIEL, Chilonium, Kirlia.
KIENSHEIM. Cunonis Villa.
KIERAZIM, Hieratis.
KIERTEMUNDE, Cartemunda.
KILBEGS, Calebachus.
KILIA, Cœla.
KILIOS, Tylæ.
KILKENNY, Cella St. Canici.
KILLALOE, Laona.
KILLIS, Ciliza.
KILMALON, Killocia.
KILMARE, Jernus.
KILMORE, Chilmorla.
KIMBOLTON, Cinnibantum.
KINGSTOWN, Regiopolis.
KINNAIRDS HEAD, Tæzalum.
KINNESRIN, Chalcis.
KIOEGE, Coagia.
KIOELEN, Juga Suevonis Montis, Sevomons.
KIOPING, Copinga.
KIRCHHEIM, Clarenna.
KIRGISKAISAKEN, Cachassæ.
KIRIAKI, Lecythus.
KIRKESIA, Carchemis.
KIRKUK, Memmis.
KIRKWALL, Carcoviaca.
KIRPE, Rhoe.
KIRPEH, Thynia.
KISIL-IRMAK, Halys.
KISIL-OSAN, Gosan.
KISIL-RUBAT, Chalonitis.
KISLAR-KALESSI, Cumania.
KISLEGG, Cassiliacum.
KISSANO, Ossa.
KISSIA, Cephisia.
KISSINGEN, Kizinga.
KISSOR, Seggo.
KISTNA, Mæsolus.
KITROS, Pydna.
KITZBUHEL, Hædicollis, Hædiopolis.
KIUSTENZA, Constantiana.
KIUTAHYA, Cotyæum.
KJOELEN. See KIOELEN.
KLADRAU, Cladrubum.
KLAGENFURT. See CLAGENFURT.
KLATTAU, Brodentia.
KLEGGAU, Latobrigicus Pagus.
KLER, Hilara.
KLITSCHOW, Klitsovia.
KNIN. Arbuda.
KOCHEL, Coveliacæ.
KOCHEL-SEE, Coveliacensis Sinus.
KOCHER, Cocharus.
KOCKEL. Coveliacus.
KOCKELBURG. Kukoliensis Comitatus.
KODSHA-SHEHR, Nacoleia.
KOEBEN. Cobena.
KOECHERSBERG, ncordia.
KOEGE. Congia.
KOEVAR, Covariensis Districtus.
KOFEL. or COVOLO. Claustrum Cubali.
KOHLENWALD, Carbonaria Sylva.
KOLA. Cola Lapporum.
KOLOKYTHIA (GULF OF), Laconicus Sinus.
KOLOSCH, Colosia.
KOLOTSCHA, Colocia.
KOLSUM. Clysma.
KONDINSK, Condora.
KONG, Mandras.
KONGEN, Condate.
KON-HISSAR, Coropassus.
KONIGINGRATZ, Gradium Reginæ, Regino gradecium.
KONIGSBERG, Mons Regius, Regiomontum.
KONIGSFELDEN, Campus Regius.
KONIGSHOF. Curia Regis ad Albim.
KONIGSHOFEN, Curia Regis Badensis.
KONIGSHOFEN-IM-GRABFELDE, Curia Regia in arvis.
KONIGSLUTTER, Luttera Regia.
KONIGSMACHERN, Machera Regis.
KONIGSSAAL, Aula Regia.
KONIGSSTADTEN, Mons Comianus.
KONIGSTEIN, Lapis Regius.
KONIGSWART, Marabodui Castrum.
KONIGSWINTER, Hiberna Regia.
KONKUN, Gogana.
KONTSHUK-TSCHEKINESCH, Regio.
KOPENHAGEN. See COPENHAGEN.
KOPING, Copinga.
KORADSCHE, Cherseus.
KORNA, Didigua.
KOROM, Cornicum.
KOROS, Chrysius.
KOROS-BANYA, Chrysii Auraria.
KOSCHING, Kesching.
KOSEL, Coselia.
KOSLOW, Eupatoria.
KOSSEIR, Berenice.
KOSSIR, Ænnum.
KOSTENDIL, Ulpianum.
KOTHEN, Cotha, Cothenæ.
KOTSCH, Canthi Colpus.
KOTTA, Cosa.
KOUKLA, Golgus.
KOUM-OMBOS, Ombri.
KRAIN, Carnia, Carniola.
KRAINBURG. See CRAINBURG.
KRAINBURG, Carrodunum.
KRAKH, Characmoba.
KRALJEWA-WELIKA, Clara.
KRASNISLAW, Crasnoslovia.
KRAVADA, Caryæ.
KRESSEL, Jastus.
KREUTZ. See CREUTZ.
KRIM, Chersonesus Taurica.

KRINK, Coriticum.
KRISNA, Nanaguna.
KROJA, Clodianæ.
KRONSTADT. See CRONSTADT.
KRUMAU, Cromena.
KTYPO-BERG, Messapium.
KUBAN, Hypanis.
KUBBET-JAMBO, Zygæna.
KUBLIS, Convallium.
KUCHEL, Cucullæ.
KUDAK, Borysthenis, Cutacum.
KUDROS, Cytorus.
KUFFHAUSEN, Kiphhusanus Mons.
KUFSNACHT, Cussenacum.
KUFSTEIN, Albianum.
KUILENBURG, Caruo.
KULELI, Nice.
KULPA, Colapis.
KUMIS, Parthia.
KUNDWIC, Poecessa.
KUNSELYSEG, Cunorum Sedes.
KUPFERBERG, Cuprimontium ad Alimonam.
KUPFERBERG, Cuprimontium.
KUPFERBERG, in Sweden, Cuprimontium Gestriciæ.
KUPFERBERG, in Bohemia, Cuprimontium Bohemiæ.
KUPFERBERG, Cuprimontium Hermionum.
KUPH, Emma.
KUR, Cyrus.
KURANKEVI, Hamaxitus.
KURDEN, Carduchi.
KURENK, Candriaces.
KURYMA, Carsidava.
KUTAIS. See CUTAIS.
KUTSCHAK, Palmata.
KUTTENBERG, Cutna.
KYLE, Coila, Covalia.
KYLL, Celbis.

L

LA, Laha.
LABIAU, Labiavia.
LABOUR (LE), Lapurdensis Tractus.
LAC-DE-LOCATA, Sordice.
LACEDOGNA. See CEDOGNA.
LACH, Lacensis Abbatia.
LACHA, Olympus.
LACHEN, Ad Lacum.
LACOSTA-RUFARIA, Rufræ.
LADENBURG, Lobdunum, Lupodunum.
LADIKIEH, Laodicea.
LADRONES, Latronum Insulæ.
LAGALNIK, Axinces.
LAGAN, Logia.
LAGER-THAL, Lagurina Vallis.
LAGHETTO, Regillus Lacus.
LAGNY, Latiniacum.
LAGO-CASTELLO, Albanus Lacus.
LAGO-DI-BAGNI, Lacus Albuneus.
LAGO-DI-BOLSENA, Volsiniensis Lacus.
LAGO-DI-BRACCIANO, Sabate.
LAGO-DI-CAMARANA, Camarina Palus.
LAGO-DI-CELANO, Fucinus Lacus.
LAGO-DI-CHIAVENNA, Comacenus Lacus.
LAGOI, Bistonis Lacus.
LAGOLIASTO, Luquido.
LAGO-MAGGIORE, Verbanus Lacus.
LAGULA, Calinda.
LAGUNA-DE-LA-JUNDA, Bælon.
LAGUNES, Venetæ Paludes.
LAGUSTA, Ladesia.
LAHEVI, Hyala.
LAHN, Lagana.
LAHOLM, Lagaholmia, Laholmiæ.
LAHOR, Bucephala, Lahorium, Acesines.
LAIBACH, Æmona, Labacum.
LAINO, Laus.
LAITTOURE, or LEICTOURE, Lectora.
LAJIURA, Adule.
LAKENO, Lagina.
LALAIN, Lalinum.
LAMBACH, Lambacum.
LAMBESE, Lambæsa.
LAMBIRIO, Erineo.
LAMBRO, Lambrus.
LAMEGO, Lama, Lameca, Lamecum.
LAMINA, Homolium.
LAMPADOSA, Lopadusa.
LAMPTA, Vobrix.
LAMUZO, Lamotis.
LANARK, Curia.
LANCAN, Dorias.
LANCASTER, Alione, Lancastria.
LANCEROTTE, Lancellotta.
LANCESTON, Fanum St. Stephani.
LANCIANO, Anxanum.
LAND (SAXON), Fundus Regius Saxonicus.
LAND-OB-DER-ENS, Austria Superior.
LAND-UNTER-DER-ENS, Austria Inferior.
LANDAFF, Fanum ad Taffum.
LANDAU, Landavia, Landavium.
LANDEN, Landæ.
LANDERNEAU, Landericiacum.
LANDES (PAY DES), Landarum Tractus, Ager Syrticus, Landæ.
LANDI, Landorum Status.
LAND'S END (THE), Belerium, Prom. Antivestæum.
LANDSHUT, Consuanetes, Landishutum.
LANDSKRONA, Coronia.
LANEBOURG, Lancioburgum, Laneburgum.
LANGEAC, Langiacum.
LANGENARGEN, Argentaria.
LANGENSALZA, Longosalissa.
LANGENZENN, Cenna.
LANGETS, Langesium.
LANGEY, Alingavia.
LANGON, Alingo, Alingonus portus.
LANGRES, Andomatunum, Andemantunum, Lingonum.
LANGUEDOC, Langedocia, Occitania.
LANGUES, Langæ.
LANNOY, Lannojum, Alnetum.
LANQUART, Langarus.
LAN-SU, Elæus.
LANZO, Axima.
LAON, Laudunum, Lugdunum Clavatum.
LAONDA, Leben, Lebena.
LAPITO, Lapathos.
LAPLAND, Lapponia, Lappia.
LAR, Laria.
LARACHE, Lixa.
LARBUSS, Lares.
LARGITZEN, Larga.
LARISTAN, Elymais.
LARIZO, Larissa Pensilis.
LARNES, Larymna.
LARRAGA, Tarraga.
LARZII, Vellanis.
LAS-CABEZAS, Ugia, Urgia.
LASCOWITZ, Budorgis.
LASTHI, Dictæus Mons.
LATARACO, Hetriculum.
LATICZOW, Latiszovia.
LATIK, Lystra.
LATZFASS, Fons Latius.
LAUDUN, Ladanum.

LAUENBURG, Cœnœnum.
LAUFEN, Redajum.
LAUFEN, Artobriga.
LAUFENBURG, Gannodurum.
LAUFFEN, Laviacum.
LAUINGEN, Lauginga, Lavinga.
LAUIS, or LUGANO, Lacus Cerusius.
LAUIS, Junianum.
LAURAGUAIS, Lauriacus, Lauriacensis Ager.
LAURANA, Lauranum.
LAUSANNE, Lausonium.
LAUSITZ, Lusatia.
LAUTREC, Lautricum.
LAUWER-SEE, Labacus Sinus.
LAVAGNA, Entella.
LAVAGNA, Lavania.
LAVAL, Vallis Guidonis.
LAVANT, Laventus.
LAVAUR, Vaurum.
LAVEDAN, Levitania.
LAVELLO, Labellum.
LAVEMUNDE, Laventina, Ostium Laventæ.
LAVIERS, Latverum.
LAVINO, Labinius.
LAWERS, Labacus.
LAYBACH. See LAIBACH.
LAYE, Ledia.
LEBAN, Lebona.
LEBEDIGLI, Lebedus.
LEBER-THAL, Leporacensis Vallis.
LEBIDA, Leptis Magna.
LEBRIJA, or LEBRIXA, Nebrissa.
LECCE, Aletium.
LECCE, Hydruntina Provincia.
LECCO, Leucum, Leccum.
LECCO (LAGO DI), Comacenus Lacus.
LECH, Lichus, Licus.
LECHENICH, Legioniacum.
LECHFELD, Lyciorum Campus.
LECHGEMUND, Licimacum.
LECK, Lecca, Fossa Corbulonis.
LECTOURE. See LAITTOURE.
LEDESMA, Bletisa.
LEE, Lea.
LEEDS, Ledesia.
LEEWARDEN, Leovardia.
LEGHORN, or LIVORNO, Liburnicus Portus.
LEGINE, Vicus Virginis.
LEGION, Legiodunum.
LEGNANO, Leoniacum, Lignanum.
LEGRAD, Legradinum.
LEGRAD, Jovia.
LEGRANO, Laurium.
LEIBNITZ, Laibnitia.
LEICESTER, Legecestria.
LEIDEN, or LEYDEN, Lugdunum Batavorum.
LEIGHLIN-BRIDGE, Lechlinia.
LEININGEN, Linanganus Principatus.
LEINSTER, Lagenia.
LEIPZIG, Lipsia.
LEIRIA, Collippo.
LEISSNIG, Leisnicium.
LEITH, Letha.
LEITHA, Litaha.
LEMBERG, Leopolis.
LEMENE, Romatinum.
LEMGO, Lemgovia.
LEMO, Lemuris.
LEMSTER, Leonis Monasterium.
LEMTA, Leptis Minor.
LEMWIG, Lemoiga, Lemvicum.
LENCZIG, Lancicia, Lancicium.
LENHAM, Durolenum.
LENNOX, Levinia, Elgovia.
LENS, Elenæ, Lentium.

LENT-EN-DOMBES, Lentulum.
LEOBEN, Leobia
LEON, Legionense Regnum.
LEON, Legio.
LEON, Cotinussa.
LEONDARI, Helisson.
LEOPOLDSTADT, Leopoldinum.
LEPANTO, Naupactus.
LEPANTO, Corinthiacus Sinus.
LEPT, Electra.
LERICE, Erycis, Portus Ericus.
LERIDA, Ilerda.
LERINIAN ISLANDS, Lerinæ.
LERIZ, Læros.
LERMA, Libarna.
LERO, Leria.
LERS, Lertius.
LESCAR, Beneharnum, Lascara Bearnensium
LESCHE, Letia, Lieta.
LESCHES, Vologatis.
LESER, Lesura.
LESIGNAN, Lesiniacum.
LESINA, Pharus, Pharia.
LESINA (LAGO DI), Pantanus.
LESUZA, Colonia Libisosanorum.
LETINES, Lestinæ, Liptinæ.
LETTE, Latera.
LETTEN, Lettia, Littia.
LETTERE, Letteranum, Lycteræ.
LEUBUS, Leobusium.
LEUCATE, Leocata.
LEUCOS, Glauchus, Glaucus.
LEUDESDORF, Landulfesdorf.
LEUK, Leucia.
LEUKERBAD, Leucenses Thermæ.
LEUSE, Letusa.
LEUTKIRCH, Ectodurum.
LEUTMERITZ, Litomericium, Litomerium.
LEUTOMISCHL, Litomislium.
LEUTSCHAU, Leuconium, Leutschovia.
LEUWARDEN, Leovardia.
LEVADIA, Lebadea.
LEVAGNA, Lebonia.
LEVANT (ISLE DU), Hypæa.
LEVANTA, Castellum Orientis.
LEVENZO, Buccina.
LEVIG, Levico.
LEVITHO, Lebinthus.
LEVROUX, Leprosium
LEWENZ, Levia.
LEWES, Lesua.
LEWIS, Leogus, Ebuda Occidentalis.
LEYE, Legia.
LEYTE, or LEYTHA, Lutia, Scarniunga.
LEZ, Lædus.
LIACA, Lacmon.
LIACURA, Lycorea.
LIAMONE, Cercidius.
LIANNE, Elna
LIBAU, Liba.
LICENZA, Digentia.
LICHSTALL, Leucostabulum.
LICOSA, Leucasia.
LIDBACH, Hlidbeki.
LIDDESDALE, Lidalia.
LIDKOEPING, Licopia.
LIEBE, Lobavia.
LIEBENTHAL, Leovallis.
LIEFLAND, Livonia.
LIENZ, Loncium.
LIER, or LIRE, Ledi, Lyra.
LIESSE, Lætiæ, Lætitiæ.
LIEUVAIN (LE), Lesuinus pagus.
LIFFEY, Avenlifnius.
LIGNE, Ligniacum.

LIGNY, Lignium, Ligniacum.
LILIENFELD, Campililium.
LILLE, Insulæ.
LILLEBONNE, Juliobona.
LILLERS, Lilertium, Lillerium.
LILLO, Lilloa.
LIMA, Forum Limicorum.
LIMA, Belio.
LIMAGNE, Limania, Alimania.
LIMBACH, Olimacum.
LIMBURG, Transmosana ditio.
LIME, Lemanis, Novus Portus.
LIMEGOS, Limodia.
LIMEUILLE, Limolium.
LIMMAT, Limaga.
LIMNE-STIVA, Hylice.
LIMOGES, Augustoritum, Lemovicum.
LIMOSIN, Lemovicensis Provincia, Lemovices.
LIMOUX, Limosum.
LINDAU, Lindavia.
LINDO, Lindus.
LINDRE, Lindrensis Lacus.
LINGEN, Lingo.
LINIERES, Linarium.
LINKOEPING, Lincopia.
LINLITHGOW, Lindum.
LINNHE-LOCH, Longus.
LINZ, Aredata, Lincium.
LINZA, Lencia.
LIONNOIS, Lugdunensis Ager.
LIONO, Pelva.
LIONS, Lionium.
LIPARI ISLANDS, Æoliæ Insulæ.
LIPESE, Liptavia.
LIPNIK, Prista.
LIPPE, Lippia, Lupia.
LIPPO, Hypia.
LIPPSPRING, Lippebrune.
LIPPSTADT, Lipstadium.
LIPSO, Lepsia.
LIPSOCATALIA, Psyttala.
LIPUDA, Aretas.
LIRIA, Edeta.
LIS. See LYS.
LISCABIANCA, Æricusa, Ericusa.
LISIEUX, Civitas Lexoviorum, Lexovium.
LISOGNO, Lisonius.
LISSA, Hissa.
LISSA, Limiosaleum.
LISSABON, Felicitas Julia, Olisipo.
LITHAUEN, Lithuania.
LIVADIA, Græcia Propria, Hellas.
LIVADIA, Copais.
LI-VAGNOLI, Trifanum.
LIVANITIS, Synos.
LIVENZA, Liquentia.
LIVERDUN, Liberdunum.
LIVINER-THAL, Lepontina Vallis.
LIVORNO. See LEGHORN.
LIVRON, Libronicum.
LIYOR, Zaba.
LLOBREGAT, Rubricatus.
LOBBES, Labieni Castra, Laubium.
LOCARNO, Lucarnum.
LOCATE (LAC DE), Sordice.
LOCHABER, Lochabria.
LOCHES, Luccæ.
LOCKNITZ, Lochenitium.
LOCKUM, Abbatia Loccensis.
LOCONI, Luguidonis Portus.
LOCRA, Eliocrata.
LODESANO, Laudensis Ager.
LODEVE, Lutevensium, Civ. Luteva.
LODI, Laudum, Laus Pompeja Nova.
LODI-VECCHIO, Laus Pompeja.
LOHAR, Calaguris Fibularensis.
LOHE, Lavus.
LOHITSCH, Longaticum.
LOING, Lupia.
LOIR (LE), Lædus.
LOIRE (LA), Liger, Ligeris.
LOIRET, Ligerula.
LOISACH, Loysa.
LOITZ, Lutitia.
LOMAGNE, Leomania.
LOMATSCH, Glomacium.
LOMBARDY, Bojus Ager, Langobardia.
LOMBEZ, Bersinum, Lombarium.
LONDON, Augusta Trinobantium, Londinum.
LONDONDERRY, Londino Deria.
LONDONDERRY, Robertum.
LONGFORD, Longofordia.
LONGO-SARDO, Tibula.
LONGWY, Longo-Vicus.
LONJUMEAU, Longum Gemellum.
LONS-LE-SAULNIER, Ledum Salarium, Ledo Salinarius.
LONSEE, Ad Lunam.
LONTEL, Luna.
LOON, Lossa.
LORA, Axalita.
LORCA, Ilorci.
LORCH, Laureacum, Lauracense Monasterium.
LORCH, Colonia Aureliana Lauracensis.
LORETTO, Fanum Mariæ Lauretanæ, Lauretum.
LORGOE, Lara.
LORGUES, Leonicæ.
LORI, Pelorus.
LORRIS, Loriacum.
LOSA, Ludosia.
LOSCHANZ, Losontium.
LOSERE, Lesora.
LOS-SANTOS, Sentice.
LOT, Oltis, Loda.
LOTHIAN, Laudania.
LOUA, Garra.
LOUAIN, Luvia.
LOUDUN, Juliodunum.
LOURDE, Lapurdum.
LOUTH, Lutum.
LOUTH, Ludum.
LOUTRA, Limnæa.
LOUVIERS, Luparia.
LOWEN, Lovania, Lovanium.
LOWENBERG, Leorinum.
LOWENSTEIN, Leostenii Comitatus.
LOWICZ, Lovitium.
LOYES, Loja.
LUBBEN, Lubena.
LUBECK, Lubecum.
LUBLAU, Lublavia.
LUC, Lucus Augusti, Tellonum.
LUCAN, Lycandus.
LUCCA, Luca.
LUCERA, Luceria.
LUCERN, Luceria, Lucerna.
LUCIENSTEIG, Clivus St. Lucii.
LUCIOL, Lophosagium.
LUCKAU, Luccavia.
LUCMANER, Adula, Lucomonis Mons.
LUÇON, Lussonia Insula.
LUÇON, Lucio.
LUCRINA (LAGO DI), Lucrinus lac.
LUDD, Diospolis.
LUDE, Lusdum.
LUDERS, Laudera.
LUDEVE, Leuteva.
LUDWIGSBURG, Ludovici Arx.
LUGANO (LAGO DI), Ceresius Lacus.
LUGDE, Luda ad Ambram.

LUGNETZERT-HAL, Leguntina Vallis.
LUGO, Lucus Augusti.
LUGODURI, Luquido.
LUGOS, Lugosium.
LUGUGNANO, Tilaventum minus.
LUKAN, Amanus Mons.
LULA, Lulea.
LUMBERIS, Ilumberis.
LUND, Londinum, Lunda Gothorum, Lundinum Scanorum.
LUNEBURG, Lunæburgum.
LUNEGIANO, Luna.
LUNEL-LA-VILLE, Lunate.
LUNEVILLE, Lunaris Villa.
LUNGA-INSEL, Sirenum Petræ.
LUPAT, Lopadium.
LURE, Lutera.
LURKAKI, Lyrcea.
LUSIGNAN, Leziniacum, Lusignanum.
LUSTENAU, Lustena.
LUTACH, Littamum.
LUTTENBERG, Lentudum.
LUTTERWORTH, Lactodurum.
LUTTICH, Leodicum, Leodium.
LUTZEN, Lucena.
LUXEMBURG, Augusta Romanduorum, Luciburgum.
LUXEVIL, or LUXEU, Luxoium, Luxovium.
LYACURA, Lycoreus.
LYCURIA, Lessa.
LYMFIORD, Sinus Lymicus.
LYNN, Lynum Regis.
LYON, Lugdunum Æduorum.
LYS, Legia, Lieva.

M

MAAN, Maon.
MAAS, or MEUSE, Mosa, Patabus fluvius.
MAASEYK, Maseca.
MAB, Ar.
MACAO, Amacaoum.
MACARMEDA, Erpis.
MACHECOU, Ratiastum.
MACHECOUL, Machicolium, Machecum.
MACON, Madascona, Matisco.
MACRAN, Cedrosia.
MACRI, Telmessus.
MACRI (GULF OF), Glaucus Sinus.
MACRIPLAI, Geranea.
MACRONISI, Helena.
MADAGASCAR, Minuthias, Hannonis Insula, Divi Laurentii Insula.
MADARE, Mons Glaber.
MADDALONI, Suessula, Trebula.
MADEBA, Medeba.
MADEIRA, Purpurariæ Insulæ.
MADRAS, Melange.
MADRID, Mantua Carpentanorum, Madritum.
MADURA, Modura.
MAEIN, Baal-Meon.
MAESTRICHT, Trajectum Tungrorum, Traj. ad Mosam.
MAGAME, Dionysopolis.
MAGDEBURG, Parthenopolis, Magdeburgum.
MAGDENAU, Augia Virginum.
MAGGIA, or MAIN, Madia.
MAGLIANO, Manliana.
MAGNE, Macna.
MAGNISA, Magnesia ad Sipylum.
MAGRA, Macra.
MAHEDIA, Hannibalis Turris.
MAHI, Mais.
MAHON, Mago, Portus Magonis.
MAHREN, Moravia.
MAIDSTONE, Madus Vagniacæ.
MAILLEZAIS, Malleaca, Malliacum.
MAILLOC, Tilomusiacum.
MAIN, Mænus, Mœnus.
MAINA, Hippola.
MAINA, Tænarum.
MAINE, Cenomania, Prov. Cenomanensis.
MAINFELD, Ambitivus pagus.
MAINGAU, Menigau.
MAINLAND, Hethlandia, Pomonia.
MAINOTTEN, Eleuthero-Lacones.
MAIN-THAL, Madiæ Vallis.
MAINZ, Magontia, Mogontiacum.
MAIRE, Marca, Merula.
MAIS, Madia.
MAITO, Madytos.
MAJORCA, Balearis Major, Majorica.
MAKERSS, Carsus.
MAKESIN, Maguda.
MALABAR, Male.
MALACCA, Aurea Chersonesus.
MALAGA, Malaca.
MALANGARA, Olmiæ.
MALATHIJA, or MALATIA, Melitene.
MALATZKERT, Philadelphia.
MALCHOF, Malchovia.
MALDON, Camalodunum.
MALEA, Jala.
MALMEDY, Malmundariæ, Malmundarium.
MALMESBURY, Maldunense Cœnobium.
MALMOE, Ellebogium.
MALO, Mallus.
MALORA, Lamellus.
MALTA, Melita.
MALVASIA, Monembasia. See NAPOLI-DI-MALVASIA.
MAMERS, Mamerciæ.
MAN (ISLE OF), Menavia, Mona.
MANAHADA, Dosaron.
MANAHADA, Tyndis.
MANCASI, Nueriola.
MANCHESTER, Manduessedum, Marcunium.
MANCHING, Vallatum.
MANCOPALDO, Delphinium.
MANDA, Manila.
MANDEURE, Epamonduodurum, Mandurïs.
MANDONIA, Maro.
MANDORVAY, Mandova.
MANGALIA, Calatis.
MANGALOR, Mangaruth, Mandagara.
MANILA, Manilia, Lusonia.
MANILLES, Lussoniæ Insulæ.
MANLIEU, Magnus locus.
MANOMUZZA, Vagedrusa.
MANOSQUE, Manesca, Manuasca.
MANPOLY, Bambana.
MANRESA, Minorissa.
MANS, Cenomania, Cenomannum.
MANSPURG, Monatium.
MANTES, Medanta.
MAQUIZ, Ossigi Laconicum.
MARAND, Morunda.
MARANO, Marianum.
MARANS, Marantium.
MARANT, Marantium.
MARARCO, Tomara.
MARASCH, Mariscum, Melitena.
MARATA, Myronos.
MARATHONA, Marathon.
MARATHONISI, Cranæ.
MARAWA AND MADURA, Pandionis regio.
MARBACH, Collis Peregrinorum.
MARBELLA, Salduba.
MARBURG, Amasia Cattorum, Mattium.

MARBURG, Marciana.
MARCELLINO, Myla.
MARCH, Marus.
MARCHE-EN-FAMINE, Marchia Famina.
MARCHENA, Colonia Marcia.
MARCHIENNES (AU-PONT), Hormum, Marchianæ.
MARCH-THAL, Martalum.
MARCIGLIANO VECCHIO, Crustumeria.
MARCIGNY, Marciniacum.
MARCILLAC, Marcilliacum.
MARDIN, Maride.
MAREB, Mariaba Baramalacum.
MARECHIA, Ariminus.
MARENNES, Marinæ.
MARETIMO, Hiera.
MARGANA, Morgyna.
MARGARET ISLAND, Margarethæ Divæ Insulæ, Fanum S. Margarethæ.
MARGHINAN, Demus.
MARGOFEST, Marcodava.
MARHES, Lacotena.
MARIA-DE-COVULTERE, Compulteria.
MARIAB, Margus.
MARIAGER, Ager Marianus, Mariæ.
MARIBOE, Habitaculum Mariæ.
MARIENBAD, Balneum Mariæ.
MARIENTHAL, Cœnobium Mariæ Vallense.
MARIENTHAL, Cœnobium B. Virginis Mariæ, Vallis Mariæ.
MARIEN-THERESIENSTADT, Maria-Theresianopolis.
MARIENWERDER, Mariæ Verda.
MARIGNANO, Merinianum, Melignanum.
MARIGNY, Marennium.
MARINES, Mons Mariorum.
MARINGUES, Maringæ.
MARINO, Bovillæ, Lamum.
MARITZA, Hebrus.
MARIUT, Mareia Lacus.
MARIUT, Marea.
MARK, Marchia, Comitatus Marchiæ.
MARK-ANCONA, Picenum.
MARKMAGEN, Marcomagus.
MARK ZATERN, Tarodunum.
MARLBOROUGH, Cunetio.
MARMARA, Hyrcania.
MARMARICE, Cressa.
MARMOLEJO, Uciense.
MARMORA, Elaphonesos.
MARMORA (SEA OF), Propontis.
MARMORE (LAGO DELLE), Reatina palus.
MARMOUTIER, Majoris Monasterium.
MARMOUTIER, Mauri Monasterium.
MARNA, Manarmanis Portus.
MARNE, Matrona.
MARO, Macrum.
MAROCCO, Maurocitanum, Maroccanum Regnum.
MAROCCO, Bocanum Hemerum, Marochium.
MAROGNO, Maronea.
MAROS, Marosiensis Sedes.
MAROSH, Marisus.
MARQUISE, Marci.
MARRO, Metaurus.
MARSALA, Lilybæum.
MARSCHLINS, Marsilinum.
MARSEILLES, Massilia.
MARSICO VECCHIO, Abellinum Marsicum.
MARTABAN, Aspithra.
MARTABAN (GULF OF), Magnus Sinus.
MARTEL, Martelli Castrum.
MARTIGNI, Civitas Vallensium.
MARTIGUES, Anatiliorum Urbs, Martigium.
MARTINACH, Octodurus (vicus) Octodurum.
MARTINSBERG, Arx St. Martini.
MARTORANA, Mamertium.
MAR-UJVAR, Patavissa.
MARVAO, Medobreja.
MARVDASCHT, Cœle-Persis.
MARVEJOLS, Marengium, Marologium.
MARVILLE, Martia Villa, Martis Villa.
MARVISAN, Phazemon.
MARZA-STROCCO, Herculis Fanum.
MARZA-SUZA, Apollonia.
MASCAT, Machorbe, Moscha.
MAS-D'ASIL, Asilium Mansum.
MASELI, Gerra.
MASEVAUX, Masonis Monasterium.
MASJEBKAN, Massiæ.
MASMA, Melsus.
MASSA-CARRARA, Herculis Fanum.
MASSA-DI-SORRENTO, Massa Lubiensis
MASSAY, Maciacum.
MASSEL, Hegetmatia.
MASSERANO, Massianum.
MASSRA-GOR-EL-SZAPHIA, Bela, Zoar.
MASULIPATAM, Cottobara, Mesolia.
MATAGI, Matisa.
MATALONE, Meta Leonis.
MATARE, Beth-Semes, Heliopolis.
MATARO, Diluron.
MATELICA, Matilica.
MATERA, Mateola.
MATRAY, Matrejum.
MATRIGA, Hermonassa.
MATZDORF, Matthæi Villa.
MAUBEUGE, Malburium Monasterium, Malobodium.
MAUDRE, Maldra, Malara.
MAULEON-DE-SOULE, Malleo, Mallosolium.
MAULEVRIER, Mallevrium.
MAULO, Hirminius.
MAUMAGUES, Mamaceæ.
MAURIENNE, Garocelia Vallis, Comitatus Maurianæ.
MAURIK, Mannaricium.
MAURO-NERO, Melas.
MAUSCHID, Muza.
MAUSPURG, Mennejanæ.
MAUTERN, Mutarensis Civitas.
MAVRE-MATIA, Messene.
MAXARA, or MOXARA, Murgis.
MAY, Sonus.
MAYEN, Magniacum.
MAYENFELD, Lupinum, Maisvilla.
MAYENNE, Meduanum.
MAYENNE, Medana, Meduana.
MAYO, Mayensis Comitatus.
MAZAGAN, Rusibis, Mazacanum.
MAZARA, Mazaræ Vallis.
MAZARAKI, Onchestus.
MAZERES-EN-FOIX, Castrum Maseria.
MAZORBO, Majorbium.
MAZUNAH, Fundus Mazucanus.
MEAUX, Civitas Meldorum, Meldæ.
MECCA, Macoraba.
MECHELN, Mechlinia, Maalinæ, Malinæ.
MECKLENBURG, Megalopolis, Mecklenburgum.
MEDEA, Lamido.
MEDELLIN, Metallinum, Castra Metellina.
MEDENBLICK, Medemelacum.
MEDENICO, Metapa.
MEDIASH, Mediensis Sedes.
MEDIASH, Mediesus.
MEDINA, Athrulla, Jatrepa.
MEDINA-CELI, Methymna Cetia, Medina Celia.
MEDINA-DEL-CAMPO, Methymna Campestris.
MEDINA-DE-LOS-TURRES, Methymna Turrium.
MEDINA-DE-RIO-SECCO, Methymna sicca, Cauca.

MEDINA-SIDONIA, Methymna Asidonia, Asindo.
MEDING, Mediana.
MEDOLO, Mutila.
MEDON, Dioclea.
MEGARISCH-UZZIS, Magusa.
MEHRERAU, Augia Brigantina.
MEHUN, Magdunum.
MEINDER, Mæander.
MEINTA-POSSEDA, Heraclea Cassiotidis.
MEISSAU, Medoslanium.
MEISSEN, Misnia.
MEISSEN, Misena, Misna.
MEJERDA, Bagrada.
MELA, Mileum.
MELASSO, Pedasum.
MELAZZO, Mylæ.
MELCHEDE, Melocavus.
MELCK, Medelicium.
MELENIK, Melenicum.
MELFA, Melpis.
MELFI, Melphia.
MELIAPUR, Maliarpha.
MELIDE, Melita.
MELILLA, Rusadir.
MELINDE, Essina, Melindum.
MELINDE, Barbaricus Sinus.
MELORA, Mænaria.
MELUN, Melodunum,
MELZO, Melpum.
MEMEL, Chronus.
MEMEL, Clupeda, Memelia.
MEMERS, Mamerciæ.
MEMMINGEN, Septemiacis, Memminga.
MENAN, Sobannus.
MENBIGZ, Menba.
MEND, Arbis.
MENDAYA, Nemetobriga.
MENDE, Mimatum.
MENDEN, Menithinna.
MENDES, Myndus.
MENDIPHILLS, Minarii Montes.
MENDRA, Bedirum.
MENDRES, Mendrisio.
MENERBINO, Minervium.
MEN-HABER, Arbis.
MENIDI, Acharna.
MENIMEN, Temnus.
MENIN, Menena.
MENSIEH, Ptolemais.
MENTANA, Nomentum.
MENUF, Momemphis.
MEPHITI, Lacus Ampsancti.
MEPPEN, Meppia.
MERAKIAH, Maraclea.
MERAKIN, Marathos.
MERAWE, Meroe.
MERBURG, Mersoburgum, Merspurgum.
MERCOEUR, Mercorium.
MERCOGLIANO, Mercuriale.
MER-DE-MARTIGUES, Avaticorum Stagnum.
MERENDA, Myrrhinus.
MERGENTHEIM, Mariæ Domus.
MERGHI, Berobe.
MERGUI, Catigara.
MERIDA, Augusta Emerita.
MERIDA-D'YUCATAN, Emerita Jucataniensis.
MERIOFITO, Myriophytus.
MERIONETHSHIRE, Mervinia, Merviniensis comitat.
MERJ-IBN-AMER, Esdraela.
MERO, Mearus, Metarus.
MERON-ERRUD, Maruca.
MEROTH, Meros.
MERSA, Carthago.
MERSEBURG, Martisburgum, Marsiburgum, Martiopolis.
MERTOLA, Julia Myrtilis.
MERU, Mervacum.
MERVILLE, Mauronti Villa, Menariacum.
MERZIG, Marcerum.
MESAGNA, Messapia.
MESEMBER, Menebria.
MESHERGA, Giuf.
MESHID-HOSEIN, Bolagasus.
MESJED-ALI, Hira.
MESSIMURI, Campsa.
MESSIN (Le), Metensis Tractus
MESSINA, Messana.
MESSINA (Faro di), Fretum Siculum.
MESSIS, Mopsuestia.
MESTRO, Mestus.
MESURATA, Misynus.
MESVRES, Massava.
METE, Mundu.
METELINO, Lesbos.
METELINO, Mitylene.
METELN, Mediolanium.
METLING, Metulum, Metlinga.
METRO, Metaurus.
METSCHNOE, Bice.
METZ, Divodurum, Metæ.
MEUDON, Modunum, Metiosedum.
MEULAN, Medlindum, Mellentum.
MEURS, Meursia, Mursia.
MEURTE (La), Murta.
MEWE, or Gniew, Gnevum.
MEXICO, Hispania Nova.
MEYNIAC, Manica.
MEZIDUN, Mansio Odonis.
MEZIERES, Maceriæ, Meseria.
MEZZO-LOMBARDO, Medium St. Petri.
MHAR, Raumathi Vicus.
MICHELSTADT, Michilinstadium.
MICILA, Myræum.
MIDDELBURG, Metelli Castrum, Medioburgum.
MIDDLAER, Meddedacum.
MIDJEH, Halmydessus.
MIDLAREN, Ad Tres Lares.
MIDOUX, Midorius fluvius.
MIERA, Sanda.
MIGNONE, Minio, Minone.
MIHALY, Fanum St. Michaelis.
MIHOLZI, Mariniana.
MIJARES, Uduba.
MILAN, Mediolanum.
MILATI, Nalata.
MILDEN, Moldunum.
MILETO, Melita.
MILETONI, Miletus.
MILHAU, Amilhanum, Æmilianum.
MILHAUD, Æmilianum.
MILIANA, Catada.
MILLAS, Millæ.
MILLY, in the French dep. of Seine-et-Oise, Mauriliacum.
MILLY, Milliacum.
MILO, Melos.
MINA-EL-DSAHA, Disahab.
MINCIO, Mincius.
MINDELHEIM, Rostrum Nemoviæ.
MINDGEN, Medianis.
MINEO, Menæ.
MINGARDO, Melpes.
MINGRELIA, Colchis.
MINHO, Minius.
MINIATO-AL-TEDESCO, Miniatum.
MINORI, Minora.
MINUGAT, Aspendus.
MIRABELLO, Olus.

MIRAKA, Olympia.
MIRANDA, Erineus.
MIRANDA-DE-DUERO, Continum.
MIRANDA-DE-EBRO, Deobriga.
MIRANDELLA, Caladunum.
MIRECOURT, Mercurii Curtis.
MIREPOIX, Mirapicum, Mirapincum.
MIREVAUX, Mira Vallis.
MIRZNO, Muziris.
MISEVRIA, Mesambria.
MISITRA, Tiasa.
MISSEN-HEAD, Notium.
MISTRETTA, Amastra.
MITTENA, Minthe.
MITTENWALD, Inutrium.
MITTENWALDE, Monosgada.
MOCHA, Pseudocelis, Moca.
MOCHO, Cambysæ Ærarium.
MOCKERN, Mokrianici.
MODENA, Mutina.
MODON, Mothone.
MODSCHEB, Arnon.
MOGELN, Mogelina.
MOGHISTAN, Carmania.
MOGLA, Alinda.
MOGLAENA, Emathia.
MOGLIA, Cresius.
MOISSAC, Mussiacum.
MOJENVIC, Mediovicus.
MOJET-EL-UHRID, Carith.
MOKOSSI, Oxynia.
MOLA, Molæ Formianæ.
MOLA, Turres Julianæ.
MOLA-DI-GAETA, Formiæ.
MOLADIVA, Moduti.
MOLDAU, Moldavia.
MOLETIO, Minoa.
MOLFETTA, Melfitum.
MOLINI, Lerna Palus.
MOLISE, Lunciana Provincia.
MOLISE, Melæ.
MOLIWA, Methymna.
MOLLICORNO, Larenusiæ Insulæ.
MOMPELGARD, Magetobriga, Mons Belligardus.
MON, Mona.
MONACO, Herculis Monœci Portus.
MONASTERACE, Mystia.
MONBASA, Itenedium Menuthesias.
MONCAJO, Caunus Mons, Cacus Mons.
MONCALE, or MONTCALVO, Castrum montis Calerii.
MONCLOVA, Obucula.
MONCORNET, Mons Cornutus.
MONDEJO, Monda.
MONDIDIER, Desiderii Mons.
MONDONEDO, Mindonia.
MONDOUBLEAU, Mons Dubelli.
MONDOVI, Mons Vici.
MONDRAGONE, Massicus Mons.
MONDSGEBIRG, Lunæ Montes.
MONEGLIA, Ad Monilia.
MONGRI, Mons Jovis.
MONLUEL, Lupelli Mons.
MONMORENCY, Maureciacus Mons.
MONMORILLON, Mons Maurilionis.
MONMOUTH, Monumethia.
MONOSQUE, Bormanicium.
MONS, Montes, Montes Hannoniæ.
MONTABAUR, Montaborium, Mons Tabor.
MONTAGNAC, Montiniacum.
MONTAIGU, Mons Acutus.
MONTAILLEU, Montala.
MONTALCINO, Mons Alcinous.
MONTALTO, Babia.
MONTALTO, Mons Altus.
MONTANOS-DE-EUROPA, Vindius.
MONTARGIS, Mons Argi.
MONTAUBAN, Mons Albanus.
MONTBAZON, Mons Basonis.
MONTBRISON, Mons Bruso, Mons Brisonis
MONTCALVO. See MONCALE.
MONT-CASSEL, Menapiorum Castellum.
MONT-COLOBRE, Colubraria.
MONTE-ALBODDO, Bodius Mons.
MONTE-ALTO, Forum Aurelii.
MONTE-ARTESINO, Montes Ærii.
MONTE-CASSINO, Cassinum.
MONTE-CHIARUGOLO, Mons Ceritus.
MONTE-CHRISTO, Oglasa, Mons Christi.
MONTE-CORVO, Coritus.
MONTE-DELLA-SIBILLA, Fiscellus.
MONTE-DI-COTRONE, Latymnus.
MONTE-DI-GIOVANNE, Canterius Mons.
MONTE-DI-GORANTE, Cragus.
MONTE-DI-LICATA, Ecnomos.
MONTE-DI-NOVE, Novana.
MONTE-DI-PALATSCHIA, Latmus.
MONTE-DI-ST.-CIRIACO, Cumerium.
MONTE-DI-ST.-JULIAN, Epomeus Mons.
MONTE-DI-ST.-MARIA-MAGGIORE, Esquiliæ.
MONTE-DI-ST.-ORESTE, Soractes.
MONTE-FIASCONE, Mons Physcon, Mons Flasconis.
MONTE-LATTARIO, Lactis Mons.
MONTE-LEONE, Leonis Mons.
MONTELESE, Mons Silicis.
MONTELIMART, Montilium Adema.
MONTELLANO, Gemella.
MONTE-MARANO, Mons Maranus.
MONTE-MARIO, Clivus Cinnæ.
MONTEMAYOR, Ulia.
MONTEMORA-VELHO, Medobriga.
MONTE-NEGRO, Scardus.
MONTE-PELLEGRINO, Ercta.
MONTE-PULCIANO, Plutium, Mons Politianus.
MONTEREAU-FAULT-YONNE, Condate, Monasteriolum ad Icaunum.
MONTE-SANTO, Athos.
MONTE-SARDO, Electris.
MONTE-ST.-GIOVANNI, Tetricus Mons.
MONTE-ST.-JULIANO, Eryx, Epomeus Mons.
MONTE-ST.-ORESTE, Faliscorum Mons.
MONTE-VISO, Vesulus Mons.
MONTE-VULCANO, Ithome.
MONTFERRAT, Mons Ferratus.
MONTGOMERY, Mons Gomericu.
MONTI, In Montibus.
MONTI-DELLA-CHIMERA, Ceraunii Montes.
MONTI-SORI, Heræi Montes.
MONTLEHERY, Leherici Mons.
MONT-LOUIS, Ludovici Mons.
MONT-MARTIN, Fanum Martis.
MONTMARTRE. Mons Martyrum.
MONTMAUR, Mons Seleuci.
MONTMEDY, Mons Medius.
MONTMIRAIL, Mons Mirabilis.
MONT-MORILLON, Mons Maurilionis.
MONTONE, Utis.
MONTPELLIER, Mons Pessulanus.
MONTPENSIER, Monpænserium.
MONTREAL, Mons Regalis,
MONTRESOR, Mons Thesauri.
MONTREUIL, Monasteriolum, Montriolum.
MONTREVIL, Mons Rebelli.
MONTRICHARD, Mons Ricardi.
MONTROSA, Mons Rosarum, Celurca.
MONT-SALEON, Seleucus Mons.
MONTSERRAT, Mons Edulius.
MONZA, Modœtia.
MONZON, Mendiculeia.

MORAISAH, Nepheris.
MORASCA, Clausula.
MORAVA-HISSAR, Horrea Margi.
MORAWA, Mora, Marus.
MOREA, Peloponnesus.
MOREE, Latameda.
MORESBY, Morbium.
MORIN, Muora.
MORLAIX, Morlocum, Mons Relaxus
MORPETH, Morstorpitum, Corstorpitum.
MORRO, Romula.
MORTAGNE, in the French dep. of Orne, Moritania.
MORTAGNE, Tamnum Moritania.
MORTAIN, Moretonium, Moritolium.
MORTALES, Murus.
MORTARA, Pulchra sylva.
MORTARO, Colentum.
MORVAN, Morundia.
MOSCOW, Moscua.
MOSCOW, Moscus.
MOSEL, Mosella.
MOSUL, Labbana, Lambana, Mausilium.
MOTRICO, Tritium Tuboricum.
MOTRIL, Firmium Julium.
MOTTA-DI-BURZANO, Locri.
MOTTER, Matra.
MOTZING, Mocenia.
MOUDON, or MILTEN, Minnidunum.
MOULINS, Gergobia, Molinæ.
MOUNSTER, Momonia.
MOURA, Arucci.
MOUSON, Mosomagum.
MOUSTIERS-EN-TARANTAISE, Tarantasia, Monasterium in Tarantesia.
MOUTIERS, Centronum Civitas.
MOWIL-GANGA, Ganges.
MOYEN, Meduantum.
MOYLAND, Mediolanum.
MOZZA, Modicia.
MUDGULL, Modogulla.
MUFITI, Mephitis Ædes.
MUGA, Tichis.
MUGELLO, Mucialla.
MUHLBERG, Moliberga.
MUHLHAUSEN, Meliodunum.
MULL (ISLE OF), Maleos.
MULTAN, Mallorum Metropolis.
MULUCHA, Molochat.
MUNCHEN, Monacum, Monachium.
MUNSTER, Miningroda, Monasterium
MURANO, Muranum.
MURAT, Muratum Alverniæ.
MURAU, Ad Pontem Muri.
MURCIA, Murgis, Murcia.
MUREK, Murœla.
MURGANA, Mergana.
MURILLO-DE-RIO-LEZA, Verela.
MURI-UCCI, Helorum.
MURO, Murus.
MUROS, Ebora.
MURRANA (LA), Crabra.
MURR, Murra.
MURRAY, Moravi Scottiæ.
MURTEN, Moratum.
MURTNERSEE, Aventicensis Lacus.
MUSCONISI, Hecatonnesi.
MUSDAG, Imaus.
MUSSELBURGH, Musselburgum.
MUSSY-SUR-SEINE, Museium Episcopale.
MUSTAPHA-PALANKA, Remisiana.
MUZON, Musonium.
MYHIE, Mophis.

N

NAABE, Nablis, Nabus.
NABAL, Neapolis.
NABBAKI, Neapolis.
NABLOUS, Mabartha, Neapolis.
NADIN, Nedinum.
NADIR, Nicii urbs.
NAERDEN, Narda, Nardenum.
NAGARAM, Nagararis.
NAGERA, Anagarum.
NAGGUR, Nicæa.
NAGY-BANJA, Paralissum, Neostadium.
NAHE, Nava.
NAHR-BEIRUT, Magoras.
NAHR-EL-ARDEN, Jordanes.
NAHR-EL-BERD, Eleutherus.
NAHR-EL-KIBIR, Eleutherus.
NAHR-EL-MUKATTUA, Kison.
NAHR-KELB, Lycus.
NAKCHIVAN, Naxuana.
NAMUR, Namon, Namurcum.
NANCY, Nancejum.
NANTERRE, Nannetodurum, Nemetodurum.
NANTES, Civitas Nannetica, Nannetes, Namnetæ.
NANTEUIL-LE-HAUDOIN, Nantogilum.
NAPOLI-DI-MALVASIA, Epidaurus Limera.
NAPOLI-DI-ROMANIA, Nauplia.
NAPOLI-DI-ROMANIA (GOLFO DI), Argolicus Sinus.
NAPOULE, Avenionis Castrum.
NARANJA, Babba.
NARBONNE, Narbo, Narbo Martius.
NARDO, Neritum.
NARENTA, Naro.
NARLA, Pons Nartiæ.
NASCARO, Syrus.
NASLI, Nysa.
NASQUE, Vindalicus fluv.
NASRA, Nazareth.
NASSENFELS, Vetonianæ.
NASSUK, Nasica.
NATOLIA, Anatolia, Asia Minor.
NATRON (VALLEY OF), Nitriotes nomos.
NAUA, Gaulana.
NAUCRIA, Nonacris.
NAUGRACUT, Batanagra.
NAUMBURG, Numburgum, Neoburgum.
NAUNI, Nonymna.
NAVARRA, Vasconia.
NAVARREINS, Navaresium.
NAY, Novum Oppidum.
NEBI-ABEL, Abila.
NEBIO, Cersuntum.
NECKAR, Nicar, Nicer.
NECKAR-GEMUND, Gemunda ad Nicrum.
NED-ROMA, Siga, Celama.
NEGA-PATNAM, Nicama.
NEGLE, Necla.
NEGOMBO, Anubingara.
NEGRO, Tanager.
NEGROPONTE (ISLAND OF), Eubœa.
NEGROPONTE, Chalcis.
NEISSE, Nissa.
NEITRA, Nitria.
NEKADEH, Maximianopolis.
NEMBROT, Quinda.
NEMI (LAGO DI), Lacus Aricius.
NEMOURS, Nemorosium, Nemosium.
NEO KASTRO, Coryphasium.
NEPI, Colonia Nepensis, Nepita.
NERIS, Aquæ Neræ.
NERMOUTIER, or NOIRMOUTIER, Nigrum Monasterium, Insula Dei.

NESIM, Europus.
NESIVO, Libethra.
NESTE, Negeta.
NETHERBY, Castra Exploratorum.
NETO, Neæthus.
NETTERBY, Æsica.
NETTUNO, Antium.
NEU-BRANDENBURG, Brandenburgum Novum.
NEU-CARLEBY, Carolina Nova.
NEUCHASTEL, Nova Castella.
NEUE-YSSEL, Drusiana Fossa.
NEUFCHATEAU, or NEUENBURG, Neocomum, Novum castrum.
NEUFCHATEAU, Nova Castella.
NEUHAUSEL, Arx Nova, Neoselium.
NEUHAUS, Henrici Hradecium.
NEUMARKT, Agropolis.
NEUPEGU, Besynga.
NEUSATZ, Neoplanta.
NEUSOHL, Neosolium.
NEUSS, Niusa castrum, Novesium.
NEUSTADT-AN-DER-HARDT, Neapolis Nemetum.
NEU-URGANTSCH, Gorgo.
NEVA, Neve.
NEVERS, Nivernium, Noviodunum.
NEW-ABERDEEN, Aberdona Nova.
NEW-CASTILE, Castella Nova.
NEWCASTLE, Gabrosentum.
NEW-GALLOWAY, Galveja.
NEW-GRANADA, Castella Aurea.
NEW-GUINEA, Guinea Nova.
NEW-HOLLAND, Hollandia Nova.
NEWTON-STEWART, Lucopibia.
NICARIA, Icaria.
NICOSIA, in Cyprus, Leucosia.
NICOSIA, in Sicily, Erbita, Herbita.
NIDROSIA, Niderus.
NIEBLA, Elepta.
NIEMEN, Nemenus, Chronus.
NIEMTSCH, Nemetzi.
NIEN-HERSE, Herisiæ Novæ.
NIENHUS, Nuæsium.
NIEULAY, Arx Niculana.
NIEVRE, Niveris.
NIGHDE, Cadyna.
NIGOLA, Misa.
NIHEIM, Niemia.
NIKDEH, Tyana.
NIKIOEPING, Neapolis Danica.
NIKOLO, Tenos.
NIKSARA, Neocæsarea.
NIKUB, Nicopolis.
NIMEGUEN, Noviomagum.
NIMES, Nemausus.
NIMIS, Nomasus.
NINFA, Nymphæus.
NIOLO, Nicæa.
NIORT, Nyrax.
NISEBIN, Nisibis.
NISI, Enisis.
NISIRO, Nisyrus.
NISITA, Nesis.
NISSA, Naissus.
NITH, Novius.
NIVELLES, Niella.
NIZZA, Nicæa, Nicia, Nizza.
NOAILLES, Noviliacum.
NOARA, Noæ.
NOCERA, Nuceria.
NOGAY STEPPE, Hylæa.
NOGENT-L'ARTAUD, Novigentum Artaldi, ad Matronam.
NOGENT-LE-ROTROU, Novigentum Retrudum, Noviodunum Diablintum.
NOGENT-SUR-SEINE, Novientum, Novigentum.
NOGUERRA-PALLERESA, Nucaria Palliarensis.
NOGUERRA-RIBAGORCANA, Nucaria Ripacurtia.
NOIREAU, Norallus.
NOJA, Netium.
NOKRA, Negra.
NOLI, Naulum.
NONA, Ænona.
NONS, Anonius.
NONSBERG, Anania.
NORCIA, Nursia.
NORDEN, Nordedi pagus.
NORDSTRAND, Glessaria.
NORIC ALPS, Alpes Noricæ.
NORMA, Norba.
NORMANDY, Normannia.
NORTEN, Nortunum.
NORTHAMPTON, Antonia septentrionalis.
NORTHFLEED, Vagniacæ.
NORTH SEA, Germanicum Mare.
NORWAY, Nerigon, Norwegia.
NORWICH, Venta Icenorum, Nordovicum.
NOSENSTADT, Nentidava.
NOTO, Neetum.
NOTRE-DAME-D'ABONDANCE, Abbatia Abundantiæ.
NOUAN, Noviodunum.
NOUS-SHEHR, Nysa.
NOVA CAPUA, Casilinum.
NOVAIA-ZEMBLIA, Carambice.
NOVARA, Novaria.
NOVI-CASTELLI, Cortyna.
NOYA, Noëla.
NOYER, Nucetum.
NOYON, Noviomagus in Veromanduis, Noviomum.
NOZEROI, Nucillum.
NUESTRA-SENORA-DA-PETTOVE, Budua.
NUITS, Vidubia, Nutium.
NUMAGEN, Noviomagus in Treviris.
NURE, or NEURE, Neorum.
NURNBERG, Norimberga, Noriberga.
NURTINGEN, Grinario.
NUSCO, Numistro.
NYMS, Nemesa.
NYNE, or NEANE, Aufona.
NYONS, Civitas Equestrium Augusta.
NYSSLOT, Arx Nova.

O

OB, Carambucis.
OBER-SEE, Brigantinus Lacus.
OBER-GLOGAU, Glogovia minor.
OBERGESTELEN, Castellio Superior.
OBER-HALBSTEIN, Impatis Ministerium.
OBERKIRCH, Hypergræcia.
OBER-LAIBACH, Hyper-Labacum.
OBERRHEIN, Obringa.
OBER-UNNACZ, Sarnade.
OBER-WELS, Lacufelicis.
OBERWESEL, Ficella-Vosalia.
OBER-WOLS, Viscellæ.
OBERYSSEL, Transisalanea.
OBRE, Ebora.
OCCA, Fluv. Aucensis.
OCHRIDA (LAKE OF), Lychnidus Lacus.
OCHRIDA, Lychnidus.
OCHSENFURT, Bosphorus, Oxonium.
OCHULA, Echetla.
OCZAKOW, Axiaca, Odessus.
ODENSEE, Ottonia.
ODENWALD, Ottonica sylva.
ODER, Viadrus.

ODERZO, Opitergium.
ODIEL, Luxia.
ODILIENBERG, Hoinborchum.
ODORNEH, Physcus.
OEDENBURG, Sempronium.
OEHRINGEN, Auriana.
OELS, Olsna.
OERDINGEN, Castra Hordeani.
OESCH, Castiodum.
OESEL, Latris, Osilia.
OETLING, Celeusum.
OETTING, Losodica, Ottinga.
OF, Ophis.
OFANTO, Aufidus.
OFEN, Buda.
OFENA, Aufina.
OFETO, Orethus.
OGGION, Eugeon.
OGLIO, Ollius.
OHM, Amana.
OIARSO, Easo.
OISE, Esia, Æsia.
OISEMONT, Avimons.
OISNITZ, Mura Pzombath.
OKA, Aucensis Fluvius.
OKEN-GATE, Uxacona.
OLAN-MUREN, Bautes.
OLD ABERDEEN, Aberdona Vetus.
OLD CARLISLE, Virosidum.
OLDENBROOK, Tecelia.
OLDENBURG, Branesia, Oldenburgum.
OLD PENRITH, Voreda.
OLD SARUM, Sorbiodunum.
OLERON (ISLAND OF), Olarion, Uliarus.
OLERON, Elorona, Iluro.
OLESA, Rubricata.
OLIVA, Obila.
OLIVENZA, Evandria, Oliventia.
OLMUTZ, Eburum, Olmucium.
OLONNE (LES SABLES D'), Arenæ Olonenses.
OMBRONE, Umbro.
OMEGNA, Eumenia.
OMISE, Almissum, Dalmissum.
OMMIRABI, Anatis.
ONCHE, Oscarus.
ONOR, Pseudostomus.
ONTARIO, Andiatrocus.
OOST-VRYE, Franconatus Orientalis.
OPH, Oplus.
OPINO, Opinum.
OPORTO, Cale.
OPPEDE, Fines.
OPPIDO, Opinum, Mamertium.
OPPIDOLO, Cossyra.
ORANGE, Arausio.
ORANGE (district of), Arausionensis Principatus.
ORANIENBURG, Arausionense Castrum.
ORBE (the river), Obris, Urbis.
ORBE, Urba.
ORBITELLO, Cosa.
ORCHIMONT, Luxemburg, Ursimontium.
ORCO (L'), Morgus.
ORDIAGO, Ordia.
ORDUNA, Dardania.
OREAJA, Variana.
ORENSE, Aquæ Calidæ, Cilinorum, Auria.
OREO, Hestiæa.
ORESZOVITZ, Oescus.
ORETO (DE), Oretum Germanorum.
ORFA, or URIA, Edessa.
ORFA, Edessena Regio.
ORFAN, Phagre.
ORGAZ, Althæa.
ORGON, Enarginum.
ORIA, Uria.
ORIENT (L'), or PORT LOUIS, Vindana.
ORIGANO, Aurelianum.
ORIHUELA, Bigastrum, Orcelis.
ORILLAC, Auriliacum.
ORIOLO, Forum Claudii.
ORISTAGNI, Arborea, Oristanum.
ORISTANO, Othoca.
ORISTANO (the river), Thyrsus.
ORLEANOIS, Aurelianensis Ager.
ORLEANS, Aurelia, Genabum.
ORLEANS (ISLAND OF), Aurelianensis Insula.
ORLON, Orolauno.
ORMEA, Ulmeta.
ORMUS, Armuza.
ORMUS (ISLAND OF), Armuzia, Ogyris.
ORN, Auriana.
OROSAI, Feronia.
OROSCHUK, Scaidava, Ad Novas.
ORSCHOWA, Clodova.
ORSERO, Ursaria.
ORTA, Horta.
ORTA (LAGO D'), Hortanus lacus.
ORTA CHAN, Noorda.
ORTENAU, Mortingia.
ORTONA, Orton.
ORTOSA, Orthosia.
ORVIETO, Herbanum, Urbs vetus.
ORZIANA, Ostra.
OSA, Ossa.
OSCHATZ, Ossitium.
OSENIK, Sucidava.
OSERO, Apsorus, Absorus.
OSIMO, Ausimum, Auximum.
OSMA, Uxama.
OSSUNA, Genua Ursorum.
OST-BOTHNIEN, Bothnia Orientalis.
OSTE, Hosta.
OSTERHOFEN, Austravia.
OSTFRIESLAND, Embdanus Comitatus, Frisia Orientalis.
OSTGOTHALAND, Gothia Orientalis, Ostrogothia.
OSTIGLIO, Hostilia.
OSTRAVEN, Tripolis.
OSTREVAND, Austerbatium.
OSTROCO, Cœnobium Insulanum.
OSTUNI, Hostunum.
OSUNA, Ossona.
OSVIA, Cherronesus.
OSWALD (SAINT), Adrante.
OSYMIANA, Oshmian.
OSZLOP, Osones.
OTOZECZ, Metubarris.
OTRANTO, Hydruntum.
OTRICOLI, Ocriculum.
OTTERTON, Othona.
OUDENARDE, Oldenarda.
OUDEWATER, Aquæ Veteres, Veraquinum.
OUESSANT, Uxantis.
OULX, Martis Stadio, Vallis Ocelana.
OUST. See AOUST.
OVIEDO, Lucus Asturum, Ovetum.
OWAR, Eburum.
OXFORD, Oxonia.

P.

PACHISTA, Selleis.
PACHSU, Ericusa.
PADERBORN, Fontes Paderæ, Paderborna.
PADOVA, or PADUA, Patavium.
PADRON, Iria Flavia.
PAGLIA, Pablia, Pallia.
PAGLIONE, Nicæa.

PALÆO-CHITRO, Chytri.
PALÆO-CHORO, Pheræ.
PALÆOPOLI, Elis.
PALÆOVOUNI, Helicon.
PALAGONIA, Palice.
PALAIOPOLI, Mantinea.
PALAMITA, Parthenicum.
PALANCA, Ophiusa.
PALATSHA, Miletus.
PALATSHA, Myus.
PALAZZO, Palatium.
PALAZZOLO, Alba Longa.
PALENCIA, Palantia.
PALEO-CASTRO, Buthrotum.
PALEO-EPISCOPI, Tegea.
PALEO-POLIS, Gytheum.
PALERMO, Colonia Augusta Panhormitanorum, Panormus.
PALESTRINA, Præneste.
PALI-KOIL, Pallura.
PALLANO, Volano.
PALLUAU, Paludellum.
PALMA, Junonia, Ombrio, Capraria.
PALMA-DI-SOLO, Solci.
PALMAROLA, Palmaria.
PALOMBARO, Palumbinum.
PALUDE-PONTINE, Pomptinæ Paludes.
PAMBUK-KALESSI, Hierapolis.
PAMIERS, Apamia.
PAMPELONA, Pompelon, Pompejopolis, Andelus.
PAN, Coli.
PANARI, Hicesia.
PANARO, Scultenna.
PANDOSIA, Molossia.
PANERMO, Placia.
PANITZA, Permessus.
PANTALEREA, Cossyra.
PANTIK, Pantichium.
PAOLO (Lago di), Saturæ palus.
PAPADORAS, Epicaria.
PAPANODIZIA, Papa-Adassi.
PAPASLI, Passa.
PAP-CASTLE, Epiacum.
PAPOSLI, Cænophrurion.
PARAY-LE-MOINEAU, Parcium Moniacum.
PARCO, Paropus.
PARENZO, Parentium.
PARIS, Lutetia Parisiorum, Parisii.
PARMA, Colonia Julia Augusta Parma.
PARO, Paros.
PARQUEROLLES, Prote.
PARTENKIRCH, Parradunum.
PARTHENAY, Partiniacum.
PASA, Pasargada.
PASAROFDSCHA, Margum.
PAS-DE-CALAIS, Fretum Britannicum, Gallicum.
PASSAU, Bacodurum, Passavium, Patavia, Castellum Batavinum.
PASSY, Paciacum.
PATAN, Balæpatna.
PATAVIONE, Patavio.
PATELPUTER, Palibothra.
PATERA, Arsinoe.
PATERNO, Hybla.
PATINO, Patmos.
PATRASSO, Colonia Augusta Aroë Patrensis, Patræ.
PATRICA, Lavinium.
PATTI, Timethus.
PAU, Epaunensis Civitas, Palum.
PAUZKE, Bugustia, Putiscum.
PAVIA, Ticinum, Papia.
PAVONARE, Cyaneæ, Pavonariæ Insulæ.
PECCIANA, Sphetzanium.
PEDERNOSO, Paterniana.
PEENE, Pana.
PEGNITZ, Pegnesus.
PEGU, Dorius, Besynga.
PEGU, Triglyptum.
PEHKELY, Peuceliotis, Pactyica Regio.
PEINA, Boynum.
PEKNO, Pephnus.
PELAGOSA, Adriæ Scopulus.
PELETIS, Peltæ.
PELUDJI, Icarus.
PEMSEY, Pevenscia.
PENNA-DI-FRANCIA, Lancia.
PENNA-DI-ORDUNA, Dardanus.
PENNAFLOR, Celsita.
PERA, Chrysoceras.
PERCHE, Perticus, Perticensis provincia.
PEREJASLAW, Marcianopolis.
PERGAMO, Pergamum.
PERIGNI, Pocrinum.
PERIGORD, Petrocoriensis tractus.
PERIGUEUX, Vesunna, Petricorium.
PERLIPE, Heraclea Pelagoniæ.
PERM, Biarmia, or Permia.
PERNALLA, Dunga.
PERNAMBUCO, Fernambocum.
PERNAU, Chesinus.
PERNES, Paternæ.
PERPIGNAN, Roscianum, Perpinianum.
PERTH, Fanum St. Joannis ad Tavum.
PERTUIS, Petronii Vicus.
PERTUS, Fauces Pertusæ.
PERUGIA, Perusia.
PERUGIA (Lago di), Trasimenus lac.
PESARO, Pisaurum.
PESCARA (the river), Aternus.
PESCARA, Aternum.
PESCHIERA, Ardelica, Piscaria.
PESCHIOLO, Plestina.
PESTH, Pessium, Contra Acincum, Pestum.
PESTO, Pæstum.
PETAU, Petobio.
PETERBOROUGH, Petuaria, Petroburgum.
PETERSBURG, Petropolis.
PETERSHAGEN, Huculbi.
PETERWARDEIN, Acimincum, Petro-Varadinum.
PETREA, Oaxes.
PETRINA, Lerna.
PEZENAS, Piscenæ.
PFEFFERS, Ad Favarias, Fabarium.
PFIN, Ad Fines.
PFIRT, Ferreta.
PFOERING, Epona.
PFORZ, Porca.
PFORZHEIM, Porta Hercyniæ.
PFULLENDORF, Bragodunum.
PFUNZ, Ponte.
PHADISA, Chadisia.
PHALACHTILA, Phalachthia.
PHANARI, Phaleria.
PHARILLON, Pharus.
PHILINE, Philea.
PHILIPPI, Colonia Augusta Julia Philippensis.
PHONEA, Pheneus.
PHOZZEYIL, Phasaëlis.
PIACENZA, Placentia.
PIANÇA, Diomedis Promontorium.
PIANOSA, Planasia.
PIAVE, Plavis.
PIAZZA, Piacus.
PICIGHETTONE, Forum Diuguntorum.
PICO (Lago di), Cyminus lac.
PICONZA, Picentia.
PICTS WALL (The), Murus Picticus.
PIDAURO, Epidaurus.

PIED, Joviacum.
PIE-DI-LUCO, Velinus lacus.
PIEDMONT, Pedemontium.
PIENZA, Corsilianum, Pientia.
PIEROLA, Andes.
PIERREPORT, Durvus Mons.
PIERTE-BRANDOMIL, Glandomirum.
PIETRA-DELLA-NAVE, Taurianus scopulus.
PIETRA-SANCTA, Fanum Feroniæ.
PIETRE-SIZZOSE, Electrides.
PIETRO, Geryonis Oraculum.
PIGNEROL, Pindrolium.
PIGNEROSO, Vibiforum.
PILATUS-BERG, Mons Fractus.
PILNIKAU, Pilingi-Villa.
PILSEN, Pelsina.
PINCK, Bynca.
PINEDA, Eluro.
PINHEL, Cepiana, Pinelum.
PIOMBINO, Plumbinum.
PIORACA, Prolaque.
PIPERNO-VECCHIO, Privernum.
PIRE, Pirina.
PIRGO, Apollonia sc. ad Mare Hadriaticum.
PIRLIPPE, Cercinas.
PISA, Pisæ.
PISATELLO, Rubicon.
PISCOPIA, Curias.
PISCOPIA, or DILOS, Telos.
PISMA, Cyane.
PISTOJA, Pistoria.
PISUERGA, Pisoraca.
PITAREVIL, Epidarum.
PITHIVIERS, Aviarium.
PITSCHEN, Bicina.
PITTY, DARRAWAY, RITSCHEL, or FETTY, Indi Ostia.
PIZZIGHITONE, Picelo.
PLAISANCE, Placentia.
PLATAMONE, Aliacmon.
PLATANI, Camicus. See CANNA.
PLAVE, Plavis, Plavia.
PLEISSE, Plissa.
PLINTENBURG, Altum Castrum.
PLOMBIERES, Plumbariæ.
PLON, Plona.
PO, Bodincus, Padus.
POCIGLIANO, Politorium.
PODHONISTA, Cephisus.
PODLACHIA, Bielcensis Palatinatus.
POESING, Basinium.
POGLION, Paulon.
POISSY, Pisciacum.
POITIERS, Limonum, Pictavium.
POITOU, Ager Pictonicus.
POIX, Pisæ.
POLCEVERA, Porcifera.
POLEGIA, Palatium.
POLEN, Polonia.
POLENZA, Carrea.
POLICANDRO, Pholegandros.
POLICASTRO, Buxentum, Palæocastrum.
POLICORO, Heraclea Lucaniæ.
POLIGNANO, Turres Aurelianæ, Polinianum.
POLIGNY, Pollemniacum, Polinianum.
POLINO, Polyægos.
POLLENZA, Pollentia.
POLLINA, Apollonia.
POLLINA, Monalus.
POLLONIA, Laons.
POLYSTILO, Abdera.
POMMERN, Pomerania.
PONCHES, Pontes.
PONDICHERRY, Ponticerium.
PONFERRADA, Flavium Interamnium, Pons ferratus.
PONIEWICZ, Upita.
PONS-DE-TOMMIERES, Pontus Thomeriæ.
PONT, Bundium.
PONTA-DI-TRIPITI, Hermæa.
PONTAFELLA, Pons Fellæ.
PONTALAREA, Paconia.
PONTA-MICHA, Miculi Mons.
PONT-A-MOUSSON, Mussipontum.
PONTARLIER, Ariorica, Pons Ælii.
PONT-AUDEMER, Breviodurum, Pons Audemari.
PONT-BEAUVOISIN, Pons Bellovicinus.
PONT-DE-CE, Pons Cæsaris.
PONT-DE-L'ARCHE, Pons Arcuensis.
PONT-DE-SORGUE, Cypresseta.
PONT-DE-VAUX, Pons Valensis.
PONTE-CENTESIMO, Forum Flaminii.
PONTECORVO, Fregellæ.
PONTE-DELLA-RIVA, Rivus.
PONTE-DI-QUATRO-CAPI, Fabricius Pons.
PONTEDURA, Portunata.
PONTE-NURA, Emporium.
PONTHIEU, Pagus Pontivus.
PONTIA, Fontia.
PONTICO, Cicynethus.
PONTICONESI, Lichades.
PONTOISE, Briva Isaræ, Pontisara.
PONTREMOLI, Aqua, Pons Tremulus.
PONTS-DE-CE, Pons Saii.
PONT-ST.-ESPRIT, Fanum Spiritus Sancti.
POPULONIA, Poplonium.
PORCUNA, Lacippo.
POROS, Calauria.
PORTA, Stratus.
PORT-A-CONDE, Coracodes Portus.
PORTAIS, MONTE-DI-MAINA and BRACCIO-DI-MAINA, Taygetum.
PORT-AMAN, Mysocaras.
PORT-DE-VENDRE, Veneris portus.
PORT-EN-BESSIN, Grannona.
PORTHLAND, Vindelis.
PORTIMAO, Hannibalis portus.
PORTO, Augusti Portus.
PORTO, Portus Cale.
PORTO-BADISCO, Veneris Portus.
PORTO-BUFALO, Porthmus.
PORTO-CESAREA, Sasina.
PORTO-CONSTANZA, Salamis.
PORTO-CROS, Mese.
PORTO-D'ATRI, Matrinum.
PORTO-DE-TORRE, Libissonis Turris.
PORTO-DI-FUORI, Cæsarea.
PORTO-DI-GORO, Carbonaria.
PORTO-DI-MAGNAVACCA, Ostium Sagis.
PORTO-DI-PALINURO, Palinurus Portus.
PORTO-DI-PALO, Portus Pachyni.
PORTO-DI-PRIMARO, Portus Vaterni.
PORTO DRACONE, Piræeus.
PORTO-FANARI, Elææ Portus.
PORTO-FAVONA, Favonii Portus.
PORTO-FERRAJO, Argous Portus, Portus Ferrajus, Ferrarius.
PORTO-FINO, Delphini Portus.
PORTOGAVEDO, Rusippir.
PORTO-INTERITO-DI-BELL'-OCCHIO, Ostium Caprasiæ.
PORTO-JERO, Hiera.
PORTO-MENDRA, Thoricus.
PORTO-PORO, Pronesus.
PORTO-RAGUSIO, Aamantia.
PORTO-RAPHTI, Panormus.
PORTO-RICO, Insula St. Joannis Portus Divitis.
PORTO-VECCHIO, Syracusanus portus.
PORTO-VECCHIO (GOLFO DI), Rubra.
PORTO-VENERE, Veneris portus.
PORT-ROYAL, Annapolis.

PORTS-DE-LONGUE, Robrica.
PORTSMOUTH, Magnus Portus, Portsmuthum.
PORT-SUR-SAONE, Portus Abucini.
PORTUGAL, Lusitania.
PORTZ, Budoris.
POSEGA, Basiana.
POSEN, Posna.
POSSEDA, Posidium.
POSSIDARIA, Pausinus.
POSTELBERG, Apostolorum Porta.
POTENZA, Potentia.
POTRINI, Bryseæ.
POTSDAM, Bostampium.
POZZI, Physcus.
POZZUOLO, Dicæarchia, Puteoli.
POZZUOLO, Puteolanus Sinus.
PRAGUE, Bojobinum, Casurgis, Praga.
PRASSA, Brasia.
PRASTIAS, Thalame.
PRATICA, Satricum.
PRAVADI, Soatra.
PREGEL, Guttalus, Chronus.
PRENZLAU, Premislavia.
PRESSBURG, Brecislaburgum, Posonium.
PREVESA, Nicopolis.
PRIEGO, Segobriga.
PRILIPO, Bora.
PRINCIPATO-ULTRA, Samnium.
PRINCES ISLANDS, Demonesos.
PROCIDA, Prochyta.
PRODANO, Prote.
PROERNA, Proarna.
PROMASENS, Viromagus.
PROMONTORE, Polaticum promontorium.
PROMONTORIO - DEL - MONTE - GARGANO, Garganum Promontorium.
PROTE, Prota.
PROVENCE, Provincia.
PROVINS, Provinum.
PRUM, Brumia.
PRUM (the river), Pronæa.
PRUSSIA, Borussia.
PRUTH, Hierasus, Poras.
PSATO, Pagæ.
PUEBLA-DE-LOS-ANGELOS, Angelopolis Americana.
PUECHE-D'ISSOLON, Uxellodunum.
PUENTE-DE-DON-GARRAY, Numantia.
PUENTE-DE-LA-REYNA, Cares.
PUENTE-DEL-ARZOBISPO, Augustobrica.
PUERTO-DE-ROBANON, Mons sacer.
PUERTO-DE-SOLON, Solauris.
PUERTO-DE-S.-MARTIN, Visci portus.
PUERTO-DE-SANTA-MARIA, Menesthei Portus.
PUICERDA, Podium Cretanum.
PULTANAH, Plithana.
PUNAH, Banavasi.
PUNTA-DELLA-LICOSA, Posidium.
PUNTA-DELLO-SPARTIMENTO, Palinurum.
PUNTA-DEL-PEZZOLO, Cænys.
PUNTA-DI-GIGANTE, Plemmyrium promontorium.
PUSCIANA (Lago di), Eupilis.
PUTILI, Phasis.
PUTRIA (Lago di), Linterna palus.
PUY (Le), Podium.
PUY-DE-DOME, Duma.
PUY-LAURENS, Podium Laurentii.
PUY-NOTRE-DAME (Le), Anicium, Urbs Vellanorum, Podium Andegavense.
PYRGO, Letrini.
PYRITZ, Periscum.
PYRMONT, Arx Pyrmontana.
PYZDRY, Peisera.
PZATE, Mases.

Q

QUANARO, Flanaticus Sinus.
QUARANTANIA, Jericho.
QUEILES, Chalybs.
QUEISS, Quissus.
QUERCY, Cadurca Terra, Tractus Cadurcensis Cadurcinus.
QUESNOY, Quercetum.
QUIETO, Quætus.
QUIMPER, Alauna, Civitas Curiosopitum, Corisopitum.
QUISCOR, Ionia.
QUITO, Fanum St. Francisci.

R

RAAB, Arabo.
RAAB, Jauriensis Comitatus.
RAAB, Jaurinum.
RABBIA (La), Asyphus.
RACCA, Constantia, Nicephorium.
RACHADA, Chrysoana.
RACKHASBURG, Alicanum, Racospurgum.
RACKSKENEY, Intercisa.
RACLIA, Heraclea.
RADOM, Carrodunum.
RAGOGNA, Reunia.
RAGUSA, Hybla Minor.
RAGUSA, in Dalmatia, Rhausium.
RAGUSA-VECCHIA, Epidaurum.
RAHABA, Rechobotir.
RAJIK, Fragiza.
RAJOO, Sadus.
RAMA, Arimathæa.
RAMANATHA COR, Cory.
RAMBERT (Saint), Ragnebertus (Sanctus).
RAMBOUILLET, Ramboletum.
RAMGONGA, Magon.
RAMPANO, Biandina.
RAMSEY, Rimnus.
RANÇON, Andecamulum.
RANDAZZO, Tissa.
RANDERS, Randrusia.
RAOZ, Partiscum.
RAPPERSWEIL, Ruperti villa.
RAS-AIN, Colonia Septimia Resainesiorum, Resæna.
RAS-AL-ANF, Pentedactylus.
RAS-AL-DAWER, Ision.
RAS-EL-ABIAD, Candidum promontorium.
RAS-EL-DOAR, Mnemium.
RAS-EL-HHAD, Didymi montes.
RAS-HEILI, Elancôn Emporium.
RAS-MOHAMMED, Pharan.
RAS-SEM, Phycus.
RASTADT, or Rachstadt, Teurnia.
RAUM, Rauranum.
RAWY, Hyarotis.
RAZEK, Eragiza.
RECCO, Ricina.
RECHBERG, Saloca.
RECULVER, Regulbium.
REDNITZ, Radiantia.
REGEN, Reginus.
REGENSBURG, Augusta Tiberii, Ratisbona
REGGIO, Rhegium, Regium Lepidi.
REGINA, Erinum.
REICHENAU, Augia Dives.
REINA, Herinum.
REMIREMONT, Avendi Castrum.
RENNES, Condate, Rhedones.
RENN-KEVI, Ophrynium.

RENO, Bononiensis Amnis.
REOL, Rigodulum.
REQUENA, Lobetum.
RESINA, Retina.
RETHA, Raphia.
RETONE, Eretenus.
RETOVINO, Retovinum.
RETTIMO, Rhitymnia.
RETZ, Cotia Silva.
REVEL, Sermyle.
RHAIN, Clarenna.
RHEIMS, Civitas Remorum, Durocortorum.
RHEIN, Rhenus.
RHEINAU, Augia Rheni.
RHEINFALL, Cataracta Rheni.
RHEINMAGEN, Rigomagum.
RHEIN-ZABERN, Tabernæ.
RHEONTES, Brasia.
RHODEZ, Segodunum.
RHODIS, Macaria.
RHONE, Rhodanus.
RHYNSBURG, Matilo.
RIBBIL, Belisama.
RIBCHESTER, Bœtonomacum.
RIBNIK, Castra Trajana.
RICHBOROUGH, Ritupæ.
RIESENBURG, Prabuctra.
RIESEN-GEBIRGE, Vandalici montes, Montes Sudeti, Sudetes.
RIETI, Reate.
RIEZ, Albece, Regium.
RIHAH, Jericho.
RIMINI, Ariminum.
RINGWOOD, Regnum.
RIO-BRAVO, Borealis Fluvius.
RIO-DE-CASTRO, Florius.
RIO-DE-LA-PUENTE, Nelo.
RIO-GUADALMEDINA, Fluvius Fœderatorum.
RION, Phasis.
RIONS, Serion.
RIO-VERDE, Salduba.
RIPA-ALBA, Hermæa Externa.
RIPE, Circidius.
RIPON, Rhidogonum.
RISANO, Rhizon.
RISANO (the river), Formio.
RISO, or CARPASSO, Caput Risum.
RISSO, Erisse.
RISSO, Larissus.
RITORBIO, Litubium.
RITZINGEN, Ricciacum.
RIVOLI, Ripula.
RIVOLTELLA, Ad Flexum.
RIWA, Rhebas.
RIZEH, Rhizus.
ROA, Rauda.
ROANNE, Rodumna.
ROBER, Erubris.
ROBRINO, Lacus Rubracensis.
ROCANELLO, Cylistarnus.
ROCCA-IMPERIALE, Lagaria.
ROCCA-LANZONE, Lanzonis Mons.
ROCHEFORT, Rupifortium.
ROCHELLE, Rupella.
ROCHESTER, Durobrivis, Roffa.
ROCROY, Rupes Regia.
RODA, Bæcula.
RODAUN, Eridanus.
RODBYE, Erythropolis.
RODEZ, Rutena.
RODIA, Garnæ Portus.
RODOSTO, Bisanthe.
ROER, Rura.
ROERMOND, Munda Ruræ.
ROGLENA, Ægæ Macedoniæ.
ROHATYN, Mœtonium.
ROMAGNA, Flaminia.
ROMAN, Augustana Prætoria sc. Daciæ.
RONCO, Bedesis.
RONDA, Acinipo.
ROPO, Oropus.
ROSAMARINA, Chydas.
ROSARNO, Medama.
ROSAS, Rhoda.
ROSELLO, Rusellæ.
ROSENBERG, Oleszno.
ROSETTE, Bolbitine.
ROSSA-JARIGATSH, Dromos Achilleos.
ROSSANO, Roscianum, Ruscianum.
ROSTOCK, Bunitium, Rostochium.
ROTH, Aurisium.
ROTIGLIANO, Rudiæ.
ROTTWEIL, Aræ Flaviæ, Rotovilla.
ROUDON, Rutunium.
ROUEN, Rotomagus.
ROUSSILLON, Ursolæ, Ruscino.
ROUWADDE, Aradus.
ROVEREDO, Roboretum.
ROVIGNO, Rivonium, Rovinum.
ROVIGO, Rhodigium.
ROW, Eractum.
ROXBURGH, Marchenium.
ROXBURGHSHIRE, Deviotia.
ROYAN, Novioregum.
RUAD, Aradus.
RUBIERA, Herberia.
RUFFACH, Aquæ Rubeæ.
RUGEN, Rugia.
RUGGA, Caraga.
RUMKALA, Urema.
RUNTERS, Contrum.
RUSSLAND, Russia.
RUSTINGEN, Hriustri.
RUVO, Rufrium.
RUVO, in the Papal states, Rubi.

S

SAADA, Saba.
SAALE, Sala.
SAAR, Saravus.
SAARBRUCK, Augusti Muri, Saræ pons.
SAARBURG, Caranusca.
SAAR-LOUIS, Arx Ludovici ad Saram.
SAATZ, Lucensis Circulus.
SABART, Colonia Sabrata.
SABATZ, Basiana.
SABBEA, Saba.
SABEIA, Jeracum Insula.
SABIONCELLA, Hyllis.
SABIOTE, Salaria.
SACCANIA, Argia.
SACCO, Casinus, Trerus.
SACLAS, Salioclita.
SADO, Catipos.
SADODELA, Edus.
SAFI, Risardir.
SAFNAS, Daphnæ Pelusiæ.
SAFURI, Diocæsarea.
SAGRIANO, Sagra.
SAGRO, Sagrus.
SAHAR, Sariphi montes.
SAHIRYE, Chebar.
SAID-GAZELLE, Synnada.
SAIN, Sena.
SAINTES, Mediolanum Santonum.
SAINT-JEAN-POUNGET, Vanesia.
SAINT-TIBERI, Araura.
SAJA, Saunium.

SAKARJA, Sangarius.
SALABRENNA, Selambina.
SALACA, Berenice Panchrysos.
SALAHIEH, Sele.
SALAMANCA, Elmatica, Ermandica, Salamantica.
SALANDRA, Acalandra.
SALANDRELLA, Acalandrus.
SALBAZAR, Halonæ.
SALÇES, Salsulæ.
SALDANNA, Eldana.
SALEHIEH, Sile.
SALEMI, Civitas Halicyensis.
SALEMJAT, Salamias.
SALEPH, Calycadnus.
SALERNO, Salernum.
SALERNO (GULF OF), Pæstanus sinus.
SALFATARA, Forum Vulcani.
SALHADJAR, Naucratis.
SALINA, Didyme, Evonymos.
SALINELLO, Helvinus.
SALINO, Sannum.
SALINS, Salinæ.
SALKHAT, Salcha.
SALM, Salmona.
SALNICH, Celidnus.
SALO, Longa Salina.
SALONA, Amphissa.
SALONA, Crissæus sinus.
SALONE, Jader.
SALONICA (GULF OF), Macedonicum mare.
SALONICHI, Thessalonica.
SALONICHI, Thermæus sinus.
SALOW-KEAH, Chidibbelensium Municipium.
SALPI, Salapia.
SALPI (LAGO-DI), Salapina palus.
SALSO (LAGO), Lacus Andurianus.
SALURN, Salurnis.
SALUZZO, Augusta Vagienorum, Salutiæ.
SALZACH, Jovavus.
SALZBURG, Juvavia, Salisburgum.
SAMALHUT, Co.
SAMALLUT, Cynopolis.
SAMARA, Panticapes.
SAMBRE, Sabis.
SAMBUL, Sambus.
SAMELAND, Lappia.
SAMMOLICO, Summus Lacus.
SAMOTHRAKI, Samothrace.
SAMSUN-KALESI, Priene.
SAMTHAURO, Seumara.
SAN-GERMANO, Casinum.
SAN-MATHEO, Incibili.
SANAGA, Nia.
SANCERRE, Sacrum Cæsaris.
ST. AGATA-DI-GOTI, Agathopolis.
ST. ALBANS, Fanum St. Albani.
ST. AMAND, Amandopolis.
ST. AMARIN, Amarinum.
ST. AMOUR, Fanum St. Amatoris.
ST. ANDEOL, Fanum St. Andeoli.
ST. ANDREE, Flavium.
ST. ANDREW'S, Andreopolis.
ST. ANGELO, Castrum St. Angelo.
ST. ANGELO, Angulum.
ST. ANGELO, Angelopolis.
ST. ANGELO-DE-LOMBARDI, Angelopolis ad Lombardum.
ST. ANGELO-DE-VADO, Angelopolis ad Metaurum, Tifernum Metaurense.
ST. ANTIOCO, Enosis.
ST. AUBIN-DU-CORMIER, Fanum St. Albini.
ST. AVO, Fanum St. Naboris.
ST. AY, Anisiacum ad Ligerum.
ST. BERTRAND, Bertranopolis.
ST. BLASIOS, Panope.
ST. BRIEUX, Briocense Oppidum.
ST. CATALINA, Insula Sanctæ Catharinæ.
ST. CATHARINA, Clunium.
ST. CATHARINA, Insula Catharinæ.
ST. CERE, Cæsareanum.
ST. CERGUE, Sanfredo.
ST. CHAMOND, Castrum St. Anemundi.
ST. CHRISTOVAL, Fortalitium St. Christophori.
ST. CLAUDE, Fanum St. Augendi.
ST. CLOUD, Fanum St. Clodoaldi.
ST. CROIX-DE-VOLVESNE, Vernosole.
ST. CRUX, Fanum St. Crucis.
ST. CRUZ-DE-LA-ZORZA, Vicus Caminarius.
ST. DAVID, Menevia.
ST. DAVID'S HEAD, Octapitarum.
ST. DENIS, Catolacum.
ST. DIDIER, Fanum St. Desiderii.
ST. DIE, Fanum St. Deodati.
ST. DIZIER, Dominicopolis.
ST. DOMINGO, Hispaniola.
ST. DONATO, Forum Appii.
ST. ESPRIT, Fanum Spiritus Sancti Vasconiæ.
ST. ETIENNE, Fanum St. Stephani.
ST. ETIENNE, Insula St. Stephani.
ST. EUFEMIA, Lameticus sinus.
ST. FANGON, Fanum St. Facundi.
ST. FE', Fanum St. Fidei.
ST. FELIX-DE-GUISOLA, Lunarium.
ST. FIORENZA, Canelata.
ST. FLORENTIN, Eburobrica.
ST. FOY-LA-GRANDE, Fanum St. Fidei.
ST. FRIQUE, Fanum St. Africani.
ST. GABINA, Lesa.
ST. GALL, Fanum St. Galli.
ST. GANNAT, Gannatum.
ST. GENAIS, Insula St. Genesii.
ST. GEORGE, Insula St. Georgi Majoris.
ST. GEORGIO, Morgentia.
ST. GERMAIN-EN-LAYE, Fanum St. Germani in Laja.
ST. GILLES, Fanum St. Ægidii Pictaviæ.
ST. GILLES, Fanum St. Ægidii.
ST. GIOAN, Arx St. Juliani.
ST. GIOVANNI, Cranii.
ST. GIOVANNI, Castrum St. Joannis.
ST. GIOVANNI-DI-BIDINI, Bidis.
ST. GOTTHARD, Alpes Lepontiæ, Alpes Summæ.
ST. GUELFO, Castrum St. Guelphi.
ST. GUILIN, Fanum St. Gisleni.
ST. HONORAT, Lerina.
ST. JAGO, Insula St. Jacobi.
ST. JAGO-DELLA-HIGUERA, Segeda Augurina.
ST. JAGO-DE-CASEM, Calipus.
ST. JAGO-DE-LEON, Fanum St. Jacobi Legionensis.
ST. JAGO-DE-VILLELA, Salaniana.
ST. JAGO-DI-CHILI, Fanum St. Jacobi Majoris.
ST. JAMES, Fanum St. Jacobi.
ST. JEAN, Stachir.
ST. JEAN-D'ANGELI, Angeriacum.
ST. JEAN-DE-BOURNAY, Turecionnum.
ST. JEAN-DE-GARGUIER, Gargarius.
ST. JEAN-DE-LAUNE, Fanum St. Joannis Laudonensis.
ST. JEAN-DE-LUZ, Fanum St. Joannis Luisii.
ST. JEAN-DE-MAURIENNE, Brennovicum.
ST. JEAN-PIE-DE-PORT, Fanum St. Joannis Peteportuensis.
ST. JOANNES-DE-LAS-BADESAS, Badesa.
ST. JOHANN-IM-TAURN, Sabatinca.
ST. JOHN'S POINT, Isamnium.
ST. JOSSE, Cella St. Jodoci.
ST. JUAN-DE-MEDUA, Nymphæum.
ST. JUAN-DE-PUERTO-RICCO, Fanum St. Joannis Portus Divitis.

ST. KIRILI, Galepsus.
ST. LEONHARD, Gabromagus.
ST. LIZIER-DE-CONSERANS, Austria.
ST. LO, Briovera.
ST. LOCAT, Castrum Octavianum.
ST. LORENZO, Graticula.
ST. LORENZO, Suasa.
ST. LUCAR-DE-BARRAMEDA, Fanum St. Luciferi.
ST. LUCAR-LA-MAYOR, Solia.
ST. LUCIÆ, Cenestum.
ST. MACAICE, Castrum St. Macarii.
ST. MAIXENT, Maxentii Fanum.
ST. MALO, Aletha Maclovia.
ST. MARCIN, Candalicas.
ST. MARCO-IN-LAMIS, Argentanum.
ST. MARGUERITA, Lero.
ST. MARIA-DE-RIBAREDONDA, Segesamunclum.
ST. MARIA-DELLE-GRACIE, Capua.
ST. MARIA-DI-LEUCA, Leuca.
ST. MARIA-FORCASSI, Forum Cassii.
ST. MARINELLA, Castrum Novum.
ST. MARTON, Fanum St. Martini.
ST. MAURA, Leucadia.
ST. MAUR-DE-FOSSEZ, Bagaudarum Castrum.
ST. MAURICE, Bergintrum.
ST. MAURICE, Acaunum.
ST. MENEHOULD, Fanum St. Menehildis.
ST. MESSIO, Ulyssis portus.
ST. MICHAEL, Castrum St. Michaelis.
ST. MICHAEL, Fanum St. Michaelis.
ST. MICHAELSTOWN, Fanum St. Michaelis.
ST. NICOLE-DU-PORT, Fanum St. Nicolai sc. Lotharingiæ.
ST. NIKLAS, Fanum St. Nicolai.
ST. OLOLA, Fanum St. Eulaliæ.
ST. OMER, Audomaropolis.
ST. OSPIZIO, Arx St. Hospitii.
ST. OUEN, Corobilium.
ST. PAPOUL, Fanum St. Papoli.
ST. PATRICK, Edrus.
ST. PAUL-DE-TROIS-CHATEAUX, Augusta Tricastinorum.
ST. PELINO, Corfinium.
ST. PHILADELFO, Haluntium.
ST. POELTEN, Fanum St. Hippolyti.
ST. PONS-DE-THOMIERES, Fanum St. Pontii Tomerianum.
ST. POURCAIN, Fanum St. Portiani.
ST. QUENTIN, Augusta Veromanduorum.
ST. REMI, Fanum St. Remigii.
ST. REMO, Fanum St. Remuli.
ST. RICQUIER, Centulum.
ST. RITI, Chalcidicus Mons.
ST. SEBASTIAN, Donastienum.
ST. SEBASTIAN, Fanum St. Sebastiani.
ST. SEBASTIAN, Morosgi.
ST. SEVERA, Cæretanorum Emporium.
ST. SEVERA, Pyrgi.
ST. SEVERINA, Siberene.
ST. SEVERINO, Septempeda.
ST. STEFANO, Cosanus Portus.
ST. THEODOR, Leuce.
ST. THOMAS, Insula St. Thomæ.
ST. THOMAS, Berzinus sinus.
ST. TROND, Fanum St. Trudonis.
ST. TROPEZ, Fanum St. Eutropii.
ST. URSANE, Fanum St. Ursicini.
ST. VINCENT-DE-BEIRA, Verurium.
ST. VITTORINO, Amiternum.
SANDARLIK, Myrina, Pitane.
SANDARLIK, Cymæus sinus.
SANGUENARES, Beleridæ Insulæ.
SANGUINARIA, Alesus.
SANSCOIN, Tinconcium.
SANTA-CATERINA, Phalarium.
SANTA-CROCE, Olympus.
SANTALOLI, Sambra.
SANTANDER, Fanum St. Andreæ, Andreopolis.
SANTAREM, Scalabis.
SANTAVER, Contrebia.
SANTEN, Castra Vetera.
SANTERNO, Badrinus, Vatrenus.
SANTILLANA, Fanum St. Julianæ, Julianopolis.
SANTIMBENIA, Nymphæus.
SANTONA, Victoriæ Juliobrigensium portus.
SANTORIN, Calliste.
SAONA, Savo.
SAONE, Segona, Arar, Saucona.
SAPAISCHE-PASSE, Symbolum.
SAPHAT, Bethulia.
SAPIGO, Sation.
SARABAT, Hermus, Pactolus.
SARACATIN, Siarum.
SARAGOSSA, Cæsaraugusta.
SARDARA, Aquæ Lesitanæ.
SARGEL, Canuccis.
SARIGOL, Rhode.
SARIONE, Lassira.
SARLIKI, Cranon.
SARNO, Sarnus.
SARRANT, Sartalis.
SARSINA, Bobium.
SART, Sardes.
SARUTHAN, Lydia.
SARVAR, Castriferrense Oppidum.
SARVIGNANO, Compitum Anagninum.
SARWITZ, Valdanus.
SAS-VAN-GEND, Agger Gandavensis.
SASAMO, Segisamo.
SASEF, Phthuris.
SASENO, Saso.
SASSOFERRATO, Juficum.
SATAL, Esepus.
SATAS-DERE, Æsepus.
SATZ, Zatecum, Satecium.
SAU, or SAVE, Savus.
SAUDRE, Sedera.
SAUER, or LA SURE, Sura.
SAULCOURT, Sodalcourt.
SAULIEU, Sidoleucum.
SAUMUR, Segora, Salmurium.
SAUVIGNY, Silviniacum.
SAVATO, Sabbatus.
SAVERNE, Sabrina.
SAVIO, Isapis, Sabis.
SAVONA, Sabata.
SAVOYEN, Sabaudia.
SAVUTO, Acheron.
SAXONY, Saxonia.
SBEKKAH, Cerbica Speculum.
SCALEMURA, Anemurium.
SCALONA, Ascalon.
SCAMANDRO, Scamandrus.
SCANDINAVIA, Scandia.
SCARDAMOULA, Cardamyle.
SCARLINO, Scabris portus, Salebro
SCARPANTO, Carpathus.
SCARPANTO, Carpathium mare.
SCATARI, Las.
SCAVEN-SHALE, Hunnum.
SCEMPSAT, Samosata.
SCHAERDING, Stanacum.
SCHAFHAUSEN, Scaphusia, Schafhusia.
SCHAGEPUR, Sageda.
SCHAIMA, Omana, Talmena.
SCHAMSER-THAL, Lapidaria Vallis.
SCHARAPANI, Sarapana.
SCHAREDSJE, Tharo.

SCHARMELY, Thilaticomum.
SCHARNITZ, Scarbia.
SCHEICH-EL-HARDY, Passalon.
SCHELDE, Scaldia, Scaldis.
SCHELLING, Actania.
SCHENET, Termessus.
SCHENIRUN, Samaria.
SCHENKENSCHANS, Aax Schenkiana.
SCHEREZUR, Siazur.
SCHERPENHEUVEL, Aspricollis.
SCHIBAM, Catabanum.
SCHINTAU, Stragona.
SCHLEGEL, Ad Plagam Beatæ Virginis.
SCHLESIEN, Silesia.
SCHLESWIG, Heideba, Sliesthorp.
SCHLETTSTADT, Selestadium.
SCHLINS, Celinum.
SCHMECHTEN, Acidula Schmechtana.
SCHONGAU, Esco.
SCHOTTWIEN, Fauces Noricorum.
SCHOUWEN, Scaldia.
SCHUSSENRIED, Abbatia Sorethana.
SCHUTT, Cituatum Insula.
SCHWAAN, Cygnea.
SCHWABEN, Suevia.
SCHWAB-MUNCHEN, Rapæ.
SCHWARZE AND ROTHE BERGE, Acabe.
SCHWARZE-HARUTSCH, Mons Ater.
SCHWARZ-WALD, Marciana, Nigra Sylva, Arnoba.
SCHWAZ, Masciacum.
SCHWEINFURT, Devona, Trajectum Suevorum.
SCHWEIZ, Helvetia.
SCHWERIN, Squirsina.
SCHWETZINGEN, Solicinium.
SCHWYZ, Suiza, Suitia.
SCIACCA, Ad Aquas.
SCICLI, Casmena.
SCILLA, Scyllæum.
SCILLY ISLANDS, Cassiterides Insulæ, Silurum Insulæ.
SCIO (the island), Chios.
SCIO, Chios.
SCOGLIO-DI-VIDO, Ptychia.
SCOPELLO, Cetaria.
SCOPELO, Peparethus.
SCOTLAND, Britannia Barbara, Scotia.
SCRIPU, Orchomenus.
SCUTARI, Chrysopolis.
SCUTARI (LAGO DI), Labeatis.
SEBAKET-BARDOIL, Sirbon.
SEBASKERES, Cnysius.
SEBBA-RUS, Tretum.
SEBEN, Sebatum.
SEBENICO, Sicum.
SEBSWAR, Hyrcania.
SECCHIA, Gabellus.
SECKAU, Secovium.
SECKINGEN, Sanctio.
SECKSSTADTE, Hexapolis.
SEDSCHESTAN, Drangiana, Anabon.
SEDSCHIDSCHIEK, Teos.
SEE-ALPEN, Alpes Maritimæ.
SEGARRA, Sigarra.
SEGNA, Senia.
SEGNI, Signia.
SEGORBE, Segobriga, Etobema.
SEGOVIA, Arevacorum Urbs.
SEGRE, Sicoris.
SEGURA, Tader.
SEGUTA, Synaos.
SEIBUSE, Rubricatus.
SEID-GAZA, Midæum.
SEIHAN, Sarus.
SEILLANS, Civitas Salinarum.
SELE, Silarus.
SELEFKIEH, Seleucia.
SELENGA, Oechardus.
SELENTI, Selinus.
SELIMBRIA, Peneus.
SELIM-EL-ASSOUAN, Meroe.
SELIVREA, Selybria.
SELLA, Salia.
SELLE, Sala.
SELTZ, Elizatium, Salatio.
SEMAGDA, Panagra.
SEMAVAT, Barsita.
SEMEH, Hierocæsarea.
SEMENDRECK, Samothracia.
SEMMENUD, Sebennytus.
SEMUR-EN-BRIONNOIS, Castrum Sinemurum.
SENAAR, Sirbitum.
SENEGAL, Daradus.
SENEZ, Civitas Saniciensium.
SENIO, Sinnius.
SENLIS, Augustomagus, Silvanectum.
SENN, Cæne.
SENS, Agedincum, Agendincum, Senones.
SEO, Sebinus Lacus.
SEO-D'URGEL, Orgellis Civitas.
SEPHARVAIM, Hipparenum.
SEPINO, Sæpinum.
SEPOLIA, Ceramis.
SERCHIO, Æsar, Anser.
SERES, Serræ.
SERETH, Ordessus.
SERFANDACAR, Epiphania Ciliciæ.
SERIA, or SAHARIDSJE, Seriane.
SERINAGUR, Cylindrina.
SERINGAPATAM, Brachmé.
SERIPHOS, Sarepta.
SERMIONE, Sirmio.
SERMONETA, Sulmo.
SERPA, Fabia Prisca Serpensis.
SERPHANT, Sarepta.
SERPHANTO, Acis.
SERPHEAT, Soriphæa.
SERRÆ, Ceres.
SERUK, Gerra.
SERVIA, Mœsia superior.
SESIA, Sessites.
SESSA, Aurunca, Suessa.
SESSE, Setia.
SESTINO, Sestinum.
SESTO, Sextium.
SESTRI, Segesta Tiguliorum.
SETENIL, Setia.
SETHIA, Cytæum.
SETIDA, Sala.
SETIF, Sitifis.
SETINES, Athenæ.
SETLEDJE, Hesydrius.
SETTENIL, Arx Septenilia.
SETUBAL, Cætobrix.
SEVENNEN, Gebenna, Gebennici montes.
SEVENNES, Cebennica sc. Regio.
SEVERN, Sabriana.
SEVEUX, Segobodium.
SEVIERSKY, Hyrgis.
SEVILLA, Hispalensis Provincia.
SEVILLA, Colonia Romulea, Hispalis.
SEVILLA-DE-L'ORO, Hispalis Aurea.
SEVILLA-LA-VIEJA, Colonia Italicensis.
SEZAME, Scingomagus.
SGIGATA, Rusicada.
SGURA, Caprus.
SHAMAKHI, Camechia.
SHANNON, Juernus, Scena, Senus.
SHAUR, Æni Insula.
SHEDUAN, Sapirene.

SHEPEY, Toliapis.
SHERBOURN, Clarus Fons.
SHETLAND ISLANDS, Hetlandicæ Insulæ, Æmodæ.
SHIBKAH-EL-LOWDIAH, Triton.
SHISSO, Naxos.
SHOA, Ammoniaca Regio.
SHOHBA, Dionysias.
SHONBRUNN, Fons Bellus.
SHONEN, Baltia, Scania, Scandia
SHREWSBURY, Salopia.
SIAM, Lestarum Regio.
SIB, Ecdippa.
SIBARI, Sybaris.
SICACOLA, Cocala.
SICKLINGEN, Sicila.
SICULIANA, Camicus.
SIDERA, Zephyrium.
SIDERO-CASTRO, Dorion.
SIDERO-KAPSA, Scydra.
SIDEROKAPSA, Bolbe.
SIDERONI, Scidrus.
SIDI-EMBARAK, Perdices.
SIDRA, Syrtis major.
SIEBENBURGEN, Dacia Mediterranea, Transylvania.
SIENA, Colonia Seniensis, Sena.
SIERRA-DE-CAZORLA, Tugiensis saltus.
SIERRA-D'OCCA, Segisama Julia.
SIERRA-DE-ORCAMO, Vasconum Saltus.
SIERRA-DE-RONDA, Arundæ Montes.
SIERRA-LEONE, Deorum Currus.
SIERRA-SEGURA, Argenteus Mons.
SIERRA-XIMIERA, Abyla.
SIFANTO, Meropia, Siphnus.
SIGEAU, Rubrensis Lacus.
SIGELLO, Helvillum.
SIGUENZA, Segontia.
SIGURO, Sycurium.
SIKI, Scylace.
SIKINO, Œnoe, Sicinus.
SIL, Silis.
SILARO, Silarum.
SILHET, Cirradia.
SILISTRIA, Dorostena, Dorostorum.
SIMANCA, Septimanca.
SIMARI, Semirus.
SIMEREN, Singas.
SIMMI, Syme.
SIMO, Doricus sinus.
SIN, Kades.
SINAAB, Tigauda.
SINANO, Megalopolis.
SIND, Sittocatis.
SINDSHAR, Singara.
SINGANFU, Sera.
SINIGAGLIA, Sena, Senogallia.
SINNO, Siris.
SINOP, Colonia Julia Augusta Sinope.
SINZIG, Sentiaca villa.
SIPICCIANO, Sepinum.
SIR, Jaxartes.
SIRA, Syros.
SIRAGOSSA, Syracusa.
SIRMAT, Armosata.
SIRMIA, Gephyra.
SIRNA, Cyrnos.
SISSECK, Siscia.
SISTERON, Civitas Segesterorum.
SITGES-VILLANUEVA AND SEGUR, Subur.
SITTART, Sitterium.
SITTEN, Castrum Sedunum, Sedunum.
SIWAS, Sebastia, Cabira.
SIWAS, Amasiæ Præfectura.
SKALA-NOVA, Neapolis.
SKANGERO, Scandile.
SKENI, Hypius.
SKIATHO, Sciathus.
SKINOSA, Schinussa.
SKIPSILAR, Scapte Hyle.
SKODAR, Scodra.
SKYE, Ebuda Orientalis.
SKYRO, Scyrus.
SLIOKAN, Staliocanus portus.
SLUIN, Stlupi.
SLUYS, Clausulæ.
SMYRNA, Meleti sinus.
SOBAD-KOH, Coronus.
SOBENIERES, Sipia.
SOCOTORA, Dioscoridis Insula.
SOEST, Susatum.
SOFIAN, Gauzania.
SOGD, Polytimetus.
SOHAR, Papiæ Insulæ.
SOHLENHOFEN, Cella Solæ.
SOISSONS, Sexoniæ, Augusta Suessonum.
SOKAR, Malichi Insulæ.
SOLFATARA, Campi Phlegræi.
SOLFELD, Flavia Solva.
SOLIA, Æpea.
SOLITO, Soletum.
SOLLIES, Solliniensium civitas.
SOLOTHURN, Salodurum.
SOLSONA, Celsona, Setelsis.
SOLTA, Olynta.
SOMME, Somena.
SOMMEPI, Fons Pigi.
SONNINO, Somnium.
SOPARIS, Compsatus.
SOPOTO, Photice.
SORIA, Numantia nova.
SORICH, Seretium.
SORISTAN, Syria.
SORRENTO, Surrentum.
SOSPELLO, Hospitellum.
SOULOSSE, Solimariaca.
SOUTHAMPTON, Clausentum.
SOUTH RONALDSHA, Ocetis.
SOUTHWOLD, Ad Pontem.
SPA, Aquæ Spadanæ, Tungrorum fons.
SPAIN, Hispania.
SPALATRO, Palatium Diocletiani, Spalatium, Aspalatos.
SPANISH SEA, Ibericum Mare.
SPELLO, Flavia Constans.
SPERLONGA, Spelunca.
SPERMADORI, Œnussæ.
SPEZIA, Tiparenus.
SPINO, Spina.
SPIRE, Augusta Nemetum, Spira.
SPIRNAZZA, Panyasus.
SPITAL, Lauriana.
SPITZBERGEN, Montes Acuti.
SPITZE v. CAMBOYA, Notium.
SPLUGEN, Cuneus Aureus.
SPLUGERBERG, Culmen Ursi.
SPREE, Spreha.
SPURNHEAD, Ocellum.
SQUILLACE, Scylaceum.
SSEIDA, Sidon.
SSUR, Tyrus.
STADTHAGEN, Civitas Indaginis.
STAEFFIS, Esteva.
STAFFORA, Iria.
STAGNO-DI-CAGLIARI, Caralitanus Sinus.
STAIN, Colatio.
STALIMENE, Lemnos.
STAMPALIA, Astypalæa.
STANCHIO, Cos.
STANDIA, Dia.

STANWICKS, Congavata.
STARIGRAD, Ortopla.
STAVRO, Stagira.
STEIN, Gaunodurum.
STEIN-AM-ANGER, Colonia Claudia Sabaria.
STENAY, Satanacum, Stanacum.
STERZINGEN, Stiriacium.
STETTIN, Sedinum.
STEYERMARK, Stiria.
STILLIDA, Phalara.
STILO, Cocintia.
STIRI, Stiris.
STIRLING, Mons Dolorosus.
STIVA, Cirphis.
STOBI, Stymbara.
STOCKHOLM, Holmia.
STOER, Sturia.
STOMBAR, Exonaba.
STRADELLA, Jella.
STRAITS OF BONIFACIO, Taphros.
STRAITS OF CONSTANTINOPLE, Bosporus Thraciæ.
STRAITS OF GIBRALTAR, Gadianum Fretum.
STRAITS OF KAFFA, or FEODOSIA, Bosporus Cimmerius.
STRALSUND, Sumonia.
STRAMA, Strymon.
STRAMULIPA, Bœotia.
STRANGFORD, Vinderius.
STRANKE, Turres.
STRASSBURG, in France, Argentina Civitas, Argentoratum.
STRASSBURG, in Prussia, Brodnica.
STRATIOTIKI, Munychia.
STRAUBINGEN, Acilia Augusta.
STRIDO, Stridon.
STROBEZ, Epetium.
STROBITZI, Lepreum.
STROFADIA, or STRIVALI, Plotæ Insulæ.
STROMBOLI, Strongyle.
STRONGOLI, Petelia.
STRONGYLA, Prepesinthus.
STRONGYLO, Apesantus.
STUHLINGEN, Juliomagus.
STUHLWEISSENBURG, Albanium, Alba Regalis.
STURA (KLEINE), Varusa.
SUAKEM, Suche.
SUBIACO, Sublaqueum.
SUCCADANA, Catigara.
SUDAG, Cytæa.
SUDER-SEE, Flevo Lacus.
SUEZ, Arsinoe.
SUGNEZ, Segni.
SUHHADJE, Crocodilopolis.
SUKUEH, Oriza.
SULCHEN, Solicinium, Sumlocennæ.
SULLY-SUR-LOIRE, Sulliacum.
SULMONE, Colonia Solmontina.
SULTAN-HISSAR, Trallis.
SULTANIEH, Batana.
SULZBACH, Salisso.
SUMPORT, Summum Pyrenæum.
SUMRE, Simyra.
SUND, Danicum Fretum.
SUNDGAU, Ferranus Comitatus.
SUNDJICK, Sindicus portus.
SUNTEL-GEBIRGE, Herculis Lucus.
SURA, Corsote.
SUSA, in Tunis, Cabar-Susis.
SUSA, Cottii Regnum.
SUSA, in Piedmont, Segusio.
SUSAM-ADASSI, Melamphylos.
SUSUGHERLI, Macestus.
SUTRI, Colonia Sutrina.
SWARTE-SLUYS, Clausulæ Nigræ.
SWEDEN, Suecia.
SYL, Rhabon, Tiarantus.
SYOUTH, Lycopolis.
SZALA and BENI-TAMER, Sala.
SZAMMERA, Hippos.
SZAMOS-UJVAR, Napoca.
SZELEPSA, Seldepa.
SZENT-BENEDICT, Fanum St. Benedicti.
SZENT-ENDRE, Ulcisia castra.
SZENT-GIORGY, Fanum St. Georgii.
SZENT-GROT, Fanum St. Gotthardi.
SZENT-JOB, Fanum St. Jobi.
SZENT-KESDY-LELEK, Fanum St. Spiritus.
SZENT-MARTIN, Fanum St. Martini.
SZENT-PETER, Fanum St. Petri.
SZERDAHELY, Hilicanum.
SZIR, Jazer.
SZOENY, Bregetio.

T

TABARCA, Tabraca.
TABARIA, Galilæum Mare.
TADCASTER, Calcaria.
TADMOR, Palmyra.
TAFA, Taphis.
TAFILETH, Talubath.
TAFNA, Siga.
TAG-AIAGHI, Zagrus.
TAGLIAMENTO, Tilaventum.
TAIBE, Tyba.
TAIMAKI, Thaumaci.
TAINEH, Colonia Ælia Augusta Mercurialis Thænit.
TAJO, Tagus.
TAKUSCH, Tacatua.
TALAVERA-DE-LA-REYNA, Elbora, Libora.
TALAVERA-LA-VIEJA, Evandria.
TALCINO, Talcinum.
TAMAR, Tamarus.
TAMARO, Tamarus.
TAMASA, Charieis.
TAMBRE, Tamara.
TAMERTON, Tamara.
TAMSWEG, In Muro.
TANARO, Tanarus.
TANASSERIM, Doanas.
TANGER, Tingis.
TANOAL, Dana.
TANTAN-VELHA, Sabanna.
TAOCHAN-ADASI, Lagussæ.
TAORMINA, Colonia Augusta Tauromenitana.
TAPTI, Goaris.
TARA, or TARO, Taras.
TARABOSAN, Trapezus.
TARANTAISE, Tarantasia.
TARBES, Castrum Bigorrense.
TARCHI-VOLICATI, Gyrton.
TARENT, or TARANTO, Tarentum.
TARIFA, Julia Jozas.
TARN, Tarnis.
TARO, Tarus.
TARRAGONA, Tarracon, Tarraco.
TARRAZONA, Turiaso.
TARSO, Tarsus.
TARTARO, Atrianus.
TARTURA, Dor.
TA-SAVA, Sabæ.
TATALIA, Talia.
TATRIA, Talaria.
TATZA, Ægissos.
TAUD, Tathis.
TAURIS, Gabris, Tigranaana, Gaza.

TAURO-CASTRO, Rhamnus.
TAVAY (the river), Ambastus.
TAVAY, Rhabana.
TAVETSCH, Ætuaticus vicus.
TAVIGNANO, Rhotanum.
TAVOLARA, Ermæa, Mercurii Insula.
TAXATA, Colonia.
TAY, Taus, Tavus.
TEANO, Teanum Sidicinum.
TECH, Tecum.
TECRIT, Birtha.
TEDELES, Saldæ.
TEFESSAD, Rusicibar.
TEGENGILL, Igenia.
TEGEWSEE, Tiges.
TEISSENDORF, Artobriga.
TEJADA, Ptucci.
TEKET, Panormus.
TEKIEH, Lycia.
TEKMAN, Hieron Oros.
TEKROVA, Phaselis.
TELAMONE, Telamon.
TELESE, Telesia.
TEL-ESSABE, Leontopolis.
TELGATE, Tollegatæ.
TELHOUE, Tarichea.
TELIGOL, Axiaces.
TELOWUNI, Hymettus.
TEMISCHBERG-AM-CUBAN, Surub.
TEMLUKE, Sigus.
TEMPI, Elephantaria.
TENASSERIM, Cotiaris.
TENERIFFE, Convallis.
TENES, Cartenna.
TENGEL, Cephalonnesos.
TER, Alba.
TERAMO, Interamna Palæstina.
TERANO, Interamna cognom. Lirinas.
TEREK, Alonta.
TERINA (GOLFO DI), Hipponiates sinus.
TERMED, Oxiana.
TERMINI, Himerenses Thermæ.
TERMOLI, Buca.
TERNI, Interamnia, Interamna.
TEROUENNE, Tarvanna.
TERPEZ, Portospana.
TERRACIA, Egara.
TERRACINA, Auxur.
TERRA-DI-BARI, Peucetia.
TERRA-DI-LAVORO, Campania, Terra Laboris.
TERRA-D'OTRANTO, Calabria, Aletina Provincia.
TERRA-NUOVA, Gela.
TERREMOTTO, Boagrius.
TERSACZ, Tarsium.
TERSAT, Tarsatica.
TERUEL, Turdetanorum urbs.
TERZA (LA), Fratruertium.
TESINO, Tesana.
TESSINO, Ticinus.
TEST, Anton.
TET, Ruscino, Telis.
TETRAGI, Lycæus.
TETUAN, Jagath.
TEUKIRA, Arsinone.
TEULADA, Tegula.
TEUTENHOF, Teutoburgium.
TEVERONE, Anienus.
TEZOTA, Arina.
TEZZUTE, Lambesa.
THABOR, Stabyrios.
THAK, Candace.
THALANDONISI, Atalanta.
THANA-WAR, Dagana.
THANET, Tanatis.
THEAKI, Ithaca.
THEIS, Pathyssus, Tibiscus
THELABBA, Thelabib.
THEMSE, Tamesis, Tamesa.
THENNIS, Hanes.
THERMIA, Cythnus.
THETEN, Matrica.
THETFORD, Sitomagus.
THIBA, Thebæ.
THIEL, Sitilla.
THIENGEN, Tenedo.
THIEVRES, Teucera.
THIN, Tegna.
THINE, Pelusium.
THIONVILLE. See DIEDENHOFEN.
THIRA, Sancterinum.
THOMASBRUCK, Aggeri pontum.
THORN, Thorunium, Thorunum.
THOUARS, Duracium, Toarcium.
THUE, Jadua.
THUR, Duria.
THURINGEN, Thuringia.
THYRSO, Caput Thyrsi.
TIEFENCASTELL, Imum Castrum.
TIEZ, Canasida.
TIFFANGE, Theiphalia.
TIFLIS, Tephlis.
TIGARI, Ticarius.
TILIOS, or TIOS, Tejum, Tium.
TILSIT, Chronopolis.
TIMAVO, Timavus.
TIMIA, Tinia.
TIMOK, Timachus.
TINEVELLY, Colchorum Regio.
TINNA, Thimonepsi.
TINO, Tenos.
TINTO, Iberus, Urium.
TINZEN, Tinnetio.
TIRAN, Phocarum Insula.
TIRCONEL, Conatia.
TIREH, Metropolis.
TIROL, Teriola castra.
TISARA, Diocæsarea.
TIVE, or TEIVE, Tuerobis.
TIVOLI, Tibur.
TIZ, Cyiza.
TOCAT, Berisa.
TODI, Tuder.
TODORIANO, Theudaria.
TOENESTEIN, Antonianæ Acidulæ.
TOLEDO, Toletum.
TOLOMETA, Ptolemais.
TOMAR, Concordia.
TOMBUCTU, Pessida.
TOMISWAR, Tomi.
TONGERN, Aduaca Tungrorum.
TONNERE, Tenodurense castrum, Tornodurum.
TONSICH, Docea.
TORBIA, Augusti Tropæa.
TORDA, Salinæ.
TORDERA, Larnum.
TORDINO, Macrinus.
TORETTO, Taurania.
TORGAU, Torgavia.
TORKOLA, Tauris Insula.
TORO (ISLAND OF), Boaris.
TORO, Sarabris.
TORQUEMADA, Augusta Nova, Porta Augusta.
TORRECILLAS, Turres.
TORRE-D'AGNAZZO, Egnatia.
TORRE-DE-PATERNO, Laurens Castrum.
TORRE-DELLE-SALINE, Salinæ.
TORRE-DEL-SAVIO, Sabis.
TORRE-DI-FIUMENICA, Paternum.
TORRE-DI-ORESTANO, Osaca.

TORRE-DI-PATRIA, Linternum.
TORRE-DI-POLLUCE, Selinus.
TORRE-DI-SENNA, Siris.
TORRE-DI-VADO, Voluterranavada.
TORRE-FORCADIZZO, Nora.
TORRE-SEGURA, Castellum Truentinum.
TORRE-TORRENTE, Turrus.
TORRE-VECCHIA, Vetulonia.
TORRES-VEDRAS, Arandis.
TORTONA, Dertona, Colonia Julia Augusta Dertona.
TORTOSA, in Catalonia, Dertosa.
TORTOSA, in the pash. of Terablus, Antaradus.
TOSA, Athiso.
TOSCANA, Etruria, Hetruriæ Magnus Ducatus.
TOSCANELLA, Ascania, Tuscana, Tuscia.
TOSTER, Susa.
TOUL, Tullum.
TOULON, Telo Martius.
TOULON-SUR-ARROUX, Telonnum.
TOULOUSE, Tolosa.
TOUR-D'ORDRE (La), Farum.
TOUR-DE-ROUSSILLON (La), Ruscino.
TOURNAY. See Dornik.
TOURNON, Tornomagensis vicus.
TOURNOUS, Tinurtium.
TOURS, Cæsarodunum, Turonium.
TOUS, Tutia.
TOVARRA, Turbula.
TOWY, Tobius.
TRACHENBERG, Dracomontanus Principatus.
TRACHENBERG, Dracomontium.
TRAEN, Drachonus.
TRAINA (the river), Cyamosorus.
TRAINA, Imachara.
TRAISMAUR, Trigisamum.
TRAJETTA, Minturnæ.
TRAJONISI, Caprarum Insula.
TRANI, Turenum.
TRAPANI, Drepanum.
TRAPANI-DEL-MONTE, Eryx.
TRASEN, Trigsamus.
TRAU, Tragurium.
TRAUN, Traunus.
TRAUN-SEE, Gemundanus Lacus.
TRAVANCORE, Cottiara.
TRAVE, Chalusus.
TRAVEMUNDE, Dragamuntina.
TRAVENDAHL, Treva.
TRAYGUERA, Teari Juliensis, Tiariula.
TREBISOND, Trapezus.
TREJA, Trea.
TREMITI, Diomedeæ Insulæ.
TRENTE, or Trient, Civitas Tridentina, Tridentum.
TRENTSIN, Singone.
TREVI, Augusta Treba, Treba.
TREVI, Trebiæ.
TREVICO, Trivicum.
TREVINNO, Beleja.
TREVISO, Tarvesium.
TREVOUX, Trevoltium, Trivultium.
TRIADITZA, Ulpia Sardica.
TRIBOLI, Tripolis.
TRIBUR, Triburia.
TRIBUXENA, Colobona.
TRICALA, Tricca.
TRICARICO, Trica.
TRICIO, Tricium Metallum.
TRIER, Augusta Trevirorum.
TRIESTE, Tergeste.
TRIEUX, Tetus.
TRIGARDON, Œniadæ.
TRIGNO, Trinium.
TRINKOMALE, Spatana.
TRIONTO, Træis.
TRIPOLI, in Asiatic Turkey, Ischopolis.
TRIPOLI, or Tarablus, Tripolis.
TRIPOLI, in Africa, Civitas Œensis.
TRIPOLITZA, Tripolis.
TRISTENA, Nemea.
TRITI, Tritæa.
TRIVENTE, Triventium.
TROIS MARIES (Les), Delphicum Templum.
TROJA, Æcæ.
TROJA, Ilium Novum.
TROJANERBERG, Cetius Mons.
TRONTE, Juvantius.
TRONTELLO, Trotilum.
TRONTO, Truentinus Fluvius.
TROPEA, Herculis Portus.
TROPPAU, Oppavia, Troppavia.
TROYES, Augustobona.
TRUXILLO, Castra Julia.
TRUZKOY, Gerrhus.
TSHALUMKA, Cucunda.
TSHATYR-DAGH, Berosus Mons.
TSHEPINI, Gargaria.
TSHERDIN, Czerdinum.
TSHERKASSEN, Sabira.
TSHERNEMBLE, Terponus.
TSHERSCHEMBI, Heracleum.
TSHERTSHEMIR-KUPRI, Rosologiacum.
TSHESHME, Zeugma.
TSHESKMEH, Erythrea.
TSHIL-MINAR, Persepolis.
TSCHONGRAD, Csongradiensis Comitatus.
TSHOROK, Isis.
TSHUMUS-DAG and Kudg-Dag, Temnus.
TSHURLU, Tzurulum.
TSHAUTSCHE-AGHISI, Posidium.
TSUCHPAREH, Caspatyrus.
TUBNAH, Tubunæ.
TUDDER, Theudurum.
TUDELA, Tulonium.
TUERTO, Astura.
TUISY, Diciacum.
TULCZE, Salsovia.
TULLEI-AUX-GROSSEILLES, Tusiacum.
TULN, Castra Catulina.
TUNIS, Tunes.
TUNSA, or Tuncza, Tonzus.
TURANO, Telonius, Tolenus.
TURCOMANA, Iran and Eriwan, Armenia Magna.
TURIN, Augusta Taurinorum.
TURKEY, Turcia.
TURKHEIM, Cælius.
TURTUKAI, Transmarisca.
TUSA, Halesa.
TUSLA, Doriscus.
TUSLAG, Tatta.
TUTTLINGEN, Dutlinga.
TUY, Tudæ.
TYBEIN, Duinum.
TYENA, Eusebia ad Taurum.
TYNA, Arimegon.
TYROS, Thyræa.
TZERNITZ, Zervæ.

U

UBEDA, Betula.
UDINE, Uthina, Utinum.
UDUBA, Turulis.
UEBERLINGEN, Iburinga.
UEBERLINGER-SEE, Acronius Lacus
UFFENTE, Ufens.
UGENTO, Uxentum.

ULM, Alcimoënnis, Ulma.
ULSTER, Hultonia, Ultonia.
ULTEN, De Ultimis.
UMBRIATICO, Brystacia.
UM-ES-SURRAB, Surratha.
UNISCH, Œnoe.
UNTERSEEN, Interlacus.
UNTERWALDEN, Subsylvania.
UNTER-ZELL, Cella Dei Inferior.
UNTER-ZELLERSEE, Venetus Lacus.
UPOIX, Epotium.
URAL MOUNTAINS, Hyperborei Montes, Norossus.
URBANIA, Urbinum.
URBINO, Urbinum.
URBISAGLIA, Pollentia.
URDINGEN, Hordeani Castra.
URK, Flevo Insula.
UR-MEZOE, Campus Dominorum.
URMI, Spauta.
URS, Ussubium.
URTZIANI, Phricium.
URUMIYA, Magrada.
USALET, Usaletum.
USBEKEN, Daæ.
USK, Castrum Oscæ.
USKUB, Scopi, Scupi.
UTENDE, Siatutanda.
UTRECHT, Trajectum ad Rhenum, Antonia.
UTTA, Susaleus vicus.
UZEN, Ozene.
UZES, Ucense Castrum, Ucetia.

V

VABRES, Vabrense Castrum.
VACCA, Bovenna.
VACUR, Cæsarea Mauretaniæ Cæsariensis.
VADO, Sabatia vada.
VADUTZ, Dulcis Vallis.
VAGNOLI (Li), Trifanum.
VAGY-VASONY, Vacontium.
VAISON, Aëria.
VALBACH, Valena.
VALDAHON, Velatudurum.
VAL-DI-CAMONICA, Camunorum Vallis.
VAL-DI-MONTIRONE, Campi Macri.
VAL-SUGAN, Euganea Vallis.
VALENCE, Civitas Valentinorum, Valentia.
VALENCIENNES, Valentianæ.
VALENTIA-DE-DON-JUAN, Palantia.
VALENZA, Forum Fulvii.
VALERA-LA-VIEJA, Nertobriga, Valoria.
VALGEN-SEE, Lacus Italicus.
VALLADOLID, Pintia, Vallisoletum.
VALLE-FREDDA, Frigida Vallis.
VALOIS, Valesia.
VALPAJELA, Vapanes.
VALPO, Jovallium.
VAN DIEMEN'S LAND, Diemeni Insula.
VANESERIS, Zalacum.
VANNES, Civitas Venetorum, Venetia.
VAR, Varum.
VARANA (Lago di), Germe.
VARAYE, Varadetum.
VARCERO, Colentum.
VARCILES, Varcilenses.
VARDAR, Axius.
VAREA, Varia.
VARESE, Baretium.
VARHELY, Augusta Dacica.
VARIGNANO, Claterna.
VARNA, Barne, Odessus.
VARS, Varcia.
VASTEO, Pitinum.
VASTO, Basta.
VASTO-D'AMMONE, Histonium.
VATISA, Polemonium.
VATISA, Phatisane.
VATKA, Bœa.
VAUCLUSE, Clusæ Vallis, Vallis Clausæ.
VAUCOULEURS, Lorium.
VAUDRE, Vodgoriacum.
VEDENE, Vindalium.
VEERE, Campiveria.
VEGEL, Besippo.
VEGJA, Baga.
VEGLIA (Island of), Curicta.
VEGLIA, Curictum.
VEINE, Davianum.
VELACHA, Voluce.
VELEY, Vellava.
VELEZ, Mænoba.
VELEZ-DE-GOMERA, Parietina.
VELEZ-MALAGA, Menoba.
VELIKA, Castra Varia.
VELINO, Velinus.
VELIZZA, Tithorea.
VELLETRI, Velitræ.
VELTLIN, Vallis Tellina.
VENAISIN, Comitatus Vindascinus, Avenionensis.
VENCE, Vintium.
VENDOME, Castrum Vindonicum, Vindocinum.
VENDRELL, Palfuriana.
VENE (Le), Sacraria.
VENEDIG, Venetiæ.
VENETICO, Lyncestius amnis.
VENLOO, Sablones.
VENOSA, Venusia.
VENTAS-DE-CAPARRA (Las), Capara.
VENTOLIERE, Bentoliera.
VENZONE, Vannia.
VERBASZ, Urpanus.
VERBICANO, Vergæ.
VERBOUSZKO, Varuaria.
VERCELLI, Vercellæ.
VERDEN, Ferda, Tuliphurdum, Fardium.
VERDON, Dila.
VERDUN, Viridunum.
VERGA, Barca.
VERGELLO, Vergellus.
VERIA, Berœa.
VEROCZE, Sorota.
VEROLI, Verulæ.
VERONA, Colonia Augusta Verona, Nova Gallieniana.
VERRES, Vitricium.
VERSIGLIA, Vesidia.
VERVIER, Verveviæ.
VERVINS, Verbinum.
VERZINA, Vertinæ.
VESCIANO, Bescia.
VESCOVATO, Episcopatus.
VESCOVIA, Episcopium.
VESRACH-REVI, Larissa.
VESUV, Vesuvius.
VETERI, Marcina, Macrina.
VEVAY, Viviscum, Bibiscum.
VEZELAY, Vizeliacum.
VEZZANO, Vitianum.
VEZZO, Vegium.
VIADANA, Vulturnia.
VIANA, Avarum Promontorium.
VIAREGGIO, Fossa Papiriana.
VICENZA, Vicentia.
VICH, Ausa.
VICHTACH, Bibacum.
VICHY, Aquæ Calidæ.
VICO, Elbii Vicus, Vicus Matrini.

VICO (Lago di), Elbii Lacus.
VICO-DI-SORENTO, Æqua
VICO-LAGO-DI, Ciminius lac.
VICTIMOLO, Ictimuli.
VID, Utum.
VIDIN, Bidinum.
VIENNA, Flaviana Castra, Vindobona, Vienna.
VIENNE, Vienna.
VIERRADEN, Ad Quatuor Rotas.
VIESTI, Apenestæ.
VIEUX, Civitas Viducassium.
VIGAN (Le), Vindomagus.
VIGANO, Vicus Serninus.
VIGNIANELLO, Junianellum.
VIGO, Vicus Spacorum.
VILCHES, Vescelia.
VILLADASO, Forum Statiellorum.
VILLA-DE-DOS-HERMANNOS, Orippo.
VILLA-DE-REYNA, Regiana.
VILLA-DEL-CARPIS, Onoba.
VILLA-DIEGO, Diegi Villa.
VILLA-FRANCA, Flavium, Bergidium Flavium.
VILLA-FRANCHE-DE-ROVERGUE, Francopolis.
VILLA-HERMOSA, Villa formosa.
VILLAINE (La), Vicinovia.
VILLA-LOYSA, Jonosia.
VILLA-NOVA-DEL-RIO, Canama.
VILLENA, Arbacala.
VILLOE, Vellejacium.
VINAY, Ventia.
VINCENNES, Ad Vicenas.
VINCENNES, Ad Vicenas Americanas.
VINDONITE, Vindenuta.
VINHAES, Veniatia.
VINKOVCZE, Cibalæ.
VINTIMIGLIA, Albium Intemelium.
VIRE, Olina.
VISAPUR, Musopale.
VISEGRAD, Cirpi.
VISO, Vesulus mons.
VISSA, Severus mons.
VISTRIZZA, Astræus.
VITERBO, Fanum Voltumnæ, Viterbium.
VITESLAU, Ægeta.
VITOLANO, Orbitanium.
VITRY-LE-FRANÇOIS, Victoriacum Francicum.
VITTORIA, Camarica, Suestasium.
VITULO, Œtylus.
VIVIERS, Albia Augusti, Vivarium.
VIZINDRUK, Barace.
VLADI-CAUCAS, Caucasiæ Portæ.
VLAERDINGEN, Flenium, Schiedamum.
VLICHA, Olpæ.
VLIE, Fossa Corbulonis.
VLIELAND, Flevolandia.
VLIESSINGEN, Flesinga.
VOBARNO, Voberna.
VOCONC, Fanum Vacunæ.
VOGELBERG, Avicula.
VOGESEN, Vogesus.
VOGHERA, Iria.
VOGTLAND, Variscia.
VOINITZA, Anactoria.
VOJUSSA, Æas.
VOLANO, Volenes.
VOLCANO, Hiera.
VOLKELMARKT, Gentiforum.
VOLTERRA, Volaterræ.
VOLTURNO, Athurnus, Volturnus.
VOMANO, Vomanus.
VONC, Vungo.
VORBURG, Forum Adriani.
VORST, Veruca castellum.
VOSCOPOLI, Uscana.
VOSFOR, Nymphæa.
VOSTIZZA, Ægium, Egea.
VOUGA, Vacua.
VRAITA, Fevus.
VRANA, Brauron
VUNO, Phanote.
VYZIA, Bizya.

W

WAAG, Cusus, Vagus.
WAAL, Nabalia, Vahalis.
WADI-EL-ARISCH, Ægypti fluvius.
WADI-EL-BERBER, Tusca.
WADI-EL-WALE, Aqua Dimonis.
WADI-QUAMA, Cinyps.
WADI-ZERKA, Jaboch.
WADY-GHARENDEL, Elim.
WADY-MEHARRAKA, Hiera Sycaminos.
WADY-SZYR, Jazer.
WAERDER, Lauri.
WAIHEND, Arachosia.
WAIHEND, Arachotoscrene.
WAIN, Viana.
WAKR, Chabinus.
WALCHEREN, Gualacra, Valachria.
WALD, Gabreta Sylva.
WALDAI MOUNTAINS, Alani montes.
WALDMUNCHEN, Seguacatum.
WALES, Cambria, Britannia Superior, Vallesia.
WALLINGFORD, Calcua.
WALLSEND, Legedunum.
WAN, Arcissa.
WAN, Mantiana Palus.
WANGEN, Vemania.
WARDEIN, or Gross-Wardein, Varadinum.
WAREN, Virunum.
WARNE, Suebus.
WARSAW, Varsovia.
WARWICK, Calunia, Præsidium Cornaviorum.
WASCH, Bascatis.
WASH (The), Metaris Æstuarium.
WASITH, Cybate.
WASSERBILLICH, Aquæbellicus.
WASSERBURG, Aquaburgum.
WASSERLEONBURG, Santicum.
WEICHSEL, Vistula.
WEICHSELMUNDE, Munda Vistulæ.
WEILHEIM, Cambodunum.
WEISSENAU, Alba Augia Naviscorum.
WEISSENBURG, Alba Selusiana.
WEISSENBURG. See Albe Julie.
WEISSENFELS, Leucopetra.
WEISSERITZ, Albula.
WEISSES MEER, Album Mare.
WEIUNG, Viana.
WELKA, Fulfinium.
WELLS, Fontes Belgæ.
WELS, Falciana, Colonia Aurelia Antoniana Ovilia.
WELTEN, Veldidena.
WELTENBURG, Attobriga.
WELZHEIM, Valentia.
WENERSBORG, Bretta.
WENSUSSEL, Burgla.
WERDEN, Moradunum.
WERI, Hwerenofeldia, Genewara.
WERMELAND, Varmia.
WERNITZ, Biriciana.
WERTACH, Vinda.
WERWICK, Viroviacum.
WESEL, Aliso, Vesalia.
WESEN, Guesta.
WESER, Visurgis.

WEST-BOTHNIA, Bothnia Occidentalis.
WESTCAPPEL, Fanum Herculis Magusani.
WESTERWALD, Burnonia Sylva.
WESTPHALIA, Guestfalia.
WEXFORD, Manapia.
WHITE, Elæusa.
WHITHORN, Candida Casa, Leucophibia.
WIBLINGEN, Ebilingua.
WICK, Ilea.
WICKEN, Vicus.
WIDDEN, Biminacium, Viminacium.
WIENER-WALD, Bojorum Deserta.
WIESBADEN, Aquæ Mattiacæ.
WIESELBURG, Flexum.
WIGHT (ISLE OF), Vecta, Vectis.
WILHELMSTADT, Guilielmostadium.
WILLERING, Hilaria.
WILLOUGHBY, Verometum.
WIMPFEN, Cornelia, Wimpina.
WINCHESTER, Venta Belgarum, Vintonio.
WINDISCH, Vindonissa.
WINDISCHGRATZ, Slavogræcium.
WINDSOR, Vindesorium.
WINTERTHUR, Vitodurum.
WINZENDORF, Alvincium.
WIPACH, Frigidus.
WIPPER, Viadus, Viadrus.
WIRM, Verna.
WISMAR, Marionis.
WISSANT, Iccius Portus.
WISSEMBURG, Selusium.
WITTENBERG, Vitemberga, Witteberga, Leucorea.
WOERINGEN, Buruncum.
WOLFENBUTTEL, Guelferbytum.
WOLGA, Rha.
WOLLIN, Fanesiorum Insula.
WOLLIN, Julinum.
WOODCOTE, Noviomagus.
WORCESTER, Brangonia, Vigornia.
WORMS, Augusta Vangionum, Wormatia.
WORMSER-BAD, Bormianæ Thermæ.
WOSTANI, Ægyptus Media.
WRIETZEN, Viritium.
WROXETER, Uriconium.
WURZBURG, Artaunum, Herbipolis, Wurceburgum.
WUTACH, Juliomagus.
WYK-DE-DUURSTEDE, Batavodurum.

X

XALON, Bilbilis.
XANTEN. See SANTEN.
XELSA, Celsa.
XERES-DE-CAVELLEROS, Soria.
XEREZ-DE-LA-FRONTERA, Asta Regia.
XIGONZA, Saguntia.
XILOCASTRO, Ægira.
XIRIA, or CHELME, Erymanthus.
XUCAR, Sucro.

Y

YARE, Garienis.
YARMOUTH, Garianonum.
YARUM, Girvium.
YEMEN, Amania.
YENNE, or HYENNE, Cauna.
YEPES, Ispinum.
YEURE-LE-CHATEAU, Euræ Castrum
YGUALADA, Aqualatum.
YIL, Hyla.
YLST, Ilostum.
YNIESTA, Egelasta.
YONNE, Icauna.
YORK, Belgium Novum.
YORK, Colonia Eboracensis.
YPERLE, Hiscar, Ypra.
YPERS, or YPERN, Ipra.
YPOL, Ibola.
YS-CAP, Glaciale Promontorium.
YSENDIK, Isendicum.
YSSEL. See ISSEL.
YSTADT, Istadium,
YUCATAN, Jucatania.
YUNTO, Tagrus.
YVOIRE, Aquaria.
YVOIX, Cariniacum.
YVOIX, Epoissus.

Z

ZABERN. See BERG-ELSASS-RHEIN-ZABERN.
ZADAICA, Meninx.
ZAFRA, Zephyrium.
ZAHARA, Lastigi.
ZAINAH, Diana.
ZALAMEA-DE-LA-SERENA, Julipa.
ZALONGOS, Charadrus.
ZAMBRONA, Hipponum Promontorium
ZAMORA, Ocellodurum.
ZANNONE, Sinonia.
ZANTE, Zacynthus.
ZARACHA, Beligrada.
ZARA-VECCHIA, Jadera.
ZARAVA, Ezra.
ZARCO, Pharcadon.
ZARI, Cambyses.
ZARNOWICE, Carrodunum.
ZARTEN, Tarodunum.
ZAXOSULTAN, Zautha.
ZEA (ISLAND OF), Cea.
ZEA, Cæa, Cea.
ZEBEN, Cibinium Minus.
ZEHNGERICHTEN-BUND, Fœdus Decem Jurisdictionum.
ZEILA, Aromata.
ZEITUN, Maliacus sinus.
ZEITZ, Citium.
ZELEBI, or SELEBY, Zonobia.
ZELL, Cella Badensis.
ZELL, Cella Franconica.
ZELLA, Cella St. Blassii.
ZELL-AM-HERMERSBACH, Cella Rudolphi.
ZELL-AM-ZELLER-SEE, Cella ad Lacum Inferiorem.
ZELL-AN-DER-PRAMB, Cella Austriæ superioris.
ZELL IM HAMM, Cella ad Mosellam.
ZELL-IM-ZILLERTHALE, Cella Tirolonsis.
ZELL-OB-DER-IPS, Cella ad Isim.
ZELL-UBER-DER-TRASEN, Cella Austriæ Inferioris.
ZELLERSEE, Lacus inferior.
ZELLETI, Cilla.
ZELSENHEIM, Elcebus.
ZEPT, Cilicenæ.
ZERBI, Meninx.
ZERBST, Servesta, Servestum.
ZERENI (GOLF DE), Laturus sinus.
ZERMANIA, Tedanius.
ZERNA, Siracella.
ZERNITZ, Zervæ.
ZETERAMBOL, Germanicopolis.
ZEUTUNI, Lamia.
ZICKA, Mogetiana.
ZIEGENRUCK, Capræ Dorsum.

ZIEL, or ZILE, Zela.
ZINARA, Cinara, Cynarus.
ZIPS, Cepusiensis Comitatus.
ZIPSERHAUS, Arx Scepusiensis, Cepusium.
ZIRL, Cireola.
ZIRO, Crimisa.
ZIZAERIS, Diacia.
ZIZERS, Ciceres.
ZOMPOR, Zurobara.
ZORBIG, Sorbiga.
ZOTRIWAR, Cithrum.
ZOTTENBERG, Sabotus mons.
ZOWAMOORE, Ægimurus.
ZOWARIN, Zama.
ZOXO-SULTAN, Gordiani Monumentum.
ZSCHINER-STUHL, Cskiensis Sedes.
ZUBA, Modin.
ZUCALA, Isthmus Tauricus.
ZUG, Tugium.
ZULCH, Cilicia.
ZULPICH, or TULPICH, Tolbiacum.
ZUNCHIO, Epitalium.
ZURICH, Tigurum.
ZURZACH, Certiacum, Forum Tiberii.
ZUYD-SCHANS, Arx Austrina.
ZWADENBURGERDAMM, Niger pullus.
ZWETL, Clara Vallis.
ZWEYBRUCKEN, Bipontium, Bipontum.
ZWICKAU, Cygnea.
ZWIEFALTEN, Duplices Aquæ.
ZYDOWO, Setidava.

ARTICLE III.

ON THE PRONUNCIATION OF MODERN GEOGRAPHICAL NAMES

*** This article is mainly a reprint of the excellent Introduction to 'Lippincott's Pronouncing Gazetteer,' a work of great merit, published at Philadelphia, U. S., in 1855. We have also employed in its compilation the remarks attached by Professor Porter of Yale college to his 'Pronouncing Vocabulary of Modern Geographical Names.'

I. GEOGRAPHICAL names may properly be divided into two great classes. The first division comprises those which, in their native language, are written either in Roman letters, or else in characters (like the German and Greek) which can readily be converted into corresponding Roman letters. The mode of spelling such names is generally uniform, and the same in each of the different European languages. Thus, ETNA or ÆTNA—the name by which the great volcanic mountain of Sicily is known to the Italians—is also the English, French, German, Dutch, Spanish, and Portuguese name for the same mountain. So LAYBACH or LAIBACH, the capital of Illyria, is not only the German, but the English, French, Italian, and Spanish name. Thus, also, YORK, a city of England, is not merely the English name, but the French, German, Dutch, Italian, Spanish, &c. The same rule holds true with regard to a large majority of the names of Western and Southern Europe.

The other division of geographical names embraces those which, in their native language, are neither written in Roman letters, nor in characters that can be converted into corresponding Roman letters: such are the Russian, Turkish, Arabic, Persian, &c. Names of this class, with few if any exceptions, are written differently in the different European languages, each nation aiming to express the *sound* or pronunciation of the foreign name by the letters of its own tongue. Thus, for example, an Englishman visiting the capital of Fezzan, in Northern Africa, and wishing to indicate the sound of the name as pronounced by the inhabitants, would naturally write it MOORZOOK, or adopting the Italian pronunciation of the vowel *u*, generally followed in foreign names in the body of this work, MURZUK; a Frenchman would write MOURZOUK; a German MURSUK; these various spellings being intended to represent precisely the same sound. Again, if an Englishman wished to represent the native pronunciation of a certain city of Persia, he would write it SHOOSTER or SHUSTER; a Frenchman would spell it CHOUSTER; a German, SCHUSTER; an Italian, SCIUSTER; a Portuguese, CHUSTER or XUSTER, &c.; these being, in fact, the correct spellings of the above name in those different languages respectively.

This mode or rule of writing African and Oriental names is, generally speaking, very strictly adhered to by the greater number of French and German writers;[1] but, unfortunately, the English geographers have, for the most part, pursued a very different course. Instead of conforming to a rule which has the double merit of being simple and easy for the writer,[2] and clear and satisfactory to the reader, they have, by sometimes writing in the French and sometimes in the German mode, and not unfrequently combining the two in the same name, involved the department of Oriental geography in a confusion which is most perplexing to all, and is absolutely inextricable to those who have not made this subject one of long and laborious study. Many of the foreign spellings found in the writings of English geographers have doubtless been taken from French and German works, the copyist neglecting to translate the names, while making a translation of the rest of the book.

II. It is obviously an immense advantage, in an English work intended for general readers, to write names in such a manner that they can without difficulty be pronounced correctly by the ordinary English scholar. Another strong argument in favour

1 It is worthy of remark that the French not only spell the names of those Asiatic countries which have not submitted to European domination—as Persia, China, &c.—according to the rule above given, but even the names of British India, although these might be considered, in some sense, as belonging to the same class as those of Western Europe. Accordingly, we have not merely

French.		English.	German.
Badakhchan,	for	Badakshan,	Badakhschan.
Beloutchistan,	„	Beluchistan,	Beludschistan.
Caboul,	„	Cabul,	Kabul.
Chiraz,	„	Sheeraz or Shiraz,	Schiras.
Kirmanchah,	„	Kermanshah,	Kermanschah.
Mechhed,	„	Meshid,	Mesched.
Recht,	„	Reshd,	Rescht.
Thian-chan,	„	Thian-shan or Tian-shan,	Thian-schan.
Chan-si,	„	Shan-see or Shan-si,	Schan-si.

but we also find in the most popular French geographical works, examples like the following:—

French.		English.
Adjmir,	instead of	Ajmir or Ajmere.
Djelalabad,	„	Jelalabad.
Djeypour,	„	Jeypur.
Djessalmir,	„	Jessulmir.
Djoudpour,	„	Jodpur or Judpur.
Hindou,	„	Hindu.
Haïderabad,	„	Hyderabad.
Hougli	„	Hoogly or Hugli.
Maïssour,	„	Mysore.
Pounah,	„	Poonah or Punah.

It may be remarked that the French names in the foregoing list, with one or two exceptions, represent the same sound as the English, and are evidently only a French translation of the English name. It should be further observed, that, while the German geographers spell the names of other parts of Asia (see the last column in the first of the foregoing lists) according to the principle above stated, (that is, representing the *sound* of the foreign name by the letters of their own language.) they generally write the names of British India after the English mode; as, HYDERABAD, JEYPUR, MYSORE, &c.

2 Of course, reference is here made to an English traveller writing the name for the first time. It certainly would be quite as easy, to say the least, for an Englishman to represent any given sound in such a name by the letters of his mother tongue as by those of a foreign language.

of the use of English orthoepy in such cases, is that it is capable of conveniently expressing or representing a greater variety of sounds than any other European language. There is no sound, we believe, that is extensively used in oriental names which we cannot express as well as the French, and there are several which we can express much better than they; while there are some sounds, perfectly familiar to our tongue, which they cannot express at all. Take, for example, the sound of our *j*—a very common one in the names both of Asia and Africa: what we express by a single letter is indicated less perfectly in the French language by two—*dj*, (as, *Djidda* for JIDDA, *Djoulamerk* for JULAMERK, &c.;) so, also, the sound of *ch*, one of continual occurrence both in the names of Asia and Eastern Europe, is represented in French by three letters—*tch*, (as *Tchanda* for CHANDA, *Tchambal* for CHUMBUL, &c.) True, the English often use (especially at the end of a syllable) *tch* in order to denote the same sound; it is, however, important to observe that this sound, as well as that of *j*, is one of the most common and familiar to the English tongue, while both are foreign to the French language, since neither of them is to be found in any genuine French word. Again, our *w* expresses a sound (common in the Oriental languages) which is not nearly so neatly nor so well expressed by the French *ou*, the latter being the proper equivalent of our *oo* or the Italian *u*. This defect is so obvious, that some of the most eminent French writers, (Pauthier, for example, in his works on China) make use of the English *w* in writing certain foreign names. Thus instead of employing *ouou* to express the sound of *wu* or *woo*, they write *wou;* instead of *ouen*, *wen*, &c. Lastly, there are sounds expressed in our tongue with the utmost facility which French orthoepy cannot represent at all; among these are the sounds of the Greek θ (*th*) and δ (*d*) common in Turkey and the Grecian islands: the former exactly corresponding to our *th* in *thin*, the latter to *th* in *this*. With respect to the German language, there is, if we mistake not, but one frequently occurring sound in oriental names (that of *kh*—represented by the German *ch*) which can be represented by it better than by the English, while there are many which can not only be expressed in English more conveniently, but more correctly, than in German. Thus, the Germans employ four letters (*dsch*) to indicate the sound of our *j*, and, after all, represent it most imperfectly; for example, they write *Dschulamerk*, *Dschidda*, *Dschilolo*, for JULAMERK, JIDDA, and JILOLO. Their four letters *tsch* do not represent correctly the sound of our *ch*, nor does their *sch* convey even a tolerable idea of our *zh*, (that of *s* in *pleasure* or *occasion*.) Like the French, they have no letter or combination of letters equivalent to our *w*, nor can they in any manner represent the sound of the modern Greek θ or δ.

Another mode of writing oriental names, now adopted by some of the best English geographers, and generally followed in the 'Journal of the Royal Geographical Society,' and in the present work, is to employ the English consonants in conjunction with the Italian vowels, those vowels which have a long or full sound being generally marked with an accent, thus: *Shapúr*, for SHAPOOR or SHAPOUR; *Tabríz*, for TABREEZ, &c. It is to be observed, however, that any general rule for writing Oriental names is, or ought to be, somewhat modified by those great practical laws of language, founded on usage, which overrule all considerations of mere theoretical propriety; and that it is not desirable to change any spelling which universal usage appears to have established. Not only ought the common English names of the great countries of Asia and Africa (as PERSIA, HINDOSTAN, CHINA, EGYPT, &c.) to be retained, but the ordinary spelling of names of far less note, when sanctioned by the universal, or almost universal, practice of the best English writers, should be adhered to. Thus, we still write OUDE, and not *Ood;* DELHI, and not *Delhee* or *Delli;* CAIRO, not *Kahira*, &c. In short, it is only when there appears to be no settled usage among geographical writers, that we may take the liberty of selecting out of many forms that which we deem preferable to the others.

III. It is scarcely necessary to dwell upon the essential importance of pronunciation, since it enforces its own claims upon "all who talk or read." In fact, the need of some uniform system of geographical pronunciation appears now to be universally felt and acknowledged; and geographers seem to be nearly unanimous in opinion, that the only rational system of geographical pronunciation is that which is based on the principle of pronouncing all names of places as nearly as possible as they are pronounced by the educated people of the respective countries to which they belong, with the exception of those few well-known foreign names which appear to have acquired a fixed English pronunciation, as PARIS, NAPLES, FLORENCE, VENICE, MUNICH, &c.; these exceptional names being pronounced according to the usage of the best English speakers.

It is admitted that cases not unfrequently occur, in which it is impossible to convey, with any great degree of precision, the native pronunciation of other countries by means of English letters; but something is undoubtedly gained by such an approximation to the true sound, as would enable one more readily to understand, and to be understood by those who are familiar with the names of places as spoken by the inhabitants themselves. Some, indeed, have maintained the propriety of pronouncing foreign names as they are written, giving to every letter its proper English sound. But such a system would obviously lead to the greatest confusion, and be attended with inextricable difficulties. What, for example, would be the proper English pronunciation of *Seine?* Should the *ei* be pronounced like *ee*, as in the words *seize*, *ceiling*, *receive*, &c.? or like *ai*, as in *vein*, *weight*, *inveigh;* or like *i* long, as in *height*, *sleight*, &c.? Should SEINE then be pronounced *seen*, *sain*, or *sine?* or should we sound the final *e*, and make it *see-nee*, *sai-nee*, or *si-nee?* This one instance, out of a multitude, may perhaps serve to show the endless diversity and confusion into which such a system, or rather want of system, must of necessity lead. But this is not all; there are innumerable cases wherein it is very difficult, if not impossible, to pronounce the names of other countries according to the English sound of the letters, *e. g.*, CZERNIGOW, CSONGRAD, HJELMÄR, HJORING, LJUSNE, SZEGEDIN, TJIRINGIN, &c., while there is no difficulty whatever in pronouncing them according to the native sound. A multitude of instances also occur in which the English manner of pronouncing names, though not difficult, is far less euphonious than that of the country to which such names belong. MINHO, (meen'yo,) a river, and BATALHA, (ba-tal'ya,) a town of Portugal, and BACCHIGLIONE, (bak-keel-yo'na,) a river of Italy, may serve as examples.

The acquisition of ten or twelve new sounds, which might be easily learned by persons of ordinary aptitude in a few hours, would enable any one, who can read correctly the English pronunciation as marked in Walker's or Worcester's Dictionary, to pronounce with tolerable correctness all the names of Portugal, Spain, Italy, France, Belgium, the Netherlands, Germany, Denmark, Sweden, Norway, and perhaps of Hungary. The difficulties in regard to the names of Poland, Bohemia, and the Slavonian countries generally, are confessedly much greater—but of this hereafter. Persons who view the different European languages separately, are apt to regard the mastering of the difficult sounds in all, as a much more formidable task than it really is. They forget that a large proportion of the most difficult sounds are common to several different languages, and after being once thoroughly learned, can, of course, present no further obstacle. Of this class is the French and Dutch (or Flemish) *u*, equivalent to the German and Hungarian *ü*, and to the Danish, Norwegian, and Swedish *y*. Here it will be seen that the same sound, intermediate to the English *ee* in *feet*, and the *oo* in *book*, occurs in SEVEN different languages. The German *œ* or *ö* is likewise found in Danish, Norwegian, Swedish, and Hungarian, and nearly corresponds to one of the most difficult Dutch sounds, that of *eeu* as in LEEUWARDEN; the German *ch* in *ach* or *noch* is almost exactly equivalent in sound to the Spanish *j* or *x*, and is essentially the same as the Polish, Dutch, Scotch, and Welsh *ch*, and the Dutch *g*.

It is worthy of remark that in all the thousands of Italian names, there is not a single sound which an Englishman cannot utter with perfect ease. We do not mean to deny that there are combinations of letters, as *cia* (pronounced *cha*,) *giu*, (pronounced *joo* or *ju*,) and *sciu*, (pronounced *shoo* or *shu*,) which, without any explanations, might appear difficult to the mere English scholar; but when these combinations are represented by their proper English equivalents, there is obviously no difficulty whatever. What has been said of Italian, is substantially true of Portuguese; the nasal sounds in such words as *alem*, *sao*, though having no exact equivalent in English, are by no means difficult for an Englishman to acquire; and, indeed, unless great nicety be called for, might very well be supplied by the English *ng*, (thus *alem* might be pronounced a-leng′, and *sao*, sowng.) In Spanish there is but one difficult sound, that of *j* or *x*, corresponding, as already remarked, to the German guttural *ch*. So that by the acquisition of one, or, at most, of two foreign sounds, an Englishman or American will be enabled (if the pronunciation be properly marked) to pronounce correctly the fifty thousand or more names of Italy, Portugal, Spain, Brazil, and Spanish America. It would, we think, be well worth while to give the pronunciation of these names, even were it impossible to give correctly those of any of the other European countries. But no such impossibility exists. With the exception of the Slavonic languages, the pronunciation of the French is by far the most difficult for the English scholar to acquire. Yet any child, with an average capacity for acquiring language, will master all the French sounds in a few lessons. The German comes next in point of difficulty, and after it scarcely any thing more remains to be done so far as regards the languages of Western and Northern Europe. As to the Slavonic languages, we admit that a number of the Polish and Bohemian names are absolutely unpronounceable; but as Poland is a part of Russia, we might with propriety adopt in such cases the Russian name and pronunciation, which, with few exceptions, are sufficiently easy for the English speaker. So, also, if we find it impossible to pronounce some of the names of Bohemia, Galicia, &c., we might in like manner substitute the German names in those instances—the German being the official language of the Austrian government, to which Bohemia and Galicia belong. With regard, however, to the languages of Russia, Turkey, &c., the names of the places in these countries, as we meet with them, being not really Russian or Turkish names, but only French, German, or English transcripts of those names, there is often much greater difficulty in determining the correct spelling and pronunciation. Even scholars the most thoroughly acquainted with those languages, often feel uncertain as to the best mode of representing names of this class in the languages of Western Europe. Hence it follows that it is next to impossible, in such cases, to attain that correctness, or exactness, in spelling, and precision in marking the pronunciation, which can be attained with comparative ease in names written in Roman or German characters. Of the topographical names of Great Britain and the United States, it may be asserted that while in the fixedness of the spelling they resemble those of Spain, France, &c., in difficulty of pronunciation they almost exceed those of Russia or Poland. This characteristic of English and American names is evidently to be attributed to the introduction of new and incongruous foreign elements into the language, after this has been to a certain extent formed and fixed. The existence in the same country of several entirely distinct nations, Welsh, Saxons, Danes, &c., each tenacious of its own natural peculiarities, which have been in a measure protected and fostered by the freedom of the government, has doubtless been the principal cause of the great irregularity and incongruity of the English language. We find, indeed, as regards English names, scarcely one general rule of pronunciation, either in relation to accent or to the sounds of the letters on which we can with confidence rely: the actual practice of the best speakers is, therefore, our only safe guide. Anomalies similar to those which prevail in England, (though modified by varying circumstances,) are not unfrequently to be met with in the United States. TERRE BONNE, (pronounced *tar bōn* or *tar bonn*,) NATCHITOCHES, (usually called *nak-e-tŭsh′*,) TERRA HAUTE, (*ter′rah hōt*, *tar hōt* or *ter′rah hŭt*,) and SAULT SAINT MARIE, (*soo sent ma′ree*,) are remarkable examples of this class. Settled originally and named by the French, although afterwards chiefly inhabited by those speaking English, the names of those places are neither English nor French, nor even a mixture of the two, but a strange corruption produced, it would seem, by each of the American settlers imagining that if he was careful to utter a sound which none could suspect of being English, he must as a matter of course speak very good French.

IV. If any might be allowed to pronounce foreign names without regard to the peculiar sounds of the letters, or to the accents, which prevail in other countries, this privilege might surely be claimed by the poets, who, in the use even of English words, are considered to enjoy a special license. It is very evident, however—as every one who will take the trouble to examine the works of the better class of them, must admit—that, as a body, they have neither claimed nor used this privilege. On the contrary, our more distinguished poets have usually exhibited a classical—we might almost say a punctilious — accuracy, in the employment of foreign names, whether of places or persons. To illustrate by examples:—GRANADA[3] and GE-

[3] "In Lithuania had he served and Russe;
No Christian man so oft of his degree.
At Algezi′ras, in GRANADA, he
Had join'd the siege;———"—CHAUCER.

"The Moorish king rides up and down
Through GRANADA's royal town"——

NOA[4] are pronounced by all the great poets who use these names, from Chaucer and Shakspeare down to the present time, with the native accentuation; that is, GRANADA has the accent on the penultima, and GENOA on the antepenultima, though the generality of speakers, who have not heard these names pronounced, but merely follow analogy, or their own notions of propriety, reverse the accentuation, making GRANADA rhyme with *Canada*, and GENOA with *boa*. No poet, perhaps, employs foreign names so frequently as Byron, and yet—though he often writes very carelessly—it would be difficult, in all the poetry he has written, to point out half a dozen instances where he has not conformed to the foreign accentuation, excepting always those few well-known names which have acquired an established English pronunciation, and in these cases he appears invariably to adopt the pronunciation of the best English speakers. The same may be said of Scott; though he writes with great freedom, he rarely, if ever, violates the strictest rules of geographical pronunciation. In the poetry of Rogers, Southey, Moore, Campbell, and Montgomery, we have met with scarcely a solitary example of departure from the native accentuation of names, which does not properly come within the exception above stated.

Among the principal languages of continental Europe, the German, in its accent and in the metre of its verse, has the nearest affinity to the English; and it is worthy of remark, that precisely the same general usage prevails with respect to foreign names in German poetry, as in that of our own tongue. It may be observed, however, that the Germans conform more exactly to the native accentuation of the names of other countries than the English. Any one may satisfy himself of the correctness of these statements, if he will take the trouble to consult the poems of Schiller, who appears to have had occasion to use foreign names far more frequently than almost any other German poet. In his drama of 'Don Carlos,' MADRID occurs near twenty times, and *always* with the accent on the last syllable. This one fact (even were there no other) may show how sparingly the "poetical license," so often alluded to, is used by the most popular, and perhaps the most careless in versification, of all the great poets of Germany. MIRANDOLA (a town in Italy) occurs twice, and in both instances has the accent on the antepenultima. This is the more remarkable, because MIRANDOLA is an exception to the general rule of Italian pronunciation, which places the accent on the penultima of words ending in a vowel. If our poet is not equally correct in regard to ALCALA, (a small town of Spain,) it was owing, doubtless, either to the difficulty of making such a name "lie smooth in rhyme," or to his being ignorant of its true accentuation. The latter is by no means improbable, inasmuch as the Spanish language is far less studied by the generality of European scholars than the Italian. That it was not the result of carelessness is shown by the fact that ALCALA is always pronounced in the poem in the same manner, and according to the general rule of Spanish accentuation; that is, with the accent on the penultima. It will be perceived that Schiller places the accent on the last syllable of PARIS, ST. DENIS, and ST. QUENTIN—in all of which he differs from the English and conforms to the French accentuation. But perhaps the most remarkable illustration of this tendency to adopt the native pronunciation of foreign names is found in his drama of 'Mary Stuart;' where the poet, with the obvious intention of obliging his countrymen to pronounce the English names correctly, invariably in his verse spells LEICESTER 'LESTER,' although in the explanatory (prose) parts of the play he as invariably writes it LEICESTER, as we do in English. For the same reason, doubtless, he writes BOLEYN 'BOULEN,' as this spelling would lead his countrymen to pronounce the name *Boo'len*, nearly as it is spoken in England. Had he written it BOLEYN, the Germans would be in danger of placing the accent on the last syllable, as it is frequently pronounced in the United States.

What has been said above respecting the usage of the poets refers principally to *accentuation*, which, for the most part, can be readily determined by the metre of the poetry. Their manner of pronouncing the letters of a foreign name is far less easily ascertained, since it can only be known when the name ends a line in rhyme; and even then it is often extremely uncertain, as they appear to consider themselves entitled, in such cases, to much greater license than in the accentuation of words. Thus we often see associated, in rhyme, words which correspond but very imperfectly in sound, as *enemy* and *lie*, *mourn* and *burn*, &c. Nevertheless, by comparing a number of examples, and especially by observing the usage of those poets who are most remarkable for the correctness of their rhymes, we shall frequently be enabled to ascertain the true pronunciation of a word or name. It will be found that while foreign names that are in familiar use in our own language, have an English pronunciation, those not very well known are generally pronounced with the foreign sound of the letters, as will be seen from the following passages:

"Not now to while an hour *away*,
Gone to the falls in *Valombré*."—

"'Tis *Jacqueline!* 'tis *Jacqueline*,"
Her little brother laughing cried;
"I know her by her kirtle *green*,
She comes along the mountain side."

"De Courcy, lord of *Argentiere!*

. . . .

Thy thirst for vengeance sought the *snare*"—ROGERS.

. . "Winding between Alpine trees;
Spiry and dark around their house of *prayer*,
Below the icy bed of bright *Argentiere*."—WORDSWORTH.

"This circumstance may serve to give a notion
Of the high talents of this new *Vauban*,[5]
But the town ditch below was deep as ocean,
The rampart higher than you'd wish to *hang*."—BYRON

"For many an age remember'd *long*
Shall live the towers of *Hougomont*,[5]
And fields of Waterloo."—SCOTT.

It would be easy to cite a multitude of such examples; but these will perhaps be sufficient to illustrate our position.

"And GRANADA must be won,
And thyself with her undone."—BYRON.

"GRANADA caught it in her Moorish hall,
Galicia bade her children fight or fall."—SCOTT.

"There was crying in GRANADA when the sun was going down."——

"Farewell, farewell, GRANADA, thou city without peer."
LOCKHART.

4 "Signior Baptista may remember me
Near twenty years ago, in GENOA, where
We were lodgers at the Pegasus."—SHAKSPEARE.

"Were GENOA'S galleys riding in the port——"—BYRON.

How quick they carved their victims and how well,
Let Saxony, let injured GENOA tell."—MOORE.

"That noble haven, where, when GENOA reign'd,
A hundred galleys shelter'd——"—ROGERS.

"My native GENOA, if with tearless eye
Prone in the dust thy beauteous form I see."
MONTGOMERY.

5 In these names the letter *n* is similar in sound to *ng*. The *t* at the end of Hougomont is silent. The rhymes, however, are not quite perfect: the *o* in the last syllable of Hougomont should be sounded like *o* in *won't*. The latter syllable of *Vauban* sounds like *bong*.

On the other hand, we shall find the poets pronounce foreign names of some celebrity, such as NILE, PARIS, LYONS, CADIZ, POITIERS or POICTIERS, &c., with the English sound of the letters, as may be seen from these and similar examples:

"Deep in those solitary woods
Where oft the genii of the floods
Dance round the cradle of their *Nile*,
And hail the new-born Giant's *smile*."—MOORE.

"Oh, never talk again to me
Of northern climes and British *ladies*;
It has not been your lot to see,
Like me, the lovely girl of *Cadiz*."—BYRON.

"And Courtenay's pride and Percy's fame
Blazed broader yet in after *years*,
At Cressy red and fell *Poitiers*."—SCOTT.

. . . . "So the shaft
Of victory mounts high, and blood is quaff'd
In fields that rival Cressy and *Poictiers*—
Pride to be wash'd away by bitter *tears*."—WORDSWORTH.

Instead of saying that the poets conform to the native accentuation of proper names, except in cases where these are well known, we might, perhaps, with more propriety, say that they merely follow the practice of the best speakers, of which their own may generally be regarded as the written representation. We sometimes hear it objected to the authority of the poets, that well-known names are occasionally pronounced, even by those who rank among our best poets, quite differently from the prevailing practice of the best speakers. Goldsmith, for example, accentuates the penultima of NIAGARA,[6] and Campbell the antepenultima of WYOMING.[7] just reversing the correct pronunciation, which places the accent on the third syllable from the end in NIAGARA, and on the second in WYOMING. But this discrepancy is readily explained by the fact, that neither of these poets was ever in America, and probably had no opportunity to learn the practice of the best speakers in the United States. It may be observed, however, that Goldsmith gave what was probably the correct accentuation of NIAGARA in his day. Scarcely a single instance can be pointed out, wherein any good English poet has misaccentuated a name with the sound of which he was familiar; we are, therefore, warranted in concluding, that when a name in a region remote or rarely visited is mispronounced, it is owing rather to the want of knowledge than to the carelessness or license of the poet.

V. The Latin names of foreign countries are nearly always to be pronounced with the English sounds of the letters. Thus, in BAVA'RIA, BULGA'RIA, LUSA'TIA, and TRANSYLVA'NIA, the accented A should have its first sound. In Prussia and Russia, however, the *u*, instead of being pronounced short, as it would unquestionably be in ancient Latin names of this kind, is usually sounded like *oo*, assuming in these instances the character of the German or Russian *u*. In like manner, the first syllable of BULGARIA is to be pronounced Bŏŏl, and not Bŭl.

In the anglicized forms of foreign names, and in most well-known names of foreign countries, the same rules of pronunciation, generally speaking, obtain, as in genuine English words, *e. g.* SPAIN, ITALY, NAPLES, &c.

At the same time, we may observe a general tendency to adopt those sounds of the English vowels which approach most nearly to the foreign sounds: thus the *a* in the first syllable of ADRIATIC and PARIS has a sound much nearer to the French and Italian *a* than its first sound, though this would probably be given to these names by the mere English scholar who should be guided by analogy solely. In like manner, the *i* in MILAN is short, so as nearly to correspond to the Italian *i*, which is like our *e* or *ee*. Again, in the pronunciation of foreign names that have become thoroughly anglicized, it is interesting to observe the tendency of our language to throw the accent as far as possible from the termination. Thus PARIS is pronounced with the accent on the first syllable, though the French appear to place it on the last; and the Germans, who have not the same aversion to the ultimate accent that we have, lay the stress of voice distinctly on the final syllable, thus, pa-riss'. HANOVER, which might be pronounced with the native accentuation (HANO'VER) without the slightest offence to the genius of our tongue, has become irrecoverably HAN'OVER. Thus, also, we pronounce ANDALU'SIA (in Spanish ANDALUCI'A,) AR'AGON, (in Spanish, ARAGŌN',) &c. So, in the United States, the old NIAGA'RA has become unalterably fixed as NIAG'ARA; and HURON' has given place to HU'RON.

VI. It may be observed that with respect to foreign names, not only in the French, Italian, and other languages that are written in the Roman letters, but also in German and Greek, (the characters of which may be readily converted into corresponding Roman letters,) it is generally customary in English to retain the literal spelling, *e. g.* ANSPACH, (German, *Anspach*,) KÖNIGSBERG, (German, *Königsberg*,) CHIOS, (Greek, Χιος,) &c., excepting a very few well-known names, as LYONS, (French, *Lyon*,) NAPLES, (Italian, *Napoli*,) MUNICH, (German, *München*,) DANTZIC, (German, *Danzig*,) LEIPSIC, (German, *Leipzig*;) we often find, however, the last two names spelled literally *Danzig* and *Leipzig*. On the other hand, names in languages of which the characters cannot be readily converted into Roman letters, or which are but little known as written languages, are usually spelled according to their sound in some well-known European tongue. Thus the name of one of the cities of Persia is written, in English, *Shooster* or *Shuster*; in German, *Schuster*; and in French, *Chouster*; precisely the same sound being expressed by these different spellings. We find, in English works of the highest character, these various modes of writing oriental and other names employed indiscriminately.

VII. In giving the pronunciation of geographical names, to represent the accent correctly is a principal difficulty to be encountered. Those acquainted with French are aware that this language has no accent in the sense in which we employ the term. The same may be said of the Hungarian, and perhaps also of the Arabic, Turkish, and Persian. At all events, those best acquainted with these languages are not unfrequently at a loss, when they wish to represent accurately in English the accentuation of Arabic, Turkish, and Persian words. Nevertheless, it will be found that here, as in the French and Hungarian, there is something analogous to our accent, which will generally serve to guide us in marking in English the accentuation of names in these languages. The different syllables of oriental names, however, like those of French words, are not unfrequently pronounced with a stress of voice so nearly equal, that, in a number of instances, the use of the secondary accents is necessary in order to indicate more precisely the true pronunciation.

(1.) It may not be improper here to remind the

[6] "Where wild Oswego spreads her swamps around,
And NIAGARA stuns with thund'ring sound."
The Traveller.

[7] "On Susquehanna's side fair WYOMING."

"And scarce had WYOMING of war and crime
Heard but in transatlantic story rung."
Gertrude of Wyoming.

reader, that the accents which we often see upon oriental names are by no means to be understood as always indicating the manner in which an Englishman should accentuate these names in pronunciation, as they are often employed to denote some particular sound in the vowels over which they are placed. Thus some authors place an accent upon *a*, when they wish merely to signify that this letter has the clear full sound of *a* in *far*. In the same manner, an accent is placed upon *u*, in order to show that it has the Italian or German sound, or, in other words, is to be pronounced like *oo*.

(2.) The sound of *a* in several of the oriental languages is often very broad, approaching nearly to that of *au* in English. Hence we often see AFGHAUN instead of AFGHAN, CAUVERY instead of CAVERY, *sultaun* instead of *sultan*, &c. In NEPAUL and BHOPAUL the French indicate the same sound by using *â* with a circumflex, *e. g.* NEPÂL, BHOPÂL, &c. This method is, perhaps, preferable to ours, the sound of â being intermediate between that in the English word *far* and that in *fall*.

(3.) The Arabic article *al* or *el* is often changed in pronunciation, so as to correspond with the initial consonant of the word to which it is prefixed; thus, EL-RASHEED, (Rashid,) EL-SHAM, (the Arabic name of Syria,) and EL-SIOOT, are pronounced, and should be written in English, *Er-Rashid*, *Esh-Sham*, *Es-Siut*. The vowel sound of the article also varies considerably, sometimes approaching that of *oo;* thus the "Country of dates," is usually pronounced Beled'ŏŏl Jer-eed'. So ES-SIOOT (or SIUT, without the article) is sometimes written *Assyout* and *Osiut*.

(4.) *Gh*, in the Arabic and some other oriental tongues, is not merely a hard *g*, as in the Italian, nor an aspirate like the German *ch*, as in the Irish language, but a harsh guttural, bearing the same relation to the German *ch* that *g* bears to *k*. It has no equivalent in any European language, but may be best represented by a hard *g*.

(5.) *Kh* is equivalent to *ch* in German, when followed by *a*, *o*, *u*, or *r;* and may be represented by the English letter *k* in orthoepy.

ARTICLE IV.

GENERAL PRINCIPLES OF PRONUNCIATION OF THE MORE IMPORTANT EUROPEAN AND ORIENTAL LANGUAGES.

I.

BOHEMIAN.

The Bohemian or Tchechian, a Slavonic dialect spoken by about 7,200,000, nearly corresponds in the sound of its letters to the Polish. See Section X. of this article.

II.

DANISH.

The Danish is a Teutonic language of the Scandinavian branch.

1. *A* is pronounced generally as in the English word *far*, though it frequently approaches the sound of *a* in *fat*.

2. *E*, at the end of an accented syllable, usually has a sound like that of *i* in *pin;* in other cases it is sometimes like *e* in *met*, and sometimes like *e* in *battery*.

3. *I* is like *ee*, or like *i* in *pin*.

4. *O* is like the English *o*.

5. *U* is like *oo*, or the German and Italian *u*.

6. *Y* is equivalent to the German *ü*.

7. *Aa* sounds like long *ō*.

8. *Ae* sounds like *a* in *fate*.

9. *Ie* sounds like *ee* in English, or *i* in Italian.

10. *Oe* or *ö* is the same as in German.

11. The consonants *b, c, f, h, k, l, m, n, p, q, s, t, x, z*, are like the English.

12. *D*, between two vowels, or at the end of a syllable in which it follows a vowel, sounds like *th* in *this;* it is not sounded after a consonant.

13. *G* is always hard; at the end of a word it is sounded very slightly, so as to resemble *h;* thus, AALBORG is pronounced nearly *ol'-bor'h*. Preceded by *n*, it gives a nasal sound to that letter.

14. *J* is like the English *y*, (consonant).

15. *R* is similar to the German.

16. *V* is usually like the English, but it appears to have a vowel sound after *a;* thus, *havn* is pronounced almost *houn*.

17. *W* is found only in words borrowed from the German, and has the sound of *v* as in that language.

III.

DUTCH.

A Teutonic language of the Low-German class, spoken by the natives of Holland.

1. The vowels *a, e, i, o*, and *u* are similar to the German, and have each at least two sounds, as in that language.

2. *Y* is like long *i* in English, as in *nigh*.

OBS. *Ij* is sometimes made use of instead of *y;* thus, OVERYSSEL is not unfrequently written *Overijssel*.

3. *Ae* is equivalent to *aa* or long *ā*.

4. *Ie* sounds like *ee* in English, or *i* in Italian.

5. *Oe* sounds like *oo*, or *u* in Italian.

6. *Oo* sounds like *ō* long.

7. *Ui* or *uy* is similar to *oi* in English.

8. The consonants *b, c, f, h, k, l, m, n, p, q, r, s, t, x, z*, are similar to the English.

9. *D*, at the end of a word, is like *t;* in other cases it is the same as in English.

10. *G* resembles in sound a strongly aspirated *h*, or the German *ch*. When united with *h*, in the form *gh*, it sounds like *g* in the word *go*.

11. *J* is equivalent to the English *y*, (consonant).

12. *V*, at the beginning of a word or name, usually sounds like *f*, as in German.

13. *W* is somewhat like the German, but softer. In the word *nieuw*, followed by a consonant, as NIEUW-POORT, (written also NIEUPOORT,) it is silent.

14. *Ch* is similar to the German *ch*. (15.) *Sch*, however, has not, as in German, the sound of the English *sh*, but the pure sound of *s*, followed by the guttural *ch*, resembling *sk* in English.

OBS. The FLEMISH is so closely allied to the Dutch, that it may be regarded as essentially the same language.

IV.

FRENCH.

A Romance language formed by the Latin engrafted on Celtic.

1. *A*, in French, is generally considered to have two sounds: the first long, as in the English word *far*, *e. g.* in *pas;* the second short, almost like *a* in *fat*, *e. g.* in *bal*. *A* circumflexed, (*â*,) however, has a sound broader than the *a* in *pas*, being intermediate between that in *far* and that in *fall*.

OBS. The French *a* would frequently seem to be intermediate between its second English sound and that of short *u*. However this may be, the French writers often employ *a* in spelling oriental names, when the English make use of *u;* thus, in CUTCH, (Fr. *Catch;*) FURRUKABAD, (Fr. *Farrakâbâd;*) MUSKAT, (Fr. *Mascate*).

2. *E* has three sounds: (1.) *close*, like *a* in *fate*, *e. g.* in *été;* (2.) *open*, nearly as in *met* or *ebb*, but more prolonged, *e. g.* in *procès* and *tête;*[1] (3.) *obscure*, as in *battery*, *e. g.* in *retour*, *devrait*.[2] At the end of polysyllables it is generally entirely mute.

3. *I* has two sounds: the first short, and nearly as in the English word *fig*, *e. g.* in *il*, *ami;* the second like *ie* in *field*, or *marine*, *e. g.* in *gîte*. This

1 In pronouncing this sound, the mouth must be freely opened, whence the name.

2 The *e* in these and similar cases is often scarcely sounded at all, and appears to pass imperceptibly into *e* mute; *retour* and *devrait* may be pronounced *r'toor* and *d'vray*.

second sound may be represented by *ee* as in English, or *i* in Italian.

4. *O* has three sounds: (1.) long, nearly as in *robe*, *e. g.* in *trône;* (2.) as in *rob*, *e. g.* in *parole;* (3.) as in *lord*, *e. g.* in *corps*.

Obs. *O* circumflexed, in French, has a deeper and fuller sound than *o* long in English. No English vowel or vowels accurately indicate this sound.

5. The sound of the French *u* has no equivalent in English. It may be said to be intermediate between *ee* and *oo;* but it can be learned from an oral instructor only.

Obs. *U*, before *n*, nasal, has its second English sound nearly, *un* being pronounced almost *ŭng*.

6. *Y* single is similar to the French *i;* but between two vowels is like double *i*.

7. *Ai* is like ĕ, or *e* open.

8. *Au* is like ō.

9. *Ei* is like ĕ in *there*.

Obs. In giving the pronunciation of French names in which *ai* or *ei* (not followed by *n* nasal) occur, *e. g.* Lorraine, Seine, &c., we sometimes represent the sound of these diphthongs, by *a* or *ai*, because this is the usual mode of anglicizing such names. But with regard to places less known, the sounds *ai* and *ei* are rendered by *e*, sounded as in *met*. In these latter instances the sound of *e* should be somewhat prolonged, the mouth at the same time being freely opened.

10. *Eu* is similar to the English *u* in *tub*, but the sound is more prolonged, nearly resembling *u* in *fur*.

Obs. *Eu*, in the different parts of the verb *avoir*, always has the sound of simple *u*.

11. *Ie* is like *ee* in English, or the Italian *i*.

12. *Oi* usually sounds like wa, *e. g. moi* is pronounced mwa or *mwŏh*.

Obs. *Oi* was formerly used in the termination of the French verbs, *e. g. avois, avoit, avoient;* also in the final syllable of a number of adjectives, as *Polonois*, and *Lyonnois*. The *oi* in these words—which are now usually written *avais, avait, avaient, Polonais, Lyonnais*—sounds like *ai*, (or ĕ.)

13. *Ou* sounds like *oo* in English, or *u* in Italian.

14. *B, d, f, k, p, t, v*, and *z*, are the same as in English.

15. *C* has the sound of the English *k* before *a*, *o*, *u*, and before consonants. It has the sound of *s* hard before *e*, *i*, and *y;* and before *a*, *o*, and *u* when written with a cedilla ç.

16. *G*, before *a*, *o*, and *u*, is hard, as in the English word *gap;* before *e*, *i*, and *y*, it is soft, having the sound of *zh*, or of *s* in *pleasure*. *Gu* sounds like *g* hard; thus, *gué, guide*, are pronounced *ga, gheed*. *Gn* is like *ni* in the English word *onion*. Final *g* is silent, except in the noun *bourg*, and in names terminating in *berg*, where it sounds nearly as *k*.

17. *H* is never pronounced in French so forcibly as in English. Some orthoepists say that *h* has no sound in French.

18. *J* sounds like soft *g* in French, or *zh* in English, or *z* in *azure*.

19. *L* has usually the same sound as in English; but when it ends a word, being preceded by *i*, or when *ll* follows *i*, in any situation, it usually has what is called its *liquid* sound. This may be said to answer nearly to the sound of *lli* in *million*, the sound of *l* in such cases being blended with that of *y*, (consonant; thus *papillon* is pronounced *pa`-peel`-yong´*; Chantilly, *shong`-teel`-yee´*, &c. According to the present practice of the more polite French speakers, the sound of *l* is scarcely heard at all in such words, the long *ee* sound being substituted for it; so that their pronunciation might rather be indicated thus—*pa`-pee`-yong´; shong`-tee`-yee´*.

20. *M* and *N*, when followed by a vowel, or when double, have the same sound as in English; but when at the end of a word, (not immediately followed by another word beginning with a vowel,) or when followed by another consonant, in the middle of a word, they have what is termed the *nasal* sound, which resembles that of *ng*, as in *long, pang*, &c., but is somewhat softer;[3] thus, *m* and *n* are nasal in such words as *comparer, contente*, but have their natural sound in such as *commune, connu*. Melun, before a consonant, or standing by itself, would be pronounced almost *m'lung;* but if followed immediately by a vowel, as in the sentence, *Melun a six mille habitans*, the final *n* is sounded distinctly like *nn*. The pronoun *sien*, when not followed by a vowel, is pronounced nearly *se-ang´;* but when it takes the feminine termination, the *n*, being doubled, has the same sound as in English, so that *sienne* is pronounced *se-enn´*.

M or *n*, nasal, when preceded by *e*, usually causes this vowel to assume the broad sound of *a:* thus, *dents, sens*, are pronounced like the French words *dans* and *sans*, almost as if written in English *dŏng* and *sŏng*.

21. *In, im, ain, aim, ein, oin*, and *en, preceded immediately by i*, when nasal, have a sound nearly resembling that of *ang* in the English word *pang*. In such cases, *in, im, ain, aim, ein*, and *en* are pronounced alike, *ang;* the *o* in *oin* has the sound of our *w*, so that *loin* and *soin* are pronounced almost *lwang, swang*.

22. In *om* and *on*, nasal, the *o* is long, as in *won't*.

23. *Q* or *qu*, in French, always sounds like *k*, *e. g. quel* is pronounced *kel; qui, kee*.

Obs. *Q*, in *French* words, (except when terminal, as in *coq* and *cinq*.) is always followed by *u*, though it is sometimes employed without this letter in writing certain foreign names. Thus Balbi and several other authors, both French and English, write *Qené* for Keneh; *Qoum* for Koom, &c. In such cases, *q* is used to denote a sound like that of *k*, but somewhat more guttural.

24. *R* is like the English, but is trilled more strongly, especially when it precedes another consonant, or stands at the end of a word, as in *vertu, punir*: in similar cases, the English *r* is but very slightly sounded.

25. *S*, when single and between two vowels, sounds like *z*, or as in the English words *rose, please*. In other cases, it has the hard sound of *s* in *sister*. At the ends of words, it is, with few exceptions, silent.

26. *X* generally has the same sound as in English, but is sometimes sounded like *s*, *e. g.* in *six*, pronounced *seece*, and *Bruxelles*, (Brussels,) pronounced *bru`sell´;* and occasionally like *z*, as in *dixième, dee`-ze-aim´*. Final *x* is silent.

27. *Ch* is like *sh* in English: followed by a consonant, and at the end of a word, it sounds like *k*. The sound of the English *th*, as in *thin*, has no equivalent in French; but is represented by *t* hard.

28. *Gn* (the same as in Italian) has a sound which blends that of *n* and *y*, (consonant,) or, in other words, is equivalent to the sound of *ni* in *minion*. Thus, Avignon is pronounced *a`-veen`-yong´*.

29. Silent Letters.—The vowel *e* at the end of a word, when not marked with an accent, is invariably mute, *e. g.* in *parle, contente*,[4] &c.

3 In uttering this sound, care should be taken not to press the back part of the tongue against the palate, as is done in pronouncing the English *ng*.

4 The particles *le, ne*, and the pronouns *je, me, te*, &c. are perhaps, strictly speaking, exceptions; but though the *e* in these words is not always absolutely mute, it is very often so; thus, the sentence *vous me trouverez le même*, is pronounced *voom troov rel*

30. The French consonants, when occurring at the end of a word, are generally not pronounced, unless they are immediately followed by a word beginning with a vowel, *e. g.* in *contenT*, *Bordeaux*, and *dents*. If, however, they are followed by a mute *e*, or any other vowel, they must always be articulated, *e. g.* in *contente*, *denté*, &c.

Obs. 1. The letters *c*, *f*, *l*, and *r* are, when final, very often pronounced; *e. g.* in *avec*, *neuf*, *il*, and *punir*.

Obs. 2. The French articulate the final consonants in almost all foreign and classical names, *e. g.* in Amsterdam, (*m* not nasal,) Ve′nus, &c.

Accent.—It may be observed that the French language has no accent in the sense in which we employ this term. The marks called *accents*, that are placed over the different vowels, serve only to indicate some particular sound of these letters, and not that peculiar impulse of the voice which characterizes an accented syllable in the English and most other European tongues. Thus, the accent over the *e* in *parlé*, serves to show that this vowel has its first French sound, and at the same time distinguishes it from *parle*, another form of the same verb, in which the *e* is mute. The circumflex imparts to the vowels over which it is placed a longer and deeper sound than ordinary; *e. g.* in *hâte*, *tempête*, *gîte*, and *apôtre*.

It is commonly said that the French pronounce all the syllables of a word with an equal stress of voice, but that they *seem*, to an English ear, to accentuate the last, because, in our language, the universal tendency is to throw the accent toward the beginning of the word. Others maintain that in pronouncing words of a number of syllables, the voice of a native French speaker almost invariably rises and dwells on the last, and that this peculiar terminal intonation is analogous, and nearly equivalent, to our accent. With very few exceptions, the principal accent may be placed on the last syllable of French geographical names; at the same time secondary accents may be used in order to prevent certain syllables from being pronounced too slightly or indistinctly, as is usually the case with unaccented syllables in English. The pronunciation of Orléans, for example, may be thus given—*or′lā-ong′*. Care should be taken not to break such names into as many isolated sounds as there are different syllables; but, while pronouncing these syllables with a stress of voice nearly equal, to let each glide smoothly into that which follows it. It may be observed, that the French, in uttering short sentences, usually make the different words run into each other, as if they were parts of the same word.

V.

GERMAN.

A Teutonic language divided into two principal branches; the High German and the Low German.

1. *A*, in German, usually sounds as in the English word *far*, though sometimes approximating to the *a* in *fat*.

2. *E*, when long, sounds like *a* in *fate;* when short, like *e* in *met:* frequently, however, it has an obscure sound, like *e* in *battery*.

3. *I*, long, sounds like *i* in *marine*, (as *i* in Italian, or *ee* in English;) *i*, short, like *i* in *pit*.

4. *O*, long, is like that in *no; o*, short, like that in *on*.

5. *U*, long, is like *oo* in *bull*, *full; u*, short, like *oo* in *good*.

6. *Y* sounds like the German *i*.

mame, the vowel in *me* and *ne* being entirely suppressed, and the consonants attached to the preceding words.

7. *Ae* or *ä*, is similar to the German *e*, or to the English *a* in *fate*, or *ai* in *air*.

8. *Oe*, or *ö*, nearly resembles the French *œu* in *œuvre*, but has no parallel sound in English. The sound nearest to it is that of *e* in *her*, or *u* in *fur*. The German poets often rhyme it with *e*, (*a* or *e*.)

9. *Ue*, or *ü*, is like the French *u*. Its sound is not found in the English language.

10. *Au* is equivalent to the English *ou* in *our* or *out;* but somewhat broader, giving predominance to the sound of *a*.

11. *Au* and *eu* resemble in sound the English *oi*, as in *oil* or *boil;* only somewhat approaching in its termination to that of the English *oo*.

12. *Ei* and *ey* have the sound of *i* in *mine*, or *ey* in *eye*.

13. *Ai* and *ay* are similar to the preceding, but somewhat broader.

Obs. It may be observed that *ai* and *au*, in German, as well as in several other languages, are *proper* diphthongs, the vowels preserving their distinct and proper sound; thus *ai* is equivalent to a′-e, and *au* to a′-oo, in English.

14. *Ui* sounds like *oo-e*.

15. *Ie* is equivalent to *ee* in English, or *i* in Italian.

16. The consonants *f*, *k*, *l*, *m*, *n*, *p*, *q*, *t*, and *x* are pronounced as in English.

17. *B* and *d*, at the beginning of a word, have the same sound as in English. At the end of a word, and after the vowel of the syllable to which it belongs, *b* is pronounced like *p;* thus *lob* is pronounced *lop; erpse, erbse*. *D*. at the end of a word is pronounced like *t;* thus *Brod*, pronounced *brot*.

18. *C*, before *a*, *o*, and *u*, sounds like *k;* before *e* *i*, and *y*, like *ts*.

19. *Ch* has a sound in German unknown in our language, and which, consequently, can be learned from an oral instructor only. It somewhat resembles that of our *h*, with a strong aspiration: after *a*, *o*, and *u*, it is guttural; for example, in the word *ach*.[5] When it follows *e*, *i*, *ä*, *ö*, *ü*, *äu*, or *eu*, it seems to be sounded more in the palate, as in *ich*.[5]

Obs. *Ch*, before *s*, *radical*, (*i. e.* forming a part of the root of the word,) has the sound of *k;* thus *Ochs* is pronounced *oks; Sachsen*, *sak′-sen*, &c.

20. *G*, at the beginning of a word, sounds as in the English word *get*. In other situations, it should be pronounced nearly like the German *ch*. In some German dialects, however, it is sounded, in all cases, nearly like *g* hard, in English.

21. *H* is pronounced only when it begins a word, or at the beginning of the radical syllable.

Obs. 1. When *g* and *h* occur in the middle of a compound word, they have the same sound as when they are initial, provided they begin any part which is a complete word in itself; thus, in the participle *gegeben*, the latter *g* has the same sound as the former, because it begins the verb *geben*, from which that participle is derived. *H*, in similar instances, is pronounced; *e. g.* in *gehabt*, *aufhalten*, &c.

Obs. 2. *G* and *h*, occurring after a vowel, lengthen its sound; *e. g.* in *Tag*, *Zahl*, *Floh*, pronounced *tag*, *tsal*, *flō*, &c.

22. *J* has the sound of the English *y*, (consonant,) in *yes* or *you*.

23. *Q* is only used before *u*, and sounds as in the English word *quit*.

5 Those who have no opportunity of acquiring this sound from a German, might perhaps learn it from a Scotchman, as the Scotch *ch*, in the word *loch*, is essentially the same with the German, though pronounced somewhat more strongly.

24. *R* is pronounced like *rr* in the English word *terror*, but somewhat more strongly.

Obs. Care should be taken to pronounce the *r*, in German, distinctly and forcibly. In such words as *berg* and *werth*, the learner should be particularly on his guard against allowing the *e* to become like short *u*, as in similar words in English. The *e*, in such cases, should have the same sound as in our word *merit*, so that *berg* should be pronounced almost as if written *bairg*, (not *burg;*) *werth*, as *wairt*, (not *wurt*,) but somewhat shorter.

25. *S*, at the beginning of a word, or between two vowels, is like *z;* in other cases it is sharp or hard, as in *this*. *Ss* is always sharp.

26. *Sch* sounds like the English *sh; sz* like *ss*.

27. *Th* is pronounced like *t*, as in most other languages.

28. *V* sounds like *f* in English, except when between two vowels; it is then usually pronounced like our *v*.

29. *W* resembles our *v*, but in pronouncing it the upper teeth should not be allowed to touch the lower lip, as is done in uttering the English *v*.

30. *Z* and *tz* sound like *ts*.

The following table exhibits the mode of representing English vowel sounds by German letters:

The English.		German.
a in *bare* is represented by *ä* or *e*, and would be written		*bär* or *ber*.
a in *all*	*ee* or *eh*	*eel* or *ehl*.
a in *far*	*a*	*far*.
a in *at*	no equivalent.	
e in *mere*	*i* or *ie*	*mir* or *mier*.
e in *met*	e	*mate*.
i in *fine*	*ei* or *ey*	*fein*.
i in *file*	*i*	*file*.
o in *hole*	*o*	*hohl*.
o in *for*	*o*	*for*.
o in *do*	*u*	*du*.
u in *bull*	*u*	*bull*.
u in *uniform*	*ju*	*juniform*.
u in *hut*	no equivalent.	
oi in *oil* nearly by	*äu*	*äul*.
ou in *out* nearly by	*au*	*aut*.

The equivalent consonantial sounds are:

English.	German.
soft *ch* represented by	*tsch*.
j and soft *g*	no equivalent.
f	*f* or *v*.
s	*ss*.
sh	*sch*.
th in think or in that	no equivalent.
v nearly by	*w*.
w	no equivalent.
x	*x* or *chs*.
y in *yes*	*j* (*jess*).
z	no equivalent.

Obs. No general rule can be gi en for the accent of German words or names: it may be remarked, however, that the penultimate accent occurs much less frequently than in the Spanish or Italian languages.

VI.

MODERN GREEK OR ROMAIC.

The modern Greek is the representative of the ancient Hellenic dialects. This modern dialect of the Greek tongue, though modern in name and form, approaches much nearer the Greek or Homer's time than does Italian, Latin; or Dutch, German. Most of its grammatical forms can still be recognised by a classical scholar; the ancient declensions are still preserved in it; and even the conjugation of the verb hardly presents any new elements. Modern scholars are now beginning to be of opinion that the best "living conception of what the spoken Greek language really was in its best days" may be obtained amongst the modern Greeks.

1. Α α (alpha) like *a* in *far*.
2. Ε ε (epsīlon) like *a* in *fate*.
3. Η η (ēta) like *ee* in English.
4. Ι ι (iōta) like *e* in *me*, or *i* in *pin*.
5. Ο ο (omicron) like *o* in English.
6. Υ υ (upsīlon) nearly like the French *u*, (or *ü*.)
7. Ω ω (ōmĕga) like *o* in English, there being no difference between this and *omicron* in prose; in poetry ω is longer.
8. Αι like *a* in *fate*.
9. Ει and οι sound like *ee* in the English *seen*, or the Italian *i*.
10. Ου like our *oo* in *boom*, or the Italian *u*.
11. Β ϐ (bēta) like *v* in English.
12. Γ γ (gamma) like *g* hard, as in *get*.
13. Δ δ (delta) like *th* in *this*.
14. Ζ ζ (zēta) like the English *z*.
15. Θ ϑ (thēta) like *th* in *thin*.
16. Κ κ (kappa) like *k*.
17. Λ λ (lambda) like *l*.
18. Μ μ (mu) like *m*.
19. Ν ν (nu) like *n*.
20. Ξ ξ (xi) like *x*.
21. Π π (pi) usually like the English *p;* but after μ (m) it is like *b;* thus ἔμπορος is pronounced *em'bo-ros*.
22. Ρ ϱ (rho) similar to the German *r*.
23. Σ σ ς (sigma) like the English *s*.
24. Τ τ (tau) usually like the English *t;* after ν, (n), however, it is sounded like *d; e. g.* ἐντὸς is pronounced *en-dos'*.
25. Υ υ, (*consonant*,) when before a vowel, or the liquids *l*, *m*, *n*, *r*, is like our *v; e. g.* αὐϱύω is pronounced *av-a-rü'o*, αὐλὸς, *av-los'*, αὔϱιον, *av'-re-on*: in other cases it is like *f; e. g.* Λευκαδία is pronounced *lef-ka-thee'a*.
26. Φ φ (phi) is equivalent to our *f*
27. Χ χ (chi) is similar to *ch* in German.
28. Ψ ψ (psi) is like *ps* in English.

Obs. Recently it has become the practice to give the islands, towns, &c. of modern Greece, their ancient names; and it is probable that at no distant period such modern corruptions as *Theaki* for *Ithaca*, *Scio* for *Chios*, may be regarded as obsolete.

VII.

HUNGARIAN.

1. *A*, unaccented, is like *o* in *not;* with an accent, (*á*,) it has the sound of *a* in *far*, and is always long; thus, Aba Uj-var, the name of a town, is pronounced *ob-oh oo-e var*.

2. *E*, unaccented, is like *e* in *met;* with an accent, (*é*,) it has a sound intermediate between *e* in *met* and *i* in *pit*, but more prolonged, somewhat like the *e* in *help*.

3. *I*, and *y* when a vowel, are similar to *e* in *me*, or *i* in *fig*.

4. *O*, without an accent, is the same as in English; when accented, (*ó*,) it has a longer and deeper sound.

5. *U*, without an accent, is like *oo* in English, or the Italian *u;* with the accent, (*ú*,) its sound is fuller and deeper.

6. *Oe*, or *ö*, and *ü;* are the same as in German.

7. The consonants *b*, *d*, *f*, *h*, *k*, *l*, *m*, *n*, *p*, *t*, *v*, *z*, are like the English.

8. *C* is not used without being joined with some other consonant; *cs* is sounded like *ch* in English; *cz* like *ts*.

9. *G*, except when followed by *j* or *y*, is always hard, as in the English word *get*. *Gh* sounds like a simple *g*. *G* followed by *j* or *y* sounds like *dj*, *dy*.

10. *J* is usually like *e* in English; *uj* is pronounced *oo-e*. *Dj* and *gj* are equivalent to *dy* and *gy*, and *tj* to *ty*.

11. *R* is like the German; in other words, is to be trilled more strongly than the English.

12. *S* is like the English *sh*.

13. *Sz* is like *s* sharp, or *ss*.

14. *Ts* is equivalent to *cs*, or *ch* in English.

15. *Tz* is like *cz*, or *ts* in English.

16. *Y*, in Hungarian, is nearly always a consonant. When it follows *d*, *g*, *l*, *n*, and *t*, it seems to be blended with these letters, so as to form but one consonant sound.

17. *Dy* and *gy* are alike. *Magyar* is pronounced *mŏd-yŏr*.

18. *Ly* is like *l* in Spanish, or *lli* in the English word *million*. Vassarhely is pronounced in three syllables—*va-shar-hel*.

19. *Ny* is like the Spanish *n*, or *ni* in *minion*. Martony is pronounced in two syllables—*mar-ton*.

20. *Ty* approximates to the sound of our *ch*, bearing the same relation to *t* that *dy* does to *d*.

21. *Zs* sounds like the French *j*, or *zh* in English.

Obs. The Hungarian language cannot be said to have any accent, in the sense in which we employ this term: the syllables of words, however, are distinguished from each other by *quantity*.

VIII.

ITALIAN.

1. *A*, in Italian, is like the English *a* in *far*, though its sound varies somewhat in different situations.[6]

2. *E* has two sounds: (1.) close, as *a* in *fate*; (2.) open, like *e* in *met*.

3. *I* is like *e* in *me*.

4. *O* has two sounds: (1.) close, as in *note*; (2.) open, similar to *o* in *not*, but rather broader.

5. *U* is like *oo* in English.

6. *Ai* and *au*, in Italian, are *proper* diphthongs. Accordingly, Cairo is to be pronounced *kī'-ro*, and Ausa, *ŏw'-sa*, &c.

7. The consonants *b*, *d*, *f*, *l*, *m*, *n*, *p*, *q*, *s*, *t*, and *v* are similar to the English.

Obs. *K*, *w*, *x*, and *y* are not used by the Italians, except in spelling foreign names.

8. *C* and *cc*, before *a*, *o*, and *u*, are sounded like *k*; before *e*, and *i*, and *y*, like *ch* in *cherry* or *chilly*.

Obs. *Cc* should be pronounced more strongly than a single *c*. This remark will apply to all double letters in Italian, as well as in most other languages.

9. As *c*, when immediately before *a*, *o*, or *u*, is never pronounced like *ch*, in order to express this sound in such cases, the vowel *i* is inserted; thus, *cia*, *cio*, *ciu*, are pronounced *cha*, *cho*, *choo*. See table below.

10. *Ch* is employed to express the sound of *k* before *e* and *i*.

11. *G*, before *a*, *o*, and *u*, is hard, as in the English word *get*; before *e*, *i*, and *y*, it sounds like the English *j*: *gia*, *gio*, *giu*, are pronounced *ja*, *jo*, *joo*. See table below.

12. *Gh* is used to express the sound of hard *g*, before *e* and *i*.

13. *Gli* has the sound of the liquid *l*, or of *lli* in *million*; thus, Boglio is pronounced *bole'-yo*, or *bol'-yo*.

14. *Gn* has the same sound as in French; or, in other words, is like the Spanish *n*; *e. g.* Bologna is pronounced *bo-lon'-ya*.

15. *H* is never sounded in Italian.

16. *J*, at the beginning of a syllable, is like the English *y*, (consonant;) at the end of a word it is equivalent to *ii* (Italian.)

17. *R* resembles the French, but is trilled somewhat more strongly.

18. *Sc*, before *e* and *i*, is like the English *sh*; thus Scio is pronounced *Shee'-o*.

19. *Z* commonly has the sound of *dz* in English, or of the *ds* in *Windsor*; *zz* is pronounced like *ts* in *benefits*.

The following table will perhaps enable the reader more readily to understand the mode in which *c* and *ch*, *g* and *gh* are employed by the Italians:

ca	is pronounced	ka.	ga	is pronounced	ga.
che	,,	ka.	ghe	,,	ga.
chi	,,	ke.	ghi	,,	ghe.
co	,,	ko.	go	,,	go
cu	,,	koo.	gu	,,	goo.
cia	,,	cha.	gia	,,	ja
ce	,,	cha.	ge	,,	ja.
ci	,,	che.	gi	,,	je.
cio	,,	cho.	gio	,,	jo.
ciu	,,	choo.	giu	,,	joo.

Obs. 1. In Italian, the accent of words ending in a vowel is usually on the penultima; but to this general rule there are many exceptions.

Obs. 2. It may be observed that, in consequence of the position of Italy, and its former extensive and intimate commercial relations with the Levant, a great number of the geographical names of Greece, Syria, and Egypt, as well as many of those along the southern shore of the Mediterranean, are written in the Italian mode, and should be pronounced according to the principles of this language; *e. g.* Corfu, Tripolizza, Scio, Jaffa, Cairo, &c.

Obs. 3. Italian words being pronounced exactly as they are written, and having no silent letter except *h*, while every vowel preserves its proper sound, independently of the consonants which accompany it, many modern geographers have adopted its vowel sounds in the writing of foreign names, as the best substitute, upon the whole, for that desideratum in a geography of true phonetic character.

IX.

NORWEGIAN.

As a written language, the Norwegian may be said to be identical with the Danish, since not only the grammar, but, with very few exceptions, the words of both are precisely the same. In pronunciation, however, the Norwegians differ widely from the Danes, while these, again, differ considerably among themselves. Under Section II. we have given the elements of Danish pronunciation, as the language is spoken by the educated classes in Copenhagen. The principal points of difference between this and the Norwegian appear to be the following: (1.) *d*, in the latter tongue, always has its proper sound, while in the Danish it is often pronounced like the English *th*; (2.) *g*, at the end of a word, in Norwegian, is to be sounded distinctly as *g* hard in English; (3.) *o*, ending a syllable, is pronounced like our *oo*; (4.) *e*, at the end of a word, always retains its distinct sound; thus, Odense would be

6 There are a number of niceties in Italian pronunciation which, however interesting to a linguist, cannot be noticed in an outline like the present. The difficulty of giving a brief, and, at the same time, a satisfactory exposition of the principles of this language, is increased by the existence of different dialects in different parts of Italy. It has been deemed sufficient, in this synopsis, merely to explain those principles of pronunciation which appear to be recognised by the Italians generally.

pronounced *oo'-den-seh*, and not *o'-den-seh*, as in Danish.

X.

POLISH.

The Polish, or Lekhian, spoken in ancient times beyond its present limits, in parts of Pomerania and Silesia, is a dialect of the Slavonic or Windic.

1. *A* sounds as *a* in the English word *far*.

2. *E*, without an accent, like *e* in *met;* with an accent, (*é*,) like *a* in *fate*.

3. *I* as in *marine*.

4. *O*, unaccented, as in *note;* with an accent, like *oo* in *good*, or *ŏŏ*.

5. *U* is like *oo* in *moon*.

6. *Y* resembles *e* in *me*, but is more guttural, being similar to *i* in *pin*.

7. The consonants *b, d, f, g*, (always hard,) *h, k, l, m, n, p, s*, (always sharp,) *t* and *z*, are essentially the same as in English.

8. *C* in all cases, even before *a* or *o*, sounds like *ts* in English; *cz* is equivalent to our *ch; ch* is like the German *ch*.

9. *J* is like the German, being equivalent to *y*, (consonant).

10. *R* is like the German.

11. *W* is similar to the German, resembling our *v*.

12. *N*, with an accent over it, (*n'*,) sounds like the Spanish *n*.

13. *S*, marked in a similar manner, (*s'*,) has a sound blending that of *s* and *y*, (consonant). *S'c'* has a sound which cannot be given in English: its nearest approximation in our language is *sts*.

14. *Sz* is equivalent to *sh* in English.

15. *Z*, with a point over it, is like the French *j* or *zh* in English.

16. *Z*, with an accent (*z'*,) is somewhat similar to the above, but has no equivalent in our language.

Obs. 1. The accent in Polish words or names of more than one syllable, is always on the penultima. In Russian, however, it is almost always on the last syllable.

Obs. 2. The sounds of the letters in Slavonian, Bohemian, and Illyrian correspond, with very slight exceptions, to those of the Polish language.

XI.

PORTUGUESE.

1. The vowels *a, e, i, o, u*, and *y*, and the diphthongs *ai, ay, au*,[7] are essentially the same as in Spanish.

2. *Ao* is pronounced almost *ŏwng*.

3. The consonants *b, d, f, l, m, n, p, s, t, v*, and *z* are similar to the English.

4. *C* is the same as in French, differing from the English only by sometimes having the cedilla.

5. *Ch* is the same is in French; or, in other words, is like our *sh*.

6. *G* and *j* are the same as in French.

7. *H*, in Portuguese, is always silent. When, however, it follows *l* or *n*, it renders these letters *liquid;* thus, *filho* is pronounced *feel'-yo*; *sen-hora*, *sane-yo'-ra*, &c.

8. *M*, frequently, and *n*, sometimes, has a nasal sound. *Sam*, like SAO, is pronounced almost *soung; alem* or *alen* sounds like *a-leng'*.

9. *Qu* is pronounced as in French, the *u* in this case not being sounded.

10. *R* is like the French.

11. *X* is sounded like *ch* in Portuguese, or *sh* in English.

7 *Ei* and *ey* are almost the same as in Spanish, but have a sound sometimes approaching to the English long *i*.

Obs. In Portuguese, the general rules of accentuation are similar to those in the Spanish language.

XII.

RUSSIAN.

The Russian language is a dialect of Slavonic; and is subdivided into three distinct dialects, that of the Great Russian, now the literary and official language of Russia; the Little Russian, a compound of the Great Russian and the Polish; and the White Russian, formerly the official language of Lithuania. Till the time of Peter the Great, the Russians employed an alphabet invented by Cyrillus, a Greek monk of the 9th cent., and taken chiefly from the Greek, with some signs added to it, to represent sounds peculiar to the Slavonic dialects, such as:

the *s* in *pleasure*, or *j* in the French *jamais*,
sh in *she*,
ts in *benefits*,
ch in *church*,
o in *work*,
i in *bird*,
on in the French *balcon*,
ea in *yea*,
in (nasal) as in the French *enfin*,

and a whole class of vowal characters to represent the initial *y*, as in *ya* in *Yarmouth*,
yu in *yule*,
yee,
ien in the French *bien*,
ion in the French *nation*.

Peter the Great struck off 9 letters of the ancient alphabet, and gave the rest a more rounded form.

Obs. As the Russians neither employ Roman characters, nor those which can be readily converted into corresponding Roman letters, in writing the geographical names of Russia, the mode adopted with respect to oriental names is perhaps the best.

XIII.

SPANISH.

1. The Spanish *a* sounds as in the English word *far; e* like *a* in *ale; i* like *e* in *mete; o* as in English; *u* like *oo;* and *y* like Spanish *i*.

2. *Ai* and *ay* are like long *i* in English. *Au* sounds like *ou* in *our*. *Ei* and *ey* are pronounced a'-e.

3. The consonants *f, l*, (single,) *m, n, p, s, t*, and *v* are pronounced nearly as in English.

4. *B*, at the beginning of a word, sounds as in English; but when between two vowels, its sound somewhat resembles that of *v*, but with this difference—*v* is pronounced with the upper teeth placed against the under lip, while the sound of the Spanish *b* is formed by bringing the lips loosely or feebly into contact. This sound seems to be between that of *b* and the English *w*.

5. *C*, before *a, o*, and *u*, is pronounced as in English; before *e* and *i*, it has the sound of *th* in the word *thin*. In the Catalan dialect it is the same as in English.

6. *Ch* has the same sound as in English, except in the dialect of Catalonia, where it is pronounced like *k*.

7. *D*, at the beginning of a word, is sounded nearly as in English, but is pronounced with the tip of the tongue against the upper teeth, while in pronouncing the English *d*, the tongue is made to touch the roof of the mouth. At the end of a syllable, or between two vowels, *d*, in Spanish, sounds like the English *th* in *this*, but is somewhat softer

8. *G*, before *a*, *o*, and *u*, is hard, as in English.

9. *G*, before *e* and *i*, and *j* before every vowel, are pronounced like a strong guttural *h*, similar to the German *ch* in *ach*, which confers a force and manliness that is far from being disagreeable. The *g*, *j*, and *x*, before vowels are generally written now with *j*. Thus XIMENEZ is now generally written JIMENEZ.

10. *Gua* and *guo* sound somewhat like *gwa*, *gwo*, but the *g* is so soft that it is scarcely perceived; so that in these cases the sound of *gu* seems to approximate very nearly to that of the English *w*. *Gu*, before *e* and *i*, is usually sounded like *g* hard; thus, GUIANA is pronounced *ghe-a'na*. When, however, the *u* is marked with a diæresis, thus, *güi*, these two letters have the same sound as when before *a* or *o*, and consequently *güi* is pronounced *gwe* or *we*. See table at the end of this section.

11. *H*, in Spanish, is never pronounced, except in words beginning with *hue*, and then very slightly.

12. *J*: for this letter, see 9 and 18 of this Section.

13. *Ll* (now sometimes written *l*) has a sound which combines that of *l* and *y*, (consonant,) and is similar to the liquid *l* in French; thus *villa* or *vila* is pronounced *veel'-ya;* LLERENA, *lya-ra'-na*.

14. *N*, in similar manner, unites the sounds of *n* and *y*, and is like *gn* in French; thus, *pena*, is pronounced *pen'ya*, or *pane'ya*.

15. *Q*, in Spanish, is always followed by *u*. *Qu*, before *a* and *o*, is sounded as in English, or, in other words, is equivalent to *kw;* before *e* and *i*, it is pronounced like *k*, unless the *u* be marked with a diæresis, in which case it is like *kw*. See table at the end of this section.

16. *R* is similar to the French, but is trilled more strongly.

17. *T* is to be pronounced by putting the tip of the tongue against the upper teeth.

18. *X* is usually sounded like the Spanish *j*, which letter, according to the present mode of spelling, has, we have remarked, been generally substituted for it; thus, instead of the old spelling, XIMENES, XUCAR, &c., we now see *Jimenes*, *Jucar*, &c. *X*, before a consonant, or before a vowel marked with this sign ∧, is sounded as in English: Example—*Exterior*, *Exâminar*.

19. *Z* is to be pronounced like *th* in *thin*.

The following table may serve to show more clearly the manner in which *c*, *g*, *j*, *q*, *x*, and *z* are used in Spanish.

ca —	is pro'ced	ka.	cua or qua	is pro'ced	qua.
que	„	ka.	cue or qüe	„	qua.
qui	„	ke.	cui or qüi	„	que.
co —	„	ko.	cuo or quo	„	quo.
cu —	„	koo.			
ga	„	ga.	gua	„	gwa or wa.
gue	„	ga.	güe	„	gwa or wa.
gui	„	ghe	gui	„	gwe or we.
go	„	go.	guo	„	gwo or wo.
gu	„	goo.			
ja or xa	„	HA.	za —	„	tha.
je, xe, or ge	„	HA.	ze or ce	„	tha.
ji, xi, or gi	„	HE.	zi or ci	„	the.
jo or xo	„	HO.	zo —	„	tho.
ju or xu	„	HOO.	zu —	„	thoo.

OBS. 1. Spanish words or names ending in a consonant have the accent almost always on the last syllable: those ending in a vowel are generally accentuated on the penultima. If a word or name be an exception to either of these rules, in correctly printed Spanish works the accent is usually marked; as, CÓRDOVA, ALCALA', JU'CAR, CA'CERES. It should be observed that the *s* in the *plural* does not change the accent; hence, *casas*, "houses," though ending in a consonant, has the penultimate accent as well as *casa*, "a house."

OBS. 2. The Spanish language, as spoken in Mexico and South America, differs in some points materially from the true Spanish. Thus, *z* (and *c*, before *e* and *i*) instead of having the sound of *th*, are generally pronounced like *s*, and *b* and *v* are used interchangeably. Among the uneducated classes, *ll* is universally sounded like *y;* thus, *gallo* is pronounced almost *ga'yo*.

XIV.

SWEDISH.

1. The vowels *a*, *e*, *i*, *ä*, and *ö* are similar to the German.

2. *A* sounds like the English *o*. LULEA is pronounced *loo'la-ō;* TORNEA, *tor'na-ō*, &c.

3. *O*, at the end of a syllable, is like our *oo;* in other cases, like *o* in *not*.

4. *U* appears to blend the sounds of short *e* and of *oo*, being somewhat similar to *e'-oo*. The English *oo* is the sound nearest to it of any in our language.

5. *Y* is the same as in Danish; or, in other words, is similar to the French *u*.

6. The Swedish consonants are, for the most part, pronounced like the English, with the exception of *j*, and *g*, (before *e*, *i*, *ä*, *ö*, and *ü*,) which are equivalent to *y* consonant, and of *z*, which sounds like *ts*.

7. *Kj* sounds like the English *ch* in *child;* thus, KJÖPING (written, also, simply *Köping*) is pronounced *chöp'-ing*.

XV.

WELSH.

The Welsh is the Cambrian branch of the old Celtic.

1. *A* is pronounced as in the English word *far*, and sometimes as in *fat*. When circumflexed, it has the same sound as in *fate*.

2. *E* sounds as in *met;* when circumflexed, as in *bear*, *fête*.

3. *I* sounds as in *fig* or *marine*.

4. *O*, with an acute accent, as *o* in *gone;* when circumflexed, as in *note*.

5. *U* has the sound of *i* in *pin;* circumflexed, as *ee* in *seen*.

6. *W* sounds like the English *oo*.

7. *Y*, in the penultima or antepenultima, has the sound of *u* in *fur* or *tub;* in the final syllable, or in a monosyllable, it sounds as in *pin;* when circumflexed, it is like *ee*. The English word *sundry* will serve to exemplify its sound both in the penultima and ultima.

8. The Welsh diphthongs may all be said to be *proper*. *Ai* sounds like long *ī* in English; *aw*, like *ow*, as in *now;* *uw*, like *e-oo* or *u*, &c.

9. The consonants *b*, *d*, *h*, *k*, *l*, *m*, *n*, *p*, *r*, *s*, *t*, are the same as in English.

10. *C* is always hard, like *k*.

11. *F* (single) is like our *v*.

12. *G* is always hard, as in *get*.

13. *Ch* is similar to the German *ch*, or to the Greek *χ*.

14. *Dd* has the sound of *th* in *this*.

15. *Ff* is like the English *f*.

16. *Ll* is *l* aspirated, and has no equivalent in English. It bears a resemblance (though somewhat remote) to the sound of *thl*. "It is pronounced by fixing the tip of the tongue to the roof of the mouth, and breathing forcibly through the jaw-teeth on

both sides, but more on the right, as if written in English *llh.*" This sound (except at the beginning of names) is usually anglicized by *thl: e. g.* LLANELLY, *lan-eth'lee,* LLANGOLLEN, *lan-goth'len,* &c.

17. *Ph* is the same as in English, or like *f.*

18. *Th,* in Welsh, is the same as in the English words *thin, mouth,* &c.

OBS. The accent, in all Welsh words, is either on the penultima or on the last syllable—never on the antepenultima.

XVI.

ORIENTAL NAMES.

The Royal Asiatic and Geographical Societies have adopted a mode of spelling Oriental names first suggested by Sir William Jones. The vowels and diphthongs in words written according to this system are to be pronounced as follows:

VOWELS.

a as in *bat* or *cat;* in India as *u* in *but* or *cut.*
á as in *father, rather, lath;* sometimes, in Western Africa, as *e* in *there.*
e as in *bet, let.*
é as in *there, where;* or *ai* in *fair; ay* in *day; ei* in *weigh; ey* in *they.*
i as in *bit, wit.*
í as in *ravine, fatigue; ie* in *mien; ee* in *seen, between.*
o as in the French words *mol, poltron.*
ó as in *bold, cold; store, sore; oa* in *roar, hoar; ou* in *soul, source; ow* in *bow, row.*
u as in *full, pull, bull; oo* in *wood, wooden.*
ú as in *lunar, lurid; oo* as in *moon, spoon; ou* in *youth, your.*
aï as in *high, stile; y* in *dry, fly; ei* in *height, sleight.*
eï as in *their, eight; ey* in *they, prey; ai* in *pail, paint; ay* in *pray, say; a* in *ale, pale.*
ou as in *foul, count; ow* in *owl, fowl.*
eu as in *feudal, feud; ew,* in *few, mew, strew.*

CONSONANTS.

b as in *bad.*
ch as in *cheese, chair.*
d as in *dot, dole.*
dh, z as *th* in *thee, thou, that;* as *z* by Indians, Persians, and Turks.
d, dh, z ditto, ditto, preceded by *d,* or as double *d;* as *z* by Indians, Persians, and Turks.
dh, z ditto, ditto, ditto.
f as in *feel.*
g as in *good, girl.*
gh as in *rain, wrist,* as pronounced in Northumberland; often as a guttural *r* by the African Arabs.
h as in *hot, hill.*
h as in *hot;* or as *h* doubled.
k as in *kill, kite.*
kh, ch as in *loch;* or *gh* in *bragh; i. e.* the Welsh, Irish, Scotch, German, and Gr. *ch;* the Spanish *j,* &c.
k as in *kind,* but deep in the throat; and by the Moors and Egyptians, as *g* in *gold.*
l as in *lamb, lore.*
m as in *mad, man.*
n as in *not, near;* as *m* before *b* or *p.*
n, g as in *ring, sing;* only used in Turkish.
p as in *purse, peal.*
r as in *rain, red.*
s as in *sun, sweet.*
sh as in *short, shade.*
s as in double *ss.*
t as in *toad, tail.*
th, s as in *thick, thigh;* as *s* by Indians, Persians, and Turks; as *t* or *ts* by the common Arabs and Moors.
t as in *tt* doubled.
v as in *vast, very.*
w as in *waste, wet.*
y as in *you, yellow.*
z as in *zeal, zest.*
zh, s as in *pleasure, measure;* the French *j.*
'a, 'e, 'i, 'o, 'u } a peculiarly deep guttural catch or stoppage in the throat, which may be expressed by a short pause.

No Arab word can begin with two consonants; thus they say *Iskander* for *Skander,* from Alexander.

No two vowels can follow each other immediately; thus *"afiyeh,* instead of *"afieh.*

The double consonant is expressed by a mark (˘) above the letter.

The *a* of the article *al* is cut off after a vowel, and its *l* takes the sound of *d, dh, d, dh, dh, r, s, sh, s, t, th, t, l, n, z,* when it immediately precedes those letters, thus: *'Abdu-r-rahmán, Shujáu-d-daulah, Khairu-d-dín, Zeïnu-n-nisá, Mantiku-t-taïr,* for *'Abdu-l-rahmán,* &c.

The *emphasis* or *stress* is laid on the long vowel by the Arabs, but on the last syllable by the Persians and Turks. In compound names and titles, such as *Káyim-makám, Yenícheri-Aghá, Khazínah-dár-Odah-Báshí*—the Turks lay the greatest stress on the last syllable of the first word, thus: *Odáh-Bashí, Mahmúd-Aghá.*

The final *b, d,* and *j,* are sounded *p, t,* and *ch,* by the Khazináhdár Turks.

XVII.

ORTHOGRAPHY OF CHINESE NAMES.

M. Edward Biot has prefixed the following useful comparative table of the different methods of representing Chinese sounds in French, Portuguese, and English, to his 'Dictionnaire des noms compris dans l'empire Chinois:'

French.	Portuguese.	English.
Cha,	xa,	sha.
Chan,	xan,	shan.
Chang,	xam,	shang.
Chao,	xao,	shaou.
Che,	xe,	shay. shĕ. shĭh.
Chen,	xen,	shen.
Cheou,	xeu,	show.
Chi,	xy-xĕ,	she.
Chin,	xin,	shin.
Ching,	xim,	shing.
Cho,	xo,	shŏ-shŭh.
Chou-chu,	xo-xu,	shoo.
Chouang,	xoam,	schwang.
Choue,	xue,	shwŏ.
Chouï,	xuy,	shwŭy.
Chun,	xun,	shun.
Fa,	fa,	fa.
Fan,	fan,	fan.
Fang,	fam,	fang.
Feï-fi,	fy,	fe or fel.
Fen,	fuen,	fun.
Feou,	feu,	fow.
Fo-fe,	foe,	fŭh.
Fou,	fu-fŏ,	foo.

French.	Portuguese.	English.
Foung,	fum,	fung.
Haï,	hay,	hae.
Han,	han,	han.
Hang,	ham,	hang.
Hao,	hao,	haou.
He,	hĕ,	hĭh.
Hen,	hen,	hăn.
Heng,	hem,	hăng.
Heou,	heu,	how.
Hi,	hy-hie,	he, hĕĭh.
Hia,	hia,	hea.
Hiaï,	hiay,	heae.
Hian, Hien,	} hien,	hëen.
Hiang,	hiam,	heang.
Hiao,	hiao,	heaou.
Hieï-hie,	hie,	hee.
Hieou,	hieu,	hew.
Hin,	hin,	hin.
Hing,	him,	hing.
Hiŏ,	hiŏ,	hĕo.
Hiou,	hio,	heu.
Hiouan, Hiouen,	} hiuen,	heuen.
Hioue,	hiue,	heuĕ.
Hioung,	hium,	heung.
Hiu,	hiu-hio,	heu.
Ho,	ho,	ho.
Hoa,	hoa,	hwa.
Hoaï,	hoay,	hwae.
Hoan,	hoan,	hwan.
Hoang,	hoam,	hwang.
Hoe,	hoe,	hwŏ.
Hoeï,	hoey,	hwuy.
Hoen,	hoen,	hwăn.
Hou,	hu-hŏ,	hoo.
Houan,	hoan,	hwan.
Houng,	hum,	hung.
Houo,	huo,	hwŭh.
I. See letter Y.		
Jang,	jam,	jang.
Jao,	jao,	jaou.
Je,	ge,	jĕ.
Jen,	gen,	jăn.
Jeng,	gem,	jăng.
Jeou,	geu,	jow.
Ji,	ge,	jĭh.
Jin,	gin,	jin.
Jo,	jo,	jŏ, jŭh.
Jou,	ju, jo,	joo.
Jouan,	juen,	juen.
Jouï,	juy,	juy.
Joung,	jum,	jung.
Joun or Jun,	jun,	jun.
Kaï,	cay,	kae.
Khaï,	cay,	kae.
Kan,	can,	kan.
Khan,	can,	kan.
Kang,	cam,	kang.
Khang,	cam,	kang.
Kao,	cao,	kaou.
Khao,	cao,	kaou.
Ke,	ke,	kĭh.
Khe,	ke,	kih.
Ken,	ken,	kăn.
Khen,	ken,	kăn.
Keng,	kem,	kăng.
Kheng,	kem,	kăng.
Keou,	keu,	kow.
Kheou,	keu,	kow.

French.	Portuguese.	English.
Ki,	ky-kie,	ke.
Khi,	ky-kie,	keĭh.
Kia,	Kia,	kea.
Khia,	kia,	kea.
Kiaï,	kiay,	keae.
Khiaï,	kiay,	keae.
Kian, Kien,	} kien,	këen.
Khian, Khien,	} kien,	këen.
Kiang,	kiam,	keang.
Khiang,	kiam,	keang.
Kiao,	kiao,	keaou.
Khiao,	kiao,	keaou.
Kie,	kie,	këe.
Khie,	kie,	këe.
Kieou,	kieu,	kew.
Khieou,	kieu,	kew.
Kin,	kin,	kin.
Khin,	kin,	kin.
King,	kim,	king.
Khing,	kim,	king.
Kio,	kio,	keŏ.
Khio,	kio,	keŭh.
Kiou,	kio,	keu.
Khiou,	kio,	keu.
Kioue,	kiue,	keue.
Khioue,	kiue,	keue.
Kiouen,	kiuen,	keuen.
Khiouen,	kiuen,	keuen.
Kioung,	kium,	keung.
Khioung,	kium,	keung.
Kiu,	kiu,	keu.
Khiu,	kiu,	keu,
Kiun,	kiun,	keun.
Ko,	co,	ko.
Kho,	co,	kŏ-kŭh.
Kou,	cu,	koo.
Khou,	cu,	koo.
Koua,	cua,	kwa.
Khoua,	cua,	kwă.
Kouaï,	cuay,	kwae.
Khouaï,	cuay,	kwae.
Kouan,	cuan,	kwan.
Khouan,	cuan,	kwan.
Kouang,	cuam,	kwang.
Khouang,	cuam,	kwang.
Koue,	cue,	kwŏ.
Koueï,	cuey,	kwei.
Khoueï,	cuey,	kwei.
Kouen,	cuen,	kwăn.
Khouen,	cuen,	kwăn.
Koueng,	cuem,	kwăng.
Khoueng,	cuem,	kwăng.
Koung,	cum,	kung.
Khoung,	cum,	kung.
Kouo-kou,	cuo,	kwo.
Kouon,	cuon,	kwan.
La,	la,	la.
Laï,	lay,	lae.
Lan,	lan,	lan.
Lang,	lam,	lang.
Lao,	lao,	laou.
Le,	le,	lĭh.
Leng,	lem,	lăng
Leou,	leu,	low.
Li,	ly-lie,	le.
Liang,	leam,	leang.
Liao,	leao,	leaou.
Lie,	lie,	lëĕ.
Lien,	lien,	lëĕn.
Lieou,	lieu,	lew.
Lin,	lin,	lin.

French.	Portuguese.	English.
Ling,	lim,	ling.
Lio,	lio,	leŏ.
Liouen,	liuen,	leuen.
Liu,	liu,	leu.
Lo,	lo,	lo.
Lou,	lu-lo,	loo.
Louan,	luon,	lwan.
Louï-leï,	luy,	luy.
Loung,	lum,	lung.
Lun,	lun,	lun.
Ma,	ma,	ma.
Maï,	may,	mae.
Man,	man,	man.
Mang,	mam,	mang.
Mao,	mao,	maou.
Me,	mĕ,	mĭh.
Meï,	moey,	mei.
Men,	muen,	mun.
Meng,	mem,	măng.
Meoü,	meu,	mow.
Mi,	my-mie,	me.
Mien,	mien,	mëen.
Miao,	miao,	meaou.
Mie,	mie,	mëĕ, meĭh.
Mieou,	mieu,	mew.
Min,	min,	min.
Ming,	mim,	ming.
Mo,	mo,	mo.
Mou,	mu-mo,	moo.
Moung,	mum,	mung.
Na,	na,	na.
Naï,	nay,	nae.
Nan,	nan,	nan.
Nang,	nam,	nang.
Nao,	nao,	naou.
Neï,	nuy,	nuy.
Neng,	nem,	năng.
Neou,	neu,	now.
Ni,	ny, nie,	ne.
Niang,	niam,	neang.
Niao,	niao,	neaou.
Nie,	niĕ,	nëĕ.
Nieï,	niĕ,	neĭh.
Nien,	nien,	nëen.
Nieou,	nieu,	new.
Nin,	nin,	nin.
Ning,	nim,	ning.
Nio,	nio,	neŏ.
Niu,	niu,	new.
No,	no,	no.
Nou,	nu-no,	noo.
Nouan,	nuon,	nwan.
Noung,	num,	nung.
Nun,	nun,	nun.
'Aï or Ngaï,	ngay,	gae.
'An or Ngan,	ngan,	gan.
'Ang or Ngang,	ngam,	gang.
'Ao or Ngao,	ngao,	gaou.
'E or Nge,	nge,	gĭh.
'En or Ngen,	ngen,	găn.
'En or Ngeng,	ngem,	găng.
'Eou or Ngeou,	ngeu,	gow.
'O or Ngo,	ngo,	gơ.
'Ou or Ngou,	ngu,	goo.
Ou,	u,	woo.
Oung,	um,	ung.
Pa,	pa,	pa.
P'a,	pa,	pa.
Paï,	pay,	pae.
P'aï,	pay,	pae.
Pan,	pan,	pan.
P'an,	pan,	pan.
Pang,	pam,	pang.
P'ang,	pam,	pang.
Pao,	pao,	paou.
P'ao,	pao,	paou.
Pe,	pe,	pĭh.
P'e,	pe,	pĭh.
Peï,	poey,	pei.
P'ei,	poey,	pei.
Pen,	puen,	pun.
P'en,	puen,	pun.
Peng,	pum,	păng.
P'eng,	pum,	păng.
P'eou,	peu,	pow.
Pi-peï,	pi-pie,	pe.
Piao,	piao,	peaou.
P'iao,	piao,	peaou.
Pien,	pien,	pëen.
P'ien,	pien,	pëen.
Pieï pie,	pie,	pëĕ, peih.
P'ieï-p'ie,	pie,	pëĕ-peĭh.
Pieou,	pieu,	pew.
Pin,	pin,	pin.
P'in,	pin,	pin.
Ping,	pim,	ping.
P'ing,	pim,	ping.
Po,	po,	pŏ.
P'o,	po,	pŏ.
Pou,	pu-po,	poo.
P'ou,	pu-po,	poo.
Poung,	pum,	pung.
Sa,	sa,	sa.
Saï,	say,	sae.
San,	san,	san.
Sang,	sam,	sang.
Sao,	sao,	saou.
Se-che,	se,	sĭh.
Sen,	sen,	săn.
Seng,	sem,	săng.
Seou,	seu,	sow.
Si,	sy-sie,	se.
Sien,	sien,	sëen.
Siang,	siam,	seang.
Siao,	siao,	seaou.
Sieï-sie,	sie,	seĭh.
Sieou,	sieu,	sew.
Sin,	sin,	sin.
Sing,	sim,	sing.
Sio,	sio,	seŏ.
Siouen,	siuen,	seuen.
Siouei,	siue,	seuĕ.
Siu,	siu,	seu.
Siun,	siun,	siun.
So,	so,	so.
Sou,	su-so,	sŭh.
Souan,	suan,	swan.
Souï,	suy,	suy.
Soung,	sum,	sung.
Sun,	sun,	sun.
Sse-chi,	su-szu,	sze.
Ta,	ta,	ta.
Tha,	ta,	ta.
Taï,	tay,	tae.
Thaï,	tay,	tae.
Tan,	tan,	tan.
Than,	tan,	tan.
Tang,	tam,	tang.
Thang,	tam,	tang.
Tao,	tao,	taou.
Thao,	tao,	taou.
Te,	te,	tĭh.

French.	Portuguese.	English.
The,	te,	tĭh.
Teng,	tem,	tăng.
Theng,	tem,	tăng.
Teou,	teu,	tow.
Theou,	teu,	tow.
Ti,	ty, tie,	te.
Thi,	ty, tie,	te.
Tiao,	tiao,	teaou.
Thiao,	tiao,	teaou.
Tieï-tie,	tie,	tëe.
Thieï-thie,	tie,	tëe.
Tien,	tien,	teĕn.
Thien,	tien,	teĕn.
Tieou,	tieu,	tew.
Ting,	tim,	ting.
Thing,	tim,	ting.
To,	to,	tŏ.
Tho,	to,	to.
Tou,	to-tu,	too.
Thou,	to-to,	tŭh.
Touan,	tuon,	twan.
Thouan,	tuon,	twan.
Touï,	tuy,	tuy.
Thouï,	tuy,	tuy.
Toung,	tum,	tung.
Thoung,	tum,	tung.
Tun,	tun,	tun.
Thun,	tun,	tun.
Tcha,	cha,	cha.
Tch'a,	cha,	cha.
Tchaï,	chay,	chae.
Tch'aï,	chay,	chae.
Tchan. See Tsan.		
Tchang,	cham,	chang.
Tch'ang,	cham,	chang.
Tchao,	chao,	chaou.
Tch'ao,	chao,	chaou.
Tche,	che,	chay.
Tch'e,	che,	chay.
Tchen,	chen,	chen.
Tch'en,	chen,	chen.
Tcheou,	cheu,	chow.
Tch'eou,	cheu,	chow.
Tchi,	chy, che,	che.
Tch'i,	chy, che,	che.
Tchin,	chin,	chin.
Tch'in,	chin,	chin.
Tcho,	cho,	chŏ.
Tch'o,	cho,	cho.
Tchou, tchu,	chu,	choo-chŭh.
Tchou, } Tch'u, }	chu,	choo-chŭh.
Tch'oua,	choa,	chwa.
Tch'oua,	choa,	chwa.
Tchouen,	chuen,	chuen.
Tch'ouen,	chuen,	chuen.
Tchouang,	choam,	chwang.
Tch'ouang,	choam,	chwang.
Tchoue,	chue,	chuĕ.
Tch'oue,	chue,	chuĕ.
Tchouï,	chuy,	chuy.
Tch'ouï,	chuy,	chuy.
Tchoung,	chum,	chung.
Tch'oung,	chum,	chung.
Tchu. See Tchou.		
Tchun,	chun,	chun.
Tch'un,	chun,	chun.
Tsa,	ça,	tsă.
Thsa,	ça,	tsă.
Tsaï,	çay,	tsae.
Thsaï,	çay,	tsae.
Tsan,	çan,	tsan.
Thsan,	çan,	tsan.
Tsang,	çam,	tsang.

French.	Portuguese.	English.
Thsang,	çam,	tsang.
Tsao,	çao,	tsaou.
Thsao,	çao,	tsaou.
Tse,	çan,	tsĭh, tsze.
Thse,	çe,	tsĭh, tsze.
Tseng,	çem,	tsăng.
Thseng,	çem,	tsĕng.
Tseou,	çeu,	tsow.
Thseou,	çeu,	tsow.
Tseu,	çu,	tsze.
Thseu,	çu,	tsze.
Tsi,	çy-cie,	tse, tseĭh.
Thsi,	çy-çie,	tse, tseĭh.
Tsien,	çien,	tsëen.
Thsien,	çien,	tsëen.
Tsiang,	çiam,	tseang.
Thsiang,	çiam,	tseang.
Tsiao,	çiao,	tseaou.
Thsiao,	çiao,	tseaou.
Tsieï-tsie,	çie,	tseay.
Thsīe-thsie,	çie,	tsëē.
Tsieou,	çieu,	tsew.
Thsieou,	çieu,	tsew.
Tsin,	çin,	tsin.
Thsin,	çin,	tsin.
Tsing,	çim,	tsing.
Thsing,	çim,	tsing.
Tsio,	çio,	tseŏ.
Tsiouen,	çiuen,	tseuen.
Thsiouen,	çiuen,	tseuen.
Tsioueï,	çiue,	tseuĕ.
Tsiu,	çiu,	tseu.
Thsiu,	çiu,	tseu.
Tsiun,	çiun,	tseun.
Tso,	ço,	tso.
Thso,	ço,	tso
Tsou,	çu-ço,	tsoo.
Thsou,	çu-çu,	tsoo, tsŭh.
Tsouan,	çuon,	tswan.
Thsouan,	çuon,	tswan.
Tsouï,	çuy,	tsuy.
Thsouï,	çuy,	tsuy.
Tsoung,	çum,	tsung.
Thsoung,	çum,	tsung.
Tsun,	çun,	tsun.
Thsun,	çun,	tsun.
Wa,	va,	wa.
Waï,	vay,	wae.
Wan,	van,	wan.
Wang,	vam,	wang.
We,	ve-voe,	wŭh.
Weï,	vy,	we.
Wen,	ven,	wăn.
Wo,	vo,	wo.
Wou,	vu-vo,	woo.
Y or I,	y-ye,	e.
Ya,	ya,	ya.
Yaï,	yay,	yae.
Yan, yen,	yen,	yen.
Yang,	yam,	yang.
Yao,	yao,	yaou.
Ye,	ye,	yay-yĭh.
Yen-yn,	yn,	yin.
Yeou,	yeu,	yew.
Yng or ing,	ym,	ying.
Yo,	yo,	yŏ-yŭh.
You,	yo,	yo.
Youeï,	yue,	yuĕ.
Youen,	yuen,	yuen.
Young,	yum,	yung.
Yu,	yu,	yu.
Yun,	yun,	yun.

XVIII.

DIGRAPHIC ALPHABET OF ORDINARY TYPES FOR APPROXIMATING TO THE SOUNDS OF ALL LANGUAGES.

Mr. Alexander J. Ellis, pending the introduction of a true phonetic character, has suggested the use of the following sets of letters, each expressing a distinct, recognised, and as nearly as possible invariable sound. "For scientifically accurate etymological and phonetical researches, for investigations of the changes which words have undergone in time or place," writes Mr. Ellis, "a very full alphabet is required. But for the general purposes of the traveller, the missionary, the historian, the ethnologist, the geographer, the newspaper correspondent, the grammarian, the educator, a much rougher instrument suffices, provided that means are furnished for refining it when required. For this purpose," Mr. Ellis continues, "the following alphabet has been constructed, entirely from letters of ordinary founts, so that it can be immediately used in any missionary station, newspaper establishment, or country printing-office. Most of the letters are employed in familiar significations, so that words thus written almost suggest their own proper sounds. The key words here affixed are chosen so as to show the different shades of meaning attached to each letter. The alphabet is termed *digraphic* from the Greek δὶς *dhis*, double, and γραφή *ghrafii*, writing, because two or more letters are systematically employed to represent single sounds. The letters *a,h* are merely diacritical or distinctive, having no independent signification. The use of the diæresis (,) prevents all confusion. The long vowel sounds are represented by reduplication, but the construction of the alphabet is such that no confusion will arise if ā ē ī ō ū are employed to represent *aa, ee, ii, oo, uu* wherever they occur. The letters *ao, oe, ue* may also be represented by ǫ, ö, ü and their long sounds *aao, ooe, uue* by ǫǫ, öö, üü respectively. It should be remembered in employing this alphabet that any alteration in the use of the letters would entirely destroy the advantages arising from its systematic connection with a more complete form. Prof. Max Mueller of Oxford having published a 'Missionary Alphabet' on similar principles, it is added in a separate column.

G. German, Gr. Greek, Fr. French, I. Italian.

			Müller.
A	*a*	man, G. m*a*nn, Fr. p*a*s m*â*t p*a*tte	a
	aa	f*a*ther, G. m*ah*nen, Fr. p*â*te	*a*
	ao	n*o*t, I. r*o*cco	*o*
	aao	n*ough*t, I. p*o*co	ou
	ai	h*igh* *aye*, G. m*ei*n h*ai*n, Fr. pa*ïe*n	ai
	aoi	h*oy* *oi*l, G. *eu*le h*äu*ser	oi
	au	h*ow* h*ou*se, G. h*au*s, Fr. r*aou*t	au
	aa	Fr. ch*a*nt ch*a*mp *an en* pa*on*	-
B	*b*	*b*e e*bb* sha*bb*y	b
C	*c*	African full-tongued clucks	-
	cc	African half-tongued clucks	-
D	*d*	*d*o o*dd* ca*dd*y, Cerebral and dental *d*	d
	dh	*th*e, Span. Ma*d*ri*d*, Dan. ve*d*, Gr. δ	dh
	dzh	*j*u*dg*e, I. *g*iglio *Gi*acomo	*g*
E	*e*	m*e*n, G. m*ä*nner, Fr. *e*lle *é è*	e
	ee	m*a*ne, G. m*äh*ne b*ee*t, Fr. *ée* b*ê*te	*e*
	ea	Fr. v*in* f*in* m*en*tor app*en*dice bi*en*	-
Ǝ	ə	n*u*t t*o*n, Fr. *un* (əa) see *a* (after *n*)	0
	əa	Fr. *un* chac*un* auc*un* H*un*, à j*eun*	
F	*f*	*f*ace sta*ff* mu*ff*in, Gr. φ	*f*
G	*g*	*g*o e*gg* sta*gg*er	g
	gh	G. ta*g* tei*g*, Dutch *g*, Ar, *gh*ain, Gr. γ	-
H	H	*h*e *h*ew w*h*ich (,*h*) may also be used, but not *h*. In writing place a dot or two strokes under *h* to indicate H.) H*l* Welsh *ll*all; H*r* Welsh *rh*ag, HW *wh*ey; H*y* *h*ue (H*yuu*)	h (ʻ)
	h	only used to form the digraphs *dh, gh, kh, sh, zh*, &c.	-
I	*i*	b*i*n b*ee*n b*e*gone, G. b*i*n, Fr. *i*l	i
	ii	b*ea*n b*ea*m f*ee*l, G. f*ie*l, Fr. *î*le	*i*
	iu	f*ew* v*iew*, G. *iu*, Fr. *iou*	-
K	*k*	*k*eep *c*ape be*ck* be*ck*on, Fr. *qu*eue	k
	kh	G. da*ch* di*ch*, Dutch *ch*, Gr. χ	*h* (ʻ*h*)
L	*l*	*l*o e*ll* sa*ll*ow	l
	ly	I. gi*gl*io, Spanish *ll*, Portuguese *lh*	-
M	*m*	*m*e ai*m* ha*mm*er	m
N	*n*	*kn*ee *n*ay pai*n* ho*n*our, Cerebral *n*	n
	ng	si*ng* si*ng*er fi*ng*er, G. sa*ng*	ng (n·)
	ɳ	Fr. a*n* (*aɳ*) vi*n* vie*n*s (*veɳ vieɳ*) o*n* (o*ɳ*) u*n* (əɳ), in writing lengthen the second stroke of an *n* like the Greek η.	
	ny	Fr. monta*gn*e, I. *gn*	ny
O	*o*	wind*ow* *o*mit, G. v*o*n F. h*o*mme	-
	oo	b*o*ne, G. m*oh*ne, Fr. Sa*ô*ne	*o*
	oe	G. st*oe*cke, Fr. j*eu*ne, *e* mute	-
	ooe	G. G*oe*the, Fr. j*eû*ne	ö
	oɳ	F. *on* chans*on*	-
P	*p*	*p*ea a*p*e ha*pp*y	p
Q	*q*	Arabic *q*aaf, Hebrew ק	-
R	*r*	*r*a*r*e va*r*y fu*r* fi*r*, G. *r*, Fr. *r*.	r
S	*s*	*s*ee *c*ea*s*e mi*ss*ile, Dental *s*	s
	sh	*sh*e, G. *sch*, Fr. *ch*, I. *sc*ena	*s*
T	*t*	*t*ea ea*t* ki*tt*en, Cerebral and dental *t*	t
	th	*th*in, Spanish *z*, *c*e, G θ	th
	tsh	*ch*eese ea*ch* ma*tch*, I. *c*e *c*iò Span. *ch*	*k*
U	*u*	p*u*ll, G. pf*u*nd, Fr. m*ou*ler	u
	uu	p*oo*l, G. pf*uh*l, Fr. *aoû*t	*u*
	ue	G. h*ue*tte, Fr. h*u*tte	-
	uue	G. h*ue*te, Fr. *û*	ü
V	*v*	*v*ie e*v*e se*v*en, G. *w*	v
W	*w*	*w*eal, Arabic *w*aa	w
Y	*y*	*y*et, G. *j*	y
Z	*z*	*z*eal ea*s*e hou*s*es, Dental *z*	z
	zh	a*z*ure vi*s*ion, Fr. *j*	*z*
	(')	indistinct murmur, *bed'* Fr. tabl*e* sabr*e*	
	(ʻ)	slight whisper, *bet*ʻ; direction to whisper the word which follows, as ʻ*bet*.	
	(,)	diæresis, to separate the parts of digraphs when each letter has a separate meaning, or to indicate letters doubled in pronunciation, as *mii*n,nes, *bu*k,kees, (meanness, bookcase). It may be used also for Arabic HAMZA and ,ain.	
	(-)	hyphen to connect letters in separate words, as Fr. *la-m'suer ki-v'ne-d toabe* (la mesure qui venait de tomber).	
	(ˊ)	over or under a letter (*shraíben, Lə́ndən*) or at the close of a syllable (*vr'tiu*), or (·) at the close (*shraib·en, Lən·dən, vr·tiu*), marks the place of the accent, and should *never* be omitted. The form (·) will also serve as a diæresis, thus *sen·no* for *sen,no* or *sén,no*.	

ARTICLE V.

OUTLINES OF GENERAL ETHNOLOGY.

The science of Ethnology—to which considerable attention has been paid in the compilation of the present Gazetteer—seeks to solve the often agitated question of the unity of the human race, or whether, as Scripture seems to inform us, the various races of mankind now existing on the globe, or whose existence in past ages is matter of authentic history, are "of one blood," the descendants of one common parent; to trace the descent and affiliation of nations and tribes; and to note the operation of external and internal influences on the physical peculiarities and geographical distribution of the different families of mankind.

LINNÆUS referred all the human family to five races: the AMERICAN, the EUROPEAN, the ASIATIC, the AFRICAN, and a fifth division embracing what he styled men of preternatural or abnormal formation.

BUFFON attempted a more philosophical arrangement of the race of mankind into the HYPERBOREAN or Lapland family, embracing the inhabitants of the Polar regions; the TARTAR family, embracing the Eastern and Central nations of Asia; the SOUTHERN-ASIATIC family, which comprises the South-Sea islanders; the EUROPEAN, the ETHIOPIAN, and the AMERICAN. He ultimately, however, reduced these six families to five, by classifying the Tartars and Laplanders together.

BLUMENBACH'S CLASSIFICATION.

BLUMENBACH recognised four great subdivisions or varieties of the human race, namely, the CAUCASIAN; the MONGOLIAN, in which he included both the Tartar and the Laplander of Buffon; the ETHIOPIAN; the MALAY, corresponding to the Southern-Asiatics of Buffon; and the AMERICAN, comprising the native races that once peopled the American continent. Of these, Blumenbach regarded the Caucasian or European to be the primary stock; and the other five to be offsets from this stock. No two persons are ever met with possessing precisely the same sort of features; yet there are certain generally characteristic countenances belonging to the different races of mankind, and even to particular nations. To enter into all the minute variations would be endless, but they have been reduced by Blumenbach to the five following, which include all the leading traits:

1. An oval countenance, with the features moderately inter-distinguished; high and broad forehead; nose aquiline, or slightly convex or prominent; cheek-bones not prominent; small mouth and lips slightly curved; chin full and well rounded. This face is most consonant with our ideas of beauty, and is characteristic in greater or less degrees of perfection of the *Caucasian* tribes.

2. Broad and flattish face; parts ill-distinguished from each other; space between the eyes flat and broad; flat nose; projecting cheeks; narrow and oblique eyelids, and chin rather prominent. This is the *Mongolian* face.

3. Broad, but not flat visage; prominent cheek bones; short-forehead; eyes deeply fixed; nose flattish, but prominent. Such is the *American* face.

4. Narrow face, projecting below; narrow, retreating, and arched forehead; prominent eyes; thick nose and lips, &c. *Negro* face.

5. Face not so narrow as the last, rather projecting downwards; bottled nose and large mouth. *Malay* countenance.

Placing these variations of the human countenance thus beside each other, the difference appears very striking; but we must repeat of the characters of countenance above described, as of those of colour, that they are by no means to be found the same in all the individuals of each respective race; on the contrary, individual deviations from the given type are numerous. Among the African and American nations, many persons are found with features as regular and as handsome as any Europeans. The same is more especially true of the South-sea islanders. Among ourselves how many have the characteristic features of other races! It may be observed, however, that rather less deviation from the characteristic model is observable in the Mongolian than in the other varieties. Though even there there is considerable difference, if not in form, at least in expression of features. As for example, between the Calmuck and the Chinese.

From the features we naturally proceed to the skull. This, in the Caucasian variety, is more fully developed in the upper and front parts, these forming a large and smooth convexity, a little flattened towards the temples. There is a general softness, harmony, and proportion in the entire contour of the head. Some differences in the formation of the skull have been observed in different nations of the Caucasian race; in the Turks, for instance, a singular globularity of this part has been remarked; and in the Poles and Russians a considerable contraction of the orbits. But we have not sufficient information on this subject to lead us to any definite conclusions.

In the Mongolian variety the head is of a square form, with the forehead low and slanting. The orbits are large and open, and the superciliary arches elevated.

In the Ethiopian variety the front of the head is laterally compressed and considerably elongated, so that the length of the whole skull from the teeth to the occiput is great. It forms a complete contrast to the globular head of some Europeans, and to the square head of the Calmucks. The forehead is wonderfully narrowed off, and the face widened below; the frontal bone is shorter, and it and the parietal less excavated than in the European. The temporal ridge is higher; the compression of the front of the skull arises from the strength of the temporal muscles; the foramen magnum is larger and farther back, and the apertures for the nerves are larger; the bony substance is hard, and the whole weight of the skull more considerable; the organs of sense are more developed, and the narrow forehead and protruded muzzle give to the Negro head the appearance of a decidedly animal character.

Some of the South African tribes vary a little from the Negro conformation of skull. In the head of a Bushman, given by Blumenbach, the cranium is less compressed, the orbits and cheek-bones are wide, and the jaws not prominent. There are other differences, but we scarcely know sufficient of these tribes to class them under any given variety. The origin of the Hottentots, Caffres, Bushmen, and their subdivisions, found at the extremity of South Africa, is quite unknown. They exhibit characters strongly approximating to, and discrepancies equally deviating from, their Negro neighbours.

The American head is thus characterized; broad cheek-bones, depressed forehead, deep orbits, and the nasal cavity generally large.

The Esquimaux and Greenlanders, who seem to form a link between the Americans and Mongolians, have broad cheek-bones, large jaws and face, flattened nose, the cranium sufficiently ample, but distinguished by a posterior elongation.

The Carib tribes are conspicuous for a most remarkable depression of the forehead, which defect, like others of the Americans, they increase by artificial means. The hinder parts of the skull greatly preponderate; the face is large and muscular; the nasal bone neither small nor flat; the cavity is large, and the jaws and teeth exhibit manifestations of great strength.

The general characters in this respect attributed to the Malay variety are, a moderately-narrowed cranium slanting at the interior and upper part; face large, and jaws prominent. But, indeed, the numerous nations comprehended, with not much philosophical precision, under this variety, exhibit very various and opposing characters; some are not distinguishable in the formation of this part from Europeans, some partake of the Mongole, and many of the Negro type. In truth, the above division of skulls is somewhat arbitrary, and though sufficient for general purposes, is by no means universally applicable.

Camper was the first anatomist who distinguished and described in an accurate manner the differences of form which have been discovered on comparing the skulls of different races; he invented a technical method by which he imagined he could display in a single measurement the essential difference of skulls as to form and capacity, not only in reference to various nations of men, but likewise to the inferior species of animals; this method he thus describes:—The basis on which the distinction of nations is founded may be displayed by two straight lines, one of which is to be drawn through the meatus auditorius to the base of the nose, and the other touching the prominent centre of the forehead, and falling thence on the most advancing part of the upper jaw-bone, the head being viewed in profile. In the angle produced by these two lines the distinctions are found to exist in different nations and species of animals. The heads of birds display the smallest angle, and it always becomes of greater extent in proportion as the animal approaches more nearly the human figure. In one species of the ape tribe, the head has a facial angle of 42 degrees; in another of the same family, the facial angle contains 50 degrees; next to this is the head of an African Negro, which, as well as the Calmuck, forms an angle of 70 degrees; whilst in the heads of Europeans it contains 80 degrees. On this difference of 10 degrees in the facial angle the superior beauty of the European depends; while the high character of beauty, in some ancient works of statuary, as in the head of Apollo and the Medusa of Sisocles, is given by an angle of 100 degrees.

CUVIER'S CLASSIFICATION.

Cuvier distinguishes only three races: the Caucasian or White; the Mongolian or Yellow; and the Ethiopian or Black; but hesitates whether to refer the Malays, Papuans, and South-Sea islanders to any of these races. The American he classifies with the Mongolian. "Although," says that distinguished naturalist, "the human species would appear to be single, since the union of any of its members produces individuals capable of propagation, there are, nevertheless, certain hereditary peculiarities of conformation observable, which constitute what are termed *races.* Three of these in particular appear eminently distinct: the *Caucasian*, or white, the *Mongolian*, or yellow, and the *Ethiopian*, or negro.

"The Caucasian, to which we belong, is distinguished by the beauty of the oval which forms the head: and it is this one which has given rise to the most civilized nations,—to those which have generally held the rest in subjection: it varies in complexion and in the colour of the hair.

"The Mongolian is known by his projecting cheek-bones, flat visage, narrow and oblique eyebrows, scanty beard, and olive complexion. Great empires have been established by this race in China and Japan, and its conquests have sometimes extended to this side of the Great Desert; but its civilization has always remained stationary.

"The Negro race is confined to the southward of the Atlas chain of mountains: its colour is black, its hair crisped, the cranium compressed, and nose flattened. The projecting muzzle and thick lips evidently approximate it to the apes: the hordes of which it is composed have always continued barbarous.

"The name *Caucasian* has been affixed to the race from which we descend, because tradition and the filiation of nations seem to refer its origin to that group of mountains situate between the Caspian and Black Seas, whence it has apparently extended by radiating all around. The nations of the Caucasus, or the Circassians and Georgians, are even now considered as the handsomest on earth. The principal ramifications of this race may be distinguished by the analogies of language. The Armenian or Syrian branch, spreading southward, produced the Assyrians, the Chaldeans, the hitherto untameable Arabs, who, after Mahomet, expected to become masters of the world; the Phœnicians, the Jews, the Abyssinians, which were Arabian colonies, and most probably the Egyptians. It is from this branch, always inclined to mysticism, that have sprung the most widely extended forms of religion. Science and literature have sometimes flourished among its nations, but always in a strange disguise and figurative style.

"The Indian, German, and Pelasgic branch is much more extended, and was much earlier divided; notwithstanding which, the most numerous affinities have been recognized between its four principal languages—the Sanscrit, the present sacred language of the Hindoos, and the parent of the greater number of the dialects of Hindostan; the ancient language of the Pelasgi, common parent of the Greek, Latin, many tongues that are extinct, and of all those of the south of Europe; the Gothic or Teutonic, from which are derived the languages of the north and north-west of Europe, such as the German, Dutch, English, Danish, Swedish, and their dialects; and finally, the Sclavonian, from which are descended those of the north-east, the Russian, Polish, Bohemian, and that of the Vandals. It is by this great and venerable branch of the Caucasian stock, that philosophy, the arts and sciences, have been carried to their present state of advancement; and it has continued to be the depository of them for thirty centuries. It was preceded in Europe by the Celts, whose tribes, once very numerous, came by the north, and are now confined to its most western extremities; and by the Cantabrians, who passed from Africa into Spain, and have become confounded with the many nations whose posterity have intermingled in that peninsula. The ancient Persians originate from the same source as the Indians, and their descendants still present a very close resemblance to the nations of Europe.

"The Scythian and Tartar branch, extending first towards the north and north-east, and always wandering over the immense plains of those countries, returned but to devastate the happier abodes of their more civilized brethren. The Scythians, who, at so remote a period, made irruptions into Upper Asia; the Parthians, who there destroyed the Greek and Roman domination; the Turks, who there subverted that of the Arabs, and subjugated in Europe the unfortunate remnant of the Grecian people, were all offsets from this branch. The Finlanders and Hungarians are tribes of the same division, which have strayed among the Sclavonic and Teutonic nations. Their original country, to the north and eastward of the Caspian Sea, still contains inhabitants who have the same origin, and speak similar languages; but these are mingled with many other petty nations, variously descended, and of different languages. The Tartars remained unmixed longer than the others throughout that extent of country included between the mouth of the Danube to beyond the Irtisch, from which they so long menaced Russia, and where they have finally been subjugated by her. The Mongoles, however, have mingled their blood with that of the nations they conquered, many traces of which may still be found among the inhabitants of Lesser Tartary.

"It is to the east of this Tartar branch of the Caucasian race that the Mongolian race begins, whence it extends to the eastern ocean. Its branches, the Calmucks and Kalkas, still wandering shepherds, traverse the great desert. Thrice did their ancestors, under Attila, Genghis, and Tamerlane, spread far the terror of their name. The Chinese are the most anciently civilized branch, not only of this race, but of all known nations. A third branch, the Mandshurs, have recently conquered and still govern China. The Japanese, Coreans, and nearly all the hordes which extend to the north-east of Siberia, subject to Russia, are also to be considered, in a great measure, as originating from this race; and such also is deemed to be the fact with regard to the original inhabitants of various islands bordering on that archipelago. With the exception of some Chinese literati, the nations of the Mongolian race pertain generally to different sects of Buddism, or the religion of Fo. The origin of this great race appears to have been in the Altai mountains, as that of ours in the Caucasus; but it is impossible to trace with the same certainty the filiation of its different branches. The history of these wandering nations is as fugitive as their establishments; and that of the Chinese, confined exclusively to their own empire, furnishes little that is satisfactory with respect to their neighbours. The affinities of their languages are also too little known to direct us in this labyrinth.

"The languages of the north of the peninsula beyond the Ganges, as well as that of Thibet, bear some relation to the Chinese, at least in their monosyllabic structure; and the people who speak them are not without resemblance to the other Mongoles: but the south of this peninsula is inhabited by Malays, whose forms approach them much nearer to the Indians, and whose race and language are distributed over the coasts of all the islands of the Indian archipelago. The innumerable small islands of the southern ocean are also peopled by a handsome race, who appear to hold a near relation to the Indians, and whose language has much affinity with the Malay: but in the interior of the larger islands, particularly in the milder portions of them, there exists another race of men with black complexions, and Negro faces, all extremely barbarous, which are named Alfourous [see article ALFORAS]; and on the coasts of New Guinea and the neighbouring islands, are other Negroes nearly similar to those of the eastern coast of Africa, which are termed Papous; to the latter are generally referred the natives of Van Diemen's Land [now rapidly approaching to extermination,] and those of New Holland to the Alfourous.

"Neither the Malays nor the Papous are easily referable to either of the three great races; but can the former be clearly distinguished from their neighbours on both sides, the Caucasian Indians and the Mongolian Chinese? We avow that we cannot discern in them sufficient traits for that purpose. Are the Papous Negroes, which may formerly have strayed into the Indian Ocean? We possess neither figures nor descriptions precise enough to enable us to reply to this question.

"The inhabitants of the north of both continents, the Samoyedes, the Laplanders, and the Esquimaux, are derived, according to some, from the Mongolian race: but others regard them as mere degenerate offsets from the Scythian and Tartar branches of the Caucasian race.

"The Americans have not yet been referred clearly to either of the races of the eastern continent; nevertheless, they have no precise or constant character, which can entitle them to be considered as a particular one. Their copper-coloured complexion is not sufficient: their general black hair and scanty beard would induce us to approximate them to the Mongoles, if their defined features, their nose as projecting as ours, their large and open eyes, did not oppose such a theory, and correspond with the features of the European. Their languages are as numberless as their tribes, and no demonstrative analogies have as yet been obtained, either with each other, or with those of the ancient world."

FISCHER'S CLASSIFICATION.

FISCHER, in his *Synopsis Mammalium*, indicates what he conceives to be seven species of *Homo* (reducing the number that had previously been assigned by Bory St. Vincent): and the numerous divisions and subdivisions of that naturalist being tolerably in accordance with the apparent value of the characters presented, whether or not they truly represent the real distinctions, or, in some instances, similarity be confounded with indentity (a problem to which philology seems to offer the only key), the outline of his arrangement may be transferred to the present work, where it may chance to prove useful to some observers. His supposed *species* are as follow:—

1. *H. Japeticus*, Bory; corresponding to the Caucasian race of Cuvier.—This is distributed under three principal varieties, termed *Caucasicus*, *Arabicus*, and *Indicus*: of these the first is arranged into five

subvarieties, named *Caucasicus* (*Orientalis*), *Pelagius* (*Meridionalis*), *Celticus* (*Occidentalis*), *Germanicus* (*Borealis*), and *Sclavonicus* (*Intermedius*), which severally comprehend the Caucasic, Pelasgic, Celtic, Teutonic, and Sclavonic (including the Sarmatic) nations; the second into two subvarieties, *Atlanticus* (*Occidentalis*), and *Adamicus* (*Orientalis*), respectively containing the Phœnicians, ancient Numidians, and Guanches, or the Punic nations, and the Abyssinians, primitive Egyptians (modern Copts), Jews, Armenians, Arabians, &c., or the Coptic and Semitic nations.

2. *H. Neptunianus*, Bory.—Ranged under three subdivisions: the first unnamed (Qu. *Malayanus?*) allied to—probably much mingled with—the Indian variety of *H. Japeticus*, and consisting of the well-known Malays, which people the coasts only of the peninsula of Malacca, the islands of the Indian ocean, Madagascar, &c., never penetrating inland; the second, *Occidentalis*, comprising the New Zealanders, and natives of the Society, Friendly, Sandwich, and other islands scattered over the Pacific ocean,—it is suggested, also, (but with due and much required hesitation,) the ancient Mexicans and Peruvians: the third, *Papuensis*, composed of certain inhabitants of part of the north coast of New Guinea, the shores of the islands Waigou, Salwaty, Gammeu, and a few others, is obviously a hybrid race, derived from the intermixture of the Malay and true Papou. Cuvier has remarked the affinity of language subsisting between the Malays and South Sea Islanders.

3. *H. Scythicus*, Bory.—The first division of this, unnamed (Qu. *Mongolensis?*) consists of the Calmucks and other Tartars; the second, *Sinicus* (*Homo sinicus* of Bory), of the Chinese, Japanese, &c.; and the third and last, *Hyperboreus* (*Homo hyperboreus*, Bory), of the Esquimaux. It corresponds to the Mongolian race of Cuvier.

4. *H. Americanus*, Bory.—"*Species*," the author writes, "*adhuc male cognita, forsan tota vel ex parte ad Scythicam reducenda*," of which the latter only is in the least probable. "*Autochthones Americæ meridionalis, in stirpes innumeras distributi; e. g. Omaguæ, Guarani, Coroadi, Atures, Otomaqui, Botucudi, Guiacæ, Cherrucæ, &c.*" A second division is designated *Patagonus*, (being the *Homo Patagonus* of Bory,) composed of the large-statured Patagonians.

5. *H. Columbicus*, Bory.—The ordinary *red Indian* of America.

6. *H. Æthiopicus*, Bory.—Divided into the true Negro, not otherwise named; *Caffer*, *Homo Caffer*, Bory, inhabiting Caffraria, and part of the coast of Madagascar; *Melanoides* (*Homo melaninus*, Bory), the Papous or indigenous inhabitants of Madagascar, the shores of New Guinea, the islands of New Britain, New Ireland, and many others, also of Van Diemen's Land; and *Hottentotus* (*Homo Hottentotus*, Bory), the Bush and other Hottentots, which, it may be remarked, have not a few analogies with the nomadic Mongoles. The last appear to have been much reduced and encroached on, till a remnant only is left near the south coast of Africa, just as the Celts are now confined to the extreme west of Europe.

7. Lastly, *H. Polynesius*, Fischer (*H. australaricus*, Bory).—The Alfourous, the lowest in the scale of human beings: comprising the inland inhabitants of the Malay peninsula, the islands of the Indian Ocean, Madagascar, New Guinea, New Holland, &c. Such is the arrangement of an able and accomplished naturalist, published in 1829.

PRICHARD'S CLASSIFICATION.

PRICHARD, in his elaborate 'Researches into the Physical History of Mankind,' contends strenuously and successfully for unity of species in the genus *Homo*. The following are the leading varieties of Man according to Dr. Prichard.

"On comparing the principal varieties of form and structure which distinguish the inhabitants of different countries, we find that there are seven classes of nations which may be separated from each other by strongly marked lines. Among their principal characteristics are peculiar forms of the skull, but these are by no means the only difference which require notice and particular description. These seven principal classes are, first, those nations which in the form of their skulls and other physical characters resemble EUROPEANS, including many nations in Asia and some in Africa; secondly, races nearly similar in figure, and in the shape of the head, to the Kalmucks, Mongoles, and Chinese. These two first classes of nations will be designated, for reasons to be explained, IRANIAN and TURANIAN nations, in preference to Caucasian and Mongolian. The third class are the native AMERICAN nations, excluding the Esquimaux and some tribes which resemble them more than the majority of inhabitants of the New World. The fourth class comprises only the HOTTENTOT and Bushman race. A fifth class are the NEGROES; the sixth, the PAPUAS, or woolly-haired nations of Polynesia; the seventh, the ALFOUROU and Australian races. The nations comprised under these departments of mankind differ so strikingly from each other, that it would be improper to include any two of them in one section, and there is no other division of the human family that is by physical traits so strongly characterized. There are, indeed, some nations that cannot be considered as falling entirely within either of these divisions, but they may be looked upon as approximating to one or another of them." Dr. Prichard affirms, of the Caucasian race of Cuvier, that, "there is no truth in the assertion that the traditions of all these nations deduce their origin from Caucasus," and states, of his Indo-Atlantic, or *Iranian* nations, that "complexion does not enter among the characters of this type, since it is of all shades, from the white and florid colour of the northern Europeans, to the jet-black of many tribes in Lybia, and southward of Mount Atlas."

LATHAM'S CLASSIFICATION.

Dr. Latham only admits three great primary divisions of the human race: 1st, the JAPETIDÆ, corresponding nearly with the Iranian race of Prichard; 2d, the MONGOLIDÆ, identical with Cuvier's group, and embracing the Malay, the American, and the Australian; and 3d, the ATLANTIDÆ or African race, embracing the Semitic nations.

PICKERING'S CLASSIFICATION.

In the 7th vol. of the magnificent report of the United States Exploring Expedition, published by the government of that country, Dr. Pickering has sketched, from personal observation chiefly, eleven races of men, under which he classifies all known varieties of the human family. An outline of his valuable observations will, it is thought, form a valuable addition to our Appendix; and serve to reunite various scattered ethnographical notices throughout the body of our Gazetteer.

Dr. Pritchard commences his enumeration of races by remarking that three races of men are familiarly known, and are admitted by general consent. "The same three physical races have been considered by eminent naturalists (who, however, have not travelled) to comprise all the varieties of the human family. Blumenbach has indicated a fourth race, the Malay; and even a fifth has been shadowed forth in the accounts of the Australian seas. It was impossible, however, from the materials furnished by books, to define the geographical boundaries of these races; a point which seemed of importance, as forming in a good degree the basis of our reasoning on the whole subject.

"This then was one of the objects of investigation I proposed to myself on joining the Exploring Expedition. At one time during the voyage, I thought my task nearly accomplished; and, after visiting Australia and New Zealand, I actually penned an opinion, that the races of men were five in number. Soon, however, I was compelled to admit three more: neither was this the limit of the productiveness of nature, in new and undreamt of combinations of feature. More careful observation than at the outset had seemed necessary was now called into requisition; and often, for a time, I experienced perplexity. One difficulty arose, in fixing in the mind, while passing from place to place, the relative shades of complexion. Fortunately for my purpose, tattooing was practised in many of the countries visited, and these markings afforded a convenient test of the depth of hue. Individuals, also, of three or more races being present among the crews of our vessels, afforded the means of making some direct comparisons. In the end all difficulties vanished, and I was enabled to arrive at satisfactory conclusions. It should be observed, that in the countries visited by the expedition, the inhabitants present among themselves great uniformity of feature and complexion: while in the Arab countries and in Western Hindostan, there is an astonishing diversity of aspect in the population; independently, to all appearance, of the great mixture of races. The mountain-region of Abyssinia is said likewise to present a seemingly heterogeneous population; but in all the countries which I have myself visited, the varieties of feature have appeared susceptible of reduction to the arrangement adopted in the present work.

"I have seen in all ELEVEN RACES OF MEN; and though I am hardly prepared to fix a positive limit to their number, I confess, after having visited so many different parts of the globe, that I am at a loss where to look for others. They may be enumerated conveniently enough in the order of complexion; and beginning with the lightest, I will add some of the more obvious distinctive characters.

a. *White.*

1. ARABIAN. The nose prominent, the lips thin, the beard abundant, and the hair straight or flowing.

2. ABYSSINIAN. The complexion hardly becoming florid; the nose prominent, and the hair crisped.

Brown.

3. MONGOLIAN. Beardless, with the hair perfectly straight and very long.

4. HOTTENTOT. Negro features, and close woolly hair; and the stature diminutive.

5. MALAY. Features not prominent in the profile; the complexion darker than in the preceding races, and the hair straight or flowing.

c. *Blackish-brown.*

6. PAPUAN. Features not prominent in the profile; the beard abundant, the skin harsh to the touch, and the hair crisped or frizzled.

7. NEGRILLO. Apparently beardless; the stature diminutive, the features approaching those of the Negro, and the hair woolly.

8. INDIAN OR TELINGAN. The features approaching those of the Arabian; and the hair, in like manner, straight or flowing.

9. ETHIOPIAN. The complexion and features intermediate between those of the Telingan and Negro; and the hair crisped.

d. *Black.*

10. AUSTRALIAN. Negro features, but combined with straight or flowing hair.

11. NEGRO. Close woolly hair; the nose much flattened, and the lips very thick.

"In an absolute sense, the terms *white* and *black* are both inapplicable to any shade of the human complexion; but they are sanctioned by general usage, and there may be some convenience in retaining the above four general divisions. Two of the races may therefore be designated as white, three as brown, four as blackish-brown, and two as black. Five of the races have the hair straight or flowing; while in the others it is more or less crisped, and in two of them it may with propriety be termed wool.

"Other modes of associating the races may also be mentioned. Maritime habits, and the part they appear to have taken in colonizing the globe, would lead us to separate the Malay, Negrillo, and Papuan, or the three island from the eight continental races.

"Again, looking to their distribution over the surface of the globe, six of the races may be regarded as Asiatic or East Indian, and four as African; the eleventh (the white race) being in common, or holding geographically an intermediate position.

"The existence of races, it should be observed, is a phenomenon independent of climate. All the physical races that occur in cold regions can be traced by continuity to the tropics, where moreover we find other races in addition. By the same evidence of geographical continuity, the population of one hemisphere can be satisfactorily derived from the other; but a difficulty arises in narrowing the circle. On the one hand, it seems quite impossible to trace the four African races to any part of Asia; and, on the other, it will be equally difficult to connect the Mongolian race with the African continent."

Colour of the skin in different races of men.—The colouring matter is understood to reside in a membranous network of greater or less density extending over the surface of the body, called the *rete mucosum.* This is situated between the chorion or true skin and the cuticle. The *rete mucosum*, or, as it is sometimes called, the cutaneous reticle, consists of a fine texture of vessels, containing fluids of different shades in the black and tawny races. It seems, however, doubtful whether any such membrane for the deposition of colouring fluid exists in white men, though the varieties of fair and dark which we observe among them would seem to require some organization of this kind; nor does this theory sufficiently illustrate the occasional instances of pied or spotted men. The human skin exhibits various shades of white, yellow, red, brown, and black. There is every possible intermediate shade between the fairest white and the deepest black, but no one gradation of colour is found in all the individuals of any nation. Generally speaking, however, we may refer all the national varieties of colour to the five following classes :—

(1.) *White*, accompanied with redness of cheeks. This characterizes all the Europeans except the Laplanders, the Western Asiatics, and the Northern Africans. Considerable variety will be found to exist in the colour generally called white. The Albino possesses a skin of a reddish or a dead white colour, with yellowish white or milk-white hair, and red or very light coloured eyes. The hair over the whole body is unusually soft and white, not of the hoary colour of age, nor the light yellow or flaxen tint of the fair-haired races. It is rather that sort of colour peculiar to a white horse. These peculiarities evidently arise from a deficiency in the colouring principle, which is much the same in the skin, hair, and eyes. The latter organs are in the Albinos peculiarly sensible to the stimulus of light, in consequence of the want of a black pigment, the office of which is to absorb its superfluous portions. Hence we find the eyelids of these people generally closed, and the eyes usually exhibiting some appearances of morbid phenomena. But in twilight, dusk, or even a close approach to darkness, they see remarkably well. This peculiarity exists from birth, never changes, and may be propagated by generation. Some would refer the Albino variety to disease, but this notion appears incorrect, inasmuch as most of the individuals thus characterized are observed to be perfectly strong and healthy. This variety was first observed in the Africans, but it is far from being peculiar to that race. It has, however, never occurred except in detached instances, for though it is observed to be more prevalent in some parts of the world than others, the notion of entire Albino tribes has been exploded. There is another race of men, with a remarkably fair complexion, yellow, flaxen, or red hair, and blue or gray eyes. In these persons the cutaneous capillaries are easily filled, and consequently they exhibit a general sanguineous tint, deeper and more florid in the face. The ancient and modern Germans belong to this variety, and generally the Danes, Dutch, Swedes, English, &c. Lastly, a race very extensive is found with skin of a brownish white, and dark brown or black hair. The Southern Europeans and Western Asiatics are of this character.

(2.) The second grand variety in human colour is *yellow* or *olive*. This characterizes all the Mongolian tribes, and, generally speaking, most of the natives of Upper Asia.

(3.) Is the *red* or *copper* colour, which in various shades is prevalent over the entire American continent, and chiefly confined to it.

(4.) *Brown* or *tawny*. This in lighter or darker variations belongs to the inhabitants of the peninsula of Malacca, and is extended through most of the islands of the Pacific ocean.

(5.) *Black*, in an amazing variety of shades, characterizes all the African continent, the northern and southern parts excepted, New Holland, Van Diemen's Land, New Guinea, New Hebrides, and some other islands of the South sea. It is mingled with the ordinary colour of the natives in Brazil, California, and India.

It is not to be supposed that these different colours, thus generally described, prevail each of them uniformly in all the individuals of the race ; on the contrary, there are considerable gradations, and even tribes have been found among the Americans, and individuals are constantly produced in each respective race, with characteristics approaching to those of some other. Children born from an intermixture of different races hold generally a middle station between the two. Thus the *Mulatto* forms a medium between the African and European. The colour will be more or less dark according to the complexion of the European father and mother. The cheeks are not ruddy, and the hair is black and curled, but less short than the Negroes. The iris is always dark.

From the Europeans and Mulattoes proceed what are called *Tercerons*. These generally resemble Europeans. The hair has nothing of the woolly curl, the skin has a slight brown tint, and the cheeks are red. The offspring of these last and the Europeans are not to be distinguished from our own race. An opposite course will reduce the Mulatto offspring to the characters of the Negro, and by intermixture with the latter, the fourth generation will be perfectly black. From the native Indians and the Europeans are born *Mestizos*. They are much lighter than the Mulattoes, and often not distinguishable in colour from Europeans. The small beard, hands, and feet, and the obliquity of the eyes, mark their Indian blood. The offspring from them and European fathers are in all respects like the Europeans. From Negroes and Americans spring *Zambos*, resembling Mulattoes, but darker. Among the dark races are sometimes found persons spotted with white.

Under this head of colour, we may briefly notice the varieties in the hair, beard, and iris.

The structure and properties of the hair are closely allied to those of the skin, and it derives the means of its growth, and probably its colouring matter, from the cutaneous vessels. Each hair may be traced through the cuticle and surface of the cutis to a bulb partly in the chorion, and partly in the cellular membrane. This bulb consists of a thick outer covering, in which the root of the hair and a vascular pulp by which the root is secreted, are contained. There is a close analogy between the skin and hair. The latter, in the Albino, as before observed, is short and white. A light complexion and thin skin, are usually accompanied with fair or red hair, and darker hair usually belongs to a dark colour and thick skin. In the coloured varieties of the human race, the hair is black and always coarser than that of Europeans. In the spotted Negroes the hair growing out of a white patch on the head, is white, a presumptive proof that the colouring matter of the skin and hair is the same. The principal differences in the hair are four. 1. Brownish, deviating into yellow or red, or into black. It is copious, long, and soft, and characterizes the natives of the temperate climates of Europe, and somewhat stronger and darker, and belongs to the Eastern Asiatics and Northern Africans, and the Celtic and Slavonic races in Europe. 2. Black, strong, straight, and thin. This character of hair belongs to the American and Mongolian varieties. 3. A softer black, thick, and curled, is found among the South-sea islanders. 4. Black and crisp, in all the Negro tribes. It may be proper to notice here that the hair of the African has been ascertained to bear no resemblance to wool except in appearance, and that it has all the characters of true hair.

The above divisions hold good generally, but we find, as in colour, so in hair, many individual exceptions in the different races. As great differences are observable in the various races in point of beard, as of hair generally. Most individuals of the dark races are remarkably deficient in this particular. The Mongolians have much less beard than the Europeans, and it grows later. The Calmucks have small and poor mustachios, and very little hair on the body. The Buriats are nearly beardless, so are the Tungooses and other hordes of Eastern Siberia. The Chinese resemble the Mongolian tribes in this particular; but contrary to their practice the latter encourage the growth of the beard. The custom of exterminating the beard and hair on the body is common among many of the dark nations; but this would not have been attempted, and could not be executed, if nature had furnished them in this article so plentifully as she has done the Europeans.

There has been much controversy whether the native Americans have beards or not; it is now completely ascertained that they have beards, but weak and imperfect, and that the practice of exterminating them is general. The genuine Negroes have very little beard or growth of hair over the body. But the South-sea islanders are by no means deficient in these excrescences.

An analogy similar to that between the hair and skin exists between the latter and the iris of the eye. New-born children in Europe have generally light eyes and hair, and both gradually darken in those of dark complexion. In old persons, as the hair turns gray, the eye loses a portion of its colour. In the Albino there is an entire deficiency of proper colouring matter, so that the iris has a reddish hue from the colour of the blood in the capillaries. The same sympathetic variations in the skin, hair, and eyes are also observable in other animals. The principal colours of the human eye are blue, passing to grayish in the lighter tints, a sort of obscure orange, a kind of middle tint between blue and orange, sometimes very green in red-haired persons; and lastly, brown, verging to hazel on the one side, and black on the other. To these the reddish eye of the Albino may be added. These varieties occur constantly in individuals of the same race and family. Sometimes they are confined to particular tribes of the same nation. The Gothlanders of Sweden are described with light hair and grayish eyes. The Findlanders with yellow hair and brown eyes, and the Laplanders with both black. Blue eyes with yellow hair have always marked the Germanic tribes. Blue eyes with black or dark hair form a combination not uncommon among the tribes of Kurdistan, and others of the Caucasian race, who inhabit elevated situations in Asia. The iris is dark in all the coloured varieties, but in the Negro it is so black as to be distinguished with difficulty from the pupil.

I.—THE MONGOLIAN RACE.

"I have thought," says Dr. Pickering, the substance of whose work we now proceed to condense, "I have thought to distinguish in the Mongolian race physical traits and a style of feature, at variance in some respects with those of the remaining series of races. One of these peculiarities consists in the occurrence of a feminine aspect in both sexes. In the absence of any striking difference in stature or dress, I have often seen the stranger at a loss to distinguish men from women; a difficulty not depending altogether on the absence of a beard, and which, so far as my observation extends, does not take place in the other races. The well-characterised Mongolian head is less compressed at the sides than is the head in the other races, so that when viewed in front, it presents a more rounded contour. The forehead recedes, but in continuation of a general curve from the chin upwards; and it frequently happens that the nose is likewise arched. The latter withal is less prominent than in the White race, and the lips are somewhat thicker. The complexion is always sufficiently light to show a flush, and in the far north, it sometimes becomes decidedly florid. This is said to be the case with the coast-tribes of North-west America; and indeed I have seen among them two females, in all probability of unmixed race, who, from their fair complexion, might very well have passed for Europeans. The Mongolian is pre-eminently a beardless race: the chin often remaining perfectly smooth, even to extreme age. In the instances where a thin beard does make its appearance, I have never seen it attain a greater length than two or three inches, and it was always perfectly straight. The hair also has appeared to me more uniformly straight, and to have a tendency to grow longer than in the other races; when left to itself, I think it will not unfrequently reach the ground. As to the 'oblique eye,' so generally spoken of as characteristic of the Chinese, I have found it among them in some instances, and also among the Chinooks of North-west America; but I have not been able to make much use of it as a distinctive character. I have moreover seen individuals of the Malay race having their eyes small, and as if half-closed, and I am induced to think some confusion may have arisen from this source. I was not more successful with the alleged 'absence of a projecting inner angle to the lids,' which has likewise been spoken of as a Chinese peculiarity. Some writers have found a want of clearness in the *sclerotica*, or 'white of the eye,' of the aboriginal American, a point I have not examined, but which seems to deserve attention.

"The Arctic regions seem exclusively possessed by the Mongolian race, which besides is diffused through a greater variety of climates than any other, and over a far larger area. This comprises about one-half of Asia, and, with a slight exception, all aboriginal America, or more than two-fifths of the land-surface of the globe. Notwithstanding the recent encroachments, the greater portion of the American continent is still inhabited by Mongolian tribes; and while some of them wander towards the North, further than civilized man has hitherto been able to follow, others are still the nearest dwellers to the Southern pole."—See the ethnographical section in article MONGOLIA.

The American Race.—The American aborigines are generally distinguished by long, straight, black hair, thin beards, prominent cheek-bones, a copper or brown-red colour, varying to lighter or darker shades, thick lips, black eyes, often obliquely placed as in other Mongolian races, and noses either flat or aquiline. "In referring to such exceptions as exist to this description," says Mr. Bradford, in his 'American Antiquities,' "we should not forget that tribal distinctions are everywhere maintained with great tenacity, and, of consequence, that to such occasional aberrations from the common standard as have originated among any particular family, a great opportunity for perpetuation has been afforded. The uniformity of the American physical appearance has, however, struck most travellers and naturalists with great force. The testimony of a few may be cited, though authorities might easily be accumulated. 'The Indians of New Spain,' says Humboldt, 'bear a general resemblance to those who inhabit Canada, Florida, Peru, and Brazil. They have the same swarthy and copper colour, flat and smooth hair, small beard, squat body, long eye, with the corner directed upwards towards the temples, prominent cheek-bones, thick lips, and an expression of gentleness in the mouth strongly contrasted with a gloomy and severe look.' 'Over a million and a half of square leagues, from the Terra del Fuego islands to the river St. Lawrence and Behring's straits, we are struck at the first glance with the general resemblance in the features of the inhabitants.' 'The Indians,' says Ulloa, 'are of a copper colour, which by the action of the sun and air grows darker. I must remark that neither heat nor cold produces any sensible change of colour, so that the Indians of the Cordilleras of Peru are easily confounded with those of the hottest plains; and those who live under the line cannot be distinguished by the colour from those who inhabit the fortieth degrees of north and south latitude.' 'I had no sooner beheld these Americans,' observes the enterprising Ledyard of the natives of Nootka, 'than I set them down for the same kind of people that inhabit the opposite side of the continent. They are rather above the middle stature, copper-coloured, and of an athletic make; they have long black hair.' 'I have been forcibly struck,' says Mr. Flint, 'with the general resemblance in the countenance, make, conformation, manners, and habits of the Indians. A savage of Canada and the Rio-del-Norte are substantially alike; they are all, in my mind, unquestionably from a common stock.' One testimony, still more explicit, if possible, may be added. Ulloa, upon his return from South America, touched at Louisburg, at which place he remarks, 'In this and the adjacent islands were a considerable number of inhabitants, born in the country, or on the main land; and, what is remarkable, these Indians not only resemble those of Peru in complexion and aspect, but there is also a considerable affinity between their manners and customs; the only visible difference is in stature, and in this the advantage lies visibly on the side of the inhabitants of these northern climates.'

"That this uniformity is universal, and applies to all the tribes, cannot be maintained, and it would be absurd to suppose that it existed. But yet no varieties have been observed which approach the Indians anywhere near the white and black races, and where an exception occurs in one particular, the other peculiarities are still retained. It is true, many statements have been made concerning the existence of white and black Indians, but, upon examination, they are found to have proceeded usually from the early travellers, who were often vague and exaggerated in their use of terms, or to have been founded upon misnomers, or to have related to tribes who had intermarried with Europeans. An instance of the erroneous conclusions which may be drawn from a misnomer is afforded in a tribe of the Caucasus, who, though called 'the Black Circassians, are of a very fair complexion.' Thus, in America, the tribes of the upper Orinoco, who have been styled 'White Indians,' according to Humboldt, who had an opportunity for personal examination, differ from other Indians only by a much less tawny skin, having, at the same time, the features, the stature, and the smooth, straight black hair of their race. The Arkansas in North America, of whom the same assertion has been made, though not of a copper colour, are dark and tawny, and possess all the Indian peculiarities of form and feature. An idea may be gathered of the inaccurate notions formerly prevailing upon this subject, by the assertion of Charlevoix, that several tribes, and among them some of the Esquimaux, have white hair; indeed he adds of the latter nation that they have a beard 'so thick up to their eyes that it is difficult to distinguish any features of their face,' that 'they are tall and pretty well shaped,' and that 'their skin is white as snow.' Frezier also says, that the Fuegians are almost as white as Europeans, a statement abundantly disproved by subsequent observations. For the purpose of showing how restricted these apparent exceptions are, and in order to indicate the general predominance of those characteristics which mark the race, it may be well to attempt a brief physical synopsis of the tribes, confirmed by the testimony of various travellers.

"The Esquimaux have generally been distinguished from the American race, in consequence of their colour, diminutive stature, and other peculiarities, but it is apprehended, that in producing these signs of difference, so far as they really exist, natural causes, such as food, mode of life and climate, have chiefly operated. The strongest evidence, however, of their affiliation to the other Indian tribes is afforded in the physical appearance of the Fuegians, who occupy a region where similar causes have existed and produced the same results. 'The general form of the Fuegians is peculiar, the head and body being particularly large, and the extremities unusually small; but the feet are broad though short. This peculiarity, no doubt, is owing to their mode of life, &c. From the same cause, want of exercise, this is the form of the Esquimaux and the Laplanders.' The Fuegians have generally straight, long, and jet black hair, scanty beard, a broad face, black, angular Chinese eyes, copper complexion and small stature. In the work just cited, there are several interesting descriptions of particular individuals of this tribe, as follows: 'The complexion of this man was dark, his skin of a copper colour, the native hue of the Fuegian tribes, the eyes and hair black,—this is universal, as far as I have seen, and predominant throughout all the aborigines of America, from the Fuegians to the Esquimaux.' And again, 'the features of this individual were rounder than they generally are among those of his nation, the form of whose countenance resembles that of the Laplanders and Esquimaux. They have broad faces with projecting cheek bones; the eyes of an oval form, and drawn towards the temples.' The stature of the Fuegians is generally from four feet ten inches to five feet six inches, and their figures are similar to those of the Esquimaux."—See article FUEGO (TIERRA-DEL).

"The Patagonians afford a striking instance of the exaggerated and gross inaccuracies which pervade many of the accounts of the Indian tribes. Indeed it is but recently that the fanciful stories which had been propagated concerning the stature of these natives have been disproved; and a careful examination of all the authorities, together with the observation of recent voyagers, has satisfactorily shown that their height has been greatly over-estimated. The medium stature of this race appears to be from five feet ten inches to six feet. Falkner, however, bore impartial and correct testimony. 'The Patagonians, or Puelches,' he says, 'are a large-bodied people, but I never heard of that gigantic race which others have mentioned, though I have seen persons of all the different tribes of southern Indians.' The colour of the Patagonians is a rich, reddish brown, rather darker than copper, the head is long, broad, and flat, the forehead low, the face of a square form, the eyes small and often obliquely placed, the nose rather flat, but sometimes aquiline, the hair long, lank, and black, and the beard thin. Proceeding to the north, the nomade Pehuenches and the Araucanians, according to Mr. Poeppig, 'belong to the same branch of the great copper-coloured, or Patagonian race.' The Araucanians are of a reddish brown or copper colour, are finely shaped and muscular, and have small black eyes, a broad face, flattish nose, coarse black hair, and no beard. One tribe it has been thought offers an exception to this description,—the Boroanes, or Borea Indians living near Valdivia, in Chile. Some of them are said to have light eyes, a fair complexion and red hair. Frezier disposes of this objection for he says, in speaking of those who are thus distinguished, 'these are descended from the women taken in the Spanish towns they destroyed; his statement is fully confirmed also by Ulloa, who traces this diversity to the same cause."—See articles ARAUCANIANS, and PATAGONIA.

"The present Peruvian Indians, who are of the same race as the ancient inhabitants, are described as of a copper colour, with high cheek bones, small black eyes set widely apart, hair coarse and black, without any inclination to curl, beard scanty, nose somewhat flattened, small stature, and the feet small: these characters are of general prevalence among all the natives. The Bolivian Indians, according to Dr. Ruschenberger, are dark copper-coloured, the nose is flattened, and the eyes are obliquely placed."—See article PERU.

"In general, the same physical description will apply to the numerous hordes of Brazil. The Mongul physiognomy is more striking in these tribes than in those of any other part of America. Prince Maximilian describes one of these Indians as 'distinguished from all the rest by his Calmuck physiognomy;' but the same characters are of almost universal prevalence. Passing further to the north, but little variation from this type can be found in the tribes inhabiting Guiana and Colombia. Humboldt has termed the complexion of the Chaymas and other natives of this region, a dull brown or brown-red, inclining to a tawny colour, and he adds, that they resemble the Monguls, by the form of the eye, their high cheek bones, their straight hair, and the almost entire absence of beard.

"The natives of the West India Islands at the discovery consisted of two classes, differing slightly in appearance, and more considerably in manners and character. The Charibs, who occupied the small islands of the southern part of this group, were of a fierce, cruel, and indomitable disposition. The Arrowauks, who resided in the larger and northernmost islands, were of a milder character, and appear to have possessed more of the arts of civilisation. The Charibs were of a tawny or dark-brown complexion, middling stature, robust, and muscular, with small black eyes, long straight black hair, scanty beard, and flat retreating foreheads. The Arrowauks were taller than the former race, and of a reddish brown colour, their foreheads were flat, though not so retreating as those of the Charibs, the hair was straight, black, and lank, the beard scanty, the eyes black, cheek bones prominent, the face broad, and the nose flat."—See article CARIBBEES.

"The Mexican Indians have generally 'a swarthy and copper colour, flat and smooth hair, small beard and squat body, long eye, with the corner directed upwards towards the temples, prominent cheek bones, and thick lips.' Clavigero says, that the moral and physical qualities of the Mexicans proper, were the same as those of the adjacent nations, so that the description of the one is equally applicable to the rest; he considers the colour of the skin, however, as olive. According to Humboldt, the Aztec and Otomite tribes have more beard than the others, and many of them wear small mustaches."—See ethnographical section of article MEXICO.

"The aboriginal inhabitants of Upper California, when first observed by the early Spanish travellers, were of the same stock as those living in the adjoining peninsula. Though the different tribes varied in some unimportant particulars, they bore a general resemblance to each other;—and though some are described as of a diminutive size, they were usually of ordinary stature. They were of a darker colour than the natives of the more southern provinces, their lips were large and projecting, noses broad and flat, foreheads low, beard scanty, and hair straight and black;—long beards, it is said, have occasionally been observed. This was one of the localities where American negroes have been placed, an idea which the following passage from Venegas may possibly dispose of. 'It is known,' he says, 'that some ships have left Mulattoes and Mestizoes at Cape San Lucas.' 'Father Juan de Torquemada tells us that the Californians showed no manner of surprise at the sight of negroes, there being some of that caste among them, the race of those who had been left by a ship from the Philippine islands.' The nations near the mouth of the Columbia river generally resemble each other in their physical appearance. Their stature is diminutive, 'the complexion is the usual copper-coloured brown of the North American tribes, though rather lighter than that of the Indians of the Missouri;' the mouth is wide and the lips are thick; the nose is wide at the extremity, and low between the eyes; the eyes are generally black, the face broad, forehead flat, and the hair straight and black. The form of the forehead arises from artificial compression of the cranium, which is carried to such an extent, that the forehead often runs in a straight line from the nose to the crown of the head. Further to the North and along the Pacific, tribes have been observed, which are said to be distinguished by a lighter complexion. The Cheyennes and other tribes on the shores of the south-western tributaries of the Missouri, though differing somewhat in features and size, 'still, in the direction of the eye, the prominence of the cheek bones, the form of the lips, chin, and retreating forehead, are precisely similar' to the Missouri Indians."—See articles CALIFORNIA and OREGON.

"The Osages are a fine race of men, tall, well made, of a tawny red colour, with aquiline features, prominent cheek bones, and straight black hair. 'They do not seem to differ in point of features and colour from the Missouri Indians; their stature is by no means inferior to the latter.' Mr. Brackenridge also remarks of this tribe, that 'they have been noted for their uncommon stature. This is somewhat exaggerated, though they are undoubtedly above the ordinary size of men. The wandering, or semi wandering nations of Louisiana may be characterized as exceeding the whites in stature.' The Arkansas are dark, but not copper-coloured, they possess fine aquiline features, scanty beards, long black hair and elongated angular eyes. The same general characteristics as those already indicated, with the exception of a greater prominence of the cheek bones, mark the Sioux, Pawnees, and other tribes inhabiting the region west of the Mississippi. The Mandans, however, appear to have a lighter colour, and though possessing the Indian form in other respects, their hair in some instances is of a light chestnut colour and the eyes are of a bluish cast. But connected as they are by affinities in language to other tribes, whose Indian physiognomy cannot be doubted, it is possible that these peculiarities have been produced by an intermixture of the race.

"'We see nothing,' says Charlevoix, 'in the outward appearance of the Natchez, that distinguishes them from the other savages of Canada and Louisiana;' the same appears to have been the case with all the southern Indians, and a description of one tribe will therefore answer for the rest. 'The Chickasaws,' says Mr. Adair, 'are a comely, pleasant looking people. Their faces are tolerably round, contrary to the visage of the Choctaws, which inclines much to flatness, as is the case of most of the other Indian Americans. The lips of the Indians in general are thin, their eyes are small, sharp and black, and their hair is lank, coarse and darkish; they pluck their beards.'

"The similarity in the physical appearance of the numerous tribes of the Algonquin-Lenape race and of the Iroquois was equally striking. Smith describes the Powhatans as generally tall and of good proportions, with a brown colour, black hair, and little beard. Loskiel says that 'the Delawares and Iroquois, and other nations in league with them, resemble each other much, both as to their bodily and mental qualifications. Their skin is of a reddish brown, nearly resembling copper, but in different shades. Some are of a brown yellow, not much differing from the mulattoes; some light brown, hardly to be known from a brown European, except by their hair and eyes. The former is jet black, stiff, lank, and coarse, and almost like horse hair.' Smith describes the Iroquois as tall, beardless, of a tawny complexion, and having black uncurled hair; and Charlevoix, as of a lofty stature, with black hair and a scanty beard. 'The colour of the savages,' he says, 'does not prove a third species between the white and black, as some people have imagined. They are very swarthy and of a dirty dark red.' The Knisteneaux, a branch of the Algonquin race, who had penetrated the farthest towards the north-east, and inhabited the territory from the Atlantic to the shores of Hudson's bay, the St. Lawrence and Churchill rivers, are described by Mr. M'Kenzie 'as of moderate stature. Their complexion is of a copper-colour, and their hair black, which is common to all the natives of North America. It is cut in various forms according to the fancy of the several tribes, and by some is left in the long lank flow of nature. They very generally extract their beards: their eyes are black, keen, and penetrating—their countenance open and agreeable.' Some individuals have been seen with full beards. The tract of country situated to the north-west of the Knisteneaux is inhabited by the Chippewyans—all the territory between 60° and 65° N lat. and long. 110° and 100° West, they consider as their lands and home. But tribes of this race border in the east on the Knisteneaux, and extend on the west to the Pacific, and on the north to the territory of the Esquimaux. At least this is to be inferred from ethnographical analogies, for tribes who speak their language are found over this vast district even as far south as lat. 52° N on the Columbia river. 'Their complexion is swarthy,' says M'Kenzie, 'their features coarse, and their hair lank—but not always of a dingy black—nor have they universally the piercing eye which generally animates the Indian countenance.' 'The men in general extract their beards, though some of them are seen to prefer a bushy black beard to a smooth chin.' The tribes situate to the westward in the vicinity of the Pacific, have been described in a similar manner, with the exception of their complexion, which is said to be of a 'light copper-colour,' accompanied with long lank hair and black eyes. Mr. Hearne remarks of these Indians, that their complexion 'is somewhat of the copper cast, inclining rather towards the dingy brown;' that their foreheads are low, cheek bones high, eyes small, and their hair black, strong and straight. 'Few of the men have any beard; this seldom makes its appearance until they arrive at middle age.' "—See ethnographical section of article AMERICA.

"The Esquimaux inhabit all the northern regions of the continent stretching along the Arctic seas from the Atlantic to the Pacific, and a considerable distance south along the shores of the two oceans. The western tribes appear to be the most assimilated to the general type of the American race, and as their language and customs connect them with those toward the east, it is possible that the varieties observed in the colour and features of the latter are owing to foreign causes. Generally upon the northern coast opposite to Asia, the inhabitants are stout and short, of a swarthy colour, with thick lips, black eyes and hair, thin beard, and high cheek bones. Further to the east, the Esquimaux met by Hearne, on the Coppermine river, are described as of small stature, and a dirty copper-coloured complexion, though some of the women are more fair. According to Crantz this race is of a dwarfish size, the face is broad and flat, the cheek bones high, the eyes black, the hair long, straight, and black, and the hands and feet small. Their colour, he says, is olive, though there are some who have a moderately white skin, but the children are born white. Captain Back, in lat. 67°, long. 94°, observed Esquimaux who had a luxuriant growth of beard; their eyes were obliquely placed; and it is remarkable that they were not tattooed. 'The male Esquimaux have rather a prepossessing physiognomy, but with very high cheek bones, broad foreheads, and small eyes, rather farther apart than those of a European. The corners of their eyelids are drawn together so close, that none of the white is to be seen; their mouths are wide and their teeth wide and regular. The complexion is a dusky yellow, but some of the young women have a little colour bursting through this dark tint. The noses of the men are rather flattened, but those of the women are rather prominent. The males are, generally speaking, between five feet five inches and five feet eight inches high, bony and broad shouldered, but do not appear to possess much muscular shape. But the most surprising peculiarity is the smallness of their hands and feet.' Captain Graah says, 'the Greenlanders inhabiting the southern part of the west coast, (and it is to be observed, this is the coast upon which the ancient Icelandic settlements were probably made,) have little in their exterior in common with genuine Esquimaux; and the inhabitants of the country about the bay of Disco in North Greenland, and the natives of the east coast seem to me to have still less. They have neither the full fleshy person, nor the prominent paunch of the Esquimaux, but are on the contrary slender and even meagre. They are moreover distinguished from the Esquimaux by a uniformity of head and cast of countenance which is handsomer and more expressive. The women and children have many of them brown hair, and a complexion scarcely less fair than that of our peasantry. But as I should not venture to conclude, that the Esquimaux about Hudson's bay have any claims to the honour of a Roman parentage from the circumstance of Sir Edward Parry's having seen many a good Roman nose among them, neither do I conceive, that the natives of the east coast of Greenland are descended from the old Icelandic colonies, because in some points they resemble Europeans. Their lank hair, their black and somewhat Chinese eyes, their disproportionally large hands and feet, their temper and disposition, their manners, customs, and language, all indicate that they are of the same stock originally with the Esquimaux. Some few of them wear beards and mustaches, but by far the greater number eradicate the beard as it appears.' Captain Parry says of some Esquimaux, N lat. 67°, long. 85°, 'their countenances at the time impressed me with the idea of Indian rather than of Esquimaux features—but this variety of physiognomy we afterwards found not to be uncommon among these people.' The same traveller, in his general description of those at Winter island

and Igloolik, observes that they are of low stature, their hands and feet are remarkably small, their faces are generally round and full, eyes small and black, nose also small and sunk far in between the cheek bones, but not much flattened. 'In the young of both sexes the complexion is clear and transparent, and the skin smooth. The colour is scarcely a shade darker than that of a deep brunette, so that the blood is plainly perceptible when it mounts into the cheeks; the eyes are not horizontal, but much lower at the end next to the nose than at the other. The hair is black, glossy, and straight. The men wear the hair on the upper lip and chin from one to one and a half inch in length,—and some were distinguished by a little tuft between the chin and lower lip.' It seems most probable, therefore, that the aberrations from the general physical standard, observed in the eastern Esquimaux, have been occasioned by a European intermixture; by intermarriages, probably, with the lost Icelandic colonists alluded to by Captain Graah. Of the western branches of the race the colour and features are essentially the same as those which characterize the aborigines generally. It thus appears that a most striking physical uniformity prevails among all the American tribes, that the variations from the predominant type are trifling and infrequent, and where they do exist, may in several cases be traced to intermarriages with individuals of the white or black race."—See articles ESQUIMAUX and TCHUKTSHES.

"It remains to inquire whether the ancient and civilized nations of the United States, Peru, and Mexico belonged to the same race. If the tribes now inhabiting Mexico and Peru are to be regarded as the descendants of those nations, the question would meet an easy solution; but the peculiar conformation of the ancient skulls found in the mounds, and the singular physiognomy observed in the human representations upon the Mexican paintings and monuments, afford some cause for hesitation. But it should be remembered that in the early stages of art, there is a tendency to delineate monstrous and exaggerated figures, which at a subsequent period, when more elegance and correctness are attained, custom, prejudice and superstition having once rendered sacred, are never abandoned. The Egyptians, says Plato, having once determined upon the forms to be exhibited in their paintings and sculptures, it was no longer lawful for painters or other imitative artists to attempt any innovation. That such an arbitrary rule prevailed in Mexico, where, from the absence of the art of writing, it became necessary to adhere to the forms which had been adopted in their hieroglyphical paintings, not only appears reasonable, but is manifest by the most cursory examination of the picture-writings. Accordingly we find in some cases, particularly in their sculpture, when it was designed to represent real individuals, that the style is changed, and the figures approach nearer to the present Indian physiognomy. In one respect the representations on the monuments and paintings are assimilated to the type of the Red race; for with the exception of some sacred figures probably intended to denote the white and bearded Quetzalcoatl, they are all delineated as beardless. The enormous size of the nose, which it is difficult to suppose ever characterized any people, was naturally exaggerated by the first artists, in consequence of the retreating forehead, with which it was accompanied. The custom of cranial compression was common to many American nations, and may be clearly traced, not only to the barbarous, but civilized races. The unusual forms thus given to the skull are not universally the same. Thus Charlevoix observes, 'there are on this continent some nations, which they call Flat-heads, which have, in fact, their foreheads very flat, and the top of their heads somewhat lengthened. This shape is not the work of nature: it is the mothers who give it to their children, as soon as they are born. For this end, they apply to their foreheads, and the back part of their heads, two masses of clay, or of some other heavy matter, which they bind by little and little, till the skull has taken the shape they desire to give it. It is quite the reverse with certain Algonquins amongst us, named Round-heads, or Bowl-heads, (Têtes de Boules), whom I have mentioned before, for they make their beauty consist in having their heads perfectly round, and mothers take care also very early to give them this shape.' Adair describes another form among the Choctaws and other southern tribes. 'The Indians flatten their heads,' he remarks, 'in divers forms, but it is chiefly the crown of the head they depress. The Choctaw Indians flatten their foreheads from the top of the head to the eyebrows, with a small bag of sand; which gives them a hideous appearance, as the forehead naturally shoots upwards, according as it is flattened; thus the rising of the nose, instead of being equi-distant from the beginning of the chin to that of the hair, is by their wild mechanism placed a great deal nearer to the one and further from the other.' The same practice, he adds, prevails among the tribes 'around South Carolina and all the way to New Mexico.' Accordingly we find, that the Waxsaws and Natchez distorted the natural shape of the head by similar artificial means. The Flat-head tribes west of the Rocky mountains and on the Columbia river have carried the same custom to a frightful extent; their heads present an inclined plane from the crown to the upper part of the nose, —a peculiarity which is produced in the following manner. 'Immediately after birth, the infant is placed in a kind of oblong cradle, formed like a trough, with moss under it. One end, on which the head reposes, is more elevated than the rest. A padding is then placed on the forehead, with a piece of cedar bark over it, and by means of cords passed through small holes, on each side of the cradle, the padding is pressed against the head. It is kept in this manner upwards of a year. When released from this inhuman process, the head is perfectly flattened, and the upper part of it seldom exceeds an inch in thickness.' The Charibs compressed the heads of their infants by boards and ligatures, and rendered the forehead so flat, 'that they could see perpendicularly when standing erect;' and the Arrowauks practised the same custom, endeavouring, however, to give the crown of the head a greater elongation. In South America, according to Condamine, 'the appellation Omaguas, in the language of Peru, as well as Cambevas in that of Brazil, given to the same people by the Portuguese of Para, signifies Flat-heads. For they have the whimsical custom of pressing between two plates the forehead of their newly born children, in order to give them this singular shape, and make them, as they say, resemble the full moon.' 'Among the variety of singular customs prevailing in these nations,' says Ulloa, 'one cannot help being surprised at the odd taste of the Omaguas, a people otherwise so sensible, who, to render their children what they call beautiful, flatten the fore and hind parts of the head, which gives them a monstrous appearance; for the forehead grows upwards, in proportion as it is flattened, so that the distance from the rising of the nose to the beginning of the hair exceeds that from the lower part of the nose to the bottom of the chin. And the same is observable in the back part of the head: the sides also are very narrow from a natural consequence of the pressure; as thus, the parts pressed, instead of spreading conformable to the common course of nature, grow upwards.' This practice is of great antiquity among them, and kept up so strictly, that they make a jest of other nations calling them calabash-heads. In order to give children this beautiful flatness, the upper part of the head is put, soon after their birth, betwixt two pieces of board, and this is repeated from time to time, till they have brought the head to the fashionable form.' The Mantas, consisting of several tribes subdued by the Incas, are described by Garcillasso de la Vega, as having their heads very much deformed. 'As soon as their children were born, they applied to the front of the head and the back of the neck two small boards, between which they compressed the head, until they had arrived to the age of five years; and by these means the head became flat and very long.' There are various authorities to show that artificial pressure of the cranium was common to many Peruvian nations, and none more satisfactory than the decree of the Synod of Lima, (passed 1585,) cited by Blumenbach, which prohibited the custom,—at the same time alluding to it as formerly universal in Peru. Dr. Morton has traced it likewise into Venezuela and Nicaragua. The same learned naturalist, in his researches into American craniology, has arrived at the conclusion, that the American nations, excepting the Polar tribes, are of one race and one species, but of two great families, which resemble each other in physical, but differ in intellectual character, and that the cranial remains discovered in the mounds, from Peru to Wisconsin, belong to the same race, and probably to the Toltecan family. The skulls from the mounds are described as being flattened on the occiput and frontal bones, 'in such manner as to give the whole head a sugar-loaf or conical form, whence also their great lateral diameter and their narrowness from back to front,' and the result of his investigation seems to be, that this peculiar configuration, as well as that of the Mexican heads represented in their sculptures and paintings, appears to result in part from the application of mechanical pressure. But there are other crania brought by Mr. Pentland from Peru, which it is supposed belong to an ancient and extinct race. They are remarkable for their unusually great length and narrowness; the face is very projecting, the forehead retreating, so that the facial angle is smaller than in any known race of men; the os frontis is continued far backwards towards the vertex, and is very long, narrow and flat. By the discovery of these skulls the interesting question has been presented to naturalists, whether they are of a natural form, or altered by art. Professor Tiedemann says, 'a careful examination of these skulls has convinced me that their peculiar shape cannot be owing to artificial pressure. The great elongation of the face, and the direction of the plane of the occipital bone are not to be reconciled with this opinion, and, therefore, we must conclude that the peculiarity of shape depends on a natural conformation. If this view of the subject be correct, it follows that these skulls belonged to a race of men now extinct, and which differed from any now existing.' But it is clear that the habit of mechanical compression of the head was common to many American nations, and prevalent in Peru. 'There is no race on the globe,' says Humboldt, 'in which the frontal bone is more depressed backwards than in the American.' 'The custom of flattening the head had its origin in the idea that beauty consists in such a form of the frontal bone as to characterize the race in a decided manner.' In fact, Waldeck saw in Yucatan profiles of the present Indian race similar to those sculptured at Palenque; those ancient profiles, he says, are at an angle of 74°, which must be attributed to the custom of flattening the head. It is true these differ in form from those of Peru under consideration, but the alteration by compression has been as great. In view of the various processes which were used, the nature of the substances which were applied, and the parts which were compressed, it does not seem improbable that the singularly shaped Peruvian skulls may have been altered by artificial means. The tendency of the cranium to lateral expansion under a pressure from the front and back, which does not appear to have taken place in these heads, may have been prevented by ligatures or other mechanical applications to the sides of the head. The testimony of Dr. Pritchard is strongly to this point; 'it is more probable,' he says, 'that the ancient skulls of Titicaca owed their strange configuration to a process, which we know is capable of explaining the phenomena, than that they constituted an original race, a circumstance of which we have no other evidence than that derived from the shape of the cranium. Professor Scouler has given the sketch of an infant skull of one of the Columbian tribes, which is as much elongated as the skulls brought by Mr. Pentland from Titicaca.' It thus appears that there are no decided and general characteristics yet clearly ascertained, which separate any of the ancient civilized nations from the great American family,—nor any peculiar cranial conformations, which might not be justly attributed to the prevalent custom of altering the head by mechanical applications, so as to produce that form, which, according to aboriginal notions, possessed the highest degree of beauty.

"In the prosecution of our inquiry as to the number of primitive races, it becomes necessary to examine, in the next place, in what points of physical appearance the American and Mongolian races are assimilated. The Mongols have exercised an important influence in the affairs of Oriental Asia, having at different periods subjected Hindostan, Siam, Thibet and China. Hordes of this race roam over the vast regions of Siberia, and are found wandering from China to the banks of the Dnieper; and it is probable that in ages far beyond the scope of authentic history they have been distributed still more widely. Considerable confusion has been occasioned by the incorrect application of the term Tartars to the Turkish race, an appellation which is now too well settled to admit of change. It appears to have been originally applied by Asiatic writers to the Mongol race, and when the Turkish tribes were subdued by them, the name passed from the princes, who were Mongols, to their subjects. 'Remusat, who, with Klaproth, had determined the original identity of the Mongols and Tartars, proposed to confine the latter appellation to the former race. But in the utter commixture of the northern tribes, to use Tartar as a generic name, would lead but to further confusion.' The Tartars, as the term is now understood, belong to the Caucasian family, and in their physical characteristics they are distinguished by an oval head, a fresh, white complexion, and great beauty and symmetry of countenance. The Mongol physiognomy is widely different, and is nearly allied to the American; the complexion is described by most physiologists as of an olive or yellow colour, but as will be seen, there are some tribes in Siberia of a regular copper colour.

"Commencing on the western shore of Behring's straits, we find the Tchutski, who are divided into two classes, the Sedentary and the Wandering

They may be clearly identified with the American family, as well by their language as by their manners, customs, and appearance, and it is thought they are of American origin. They speak the same language, however, as the Koriacks, or Korœiki, a tribe in their vicinity. The resemblance in their customs to those of the Americans is not peculiar to this nation, but is common to many others of Siberia, and no argument, therefore, in favour of their American origin can be drawn from that circumstance. 'From what I have seen of the Koriacks,' says Cochrane, 'I have no doubt of their being of the same tribe as the Tchutski; they have the same features, manners, and customs, and the same language.' According to a Russian author of great authority, 'the language of the Tchutski is derived from that of the Koreki, and differs from it in dialect only;' 'they agree in most of their customs and habits with the Kamtschatdales,' and 'the Tchutski should be accounted a race of the Koreki.' The Koriacks are also divided into two tribes; the wandering family are of smaller stature. They are described as having long black hair, small eyes, a short nose, and large mouth. Strahlenburgh says that they have no beards, but only a few loose hairs scattered over their chins. 'The Lutorzi,' he adds, (the Tchutski?) 'who live eastward of the former, and towards the coast of the main ocean, were, as to shape, customs, and language, the very same with the Korœiki, except that they made their habitations under ground.' 'They are beardless like the Laplanders, Samoides, and Ostiacs; for, in the first place, they have naturally very little hair about the mouth, and what little they have, they pluck out, as do also the Yakuti, Tungusi, and Kalmucks.'" —See articles MONGOLIA, TARTARS, and LAPLAND.

"The Kamtschatdales are swarthy, of a small stature, have straight black hair, high cheek bones, oblique eyes, and scanty beard. It has been supposed of this people, as well as of the Tchutski, that they were of American origin; an idea which Captain Cochrane pronounces ridiculous.

"The Yakuts are of a low stature, with long black hair, but little beard, and 'their complexion is a light copper colour.' The Mantchoos belong to the Tongoo, or Tungusi race, and resemble them in appearance and features; and the latter, besides the oblique eye, black hair, and slender beard of the Mongols, are described as 'copper-coloured.'" —See article JAKUTS.

"The Burretti, the Calmucks, and the Mongols near Bogdo, are all described as of a complexion varying from a yellow or swarthy hue to a brownish red or copper-colour, with small beards, black hair, and oblique eyes.

"The Chinese, Japanese, Siamese, and other inhabitants of Indo-China, all present the same general physical type, modified in various degrees, but conforming in a great measure to the characters of the Mongols; and the same race appears to have penetrated into Thibet, Bootan, and Nepaul. The close analogy which exists between the Mongol and American families, notwithstanding certain diversities which have been observed, cannot be better illustrated than by the testimony of travellers, and of those who have had personal opportunities for direct comparison. 'There is a great resemblance in feature,' says Mr. Andrews, 'between these Indians along the banks of the Rio Chico in South America, and some of the people I have seen in the East, especially the mixed breed of Chinese and Malay in Java. The high cheek-bone, sharp angular eyes, and small beards, agree.' 'I shall only remark further,' observes Mr. Bell, 'that from all the accounts I have heard and read of the natives of Canada, there is no nation in the world which they so much resemble as the Tongoos.' 'The American race,' says Humboldt, 'has a striking resemblance to the Mongol nations, which include those formerly called Huns, Kulans, and Kalmucks.' 'The viceroy of Brazil retains a dozen of the native Indians in his service as rowers of his barge. We observed the Tartar or Chinese features, particularly the eye, strongly marked in the countenances of these Indians. The copper tinge was rather deeper than the darkest of the Chinese, but their beards, being mostly confined to the upper lip and the point of the chin, together with their strong black hair, bore a very near resemblance.'

"Chinese colonists have been imported into Brazil, and afford a valuable opportunity of contrasting their appearance with the native Americans. 'The physiognomy of the Chinese colonists,' says Spix and Martius, 'was particularly interesting to us, and was in the sequel still more so, because we thought we could perceive in them the fundamental lines which are remarked in the Indians. The figure of the Chinese is, indeed, rather more slender, the forehead broader, the lips thinner and more alike, and the features in general more delicate and mild than those of the American who lives in the woods; yet the small, not oblong, but roundish, angular, rather pointed head, the broad crown, the prominent sinus frontales, the low forehead, the pointed and projecting cheek-bones, the oblique position of the small narrow eyes, the blunt, proportionally small, broad flat nose, the thinness of the hair on the chin and the other parts of the body, the long smooth black hair of the head, the yellowish or *bright reddish tint* of the skin, are all characteristics common to the physiognomy of both races. The mistrustful, cunning, and, as it is said, often thievish character, and the expression of a mean way of thinking, and mechanical disposition, appear in both in the same manner. In comparing the Mongol physiognomy with the American, the observer has opportunity enough to find traces of the series of developments through which the eastern Asiatic had to pass, under the influence of the climate, in order to be transformed into an American. In these anthropological investigations we arrive at the remarkable result, that certain characteristics, which constitute the principal difference of the races, do not easily pass into others, whereas, those which depend only upon *more* or *less*, gradually vanish or degenerate, through a series of different gradations.'

"Of the Chiriguanos, a Peruvian tribe, Mr. Temple says, 'They are of a copper colour, approaching to sallowness, with long shining black hair, and as the Indians of South America generally are, without beards. Had I seen them in Europe, I should have supposed them to be Chinese, so closely do they resemble those people in their features.'

"'In some points of physiology,' remarks Mr. Davis of the Chinese, 'the people whom we describe bear a considerable resemblance to the North American Indians. There is the same lank, black, and shining hair, the same obliquity of the eyes, and eyebrows turned upwards at the outer extremities, and a corresponding thinness and tufty growth of beard. The Chinese, too, is distinguished by a nearly total absence of hair from the surface of the body. We may remark here that the Esquimaux, as represented in the plates to Captain Lyon's voyage, bear a very striking resemblance to the Taukea, or 'boat-people' of the coast of China, who are treated by the government as a different race from those on shore, and not allowed to intermarry with them. Whether the miserable inhabitants of the cold regions to the north have thus migrated southward, along the coast, at some former periods, in search of a more genial climate, must be a mere matter of conjecture in the absence of positive proof.'" — See ethnographical section of article CHINA.

"Mr. Ledyard, who had personal opportunities of observing the peculiar physiognomy of the American Indians, and who had travelled through Siberia, is still more positive in his assertions, as to the resemblance between the Americans and Mongols. His testimony being of the highest kind, deduced from his own personal examination, we shall quote extensively from his remarks upon this subject, premising that in his use of the term Tartar, he applied it to all those tribes possessing the Mongol physical characteristics. In a letter to Mr. Jefferson from Siberia, he says, 'I shall never be able, without seeing you in person, and perhaps not then, to inform you how universally and circumstantially the Tartars resemble the aborigines of America. They are the same people—the most ancient and the most numerous of any other; and had not a small sea divided them, they would all have been still known by the same name.' 'I know of no people among whom there is such a uniformity of features (except the Chinese, the Jews, and the Negroes) as among the Asiatic Tartars. They are distinguished indeed by different tribes, but this is only nominal. Nature has not acknowledged the distinction, but, on the contrary, marked them, wherever found, with the indisputable stamp of Tartars. Whether in Nova Zembla, Mongolia, Greenland, or on the banks of the Mississippi, they are the same people, forming the most numerous, and, if we must except the Chinese, the most ancient nation of the globe; but I, for myself, do not except the Chinese, because I have no doubt of their being of the same family.' 'I am certain that all the people you call *red* people on the continent of America, and on the continents of Europe and Asia, as far south as the southern parts of China, are all one people, by whatever names distinguished, and that the best general name would be Tartar. I suspect that all *red* people are of the same family.' And again: 'With respect to the national or genealogical connection which the remarkable affinity of person and manners bespeaks between the Indians on this and on the American continent, I declare my opinion to be, without the least scruple and with the most absolute conviction, that the Indians on the one and on the other are the same people.'"

II.—THE MALAY RACE.

"If the Mongolian occupies a larger portion of the surface of the globe, the Malay," says Dr. Pickering, "is yet the most widely-scattered race, and, in some respects, it is also the most remarkable. In institutions and social condition it exhibits, perhaps, greater variety than all the other races combined; and, from a universal instinctive attachment to the water, it almost merits the appellation of 'amphibious.' Endowed, too, with a wandering disposition, less regardful of a home than the rest of mankind, its march has been truly 'upon the waves;' and, beyond the Atlantic, it has reached almost every islet in the ocean that affords the least means of subsistence."—See article MALAY PENINSULA.

"The Malay complexion is very uniform, and is always decidedly darker than the Mongolian. I have never seen it light enough to show the least trace of a flush, nor, on the other hand, so dark but that the marks of tattooing were conspicuously visible. The colour may be termed reddish-brown, more nearly than in the Mongolian race, approaching the hue of tarnished copper. The hair seems in greater quantity than in the other races, the Papuan, perhaps, excepted; and it is straight, or at most wavy, and usually raven-black. When cropped within about two inches, I have observed that it will generally stand erect, owing, apparently, to a coarser texture than in the Telingan and White races. The beard grows long, but is almost always thin, though some variety prevails in different countries. The East Indian tribes are nearly beardless, while among the Polynesians a beard is not unusual, though it does not seem to get strong till late in life. I have occasionally seen Polynesians, in whom the beard was nearly thick enough to conceal the skin. The practice, however, of eradicating it prevails very generally in the countries inhabited by the Malay race.

"In style of feature there is often no striking dissimilarity from Europeans, especially in middle-aged and elderly men, many of whom have the nose aquiline; while in females and young men it is almost always flattened. Nevertheless, I think it will be found that there is less prominence of profile in the Malay than in any other race. This appears to be owing, in part, to the absence of rigidity in the cartilage of the nose, as with the Negro. The lips are likewise thicker than among Europeans. The profile has appeared to me usually more vertical than in the white race, but this may be owing in part to the mode of carriage, for the skull does not show a superior facial angle. A more marked peculiarity, and one very generally observable, is the elevated occiput, and its slight projection beyond the line of the neck. The face, in consequence, when seen in front, appears broader than among Europeans, as is the case with the Mongolian, though for a different reason. In the Mongolian the front is depressed, or the cranium inclines backwards, while in the Malay it is elevated or brought forwards. The Mongolian traits are heightened artificially by the Chinooks; but it is less generally known that a slight pressure is often applied to the occiput by the Polynesians, in conformity with the Malay standard. A peculiarity in the Malay skull has been pointed out to me by Dr. Morton, in the tendency to unusual prolongation and projection of the upper maxilla. This character, though not universal, is strikingly exemplified in several East Indian skulls; and I have found traces of the same in the Hawaiian skulls obtained by the Expedition. It accords with the remark of Mr. Hale, 'that the upper lip is very often, among Polynesians, slightly turned up.'

"A remarkable variation in stature occurs," says Dr. Pickering, "in the Malay race. The Polynesians (particularly the Taheitians, Samoans, and Tonga islanders), appear to exceed in size the rest of mankind, while the East Indian tribes, and the inhabitants of the Indo-Chinese countries, fall decidedly below the general average. I will not undertake to offer an explanation, but there are facts connected with the nature of the food that have appeared worthy of notice. Both divisions of the race live principally on vegetable food, but there is this striking difference; the food of the East Indian consists almost exclusively of rice, while the Polynesian (to whom grain of every kind is unknown) draws his subsistence mainly from farinaceous roots, and from certain fruits of similar consistence. The further superiority in the stature of the Polynesian chiefs seems a remarkable circumstance, when we consider that they are hereditary. At the Hawaiian islands, residents declared that it depended very much on 'the greater quantity of food they obtained while young,' and I have reason to believe that the remark is not without foundation. We are reminded here of the process of forming a queen among bees. At the same time, certain kinds of food are undoubtedly more favourable than others for fully developing the human frame."—See ethnographical sections of articles HAWAIIAN ISLANDS, MALACCA PENINSULA, MADAGASCAR, and NEW ZEALAND.

Mr. Bradford remarks, that, "in the opinion of Cuvier, the Malays are not easily referrible to either of the great races, but adds Cuvier, 'Can they be clearly distinguished from their neighbours, the Caucasian Hindoos and the Mongolian Chinese? As for us, we confess we cannot discover any sufficient characteristics in them for that purpose.' M. Lesson, also, who has bestowed great research upon these insular nations, has concluded that the Malays are a mixed race of Mongols and Indo-Caucasians. The complexion of the Malays is brown, from a light tawny or yellow hue to a deep bronze; the hair is long, lank, and black, the beard weak; their eyes are black and oblique, the nose full and broad towards the apex, the mouth large, the bones of the face large and prominent, and the head narrow and compressed. Their persons are generally below the middle size and somewhat robust. The real Polynesian nations are described generally as of a dark complexion, varying from olive through shades of reddish brown to a copper-colour, with long black hair, straight or curling, and scanty beards. 'The general complexion of both men and women (of the Polynesian tribes) is a dark coppery brown, but it varies from the lightest hue of copper to a rich mahogany or chocolate, and in some cases almost to black.' Sometimes features are observed which approach to the Caucasian variety.

"The natives of the Sandwich islands are described by Mr. Ellis as 'in general rather above the middle stature, well formed, with fine muscular limbs, open countenances, and features frequently resembling those of Europeans.' 'Their hair is black or brown, strong, and frequently curly. Their complexion is neither yellow like the Malays, nor red like the American Indians, but a kind of olive and sometimes reddish brown.' Mr. Ledyard remarks of the inhabitants of the Society islands, that 'they are tall, strong, well limbed and fairly shaped.' 'Their complexion is a clear olive or brunette, and the whole contour of the face quite handsome, except the nose, which is generally a little inclined to be flat. Their hair is black and coarse. The men have beards, but pluck the greatest part of them out.' Of the Otaheitans, Captain Fitzroy says, 'To my eye they differed from the aborigines of southern South America, in the form of their heads, in the width or height of their cheek bones, in their eyebrows, in their colour, and most essentially in the expression of their countenances. High foreheads, defined and prominent eyebrows, with a rich bronze colour, give an Asiatic expression to the upper part of their faces; but the flat noses (carefully flattened in infancy) and thick lips, are like those of the South Americans.' The natives of the Pelew islands are of a deep copper-colour with long black hair and scanty beard. They are well made and of middle stature. The inhabitants of Easter island were a handsome race with oval countenances, jet black hair, scanty beard, and black eyes. Here again, the resemblance between the Malays and Polynesians and Americans has attracted the attention of those who have possessed the opportunity of comparing their physical appearance. Of the Indians of Acapulco, in Mexico, Captain Basil Hall observes, 'their features and colour partake somewhat of the Malay character; their foreheads are broad and square; their eyes small and not deep seated; their cheek bones prominent, and their heads covered with black straight hair; their stature about the medium standard, their frame compact and well made.' 'I had not long since,' says Mr. Smith, 'a striking proof of the visible resemblance between the figure, countenance, and whole appearance of the Malay and the American Indian. Mr. Van Polanen, late minister from the late Republic of Holland to the United States, and afterwards holding a high office at the Cape of Good Hope, and in the island of Java, on his return from the East, fixed his residence in Princeton. He brought with him two Malay servants. As they were one day standing in his door, there happened to pass by two or three Indians belonging to a small tribe, which still holds some lands in the state of New Jersey. When they approached the door, the attention of each party was strongly arrested by the appearance of the other. They contemplated one another with evident marks of surprise; and by their signs and gestures discovered their mutual astonishment at seeing such a likeness to themselves. Every person, indeed, who sees these Malays, and is acquainted with the countenance of our native Indians, is forcibly struck with the resemblance. The chief difference between them is, that the features of the Malays are more soft, the cheek-bone not quite so much raised, and the outline of the face somewhat more circular.' 'My first impression,' says Mr. King, 'upon seeing several New Zealanders in their native dress and dirtiness, was that they were a race intermediate between the Otaheitans and Fuegians, and I afterwards found that Mr. Stokes and others saw many precise resemblances to the Fuegians; while every one admitted their likeness to the Otaheitans. To me they all seem to be one and the same race of men, altered by climate, habits, and food, but descended from the same original stock.' And again, 'a word about the inhabitants and I leave the Keelings (islands). No material difference was detected by me between the Malays on these islands, and the natives of Otaheite and New Zealand. I do not mean to assert that there were not numbers of men at each of those islands to whom I could not trace resemblances, (setting individual features aside) at the Keelings. I merely say that there was not one individual among the two hundred Malays I saw there, whom I could have distinguished from a Polynesian islander, had I seen him in the Pacific."

III.—THE AUSTRALIAN RACE.

"The Australian," says Mr. Pickering, "may be characterised as having the complexion and features of the Negro, with hair in the place of wool. On closer examination, however, other points of diversity are remarked; and I think it will be very generally found that the forehead does not recede as in the Negro, an unusually deep-sunk eye giving it rather the appearance of projecting. The eye, at the same time, though uniformly small, is uncommonly piercing. With regard to other races somewhat approximating in personal appearance, the genuine hair will at all times distinguish the Australian.

"About thirty Australians came under my own observation, who neither had the lips so uniformly thick, nor the nose so much depressed, as in the Negro; but in certain instances both nose and mouth were wider. Some individuals were of surpassing ugliness; while others, contrary to all anticipation, had the face decidedly fine; and several of the young women had a very pleasing expression of countenance. The general form, though sometimes defective, seemed, on the average, better than that of the Negro; and I did not find the undue slenderness of limb which has been commonly attributed to the Australians. Strange as it may appear, I would refer to an Australian as the finest model of the human proportions I have ever met with; in muscular development combining perfect symmetry, activity, and strength; while his head might have compared with an antique bust of a philosopher.

"The Australian complexion appeared to me fully as dark as that of the Negro; but I did not institute a critical comparison, neither did I see the two races in company. The hair seemed rather coarse, and instead of being perfectly straight was usually undulating, or even curling in ringlets. The head was by no means deficient, though less abundant than in the Feejeean.

"Although the range or territorial limits of the different Australian communities are carefully observed, the idea of any ownership in the soil does not appear to have occurred to these people. They neither at any time resisted the new-comers from Europe, nor did they withdraw from the contact; but they continue to ramble over their old domain, complaining only of the disappearance of the kangaroo. For this loss, which they attribute to the 'White man's kangaroo (the sheep and bullock) having driven away their own,' they consider themselves entitled to indemnity. On the frontiers, the natives doubtless occasionally spear the cattle; though it was evident that they have been far more sinned against than sinning. They have further been peculiarly unfortunate in being exposed to the contact of such a class of people as the escaped convicts. According to common report, 'the natives cannot be induced to cultivate the soil, and are equally averse to all other kinds of labour;' but in the towns I met with one or two individuals who appeared to be engaged in some occupation. I am not aware, however, that any instance is on record of an Australian being employed as a mariner.

"A striking difference from the aboriginal Americans, and perhaps from all other known people, is found in the absolute rejection of all innovation by the wild communities. European customs or articles of manufacture do not make the slightest progress among them, and relations of commercial intercourse continue altogether unknown. In a region which has not added to the list of esculent plants, it does not appear surprising that cultivation should be unknown; especially as the soil rarely affords the capability. Moreover, such is the forbidding character of the natural vegetation, that this contributes very slightly to the sustenance of man; and I have sometimes doubted whether any different branch of the human family could have maintained its existence on the slender natural resources of Interior Australia. The original population, notwithstanding its sparseness, was yet in proportion to the means of subsistence; and the European accession has been owing solely to the introduction of sources of supply previously unknown. The Australians are strictly in 'the hunter state,' deriving their subsistence from game, and from the spontaneous productions of the country; and they can only be compared with the hunting tribes of America. The parallel, however, is not very obvious; and the Australians appear almost as anomalous as the inferior animals and plants of the same quarter of the globe. They have commonly been placed at the bottom of the scale of civilisation; yet in the few arts in their possession they have displayed singular ingenuity: for instance, in the method of ascending trees by making successive notches; the 'throwing-stick' for the light javelin; and especially the boomerang, the curved missile whose devious course remained so long the subject of incredulity. Their manufactures, indeed, are chiefly confined to their peculiar weapons, which serve both for warfare and the chase. They also use defensive armour, the light oblong shield, which is rendered efficient only by the height of skill and activity."

Dr. Thomson, in a paper recently read before the Ethnological society, after remarking that the experience and observation of all travellers in New Holland during the last thirty years confirm the opinion that the aboriginal races of that country are gradually, yet certainly, in course of disappearance from the face of the globe, and that, from whatever cause it may be, the rude inhabitants of these extensive territories, over which we now claim jurisdiction, are on all sides steadily giving way before the rapid advances of the white man, says, "Of many tribes which not sixty years ago existed in the neighbourhood of Sydney, each numbering from 200 and upwards, several have already disappeared, as the Botany Bay tribe, the Five Islands tribes; and of others only a trace exists, to be seen occasionally wandering about the streets of the metropolis of New South Wales." Dr. Thomson considers it probable that in their physical characters these tribes do not differ greatly from each other; that such differences as are observed to exist may be easily accounted for by the climate and the supply of food of the districts in which each is to be found. He thinks, from a careful examination of the natives of various parts of New Holland, at points tolerably distant from each other, that they have had one common origin. His paper contained observations on some of the tribes inhabiting the neighbourhood of the Castlereagh and Lochlans branches of the river Macquarie; some of the tribes surrounding Moreton Bay, the Broken Bay tribe near Sidney, and a tribe inhabiting Albany island. Of the tribes which once inhabited the Castlereagh and Lochlans branches of the Macquarie river, several are said to be at the present time quite extinct. The Bogan blacks number not more than 100. These natives are generally about the middle height, body spare; hands and feet of tolerably good proportion; arms and legs, especially the latter, long; head rather round and oval, forehead rather broad, but low; the cheek bones rather high, the eyes sunken, the eyebrows arched; they have but little whiskers, but some of them have a good deal of beard on the chin, which is rounded off. The lips are thick and prominent, especially the upper one, the mouth being expressive of determination. The skin is dark brown, approaching to black, and very smooth. The hair is inclined to be long, soft, and silky, very glossy, and always curly. The women are well-proportioned in their youth, and the hands and feet of many of them are very small. In their manner and speech the aborigines of Australia are quick and animated, and easily excited. They are very revengeful, as many of the European settlers have found to their cost. Dr. Thomson says that "all the tribes have distinct markings, caused by longitudinal or tranverse gashes on the breast, back, shoulders, and arms." He says all the heads he has examined are remarkably thick and heavy in proportion to the size, and he remarks that "nearly all of them had marks of waddy-blows and other injuries, which would have killed anything but a New Hollander."

Coming from among the Pacific islands, Mr. Pickering "was surprised at the facility with which the Australians acquire the English language, and at the correctness of their pronunciation. The latter circumstance is doubtless attributable chiefly to the frequent occurrence of terminal consonants in their own languages. 'A child educated at a school in Sydney showed intellectual capacity equal in every respect to that of his European companions.' A difficulty, however, had been experienced in undertakings of this sort, in consequence of the mothers 'returning to claim and carry away their children.' Half-breeds must be rare, as I did not meet with an instance; but some were spoken of; and native women were occasionally seen accompanying the wagons from the interior.

"A great uniformity in the habits of the people appears to prevail throughout Australia; and it is only in the extreme north-east, in the immediate vicinity of Torres straits, that any important variation is known to take place. In this quarter Flinders observed the use of 'the bow; and he met with some individuals in the gulf of Carpentaria, who appeared to have been circumcised; but further west the usual Australian weapons again made their appearance.' A missile club, analogous to the Feejeean, is represented in the '*Voyage aux Terres Australes*,' but belonging, I think, to a different part of the continent. The source of these innovations may readily be divined: but a greater one was found by Flinders at the Murray islands, immediately within the straits; 'the cocoa-nuts, bananas, and joints of bamboo filled with water, that were brought off by the natives,' showing a knowledge of agriculture; while at the same time, no physical difference from the continental Australians appears to have been remarked. There are other reasons for supposing that the Australian race does actually extend among the little-known islands of the same vicinity."

Mr. Earl, in a Memoir on the Tribes of the North coast of Australia, remarks that the natives of the north coast are not only less known to us than those of the other coasts of the island, but possess a peculiar interest from their proximity to the Indian Archipelago. "A circle," says Mr. Earl, "drawn round Port Essington, at a distance of 500 miles, would enclose almost an equal number of distinct tribes." The good understanding which existed between the colony and the natives in its vicinity induced parties of warriors and their chiefs to come from the remote interior to visit the whites; and in the month of April, when the Macassar prahus congregate at Port Essington, there may be seen, besides the tribes of Australia, people from many of the islands of the Archipelago. Mr. Earl, however, confines his observations to the natives of Australia who inhabit the coast from the Cobourg peninsula towards the east. Certain general characteristics, he says, are observable among all the tribes of this part of Australia. Thus their skin is invariably embossed with raised cicatrices, the septum of the nose is generally pierced among the men; and clothing is disregarded but by way of ornament; they paint their bodies. Their mode of treating the dead differs, however, from that in use among the tribes of other parts of the country. They leave the bodies wrapped in the bark of the tea-tree till nothing remains but the skeleton, which is then removed to the general cemetery, or placed in the hollowed trunk of a decayed tree. They are divided into three castes. They have some superstitions similar to those of the Indian islanders, while others are like those which pervade Australia. In the Cobourg peninsula there are four distinct communities, of which the most powerful occupies the southern coast and the islands of Van Diemen's Gulf; the upper part of the harbour of Port Essington being also in their possession. In the mountain-range there is a very numerous tribe—Marigianbirik. Beyond these we know nothing; nor have we much accurate knowledge respecting the tribes distributed between Jalakuru and the gulf of Carpentaria. With the Yaaks, or Croker island tribe, our acquaintance is of older date than with the others; the Raffles Bay settlement in 1827 was within their territory. These are described as most forbidding in appearance, and little better in disposition. Fear of the whites, however, secured their good behaviour. The Yarto and Syi tribes resemble each other in general characteristics, though their dialects are totally dissimilar. They are a better-looking people than the Croker islanders, and from the beginning showed great partiality for the English. The Bijna-lumbo tribe possess a superior physical organisation; and, indeed, Mr. Earl is inclined to think there has been some infusion of Polynesian blood among the aborigines of this part of Australia. Of the four dialects spoken by the tribes of the Cobourg peninsula, one only appears to differ from those spoken in other parts of New Holland; and this difference consists only in the words almost invariably ending in a vowel. This, however, seems only accidental. The consonants *f* and *s* are rejected throughout the Australian dialects; as is also the *h* aspirate. With the single exception mentioned above, two-thirds at least of the words end in a consonant, and often a double consonant, as *-alk*, *-irt*, &c.; the nasal *-ng* is very common. The natives of the coast, from frequent intercourse with the Macassar trepang-fishers, have picked up a good deal of their language, which is a dialect of the Polynesian; and as they spoke in this *patois* to the English on their first arrival, these latter, from ignorance, made vocabularies of it, taking it for the native language of the tribes. Mr. Earl, proceeding to treat of the great inland tribes, says that the communities in the interior are much more numerous and better organised than the coast tribes. One great chief, dignified by the title of rajah, has control over several large communities, each having its own chief. They live on the spontaneous produce of the country, which is abundant. Their hair, which is fine, is adorned with parrot's feathers or opossum's fur, and makes a very neat appearance. They are treated with great deference and attention by the coast natives. Mr. Earl looks upon these people as a kind of mixed race between the Polynesian and aborigines, who much resemble the Papuas of New Guinea. The people of Arnheim-bay are described as numerous and powerful, and very formidable when hostile: so say the Macassars. They are remarkable for their bulky forms and fine chests; their lower extremities, however, are not well formed, having the curved shin; their features are coarse, and the nose broad and flat, but the general expression is pleasing. All the males above twelve years old were circumcised. The western side of the gulf of Carpentaria is well-peopled; but beyond this, as far as the head of the gulf, the natives are few and scattered. Mr. Earl says he has observed that on the north coast of Australia the population bears a striking proportion to the quantity of vegetable food, to procure which the natives take great trouble, while they do not seem to be nearly so partial to animal diet.

"All visitors," resumes Mr. Pickering, "speak of seeing straight-haired individuals among the natives of the Louisiade; and one of the figures of the natives of Eastern New Guinea, contained in Belcher's Voyage, seems to correspond with the Australian. It seems also probable that, as we become better acquainted with this portion of the globe, there will appear less isolation in the customs of the Australians. We can at present refer to various connecting circumstances: to the use of a throwing-stick for the javelin, in Eastern New Guinea, in New Britain, and even in New Caledonia; to the oblong shield or buckler, worn in the Louisiade and around Dampier's straits; and also to some corresponding methods of dressing the hair. Some words, too, appear to be in common: thus, *dundu* (the Australian name for the black swan) occurs in New Britain, where, according to Morrell and Jacobs, it is applied to a species of emu or cassowary."—See ethnographical section of the article AUSTRALIA.

"We come now to the controverted point of the physical race of the natives of Van Diemen's Land, situated to the southward of the continent. They were spoken of at Sydney as really distinguishable from the Australians by their 'woolly hair; peculiar, however, in its texture;' and, at the same time, all idea of affinity with the Negro race was rejected. Mitchell has some observations on the Tasmannians, and concludes in favour of associating them with the continental Australians; and this opinion seems the most probable. Indeed, the portraits of Labillardiere would have much weight with me, as they exhibit very decidedly the projecting forehead of the Australian. Not being aware of the presence of a Tasmannian in Sydney, I lost the opportunity of seeing him: those who were more fortunate were unable to recognise any resemblance in the Feejeeans, whom we shortly afterwards visited."

IV.—THE PAPUAN RACE.

"The term 'Papuan,' notwithstanding some ambiguity, may be conveniently applied to a race of robust blacks, of whom the only examples I have seen (with probably one exception), have been the natives of the Feejee Islands. So far as my observation extends, the Papuan race differs from the rest of mankind in one remarkable physical peculiarity, the hardness or harshness of the skin. This point long since attracted the attention of those Tonga people, from whom Mariner derived his accurate notices of the Feejee islands. It is proper to add, that I have not examined the quality of the skin in the Negrillo race.

"The hair of the Papuan is in great quantity, is naturally frizzled and bushy, and so coarse as to be rather wiry than woolly. When dressed according to the Feejee fashion, it forms a resisting mass, and offers no slight protection against the blow of a club. I have had occasion to remark that it actually incommoded the wearer when lying down; and to this circumstance, rather than to any foppery, I am disposed to attribute the origin of the wooden neck-pillow. The beard does not appear to grow so long, or to cover so large a portion of the face as in the White race; but the Papuan exceeds the remaining races, in the quantity of beard. The complexion is of a deeper shade than in the Malay race, but is much the same as in the Bengalee or Telingan. The features in many respects resemble those of the Negro, but the lips are not quite so thick, and the nose is somewhat more prominent; while a very general elongation of the face imparts a different aspect. I did not meet with a really fine head among the Feejeeans, but the countenance was often grave and peculiarly impressive; and I had frequent occasion to remark, that strangers did not readily forget the features of Veindovi. In average stature, the Feejeeans were found to exceed the White race; but they fell below the men of Tonga and Samoa. The profile in general appeared to be as vertical, if not more so, than in the White race; but this, I find, is not confirmed by the facial angle of the skull, and it may possibly be accounted for by some difference in the carriage of the head. The Feejeean skulls brought home by the Expedition, will not readily be mistaken for Malayan; they bear rather the Negro outline; but they are much compressed, and differ materially from all other skulls that I have seen."

The Papuan race may be traced with tolerable certainty from the Feejee Islands to the shores of New Guinea. The population appears to be everywhere dense, and to be divided, as at the Feejee Islands, into two classes that are politically independent, the fishermen or people of the coast, and the mountaineers. The practice of cannibalism seems also co-extensive with the race, while the surrounding islanders, though often in a less advanced state of society, as generally hold it in horror. The Papuan race appears also to extend through a portion of the East India islands; but it probably does not occur to the westward of Floris, or Ende. Notwithstanding the various remarkable coincidences in customs, as the use of the neck pillow, circumcision, similar modes of dressing the hair, even to the staining of it of a flaxen hue, the Papuan race does not appear to exist in Eastern Africa. MM. Lesson and Garnot, describing the genuine Papuan race, say, "the colour of the skin is black, mixed with an eighth part of yellow, which imparts to it a clear tint of various intensity. Their hair is black, very thick, and moderately woolly. They wear it frizzled out in a very remarkable manner, or let it fall upon their necks in long and twisted masses. Their countenance and features are regular, except their noses, which are somewhat flattened, with the nostrils enlarged in the transverse direction. Their chins are small and well formed; their cheek bones are prominent, their foreheads elevated, their eyebrows thick and long. Their beards are thin; they let them grow upon the upper lip and chin like many African nations." In this description it will be perceived, says Bradford, that there is "just that degree of diversity from the Ethiopian or Negro, which a slight mixture with the Malay islanders of the Pacific might produce. This conclusion is strengthened by the circumstance, that as we proceed towards the Malayan islands, where the races may be mixed more equally, we find the Papuan complexion becoming lighter and approaching that of the Oceanic nations. Mr. Lesson, who supposes them to have migrated into the islands of the Pacific subsequently to the Oceanic tribes, traces a close resemblance between them and the dark-coloured tribes of Madagascar, a fact which adds to the force of our conjecture, for in that island, as will hereafter appear, there is also a race not aboriginal, distinguished by an olive colour, straight black hair and thin beard, and similar to the Malay race in their leading characteristics; and from which, by intermarriage with the Negro, has probably originated the very variety resembling the Papuas. It is curious, that in America, we find the same consequences attending the mingling of the two races. 'In this part' (of Brazil), say MM. Von Spix and Von Martius, 'we met with several families of the people called Cafusos, who are a mixture of blacks and Indians. Their external appearance is one of the strangest that a European can meet with. They are slender and muscular, in particular the muscles of the breast and arms are very strong; the feet, on the contrary, in proportion weaker. Their colour is a dark copper or coffee brown. Their features, on the whole, have more of the Ethiopic than of the American race. The countenance is oval, the cheek-bones high, but not so broad as in the Indians; the nose broad and flattened, but neither turned up nor much bent; the mouth broad, with thick but equal lips, which, as well as the lower jaw, project but little; the black eyes have a more open and freer look than in the Indians, yet are still a little oblique, if not standing so much inward as in them, on the other hand not turning outwards as in the Ethiopians. But what gives these Mestizoes a peculiarly striking appearance, is the excessively long hair of the head, which, especially at the end, is half curled, and rises almost perpendicularly from the forehead to the height of a foot, or a foot and a half, thus forming a prodigious and very ugly kind of peruke. This strange head of hair, which, at first sight, seems more artificial than natural, and almost puts one in mind of the plica polonica, is not a disease, but merely a consequence of their mixed descent, and the mean between the wool of the Negro and the long stiff hair of the American. This conformation of the hair gives the Cafusos a resemblance with the Papuas in New Guinea, and we therefore thought it interesting to give the representation of a woman of that race in her peculiar costume. To this it needs only to add, that with the Papuas of New Guinea 'the hair is long and woolly, and frequently forms a huge peruke three feet in diameter.' Thus it appears, that in three of the quarters of the globe, where the Negro and Mongolian races have intermarried, the physical result is nearly the same, a circumstance appearing to justify the inference, that the Papuas are a mixed race."—See articles FIJI ISLANDS and NEW GUINEA.

V.—THE NEGRILLO RACE.

"The Negrillo race has much the same complexion as the Papuan; but differs in the diminutive stature, the general absence of a beard, the projecting of the lower part of the face or the inclined profile, and the exaggerated Negro features. The hair also is more woolly than in the Papuan, though far from equalling in knotty closeness that of the Negro. On a direct comparison with the Negro, I have observed that the complexion appears to be rather red than black. Geographical distribution," Mr. Pickering continues, "sufficiently indicates that the Negrillo race once occupied more space than it does at this time, and that in many instances it has preceded the dissemination of other races."—See articles ANDAMAN ISLANDS, NICOBAR ISLANDS.

VI.—THE TELINGAN, OR INDIAN RACE.

"The Eastern Hindoos, those at least who make visits to the East India islands, present great uniformity in their personal appearance; and in this respect they agree with the tribes and nations situated to the eastward of Hindostan. The complexion is much the same as in the two preceding races, and is so decidedly darker than in the Malayan, that by common consent it is called black; although, on comparison, the hue differs widely from that of the unmixed Negro. The true colour may be formed by mixing red and black; and in reference to the use of the term of 'purple-brown' and that of 'olive,' it should be observed that neither blue nor green enter into any variety of human complexion. The features approximate very closely to those of the White race; but, in general, the mouth appeared to be wider, the nose rather less prominent, and the lips sensibly thicker. The profile was observed to be less vertical than in the surrounding Malays; the lower part of the face projecting with a regular arch, as in the Mongolian; and there was a further correspondence with the latter race, in the frequent instances of the arched nose. The skin was ascertained to be very soft. The beard occurred more frequently, and was decidedly more copious than in the Malayan race. The hair was straight and fine, and I have never seen it of any other colour than black. I have not met with Albinoes in the Telingan race."

VII.—THE NEGRO RACE.

"At the present day, the personal appearance of the Negro is generally familiar; and the thick lips, flattened nose, retreating forehead, close woolly hair, and dark complexion, have become proverbial. The Negro appears to exceed all other races in depth of hue, and in the close woolly texture of the hair is rivalled only by the Hottentot. The absence of rigidity and of a divided apex in the cartilage of the nose is a character common equally to the Malayan, and, probably, to some of the other races. In the case of two Albino children, the Negro aspect had so entirely disappeared, that they might have passed for the children of Europeans, but for the remarkable appearance of the hair, which I could only compare to a white fleece. The Negro race seems to occupy about one-half of Africa, and, excluding the northern and southern extremes with the table-land of Abyssinia, it holds all the more temperate and fertile parts of the continent. These limits, to all appearance, would not have been exceeded to this day, aside from foreign interference; but, as one consequence of the events of the last two centuries, the Negro race seems destined to fill hereafter an important place in general history."—See article NEGROES.

a. *Europeanised Negroes.*

"Negroes are now to be found in most parts of the globe where Europeans have established themselves; and although separate notices seem hardly to be required, I have selected the following. During a week spent in different parts of the island of Madeira, I met with no Negroes, except at the port of Funchal; and the few individuals seen at this place may have been connected with the foreign shipping. On the other hand, the population of the Cape Verd islands, judging from our hasty visit, appeared to be principally composed of Negroes; and they were living in a very rude state for the subjects of a civilised government. They, however, exhibited a certain air of independence, not seen in the same race in the United States. Brazil had more of the aspect of a European colony, although Negroes formed the most numerous class of the population of Rio Janeiro and its environs; and instances of even mixed descent were comparatively rare. Most of these Negroes were slaves; but very many were employed as soldiers in the military police. Others held places of honour and trust under the government; and one man was spoken of, as being distinguished for his abilities as an advocate. Several of the women too moved, with their Portuguese husbands, in the first circles of society. I was careful to look among the imported Negroes for traces of some different race; but I was unable to detect any; neither have I met with more success in other parts of America. A second race may have been sometimes included in the importations from Madagascar; but otherwise, I think, instances have been extremely rare; prior at least to the recent introduction by the English of people of Hindostan. Some Negroes were seen among the 'guachos' or mounted herdsmen of the Rio Negro, in North Patagonia. 'They had come by land from Buenos Ayres,' and they hardly appeared to be on the footing of slaves. There are, perhaps, fewer Negroes in Chili than in any other

Europeanised portion of America. I do not remember seeing more than ten or twelve, during the eighteen days I spent in Chili, chiefly at the cities of Valparaiso and Santiago. In Peru, Negroes are common, but they are chiefly confined to the vicinity of the coast, where many of them are held as slaves, and where the practice of merely 'requiring them to furnish their masters with stated sums of money, had contributed essentially to the frequency of robberies.' Some of the free Negroes seen at Lima had reached the middle class of society, but I did not hear of any one who had acquired eminence, or who held an important office. In the churches, however, some deceased Negro priests had been canonised. Among the military, Negroes were sometimes observed in the ranks; together with a much larger proportion of persons of mixed origin, 'zamboes and mulattoes.' In the region of the Andes I saw but one Negro, who held some subordinate office in a small village.

"In New South Wales I met with two or three Negroes in the streets of Sydney; but I saw none on my visit to the district of the Hunter river. Free Negroes were occasionally seen on such of the Polynesian islands as have resident Whites; having been derived, as in various other parts of the world, chiefly from the American shipping. I also met with a few half-breeds, between the Negro and Polynesian. Up to the time of our visit, no Negroes had taken up their abode at the Feejee group. Among the East India islands, Negroes seem to be very nearly wanting; and, indeed, we read of their being carried to the Indo-Chinese countries as curiosities. At Singapore, situated in respect to the route of travel at the entrance of the whole region, I met with three Negroes; and one of them had been introduced partly through the agency of Europeans. This man was large and muscular in comparison with the surrounding population, and was called a 'Koffri,' a class of persons said to have been originally imported from Mozambique into Southern India, and who bear a high character for fidelity. It seemed singular that their identity with the Negro was not recognised even by American residents. At Cape Town the lower class of the population appeared to be composed chiefly of Negroes; and, according to report, 'they had been derived principally from Mozambique and Madagascar.' They appeared to be more cleanly, and to live on a better footing with their White neighbours, than their brethren in our northern cities.

"At St. Helena, the Negro portion of the population, as already stated, 'had been derived exclusively from Madagascar.' A remarkable deviation from direct commercial intercourse may be here noted; the circumstance 'that slaves from Africa are now carried across the island of Madagascar, to be sold to Europeans from its eastern coast.' One or more Negroes from the United States had taken up their residence at Zanzibar, and others again at Bombay; conforming to the customs of their several adopted countries, and having no wish to return to the land of their nativity.

"In passing through the Mediterranean countries, I observed two or three Negroes at Malta, and as many at Marseilles; but I met with none at the Sicilian and Italian ports, nor in the interior of France. The general excellence of the Negro ear for music is a subject of common remark in the United States, and is manifested in many of the ordinary occurrences of life. Indeed it has been stated, apparently on sufficient grounds, that much of our popular music can be traced to a Negro origin. Observations on my recent journey have led me to suspect that some of this music may have a yet more distant source, and one perhaps more ancient than is commonly supposed."

b. *Arabised Negroes.*

"With the exception of the two individuals at Singapore, I first saw Muslim Negroes in Egypt. In this country, the Negroes seem to be in a good measure confined to the two cities of Cairo and Alexandria; and, excepting a large body of soldiers, are principally employed as house-servants. It sometimes happens that one of them 'inherits his master's property together with the charge of his family;' in accordance apparently with ancient usage, with Abraham's selection of an heir from among the members of his household. Negroes are occasionally purchased and educated both by Copts and by resident Europeans; but the influences of locality prevailing, they do not form a distinct class from the Muslim Negroes. I do not remember to have seen Negro servants among the Jews, either at Cairo or anywhere in the East.

"I did not in Egypt observe Negroes engaged in the labours of agriculture. Neither, indeed, are they so represented on the ancient monuments; where Whites, distinguished by their profile from the proper Egyptians, are often serving in the capacity of slaves. Negroes are figured principally in connexion with and as illustrating the military campaigns of the eighteenth dynasty; and there is evidence, moreover, that one of the Pharaohs of this dynasty (Thouthmosis IV.) selected a negress for his queen. I do not remember seeing Negroes represented on the anterior monuments, nor indeed on those of a much later date; though I did not, as carefully as the case requires, search the records of the Ptolemaic conquests.

"I am not aware of any fact contravening the assumption that Negro slavery may have been of modern origin; and the race even seems to have been very little known to the ancient Greeks and Romans. I have seen the Negro skull which was exhumed in the island of Malta, among the ruins of Hadjerkim; but, in the absence of inscriptions, I am not aware that any date has been assigned to these anomalous constructions, other than referring them indefinitely to idolatrous times. In regard, however, to the general subject of the ancient Oracles, I will mention, on the authority of Dr. Schembri, that 'the name of 'dodon' is to this day applied by the Maltese to a particular variety of black pigeon.'

"At Suez I remarked one Negro among a party of 'Arabs of the Hedjaz,' and this man wore the same costume, and appeared to be living on a footing of equality with his companions. At Mocha Negroes were numerous, and they had been derived from two distinct portions of Africa. Some had come, like those of Egypt, from the White River of the Nile, and were called 'Nuba.' These were 'highly esteemed as soldiers, and, as such, were kept in numbers by some of the more powerful chiefs of Southern Arabia.' The majority, however, of the Negroes I saw at Mocha had been brought from Zanzibar. At Mocha I also met with a class of persons more or less mixed, who were called 'HADEM,' a term which appeared to be equivalent to that of 'low people,' and which was used in other instances to be mentioned hereafter. They were characterised as 'a sort of Arabian Negro coming from the Interior or the Desert, who, among other usages, practise eating dead animals.' Of the individuals pointed out to me, some were not distinguishable from the Arabs of the country; but others presented traces of Negro descent, and one man had the aspect of an unmixed Negro

"At Muscat the influx of Negroes appeared to be exclusively from Zanzibar, and, excepting those brought as slaves, they consisted chiefly of SOAHILI. The latter were perhaps only visitors, since they compose the principal part of the crews of the numerous dows trading to the African coast. The Soahili are all Muslims, and I did not hear of any who were held in slavery in this part of Arabia; but at Mocha I met with slaves who called themselves Soahili. In the streets of Bombay I frequently fell in with Soahili, who belonged, as before, to the Arab shipping. But at Singapore I saw none of these people, and their absence was confirmed by the testimony of the Abyssinian Arab. At Zanzibar the Soahili were found to assume the exclusive management of the small dows trading along the coast as far south as Mozambique and Madagascar; while another, and, in their own estimation, a more responsible class of navigators, direct the larger vessels to Arabia and North-western Hindostan. The Soahili are a mixed nation, consisting principally of Negroes, but embodying the remains of an infusion of Whites; and many individuals among them would not be distinguished from Arabs. The inhabitants of the fertile islands of Zanzibar and Pemba are chiefly Soahili, together with, in all probability, the mass of the population at Mombas and in the Arab towns in the delta of the Juba. Soahili, however, are scattered along the coast beyond these limits, and adventurers sometimes join the caravans of the interior Negroes, and make visits to the Great Lake, or, in conjunction with Comoro men, form independent trading expeditions for minor distances. The Soahili language has, in consequence, become the medium of commercial intercourse along the eastern coast of Africa. It contains numerous foreign words, derived principally from the countries around the Persian Gulf; but it is properly a Negro language. At the same time, instances may probably be found of the transmission of words in the opposite direction."

c. *Malayised Negroes.*

"This class of people is confined to the immediate vicinity of Eastern Africa, and constitutes the chief part of the population of the Comoro islands and Madagascar. The languages and customs of these islanders are not Arab, but correspond with those of the East Indies; and most of the plants cultivated have evidently been derived from the same quarter. There are no records and no existing intercourse. Aboriginal Negroes may, or may not, have once dwelt upon these African islands, but the state of things now prevailing is clearly the result of East Indian colonisation. Indeed, the Malayan race is actually present in Madagascar, and is still dominant there; and as to the time of arrival, it is sufficiently evident that Malay influence has preceded the visits of the Arabs. The Madagascar people seen at Zanzibar, with the exception of the two Ovahs already noticed, were all Negroes, who did not even present any distinguishable traces of Malayan admixture. I did not hear that any of them had voluntarily left their native country, but many appeared to be free; and a cluster of their houses was pointed out to me as an example of the Madagascar style of building. All the Madagascar people whose origin I ascertained, belonged to the 'Sakalava' nation. Comoro men were found to be yet more numerous at Zanzibar, where they form a very considerable portion of the population of the city. I was equally at a loss to discover in the Comoro men any traces of the physical Malayan; and some of them assured me that "there are no long-haired people upon their native islands."

d. *Aboriginal Eastern Africa.*

"On one of my visits to the slave-market at Zanzibar, a Negro among the bystanders attracted my attention on account of the rudeness of his dress and his whole appearance. The race was evidently pure; and if there was any peculiarity, it was only in the beard being nearly wanting. On inquiry, I learned with surprise that he was an aboriginal of the island; and further, that having come from an unfrequented district, he was acquainted with the original language. He searched out two other 'Hadem,' in order that I might hear him converse in it; but these, it appeared, could only speak Soahili. A vocabulary would doubtless have been interesting, and I regret being unable to give more than a single word, *monakela*, which signifies 'a small boy.' Zanzibar, being situated nearly 20 miles from the main land, these circumstances have a bearing on the question of the original population of the Comoro Islands and Madagascar.

"The following details respecting the tribes which inhabit the main land of Eastern Africa were collected at Zanzibar.

"The WANIKA possess the country around Mombas, and their towns were found by Dr. Krapf to be 'protected by being built in clumps of thick woods;' it is presumed, against the incursions of the Ethiopian tribes who border on the north and partly on the west. The Wanika are known at Zanzibar only as slaves, and, as such, are in request for the special purpose of 'procuring toddy;' for, from living on the coast, they have been accustomed to climb cocoa-palms. Notwithstanding that they were said to be common on the plantations, I fell in with but a single individual. This man bore the marks of national designation, consisting of a single notch filed between the two upper front teeth, with numerous small scars on the breast.

"A detached tribe of Negroes are found upon the African coast north of the Wanika in the delta of the Juba. The Sultan of Patta termed them 'POKOMO,' and also 'Hadem;' and he spoke of them as 'rather a good sort of people, who will pull a boat up the Oozy, being relieved at each town by a fresh set of hands. They live on one side of the river, the other side belonging to the Galla.'

"The CHAGA dwell in the interior, to the southwest of the Wanika, and on the upper part of the Pungany river. All accounts of the Chaga agree in representing them to be an agricultural and a Negro nation. Individuals were said to be common at Zanzibar, but I did not succeed in finding any. In all my inquiries respecting the people of Eastern Africa, I could not hear of pastoral Negroes, nor of Ethiopian cultivators; but there are some undetermined tribes holding an intermediate position along the borders of the cultivable soil, who may offer trifling exceptions. These minor tribes are situated to the westward of the Wanika, and apparently to the northward of the Chaga, and they may be conveniently noticed in this place.

"The WAKAMBA, already alluded to, have, by some writers, been termed 'Merremengo.' Sadik stated that 'they are not bad people, but all the same with the Monomoisy; they trade and do not fight. They do not bring slaves, except a few, but trade in ivory.' Dr. Krapf, writing from Mombas, considers them a kindred nation to the Wanika and M'Sigua, but as having no affinity with the M'Kuafy, and it may therefore be inferred that they use a Negro language. He further states, that they 'live partly among the Wanika and partly extended into the interior, that they throw their dead in the bush, and that some lead a pastoral life, and possess a fine breed of cattle.' I did not meet with any of the Wakamba, although they were spoken of as not rare at Zanzibar.

"The WATAITA seem also to be borderers, their country 'commencing not more than a five days' journey from the coast.' They were spoken of as 'good people, who do not fight unless the Wakamba come out against them.' Sadik called them 'Wanika, but a different nation.' They were enumerated among the tribes pillaged by the M'Kuafy.

"The TAVAITI dwell to the westward of the preceding: and a girl belonging to this tribe was seen at Zanzibar. She was apparently in the condition of a slave, though not of the lowest class; and she spoke, it was said, neither the Chaga nor the M'Kuafy language. Her hair was cropped close; and from the hasty glance I obtained, I was unable to decide whether the Negro race was entirely pure.

"The WAMPUGO appear to be quite a local tribe; dwelling, according to the M'Kuafy lad, 'on the river Ruvu, which comes into the Pungany, and not more than half a day's journey from his own country.' Sadik said that 'they brought ivory, and not slaves.'

"The MONOMOISY, on the other hand, occupy an extensive territory, which is situated far in the interior of the continent, and is bounded on the west by the 'Great Lake.' An Arab informed me, 'that on his journey to the Monomoisy country he crossed numerous rivers coming from the mountains; none of them, however, of greater depth than mid-leg, and donkeys had been taken the whole distance. He was four and a-half months in going, and was absent in all three years; but he did not visit the Lake. He returned with a party of 700 persons, and was only 76 days in reaching the coast.' I did not inquire of him the point of arrival; but I would observe, that, according to some authorities, Kilwa (Quiloa) is nearer than Zanzibar to the Monomoisy country. In some memoranda communicated to me at Zanzibar, I find the following statement of a Soahili, who 'was eleven months absent on a similar journey. He saw the Great Lake, which has many islands. The canoes carry from five to fifteen men. He did not go armed while he remained among the Monomoisy. The kings fight much among themselves. The Monomoisy have no money, but they are fond of beads and brass.' I learned, moreover, that the imaum has some sort of patriarchal influence with the Monomoisy, and that the great caravans, which come to Kilwa and to the coast opposite Zanzibar, 'bring him a present, and look to him for protection during their stay.' Beasts of burden appear to be entirely unknown to the Monomoisy; and the ponderous elephant tusks, when brought to the coast, have been sometimes found to contain a little box or other article, deposited under the fancied idea that 'it will lessen the weight.' According to Sadik, 'the Monomoisy trade only in ivory, and do not bring slaves; but sometimes a bad fellow among them will manage to sell his friend.'

"The WABIZA (Moviza of some writers), who have been alluded to above, appear to be the neighbours of the Monomoisy on the south. According to another account, 'they dwell at the Great Lake, and generally come to Kilwa.'

"The N'YASA, who inhabit the islands and perhaps the further shores of the Great Lake, seemed to be the most distant tribe known at Zanzibar. Individuals in the state of slavery were not uncommon; and one of them stated 'that he was one day in reaching the Great Lake, and spent another in crossing it, making two days to the Monomoisy country.'

"The M'SIGUA dwell on the coast, and appear to be much scattered among the Wanika, although their proper country, I believe, is south of the Pungany river. Individuals were commonly seen in the slave-market at Zanzibar, where they were recognized by the numerous small scars on the temples.

"Dr. Krapf speaks of the M'SAMBARA, who dwell somewhat further in the interior, as 'a kindred nation to the M'Sigua, Wanika, Wakamba, and Soahili;' grounding his opinion, it is presumed, principally upon the affinities of language. With the exception of the M'Sigua, almost all the slaves offered for sale at Zanzibar had been brought from the southward, and principally from Kilwa Small dows arriving from that place were crowded with the living freight; and the influx was represented as being much greater at a different season of the year.

"The MAKUA occupy the coast immediately to the northward of Mozambique, and trade with that city. They were spoken of as 'bad people, who kill the crews of vessels that happen to get on shore; and dows, in consequence, do not visit that part of the coast, unless several are in company.' A Makua girl, pointed out to me in the slave-market, bore the national mark of a crescent-shaped scar in the middle of the forehead.

"The district of the M'KINDO is back of Kilwa; and a man of this tribe, who was seen in the slave-market, was only 'two days in reaching Kilwa.' A slave girl, who likewise belonged to a tribe near Kilwa, had a thread-like scar a little below the hair, quite encircling the head.

"The country of the M'HIAO is much further in the interior, and must be extensive, since it furnishes a large portion of the slaves offered for sale at Zanzibar. The markings vary in different individuals, but often consist of raised scars or welts, crossing each other like stars. Many of the females have the upper lip perforated; and the opening enlarging with age, at last forms a false gaping mouth, which destroys all natural expression of countenance, and singularly disfigures them. I once observed two M'Hiao women, free residents of Zanzibar, in the act of purchasing one of these elderly deformed persons.

"On comparing the various circumstances already mentioned, it may, I think, be inferred that if the outlet of the Monomoisy river or lake is upon the eastern coast, it is to be looked for only in the Juba. It may also be observed, that the mouth of the Juba being situated in barren territory, its waters must be derived from a distance, either from the mountains of Abyssinia or from the rainy region to the southward. Some of the Negroes imported from Mozambique into Rio Janeiro had come from the southern part of the above-described slave-district; and Mr. Hale met with Wakondy, M'Hiao, and Makua. Others, however, had been brought from further south, as the TAKWANI, from the Zambezi river, and the N'YAMBANA, from Inhambane. The latter were observed to be marked in a singular and unique manner with a row of fleshy knobs like buttons, which had been artificially formed upon the forehead.

c. *South Africa.*

"There were no KAFFERS at Cape Town at the time of the visit of the Vincennes; but from statuary and other sources I obtained abundant evidence that these tribes belong physically to the Negro race. They appear, however, to be more familiar with pastoral life than their brethren on the north; but they agree in their dialects, which, according to Mr. Hale, form a subdivision of the same general language. A fortnight previous to our arrival, some families of the FINGO tribe were encamped in the vicinity of Cape Town; but on visiting the spot designated, I found that they had departed. The owner of a neighbouring plantation stated that 'a large body of Fingoes once remained there for several weeks, and formed a regular town or "kraal." Their hair was exactly like that of the Negro; but the complexion varied, in some individuals being nearly as light as that of the Hottentot, and in others quite black. The men marry as many wives as possible, as by this means they secure at all times a support.'"—See ethnographical section of articles CAFFRARIA and CAPE OF GOOD HOPE.

f. *Western Africa.*

"I have seen aboriginal Negroes of Western Africa only at Rio Janeiro, where I enjoyed the advantage of being present during many of Mr. Hale's examinations, and of thus learning the preci[illegible] place of origin of the different individuals. T[illegible] mostly belonged to the countries south of the e[illegible]r, to tribes speaking dialects 'of the Congo-M[illegible]an subdivision of the general language, such a[illegible]e BENGUELA, KASANGI, and CONGO; the KAMB[illegible]A, from the Congo river; and the MUNDJOLA, from the country back of Loango.' The Negroes from the coast north of the equator, or from Guinea, had been brought principally from Mina, a port in the Ashantee country. They as a class, and particularly individuals from the far interior, were much dreaded by the planters. I saw one man who spoke a little Arabic, and 'in the instance of a former insurrection, the slaves were said to have corresponded in this language.' I would remark in reference to [illegible] condition of Western Africa, that the descrip[illegible]s I have read exhibit, on the part of the natives, [illegible]efi[illegible]ncy in shrewdness and natural good sense, [illegible]mpanied, however, by a grossness of manners [illegible]ound in those parts of the world which I have [illegible]d.

g. *Central Africa.*

[illegible]tives of Central Africa were chiefly seen in [illegible] where, through the assistance of Dr. Gas-[illegible] I was enabled to learn the history of a [illegible] of individuals selected from the Negro regi-[illegible] On inquiring for SHILLUK, several were at [illegible]inted out, but it appeared that the term, as [illegible]d, had reference to height of stature, rather [illegible] the river tribe bearing the name. The per-[illegible]ignated called themselves JANGA; and one [illegible] man 'came from the mountains, and spoke [illegible]kawi language.' Three other Janga, one [illegible] was short, were from the plains. Three [illegible]ls called themselves NUBAWI, the name of the class of Negroes who have been already mentioned as well known at Mocha and in Southern Arabia.

"One Negro from Kordofan termed himself an Arab, and said that his language was the Arabic, and the same account was given by three other natives of that country, although they were all manifestly of unmixed race. Another Negro styled his nation HAMAJAUWI. And two individuals were from Fazoglo. It appeared that all present, without exception, came from the Bahr-el-Abiad, or the White River; and as late accounts place Darfour within the valley, it is quite possible that no Negroes may reach Egypt from countries beyond the waters of the Nile. I did not find in Egypt the scars and marks on the skin, or the filed teeth, which usually distinguish aboriginal Negroes. In the slave-market I remarked among the females considerable variety in the mode of dressing the hair, and much pains had been often taken to form it into rolls, or, in other instances, into numerous slender braids. I once observed the eyelids coloured with blue paint, as a substitute for 'kohhl,' which latter material would hardly in this instance have been conspicuous on the complexion. The Negroes seen in Egypt appeared to me unusually slender, even after making every allowance for the style of dress.

"At the present day, three important Muslim-Negro kingdoms are found along the southern border of the Desert—Darfour, Bornou, and Soudan. An account of the two last may be found in the work of Denham and Clapperton. Darfour is less known, although, in the eye of the Egyptian Arab, it ranks as one of the three great political powers which divide the earth, the two others being Persia and Constantinople. At Singapore I saw two Negroes in the shop of the Abyssinian Arab, who, in their deportment, in their air of resolution and self-respect, seemed a different class of beings from those of their race I had previously met with. On inquiry, I was informed that 'the black Arabs come chiefly from Sennaar,' and on regarding the two individuals more closely, I thought I could distinguish the costume of Bornou, as figured by Denham and Clapperton. It appears, however, from Burckhardt, that pilgrims very rarely come from as far west as Bornou, and the probable origin of the above individuals may have been Kordofan or Darfour. The head was habitually uncovered, and one had the wool so close as fairly to justify the appellation of 'pepper-haired,' which is so universally applied to the race in Eastern countries. They had doubtless come by the way of Mocha. But I did not meet with the same costume on my recent journey."—See articles CAFFRES, NEGROES, SUDAN.

VIII.—THE ETHIOPIAN RACE.

"The Ethiopian race is in some measure intermediate in personal appearance between the Telingan and the Negro. The complexion, too, seems generally darker than in the Telingan race, holding the third rank in depth of hue. The hair is crisped, but fine in its texture, and I have never seen it wiry, as in the Papuan, from which latter race the Ethiopian differs, in having a soft skin and European-like features. Various modern travellers have been struck with the resemblance of the Ethiopian to the Telingan race. The same circumstance was noticed in ancient times by Herodotus, who also speaks of the remarkable beauty of those living southwest of Arabia, an opinion in which, from actual observation, I am now prepared to concur.

"The Ethiopian race occupies the hottest countries of Africa. Most of its tribes are purely pastoral; and some of them are, perhaps, alone acquainted with all the recesses of the Great Desert. The only portions of the race that are known to lead an agricultural life are the Nubians of the Nile, and some of the tribes bordering on the table-land of Abyssinia. The Ethiopian race was not seen in its purity during the voyage of the exploring expedition, but it was doubtless present among some of the mixed Arabs of Singapore. My acquaintance with it commenced on the day of my landing in Egypt.

NUBIANS.

a. *Barabra of the Nile.*

"The Barabra found in Lower Egypt are principally adventurers from the banks of the Nile in Nubia, who seek employment for hire, and after a series of years return to their own country. They are favoured in certain particulars by the government, being 'exempted from military duty, and being sometimes employed as taskmasters, an occupation in which they particularly excel.' They are extremely clannish, and strong antipathies exist between them and the common Egyptian Arabs. Old residents have assured me that they have never known in Lower Egypt of an instance of mixed offspring between these two classes. Some of the boats on the Nile are manned and commanded by Barabra, who, however, hardly form mixed crews, or serve under Arab commanders. These Barabra boatmen appeared to belong principally to the immediate vicinity of the Cataract, where a fondness for the water is acquired in early life; and they perhaps correspond to the 'Ichthyophagi of Elephantine' mentioned by Herodotus. A third class of Barabra are the 'Jellab,' who procure slaves on the upper parts of the Nile, and conduct them in slave caravans to Cairo and Alexandria. A fourth class consists of a large portion of the slaves, those who in Egypt are termed 'Abyssinians.' Such as I saw offered for sale were nearly all children and half-grown persons, having their hair dressed with much pains, according, apparently, to their respective national fashions. Some of the children proved to be 'Galla,' but through the diversity of dialects, their place of origin could rarely be ascertained, and they, perhaps, had not all been derived from the southern border of Abyssinia. On ascending the Nile, I met with the first instances of mixed descent at Kenneh, the modern capital of the Thebaid, about 30 m. below the site of ancient Thebes. Market-women of the Ethiopian race likewise made their appearance at Kenneh, where a change took place in the weights and measures, and cowries were seen used as money, and where articles from 'Hindi' were commonly offered for sale, such as black pepper, which had been brought, together with dates and coffee, from Mecca. Above Kenneh, men walking with spears were occasionally met with, and, according to the words of my Barabra servant, 'thief had finished.' The true boundary, however, between the two races, appears to be at Silsilis, about 100 m. above Thebes, and 40 m. below the Cataract. The KENOOS begin at this point, and with them a change takes place in the mode of raising water for irrigation, a task now accomplished by means of the wheel. I saw numbers of Kenoos on landing at different places, and particularly at Assouan and Philæ, the limit of my journey. The Barabra boatmen above mentioned, hardly appeared to form a distinct class. Contrary to the practice of the Arabs, the women were observed to take much pains in braiding and dressing their hair in a variety of modes; but they had other customs which they had evidently borrowed from the Arabs, as the use of kohhl or eye-paint, and of tattooing, both continued, notwithstanding the deep hue of the complexion. The Kenoos are all Muslims, but they do not appear to regard very rigidly some of the observances of their religion, and swine were seen at Assouan. At the same place I fell in with a body of slaves on their way down the river, and also with an Arab merchant, proceeding with bales of goods to Sennaar.

b. *Barabra of the Desert.*

"The ABABDEH inhabit the eastern desert of the Thebaid, between the Nile and the Red Sea; and I met with individuals at Kenneh, and at other towns higher up the river, but principally at Assouan. The Ababdeh use the Arabic language, and they do not understand that of the agricultural Nubians. An opinion, however, prevailed at Assouan, that they once had a language of their own, as the Bishareen now have, and that it was displaced by the Arabic. The Ababdeh 'bring charcoal to Assouan, procuring it within two, three, or more days' journey in the desert, and they occasionally bring ostrich eggs, obtained at a distance of not more than 200 m.' The BISHAREEN dwell in the district to the southward of the Ababdeh, and situated in like manner eastward of the Nile. Two individuals seen at Assouan enabled me to identify the physical race. One of them wore a pin for dressing the hair, similar to the Feejeean, but shorter and curved. The Ababdeh and Bishareen do not mix with their hair any foreign substance except ghee or butter, which in this climate is entirely fluid, and is an effectual safeguard against vermin. The [illegible]ved club which these tribes have been accus[illegible] to use, from remote antiquity, is small an[illegible] in comparison with that of the Feejeeans."—[illegible]ticles ABABDEH, BERBERS, and BISHARI.

"I could hear nothing in Egypt of the [illegible] of the central Sahara; and it is possible that [illegible]y have no communication with the Egyptia[illegible]s. Herodotus seems to allude to these people under the name of the 'Ethiopian Troglodytes;' and they are probably to be regarded as the proper aboriginals of the Desert. According to Horneman and others, the Tibboo continue in part pagan; and they hold their flocks about the widely separated wells and Oases, some of which are perhaps known only to themselves; yet in their trackless solitudes they are not altogether free from the depredatio[illegible] the Tuarick and Arab. Some points i[illegible] th[illegible]tory present unexpected analogy to the [illegible] the Australians; as their use of a curved [illegible] and their counterfeiting a lump of black basalt [illegible]order to elude observation, as the Australians counterfeit a burned stump. Denham and Clapperton moreover state, that the Tibboo couriers 'are the only persons who will undertake singly to cross the Sahara.' On my journey from Cairo to Suez I met the caravan returning from Mecca; but it was unfortunately at night, and I obtained a view of the travellers only by the light of their own torches. There seemed to be a large proportion of Ethiopians, many of them doubtless, from Nubia; but some, who made inquiries about Alexandria, and maintained a more independent bearing than the Egyptian Barabra may have belonged to the far west. The Ethiopian profile has appeared to me to have furnished the model for the Egyptian features of the earliest monuments, as far down as the conclusion of the eighteenth dynasty; and we may otherwise look in vain for representations of a people who at least must have been intimately known to the ancient inhabitants of the Thebaid. I am aware that the Desert tribes have been separately and distinctly delineated, though even here I would appeal to the profile; and after making every allowance for a conventional style of art, those highly finished portraits of Egyptians, in which the hair is disposed in numerous slender braids, have appeared to me intended for men of the Ethiopian race. The same rule respecting the hair will, I think, be found to prevail in the mummies (though few of these have proved to be of greater antiquity than the Greek or Ptolemaic period); and the wooden neck-pillow accompanying the mummies was certainly not an invention of men of the White race. On the monuments of the eighteenth dynasty, men of the White race are frequently represented, but chiefly as slaves, or as foreigners; and they may always be distinguished by the profile from the Egyptians. The earliest unequivocal representations of the White race occur in the grottoes of Beni Hassan, excavated during the twelfth dynasty, or towards the conclusion of 'the time of the pyramids;' and it should also be observed, that some of the transactions here recorded seem connected with the first appearance in Egypt of the Hyksos or Shepherds."—See articles TUARICKS and TIBBUS.

SOMALI.

"The Somali inhabit that projecting portion of the African continent which is opposite to, and hems in, the southern angle of Arabia; and they extend along the coast and more or less into the interior, from the Straits of Babel-el-Mandeb to the river Juba under the equator. Their country, although arid and intensely hot, affords pasturage, and seems eminently adapted for rearing a peculiar close-haired variety of sheep. The Somali also engage in maritime pursuits; and many of them are to be found established in separate communities, at the principal ports on the lower portion of the Red Sea. The Somali seem to be very much confined to the above limits."—See article SOMALI TERRITORY.

DENKALI.

"The Danakil tribes inhabit the district north of the Somali, and fronting on the lower portion of the Red Sea; and on a line with Zeyla, they appear to be much extended into the interior. They 'speak a different language from the Somali; and individuals visit Mocha for the purpose of bringing slaves, which they obtain in the interior,' and which, it is presumed, are chiefly Galla. The Denkali were said to 'conduct themselves well while at Mocha, but to come only a few at a time.' The first individual I met with happened to be of short stature and short-necked; and he might readily have been mistaken for a Feejeean, agreeing even in the disposal of his hair. Points of difference, however, were perceived on a closer inspection; and the other Denkali seen, did not differ physically from the Somali and Barabra.

GALLA.

"The Galla tribes have long been regarded as the terror of East Africa, and the extensive region they inhabit has been very little visited by strangers. The Galla are, for the most part, purely pastoral in their habits; but, according to Mr. Isenberg, the tribes bordering on the highlands of Abyssinia cultivate the ground.

"The NORTHERN GALLA furnish great numbers of slaves, principally women and children, who are sent by the way of Abyssinia into Egypt and Arabia. In the latter country many of the town Arabs are of partial Galla descent. I did not meet with like instances at Cairo and Alexandria, where, however, the Galla women hold a similar position with those in Arabia. Some of them, moreover, were married to European residents of the first respectability: having been carefully educated, they proved elegant and accomplished companions. At Zanzibar I met with but one of the Northern Galla, a slave-woman, who called her nation 'Chory,' and who, according to her own account, came 'by the way of Mocha from Massowa.' The point of embarkation interposes an objection to affinity with the Bishareen, and she was expressely termed a Galla by the Arabs in company.

"The SOUTHERN GALLA are rarely slaves, but I heard of an instance in a 'Musagna' woman, who was the only representative of these tribes at Zanzibar. It was said, that in times of scarcity the Southern Galla will sell some of their number, who, however, are not esteemed; for when directed to perform any service, they will ask questions, or reply 'Go and do it yourself.' The nation is well known by report at Zanzibar, as frequenting the western bank of the Oozy, and intercepting the land-route from Bombas to Lamo, and as visiting the settlements on the coast for the purposes either of plunder or traffic. The Southern Galla refuse vegetable food, and live exclusively on milk and meat; the only semblance of an exception that I could hear of, being the fact, that they have been known to 'steal corn from the people of Brava.' The sultan of Patta and an Arab who was present at the interview, gave me the following account of the Southern Galla: 'Their country is called Bararata, and is situated eight days' journey from Patta;' perhaps not including the 'two days to the river Oozy, where [illegible] trade. They bring cattle, which ha[illegible] which are greatly superior to the In[illegible]nzibar. They have camels, which th[illegible] for carrying burdens, but not for riding. They elect a king every eight years, and the new king circumcises his subjects. They shave the head, preserving a lock of hair on it for every man they have killed. Their king, when sent for, will come to Oozy; and, if a present were made, he would be willing to take an unattended European into the country, and would bring him back in safety; but if several persons wished to go, the visit would be impossible. For the last eighty-five years, or since the time of his grandfather, the Patta people have not been able to go into the Galla country. In person, the Galla are tall and good-looking, and resemble the M'Kuafi, with whom they often fight; both nations having precisely the same description of hair. The Galla have strong beards, but these are habitually clipped. They have no bows; and in warfare they use only the spear (which is held, and is not thrown), protecting themselves with shields. They are very much afraid of fire-arms.'"—See article GALLAS.

M'KUAFI.

"The M'Kuafi dwell in the interior, to the west of Mombas, and south of the proper Galla country. The following information respecting them was derived from young persons seen at Zanzibar; where, according to the Arabs, slaves of this class were formerly cheap and not much esteemed, but now bring high prices. A M'Kuafi girl stated, that 'she had been captured by the Mussai, who killed her father and mother, and who sold her to the Chaga. She was twenty-five days in reaching the coast. Formerly, her nation was powerful above others; so that one woman with a stick would stop a thousand persons from passing through the country, unless a present was first made; but her people are broken, and at present they would not fight the M'Sigua. Her people do not cultivate the ground, and they eat only milk and meat. Children when hungry help themselves by direct application to the cow. Cattle are killed by piercing the spine; numbers of them every day, until each family is supplied. The M'Kuafi have no fixed meal-times; but they eat whenever they feel inclined, inviting their neighbours of the same village to partake with them. Each family has its own cattle, which all go to pasture together, and outside the town is a place to receive them at night. The men marry as many women as they please, and each wife has a separate house. These habitations are tents of bullock-skin, supported by poles set around. The men decorate themselves with large beads, and their dress is made of skin, and consists of a painted cincture full of openings and hanging stripes, and of a long cloak worn over one shoulder. Cloth, however, is now brought by traders. The women, by way of ornament, coil brass wire about the arm as far as the elbow. The beads and brass wire are procured at Pemba, by selling ivory obtained from elephants, some of which are found dead, while others are purposely killed.'

MUSSAI.

"The Mussai hold the country west of the M'Kuafi; and they are the most distant of the pastoral tribes known at Zanzibar. According to the girl above-mentioned, they speak the same language as the M'Kuafi, with whom they agree in not cultivating the soil, and in living exclusively on milk and meat. Indeed, all the accounts show them to be a kindred nation. My interpreter, Sadik, had conversed with persons who, on a trading expedition to Chaga, saw there some of the Mussai. These were described as being 'like the M'Kuafi, but stronger; and as having shields that reach to the breast, together with long-pointed spears and round-headed clubs, neither of which are missile. The young men only go out to fight, sometimes five hundred of them in a body; and when they capture a town, they put to death all except the handsome women, remaining afterwards for two or three days upon the ground, lest they should seem to run away.'

OTHER ETHIOPIANS.

"I have felt much uncertainty respecting the physical race of the Fellata of Central Africa, especially as the descriptions of them by Denham and Clapperton seem to accord entirely with the Ethiopian. The Foula, who dwell near the Western coast, are said to be connected in language and customs with the Fellata, except that they are not Muslims; and to differ from the Negro tribes among which they are scattered, in leading a pastoral life.

IX.—THE HOTTENTOT RACE.

"I am not sure that I have seen Hottentots of pure race; and in the following account I am obliged to rely in part on published figures and descriptions of them. Eye-witnesses uniformly agree in regarding the Hottentot as differing in physical race from the Negro, and they refer to the diminutive stature and the light complexion; the latter, in some well authenticated instances, having even been known to exhibit a flush. Both races seem to be alike in the texture of the hair, which is more closely woolly than in the rest of mankind; and to agree also in the general style of feature. But many travellers describe the Hottentot countenance as being to a marked degree peculiar. The three days I spent at the Cape of Good Hope were in great part occupied in an unsuccessful search for examples of the Hottentot race. Some individuals of mixed descent were pointed out to me, and I thought I could myself distinguish others. A waggon from the Interior was accompanied by several individuals who differed in their personal appearance from mulattoes; and who, like the other mixed Hottentots I saw, were rather good-looking. Hottentots were said to be extremely serviceable, and indeed indispensable, to military expeditions; as well from their knowledge of the country, as from their intimate acquaintance with the habits of the Kaffer tribes. I have found many points of interest in the Hottentot character, as portrayed by travellers, who universally bear testimony to the faithfulness, efficiency, and courage of their guides in trying situations, amid the dangers of this difficult and desolate country. Unlike many wild tribes, the Hottentot did not shrink before the advance of Europeans; but readily adopting the habits of civilisation, these people have ever proved active and useful assistants of the colonist, their history in this respect contrasting strongly with that of the neighbouring Kaffer tribes. Indeed, the advantage of the Hottentot character appears to have contributed essentially to the admission of Europeans into this, the only part of Africa that has proved accessible to foreigners; and the benefit deserved a better return than unequal legislation. The following quotation is descriptive of the wild life led by the Bosjesmans of the frontier, who furnish abundant proof that man in his natural state is by no means a helpless animal. 'They live among rocks and woods; have a keen, vivid eye, always on the alert; will spring from rock to rock like the antelope; sleep in nests which they form in the bushes, and seldom pass two nights in the same place, supporting themselves by robbery, or by catching wild animals, as reptiles and insects.'

"Recent discoveries have shown that the Hottentot race is more widely diffused than was previously supposed, and that it extends in a scattered manner for an unknown distance towards the heart of Africa. According to the testimony of Owen, Morrell, and others, Hottentot tribes occur along the western coast as far as latitude 20° south, and remain, in some instances, still unchanged by contact with Europeans."—See articles BUSHMEN, CAPE OF GOOD HOPE, and HOTTENTOTS.

X.—THE ABYSSINIAN RACE.

"I have seen but few genuine Abyssinians, and these few have not presented among themselves a very uniform personal appearance. In a general way the race may be said to possess European features in combination with crisped or frizzled hair. The complexion, however, though it is often very light, does not appear ever to become florid. The race seems to be confined to the table-land of Abyssinia, and to its prolongation in a mountainous tract that extends towards the interior of the continent; but even within these limits there appears to be a large infusion of Ethiopians belonging to the Galla tribes. The Abyssinians, having been converted during the early ages of Christianity, continue to make pilgrimages to Jerusalem; but otherwise, their presence in foreign countries is, for the most part, involuntary. My first interview with Abyssinians took place at Singapore; where, on entering one of the principal shops, I was greatly surprised to find its three occupants differing in physical race from all the men I had previously met with. The wonder increased on perceiving their superiority in refinement and intelligence over the other Orientals around; and that although so unlike, they did not, on these points, seem to fall below Europeans. The hair was much alike in all three, and was crisped and fine, neither coarse enough nor in sufficient quantity to form a resisting mass. The beard of one individual was in pellets, absolutely like the close wool of the Negro; but the prominence of nose, greater even than usually occurs in the White race, bore sufficient testimony to his purity of descent. The second individual had the face very much elongated, but the nose was not particularly prominent. The third individual had a straighter beard, which was black and grey in regular stripes, as in another instance already mentioned. The complexion was the same in all three, and though very light, was by no means of a sickly hue; and, indeed, these persons might readily have been passed in the street as belonging to the White race.

"I will here insert the opinions of two travellers in Abyssinia, communicated to me orally. Mr. Rochon regarded the Galla as differing physically from the Abyssinians, and spoke of the latter as being 'a fine set of people, men absolutely such as ourselves, and capable of doing anything that we can do; except only that they have been secluded from the remainder of the world.' Mr. Isenberg, whom I saw at Bombay, thought that 'under the same advantages, Abyssinia might rise to an equality with a European nation. In times of peace there is tolerable security for person and property, and the present condition of Abyssinia is not very unlike that of a European nation when distracted with civil war.' The Abyssinian is the third physical race, which will enter into the question of the primitive Egyptians. It seems, however, that the true Abyssinian, (as first pointed out to me by Mr. Gliddon,) has been separately and distinctly figured on the Egyptian monuments, in the two men leading a camelopard in the tribute-procession to Thouthmosis III.; and this opinion was confirmed by an examination of the original painting at Thebes."—See ethnographical section of article ABYSSINIA.

XI.—THE WHITE OR ARABIAN RACE.

"Europeans and European colonists are comparatively uniform in their complexion and personal appearance, and they can hardly conceal their origin by dress, even amid the population of Northern Africa and North-western Asia. The inhabitants of the two last-named countries present among themselves more diversity, although at the same time there is rarely any difficulty in recognising the physical race. The White race then, as it exists in northern climates, may be characterised by its superiority in lightness of complexion, in thinness of lip, in prominence of nose, and in length and copiousness of beard. No one of these tests is of itself sufficient to distinguish the race, for Abyssinians in some instances, rival it in prominence of nose; Telingans, or even Ethiopians, in thinness of lip; many Papuans have as copious a beard; and I have myself seen the florid complexion among Mongolians of high northern latitudes. So far, however, as my observation has extended, flaxen hair, red hair, and blue eyes, (albinoes being excepted,) are found only in the White race. The hottest portion of the globe appears to be about seventeen degrees in width, counting from lat. 27° N, and extends from the Atlantic ocean to the Ganges. One-third, perhaps, of this immense tract is inhabited by the White race, although often under a physical aspect that would not readily be acknowledged by Europeans. The complexion, always dark, is in frequent instances sufficiently so to conceal a flush; indeed the Malay-brown complexion seems rather to predominate; and I have seen Arabs of a deeper hue who yet were apparently of unmixed descent: moreover, a considerable proportion of the inhabitants of Southern Arabia are nearly or quite beardless. In short, the White race is here protean or polymorphous, and exhibits a diversity in feature and complexion that I have not found in the other races. The prominence of profile is, however, for the most part, permanent; and I further remarked that the various series of expressions of countenance, which pertain respectively to the other races, appeared to be absent. It should be observed, however, that actual mixtures of race have been more frequent and more complicated in the southern Arab countries, and have been taking place there for a longer period, than in the other parts of the globe. While acknowledging that we are only beginning to be acquainted with the countries and people of the East, I have been impressed with one view of the circumstances, in correspondence apparently with an ancient condition of the human family. In the course of my recent tour, I was continually hearing from the lips of Orientals the words of different ancient and modern European languages, until at last the whole class of these languages seemed as if merely recomposed from fragments of Arabic and Sanscrit. Of fragments indeed, which have been disguised more or less by interchanges during some thousands of years; and if any European words can be traced to a different source, they at least remain to be pointed out. The same state of things appears to exist in the Malay class of languages, and instances have been discovered of English words which can be traced through the Sanscrit to the remotest islands of the Pacific. The rule may, perhaps, be further extended to the languages of the African continent; but whether applicable in any degree to the Chinese, or to the aboriginal American languages, I am uninformed. At the present day the White race may be conveniently disposed in two divisions, as well geographical, as differing in institutions and habits of life: the Frank or European, and the Oriental. The extraordinary intensity of heat, in some of the countries inhabited by the White race, may help to explain a portion of these differences. One of the above divisions may in some measure be said to rule the land as the other rules the sea; for the extent of the caravan routes is almost an equivalent to the universal maritime intercourse now attained by Europeans.

EUROPEAN COLONISTS.

"At the present day, Europeans and their descendants are to be found in most parts of the globe that are accessible by sea; the eastern coast of Asia, together with New Guinea and the neighbouring islands, forming the only considerable exceptions.

"Madeira, having been discovered in modern times, and having been found uninhabited, must be regarded as a colony; although in the character of its population it seems rather to form a part of Europe. A week spent at this island, chiefly in a journey with horses, of nearly a hundred miles, afforded me a fair opportunity of seeing the inhabitants. According to an idea commonly entertained in regard to the Spanish and Portuguese, I had anticipated universal swarthiness, with the hair and eyes invariably black; but I found as much variety in complexion and feature as at home, and many instances even of flaxen hair. The singularly rugged character of the surface (the roads intersecting deep mountain ravines and leading along the brink of frightful precipices), tends greatly to impede intercourse. At San Vincente, on the northern coast we were told that but three travellers had visited the place during the preceding four months: and one of the Catholic Fathers residing there, had grown grey in that single ravine, which he had never once left since his first arrival. The peasantry of Madeira appeared to be laborious, civil, and sober; but the idea of political rights, in all probability, seldom enters their heads; it may be doubted, whether the news of the great world around ever reaches them. In secluded situations we sometimes met with young persons in a state of nudity, and on one occasion, with several who were nearly full-grown. The strange tone of voice everywhere heard, was explained by Mr. Drayton; for it appeared that 'the peasantry in their conversation and music, habitually use the minor key.'

"During the few hours spent on shore at the Cape Verd Islands, I saw only three or four Whites; who were residing in the town of Porto Praya. They were sallow and languid; having evidently suffered in their health, from the powerful influence of the sun.

"At Rio Janeiro I did not find the like marks of a deteriorating influence in the climate; and the popular prejudice against the general healthiness of the tropics was soon perceived to be without foundation. In the city, artisans direct from Europe, were seen working steadily at their various vocations; some of which, such as that of the blacksmith, were peculiarly oppressive. In the course of an excursion beyond the Organ Mountains, I had an opportunity of witnessing some of the hardships of a life in Interior Brazil. I had previously imagined with many in the North, that the bane of tropical climates consisted in the absence of a stimulus to industry; and that the soil poured forth, almost spontaneously, a profusion of dainties in addition to the necessaries of life; I was, therefore, unprepared to find the stigma of luxury reflected back upon my own countrymen, even upon those undergoing the reputed privations of our frontiers. Among other peculiarities in manners, derived to all appearance from the parent country, the seclusion of the women may be mentioned; for their existence in certain families appeared to be known chiefly through report.

"During our short stay at the mouth of the Rio Negro, in North Patagonia, I did not come in contact with many of the inhabitants. One of the pilots was a Frenchman, and the other an Englishman, and I met with three individuals from the United States. The remainder, about twenty in number, were 'guachos,' or mounted herdsmen; who (with the exception of some Negroes) were of Spanish descent, and who were rendered conspicuous in the distance by the scarlet and white of their fantastic dress. It has been said that the guachos, 'owing to their being perpetually on horseback, appear personally to disadvantage when dismounted;' I did not remark this, but I saw some well-formed men, who, moreover, did not appear to be deficient in either intelligence or general information. I observed, however, an evident disinclination to walking; and it must be a rare circumstance in this region to meet a man on foot. The country has such a desert aspect, that it is a matter of astonishment how the great numbers of horses and bullocks are enabled to procure subsistence. We here began to perceive an influence in the surface of a country, in modifying the habits of its population. In an open accessible region, men in the course of their ordinary pursuits, become accustomed to rove widely, tenfold more than in agricultural districts, or even a hundred fold more than where hemmed in by close woods.

"Northern Chili, although less barren in aspect than the country on the Rio Negro, supports comparatively fewer cattle; but being broken and mountainous, irrigation, for agricultural purposes, is sometimes available. The climate seems extremely healthy, as is shown, among other circumstances, by the robust forms of the women of the lower orders. The hospitality and obliging disposition of the country people, formed a strong contrast to the habits of the Brazilians; and extended even to annoyance on our botanical excursions by the offer of horses. At a ball, given in honour of a victory, I had occasion to admire the good taste shown in the arrangements, and the dancing; an accomplishment in which the Spanish ladies are acknowledged to excel; and I was also struck with the similarity between such fêtes in this remote quarter of the globe, and those I had attended at home. It will be difficult to find another continental nation that is so completely isolated by natural boundaries; the Andes forming a barrier to the extension of population on the east, and the Atacaman desert on the north. Chili is generally conceded to be in advance of the other Spanish American countries; among which it alone has enjoyed internal tranquillity. It should also be observed, that only in one direction can Chili conveniently carry on foreign war. The English language was found to be making progress in Chili. I often heard it spoken, and I observed translations on many of the sign-boards, both at Valparaiso and in the Interior; showing the numerical importance of customers from England and the United States. Of other foreign residents, those from Germany and France appeared to be the most numerous.

"California, the only other Spanish-American country I have visited, presented a third variation in the costume, which was again fantastic, and unlike anything European. The Californians are mounted herdsmen, skilled in the use of the lasso; and they are as expert in feats of horsemanship as their brethren of the far south. Great ignorance of geography and of foreign countries was found to prevail even among the upper ranks, as we had remarked at Lima. In the other countries visited by the Expedition, the European emigrants had not acquired separate nationality. They were rather residents; or if born on the spot, they nevertheless did not differ from the population of Europe and the United States.

"The Hawaiian is the only Polynesian group on which foreigners are established in sufficient numbers to form anything like society. Some hundreds of them, chiefly from the United States, were residing at the commercial centre, the town of Honolulu. The island of Taheiti has also become an important place of resort for shipping; but its commercial advantages are inferior to those of the Hawaiian group, and we met with but few European residents, who were unconnected with the mission. Europeans had but recently established themselves at the Samoan or Navigator Islands; and, excepting the missionaries, they consisted exclusively of runaway sailors. I met with seven on Savaii; and nearly a hundred were supposed to be scattered about the group,—some following the occupation of pilots, blacksmiths, or interpreters; but the majority content to obtain without labour the means of subsistence; and certain individuals had even adopted the Samoan costume. In various native families, White men were kept as a sort of pet, although 'they were at liberty to leave at any moment, provided that they took nothing away.' The value, however, of this description of property within a short time, had much depreciated in the estimation of the natives. Although Tongataboo has been so much frequented

Europeans had not exerted much influence over the minds of the natives. Indeed, a low island like this, which produces chiefly cocoa-nuts and bananas, and is otherwise not an agreeable place of residence, may, probably, for years to come, offer a point of resistance.

"There were many European residents round the Bay of Islands, in New Zealand; and in the course of our excursions it was ascertained that runaway sailors were living in some of the native villages on much the same footing as at the Samoan Islands.

"In the East Indies we met with but two or three Europeans of the lower class, and these expressed dissatisfaction with their situation, and they evidently could not maintain themselves among the surrounding population. The European residents continue to be confined to those portions of the Archipelago which are under European rule; and they consist of the officers of government, civil, military, and ecclesiastic, and of a few persons engaged in foreign commerce. Persons of European descent, more or less mixed, now form an intermediate class of some importance. European power in Hindostan amounts rather to the possession of the government than to a proper colony; for, should it now cease, there would scarcely be a separate class of population left behind, except only in the instance of the mixed Portuguese. Of mixed European colonies, the most complicated, perhaps, is the recent establishment of the English at Aden, where, without counting the troops, the majority of the inhabitants appear to be voluntary emigrants from Hindostan.

HEBREWS.

"The White men met with previous to arriving at Singapore, had been exclusively of the European stock; and only at that place did the Exploring Expedition come in contact with Orientals—with men physically like ourselves, who yet differed from us in their customs, and who had not derived their arts and acquirements from Europe. The Jews, indeed, at the present day, are common to both divisions of the race, being spread abroad like a net throughout all the countries possessed by Whites, and even to a short distance beyond, in the instances of Abyssinia and India. Their various adopted countries having for ages exerted modifying influences, four classes may now be distinguished in the Jewish family: the Syrian, Indian, Arabian, and European Jews.

"I first saw SYRIAN JEWS at Singapore; some ten or twelve individuals, who were robed and turbaned, and who in person appeared to advantage among the surrounding population. The complexion was fair, and in a different costume they probably would not have been distinguished from Europeans. One of them who spoke English invited me into his shop, and stated, in the course of conversation, that 'he came from Aleppo when quite young, travelling by land with his father to India.' He knew nothing of the Malabar Jews, and, in respect to the origin of Black Jews, he showed us one in his service, who was of mixed Telingan descent. He was unwilling to acknowledge the fellowship of European Jews, some of whom he had seen at Calcutta, objecting that they shaved the beard, and in various other particulars he thought they had departed from the law. He spoke of a European Jew then residing at Singapore, but said that 'he should never think of questioning him on the subject of religion, unless on the occasion of a proposal to marry a daughter.' At Bombay I saw a Jew who, about twenty-five years previously, came by land from Jerusalem. He stated, in regard to his journey through so many foreign countries, 'that he found Jews along the whole route, with whom he could communicate in the Hebrew language.' The agent of the English government at Muscat, a Jew, probably of the Syrian class, died a short time previous to my arrival in that city. I saw there, however, a young Jew from Bagdad, whose scanty turban included the Egyptian *turboosh*, and who wore a side-lock, differing, however, in some respects, from that of the Arabian Jews. He carried a rosary, or string of prayer-beads, like the Catholics and Muslims—a custom of high antiquity. A Jew had resided for several years at Zanzibar, to which place 'he came in a dow,' it is presumed, for purposes of commerce. I did not learn his further history; but, from his allowing the whole of his beard to grow without regard to side-lock, I supposed him to be a Syrian Jew. I did not observe the side-lock among the Jews of Egypt, who likewise belong to the Syrian class. They inhabit a distinct quarter of the city of Cairo, where, as in Europe, their principal occupation appeared to be dealing in money. Their traditionary accounts of Egyptian matters have not proved to be of much value. On viewing, however, the scenes of their early history, one can better realize the wonderful and unique development of a whole nation from a single family.

"At Bombay the INDIAN JEWS form a considerable body of artisans among the lower orders of the population. The greater proportion of them distinguish themselves as Beni-Israel, and, in the opinion of competent inquirers, the title had not been lightly assumed. Polygamy, I was informed, 'exists among the Beni-Israel, but principally when the first wife proves barren; and the ancient custom of marrying a brother's widow is not continued.' In such of the Beni-Israel as were pointed out to me, I did not perceive any peculiarity in personal appearance distinguishing them from the rest of this Jewish community, which presented an unexpected variety in feature, complexion, and even in costume. I observed, however, one point of agreement, in the head being shaved—entirely, in some instances, but usually in such a manner as to leave traces of a side-lock. The head-mantle, apparently a relic of the life of their forefathers in the desert, was used in the synagogue; and their religious service was the more impressive as being practised in the midst of an idolatrous population. 'A few years ago, the Bombay Jews were in a very low and degraded state, being dependent for the little Hebrew they possessed on the Arabian Jews, by whom they were held in very little consideration. The attention of the missionaries having been turned in this direction, the Bombay Jews were supplied with printed copies of the Scriptures, since which time their knowledge of Hebrew had become more general, and their condition in other respects had improved. They had expressed much gratitude for the benefit, saying, that the sun then rose upon them for the first time.'

"I saw numbers of ARABIAN JEWS at Aden, which place seems to have been principally inhabited by them prior to the English occupancy. They appeared to have preserved more of the primitive Hebrew customs than I have seen among their brethren in other countries. Their style of architecture offered unexpected analogy to the ancient Arabic, as depicted on the Egyptian monuments, the buildings being devoid of arches and of all curved lines, but having square towers with battlements, and at each story a projecting ledge. The men wore long side-locks, reaching from above the ear to the shoulder, and these may perhaps constitute the 'corners of the beard' in the passage of Scripture which has sometimes been differently interpreted. The women had their eyelids and eyebrows painted, and they wore a large silver ring in the nose, and numbers of similar ones in the ears; children sometimes had their ears folded downwards with the weight of the rings inserted around the margin. Education, as among the southern Arabs, was general, and all the children attended school. In this scorching climate, the Jewish complexion, instead of being in any degree florid, presented a universal tinge of yellow, but it was obviously lighter than the hue of the common Arabs of the country. Some of the boys had a coarse expression of countenance, combined with flaxen hair, reminding me of faces seen occasionally in Northern climates, and at variance with the usual Jewish physiognomy. I do not remember, on any other occasion, meeting with flaxen hair among the orientals.

"It would seem that the Hebrew is not altogether a dead language with the Jews of Aden; but they commonly use the Arabic, which they write with the Hebrew character. I obtained from them a copy of the Pentateuch, 'written at Sanaa, in the interior,' and I saw another work in the Hebrew character, which 'had been brought from Bagdad.' It was said that the Aden Jews refused to hold communion with their Indian brethren, who had been brought to the place by the English, and that they worshipped apart. The Jews at Mocha lived outside the walls of the town in a village by themselves, and they consisted of artizans, especially builders, as at Aden. Notwithstanding a residence for many successive generations, the complexion remained lighter, and the beard longer than in the Arabs of Mocha. With respect to the Jews of Abyssinia, geographical considerations would lead us to suppose that they were originally derived from Southern Arabia, rather than by the route of Egypt and the Nile.

ARMENIANS.

"Of all Orientals, the Armenians, next to the Jews, are most frequently to be met with in the society of Europeans. They do not occupy much space in history, but it seems a remarkable circumstance that, without any decided distinctive traits, they should have preserved their nationality from remote antiquity. The range of Armenian emigration is chiefly a northern one, extending from the Mediterranean, by the way of Constantinople, to Northern Hindostan and to Calcutta. I heard nothing of Armenians in the Arab countries, nor in South-western Hindostan, except at the city of Bombay. Here I saw several of their tombs in the European burying-ground, and I remarked that they were provided with a niche for burning incense, like those of the Indian Muslims. In the Mediterranean, Armenians are now frequent passengers in the steamboats to and from the Levant, a circumstance attributable, in a good measure, to the extent of their commercial relations. Many Armenians are doubtless settled in the principal cities of the United States, although their presence may not be generally known."—See article ARMENIA.

ASSYRIANS.

"I have not heard of any modern distinctive name for the inhabitants of the valley of the Euphrates, and I have therefore had recourse to an ancient appellation. The district is well known on the Indian ocean as belonging to Turkey, its southern angle being interposed between Persia and the territory of the Arabs. Formerly it was the seat of the power of the caliphs, and, although the lower country has been much overrun by Arab tribes, I believe there is some separate nationality remaining. From the geographical position, it seems probable that many natives of the valley of the Euphrates visit Hindostan and the shores of the Persian gulf; but none were pointed out to me on my recent journey, and, perhaps they are not generally distinguished from other Muslims.

MOGULS.

"I have not found a more convenient name for the modern Persians than that of Moguls, which I sometimes heard applied to them at Bombay, but the term, I believe, is also used in reference to rank. Numbers of these Persian emigrants were settled at Bombay, where they were distinguished by wearing, in place of a turban, a high pointed cap, made apparently of black lambskin. They seemed to be a rude, hardy, able-bodied set of men, and they were, perhaps, all of them, from the northern section of the country; for Persians wearing a different costume will be spoken of hereafter. They belonged among Muslims to the sect of Ali. I saw none of the above class of people at Singapore. But in the streets of Cairo I once met with several strangers who wore a somewhat similar dress, and who had, doubtless, come either from Persia or from a neighbouring country.

AFGHANS.

"Afghans were spoken of as being by no means rare at Bombay, and I met with two or three individuals wearing the Afghan dress, as this was described to me; but, in the absence of my interpreter, I could not ascertain their place of origin. Some 'Cabul people' were once pointed out to me at Bombay. These differed from the preceding individuals in their costume, and they strongly resembled the Sindians and Bedouins of Northwestern India. They wore a hemispherical cap on the top of the head."—See ethnographical section of article AFGHANISTAN.

SIKHS.

"Sikhs were well known at Singapore, which place they had reached by the route of the Ganges and Calcutta. An individual was pointed out to me, who was perhaps not a fair example of the nation, and who was obviously of mixed descent. His complexion was as dark as that of the Telinga people, but he presented points of resemblance to the Ethiopian, and even to the Negro race. Notwithstanding that the Punjab, or country of the Sikhs, is situated on the waters of the Indus, I did not meet with any of these people at Bombay.

PARSEES.

"The Parsees are the living representatives and remnant of the Persians of antiquity. And they continue, as in former times, to reject image-worship, and to pay their adorations to the elements, holding that of fire in more special reverence. They agree even in various minor particulars which are related by Herodotus of their forefathers, as in 'their names ending without exception alike; in those words which are expressive of personal or of any other distinction, terminating in the Doric san, which is the same with the Ionic sigma; and in their dead being never interred (placed in the cemetery) till some bird or dog has discovered a pro pensity to prey on them.' When compared with the Parthians of a somewhat later period, the points of resemblance will be found to be yet more numerous. Resident Parsees were spoken of at Singapore, but I first met with individuals at Aden. Their settlement here was in consequence of the English establishment; and, unlike the Banians, they are not found in other parts of Arabia. At Zanzibar, indeed, I met with two or three Parsee artizans, who had been engaged by the Imaum for temporary employment. The servants in some of the Bombay steam-boats were Parsees; and one of them, seen in full costume on the road to Cairo, reminded me of the strangely different auspices under which this people, after the lapse of ages, were revisiting the Valley of the Nile. The Parsees shave the head, and wear a round inner cap like that of the Arabs. They usually preserve the mustache, but they always shave the chin. Sometimes a portion of the hair is left on the sides of the head, in continuation of the whiskers; in accordance, in some measure, with the practice of the Pharaonic princes. The dress of the Bombay Parsees has varied perceptibly within two or three generations, but always within certain limits, as in respect to the fulness of the robe. The costume of the women throughout the East has a general similarity, whether among the Arabs, Hindoos, or Parsees. Unlike, however, the practice with the higher class of Arabs, the Parsee women do not conceal the face." See article PARSIS.

THE ARAB COUNTRIES.

a. *Arab Colonists.*

"Numerous adventurers from Arabia are scattered around the shores of the Indian Ocean and among the East India islands, seeking often a permanent residence in countries more desirable in every respect than their own. The Exploring Expedition first met with an Arab at Soolo, who has been already noticed while speaking of the Malayan race. At Singapore various individuals of dissimilar personal appearance, and even differing in physical race, laid claim to the title of Arabs. Moreover, we were here introduced to the extraordinary diversity of aspect which, independently of mixtures of race, prevails in the southern Arab countries.

b. *Arabs of the Mediterranean.*

"The little intercourse between the inhabitants of the northern and southern shores of the Mediterranean seems a surprising circumstance to a stranger from the United States; especially when it is considered that this little is chiefly carried on through Gibraltar and Malta. The identity in language has contributed to throw the traffic into the hands of the Maltese; numbers of whom are now settled around the southern and eastern borders of the Mediterranean. On my subsequent visit to Malta I was better prepared to distinguish the many coincidences with the customs of the Arabs, as in the style of architecture, in the female dress, and in the fashion of early marriages. Also, in the four-pronged anchors, such as I had seen used in the Indian Ocean; where, moreover, I had been reminded by the practice of towing a boat under sail, of some of the incidents narrated in the Voyage of St. Paul. The painted row-galleys of Gozo offer a decided approximation to the antique Mediterranean model; and the inhabitants of this neighbouring island, which from being without harbours is very little visited, have retained more of their ancient usages than the Maltese. I sometimes witnessed the transhipment of Alexandria wheat for distribution to surrounding ports; Egypt being, as in the time of the Romans, the granary of this portion of the Mediterranean. The antiquities of Malta, although consisting principally of fragments, are extremely interesting; especially on account of the intimate relations between this island and ancient Carthage. The variety is unusual, including Phœnician or it may be Carthaginian, early Greek, and Etruscan, Egyptian, Roman, Saracenic or Muslim-Arabic; and another class already alluded to, which may receive the provisional name of Mauritanian.

"On the 8th of December, I left Malta in the French steam-boat, and on the second day after we were close in with the rugged mountainous land of Greece. Although forewarned by an ancient writer, that 'Greece was always the child of poverty,' I did not anticipate the aspect of sterility. Among other undulations of surface offered to our view, was a valley, remarkable only for having existed in the reminiscences of three thousand years; a young Turk, who happened to be looking in that direction, pronounced its name, 'Sparda.' Astonishment may well be excited, that such a district should have brought Xerxes out of the depths of Asia. Even in modern times this further peninsula is said 'to have been never actually conquered; the Turks contenting themselves with requiring only some trifling tribute.' On the 11th we arrived at Syra, where, with other passengers, I was immediately transferred to the Alexandria steam-boat. I counted at Syra, no less than eleven vessels in the course of building; for the Greeks, as in former ages, still hold the navigation of the eastern division of the Mediterranean; their vessels being comparatively rare to the westward of Malta. The Greek costume is in many respects half oriental, like the position of the country; and, indeed, the people themselves, according to Eastern classification, are hardly ranked among Franks or Europeans.

"We obtained a distant view of Crete and of Mount Ida; and on the 14th we reached the low shores of Egypt. The first glance on landing at Alexandria sufficed to dispel all ideas of romance as connected with Oriental life. Excepting my slight introduction at Malta, I was now for the first time within the territorial domain of history; and passing the modern village, I looked around, expecting to see ruins. What had become of the great city of Alexander, for so many centuries the emporium of the world? The ground was only hilly and uneven; and a cultivator complained of its being 'stony,' meaning by the expression the fragments of pottery. Two objects, however, still mark the site of the ancient city; and towards one of them, on the following day, I directed my steps. The cabins of the South Seas were palaces in comparison with the low mud hovels on the way, and the whole neighbourhood was disgusting. On reaching the base of the obelisk my attention was principally engaged by the size and the bold sculpture of the characters, seemingly designed to transmit a record through ages of barbarism. The dimensions of the stone were sufficiently astonishing; and I next considered the intellectual attainments of the existing population. Turning to my Barabra guide, 'Ali,' said I, 'who do you think placed the stone here?' His reply, coming from an unexpected source, startled me; 'Pharoon.' Richly carved capitals and columns, some of marble, some of polished porphyry, and some of granite, had been here and there dug out of the ground, abundantly vindicating the splendour of the ancient

Grecian city; and I occasionally remarked a fragment which had proved too massive to be of use to modern Egyptian architects. My Barabra guide pointed out these various antique relics, referring them always to 'Sooltan Iskander,' and only in a single instance did I find him at fault—in that of the unmeaning Roman column which he persisted in attributing to Pharoon. On a subsequent occasion the accuracy of the popular tradition was most unexpected, for Ali declared that 'he did not know the builder of the Pyramids, who, however, was certainly not Pharoon.'

"On the 20th of December I started for Cairo in the steam-boat, by canal and river. Owing principally to the groves of date-palms planted around the villages, the character of the scenery is rather Indian than European. In ancient times the people of the Mediterranean knew of the tropics only by the valley of the Nile; which, notwithstanding its two foreign palms, affords but a very slight insight. At Cairo, I had the good fortune to meet with Mr. Bonomi, and I was thus at once initiated into the subject of the Egyptian antiquities. I visited the Pyramids, placed like a rock in the current of time: a spot where the mind is directed to the roll of empires. I witnessed the perfection of masonry in the workmanship around the entrance of the great Pyramid; and the fact also that the principle of the arch is carried out in a neighbouring coeval tomb. While at Saccara, the beauty of the sculptures in a tomb of the sixth dynasty seemed to justify Herodotus in his exalted commendation of the Labyrinth. Of human works, the most ancient not only promise to endure the longest, but they bear the impress of superior skill and of extreme purity of taste. And the decline of art in Egypt is clearly referable to the Pharaonic ages; to the accession of the eighteenth dynasty. A point of higher interest is the fact, that the earliest Egyptian monuments appear to have preceded the origin of idolatry; but this question, with many others relating to the same period, may probably be decided by the researches of Lepsius. At Thebes, once the mistress of the world, the remains are nearly all Pharaonic, and in a corrupted style of art; yet they present a greater variety of subjects of interest than the antiquities of Lower Egypt, or those of any other country. In the tombs of the Pharaohs such is the freshness of the unfinished paintings, that one can hardly avoid expecting the return of the artist to his task. The excavations are in a material which resembles lithographic stone, and which is admirably adapted for painting in a climate devoid of the elements of destruction. But it must not be supposed that Egypt is mainly indebted to this circumstance for the abundance of antiquities, since the climate of various other countries is equally favourable for the preservation of the works of human hands. Although I may appear singular, I confess being interested in the Egyptian colossi. These are by no means simple enlarged figures of men, a kind of work that does not require genius; but they have lines not strictly anatomical, which yet appeared peculiarly expressive of physical strength. Among the numerous colossi at Thebes, the largest excites astonishment, not merely from the size and the freshness of polish, but from having been subjected to a mechanical force capable of rending it. The mode of warfare described by Homer is abundantly illustrated on the walls of the Theban temples. Here are portraits of individuals, costumes of foreign nations, military campaigns in distant lands, and other evidence of great international events, most of them anterior to written history. In fearful pre-eminence among the records of carnage, the most recent of the larger temples bears the battles of Sethosis. This Pharaoh preceded Alexander by about a thousand years, and a narration of his career is given by Manetho with much simplicity and dignity in the following words:—'Sethosis then made an expedition against Cyprus and Phœnicia, and besides, against the Assyrians and the Medes; and he subdued them all, some by his arms, some without fighting, and some by the terror of his great army; and, being elated with his success, he went onward the more confidently, and overthrew the cities and the kingdoms of the East.'

"While ascending the Nile, I remarked, among much individual variation, that the complexion sensibly increases in depth, until in the Thebaid the Malay-brown hue is not uncommon. This gradual change takes place in both Copt and Arab, and, to all appearance, is independent of mixture of physical race; although individuals might readily be selected who, in Europe and the United States, would not be recognized as Whites. With the exception of the Copts, the modern Egyptians all rank themselves as Arabs. I found, however, that the people living near the Pyramids, on the western margin of the valley, equally claimed Berber or Libyan descent. The circumstance seems to favour the suggestion of Hodgson, that the Libyan was once the spoken language of Egypt. The coincidences between Libyan and Sanscrit words are numerous; and the point might be worth ascertaining, how far a Mahratta from the Dekkan and a Kabyle from Mount Atlas, would be able to understand each other's language.

"I frequently met with Copts, a class of the population noted for having maintained their religion throughout the Muslim conquest, and for having preserved the sacred language of ancient Egypt. In the upper country, where the Copts have monasteries and villages, and where they cultivate the soil, the complexion is dark; and, in the absence of my interpreter, I should not have distinguished them from the rest of the peasantry. The Copts of Cairo and Alexandria have the fair European complexion, accompanied, frequently, with a Greek or an Italian cast of countenance, that suggested doubts as to their exclusive Oriental descent.

"Although Egypt at present may be regarded as politically independent, it is not the less governed by Turks; and these have appropriated most of the offices of honour and profit. In appearance and manners the Turks of Egypt are not very distinguishable by strangers; but they experience the same difficulty with Europeans in rearing their children in the climate of Egypt. The few Turks I saw in other parts of the Mediterranean, were more readily recognized from their wearing a distinctive dress. As a class, the Turks are manifestly less inclined than the Armenians to assimilate with Europeans.

"Muslim Syrians were said to be numerous at Cairo, although I was unable to distinguish them. I do not know whether Christian Syrians are in the habit of visiting this city; but I saw one of them at Malta, a person who had acquired eminence by his learning.

"Among the most interesting visitors at Cairo were a party of Bedouins from Petra, near Palestine, who had accompanied some travellers, and who were making their journey subserve the additional purpose of buying corn. They were six or eight in number, and from their lineaments it was evident that they belonged to one family. The hair was straight, and the features were those of the White race; but the depth of hue surprised me, as it fully equalled that of my Barabra servant, who chanced to be standing near. They were warmly clothed, in which respect they differed from the southern Bedouins.

"I once met with three or four ARABS OF THE HEDJAZ, in the streets of Cairo, and I was struck with their peculiarly dignified deportment. Their complexion was not particularly dark, and the physical race could not be mistaken. At Suez, however, where this class of Arabs are frequent visitors, I observed signs of mixed descent; and in an instance already mentioned, a Negro was in company. In place of a turban, all wore a mantle striped with yellow, an article of dress, perhaps, general in the Hedjaz, but which I have occasionally seen in Southern Arabia, and among Arab visitors at Bombay and Zanzibar. At the last-named place, a man wearing it was pointed out to me as a 'Wahabi,' who had fought against the Pasha of Egypt; and I have elsewhere heard this mantle termed the badge of the Wahabi. At Muscat I saw it worn by Negroes, whose filed teeth designated an African birth-place.

c. *Southern Arabs.*

"On the 24th of March I left Suez in the English steam-boat, and on the 30th I landed at Aden, where I remained for three weeks, besides making two subsequent visits. Egress was here prohibited to the new comers; but the people of the country were often seen bringing upon camels articles of produce for the supply of the market. A body of these Arabs, hired by the English authorities, were at work on the road, and a few had become permanent residents. In some individuals the complexion was yet darker than I had seen it among the fellahs of the Thebaid; while in other respects the affinity with the White race was hardly distinguishable. I remarked nothing peculiar in the costume, except on one occasion, in the instance of three Arabs wearing caps, which were not padded like those of the Banians, but which equally resembled in outline the common monumental pattern of Egypt.

"On the 22d of April I left Aden in the brig Cherokee, Captain Mansfield, of Salem; and we arrived at Mocha on the following day. A boat soon came off to the anchorage, bringing the Interpreter, who, with three other persons employed in the intercourse with American vessels, were the only natives of the place who could speak a European language. The interpreter stated that 'Mocha had now a good governor, and that all was right;' and he proceeded to descant, with great simplicity, upon the manifold advantages of good government. On landing, we visited the governor, who was called a Bedouin, but who resembled, in personal appearance, the other principal Arabs of the place. These seemed to be a superior set of men to the Muslims of Cairo. Many of them were tall and of large stature; and although their physical race would anywhere be recognised, I thought I could distinguish a local cast of countenance, independent of the unusual shortness and scantiness of beard. Indeed, the only long beards I saw among the native population of Mocha were those of the Jews. The common Arabs of the town varied more than the upper classes from the usual aspect of the White race. Their average stature was decidedly inferior; the beard was yet more scanty, and instances occurred of its entire absence, in combination, too, with the deep Malay-brown complexion. Polygamy seemed hardly to exist among the mass of the population, notwithstanding the frequency of unfruitful marriages; neither did I hear of any undue facility of divorce. I remarked many other points of difference from the Egyptian Arabs: as in the costume (the head-dress usually consisting of a blue turban more or less covering the ears); in the eating of camel-meat; and especially in the universal diffusion of education. The country immediately behind Mocha is level, and, on a general view, appears to be utterly desert. Here and there, however, I met with small patches of cultivated soil that are irrigated from wells, a discouraging task in such a burning climate. On inquiring as to the safety of travelling, I was told that 'it would be better not to go far from town, for, being a stranger, if I should happen to be robbed, I should be unable to tell by whom.' I asked my informant if he would be willing to make the journey by land to Aden? 'Nothing would induce him to go among people with whom he was unacquainted.' I then asked whether the people of Mocha would fight for their governor? This question elicited surprise, and he rejoined,—'Why should we fight for the governor, who only taxes us?' In addition to the slight tattooing usual with Arab women, I observed, at Mocha, ornamental markings with a black pigment; but these occurred more especially upon children. Similar markings with pigment, rather than true tattooing, have appeared to me to be represented on the Pharaonic monuments, on the bodies of certain foreigners. A cane which I obtained at Suez, having the oblique monumental handle, was recognised as being made after the pattern of Western Arabia; while the fashion of the rounded hook was referred to the cane of Muscat."—See ethnographical section of article ARABIA.

"The few BEDOUINS pointed out to me were hardly distinguishable from the lower class of town Arabs, except by the absence of a turban. They were all young men, having uniformly the Malay-brown complexion, and the hair curling in ringlets all over the head. A powerful tribe of Bedouins, who sometimes rob and kill people, even within a few hours of Mocha, interrupted the land-route to Aden. Couriers, however, are occasionally dispatched, and we transmitted letters by one of them. 'About eleven years prior to our visit, Mocha was seized by the Turks, (as the Egyptian forces were here called,) who left a garrison of some five hundred men.' This state of things afforded a pretext to the Bedouins of the vicinity to pillage the place. 'Assembling to the number of some thousands, they approached the town, regardless of the artillery, and scaled the walls with ladders. They afterwards robbed every house, and left the inhabitants almost literally naked.' Mocha, since that event, has been ruled by a Bedouin. Among other results of the affair, it was found that Egyptian money, which had been in use during the short accession of Turkish power, gave place to the former currency. I sometimes heard references to the palmy days when 'Mocha belonged to the king of Sanaa;' when her staple was more in demand, and especially when it was paid for in dollars. The recent death of the last-named potentate had added to the usual disturbances of the country, and a battle was daily anticipated, in consequence of the revolt of some of the villages belonging to his soldiers.'

"Bedouins are often seen at Zanzibar—a circumstance the more unexpected, as I had ascertained that they do not visit Singapore. Besides the individual adventurers at Zanzibar, the Imaum sometimes imports bodies of Bedouins, to serve as soldiers in his military campaigns. On the 3d of October I left Zanzibar in the barque, Wm. Schroeder, Captain Jackson of Salem; and, after making an unsuccessful attempt to land at Mombas, we arrived on the 18th at Muscat. The people of this part of Arabia, like those of the Hedjaz, have long beards, and they may be recognised by the checkered and banded turban, which I first saw at Zanzibar. Some individuals were found to be more polished and liberal-minded than any other Arabs I have met with; and a more advanced tone of society pervaded their principal town. The name of Ishmael was well known at Muscat; and two classes of Arabs were recognised 'who are fighting to this day. The Kahtan Arabs ruled formerly for a century and a half; and they are many of them a fine-looking set of men, having, unlike the Abyssinians, regular hair. They are all Muslims, and are not rare in the interior, but there were none of them residing in Muscat. The Imaum's family came from the town of Adam in Uzjd, and belonged to the other class of Arabs, the Hameer, as did, likewise, most of the people of Muscat, and many of the Bedouins. Inscriptions on the rocks are unknown in Eastern Arabia.' The Bedouins visiting Muscat differed in personal appearance from those seen at Mocha; and they were more readily recognised as belonging to the White race. Their complexion, however, in most instances seemed to be yet darker than in the Mongolian race. Their long hair was much undulated; and they usually wore a fillet around the head, such as is seen on the Pharaonic monuments in representations of captive monarchs. One of the Bedouins, who bore no outward marks of distinction, was pointed out to me as being chief 'over a thousand men; all to get angry whenever he is angry and at no expense.' I was informed that no one came to Muscat from a greater distance inland than two or three hundred miles, and that the great interior of Arabia is unknown even to the Bedouins, who, however, suppose it to be inhabited. Persons sometimes go by land to Mecca, keeping near the Persian Gulf; but this is not often done, the route being considered dangerous from the scarcity of water; and the usual course is by sea to Bahrein, and thence through the Interior. The communication along the outer coast of Arabia is carried on altogether by sea."—See article BEDOUINS.

d. *Indian Muslims.*

"I have spoken in another chapter of the Indian Muslims who were seen at Singapore, and who came chiefly from Eastern Hindostan. Those met with on my second Voyage belong to the North and West; and, for the most part, unequivocally to the White race. As the pilgrim vessels usually touch at Mocha, numbers of Indian Muslims are to be met with in that city; where I soon learned to distinguish them by the straightness of the beard; a point in which they differ from Europeans, and which is expressed in certain Greek and Roman representations of Orientals. Some of these Indian Muslims dealt in precious stones, while others had become residents, and had opened shops. A Bombay ship at the anchorage, was laden principally with Surat cloth; and had on board about forty merchants, each of them, as in the Chinese junks, having the charge of his own wares.

"The larger edifices at Mocha are built in much the same style as those of Cairo, except that there is greater refinement of architectural ornament. This was traced to the workmen of India: and I soon perceived, that the Arabs here, as everywhere, are a rude people; and that all the arts requiring skill and ingenuity, belong properly to Hindostan. On proposing to have an umbrella covered, I was told that this 'could not be done; because the people of Mocha are not India people.'

"I met with a class of individuals called PERSIANS, who are noted as the fairest Orientals that come to Mocha; and who appear to me to have close affinity with the Indian Muslims. One of them accosted me in English, and among other matters, stated that 'he was from Iraun, and that the English and his countrymen were friends:' although a person of much intelligence, he had never before heard of America, and he was evidently at a loss to comprehend where it could be situated. I met with another of these Persians, immediately after he had performed a surgical operation; and, to my surprise, he described the 'couching of the cataract;' but I did not undo the bandages to verify his account. At Bombay, and throughout my journey in the interior of India, I found myself always at a distance from the seat of arts above alluded to; and my inquiries now placed it to the northward, and in the direction of the Indus. Other attendant circumstances, taken in connexion with the geographical position of the Upper Indus, seem to point yet further east. Bombay appears to be the head-quarters of the Indian Muslims, and I remarked among them various evidence of Persian descent, as in the custom of child marriages. Polygamy is rare, even among the wealthy; but their religion has been much corrupted; 'they have saints, and they address prayers to Mohammed, and even to Ali.' I was informed, however, that four castes of Muslims are recognised at Bombay, who all eat together, but who will not intermarry. It has been sometimes asserted that the boatmen of the western coast of Hindostan are all Muslims. Exceptions, however, must be made in regard to Cutch; and a Rajpoot sailor was serving in the Bombay ship at Mocha under a Muslim commander. Indian Muslims are likewise common at Zanzibar, and it was said of them, that 'after getting married and mixed up a little, they wanted to pass for Arabs.' Most of the White women I saw at Zanzibar belonged to the class of Indian Muslims.

"Two or three BELOOCHES were pointed out to me at Zanzibar, and others again at Bombay. They seemed to have affinity with the Indian Muslims, rather than with the Sindians and Bedouins; but I am hardly prepared to decide from these few examples. Belooche women have been sometimes imported for the harems of Zanzibar."—See article BELUCHISTAN.

e. *Western Hindoos.*

"In the various ports of Arabia and Eastern Africa, individuals may occasionally be met with, who differ remarkably in their bearing and costume from the surrounding population. An equal difference will be found in the qualities of the mind, in a surprising amount of intelligence and general information. These people have influence, not, however, derived from battles; their empire is not of the sword; neither is it based on their religion. They are the idolatrous merchants of India, who, for ages, have mainly held the commerce of the above-named countries.

"My acquaintance with the CUTCH BANIANS commenced at Aden: and at first I mistook them for Parsees, whom they resemble in many striking particulars; as in their names, in their general style of dress, and in their shaving the beard with the exception of the mustache (a circumstance possibly connected with the prevalence of the same fashion at Cairo), but on further acquaintance, I was soon able to distinguish a peculiar style of countenance. The Banians visiting Mocha and the neighbouring parts of Arabia, belong to a different caste from those usually seen at Zanzibar, Muscat and Bombay. They may be distinguished by the more flattened turban; the shape of which, as is shown by Niebuhr's figure, has not changed for the last eighty years. At Zanzibar I was more particularly struck with the analogy to the ancient Egyptian priests; for in like manner the Banians seemed to be the depositaries of knowledge and of the relics of civilisation in the midst of barbarians. Herodotus says, 'that the Egyptians venerate the cow;' and this is a fundamental principle of the Hindoo religion. Also, in minor particulars, the undress Banian cap entirely resembles the pattern most frequently represented on the Egyptian monuments. One of the principal Banians at Zanzibar spoke English, and he repeatedly complained to me of the 'hard life required in being a Banian. There are eighty-four castes of Banians;' but, said he, 'God made all men free. The Banians who go to sea do not associate with those who remain in India; and the same rule prevails among the Lascars, who are at Sudras, and are allowed to eat fish.

"The mass of the population of Bombay is composed of Hindoos who, by way of territorial distinction, are usually termed MAHRATTAS. From my first view of them I could only regard them as Arabs; as the unconverted idolatrous Arabs, such as we are familiar with in the history of the Mediterranean countries; and I soon perceived that it was possible at the present day, to obtain an insight into the influence on the mind of the ancient system of image-worship. I perceived also that in various points of difference from the modern Arabs, the Mahrattas accord with the ancient Egyptians. Thus polygamy is rare; the women do not conceal the face; the men carry burdens (with the balance-beam) on the shoulders; and the drinking-cups and water-jars are universally made either of copper or brass. Other customs witnessed have not yet become obsolete in Egypt; as that of the women carrying burdens on the head, and their collecting manure (for fuel) with the hands. It is true, the analogy in surrounding circumstances (periodical rains compensating in reservoirs for the inundations of the Nile) tends doubtless to assimilate the population; yet with every allowance, the Mahrattas appeared to me to be more than a counterpart to the Egyptians of Herodotus. The fact, too, should be noted, that the Sanscrit language, which has been inherited by the Mahrattas, is regarded by writers as originally foreign to India. In former times the nations of India were principally four, corresponding with the four languages—the Mahratta, Tolugu, Canara, and Tamul. The Mahrattas have retained the Sanscrit alphabet, but they possess very little literature of their own, and the Tamul language is the only one of the four that is rich in literary treasure."—See article MAHRATTA TERRITORY.

f. *Irregular Castes of India.*

"In India the term 'caste' is not restricted to the followers of the Braminical religion, but it is used in a general sense, and Parsees, Muslims, Jews, and Europeans are commonly spoken of as so many castes. The population is heterogeneous, made up of the remnants or wrecks of a variety of nations, some of which are perhaps extinct at their place of origin; for it would seem that amid 'the violence with which the earth has been filled,' toleration has existed only in India.

"The MAHARS are commonly supposed to be the original inhabitants of the Dekkan. Three or four of them were pointed out to me at Ahmednuggur, but in costume and personal appearance they did not present any marked difference from the surrounding population. They may have been a little darker, and one individual who accompanied me to Adjunta and back, seemed to be as dark as a Telingan.

"I met with a party of BHILLS on the road, near the limit of my excursion. They presented very distinctly the lineaments of the white race, but I did not remark any thing peculiar in their dress or personal appearance.

"The BRINJARRY were principally met with on first entering the Dekkan, their immense herds of bullocks sometimes occupying the road for nearly a mile. At regular intervals, a woman in the midst was spinning or sewing as she walked along, and one was carrying a dog, an act which, according to my interpreter, 'neither Hindoos nor Parsees would do.' The dress more resembled the European female costume than the Hindoo, but the Brinjarry women had large anklets of fantastic pattern, with other brass ornaments. The men carried spears having the broad blades covered with leathern sheaths. In parties returning towards the interior, the bullocks were all heavily laden with merchandise. Except in the masculine forms of the women, I could find no physical difference from the general population of the country. The lineaments of the white race were readily distinguishable, except in an individual already noticed as resembling an Australian. The herds were spoken of as the property of the women, who, indeed, had the air of being the heads of the community; but whether the system of polyandry prevails, as among certain classes in the south of India, I did not ascertain.

"The DUNGUR were encamped near every village of importance, in small huts formed of rush-mats. I have seen these huts not only hemispherical in shape, but disposed in a circle, with their entrances towards the centre, thus presenting an unexpected resemblance to the 'kraals' of Africa. The Dungur have drays of a very rude and primitive construction, formed of three cross timbers resting on an axle, the wheels being merely sections of the trunk of a tree.

"The SIKLIGUR have similar wandering habits, but they hold themselves aloof; they are less numerous, and are said to exercise the profession of 'chatty-makers,' or tin men. I fell in with one of their encampments, which consisted of low woollen tents. This, I think, is the class of people to which the origin of the Gipsies has been sometimes referred.

"Some of the numerous irregular castes of India are supposed to be the remains of the aboriginal population. I did not meet with any direct evidence of this, but I am inclined to think there may be something in the point of view. Some parallelism in social condition seems wanting in the white race, and the required connection may possibly be traced in the hill-people of Nepaul, the Karens of Birmah, and even tribes situated further to the south-east.

XII. ASSOCIATION.—NUMERICAL PROPORTIONS.

In order to complete the foregoing account, Mr. Pickering furnishes a specification of the several races occurring at each place visited, or, in other words, an analysis of the population.

"In the United States, three races are familiarly known—the White, the Negro, and the Mongolian, the latter represented both by the aboriginal population and by a few Chinese settled in our principal cities. A fourth race, the Malayan (represented chiefly by Polynesians), is not uncommon in the ports frequented by whale-ships; and, in some few instances, I have met with a fifth race, the Telingan, or true Indian.

"At Madeira I met with two races—the White and the Negro; but the latter race occurred in only a few instances, and exclusively at the principal sea-port town.

"At the Cape Verde Islands I met with the same two races, but they occurred in inverse proportions, and the White race was here comparatively rare.

"At Rio Janeiro and the vicinity, I fell in with only the same two races—the White and the Negro, and they appeared to be here in nearly equal proportions.

"The same two races were again seen at the mouth of the Rio Negro, in North Patagonia; but a third race—the Mongolian or the aboriginal—was said to be present among the population of the neighbouring village.

"In Terra del Fuego there occurred but a single race—the Mongolian, or the aboriginal. I have never heard of runaway sailors resorting to the southern side of the straits of Magellan.

"In Northern Chili, at Valparaiso and Santiago, I met with the three races last mentioned—the White, the Mongolian or the aboriginal, and the Negro, but the latter race was decidedly rare.

"In Peru, at Lima and in the vicinity, I met with the same three races, here, however, interspersed in more equal proportions; and a variety of complicated mixtures had sprung up, which were distinguished by different names. On the Andes the aboriginal race was found to preponderate, and the Negro was rare.

"At San Francisco, in North California, I met with three races—the Malayan—which is the aboriginal, and is also represented by the Polynesian crews of trading vessels—the White race, and, in a single instance, the Negro.

"Three races also occurred in Oregon—the Mongolian, here the aboriginal, and far predominating at the time of our visit; the White race, consisting principally of the missionaries, and of persons connected with the Hudson's Bay Company; and the Malayan, represented by a few Polynesians. A Negro, I believe, effected his escape from one of the vessels of the expedition, and, in this manner, a fourth race was added to the foregoing.

"Leaving now the shores of America, at the Hawaiian Islands, I met with five races—the Malayan, here the aboriginal; the White; the Negro; the Mongolian—represented by several imported Chinese—and the fifth race, by an individual from Hindostan.

"A single race occurred among the eastern Paumotus—the Malayan, or the aboriginal. In the western part of the same coral archipelago we received on board a shipwrecked European.

"At Taheiti, notwithstanding it is so much frequented by trading vessels, I observed but two races—the Malayan, or the aboriginal, and the White race. The same two races occurred at the Samoan, or Navigator islands, where the resident Whites consisted of the missionaries and of numerous deserters from trading vessels.

"The island of Tongataboo was the first place visited where two races had been brought in contact without European intervention; these were, the Malayan, or the aboriginal, and the Papuan from the Feejee islands. Three additional races had arrived in trading vessels—the White, the Negro—represented by a single resident—and the Negrillo, by a lad from Aramanga.

"At the Feejee Islands the Papuan and the Malayan races had again aboriginally met together, but Polynesians had likewise been introduced by trading vessels, as had the resident Whites, and a native of Hindostan, making in all, four races.

"At the Bay of Islands, in New Zealand, I met with but two races—the Malayan, or the aboriginal, and the White, the latter already settled in considerable numbers, before obtaining a cession of the sovereignty of the group.

"At Sydney, and in the vicinity, I met with five races—the Australian, or the aboriginal; the White, far preponderating; the Negro, in a few instances; the Malayan, represented by Polynesians, chiefly from New Zealand; and the Telingan, by some natives of Hindostan.

"In Luzon there occurred five races, two of them aboriginal, the Malayan and the Negrillo. The Mongolian race was represented by numerous Chinese residents; the Telingan, by a few Lascars from the shipping; and the White race by Europeans, and persons of European descent.

"At Caldera, on Mindanao, I met with but two races—the Malayan and the White; the latter represented by the commander of the post, and possibly, also, by some traces among the Muslim population.

"At Sooloo I met with four races—the Malayan; the Negrillo, in an individual, perhaps aboriginal; the Mongolian, represented by a single resident Chinese; and the Telingan, by two captive Lascars. There was, besides, evidence of some mixture of a fifth race, in the persons of the chiefs and the Muslim priest.

"At Singapore, individuals belonging to no less than eight physical races were found to be congregated together. The White race was represented by Europeans and by numerous Orientals; the Mongolian by Chinese; the Malayan, by the native population of the East Indies; the Telingan, by adventurers from Hindostan; the Negrillo, by slaves from New Guinea; the Negro, too, was present, in a few instances, as was likewise the true Abyssinian, and, in all probability, the Ethiopian, in some of the mixed Arabs. Indeed, as we had brought with us the Papuan, in the person of Veindovi, the Australian and the Hottentot were the only races remaining unrepresented at Singapore. The existence of such a spot on the globe is a fact deserving attention, especially when it is considered that this gathering is chiefly independent of European shipping, and of the modern town of Singapore; the same causes, and the same modes of conveyance, having for many centuries directed emigration from various quarters to the Straits of Malacca.

"At the Cape of Good Hope the Hottentot race is the aboriginal; but I did not meet with individuals who were unmixed. The four additional races introduced through European intercourse have also become much intermixed. At Cape Town the Negro appeared rather to preponderate in numbers over the White race, while the Malayan and the Telingan were comparatively rare.

"At St. Helena, in a population of only a few hundreds, introduced in European shipping, I observed no less than four races: the White; the Negro; the Mongolian, represented by a Chinese; and, to all appearance, the Telingan was present, in the descendants of individuals derived from Madagascar.

"At Zanzibar, on the east coast of Africa, I met with six races: the Negro, here preponderating; the White, derived either directly or originally from Arabia, Persia, and Western Hindostan; the Malayan, represented by three or four Ambolambo from Madagascar; the Mongolian, by one resident Chinese; the true Abyssinian, observed in a single instance; and the Ethiopian, which, during the monsoon then prevailing, was somewhat rare.

"At Bombay I observed five races: the White; the Telingan, rare in a state of purity; the Negro, chiefly among the crews of Arab vessels arriving from Zanzibar; the Mongolian, represented by a few Chinese residents; and the Ethiopian, which, in a single instance, was seen unmixed.

"At Muscat I met with four races: the White, represented exclusively by Orientals; the Telingan, apparently present in two individuals from the shores of the Persian Gulf; the Negro, chiefly derived through Zanzibar; and the Ethiopian, seen in individuals of mixed race, and in a single Somali.

"The same four races occurred at Aden. The Ethiopian, however, was here represented by numerous Somali visitors; and the Telingan, by individuals among the Sepoy regiments.

"At Mocha I fell in with five races: the White; the Mongolian, represented in a half-breed Chinese; the Abyssinian, seen in several instances; the Ethiopian, which is here numerous; as is, likewise, the Negro race.

"In Egypt I met with but three races—the same known there from remote antiquity: the Ethiopian, the Negro, and the White. I learned, however, that individuals belonging to a fourth race, the true Abyssinian, are sometimes seen at Cairo.

"I observed two races at Malta: the White race; and, in the instance of two lads, who were probably derived from the neighbouring Muslim countries, the Negro.

"On the continent of Europe I met with the same two races; but with the Negro only at Marseilles, and in the instance of two or three resident individuals. Of isolated districts, Africa appears to contain the greatest number of aboriginal races; and, next in order, will probably follow either Madagascar or New Guinea.

"Density of population being independent of territorial extent, the races, when compared by the number of individuals, rank differently from the appearance of things on the map. The usual estimates of the population of the globe vary from eight hundred to a thousand millions; and, taking the mean, the human family would seem to be distributed among the races in something like the following proportions:—

The White, . .	350,000,000	The Abyssinian,	3,000,000
The Mongolian,	300,000,000	The Papuan, .	3,000,000
The Malayan, .	120,000,000	The Negrillo, .	2,000,000
The Telingan, .	60,000,000	The Australian,	500,000
The Negro, . .	55,000,000	The Hottentot,	500,000
The Ethiopian, .	5,000,000		

XIII.—RELATIONS BETWEEN THE RACES.

"Although languages indicate national affiliation, their actual distribution is, to a certain extent, independent of physical race. Confusion has sometimes arisen from not giving due attention to this circumstance; and, indeed, the extension or the imparting of languages is a subject which has received very little attention. Writers sometimes reason as if nations went about in masses, the strong overcoming the weak, and imposing at once their customs, religion, and languages on the vanquished. But a careful examination of the present being regarded as the best guide to the history of the past, the following examples, selected from among others which I have noted, may be found deserving of attention:—

"In the mining towns of the Peruvian Andes, the Spanish language is not commonly spoken; but Spaniards and other foreigners learn the Quichua, or the native Peruvian.

"In Luzon, though a 'conquered country' for more than two centuries, the Tagala is still the general language, and the Spanish makes no progress in the interior towns and districts.

"At Singapore, where the native population is lost amid the influx of foreigners, the current language is, nevertheless, the proper Malay.

"Were foreign intercourse cut off from the Hawaiian Islands, I think I shall be supported in the opinion that, notwithstanding the large number of resident Whites, the English language would soon become obsolete.

"The English language makes no progress in Hindostan, neither do the Roman letters; and among the native population, newspapers are printed in five or six different alphabets, and a yet greater variety of languages.

"The language most likely to prevail at Aden, in case the English establishment becomes permanent, is the Hindoostanee.

"On the East Coast of Africa, notwithstanding the large number of resident Arabs, the general medium of communication is the Soahili, a Negro language.

"In Malta, a Catholic and European island, Arabic is alone spoken by the mass of the population; moreover, it is not here a written language.

"The adoption of a language seems to be very much a matter of convenience, depending often on the numerical majority. A stranger learns the language of the community in which he may happen to fix his abode; and his children often know no other. On the supposition, then, that POLYNESIANS in former times may have reached the shores of America, it does not necessarily follow that we ought to meet with traces of their language. Some combined association of individuals, some clannism, seems requisite to the preservation of a language in the midst of a foreign nation.—Thus it may be inferred, that the HEBREWS retained their language during their sojourn in Egypt, while in their present dispersed condition they speak the languages of their several adopted countries, to the very common neglect of their own.—The GYPSIES, who are almost as widely, though less individually dispersed, retain everywhere something of their original language.—But a more remarkable instance is found in the YUCHI, a tribe 'long incorporated in the Creek Nation; their peculiar language abounding in low, difficult, guttural sounds, so that with the exception of a few words, no one but the children of the Yuchi has ever been able to acquire it.'

"The following are instances of more complicated relations:—A foreigner might naturally suppose, that he could learn, in the United States, something of African languages and customs; but he might traverse the country without hearing a word of any Negro language, a single generation being sufficient to efface all knowledge of Africa.—Hayti would furnish him a more perplexing case; and perhaps nothing short of written history could solve the enigma of French being the language of that island.—Moreover, his search for African institutions would not be more successful among the St. Vincent CARIBS.—And thus we have three examples where one physical race of men has succeeded to the language and institutions of another.

"Turning now to the Pacific and East India Islands, we find languages of the Malay class spoken by three distinct physical races; and all analogy indicating that this type of language belonged originally to a single race, the particular source becomes a fair subject of inquiry. In determining this point, aid may be obtained from the geographical distribution of the three island races; from the well-known composition of the population of the East Indies, where the Malayan is an invading race, intruding and encroaching on the territories of the Negrillo; and from the fact, that the Negro race uses the same type of language at the Comoro Islands and Madagascar; while at the last-named island, the physical race of the Vazimba remains undetermined.

"The relations between the three island races are further illustrated by the state of the diffusion of knowledge in the Pacific, where almost every Polynesian art can be distinctly traced to the Feejee Islands. Agriculture and civilisation are very commonly regarded as inseparable; but the Feejeeans, notwithstanding their ferocious and barbarous manners, are found to possess a high degree of skill in cultivating the soil. A little further west, at Vanikoro, the trunk of a tree, rudely fashioned, is simply grooved to afford foothold, while an outrigger is attached, and a mat is spread for a sail. In our part of the globe, few would be willing, thus on a simple log, to launch out into an unknown ocean. The pattern, however, seems to shadow forth the Feejee canoe; and it by no means follows, that the race which elicited the spark of civilisation should have most profited by the advantage.

"The inventions of the Negrillo, a race so averse to, and almost refusing voluntary intercourse with strangers, were not likely to benefit the rest of mankind. If acquired, elaborated, and improved by the Feejeeans, still, from the social condition of this people, their dislike of change, and unwillingness to leave home, the knowledge might not have extended further. A third race is called into requisition, one of a roving disposition, the proper children of the sea; who, naturally the most amiable of mankind, are free from antipathies of race, are fond of novelty, inclined rather to follow than to lead; and who, in every respect, seem qualified to become a medium of communication between the different branches of the human family. Such an office is even indicated in the geographical distribution; for the Malayan alone makes a near approach to universal contact with the other races.

"While admitting the general truth, that mankind are essentially alike, no one doubts the existence of character, distinguishing not only individuals, but communities and nations. I am persuaded that there is, besides, a character of race. It would not be difficult to select epithets, such as 'amphibious, enduring, insititious;' or to point out, as accomplished by one race of men, that which seemed beyond the powers of another. Each race possessing its peculiar points of excellence, and at the same time counterbalancing defects, it may be, that union was required to attain the full measure of civilisation.

"In the organic world, each new field requires a new creation; each change in circumstances going beyond the constitution of a plant or animal, is met by a new adaptation, until the universe is full; while among the immense variety of created beings, two kinds are hardly found fulfilling the same precise purpose. Some analogy may possibly exist in the human family; and it may even be questioned whether any one of the races existing singly would, up to the present day, have extended itself over the whole surface of the globe."

XIV.—GEOGRAPHICAL PROGRESS OF KNOWLEDGE.

Leaving now the physical diversity of the human family, and recurring to the order of the voyage of the Expedition: in proceeding always west, the American tribes first engaged our attention, and next the widely scattered Polynesians. Among the latter people it soon became evident that their customs, arts, and attainments, were not of independent growth. And further, that the supposition of the Polynesians being the descendants—the scattered remnants of a highly civilised people now relapsed into barbarism, did not altogether meet the case. The East Indies being regarded by general consent as the place of origin of the Polynesians, it was natural to seek there for some corresponding tribe, for something analogous to the relation among European nations, of England to the United States. But on the contrary, surprise arose at the want of resemblance, and the term 'Malay origin,' if used in a national sense, was discovered to be incorrect. So, in looking to Asia, to China, Japan, or the neighbouring countries, for some tribe corresponding with the aboriginal Americans, we shall equally find ourselves at fault. For the remote environs of Behring's Straits do not offer a fair exception. After being much perplexed for an explanation, it was only on taking leave of the East Indies that the following simple reflection occurred to me:—If the human family has had a central origin, and has gradually and regularly diffused itself, followed by the principal inventions and discoveries, the history of man would then be inscribed on the globe itself; and each new revolution obliterating more or less of the preceding, his primitive condition should be found in the furthest remove from the geographic centre: as, in the case of a pebble dropped into the water, the earliest wave keeps most distant from the point of origin.

"If then we could go back into the early history of the East Indies, we might find there a condition of society approximating to that of the Polynesian Islands. And we can readily understand how customs may continue in remote situations, after having become obsolete at the place where they originated.

"They who send ships over every sea, and who live where inventions from different sources become common property at once, may not readily appreciate the state of things in the absence of such universal intercourse. But on the other side of the globe, in the vast space between Arabia and the coast of America, traces of successive waves in society seem actually to exist.

a. *The East Indies.*

"Thus in the East Indies, in the north-eastern or the heathen part of Celebes, we observe many striking coincidences with the state of society among the ancient Greeks. We have the spear, the shield, and the sword; and these weapons are even almost identical in pattern. Altars and sacrifices, so generally obsolete at the present day, are here extant. Auguries continue to be drawn from the entrails of offered victims, and also from the flight of birds. A separate literature has become an object of national pride, while the proa of the surrounding seas approaches in many respects the ancient rowgalley of the Mediterranean. A further analogy may even be found in the custom of putting to death strangers arriving by sea, which, it appears, was not unknown among the ancient Greeks. A little to the west of Celebes, the literature of the Malay nation contains a translation of the Fables of Æsop; who, according to the unsatisfactory accounts we have of him, was one of the earliest of the Greek writers. And further, the fact may be noted, that the Æsopian style of composition is still in vogue at Madagascar. Superstitions also appear to be subject to the same laws of progression with communicated knowledge; and the belief in ghosts, evil spirits, and sorcery, current among the ruder East

Indian tribes, in Madagascar, and in a great part of Africa, seems to indicate that such ideas may have elsewhere preceded a regular form of mythology. In the East, Luzon is one of the most distant points reached by the invention of letters, or having an aboriginal alphabet. I was, therefore, much struck with some coincidences in the forms of the letters, between the obsolete Tagala alphabet and the ancient Geez of Abyssinia; while in the intervening countries the alphabets, although various, are altogether unlike. It is true the connexion is not entirely obliterated, but may be traced in the mode of marking the termination of words among the Bugis and the Siamese.

"The multiplicity of alphabets in the East Indies seems to offer a parallel to the multiplicity of languages in America; and I have found nothing, in all this, contravening the idea of a single source to the invention. Many of the East India alphabets are plainly derived one from another, the form of the characters often being merely modified and new ones superadded; but I shall assume only that the knowledge of the existence of the art was derived from abroad. Much of the difference, between alphabetic writing in the East Indies and our own more perfect system, seems attributable to the circumstance, that the alphabet transmitted to the West by Cadmus had been already elaborated.

"Let the reader, in thought, divest himself of his education, and suppose that his mother tongue had never been reduced to writing; and, further, that the possibility of representing sounds by signs had simply arisen in his mind. The attempt to carry this idea into practice will be found by no means so easy as it may at first appear; and, by most persons, would probably be abandoned. With the aid of example, there would be a greater prospect of success; but even, with several characters communicated, their forms would probably be modified, and new letters invented, for sounds difficult to be reconciled, or that might be unrepresented. Where the model has been imperfect, it has appeared to me that the system founded upon it would naturally be complicated; one requiring long study, and which might serve for a profession, a means of procuring support; in short, more resembling the inconvenient alphabets of the East India islands, than the simple analysis of sounds which we have in the Roman letters.

b. *The Feejee Islands.*

"Eastward from Celebes, and nearly equidistant with Madagascar, we have the Feejee Islands, situated entirely beyond the direct influence of the East Indies. We find, nevertheless, a regular system of mythology; oracles; the javelin, or throwing-spear; the patera, or shallow drinking-bowl; the 'pipe of Pan;' and the 'alabastron,' with the practice of anointing. We have likewise the wooden neck-pillow, and the practice of circumcision, both common to the ancient inhabitants of the Thebaid, and also to the modern Abyssinians and their neighbours on the south. We have, further, the legend of an enormous bird, the 'Rok' of the Arabian Tales; and a similar story is said to be likewise current in Madagascar and in the neighbouring parts of Africa. Among the subjects which promise to throw light on the affiliation of nations, none more deserves attention than the calendar and state of astronomical knowledge. These points have been much neglected in the accounts of the East Indian and Pacific islands. The Feejeeans have a regular calendar; and, I may add, that Veindovi kept a strict account of the number of the moons, during the whole time he was on board the Vincennes.

c. *Polynesian.*

"The Polynesians, notwithstanding their more eastern geographical position, have most of their customs, arts, and acquired knowledge, in common with the Feejeeans; but there are certain points which require a separate notice. The eagerness with which whole communities, young and old, devoted themselves to reading and writing, when these arts were introduced, and the care previously taken to preserve and transmit poetry and history by means of living records, are facts of interest, under more than one point of view. Whoever has witnessed something of this, will hardly be willing to admit that such a people would ever have lost the art of writing. It should seem, then, as has been remarked by others, that the Polynesians belong to a wave of migration that preceded the invention of letters. Corresponding results may be deduced: from the ignorance of money, or of a medium of exchange; although property is acknowledged, inherited, and guarded by regular laws, even to a division of the soil. And again, from the absence of all kinds of 'grain;' notwithstanding the advanced state of the art of cultivation. The strict adherence to truth among portions of the Polynesian family, stands opposed to the treachery of the Feejeeans; and further, it seems to result mainly from the simplicity of the Polynesian character, the contrary practice being looked upon rather in the light of an error of judgment. The pseudo-accomplishment, however, proved to be not unknown to the chiefs. Other traits in the moral man might be adduced; for the 'golden age,' in accordance almost literally with the ideas of the ancient poets, still exists on some of the secluded Polynesian isles.

d. *America.*

"I hardly know whether to refer here to America; though in general the aboriginal Americans are to be regarded as in a less advanced state of society than the Polynesians. We note throughout Aboriginal America the entire absence of poetry and literature. Records, indeed, of a certain description were preserved by the Mexicans; but among the tribes of the United States, the tradition of remarkable events is said to disappear in very few generations. We note also in extensive and varied districts, a complete ignorance of agriculture; and where it is otherwise there is usually no division of the soil. A further difference from the islanders is found in the absence, in most instances, of hereditary governments. The custom of scalping, so well known in North America, may be noticed in this place, on account of its having been described by Herodotus; while it is not, that I am aware of, in vogue among any Asiatic tribe of the present day."

The 'zoological deductions' of Mr. Pickering are thus summed up by him. "Man does not belong to the cold and variable climates; his original birthplace has been in a region of perpetual summer, where the unprotected skin bears without suffering the slight fluctuations of temperature; and there has been a time when the human family had not strayed beyond these geographical limits.

"According to another principle, established by the observations of naturalists, species are found to have a certain geographical range; and notwithstanding a few instances of wide diffusion, Nature has not reproduced a species in different quarters of the globe.

"Oriental tradition, in placing the origin of mankind in Ceylon, reconciles in many respects the requirements of zoology. But there are certain coincidences which point in a different direction. Of all parts of the globe, save only the very poles, the equatorial countries continue the least known, owing, in some measure, to the forests with which they are generally invested, to the rains, to ferocious beasts of prey, and to the unhealthiness of the coasts. The portions which I have myself visited, have not offered any striking variation in natural productions from the adjoining countries: and I only remarked a cooler climate than in the vicinity of the northern tropic. It appears, however, that some of the most remarkable objects in the vegetable and animal creation have their home precisely under the equator. The races of men could all be conveniently derived from the same two centres, the one in the East Indies and the other in Africa; and if we could suppose separate species, this, agreeably to known laws, would reconcile the geographical portion of the subject. The objection of the slight external differences would not of itself be sufficient, for it is found in the organic world, that species differ by peculiarities which tend merely to change the aspect, while any radical change of structure involves the habits of life, and marks out a genus or group; in other words, that specific characters are external, while generic characters are anatomical. On the other hand, a very decided change may take place in the aspect, without a specific difference, as is more particularly the case in domestic animals and plants. Among these, moreover, we occasionally observe new varieties springing up at a distance from the original locality. Thus, new varieties of plums, apples, and pears have made their appearance in America since the introduction from abroad of the parent stocks. Not so, however, with the human family. Notwithstanding the mixtures of race during two centuries, no one has remarked a tendency to the development of a new race in the United States. In Arabia, where the mixtures are more complicated, and have been going on from time immemorial, the result does not appear to have been different. On the Egyptian monuments I was unable to detect a change in the races of the human family. Neither does written history afford evidence of the extinction of one physical race of men, or of the development of another previously unknown. It is true, varieties appear to have more permanence in certain kinds of animals than in others. Thus, the breed of sheep, figured on the most ancient Egyptian monuments, is unknown at the present day, and at least one variety of the bullock formerly known in Egypt, has in like manner become extinct. On the other hand, among the many varieties of the dog figured at Beni-Hassan (anterior to the Pharaonic ages), two have been pronounced identical with the greyhound and the turnspit now existing. Within my own observation, I have found no tendency in varieties to revert, in the course of successive generations, to the original type. But this whole subject of varieties has received from naturalists very little attention, and when it shall be better understood, it may, perhaps, throw light upon the origin of the races of the human family. There is, I conceive, no middle ground between the admission of eleven distinct species in the human family, and the reduction to one. The latter opinion, from analogy with the rest of the organic world, implies a central point of origin. Further, zoological considerations, though they do not absolutely require it, seem most to favour a centre on the African continent. Confirmatory circumstances of a different character are not wanting, some of which may be worth enumerating. In the first place, we have no less than five races actually inhabiting Tropical Africa, while the diversity diminishes as we recede, until districts of greater territorial extent are inhabited by but a single race. Again, some accordance may be found in the history and habits of the Arab tribes, holding, as they do, the country interposed between the heart of Africa and extensive and important seats of population; while these are occupied in regular succession by pastoral and agricultural nations. The disowning too, by the Desert tribes, of such individuals as take up agricultural pursuits, looks very like the incipient extension of population. There is something, also, in the fixed condition of the African tribes, on which Time, in some instances, seems scarcely to make an impression; something in the near approach to the absence of invention, or of the spontaneous evolution of knowledge; something in the seeming yielding of civilisation before barbarism; and in the bounds prescribed to ambition, from the general failure of attempts to colonise this continent; Africa, under various guise, having been always sending forth streams of population, and at the same time prohibiting a return." These views, treating the question of man's origin purely as one of natural history, are moderately stated; but the main deduction at present to be made from them seems to be this, that they only add fresh confirmation to the theory of man's unity, by showing that no well-defined groups of the human family have yet been found, which can be fairly and philosophically regarded as representing separate species, distinguished from each other by such wide and impassable diversities as to establish a different origin, or invalidate the scriptural doctrine of the original unity of all the families of mankind.

THE END.